More Than Just a Textbook

Internet Resources

StudentWorks™ *Plus* Online This interactive **eBook** includes the complete Student Edition with audio, Math in Motion, Personal Tutor, Self-Check Quizzes, and much more – all at point of use!

Step 1 **Connect to**  glencoe.com

Step 2 **Connect to resources by using simple and convenient QuickPass codes.**

PA5150c1

"PA" for "Pre-Algebra"

This edition, ISBN 978-0-07-888515-0

Enter the appropriate chapter number. c1 = Chapter 1

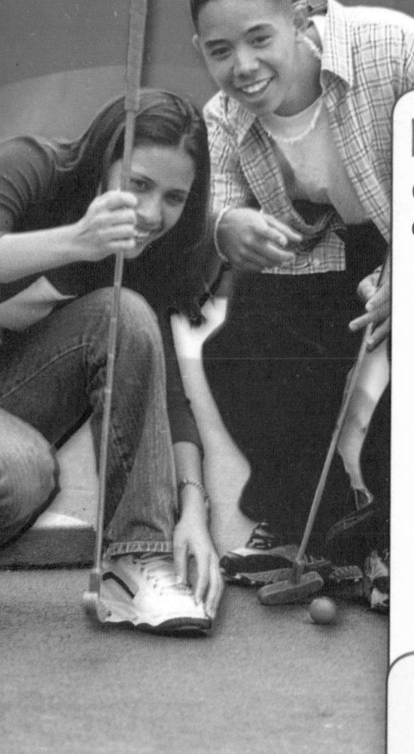

For Students

Connect to the Student Edition **eBook** that contains all of the following online resources. You don't need to take your textbook home every night.

- Personal Tutor
- Self-Check Quizzes
- Chapter Readiness Quizzes
- Math in Motion: Animation
- Math in Motion: BrainPOP®
- Math in Motion: Interactive Lab
- Extra Examples
- Chapter Test Practice
- Standardized Test Practice
- Study to Go
- Vocabulary Review Games
- Graphing Calculator Keystrokes
- Multilingual eGlossary
- Scavenger Hunts
- Workbooks
- Hotmath Math Homework Help Homework Help

For Teachers

- Teaching Today
- **Advance Tracker**
  - Diagnostic, formative, and summative assessment
  - Progress reports
  - Differentiated instruction
- State Resources
- Professional Development at www.mhpd.com
  - Video Clips
  - Online Credit Courses
- Research
  - White Papers
  - Efficacy Studies

For Parents

Connect to www.glencoe.com to access **StudentWorks *Plus* Online** and all of the resources for students and teachers listed above.

Glencoe McGraw-Hill

Pre-Algebra

Authors

Carter • Cuevas • Day • Malloy • Molix-Bailey • Price • Willard

McGraw Hill Glencoe

About the Cover

Nothing beats a hole-in-one in miniature golf! Next time you are playing miniature golf, think about the math you use—from keeping score to estimating distance. But there is more math than that! When you putt the ball in a straight line, the line can represent a linear function. A linear function is an equation whose graph is a line. You will learn more about linear functions in Chapter 8.

The McGraw-Hill Companies

Glencoe

Send all inquiries to:
Glencoe/McGraw-Hill
8787 Orion Place
Columbus, OH 43240-4027

ISBN: 978-0-07-888515-0
MHID: 0-07-888515-9

Printed in the United States of America.

7 8 9 10 DOW 17 16 15 14 13 12 11

CONTENTS IN BRIEF

Focal Points and Connections
See page iv for key.

The Curriculum Focal Points identify key mathematical ideas for this grade. They are not discrete topics or a checklist to be mastered; rather, they provide a framework for the majority of instruction at a particular grade level and the foundation for future mathematics study. The complete document may be viewed at www.nctm.org/focalpoints.

KEY

G8-FP1 Grade 8 Focal Point 1

G8-FP2 Grade 8 Focal Point 2

G8-FP3 Grade 8 Focal Point 3

G8-FP4C Grade 8 Focal Point 4 Connection

G8-FP5C Grade 8 Focal Point 5 Connection

G8-FP6C Grade 8 Focal Point 6 Connection

G8-FP7C Grade 8 Focal Point 7 Connection

G8-FP1 Algebra: **Analyzing and representing linear functions and solving linear equations and systems of linear equations**

Students use linear functions, linear equations, and systems of linear equations to represent, analyze, and solve a variety of problems. They recognize a proportion ($y/x = k$, or $y = kx$) as a special case of a linear equation of the form $y = mx + b$, understanding that the constant of proportionality (k) is the slope and the resulting graph is a line through the origin. Students understand that the slope (m) of a line is a constant rate of change, so if the input, or x-coordinate, changes by a specific amount, a, the output, or y-coordinate, changes by the amount ma. Students translate among verbal, tabular, graphical, and algebraic representations of functions (recognizing that tabular and graphical representations are usually only partial representations), and they describe how such aspects of a function as slope and y-intercept appear in different representations. Students solve systems of two linear equations in two variables and relate the systems to pairs of lines that intersect, are parallel, or are the same line, in the plane. Students use linear equations, systems of linear equations, linear functions, and their understanding of the slope of a line to analyze situations and solve problems.

G8-FP2 Geometry and Measurement: **Analyzing two- and three-dimensional space and figures by using distance and angle**

Students use fundamental facts about distance and angles to describe and analyze figures and situations in two- and three-dimensional space and to solve problems, including those with multiple steps. They prove that particular configurations of lines give rise to similar triangles because of the congruent angles created when a transversal cuts parallel lines. Students apply this reasoning about similar triangles to solve a variety of problems, including those that ask them to find heights and distances. They use facts about the angles that are created when a transversal cuts parallel lines to explain why the sum of the measures of the angles in a triangle is 180 degrees, and they apply this fact about triangles to find unknown measures of angles. Students explain why the Pythagorean theorem is valid by using a variety of methods—for example, by decomposing a square in two different ways. They apply the Pythagorean theorem to find distances between points in the Cartesian coordinate plane to measure lengths and analyze polygons and polyhedra.

G8-FP3 Data Analysis and Number and Operations and Algebra: **Analyzing and summarizing data sets**

Students use descriptive statistics, including mean, median, and range, to summarize and compare data sets, and they organize and display data to pose and answer questions. They compare the information provided by the mean and the median and investigate the different effects that changes in data values have on these measures of center. They understand that a measure of center alone does not thoroughly describe a data set because very different data sets can share the same measure of center. Students select the mean or the median as the appropriate measure of center for a given purpose.

Connections to the Focal Points

G8-FP4C **Algebra:** Students encounter some nonlinear functions (such as the inverse proportions that they studied in grade 7 as well as basic quadratic and exponential functions) whose rates of change contrast with the constant rate of change of linear functions. They view arithmetic sequences, including those arising from patterns or problems, as linear functions whose inputs are counting numbers. They apply ideas about linear functions to solve problems involving rates such as motion at a constant speed.

G8-FP5C **Geometry:** Given a line in a coordinate plane, students understand that all "slope triangles"—triangles created by a vertical "rise" line segment (showing the change in y), a horizontal "run" line segment (showing the change in x), and a segment of the line itself—are similar. They also understand the relationship of these similar triangles to the constant slope of a line.

G8-FP6C **Data Analysis:** Building on their work in previous grades to organize and display data to pose and answer questions, students now see numerical data as an aggregate, which they can often summarize with one or several numbers. In addition to the median, students determine the 25th and 75th percentiles (1st and 3rd quartiles) to obtain information about the spread of data. They may use box-and-whisker plots to convey this information. Students make scatterplots to display bivariate data, and they informally estimate lines of best fit to make and test conjectures.

G8-FP7C **Number and Operations:** Students use exponents and scientific notation to describe very large and very small numbers. They use square roots when they apply the Pythagorean theorem.

About the Authors

Macmillan/McGraw-Hill and Glencoe/McGraw-Hill K–12 Mathematics Lead Authors

Our lead authors ensure that the Macmillan/McGraw-Hill and Glencoe/McGraw-Hill mathematics programs are truly vertically aligned by beginning with the end in mind–success in Algebra 1 and beyond. By "backmapping" the content from the high school programs, all of our mathematics programs are well articulated in their scope and sequence, ensuring that the content in each program provides a solid foundation for moving forward. These authors also worked closely with the entire K–12 author team to ensure vertical alignment of the instructional approach and visual design.

Dr. John A. Carter, Ph.D.
Assistant Principal for Teaching and Learning
Adlai E. Stevenson High School
Lincolnshire, Illinois

Areas of Expertise: Using technology and manipulatives to visualize concepts; Mathematics Achievement of English-Language Learners

Dr. Gilbert J. Cuevas, Ph.D.
Professor of Mathematics Education
Texas State University–San Marcos
San Marcos, Texas

Areas of Expertise: Applying concepts and skills in mathematically rich contexts; Mathematical Representations

Dr. Roger Day, Ph.D., NBCT
Mathematics Department Chairperson
Pontiac Township High School
Pontiac, Illinois

Areas of Expertise: Understanding and applying probability and statistics; Mathematics Teacher Education

Dr. Carol Malloy, Ph.D.
Associate Professor
University of North Carolina at Chapel Hill
Chapel Hill, NC

Areas of Expertise: Representations and critical thinking; Student Success in Algebra 1

Additional Pre-Algebra Authors

The entire Pre-Algebra author team strives to create a program that can be used by all types of Pre-Algebra teachers with all types of Pre-Algebra students. Each author brings their special expertise to making a program that will contribute to the success of every student who uses this instructional resource.

Rhonda J. Molix-Bailey
Mathematics Consultant
Mathematics by Design
DeSoto, Texas

Areas of Expertise: Geometry; Geometry in grades K–12 mathematics

Jack Price, Ed.D.
Professor Emeritus
California State Polytechnic University
Pomona, California

Areas of Expertise: Proportionality; Probability, statistics, and geometry in K–12 mathematics; Pre-Service Teacher Education

Teri Willard, Ed.D.
Assistant Professor
Department of Mathematics
Central Washington University
Ellensburg, Washington

Areas of Expertise: Functions; Functions in grades 6–12 mathematics

Contributing Author

This program is the beneficiary of the imagination of Dinah Zike through the contribution of the Foldables Study Organizers.

Dinah Zike
Educational Consultant
Dinah-Might Activities, Inc.
San Antonio, Texas

Math Online Meet the Authors at glencoe.com

Consultants

Glencoe/McGraw-Hill wishes to thank the following professionals for their feedback. They were instrumental in providing valuable input toward the development of this program in these specific areas.

Mathematical Content

Viken Hovsepian
Professor of Mathematics
Rio Hondo College
Whittier, California

Grant A. Fraser, Ph.D.
Professor of Mathematics
California State University, Los Angeles
Los Angeles, California

Arthur K. Wayman, Ph.D.
Professor of Mathematics Emeritus
California State University, Long Beach
Long Beach, California

Gifted and Talented

Shelbi K. Cole
Research Assistant
University of Connecticut
Storrs, Connecticut

College Readiness

Robert Lee Kimball, Jr.
Department Head, Math and Physics
Wake Technical Community College
Raleigh, North Carolina

Differentiation for English-Language Learners

Susana Davidenko
State University of New York
Cortland, New York

Alfredo Gómez
Mathematics/ESL teacher
Fowler High School
Syracuse, New York

Graphing Calculator

Ruth M. Casey
T^3 National Instructor
Frankfort, Kentucky

Jerry Cummins
Former President
National Council of Supervisors of Mathematics
Western Springs, Illinois

Mathematical Fluency

Robert M. Capraro
Associate Professor
Texas A&M University
College Station, Texas

Pre-AP

Dixie Ross
Mathematics Teacher
Pflugerville High School
Pflugerville, Texas

Reading and Writing

ReLeah Cossett Lent
Author and Educational Consultant
Morganton, Georgia

Lynn T. Havens
Director of Project CRISS
Kalispell, Montana

Teacher Reviewers

Each Reviewer reviewed at least two chapters of the Student Edition, giving feedback and suggestions for improving the effectiveness of the mathematics instruction.

Sherri Abel
Mathematics Teacher
Eastside High School
Taylors, South Carolina

Kelli Ball, NBCT
Mathematics Teacher
Owasso 7th Grade Center
Owasso, Oklahoma

Cynthia A. Burke
Mathematics Teacher
Sherrard Junior High School
Wheeling, West Virginia

Patrick M. Cain, Sr.
Assistant Principal
Stanhope Elmore High School
Millbrook, Alabama

Robert D. Cherry
Mathematics Instructor
Wheaton Warrenville South High School
Wheaton, Illinois

Tammy Cisco
8th Grade Mathematics/ Algebra Teacher
Celina Middle School
Celina, Ohio

Amber L. Contrano
High School Teacher
Naperville Central High School
Naperville, Illinois

Catherine Creteau
Mathematics Department
Delaware Valley Regional High School
Frenchtown, New Jersey

Glenna L. Crockett
Mathematics Department Chair
Fairland High School
Fairland, Oklahoma

Jami L. Cullen
Mathematics Teacher/Leader
Hilltonia Middle School
Columbus, Ohio

Kendrick Fearson
Mathematics Department Chair
Amos P. Godby High School
Tallahassee, Florida

Lisa K. Gleason
Mathematics Teacher
Gaylord High School
Gaylord, Michigan

Tracie A. Harwood
Mathematics Teacher
Braden River High School
Bradenton, Florida

Bonnie C. Hill
Mathematics Department Chair
Triad High School
Troy, Illinois

Clayton Hutsler
Teacher
Goodwyn Junior High School
Montgomery, Alabama

Gureet Kaur
7th Grade Mathematics Teacher
Quail Hollow Middle School
Charlotte, North Carolina

Rima Seals Kelley, NBCT
Mathematics Teacher/ Department Chair
Deerlake Middle School
Tallahassee, Florida

Holly W. Loftis
8th Grade Mathematics Teacher
Greer Middle School
Lyman, South Carolina

Carol Y. Lumpkin
Mathematics Educator
Crayton Middle School
Columbia, South Carolina

Ron Mezzadri
Supervisor of Mathematics K-12
Fair Lawn Public Schools
Fair Lawn, New Jersey

Bonnye C. Newton
SOL Resource Specialist
Amherst County Public Schools
Amherst, Virginia

Kevin Olsen
Mathematics Teacher
River Ridge High School
New Port Richey, Florida

Kara Painter
Mathematics Teacher
Downers Grove South High School
Downers Grove, Illinois

Sheila L. Ruddle, NBCT
Mathematics Teacher, Grades 7 and 8
Pendleton County Middle/High School
Franklin, West Virginia

Angela H. Slate
Mathematics Teacher/Grade 7, Pre-Algebra, Algebra
LeRoy Martin Middle School
Raleigh, North Carolina

Cathy Stellern
Mathematics Teacher
West High School
Knoxville, Tennessee

Dr. Maria J. Vlahos
Mathematics Division Head for Grades 6-12
Barrington High School
Barrington, Illinois

Susan S. Wesson
Mathematics Consultant/ Teacher (Retired)
Pilot Butte Middle School
Bend, Oregon

Mary Beth Zinn
High School Mathematics Teacher
Chippewa Valley High Schools
Clinton Township, Michigan

Each Reviewer reviewed at least two chapters of the Student Edition, giving feedback and suggestions for improving the effectiveness of the [illegible]

CHAPTER 0

Start Smart: Preparing for Pre-Algebra

Jack Sullivan/Alamy

Chapter 0 Support

Helping You Learn

Math Online

Unit 1
Rational Numbers and Equations

CHAPTER 1

The Tools of Algebra

Chapter 1 Support

Helping You Learn

Math Online

Preparing for Testing

Unit 1
Rational Numbers and Equations

CHAPTER 2

Operations with Integers

VisionsofAmerica/Joe Sohm/Getty Images

Chapter 2 Support

Helping You Study

Math Online

Preparing for Testing

Unit 1
Rational Numbers and Equations

CHAPTER 3

Operations with Rational Numbers

Chapter 3 Support

Helping You Study

- **New Vocabulary** 121, 128, 141, 147, 153
- **Key Concepts** 122, 130, 134, 141, 142, 147, 148, 153, 154
- **Check Your Progress** 121, 122, 123, 124, 128, 129, 130, 134, 135, 136, 141, 142, 143, 147, 148, 149, 150, 153, 154, 155
- **Check Your Understanding** 124, 130, 136, 144, 150, 155
- **Multiple Representations** 126, 132, 156
- **H.O.T. Problems** 126, 132, 138, 145, 151, 157
- **Skills Review** 127, 133, 139, 146, 152, 158

Math Online

- **Math in Motion: Animation** 117, 119, 128
- **Personal Tutor** 121, 122, 123, 124, 128, 129, 130, 134, 135, 136, 141, 142, 143, 147, 148, 149, 150, 153, 154, 155
- **Self-Check Quizzes** 121, 128, 134, 141, 147, 153
- **Extra Examples** 121, 128, 134, 141, 147, 153
- **Homework Help** 121, 128, 134, 141, 147, 153

Preparing for Testing

- **Extended Response** 133, 139, 167
- **Multiple Choice** 127, 133, 139, 140, 144, 146, 152, 158, 163, 166
- **Short/Gridded Response** 127, 146, 152, 158, 165, 167
- **Worked-Out Example** 143

Unit 1
Rational Numbers and Equations

CHAPTER 4

Expressions and Equations

Howie Garber/Photographer's Choice/Getty Images

Chapter 4 Support

Helping You Study

- **New Vocabulary** 171, 178, 184, 199
- **Key Concepts** 171, 184, 185, 191, 193
- **Check Your Progress** 171, 172, 173, 178, 179, 180, 185, 186, 191, 192, 193, 199, 200, 201, 205, 206
- **Check Your Understanding** 173, 180, 187, 194, 202, 207
- **Multiple Representations** 175, 182, 188, 195, 203, 208
- **H.O.T. Problems** 175, 182, 188, 195, 203, 208
- **Skills Review** 176, 183, 189, 196, 204, 209

Math Online

- **Math in Motion: Animation** 169, 191
- **Math in Motion: BrainPOPs** 171
- **Math in Motion: Interactive Labs** 191
- **Personal Tutor** 171, 172, 173, 178, 179, 180, 185, 186, 191, 192, 193, 199, 200, 201, 205
- **Self-Check Quizzes** 171, 184, 191, 199, 205, 206
- **Extra Examples** 171, 184, 191, 199, 205, 206,
- **Homework Help** 171, 184, 191, 199, 205, 206

Preparing for Testing

- **Extended Response** 176, 196, 209, 217
- **Multiple Choice** 176, 183, 189, 190, 196, 202, 204, 209, 213, 215, 216
- **Short/Gridded Response** 183, 189, 204, 217
- **Worked-Out Example** 201

Unit 1
Rational Numbers and Equations

CHAPTER 5

Multi-Step Equations and Inequalities

Chapter 5 Support

Unit 2
Proportions and Similarity

CHAPTER 6

Ratio, Proportion, and Similar Figures

Steffen Schmidt/epa/CORBIS

Chapter 6 Support

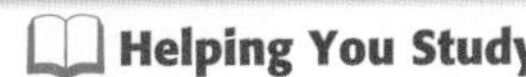

Helping You Study

Math Online

Preparing for Testing

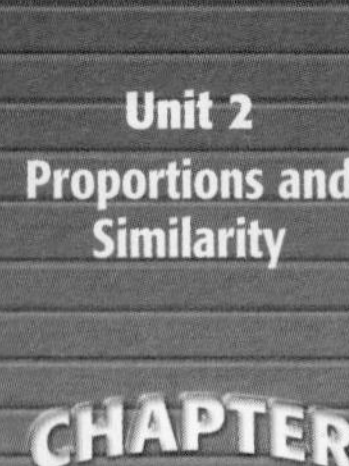

Unit 2
Proportions and Similarity

CHAPTER 7

Percent

PhotoLink/Getty Images

Chapter 7 Support

Helping You Study

- **New Vocabulary** 331, 345, 357, 364, 370, 376
- **Key Concepts** 331, 337, 339, 345, 346, 351, 364
- **Check Your Progress** 331, 332, 333, 337, 338, 339, 345, 346, 347, 351, 352, 353, 357, 358, 359, 364, 365, 366, 370, 371, 376, 377, 378
- **Check Your Understanding** 334, 340, 347, 353, 360, 366, 372, 378
- **Multiple Representations** 341, 361, 368, 373, 380
- **H.O.T. Problems** 335, 341, 349, 354, 361, 368, 373, 380
- **Skills Review** 336, 342, 350, 355, 362, 369, 374, 384

Math Online

- **Math in Motion: Animation** 329, 343
- **Personal Tutor** 331, 332, 333, 337, 338, 339, 345, 346, 347, 351, 352, 353, 357, 358, 359, 364, 365, 366, 370, 371, 375, 376, 377, 378
- **Self-Check Quizzes** 331, 337, 345, 351, 357, 364, 370, 376
- **Extra Examples** 331, 337, 345, 351, 357, 364, 370, 376
- **Homework Help** 331, 337, 345, 351, 357, 364, 370, 376

Preparing for Testing

- **Extended Response** 355, 362, 381, 391
- **Multiple Choice** 336 , 342, 350, 355, 356, 360, 362, 369, 374, 381, 387, 388, 389, 390
- **Short/Gridded Response In** 336, 342, 350, 369, 374, 391
- **Worked-Out Example** 359

Unit 3
Linear and Nonlinear Functions

CHAPTER 8

Linear Functions and Graphing

STAN HONDA/AFP/Getty Images

Chapter 8 Support

Unit 3
Linear and Nonlinear Functions

CHAPTER 9

Powers and Nonlinear Functions

Chapter 9 Support

Unit 4
Two- and Three-Dimensional Space

CHAPTER 10

Real Numbers and Right Triangles

Chapter 10 Support

Helping You Study

Math Online

Preparing for Testing

Unit 4
Two- and Three-Dimensional Space

CHAPTER 11

Distance and Angle

Chapter 11 Support

Helping You Study

- **New Vocabulary** 589, 598, 605, 611, 617, 624, 631, 636, 644
- **Key Concepts** 589, 590, 598, 611, 618, 624, 625, 626, 631, 636, 638, 645
- **Check Your Progress** 590, 591, 599, 600, 605, 606, 607, 612, 617, 618, 619, 624, 625, 626, 631, 632, 637, 638, 644, 645
- **Check Your Understanding** 592, 601, 608, 613, 620, 627, 633, 638, 646
- **Multiple Representations** 594, 609, 615, 629, 634, 640
- **H.O.T. Problems** 594, 603, 609, 614, 621, 629, 634, 640, 648

Math Online

- **Math in Motion: Animation** 587, 611
- **Math in Motion: Interactive Labs** 631
- **Personal Tutor** 590, 591, 599, 600, 605, 606, 607, 612, 617, 618, 619, 624, 625, 626, 631, 632, 637, 638, 644, 645,
- **Self-Check Quizzes** 589, 598, 605, 611, 617, 624, 631, 636, 644
- **Extra Examples** 589, 598, 605, 611, 617, 624, 631, 636, 644
- **Homework Help** 589, 598, 605, 611, 617, 624, 631, 636, 644

Preparing for Testing

- **Extended Response** 595, 604, 610, 630, 635, 641, 659
- **Multiple Choice** 595, 604, 610, 615, 620, 622, 623, 630, 635, 641, 649, 655, 658
- **Short/Gridded Response** 615, 622, 649, 656, 657, 659
- **Worked-Out Example 618**

Unit 4
Two- and Three-Dimensional Space

CHAPTER 12

Surface Area and Volume

Duomo/CORBIS

Chapter 12 Support

Helping You Study

Math Online

Preparing for Testing

Unit 5
Data Sets

CHAPTER 13

Statistics and Probability

Chapter 13 Support

Helping You Study

- **New Vocabulary** 730, 737, 743, 750, 757, 765, 771, 777, 783, 790
- **Key Concepts** 730, 731, 744, 765, 771, 772, 777, 790, 791, 792
- **Check Your Progress** 730, 731, 732, 737, 738, 743, 744, 745, 746, 750, 751, 752, 758, 765, 766, 767, 772, 773, 777, 778, 779, 783, 784, 785, 791
- **Multiple Representations** 741, 787
- **H.O.T. Problems** 734, 741, 748, 754, 761, 769, 775, 780, 787, 794

Math Online

- **Math in Motion: Animation** 727
- **Math in Motion: BrainPOPs** 791
- **Math in Motion: Interactive Labs** 765
- **Personal Tutor** 730, 731, 732, 736, 737, 738, 743, 744, 745, 746, 750, 751, 752, 756, 757, 758, 763, 765, 766, 767, 772, 773, 777, 778, 779, 783, 784, 785, 789, 791, 792
- **Self-Check Quizzes** 730, 737, 743, 750, 757, 765, 771, 777, 783, 790
- **Extra Examples** 730, 737, 743, 750, 757, 765, 771, 777, 783, 790

Preparing for Testing

- **Extended Response** 735, 770, 807
- **Multiple Choice** 733, 735, 742, 749, 755, 762, 764, 770, 776, 781, 788, 795, 803, 804, 805, 806
- **Short/Gridded Response** 742, 749, 755, 762, 776, 781, 788, 795, 807
- **Worked-Out Example** 731

Christian Liewig/Liewig Media Sports/CORBIS

Looking Ahead to Algebra 1

Student Handbook

Jack Sullivan/Alamy

Start Smart: Preparing for Pre-Algebra

Chapter 0 contains lessons on topics from previous courses. You can use this chapter in various ways.

- Begin the school year by taking the Pretest. If you need additional review, complete the lessons in this chapter. To verify that you have successfully reviewed the topics, take the Posttest.
- As you work through the text, you may find that there are topics you need to review. When this happens, complete the individual lessons that you need.
- Use this chapter for reference. When you have questions about any of these topics, flip back to this chapter to review definitions or key concepts.

CHAPTER 0 Start Smart: Preparing for Pre-Algebra

Then

You have already written, interpreted, and used mathematical expressions and equations. (Previous Course)

Now

- Use the four-step plan to solve problems.
- Add, subtract, multiply, and divide decimals.
- Write a verbal rule to represent a pattern or sequence.

Why?

BEGINNINGS You are getting ready to start Pre-Algebra. Like the beginning of a race, you need to be prepared for what is ahead. This chapter will "rev-up" your skills!

Get Started on Chapter 0

You will review several concepts, skills, and vocabulary terms as you study this chapter. To get ready, identify important terms and organize your resources.

Study Organizer

Start Smart Make this Foldable to help you organize your Chapter 0 review notes. Begin with a piece of 11″ by 17″ paper.

1. **Fold** a 2″ tab along the long side of the paper.

2. **Unfold** the paper and fold in thirds widthwise.

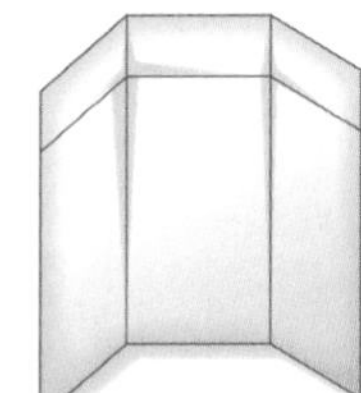

3. **Open** and draw lines along the folds. Label the head of each column as shown. Label the front of the folded table with the chapter title.

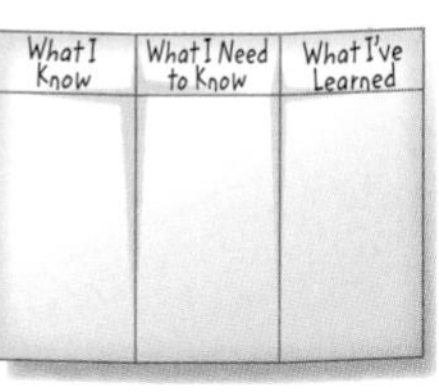

New Vocabulary

English		Español
look for a pattern	p. P8	buscar un patrón
guess and check	p. P8	adivina y verifica
make a table	p. P9	trabajar a la inverse
work backward	p. P9	hacer una tabla
decimal	p. P12	decimal
annex	p. P12	anexionar
customary system	p. P18	sistema inglés
metric system	p. P18	sistema métrico
data	p. P20	datos
pictograph	p. P20	pictograma
line graph	p. P20	gráfica lineal
bar graph	p. P21	gráfica de barras

Review Vocabulary

centimeter • centímetro a metric unit of length that is equal to one-hundredth of a meter

inch • pulgada customary unit of length, there are 12 inches in one foot

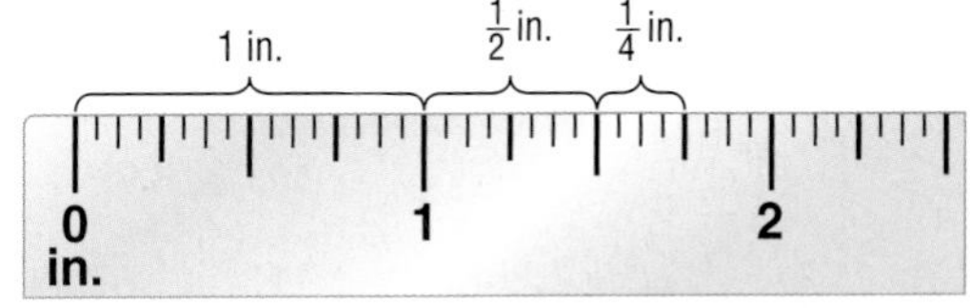

millimeter • milímetro a metric unit of length that is equal to one-thousandth of a meter

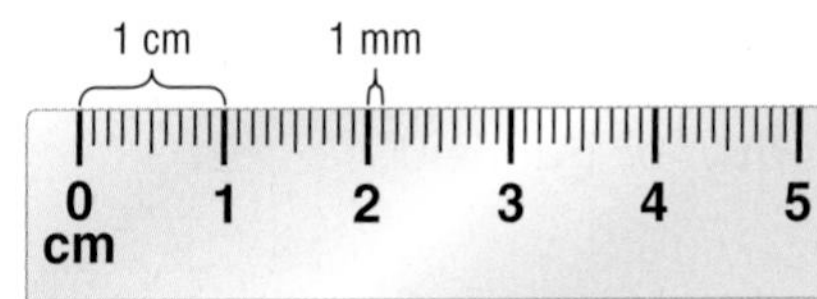

Multilingual eGlossary glencoe.com

Math Online

- Study the chapter online
- Explore **Math in Motion**
- Get extra help from your own **Personal Tutor**
- Use **Extra Examples** from additional help
- Take a **Self-Check Quiz**
- **Review Vocabulary** in fun ways

CHAPTER 0

Pretest

Use any problem-solving strategy to solve each problem.

1. **STAMPS** Suppose stamps for postcards cost $0.26 and stamps for first-class letters cost $0.41. Diego wanted to send postcards and letters to 10 friends. If he had $3.35 for stamps and spends all of it, how many postcards and how many letters could he send?

2. **PATTERNS** If the pattern below continues, where will the heart be in the 20th figure?

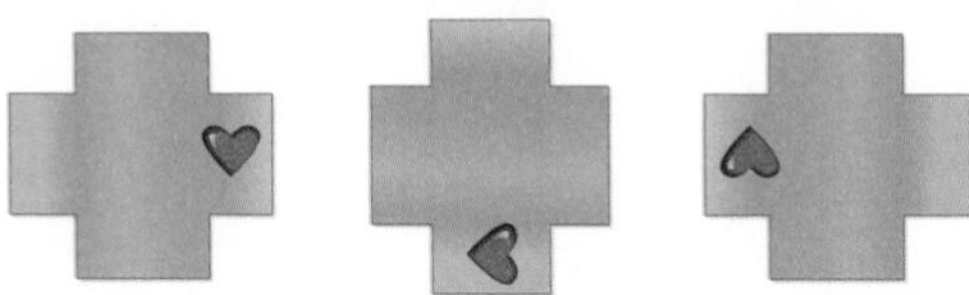

3. **WATER** A 500-gallon bathtub is being filled with water. Eighty gallons of water are in the bathtub after 4 minutes. How long will it take to fill the bathtub?

4. **GEOGRAPHY** The table shows the areas of the five Great Lakes.
   a. About how many times larger is Lake Superior than Lake Ontario?
   b. How many square miles do the Great Lakes cover altogether?

Great Lakes

Name	Area (sq mi)
Lake Superior	31,698
Lake Huron	23,011
Lake Michigan	22,316
Lake Erie	9922
Lake Ontario	7320

5. **WORK** Two workers can make two chairs in two days. How many chairs can 8 workers working at the same rate make in 20 days?

Find each sum or difference.

6. 9.5 + 8.6
7. 11.25 + 13.46
8. 48.64 + 52.91
9. 39.45 + 42.4
10. 178.96 + 201.841
11. 3054.2 + 514.67
12. 8.7 − 5.4
13. 142.67 − 98.59
14. 267.84 − 192.41
15. 58.6 − 31.65
16. 237.42 − 98.6
17. 982 − 457.28

18. **SALES** An MP3 player is on sale for $149.98. If the original price of the MP3 player was $225.49, how much money will you save?

Find each product or quotient.

19. 9.4×5
20. 7.5×4.3
21. 12.6×18.2
22. 24×6.9
23. 42.3×5.24
24. 9.865×4.3
25. $12.4 \div 5$
26. $36.98 \div 8.6$
27. $38.7828 \div 6.84$
28. $840.9744 \div 12.96$
29. $1877.9904 \div 6.24$
30. $782.8886 \div 42.41$

31. **BOWLING** Xavier's bowling average is 185.4 and Lola's average is 122.9. About how many times greater is Xavier's average than Lola's?

Write a verbal rule to describe each pattern or sequence. Then find the tenth term.

32.

Number of Tanks	Number of Fish
1	7
2	14
3	21

33.

Number of Cars Washed	Money Earned ($)
1	10
2	15
3	20

34. 8, 9, 10, 11, ...

35. 4, 9, 14, 19, ...

Find the perimeter of each figure.

36.

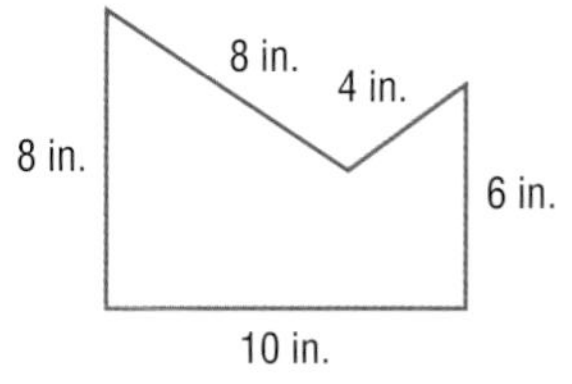

37.

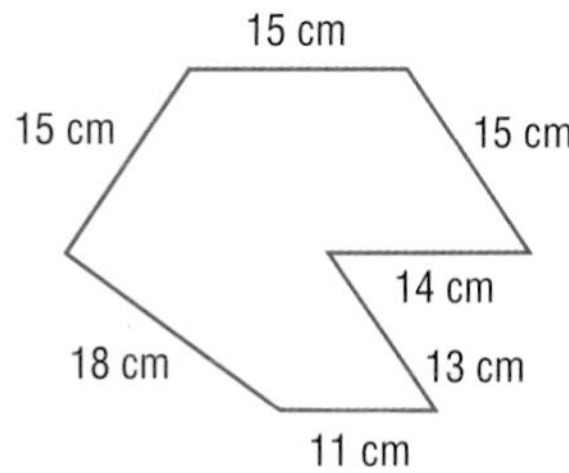

38.

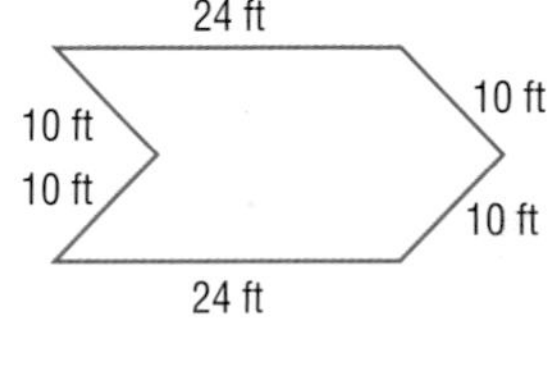

Write the customary unit that you would use to measure each length.

39. width of a video game system

40. length of a car

41. distance between cities

42. height of a door

Complete.

43. 6160 yd = ■ mi

44. ■ in. = 6 ft

45. ■ ft = 45 yd

46. 2 mi = ■ yd

47. ■ yd = 20 ft

48. 2784 mm = ■ cm

49. 6.4 km = ■ m

50. ■ m = 210 cm

51. ■ cm = 4.3 m

52. **CELL PHONES** The pictograph shows the number of students in each class who have a cell phone. How many times as many students have a cell phone in Mr. Watkins' class as in Mr. Hernandez' class?

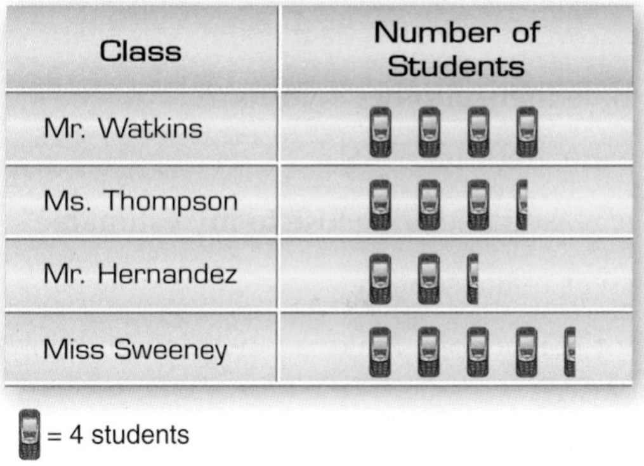

Class	Number of Students
Mr. Watkins	
Ms. Thompson	
Mr. Hernandez	
Miss Sweeney	

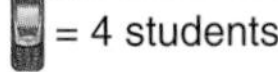 = 4 students

53. **TICKETS** The Drama Club is selling tickets for the school play. The bar graph shows the number of tickets sold each day for one week. How many tickets were sold Wednesday through Friday?

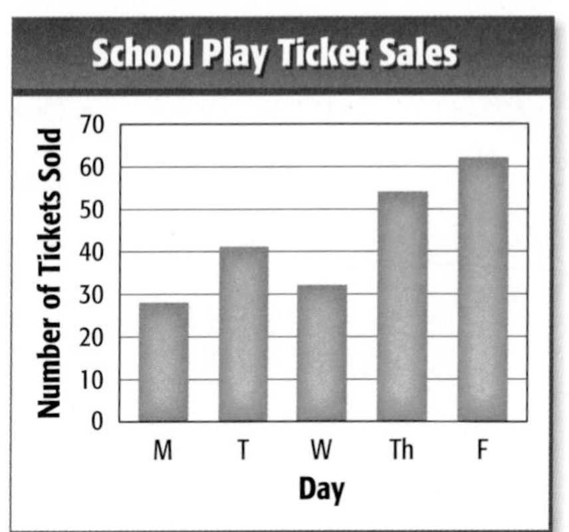

0-1 A Plan for Problem Solving

Now

- Use the four-step plan to solve problems.

Math Online

glencoe.com

- Extra Examples
- Personal Tutor
- Self-Check Quiz
- Homework Help

INSTANT MESSAGING The graphic shows the results of a class survey about which instant messaging abbreviations students use most. About how many times as great is the number of students who use TTYL as the number of students who use WTG?

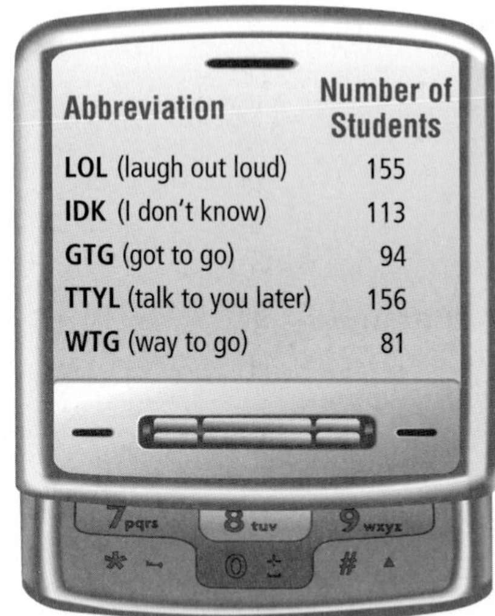

Abbreviation	Number of Students
LOL (laugh out loud)	155
IDK (I don't know)	113
GTG (got to go)	94
TTYL (talk to you later)	156
WTG (way to go)	81

It is often helpful to have an organized plan to solve math problems. The following four steps can be used to solve any math problem.

Understand

- Read the problem quickly to gain a general understanding of it.
- Ask, "What facts do I know?"
- Ask, "What do I need to find out?"
- Ask, "Is there enough information to solve the problem? Is there extra information?"

Plan

- Reread the problem to identify relevant facts.
- Determine how the facts relate to one another.
- Make a plan and choose a strategy for solving it. There may be several strategies that you can use.
- Estimate what you think the answer should be.

Solve

- Use your plan to solve the problem.
- If your plan does not work, revise it or make a new plan.

Check

- Reread the problem. Is there another solution?
- Examine your answer carefully.
- Ask, "Is my answer reasonable and close to my estimate?"
- Ask, "Does my answer make sense?"
- If your answer is not reasonable, make a new plan and solve the problem another way.
- You may also want to check your answer by solving the problem again in a different way.

Real-World Link

According to a recent survey, 75% of all online teens use instant messaging. Of these, 48% say that they send more than one instant message per day.

Source: Pew Internet and American Life Project

Real-World EXAMPLE 1 The Four-Step Plan

INSTANT MESSAGING Refer to the information at the top of the previous page.

Use the four-step plan to find *about* how many times as many students use TTYL than WTG.

Understand You know that 156 students use TTYL and 81 students use WTG. You need to find *about* how many times as great the number who use TTYL is as the number who use WTG.

Plan Divide the number of students that use TTYL by the number that use WTG. Since the question asks for *about* how many, you can estimate.

Solve $160 \div 80 = 2$

So, about twice as many students use TTYL as use WTG.

Check Multiply 81 by 2. Since $162 \approx 160$, the answer is reasonable. ✓

Personal Tutor glencoe.com

Exercises

Use the four-step plan to solve each problem.

1. **PIZZA** For today's lunch, the school cafeteria is offering make your own pizzas with the options shown at the right. How many different pizzas can be ordered with one cheese and one topping?

Pizza Options		
Crust	**Cheese**	**Toppings**
thin	mozzarella	pepperoni
deep dish	cheddar	sausage
		onions
		mushrooms

2. **MONEY** Antonia bought a video game system for $323.96. She paid in 12 equal installments. About how much did she pay each month?

3. **GEOMETRY** Megan has 1200 sugar cubes. What is the largest cube she could build with the sugar cubes?

4. **COFFEE** The sales of Café Mocha's coffee of the month are shown.

   a. How many more cups of this kind of coffee did Café Mocha sell in January than in April?

   b. If Café Mocha charges $1.98 for each cup of coffee, about how much money did Café Mocha take in on coffee of the month in March?

   c. In May, Café Mocha raised their price for each cup of coffee to $2.25. How much money did Café Mocha take in for coffee of the month in May?

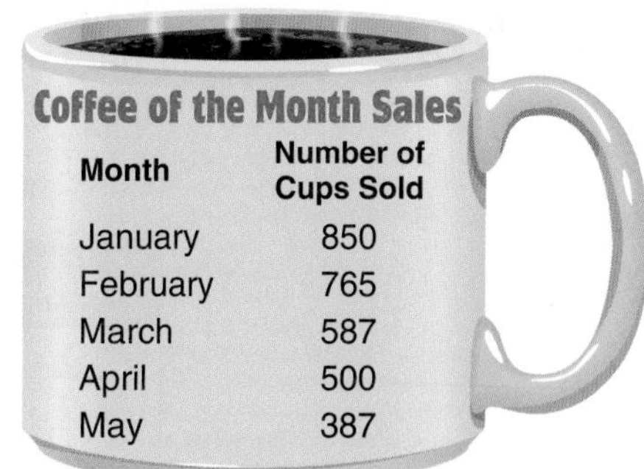

Coffee of the Month Sales

Month	Number of Cups Sold
January	850
February	765
March	587
April	500
May	387

0-2 Problem-Solving Strategies

Now

- Use problem-solving strategies to solve nonroutine problems.
- Select an appropriate strategy.

Vocabulary

look for a pattern
guess and check
make a table
work backward

Math Online

glencoe.com

- Extra Examples
- Personal Tutor
- Self-Check Quiz
- Homework Help

There are many problem-solving strategies in mathematics. One common strategy is to **look for a pattern**. To use this strategy, analyze the first few numbers in a pattern and identify a rule that is used to go from one number to the next. Then use the rule to extend the pattern and find a solution.

Real-World EXAMPLE 1 Look for a Pattern

E-MAIL Ramon got an E-mail from Angela. After 10 minutes, he forwarded it to 2 of his friends. After 10 more minutes, those 2 friends forwarded it to 2 more friends. If the E-mail was forwarded like this every 10 minutes, how many people received Angela's E-mail after 40 minutes?

Understand You know how long it takes to forward the E-mail. You need to find the total number of people who received the E-mail.

Plan Organize the data in a table. Look for a pattern in the data and extend the pattern.

Time (min)	People Receiving Message
0	1
10	2
20	4
30	■
40	■

Solve 1, 2, 4, ■, ■

×2 ×2 ×2 ×2

To continue the pattern, multiply each term by 2.

$4 \times 2 = 8$ $\quad$ $8 \times 2 = 16$

So, $1 + 2 + 4 + 8 + 16$ or 31 people got the E-mail.

Check In the 40th minute, 16 people received the E-mail. Half as many received it each time before that. So, it is reasonable that the total will be less than 16×2 or 32. ✓

Personal Tutor glencoe.com

To solve other problems, you can make a reasonable guess and then check it in the problem. You can then use the results to improve your guess until you find the solution. This strategy is called **guess and check**.

Real-World EXAMPLE 2 Guess and Check

LIFE SCIENCE Each hand in the human hand has 27 bones. There are 6 more bones in the fingers than in the wrist. There are 3 fewer bones in the palm than in the wrist. How many bones are in each part of the hand?

Make a guess to find the bones in each part of the hand.

Wrist	Palm wrist − 3	Fingers wrist + 6	Total Bones (27)	Correct?
5	$5 - 3 = 2$	$5 + 6 = 11$	$5 + 2 + 11 = 18$	This is too low.
7	$7 - 3 = 4$	$7 + 6 = 13$	$7 + 4 + 13 = 24$	This is too low.
9	$9 - 3 = 6$	$9 + 6 = 15$	$9 + 6 + 15 = 30$	This is too high.
8	$8 - 3 = 5$	$8 + 6 = 14$	$8 + 5 + 14 = 27$	This is correct. ✓

There are 8 bones in the wrist, 5 bones in the palm, and 14 bones in the fingers.

Personal Tutor glencoe.com

Another strategy for solving problems is to **make a table**. A table allows you to organize information in an understandable way.

Real-World EXAMPLE 3 Make a Table

SNACKS A vending machine accepts dollars, and each item in the machine costs 65 cents. If the machine gives only nickels, dimes, and quarters, what combinations of those coins are possible as change for one dollar?

The machine will give back $1.00 – $0.65 or 35 cents in change in a combinations of nickels, dimes, and quarters.

Make a table showing different combinations of nickels, dimes, and quarters that total 35 cents. Organize the table by starting with the combinations that include the most quarters.

quarters	dimes	nickels
1	1	0
1	0	2
0	3	1
0	2	3
0	1	5
0	0	7

The total for each combination of these coins is 35 cents. There are 6 combinations possible.

Personal Tutor glencoe.com

Real-World Link

More than 50% of the vending machines in the United States sell cold drinks and about 20% of the machines sell snacks.

Source: Vencoa

In most problems, a set of conditions or facts is given and an end result must be found. However, some problems start with the result and ask for something that happened earlier. The **work backward** strategy can be used to solve problems like this.

To use the work backward strategy, start with the end result and *undo* each step.

Real-World EXAMPLE 4 Work Backward

MONEY Kendrick spent half of the money he had this morning on lunch. After lunch, he loaned his friend a dollar. Now he has $1.50. How much money did Kendrick start with in the morning?

Start with the end result, $1.50, and work backward to find Kendrick's starting amount.

Kendrick now has $1.50. ⟶ $1.50

Undo the $1 he loaned to his friend. ⟶ +1.00

$2.50

Undo the half he spent for lunch. ⟶ × 2

$5.00

The amount Kendrick started with was $5.00.

Check Kendrick started with $5. If he spent half of that on lunch and loaned his friend $1.00, he would have $1.50 left. The solution is correct. ✓

Personal Tutor glencoe.com

Exercises

Solve each problem by looking for a pattern.

1. **MEASUREMENT** What is the perimeter of the twelfth figure?

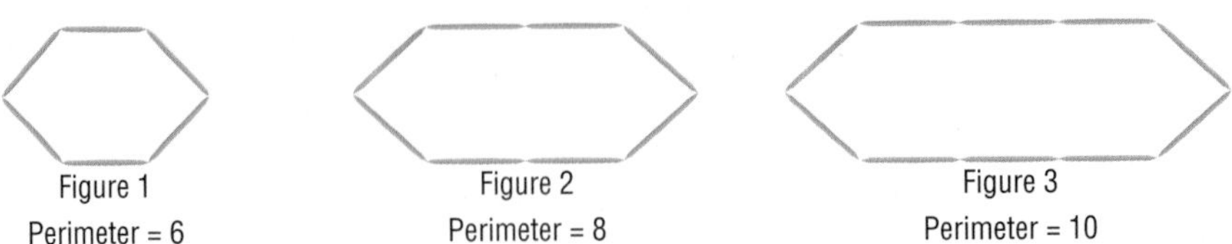

2. **SCIENCE** A ball bounces back 0.6 of its height on every bounce. If a ball is dropped from 200 feet, how high does it bounce on the fifth bounce? Round to the nearest tenth.

Use the guess and check strategy to solve each problem.

3. **NUMBER SENSE** The product of two consecutive odd integers is 783. What are the integers?

4. **MUSIC** Rafael is burning a CD for Selma. The CD will hold 35 minutes of music. Which songs should he select from the list to record the maximum time on the CD without going over?

Song	A	B	C	D	E
Time	5 min 4 s	9 min 10 s	4 min 12 s	3 min 9 s	3 min 44 s
Song	F	G	H	I	J
Time	4 min 30 s	5 min 0 s	7 min 21 s	4 min 33 s	5 min 58 s

Solve each problem by making a table.

5. **CARDS** Jorge had 55 football cards. He traded 8 cards for 5 from Elise. He traded 6 more for 4 from Leon and 5 for 3 from Bret. Finally, he traded 12 cards for 9 from Ginger. How many cards does Jorge have now?

Real-World Link

The first trading cards were made around 1887 and were made from a cloth-like material and measured 1.5 inches by 2 inches. Today, these are among the rarest and most valuable trading cards.

Source: Trading Card Central

6. **GAMES** The cubes at the right are each numbered 1 through 6. During a game, both are rolled and the faces landing up are added. How many ways can a person playing the game roll a sum less than 8?

Solve each problem by working backward.

7. **MONEY** Tia used half of her allowance to buy a ticket to the class play. Then she spent $1.75 for ice cream. Now she has $2.25 left. How much is her allowance?

8. **TIME** To catch a 7:30 A.M. bus, Don needs 30 minutes to shower and dress, 15 minutes for breakfast, and 10 minutes to walk to the bus stop. To catch the bus, what is his latest possible wake-up time?

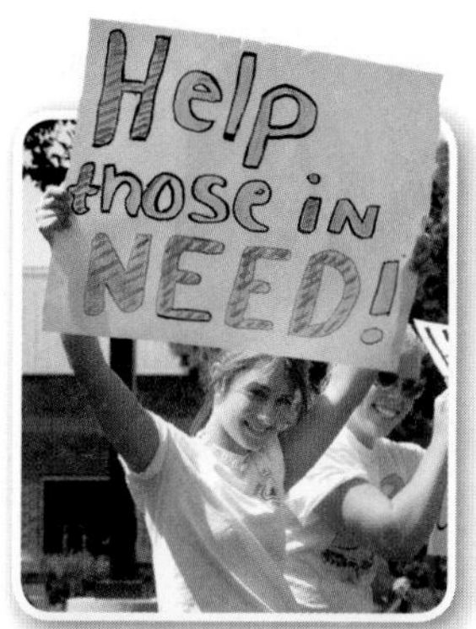

Real-World Link

The top five fundraising activities for schools are
1) candy bars
2) catalogs
3) cookie dough
4) discount cards
5) scratch-off tickets

Source: Fundraiser Help

Use any strategy to solve each problem.

9. **FUNDRAISING** The Science Club sold candy bars and soft pretzels to raise money for an animal shelter. They raised a total of \$62.75. They made 25¢ profit on each candy bar and 30¢ profit on each pretzel sold. How many of each did they sell?

10. **STOCKS** Mr. and Mrs. Delgado each own an equal number of shares of a stock. Mr. Delgado sells one-third of his shares for \$2700. What was the total value of Mr. and Mrs. Delgado's stock before the sale?

11. **MONEY** Odell has the same number of quarters, dimes, and nickels. In all he has \$4 in change. How many of each coin does he have?

12. **GAMES** Three counters are used for a board game. If the counters are tossed, how many ways can at least one counter with Side A turn up?

Counters	Side 1	Side 2
Counter 1	A	B
Counter 2	A	C
Counter 3	B	C

13. **NUMBER SENSE** A certain number is multiplied by 3, and then 5 is added to the result. The final answer is 41. What is the number?

14. **TRAVEL** Courtney travels south on her bicycle riding 8 miles per hour. One hour later, her friend Horacio starts riding his bicycle from the same location. If he travels south at 10 miles per hour, how long will it take him to catch Courtney?

StudyTip

Strategies Other problem solving strategies include
- draw a diagram
- make a model
- solve a simpler problem
- use logical reasoning

15. **GAMES** The spinner is used to play a certain game. On your turn, you must spin the spinner twice. How many different combinations of colors could you spin? List all possible combinations.

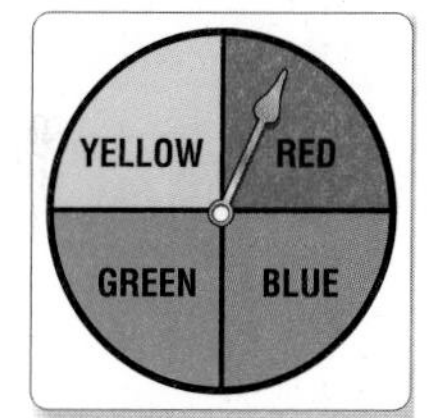

16. **FIRES** A forest fire spread to 41 acres in 10 hours. Each hour the fire spread to four more acres than the previous hour. How many acres were consumed during each hour of the fire?

17. **MONEY** Lawanda put \$15 of her paycheck into her savings account. Then she spent one-half of what was left on clothes. She paid \$24 for a concert ticket and later spent one-half of what was then left on a book. When she got home, she had \$14 left. What was the amount of Lawanda's paycheck?

18. **AGE** Brianne is three times as old as Camila. Four years from now she will be just two times as old as Camila. How old are Brianne and Camila now?

19. **MONEY** Anita sold tickets to the school musical. At the end of the day, she filled out a register like the one at the right to make her deposit. If she deposited a total of 12 bills, how many of each bill did she deposit?

Deposit	
Type of Bill	Number Deposited
\$1	0
\$5	■
\$10	■
\$20	■
Total	\$175

0-3 Number and Operations

A **decimal** is a number that has a digit in the tenths place, hundredths place, and beyond. To add or subtract decimals, line up the decimal points first.

Now
- Add and subtract decimals.
- Multiply and divide decimals.

Review Vocabulary
decimal
annex

Math Online
glencoe.com
- Extra Examples
- Personal Tutor
- Self-Check Quiz
- Homework Help

Real-World EXAMPLE 1 Add and Subtract Decimals

SPORTS The table shows the scores for each event for the two champions of the 2007 Southeastern Conference (SEC) Gymnastics Championships.

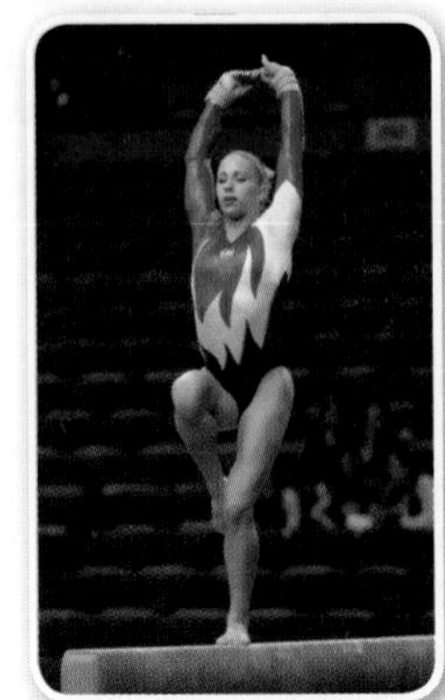

2007 SEC Gymnastics Championships		
Event	Courtney Kupets	Katie Heenan
Bars	9.925	9.85
Beam	9.825	9.9
Floor	9.875	9.9
Vault	9.975	9.95

a. Find the all-around score for each champion.

The all-around score is found by adding together the four event scores.

Courtney Kupets		Katie Heenan
9.925	**Line up the**	9.85
9.825	**decimal points.**	9.90
9.875	**Add as with**	9.90
+ 9.975	**whole numbers.**	+ 9.95
39.600		39.60

Annex a zero to align the columns.

So, both women had an all-around score of 39.6.

b. How much greater was Katie's beam score than Courtney's?

Find the difference between the scores.

$$\begin{array}{r} 9.900 \\ -\ 9.825 \\ \hline 0.075 \end{array}$$

Annex zeros.

Subtract as with whole numbers.

So, Katie's beam score was 0.075 more than Courtney's.

Personal Tutor glencoe.com

EXAMPLE 2 Multiply Decimals

Find 4.5 × 2.7.

Multiply as with whole numbers. To place the decimal point, find the sum of the number of decimal places in each factor. The product has the same number of decimal places.

$$\begin{array}{r} 4.5 \\ \times\ 2.7 \\ \hline 315 \\ +\ 90 \\ \hline 12.15 \end{array}$$

4.5 ← **one decimal place**
2.7 ← **one decimal place**
12.15 ← **two decimal places**

Personal Tutor glencoe.com

EXAMPLE 3 Divide Decimals

Divide 0.086 ÷ 0.04.

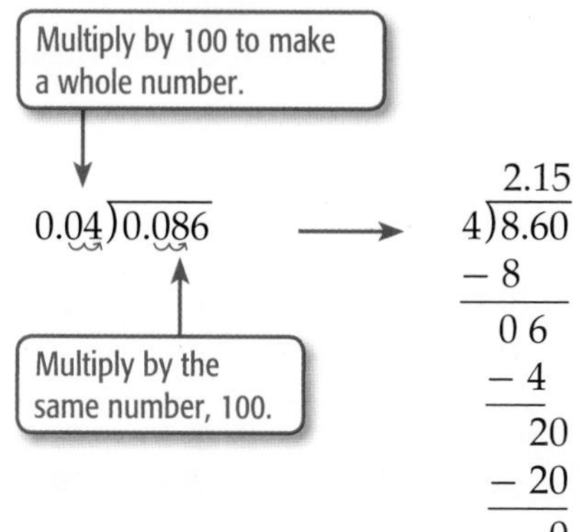

Multiply by 100 to make a whole number.
0.04)0.086
Multiply by the same number, 100.

$$\begin{array}{r} 2.15 \\ 4\overline{)8.60} \\ -8 \\ \hline 0\,6 \\ -4 \\ \hline 20 \\ -20 \\ \hline 0 \end{array}$$

Place the decimal point.
Divide as with whole numbers.

Annex a zero to continue.

The quotient is 2.15.

StudyTip

Powers of 10 To multiply by 100, move the decimal point two places to the right.

Personal Tutor glencoe.com

Exercises

Find each sum or difference.

1. 17.8 + 22.29 **2.** 32.34 + 15.19 **3.** 6.651 + 13.76

4. 45 − 31.52 **5.** 113.6 − 41.8 **6.** 178.4 − 147.3

7. 0.37 + 2.548 **8.** \$8.74 + \$3.15 **9.** \$92.71 + \$115.23

10. 36.17 − 20 **11.** \$17.00 − \$4.61 **12.** \$98.10 − \$20.10

Find each product or quotient.

13. 5.7 × 9.4 **14.** 7.34 × 7.6 **15.** 21.5 × 4.9

16. 13.02 ÷ 4.2 **17.** 5.082 ÷ 2.42 **18.** 0.1485 ÷ 1.35

19. 145.2 × 27.5 **20.** 1.723 × 2.49 **21.** 8.42 × 0.137

22. 25.334 ÷ 5.3 **23.** 100.1 ÷ 14.3 **24.** 1864.3 ÷ 3.62

25. WEATHER The average annual precipitation for Evansville, Indiana, is 44.27 inches. Fort Wayne has an average annual precipitation of 36.55 inches. How much greater is the precipitation average for Evansville than for Fort Wayne?

EN
H

Real-World Link

The 400-meter medley relay was introduced as an Olympic sport in 1960. The gold medal winning time that year was 245.4 seconds.

26. SWIMMING A record time for the 400-meter medley relay was set by the four swimmers shown. The times for each leg that they swam are shown.

Swimmer	Leg	Time (s)
Aaron Peirsol	backstroke	53.45
Brendan Hansen	breaststroke	59.37
Ian Crocker	butterfly	50.28
Jason Lezak	freestyle	47.58

a. What was the total time of the world record?

b. How much faster did Jason Lezak swim his leg than Aaron Peirsol?

c. Refer to the information at the left. How much faster is the world record time than the gold medal winning time in the 1960 Olympics?

0-4 Algebra

The world around you is made up of patterns and sequences. Tables, rules, formulas, and algebraic expressions can model these relationships.

Now

- Write a verbal rule to represent a pattern or sequence.
- Find any term value in a pattern or sequence.

Math Online

glencoe.com

- Extra Examples
- Personal Tutor
- Self-Check Quiz
- Homework Help

EXAMPLE 1 Write a Rule

Write a verbal rule to find the number of dots in any figure below. Then find the number of dots in the 15th figure.

Figure	● ●●	●● ●●●	●●● ●●●●	●●●● ●●●●●
Figure Number	1	2	3	4
Number of Dots	3	5	7	9

You can draw each figure or you can look for a pattern and write a rule.

Make a table to organize the information.

Figure Number	1	2	3	...
Number of Dots	3	5	7	...

$2 \cdot 3 + 1 = 7$

The number of dots is *one more than the twice the figure number.*

So, the 15th figure would have $2 \cdot 15 + 1$ or 31 dots.

Personal Tutor glencoe.com

EXAMPLE 2 Find Terms

Find the 25th term in the sequence 4, 8, 12, 16, ...

Position	1	2	3	4	...
Value of Term	4	8	12	16	...

The value of each term is 4 times its position in the sequence.

So, the 25th term would have a value of $4 \cdot 25$ or 100.

Personal Tutor glencoe.com

EXAMPLE 3 Find a Rule

Describe how to find the number of diagonals that can be drawn from any one vertex in the polygons shown. Then find the number of diagonals from one vertex in a 10-sided polygon.

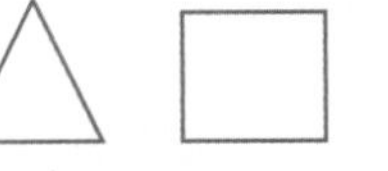

Draw the diagonals from a single vertex.

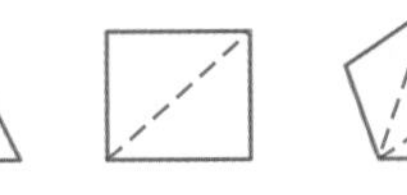

Number of Sides	3	4	5	...
Number of Diagonals	0	1	2	...

The number of diagonals is *three less than the number of sides.*

So, there are $10 - 3$ or 7 diagonals that can be drawn.

Personal Tutor glencoe.com

Exercises

Write a verbal rule to describe each pattern or sequence. Then find the 20th term in each.

1.

Time (hours)	Bracelets Made
1	25
2	50
3	75

2.

Number of Lawns	Dollars Paid
1	15
2	25
3	35

3. 4, 5, 6, 7, ...

4. 3, 8, 13, 18, ...

Real-World Link

The average ticket price for a movie in the United States is about $6.60 per ticket.

Source: National Association of Theatre Owners

5. MOVIES Tickets at the local movie theater cost $6.50 per ticket.

a. Copy and complete the table that shows the relationship between the number of tickets purchased and the total cost.

Number of Tickets	1	2	3	4	5
Total Cost ($)	6.50	■	■	■	■

b. Write a rule that describes the relationship between the number of tickets and the total cost.

c. How much will it cost to purchase 8 tickets?

6. CONSTRUCTION Micah is tiling his kitchen floor with blue and white tiles. He surrounds each pair of blue tiles with white tiles. Four copies of this design are shown below.

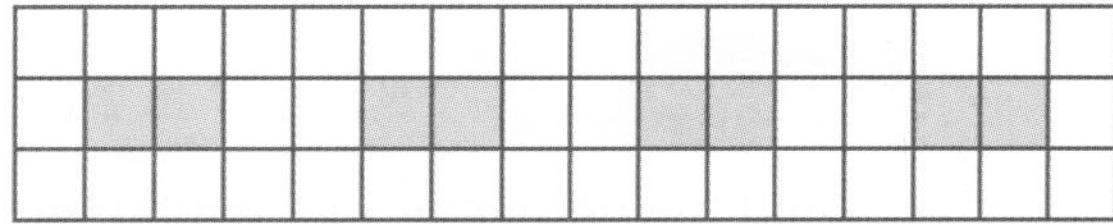

a. Copy and complete the table to show the number of white tiles needed for each given amount of blue tiles.

Blue Tiles	2	4	6	8	20	100
White Tiles	10	■	■	■	■	■

b. Write a rule that describes the relationship between the number of white tiles and the number of blue tiles.

c. How many blue tiles were used if 800 white tiles were used?

StudyTip

Patterns In these Exercises, you are using both the Make a Table and Look for a Pattern problem solving strategies.

7. PATTERNS Look at the sequence of cubes below.

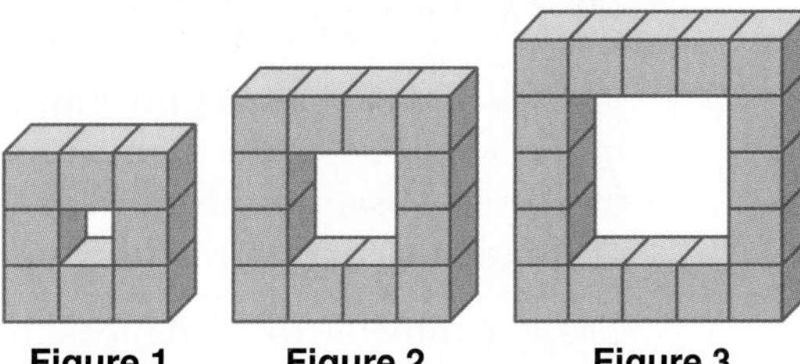

a. Make a table that shows the number of cubes in each figure.

b. Write a rule for the number of cubes in each figure.

c. How many cubes would be in the tenth figure of the pattern?

0-5 Geometry

Now
- Find the perimeter of a figure.
- Measure sides of figures to find the perimeter.

Math Online
glencoe.com
- Extra Examples
- Personal Tutor
- Self-Check Quiz
- Homework Help

Perimeter is the distance around any closed figure. While the figure itself is two-dimensional, the perimeter is only one dimension: length. You can find the perimeter by adding the measures of all of the sides of the figure.

Real-World EXAMPLE 1 Find Perimeter

BASEBALL The distance between each base on a Little League baseball diamond is 60 feet. Find the total distance you will run if you hit a home run.

To find the total distance run, find the perimeter of the baseball diamond.

60 feet + 60 feet + 60 feet + 60 feet = 240 feet

You will run a total of 240 feet.

Personal Tutor glencoe.com

A figure like a baseball diamond that has equal sides and equal angles is a regular figure. You can find the perimeter of a regular figure by multiplying the side length by the number of sides.

EXAMPLE 2 Perimeter of a Regular Polygon

Find the perimeter of a regular octagon with side lengths of 2 centimeters.

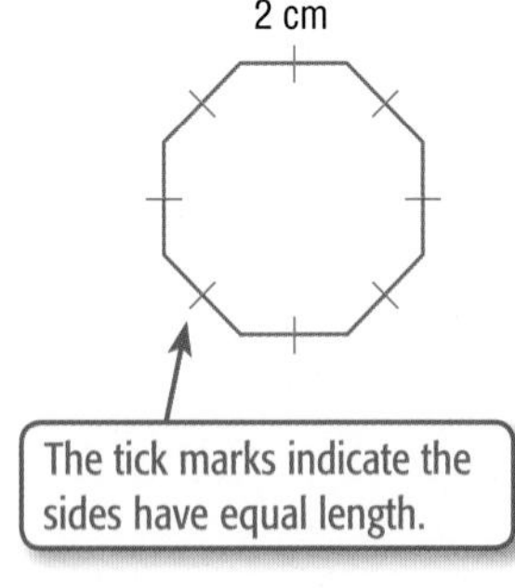
2 cm
The tick marks indicate the sides have equal length.

A regular octagon has 8 sides of equal length.

Multiply 8 by 2 centimeters to get the perimeter.

$8 \cdot 2 = 16$

So, the octagon has a perimeter of 16 centimeters.

Personal Tutor glencoe.com

Real-World EXAMPLE 3 Find the Perimeter

SCRAPBOOKING Laila is putting the letter shown at the right into a scrapbook. She wants to put a decorative trim around the letter. Measure the sides of the letter to the nearest millimeter to determine how much trim she will need.

Use a centimeter ruler to measure each side of the letter to the nearest millimeter. Then add the six lengths.

6 mm + 21 mm + 16 mm + 6 mm + 10 mm + 15 mm = 74 mm

Laila will need 74 millimeters of trim for the letter.

Personal Tutor glencoe.com

Exercises

Find the perimeter of each figure.

1.

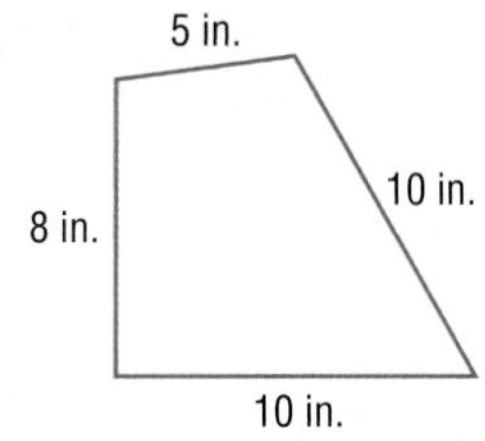

2.

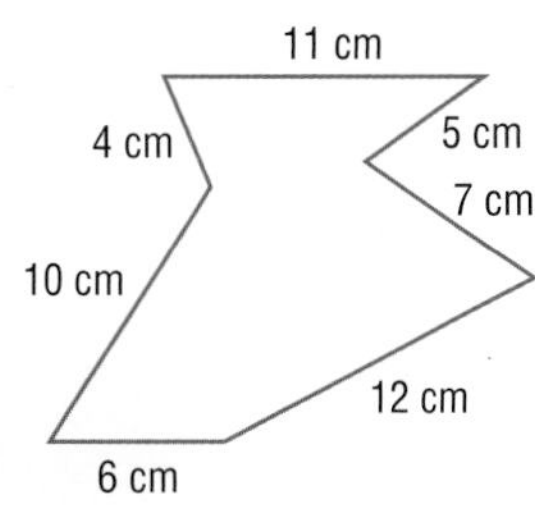

3.

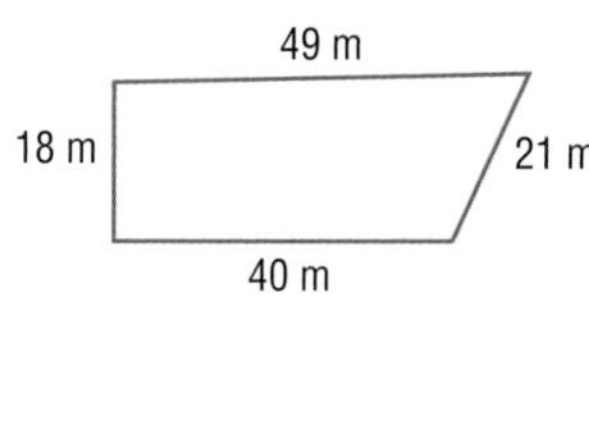

Find the perimeter of each regular polygon.

4.

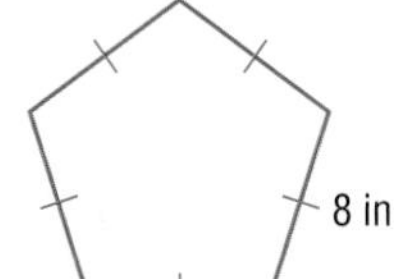

5.

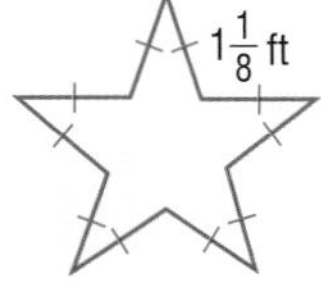

6.

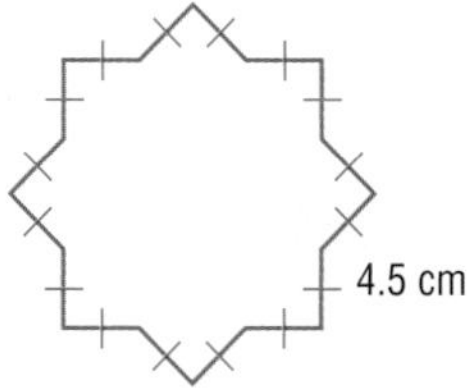

> ***Watch Out!***
>
> **Rulers** Be sure to use the correct ruler when measuring length. Use the metric side for Exercises 7–9 and the customary side for Exercises 10–12.

Measure the sides of each figure to the nearest millimeter. Then find the perimeter.

7.

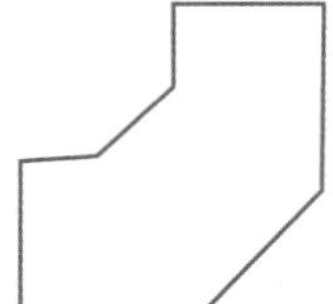

8.

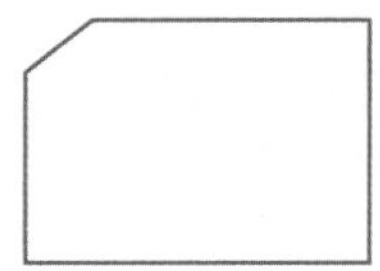

9.

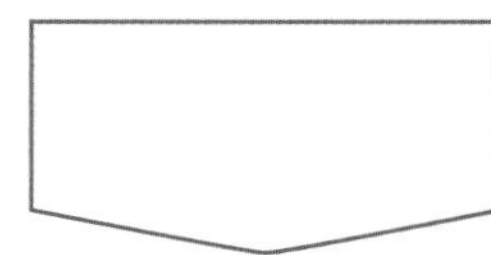

Measure the sides of each figure to the nearest $\frac{1}{16}$ inch. Then find the perimeter.

10.

11.

12.

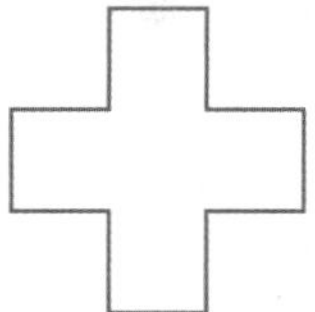

13. GEOMETRY Measure one side of the square to the nearest inch.

a. Draw a square that has sides that are twice the length of the square.

b. Draw two more squares, one with side lengths that are three times as great as the original and one with side lengths that are four times great as the original. Record the side length and perimeter of each square in a table like the one shown below.

Side Length	Perimeter
■	■
■	■
■	■
■	■

c. What happens to the perimeter of a square if the side length is doubled? tripled? quadrupled?

d. Predict the perimeter if the original side length is multiplied by 8.

0-6 Measurement

Now
- Convert within measurement systems.
- Estimate and measure objects.

Vocabulary
customary system
metric system

Math Online
glencoe.com
- Extra Examples
- Personal Tutor
- Self-Check Quiz
- Homework Help

A regulation baseball bat is about 3 feet long. But the world's largest bat is 40 yards long. Yards and feet are measurements of length. There are two common systems of measurement, the customary system and the metric system. The most common units for measuring length are shown.

Customary System	
Unit	**Model**
1 inch (in.)	width of a quarter
1 foot (ft) = 12 in.	height of a textbook
1 yard (yd) = 3 ft	length from nose to fingertip
1 mile (mi) = 1760 yd	10 city blocks
Metric System	
Unit	**Model**
millimeter (mm)	thickness of a coin
centimeter (cm) = 10 mm	half the width of a penny
meter (m) = 100 cm	width of a doorway
kilometer (km) = 1000 m	six city blocks

EXAMPLE 1 Estimate and Measure Length

Estimate the customary length of the key. Then measure to find the actual length.

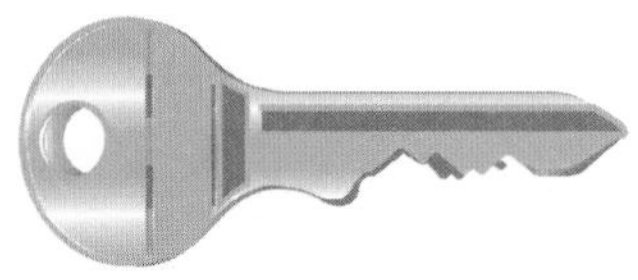

The length of the key is about the width of two quarters. An appropriate estimate is about 2 inches long.

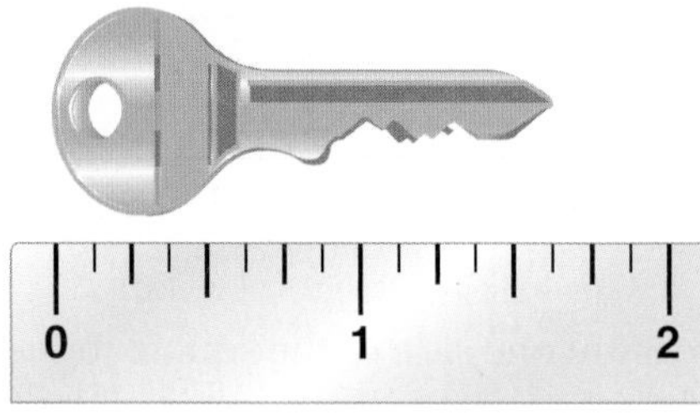

The key is $1\frac{5}{8}$ inches long.

Personal Tutor glencoe.com

StudyTip

Conversions When converting a larger unit to a smaller unit, multiply. When converting a smaller unit to a larger unit, divide.

EXAMPLE 2 Estimate and Measure Length

Estimate the metric length of the caterpillar. Then measure to find the actual length.

The caterpillar is about $2\frac{1}{2}$ pennies wide. An appropriate estimate is about 5 centimeters long.

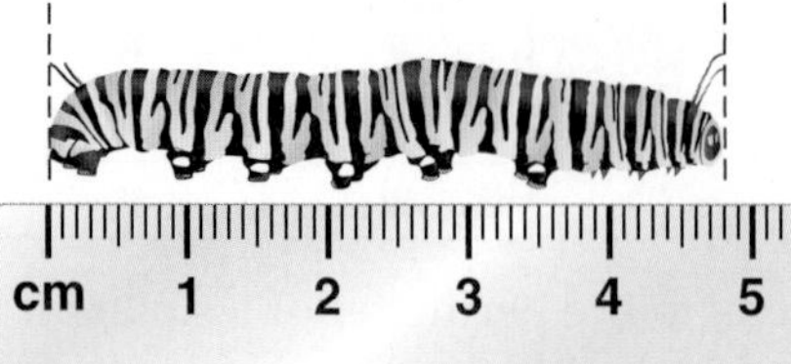

The caterpillar is 48 millimeters or 4.8 centimeters long.

Personal Tutor glencoe.com

EXAMPLE 3 Convert Measurements

Complete each conversion.

a. ■ **yd = 129 ft**

Since 1 yard = 3 feet, divide 129 by 3.

$129 \div 3 = 43$

So, 43 yard = 129 feet.

b. ■ **in. = 4.5 ft**

Since 12 inches = 1 foot, multiply 12 by 4.5.

$12 \times 4.5 = 54$

So, 54 inches = 4.5 feet

c. ■ **mm = 13 cm**

Since 10 millimeter = 1 centimeter, multiply 13 by 10.

$13 \times 10 = 130$

So, 130 millimeter = 13 centimeter.

d. ■ **m = 175 cm**

Since 1 meter = 100 centimeter, divide 175 by 100.

$175 \div 100 = 1.75$

So, 1.75 meter = 175 centimeter.

Personal Tutor glencoe.com

Exercises

Write the customary unit of length that you would use to measure each of the following.

1. height of a box of cereal
2. length of a truck
3. distance across Tennessee
4. height of a classroom

Watch Out!

Rulers Be sure to use the correct ruler when measuring length. Use the customary side for Exercise 9 and the metric side for Exercise 10.

Write the metric unit of length that you would use to measure each of the following.

5. length of a toothbrush
6. thickness of a dime
7. length of the Mississippi River
8. depth of a pond
9. Estimate the customary length of the paperclip. Then measure to find actual length.
10. Estimate the metric length of the fish. Then measure to find the actual length.

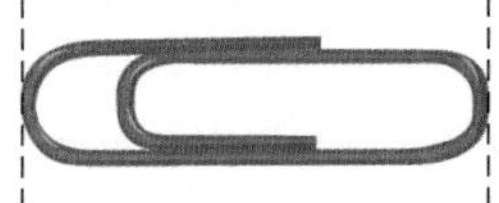

Complete.

11. ■ mi = 3080 yd
12. 12 ft = ■ in.
13. ■ in. = 5.25 ft
14. 4400 yd = ■ mi
15. ■ ft = 51 yd
16. 18 in. = ■ ft
17. ■ cm = 7.5 m
18. 5.7 km = ■ m
19. ■ mm = 54 cm
20. 4.75 m = ■ cm
21. ■ m = 740 cm
22. 1675 mm = ■ cm

0-7 Data Analysis

Now

- Read and interpret pictographs, line graphs, and bar graphs.
- Make pictographs, line graphs, and bar graphs.

Review Vocabulary

data
pictograph
line graph
bar graph

Math Online

glencoe.com

- Extra Examples
- Personal Tutor
- Self-Check Quiz
- Homework Help
- Math in Motion

Data are pieces of information that are often numerical and can be displayed using tables and graphs.

A **pictograph** compares data by using picture symbols.

Real-World EXAMPLE 1 Pictographs

DRIVER PERMITS The pictograph shows the number of students in each homeroom that have their driving permit. How many times as great is the number of students who have their driver's permits in homeroom 16 as in homeroom 14?

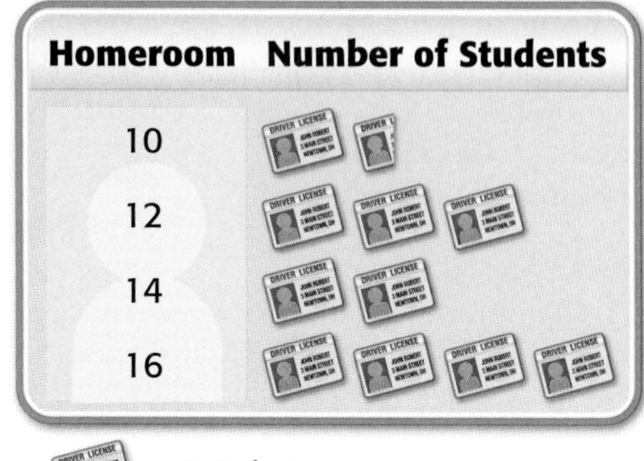

In homeroom 16, 8 students have their permits. In homeroom 14, 4 students have their permits.

To find how many times as many students in homeroom 16 have their permits as homeroom 14, divide.

$8 \div 4 = 2$

So, two times as many students in homeroom 16 have their permits as in homeroom 14.

Personal Tutor glencoe.com

Line graphs show how a data set changes over a period of time. By examining the direction of the connected lines, you can describe the trend in the data.

Real-World EXAMPLE 2 Line Graphs

ANIMALS The line graph shows the growth of a baby Asian elephant at the National Zoo in Washington, D.C. Find about how much weight the elephant gained from 10 months to 12 months.

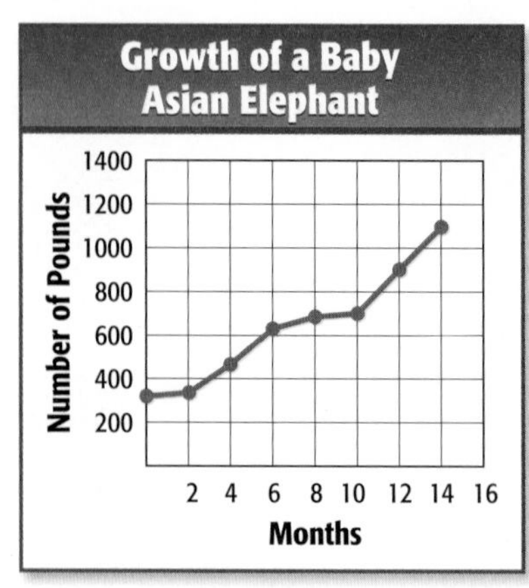

Look at the graph. At ten months, the elephant weighed about 700 pounds. At twelve months, the elephant weighed about 900 pounds.

$900 - 700 = 200$

So, the elephant gained about 200 pounds between the ages of 10 months and 12 months.

Personal Tutor glencoe.com

Real-World Link

There are almost 50,000 shopping malls nationwide. Connecticut has the greatest number of malls per person. For every 1000 people in Connecticut, there are 0.238 malls.

Source: National Research Bureau

Math *in Motion,* Interactive Lab glencoe.com

Bar graphs are used to compare categories of data. The height of each bar represents the frequency of each category of data.

Real-World EXAMPLE 3 Bar Graphs

SHOPPING The bar graph shows the approximate number of shopping malls in the states with the most shopping malls. About how many more malls are in California than in Illinois?

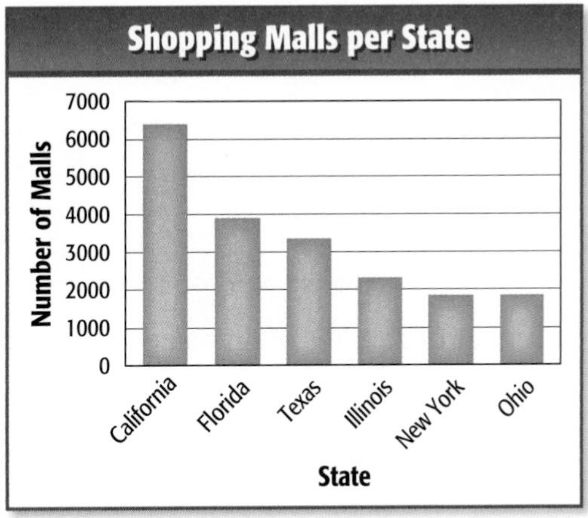

Source: National Research Bureau

There are about 6500 malls in California and about 2250 malls in Illinois.

$6500 - 2250 = 4250$

So, there are about 4250 more malls in California than in Illinois.

Personal Tutor glencoe.com

Exercises

For Exercises 1–3, refer to the pictograph in Example 1.

1. How many more students would have to have their driver's permit in homeroom 10 to exceed the number of students who have their permit in homeroom 12?
2. What is the total number of students who have their driver's permit shown in the graph?
3. Write a question that could be answered using the pictograph. Have a classmate answer your question.

For Exercises 4–6, refer to the line graph in Example 2.

4. Describe the change in the elephant's weight from 2 months to 14 months.
5. Predict the elephant's weight by the 16th month.
6. For which two-month period did the elephant have the greatest weight change?

For Exercises 7–9, refer to the bar graph in Example 3.

7. What can be determined about the number of malls in New York and Ohio?
8. About how many times as many malls are in California as in Ohio?
9. Use the Internet or another source to find the number of malls in your state. Then write a question that compares the data in the bar graph to the data you found.

CHAPTER 0 Posttest

Use any problem-solving strategy to solve each problem.

1. **GO-KARTS** The length of Fun Center's go-kart track is 843 feet. If Nadia circled the track 9 times, how many feet did she travel?

2. **PATTERNS** Draw the fifth figure in the pattern below.

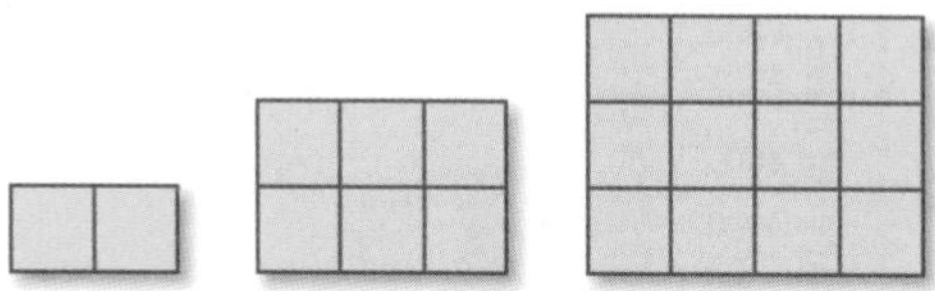

3. **GAMES** Find the number of squares of any size in the game board shown at the right.

4. **SUPPLIES** An art supply store sells 5 different sized canvases. The surface area of the middle-size canvas is 3.5 times larger than the surface area of the extra-small canvas. If the surface area of the extra-small canvas is 81 square inches, what is the surface area of the middle-sized canvas?

5. **MONEY** How many ways can you make change for a dollar using only quarters, dimes, and nickels?

6. **NUMBER SENSE** The product of two consecutive even integers is 1088. What are the integers?

Find each sum or difference.

7. 6.5 + 2.3
8. 25.72 + 18.67
9. 59.04 + 77.84
10. 25.54 + 36.7
11. 257.76 + 345.487
12. 4865.7 + 705.76
13. 5.3 – 4.7
14. 458.07 – 67.75
15. 851.79 – 570.85
16. 73.2 – 42.86
17. 569.43 – 97.1
18. 852 – 469.72

19. **SALES** A digital camera is on sale for $139.99. If the original price of the camera was $175.59, how much money will you save?

20. **TEMPERATURE** A city's average high temperature during the month of May is 64.5°F. Its average low for the same month is 35.9°F. Find the difference between the average high and average low temperature.

Find each product or quotient.

21. 4.7×6
22. 6.2×3.9
23. 43.8×76.3
24. 49×3.7
25. 82.7×6.72
26. 6.535×3.7
27. $28.8 \div 4$
28. $92.4 \div 5.5$
29. $156.3408 \div 9.87$
30. $382.1565 \div 36.57$
31. $2542 \div 8.64$
32. $6482.84 \div 46.22$

33. **MOVIES** The all-time top grossing movie earned approximately 1.835 billion dollars. The tenth all-time grossing movie earned about 0.915 billion dollars. About how many times as much did the top movie make as the tenth?

Write a verbal rule to describe each pattern or sequence. Then find the 10th term in each.

34.

Number of Teams	Number of Players
1	7
2	14
3	21

35.

Number of People	Bracelets Made
1	15
2	25
3	35

36. 6, 7, 8, 9, ...

37. 6, 11, 16, 21, ...

Find the perimeter of each figure.

38.

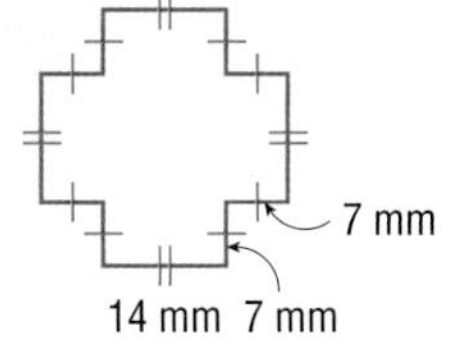

39.

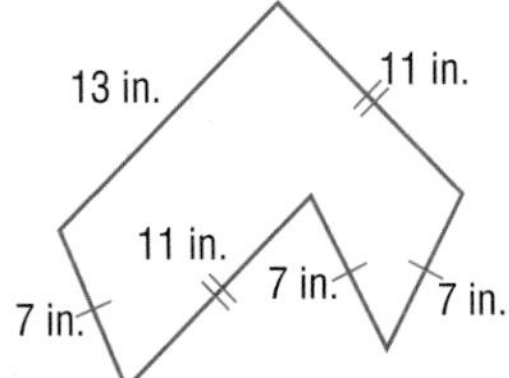

40.

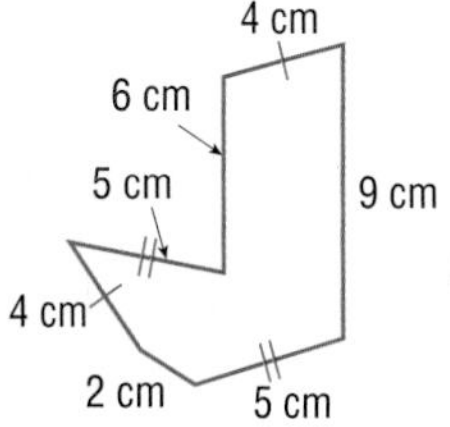

Write the metric unit of length that you would use to measure each of the following.

41. width of a DVD player

42. length of a school bus

43. distance between state capitals

44. height of a room

Complete.

45. 7040 yd = ■ mi

46. ■ in. = 12 ft

47. ■ ft = 72 yd

48. 10 ft = ■ in.

49. 3.5 mi = ■ yd

50. ■ yd = 30 ft

51. 1656 mm = ■ cm

52. 4.8 km = ■ m

53. ■ m = 620 cm

54. 5.72 m = ■ cm

55. ■ mm = 43 cm

56. ■ cm = 3.9 m

57. VIDEO GAMES The pictograph shows the number of students in each grade that have a video game system. About how many times as great is the number of students who have a video game system in eighth grade as in seventh grade?

Grade	Number of Students
Sixth	
Seventh	
Eighth	

= 50 students

58. CELL PHONES The local cell phone store is having a sale on cell phones. The graph shows the number of cell phones sold each day for one week. How does the number of phones sold Sunday through Thursday compare to the number of phones sold Friday and Saturday?

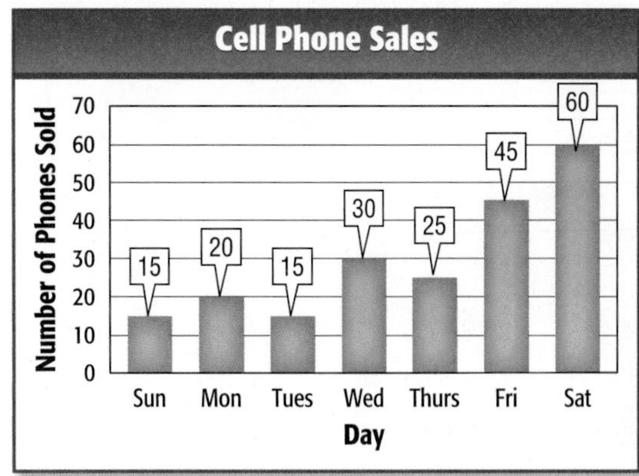

CHAPTER 1 The Tools of Algebra

Then

In previous courses, you have performed mathematical operations and evaluated numerical expressions.

Now

In Chapter 1, you will:

- Translate verbal phrases into numerical expressions.
- Evaluate expressions and use the order of operations.
- Identify and use properties of addition and multiplication.
- Use words, tables, equations, and graphs to represent relations and functions.

Why?

TECHNOLOGY

Geocaching is a new adventure game where participants bury a container, or *cache*, somewhere in the world. They then list the coordinates of their cache on a website. Other participants use a Global Positioning System, or GPS, to find hidden caches. There are nearly 500,000 active caches hidden worldwide.

The Tools of Algebra

Activity

If you travel from the Samoa Peninsula east on SR 255 and then south on Hwy 101, you will be in the Humboldt Redwoods State Forest, home of the Avenue of the Giants. You should be looking for a famous tree with another plaque. Soon enough, you see the plaque of the Immortal Tree.

IMMORTAL TREE

Eureka

Bayview

4/5

Math *in Motion*, Animation glencoe.com

Get Ready for Chapter 1

Diagnose Readiness You have two options for checking Prerequisite Skills.

Text Option

Take the Quick Check below. Refer to the Quick Review for help.

*Quick*Check

Find each sum or difference. (Previous Course)

1. 5.8 + 4.2 **2.** 7.1 + 6.5 **3.** 1.3 + 2.6

4. 9.4 – 4.2 **5.** 8.6 – 1.1 **6.** 5.7 – 3.5

7. LUNCH Calvin has $7.80. He spends $3.30 on lunch. How much money does Calvin have left?

8. MONEY Phil has $9.50. His sister gives him the $3.75 she owes him. How much money does Phil have now?

Estimate each sum, difference, product, or quotient. (Previous Course)

9. 1600 + 192 **10.** 524 – 349

11. 119 × 63 **12.** 1210 ÷ 398

13. ANIMALS The Siberian Tiger is the largest cat in the world. It can run as fast as 50 miles per hour. At that rate, about how far can it travel in 4 hours?

14. BABYSITTING Ginny earns $5 per hour babysitting. Estimate how much she will earn in 4.75 hours.

Write the number that represents each given point on the number line. (Previous Course)

A B C D
0 1 2 3 4 5 6 7 8 9 10

15. *A* **16.** *B* **17.** *C* **18.** *D*

19. TEMPERATURE The low temperature yesterday was 32°F. Graph this temperature on a number line.

*Quick*Review

EXAMPLE 1

Find 11.9 – 2.15.

$$\begin{array}{r} \overset{8\,10}{11.\not{9}\not{0}} \\ -\ 2.15 \\ \hline 9.75 \end{array}$$

Annex a zero to align the decimal points.
Subtract.

11.9 – 2.15 = 9.75

EXAMPLE 2

Estimate 117 + 51.

$$\begin{array}{r} 117 \\ +\ 51 \\ \hline \end{array} \longrightarrow \begin{array}{r} 120 \\ +\ 50 \\ \hline 170 \end{array}$$

Round to the nearest ten.
Add.

117 + 51 ≈ 170

EXAMPLE 3

Write the number that represents point *A* on the number line.

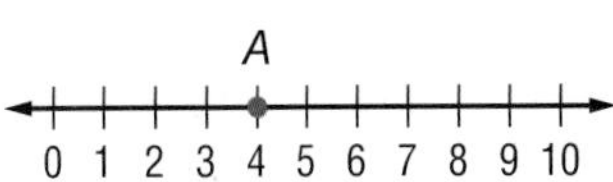

The number line starts at 0 and increases by 1. So point *A* is at 4.

Online Option

Math Online Take a self-check Chapter Readiness Quiz at glencoe.com.

Get Started on Chapter 1

You will learn several new concepts, skills, and vocabulary terms as you study Chapter 1. To get ready, identify important terms and organize your resources. You may wish to refer to **Chapter 0** to review prerequisite skills.

FOLDABLES® Study Organizer

The Tools of Algebra Make this Foldable to help you organize your Chapter 1 notes about algebra. Begin with seven sheets of notebook paper.

1. **Staple** the seven sheets together to form a booklet.

2. **Cut** a tab on the second page the width of the white space. On the third page, make the tab 2 lines longer, and so on.

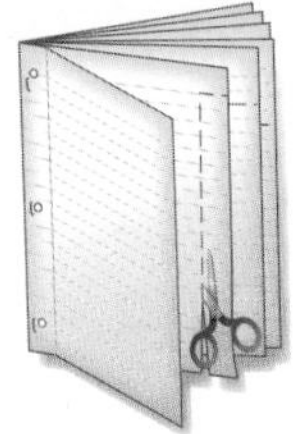

3. **Write** the chapter title on the cover and label each tab with the lesson number.

New Vocabulary

English		Español
numerical expression	• p. 5 •	expresión numérica
order of operations	• p. 6 •	orden de las operaciones
algebra	• p. 11 •	álgebra
variable	• p. 11 •	variable
algebraic expression	• p. 11 •	expresión algebraica
counterexample	• p. 19 •	contraejemplo
simplify	• p. 20 •	reducir
y-axis	• p. 25 •	efe *y*
coordinate plane	• p. 25 •	plano de coordenadas
origin	• p. 25 •	origen
x-axis	• p. 25 •	eje *x*
ordered pair	• p. 25 •	par ordenado
function	• p. 33 •	función
function rule	• p. 33 •	regla de funciones
function table	• p. 33 •	tabla de funciones
equation	• p. 34 •	ecuación

Review Vocabulary

difference • (Previous Course) • diferencia the answer to a subtraction problem

operations • (Previous Course) • operación common words that indicate different operations

Operations			
+	−	×	÷
plus	minus	times	divide
the sum of	difference between	the product of	quotient
increased by	decreased by	of	divided by
more than	less than		among

quotient • (Previous Course) • cociente the answer to a division problem

Multilingual eGlossary glencoe.com

Math Online glencoe.com

- Study the chapter online
- Explore **Math in Motion**
- Get extra help from your own **Personal Tutor**
- Use **Extra Examples** for additional help
- Take a **Self-Check Quiz**
- **Review Vocabulary** in fun ways

1-1

Words and Expressions

Then
You have already performed mathematical operations. (Previous Course)

Now
- Translate verbal phrases into numerical expressions.
- Use the order of operations to evaluate expressions.

New Vocabulary
numerical expression
evaluate
order of operations

Math Online
glencoe.com
- Extra Examples
- Personal Tutor
- Self-Check Quiz
- Homework Help

Why?

Lisa and Patty want their friends to make ice cream at their birthday party. Each friend will need the following ingredients for one small batch of ice cream.

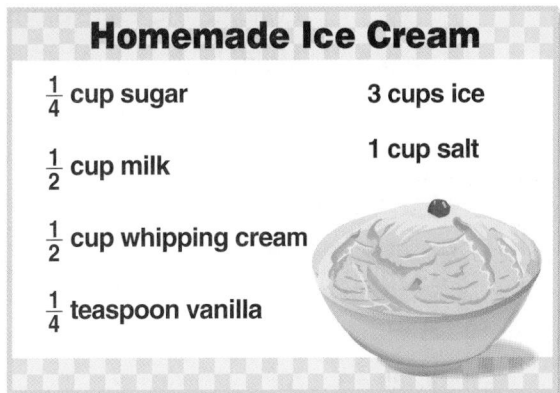

Homemade Ice Cream

$\frac{1}{4}$ cup sugar
$\frac{1}{2}$ cup milk
$\frac{1}{2}$ cup whipping cream
$\frac{1}{4}$ teaspoon vanilla
3 cups ice
1 cup salt

a. Determine how much ice is needed to make one batch of ice cream.

b. How many cups of ice will be needed for 9 people at the party?

c. If there are 12 people at the party, how many cups of ice are needed?

d. Describe how you will determine the number of cups of ice needed for any number of friends.

Translate Verbal Phrases into Expressions To find the number of cups of ice for 9 friends, Lisa and Patty can use the numerical expression 9×3. **Numerical expressions** contain a combination of numbers and operations such as addition, subtraction, multiplication, and division.

EXAMPLE 1 Translate Phrases into Expressions

Write a numerical expression for each verbal phrase.

a. the total amount of money if you have nine dollars and twelve dollars

Phrase the sum of nine and twelve

Expression $9 + 12$

b. the age difference between fifteen years old and ten years old

Phrase the difference of fifteen and 10

Expression $15 - 10$

Check Your Progress

1A. the cost of ten yo-yos if each costs three dollars

1B. the number of students in each group if fifteen students are divided into five equal groups

Personal Tutor glencoe.com

Order of Operations To **evaluate** an expression, you find its numerical value. If a numerical expression has more than one operation, use the order of operations. The **order of operations** are the rules to follow when evaluating an expression with more than one operation.

StudyTip

Grouping Symbols Grouping symbols include:

- parentheses ( ),
- brackets [ ], and
- fraction bars, as in $\frac{6+4}{2}$, which means $(6+4) \div 2$.

Key Concept **Order of Operations** *For Your FOLDABLE*

Step 1 Evaluate the expressions inside grouping symbols.

Step 2 Multiply and/or divide in order from left to right.

Step 3 Add and/or subtract in order from left to right.

Watch Out!

In Example 2b, you divide before you multiply because division comes first in order from left to right.

EXAMPLE 2 Evaluate Expressions

Evaluate each expression.

a. $20 - 3 \times 5$

$20 - 3 \times 5 = 20 - 15$ — Multiply 3 and 5 first.

$= 5$ — Subtract 15 from 20.

b. $30 \div 5 \times 3$

$30 \div 5 \times 3 = 6 \times 3$ — Divide 30 by 5.

$= 18$ — Multiply 6 and 3.

c. $4(10 - 7) + 2 \cdot 3$

$4(10 - 7) + 2 \cdot 3 = 4(3) + 2 \cdot 3$ — Evaluate (10 – 7) first.

$= 12 + 2 \cdot 3$ — 4(3) means 4×3 or 12.

$= 12 + 6$ — $2 \cdot 3$ means 2×3 or 6.

$= 18$ — Add 12 and 6.

d. $5[11 - (7 + 5) \div 4]$

$5[11 - (7 + 5) \div 4] = 5[11 - 12 \div 4]$ — Evaluate (7 + 5).

$= 5(11 - 3)$ — Divide 12 by 4.

$= 5(8)$ — Subtract 3 from 11.

$= 40$ — Multiply 5 and 8.

e. $\frac{60 - 15}{2 + 7}$

$\frac{60 - 15}{2 + 7} = (60 - 15) \div (2 + 7)$ — Rewrite as a division expression.

$= 45 \div 9$ — Evaluate (60 – 15) and (2 + 7).

$= 5$ — Divide 45 by 9.

Check Your Progress

2A. $6 - 3 + 5$

2B. $24 \div 3 \times 9$

2C. $2[(10 - 3) + 6(5)]$

2D. $\frac{19 - 7}{25 - 22}$

Personal Tutor glencoe.com

Real-World Link

The first call on a portable cellular phone was made in 1973. The phone weighed 28 ounces and was 13 inches by 1.75 inches by 3.5 inches.

Real-World EXAMPLE 3 Write Expressions to Solve Problems

CELL PHONES A cell phone company charges $20 per month and $0.15 for each call made or received. Write and evaluate an expression to find the cost for 40 calls. Then make a table showing the cost for 40, 50, 60, and 70 calls.

Understand You know how much the company charges per month. You need to find how much it will cost for 40, 50, 60, and 70 calls.

Plan Write an expression to find each cost. Organize the results in a table.

Solve First, write an expression for 40 calls.

Words	$20 per month	and	$0.15 for each call made or received
Expression	20	+	0.15 • 40

$20 + 0.15 \cdot 40 = 20 + 6$ **Multiply.**

$= 26$ **Add.**

So, 40 calls will cost $26. Make a table showing the costs for 40, 50, 60, and 70 calls.

Number of Calls	Expression	Cost ($)
40	20 + 0.15 • 40	26.00
50	20 + 0.15 • 50	27.50
60	20 + 0.15 • 60	29.00
70	20 + 0.15 • 70	30.50

Check Look for a pattern in the costs. As the number of calls increases by 10, the cost increases by $1.50. ✓

Check Your Progress

3. **CELL PHONES** The same phone company offers another plan where they charge $15 per month and $0.25 for each call made or received. Write and then evaluate an expression to find the cost for 40 calls during one month. Then make a table showing the cost for 40, 50, 60, and 70 calls.

Personal Tutor glencoe.com

Check Your Understanding

Example 1 p. 5

Write a numerical expression for each verbal phrase.

1. the cost of six electronic hand-held games if each costs eight dollars
2. the cost of one box of cereal if four boxes cost twelve dollars

Example 2 p. 6

Evaluate each expression.

3. $18 + 2 \times 4$
4. $2 \times 9 \div 3$
5. $4(6) + 9$
6. $6(17 - 8)$
7. $4[6(2) - 3]$
8. $3[(20 - 7) + 1]$
9. $\frac{15 - 5}{6 - 4}$
10. $\frac{34 + 18}{27 - 14}$

Example 3 p. 7

11. **TOURS** A tour bus costs $75 plus $6 for each passenger. Write and evaluate an expression to find the total cost for 25 passengers. Then make a table showing the cost for 25, 30, 35, and 40 passengers.

Practice and Problem Solving

● = **Step-by-Step Solutions** begin on page R11.
Extra Practice begins on page 810.

Example 1 p. 5

Write a numerical expression for each verbal phrase.

12. the height difference between fourteen and nine inches

13. the total number of fish if you had eight and bought four more

14. the number of weeks until vacation if vacation is twenty-eight days away

15. the total length of songs on a CD if each of nine songs is three minutes long

16. the total money earned if Lee sold four scarves at twenty dollars each

17. the number of people left to play if ten of fifteen have played

Example 2 p. 6

Evaluate each expression.

18. $3 \cdot 6 - 4$

19. $18 - 4 \times 2$

20. $16 \div 4 + 15$

21 $2[3 + 7(4)]$

22. $6 + 4(3)$

23. $12(11) - 56$

24. $8[(12 - 5) + 4]$

25. $4(8) \div (8 - 6)$

26. $\frac{28 + 12}{13 - 5}$

27. $7[8(3) \div (15 - 9)]$

28. $\frac{5 + 9}{10 - 3}$

29. $\frac{16 - 8}{15 - 11}$

Example 3 p. 7

30. TILING A decorative floor pattern has one red square tile surrounded by 12 blue tiles. Write and evaluate an expression to show how many total tiles (red and blue) are needed if there are 15 red tiles. Then make a table showing the total number of tiles if there are 15, 20, 25, or 30 red tiles.

31 FINANCIAL LITERACY To place an ad in a newspaper, it costs $8 plus $0.75 for each line. Write and evaluate an expression to find the total cost for an ad that has 6 lines. Then make a table showing the cost if there are 6, 10, 14, and 18 lines in the ad.

32. ZOO The table shows the prices of admission to the local zoo.

a. Write an expression that can be used to find the total cost of admission for 4 adults, 3 children, and 1 senior.

b. Find the total cost.

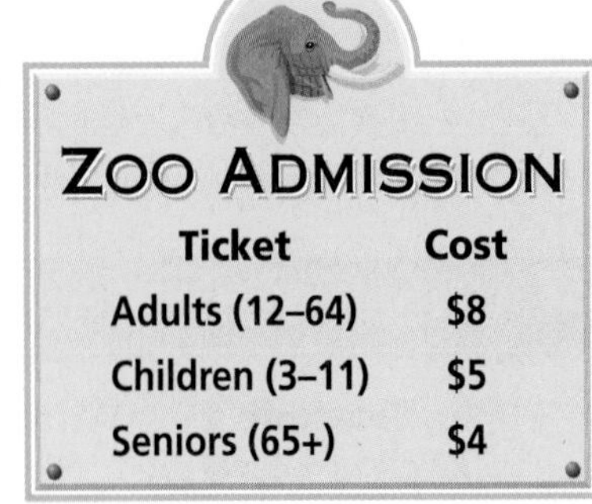

ZOO ADMISSION

Ticket	Cost
Adults (12–64)	$8
Children (3–11)	$5
Seniors (65+)	$4

33. MULTIPLE REPRESENTATIONS In this problem, you will investigate expressions using a toothpick sequence.

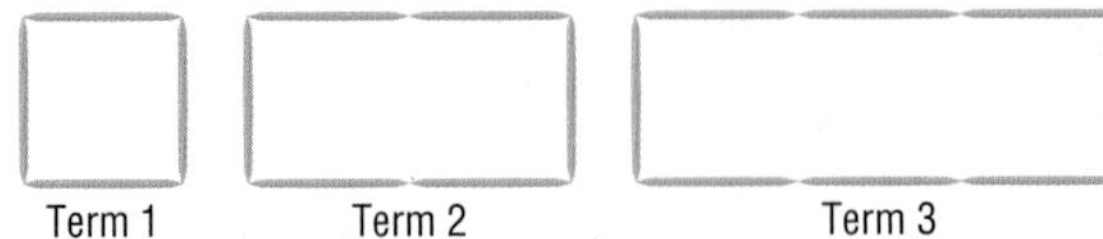

a. TABULAR Make a table showing term number and number of toothpicks.

b. VERBAL Write a verbal rule to find the number of toothpicks for any term.

H.O.T. Problems

Use Higher-Order Thinking Skills

34. OPEN ENDED Write two different expressions, each with more than one operation, that have a value of 20.

35. REASONING Use the expression $64 - 20 \div 4 + 6$.

a. Where should the parentheses be placed so that the value of the expression is 17?

b. Place different parentheses in the expression to find an answer other than 17. Then evaluate the new expression.

36. WRITING IN MATH The numerical expression 3^2 means to write 2 factors of 3 and multiply. So, $3^2 = 3 \cdot 3$ or 9. Likewise, $2^3 = 2 \cdot 2 \cdot 2$ or 8. Using this information, how would you evaluate the expression $(1 + 3)^2 + 4 \div 2$? Explain your reasoning.

37. CHALLENGE Leah bought two packages of special decorative trim for a craft project. One package contains 350 centimeters of trim. The second package contains 200 inches of trim.

a. If one inch $\approx$ 2.54 centimeters, write and evaluate an expression to find the number of centimeters in 200 inches.

b. Find the total number of centimeters of trim in the two packages.

c. How many packages of trim containing 90 inches each will Leah need to buy to have about the same amount of trim from part **b**? Explain.

38. WRITING IN MATH Explain why the rules for the order of operations are important. Support your answer with two numerical examples.

Standardized Test Practice

39. Rusty is evaluating $96 \div 3 \times 4 + 7$ as shown below.

$$96 \div 3 \times 4 + 7$$
$$3 \times 4 = 12$$
$$96 \div 12 = 8$$
$$8 + 7 = 15$$

What should Rusty have done differently in order to evaluate the expression correctly?

A multiplied $(96 \div 3)$ by $(4 + 7)$

B divided 96 by (3×11)

C multiplied $(96 \div 3)$ by 4 and added 7

D divided 96 by $(4 + 7)$

40. Evaluate $7[12 - (6 - 2) \div 4]$.

F 7

G 14

H 77

J 83

41. Aman is a musician. He charges \$50 for each of the first three hours he plays and \$24.95 for each additional hour. Which expression *cannot* be used to find the total amount he charges if he plays for 7 hours?

A 4 × \$24.95 + 3 × \$50

B 7(\$50 + \$24.95)

C \$50 + \$50 + \$50 + 4(\$24.95)

D 3(\$50) + 4 (\$24.95)

42. SHORT RESPONSE Max wants to buy 5 hats and 4 T-shirts. Write an expression to find the total cost of 5 hats and 4 T-shirts.

Hats	\$15
T-shirts	\$20

Algebra Lab
Rules and Expressions

Math Online glencoe.com
Math *in Motion,* Animation

Objective
Determine rules for a given pattern.

Numerical patterns can often be defined by a rule. You can then use the rule to find any term in the pattern.

ACTIVITY 1

The pattern below is made from toothpicks. The first term uses 3 toothpicks, the second term uses 5 toothpicks, and the third term uses 7 toothpicks.

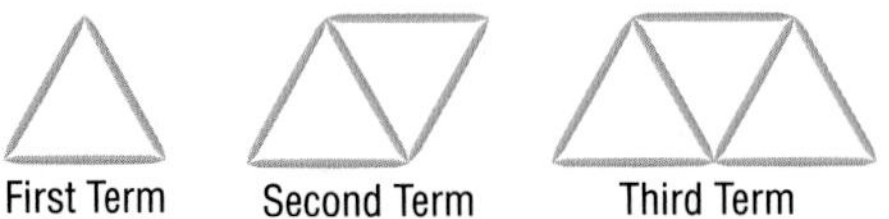

Step 1 Draw the next two terms in the pattern.

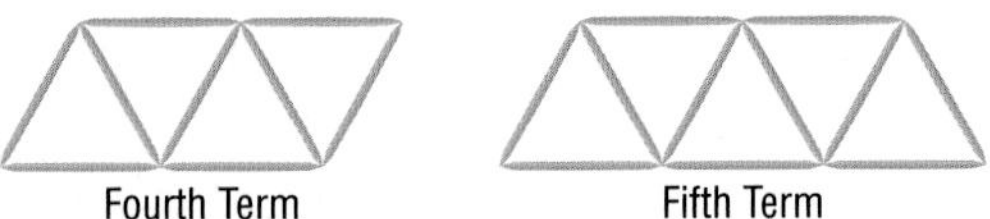

Step 2 Copy and complete the table to show the number of toothpicks needed if the pattern continues.

Term Number	1	2	3	4	5	6	7
Number of Toothpicks	3	5	7	■	■	■	■

+2 +2 +2

Step 3 Describe a rule you could use to find the number of toothpicks for any figure.

Since the number of toothpicks increases by 2 for each term, multiply the term number by 2. Then add 1 to get the number of toothpicks in the figure.

Analyze the Results

1. Using your rule, how many toothpicks would be needed for the twelfth term? the thirtieth term?

For Exercises 2 and 3, use the tile pattern below.

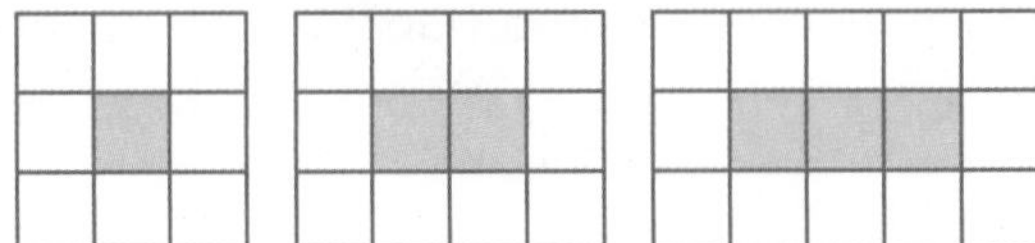

2. Make a table to show how many white tiles are needed for each blue tile.

3. Describe a rule that could be used to find the number of white tiles needed for any figure.

4. Create a pattern of your own. Share your pattern with another student and ask him or her to write a rule to describe your pattern.

1-2 Variables and Expressions

Then

You have already written numerical expressions. (Lesson 1-1)

Now

- Translate verbal phrases into algebraic expressions.
- Evaluate expressions containing variables.

New Vocabulary

algebra
variable
algebraic expression
defining a variable

Math Online

glencoe.com

- Extra Examples
- Personal Tutor
- Self-Check Quiz
- Homework Help

Why?

Marin received $100 from her grandfather for her birthday. She wants to use the money to buy an MP3 player for $69 and download some songs that are $0.88 each from the Internet with the rest of the money.

Total Cost of MP3 Player and Songs

Number of Songs	Cost to Download Songs	Total Cost
0	0×0.88 or $0	$69 + 0$ or $69
1	1×0.88 or $0.88	$69 + 0.88$ or $69.88
5	5×0.88 or $4.40	$69 + 4.40$ or $73.40
10	10×0.88 or $8.80	$69 + 8.80$ or $77.80

a. Suppose Marin wants to download 15 songs. What would be the total cost of the player and the songs?

b. How can you find the cost of 20 songs?

c. If we used the letter s to represent *any number of songs*, what expression could be used to represent the total cost of the MP3 player and the songs?

Algebraic Expressions and Verbal Phrases **Algebra** is a branch of mathematics that uses symbols. A variable is often used in algebra. A **variable** is a letter or symbol used to represent an unknown value. In the example above, s was used to represent *any number of songs*. Any letter can be used as a variable.

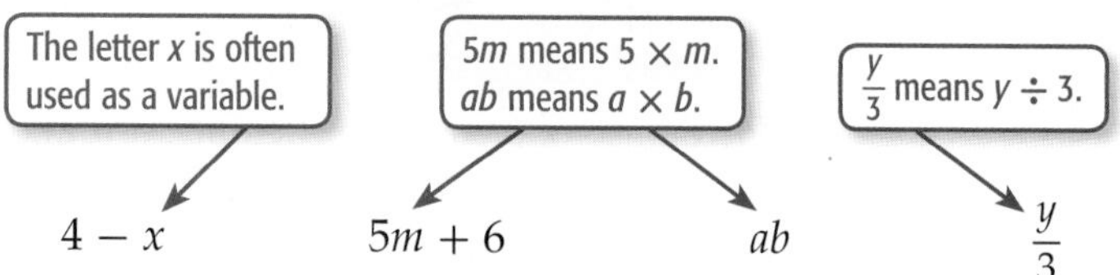

An expression like $5m + 6$ is an **algebraic expression** because it contains at least one variable and at least one mathematical operation.

The first step in translating verbal phrases into algebraic expressions is to choose a variable and a quantity for the variable to represent. This is called **defining a variable**. All of the steps involved in writing algebraic expressions are shown below.

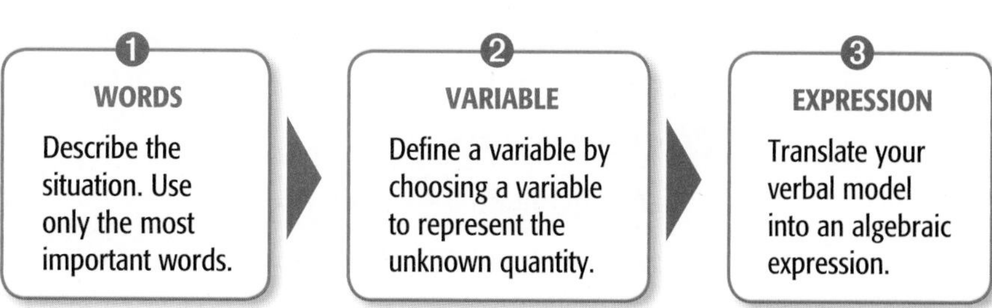

1
WORDS
Describe the situation. Use only the most important words.
2
VARIABLE
Define a variable by choosing a variable to represent the unknown quantity.
3
EXPRESSION
Translate your verbal model into an algebraic expression.

Math History Link

Al-Khwarizmi (c 780–850) Al-Khwarizmi is considered to be the father of algebra. He wrote a book titled *Al-Kitab al-Jabr wa-l-Muqabala* in approximately 820 A.D. The word *algebra* is derived from this title.

EXAMPLE 1 Translate Phrases into Algebraic Expressions

Translate each phrase into an algebraic expression.

a. three dollars more than the cost of a sandwich

Words	three dollars more than the cost of a sandwich
Variable	Let c represent the cost of the sandwich.
Expression	$3 + c$

b. Mari had \$2 and made \$6 an hour babysitting.

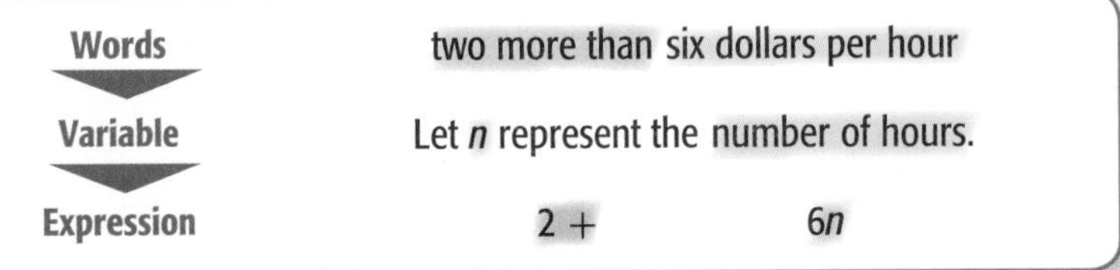

Words
two more than six dollars per hour
Variable
Let n represent the number of hours.
Expression
2 +
6n

Check Your Progress

1A. two miles less than the athlete ran

1B. five points more than the points scored by field goals if each field goal is worth 3 points

Personal Tutor glencoe.com

Evaluate Expressions To evaluate an algebraic expression, replace the variable(s) with known values and then use the order of operations. When you replace a variable with a number, you are using the **Substitution Property of Equality.**

Key Concept Substitution Property of Equality

For Your FOLDABLE

Words If two quantities are equal, then one quantity can be replaced by the other.

Symbols For all numbers a and b, if $a = b$, then a may be replaced by b.

EXAMPLE 2 Evaluate Expressions

Evaluate $d + 5 - f$ if $d = 16$ and $f = 18$.

$d + 5 - f = 16 + 5 - 18$	**Replace d with 16 and f with 18.**
$= 21 - 18$	**Add 16 and 5.**
$= 3$	**Subtract 18 from 21.**

Check Your Progress

2A. Evaluate $6 - e + f$ if $e = 3$ and $f = 9$.

2B. Evaluate $7k + h$ if $k = 4$ and $h = 10$.

Personal Tutor glencoe.com

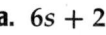

EXAMPLE 3 Evaluate Expressions

Evaluate each expression if $r = 1$, $s = 5$, and $t = 8$.

a. $6s + 2t$

$6s + 2t = 6(5) + 2(8)$ — Replace s with 5 and t with 8.

$= 30 + 16$ or 46 — Multiply. Then add.

b. $\frac{st}{20}$

$\frac{st}{20} = st \div 20$ — Rewrite as a division expression.

$= (5 \cdot 8) \div 20$ — Replace s with 5 and t with 8.

$= 40 \div 20$ or 2 — Multiply. Then divide.

c. $r + (40 - 3t)$

$r + (40 - 3t) = 1 + (40 - 3 \cdot 8)$ — Replace r with 1 and t with 8.

$= 1 + (40 - 24)$ — Multiply 3 and 8.

$= 1 + 16$ or 17 — Subtract 24 from 40. Then add 1 and 16.

Check Your Progress

Evaluate each expression if $a = 4$, $b = 8$, and $c = 12$.

3A. $3a + 2c$ **3B.** $\frac{ab}{16}$ **3C.** $c + (5b - 2a)$

Personal Tutor glencoe.com

Real-World EXAMPLE 4 Use Expressions to Solve Problems

BOATS **A company rents a house boat for \$200 plus an extra \$30 per day.**

a. **Write an expression that can be used to find the total cost to rent a house boat.**

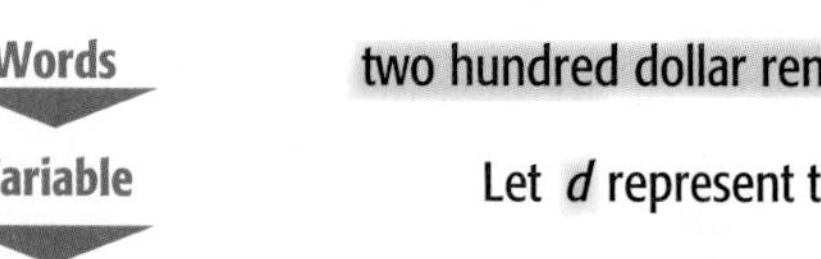

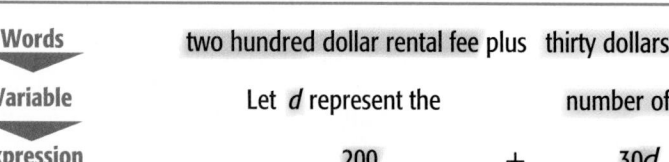

Words	two hundred dollar rental fee	plus	thirty dollars per day
Variable	Let d represent the		number of days.
Expression	200	+	$30d$

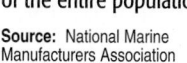

The expression is $200 + 30d$.

b. **Suppose the Gregoran family wants to rent a house boat for six days. What will be the total cost?**

$200 + 30d = 200 + 6(30)$ — Replace d with 6.

$= 200 + 180$ or 380 — Multiply. Then add.

The total cost will be \$380.

Real-World Link

In a recent year, 72.6 million adults in the United States went boating. This is about 25% of the entire population.

Source: National Marine Manufacturers Association

Check Your Progress

4. **SALES** At a garage sale, Georgia found some used DVDs and CDs that she wanted to buy. Each DVD was marked at \$5 and each CD was marked at \$3. Write an expression to find the total cost to buy some DVDs and CDs. Then find the cost of buying 4 DVDs and 7 CDs.

Personal Tutor glencoe.com

Check Your Understanding

Example 1 p. 12

Translate each phrase into an algebraic expression.

1. four dollars less than the cost of the sweater
2. 13 more students than teachers
3. money earned babysitting at $10 per hour
4. 30 pencils divided among some students

Examples 2 and 3 pp. 12–13

ALGEBRA Evaluate each expression if $g = 6$, $h = 10$, and $j = 5$.

5. $h + 15$
6. $g - 3$
7. $20 - h + g$
8. $22 - 3j$
9. $\frac{gh}{j}$
10. $4g + (3h - 4j)$

Example 4 p. 13

11. **CAPACITY** One pint of liquid is the same as 16 fluid ounces.
    a. Suppose the number of pints of liquid is represented by p. Write an expression to find the number of fluid ounces.
    b. How many fluid ounces is 5 pints?

 = **Step-by-Step Solutions** begin on page R11.
Extra Practice begins on page 810.

Practice and Problem Solving

Example 1 p. 12

Translate each phrase into an algebraic expression.

12. three times as many balloons
13. twenty-four pieces of candy divided among some students
14. the number of people increased by thirteen
15. the number of inches in any number of feet
16. four more than the number of weeks in a group of days
17. four less than the amount of cents in a number of dimes

Examples 2 and 3 pp. 12–13

ALGEBRA Evaluate each expression if $a = 9$, $b = 4$, and $c = 11$.

18. $b + 9$
19. $13 - a$
20. $2c - 5$
21. $18 + 4b$
22. $\frac{ab}{6}$
23. $\frac{8a}{b}$
24. $5c - 4a$
25. $7b - 2c$
26. $45 - \frac{bc}{2}$
27. $\frac{ac}{3} - 15$
28. $4b + 3c - 5a$
29. $6c - 2a + 6b$

Example 4 p. 13

30. **PHOTOGRAPHY** A studio charges a sitting fee of $25 plus $7 for each portrait sheet ordered. Write an expression that can be used to find the total cost to have photographs taken. Then find the cost of purchasing twelve portrait sheets.

31. **MEASUREMENT** One gallon of water is equal to 231 cubic inches. Write an expression for the number of gallons of water in any number of cubic inches of water.

ALGEBRA Evaluate each expression if $x = 9$, $y = 4$, and $z = 12$.

32. $7z - (y + x)$
33. $(8y + 5) - 2z$
34. $(5z - 4x) + 3y$
35. $6x - (z - 2y)$
36. $2x + (4z - 13) - 5$
37. $(29 - 3y) + 4z - 7$

38. **FINANCIAL LITERACY** After the included minutes have been used, a cell phone company charges an additional $0.08 per minute. Plan A uses a flat rate of $0.10 per minute for all calls. Which plan is the least costly if a person uses 750 minutes per month? Explain.

Plan	Monthly Fee	Included Minutes
A	$0	None
B	$29.99	500
C	$39.99	1000
D	$49.99	1500

Real-World Link

The largest apple crop in the United States was in 1998. Approximately 277.3 million bushels were harvested.

Source: U.S. Apple Association

39. **FOOD** One bushel of apples from a dwarf apple tree is equal to 42 pounds. Write an expression to find the number of pounds of apples in any number of bushels. If one tree can produce 6 bushels, how many pounds of apples will an orchard of 100 trees produce?

40. **MULTIPLE REPRESENTATIONS** In this problem, you will use algebra to describe a relationship. Jacinda used the table below to help convert measurements while she was cooking.

Number of Cups (c)	4	8	12	16
Number of Quarts (q)	1	2	3	4

a. **VERBAL** Write an expression in words that describes the relationship between the number of quarts and the number of cups.

b. **ALGEBRAIC** Write an algebraic expression that represents the number of quarts in c cups.

c. **NUMERICAL** Use the expression in part **b** to find the number of quarts in 100 cups.

H.O.T. Problems

Use Higher-Order Thinking Skills

41. **OPEN ENDED** Write an algebraic expression that has two different variables and two different operations: addition, subtraction, multiplication, or division. Then write a real-world problem that uses the expression.

42. **FIND THE ERROR** John and Ramon are writing an algebraic expression for the phrase *five less than a number*. Is either of them correct? Explain.

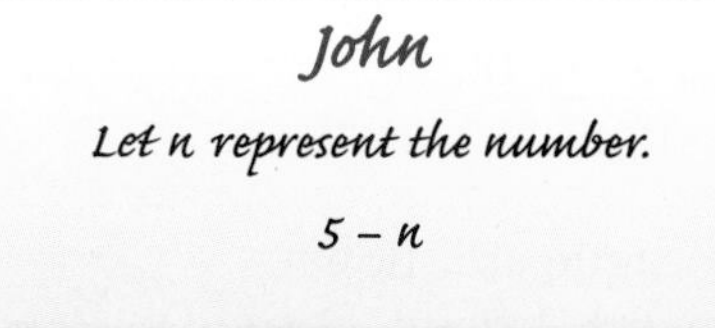
John
Let n represent the number.
5 − n

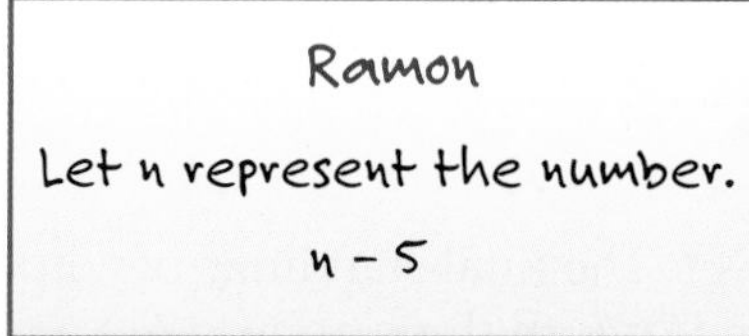
Ramon
Let n represent the number.
n − 5

Problem-Solving Tip

Make a Table You can make a table to organize the information in Exercise 43. You can find the pattern involved by comparing the number of sides to the number of toothpicks.

43. **CHALLENGE** Franco constructed the objects below using toothpicks.

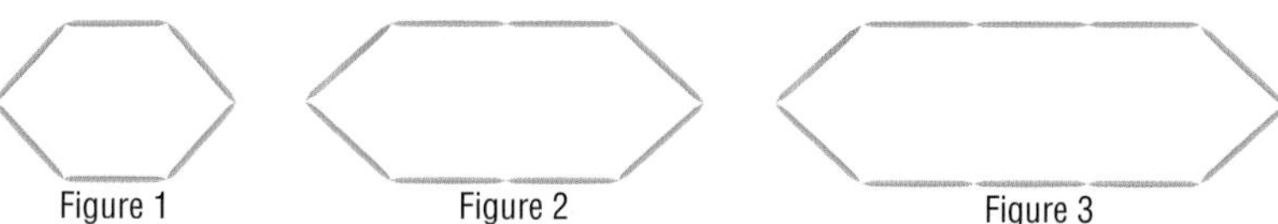
Figure 1
Figure 2
Figure 3

Write two different rules that relate the figure number to the number of toothpicks in each figure. Explain how you arrived at your answers.

44. **WRITING IN MATH** Cassandra needs to evaluate the expression $a(x + y)$. After she replaces the variables with numerical values, in which order should she perform the operations of addition and multiplication? Explain.

Standardized Test Practice

45. What word phrase is equivalent to the expression $5x + 9$?

A I have five cents more than nine nickels

B I have nine cents plus five cents

C I have nine cents more than five nickels

D I have nine cents less than five nickels

46. Which rule describes the ordered pairs in this table?

x	y
1	1
2	4
3	7
4	10

F $y = x$

G $y = 2x$

H $y = 3x - 2$

J $y = 2x + 2$

47. What is "9 more than the product of five and a number n" written as an algebraic expression?

A $5 + 9n$

B $9 + 5n$

C $5 + 9 \div n$

D $9 + n \div 5$

48. **EXTENDED RESPONSE** At a video store, DVDs cost $4 to rent for two days, and games cost $3 to rent for one day.

a. Complete the table below to show how much Ava will pay to rent one DVD and one game for the number of days given.

Number of Days	Total Cost ($)
2	■
4	■
6	■

b. How much will Ava pay to rent 4 DVDs and 3 games for 10 days?

Spiral Review

Find the value of each expression. (Lesson 1-1)

49. $3 \cdot 6 - 4$

50. $12 - 3 \times 3$

51. $9 + 18 \div 3$

52. $56 \div (7 \cdot 2) \times 6$

53. $75 \div (7 + 8) - 3$

54. $70 - (16 \div 2 + 21)$

55. $\dfrac{45 - 18}{9 \div 3}$

56. $\dfrac{8 \div 8 + 11}{15 - 4(3)}$

57. $4(20 - 13) + 4 \times 5$

HOCKEY The final standings of a hockey league are shown. A win is worth three points, and a tie is worth 1 point. Zero points are given for a loss. (Lesson 1-1)

Team	Wins	Losses	Ties
Knights	14	9	7
Huskies	11	9	10
Wildcats	10	9	11
Mustangs	9	10	11
Panthers	10	14	6

58. How many points do the Wildcats have?

59. How many points do the Huskies have?

60. How many more points do the Knights have than the Panthers?

Skills Review

Find each difference. (Previous Course)

61. $47 - 24$

62. $58 - 32$

63. $93 - 61$

64. $154 - 107$

EXTEND 1-2

Spreadsheet Lab
Expressions and Spreadsheets

Math Online glencoe.com
- Graphing Technology Personal Tutor

One of the most common computer applications is a spreadsheet program. A **spreadsheet** is a table that performs calculations. It is organized into boxes called **cells**, which are named by a letter and a number. In the spreadsheet below, cell B2 is highlighted.

An advantage of using a spreadsheet is that values in the spreadsheet are recalculated when a number is changed. You can use a spreadsheet to investigate patterns in data.

ACTIVITY 1

Here's a number trick! Pick any number. Double it. Then add four and divide by two. Then subtract the original number. I guarantee you the answer is 2.

You can use a spreadsheet to test different numbers. Suppose we start with the number 15.

I Know The Answer!

◇	A	B	C
1	Think of any number.	15	15
2	Double it.	=B1*2	30
3	Add 4.	=B2+4	34
4	Divide by 2.	=B3/2	17
5	Subtract the original number.	=B4-B1	2
6			

Sheet 1 | Sheet 2 | Sheet 3

The spreadsheet takes the value in B1,doubles it, and enters the value in B2. Note the * is the symbol for multiplication.

The spreadsheet takes the value in B3, divides by 2, and enters the value in B4. Note that / is the symbol for division.

The result is 2.

Exercises

To change information in a spreadsheet, move the cursor to the cell you want to access and click the mouse. Then type in the information and press Enter. Find the result when each value is entered in B1.

1. 7 **2.** 10 **3.** 50

4. 115 **5.** 1200 **6.** 2132

7. **MAKE A CONJECTURE** What is the result if a decimal is entered in B1? a negative number?

8. **WRITING IN MATH** Explain why the result is always 2. Write an expression that describes your answer.

9. Make up your own mind-reading trick. Enter it into a spreadsheet to show that it works. Write an expression to describe the trick.

1-3 Properties

Then

You have already evaluated numerical and algebraic expressions. (Lessons 1-1 and 1-2)

Now

- Identify and use properties of addition and multiplication.
- Use properties of addition and multiplication to simplify algebraic expressions.

New Vocabulary

properties
counterexample
simplify
deductive reasoning

Math Online

glencoe.com

- Extra Examples
- Personal Tutor
- Self-Check Quiz
- Homework Help

Why?

When you make a peanut butter and jelly sandwich, do you spread the peanut butter first or the jelly? The order does not matter because in the end you have a tasty sandwich.

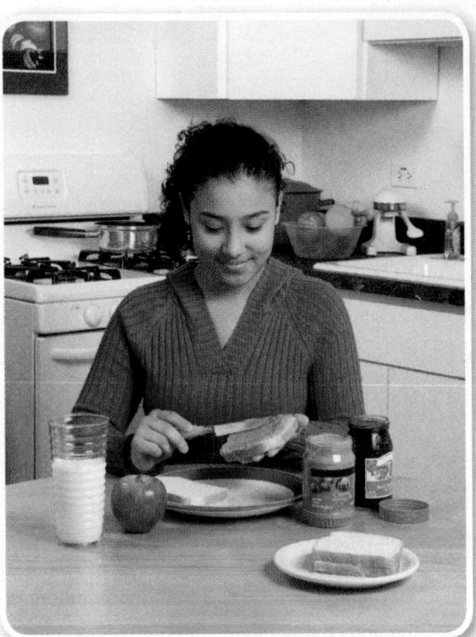

a. Name two other activities where order does not matter.

b. Name two activities where order *does* matter.

c. Name a mathematical operation in which you can switch the numbers and still have the same value.

Properties of Addition and Multiplication In algebra, **properties** are statements that are true for any numbers. For example, the expression $30 + 10$ and $10 + 30$ have the same value, 40. This illustrates the **Commutative Property of Addition**.

Likewise, $30 \cdot 10$ and $10 \cdot 30$ have the same value, 300. This illustrates the **Commutative Property of Multiplication**.

Key Concept — Commutative Properties (For Your FOLDABLE)

Words The order in which numbers are added or multiplied does not change the sum or product.

Symbols For any numbers a and b, $a + b = b + a$.

For any numbers a and b, $a \cdot b = b \cdot a$.

Examples $6 + 9 = 9 + 6$, $15 = 15$ $\qquad$ $4 \cdot 7 = 7 \cdot 4$, $28 = 28$

Suppose you want to evaluate the expression $16 + (14 + 58)$. Since $4 + 6$ is 10, it is easier to use mental math by grouping the numbers as $(16 + 14) + 58$. This illustrates the **Associative Property of Addition**. There is also an **Associative Property of Multiplication**.

Key Concept — Associative Properties (For Your FOLDABLE)

Words The order in which numbers are grouped when added or multiplied does not change the sum or product.

Symbols For any numbers a, b, and c, $(a + b) + c = a + (b + c)$.

For any numbers a, b, and c, $(a \cdot b) \cdot c = a \cdot (b \cdot c)$.

Examples

$(3 + 6) + 1 = 3 + (6 + 1)$
$9 + 1 = 3 + 7$
$10 = 10$

$(5 \cdot 9) \cdot 2 = 5 \cdot (9 \cdot 2)$
$45 \cdot 2 = 5 \cdot 18$
$90 = 90$

In addition to the Commutative and Associative Properties, the following properties are also true for any numbers.

Key Concept **Number Properties** For Your FOLDABLE

Property	Words	Symbols	Examples
Additive Identity	When 0 is added to any number, the sum is the number.	For any number a, $a + 0 = 0 + a = a$	$5 + 0 = 5$ $0 + 5 = 5$
Multiplicative Identity	When any number is multiplied by 1, the product is the number.	For any number a, $a \cdot 1 = 1 \cdot a = a$.	$8 \cdot 1 = 8$ $1 \cdot 8 = 8$
Multiplicative Property of Zero	When any number is multiplied by 0, the product is 0.	For any number a, $a \cdot 0 = 0 \cdot a = 0$	$3 \cdot 0 = 0$ $0 \cdot 3 = 0$

ReadingMath

Identity The word identity means *sameness of essential character*. The additive identity, 0, and multiplicative identity, 1, allow the original number to remain the same.

Do these properties apply to subtraction or division? One way to find out is to look for a counterexample. A **counterexample** is an example that shows a conjecture is not true.

EXAMPLE 1 Find a Counterexample

Is division of whole numbers associative? If not, give a counterexample.

The Associative Property of Multiplication states $(a \cdot b) \cdot c = a \cdot (b \cdot c)$. Check $(a \div b) \div c \stackrel{?}{=} a \div (b \div c)$.

$(27 \div 9) \div 3 \stackrel{?}{=} 27 \div (9 \div 3)$ **Pick values for *a*, *b*, and *c*.**

$(3) \div 3 \stackrel{?}{=} 27 \div (3)$ **Simplify.**

$1 \neq 9$ **Simplify.**

We found a counterexample. So, division of whole numbers is not associative.

StudyTip

Counterexample You can disprove a statement by finding only one counterexample.

Check Your Progress

1. Is subtraction of decimals associative? If not, give a counterexample.

Personal Tutor glencoe.com

EXAMPLE 2 Identify Properties

Name the property shown by each statement.

a. $4 + (a + 3) = (a + 3) + 4$

The order of the numbers and variables changed. This is the Commutative Property of Addition.

b. $1 \cdot (3c) = 3c$

The expression was multiplied by 1 and remained the same. This is the Multiplicative Identity Property.

Check Your Progress

2A. $d + 0 = d$

2B. $8 \cdot 1 = 8$

2C. $14 + (9 + 10) = (14 + 9) + 10$

2D. $5 \times 7 \times 2 = 7 \times 2 \times 5$

Personal Tutor glencoe.com

Simplify Algebraic Expressions To **simplify** an algebraic expression, perform all possible operations. You can use the properties you learned in this lesson. Using facts, properties, or rules to reach valid conclusions is called **deductive reasoning**.

EXAMPLE 3 Simplify Algebraic Expressions

Simplify each expression.

a. $(3 + e) + 7$

$(3 + e) + 7 = (e + 3) + 7$	Commutative Property of Addition
$= e + (3 + 7)$	Associative Property of Addition
$= e + 10$	Simplify.

b. $8 \cdot (x \cdot 5)$

$8 \cdot (x \cdot 5) = 8 \cdot (5 \cdot x)$	Commutative Property of Multiplication
$= (8 \cdot 5) \cdot x$	Associative Property of Multiplication
$= 40x$	Simplify.

Check Your Progress

3A. $12 \cdot (10 \cdot z)$

3B. $10 + (p + 18)$

Personal Tutor glencoe.com

STANDARDIZED TEST EXAMPLE 4

Test-TakingTip

Multiple Choice Eliminate the options you know are incorrect. If possible, circle the word, phrase or number that makes the option incorrect.

Which of the following is an example of the Commutative Property of Addition?

A $(3 \cdot 4) + 5 = 5 + (3 \cdot 4)$

B $(7 + 8) + 2 = 7 + (8 + 2)$

C $8 \cdot 9 = 9 \cdot 8$

D $1 + 0 = 1$

Read the Test Item

You need to identify the correct expression.

Solve the Test Item

The Identity Property of Addition shows that any number added to zero is equal to that number. So, option D can be eliminated.

The Associative Property of Addition deals with grouping of individual terms. The factors are grouped in different ways on each side of the statement. So, option B can be eliminated.

Option C can be eliminated because it shows the Commutative Property of Multiplication.

The answer is A.

Check Your Progress

4. Which of the following is an example of the Identity Property of Multiplication?

F $10 \cdot 1 = 1 \cdot 10$

G $(5 \cdot 6) \cdot 3 = 3 \cdot (5 \cdot 6)$

H $8 \cdot 1 = 8$

J $4 \cdot 4 = 16$

Personal Tutor glencoe.com

Check Your Understanding

Example 1 p. 19

1. Is subtraction of whole numbers commutative? If not, give a counterexample.

Example 2 p. 19

Name the property shown by each statement.

2. $8 \cdot 4 = 4 \cdot 8$
3. $6 \cdot 1 = 6$
4. $9 + 3 + 20 = 3 + 9 + 20$
5. $7 + 0 = 7$
6. $13 + 12 = 12 + 13$
7. $6 \times (1 \times 9) = (6 \times 1) \times 9$

Example 3 p. 20

ALGEBRA Simplify each expression.

8. $(12 + m) + 4$
9. $3 + (k + 8)$
10. $(15 + s) + 4$
11. $8 \cdot (x \cdot 5)$
12. $(10 \cdot r) \cdot 5$
13. $(12 \cdot a) \cdot 6$

Example 4 p. 20

14. **MULTIPLE CHOICE** Which of the following is an example of the Identity Property of Addition?

A $3 + 4 = 4 + 3$

B $7 + 7 = 14$

C $(5 + 2) + 6 = 5 + (2 + 6)$

D $12 + 0 = 12$

Practice and Problem Solving

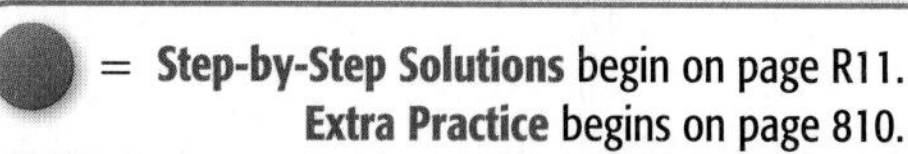
= **Step-by-Step Solutions** begin on page R11. **Extra Practice** begins on page 810.

Example 1 p. 19

State whether each conjecture is true. If not, give a counterexample.

15. The sum of two odd numbers is always odd.
16. The product of odd numbers is always even.

Example 2 p. 19

Name the property shown by each statement.

17. $0 + 14 = 14$
18. $8 \cdot 1 = 8$
19. $15 + 17 = 17 + 15$
20. $(2 \cdot 8) \cdot 5 = 2 \cdot (8 \cdot 5)$
21. $14 \times 0 \times 3 = 0$
22. $4 + (9 + 2) = (4 + 9) + 2$
23. $7 + x + 11 = x + 7 + 11$
24. $5k \times 1 = 5k$

Example 3 p. 20

ALGEBRA Simplify each expression.

25. $(d + 12) + 16$
26. $14 + (27 + m)$
27. $(54 + p) + 16$
28. $(r + 32) + 24$
29. $(8 \cdot s) \cdot 9$
30. $g \cdot (5 \cdot 7)$
31. $11 \cdot (t \cdot 4)$
32. $15b(5)$
33. $6(12c)$
34. $(7 + p) + 13$
35. $29 + (1 + t)$
36. $4 \cdot (x \cdot 2)$

37. **BASKETBALL** Use the table to write an expression that shows how many total baskets the Cavaliers made during the season. Simplify the expression.

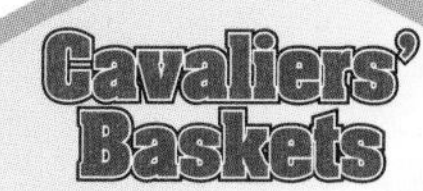

Cavaliers' Baskets	
Free Throws	1484
2-Point Field Goals	f
3-Point Field Goals	494

38. **HOMEWORK** Moreno likes to do her social studies homework before she does her math homework. Is doing social studies homework and math homework commutative? Explain.

ALGEBRA Translate each verbal expression into an algebraic expression. Then simplify the expression.

39 the sum of two times a number and five added to six times a number

40. the product of seven and four times a number multiplied by three

41. eight more than the sum of six times a number and nine times a number added to one

42. the product of eleven and five times a number multiplied by four

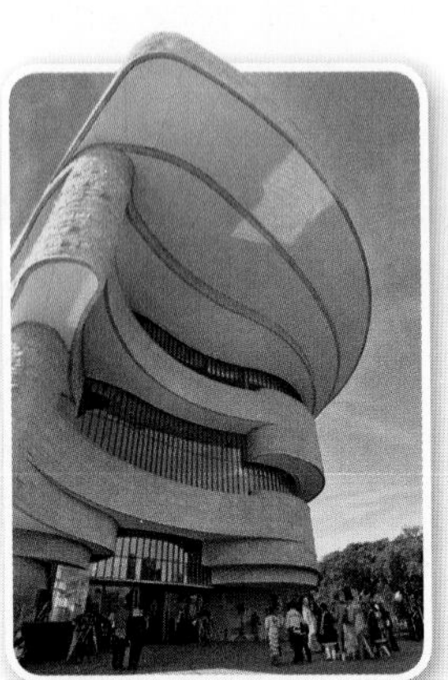

Real-World Link

The Smithsonian Institute opened in 1846 as a national museum. It currently has 19 museums and 9 research centers. In 2006, 23.2 million people visited the museums. Admission to most of the museums is free.

43. **FINANCIAL LITERACY** The Center of Wonders science center has the rates shown.

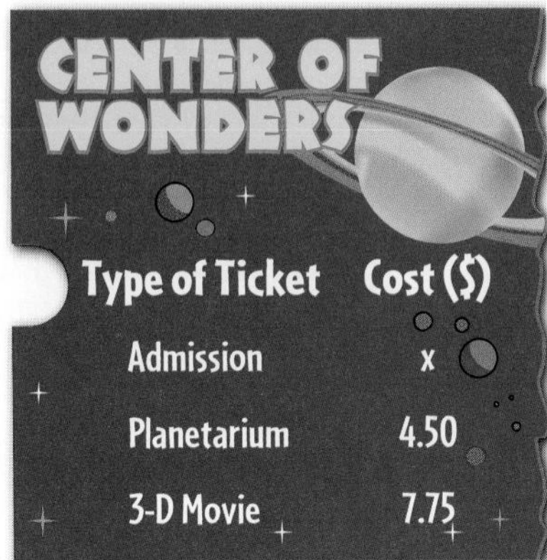

Type of Ticket	Cost ($)
Admission	x
Planetarium	4.50
3-D Movie	7.75

a. Write an algebraic expression that could be used to find the total cost for five people to get into the center, visit the planetarium, and watch a 3-D movie.

b. If the cost of admission to the center is \$12, how much will it cost for four people to get into the center and watch a 3-D movie?

H.O.T. Problems
Use Higher-Order Thinking Skills

44. **OPEN ENDED** Write an algebraic expression that can be simplified using at least two different properties. Simplify the expression showing each step and justify for each step.

45. **WRITING IN MATH** Is the following statement *true* or *false*? Explain your reasoning.

$$15 + (4 \cdot 6) = (15 + 4) \cdot 6$$

46. **FIND THE ERROR** Meghan and Alejandro are simplifying the expression $8x \cdot 4 \cdot 2x \cdot 3$. Is either of them correct? Explain your reasoning.

Meghan	Alejandro
$8x \cdot 4 \cdot 2x \cdot 3 = 16x \cdot 12$	$8x \cdot 4 \cdot 2x \cdot 3 = 192x$

47. **CHALLENGE** If you take any two whole numbers and add them together, the sum is always a whole number. This is the Closure Property for Addition. The set of whole numbers is *closed* under addition.

a. Is the set of whole numbers closed under subtraction? If not, give a counterexample.

b. Suppose you had a very small set of numbers that contained only 0 and 1. Would this set be closed under addition? If not, give a counterexample.

c. There is also a Closure Property for Multiplication of Whole Numbers. State this property using the addition property above as a guideline.

d. Is the set {0, 1} closed under multiplication? Explain.

48. **WRITING IN MATH** The number 1 is the identity for multiplication. Do you think that division has an identity? Explain your reasoning.

Standardized Test Practice

49. Which statement is an example of the Identity Property?

A $3 \cdot x \cdot 0 = 0$
B $7(4x) = (4 \cdot 7)x$
C $5 + (4 + x) = (5 + 4) + x$
D $4x + 0 = 4x$

50. Which expression can be used to find the perimeter of the rectangle below?

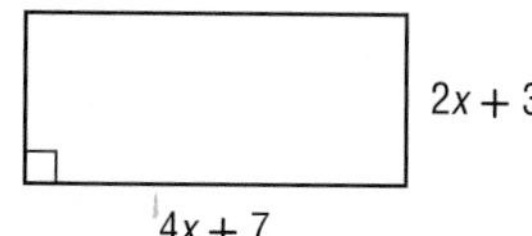

F $8x + 21$
G $12x + 20$
H $6x + 10$
J $8x + 10$

51. Which property is illustrated by the statement below?

$$12 \cdot (n \cdot 5) = (12 \cdot n) \cdot 5$$

A Commutative Property
B Associative Property
C Identity Property
D Zero Property

52. **SHORT RESPONSE** Simplify the expression, show and justify each step.

$$10 \cdot (x \cdot 3)$$

Spiral Review

ALGEBRA Translate each phrase into an algebraic expression. (Lesson 1-2)

53. Bianca's salary plus a $200 bonus

54. three more than the number of cakes baked

55. six feet shorter than the mountain's height

56. eight less than the quotient of the number of quarters and four

57. **SCIENCE** The number of times a cricket chirps can be used to estimate the temperature in degrees Fahrenheit. Use $c \div 4 + 37$, where c is the number of chirps in 1 minute. (Lesson 1-2)

a. Find the approximate temperature if a cricket chirps 136 times a minute.
b. What is the temperature if a cricket chirps 100 times in a minute?

Evaluate each expression. (Lesson 1-1)

58. $50 \div 2 \times 5$

59. $6(8 - 4) + 3 \cdot 7$

60. $16 - 2 \cdot 4$

61. $18 + 2 \cdot 3$

62. $49 - 25 + 5$

63. $3(7 \cdot 5) \cdot 2$

Skills Review

Find each product. (Previous Course)

64. 22×7

65. 9×45

66. 8×34

67. 18×3

68. 15×13

69. 8×42

70. 109×21

71. 43×119

CHAPTER 1

Mid-Chapter Quiz

Lessons 1-1 through 1-3

Write a numerical expression for each verbal phrase. (Lesson 1-1)

1. the total number of video games if Tonya has 26 and Mary has 38

2. the number of years until the Olympics if the Olympics are 36 months away

Evaluate each expression. (Lesson 1-1)

3. $3 \cdot 4 + 2$

4. $45 \div 15 \times 5$

5. $(18 - 6) \div 4$

6. $3[(4 + 6) \div 2]$

7. **POPCORN** To rent a popcorn machine, it costs \$35 plus \$6.95 for each hour. Write and evaluate an expression to find the total cost of renting the popcorn machine for 5 hours. Then make a table showing the cost for 5, 6, 7, and 8 hours. (Lesson 1-1)

8. **SWIMMING** The table shows the prices of admission to a pool.

Age	Cost
Adults (18–62)	\$6
Children (3–17)	\$4
Seniors (63+)	\$3

Write an expression that can be used to find the total cost of admission for 2 adults, 3 children, and 2 seniors. Then find the total cost. (Lesson 1-1)

Evaluate each expression if $a = 5$, $b = 7$, and $c = 9$. (Lesson 1-2)

9. $a + 7$

10. ab

11. $(c + 6) \times a$

12. $(a + c) \div b$

13. **MULTIPLE CHOICE** Carlos does lawn work on the weekends. He charges \$12 per lawn and \$5 per hour trimming bushes. Which expression represents the total amount of money Carlos charges to mow 2 lawns and trim bushes for h hours? (Lesson 1-2)

A $12 + 2 + 5 + h$

B $2h + 5 \cdot 12$

C $2 \cdot 12 + h$

D $2 \cdot 12 + 5h$

14. **MEASUREMENT** There are 12 inches in 1 foot. Write an algebraic expression that represents the number of inches in f feet. Then complete the table. (Lesson 1-2)

Number of Feet	Number of Inches
2	24
4	■
6	■
8	96

15. **MULTIPLE CHOICE** Which of the following phrases represents the expression $3n - 4$? (Lesson 1-2)

F The cost of a new DVD is \$4 more than three times the price of a used DVD.

G The cost of a new DVD is \$4 less than three times the price of a used DVD.

H The cost of a new DVD is \$3 more than four times the price of a used DVD.

J The cost of a new DVD is \$3 less than four times the price of a used DVD.

Name the property shown by each statement. (Lesson 1-3)

16. $35 + 0 = 35$

17. $(3 + y) + 11 = 3 + (y + 11)$

18. $9 \cdot 3 = 3 \cdot 9$

Simplify each expression. (Lesson 1-3)

19. $13 + (27 + a)$

20. $x \cdot (6 \cdot 9)$

21. $(z + 4) + 12$

22. $10a(6)$

23. $18 + (h + 15)$

24. **REASONING** *True* or *False*? Is the difference of two whole numbers always a whole number? If false, give a counterexample. (Lesson 1-3)

25. **TECHNOLOGY** When you listen to music on an MP3 player, you turn it on and then program the player. Is programming the MP3 player and turning it on commutative? Explain. (Lesson 1-3)

1-4 Ordered Pairs and Relations

Then

You have already graphed numbers on number lines. (Previous Course)

Now

- Use ordered pairs to locate points.
- Use graphs to represent relations.

New Vocabulary

coordinate system
coordinate plane
y-axis
origin
x-axis
ordered pair
x-coordinate
y-coordinate
graph
relation
domain
range

Math Online

glencoe.com

- Extra Examples
- Personal Tutor
- Self-Check Quiz
- Homework Help
- Math in Motion

Why?

Brenna is planning a treasure hunt for her little brother in their yard. She has drawn an imaginary grid over the area for the hunt.

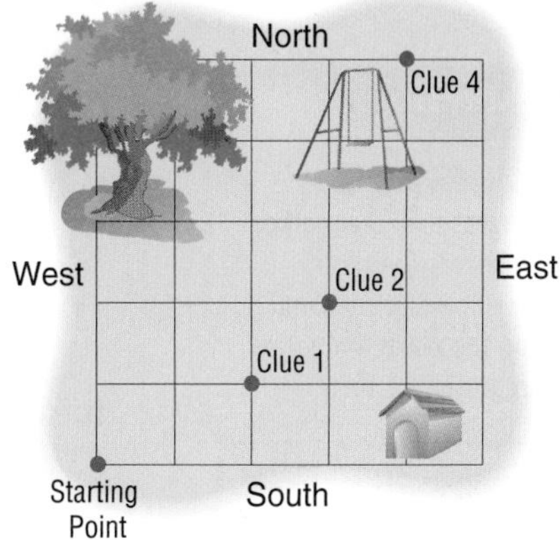

Clue 1 is located 2 units over (east) and 1 unit up (north) on the grid from the starting point named (0, 0). This location will be named (2, 1) on the list of clues. Use this information to answer the questions below.

a. Find Clue 2 on the grid. Name this location using the method used for Clue 1.

b. Clue 3 is located at (2, 4). Describe its location on the grid.

c. Describe how to get from Clue 3 to Clue 4 using units and directions. Give the final location of Clue 4.

d. To get to Clue 5, you can go 2 units south and 1 unit west from Clue 4. Name the location.

e. Clue 5 tells you that Clue 6 is 3 units west and 2 units south from Clue 5. Name the location.

f. Clue 6 says that the treasure is located halfway between Clue 1 and Clue 2. What is the location of the treasure?

Ordered Pairs In mathematics, a **coordinate system** or **coordinate plane** is used to locate points. The coordinate system is formed by the intersection of two number lines that meet at right angles at their zero points.

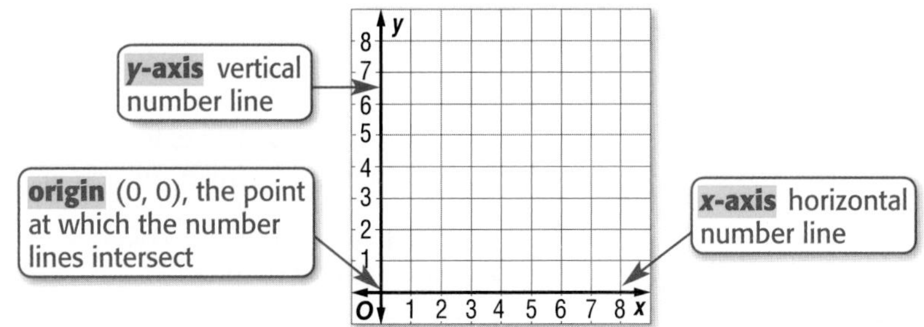

An **ordered pair** of numbers is used to locate any point on a coordinate plane. The first number is called the **x-coordinate** and the second number is called the **y-coordinate.**

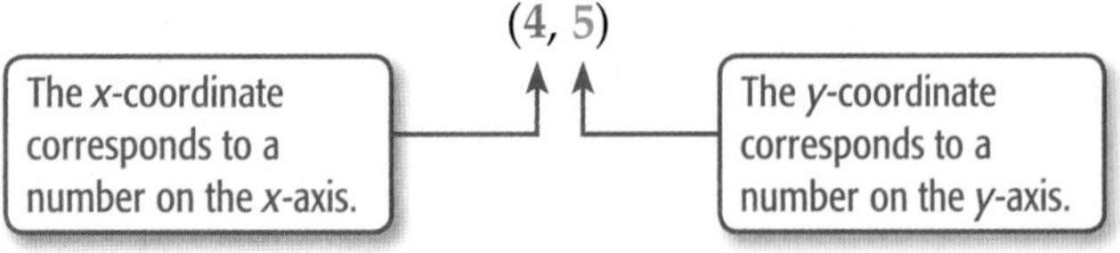

To **graph** an ordered pair, draw a dot at the point that corresponds to the ordered pair. The coordinates are your directions to locate the point.

EXAMPLE 1 Graph Ordered Pairs

Graph each ordered pair on a coordinate plane.

a. (5, 3)

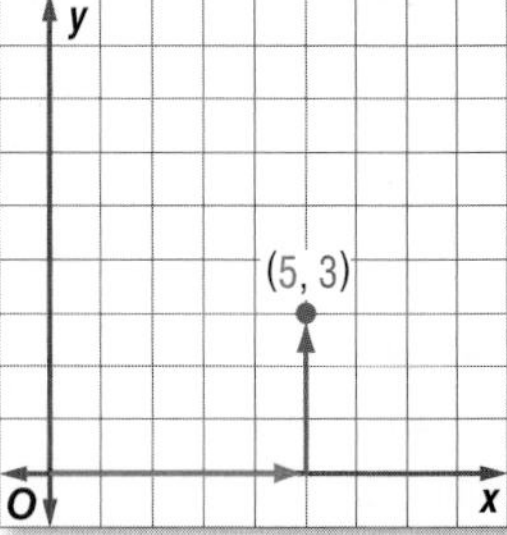

Step 1 Start at the origin.

Step 2 Since the x-coordinate is 5, move 5 units to the right.

Step 3 Since the y-coordinate is 3, move 3 units up. Draw a dot.

StudyTip

Coordinate Planes Unless they are marked otherwise, you can assume that each unit on the x-axis and y-axis represents 1 unit.

b. (0, 4)

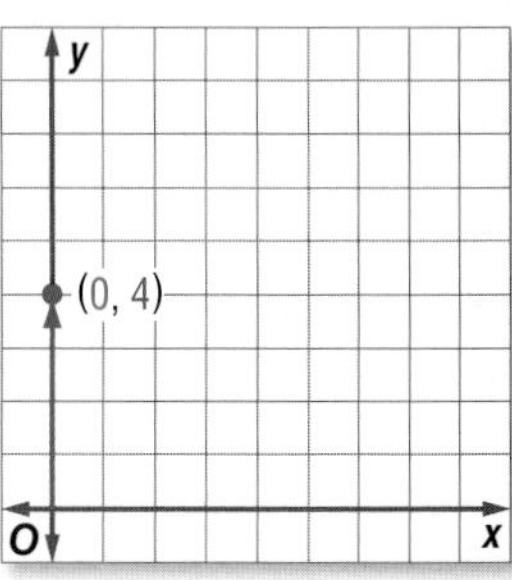

Step 1 Start at the origin.

Step 2 Since the x-coordinate is 0, you do not need to move right.

Step 3 Since the y-coordinate is 4, move 4 units up. Place the dot on the axis.

Check Your Progress

1A. (2, 3) **1B.** (5, 0) **1C.** $\left(3, 1\frac{1}{2}\right)$ **1D.** $\left(6\frac{1}{2}, 5\frac{1}{2}\right)$

Personal Tutor glencoe.com

Sometimes a point on a graph is named by using a letter. To identify its location, you can write the ordered pair that represents the point.

EXAMPLE 2 Identify Ordered Pairs

Write the ordered pair that names each point.

a. A

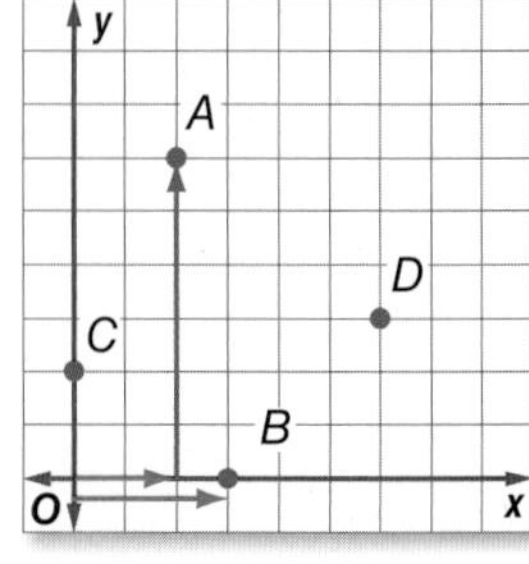

Step 1 Start at the origin.

Step 2 Move right on the x-axis to find the x-coordinate of point A, which is 2.

Step 3 Move up the y-axis to find the y-coordinate, which is 6.

The ordered pair for point A is (2, 6).

b. B

The x-coordinate of point B is 3, and the y-coordinate is 0.

The ordered pair for point B is (3, 0).

Check Your Progress

2A. C **2B.** D

Personal Tutor glencoe.com

ReadingMath

Domain and Range The domain of a relation is also called the *input*. The range of a relation is also called the *output*.

Relations A set of ordered pairs such as {(2, 3), (3, 5), (4, 1), (5, 6)} is a **relation**. A relation can also be shown in a table or a graph. The **domain** of the relation is the set of *x*-coordinates. The **range** of the relation is the set of *y*-coordinates.

Ordered Pairs

(2, 3)
(3, 5)
(4, 1)
(5, 6)

The domain is {2, 3, 4, 5}.

The range is {3, 5, 1, 6}.

Table

x	y
2	3
3	5
4	1
5	6

Graph

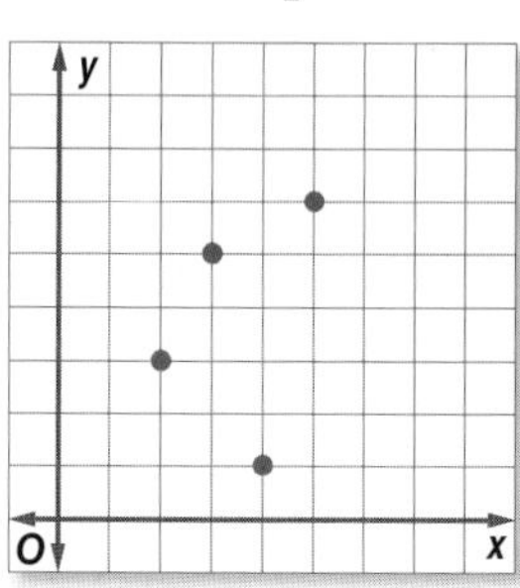

Math *in Motion*, Animation glencoe.com

EXAMPLE 3 Relations as Tables

Express the relation {(0, 2), (1, 4), (2, 5), (3, 8)} as a table. Then determine the domain and range.

x	0	1	2	3
y	2	4	5	8

The domain is {0, 1, 2, 3}, and the range is (2, 4, 5, 8}.

Check Your Progress

3. Express the relation {(2, 4), (0, 3), (1, 4), (1, 1)} as a table. Then determine the domain and range.

Personal Tutor glencoe.com

Real-World EXAMPLE 4 Relations as Graphs

SEAHORSES **A seahorse swims at a rate of about 5 feet per hour.**

a. Make a table of ordered pairs in which the *x*-coordinate represents the hours and the *y*-coordinate represents the number of feet for 1, 2, and 4 hours.

x	y
1	5
2	10
4	20

b. Graph the ordered pairs and describe the graph.

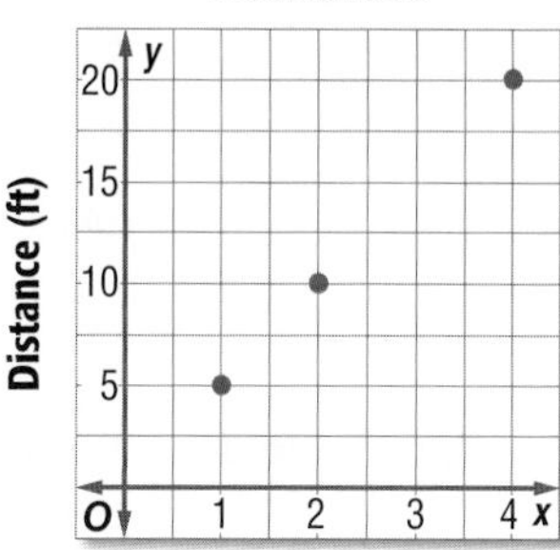

The points appear to lie in a line.

Real-World Link

Seahorses are the slowest swimming fish due to their unusual body shape. To move, they flutter their back fin up to 35 times per second.

Check Your Progress

4. MEASURMENT One square mile is equal to six hundred forty acres.

A. Make a table of ordered pairs in which the *x*-coordinate represents the number of square miles and the *y*-coordinate represents the number of acres in 1, 2, and 3 square miles.

B. Graph the ordered pairs. Then describe the graph.

Personal Tutor glencoe.com

Check Your Understanding

Example 1 p. 26

Graph each ordered pair on a coordinate plane.

1. $F(6, 0)$
2. $A(2, 5)$
3. $W(4, 1)$
4. $Z(0, 1)$

Example 2 p. 26

Refer to the coordinate plane shown at the right. Write the ordered pair that names each point.

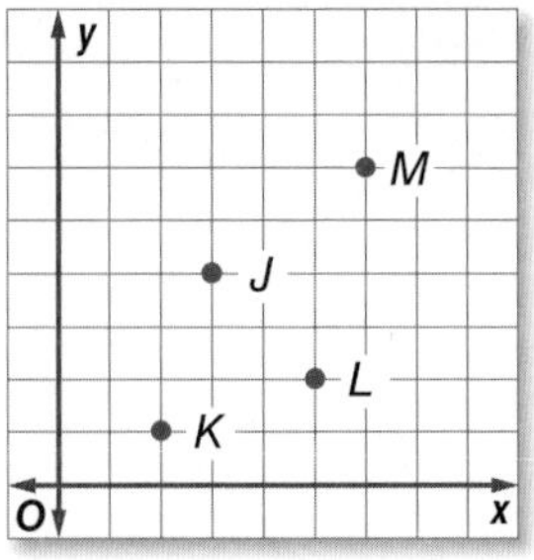

5. J
6. K
7. L
8. M

Example 3 p. 27

Express each relation as a table. Then determine the domain and range.

9. $\{(3, 4), (1, 5), (4, 2)\}$
10. $\{(1, 3), (2, 6), (3, 3), (4, 7)\}$

Example 4 p. 27

11. **CAPACITY** One quart is the same as two pints.
    a. Make a table of ordered pairs in which the x-coordinate represents the number of quarts and the y-coordinate represents the number of pints in 1, 2, 3, and 4 quarts.
    b. Graph the ordered pairs. Then describe the data.

Practice and Problem Solving

 = **Step-by-Step Solutions** begin on page R11.
Extra Practice begins on page 810.

Example 1 p. 26

Graph each ordered pair on a coordinate plane.

12. $A(4, 7)$
13. $B(0, 4)$
14. $C(7, 3)$
15. $D(3, 4)$
16. $F(6, 1)$
17. $G(6, 5)$
18. $H(3, 0)$
19. $J(2, 2)$

Example 2 p. 26

Refer to the coordinate plane shown at the right. Write the ordered pair that names each point.

20. L
21. M
22. N
23. P
24. Q
25. R
26. S
27. T

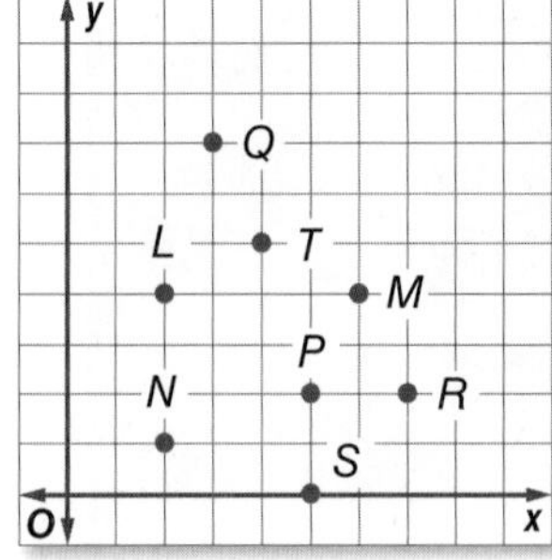

Example 3 p. 27

Express each relation as a table. Then determine the domain and range.

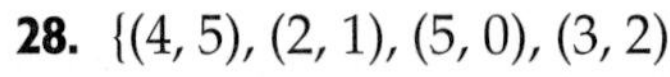

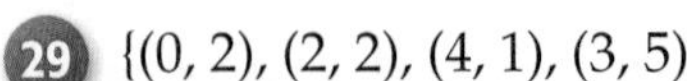

28. $\{(4, 5), (2, 1), (5, 0), (3, 2)\}$
29. $\{(0, 2), (2, 2), (4, 1), (3, 5)\}$
30. $\{(6, 0), (4, 5), (2, 1), (3, 1)\}$
31. $\{(5, 1), (3, 7), (4, 8), (5, 7)\}$

Example 4 p. 27

32. **PIZZA** The cost of a mini pizza is \$7 at Pizza Pizza.
    a. Make a table of ordered pairs in which the x-coordinate represents the number of mini pizzas and the y-coordinate represents the cost of 1, 3, 5, and 7 mini pizzas at Pizza Pizza.
    b. Graph the ordered pairs. Then describe the graph.

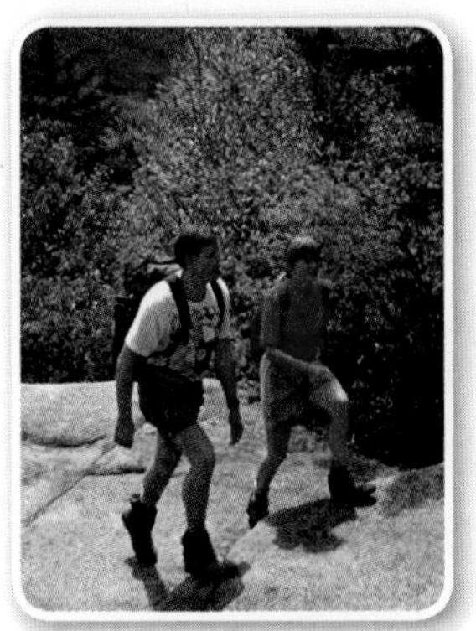

Real-World Link

The Appalachian Trail is one of the most hiked trails in the United States. It winds through 14 states with an estimated length of 2170 miles. Almost 10,000 people have hiked the entire trail.

Source: Appalachian Trail Conservancy

33. HIKING Aaron is hiking in a state park. He averages 3 miles per hour.

a. Make a table of ordered pairs in which the x-coordinate represents the number hours and the y-coordinate represents the number of miles hiked in 1, 2, 4, and 6 hours.

b. Graph the ordered pairs.

34. MULTIPLE REPRESENTATIONS In this problem, you will explore more about relations. Suppose Jamal has only 30 minutes to practice the piano and study for a science test.

a. TABULAR Make a table of ordered pairs showing at least 6 ways Jamal can split the time between the two activities. Let the x-coordinate represent the number of minutes spent playing the piano and the y-coordinate represent the number of minutes spent studying.

b. GRAPHICAL Graph the ordered pairs.

c. VERBAL Describe the general pattern of points of your graph.

d. GRAPHICAL Connect the points on your graph with line segments. Then choose a point on the graph that is *not* one of the points you plotted. Use the coordinates to predict a pair of values for the piano time and study time.

35. SCHOOL Six students in Mr. Maloney's class made a table of ordered pairs for their height in inches x and their shoe size y.

Height (in.)	58	56	62	60	59	61
Shoe Size	6	$5\frac{1}{2}$	$8\frac{1}{2}$	8	7	$7\frac{1}{2}$

a. Graph the ordered pairs.

b. Compare this graph to the graph in Example 4.

H.O.T. Problems Use Higher-Order Thinking Skills

36. OPEN ENDED Write a set of four ordered pairs. Then create a table and graph of the pairs and state the domain and range.

37. REASONING The numbers 4, 7, 10, 13, … form an *arithmetic sequence* because each term can be found by adding the same number to the previous term.

Term Number	1	2	3	4
Term	4	7	10	13

a. Write the set of ordered pairs (term number, term).

b. Graph the ordered pairs.

c. Describe the shape of the graph.

d. If possible, write a rule to find the twentieth term. Explain how you found the rule or why you cannot write a rule.

38. CHALLENGE Describe all of the possible locations for the graph of (x, y) if $x = 2$.

39. WRITING IN MATH Refer to Exercises 32 and 33. For which graph would it make more sense to connect the points with line segments? Explain.

40. WRITING IN MATH Explain why the point $M(4, 3)$ is different from the point $N(3, 4)$.

Standardized Test Practice

41. On the map of a campsite shown below, the tent is located at (3, 7). Which point represents the location of the tent?

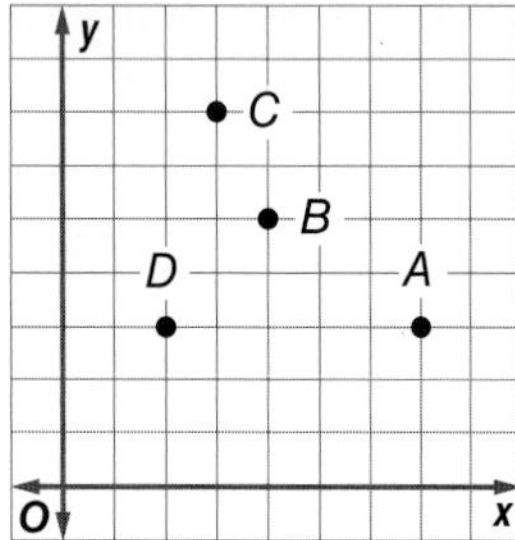

A point A **C** point C

B point B **D** point D

42. Rectangle $ABCD$ has vertices $A(1, 3)$, $B(1, 6)$, and $C(5, 6)$. What are the coordinates of point D?

F (6, 5) **H** (5, 3)

G (5, 1) **J** (6, 1)

43. What is the domain of the relation below?

x	y
1	3
2	4
4	8
6	1
7	4

A {1, 2, 4, 6, 7}

B {1, 3}

C {3, 4, 8, 1, 4}

D {(1, 3), (2, 4), (4, 8), (6, 1), (7, 4)}

44. GRIDDED RESPONSE Point Z is located at (4, 7) on a coordinate plane. Point T is located 3 units to the right and 4 units down from point Z. What is the x-coordinate of point T?

Spiral Review

Name the property shown by each statement. (Lesson 1-3)

45. $5 \cdot 3 = 3 \cdot 5$

46. $6 \cdot 2 \cdot 0 = 0$

47. $0 + 13 = 13$

48. $(5 + x) + 6 = 5 + (x + 6)$

State whether each conjecture is true. If not, give a counterexample. (Lesson 1-3)

49. Division of whole numbers is associative.

50. Subtraction of whole numbers is commutative.

51. SHOPPING Melinda purchased the items shown in the table. (Lesson 1-2)

a. Write an expression to show the total cost of the items.

b. Suppose the cost of the sweater is $25. How much did she spend in all?

Item	Price ($)
sweater	s
purse	$s + 12$
belt	$s + 4$

Skills Review

Find the value of each expression. (Lesson 1-2)

52. $x + 7$ if $x = 8$

53. $d - 5$ if $d = 12$

54. $18 + m$ if $m = 4$

55. $f - 12$ if $f = 28$

56. $s + 14$ if $s = 32$

57. $19 - b$ if $b = 5$

Algebra Lab: Relations and Functions

Math Online glencoe.com
Math in Motion, Animation

A **relation** is any set of ordered pairs. A **function** is a special relation in which each member of the domain is paired with *exactly* one member in the range.

Here is an example to help you remember how to identify functions. Suppose three students are asked to choose their favorite pet. The mapping diagrams below show some possible results.

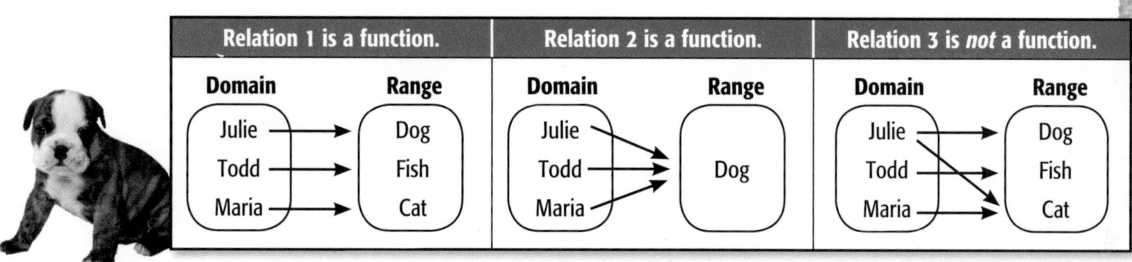

In the example above, the first two relations are functions, because each person chose only one favorite pet. The third relation is *not* a function, because Julie chose two favorite pets, a dog and a cat.

ACTIVITY 1

Step 1 Three students reported the number of cell phone minutes they had every month. Copy and complete the mapping diagram shown below.

Student	1	2	3
Number of Minutes	400	400	750

Domain: 1, 2, 3 — Range: (empty)

Step 2 Student 1 added a phone in the middle of the month and reported two sets of times. Copy and complete the mapping diagram with the new information.

Student	1	1	2	3
Number of Minutes	600	400	400	750

Domain: 1, 2, 3 — Range: (empty)

Analyze the Results

1. A relation can be written as a set of ordered pairs, with the input as the x-coordinate and the output as the y-coordinate. For each relation diagram you drew in the Activity above, write the relation as a set of ordered pairs.
2. Is each relation above a function? Explain your reasoning in terms of the ordered pairs.

Make a mapping diagram for each relation. Then determine whether each relation is a function. Explain.

3. $\{(2, 5), (4, 5), (6, 6), (7, 8)\}$
4. $\{(12, 18), (16, 21), (16, 25), (20, 30)\}$
5. $\{(1, 3), (9, 15), (6, 10), (9, 8)\}$
6. $\{(5, 6), (10, 11), (8, 13), (0, 7)\}$

A *function rule* is the operation(s) performed on the domain value to get the range value.

ACTIVITY 2

Step 1 Use centimeter cubes to build the figures below.

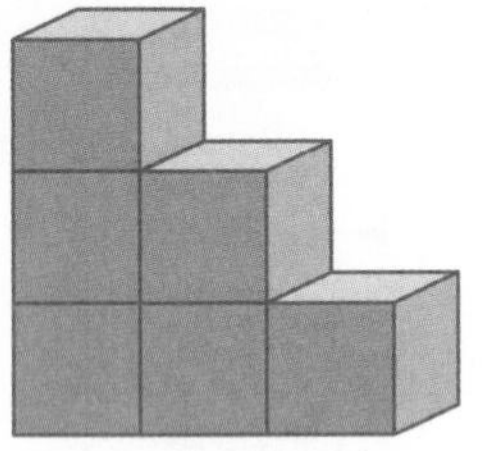

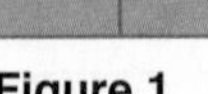

Figure 1

Figure 2

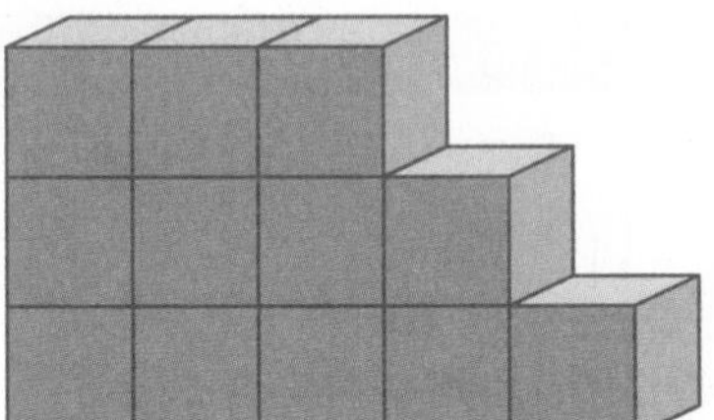

Figure 3

Step 2 Make a table like the one shown and record the figure number and number of cubes used in each figure.

Step 3 Construct the next figure in this pattern. Record your results.

Step 4 Repeat Step 3 until you have found the next four figures in the pattern shown above.

Figure Number	Number of Cubes
1	6
2	9
3	12
⋮	⋮

Analyze the Results

7. Suppose x represents the figure number and y represents the number of cubes. Graph the data from your table on a coordinate plane.

8. Look at the results in the table and graph. Does this data represent a function? Explain your reasoning.

9. Write a rule to determine the number of cubes for any figure number. Use x for the figure number and y for the number of cubes.

10. *Perimeter* is the distance around a figure. Write a rule that describes the relationship between the figure number x and the perimeter y for the figures shown below. Then make a table and graph the data on a coordinate plane.

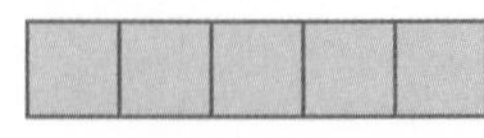

Figure 1

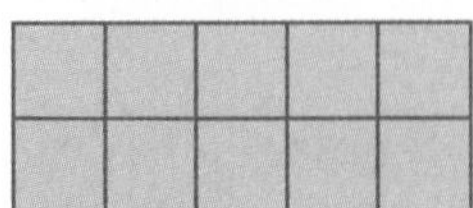

Figure 2

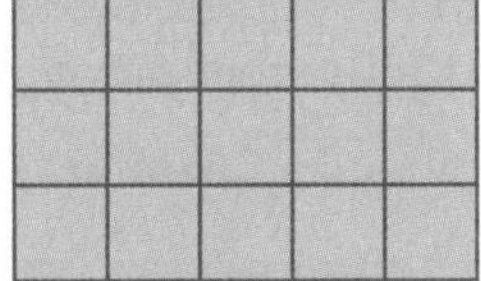

Figure 3

11. **WRITING IN MATH** Look at the results from Exercise 10. Does this data represent a function? Explain your reasoning.

1-5

Words, Equations, Tables, and Graphs

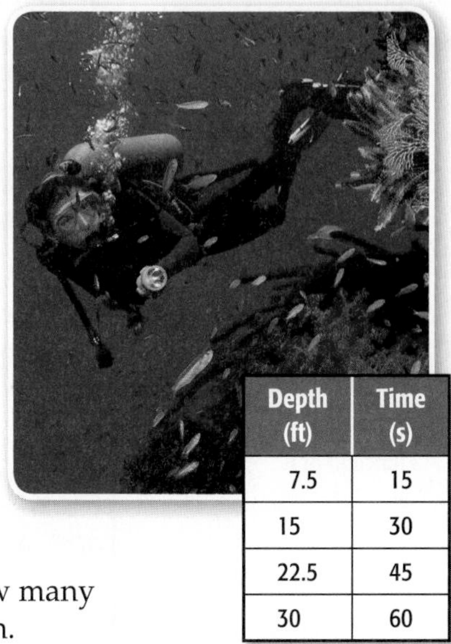

Then

You have already learned how to represent relations as tables and graphs. (Lesson 1-4)

Now

- Use multiple representations to represent functions.
- Translate among different verbal, tabular, graphical, and algebraic representations of functions.

New Vocabulary

function
function rule
function table
equation

Math Online

glencoe.com
- Extra Examples
- Personal Tutor
- Self-Check Quiz
- Homework Help

Why?

The table shows the time it should take a scuba diver to ascend to the surface from several depths to prevent sickness.

Depth (ft)	Time (s)
7.5	15
15	30
22.5	45
30	60

a. On grid paper, graph the data as ordered pairs (depth, time).

b. Write a rule to describe the relationship between the sets of numbers.

c. If a scuba diver is at a depth of 45 feet, how many seconds should she take to ascend? Explain.

Represent Functions A **function** is a relation in which each member of the domain is paired with *exactly* one member in the range. A **function rule** is the operation(s) performed on the domain value to get the range value. A **function table** is a table that lists the x-coordinates (input), rule, and y-coordinates (output).

EXAMPLE 1 Make a Function Table

In a game of *What's My Rule?* Kinna picked the card shown at the right. Make a function table for four different input values and write an algebraic expression for the rule. Then state the domain and range of the function.

What's My Rule?
double a number, then add three

Step 1 Create a function table showing the input, rule, and output. Enter four different input values.

Step 2 The rule "double a number, then add three" translates to $2x + 3$. Use the rule to complete the table.

Step 3 The domain is {1, 2, 3, 4}. The range is {5, 7, 9, 11}.

Input (x)	Rule: $2x + 3$	Output (y)
1	$2(1) + 3$	5
2	$2(2) + 3$	7
3	$2(3) + 3$	9
4	$2(4) + 3$	11

Check Your Progress

1. Jenna picked a game card with the function shown. Make a function table for four different input values and write an algebraic expression for the rule. Then state the domain and range of the function.

What's My Rule?
triple the number and subtract one

Personal Tutor glencoe.com

Vocabulary Link

function
Everyday Use a relationship in which one quality or trait depends on another. Height is usually a function of age.
Math Use a relationship in which a range value depends on a domain value, y is a function of x.

Multiple Representations Words, equations, tables, and graphs can be used to represent functions. An **equation** is a mathematical sentence stating that two quantities are equal. Functions are often written as equations with two variables—one to represent the input and one to represent the output.

Concept Summary — Multiple Representations

For Your FOLDABLE

Words
Distance is equal to 60 miles per hour times the number of hours.

Equation
$d = 60t$

Table

Time (h)	Distance (mi)
1	60
2	120
3	180
4	240

Graph

d
Distance (mi)
240
200
160
120
80
40
0
1 2 3 4 5 6
t
Time (h)

Real-World Career

Air Traffic Controller
Air traffic controllers monitor the movement of objects in the air. They must have good spatial awareness and computational skills.

There are 14 schools that offer programs to become an air traffic controller. After graduation, students must then enroll in the FAA Air Traffic Controller Academy.

EXAMPLE 2 Use Multiple Representations

TECHNOLOGY The navigation message from a satellite to a GPS in an airplane is sent once every 12 minutes.

a. Write an equation to find the number of messages sent in any number of minutes.

Let t represent the time, and n represent the number of messages. The equation is $n = t \div 12$.

b. Make a function table to find the number of messages in 120, 180, 240, and 300 minutes. Then graph the ordered pairs.

Input (t)	$t \div 12$	Output (n)
120	$120 \div 12$	10
180	$180 \div 12$	15
240	$240 \div 12$	20
300	$300 \div 12$	25

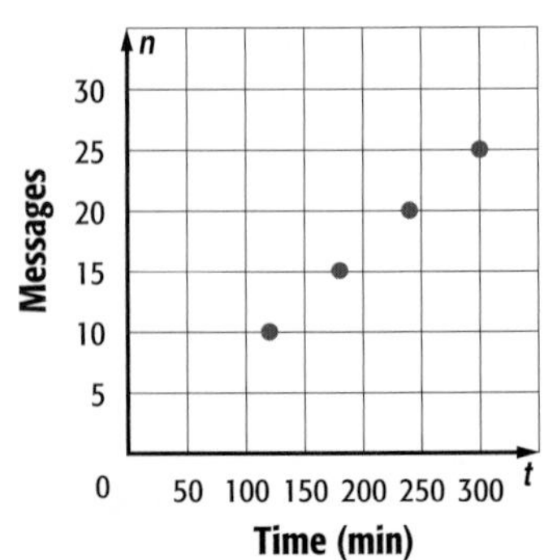

Check Your Progress

2. **MULTIPLE REPRESENTATIONS** The speed of sound is about 1088 feet per second at 32°F in dry air at sea level.

   A. **ALGEBRAIC** Write an equation to find the distance traveled by sound for any number of seconds.

   B. **TABULAR** Make a function table to find the distance sound travels in 0, 1, 2, and 3 seconds. Then graph the ordered pairs for the function.

Personal Tutor glencoe.com

Check Your Understanding

Example 1 p. 33

Copy and complete each function table. Then state the domain and range of the function.

1. The team scores 6 points for each touchdown.

Number of Touchdowns	Number of Points
Input (x)	Output (y)
1	■
2	■
5	■
7	■

2. Bob spent 5 more than 3 times what Anna spent.

Anna's Spending ($)	Bob's Spending ($)
Input (x)	Output (y)
2	■
4	■
6	■
8	■

Example 2 p. 34

3. **MULTIPLE REPRESENTATIONS** There are sixteen ounces in one pound.
   a. **ALGEBRAIC** Write an equation that can be used to find the number of ounces in any number of pounds.
   b. **TABULAR** Make a function table to find the number of ounces in 5, 8, 11, and 13 pounds.
   c. **GRAPHICAL** Graph the ordered pairs for the function.

Practice and Problem Solving

= **Step-by-Step Solutions** begin on page R11. **Extra Practice** begins on page 810.

Example 1 p. 33

Copy and complete each function table. Then state the domain and range of the function.

4. Each ticket to the school musical costs $8.

Number of Tickets	Total Cost ($)
Input (x)	Output (y)
4	■
7	■
9	■
12	■

5. The dog weighs 4 pounds more than the cat.

Weight of Cat (lb)	Weight of Dog (lb)
Input (x)	Output (y)
3	■
6	■
9	■
12	■

6. Today's attendance is four less than half of yesterday's attendance.

Yesterday's Attendance	Today's Attendance
Input (x)	Output (y)
14	■
18	■
22	■
26	■

7. Casey has 5 less than 4 times as many baseball cards than Ben.

Ben's Cards	Casey's Cards
Input (x)	Output (y)
3	■
7	■
11	■
15	■

Example 2 p. 34

8. **MULTIPLE REPRESENTATIONS** One roll of quarters contains 40 quarters.

   a. **ALGEBRAIC** Write an equation that can be used to find the number of quarters q in any number of rolls of quarters r.

   b. **TABULAR** Make a function table to find the number of quarters in 3, 4, 5, and 6 rolls.

   c. **GRAPHICAL** Graph the ordered pairs for the function.

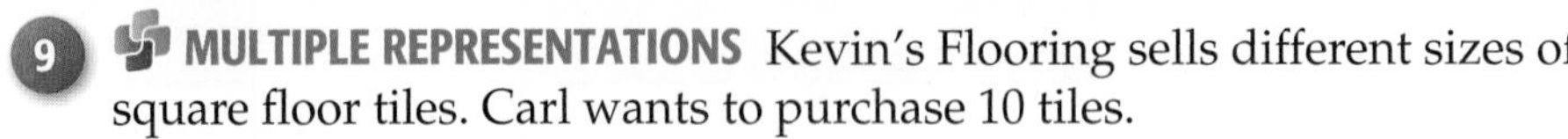

9. **MULTIPLE REPRESENTATIONS** Kevin's Flooring sells different sizes of square floor tiles. Carl wants to purchase 10 tiles.

   a. **ALGEBRAIC** Write an equation that can be used to find the area of any ten square floor tiles. (*Hint:* area = side × side)

   b. **TABULAR** Make a function table to find the area covered by ten tiles that measure 6, 12, 15, and 24 inches on one side.

   c. **GRAPHICAL** Graph the ordered pairs for the function.

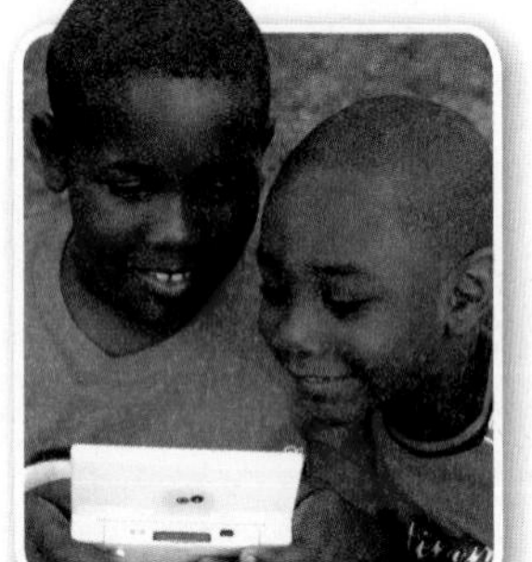

Real-World Link

In 2006, sales for computer and video games totaled about $7,400,000,000. The average age of game buyers was 38.

Source: Entertainment Software Association

10. **MULTIPLE REPRESENTATIONS** Sales for a new video game offered by Technogames is shown at the right.

    a. **TABULAR** Make a function table showing the input (month) and the output (video games sold).

    b. **ALGEBRAIC** Write an equation that can be used to find the number of games sold g for any month m.

    c. **ANALYTICAL** Is the set of ordered pairs (month, games sold) a function? Explain.

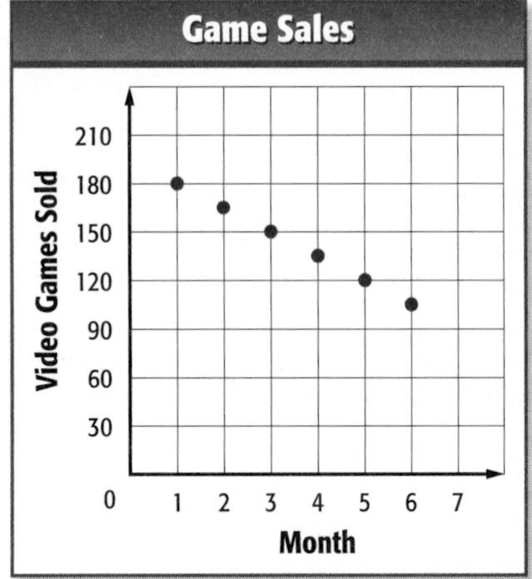

11. **MULTIPLE REPRESENTATIONS** The table shows the temperatures at various depths in a lake.

    a. **GRAPHICAL** Graph the ordered pairs on a coordinate plane.

    b. **ALGEBRAIC** Can you write one equation that can be used to find the temperature t based on the depth in the lake d? Explain.

    c. **ANALYTICAL** Is the relation a function? Explain.

Depth (ft)	Temperature (°F)
0	74
10	72
20	71
30	61
40	55
50	53

H.O.T. Problems

Use Higher-Order Thinking Skills

12. **OPEN ENDED** Write about a real-world situation that can be represented by the equation $y = 4x$.

13. **CHALLENGE** Write a rule for the function shown in the table.

Input (x)	1	2	3	4
Output (y)	10	12	14	16

14. **WRITING IN MATH** Give a set of data that does *not* represent a function. Explain your reasoning.

Standardized Test Practice

15. EXTENDED RESPONSE The function table below follows a function rule.

x	y
1	5
3	11
5	17
7	23
8	■
10	■

a. Complete the table by filling in the two missing numbers.

b. Based on the table, write a function rule that represents the relationship between x and y.

16. Walik is buying CDs from an online store. Each CD costs $12.99. There is a flat shipping charge of $4.95. Which expression represents the cost of purchasing m CDs?

A $m(12.99 + 4.95)$

B $4.95m + 12.99$

C $12.99m + 4.95$

D $(12.99 - 4.95)m$

17. Which of the following equations would best describe the graph shown below?

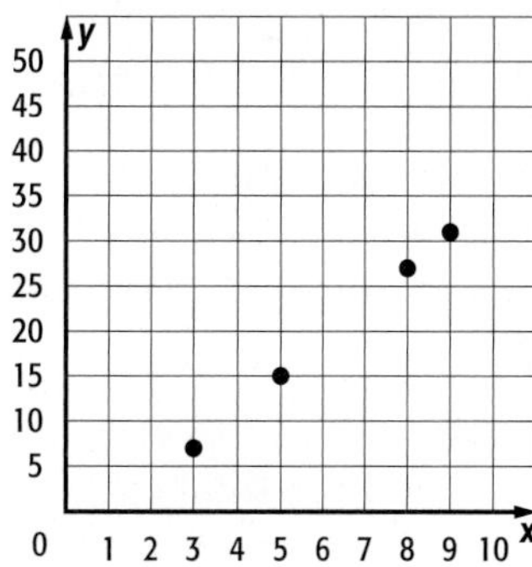

F $y = 2x + 1$

G $y = 3x - 2$

H $y = 5x - 10$

J $y = 4x - 5$

18. Which of the following sets of ordered pairs is a function?

A {(0, 0), (1, 0), (0, 1), (1, 1)}

B {(0, 3), (1, 3), (2, 3), (3, 3)}

C {(0,1), (0, 2), (0, 3), (0, 4)}

D {(0,0), (0, 4), (4, 0), 4, 4)}

Spiral Review

Refer to the coordinate plane at the right. Write the ordered pair that names each point. (Lesson 1-4)

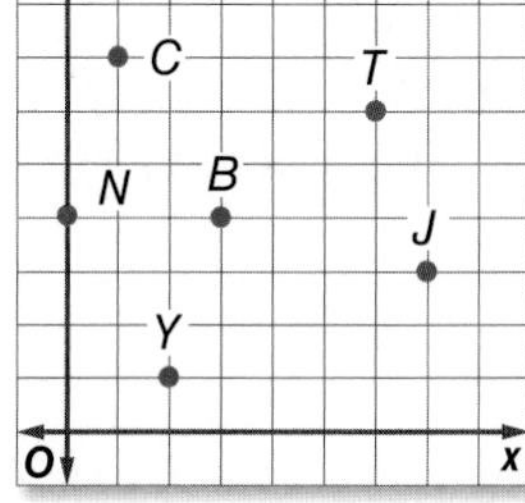

19. C **20.** J **21.** N

22. T **23.** Y **24.** B

ALGEBRA Simplify each expression (Lesson 1-3)

25. $(m + 8) + 4$ **26.** $(17 + p) + 9$ **27.** $21 + (k + 16)$

28. $(6 \cdot c) \cdot 8$ **29.** $8 \cdot (y \cdot 2)$ **30.** $25s\,(3)$

31. MONEY There are 20 nickels in one dollar.

a. Write an algebraic expression that can be used to find the number of nickels in any number of dollars n.

b. How many nickels are in $7.00?

Skills Review

Find each quotient. (Previous Course)

32. $68 \div 4$ **33.** $84 \div 6$ **34.** $126 \div 9$ **35.** $135 \div 9$

EXTEND 1-5

Graphing Technology Lab
Function Tables

Math Online glencoe.com
- Other Calculator Keystrokes
- Graphing Technology Personal Tutor

You can use a TI-Nspire graphing calculator to create function tables. If you enter a function and the domain values, the calculator will give you the corresponding range values.

ACTIVITY

Packages of batteries cost $4 each at a store. Bridgett has a coupon for $2 off her total purchase. Find the total cost y of buying x packages of batteries. Use a function table to find the range of $y = 4x - 2$ if the domain is {2, 5, 6, 8, 10}.

Step 1 Enter the function.

- The graphing calculator uses x for the domain values and *f1(x)* for the range values. So, $f1(x) = 4x - 2$ represents $y = 4x - 2$.
- Access Graphs and Geometry.

KEYSTROKES: on [home] 2

- Enter $f1(x) = 4x - 2$.

KEYSTROKES: 4 [X] [–] 2 [enter]

Step 2 Format the table.

- Add Function Table.

KEYSTROKES: [menu] 2 8

- Edit Function Table Settings.

KEYSTROKES: [menu] 5 3 ▸▸▸▸ [click] ▾ [click] ▸▸ [click]

- You can use [ctrl] + [tab] to switch between the graph and the table.

Step 3 Find the range by entering the domain values.

- Enter the domain values given above. The range values will appear automatically.

KEYSTROKES:

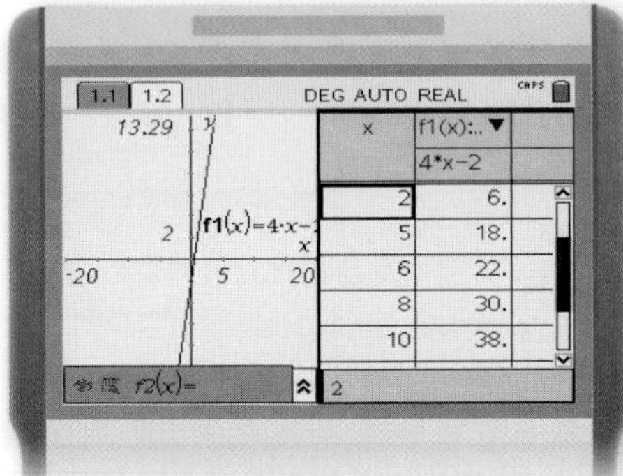

- Note the graph is a different representation of the same data.
- The *trace* function could be used to find domain and range values that were not requested.

Analyze the Results

Use the Function Table option on a TI-Nspire to complete each exercise.

1. **MULTIPLE REPRESENTATIONS** Suppose you are using the formula $d = rt$ to find the distance d a car travels for the times t in hours given by {0, 1, 3.5, 10}.

   a. **ALGEBRAIC** If the rate is 60 miles per hour, what function should be entered in the $f(x) =$ list?

   b. **GRAPHICAL** Make a graph and function table for the given domain. Make sure to use an appropriate window.

   c. **NUMERICAL** Between which two times in the domain does the car travel 150 miles?

   d. **NUMERICAL** How many miles will the car have traveled after 12 hours?

   e. **VERBAL** Describe how a function table and graph can be used to estimate the time it takes to drive 150 miles.

EXPLORE
1-6 Algebra Lab Scatter Plots

Math Online glencoe.com
Math *in Motion*, Animation

Objective
Use a scatter plot to investigate the relationship between two sets of data.

Sometimes it is difficult to determine whether a relationship exists between two sets of data by simply looking at them. To determine whether a relationship exists, you can write the data as a set of ordered pairs and then graph the ordered pairs on a coordinate plane. This kind of graph is called a **scatter plot**.

ACTIVITY 1

Collect data to investigate whether a relationship exists between height and arm span.

Step 1 Have a classmate measure your height and the length of your arm span with a yardstick to the nearest half inch. Then write your height and arm span as an ordered pair.

Step 2 Combine your data with that of your classmates.

Step 3 Make a list of ordered pairs in which the x-coordinate represents height and the y-coordinate represents arm span.

Step 4 Draw a coordinate plane like the one shown and graph the ordered pairs (height, arm span).

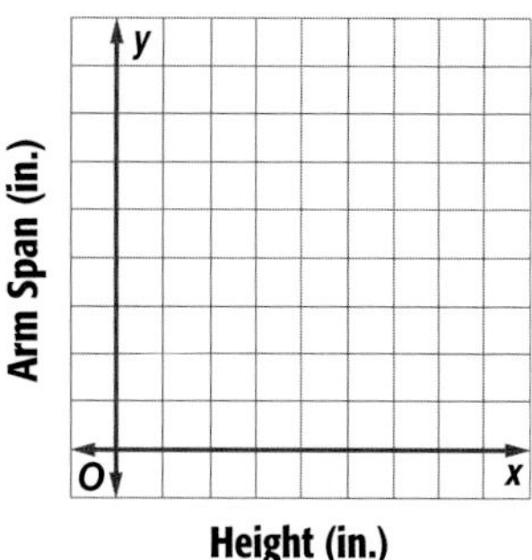

StudyTip
A scatter plot is a collection of points that may or may not show a relationship between sets of data.

Analyze the Results

1. Does there appear to be a trend in the data? If so, describe the trend.
2. Using your graph, estimate the arm span of a person whose height is 60 inches. 72 inches.
3. How does your arm span compare with your height?
4. **MAKE A CONJECTURE** Suppose the variable x represents height and the variable y represents arm span. Write an equation relating x to the arm span y.
5. **COLLECT DATA** Collect and graph data to determine whether a relationship exists between height and foot length. Explain your results.

1-6 Scatter Plots

Then
You have already graphed ordered pairs and relations on a coordinate system. (Lesson 1-4)

Now
- Construct scatter plots.
- Analyze trends in scatter plots.

New Vocabulary
scatter plot

Math Online
glencoe.com
- Extra Examples
- Personal Tutor
- Self-Check Quiz
- Homework Help

Why?

The graph shows the percent of the total population that are Internet users in the United States over the years.

a. Do you see a trend in the data?

b. Predict the percent of the total population that are internet users in the U. S. in 2012.

c. Do you think this trend will continue? Explain your reasoning.

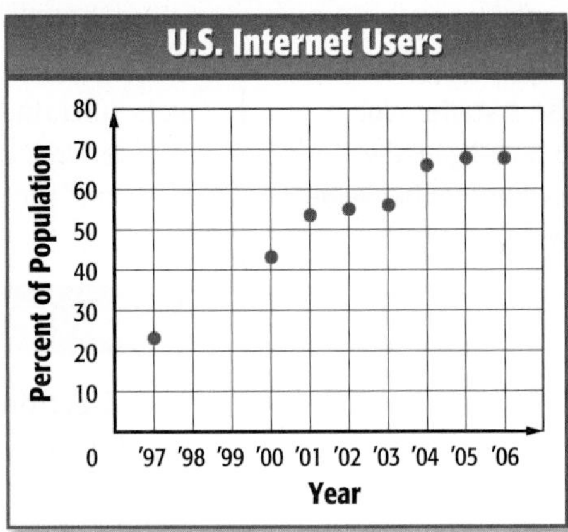

Source: NY Times Almanac

Construct Scatter Plots A **scatter plot** shows the relationship between a set of data with two variables, graphed as ordered pairs on a coordinate plane.

EXAMPLE 1 Construct a Scatter Plot

POPULATION Make a scatter plot of the approximate population of Tampa, Florida.

Year	Population (thousands)
1940	108
1950	125
1960	275
1970	278
1980	272
1990	280
2000	303

Source: U.S. Census Bureau

Let the horizontal axis, or x-axis, represent the year. Let the vertical axis, or y-axis, represent the population. Then graph ordered pairs (year, population).

Population of Tampa, Florida 1940–2000

Population (thousands): 0, 50, 100, 150, 200, 250, 300, 350

Year: '40 '50 '60 '70 '80 '90 '00

Check Your Progress

1. **BASKETBALL** Make a scatter plot of the number of field goals made by a WNBA player from 1997–2006.

Season	'00	'01	'02	'03	'04	'05	'06	'07	'08	'09
Number of Field Goals Made	160	202	182	197	221	189	165	223	204	257

Personal Tutor glencoe.com

StudyTip

Scatter Plots Data that appear to go uphill from left to right show a positive relationship. Data that appear to go downhill from left to right show a negative relationship.

Analyze Scatter Plots The following scatter plots show the types of relationships or patterns of two sets of data.

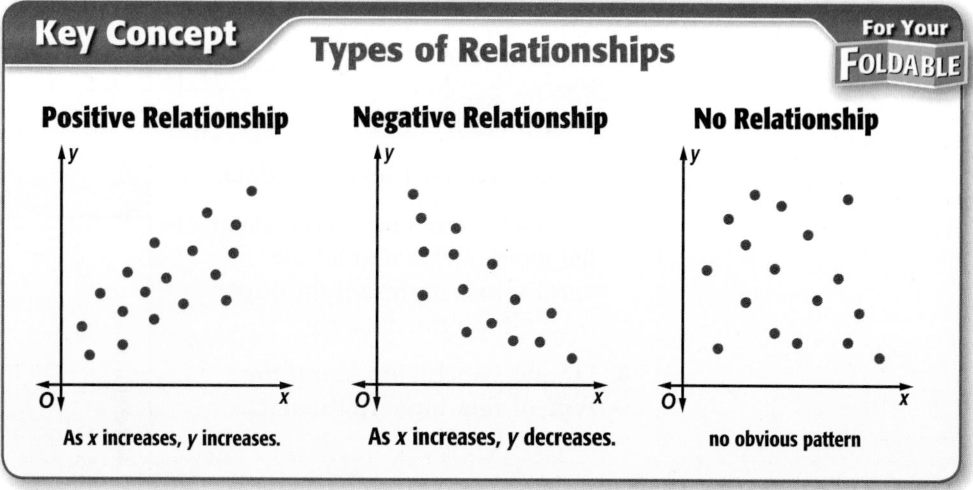

Key Concept
Types of Relationships
For Your FOLDABLE
Positive Relationship
Negative Relationship
No Relationship
y
O
x
As x increases, y increases.
As x increases, y decreases.
no obvious pattern

Real-World Link

Over 5 million people celebrate their birthday each week in the United States. On average, 700,000 people celebrate their birthday every day.

Source: Hallmark

EXAMPLE 2 Interpret Scatter Plots

PEOPLE Determine whether the scatter plot of the birth month and age of people in a park might show a *positive*, *negative*, or *no* relationship. Explain your answer.

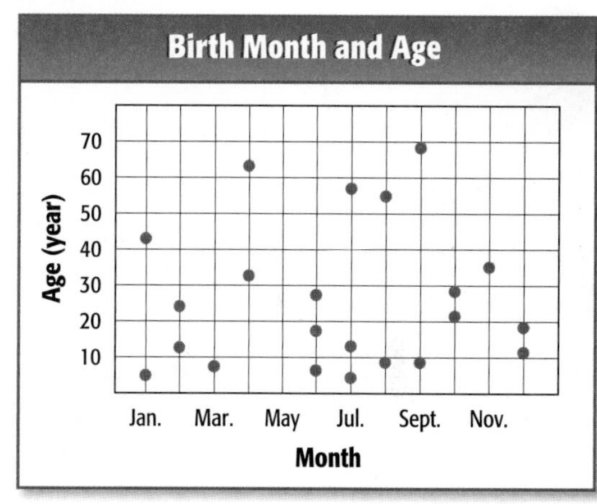

A person's age is not affected by their birth month. Therefore, the scatter plot of the data would show no relationship.

Check Your Progress

2. **TEST SCORES** Determine whether the scatter plot of the study time and resulting test score might show a *positive*, *negative*, or *no* relationship. Explain your answer.

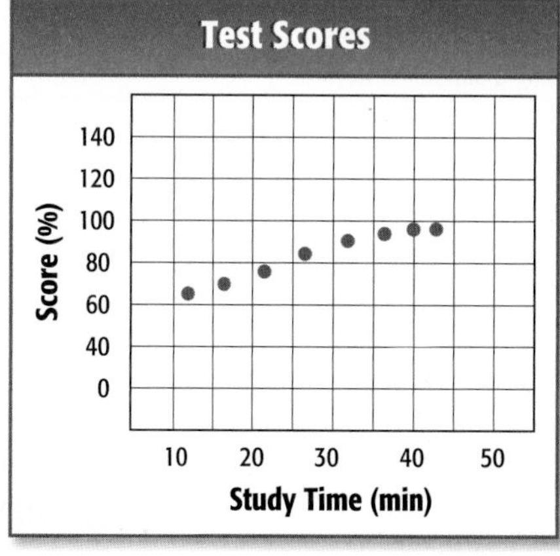

Personal Tutor glencoe.com

You can also use scatter plots to spot trends, draw conclusions, and make predictions about the data.

Real-World EXAMPLE 3 Analyze Scatter Plots

SALES **The table shows how many cups of coffee were sold during outdoor games at certain temperatures.**

Temperature (°F)	34	36	43	45	48	54	59	62	65	71	75
Number of Cups	42	39	34	33	37	18	18	13	12	9	8

a. Make a scatter plot of the data.

Let the horizontal axis represent the temperature and let the vertical axis represent the cups sold. Graph the data.

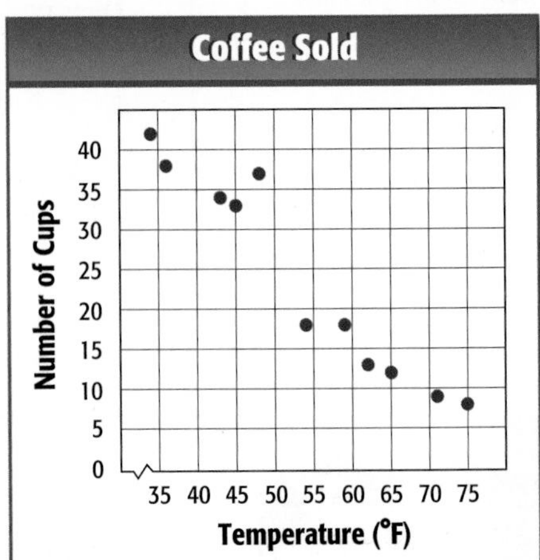

b. Draw a conclusion about the type of relationship the data shows. Explain.

As the temperature increases, the number of cups sold decreases. So, the scatter plot shows a negative relationship.

c. Predict the number of cups sold if the temperature is 30° Fahrenheit.

By looking at the pattern in the graph, you can predict that between 40 and 45 cups will be sold.

Check Your Progress

3. **KEYBOARDING** The table shows keyboarding speeds in words per minute (wpm) of 12 students.

Experience (weeks)	4	7	8	1	6	3	5	2	9	6	7	10
Speed (wpm)	38	46	48	20	40	30	38	22	52	44	42	55

A. Make a scatter plot of the data.

B. Draw a conclusion about the type of relationship the data shows.

C. Predict the keyboarding speed of a student with 12 weeks of experience.

Personal Tutor glencoe.com

Real-World Link

Average typing speed is about 50 to 70 words per minute. The world's fastest typer can maintain a typing speed of 150 words per minute for 50 minutes using a simplified keyboard.

Check Your Understanding

Examples 1–3 pp. 40–42

1. **TEXT MESSAGING** The table shows the number of friends per group in a cell phone network and the average number of text messages made per day.

Number of Friends	5	4	2	4	6	3	4	3	5	2
Number of Text Messages	131	86	39	98	234	43	100	85	190	55

a. Make a scatter plot of the data.

b. Draw a conclusion about the type of relationship the data shows. Explain.

c. If a relationship exists, predict the number of text messages made during the week for a group of 8.

Practice and Problem Solving

● = **Step-by-Step Solutions** begin on page R11.
Extra Practice begins on page 810.

Example 1 p. 40

2. **MUSIC** The table shows the number of songs and the total number of minutes on different CDs. Make a scatter plot of the data.

Number of Songs	15	20	13	12	15	16	17	18	20	19	11	14
Total Minutes	64	63	70	59	61	77	75	71	78	75	63	69

3. **OLYMPICS** The table shows the winning times for the women's Olympic 100-meter run. Make a scatter plot of the data.

Year	'28	'32	'36	'48	'52	'56	'60	'64	'68
Winning Time(s)	12.2	11.9	11.5	11.9	11.5	11.5	11.0	11.4	11.08
Year	'72	'76	'80	'84	'88	'92	'96	'00	'04
Winning Time(s)	11.07	11.08	11.06	10.97	10.54	10.82	10.94	10.75	10.93

Example 2 p. 41

Determine whether the data for the following might show a *positive*, *negative*, or *no* relationship. Explain your answer.

4.

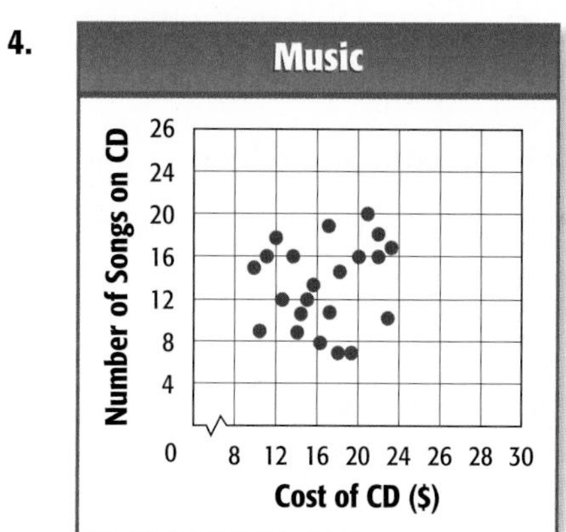

5.

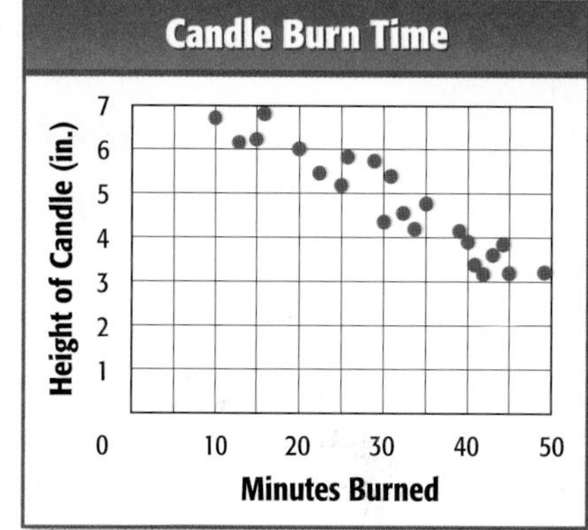

6. size of household and amount of water bill
7. distance driven and gallons of gasoline used
8. age of adults and number of pets currently owned
9. size of lawn and amount of water used to water the lawn

Example 3 p. 42

10. **MULTIPLE REPRESENTATIONS** The table shows the approximate number of students, in millions, that attended public schools in the U.S. in various years.

Year	1900	1920	1940	1960	1980	2000
Number of Students	16	22	25	35	42	47

a. **GRAPHICAL** Make a scatter plot of the data.

b. **VERBAL** Draw a conclusion about the type of relationship the data shows. Explain your reasoning.

c. **ANALYTICAL** If a relationship exists, predict the number of students in public school in the year 2020.

Real-World Link

The highest temperature recorded in the United States was 134°F on July 10, 1913, in Death Valley, California. This is just 2 degrees below the world temperature high.

Source: National Weather Service

11. **MULTIPLE REPRESENTATIONS** The table shows the average high temperatures for Louisville, Kentucky.

Month	Temperature (°F)	Month	Temperature (°F)
January	41	July	87
February	47	August	86
March	57	September	79
April	67	October	68
May	75	November	56
June	83	December	45

a. **GRAPHICAL** Make a scatter plot of the data.

b. **ANALYTICAL** Draw a conclusion about the type of relationship the data shows. Explain.

12. **MULTIPLE REPRESENTATIONS** A person breathes about 20 times per minute.

a. **TABULAR** Let x represent minutes and y represent the number of times a person breathes. Make a table using the x-values of 1, 2, 4, 8, and 10.

b. **GRAPHICAL** Make a scatter plot of the data.

c. **ALGEBRAIC** Write an equation relating x and y.

d. **ANALYTICAL** Predict how many times a person would breathe in 25 minutes. Explain how you arrived at your answer.

13. **MULTIPLE REPRESENTATIONS** The scatter plot at the right shows a relationship between two variables, x and y.

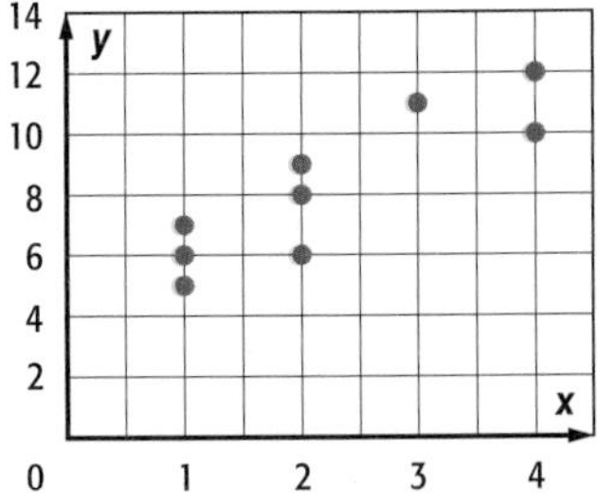

a. **TABULAR** Make a table showing the x-values and y-values.

b. **VERBAL** Write about a situation that could produce the data values shown.

c. **ANALYTICAL** Draw a conclusion about the type of relationship the data shows. Explain.

14. **DVDS** The scatter plot at the right shows the number of DVD players sold from 1996 to 2006.

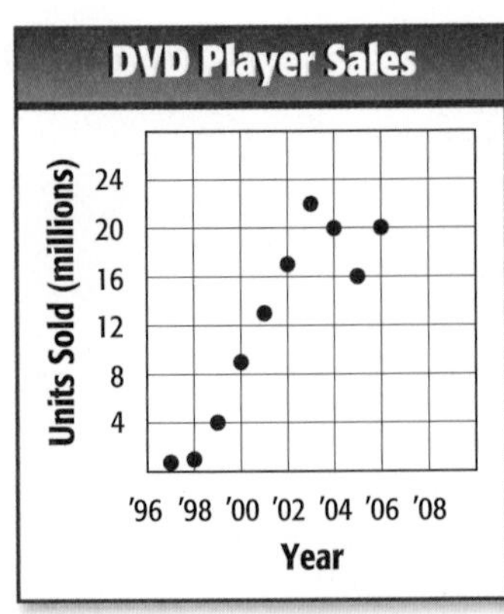

a. Do the data show a *positive*, *negative*, or *no* relationship between the year and the number of players sold?

b. What appears to be the trend between 1997 and 2003? 2003 and 2006?

c. What factors could contribute to the trend displayed in the scatter plot?

15. **AREA** The area of a rectangle can be found using the expression $\ell \cdot w$ where ℓ is the length and w is the width of the rectangle.

   a. Make a table of values showing possible lengths and widths of a rectangle with an area of 36 square inches.

   b. Graph the ordered pairs (length, width). Connect the points.

   c. Predict the width of a rectangle with a length of 7 inches.

H.O.T. Problems

Use Higher-Order Thinking Skills

16. **OPEN ENDED** Describe an example of two sets of data that would show a positive relationship.

NUMBER SENSE What type of relationship is shown on a graph that shows the following values.

17. As x increases, y decreases.

18. As x decreases, y decreases.

19. **WRITING IN MATH** Explain why a scatter plot of swimming pool attendance and ice cream sales might show a positive relationship. Does this mean that one factor caused the other?

Real-World Link

A recent survey showed that America's favorite ice cream flavors are vanilla, chocolate, neopolitan, strawberry, and cookies and cream.

Source: National Eating Trends Services

20. **CHALLENGE** The number 3 is considered the *principal square root* of 9 because 3×3 is equal to 9. Some numbers have whole number square roots and others, like 5, do not.

   a. **TABULAR** Let x represent numbers with whole number square roots and y represent the positive square root of x. Copy and complete the table. Is the data in the table a function? Explain.

x	y
1	1
4	■
9	■
16	■
■	5
36	■

   b. **GRAPHICAL** Make a scatter plot of the data. Describe the relationship shown in the graph.

   c. **TABULAR** Reverse the x- and y-coordinates and write the new set of ordered pairs in a table. Is the set of ordered pairs a function? Explain.

   d. **GRAPHICAL** Make a scatter plot of the data in part **c**. Describe the relationship shown in the graph.

21. **FIND THE ERROR** Sonya and Melisa were asked to determine if the table represents a function. Is either of them correct? Explain your reasoning.

x	y
4	1
3	3
1	5
2	5

> Sonya
>
> It is not a function because it shows a negative relationship.

> Melisa
>
> It is not a function becauses the y-value of 5 is paired with two x-values

22. **WRITING IN MATH** The relationship between variables can be strong or weak. Describe a situation that would have a *strong* positive relationship. What makes it a strong positive relationship?

Standardized Test Practice

23. The scatter plot below shows the relationship between the age of a car and the average number of visits to a repair shop.

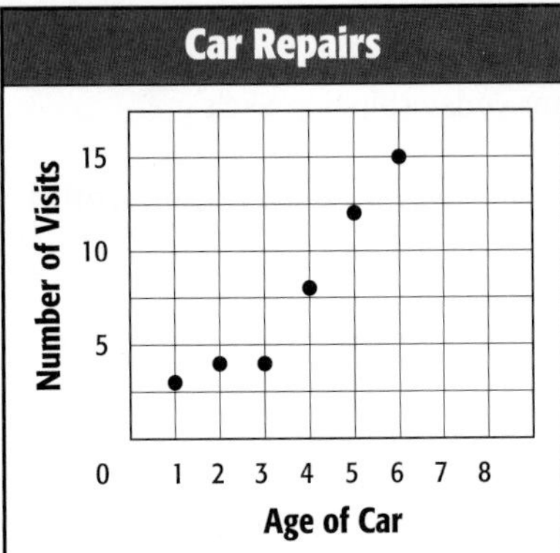

Which statement below is best supported by this scatter plot?

A As the age of the car increases, the number of visits doubles.

B As the age of the car increases, the number of visits decreases.

C As the age of the car increases, the number of visits increases.

D As the age of the car increases, the number of visits stays the same.

24. Refer to the graph in Exercise 23. Which kind of relationship exists between the age of the car and the number of visits to the repair shop?

F positive **H** neutral

G negative **J** none

25. The table shows the number of people that attended a new movie over the course of a week.

Day	Attendance
1	13,400
3	13,000
5	12,600
7	12,200

Which answer choice below is the *best* prediction for attendance on the 8th day?

A 11,800 **C** 12,100

B 12,000 **D** 11,700

26. SHORT RESPONSE Based on the table in Exercise 25, which kind of relationship exists between the length of time the movie is at the theatre and the number people that attend the movie?

Spiral Review

27. MULTIPLE REPRESENTATIONS Cornet Cable charges $32.50 a month for basic cable television. Each premium channel selected costs an additional $4.95 per month. (Lesson 1-5)

a. ALGEBRAIC Write an expression to find the cost of a month of cable service.

b. TABULAR Make a table to show the monthly cost for 0, 1, 2, 3, and 4 premium channels.

c. GRAPHICAL Graph the ordered pairs from the table.

Graph each ordered pair on a coordinate system. (Lesson 1-4)

28. $A(3, 3)$ **29.** $D(1, 8)$ **30.** $G(2.5, 7)$

31. $X(7, 2)$ **32.** $P(0, 6)$ **33.** $N(4\frac{1}{2}, 0)$

Skills Review

Replace each ● with <, >, or = to make a true sentence.

34. 6 ● 8 **35.** 12 ● 9 **36.** 14 ● 15

37. 3 ● 3 **38.** 17 ● 15 **39.** 29 ● 19

Graphing Technology Lab
Scatter Plots

Math Online glencoe.com
- Other Calculator Keystrokes
- Graphing Technology Personal Tutor

You have learned that graphing ordered pairs as a scatter plot on a coordinate plane is one way to make it easier to "see" if there is a relationship. You can use a graphing calculator to create scatter plots.

ACTIVITY 1

GEOGRAPHY The U.S. Census Bureau estimated the population and area of certain countries. Mr. Henderson's geography class wanted to see if there was a correlation between a country's size and population. Use the table of data below to make a scatter plot.

Country	Hemisphere	Area (million square miles)	Population (millions)
Australia	Eastern	2.97	20.3
Brazil	Western	3.29	188.1
Canada	Western	3.86	33.1
Chile	Western	0.29	16.1
Egypt	Eastern	0.39	78.9
India	Eastern	1.27	1095.4
Japan	Eastern	0.15	127.5
Mexico	Western	0.76	107.5
Russia	Eastern	6.60	142.9
United States	Western	3.72	298.4

Step 1 Enter the data.

- Clear any existing list.

KEYSTROKES: STAT ENTER ▲ CLEAR ENTER

- Enter the area as L1 and population as L2.

KEYSTROKES: STAT ENTER 2.97 ENTER 3.29 ENTER … 3.72 ENTER ▶ 20.3 ENTER 188.1 ENTER … 298.4 ENTER

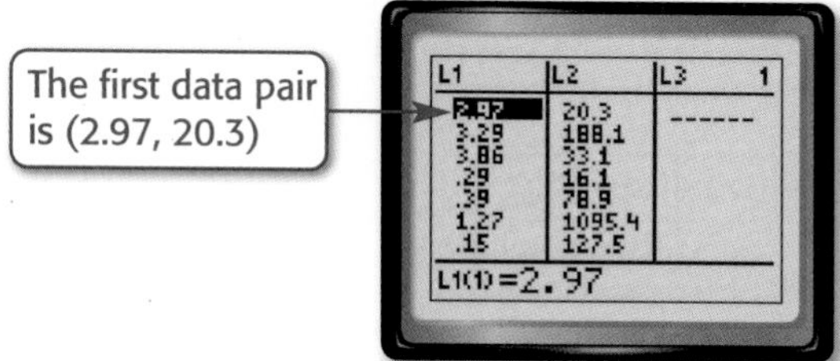

Step 2 Format the graph.

- Turn on the statistical plot.

KEYSTROKES: 2nd STAT PLOT ENTER ENTER

- Select the scatter plot, L1 as the Xlist and L2 as the Ylist.

KEYSTROKES: ▼ ENTER ▼ 2nd L1 ENTER

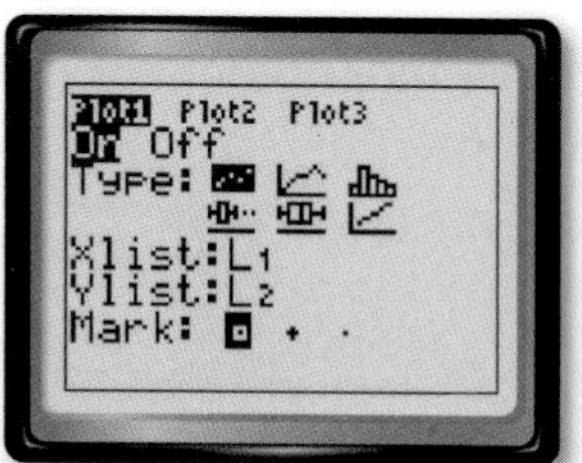

Step 3 Graph the data.

- Display the scatter plot.

  KEYSTROKES: ZOOM 9

- Use the TRACE feature and the left and right arrow keys to move from one point to another.

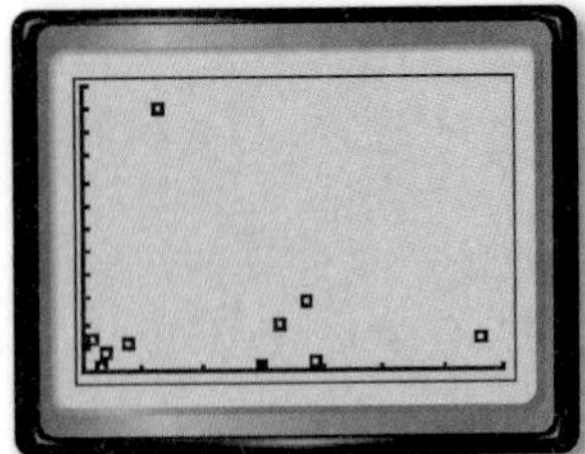

Analyze the Results

1. Press TRACE. Use the left and right arrow keys to move from one point to another. What do the coordinates of each data point represent?
2. Describe the scatter plot.
3. Is there a relationship between the area of a country and its population? If so, write a sentence or two that describes the relationship.
4. Based on the scatter plot, does the area of a country affect its population?
5. Separate the data by hemisphere. Enter the area and population for the western hemisphere as lists L1 and L2 and for the eastern hemisphere as lists L3 and L4. Use the graphing calculator to make a scatter plot with different marks for the western and eastern hemispheres. Do the different scatter plots agree with your answer in Exercise 3? Explain.
6. **SCIENCE** A zoologist studied the extinction times (in years) of island birds. The zoologist wanted to see if there was a relationship between the average number of nests and the time needed for each bird to become extinct on the islands. The results are shown in the table.

Bird Name	Bird Size	Average Number of Nests	Extinction Time
Buzzard	large	2.0	5.5
Quail	large	1.0	1.5
Curlew	large	2.8	3.1
Cuckoo	large	1.4	2.5
Magpie	large	4.5	10.0
Swallow	small	3.8	2.6
Robin	small	3.3	4.0
Stonechat	small	3.6	2.4
Blackbird	small	4.7	3.3
Tree-Sparrow	small	2.2	1.9

   a. Use your graphing calculator to make a scatter plot of the data.

   b. Is there a relationship between the average number of nests and extinction times? If so, write a sentence or two that describes the relationship.

   c. Are there any differences between the extinction times of large birds versus small birds?

7. Make a scatter plot of the data and describe the relationship, if any, between the x- and y-values.

x	3.2	3.8	4.3	4.7	5.5	5.9	7.2	7.8	8.2
y	15.7	13.2	13.9	11.1	12.8	11.4	10.7	11.3	10.4

8. **RESEARCH** Find two sets of data on your own. Then determine whether a relationship exists between the data.

CHAPTER 1 Study Guide and Review

Math Online glencoe.com
• STUDY TO GO
• Vocabulary Review

Chapter Summary

Key Concepts

Order of Operations (Lesson 1-1)

Step 1 Evaluate the expressions inside grouping symbols.

Step 2 Multiply and/or divide in order from left to right.

Step 3 Add and/or subtract in order from left to right.

Expressions (Lessons 1-1 and 1-2)

- A numerical expression contains a combination of numbers and operations.
- An algebraic expression is a numerical expression that contains at least one variable.

Properties (Lesson 1-3)

- Commutative Property
  $4 + 5 = 5 + 4$
  $2 \cdot 7 = 7 \cdot 2$
- Identity Property
  $3 + 0 = 3$
  $3 \cdot 1 = 3$
- Associative Property
  $10 + (8 + 3) = (10 + 8) + 3$
  $6 \cdot (5 \cdot 9) = (6 \cdot 5) \cdot 9$

Coordinate Plane (Lesson 1-4)

- *x*- and *y*- coordinates are used to indicate a point's position in a coordinate system.
- The domain of a relation is the set of *x*-coordinates and the range of a relation is the set of *y*-coordinates.

Functions (Lesson 1-5)

- A function is a relation that assigns exactly one *domain* value for each *range* value.

FOLDABLES® Study Organizer

Be sure the Key Concepts are noted in your Foldable.

Key Vocabulary

algebra (p. 11)
algebraic expression (p. 11)
coordinate plane (p. 25)
coordinate system (p. 25)
counterexample (p. 19)
deductive reasoning (p. 20)
defining a variable (p. 11)
domain (p. 27)
equation (p. 34)
evaluate (p. 6)
function (p. 33)
function rule (p. 33)
function table (p. 33)
graph (p. 26)
numerical expression (p. 5)
order of operations (p. 6)
ordered pair (p. 25)
origin (p. 25)
properties (p. 18)
range (p. 27)
relation (p. 27)
scatter plot (p. 40)
simplify (p. 20)
variable (p. 11)
***x*-axis** (p. 25)
***x*-coordinate** (p. 25)
***y*-axis** (p. 25)
***y*-coordinate** (p. 25)

Vocabulary Check

State whether each sentence is *true* or *false*. If *false*, replace the underlined term to make a true sentence.

1. An example of a(n) <u>algebraic expression</u> is $a + 2c + 6$.
2. The set of all *y*-coordinates is called the <u>domain</u>.
3. A <u>variable</u> is a letter used to represent a value.
4. The <u>*x*-axis</u> is formed by the intersection of two number lines at right angles at their zero points.
5. An <u>ordered pair</u> names a point on a coordinate system.
6. To find the value of a numerical expression, you <u>evaluate</u> it.
7. A <u>numerical expression</u> contains a combination of numbers and operations.
8. The set of all *x*-values is called the <u>domain</u>.
9. A <u>coordinate plane</u> is a graph that shows the relationship between a set of data with two variables.

CHAPTER 1

Study Guide and Review

Lesson-by-Lesson Review

1-1 Words and Expressions (pp. 5–9)

10. Write a numerical expression for the verbal phrase *the total number of students if there are nineteen in one class and thirteen in another.*

Evaluate each expression.

11. $6(7) + 3$

12. $4[(12 - 4) + 2]$

13. **PROFITS** Yu, Collin, and Sydney spent \$284 to make bracelets. They sold the bracelets for \$674. If they split the profits evenly, how much did each person earn?

EXAMPLE 1

Write the phrase *the total number of post cards if eight people each buy five* as a numerical expression.

Phrase	the product of 8 and 5
Expression	8×5

1-2 Variables and Expressions (pp. 11–16)

14. Translate into an algebraic expression *the quotient of the number of points and three.*

ALGEBRA **Evaluate each expression if $a = 4$, $b = 8$, and $c = 11$.**

15. $3a + b$

16. $16 - c$

17. $3a + 4b - c$

18. $\frac{3b}{a} + ac$

19. **MEASUREMENT** There are 36 inches in a yard. Write an expression to find the number of yards in x inches.

EXAMPLE 2

Evaluate $x - 5 + 2y$ if $x = 6$ and $y = 4$.

$x - 5 + 2y = 6 - 5 + 2(4)$	Replace x with 6 and y with 4.
$= 6 - 5 + 8$	Multiply 2 and 4.
$= 1 + 8$	Subtract 5 from 6.
$= 9$	Add 1 and 8.

1-3 Properties (pp. 18–23)

Name the property shown by each statement.

20. $12 \times 0 = 0$

21. $(3 \cdot 5) \cdot 2 = 3 \cdot (5 \cdot 2)$

ALGEBRA **Simplify each expression.**

22. $3 \cdot (2 \cdot x)$

23. $(5 + v) + 7$

24. **TOYS** Gloria has 58 dolls. If she does not add any dolls to her collection, write a number sentence that represents the situation. Then name the property that is illustrated.

EXAMPLE 3

Simplify $(4 + x) + 6$.

$(4 + x) + 6 = (x + 4) + 6$	Commutative (+)
$= x + (4 + 6)$	Associative (+)
$= x + 10$	Simplify.

MIXED PROBLEM SOLVING
For mixed problem-solving practice, see page 843.

1-4 Ordered Pairs and Relations (pp. 25–30)

Express each relation as a table. Then determine the domain and range.

25. {(2, 3), (2, 6), (2, 5)}

26. {(1, 4), (2, 8), (3, 12), (4, 16)}

27. **FAIRS** It costs $2 per person to ride the Ferris wheel.

a. Make a table of ordered pairs in which the x-coordinate represents the number of people and the y-coordinate represents the cost for 4, 8, 12, and 16 people.

b. Graph the ordered pairs and then describe the graph.

EXAMPLE 4

Express the relation {(2, 1), (5, 6), (2, 7), (6, 1)} as a table. Then determine the domain and range.

x	2	5	2	6
y	1	6	7	1

The domain is {2, 5, 6} and the range is {1, 6, 7}.

1-5 Word, Equations, Tables, and Graphs (pp. 33–37)

28. **MULTIPLE REPRESENTATIONS** Shanna downloaded 5 more songs than videos.

a. **ALGEBRAIC** Write an equation that can be used to find the number of songs downloaded given the number of videos downloaded.

b. **TABULAR** Complete the function table for the number of songs downloaded when the number of videos downloaded is 2, 4, 6, and 8.

Number of Videos	Number of Songs
Input (v)	Output (s)
2	■
4	■
6	■
8	■

c. **GRAPHICAL** Graph the ordered pairs for the function.

EXAMPLE 5

MULTIPLE REPRESENTATIONS There are 3 apples for each horse.

a. **ALGEBRAIC** Write an equation that can be used to find the number of apples needed for any number of horses.
$w = 3h$

b. **TABULAR** Make a function table for 3, 5, 7, and 11 horses.

Number of Horses	Number of Apples
Input (x)	Output (y)
3	9
5	15
7	21
11	33

c. **GRAPHICAL** Graph the ordered pairs for the function.

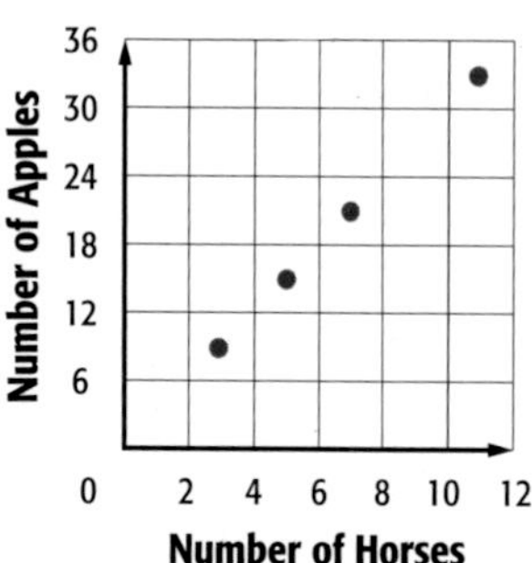

CHAPTER 1

Study Guide and Review

1-6 Scatter Plots (pp. 40–46)

29. HEALTH CARE The table shows the number of physicians and hospital beds for nine rural counties. Make a scatter plot of the data.

Physicians	11	26	10	19	22	9	15	7	1
Hospital Beds	85	67	32	69	49	43	90	49	18

30. SLEEP The table shows the amount of sleep students received the night before a standardized test and their score on the test.

Number of Sleep Hours	8	9.5	8	5	9	7
Score	89	91	94	68	81	77

a. Make a scatter plot of the data.

b. Draw a conclusion about the type of relationship the data shows. Explain.

c. Predict the score of a person who got 4 hours of sleep the night before the test.

EXAMPLE 6

The table shows the number of cups of lemonade two girls sold at their lemonade stand at certain temperatures.

Number of Cups	5	7	9	6	8
Temperature (F°)	75	85	90	86	88

a. Make a scatter plot of the data.

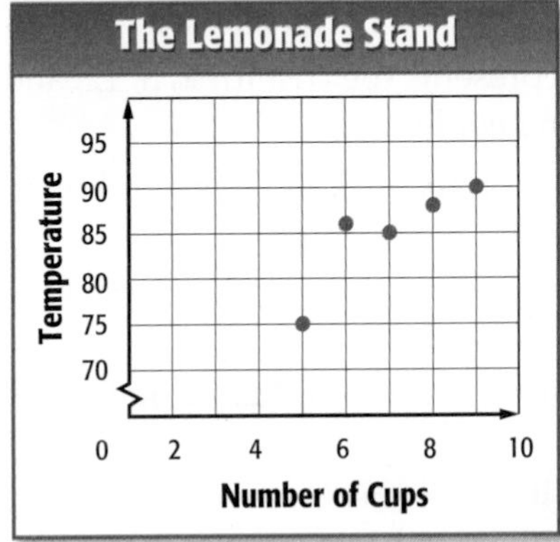

b. Draw a conclusion about the type of relationship the data shows. Explain.
The scatter plot shows a positive relationship since the number of cups sold increases as the temperature increases.

c. Predict the number of cups sold when the temperature is 95°.
By looking at the scatter plot of the data, it appears that the girls will sell about 10 cups of lemonade.

Practice Test

Math Online glencoe.com
Chapter Test

Write a numerical expression for each verbal phrase.

1. two people share a lunch bill of \$10
2. the new price if \$6 is taken off a \$15 item

Evaluate each expression.

3. $8 \div 2 + 11$
4. $(12 - 4) + 4$
5. $15 + 7 \cdot 9$
6. **ALGEBRA** Evaluate $3x - y$ if $x = 10$ and $y = 7$.
7. **MULTIPLE CHOICE** What is the value of $(a + b) \div c$ if $a = 17$, $b = 7$ and $c = 12$?

   A 24　　C 10

   B 19　　D 2

8. **GOLF** The table below shows the cost to play miniature golf. Write and evaluate an expression that can be used to find the total cost for 2 adults, 4 children, and 1 senior.

Age	Cost (\$)
Adults (16–62)	\$5
Children (3–15)	\$3
Seniors (63+)	\$4

9. Translate the phrase *the number of books increased by thirteen* into an algebraic expression.
10. **SPORTS** The table shows the point values of different scoring plays in football.

Scoring Play	Points
touchdown	6
extra point	1
field goal	3

   a. During the championship game, the winning team scored t touchdowns, e extra points, and f field goals. Write an algebraic expression to show the total points scored.

   b. If the team scored 27 total points, give one possible combination for touchdowns, extra points, and field goals.

Name the property shown by each statement.

11. $x + y = y + x$
12. $20 \cdot 1 = 20$
13. $(2 \cdot 3) \cdot 5 = 2 \cdot (3 \cdot 5)$
14. $12 + 0 = 12$

Simplify each expression.

15. $23 + (z + 47)$
16. $5a(3)$
17. $(8 \cdot b) \cdot 4$
18. $(17 + m) + 31$

Refer to the coordinate system at the right. Write the ordered pair that names each point.

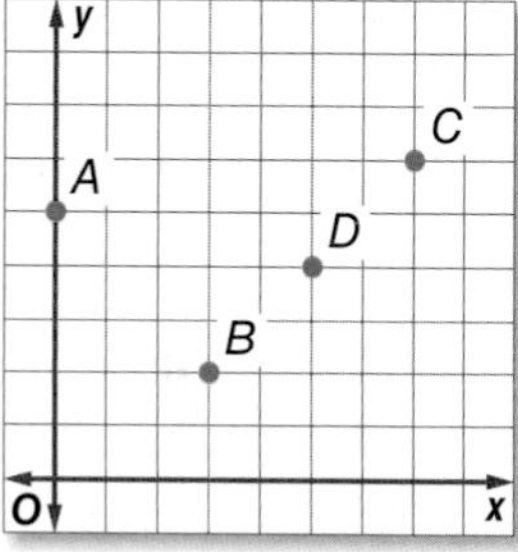

19. A
20. B
21. C
22. D
23. **MULTIPLE REPRESENTATIONS** Luz is three times as old as Malik.

   a. **ALGEBRAIC** Write an equation that can be used to find Luz's age given Malik's age.

   b. **TABULAR** Make a function table to show Luz's age if Malik is 1, 2, 3, 4, or 5.

   c. **GRAPHICAL** Graph the ordered pairs.

24. Express the relation {(1, 3), (2, 7), (4, 13), and (5, 15) as a table. Then determine the domain and range.
25. **MULTIPLE CHOICE** The scatter plot shows semester grades and school days missed for the students in Mr. Hernandez's math class. Which of the following is a reasonable score for a student who missed three days?

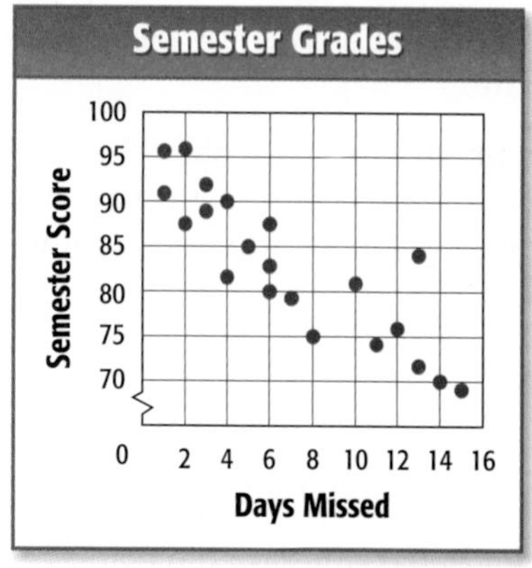

   F 80　　G 85　　H 90　　J 95

CHAPTER 1 **Preparing for Standardized Tests**

Reading to Solve Problems

The first step to solving any math problem is to read the problem. When reading a math problem to get the information you need to solve, it is helpful to use special reading strategies.

Strategies for Reading Math Problems

Step 1

Read the problem quickly to gain a general understanding of it.

- **Ask yourself:** "What do I know?" "What do I need to find out?"
- **Think:** "Is there enough information to solve the problem? Is there extra information?"
- **Highlight:** If you are allowed to write in your test booklet, underline or highlight important information. Cross out any information you don't need.

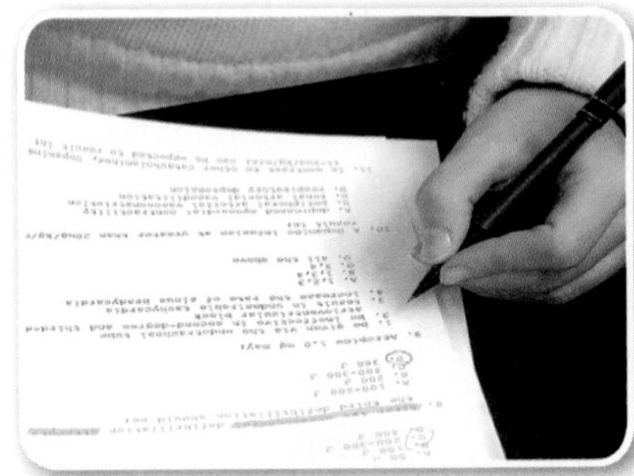

Step 2

Reread the problem to identify relevant facts.

- **Analyze:** Determine how the facts are related.
- **Key Words:** Look for keywords to solve the problem.
- **Vocabulary:** Identify mathematical terms. Think about the concepts and how they are related.
- **Plan:** Make a plan to solve the problem.
- **Estimate:** Quickly estimate the answer.

Step 3

Identify any obvious wrong answers.

- **Eliminate:** Eliminate any choices that are very different from your estimate.
- **Units of Measure:** Identify choices that are possible answers based on the units of measure in the question. For example, if the question asks for area, only answers in square units will work.

Step 4

Look back after solving the problem.

- **Check:** Make sure you have answered the question.

EXAMPLE

Read the problem. Identify what you need to know. Then use the information in the problem to solve.

The 8th graders are going on a field trip. They will rent a bus that costs $40 plus $3 for each passenger. Each student must bring $6.50 for the field trip. How much will it cost to rent the bus if there will be 42 students, 4 teachers, and 5 parent volunteers going on the field trip?

A $40 **C** 42 people

B $193 **D** 51 people

Read the problem. Highlight important information. Cross out information you don't need.

~~**The 8th graders are going on a field trip**~~**. They will rent a bus that costs $40 plus $3 for each passenger.** ~~**Each student must bring $6.50 for the field trip**~~**. How much will it cost to rent the bus if there will be 42 students, 4 teachers, and 5 parent volunteers going on the field trip?**

Answer choices C and D are numbers of people, not dollar amounts. Since they do not have the correct units, they can be eliminated. You know the correct answer will be choice A or B.

Answer choice A gives only the base price for renting the bus, $40. It does not include the cost of the passengers, so it can be eliminated. The correct answer is B.

Exercises

Read each problem. Identify what you need to know. Then use the information in the problem to solve.

1. An electrician charges a fee of $35 plus $20 per hour to work on customers' homes. The electrician works 5 days per week. How much would it cost to hire the electrician for 4 hours?

A $80 **B** $115 **C** $145 **D** $160

2. Which algebraic expression represents sixteen stickers shared equally by some students?

F $s \div 16$ **G** $16 \div s$ **H** $16s$ **J** $s + 16$

3. A toy manufacturer can produce 500 toys in one hour. Each toy weighs 24 ounces. If the manufacturer operates for 8 hours per day, how many toys can be produced in one day?

A 96,000 ounces **C** 4000 toys

B 12,000 ounces **D** 192 toys

4. Danika bought an MP3 player for $197.78. She paid $50 then made 6 equal monthly payments. What was her monthly payment amount?

F $39.55 **H** $24.63

G $83.33 **J** $32.96

CHAPTER 1 Standardized Test Practice

Chapter 1

Multiple Choice

Read each question. Then fill in the correct answer on the answer document provided by your teacher or on a sheet of paper.

1. The table shows the prices of admission to an amusement park.

Amusement Park Admission	
Ticket	**Cost ($)**
Children (up to 12)	8
Adults (13–64)	20
Seniors (65+)	15

Which expression can be used to find the total cost of admission for 4 children, 2 adults, and 3 seniors?

A $4(8) + 2(20) + 3(15)$

B $(4 + 8) + (2 + 20) + (3 + 15)$

C $(4 + 2 + 3) \times (8 + 20 + 15)$

D $9 \times (8 + 20 + 15)$

2. In Exercise 1, what is the total cost of admission for 4 children, 2 adults, and 3 seniors?

F $88

G $104

H $117

J $125

3. One gallon is equivalent to 16 cups. Write an expression to find the number of cups in n gallons.

A $n + 16$

B $16n$

C $n - 16$

D $\frac{n}{16}$

Test-TakingTip

Question 3 You can check your answer by replacing n with 1 gallon. The result should be equal to 16 cups.

4. Jeremy has five more than twice the number of baseball cards that José has. If José has c cards, which expression represents the number of baseball cards that Jeremy has?

F $2(c + 5)$

G $5c + 2$

H $2c - 5$

J $2c + 5$

5. Use the table to write an expression that shows Marie's total combined score from the three judges.

Marie's Dance Routine	
Judge	**Score**
Smitherman	8
Rodriguez	s
Hampton	7

A $15s$

B $s + 15$

C $s + 8$

D $s + 7$

6. What are the coordinates of point K on the coordinate grid below?

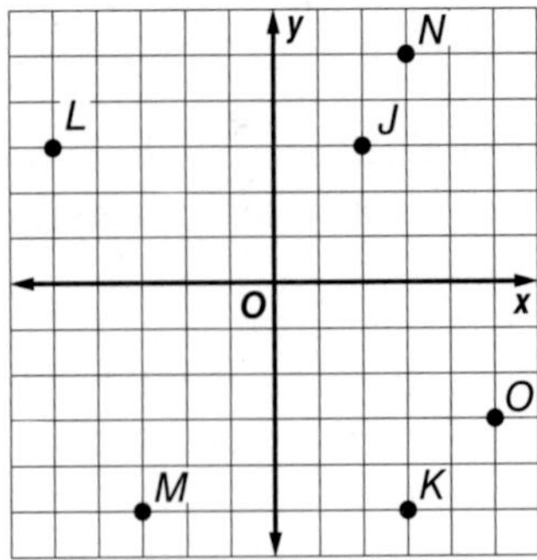

F $(-5, -3)$

G $(-3, 5)$

H $(-5, 3)$

J $(3, -5)$

Short Response/Gridded Response

Record your answers on the answer sheet provided by your teacher or on a sheet of paper.

7. What mathematical property is illustrated below?

$$5 + 3 = 3 + 5$$

8. GRIDDED RESPONSE Rachel had $80 in a savings account before making several withdrawals and deposits as shown in the table. How much money in dollars did Rachel have in the account after the activity shown?

Rachel's Savings Account	
Activity	**Amount ($)**
Starting Balance	80
Deposit	10
Deposit	5
Withdrawal	20
Deposit	10
Withdrawal	15

9. State the domain and range of the relation graphed below.

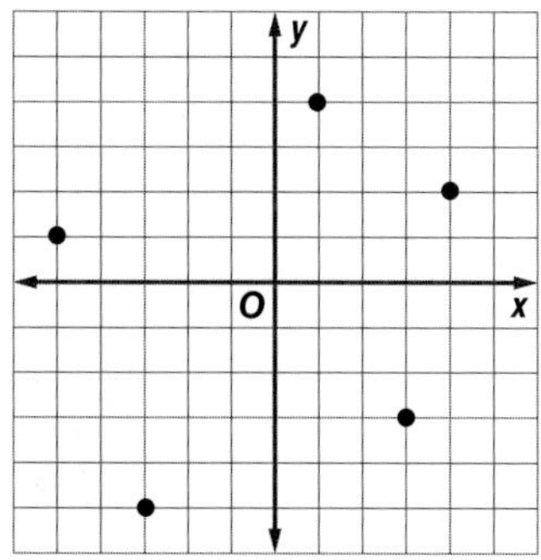

10. GRIDDED RESPONSE Evaluate the expression $\frac{2a + 4}{17 - 3b}$ for $a = 4$ and $b = 5$.

Extended Response

Record your answers on a sheet of paper. Show your work.

11. The scatter plot below shows the relationship between the value of several cars and the ages of the cars, in years.

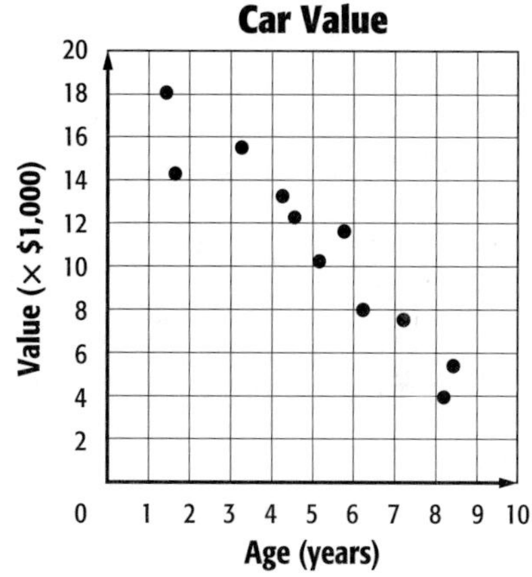

a. Does the scatter plot show a *positive, negative,* or *no* relationship? Explain.

b. Draw a conclusion about the relationship between the value of a car and the car's age.

c. About how much would you expect a car that is 5 years old to be worth? Explain.

Need Extra Help?											
If you missed Question...	1	2	3	4	5	6	7	8	9	10	11
Go to Lesson or Page...	1-1	1-1	1-2	1-2	1-3	1-4	1-3	1-5	1-4	1-2	1-6

CHAPTER 2 Operations with Integers

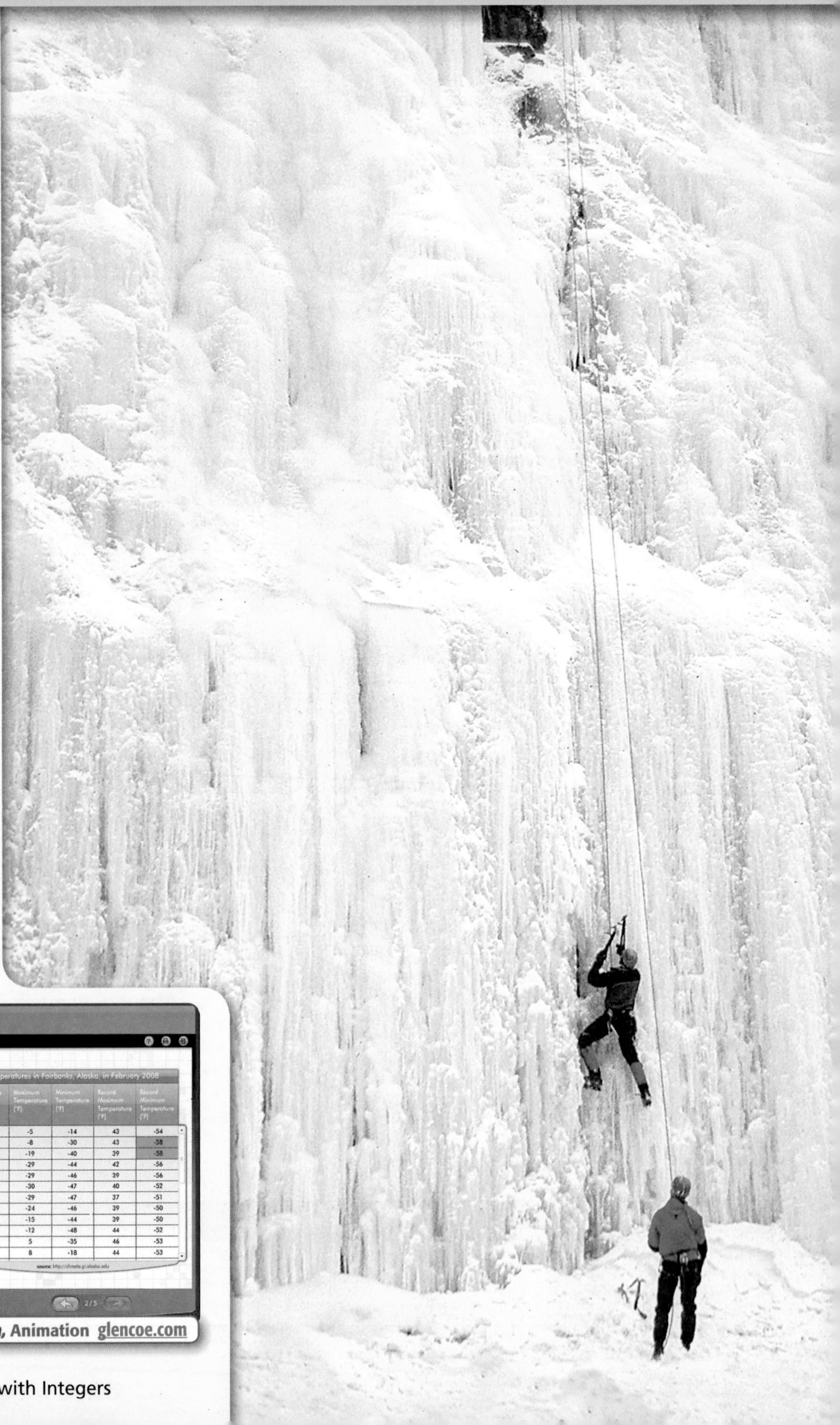

Then

In Chapter 1, you solved problems using the tools of algebra.

Now

In Chapter 2, you will:

- Compare, order, add, subtract, multiply, and divide integers.
- Graph points and algebraic relationships on a coordinate plane.
- Define, identify, and draw transformations.

Why?

WEATHER Rapid temperature changes, ice fog, and summer thunderstorms with hail, lightning, and snow are normal for Fairbanks, Alaska. Fairbanks experiences some of the most extreme weather in the world. Winter spans eight months and the temperature is often below zero for entire months.

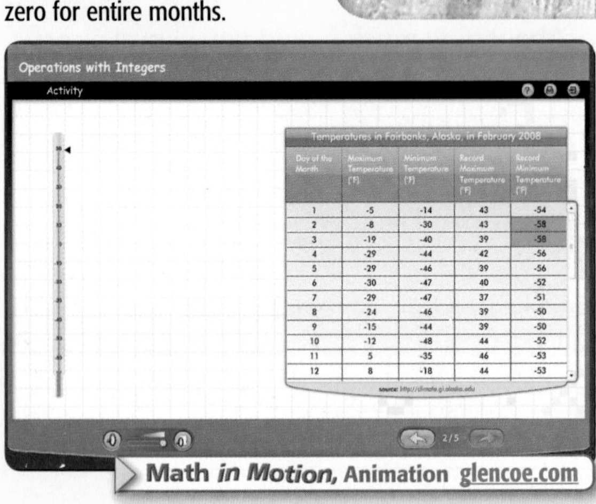

Temperatures in Fairbanks, Alaska, in February 2008

Day of the Month	Maximum Temperature (°F)	Minimum Temperature (°F)	Record Maximum Temperature (°F)	Record Minimum Temperature (°F)
1	-5	-14	43	-54
2	-8	-30	43	-58
3	-19	-40	39	-58
4	-29	-44	42	-56
5	-29	-46	39	-56
6	-30	-47	40	-52
7	-29	-47	37	-51
8	-24	-46	39	-50
9	-15	-44	39	-50
10	-12	-48	44	-52
11	5	-35	46	-53
12	8	-18	44	-53

source: http://climate.gi.alaska.edu

Math *in Motion*, Animation glencoe.com

Get Ready for Chapter 2

Diagnose Readiness You have two options for checking Prerequisite Skills.

Text Option

Take the Quick Check below. Refer to the Quick Review for help.

QuickCheck

Evaluate each expression if $x = 2$, $y = 11$, and $z = 5$. (Lesson 1-2)

1. $x + y + z$

2. $yz - xy$

3. $y + xz$

4. $4z + 3y$

5. **SALES** Flora sold three times as many bottles of water on Sunday than on Saturday. How many bottles of water did she sell Saturday if she sold 120 bottles on Sunday?

Find the next term in each list. (Previous Course)

6. 28, 34, 40, 46, 52, …

7. 135, 120, 105, 90, 75, …

8. **PHONE** The telephone company charges \$0.30 for the first minute and \$0.15 for each additional minute. How much would it cost to talk for 10 minutes?

Use the graph to name the coordinates of each point. (Lesson 1-4)

9. *D*

10. *B*

11. *A*

12. *J*

13. *C*

14. *G*

15. *F*

16. *H*

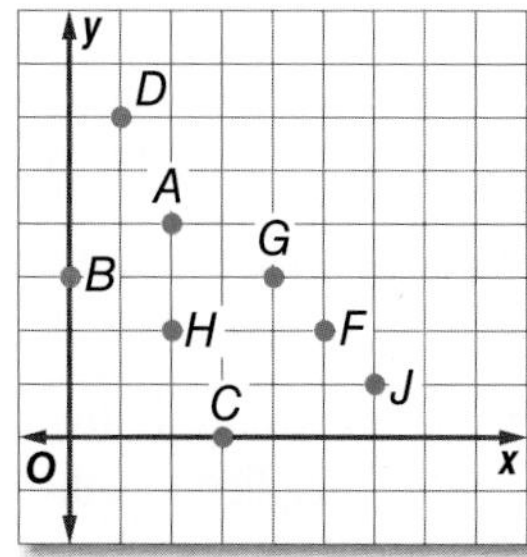

QuickReview

EXAMPLE 1

Evaluate $2b + 3c$ if $b = 2$ and $c = 3$.

$2b + 3c = 2(2) + 3(3)$ — **Replace *b* with 2 and *c* with 3.**

$= 4 + 9$ — **Multiply.**

$= 13$ — **Add.**

EXAMPLE 2

Find the next term in the list.

6, 11, 16, 21, …

6, 11, 16, 21, ■

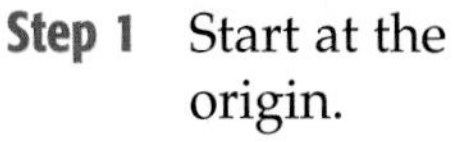

+5 +5 +5 +5

The next term is 21 + 5 or 26.

EXAMPLE 3

Use the coordinate plane to write the ordered pair that names point *A*.

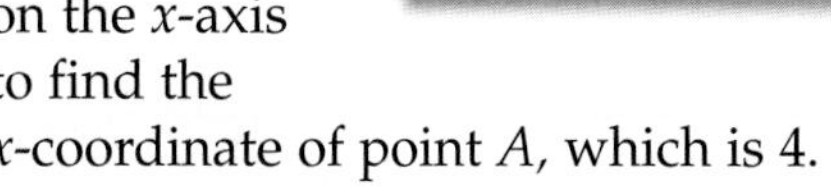

Step 1 Start at the origin.

Step 2 Move right on the *x*-axis to find the *x*-coordinate of point *A*, which is 4.

Step 3 Move up the *y*-axis to find the *y*-coordinate, which is 1.

The ordered pair for point *A* is (4, 1).

Online Option

Math Online Take a self-check Chapter Readiness Quiz at glencoe.com.

Get Started on Chapter 2

You will learn several new concepts, skills, and vocabulary terms as you study Chapter 2. To get ready, identify important terms and organize your resources. You may wish to refer to **Chapter 0** to review prerequisite skills.

FOLDABLES® Study Organizer

Operations With Integers Make this Foldable to help you organize your notes about operations with integers. Begin with a sheet of 11″ by 17″ paper.

1. **Fold** the short sides toward the middle.

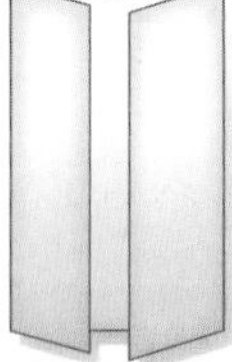

2. **Fold** the top to the bottom.

3. **Open.** Cut along the second fold to make four tabs.

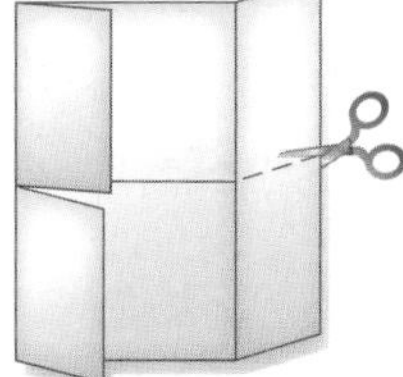

4. **Label** each of the tabs as shown.

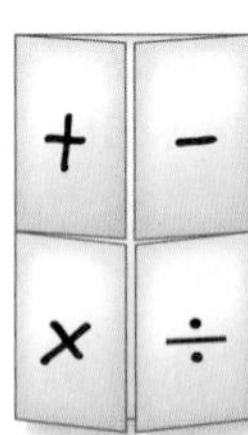

Math Online glencoe.com

- Study the chapter online
- Explore **Math in Motion**
- Get extra help from your own **Personal Tutor**
- Use **Extra Examples** for additional help
- Take a **Self-Check Quiz**
- **Review Vocabulary** in fun ways

New Vocabulary

English		Español
integer	p. 61	entero
coordinate	p. 62	coordine
zero pair	p. 67	par nulo
additive inverse	p. 71	inverso de añadidura
quadrants	p. 97	cuadrantes
reflection	p. 101	reflejo
transformation	p. 101	transformación
translation	p. 101	traducción

Review Vocabulary

addends • p. 69 • sumandos numbers that are added together

coordinate plane • p. 96 • plano de coordenadas
a coordinate plane is formed by the intersection of two number lines that meet at right angles at their zero points

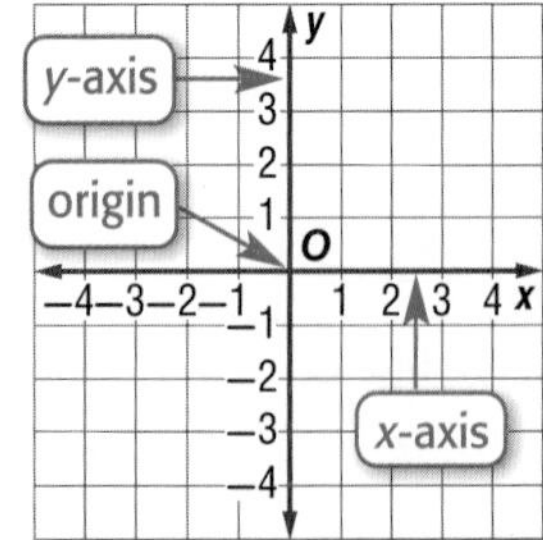

expression • p. 63 • expresión a combination of numbers, variables, and at least one operation

factors • p. 85 • factors numbers that are multiplied together

product • p. 83 • producto the result when two or more numbers are multiplied together

sum • p. 69 • suma the result when two or more numbers are added together

Multilingual eGlossary glencoe.com

2-1 Integers and Absolute Value

Then

You have already compared and ordered whole numbers. (Previous Course)

Now

- Compare and order integers.
- Find the absolute value of an expression.

New Vocabulary

negative number
positive number
integers
coordinate
inequality
absolute value

Math Online

glencoe.com

- Extra Examples
- Personal Tutor
- Self-Check Quiz
- Homework Help

Why?

In miniature golf, *par* is the number of strokes a golfer should take to complete a hole. In the graph, a score of −3 represents 3 strokes under par.

a. What does a value of −2 represent?

b. Which golfer's score was farthest from par?

c. How would you represent 2 strokes above par?

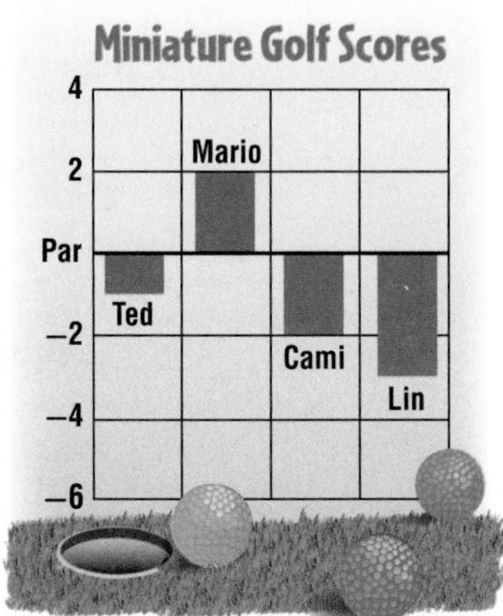

Compare and Order Integers A **negative number** is a number less than zero. A **positive number** is a number greater than zero.

Negative numbers like −3 and positive numbers like +3, are members of the set of integers. An **integer** is any number from the set {…, −3, −2, −1, 0, 1, 2, 3, …}, where … means continues indefinitely.

Integers can be represented as points on a number line.

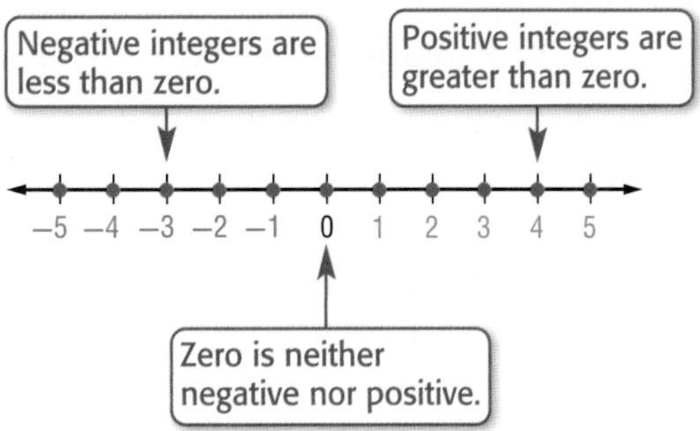

You can write integers to represent real-world situations.

EXAMPLE 1 Write Integers for Real-World Situations

Write an integer for each situation.

a. 23°F *below* zero

Because it is below zero, the integer is −23.

b. 11 inches more than normal

Because it is *more than* normal, the integer is +11 or 11.

Check Your Progress

1A. a loss of 8 yards

1B. a deposit of $15

Personal Tutor glencoe.com

To graph an integer, locate the point named by the integer on a number line. The **coordinate** is the number that corresponds to the point on a number line.

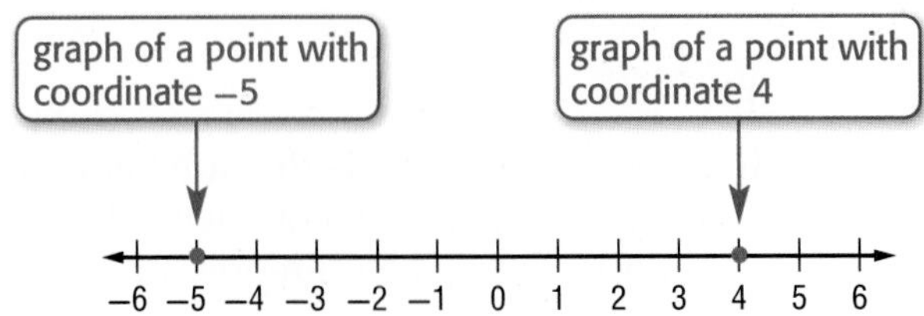

Any mathematical sentence containing $<$ or $>$ is called an inequality. An **inequality** compares numbers or quantities. When two numbers are graphed on a number line, the number to the left is always less than the number to the right.

> **ReadingMath**
>
> **Inequalities** The inequality symbol always points to the lesser number.

EXAMPLE 2 Compare Two Integers

Use the integers graphed on the number line below.

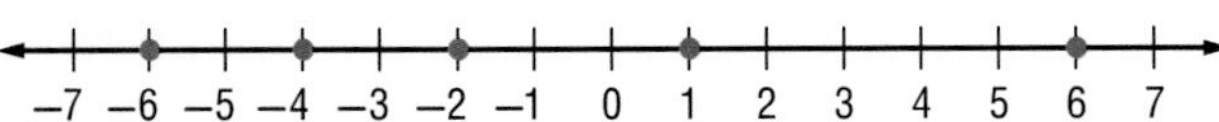

a. Write two inequalities involving 1 and –2.

Since 1 is to the right of –2, 1 is greater than –2. So, $1 > -2$.

Since –2 is to the left of 1, –2 is less than 1. So, $-2 < 1$.

b. Replace the ● with $<$, $>$, or $=$ in –4 ● –6 to make a true sentence.

Since –4 is to the right of –6, –4 is greater. So, $-4 > -6$.

Check Your Progress

2A. Write two inequalities involving –7 and –3.

2B. Replace the ● with $<$, $>$, or $=$ in –1 ● 2 to make a true sentence.

Personal Tutor glencoe.com

> **Real-World Link**
>
> The American Falls, one of the falls that makes up the Niagara Falls, has frozen six times since recorded history. In February 1936, the American Falls was frozen for 15 days.
>
> **Source:** Niagara Parks

Real-World EXAMPLE 3 Order Integers

VIDEO GAMES Bethany and her friends played a question and answer video game. Their scores at the end of the game were 1, –5, 0, –1, 2, and 4. Order the scores from least to greatest.

Graph each integer on a number line.

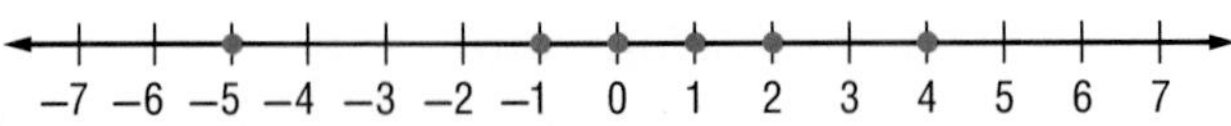

Write the numbers as they appear from left to right. The scores –5, –1, 0, 1, 2, and 4 are in order from least to greatest.

3. WEATHER The recorded highs in degrees Celsius in Niagara Falls from February 21–28 of a recent year are 4, 2, 3, –6, –5, –1, 0, and 1. Order the temperatures from greatest to least.

Absolute Value Notice on the number line that −6 and 6 are each 6 units from 0, even though they are on opposite sides of 0. The **absolute value** of a number is the distance the number is from zero on a number line. So, −6 and 6 have the same absolute value.

Key Concept Absolute Value

For Your FOLDABLE

Words The absolute value of a number is the distance the number is from zero on the number line. The absolute value of a number is always greater than or equal to zero.

Example $|6|$ and $|-6|$

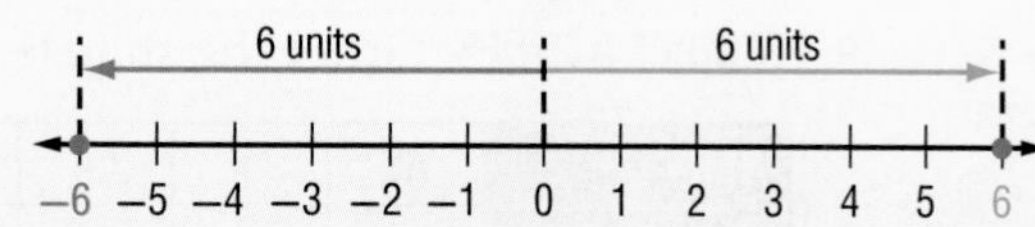

Symbols $|6| = 6$ The absolute value of 6 is 6.

$|-6| = 6$ The absolute value of −6 is 6.

Vocabulary Review

expression A combination of numbers, variables, and a least one operation. (Lesson 1-2)

EXAMPLE 4 Expressions with Absolute Value

Evaluate each expression.

a. $|-4|$

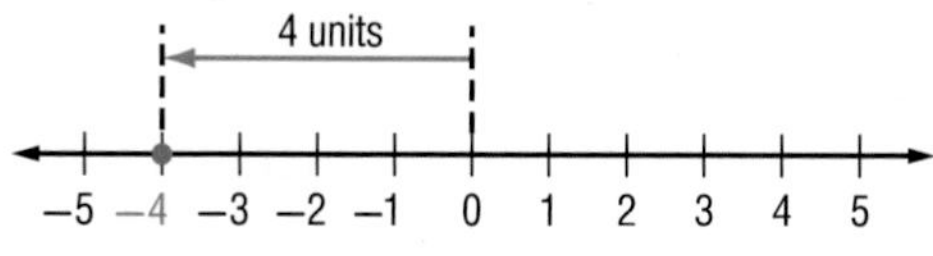

On the number line, the graph of −4 is 4 units from 0.

$|-4| = 4$

b. $|-8| - |5|$ The absolute value of −8 is 8.

$|-8| - |5| = 8 - 5$ The absolute value of 5 is 5.

$= 3$ Simplify.

Check Your Progress

4A. $|-3|$ **4B.** $|-4| - |3|$

Personal Tutor glencoe.com

You can use absolute value notation with algebraic expressions involving variables since variables represent numbers.

EXAMPLE 5 Algebraic Expressions with Absolute Value

ALGEBRA **Evaluate $6 + |x|$ if $x = -2$.**

$6 + |x| = 6 + |-2|$ Replace *x* with −2.

$= 6 + 2$ The absolute value of −2 is 2.

$= 8$ Simplify.

Check Your Progress

5. Evaluate $|y| + 8$ if $y = -7$.

Personal Tutor glencoe.com

Check Your Understanding

Example 1 p. 61

Write an integer for each situation. Then graph on a number line.

1. a bank withdrawal of \$500
2. a gain of 4 pounds

Example 2 p. 62

Write two inequalities using the number pairs. Use the symbols < or >.

3. 2 and −5
4. −4 and −8
5. −1 and 1

Replace each ● with <, >, or = to make a true sentence.

6. −9 ● −16
7. −7 ● 7
8. −6 ● 0

Example 3 p. 62

9. **TEMPERATURES** Order the state temperatures from least to greatest.

State	AL	AK	CA	FL	HI	ME	NJ	OH	TX
Temperature	−27	−80	−45	−2	12	−48	−34	−39	−23

Example 4 p. 63

Evaluate each expression.

10. $|-12|$
11. $|-14| + |3|$
12. $|18| - |-5|$

Example 5 p. 63

ALGEBRA Evaluate each expression if $x = 7$ and $y = -6$.

13. $15 - |y|$
14. $|y| + x$
15. $3|y|$

Practice and Problem Solving

= **Step-by-Step Solutions** begin on page R11.
Extra Practice begins on page 810.

Example 1 p. 61

Write an integer for each situation. Then graph on a number line.

16. 5 strokes above par
17. 200 feet below sea level
18. an elevator descends 18 floors
19. no gain on fourth down

Example 2 p. 62

Write two inequalities using the number pairs. Use the symbols < or >.

20. 5 and −11
21. −8 and 14
22. −6 and −1
23. −12 and −11
24. 0 and −4
25. 4 and 0
26. $|55|$ and $|50|$
27. $|-27|$ and $|-30|$

Replace each ● with <, >, or = to make a true sentence.

28. −11 ● −9
29. −14 ● −17
30. 15 ● −6
31. −2 ● 16
32. 21 ● 0
33. 0 ● −35
34. $|13|$ ● $|-13|$
35. $|-27|$ ● $|-27|$

Example 3 p. 62

36. **CAR RACING** In a recent year, Jimmy Johnson was the point leader in NASCAR's Chase to the Cup. Other drivers' standings are shown.
    a. Write an integer to describe each driver's standing with respect to the leader.
    b. Order the integers from least to greatest.

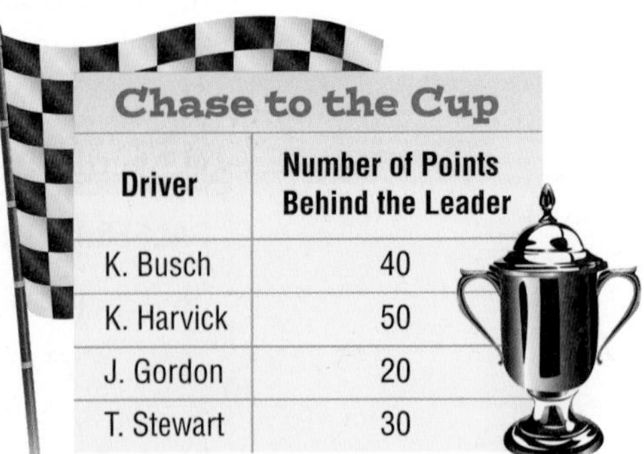

Chase to the Cup

Driver	Number of Points Behind the Leader
K. Busch	40
K. Harvick	50
J. Gordon	20
T. Stewart	30

37. **GOLF** The top fourth round scores of a recent PGA Championship were +4, −2, +6, +1, −4, −3, +5, −1, +2, and +3. Order the scores from least to greatest.

Example 4 p. 63

Evaluate each expression.

38. $|8|$ **39.** $|-17|$ **40.** $|-21|$

41. $-|15|$ **42.** $|0| + -|4|$ **43.** $-|-7| + |12|$

44. $|12| - |-2|$ **45.** $|-32| - |-6|$ **46.** $|18 - 4| - |9 - 4|$

Example 5 p. 63

ALGEBRA Evaluate each expression if $x = -3$, $y = 4$, and $z = 2$.

47. $10 - |x|$ **48.** $2y - |x|$ **49.** $|z| + 19$

50. $3y + 3z + |x|$ **51.** $|4yz| - 3|x|$ **52.** $2(z + y) - |x|$

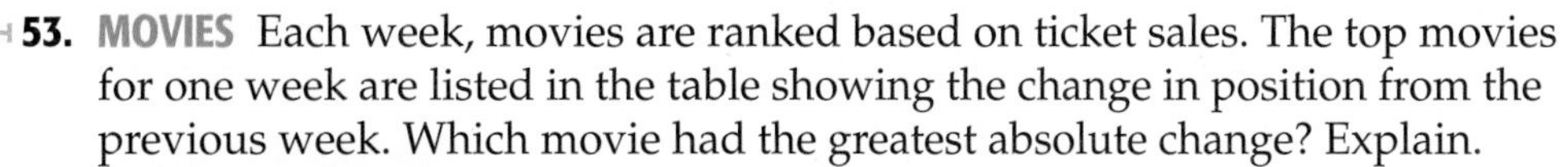

53. **MOVIES** Each week, movies are ranked based on ticket sales. The top movies for one week are listed in the table showing the change in position from the previous week. Which movie had the greatest absolute change? Explain.

Movie	A	B	C	D	E	F	G	H
Change in Position	−2	−7	+1	−3	+2	−8	−4	0

Real-World Link

In 2006, 1.4 billion people attended movie theatres. This was more than theme park and sporting event attendance combined.

Source: Motion Picture Association

SCIENCE The table below shows the freezing point of various elements.

Element	Freezing Point (°C)
chlorine	−101
helium	−272
krypton	−157
neon	−249
nitrogen	−201

54. Write two inequalities using the freezing point of neon and helium.

55. Order the temperatures from least to greatest using a number line.

56. Is the absolute value of the freezing point of chlorine greater than or less than the absolute value of the freezing point of nitrogen?

57. **SOLAR SYSTEM** The average temperature of Saturn is −218°F while the average temperature of Jupiter is −162°F. Which planet has the lower average temperature? Explain.

Order the integers in each set from greatest to least.

58. $\{4, -2, -10, 3\}$ **59.** $\{-13, 5, 0, -5\}$

60. $\{7, -26, -15, 32, -19\}$ **61.** $\{-28, 62, -35, 20, -59\}$

62. $\{-42, 1, -6, 74, 0, -11\}$ **63.** $\{88, -72, -83, 232, -165, -94\}$

H.O.T. Problems

Use **H**igher-**O**rder **T**hinking Skills

64. **OPEN ENDED** Write a real-world situation in which you compare two negative integers.

CHALLENGE Determine whether each statement is *always, sometimes,* or *never* true. Explain your reasoning.

65. $|x| = |-x|$ **66.** $|x| = -|x|$ **67.** $|-x| = -|x|$

68. **REASONING** Find all the values of x that make the statement $|x| = 7$ true.

69. **CHALLENGE** What is the least integer value of n such that $n > 0$?

70. **WRITING IN MATH** Order the integers −12, −5, −15, −10, −3 from least to greatest without using a number line. Explain your method.

Standardized Test Practice

71. Which of the following statements is false if $a = 3$ and $b = -3$?

A $|b| = a$
B $|b| > 0$
C $|b| = |a|$
D $|b| < 0$

72. If $|x| = 1$, what is the value of x?

F 1 and 2
G 1 and 0
H −1 and 0
J 1 and −1

73. EXTENDED RESPONSE Order the integers $|-5|, -|9|, -4, 0, -|10|, 7$ from least to greatest. Explain how you determined the order.

74. The table shows the number of points selected players have at the end of a game.

Player	Points
A	−10
B	−50
C	−5
D	0
E	−15

Which list shows the finishing order of the players from first to fifth?

A D, C, A, E, B
B B, E, A, C, D
C D, C, A, B, E
D B, C, A, D, E

Spiral Review

Determine whether a scatter plot of the data for the following might show a *positive, negative,* or *no* relationship. Explain your answer. (Lesson 1-6)

75.

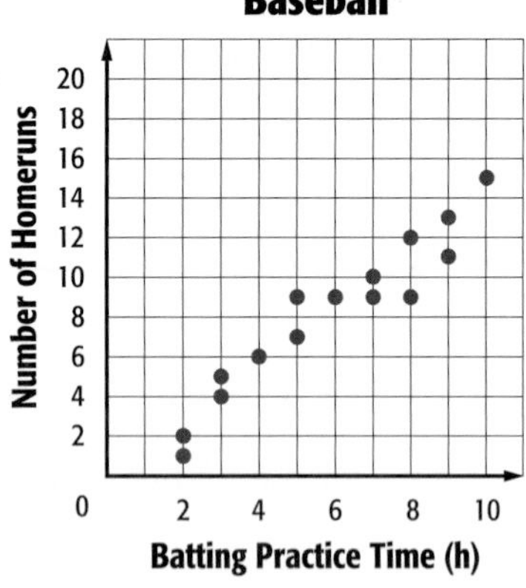

76.

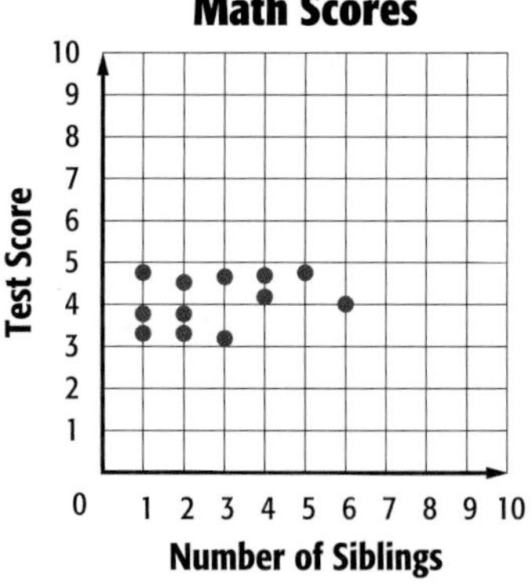

77. MULTIPLE REPRESENTATIONS A roll of wrapping paper costs $3. (Lesson 1-5)

a. ALGEBRAIC Write an equation that can be used to find the cost y of buying x number of rolls of wrapping paper.

b. TABULAR Make a function table to find the cost of 3, 4, 5, and 6 rolls.

c. GRAPHICAL Graph the ordered pairs.

Skills Review

Find each sum or difference. (Previous Course)

78. 121 − 56
79. 381 + 57
80. 743 − 259
81. 427 + 598
82. 892 − 645
83. 63 + 97 + 21

EXPLORE 2-2 Algebra Lab
Adding Integers

Math Online glencoe.com
Math *in Motion*, Animation

You can use algebra tiles and an integer mat to model operations with integers. In a set of algebra tiles, [1] represents the integer 1, and [−1] represents the integer −1.

ACTIVITY 1

Find the sum −2 + (−4) using algebra tiles.

Recall that addition means *to combine*. The expression, −2 + (−4) tells you to combine a set of 2 negative tiles with a set of 4 negative tiles.

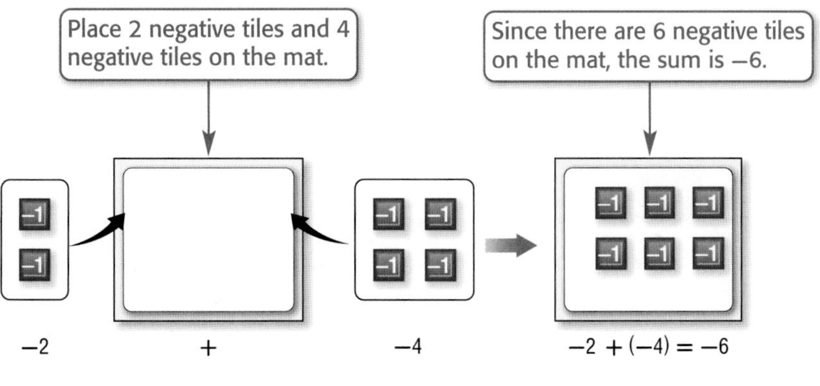

Place 2 negative tiles and 4 negative tiles on the mat.
Since there are 6 negative tiles on the mat, the sum is −6.
−1
−2
+
−4
−2 + (−4) = −6

So, −2 + (−4) = −6.

One positive tile paired with one negative tile is called a **zero pair.** You can add or remove zero pairs from a mat because removing or adding zero does not change the value of the tiles on the mat.

ACTIVITY 2

Find the sum −3 + 2 using algebra tiles.

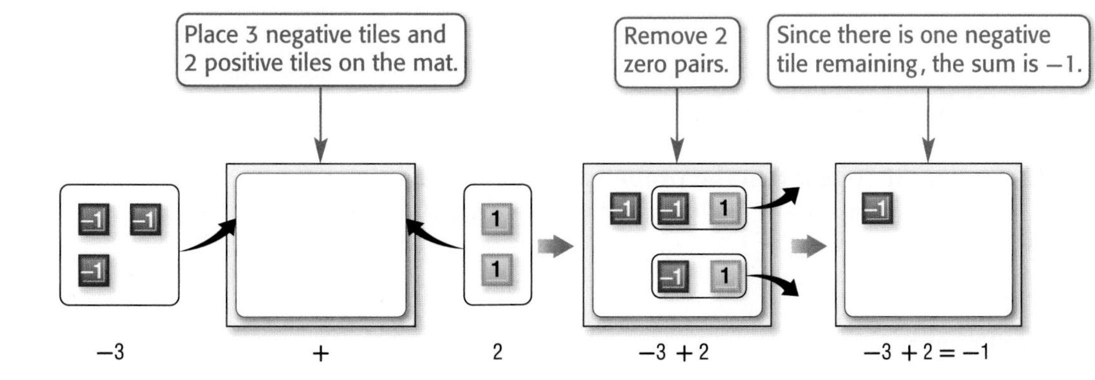

Place 3 negative tiles and 2 positive tiles on the mat.
Remove 2 zero pairs.
Since there is one negative tile remaining, the sum is −1.
−1
1
−3
+
2
−3 + 2
−3 + 2 = −1

So, −3 + 2 = −1.

ACTIVITY 3

Complete the Addition Table below using algebra tiles.

In the highlighted portion of the table, the addends are -4 and 3, and the sum is -1. So, $-4 + 3 = -1$.

Addition Table

+	4	3	2	1	0	−1	−2	−3	−4
4	8	7	6	5	4	3	2	1	0
3	7	6	5	4	3	2	1	0	−1
2	6	5							
1	5	4							
0	4	3							
−1	3	2							
−2	2	1							
−3	1	0							
−4	0	−1							

addends (top row) · sums (body) · addends (left column)

Analyze the Results

Model and find each sum using algebra tiles.

1. $-2 + (-1)$
2. $-4 + (-4)$
3. $-3 + (-4)$
4. $-6 + (-3)$
5. $1 + (-2)$
6. $(-5) + 3$
7. $(-2) + 2$
8. $2 + (-6)$

For Exercises 9–11, use the Addition Table above.

9. **MAKE A CONJECTURE** Look at all of the positive sums in the table. What is true about the addends that result in a positive sum?
10. **MAKE A CONJECTURE** Look at all of the negative sums in the table. What is true about the addends that result in a negative sum?
11. **MAKE A CONJECTURE** Look at all of the sums that are zero. What is true about the addends that result in a sum of zero?

For Exercises 12–14, does it appear that the property is true for addition of integers? If so, write two examples that illustrate the property. If not, give a counterexample.

12. Identity Property
13. Commutative Property
14. Associative Property
15. **WRITING IN MATH** Write a rule to help you determine the sum of any two integers.

2-2 Adding Integers

Then

You added integers using algebra tiles. (Explore 2-2)

Now

- Add two integers.
- Add more than two integers.

New Vocabulary

opposites
additive inverse

Math Online

glencoe.com

- Extra Examples
- Personal Tutor
- Self-Check Quiz
- Homework Help

Why?

An increase in temperature is represented by a positive integer. A decrease in temperature is represented by a negative integer. In one hour, the temperature dropped 2 degrees. In the next hour, the temperature dropped an additional 3 degrees.

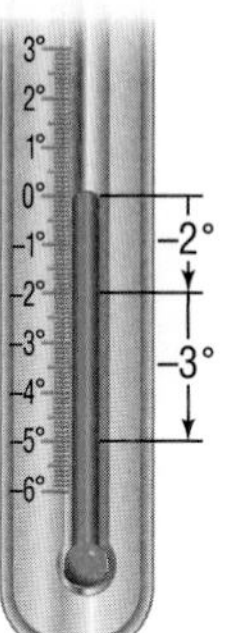

a. What integer represents the total temperature change over these two hours?

b. Write an addition expression that describes this situation.

c. Suppose the temperature rose 4 degrees in the third hour and dropped 1 degree in the fourth hour. Write an addition expression to represent the temperature change over all four hours. What integer represents that change?

Add Integers The equation $-2 + (-3) = -5$ is an example of adding two integers with the same sign. Notice the sign of the sum is the same as the sign of the addends.

EXAMPLE 1 Add Integers Using a Number Line

Find $-1 + (-5)$.

Use a number line.

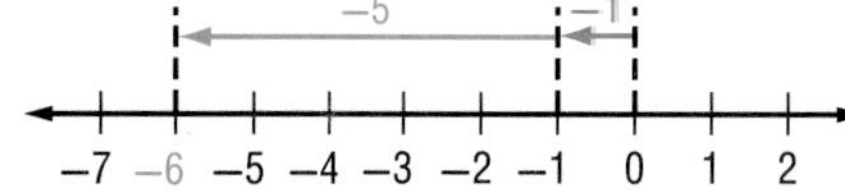

Step 1 Start at zero.

Step 2 Move 1 unit to the left.

Step 3 From there, move 5 more units to the left.

So, $-1 + (-5) = -6$.

Check Your Progress

Find each sum.

1A. $-3 + (-4)$ **1B.** $-6 + (-14)$ **1C.** $-7 + (-2)$

Personal Tutor glencoe.com

These and other similar examples suggest a rule for adding integers with the same sign.

Key Concept: Adding Integers with the Same Sign

For Your FOLDABLE

Words To add integers with the same sign, add their absolute values. The sum is:

- positive if both integers are positive.
- negative if both integers are negative.

Examples $-2 + (-4) = -6$ $8 + 1 = 9$

EXAMPLE 2 Add Integers with the Same Sign

Find $-3 + (-6)$.

$-3 + (-6) = -9$ Add $|-3|$ and $|-6|$. Both numbers are negative, so the sum is negative.

Check Your Progress

Find each sum.

2A. $-8 + (-2)$

2B. $-1 + (-12)$

Personal Tutor glencoe.com

To add integers with different signs you can also use a number line.

StudyTip

Adding Integers on a Number Line Always start at zero. Move right to model a positive integer. Move left to model a negative integer.

EXAMPLE 3 Add Integers Using a Number Line

Find each sum.

a. $6 + (-3)$

Use a number line.

Step 1 Start at zero.

Step 2 Move 6 units to the right.

Step 3 From there, move 3 units to the left.

So, $6 + (-3) = 3$.

b. $1 + (-4)$

Use a number line.

Step 1 Start at zero.

Step 2 Move 1 unit to the right.

Step 3 From there, move 4 units to the left.

So, $1 + (-4) = -3$.

Check Your Progress

3A. $5 + (-2)$

3B. $4 + (-8)$

Personal Tutor glencoe.com

These examples suggest a rule for adding integers with different signs.

Key Concept **Adding Integers with Different Signs** *For Your FOLDABLE*

Words To add integers with different signs, subtract their absolute values. The sum is:

- positive if the positive integer's absolute value is greater.
- negative if the negative integer's absolute value is greater.

Examples $9 + (-4) = 5$ $-9 + 4 = -5$

EXAMPLE 4 Add Integers with Different Signs

Find each sum.

a. $12 + (-6)$

$12 + (-6) = 6$ — To find $12 + (-6)$, subtract $|-6|$ from $|12|$. The sum is positive because $|12| > |-6|$.

b. $-7 + 5$

$-7 + 5 = -2$ — To find $-7 + 5$, subtract $|5|$ from $|-7|$. The sum is negative because $|-7| > |2|$.

Check Your Progress

4A. $-20 + 4$

4B. $16 + (-5)$

Personal Tutor glencoe.com

Real-World EXAMPLE 5 Solve Equations

WHALES A blue whale was at a depth of 275 feet below the surface of the water. After 10 minutes, it rose 194 feet. What is the current depth of the blue whale? Write an addition equation and then solve.

Words	Beginning depth	plus	increase after 10 minutes	equals	current depth
Variable	Let d = the current depth.				
Equation	-275	$+$	194	$=$	d

Solve the equation. — Estimate $-275 + 200 = -75$.

$-275 + 194 = d$ — To find the sum, subtract 194 from $|-275|$.

$-81 = d$ — The sum is negative because $|-275| > |194|$.

The current depth is -81 feet. — Check for Reasonableness $-81 \approx -75$ ✓

Real-World Link

The blue whale is the largest mammal on Earth. It can grow up to 100 feet long and can weigh as much as 300,000 pounds. There's enough room on its tongue for 50 people!

Source: National Geographic

Check Your Progress

5. SCUBA DIVING A scuba diver is 120 feet below the water's surface. She then ascends 83 feet. What is her current depth? Write an addition equation and then solve.

Personal Tutor glencoe.com

Add More Than Two Integers The integers -4 and 4 are an example of opposites. **Opposites** are two numbers with the same absolute value but different signs. An integer and its opposite are also called **additive inverses**.

Key Concept Additive Inverse Property

For Your FOLDABLE

Words The sum of any number and its additive inverse is zero.

Examples $2 + (-2) = 0$

Symbols $a + (-a) = 0$

This property will be useful when adding 2 or more integers.

StudyTip

Adding Mentally

- One way to add a group of integers mentally is to look for addends that are opposites.
- Another way is to group the positive addends together and the negative addends together. Then add.

EXAMPLE 6 Add More Than Two Integers

Find each sum.

a. $-6 + (-15) + 6$

$-6 + (-15) + 6 = -6 + 6 + (-15)$ — Commutative Property

$= 0 + (-15)$ — Additive Inverse Property

$= -15$ — Identity Property of Addition

b. $7 + (-1) + 26 + (-13)$

$7 + (-1) + 26 + (-13) = 7 + 26 + (-1) + (-13)$ — Commutative Property

$= (7 + 26) + [-1 + (-13)]$ — Associative Property

$= 33 + (-14)$ or 19 — Simplify.

Check Your Progress

6A. $4 + (-2) + (-7)$

6B. $-10 + 3 + (-7) + 12$

Personal Tutor glencoe.com

Check Your Understanding

Examples 1–4 pp. 69–71

Find each sum.

1. $-5 + (-6)$ **2.** $14 + (-5)$ **3.** $-18 + 11$ **4.** $16 + (-13)$

Example 5 p. 71

5. **GAME SHOWS** A contestant has –1500 points. He loses another 1250 points. What is his new score? Write an addition equation and then solve.

Example 6 p. 72

Find each sum.

6. $-7 + 14 + 7$ **7.** $11 + (-2) + (-10)$ **8.** $-5 + 4 + (-5) + 3$

Practice and Problem Solving

● = Step-by-Step Solutions begin on page R11. Extra Practice begins on page 810.

Examples 1–4 pp. 69–71

Find each sum.

9. $-7 + (-3)$ **10.** $-6 + (-14)$ **11.** $17 + (-8)$ **12.** $21 + (-11)$

13. $8 + (-13)$ **14.** $12 + (-16)$ **15.** $11 + (-5)$ **16.** $13 + (-2)$

Example 5 p. 71

For Exercises 17 and 18, write an addition equation. Then solve.

17. **FOOTBALL** A football team gained 6 yards on a play. It then lost 10 yards on the next play. What was the total change of yardage for the team?

18. **TEMPERATURE** The temperature in Rockford, Illinois, was –3°F. The temperature rose 7 degrees. What is the current temperature?

Example 6 p. 72

Find each sum.

19. $10 + (-3) + 3$ **20.** $-7 + (-1) + 7$

21. $-15 + 8 + (-9)$ **22.** $-6 + (-2) + 14$

23. $8 + (-11) + (-19) + 11$ **24.** $13 + 20 + (-17) + (-13)$

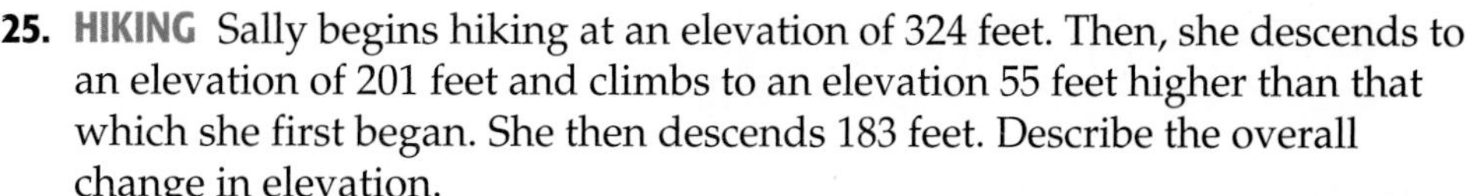

25. **HIKING** Sally begins hiking at an elevation of 324 feet. Then, she descends to an elevation of 201 feet and climbs to an elevation 55 feet higher than that which she first began. She then descends 183 feet. Describe the overall change in elevation.

Problem-Solving Tip

Draw a Diagram To find the overall change in elevation, draw a diagram.

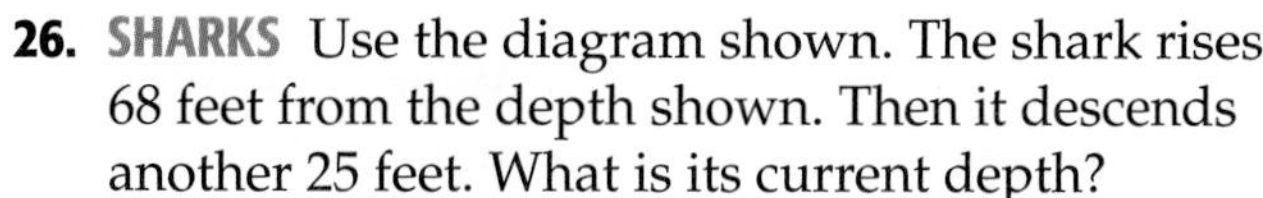

26. **SHARKS** Use the diagram shown. The shark rises 68 feet from the depth shown. Then it descends another 25 feet. What is its current depth?

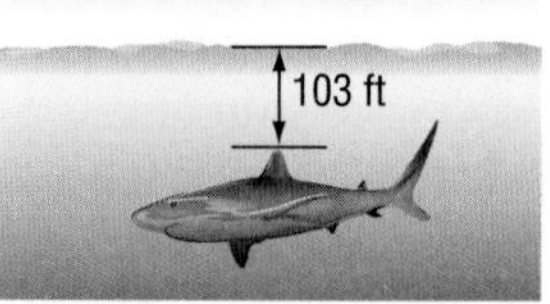

Find each sum.

27. $|-4 + 15|$
28. $|-2 + 11|$
29. $|17 - 25|$

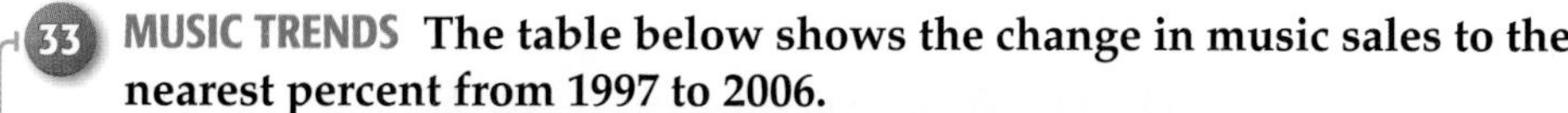

30. $|21 - 42|$
31. $|-13 + (-22)|$
32. $|-17 + (-39)|$

33. **MUSIC TRENDS The table below shows the change in music sales to the nearest percent from 1997 to 2006.**

Style of Music	Percent of Music Sold in 1997	Percent Change as of 2006
Rock	33	+1
Rap/Hip Hop	10	+2
Pop	9	−2
Country	14	−1

a. What is the percent of music sold in 2006 for each of these music categories?

b. What was the total percent change in the sale of these types of music?

Real-World Link

With 38%, the United States is the leader of album sales. Japan is second with 10%.

Source: Nation Master

ALGEBRA Evaluate each sum if $x = -6$ and $y = -4$.

34. $-35 + x + y + (-15)$
35. $x + y + x + y$

H.O.T. Problems

Use **H**igher-**O**rder **T**hinking Skills

36. **OPEN ENDED** Give an example of two negative integers and one positive integer that has a positive sum. Then find the sum.

37. **WRITING IN MATH** Write a real-world problem that can be solved by using the number line shown at the right.

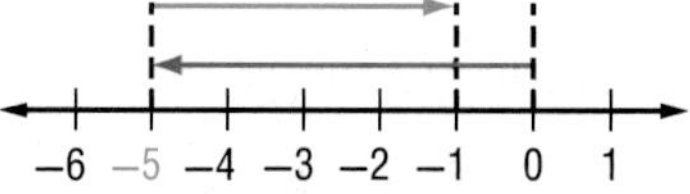

38. **CHALLENGE** *True* or *false?* $-n$ always names a negative number. If false, give a counterexample.

CHALLENGE Name the property illustrated by each of the following.

39. $a(b + (-b)) = (b + (-b))a$
40. $a(b + (-b)) = 0$

41. **WHICH ONE DOESN'T BELONG?** Identify the expression that does not belong with the other three. Explain your reasoning.

$7 + (-13)$	$-35 + 29$	$-22 + (28)$	$-2 + (-19) + 15$

CHALLENGE Simplify each expression.

42. $-5x + (-2) + 3x + 9$
43. $7 + (-4y) + y + (-3)$

44. **WRITING IN MATH** Explain how you know the sum of 5, 4, and -5 is positive without actually adding.

Standardized Test Practice

45. The table shows the low temperature and the high temperature on Monday for a certain city.

High Temperature	4°F
Low Temperature	−10°F

How much did the temperature rise?

A 14°F
B 10°F
C 6°F
D 4°F

46. **SHORT RESPONSE** On a drive, a football team gained 3 yards, lost 11 yards, and gained 15 yards. What was their total yardage for that drive?

47. A bird is flying at an altitude of 100 feet. It descends 30 feet then ascends 50 feet. Which of the following expressions best represents this situation?

F $100 + 30 + (-50)$
G $100 + (-30) + (-50)$
H $100 + (-30) + 50$
J $100 + 30 + 50$

48. **EXTENDED RESPONSE** A team started with 4 points. They gained 3 points, lost 6 points, and gained 2 points.

a. Write an expression to find how many points the team has now.
b. How many points does the team have now?

Spiral Review

49. **WEATHER** The record low temperature for Wisconsin is 54°F below zero. Write an integer to represent this situation. (Lesson 2-1)

Determine whether a scatter plot of the data for the following might show a *positive, negative,* or *no* relationship. (Lesson 1-6)

50. speed of a car and miles traveled in four hours

51. eye color and weight

52. car value and age of car

Name the property shown by each statement. (Lesson 1-3)

53. $12 \cdot 7 = 7 \cdot 12$
54. $1 \cdot 4 \cdot 0 = 0$
55. $4xy \cdot 1 = 4xy$

56. **SOCCER** A soccer league ranks each team in their league using points. A team gets three points for a win, one point for a tie, and zero points for a loss. Write an expression that can be used to find the total number of points a team receives. (Lesson 1-2)

Evaluate each expression. (Lesson 1-1)

57. $9 \div 3 \cdot 6$
58. $25 - 12 \div 4$
59. $(7 \cdot 5) + (8 \cdot 2)$
60. $(34 \div 2) - (8 \cdot 2)$

Skills Review

ALGEBRA Evaluate each expression if $a = 5$, $b = 12$, and $c = 7$. (Lesson 1-2)

61. $2c - 11$
62. $4a - 16$
63. $bc - ac$
64. $(3b - c) - 4c$

EXPLORE

2-3 Algebra Lab
Subtracting Integers

Math Online glencoe.com
Math *in Motion*, Animation

You can also use algebra tiles to model subtraction of integers. Remember, one meaning of subtraction is to *take away*.

ACTIVITY 1 Find 7 − 4.

Place 7 positive tiles on the mat. Remove 4 positive tiles.

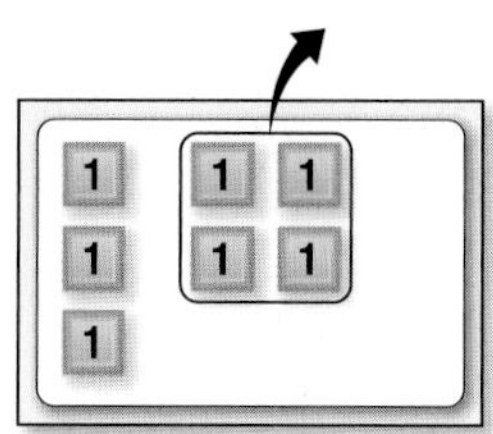

So, 7 − 4 = 3.

ACTIVITY 2 Find −8 − (−3).

Place 8 negative tiles on the mat. Remove 3 negative tiles.

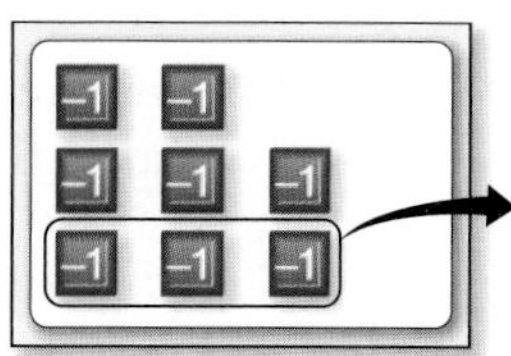

So, −8 − (−3) = −5.

ACTIVITY 3 Find 5 − (−2).

Place 5 positive tiles on the mat. There are no negative tiles to remove. Add 2 zero pairs to the set.

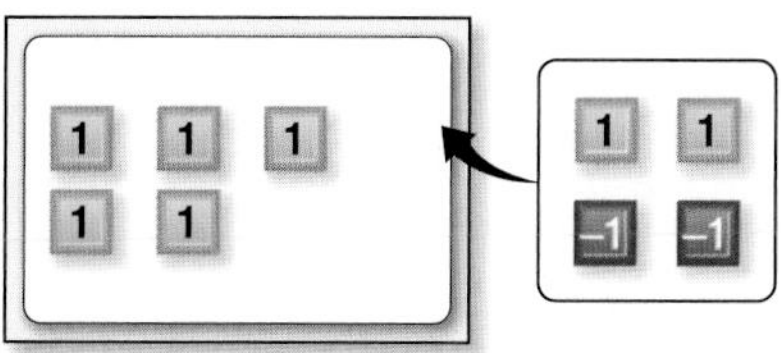

Then remove the 2 negative tiles.

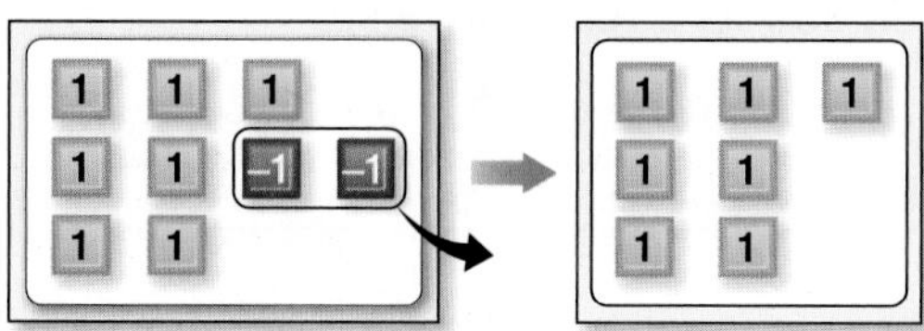

So, 5 − (−2) = 7.

ACTIVITY 4 Find −6 − 2.

Place 6 negative tiles on the mat. Since there are no positive tiles, add 2 zero pairs to the mat.

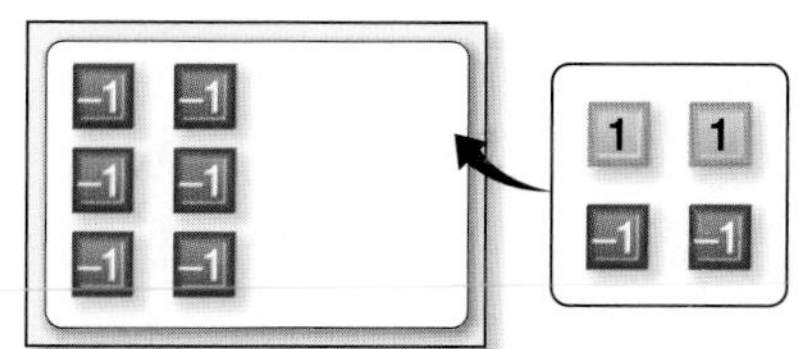

Then remove the 2 positive tiles.

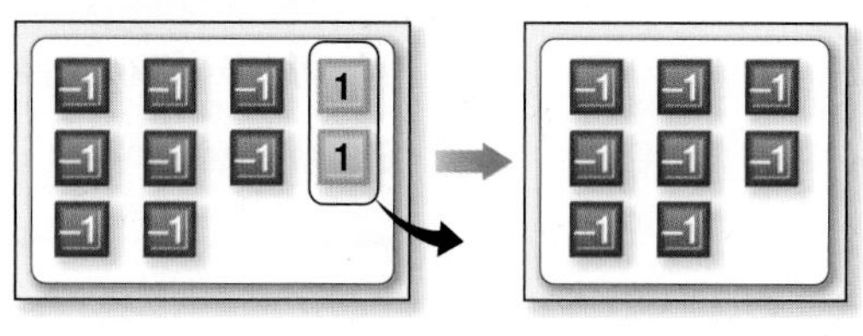

So, −6 − 2 = −8.

Analyze the Results

Model and find each difference using algebra tiles.

1. 9 − 7
2. 5 − (−3)
3. 6 − (−3)
4. 1 − (−5)
5. 3 − (−9)
6. −8 − 3
7. −8 − (−1)
8. −1 − 4

9. **MAKE A CONJECTURE** Write a rule that will help you determine the sign of the difference of two integers.

2-3 Subtracting Integers

Then
You have already learned to subtract integers using algebra tiles. (Explore 2-3)

Now
- Subtract integers.
- Evaluate expressions containing variables.

Math Online
glencoe.com
- Extra Examples
- Personal Tutor
- Self-Check Quiz
- Homework Help
- Math in Motion

Why?

The table shows the amount of money remaining in the monthly allowance of the Coughlin children at the middle of the month. A negative amount means the child overspent the monthly allowance and owes his or her parents.

Children	Monthly Allowance Balance ($)
Elizabeth	−5
Ginny	12
James	4

a. Suppose James spends $9 between the middle of the month and the end of the month. Write an expression to find his allowance balance at the end of the month.

b. How much does James owe his parents at the end of the month? Write this amount as an integer.

Subtract Integers The number line shows $4 - 9 = -5$.

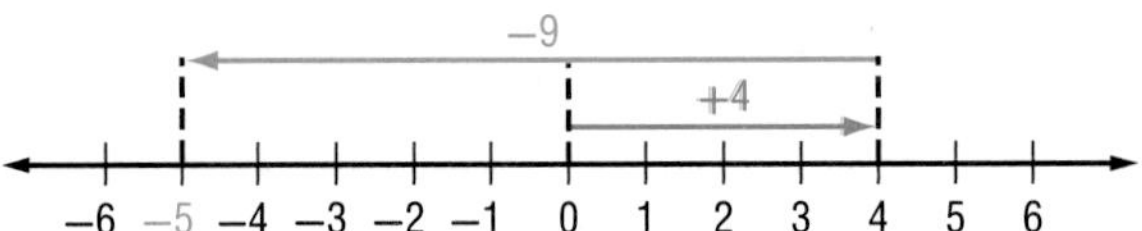

When you subtract 9 on the number line, the result is the same as adding −9.

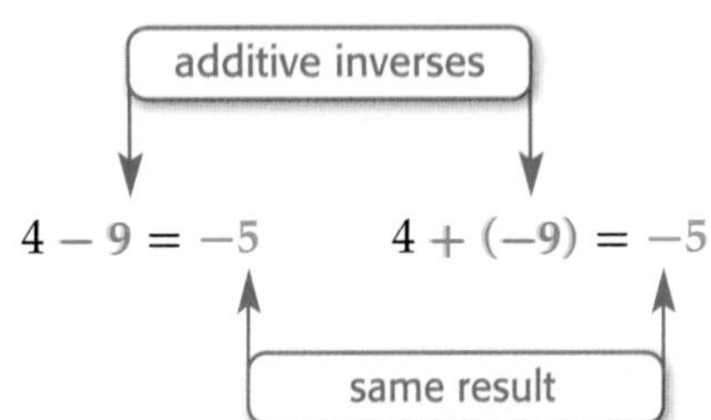

Key Concept Subtracting Integers — For Your FOLDABLE

Words To subtract an integer, add its additive inverse.

Examples $2 - 7 = 2 + (-7)$ **Symbols** $a - b = a + (-b)$

Math *in Motion,* BrainPOP® glencoe.com

EXAMPLE 1 Subtract a Positive Integer

Find each difference.

a. $6 - 15$

$6 - 15 = 6 + (-15)$ To subtract 15, add −15.

$= -9$ Simplify.

b. $-7 - 8$

$-7 - 8 = -7 + (-8)$ To subtract 8 add −8.

$= -15$ Simplify.

Check Your Progress

1A. $4 - 15$

1B. $-3 - 12$

Personal Tutor glencoe.com

StudyTip

Inductive Reasoning Making a conjecture based on a pattern of examples or past events

In Example 1, you subtracted a positive integer by adding its additive inverse. Use inductive reasoning to see if the method also applies to subtracting a negative integer.

Adding the Additive Inverse		
Input	rule: $4 + (-x)$	Output
2	$4 + (-2)$	2
1	$4 + (-1)$	3
0	$4 + 0$	4
−1	$4 + 1$	5

Subtracting an Integer		
Input	rule: $4 - x$	Output
2	$4 - 2$	2
1	$4 - 1$	3
0	$4 - 0$	4
−1	$4 - (-1)$	■

Continuing the pattern in the first column, $4 - (-1) = 5$. The result is the same as when you add the additive inverse.

EXAMPLE 2 Subtract a Negative Integer

Find each difference.

a. $9 - (-2)$

$9 - (-2) = 9 + 2$ To subtract −2, add 2.
$= 11$

b. $3 - (-5)$

$3 - (-5) = 3 + 5$ To subtract −5, add 5.
$= 8$

2A. $18 - (-2)$

2B. $-5 - (-11)$

Personal Tutor glencoe.com

Real-World Career

Climatologist
A climatologist studies the long term trends in the climate, collects climate data, investigates climate indicators, and makes predictions regarding climate patterns.

To be a climatologist, you need a bachelor's degree in meteorology or atmospheric sciences.

Source: NASA

Real-World EXAMPLE 3 Subtract a Negative Integer

WIND CHILL **The wind makes the outside temperature feel colder than the actual temperature. How much colder does a temperature of 10°F with a 30-mile-per-hour wind feel than the same temperature with a 20-mile-per-hour wind?**

Wind Chill Temperature				
	Wind Speed (miles per hour)			
Temperature (°F)	Calm	10	20	30
	20°	9	4	1
	10°	−4	−9	−12
	0°	−16	−22	−26
	−10°	−28	−35	−39

Understand You need to find how much colder 10°F feels with a 30-mile-per-hour wind than with a 20-mile-per-hour wind.

Plan Subtract the wind chill temperature at 20 miles per hour from the temperature at 30 miles per hour.

Solve $-12 - (-9) = -12 + 9$ To subtract −9, add 9.

$= -3$ Add −12 and 9.

It feels 3°F colder.

Check Since $-12 < -9$, it makes sense that the temperature is colder.

Check Your Progress

3. SPACE On Mars, the temperature ranges from 68°F during the day to −220°F at night. Find the change from the day temperature to the night temperature.

Personal Tutor glencoe.com

Evaluate Expressions Use the rule for subtracting integers to evaluate expressions.

EXAMPLE 4 Evaluate Algebraic Expressions

Evaluate.

$a - b - c$ if $a = 4$, $b = 3$, and $c = -10$.

$a - b - c = 4 - 3 - (-10)$	Replace a with 4, b with 3, and c with -10.
$= 1 - (-10)$	Use order of operations.
$= 1 + 10$	To subtract -10, add its additive inverse, 10.
$= 11$	Add 1 and 10.

Evaluate each expression if $\ell = 7$, $m = -3$, and $n = -10$.

4A. $n - \ell$ **4B.** $\ell - m + n$

Personal Tutor glencoe.com

Check Your Understanding

Examples 1 and 2 pp. 76–77

Find each difference.

1. $3 - 5$ **2.** $10 - 15$ **3.** $-10 - 14$ **4.** $-8 - 9$

5. $17 - (-14)$ **6.** $16 - (-12)$ **7.** $-7 - (-11)$ **8.** $-4 - (-3)$

Example 3 p. 77

9. **ANIMALS** A gopher begins at 7 inches below the surface of a garden and digs another 9 inches. Find an integer that represents the gopher's position in relation to the surface of the garden.

Example 4 p. 78

ALGEBRA Evaluate each expression if $a = 5$, $b = -7$, and $c = -16$.

10. $2a - (-11)$ **11.** $c - b$ **12.** $a + b - c$

Practice and Problem Solving

● = **Step-by-Step Solutions** begin on page R11.
Extra Practice begins on page 810.

Examples 1 and 2 pp. 76–77

Find each difference.

13. $6 - 7$ **14.** $4 - 8$ **15.** $-5 - 2$ **16.** $-9 - 3$

17. $5 - (-10)$ **18.** $1 - (-18)$ **19** $-12 - (-11)$ **20.** $-15 - (-14)$

21. $-20 - (-30)$ **22.** $-38 - (-40)$ **23.** $-32 - 28$ **24.** $-47 - 34$

Example 3 p. 77

25. **FINANCIAL LITERACY** Suppose you deposited \$25 into your checking account and wrote a check for \$38. What was the change in your account balance?

26. **SCORES** At the end of the first round of a game show, Jillian had a score of 40 points and Marty had a score of -50 points. Find the difference between their two scores.

Example 4 p. 78

ALGEBRA Evaluate each expression if $a = -3$, $b = 8$, and $c = -12$.

27. $7 - b$ **28.** $10 - c$ **29.** $a - 9$ **30.** $b - 5$

31. $c - b$ **32.** $c - a$ **33.** $a + b - c$ **34.** $a - b - c$

Real-World Link

In 2006, the number of girls playing high school sports hit an all-time high of 2,953,355 participants.

Source: NFHS

35. **SPORTS** The table shows the approximate participation numbers for different high school sports.

a. Write an integer to represent the change in the participation related to each sport from 2005 to 2006.

b. What was the total change in participation related to these sports from 2005 to 2006?

U.S. High School Sports Participation (thousands)		
Sport	2005	2006
Baseball	461	472
Basketball	1002	999
Softball	366	370
Football	1047	1073
Gymnastics	21	19
Hockey	44	43
Tennis	318	327

Source: National Federation of State High School Associations

Find each difference.

36. $125 - (-114)$ **37.** $-320 - (-106)$ **38.** $-2200 - (-3500)$

39 **STOCKS** The daily closing prices for a company's stock are shown.

Date	May 3	May 4	May 5	May 6	May 7
Closing Price	\$33.30	\$30.59	\$31.04	\$31.97	\$30.15
Change	—	■	■	■	■

a. Find the change in the closing price since the previous day.

b. What is the difference between the highest and lowest changes?

40. **MULTIPLE REPRESENTATIONS** In this problem, you will apply subtraction of integers to a real-world situation. An underwater video camera is 7 feet below the surface. It will be lowered an additional f feet.

a. **ALGEBRAIC** Write a function rule to show how many total feet below the surface the camera will be after it is lowered.

b. **TABULAR** Make a function table to show the depth of the camera if it is lowered 5, 8, 10, or 12 feet.

H.O.T. Problems

Use Higher-Order Thinking Skills

41. **OPEN ENDED** Write a subtraction expression with a positive integer and a negative integer whose difference is positive. Then find the difference.

42. **FIND THE ERROR** Rick and Michael are finding $-6 - (-2)$. Is either of them correct? Explain your reasoning.

Rick	Michael
$-6 - (-2) = 6 - 2$	$-6 - (-2) = -6 - 2$
$= 4$	$= -8$

43. **CHALLENGE** *True* or *false*? A subtraction expression with a positive integer and a negative can have a difference of zero. If *false*, give a counterexample.

44. **WRITING IN MATH** Write an expression involving the subtraction of a negative integer. Then write an equivalent addition expression. Explain why the result is the same.

Standardized Test Practice

45. The melting point of mercury is −39°C. The freezing point of alcohol is −114°C. How much warmer is the melting point of mercury than the freezing point of alcohol?

A −153°C
B −75°C
C 75°C
D 153°C

46. Which expression is modeled below?

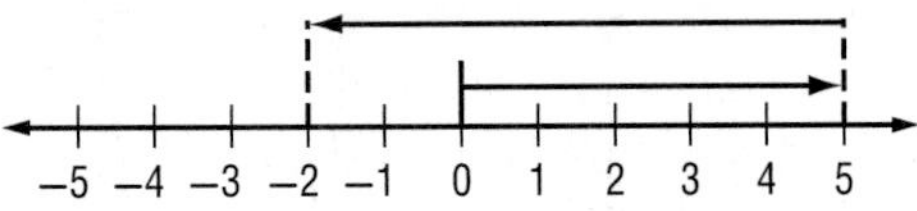

F 5 − 7
G 5 − 2
H 0 − 2
J 5 + 2

47. GRIDDED RESPONSE The crest of a mountain is 5740 feet above sea level. The base of the mountain is 25 feet below sea level. What is the difference in feet between the crest and base of the mountain?

48. Which statement about subtracting integers is always true?

A positive − negative = positive
B negative − positive = positive
C negative − negative = positive
D positive − positive = positive

Spiral Review

49. FOOTBALL A team gained 4 yards on one play. On the next play, they lost 5 yards. Write an addition sentence to find the change in yardage. (Lesson 2-2)

Replace each ● with <, >, or = to make a true sentence. (Lesson 2-1)

50. −18 ● −8
51. 0 ● −3
52. 9 ● −9

53. MULTIPLE REPRESENTATIONS In this problem, you will work with functions. It costs $6 to buy a student ticket to the movies. (Lesson 1-5)

a. **ALGEBRAIC** Write an equation that can be used to find the cost of any number of student tickets.

b. **TABULAR** Make a function table to find the cost of 2, 4, 5, and 7 tickets.

c. **GRAPHICAL** Graph the ordered pairs for the function.

ALGEBRA Translate each phrase into an algebraic expression. (Lesson 1-2)

54. eight more than the amount Kira saved
55. five runs fewer than the Pirates scored
56. the quotient of a number and four, minus five
57. seven increased by the quotient of a number and eight

Skills Review

Find each product. (Previous Course)

58. 9 × 8
59. 14 × 4
60. 3 × 2 × 8
61. 6 × 7 × 10

CHAPTER 2

Mid-Chapter Quiz

Lessons 2-1 through 2-3

Write an integer for each situation. Then graph on a number line. (Lesson 2-1)

1. 300 feet below sea level
2. a profit of $90

Replace each ● with <, >, or = to make a true sentence. (Lesson 2-1)

3. 9 ● −5
4. −3 ● 0
5. −8 ● −6
6. 2 ● −4

7. **MULTIPLE CHOICE** Refer to the number line. Which statement is true? (Lesson 2-1)

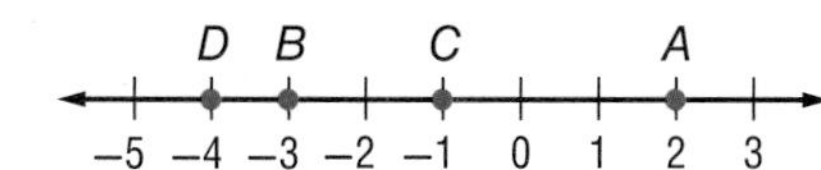

A $|B| < |C|$
C $C > A$
B $B > C$
D $|D| > |A|$

8. **GEOGRAPHY** The table shows the elevations of geographic areas in relation to sea level. Order the elevations from least to greatest. (Lesson 2-1)

Area	Elevation (m)
Dead Sea	−418
Death Valley	−86
Lake Eyre	−15
Salton Sea	−66

Find each sum. (Lesson 2-2)

9. −6 + (−15)
10. −4 + 12
11. −7 + 9 + (−8)
12. 12 + (−6) + (−15)
13. |−33 + 19|
14. |−23 + −20|

15. **MULTIPLE CHOICE** Which day had the greatest change in stock price? (Lesson 2-2)

Day	Open Price	Close Price
Monday	$43.29	$48.55
Tuesday	$48.55	$46.65
Wednesday	$46.65	$41.30
Thursday	$41.30	$45.99

F Monday
H Wednesday
G Tuesday
J Thursday

16. **WEATHER** The highest recorded temperature on Earth was 136°F in Libya. The lowest recorded temperature on Earth was −128.6°F in Antarctica. What is the average of the two temperatures? (Lesson 2-2)

17. **ACCOUNTING** A company showed the following earnings for a three-month period. How much did the company earn during this time period? (Lesson 2-2)

Month	Earnings ($)
January	−$3674
February	$4013
March	−$1729

Find each difference. (Lesson 2-3)

18. 15 − 21
19. −16 − 9
20. 35 − (−7)
21. −16 − (−11)

ALGEBRA Evaluate each expression if $x = 3$, $y = -2$, and $z = 6$. (Lesson 2-3)

22. $x - y$
23. $x - z - y$

24. **WEATHER** If the temperature is −1°F and it drops 5°F overnight, what is the new temperature? (Lesson 2-3)

25. **ASTRONOMY** Use the graph below to find the difference between the highest and lowest points on each planet. Which planet has the greatest difference? (Lesson 2-3)

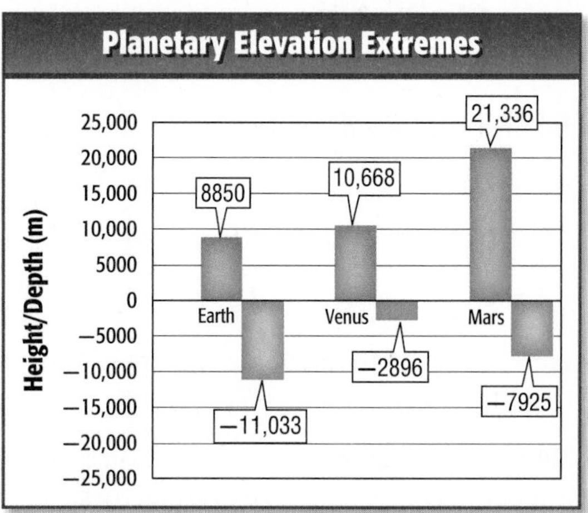

EXPLORE 2-4

Algebra Lab
Multiplying Integers

Math Online glencoe.com
Math *in Motion*, Animation

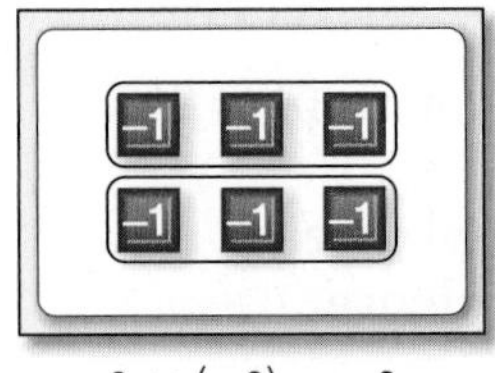
2 × (−3) = −6

You can also use algebra tiles to model multiplication of integers. Remember that 2 × 3 means *two sets of three items.* So, you can show 2 × 3 by placing 2 sets of 3 positive tiles on a mat.

Similarly, you can model 2 × (−3) by placing 2 sets of 3 negative tiles on the mat, as shown at the right.

If the first factor is negative, you will need to *remove* tiles from the mat.

ACTIVITY

Find −2 × (−3) using algebra tiles.

Step 1 The expression −2 × (−3) means to *remove* 2 sets of 3 negative tiles. To do this, first place 2 × 3 or 6 zero pairs on the mat.

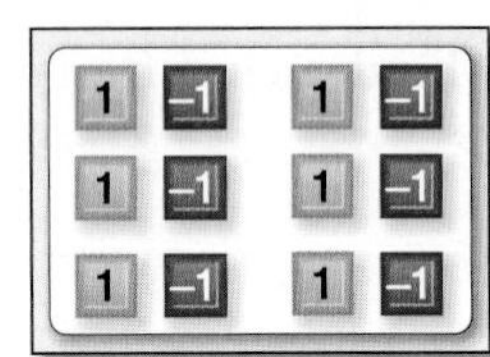

Step 2 Then remove 2 sets of 3 negative tiles from the mat. There are 6 positive tiles remaining.

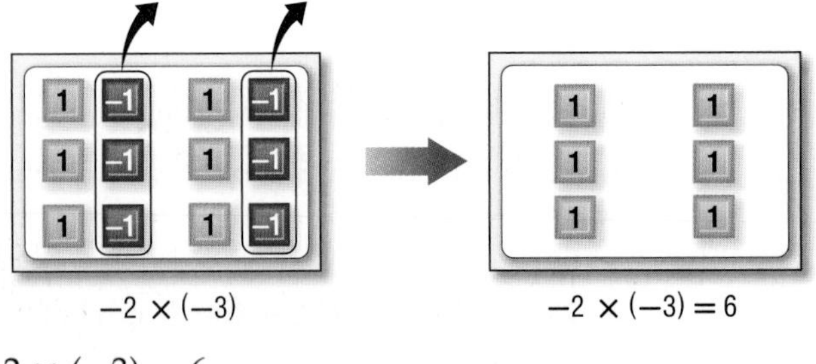
−2 × (−3)   −2 × (−3) = 6

So, −2 × (−3) = 6.

Analyze the Results

1. Explain the meaning of −2 × 3. Then find the product using algebra tiles.

Model and find each product using algebra tiles.

2. 6 × (−2)
3. 3 × (−5)
4. 3 × (−4)
5. 1 × (−8)
6. −4 × (−2)
7. −5 × (−2)
8. −7 × (−1)
9. −2 × (−2)
10. −4 × 2
11. −3 × 5
12. −2 × 6
13. −1 × 3
14. **WRITING IN MATH** How are the operations −3 × 4 and 4 × (−3) the same? How do they differ?
15. **MAKE A CONJECTURE** Find a rule you can use to find the sign of the product of two integers given the sign of both factors.

2-4 Multiplying Integers

Then

You multiplied integers using algebra tiles. (Explore 2-4)

Now

- Multiply integers.
- Simplify algebraic expressions.

Math Online

glencoe.com

- Extra Examples
- Personal Tutor
- Self-Check Quiz
- Homework Help

Why?

Time (min)	Depth (ft)
1	−40
2	−80
⋮	⋮

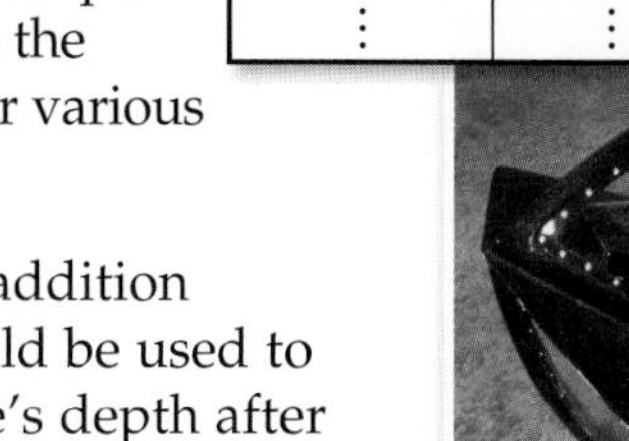

A one passenger submersible descends at a rate of 40 feet per minute. The table shows the submersible's depth after various minutes.

a. Write two different addition expressions that could be used to find the submersible's depth after 3 minutes. Then find their sum.

b. Write a multiplication expression that could be used to find this same depth. Explain your reasoning.

Multiply Integers Multiplication is repeated addition. So, $3(-40)$ means that -40 is used as an addend 3 times.

$$3(-40) = (-40) + (-40) + (-40)$$
$$= -120$$

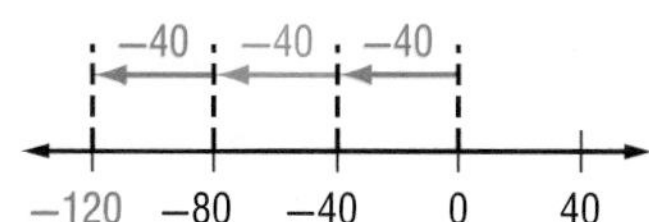

By the Commutative Property of Multiplication, $3(-40) = -40(3)$. This and other similar examples suggests the following rule for multiplying integers.

Key Concept | **Multiplying Two Integers with Different Signs** | For Your FOLDABLE

Words The product of two integers with different signs is negative.

Examples $2(-6) = -12$ $\quad -2(6) = -12$

EXAMPLE 1 Multiply Integers with Different Signs

Find each product.

a. $-3 \cdot 12$

$-3(12) = -36$ The factors have different signs. The product is negative.

b. $4(-7)$

$4(-7) = -28$ The factors have different signs. The product is negative.

Check Your Progress

1A. $7(-8)$ **1B.** $-6 \cdot 12$

Personal Tutor glencoe.com

The product of two positive integers is positive. What is the sign of the product of two negative integers? Look at the pattern below.

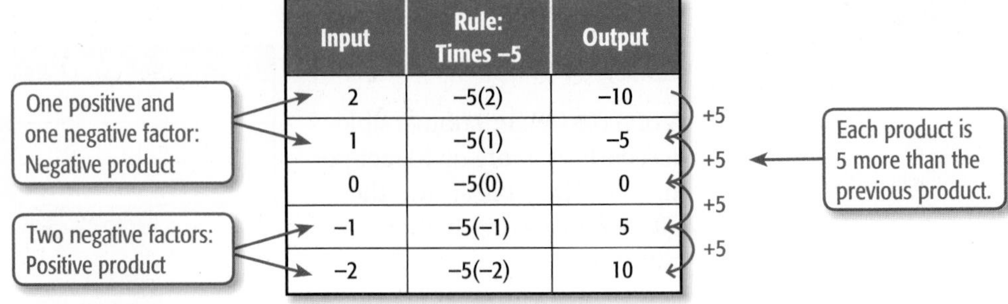

Input	Rule: Times –5	Output
2	–5(2)	–10
1	–5(1)	–5
0	–5(0)	0
–1	–5(–1)	5
–2	–5(–2)	10

One positive and one negative factor: Negative product
Two negative factors: Positive product
+5
+5
+5
+5
Each product is 5 more than the previous product.

Key Concept: Multiplying Two Integers with Same Signs

For Your FOLDABLE

Words The product of two integers with the same sign is positive.

Example $4 \cdot 6 = 24$ $-4(-6) = 24$

EXAMPLE 2 Multiply Integers with the Same Sign

Find each product.

a. $-5(-7)$

$-5(-7) = 35$ **The product is positive.**

b. $-8(-14)$

$-8(-14) = 112$ **The product is positive.**

Check Your Progress

2A. $-5(-11)$

2B. $-13(-4)$

Personal Tutor glencoe.com

Real-World EXAMPLE 3 Multiply Integers with Different Signs

AIRPLANES An airplane descends at a rate of 175 feet per minute. What is the airplane's change in altitude after 5 minutes?

Understand You need to find how many feet the airplane descended.

Plan The word *descends* means *downward,* so the rate per minute is represented by -175. You could make a function table like the one at the right. You could also multiply 5 and -175 to find the change in altitude after 5 minutes.

x	y
1	–175
2	–350
3	–525
⋮	⋮

Solve $5(-175) = -875$ **The product is negative.**

So, the change in altitude is -875 feet.

Check $5(-200)$ is -1000 and -875 is close to -1000. ✓

Check Your Progress

3. DEPTH A scuba diver descends from the surface at a rate of 7 feet per minute. What was the scuba diver's depth at 15 minutes?

Personal Tutor glencoe.com

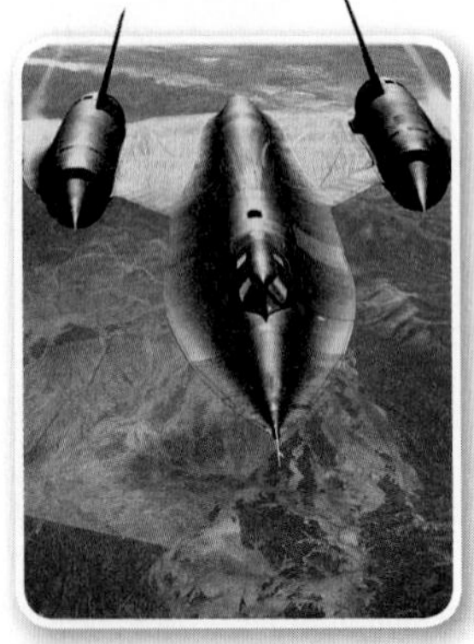

Real-World Link

The fastest airplane is the Lockheed SR-71 Blackbird. It is capable of speeds over 2200 miles per hour, which is more than 3 times the speed of sound.

Use the Commutative and Associative Properties of Multiplication to multiply more than two integers.

EXAMPLE 4 Multiply More than Two Integers

Find −4(12)(−5).

Method 1 Use the Associative Property

$-4(12)(-5) = [-4(12)](-5)$	**Associative Property**
$= -48(-5)$	**$-4(12) = -48$**
$= 240$	**$-48(-5) = 240$**

Method 2 Use the Commutative Property

$-4(12)(-5) = -4(-5)(12)$	**Commutative Property**
$= 20(12)$	**$-4(-5) = 20$**
$= 240$	**$20(12) = 240$**

StudyTip

Mental Math Look for products that are multiples of ten to make the multiplication simpler.

Check Your Progress

4A. Find $-7(9)(-6)$.

4B. Find $-3(-4)(-5)$.

Personal Tutor glencoe.com

Algebraic Expressions You can use the rules for multiplying integers to simplify and evaluate algebraic expressions.

EXAMPLE 5 Simplify Algebraic Expressions

Simplify −7a(4b).

$-7a(4b) = (-7)(a)(4)(b)$	**$-7a = (-7)(a)$, $4b = (4)(b)$**
$= (-7 \cdot 4)(a \cdot b)$	**Commutative Property of Multiplication**
$= -28ab$	**$-7 \cdot 4 = -28$, $a \cdot b = ab$**

Check Your Progress

5A. Simplify $-3(6y)$.

5B. Simplify $-9x(3y)$.

Personal Tutor glencoe.com

EXAMPLE 6 Evaluate Algebraic Expressions

Evaluate $3xy$ if $x = 2$ and $y = -6$.

$3xy = 3(2)(-6)$	**Replace x with 2 and y with −6.**
$= [3(2)](-6)$	**Associative Property of Multiplication**
$= 6(-6)$	**The product of 3 and 2 is positive.**
$= -36$	**The product of 6 and −6 is negative.**

Check Your Progress

6A. Evaluate $2rs$ if $r = 5$ and $s -10$.

6B. Evaluate $4ab$ if $a = -8$ and $b = -4$.

Personal Tutor glencoe.com

Check Your Understanding

Examples 1, 2, and 4 pp. 83–85

Find each product.

1. $-6 \cdot 7$
2. $-5(-8)$
3. $8(-3)(-5)$
4. $-2(-9)(-5)$

Example 3 p. 84

5. **MONEY** Mr. Heppner bought lunch with his debit card every day for 5 days. Each day he spent $8. If these were his only transactions this week, what was the change in his account balance?

Example 5 p. 85

ALGEBRA Simplify each expression.

6. $-2 \cdot 9m$
7. $-3a(7b)$
8. $-6e(-4f)$

Example 6 p. 85

ALGEBRA Evaluate each expression.

9. $8j$, if $j = -11$
10. $-9cd$, if $c = -3$ and $d = -7$

Practice and Problem Solving

 = **Step-by-Step Solutions** begin on page R11. **Extra Practice** begins on page 810.

Examples 1, 2, and 4 pp. 83–85

Find each product.

11. $3(-9)$
12. $8 \cdot -9$
13. $25 \cdot 3$
14. $-4(-8)$
15. $-7 \cdot -7$
16. $2(-11)(5)$
17. $-8(-7)(-6)$
18. $-8(-20)(5)$

Example 3 p. 84

19. **TEMPERATURE** The temperature dropped 2°F every hour for the last 6 hours. What is the total change in temperature?

20. **ELEVATORS** An elevator takes passengers from the ground floor down to an underground parking garage. Where will the elevator be in relation to the ground floor after 5 seconds if it travels at a rate of 3 feet per second?

Example 5 p. 85

ALGEBRA Simplify each expression.

21. $5(-6m)$
22. $-5 \cdot 10s$
23. $-9m(-9n)$
24. $11a(7c)$
25. $4a(b)(-9)$
26. $-12(-j)(-3k)$
27. $3e(-2f)(9g)$
28. $3r(7s)(5t)$

Example 6 p. 85

ALGEBRA Evaluate each expression.

29. $10n$, if $n = -10$
30. $8m$, if $m = -5$
31. $4xy$ if $x = -6$ and $y = 3$
32. $-15st$, if $s = 4$ and $t = -9$

FUNCTION TABLES Identify the function rule for each table.

33.

x	y
1	−2
2	−4
3	−6

34.

x	y
2	−10
0	0
−2	10

35.

x	y
−3	12
−6	24
−9	36

36. **FITNESS** The table shows the number of Calories burned per minute for a 120-pound person during different activities. What is the change in the number of Calories in a 120-pound person's body if he runs for 20 minutes and swims for 25 minutes?

Activity	Calories per Minute
Ballet Dancing	6
Bicycling	12
Running	18
Swimming	8

Replace each ● with <, >, or = to make a true sentence.

37. $-4(6)$ ● $3(8)$

38. $(-5)(-2)(9)$ ● $(4)(10)(-9)$

39. $(-11)(-3)(6)(-2)$ ● $-24(18)$

40. INVESTMENTS The price of stock fell \$2 each day for 14 consecutive days. The original price of the stock was \$41. Write an expression that you could use to find the price of the stock on any day.

41. TRIVIA Two groups were playing a trivia game. Each team earned 5 points for every correct answer, lost 8 points for every incorrect answer, and lost 2 points for every passed question. For each team, write an expression to determine the number of points they earned. Which team won?

Team	Correct	Incorrect	Passed
1	12	3	1
2	13	2	7

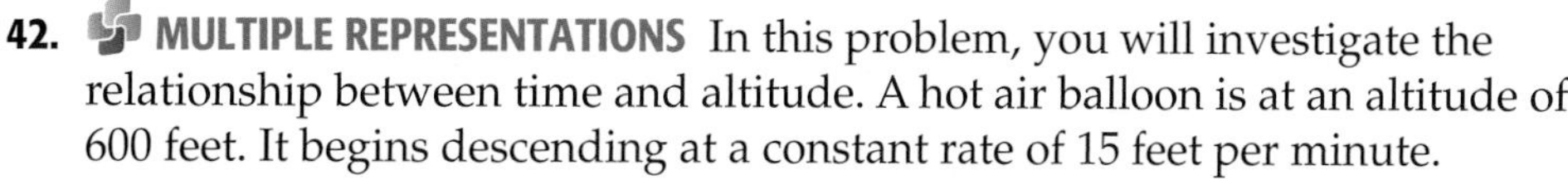

42. MULTIPLE REPRESENTATIONS In this problem, you will investigate the relationship between time and altitude. A hot air balloon is at an altitude of 600 feet. It begins descending at a constant rate of 15 feet per minute.

a. TABULAR Complete the table below that shows the altitude of the balloon every 5 minutes.

Input x (minutes)	$y = -15x + 600$	Output y (altitude)
0	$-15(0) + 600$	600
5	■	■
10	■	■
15	■	■

b. VERBAL What does the input value of 0 represent?

c. GRAPHICAL Graph the ordered pairs.

d. ALGEBRAIC Determine the time it will take for the balloon to reach the ground. Explain how you solved.

Real-World Link

Hot-air balloons can lift off when the air inside the balloon is heated. This makes the air inside the balloon less dense than the air outside the balloon. It takes about 65,000 cubic feet of air to lift 1000 pounds.

Source: How Stuff Works

H.O.T. Problems Use Higher-Order Thinking Skills

43. OPEN ENDED Name two integers that have a product between -10 and -15.

44. REASONING Name all of the values of x if $7|x| = 63$.

45. CHALLENGE Positive integers A and C satisfy $A(A - C) = 23$. What is the value of C?

46. REASONING Calculate $(-10)(5)(18)[7 + (-7)]$ mentally. Justify your answer.

CHALLENGE Determine whether each of the following is *true* or *false*. If *false*, give a counterexample.

47. The product of three negative integers is positive.

48. The product of four negative integers is positive.

49. WRITING IN MATH When multiplying more than two integers, how can you determine the sign of the product?

Standardized Test Practice

50. A submarine descends at a constant rate of 300 feet per minute. Which of the following equations could be used to find the altitude of the submarine after 5 minutes?

A $5(-300) = -1500$
B $5(300) = 1500$
C $-5(300) = -1500$
D $-5(-300) = 1500$

51. Which of the following equations is modeled by the number line below?

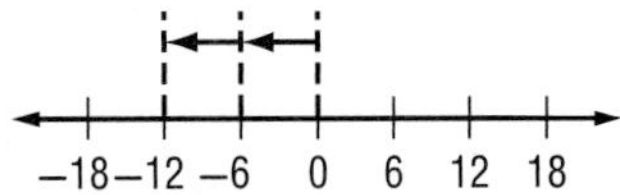

F $-2(-6) = -12$
G $-2(6) = -12$
H $2(6) = -12$
J $2(-6) = -12$

52. Simplify $-14(-2)(-12)$.

A -336
B -168
C 168
D 288

53. GRIDDED RESPONSE The distance from the water line to the bottom of a cargo ship changes based on the weight of the cargo.

Weight of Cargo (tons)	Depth of Ship (feet)
35	30
25	25
15	20

If the pattern in the table continues, find the depth of the ship, in feet, if the ship is carrying 100 tons of cargo.

Spiral Review

54. GEOGRAPHY The highest point in California is Mount Whitney, with an elevation of 14,494 feet. The lowest point is Death Valley, with an elevation of −282 feet. How much greater is the elevation of Mount Whitney than Death Valley? (Lesson 2-3)

Find each sum. (Lesson 2-2)

55. $6 + (-9) + 9$ **56.** $-7 + (-13) + 4$ **57.** $-9 + 16 + (-10)$ **58.** $-12 + 18 + (-12)$

59. WEATHER The table shows the lowest recorded temperatures for certain states.

Record Lowest Temperatures by State

State	Station	Date	Temperature (°F)
Alaska	Prospect Creek Camp	Jan. 23, 1971	−80
Montana	Rogers Pass	Jan. 20, 1954	−70
Wisconsin	Danbury	Jan. 24, 1922	−54

a. Compare the lowest temperature in Montana and the lowest temperature in Wisconsin using an inequality.

b. Write the temperatures in order from greatest to least.

Skills Review

Find each quotient. (Previous Course)

60. $72 \div 9$ **61.** $108 \div 12$ **62.** $84 \div 7$ **63.** $52 \div 4$

EXPLORE 2-5

Algebra Lab
Dividing Integers

Math Online glencoe.com
Math *in Motion*, Animation

You can model division by separating algebra tiles into equal-sized groups.

ACTIVITY 1

Find 10 ÷ 2.

Place 10 positive tiles on the mat to represent 10.

Separate the tiles into 2 equal-sized groups.

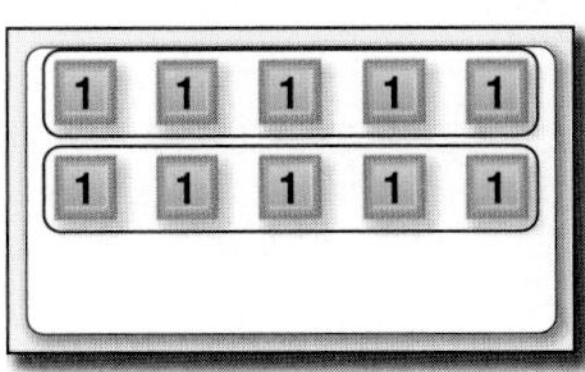

$10 \div 2 = 5$

There are 5 positive tiles in each of the 2 groups.

So, $10 \div 2 = 5$.

ACTIVITY 2

Find −12 ÷ 2.

Place 12 negative tiles on the mat to represent −12.

Separate the tiles into 2 equal-sized groups.

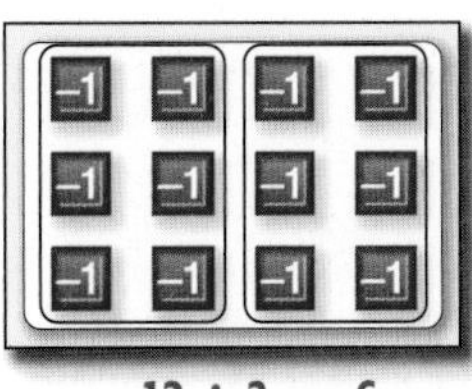

$-12 \div 2 = -6$

There are 6 negative tiles in each of the 2 groups.

So, $-12 \div 2 = -6$.

Analyze the Results

Model each quotient using algebra tiles.

1. $12 \div 6$

2. $16 \div 2$

3. $14 \div 7$

4. $-8 \div 2$

5. $-9 \div 3$

6. $-6 \div 2$

7. $-16 \div 4$

8. $-5 \div 5$

9. $-10 \div 2$

For Exercises 10–12, study the quotients in Exercises 1–9.

10. When the dividend and the divisor are both positive, is the quotient positive or negative? How does this compare to the sign of a product when both factors are positive?

11. When the dividend is negative and the divisor is positive, is the quotient positive or negative? How does this compare to the sign of a product when one factor is positive and one is negative?

12. **MAKE A CONJECTURE** Write a rule that will help you determine the sign of the quotient of two integers.

2-5 Dividing Integers

Why?

Rappelling is one way climbers descend rocks and mountains. A climber was at an elevation of 800 feet above sea level. Five minutes later, the climber was at an elevation of 755 feet above sea level.

a. What was the climber's change in altitude?

b. Suppose the climber descended the same number of feet each minute. Write an expression to determine the number of feet the climber descended each minute.

Then
You divided integers using algebra tiles. (Explore 2-5)

Now
- Divide integers.
- Find the mean (average) of a set of data.

New Vocabulary
mean

Math Online
glencoe.com
- Extra Examples
- Personal Tutor
- Self-Check Quiz
- Homework Help

Divide Integers The expression $-45 \div 5$ is an example of dividing integers. Division of integers is related to multiplication of integers. So, one way to find the quotient is by using related multiplication sentences.

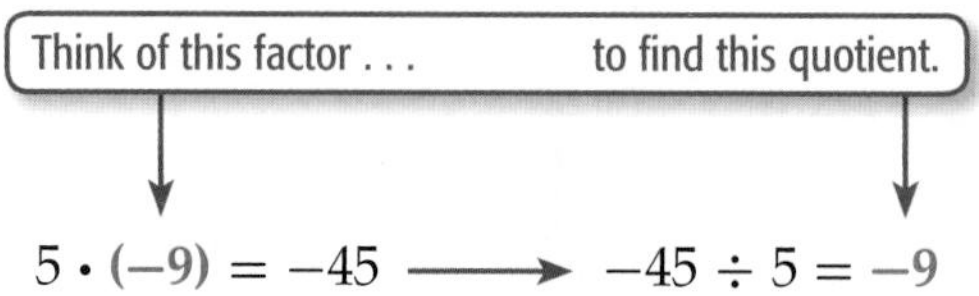

Think of this factor . . .
to find this quotient.

$$5 \cdot (-9) = -45 \longrightarrow -45 \div 5 = -9$$

In the division sentence $-45 \div 5 = -9$, notice that the dividend and the divisor have different signs. So, the quotient is negative. This suggests a rule for dividing integers with different signs.

Key Concept **Dividing Integers with Different Signs** *For Your FOLDABLE*

Words The quotient of two integers with different signs is negative.

Example $-10 \div 5 = -2$ $\quad$ $10 \div (-5) = -2$

EXAMPLE 1 Divide Integers with Different Signs

Find each quotient.

a. $-35 \div (7)$

$-35 \div (7) = -5$ **The quotient is negative.**

b. $\frac{64}{-8}$

$\frac{64}{-8} = -8$ **The quotient is negative.**

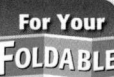

Check Your Progress

1A. $-63 \div 7$

1B. $\frac{110}{-10}$

Personal Tutor glencoe.com

You can also use multiplication and division sentences to find the quotient of integers with the same sign.

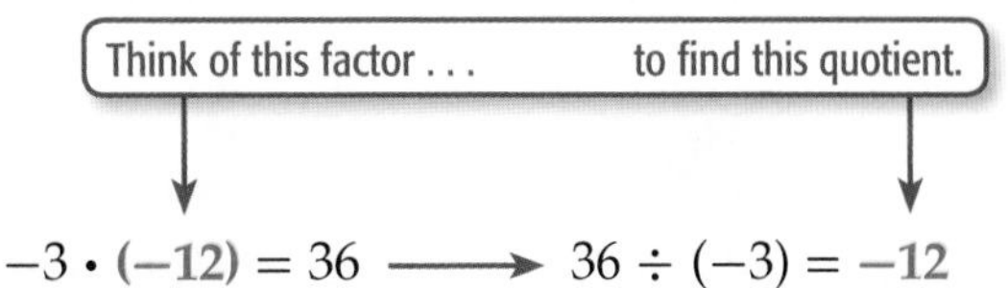
Think of this factor . . .
to find this quotient.

$-3 \cdot (-12) = 36 \longrightarrow 36 \div (-3) = -12$

This suggests a rule for dividing integers with the same sign.

Key Concept **Dividing Integers with the Same Signs** — For Your FOLDABLE

Words The quotient of two integers with the same sign is positive.

Example $-10 \div (-5) = 2$ $\quad 10 \div 5 = 2$

EXAMPLE 2 Divide Integers with the Same Signs

Find each quotient.

a. $48 \div 12$

$48 \div 12 = 4$ — The quotient is positive.

b. $\frac{-56}{-8}$

$\frac{-56}{-8} = 7$ — The quotient is positive.

StudyTip

Check Your Work Always check your work after finding an answer. Does $-8 \times 7 = -56$?

Check Your Progress

2A. $-35 \div (-5)$

2B. $\frac{39}{3}$

Personal Tutor glencoe.com

You can use the rules for dividing integers to evaluate algebraic expressions.

EXAMPLE 3 Evaluate Algebraic Expressions

Evaluate each expression if $x = -6$ and $y = -3$.

a. $12y \div x$

$12y \div (x) = 12(-3) \div (-6)$ — Replace y with -3 and x with -6.

$= -36 \div (-6)$ — The product of 12 and -3 is negative.

$= 6$ — The quotient of -36 and -6 is positive.

b. $\frac{-5x}{y}$

$\frac{-5x}{y} = \frac{-5(-6)}{-3}$ — Replace x with -6 and y with -3.

$= \frac{30}{-3}$ — The product of -5 and -6 is positive.

$= -10$ — The quotient of 30 and -3 is negative.

Check Your Progress

3A. $4b \div a$ if $a = -2$ and $b = -5$

3B. $4y \div (2x)$ if $x = 14$ and $y = -7$

Personal Tutor glencoe.com

Mean (Average) Division is used in statistics to find the average, or mean, of a set of data. To find the **mean** of a set of numbers, find the sum of the numbers and then divide by the number of items in the set.

Real-World EXAMPLE 4 Find the Mean

WEATHER The wind chill temperatures in degrees Fahrenheit for 7 days were −6, −5, 2, −10, 1, −9 and 6. Find the mean temperature.

$$\frac{-6 + (-5) + 2 + (-10) + 1 + (-9) + 6}{7} = \frac{-21}{7}$$ Find the sum of the temperatures. Divide by the number of days.

$$= -3$$ Simplify.

The mean wind chill temperature is −3°F.

Check Your Progress

4. **GOLF** Linda has scores of −3, −2, 1, and 0 during 4 rounds of golf. Find the mean of her golf scores.

Personal Tutor glencoe.com

Concept Summary Operations with Integers

For Your FOLDABLE

Words	Examples
Adding Integers **Same Signs:** Add absolute values. The sum has the same sign as the integers. **Different Signs:** Subtract absolute values. The sum has the same sign as the integer with the greater absolute value.	$3 + 2 = 5$ $-3 + (-2) = -5$ $-3 + 2 = -1$ $3 + (-2) = 1$
Subtracting Integers To subtract an integer, add its additive inverse.	$3 - 5 = 3 + (-5)$ or -2 $3 - (-5) = 3 + 5$ or 8
Multiplying and Dividing Two Integers **Same Signs:** The product or quotient is positive. **Different Signs:** The product or quotient is negative.	$3 \cdot 2 = 6$ $6 \div 3 = 2$ $-3(-2) = 6$ $-6 \div (-3) = 2$ $-3 \cdot 2 = -6$ $-6 \div 3 = -2$ $3(-2) = -6$ $6 \div (-3) = -2$

Check Your Understanding

Examples 1 and 2 pp. 90–91

Find each quotient.

1. $40 \div -10$
2. $\frac{39}{13}$
3. $-26 \div (-3)$
4. $\frac{-54}{6}$
5. $-48 \div 3$
6. $\frac{72}{-18}$
7. $36 \div (-4)$
8. $\frac{-72}{-9}$

Example 3 p. 91

ALGEBRA Evaluate each expression if $s = -2$ and $t = 7$.

9. $14s \div t$
10. $\frac{-10t}{s}$
11. $4t \div (2s)$

Example 4 p. 92

12. **MONEY** The following are the changes of a value of a certain stock over the last 5 days: −\$7, +\$3, +\$6, −\$2, −\$5. Find the mean change.

Practice and Problem Solving

● = **Step-by-Step Solutions** begin on page R11.
Extra Practice begins on page 810.

Examples 1 and 2 pp. 90–91

Find each quotient.

13. $-33 \div 11$
14. $28 \div -14$
15. $-36 \div (-2)$
16. $-60 \div (-5)$
17. $\frac{-150}{10}$
18. $\frac{600}{-20}$
19. $126 \div 9$
20. $750 \div 15$
21. $-770 \div 7$
22. $-560 \div 8$
23. $\frac{-350}{-70}$
24. $\frac{-480}{-16}$

Example 3 p. 91

ALGEBRA Evaluate each expression.

25. $\frac{n}{-13}$, if $n = -182$
26. $252 \div k$, if $k = 9$
27. $\frac{-6a}{b}$, if $a = -24$ and $b = -4$
28. $\frac{9y}{x}$, if $x = -21$ and $y = -35$
29. $-2st \div (-3t)$, if $s = 18$ and $t = -14$
30. $4qr \div (2r)$ if $q = -16$ and $r = -8$

Example 4 p. 92

31. **FINANCIAL LITERACY** The last 5 transactions at Mr. Brigham's ATM were \$250, −\$60, −\$94, \$300, and −\$186. Find the mean transaction amount.

32. **GAMES** The final scores of contestants on a game show were −14, 0, 78, −12, 46, and 64. What is the mean score?

TEMPERATURE The expression $\frac{5(F-32)}{9}$, where F represents the temperature in degrees Fahrenheit, can be used to convert temperatures from degrees Fahrenheit to degrees Celsius.

33. The surface temperature on Mercury at night can fall to −300°F. Convert this temperature to degrees Celsius. Round to the nearest tenth.

34. The extreme high and low temperatures for different states are shown in the table.
    a. Find the extreme high and low temperatures for each state in degrees Celsius. Round to the nearest tenth.
    b. The difference between the extreme high and low temperatures is called the range. Find the range of the temperatures in degrees Celsius for each state.
    c. List the states in order from least to greatest ranges.

Extreme Temperature

State	Extreme Low (°F)	Extreme High (°F)
Arizona	−40	128
Florida	−2	109
Kentucky	−34	114
Michigan	−51	112
New York	−57	108

35. **MONEY** Last year, a small clothing company's total income was \$64,000, while its total expenses were \$67,600. Use the expression $\frac{I-E}{12}$, where I represents total income and E represents total expenses, to find the average difference between the company's income and expenses each month. Explain what the answer means.

Replace each ● with <, >, or = to make a true sentence.

36. $-80 \div (-2)$ ● $120 \div 3$
37. $-1750 \div 70$ ● $-1008 \div 48$
38. $\frac{240}{-80}$ ● $\frac{-150}{-50}$
39. $\frac{675}{-45}$ ● $\frac{867}{-51}$

Problem-Solving Tip

Guess and Check To find the value of x, try different values until you find the one that makes each sentence true.

ALGEBRA Find the value of x that makes each statement true.

40. $-375 \div x = -15$

41. $22 = x \div (-34)$

42. $x \div (-17) = -35$

43. $-689 \div x = 53$

For Exercises 44–46, write an expression to represent each real-world situation. Then evaluate the expression and interpret the meaning of the solution.

44. Jean owes her parents \$90, to be paid in 5 equal installments. How much is each installment?

45. The temperature dropped a total of 40°F over an 8 hour period. What was the mean hourly temperature drop?

46. In October, the full price of a television is \$695. Over the next seven months, the price of the television drops a total of \$140. How much did the price of the television drop each month on average?

47. **GEOGRAPHY** The table shows the deepest point of each of the Great Lakes.

a. What is the mean of the deepest points of the Great Lakes?

b. Suppose each of the deepest points were 10 meters higher. Find the mean. Compare the new mean to the original.

Great Lake	Deepest Point (m)
Erie	−64
Huron	−229
Michigan	−281
Ontario	−244
Superior	−406

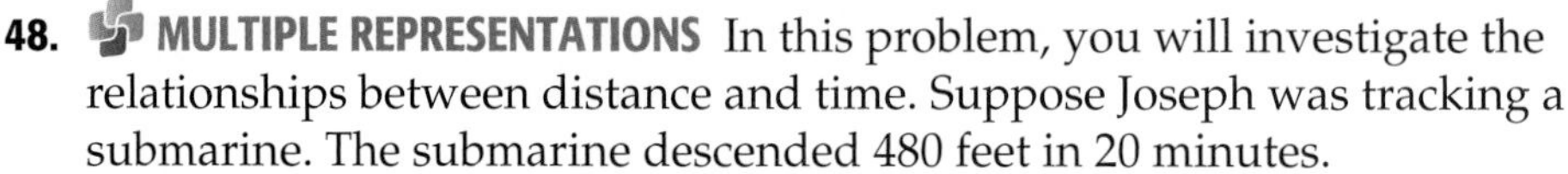

48. **MULTIPLE REPRESENTATIONS** In this problem, you will investigate the relationships between distance and time. Suppose Joseph was tracking a submarine. The submarine descended 480 feet in 20 minutes.

a. NUMERICAL How many feet did the submarine descend per minute?

b. ALGEBRAIC Write a function rule to determine how many feet the submarine will descend after any number of minutes.

c. TABULAR Create a function table to show how far the submarine descended after 5, 10, 15, and 18 minutes.

Real-World Link

There is enough water in the five Great Lakes to cover the entire continental United States with 9.5 feet of water.

Source: Alliance for the Great Lakes

H.O.T. Problems

Use Higher-Order Thinking Skills

49. **OPEN ENDED** Write a division expression with a quotient between −20 and −25.

50. **CHALLENGE** The mean temperature during 5 days was −10°F. Give a sample set of what the temperatures might have been for the 5 days.

51. **REASONING** Find the next two numbers in the pattern 1024, −256, 64, −16,. . . . Explain your reasoning.

52. **CHALLENGE** Addition and multiplication are said to be *closed* for whole numbers, but subtraction and division are not. That is, when you add or multiply any two whole numbers, the result is a whole number. Which operations are closed for integers?

53. **WRITING IN MATH** Explain whether the Associative Property and Commutative Property are true for division of integers. Support your reasoning with an example.

Standardized Test Practice

54. Yesterday's low temperature was 24 degrees Fahrenheit. Use the expression $\frac{5(F-32)}{9}$, where F represents the temperature in degrees Fahrenheit, to find the approximate low temperature in degrees Celsius.

A −9.8°C
B −4.4°C
C 4.4°C
D 9.8°C

55. Miss Washer recorded the low temperature each day for a week. What is the mean low temperature?

Day	M	T	W	Th	F
Temperature (°F)	−18	12	−7	9	−2

F 9.6°F
G 1.2°F
H −9.6°F
J −1.2°F

56. The depth of a reservoir decreased 84 inches in two weeks. If the water depth changed by the same amount each day, how much did the depth of the water change per day?

A −7 inches
B −6 inches
C 6 inches
D 7 inches

57. EXTENDED RESPONSE The temperature dropped 30°F in a 6 hour period.

a. What was the mean hourly temperature change?
b. Write a function rule to represent the situation.
c. Create a function table to show how much the temperature changed in 1, 3, and 5 hours.

Spiral Review

58. TIDES During low tide, in Wrightsville, North Carolina, the beachfront in some places is about 350 feet from the ocean to the homes. High tide can change the width of a beach at a rate of −17 feet an hour. It takes 6 hours for the ocean to move from low to high tide. (Lesson 2-4)

a. What is the change in the width of the beachfront from low to high tide?
b. What is the distance from the ocean to the homes at high tide?

Find each difference. (Lesson 2-3)

59. 3 − 8
60. 4 − 5
61. 2 − 9
62. −9 − (−7)
63. −7 − (−10)
64. −11 − (−12)

65. MONEY The starting balance in a checking account was $50. What was the balance after checks were written for $25 and for $32? (Lesson 2-2)

Skills Review

Use the coordinate plane to name the ordered pairs for each point. (Lesson 1-6)

66. Point A
67. Point G
68. Point C
69. Point E

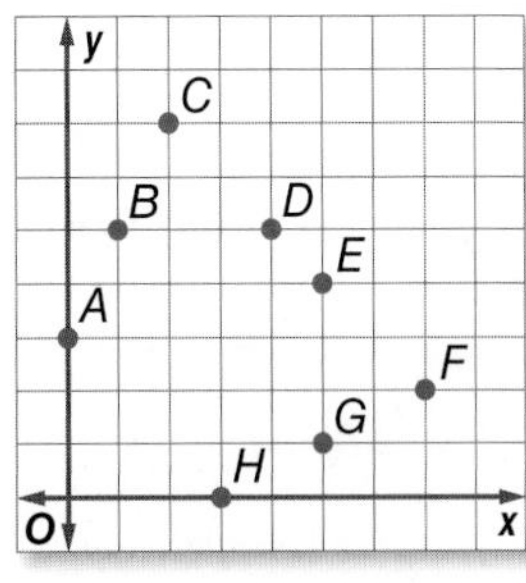

2-6 Graphing in Four Quadrants

Then

You have already used ordered pairs to name and locate points. (Lesson 1-4)

Now

- Graph points on a coordinate plane.
- Graph algebraic relationships.

New Vocabulary

quadrants

Math Online

glencoe.com

- Extra Examples
- Personal Tutor
- Self-Check Quiz
- Homework Help
- Math in Motion

Why?

Duncan and his friends have agreed to meet at the community swimming pool. The map shows where each person lives in relationship to the swimming pool.

a. Explain how Kacy would go to the swimming pool.

b. Which two friends live the same distance from the swimming pool? Why are they in different locations?

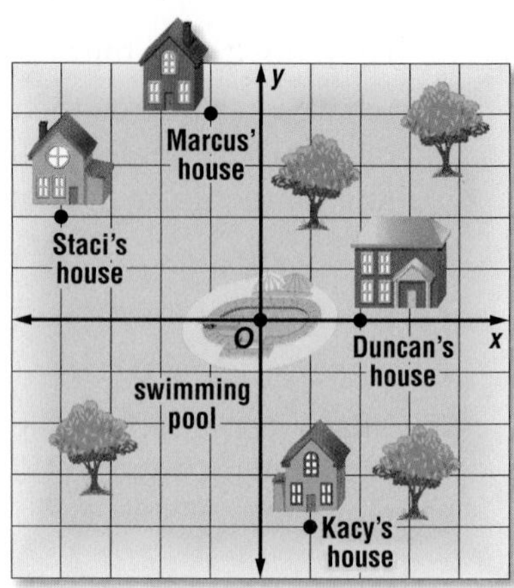

Graph Points Street maps use a coordinate system. The coordinate system you used in Lesson 1-4 can be extended to include points below and to the left of the origin.

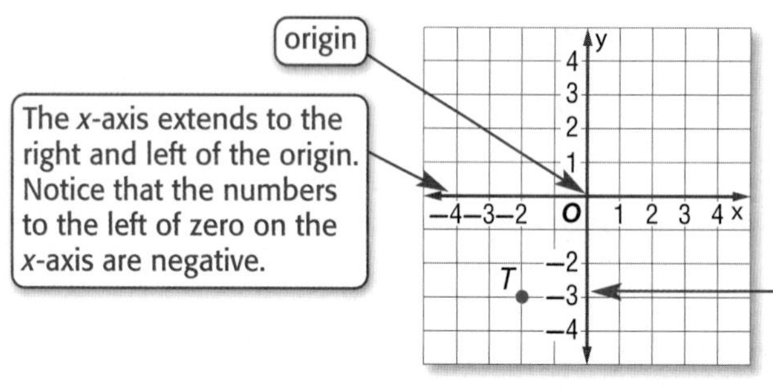

Recall that a point graphed on the coordinate system has an x-coordinate and a y-coordinate. The dot at the ordered pair $(-2, -3)$ is the graph of point T.

x-coordinate → $(-2, -3)$ ← y-coordinate

EXAMPLE 1 Write Ordered Pairs

Write the ordered pair that names each point.

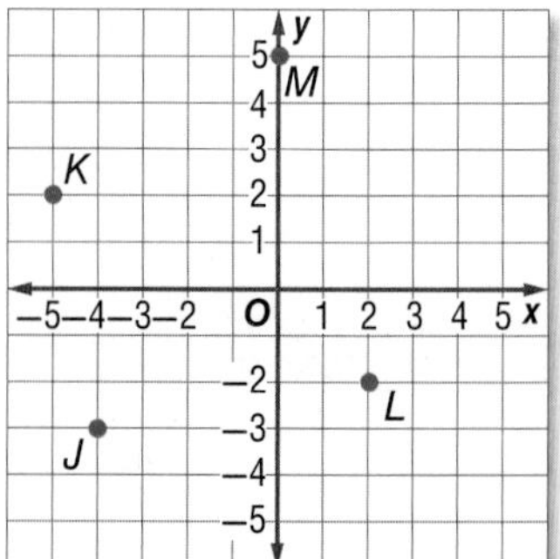

a. J

The x-coordinate is -4.
The y-coordinate is -3.
The ordered pair is $(-4, -3)$.

b. L

The x-coordinate is 2.
The y-coordinate is -2.
The ordered pair is $(2, -2)$.

Check Your Progress

1A. M **1B.** K

Personal Tutor glencoe.com

Math *in Motion*, Interactive Lab glencoe.com

The x-axis and the y-axis separate the coordinate plane into four regions, called **quadrants**. The quadrants are named I, II, III, and IV.

The axes and points on the axes are not located in any of the quadrants.

EXAMPLE 2 Graph Points and Name the Quadrant

Graph and label each point on a coordinate plane. Name the quadrant in which each point lies.

a. $A(-2, -4)$

Start at the origin. Move 2 units left. Then move 4 units down and draw a dot. Point $A(-2, -4)$ is in quadrant III.

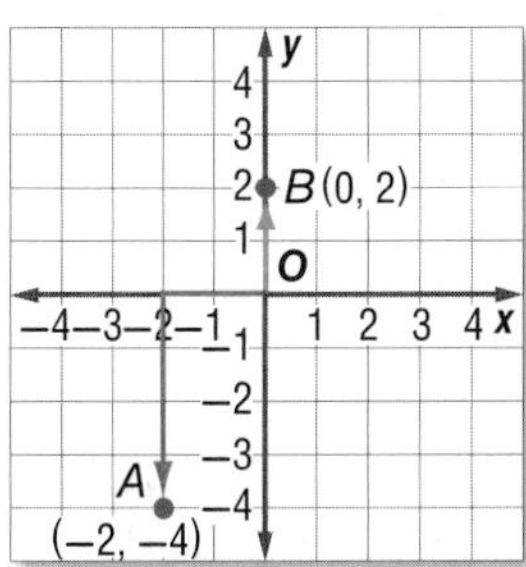

b. $B(0, 2)$

Start at the origin. Since the x-coordinate is 0, the point will lie on the y-axis. So, move 2 units up. Point $B(0, 2)$ is not in a quadrant.

Check Your Progress

2A. $H(4, -3)$ **2B.** $I(-1, 4)$ **2C.** $J(0, -2)$

Personal Tutor glencoe.com

Math History Link

René Descartes (1596–1650)
The coordinate plane can also be called the *Cartesian* plane. The word *Cartesian* comes from the name Descartes. René Descartes was a French mathematician who is credited with developing the coordinate plane.

Graph Algebraic Relationships You can use a coordinate graph to show relationships between two numbers.

Real-World EXAMPLE 3 Graph an Algebraic Relationship

GOLF The difference between John and Tarie's golf score is 2. If x represents John's score and y represents Tarie's score, make a function table of possible values for x and y. Graph the ordered pairs and describe the graph.

Choose values for x and y that have a difference of 2. Then graph the ordered pairs.

The points are along a diagonal line that crosses the x-axis at $x = 2$.

$x - y = 2$		
x	y	(x, y)
2	0	(2, 0)
1	−1	(1, −1)
0	−2	(0, −2)
−1	−3	(−1, −3)
−2	−4	(−2, −4)

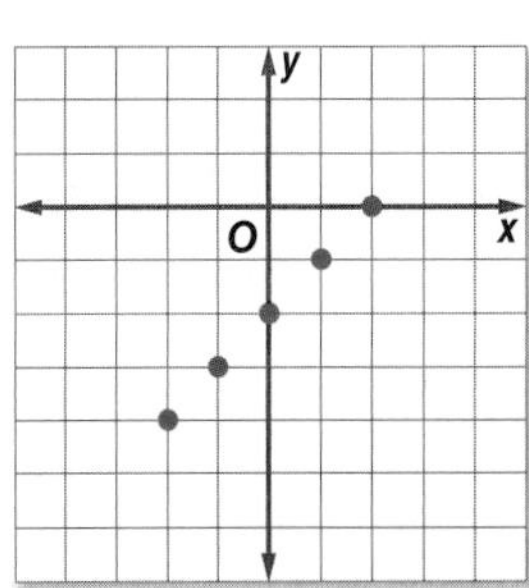

StudyTip

Scale When no numbers are shown on the x- or y-axis, you can assume that each square is one unit long on each side.

Check Your Progress

3. GOLF The sum of two golf scores is 3. If x represents one score and y represents the other score, make a function table of possible values for x and y. Graph the ordered pairs and describe the graph.

Personal Tutor glencoe.com

Check Your Understanding

Example 1 p. 96 **Name the ordered pair for each point graphed at the right.**

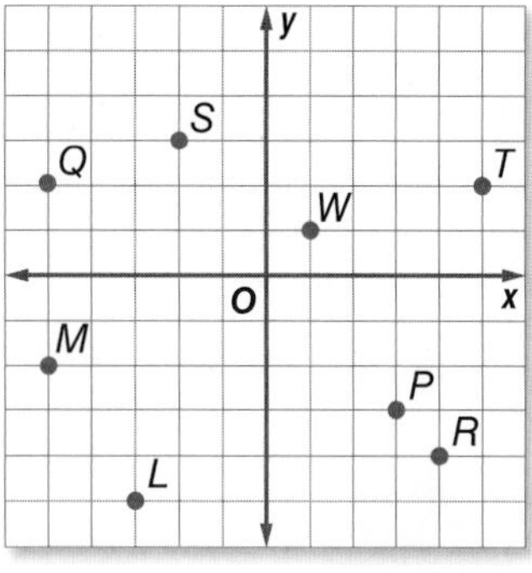

1. Q
2. P
3. T
4. M

Example 2 p. 97 **Graph and label each point on a coordinate plane. Name the quadrant in which each point is located.**

5. $A(-2, 3)$
6. $B(4, -1)$
7. $C(-3, -2)$
8. $D(0, -5)$

Example 3 p. 97

9. **TEMPERATURE** The difference of two temperatures is 4°F. If x represents the first temperature and y represents the second temperature, make a function table of possible values for x and y. Graph the ordered pairs and describe the graph.

Practice and Problem Solving

● = **Step-by-Step Solutions** begin on page R11.
Extra Practice begins on page 810.

Example 1 p. 96 **Name the ordered pair for each point graphed at the right.**

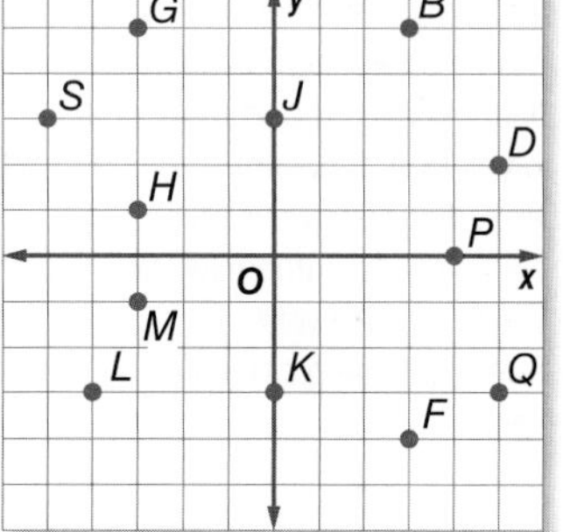

10. S
11. H
12. D
13. B
14. M
15. L
16. F
17. Q
18. K
19. J

Example 2 p. 97 **Graph and label each point on a coordinate plane. Name the quadrant in which each point is located.**

20. $Z(-1, 1)$
21. $Y(-2, 3)$
22. $X(5, 6)$
23. $W(6, 2)$
24. $V(-1, -6)$
25. $S(2, -1)$
26. $T(-5, 0)$
27. $R(0, -4)$
28. $P(-4, 5)$
29. $Q(-3, 3)$
30. $N(1, -1)$
31. $K(5, -3)$

Example 3 p. 97

32. **FOOTBALL** After two plays, the Wildcats gained a total of 16 yards. If x represents the number of yards for play one, and y represents the number of yards for play two, make a function table of possible values for x and y. Graph the ordered pairs and describe the graph.

33. **SCUBA DIVING** The difference in depth between two scuba divers is 10 feet. If x represents the depth of one scuba diver and y represents the depth of the second scuba diver, make a function table of possible values for x and y. Graph the ordered pairs and describe the graph.

Name the quadrant in which each point lies.

34. $A(5, |-6|)$
35. $E(|-5|, -3)$
36. $I(x, y)$ if $x < 0, y > 0$
37. $U(x, y)$ if $x > 0, y < 0$

38. GEOMETRY Graph points $A(-4, 3)$, $B(1, 3)$, $C(1, 2)$, and $D(-4, 2)$ on a coordinate plane and connect them to form a rectangle.

a. Add 4 to the x-coordinate of each ordered pair and re-draw the figure.

b. Compare the two rectangles.

39 **TEMPERATURE** The function table shows temperatures in Celsius and the corresponding temperatures in Fahrenheit. Graph the ordered pairs (Celsius, Fahrenheit) to show the relationship between Celsius and Fahrenheit.

Celsius	−10	−5	0	5	10
Fahrenheit	14	23	32	41	50

40. FINANCIAL LITERACY The function table shows the balance on a $50 music card after a certain number of songs have been downloaded. Make a graph to show how the number of songs downloaded and the remaining balance are related.

Songs Downloaded	Balance ($)
0	50
5	45
10	40
15	35

Real-World Link

The first MP3 player to become widely available was released in 1999 and cost around $200. It weighed about 2 ounces and held 32 minutes of music. In contrast, an MP3 player that holds 8 hours of music costs less than $50 in 2007.

For each graph, create a function table showing the input, rule, and output.

41.

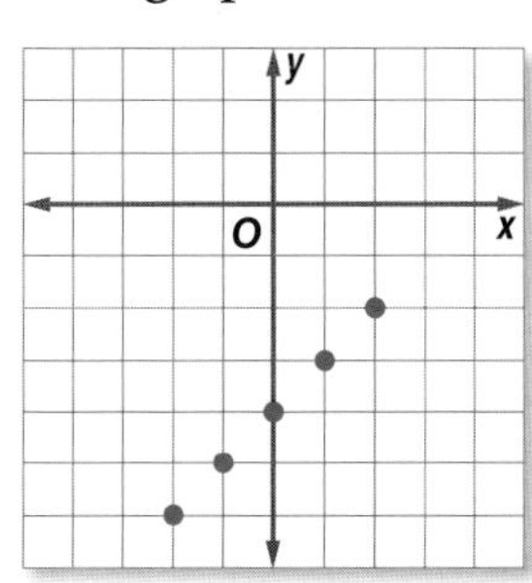

42.

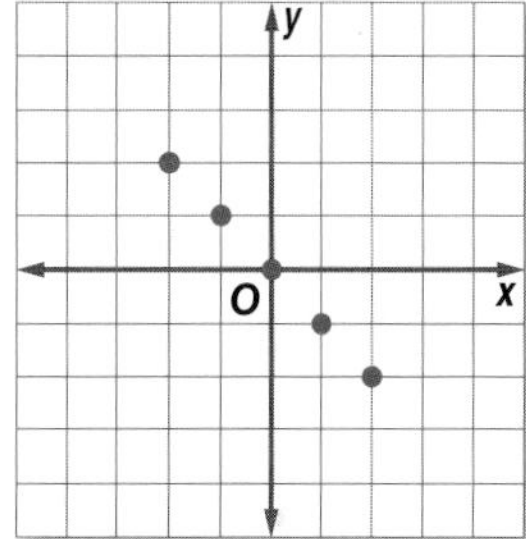

Graph and label each point on a coordinate plane.

43. $A(-6.5, 3)$ **44.** $B(-2, -5.75)$ **45.** $C(4.1, -1)$ **46.** $D(-3.4, 1.5)$

H.O.T. Problems Use Higher-Order Thinking Skills

47. OPEN ENDED Write the coordinates of a point located in quadrant II.

48. CHALLENGE The product of two numbers is 12.

a. Make a function table using −3, −2, −1, 1, 2, and 3 as input values.

b. Graph the ordered pairs. Compare and contrast your graph with the graph in Example 3.

49. CHALLENGE **Determine whether each statement is *always, sometimes,* or *never* true. Explain or give a counterexample to support your answer.**

a. Both x- and y-coordinates of a point in quadrant I are negative.

b. The x-coordinate of a point that lies on the x-axis is negative.

50. WRITING IN MATH Are the points at $(-7, 8)$ and $(8, -7)$ in the same location? Explain your reasoning.

Standardized Test Practice

51. Which point on the graph best represents the location of the library?

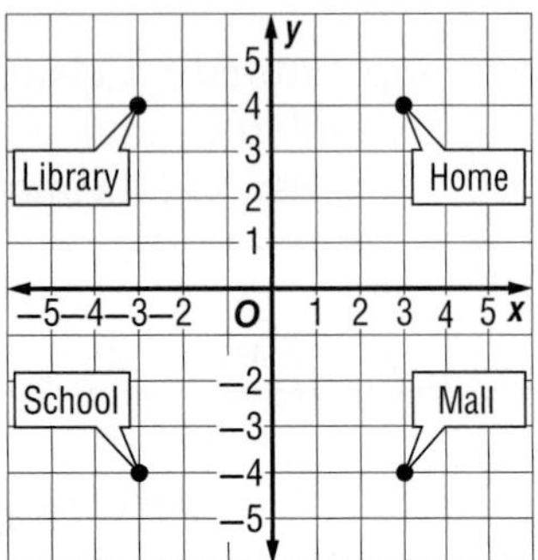

A (3, 4)
B (−3, −4)
C (−3, 4)
D (3, −4)

52. What building is located at point (−3, −4) on the graph above?

F School
G Mall
H Library
J Home

53. In which quadrant on the coordinate plane is point (2, −3)?

A quadrant I
B quadrant II
C quadrant III
D quadrant IV

54. EXTENDED RESPONSE Juan wants to rent 4 DVDs. Each DVD costs $3 for 2 days.

a. Complete the table to show his total cost for the number of days given.

Number of Days	Total Cost
2	■
4	■
6	■

b. How much will Juan have to pay if he wants to keep the DVDs for 10 days?

Spiral Review

Find each quotient (Lesson 2-5)

55. $-27 \div (-9)$

56. $-77 \div 7$

57. $-300 \div 6$

58. GLACIERS A glacier was receding at a rate of 300 feet per day. What is the glacier's movement in 5 days? (*Hint:* The word *receding* means moving backward.) (Lesson 2-4)

59. SWIMMING Lincoln High School's swim team finished the 4 × 100-meter freestyle relay in 5 minutes 18 seconds. Prospect High School's swim team finished the race in 5 minutes 7 seconds. Write an integer that represents Lincoln's finish compared to Prospect's finish. (Lesson 2-3)

Evaluate each expression. (Lessons 2-1 through 2-3)

60. $|-10|$

61. $|10| - |-4|$

62. $|16| + |-5|$

Skills Review

Use the grid to name the point for each ordered pair. (Lesson 1-4)

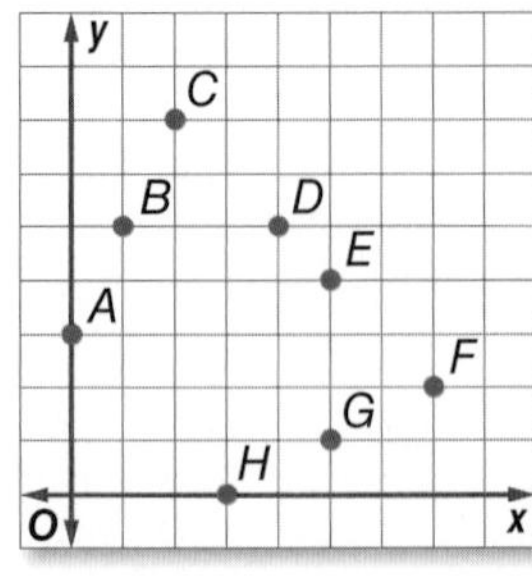

63. (1, 5)

64. (7, 2)

65. (4, 5)

66. (0, 3)

67. (2, 7)

68. (5, 4)

2-7

Translations and Reflections on the Coordinate Plane

Then

You have already graphed points on a coordinate plane. (Lesson 2-6)

Now

- Define and identify transformations.
- Draw translations and reflections on a coordinate plane.

New Vocabulary

transformation
image
translation
reflection
line of symmetry

Math Online

glencoe.com

- Extra Examples
- Personal Tutor
- Self-Check Quiz
- Homework Help

Why?

Suppose a line of tuba players marched across the field for 20 yards, then turned left and marched another 5 yards.

a. Compare and contrast the original position and resulting position of the tubas.

b. Suppose that a line of bass drums is on the 50-yard line. One half of the drummers turned around and marched 10 yards back, while the other half kept marching 10 yards forward. Compare and contrast these motions.

Transformations A **transformation** is an operation that maps an original geometric figure onto a new figure called the **image**. Two common transformations on the coordinate plane are shown.

A **translation** is when you slide a figure from one position to another without turning it.

Translation

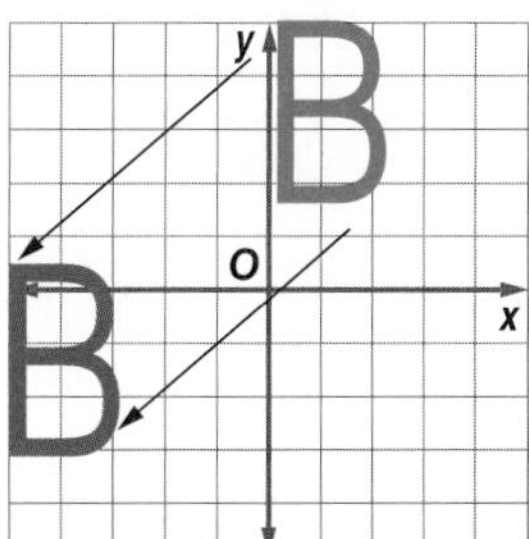

A **reflection** is when you flip a figure over a line. This line is called the **line of symmetry**.

Reflection

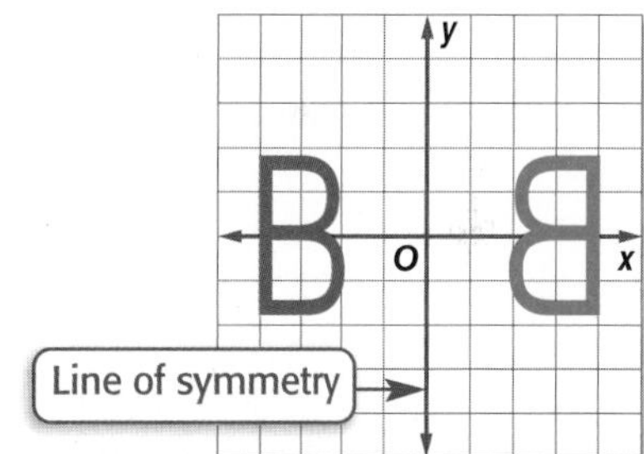

Key Concept: Translations and Reflections

For Your FOLDABLE

Translation

- called a *slide*
- image is the same shape and the same size as original figure
- orientation is the *same* as the original figure

Reflection

- called a *flip*
- figures are mirror images of each other
- image is the same shape and same size as original figure
- orientation is *different* from the original figure

StudyTip

Notation The notation M' is read *M prime*. It corresponds to point *M*.

Translations When translating a figure, every point of the original figure is moved the same distance and in the same direction.

Translation 5 units right

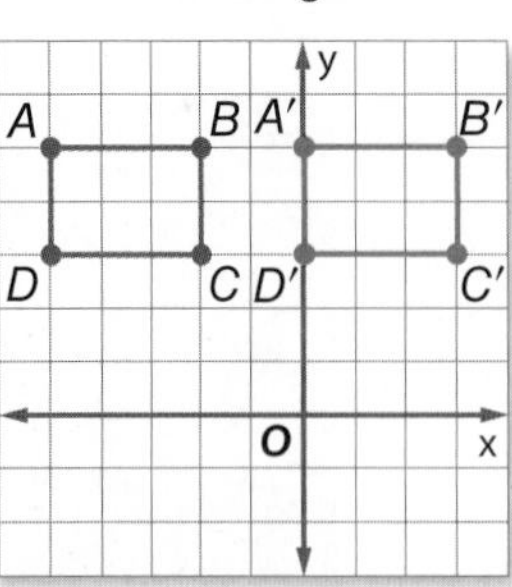

Translation 4 units up

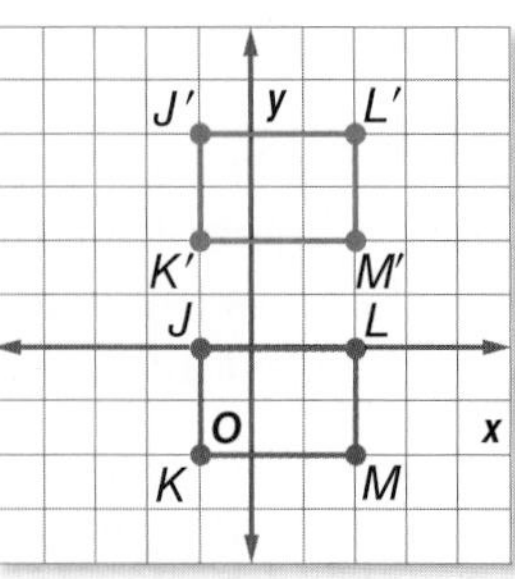

Translation 3 units left, 5 units down

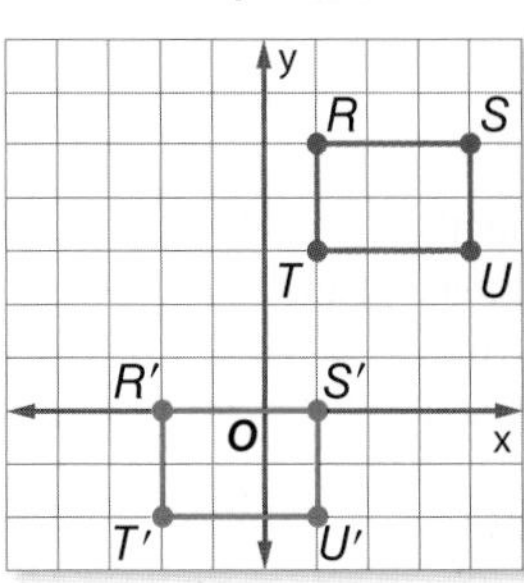

You can describe the translation using an ordered pair (a, b). For example, a translation of 3 units left and 5 units down corresponds to $(-3, -5)$. A translation moves every point $P(x, y)$ to an image $P'(x + a, y + b)$.

STANDARDIZED TEST EXAMPLE 1

ReadingMath

Vertex, Vertices A *vertex* of a figure is a point where two sides of the figure meet. *Vertices* is the plural of *vertex*.

Rectangle *JKLM* is shown. If it is translated 6 units to the right and 4 units down, find the coordinates of the vertices of the image.

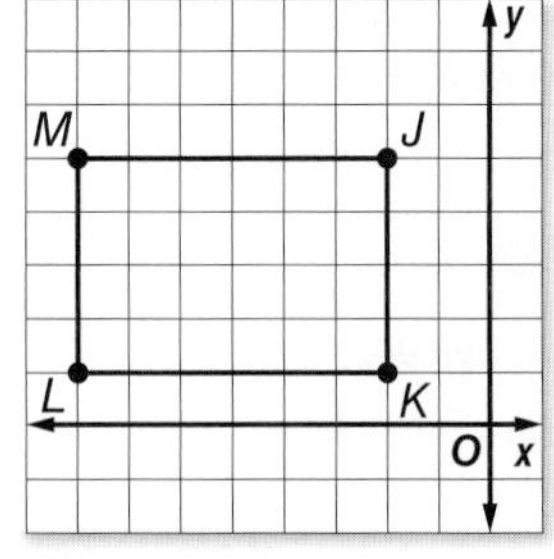

A $J'(-8, 1)$, $K'(-8, -3)$, $L'(-14, -3)$, $M'(-14, -3)$

B $J'(4, 1)$, $K'(4, 0)$, $L'(-2, -3)$, $M'(-2, -1)$

C $J'(4, 1)$, $K'(4, -3)$, $L'(-2, -3)$, $M'(-2, 1)$

D $J'(-2, -1)$, $K'(-2, -3)$, $L'(4, -3)$, $M'(4, -1)$

Read the Test Item

This translation can be written as $(6, -4)$. To find the coordinates of the translated image, add 6 to each x-coordinate and add -4 to each y-coordinate.

Solve the Test Item

original		translation		image
$J(-2, 5)$	+	$(6, -4)$	→	$J'(4, 1)$
$K(-2, 1)$	+	$(6, -4)$	→	$K'(4, -3)$
$L(-8, 1)$	+	$(6, -4)$	→	$L'(-2, -3)$
$M(-8, 5)$	+	$(6, -4)$	→	$M'(-2, 1)$

The coordinates of the vertices of rectangle $J'K'L'M'$ are $(4, 1)$, $(4, -3)$, $(-2, -3)$, and $(-2, 1)$. So the answer is C.

Check Your Progress

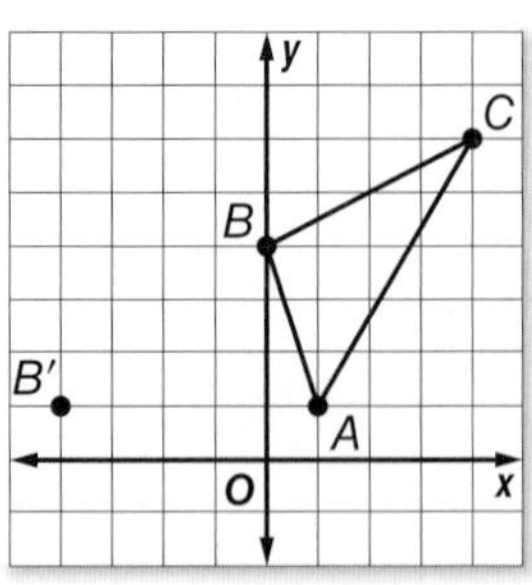

1. Triangle *ABC* is translated so that *B* is mapped to B'. Which coordinate pair represents C'?

F $(-4, 1)$

G $(0, 3)$

H $(-1, 1)$

J $(1, 3)$

Personal Tutor glencoe.com

Reflections When reflecting a figure, every point of the original figure has a corresponding point on the other side of the line of symmetry. Corresponding points are the same distance from the line of symmetry.

Opposites
Opposites are two numbers with the same absolute value but different signs. (Lesson 2-2)

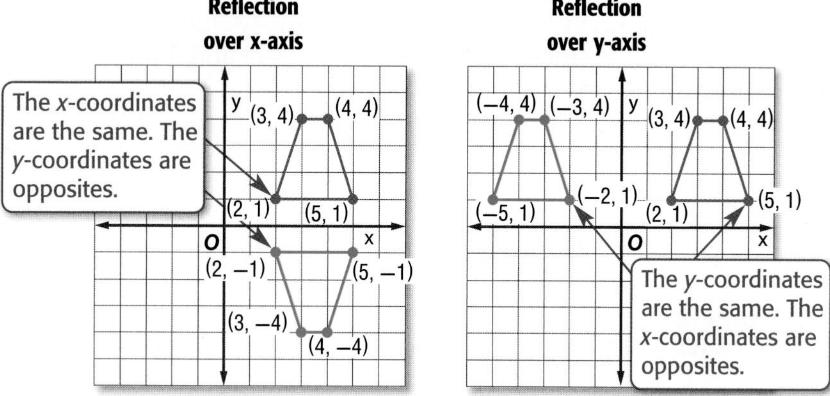

To reflect a point over the x-axis, use the same x-coordinate and multiply the y-coordinate by -1. To reflect a point over the y-axis, use the same y-coordinate and multiply the x-coordinate by -1.

EXAMPLE 2 Reflections on a Coordinate Plane

The vertices of figure *DEFG* are *D*(4, −2), *E*(5, −5), *F*(2, −4), and *G*(1, −1). Graph the figure and its image after a reflection over the *y*-axis.

To find the coordinates of the vertices of the image after a reflection over the y-axis, use the same y-coordinate. Replace the x-coordinate with its opposite.

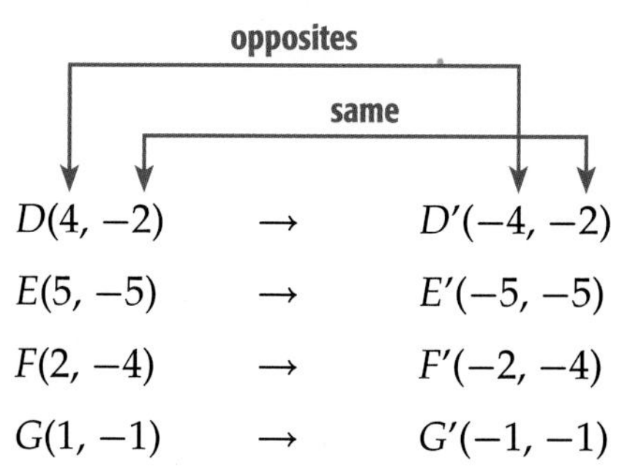

$D(4, -2)$ → $D'(-4, -2)$

$E(5, -5)$ → $E'(-5, -5)$

$F(2, -4)$ → $F'(-2, -4)$

$G(1, -1)$ → $G'(-1, -1)$

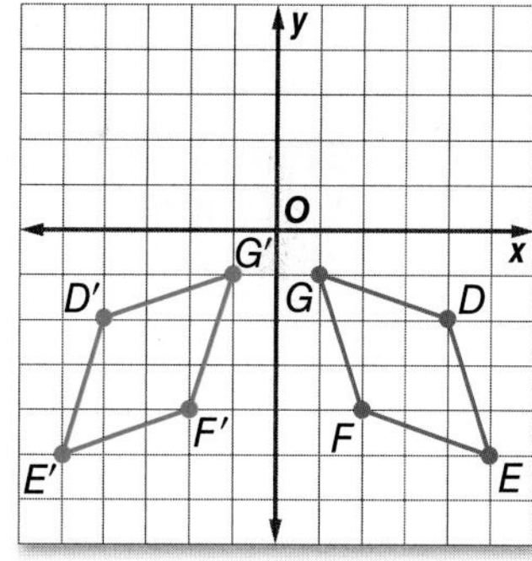

The coordinates of the vertices of the image are $D'(-4, -2)$, $E'(-5, -5)$, $F'(-2, -4)$, and $G'(-1, -1)$.

Check Your Progress

2A. The vertices of $\triangle ABC$ are $A(4, -2)$, $B(0, 2)$, and $C(5, 2)$. Graph the triangle and its image after a reflection over the x-axis.

2B. The vertices of rectangle $WXYZ$ are $W(-3,-3)$, $X(-3, 4)$, $Y(2, 4)$ and $Z(2, -3)$. Graph the rectangle and its image after a reflection over the y-axis.

Personal Tutor glencoe.com

Check Your Understanding

Example 1
p. 102

1. **MULTIPLE CHOICE** Triangle *MNP* is shown on the coordinate plane. Find the coordinates of the vertices of the image of the triangle *MNP* translated 5 units to the right and 3 units up.

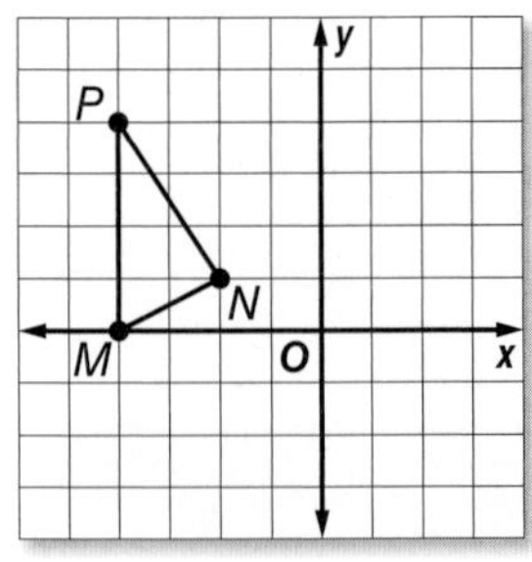

A $M'(1, 3)$, $N'(3, 4)$, $P'(1, 7)$

B $M'(3, 1)$, $N'(3, 4)$, $P'(1, 7)$

C $M'(1, 3)$, $N'(4, 3)$, $P'(7, 1)$

D $M'(1, 3)$, $N'(3, -4)$, $P'(1, 7)$

Example 2
p. 103

2. The vertices of $\triangle LMN$ are $L(2, 1)$, $M(5, 2)$, and $N(-1, 4)$. Graph the triangle and its image after a reflection over the x-axis.

Practice and Problem Solving

 = **Step-by-Step Solutions** begin on page R11.
Extra Practice begins on page 810.

Example 1
p. 102

For Exercises 3 and 4, use the coordinate plane at the right. Triangle *XYZ* is shown.

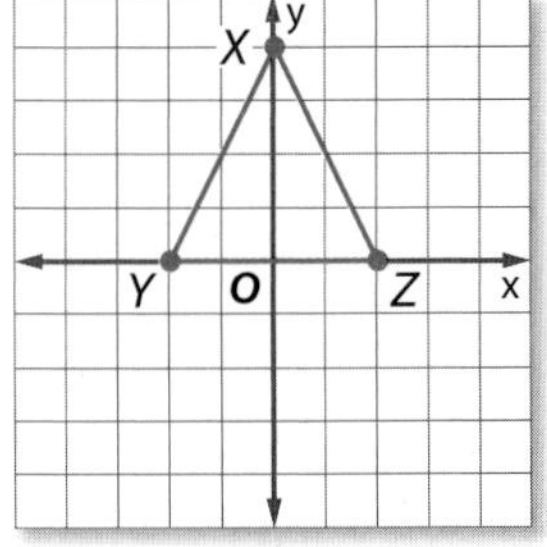

3. Find the coordinates of the vertices of the image of $\triangle XYZ$ translated 4 units to the right and 5 units down.

4. Find the coordinates of the vertices of the image of $\triangle XYZ$ translated 2 units to the left and 3 units up.

Example 2
p. 103

5. The vertices of $\triangle FGH$ are $F(-3, 4)$, $G(0, 5)$, and $H(3, 2)$. Graph the triangle and its image after a reflection over the y-axis.

6. The vertices of figure *RSTV* are $R(2, 4)$, $S(4, 3)$, $T(4, -2)$, and $V(1, -2)$. Graph the figure and its image after a reflection over the y-axis.

7. The vertices of figure *ABCD* are $A(-3, -1)$, $B(-5, -1)$, $C(-5, -6)$, and $D(-3, -3)$. Graph the figure and its image after a reflection over the x-axis.

8. **ART** Reflect the figure below over the x-axis. Sketch the figure and its image on grid paper. What is the animal?

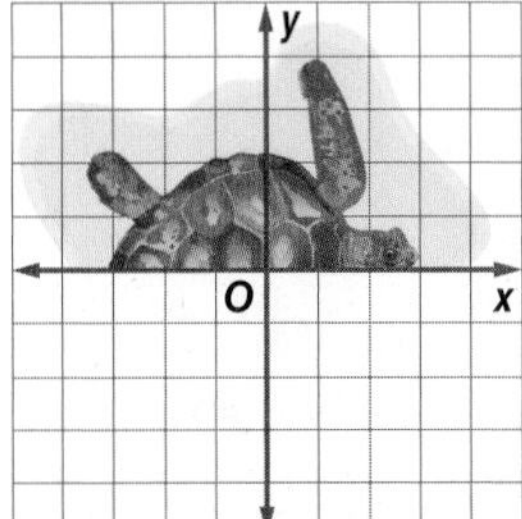

9. **CHESS** In chess, the rook can only move vertically or horizontally across the board. The chessboard below shows the movement of a rook after two turns. Describe this translation in words.

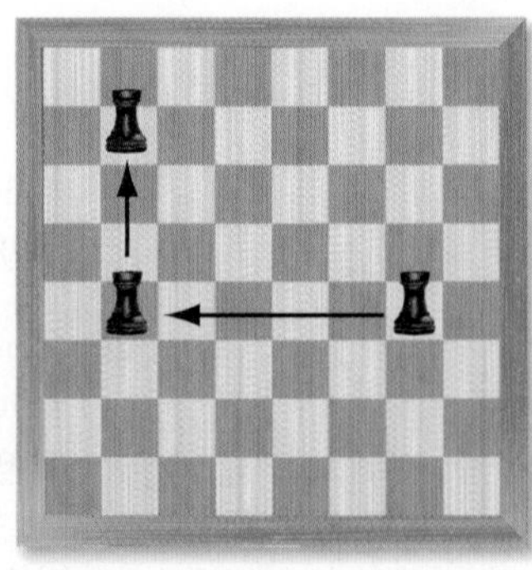

For Exercises 10–14, identify each transformation as a *translation* or a *reflection*. If the green image is the original, describe each transformation.

10.

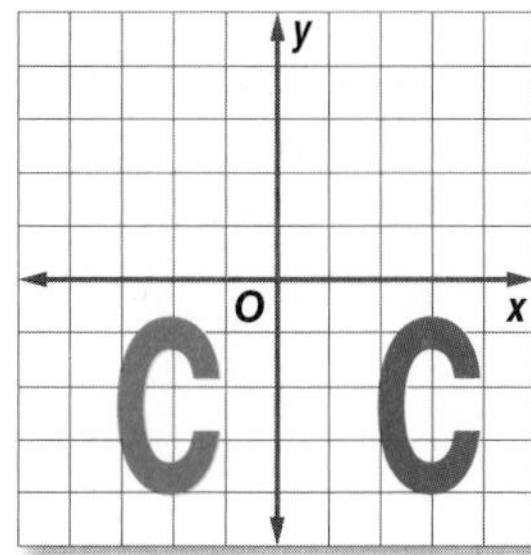

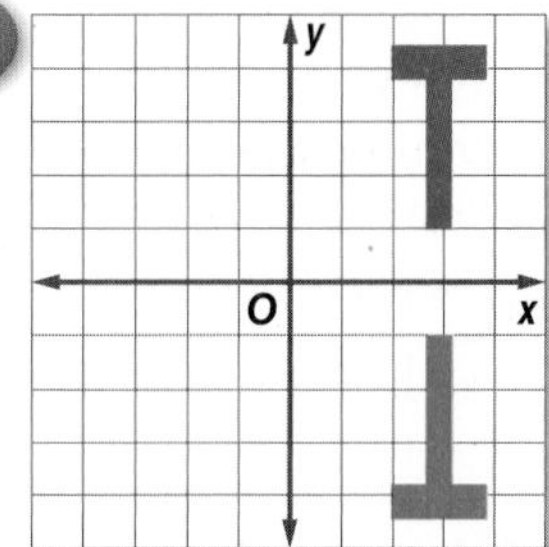

12.

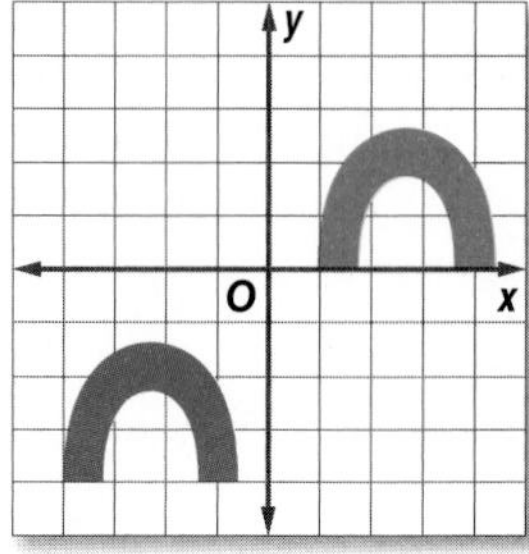

Real-World Link

The world's largest rubber stamp, located in Cleveland, Ohio has the word "FREE" on it. The stamp, is 28 feet tall and 48 feet long.

Source: Roadside America

13. A figure has vertices $H(1, -1)$, $J(1, 5)$, $K(3, -1)$, and $L(3, 5)$. The image's vertices are $H'(-4, -5)$, $J'(-4, 1)$, $K'(-2, -5)$, and $L'(-2, 1)$.

14. Triangle QRS has vertices $Q(1, -1)$, $R(5, -3)$, and $S(3, 2)$. The vertices of the image are $Q'(1, 1)$, $R'(5, 3)$, and $S'(3, -2)$.

15. **GEOMETRY** Triangle RST has vertices $R(4, 2)$, $S(-8, 0)$, and $T(6, 7)$. When translated, R' has coordinates $(-2, 4)$. Find the coordinates of S' and T'. Then describe the translation of triangle RST onto triangle $R'S'T'$.

16. **ART** A mosaic is a type of art created using glass, stone, tile, or other materials. Describe the transformation that maps the red outlined tile to the purple outlined tile.

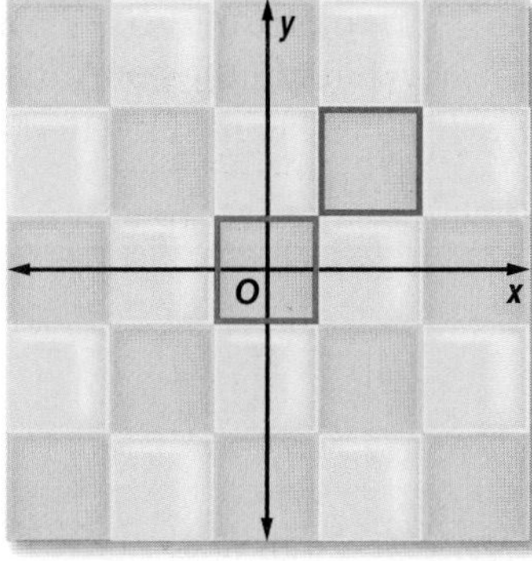

17. **STAMPS** The image the ink makes on a page is a reflection of the rubber stamp. Suppose you create a rubber stamp that would print the word MATH. Draw the rubber stamp. Is the image a reflection over the x-axis or y-axis?

H.O.T. Problems

Use **H**igher-**O**rder **T**hinking Skills

18. **OPEN ENDED** Draw a figure on the coordinate plane. Then reflect the figure over the y-axis.

19. **WRITING IN MATH** Suppose you reflect a figure over the x-axis and then you reflect the figure over the y-axis. Is there a single transformation using reflections or translations that maps the original figure to its image? If so, name it. Explain your reasoning.

20. **WHICH ONE DOESN'T BELONG?** Without graphing, identify the pair of points that does not represent a reflection over the y-axis. Justify your reasoning.

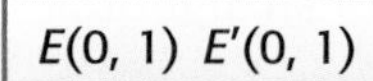

$E(0, 1)$ $E'(0, 1)$	$F(-2, 5)$ $F'(2, 5)$	$G(-3, -4)$ $G'(-3, 4)$	$H(5, 0)$ $H'(-5, 0)$

21. **CHALLENGE** Discuss how an image compares to the original figure if you reflect a triangle in Quadrant I over the x-axis, then translate the image 4 units right and 3 units up. Determine if a single transformation can map the original figure to the final image.

22. **WRITING IN MATH** A figure is translated by $(2, -3)$ and then the image is translated by $(-2, 3)$. Without graphing, describe the final position of the figure. Explain your reasoning.

Standardized Test Practice

23. Which of the following is a vertex of the figure shown below after a translation of 2 units right and 2 units up?

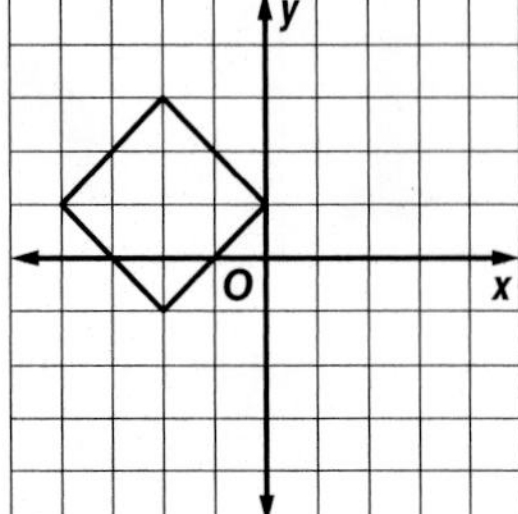

A (5, 0)
B (0, 2)
C (2, 0)
D (0, 5)

24. SHORT RESPONSE The coordinates of a triangle are $A(0, -1)$, $B(-2, -1)$, and $C(3, 5)$. What are the coordinates of the triangle after it has been translated 3 units left and 4 units down?

25. Which of the following *best* represents a reflection over the vertical line segment in the center of the rectangle?

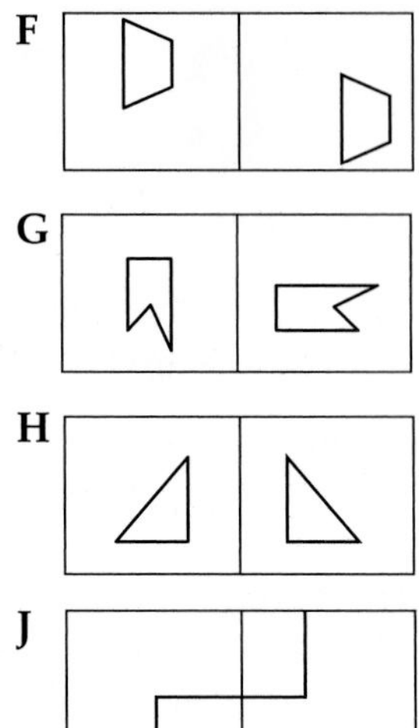

F

G

H

J

26. SHORT RESPONSE What are the coordinates of the point $(-3, 5)$ after it has been reflected over the y-axis?

Spiral Review

Name the ordered pair for each point graphed at the right. (Lesson 2-6)

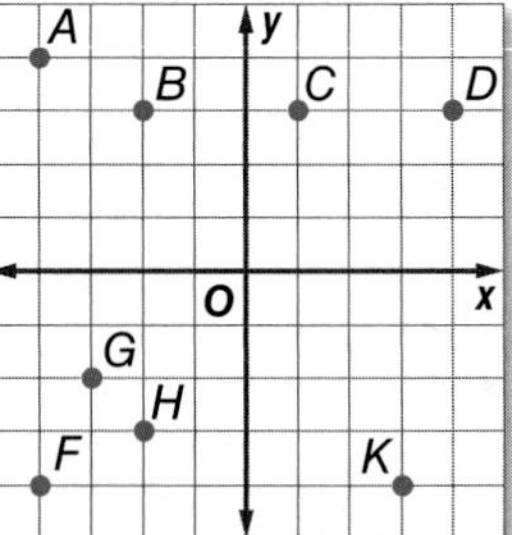

27. *A*

28. *C*

29. *G*

30. *K*

31. BASKETBALL In their first five games, the Jefferson Middle School basketball team scored 46, 52, 49, 53, and 45 points. What was their average number of points per game? (Lesson 2-5)

32. MONEY The starting balance in a checking account was −\$50. What was the balance after a \$100 deposit was made and checks were written for \$25 and for \$32? (Lesson 2-2)

33. Translate the phrase *three times as many cards as Neville has* into an algebraic expression. (Lesson 1-2)

Skills Practice

Divide. (Previous Course)

34. $4\overline{)2.0}$

35. $5\overline{)1.0}$

36. $12\overline{)6.0}$

37. $8\overline{)5.000}$

CHAPTER 2

Study Guide and Review

Math Online glencoe.com
- STUDY TO GO
- Vocabulary Review

Chapter Summary

Key Concepts

Integers and Absolute Value (Lesson 2-1)

- Numbers on a number line increase as you move from left to right.
- The absolute value of a number is the distance the number is from zero on the number line.

Adding and Subtracting Integers (Lessons 2-2 and 2-3)

- To add integers with the same sign, add their absolute values. Give the result the same sign as the integers.
- To add integers with different signs subtract their absolute values. Give the result the same sign as the integer with the greater absolute value.
- To subtract an integer, add its additive inverse.

Multiplying and Dividing Integers (Lessons 2-4 and 2-5)

- The product or quotient of two integers with the same sign is positive.
- The product or quotient of two integers with different signs is negative.

The Coordinate Plane (Lesson 2-6)

- The *x*-axis and the *y*-axis separate the coordinate plane into four quadrants.
- The axes and points on the axes are not located in any of the quadrants.

Transformations (Lesson 2-7)

- Sliding a figure from one position to another without turning it is called a translation.
- Flipping a figure over a line is called a reflection.

Be sure the Key Concepts are noted in your Foldable.

Key Vocabulary

absolute value (p. 63)
additive inverse (p. 71)
coordinate (p. 62)
inequality (p. 62)
integer (p. 61)
mean (p. 92)
negative number (p. 61)
opposites (p. 71)
positive number (p. 61)
quadrant (p. 97)
reflection (p. 101)
transformation (p. 101)
translation (p. 101)

Vocabulary Check

Determine whether each statement is *true* or *false*. If *false*, replace the underlined word or number to make a true statement.

1. Two numbers with the same absolute values but different signs are <u>opposites.</u>
2. A positive number is a number <u>less</u> than zero.
3. A reflection occurs when you <u>turn</u> a figure over a line.
4. The number that corresponds to a point on the number line is called a(n) <u>integer.</u>
5. An integer and its <u>opposite</u> are also called additive inverses of each other.
6. The set of <u>quadrants</u> includes positive whole numbers, their opposites, and zero.
7. The <u>absolute value</u> of a number is the distance the number is from zero on the number line.
8. A point located in quadrant <u>IV</u> has an *x*-coordinate that is positive and a *y*-coordinate that is negative.
9. To find the <u>additive inverse</u> of a set of numbers, find the sum of the numbers and then divide by the number of items in the set.
10. Numbers like <u>−6</u> and $\underline{-\frac{3}{5}}$ are examples of integers.
11. You can write an inequality when two quantities are <u>not equal</u>.

CHAPTER 2 Study Guide and Review

Lesson-by-Lesson Review

2-1 Integers and Absolute Value (pp. 61–66)

Write two inequalities using the number pairs. Use the symbols < or >.

12. -20 and -18

13. 0 and -5

Replace each ● with <, >, or = to make a true sentence.

14. $5 \bullet -5$

15. $7 \bullet 7$

16. $-3 \bullet 1$

17. $-14 \bullet -22$

Evaluate each expression.

18. $|-16|$

19. $|4|$

20. $-|34|$

21. $|-2| + |-11|$

22. **BASEBALL CARDS** Jamal traded away 7 shortstop cards for 5 pitcher cards. Find an integer that represents the change in the number of cards Jamal had after the trade.

EXAMPLE 1

Write two inequalities comparing -5 and -4. Use the symbols < or >.

−6 −5 −4 −3 −2 −1 0 1 2 3

Since -4 is to the right of -5, $-4 > -5$.

Since -5 is to the left of -4, $-5 < -4$.

EXAMPLE 2

Evaluate $|-5|$.

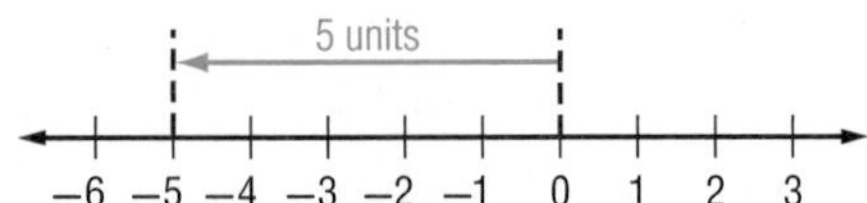

The graph of -5 is 5 units from 0.

So, $|-5| = 5$.

2-2 Adding Integers (pp. 69–74)

Find each sum.

23. $-5 + (-1)$

24. $-3 + (-7)$

25. $-6 + 10$

26. $4 + (-9)$

27. $7 + (-2)$

28. $2 + 8 + (-3)$

29. $-12 + 5 + (-6)$

30. $-7 + 5 + (-4)$

31. **GAME SHOWS** A contestant on a quiz game show has -25 points. If she loses an additional 50 points, what is her score? Write an addition equation and then solve.

32. **GOLF** A golfer's scores for the last five weeks are -3, $+5$, -1, -2, and $+4$. What is the sum of his scores?

EXAMPLE 3

Find $-2 + (-3)$.

Use a number line.

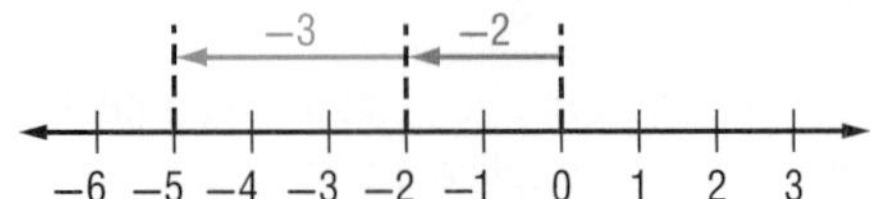

Start at zero. Move 2 units to the left. From there, move 3 more units to the left.

So, $-2 + (-3) = -5$.

EXAMPLE 4

Find $9 + (-4)$.

$9 + (-4) = 5$ — **Subtract $|-4|$ from $|9|$. The sum is positive.**

MIXED PROBLEM SOLVING
For mixed problem-solving practice, see page 844.

2-3 Subtracting Integers (pp. 76–80)

Find each difference.

33. $13 - 7$
34. $-2 - 5$
35. $8 - (-3)$
36. $-1 - (-4)$
37. $-4 - 6$
38. $3 - 5$

39. **ELEVATION** The table shows the highest and lowest elevations for North America. Find the difference between the highest and lowest elevations.

Lowest Elevation (feet)	Highest Elevation (feet)
−282	20,320

EXAMPLE 5

Find −13 − 4.

$-13 - 4 = -13 + (-4)$ To subtract 4, add −4.
$= -17$

So, $-13 - 4 = -17$.

EXAMPLE 6

Find 10 − (−2).

$10 - (-2) = 10 + 2$ To subtract −2, add 2.
$= 12$

So, $10 - (-2) = 12$.

2-4 Multiplying Integers (pp. 83–88)

Find each product.

40. $-2(3)$
41. $-5(6)$
42. $-7(-9)$
43. $-12(-4)$
44. $11(-7)$
45. $-10(14)$

46. **ICE SKATING** For each jump she completes incorrectly in the competition, Dawn receives −2 points. If Dawn completes six jumps incorrectly, what is her score?

EXAMPLE 7

Find 3(−7).

$3(-7) = -21$ The factors have different signs. The product is negative.

EXAMPLE 8

Find −5(−4).

$-5(-4) = 20$ The factors have the same sign. The product is positive.

2-5 Dividing Integers (pp. 90–95)

Find each quotient.

47. $-16 \div (-4)$
48. $-56 \div (-8)$
49. $-30 \div 5$
50. $15 \div (-3)$
51. $-88 \div -11$
52. $170 \div (-10)$

53. **RACING** For the first five legs of a bicycle race, Elena was +32 seconds, +5 seconds, +10 seconds, +8 seconds, and +12 seconds behind the leader. What was the average time she was behind the leader?

EXAMPLE 9

Find −24 ÷ (−6).

$-24 \div (-6) = 4$ The quotient is positive.

EXAMPLE 10

Find 15 ÷ (−3).

$15 \div (-3) = -5$ The quotient is negative.

CHAPTER 2 Study Guide and Review

2-6 Graphing in Four Quadrants (pp. 96–100)

Graph and label each point on a coordinate plane. Name the quadrant in which each point is located.

54. $X(5, 2)$

55. $Y(-3, 3)$

56. $W(2, 0)$

57. $Z(-1, -4)$

58. GAMES The coordinate plane shown represents a board game. Name the quadrant in which each player's game piece is located.

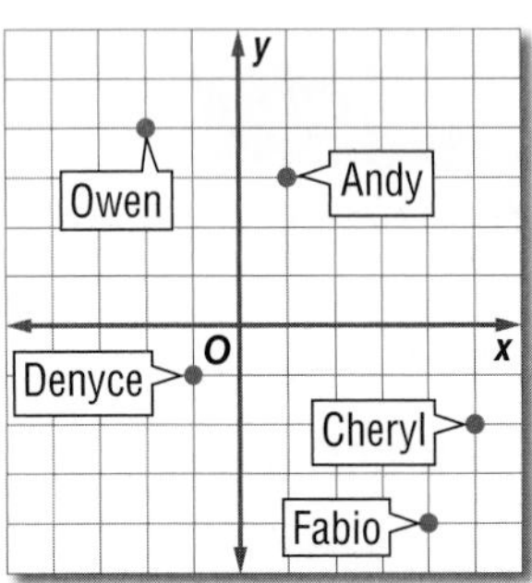

EXAMPLE 11

Graph and label $J(-3, 5)$ point on a coordinate plane. Name the quadrant in which the point is located.

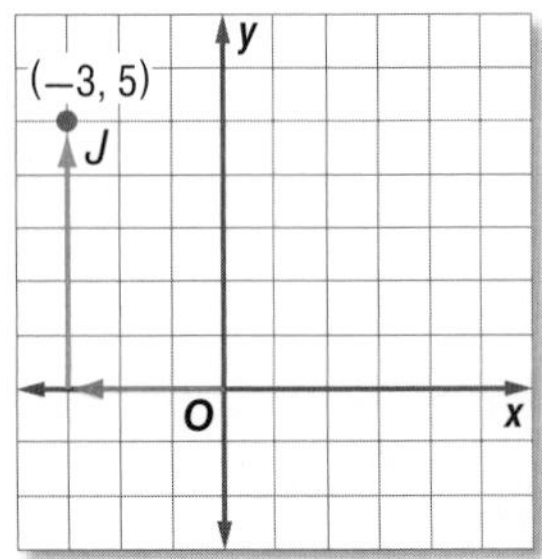

Point $J(-3, 5)$ is in quadrant II.

2-7 Translations and Reflections on the Coordinate Plane (pp. 101–106)

59. The vertices of figure *ABCD* are $A(1, -3)$, $B(4, -3)$, $C(1, -1)$, and $D(-2, -1)$. Find the vertices after a reflection over the *x*-axis.

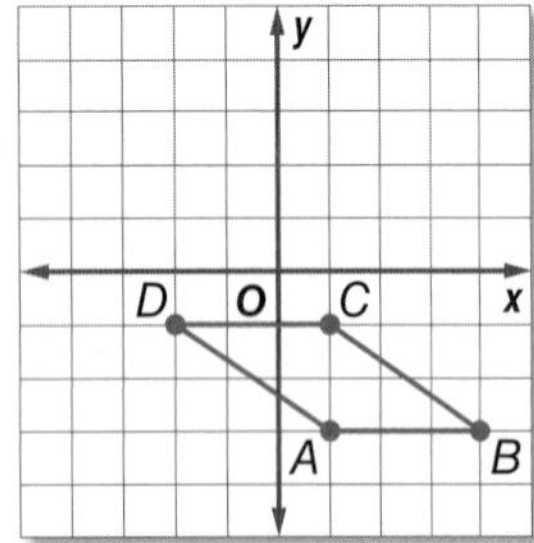

60. A triangle has vertices $N(6, 3)$, $P(3, 9)$, and $Q(9, 6)$. The triangle is translated 2 units right and 2 units down. Graph the figure and its image.

61. ESCALATORS What type of transformation is used when moving up an escalator?

EXAMPLE 12

The vertices of $\triangle JKL$ are $J(1, 2)$, $K(3, 2)$, and $L(1, -1)$. Find the vertices of the image after a translation 3 units left and 2 units up. Then find the vertices of the image after $\triangle JKL$ is reflected over the *x*-axis.

Translation This translation can be written as $(-3, 2)$.

original		translation		image
$J(1, 2)$	+	$(-3, 2)$	→	$J'(-2, 4)$
$K(3, 2)$	+	$(-3, 2)$	→	$K'(0, 4)$
$L(1, -1)$	+	$(-3, 2)$	→	$L'(-2, 1)$

Reflection Use the same *x*-coordinate and replace the *y*-coordinate with its opposite.

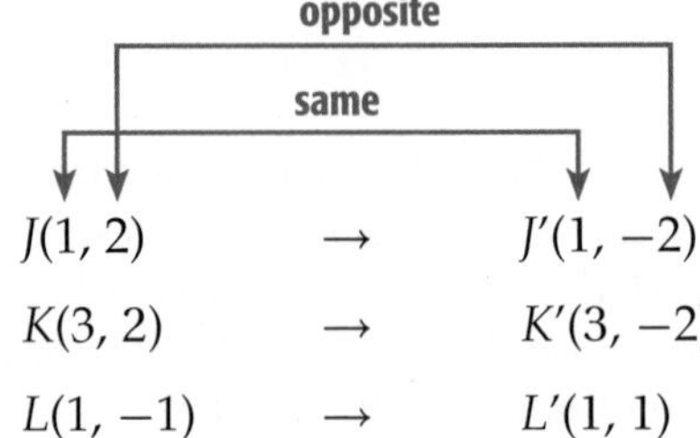

CHAPTER 2 Practice Test

Math Online glencoe.com
Chapter Test

Write two inequalities using each pair of numbers. Use the symbols $<$ and $>$.

1. 7 and -5
2. -1 and 0
3. -21 and -22

4. **MULTIPLE CHOICE** A scuba diver records her depth in the lake every minute. Choose the group of depths that is listed in order from least to greatest.

   A -13 ft, -12 ft, -9 ft, -3 ft, -1 ft, -5 ft

   B -5 ft, -3 ft, -1 ft, -9 ft, -12 ft, -13 ft

   C -12 ft, -13 ft, -3 ft, -1 ft, -9 ft, -5 ft

   D -13 ft, -12 ft, -9 ft, -5 ft, -3 ft, -1 ft

Find each sum or difference.

5. $(-3) + (-7)$
6. $7-(-5)$
7. $-4 + 11$
8. $11 + (-13)$
9. $3 - 11$
10. $-21 - (-6)$
11. $13 + (-2) + (-9)$
12. $-4 - (-8)$

13. **GOLF** Jack played in a 4-day golf tournament. His score at the end of each day is shown in the table. What was his final score at the end of the tournament?

Day	Score
Monday	-3
Tuesday	$+1$
Wednesday	$+4$
Thursday	-1

Find each product or quotient.

14. $7(-6)$
15. $-36 \div (-6)$
16. $54 \div (-6)$
17. $-5(-9)$
18. $-4(7)$
19. $-95 \div 5$
20. $2(-3)10$
21. $-132 \div 11$

22. **SWIMMING POOL** The water in the swimming pool drains at a rate of 24 gallons per minute. Describe the change in the amount of water in the swimming pool after 1 hour.

ALGEBRA Evaluate each expression if $a = -3$, $b = 6$, and $c = -9$.

23. $ca - b$
24. $|bc| \div (-|a|)$
25. $4b + |a|$
26. $\frac{ab}{c} - 6$

27. **AIRPLANES** An airplane descends 500 feet each minute when beginning to land.

   a. **VERBAL** Write an expression to find how many minutes an airplane has been descending if it has descended x feet.

   b. **ALGEBRAIC** Find the total number of minutes an airplane has been descending if the plane has descended 9000 feet.

Graph and label each point on a coordinate plane. Name the quadrant in which each point is located.

28. $A(-4, 3)$
29. $B(1, -3)$
30. $C(-2, -4)$
31. $D(5, 6)$

32. **MULTIPLE CHOICE** Suppose Elan's home represents the origin on a coordinate plane. If Elan leaves his home and walks two miles west and then four miles north, what is the location of his destination as an ordered pair?

   F $(-2, 4)$ H $(-2, -4)$

   G $(2, 4)$ J $(4, -2)$

33. Triangle ABC is graphed on the coordinate plane shown below. Find the coordinates of the vertices of the image of ΔABC translated 3 units to the left and 2 units up.

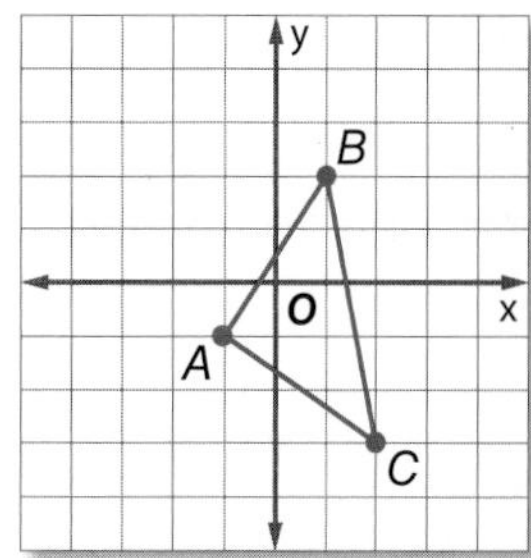

Preparing for Standardized Tests

Gridded Response Questions

In addition to multiple choice, short answer, and extended response questions, you will likely encounter gridded response questions on standardized tests. For gridded response questions, print your answer on an answer sheet and mark in the correct circles on the grid to match your answer.

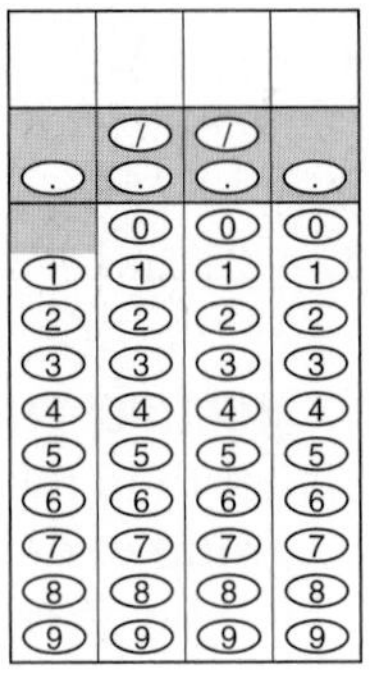

Strategies for Solving Gridded Response Questions

Step 1

Read the problem carefully.

- **Ask yourself:** "What information is given?" "What do I need to find?" "How do I solve this type of problem?"
- **Solve the Problem:** Use the information given in the problem to solve.
- **Check your answer:** If time permits, check your answer to make sure you have solved the problem correctly.

Step 2

Print your answer in the answer boxes.

- Print only one digit or symbol in each answer box.
- Do not write any digits or symbols outside the answer boxes.
- You may print your answer with the first digit in the left answer box, or with the last digit in the right answer box. You may leave blank any boxes you do not need on the right or the left side of your answer.

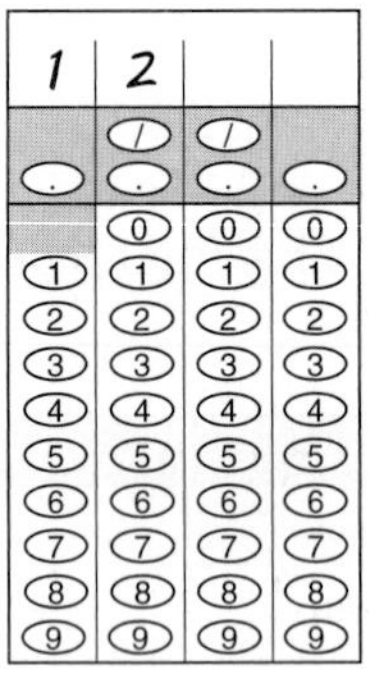

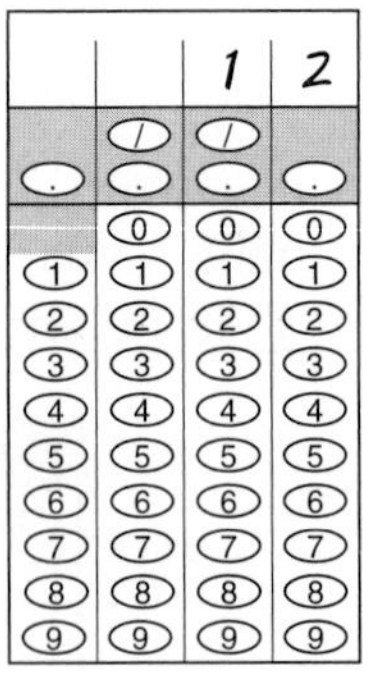

Step 3

Fill in the grid.

- Fill in only one bubble for every answer box that you have written in. Be sure not to fill in a bubble under a blank answer box.
- Fill in each bubble completely and clearly.

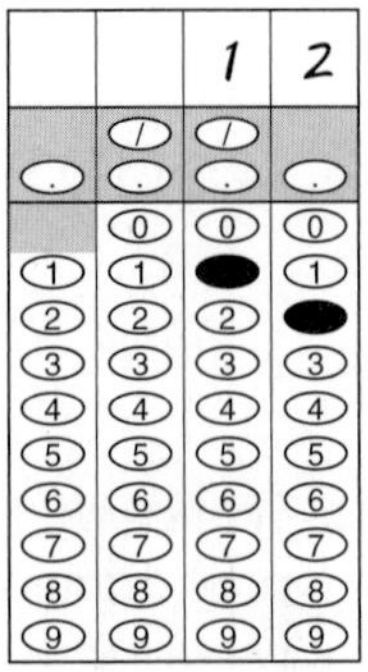

EXAMPLE

Read the problem. Identify what you need to know. Then use the information in the problem to solve.

> **GRIDDED RESPONSE** Manuel rode his bike for 15 miles. The trip took him 2 hours. Find Manuel's rate of speed.

Read the problem carefully. You are given the distance and time that Manuel rode his bike. You are asked to find his rate of speed.

Use the formula $d = rt$. Replace d with 15 and t with 2. Solve for r.

Solve the Problem	Fill in the Grid
$d = rt$ $15 = r \cdot 2$ $\frac{15}{2} = \frac{r \cdot 2}{2}$ $7\frac{1}{2} = r$ You cannot grid in the mixed number $7\frac{1}{2}$. Either grid the fraction $\frac{15}{2}$ or the decimal 7.5.	1 5 / 2 7 . 5 7 . 5

Exercises

Read each problem. Identify what you need to know. Then use the information in the problem to solve. Copy and complete an answer grid on your paper.

1. **GRIDDED RESPONSE** Last year, Michael was 61 inches tall. During the past year, he grew a total of 4 inches. How tall is Michael now? Express your answer in inches.

2. **GRIDDED RESPONSE** Jamal scored 14, 12, 9, 17, 15, and 11 points in his last 6 basketball games. What is the mean number of points per game that he scored?

3. **GRIDDED RESPONSE** Find the sum of the integers below.

$$14 + (-19) + (-3) + 35$$

4. **GRIDDED RESPONSE** The table shows the number of Calories burned per minute for a 130-pound person during different activities.

Activity	Calories Per Minute
Basketball	11
Bicycling (10 mph)	8
Jogging (6 mph)	9
Swimming	6
Walking (3 mph)	4

How many Calories would Colleen burn if she rides her bicycle for 15 minutes and jogs for 30 minutes?

5. **GRIDDED RESPONSE** Point D has coordinates $D(8, -2)$. What will the y-coordinate of the image be after a reflection over the x-axis?

CHAPTER 2

Standardized Test Practice

Cumulative, Chapters 1 through 2

Multiple Choice

Read each question. Then fill in the correct answer on the answer document provided by your teacher or on a sheet of paper.

1. The record low temperatures of four U.S. cities are shown in the table.

Record Low Temperatures	
City	**Temperature (°F)**
Denver, CO	−30
Hartford, CT	−26
Atlanta, GA	−8
Indianapolis, IN	−27

Which city has the coldest record low temperature?

A Atlanta **C** Hartford

B Denver **D** Indianapolis

2. Lucinda received $35 for her birthday and deposited it into her checking account. Later in the week, she wrote a check for $52 to pay for a DVD player. Which of the following represents the change in Lucinda's checking account balance?

F $87 **H** −$17

G $17 **J** −$87

3. Write a function rule for the input and output values shown in the table below.

x	−4	−2	0	3
y	12	6	0	−9

A $y = 3x$ **C** $y = -2x$

B $y = 2x$ **D** $y = -3x$

Test-TakingTip

Question 2 After the deposit and check, would Lucinda's balance increase or decrease? This will tell you what sign the correct answer should be.

4. Find the quotient of the expression shown below.

$$\frac{-360}{30}$$

F −12 **H** 8

G −8 **J** 12

5. The total cost of Jason's guitar is $180. If he pays for the guitar in 12 equal payments, how much is each payment?

A $12 **C** $15

B $14 **D** $16

6. Which of the following equations describes the rule shown in the table?

x	1	2	3	4
y	5	6	7	8

F $y = 5x$

G $y = 4x$

H $y = x + 4$

J $y = x + 5$

7. Which term *best* describes flipping an object to create a mirror image?

A movement **C** slide

B reflection **D** translation

8. Sholanda is 3 years older than twice her sister Lakita's age. If Lakita is *a* years old, which expression represents Sholanda's age?

F $3a + 2$

G $3a - 2$

H $2a + 3$

J $2a - 3$

Short Response/Gridded Response

Record your answers on the answer sheet provided by your teacher or on a sheet of paper.

9. GRIDDED RESPONSE A scuba diver is 83 feet below the surface of the water when she dives to a depth of 114 feet below the surface. How many feet did she dive from her previous depth to arrive at her current depth? Express your answer in feet.

10. GRIDDED RESPONSE Marco and Monica are playing a board game. At the end of the first round, Marco had a score of 30 points and Monica had a score of −20 points. Find the difference between the two scores.

11. Suppose triangle ABC is reflected over the y-axis.

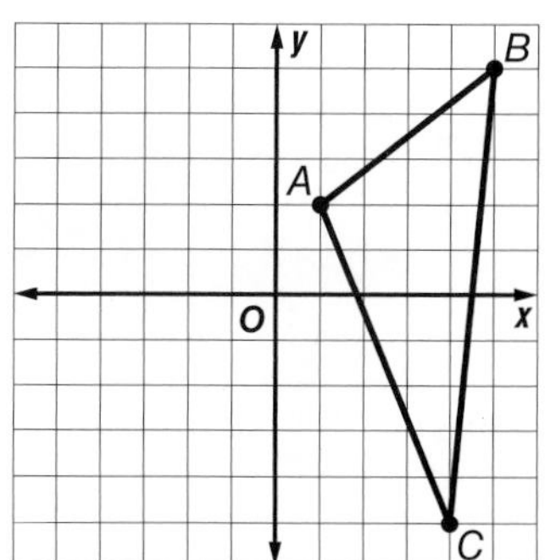

What are the coordinates of the vertices of the image?

12. Suppose the mean low temperature over a 5-day period in Chicago was –12°F. Give a sample set of what the low temperatures might have been for the 5 days.

13. Alexandria recently had a birthday. Her age is now four less than twice her younger sister Sara's age. Let s represent Sara's age in years.

a. Write an expression that represents Alexandria's age in terms of Sara's age.

b. Suppose Sara is 6 years old. Use the expression from part **a** to find Alexandria's age.

Extended Response

Record your answers on a sheet of paper. Show your work.

14. After three rounds of a 4-round tournament, Reggie is the leader. Other players' standings in relation to Reggie are shown in the table.

Golf Standings	
Player	**Number of Strokes Behind the Leader**
Reggie	0
Benjamin	3
Alejandro	7
Christopher	4
Thomas	1

a. Write an integer to describe each golfer's standing with respect to the leader after 3 rounds.

b. Order the integers from least to greatest.

c. Which player is currently in second place?

Need Extra Help?

If you missed Question...	1	2	3	4	5	6	7	8	9	10	11	12	13	14
Go to Lesson or Page...	2-1	2-3	2-4	2-5	2-5	1-5	2-7	1-2	2-2	2-3	2-7	2-5	1-3	2-1

CHAPTER 3 Operations With Rational Numbers

Then

In Chapter 2, you learned to add, subtract, multiply, and divide integers.

Now

In Chapter 3, you will:

- Write fractions as terminating or repeating decimals.
- Identify, add, subtract, multiply, and divide rational numbers.
- Evaluate algebraic expressions with fractions.

Why?

COOKING

Cookbooks are among the top selling books each year. Recipes in the books give a set of instructions to make a dish. In most recipes, fractions and mixed numbers appear. Being able to understand and use rational numbers will help make a more delicious meal.

Operations with Rational Numbers

Activity

One of your friends wants you to make Brownie Fruit Pizza for a backyard party. Use the recipe for to find out how much of each ingredient is needed to double the recipe.

Brownie Fruit Pizza

Ingredients

2 rolls (16.5 oz each) fudge brownie batter

1 package (8 oz) cream cheese, softened

$\frac{1}{3}$ cup sugar

$1\frac{1}{2}$ cups sliced fresh strawberries

1 cup fresh blueberries

$1\frac{1}{4}$ cups fresh raspberries

Directions

1. Heat oven to 350°F. Grease a 12-inch pan. Spread batter evenly in the bottom of the pan to form the crust.
2. Bake 15 to 20 minutes. Cool completely.
3. In a small bowl, beat cream, cheese and sugar. Spread mixture over cooled crust. Arrange fruit over creamed cheese. Refrigerate until chilled.

Makes 1 fruit pizza.
12 servings.

2/6

Math *in Motion*, Animation glencoe.com

Get Ready for Chapter 3

Diagnose Readiness You have two options for checking Prerequisite Skills.

Text Option

Take the Quick Check below. Refer to the Quick Review for help.

QuickCheck

Find each quotient. Round to the nearest tenth, if necessary. (Previous Course)

1. $4 \div 7$
2. $-1 \div 6$
3. $3 \div 15$
4. $-18 \div 3$
5. $-5 \div (-11)$
6. $6 \div (-19)$
7. $-28 \div 16$
8. $-63 \div (-7)$
9. **LANDSCAPING** How many bricks, each 0.25 meter long, are needed to make a row that is 7.25 meters long?

Write each fraction in simplest form. If the fraction is already in simplest form, write *simplified*. (Previous Course)

10. $\frac{6}{30}$
11. $\frac{40}{50}$
12. $\frac{17}{36}$
13. $\frac{12}{80}$
14. $\frac{32}{64}$
15. $\frac{56}{71}$
16. **SURVEY** Twelve of the 28 students in math class have blonde hair. In simplest form, what fraction of the students in math class does *not* have blonde hair?

Find each sum or difference. (Lessons 2-2 and 2-3)

17. $3 + (-7)$
18. $-11 + 19$
19. $(-2) + (-5)$
20. $-6 - (-12)$
21. **TRAVEL** A family drives 26 miles east from their house. Then they drive 17 miles west. Find an integer that represents the family's position in relation to their house.

QuickReview

EXAMPLE 1

Find $5 \div 9$. Round to the nearest tenth.

$5 \div 9 = 0.555...$ Find the quotient.
≈ 0.6 Round to the nearest tenth.

EXAMPLE 2

Write $\frac{12}{36}$ in simplest form.

Factors of 12: 1, 2, 3, 4, 6, 12
Factors of 36: 1, 2, 3, 4, 6, 9, 12, 18, 36
The GCF of 12 and 36 is 12.

$\frac{12}{36} = \frac{12 \div 12}{36 \div 12}$ Divide the numerator and the denominator by the GCF.
$= \frac{1}{3}$ Simplest form

EXAMPLE 3

Find $10 - 16$.

$10 - 16 = 10 + (-16)$ To subtract 16, add -16.
$= -6$ Simplify.

Online Option

Math Online Take a self-check Chapter Readiness Quiz at glencoe.com.

Get Started on Chapter 3

You will learn several new concepts, skills, and vocabulary terms as you study Chapter 3. To get ready, identify important terms and organize your resources. You may wish to refer to **Chapter 0** to review prerequisite skills.

FOLDABLES® Study Organizer

Applying Rational Numbers Make this Foldable to help you record information about rational numbers. Begin with a sheet of notebook paper.

1. **Fold** lengthwise to the holes.

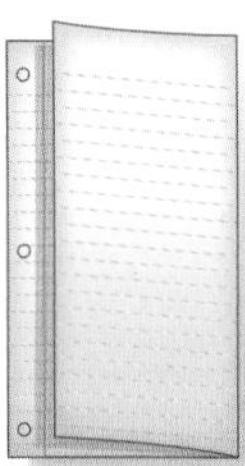

2. **Cut** along the top line and then make equal cuts to form 6 tabs.

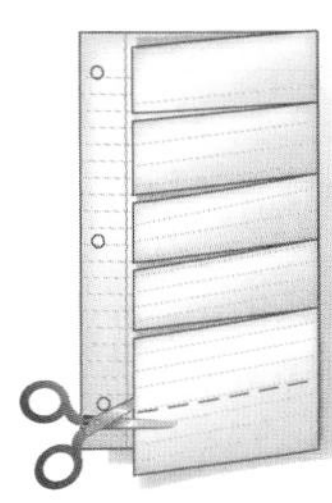

3. **Label** the major topics as shown.

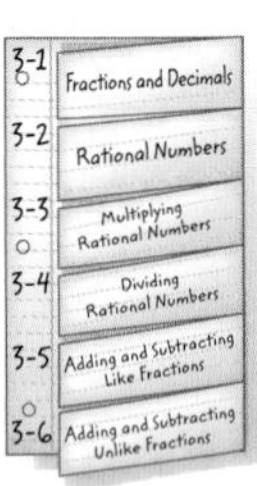

Math Online glencoe.com

- Study the chapter online
- Explore **Math in Motion**
- Get extra help from your own **Personal Tutor**
- Use **Extra Examples** for additional help
- Take a **Self-Check Quiz**
- **Review Vocabulary** in fun ways

New Vocabulary

English		Español
terminating decimal	p. 121	decimal terminal
repeating decimal	p. 122	decimal periódico
bar notation	p. 122	notación de barra
rational number	p. 128	número racional
multiplicative inverse	p. 141	inversos multiplicativos
reciprocal	p. 141	recíproco
like fractions	p. 147	fracciones semejantes
unlike fractions	p. 153	fracciones con distintos denominadores

Review Vocabulary

GCF (greatest common factor) • p. 858 • máximo común divisor (MCD) the greatest number that is a factor of two or more numbers; Example: the GCF of 14 and 35 is 7

LCM (least common multiple) • p. 860 • mínimo común múltiplo (MCM) the least of the nonzero common multiples of two or more numbers; Example: the LCM of 4 and 6 is 12

opposites • p. 71 • opuestos two numbers with the same absolute value but different signs

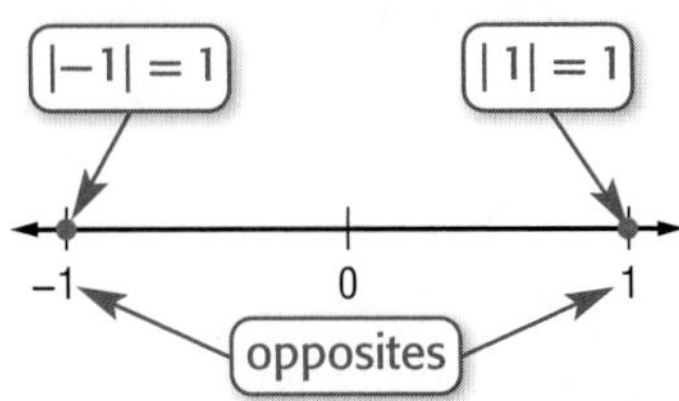

Multilingual eGlossary glencoe.com

EXPLORE **3-1**

Algebra Lab
Fractions and Decimals on the Number Line

Math Online glencoe.com
Math *in Motion,* Animation

Objective
Represent positive and negative fractions and decimals on a number line.

You have already used a number line to graph integers. The number line below shows the graph of the integers 1 and –1.

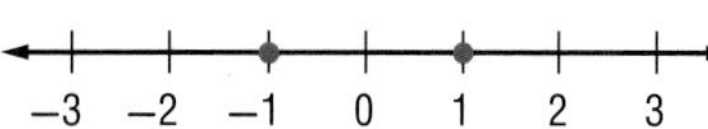

In this lab, you will graph positive and negative fractions and decimals on a number line.

ACTIVITY 1

Use a number line to graph $\frac{5}{8}$.

Step 1 Draw a number line showing 0 and 1.

Step 2 Since the denominator is eighths, divide the number line between 0 and 1 into 8 equal parts.
Label the number line with $\frac{1}{8}$, $\frac{2}{8}$, $\frac{3}{8}$, and so on.

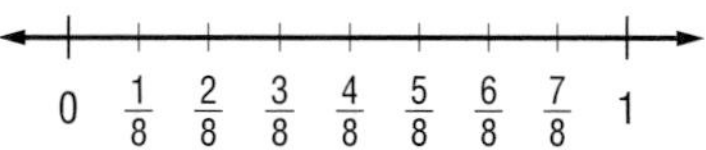

Step 3 Draw a dot on the number line above the $\frac{5}{8}$ mark.

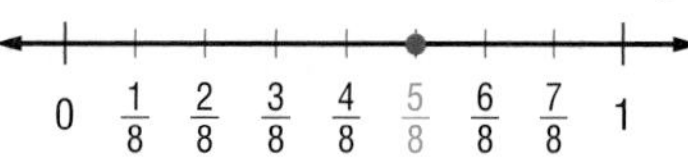

ACTIVITY 2

Use a number line to graph $-2\frac{4}{5}$.

Step 1 Draw a number line showing –2 and –3.

Step 2 Since the denominator is fifths, divide the number line between –2 and –3 into 5 equal parts.
On the number line, $-2\frac{1}{5}$ lies just to the left of –2. Continue in this manner until you reach –3.

$-3 \quad -2\frac{4}{5} \quad -2\frac{3}{5} \quad -2\frac{2}{5} \quad -2\frac{1}{5} \quad -2$

Step 3 Draw a dot on the number line above the $-2\frac{4}{5}$ mark.

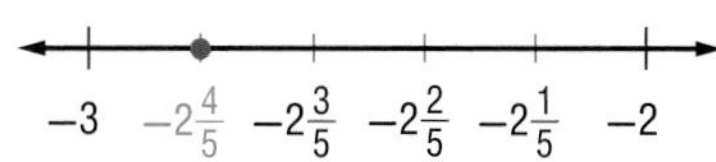

Exercises

Graph each fraction on a number line.

a. $\frac{2}{3}$ **b.** $-\frac{3}{4}$ **c.** $-2\frac{5}{6}$ **d.** $3\frac{1}{5}$

StudyTip

Scale When deciding on the scale for the number line, use the first integer less than −1.3 and the first integer greater than 1.3. So, the number line should range from −2 to 2.

ACTIVITY 3

Use a number line to graph 1.3 and −1.3.

Step 1 The decimals are expressed in tenths. So, divide the number line between −2 and 2 so there are 10 sections between each integer.

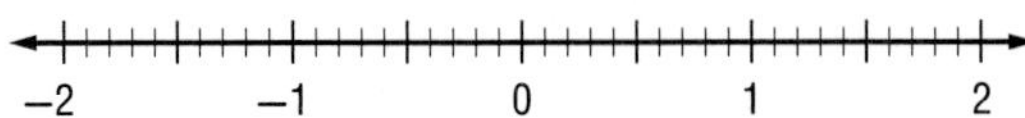

Step 2 Locate 1.3 on the number line. Draw a dot on the number line above the 1.3 mark. Draw another dot on the number line above the −1.3 mark.

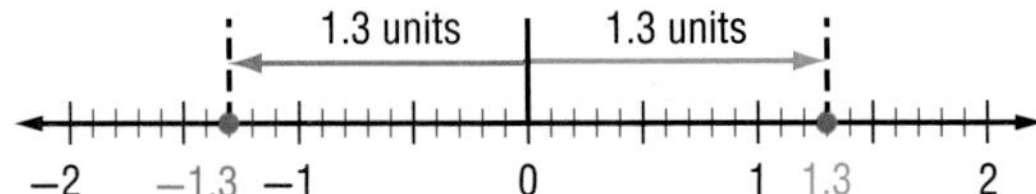

The two points graphed on the number line are the same distance away from zero. The numbers 1.3 and −1.3 are opposites. Recall that opposites are two numbers with the same absolute value but different signs.

Exercises

Graph each decimal and its opposite on a number line.

e. 0.5 **f.** −2.8 **g.** 1.35 **h.** −2.24

Analyze the Results

1. In Activity 2, you graphed $-2\frac{4}{5}$ on a number line. From that point, in what direction on the number line would you move if you wanted to graph $-3\frac{1}{5}$?
2. Equivalent numbers are graphed in the same place on a number line. Use a number line to determine if the following pairs of numbers are equivalent.
   a. -1.6 and $-1\frac{3}{5}$ **b.** $\frac{2}{3}$ and 0.8 **c.** -1.75 and $-1\frac{3}{4}$
3. Use a number line to determine which number is greater in each pair of numbers. Justify your reasoning.
   a. $-2\frac{5}{8}$ or -2.75 **b.** -1.4 or -1.7 **c.** $\frac{1}{5}$ or $-1\frac{1}{5}$
4. When graphing fractions on a number line, explain how to divide the number line into equal parts.
5. **OPEN ENDED** Name a fraction and its opposite. Then graph the fractions on a number line.
6. **WRITING IN MATH** Explain why $\frac{3}{4}$ is graphed to the right of $\frac{1}{4}$ but $-\frac{3}{4}$ is graphed to the left of $-\frac{1}{4}$.

3-1 Fractions and Decimals

Then

You have used a number line to graph integers. (Lesson 2-1)

Now

- Write fractions as terminating or repeating decimals.
- Compare fractions and decimals.

New Vocabulary

terminating decimal
repeating decimal
bar notation

Math Online

glencoe.com

- Extra Examples
- Personal Tutor
- Self-Check Quiz
- Homework Help
- Math in Motion

Why?

Tara was making tacos for her friends. She bought $\frac{1}{2}$ pound of chicken. The scale showed 0.5 pound.

a. Explain why $\frac{1}{2}$ is the same as 0.5.

b. She needed $\frac{3}{4}$ pound of cheese. What decimal is equivalent to $\frac{3}{4}$? Explain.

Write Fractions as Decimals Some fractions like $\frac{1}{2}$ and $\frac{3}{4}$ can be written as a decimal by making equivalent fractions with denominators of 10, 100, or 1000. However, any fraction $\frac{a}{b}$, where $b \neq 0$, can be written as a decimal by dividing the numerator by the denominator. So, $\frac{a}{b} = a \div b$.

If the division ends, or terminates, when the remainder is zero, the decimal is a **terminating decimal**.

EXAMPLE 1 Write a Fraction as a Terminating Decimal

Write $\frac{7}{8}$ as a decimal.

Method 1 Use paper and pencil.

$$\begin{array}{r} 0.875 \\ 8\overline{)7.000} \\ \underline{-6\,4} \\ 60 \\ \underline{-56} \\ 40 \\ \underline{-40} \\ 0 \end{array}$$

Place the decimal point.
Annex zeros and divide as with whole numbers.

Division ends when the remainder is 0.

So, $\frac{7}{8} = 0.875$.

Method 2 Use a calculator.

7 [÷] 8 [ENTER] 0.875

Using either method, $\frac{7}{8} = 0.875$.

Check Your Progress

Write each fraction as a decimal.

1A. $\frac{4}{5}$

1B. $\frac{3}{16}$

Personal Tutor glencoe.com

Vocabulary Link

Terminating
Everyday Use bringing to an end
Math Use a decimal whose digits end

Not all fractions can be written as terminating decimals. Sometimes a digit or group of digits repeats without end in the quotient.

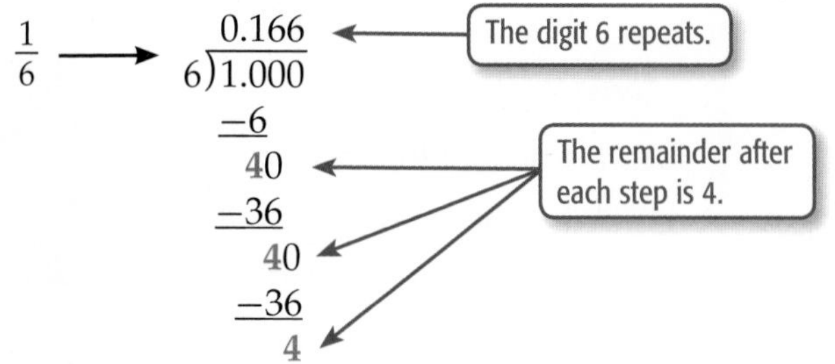

The digit 6 repeats.
The remainder after each step is 4.

Check 1 ÷ 6 ENTER 0.1666666667 ✓ **The last digit is rounded.**

You can indicate that the digit 6 repeats by annexing dots. So, $\frac{1}{6} = 0.1666666666\ldots$. This decimal is called a repeating decimal.

Repeating decimals have a pattern in their digits that repeats without end. **Bar notation** is a bar or line placed over the digit(s) that repeats. The table shows some examples of repeating decimals and their bar notations.

Decimal	Bar Notation
0.166666...	$0.1\overline{6}$
0.353535...	$0.\overline{35}$
12.6888888...	$12.6\overline{8}$
5.71428571...	$5.\overline{714285}$

EXAMPLE 2 Write Fractions as Repeating Decimals

Write each fraction as a decimal. Use a bar to show a repeating decimal.

a. $\frac{5}{12}$

$\frac{5}{12} \rightarrow 12\overline{)5.0000\ldots}$ with quotient $0.4166\ldots$ **The digit 6 repeats.**

So, $\frac{5}{12} = 0.41\overline{6}$.

b. $-\frac{2}{11}$

$-\frac{2}{11} \rightarrow 11\overline{)2.0000\ldots}$ with quotient $0.1818\ldots$ **The digits 18 repeat.**

So, $-\frac{2}{11} = -0.\overline{18}$.

Check Your Progress

2A. $-\frac{5}{6}$

2B. $\frac{7}{9}$

Personal Tutor glencoe.com

It is helpful to memorize these fraction-decimal equivalents.

Concept Summary **Fraction-Decimal Equivalents** For Your FOLDABLE

$\frac{1}{2} = 0.5$ $\frac{1}{3} = 0.\overline{3}$ $\frac{1}{4} = 0.25$ $\frac{1}{5} = 0.2$ $\frac{1}{10} = 0.1$ $\frac{1}{100} = 0.01$

$\frac{2}{3} = 0.\overline{6}$ $\frac{3}{4} = 0.75$ $\frac{2}{5} = 0.4$ $\frac{3}{5} = 0.6$ $\frac{4}{5} = 0.8$ $\frac{5}{6} = 0.8\overline{3}$

Real-World Link

The new food pyramid was released in 2005 by the United States Department of Agriculture (USDA). It provides guidelines for a balanced diet by recommending amounts of different foods based on percents of total Calorie intake. Male and female teens should consume 1300 grams of Calcium each day.

Source: USDA

Real-World EXAMPLE 3 Write a Fraction as a Decimal

FOOD According to the USDA, teenage boys should consume an average of 2700 Calories per day. About 360 Calories should come from milk. To the nearest hundredth, what part of a teenage boy's total Calories should come from milk?

Divide the number of Calories that should come from milk, 360, by the number of total Calories, 2700.

360 [÷] 2700 [ENTER] **0.133...** or $0.1\overline{3}$

Look at the digit to the right of the thousandths place. Round down since $3 < 5$.

Milk should be 0.13 of the daily Calories consumed by a teenage boy.

Check Your Progress

3. **GOLF** In a recent Masters Tournament, Zach Johnson's first shot landed on the fairway 45 out of 56 times. To the nearest thousandth, what part of the time did his shot land on the fairway?

Personal Tutor glencoe.com

Compare Fractions and Decimals It may be easier to compare numbers when they are written as decimals.

EXAMPLE 4 Compare Fractions and Decimals

Replace each ● with <, >, or = to make a true sentence.

a. $\frac{1}{4}$ ● 0.2

$\frac{1}{4}$ ● 0.2	**Write the sentence.**
0.25 ● 0.20	**Write $\frac{1}{4}$ as a decimal. Annex a zero to 0.2.**
$0.25 > 0.20$	**In the hundredths place, $5 > 0$.**

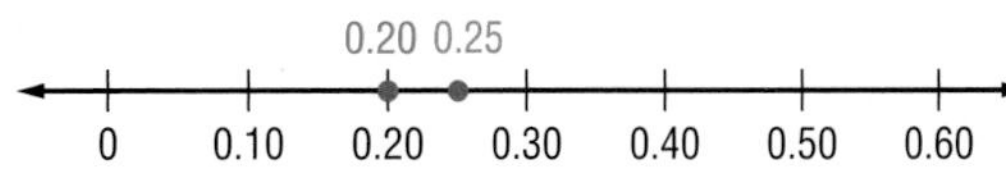

Check Since 0.20 is to the left of 0.25 on the number line, $\frac{1}{4} > 0.2$.

b. $-\frac{5}{8}$ ● $-\frac{6}{9}$

Write the fractions as decimals and then compare the decimals.

$-\frac{5}{8} = -0.625$ $\quad -\frac{6}{9} = -0.666\ldots$ or $-0.\overline{6}$

−0.666... −0.625

−0.68 −0.67 −0.66 −0.65 −0.64 −0.63 −0.62 −0.61

Since −0.625 is to the right of $-0.\overline{6}$ on the number line, $-\frac{5}{8} > -\frac{6}{9}$.

Check Your Progress

4A. $\frac{7}{8}$ ● 0.87

4B. $-\frac{7}{15}$ ● $-\frac{5}{12}$

Personal Tutor glencoe.com

Problem-Solving Tip

Use a Graph You can use a graph to visualize data, analyze trends, and make predictions. In this example, you can compare the decimals on a number line.

Real-World EXAMPLE 5 Compare Fractions Using Decimals

FUNDRAISING Thirty out of 36 freshmen and 34 out of 40 sophomores participated in a marathon for charity. Which class had a greater fraction participating in the marathon?

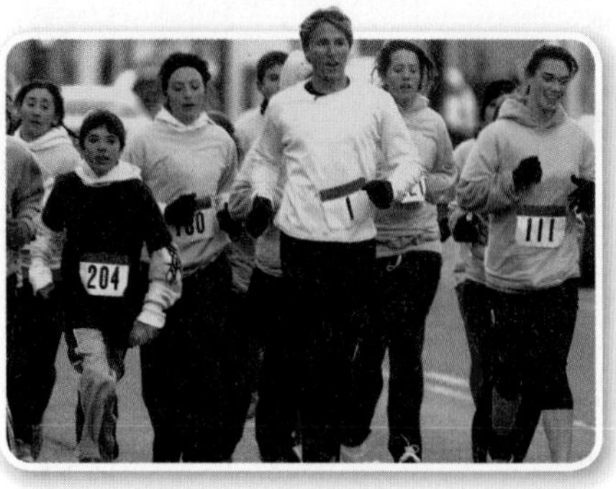

Write each fraction as a decimal. Then compare the decimals.

freshmen: $\frac{30}{36} = 0.8\overline{3}$

sophomores: $\frac{34}{40} = 0.85$

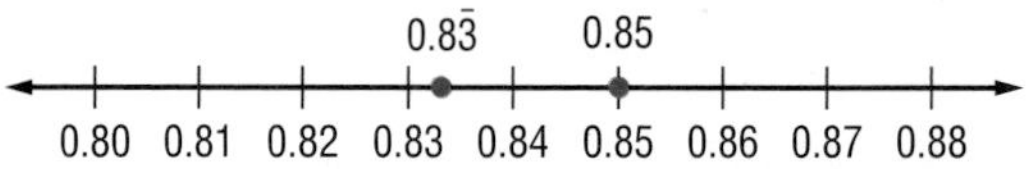

On a number line, $0.8\overline{3}$ is to the left of 0.85. Since $0.8\overline{3} < 0.85$, $\frac{30}{36} < \frac{34}{40}$. So, a greater fraction of sophomores participated in the marathon.

Check Your Progress

5. **MOVIES** Over the weekend, $\frac{16}{28}$ of the 8th-grade girls and $\frac{19}{30}$ of the 8th-grade boys went to see a new comedy movie. Did a greater fraction of girls or boys see the movie?

Personal Tutor glencoe.com

Check Your Understanding

Examples 1 and 2 pp. 121–122

Write each fraction as a decimal. Use a bar to show a repeating decimal.

1. $\frac{3}{5}$
2. $\frac{5}{16}$
3. $-\frac{3}{20}$
4. $\frac{5}{8}$
5. $-\frac{2}{3}$
6. $-\frac{7}{9}$

Example 3 p. 123

7. **FOOTBALL** In one season, the New England Patriots converted 16 of 20 fourth downs. What part of the time did the Patriots convert on fourth down?

Example 4 p. 123

Replace each ● with <, >, or = to make a true sentence.

8. $0.89 \bullet \frac{11}{13}$
9. $-\frac{2}{3} \bullet -\frac{3}{5}$
10. $-0.21 \bullet \frac{1}{5}$
11. $\frac{5}{9} \bullet \frac{6}{11}$
12. $-\frac{9}{15} \bullet -0.61$
13. $\frac{3}{4} \bullet \frac{7}{9}$

Example 5 p. 124

14. **WATER USAGE** Of Nikki's home water usage, $\frac{7}{50}$ comes from lawn watering, and $\frac{3}{20}$ comes from cooking. Does a greater fraction of water usage come from lawn watering or from cooking?

15. **SCHOOL** On his first reading test, Tre answered $\frac{26}{30}$ questions correctly. On his second reading test, he answered $\frac{34}{40}$ questions correctly. On which test did Tre have the better score?

Practice and Problem Solving

● = **Step-by-Step Solutions** begin on page R11.
Extra Practice begins on page 810.

Examples 1 and 2 pp. 121–122

Write each fraction as a decimal. Use a bar to show a repeating decimal.

16. $\frac{3}{8}$ **17.** $\frac{7}{20}$ **18.** $-\frac{8}{25}$ **19.** $-\frac{3}{16}$

20. $\frac{4}{5}$ **21.** $\frac{9}{25}$ **22.** $-\frac{1}{8}$ **23.** $-\frac{7}{16}$

24. $\frac{3}{11}$ **25.** $\frac{33}{45}$ **26.** $-\frac{5}{11}$ **27.** $-\frac{2}{9}$

Example 3 p. 123

28. **BUSINESS** The customer service department resolved 106 of 120 customer complaints in a one-hour time span. To the nearest thousandth, find the resolve rate of the customer service department.

29 **HOCKEY** In a recent season, Niklas Backstrom of the Minnesota Wild saved 955 out of 1028 shots on goal. To the nearest thousandth, what part of the time did Backstrom save shots on goal?

Example 4 p. 123

Replace each ● with <, >, or = to make a true sentence.

30. $\frac{6}{15}$ ● 0.4 **31.** 0.7 ● $\frac{17}{20}$ **32.** $\frac{5}{6}$ ● $\frac{7}{8}$ **33.** $\frac{5}{7}$ ● $\frac{10}{14}$

34. $-\frac{2}{9}$ ● $-\frac{1}{4}$ **35.** $-\frac{1}{8}$ ● $-\frac{1}{10}$ **36.** $0.\overline{6}$ ● $\frac{5}{9}$ **37.** $\frac{1}{2}$ ● 0.67

Example 5 p. 124

WEATHER **The graph at the right shows the amount of rain, in inches, that fell in a 5-day period.**

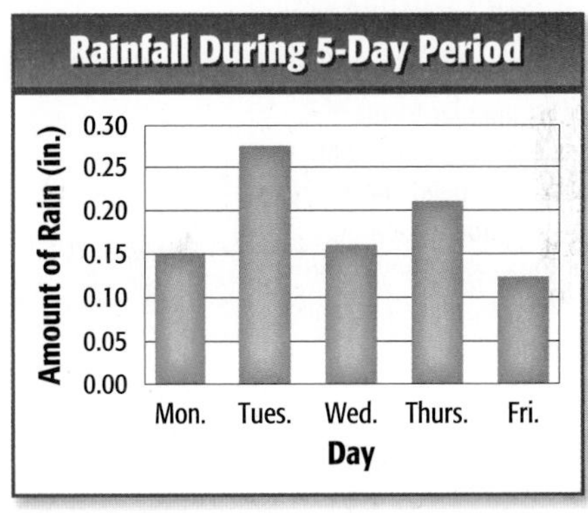

38. On which days did it rain less than one fifth of an inch?

39. Did more or less than one-fourth inch of rain fall on Tuesday? Explain.

40. Suppose it rained $\frac{9}{10}$ inch on Saturday. How does this compare to the previous five days?

Replace each ● with <, >, or = to make a true sentence.

41. $-\frac{5}{13}$ ● $-0.\overline{36}$ **42.** $0.\overline{54}$ ● $\frac{6}{11}$ **43.** $-5.\overline{42}$ ● $-5\frac{3}{7}$

44. $-\frac{5}{16}$ ● $-\frac{8}{25}$ **45.** -2.2 ● $-2\frac{2}{7}$ **46.** $-5\frac{1}{3}$ ● $-5\frac{3}{10}$

47. **CARPENTRY** A carpenter has some bolts that are marked $\frac{1}{2}$, $\frac{5}{16}$, $\frac{3}{32}$, $\frac{3}{4}$, and $\frac{3}{8}$. If all measurements are in inches, how should these bolts be arranged from least to greatest?

Order each group of numbers from least to greatest.

48. $-0.29, -\frac{3}{11}, -\frac{2}{7}$ **49.** $2\frac{3}{5}, 2.67, 2\frac{2}{3}$

50. $-1.\overline{1}, -1\frac{1}{8}, -1\frac{1}{10}$ **51.** $\frac{2}{25}, \frac{1}{13}, 0.089$

52. **SOFTBALL** The table shows the number of times at bat and hits that players on Rawson Middle School team had last season. Order the players based on their batting averages from greatest to least. (*Hint:* Divide the number of hits by the number of at bats.)

Player	Hits	At Bats
Kristen	35	47
Cho	51	73
Brooke	36	50
Alma	49	65
Jessica	46	60

Write each decimal using bar notation.

53. 0.99999... 54. 4.636363... 55. −10.3444... 56. −22.8151515...

57. **MULTIPLE REPRESENTATIONS** Use the number line shown.

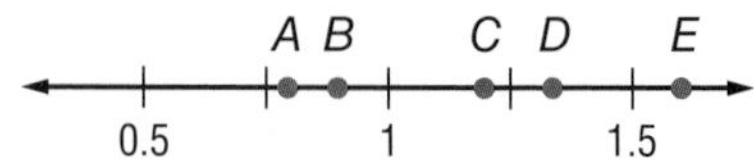

a. **NUMERICAL** Find a fraction or mixed number that might represent each point on the graph.

b. **ALGEBRAIC** Write an inequality using two of your values.

Real-World Link

Glass beads are one of the most popular forms of jewelry. But, did you know that glass beads have been made for about 9,000 years? They come in all shapes, colors, and finishes.

58. **JEWELRY** The table shows the number of each type of bead on 3 bracelets that Mrs. Fraser made for a craft show. Which bracelet has the greatest fraction of glass beads? the least?

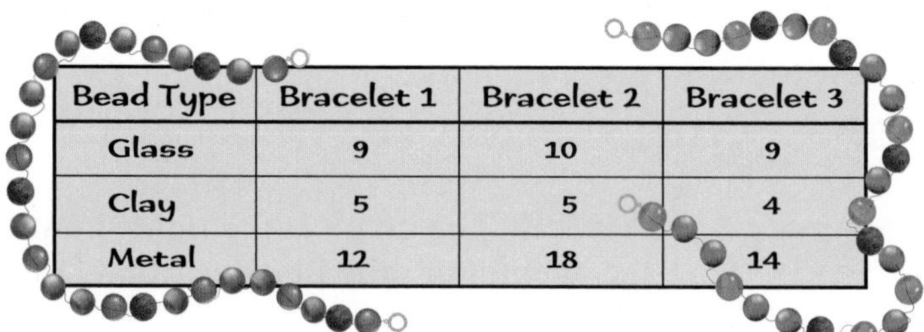

Bead Type	Bracelet 1	Bracelet 2	Bracelet 3
Glass	9	10	9
Clay	5	5	4
Metal	12	18	14

H.O.T. Problems

Use Higher-Order Thinking Skills

59. **OPEN ENDED** Give one example each of real-world situations where it is most appropriate to give a response in fractional form and in decimal form.

60. **CHALLENGE** Are there any rational numbers between $0.\overline{4}$ and $\frac{4}{9}$? Explain.

61. **CHALLENGE** A *unit fraction* is a fraction that has 1 as its numerator. Write the four greatest unit fractions that are repeating decimals. Write each fraction as a decimal.

62. **WRITING IN MATH** Luke is making lasagna that calls for $\frac{4}{5}$ pound of mozzarella cheese. The store only has packages that contain 0.75- and 0.85-pound of mozzarella cheese. Which of the following techniques might Luke use to determine which package to buy? Justify your selection(s). Then use the technique(s) to solve the problem.

mental math	number sense	estimation

63. **CHALLENGE** Write the following fractions as decimals: $\frac{1}{9}$, $\frac{23}{99}$, and $\frac{75}{99}$. Make a conjecture about how to express these kinds of fractions as decimals.

64. **WRITING IN MATH** Explain how 0.5 and $0.\overline{5}$ are different. Which is greater?

Standardized Test Practice

65. Sherman answered $\frac{4}{5}$ of the multiple-choice questions on his science test correctly. Write this fraction as a decimal.

A 0.4

B 0.45

C 0.8

D 4.5

66. Which of the following show the fractions $\frac{2}{5}, \frac{3}{8}, \frac{1}{3}, \frac{1}{2}$, and $\frac{5}{12}$ in order from least to greatest?

F $\frac{1}{3}, \frac{3}{8}, \frac{2}{5}, \frac{1}{2}, \frac{5}{12}$

G $\frac{1}{2}, \frac{1}{3}, \frac{2}{5}, \frac{3}{8}, \frac{5}{12}$

H $\frac{1}{3}, \frac{3}{8}, \frac{2}{5}, \frac{5}{12}, \frac{1}{2}$

J $\frac{1}{2}, \frac{5}{12}, \frac{2}{5}, \frac{3}{8}, \frac{1}{3}$

67. The fraction $\frac{7}{9}$ is found between which pair of fractions on a number line?

A $\frac{3}{5}$ and $\frac{3}{4}$ **C** $\frac{7}{10}$ and $\frac{3}{4}$

B $\frac{7}{10}$ and $\frac{4}{5}$ **D** $\frac{3}{5}$ and $\frac{2}{3}$

68. SHORT RESPONSE Which items shown in the table have a recycle rate less than one half?

Material	Fraction Recycled
Paper	$\frac{5}{11}$
Aluminum Cans	$\frac{5}{8}$
Glass	$\frac{2}{5}$

Spiral Review

Find each product or quotient. (Lessons 2-4 and 2-5)

69. $4(-12)(-5)$ **70.** $-2(42)(3)$ **71.** $-54 \div (-6)$ **72.** $72 \div (-9)$

73. SCUBA DIVING A scuba diver descends from the surface of the lake at a rate of 6 meters per minute. Where will the diver be in relation to the lake's surface after 4 minutes? (Lesson 2-4)

74. EMPLOYMENT The scatter plot shows the years of experience and salaries of twenty people. Does the data show a *positive*, *negative*, or *no* relationship? Explain. (Lesson 1-6)

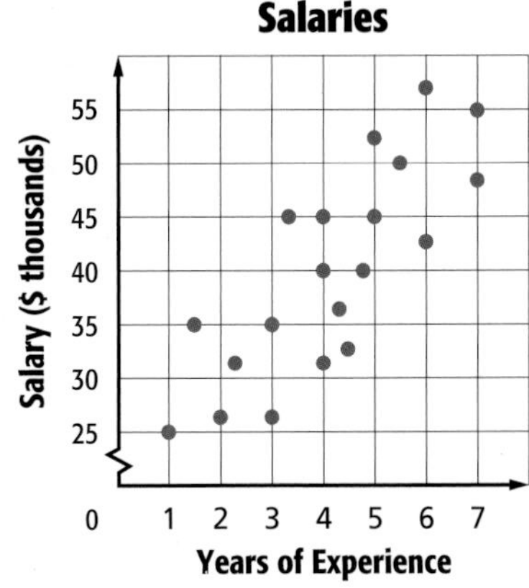

Evaluate each expression if $x = 7$, $y = 3$, and $z = 5$. (Lesson 1-2)

75. $x + y + z$ **76.** $4x - z$

77. $6y - z$ **78.** $9x + 8y$

Skills Review

Write each decimal in word form. (Previous Course)

79. 0.34 **80.** 5.836 **81.** 0.3 **82.** 2.875

3-2 Rational Numbers

Why?

Mr. Marsh's banana waffle recipe is shown at the right.

a. What measures are written as whole numbers?

b. What measures are written as fractions or mixed numbers?

c. Which measures are written as integers?

Banana Waffles
2 Servings

- $1\frac{3}{4}$ cups flour
- $\frac{3}{4}$ cup mashed banana
- 2 egg whites, whipped
- $1\frac{1}{2}$ cups skim milk
- $\frac{1}{2}$ cup applesauce
- 1 teaspoon cinnamon
- 1 teaspoon baking powder
- $\frac{1}{4}$ teaspoon salt

Then

You have already studied natural numbers, whole numbers, and integers. (Lesson 2-1)

Now

- Write rational numbers as fractions.
- Identify and classify rational numbers.

New Vocabulary

rational numbers

Math Online

glencoe.com

- Extra Examples
- Personal Tutor
- Self-Check Quiz
- Homework Help
- Math in Motion

Rational Numbers Numbers like 1, 0, −3, and $1\frac{1}{2}$ can be organized into sets.

When you first learned to count using the numbers 1, 2, 3, …, you were using members of the set of *natural numbers*, N = {1, 2, 3, …}.

If you add zero to the set of natural numbers, the result is the set of *whole numbers*, W = {0, 1, 2, 3, …}.

Whole numbers and their opposites make up the set of *integers*, Z = {…, −3, −2, −1, 0, 1, 2, 3, …}.

Any number that can be written as a fraction is part of the set of **rational numbers**, Q. Some examples of rational numbers are shown below.

$$0.87 \quad -23 \quad \frac{2}{3} \quad -2.\overline{56} \quad 1\frac{1}{2}$$

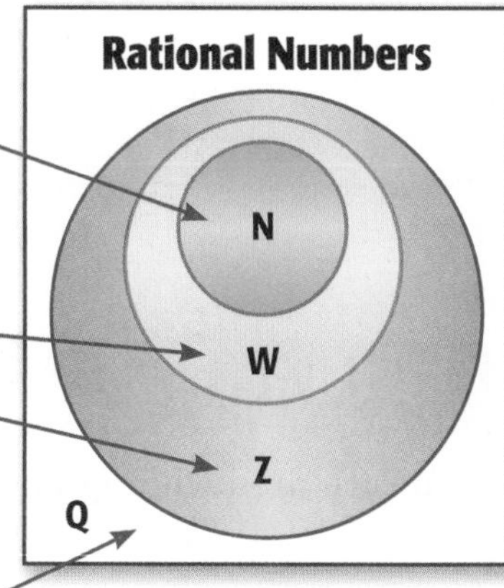

EXAMPLE 1 Write Mixed Numbers and Integers as Fractions

Write each rational number as a fraction.

a. $6\frac{1}{6}$

$6\frac{1}{6} = \frac{37}{6}$ Write $6\frac{1}{6}$ as an improper fraction.

b. −23

$-23 = \frac{-23}{1}$ or $-\frac{23}{1}$

Check Your Progress

1A. $4\frac{2}{3}$

1B. 7

Personal Tutor glencoe.com

Fractions, mixed numbers, and integers are all rational numbers. Terminating decimals are also rational numbers because they can be written as a fraction with a denominator of 10, 100, 1000, and so on.

ReadingMath

Decimal Point Use the word *and* to represent the decimal point.

- Read 0.625 as *six hundred twenty-five thousandths.*
- Read 20.005 as *twenty and five thousandths.*

Real-World EXAMPLE 2 Write Terminating Decimals as Fractions

a. Write 0.64 as a fraction in simplest form.

$0.64 = \frac{64}{100}$ **0.64 is 64 hundredths.**

$= \frac{16}{25}$ **The GCF of 64 and 100 is 4.**

thousands	hundreds	tens	ones	tenths	hundredths	thousandths	ten-thousandths
0	0	0	0.	6	4	0	0

b. GAMES A handheld video game system weighs 9.675 ounces. Write this decimal as a mixed number in simplest form.

$9.675 = 9\frac{675}{1000}$ **0.675 is 675 thousandths.**

$= 9\frac{27}{40}$ **The GCF of 675 and 1000 is 25.**

thousands	hundreds	tens	ones	tenths	hundredths	thousandths	ten-thousandths
0	0	0	9.	6	7	5	0

Check Your Progress

Write each decimal as a fraction or mixed number in simplest form.

2A. 0.84

2B. 5.875

2C. MUSIC Rock music accounted for 0.35 of the total music sales in a recent year. Write this decimal as a fraction in simplest form.

Personal Tutor glencoe.com

Any repeating decimal can be written as a fraction, so repeating decimals are also rational numbers.

EXAMPLE 3 Write Repeating Decimals as Fractions

Write $0.\overline{6}$ as a fraction in simplest form.

$N = 0.6666...$ **Let *N* represent the number.**

$10N = 10(0.6666...)$ **Multiply each side by 10 because one digit repeats.**

$10N = 6.666...$

Subtract N from $10N$ to eliminate the repeating part, 0.666...

$$\begin{aligned} 10N &= 6.666... \\ -N &= 0.666... \\ \hline 9N &= 6 \end{aligned}$$

$10N - N = 10N - 1N$ or $9N$

$\frac{9N}{9} = \frac{6}{9}$ **Divide each side by 9.**

$N = \frac{6}{9}$ or $\frac{2}{3}$

Check 6 [÷] 9 [ENTER] 0.666666667 ✓

StudyTip

Repeating Decimals When *two* digits repeat, multiply each side by 100. Then subtract *N* from 100*N* to eliminate the repeating part.

Check Your Progress

3. Write $0.\overline{42}$ as a fraction in simplest form.

Personal Tutor glencoe.com

ReadingMath

Ratios Rational comes from the word ratio. A *ratio* is the comparison of two quantities by division.
Recall that $\frac{a}{b} = a \div b$, where $b \neq 0$.

Identify and Classify Rational Numbers All rational numbers can be written as terminating or repeating decimals. Decimals that neither terminate nor repeat, such as the numbers below, are called *irrational numbers*. ***You will learn more about irrational numbers in Chapter 10.***

$\pi = 3.141592...$	→	**The digits do not repeat.**
$8.787787778...$	→	**The same block of digits does not repeat.**

Concept Summary — Rational Numbers

For Your FOLDABLE

A rational number is any number that can be expressed as the quotient $\frac{a}{b}$, where a and b are integers and $b \neq 0$.

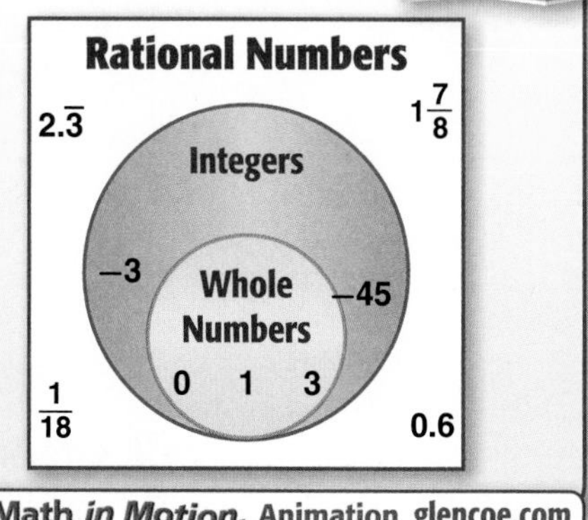

Math *in Motion,* Animation glencoe.com

EXAMPLE 4 Classify Numbers

Identify all sets to which each number belongs.

a. $-2\frac{6}{11}$

Since $-2\frac{6}{11}$ can be written as $-\frac{28}{11}$, it is rational.

b. **1.313313331...**

This is a nonterminating and nonrepeating decimal. So, it is irrational.

c. **45**

45 is a whole number, an integer, and a rational number.

Check Your Progress

5A. 0 **5B.** $1\frac{4}{5}$ **5C.** 1.414213562...

Personal Tutor glencoe.com

Check Your Understanding

Example 1 p. 128

Write each number as a fraction.

1. $3\frac{3}{4}$ **2.** -9 **3.** $-1\frac{3}{4}$

Examples 2 and 3 p. 129

Write each decimal as a fraction or mixed number in simplest form.

4. 0.07 **5.** $-3.\overline{85}$ **6.** $0.\overline{78}$

7. **MEASUREMENT** There are approximately 2.54 centimeters in 1 inch. Express 2.54 as a mixed number.

Example 4 p. 130

Identify all sets to which each number belongs.

8. -632 **9.** $0.\overline{56}$ **10.** 21

Practice and Problem Solving

● = **Step-by-Step Solutions** begin on page R11.
Extra Practice begins on page 810.

Example 1 p. 128

Write each number as a fraction.

11. $1\frac{5}{6}$ **12.** -12 **13.** $-10\frac{7}{8}$ **14.** 49

Example 2 p. 129

Write each decimal as a fraction or mixed number in simplest form.

15. 3.625 **16.** 0.55 **17.** -5.36

18. -0.265 **19.** -1.3 **20.** 0.9

Example 2 p. 129

21. **FINANCIAL LITERACY** Recently, one U.S. dollar was equal to 0.506 British pounds. Express 0.506 as a fraction in simplest form.

22. **POPULATION** The estimated portions for various age groups of the population for 2010 are shown in the table.

a. Find the fraction in simplest form of the population that is 19 years of age or younger.

b. Find the fraction in simplest form of the population that is 20 to 64 years of age.

Age Group	Portion of Population
19 years and under	0.27
20 to 64 years	0.60
65 years and over	0.13

Source: United States Census Bureau

Example 3 p. 129

Write each decimal as a fraction or mixed number in simplest form.

23. $-2.\overline{5}$ **24.** $0.\overline{36}$ **25.** 0.161616...

26. $9.\overline{27}$ **27.** $-0.\overline{09}$ **28.** $-10.\overline{74}$

Example 4 p. 130

Identify all sets to which each number belongs.

29. -8 **30.** 14 **31.** 9.23

32. $1\frac{5}{9}$ **33.** 0.323322333... **34.** 3.141516...

35. **JEWELRY** Maria has a bead that is 0.6 inch long. She wants to use the bead to fill a space that is $\frac{5}{8}$ inch long. Will the bead fit? Explain.

36. **FOOD** All of the Calories in one cup of milk come from fat, protein, and carbohydrates. Use the table to find the fraction of Calories that comes from protein. Write the fraction in simplest form.

Nutrient	Decimal Part of Calories
Fat	0.03
Protein	■
Carbohydrates	0.53

Source: *Nutrition Data*

Replace each ● with <, >, or = to make a true sentence.

37. -0.23 ● -0.3 **38.** $\frac{8}{9}$ ● 0.888… **39.** 0.714 ● $\frac{5}{7}$

40. $-1\frac{1}{11}$ ● -0.9 **41.** $4.\overline{63}$ ● $4\frac{5}{8}$ **42.** $-5.\overline{3}$ ● 5.333…

Write each decimal as a fraction or mixed number in simplest form.

43. $0.\overline{652}$ **44.** $0.1\overline{8}$ **45.** $0.72\overline{4}$

46. $3.\overline{596}$ **47.** $9.2\overline{43}$ **48.** $0.24\overline{67}$

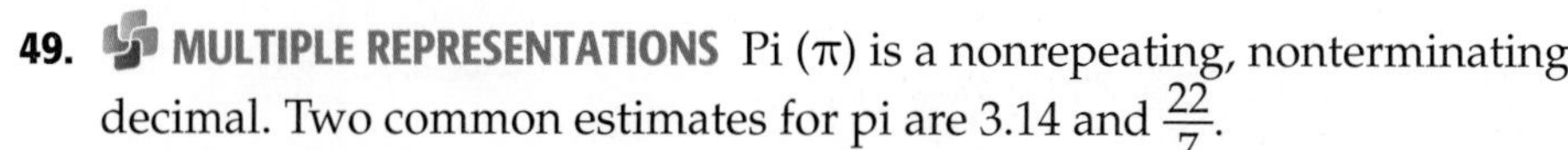

49. **MULTIPLE REPRESENTATIONS** Pi (π) is a nonrepeating, nonterminating decimal. Two common estimates for pi are 3.14 and $\frac{22}{7}$.

a. **GRAPHICAL** Use a calculator to find the value of π to seven decimal places. Then graph the three values on a number line.

b. **SYMBOLIC** Write an inequality comparing the values.

c. **VERBAL** To find the circumference of a circle, you multiply pi by the diameter *d* of the circle. Explain when you might use 3.14 to find the circumference and when you might use $\frac{22}{7}$ to find the circumference.

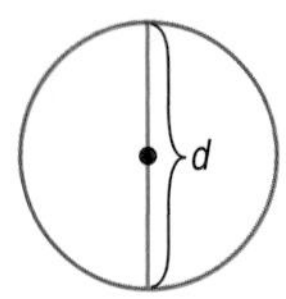

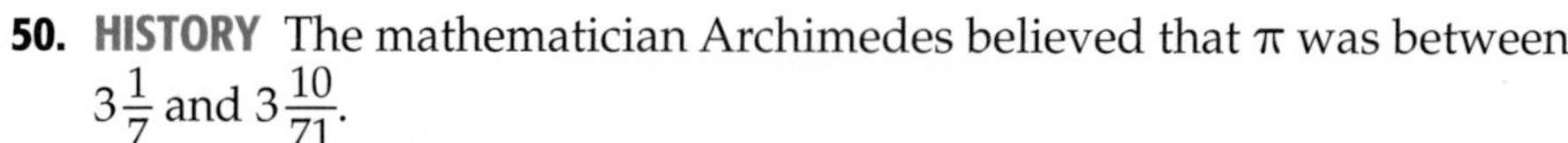

50. **HISTORY** The mathematician Archimedes believed that π was between $3\frac{1}{7}$ and $3\frac{10}{71}$.

a. Express each mixed number as a decimal rounded to the nearest thousandth. Was Archimedes' theory correct? Explain.

b. The Rhind Papyrus records that the Egyptians used $\frac{256}{81}$ for π. Express the fraction as a decimal rounded to the nearest thousandth. Which value is closer to the actual value of π, Arcihimedes' or the Egyptians' value?

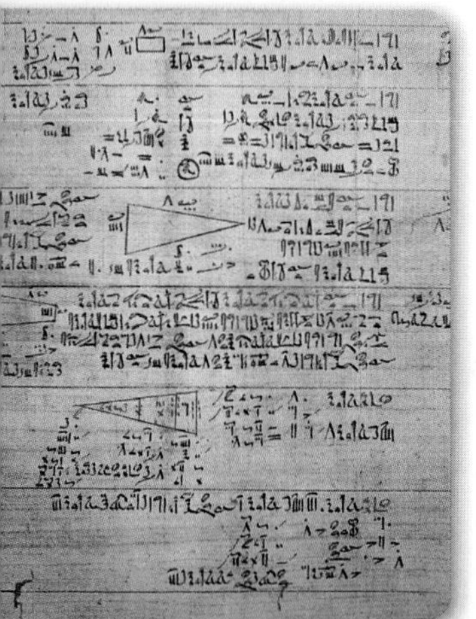

Math History Link

Rhind or Ahmes Papyrus (c1600 B.C.)
The Rhind Papyrus is the chief source of information on Egyptian mathematics. It was originally written by the Egyptian, Ahmes, around 1600 B.C. and contains 15 problems that show the first references to Egyptians using fractions.

Order each set of rational numbers from least to greatest.

51. $-3.4, 3\frac{4}{11}, -3.\overline{42}, 3.38$

52. $\frac{1}{3}, 0.\overline{13}, \frac{5}{13}, 0.32$

53. $-1\frac{13}{14}, -1.9, -1\frac{9}{11}, -1.95$

54. $9\frac{4}{5}, 9.\overline{79}, 9\frac{11}{13}, 9.82$

55. **ANIMALS** A lion's speed is $\frac{5}{7}$ the speed of a cheetah. Find the least rational number with a denominator of 9 that is greater than $\frac{5}{7}$. Find the greatest rational number with a denominator of 8 that is less than $\frac{5}{7}$. Write an inequality comparing the three numbers.

H.O.T. Problems

Use Higher-Order Thinking Skills

56. **OPEN ENDED** Choose a repeating decimal in which three digits repeat. Write the number as a fraction or mixed number in simplest form.

57. **WRITING IN MATH** Explain why $0.\overline{76}$ is greater than 0.76.

58. **CHALLENGE** Antonio stated that $0.\overline{9} = 1$. Show that he is correct.

59. **REASONING** Determine whether the following statements are *true* or *false*. If true, explain your reasoning. If false, give a counterexample.

a. All integers are rational numbers.

b. All whole numbers are integers.

c. A rational number is always an integer.

d. All natural numbers are rational.

60. **WRITING IN MATH** How are repeating decimals usually represented in real-world situations? Give an example to explain your reasoning.

Standardized Test Practice

61. Which fraction is between 0.12 and 0.15?

A $\frac{3}{25}$ C $\frac{3}{20}$

B $\frac{1}{8}$ D $\frac{1}{5}$

62. Which of the following is *not* a rational number?

F $\frac{4}{9}$ H $0.\overline{62}$

G -4.27 J $-3.131131113\ldots$

63. Last football season, Jason made 0.85 of his field goal attempts. Write this decimal as a fraction in simplest form.

A $\frac{85}{100}$ C $\frac{17}{20}$

B $\frac{20}{17}$ D $\frac{100}{85}$

64. EXTENDED RESPONSE The table shows the results of a survey about how students get to school.

Method of Transportation	Portion of Students
bus	0.40
walk	0.18
car	0.36
bicycle	0.04
other	0.02

a. Write each decimal in the table as a fraction in simplest form.

b. List the methods of transportation in order from least to greatest.

c. Which method of transportation do most students use to get to school?

Spiral Review

Write each fraction as a decimal. Use a bar to show a repeating decimal. (Lesson 3-1)

65. $-\frac{5}{8}$ **66.** $\frac{1}{6}$ **67.** $-\frac{2}{10}$ **68.** $\frac{4}{7}$

Graph the figure at the right and its image after the transformation indicated. (Lesson 2-7)

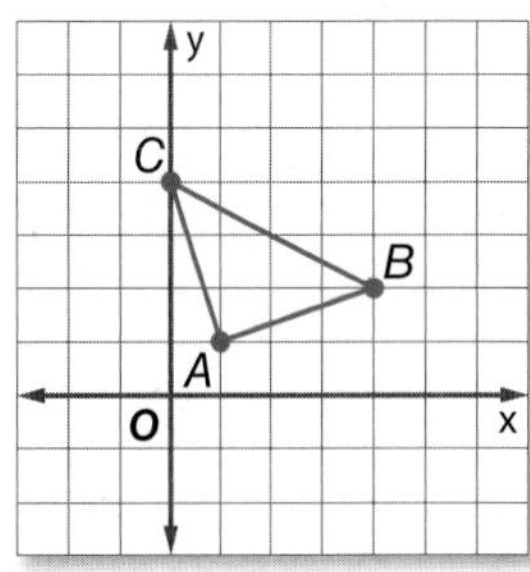

69. translation 3 units down and 2 units left

70. translation 4 units up and 1 unit right

71. reflection across the x-axis

72. reflection across the y-axis

State the domain and range for each relation.

73. {(0, 0), (3, 2), (4, 6), (8, 12)}. (Lesson 1-4)

74. {(1, 2), (3, 4), (5, 6), (7, 8)}.

75. Mount Kilimanjaro's altitude is 5895 meters. Lake Assal's altitude is –155 meters. Find the difference between these altitudes. (Lesson 2-3)

Skills Review

Find each product. (Lesson 2-4)

76. $-6(-12)$ **77.** $15(-3)(-4)(0)$ **78.** $-3(5)(-9)$ **79.** $14(-20)$

3-3

Multiplying Rational Numbers

Then

You have already multiplied positive fractions. (Previous Course)

Now

- Multiply positive and negative fractions.
- Evaluate algebraic expressions with fractions.

Math Online

glencoe.com

- Extra Examples
- Personal Tutor
- Self-Check Quiz
- Homework Help

Why?

The students in Genoa Middle School were surveyed about their favorite school lunch. One half of the students chose pizza. Of those students, one third chose plain cheese. What part of the students in the school chose plain cheese pizza as their favorite?

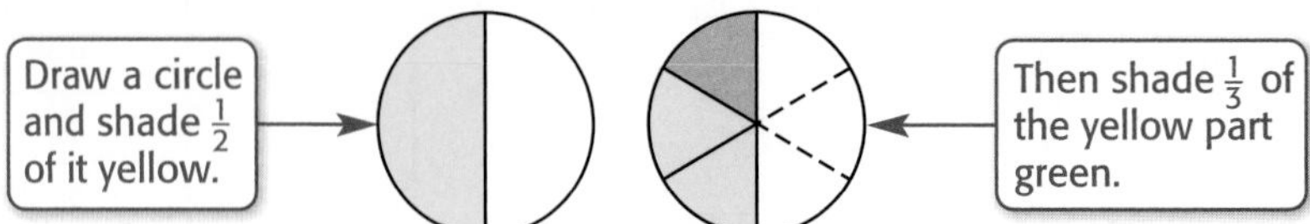

Draw a circle and shade $\frac{1}{2}$ of it yellow.
Then shade $\frac{1}{3}$ of the yellow part green.

The green area represents the product of $\frac{1}{3}$ and $\frac{1}{2}$. One sixth of the circle is shaded green. So $\frac{1}{3} \times \frac{1}{2} = \frac{1}{6}$. One sixth of the students chose plain cheese.

Use an area model to find each product.

a. $\frac{1}{2} \cdot \frac{1}{4}$ b. $\frac{1}{3} \cdot \frac{1}{3}$ c. $\frac{2}{5} \cdot \frac{3}{4}$

d. **MAKE A CONJECTURE** What is the relationship between the numerators and denominators of the factors and the numerator and denominator of the product?

Multiply Fractions These models suggest a rule for multiplying fractions.

Key Concept — Multiplying Fractions

For Your FOLDABLE

Words To multiply fractions, multiply the numerators and multiply the denominators.

Symbols $\frac{a}{b} \cdot \frac{c}{d} = \frac{a \cdot c}{b \cdot d}$, where $b, d \neq 0$

Example $\frac{3}{4} \cdot \frac{1}{2} = \frac{3 \cdot 1}{4 \cdot 2}$ or $\frac{3}{8}$

EXAMPLE 1 Multiply Fractions

Find $\frac{1}{6} \cdot \frac{2}{3}$. Write the product in simplest form.

$\frac{1}{6} \cdot \frac{2}{3} = \frac{1 \cdot 2}{6 \cdot 3}$ ← **Multiply the numerators.** ← **Multiply the denominators.**

$= \frac{2}{18}$ or $\frac{1}{9}$ **Simplify. The GCF of 2 and 18 is 2.**

Check Your Progress

Find each product. Write in simplest form.

1A. $\frac{1}{2} \cdot \frac{4}{10}$ **1B.** $\frac{5}{12} \cdot \frac{6}{10}$

Personal Tutor glencoe.com

If the fractions have common factors in the numerators and denominators, you can simplify before you multiply.

StudyTip

Negative Signs When working with negative fractions, assign the negative sign to the numerator.
Example: Write $-\frac{7}{9}$ as $\frac{-7}{9}$.

EXAMPLE 2 Multiply Negative Fractions and Mixed Numbers

Find each product. Write in simplest form.

a. $\frac{3}{4}\left(-\frac{7}{9}\right)$

$\frac{3}{4}\left(-\frac{7}{9}\right) = \frac{\overset{1}{\cancel{3}}}{4}\left(\frac{-7}{\underset{3}{\cancel{9}}}\right)$ Divide 3 and 9 by their GCF, 3.

$= \frac{1 \cdot -7}{4 \cdot 3}$ Multiply the numerators and multiply the denominators.

$= -\frac{7}{12}$ Simplify.

b. $2\frac{1}{3} \cdot 2\frac{5}{7}$

Estimate $2 \cdot 3 = 6$

$2\frac{1}{3} \cdot 2\frac{5}{7} = \frac{7}{3} \cdot \frac{19}{7}$ Rename $2\frac{1}{3}$ as $\frac{7}{3}$ and $2\frac{5}{7}$ as $\frac{19}{7}$.

$= \frac{\overset{1}{\cancel{7}}}{3} \cdot \frac{19}{\underset{1}{\cancel{7}}}$ Divide by the GCF, 7.

$= \frac{1 \cdot 19}{3 \cdot 1}$ Multiply.

$= \frac{19}{3}$ or $6\frac{1}{3}$ Simplify.

Check The solution is close to the estimate. ✓

Check Your Progress

2A. $-\frac{9}{12} \cdot -\frac{2}{3}$

2B. $\frac{6}{9} \cdot -\frac{3}{11}$

2C. $3\frac{3}{8} \cdot 2\frac{1}{3}$

2D. $-1\frac{5}{6} \cdot 5\frac{1}{7}$

Personal Tutor glencoe.com

Evaluate Expressions with Fractions Variables can represent fractions in algebraic expressions.

EXAMPLE 3 Evaluate Rational Expressions Using Multiplication

Evaluate $\frac{1}{2}ab$ if $a = \frac{6}{7}$ and $b = -\frac{4}{9}$. Write in simplest form.

$\frac{1}{2}ab = \frac{1}{2}\left(\frac{6}{7}\right)\left(-\frac{4}{9}\right)$ Replace a with $\frac{6}{7}$ and b with $-\frac{4}{9}$.

$= \frac{1}{\underset{1}{\cancel{2}}}\left(\frac{\overset{2}{\cancel{6}}}{7}\right)\left(-\frac{\overset{2}{\cancel{4}}}{\underset{3}{\cancel{9}}}\right)$ The GCF of 6 and 9 is 3. The GCF of 2 and 4 is 2.

$= -\frac{4}{21}$ Simplify.

Check Your Progress

Evaluate each expression if $x = \frac{3}{8}$, $y = -2\frac{2}{9}$, and $z = -\frac{7}{10}$. Write in simplest form.

3A. xy

3B. $5x$

3C. yz

Personal Tutor glencoe.com

Real-World Link

The world's tallest tower is the KVLY-TV Tower located in Fargo, North Dakota. The 2063-foot tower is supported by 7.5 miles of steel wires.

Real-World EXAMPLE 4 Multiply Fractions by Whole Numbers

ROLLER COASTERS The first drop on a certain roller coaster at a theme park is 255 feet. The first drop on another roller coaster is about $\frac{11}{20}$ as high. Find the height of the drop on the second roller coaster.

To find the height of the drop on the second roller coaster, multiply $\frac{11}{20}$ by 255.

$\frac{11}{20} \cdot 255 = \frac{11}{20} \cdot \frac{255}{1}$ — Rename 255 as $\frac{255}{1}$.

$= \frac{11}{\cancel{20}_{4}} \cdot \frac{\cancel{255}^{51}}{1}$ — Divide by the GCF, 5.

$= \frac{11 \cdot 51}{4 \cdot 1}$ — Multiply.

$= \frac{561}{4}$ or $140\frac{1}{4}$ — Simplify.

So, the height of the drop is about 140 feet.

Check Your Progress

4. **SKYSCRAPERS** The Sears Tower in Chicago is about 1450 feet. The Empire State Building in New York City is about $\frac{4}{5}$ as tall. About how tall is the Empire State Building?

Personal Tutor glencoe.com

Check Your Understanding

Examples 1 and 2 pp. 134–135

Find each product. Write in simplest form.

1. $\frac{7}{8} \cdot \frac{1}{2}$
2. $\frac{1}{3} \cdot \frac{2}{5}$
3. $-\frac{2}{3} \cdot \frac{3}{16}$
4. $-\frac{3}{5} \cdot -\frac{10}{21}$
5. $-4\frac{1}{2} \cdot -1\frac{1}{9}$
6. $-2\frac{1}{2} \cdot 5\frac{2}{3}$

Example 3 p. 135

ALGEBRA Evaluate each expression if $x = \frac{14}{15}$, $y = -1\frac{2}{5}$, and $z = -\frac{3}{7}$. Write the product in simplest form.

7. xy
8. $z \cdot z$
9. xz
10. $\frac{3}{4}xz$
11. $4y$
12. $2\frac{1}{3}z$

Example 4 p. 136

13. **GEOGRAPHY** "Midway" is the name of 252 towns in the United States. "Pleasant Hill" occurs $\frac{5}{9}$ as many times. How many towns named "Pleasant Hill" are there in the United States?

14. **SCHOOL SPORTS** Of the 480 students at Pleasantview Middle School, $\frac{13}{20}$ play a school sport. How many students play a sport?

Practice and Problem Solving

● = Step-by-Step Solutions begin on page R11.
Extra Practice begins on page 810.

Examples 1 and 2 pp. 134–135

Find each product. Write in simplest form.

15. $\frac{3}{4} \cdot \frac{1}{8}$
16. $\frac{3}{7} \cdot \frac{1}{6}$
17. $\frac{2}{3} \cdot \frac{4}{9}$
18. $\frac{1}{12} \cdot \frac{3}{8}$
19. $\frac{5}{10} \cdot \frac{2}{9}$
20. $\frac{4}{5} \cdot \frac{5}{8}$
21. $-\frac{1}{15} \cdot -\frac{10}{13}$
22. $-\frac{6}{10} \cdot -\frac{1}{8}$
23. $3\frac{1}{3} \cdot -\frac{1}{5}$
24. $\frac{12}{45} \cdot -\frac{9}{16}$
25. $-1\frac{1}{2} \cdot \frac{2}{3}$
26. $4\frac{3}{8} \cdot -3\frac{3}{7}$

Example 3 p. 135

ALGEBRA **Evaluate each expression if $a = \frac{10}{24}$, $b = -3\frac{1}{8}$, and $c = -\frac{4}{5}$. Write the product in simplest form.**

27. bc
28. ab
29. $2c$
30. $\frac{2}{3}abc$
31. $-4bc$
32. $-3\frac{4}{5}ac$

Example 4 p. 136

33. **BEEF** The average person living in Argentina consumes about 145 pounds of beef per year. The average person living in the United States consumes about $\frac{3}{5}$ as much. How many pounds of beef does the average American consume every year?

34. **BRIDGES** The Golden Gate Bridge in San Francisco is 4200 feet long. The Brooklyn Bridge in New York City is $\frac{19}{50}$ as long. How long is the Brooklyn Bridge?

Find each product. Write in simplest form.

35. $\frac{3}{5} \cdot \frac{10}{28} \cdot \frac{2}{9}$
36. $\frac{2}{3} \cdot \frac{1}{4} \cdot \frac{6}{13}$
37. $3\frac{1}{2} \cdot \left(-1\frac{1}{14}\right) \cdot \frac{4}{5}$
38. $4\frac{1}{5} \cdot -1\frac{3}{7} \cdot \frac{6}{11}$
39. $-\frac{6}{11} \cdot -4 \cdot -2\frac{3}{4} \cdot \frac{1}{3}$
40. $-\frac{9}{10} \cdot 7 \cdot 2\frac{1}{3} \cdot \frac{1}{21}$

41. **LAWN CARE** Dexter's lawn is $\frac{2}{3}$ of an acre. If $7\frac{1}{2}$ bags of fertilizer are needed for 1 acre, how much will he need to fertilize his lawn?

Real-World Link

On April 1, 2004, the first American-made hybrid SUV was released. The hybrid averaged 40 miles to the gallon and saved the consumer an average of $4000 in gasoline costs.

42. **HYBRID CARS** A certain hybrid car can travel $1\frac{4}{11}$ times as far as a similar nonhybrid car with one gallon of gasoline. If the nonhybrid car can travel 33 miles per gallon of gasoline, how far can the hybrid travel on $\frac{4}{5}$ gallon of gasoline?

MEASUREMENT **Complete.**

43. ■ ounces = $\frac{3}{4}$ pound (*Hint:* 1 pound = 16 ounces)
44. ■ feet = $\frac{2}{3}$ mile (*Hint:* 1 mile = 5280 feet)
45. $\frac{5}{6}$ foot = ■ inches
46. $\frac{1}{4}$ minute = ■ seconds
47. ■ cups = $\frac{1}{4}$ gallon (*Hint:* 1 gallon = 16 cups)
48. $\frac{3}{4}$ year = ■ weeks

49. **RESEARCH** Use a cookbook to find a recipe for guacamole. Change the recipe to make $2\frac{1}{4}$ times the original amount.

50. ANALYZE TABLES Use the table that shows statistics from the last election for 8th grade class president. There are 540 students in the 8th grade.

Class Elections	
Fraction of class that voted	$\frac{3}{4}$
Fraction of votes for Hector	$\frac{3}{5}$
Fraction of votes for Nora	$\frac{1}{3}$

a. How many students voted for Hector?

b. How many students voted for Nora?

c. Were there other candidates for class president? How do you know? Explain your reasoning. If there were other candidates, what fraction of the student body voted for them?

StudyTip

Area Models To find each product, you could draw an area model.

51. NUMBER SENSE The expression $\frac{1}{2} \times 4$ means $\frac{1}{2}$ of 4. The number line shows that the product of $\frac{1}{2}$ and 4 is 2. Find each product using a number line.

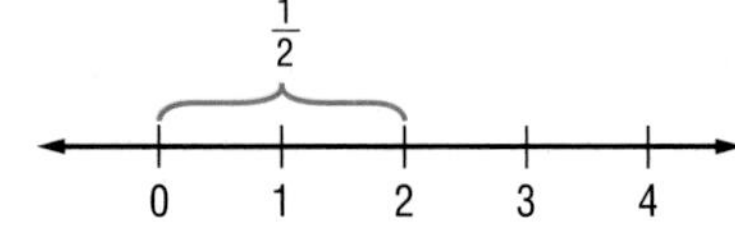

a. $\frac{2}{3}$ of 6

b. $\frac{3}{4}$ of 8

c. $\frac{1}{2}$ of $\frac{2}{3}$

d. $\frac{1}{2}$ of 2

e. $\frac{2}{3}$ of $\frac{3}{2}$

f. $\frac{3}{7}$ of $\frac{7}{3}$

g. Look back at the solutions for Exercises d–f. What pattern do you notice?

h. What is the product of $\frac{a}{b} \cdot \frac{b}{a}$ where $a, b \neq 0$?

H.O.T. Problems

Use Higher-Order Thinking Skills

52. OPEN ENDED Find two rational numbers greater than $\frac{1}{3}$ whose product is less than $\frac{1}{3}$.

53. FIND THE ERROR Kelly and Marina are finding $-4\frac{1}{6} \cdot 2\frac{2}{9}$. Is either of them correct? Explain your reasoning.

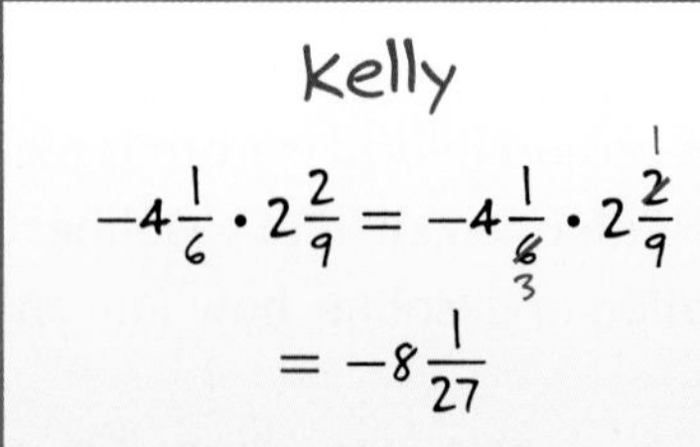

Kelly
$-4\frac{1}{6} \cdot 2\frac{2}{9} = -4\frac{1}{\not{6}_{3}} \cdot 2\frac{\not{2}^{1}}{9}$
$= -8\frac{1}{27}$

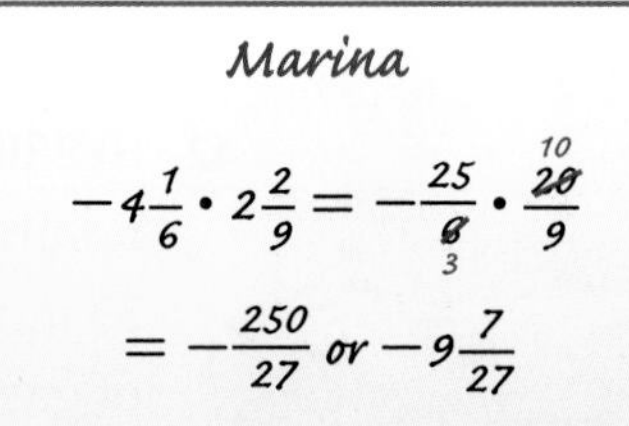

Marina
$-4\frac{1}{6} \cdot 2\frac{2}{9} = -\frac{25}{\not{6}_{3}} \cdot \frac{\not{20}^{10}}{9}$
$= -\frac{250}{27}$ or $-9\frac{7}{27}$

CHALLENGE **Find each missing fraction.**

54. $\frac{2}{3} \cdot \frac{x}{y} = -\frac{3}{8}$

55. $\frac{a}{b} \cdot -\frac{3}{4} = \frac{5}{8}$

56. $\frac{8}{9} \cdot \frac{m}{n} = \frac{14}{27}$

57. REASONING Investigate the product of a fraction between 0 and 1 and a whole number or mixed number. Is the product *always*, *sometimes*, or *never* less than the whole number or mixed number? Explain.

58. WRITING IN MATH Estimate $3\frac{3}{5} \cdot 4\frac{2}{3}$. Then find the actual product. Explain why the estimate and the product are different. What could you do to make your estimate closer to the actual product?

Standardized Test Practice

59. Of the students in Mr. Boggs' class, $\frac{3}{5}$ participate in an after school sport. Of these, $\frac{1}{3}$ participate in track and field. What fraction of the students participates in track and field?

A $\frac{1}{5}$

B $\frac{1}{3}$

C $\frac{3}{5}$

D $\frac{14}{15}$

60. Which statement is shown on the number line below?

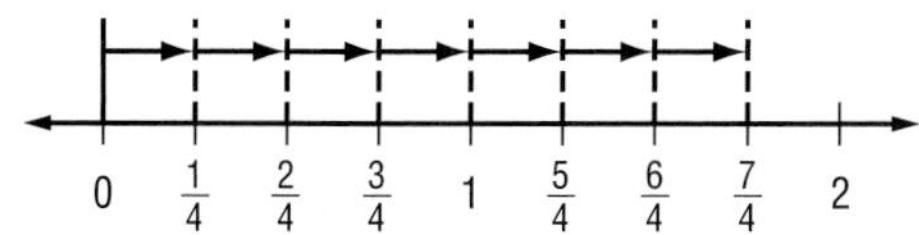

F $\frac{1}{4} + 7 = \frac{7}{4}$

G $\frac{1}{4} \cdot 7 = \frac{7}{4}$

H $\frac{7}{4} \cdot 7 = \frac{1}{4}$

J $\frac{7}{4} + 7 = \frac{1}{4}$

61. What is the value of the expression $2ab$ if $a = \frac{5}{7}$ and $b = -\frac{3}{8}$?

A $-2\frac{15}{56}$

B $-\frac{15}{28}$

C $\frac{15}{28}$

D $2\frac{15}{56}$

62. EXTENDED RESPONSE The length of one side of a square garden tile is $1\frac{2}{3}$ feet.

a. Write a mixed number to represent the perimeter of the tile.

b. Write a mixed number to represent the area of the tile.

c. What is the perimeter of the tile in inches?

d. What is the area of the tile in square inches?

Spiral Review

Write each decimal as a fraction or mixed number in simplest form. (Lesson 3-2)

63. 4.02 **64.** 0.215 **65.** −5.125

66. $-0.\overline{3}$ **67.** $4.\overline{5}$ **68.** $-2.\overline{05}$

Replace each ● with <, >, or = to make a true sentence. (Lesson 3-1)

69. 0.3 ● $\frac{1}{4}$ **70.** $\frac{5}{8}$ ● 0.65 **71.** $\frac{2}{5}$ ● 0.4

72. $\frac{7}{8}$ ● $\frac{8}{9}$ **73.** $\frac{1}{5}$ ● $0.\overline{5}$ **74.** $3\frac{4}{9}$ ● $3.\overline{4}$

75. SLEEP In an online survey, about $\frac{1}{4}$ of teenagers go to sleep between 9 and 10 P.M., while $\frac{13}{50}$ of teenagers go to sleep at 12 A.M. or later. Which group is larger? (Lesson 3-1)

Find each product. (Lesson 2-4)

76. 14(−5) **77.** −8(−11) **78.** −7(−8)(−3) **79.** 2(−8)(−9)(10)

Skills Review

Multiply. (Lesson 2-4)

80. −50(−5) **81.** (12)(−2)(8) **82.** (−1)(16)(−2) **83.** 14(−2)(−3)

CHAPTER 3

Mid-Chapter Quiz

Lessons 3-1 through 3-3

Write each fraction as a decimal. Use a bar to show a repeating decimal. (Lesson 3-1)

1. $\frac{9}{20}$ **2.** $-\frac{3}{11}$

3. $\frac{3}{4}$ **4.** $-\frac{4}{7}$

5. MULTIPLE CHOICE In a recent year, a baseball team won 36 of their 42 games. Which of the following shows the part of games they won to the nearest thousandth? (Lesson 3-1)

A 0.857 **C** 1.17

B 0.86 **D** 1.167

6. SHOPPING A store estimates that 14 out of 120 people return items to the store. To the nearest thousandth, find the rate of customer returns. (Lesson 3-1)

Replace each ● with <, >, or = to make a true sentence. (Lesson 3-1)

7. $\frac{3}{9}$ ● $0.\overline{3}$ **8.** $-\frac{3}{8}$ ● -0.5

9. $1\frac{5}{6}$ ● 1.8 **10.** $4.\overline{25}$ ● $\frac{17}{4}$

11. MANUFACTURING A garbage bag has a thickness of 0.8 mil, which is equal to 0.0008 inch. What fraction of an inch is this? (Lesson 3-2)

Write each decimal as a fraction or mixed number in simplest form. (Lesson 3-2)

12. −4.075 **13.** −1.3636...

14. 0.42 **15.** $3.08\overline{3}$

16. GEOGRAPHY Africa makes up $\frac{1}{5}$ of Earth's entire land surface. Use the table to find the fraction of Earth's land surface that is made up by each of the other continents. Write each fraction in simplest form. (Lesson 3-2)

Continent	Decimal Portion of Earth's Land
Antarctica	0.095
Asia	0.295
Europe	0.07
North America	0.16

Source: *Incredible Comparisons*

17. TRAVEL One of the fastest commuter trains is the Japanese Nozomi, which averages 162 miles per hour. About how many minutes would it take to travel 119 miles from Hiroshima to Kokura on the train? (Lesson 3-3)

Find each product. Write in simplest form. (Lesson 3-3)

18. $\frac{5}{18} \cdot \frac{4}{15}$ **19.** $-2\frac{1}{3} \cdot 2\frac{1}{7}$

20. $-1\frac{1}{2} \cdot \frac{2}{3}$ **21.** $-\frac{3}{16} \cdot (-3\frac{5}{9})$

22. MULTIPLE CHOICE The table shows the number of sports films created with different themes.

Sport Theme	Films
boxing	204
horse racing	139
football	123
baseball	85

Which theme occurs $\frac{5}{12}$ as many times as boxing? (Lesson 3-3)

F horse racing

G football

H baseball

J none of the above

ALGEBRA Evaluate each expression if $w = -3$, $x = \frac{3}{4}$, $y = -\frac{4}{5}$, **and** $z = -2\frac{2}{9}$. (Lesson 3-3)

23. $-wyz$ **24.** $\frac{2}{3}xy$

25. $5wxyz$ **26.** $xy \cdot xy$

27. $-\frac{1}{2}wx$ **28.** $-\frac{3}{2}yz$

29. RIVERS The Nile River is 4160 miles long. The Amazon River is $\frac{25}{26}$ as long. How long is the Amazon River? (Lesson 3-3)

30. JEWELRY Magda is making five necklaces. She uses $20\frac{3}{4}$ inches of wire for each necklace. How much wire will Magda use? (Lesson 3-3)

3-4 Dividing Rational Numbers

Then

You have already divided positive fractions and multiplied rational numbers. (Lesson 3-3)

Now

- Divide positive and negative fractions using multiplicative inverses.
- Divide algebraic fractions.

New Vocabulary

multiplicative inverse
reciprocal

Math Online

glencoe.com

- Extra Examples
- Personal Tutor
- Self-Check Quiz
- Homework Help

Why?

Mrs. Hollern had 3 apples. She cut each of them in half.

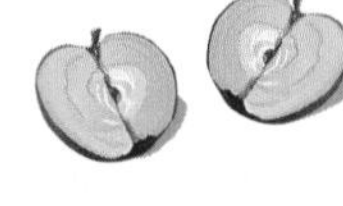

There are six half-pieces in 3 apples, so $3 \div \frac{1}{2} = 6$.

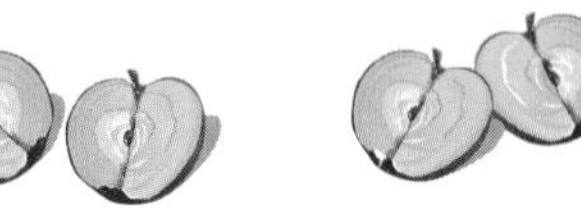

Another way to find the number of sections is to multiply $3 \times 2 = 6$.

Use an area model or another model to find each quotient. Explain how the model shows the quotient.

a. $5 \div \frac{1}{2}$ **b.** $2 \div \frac{1}{4}$ **c.** $3 \div \frac{1}{3}$

d. **MAKE A CONJECTURE** Write about how dividing by a fraction is related to multiplying.

Divide Fractions All of the properties of integers also apply to rational numbers. The statement $\frac{1}{4} \cdot 4 = 1$ demonstrates another property. Two numbers whose product is 1 are called **multiplicative inverses** or **reciprocals**.

Key Concept — Inverse Property of Multiplication (For Your FOLDABLE)

Words The product of a number and its multiplicative inverse is 1.

Symbols For every number $\frac{a}{b}$, where $a, b \neq 0$, there is exactly one number $\frac{b}{a}$ such that $\frac{a}{b} \cdot \frac{b}{a} = 1$.

Example $\frac{2}{3} \cdot \frac{3}{2} = 1$

EXAMPLE 1 Find Multiplicative Inverses

Find the multiplicative inverse of each number.

a. $\frac{7}{16}$

$\frac{7}{16}\left(\frac{16}{7}\right) = 1$ **The product is 1.**

The multiplicative inverse or reciprocal of $\frac{7}{16}$ is $\frac{16}{7}$.

b. $-6\frac{1}{3}$

$-6\frac{1}{3} = -\frac{19}{3}$ **Write $-6\frac{1}{3}$ as an improper fraction.**

$-\frac{19}{3}\left(-\frac{3}{19}\right) = 1$ **The product is 1.**

The multiplicative inverse or reciprocal of $-6\frac{1}{3}$ is $-\frac{3}{19}$.

Check Your Progress

1A. $-\frac{7}{9}$ **1B.** $2\frac{1}{12}$

Personal Tutor glencoe.com

Multiplicative inverses are used in division. Consider $\frac{4}{9} \div \frac{3}{5}$ and $\frac{a}{b} \div \frac{c}{d}$.

$$\frac{\frac{4}{9}}{\frac{3}{5}} = \frac{\frac{4}{9} \cdot \frac{5}{3}}{\frac{3}{5} \cdot \frac{5}{3}}$$ **Multiply the numerator and denominator by $\frac{5}{3}$, the multiplicative inverse of $\frac{3}{5}$.**

$$= \frac{\frac{4}{9} \cdot \frac{5}{3}}{1}$$ **$\frac{3}{5} \cdot \frac{5}{3} = 1$**

$$= \frac{4}{9} \cdot \frac{5}{3}$$

$$\frac{\frac{a}{b}}{\frac{c}{d}} = \frac{\frac{a}{b} \cdot \frac{d}{c}}{\frac{c}{d} \cdot \frac{d}{c}}$$ **Multiply the numerator and denominator by $\frac{d}{c}$, the multiplicative inverse of $\frac{c}{d}$.**

$$= \frac{\frac{a}{b} \cdot \frac{d}{c}}{1}$$ **$\frac{c}{d} \cdot \frac{d}{c} = 1$**

$$= \frac{a}{b} \cdot \frac{d}{c}$$

These examples suggest the following rule for dividing fractions.

Key Concept Dividing Fractions

For Your FOLDABLE

Words To divide by a fraction, multiply by its multiplicative inverse.

Examples $\frac{4}{9} \div \frac{3}{5} = \frac{4}{9} \cdot \frac{5}{3}$ $\quad \frac{a}{b} \div \frac{c}{d} = \frac{a}{b} \cdot \frac{d}{c}$, where b, c, and $d \neq 0$

StudyTip

Dividing By a Whole Number When dividing by a whole number, always rename it as an improper fraction first. Then multiply by its reciprocal.

EXAMPLE 2 Divide by a Fraction or Whole Number

Find each quotient. Write in simplest form.

a. $\frac{1}{9} \div \frac{5}{12}$

$\frac{1}{9} \div \frac{5}{12} = \frac{1}{9} \cdot \frac{12}{5}$ **Multiply by the reciprocal of $\frac{5}{12}$, $\frac{12}{5}$.**

$= \frac{1}{\cancel{9}_3} \cdot \frac{\cancel{12}^4}{5}$ **Divide by the GCF, 3.**

$= \frac{4}{15}$ **Simplify.**

b. $\frac{3}{7} \div 8$

$\frac{3}{7} \div 8 = \frac{3}{7} \div \frac{8}{1}$ **Write 8 as $\frac{8}{1}$.**

$= \frac{3}{7} \cdot \frac{1}{8}$ **Multiply by the reciprocal of $\frac{8}{1}$, $\frac{1}{8}$.**

$= \frac{3}{56}$ **Simplify.**

Check Your Progress

2A. $\frac{1}{3} \div \frac{7}{15}$ **2B.** $\frac{5}{8} \div \left(-\frac{3}{4}\right)$ **2C.** $\frac{3}{4} \div 11$ **2D.** $-\frac{6}{7} \div 12$

Personal Tutor glencoe.com

EXAMPLE 3 Divide by a Mixed Number

Find $-4\frac{2}{3} \div 3\frac{1}{9}$.

$-4\frac{2}{3} \div 3\frac{1}{9} = -\frac{14}{3} \div \frac{28}{9}$ **Rename the mixed numbers as improper fractions.**

$= -\frac{14}{3} \cdot \frac{9}{28}$ **Multiply by the reciprocal, $\frac{9}{28}$.**

$= -\frac{\cancel{14}^1}{\cancel{3}_1} \cdot \frac{\cancel{9}^3}{\cancel{28}_2}$ **Divide out common factors.**

$= -\frac{3}{2}$ or $-1\frac{1}{2}$ **Simplify.**

Check Your Progress

3A. Find $6\frac{3}{8} \div \left(-4\frac{1}{4}\right)$. **3B.** Find $-6\frac{4}{5} \div \left(-2\frac{2}{5}\right)$.

Personal Tutor glencoe.com

Division can be used to find the number of equal size groups in a real-world situation.

Test-Taking Tip

Underlining Underline key words and identify what you are being asked to do in the problem.

STANDARDIZED TEST EXAMPLE 4

Tessa feeds her dog Roscoe $3\frac{3}{4}$ cups of dog food per day. If she buys a bag of food that contains 165 cups, how many days will the bag of food last?

A 600 days **B** 480 days **C** 90 days **D** 44 days

Read the Test Item

You need to find how many days the bag of food will last.

Solve the Test Item

To find how many days, divide. $165 \div 3\frac{3}{4}$ **Think** **How many $3\frac{3}{4}$s are in 165?**

$165 \div 3\frac{3}{4} = \frac{165}{1} \div \frac{15}{4}$ **Rewrite 165 and $3\frac{3}{4}$ as improper fractions.**

$= \frac{165}{1} \cdot \frac{4}{15}$ **Multiply by the reciprocal of $\frac{15}{4}$, $\frac{4}{15}$.**

$= \frac{\overset{11}{\cancel{165}}}{\cancel{1}} \cdot \frac{\cancel{4}}{\underset{1}{\cancel{15}}}$ **Divide out common factors.**

$= 44$ **Simplify.**

So, the correct choice is D.

Check Your Progress

4. A box of cereal contains $15\frac{3}{5}$ ounces. If one bowl holds $2\frac{2}{5}$ ounces of cereal, how many bowls of cereal are in one box?

F $6\frac{1}{2}$ **G** $13\frac{1}{5}$ **H** 18 **J** $37\frac{11}{25}$

Personal Tutor glencoe.com

Real-World Link

About 50% of Americans start their day with a bowl of cereal. So, it is not surprising that 2.7 billion packages of cereal are sold each year. That is enough cardboard to wrap around Earth 13 times.

Source: American Cereal Council

Divide Algebraic Expressions You can divide algebraic fractions in the same way that you divide numerical fractions.

EXAMPLE 5 Divide Algebraic Fractions

Find $\frac{5}{3ab} \div \frac{15}{abc}$. Write the quotient in simplest form.

$\frac{5}{3ab} \div \frac{15}{abc} = \frac{5}{3ab} \cdot \frac{abc}{15}$ **Multiply by the reciprocal of $\frac{15}{abc}$, $\frac{abc}{15}$.**

$= \frac{\overset{1}{\cancel{5}}}{\underset{1}{3\cancel{ab}}} \cdot \frac{\overset{1}{\cancel{ab}c}}{\underset{3}{\cancel{15}}}$ **Divide out common factors.**

$= \frac{c}{9}$ **Simplify.**

Check Your Progress

Find each quotient. Write in simplest form.

5A. $\frac{5ab}{6} \div \frac{10b}{7}$

5B. $\frac{mn}{4} \div \frac{m}{8}$

Personal Tutor glencoe.com

Check Your Understanding

Example 1 p. 141

Find the multiplicative inverse of each number.

1. $\frac{6}{7}$
2. $-5\frac{1}{2}$
3. -63

Examples 2 and 3 p. 142

Find each quotient. Write in simplest form.

4. $-\frac{4}{5} \div \frac{8}{9}$
5. $-\frac{5}{7} \div \frac{2}{35}$
6. $\frac{4}{9} \div (-2)$
7. $\frac{7}{9} \div (-14)$
8. $-2\frac{1}{5} \div \left(-3\frac{2}{3}\right)$
9. $7\frac{1}{9} \div \left(-1\frac{1}{3}\right)$

Example 4 p. 143

10. **MULTIPLE CHOICE** Sonia is making a quilted wall hanging that is 38 inches wide. If each quilt square is $4\frac{3}{4}$ inches wide, how many squares will she need to complete one row of the wall hanging?

A $6\frac{1}{2}$

B 8

C $42\frac{3}{4}$

D 190

Example 5 p. 143

ALGEBRA Find each quotient. Write in simplest form.

11. $\frac{4ab}{c} \div \frac{3a}{2c}$
12. $\frac{mn}{6} \div \frac{3m}{p}$
13. $\frac{3xy}{yz} \div \frac{6y}{5}$

Practice and Problem Solving

 = **Step-by-Step Solutions** begin on page R11.
Extra Practice begins on page 810.

Example 1 p. 141

Find the multiplicative inverse of each number.

14. $-\frac{4}{5}$
15. $\frac{10}{19}$
16. $6\frac{1}{8}$
17. $-4\frac{2}{7}$
18. 19
19. -54

Examples 2 and 3 p. 142

Find each quotient. Write in simplest form.

20. $-\frac{1}{8} \div \frac{2}{5}$
21. $-\frac{5}{12} \div \frac{2}{3}$
22. $-\frac{6}{7} \div \left(-\frac{16}{21}\right)$
23. $-\frac{4}{9} \div (-24)$
24. $-\frac{9}{10} \div (-21)$
25. $-6\frac{1}{9} \div 3\frac{2}{3}$
26. $-10\frac{3}{5} \div \left(-2\frac{2}{5}\right)$
27. $2\frac{3}{8} \div 1\frac{1}{6}$

Example 4 p. 143

28. **COOKING** Hannah is making chocolate chip cookies. How many batches of cookies can she make if she has $7\frac{1}{2}$ cups of brown sugar? Use the recipe card.

29. **DRAMA CLUB** How many play costumes can be made with $49\frac{1}{2}$ yards of fabric if each costume requires $4\frac{1}{8}$ yards?

Chocolate Chip Cookies

- 1 cup softened butter (2 sticks)
- $\frac{1}{2}$ cup granulated sugar
- $1\frac{1}{2}$ cups packed brown sugar
- 2 eggs
- $2\frac{1}{2}$ cups all-purpose flour
- $\frac{3}{4}$ teaspoon salt
- 1 teaspoon baking powder
- 1 teaspoon baking soda
- 18 ounces chocolate chips

Example 5 p. 143

ALGEBRA Find each quotient. Write in simplest form.

30. $\frac{x}{20} \div \frac{x}{5}$
31. $\frac{m}{6n} \div \frac{7m}{3n}$
32. $\frac{m}{np} \div \frac{3m}{2p}$
33. $\frac{5a}{3bc} \div \frac{2a}{9bc}$

34. **BABYSITTING** Barbara babysat for $3\frac{1}{4}$ hours and earned \$19.50. What was her hourly rate?

35. **TRAINS** A train traveled 405 miles in $4\frac{1}{2}$ hours. How fast was the train traveling on average? (*Hint:* distance equals the rate multiplied by the time.)

36. **PHOTOS** Sydney reduced her favorite photograph to put in a scrapbook. How many times as wide is the actual photo than the reduced photo?

4 in.

3 in.

ALGEBRA Evaluate each expression if $m = 2\frac{2}{5}$, $n = -\frac{3}{10}$, and $p = 6$.

37. $mn \div p$

38. $\frac{m}{n}$

39. $np \div m$

Real-World Link

One of the more common tie dyeing techniques is the swirl. To produce the swirl effect on a T-shirt, pinch the center of the shirt and twist into a spiral. Then tie the shape with 2 rubberbands to form 4 sections.

40. **TIE DYE** Ms. Augello is making tie dye shirts with her students. Each gallon of hot water needs $\frac{2}{3}$ cup of tie dye. If Ms. Augello has $5\frac{1}{4}$ cups of tie dye, how many batches of solution will she be able to make?

41. The model at the left shows $\frac{3}{4} \div \frac{1}{2}$. The model at the right shows $\frac{3}{4} \div \frac{1}{4}$.

How many $\frac{1}{2}$s are in $\frac{3}{4}$?

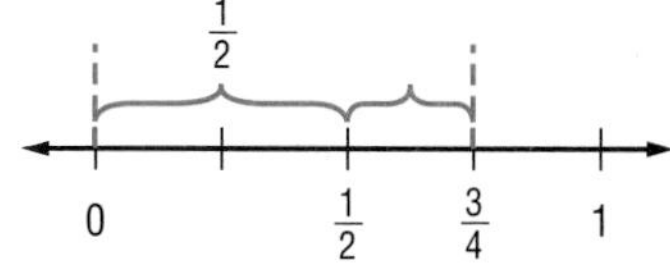

There are $1\frac{1}{2}$ $\frac{1}{2}$s in $\frac{3}{4}$.

How many $\frac{1}{4}$s are in $\frac{3}{4}$?

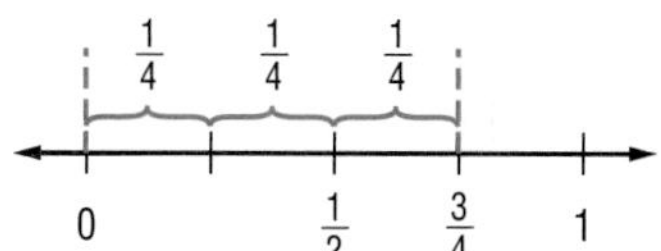

There are three $\frac{1}{4}$s in $\frac{3}{4}$.

Make a conjecture about what happens to the quotient as the value of the divisor increases. Test your conjecture.

H.O.T. Problems

Use **H**igher-**O**rder **T**hinking Skills

42. **OPEN ENDED** Choose two fractions and use an area model or number line to show that division of rational numbers is not commutative.

43. **CHALLENGE** Give a counterexample to this statement. *The quotient of two fractions between 0 and 1 is always a whole number.*

44. **WRITING IN MATH** Which is greater $40 \cdot \frac{1}{4}$ or $40 \div \frac{1}{4}$? Explain.

45. **REASONING** Is a whole number divided by a proper fraction *always*, *sometimes*, or *never* greater than the whole number?

46. **WRITING IN MATH** Explain why, for a positive number n, $n \div \frac{1}{2} > n$.

Standardized Test Practice

47. Heidi is having a party. She is planning that each of her 16 guests will have $\frac{3}{4}$ cup of snack mix. She has made 12 cups of snack mix. Which expression could Heidi use to determine if she has made enough snack mix for each of her guests?

A $16 \div \frac{3}{4}$ **C** $12 \div \frac{3}{4}$

B $12 \div 16$ **D** $\frac{3}{4}(12)$

48. A bag of potting soil contains $4\frac{1}{4}$ pounds of soil. Each flower that Mr. Henderson plants will need $\frac{1}{8}$ pound of soil. How many flowers will he be able to plant?

F 16 **H** 32

G 28 **J** 34

49. A recipe for one batch of soft pretzels calls for $\frac{1}{4}$ cup of salt and $\frac{2}{3}$ cup of sugar. If Mrs. Valdez used $\frac{7}{8}$ cup of salt and $2\frac{1}{3}$ cups of sugar, how many batches of pretzels has she made?

A $3\frac{1}{2}$ **C** $2\frac{1}{4}$

B 3 **D** 2

50. SHORT RESPONSE Popcorn is sold in a variety of sizes. Use the table to find how many times as large the regular bag of popcorn is than the snack bag.

Size	Amount (cups)
Snack	$3\frac{1}{2}$
Regular	$8\frac{3}{4}$
Large	12

Spiral Review

Find each product. Write in simplest form. (Lesson 3-3)

51. $2 \cdot \frac{9}{16}$ **52.** $-4\frac{4}{7} \cdot 2\frac{5}{8}$ **53.** $\frac{3}{20} \cdot \left(-\frac{10}{11}\right)$

54. $-6\frac{1}{2} \cdot \left(-3\frac{1}{4}\right)$ **55.** $-\frac{5}{6} \cdot \left(-1\frac{7}{35}\right)$ **56.** $1\frac{1}{8} \cdot 1\frac{1}{3}$

57. WHITE HOUSE The White House covers an area of 0.028 square mile. What fraction of a square mile is this? (Lesson 3-2)

58. SPORTS The Wildcat football team was penalized the same amount four times during the third quarter. The total of the four penalties was 60 yards. If –60 represents a loss of 60 yards, write a division sentence to represent this situation. Then express the number of yards of each penalty as an integer. (Lesson 2-5)

Find each product. (Lesson 2-4)

59. $12(-6)$ **60.** $-12(-11)$ **61.** $4(-2)(-6)$

Skills Review

Find each sum or difference. (Lessons 2-2 and 2-3)

62. $23 - (-13)$ **63.** $-42 + (-26)$ **64.** $-80 - (-80)$

65. $n + 2n$ **66.** $-4x - (-3x)$ **67.** $5n - 10n$

3-5 Adding and Subtracting Like Fractions

Then
You have already added and subtracted positive fractions with like denominators.
(Previous Course)

Now
- Add rational numbers with common denominators.
- Subtract rational numbers with common denominators.

New Vocabulary
like fractions

Math Online
glencoe.com
- Extra Examples
- Personal Tutor
- Self-Check Quiz
- Homework Help

Why?

Javier is making a smoothie that uses $\frac{1}{8}$ cup of milk and $\frac{3}{8}$ cup of pineapple juice. Javier will use $\frac{4}{8}$ cup of liquid in his smoothie. Use the measuring cup to find each of the following measures.

a. $\frac{1}{8}c + \frac{1}{8}c$ **b.** $\frac{3}{8}c + \frac{2}{8}c$

c. $\frac{5}{8}c + \frac{2}{8}c$ **d.** $\frac{8}{8}c - \frac{4}{8}c$

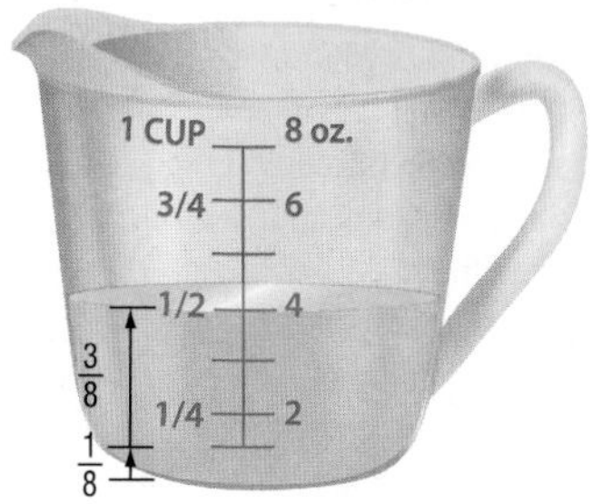

Add Like Fractions **Like fractions** are fractions with the same denominator.

Key Concept: Adding Like Fractions
For Your FOLDABLE

Words To add fractions with like denominators, add the numerators and write the sum over the denominator.

Symbols $\frac{a}{c} + \frac{b}{c} = \frac{a+b}{c}$, where $c \neq 0$

Example $\frac{2}{8} + \frac{3}{8} = \frac{2+3}{8}$ or $\frac{5}{8}$

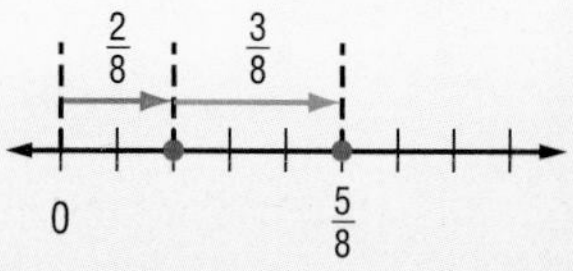

EXAMPLE 1 Add Fractions

Find each sum. Write in simplest form.

a. $\frac{7}{10} + \frac{6}{10}$

Estimate $1 + \frac{1}{2} = 1\frac{1}{2}$

$\frac{7}{10} + \frac{6}{10} = \frac{7+6}{10}$ The denominators are the same. Add the numerators.

$= \frac{13}{10}$ or $1\frac{3}{10}$ Simplify and rename as a mixed number. Is the answer reasonable?

b. $\frac{5}{8} + \left(-\frac{7}{8}\right)$

Estimate $\frac{1}{2} + (-1) = -\frac{1}{2}$

$\frac{5}{8} + \left(-\frac{7}{8}\right) = \frac{5+(-7)}{8}$ The denominators are the same. Add the numerators.

$= \frac{-2}{8}$ or $-\frac{1}{4}$ Simplify. Compare to the estimate. Is it reasonable?

Check Your Progress

1A. $\frac{5}{6} + \frac{4}{6}$ **1B.** $\frac{4}{7} + \left(-\frac{6}{7}\right)$

Personal Tutor glencoe.com

EXAMPLE 2 Add Mixed Numbers

Find $2\frac{3}{8} + 3\frac{7}{8}$. Write in simplest form.

Estimate $2 + 4 = 6$

$2\frac{3}{8} + 3\frac{7}{8} = (2 + 3) + \left(\frac{3}{8} + \frac{7}{8}\right)$ Add the whole numbers and fractions separately.

$= 5 + \frac{10}{8}$ Add the numerators.

$= 5\frac{10}{8}$ or $6\frac{1}{4}$ Simplify. Rename $5\frac{10}{8}$ as $6\frac{2}{8}$ or $6\frac{1}{4}$.

Check for Reasonableness $6\frac{1}{4} \approx 6$ ✓

Check Your Progress

Find each sum. Write in simplest form.

2A. $1\frac{3}{4} + 4\frac{3}{4}$ **2B.** $3\frac{2}{5} + 8\frac{1}{5}$ **2C.** $-2\frac{3}{7} + \left(-4\frac{5}{7}\right)$

Personal Tutor glencoe.com

StudyTip

Alternative Method When adding or subtracting mixed numbers, you can write them as improper fractions before adding or subtracting. If any of the numbers are negative, it is easier to use this method.

$2\frac{3}{8} + 3\frac{7}{8} = \frac{19}{8} + \frac{31}{8}$

$= \frac{50}{8}$ or $6\frac{1}{4}$

Subtract Like Fractions The rule for subtracting fractions with like denominators is similar to the rule for addition.

Key Concept Subtracting Like Fractions

For Your FOLDABLE

Words To subtract fractions with like denominators, subtract the numerators and write the difference over the denominator.

Symbols $\frac{a}{c} - \frac{b}{c} = \frac{a - b}{c}$, where $c \neq 0$

Example $\frac{4}{9} - \frac{3}{9} = \frac{4 - 3}{9}$ or $\frac{1}{9}$

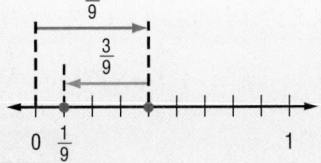

EXAMPLE 3 Subtract Fractions

Find $\frac{3}{10} - \frac{9}{10}$. Write in simplest form.

Estimate $\frac{1}{2} - 1 = -\frac{1}{2}$

$\frac{3}{10} - \frac{9}{10} = \frac{3 - 9}{10}$ The denominators are the same. Subtract the numerators.

$= \frac{-6}{10}$ or $-\frac{3}{5}$ Simplify.

Check for Reasonableness $-\frac{3}{5} \approx -\frac{1}{2}$ ✓

Check Your Progress

Find each difference. Write in simplest form.

3A. $\frac{5}{15} - \frac{10}{15}$ **3B.** $\frac{3}{9} - \frac{4}{9}$ **3C.** $\frac{7}{8} - \frac{3}{8}$

Personal Tutor glencoe.com

EXAMPLE 4 Subtract Mixed Numbers with Regrouping

Evaluate $x - y$ when $x = 3\frac{1}{4}$ and $y = 1\frac{3}{4}$.

$x - y = 3\frac{1}{4} - 1\frac{3}{4}$ **Replace x with $3\frac{1}{4}$ and y with $1\frac{3}{4}$.**

Since $\frac{1}{4}$ is less than $\frac{3}{4}$, rename $3\frac{1}{4}$ before subtracting.

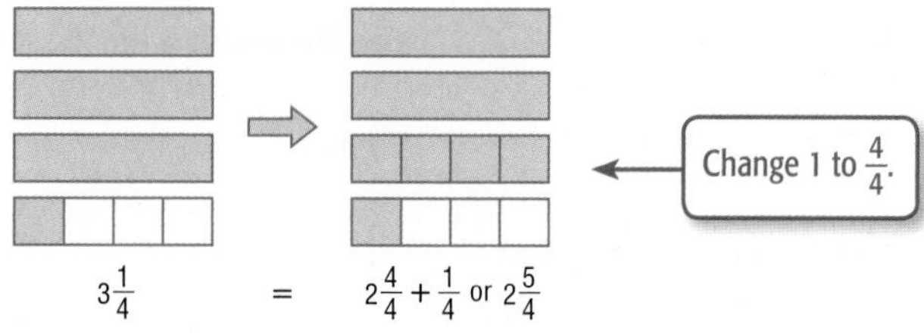

$3\frac{1}{4} - 1\frac{3}{4} = 2\frac{5}{4} - 1\frac{3}{4}$ **Rename $3\frac{1}{4}$ as $2\frac{5}{4}$.**

$= 1\frac{2}{4}$ **Subtract the whole numbers and then the fractions.**

$= 1\frac{1}{2}$ **Simplify.**

Check Your Progress

4. Evaluate $x - y$ when $x = 9\frac{3}{8}$ and $y = 5\frac{5}{8}$.

Personal Tutor glencoe.com

Real-World EXAMPLE 5 Subtract Mixed Numbers

CRAFTS LaShaun has $5\frac{1}{8}$ yards of ribbon to border scrapbook pages. If she uses $1\frac{7}{8}$ yards on one page, how much ribbon is left?

Understand You know how much ribbon she has and how much ribbon she will use.

Plan Subtract the amount of ribbon she will use from the total amount of ribbon.

Estimate $5\frac{1}{8} - 1\frac{7}{8} \approx 5 - 2$ or 3 yards

Solve $5\frac{1}{8} - 1\frac{7}{8} = 4\frac{9}{8} - 1\frac{7}{8}$ **Rename**

$= 3\frac{2}{8}$ **Subtract the whole numbers and then the fractions.**

$= 3\frac{2}{8}$ or $3\frac{1}{4}$ **Simplify.**

LaShaun has $3\frac{1}{4}$ yards of ribbon remaining.

Check Since $3\frac{1}{2}$ is close to 3, the answer is reasonable. ✔

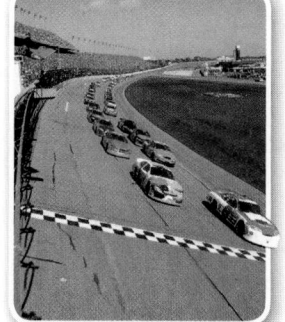

Real-World Link

The Daytona 500 is the first Nextel Cup Race of the annual NASCAR season. In 2008, the race celebrated its 50th anniversary.

Check Your Progress

5. CAR RACING The Daytona International Speedway is one of the longest tracks used in NASCAR races. It is $2\frac{2}{4}$ miles long. Richmond International Speedway is $\frac{3}{4}$ mile long. How much longer is the Daytona Speedway than the Richmond Speedway?

Personal Tutor glencoe.com

You can use the same rules for adding or subtracting like algebraic fractions as you did for adding or subtracting like numerical fractions.

EXAMPLE 6 **Add or Subtract Algebraic Fractions**

Find $\frac{2a}{10} + \frac{4a}{10}$. Write in simplest form.

$\frac{2a}{10} + \frac{4a}{10} = \frac{2a + 4a}{10}$ — **The denominators are the same. Add the numerators.**

$= \frac{6a}{10}$ or $\frac{3a}{5}$ — **Simplify.**

Check Your Progress

Find each sum or difference. Write in simplest form.

6A. $\frac{y}{8} + \frac{5y}{8}$ **6B.** $\frac{3x}{7} - \frac{5x}{7}$ **6C.** $\frac{x}{5} + \frac{4x}{5}$

Personal Tutor glencoe.com

Check Your Understanding

Examples 1–3 pp. 147–148

Find each sum or difference. Write in simplest form.

1. $\frac{3}{6} + \frac{5}{6}$ **2.** $\frac{2}{9} + \left(-\frac{4}{9}\right)$ **3.** $\frac{4}{12} - \frac{10}{12}$

4. $\frac{8}{15} - \frac{11}{15}$ **5.** $3\frac{3}{8} + 6\frac{5}{8}$ **6.** $2\frac{1}{6} + 8\frac{3}{6}$

Example 4 p. 149

ALGEBRA **Evaluate each expression if $s = 7\frac{1}{7}$ and $t = 6\frac{3}{7}$.**

7. $s - t$ **8.** $t - s$ **9.** $2s - t$

Example 5 p. 149

10. **WOODWORKING** Mia is making a bookcase and has $92\frac{5}{8}$ inches of wood. If she uses $30\frac{7}{8}$ inches of wood for the top and bottom, find the amount of wood she has left for the sides.

Example 6 p. 150

ALGEBRA **Find each sum or difference. Write in simplest form.**

11. $\frac{5x}{y} - \frac{7x}{y}, y \neq 0$ **12.** $\frac{3a}{13} + \left(-\frac{8a}{13}\right)$ **13.** $-\frac{4c}{ab} - \frac{8c}{ab}, a, b \neq 0$

Practice and Problem Solving

= **Step-by-Step Solutions** begin on page R11.
Extra Practice begins on page 810.

Examples 1–3 pp. 147–148

Find each sum or difference. Write in simplest form.

14. $\frac{5}{6} + \left(-\frac{4}{6}\right)$ **15.** $-\frac{11}{12} + \frac{7}{12}$ **16.** $\frac{3}{14} + \left(-\frac{5}{14}\right)$ **17.** $-\frac{3}{7} + \frac{6}{7}$

18. $5\frac{1}{4} + \left(5\frac{1}{4}\right)$ **19.** $12\frac{5}{9} + 1\frac{1}{9}$ **20.** $2\frac{12}{13} + \left(-7\frac{10}{13}\right)$ **21.** $-8\frac{3}{10} + 4\frac{9}{10}$

22. $\frac{2}{15} - \frac{7}{15}$ **23.** $\frac{5}{11} - \frac{7}{11}$ **24.** $-\frac{1}{5} - \frac{4}{5}$ **25** $-\frac{7}{20} - \frac{7}{20}$

Example 4 p. 149

ALGEBRA **Evaluate each expression if $a = 4\frac{4}{9}$, $b = 6\frac{5}{9}$, and $c = \frac{1}{9}$.**

26. $b - a$ **27.** $a - b$ **28.** $c - a$ **29.** $c - b$

Example 5
p. 149

30. **MEASUREMENT** Nan was $59\frac{7}{8}$ inches tall at the end of the summer. She was $62\frac{1}{8}$ inches by her birthday in March. How much did she grow during that time?

31. **FOOD** Yahto needs $3\frac{3}{4}$ cups of sugar to make cookies. He needs an additional $\frac{3}{4}$ cup for bread. Find the total amount of sugar that Yahto needs.

Example 6
p. 150

ALGEBRA Find each sum or difference. Write in simplest form.

32. $\frac{6x}{15} + \frac{5x}{15}$ 33. $\frac{2m}{5} + \frac{7m}{5}$ 34. $\frac{7a}{10} + \frac{4a}{10}$ 35. $-\frac{p}{14} + \frac{6p}{14}$

Find each sum or difference. Write in simplest form.

36. $-2\frac{9}{10} + \left(-9\frac{9}{10}\right) + \left(-6\frac{9}{10}\right)$ 37. $\frac{1}{9} - 2\frac{4}{9} - \frac{5}{9}$

38. **SPORTS** A triathlon consists of three races: swimming, biking and hiking. If an athlete swims for $18\frac{2}{4}$ minutes, runs for $37\frac{3}{4}$ minutes, and bikes for $59\frac{1}{4}$ minutes, what was his total time?

39. **PETS** The table shows the weight of Leon's dog during its first 5 years.

Age (years)	1	2	3	4	5
Weight (pounds)	$17\frac{2}{8}$	$18\frac{5}{8}$	$19\frac{4}{8}$	$18\frac{3}{8}$	$20\frac{7}{8}$

a. How much weight did Leon's dog gain or lose between ages 3 and 4? between years 1 and 5?

b. If Leon's dog gains $1\frac{3}{8}$ pounds each year between years 5 and 7, how much will his dog weigh?

Real-World Career

Chef
Chefs frequently work with fractional measurements when preparing dishes and when creating their own recipes.

Education and training for chefs range from on the job training to obtaining a 4-year degree. Many cities have culinary institutes which offer associate degree programs with restaurant experience.

40. **COOKING** Chad is making lasagna for a party. The recipe uses $1\frac{2}{4}$-teaspoons of basil, $\frac{2}{4}$ teaspoon of salt, $\frac{1}{4}$ teaspoon of pepper, and 4 teaspoons of parsley. If he needs to double the recipe, how many teaspoons will he use?

ALGEBRA Find each sum or difference. Write in simplest form.

41. $\frac{3pr}{2n} + \frac{7pr}{2n} - \frac{pr}{2n}$ 42. $-\frac{8x}{y} + \frac{6x}{y} + \left(-\frac{3x}{y}\right)$

H.O.T. Problems

Use Higher-Order Thinking Skills

43. **OPEN ENDED** Write a subtraction problem with a difference of $-\frac{2}{3}$.

44. **CHALLENGE** Lopez Construction is replacing a window in a house. The window is currently 3 feet wide by 4 feet tall. The homeowner wants to add 9 inches to each side of the window. What is the new perimeter of the window in feet? Support your answer with a model.

45. **FIND THE ERROR** Xavier said the sum of $-4\frac{1}{9}$ and $1\frac{7}{9}$ is $-3\frac{8}{9}$. Is he correct? Explain your reasoning.

46. **CHALLENGE** Explain how you could use mental math to find the following sum. Then find the sum. Support your answer with a model.

$$1\frac{1}{4} + 2\frac{1}{3} + 3\frac{2}{3} + 4\frac{1}{2} + 5\frac{1}{2} + 6\frac{3}{4}$$

47. **WRITING IN MATH** Write a real-world problem about cooking that can be solved by adding or subtracting fractions. Then solve the problem.

Standardized Test Practice

48. The average times it takes Miguel to cut his lawn and his neighbor's lawn are given in the table.

Lawn	Time to Cut (h)
Miguel's	$\frac{3}{4}$
Neighbor's	$1\frac{1}{4}$

Last summer, he cut his lawn 10 times and his neighbor's 6 times. How many hours did he spend cutting both lawns?

A $13\frac{1}{2}$ h **C** $14\frac{1}{2}$ h

B 14 h **D** 15 h

49. A piece of wood is $1\frac{9}{16}$ inches thick. A layer of padding $\frac{15}{16}$ inch thick is placed on top of the wood. What is the total thickness of the wood and the padding?

F $1\frac{3}{8}$ in. **H** $1\frac{15}{16}$ in.

G $1\frac{1}{2}$ in. **J** $2\frac{1}{2}$ in.

50. Ronata is putting lace around the tablecloth shown below. How much lace will she need to cover all 4 sides?

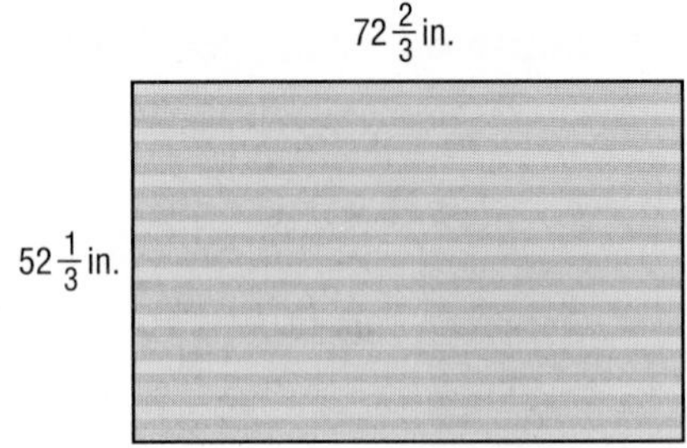

A $20\frac{1}{2}$ in. **C** 125 in.

B 41 in. **D** 250 in.

51. **SHORT RESPONSE** Simplify the expression below.

$$-5\frac{7}{9} - 2\frac{4}{9} + 1\frac{8}{9}$$

Spiral Review

Find each quotient. Write in simplest form. (Lesson 3-4)

52. $\frac{2}{7} \div \frac{5}{14}$

53. $-3\frac{1}{8} \div \frac{5}{16}$

54. $4\frac{2}{3} \div \left(-3\frac{1}{9}\right)$

Find each product. Write in simplest form. (Lesson 3-3)

55. $\frac{3}{10} \cdot \frac{4}{21}$

56. $\frac{3}{8}(-6)$

57. $\frac{5}{19} \cdot 2\frac{2}{18}$

58. **FURNITURE** A shelf $16\frac{5}{8}$ inches wide is to be placed in a space that is $16\frac{3}{4}$ inches wide. Will the shelf fit in the space? Explain. (Lesson 3-1)

59. **FOOTBALL** The Hawks started a play on their own 31-yard line. They lost 9 yards on one play and another 5 yards on the next play. Find the team's field location after the two plays. (Lesson 2-3)

Skills Review

Find the LCM of each pair of numbers or monomials.
(Concepts and Skills Bank pp. 860–861)

60. 6, 8

61. 12, 15

62. 3, 7

63. 15, 45

64. $2a, 2b$

65. x, x^2y

3-6 Adding and Subtracting Unlike Fractions

Then
You have already added and subtracted rational numbers with like denominators. (Lesson 3-5)

Now
- Add unlike fractions.
- Subtract unlike fractions.

New Vocabulary
unlike fractions

Math Online
glencoe.com
- Extra Examples
- Personal Tutor
- Self-Check Quiz
- Homework Help

Why?

Tasha feeds her cats $\frac{2}{3}$ cup of cat food in the morning and $\frac{1}{2}$ cup in the evening. You can use the least common multiple, or LCM, to find how much food her cats eat each day.

a. What is the LCM of the denominators?

b. Each model is divided into six parts. What parts of each model are shaded?

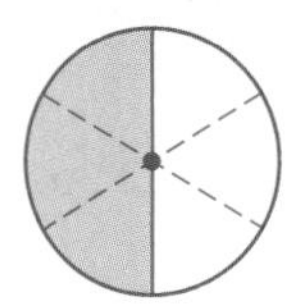
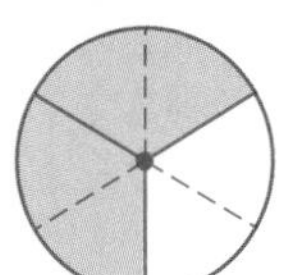

c. How many parts are in the sum $\frac{2}{3} + \frac{1}{2}$?

Add Unlike Fractions **Unlike fractions** are fractions with different denominators. Use the least common multiple of the denominators to rename the fractions before adding them.

Key Concept — Adding Unlike Fractions (For Your FOLDABLE)

Words To add fractions with unlike denominators, rename the fractions with a common denominator. Then add and simplify as with like fractions.

Example

$$\frac{2}{3} + \frac{1}{2} = \frac{2}{3} \cdot \frac{2}{2} + \frac{1}{2} \cdot \frac{3}{3}$$
$$= \frac{4}{6} + \frac{3}{6}$$
$$= \frac{7}{6} \text{ or } 1\frac{1}{6}$$

EXAMPLE 1 Adding Unlike Fractions

Find $\frac{3}{5} + \frac{1}{3}$. Write in simplest form. Estimate $1 + 0 = 1$

$\frac{3}{5} + \frac{1}{3} = \frac{3}{5} \cdot \frac{3}{3} + \frac{1}{3} \cdot \frac{5}{5}$ Use 3 • 5 or 15 as the common denominator.

$= \frac{9}{15} + \frac{5}{15}$ Rename each fraction with the common denominator.

$= \frac{14}{15}$ Add the numerators.

Check for Reasonableness $\frac{14}{15} \approx 1$ ✓

Check Your Progress

Find each sum. Write in simplest form.

1A. $\frac{1}{6} + \frac{3}{4}$

1B. $\frac{2}{7} + \frac{3}{14}$

Personal Tutor glencoe.com

EXAMPLE 2 Add Fractions and Mixed Numbers

Find each sum. Write in simplest form.

a. $-\frac{5}{6} + \frac{1}{8}$ — Estimate $-1 + 0 = -1$

$$-\frac{5}{6} + \frac{1}{8} = \frac{-5}{6} \cdot \frac{4}{4} + \frac{1}{8} \cdot \frac{3}{3}$$ The LCD of 6 and 8 is 24.

$$= \frac{-20}{24} + \frac{3}{24}$$ Rename each fraction using the LCD, 24.

$$= \frac{-17}{24}$$ Simplify.

StudyTip

LCD You can rename fractions using any common denominator. However, using the least common denominator will make simplifying the solution easier.

b. $-\frac{3}{10} + \left(-5\frac{3}{4}\right)$ — Estimate $-\frac{1}{2} + -6 = -6\frac{1}{2}$

$$-\frac{3}{10} + \left(-5\frac{3}{4}\right) = \frac{-3}{10} + \frac{-23}{4}$$ Write $-5\frac{3}{4}$ as an improper fraction.

$$= \frac{-3}{10} \cdot \frac{2}{2} + \frac{-23}{4} \cdot \frac{5}{5}$$ The LCD of 10 and 4 is 20.

$$= \frac{-6}{20} + \frac{-115}{20}$$ Rename each fraction using the LCD, 20.

$$= \frac{-121}{20} \text{ or } -6\frac{1}{20}$$ Simplify.

Check Your Progress

2A. $3\frac{3}{4} + \frac{5}{14}$ **2B.** $-6\frac{8}{9} + 7\frac{5}{12}$ **2C.** $3\frac{3}{5} + \left(-4\frac{5}{6}\right)$

Personal Tutor glencoe.com

Subtract Unlike Fractions The rule for subtracting fractions with unlike denominators is similar to the rule for addition.

Key Concept Subtracting Unlike Fractions

For Your FOLDABLE

To subtract fractions with unlike denominators, rename the fractions with a common denominator. Then subtract and simplify as with like fractions.

EXAMPLE 3 Subtract Fractions and Mixed Numbers

Find each difference. Write in simplest form.

a. $\frac{3}{8} - \frac{3}{4}$

$$\frac{3}{8} - \frac{3}{4} = \frac{3}{8} - \frac{3}{4} \cdot \frac{2}{2}$$ The LCD is 8.

$$= \frac{3}{8} - \frac{6}{8}$$ Rename using the LCD.

$$= -\frac{3}{8}$$ Simplify.

b. $9\frac{3}{5} - 7\frac{2}{3}$

$$9\frac{3}{5} - 7\frac{2}{3} = \frac{48}{5} - \frac{23}{3}$$

$$= \frac{48}{5} \cdot \frac{3}{3} - \frac{23}{3} \cdot \frac{5}{5}$$ The LCD is 15.

$$= \frac{144}{15} - \frac{115}{15}$$ Rename using the LCD.

$$= \frac{29}{15} \text{ or } 1\frac{14}{15}$$ Simplify.

StudyTip

Reasonableness Use estimation to check whether your answer is reasonable.

$$\frac{3}{8} - \frac{3}{4} \approx \frac{1}{2} - 1$$

$$\approx -\frac{1}{2}$$

$-\frac{3}{8}$ is close to $-\frac{1}{2}$.

Check Your Progress

3A. $\frac{3}{4} - \frac{8}{9}$ **3B.** $7\frac{1}{6} - 6\frac{5}{8}$ **3C.** $5\frac{1}{3} - \left(-4\frac{5}{9}\right)$

Personal Tutor glencoe.com

Real-World Link

Computers can connect to networks without cables by using wireless cards. Wireless networks, or WiFi, use radio waves to transmit the information. A typical indoor range for a wireless connection is 150–300 feet from the access point.

Source: How Stuff Works

Real-World EXAMPLE 4 Add and Subtract Mixed Numbers

COMPUTERS **To set up a computer network in an office, a 100-foot cable is cut and used to connect three computers to the server as shown. How much cable is left to connect the third computer?**

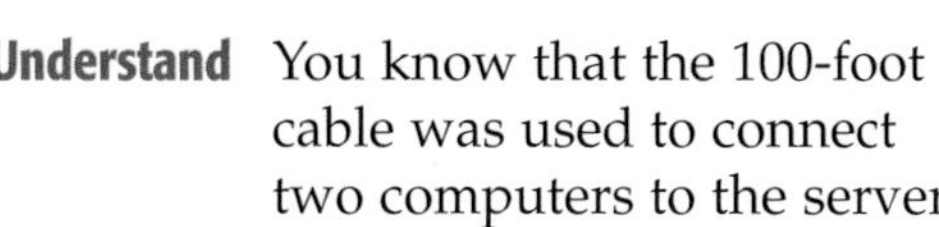

Understand You know that the 100-foot cable was used to connect two computers to the server.

Plan Add the measures of the cables that were already used and subtract that sum from 100.

Estimate $100 - (19 + 41) \approx 100 - 60$ or 40 feet

Solve

$19\frac{1}{8} + 40\frac{3}{4} = 19\frac{1}{8} + 40\frac{6}{8}$ — Rename $40\frac{3}{4}$ using the LCD, 8.

$= 59\frac{7}{8}$ — Simplify.

$100 - 59\frac{7}{8} = 99\frac{8}{8} - 59\frac{7}{8}$ — Rename 100 as $99\frac{8}{8}$.

$= 40\frac{1}{8}$ — Simplify.

There is $40\frac{1}{8}$ feet of cable left to connect the third computer.

Check Since $40\frac{1}{8}$ is close to 40, the answer is reasonable. ✓

Check Your Progress

4. **FROGS** At a recent frog jumping contest, the winning frog jumped $21\frac{1}{3}$ feet. The second place frog jumped $20\frac{1}{2}$ feet. How much farther did the first place frog jump?

Personal Tutor glencoe.com

Check Your Understanding

Examples 1 and 2 (pp. 153–154)

Find each sum. Write in simplest form.

1. $\frac{1}{15} + \frac{3}{5}$
2. $-\frac{5}{9} + \frac{1}{6}$
3. $\frac{7}{8} + \left(-\frac{2}{7}\right)$
4. $8\frac{5}{12} + 11\frac{1}{4}$
5. $-2\frac{1}{3} + \left(-7\frac{1}{2}\right)$
6. $4\frac{3}{8} + 10\frac{5}{12}$

Example 3 (p. 154)

Find each difference. Write in simplest form.

7. $-\frac{1}{4} - \frac{7}{9}$
8. $\frac{3}{5} - \frac{9}{10}$
9. $\frac{5}{8} - \frac{7}{12}$
10. $-1\frac{1}{3} - 4\frac{2}{7}$
11. $5\frac{5}{6} - \left(-2\frac{1}{4}\right)$
12. $12\frac{1}{2} - 6\frac{3}{8}$

Example 4 (p. 155)

13. **COOKING** Dwayne needs $1\frac{2}{3}$ cups of shredded cheese to put in his enchilada casserole and $\frac{3}{4}$ cup of cheese for the top. If he has 3 cups of cheese in all, how much cheese will he have left?

Practice and Problem Solving

● = **Step-by-Step Solutions** begin on page R11. **Extra Practice** begins on page 810.

Examples 1 and 3 pp. 153–154

Find each sum or difference. Write in simplest form.

14. $-\frac{8}{35}+\frac{2}{5}$ **15.** $\frac{5}{7}+\left(-\frac{10}{21}\right)$ **16.** $-\frac{5}{8}+\left(-\frac{8}{9}\right)$

17. $-\frac{1}{3}+\left(-\frac{10}{11}\right)$ **18.** $\frac{7}{8}-\frac{2}{5}$ **19.** $\frac{1}{6}-\frac{5}{7}$

20. $-\frac{2}{5}-\frac{1}{3}$ **21.** $-\frac{3}{10}-\frac{1}{15}$ **22.** $-\frac{3}{4}-\frac{5}{6}$

23 **BASEBALL** During spring training, the Detroit Tigers won about $\frac{2}{3}$ of the games they played while the Cleveland Indians won $\frac{7}{15}$ of the games they played. What fraction more of the games did Detroit win than Cleveland?

24. **COLLEGE** In a college dormitory, $\frac{1}{10}$ of the residents are juniors and $\frac{2}{5}$ of the residents are sophomores. What fraction of the students at the dormitory are juniors and sophomores?

Examples 2 and 3 p. 154

Find each sum or difference. Write in simplest form.

25. $8\frac{1}{2}+3\frac{4}{5}$ **26.** $-10\frac{2}{3}+9\frac{7}{12}$ **27.** $16\frac{5}{6}-12\frac{1}{3}$

28. $6\frac{6}{7}-11\frac{7}{8}$ **29.** $-4\frac{1}{9}-7\frac{2}{3}$ **30.** $-\frac{5}{6}+8\frac{1}{4}$

31. $\frac{9}{16}+3\frac{5}{6}$ **32.** $-5\frac{3}{5}+\left(-7\frac{1}{6}\right)$ **33.** $-10\frac{1}{2}-6\frac{5}{7}$

Example 4 p. 155

34. **JEWELRY** Sybrina wants to make a 17-inch necklace with a $\frac{3}{4}$-inch bead, a $1\frac{1}{2}$-inch bead and another $\frac{3}{4}$-inch bead on it. What is the length of the remaining part of the necklace?

35. **BAKING** Kenzie is making three desserts for a party. The recipes call for $\frac{2}{3}$ cup of sugar, $1\frac{5}{6}$ cups of sugar, and $2\frac{3}{4}$ cups of sugar. If she has 6 cups of sugar, how much sugar will she have left over?

36. **YEARBOOKS** The length of a page in a yearbook is 10 inches. The top margin is $\frac{1}{2}$ inch, and the bottom margin is $\frac{3}{4}$ inch. What is the length of the page inside the margins?

37. **MULTIPLE REPRESENTATIONS** The perimeter of a geometric figure is the distance around the figure. You can find the perimeter of a rectangle by adding the measures of all four sides.

a. **TABULAR** Copy and complete the table at the right by listing the lengths and widths of three additional rectangles that have a perimeter of 20.

Perimeter of 20	
Length	Width
8	2
■	■
■	■
■	■

b. **GRAPHICAL** Write the values from the table as ordered pairs (ℓ, w). Graph the ordered pairs on the coordinate plane.

c. **NUMERICAL** Use the graph to predict the length of a rectangle with a perimeter of 20 inches and a width of $5\frac{1}{2}$ inches. Check the prediction by finding the actual length of the rectangle.

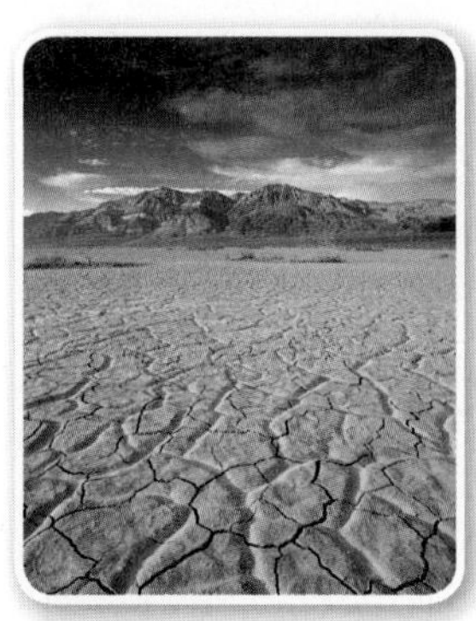

Real-World Link

A shortage of rainfall can cause major problems in forestry, agriculture, tourism, and city water supplies. The U.S. Federal Emergency Management Agency estimates that droughts cost the country $6–$8 billion each year.

38. ANALYZE TABLES Use the table to find the total average precipitation that falls in August, September, and October.

Average Precipitation	
Month	**Amount (in.)**
Aug.	$2\frac{47}{50}$
Sept.	$1\frac{22}{25}$
Oct.	$1\frac{1}{2}$

39. RESEARCH Use the Internet or another source to find out the monthly rainfall totals in your community during the past year. How much rain fell in August, September, and October?

Find each difference. Write in simplest form.

40. $-3\frac{2}{5} - (-2\frac{4}{7})$ **41.** $-19\frac{3}{8} - (-4\frac{3}{4})$ **42.** $8\frac{5}{12} - (-12\frac{13}{18})$

43. $-35\frac{5}{6} - 23.\overline{3}$ **44.** $-17\frac{7}{8} - (-17.\overline{9})$ **45.** $24.\overline{56} - (-12.\overline{1})$

46. GEOMETRY The length of a rectangle is $3\frac{1}{3}$ inches. The width is $\frac{1}{5}$ of the length. Find the width and the perimeter of the rectangle. Support your answer with a drawing.

H.O.T. Problems Use Higher-Order Thinking Skills

47. OPEN ENDED Write a subtraction problem using unlike fractions with a least common denominator of 24. Find the difference.

48. REASONING Is the difference between a positive mixed number and a negative mixed number *always*, *sometimes*, or *never* positive? Justify your answer with an example.

49. WRITING IN MATH Explain why you cannot add or subtract fractions with unlike denominators without renaming the fractions. You may use a diagram to illustrate your answer.

50. FIND THE ERROR Cooper and Yu are adding the fractions $\frac{1}{3}$, $\frac{7}{9}$, and $\frac{4}{15}$. Their first step is to find the least common denominator of 3, 9, and 15. Is either of them correct? Explain your reasoning.

Cooper	Yu
The least common denominator of 3, 9, and 15 is 3 because 3 divides into all these numbers evenly.	*The least common denominator of 5, 9, and 15 is 90 because you can divide 90 by all of those numbers without getting a remainder.*

51. CHALLENGE A set of measuring cups has measures of 1 cup, $\frac{3}{4}$ cup, $\frac{1}{2}$ cup, $\frac{1}{3}$ cup, and $\frac{1}{4}$ cup. How could you measure $\frac{1}{6}$ cup of milk by using these measuring cups?

52. WRITING IN MATH Suppose you use 24 instead of 12 as a common denominator when finding $2\frac{3}{4} - 5\frac{5}{6}$. Will you get the correct answer? Explain.

Standardized Test Practice

53. A recipe for snack mix contains $2\frac{1}{3}$ cups of mixed nuts, $3\frac{1}{2}$ cups of granola, and $\frac{3}{4}$ cup raisins. What is the total amount of snack mix?

A $5\frac{2}{3}$ c

B $5\frac{7}{12}$ c

C $6\frac{2}{3}$ c

D $6\frac{7}{12}$ c

54. SHORT RESPONSE The graph shows the results of an election for class president. What fraction of the votes did Michaela receive?

Class Election Results

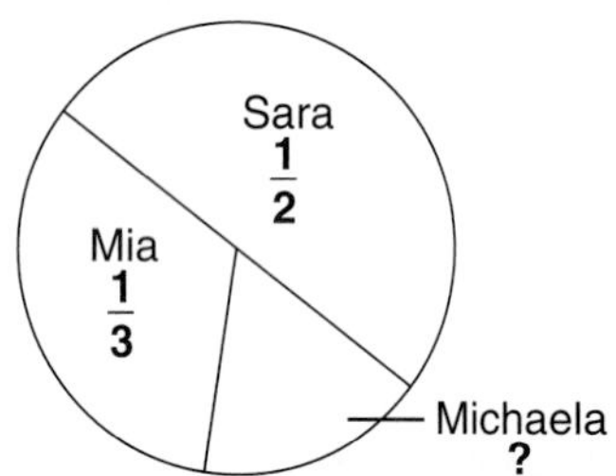

55. The results of a grocery store survey are listed in the table. Find the fraction of families who grill out 2, 3, or 4 or more times per month.

How Often Do You Grill Out?

Times per Month	Fraction of People
Less than 1	$\frac{11}{50}$
1	$\frac{2}{25}$
2–3	$\frac{4}{25}$
4 or more	$\frac{27}{50}$

F $\frac{2}{25}$

G $\frac{23}{100}$

H $\frac{7}{10}$

J $\frac{39}{50}$

56. Marco spent $\frac{4}{5}$ hour doing homework on Monday. On Tuesday, he spent $1\frac{1}{3}$ hours doing homework. How long did he spend working on homework for those two days?

A $\frac{8}{15}$ h

B $1\frac{5}{8}$ h

C $1\frac{16}{15}$ h

D $2\frac{2}{15}$ h

Spiral Review

Find each sum or difference. Write in simplest form. (Lesson 3-5)

57. $6\frac{1}{12} - \left(-8\frac{5}{12}\right)$

58. $2\frac{5}{12} + \left(2\frac{7}{12}\right)$

59. $2\frac{3}{8} - 1\frac{5}{8}$

60. CARPENTRY A 3-foot-long shelf is to be installed between two walls that are $32\frac{5}{8}$ inches apart. How much of the shelf must be cut off so that it fits between the walls? (Lesson 3-5)

Find each quotient. Write in simplest form. (Lesson 3-4)

61. $\frac{8}{9} \div \frac{4}{3}$

62. $12 \div \frac{4}{9}$

63. $\frac{a}{7} \div \frac{a}{42}$

64. FOOD How many $\frac{1}{4}$-pound hamburgers can be made from $2\frac{3}{4}$ pounds of ground beef? (Lesson 3-4)

65. MEASUREMENT A *micron* is a unit of measure that is approximately 0.000039 inch. Express this decimal as a fraction. (Lesson 3-2)

Skills Review

Multiply. (Previous Course)

66. $2 \cdot (5 \cdot 10)$

67. $(1 \cdot 5) \cdot 10$

68. $(15 + 7) \cdot 11$

CHAPTER 3

Study Guide and Review

Math Online glencoe.com

- STUDY TO GO
- Vocabulary Review

Chapter Summary

Key Concepts

Comparing Rational Numbers (Lesson 3-1)

- To compare fractions with unlike denominators, write the fractions as decimals and compare.

Fractions and Decimals (Lessons 3-1 and 3-2)

- Any number that can be written as a fraction is a rational number.
- Decimals that are terminating or repeating are rational numbers.

Multiplying and Dividing Rational Numbers (Lessons 3-3 and 3-4)

- To multiply fractions, multiply the numerators and then multiply the denominators.
- The product of a number and its multiplicative inverse, or reciprocal, is 1.
- To divide by a fraction, multiply by its multiplicative inverse.

Adding and Subtracting Rational Numbers (Lessons 3-5 and 3-6)

- To add like fractions, add the numerators and write the sum over the denominator.
- To subtract like fractions, subtract the numerators and write the difference over the denominator.
- To add or subtract fractions with unlike denominators, rename the fractions with the LCD. Then add or subtract.

FOLDABLES® Study Organizer

Be sure the Key Concepts are noted in your Foldable.

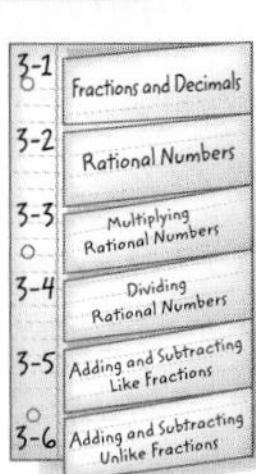

Key Vocabulary

bar notation (p. 122)
like fractions (p. 147)
multiplicative inverse (p. 141)
rational number (p. 128)
reciprocal (p. 141)
repeating decimal (p. 122)
terminating decimal (p. 121)
unlike fractions (p. 153)

Vocabulary Check

State whether each sentence is *true* or *false*. If *false*, replace the underlined term to make a true sentence.

1. Numbers that can be written as fractions are called reciprocals.
2. The decimal 4.7 is a terminating decimal.
3. The fractions $\frac{4}{6}$ and $\frac{1}{3}$ are like fractions.
4. To add unlike fractions, rename the fractions using the GCF.
5. The product of a number and its multiplicative inverse is 1.
6. A mixed number is another name for the multiplicative inverse.
7. Like fractions are fractions that have the same numerator.
8. Terminating decimals use bar notation to show which digits terminate.
9. You need a common denominator to divide fractions.
10. Decimals that repeat or terminate are rational numbers.

CHAPTER 3 Study Guide and Review

Lesson-by-Lesson Review

3-1 Writing Fractions as Decimals (pp. 121–127)

Write each fraction or mixed number as a decimal. Use a bar to show a repeating decimal.

11. $\frac{3}{10}$
12. $\frac{2}{5}$
13. $-\frac{5}{6}$
14. $-7\frac{4}{9}$
15. $\frac{5}{8}$
16. $1\frac{4}{15}$

Replace each ● with <, >, or = to make a true sentence.

17. $2\frac{1}{2}$ ● $2\frac{5}{12}$
18. $\frac{5}{8}$ ● 0.625
19. 10.74 ● $10\frac{7}{10}$
20. $-\frac{5}{6}$ ● −0.83
21. $4.\overline{37}$ ● $4\frac{19}{50}$
22. −2.54 ● $-2\frac{27}{50}$

23. **CARPENTRY** Antoine is cutting a $5\frac{5}{16}$-inch board for a project. Write $5\frac{5}{16}$ as a decimal.

EXAMPLE 1

Write $\frac{3}{4}$ as a decimal.

$$\begin{array}{r} 0.75 \\ 4\overline{)3.00} \\ \underline{2\,8} \\ 20 \\ \underline{-20} \\ 0 \end{array}$$

Divide 3 by 4.

Divide until the remainder is zero or until a sequence of numbers repeats.

EXAMPLE 2

Replace the ● with <, >, or = to make $\frac{4}{5}$ ● 0.75 a true sentence.

$\frac{4}{5}$ ● 0.75 — **Write the sentence.**

0.8 ● 0.75 — **Write $\frac{4}{5}$ as a decimal.**

$0.8 > 0.75$ — **In the tenths place, $8 > 7$.**

3-2 Rational Numbers (pp. 128–133)

Write each decimal as a fraction or mixed number in simplest form.

24. 2.08
25. −0.45
26. 0.875
27. −0.56
28. $0.\overline{1}$
29. $-2.\overline{03}$
30. $0.\overline{5}$
31. $10.\overline{27}$

Identify all sets to which each number belongs.

32. −4
33. $3\frac{1}{3}$
34. 1.151551555…
35. $-0.\overline{67}$

36. **MUSIC** Suzanne practiced playing the piano for $1.\overline{6}$ hours after school. Write $1.\overline{6}$ as a mixed number.

EXAMPLE 3

Write 1.25 as a fraction in simplest form.

$1.25 = 1\frac{25}{100}$ — **1.25 is *1 and 25 hundredths*.**

$= 1\frac{1}{4}$ — **Simplify. The GCF of 25 and 100 is 25.**

EXAMPLE 4

Write $0.\overline{7}$ as a fraction in simplest form.

$N = 0.777\ldots$

$10N = 10(0.777\ldots)$ — **Multiply each side by 10.**

$10N = 7.777\ldots$

$-N = 0.777\ldots$ — **Subtract N from $10N$.**

$9N = 7$ — **Simplify.**

$N = \frac{7}{9}$ — **Divide each side by 9.**

MIXED PROBLEM SOLVING
For mixed problem-solving practice, see page 845.

3-3 Multiplying Rational Numbers (pp. 134–139)

Find each product. Write in simplest form.

37. $\frac{1}{5} \cdot \frac{3}{4}$

38. $-\frac{3}{7} \cdot \frac{4}{9}$

39. $-\frac{2}{3} \cdot (-5)$

40. $-3\frac{1}{2} \cdot \left(-5\frac{1}{5}\right)$

41. CRAFTS Mireille has a piece of ribbon that is 10 inches long. Abi's ribbon is $\frac{5}{8}$ as long. How long is Abi's ribbon?

42. BACKPACKING A liter of water weighs approximately $2\frac{1}{5}$ pounds. While backpacking, Enrique wants to carry $3\frac{1}{2}$ liters of water with him. Find the weight of the water that Enrique is taking with him.

EXAMPLE 5

Find $\frac{3}{8} \cdot \frac{20}{27}$. Write in simplest form.

$\frac{3}{8} \cdot \frac{20}{27} = \frac{3 \cdot 20}{8 \cdot 27}$ — Multiply the numerators. Multiply the denominators.

$= \frac{60}{216}$ or $\frac{5}{18}$ — Simplify. The GCF of 60 and 216 is 12.

EXAMPLE 6

Find $-4\frac{1}{6} \cdot \frac{3}{5}$. Write in simplest form.

$-4\frac{1}{6} \cdot \frac{3}{5} = -\frac{25}{6} \cdot \frac{3}{5}$ — Rename $-4\frac{1}{6}$ as an improper fraction.

$= -\frac{\overset{5}{\cancel{25}}}{\underset{2}{\cancel{6}}} \cdot \frac{\overset{1}{\cancel{3}}}{\underset{1}{\cancel{5}}}$ — Divide by the GCFs, 5 and 3.

$= -\frac{5}{2}$ or $-2\frac{1}{2}$ — Multiply. Then simplify.

3-4 Dividing Rational Numbers (pp. 141–146)

Find the multiplicative inverse of each number.

43. -16

44. $\frac{7}{9}$

45. $3\frac{4}{5}$

46. $-4\frac{1}{3}$

Find each quotient. Write in simplest form.

47. $\frac{7}{9} \div \left(-\frac{4}{15}\right)$

48. $-2\frac{2}{3} \div 2\frac{2}{7}$

49. $\frac{3}{5} \div \frac{9}{10}$

50. $3\frac{1}{9} \div \left(-1\frac{1}{6}\right)$

51. FOOD Pilar drinks $1\frac{3}{4}$ glasses of milk each day. At this rate, how many days will it take her to drink a total of 14 glasses?

EXAMPLE 7

Find the multiplicative inverse of $2\frac{3}{4}$.

$2\frac{3}{4} = \frac{11}{4}$ — Rename $2\frac{3}{4}$ as an improper fraction.

$\frac{11}{4} \cdot \frac{4}{11} = 1$ — The product is 1.

The multiplicative inverse of $2\frac{3}{4}$ is $\frac{4}{11}$.

EXAMPLE 8

Find $\frac{4}{9} \div \frac{2}{15}$. Write in simplest form.

$\frac{4}{9} \div \frac{2}{15} = \frac{4}{9} \cdot \frac{15}{2}$ — Multiply by the reciprocal of $\frac{2}{15}, \frac{15}{2}$.

$= \frac{\overset{2}{\cancel{4}}}{\underset{3}{\cancel{9}}} \cdot \frac{\overset{5}{\cancel{15}}}{\underset{1}{\cancel{2}}}$ — Divide out common factors.

$= \frac{10}{3}$ or $3\frac{1}{3}$ — Simplify.

CHAPTER 3

Study Guide and Review

3-5 Adding and Subtracting Like Fractions (pp. 147–152)

Find each sum or difference. Write in simplest form.

52. $\frac{8}{15} + \left(-\frac{2}{15}\right)$

53. $\frac{6}{12} - \frac{11}{12}$

54. $2\frac{5}{12} - \left(-8\frac{7}{12}\right)$

55. $5\frac{3}{7} + 2\frac{6}{7}$

56. **EXERCISE** Samantha is going to walk $3\frac{5}{16}$ miles today and $2\frac{3}{16}$ miles tomorrow. What is the total distance she will walk?

57. **PETS** Last week, Douglas fed his puppy $10\frac{1}{4}$ cups of food. This week the puppy will be fed an additional $1\frac{1}{4}$ cups of food. Find the total amount of food the puppy will be fed this week.

EXAMPLE 9

Find $\frac{3}{4} - \left(-\frac{3}{4}\right)$. Write in simplest form.

$\frac{3}{4} - \left(-\frac{3}{4}\right) = \frac{3}{4} + \frac{3}{4}$ — **To subtract $-\frac{3}{4}$, add $\frac{3}{4}$.**

$= \frac{3+3}{4}$ — **The denominators are the same. Add the numerators.**

$= \frac{6}{4}$ — **Simplify.**

$= 1\frac{1}{2}$ — **Simplify.**

3-6 Adding and Subtracting Unlike Fractions (pp. 153–158)

Find each sum or difference. Write in simplest form.

58. $\frac{2}{5} + \frac{1}{15}$

59. $-3\frac{5}{6} - 2\frac{1}{2}$

60. $\frac{4}{7} + \left(-1\frac{1}{3}\right)$

61. $\frac{3}{10} - \left(-\frac{1}{8}\right)$

62. $25\frac{1}{3} - 14\frac{2}{5}$

63. $7\frac{3}{4} + 1\frac{3}{8}$

64. $-\frac{5}{9} - 3\frac{2}{3}$

65. $-4\frac{1}{6} + \frac{3}{4}$

66. **COOKING** Monica needs $2\frac{3}{4}$ cups of flour for a batch of cookies and $3\frac{1}{3}$ cups of flour for a dozen muffins. How many cups of flour does Monica need altogether?

67. **TRAVEL** Dane and his family drove 357.9 miles in one day. If their trip is a total of $524\frac{3}{4}$ miles, how much farther do they need to drive?

EXAMPLE 10

Find $-\frac{3}{8} + \frac{5}{6}$. Write in simplest form.

$-\frac{3}{8} + \frac{5}{6} = -\frac{3}{8} \cdot \frac{3}{3} + \frac{5}{6} \cdot \frac{4}{4}$ — **The LCD is 24. Rename the fractions using the LCD.**

$= -\frac{9}{24} + \frac{20}{24}$ — **Simplify.**

$= \frac{-9+20}{24}$ — **Add the numerators.**

$= \frac{11}{24}$ — **Simplify.**

EXAMPLE 11

Find $6\frac{5}{9} - 4\frac{11}{12}$. Write in simplest form.

$6\frac{5}{9} - 4\frac{11}{12} = 6\frac{20}{36} - 4\frac{33}{36}$ — **The LCD is 36. Rename the fractions using the LCD.**

$= 5\frac{56}{36} - 4\frac{33}{36}$ — **Since $\frac{20}{36}$ is less than $\frac{33}{36}$, rename $6\frac{20}{36}$.**

$= 1\frac{23}{36}$ — **Subtract the whole numbers and then the fractions.**

CHAPTER 3

Practice Test

Math Online glencoe.com
Chapter Test

Write each fraction as a decimal. Use a bar to show a repeating decimal.

1. $\frac{3}{9}$
2. $-\frac{3}{25}$
3. $\frac{1}{8}$
4. $\frac{2}{7}$

Write each decimal as a fraction or mixed number in simplest form.

5. 0.38
6. 10.17
7. $-5.\overline{5}$
8. -1.44

Replace each ● with <, >, or = to make a true sentence.

9. $\frac{3}{10}$ ● $0.\overline{3}$
10. -0.58 ● $-\frac{1}{4}$
11. $-6\frac{7}{8}$ ● -6.8
12. $1\frac{2}{3}$ ● $1\frac{5}{7}$

13. **FOOTBALL** The table shows the average number of points scored per game for four NFL teams. Denver scored $\frac{3}{4}$ as many points as Dallas scored. How many average points per game did Denver score?

Team	Average Points per Game
San Diego	$30\frac{4}{5}$
Indianapolis	$26\frac{7}{10}$
Chicago	$26\frac{7}{10}$
Dallas	$26\frac{3}{5}$

Find each product or quotient. Write in simplest form.

14. $-\frac{2}{9} \cdot \frac{3}{14}$
15. $4\frac{4}{7} \cdot 9\frac{1}{3}$
16. $\frac{5}{6} \div \left(-\frac{7}{18}\right)$
17. $-8\frac{4}{9} \div 2\frac{1}{9}$

18. **MONEY** A dollar bill remains in circulation about $1\frac{1}{4}$ years. A coin lasts about $22\frac{1}{2}$ times longer. How long is a coin in circulation?

19. **FOOD** If each guest at a party eats two-thirds of a small pizza, how many guests would finish 12 small pizzas?

Find each sum or difference. Write in simplest form.

20. $-\frac{3}{7} + \frac{5}{14}$
21. $1\frac{1}{5} - \left(-\frac{7}{15}\right)$
22. $6\frac{3}{4} - 2\frac{1}{6}$
23. $9\frac{1}{3} + \left(-7\frac{5}{6}\right)$

24. **MULTIPLE CHOICE** Use the table to find the fraction of people who voted for Collins, Johnson, and Shaw in the election for freshman class president.

Candidate	Fraction of People
Collins	$\frac{3}{20}$
Johnson	$\frac{21}{50}$
Juarez	$\frac{3}{10}$
Shaw	$\frac{3}{100}$

A $\frac{3}{10}$
C $\frac{3}{5}$
B $\frac{2}{5}$
D $\frac{7}{10}$

Evaluate each expression if $p = \frac{1}{3}$, $r = \frac{5}{8}$, $a = 2\frac{1}{2}$, and $c = 6$.

25. pr
26. $r \div a$
27. $c - p$
28. $ac + p$

29. **MULTIPLE CHOICE** A pipe that is $12\frac{3}{4}$ feet long is cut into pieces that are each $2\frac{2}{3}$ feet long. Which step below would give the number of pieces into which the pipe is cut?

F Subtract $2\frac{2}{3}$ from $12\frac{3}{4}$.

G Divide $12\frac{3}{4}$ by $2\frac{2}{3}$.

H Multiply $12\frac{3}{4}$ by $2\frac{2}{3}$.

J Add $2\frac{2}{3}$ to $12\frac{3}{4}$.

30. **COOKING** Reena is making $2\frac{1}{2}$ times a recipe. If the recipe calls for $1\frac{2}{3}$ cups of milk, how much milk will Reena need?

CHAPTER 3

Preparing for Standardized Tests

Short-Response Questions

Short-response questions require you to provide a solution to the problem, as well as any method, explanation, and/or justification you used to arrive at the solution.

These questions are sometimes called *constructed-response, open-response, open-ended, free-response,* or *student-produced* questions.

The following is sample rubric, or scoring guide.

Credit	Score	Criteria
Full	2	The answer is correct and a full explanation is provided that shows each step in arriving at the final answer.
Partial	1	There are two different ways to receive partial credit. • The answer is correct, but the explanation is incomplete. • The answer is incorrect, but the explanation is correct.
None	0	Either an answer is not provided or the answer does not make sense.

On some standardized tests, no credit is given for a correct answer if your work is not shown.

In solving short-response questions, remember to...

- explain your reasoning or state your approach to solving the problem.
- show all of your work or steps.
- check your answer if time permits.

EXAMPLE

Read the problem. Identify what you need to know. Then use the information in the problem to solve. Show your work.

The table shows the number of at-bats and hits of three teammates. Which player had the greatest fraction of at-bats that were hits? Show your work.

Softball Statistics

Player	At-Bats	Hits
Umeko	84	35
Melanie	75	30
Olivia	64	28

Read the problem statement carefully. You are given the number of at-bats for 3 players and the number of hits they each had. You need to find the player with the greatest fraction of at-bats that were hits.

Full Credit Solution

In this sample solution, the student gave a clear explanation of the process, showed work, and arrived at the correct answer.

In order for me to compare the fractions, I need to write them in the same form. I'll use a calculator to divide the number of hits by the number of at-bats for each player to write the fractions as decimals.

Umeko	Melanie	Olivia
35 hits out of 84 at-bats	30 hits out of 75 at-bats	28 hits out of 64 at-bats
35 [÷] 84 [ENTER] .41666667	30 [÷] 75 [ENTER] 0.4	28 [÷] 64 [ENTER] .4375
35 out of 84 = $0.41\overline{6}$	30 out of 75 = 0.4	28 out of 84 = 0.4375

The steps, calculations, and reasoning are clearly stated.

Since $0.4 < 0.41\overline{6} < 0.4375$, Olivia had the greatest fraction of at-bats that were hits.

Be sure to complete this final step to answer the question asked.

Partial Credit Solution

In this sample solution, there are no explanations for finding the calculations.

Umeko	Melanie	Olivia
35 hits out of 84 at-bats = $0.41\overline{6}$	30 hits out of 75 at-bats = 0.4	28 hits out of 64 at-bats = 0.4375

Since $0.4 < 0.41\overline{6} < 0.4375$, Olivia had the greatest fraction of at-bats that were hits.

No Credit Solution

In this sample solution, there are no explanations for finding the calculations, and the calculations are wrong.

Umeko	Melanie	Olivia
35 hits out of 84 at-bats = 2.4	30 hits out of 75 at-bats = 2.5	28 hits out of 64 at-bats = 2.2

Melanie had the greatest fraction of at-bats that were hits.

Exercises

Read each problem. Identify what you need to know. Then use the information in the problem to solve. Show your work.

1. A customer service department satisfactorily resolved 55 out of 60 customer complaints over the weekend. To the nearest thousandth, find the satisfaction rate of the customer service department.

2. A zebra's top running speed is $\frac{5}{4}$ the top running speed of a giraffe. If a zebra can run up to 40 miles per hour, how fast can a giraffe run?

3. Find the sum of the expression shown below.

$$-3\frac{2}{7} + 8\frac{3}{4}$$

CHAPTER 3

Standardized Test Practice

Cumulative, Chapters 1 through 3

Multiple Choice

Read each question. Then fill in the correct answer on the answer document provided by your teacher or on a sheet of paper.

1. The table shows the number of field goal attempts and successes for four place kickers this season.

Kicker	Attempts	Success
Roland	39	30
Michael	45	36
Cameron	32	25
Jorge	48	41

Which kicker had the highest success rate?

A Cameron **C** Michael

B Jorge **D** Roland

2. The portion of fish caught at East Fork Lake yesterday that were large mouth bass was 0.62. Which of the following represents the number of large mouth bass caught as a fraction?

F $\frac{15}{26}$ **H** $\frac{31}{50}$

G $\frac{29}{50}$ **J** $\frac{8}{25}$

3. Rachael's new printer can print $1\frac{2}{3}$ times as many pages per minute as her old printer. Her old printer could print 9 pages per minute. How many pages per minute does Rachael's new printer print?

A 12 pages per minute

B 13 pages per minute

C 14 pages per minute

D 15 pages per minute

4. What are the coordinates of point A (4, −1) after it has been reflected over the y-axis?

F (−4, −1) **H** (4, −1)

G (−4, 1) **J** (4, 1)

5. What are the coordinates of point Z on the coordinate plane?

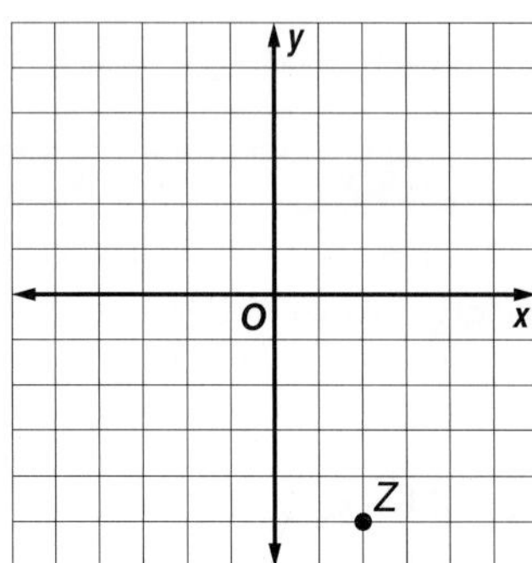

A (−5, 2) **C** (−2, 5)

B (5, −2) **D** (2, −5)

6. Which property is illustrated in the statement shown below?

$$23 \cdot 1 = 23$$

F Commutative Property

G Associative Property

H Identity Property

J Zero Property

7. Find the sum of the expression shown below.

$$-\frac{8}{15} + \frac{12}{15}$$

A $\frac{3}{15}$

B $\frac{4}{15}$

C $-\frac{5}{15}$

D $-\frac{4}{15}$

Test-TakingTip

Question 7 The fractions have the same denominator, so the sign of the answer will match the numerator with the greater absolute value.

Short Response/Gridded Response

Record your answers on the answer sheet provided by your teacher or on a sheet of paper.

8. Use the number line below to answer each of the following.

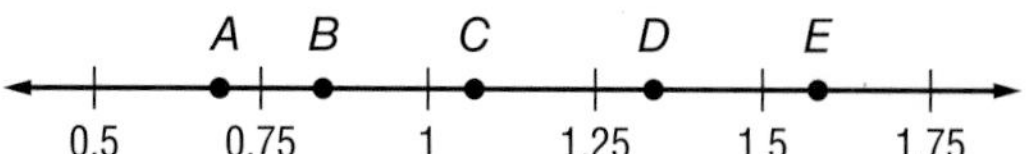

a. Write a fraction or mixed number that could represent each point on the graph.

b. Write an inequality comparing two of the points.

9. **GRIDDED RESPONSE** Mario made 27 out of 40 penalty kicks last season. Write the fraction of penalty kicks that he did not make as a decimal.

10. Jamie claims that in order for the sum of two integers to be positive, both of the addends must be positive. Do you agree or disagree with this claim? Justify your answer.

11. **GRIDDED RESPONSE** Nicole drove 156 miles to her grandmother's house last week for a family reunion. Her sister drove $\frac{5}{4}$ of this distance to get to the reunion. How many more miles did Nicole's sister drive than Nicole?

12. There are 60 minutes in 1 hour.

a. Write an equation to find the number of hours h in any number of minutes m.

b. How many hours are in 210 minutes?

13. The area model shows the product of two rational numbers. Write an expression that represents the model. Then state the product.

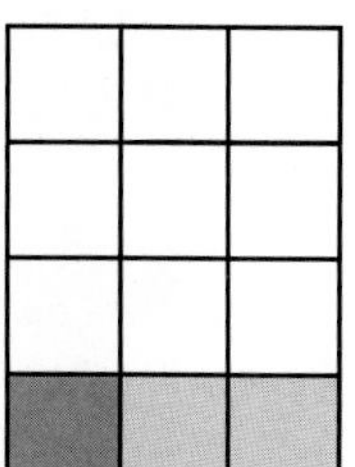

Extended Response

Record your answers on a sheet of paper. Show your work.

14. Mr. Lombardo has five wrenches in his toolbox that are labeled $\frac{3}{8}$ in., $\frac{1}{4}$ in., $\frac{5}{16}$ in., $\frac{7}{8}$ in., and $\frac{9}{32}$ in.

a. Write each fraction as a decimal. Use a bar to show a repeating decimal.

b. Order the sizes of the wrenches from smallest to largest.

c. How else could you order the wrenches from smallest to largest? Explain your reasoning.

Need Extra Help?														
If you missed Question...	1	2	3	4	5	6	7	8	9	10	11	12	13	14
Go to Lesson or Page...	3-1	3-2	3-3	2-7	1-4	1-3	3-5	3-1	3-2	2-2	3-3	1-5	3-3	3-1

CHAPTER 4 Expressions and Equations

Then

In Chapters 2 and 3, you worked with integers and rational numbers.

Now

In Chapter 4, you will:

- Use the Distributive Property.
- Solve equations by using properties of equality.
- Write equations to solve problems.

Why?

ENTERTAINMENT
The entertainment field uses equations to balance the costs and profits of a project. These situations can be represented using variables, coefficients, and constants. The variable represents something that changes, such as the cost of costumes.

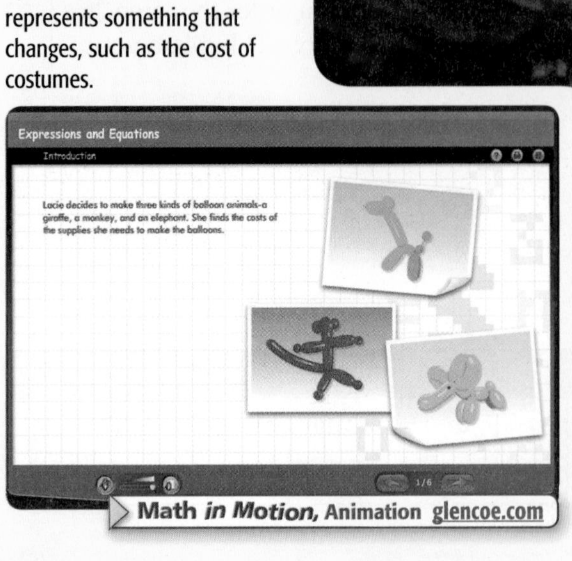

Expressions and Equations
Introduction
Lacie decides to make three kinds of balloon animals–a giraffe, a monkey, and an elephant. She finds the costs of the supplies she needs to make the balloons.
1/6

Math *in Motion*, Animation glencoe.com

Get Ready for Chapter 4

Diagnose Readiness You have two options for checking Prerequisite Skills.

Text Option Take the Quick Check below. Refer to the Quick Review for help.

QuickCheck

Find each product. (Lesson 2-4)

1. $3(-3)$
2. $-4(2)$
3. $-7(-4)$
4. $-4 \cdot 5$
5. $-11(-8)$
6. $9(-4)$

7. **STOCK MARKET** The price of a stock decreased \$2.05 each day for five consecutive days. What was the total change in value of the stock over the five day period? (Lesson 2-4)

Write each subtraction expression as an addition expression. (Lesson 2-3)

8. $5 - 9$
9. $4 - 10$
10. $-11 - 9$
11. $-19 - 10$

12. **MONEY** Student Council spent \$178 on decorations and \$110 on snacks for the dance. Write an addition expression for the amount remaining in the dance budget if Student Council initially had \$593.

Find each sum. (Lesson 2-2)

13. $6 + (-7)$
14. $-8 + 6$
15. $3 + (-3)$
16. $4 + (-10)$
17. $-13 + (-8)$
18. $-11 + 12$

19. **CAVERNS** A tour group began 26 feet underground. During their tour, they went down 15 feet more and then went up 19 feet. Express their current depth as an integer.

QuickReview

EXAMPLE 1

Find 7(−2).

$7(-2) = -14$ The factors have different signs, so the product is negative.

EXAMPLE 2

Find −5(−9).

$-5(-9) = 45$ The factors have the same sign, so the product is positive.

EXAMPLE 3

Write 8 − 12 as an addition expression.

$8 - 12 = 8 + (-12)$ To subtract 12, add −12.

EXAMPLE 4

Find −5 + 7.

$-5 + 7 = 2$ Subtract $|-5|$ from $|7|$. The sum is positive because $|7| > |-5|$.

Online Option Math Online Take a self-check Chapter Readiness Quiz at glencoe.com.

Get Started on Chapter 4

You will learn several new concepts, skills, and vocabulary terms as you study Chapter 4. To get ready, identify important terms and organize your resources. You may wish to refer to **Chapter 0** to review prerequisite skills.

FOLDABLES® Study Organizer

Solving Equations Make this Foldable to help you organize information about expressions and equations. Begin with a plain sheet of $8\frac{1}{2}'' \times 11''$ paper.

1. **Fold** in half lengthwise.

2. **Fold** in thirds and then fold each third in half.

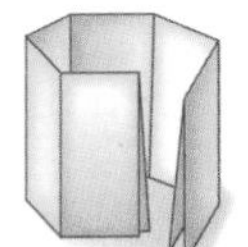

3. **Open.** Cut one side along the folds to make tabs.

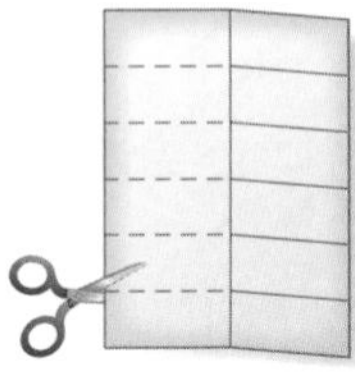

4. **Label** each tab with a lesson number as shown.

4-1
4-2
4-3
4-4
4-5
4-6

Math Online glencoe.com

- Study the chapter online
- Explore **Math in Motion**
- Get extra help from your own **Personal Tutor**
- Use **Extra Examples** for additional help
- Take a **Self-Check Quiz**
- **Review Vocabulary** in fun ways

New Vocabulary

English		Español
Distributive Property	p. 171	Propiedad distributiva
equivalent expressions	p. 171	expresiones equivalentes
coefficient	p. 178	coeficiente
constant	p. 178	constante
like terms	p. 178	terminos semejantes
term	p. 178	término
simplest form	p. 179	forma reducida
simplifying the expression	p. 179	reducir la expresión
equivalent equations	p. 184	ecuaciones equivalentes
inverse operations	p. 184	operaciones inversas
solution	p. 184	solución
solving the equation	p. 184	resolver la ecuación
two-step equation	p. 199	ecuación de dos pasos

Review Vocabulary

algebraic expression • p. 11 • expresión algebraica any combination of terms and operations

$5x + 3$ (Expression) $\quad 5x + 3 = 13$ (Equation)

equation • p. 34 • ecuación a mathematical sentence stating that two quantities are equal

Multilingual eGlossary glencoe.com

4-1

The Distributive Property

Then

You have already evaluated numerical expressions using addition and multiplication.

Now

- Use the Distributive Property to write equivalent numerical expressions.
- Use the Distributive Property to write equivalent algebraic expressions.

New Vocabulary

equivalent expressions
Distributive Property

Math Online

glencoe.com

- Extra Examples
- Personal Tutor
- Self-Check Quiz
- Homework Help

Why?

Lita's mother is paying for her and two friends to go to the movies. She will buy each of them a snack and a drink.

a. Find the total amount that Lita's mother will need to pay.

b. Describe the method you used to find the total cost.

Here are two ways to find the total cost.

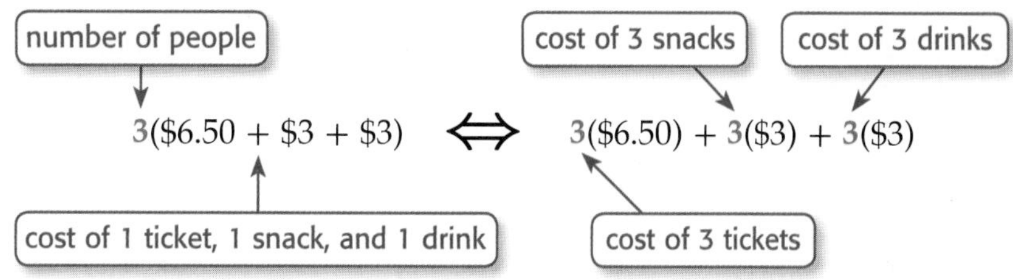

Numerical Expressions The expressions above are **equivalent expressions** because they have the same value, $37.50. This example also shows the **Distributive Property**.

Key Concept Distributive Property — For Your FOLDABLE

Words To multiply a sum or difference by a number, multiply each term inside the parentheses by the number outside the parentheses.

Symbols $a(b + c) = ab + ac$ $\quad a(b - c) = ab - ac$

Example $5(6 + 7) = 5 \cdot 6 + 5 \cdot 7$ $\quad (9 - 3)8 = 9 \cdot 8 - 3 \cdot 8$

Math *in Motion*, BrainPOP® glencoe.com

EXAMPLE 1 Evaluate Numerical Expressions

Use the Distributive Property to write each expression as an equivalent expression. Then evaluate the expression.

a. $5(12 + 4)$

$5(12 + 4) = 5 \cdot 12 + 5 \cdot 4$

$= 60 + 20$ **Multiply.**

$= 80$ **Add.**

b. $(20 - 3)8$

$(20 - 3)8 = 20 \cdot 8 - 3 \cdot 8$

$= 160 - 24$ **Multiply.**

$= 136$ **Subtract.**

Check Your Progress

1A. $(6 + 3)4$ **1B.** $4(9 - 2)$

Personal Tutor glencoe.com

The Distributive Property allows you to find some products mentally. For example, you can find $7 \cdot 34$ mentally by evaluating $7 \cdot (30 + 4)$.

$$7 \cdot (30 + 4) = 7 \cdot 30 + 7 \cdot 4$$
$$= 210 + 28 \quad \text{Think } 7 \cdot 30 = 210$$
$$= 238 \quad \text{Think } 210 + 28 = 238$$

Real-World EXAMPLE 2 Use the Distributive Property

FINANCIAL LITERACY On a school visit to Washington, D.C., Dichali and his class visited the Smithsonian Air and Space Museum. Tickets to the IMAX movie cost $8.99. Find the total cost for 20 students to see the IMAX movie.

Understand You know how many students will be attending the movie and how much the movie costs. You need to find the total cost for the group to see the IMAX movie.

Plan You can use the Distributive Property and mental math to find the total cost for the movie. To find the total cost mentally, find 20($9.00 – $0.01).

Solve 20($9.00 – $0.01) = 20($9.00) – 20($0.01) **Distributive Property**

= $180 – $0.20 **Multiply.**

= $179.80 **Subtract.**

The total cost is $179.80.

Check You can check your result by multiplying 20 • $9 to get $180. The answer seems reasonable. ✓

Real-World Link

The National Air and Space Museum opened in Washington, D.C., in 1976. It contains the world's largest collection of aircraft and spacecraft, including the 1903 Wright brothers' *Kitty Hawk Flyer.*

Source: Smithsonian Museum

Check Your Progress

2. **FOOD** A spaghetti dinner at the Italian Village restaurant costs $10.25. Use the Distributive Property and mental math to find the total cost of the dinner for Sherita, her brother, and her parents.

Personal Tutor glencoe.com

Algebraic Expressions You can model the Distributive Property by using algebra tiles and variables.

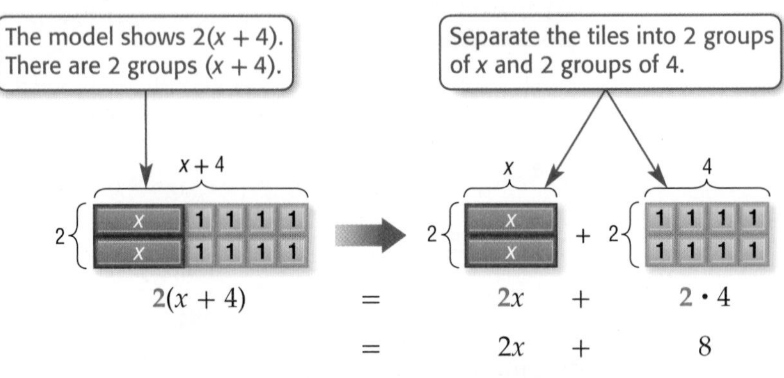

The expressions $2(x + 4)$ and $2x + 8$ are equivalent expressions because no matter what the value of x is, these expressions have the same value.

Vocabulary Link

Distribute
Everyday Use to deliver to each member of a group

Distributive
Math Use a property that allows you to multiply each member of a sum by a number

EXAMPLE 3 Simplify Algebraic Expressions

Use the Distributive Property to write each expression as an equivalent algebraic expression.

Watch Out!

Distributive Property In Example 3a, remember to distribute the 4 to both values inside the parentheses.

a. $4(x + 5)$

$4(x + 5) = 4x + 4 \cdot 5$

$= 4x + 20$ **Simplify.**

b. $(y + 10)6$

$(y + 10)6 = y \cdot 6 + 10 \cdot 6$

$= 6y + 60$ **Simplify.**

Check Your Progress

3A. $2(a + 5)$

3B. $(b + 6)3$

Personal Tutor glencoe.com

EXAMPLE 4 Simplify Expressions with Subtraction

Use the Distributive Property to write each expression as an equivalent algebraic expression.

a. $3(m - 4)$

$3(m - 4) = 3[m + (-4)]$ **Rewrite $m - 4$ as $m + (-4)$.**

$= 3 \cdot m + 3 \cdot (-4)$ **Distributive Property**

$= 3m + (-12)$ **Simplify.**

$= 3m - 12$ **Definition of subtraction**

b. $-9(n - 7)$

$-9(n - 7) = -9[n + (-7)]$ **Rewrite $n - 7$ as $n + (-7)$.**

$= -9 \cdot n + (-9)(-7)$ **Distributive Property**

$= -9n + 63$ **Simplify.**

Check Your Progress

4A. $4(d - 3)$

4B. $-7(e - 4)$

Personal Tutor glencoe.com

Check Your Understanding

Example 1 p. 171

Use the Distributive Property to write each expression as an equivalent expression. Then evaluate the expression.

1. $7(9 + 3)$

2. $4(3 + 5)$

3. $(7 + 8)2$

4. $(5 + 6)8$

Example 2 p. 172

5. **SCHOOL SUPPLIES** You purchase 3 blue notebooks and 2 red notebooks. Each notebook costs \$1.30. Use mental math to find the total cost of the notebooks. Justify your answer by using the Distributive Property.

Examples 3 and 4 p. 173

Use the Distributive Property to write each expression as an equivalent algebraic expression.

6. $8(m + 4)$

7. $(p + 4)5$

8. $-6(b - 5)$

9. $9(a - 10)$

Practice and Problem Solving

= **Step-by-Step Solutions** begin on page R11.
Extra Practice begins on page 810.

Example 1 p. 171

Use the Distributive Property to write each expression as an equivalent expression. Then evaluate the expression.

10. $6(3 + 9)$ **11.** $8(8 + 5)$ **12.** $(10 + 9)3$

13. $(12 + 7)5$ **14.** $9(12 - 3)$ **15.** $3(15 - 5)$

16. $-4(8 - 5)$ **17.** $-7(16 - 8)$ **18.** $14(20 - 4)$

Example 2 p. 172

19. **SHOPPING** Martine bought two pairs of jeans that are on sale for $32.85 each. Use mental math to find the total cost of the jeans. Justify your answer by using the Distributive Property.

20. **EXERCISE** Tionne can ride 6 miles on her bike in one hour. If she rode for 1.5 hours on Saturday and 2 hours on Sunday, use mental math to find the total distance she rode that weekend. Justify your answer by using the Distributive Property.

Examples 3 and 4 p. 173

Use the Distributive Property to write each expression as an equivalent algebraic expression.

21. $4(y + 7)$ **22.** $8(x + 2)$ **23.** $(a + 9)6$

24. $(b + 4)12$ **25.** $5(t - 6)$ **26.** $3(r - 1)$

27. $-1(d - 10)$ **28.** $-5(f - 5)$ **29.** $(x - 3)(-7)$

30. **CELL PHONE** **The double bar graph shows average monthly cell phone usage by age.**

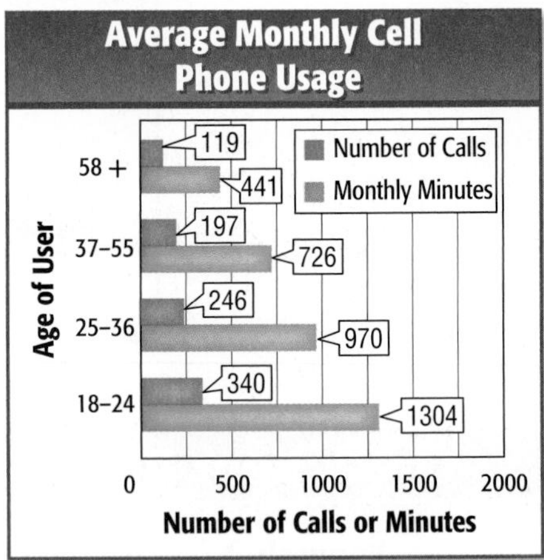

a. Find the total number of minutes used in a month by a family with an 18-year old, a 19-year old, and two 37–55-year old members.

b. Find the total number of calls made in 3 months by two 20-year olds and two 30-year olds.

c. Which group placed the most monthly calls? How do you know?

MENTAL MATH **Find each product mentally. Justify your answer.**

31. $8 \cdot 22$ **32.** $13 \cdot 39$ **33.** $19 \cdot 41$

34. $29 \cdot 13$ **35.** $75 \cdot 40$ **36.** $95 \cdot 38$

37. $9 \cdot 49$ **38.** $31 \cdot 11$ **39.** $121 \cdot 15$

40. **MONEY** Sarah charges $6.50 per hour to babysit. She babysat for 3 hours on Friday, and 5 hours on Saturday. Write two equivalent expressions for her total wages. Then find her total wage.

41. **ENTERTAINMENT** Admission to the state fair is $8 for adults and $7 for students. Write two equivalent expressions if two adults and two students go to the fair. Then find the total admission cost.

42. **MULTIPLE REPRESENTATIONS** In this problem, you will use the Distributive Property. The volume of seed in a bird feeder is represented by the equation $V = 12(24 - h)$.

a. **TABULAR** Make a table of ordered pairs (h, V).

b. **GRAPHICAL** Graph the ordered pairs on the coordinate plane.

c. **VERBAL** Explain what happens to the volume as the height increases.

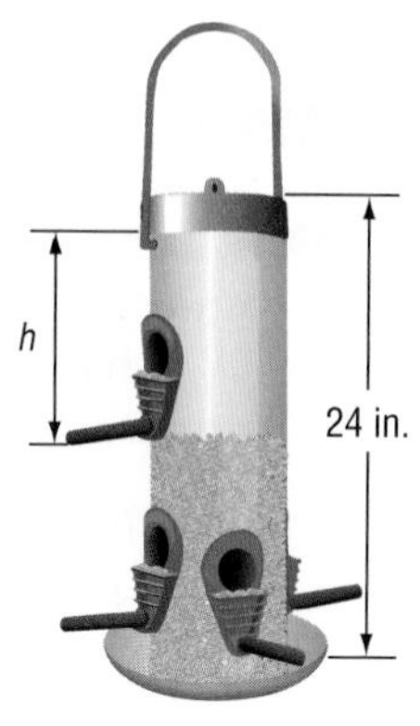

Use the Distributive Property to write each expression as an equivalent expression. Then evaluate the expression.
(*Hint*: $3\frac{1}{4}$ can be written as the sum of $3 + \frac{1}{4}$.)

43. $4\frac{1}{5} \cdot 5$

44. $10 \cdot 5\frac{1}{2}$

45. $6 \cdot 4\frac{2}{3}$

46. $2\frac{2}{7} \cdot 14$

47. **COSTUMES** Aiko uses $2\frac{1}{3}$ yards of fabric to make costumes for a play. Use the Distributive Property to find how much fabric she will need if she makes 9 costumes.

Use the Distributive Property to write each expression as an equivalent algebraic expression.

48. $3(a + b)$

49. $(e + f)(-5)$

50. $-6(x - y)$

51. $-4(j - k)$

52. $10(r - s)$

53. $(u - w)(8)$

54. **ENTERTAINMENT** Admission to Hersheypark Theme Park in Hershey, Pennsylvania, is \$45.95 for an adult and \$26.95 for children. The Diego family has a coupon for \$10 off each admission ticket. Write an expression to find the cost for x adults and y children.

Real-World Link

HersheyPark in Hershey, PA, began as a park for people who worked in Milton Hershey's chocolate company. In the 1970s, it became one the country's top theme parks. Today, the park contains 11 roller coasters and 13 water rides.

H.O.T. Problems

Use Higher-Order Thinking Skills

55. **OPEN ENDED** Write an equation using three integers that is an example of the Distributive Property.

56. **FIND THE ERROR** Julia and Catelyn are using the Distributive Property to simplify $3(x + 2)$. Is either of them correct? Explain your reasoning.

Julia	Catelyn
$3(x + 2) = 3x + 2$	$3(x + 2) = 3x + 6$

57. **CHALLENGE** Is $3 + (x \cdot y) = (3 + x) \cdot (3 + y)$ a true statement? If so, explain your reasoning. If not, give a counterexample.

58. **WRITING IN MATH** Explain how you can use the Distributive Property and mental math to simplify $2\frac{1}{2} \cdot 4\frac{1}{2}$.

Standardized Test Practice

59. Admission to a science museum is d dollars and a ticket for the 3-D movie is t dollars. Which expression represents the total cost of admission and a movie for p people?

A dtp
B $p + (dt)$
C $p(d + t)$
D $d(p + t)$

60. Which expression represents the total areas of the rectangles?

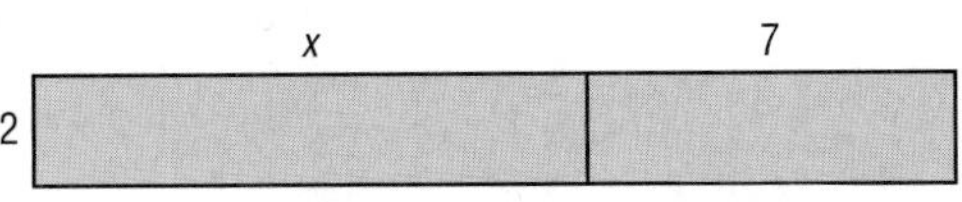

F $2 + x + 7 + 2 + x + 7$
G $2x + 14$
H $2x + 7$
J $14x$

61. Which expression can be written as $7(c + d)$?

A $7c \cdot 7d$
B $(7 + c) \cdot (7 + d)$
C $7c + 7d$
D $(7 + c) + (7 + d)$

62. EXTENDED RESPONSE A car rental company charges \$45 per day to rent a car.

a. Write an equation to show the total cost c of renting a car for d days.

b. If you rent the car for more than seven days, the cost will be reduced by \$10 per day. Write an equation to show the total cost c of renting a car for d days if you rent the car for more than seven days.

c. How much will it cost to rent the car for 4 days? 9 days?

d. If you have \$300, for how many days can you rent the car?

Spiral Review

Find each sum or difference. Write in simplest form. (Lesson 3-6)

63. $-\frac{5}{8} + \frac{3}{4}$ **64.** $-2\frac{1}{2} - \frac{2}{3}$ **65.** $\frac{2}{5} + \frac{1}{6}$ **66.** $-5\frac{6}{7} + \frac{1}{9}$

67. SEWING Jessica needs $5\frac{5}{8}$ yards of fabric to make a skirt and $14\frac{1}{2}$ yards to make a coat. How much fabric does she need in all? (Lesson 3-6)

68. GARDENING Tate's flower garden has a perimeter of 25 feet. He plans to add 2 feet 9 inches to the width and 3 feet 9 inches to the length. What is the new perimeter in feet? (Lesson 3-6)

ALGEBRA Evaluate each expression. (Lessons 2-4 and 2-5)

69. $-6h$, if $h = -20$

70. $-4st$, if $s = -9$ and $t = 3$

71. $\frac{x}{-5}$, if $x = -85$

72. $\frac{108}{m}$, if $m = -9$

Skills Review

Write each subtraction expression as an addition expression. (Lesson 2-3)

73. $9 - 12$

74. $-2 - 6$

75. $-10 - (-3)$

76. $-12 - 14$

EXPLORE 4-2

Algebra Lab
Simplifying Algebraic Expressions

Math Online glencoe.com
Math *in Motion*, Animation

In a set of algebra tiles, [x] represents the variable x, [1] represents the integer 1, and [–1] represents the integer -1. You can use algebra tiles to represent and simplify algebraic expressions.

ACTIVITY 1

Simplify $2x + 4 + 4x + 1$.

Step 1

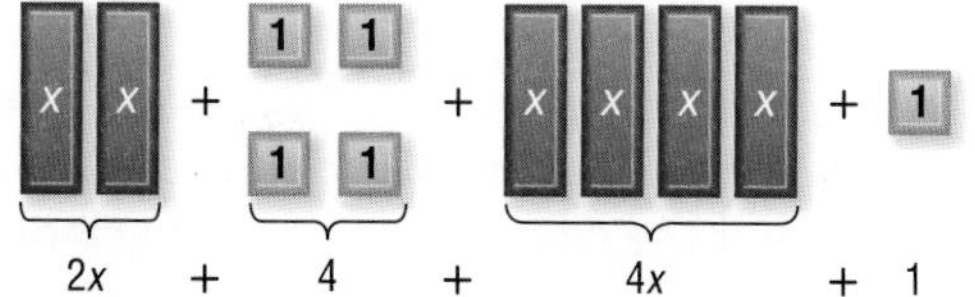

Model the expression.

Step 2

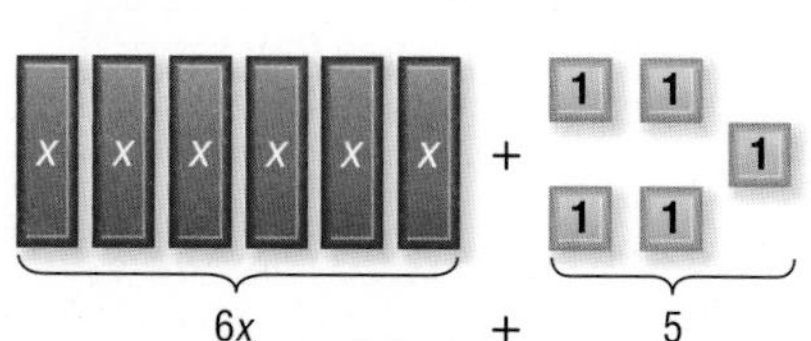

Group like tiles together. There are 6 x-tiles and 5 1-tiles.

So, $2x + 4 + 4x + 1 = 6x + 5$.

ACTIVITY 2

Simplify $x + 6 + 3x - 3$.

Step 1

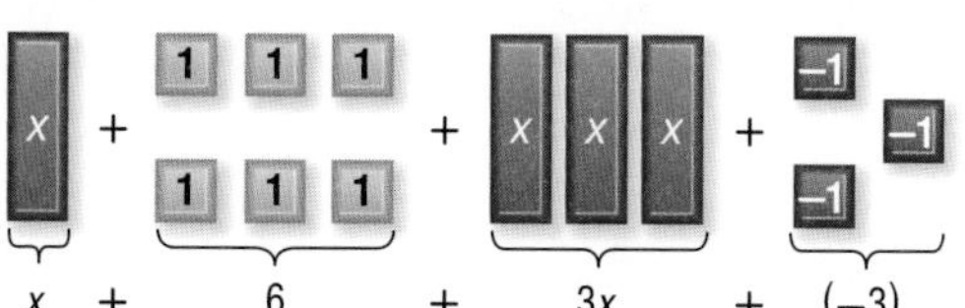

Model the expression.

Step 2

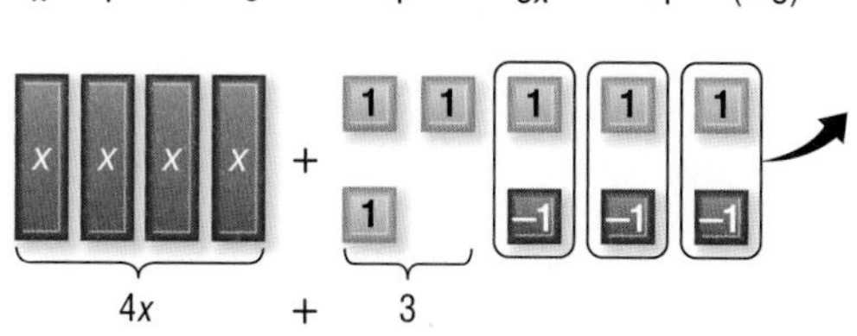

Group like tiles together. Remove zero pairs.

So, $x + 6 + 3x - 3 = 4x + 3$.

Analyze the Results

Model and simplify each expression using algebra tiles.

1. $3x + 4 + x + 3$ **2.** $2x + 3 + 2x + 3$ **3.** $x + 7 + 5x$

4. $4x - 1 + 2x + 5$ **5.** $3x + 2x - 4$ **6.** $2x + 2 + 2x - 2$

7. What mathematical properties allow you to sort the algebra tiles by their shapes?

8. What mathematical property allows you to remove zero pairs?

4-2 Simplifying Algebraic Expressions

Why?

Sandra needs soccer shoes and socks. Shoes cost \$45, a pair of white socks costs \$5, and a pair of red socks costs \$7.50.

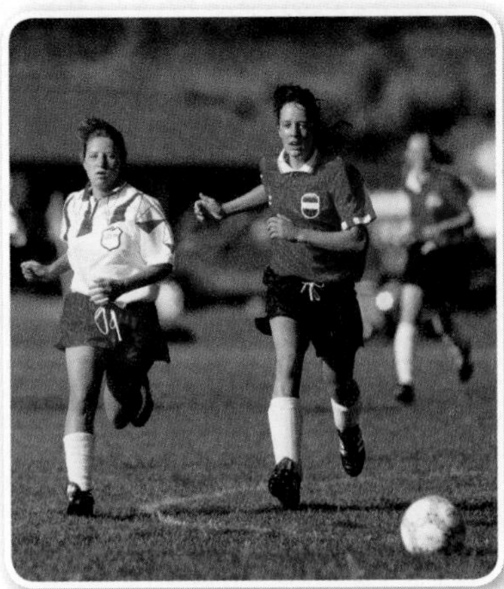

a. Find the total cost of one pair of shoes and 2 pairs of white socks.

b. If she buys 2 pairs of white socks and 2 pairs of red socks, what is the total cost?

c. Let x represent the number of pairs of socks she buys. Write an expression to represent the cost of one pair of shoes, x pairs of white socks, and x pairs of red socks.

Then
You already know how to write algebraic expressions. (Lesson 1-2)

Now
- Identify parts of an algebraic expression.
- Use the Distributive Property to simplify algebraic expressions.

New Vocabulary
term
coefficient
like terms
constant
simplest form
simplifying the expression

Math Online
glencoe.com
- Extra Examples
- Personal Tutor
- Self-Check Quiz
- Homework Help

Parts of Algebraic Expressions When addition or subtraction signs separate an algebraic expression into parts, each part is a **term**. The numerical part of a term that contains a variable is called the **coefficient** of the variable.

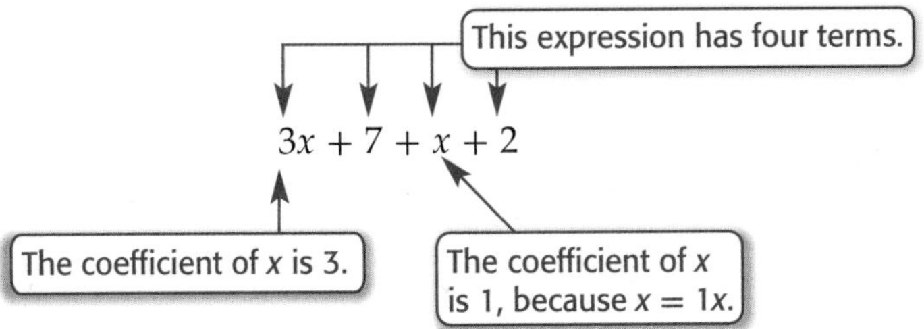

This expression has four terms.
3x + 7 + x + 2
The coefficient of x is 3.
The coefficient of x is 1, because x = 1x.

In this chapter, we will work only with terms with an exponent of 1. In this case, **like terms** are terms that contain the same variables, such as $2n$ and $5n$ or $6xy$ and $4xy$. A term without a variable is called a **constant**.

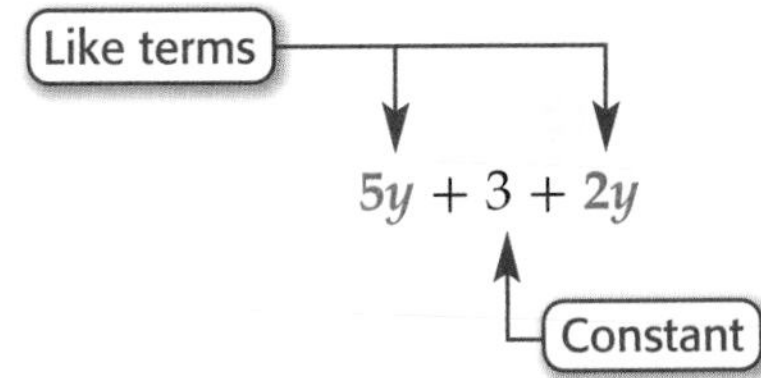

Like terms
5y + 3 + 2y
Constant

EXAMPLE 1 Identify Like Terms

Identify the like terms in the following expressions.

a. $3x + 4y + 4x$

$3x$ and $4x$ are like terms since the variables are the same.

b. $5x + 3 + 7x + 4$

$5x$ and $7x$ are like terms since the variables are the same. Constant terms 3 and 4 are also like terms.

Check Your Progress

1. Identify the like terms in the expression $-4x + 2y + 3y + 2x$.

Personal Tutor glencoe.com

Rewriting a subtraction expression using addition will help you identify the terms of an expression.

EXAMPLE 2 **Identify Parts of an Expression**

Identify the terms, like terms, coefficients, and constants in the expression $6x - 2y + x - 5$.

$6x - 2y + x - 5 = 6x + (-2y) + x + (-5)$ **Definition of subtraction**

$= 6x + (-2y) + 1x + (-5)$ **Identity Property**

The terms are $6x$, $-2y$, x, and -5. The like terms are $6x$ and x. The coefficients are 6, −2, and 1. The constant is −5.

Check Your Progress

2. Identify the terms, like terms, coefficients, and constants in the expression $3n + 5m - 6m + 2$.

Personal Tutor glencoe.com

Math History Link

Emmy Noether (1882–1935) Emmy Noether was known as the "mother of modern abstract algebra." Her later works focused on noncommutative algebra, which is where the order in which elements are multiplied *does* affect the solution.

Source: *Encyclopaedia Britannica*

Simplify Algebraic Expressions An algebraic expression is in **simplest form** if it has no like terms and no parentheses. When you use the Distributive Property to combine like terms, you are **simplifying the expression**.

EXAMPLE 3 **Simplify Algebraic Expressions**

Simplify each expression.

a. $4x + 6 + 2x$

$4x + 6 + 2x = 4x + 2x + 6$ **Commutative Property**

$= (4 + 2)x + 6$ **Distributive Property**

$= 6x + 6$ **Simplify.**

b. $5n + 2 - n - 6$

$5n + 2 - n - 6 = 5n + 2 + (-n) + (-6)$ **Definition of Subtraction**

$= 5n + 2 + (-1n) + (-6)$ **Identity Property**

$= 5n + (-1n) + 2 + (-6)$ **Commutative Property**

$= [5 + (-1)]n + 2 + (-6)$ **Distributive Property**

$= 4n + (-4)$ or $4n - 4$ **Simplify.**

c. $6y - 3(x - 2y)$

$6y - 3(x - 2y) = 6y + (-3)[x + (-2y)]$ **Definition of Subtraction**

$= 6y + (-3x) + (-3 \cdot -2)y$ **Distributive Property**

$= 6y + (-3x) + 6y$ **Simplify.**

$= 6y + 6y + (-3x)$ **Commutative Property**

$= (6 + 6)y + (-3x)$ **Distributive Property**

$= 12y + (-3x)$ or $12y - 3x$ **Simplify.**

Watch Out!

Distributive Property In Example 3c, remember to distribute −3, not +3, to the terms in the parentheses.

Check Your Progress

3A. $4x + 6 - 3x$ **3B.** $2m + 3 - 7m - 4$ **3C.** $4(q + 8p) + p$

Personal Tutor glencoe.com

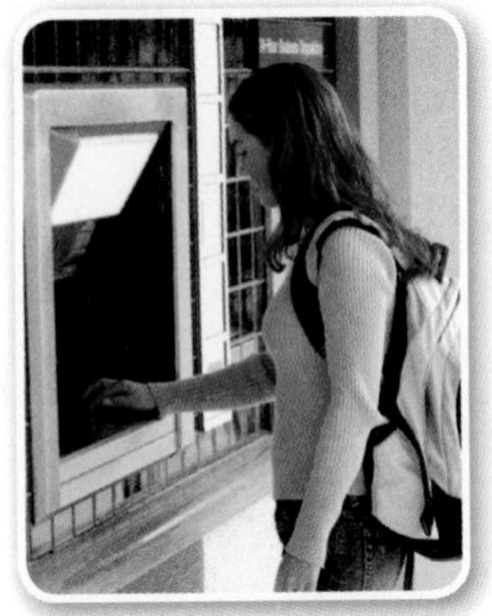

Real-World Link

According to FDIC, there are three main goals for teens to save money. First, set goals for how much you want to save/spend each month. Second, have a plan for saving money. Third, cut back on spending but don't stop spending.

Source: Federal Deposit Insurance Corporation

Real-World EXAMPLE 4 Write and Simplify Algebraic Expressions

MONEY You have some money in a savings account. Your sister has $25 more than you have in her account. Write an expression in simplest form that represents the total amount of money in both accounts.

Words	amount of your money plus amount of your sister's money
Variables	Let x = amount of your money. Let $x + 25$ = amount of your sister's money.
Expression	$x \quad + \quad (x + 25)$

$x + (x + 25) = (x + x) + 25$	**Associative Property**
$= (1x + 1x) + 25$	**Identity Property**
$= (1 + 1)x + 25$	**Distributive Property**
$= 2x + 25$	**Simplify.**

The expression $2x + 25$ represents the total amount of money you and your sister have in your accounts.

Check Your Progress

4. **STAMPS** Mato and Lola both collect stamps. Lola has 16 more stamps in her collection than Mato has. Write an expression in simplest form that represents the total number of stamps in both collections.

Personal Tutor glencoe.com

Check Your Understanding

Examples 1 and 2 pp. 178–179

Identify the terms, like terms, coefficients, and constants in each expression.

1. $-2a + 3a + 5b$
2. $2x + 3x + 4 + 4x$
3. $mn + 4m + 6n + 2mn$
4. $3a + 5b + 4 + 6a$
5. $3x + 4x + 5y$
6. $-4p - 6q - 5$

Example 3 p. 179

Simplify each expression.

7. $6x + 2x + 3$
8. $-2a + 3a + 6$
9. $7x + 4 - 5x - 8$
10. $5a - 2 - 3a + 7$
11. $-3(m - 1) + 4m + 2$
12. $4a - 6 - 2(a - 1)$

Example 4 p. 180

13. **JEWELRY** Marena is using a certain number of blue beads in a bracelet design. She will use 7 more red beads than blue beads. Write an expression in simplest form that represents the total number of beads in her bracelet design.

14. **ENTERTAINMENT** Kyung bought 3 CDs that cost x dollars each, 2 DVDs that cost $10 each; and a book that costs $15. Write an expression in simplest form that represents the total amount that Kyung spent.

Practice and Problem Solving

= **Step-by-Step Solutions** begin on page R11.
Extra Practice begins on page 810.

Examples 1 and 2 pp. 178–179

Identify the terms, like terms, coefficients, and constants in each expression.

15. $3a + 2 + 3a + 7$

16. $4m + 3 + m + 1$

17. $3c + 4d + 5c + 8$

18. $7j + 11jk + k + 9$

19. $4x + 4y + 4z + 4$

20. $3m + 3n + 2p + 4r$

Example 3 p. 179

Simplify each expression.

21. $4a + 3a$

22. $9x + 2x$

23. $-5m + m + 5$

24. $6x - x + 3$

25. $7p + 3 + 4p + 5$

26. $2a + 4 + 2a + 9$

27. $4a - 3b - 7a - 3b$

28. $-x - 2y - 8x - 2y$

29. $x + 5(6 + x)$

30. $2a + 3(2 + a)$

31. $-3(6 - 2r) - 3r$

32. $-2(2x - 5) - 4x$

Example 4 p. 180

For each situation write an expression in simplest form that represents the total amount.

33. **FASHION** Mateo has y pairs of shoes. His brother has 5 fewer pairs.

34. **CELL PHONES** You used p minutes one month on your cell phone. The next month you used 75 fewer minutes.

35. **SPORTS** Nathan scored x points in his first basketball game. He scored three times as many points in his second game. In his third game, he scored 6 more than the second game.

36. **MONEY** On Monday, Rebekah spent d dollars on lunch. She spent \$0.50 more on Tuesday than she did on Monday. On Wednesday, she spent twice as much as she did on Tuesday.

Simplify each expression.

37. $2(x - y) + 3x$

38. $-3(a - 2b) - 4b$

39. $-4(3m + 2n) - 5m + y$

40. $\frac{2}{3}(6a + 3b) - \frac{1}{2}(a - 2b)$

41. $\frac{1}{4}(m + 2n) - \frac{1}{3}(3m - 3n)$

42. $2(x - y) - (x + y)$

43. $\frac{2}{5}(2a - b) + \frac{2}{3}(a + 2b)$

44. $-\frac{3}{4}(3x + 2y) - \frac{3}{8}(x - 3y)$

45. **ALGEBRA** **Write an expression to represent each model. Then simplify the expression using algebra tiles.**

a. 1 1 1 1 1 1 1 + x x x x + −1 −1 −1 −1 + 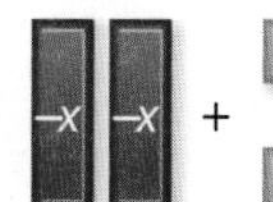 + 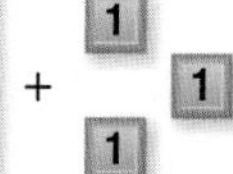

b. −1 −1 −1 −1 −1 −1 −1 −1 +  + 1 1 + 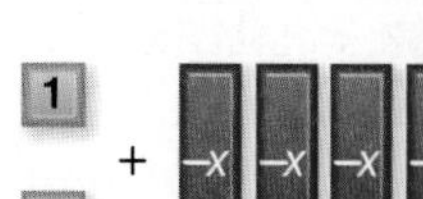 +  + 

c. Use algebra tiles to write and simplify your own expression.

StudyTip

Perimeter The perimeter of a figure is the distance around the figure. You can find the perimeter of a rectangle using the expression $2\ell + 2w$ where ℓ is the length and w is the width of the rectangle.

46. **MULTIPLE REPRESENTATIONS** In this problem, you will investigate the perimeter of a rectangle. Consider a rectangle that has a length that is twice its width.

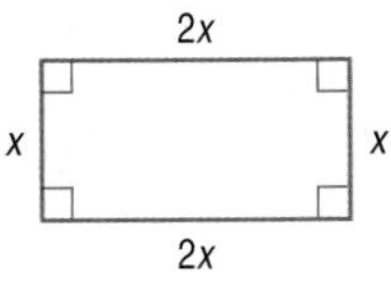

a. **TABULAR** Make a table that shows the width of a rectangle and its perimeter for widths of 1, 2, 3, 4, 5, and 6 units.

b. **GRAPHICAL** Graph the ordered pairs (width, perimeter).

c. **ALGEBRAIC** Write an expression in simplest form for the perimeter of a rectangle if the width is w units.

d. **VERBAL** If you double the width, what happens to the perimeter? Justify your reasoning.

GEOMETRY Write an expression in simplest form for the perimeter of each rectangle.

47. 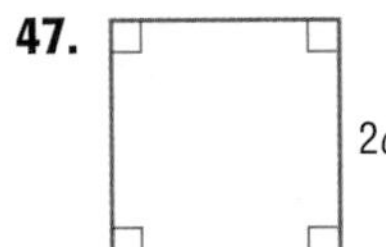

48. 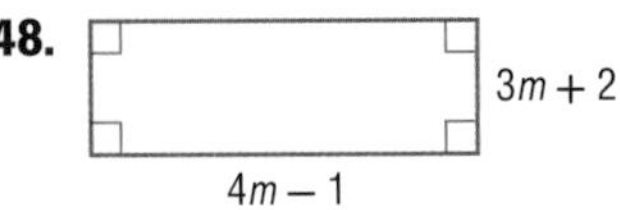

Simplify. Identify the properties you used in each step of your calculation.

49. $16 \cdot (-31) + 16 \cdot 32$

50. $72(38) + (-72)(18)$

51. $24 \cdot (-15) + 36 \cdot 15$

52. $22(-18) - 22\ (24)$

53. **AGE** This year Ana's mother is 2 years more than 3 times Ana's age. Write an expression in simplest form for the total of their ages.

H.O.T. Problems Use Higher-Order Thinking Skills

54. **OPEN ENDED** Write an expression containing at least 2 unlike terms. Then simplify the expression.

55. **CHALLENGE** Simplify $(2 + x)(y + 5)$.

56. **REASONING** Classify the following statement as *sometimes*, *always*, or *never* true. Explain your reasoning.

When using the Distributive Property, if the term outside the parentheses is negative, then the sign of each term inside the parentheses will change.

57. **WHICH ONE DOESN'T BELONG?** Identify the algebraic expression that does not belong with the other three. Explain your reasoning.

$-6(x - 2)$ | $x + 12 - 7x$ | $-x - 5x + 12$ | 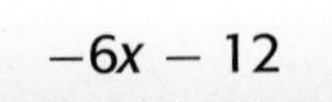

58. **CHALLENGE** In a three-digit number, the second and third digits are the same. The first digit is 4 more than the sum of the second and third digits. Write an expression in simplest form for the total sum of all three digits.

59. **WRITING IN MATH** Suppose your friend simplifies $4x - 2(x + 5)$ as $2x + 10$. Identify the error and correct it.

Standardized Test Practice

60. The perimeter of $\triangle DEF$ is $4x + 3y$. What is the measure of the third side of the triangle?

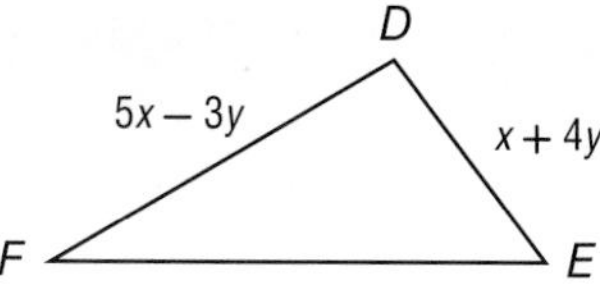

A $-2x + 2y$
B $2x + 2y$
C $x - y$
D $-x + 2y$

61. Which of the following expressions is equivalent to $4x + 4y$?

F $4xy$
G $4(x + y)$
H $4x + y$
J $x + 4y$

62. Adriana spent m minutes on her homework on Monday. She spent 45 more minutes doing her homework on Tuesday than on Monday. Which expression represents the total amount of time she spent on her homework on Monday and Tuesday?

A $m + 45$
B $m - 45$
C $2m + 45$
D $2m - 45$

63. SHORT RESPONSE Simplify the following expression.

$$7(3a - 2b) + 5b - 3(4a + 2)$$

Spiral Review

Use the Distributive Property to write each expression as an equivalent expression. (Lesson 4-1)

64. $8(z - 3)$

65. $(a - 6)(-5)$

66. $15(s + 2)$

67. ENTERTAINMENT The table shows the cost of different items at a movie theater. Write two equivalent expressions for the total cost of four movie tickets and four bags of popcorn. Then find the total cost. (Lesson 4-1)

Item	Cost ($)
ticket	7.00
small popcorn	3.00
small drink	2.50
candy bar	1.75

68. COOKING Simon has $1\frac{1}{4}$ cups of margarine. He needs $\frac{1}{2}$ cup for a cake and another $\frac{1}{3}$ cup for the icing. How much margarine will he have left? (Lesson 3-6)

Write two inequalities using the number pairs. Use the symbols $<$ or $>$. (Lesson 2-1)

69. -6 and -2

70. -10 and -13

71. 0 and -9

72. $|-11|$ and $|-7|$

73. $|15|$ and $|18|$

74. $|-12|$ and $|14|$

ALGEBRA Find the value of each expression if $a = 6$ and $b = 7$. (Lesson 1-2)

75. $\dfrac{4b + 3a}{b - 5}$

76. $\dfrac{6a - 2ab}{a + 2}$

77. $\dfrac{3(4a - 3b)}{b - 4}$

Skills Review

Find each sum or difference. (Lessons 2-2 and 2-3)

78. $-21 - 6$

79. $62 - (-12)$

80. $-32 + 26$

4-3

Solving Equations by Adding or Subtracting

Then

You have already worked with the additive inverse of a number when you subtracted integers. (Lesson 2-3)

Now

- Solve equations by using the Addition and Subtraction Properties of Equality.
- Translate verbal sentences into equations.

New Vocabulary

equation
solution
solving the equation
inverse operation
equivalent equation

Math Online

glencoe.com

- Extra Examples
- Personal Tutor
- Self-Check Quiz
- Homework Help
- Math in Motion

Why?

Kareem wants to determine how much his dog weighs. When he weighs himself, he weighs 120 pounds. When he and his dog are both on the scale, they weigh 168 pounds.

a. Write an addition sentence that relates Kareem's weight and his dog's weight to the total weight.

b. How much does Kareem's dog weigh?

Solve Equations by Adding A mathematical sentence that contains an equals sign, (=), showing that two expressions are equal is called an **equation**. You can write the equation $120 + x = 168$ to model the situation above. A value for the variable that makes an equation true is called a **solution**.

$120 + x = 168$		
$120 + 40 \stackrel{?}{=} 168$	**Replace *x* with 40.**	
$160 \neq 168$	**False**	

$120 + x = 168$		
$120 + 48 \stackrel{?}{=} 168$	**Replace *x* with 48.**	
$168 = 168$	**True ✓**	

For $120 + x = 168$, the solution is 48. The process of finding a solution is called **solving the equation**.

You can use inverse operations to solve an equation. **Inverse operations** "undo" each other. For example, to undo the subtraction of 5 in the equation $y - 5 = 10$, you could add 5 to each side of the equation.

Key Concept Addition Property of Equality — For Your FOLDABLE

Words If you add the same number to each side of an equation, the two sides remain equal.

Symbols For any numbers a, b, and c, if $a = b$, then $a + c = b + c$.

Examples

$$4 = 4 \qquad 4 + 9 = 4 + 9 \qquad 13 = 13$$

$$y - 5 = 10 \qquad +5 = +5 \qquad y = 15$$

Equivalent equations are equations that have the same solution. Because the equations $y - 5 = 10$ and $y = 15$ have the same solution, 15, they are equivalent equations.

StudyTip

Solving Equations When you solve an equation, you should always check to be sure that the first and last equations are equivalent.

EXAMPLE 1 Solve Equations by Adding

Solve each equation. Check your solution and graph it on a number line.

a. $x - 7 = -4$

$x - 7 = -4$	Write the equation.
$\underline{+7 = +7}$	Addition Property of Equality
$x + 0 = 3$	Additive Inverse Property; $-7 + 7 = 0$
$x = 3$	Identity Property; $x + 0 = x$

To check that 3 is the solution, replace x with 3 in the original equation.

Check	$x - 7 = -4$	Write the equation.
	$3 - 7 \stackrel{?}{=} -4$	Check to see whether this sentence is true.
	$-4 = -4$ ✓	The sentence is true.

The solution is 3. To graph 3, draw a dot at 3 on a number line.

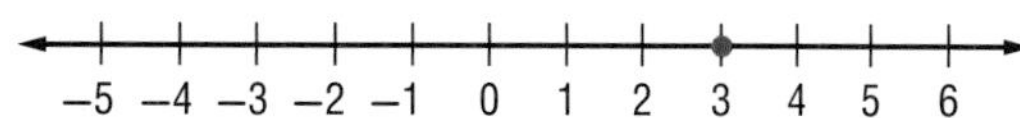

StudyTip

Position of the Variable You could begin solving Example 1b by rewriting the equation so that the expression containing the variable is to the left of the equals sign.

$-13.9 = n - 9.7$
↓
$n - 9.7 = -13.9$

b. $-13.9 = n - 9.7$

$-13.9 = n - 9.7$	Write the equation.
$\underline{+9.7 = \quad +9.7}$	Addition Property of Equality
$-4.2 = n + 0$	Additive Inverse Property; $-9.7 + 9.7 = 0$
$-4.2 = n$	Identity Property; $n + 0 = n$

The solution is –4.2. **Check your solution.**

To graph –4.2, draw a dot at –4.2 on a number line.

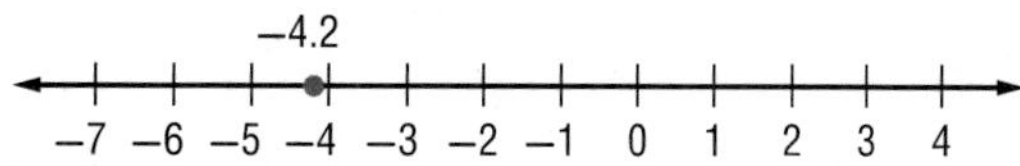

Check Your Progress

1A. $x - 5 = 20$

1B. $y - 6.4 = 10.7$

Personal Tutor glencoe.com

Solve Equations by Subtracting Some equations can be solved by subtracting the same number from each side.

Key Concept — Subtraction Property of Equality

For Your FOLDABLE

Words If you subtract the same number from each side of an equation, the two sides remain equal.

Symbols For any numbers a, b, and c, if $a = b$, then $a - c = b - c$.

Examples

$7 = 7$
$7 - 3 = 7 - 3$
$4 = 4$

$n + 4 = 9$
$\underline{-4 = -4}$
$n = 5$

Math *in Motion,* Animation glencoe.com

StudyTip

Isolating the Variable When trying to decide which value to subtract from each side of an equation, remember that your goal is to get the variable by itself on one side of the equal sign. This is called *isolating the variable.*

EXAMPLE 2 Solve by Subtracting

Solve each equation. Check your solution.

a. $23 = y + 10$

$23 = y + 10$	**Write the equation.**
$-10 = \quad -10$	**Subtraction Property of Equality**
$13 = y$	**Additive Inverse and Identity Properties**

To check that 13 is the solution, replace y with 13 in the original equation.

Check	$23 = y + 10$	**Write the equation.**
	$23 \stackrel{?}{=} 13 + 10$	**Check to see whether this sentence is true.**
	$23 = 23$ ✓	**The sentence is true.**

b. $m + 6.7 = 3.4$

$m + 6.7 = 3.4$	**Write the equation.**
$-6.7 = -6.7$	**Subtraction Property of Equality**
$m = -3.3$	**Additive Inverse and Identity Properties**
The solution is -3.3.	**Check your solution.**

Check Your Progress

2A. $16 + z = 14$

2B. $0.7 + a = 0.4$

Personal Tutor glencoe.com

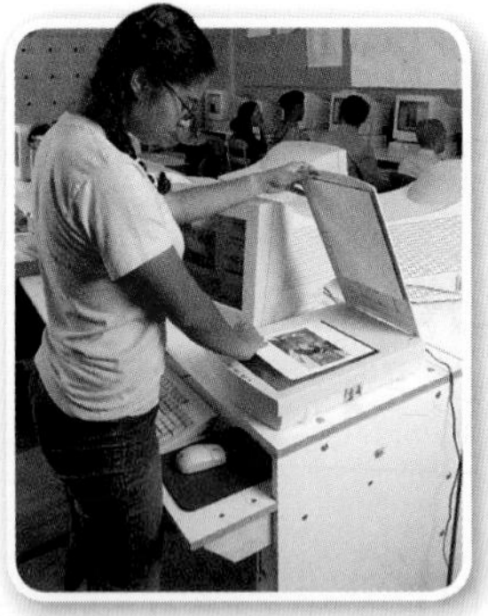

Real-World Link

Digital scrapbooking is becoming more and more popular. The advantages to digital scrapbooking include being able to enhance your photos, printing multiple copies of pages, and being able to undo mistakes.

Source: Kaboose Scrapbook LLC

Real-World EXAMPLE 3 Solve by Subtracting

PHOTO Pilar is making digital scrapbooks as gifts for her family. After saving her current work onto a CD, the CD is $\frac{5}{6}$ full. If the CD was $\frac{1}{3}$ full before she started saving, what fraction of the space on the CD do the new pages occupy?

Words	Starting amount	+	amount for new pages	is	new amount.
Variable	Let x = amount for the new pages.				
Equation	$\frac{1}{3}$	+	x	=	$\frac{5}{6}$

$\frac{1}{3} + x = \frac{5}{6}$	**Write the equation.**
$\frac{1}{3} - \frac{1}{3} + x = \frac{5}{6} - \frac{1}{3}$	**Subtraction Property of Equality**
$x = \frac{1}{2}$	**Additive Inverse and Identity Properties**

The new scrapbook pages take up $\frac{1}{2}$ of the space on the CD.

Check Your Progress

3. BUILDINGS The Jefferson Memorial in Washington, D.C., is 129 feet tall. This is 30 feet taller than the Lincoln Memorial. Write and solve an equation to find the height of the Lincoln Memorial.

Personal Tutor glencoe.com

Check Your Understanding

Examples 1 and 2 pp. 185–186

ALGEBRA Solve each equation. Check your solution and graph it on a number line.

1. $25 = y - 14$
2. $67 = m - 29$
3. $d - \frac{1}{3} = \frac{1}{6}$
4. $x + 24 = 72$
5. $p - 13 = -45$
6. $x - 36 = -2$
7. $0.53 + a = 1.97$
8. $1\frac{3}{4} = b + \frac{5}{8}$
9. $a + 5.7 = 9.2$

Example 3 p. 186

10. **FUNDRAISING** Joaquim sold 43 magazine subscriptions to raise money for a class trip. This is 15 less than the number Den sold. Write and solve a subtraction equation to find the number of subscriptions Den sold.

11. **WEATHER** The air pressure before a storm was 29.15 inches. After the storm, the pressure was 28.79 inches. Write and solve an addition equation to find how much the pressure changed during the storm.

Practice and Problem Solving

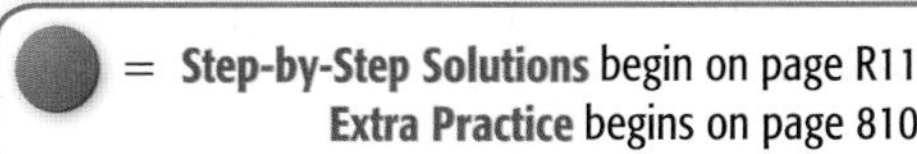
= **Step-by-Step Solutions** begin on page R11.
Extra Practice begins on page 810.

Examples 1 and 2 pp. 185–186

ALGEBRA Solve each equation. Check your solution and graph it on a number line.

12. $p - 12 = 20$
13. $x - 24 = 73$
14. $-14 = y - 16$
15. $-31 = r - 36$
16. $y + 14 = 72$
17. $m + 21 = 60$
18. $m + 1\frac{3}{8} = 5$
19. $y + \frac{3}{4} = -\frac{1}{2}$
20. $-6.5 = x - 0.54$
21. $56 = -78.9 + p$
22. $1.4 + t = 3.6$
23. $2.9 + z = -1.2$
24. $0.97 + a = 2.6$
25. $-1\frac{3}{8} = x - \frac{1}{2}$
26. $8\frac{3}{8} = r - 2\frac{1}{3}$

Example 3 p. 186

27. **FINANCIAL LITERACY** Amado budgets \$65 for his monthly cell phone bill. This is \$25 less than his monthly savings deposit. Write and solve a subtraction equation to find how much money Amado saves each month.

28. **PETS** Keisha feeds her dog $\frac{2}{3}$ cup of food in the morning. She feeds the dog a total of $1\frac{1}{2}$ cups of food every day. Write an addition equation to find how much food she gives the dog the rest of the day.

ALGEBRA Solve each equation. Check your solution.

29. $e - (-36) = -5$
30. $f - (-40) = -12$
31. $-2.5 + g = -1.3 + -1.1$
32. $-1.7 + h = -2.2 - 3.4$
33. $j + 17 - 23 = -7$
34. $k - 32 - (-16) = -9$
35. $1\frac{1}{4} + b = 1.6$
36. $y + 5.8 = \frac{3}{20}$

37. **SPORTS** Damon scored 13 points more than Wes and 9 points less than Ross. Ross scored 97. Write and solve equations to find the scores of Damon and Wes.

38. **GEOMETRY** The perimeter of a triangle is 27.1 feet. The sides of the triangle measure 9.8 feet, 10.9 feet, and x feet. Write and solve an equation to find the length of the missing side.

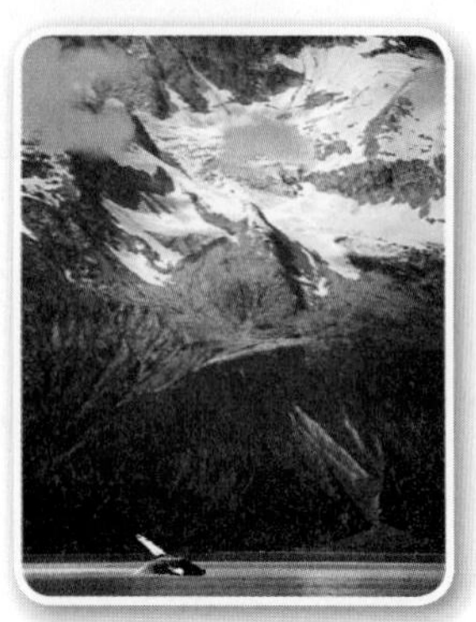

Real-World Link

In a recent year, annual visitation to the national parks reached an all-time high of 438,392,184 visitors.

Source: NPS

NATIONAL PARKS The graph shows the areas of the six largest national parks in the United States.

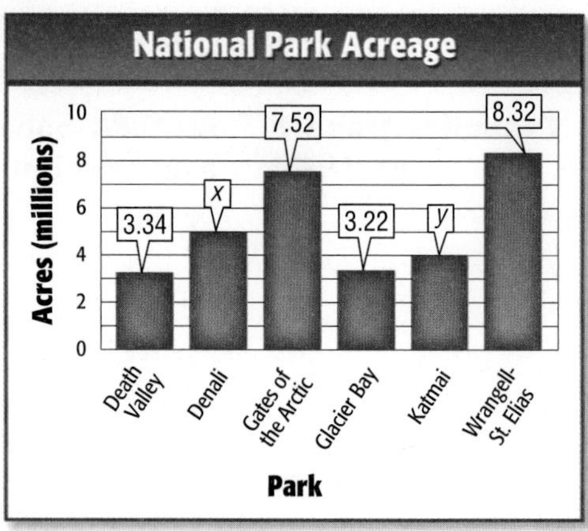

39 The Gates of the Arctic National Park is 2.78 million acres larger than Denali National Park. Write and solve an equation to find the size of Denali National Park.

40. Death Valley National Park is 0.75 million acres smaller than Katmai National Park. Write and solve an equation to find the size of Katmai National Park.

41. **MULTIPLE REPRESENTATIONS** In this problem, you will investigate more about functions.

x	y
5	3
4	4
0	8
−1	9

a. **ALGEBRAIC** Find the function rule.

b. **GRAPHICAL** Graph the ordered pairs.

c. **VERBAL** Write a real-world situation for the rule.

d. **VERBAL** How many solutions does $x + y = 8$ have? If $y = 3$, how many solutions does $x + 3 = 8$ have?

42. **PUBLISHING** A newspaper is $12\frac{1}{4}$ inches wide and 22 inches long. This is $1\frac{1}{4}$ inches narrower and half an inch longer than the old edition. What were the dimensions of the old edition?

H.O.T. Problems

Use **H**igher-**O**rder **T**hinking Skills

43. **OPEN ENDED** As shown in Example 2, $23 = y + 10$ is equivalent to the equation $y = 13$. Write an equation that is equivalent to $x = -2.4$.

44. **FIND THE ERROR** Liam and Marcus are solving the equation $x - (-2) = 4$. Is either of them correct? Explain your reasoning.

Liam

$$\begin{aligned} x - (-2) &= 4 \\ +(-2) &= +(-2) \\ \hline x &= 2 \end{aligned}$$

Marcus

$$\begin{aligned} x - (-2) &= 4 \\ +2 &= +2 \\ \hline x &= 6 \end{aligned}$$

45. **WRITING IN MATH** Jaime's golf score after a round of golf was −9. She had decreased her score by 5 strokes from the day before. Explain how you could use a model to determine her golf score on the previous day. Then use the model to solve the problem.

46. **CHALLENGE** Is the following statement *always*, *sometimes*, or *never* true? Explain your reasoning.

If $a + x = 100$, then x is less than 100.

47. **WRITING IN MATH** Write a real-word problem that can be modeled by the equation $p - 2.70 = 3.25$.

Standardized Test Practice

48. What situation can be represented by the equation $p = c + 12$?

A The Cardinals scored 12 more runs last season than the Panthers.

B The Panthers scored 12 more runs last season than the Cardinals.

C Together, the Panthers and the Cardinals scored 12 runs last season.

D The Panthers scored 12 runs last season.

49. What value of x makes the equation true?

$$5.47 - (-x) = 9.24$$

F -14.71 H 3.77

G -3.77 J 14.71

50. **GRIDDED RESPONSE** Refer to the table in Exercise 51. How many light years closer is Alpha Centauri B to Earth than Wolf 359?

51. The table shows the five nearest stars to Earth, excluding the Sun.

Star	Distance (light-years)
Proxima Centauri	4.22
Alpha Centauri A	4.40
Alpha Centauri B	4.40
Barnard's Star	5.94
Wolf 359	7.79

Which equation will best help you find how much closer Proxima Centauri is to Earth than Barnard's Star?

A $x - 5.94 = 4.22$

B $x + 4.22 = 5.94$

C $5.94 + x = 4.22$

D $5.94 + 4.22 = x$

Spiral Review

Simplify each expression. (Lessons 4-1 and 4-2)

52. $-2(z - 4)$

53. $(r - 5)6$

54. $8e - 4(2f + 5e)$

55. $4 + 3(c - 12)$

56. $-3(a + 2) - a$

57. $8 + x - 5x$

58. **MONEY** Suppose you work in a grocery store 4 hours on Friday and 5 hours on Saturday. You earn $7.25 an hour. Write two different expressions to find your wages. Then find the total wages for that weekend. (Lesson 4-1)

59. **POPULATION** The city of Heath makes up $\frac{1}{10}$ of the population in Rockwall County. Use the table to find the fraction of Rockwall County's population that lives in other cities. Write each fraction in simplest form. (Lesson 3-2)

City	Decimal Part of Rockwall County's Population
Fate	0.018
McLendon-Chisholm	0.02
Rockwall	0.42
Royse City	0.07

Skills Review

Find each product or quotient. (Lessons 2-4 and 2-5)

60. $-3(-15)$

61. $42 \div (-6)$

62. $-4 \cdot 8$

63. $\frac{-12}{-3}$

64. $\frac{-27}{9}$

65. $(-25)(-5)$

Mid-Chapter Quiz

Lessons 4-1 through 4-3

1. **MULTIPLE CHOICE** Lucita works at a bookstore and earns \$5.50 per hour. She works 3 hours on Friday and 7 hours on Saturday. Which expression does *not* represent her wages for those days? (Lesson 4-1)

   A $5.50(3 + 7)$ **C** $5.50(3) + 5.50(7)$

   B $10(5.50)$ **D** $7(5.50 + 3)$

2. **BUSINESS** A local newspaper can be ordered for delivery on weekdays or Sundays. A weekday paper is 35¢ and the Sunday Edition is \$1.50. The Stadlers ordered delivery of the weekday papers. The month of March had 23 weekdays and April had 20. How much should the carrier charge the Stadlers for those two months? (Lesson 4-1)

Simplify each expression. (Lessons 4-1 and 4-2)

3. $8(x + 3)$
4. $4(x - 5)$
5. $9y + 3 - y$
6. $6(m + 2) - 2m$
7. $5 - 4(12 - 3y)$
8. $2p - 7 - 6p - 8$
9. $8 - 12r - 5r - 3$
10. $-5t + 12 + 8 - 7t$

11. **CONSTRUCTION** A paving brick is shown. Find the perimeter of five bricks. (Lesson 4-2)

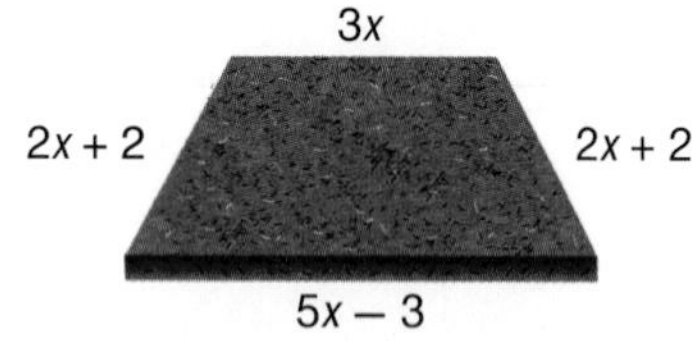

12. **SHOPPING** You buy x pairs of shoes that each cost \$24.95, the same number of socks that each cost \$4.75, and a pair of pants that costs \$29.99. Write an expression in simplest form that represents the total amount of money spent. (Lesson 4-2)

13. **GEOMETRY** Write an expression in simplest form that represents the total distance around the figure below. (Lesson 4-2)

2x – 5 in.
6 in. 6 in.
2x – 5 in.

14. **AVIATION** On December 17, 1903, the Wright brothers made the first flights in a power-driven airplane. Orville's flight covered 120 feet, which was 732 feet shorter than Wilbur's. Write and solve a subtraction equation to find the length of Wilbur Wright's flight. (Lesson 4-3)

ALGEBRA Solve each equation. (Lesson 4-3)

15. $p + 15 = 34$
16. $\frac{2}{3} + y = 2\frac{1}{2}$
17. $a + 12 = -7$
18. $45 = b + 18$
19. $-33 = x - 14$
20. $-3.6 = t - 6.8$
21. $-\frac{3}{8} = \frac{1}{5} + r$
22. $w - 0.87 = -2.4$

23. **MULTIPLE CHOICE** The table shows the five nearest train stops along the route from Main Street to Peach Court. Which equation will best help you find how much farther Peach Court is from Main Street than it is from City Center? (Lesson 4-3)

Train Stop	Distance to Main Street (mi)
City Center	4
14th Street	6
Grand Hotel	7
Stadium	12
Peach Court	17

   F $x - 17 = 4$ **H** $x - 4 = 17$

   G $x + 17 = 4$ **J** $x + 4 = 17$

24. **MONEY** Ricardo spent \$37 for a jacket, which included \$2.38 in sales tax. Write and solve an addition equation to find the price of the jacket before tax. (Lesson 4-3)

25. **DRAMA** Last school year, the fall play sold 45 more tickets than the spring play. Write and solve an equation to find the number of tickets sold at the spring play. (Lesson 4-3)

Play	Number of Tickets
Fall	214
Spring	■

4-4

Solving Equations by Multiplying or Dividing

Then

You have already solved equations using addition and subtraction. (Lesson 4-3)

Now

- Solve equations by using the Division Property of Equality.
- Solve equations by using the Multiplication Property of Equality.

Math Online

glencoe.com

- Extra Examples
- Personal Tutor
- Self-Check Quiz
- Homework Help
- Math in Motion

Why?

The Spirit Club at Westown Middle School is sponsoring a car wash. They charge $5 to wash each car.

Let c represent the number of cars the club washes and m represent the money the club raises. Then $5c = m$.

a. Suppose the Spirit Club wants to raise $120. Write an equation to find the number of cars they need to wash.

b. How can you find the number of cars?

Number of Cars	Money Raised
1	$5(1) = $5
2	$5(2) = $10
3	$5(3) = $15

Solve Equations by Dividing The equation $5c = 120$ represents the relationship described above. The operation involved in the equation is multiplication. To undo multiplication, use division.

Key Concept — Division Property of Equality

For Your FOLDABLE

Words When you divide each side of an equation by the same nonzero number, the two sides remain equal.

Symbols For numbers a, b, and c, where $c \neq 0$, if $a = b$, then $\frac{a}{c} = \frac{b}{c}$.

Examples

$$1.8 = 1.8 \qquad 4x = 24$$
$$\frac{1.8}{2} = \frac{1.8}{2} \qquad \frac{4x}{4} = \frac{24}{4}$$
$$0.9 = 0.9 \qquad x = 6$$

EXAMPLE 1 Solve Equations by Dividing

Solve $5c = 120$.

$5c = 120$	**Write the equation.**
$\frac{5c}{5} = \frac{120}{5}$	**Division Property of Equality**
$1c = 24$	**$5 \div 5 = 1$, $120 \div 5 = 24$**
$c = 24$	**Identity Property; $1c = c$**

The solution is 24.

Check Your Progress

Solve each equation.

1A. $-54 = 6x$

1B. $7a = 63$

Personal Tutor glencoe.com

EXAMPLE 2 Solve Equations by Dividing

Solve $4x = -48$. Check your solution and graph it on a number line.

$4x = -48$	**Write the equation.**
$\frac{4x}{4} = \frac{-48}{4}$	**Division Property of Equality**
$1x = -12$	**$4 \div 4 = 1, -48 \div 4 = -12$**
$x = -12$	**Identity Property; $1x = x$**

To check your solution, replace x with -12 in the original equation.

Check $4x = -48$	**Write the equation.**
$4(-12) \stackrel{?}{=} -48$	**Replace x with -12.**
$-48 = -48$ ✓	**The sentence is true.**

The solution is -12. To graph it, draw a dot at -12 on the number line.

-12

$-14 \quad -12 \quad -10 \quad -8 \quad -6 \quad -4$

Check Your Progress

2A. Solve $-121 = 11x$.

2B. Solve $-15x = -105$.

Personal Tutor glencoe.com

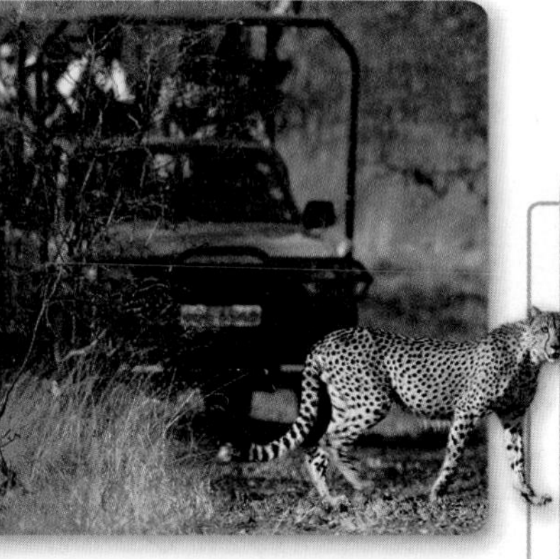

Real-World Link

Safari zoos allow animals the chance to roam while visitors tour in their cars. The first one in the United States opened in Florida in 1967.

Source: Lion Country Safari

Real-World EXAMPLE 3 Solve an Equation by Dividing

ZOOS A drive-through safari zoo charges $12.50 per person for admission. In one hour, the park raised $675 in admission fees. Write and solve an equation to find how many people visited that hour.

Words	Admission fee	times	the number of visitors	equals	the total money raised.
Variable	Let v = the number of zoo visitors.				
Equation	12.50	$\cdot$	v	=	675

$12.5v = 675$	**Write the equation.**
$\frac{12.5v}{12.5} = \frac{675}{12.5}$	**Division Property of Equality**
$v = 54$	**Simplify. Check this solution.**

The zoo admitted 54 people in one hour.

Check Your Progress

3. PARKS An in-state one year camping permit for New Mexico State Parks costs $180. If the total income from the camping permits is $8280 during the first day of sales, how many permits were purchased?

Personal Tutor glencoe.com

Solve Equations by Multiplying Equations in which a variable is divided can be solved by multiplying each side by the same number.

Key Concept **Multiplication Property of Equality** For Your FOLDABLE

Words When you multiply each side of an equation by the same nonzero number, the two sides remain equivalent.

Symbols For any numbers a, b, and c, if $a = b$, then $ca = cb$.

Examples

$$-3 = -3 \qquad \frac{1}{5}x = -3$$

$$4(-3) = 4(-3) \qquad 5\left(\frac{1}{5}x\right) = 5(-3)$$

$$-12 = -12 \qquad x = -15$$

Math *in Motion*, Interactive Lab glencoe.com

EXAMPLE 4 **Solve Equations by Multiplying**

Solve $\frac{y}{4} = -8$. Check your solution.

$\frac{y}{4} = -8.$	Write the equation.
$4 \cdot \frac{y}{4} = 4 \cdot (-8)$	Multiplication Property of Equality
$1y = -32$	Multiplicative Inverse Property; $4 \cdot \frac{1}{4} = 1$
$y = -32$	Identity Property. Check your solution.

StudyTip

Division In Example 4, $\frac{y}{4}$ means y divided by 4.

Check Your Progress

Solve each equation. Check your solution.

4A. $7 = \frac{x}{-2}$

4B. $\frac{a}{6} = 12$

Personal Tutor glencoe.com

To solve an equation such as $-\frac{3}{5}x = -6$, you can divide each side by $-\frac{3}{5}$ or multiply each side by $-\frac{5}{3}$.

EXAMPLE 5 **Solve Equations by Multiplying by the Reciprocal**

Solve $-\frac{3}{5}x = -6$. Check your solution.

$-\frac{3}{5}x = -6$	Write the equation.
$-\frac{5}{3}\left(-\frac{3}{5}\right)x = -\frac{5}{3}\left(-\frac{6}{1}\right)$	Multiply each side by $-\frac{5}{3}$.
$1x = 10$	Multiplicative Inverse Property; $-\frac{5}{3}\left(-\frac{3}{5}\right) = 1$
$x = 10$	Identity Property. Check your solution.

StudyTip

Multiplicative Inverse Remember that the product of a number and its multiplicative inverse is 1. Use this property when the coefficient of x is a fraction.

Check Your Progress

Solve each equation. Check your solution.

5A. $\frac{6}{7}m = -24$

5B. $5 = -\frac{5}{9}x$

Personal Tutor glencoe.com

Check Your Understanding

Examples 1, 2, 4 and 5 pp. 191–193

Solve each equation. Check your solution.

1. $5c = -65$
2. $-42 = -7m$
3. $8p = 96$
4. $\frac{n}{12} = 12$
5. $18 = \frac{t}{-2}$
6. $0.6h = 1.8$
7. $-3.4 = 0.4j$
8. $-\frac{3}{4}k = 12$
9. $36 = \frac{3}{5}m$

Example 3 p. 192

10. BOATING A forest preserve rents canoes for $18 per hour. Corey has $90 to spend. Write and solve a multiplication equation to find how many hours he can rent a canoe.

11. SPACE The weight of an object on the Moon is one-sixth its weight on Earth. If an object weighs 54 pounds on the Moon, write and solve a division equation to find how much it weighs on Earth.

Practice and Problem Solving

 = **Step-by-Step Solutions** begin on page R11. **Extra Practice** begins on page 810.

Examples 1, 2, 4 and 5 pp. 191–193

Solve each equation. Check your solution.

12. $9x = 54$
13. $5s = -60$
14. $64 = -4r$
15. $-72 = 3y$
16. $0.3x = -4.5$
17. $4.95 = 0.3t$
18. $-8.4 = -6g$
19. $-28 = \frac{d}{-14}$
20. $\frac{b}{9} = -108$
21. $16 = -\frac{b}{4}$
22. $\frac{x}{-8} = -4$
23. $-32 = -\frac{4}{3}s$
24. $-25 = -\frac{5}{6}r$
25. $-\frac{9}{10}k = 72$
26. $\frac{2}{3}n = -22$

Example 3 p. 192

27. FRUIT Rashid picked a total of 420 strawberries in $\frac{5}{6}$ hour. Write and solve a multiplication equation to find how many strawberries Rashid could pick in 1 hour.

28. ATHLETICS Marcus ran every day for 14 weeks to train for a marathon. Write and solve a division equation to find how many days he trained.

Solve each equation. Check your solution.

29. $5p - 2p = -12$
30. $42 = 4x + 3x$
31. $-2(6y) = 144$
32. $72 = -12(-3x)$
33. $\frac{r}{4} = -25 + 9$
34. $\frac{m}{-3} = -5 - 18$
35. $\frac{1}{3}n = \frac{2}{9}$
36. $\frac{5}{8} = -\frac{1}{2}x$
37. $-0.7 = -\frac{7}{9}z$
38. $1\frac{7}{8}y = 4\frac{1}{2}$
39. $2\frac{1}{3} = -9m$
40. $-\frac{7}{9}t = -\frac{28}{36}$

41. ANIMALS The sleeping heart rate of a black bear during hibernation is about $\frac{2}{5}$ of its summer rate. If the sleeping heart rate of a bear is 28 beats per minute during hibernation, find the summer sleeping heart rate.

42. GEOMETRY The formula for finding the area of a triangle is $A = \frac{1}{2}bh$, where A represents the area, b represents the length of the base of the triangle, and h represents the height of the triangle. Write and solve equations to complete the table of values.

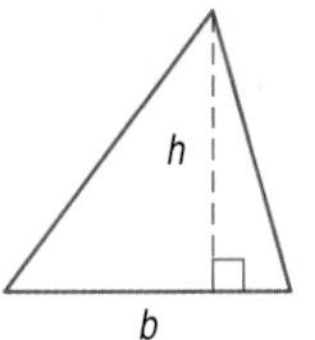

Area (A)	15	15	15	15	15
base (b)	1	2	3	4	5
height (h)	■	■	■	■	■

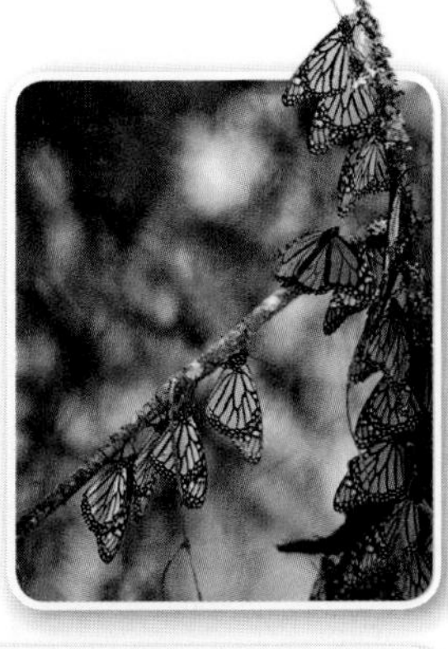

Real-World Link

Monarchs east of the Rocky Mountains migrate to small forest groves in Mexico's mountains. As many as 500 million butterflies have migrated in a given year.

Source: Monarch Watch

43 **MULTIPLE REPRESENTATIONS** Every autumn, the North American Monarch butterfly migrates up to 3000 miles to California and Mexico where it hibernates until early spring. The butterfly travels on average 50 miles per day.

a. ALGEBRAIC Write an equation that represents the distance d a butterfly will travel in t days.

b. TABULAR Use the equation to complete the table at the right.

Time (days)	1	2	3	4	5	6
Distance (miles)	■	■	■	■	■	■

c. GRAPHICAL Graph the points from the table on the coordinate plane. Graph time on the x-axis and distance on the y-axis.

d. GRAPHICAL Using the graph, estimate the number of days it will take the butterfly to travel 450 miles.

e. VERBAL How many days will it take a butterfly to travel 2500 miles? Which method did you use to solve the problem?

H.O.T. Problems

Use Higher-Order Thinking Skills

44. OPEN ENDED Write a multiplication equation and a division equation which has a solution of −5.

45. WRITING IN MATH Write a real-world example that uses an equation containing a decimal and a fraction. Then find the solution.

46. REASONING *True* or *false*: $\frac{x}{4}$ is equivalent to $\frac{1}{4}x$. Explain your reasoning.

47. FIND THE ERROR Sam and Rachel are solving $\frac{x}{4} = -20$. Is either of them correct? Explain your reasoning.

Rachel

$\frac{x}{4} \times 4 = -20 \times 4$

$x = -80$

Sam

$\frac{x}{4} \div 4 = -20 \div 4$

$x = -5$

48. CHALLENGE If $\frac{3}{10}x = 3$, what is the value of $7x + 13$?

49. WRITING IN MATH Suppose your friend says he can solve $3x = 15$ by using the Multiplication Property of Equality. Is he correct? Justify your response.

Standardized Test Practice

50. During a vacation, the Morales family drove 63 miles in 1 hour. If they averaged the same speed during their trip, which equation can be used to find how far the Morales family drove in 6 hours?

A $\frac{63}{x} = 6$

B $\frac{x}{6} = 63$

C $6x = 63$

D $63x = 6$

51. The solution of which equation is *not* graphed on the number line below?

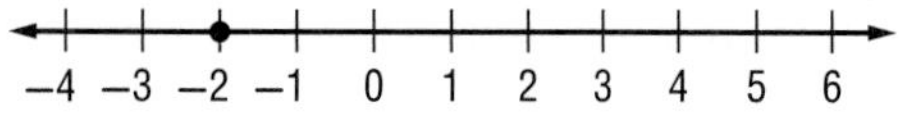

F $12 = -6x$

G $8x = -16$

H $-14 = 7x$

J $-18x = -36$

52. Ella paid $11.85 for 3 magazines. If each magazine was the same price, how much did each magazine cost?

A $4.59

B $4.00

C $3.95

D $3.59

53. EXTENDED RESPONSE Stanley paid half of what Royce paid for his baseball glove. Royce paid $64 for his glove.

a. Write an equation to find how much Stanley paid for his glove.

b. How much did Stanley pay for his glove?

Spiral Review

Solve each equation. Check your solution. (Lesson 4-3)

54. $x - 5 = -22$

55. $4 = 7 + p$

56. $-40 = y - 9$

57. $2.3 + r = 1.6$

58. $d - 2.7 = -1.4$

59. $t + (-16) = -24$

60. $p + \frac{1}{10} = -\frac{3}{4}$

61. $\frac{2}{3} + k = \frac{1}{6}$

62. $d - \frac{4}{9} = -\frac{1}{12}$

Simplify each expression. (Lesson 4-2)

63. $5(t + 3)$

64. $7x - 12x$

65. $9p + 4 + 3p$

66. $3w + 4s - w + 5s$

67. $7 - 4(x + 3)$

68. $3(2 + 3x) + 21x$

69. ALGEBRA Find the values that complete the table at the right for $y = -4x$. (Lesson 2-4)

x	−2	−1	0	1
y	■	■	■	■

70. AGE Gabriel is 12 years old, and his younger brother Elias is 2 years old. How old will each of them be when Gabriel is twice as old as Elias? (Lesson 1-1)

Skills Review

Find the value of each expression. (Lessons 2-2, 2-3, and 2-4)

71. $-3 + 7(4)$

72. $\frac{6 - 9}{8 + 4}$

73. $5 - 3(6 + 2)$

74. $14 - 24 + 6 \cdot 8$

75. $9 \cdot 7 - 4 \cdot 5$

76. $3[15 - (-9)]$

EXPLORE 4-5

Algebra Lab
Solving Two-Step Equations Using Algebra Tiles

Math Online glencoe.com
Math *in Motion,* Interactive Lab

Like one-step equations, two-step equations can be solved using algebra tiles. I'm thinking of a number. If you multiply it by 3 and add 2, the result is 8. To solve the problem, you can use the *work backward strategy*. Undo each operation in reverse order.

ACTIVITY 1

Model and solve $3x + 2 = 8$ using algebra tiles.

Step 1

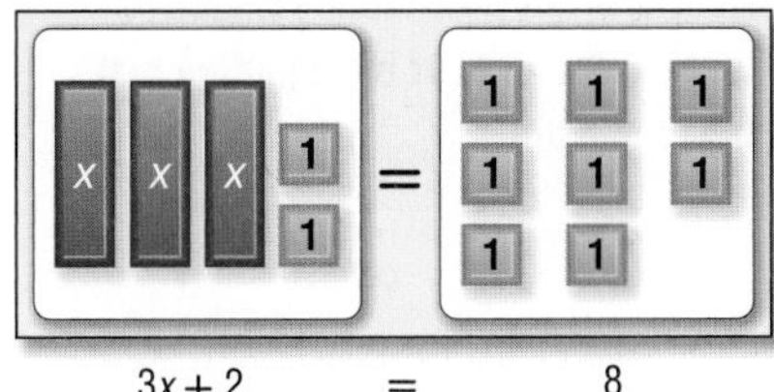

Model the equation.

Step 2

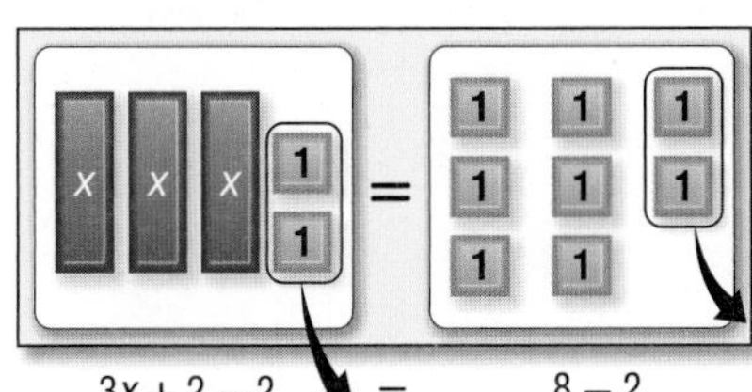

Remove two 1-tiles from each side of the mat.

Step 3

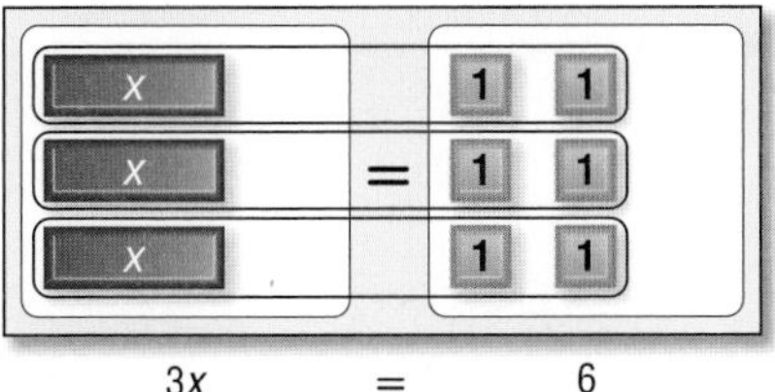

Separate the remaining tiles into three equal groups.

So, $x = 2$.

Step 4 Check your solution by replacing x with 2 in the original equation.

$3x + 2 = 8$ **Write the equation.**

$3(2) + 2 = 8$ **Replace x with 2**

$6 + 2 = 8$ **Multiply.**

$8 = 8$ ✓ **The statement is true.**

Analyze the Results

Model and solve each equation using algebra tiles.

1. $2x + 2 = 12$ **2.** $9 = 4 + 5x$ **3.** $3x + 6 = 15$

Some equations are solved by using zero pairs. One +1 tile and one −1 tile make a zero pair. Since $1 + (-1) = 0$, you can add or subtract a zero pair from either side of an equation mat without changing its value.

ACTIVITY 2

Model and solve $2x + 4 = 2$ using algebra tiles.

Step 1

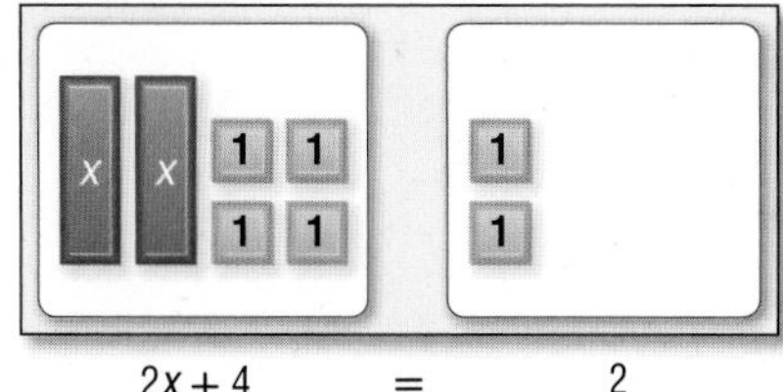

Model the equation. Notice it is not possible to remove four positive 1-tiles from the right side of the equation mat.

Step 2

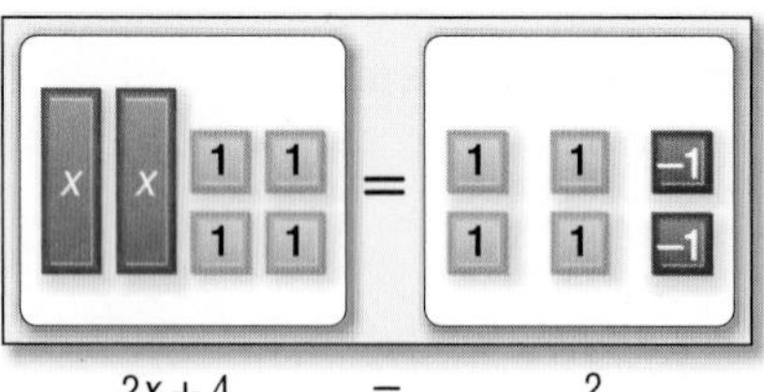

Add 2 zero pairs to the right side of the mat so you have enough positive 1-tiles.

Step 3

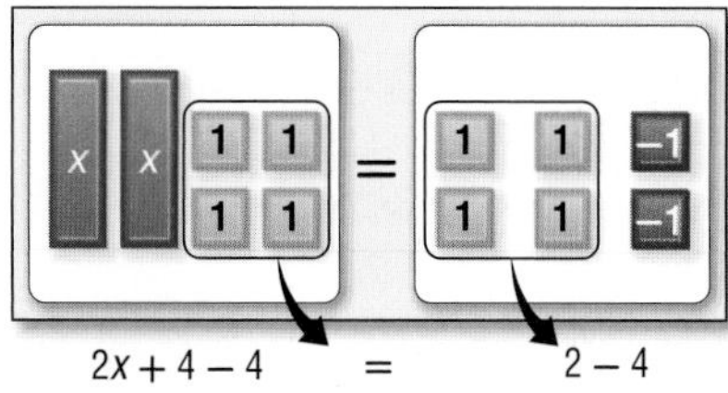

Remove the same number of 1-tiles from each side of the mat until the *x*-tile is alone on one side.

Step 4

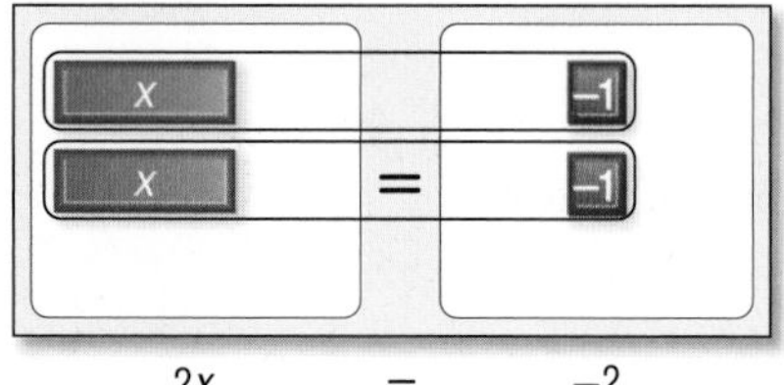

Separate the remaining tiles into two equal groups.

So, $x = -1$.

Analyze the Results

Model and solve each equation using algebra tiles.

4. $4x + 3 = -9$ **5.** $3x - 3 = 6$ **6.** $-6 = 2x + 4$

7. What property is shown by removing tiles from each side?

8. What property is shown by separating the tiles into groups?

4-5 Solving Two-Step Equations

Then

You have already solved one-step equations. (Lessons 4-3 and 4-4)

Now

- Solve two-step equations.
- Solve real-world problems involving two-step equations.

New Vocabulary

two-step equation

Math Online

glencoe.com

- Extra Examples
- Personal Tutor
- Self-Check Quiz
- Homework Help
- Math in Motion

Why?

Santos is planning a hot air balloon ride. The cost of the ride is \$125 plus \$50 for each hour in the air. The equation $c = 50h + 125$ represents the total cost c to ride a hot air balloon for h hours.

a. Find the total cost if Santos rides for 2 hours.

b. How long could Santos ride if he had \$325?

Solve Two-Step Equations A **two-step equation** like the one above contains two operations. To solve a two-step equation, use inverse operations to undo each operation in reverse order of the order of operations.

EXAMPLE 1 Solve a Two-Step Equation

Solve $3a + 9 = 33$. Check your solution.

Method 1 The Vertical Method

$3a + 9 = 33$ Write the equation.

$$\begin{array}{r} 3a + 9 = 33 \\ -9 = -9 \\ \hline 3a = 24 \end{array}$$

Subtraction Property of Equality

Simplify.

$\frac{3a}{3} = \frac{24}{3}$ Division Property of Equality

$a = 8$ Simplify.

Method 2 The Horizontal Method

$3a + 9 = 33$ Write the equation.

$3a + 9 - 9 = 33 - 9$ Subtraction Property of Equality

$3a = 24$ Simplify.

$\frac{3a}{3} = \frac{24}{3}$ Division Property of Equality

$a = 8$ Simplify.

Using either method, the solution is 8.

Check $3a + 9 = 33$ Write the equation.

$3(8) + 9 \stackrel{?}{=} 33$ Replace a with 8.

$24 + 9 \stackrel{?}{=} 33$ Multiply.

$33 = 33$ ✓ The sentence is true.

Check Your Progress

Solve each equation. Check your solution.

1A. $6x + 1 = 25$

1B. $4x - 5 = -33$

Personal Tutor glencoe.com

StudyTip

Vertical Method You can also use the vertical method when solving equations.

EXAMPLE 2 Solve a Two-Step Equation

Solve $\frac{p}{5} - 12 = 20$.

$\frac{p}{5} - 12 = 20$	Write the equation.
$\frac{p}{5} - 12 + 12 = 20 + 12$	Addition Property of Equality
$\frac{p}{5} = 32$	Simplify.
$5 \cdot \frac{p}{5} = 5 \cdot 32$	Multiplication Property of Equality
$p = 160$	Simplify. Check your solution.

Check Your Progress

2A. $8 = 15 + \frac{n}{3}$

2B. $-\frac{1}{6}x - 3 = 2$

Personal Tutor glencoe.com

Math *in Motion*, BrainPOP® glencoe.com

EXAMPLE 3 Equations with Negative Coefficients

Solve $9 - t = -34$.

$9 - t = -34$	Write the equation.
$9 - 1t = -34$	Identity Property: $t = 1t$
$9 + (-1t) = -34$	Definition of Subtraction
$-9 + 9 + (-1t) = -9 + (-34)$	Addition Property of Equality
$-1t = -43$	Simplify.
$\frac{-1t}{-1} = \frac{-43}{-1}$	Division Property of Equality
$t = 43$	Simplify. Check your solution.

Check Your Progress

3A. $-15 - b = 44$

3B. $-6.5 = -4.3 - n$

Personal Tutor glencoe.com

StudyTip

Distributive Property You use the Distributive Property to mentally simplify $2x + x$.

$$2x + 1x = (2 + 1)x$$
$$= 3x$$

EXAMPLE 4 Combine Like Terms Before Solving

Solve $2x + x - 27 = 3$. Check your solution.

$2x + x - 27 = 3$	Write the equation.
$2x + 1x - 27 = 3$	Identity Property; $x = 1x$
$3x - 27 = 3$	Distributive Property; $2x + x = 3x$
$3x - 27 + 27 = 3 + 27$	Addition Property of Equality
$3x = 30$	Simplify.
$\frac{3x}{3} = \frac{30}{3}$	Division Property of Equality
$x = 10$	Simplify. Check your solution.

Check Your Progress

4A. $4 - 9c + 3c = 58$

4B. $3.4 = 0.4m - 2 + 0.2m$

Personal Tutor glencoe.com

Solve Real-World Problems You can write and solve two-step equations to solve many real-world problems.

STANDARDIZED TEST EXAMPLE 5

Problem-SolvingTip

Eliminating Choices Sometimes, you can save time by looking at the possible answers and eliminating choices. In this case, you can eliminate answer choice D. You would not undo multiplying 18 by subtracting 18.

Deon wants to go on a camping trip with his hiking club. The trip costs \$199. He paid a deposit of \$55 and will save an additional \$18 per week to pay for the trip. The equation $55 + 18w = 199$ can be used to find how many weeks Deon will need to save. Which series of steps can be used to solve the equation?

A Divide 199 by 18. Then subtract 55.

B Subtract 55 from 199. Then divide by 18.

C Subtract 199 from 55. Then multiply by 18.

D Subtract 18 from 199. Then divide by 55.

Read the Test Item

Solve the equation so you can write the steps in the correct order that are necessary to solve the problem.

Solve the Test Item

$55 + 18w = 199$	Write the equation.
$55 - 55 + 18w = 199 - 55$	Subtraction Property of Equality
$18w = 144$	Simplify.
$\frac{18w}{18} = \frac{144}{18}$	Division Property of Equality
$w = 8$	Simplify.

To solve the equation, you first subtract 55 and then divide by 18. Choice B is the correct answer.

Check

When you solve an equation, you undo the steps in evaluating an expression in *reverse* order of the order of operations. In this equation, you would first undo adding 55 by subtracting 55 then undo multiplying by 18 by dividing by 18.

Real-World Link

The average price of a popular computer in 2005 was \$963. This is about half the cost of a computer in 1995.

Source: CNET Networks

Check Your Progress

5. Salvatore purchased a computer for \$650. He paid \$105 initially and will pay \$20 per month until the computer is paid off. The equation $105 + 20x = 650$ can be used to find how many months he will make payments. Which series of steps can be used to solve the equation?

F Add 105 to 650. Then divide by 20.

G Multiply 650 by 20. Then subtract 105.

H Subtract 105 from 650. Then divide by 20.

J Subtract 20 from 650. Then divide by 105.

Personal Tutor glencoe.com

Check Your Understanding

Examples 1 and 2 pp. 199–200

Solve each equation. Check your solution.

1. $4p + 9 = 25$ **2.** $-2x + 1 = 7$ **3.** $5y - 3 = -23$

4. $\frac{p}{4} - 6 = -8$ **5.** $\frac{t}{-6} + 1 = 3$ **6.** $\frac{r}{-2} - 12 = -27$

Examples 3 and 4 p. 200

Solve each equation. Check your solution.

7. $-7 - 8d = 17$ **8.** $23 - 2c = 41$ **9.** $1 - 2k = -9$

10. $-4 = 8y - 9y + 6$ **11.** $-1.3j + 0.4 = -1.16$ **12.** $1.1 - t + 2.2t = 5.9$

Example 5 p. 201

13. **MULTIPLE CHOICE** Kaleigh has \$25. She plans to save \$5 each week. The equation $\$25 + \$5w = \$150$ represents how long it will take her to save \$150. Which series of steps could be used to solve the equation?

A Add 25 to 150. Then divide by 5.

B Divide 150 by 5. Then add 25.

C Subtract 25 from 150. Then divide by 5.

D Divide 150 by 5. Then subtract 25.

Practice and Problem Solving

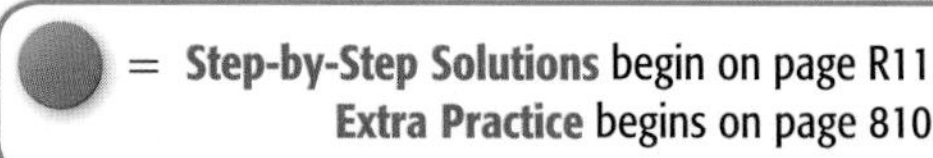

Examples 1 and 2 pp. 199–200

Solve each equation. Check your solution.

14. $5a + 3 = 28$ **15.** $3b + 15 = 27$ **16.** $25 = 2c - 9$

17. $4d - 18 = -34$ **18.** $\frac{g}{3} + 4 = 2$ **19.** $\frac{h}{9} - 3 = 2$

20. $-16 = \frac{k}{2} - 7$ **21.** $20 = \frac{m}{5} + 12$ **22.** $\frac{n}{4} - 20 = -1$

Examples 3 and 4 p. 200

Solve each equation. Check your solution.

23. $46 - 8x = -18$ **24.** $y - 7y + 6 = 30$ **25.** $-7 = \frac{p}{-5} - 1$

26. $14 = \frac{s}{-3} - 8$ **27.** $x + 7 - 2x = 18$ **28.** $46 - 3n = -23$

Example 5 p. 201

29. **FINANCIAL LITERACY** The cost of a family membership at a health club is shown at the right. The Johnson family budgets \$800 to use the health club. Solve $125 + 45f = 800$ to find the number of months the family can use the club.

30. **ENTERTAINMENT** The second book in a fantasy series is 112 pages longer than the first book. The total number of pages in both books is 524. Solve the equation $b + b + 112 = 524$ to find the number of pages b in the first book.

Solve each equation. Check your solution.

31. $6.1e + 1.07 = 9$ **32.** $-2.5c + 6.7 = -1.3$ **33.** $\frac{2}{3} - 6y = -1\frac{5}{6}$

34. $\frac{3}{4}x + 1.5 = 2.7$ **35.** $\frac{f}{-4} + 20.5 = 12.9$ **36.** $54.8 - \frac{d}{5} = 60.1$

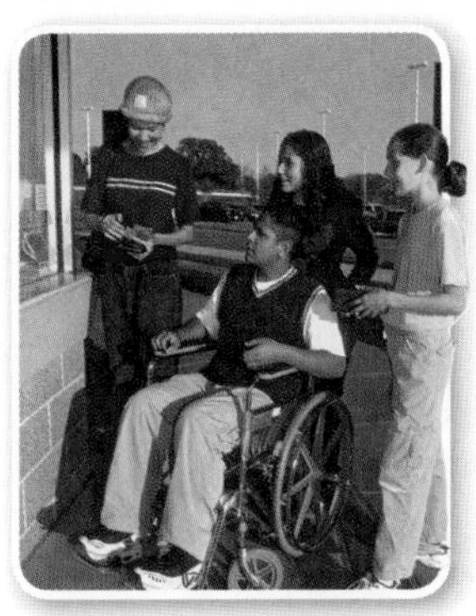

Real-World Link

In 2006, about 1.45 billion movie tickets were sold at an average cost of $6.55. This generated a revenue of $9.49 billion which was up 5.5% from the previous year.

Source: Variety Weekly

37 **ENTERTAINMENT** Janelle and some of her friends went to the movies. Tickets cost $6 per person, and they each received a $1.50 student discount. Each girl also purchased a snack for $2.25. The total cost was $40.50. Solve the equation $6s - 1.5s + 2.25s = 40.50$ to find how many girls went to the movies.

Solve each equation. Check your solution.

38. $\frac{3x}{2} + 4x = 22$

39. $40.77 = \frac{y}{5} + 2.4y + \frac{y}{10}$

40. $\frac{x}{2} + \frac{5x}{6} + \frac{x}{4} = 380$

41. $\frac{-2x + 5}{2} = 17$

42. **MULTIPLE REPRESENTATIONS** In this problem, you will investigate a function. Tia's family is installing a fence around three sides of her backyard as shown at the right. The equation $2w + 24 = f$ represents the relationship between the width of the fenced area and the total amount of fencing needed.

a. **TABULAR** Make a function table to show the amount of fencing needed for widths of 12, 15, and 18 feet.

b. **ALGEBRAIC** Find the width of the fenced area if Tia has 92 feet of fencing.

H.O.T. Problems

Use **H**igher-**O**rder **T**hinking

43. **OPEN ENDED** Write a real-world example that could be solved by using the equation $2x + 7 = 15$. Then solve the equation.

44. **CHALLENGE** The model at the right represents the equation $6y + 1 = 3x + 1$. What is the value of x?

45. **WRITING IN MATH** Describe why knowing how to simplify expressions is important when solving equations.

46. **FIND THE ERROR** Toshiro and Evelina are solving the equation $7 - 2x = -51$. Is either of them correct? Explain your reasoning.

Toshiro

$$7 - 2x = -51$$
$$7 + 7 - 2x = -51 + 7$$
$$2x = -44$$
$$\frac{2x}{2} = \frac{-44}{2}$$
$$x = -22$$

Evelina

$$7 - 2x = -51$$
$$-7 + 7 - 2x = -51 + (-7)$$
$$-2x = -58$$
$$\frac{-2x}{-2} = \frac{-58}{-2}$$
$$x = -29$$

47. **WRITING IN MATH** Evaluate $3(2) + 5$. Then solve the equation $3x + 5 = 11$. How are the problems and solutions similar? How are they different?

Standardized Test Practice

48. The results of a student council fundraiser are shown in the table.

Purchase Price for 144 Pens	Profit for 144 Pens
\$309.60	\$50.40

Use the equation below to find the selling price p of one pen.

$$144p - 309.60 = 50.40$$

A \$1.80
B \$2.15
C \$2.50
D \$2.72

49. Ms. Fraser's total monthly cell phone bill b can be found using the equation $b = 45.60 + 0.10t$, where t represents the number of text messages she made. Find the number of text messages she made in a month in which the total charge was \$56.70.

F 101
G 111
H 125
J 131

50. The distance d that Maxie can run in her first training run is represented by the equation $d = \frac{1}{2}m - 2$. What is the maximum distance m that she can run if her first training run is 3 miles?

A 10 miles
B 8 miles
C 6 miles
D 4 miles

51. **GRIDDED RESPONSE** Jody bought two pairs of jeans. The first pair costs \$12 less than 3 times the cost c of the second pair. The first pair of jeans costs \$45. The equation below can be used to find the cost in dollars of the second pair of jeans.

$$3c - 12 = 45$$

Solve the equation to find the cost of the second pair of jeans.

Spiral Review

Solve each equation. (Lessons 4-3 and 4-4)

52. $36 = -12y$

53. $4 = \frac{x}{14}$

54. $5y = \frac{3}{2}$

55. $x - 13 = -45$

56. $\frac{2}{3} + p = 1$

57. $t + 12.4 = 16.23$

58. **WEATHER** The difference between the record high and low temperatures in Columbus, Ohio, is 128° F. The record high temperature is 106° F. Write and solve an equation to find the record low temperature. (Lesson 4-3)

59. **MONEY** You have saved some money. Your friend has saved \$40 more than you. Write an expression in simplest form that represents the total amount of money you and your friend have saved. (Lesson 2-5)

Skills Review

Evaluate each expression if $x = 4$, $y = -10$, and $z = 14$. (Lessons 2-2, 2-3, and 2-4)

60. xy

61. $y + z$

62. $2x - y$

63. $2z + 2y$

64. xyz

65. $z - 3x + y$

4-6 Writing Equations

Why?

Marisol and Ivy spent a total of \$22 for lunch. Ivy's lunch cost \$5 more than Marisol's. How much did each girl spend for lunch?

a. Whose lunch cost more?

b. How much more?

c. If m represents the cost of Marisol's lunch, how much did Ivy's lunch cost?

d. Write an expression that represents *the sum of Marisol's and Ivy's lunches.*

Then

You have already translated verbal phrases into algebraic expressions. (Lessons 1-2)

Now

- Write two-step equations.
- Solve verbal problems by writing and solving two-step equations.

Math Online

glencoe.com

- Extra Examples
- Personal Tutor
- Self-Check Quiz
- Homework Help

Write Two-Step Equations You can summarize this information by writing an equation. Suppose the total cost of the lunches is \$22.

Words	Marisol's lunch	plus	Ivy's lunch	costs	\$22.
Symbols	Let m = the cost of Marisol's lunch. Let $m + 5$ = the cost of Ivy's lunch.				
Equation	m	$+$	$m + 5$	$=$	22

EXAMPLE 1 Translate Sentences into Equations

Translate each sentence into an equation.

a. Zack has 6 shirts. This is 4 less than twice the number of shirts n that Xavier has.

$6 = 2n - 4$

b. Eight more than the quotient of a number y and -3 is -24.

$8 + \frac{y}{-3} = -24$

c. Jeremy has 13 baseball cards, which is 7 more than 3 times the number m Michael has.

$13 = 7 + 3m$

Check Your Progress

1A. Four more than three times a number x is -26.

1B. Hannah has 24 stickers. This is 6 less than twice the number of stickers n Molly has.

1C. The quotient of a number n and 7, increased by 6, is equal to 12.

Personal Tutor glencoe.com

EXAMPLE 2 Write and Solve an Equation

Juan's father was 29 years old when Juan was born. This year, the sum of their ages is 53. Find their ages.

Let x = Juan's age. Then, $x + 29$ = Juan's father's age.

$x + x + 29 = 53$	**Write the equation.**
$2x + 29 = 53$	**Distributive Property**
$2x + 29 - 29 = 53 - 29$	**Subtraction Property of Equality**
$2x = 24$	**Simplify.**
$x = 12$	**Mentally divide each side by 2.**

Juan is 12 years old. His father is 12 + 29 or 41 years old.

Check Your Progress

2. Deisha saved d dollars last month. This month she saved $8 more than 3 times the amount she saved last month. She saved a total of $141. Write and solve an equation to find how much she saved last month.

Personal Tutor glencoe.com

StudyTip

Equations Look for the words *is, total, equals,* or *is equal to* when you translate sentences into equations.

Two-Step Verbal Problems In some real-world situations, you start with a given amount and then increase it at a certain rate.

Real-World EXAMPLE 3 Solve a Two-Step Verbal Problem

FUNDRAISING Logan collected pledges for the charity walk-a-thon. He will receive total contributions of $68 plus $20 for every mile that he walks. How many miles will he need to walk to raise $348?

Understand He has already raised $68. He can raise another $20 per mile until he reaches $348. You need to find how many miles he needs to walk.

Plan Organize the data for the first few miles in a table. Notice the pattern. Then write an equation to represent the situation.

Number of Miles	Total Amount Raised
0	20(0) + 68 = $68
1	20(1) + 68 = $88
2	20(2) + 68 = $108

Solve Let m = the number of miles.

Then, $20m + 68 = 348$.

$20m + 68 = 348$	**Write the equation.**
$20m + 68 - 68 = 348 - 68$	**Subtraction Property of Equality**
$20m = 280$	**Simplify.**
$m = 14$	**Mentally divide each side by 20.**

Logan needs to walk 14 miles to raise $348.

Check If he walks 14 miles he will have $20 • 14 or an additional $280. The answer seems reasonable. ✓

Check Your Progress

3. **SHOPPING** Jasmine bought 6 CDs, all at the same price. The tax on her purchase was $5.04, and the total was $85.74. What was the price of each CD?

Personal Tutor glencoe.com

Check Your Understanding

Example 1
p. 205

Translate each sentence into an equation.

1. The quotient of a number and 3, less 8, is 16.
2. Tiffani spent $95 for clothes. This is $15 more than 4 times the amount her sister spent for school supplies.
3. Morgan has 98 baseball cards in his collection, which is twelve less than the product of 5 and the number of cards Tyler has.

Example 2
p. 206

Solve each problem by writing and solving an equation.

4. **SHOPPING** Kendra pays $132 for shoes and clothes. The clothes cost $54 more than the shoes. How much do the shoes cost?
5. **CAR WASH** During the spring car wash, the Activities Club washed 14 fewer cars than during the summer car wash. They washed a total of 96 cars during both car washes. How many cars did they wash during the spring car wash?

Example 3
p. 206

6. **FITNESS** A gym charges a $50 activation fee and $17 per month for a membership. If you spend $356, for how many months do you have a gym membership?

Practice and Problem Solving

 = **Step-by-Step Solutions** begin on page R11.
Extra Practice begins on page 810.

Example 1
p. 205

Translate each sentence into an equation.

7. Eighteen more than twice a number is 8.
8. The product of a number and 9, less 20 is 7.
9. There are 48 soccer teams in the Springtown Association. This is three less than three times the number of teams in the Lyon Association.
10. Eileen swam for 85 minutes. This is 21 more minutes than 4 times the number of minutes Ethan swam.

Examples 2 and 3
p. 206

Solve each problem by writing and solving an equation.

11. **BASKETBALL** In 2007, Candace Parker, from the University of Tennessee, made 37 more field goals than she did in 2006. She had a total of 497 field goals for those years. How many field goals did she make in 2006?
12. **CARS** The Marsh family took a vacation that covered a total distance of 1356 miles. The return trip was 284 miles shorter than the first part of the trip. How long was the return trip?
13. **VIDEO GAMES** Three friends share the cost of renting a game system. Each person also rents one game for $8.50. If each person pays $13.25, what is the cost of renting the system?
14. **VACATION** Suppose you purchase 3 identical T-shirts and a hat. The hat cost $21 and you spend $60 in all. How much does each T-shirt cost?

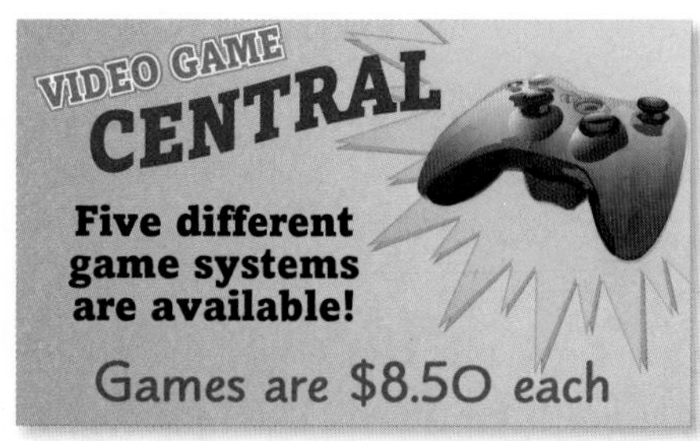

VIDEO GAME
CENTRAL
Five different game systems are available!
Games are $8.50 each

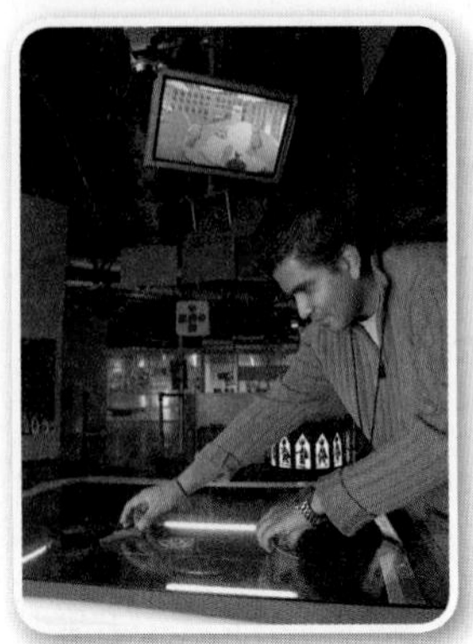

Real-World Career

Computer animation involves changing still images to create the illusion of movement. Animators created 2D and 3D digital images that are found in movies, commercials, and on television.

Excellent drawing skills and formal training in graphic art and a degree in multimedia technology are strong requirements.

Source: WestOne Services

15. BOOK FINES You return a book that was 6 days overdue. Including a previous unpaid balance of \$0.90, your new balance is \$2.40. How much is the daily fine for an overdue book?

16. CONSTRUCTION Tao is building a door. The height of the door is 1 foot more than twice its width. If the door is 7 feet high, what is its width?

17. ENTERTAINMENT In his DVD collection, Domingo has eight more than twice as many animated movies as action movies. If he has 24 animated movies, write and solve an equation to find how many action movies are in his collection.

18. TRAVEL At the start of the school trip to Washington D.C., the tour bus has 40 gallons of gasoline in the fuel tank. Each hour, the bus uses 7 gallons of gasoline. The bus will stop for gas when there are 10 gallons left.

a. Make a table to show how many gallons of gasoline are remaining in the tank after each hour.

b. Write and solve an equation to find how many hours will pass before the bus will have to stop for gasoline.

19. MULTIPLE REPRESENTATIONS In this problem, you will use tables, graphs, and equations to solve a problem. Misty is saving money to buy an MP3 player that costs \$212. She has already saved \$47 and plans to save an additional \$15 per week.

Number of Weeks	Amount of Savings (\$)
1	62
2	■
3	■
4	■

a. ALGEBRAIC Write a variable expression to represent the amount of money saved after w weeks. Then use the expression to complete the table at the right.

b. GRAPHICAL Make a scatter plot of the data in the table. How can you use the graph to find the number of weeks it will take her to save enough money for the MP3 player?

c. ALGEBRAIC Write and solve an equation to find the number of weeks it will take her to save the money.

d. VERBAL Compare the methods for finding the solution that you used in parts **b** and **c.**

H.O.T. Problems

Use Higher-Order Thinking Skills

20. OPEN ENDED Write a two-step equation with a solution of 6. Write the equation using both words and symbols.

21. NUMBER SENSE An example of two consecutive even numbers is 4 and 6. They can be represented by n and $n + 2$. Find 3 consecutive even numbers whose sum is 30.

22. REASONING The equations $\frac{x+4}{5} = 20$ and $\frac{x}{5} + 4 = 20$ are both two-step equations. Compare and contrast how to solve them.

23. CHALLENGE Emelia discovered that if she takes three-fourths of her age and adds 9, it produces the same result as when she takes one-fourth of her age and adds 21. How old is Emelia?

24. WRITING IN MATH Explain how two-step equations are used to represent real-world problems.

Standardized Test Practice

25. An electrician charges \$35 for a house call and \$80 per hour for each hour worked. If the total charge was \$915, which equation would you use to find the number of hours n that the electrician worked?

A $35n + 2n(80) = 915$

B $80 + 35n = 915$

C $35 + (80 - n) = 915$

D $35 + 80n = 915$

26. Belinda scored 16 goals this season. This is 4 more than three times the number she scored last season. Which equation could you use to find how many goals she scored last season?

F $4n + 3 = 16$

G $3n + 4 = 16$

H $4n - 3 = 16$

J $3n - 4 = 16$

27. EXTENDED RESPONSE Sheila wants to buy a mountain bike for \$180. She has already saved \$60. For the next six months, she wants to save an equal amount each month in order to save the total amount.

a. Write an equation to represent the situation.

b. How much money will Sheila need to save each month?

28. You and your friend spent a total of \$15 for lunch. Your friend's lunch cost \$3 more than yours did. How much did you spend for lunch?

A \$6

B \$7

C \$8

D \$9

Spiral Review

Solve each equation. Check your solution. (Lessons 4-3, 4-4, and 4-5)

29. $x + 12 = -10$

30. $-\frac{y}{6} = -2$

31. $2p + 13 = -7$

32. $7y + 3 = -11$

33. $-8t - 9 = -41$

34. $8z = 14$

35. CONCERTS A concert ticket costs t dollars, a hamburger costs h dollars, and soda costs s dollars. Write an expression that represents the total cost of a ticket, hamburger, and soda for n people. (Lesson 4-1)

36. AIR PRESSURE The air pressure decreases as the distance from Earth increases. The table shows the air pressure for certain distances. (Lesson 2-6)

a. Write a set of ordered pairs for the data.

b. Graph the data.

c. State the domain and the range of the relationship shown.

Air Pressure

Height (mi)	Pressure (lb/in^2)
0 (sea level)	14.7
1	10.2
2	6.4
3	4.3
4	2.7
5	1.6

37. FOOD The SubShop had 36, 45, 41, and 38 customers during the lunch hour the last four days. Find the mean of the number of customers. (Lesson 2-5)

Skills Review

Use the Distributive Property to write each expression as an equivalent algebraic expression. (Lesson 4-1)

38. $4(x + 3)$

39. $8(y - 2)$

40. $-6(z - 7)$

41. $-2(-9 - p)$

Study Guide and Review

Math Online glencoe.com
- STUDY TO GO
- Vocabulary Review

Chapter Summary

Key Concepts

Distributive Property (Lesson 4-1)

- For any numbers a, b, and c, $a(b + c) = ab + ac$.

Properties of Equality (Lessons 4-3 through 4-4)

Addition Property of Equality
- For any numbers a, b, and c, if $a = b$, then $a + c = b + c$.

Subtraction Property of Equality
- For any numbers a, b, and c, if $a = b$, then $a - c = b - c$.

Division Property of Equality
- For any numbers a, b, and c, where $c \neq 0$, if $a = b$, then $\frac{a}{c} = \frac{b}{c}$.

Multiplication Property of Equality
- For any nonzero numbers a, b, and c, if $a = b$, then $ca = cb$.

Solving Equations (Lessons 4-3 through 4-5)

- To solve a single-step equation, use the Properties of Equality and inverse operations.
- To solve a two-step equation, undo operations in reverse order of the order of operations.

Writing Equations (Lesson 4–6)

- You can write verbal sentences as equations.
- Some real-world situations start with a given amount and then increase it at a certain rate. These situations can be represented by two-step equations.

FOLDABLES® Study Organizer

Be sure the Key Concepts are noted in your Foldable.

4-1
4-2
4-3
4-4
4-5
4-6

Key Vocabulary

coefficient (p. 178)
constant (p. 178)
Distributive Property (p. 171)
equation (p. 184)
equivalent equation (p. 184)
equivalent expressions (p. 171)
inverse operation (p. 184)
like terms (p. 178)
simplest form (p. 179)
simplifying the expression (p. 179)
solution (p. 184)
solving the equation (p. 184)
term (p. 178)
two-step equation (p. 199)

Vocabulary Check

Complete each sentence with the correct term. Choose from the list above.

1. Expressions that have the same value are called ______.
2. An algebraic expression is in ______ if it has no like terms and no parentheses.
3. A term without a variable is called a(n) ______.
4. The numerical part of a term that contains a variable is called the ______.
5. A value for the variable that makes an equation true is called a(n) ______.
6. Two ______ have the same solution.
7. The ______ allows you to multiply a sum or difference by a number.
8. An equation that contains two steps is called a(n) ______.
9. Addition and subtraction are examples of ______.
10. ______ contain the same variable.

MIXED PROBLEM SOLVING
For mixed problem-solving practice, see page 846.

Lesson-by-Lesson Review

4-1 The Distributive Property (pp. 171–176)

Use the Distributive Property to write each expression as an equivalent algebraic expression.

11. $(y + 3)7$

12. $-2(a - 7)$

13. $-1(b - 9)$

14. $(8m - 4)(-5)$

15. **FOOD** The Stuart family has 5 members. They each purchase a soda at \$2.50 each and a hotdog at \$3.50 each. Use mental math to find the total cost of the food. Justify your answer by using the Distributive Property.

EXAMPLE 1

Use the Distributive Property to write $3(x - 6)$ as an equivalent algebraic expression.

$3(x - 6) = 3x - 3 \cdot 6$

$= 3x - 18$ Simplify.

4-2 Simplifying Algebraic Expressions (pp. 178–183)

Simplify each expression.

16. $6a + 5a$

17. $3x + 6x$

18. $7m - 2m + 3$

19. $6x - 3 + 2x + 5$

20. $a + 6(a + 3)$

21. $2(b + 3) + 3b$

22. **BASKETBALL** Karen made 5 less than 4 times the number of free throws that Kimi made. Write an expression in simplest form that represents the total number of free throws.

EXAMPLE 2

Simplify $-6x + 5 + x$.

$-6x + 5 + x$

$= -6x + x + 5$ Commutative Property

$= [(-6) + 1]x + 5$ Distributive Property

$= -5x + 5$ Simplify.

4-3 Solving Equations by Adding or Subtracting (pp. 184–189)

Solve each equation. Check your solution.

23. $x + 4 = 10$

24. $a - 9.45 = -10.6$

25. $x + 3\frac{1}{4} = 2\frac{1}{5}$

26. $-5.3 = m + 4.1$

27. $p - 6 = 12$

28. $s - \frac{2}{9} = \frac{2}{3}$

29. **REPORTS** Sonia needs to add 13 more pages to complete an assignment that is supposed to be 37 pages long. Write and solve an addition equation to find how many pages she has already completed.

EXAMPLE 3

Solve $x + 6 = 11$.

$x + 6 = 11$ Write the equation.

$x + 6 - 6 = 11 - 6$ Subtraction Property of Equality

$x = 5$ Simplify.

EXAMPLE 4

Solve $a - 4 = -3$.

$a - 4 = -3$ Write the equation.

$a - 4 + 4 = -3 + 4$ Addition Property of Equality

$a = 1$ Simplify.

CHAPTER 4

Study Guide and Review

4-4 Solving Equations by Multiplying or Dividing (pp. 191–196)

Solve each equation. Check your solution.

30. $12m = 24$

31. $\frac{x}{5} = 4$

32. $-2x = 22$

33. $5x = 25$

34. $\frac{x}{-4} = 16$

35. $\frac{1}{6}x = -4$

36. FASHION Rosa is making scarves for her friends. Each scarf requires 48 inches of material. Write and solve a multiplication equation to find how many scarves Rosa can make if she has 336 inches of material.

EXAMPLE 5

Solve $-4x = -32$.

$-4x = -32$ Write the equation.

$\frac{-4x}{-4} = \frac{-32}{-4}$ Division Property of Equality

$x = 8$ Simplify.

EXAMPLE 6

Solve $\frac{a}{-2} = 5$.

$\frac{a}{-2} = 5$ Write the equation.

$-2\left(\frac{a}{-2}\right) = -2(5)$ Multiplication Property of Equality

$a = -10$ Simplify.

4-5 Solving Two-Step Equations (pp. 199–204)

37. $3 + 4c = 15$

38. $2.1n - 5.31 = 18$

39. $\frac{a}{3} + 2 = 5$

40. $\frac{x}{5} - 3 = 7$

41. $\frac{4}{7} + 2p = \frac{2}{7}$

42. $0.12t - 0.6 = -0.06$

43. BOOKS Nate read 10 more books than Maren for the summer reading program. The total number of books they read is 60. Solve $x + x + 10 = 60$ to find the number of books Nate read.

EXAMPLE 7

Solve $3x + 5 = 29$.

$3x + 5 = 29$ Write the equation.

$3x + 5 - 5 = 29 - 5$ Subtraction Property of Equality

$3x = 24$ Simplify.

$\frac{3x}{3} = \frac{24}{3}$ Division Property of Equality

$x = 8$ Simplify.

4-6 Writing Equations (pp. 205–209)

Translate each sentence into an equation. Then find each number.

44. Toya bought some fruit for $5 and 3 boxes of cereal and spent a total of $17.

45. Six less than twice a number is −22.

46. MONEY Noelle spent $36 on books and pens. She spent $12 more on books than she did on pens. How much did she spend on books?

EXAMPLE 8

The product of a number and 6 is −36. Write and solve an equation to find the number.

$6n = -36$ Write the equation.

$\frac{6n}{6} = \frac{-36}{6}$ Division Property of Equality

$n = -6$ Simplify.

CHAPTER 4

Practice Test

Math Online glencoe.com
Chapter Test

Use the Distributive Property to write each expression as an equivalent algebraic expression.

1. $6(s + 10)$ **2.** $9(a - 4)$

3. $-5(3 - b)$ **4.** $11(m + 7)$

5. **ENTERTAINMENT** Suppose you pay $15 per hour to go horseback riding. You ride 2 hours today and plan to ride 4 more hours this weekend.

a. Write two different expressions to find the total cost of horseback riding.

b. Find the total cost.

Simplify each expression.

6. $x + 3x$ **7.** $10 + 6x - 11 + 7x$

8. $14n - 3(n + 8)$ **9.** $-7b - 5(b - 4)$

10. **MUSIC** Omar and Deb each have a digital music player. Deb has 37 more songs on her player than Omar has on his player. Write an expression in simplest form that represents the total number of songs on both players.

11. **MULTIPLE CHOICE** The table shows the prices of different items at a snack bar.

Item	Cost ($)
hot dog	2.50
drink	1.75
fries	2.00
hamburger	3.00
chips	0.75

After buying some items, you receive $3.75 in change. If you paid with a $10 bill, which equation could *not* be used to determine the amount of money you spent?

A $3.75 + n = 10$ **C** $10 - 3.75 = n$

B $10 - n = 3.75$ **D** $10 + n = 3.75$

Solve each equation. Check your solution.

12. $a - (-12) = 6$ **13.** $16 + t = -9$

14. $g + (-18) = -36$ **15.** $-b - 15 = -21$

16. $m + 3.75 = -4.15$ **17.** $\frac{7}{9} - d = \frac{1}{3}$

Solve each equation. Check your solution.

18. $\frac{r}{13} = -4$ **19.** $-75t = 300$

20. $-8t = -72$ **21.** $\frac{c}{3} = -42$

22. $-\frac{m}{12} = \frac{3}{5}$ **23.** $0.4d = 3.6$

24. **GENETICS** Approximately one-seventh of the people in the United States are left-handed. The population of the United States is about 300 million. Write and solve an equation to estimate how many people in the United States are left-handed.

MEASUREMENT **Use the table to write and solve an equation to find each quantity.**

Customary System (capacity)
1 cup = 8 fluid ounces
1 pint = 2 cups
1 quart = 2 pints
1 gallon = 4 quarts

25. the number of cups in 7 pints

26. the number of quarts in 16 pints

27. the number of cups in 2 gallons

Solve each equation. Check your solution.

28. $16 + 5w = 31$ **29.** $28 = 4g - 4$

30. $211 - \frac{k}{3} = 111$ **31.** $7.2j - 1.9 = 3.5$

32. **TRAVEL** Mr. Carter is renting a car from an agency that charges $20 per day plus $0.15 per mile. He has a budget of $80 per day. Write and solve an equation to find the maximum number of miles he can drive each day.

33. **MULTIPLE CHOICE** Kenneth signed up to receive Internet service for $13 per month plus a $30 start-up fee. Which equation could be used to find the number of months he can receive Internet service for $134?

F $\$134 + 30 = \$13m$

G $\$30 - \$13m = \$134$

H $\$13 + 30m = \$134$

J $\$30 + \$13m = \$134$

CHAPTER 4 Preparing for Standardized Tests

Write and Solve an Equation

Many standardized test questions can be solved by writing and solving an equation. Follow the steps below to help you successfully solve these types of problems.

Strategies for Writing and Solving Equations

Step 1

Read the problem statement carefully.

Ask yourself:

- What am I being asked to solve?
- What information is given in the problem?
- What is the unknown quantity that I need to find?

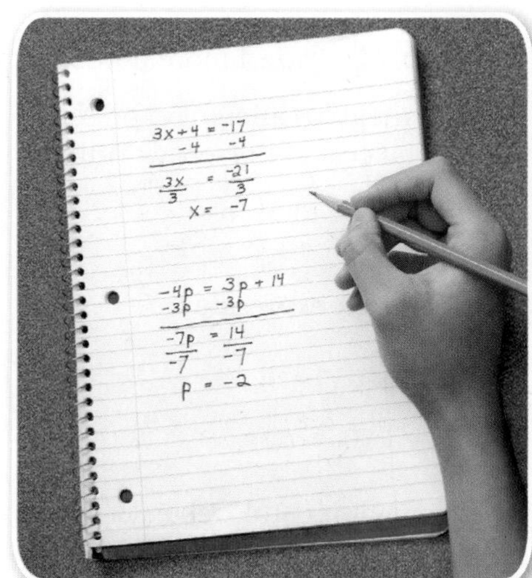

Step 2

Translate the problem statement into an equation.

- Assign a variable to the unknown quantity.
- Write the word sentence as a mathematical number sentence.
- Look for keywords such as *is, is the same as, is equal to,* or *is identical to* that indicate where to place the equal sign.

Step 3

Solve the equation.

- Solve for the unknown in the equation.
- Check your answer to be sure it makes sense.

EXAMPLE

Read the problem. Identify what you need to know. Then use the information in the problem to solve.

Lisa baked 48 cookies for the school bake sale. This is three times the number of cookies that Kyle baked. How many cookies did Kyle bake for the sale?

A 16 cookies

B 21 cookies

C 24 cookies

D 28 cookies

Read the problem carefully. You know how many cookies Lisa baked. You also know that this is three times the number of cookies Kyle baked.

The unknown quantity that you need to find is the number of cookies that Kyle baked.

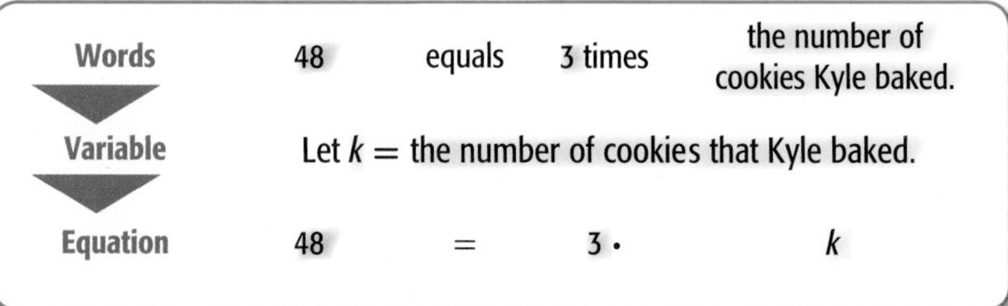

Words
48
equals
3 times
the number of cookies Kyle baked.
Variable
Let k = the number of cookies that Kyle baked.
Equation
48
=
3 ·
k

Solve the equation for k.

$48 = 3k$ **Write the equation.**

$\frac{48}{3} = \frac{3k}{3}$ **Divide each side by 3.**

$16 = k$ **Simplify.**

So, Kyle baked 16 cookies for the bake sale. The correct answer is A.

Exercises

Read each problem. Identify what you need to know. Then use the information in the problem to solve.

1. Angelica is 4 years older than her sister, Suzie. Angelica is 26 years old. How old is Suzie?

A 19 years old

B 20 years old

C 21 years old

D 22 years old

2. The quotient of a number and 5 is equal to 19. What is the number?

F 60

G 75

H 80

J 95

3. Mr. Lombardi is installing rectangular tiles like the one shown in his bathroom shower. The perimeter of one tile is 38 inches.

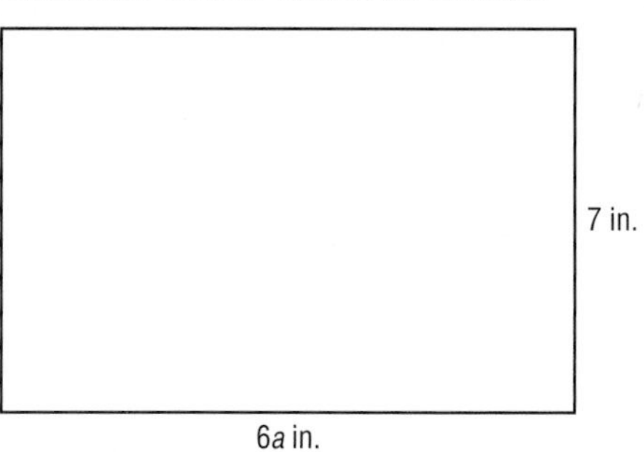

What is the length of the tile?

A 11 in.

B 12 in.

C 13 in.

D 14 in.

CHAPTER 4

Standardized Test Practice

Cumulative, Chapters 1 through 4

Multiple Choice

Read each question. Then fill in the correct answer on the answer document provided by your teacher or on a sheet of paper.

1. The 7th graders have collected 86 cans for a food drive so far. This is fifteen less than the number of cans collected by the 8th graders. Which of the following equations could be used to find the number of cans collected by the 8th graders?

   A $n - 15 = 86$ **C** $n - 86 = 15$

   B $n + 15 = 86$ **D** $n + 86 = 15$

2. Refer to Exercise 1. How many cans have the 8th graders collected so far?

   F 71 **H** 101

   G 91 **J** 105

3. The recipe shown makes one batch of chocolate chip cookies. How many batches can you make if you have $12\frac{1}{2}$ cups of flour?

   Recipe

   1 cup softened butter (2 sticks)
   $\frac{1}{2}$ cup granulated sugar
   $1\frac{1}{2}$ cups packed brown sugar
   2 eggs
   $2\frac{1}{2}$ cups all-purpose flour
   $\frac{3}{4}$ teaspoon salt
   1 teaspoon baking powder
   1 teaspoon baking soda
   18 ounces chocolate chips

   A 8 **C** 6

   B 7 **D** 5

4. Write an expression in simplest form for the perimeter of the rectangle shown.

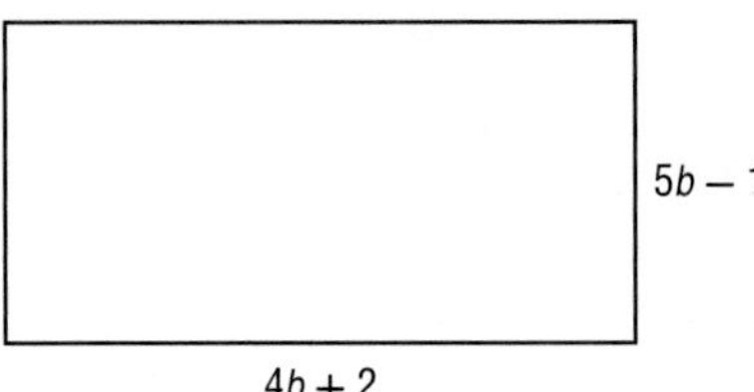

   F $18b + 4$ **H** $16b + 2$

   G $18b + 2$ **J** $9b + 1$

5. What is the least common denominator of $\frac{7}{12}$ and $\frac{5}{8}$?

   A 12 **C** 24

   B 16 **D** 48

6. Myriah's long distance telephone plan is shown. If she only wants to spend \$10 per month on long distance, how many minutes can she use?

   Long Distance Charges
   - \$3.95 per month
   - \$0.05 per minute

   F 121 minutes **H** 135 minutes

   G 126 minutes **J** 142 minutes

7. Rectangle *WXYZ* has vertices $W(0, -5)$, $X(5, -5)$, and $Y(5, 3)$. What are the coordinates of point *Z*?

   A (0, 5) **C** (0, 3)

   B (5, 0) **D** (3, 0)

Test-TakingTip

Question 6 To find how much money she can spend on minutes, find \$10 − \$3.95. Then write a multiplication equation to find the total number of minutes.

Short Response/Gridded Response

Record your answers on the answer sheet provided by your teacher or on a sheet of paper.

8. **GRIDDED RESPONSE** Josiah put \$45 in a savings account. If he saves an additional \$10 per week, the equation $\$10w = \$175 - \$45$ represents how long it will take him to save \$175. Find the number of weeks Josiah will need to save in order to reach \$175.

9. Douglas charges \$18 per yard to mow lawns in his neighborhood. Yesterday he mowed 3 lawns. Today he mowed 4 lawns.

a. Write two equivalent expressions for the total amount of money Douglas earned mowing lawns over the two days.

b. How much did he earn altogether?

10. **GRIDDED RESPONSE** What is the least common denominator of $\frac{3}{16}$ and $\frac{1}{3}$?

11. At the start of a long trip, Jasmine has 19 gallons of gasoline in her car. Every 30 minutes, her car uses a gallon of gasoline. Jasmine plans to stop and refuel when there are 3 gallons of gasoline left in her tank.

a. Write an equation to represent the amount of gasoline in her car's tank after h half-hours.

b. Solve the equation from part **a** to find how long Jasmine will be able to drive before she needs to refuel.

12. Evaluate $x \cdot y \div 4$ when $x = 10$ and $y = -3$.

13. **GRIDDED RESPONSE** Evaluate abc if $a = -\frac{2}{5}$, $b = \frac{20}{32}$, and $c = -\frac{4}{10}$.

14. What mathematical property is illustrated below?

$$(p + 5) + 13 = p + (5 + 13)$$

Extended Response

Record your answers on a sheet of paper. Show your work.

15. On Wednesday, Carlos spent d dollars for lunch. He spent \$0.80 less than this amount for lunch on Thursday. Friday, he spent twice as much as he did on Thursday.

a. Write an expression for the amount of money Carlos spent for lunch on Thursday.

b. Write an expression for the amount Carlos spent for lunch on Friday.

c. Write an expression for the total amount of money Carlos spent for lunch on all three days.

d. If Carlos spent \$8 for lunch on Wednesday, how much money did he spend all three days?

Need Extra Help?															
If you missed Question...	1	2	3	4	5	6	7	8	9	10	11	12	13	14	15
Go to Lesson or Page...	4-3	4-3	3-4	4-2	3-6	4-5	1-4	4-5	4-1	3-6	4-6	2-5	3-3	1-3	4-2

CHAPTER 5 Multi-Step Equations and Inequalities

Then

In Chapter 4, you learned how to solve simple equations.

Now

In Chapter 5, you will:

- Use the Distributive Property to solve equations and inequalities.
- Select and use appropriate operations to solve problems and justify solutions.

Why?

BUDGETS Many students receive a monthly allowance. A budget can help you determine how much of your allowance you can spend or save every month. You can use an inequality to calculate the number of months that it will take to save enough money to purchase an item.

Multi-Step Equations and Inequalities
Activity
Write an equation to determine the cost of purchasing the bicycle and the helmet if the sales tax is 6 cents (0.06) on every dollar.
Move the prices from the items to complete the expression.
Cost (C) = [ ] + 26 + (180 + [ ])0.06
$180
$26
2/5

Math *in Motion*, Animation glencoe.com

Get Ready for Chapter 5

Diagnose Readiness You have two options for checking Prerequisite Skills.

Text Option Take the Quick Check below. Refer to the Quick Review for help.

QuickCheck

Solve each equation. Check your solution. (Lesson 4-5)

1. $-3m - 8 = 10$
2. $5n - 9 = 6$
3. $\frac{d}{9} + 5 = 15$
4. $-3 = 7 + \frac{f}{-2}$
5. $-18 = 4b + 10$
6. $-9 - \frac{h}{4} = 5$
7. **SALES** Suppose a computer costs \$600. Tony pays a down payment of \$150 and plans to pay the balance in 6 equal installments. How much will each installment be?

QuickReview

EXAMPLE 1

Solve $\frac{r}{4} + 6 = 5$.

$\frac{r}{4} + 6 = 5$	Write the equation.
$\frac{r}{4} + 6 - 6 = 5 - 6$	Subtract 6 from each side.
$\frac{r}{4} = -1$	Simplify.
$4\left(\frac{r}{4}\right) = 4(-1)$	Undo division. Multiply each side by 4.
$r = -4$	Simplify.

Find each sum or difference. (Lessons 2-2 and 2-3)

8. $-42 + (-23)$
9. $42 + (-79)$
10. $-27 + 5$
11. $15 + (-88)$
12. $12 - 60$
13. $-35 - (-35)$
14. **STOCKS** The stock market fell 507.99 points on October 19, 1987. If the stock market was 2246.73 points at the beginning of the day, what was its value at the end of the day?

EXAMPLE 2

Find $-30 - (-42)$.

$-30 - (-42) = -30 + 42$	To subtract -42, add 42.
$= 12$	Simplify.

Find each product or quotient. (Lessons 2-4 and 2-5)

15. $5(-8)$
16. $-10(-12)$
17. $-3(5)(-8)$
18. $54 \div (-2)$
19. $-20 \div 5$
20. $-27 \div (-3)$
21. **CHEMISTRY** A solution cooled at a rate of 6°F every 5 minutes. What integer represents the change in the solution's temperature in $\frac{1}{2}$ hour?

EXAMPLE 3

Find $6(-15)$.

$6(-15) = -90$	The factors have different signs so the product is negative.

Online Option **Math Online** Take a self-check Chapter Readiness Quiz at glencoe.com.

Get Started on Chapter 5

You will learn several new concepts, skills, and vocabulary terms as you study Chapter 5. To get ready, identify important terms and organize your resources. You may wish to refer to **Chapter 0** to review prerequisite skills.

FOLDABLES® Study Organizer

Solving Equations Make this Foldable to help you organize information about equations and inequalities. Begin with three sheets of $8\frac{1}{2}'' \times 11''$ paper.

1. **Stack** 3 sheets of paper $\frac{3}{4}$ inch apart.

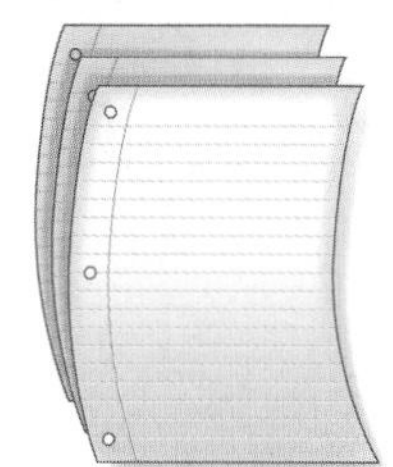

2. **Roll** up the bottom edges. All tabs should be the same size.

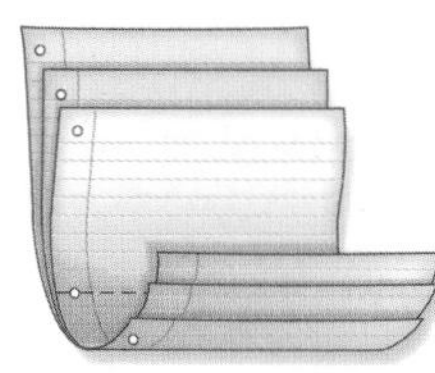

3. **Crease** and staple along the fold.

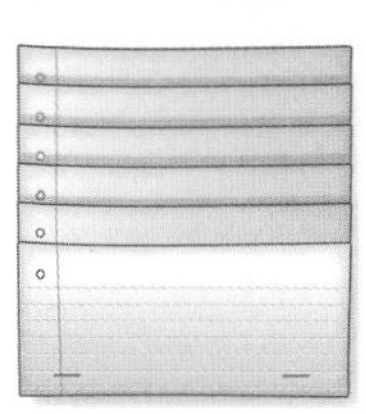

4. **Label** the tabs with topics from the chapter.

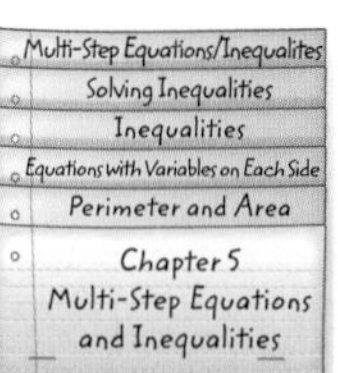

Math Online glencoe.com

- Study the chapter online
- Explore **Math in Motion**
- Get extra help from your own **Personal Tutor**
- Use **Extra Examples** for additional help
- Take a **Self-Check Quiz**
- **Review Vocabulary** in fun ways

New Vocabulary

English		Español
formula	p. 221	fórmula
perimeter	p. 221	perímetro
area	p. 222	área
inequality	p. 234	desigualdad
null Set or empty Set	p. 250	conjunto vacío
identity	p. 250	identidad

Review Vocabulary

Distributive Property • p. 171 • Propiedad distributíva
To multiply a sum by a number, multiply each number in parentheses by the number outside the parentheses

Properties of Equality • Propiedad de igualdad	
Addition (p. 184)	For any numbers a, b, and c, if a $= b$, then $a + c = b + c$.
Subtraction (p. 185)	For any numbers a, b, and c, if $a = b$, then $a - c = b - c$.
Multiplication (p. 193)	For any numbers a, b, and c, if $a = b$, then $ca = cb$.
Division (p. 191)	For any numbers a, b, and c, where $c \neq 0$, if $a = b$, then $\frac{a}{c} = \frac{b}{c}$.

Multilingual eGlossary glencoe.com

5-1

Perimeter and Area

Then

You have already found values of algebraic expressions by substituting values for the variables. (Lesson 1-2)

Now

- Solve problems involving the perimeters of triangles and rectangles.
- Solve problems involving the areas of triangles and rectangles.

New Vocabulary

formula
perimeter
area

Math Online

glencoe.com

- Extra Examples
- Personal Tutor
- Self-Check Quiz
- Homework Help

Why?

In professional baseball, the infield diamond is actually a square that measures 90 feet on each side.

a. How far does a player run if he runs from home plate to second base?

b. How far does a player run if he hits a home run?

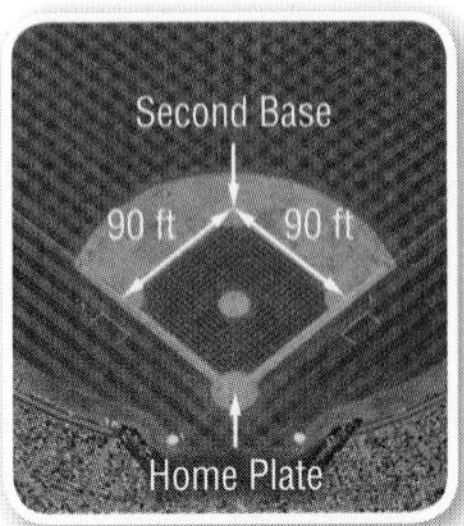

Perimeter A **formula** is an equation that shows a relationship among certain quantities. Formulas are commonly used in measurement. For example, the distance around a geometric figure is called the **perimeter**.

The formula $P = 4s$, where s represents the distance from home plate to first base, can be used to find the perimeter of the infield.

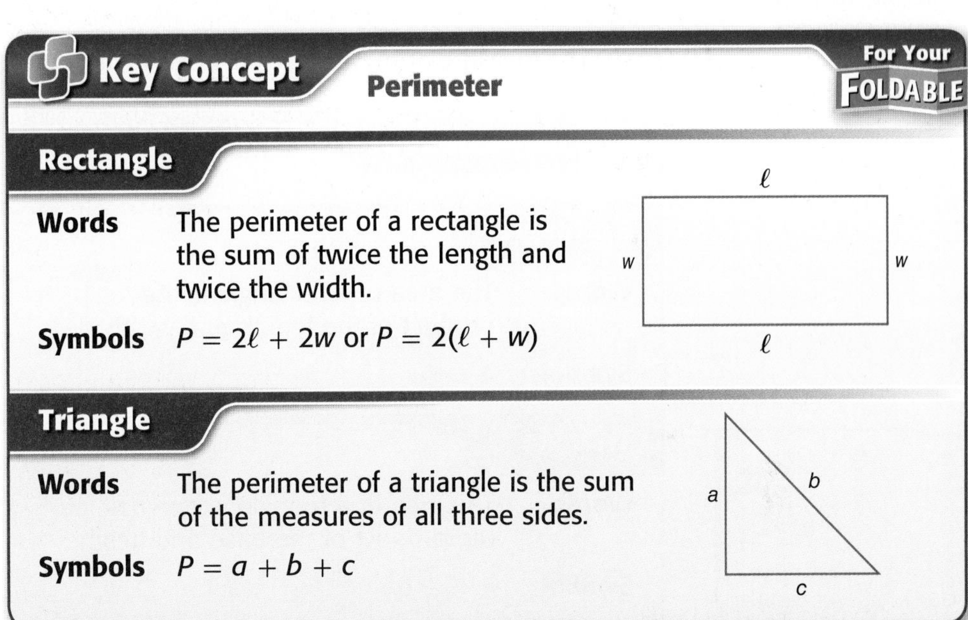

Key Concept Perimeter — For Your FOLDABLE

Rectangle

Words The perimeter of a rectangle is the sum of twice the length and twice the width.

Symbols $P = 2\ell + 2w$ or $P = 2(\ell + w)$

Triangle

Words The perimeter of a triangle is the sum of the measures of all three sides.

Symbols $P = a + b + c$

EXAMPLE 1 Find the Perimeter

Find the perimeter of the triangle.

$P = a + b + c$	**Write the formula for perimeter.**
$P = 12.8 + 28.5 + 17$	**Replace *a* with 12.8, *b* with 28.5, and *c* with 17.**
$P = 58.3$ cm	**Simplify.**

The perimeter is 58.3 centimeters.

Check Your Progress

1. Find the perimeter of a rectangle with length 15.2 meters and width 10.5 meters.

Personal Tutor glencoe.com

EXAMPLE 2 Find the Length

The perimeter of a rectangle is 42 feet. Its width is 10 feet. Find the length.

$P = 2\ell + 2w$	Write the formula for perimeter.
$42 = 2\ell + 2(10)$	Replace P with 42 and w with 10.
$42 = 2\ell + 20$	Simplify.
$42 - 20 = 2\ell + 20 - 20$	Subtraction Property of Equality
$22 = 2\ell$	Simplify.
$\frac{22}{2} = \frac{2\ell}{2}$	Division Property of Equality
$11 = \ell$	Simplify.

The length of the rectangle is 11 feet.

Check Your Progress

2. The perimeter of a rectangle is 26 yards. Its length is 8 yards. Find the width.

Personal Tutor glencoe.com

StudyTip

Units of Measure When finding perimeter, use linear units such as feet, inches, or centimeters. When finding area, use square units such as square feet, square inches, or square centimeters.

Area The measure of the surface enclosed by a figure is its **area**.

Key Concept Area

For Your FOLDABLE

Rectangle

Words The area of a rectangle is the product of the length and width.

Symbols $A = \ell w$

w ℓ

Triangle

Words The area of a triangle is one-half the product of the base and height.

Symbols $A = \frac{1}{2}bh$

h b

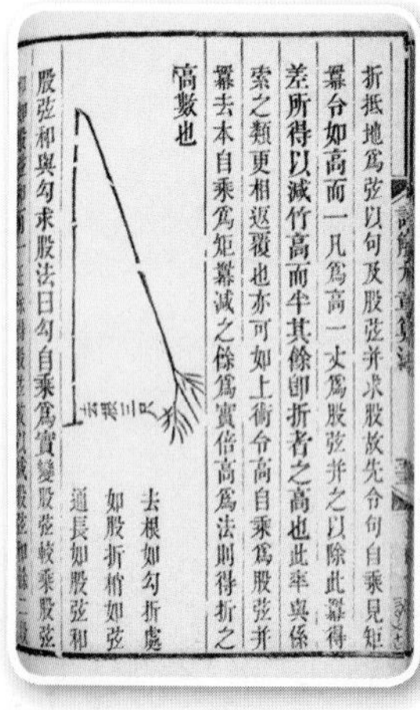

Math History Link

Jiuzhang Suanshu (c. 100 B.C.)
The Jiuzhang Suanshu or Nine Chapters on the Mathematical Art is a Chinese handbook of mathematics. The first chapter deals with areas of shapes.

Source: Clark University

EXAMPLE 3 Find the Area

The base of a triangle is 12 inches, and its height is $5\frac{1}{4}$ inches. Find the area.

$A = \frac{1}{2}bh$	Write the formula for area.
$A = \frac{1}{2} \cdot 12 \cdot 5\frac{1}{4}$	Replace b with 12 and h with $5\frac{1}{4}$.
$A = 31\frac{1}{2}$	Simplify. This can also be written as 31.5 square inches.

The area is 31.5 square inches.

Check Your Progress

3. The width of a rectangle is 11.4 meters and its length is 18.9 meters. Find the area.

Personal Tutor glencoe.com

Real-World EXAMPLE 4 Find the Width

DECORATING Tyler has enough blue paint to cover 800 square inches. He wants to paint a rectangular stripe on his bedroom wall that is 120 inches long. Find the width of the section he can paint.

Understand You know Tyler wants to paint a rectangle that is 120 inches long. He has enough paint to cover 800 square inches. You need to find the width of the rectangle.

Plan Use the formula $A = \ell w$ to find the width.

Solve Solve for w. Then substitute.

$A = \ell w$	**Write the formula for area.**
$\frac{A}{\ell} = \frac{\ell w}{\ell}$	**Divide each side by ℓ.**
$\frac{A}{\ell} = w$	**Simplify.**
$\frac{800}{120} = w$	**Replace *A* with 800 and ℓ with 120.**
$6\frac{2}{3} = w$	**Simplify.**

The stripe should be $6\frac{2}{3}$ inches wide.

Check Check by multiplying. $120 \cdot 6\frac{2}{3} = 800$ ✓

> **StudyTip**
>
> **Literal Equations** In Example 4, the area formula contains 3 variables. When an equation or formula has more than one variable, it is called a *literal equation.* You can apply the properties of equality to solve for any of the variables.

Check Your Progress

4. **FLAGS** The area of the triangular flag case shown is 69 square inches. Find the height of the case.

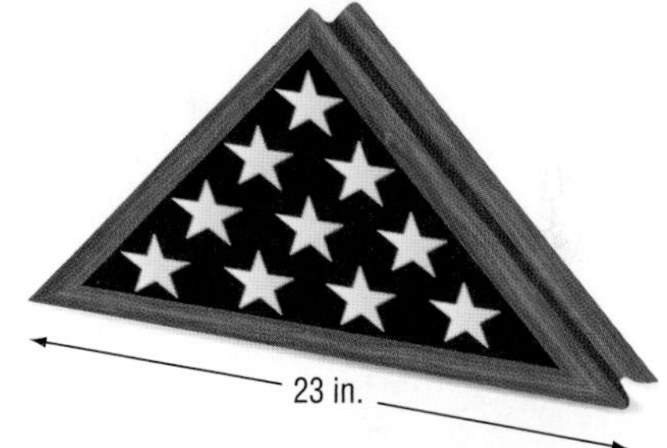

Personal Tutor glencoe.com

Check Your Understanding

Examples 1 and 3 pp. 221–222

1. Find the perimeter and area of the rectangle.

2 cm
12 cm

Example 2 p. 222

2. The perimeter of a rectangle is 45 meters. Its width is 6 meters. What is the length of the rectangle?

Example 4 p. 223

3. **CRAFTS** Dena's mother is making a rectangular quilt. She has 117 squares and wants the quilt to be nine squares wide. How many squares will make up the length of the quilt?

Practice and Problem Solving

● = **Step-by-Step Solutions** begin on page R11.
Extra Practice begins on page 810.

Examples 1 and 3 pp. 221–222

Find the perimeter and area for each figure.

4.

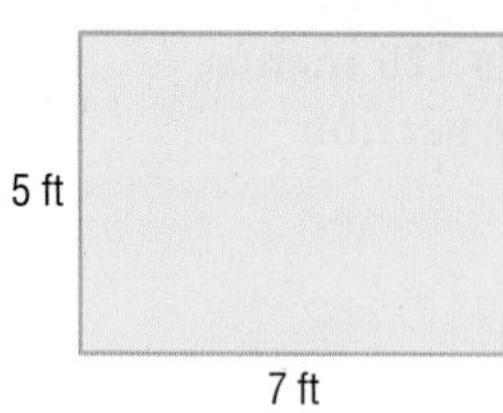

5.

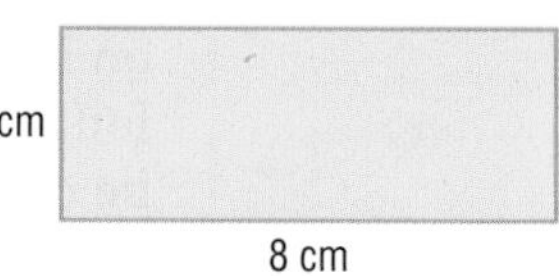

6.

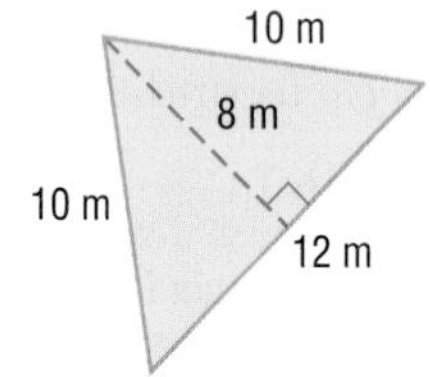

7.

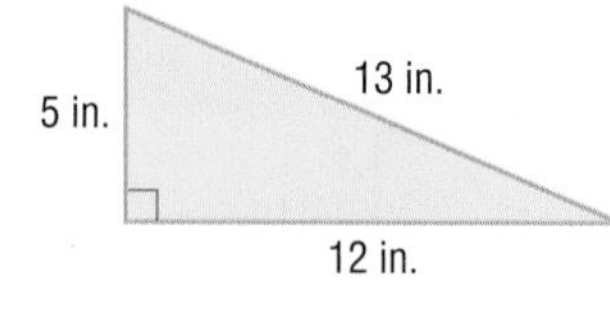

Example 2 p. 222

8. A rectangle has a length of 16.3 meters and a perimeter of 80.6 meters. What is the width of the rectangle?

9. Find the third side of a triangle if the perimeter is $124\frac{1}{4}$ feet and two of the sides each measure $36\frac{3}{4}$ feet.

Example 4 p. 223

10. PARKS The city park shown at the right has an area of 6889 square meters. The park's length along Front Street is 83 meters. Find the length of the park along Second Avenue.

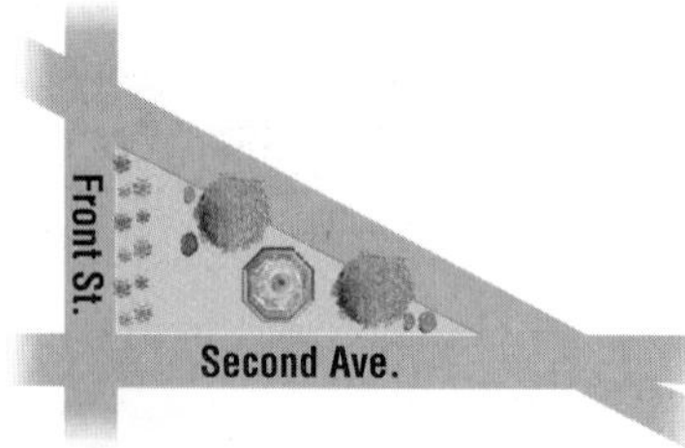

Find the missing dimension for each figure.

11.

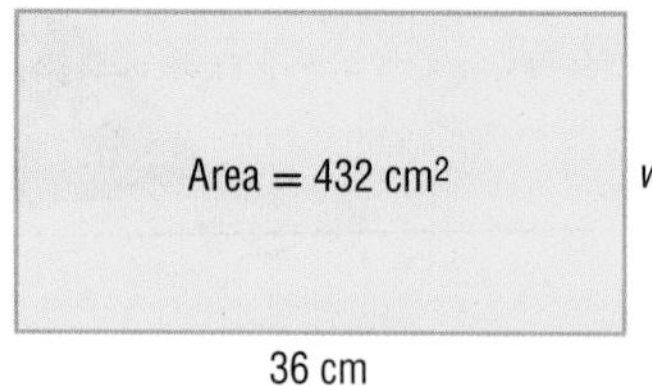

12.

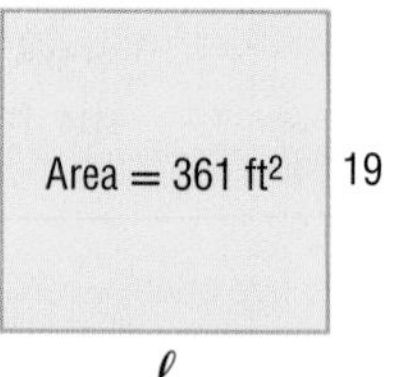

Real-World Link

A pumpkin pie with a 14-foot diameter holds the record for the world's largest pumpkin pie. It took over 10 hours to bake.

Source: American City and Country

13. FOOD The circumference of a circle is the distance around the circle. It can be found by using the formula $C \approx 3.14\,d$, where d represents the diameter. Find the circumference of the pie described at the left in inches.

14. GEOMETRY A *trapezoid* is a four-sided figure with exactly one set of parallel sides.

a. The area of a trapezoid is one half the product of the height h and the sum of the two bases b_1 and b_2. Translate this relationship into a formula. Use the formula to find the area of each trapezoid shown below.

b.

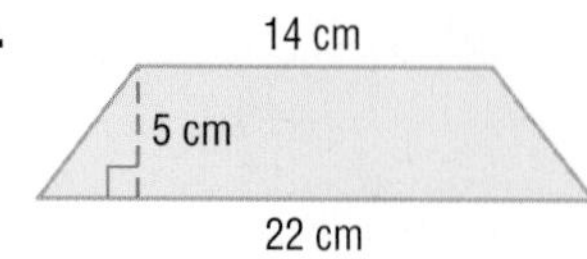

c.

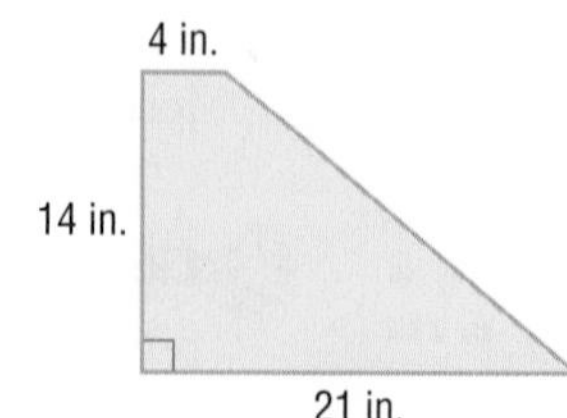

Problem-Solving Tip

Solve a Simpler Problem Sometimes it is easier to break a problem into parts, solve each part, then combine the solutions of the parts. In Exercise 15b, you know one bag covers 150 square feet. The width of the garden is 15 feet so you will use one bag for every 10 feet of length.

15. **GARDENING** Use the diagram of a vegetable garden.

    a. You want to put fencing around the garden to keep animals out. How much fencing will you need?

    b. If one bag of fertilizer covers 150 square feet, how many bags would you need for the garden?

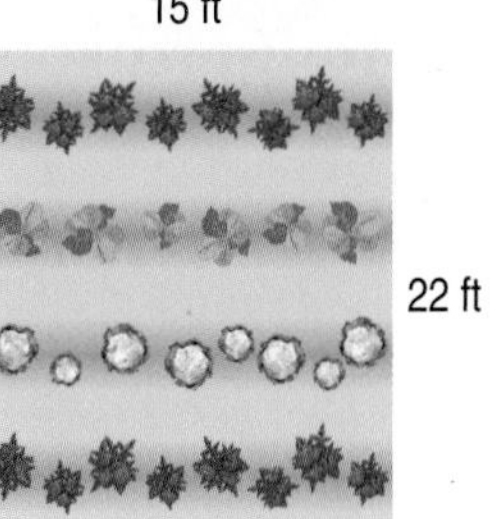

Find the area of each rectangle.

16. 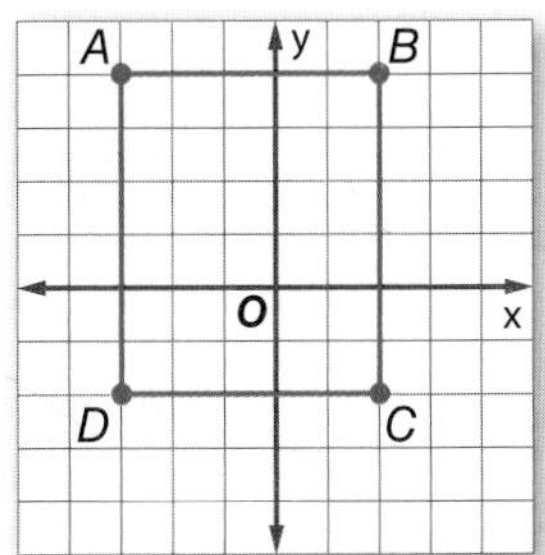

17. 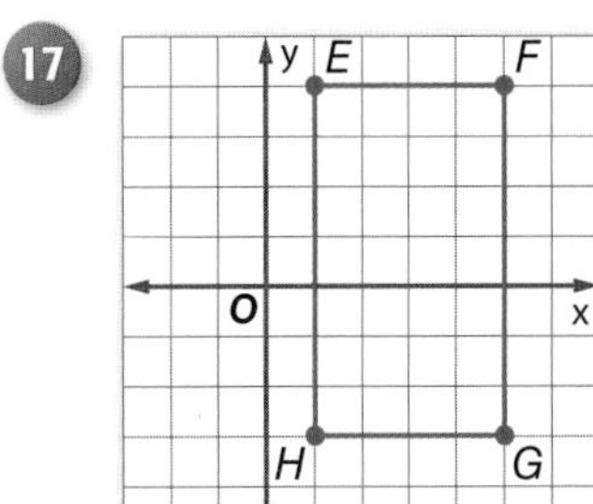

Solve for the indicated variable.

18. $d = rt$

19. $V = Ah$

20. $2p = s - t$

21. $r = st + p$

22. $P = 2(\ell + w)$

23. $A = \frac{1}{2}bh$

24. **MULTIPLE REPRESENTATIONS** In this problem, you will explore a rectangle that has a perimeter of 20 inches.

    a. **TABULAR** Copy and complete the table at the right using only whole number values for the length and the width.

    b. **GRAPHICAL** Graph the points (w, A) on a coordinate plane. Describe the graph.

Perimeter = 20 inches		
Length ℓ	Width w	Area A
9	1	9
8	2	16
7	■	■
⋮	■	■
1	■	■

H.O.T. Problems

Use Higher-Order Thinking Skills

25. **OPEN ENDED** Write and solve a real-world problem in which you would use the perimeter or area formula.

26. **REASONING** Classify the following statements as *true* or *false*. Explain your reasoning and provide examples.

    a. The area of a larger rectangle that can be divided into smaller rectangles is the sum of the areas of the smaller rectangles.

    b. The perimeter of a larger rectangle that can be divided into smaller rectangles is the sum of the perimeters of the smaller rectangles.

27. **CHALLENGE** Find the dimensions of a rectangle with the largest area if the perimeter of the rectangle is 40 feet.

28. **WRITING IN MATH** Describe the effect on the perimeter and area of a rectangle if its length and width are doubled.

Standardized Test Practice

29. The rectangle below has a perimeter of P yards.

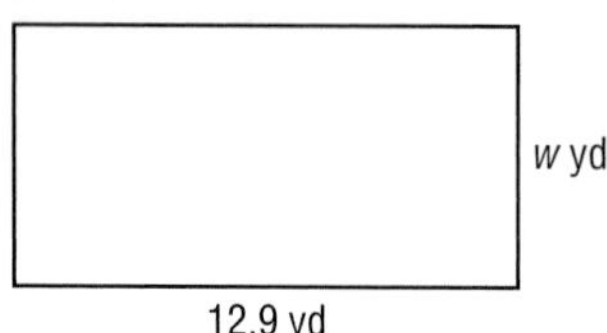

Which of the following could be used to find the width of the rectangle?

A $w = P - 12.9$

B $w = P - 25.8$

C $w = \frac{P}{2} - 12.9$

D $w = \frac{P}{2} - 25.8$

30. If the width of a rectangle is doubled but its length remains the same, what will happen to its area?

F The area will be four times as large.

G The area will be twice as large.

H The area will not change.

J The area will be three times as large.

31. An architect designed a great room for the Martes family. The small carpeted areas are 8 foot squares. The rest of the floor is wood.

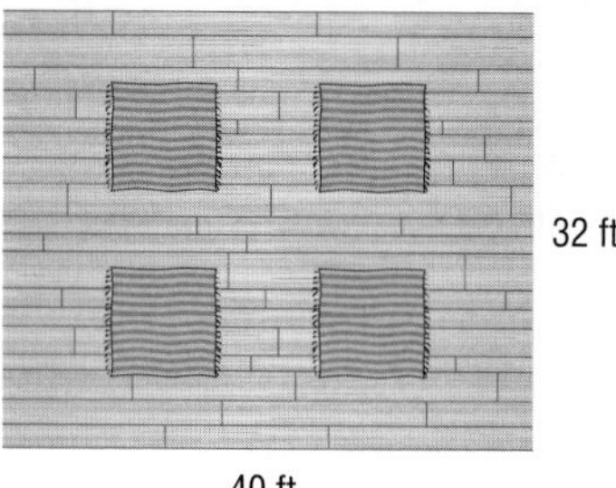

What is the area in square feet of the floor that is *not* carpeted?

A 64

B 256

C 1024

D 1280

32. **GRIDDED RESPONSE** A rectangular desk in a classroom is 6 inches longer than it is wide. The perimeter of the desk is 84 inches. What is its length in inches?

Spiral Review

33. **TEMPERATURE** Suppose the current temperature is 17°F. It is expected to rise 3°F each hour for the next several hours. Write and solve an equation to find in how many hours the temperature will be 32°F. (Lesson 4-6)

Solve each equation. Check your solution. (Lesson 4-5)

34. $22 = 8k - 18$

35. $8 = \frac{c}{-3} + 15$

36. $5r + 3r - 6 = 10$

MEASUREMENT **The table shows several conversions in the customary system. Write and solve an equation to find each quantity.** (Lesson 4-4)

Customary System (length)		
1 mile	=	5280 feet
1 mile	=	1760 yards
1 yard	=	3 feet
1 foot	=	12 inches
1 yard	=	36 inches

37. the number of feet in 132 inches

38. the number of yards in 15 feet

39. the number of miles in 10,560 feet

Find each quotient. (Lesson 2-5)

40. $-64 \div (-8)$

41. $72 \div (-9)$

Skills Review

Evaluate each expression. (Lesson 1-2)

42. $6w - 4$ if $w = 9$

43. $45 - 4x$ if $x = 12$

44. $-7b - 9$ if $b = -6$

EXTEND 5-1

Spreadsheet Lab: Perimeter and Area

Math Online glencoe.com
• Other Calculator Keystrokes
• Graphing Technology Personal Tutor

A spreadsheet allows you to use formulas to investigate problems. When you change a numerical value in a cell, the spreadsheet recalculates the formula and automatically updates the results.

ACTIVITY

Kina wants to build a kennel for her dog using 50 feet of fencing. She wants the dog to have the largest possible play area. Find the whole number dimensions of the kennel.

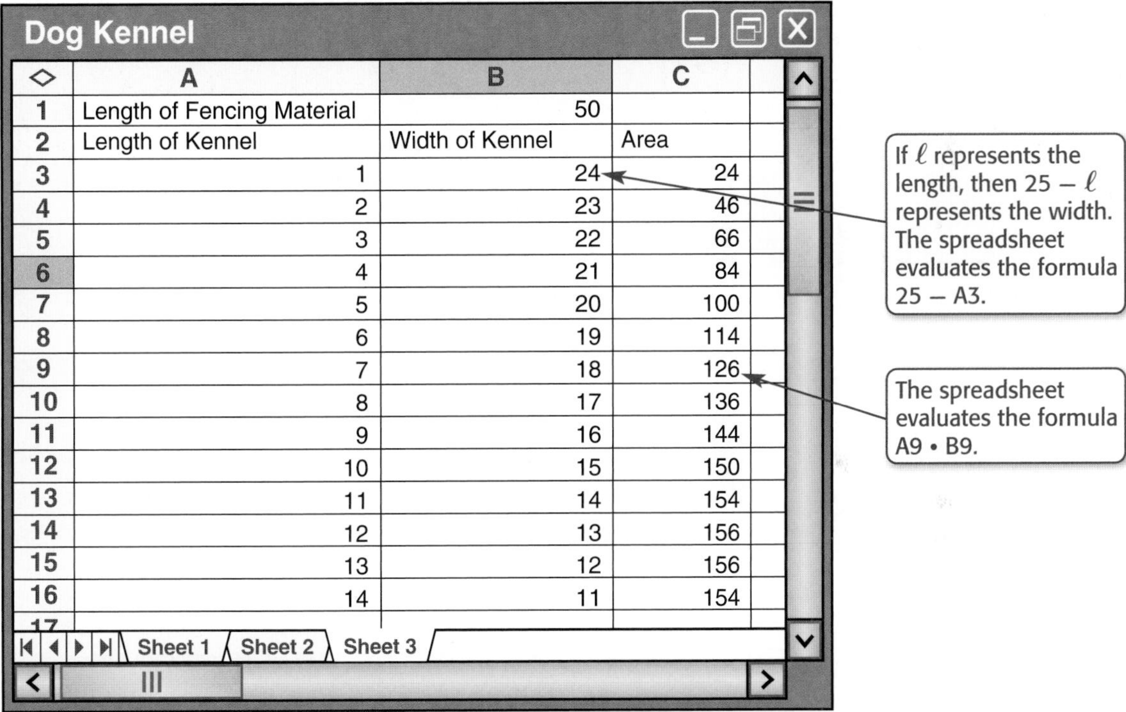
Dog Kennel
Sheet 1
Sheet 2
Sheet 3
If ℓ represents the length, then 25 – ℓ represents the width. The spreadsheet evaluates the formula 25 – A3.
The spreadsheet evaluates the formula A9 • B9.

◇	A	B	C
1	Length of Fencing Material	50	
2	Length of Kennel	Width of Kennel	Area
3	1	24	24
4	2	23	46
5	3	22	66
6	4	21	84
7	5	20	100
8	6	19	114
9	7	18	126
10	8	17	136
11	9	16	144
12	10	15	150
13	11	14	154
14	12	13	156
15	13	12	156
16	14	11	154

The largest area is 156 square feet. It occurs when the length is 12 feet and the width is 13 feet or the length is 13 feet and the width is 12 feet.

Analyze the Results

1. What happens to the area as the length of the kennel increases beyond 14 feet?
2. Suppose you want to find the largest area you can enclose with 70 feet of fencing. Which cell should you modify to solve this problem?
3. Use a spreadsheet to find the whole number dimensions of the greatest area you can enclose with 60 feet, 70 feet, and 80 feet.
4. **MAKE A CONJECTURE** Use any pattern you may have observed in your answers to find the dimensions of the greatest area you can enclose with 100 feet of fencing.

EXPLORE **5-2** Algebra Lab

Equations with Variables on Each Side

Math Online glencoe.com
Math *in Motion,* Animation

Use algebra tiles to solve equations with variables on each side of the equation.

ACTIVITY

Solve $2x - 5 = 3x + 3$ using algebra tiles.

Step 1

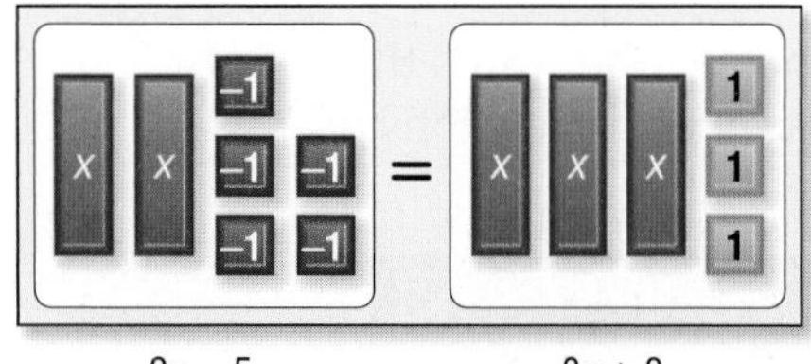

$2x - 5 = 3x + 3$

Model the equation.

Step 2

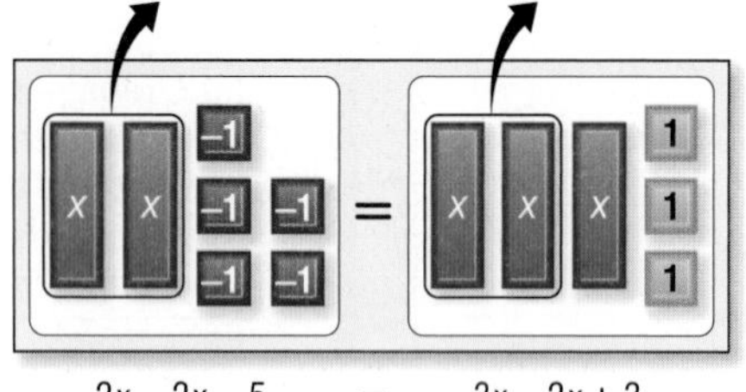

$2x - 2x - 5 = 3x - 2x + 3$

Remove two x-tiles from each side of the mat. All of the x-tiles are on one side of the mat.

Step 3

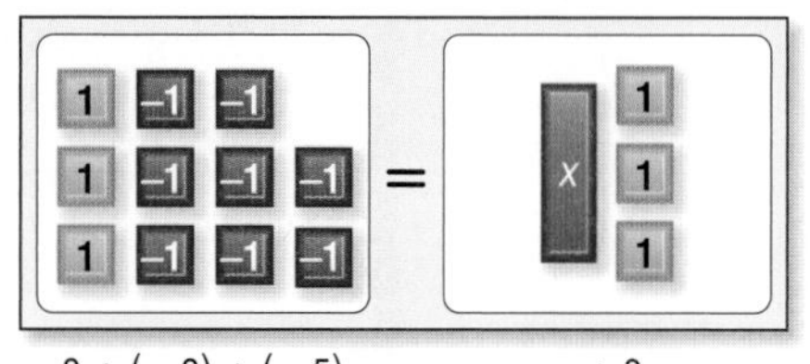

$3 + (-3) + (-5) = x + 3$

It is not possible to remove three 1-tiles from each side of the mat. So, add 3 zero pairs to the left side of the mat.

Step 4

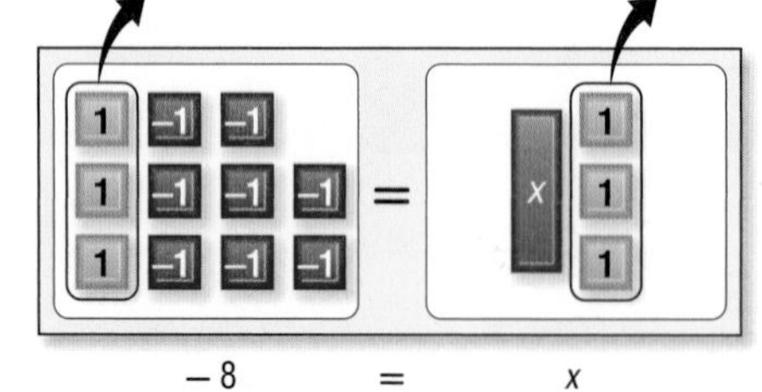

$-8 = x$

Now you can remove three 1-tiles from each side. There are 8 negative tiles on the left side of the mat.

The solution is -8.

Analyze the Results

Use algebra tiles to model and solve each equation.

1. $4x - 3 = 3x + 1$ **2.** $2x + 6 = 5x - 3$ **3.** $x + 8 = 3x - 4$ **4.** $2x - 5 = 5x + 4$

5. Does it matter whether you remove x-tiles or 1-tiles first? Explain.

6. What property of equality allows you to remove an x-tile from each side of the mat?

5-2

Solving Equations with Variables on Each Side

Then

You have already solved two-step equations. (Lesson 4-5)

Now

- Solve equations with variables on each side.
- Solve equations that involve grouping symbols.

Math Online

glencoe.com

- Extra Examples
- Personal Tutor
- Self-Check Quiz
- Homework Help

Why?

Antelopes and zebras are racing to a watering hole. The antelope can travel 90 feet per second while the zebra travels 60 feet per second. The zebra has a 120-foot head start.

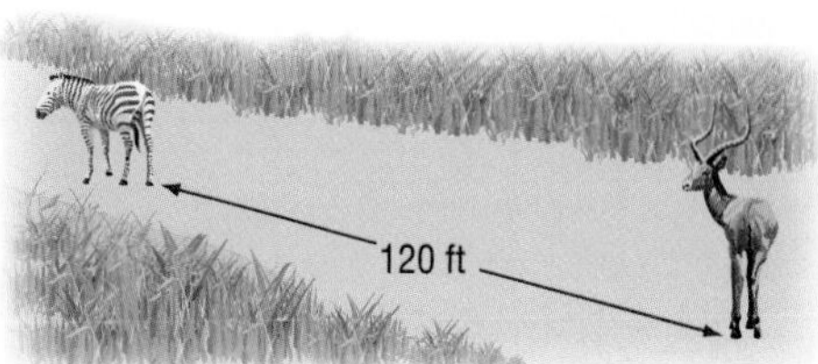

a. Write an expression for the distance covered by the antelope in t seconds.

b. Write an expression for the total distance covered by the zebra in t seconds.

c. Are the two expressions equal? How could you show they are equal?

Equations with Variables on Each Side The equation $90t = 60t + 120$ represents the point in the race when the antelope catches up to the zebra. To solve equations with variables on each side, use the Addition or Subtraction Property of Equality to write an equivalent equation with the variables on one side. Then solve the equation.

EXAMPLE 1 Equations with Variables on Each Side

Solve $3x = x + 8$.

$3x = x + 8$ **Write the equation.**

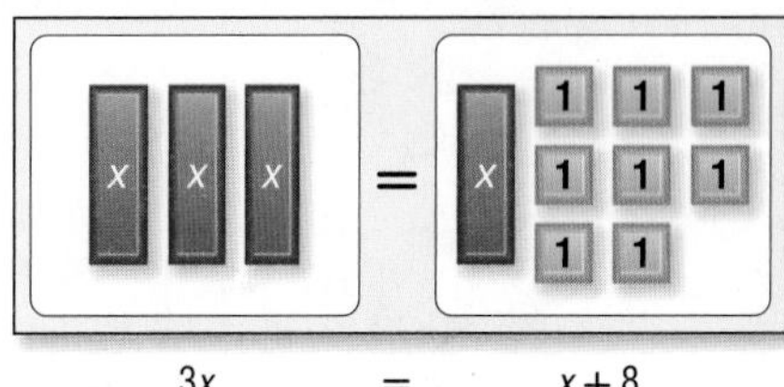

$$\begin{aligned} 3x &= x + 8 \\ -x &= -x \\ \hline 2x &= 8 \end{aligned}$$

Subtact x from each side.

Simplify.

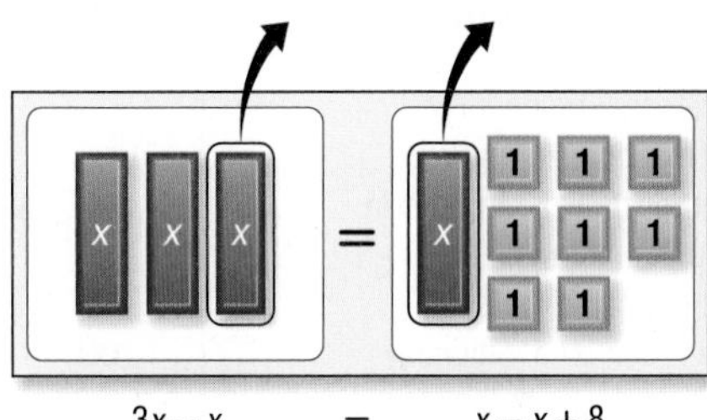

$\frac{2x}{2} = \frac{8}{2}$ **Divide each side by 2.**

$x = 4$ **Simplify.**

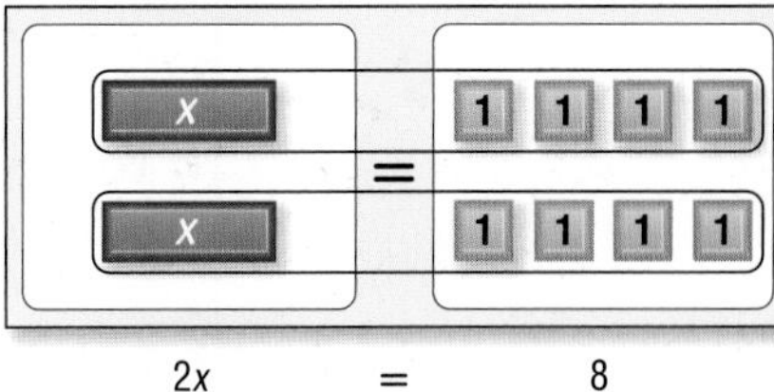

The solution is 4.

Check Your Progress

Solve each equation.

1A. $7x = 5x + 4$

1B. $3x - 2 = x + 4$

Personal Tutor glencoe.com

StudyTip

Solve a Simpler Equation Notice that the equations in the third step of each solution are similar to the equations that you solved in Lesson 4-5.

EXAMPLE 2 Equations with Variables on Each Side

Solve $7y + 8 = 4y - 10$. Check your solution.

$7y + 8 = 4y - 10$	**Write the equation.**
$7y - 4y + 8 = 4y - 4y - 10$	**Subtract $4y$ from each side.**
$3y + 8 = -10$	**Simplify.**
$3y + 8 - 8 = -10 - 8$	**Subtract 8 from each side.**
$3y = -18$	**Simplify.**
$y = -6$	**Mentally divide each side by 3.**

Check	$7y + 8 = 4y - 10$	**Write the equation.**
	$7(-6) + 8 \stackrel{?}{=} 4(-6) - 10$	**Substitute –6 for y.**
	$-42 + 8 \stackrel{?}{=} -24 - 10$	**Check to see whether this sentence is true.**
	$-34 = -34$ ✓	**The sentence is true.**

Check Your Progress

Solve each equation. Check your solution.

2A. $2x + 3 = 3x - 2$ **2B.** $5p + 15 = p - 49$

Personal Tutor glencoe.com

Real-World Link

Personal trainers can help teens build strength in order to avoid injuries in sports. Trainers for teenagers should be certified by a nationally recognized organization.

Source: IDEA Health & Fitness Association

Real-World EXAMPLE 3 Solve Equations with Variables on Each Side

EXERCISE **A personal trainer charges a one time fee of \$60 plus \$25 for each individual session. A new fitness club charges a yearly fee of \$450 plus \$10 for each session with a personal trainer. Write and solve an equation to determine for what number of sessions will the costs be equal.**

Words	one time fee	+	cost per session	×	number of sessions	=	yearly fee	+	cost per session	× number of sessions
Variable	Let s = number of sessions.									
Equation	60	+		$25s$		=	450	+		$10s$

$60 + 25s = 450 + 10s$	**Write the equation.**
$60 + 25s - 10s = 450 + 10s - 10s$	**Subtract $10s$ from each side.**
$60 + 15s = 450$	**Simplify.**
$60 - 60 + 15s = 450 - 60$	**Subtract 60 from each side.**
$15s = 390$	**Simplify.**
$\frac{15s}{15} = \frac{390}{15}$	**Divide each side by 15.**
$26 = s$	**Simplify.**

You would need to have 26 trainer sessions in order for the costs to be equal.

Check Your Progress

3. **CRUISES** Red Bird Cruises charges \$85 per day plus a one-time fee of \$75 for taxes and gratuities. King Cruises charges \$100 per day plus a fee of \$30. Write and solve an equation to determine for what number of days the charge for the cruises will be the same.

Personal Tutor glencoe.com

Check Your Understanding

Examples 1 and 2 pp. 229–230

Solve each equation. Check your solution.

1. $x + 6 = 3x$

2. $4y = 2y - 10$

3. $7z - 4 = 12 + 3z$

4. $10t - 3 = t + 15$

5. $28 - x = 3x - 84$

6. $2p - 7 = 13 - 8p$

Example 3 p. 230

7. **MONEY** An Internet movie rental company charges a yearly membership fee of \$50 plus \$1.99 per DVD rental. Your neighborhood rental store has no membership fee and charges \$3.99 per DVD rental. Write and solve an equation to find the number of DVDs so the cost for each will be the same.

Practice and Problem Solving

= **Step-by-Step Solutions** begin on page R11.
Extra Practice begins on page 810.

Examples 1 and 2 pp. 229–230

Solve each equation. Check your solution.

8. $2x + 3 = x$

9. $8 - v = 7v$

10. $8 - 2c = 2c$

11. $q - 2 = -q + 1$

12. $-2 + x = -2 + 2x$

13. $14 + 3a = -2a - 1$

14. $12p = p + 14 + p$

15. $5b - 4b = 6b - 2$

Example 3 p. 230

16. **CAR RENTALS** Use the table at the right to write and solve an equation to find the number of miles a rental car must be driven for each option to cost the same for one day.

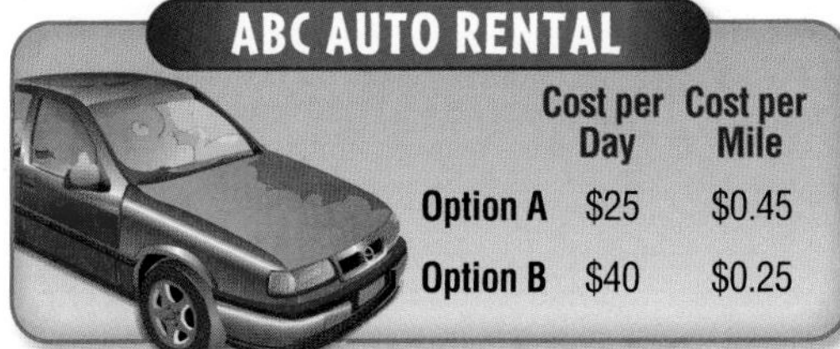

ABC AUTO RENTAL

	Cost per Day	Cost per Mile
Option A	\$25	\$0.45
Option B	\$40	\$0.25

17. **MUSIC DOWNLOADS** Denzel is comparing websites for downloading music. One charges a \$5 membership fee plus \$0.50 per song. Another charges \$1.00 per song, but has no membership fee. Write and solve an equation to find how many songs Denzel would have to buy to spend the same amount at both websites.

Solve each equation. Check your solution.

18. $3.2 + 0.3x = 0.2x + 1.4$

19. $0.4x = 2x + 1.2$

20. $7.2 - 3c = 2c - 2$

21. $3 - 3.7b = 10.3b + 10$

22. $-\frac{1}{4}x + 6 = \frac{2}{3}x + 28$

23. $\frac{2}{5}x - 8 = 20 + \frac{3}{4}x$

24. **SHOPPING** Gabriella bought some school supplies for \$48 and then bought 3 CDs. Min did not buy any school supplies but bought 7 CDs. All the CDs cost the same amount and they both spent the same amount of money. Write and solve an equation to find the cost of one CD.

25. **FINANCIAL LITERACY** One cell phone company charges \$19.95 a month plus \$0.21 per text message, and a second company charges \$24.95 per month plus \$0.16 per text message. For how many text messages is the cost of the plans the same?

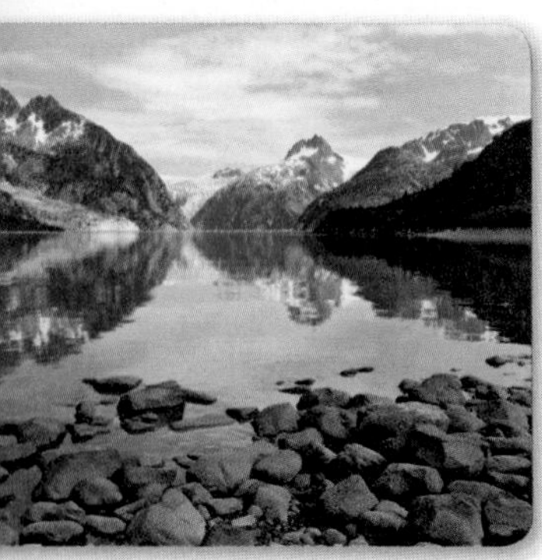

Real-World Link

Twenty-two states have borders on at least one of three oceans for a total of 12,383 miles of ocean coastline. Alaska alone accounts for 6640 miles of coast.

26. AGES Five years ago Ang was $\frac{1}{2}$ as old as Paula. Now he is $\frac{3}{5}$ as old as Paula. Complete the table shown below. Then use the table to write and solve an equation to find their current ages.

Student	5 years ago	Now
Ang	$\frac{1}{2}a$	■
Paula	a	$a + 5$

27 COASTLINE Florida's coastline is 118 miles shorter than four times the coastline of Texas. It is also 983 miles longer than the coastline of Texas. Find the lengths of the coastlines of Florida and Texas.

28. GEOMETRY Use the square shown at the right.

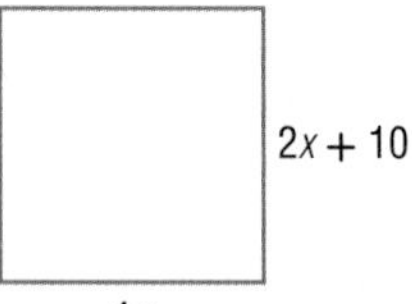

a. What is the value of x?

b. Find the length of each side of the square.

29. LANDSCAPE Jamie is going to fence the rectangular and triangular sections of grass shown below. The perimeters of the two sections are now equal. If w represents the width of the rectangle, how could you find the lengths of the sides of the rectangle and of the triangle? Justify your response and use your method to solve the problem.

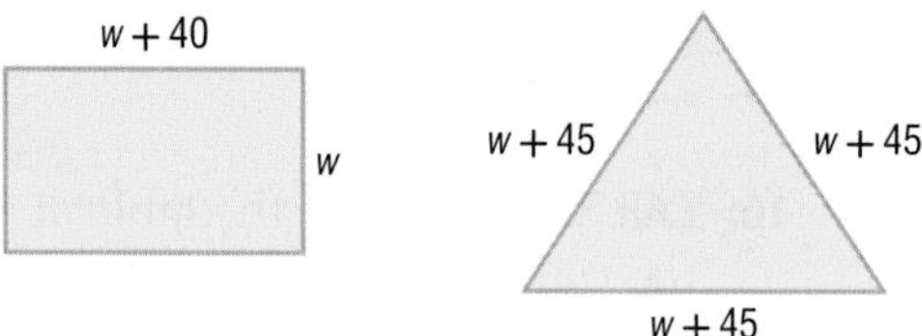

H.O.T. Problems Use Higher-Order Thinking Skills

30. OPEN ENDED Write an equation that has variables on each side and has a solution of -2.

31. NUMBER SENSE Three times the sum of three consecutive integers x, $x + 1$, and $x + 2$, is 72. What are the integers?

32. CHALLENGE The formula $F = 1.8C + 32$ can be used to find the temperature in degrees Fahrenheit F when the temperature is given in degrees Celsius C. For what value is the temperature in degrees Fahrenheit equal to the temperature in degrees Celsius? Justify your reasoning by writing and solving an equation. (*Hint:* If Fahrenheit and Celsius are equal, they can be assigned the same variable.)

33. FIND THE ERROR Mykia solved the equation $10x + 6 = 8x - 4$. Her steps are shown at the right. Is Mykia correct? If not, what error did she make? Then correct Mykia's error.

$$10x + 6 = 8x - 4$$
$$10x - 10x + 6 = 8x - 4 - 10x$$
$$6 = 4 - 2x$$
$$2 = -2x$$
$$-1 = x$$

34. WRITING IN MATH Write a real-world problem that could be solved by using the equation $54 + 3.5x = 8x$. Then solve the equation and interpret your solution.

Standardized Test Practice

35. Yesterday, the math club had 1 less than 3 times their average attendance. Last week they had 3 more than their average attendance. If the attendance for both weeks were equal, what is the average attendance?

A 1 **C** 3

B 2 **D** 4

36. For which of the following is −8 a solution?

F $-2c + 18 = 10c + 12$

G $6m - 15 = 9m + 9$

H $4 + 7s = 5s + 20$

J $5d - 13 = 19 - 3d$

37. GRIDDED RESPONSE What is the solution of the equation $12x + 4 = 2x - 16$?

38. A cellular company has the following options for text messaging plans.

Text Plans	Monthly Fee	Cost per Message
Plan A	$10	$0.15
Plan B	$20	$0.05

Which equation shows how many text messages would need to be sent in order for the costs for one month to be the same?

A $10 + 0.05m = 20 + 0.15m$

B $10m + 0.15 = 20m + 0.05$

C $10 + 0.15m = 20 + 0.05m$

D $10(m + 0.15) = 20(m + 0.05)$

Spiral Review

Find the perimeter and area of each figure (Lesson 5-1)

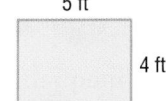

39. 2 m, 6 m

40.

41. SPORTS Carla paid $45 to join a golf camp for the summer. She will also pay $15 for every private lesson that she takes. If she has budgeted $225 for the camp, how many private lessons can she take? (Lesson 4-6)

Simplify each expression. (Lesson 4-2)

42. $2x + 5x$ **43.** $7b + 2b$ **44.** $y + 10y$

45. WEATHER The table shows the number of tornadoes that occurred in Nebraska in May and the total number of tornadoes for selected years. What decimal part, rounded to the nearest hundredth, of the annual tornadoes occurred in May for each year? (Lesson 3-1)

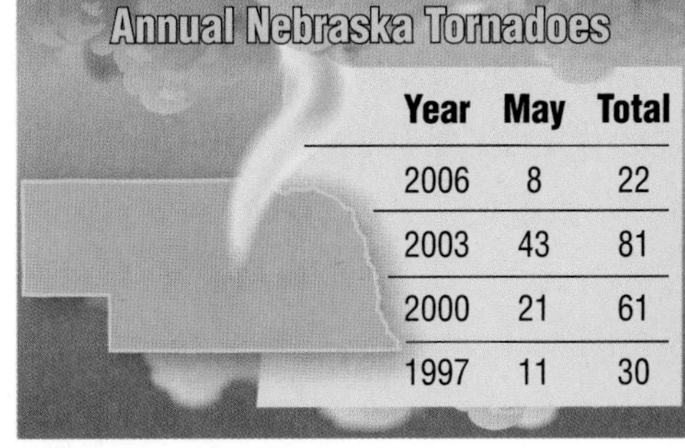
Annual Nebraska Tornadoes

Year	May	Total
2006	8	22
2003	43	81
2000	21	61
1997	11	30

Source: Nebraska Severe Weather

46. WEATHER The low temperatures for 7 days in January were −2°F, 0°F, 5°F, −1°F, −4°F, 2°F, and 0°F. Find the average for the 7-day period. (Lesson 2-5)

Skills Review

Evaluate each expression. (Lesson 1-2)

47. $8c + 5$, if $c = 6$ **48.** $22 - 3h$, if $h = 4$ **49.** $36 - (-6g)$, if $g = -2$

5-3

Inequalities

Then

You have already solved equations and graphed the solution on a number line. (Lessons 4-3 and 4-4)

Now

- Write inequalities.
- Graph inequalities on a number line.

New Vocabulary

inequality

Math Online

glencoe.com

- Extra Examples
- Personal Tutor
- Self-Check Quiz
- Homework Help

Why?

The table shows the admission prices at a water park.

Water Park Tickets

Ticket	Price	Price after 3 P.M.
Children under 2	Free	Free
Children ages 2 - 11	\$12	\$8
Adults ages 12 - 59	\$14	\$9
Seniors ages 60 & up	\$10	\$7

a. Would a 2-year-old child be able to attend the water park free?

b. When would an adult age 34 pay less than \$14 for admission?

c. Find the difference between the admission prices before 3 P.M. for a person age 59 and a person age 60.

Write Inequalities An **inequality** is a mathematical sentence that compares quantities that are not equal. Inequalities contain the symbols $<$, $>$, $\leq$, or $\geq$.

EXAMPLE 1 Write an Inequality

Write an inequality for each sentence.

a. The DVD costs more than \$15.

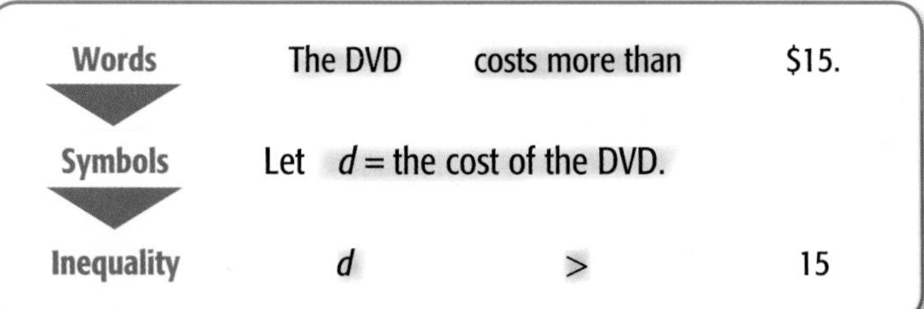

Words
The DVD
costs more than
$15.
Symbols
Let d = the cost of the DVD.
Inequality
d
>
15

b. A dog weighs less than 50 pounds.

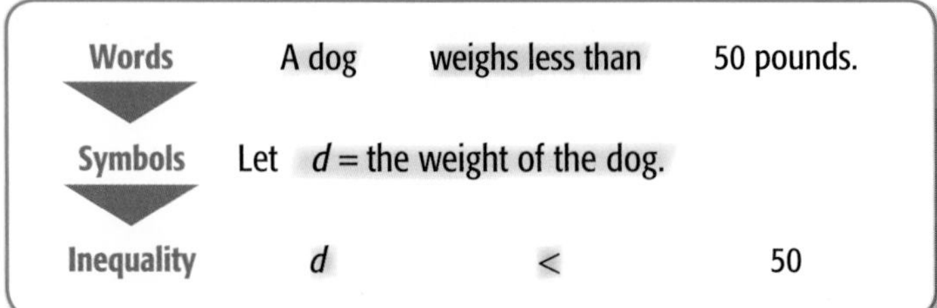

Words
A dog
weighs less than
50 pounds.
Symbols
Let d = the weight of the dog.
Inequality
d
<
50

Check Your Progress

1A. The height requirement is greater than or equal to 40 inches.

1B. The speed limit is less than or equal to 35 miles per hour.

Personal Tutor glencoe.com

ReadingMath

Inequalities Notice that $\leq$ and $\geq$ are a combination of the $<$ or $>$ symbol with part of the symbol for equals, $=$.

The table below shows some common verbal phrases and the corresponding mathematical inequalities.

Concept Summary **Inequalities** *For Your FOLDABLE*

$<$	$>$	$\leq$	$\geq$
• is less than • is fewer than	• is greater than • is more than • exceeds	• is less than or equal to • is no more than • is at most	• is greater than or equal to • is no less than • is at least

Real-World EXAMPLE 2 Write an Inequality

CIVICS You must be at least 18 years old to vote. Write an inequality to describe this situation.

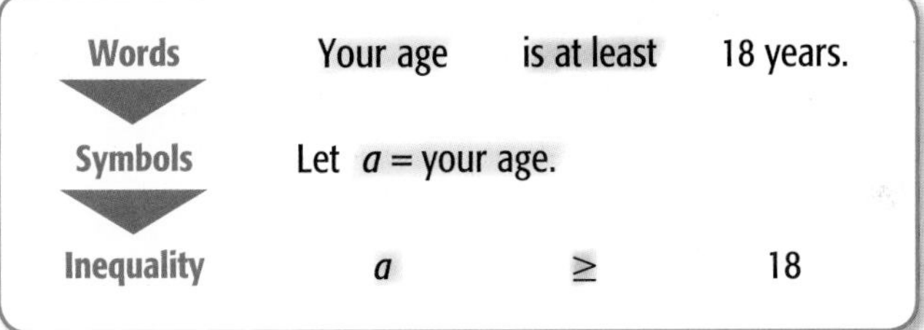

Words
Your age
is at least
18 years.
Symbols
Let a = your age.
Inequality
a
$\geq$
18

The inequality is $a \geq 18$.

Check Your Progress

2. **DRIVER'S EDUCATION** A student must have at least 10 hours of instructor assisted driving time to pass the course. Write an inequality to describe this situation.

Personal Tutor glencoe.com

Real-World Link

The right to vote has changed since the Constitution was written.

1870–The 15th Amendment included men of color.

1920–The 19th Amendment gave women the right to vote.

1971–The 26th Amendment lowered the voting age from 21 years to 18 years.

Inequalities with variables are open sentences. When the variable in an open sentence is replaced with a number, the inequality may be true or false.

EXAMPLE 3 Determine Truth of an Inequality

For the given value, state whether each inequality is *true* or *false*.

a. $2t + 8 > 7; t = -1$

$2t + 8 > 7$ **Write the inequality.**

$2(-1) + 8 \overset{?}{>} 7$ **Replace t with -1.**

$6 \not> 7$ **Simplify.**

This sentence is false.

b. $p - 42 \leq -2; p = 40$

$p - 42 \leq -2$ **Write the inequality.**

$40 - 42 \overset{?}{\leq} -2$ **Replace p with 40.**

$-2 \leq -2$ **Simplify.**

Although the inequality $-2 < -2$ is false, the equation $-2 = -2$ is true. So, this sentence is true.

Check Your Progress

3A. $3 + x \leq 12, x = 6$

3B. $y - 7 < 10, y = 17$

Personal Tutor glencoe.com

Graph Inequalities Inequalities can be graphed on a number line. The graph helps you visualize the values that make the inequality true.

StudyTip

Graphing Inequalities When inequalities are graphed, an open dot means the number is *not* included (< or >) and a closed dot means it is included (≤ or ≥).

EXAMPLE 4 Graph an Inequality

Graph each inequality on a number line.

a. $a > 6$

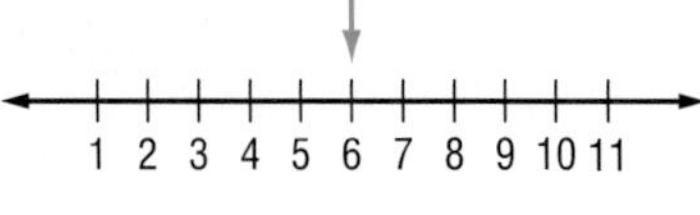

Locate 6 on the number line. It is a key point in the inequality.

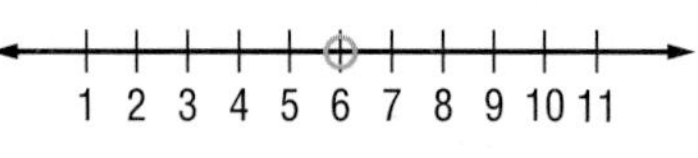

Draw an *open* dot on 6 because 6 is *not* included.

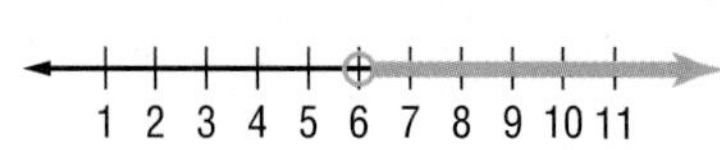

The inequality $a > 6$ means that all numbers *greater than* 6 will make the sentence true. Draw an arrow from the dot pointing to the right.

b. $x \leq -1$

–6 –5 –4 –3 –2 –1 0 1 2 3 4

Locate –1 on the number line. It is a key point in the inequality.

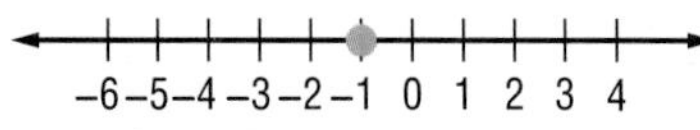

Draw a *closed* dot on –1 because –1 *is* included.

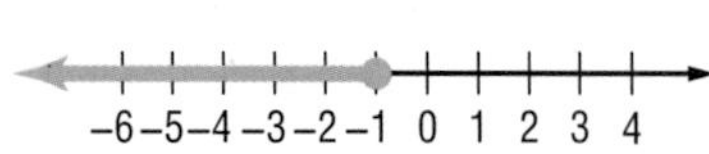

The inequality $x \leq -1$ means that all numbers *less than or equal to* –1 will make the sentence true. Draw an arrow from the dot pointing to the left.

Check Your Progress

4A. $x < 5$ **4B.** $x \geq -2$ **4C.** $x > 0$

Personal Tutor glencoe.com

EXAMPLE 5 Write an Inequality

Write an inequality for the graph.

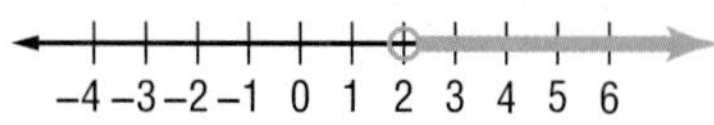

An open circle is on 2, so the point 2 is *not* included in the graph. The arrow points to the right, so the graph includes all numbers greater than 2. The inequality is $x > 2$.

Check Your Progress

Write an inequality for each graph.

5A. –5 –4 –3 –2 –1 0 1 2 3 4 5

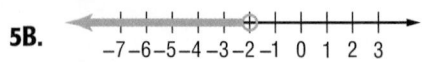

5B. –7 –6 –5 –4 –3 –2 –1 0 1 2 3

Personal Tutor glencoe.com

Check Your Understanding

Example 1 p. 234

Write an inequality for each sentence.

1. Lacrosse practice will be no more than 45 minutes.
2. Mario is more than 60 inches tall.

Example 2 p. 235

3. **SOCCER** More than 8000 fans attended the Wizards' opening soccer game at Arrowhead Stadium in Kansas City, Missouri. Write an inequality to describe the attendance.

Example 3 p. 235

For the given value, state whether the inequality is *true* or *false*.

4. $13 - x < 4; x = 9$
5. $45 > 2x - 5; x = 20$

Example 4 p. 236

Graph each inequality on a number line.

6. $x < -1$
7. $y \geq 5$
8. $w > 9$
9. $z \leq 2$

Example 5 p. 236

Write an inequality for each graph.

10. 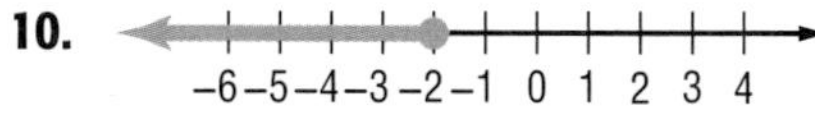

11. 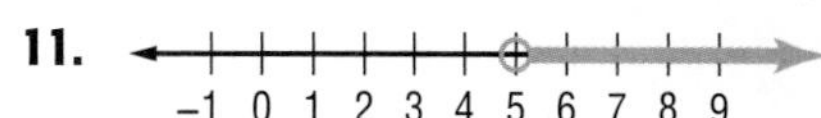

Practice and Problem Solving

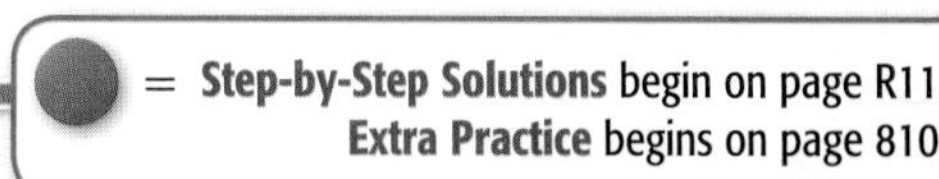

Example 1 p. 234

Write an inequality for each sentence.

12. The elevators in an office building have been approved for a maximum load of 3600 pounds.
13. Children under the age of 2 fly free.
14. An assignment requires at least 45 minutes.
15. While shopping, Abby spent no more than $50.

Example 2 p. 235

ANALYZE GRAPHS The graph shows the average life span of various animals.

16. The average lifespan of a Galapagos tortoise is at least 4 times that of a chimpanzee. Write an inequality for the lifespan of a tortoise.
17. The average lifespan of a lobster is at most the lifespan of a cat. Write an inequality for the lifespan of a lobster.

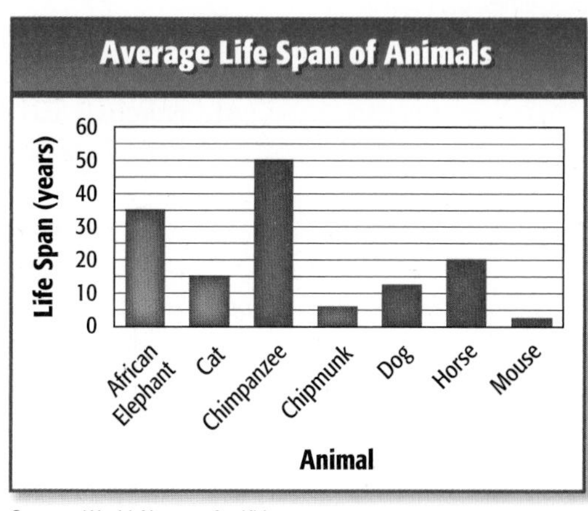

Source: *World Almanac for Kids*

Example 3 p. 235

For the given value, state whether the inequality is *true* or *false*.

18. $13 - a < 29; a = -30$
19. $4b \geq -12; b = -4$
20. $2c + 18 \leq 50; c = 15$
21. $\frac{120}{d} > 40; d = 3$
22. $\frac{55}{f} > -22; f = -5$
23. $c + 19 < 2c; c = 20$

Example 4 p. 236

Graph each inequality on a number line.

24. $x < 0$ **25.** $y \geq 3$ **26.** $p > -4$

27. $t > 6$ **28.** $s \geq -2$ **29.** $r \leq -4$

Example 5 p. 236

Write an inequality for each graph.

30. −5 −3 −1 1 3 5

31. −2 0 2 4 6

32. −14 −12 −10 −8 −6

33. −4 −2 0 2 4

34. 5 7 9 11 13

35.

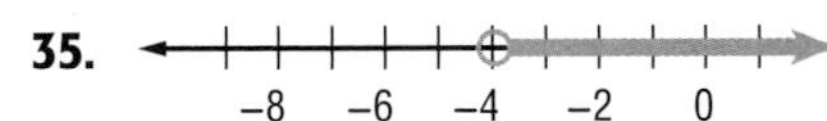

36. BUDGET Madison Middle School spends $750 per year on club activities. They spend at least twice that amount on after-school activities. Write an inequality that represents how much they spend on after-school activities.

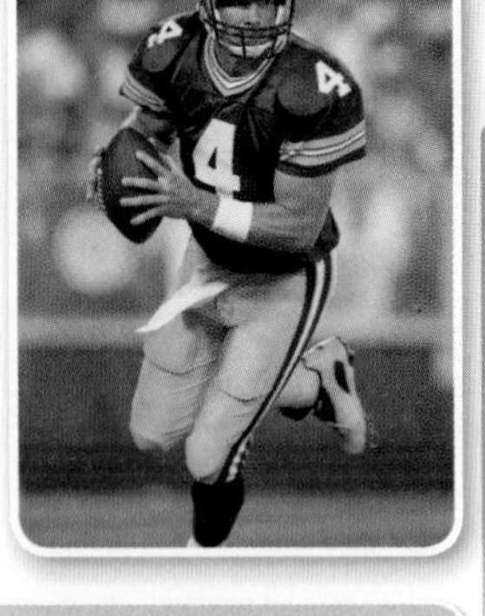

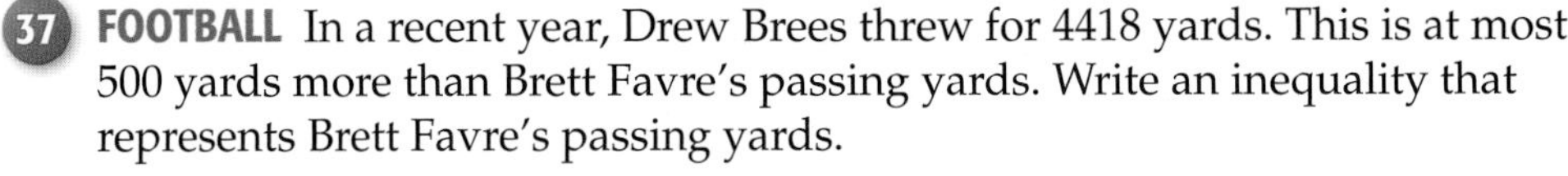

37. FOOTBALL In a recent year, Drew Brees threw for 4418 yards. This is at most 500 yards more than Brett Favre's passing yards. Write an inequality that represents Brett Favre's passing yards.

38. RESEARCH Use the Internet or another source to find the state or national spending limits on certain government branches, organizations, or projects. Write an inequality to express one or more of these limits.

Real-World Link

In 2007, Brett Favre broke an NFL record for the number of career completed passes. He has also won the Most Valuable Player award three times, the most in NFL history.

Source: National Football League

H.O.T. Problems

Use Higher-Order Thinking Skills

39. OPEN ENDED State three numbers that could be solutions of the inequality $h \leq -12$. Then justify your response by using a number line.

40. WRITING IN MATH Write an email to a friend who missed class today explaining how to tell the difference between graphing an inequality with a closed dot and one with an open dot. Use examples to clarify your explanation.

41. NUMBER SENSE Provide a counterexample to the statement *All numbers less than 0 are negative integers.*

42. FIND THE ERROR Alex and Carlos are graphing the inequality $p < 14$. Is either of them correct? Explain.

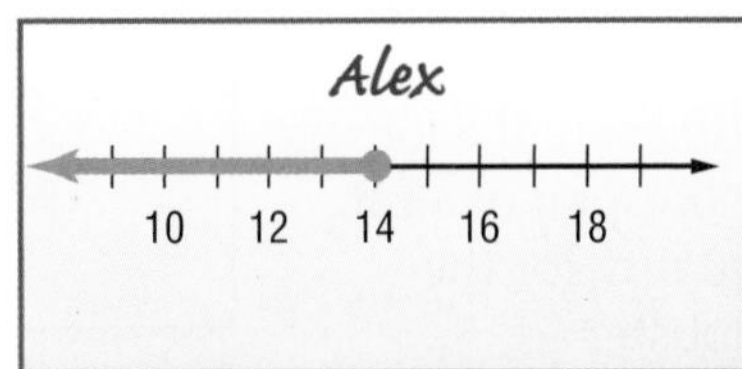

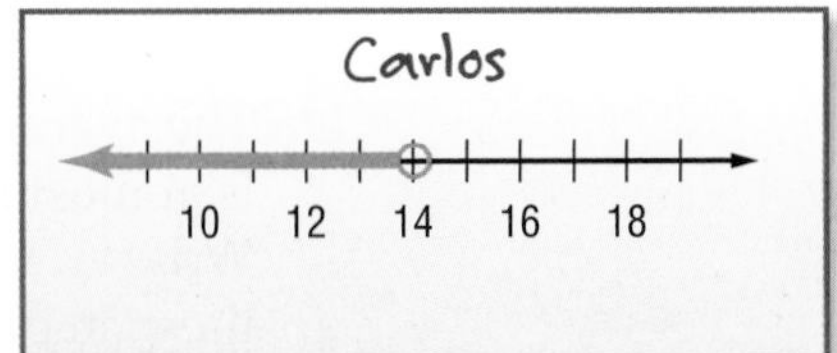

43. WRITING IN MATH Write a real-world example for the inequality below.

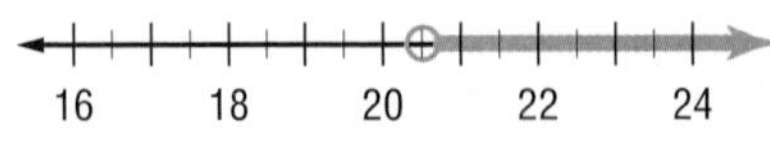

Standardized Test Practice

44. Which of the following best represents the sign shown?

Must be
over 48
inches tall
to ride.

A $t > 48$ **C** $t \geq 48$

B $t < 48$ **D** $t \leq 48$

45. SHORT RESPONSE Graph the following inequality on a number line.

$$p < 3$$

46. An elevator's maximum load is 3400 pounds. Which of the following best represents that sentence?

F $\ell > 3400$ **H** $\ell \geq 3400$

G $\ell < 3400$ **J** $\ell \leq 3400$

47. The Chess Club is having a bake sale to raise money for a tournament. The club must raise at least three times what they raised at the last bake sale. What is the minimum amount they must raise if the last bake sale raised $45?

A $15 **C** $90

B $45 **D** $135

Spiral Review

Solve each equation. (Lesson 5-2)

48. $3t - 6 = 6t + 30$

49. $2y + 14 = 42 - 5y$

50. $3x = 12 - 3x$

51. $4p = -p - 20$

52. LANDSCAPING Jordan is paving a rectangular patio with 308 bricks. If there are 22 bricks running along the length of the patio, how many bricks run along the width of the patio? (Lesson 5-1)

53. CELLULAR PHONE A cell phone plan charges $7 per month and $0.10 per minute. If the monthly cost is $25, solve $10m + 700 = 2500$ to find the number of minutes you can talk that month. (Lesson 4-5)

Find each sum or difference. (Lesson 3-5)

54. $\frac{2}{5} + \frac{1}{5}$

55. $\frac{3}{10} + \frac{7}{10}$

56. $-\frac{3}{4} + \left(-\frac{3}{4}\right)$

57. $-\frac{13}{16} + \left(-\frac{9}{16}\right)$

58. $7\frac{2}{5} + 4\frac{2}{5}$

59. $5\frac{17}{20} + 5\frac{9}{20}$

60. TEMPERATURE The low temperatures in degrees Fahrenheit for each day last week are $-3°$, $9°$, $4°$, $-7°$, $5°$, $0°$, and $-1°$. Find the average low temperature. (Lesson 2-5)

Skills Review

Solve each equation. (Lessons 4-3 and 4-4)

61. $2y = -42$

62. $b + 5 = -10$

63. $90 = -6t$

64. $56 = c + 24$

65. $f - 4.2 = -6$

66. $-2.1m = 8.4$

CHAPTER 5

Mid-Chapter Quiz

Lessons 5-1 through 5-3

Find the perimeter and area of each figure. (Lesson 5-1)

1.

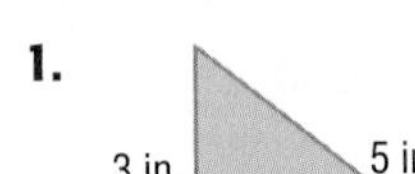

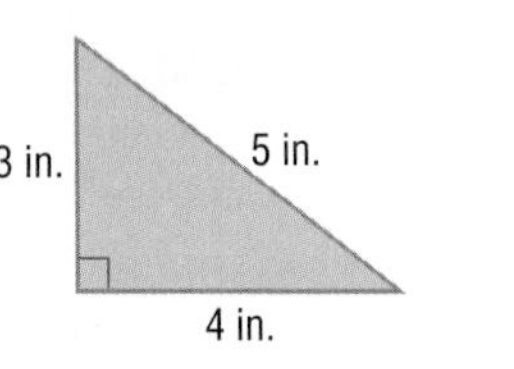

2.

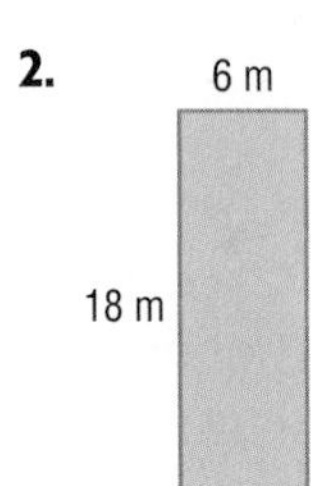

3. MEASUREMENT Find the base of a triangle if the height is 9.75 feet and its area is 58.5 square feet. (Lesson 5-1)

4. MEASUREMENT The perimeter of a rectangle is 28 inches. If the width is 5 inches, what is the length of the rectangle? (Lesson 5-1)

5. CAKES Mrs. Barsch had a cake made for her daughter's graduation party. The shaded area is where a photo will be placed and the rest of the cake will be covered with frosting.

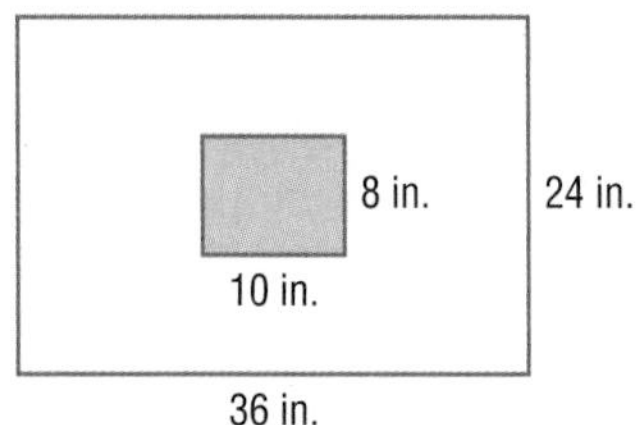

a. What is the perimeter of the entire cake?

b. What is the area of the photograph? (Lesson 5-1)

Solve each equation. Check your solution. (Lesson 5-2)

6. $-2w + 5 = 26 + w$

7. $6z + 2 = 4z - 6$

8. $12x - 19 = 3x + 8$

9. MULTIPLE CHOICE An online game site has two membership plans. The first plan gives you unlimited play for \$40 a month. The second plan charges a monthly access fee of \$4.25 plus \$2.75 for each hour you play. After how many hours do the two plans cost the same? (Lesson 5-2)

A 6.6 h **C** 11.2 h

B 9.0 h **D** 13.0 h

10. TESTS Carolina's score is 5 less than twice Aiko's score. It is also 45 points greater than Aiko's score. What score did the two girls receive? (Lesson 5-2)

Write an inequality for each sentence. (Lesson 5-3)

11. More than 35,000 people attended a college football game.

12. Toby wants to spend no more than 3 hours working on his model car.

13. Jameson's new guitar will cost at least \$450.

14. MULTIPLE CHOICE Which of the following represents the inequality $-\frac{17}{5} \leq y$? (Lesson 5-3)

F −5 −4 −3 −2 −1 0 1 2 3 4 5

G

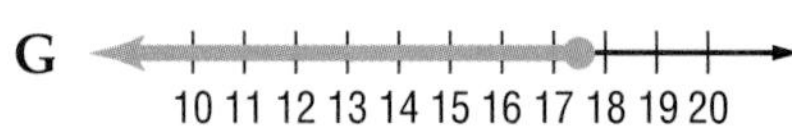

H −5 −4 −3 −2 −1 0 1 2 3 4 5

J −5 −4 −3 −2 −1 0 1 2 3 4 5

For the given value, state whether each inequality is *true* or *false*. (Lesson 5-3)

15. $4a + 5 \geq 16; a = 4$

16. $3m - 12 < -40; m = 9$

17. $5s + 8 > -10; s = -3$

Graph each inequality on a number line. (Lesson 5-3)

18. $8 > d$

19. $f < 11$

20. FITNESS The table shows a gym class's average results for boys and girls participating in the long jump.

Gender	Distance
Male	17 feet 5 inches
Female	14 feet 3 inches

Ching-Li could jump no farther than 12 inches more than the average distance for males. Write an inequality that gives the possible distances that Ching-Li could jump. (Lesson 5-3)

5-4 Solving Inequalities

Then

You have already used the Properties of Equality to solve equations. (Lessons 4-3 and 4-4)

Now

- Solve inequalities by using the Addition and Subtraction Properties of Inequality.
- Solve inequalities by multiplying or dividing by a positive or negative number.

Math Online

glencoe.com

- Extra Examples
- Personal Tutor
- Self-Check Quiz
- Homework Help
- Math in Motion

Why?

The bar graph shows the number of touchdowns the starting quarterbacks for different high school football teams threw last season.

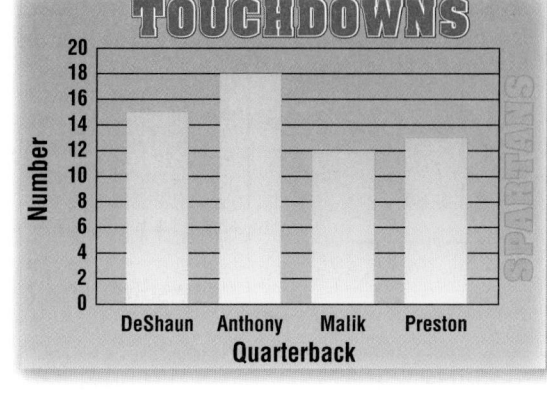

a. Write an inequality that compares the number of touchdowns thrown by Anthony and Malik.

b. Suppose each quarterback threw 3 more touchdowns. Write a new inequality that compares the number thrown by Anthony and Malik.

Solve Inequalities by Adding or Subtracting In this situation you added 3 to each side of the inequality. The inequality was still true. This and other similar examples suggest the following properties.

Key Concept: Addition and Subtraction Properties

For Your FOLDABLE

Words When you add or subtract the same number from each side of an inequality, the inequality remains true.

Symbols For all numbers a, b, and c,

1. if $a < b$, then $a + c < b + c$ and $a - c < b - c$.
2. if $a > b$, then $a + c > b + c$ and $a - c > b - c$.

Example

$$5 < 9 \qquad\qquad 11 > 6$$
$$5 + 4 < 9 + 4 \qquad\qquad 11 - 3 > 6 - 3$$
$$9 < 13 \qquad\qquad 8 > 3$$

These properties are also true for $a \le b$ and $a \ge b$.

EXAMPLE 1 Solve an Inequality

Solve $x + 5 > 12$. Check your solution.

$x + 5 > 12$	Write the inequality.
$-5 \quad -5$	Subtraction Property of Inequality
$x > 7$	Simplify.

Check

$x + 5 > 12$	Write the inequality.
$9 + 5 \overset{?}{>} 12$	Replace x with any number greater than 7.
$14 > 12$ ✓	The statement is true.

Check Your Progress

1A. $y + 10 < 3$

1B. $x + 7 \ge 10$

Personal Tutor glencoe.com

StudyTip

Equations and Inequalities In an equation, if $a = b$, then $b = a$. In an inequality, if $a < b$, then $b > a$.

Example: $7 = 2 + 5$ and $2 + 5 = 7$
$2 + 8 > 7$
but $7 < 2 + 8$

When graphing inequalities, it is often easier to visualize the solution when the variable is on the left side of the inequality symbol.

EXAMPLE 2 Graph the Solution of an Inequality

Solve $3 \le b - 1\frac{1}{3}$. Graph the solution on a number line.

$3 \le b - 1\frac{1}{3}$	**Write the inequality.**
$3 + 1\frac{1}{3} \le b - 1\frac{1}{3} + 1\frac{1}{3}$	**Addition Property of Inequality**
$4\frac{1}{3} \le b$ or $b \ge 4\frac{1}{3}$	**Simplify.**

The solution is $b \ge 4\frac{1}{3}$.

Check	$3 \le b - \frac{1}{3}$	**Write the inequality.**
	$3 \overset{?}{\le} 4\frac{1}{3} - 1\frac{1}{3}$	**Replace b with $4\frac{1}{3}$.**
	$3 \le 3$ ✓	**The statement is true.**

Graph the solution.

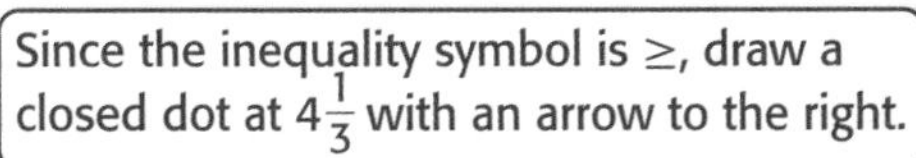

Since the inequality symbol is ≥, draw a closed dot at $4\frac{1}{3}$ with an arrow to the right.

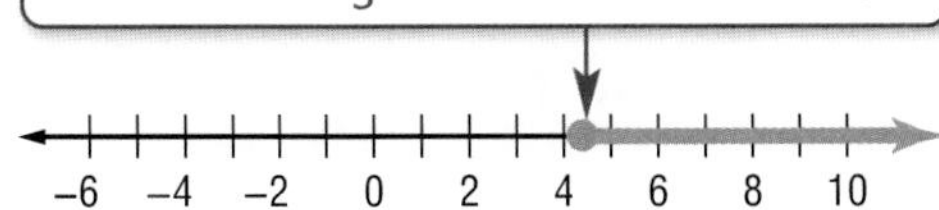

Check Your Progress

Solve each equation. Graph the solution on a number line.

2A. $3 \ge g + 7$

2B. $b + \frac{5}{7} > 2$

Personal Tutor glencoe.com

Solve Inequalities by Multiplying or Dividing by a Positive Number Some inequalities like $4x > 8$ are solved by multiplication or division. You can multiply or divide each side of an inequality by a positive number and the inequality is still true.

StudyTip

Positive Number The statement $c > 0$ means that c is a positive number.

Key Concept Multiplication and Division Properties

For Your FOLDABLE

Words When you multiply or divide each side of an inequality by the same *positive* number, the inequality remains true.

Symbols For all numbers a, b, and c, where $c > 0$,

1. if $a < b$, then $ac < bc$ and $\frac{a}{c} < \frac{b}{c}$.
2. if $a > b$, then $ac > bc$ and $\frac{a}{c} > \frac{b}{c}$.

Example

$-6 < 10$	$20 > 16$
$-6 \cdot 2 < 10 \cdot 2$	$\frac{20}{4} > \frac{16}{4}$
$-12 < 20$	$5 > 4$

These properties are also true for $a \le b$ and $a \ge b$.

StudyTip

Words to Symbols Remember, *at most* translates to $\leq$, while *at least* translates to $\geq$.

Real-World Link

Jewelry and other artifacts made of gold have been discovered in archaeological sites that date back to 3000 B.C.

Source: Smithsonian

Real-World EXAMPLE 3 Divide by a Positive Number

JEWELRY Macy is making each of her 7 friends a bracelet. She does not want to spend more than \$40 on the bracelets. Find the maximum cost for each bracelet.

Understand You know that Macy wants to make 7 bracelets and does not want to spend more than \$40. You need to determine the most she can spend per bracelet.

Plan Since Macy wants to spend at most \$40, write and solve an inequality using the symbol $\leq$.

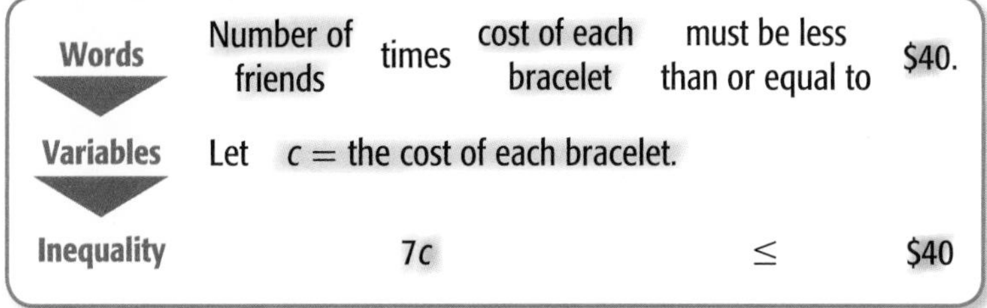

Words	Number of friends	times	cost of each bracelet	must be less than or equal to	\$40.
Variables	Let c = the cost of each bracelet.				
Inequality		$7c$		$\leq$	\$40

Solve $7c \leq 40$ — **Write the inequality.**

$\frac{7c}{7} \leq \frac{40}{7}$ — **Division Property of Inequality**

$c \leq 5\frac{5}{7}$ or $5.\overline{714285}$ — **Simplify.**

Macy can spend no more than \$5.71 per bracelet.

Check If she spends \$5.71 per bracelet, then the total cost is \$5.71 × 7 or \$39.97. If she spends \$5.72 per bracelet, then the total cost is \$5.72 × 7 or \$40.04. So, the answer is correct. ✓

Check Your Progress

3. WORK It takes Alfonzo $\frac{3}{4}$ hour to mow a lawn. Write and solve an inequality to find the number of lawns he can mow if he works at least 15 hours.

Personal Tutor glencoe.com

Multiply or Divide an Inequality by a Negative Number What happens when each side of an inequality is multiplied or divided by a negative number?

Graph 2 and 4 on a number line.

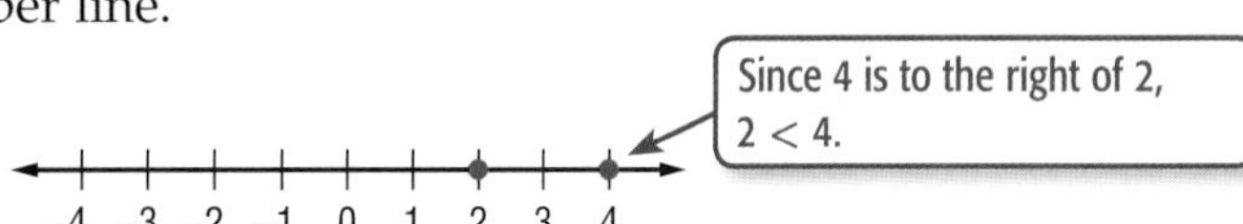

Now, multiply each number by −1.

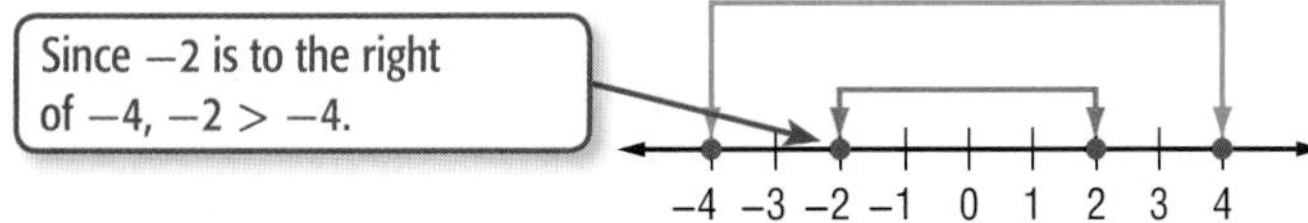

Notice that the numbers being compared switched positions as a result of being multiplied by a negative number. In other words, their order reversed. This suggests the properties shown on the next page.

StudyTip

Negative Number The statement $c < 0$ means that c is a negative number.

Key Concept — Multiplication and Division Properties

For Your FOLDABLE

Words When you multiply or divide each side of an inequality by the same *negative* number, the inequality symbol must be reversed for the inequality to remain true.

Symbols For all numbers a, b, and c, where $c < 0$,

1. if $a < b$, then $ac > bc$ and $\frac{a}{c} > \frac{b}{c}$.
2. if $a > b$, then $ac < bc$ and $\frac{a}{c} < \frac{b}{c}$.

Examples

$$-4 < 5 \qquad 18 > -12$$
$$-4 \cdot -3 > 5 \cdot -3 \qquad \frac{18}{-3} < \frac{-12}{-3}$$
$$12 > -15 \qquad -6 < 4$$

Math *in Motion,* BrainPOP® glencoe.com

These properties are also true for $a \le b$ and $a \ge b$.

EXAMPLE 4 Multiply or Divide by a Negative Number

Solve each inequality. Then graph the solution on a number line.

a. $-5x < 45$

$-5x < 45$	Write the inequality.
$\frac{-5x}{-5} > \frac{45}{-5}$	Division Property of Inequality
$x > -9$	Simplify.

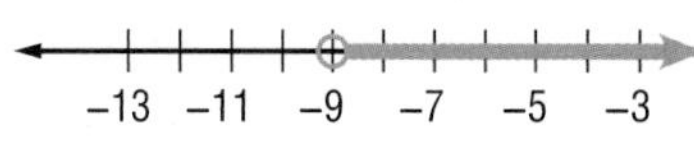

b. $\frac{b}{-8} \ge -6$

$\frac{b}{-8} \ge -6$	Write the inequality.
$(-8)\,\frac{b}{-8} \le -6\,(-8)$	Multiplication Property of Inequality
$b \le 48$	Simplify.

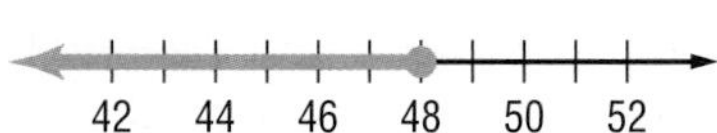

StudyTip

Checking Solutions For Example 4a, try a number that is less than −9 to show it is *not* a solution.

$$-5x < 45$$
$$-5(-10) \overset{?}{<} 45$$
$$50 \not< 45$$

Check Your Progress

4A. $-\frac{y}{4} < 3.$

4B. $7 \ge -2f$

Personal Tutor glencoe.com

Check Your Understanding

Example 1 p. 241

Solve each inequality. Check your solution.

1. $y + 7 \le 12$

2. $b + 20 > -13$

3 $-7 < x + (-3)$

Example 2 p. 242

Solve each inequality. Graph the solution on a number line.

4. $d - 9.3 \ge 12.5$

5. $3\frac{1}{5} > f - \frac{4}{5}$

6. $g - 22 \le -40$

Example 3 p. 243

7. JOBS Isabel earns $7.50 per hour on the weekends. Write and solve an inequality to find how many hours she needs to work to earn at least $120.

Example 4 p. 244

Solve each inequality. Graph the solution on a number line.

8. $-20 < -4t$

9. $-8z \le -24$

10. $18 > -\frac{2}{3}g$

Practice and Problem Solving

● = **Step-by-Step Solutions** begin on page R11.
Extra Practice begins on page 810.

Examples 1 and 2 pp. 241–242

Solve each inequality. Check your solution.

11. $a + 18 < 40$ **12.** $h - 12 > 52$ **13.** $y - 4.2 \leq 6.5$

14. $g + 5.9 \geq 10$ **15.** $p - 14 > 12$ **16.** $x + 3.75 < 5$

17. $n - 0.1 \leq 1.4$ **18.** $7 > z + \frac{2}{3}$ **19.** $22 \geq c - 2.1$

20. $13 \geq 9 + b$ **21.** $14\frac{1}{2} < b - 1\frac{1}{4}$ **22.** $t + \frac{1}{5} < 2\frac{7}{10}$

Example 3 p. 243

23. **ARCADES** Montel spends \$0.75 every time he plays his favorite video game. If he has \$10, write and solve an inequality to find how many video games he can play.

24. **ANIMALS** Write and solve an inequality to find the swimming rate of a manatee that swims at least 15 miles in 3 hours.

Example 4 p. 244

Solve each inequality. Graph the solution on a number line.

25. $4x > -36$ **26.** $7y \leq -49$ **27.** $-5m \geq 15$

28. $-2p < -3$ **29.** $45 \leq -10r$ **30.** $-66 \geq -11t$

31. $\frac{7}{2}y > 63$ **32.** $\frac{3}{4} \leq -\frac{5}{7}m$ **33.** $-\frac{3}{24}b \geq -\frac{1}{4}$

34. $-\frac{x}{4} > 3$ **35.** $\frac{4}{9}c \leq -\frac{4}{5}$ **36.** $-\frac{3}{4} \geq -\frac{6}{10}a$

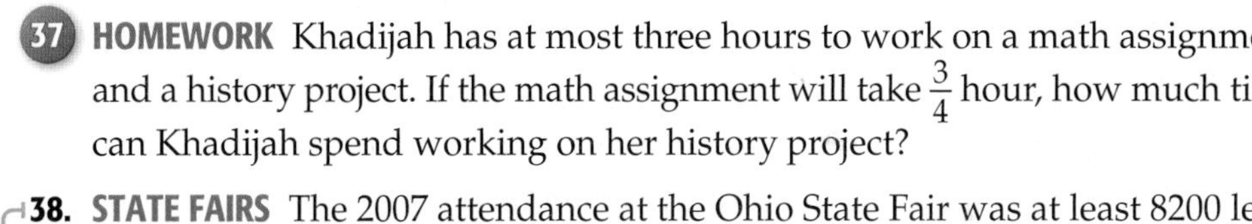

37 **HOMEWORK** Khadijah has at most three hours to work on a math assignment and a history project. If the math assignment will take $\frac{3}{4}$ hour, how much time can Khadijah spend working on her history project?

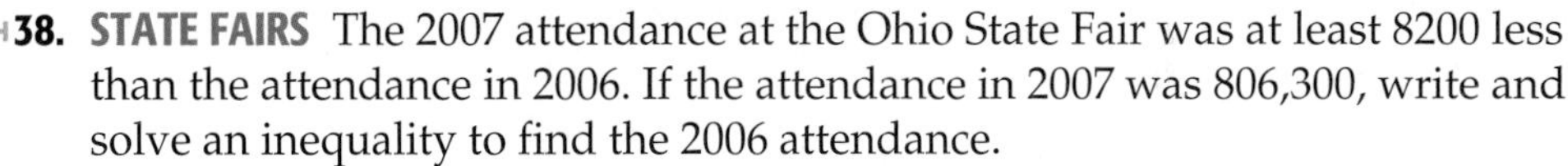

38. **STATE FAIRS** The 2007 attendance at the Ohio State Fair was at least 8200 less than the attendance in 2006. If the attendance in 2007 was 806,300, write and solve an inequality to find the 2006 attendance.

Real-World Link

The Ohio State Fair began in 1850. One of its most popular attractions is a butter cow, which is sculpted from 2000 pounds of real butter.

Source: Ohio State Fair

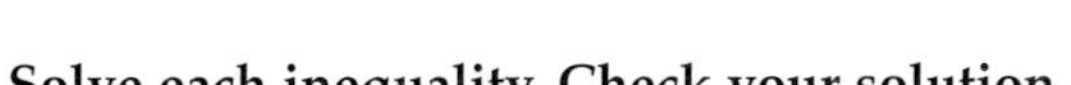

Solve each inequality. Check your solution.

39. $a - 3.5 < \frac{2}{5}$ **40.** $b + 4\frac{1}{2} \geq 0.4$ **41.** $26 \leq 5 - c$

42. $12 - d > 40$ **43.** $4.5 \geq \frac{3}{2}r$ **44.** $f - 8 < -1.1$

45. $t - 6 \leq 2.5$ **46.** $-9.6 \geq \frac{1}{3}y$ **47.** $-\frac{7}{11} \geq -\frac{14}{33}g$

Write an inequality to represent each situation. Then solve the inequality.

48. Seven more than a number is at most 24.

49. The quotient of a number and −3 is greater than the quotient of 5 and 6.

50. 18 is at least the product of −6 and a number.

51. The difference of a number and 15 is no more than −8.

52. **FINANCIAL LITERACY** Brian is saving money to buy a new mountain bike. The bike that he likes costs \$375.95, and he has already saved \$285.50. Write and solve an inequality to find the amount he must still save.

Solve each inequality. Check your solution.

53. $4a - 7 \geq 21$

54. $8 < -\frac{m}{6} - 2$

55. $3b + 15 \leq 8b - 5$

56. $\frac{m}{5} - 12 > -18$

57. $-\frac{3}{4}d - 6 \geq 9$

58. $13 - 3n < -8$

59. $11 - 2g > -7 - 8g$

60. $15 \leq -\frac{k}{4} - 9$

61. $-12 \geq -3q - 18$

62. **MULTIPLE REPRESENTATIONS** Consider the inequalities $b \geq 4$ and $b \leq 13$.

a. GRAPHICAL Graph each inequality on the same number line.

b. VERBAL Do the solution sets of the two inequalities overlap? If so, what does this overlapping area represent?

c. ALGEBRAIC A *compound inequality* is an inequality that combines two inequalities. Write a compound inequality for the situation.

d. GRAPHICAL Look back at the graph of the solutions for both inequalities. Make another graph that shows only the solution of the compound inequality.

Graph each compound inequality on a number line.

63. $-3 < n < 5$

64. $4 \geq m > -2$

65. $8 \leq g < 14$

Real-World Link

Throughout the day, the body temperature of a healthy person can fluctuate by 1°F or more depending on their physical activity. A person's body temperature is usually lowest just after waking up.

Source: WebMD

66. **SCIENCE** Use the body temperature scale shown.

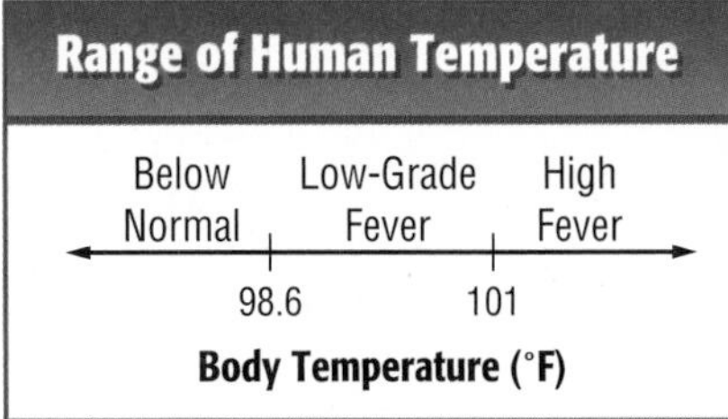

a. Suppose Malia has a temperature of 99.2°. Write and solve an inequality to find how much her temperature must increase before she is considered to have a high fever.

b. Hypothermia occurs when a person's body temperature falls below 95°F. Write and solve an inequality that describes how much lower the body temperature of a person with hypothermia will be than a person with a normal body temperature of 98.6°F.

H.O.T. Problems

Use Higher-Order Thinking Skills

67. **OPEN ENDED** Write an inequality for the following sentence.

The quotient of a number and −5 increased by 4 is at most 8.

Name three numbers that are possible solutions. Explain.

68. **WRITING IN MATH** Write a real-world problem involving an inequality and negative numbers where the inequality symbol would *not* be reversed when finding the solution.

69. **CHALLENGE** Twenty more than half a number is at least 45. Find the least number that meets this condition.

70. **REASONING** Is the following statement *true* or *false*? If false, provide a counterexample.

For all values of x, two times x is greater than x.

71. **REASONING** Is it *always, sometimes,* or *never* true that if $x \leq y$, then $y > x$? Explain your reasoning.

72. **WRITING IN MATH** Explain to a friend who was not in class today how to solve inequalities that involve multiplication and division.

Standardized Test Practice

73. The length of the rectangle is greater than its width. Which inequality represents the possible values of x?

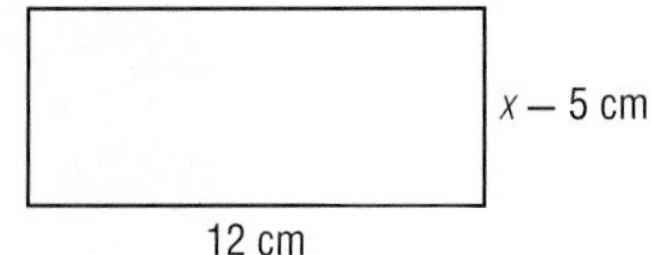

A $x \leq 17$

B $x < 17$

C $x \geq 17$

D $x > 17$

74. If $n + 15 > 4$, then n could be which of the following values?

F -13

G -12

H -11

J -10

75. The solutions for which inequality are represented by the following graph?

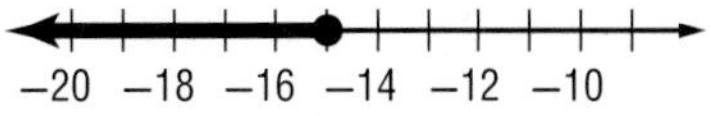

−20 −18 −16 −14 −12 −10

A $\frac{x}{-3} < 5$

B $\frac{x}{-3} \leq 5$

C $\frac{x}{3} > -5$

D $\frac{x}{-3} \geq 5$

76. EXTENDED RESPONSE The product of a number and four is at most thirty.

a. Write an inequality for the sentence.

b. Solve the inequality.

c. Graph the solution on a number line.

Spiral Review

Write an inequality for each sentence. (Lesson 5-3)

77. Leticia made at least $45 babysitting last weekend.

78. Marc could pay no more than $8500 for his car.

79. Adrienne needs an 86% or better to get a B in the class.

Solve each equation. Check your solution. (Lesson 5-2)

80. $4h + 5 = -6h - 19$

81. $7d - 13 = 17 + 3d$

82. $n - 14 = 3n$

83. $2g + 12 = 3g - 1$

84. $8y + 5 = 5y - 5 + 2y$

85. $4t = 2t - 26$

86. PLUMBING A standard showerhead uses about 6 gallons of water per minute. The table shows the relationship between time in minutes and the number of gallons of water used. (Lesson 1-5)

a. Given m, the number of minutes, write an equation that can be used to find g, the number of gallons used.

b. How many minutes elapsed if 72 gallons of water were used?

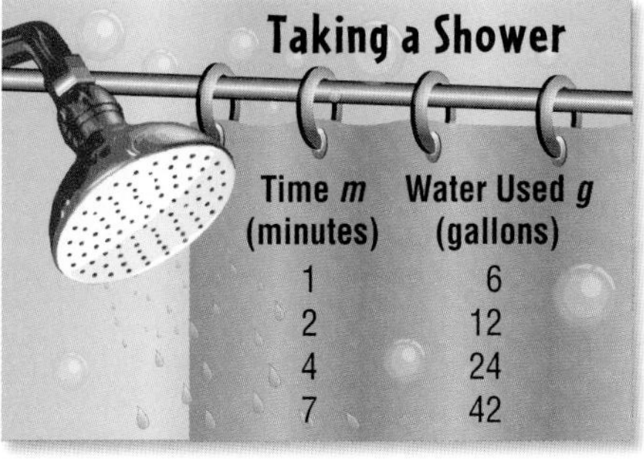

Taking a Shower

Time m (minutes)	Water Used g (gallons)
1	6
2	12
4	24
7	42

Skills Review

Solve each equation. (Lesson 4-5)

87. $3x + 1 = 7$

88. $5x - 4 = 11$

89. $4h + 6 = 22$

90. $8n + 3 = -5$

91. $37 = 4d + 5$

92. $9 = 15 + 2p$

5-5

Solving Multi-Step Equations and Inequalities

Then
You have already solved two-step equations and inequalities.

Now
- Solve multi-step equations.
- Solve multi-step inequalities.

New Vocabulary
null or empty set
identity

Math Online
glencoe.com
- Extra Examples
- Personal Tutor
- Self-Check Quiz
- Homework Help

Why?

A good rule to know when training for a marathon is you will generally have enough endurance to finish a race that is 3 times your average daily distance.

a. Write an equation that represents the relationship between daily average distance d and possible race lengths l.

b. Suppose your average daily run is 2 kilometers. Write an equation that represents the amount that you need to increase your daily run by d to have enough endurance for a 12-kilometer race.

Solve Equations and Inequalities with Grouping Symbols To find the minimum number of kilometers you need to increase your daily run, you can solve the equation $3(d + 2) = 12$. First, use the Distributive Property to remove the grouping symbols.

EXAMPLE 1 Solve Equations and Inequalities with Parentheses

a. Solve $3(d + 2) = 12$. Check your solution.

$3(d + 2) = 12$	Write the equation.
$3d + 6 = 12$	Distributive Property
$-6 \quad -6$	Subtraction Property of Equality
$3d = 6$	Simplify.
$\frac{3d}{3} = \frac{6}{3}$	Division Property of Equality
$d = 2$	Simplify.

b. Solve $4(x - 3) > 6$.

$4(x - 3) > 6$	Write the inequality.
$4x - 12 > 6$	Distributive Property
$+12 \quad +12$	Subtraction Property of Inequality
$4x > 18$	Simplify.
$\frac{4x}{4} > \frac{18}{4}$	Division Property of Inequality
$x > 4.5$	Simplify.

Check Your Progress

Solve. Check your solution.

1A. $3 = 4(x + 2)$

1B. $4(b - 3) \leq 72$

Personal Tutor glencoe.com

StudyTip

Alternative Method You can also solve the equation using the vertical method shown on p. 248.

EXAMPLE 2 Solve Multi-Step Equations

Solve $4(x + 5) = 3(2x + 4)$. Check your solution.

$4(x + 5) = 3(2x + 4)$	Write the equation.
$4x + 20 = 6x + 12$	Distributive Property
$4x - 4x + 20 = 6x - 4x + 12$	Subtraction Property of Equality
$20 = 2x + 12$	Simplify.
$20 - 12 = 2x + 12 - 12$	Subtraction Property of Equality
$8 = 2x$	Simplify.
$4 = x$	Division Property of Equality

Check Your Progress

Solve each equation. Check your solution.

2A. $12m + 12 = 6(3m + 3)$

2B. $5(n - 3) = 3(n + 7)$

Personal Tutor glencoe.com

STANDARDIZED TEST EXAMPLE 3

Mariella's parents have budgeted $575 for her quinceañera. The cost of the party room is $75. How much can the family spend per guest on food if each of the 40 guests receives a $5 favor?

A $5.00 **B** $7.50 **C** $8.00 **D** $9.50

Real-World Career

Event Planner
Event planners plan, budget , and organize events including parties, weddings, and business conferences.

Education and training for event planners can include a college degree in hospitality at either a two-year or a four-year program.

Read the Test Item

You need to find the amount of money the family can spend per guest on food.

Solve the Test Item

Words Party cost = room cost + the number of guests × the cost per guest.

Symbols Let c = the food cost per guest so $c + 5$ = the total cost per guest.

Equation $575 = 75 + 40(c + 5)$

$575 = 75 + 40(c + 5)$	Write the equation.
$575 = 75 + 40c + 200$	Distributive Property
$575 = 40c + 275$	Simplify.
$575 - 275 = 40c + 275 - 275$	Subtraction Property of Equality
$300 = 40c$	Simplify.
$7.5 = c$	Division Property of Equality

The answer is B.

Check Your Progress

3. Sofia recycled 3 pounds less than the amount that James recycled. Hannah recycled 3 times the amount that Sofia recycled. If they recycled a total of 53 pounds, how many pounds did Sofia recycle?

F 10 pounds **G** 13 pounds **H** 30 pounds **J** 35 pounds

Personal Tutor glencoe.com

EXAMPLE 4 Solve Multi-Step Inequalities

Solve $5a - 8 \geq 4(a - 3)$. Graph the solution on a number line.

$5a - 8 \geq 4(a - 3)$	Write the inequality.
$5a - 8 \geq 4a - 12$	Distributive Property.
$5a - 8 - 4a \geq 4a - 12 - 4a$	Subtraction Property of Inequality
$a - 8 \geq -12$	Simplify.
$a - 8 + 8 \geq -12 + 8$	Addition Property of Inequality
$a \geq -4$	Simplify.
−5 −4 −3 −2 −1 0 1 2 3 4 5	Graph the solution on a number line.

Watch Out!

Do not reverse the inequality sign just because there is a negative sign in the inequality. Only reverse the sign when you multiply or divide by a negative number.

Check Your Progress

Solve each inequality. Graph the solution on a number line.

4A. $-2(k + 1) > -16 + 5k$

4B. $2p + 5 \geq 3(p - 6)$

Personal Tutor glencoe.com

No Solution or All Numbers as Solutions Some equations have *no* solution. When this occurs, the solution is the **null or empty set**, shown by the symbol Ø or { }. Other equations may have every number as their solution. An equation that is true for every value of the variable is called an **identity**.

EXAMPLE 5 Null Set and Identities

Solve each equation.

a. $3(y - 5) + 25 = 3y + 10$

$3(y - 5) + 25 = 3y + 10$	Write the equation.
$3y - 15 + 25 = 3y + 10$	Use the Distributive Property.
$3y + 10 = 3y + 10$	Simplify.
$3y + 10 - 3y = 3y + 10 - 3y$	Subtraction Property of Equality
$10 = 10$	Simplify.

The statement $10 = 10$ is *always* true. The equation is an identity and the solution set is all numbers.

b. $-5s - 14 = 2(2s + 3) - 9s$

$-5s - 14 = 2(2s + 3) - 9s$	Write the equation.
$-5s - 14 = 4s + 6 - 9s$	Use the Distributive Property.
$-5s - 14 = 6 - 5s$	Simplify.
$-5s - 14 + 5s = 6 - 5s + 5s$	Addition Property of Equality
$-14 = 6$	Simplify.

The statement $-14 = 6$ is *never* true. The equation has no solutions and the solution set is Ø.

Review Vocabulary

Identities An identity is an equation that shows that a number or expression is equivalent to itself.

Additive Identity
$a + 0 = a$

Multiplicative Identity
$a \cdot 1 = a$

Check Your Progress

5A. $-2(3r + 4) = -5r - 8 - r$

5B. $14 + 8w = 4(8 + 2w)$

Personal Tutor glencoe.com

Check Your Understanding

Examples 1, 2, and 5 pp. 248–250

Solve. Check your solution.

1. $4(x + 1) = 28$
2. $35 = 7(2p - 1)$
3. $2(a - 2) = 3(a - 5)$
4. $16(z + 3) = 4(z + 9)$
5. $7(x + 2) = 2(x + 2)$
6. $3(d - 2) = 5(d + 8)$
7. $6x + 4 = 2(3x - 5)$
8. $20f + (-8f - 15) = 3(4f - 5)$
9. $3(1 + 2f) - 5 = 6f - 2$
10. $7x + 5 = 10(x - 7) - 3x$

Example 3 p. 249

11. **MULTIPLE CHOICE** You and three friends are going to the fair. The cost for parking is \$5 per car and admission to the fair is \$19 per person. If you have a total of \$113, how much can each person spend on food?

    A \$4 **C** \$12

    B \$8 **D** \$24

Examples 1 and 4 pp. 248 and 250

Solve. Graph the solution on a number line.

12. $-2(k - 2) \geq -20$
13. $(3r + 7)2 \leq -34$
14. $-2(g - 1) > g - 4$
15. $5p + 8 \geq 3(p + 6)$
16. $6(-2z + 5) < -19z + 16$
17. $10p \leq 7(2p - 4)$

Practice and Problem Solving

= **Step-by-Step Solutions** begin on page R11.
Extra Practice begins on page 810.

Examples 1, 2, and 5 pp. 248–250

Solve. Check your solution.

18. $6n - 18 = 4(n + 2)$
19. $12y + 5(y - 6) = 4$
20. $12z + 4 = 2(5z + 8) - 12$
21. $d - 12 = 4(d - 6)$
22. $3x + 2 = 2(2x - 7)$
23. $6(y - 5) = 2(10 + 3y)$
24. $4(2c + 8) = 5(c + 4)$
25. $10 + 12p = 3(3 + 4p)$
26. $3x + 2 + 5(x - 1) = 8x + 17$
27. $10z + 4 = 2(5z + 8) - 12$

Example 3 p. 249

28. **GEOMETRY** The perimeter of a rectangle is 50 centimeters. The length of the rectangle is one more than 3 times the width of the rectangle. What are the dimensions of the rectangle?

29. **FINANCIAL LITERACY** Tim is taking the train to Seattle to visit his grandparents. He was given \$15 to spend on snacks and reading material. Granola bars cost \$1.15 each, and magazines cost \$1.25. If Tim buys the same number of granola bars and magazines, how many can he buy?

Examples 1 and 4 pp. 248 and 250

Solve. Graph the solution on a number line.

30. $20 > 5(w + 3)$
31. $-32 \leq 9(3h + 2) + 4$
32. $3(6m - 4) \geq 24$
33. $10(3 + s) < 4s$
34. $3y - 6 > 4(y - 3)$
35. $8(2h + 6) \leq 12h + 20$
36. $3(3r + 5) \geq 24 + 10r$
37. $14t - 28 < 7(t + 6)$

38. **SCHOOL** Nomar has earned scores of 73, 85, 91, and 82 on the first four of five math tests for the grading period. He would like to finish the grading period with a test average of at least 82. What is the minimum score Nomar needs to earn on the fifth test in order to achieve his goal?

Solve.

39. $-0.2(3c + 15) = 3(0.8c - 8)$

40. $2(t + 12) - 6(2t - 3) = 14$

41. $5 - \frac{1}{2}(x - 6) < 4$

42. $6n - 18 \geq 4(n + 2.1)$

43. $2.01c - 6 = -0.15c + 6.96$

44. $\frac{1}{4}x + 13 > 0.25(2x - 32)$

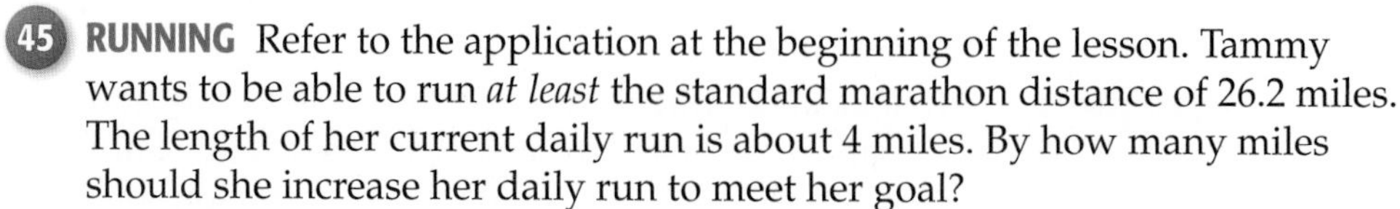

45. **RUNNING** Refer to the application at the beginning of the lesson. Tammy wants to be able to run *at least* the standard marathon distance of 26.2 miles. The length of her current daily run is about 4 miles. By how many miles should she increase her daily run to meet her goal?

Real-World Link

One of the most popular races in America is the Chicago marathon. Usually about 36,000 people participate in the 26.2 mile race.

Source: Chicago Marathon

46. **REPAIRS** Cole is having his car repaired. The mechanic said it would cost at least \$375 for parts and labor. If the cost of the parts was \$150, and the mechanic charges \$60 an hour, how many hours is the mechanic planning to work on the car?

Solve. Justify each step in the solution. Use a Property of Equality or Inequality when necessary.

47. $4(y - 3) = 2(3y + 10)$

48. $5(2f - 1) = 3(f + 3)$

49. $-1.2(w + 1.1) \leq 6.18$

50. $p > \frac{2}{3}\left(p - \frac{1}{2}\right)$

H.O.T. Problems

Use Higher-Order Thinking Skills

51. **OPEN ENDED** Write a multi-step inequality that can be solved by first adding 3 to each side.

52. **WRITING IN MATH** Explain how you can solve $45 > -6x + 3$ without multiplying or dividing each side by a negative number.

53. **FIND THE ERROR** Jada and Liu are solving $3x - 9 \leq 5(x + 10)$. Is either of them correct? Explain your reasoning.

Jada

$3x - 9 \leq 5(x + 10)$

$3x - 9 \leq 5x + 10$

$-2x \leq 19$

$x \geq -9.5$

Liu

$3x - 9 \leq 5(x + 10)$

$3x - 9 \leq 5x + 50$

$-2x \leq 59$

$x \leq -29.5$

54. **CHALLENGE** Use the information in Example 5 about equations that have no solutions or those that are identities to solve the following inequalities. Justify each step in the solution.

a. $5x - 6 > 3(x - 2) + 2x$

b. $12p + 17 \leq 3(4p - 8)$

55. **WRITING IN MATH** Explain how to determine if an equation has no solution, one solution, or if all numbers are solutions. Use examples with your explanation.

Standardized Test Practice

56. Find the value of x so that the polygons have the same perimeter.

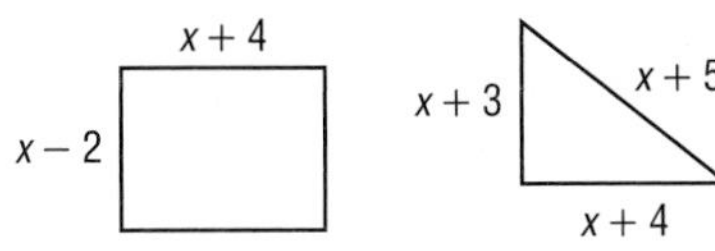

A 3
B 6
C 8
D 12

57. GRIDDED RESPONSE Damon spent $42, including tax, on shelves he bought at a home improvement store. The price of the shelves, including tax, was $7 for two shelves. In the equation below, s represents the number of shelves Damon bought.

$$42 = s(7 \div 2)$$

How many shelves s did Damon buy?

58. What is the solution of this inequality?

$$-4x + 16 \geq -4$$

F $x \geq 3$
G $x \leq 3$
H $x \geq 5$
J $x \leq 5$

59. Sandra's scores on the first five science tests are shown. Which inequality represents the score she must receive on the sixth test to have an average score of more than 88?

Test	Score
1	85
2	84
3	90
4	95
5	88

A $s \geq 86$
B $s \leq 88$
C $s < 88$
D $s > 86$

Spiral Review

Solve each inequality. Graph the solution on a number line. (Lesson 5-4)

60. $-25t \leq 400$
61. $8 > \frac{q}{3}$
62. $14 \geq 7 + a$
63. $-13 \geq x - 8$
64. $-\frac{3}{4} < w - 1$
65. $3 \leq \frac{1}{2} + a$

Write an inequality for each sentence. (Lesson 5-3)

66. Kyle's earnings were no more than $60.

67. The 10 kilometer race time of 86 minutes was greater than the winner's time.

Write each fraction or mixed number as a decimal. Use a bar to show a repeating decimal. (Lesson 3-1)

68. $\frac{1}{5}$
69. $-\frac{5}{8}$
70. $7\frac{3}{10}$
71. $\frac{1}{9}$
72. $-3\frac{3}{4}$
73. $-\frac{5}{11}$

74. BUSINESS The formula $P = I - E$ is used to find the profit P when income I and expenses E are known. One month a small business has an income of $19,592 and expenses of $20,345. (Lesson 2-3)

a. What is the profit for the month?

b. What does a negative profit mean?

Skills Review

Divide. (Lesson 0-3)

75. $7.2 \div 2$
76. $\$3.75 \div 5$
77. $\$25.90 \div 3.5$
78. $29.14 \div 4.7$

CHAPTER 5

Study Guide and Review

Math Online glencoe.com

- STUDY TO GO
- Vocabulary Review

Chapter Summary

Key Concepts

Perimeter and Area (Lesson 5-1)

- Perimeter is the distance around a geometric figure.
  Rectangle: $P = 2(\ell + w)$
  Triangle: $P = a + b + c$
- Area is the measure of the surface enclosed by a figure.
  Rectangle: $A = \ell w$
  Triangle: $A = \frac{1}{2}bh$

Solving Equations (Lessons 5-2 and 5-5)

- Use the Addition or Subtraction Property of Equality to isolate the variables on one side of an equation.
- Use the Distributive Property to remove the grouping symbols.

Inequalities (Lessons 5-3 to 5-5)

- An inequality is a mathematical sentence that contains $<$, $>$, $\leq$, or $\geq$.
- Solving an inequality means finding values for the variable that make the inequality true.
- When you multiply or divide each side of an inequality by a positive number, the inequality symbol remains the same.
- When you multiply or divide each side of an inequality by a negative number, the inequality symbol must be reversed.
- To solve an inequality that involves more than one operation, work backward to undo the operations.

FOLDABLES® Study Organizer

Be sure the Key Concepts are noted in your Foldable.

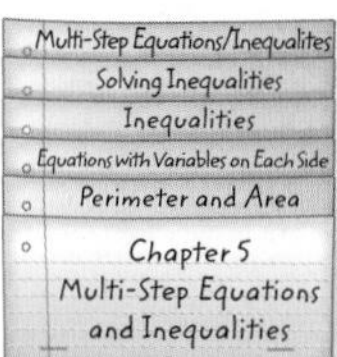

Key Vocabulary

area (p. 222)

formula (p. 221)

identity (p. 250)

inequality (p. 234)

null or empty set (p. 250)

perimeter (p. 221)

Vocabulary Check

Choose the correct term to complete each sentence.

1. An (equation, inequality) is a mathematical sentence that contains a less than or greater than symbol.
2. (Perimeter, Area) is the distance around a geometric figure.
3. The inequality symbol must be reversed when you multiply or divide by a (positive, negative) number.
4. When finding the area of a triangle multiply $\left(2, \frac{1}{2}\right)$ by the product of the base and height.
5. The symbol ≥ means (less than, greater than) or equal to.
6. The statement $d + 0 = d$ is an example of the (additive, multiplicative) identity.
7. The area of a rectangle is equal to the (sum, product) of the length and width.
8. The (Addition, Distributive) Property can be used to remove grouping symbols.
9. The measure of the amount of space in a figure is the (perimeter, area).
10. A null or empty set is shown by the symbol (Ø, λ).

MIXED PROBLEM SOLVING
For mixed problem-solving practice, see page 847.

Lesson-by-Lesson Review

5-1 Perimeter and Area (pp. 221–226)

Find the perimeter and area of each figure.

11.

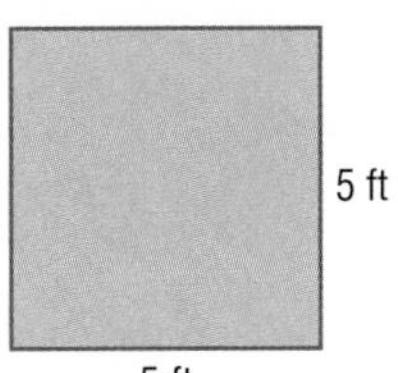

12.

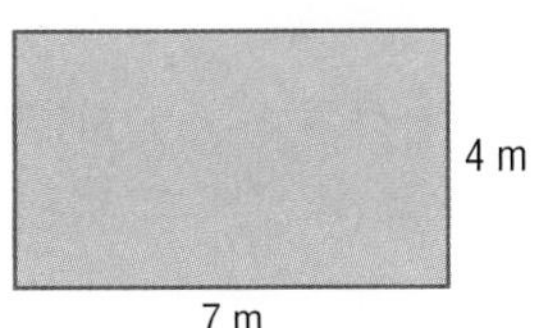

13.

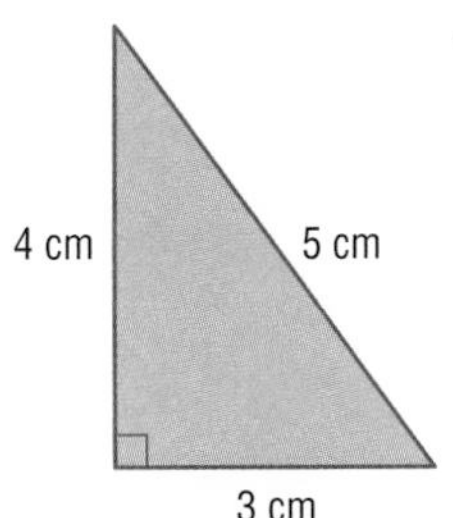

14.

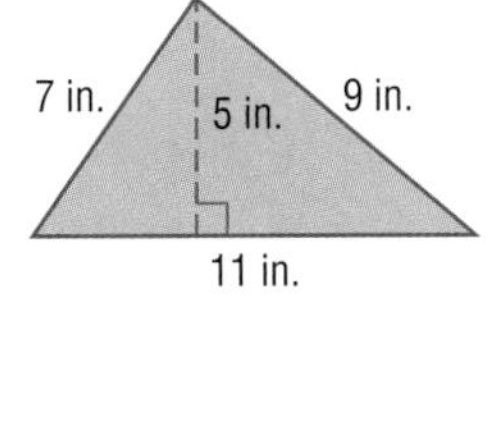

15. **GYMNASTICS** The area of a gymnastics mat is 142.5 square feet. If the width of the mat is 9.5 feet, what is the length?

EXAMPLE 1

Find the perimeter and area of a 12 inch by 6 inch rectangle.

$P = 2(\ell + w)$	**Perimeter of a rectangle**
$= 2(12 + 6)$	**Replace ℓ with 12 and w with 6.**
$= 36$	**Simplify.**

The perimeter is 36 inches.

$A = \ell w$	**Area of a rectangle**
$= 12(6)$	**Replace ℓ with 12 and w with 6.**
$= 72$	**Simplify.**

The area is 72 square inches.

5-2 Solving Equations with Variables on Each Side (pp. 229–233)

Solve each equation. Check your solution.

16. $3a + 6 = 2a$

17. $10 - x = 9x$

18. $b - 6 = -b + 2$

19. $5 + 2y - 12 = y + 9$

20. $9q + 6 = 6q - 9$

21. $c - 12 = 4c - 12$

22. **MOVIES** An online DVD rental club has two membership plans as shown. Write and solve an equation to find how many months it would take for the total cost of the two plans to be the same.

Plan	Membership Fee	Cost Per Month
A	\$20	\$5
B	\$30	\$3

EXAMPLE 2

Solve $8x + 6 = 4x - 10$. Check your solution.

$8x + 6 = 4x - 10$	**Write the equation.**
$8x - 4x + 6 = 4x - 4x - 10$	**Subtraction Property of Equality**
$4x + 6 = -10$	**Simplify.**
$4x + 6 - 6 = -10 - 6$	**Subtraction Property of Equality**
$4x = -16$	**Simplify.**
$x = -4$	**Mentally divide each side by 4.**

Check $8x + 6 = 4x - 10$

$8(-4) + 6 \stackrel{?}{=} 4(-4) - 10$

$-32 + 6 \stackrel{?}{=} -16 - 10$

$-26 = -26$ ✓

CHAPTER 5

Study Guide and Review

5-3 Inequalities (pp. 234–239)

Write an inequality for each sentence.

23. Jeremiah can spend at most \$15 at the store.

24. There are more than 35 students in the band.

For the given value, state whether each inequality is *true* or *false*.

25. $x + 6 > 7, x = 2$

26. $13 - a < 9, a = 10$

27. $16 \leq 4a, a = 4$

28. $3m + 4 \geq 12, m = 2$

29. $6x > 18, x = 3$

30. $6b + 4 > 12, b = 2$

EXAMPLE 3

State whether $x - 6 > 12$ is *true* or *false* for $x = 15$.

$x - 6 > 12$ — **Write the inequality.**

$15 - 6 \overset{?}{>} 12$ — **Replace x with 15.**

$9 \overset{?}{>} 12$ — **Simplify.**

The sentence is false. So, $9 \not> 12$.

5-4 Solving Inequalities (pp. 241–247)

Solve each inequality. Graph the solution on a number line.

31. $x - 4 < 8$

32. $y + 3 \geq 11$

33. $a - 15 \leq 3$

34. $x + 13 > -22$

35. $-3z \leq -24$

36. $6h > 42$

37. $-5x < -13$

38. $\frac{2}{3}x > 6$

39. MUSIC Jose can spend at most \$120 for CDs. If each CD costs \$20, write and solve an inequality to show the maximum number of CDs Jose can buy.

EXAMPLE 4

Solve $x - 3 < 10$. Then graph the solution on a number line.

$x - 3 < 10$ — **Write the inequality.**

$x - 3 + 3 < 10 + 3$ — **Addition Property of Inequality**

$x < 13$ — **Simplify.**

Graph the solution.

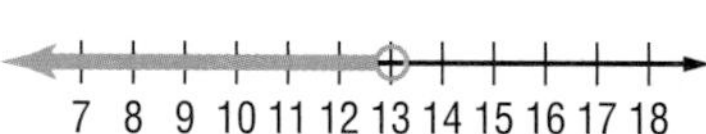

5-5 Solving Multi-Step Equations and Inequalities (pp. 248–253)

Solve. Check your solution.

40. $8b + 5 = 21$

41. $15 - 4n = -13$

42. $12 = 6(z - 4)$

43. $\frac{3}{4}(12 + 4a) = 21$

44. $-4g - 5 \geq -17$

45. $18 > -12 + 6m$

46. $24 - 3c \leq 15$

47. $\frac{2}{3}k + 9 < 5$

48. SALES A car sales associate receives a monthly salary of \$1700 a month plus \$140 for every car he sells. How many cars must he sell monthly to earn at least \$4500?

EXAMPLE 5

Solve $-5m + 8 \geq 23$.

$-5m + 8 \geq 23$ — **Write the inequality.**

$-5m + 8 - 8 \geq 23 - 8$ — **Subtraction Property of Inequality**

$-5m \geq 15$ — **Simplify.**

$\frac{-5m}{-5} \leq \frac{15}{-5}$ — **Division Property of Inequality**

$m \leq -3$ — **Simplify.**

CHAPTER 5

Practice Test

Math Online glencoe.com
Chapter Test

1. **MULTIPLE CHOICE** The rectangle below has a length of 20 centimeters and a perimeter of P centimeters. Which equation could be used to find the width of the rectangle?

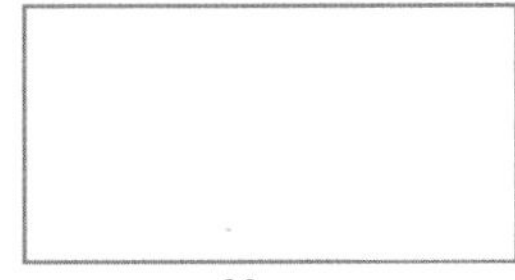

A $P = 40 + \frac{w}{2}$ C $P = 20 + w$

B $P = 40 + 2w$ D $P = 20 + 2w$

2. **MEASUREMENT** If the length of the side of each square is 6 units, what is the perimeter and area of the rectangle?

Solve each equation. Check your solution.

3. $61 = 3b + 7$
4. $-27 + 15x = 33 - 9x$
5. $16y + 24 = -18 + 9y$
6. $7(m + 6) = 105$
7. $s - 17 = 7(s - 5)$
8. $9(12 + 4z) = 3(6z - 18)$

9. **PROJECTS** The eighth grade class is making digital yearbooks on a DVD. One company charges a rate of $235 plus an additional $0.75 per DVD. Another company charges $125 plus $1.63 per DVD. Write and solve an equation to determine for what number of DVDs the costs will be equal.

For the given value, state whether the inequality is *true* or *false*.

10. $3x + 8 > 56; x = 15$
11. $72 < 5x - 3; x = 16$
12. $\frac{124}{x} \geq x + 27; x = 4$

Graph each inequality on a number line.

13. $-5 \leq a$ 14. $w \leq 12$

Write an inequality for each graph.

15. 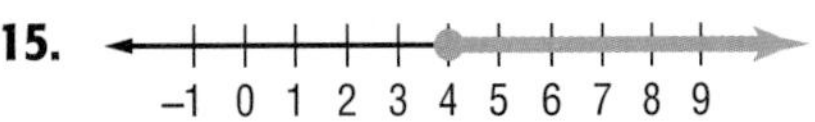

16. 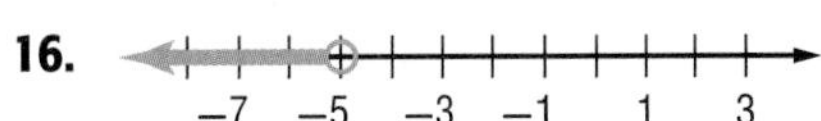

17. **MULTIPLE CHOICE** A community wants to raise at least $4000 for a new skateboarding park. They have been given a $150 donation, and are selling canvas bags for $55 each to raise the rest of the money. Which inequality describes how many bags they need to sell in order to reach this goal?

F $x \geq 35$ H $x \leq 70$

G $x \leq 35$ J $x \geq 70$

Solve each inequality. Graph the solution on a number line.

18. $-9 \geq a + (-3)$
19. $54 < 6m$
20. $-5t \geq -60$
21. $10 + 4x \geq 3(x - 6)$
22. $-7(k - 9) < -21$
23. $14n - 8 \leq (4n - 5)$

24. **TEXT MESSAGES** A cell phone company charges $0.28 for each text message. Paula plans to spend no more than $5.00 on text messages next month. Write and solve an inequality to find how many text messages she will be able to send.

25. **TRANSPORTATION** The minimum amount you can spend for renting a motor scooter is $50. The rental fee is $12 and the cost per hour is $9.50. What is the minimum number of hours you can rent the scooter?

CHAPTER 5

Preparing for Standardized Tests

Write and Solve an Inequality

Some standardized test questions will require you to be able to write and solve inequalities. The steps below can help you solve these problems.

Strategies for Writing and Solving an Inequality

Step 1

Read the problem statement carefully.

Ask yourself:

- What am I being asked to solve?
- What information is given in the problem?
- What is the unknown quantity that I need to solve for?

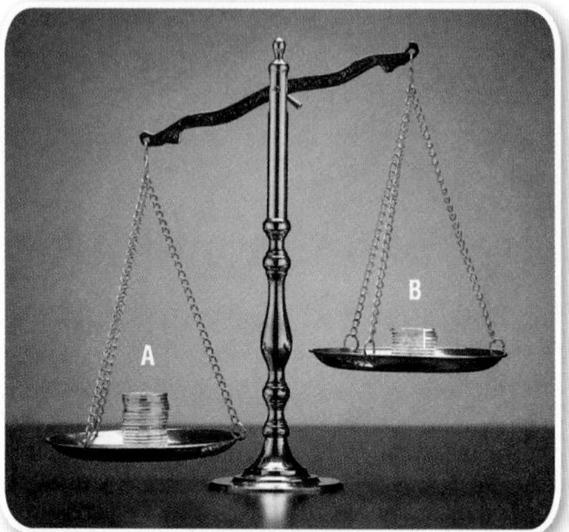

A > B

Step 2

Translate the problem statement into an inequality.

- Assign a variable to the unknown quantity. Then write the word sentence as a mathematical number sentence.
- Look for keywords such as *greater than, less than, no more than, up to,* or *at least* to indicate the type of inequality as well as where to place the inequality sign.

Step 3

Solve the inequality.

- Solve for the unknown in the inequality.
- Multiplying or dividing by a negative reverses the direction of the inequality.
- Check your answer to be sure it makes sense.

EXAMPLE

Read the problem. Identify what you need to know. Then use the information in the problem to solve.

Lorenzo can spend no more than \$120 to buy a CD player and some CDs. The CD player costs \$65, and each CD costs \$11. How many CDs can he buy?

A up to 6 **C** up to 5

B more than 6 **D** more than 5

Read the problem carefully. You know that Lorenzo can spend no more than $120 for a CD player that costs $65 and CDs that cost $11 each. You want to know how many CDs he can buy.

The unknown quantity that you need to find is the number of CDs Lorenzo can buy.

Words	$65 plus $11 per CD is *no more than* $120
Variable	Let n represent the number of CDs Lorenzo can buy.
Inequality	$\$65 + 11n \leq 120$

Solve the equation for n.

$65 + 11n \leq 120$	**Write the inequality.**
$65 + 11n - 65 \leq 120 - 65$	**Subtract 65 from each side.**
$11n \leq 55$	**Simplify.**
$n \leq 5$	**Divide each side by 11. Simplify.**

The number of CDs that Lorenzo can buy with his gift card is less than or equal to 5, so he can buy up to 5 CDs. The correct answer is C.

Exercises

Read each problem. Identify what you need to know. Then use the information in the problem to solve.

1. Nina solved the problem below, then graphed the solution on a number line.

   Two thirds of a number plus five is greater than eleven.

   Which of the following most appropriately describes the unknown number?

   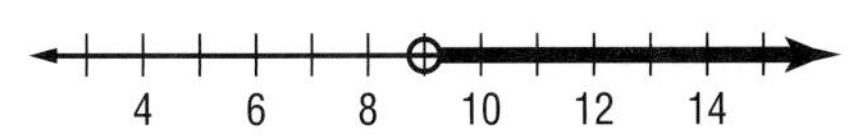

   A more than 9 C at least 9

   B up to 9 D 9 or more

2. Amy added 15 songs to her MP3 player making the total number of songs more than 100. How many songs were originally on the MP3 player?

   F more than 115 H less than 85

   G less than 115 J more than 85

3. Kara's golden retriever weighs 80 pounds. Her veterinarian told her that a healthy weight for the dog would be less than 62 pounds. If Kara's dog can lose an average of 1.5 pounds per week, how long will it take her dog to reach this goal?

   A at most 12 weeks C at most 13 weeks

   B more than 12 weeks D more than 13 weeks

CHAPTER 5

Standardized Test Practice

Cumulative, Chapters 1 through 5

Multiple Choice

Read each question. Then fill in the correct answer on the answer document provided by your teacher or on a sheet of paper.

1. The table below shows the membership cost for two CD clubs. How many CDs would you need to buy in a year in order for the total cost of both memberships to be the same?

CD Club Membership		
	Annual Fee	Cost per CD
Club #1	$50	$7.50
Club #2	$35	$8.00

A 30

B 25

C 20

D 15

2. Georgina wants to practice the piano for at least 45 minutes tonight. Which of the following inequalities represents this situation?

F $t \leq 45$

G $t < 45$

H $t \geq 45$

J $t > 45$

3. On Monday, the price of a share of stock was $79. It fell $3 each day for 11 consecutive days. Which of the following expressions could you use to find the price of the stock on any one of those days?

A $3d + 79$

B $3d - 79$

C $-3d + 79$

D $-3d - 79$

Test-TakingTip

Question 2 Read the problem carefully and look for keywords to help you choose the correct inequality. The words *at least* suggest greater than or equal to.

4. Jonathan has a gift card worth $75. If each DVD costs $12.50, how many DVDs can he buy if he spends no more than $75?

F up to 6 DVDs

G more than 6 DVDs

H up to 7 DVDs

J more than 7 DVDs

5. Solve the following equation.

$$\frac{5}{6}x - 7 = x - 10$$

A $x = 9$

B $x = 18$

C $x = 21$

D $x = 24$

6. What is the perimeter of the triangle below?

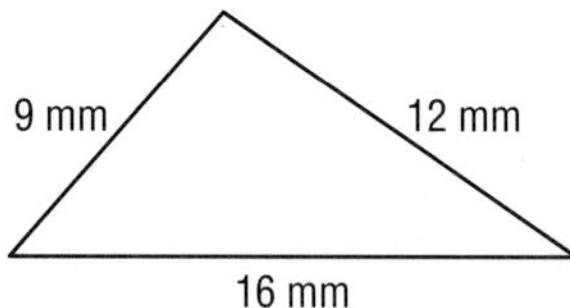

F 32 millimeters

G 35 millimeters

H 37 millimeters

J 41 millimeters

7. The top running speed of a lion is $\frac{5}{7}$ that of a cheetah. If a cheetah can run up to 70 miles per hour, how fast can a lion run?

A 45 mph

B 50 mph

C 55 mph

D 60 mph

Short Response/Gridded Response

Record your answers on the answer sheet provided by your teacher or on a sheet of paper.

8. Write an inequality to represent the number line below.

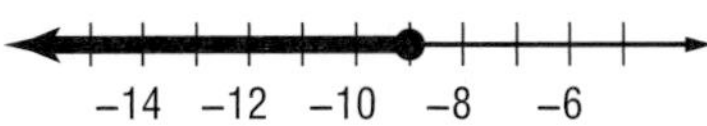

9. GRIDDED RESPONSE What is the area, in square meters, of the right triangle shown below?

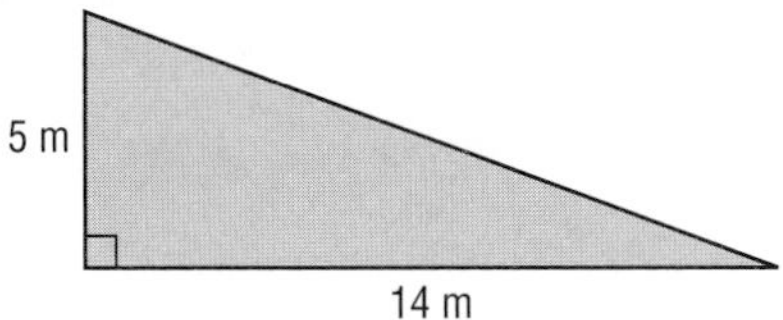

10. GRIDDED RESPONSE Solve the equation.

$$7n - 11 = 3(12 - 4n) + 10$$

11. Describe when to flip the direction of an inequality symbol when solving inequalities. Give an example.

12. Forty out of 48 freshmen and 42 out of 50 sophomores had perfect attendance last quarter.

a. What fraction of each class had perfect attendance last quarter? Write your answers in lowest terms.

b. Write each fraction from part **a.** as a decimal.

c. Which class had a greater portion with perfect attendance last quarter?

13. The equation $2w + 36 = 88$ represents the relationship between the width w of Daniel's yard and the total amount of fencing he has.

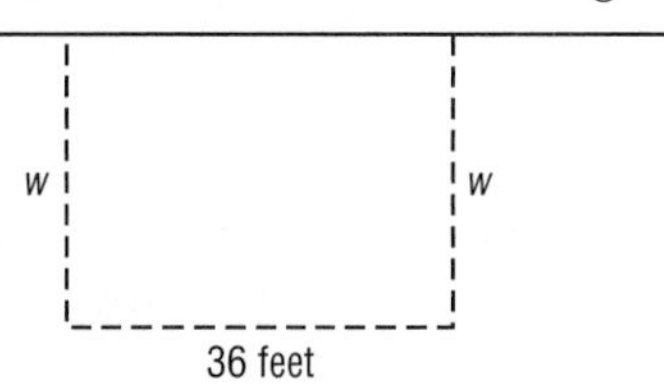

a. Solve the equation for w, the width of the yard.

b. If Daniel increased each side by 12 feet, how much total fencing would he need?

Extended Response

Record your answers on a sheet of paper. Show your work.

14. The formula $C = \frac{5}{9}(F - 32)$ can be used to convert temperatures in degrees Fahrenheit, F, to degrees Celsius, C.

a. Water boils when it reaches a temperature of 212° Fahrenheit. Write this temperature in degrees Celsius.

b. Water freezes at a temperature of 0° Celsius. Write this temperature in degrees Fahrenheit.

c. Solve the formula for F to find a new formula that converts temperatures from Celsius to Fahrenheit.

Need Extra Help?

If you missed Question...	1	2	3	4	5	6	7	8	9	10	11	12	13	14
Go to Lesson or Page...	5-2	5-3	2-4	5-4	5-5	5-1	4-4	5-3	5-1	5-5	5-4	3-1	4-5	5-2

CHAPTER 6

Ratio, Proportion, and Similar Figures

Then

In Chapter 4, you solved equations using multiplication and division.

Now

In Chapter 6, you will:

- Write ratios as fractions in simplest form.
- Find and compare unit rates.
- Use and solve proportions.

Why?

ARCHITECTURE The art and science of designing structures is detailed work. Architects produce scale models of their structures for clients. They use proportions and scale factors to draw blueprints for their structures.

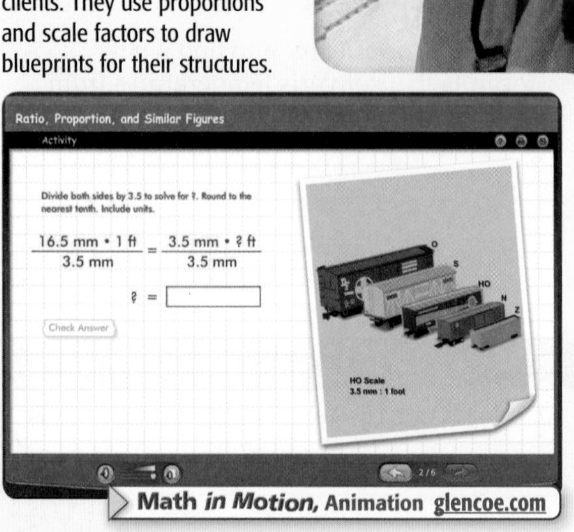

Ratio, Proportion, and Similar Figures
Activity
Divide both sides by 3.5 to solve for ?. Round to the nearest tenth. Include units.
$\frac{16.5 \text{ mm} \cdot 1 \text{ ft}}{3.5 \text{ mm}} = \frac{3.5 \text{ mm} \cdot ? \text{ ft}}{3.5 \text{ mm}}$
? =
Check Answer
O
S
HO
N
Z
HO Scale
3.5 mm : 1 foot
2/6

Math *in Motion,* Animation glencoe.com

Get Ready for Chapter 6

Diagnose Readiness You have two options for checking Prerequisite Skills.

Text Option Take the Quick Check below. Refer to the Quick Review for help.

QuickCheck

Simplify each fraction. If the fraction cannot be simplified, write *simplest form.* (Previous Course)

1. $\frac{12}{15}$

2. $\frac{24}{32}$

3. $\frac{30}{45}$

4. $\frac{5}{11}$

5. $\frac{48}{72}$

6. $\frac{16}{43}$

7. **PETS** Of the 20 students in Mr. Masuru's class, 15 have pets. Write the fraction of students that own pets in simplest form.

Complete each sentence. (Lesson 0-6)

8. 3 ft = ■ in.

9. 2 h = ■ min

10. 4 lb = ■ oz

11. 2 pt = ■ c

12. 3 m = ■ cm

13. 4 km = ■ m

14. 1.8 L = ■ mL

15. 13 kg = ■ g

16. **RUNNING** The high school cross country course is 5-kilometers long. How many meters is this?

Solve each equation. Check your solution. (Lesson 4-4)

17. $60x = 12$

18. $42 = 7n$

19. $4y = 10$

20. $9m = 18$

21. $7 = 3.5a$

22. $11 = 5.5d$

23. **BOOKS** Dyani has 3 times as many books as Mario. If Mario has 27 books, how many books does Dyani have?

QuickReview

EXAMPLE 1

Write $\frac{28}{36}$ as a fraction in simplest form.

Find the GCF of 28 and 36.

factors of 28: 1, 2, 4, 7, 14, 28

factors of 36: 1, 2, 3, 4, 6, 9, 12, 18, 36

The GCF of 28 and 36 is 4.

$\frac{28}{36} = \frac{28 \div 4}{36 \div 4}$ **Divide the numerator and denominator by the GCF, 4.**

$= \frac{7}{9}$ **Simplest form**

EXAMPLE 2

Complete 3 T = ■ oz.

$3\,\cancel{T} \times \frac{2000 \text{ lb}}{1\,\cancel{T}} = 6000 \text{ lb}$ **To convert tons to pounds, multiply by 2000.**

$6000\,\cancel{\text{lb}} \times \frac{16 \text{ oz}}{1\,\cancel{\text{lb}}} = 96{,}000 \text{ oz}$ **To convert pounds to ounces, multiply by 16.**

So, 3 T = 96,000 oz.

EXAMPLE 3

Solve $9.5x = 38$.

$9.5x = 38$ **Write the equation.**

$\frac{9.5x}{9.5} = \frac{38}{9.5}$ **Divide each side by 9.5.**

$x = 4$ **Simplify.**

Online Option Math Online Take a self-check Chapter Readiness Quiz at glencoe.com.

Get Started on Chapter 6

You will learn several new concepts, skills, and vocabulary terms as you study Chapter 6. To get ready, identify important terms and organize your resources. You may wish to refer to **Chapter 0** to review prerequisite skills.

FOLDABLES® Study Organizer

Ratio, Proportion, and Similar Figures Make this Foldable to help you organize your Chapter 6 notes about ratios, proportions, and similar figures. Begin with a piece of 11″ by 17″ paper.

1. **Fold** a two inch tab along the bottom.

2. **Fold** in thirds lengthwise.

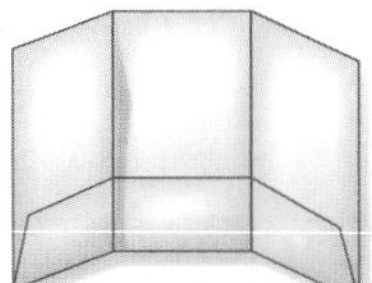

3. **Staple** the bottom fold to make a pocket.

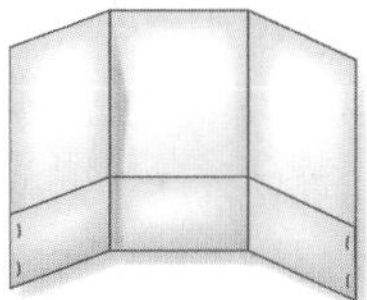

4. **Label** each pocket as shown. Place several index cards in each pocket.

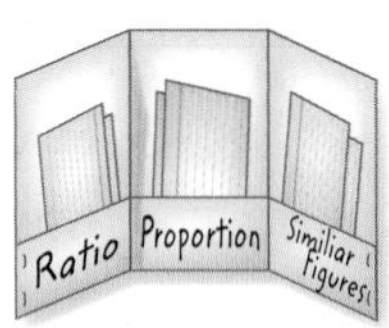

Math Online glencoe.com

- Study the chapter online
- Explore **Math in Motion**
- Get extra help from your own **Personal Tutor**
- Use **Extra Examples** for additional help
- Take a **Self-Check Quiz**
- **Review Vocabulary** in fun ways

New Vocabulary

English		Español
ratio	p. 265	razón
rate	p. 270	tasa
unit rate	p. 270	tasa unitaria
dimensional analysis	p. 275	análisis dimensional
proportion	p. 287	proporción
cross products	p. 287	productos cruzados
scale drawing	p. 294	dibujo a escala
scale model	p. 294	modelo a escala
scale	p. 294	escala
scale factor	p. 295	factor de escala
similar figures	p. 301	figures semejantes
congruent	p. 301	congruentes
dilation	p. 307	homotecia
indirect measurement	p. 313	medición indirecta

Review Vocabulary

figure • figura a two-dimensional shape that lies entirely within one plane

simplest form • forma reducida a fraction is in simplest form when the GCF of the numerator and denominator is 1

$\frac{12}{28} = \frac{12 \div 4}{28 \div 4}$ **Divide the numerator and denominator by the GCF, 4.**

$= \frac{3}{7}$ **Simplest form**

Multilingual eGlossary glencoe.com

6-1

Ratios

Why?

In mathematics, there are many different ways to compare numbers. Consider the data in the table.

Animal	Life Span (years)
gray whale	70
bottlenose dolphin	30
kangaroo	7
mouse	3

a. On average, how many times as long does a gray whale live as a kangaroo?

b. For each year a kangaroo lives, how many years does a gray whale live on average?

c. What operation did you use to find parts **a** and **b**?

Then
You have already learned operations with fractions. (Lessons 3-3 and 3-4)

Now
- Write ratios as fractions in simplest form.
- Simplify ratios involving measurements.

New Vocabulary
ratio

Math Online

glencoe.com
- Extra Examples
- Personal Tutor
- Self-Check Quiz
- Homework Help

Write Ratios as Fractions in Simplest Form A **ratio** is a comparison of two quantities by division. If the first number being compared is less than the second, the ratio is usually written as a fraction in simplest form.

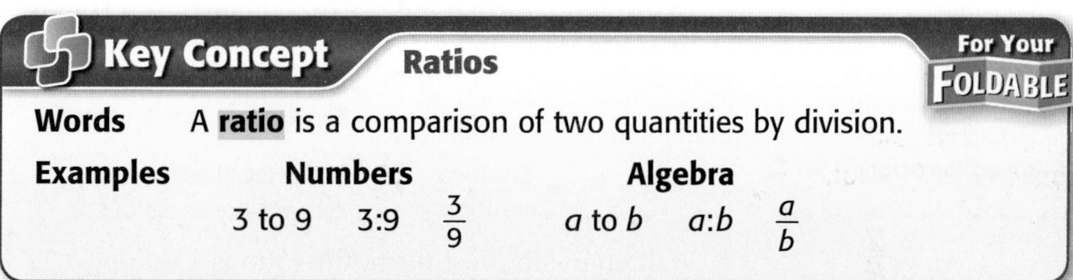

Key Concept Ratios — For Your FOLDABLE

Words A **ratio** is a comparison of two quantities by division.

Examples

	Numbers	Algebra
	3 to 9 3:9 $\frac{3}{9}$	a to b a:b $\frac{a}{b}$

Ratios can express part to part, part to whole, or whole to part relationships.

EXAMPLE 1 Write Ratios in Simplest Form

Express the ratio *12 baskets in 18 attempts* as a fraction in simplest form. Explain its meaning.

$$\frac{12}{18} = \frac{2}{3} \quad (\div 6)$$

Divide the numerator and denominator by the GCF, 6.

The ratio of baskets to shots attempted is 2 to 3. This means that for every 3 shots attempted, 2 were made. Also, $\frac{2}{3}$ of the shots attempted were baskets.

Check Your Progress

1. Refer to the table above. Express the ratio of the life span of a bottlenose dolphin to the life span of a mouse as a ratio in simplest form. Explain its meaning.

Personal Tutor glencoe.com

Real-World EXAMPLE 2 Write Ratios as Fractions

MUSIC Ten out of every 30 Americans own a portable MP3 player. Express this ratio as a fraction in simplest form. Explain its meaning.

$\frac{10}{30} = \frac{1}{3}$ (÷ 10 numerator and denominator) **Divide the numerator and denominator by the GCF, 10.**

The ratio of Americans who own a portable MP3 player is 1 to 3. This means that for every 3 Americans, 1 owns a portable MP3 player. Also, $\frac{1}{3}$ of Americans own a portable MP3 player.

Check Your Progress

2. **CAMPING** 15 out of 100 campsites at a campground are reserved for campers with pets. Express this ratio as a fraction in simplest form. Explain its meaning.

Personal Tutor glencoe.com

Simplify Ratios Involving Measurements When writing a ratio involving measurements, both quantities should have the same unit of measure.

> **Watch Out!**
>
> **Units** Always double check that the units in a ratio match. If the units do not match, you will not get the correct comparison.

EXAMPLE 3 Write Ratios as Fractions

Express the ratio *8 ounces to 3 pounds* as a fraction in simplest form.

$\frac{8 \text{ ounces}}{3 \text{ pounds}} = \frac{8 \text{ ounces}}{48 \text{ ounces}}$ **Convert 3 pounds to ounces.**

$= \frac{1 \text{ ounce}}{6 \text{ ounces}}$ **Divide the numerator and denominator by the GCF, 6.**

Written in simplest form, the ratio is 1 to 6.

Check Your Progress

3. Express the ratio *15 inches to 1 foot* as a fraction in simplest form.

Personal Tutor glencoe.com

Check Your Understanding

Example 1 p. 265

Express each ratio as a fraction in simplest form.

1. 12 boys to 16 girls
2. 24 out of 60 light bulbs
3. 36 DVDs out of 84 DVDs
4. 50 tiles to 25 tiles

Example 2 p. 266

5. **ACTIVITIES** In Mr. Blackwell's class, 15 out of 24 students play sports. Express this ratio as a fraction in simplest form. Explain its meaning.

Example 3 p. 266

Express each ratio as a fraction in simplest form.

6. 3 pints to 4 quarts
7. 2 pounds to 6 ounces
8. 9 inches to 1 yard
9. 6 gallons to 3 quarts

Practice and Problem Solving

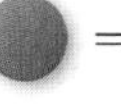 = **Step-by-Step Solutions** begin on page R11.
Extra Practice begins on page 810.

Example 1 p. 266

Express each ratio as a fraction in simplest form.

10. 9 out of 15 pets
11. 20 wins out of 36 games
12. 4 players to 52 cards
13. 45 out of 60 days
14. 16 pens to 10 pencils
15. 96 people to 3 buses

Example 2 p. 267

16. **PIANOS** On a full size piano, there are 36 black keys and 52 white keys. Express the ratio of black keys to white keys as a fraction in simplest form. Explain its meaning.

17. **RESTAURANTS** In a restaurant, 72 out of 108 tables are booths. Express the ratio of tables that are booths to the total number of tables as a fraction in simplest form. Explain its meaning.

Example 3 p. 267

Express each ratio as a fraction in simplest form.

18. 10 yards to 10 feet
19. 4 ounces to 2 pounds
20. 18 quarts to 4 gallons
21. 6 feet to 14 inches

22. **STORES** A department store conducted a study to determine what age group shops in their store.

Age Group	Number
0–17	25
18–30	75
31–45	54
46+	26

 a. Express the ratio of people ages 0–17 to people ages 18–30 as a fraction in simplest form.

 b. Express the ratio of people thirty or under to people over the age of 30 as a fraction in simplest form.

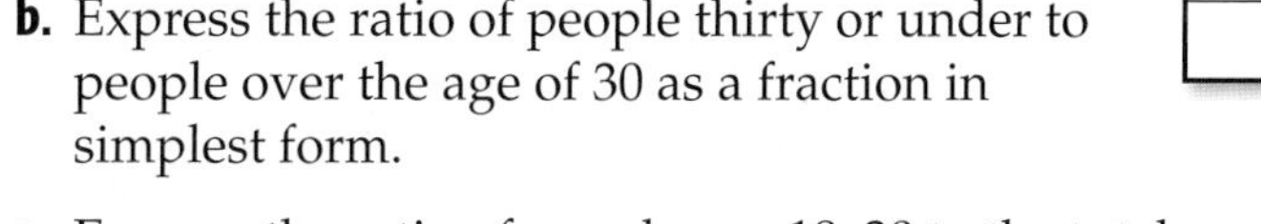

 c. Express the ratio of people age 18–30 to the total number of people as a fraction in simplest form.

Real-World Link

Wisconsin Dells, Wisconsin, is recognized as the "Water Park Capital of the World." It is home to both the largest indoor and outdoor water parks in the world.

Source: Wisconsin Dells

23. **WATER PARKS** A water park has 14 body slides, 8 tube slides, 2 types of swimming pools, and 6 water play areas. Use this information to write each ratio as a fraction in simplest form.

 a. body slides : tube slides
 b. play areas : tube slides
 c. slides : all attractions
 d. slides : not slides

24. **PHONES** A cell phone store displayed the phone at the right on a poster. The length of the phone on the poster is 3 feet 4 inches. Write a ratio comparing the length of the actual cell phone to length of the cell phone on the poster as a fraction in simplest form.

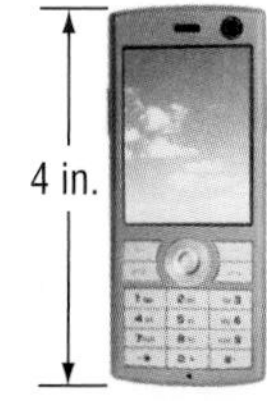

25. **ANIMALS** The table shows the heart rates of different animals.

Animal	Heart Rate (beats/min)	Mass (g)
cat	150	2000
cow	65	800,000
hamster	450	60
horse	44	1,200,000

 a. Order the animals from greatest mass to heart rate ratio to least mass to heart rate ratio.

 b. Which animal had the greatest ratio? Explain your reasoning.

Real-World Link

In the 2007 NFL season, Tom Brady threw for 50 touchdowns, making him the record holder for most touchdown passes in a single season.

Source: NFL

26. SALES Bonnie's Boutique had a T-shirt sale to make room for new inventory. At the end of the sale, 16 T-shirts were left. Of these, 6 of them were blue. Write the ratio of blue shirts to shirts left as a fraction in simplest form.

27. BAKING When cooking a turkey, you should bake it for about 1 hour for every four pounds of meat. If an 18 pound turkey is cooked for 4 hours, was it cooked long enough? If not, how long should the turkey have been cooked?

28. FOOTBALL The table shows the number of touchdowns and interceptions each NFL quarterback had in a recent season. Which quarterback had the best touchdown to interception ratio? Explain its meaning.

Player	Touchdowns	Interceptions
Drew Brees	26	11
Carson Palmer	28	13
Tom Brady	24	12
Philip Rivers	22	9

Compare each pair of ratios using <, >, or =.

29. \$27 for 9 key chains, \$45 for 15 key chains

30. 8 girls out of 18 students, 12 girls out of 22 students

31. 6 cases for \$48, 14 cases for \$88

32. 24 thriller movies out of 36 DVDs, 10 thriller movies out of 15 DVDs

H.O.T. Problems

Use **H**igher-**O**rder **T**hinking Skills

33. OPEN ENDED Give three different examples of ratios that might occur in a real-world situation.

34. SELECT A TECHNIQUE Which of the technique(s) listed below might you use to determine which of the following ratios is the greatest: 440:1200, 1200: 3750, or 350:450? Justify your selection(s). Then use the technique(s) to solve the problem.

mental math | a calculator | estimation

35. CHALLENGE The ratio of Mieko's age to his brother Sado's age is 2:3. In ten years, the ratio will be 4:5. How old is Mieko?

36. WHICH ONE DOESN'T BELONG? Select the ratio that does not have the same value as the other three. Explain your reasoning.

2 boys: 3 girls | 2 qt: 3 gal | 2 spoons: 3 utensils | 2 ft: 3 ft

37. CHALLENGE Students in Mrs. Miller's class are measuring the length and width of a table using nonstandard materials like pencils and pieces of paper. No matter which tools they use, will the ratio of the length to the width *always*, *sometimes*, or *never* be the same? Explain.

38. WRITING IN MATH In a recent year, the Jacksonville Jaguars took the ball away from their opponents 21 times. They gave the ball up to their opponents 22 times. A sports writer claims the takeaway/giveaway ratio is 21 – 22 or –1. Is this statement correct? Explain.

Standardized Test Practice

39. Of 48 orchestra members, 30 are girls. What ratio compares the number of boys to girls in the orchestra?

A 3:8 C 3:5

B 8:3 D 5:3

40. Which of the following ratios does *not* describe a relationship between the squares shown?

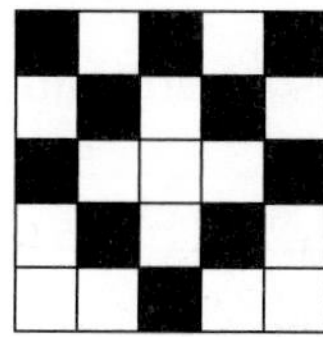

F 2 black:3 white H 2 black:5 total

G 2 black:5 white J 3 white:5 total

41. Of newly manufactured volleyballs, 12 were defective and 56 passed inspection. What ratio compares the number of defective volleyballs to the total number of volleyballs manufactured?

A 3:14 C 4:14

B 3:17 D 4:17

42. EXTENDED RESPONSE Marisol counted the number of coins she had in her piggy bank. The table shows her results.

pennies	nickels	dimes	quarters
47	14	18	21

a. Write a ratio that compares the number of nickels to the number of quarters.

b. Write a ratio that compares the number of dimes to the number of total coins.

Spiral Review

Solve. Graph your solution on a number line. (Lesson 5-5)

43. $3x + 4 \leq 31$

44. $2n + 5 > 11 - n$

45. $y + 1 \geq 4y + 4$

46. $16 - 2c < 14$

47. $18 \leq 12 - 2n$

48. $-3(b - 1) > 18$

49. APPLES More than 2500 varieties of apples are grown in the United States. Write an inequality to describe the apple varieties. (Lesson 5-3)

Name the ordered pair for each point graphed at the right. (Lesson 2-6)

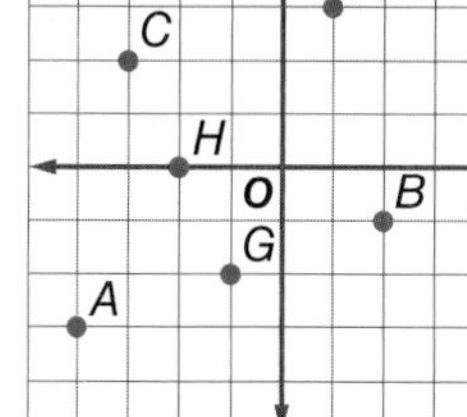

50. A

51. B

52. C

53. D

54. E

55. F

56. AEROSPACE To simulate space travel, NASA's Lewis Research Center in Cleveland uses a 430-foot shaft. The free fall of an object in the shaft takes 5 seconds to travel the 430 feet. (Lesson 2-5)

a. Write an integer to represent the change in the height of the object.

b. On average, how far does the object travel each second?

Skills Review

Divide. Round to the nearest cent. (Previous Course)

57. $5.75 ÷ 4

58. $8.30 ÷ 6

59. $2.27 ÷ 8

60. $11.50 ÷ 5

6-2 Unit Rates

Then

You have already learned how to write ratios. (Lesson 6-1)

Now

- Find unit rates.
- Compare and use unit rates to solve problems.

New Vocabulary

rate
unit rate

Math Online

glencoe.com

- Extra Examples
- Personal Tutor
- Self-Check Quiz
- Homework Help

Why?

Eighteen year old Ben Cook set a world record for text messaging by typing 160 characters in a little more than 40 seconds.

a. If he typed at a constant rate, how many characters did he text each second?

b. It actually took him 42.2 seconds to set the record. Was his rate per second greater or less than the rate in part **a**? Explain.

Find Unit Rates A **rate** is a ratio of two quantities having different kinds of units.

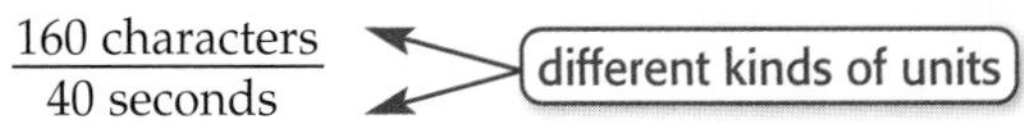

When a rate is simplified so that it has a denominator of 1, it is called a **unit rate**. To write a rate as a unit rate, divide the numerator and denominator of the rate by the denominator.

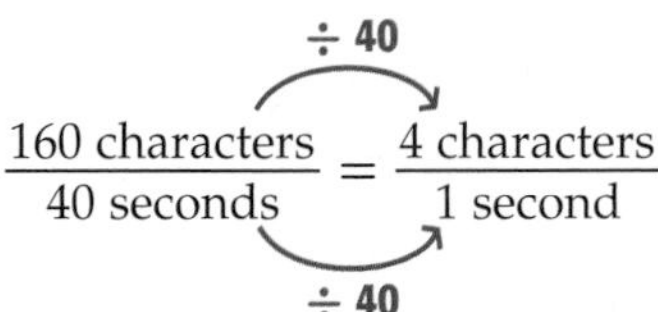

EXAMPLE 1 Find Unit Rates

Express each rate as a unit rate. Round to the nearest tenth, if necessary.

a. $11 for 4 boxes of cereal

Write the rate that compares the cost to the number of boxes.

$$\frac{\$11}{4 \text{ boxes}} = \frac{\$2.75}{1 \text{ box}} \quad (\div 4)$$

Divide the numerator and denominator by 4.

So, the cost is $2.75 per box of cereal.

b. 400 miles on 14 gallons of gasoline

Write the rate that compares the number of miles to the number of gallons.

$$\frac{400 \text{ miles}}{14 \text{ gallons}} = \frac{28.6 \text{ miles}}{1 \text{ gallon}} \quad (\div 14)$$

Divide the numerator and denominator by 14.

So, the car traveled 28.6 miles per gallon of gasoline.

Check Your Progress

1A. $18.50 for 5 pounds

1B. 100 meters in 14 seconds

Personal Tutor glencoe.com

Compare Unit Rates You can also compare unit rates to solve problems.

Real-World EXAMPLE 2 Compare Unit Rates

FINANCIAL LITERACY An online music store sells 15 songs for $12. Another online music store sells 10 songs for $9. Which online store has the lower cost per song?

Step 1 Find the unit rates of the two stores.

$$\frac{12 \text{ dollars}}{15 \text{ songs}} = \frac{0.8 \text{ dollar}}{1 \text{ song}} \quad (\div 15)$$

Divide the numerator and denominator by 15.

For the 15 songs, the unit rate is $0.80 per song.

$$\frac{9 \text{ dollars}}{10 \text{ songs}} = \frac{0.9 \text{ dollar}}{1 \text{ song}} \quad (\div 10)$$

Divide the numerator and denominator by 10.

For the 10 songs, the unit rate is $0.90 per song.

Step 2 Compare the rates. Since $0.80 < $0.90, the first store has a better rate per song.

> **StudyTip**
>
> **Alternative Method** Use the LCM. In the first case, 30 songs cost $12 × 2 or $24. In the second, 30 songs cost $9 × 3 or $27. The first has the lower cost per song.

Check Your Progress

2. **SUN SCREEN** A store sells two different sizes of the same brand of sun screen, an 8-ounce bottle for $5.76 and a 12-ounce bottle for $8.88. Which size bottle is the better buy per ounce? Explain.

Personal Tutor glencoe.com

Once you know the unit rate, you can use it to solve problems involving any amount.

Real-World Link

Bottlenose dolphins breathe through a nasal opening called a blowhole. A typical dolphin will take about 34 breaths in 4 hours.

Source: Dolphins Plus

EXAMPLE 3 Use Unit Rates

DOLPHINS Use the information at the left to find how many breaths a dolphin will take in 7 hours.

Step 1 Find the unit rate.

$$34 \text{ breaths in } 4 \text{ hours} = \frac{34 \text{ breaths} \div 4}{4 \text{ hours} \div 4} \text{ or } \frac{8.5 \text{ breaths}}{1 \text{ hour}}$$

Step 2 Multiply this unit rate by 7 to find the number of breaths a dolphin will take in 7 hours.

$$\frac{8.5 \text{ breaths}}{1 \cancel{\text{hour}}} \cdot 7 \cancel{\text{hours}} = 59.5 \text{ breaths}$$

Divide out the common units.

A bottlenose dolphin will take 59.5 breaths in 7 hours.

Check Your Progress

3. **BAKING** A bakery can make 195 doughnuts in 3 hours. At this rate, how many doughnuts can the bakery make in 8 hours?

Personal Tutor glencoe.com

Check Your Understanding

Example 1 p. 270

Express each rate as a unit rate. Round to the nearest tenth, if necessary.

1. \$120 for 5 days of work
2. 275 miles on 14 gallons
3. 338 points in 16 games
4. 6 pounds for \$19.49
5. 17 gallons in 4 minutes
6. 180 feet in 19 seconds

Example 2 p. 271

7. **PEANUTS** Jamal is comparing prices of several different brands of peanuts. Which brand is the best buy? Explain.

Brand	Size (oz)	Price
Barrel	10	\$3.39
Mr. Nut	14	\$4.54
Chip's	18	\$6.26

Example 3 p. 271

8. **DRIVING** Aisha drove 170 miles in 2.5 hours. At this same rate, how far will she drive in 4 hours?

Practice and Problem Solving

 = **Step-by-Step Solutions** begin on page R11.
Extra Practice begins on page 810.

Example 1 p. 270

Express each rate as a unit rate. Round to the nearest tenth, if necessary.

9. 156 students in 6 classes
10. 424 Calories in 3 servings
11. 147.5 miles in 2.5 hours
12. \$29.95 for 4 DVDs
13. \$231 for 3 game tickets
14. 5 tablespoons in 4 quarts
15. \$97.50 for 15 pizzas
16. 400 meters in 58 seconds

Example 2 p. 271

17. **PARTIES** The Party Planner sells 10 paper plates for \$2.50. Use the table at the right to determine which company sells paper plates for the same price per plate.

Store	Number of Plates	Price
Party Time	15	\$3.75
Good Times	20	\$6.00
Birthday, Inc.	25	\$7.50

18. **ARCHITECTURE** Building A has 7500 square feet of office space for 320 employees. Building B has 9500 square feet of office space for 370 employees. Which building has more square feet of space per employee?

Example 3 p. 271

19. **PRODUCE** A farmers market sells ears of sweet corn. At this same rate, how much will it cost to buy 28 ears of sweet corn?

20. **RECIPES** A recipe that makes 2 dozen cookies calls for $\frac{3}{4}$ cup sugar. How much sugar is needed to make $4\frac{1}{2}$ dozen cookies?

21. **SWIMMING** Which swimmer shown in the table has the fastest rate?

Swimmer	Jenny	Dana	Kaitlin
Event	50 m	100 m	200 m
Time (s)	25.02	119.2	248.07

22. **ESTIMATION** A gallon of milk sells for \$3.18, and a quart of milk sells for \$0.76. Which item has the better unit price? Justify your answer using two different methods.

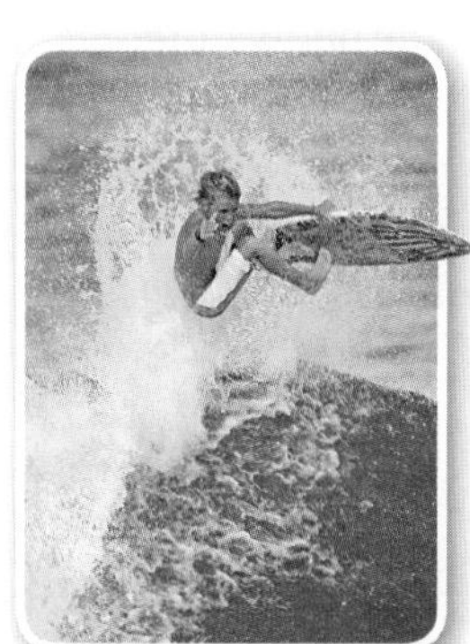

Real-World Link

The term *salinity* refers to the amount of salt in water. The salinity of Earth's oceans is about 525 grams of dissolved salt for every 15 kilograms of water.

Source: Dolphins Plus

23. **SPORTS** The line graph shows Alicia's and Jermaine's average rates in a race.

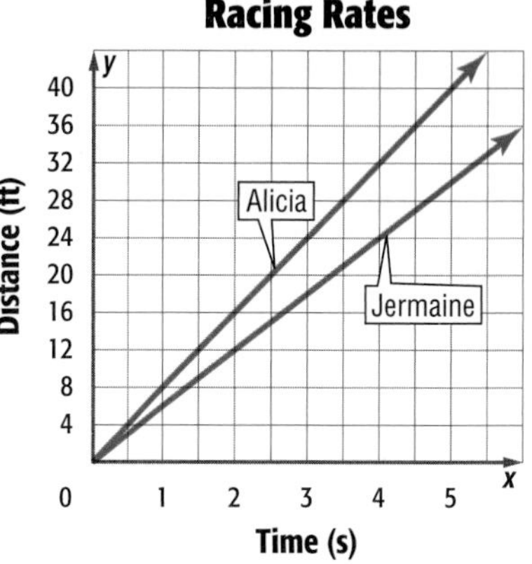

a. Express each person's speed as a unit rate.

b. How long would it take Alicia and Jermaine each to run 1 mile? (*Hint:* One mile = 5280 feet)

c. Suppose Max runs at a rate of 5 feet per second. Predict where the line representing his speed would be graphed. Explain.

24. **MULTIPLE REPRESENTATIONS** In this problem, you will explore rates. Abigail left Charlotte, North Carolina, 30 minutes before Juan to travel to Greenville, South Carolina. Two and a half hours after Abigail left, she had traveled 155 miles and Juan had traveled 144 miles.

a. **TABULAR** Make a function table to show how far each person has traveled after 1, 2, 3, and 4 hours.

b. **GRAPHICAL** Make a graph of the data. Do the two lines intersect? If so, what does this intersection point represent?

25. **EARTH SCIENCE** Use the information at the left to find the grams of salt for every gram of ocean water.

H.O.T. Problems

Use Higher-Order Thinking Skills

26. **OPEN ENDED** Give a real-world example of a rate for a unit rate of 40 miles per hour.

27. **REASONING** In which situation will the rate $\frac{x \text{ miles}}{y \text{ hours}}$ decrease? Give an example to explain your reasoning.

a. x increases, y is unchanged

b. x is unchanged, y increases

28. **FIND THE ERROR** Nadia and Josef are writing the rate \$15.75 for 4 pounds as a unit rate. Is either of them correct? Explain your reasoning.

Nadia

$\frac{\$15.75}{4 \text{ pounds}} = \frac{\$7.88}{2 \text{ pounds}}$

Josef

$\frac{\$15.75}{4 \text{ pounds}} = \frac{\$3.94}{1 \text{ pound}}$

29. **CHALLENGE** A 96-ounce container of orange juice costs \$4.80. At what price should a 128-ounce container be sold in order for the unit rate for both containers to be the same? Explain your reasoning.

30. **WRITING IN MATH** Explain why a horse that runs $\frac{3}{4}$ mile in 1 minute and 9 seconds is faster than a sprinter who runs a 100-yard dash in 12 seconds.

Standardized Test Practice

31. On Monday, Ms. Moseley drove 340 miles in 5 hours. On Tuesday, she drove 198 miles in 3 hours. Based on these rates, which statement is true?

A Her rate on Monday was 2 miles per hour slower than her rate on Tuesday.

B Her rate on Tuesday was 2 miles per hour slower than her rate on Monday.

C Her rate on Monday was the same as her rate on Tuesday.

D Her rate on Tuesday was 2 miles per hour faster than her rate on Monday.

32. Kalyin keyboards at a rate of 60 words per minute for 5 minutes and 45 words per minute for 10 minutes. How many words did she type in all?

F 105 H 650

G 510 J 750

33. EXTENDED RESPONSE The table shows the costs of different sized bags of snack mix.

Snack Mix	
Size	**Price**
x-small	\$1.96 for 12 oz
small	\$2.20 for 16 oz
large	\$4.00 for 36 oz
x-large	\$5.40 for 48 oz

a. Which size is the most expensive per ounce?

b. Pilar wants to buy the one that costs the least per ounce. What size should she buy? Explain.

34. Sal paid \$2.79 for a gallon of milk. Find the cost per quart of milk at this rate.

A \$0.55 C \$0.93

B \$0.70 D \$1.40

Spiral Review

Express each ratio as a fraction in simplest form. (Lesson 6-1)

35. 155 apples to 75 oranges

36. 7 cups to 9 pints

37. 11 gallons to 11 quarts

38. 900 pounds to 16 tons

39. WEB SITES A company pays Dante to advertise on his web site. The web site earns \$10 per month plus \$0.05 each time a visitor to the site clicks on the advertisement. What is the least number of clicks he needs to make \$45 per month or more from this advertiser? (Lesson 5-5)

40. SUBMARINES The research submarine *Alvin*, used to locate the wreck of the *Titanic*, descended at a rate of about 100 feet per minute. Write an integer to represent the distance *Alvin* traveled in 5 minutes. (Lesson 2-4)

41. Evaluate $a - b + c$ if $a = 5$, $b = 7$, and $c = 10$. (Lesson 2-3)

Skills Review

Complete. (Previous Course)

42. 36 in. = ■ ft

43. 4 yd = ■ in.

44. ■ fl oz = 6 c

45. ■ m = 4 cm

46. 476 mL = ■ L

47. 3.5 km = ■ m

6-3 Converting Rates and Measurements

Then

You have already learned to find unit rates. (Lesson 6-2)

Now

- Convert rates using dimensional analysis.
- Convert between systems of measurement.

New Vocabulary

dimensional analysis

Math Online

glencoe.com

- Extra Examples
- Personal Tutor
- Self-Check Quiz
- Homework Help

Why?

The record for the fastest land car speed was set at about 760 miles per hour in Black Rock Desert, Nevada, by a thrust powered race car.

a. How many minutes are in 1 hour?

b. Write a ratio that compares 1 hour to the number of minutes in 1 hour.

Dimensional Analysis **Dimensional analysis** is the process of including units of measurement as factors when you compute. For example, you know that 1 hour = 60 minutes. You can write conversion factors $\frac{1 \text{ hour}}{60 \text{ minutes}}$ or $\frac{60 \text{ minutes}}{1 \text{ hour}}$. Each ratio is equivalent to 1 because the numerator and denominator represent the same amount.

Real-World EXAMPLE 1 Use Dimensional Analysis

Convert 760 miles per hour to miles per minute.

Step 1 You need to convert miles per hour to miles per minute. Choose a conversion factor that converts hours to minutes, with minutes in the denominator.

Convert miles per hour... ... to miles per minute.

$$\frac{\text{miles}}{\cancel{\text{hour}}} \cdot \frac{\cancel{\text{hour}}}{\text{minute}} = \frac{\text{miles}}{\text{minute}}$$

Conversion factor

So, use $\frac{1 \text{ h}}{60 \text{ min}}$.

Step 2 Multiply.

$$\frac{760 \text{ mi}}{1 \text{ h}} = \frac{760 \text{ mi}}{1 \text{ h}} \cdot \frac{1 \text{ h}}{60 \text{ min}}$$ **Multiply by $\frac{1 \text{ h}}{60 \text{ min}}$.**

$$= \frac{760 \text{ mi}}{\cancel{1 \text{ h}}_1} \cdot \frac{\cancel{1 \text{ h}}^1}{60 \text{ min}}$$ **Divide out common units.**

$$= \frac{760 \text{ mi}}{60 \text{ min}}$$ **Multiply.**

$$\approx 12.7$$ **Divide.**

So, the car traveled about 12.7 miles per minute.

Check Your Progress

1. MONEY The average teenager spends $1742 per year on fashion related items. How much is this per week?

Personal Tutor glencoe.com

Watch Out!

Common Units Make sure that you choose conversion factors that allow you to divide out the common units.

Real-World EXAMPLE 2 Convert Rates

SPORTS Tyree and three friends attend skydiving class before their first jump. The instructor tells them they will travel at about 176 feet per second. How many miles per hour is this?

You need to convert feet per second to miles per hour.

Use 1 mi = 5280 ft and 1 h = 3600 s.

$$\frac{176\text{ ft}}{1\text{ s}} = \frac{176\text{ ft}}{1\text{ s}} \cdot \frac{1\text{ mi}}{5280\text{ ft}} \cdot \frac{3600\text{ s}}{1\text{ h}}$$ **Multiply by $\frac{1\text{ mi}}{5280\text{ ft}}$ and $\frac{3600\text{ s}}{1\text{ h}}$.**

$$= \frac{\overset{1}{\cancel{176}\cancel{\text{ ft}}}}{1\cancel{\text{ s}}} \cdot \frac{1\text{ mi}}{\underset{\underset{1}{\cancel{30}}}{\cancel{5280}\cancel{\text{ ft}}}} \cdot \frac{\overset{120}{\cancel{3600}\cancel{\text{ s}}}}{1\text{ h}}$$ **Divide the common factors and units.**

$$= \frac{120\text{ mi}}{1\text{ h}}$$ **Simplify.**

So, 176 feet per second is equivalent to 120 miles per hour.

Check Your Progress

2. **TRAINS** The TGV is a high speed rail train in France. At top speed, it runs at an average of 320 kilometers per hour. How many meters per second is this?

Personal Tutor glencoe.com

Convert Between Systems The table shows conversion factors between the Customary and Metric systems for units of length, capacity, and mass or weight.

Key Concept Measurement Conversions

For Your FOLDABLE

Length	
Customary to Metric	**Metric to Customary**
1 in. ≈ 2.540 cm	1 cm ≈ 0.394 in.
1 ft ≈ 0.305 m	1 m ≈ 3.279 ft
1 yd ≈ 0.914 m	1 m ≈ 1.094 yd
1 mi ≈ 1.609 km	1 km ≈ 0.621 mi
Capacity	
Customary to Metric	**Metric to Customary**
1 fl oz ≈ 29.574 mL	1 mL ≈ 0.034 fl oz
1 pt ≈ 0.473 L	1 L ≈ 2.114 pt
1 qt ≈ 0.946 L	1 L ≈ 1.057 qt
1 gal ≈ 3.785 L	1 L ≈ 0.264 gal
Mass or Weight	
Customary to Metric	**Metric to Customary**
1 oz ≈ 28.350 g	1 g ≈ 0.035 oz
1 lb ≈ 0.454 kg	1 kg ≈ 2.203 lb

You can also use dimensional analysis to convert between measurement systems. The two conversion factors $\frac{1\text{ ft}}{0.305\text{ m}}$ and $\frac{0.305\text{ m}}{1\text{ ft}}$ use the same conversion. Use the factor that will correctly divide out the appropriate common unit.

EXAMPLE 3 Convert Measurements Between Systems

Complete each conversion. Round to the nearest hundredth.

a. 12 centimeters to inches

Use 1 in. ≈ 2.54 centimeters.

$12 \text{ cm} \approx 12 \text{ cm} \cdot \frac{1 \text{ in.}}{2.54 \text{ cm}}$ — Multiply by $\frac{1 \text{ in.}}{2.54 \text{ cm}}$.

$\approx 12 \cancel{\text{cm}} \cdot \frac{1 \text{ in.}}{2.54 \cancel{\text{cm}}}$ — Divide out common units, leaving the desired unit, inch.

$\approx \frac{12 \text{ in.}}{2.54}$ or 4.72 in. — Simplify.

So, 12 centimeters is approximately 4.72 inches.

b. 4 quarts to liters

Use 1 qt ≈ 0.946 L.

$4 \text{ qt} \approx 4 \text{ qt} \cdot \frac{0.946 \text{ L}}{1 \text{ qt}}$ — Multiply by $\frac{0.946 \text{ L}}{1 \text{ qt}}$.

$\approx 4 \cancel{\text{qt}} \cdot \frac{0.946 \text{ L}}{1 \cancel{\text{qt}}}$ — Divide out common units, leaving the desired unit, quart.

$\approx 4 \cdot 0.946$ L or 3.784 L — Simplify.

So, 4 quarts is approximately 3.78 liters.

Check Your Progress

3A. 6 mi ≈ ■ km **3B.** 12 oz ≈ ■ g **3C.** 11 yd ≈ ■ m

Personal Tutor glencoe.com

Real-World Link

At top speed, a giant tortoise can travel about 900 feet per hour.

Source: San Diego Zoo

Real-World EXAMPLE 4 Convert Rates Between Systems

ANIMALS Use the information at the left to determine how many centimeters per second a giant tortoise travels at top speed.

To convert feet to centimeters, use 1 ft = 12 in. and 1 in. ≈ 2.54 cm.

To convert hours to seconds, use 1 h = 60 min and 1 min = 60 s.

$$\frac{900 \text{ ft}}{1 \text{ h}} \cdot \frac{12 \text{ in.}}{1 \text{ ft}} \cdot \frac{2.54 \text{ cm}}{1 \text{ in.}} \cdot \frac{1 \text{ h}}{60 \text{ min}} \cdot \frac{1 \text{min}}{60 \text{ s}}$$

$= \frac{900 \cancel{\text{ft}}}{1 \cancel{\text{h}}} \cdot \frac{12 \cancel{\text{in.}}}{1 \cancel{\text{ft}}} \cdot \frac{2.54 \text{ cm}}{1 \cancel{\text{in.}}} \cdot \frac{1 \cancel{\text{h}}}{60 \cancel{\text{min}}} \cdot \frac{1 \cancel{\text{min}}}{60 \text{ s}}$ — Divide out common units.

$= \frac{27{,}432 \text{ cm}}{3600 \text{ s}}$ — Multiply.

$= \frac{7.62 \text{ cm}}{1 \text{ s}}$ — Divide.

At top speed, a giant tortoise will travel 7.62 centimeters per second.

Check Your Progress

4. SPORTS At a recent Winter Olympics, USA short track speed skater Apolo Ohno won a gold medal by skating about 12 meters per second. Rounded to the nearest hundredth, how many miles per hour is this?

Personal Tutor glencoe.com

Check Your Understanding

Example 1 p. 275

1. **RAINFORESTS** In Brazil, about 20 acres of rainforest are destroyed each minute. At this rate, how much rainforest is destroyed per day?

Example 2 p. 276

2. **FENCES** Lexi can paint 5 yards of fencing in one hour. At this rate, how many inches does she paint per minute?

Example 3 p. 277

Complete each conversion. Round to the nearest hundredth.

3. 8 in. ≈ ■ cm
4. 5 L ≈ ■ gal
5. 15 oz ≈ ■ g
6. 24 cm ≈ ■ in.
7. 9 pt ≈ ■ L
8. 3 m ≈ ■ ft

Example 4 p. 277

9. **ELEPHANTS** An elephant can eat up to 440 pounds of vegetation every day. How many grams per minute is this?

Practice and Problem Solving

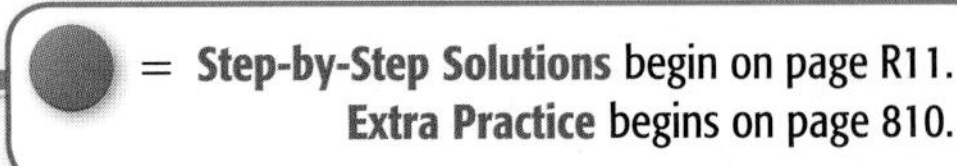
= **Step-by-Step Solutions** begin on page R11.
Extra Practice begins on page 810.

Example 1 p. 275

10. **CANDY** A candy company can produce 4800 sour lemon candies per minute. How many candies can they produce each hour?
11. **RECYCLING** In a recent year, 51.9 billion aluminum cans were recycled. About how many cans per week is this?

Example 2 p. 276

12. **TELEVISION** The average American student spends almost 1500 hours per year watching television. How many minutes per day is this?
13. **AMUSEMENT PARK** A thrill ride at an amusement park travels 55 miles per hour. How many feet per second is this?

Example 3 p. 277

Complete each conversion. Round to the nearest hundredth.

14. 4 L ≈ ■ qt
15. 16 in. ≈ ■ cm
16. 13 m ≈ ■ ft
17. 8 yd ≈ ■ m
18. 18 lb ≈ ■ kg
19. 7 L ≈ ■ gal
20. 1500 g ≈ ■ oz
21. 15 ft ≈ ■ m
22. 28 fl oz ≈ ■ mL

Example 4 p. 277

23. **SCIENCE** The velocity of sound through wood at 0° Celsius is 1454 meters per second. How many miles is this per hour?
24. **CARS** A certain car in Canada can travel 15 kilometers per 1 liter of gasoline. How many miles per gallon is this?

Complete each conversion. Round to the nearest hundredth.

25. 8 in. ≈ ■ mm
26. 16 L ≈ ■ c
27. 2 km ≈ ■ yd
28. 250 fl oz ≈ ■ L
29. 2750 g ≈ ■ lb
30. 5 gal ≈ ■ mL

31. **SPORTS** Crystal's times for each portion of a triathlon are shown in the table.

	Swim	Bike	Run
Distance (km)	1.5	40	10
Time (min)	40	86	64

a. How many meters per second did she run?

b. What was her speed in miles per hour for the aquabike portion (swimming and biking)?

Real-World Link

In the event of a fire, the Willis Tower has a sprinkler system to help extinguish the fire. The pumps that operate the system can pump up to 1500 gallons of water per minute.

Source: Willis Tower

Order each group of rates from least to greatest.

32. 100 oz/min, 2500 g/min, 10 lb/min

33. 500 m/h, 7 yd/min, 6 in./s

34. 32 mi/gal, 15 m/mL, 6600 yd/qt

35. 500 kg/h, 5 oz/s, 18 lb/min

36. **ARCHITECTURE** Use the information at the left to determine how many liters of water the system in the Willis Tower could pump in $\frac{1}{4}$ minute. Round to the nearest hundredth.

37. **FOOD** The average American consumes 20 gallons of ice cream in one year. At this rate, how many liters of ice cream will 50 Americans consume in one week? Round to the nearest hundredth.

Replace each ● with <, >, or = to make a true sentence.

38. 10 m ● 390 in.

39. 520 oz ● 15 kg

40. 14 pt ● 6622 mL

FINANCIAL LITERACY The table shows the exchange rate between the U.S. dollar and other currencies. Use dimensional analysis to make each conversion.

Exchange Rates Per 1 U.S. Dollar		
Country	**Currency**	**Rate**
European Union	euro	0.729
France	franc	4.784
Japan	yen	122.198
United Kingdom	pound	0.494

41. 150 dollars to euros

42. 275 dollars to pounds

43. 90 dollars to yen

44. 500 francs to dollars

H.O.T. Problems

Use Higher-Order Thinking Skills

45. **OPEN ENDED** Give two examples of different measurements that are equivalent to 10 centimeters per second.

46. **WRITING IN MATH** Macha needs 120 square feet of carpeting for her bedroom. Explain how to use dimensional analysis to convert 120 square feet to square yards if 1 square yard is equal to 9 square feet.

47. **WHICH ONE DOESN'T BELONG?** Select the rate that does not have the same value as the other three. Explain your reasoning.

60 mi/h	88 ft/s	500 ft/min	1440 mi/day

48. **CHALLENGE** A recipe for fruit punch uses the ingredients shown. About how many cups of each ingredient are needed? Round to the nearest tenth.

Fruit Punch	
900 mL	cranberry juice
700 mL	apple juice
300 mL	pineapple juice
150 mL	lemon juice
900 mL	club soda

49. **REASONING** What property of multiplication allows you to multiply a rate by a conversion factor without changing its value? Explain.

50. **WRITING IN MATH** Explain how you would convert 10 miles per hour to meters per second.

Standardized Test Practice

51. 55 miles per hour is the same rate as which of the following?

A 34 kilometers per hour
B 50 kilometers per hour
C 88 kilometers per hour
D 98 kilometers per hour

52. A piece of notebook paper measures $8\frac{1}{2}$ inches by 11 inches. Which of the following metric approximations is the same?

F 2 m by 2.8 m
G 3 cm by 4 cm
H 22 cm by 28 cm
J 30 m by 40 m

53. A car's mileage is registered at 29,345.5 miles. The driver sees a sign that warns of road work in 1000 feet. What will be the car's mileage when the road work begins?

A 29,345.7
B 29,345.9
C 29,356.2
D 29,356.5

54. **SHORT RESPONSE** Convert 565 miles per hour to feet per second. Show the procedure you used.

Spiral Review

Express each rate as a unit rate. Round to the nearest tenth, if necessary. (Lesson 6-2)

55. \$183 for 4 concert tickets

56. 100 feet in 14.5 seconds

57. 254.1 miles on 10.5 gallons

58. 9 inches of snow in 12 hours

59. **SHOPPING** Mrs. Gallagher wants to buy the package of soda that is less expensive per can. Which pack of sodas shown should she buy? Explain your reasoning. (Lesson 6-2)

Express each ratio as a fraction in simplest form. (Lesson 6-1)

60. 12 cars out of 30 vehicles

61. 5 cups to 5 quarts

62. 15 soccer balls out of 35 balls

63. 8 pencils to 20 crayons

ALGEBRA **Solve each equation. Check your solution.** (Lessons 4-3, 4-4)

64. $m - \frac{4}{5} = \frac{7}{10}$

65. $-7.2 = 9b$

66. $18 = \frac{2}{3}s$

67. $6\frac{4}{7} = d - \frac{11}{14}$

68. $-8.37 = c + (-5.28)$

69. $\frac{2}{3}t = \frac{1}{6}$

70. **HAMBURGERS** Clive is making hamburgers for a cookout. How many $\frac{1}{4}$-pound hamburgers can he make from $2\frac{3}{4}$ pounds of ground beef? (Lesson 3-4)

Skills Review

Write each fraction in simplest form. (Previous Course)

71. $\frac{12}{15}$

72. $\frac{18}{24}$

73. $\frac{6}{26}$

74. $\frac{10}{25}$

6-4 Proportional and Nonproportional Relationships

Then

You have already used unit rates to convert measurements. (Lesson 6-3)

Now

- Identify proportional and nonproportional relationships in tables and graphs.
- Describe a proportional relationship using an equation.

New Vocabulary

proportional
nonproportional
constant of proportionality

Math Online

glencoe.com

- Extra Examples
- Personal Tutor
- Self-Check Quiz
- Homework Help

Why?

Tom and Jenna are running laps around the gym. The number of laps each runner has completed is shown.

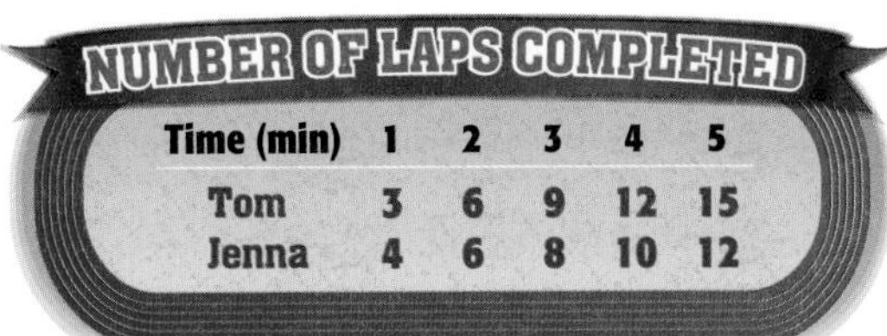

NUMBER OF LAPS COMPLETED

Time (min)	1	2	3	4	5
Tom	3	6	9	12	15
Jenna	4	6	8	10	12

a. For each minute, write a rate in simplest form that compares the number of laps completed by Tom to the time.

b. Repeat for Jenna's times.

c. What pattern(s) do you notice?

Identify Proportions Two quantities are **proportional** if they have a constant ratio or rate.

Tom's rates: $\frac{3}{1} = \frac{6}{2} = \frac{9}{3} = \frac{12}{4} = \frac{15}{5}$ Jenna's rates: $\frac{4}{1} \neq \frac{6}{2} \neq \frac{8}{3} \neq \frac{10}{4} \neq \frac{12}{5}$

The number of laps completed by Tom is proportional to the time. However, Jenna's rates are not constant. For relationships in which the ratios or rates are *not* constant, the two terms are said to be **nonproportional**.

EXAMPLE 1 Identify Proportional Relationships

Determine whether the cost of coffee is proportional to the number of pounds. Explain your reasoning.

Coffee (pounds)	1	2	3	4
Cost (dollars)	3	6	9	12

Write the rate of coffee to cost for each column in the table. Simplify each fraction.

$\frac{1}{3}$ $\frac{2}{6} = \frac{1}{3}$ $\frac{3}{9} = \frac{1}{3}$ $\frac{4}{12} = \frac{1}{3}$ **All the rates are equal.**

The number of pounds is proportional to the cost.

Check Your Progress

1. Determine whether the number of legs are proportional to the number of spiders. Explain your reasoning.

Number of Spiders	1	2	3	4
Number of Legs	8	16	24	32

Personal Tutor glencoe.com

EXAMPLE 2 Identify Proportional Relationships

Determine whether the number of miles is proportional to the number of hours. Explain your reasoning.

Time (hours)	1	2	3	4
Distance (miles)	50	70	90	110

Write the rate of time to distance for each hour in simplest form.

$\frac{1}{50}$ $\frac{2}{70} = \frac{1}{35}$ $\frac{3}{90} = \frac{1}{30}$ $\frac{4}{100} = \frac{2}{55}$ **The rates are not equal.**

The distance is *not* proportional to the time.

Check Your Progress

2. Determine whether the number of ice cubes is proportional to the number of drinks. Explain your reasoning.

Number of Drinks	1	2	3	4
Number of Ice Cubes	6	14	22	30

Personal Tutor glencoe.com

ReadingMath

Constant of Proportionality The constant of proportionality is also called the unit rate.

Describe Proportional Relationships Proportional relationships can also be described using equations of the form $y = kx$, where k is the constant ratio. The constant ratio is called the **constant of proportionality**.

Real-World EXAMPLE 3 Describe Proportional Relationships

GEOMETRY A circle's circumference is proportional to its diameter. The circle shown has a circumference of 12.56 meters. Write an equation relating the circumference of the circle to its diameter. What would be the circumference of a circle with a 6-inch diameter?

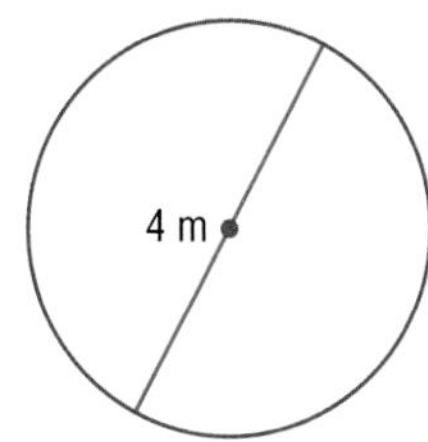

Find the constant of proportionality.

$\frac{\text{circumference}}{\text{diameter}} = \frac{12.56}{4}$ or 3.14

Words	The circumference is about 3.14 times the diameter.
Variable	Let C = circumference and d = diameter.
Equation	$C = 3.14d$

$C = 3.14d$ **Write the equation.**

$C = 3.14(6)$ **Replace *d* with 6.**

$C = 18.84$ **Multiply.**

The circumference is about 18.84 inches.

Check Your Progress

3. FOOD The cost for $2\frac{1}{2}$ pounds of meat is $7.20. Write an equation relating cost to the pounds of meat. How much will 4 pounds cost?

Personal Tutor glencoe.com

Check Your Understanding

Examples 1 and 2 pp. 281–282

Determine whether the set of numbers in each table is proportional. Explain.

1.

Blue Paint (quarts)	1	2	3	4
Yellow Paint (quarts)	5	6	7	8

2.

Ice Tea Mix (cups)	1	2	3	4
Sugar (cups)	2	4	6	8

Example 3 p. 282

3. **GASOLINE** The cost of 13 gallons of gasoline is $41.47. Write an equation relating cost to the number of gallons of gasoline. How much does 18.5 gallons of gasoline cost?

Practice and Problem Solving

● = **Step-by-Step Solutions** begin on page R11. **Extra Practice** begins on page 810.

Examples 1 and 2 pp. 281–282

Determine whether the set of numbers in each table is proportional. Explain.

4.

Cans of Concentrate	1	2	3	4
Cans of Water	4	8	12	16

5.

Shaded Squares	1	2	3	4
Total Squares	8	15	30	42

6.

Junk E-mails	10	20	30	40
Total E-mails	15	30	45	60

7.

Weeks	5	6	7	8
Days	35	42	49	56

8.

Heat Index (°F)*	67	73	78	84
Air Temperature (°F)	70	75	80	85

*at 30% Relative Humidity

9.

Jars	3	9	12	15
Jelly Beans	18	36	54	72

Example 3 p. 282

For Exercises 10 and 11, write and solve an equation.

10. **FINANCIAL LITERACY** A store is having a sale where all jeans are $\frac{1}{4}$ off the regular price. Write an equation relating the sale price to the regular price. How much would a pair of $29 jeans cost on sale?

11. **LAWN SERVICE** Luke earned $54 after mowing 3 lawns. Write an equation comparing earnings to lawns mowed. How much would Luke earn after mowing 7 lawns?

Copy and complete each table. Determine whether the pattern forms a proportion.

12. **PIZZA** Ms. Rollins had an end-of-year pizza party for the chess team. At the party, every 2 students had 5 slices of pizza.

Number of Students	2	4	6	8	10
Slices of Pizza	■	■	■	■	■

Real-World Link

There are more than 61,000 pizzerias in the United States.

Source: Pizzaware

13. **AMUSEMENT PARKS** Admission to an amusement park is $4 plus $1.50 per ride.

Number of Rides	1	2	3	4	5
Cost	■	■	■	■	■

14. **PARTY PLANNING** It will cost $7 per person to hold a birthday party at the recreation center.

Number of Guests	6	7	8	9	10
Cost	■	■	■	■	■

15 **FOOD** Eight hot dogs and ten hot dog buns come in separate packages.

a. Is the number of packages of hot dogs proportional to the number of hot dogs? Explain.

b. Is the number of hot dogs per package proportional to the number of hot dog buns per package? Explain.

16. **MULTIPLE REPRESENTATIONS** In this problem, you will investigate the relationship between graphs and their corresponding equations. Suppose Isabel decides to save $20 each week for her family vacation. Her sister already has $10 and wants to save an additional $20 each week for the vacation. These situations are modeled in the graphs below.

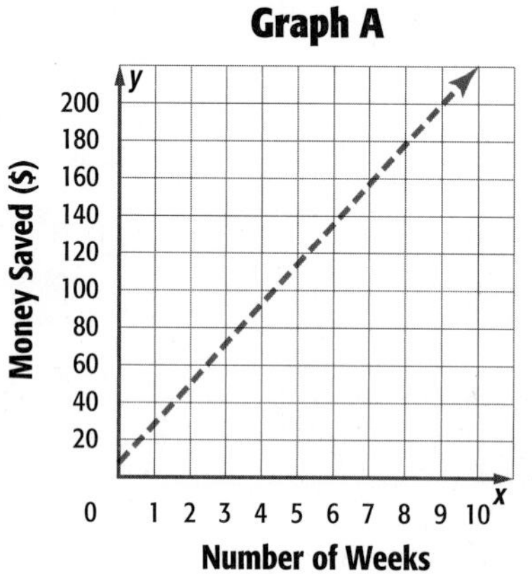

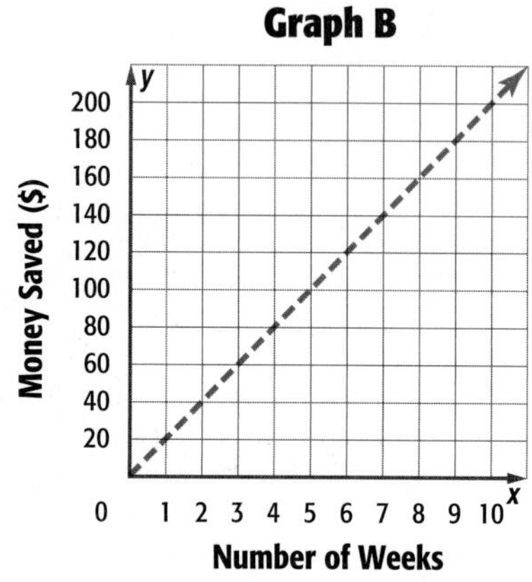

a. ANALYTICAL Which situation is proportional? Explain.

b. GRAPHICAL Compare and contrast the graphs. Which graph represents the proportional situation? Explain.

c. TABULAR Make a table showing the first six weeks of savings for each girl.

d. ALGEBRAIC Write an equation to represent each situation.

H.O.T. Problems

Use Higher-Order Thinking Skills

17. **OPEN ENDED** Give examples of two similar situations in which one is a proportional relationship and the second one is nonproportional. Then write equations that describe them.

18. **REASONING** A recipe for paper maché paste includes $\frac{1}{4}$ cup of flour for every cup of water. If there are 6 cups of flour, how many gallons of water are needed? Explain your reasoning.

19. **CHALLENGE** Many objects such as credit cards are shaped like golden rectangles. A *golden rectangle* is a rectangle in which the ratio of the length to the width is approximately 1.618 to 1. This ratio is called the *golden ratio.*

a. Find three different objects that are close to a golden rectangle. Make a table to display the dimensions and the ratio found in each object.

b. Describe how each ratio compares to the golden ratio.

c. Use the Internet or another source to find three examples of where the golden rectangle is used in architecture.

20. **WRITING IN MATH** This year Monica is 12 years old, and her little sister Patrice is 6 years old. Is Monica's age proportional to Patrice's age? Explain your reasoning using a table of values.

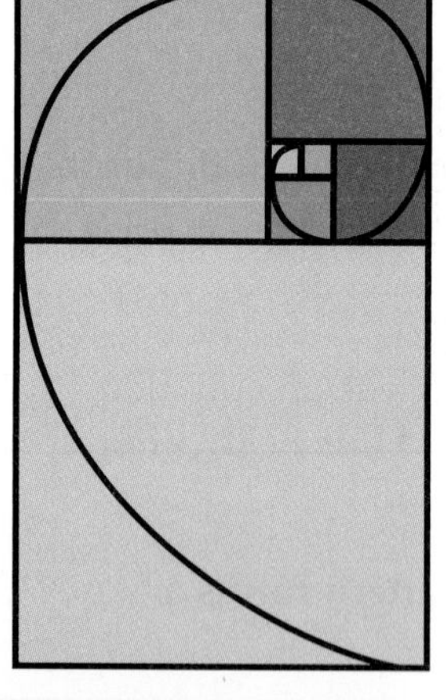

Math History Link

Golden Rectangle When a square is removed from a golden rectangle, the result is that the remaining rectangle will have the same proportions as the original rectangle. If squares are continually removed from the smaller rectangles, a spiral pattern will result. Greek mathematicians first studied this idea around 500 B.C.

Source: MathWorld

Standardized Test Practice

21. A bicycle wheel makes 30 revolutions in 45 feet. Which of these represents an equivalent rate of bicycle wheel revolutions?

A 10 revolutions in 15 ft

B 60 revolutions in 100 ft

C 15 revolutions in 10 ft

D 100 revolutions in 60 ft

22. The amount of sales tax paid on a purchase is proportional to the price of the item. Suppose the sales tax rate is 6.25%. If p is the price of the item and t is the sales tax, which equation represents this?

F $p = 0.0625t$

G $p = 6.25t$

H $t = 6.25p$

J $t = 0.0625p$

23. The cost of renting a boat for 4 hours is $51. If the cost of renting a boat is proportional to the number of rental hours, which of the following is *not* an equivalent rate?

A 6 hours for $76.50

B 3 hours for $38.25

C 7 hours for $89.25

D 5 hours for $63.00

24. **EXTENDED RESPONSE** The prices of different sized smoothies at an ice cream shop are shown below. Is the pricing guide based on a constant unit price? Explain.

Size (oz)	16	20	24
Cost ($)	3.25	3.75	4.25

Spiral Review

Complete each conversion. Round to the nearest hundredth, if necessary. (Lesson 6-3)

25. 4 in. ≈ ■ cm

26. 5 L ≈ ■ gal

27. 1500 lb ≈ ■ kg

28. 14 yd ≈ ■ m

Express each rate as a unit rate. Round to the nearest tenth, if necessary. (Lesson 6-2)

29. 140 miles on 6 gallons

30. 19 yards in 2.5 minutes

31. 236.7 miles in 4.5 days

32. 331.5 pages in 8.5 weeks

33. **SPORTS** A volleyball uniform costs $15 for the shirt, $10 for the pants, and $8 for the socks. Write two equivalent expressions for the total cost of 12 uniforms. Then find the cost. (Lesson 4-1)

34. **WEATHER** The table shows the record high and low temperatures for selected states. What is the difference between the highest and lowest temperatures for Kentucky? Massachusetts? (Lesson 2-3)

State	Lowest Temperature (°F)	Highest Temperature (°F)
Kentucky	−37	114
Massachusetts	−35	107

Skills Review

Solve each equation. (Lessons 4-3 and 4-4)

35. $7x = 126$

36. $150 = 15n$

37. $42 = 6m$

38. $12p = 78$

39. $22s = 154$

40. $245 = 7t$

CHAPTER 6

Mid-Chapter Quiz

Lessons 6-1 through 6-4

Express each ratio as a fraction in simplest form. (Lesson 6-1)

1. $16 for 6 gallons
2. 200 miles in 4 hours
3. 35 people in 7 events

Express each rate as a unit rate. Round to the nearest tenth, if necessary. (Lesson 6-2)

4. 875 customers in 7 days
5. 12 blocks in 8 minutes
6. 7 cars washed in 2 hours

7. **GROCERIES** Determine which is less expensive per ounce, 16 ounces of Swiss cheese for $2.75 or 24 ounces of cheddar cheese for $3.25. Explain. (Lesson 6-2)

8. **CONCERTS** Suppose it costs $45 to attend a concert where the band plays for 1.5 hours. (Lesson 6-2)
   a. What was the cost to attend the concert per hour?
   b. What was the cost to attend the concert per minute?

9. **MULTIPLE CHOICE** Which of the following has the same unit rate as $5 for 2 pounds? (Lesson 6-2)

   A $20 for 8 lb C $24 for 8 lb
   B $16 for 6 lb D $12 for 4 lb

Complete each conversion. Round to the nearest hundredth. (Lesson 6-3)

10. 25 lb ≈ ■ kg
11. 4 gal ≈ ■ L
12. 42 fl oz ≈ ■ mL
13. 36 in. ≈ ■ cm

14. **SWIMMING** The Olympic record for the men's 50-meter freestyle is 21.91 seconds. (Lesson 6-3)
    a. Express this speed in meters per second.
    b. At this rate, how many meters could be swum in 1 minute?

15. **MOUNTAINS** The table shows the heights of tall mountains in the U. S. (Lesson 6-3)

Mountain	Height (ft)
McKinley	20,320
Saint Elias	18,008
Foraker	17,400

    a. What is the height of Mt. McKinley in meters? Round to the nearest hundredth.
    b. What is the total height of the three mountains in yards?

16. **MULTIPLE CHOICE** An airplane flew 1500 miles in 2.5 hours. How many kilometers is this per hour? (Lesson 6-3)

    F 3750 H 750
    G 966 J 600

17. **ANIMALS** A blue whale's heart beats only 45 times in 5 minutes. Write an equation comparing the number of beats to the number of minutes. At that rate, how many times would the blue whale's heart beat in 20 minutes? (Lesson 6-4)

Copy and complete each table. Determine whether the pattern forms a proportion. Explain. (Lesson 6-4)

18. **ART MUSEUM** Mr. Dixon takes his fine arts class to the museum. The class has 25 students and museum admission is $2 per student.

Number of Students	5	10	15	20	25
Cost ($)	■	■	■	■	■

19. **FRUIT** A local apple orchard charges a $3 entrance fee plus $2 per barrel of apples.

Number of Barrels	1	2	3	4	5
Cost ($)	■	■	■	■	■

20. **DOGS** You earned $27 for walking 4 dogs. Write an equation comparing earnings to dogs walked. How much would you earn after walking 9 dogs? (Lesson 6-4)

6-5 Solving Proportions

Then
You have already used dimensional analysis to convert measurements. (Lesson 6-3)

Now
- Solve proportions.
- Use proportions to solve real-world problems.

New Vocabulary
proportion
cross products

Math Online
glencoe.com
- Extra Examples
- Personal Tutor
- Self-Check Quiz
- Homework Help

Why?

Bianca wants to make a salt map of the United States for her social studies class. The recipe for the dough is shown.

Salt Map Ingredients

Ingredient	Amount
flour	4 c
salt	2 c
water	1 c

a. Write the ratio of cups of flour to cups of salt as a fraction in simplest form.

b. Bianca needs to double the recipe to have enough dough for her map. What is the ratio of cups of flour to cups of salt in simplest form?

c. Are the ratios in parts **a** and **b** equal? Explain

Proportions In the example above, the ratios of cups of flour to cups of salt for a single batch of dough, $\frac{4}{2}$, or a double batch of dough, $\frac{8}{4}$, are equal. They both simplify to $\frac{2}{1}$.

$$\frac{4}{2} = \frac{2}{1} \qquad \frac{8}{4} = \frac{2}{1}$$

A **proportion** is an equation stating that two ratios or rates are equal. You can use the Multiplication Property of Equality to illustrate an important property of proportions.

	$\frac{4}{2} = \frac{8}{4}$	$\frac{a}{b} = \frac{c}{d}$	
Multiply each side by 2 • 4.	$\frac{4}{\cancel{2}_1} \cdot (\cancel{2}^1 \cdot 4) = \frac{8}{\cancel{4}_1} \cdot (2 \cdot \cancel{4}^1)$	$\frac{a}{\cancel{b}_1} \cdot \cancel{b}^1 d = \frac{c}{\cancel{d}_1} \cdot b\cancel{d}^1$	**Multiply each side by *bd*.**
Simplify.	$4 \cdot 4 = 8 \cdot 2$	$ad = cb$	**Simplify.**

The products 4 • 4 and 8 • 2, and *ad* and *cb* are called **cross products** of the proportion. The cross products of any proportion are equal.

$$\frac{4}{2} = \frac{8}{4} \quad 4 \cdot 4 = 8 \cdot 2$$

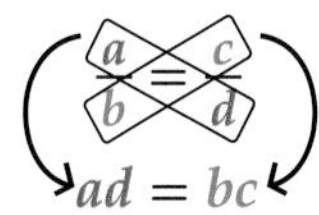

$$ad = bc$$

Key Concept Property of Proportions — For Your FOLDABLE

Words The cross products of a proportion are equal.

Symbols If $\frac{a}{b} = \frac{c}{d}$, then $ad = cb$.
If $ad = cb$, then $\frac{a}{b} = \frac{c}{d}$ if $b \neq 0$ and $d \neq 0$.

Cross products will help when you use proportional reasoning to solve a problem.

Just as in solving an equation, solving a proportion means finding the value of the variable that makes a true statement. You can use cross products to solve a proportion in which one of the quantities is not known.

StudyTip

Solve Mentally You can also solve Example 1a mentally.

Think: $90 \div 15 = 6$ and $66 \div 6 = 11$.

EXAMPLE 1 Solve Proportions

Solve each proportion.

a. $\frac{b}{15} = \frac{66}{90}$

$\frac{b}{15} = \frac{66}{90}$

$b \cdot 90 = 15 \cdot 66$ Cross products

$90b = 990$ Multiply.

$\frac{90b}{90} = \frac{990}{90}$ Divide.

$b = 11$ Simplify.

b. $\frac{3.2}{9} = \frac{n}{36}$

$\frac{3.2}{9} = \frac{n}{36}$

$3.2 \cdot 36 = 9 \cdot n$ Cross products

$115.2 = 9n$ Multiply.

$\frac{115.2}{9} = \frac{9n}{9}$ Divide.

$12.8 = n$ Simplify.

Check Your Progress

1A. $\frac{x}{4} = \frac{7}{20}$ **1B.** $\frac{7}{14} = \frac{c}{12}$ **1C.** $\frac{6.8}{t} = \frac{34}{50}$ **1D.** $\frac{m}{8.5} = \frac{42}{51}$

Personal Tutor glencoe.com

Use Proportions to Solve Problems When you solve a real-world problem using a proportion, be sure to compare the quantities in the same order.

Real-World EXAMPLE 2 Use Proportions to Solve Problems

AMUSEMENT PARKS The wait time to ride a roller coaster is 20 minutes when 160 people are in line. At this rate, how long is the wait time when 220 people are in line?

Understand You know how long the wait time is for 160 people. You need to find how long the wait time will be for 220 people.

Plan Write and solve a proportion using ratios that compare people to wait time. Let w represent the wait time for 220 people.

Solve

$\frac{20}{160} = \frac{w}{220}$ ← wait time / ← number of people

$20 \cdot 220 = 160 \cdot w$ Cross products

$4400 = 160w$ Multiply.

$\frac{4400}{160} = \frac{160w}{160}$ Divide each side by 160.

$27.5 = w$ Simplify.

Check Check the cross products. Since $20 \cdot 220 = 4400$ and $160 \cdot 27.5 = 4400$, the answer is correct. ✓

The wait time is 27.5 minutes.

Check Your Progress

2. COMMUNITY SERVICE Alicia's class is making care packages for a local shelter. They can make 8 care packages with 240 food items. How many care packages can they make with 500 food items?

Personal Tutor glencoe.com

StudyTip

Proportions You can set up a proportion in more than one way. In Example 3, the proportion can also be set up as $\frac{4}{20} = \frac{30}{x}$.

EXAMPLE 3 Use Proportions to Solve Problems

MUSEUMS Mrs. Hidalgo paid $30 for 4 students to visit an art museum. Find the cost for 20 students.

Method 1 Write and solve a proportion.

Let x represent the cost for 20 students.

$\frac{30}{4} = \frac{x}{20}$	← **cost** / ← **number of students**
$30 \cdot 20 = 4 \cdot x$	**Cross products**
$600 = 4x$	**Multiply.**
$\frac{600}{4} = \frac{4x}{4}$	**Divide each side by 4.**
$150 = x$	**Simplify.**

Method 2 Write and solve an equation.

Find the constant of proportionality, or unit cost, for each student.

$\frac{\text{cost in dollars}}{\text{number of students}} = \frac{30}{4}$ or 7.50 The cost is $7.50 per student.

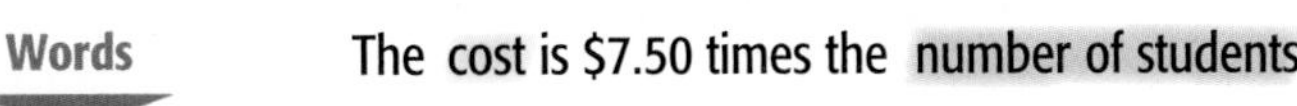

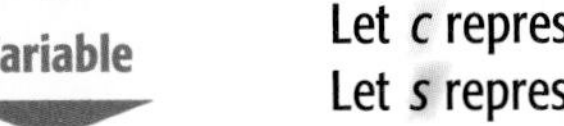

Words	The cost is $7.50 times the number of students.
Variable	Let c represent the cost. Let s represent the number of students.
Equation	$c = 7.5s$

Use this equation to find the cost for 20 students at the same rate.

$c = 7.5s$	**Write the equation.**
$c = 7.5(20)$	**Replace s with 20.**
$c = 150$	**Multiply.**

So, the cost for 20 students to visit the art museum is $150.

Check Your Progress

3. DVDS Matthew paid $49.45 for 5 DVDs at a sale.

A. Write and solve a proportion to find the cost for 8 DVDs at the same rate.

B. Write an equation relating the cost c to the number of DVDs d. How much would it cost for 11 DVDs at the same rate?

Personal Tutor glencoe.com

Real-World Link

Nearly 150 million DVD players have been sold in the United States since they went on the market in 1997.

Source: Consumer Electronics Association

Concept Summary Solve Proportions

Ways to Solve Proportions

- Mental Math
- Cross Products
- Constant of Proportionality

Check Your Understanding

Example 1 p. 288

Solve each proportion.

1. $\frac{18}{m} = \frac{27}{36}$
2. $\frac{t}{21} = \frac{9}{15}$
3. $\frac{8}{17} = \frac{16}{x}$
4. $\frac{12}{7.2} = \frac{4}{p}$
5. $\frac{n}{13} = \frac{5.8}{2.6}$
6. $\frac{4.4}{2} = \frac{c}{25}$

Example 2 p. 288

7. **TELEVISION** *Aspect ratio* is the ratio of width to height of a television screen. A widescreen television screen has an aspect ratio of 16 inches wide to 9 inches high. If a television screen is 48 inches wide, how many inches high is it?

Example 3 p. 289

8. **TIME** Joaquin has a total of 12.5 hours of football practice after school five days a week. Write an equation relating the number of days d to the number of hours h. How many hours of practice will he have for 15 school days? 22 school days?

Practice and Problem Solving

● = **Step-by-Step Solutions** begin on page R11. **Extra Practice** begins on page 810.

Example 1 p. 288

Solve each proportion.

9. $\frac{6}{8} = \frac{z}{48}$
10. $\frac{8}{12} = \frac{28}{m}$
11. $\frac{b}{30} = \frac{4}{5}$
12. $\frac{18}{15} = \frac{9}{s}$
13. $\frac{7}{c} = \frac{35}{60}$
14. $\frac{k}{56} = \frac{12}{7}$
15. $\frac{32}{a} = \frac{12.8}{5.6}$
16. $\frac{11.5}{6} = \frac{n}{22.8}$
17. $\frac{9.6}{3} = \frac{p}{0.3}$
18. $\frac{14}{w} = \frac{8.4}{4.5}$
19. $\frac{v}{6} = \frac{20.7}{5.4}$
20. $\frac{10.2}{4} = \frac{h}{12}$

Example 2 p. 288

21. **PAINTING** The classrooms at Lincoln Middle School are painted every summer. If 7 gallons of paint are needed to paint 4 classrooms, how many gallons of paint are needed to paint 16 classrooms?

22. **PIZZA** A principal is ordering pizzas for a school pizza party. He knows that 9 pizzas will feed 25 students. If there are 300 students in the school, how many pizzas will he need to order?

Example 3 p. 289

23. **BOATS** A boat traveled 150 feet in 9.7 seconds. Write an equation relating the time t to the distance d. How far would the boat travel in 1 minute? 1 minute 30 seconds?

24. **RAINFALL** The record for the most amount of rain in the shortest amount of time in the United States was 12 inches in 42 minutes. Write an equation relating the time t to the amount of rain a. How much rain fell in 15 minutes? 28 minutes?

25. **PUNCH** The table shows the amount of each ingredient in 52 ounces of punch.

Ingredient	Amount (ounces)
water	12
lime juice	4
cranberry concentrate	12
sparkling lemon water	24

   a. If you have 130 ounces of punch, how much lime juice does the punch contain?

   b. If there are 54 ounces of sparkling lemon water in the punch, how many ounces of cranberry concentrate does the punch contain?

   c. If the punch contains 44 ounces of water, how many ounces of punch do you have?

Real-World Link

The Eiffel Tower in Paris, France, was completed in 1889. It has a total weight of 10,100 tons and is 324 meters tall.

Source: SETE

Write a proportion that could be used to solve for each variable. Then solve.

26. 6 goals in 14 games
9 goals in g games

27. s inches in 0.54 hour
4.55 inches in 1.89 hours

28. 14 gallons for d dollars
8 gallons for \$24.72

29. 20 boxes on 4 shelves
b boxes on 20 shelves

30. **ART** A 14-inch-wide by 20-inch-long print of the Eiffel Tower is also available as a postcard 6 inches long. What is the width of the postcard?

Solve each proportion.

31. $\frac{a}{0.28} = \frac{4}{1.4}$

32. $\frac{3}{14} = \frac{15}{m-3}$

33. $\frac{16}{x+5} = \frac{4}{5}$

34. $\frac{x-18}{24} = \frac{15}{8}$

35. $\frac{9}{7} = \frac{b+11}{14}$

36. $\frac{15-d}{12} = \frac{37.5}{75}$

37. **MULTIPLE REPRESENTATIONS** In this problem, you will explore proportions. A craft store is offering the specials shown for different materials.

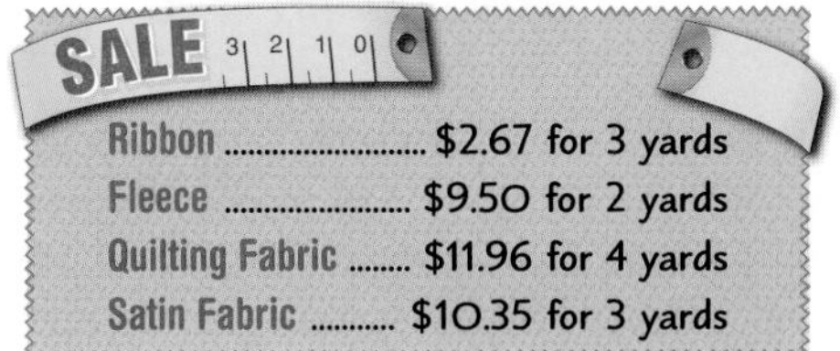

SALE
Ribbon .... $2.67 for 3 yards
Fleece .... $9.50 for 2 yards
Quilting Fabric .... $11.96 for 4 yards
Satin Fabric .... $10.35 for 3 yards

a. ALGEBRAIC Write an equation relating the cost c to the number of yards n for each material.

b. GRAPHICAL Graph the cost of each material per yard on a coordinate plane. Which item costs the least per yard? the most? How is this shown on the graph?

c. NUMERICAL How much would it cost to buy 18 inches of ribbon? How much would it cost to buy 10 meters of fleece?

H.O.T. Problems

Use Higher-Order Thinking Skills

38. **OPEN ENDED** Give two examples that are proportional to $\frac{6 \text{ hits}}{8 \text{ at bats}}$.

39. **REASONING** Suppose $a{:}b = 2{:}4$ and $b{:}c = 4{:}9$. If $a = 10$, find the value of c.

40. **CHALLENGE** Solve each proportion.

a. $\frac{4}{x} = \frac{x}{9}$ **b.** $\frac{2}{x} = \frac{x}{8}$ **c.** $\frac{4}{x} = \frac{x}{25}$ **d.** $\frac{9}{x} = \frac{x}{16}$

41. **FIND THE ERROR** Trey and Morgan are solving the proportion $\frac{x}{36} = \frac{4}{9}$. Is either of them correct? Explain your reasoning.

Trey	Morgan
$\frac{x}{36} = \frac{4}{9}$	$\frac{x}{36} = \frac{4}{9}$
$x(9) = 36(4)$	$x(4) = 36(9)$
$x = 16$	$x = 81$

42. **CHALLENGE** Rectangle $ABCD$ has a fixed area. As the length ℓ and the width w change, what do you know about their product? Is the length proportional to the width? Justify your reasoning.

43. **WRITING IN MATH** Describe a situation in which it may be easier to solve a proportion using the constant of proportionality. Explain your reasoning. Does it matter which method you use to solve a proportion? Why or why not?

Standardized Test Practice

44. Which equation could be used to find the total cost c if Fernando wanted to buy 8 pencils from the school store?

School Store Sale
3 pencils for $0.45
2 pens for $0.75
4 highlighters for $1.25

A $c = 0.15 \cdot 8$ C $c = 24 \cdot 0.45$
B $c = 0.45 \cdot 8$ D $c = 0.15 \cdot 24$

45. A line to purchase concert tickets is moving at a rate of 5 feet every 20 minutes. At this rate, how long will a person have been in line if they have moved 30 feet?

F 30 min H 1 h 30 min
G 1 h J 2 h

46. During her keyboarding test, Luisa typed 145 words in three minutes. Which of the following could *not* be used to determine the number of minutes it would take her to type 550 words?

A $\frac{145}{3} = \frac{550}{w}$ C $\frac{145}{550} = \frac{3}{w}$
B $\frac{3}{145} = \frac{w}{550}$ D $\frac{145}{550} = \frac{w}{3}$

47. EXTENDED RESPONSE There are 258 eighth graders at Henderson Middle School. The graph shows how many of the eighth graders participate in each sport.

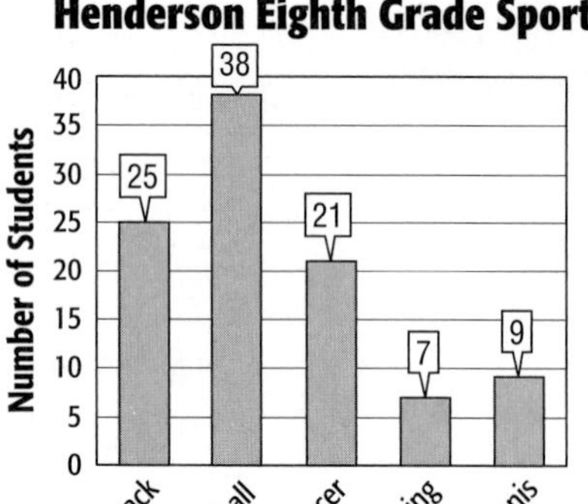

a. If each student participates in only one sport, write a proportion that could be used to predict how many students at Henderson Middle School participate in sports if there are 645 total students.

b. Solve the proportion in part **a.**

c. Write and solve a proportion that could be used to predict the number of eighth graders that participate in track if there are 220 total eighth graders.

Spiral Review

48. Determine whether the rental charge is proportional to the time. Explain your reasoning. (Lesson 6-4)

Time (hours)	1	2	3	4
Rental Charge	$13	$23	$33	$43

Use dimensional analysis to complete each conversion. Round to the nearest hundredth. (Lesson 6-3)

49. 5 in. ≈ ■ cm
(*Hint:* 1 in. ≈ 2.54 cm)

50. 10 km ≈ ■ mi
(*Hint:* 1 km ≈ 0.621 mi)

51. 26.3 cm ≈ ■ in.
(*Hint:* 1 cm ≈ 0.394 in.)

52. PLANES How long will it take an Air Force fighter jet to fly 5200 miles at 650 miles per hour? (Lesson 4-6)

Skills Review

Find each product. (Lesson 3-3)

53. $\frac{4}{9} \cdot \frac{2}{3}$

54. $\frac{1}{5} \cdot \frac{1}{8}$

55. $\frac{3}{4} \cdot \frac{3}{5}$

56. $\frac{2}{5} \cdot \frac{5}{6}$

Algebra Lab
Inverse Proportions

Math Online glencoe.com
Math in Motion, Animation

Jon earns money for a class trip by selling T-shirts at a track meet with the Heritage Middle School Service Club. The club earned \$220 from the T-shirt sale. The money will be equally divided among all the students who participated in the sale. The equation $d = \frac{220}{s}$ or $sd = 220$ represents this situation, where d is the number of dollars earned per student and s is the number of students.

ACTIVITY

Step 1 Copy and complete the table for the equation $sd = 220$.

Number of Students, s	Dollars Earned per Student, d
2	110
4	55
6	■
8	■
10	■

Step 2 Copy the blank grid below. Graph the ordered pairs from Step 1. Then connect the points with a smooth curve.

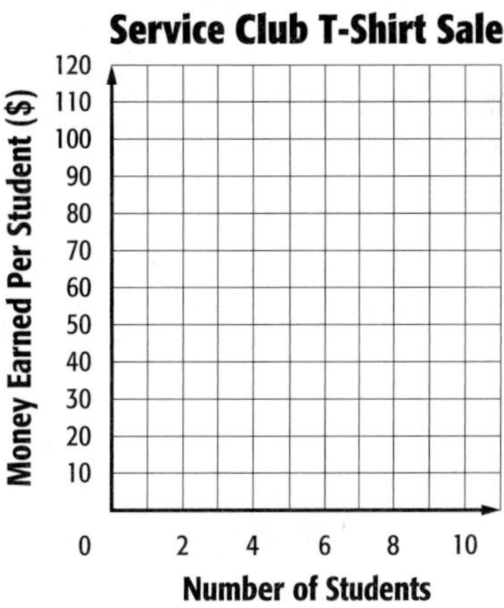

Analyze the Results

1. Is the number of students proportional to the dollars earned per student? Explain.
2. What is different about the graph in Step 2 than other functions you have seen?
3. An **inverse proportion** is a relationship formed when the product of two variables is a constant. Which situation is an inverse proportion: each student earning \$3 per T-shirt sold or the \$220 in earnings being equally divided among the total number of students? Explain.
4. **GARDENS** A rectangular herb garden has an area of 36 square feet.
   a. Make a table of ordered pairs to represent possible dimensions for the garden. Then graph the ordered pairs (length, width).
   b. Is the relationship in the graph proportional or is it an inverse proportion? Explain.
5. **JOGGING** On Monday, Lola jogs at an average rate of 8 miles per hour around a track. On Wednesday, she jogs 16 miles through a cross-country course at an unknown rate. Make a table of ordered pairs and a graph for each situation. Then decide whether each relationship is a proportion or an inverse proportion.

6-6 Scale Drawings and Models

Then

You have already written and solved proportions. (Lesson 6-5)

Now

- Use scale drawings.
- Construct scale drawings.

New Vocabulary

scale drawing
scale model
scale
scale factor

Math Online

glencoe.com

- Extra Examples
- Personal Tutor
- Self-Check Quiz
- Homework Help

Why?

A scale drawing of the White House is drawn on grid paper. The length of one square on the grid paper represents 7 feet.

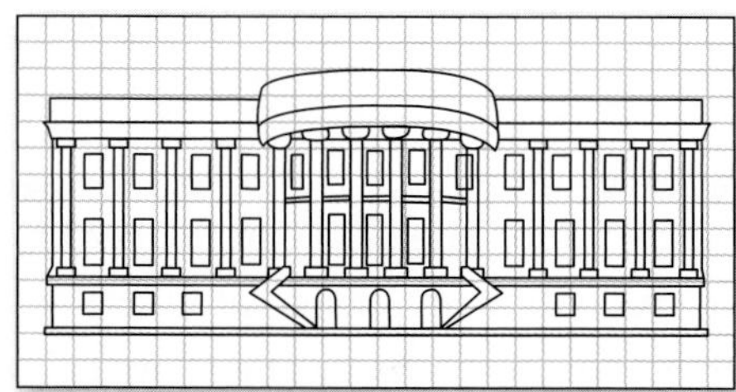

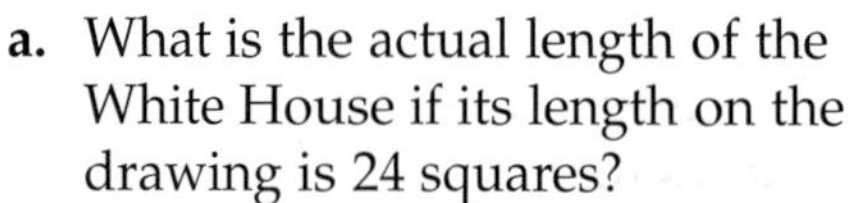

a. What is the actual length of the White House if its length on the drawing is 24 squares?

b. What is the actual height of the White House if its height on the drawing is 10 squares?

Use Scale Drawings and Models A **scale drawing** or a **scale model** is used to represent an object that is too large or too small to be drawn or built at actual size. The lengths and widths of objects on a scale drawing or model are proportional to the lengths and widths of the actual object.

The **scale** is determined by the ratio of a given length on the drawing or model to its corresponding length on the actual object. Consider the following scales.

1 in. = 3 ft	**1 inch represents an actual distance of 3 feet.**
1 cm = 2 mm	**1 centimeter represents an actual distance of 2 millimeters.**

Scales are written so that a unit length on the drawing or model is listed first.

Real-World EXAMPLE 1 Determine the Scale

INSECTS Suppose a model of a dragonfly has a wing length of 4 centimeters. If the length of the insect's actual wing is 6 centimeters, what is the scale of the model?

Let x represent the actual length.

Write and solve a proportion.

model length → / actual length → $\frac{4 \text{ cm}}{6 \text{ cm}} = \frac{1 \text{ cm}}{x \text{ cm}}$ ← model length / ← actual length

$4 \cdot x = 6 \cdot 1$ **Find the cross products.**

$4x = 6$ **Simplify.**

$x = 1.5$ **Divide each side by 4.**

So, the scale is 1 centimeter = 1.5 centimeters.

Check Your Progress

1. **ARCHITECTURE** The pillars of the World War II memorial in Washington, D.C., are 17 feet tall. A scale model of the memorial has pillars that are 5 inches tall. What is the scale of the model?

Personal Tutor glencoe.com

If the scale drawing and model have the same unit of measure, the scale can be written without units. This is called the **scale factor.** Suppose a scale model has a scale of 1 inch = 2 feet.

scale → 1 inch = 2 feet → $\frac{1 \text{ inch}}{2 \text{ feet}}$ → $\frac{1 \text{ inch}}{24 \text{ inches}}$ → 1:24 ← scale factor

One unit on the model represents an actual distance of 24 units. So, the model is $\frac{1}{24}$ the size of the actual object.

Real-World EXAMPLE 2 Find Actual Measurements

ARCHITECTURE The blueprint of a skateboard ramp shows that its length is 11.4 inches. If the scale on the blueprint is 1 inch = 6 feet, what is the length of the actual skateboard ramp?

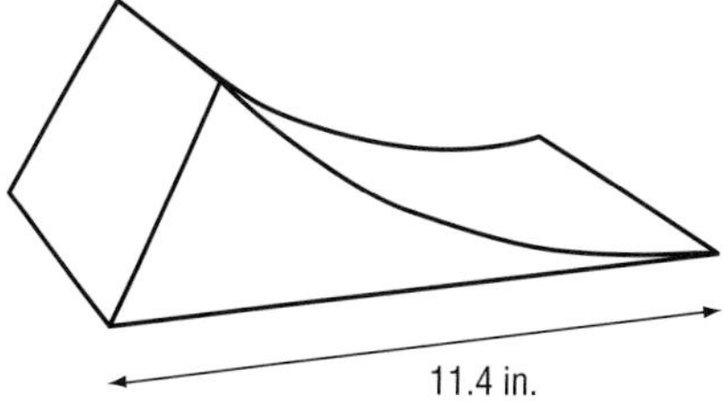

Method 1 Use a proportion.

Let x represent the actual length of the ramp. Write and solve a proportion.

plan length → / actual length → $\frac{1 \text{ inch}}{6 \text{ feet}} = \frac{11.4 \text{ inches}}{x \text{ feet}}$ ← plan length / ← actual length

$1 \cdot x = 6 \cdot 11.4$ **Find the cross products.**

$x = 68.4$ **Simplify.**

StudyTip

Alternative Method You can also solve the proportion.

plan actual

$\frac{1 \text{ in.}}{11.4 \text{ in.}} = \frac{6 \text{ ft}}{x \text{ ft}}$

Method 2 Use the scale factor.

The actual length is proportional to the length on the scale drawing with a ratio of $\frac{1 \text{ inch}}{6 \text{ feet}}$.

Find the scale factor.

$\frac{1 \text{ inch}}{6 \text{ feet}} = \frac{1 \text{ inch}}{72 \text{ inches}}$ or $\frac{1}{72}$ **Convert 6 feet to inches and divide out units.**

The scale factor is $\frac{1}{72}$.

So, the actual length is 72 times the blueprint length.

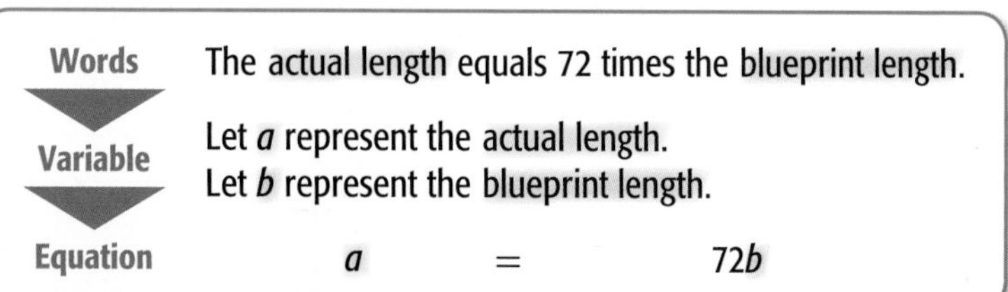

Words The actual length equals 72 times the blueprint length.
Variable Let a represent the actual length.
Let b represent the blueprint length.
Equation a = 72b

$a = 72b$ **Write the equation.**

$= 72(11.4)$ **Replace b with 11.4.**

$= 820.8$ **Simplify.**

The actual length of the ramp is 820.8 inches or 68.4 feet.

Check Your Progress

2. **EXHIBITS** A map of a natural history museum shows that the dinosaur exhibit room is 7.25 inches wide. If the scale on the map is 1 inch = 8 feet, what is the width of the actual exhibit room?

Personal Tutor glencoe.com

Construct Scale Drawings To construct a scale drawing of an object, use the actual measurements of the object and the scale to which the object is to be drawn.

Real-World EXAMPLE 3 Construct a Scale Drawing

ART Lila is painting a mural on a wall at the community center that measures 18 feet long and 12 feet tall. Make a scale drawing of the mural. Use a scale of $\frac{1}{4}$ inch = 3 feet. Use $\frac{1}{4}$-inch grid paper.

Step 1 Find the measure of the wall's length on the drawing. Let ℓ represent the length.

drawing length / actual length → $\frac{\frac{1}{4}\text{ inch}}{3\text{ feet}} = \frac{\ell\text{ inches}}{18\text{ feet}}$ ← drawing length / actual length

$\frac{1}{4} \cdot 18 = 3 \cdot \ell$ — **Find the cross products.**

$4.5 = 3\ell$ — **Simplify.**

$\frac{4.5}{3} = \frac{3\ell}{3}$ — **Divide each side by 3.**

$1.5 = \ell$ — **Simplify.**

On the drawing, the length is 1.5 or $1\frac{1}{2}$ inches.

Step 2 Find the measure of the wall's height on the drawing. Let w represent the width.

drawing width / actual width → $\frac{\frac{1}{4}\text{ inch}}{3\text{ feet}} = \frac{w\text{ inches}}{12\text{ feet}}$ ← drawing width / actual width

$\frac{1}{4} \cdot 12 = 3 \cdot w$ — **Find the cross products.**

$3 = 3w$ — **Simplify.**

$\frac{3}{3} = \frac{3w}{3}$ — **Divide each side by 3.**

$1 = w$ — **Simplify.**

On the drawing, the height is 1 inch.

Step 3 Make the scale drawing.

Use $\frac{1}{4}$-inch grid paper. Since $1\frac{1}{2}$ inches = 6 squares and 1 inch = 4 squares, draw a rectangle that is 4 squares by 6 squares.

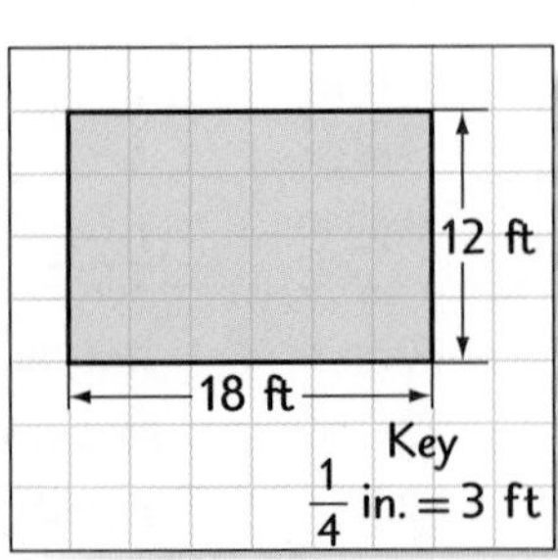

Real-World Career

Landscape Architect Landscape architects use computer sketches and models to design and show outdoor spaces.

A job in this profession requires a degree in landscape architecture which takes 4–5 years to complete.

Source: Bureau of Labor Statistics and ASLA

Check Your Progress

3. **SCHOOL BUILDINGS** An architect is designing a school courtyard that is 45 feet long and 30 feet wide. Make a scale drawing of the courtyard. Use a scale of 0.5 inch = 10 feet. Use $\frac{1}{4}$-inch grid paper.

Personal Tutor glencoe.com

Check Your Understanding

Example 1 p. 294

1. **CARS** The model of a car is shown at the right. The actual car is $14\frac{1}{2}$ feet long. What is the scale of the model car?

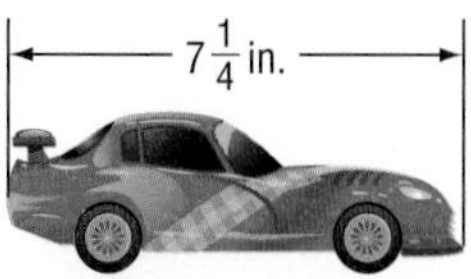

Example 2 p. 295

2. **MAPS** On the map, the scale is 1 inch = 20 miles. What is the actual distance between Kansas City and St. Louis?

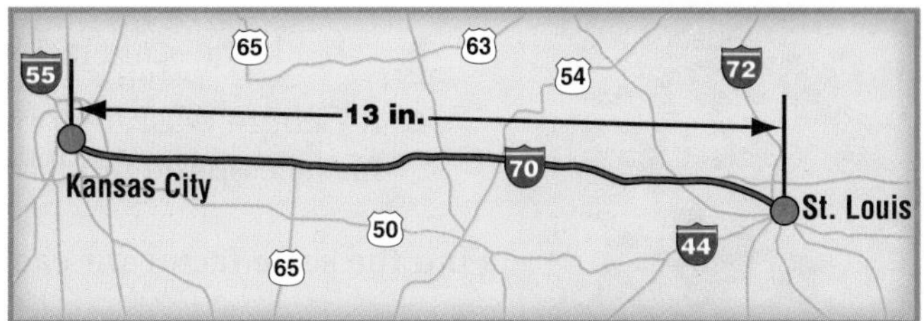

Example 3 p. 296

3. **GARDENS** Marco is designing a flower garden in his backyard that is 12 feet long and 10 feet wide. Make a scale drawing of the room. Use a scale of 0.5 inch = 2 feet. Use $\frac{1}{4}$-inch grid paper.

Practice and Problem Solving

= **Step-by-Step Solutions** begin on page R11. **Extra Practice** begins on page 810.

Example 1 p. 294

4. **AIRPLANES** A model airplane is built with a wing span of 23 inches. The actual wing span is 92 feet. Find the scale.

5. **FLAGS** The largest American flag in existence measures 255 feet wide. In an advertisement for renting this flag, the image of the flag is 4 inches wide. What is the scale of the flag?

Example 2 p. 295

6. **ARCHITECTURE** A floor plan is shown for the first floor of a new house. If one inch represents 24 feet, what are the actual dimensions of each of the rooms listed?

a. living room **b.** deck **c.** kitchen

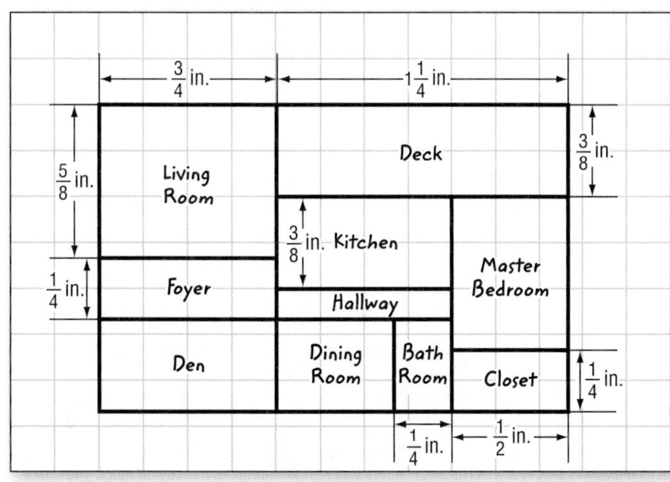

7. **ARCHITECTURE** The actual measurements for rooms are given. Using the floor plan and scale above, find the measurements on the floor plan.

a. master bedroom 12 feet by 15 feet **b.** den 18 feet by 9 feet **c.** dining room 12 feet by 9 feet

Example 3
p. 296

8. **SKATEPARKS** A skatepark is 24 yards wide by 48 yards long. Make a scale drawing of the skatepark that has a scale of $\frac{1}{4}$ in. = 8 yd.

9. **BEDROOMS** Create a scale factor that relates inches to feet and use it to make a scale drawing of your bedroom.

Real-World Link

The actual Mount Rushmore carving was made from a scale model with a scale of 1 in. = 1 ft. On the model, Teddy Roosevelt's moustache was 1 foot 8 inches long.

Source: Bureau of Labor Statistics

10. **MONUMENTS** **Use the information at the left to answer the following.**
    a. Find the length of Roosevelt's moustache on the monument.
    b. What is the scale factor?
    c. If George Washington's face is 60 feet tall on the monument, how tall is his face on the model?

Find the scale factor for each scale.

11. 6 in. = 10 ft	**12.** 10 cm = 5 m	**13.** 0.5 in. = 3 ft
14. 5 ft = 15 yd	**15.** 4 cm = 2.5 mm	**16.** 8 in. = 200 mi

17. **PLANETS** The 8 planets' distance from the Sun is shown in the table at the right.
    a. What scale would you use to create a scale drawing?
    b. Use centimeter grid paper and your scale from part **a** to create a map of the six planets closest to the Sun.

Planet	Distance from the Sun (mi)
Mercury	3.6×10^7
Venus	6.7×10^7
Earth	9.3×10^7
Mars	1.42×10^8
Jupiter	4.84×10^8
Saturn	8.87×10^8
Uranus	1.8×10^9
Neptune	2.8×10^9

18. You build a model with a scale of 1:25. Your friend builds a model of the same object with a scale of 1:50. Which model is bigger? Explain.

H.O.T. Problems

Use **H**igher-**O**rder **T**hinking Skills

19. **OPEN ENDED** Find a small rectangular item you use on a daily basis. Make a scale drawing of that item. Then write a problem based on your scale drawing.

20. **CHALLENGE** Rectangle *ABCD* is reduced by a scale factor $\frac{1}{2}$. What is the area of the new rectangle?

A B
Area = 48 in^2
D C

21. **REASONING** Determine whether the following statement is *always*, *sometimes*, or *never* true. Justify your reasoning. *If the scale factor of a scale drawing is greater than one, the scale drawing is larger than the actual object.*

22. **WHICH ONE DOESN'T BELONG?** Identify the scale that does not have the same scale factor. Explain your reasoning.

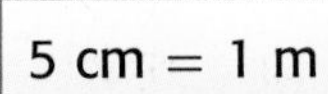

10 mm = 20 cm

10 cm = 10 m

25 mm = 0.5 m

23. **CHALLENGE** A model of an insect has a scale of 0.25 cm = 1 mm. Is the model *smaller* or *larger* than the actual insect? Justify your reasoning by using the scale factor.

24. **WRITING IN MATH** Compare and contrast *scale* and *scale factor*.

Standardized Test Practice

25. Desiree is drawing a model of the Washington Monument which has an actual height of 555.5 feet.

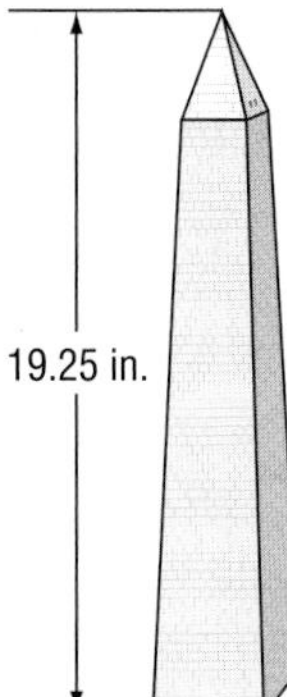

What other information is needed to find the length of the model's sides on the square base?

A the height of the top of the tower
B the length of the actual base side
C the age of the tower
D the height of the tower's first story

26. **GRIDDED RESPONSE** A blueprint has a scale of 2 inches = 2 feet. What is the scale factor of the blueprint?

27. A scale drawing of a swimming pool is shown.

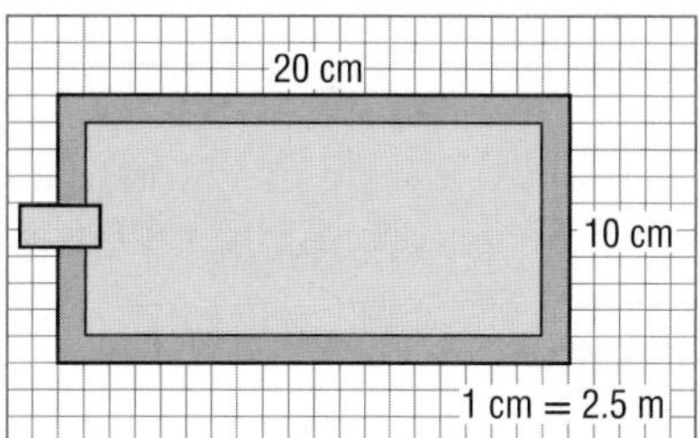

What are the actual dimensions of the swimming pool?

F 8 meters by 4 meters
G 20 meters by 10 meters
H 50 meters by 25 meters
J 80 meters by 40 meters

28. A map has a scale of 1.5 inches = 500 miles. How many inches on the map would represent 850 miles? Round to the nearest tenth.

A 2.2 inches
B 2.4 inches
C 2.6 inches
D 2.8 inches

Spiral Review

ALGEBRA **Solve each proportion.** (Lesson 6-5)

29. $\frac{p}{6} = \frac{24}{36}$
30. $\frac{4}{10} = \frac{8}{a}$
31. $\frac{18}{12} = \frac{24}{q}$
32. $\frac{5}{h} = \frac{10}{30}$
33. $\frac{7}{45} = \frac{x}{9}$
34. $\frac{7}{5} = \frac{10.5}{b}$

35. **PLANTS** Some species of bamboo can grow 245 inches in a week. Write an equation relating the height of the bamboo to the number of days. How much would a bamboo plant have grown after 3 days? (Lesson 6-4)

36. **OLYMPICS** The conversion factor for changing meters to feet is 1 meter ≈ 3.28 feet. Find the approximate distance in feet of the 110-meter dash. (Lesson 6-3)

Skills Review

Solve each equation. (Lesson 4-4)

37. $4t = 36$
38. $52 = 13n$
39. $6.2m = 114.08$
40. $5x = 3.5$
41. $7.8 = 3b$
42. $9.4 = 4g$

EXPLORE
6-7 Geometry Lab: Similar Figures

Math Online glencoe.com
Math *in Motion*, Animation

Have you ever used a copy machine to make an enlargement or reduction of a drawing? In these activities, you will draw enlargements and reductions.

ACTIVITY 1

Step 1 On grid paper, draw a rectangle with a length of 5 inches and a width of 2 inches. This is the original figure.

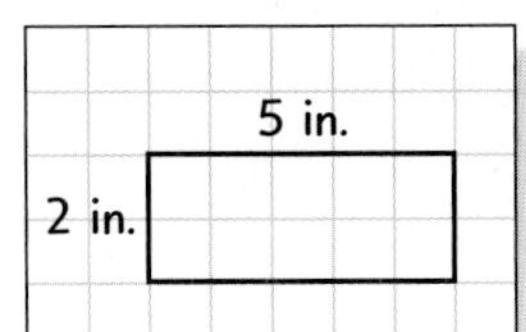

Step 2 Use a scale factor of 1.5. Draw a new rectangle with a length that is 1.5×5 inches and a width that is 1.5×2 inches.

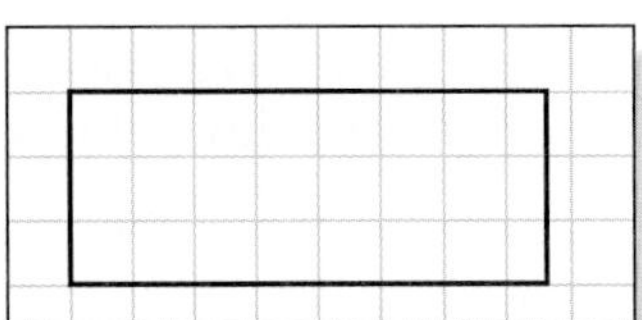

ACTIVITY 2

Step 1 On grid paper, draw a right triangle with legs that measure 3 inches and 4 inches. This is the original figure.

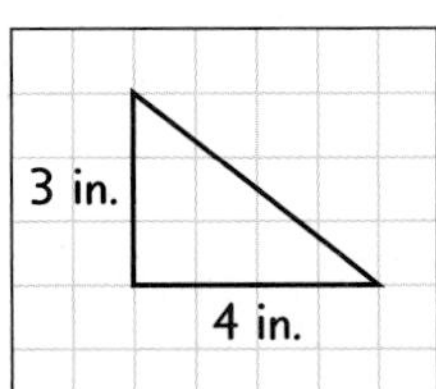

Step 2 Use a scale factor of $\frac{1}{2}$. Draw a new triangle with legs that measure $\frac{1}{2} \times 3$ inches and $\frac{1}{2} \times 4$ inches.

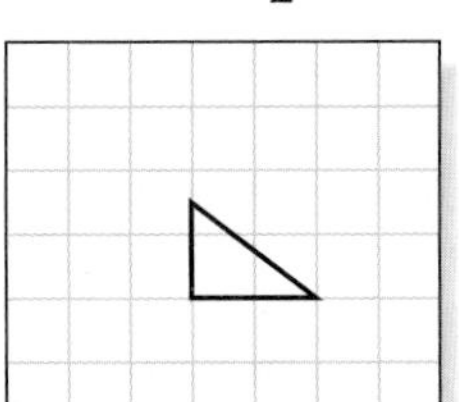

Analyze the Results

1. What are the dimensions of the new rectangle in Activity 1? Is the new rectangle an enlargement or a reduction?
2. What are the dimensions of the new triangle in Activity 2? Is the new triangle an enlargement or a reduction?
3. How do the sizes of the angles appear to compare in each pair of figures? Do you notice any patterns?
4. How do the lengths of the sides of the figures compare?
5. **MAKE A CONJECTURE** Repeat Steps 1 and 2 using different figures and different scale factors. What kinds of scale factors result in an enlargement? reduction?

6-7 Similar Figures

Then

You have already used scales and scale factors to make scale drawings of objects. (Lesson 6-6)

Now

- Find missing measures of similar figures.
- Use scale factors to solve problems.

New Vocabulary

similar figures
corresponding parts
congruent

Math Online

glencoe.com

- Extra Examples
- Personal Tutor
- Self-Check Quiz
- Homework Help

Why?

A fractal is a geometric image that can be divided into parts that are smaller copies of the whole.

a. The image repeats itself infinitely many times. Research the definition of infinity. Does the image represent an infinite pattern?

b. A fractal is considered to be *self-similar*. Using the image, create your own definition of the term *self-similar*.

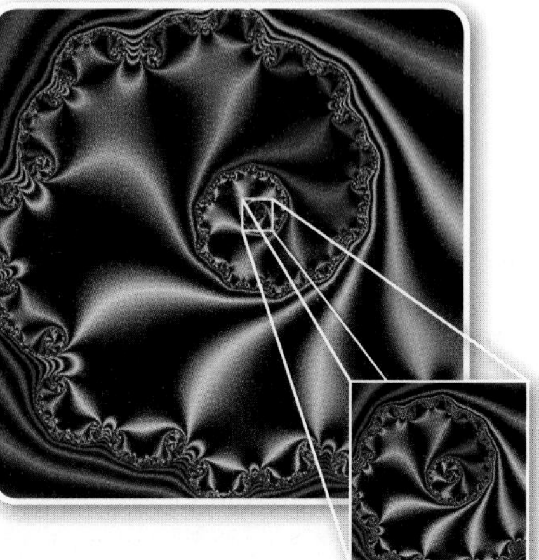

Corresponding Parts of Similar Figures **Similar figures** are figures that have the same shape but not necessarily the same size. Figure *ABCD* is similar to figure *EFGH*. In symbols, $ABCD \sim EFGH$.

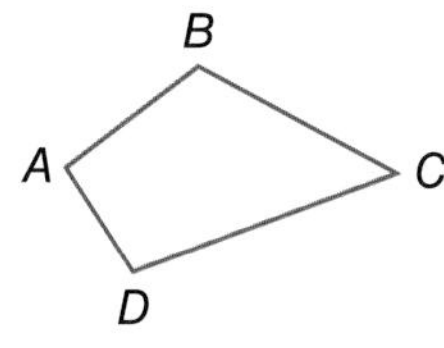

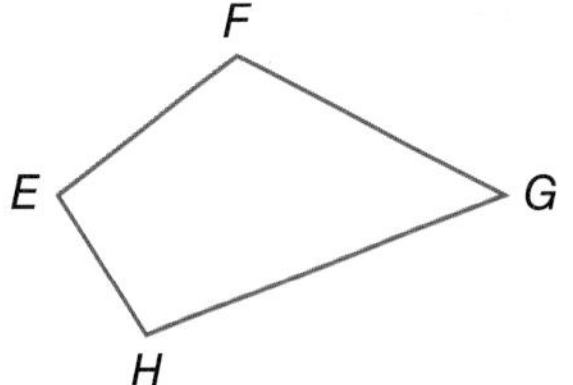

Similar figures have **corresponding parts**. These are angles and sides in the same position.

Corresponding Angles		Corresponding Sides	
$\angle A \leftrightarrow \angle E$	$\angle C \leftrightarrow \angle G$	$\overline{AB} \leftrightarrow \overline{EF}$	$\overline{CD} \leftrightarrow \overline{GH}$
$\angle B \leftrightarrow \angle F$	$\angle D \leftrightarrow \angle H$	$\overline{BC} \leftrightarrow \overline{FG}$	$\overline{DA} \leftrightarrow \overline{HE}$

Key Concept Similar Triangles — For Your FOLDABLE

Words If two figures are similar, then

- the corresponding angles are **congruent**, or have the same measure, and
- the corresponding sides are proportional and opposite corresponding angles

Model

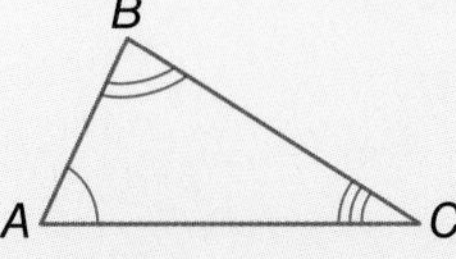

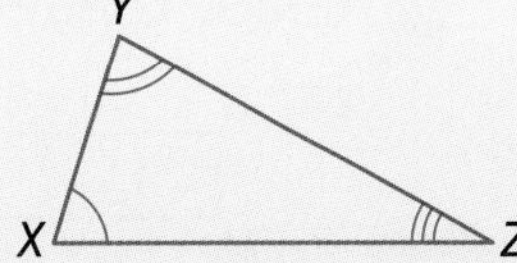

Symbols $\triangle ABC \sim \triangle XYZ$

$\angle A \cong \angle X$, $\angle B \cong \angle Y$, $\angle C \cong \angle Z$ and $\frac{AB}{XY} = \frac{BC}{YZ} = \frac{AC}{XZ}$

The symbol $\cong$ is read *is congruent to*. Arcs are used to show congruent angles.

Since corresponding sides are proportional, you can use a proportion or the scale factor to determine the measures of the sides of similar figures when some measures are known.

EXAMPLE 1 Find Measures of Similar Figures

The figures are similar. Find each missing measure.

a.

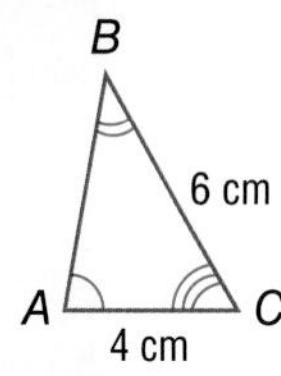

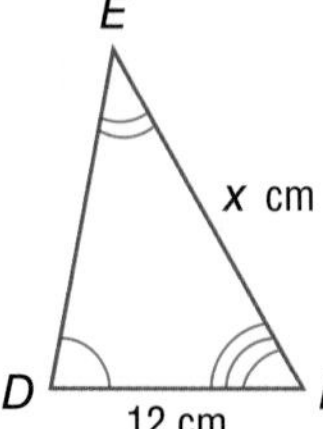

Since $\triangle ABC \sim \triangle DEF$, the corresponding angles are congruent and the corresponding sides are proportional.

$\frac{BC}{EF} = \frac{AC}{DF}$ **Write a proportion.**

$\frac{6}{x} = \frac{4}{12}$ **Replace *BC* with 6, *EF* with *x*, *AC* with 4, and *DF* with 12.**

$6 \cdot 12 = x \cdot 4$ **Find the cross products.**

$72 = 4x$ **Simplify.**

$18 = x$ **Mentally divide each side by 4.**

The length of $\overline{EF}$ is 18 centimeters.

StudyTip

Alternative Method You can also use ratios that compare sides within the figures. In Example 1a, you can use $\frac{6}{4} = \frac{x}{12}$.

b.

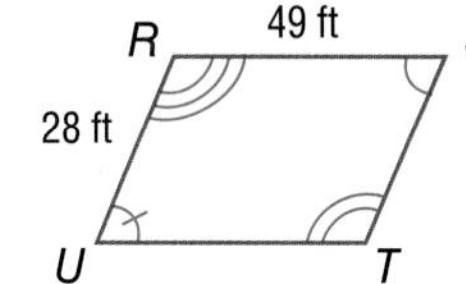

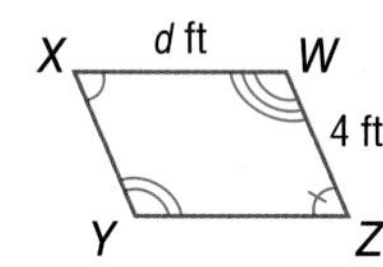

Figure *RSTU* ~ figure *WXYZ*. The corresponding sides are proportional.

$\frac{RS}{WX} = \frac{RU}{WZ}$ **Write a proportion.**

$\frac{49}{d} = \frac{28}{4}$ **Replace *RS* with 49, *WX* with *d*, *RU* with 28, and *WZ* with 4.**

$49 \cdot 4 = d \cdot 28$ **Find the cross products.**

$196 = 28d$ **Simplify.**

$7 = d$ **Divide each side by 28.**

The length of $\overline{WX}$ is 7 feet.

StudyTip

Segment Notation $\overline{RS}$ refers to the actual line segment. *RS* refers to the length of $\overline{RS}$.

Check Your Progress

1A.

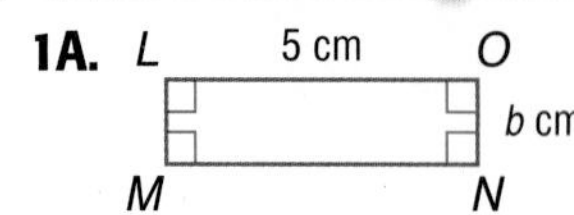

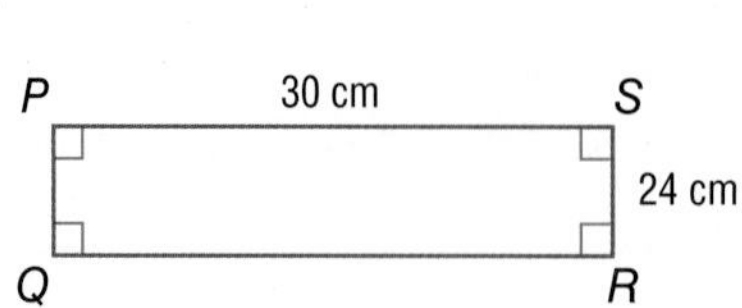

1B.

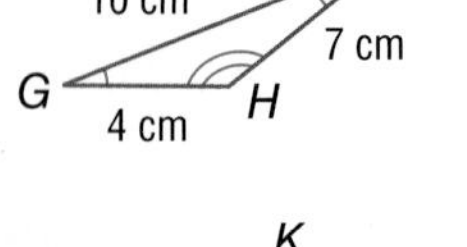

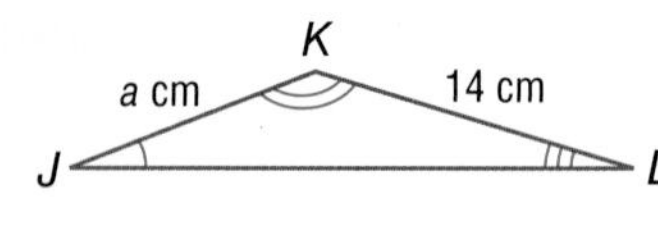

Personal Tutor glencoe.com

StudyTip

Scale Factor The scale factor is the constant of proportionality. In Lesson 6-4, you learned that proportional relationships can be described using equations of the form $y = kx$, where k is the constant of proportionality.

Scale Factors Recall that the scale factor is the ratio of a length on a scale drawing to the corresponding length on the real object. It is also the ratio of corresponding sides in similar figures.

Real-World EXAMPLE 2 Find Measures of Similar Figures

ARCHITECTURE An architect is designing a decorative window for the entrance of a new office building using similar triangles. If $\triangle ABC \sim \triangle DEF$, find the length of segment DF.

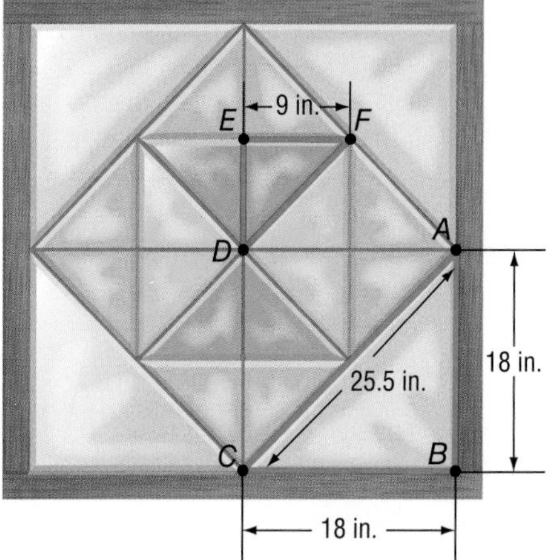

Find the scale factor from $\triangle DEF$ to $\triangle ABC$ by finding the ratio of corresponding sides with known lengths.

scale factor: $\frac{BC}{EF} = \frac{18}{9}$ or 2

Words	2 times a length on triangle *DEF*	is	a corresponding length on triangle *ABC*.
Variable	Let m represent the measure of $\overline{DF}$.		
Expression	$2m$	$=$	25.5

$2m = 25.5$ **Write the equation.**

$m = 12.75$ **Divide each side by 2.**

So, the length of $\overline{DF}$ is 12.75 inches.

Check Your Progress

2. **TILES** A rectangular blue tile has a length of 4.25 inches and a width of 6.75 inches. A similar red tile has a length of 12.75 inches. What is the width of the red tile?

Personal Tutor glencoe.com

Check Your Understanding

Example 1 p. 302

The figures are similar. Find each missing measure.

1. 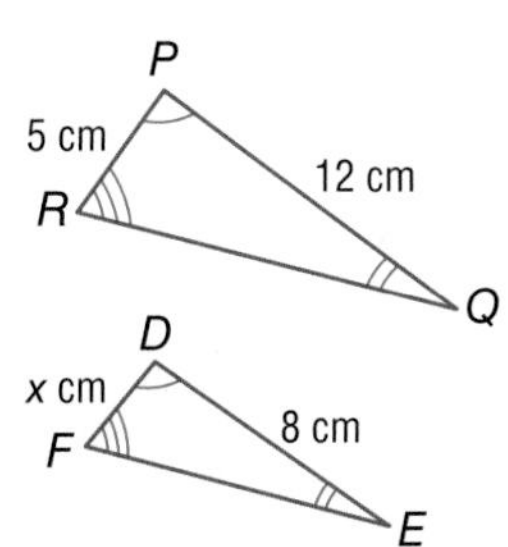

2. 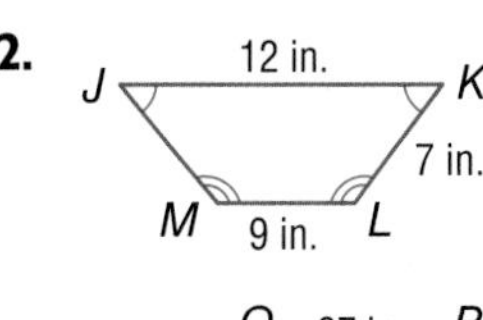

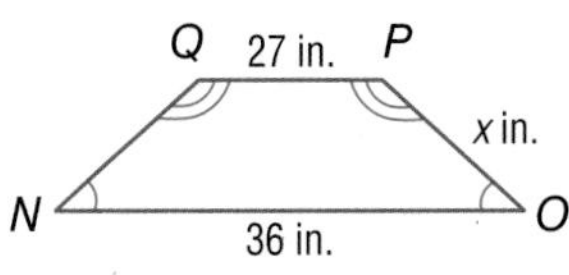

Example 2 p. 303

3. **LOGOS** The logo for an electronics store is made from similar trapezoids as shown. What is the length of the missing measure?

Practice and Problem Solving

= **Step-by-Step Solutions** begin on page R11.
Extra Practice begins on page 810.

Example 1 p. 302

The figures are similar. Find each missing measure.

4.

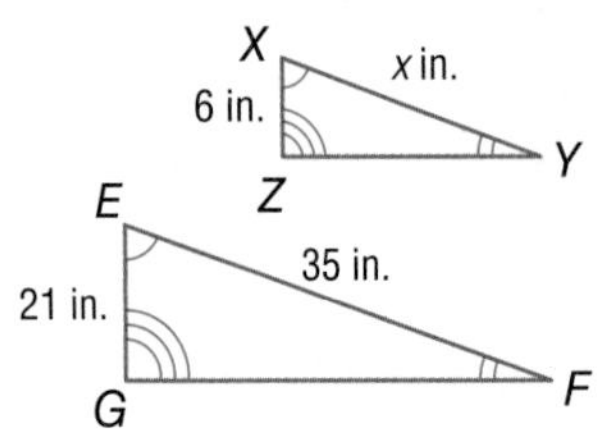

5.

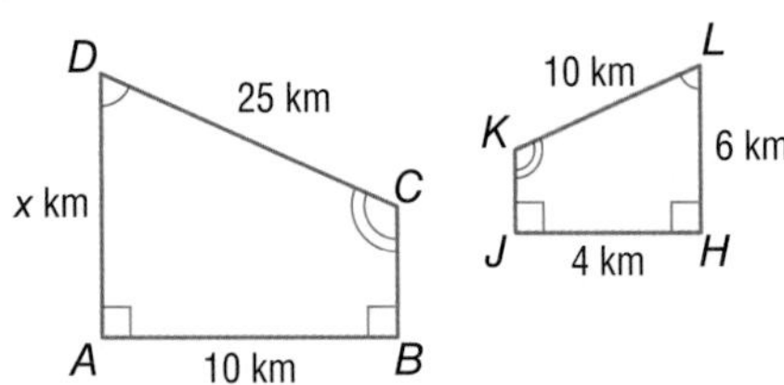

6.

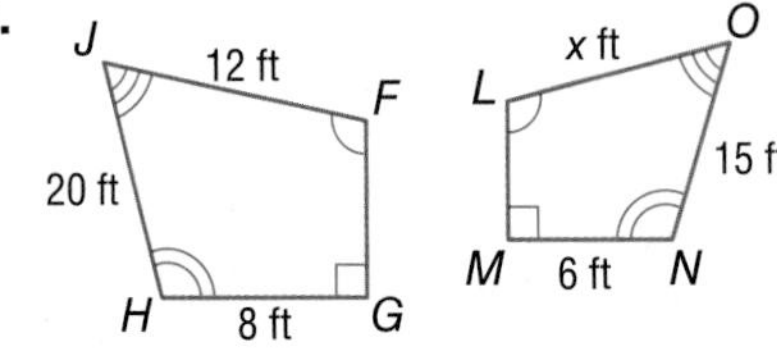

7.

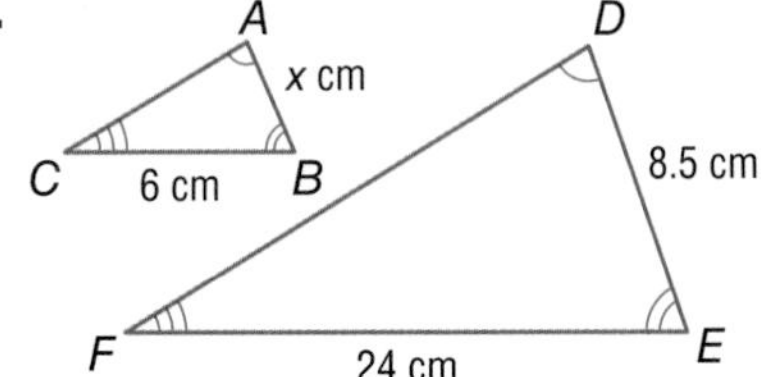

Example 2 p. 303

8. ART The design shown is made using similar triangles and quadrilaterals. triangle A ~ triangle B and quadrilateral C ~ quadrilateral D.

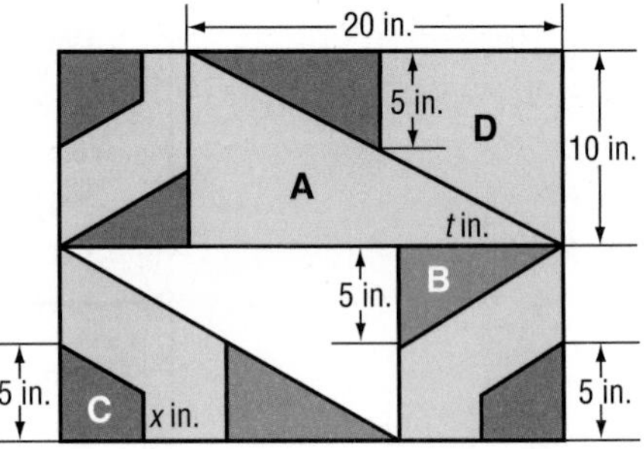

a. Find the missing measure in triangle B.

b. Find the missing measure in quadrilateral C.

9. GEOMETRY Triangle *LMN* is similar to $\triangle RST$. What is the value of *LN* if *RT* is 9 inches, *MN* is 21 inches, and *ST* is 7 inches?

10. GEOMETRY Quadrilateral *ABCD* is similar to quadrilateral *WXYZ*. What is the value of *WZ* if *AD* is 18 feet, *CD* is 27 feet, and *YZ* is 10.8 feet?

11. Using a scale factor of $\frac{2}{3}$, draw and label a rectangle similar to rectangle *ABCD*.

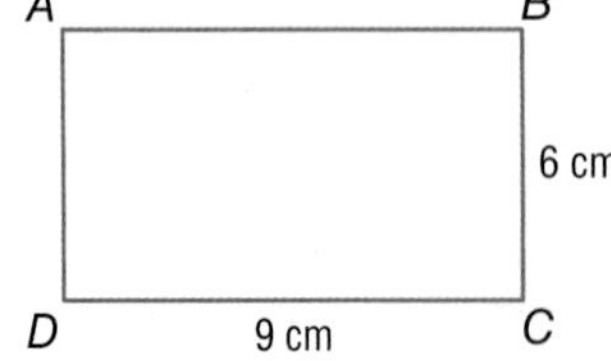

12. Using a scale factor of $\frac{4}{3}$, draw and label a triangle similar to $\triangle XYZ$.

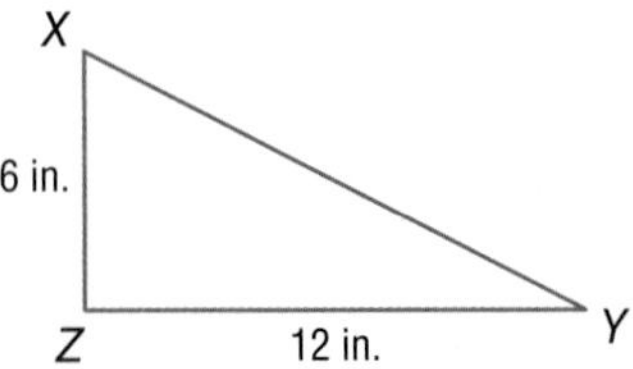

13. PERIMETER Figure *FGHJK* ~ figure *LMNPQ*. The scale factor from figure *FGHJK* to figure *LMNPQ* is $\frac{2}{3}$. What is the perimeter of figure *LMNPQ*?

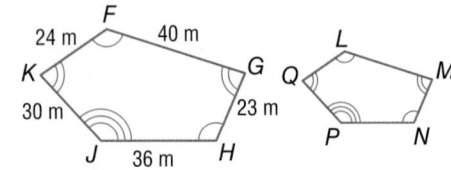

Real-World Link

The film most commonly used in movie productions is called 35 millimeter film because it is 35 millimeters wide. It was first introduced in 1892.

14. MOVIES An image that is projected onto a movie screen measures 8.8 meters by 6.4 meters. The projection is similar to the individual frame on the movie reel. If the projection has a scale factor of 400, what are the original dimensions of a frame on a movie reel?

15 **MULTIPLE REPRESENTATIONS** In this problem, you will investigate the relationship between the perimeters of similar figures. In the figures at the right, $\triangle ABC \sim \triangle XYZ$.

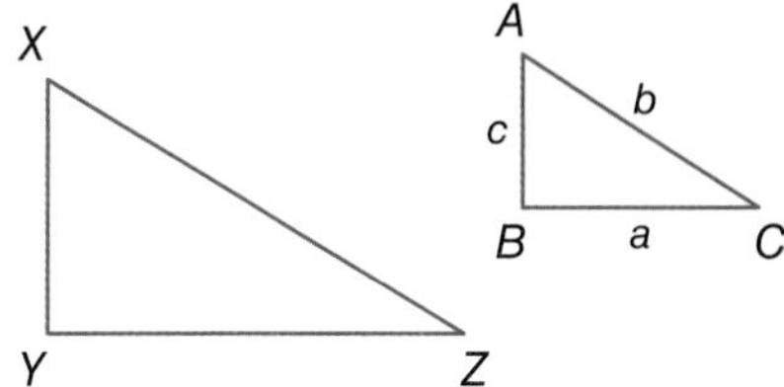

a. GEOMETRIC Write an expression for the perimeter of $\triangle ABC$.

b. SYMBOLIC If the scale factor is represented by d, write algebraic expressions for the measures of the sides of $\triangle XYZ$.

c. GEOMETRIC Write an expression for the perimeter of $\triangle XYZ$.

d. SYMBOLIC Use the Distributive Property to factor the expression from part **c**. Explain the meaning of the expression.

e. MAKE A PREDICTION Suppose $AB = 3$ inches, $BC = 4$ inches, $AC = 5$ inches, and the scale factor from $\triangle ABC$ to $\triangle XYZ$ is 2. Find the perimeter of $\triangle XYZ$ without calculating the lengths of $\overline{XY}$, $\overline{YZ}$, and $\overline{XZ}$. Justify your procedure.

f. VERBAL Explain how the perimeters of similar figures are related to the scale factor.

H.O.T. Problems

Use Higher-Order Thinking Skills

16. OPEN ENDED Draw two similar triangles whose scale factor is 1:3. Justify your answer.

REASONING Determine whether each statement is *always*, *sometimes*, or *never* true. Explain your reasoning.

17. All rectangles are similar.

18. All squares are similar.

19. FIND THE ERROR Carla and Tony are finding the length of $\overline{AB}$ where $\triangle ABC \sim \triangle DEF$, $BC = 16$ feet, $EF = 12$ feet, and $DE = 18$ feet. Is either of them correct? Explain your reasoning.

Carla

$$\frac{16}{18} = \frac{12}{x}$$

$x = 13.5$ ft

Tony

$$\frac{16}{12} = \frac{18}{x}$$

$x = 24$ ft

20. CHALLENGE *True* or *false*? If $\triangle XYZ \sim \triangle RST$, then $\frac{x}{z} = \frac{r}{t}$, where x is the side opposite $\angle X$, z is the side opposite $\angle Z$, and so on. Justify your answer. If false, provide a counterexample.

21. WRITING IN MATH Suppose you have two triangles. Triangle A is similar to triangle B, and the measures of the sides of triangle A are less than the measures of the sides of triangle B. The scale factor is 0.25. Which is the original triangle? Explain.

Standardized Test Practice

22. Quadrilateral $ABCD$ is similar to quadrilateral $EFGH$. What is the length of $\overline{FG}$?

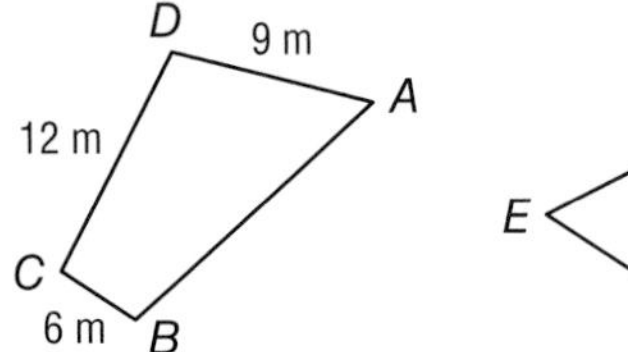

A 1.5 m **C** 4 m

B 3.8 m **D** 5.3 m

23. If polygon $ABCDE$ is similar to polygon $FGHIJ$, which of the following is *not* true?

F $\angle ABC \cong \angle FGH$

G $\angle EDC \cong \angle JIH$

H $\overline{AE}$ corresponds to $\overline{FJ}$

J $\overline{BC}$ corresponds to $\overline{HI}$

24. Triangle RST is similar to triangle XYZ.

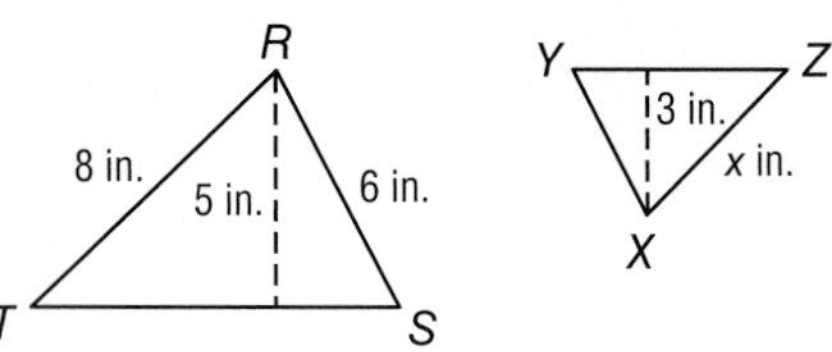

What is the length of $\overline{XZ}$?

A 1.875 in. **C** 4.8 in.

B 3.75 in. **D** 9.6 in.

25. GRIDDED RESPONSE Triangle LMN has a perimeter of 24 centimeters and is similar to triangle DEF. If the scale factor relating $\triangle LMN$ to $\triangle DEF$ is $\frac{1}{3}$, what is the perimeter, in centimeters of $\triangle DEF$?

Spiral Review

26. STATUES The Statue of Zeus at Olympia is one of the Seven Wonders of the World. On a scale model of the statue, the height of Zeus is 8 inches. (Lesson 6-6)

a. If the actual height of the statue is 40 feet, what is the scale?

b. What is the scale factor?

27. JEWELRY Flor is making bracelets. She knows that 4 bags of beads will make 14 bracelets. If she wants to make 56 bracelets, how many bags of beads will she need? (Lesson 6-5)

28. APPLE CIDER A bushel of apples will make approximately 3 gallons of apple cider. The table shows the relationship between the number of bushels of apples and the number of gallons of apple cider. (Lesson 1-5)

a. Given b, the number of bushels needed, write an equation that can be used to find g, the number of gallons of apple cider.

b. How many bushels are needed to make 54 gallons of cider?

Apple Bushels, b	Gallons of Apple Cider, g
1	3
2	6
5	15
8	24

Skills Review

Multiply. (Lesson 3-3)

29. $\frac{4}{5} \times 5$ **30.** $\frac{7}{4} \times 3$ **31.** $\frac{2}{3} \times 6$ **32.** $\frac{8}{5} \times 9$

6-8 Dilations

Then

You have already used transformations such as translations and reflections in the coordinate plane. (Lesson 2-7)

Now

- Graph dilations on a coordinate plane.
- Find the scale factor of a dilation.

New Vocabulary

dilation

Math Online

glencoe.com

- Extra Examples
- Personal Tutor
- Self-Check Quiz
- Homework Help

Why?

As a gift for her grandparents, Sierra is enlarging a favorite 4" inch by 5" inch family photo on her computer.

a. Suppose she enlarges it to 8 by 10 inches. Compare and contrast the original photo and the enlargement.

b. What is the scale factor from the original photo to the enlargement?

Dilations Recall that a transformation maps an original figure onto an image. Translations and reflections are two transformations that change the position of the object. A **dilation** is a transformation that enlarges or reduces a figure by a scale factor.

TRANSLATION

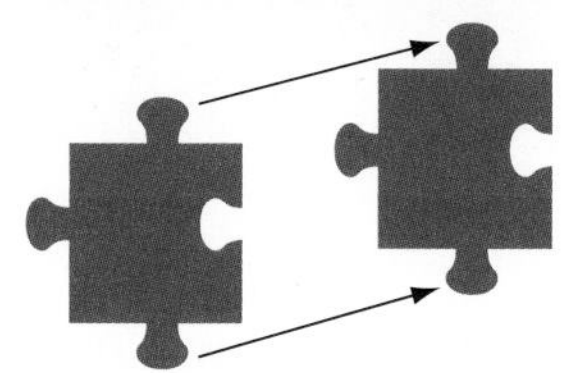

REFLECTION

DILATION

When the center of a dilation is the origin, you can find the coordinates of the image by multiplying each coordinate of the figure by the scale factor. Use the notation $(x, y) \rightarrow (kx, ky)$ to describe the dilation.

EXAMPLE 1 Dilation in a Coordinate Plane

A figure has vertices $J(2, 4)$, $K(2, 6)$, $M(8, 6)$, and $N(8, 2)$. Graph the figure and the image of the polygon after a dilation with a scale factor of $\frac{1}{2}$.

The dilation is $(x, y) \rightarrow \left(\frac{1}{2}x, \frac{1}{2}y\right)$. Multiply the coordinates of each vertex by $\frac{1}{2}$. Then graph both figures on the same coordinate plane.

$J(2, 4) \rightarrow J'\left(\frac{1}{2} \cdot 2, \frac{1}{2} \cdot 4\right) \rightarrow J'(1, 2)$

$K(2, 6) \rightarrow K'\left(\frac{1}{2} \cdot 2, \frac{1}{2} \cdot 6\right) \rightarrow K'(1, 3)$

$M(8, 6) \rightarrow M'\left(\frac{1}{2} \cdot 8, \frac{1}{2} \cdot 6\right) \rightarrow M'(4, 3)$

$N(8, 2) \rightarrow N'\left(\frac{1}{2} \cdot 8, \frac{1}{2} \cdot 2\right) \rightarrow N'(4, 1)$

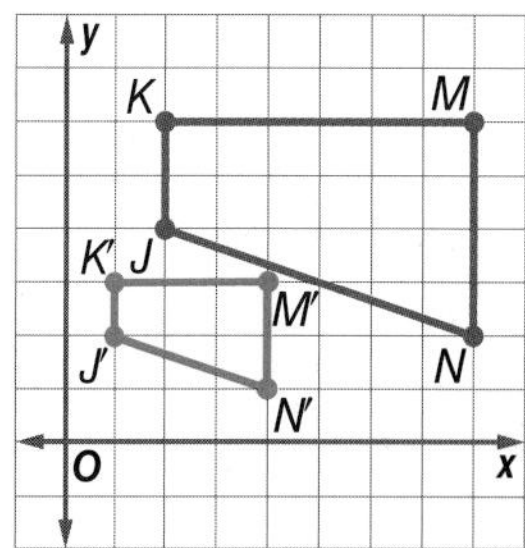

Check Your Progress

1. A figure has vertices $R(-3, 6)$, $S(3, 12)$, and $T(3, 3)$. Graph the figure and the image of the figure after a dilation with a scale factor of $\frac{1}{3}$.

Personal Tutor glencoe.com

Vocabulary Link

Dilate
Everyday Use to expand or widen
Math Use to enlarge or reduce by a scale factor

STANDARDIZED TEST EXAMPLE 2

A triangle has vertices $A(1, -2)$, $B(0, 2)$, and $C(-2, -1)$. Find the coordinates of the triangle after a dilation with a scale factor of 2.

A $A'\left(\frac{1}{2}, -1\right), B'(0, 1), C'\left(-1, -\frac{1}{2}\right)$ **C** $A'(2, -4), B'(0, 4), C'(-4, -2)$

B $A'(1, -1), B'(0, 1), C'(-1, -1)$ **D** $A'(2, -4), B'(0, 2), C'(-4, -2)$

Read the Test Item

You are asked to find the coordinates of the image after the dilation. Multiply the coordinates of each vertex by the scale factor.

Solve the Test Item

The dilation is $(x, y) \rightarrow (2x, 2y)$. To dilate the figure, multiply the coordinates of each vertex by 2.

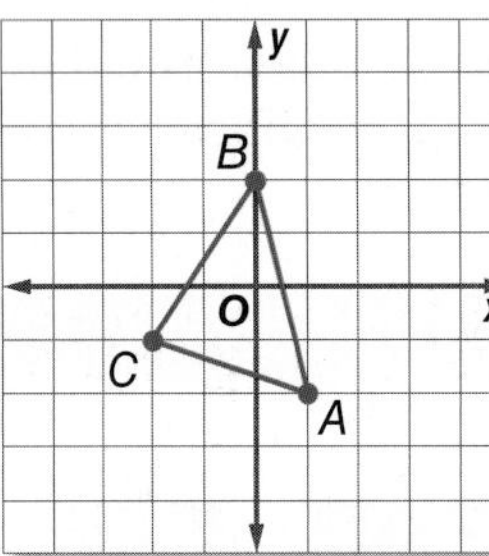

$A(1, -2) \rightarrow (2 \cdot 1, 2 \cdot -2) \rightarrow A'(2, -4)$

$B(0, 2) \rightarrow (2 \cdot 0, 2 \cdot 2) \rightarrow B'(0, 4)$

$C(-2, -1) \rightarrow (2 \cdot -2, 2 \cdot -1) \rightarrow C'(-4, -2)$

The answer is C.

Check Graph the triangle and its image on the coordinate plane.

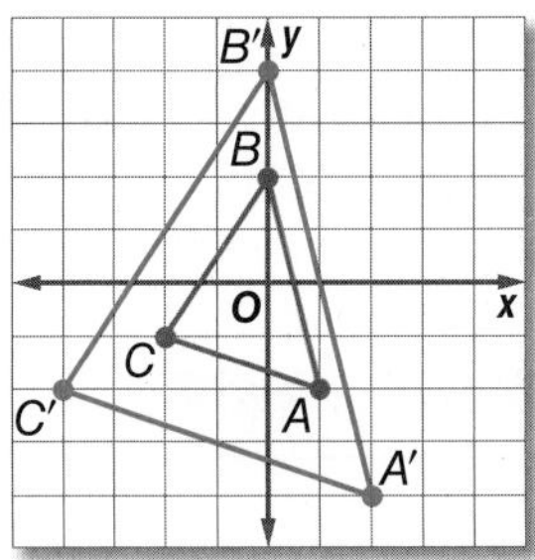

Problem-Solving Tip

Draw a Diagram Draw a diagram when the problem situation involves spatial reasoning or geometric figures. In this example, drawing a diagram helps you check that the coordinates are correct.

Check Your Progress

2. A figure has vertices $X(-3, 6)$, $Y(3, 0)$, and $Z(3, 3)$. Find the coordinates of the figure after a dilation with a scale factor of 3.

F $X'(0, 9), Y'(6, 3), Z'(6, 6)$ **H** $X'(-1, 2), Y'(1, 0), Z'(1, 1)$

G $X'(-6, -3), Y'(0, -3), Z'(0, 0)$ **J** $X'(-9, 18), Y'(9, 0), Z'(9, 9)$

Personal Tutor glencoe.com

Concept Summary — Scale Factor

For Your FOLDABLE

A dilation with a scale factor of k will be:

- an enlargement if $k > 1$,
- a reduction if $0 < k < 1$,
- the same as the original figure if $k = 1$.

StudyTip

Similarity Dilations produce images that are similar (same shape, but not same size).

EXAMPLE 3 Find a Scale Factor

MASCOTS A drawing of a school mascot is to be increased as shown at the right so it can be painted on the gymnasium doors. What is the scale factor of the dilation?

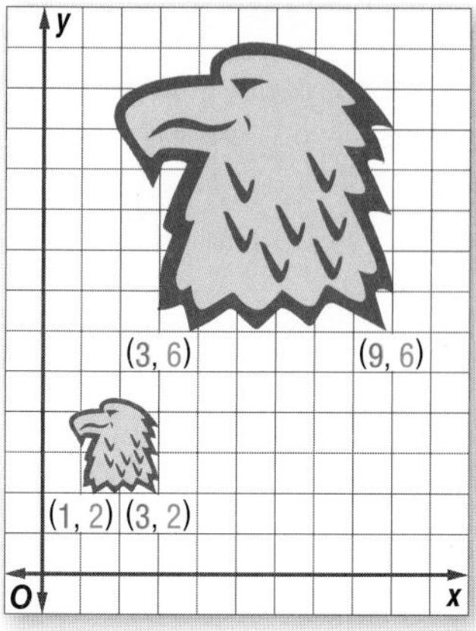

Write a ratio comparing the lengths of the sides of the two images. Subtract the x-coordinates to find the lengths.

$$\frac{\text{length on dilation}}{\text{length on original}} = \frac{9-3}{3-1}$$

$$= \frac{6}{2} \text{ or } 3$$

So, the scale factor of the dilation is 3.

Check Your Progress

3. **PHOTOS** Megan wants to reduce an 8-by-10-inch photo to a 2-by-$2\frac{1}{2}$-inch photo. What is the scale factor of the dilation?

Personal Tutor glencoe.com

Check Your Understanding

Example 1 p. 307

Find the vertices of each figure after a dilation with the given scale factor k. Then graph the image.

1. $k = \frac{1}{2}$

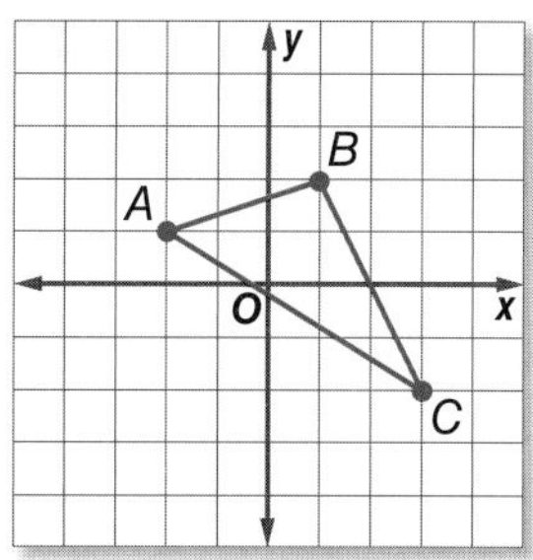

2. $k = \frac{1}{4}$

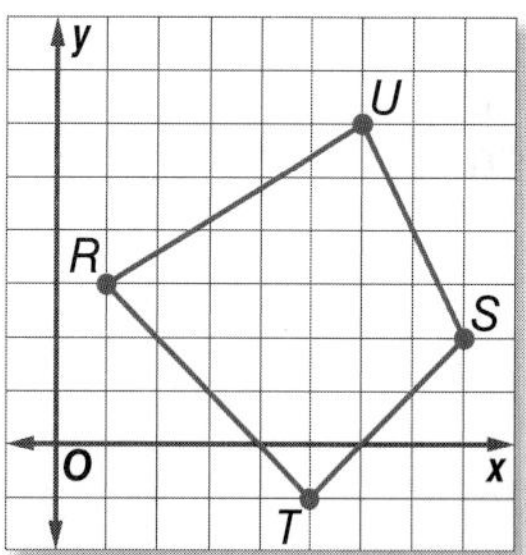

Example 2 p. 308

3. **MULTIPLE CHOICE** A square has vertices $M(0, 0)$, $N(3, -3)$, $O(0, -6)$ and $P(-3, -3)$. Find the coordinates of the square after a dilation with a scale factor of 2.5.

A $M'(0, 0)$, $N'(7.5, -7.5)$, $O'(0, -15)$, $P'(-7.5, -7.5)$

B $M'(0, 0)$, $N'(2, -2)$, $O'(0, -4)$, $P'(-2, -2)$

C $M'(2.5, 2.5)$, $N'(7.5, -7.5)$, $O'(0, -15)$, $P'(-7.5, -7.5)$

D $M'(2.5, 2.5)$, $N'(2, -2)$, $O'(0, -4)$, $P'(-2, -2)$

Example 3 p. 309

4. **IMAGES** Jorge is using a photocopier to reduce a poster that is 8 inches by 14 inches to 5 inches by $8\frac{3}{4}$ inches. What is the scale factor of the dilation?

Practice and Problem Solving

● = **Step-by-Step Solutions** begin on page R11.
Extra Practice begins on page 810.

Examples 1 and 2
pp. 307–308

Find the vertices of each figure after a dilation with the given scale factor *k*. Then graph the image.

5. $k = 4$

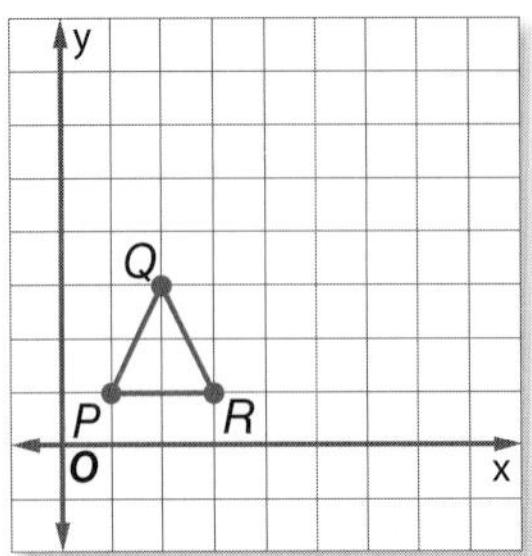

6. $k = 1.5$

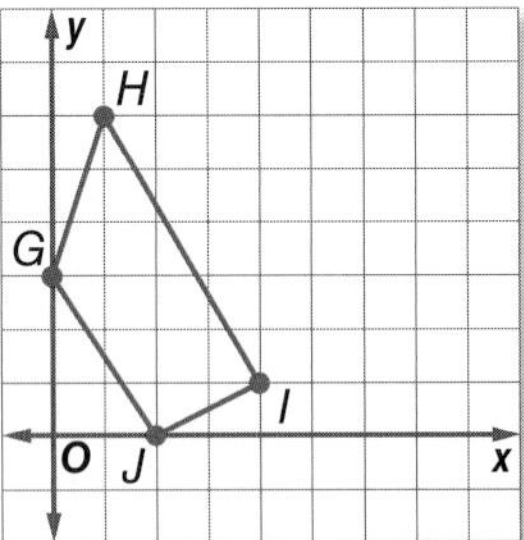

7. $k = \frac{3}{4}$

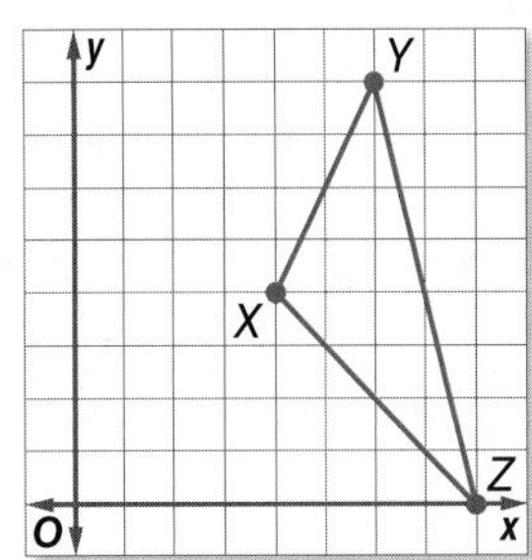

8. $k = \frac{1}{3}$

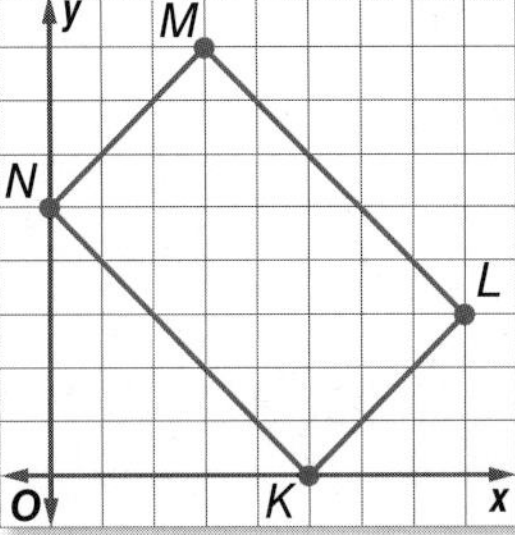

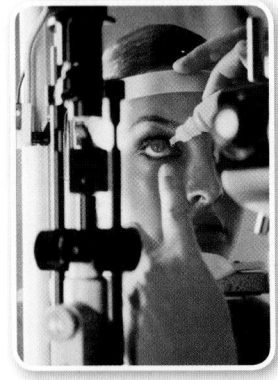

Real-World Link

Optometrists dilate eyes so that they can get a better look at the retina, optic nerve, and blood vessels in the back of your eye.

Source: St. Luke's Cataract and Laser Institute.

Find the vertices of each figure after a dilation with the given scale factor, *k*. Then graph the original image and the dilation.

9. $G(-1, 1), H(2, 1), J(3, -2), K(-2, -2); k = 2$

10. $W(0, 0), X(5, -5), Y(5, 10); k = \frac{2}{5}$

11. $R(-1, 2), S(1, 4), T(1, 1); k = 3$

12. $A(-3, -5), B(0, 2), C(3, -1), D(0, -4); k = \frac{1}{2}$

Example 3
p. 309

13. **LIFE SCIENCE** In a microscope, the image of a 0.16-millimeter paramecium appears to be 32 millimeters long. What is the scale factor of the dilation?

14. **EYES** During an eye exam, an optometrist dilates her patient's pupils to 7 millimeters. If the diameter of the pupil before dilation was 4 millimeters, what is the scale factor of the dilation?

15 **PHOTOGRAPHY** Jung is editing a digital photograph that is 640 pixels wide and 480 pixels high on his computer monitor.

a. If Jung zooms the image on his monitor 150%, what are the dimensions of the image?

b. Suppose that Jung is going to use the photograph in a design and wants the image to be 32 pixels wide. What scale factor should he use?

c. Jung resizes the photograph so it is 600 pixels high. What scale factor did he use?

16. GEOMETRY Triangle $L'M'N'$ is a dilation of triangle LMN. Find the scale factor of the dilation and classify it as an enlargement or a reduction.

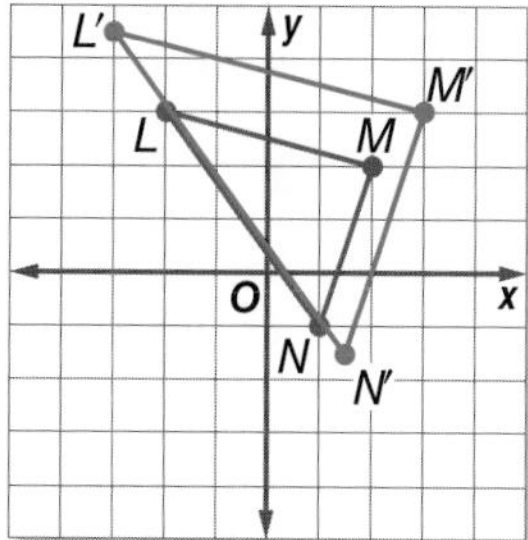

17. GEOMETRY Quadrilateral $A'B'C'D'$ is a dilation of quadrilateral $ABCD$. Find the scale factor of the dilation and classify it as an enlargement or a reduction.

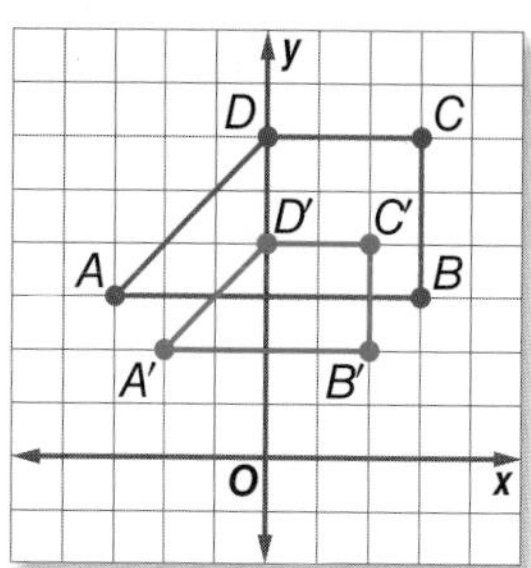

18. PRESENTATIONS Felicia wants to project a 2-inch by 2-inch slide on to a wall to create an image 128 inches by 128 inches. If the slide projector makes the dimensions of the image twice as large for each yard that it is moved away from the wall, how far away should Felicia place the projector?

H.O.T. Problems

Use Higher-Order Thinking Skills

19. OPEN ENDED Draw a triangle on grid paper. Then draw the image of the triangle after it is moved 5 units right and then dilated by a scale factor of $\frac{1}{3}$.

20. CHALLENGE Suppose a figure $ABCD$ is dilated by a scale factor of $\frac{1}{2}$ and then reflected over the x-axis and the y-axis. The final image is shown on the graph at the right. Graph the original image and list the coordinates of points A, B, C, and D.

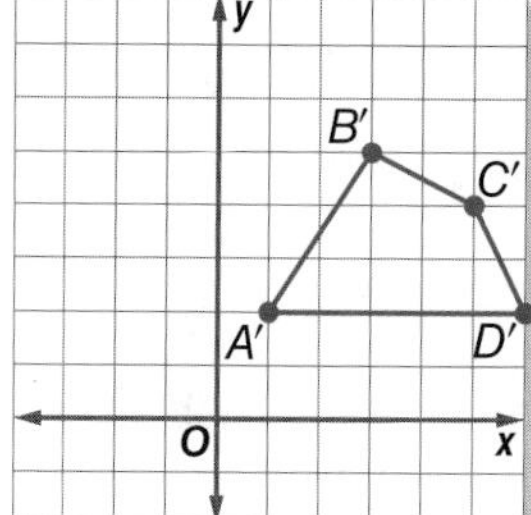

21. REASONING A triangle has coordinates $A(0, 5)$, $B(0, 0)$ and $C(10, 5)$. Find the coordinates of the final image after each set of dilations.

a. $k = 2$ followed by $k = \frac{2}{5}$

b. $k = \frac{2}{5}$ followed by $k = 2$

c. $k = \frac{1}{4}$ followed by $k = 2$

d. $k = 2$ followed by $k = \frac{1}{4}$

22. REASONING Refer to Exercise 21. Does the order in which you perform multiple dilations *always*, *sometimes*, or *never* result in the same image? Explain.

23. WHICH ONE DOESN'T BELONG? Which pair of points does not represent a dilation with center at the origin? Explain.

$A(2, 3), A'(4, 6)$	$B(4, 6), B'(2, 3)$	$C(2, 5), C'(4, 7)$	$D(-2, 4), D'(-1, 2)$

24. WRITING IN MATH A triangle has one vertex at point (3, 6). It is dilated with a center at the origin by a scale factor of 3. The resulting image is then dilated with a scale factor of $\frac{1}{3}$. What are the coordinates of that vertex after both dilations? Explain your reasoning.

Problem-Solving Tip

Draw a Picture In Exercise 24, you can draw a picture of the situation to help see each transformation. This can help you write a more concise response.

Standardized Test Practice

25. Quadrilateral $PQRS$ was dilated to form quadrilateral $P'Q'R'S'$.

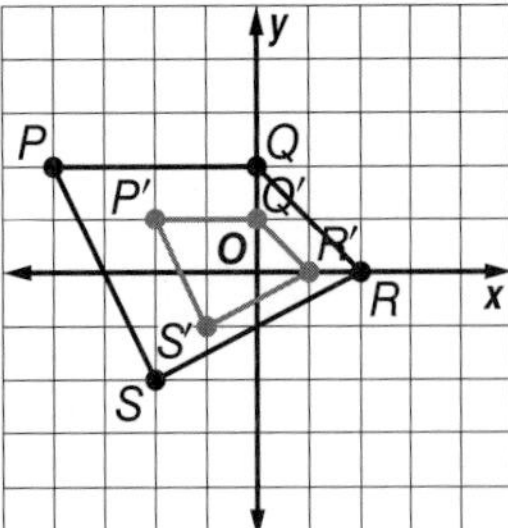

Which number *best* represents the scale factor used to dilate quadrilateral $PQRS$ to quadrilateral $P'Q'R'S'$?

A -2 **C** $\frac{1}{2}$

B $-\frac{1}{2}$ **D** 2

26. A triangle has vertices (1, −1), (−2, 6), and (4, 1). If the triangle undergoes a dilation with a scale factor of 3, what will be the vertices of the image?

F (3, −3), (−6, 18), (12, 3)

G (3, 3), (6, 18), (12, 3)

H (3, 3), (−6, 18), (12, −3)

J (3, −3), (−6, −18), (−12, 3)

27. Let $G(6, -8)$ be a point on triangle FGH. What are the coordinates of G' if the triangle is dilated by a scale factor of $\frac{1}{2}$?

A (3, 4) **C** (−3, 4)

B (3, −4) **D** (−3, −4)

28. **GRIDDED RESPONSE** Point $P(-10, 5)$ is on rectangle $PQRS$. If the coordinates of P after a dilation are $(-4, 2)$, what is the scale factor of the dilation?

Spiral Review

29. **GEOMETRY** Triangle RSK is similar to $\triangle RUV$. What is the value of x, rounded to the nearest tenth? (Lesson 6-7)

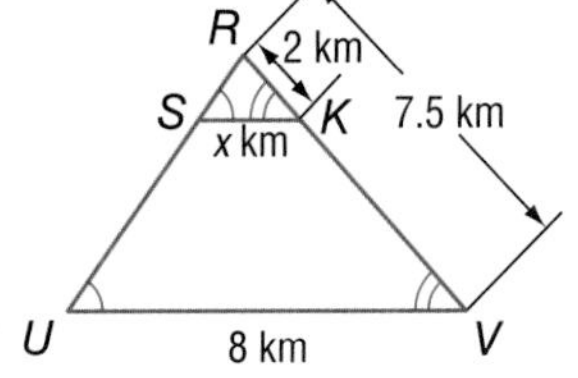

MAPS On a map of South Carolina, the scale is 1 inch = 20 miles. Find the actual distance for each map distance. (Lesson 6-6)

	From	To	Map Distance
30.	Columbia	Florence	4 inches
31.	Myrtle Beach	Greenville	12 inches

32. **CURRENCY** The table at the right shows the exchange rates for certain countries compared to the U.S. dollar on a given day. (Lesson 6-3)

Country	Rate per $1 (U.S.)
Great Britain	0.480 pounds
South Africa	6.568 rands

a. What is the cost of an item in U.S. dollars if it costs 14.99 in British pounds?

b. Find the cost of an item in South African rands if it costs 12.50 in U.S. dollars.

Skills Review

Solve each proportion. (Lesson 6-5)

33. $\frac{2}{15} = \frac{c}{72}$ **34.** $\frac{16}{7} = \frac{4.8}{h}$ **35.** $\frac{2}{9.4} = \frac{0.2}{v}$ **36.** $\frac{9}{7.2} = \frac{3.5}{k}$

6-9 Indirect Measurement

Why?

On a sunny day, you can see your shadow. The lengths of the shadows around you are proportional to the heights of the objects casting the shadows.

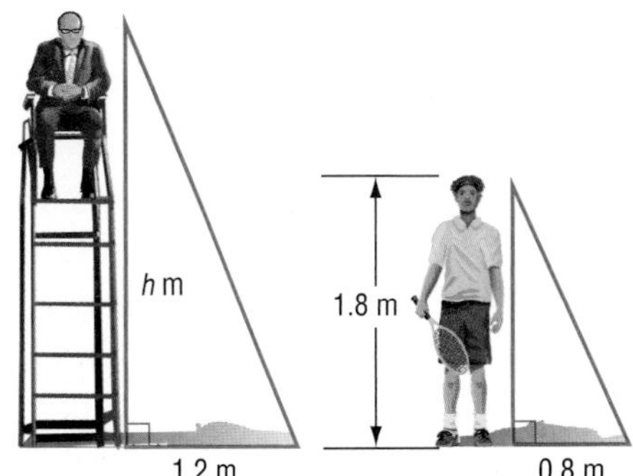

a. What kind of triangles are formed by the objects and their shadows?

b. **MAKE A CONJECTURE** How could you use your shadow to determine the height of another object?

Then

You have already used similar figures to determine missing measures. (Lesson 6-7)

Now

- Solve problems involving indirect measurement using shadow reckoning.
- Solve problems using surveying methods.

New Vocabulary

indirect measurement

Math Online

glencoe.com

- Extra Examples
- Personal Tutor
- Self-Check Quiz
- Homework Help
- Math in Motion

Indirect Measurement **Indirect measurement** allows you to use the properties of similar triangles to find measurements that are difficult to measure directly.

One type of indirect measurement is called *shadow reckoning*. Two objects and their shadows form two sides of similar triangles.

Math *in Motion*, BrainPOP® glencoe.com

Real-World EXAMPLE 1 Use Shadow Reckoning

MEMORIALS The lead statue of the Korean War Memorial in Washington, D.C., casts a 43.5-inch shadow at the same time a nearby tourist casts a 32-inch shadow. If the tourist is 64 inches tall, how tall is the lead statue?

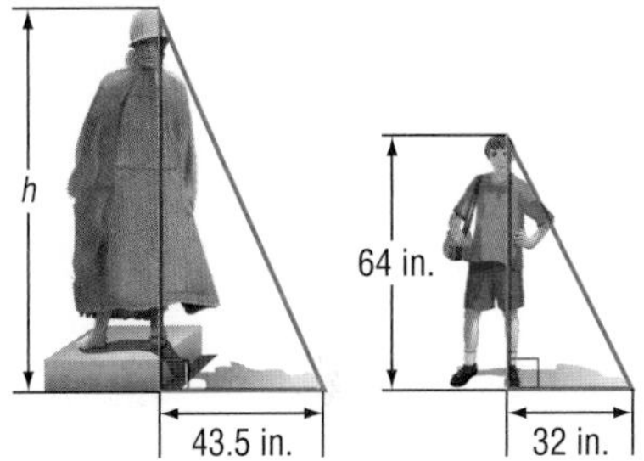

Understand You know the lengths of the shadows and the height of the tourist. You need to find the statue's height.

Plan To find the height of the statue, set up a proportion comparing the tourist's shadow to the statue's shadow. Then solve.

Solve tourist's height → / statue's height → $\frac{64}{h} = \frac{32}{43.5}$ ← tourist's shadow / ← statue's shadow

$64 \cdot 43.5 = 32 \cdot h$ **Find the cross products.**

$2784 = 32h$ **Multiply.**

$87 = h$ **Divide each side by 32.**

Check The tourist's height is 2 times the length of his or her shadow. The statue should be 2 times its shadow, or $2 \cdot 43.5$, which is 87 inches.

✓ Check Your Progress

1. **MONUMENTS** Suppose a bell tower casts a 27.6-foot shadow at the same time a nearby tourist casts a 1.2-foot shadow. If the tourist is 6 feet tall, how tall is the tower?

Personal Tutor glencoe.com

Surveying Methods Surveyors also use similar triangles, but they do not involve shadows. Notice in Example 2 that it is possible to measure three sides of the triangles directly.

EXAMPLE 2 Find Missing Measures

MAPS In the figure, $\triangle STU \sim \triangle VQU$. Find the distance across the pond.

Since the figures are similar, corresponding sides are proportional.

$\frac{UQ}{UT} = \frac{VQ}{ST}$	**Write a proportion.**
$\frac{16}{10} = \frac{x}{20}$	**$UQ = 16$, $UT = 10$, $VQ = x$, and $ST = 20$**
$10 \cdot x = 16 \cdot 20$	**Cross products**
$10x = 320$	**Multiply.**
$\frac{10x}{10} = \frac{320}{10}$	**Divide each side by 10.**
$x = 32$	**Simplify.**

So, the distance across the pond is 32 meters.

2. **CAMP** In the figure, $\triangle QRS \sim \triangle URT$. Find the distance from the boys' cabins to the Mess Hall.

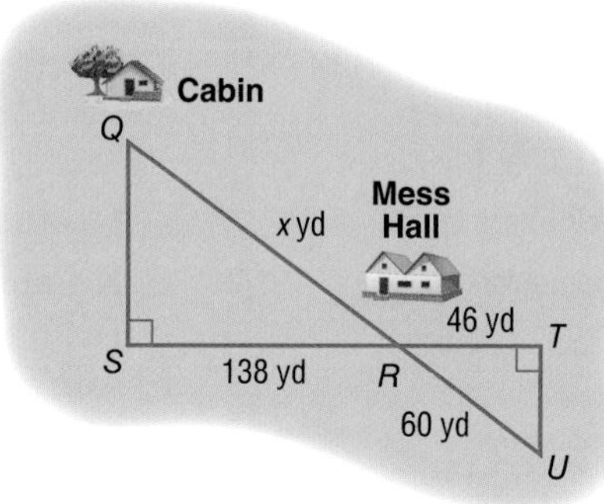

Personal Tutor glencoe.com

Check Your Understanding

Examples 1 and 2 pp. 313–314

1. **SPORTS** A basketball hoop in Miguel's backyard casts a shadow that is 8 feet long. At the same time, Miguel casts a shadow that is 4.5 feet long. If Miguel is 5.5 feet tall, how tall is the basketball hoop? Round to the nearest tenth.

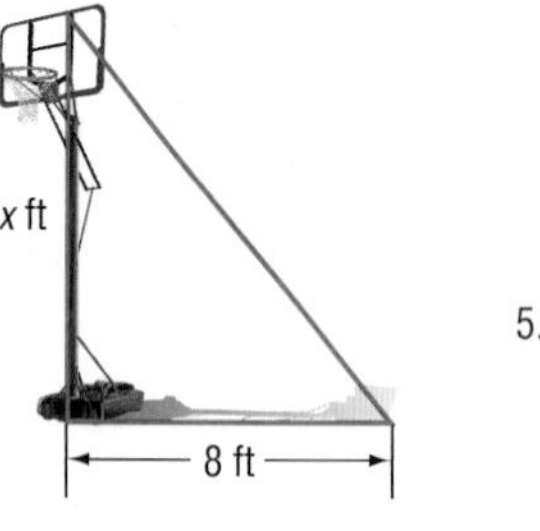

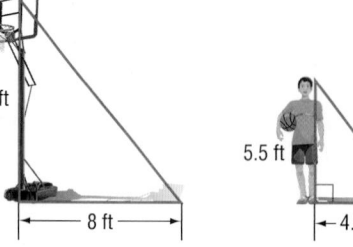

2. **SURVEYING** In the figure, $\triangle ABC \sim \triangle EBD$. Find the distance across Stallion Ravine.

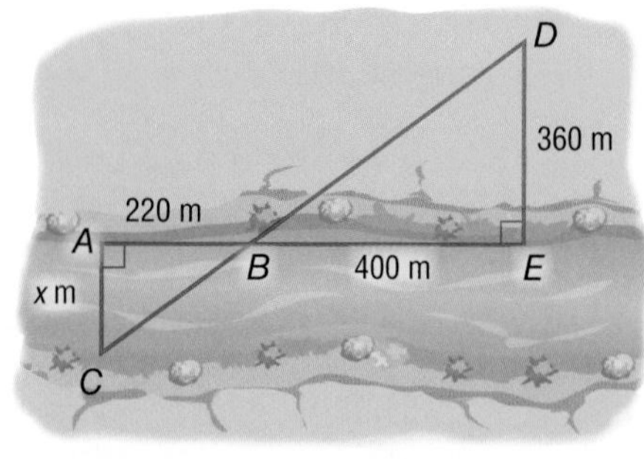

Practice and Problem Solving

Example 1 p. 313

3. **FLAGS** A flagpole is 30 feet high and a mailbox is 3.5 feet high. The mailbox casts a shadow that is 5.25 feet long. How long is the flagpole's shadow at the same time?

4. **ARCHITECTURE** The height of Medina Middle School is 25 feet tall. A mail service drop box outside the school is 4 feet tall. The drop box casts a shadow that is 6 feet long. At the same time, what is the length of the shadow of the school building?

Example 2 p. 314

5. **BRIDGES** The triangles below are similar. Find x.

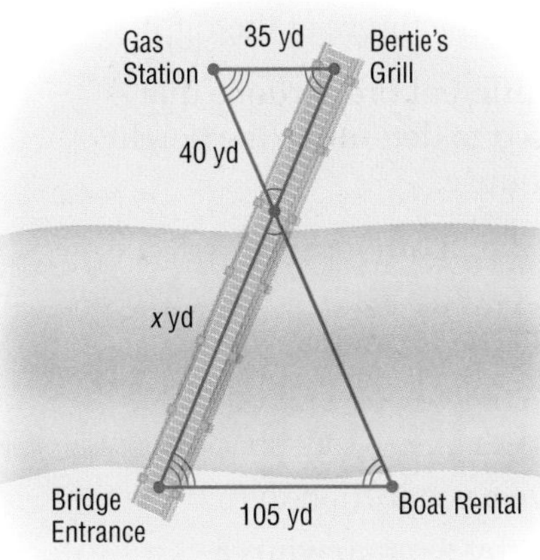

6. **MAPS** The triangles below are similar. How far is it from Athens to Yukon?

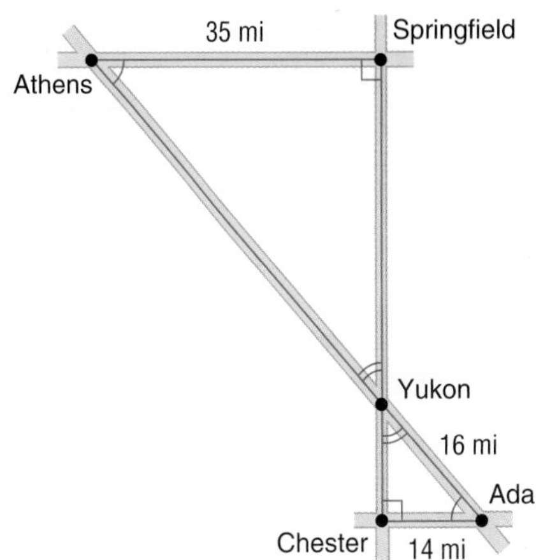

7. **ROLLER COASTERS** The height of a roller coaster is 157.5 feet. If the roller coaster's shadow is 60 feet long, how long will a person's shadow be if the person is 5 feet 3 inches tall?

Real-World Link

On July 30, 1909, the U.S. Government bought its first airplane. It was a Wright Brothers biplane. The selling price was $25,000 plus a bonus of $5,000 because it could exceed 40 miles per hour.

Source: Smithsonian

8. **GEOMETRY** All of the triangles in the figure at the right are similar.
   a. Find the measure of segment GD.
   b. If segment GF is congruent to segment FE, find the measure of segment BF.
   c. If the length of segment AD is 15 meters, what are the lengths of segments BC and CD?

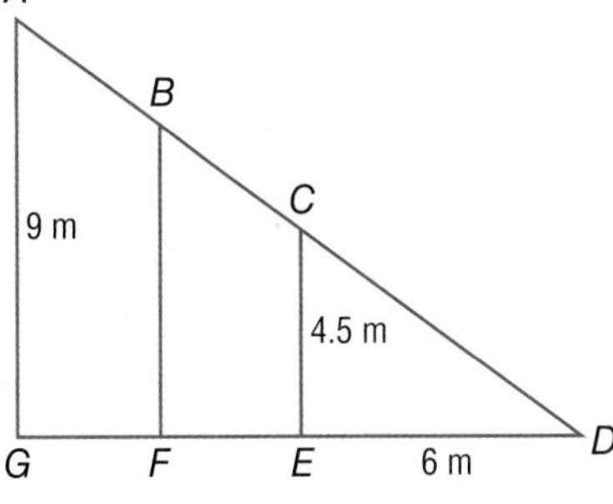

9. **MULTIPLE REPRESENTATIONS** In this problem, you will investigate similar triangles. Consider the following situation. A biplane starts to take-off from the beginning of a runway. When the plane is level with the end of the runway, it is 500 feet above the ground. A bird is flying in the same direction. It is 8 feet above the ground and 15 feet from the beginning of the runway.
   a. **MODEL** Draw a diagram of the situation.
   b. **ALGEBRAIC** Write and solve a proportion to find how far the plane is from the beginning of the runway.

10. **POLES** Electrical poles that carry electrical wire seem to get smaller the farther away they are. Find the apparent height of each pole if the tallest pole is 50 feet tall and there is 100 feet between each pole.

11. **BASKETBALL** During a basketball game, Josh, Devon, and Marco are in the following positions. Josh is 16 feet from Devon, and Devon is $5\frac{1}{3}$ feet from Marco. If Marco is 4 feet from both A and B, how wide is the key?

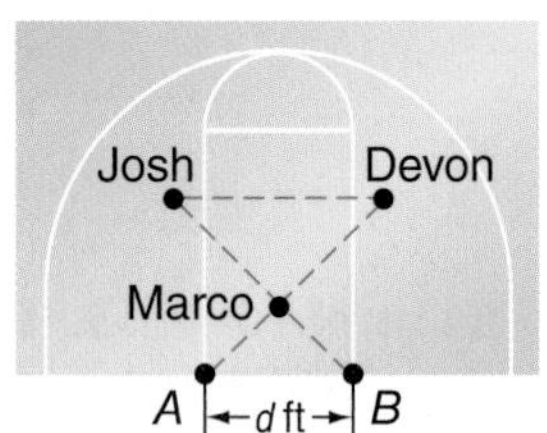

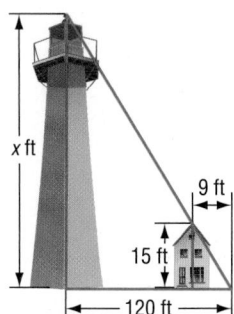

12. **LIGHTHOUSE** Use the figure at the right.
    a. Write two different proportions that could be used to determine the height of the lighthouse.
    b. How tall is the lighthouse?

Problem-Solving Tip

To solve Exercise 13, you could draw a diagram.

13. **FERRIS WHEELS** The Navy Pier Ferris Wheel in Chicago is 150 feet tall. If the Ferris wheel casts a $37\frac{1}{2}$-foot shadow, write and solve a proportion to find the height of a nearby woman who casts a $1\frac{1}{2}$-foot shadow.

14. **TREE HOUSE** A tree house casts a shadow of 18 feet while Jenet casts a shadow 9 feet. If Jenet is 5 feet tall, how tall is the tree house?

H.O.T. Problems

Use **H**igher-**O**rder **T**hinking Skills

15. **OPEN ENDED** Determine the height of a local landmark or statue in your community using shadow reckoning.

16. **CHALLENGE** In the diagram shown at the right, $\triangle ABC \sim \triangle EDC$.
    a. Write a proportion that could be used to solve for the height h of the flag pole.
    b. What information would you need to know in order to solve this proportion?

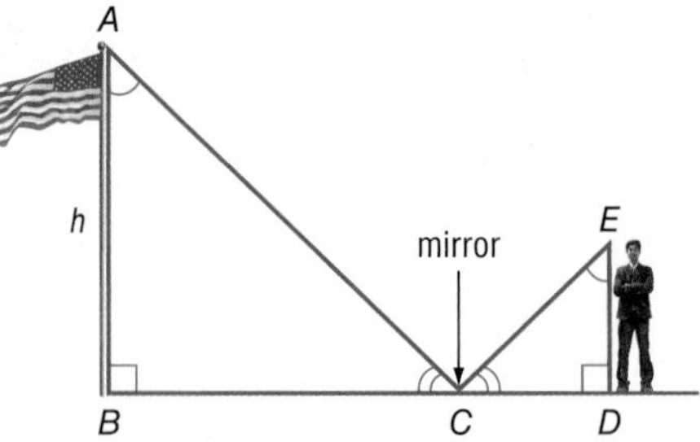

17. **REASONING** *True* or *false?* If two pairs of corresponding sides of two triangles are proportional, then you can use indirect measurement to determine the length of a missing side. Explain your reasoning.

18. **WRITING IN MATH** Give a real-world example of when you might need to use indirect measurement. Explain how you would solve the problem.

Standardized Test Practice

19. A bell tower casts a 60-inch shadow. At the same time, a statue that is 4.5 feet tall casts a 15-inch shadow. How tall is the bell tower?

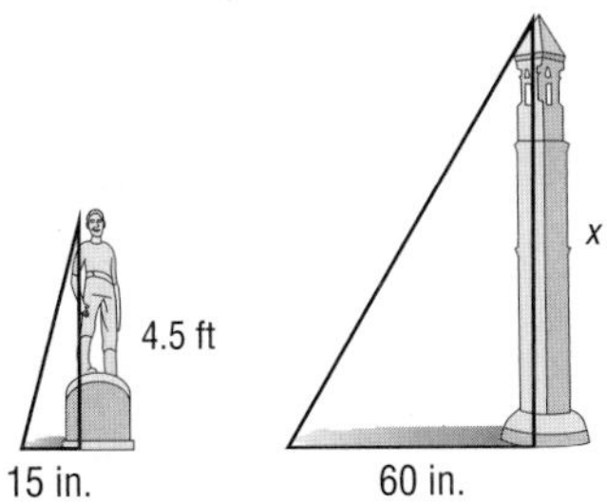

A 200 ft
B 30 ft
C 27 ft
D 18 ft

20. In the figure, $\triangle PRT \sim \triangle SRQ$. Find the distance across the golf green.

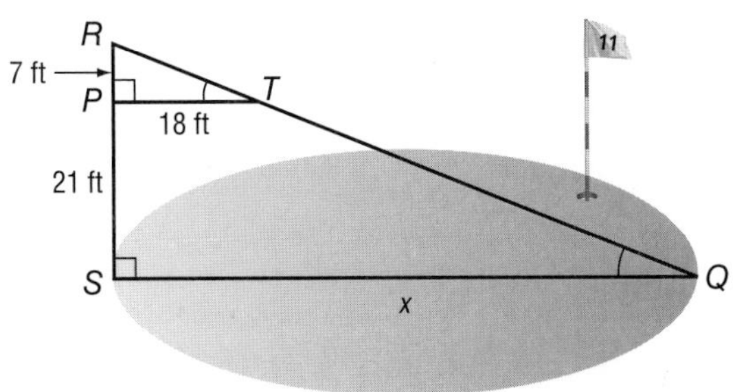

F 4.5 ft
G 35 ft
H 48 ft
J 72 ft

21. SHORT RESPONSE Find the length in kilometers of Beechwold Boulevard.

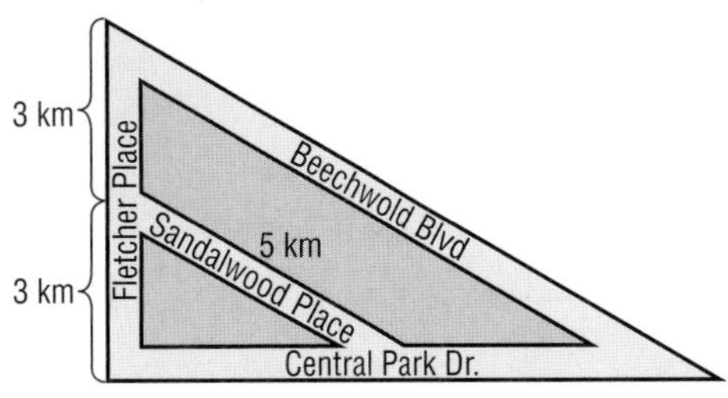

22. How tall is the street sign?

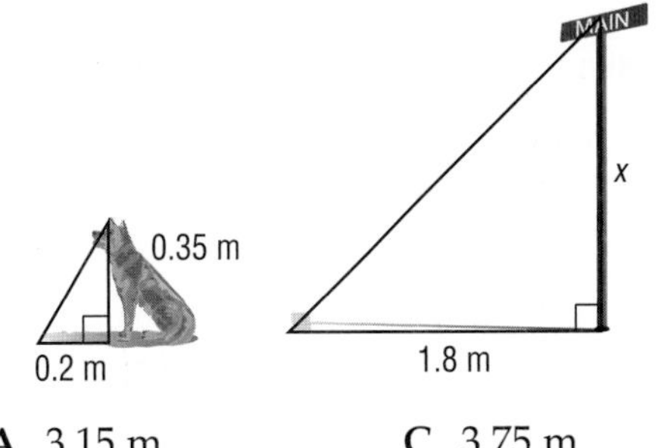

A 3.15 m
B 3.5 m
C 3.75 m
D 3.8 m

Spiral Review

23. A figure has vertices $A(-2, 1)$, $B(1, 2)$, and $C(3, -2)$. Graph the figure and its image after a dilation centered at the origin with a scale factor of 3. (Lesson 6-7)

24. Triangle *JMK* is similar to triangle *PRO*. What is the value of *x*? (Lesson 6-7)

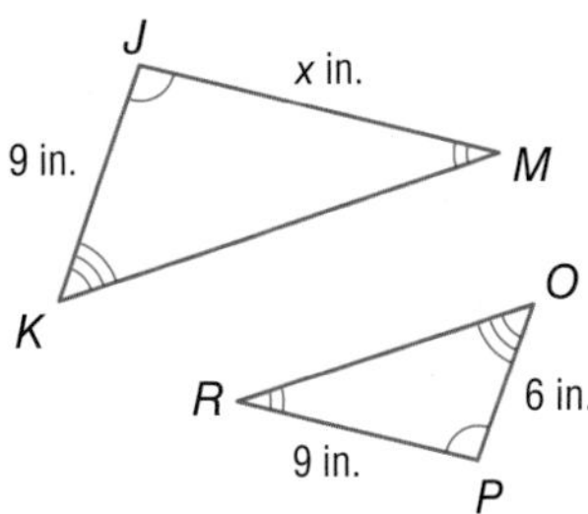

25. CRAFTS How many 6-inch long pieces of ribbon can be cut from a piece that is 3 yards long? (Lesson 6-3)

Skills Review

Write each fraction as a decimal. (Lesson 3-1)

26. $\frac{5}{8}$

27. $\frac{7}{12}$

28. $\frac{8}{12}$

29. $\frac{3}{5}$

CHAPTER 6

Study Guide and Review

Math Online glencoe.com
- STUDY TO GO
- Vocabulary Review

Chapter Summary

Key Concepts

Ratios and Rates (Lessons 6-1 and 6-2)

- A ratio is a comparison of numbers by division.
- A unit rate is a simplified rate whose denominator is 1.

Proportions (Lessons 6-4 and 6-5)

- A proportion is an equation stating two ratios or rates are equal. So, if $\frac{a}{b} = \frac{c}{d}$, then $ad = bc$.
- A proportional relationship exists when the ratios of related terms are equal.

Scale Drawings and Models (Lesson 6-6)

- A scale drawing or model represents an object that is too large or too small to be drawn or built at actual size.
- The ratio of a length on a scale drawing or model to the corresponding length on the real object is the scale.

Similar Figures and Dilations (Lessons 6-7 and 6-8)

- If two figures are similar, then the corresponding angles are congruent and the corresponding sides are proportional.
- A dilation with a scale factor:

  $k > 1$ will result in an enlargement.

  $0 < k < 1$ will result in a reduction.

  $k = 1$ will result in the same figure.

FOLDABLES® Study Organizer

Be sure the Key Concepts are noted in your Foldable.

Key Vocabulary

congruent (p. 301)
constant of proportionality (p. 282)
corresponding parts (p. 301)
cross products (p. 287)
dilation (p. 307)
dimensional analysis (p. 275)
indirect measurement (p. 313)
inverse proportion (p. 293)
nonproportional (p. 281)
proportion (p. 287)
proportional (p. 281)
rate (p. 270)
ratio (p. 265)
scale (p. 294)
scale drawing (p. 294)
scale factor (p. 295)
scale model (p. 294)
similar figures (p. 301)
unit rate (p. 270)

Vocabulary Check

Complete each sentence with the correct term. Choose from the list above.

1. A statement of equality of two ratios or rates is called a(n) __________.
2. A(n) __________ is a ratio of two measurements having different units.
3. Proportional relationships can be described by using the equation $y = kx$, where k is the __________.
4. The ratio of the length on a scale drawing to the corresponding length on the real object is called the __________.
5. Figures that have the same shape but not necessarily the same size are called __________.
6. A (n) __________ is a comparison of two quantities by division.
7. The process of including units of measurement as factors when computing is called __________.
8. Two quantities are __________ if they have a constant ratio or rate.

Lesson-by-Lesson Review

6-1 Ratios (pp. 265–269)

Express each ratio as a fraction in simplest form.

9. 10 girls out of 24 students
10. 6 red cars to 4 blue cars
11. 10 yards to 8 inches
12. 18 ounces to 3 cups
13. **BASEBALL** Jean got 12 hits out of 16 times at bat. Express this rate as a fraction in simplest form. Explain its meaning.

EXAMPLE 1

Express the ratio *2 feet to 18 inches* as a fraction in simplest form.

First, convert feet to inches.

$$\frac{2 \text{ ft}}{18 \text{ in.}} = \frac{24 \text{ in.}}{18 \text{ in.}}$$

Next, divide the numerator and denominator by the GCF, 6.

$$\frac{24 \text{ in.} \div 6}{18 \text{ in.} \div 6} = \frac{4 \text{ in.}}{3 \text{ in.}} \text{ or } \frac{4}{3}$$

6-2 Unit Rates (pp. 270–274)

Express each rate as a unit rate. Round to the nearest tenth, if necessary.

14. \$25.97 for 8 boxes
15. 400 meters in 5 minutes
16. \$175 for 4 concert tickets
17. 125 miles in 200 minutes
18. **SHOPPING** An eight pack of juice boxes costs \$4.79, and a twelve pack of juice boxes costs \$6.59. Which is a better buy? Explain.

EXAMPLE 2

Express *274 miles in 14 gallons of gasoline* as a unit rate. Round to the nearest tenth of a mile if necessary.

Write the rate that compares the miles to the number of gallons. Then divide to find the unit rate.

$$\frac{274 \text{ miles}}{14 \text{ gallons}} \overset{\div 14}{=} \frac{19.6 \text{ miles}}{1 \text{ gallon}}$$

So, the car traveled 19.6 miles on 1 gallon of gasoline.

6-3 Converting Rates and Measurements (pp. 275–280)

Complete each conversion. Round to the nearest hundredth.

19. 7 in. ≈ ■ cm
20. 20 m ≈ ■ yd
21. 25 fl oz ≈ ■ mL
22. 4 L ≈ ■ gal
23. 18 pt ≈ ■ L
24. 12 oz ≈ ■ g
25. **PLANES** A plane is flying at a speed of 425 miles per hour. How far will the plane travel in 0.75 hour?
26. **WATER** A swimming pool is being drained at a rate of 50 gallons per hour. How many milliliters per second is this?

EXAMPLE 3

A peregrine falcon can fly at a top speed of 200 miles per hour. How many feet per second is this?

First, convert miles to feet and hours to seconds.

$$\frac{200 \text{ mi}}{1 \text{ h}} = \frac{200 \text{ mi}}{1 \text{ h}} \cdot \frac{5280 \text{ ft}}{1 \text{ mi}} \cdot \frac{1 \text{ h}}{3600 \text{ s}}$$

Next, divide out the common factors.

$$= \frac{\overset{1}{\cancel{200}} \cancel{\text{mi}}}{1 \cancel{\text{h}}} \cdot \frac{\overset{176}{\cancel{5280}} \text{ ft}}{1 \cancel{\text{mi}}} \cdot \frac{1 \cancel{\text{h}}}{\underset{\underset{1}{\cancel{30}}}{\cancel{3600}} \text{ s}}$$

$$= \frac{293.3 \text{ ft}}{1 \text{ s}}$$

The falcon can fly about 293 feet per second.

CHAPTER 6

Study Guide and Review

6-4 Proportional and Nonproportional Relationships (pp. 281–285)

Determine whether the set of numbers in each table is proportional. Explain.

27.

Boxes	1	2	3	4
Pens	8	16	24	32

28.

Number of People	2	4	6	8
Brownies Eaten	2	5	7	10

29. FESTIVALS A customer at the ring toss booth gets 8 rings for \$2. Write an equation relating the cost to the number of rings. At this same rate, how much would a customer pay for 11 rings? for 20 rings?

EXAMPLE 4

Determine whether the distance is proportional to the time. Explain your reasoning.

Distance (meters)	30	56	69	80
Time (minutes)	1	2	3	4

Write the rate of distance to time for each minute in simplest form.

$\frac{30}{1}$ $\quad \frac{56}{2} = \frac{28}{1}$ $\quad \frac{69}{3} = \frac{23}{1}$ $\quad \frac{80}{4} = \frac{20}{1}$

Since the rates are not equal, the distance is not proportional to the time.

6-5 Solving Proportions (pp. 287–292)

Solve each proportion.

30. $\frac{15}{a} = \frac{5}{4}$ **31.** $\frac{m}{6} = \frac{18}{15}$

32. $\frac{28}{24} = \frac{d}{12}$ **33.** $\frac{16.5}{21} = \frac{5.5}{t}$

34. REAL ESTATE A homeowner whose house is assessed for \$120,000 pays \$1800 in taxes. At the same rate, what is the tax on a house assessed at \$135,000?

EXAMPLE 5

Solve $\frac{4}{9} = \frac{9}{x}$.

$\frac{4}{9} = \frac{9}{x}$	**Write the proportion.**
$4 \cdot x = 9 \cdot 9$	**Cross products**
$4x = 81$	**Multiply.**
$\frac{4x}{4} = \frac{81}{4}$	**Divide each side by 4.**
$x = 20.25$	**Simplify.**

6-6 Scale Drawing and Models (pp. 294–299)

On the scale drawing of a museum, the scale is 0.5 inch = 10 feet. Find the actual length of each gallery.

	Gallery	Drawing Length
35.	Modern Art	6 in.
36.	Renaissance	4.25 in.
37.	Egypt	7.5 in.

38. MAPS The length of a highway is 900 miles. If 0.5 inch on a map represents 50 miles, what is the length of the highway on the map?

EXAMPLE 6

A scale model of a car has a bumper that is 3.5 inches long. The scale on the model is 1 inch = 2 feet. What is the length of the actual car bumper?

model length → / actual length → $\frac{1 \text{ in.}}{2 \text{ ft}} = \frac{3.5 \text{ in.}}{x \text{ ft}}$ ← model length / ← actual length

$1 \cdot x = 2 \cdot 3.5$

$x = 7$

The actual length of the car bumper is 7 feet.

MIXED PROBLEM SOLVING
For mixed problem-solving practice, see page 848.

6-7 Similar Figures (pp. 301–306)

The figures are similar. Determine each missing measure.

39.

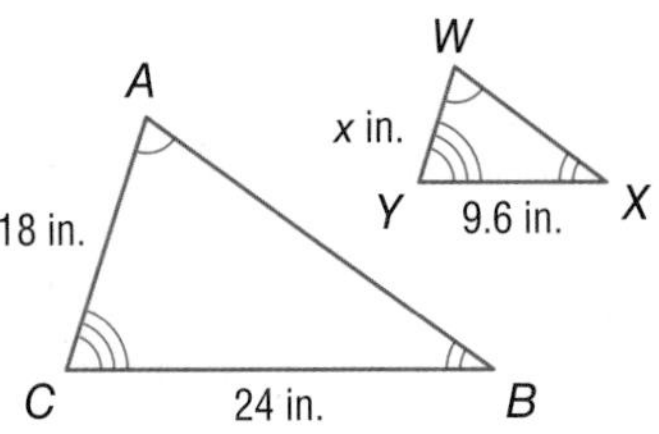

40.

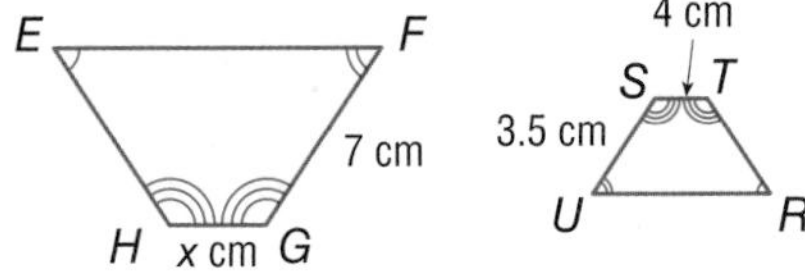

41. MOSAIC A mosaic is created using rectangular blocks. Block *A* has a length of 5 centimeters and a width of 2.5 centimeters. Block *B* is similar to block *A* and has a length of 7 centimeters. What is the width of block *B*?

EXAMPLE 7

If $\triangle LMN \approx \triangle PQR$, what is the value of x?

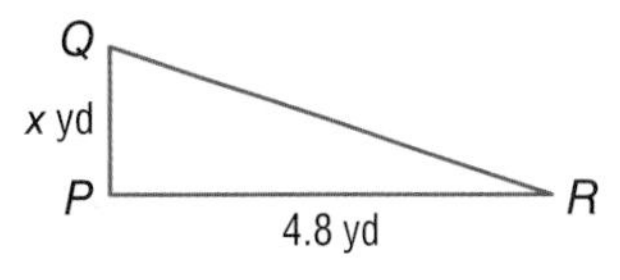

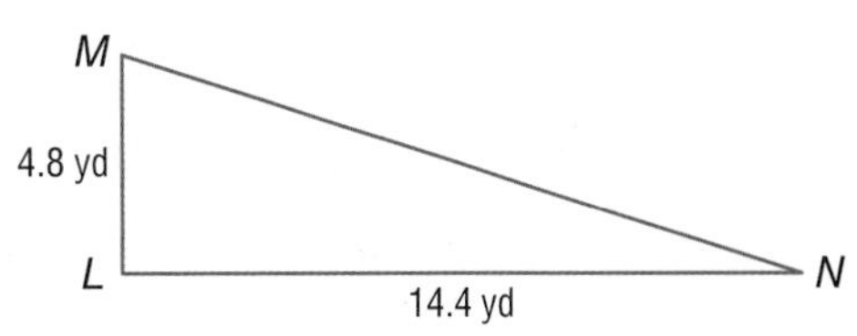

$\frac{LM}{PQ} = \frac{LN}{PR}$ **Write a proportion.**

$\frac{4.8}{x} = \frac{14.4}{4.8}$ **Replace $\overline{LM}$ with 4.8, $\overline{PQ}$ with x, $\overline{LN}$ with 14.4, and $\overline{PR}$ with 4.8.**

$x = 1.6$ **Find cross products and simplify.**

6-8 Dilations (pp. 307–312)

GEOMETRY Find the vertices of each figure after a dilation with the given scale factor, *k* centered about the origin. Then graph the original image and the dilation.

42. $W(-3, 0), X(2, 6), Y(6, 2), Z(2, -2); k = \frac{2}{3}$

43. $Q(-2, 3), R(1, 2), S(3, -1), T(-2, -2); k = 3$

44. $L(-4, -2), M(-2, 4), N(4, 0); k = \frac{1}{4}$

45. $F(1, 3), G(3, 4), H(2, 1); k = 2.5$

46. PHOTOGRAPHS Percy wants to increase the dimensions of his 5-inch by 7-inch photograph by a scale factor of 1.5 on his computer. What is the new size of the photograph?

EXAMPLE 8

A triangle has vertices $A(-2, -1)$, $B(1, 1)$ and $C(3, -3)$. Find the coordinates of the triangle after a dilation centered at the origin with a scale factor of 2.

To dilate the triangle, multiply the coordinates of each vertex by 2.

$A(-2, -1) \rightarrow A'(-2 \cdot 2, -1 \cdot 2) \rightarrow A'(-4, -2)$

$B(1, 1) \rightarrow B'(1 \cdot 2, 1 \cdot 2) \rightarrow B'(2, 2)$

$C(3, -3) \rightarrow C'(3 \cdot 2, -3 \cdot 2) \rightarrow C'(6, -6)$

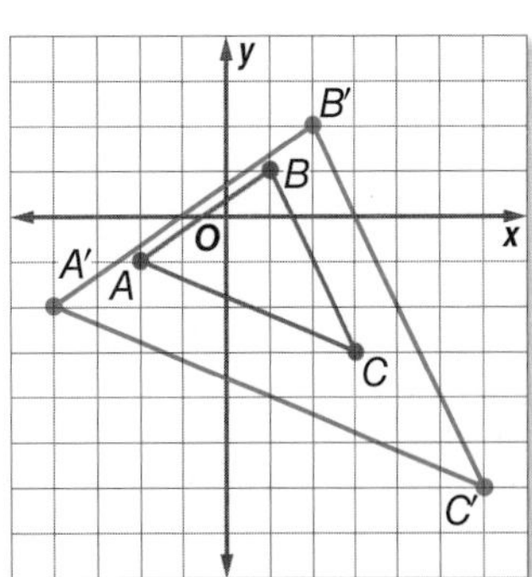

6-9 Indirect Measurement (pp. 313–317)

47. **WORLD RECORDS** At 7 feet 8 inches, the world's tallest woman casts a 46-inch shadow. At the same time, the world's shortest woman casts a 15.5-inch shadow. How tall is the world's shortest woman?

48. **HISTORY** The largest known pyramid is the Pyramid of Khufu. At a certain time of day, a vertical yard stick casts a shadow 1.5 feet long, and the pyramid casts a shadow 241 feet long. How tall is the pyramid?

49. In the figure below, $\triangle ABE \approx \triangle ACD$. What is the distance across the pond?

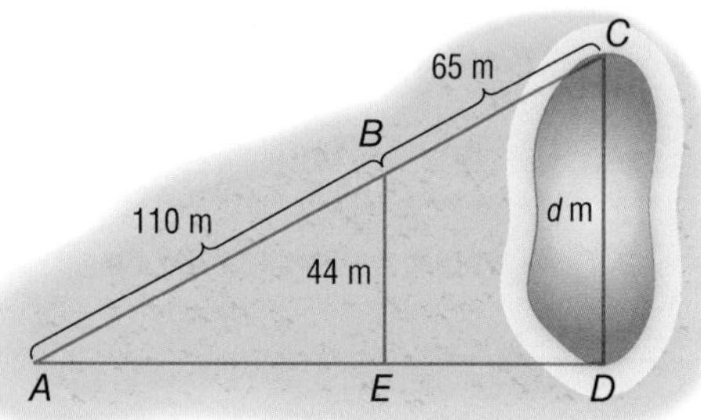

EXAMPLE 9

The Washington Monument casts a 185-foot shadow at the same time a nearby flagpole casts a 3-foot shadow. If the flagpole is 9 feet tall, how tall is the Washington Monument?

Write and solve a proportion.

flagpole height → 9 ft, monument height → x ft; flagpole's shadow ← 3 ft, monument's shadow ← 185 ft

$$\frac{9 \text{ ft}}{x \text{ ft}} = \frac{3 \text{ ft}}{185 \text{ ft}}$$

$$9 \cdot 185 = x \cdot 3$$ **Cross products**

$$1665 = 3x$$ **Multiply.**

$$\frac{1665}{3} = \frac{3x}{3}$$ **Divide each side by 3.**

$$555 = x$$

The Washington Monument is 555 feet tall.

CHAPTER 6

Practice Test

Math Online glencoe.com
Chapter Test

1. **ZOOS** The table shows how many of each type of animal are at a zoo. Express each ratio as a fraction in simplest form.

Animal	Types
Mammals	25
Birds	10
Reptiles	8
Arthropods	5

   a. birds to mammals

   b. reptiles to all animals

Express each rate as a unit rate.

2. \$75 for 5 DVDs

3. 300 words in 5 minutes

4. **WATER** Which bottle of water costs more per ounce: a 12 ounce bottle for \$1.25 or a 16 ounce bottle for \$1.50? Explain.

Complete each conversion. Round to the nearest hundredth.

5. 15 ft ≈ ■ m

6. 11 gal ≈ ■ L

7. 50 kg ≈ ■ lb

8. 18 oz ≈ ■ g

9. **TAXI** The taxi cab company charges a \$2 fee plus \$1.10 for each mile driven. Complete the table and determine whether the pattern forms a proportion. Explain.

Miles Driven	1	2	3	4	5
Cab Fare (\$)	■	■	■	■	■

10. **SALES** Jhan bought an 8-ounce fruit drink for \$1.50. Write and solve a proportion that could be used to find the cost of a gallon of fruit drink. (*Hint*: 1 gal = 128 oz)

11. **DRAWING** A scale drawing of a house that is 12 meters tall is being drawn with a scale of 2 centimeters = 3 meters. How many centimeters tall is the scale drawing?

12. **MULTIPLE CHOICE** Brandon is watching a movie clip on his computer in a window that is 6 inches by 3 inches. What is length and width of the window when dilated by a scale factor of $\frac{2}{3}$?

   A $\ell = 12$ in., $w = 9$ in.

   B $\ell = 2$ in., $w = 3$ in.

   C $\ell = 3$ in., $w = 6$ in.

   D $\ell = 4$ in., $w = 2$ in.

13. **GEOMETRY** The figures are similar. Find the missing measure.

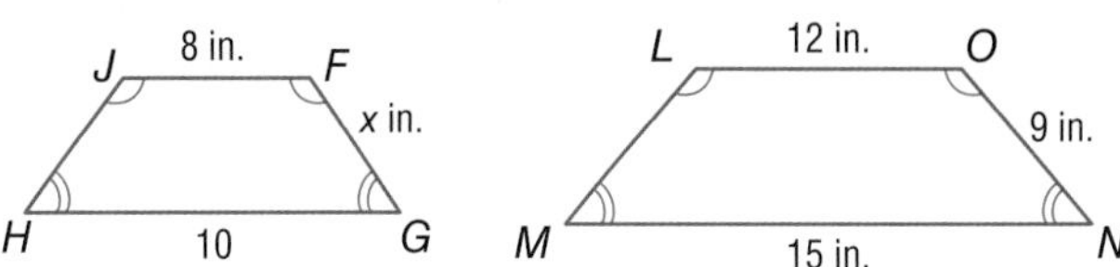

14. **GEOMETRY** Triangle *KLM* is shown. Graph the image of the triangle after a dilation centered at the origin with a scale factor of 2.

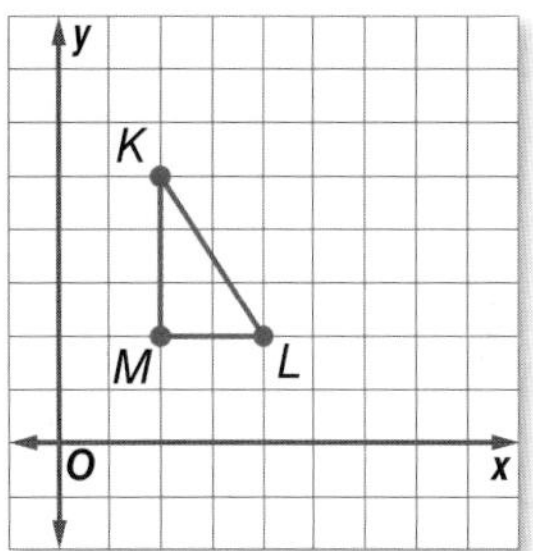

15. **MULTIPLE CHOICE** Triangle *DEF* has vertices *D*(4, 16), *E*(12, −12), and *F*(4, −8). Which of the following gives the vertices of triangle *D′E′F′* after a dilation of $\frac{1}{2}$?

   F *D′*(2, 4), *E′*(6, −3), *F′*(2, −2)

   G *D′*(16, 32), *E′*(24, −24), *F′*(8, −16)

   H *D′*(2, 8), *E′*(6, −6), *F′*(2, −4)

   J *D′*(1, 4), *E′*(3, −3), *F′*(1, −2)

16. **GAMES** In the map of Field Games, the triangles are similar. How far is the walk from Gone Fishin′ to the Cake Walk?

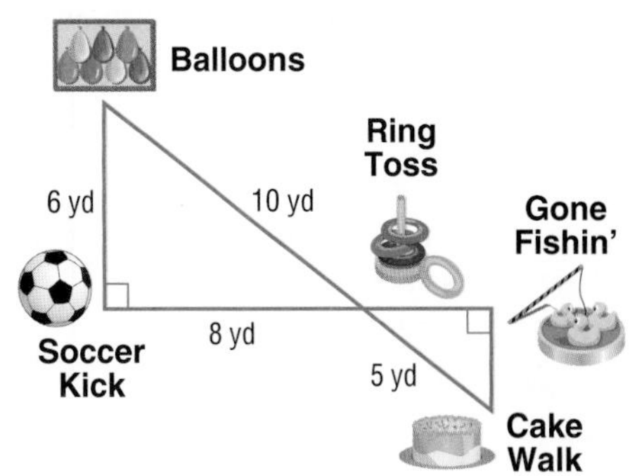

CHAPTER 6

Preparing for Standardized Tests

Draw a Diagram

Drawing a diagram can be a helpful way for you to visualize how to solve a problem. You can sketch your diagram on scrap paper or in your test booklet (if allowed). Some questions will require you to include a diagram in your answer. Do *not* make any marks however on your gridded response sheet other than your answers.

Strategies for Drawing a Diagram

Step 1

Read the problem carefully.

Ask yourself:

- What am I being asked to solve?
- What information is given in the problem?

Step 2

Sketch and label your diagram.

- Draw your diagram as clearly and accurately as possible.
- Label the diagram carefully. Be sure to include all of the given information.

Step 3

Solve the problem.

- Use your diagram to help you model the problem situation with an equation. Solve the equation.
- Check to be sure your answer makes sense.

EXAMPLE

Read the problem. Identify what you need to know. Then use the information in the problem to solve. Show your work.

SHORT RESPONSE A flagpole casts a shadow that is 26.25 feet long at the same time Jerome casts a shadow that is 3.5 feet long. If Jerome is 5.6 feet tall, what is the height of the flagpole?

Scoring Rubric	
Criteria	**Score**
Full Credit: The answer is correct and a full explanation is provided that shows each step.	2
Partial Credit: • The answer is correct, but the explanation is incomplete. • The answer is incorrect, but the explanation is correct.	1
No Credit: Either an answer is not provided or the answer does not make sense.	0

Read the problem carefully. You are given information about heights and shadows. Draw a diagram to help you solve the problem.

Example of a 2-point response:

Draw a diagram and use similar triangles to find the height of the flagpole.

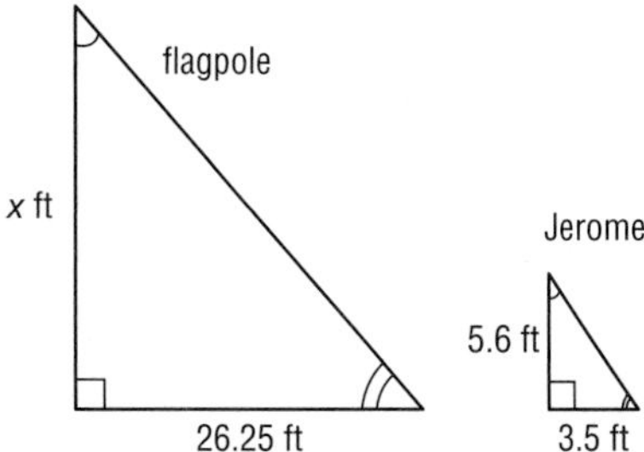

Use the similar triangles to set up and solve a proportion.

$$\text{flagpole}\left\{\frac{x}{26.25} = \frac{5.6}{3.5}\right\}\text{Jerome}$$

$$3.5x = 147$$

$$\frac{3.5x}{3.5} = \frac{147}{3.5}$$

$$x = 42$$

The height of the flagpole is 42 feet.

The steps, calculations, and reasoning are clearly stated. The student also arrives at the correct answer. So, this response is worth the full 2 points.

Exercises

Read each problem. Identify what you need to know. Then use the information in the problem to solve. Show your work.

1. The height of a roller coaster is 155 feet. The roller coaster casts a shadow that is 62 feet long. At the same time, Alexandria casts a shadow that is $2\frac{1}{5}$ feet long. How tall is Alexandria? Round your answer to the nearest inch, if necessary.

2. **GRIDDED RESPONSE** A tower casts a 24-foot shadow while a mailbox in the same location casts a 1.2-foot shadow. If the mailbox is 4 feet high, how high is the tower in feet?

3. A surveyor needs to find the distance across the river. Find the distance across the river.

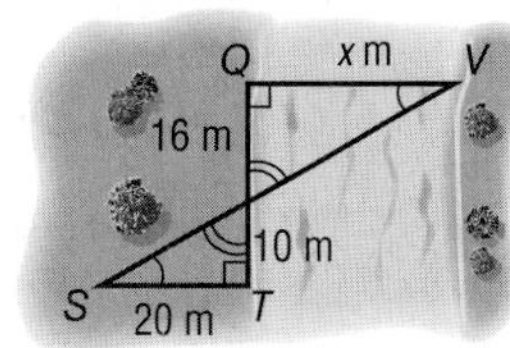

4. In triangle *ABC*, segment *AB* is 5 centimeters long, segment *BC* is 3 centimeters long, and segment *AC* is 6 centimeters long. Triangle *DEF* is similar to triangle *ABC*. If segment *DF* is 7.5 centimeters long, what is the length of segment *DE*?

CHAPTER 6

Standardized Test Practice

Cumulative, Chapters 1 through 6

Multiple Choice

Read each question. Then fill in the correct answer on the answer document provided by your teacher or on a sheet of paper.

1. Of the 60 students in the eighth grade, 24 participate in a school sport. Express this ratio as a fraction in simplest form.

A $\frac{1}{4}$ C $\frac{2}{5}$

B $\frac{3}{4}$ D $\frac{3}{5}$

2. A 128-ounce container of orange juice costs $5.12 at the local supermarket. At what price should the supermarket sell a 96-ounce container of orange juice so that the unit rate of the containers is the same?

Orange Juice Prices	
Size	Price
128 oz	$5.12
96 oz	■

F $3.76 H $3.92

G $3.84 J $4.08

3. Miguel spends $0.25 for each game ticket at the school carnival. If he has $4.50, how many game tickets can he buy?

A 12 C 16

B 15 D 18

4. Seven more than three times a number is 58. What is the number?

F 15 H 17

G 16 J 18

Test-Taking Tip

Question 1 Use estimation to eliminate unreasonable answers. Half of 60 is 30 so the fraction should be less than $\frac{1}{2}$.

5. On a map of Terrance's hometown, the distance from the stadium to a park is 3.5 inches. If the actual distance is 7 miles, what is the scale of the map?

A 1 inch = 0.5 mile

B 1 inch = 0.25 mile

C 1 mile = 0.5 inch

D 1 mile = 0.25 inch

6. Use the table shown below. What number of printer cartridges would result in a proportional relationship?

Boxes	Printer Cartridges
1	16
2	32
3	48
4	■

F 56 H 60

G 58 J 64

7. Mrs. Connors is organizing the seventh grade field trip to the museum of natural history. She knows that 3 vans will accommodate 24 passengers. How many vans will be needed for 64 passengers?

A 7 vans C 9 vans

B 8 vans D 10 vans

8. Rectangle *ABCD* is similar to rectangle *HIJK*. What is the length of $\overline{IK}$ if $BD =$ 9 centimeters, $AB =$ 12 centimeters, and $HI =$ 8 centimeters?

F 5.6 cm

G 6 cm

H 6.5 cm

J 7 cm

Short Response/Gridded Response

Record your answers on the answer sheet provided by your teacher or on a sheet of paper.

9. The right triangles below are similar.

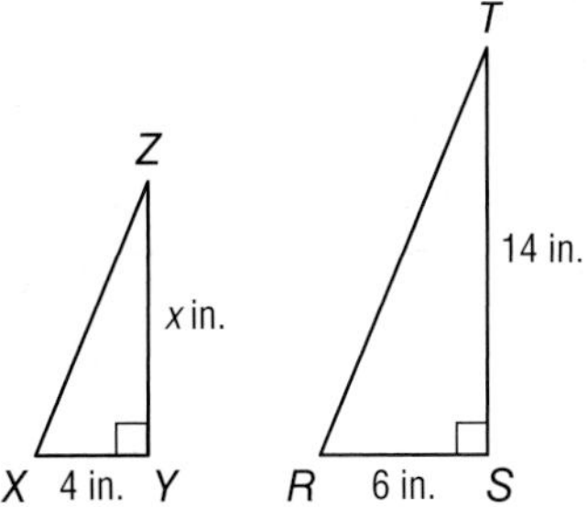

a. Find the scale factor from $\triangle XYZ$ to $\triangle RST$.

b. What is the value of x in $\triangle XYZ$?

10. GRIDDED RESPONSE In a microscope, the image of a 0.16-millimeter paramecium appears to be 8 millimeters long. What is the scale factor of the dilation?

11. Miley needs to use $\frac{1}{4}$ cup of flour in one recipe and $\frac{2}{3}$ cup in a second recipe. If she doubles both recipes, how much flour does she need in all?

12. A tree casts a shadow that is 32 feet long at the same time as a stop sign casts a shadow that is 4 feet long. The stop sign is 7 feet tall.

a. Sketch a drawing of the situation.

b. How tall is the tree?

13. GRIDDED RESPONSE The length of a rectangle is 5 centimeters longer than twice the width. If the width of the rectangle is 6 centimeters, find the area of the figure in square centimeters.

Extended Response

Record your answers on a sheet of paper. Show your work.

14. The graph shows Jeremy's and Eduardo's average rates in a race.

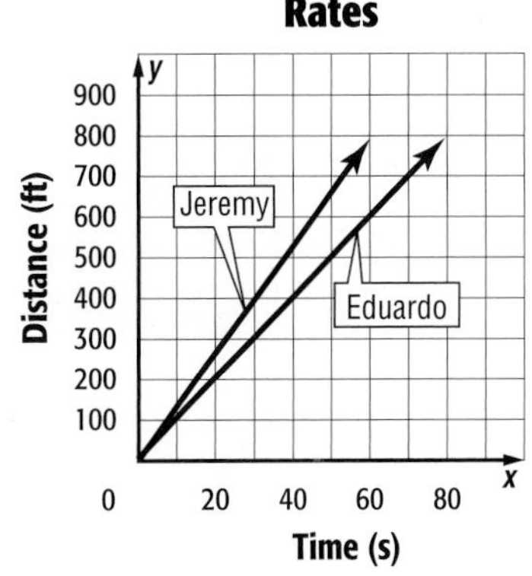

a. Express each person's average speed as a unit rate.

b. At these rates, how long would it take Jeremy and Eduardo each to complete a race of 1800 feet?

c. Suppose Albert races at an average speed of 12.5 feet per second. Predict where the line representing his speed would be graphed. Explain your reasoning.

Need Extra Help?														
If you missed Question...	1	2	3	4	5	6	7	8	9	10	11	12	13	14
Go to Lesson or Page...	6-1	6-2	5-4	4-6	6-6	6-4	6-5	6-7	6-7	6-8	3-6	6-9	5-1	6-2

CHAPTER 7 Percent

Then

In Chapter 3, you converted fractions and decimals. In Chapter 6, you solved proportions.

Now

In Chapter 7, you will:

- Express percents as fractions and decimals, and fractions as percents.
- Use the percent proportion and percent equations to solve problems.
- Construct and interpret circle graphs.

Why?

SPORTS In baseball, percents are often used to represent statistics of players. In a recent season, Jimmy Rollins, of the Philadelphia Phillies, led the major leagues with the highest on base percent with 0.344. This means he got on base 34.4% of the times he was at bat.

Percent
Introduction
In baseball, percents are often used to represent the statistics of a player. In 2007, Jimmy Rollins led the major leagues with the highest on base percentage with 0.344. This means he got on base 34.4% of the times he was at bat. To calculate these percentages, a percent proportion is often used.

Math *in Motion*, Animation glencoe.com

Get Ready for Chapter 7

Diagnose Readiness You have two options for checking Prerequisite Skills.

Text Option

Take the Quick Check below. Refer to the Quick Review for help.

QuickCheck

Convert each fraction to a decimal. (Lesson 3-1)

1. $\frac{1}{4}$

2. $\frac{1}{10}$

3. $\frac{3}{8}$

4. $\frac{3}{4}$

5. RECIPES A recipe calls for $\frac{2}{3}$ cup of sugar. Write this amount as a decimal rounded to the nearest hundredth.

QuickReview

EXAMPLE 1

Write $\frac{1}{4}$ as a decimal.

$$\begin{array}{r} 0.25 \\ 4\overline{)1.00} \\ \underline{-80} \\ 20 \\ \underline{-20} \\ 0 \end{array}$$

So, $\frac{1}{4} = 0.25$.

QuickCheck

Solve each proportion. (Lesson 6-5)

6. $\frac{2}{5} = \frac{n}{15}$

7. $\frac{3}{2} = \frac{n}{10}$

8. $\frac{3}{20} = \frac{n}{100}$

9. $\frac{11}{50} = \frac{n}{200}$

10. FERRIS WHEEL The wait time to ride the Ferris wheel at a state fair is 15 minutes when 75 people are in line. At this rate, how long is the wait time when 200 people are in line?

QuickReview

EXAMPLE 2

Solve $\frac{3}{5} = \frac{n}{20}$.

$\frac{3}{5} = \frac{n}{20}$

$3 \cdot 20 = 5n$ **Cross products**

$60 = 5n$ **Multiply.**

$\frac{60}{5} = \frac{5n}{5}$ **Divide each side by 5.**

$12 = n$ **Simplify.**

So, $n = 12$.

QuickCheck

Solve each equation. Check your solution. (Lesson 4-4)

11. $25x = 50$

12. $8a = 28$

13. $0.1x = 0.5$

14. $6n = 12$

15. MUSIC Jan has 4 times as many songs on her computer as Ian has on his computer. If Jan has 516 songs, how many songs does Ian have on his computer?

QuickReview

EXAMPLE 3

Solve $7.5x = 30$.

$7.5x = 30$ **Write the equation.**

$\frac{7.5x}{7.5} = \frac{30}{7.5}$ **Divide each side by 7.5.**

$x = 4$ **Simplify.**

So, $x = 4$.

Online Option

Math Online Take a self-check Chapter Readiness Quiz at glencoe.com.

Get Started on Chapter 7

You will learn several new concepts, skills, and vocabulary terms as you study Chapter 7. To get ready, identify important terms and organize your resources. You may wish to refer to **Chapter 0** to review prerequisite skills.

FOLDABLES® Study Organizer

Linear Functions Make this Foldable to help you organize your Chapter 7 notes about percents. Begin with 2 sheets of $8\frac{1}{2}$" × 11" notebook paper.

1. **Fold** the first two sheets in half from top to bottom. Cut along the fold from edges to margin.

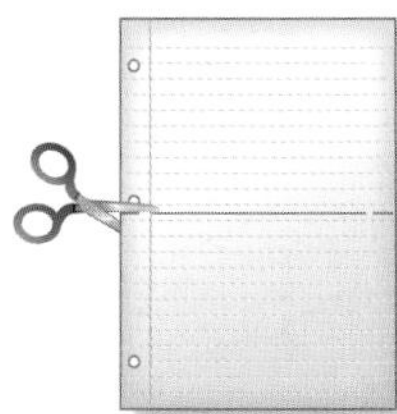

2. **Fold** the third sheet in half from top to bottom. Cut along the fold from margin to edge.

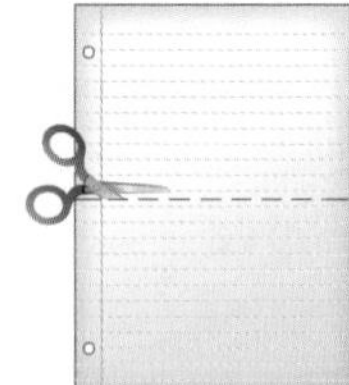

3. **Insert** the first two sheets through the third sheets and align the folds.

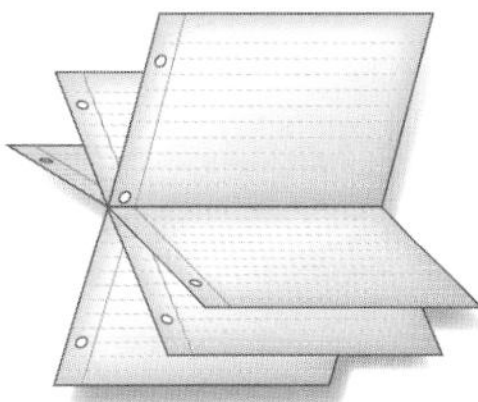

4. **Label** each page with a lesson number and a title.

Chapter 7
Percent

Math Online glencoe.com

- Study the chapter online
- Explore **Math in Motion**
- Get extra help from your own **Personal Tutor**
- Use **Extra Examples** for additional help
- Take a **Self-Check Quiz**
- **Review Vocabulary** in fun ways

New Vocabulary

English		Español
percent	• p. 331 •	por ciento
percent proportion	• p. 345 •	proporción porcental
percent equation	• p. 357 •	ecuación porcental
percent of change	• p. 364 •	porcentaje de cambio
percent of increase	• p. 364 •	porcentaje de aumento
percent of decrease	• p. 364 •	porcentaje de disminución
markup	• p. 365 •	margen de utilidad
selling price	• p. 365 •	precio de venta
discount	• p. 366 •	descuento
interest	• p. 370 •	interés
simple interest	• p. 370 •	interés simple
principal	• p. 370 •	capital
compound interest	• p. 371 •	interés compuetso
circle graph	• p. 376 •	gráfica circular

Review Vocabulary

cross products • p. 287 • productos cruzados if $\frac{a}{c} = \frac{b}{d}$, then $ad = bc$

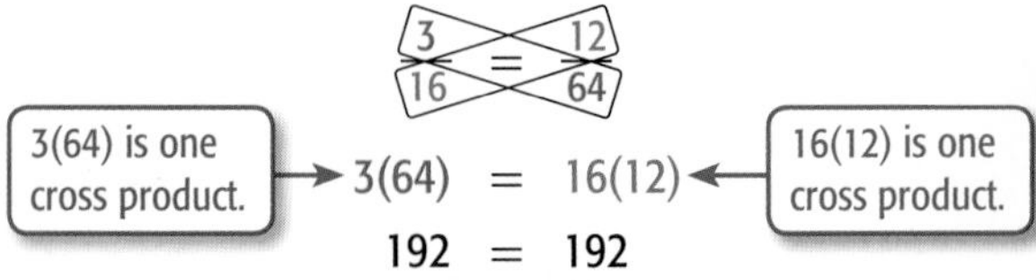
3 / 16 = 12 / 64
3(64) is one cross product.
3(64) = 16(12)
16(12) is one cross product.
192 = 192

The cross products are equal.

proportion • p. 287 • proporción a statement of equality of two or more ratios

rational number • p. 128 • número racional a number that can be written as a fraction in the form $\frac{a}{b}$, where a and b are integers and $b \neq 0$

Multilingual eGlossary glencoe.com

7-1 Fractions and Percents

Then

You have already converted fractions and decimals. (Lesson 3-1)

Now

- Express percents as fractions.
- Express fractions as percents.

New Vocabulary

percent

Math Online

glencoe.com

- Extra Examples
- Personal Tutor
- Self-Check Quiz
- Homework Help

Why?

Three players were comparing their basketball free throws.

Basketball Statistics		
Player	Number of Free Throws Made	Number Attempted
Daniel	2	4
Frank	8	10
Dale	15	20

a. For each player, write a ratio that compares the number of free throws made to the number attempted as a fraction in simplest form.

b. Rewrite the fractions in part **a** using a denominator of 100.

c. Which player has the greatest ratio of number of free throws made to the number attempted?

d. Was it easier to compare the fractions in parts **a** or **b**? Explain.

Percents as Fractions A **percent** is a ratio that compares a number to 100. To write a percent as a fraction, express the ratio as a fraction with a denominator of 100. Then simplify if possible.

Key Concept — Percent

For Your FOLDABLE

Words A percent is a part to whole ratio that compares a number to 100.

Examples 80% 80 out of 100 $\frac{80}{100}$

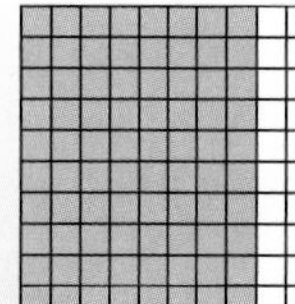

EXAMPLE 1 Percents as Fractions

Write each percent as a fraction in simplest form.

a. 60%

$60\% = \frac{60}{100}$ **Definition of percent**

$= \frac{3}{5}$ **Simplify.**

b. $12\frac{1}{2}\%$

$12\frac{1}{2}\% = \frac{12\frac{1}{2}}{100}$ **Definition of percent**

$= \frac{25}{2} \div 100$ **Write $12\frac{1}{2}$ as an improper fraction.**

$= \frac{\overset{1}{\cancel{25}}}{2} \cdot \frac{1}{\underset{4}{\cancel{100}}}$ or $\frac{1}{8}$ **Simplify.**

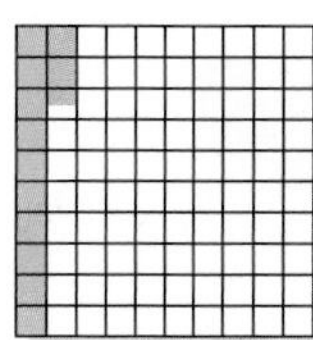

Check Your Progress

1A. 40%

1B. $20\frac{3}{4}\%$

Personal Tutor glencoe.com

Notice that a percent can be less than 1% or greater than 100%.

EXAMPLE 2 Less Than 1% or Greater Than 100%

Write each percent as a fraction in simplest form.

a. 0.8%

$0.8\% = \frac{0.8}{100}$ **Definition of percent**

$= \frac{0.8}{100} \cdot \frac{10}{10}$ **Multiply by $\frac{10}{10}$ to eliminate the decimal in the numerator.**

$= \frac{8}{1000}$ or $\frac{1}{125}$ **Simplify.**

b. 175%

$175\% = \frac{175}{100}$ **Definition of percent**

$= \frac{7}{4}$ or $1\frac{3}{4}$ **Simplify.**

Math History Link

Augustus (27 B.C.–14 A.D.) Percents were first used by Roman emperor Augustus. He levied a tax of $\frac{1}{100}$ on all goods sold at auctions.

Check Your Progress

2A. 0.2% **2B.** 150%

Personal Tutor glencoe.com

Fractions as Percents To write a fraction as a percent, write an equivalent fraction with a denominator of 100.

EXAMPLE 3 Fractions as Percents

Write each fraction as a percent.

a. $\frac{1}{4}$

First, find the equivalent fraction with a denominator of 100. Then write the fraction as a percent.

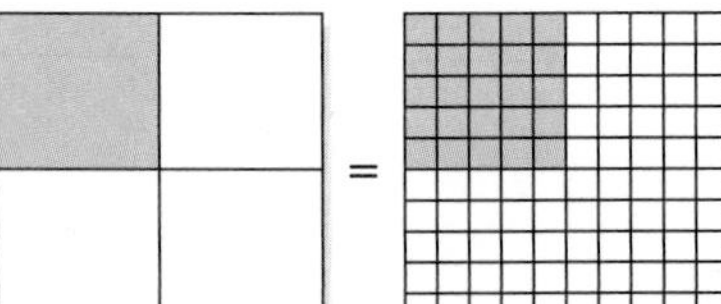

$\frac{1}{4} = \frac{1 \times 25}{4 \times 25} = \frac{25}{100}$ or 25%

So, $\frac{1}{4} = 25\%$.

b. $\frac{6}{5}$

$\frac{6}{5} = \frac{6 \times 20}{5 \times 20}$ or $\frac{120}{100}$ **Write an equivalent fraction with a denominator of 100.**

$= 120\%$ **$\frac{120}{100} = 120\%$**

So, $\frac{6}{5} = 120\%$.

ReadingMath
Percent

Root Word Cent; comes from the Latin word *centum* which means hundred. There are 100 *cents* in one dollar. *Percent* means *per hundred* or *hundredths*.

Check Your Progress

3A. $\frac{3}{10}$ **3B.** $\frac{7}{2}$

Personal Tutor glencoe.com

StudyTip

Choose the Method To write a fraction as a percent,

- use **multiplication** when a fraction has a denominator that is a factor of 100,
- use a **proportion** for any type of fraction.

If the denominator is not a factor of 100, you can write fractions as percents by using a proportion.

Real-World EXAMPLE 4 Fractions as Percents

QUIZZES On her science quiz, Jacinda got 14 questions correct out of 16. Find Jacinda's grade as a percent.

To find her grade, write $\frac{14}{16}$ as a percent.

Estimate $\frac{14}{16}$ is greater than $\frac{12}{16}$ or $\frac{3}{4}$. So, $\frac{14}{16}$ is greater than 75%.

$\frac{14}{16} = \frac{n}{100}$ **Write a proportion using $\frac{n}{100}$.**

$14 \cdot 100 = 16n$ **Cross products**

$1400 = 16n$ **Multiply.**

$\frac{1400}{16} = \frac{16n}{16}$ **Divide each side by 16.**

$87\frac{1}{2} = n$ **Simplify.**

So, $\frac{14}{16} = 87\frac{1}{2}\%$ or 87.5%.

Check for Reasonableness $87.5\% > 75\%$ ✓

Check Your Progress

4. **MUSIC** Caitlyn has practiced playing the piano 5 out of the last 8 days. What percent is this?

Personal Tutor glencoe.com

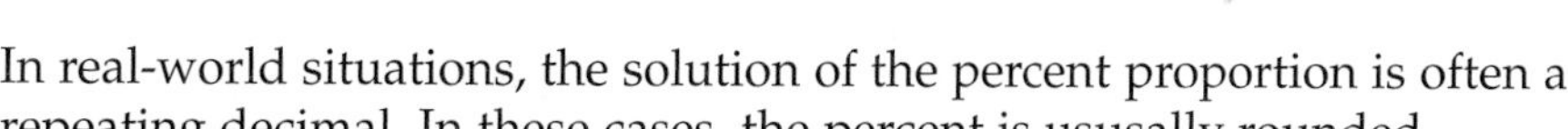

In real-world situations, the solution of the percent proportion is often a repeating decimal. In these cases, the percent is ususally rounded.

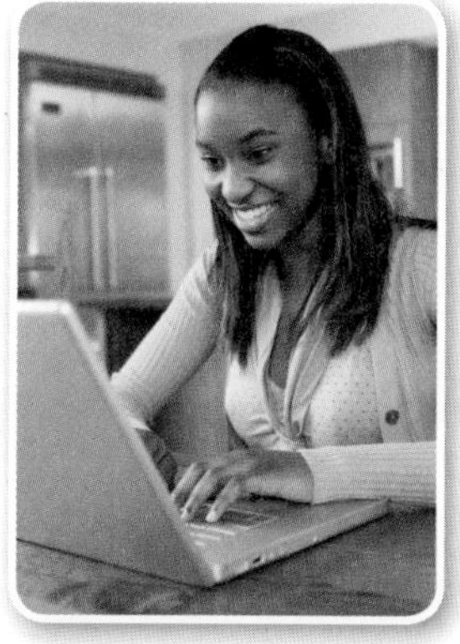

Real-World Link

A recent survey showed that 35% of Internet users spend their time playing games.

Source: Pew Internet & American Life Project

Real-World EXAMPLE 5 Rounding Percents

GAMES After a survey, Ling discovered that 4 out of every 11 eighth graders at his school play games online. Find the percent of eighth graders that play games online. Round to the nearest hundredth.

Estimate $\frac{4}{11}$ is about $\frac{4}{12}$, which equals $\frac{1}{3}$ or about 33%.

$\frac{4}{11} = \frac{n}{100}$ **Write a proportion using $\frac{n}{100}$.**

$4 \cdot 100 = 11n$ **Cross products**

$400 = 11n$ **Multiply.**

400 [÷] 11 [ENTER] 36.36363636 **Use a calculator.**

So, about 36.36% of the eighth graders play games online.

Check for Reasonableness $36.36\% \approx 33\%$ ✓

Check Your Progress

5. **SPORTS** During baseball season, Theo had 34 hits out of 55 at bats. Find Theo's at-bat percent. Round to the nearest hundredth.

Personal Tutor glencoe.com

Check Your Understanding

Examples 1 and 2 pp. 331–332

Write each percent as a fraction or mixed number in simplest form.

1. 40% **2.** $14\frac{1}{2}\%$ **3.** 150% **4.** 0.9%

Example 3 p. 332

Write each fraction as a percent. Round to the nearest hundredth.

5. $\frac{19}{20}$ **6.** $\frac{10}{4}$ **7.** $\frac{17}{18}$ **8.** $\frac{21}{24}$

Examples 4 and 5 p. 333

9. **VACATIONS** Of the students in Ms. Villata's classes, 56 out of 128 are planning to go on a summer vacation. What percent is this?

10. **TESTS** On a recent social studies test, Miguel earned 80 out of 90 points. Find Miguel's grade as a percent. Round to the nearest hundredth.

Practice and Problem Solving

 = **Step-by-Step Solutions** begin on page R11. **Extra Practice** begins on page 810.

Examples 1 and 2 pp. 331–332

Write each percent as a fraction or mixed number in simplest form.

11. 61% **12.** 28% **13.** $20\frac{2}{3}\%$ **14.** $58\frac{1}{2}\%$

15. 116% **16.** 425% **17.** 0.6% **18.** 0.33%

19. $62\frac{1}{2}\%$ **20.** $33\frac{1}{3}\%$ **21.** 1.2% **22.** 2.3%

23. **GEOGRAPHY** The tallest mountain in the world is Mt. Everest. It is about 140% as tall as the tallest mountain in the United States, Mt. McKinley. Write 140% as a mixed number in simplest form.

24. **POPULATION** According to the U.S. Census Bureau, 0.4% of the population of South Carolina is Native American. Write 0.4% as a fraction in simplest form.

Example 3 p. 332

Write each fraction as a percent. Round to the nearest hundredth.

25. $\frac{8}{32}$ **26.** $\frac{6}{25}$ **27.** $\frac{18}{4}$ **28.** $\frac{11}{5}$

29. $\frac{7}{8}$ **30.** $\frac{1}{60}$ **31.** $\frac{10}{14}$ **32.** $\frac{6}{13}$

Examples 4 and 5 p. 333

33. **FRIENDS** Of Mark's friends, 5 out of 8 are planning to go to the concert on Friday night. What percent of Mark's friend are planning to go to the concert?

34. **SPORTS** There are 18 members on the varsity baseball team. Of these players, 7 are pitchers. What percent of the team are pitchers? Round to the nearest hundredth.

35. **ANIMALS** A cheetah can run about $5\frac{1}{2}$ times as fast as a squirrel. What percent is this?

36. **GEOMETRY** The Martin's patio is shown at the right. Part of their patio is covered with a mat. What percent of their patio is covered by the mat? (*Hint:* Area is equal to length times width.)

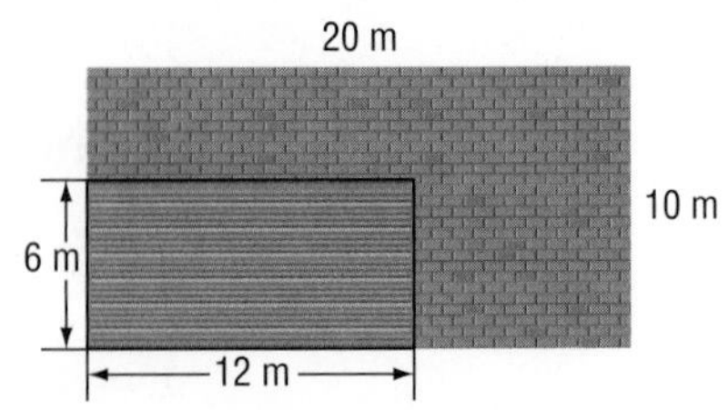

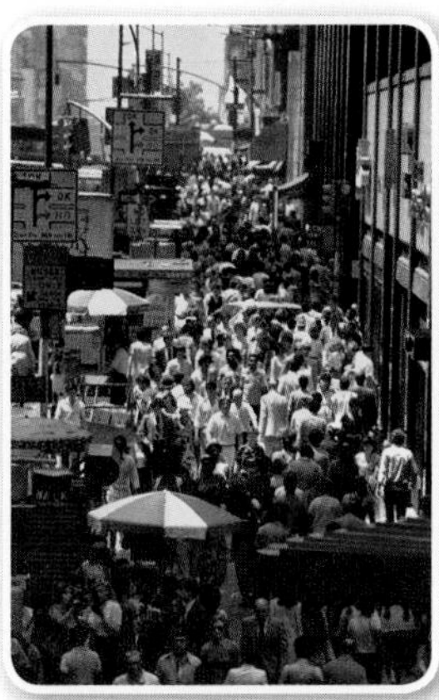

Real-World Link

On October 17, 2006, the population of the United States hit the 300 million mark. The population is estimated to be 400 million in 2043.

Source: *USA Today*

37. GEOMETRY For each model below, write a fraction in simplest form that compares the shaded portion of each figure to its total area. Then write each fraction as a percent. Which figure has the greatest part shaded?

Figure A

Figure B

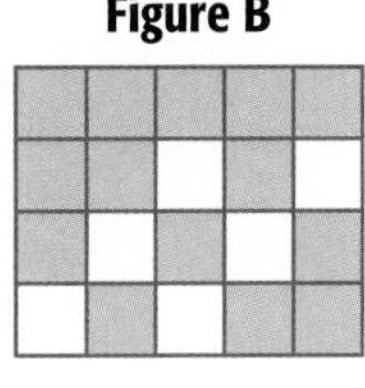

Figure C

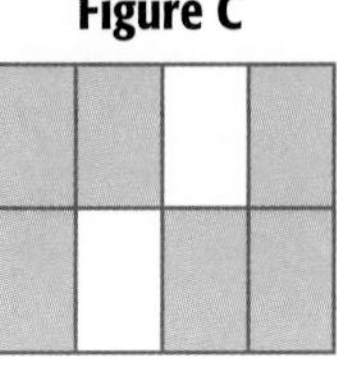

Replace each ● with <, >, or = to make a true sentence.

38. $\frac{8}{20}$ ● 45% **39.** 75% ● $\frac{36}{48}$ **40.** 120% ● $\frac{84}{72}$

41. $\frac{7}{10}$ ● $70\frac{1}{4}\%$ **42.** $\frac{1}{50}$ ● 0.002 % **43.** 300% ● $\frac{3}{10}$

44. POPULATION The circle graph shows the percent of the world's population on each continent.

a. Write each percent as a fraction.

b. What fraction of the population lives in the Americas?

c. If the world population is about 6.5 billion people, explain how you might determine the number of people who live in Asia.

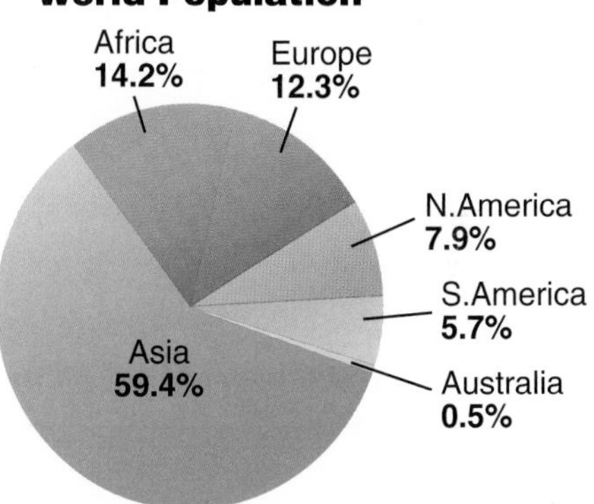

H.O.T. Problems

Use Higher-Order Thinking Skills

45. OPEN ENDED Give an example of a percent that is between $\frac{3}{4}$ and $\frac{7}{9}$. Explain your reasoning.

46. CHALLENGE What percent of the larger triangle shown is *not* shaded? Round to the nearest hundredth.

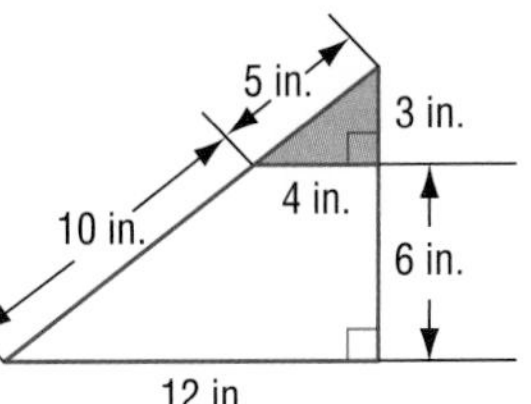

47. NUMBER SENSE Model the fraction $\frac{15}{2}$. Then write the fraction as a percent.

48. FIND THE ERROR Emma and Anna are changing 0.5% to a fraction. Is either of them correct? Explain.

Emma

$0.5\% = \frac{5}{100}$

$= \frac{1}{20}$

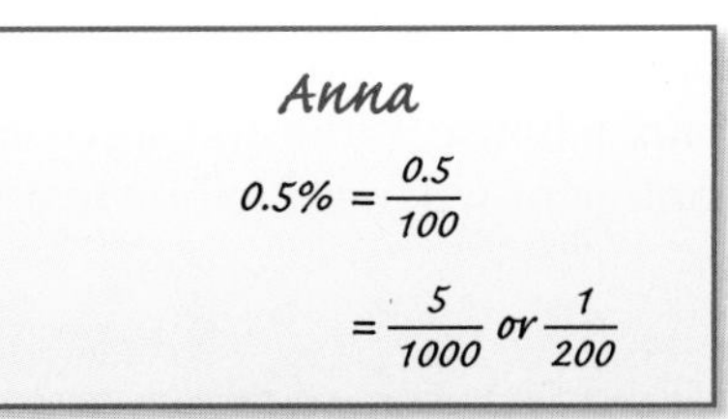

49. CHALLENGE Find the whole number value of x such that $\frac{x}{x+5}$ equals 50%. Explain your reasoning.

50. WRITING IN MATH Describe two ways to write $\frac{4}{5}$ as a percent.

Standardized Test Practice

51. Which of the following represents the shaded portion of the figure below?

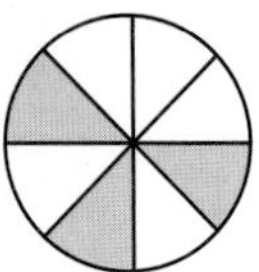

A 25%
B 37.5%
C 50%
D 62.5%

52. During her last basketball game, Carmen made 14 field goals out of 30 attempts. What percent of field goals did she make?

F 45%
G 47%
H 50%
J 51%

53. If 95% of the seventh grade students are going on the trip to the museum, what fraction of the students are *not* going on the field trip?

A $\frac{1}{20}$
B $\frac{1}{10}$
C $\frac{9}{10}$
D $\frac{19}{20}$

54. **SHORT RESPONSE** Helio is installing some software onto his computer. The screen shows that it is 45% complete. Write 45% as a fraction in simplest form. Show all of your work.

Spiral Review

55. **RIDES** Suppose a roller coaster casts a shadow of 31.5 feet. At the same time, a nearby Ferris wheel casts a 19-foot shadow. If the roller coaster is 126 feet tall, how tall is the Ferris wheel? (Lesson 6-9)

For Exercises 56 and 57 use the figure at the right.

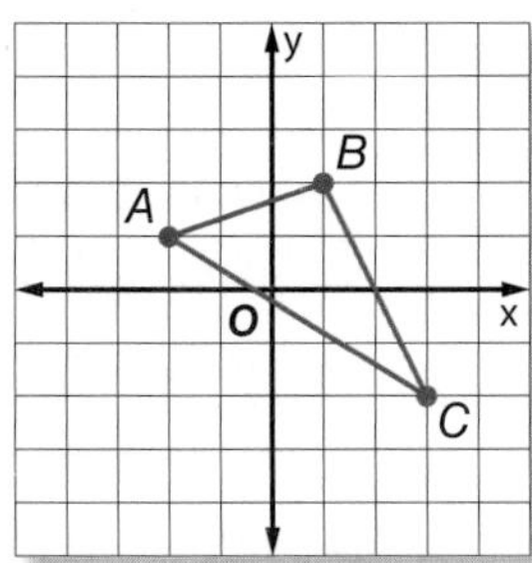

56. Find the vertices of $\triangle ABC$ after a dilation centered at the origin with a scale factor of 3. Then graph the original image and the dilation. (Lesson 6-8)

57. Find the vertices of $\triangle ABC$ after a dilation centered at the origin with a scale factor of $\frac{1}{2}$. Then graph the original image and the dilation. (Lesson 6-8)

Solve each inequality. Check your solution. (Lesson 5-4)

58. $x + 3 < 8$

59. $14 + y \geq 7$

60. $-13 \geq 9 + b$

61. $a - 5 > 6$

62. **PAINTING** A person-day is a unit of measure that represents one person working for one day. A painting contractor estimates that it will take 24 person-days to paint a house. Write and solve an equation to find how many painters the contractor will need to hire to paint the house in 6 days. (Lesson 4-4)

Skills Review

Multiply or divide. (Previous Course)

63. 18.2×100

64. 0.04×100

65. $33.3 \div 100$

66. $0.9 \div 100$

7-2 Fractions, Decimals, and Percents

Then

You have already expressed fractions as percents and percents as fractions. (Lesson 7-1)

Now

- Express percents as decimals and decimals as percents.
- Compare and order fractions, decimals, and percents.

Math Online

glencoe.com

- Extra Examples
- Personal Tutor
- Self-Check Quiz
- Homework Help

Why?

PETS The table shows the percent of spending on pets in the United States in a recent year.

Category	Percent of Sales
Food	40%
Live animal purchases	5%
Grooming & Boarding	7%
Supplies/Medicine	24%
Vet Care	24%

U.S. Spending on Pets

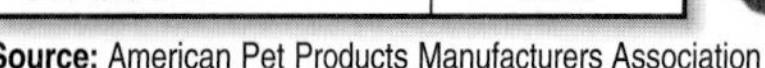

Source: American Pet Products Manufacturers Association

a. Write each percent as a fraction. Do not simplify the fractions.

b. Write each fraction in part **a** as a decimal.

c. How could you write a percent as a decimal without writing the percent as a fraction first?

Percents and Decimals You have learned to write percents as fractions and then as decimals.

$$24\% = \frac{24}{100} = 0.24 \qquad 7\% = \frac{7}{100} = 0.07$$

You can also write decimals as fractions and then as percents.

$$0.40 = \frac{40}{100} = 40\% \qquad 0.05 = \frac{5}{100} = 5\%$$

These examples suggest the following rules.

Key Concept — Percents and Decimals

For Your FOLDABLE

- To write a percent as a decimal, divide by 100 and remove the percent symbol.
- To write a decimal as a percent, multiply by 100 and add the percent symbol.

EXAMPLE 1 Percents as Decimals

Write each percent as a decimal.

a. 16%

$16\% = .16$ — **Remove the % symbol and divide by 100.**

$= 0.16$ — **Add a zero in the ones place.**

b. 9%

$9\% = .09$ — **Remove the % symbol and divide by 100. Add a placeholder zero.**

$= 0.09$ — **Add a zero in the ones place.**

Check Your Progress

1A. 78% **1B.** 2%

Personal Tutor glencoe.com

StudyTip

Mental Math To divide a number by 100, move the decimal point two places to the left. To multiply a number by 100, move the decimal point two places to the right.

StudyTip

Fractions When the numerator of a fraction is less than the denominator, the fraction is less than 100%. When the numerator is greater than the denominator, the fraction is greater than 100%.

EXAMPLE 2 Percents Less than 1 or Greater than 100

Write each percent as a decimal.

a. 0.7%

$0.7 = .007$	Remove the % symbol and divide by 100. Add placeholder zeros.
$= 0.007$	Add a zero in the ones place.

b. 537%

$537\% = 5.37$	Remove the % symbol and divide by 100.
$= 5.37$	

Check Your Progress

2A. 126% **2B.** 0.9%

Personal Tutor glencoe.com

EXAMPLE 3 Decimals as Percents

Write each decimal as a percent.

a. 0.2

$0.2 = 0.20$	Multiply by 100. Add a placeholder zero.
$= 20\%$	Add the % symbol.

b. 3.84

$3.84 = 3.84$	Multiply by 100.
$= 384\%$	Add the % symbol.

c. 0.005

$0.005 = 0.005$	Multiply by 100.
$= 0.5\%$	Add the % symbol.

Check Your Progress

3A. 0.01 **3B.** 3.91 **3C.** 0.002

Personal Tutor glencoe.com

You have expressed fractions as decimals and decimals as percents. Fractions, decimals, and percents are all different names that represent the same number. The model represents these forms.

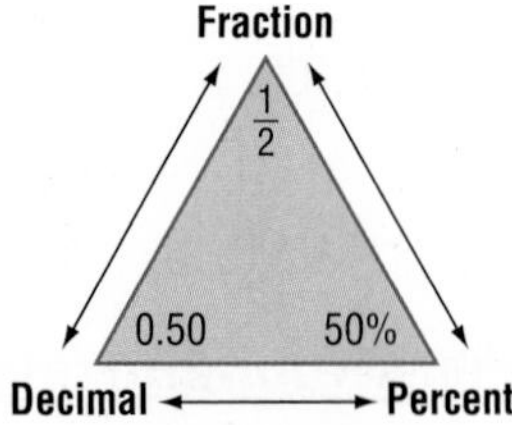

You can express a fraction as a percent by first expressing the fraction as a decimal and then expressing the decimal as a percent.

EXAMPLE 4 Fractions as Percents

Express each fraction as a percent. Round to the nearest tenth, if necessary.

a. $\frac{5}{8}$

$\frac{5}{8} = 0.625$

$= 62.5\%$

b. $\frac{7}{9}$

$\frac{7}{9} = 0.7777777\ldots$

$\approx 77.8\%$

c. $\frac{27}{13}$

$\frac{27}{13} \approx 2.076923077$

$\approx 207.7\%$

d. $\frac{3}{400}$

$\frac{3}{400} = 0.0075$

$= 0.75\%$

Check Your Progress

4A. $\frac{27}{40}$ **4B.** $\frac{1}{9}$ **4C.** $\frac{1}{250}$ **4D.** $\frac{17}{11}$

Personal Tutor glencoe.com

Compare Fractions, Decimals, and Percents You can compare fractions, decimals, and percents by writing them in the same format.

Real-World Link

Of wireless cell phone subscribers, 45% use text messaging, 44% use the camera, 17% play games, and 11% access e-mail.

Source: *ZDNet Research*

Real-World EXAMPLE 5 Compare Numbers

COMMUNICATION In an online survey, one-fifth of teenagers said that their favorite way of communicating with friends is by e-mail, 21% preferred using their cell phones, and 0.15 favored instant messaging. Which of these groups is the largest?

Write one-fifth and 0.15 as percents. Then compare with 21%.

$\frac{1}{5} = 0.20$ or 20% $\quad$ $0.15 = 15\%$

Since $21\% > 20\%$ and $21\% > 15\%$, the group that preferred cell phones is the largest.

Check Your Progress

5. **FOOD** At a restaurant, one-eighth of customers prefer fish sandwiches, 14% prefer rib sandwiches, and 0.12 prefer chicken sandwiches. Which of these groups is the largest?

Personal Tutor glencoe.com

Concept Summary — Common Equivalents

For Your FOLDABLE

$\frac{1}{5} = 0.20 = 20\%$	$\frac{1}{3} = 0.\overline{3} = 33.\overline{3}\%$	$\frac{1}{8} = 0.125 = 12.5\%$
$\frac{2}{5} = 0.40 = 40\%$	$\frac{2}{3} = 0.\overline{6} = 66.\overline{6}\%$	$\frac{3}{8} = 0.375 = 37.5\%$
$\frac{3}{5} = 0.60 = 60\%$	$\frac{1}{6} = 0.1\overline{6} = 16.\overline{6}\%$	$\frac{5}{8} = 0.625 = 62.5\%$
$\frac{4}{5} = 0.80 = 80\%$	$\frac{5}{6} = 0.8\overline{3} = 83.\overline{3}\%$	$\frac{7}{8} = 0.875 = 87.5\%$

Check Your Understanding

Examples 1 and 2 pp. 337–338

Write each percent as a decimal.

1. 45% **2.** 8% **3.** 0.6% **4.** 455%

Examples 3 and 4 pp. 338–339

Express each decimal or fraction as a percent. Round to the nearest tenth, if necessary.

5. $\frac{15}{500}$ **6.** $\frac{4}{200}$ **7.** 0.6 **8.** 1.45

Example 5 p. 339

9. **PETS** Of the students in Miss Han's class 32% own a cat, $\frac{3}{8}$ own a dog, and 0.29 own a bird. Which type of pet do most students own?

Practice and Problem Solving

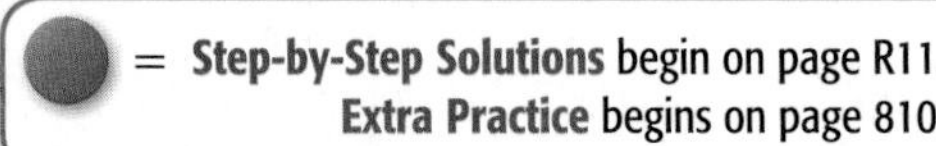
= **Step-by-Step Solutions** begin on page R11.
Extra Practice begins on page 810.

Examples 1 and 2 pp. 337–338

Write each percent as a decimal.

10. 19% **11.** 75% **12.** 2% **13.** 3%

14. 0.51% **15.** 0.91% **16.** 166% **17.** 284%

Examples 3 and 4 pp. 338–339

Write each decimal or fraction as a percent. Round to the nearest tenth, if necessary.

18. 0.57 **19** 0.43 **20.** $\frac{1}{3}$ **21.** $\frac{2}{7}$

22. $\frac{12}{800}$ **23.** $\frac{7}{200}$ **24.** 0.2 **25.** 0.8

26. 5.43 **27.** 6.20 **28.** $\frac{22}{9}$ **29.** $\frac{11}{3}$

30. $\frac{10}{2000}$ **31.** $\frac{2}{1000}$ **32.** 0.0093 **33.** 0.0002

Example 5 p. 339

34. **FRUIT** Thirty percent of high school freshman chose apples as their favorite fruit in a survey. Strawberries were chosen by 0.03 of students, and $\frac{1}{3}$ chose watermelon. Which group is the largest? Explain.

35. **MOVIES** Neka was organizing his DVDs. Of his DVDs, $\frac{1}{5}$ were comedies, 18% were drama, and 0.24 were action. Which type of movie makes up most of his collection?

Write each list of numbers in order from least to greatest.

36. 18%, $\frac{1}{8}$, 0.08 **37.** 2.2, 227%, $\frac{5}{22}$ **38.** 0.23, 28%, $\frac{4}{15}$

39. **SPORTS** After surveying 200 middle school students about their favorite sports, Antoine and Bartoli made the graph at the right.

a. Estimate the number of students that ranked basketball as their favorite sport.

b. About how many students ranked football or soccer as their favorite sport?

Favorite Sports

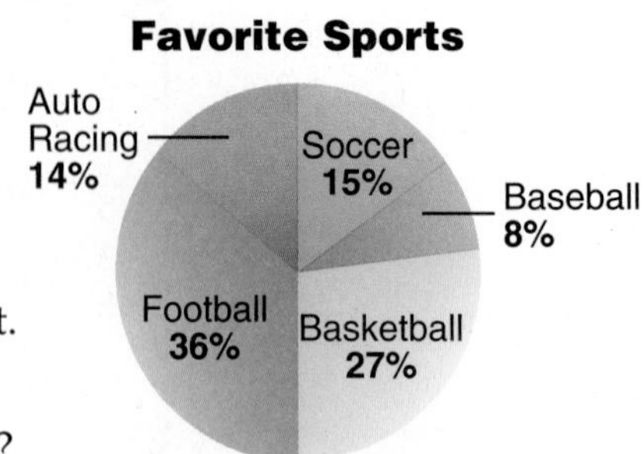

40. **SOCCER** Use the table to rank the goalies in order from the greatest percent of saves to the least percent of saves.

Player	Saves
Christina Reuter	$\frac{64}{71}$
Chantel Jones	0.906
Amber Campbell	90.4%

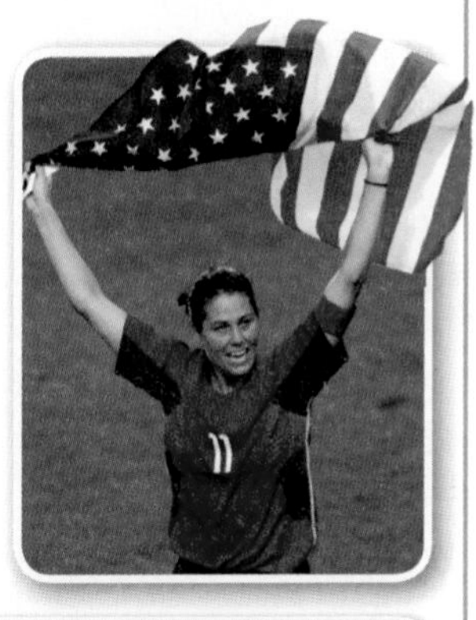

Real-World Link

The U.S. Women's National Soccer team is the only women's team to win the World Cup two times. The team won Olympic gold medals in 1996, 2004, and 2008.

Source: National Soccer Hall of Fame

41. **TESTS** On his last math test, Luther answered 1 question or 4% of the questions incorrectly. How many questions did Luther answer correctly?

42. **MULTIPLE REPRESENTATIONS** In this problem, you will investigate bar graphs and percents. The graph shows the results of a recent survey.

a. **TABULAR** Create a table that shows the fraction of total students that prefer each ride. Write each fraction in simplest form, then as a decimal and percent.

b. **NUMERICAL** Write each fraction as an equivalent fraction with a denominator of 100.

c. **ANALYTICAL** Compare and contrast the results of parts **a** and **b**.

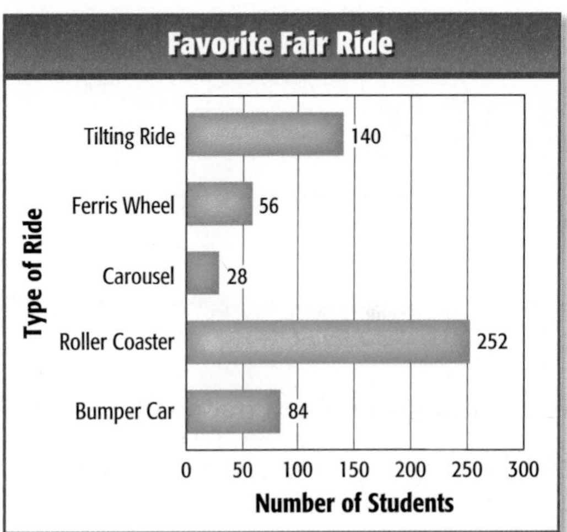

H.O.T. Problems

Use **H**igher-**O**rder **T**hinking Skills

43. **OPEN ENDED** Name a percent that is between 0.10 and 0.01. Write it as a decimal and as a fraction.

44. **WHICH ONE DOESN'T BELONG?** Identify the ratio that does not have the same value as the other three. Explain your reasoning.

$\frac{6}{25}$	2.4	24%	6 out of 25

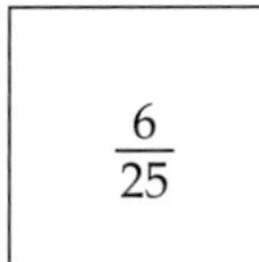

45. **FIND THE ERROR** Carlita and Len are writing 1.5 as a percent. Is either of them correct? Explain your reasoning.

Carlita	Len
1.5 = 1.5 ÷ 100 or 0.015%	1.5 = 1.5 × 100 or 150%

46. **CHALLENGE** Suppose you have two squares. The side length of the smaller square is 50% of the side length of the larger square. Is the area of the smaller square 50% of the area of the larger square? Explain your reasoning.

47. **WRITING IN MATH** Is 0.5% less than or greater than 0.0005? Explain.

Standardized Test Practice

48. Anoki's math teacher drops the lowest quiz score when averaging grades. Which score should his teacher drop?

A 12 correct out of 15

B 79%

C $\frac{56}{70}$

D 0.75

49. A toy company claims that 0.04% of its toys are defective. Which number is *not* equivalent to 0.04%?

F 0.0004

G $\frac{4}{10000}$

H $\frac{1}{2500}$

J 4.0

50. The table shows how Silvia spent her earnings from a paper route. Which represents the part of her money spent at the amusement park?

Item or Activity	Amount
Amusement park	$65
Bicycle	$105
Clothing	$30

A three-eighths

B 32.5%

C 0.65

D $\frac{2}{5}$

51. GRIDDED RESPONSE If 36 out of 80 students voted for Sophia for class president, what percent of the students voted for her?

Spiral Review

Write each percent as a fraction in simplest form. (Lesson 7-1)

52. 30%

53. $12\frac{1}{2}\%$

54. 125%

55. ZOO Refer to the graphic shown. If the triangles are similar, how far are the gorillas from the cheetahs? (Lesson 6-8)

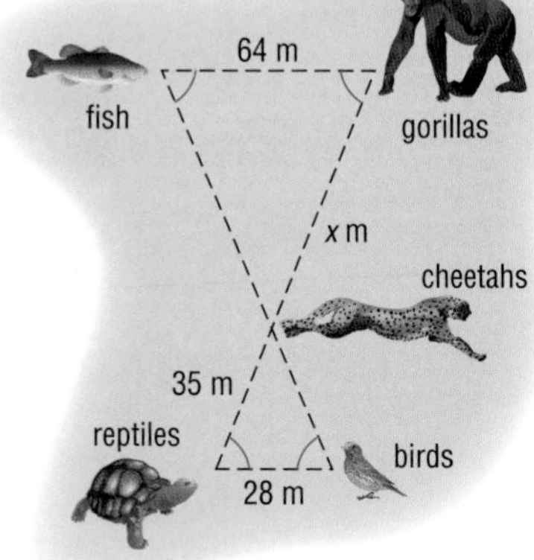

56. MEDICINE For Jillian's cough, her doctor says that she should take eight tablets the first day and then four tablets each day until her prescription runs out. There are 36 tablets. Solve $8 + 4d = 36$ to find the number of days she will take only four tablets. (Lesson 4-5)

Simplify each expression. (Lesson 4-2)

57. $6a + 4 + 2a$

58. $x + 9x + 3$

59. $3x + 2y + 4y$

60. $6c + 4 + c + 8$

61. TRAVEL Joshua's family is packing for a trip. The total weight of their luggage cannot exceed 100 pounds. They have 3 suitcases that weigh 24 pounds each and 2 sport bags that weigh 12 pounds each. Is Joshua's family's luggage within the 100-pound limit? Explain. (Lesson 1-1)

Skills Review

Solve each proportion. (Lesson 6-5)

62. $\frac{20}{4} = \frac{x}{100}$

63. $\frac{63}{7} = \frac{y}{100}$

64. $\frac{65}{8} = \frac{n}{100}$

65. $\frac{m}{10} = \frac{4.2}{100}$

66. $\frac{h}{350} = \frac{26}{100}$

67. $\frac{86.4}{k} = \frac{36}{100}$

EXPLORE **7-3** Algebra Lab

Using a Percent Model

Math Online glencoe.com

Math *in Motion,* Animation

You can use a percent model to represent many real-world situations.

ACTIVITY 1 Finding a Percent

An Internet advertisement claims that 3 out of 5 dentists prefer a certain toothpaste. What percent does this represent?

You can find the *percent* by using a model.

Step 1

Draw a 10-unit by 1-unit rectangle on grid paper. Label the units on the right from 0 to 100, because percent is a ratio that compares a number to 100.

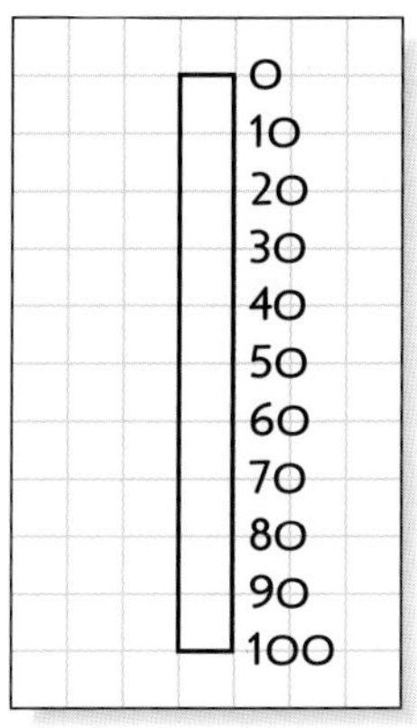

Step 2

On the left side, mark equal units from 0 to 5, because 5 represents the whole quantity. Locate 3 on this scale.

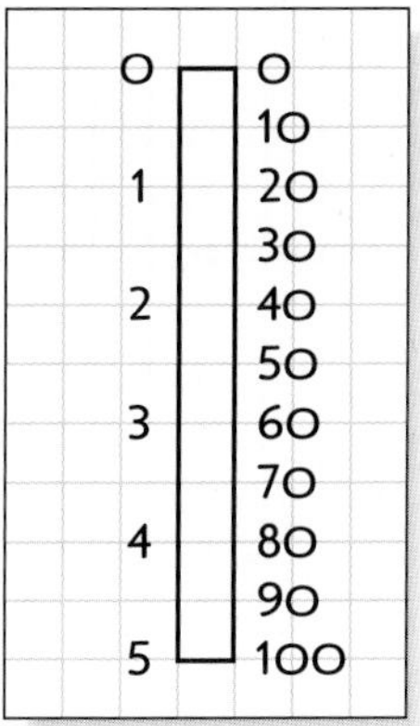

Step 3

Draw a horizontal line from 3 on the left side to the right side of the model. The number on the right side is the percent. Label the model as shown.

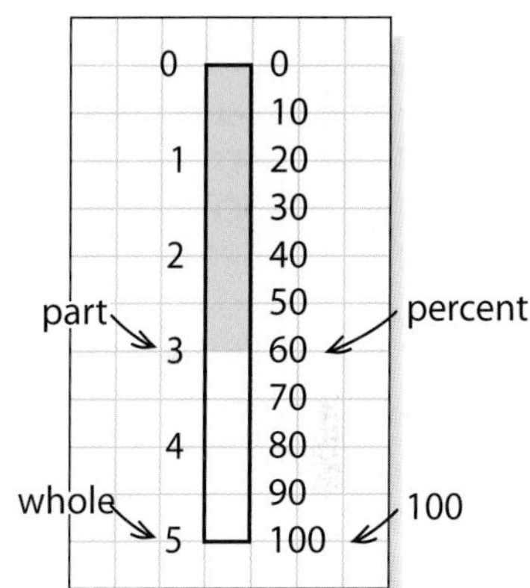

Using the model, you can see that the ratio *3 out of 5* is the same as 60%. So, according to this claim, 60% of dentists prefer this toothpaste.

Exercises

Draw a model and find the percent that is represented by each ratio. If it is not possible to find the exact percent using the model, estimate.

1. 4 out of 10

2. 7 out of 10

3. 4 out of 5

4. 2 out of 4

5. 6 out of 20

6. 5 out of 25

7. 8 out of 40

8. 6 out of 24

9. 1 out of 3

10. 4 out of 9

ACTIVITY 2 Finding a Percent

Suppose a store advertises a sale in which all merchandise is 30% off its price. If the price of a cell phone case is $20, how much will you save?

In this case, you know the percent. You need to find what part of the original price you'll save.

You can find the *part* by using a similar model.

Step 1

Draw a 10-unit by 1-unit rectangle on grid paper. Label the units on the right from 0 to 100 because percent is a ratio that compares a number to 100.

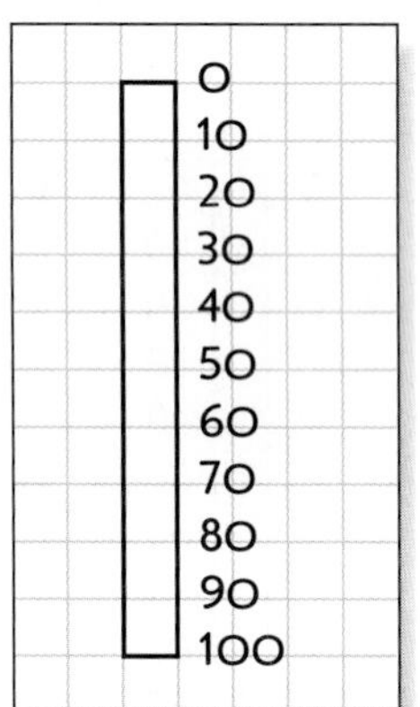

Step 2

On the left side, mark equal units from 0 to 20, because 20 represents the whole quantity.

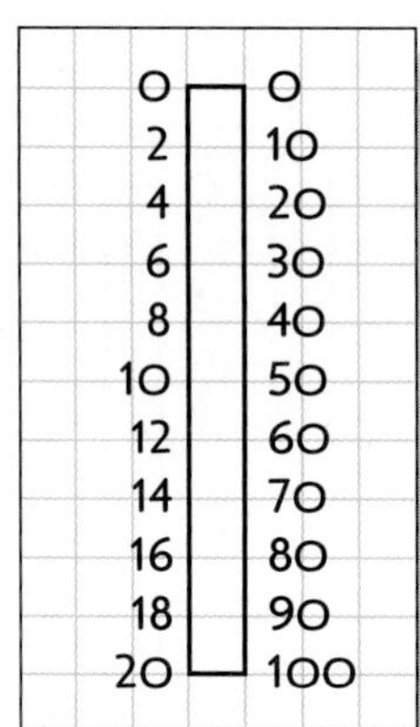

Step 3

Draw a horizontal line from 30% on the right side to the left side of the model. The number on the left side is the part. Label the model as shown.

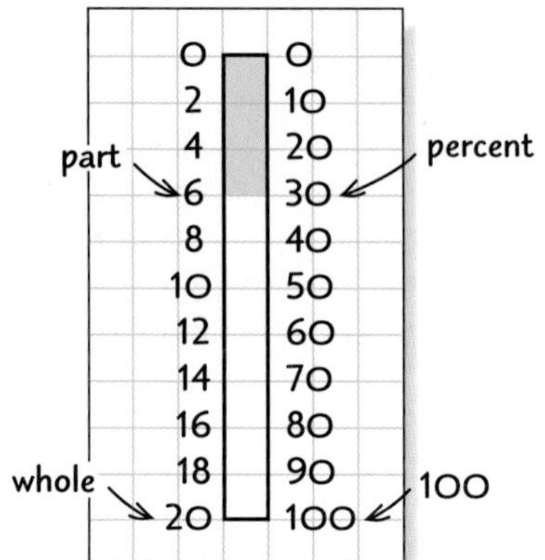

Using the model, you can see that 30% of 20 is 6. So, you will save $6 if you buy the cell phone case.

Exercises

Draw a model and find the part that is represented. If it is not possible to find an exact answer from the model, estimate.

11. 20% of 80

12. 60% of 15

13. 80% of 40

14. 30% of 50

15. 25% of 50

16. 75% of 60

17. 10% of 150

18. 45% of 500

19. $33\frac{1}{3}$% of 20

20. 90% of 20

21. REASONING Suppose you know that 60% of some number is 18. Use the model below to find the number x. Explain your reasoning.

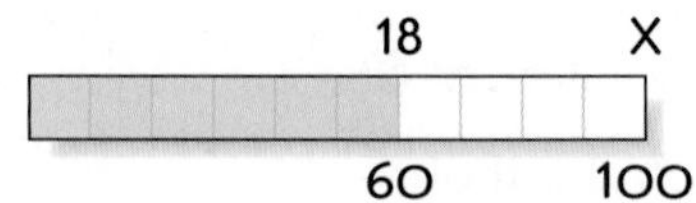

7-3 Using the Percent Proportion

Then
You have already used models to find percents. (Explore 7-3)

Now
- Use the percent proportion to solve problems.
- Apply the percent proportion to real-world problems.

New Vocabulary
percent proportion

Math Online
glencoe.com
- Extra Examples
- Personal Tutor
- Self-Check Quiz
- Homework Help

Why?

A recipe for strawberry lemonade is shown.

a. Write a ratio that compares the number of cups of water to the total number of cups in the recipe.

b. Write the ratio as a fraction in simplest form and as a percent.

Strawberry Lemonade
2 cups sugar
1 cup lemon juice
6 cups water
3 cups strawberries

The Percent Proportion From previous lessons, you know that 6 out of 12 is 50%. These numbers form a percent proportion. In a **percent proportion**, one ratio compares *part* of a quantity to the *whole* quantity. The other ratio is the equivalent percent written as a fraction with a denominator of 100.

Key Concept — Percent Proportion

For Your FOLDABLE

Words $\frac{\text{part}}{\text{whole}} = \frac{\text{percent}}{100}$

Symbols $\frac{a}{b} = \frac{p}{100}$, where a is the part, b is the whole and p is the percent written as a fraction.

EXAMPLE 1 Find the Percent

Twelve is what percent of 32?

Twelve is being compared to 32. So, 12 is the part and 32 is the whole.

Words	Twelve is what percent of 32?
Variable	Let p represent the percent.
Proportion	part → $\frac{12}{32} = \frac{p}{100}$ } percent ← whole

$\frac{12}{32} = \frac{p}{100}$ — **Write the percent proportion.**

$12 \cdot 100 = 32 \cdot p$ — **Find the cross products.**

$1200 = 32p$ — **Multiply.**

$\frac{1200}{32} = \frac{32p}{32}$ — **Divide each side by 32.**

$37.5 = p$ — **Simplify.**

So, 12 is 37.5% of 32.

Check Your Progress

1A. Fifteen is what percent of 20?

1B. What percent of 5 is 12?

Personal Tutor glencoe.com

StudyTip

Check for Reasonableness Ten percent of 450 is 45. So, the answer should be more than 45.

EXAMPLE 2 Find the Part

What number is 15.5% of 450?

The percent is 15.5, and the base is 450. Let a represent the part.

$\frac{a}{450} = \frac{15.5}{100}$ **Write the percent proportion.**

$a \cdot 100 = 450 \cdot 15.5$ **Find the cross products.**

$100a = 6975$ **Multiply.**

$a = 69.75$ **Mentally divide each side by 100.**

So, 69.75 is 15.5% of 450.

Check Your Progress

2A. What number is 11.4% of 330?

2B. Find 15.3% of 425.

Personal Tutor glencoe.com

StudyTip

Whole In percent problems, the whole usually follows the word *of*.

EXAMPLE 3 Find the Whole

Seventy-eight is 60% of what number?

The percent is 60%, and the part is 78. Let b represent the whole.

$\frac{78}{b} = \frac{60}{100}$ **Write the percent proportion.**

$78 \cdot 100 = b \cdot 60$ **Find the cross products.**

$7800b = 60b$ **Multiply.**

$\frac{7800}{60} = \frac{60b}{60}$ **Divide each side by 60.**

$130 = b$ **Simplify.**

So, 78 is 60% of 130.

Check Your Progress

3A. Thirty percent of what number is 63?

3B. Forty-five is 3% of what number?

Personal Tutor glencoe.com

Concept Summary Types of Percent Problems

For Your FOLDABLE

Type	Example	Proportion
Find the Percent	1 is what percent of 5? or What percent of 5 is 1?	$\frac{1}{5} = \frac{p}{100}$
Find the Part	What number is 20% of 5?	$\frac{a}{5} = \frac{20}{100}$
Find the Whole	1 is 20% of what number?	$\frac{1}{b} = \frac{20}{100}$

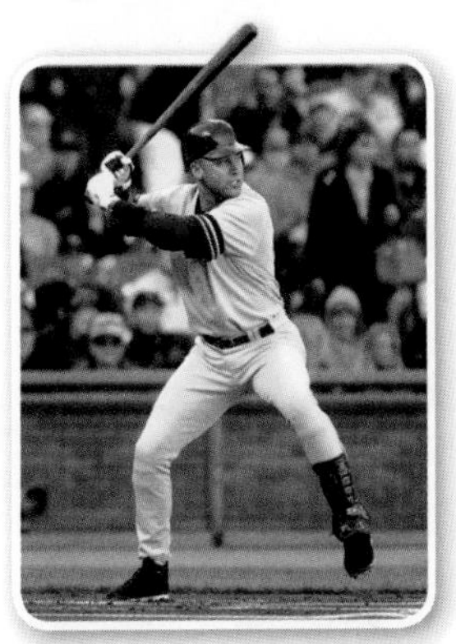

Real-World Link

Derek Jeter's career batting average through the 2007 season was 0.317. This ranks him with the 5th highest lifetime batting average of all active baseball players.

Source: MLB

Real-World EXAMPLE 4 Apply the Percent Proportion

BASEBALL The table shows the batting statistics for one season for Derek Jeter of the New York Yankees. If he had 639 at bats, what percent of his at bats were singles?

Stat	Number
single	151
double	39
triple	4
home run	12
walk	56
strikeout	100

Compare the number of singles, 151, to the total number of at bats, 639. Let p represent the percent.

$\frac{151}{639} = \frac{p}{100}$ **Write the percent proportion.**

$151 \cdot 100 = 639 \cdot p$ **Find the cross products.**

$15{,}100 = 639p$ **Simplify.**

$\frac{15{,}100}{639} = \frac{639p}{639}$ **Divide each side by 639.**

$23.6 \approx p$ **Simplify.**

So, about 23.6% of time Derek's at bats were singles.

Check Your Progress

4. **BASEBALL** What percent of his at bats were strikeouts?

Personal Tutor glencoe.com

Check Your Understanding

Examples 1–3 pp. 345–346

Use the percent proportion to solve each problem.

1. 18 is what percent of 72?
2. What percent of 8 is 20?
3. What is 74% of 56?
4. 9 is 20% of what number?
5. What percent of 2 is 8?
6. Find 6% of 300.

Example 4 p. 347

7. **TEST SCORES** Of the 120 math tests, 47 were Bs. What percent of the math tests were Bs?

Practice and Problem Solving

= **Step-by-Step Solutions** begin on page R11. **Extra Practice** begins on page 810.

Examples 1–3 pp. 345–346

Use the percent proportion to solve each problem.

8. 16 is what percent of 64?
9. 21 is what percent of 50?
10. What percent of 145 is 52.2?
11. What percent of 36 is 19.8?
12. What is 60% of 120?
13. What is 80% of 125?
14. Find 65% of 440.
15. Find 83% of 200.
16. 12 is 40% of what number?
17. 34 is 20% of what number?
18. 80% of what number is 12?
19. 4% of what number is 15?

Example 4 p. 347

20. **DOGS** Sixteen of the 80 dogs at a kennel are golden retrievers. What percent of the dogs at the kennel are golden retrievers?
21. **FLAVORS** The number of lime-flavored gumballs in a gumball machine is 85. If this is 17% of the number of gumballs in the machine, how many gumballs are in the machine?

22. **RESEARCH** Use the Internet or another source to find the percent of states that begin with *A*, *I*, *O*, or *U*.

23. **SURVEYS** Use the circle graph that shows the results of a survey about New Year's Resolutions.

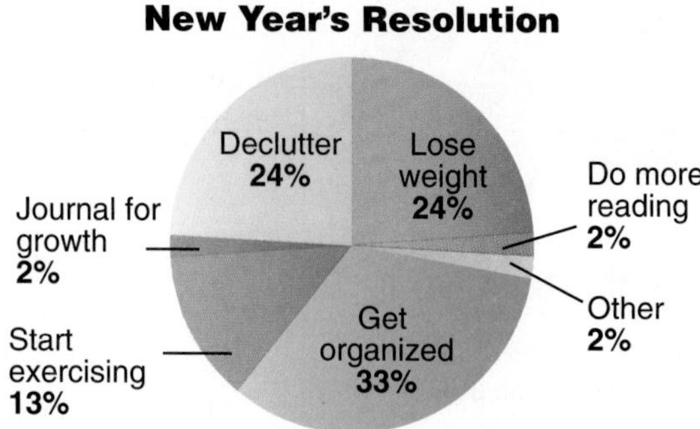

a. Determine about how many of the 2947 people surveyed said their most important New Year's resolution was to get organized.

b. About how many said their most important New Year's resolution was to do more reading or to declutter?

24. **FINANCIAL LITERACY** Accessories Central is having a summer clearance on sunglasses. Maria wants to buy a pair of sunglasses that cost \$48 with a 65% discount. The same pair of sunglasses costs \$38 with a 55% discount at Shades Inc. Which store has the better price for the pair of sunglasses? Explain your reasoning.

Real-World Link

An online survey showed that 47% of the 15- to 25-year-olds surveyed voted education as their top concern.

25. **PATTERNS** A pattern of equations is shown.

$$2\% \text{ of } 100 = 2$$
$$4\% \text{ of } 50 = 2$$
$$8\% \text{ of } 25 = 2$$
$$16\% \text{ of } 12.5 = 2$$

a. Describe the pattern.

b. Find the next equation in the pattern.

26. **SOCIAL STUDIES** The bar graph shows the results of an online survey of 1242 people aged 15–25 about their political involvement over the last 12 months.

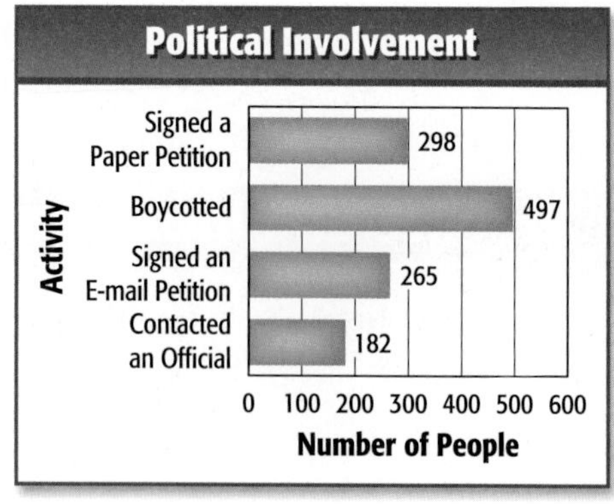

a. What percent of the people boycotted? Round to the nearest percent.

b. Of the people who signed an e-mail petition, 20% were 18 years old. How many 18 year-olds signed an e-mail petition?

c. Based on the results of the survey, predict about how many people out of 2000 would contact an official.

Watch Out!

Wholes The whole is not always the larger number.

Use the percent proportion to solve each problem. Round to the nearest tenth if necessary.

27. 45 is what percent of 15?

28. 13 is 25% of what number?

29. What is 58% of 7?

30. 8 is what percent of 2000?

31. What is 0.6% of 360?

32. 41 is $5\frac{1}{3}$% of what number?

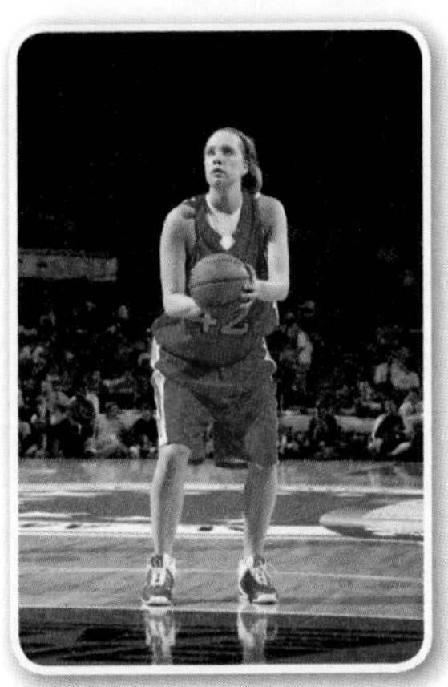

Real-World Link

The free-throw line in basketball is 2 inches wide, 12 feet long, and 15 feet from the backboard.

Source: NBA

33. BASKETBALL A professional basketball player made 465 out of 520 free throws in one season. The next season the player made 386 out of 437. For both seasons combined, what percent of free throws did the player make? Round to the nearest tenth, if necessary.

34. TALENT SHOWS The table shows the results of a student survey at Crestview Middle School.

Favorite Type of Talent Show Act	
Act	**Number of Students**
stand-up comedy	198
singing	150
dancing	212
playing instruments	80

a. What percent of the students surveyed said their favorite act was playing instruments?

b. If 10 more students vote for stand-up comedy, will the percent be greater than, less than, or equal to the current percent? Explain.

35 CELL PHONES A cell phone store has 120 cell phones in stock. Of these, 45 have keyboards. The manager of the store wants to add more cell phones with keyboards so that 40% of the stock has keyboards.

a. Write and solve a proportion to find the number of cell phones with keyboards that should be added to the store's inventory.

b. What will be the total number of cell phones in stock?

H.O.T. Problems

Use Higher-Order Thinking Skills

36. OPEN ENDED Write and solve a real-world problem involving percents.

37. CHALLENGE Without calculating, arrange the following from least to greatest value. Justify your reasoning.

25% of 160, 5% of 80, 25% of 80

38. REASONING Sabrina spelled 82% of her spelling words correctly on her spelling tests this year. If she spells 13 out of 15 words correctly on her next test, will this help or hurt her average? Explain.

39. CHALLENGE Find the value of y so that $y\% = \frac{3y + 9}{600}$.

40. FIND THE ERROR Bethany and Mei are finding what percent of 32 is 18. Is either of them correct? Explain your reasoning.

Bethany	Mei
$\frac{18}{32} = \frac{x}{100}$	$\frac{32}{18} = \frac{x}{100}$
$x \approx 56$	$x \approx 178$
So, 32 is about 56% of 18.	So, 32 is about 178% of 18.

41. REASONING Is x% of y and y% of x *always*, *sometimes*, or *never* equivalent? Explain your reasoning.

42. WRITING IN MATH Kenji states that 5.9 is 125% of 47. Is his answer reasonable? Explain your reasoning.

Standardized Test Practice

43. The table shows the results of a survey of middle school students about their favorite school mascots.

Mascot	Number of Students
Falcon	60
Stallion	123
Ram	86
Tiger	131

Based on the data, predict how many out of 2000 students would vote for the falcon.

A 600
B 400
C 300
D 200

44. If 40% of a number is 32, what is 35% of the number?

F 8
G 20
H 24
J 28

45. A place kicker expects to make 75% of his field goal attempts this season. If he attempts 36 field goals this season, which of the following statements does *not* represent the place kicker's expectation?

A The place kicker will make 27 field goals.
B The place kicker will miss 9 field goals.
C Less than $\frac{1}{4}$ of the field goal attempts will be missed.
D The place kicker will make more than $\frac{1}{2}$ of his field goal attempts.

46. SHORT RESPONSE If 68 is 25% of a number, what is 600% of the number?

Spiral Review

47. MEDIA In a survey, 35% of those surveyed said that they get the news from their local television station while three-fifths said that they get the news from a daily newspaper. From which source do more people get their news? (Lesson 7-2)

Write each percent as a fraction in simplest form. (Lesson 7-1)

48. 45%
49. 120%
50. 0.5%
51. $83\frac{1}{3}\%$

52. INSECTS In a drawing of a honeybee, the bee is 4.8 centimeters long. The actual size of the honeybee is 1.2 centimeters. What is the scale of the drawing? (Lesson 5-6)

Find each sum or difference. Write in simplest form. (Lesson 3-5)

53. $\frac{17}{18} - \frac{5}{18}$
54. $\frac{3}{10} + \frac{7}{10}$
55. $\frac{1}{2} - \frac{4}{5}$
56. $\frac{7}{15} + \frac{1}{6}$
57. $\frac{3}{4} - \frac{4}{9}$
58. $\frac{1}{2} - \frac{7}{8}$

Skills Review

Find each product. (Lesson 3-3)

59. $\frac{1}{4} \times 12$
60. $\frac{3}{4} \times 24$
61. $38 \times \frac{1}{2}$
62. $15 \times \frac{1}{3}$

7-4 Find Percent of a Number Mentally

Then

You have already found percents using the percent proportion. (Lesson 7-3)

Now

- Compute mentally with percents.
- Estimate with percents.

Math Online

glencoe.com

- Extra Examples
- Personal Tutor
- Self-Check Quiz
- Homework Help

Why?

The table shows the final standing of the first, second, and third place finishers in a recent Women's World Cup soccer tournament.

Team	Number of Wins
Germany	5
United States	4
Norway	3

a. If the team from Norway won 50% of their games, use mental math to find the total number of games they played.

b. If the team from Germany scored 2 goals in 40% of their winning games, use mental math to find the number of winning games in which they scored 2 goals.

Find Percent of a Number Mentally When you compute with common percents like 40% or 50%, it may be easier to use the fraction form of the percent. The number line shows some common percent-fraction equivalents.

Concept Summary — Percent-Fraction Equivalents — For Your FOLDABLE

$25\% = \frac{1}{4}$	$20\% = \frac{1}{5}$	$10\% = \frac{1}{10}$	$12\frac{1}{2}\% = \frac{1}{8}$	$16\frac{2}{3}\% = \frac{1}{6}$
$50\% = \frac{1}{2}$	$40\% = \frac{2}{5}$	$30\% = \frac{3}{10}$	$37\frac{1}{2}\% = \frac{3}{8}$	$33\frac{1}{3}\% = \frac{1}{3}$
$75\% = \frac{3}{4}$	$60\% = \frac{3}{5}$	$70\% = \frac{7}{10}$	$62\frac{1}{2}\% = \frac{5}{8}$	$66\frac{2}{3}\% = \frac{2}{3}$
$100\% = \frac{1}{1}$	$80\% = \frac{4}{5}$	$90\% = \frac{9}{10}$	$87\frac{1}{2}\% = \frac{7}{8}$	$83\frac{1}{3}\% = \frac{5}{6}$

EXAMPLE 1 Use a Fraction to Compute Mentally

Find the percent of each number mentally.

a. 75% of 24

$75\% \text{ of } 24 = \frac{3}{4} \text{ of } 24$ **THINK** $75\% = \frac{3}{4}$

$= 18$ **THINK** $\frac{3}{4}$ **of 24 is 18.**

b. 80% of 60

$80\% \text{ of } 60 = \frac{4}{5} \text{ of } 60$ **THINK** $80\% = \frac{4}{5}$

$= 48$ **THINK** $\frac{4}{5}$ **of 60 is 48.**

Check Your Progress

1A. 40% of 50 **1B.** 30% of 70

Personal Tutor glencoe.com

EXAMPLE 2 Use Decimals to Compute Mentally

Compute mentally.

a. 10% of 76

10% of 76 = 0.1 • 76 or 7.6

b. 1% of 122

1% of 122 = 0.01 • 122 or 1.22

Check Your Progress

2A. 10% of 42

2B. 1% of 264

Personal Tutor glencoe.com

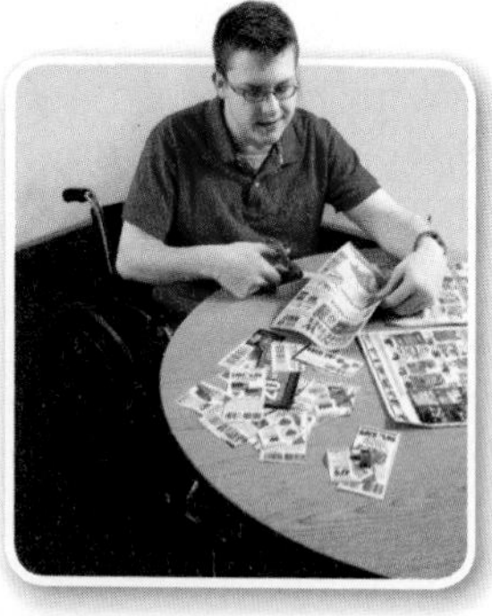

Real-World Link

Every year nearly 300 billion coupons are distributed nationwide. About 90% of these coupons are distributed in Sunday newspapers.

Source: CMS, Inc.

Real-World EXAMPLE 3 Compute Mentally

SALES Hannah is shopping for school clothes. She has a coupon that will give her 20% off her entire purchase. If the items she buys cost $110 originally, how much will she save with her coupon?

You need to find 20% of the total cost. First, find 10% of 110.

10% of 110 = 0.1 • 11.0 — **Move the decimal point one place to the left.**

= 11

20% is the same as 2 • 10%.

2 • 10% = 2 • 11 or 22 — **Replace 10% with 11.**

So, Hannah will save $22 on her purchase.

Check Your Progress

3. SALES A television that costs $750 is on sale for 15% off. What is the total discount on the television?

Personal Tutor glencoe.com

Estimate With Percents You can estimate when an exact answer is not needed.

Problem-SolvingTip

Determine Reasonable Answers Deciding whether an answer is reasonable is useful when an exact answer is not necessary.

EXAMPLE 4 Estimate Percent of a Number

Estimate.

a. 26% of 64

26% is about 25% or $\frac{1}{4}$.

$\frac{1}{4}$ of 64 is 16.

So, 26% of 64 is about 16.

b. $\frac{2}{3}$% of 891

$\frac{2}{3}\% = \frac{2}{3} \times 1\%$. 1% of 900 is 9.

891 is almost 900.

So, $\frac{2}{3}$% of 891 is about $\frac{2}{3} \times 9$ or 6.

c. 39% of 81

39% is about 40% or $\frac{2}{5}$.

81 is about 80.

$\frac{2}{5}$ of 80 is 32.

So, 39% of 81 is about 32.

d. 120% of 51

100% of 50 is 50.

20% of 50 is 10.

So, 120% of 51 is about 50 + 10 or 60.

Check Your Progress

4A. 92% of 50 **4B.** 63% of 205 **4C.** 75% of 84 **4D.** 130% of 91

Personal Tutor glencoe.com

Real-World Link

About 3 billion pizzas are sold each year in the United States. The ratio of meat toppings to vegetables toppings preferred by Americans is 31 to 50.

Real-World EXAMPLE 5 Estimate Percent of a Number

PIZZA Mr. Williams ordered 4 pizzas for a birthday party. The cost of the pizzas was \$57.96. He wants to tip the delivery person about 15%. What is a reasonable amount for the tip?

Understand You need to find the tip for the delivery person. You know the cost of the pizzas.

Plan Estimate the price of the pizzas. Find 15% of the estimated price.

Solve \$57.96 is about \$60, and 15% = 10% + 5%.

10% of \$60 is \$6.00. **Move the decimal point 1 place to the left.**

5% of \$60 is \$3.00 **5% is one half of 10%.**

So, 15% is about \$6.00 + \$3.00 or \$9.00.

A reasonable amount for the tip is \$9.

Check 10% of \$58 is \$5.80 and 20% of \$58 is \$11.60.
Since \$5.80 < \$9 < \$11.60, the answer is reasonable. ✓

Check Your Progress

5. **RESTAURANT** Haley went to dinner with her friends. Their bill was \$48.61. They want to leave their server a 15% tip. What would be a reasonable amount for the tip? Explain your reasoning.

Personal Tutor glencoe.com

Check Your Understanding

Examples 1 and 2 pp. 351–352

Find the percent of each number mentally.

1. 75% of 16
2. 25% of 32
3. 10% of 37
4. 10% of 115
5. 1% of 72
6. 1% of 231

Example 3 p. 352

7. **HOMEWORK** Jasmine has finished 30% of the exercises on her homework. If there are 40 exercises in all, how many has Jasmine completed?

Example 4 p. 352

Estimate.

8. 11% of 70
9. 53% of 20
10. 40% of 19
11. 87% of 42
12. $\frac{1}{3}$% of 598
13. 110% of 39

Example 5 p. 353

14. **SPORTS** Last basketball season, Carlos made 38% of the baskets he attempted. At this rate, about how many baskets will he make if he attempts 30 baskets?

Practice and Problem Solving

 = **Step-by-Step Solutions** begin on page R11.
Extra Practice begins on page 810.

Examples 1 and 2 pp. 351–352

Find the percent of each number mentally.

15. 40% of 80
16. 20% of 50
17. 25% of 280
18. 75% of 96
19. $33\frac{1}{3}$% of 27
20. $12\frac{1}{2}$% of 48
21. $8\frac{1}{3}$% of 72
22. $87\frac{1}{2}$% of 32
23. 10% of 125
24. 10% of 259
25. 1% of 30
26. 1% of 400

Example 3 p. 352

27. **SALES** A store is having a sale where everything is 15% off. If Jeremy wants to buy items that originally cost \$50, how much will he save?

Example 4 p. 352

Estimate.

28. 16% of 20
29. 73% of 84
30. 46% of 88
31. 25% of 49
32. $\frac{1}{2}$% of 507
33. $\frac{1}{6}$% of 295
34. 148% of 30
35. 276% of 8
36. $\frac{3}{4}$% of 801
37. $\frac{4}{5}$% of 30
38. 117% of 50
39. 194% of 15

Example 5 p. 353

40. **FINANCIAL LITERACY** The total cost for Soledad's manicure was \$32.99. She wants to give the manicurist a 20% tip. What would be a reasonable amount for the tip?

41. **INTERNET** In a national survey of 6700 teens, 81% of teens between the ages of 12 and 17 said they use the Internet to e-mail friends or relatives. About how many teens is this?

42. **MUSIC** The bar graph shows the percent of each age group that owns a portable MP3 player. Suppose there are 825 12–17 year olds in the Louisville School District. About how many of them are likely to own a portable digital music player?

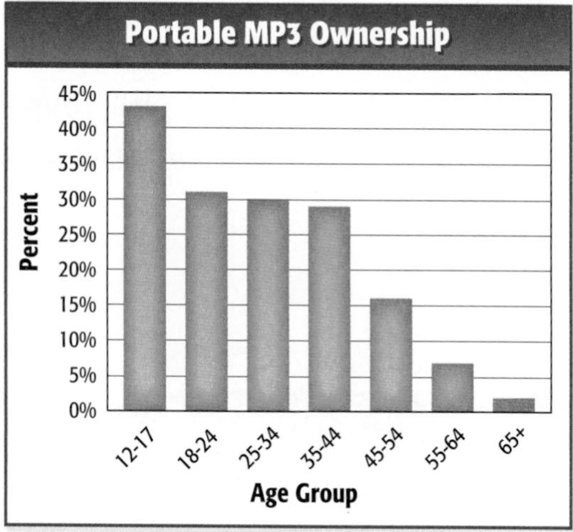

Source: Edison Media Research

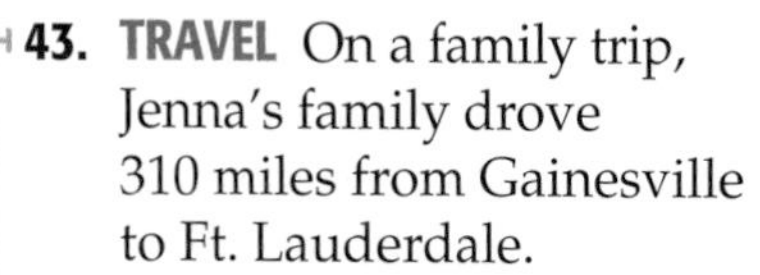

43. **TRAVEL** On a family trip, Jenna's family drove 310 miles from Gainesville to Ft. Lauderdale.
   a. Her dad drove 52% of the way. About how many miles did he drive?
   b. Jenna's mom drove 58% of the distance her dad drove. About how far did she drive?
   c. Her older brother drove the remaining miles. About how many miles did he drive?

Real-World Link

Florida is the number one tourist destination in the world. The Florida tourism industry has an annual economic impact of about \$57 billion.

Source: Visit Florida

44. **ARTS** About 41% of twelfth graders participated in school performing arts last year. A high school had 1800 students, one-fourth of which were twelfth graders. About how many twelfth graders participated in school performing arts?

H.O.T. Problems

Use **H**igher-**O**rder **T**hinking Skills

45. **OPEN ENDED** Suppose you want to find $66\frac{2}{3}$% of a. List two values of a for which you could do the computation mentally. Explain your reasoning.

46. **CHALLENGE** Find two numbers, x and y, such that 10% of x is the same as 40% of y. Explain your reasoning.

47. **WRITING IN MATH** Describe two different ways you could find 20% of 60 mentally.

Standardized Test Practice

48. Jerome bought the items listed with the original prices shown on the receipt. If he saved 20% on each item, what is the *best* estimate of how much he saved?

RECEIPT

Qty.	Item	Amount
1	Network Cable	\$50.00
1	CD 10-pack	\$12.00
1	Mouse	\$24.00
1	Memory	\$55.00

—Thank you, come again.—

A \$28
B \$32
C \$36
D \$40

49. Which fraction is between 85% and 90%?

F $\frac{5}{6}$
G $\frac{7}{8}$
H $\frac{9}{10}$
J $\frac{10}{11}$

50. Lorena, Julian, and Cho completed a group assignment that had 84 questions. Lorena answered $\frac{1}{3}$ of the questions, Julian answered 25% of the questions, and Cho answered the rest. How many questions were answered by the person who answered the greatest number of questions?

A 18
B 21
C 28
D 35

51. EXTENDED RESPONSE The price of Jamila's haircut is \$28. There is also a 7% tax added to the bill and Jamila wants to tip the stylist 15% of the total bill. Justify each solution.

a. About how much is the tax on the bill?

b. Find the approximate cost including tax.

c. About how much of a tip will Jamila give the stylist?

d. If she only has \$35, does she have enough for tax and tip? Explain.

Spiral Review

52. ACTIVITIES According to a survey about family activities, 35% of people said they enjoy playing games, while three-fifths enjoy watching movies, and $\frac{3}{8}$ enjoy sports. Which group is the largest? Explain. (Lesson 7-3)

53. GEOGRAPHY The Arctic Ocean contains 3.7% of the world's water. What fraction is this? (Lesson 7-2)

ALGEBRA **Solve each proportion.** (Lesson 6-5)

54. $\frac{k}{35} = \frac{3}{7}$

55. $\frac{3}{t} = \frac{18}{24}$

56. $\frac{10}{8.4} = \frac{5}{m}$

57. GEOMETRY The area A of the triangle is 33.75 square inches. Use the formula $A = \frac{1}{2}bh$ to find the height h of the triangle. (Lesson 5-1)

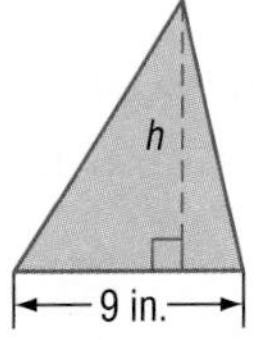

Skills Review

Solve each equation. Check your solution. (Lesson 4-4)

58. $5 = 30b$

59. $40n = 10$

60. $20 = 100k$

61. $34g = 1.7$

62. $3.6 = 90a$

63. $200j = 70$

CHAPTER 7

Mid-Chapter Quiz

Lessons 7-1 through 7-4

Write each percent as a fraction or mixed number in simplest form. (Lesson 7-1)

1. $33\frac{1}{3}\%$ **2.** 0.25% **3.** 175%

Write each fraction as a percent. Round to the nearest hundredth if necessary. (Lesson 7-1)

4. $\frac{1}{6}$ **5.** $\frac{50}{40}$ **6.** $\frac{7}{8}$

7. ANIMALS A male lion weighs $\frac{7}{10}$ more than a female lion that weighs 250 pounds. What percent is this? (Lesson 7-1)

8. BOXES What percent of the squares are dark gray? (Lesson 7-1)

9. MULTIPLE CHOICE Monday's average temperature was 70°F. Tuesday's average temperature was 20% hotter than Monday. Write the percent as a fraction. (Lesson 7-1)

A $\frac{2}{7}$ C $\frac{1}{5}$

B $\frac{1}{4}$ D $\frac{1}{7}$

Write each percent as a decimal (Lesson 7-2)

10. 54% **11.** 87.5%

12. 0.163% **13.** 225%

Express each decimal or fraction as a percent. Round to the nearest tenth, if necessary. (Lesson 7-2)

14. $\frac{8}{64}$ **15.** $\frac{56}{32}$

16. 0.14 **17.** 1.3

18. STUDENTS Of the students in Ms. Hillier's class, 27% have blonde hair, 0.33 have black hair, and $\frac{2}{5}$ have brown hair. Which hair color is most common? (Lesson 7-2)

19. MULTIPLE CHOICE Jason, Marita, Alice, and Marcus were on a road trip. Jason drove 400 miles, Alice drove $\frac{1}{6}$ of the way, Marita and Marcus each drove 0.25 of the way. If they drove a total of 1200 miles, who drove the longest? (Lesson 7-2)

F Jason H Marita

G Alice J Marcus

Use the percent proportion to solve each problem. (Lesson 7-3)

20. 15 is what percent of 45?

21. What percent of 7 is 21?

22. What is 83% of 16?

23. 30 is what percent of 75?

24. Find 8% of 145.

25. POPCORN Jaime popped a bag of popcorn in the microwave. Thirty percent of the kernels did *not* pop. If there were 120 original kernels, how many did pop? (Lesson 7-3)

26. FOOD The Dragons Soccer Team is having a pizza party and voted on what toppings the players preferred. Of the 25 players, what percent like cheese? (Lesson 7-3)

Toppings	Number of Players
Cheese	9
Pepperoni	8
Sausage	3
Hawaiian	5

Find the percent of each number mentally. (Lesson 7-4)

27. 1% of 80 **28.** 25% of 160

29. $87\frac{1}{2}\%$ of 56 **30.** 175% of 200

31. 20% of 105 **32.** $16\frac{2}{3}\%$ of 42

33. PLANETS Mercury's radius is about 40% of Venus' radius. If Venus has a radius of 3800 miles, what is the approximate radius of Mercury? (Lesson 7-4)

7-5

Using Percent Equations

Then

You have already used the percent proportion to solve problems. (Lesson 7-3)

Now

- Solve percent problems using percent equations.
- Apply the percent equation to real-world problems.

New Vocabulary

percent equation

Math Online

glencoe.com

- Extra Examples
- Personal Tutor
- Self-Check Quiz
- Homework Help

Why?

The winner of the New York City Marathon in a recent year was Jelena Prokopcuka. Her total prize money was $130,000.

a. Of her prize money, about 23% was a bonus. Use a proportion to find how much of her prize money was a bonus to the nearest dollar.

b. Express the percent as a decimal. Multiply the total prize money by the decimal. Round to the nearest dollar.

c. Describe how the answers from parts **a** and **b** are related.

Percent Equations A **percent equation** is an equivalent form of the percent proportion in which the percent is written as a decimal.

$\frac{\text{Part}}{\text{Whole}} = \text{Percent}$ ← The percent is written as a decimal.

$\frac{\text{Part}}{\cancel{\text{Whole}}} \cdot \cancel{\text{Whole}} = \text{Percent} \cdot \text{Whole}$ **Multiply each side by the whole.**

$\text{Part} = \text{Percent} \cdot \text{Whole}$ ← This form is called the percent equation.

EXAMPLE 1 Find the Part

Find 62% of 75.

Estimate $\frac{3}{5}$ **of 75 is 45.**

The percent is 62 and the whole is 75. You need to find the part.

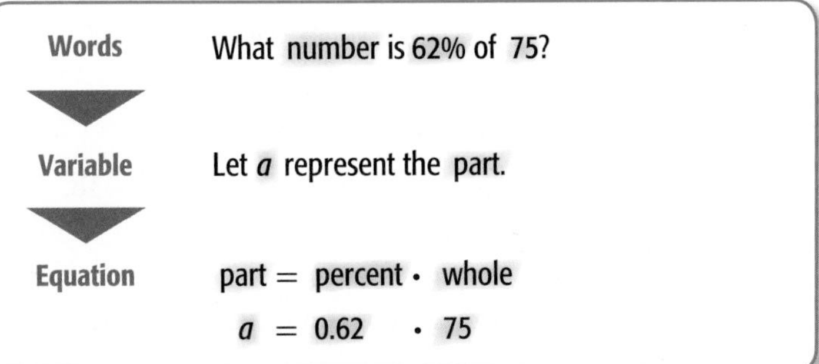

Words
What number is 62% of 75?
Variable
Let a represent the part.
Equation
part = percent · whole
$a = 0.62 \cdot 75$

$a = 0.62 \cdot 75$ **Write the percent equation.**

$= 46.5$ **Multiply.**

Check for Reasonableness $46.5 \approx 45$ ✓

1A. Find 60% of 96.

1B. Find 45% of 70.

Personal Tutor glencoe.com

StudyTip

Estimation To determine whether your answer is reasonable, estimate before finding the exact answer.

EXAMPLE 2 Find the Percent

287 is what percent of 410? **Estimate** $\frac{287}{410} \approx \frac{300}{400}$ or $\frac{3}{4}$, which is 75%.

The whole is 410 and the part is 287. Let p represent the percent.

$\underbrace{\text{part}} = \underbrace{\text{percent}} \cdot \underbrace{\text{whole}}$

$287 = p \cdot 410$ **Write the percent equation.**

$\frac{287}{410} = \frac{p \cdot 410}{410}$ **Divide each side by 410.**

$0.7 = p$ **Simplify.**

By definition, the percent is expressed as a decimal. Convert 0.7 to a percent.

Since 0.7 = 70%, 287 is 70% of 410.

Check for Reasonableness 70 ≈ 75% ✓

Check Your Progress

2A. 15 is what percent of 125?

2B. 20 is what percent of 400?

Personal Tutor glencoe.com

EXAMPLE 3 Find the Whole

33 is 55% of what number? **Estimate** 33 is 50% of 66.

The part is 33, and the percent is 55%. Let b represent the whole.

$\underbrace{\text{part}} = \underbrace{\text{percent}} \cdot \underbrace{\text{whole}}$

$33 = 0.55 \cdot b$ **Write the percent equation.**

$\frac{33}{0.55} = \frac{0.55b}{0.55}$ **Divide each side by 0.55.**

$60 = b$ **Simplify.**

So, 33 is 55% of 60.

Check for Reasonableness 60 ≈ 66 ✓

StudyTip

Percents Remember that the percent is written as a decimal in the percent equation. So, use 0.55, not 55%.

Check Your Progress

3A. 18 is 30% of what number?

3B. 79 is 80% of what number?

Personal Tutor glencoe.com

The table summarizes the three types of percent problems.

Concept Summary **The Percent Equation** For Your FOLDABLE

Type	Example	Equation
Find the Percent	15 is what percent of 60?	$15 = p(60)$
Find the Part	What number is 25% of 60?	$a = 0.25(60)$
Find the Whole	15 is 25% of what number?	$15 = 0.25b$

Real-World Link

Some states, such as Oregon, Alaska, and New Hampshire, do not have sales tax. California has the highest sales tax average with 7.25%, while the median U.S. sales tax average is 5.5%.

Source: National Tax Administration

Solve Problems The percent equation can be used to solve real-world problems.

Real-World EXAMPLE 4 Use the Percent Equation

SALES TAX Scott wants to buy a digital video recorder that costs $250. If a 6% sales tax is added, what is the total cost?

Method 1 Find the tax first. Then add.

The whole is $250. The percent is 6%. You need to find the amount of the tax, or the part. Let t represent the amount of tax.

$t = 0.06 \cdot 250$ **Write the percent equation, writing 6% as a decimal.**

$t = 15$ **Multiply.**

The tax is $15. The total cost is $250 + $15 or $265.

Method 2 Find the total percent first.

Find 100% + 6% or 106% of $250 to find the total cost, including tax.

$t = 1.06 \cdot 250$ **Write the percent equation, writing 106% as a decimal.**

$t = 1.06 \cdot 250$ or 265 **Multiply.**

Using either method, the total cost is $265.

Check Your Progress

4. **PROFIT** Last summer, Mr. Potter bought a house for $175,000. Five years later, he sold it for a 24% profit. What was the sale price of the house?

Personal Tutor glencoe.com

STANDARDIZED TEST EXAMPLE 5

Mr. Li bought a memory card for $137.46 including tax. The card had a sticker price of $129.99. About what percent sales tax did he pay?

A 5% **B** 6% **C** 7% **D** 8%

Read the Test Item

You are asked to find the estimated percent of sales tax.

Solve the Test Item

The tax is $137.46 – $129.99 or $7.47.

$7.47 = p \cdot 129.99$ **Write the percent equation.**

$\frac{7.47}{129.99} = \frac{p \cdot 129.99}{129.99}$ **Divide each side by 129.99.**

$0.057 \approx p$ **Simplify.**

So, 0.057 = 5.7%, and 5.7%, ≈ 6%. The correct choice is B.

Check Your Progress

5. The total cost of a mixer including tax was $47.70. If the original price of the mixer was $45, what was the percent of sales tax?

F 5.75% **G** 6% **H** 6.5% **J** 7%

Personal Tutor glencoe.com

Check Your Understanding

Examples 1–3 pp. 357–358

Solve each problem using a percent equation.

1. What is 40% of 75?
2. Find 13% of 27.
3. 30 is what percent of 90?
4. 15 is what percent of 300?
5. 55 is 20% of what number?
6. 24 is 80% of what number?

Examples 4 and 5 p. 359

7. **FUNDRAISERS** Last year, Kimberly sold 95 boxes of cookies. This year she wants to sell 20% more boxes than she sold last year. How many boxes will Kimberly have to sell this year to reach her goal?

8. **MULTIPLE CHOICE** Martin wants to buy a motor scooter. The cost of a motor scooter is $4968. If the total, including tax, is $5290.92, what is the percent of sales tax?

   A 5.5% B 6% C 6.5% D 7%

Practice and Problem Solving

 = **Step-by-Step Solutions** begin on page R11. **Extra Practice** begins on page 810.

Examples 1–3 pp. 357–358

Solve each problem using a percent equation.

9. Find 16% of 64.
10. What is 36% of 50?
11. 8 is what percent of 40?
12. 54 is what percent of 60?
13. 16 is 25% of what number?
14. 64 is 32% of what number?
15. 39 is 50% of what number?
16. 27 is 10% of what number?

Examples 4 and 5 p. 359

17. **SKI JACKETS** Roberto wants to buy a new ski jacket that costs $96. If the total cost, including tax, is $101.28, what is the percent of sales tax?

18. **SHOPPING** A commission is a fee paid to a salesperson based on a percent of sales. Suppose a salesperson at a jewelry store earns a 6% commission. What commission would be earned for selling a ring that costs $1300 dollars?

Solve each problem using a percent equation.

19. Find 52.5% of 76.
20. Find 23.6% of 90.
21. 33.8 is what percent of 130?
22. 79.8 is what percent of 114?

23. **FINANCIAL LITERACY** The cost, including a 6.75% sales tax, of a digital home theater system with a 40-inch high definition television is $2668.75. What is the original cost of the television and theater system?

24. **TENNIS** The results of a Wimbledon Women's Championship match is shown in the table.

   a. What was Bartoli's percent of receiving points won?

   b. Which player had a greater percent of their first serves in?

   c. Suppose in Williams' next match she has 16 break point opportunities. Based on this match, how many times will she convert on break point opportunities?

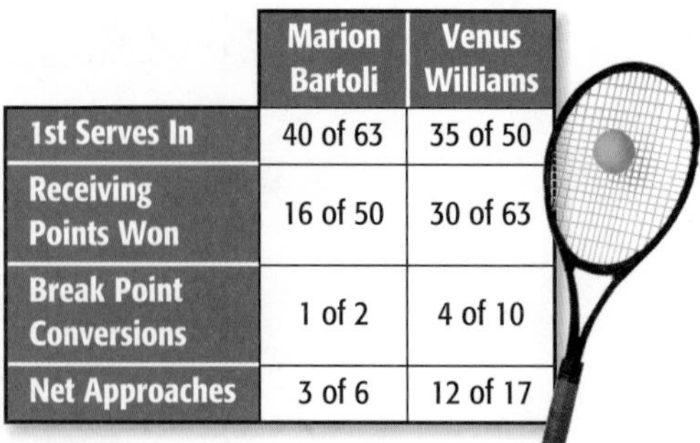

	Marion Bartoli	Venus Williams
1st Serves In	40 of 63	35 of 50
Receiving Points Won	16 of 50	30 of 63
Break Point Conversions	1 of 2	4 of 10
Net Approaches	3 of 6	12 of 17

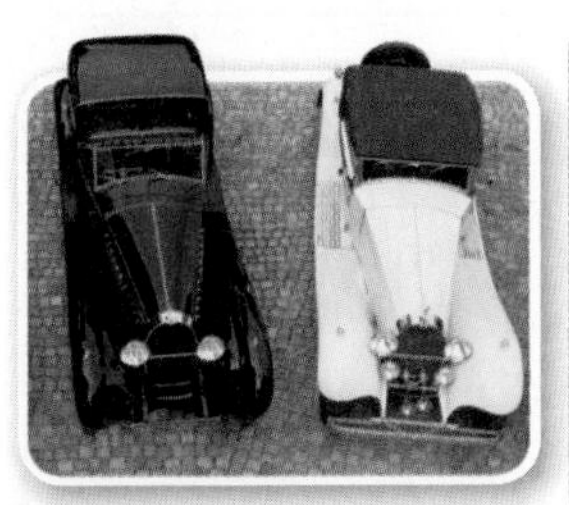

Real-World Link

The National Automobile Museum in Reno, Nevada, features over 200 antique automobiles from 1892 to present.

Source: National Automobile Museum

25 **MUSEUMS** A car museum wants to increase their collection by 20% over the next year. Currently, the museum has 120 cars in its collection.

a. Write and solve a multiplication equation to find how many cars the museum will have in the next year. How many cars will the museum need to add over the next year to meet its goal?

b. Make a table to find the number of cars in the museum collection if they increase their collection by 5%, 15%, 25%, and 35%.

26. **AREA** The table shows the area of the Great Lakes.

LAKE	AREA (square miles)
Erie	9922
Huron	23,011
Michigan	22,316
Ontario	7320
Superior	31,698

a. About what percent of the Great Lakes is covered by Lake Erie?

b. About what percent of the Great Lakes is covered by Lake Huron?

c. Suppose the area of Lake Michigan was decreased by 8%. Find its new area.

27. **MULTIPLE REPRESENTATIONS** In this problem, you will investigate percent relationships. In 2010, Aida saved $500. She plans to save 6% more than her previous' years savings for the next several years.

a. **ALGEBRAIC** Write and solve a multiplication equation to find how much money she will save next year.

b. **TABULAR** Let x represent the year and y represent the amount of money she has saved. Make a table using the x-values for 2010–2015.

c. **ANALYTICAL** Does Aida's savings increase by a constant amount each year? Explain.

Use the percent equation to solve each problem if $x = 10$.

28. $(2x)$ is 4% of what number?

29. Find $(4x)$% of 240.

H.O.T. Problems

Use **H**igher-**O**rder **T**hinking Skills

30. **OPEN ENDED** Write two percent problems in which the solution is 30%.

31. **CHALLENGE** If you found the percent of a number and the part is greater than the number, what do you know about the percent? Explain.

32. **FIND THE ERROR** Todd and Jon are finding what percent of 80 is 28. Is either of them correct? Explain your reasoning.

Todd	Jon
$28 = p \cdot 80$	$p = 80 \cdot 0.28$
$p = 0.35$	$p = 22.4$
So, 28 is 35% of 80.	So, 28 is 22.4% of 80.

33. **REASONING** Does taking a 10% discount on an item then adding a 10% sales tax result in the original price of the item? Support your answer with an example.

34. **WRITING IN MATH** Explain whether you would change the percent to a fraction or a decimal when finding 25% of 128.

Standardized Test Practice

35. Interest on a savings account is calculated every quarter of a year. During the first quarter, Alejandra earned \$54.84 in interest. This was 2% of her savings. How much was Alejandra's savings?

A \$274.20
B \$2742
C \$5484
D \$5593.68

36. A lawyer earns an annual salary of \$65,490 and receives a raise. The lawyer's new annual salary is \$68,109.60. About what percent of a raise did the lawyer receive?

F 3%
G 4%
H 5%
J 6%

37. Nate and some friends went to dinner. The total cost of dinner including a 15% tip was \$43.70. What was the cost of dinner alone?

A \$38
B \$37.50
C \$5.70
D \$4.30

38. EXTENDED RESPONSE The table shows the capacity of two collegiate football stadiums.

Stadium	Capacity
L.A. Coliseum	91,000
Ben-Hill Griffin	88,548

a. Suppose 75% of Ben-Hill Griffin Stadium is filled, and 73% of L.A. Coliseum is filled. Which stadium has a greater number of people in it?

b. How many more people are in that stadium?

Spiral Review

Find the percent of each number mentally. (Lesson 7-4)

39. 75% of 64
40. 25% of 52
41. $33\frac{1}{3}$% of 27

42. LIFE SCIENCE Carbon makes up 18.5% of the human body by weight. Determine the amount of carbon in a person who weighs 145 pounds. Round to the nearest tenth. (Lesson 7-3)

43. SNACKS The Skyway Snack Company makes a snack mix that contains raisins, peanuts, and chocolate pieces. The ingredients are shown at the right. Suppose the company wants to sell a larger-sized bag that contains 6 cups of raisins. How many cups of chocolate pieces and peanuts should be added? (Lesson 6-5)

Convert each rate using dimensional analysis. (Lesson 6-3)

44. 45 mi/h = ■ ft/s
45. 18 mi/h = ■ ft/s
46. 26 cm/s = ■ m/min
47. 32 cm/s = ■ m/min

Skills Review

Write each decimal as a percent. (Lesson 7-1)

48. 0.44
49. 0.37
50. 2.06
51. 1.82
52. 0.03
53. 0.05
54. 0.004
55. 0.007

EXPLORE 7-6

Algebra Lab
Modeling Percents and Change

Math Online glencoe.com
Math *in Motion,* Animation

You can use percents to describe a change when a number increases or decreases.

ACTIVITY 1

Suppose the width of rectangle B is increased from 2 units to 3 units, but the length is unchanged. How is the area affected?

Step 1

Draw a 4 by 2 unit rectangle on grid paper.

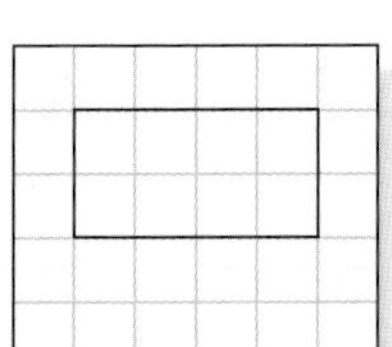

The rectangle has an area of 8 square units.

Step 2

Increase the width to 3 units. Shade the new units on the rectangle.

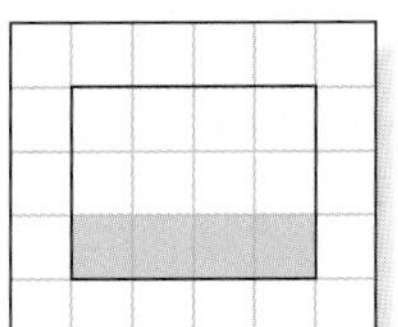

Step 3

Write a ratio comparing the shaded portion to the unshaded portion of the rectangle. Write the ratio as a percent.

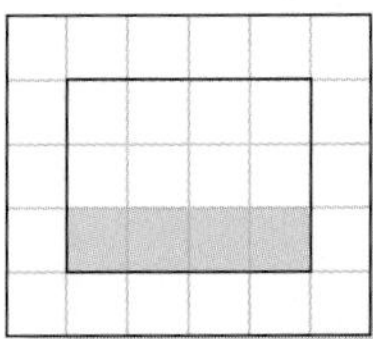

$\frac{\text{change in area}}{\text{original area}} = \frac{4}{8}$ or 50%

Compared to the original area, the new area increased by 50%.

ACTIVITY 2

Draw a rectangle with dimensions of 5 units by 4 units. Using a percent, describe the change if the width decreases from 4 units to 1 unit.

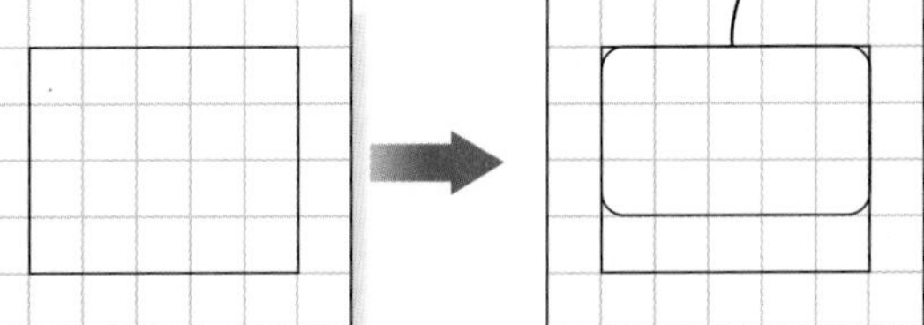

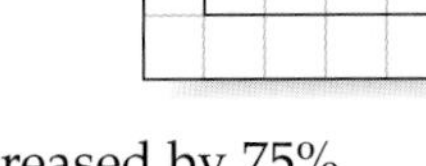

$\frac{\text{change in area}}{\text{original area}} = \frac{15}{20}$ or 75%.

Compared to the original area, the new area decreased by 75%.

Exercises

Draw each rectangle. Using a percent, describe the change for each set of rectangles.

1. A: 4 units by 1 unit
   B: 4 units by 2 units
2. C: 4 units by 3 unit
   D: 4 units by 4 units
3. W: 4 units by 4 units
   X: 4 units by 5 units
4. A: 5 units by 4 units
   B: 4 units by 4 units
5. C: 4 units by 4 units
   D: 4 units by 3 units
6. W: 2 units by 3 units
   X: 2 units by 1 unit
7. **WRITING IN MATH** For each pair of rectangles in Exercises 1–6, the change in area is 4 square units. Explain why the percent of change is different.

7-6 Percent of Change

Then

You have already solved real-world problems using the percent equations. (Lesson 7-5)

Now

- Find percent of increase and decrease.
- Solve real-world problems involving markup and discount.

New Vocabulary

percent of change
percent of increase
percent of decrease
markup
selling price
discount

Math Online

glencoe.com

- Extra Examples
- Personal Tutor
- Self-Check Quiz
- Homework Help

Why?

The table shows the number of California sea otters in recent years.

a. How many more sea otters were there in 2007 than in 2006?

b. Write the ratio $\frac{\text{amount of increase}}{\text{number of otters in 2006}}$. Then write the ratio as a percent. Round to the nearest tenth.

Year	Number of California Sea Otters
2006	2692
2007	3026

Find Percent of Change When you subtracted the original amount from the final amount, you found the *amount* of change. When you compared the change to the original amount, you found the *percent* of change.

Key Concept — Percent of Change (For Your FOLDABLE)

Words A **percent of change** is a ratio that compares the change in quantity to the original amount.

Symbols $\text{percent of change} = \frac{\text{amount of change}}{\text{original amount}}$

If the percent is positive, the percent of change is a **percent of increase**. If the percent is negative, the percent of change is called a **percent of decrease**.

EXAMPLE 1 Find the Percent of Change

Find the percent of change from 60°F to 84°F. Then state whether the percent of change is an *increase* or *decrease*.

Step 1 Subtract to find the amount of change.

$84 - 60 = 24$ **final amount – original amount**

Step 2 Write a ratio that compares the amount of change to the original amount. Express the ratio as a percent.

$\text{percent of change} = \frac{\text{amount of change}}{\text{original measurement}}$

$= \frac{24}{60}$ **Substitution**

$= \frac{2}{5}$ or 0.4 **Simplify.**

Step 3 The decimal 0.4 is written as 40%. So, the percent of change is 40%. Since the percent of change is positive, it is a percent of increase.

Check Your Progress

1. COMICS Ty had 52 comic books. Now he has 61 books. Find the percent of change. Then state whether the percent of change is an *increase* or *decrease*. Round to the nearest tenth, if necessary.

Personal Tutor glencoe.com

Watch Out!

When finding percent of change, don't assume the smaller number is the whole. When the percent of change is a decrease, the original amount will be larger than the new amount.

Real-World EXAMPLE 2 Find the Percent of Change

STAMPS McKenna had 318 stamps. Now she has 273 stamps. Find the percent of change. Round to the nearest tenth, if necessary. Then state whether the percent of change is an *increase* or *decrease*.

$$\text{percent of change} = \frac{\text{amount of change}}{\text{original amount}}$$

$$= \frac{273 - 318}{318} \qquad \frac{\text{final amount} - \text{original amount}}{\text{original amount}}$$

$$= \frac{-45}{318} \qquad \text{Simplify.}$$

$$\approx -0.141509 \qquad \text{Divide. Use a calculator.}$$

To the nearest tenth, the percent of change is −14.2%. Since the percent of change is negative, it is a percent of decrease.

Check Your Progress

2. Find the percent of change from 24 points to 18 points. Then state whether the percent of change is an *increase* or *decrease*.

Personal Tutor glencoe.com

Using Markup and Discount A store sells items for more than it pays for those items. The amount of increase is called the **markup**. The percent of markup is a percent of increase. The **selling price** is the amount the customer pays for an item.

EXAMPLE 3 Find the Selling Price

Find the selling price if a store pays $42 for a pair of roller blades and the markup is 25%.

Method 1 Find the amount of the markup first.

The whole is $42. The percent is 25. You need to find the amount of the markup, or the part. Let m represent the amount of the markup.

$m = 0.25 \cdot 42$ **part = percent • whole**

$m = 10.5$ **Multiply.**

Then add the markup to the cost. So, $42 + $10.50 = $52.50.

Method 2 Find the total percent first.

Use the percent equation to find 100% + 25% or 125% of the price. Let p represent the price.

$p = 1.25(42)$ **part = percent • whole**

$p = 52.50$ **Multiply.**

Using either method, the selling price is $52.50.

StudyTip

Total Percent In Method 2, the customer will pay 100% of the price plus an extra 25% of the price.

Check Your Progress

3. Find the selling price if a store pays $75 for a bike and the markup is 40%.

Personal Tutor glencoe.com

A **discount** is the amount by which the regular price is reduced. The percent of discount is a percent of decrease.

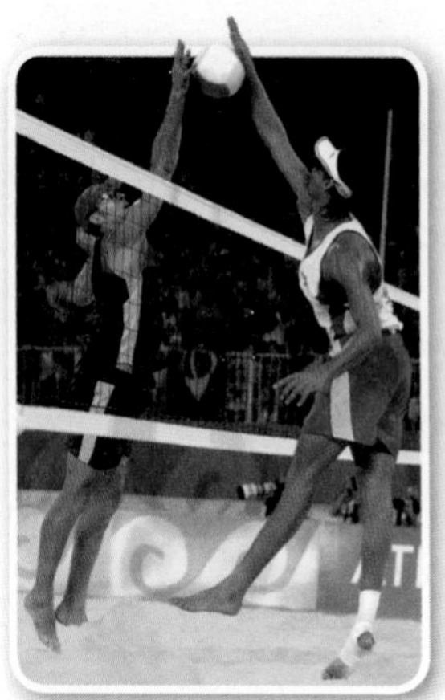

Real-World Link

Beach volleyball became an official Olympic sport in 1996. It is played on a sand court, and each half of the court measures 8 meters by 8 meters.

Source: USA Volleyball

Real-World EXAMPLE 4 Find the Sale Price

VOLLEYBALL Summer Sports is having a sale. A volleyball has an original price of $59. It is on sale for 65% off the original price. Find the sale price of the volleyball.

Method 1 Find the amount of the discount.

The percent is 65 and the whole is 59. You need to find the amount of the discount, or the part.

Let d represent the amount of the discount.

$d = 0.65 \cdot 59$ **part = percent • whole**

$d = 38.35$ **Multiply.**

Subtract the discount from the original cost to find the sale price.
So, $\$59 - \$38.35 = \$20.65$.

Method 2 Find the total percent first.

If the amount of the discount is 65%, the percent the customer will pay is 100% − 65% or 35%. Find 35% of $59.

Let s represent the sale price.

$s = 0.35(59)$ **part = percent • whole**

$s = 20.65$ **Multiply.**

Using either method, the sale price of the volleyball is $20.65.

Check Your Progress

4. **MAGAZINES** A magazine subscription has a cover price of $35. It is on sale for 67% off the original price. Find the sale price of the magazine subscription.

Personal Tutor glencoe.com

Check Your Understanding

Example 1 p. 364

Find the percent of change. Then state whether the percent of change is an *increase* or *decrease*. Round to the nearest tenth, if necessary.

1. From $40 to $32
2. From 56 inches to 63 inches

Example 2 p. 365

3. **FINANCIAL LITERACY** On Saturday, Smoothie Central made $1300 in sales. On Sunday, they made $900 in sales. What is the percent of change from Saturday to Sunday?

Example 3 p. 365

Find the selling price for each item given the cost and the percent of markup.

4. shoes: $30, 25% markup
5. CD player; $45, 31% markup

Example 4 p. 366

6. **BIKES** Find the sale price of a bike that is regularly $110 and is on sale for 45% off the original price.

Practice and Problem Solving

 = **Step-by-Step Solutions** begin on page R11.
Extra Practice begins on page 810.

Example 1 p. 364

Find the percent of change. Round to the nearest tenth, if necessary. Then state whether the percent of change is an *increase* or *decrease*.

7. From 14 inches to 26 inches
8. From \$36 to \$48
9. From 82 feet to 74 feet
10. From 16 kilograms to 5 kilograms
11. From \$128 to \$112
12. From 90 yards to 72 yards
13. From 191 ounces to 270 ounces
14. From 150 minutes to 172 minutes

Example 2 p. 365

15. **GASOLINE** A survey of gas prices in January showed that the cost per gallon one year was \$2.649. The following January, the cost per gallon was \$2.999. Find the percent change in gas prices from one year to the next to the nearest tenth.

16. **FOOTBALL** Jerome High School's football team scored 38 points in their first game. The next week they only scored 17 points. Find the percent change in the number of points scored by the football team to the nearest tenth.

Example 3 p. 365

Find the selling price for each item given the cost and the percent of markup.

17. video game: \$60; 28% markup
18. bracelet: \$26.50, 35% markup
19. jacket: \$25; 32% markup
20. stereo: \$55, 40% markup
21. wallet: \$14.50, 30% markup
22. phone: \$34, 36% markup
23. television; \$499, 20% markup
24. mountain bike: \$255, 34% mark up

Example 4 p. 366

25. **SALONS** A salon is having a sale on their hair products. Find the sale price of the shampoo and conditioner set shown at the right that regularly costs \$10.

EMME
CONDITIONER
EMME
SHAMPOO
SALE!
25% OFF

26. **ENTERTAINMENT** The unlimited rental plan at a video store costs \$30 a month. It is on sale for 35% off the original price. What is the sale price of the plan?

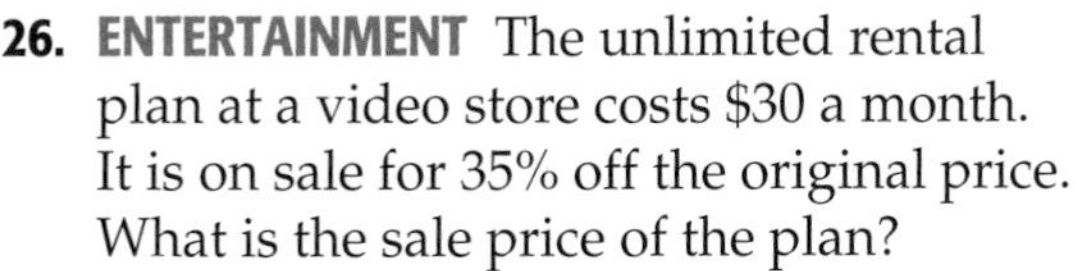

27. **TELEVISION** For the local telethon 3860 viewers called in and donated money on the first night. The next night, there was a 20% decrease in the number of calls from the first night. How many calls did the telethon receive on the second night?

Real-World Link

The first modern Olympic games were held in Athens, Greece in 1896. The Olympic rings, designed in 1913, represent the continents Africa, America, Asia, Australia, and Europe and appear on the Olympic flag.

Source: US Olympic Committee

28. **OLYMPICS** There were 10,651 athletes who participated in the 2000 Summer Olympics in Sydney, Australia. In the 2004 Summer Olympics in Athens, Greece, 11,099 athletes participated. What was the percent of change in the number of athletes participating from 2000 to 2004?

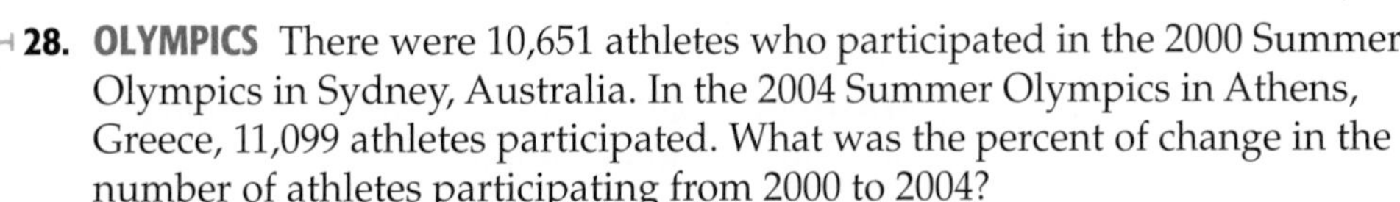

29. **BICYCLING** According to the National Sporting Goods Association, the number of people who participated in bicycle riding in 2004 was 40.3 million. In 2006, the number was 35.6 million. Find the percent of change from 2004 to 2006.

30. **DOGS** Kyla's boxer weighs about 62 pounds. When it was a puppy it only weighed 23 pounds. What is the percent of change in weight?

Problem-Solving Tip

Make a Table In Exercise 32, the data are displayed in a table. This makes finding the percent of change easier.

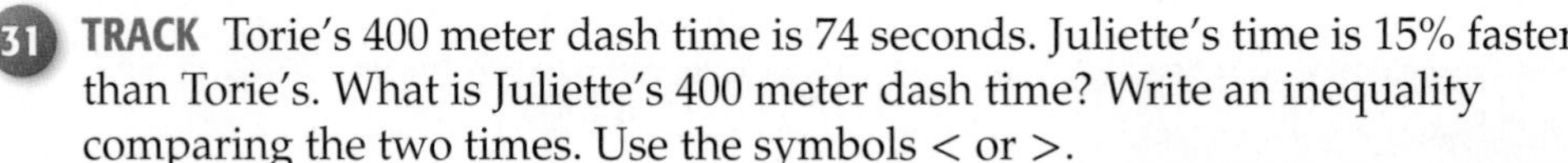

31. **TRACK** Torie's 400 meter dash time is 74 seconds. Juliette's time is 15% faster than Torie's. What is Juliette's 400 meter dash time? Write an inequality comparing the two times. Use the symbols < or >.

32. **BUSINESS** The first and second quarter earnings of two restaurants are shown at the right. Which restaurant had the greater percent of change in the second quarter?

Earnings ($)		
	A	B
Quarter 1	17,821	8112
Quarter 2	18,331	9920

33. **MULTIPLE REPRESENTATIONS** In this problem, you will compare percents of change. The table shows the population of capital cities for four different states.

City	Population 2000	Population 2006	Amount of Change	% of change
Raleigh, NC	276,093	356,321	■	■
Columbia, SC	116,278	119,961	■	■
Frankfort, KY	27,741	27,077	■	■
Columbus, OH	711,470	733,203	■	■

a. **TABULAR** Copy and complete the table. Round to the nearest whole percent.

b. **ANALYTICAL** Compare the amounts of change and the percents of change for Columbia and Columbus. Explain the differences and similarities between the two.

34. **MULTIPLE REPRESENTATIONS** In this problem, you will examine percent of change over time. The table gives the price of milk for various years.

One Gallon of Milk

Year	Price ($)
1970	1.23
1980	1.60
1990	2.15
2000	2.60

a. **ALGEBRAIC** Write two inequalities comparing the percent of increase from 1970 to 1980 to the percent of increase from 1980 to 1990. Round to the nearest percent, if necessary.

b. **GRAPHICAL** Graph the prices of milk over the years. Use the x-axis as the years and y-axis as the prices.

c. **GRAPHICAL** Use the graph to determine which decade had the greatest percent of increase in the price of milk. Explain your reasoning.

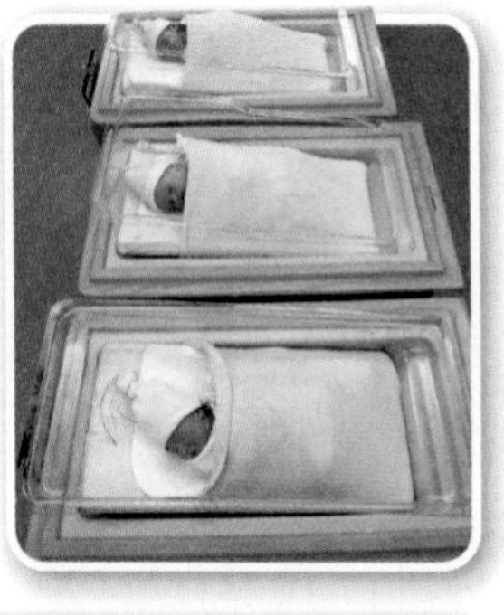

Real-World Link

The largest annual population change in the United States in the 20th century was 3,083,287, which occurred from 1949 to 1950. The largest percent of change occurred from 1909 to 1910. It was a 2.1% increase.

Source: U.S. Census Bureau

H.O.T. Problems

Use Higher-Order Thinking Skills

35. **OPEN ENDED** Give an example of a percent of increase.

36. **CHALLENGE** An item at a consignment shop is marked down 10% each week until it sells. If a bicycle was originally priced at $150, what is the cost after 3 weeks? 6 weeks?

37. **REASONING** Determine whether each statement is *true* or *false*. If false, provide a counterexample.

a. It is impossible to increase the cost of an item by more than 100%.

b. It is possible to decrease the cost of an item by less than 1%.

38. **CHALLENGE** Suppose a store has an item on sale for 25% off the original amount. By what percent does the store have to increase the price of the item in order to sell the item for the original amount? Explain.

39. **WRITING IN MATH** Write and solve a real-world problem involving a discount of an item.

Standardized Test Practice

40. If each dimension of the rectangle is tripled, what is the percent of increase in the area?

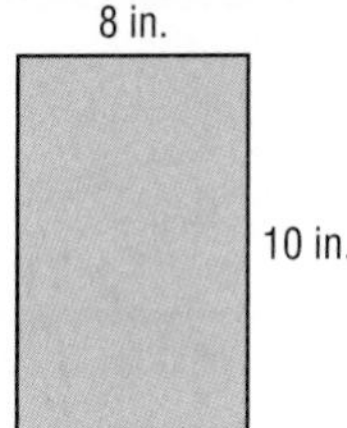

A 300%
B 600%
C 800%
D 900%

41. Which of the following represents the greatest percent of change?

F Boots that were originally priced at $90 are on sale for $63.

G A baby that weighed 7 pounds at birth now weighs 10 pounds.

H A bracelet that costs $12 to make is sold for $28.

J A savings account increased from $500 to $600 in 1 year.

42. The table shows the budget of a city.

Annual Budget	
Year	Budget (millions of $)
2005	45.6
2006	48.3
2007	45.9
2008	55.1

Which statement is supported by the table?

A The budget decreased and then increased.

B The greatest percent of change occurred from 2005 to 2006.

C The budget increased 20% from 2007 to 2008.

D The percent of change from 2005 to 2006 was the same as from 2006 to 2007.

43. **GRIDDED RESPONSE** The price of a television was $900 on Tuesday. On Wednesday the manager reduced the price 5%. What was the price in dollars of the television after the reduction?

Spiral Review

Solve each problem using the percent equation. (Lesson 7-5)

44. Find 12% of 72.

45. Find 42% of 150.

46. What is 37.5% of 89?

47. What is 24.2% of 60?

48. **FOOD** Suppose fifty-six percent of the Calories in corn chips are from fat. If one serving contains 160 Calories, estimate the number of Calories from fat in one serving of corn chips. (Lesson 7-4)

49. **FISH** Of the fish in an aquarium, 26% are angelfish. If the aquarium contains 50 fish, how many are angelfish? (Lesson 7-3)

Find each sum. Write in simplest form. (Lesson 3-6)

50. $\frac{1}{10} + \frac{1}{3}$

51. $-\frac{1}{6} + \frac{7}{18}$

52. $6\frac{4}{5} + \left(-1\frac{3}{4}\right)$

Skills Review

Write each fraction as a decimal. (Lesson 3-1)

53. $\frac{7}{8}$

54. $\frac{1}{5}$

55. $\frac{6}{20}$

56. $\frac{45}{50}$

7-7 Simple and Compound Interest

Then

You have already solved problems using the percent equation. (Lesson 7-5)

Now

- Solve simple interest problems and apply the simple interest equation to real-world problems.
- Solve compound interest problems.

New Vocabulary

interest
simple interest
principal
compound interest

Math Online

glencoe.com

- Extra Examples
- Personal Tutor
- Self-Check Quiz
- Homework Help

Why?

Stephanie received $500 for graduation. She plans to save it for college. The table shows rates for various investments for one year.

Type of Investment	Rate
Certificate of Deposit (CD)	3.75%
Money Market	4.9%
Savings	2.0%

a. If Stephanie puts her money in a savings account, she will receive 2% of $500 in interest for one year. Find the interest Stephanie will receive if she puts her money in a savings account for one year.

b. Compare the interest Stephanie will receive in one year from a money market and from a certificate of deposit for one year.

Simple Interest **Interest** is the amount of money paid or earned for the use of money by a bank or other financial institution. **Simple interest** is paid only on the initial principal of a savings account or a loan. To solve problems involving simple interest, use the following formula.

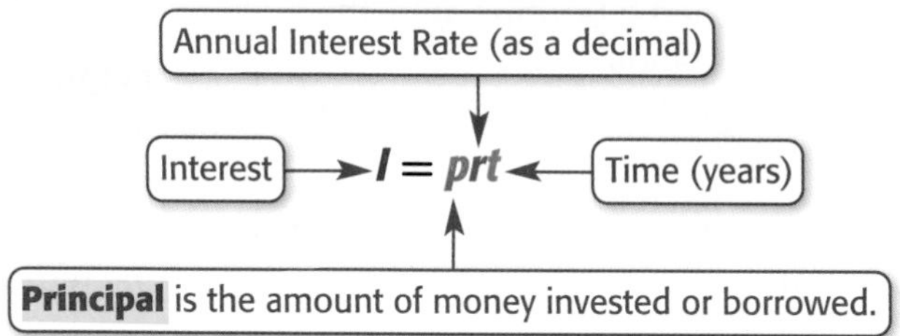

EXAMPLE 1 Find Simple Interest

Find the simple interest to the nearest cent.

a. $1000 at 4.5% for 2 years

$I = prt$	**Write the simple interest formula.**
$I = 1000 \cdot 0.045 \cdot 2$	**Replace *p* with 1000, *r* with 0.045, and *t* with 2.**
$I = 90$	**Simplify.**

The simple interest is $90.

b. $2500 at 6.75% for 3 years

$I = prt$	**Write the simple interest formula.**
$I = 2500 \cdot 0.0675 \cdot 3$	**Replace *p* with 2500, *r* with 0.0675, and *t* with 3.**
$I = 506.25$	**Simplify.**

The simple interest is $506.25.

Check Your Progress

1A. $2250 at 6% for 4 years

1B. $4000 at 4.25% for 1 year

Personal Tutor glencoe.com

Watch Out!

Converting Units When using the formula $I = prt$, remember the time is expressed in years. For example, 6 months is 0.5 year.

Real-World EXAMPLE 2 Find the Interest Rate

COMPUTERS Mr. Gabel borrowed $1860 to buy a computer. He will pay $71.30 per month for 30 months. Find the simple interest rate for his loan.

Understand You need to find the simple interest rate.

Plan Use the formula $I = prt$.

Solve First find the amount of interest he will pay.

$71.30 • 30 = $2139	**Multiply to find total amount.**
$2139 – $1860 = $279	**Subtract to find the interest.**
So, I = $279.	

The principal is $1860. So, $p = 1860$.
The loan will be for 30 months or 2.5 years. So, $t = 2.5$.

$I = prt$	**Write the simple interest formula.**
$279 = 1860 \cdot r \cdot 2.5$	**Replace I with 279, p with 1860, and t with 2.5.**
$279 = 4650r$	**Simplify.**
$\frac{279}{4650} = \frac{4650r}{4650}$	**Divide each side by 4650.**
$0.06 = r$	The simple interest rate is 0.06 or 6%.

Check Use the formula $I = prt$. $1860 × 0.06 × 2.5 or $279. ✓

Check Your Progress

2. **SAVINGS** Suppose Nantai placed $2400 in the bank for 5 years. He makes $9.20 in interest each month. Find the annual interest rate.

Personal Tutor glencoe.com

Compound Interest **Compound interest** is paid on the initial principal and on interest earned in the past.

StudyTip

Compound Interest Because you are finding the interest after the first year, substitute 1 for t instead of 2.

EXAMPLE 3 Find the Total Amount

What is the total amount of money in an account where $600 is invested at an interest rate of 8.75% compounded annually for 2 years?

Step 1 Find the amount of money in the account at the end of the first year.

$I = prt$	**Write the simple interest formula.**
$I = 600 \cdot 0.0875 \cdot 1$	**Replace p with 600, r with 0.0875, and t with 1.**
$I = 52.5$	**Simplify.**
$600 + 52.5 = 652.50$	**Add the amount invested and the interest.**

At the end of the first year, there is $652.50 in the account.

Step 2 Find the amount of money in the account at the end of the second year.

$I = prt$	**Write the simple interest formula.**
$I = 652.50 \cdot 0.0875 \cdot 1$	**Replace p with 652.50, r with 0.0875, and t with 1.**
$I = 57.09$	**Simplify.**

So, the amount in the account after 2 years is $652.50 + $57.09 or $709.59.

Check Your Progress

3. What is the total amount of money in an account where $5000 is invested at an interest rate of 5% compounded annually after 3 years?

Personal Tutor glencoe.com

Check Your Understanding

Example 1 p. 370

Find the simple interest to the nearest cent.

1. $1350 at 6% for 7 years
2. $240 at 8% for 9 months
3. $725 at 3.25% for 5 years
4. $3750 at 5.75% for 42 months

Example 2 p. 371

5. **LOANS** Mateo's sister paid off her student loan of $5000 in 3 years. If she made a payment of $152.35 each month, what was her simple interest rate for her loan? Round to the nearest hundredth.

Example 3 p. 371

Find the total amount in each account to the nearest cent if the interest is compounded annually.

6. $480 at 5% for 3 years
7. $515 at 11.8% for 2 years
8. $6525 at 6.25% for 4 years
9. $2750 at 8.5% for 3 years

Practice and Problem Solving

● = **Step-by-Step Solutions** begin on page R11.
Extra Practice begins on page 810

Example 1 p. 370

Find the simple interest to the nearest cent.

10. $275 at 7.5% for 4 years

11. $620 at 6.25% for 5 years
12. $734 at 12% for 3 months
13. $2020 at 8% for 18 months
14. $1200 at 6% for 36 months
15. $4380 at 10.5% for 2 years

Example 2 p. 371

16. **CARS** Thomas borrowed $4800 to buy a new car. He will be paying $96 each month for the next 60 months. Find the simple interest rate for his car loan.

Example 3 p. 371

Find the total amount in each account to the nearest cent if the interest is compounded annually.

17. $3850 at 5.25% for 2 years
18. $4025 at 6.8% for 6 years
19. $595 at 4.75% for 3 years
20. $840 at 7% for 4 years
21. $12,000 at 6.95% for 4 years
22. $8750 at 12.25% for 2 years

23. **CARS** Denise has a car loan of $8000. Over the course of the loan, she paid a total of $1680 in interest at a simple interest rate of 6%. How many months was the loan?

24. **INVESTMENTS** A certificate of deposit has an annual simple interest rate of 5.25%. If $567 in interest is earned over a 6 year period, how much was invested?

25. **FINANCIAL LITERACY** A bank offers the options shown for interest rates on their savings accounts. Which option will yield more money after 3 years with an initial deposit of $1500? Explain.

Kingman Bank		
Option	Rate	Type of Interest
A	6.25%	simple
B	5.75%	compounded annually

Find the total amount in each account to the nearest cent if the interest is compounded twice a year.

26. $2500 at 6.75% for 1 year
27. $14,750 at 5% for 1 year
28. $3750 at 10.25% for 2 years
29. $975 at 7.2% for 2 years

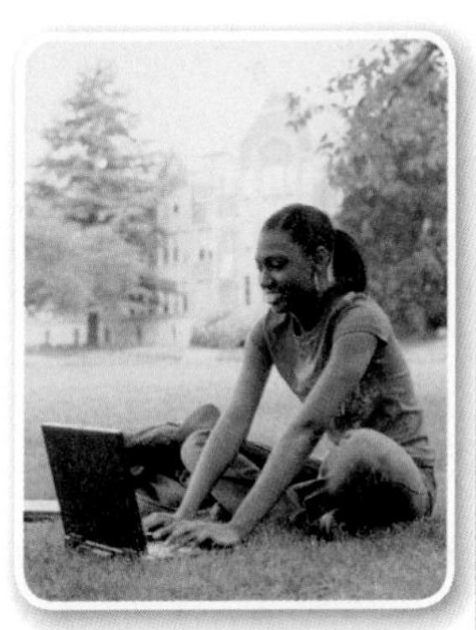

Real-World Link

College tuition continues to rise each year. The average yearly cost of a four-year private college or university is $22,218. The average yearly cost of a four-year pubilc college or university is $5836.

Source: College Board

30. **COLLEGE** Mrs. Glover placed $15,000 in a certificate of deposit for 18 months for her children's college funds. Each month she makes $56.50 in interest. Find the annual simple interest rate for the certificate of deposit.

31. **CREDIT** Jameson received his first credit card bill for a total of $325.42. Each month he makes a $50 payment and the remaining balance is charged an interest rate of 1.5%. The register at the right shows his first three monthly bills. If he does not make any more charges, what will be the amount of the fifth bill? the seventh bill?

Bill Number	Bill Amount	Payment	New Balance
1	$325.42	$50	$275.42
2	$279.55	$50	$229.55
3	$232.99	$50	$182.99

32. **MULTIPLE REPRESENTATIONS** In this problem, you will compare simple and compound interest. Consider the following situation. Ben deposits $550 at a 6% simple interest rate and Anica deposits $550 at a 6% interest rate that is compounded annually.

a. **TABULAR** Copy and complete the table.

Total Interest Earned ($)

Years	Ben	Anica
2	■	■
4	■	■
6	■	■
8	■	■
10	■	■

b. **GRAPHICAL** Graph the data on the coordinate plane. Show the time in years on the x-axis and the total interest earned in dollars on the y-axis. Plot Ben's interest in blue and Anica's interest in red. Then connect the points.

c. **ANALYTICAL** Compare the graphs of the two functions.

H.O.T. Problems

Use Higher-Order Thinking Skills

33. **OPEN ENDED** Give a principal and interest rate where the amount of simple interest earned in four years would be $80. Justify your answer.

34. **REASONING** Kai-Yo deposits $500 into an account that earns 2% simple interest. Marcos deposits $250 into an account that earns 4% simple interest. How much money does each have after 10 years? Who will have more money over the long run? Explain your reasoning.

35. **FIND THE ERROR** Sabino and Mya are finding the simple interest on a $2500 investment at a simple interest rate of 5.75% for 18 months. Is either of them correct? Explain your reasoning.

Sabino

$I = prt$

$I = 2500 \cdot 0.0575 \cdot 18$

$I = \$2587.50$

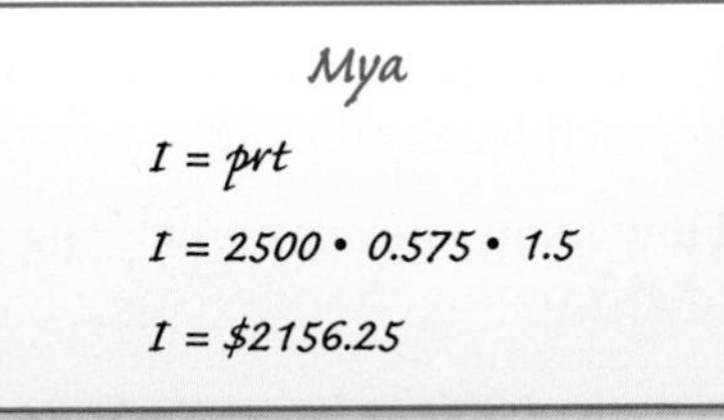

36. **CHALLENGE** Determine the length of time it will take to double a principal of $100 if deposited into an account that earns 10% simple annual interest.

37. **WRITING IN MATH** Compare simple and compound interest.

Standardized Test Practice

38. A $500 certificate of deposit has a simple interest rate of 7.25%. What is the value of the certificate after 8 years?

A $290 C $790
B $500 D $2900

39. Beatriz borrowed $1500 for student loans. She will make 30 equal monthly payments of $62.50 to pay off the loan. What is the simple interest rate for the loan?

F 4% H 8.5%
G 7% J 10%

40. A savings account with $2250 has an interest rate of 5%. If the interest is compounded annually, how much will be in the account after 2 years?

A $230.63 C $2480.63
B $337.50 D $2587.50

41. **EXTENDED RESPONSE** Which of the following plans will produce the greater earnings for an investment of $500 over 5 years? Explain.

Plan A	simple interest rate of 6.75%
Plan B	rate of 6.5% compounded annually

Spiral Review

42. **ANIMALS** In 2000, there were 356 endangered species. Five years later, 389 species were considered endangered. What was the percent of change? (Lesson 7-6)

Solve each problem using the percent equation. (Lesson 7-5)

43. 12 is what percent of 400?

44. 30 is 60% of what number?

45. **MONEY** In a recent year, the number of $1 bills in circulation in the United States was about 7 billion. (Lesson 7-4)

a. Suppose the number of $5 bills in circulation was 25% of the number of $1 bills. About how many $5 bills were in circulation?

b. If the number of $10 bills was 20% of the number of $1 bills, about how many $10 bills were in circulation?

ALGEBRA Find each product. Write in simplest form. (Lesson 3-5)

46. $\frac{2}{x} \cdot \frac{3x}{7}$

47. $\frac{a}{b} \cdot \frac{5b}{c}$

48. $\frac{4t}{9r} \cdot \frac{18r}{t^2}$

49. **EXERCISE** The table shows the amount of time Craig spends jogging every day. He increases the time he jogs every week. (Lesson 1-5)

a. Write an equation to show the number of minutes spent jogging m for each week w.

b. How many minutes will Craig jog during week 9?

Week	Time Jogging (minutes)
1	7
2	15
3	23
4	31
5	39

Skills Review

Solve each problem. (Lesson 7-5)

50. Find 66% of 90.

51. What is 0.2% of 735?

52. Find 250% of 7000.

EXTEND
7-7
Spreadsheet Lab
Compound Interest

Math Online glencoe.com
- Graphing Technology Personal Tutor

You can use a spreadsheet to investigate the impact of compound interest.

ACTIVITY

George deposits \$1600 into an account that earns 8% interest compounded semiannually. What is the value of the account after 5 years?

An 8% interest compounded semiannually means that the interest is paid twice a year, or every 6 months. The interest rate is 8% ÷ 2 or 4%.

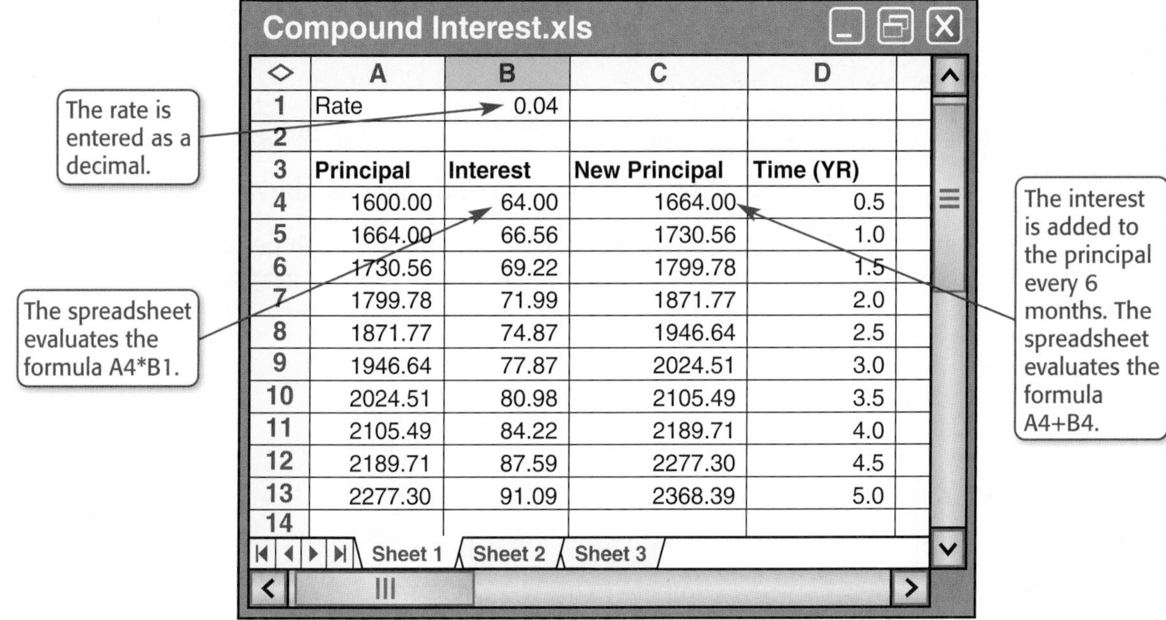

Compound Interest.xls
The rate is entered as a decimal.
The spreadsheet evaluates the formula A4*B1.
The interest is added to the principal every 6 months. The spreadsheet evaluates the formula A4+B4.
Sheet 1
Sheet 2
Sheet 3

	A	B	C	D
1	Rate	0.04		
2				
3	Principal	Interest	New Principal	Time (YR)
4	1600.00	64.00	1664.00	0.5
5	1664.00	66.56	1730.56	1.0
6	1730.56	69.22	1799.78	1.5
7	1799.78	71.99	1871.77	2.0
8	1871.77	74.87	1946.64	2.5
9	1946.64	77.87	2024.51	3.0
10	2024.51	80.98	2105.49	3.5
11	2105.49	84.22	2189.71	4.0
12	2189.71	87.59	2277.30	4.5
13	2277.30	91.09	2368.39	5.0
14				

The value of the savings account after five years is \$2368.39.

Analyze the Results

1. Suppose you invest \$1600 for five years at 8% simple interest. How does the simple interest compare to the compound interest shown above?
2. Use a spreadsheet to find the amount of money in a savings account if \$1600 is invested for five years at 8% interest compounded quarterly.
3. Suppose you leave \$150 in each of three bank accounts paying 6% interest per year. One account pays simple interest, one pays interest compounded semiannually, and one pays interest compounded quarterly. Use a spreadsheet to find the amount of money in each account after three years.
4. **MAKE A CONJECTURE** If the compounding occurs less frequently, how does the amount of interest change?
5. **MAKE A CONJECTURE** If the compounding occurs more frequently, how does the amount of interest change?

7-8 Circle Graphs

Then

You have already expressed percents as decimals. (Lesson 7-2)

Now

- Construct circle graphs.
- Analyze circle graphs to solve real-world problems.

New Vocabulary

circle graph

Math Online

glencoe.com

- Extra Examples
- Personal Tutor
- Self-Check Quiz
- Homework Help

Why?

The graphic shows the results of a recent online survey about what part teens would want to be in a band.

a. Which part was the least favorite?

b. Are all the parts accounted for in the graphic? How can you tell?

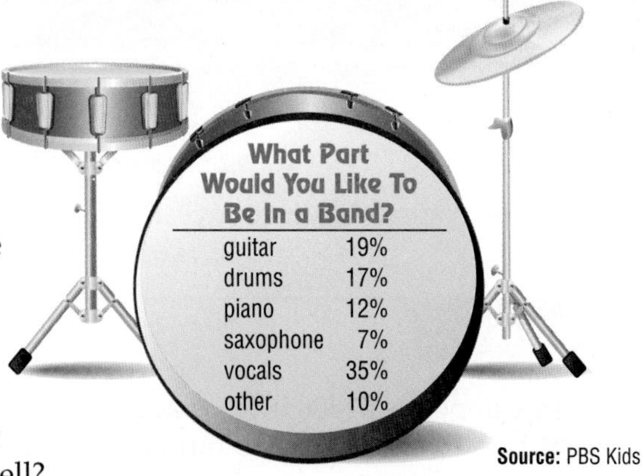

What Part Would You Like To Be In a Band?
guitar 19%
drums 17%
piano 12%
saxophone 7%
vocals 35%
other 10%
Source: PBS Kids

Circle Graphs A **circle graph** can be used to compare parts of a data set to the whole set of data. The percents in a circle graph add up to 100 since the entire circle represents the whole set.

EXAMPLE 1 Construct a Circle Graph from Percents

Construct a circle graph using the information above.

Step 1 There are 360° in a circle. So, multiply each percent by 360 to find the number of degrees for each section of the graph.

guitar: 19% of $360 = 0.19 \cdot 360 \approx 68$

drums: 17% of $360 = 0.17 \cdot 360 \approx 61$

piano: 12% of $360 = 0.12 \cdot 360 \approx 43$

saxophone: 7% of $360 = 0.07 \cdot 360 \approx 25$

vocals: 35% of $360 = 0.36 \cdot 360 = 126$

other: 10% of $360 = 0.10 \cdot 360 = 36$

Step 2 Use a compass to draw a circle and a radius. Then use a protractor to draw a 43° angle. This section represents the number of people who want to play the piano.

Step 3 From the new radius, draw the next angle. Repeat for each of the remaining angles.

Step 4 Label each section. Then give the graph a title.

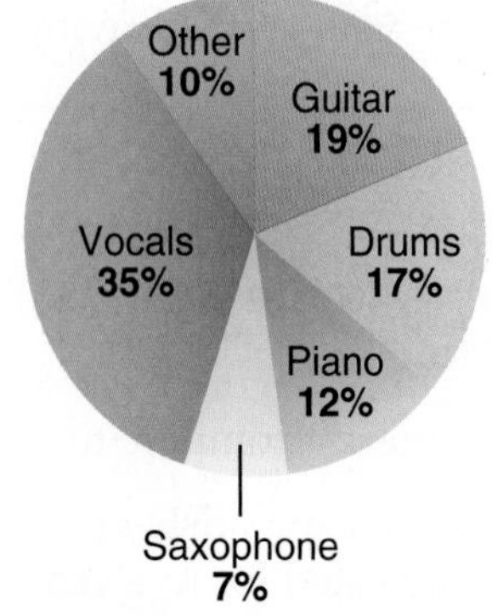

Check Your Progress

1. PIZZA The table gives the percent of favorite single pizza toppings. Construct a circle graph using the information in the table.

Topping	Percent
Cheese	34
Pepperoni	28
Mushroom	21
Other	17

Personal Tutor glencoe.com

When percents are not known, you must first determine what part of the whole each item represents.

Real-World Link

The United States contributes about 17% of the world's energy supply which makes it the leading producer. However, it is also the leading country in energy usage. The United States uses about 24% of the world's energy supply.

Source: *World Almanac*

EXAMPLE 2 Construct a Circle Graph From Data

Construct a circle graph using the information in the table at the right.

Home Heating	
Type of Fuel	**Number of Homes**
Bottled Gas	12
Electricity	64
Fuel Oil	18
Piped Gas	100
Wood	4
Other	2

Source: U.S. Census Bureau

Step 1 Find the total number of homes.

$12 + 64 + 18 + 100 + 4 + 2 = 200$

Step 2 Find the ratio that compares the number of homes in each group to the total number of homes.

bottled gas: $12 \div 200 = 0.06$
electricity: $64 \div 200 = 0.32$
fuel oil: $18 \div 200 \approx 0.09$
piped gas: $100 \div 200 = 0.50$
wood: $4 \div 200 = 0.02$
other: $2 \div 200 = 0.01$

Step 3 Use these ratios to find the number of degrees of each section.

bottled gas: $0.06 \cdot 360 = 21.6$ or about 22
electricity: $0.32 \cdot 360 = 115.2$ or about 115
fuel oil: $0.09 \cdot 360 = 32.4$ or about 32
piped gas: $0.5 \cdot 360 = 180$
wood: $0.02 \cdot 360 = 7.2$ or about 7
other: $0.01 \cdot 360 = 3.6$ or about 4

Step 4 Use a compass and a protractor to draw a circle and the appropriate sections.

Step 5 Label each section and give the graph a title. Write the ratios as percents.

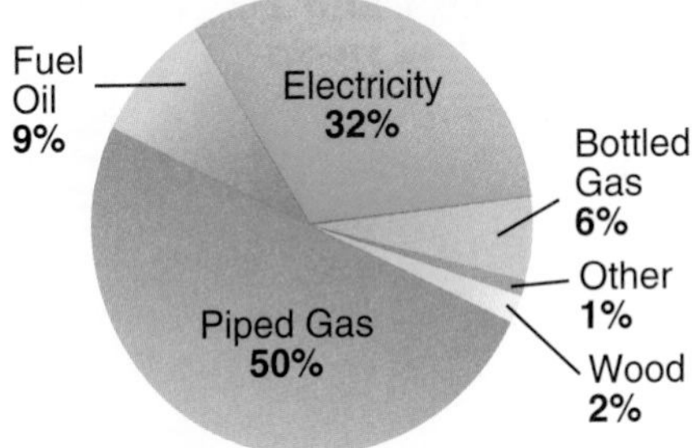

StudyTip

Alternative Method You can also find the angle measure using a proportion.

$\frac{12}{200} = \frac{x}{360}$

$4320 = 200x$

$21.6 = x$

Check Your Progress

2. Make a circle graph using the information in the table at the right that shows the major influences on teens for music choices.

Influences on Music Choice	
Influence	**Number of Teens**
Radio	860
Friends	600
Television	320
Parents	140
Other	80

Personal Tutor glencoe.com

Analyze Circle Graphs You can use the percents and central angle measures in a circle graph to solve real-world problems.

> **StudyTip**
>
> **Another Method** To check Example 3, you can solve the problem another way.
> 22% − 20% = 2%
> 2% of $10 billion is $200 million.

Real-World EXAMPLE 3 Analyze Circle Graphs

GOVERNMENT The circle graph at the right shows how the U.S. Government spends its money. Suppose the U.S. Government's budget this year is $10 billion. How much more money is spent on Health than Defense?

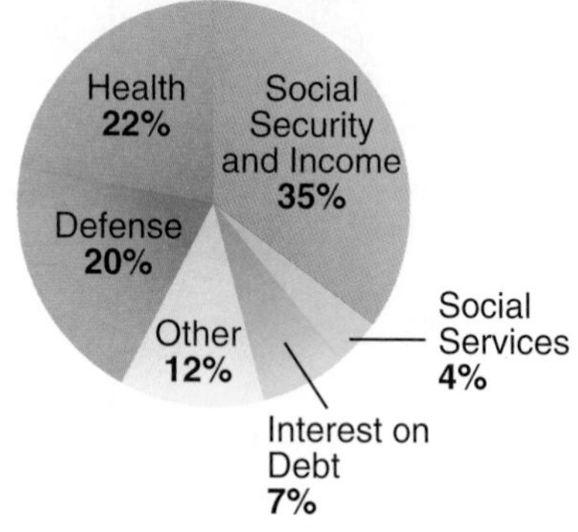

Health
22% of $10.0 billion = 0.22 × $10.0
or $2.2 billion

Defense
20% of $10.0 billion = 0.20 × $10.0
or $2.0 billion

The U.S. Government would spend $2.2 billion − $2.0 billion or $200 million more on Health than Defense.

 **Check Your Progress**

3. Suppose the U.S. Government's budget is $20 billion. How much more money is spent on Interest than on Social Services?

Personal Tutor glencoe.com

Check Your Understanding

Examples 1 and 2 pp. 376–377

Construct a circle graph for each set of data.

1.

Atmospheric Composition	
Element	**Percent**
Nitrogen	78
Oxygen	21
Other	1

Source: NASA

2.

Kingdoms of Life	
Kingdom	**Number of Species**
Bacteria	4,000
Protists	80,000
Animals	1,324,000
Fungi	72,000
Plants	270,000

Source: PBS Kids

Example 3 p. 378

3 **BOOKS** The circle graph at the right shows the results of a survey about favorite kinds of books. If 600 people were surveyed, how many more people prefer mystery books than historical fiction books?

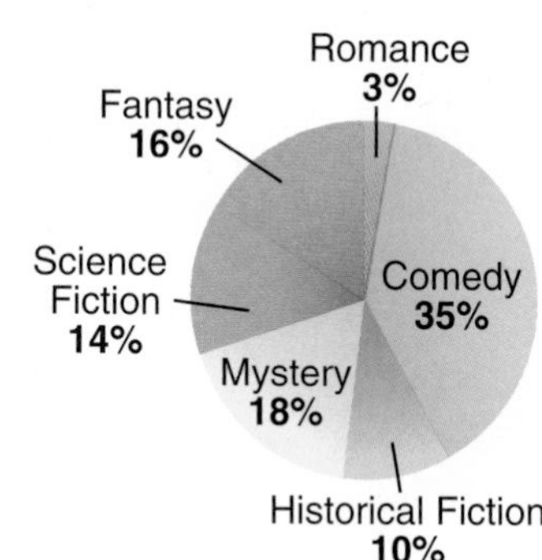

Practice and Problem Solving

● = **Step-by-Step Solutions** begin on page R11. **Extra Practice** begins on page 810.

Examples 1 and 2 pp. 376–377

Construct a circle graph for each set of data.

4.

Number of U.S. States Visited	
Number of States	**Percent of People**
0	14%
1–5	38%
6–15	26%
16–25	7%
26–50	15%

Source: PBS Kids

5.

Layers of Earth's Atmosphere	
Layer	**Depth (kilometers)**
Troposphere	11.5
Stratosphere	50
Mesosphere	85
Thermosphere	600

Source: *World Almanac*

6.

Athletic Shoe Purchases	
Age	**Number**
Under 14	95
14–17	30
18–24	50
25–34	70
35–44	75
45–64	120
65 & Older	60

7.

U.S. Landfill Composition	
Type	**Percent**
Metal	8%
Plastic	24%
Food and Yard Waste	11%
Rubber and Leather	6%
Other	21%
Paper	30%

Example 3 p. 378

8. **TRAVEL** The circle graph below shows the results of a middle school survey about favorite states to visit on vacation. If 400 students were surveyed, how many more students favor visiting Florida than Colorado?

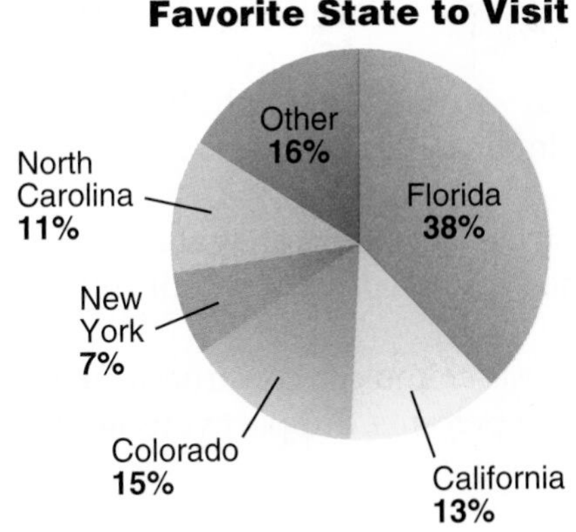

9. **TELEVISION** The circle graph below shows the results of a survey about the number of televisions per household. Suppose 250 households were surveyed. How many more households have three televisions than five or more televisions?

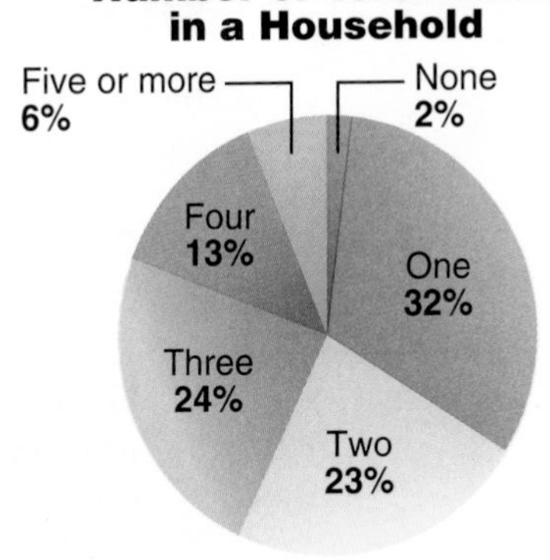

10. **PETS** Samuel surveyed his classmates about their favorite type of pet. He made a circle graph to show the results. The central angle for the section of the graph that represents cats measures 108°. If 45 classmates chose cats as their favorite type of pet, how many classmates did Samuel survey?

11. **FLOWERS** Yelina surveyed people to find their favorite flower. She made a circle graph of the data. The central angle for the portion of the graph that represents tulips measures 54°. If 75 people chose tulips as their favorite flower, how many people did Yelina survey?

12. **SPORTS** Kenard worked at a sporting goods store. To determine trends in footwear, he charted sales for a year. Then he constructed a circle graph of the data. The sales in March were double the sales in May. If the central angle in the graph for March measured 47.5°, what percent of the sales occurred in May?

Real-World Career

Sports Statisticians
Sports statisticians gather and analyze sports data. Statisticians may work for sports teams, associations, or private sports statistics companies. Most jobs require a 4-year degree in computer science or statistics.
Source: Career Prospect

13. **MULTIPLE REPRESENTATIONS** In this problem, you will investigate circle graphs. The table shows the number of grand slam wins for two different tennis players.

Grand Slam Tournament	Number of Grand Slam Wins for Steffi Graf	Number of Grand Slam Wins for Pete Sampras
Australian Open	4	2
French Open	6	0
Wimbledon	7	7
U.S. Open	5	5

a. **GRAPHICAL** Construct a circle graph for each player's Grand Slam wins.

b. **NUMERICAL** Do both players have the same percentage of U.S. Grand Slam wins? Explain your reasoning.

c. **ANALYTICAL** Compare the circle graphs. Describe the similarities and differences between the two graphs.

14. **COLLECT DATA** Design a survey to give to your classmates.

a. Display the data in a table.

b. Create a circle graph of the data.

c. Write two questions based on your data.

H.O.T. Problems

Use Higher-Order Thinking Skills

15. **CHALLENGE** Explain why a circle graph should *not* be made of the data in the table at the right.

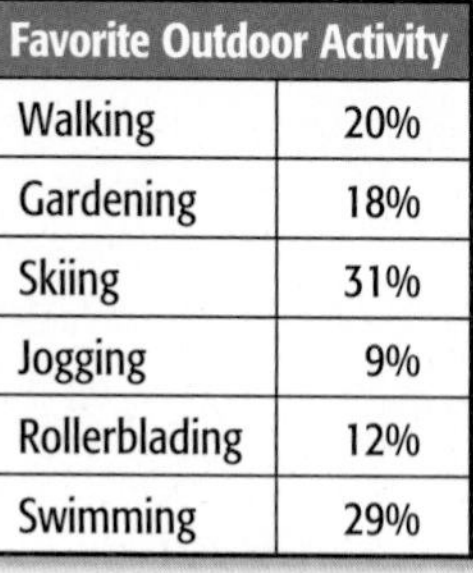

Favorite Outdoor Activity	
Walking	20%
Gardening	18%
Skiing	31%
Jogging	9%
Rollerblading	12%
Swimming	29%

16. **OPEN ENDED** Construct a circle graph with four categories that shows how you spend your free time.

17. **REASONING** *True* or *false*? You can construct a circle graph without using percents. Support your answer with examples.

18. **NUMBER SENSE** Estimate the percent of the circle graph shown at the right that is represented by Section A and Section B combined. Section C. Section D, Section E and Section F combined.

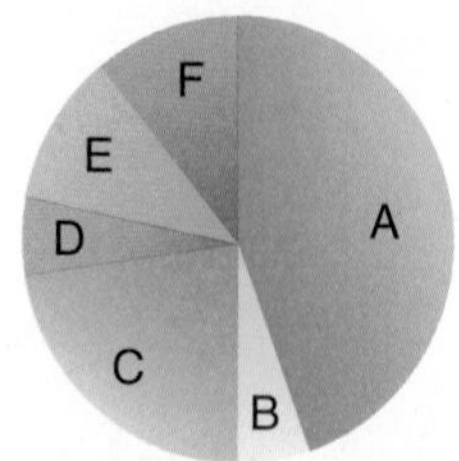

19. **WRITING IN MATH** Explain the steps you would take to create a circle graph if the data are given as percents.

Standardized Test Practice

For Exercises 20 and 21 refer to the circle graph below.

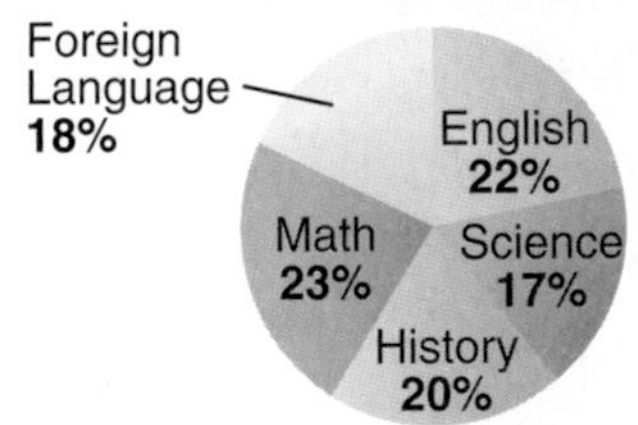

20. Which of the following is supported by the graph?

A Twice as many students prefer math to science.

B The favorite subject of most students is math.

C More students list history as a favorite than English.

D Less than half the students surveyed list Science, Foreign Language and History as their favorite subject.

21. If 800 students were surveyed, what number of students would have math as a favorite subject?

F 83 **H** 184

G 180 **J** 221

22. The table shows the results of a class survey.

Type of Movie	Number of Students
Comedy	165
Action	160
Drama	92
Science Fiction	83

In a circle graph of the data, how many degrees would make up the central angle of the section titled Action?

A 32° **C** 160°

B 115° **D** 180°

23. **EXTENDED RESPONSE** The table below shows data about automobile colors. Calculate the degree measures of the central angles of a circle graph for each category.

Car Color	Number of Cars
Red	5
White	10
Green	15
Blue	30
Other	40

Spiral Review

Find the simple interest to the nearest cent. (Lesson 7-7)

24. \$4500 at 5.5% for $4\frac{1}{2}$ years

25. \$3680 at 6.75% for $2\frac{1}{4}$ years

26. **SCHOOL** Jiliana is using a copy machine to increase the size of a 2-inch by 3-inch picture of a spider. The enlarged picture needs to measure 3 inches by 4.5 inches. What enlargement setting on the copy machine should she use? (Lesson 7-6)

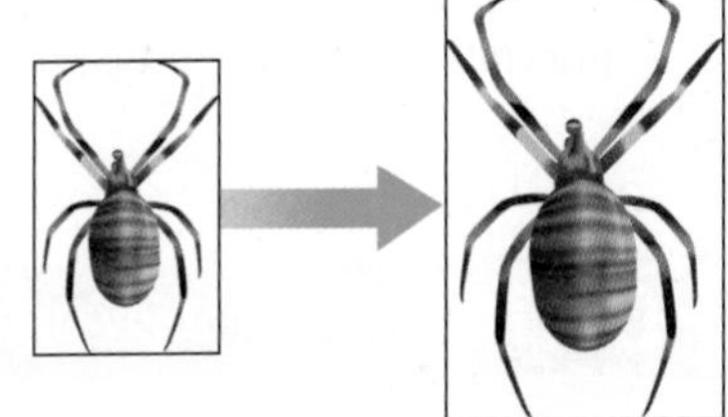

Skills Review

Determine the domain and range of each function (Lesson 1-4)

27. (4, 2), (−7, 9), (0, −1)

28. (−2, 12), (−2, 11), (−2, 10)

29. (18, 0), (14, −9), (−6, 6)

Study Guide and Review

Math Online glencoe.com
- STUDY TO GO
- Vocabulary Review

Chapter Summary

Key Concepts

Fractions, Decimals, and Percents (Lessons 7-1 and 7-2)

- A percent is a ratio that compares a number to 100.
- Fractions, decimals, and percents are all different ways to represent the same number.
- Common fraction, decimal, percent equivalents:

$\frac{1}{5} = 0.2 = 20\%$ | $\frac{1}{2} = 0.5 = 50\%$
$\frac{4}{5} = 0.8 = 80\%$ | $\frac{1}{4} = 0.25 = 25\%$
$\frac{1}{3} = 0.\overline{3} = 33\frac{1}{3}\%$ | $\frac{3}{10} = 0.3 = 30\%$
$\frac{3}{5} = 0.6 = 60\%$ | $\frac{9}{10} = 0.9 = 90\%$
$\frac{3}{4} = 0.75 = 75\%$ | $\frac{2}{3} = 0.\overline{6} = 66\frac{2}{3}\%$
$\frac{2}{5} = 0.4 = 40\%$ | $\frac{7}{10} = 0.7 = 70\%$

Percents (Lessons 7-3 through 7-6)

- A percent proportion is $\frac{\text{part}}{\text{whole}}$ = percent, where the percent is written as a fraction.
- A percent of increase, or decrease, tells how much an amount has increased, or decreased, in relation to the original amount.

Interest (Lesson 7-7)

- Interest is the amount of money paid or earned for the use of money.
- Simple interest can be found using the formula $I = prt$, where I is the interest paid or earned, p is the principal money invested or borrowed, r is the interest rate, and t is the time in years.
- Compound interest is interest paid on the initial principal and on interest earned in the past.

FOLDABLES® Study Organizer

Be sure the Key Concepts are noted in your Foldable.

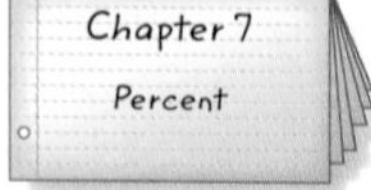

Key Vocabulary

circle graph (p. 376)
compound interest (p. 371)
discount (p. 366)
interest (p. 370)
markup (p. 365)
percent (p. 331)
percent equation (p. 357)
percent of change (p. 364)
percent of decrease (p. 364)
percent of increase (p. 364)
percent proportion (p. 345)
principal (p. 370)
selling price (p. 365)
simple interest (p. 370)

Vocabulary Check

Complete each sentence with the correct term. Choose from the list above.

1. A(n) ________ is an increase in price.
2. ________ is the amount of money paid or earned for the use of money.
3. The ________ is a ratio that compares the change in quantity to the original amount.
4. ________ is the amount by which the regular price of an item is reduced.
5. The money invested or borrowed is called the ________.
6. The ________ is the amount of money a customer pays for an item.
7. A(n) ________ is a ratio that compares a number to 100.
8. A(n) ________ is a visual representation that can be used to compare parts of a data set to the whole set of data.

Lesson-by-Lesson Review

7-1 Fractions and Percents (pp. 336–336)

Write each percent as a fraction in simplest form.

9. 30% **10.** 16% **11.** 92% **12.** 65%

13. 0.6% **14.** 0.45% **15.** 140% **16.** 212%

Write each fraction as a percent.

17. $\frac{2}{8}$ **18.** $\frac{4}{15}$ **19.** $\frac{3}{5}$ **20.** $\frac{7}{56}$

21. $\frac{11}{33}$ **22.** $\frac{12}{8}$ **23.** $\frac{9}{25}$ **24.** $\frac{36}{27}$

25. **TESTS** In Mr. Henderson's math class, 20 out of 25 students earned a grade of A or B. What percent is this?

EXAMPLE 1

Write 70% as a fraction in simplest form.

$70\% = \frac{70}{100}$ Definition of percent

$= \frac{7}{10}$ Simplify.

So, $70\% = \frac{7}{10}$.

EXAMPLE 2

Write $\frac{30}{20}$ as a percent.

$\frac{30}{20} = \frac{150}{100}$ Write an equivalent fraction with a denominator of 100.

$= 150\%$ $\frac{150}{100} = 150\%$

7-2 Fractions, Decimals, and Percents (pp. 337–342)

Write each percent as a decimal.

26. 43% **27.** 7.2% **28.** 115% **29.** 0.48%

Write each decimal or fraction as a percent. Round to the nearest tenth if necessary.

30. 0.37 **31.** 2.4 **32.** $\frac{1}{7}$ **33.** $\frac{62}{80}$

34. **PETS** In a survey, 0.2 of American households own a dog, one-fourth own cats, and 7% own a bird. Which group is the largest? Explain.

EXAMPLE 3

Write 8% as a decimal.

$8\% = 0.08$ Remove the % symbol and divide by 100. Add placeholder zero.

$= 0.08$ Add leading zero.

EXAMPLE 4

Write 0.36 as a percent.

$0.36 = 0.36$ Multiply by 100.

$= 36\%$ Add the % symbol.

7-3 Using the Percent Proportion (pp. 345–350)

Use the percent proportion to solve each problem.

35. 12 is what percent of 60?

36. What is 63% of 130?

37. 28 is 80% of what number?

38. **MUSIC** Thirty percent of the CDs that Monique owns are classical. If Monique owns 120 CDs, how many are classical?

EXAMPLE 5

Thirty six is 24% of what number?

$\frac{36}{b} = \frac{24}{100}$ Write the percent proportion.

$36 \cdot 100 = b \cdot 24$ Find the cross products.

$3600 = 24b$ Simplify.

$150 = b$ Divide each side by 24.

So, 36 is 24% of 150.

CHAPTER 7

Study Guide and Review

7-4 Find Percent of a Number Mentally (pp. 351–355)

Find the percent of each number mentally.

39. 50% of 36

40. 40% of 55

41. $33\frac{1}{3}$% of 27

42. 1% of 167

Estimate.

43. 24% of 40

44. 62% of 90

45. $\frac{1}{6}$% of 298

46. 130% of 250

47. **SPORTS** Tito had 244 free throw attempts in his high school career. If he was successful 77% of the time, about how many free throws did he make?

EXAMPLE 6

Find 40% of 90 mentally.

$40\% \text{ of } 90 = \frac{2}{5} \text{ of } 90$ **Think: 40% = $\frac{2}{5}$.**

$= 36$ **Think: $\frac{2}{5}$ of 90 is 36.**

So, 40% of 90 is 36.

EXAMPLE 7

Estimate 78% of 112.

78% is about 75% or $\frac{3}{4}$.

$\frac{3}{4}$ of 112 is 84.

So, 78% of 112 is about 84.

7-5 Using Percent Equations (pp. 357–362)

Solve each problem using a percent equation.

48. 17 is what percent of 68?

49. What is $16\frac{2}{3}$% of 24?

50. 55 is 20% of what number?

51. 48 is what percent of 32?

52. **SOUVENIRS** The items in a souvenir shop are on sale for the prices shown. What percent of the original price is the sale price for each item?

Item	Original Price	Sale Price
hat	$14.00	$10.50
beach towel	$17.50	$14.00
tote bag	$9.00	$6.30

53. **SHOPPING** A jersey is on sale for 50% off the original price. A week later, the manager takes another 50% off. Is the jersey now free? Explain.

EXAMPLE 8

84 is 60% of what number?

The part is 84 and the percent is 60%. Let w represent the whole.

part = percent · whole

$84 = 0.6 \cdot w$ **Write the percent equation.**

$\frac{84}{0.6} = \frac{0.6w}{0.6}$ **Divide each side by 0.6.**

$140 = w$ **Simplify.**

So, 84 is 60% of 140.

EXAMPLE 9

18 is what percent of 25?

The part is 18 and the whole is 25. Let p represent the percent.

part = percent · whole

$18 = p \cdot 25$ **Write the percent equation.**

$\frac{18}{25} = \frac{25p}{25}$ **Divide each side by 25.**

$0.72 = p$ **Simplify.**

Since 0.72 = 72%, 18 is 72% of 25.

MIXED PROBLEM SOLVING
For mixed problem-solving practice, see page 849.

7-6 Percent of Change (pp. 364–369)

Find the percent of change. Round to the nearest tenth, if necessary. Then state whether the percent of change is an *increase* or *decrease*.

54. From 55 lb to 24 lb

55. From \$55.75 to \$75.00

Find the selling price for each item given the cost and the percent of markup or discount.

56. tennis shoes: \$85; 24% discount

57. portable MP3 player: \$150; 36% markup

58. CLUBS The number of members in the recycling club increased by 15 people. If the club had 12 members previously, what was the percent of increase of the members in the club?

59. ICE CREAM The number of pints of mint chocolate chip sold last week was 88. If this week 110 pints were sold, what was the percent of increase?

EXAMPLE 10

Find the percent of change from 64 minutes to 16 minutes.

$$\text{percent of change} = \frac{\text{amount of change}}{\text{original amount}}$$

$$= \frac{16 - 64}{64}$$

$$= \frac{-48}{64}$$

$$= -\frac{3}{4} \text{ or } -0.75$$

The decimal −0.75 is written as −75%. So, the percent of change is −75%.

Since the percent of change is negative, it is a percent of decrease.

EXAMPLE 11

Find the selling price if a store pays \$37 dollars for a video game and the markup is 25%.

$m = 0.25 \cdot 37$ **part = percent • whole**

$m = 9.25$ **Multiply.**

Add the markup and the cost. The selling price is \$37 + \$9.25 or \$46.25.

7-7 Simple and Compound Interest (pp. 370–374)

Find the simple interest to the nearest cent.

60. \$575 at 6.25% for 7 years

61. \$12,750 at 5% for 10 years

Find the total amount in each account to the nearest cent if the interest is compounded annually.

62. \$2750 at 8% for 3 years

63. \$1500 at 12.5% for 2 years

64. BOATS Lucas borrowed \$10,500 to buy a boat. He will pay \$276.50 each month for the next 48 months. Find the simple interest rate for his loan.

EXAMPLE 12

Find the simple interest for \$2500 invested at 3.85% for 4 years.

$I = prt$ **Write the simple interest formula.**

$I = 2500 \cdot 0.0385 \cdot 4$ **Substitute**

$I = 385$ **Simplify.**

The simple interest is \$385.

7-8 Circle Graphs (pp. 376–381)

Construct a circle graph for each set of data.

65.

Daily Nutrition	
Fluids	25%
Breads	19%
Vegetables	13%
Fruits	9%
Dairy	9%
Meat/Fish/ Poultry	6%
Fats & Oils	19%

66.

Age of U.S. Senators	
Age	Number of Senators
41–50	13
51–60	33
61–70	33
71–80	16
81+	5

67. **SPORTS** The circle graph below shows the winners of the NCAA Women's basketball championship by conference. Suppose there have been 25 total championships. How many championships were won by the ACC?

NCAA Women's Basketball Winners

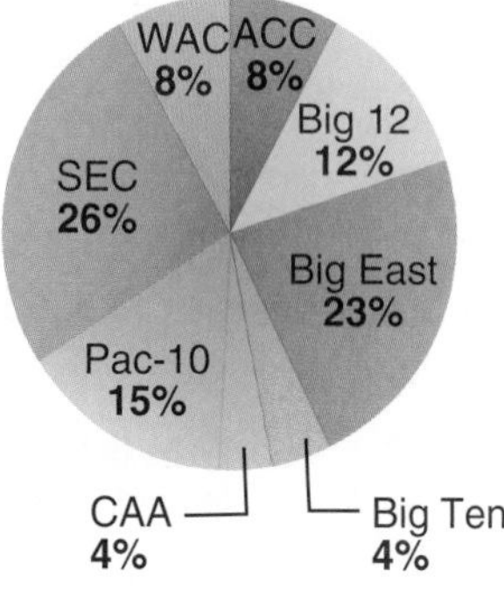

EXAMPLE 13

Construct a circle graph using the information in the table.

Earth's Oceans	
Ocean	Percent of Total
Pacific	47%
Atlantic	23%
Indian	20%
Southern	6%
Arctic	4%

Multiply each percent by 360 to find the number of degrees for each section of graph.

Pacific: 47% of 360 $\approx$ 169
Atlantic: 23% of 360 $\approx$ 83
Indian: 20% of 360 = 72
Southern: 6% of 360 $\approx$ 22
Arctic: 4% of 360 $\approx$ 14

Use a compass to draw a circle and a radius. Then use a protractor to draw a 169° angle to represent the Pacific Ocean.

From the new radius, draw the next angle. Repeat for each of the remaining angles.

Label each section. Then give the graph a title.

Earth's Oceans

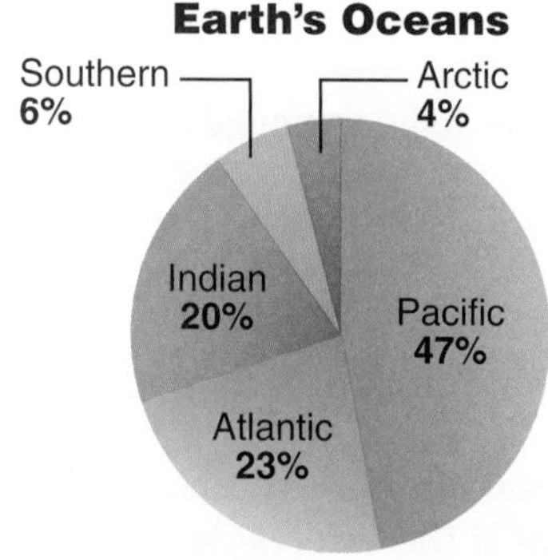

Source: World Atlas Travel

CHAPTER 7 Practice Test

Math Online glencoe.com
Chapter Test

Write each percent as a fraction or mixed number in simplest form.

1. 30%
2. 75%
3. 62%
4. 450%
5. 0.5%
6. 120%

Express each decimal or fraction as a percent. Round to the nearest tenth, if necessary.

7. $\frac{21}{25}$
8. 0.81
9. $\frac{30}{75}$
10. $\frac{11}{10}$
11. 12.54
12. 0.002

13. **TRANSPORTATION** Of the players on a football team, 32% ride their bikes to practice, $\frac{6}{20}$ walk to practice, and 0.29 ride the bus to practice. Order the types of transportation used by the football players from least to greatest.

Use the percent proportion to solve each problem.

14. Find 35% of 300.
15. What percent of 6 is 9?
16. 34 is 68% of what number?
17. What percent of 50 is 32?

18. **CAPACITY** The capacity of a fish tank is 44 gallons. If the aquarium is 23% filled, about how many gallons are in the tank?

Solve each problem using the percent equation.

19. What is 30% of 80?
20. 60 is what percent of 180?
21. 21 is 70% of what number?

22. **MULTIPLE CHOICE** Owen wants to buy a snowboard that costs $525. If there is a 5% sales tax added, what is the total cost of the snowboard?

    A $26.25
    C $530
    B $525
    D $551.25

Find the percent of change. Round to the nearest tenth, if necessary. Then state whether the percent of change is an *increase* or *decrease*.

23. From 12 h to 18 h
24. From 87 ft to 21 ft

Find the selling price for each item given the cost and the percent of the markup or discount.

25. shirt: $7, 50% discount
26. jeans: $32, 40% markup
27. sweater: $35; 28% discount
28. DVD: $15; 33% markup

29. **MULTIPLE CHOICE** Mrs. Olsen wants to buy a DVD player that regularly costs $120 and is on sale for 30% off the original price. What is the sale price of the DVD player?

    F $120
    H $84
    G $90
    J $36

30. What is the simple interest to the nearest cent of $1200 invested at 3% for 5 years?

31. Find the total amount in an account to the nearest cent if $15,000 is compounded annually at 6% for 2 years.

32. **SUBJECTS** The student council conducted a survey about students' favorite school subjects. Make a circle graph of the data in the table at the right.

Favorite Subject	Percent of Students
Math	12
Language Arts	25
Science	30
Physical Education	33

33. **SIBLINGS** The table at the right shows the number of siblings each student in the Science Club has in their family. Make a circle graph of the data.

Number of Siblings	Number of Students
0	2
1	9
2	7
3	4
4	1
5	1
6	1

CHAPTER 7

Preparing for Standardized Tests

Use Estimation

Estimating the solution to a test question before you solve it can be a useful strategy. In some cases, estimation alone might even be enough to identify the correct answer choice.

Strategies for Using Estimation

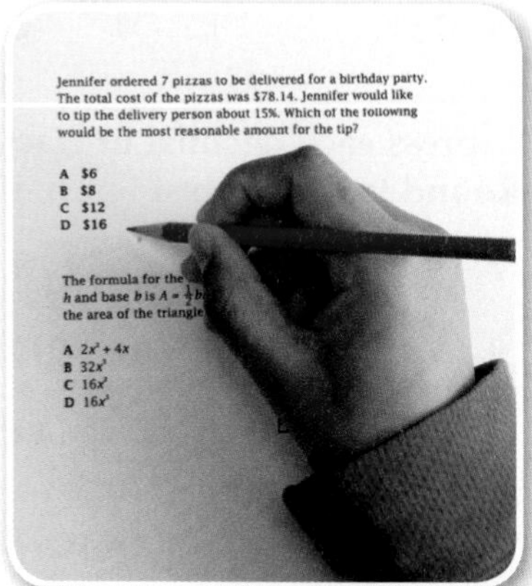

Step 1

Read the question carefully so that you understand what is being asked.

Ask yourself:

- Is an exact answer required to identify the correct answer choice?
- Could I eliminate any of the answer choices as unreasonable by quickly estimating the solution?
- Can I use estimation to check my solution?

Step 2

Use estimation when appropriate. A few examples of when estimation can be used are listed below.

- **Solving the Problem** Use the facts given in the problem to estimate the solution. Then use your estimate to choose the correct answer choice.
- **Eliminating Unreasonable Answers** Estimate the solution to the problem. If any of the answer choices are clearly different than your estimate, eliminate them as unreasonable.
- **Checking Your Answer** If time permits, check to see that your answer is reasonable using estimation.

EXAMPLE

Read the problem. Identify what you need to know. Then use the information in the problem to solve.

Jennifer ordered 7 pizzas to be delivered for a birthday party. The total cost of the pizzas was $78.14. Jennifer would like to tip the delivery person about 15%. Which of the following would be the most reasonable amount for the tip?

A $6	**C** $12
B $8	**D** $16

Read the problem carefully.

Understand You know the total cost of the pizzas. You need to determine the estimated tip for the delivery person.

Plan Estimate the tip by rounding the cost of the pizzas and using mental math.

Solve

$78.14 ≈ $80	**Round the total cost to the nearest $10.**
15% = 10% + 5%	**Break up 15% to make the problem easier to solve.**
10% of $80 is $8.	**Use mental math.**
5% of $80 is $4	**5% is $\frac{1}{2}$ of 10%. $4 is $\frac{1}{2}$ of $8.**
$8 + $4 = $12	**Add.**

So, 15% of the total cost is about $8 + $4, or $12. A reasonable amount for the tip would be $12.

The correct answer is C.

Check 10% of $78 is $7.80, and 20% of $78 is $15.60. Since $12 is somewhere in the middle between these two values, the answer seems reasonable. ✔

Exercises

Read each question. Identify what you need to know. Then use the information in the question to solve.

1. Orlando has made 69% of his free throws this basketball season. At this rate, about how many free throws would you expect him to make in his next 40 attempts?

   A about 22

   B about 24

   C about 26

   D about 28

2. Heather spent $48.85 on her haircut. About how much should she leave for a 20% tip?

   F about $10

   G about $12

   H about $14

   J about $15

3. Gregory is thinking about investing some money in a savings account.

Bank Savings Account		
Option	Rate	Type of Interest
A	5.00%	simple
B	4.25%	compound

   If Gregory has $1985 to invest, about how much interest would he earn if he invests it for 3 years using option A?

   A about $150 C about $400

   B about $300 D about $450

4. The total area of the state of Michigan is 96,716 square miles. Of that, about 41% of the area is water. Which of the following would be the most reasonable water area for the state?

   F 4000 square miles H 40,000 square miles

   G 4500 square miles J 45,000 square miles

CHAPTER 7

Standardized Test Practice

Cumulative, Chapters 1 through 7

Multiple Choice

Read each question. Then fill in the correct answer on the answer document provided by your teacher or on a sheet of paper.

1. On a recent social studies exam, Juanita answered 34 out of 40 questions correctly. Find Juanita's grade as a percent. Round to the nearest whole percent if necessary.

A 77%
B 80%
C 82%
D 85%

2. Let $X(-12, 6)$ be a point on rectangle $WXYZ$. What are the coordinates of X' if the rectangle is dilated by a scale factor of $\frac{2}{3}$?

F (8, 4)
G (−8, 4)
H (−8, −4)
J (8, −4)

3. The table shows the favorite activity of campers at a summer camp. What percent of the campers voted for horseback riding as their favorite activity? Round to the nearest tenth if necessary.

Activity	Votes
Canoeing	17
Horseback Riding	12
Hiking	5
Swimming	9

A 24.1%
B 25.2%
C 27.9%
D 33.4%

4. Lunch for the Carlton family was $39.43. They want to leave the server a 15% tip. Which of the following is the most reasonable amount for the tip?

F $6.00
G $6.75
H $7.50
J $8.00

5. Heather is riding her bike at a speed of 15 miles per hour. How many feet per second is this?

A 17 feet per second
B 18 feet per second
C 20 feet per second
D 22 feet per second

6. The circle graph below shows the results of a survey about students' favorite types of movies. If 400 students were surveyed, how many of them chose comedy as their favorite?

F 38
G 45
H 146
J 152

Favorite Type of Movie

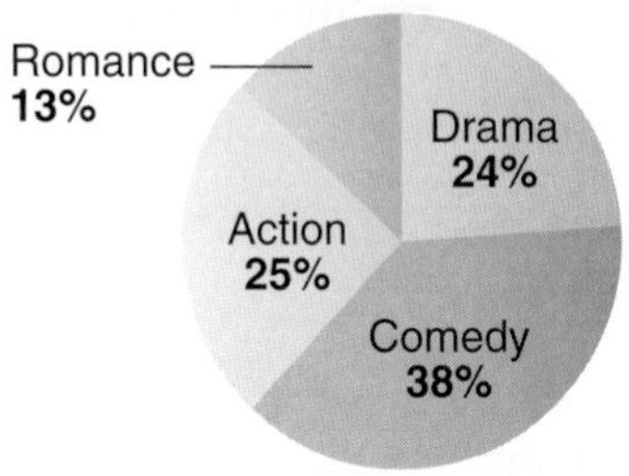

7. Craig played basketball for 56 minutes. This is 10 minutes less than twice the number of minutes that Caleb played. How many minutes did Caleb play basketball?

A 27 minutes
B 29 minutes
C 31 minutes
D 33 minutes

8. A washing machine that regularly sells for $329 is on sale for $279. What is the percent discount? Round to the nearest tenth if necessary.

F 15.2%
G 16.5%
H 17.4%
J 18.8%

Test-TakingTip

Question 4 Use mental math. Round $39.43 to $40. You know that 10% of $40 is $4, so 5% of $40 must be $2.

Short Response/Gridded Response

9. **GRIDDED RESPONSE** Dr. Bade has 24 patients to see today at his veterinarian's office. Of them, 15 are dogs. What portion of Dr. Bade's patients today are dogs?

10. Meredith wants to go on a trip this summer that costs $375. She paid a deposit of $125 and will save an additional $25 per week to pay for the trip. The equation $125 + 25w = 375$ can be used to find the number of weeks Meredith will need to save.

    a. Describe the steps needed to solve the equation.

    b. How many weeks will Meredith need to save to pay for the trip?

11. The bar graph below shows the number of votes each candidate received for class president.

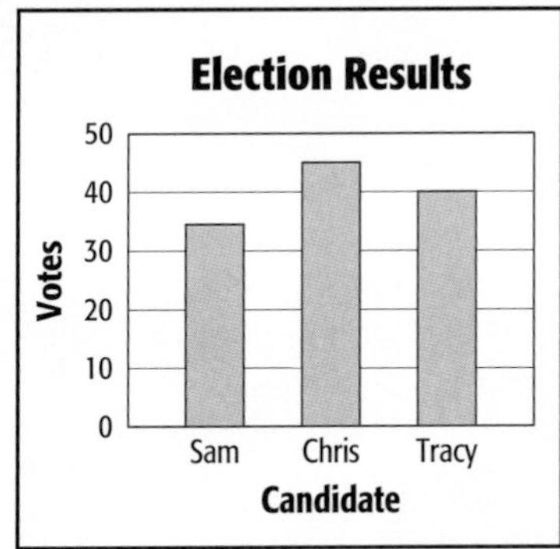

    a. How many students voted in the election?

    b. Which candidate received the most votes? What percent of the total votes did he or she receive?

12. Find and graph the solution to the following inequality.

$$5x - 3 \geq 9x + 9$$

13. **GRIDDED RESPONSE** Naomi bought the items shown below. She saved 20% on each item by using a store coupon. What was the total amount of money she saved?

The Movie Source Sales Receipt	
Item	**Original Price**
DVD Player	$85
DVD	$10
DVD	$15
Poster	$10

Extended Response

Record your answers on a sheet of paper. Show your work.

14. A retail store currently has annual sales of $160,000. The sales manager's goal is to increase sales by 15% for the following year.

    a. Write and solve an equation to find how much the store manager wants to do in sales volume next year.

    b. How much will the store's sales volume need to increase next year to meet this goal?

    c. Find the amount of sales volume next year if it is increased by 5%, 10%, 15%, and 20%.

Need Extra Help?														
If you missed Question...	1	2	3	4	5	6	7	8	9	10	11	12	13	14
Go to Lesson or Page...	7-1	6-8	7-3	7-4	6-3	7-8	4-6	7-6	7-2	4-5	7-3	5-5	7-3	7-5

CHAPTER 8

Linear Functions and Graphing

Then

In Chapter 1, you learned how to use multiple representations to represent functions.

Now

In Chapter 8, you will:

- Solve and graph linear equations with two variables.
- Write and graph linear equations using the slope and y-intercept.
- Solve systems of equations by graphing and substitution.

Why?

ANIMALS Did you know that giraffes on average eat about 100 pounds of leaves a day? They spend a majority of their day eating because it takes a lot of leaves to energize these large animals. The amount of food a giraffe eats each day can be represented by a function.

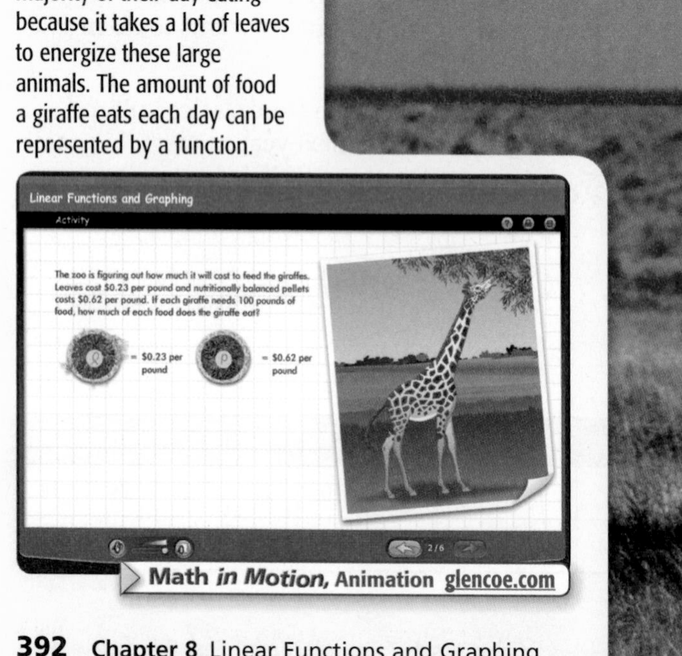

Math *in Motion,* Animation glencoe.com

Get Ready for Chapter 8

Diagnose Readiness You have two options for checking Prerequisite Skills.

Text Option

Take the Quick Check below. Refer to the Quick Review for help.

QuickCheck

Find the next term in each sequence. (Previous Course)

1. 1, 5, 9, 13, …

2. 5, 10, 20, 40, …

3. 32, 29, 26, 23, …

4. 50, 48, 46, 44, …

5. **EXERCISE** Becky started an exercise program that calls for 12 minutes of jogging each day during the first week. Each week after that, Becky increases the time she jogs by 5 minutes. In which week will she first jog more than 30 minutes?

Use the coordinate plane to name the point for each ordered pair. (Lesson 1-4)

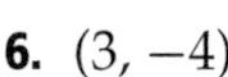

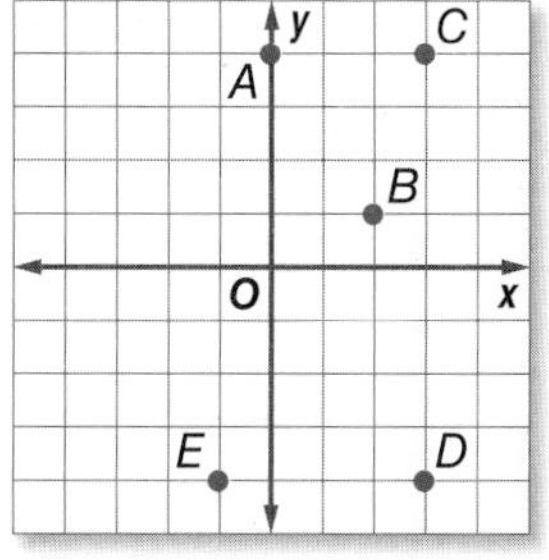

6. (3, −4)

7. (0, 4)

8. (−1, −4)

9. **DIRECTIONS** On a coordinate plane, the library is located at (4, −3) and the mall is located at (−1, 7). Using north, south, west, and east, write directions on how to walk from the library to the mall.

Write each rate as a unit rate. (Lesson 6-2)

10. 180 gallons in 10 minutes

11. 455 miles in 7 hours

12. **SHOPPING** Christa spent \$50 for 10 books at the bookstore. What was the unit rate price for each book?

QuickReview

EXAMPLE 1

Find the next term in the sequence.

4, 7, 10, 13, …

Each term is 3 more than the one before it. So, the next term is 13 + 3 or 16.

EXAMPLE 2

Use the coordinate plane to name the point for (−1, 2).

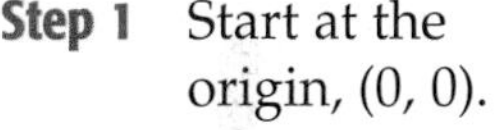

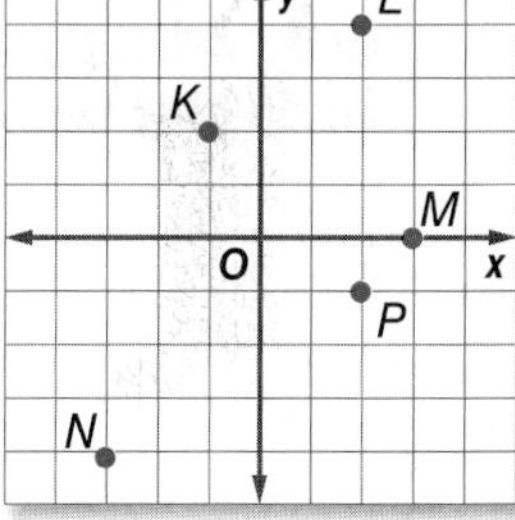

Step 1 Start at the origin, (0, 0).

Step 2 Move 1 unit left.

Step 3 Move 2 units up.

So, point *K* is at (−1, 2).

EXAMPLE 3

Write 150 miles in 2 hours as a unit rate.

To write this as a unit rate, divide 150 by 2.

150 ÷ 2 = 75

So, the unit rate is 75 miles per hour.

Online Option

Math Online Take a self-check Chapter Readiness Quiz at glencoe.com.

Get Started on Chapter 8

You will learn several new concepts, skills, and vocabulary terms as you study Chapter 8. To get ready, identify important terms and organize your resources. You may wish to refer to **Chapter 0** to review prerequisite skills.

FOLDABLES® Study Organizer

Linear Functions Make this Foldable to help you organize your Chapter 8 notes about functions and graphs. Begin with an 11" × 17" sheet of paper.

1. **Fold** the short sides so they meet in the middle.

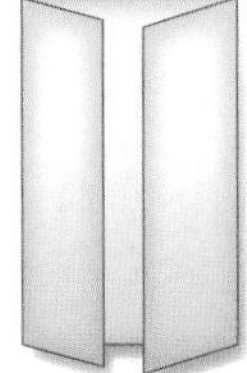

2. **Fold** the top to the bottom.

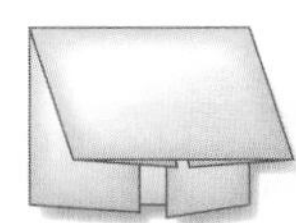

3. **Open and Cut** along the second fold to make four tabs.

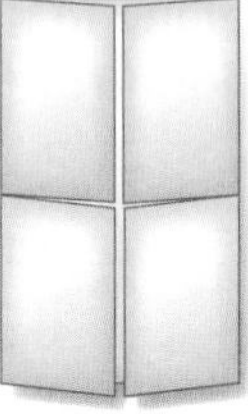

4. **Add** labels as shown.

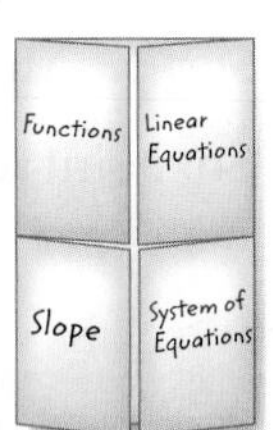

New Vocabulary

English		Español
dependent variable	• p. 395 •	variable dependiente
independent variable	• p. 395 •	variable independiente
sequence	• p. 401 •	sucesión
linear equation	• p. 406 •	ecuación lineal
rate of change	• p. 412 •	tasa de cambio
constant rate of change	• p. 418 •	tasa constante de cambio
linear relationship	• p. 418 •	relación lineal
constant of variation	• p. 420 •	constante de variación
direct variation	• p. 420 •	variación directa
slope	• p. 427 •	pendiente
slope-intercept form	• p. 433 •	forma pendiente-intersección
line of fit	• p. 448 •	recta de ajuste
system of equations	• p. 453 •	sistema de ecuaciones

Review Vocabulary

function • p. 33 • función function is a special relation in which each element of the domain is paired with exactly one element in the range

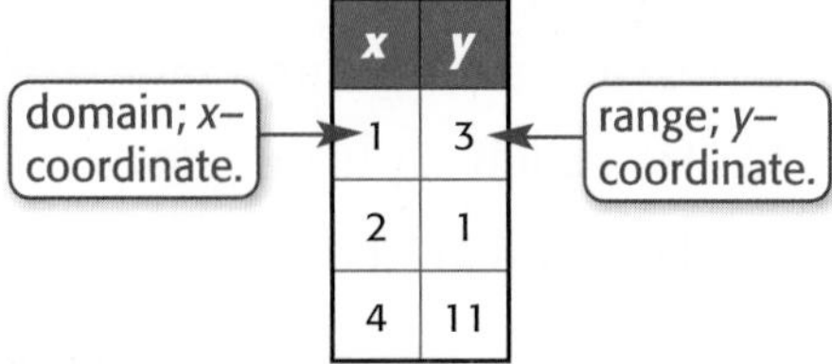

x	y
1	3
2	1
4	11

proportional relationship • p. 281 • relación proporcional the ratios of related terms are equal

rate • p. 270 • tasa a ratio of two measurements having different units

Multilingual eGlossary glencoe.com

Math Online glencoe.com

- Study the chapter online
- Explore **Math in Motion**
- Get extra help from your own **Personal Tutor**
- Use **Extra Examples** for additional help
- Take a **Self-Check Quiz**
- **Review Vocabulary** in fun ways

8-1 Functions

Then

You have already learned how to find function rules and create function tables. (Lesson 1-5)

Now

- Determine whether a relation is a function.
- Write a function using function notation.

New Vocabulary

independent variable
dependent variable
vertical line test
function notation

Math Online

glencoe.com

- Extra Examples
- Personal Tutor
- Self-Check Quiz
- Homework Help

Why?

The Thompson Middle School student council is selling school pennants to raise money for a new school welcome sign.

a. Copy and complete the table.

b. If they sell 70 pennants, how much money will they earn?

c. Explain how to find the total earnings if they sell 100 pennants.

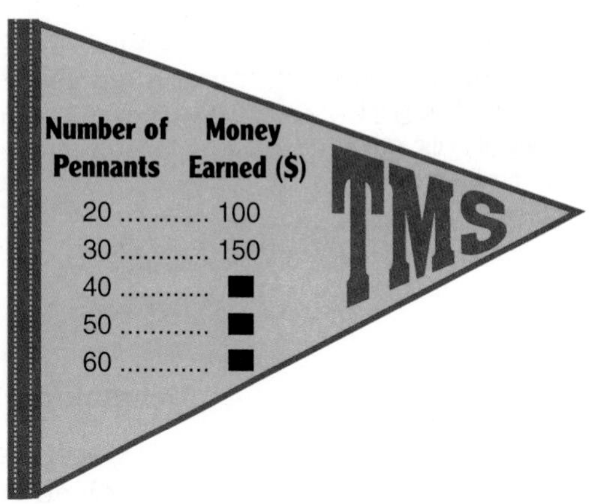

Relations and Functions The total earnings (output) of the student council depends on, or is a function of, the number of pennants sold (input). The number of pennants sold is called the **independent variable** because the values are chosen and do not depend upon the other variable. The total money earned is called the **dependent variable** because it depends on the input value.

Recall that functions are relations in which each element of the domain is paired with *exactly* one element of the range.

Function

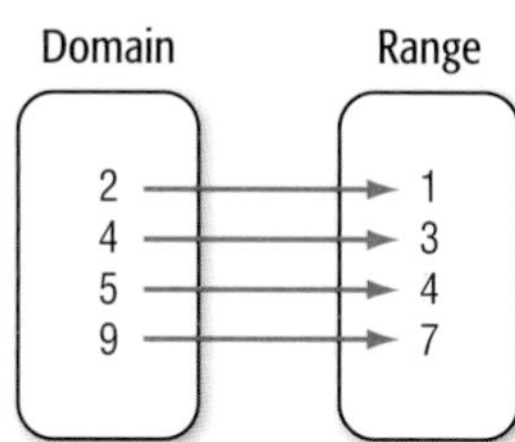

This is a function because each domain value is paired with exactly one range value.

Not a Function

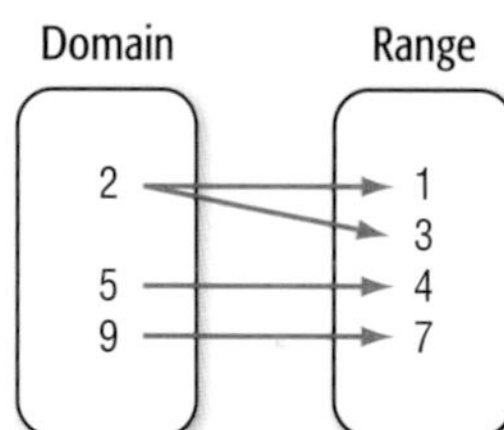

This is *not* a function because 2 in the domain is paired with two range values, 1 and 3.

EXAMPLE 1 Determine Whether a Relation is a Function

Determine whether each relation is a function. Explain.

a. {(54, 112), (56, 130), (55, 145), (54, 123), (56, 128)}

This is not a function because 54 and 56 in the domain are paired with two range values.

b.

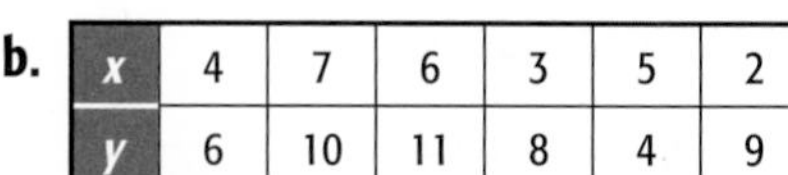

x	4	7	6	3	5	2
y	6	10	11	8	4	9

This a function. Each domain value is paired with one range value.

Check Your Progress

1A. {(5, 1), (6, 3), (7, 5), (8, 0)}

1B.

x	1	6	3	1	5	2
y	7	6	2	8	2	1

Personal Tutor glencoe.com

StudyTip

Vertical Line Test Use a pencil or straightedge to represent a vertical line. Place the pencil at the left of the graph. Move it to the right across the graph.

Another way to determine whether a relation is a function is to apply the **vertical line test** to the graph of the relation. If, for each value of x in the domain, a vertical line passes through no more than one point on the graph, then the graph represents a function. If the line passes through more than one point on the graph, it is *not* a function.

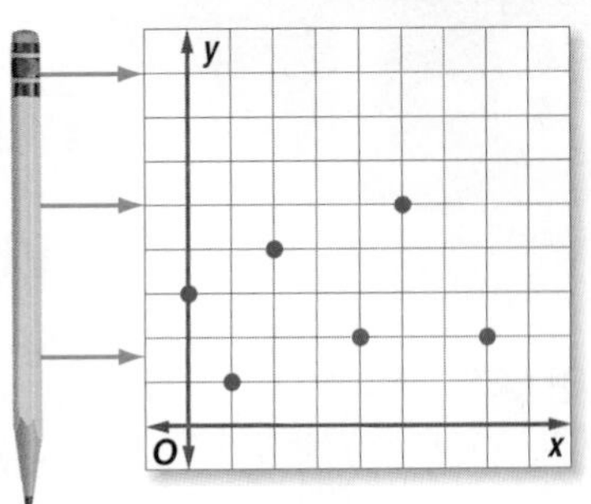

EXAMPLE 2 Use a Graph to Identify Functions

Determine whether the graph above is a function. Explain your answer.

The graph is a function because the vertical line test shows that it passes through no more than one point on the graph for each value of x.

Check Your Progress

2. Determine whether the graph at the right is a function. Explain your answer.

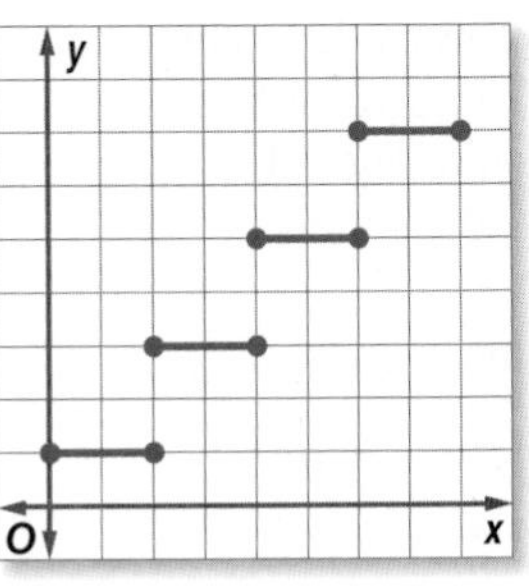

Personal Tutor glencoe.com

Function Notation A function that is written as an equation can also be written in a form called **function notation**. Consider the equation $y = 2x + 3$.

Equation	Function Notation
$y = 2x + 3$	$f(x) = 2x + 3$

The variable y and $f(x)$ both represent the dependent variable. In the example above, when $x = 4$, $f(x) = 11$. In function notation, $f(x)$ is read "f of x" and is equal to the value of the function at x.

Math History Link

Leonhard Euler (1707–1783)
By 1766, Euler lost sight in both of his eyes. Despite being blind, he continued to publish his mathematical results. Euler is credited with being the person who introduced function notation.

EXAMPLE 3 Find a Function Value

If $f(x) = 4x - 7$, find each function value.

a. $f(5)$

$f(x) = 4x - 7$	Write the function.
$f(5) = 4(5) - 7$	Replace x with 5.
$f(5) = 13$	Simplify.

b. $f(-6)$

$f(x) = 4x - 7$	Write the function.
$f(-6) = 4(-6) - 7$	Replace x with -6.
$f(-6) = -31$	Simplify.

Check Your Progress

If $f(x) = 14 + 3x$, find each function value.

3A. $f(4)$

3B. $f(-7)$

Personal Tutor glencoe.com

Describe Relationships A function can also describe the relationship between two quantities. For example, the distance you travel in a car depends on how long you are in the car. In other words, *distance is a function of time* or $d(t)$.

StudyTip

Functions A function does not always have to be in the form of $f(x)$. If a function is representing a real-world situation, any variables representing the independent and dependent variables can be used.

Real-World EXAMPLE 4 Use Function Notation

MUSIC Mariah spent $22.50 downloading songs to her digital music player from an online music store for $0.90 each.

a. Use function notation to write an equation that gives the total cost as a function of the number of songs purchased.

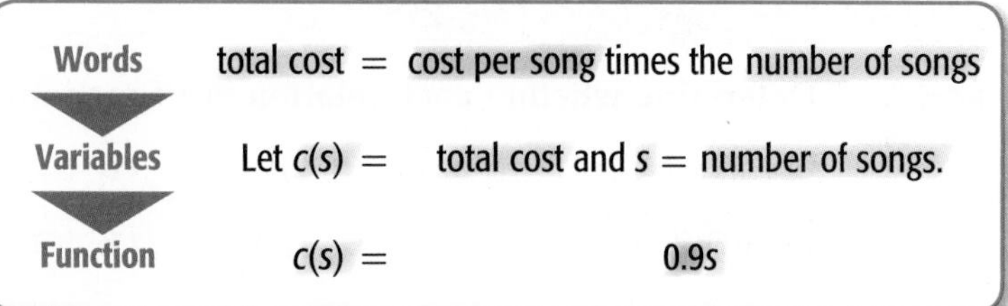

Words	total cost	=	cost per song times the number of songs
Variables	Let $c(s)$	=	total cost and s = number of songs.
Function	$c(s)$	=	$0.9s$

The function is $c(s) = 0.9s$.

b. Use the equation to determine the number of songs that Mariah downloaded.

$c(s) = 0.9s$	**Write the function.**
$22.5 = 0.9s$	**Substitute 22.5 for $c(s)$.**
$25 = s$	**Divide each side by 0.9.**

So, Mariah downloaded 25 songs.

Check Your Progress

4. **WHALES** A whale watching boat traveled at a speed of 5.5 miles per hour.

   A. Use function notation to write a function that gives the total distance traveled as a function of the time in hours spent whale watching.

   B. Use the function to find how long it took to travel 25 miles.

Personal Tutor glencoe.com

Check Your Understanding

Examples 1 and 2 pp. 395–396

Determine whether each relation is a function. Explain.

1. {(8, 2), (4, 3), (6, 5), (1, 5)}

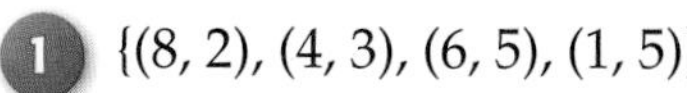

2.

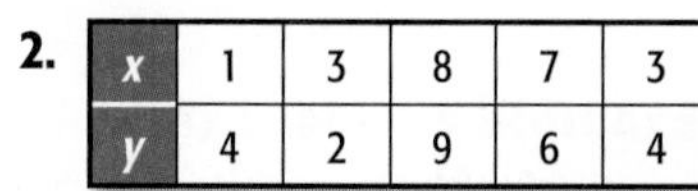

x	1	3	8	7	3
y	4	2	9	6	4

3.

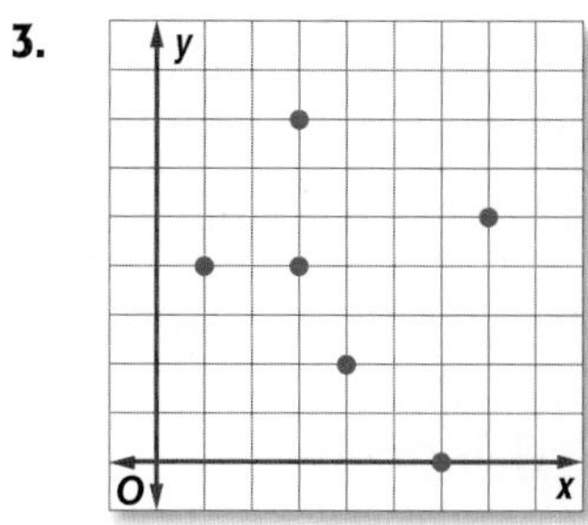

4.

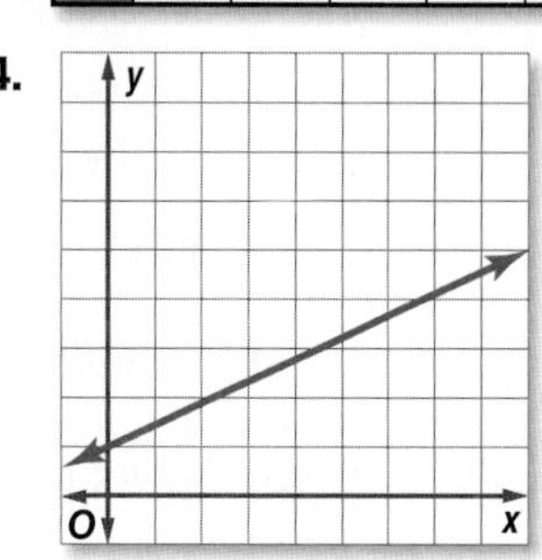

Example 3 p. 396

If $f(x) = 6x - 4$, find each function value.

5. $f(3)$
6. $f(-5)$
7. $f(8)$
8. $f(-9)$

Example 4 p. 397

9. **PORTRAITS** The Milligan family spent \$215 to have their family portrait taken. The portrait package they would like to purchase costs \$125. In addition, the photographer charges a \$15 sitting fee per person in the portrait.

   a. Use function notation to write an equation that gives the total cost as a function of the number of people in the portrait.

   b. Use the equation to find the number of people in the portrait.

Practice and Problem Solving

= Step-by-Step Solutions begin on page R11.
Extra Practice begins on page 810.

Examples 1 and 2 pp. 395–396

Determine whether each relation is a function. Explain.

10. {(10, 8), (12, 4), (15, 15), (9, 4)}

11. {(24, 16), (25, 16), (24, 17), (26, 17)}

12. {(8, 9), (9, 10), (10, 11), (11, 12)}

13. {(3, 1), (3, 3), (3, 7), (3, 9)}

14.
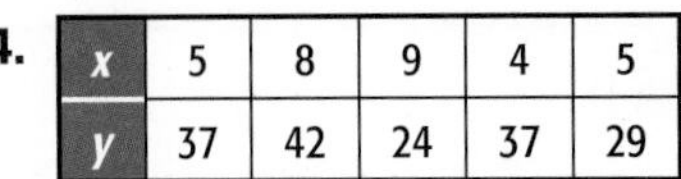

x	5	8	9	4	5
y	37	42	24	37	29

15.
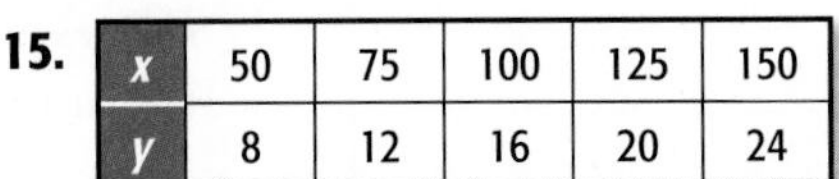

x	50	75	100	125	150
y	8	12	16	20	24

16.
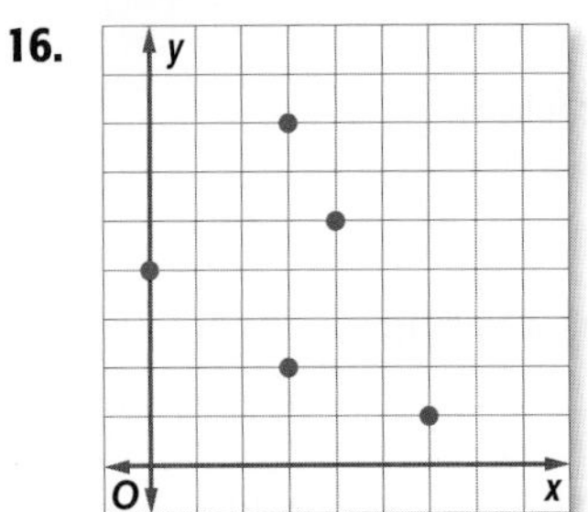

17.
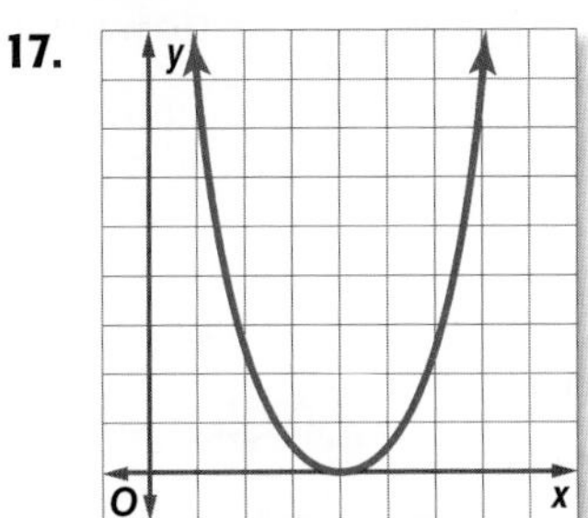

Example 3 p. 396

If $f(x) = 3x - 9$, find each function value.

18. $f(12)$
19. $f(9)$
20. $f(-8)$
21. $f(-15)$
22. $f(22)$
23. $f(3)$
24. $f(-4)$
25. $f(-11)$

If $g(x) = 5x + 4$, find each function value.

26. $g(7)$
27. $g(14)$
28. $g(-10)$
29. $g(-18)$
30. $g(24)$
31. $g(50)$
32. $g(-7)$
33. $g(-14)$

Example 4 p. 397

34. **EXPLORATION** A submersible descends to a depth of 1500 feet at a rate of 40 feet per minute to explore an underwater shipwreck.

   a. Use function notation to write an equation that gives the total depth of the submersible as a function of the total time (in minutes).

   b. Use the equation to find the total amount of time it took to reach 1500 feet.

35. **TRUCKS** Logan spends \$128.10 to rent a moving truck from a rental company. The rental company charges him \$50 for the truck plus an additional fee of \$0.55 per mile that the truck is driven.

   a. Use function notation to write an equation that gives the total cost as a function of the number of miles driven.

   b. Use the equation to find the number of miles that he drove the truck.

36. **MILK** The table shows the sales of milk in the U.S. in different years.

U.S. Milk Sales, 1960–2000 (millions of pounds)		
Year	Whole Milk	Reduced and Lowfat Milk
1960	42,999	389
1970	41,512	6,163
1980	31,253	15,918
1990	21,333	24,509
2000	18,448	23,649

Source: USDA

a. Is the set of ordered pairs (year, whole milk) a function? Explain.

b. Is the set of ordered pairs (year, reduced and lowfat milk) a function? Explain.

c. Describe the relationship between year and million pounds of whole milk purchased. Describe the relationship between year and million pounds of reduced fat and lowfat milk production. Compare the information provided by the two sets of data.

Real-World Link

Twenty-one percent of the United States' milk supply comes from California. The number two state is Wisconsin with 13%.

Source: USDA

37. **PRODUCTION** The table shows the average number of hours worked weekly by U.S. production workers for various years.

Year	Hours
1970	37
1980	35
1990	34
2000	34

a. Write the values in the table as a set of ordered pairs. Do the data represent a function? Explain.

b. The *inverse* of a relation is obtained by switching the order of the numbers in each ordered pair. For example, (1970, 37) would be (37, 1970). Write the inverses of the ordered pairs of the data. Do the data represent a function? Explain.

H.O.T. Problems

Use Higher-Order Thinking Skills

38. **OPEN ENDED** Draw two graphs, one that represents a relation that is a function and one the represents a relation that is not a function. Explain why each graph is or is not a function.

39. **WRITING IN MATH** In an arithmetic sequence, a term is found by adding a constant value to the previous term. Consider the terms in this arithmetic sequence: 36, 33, 30, 27, …

Term Number	Term
1	36
2	33
3	30
⋮	⋮
7	■

a. What value was added to each term in the sequence?

b. In a sequence, each number is assigned a term number. Copy and complete the table. Is the set of ordered pairs (term number, term) a function? Explain.

c. Graph the set of ordered pairs. Describe the relationship between term number and term shown in the graph.

CHALLENGE If $f(x) = 4x - 3$ and $g(x) = 8x + 2$, find each function value.

40. $f[g(3)]$
41. $g[f(5)]$
42. $f[g(-8)]$
43. $g[f(12)]$
44. $f\{g[f(2)]\}$
45. $g\{f[g(-4)]\}$

46. **WRITING IN MATH** How can the relationship between water depth and time to ascend to the water's surface be a function? Explain how the two variables are related. Discuss whether water depth can ever correspond to two different times.

Standardized Test Practice

47. The equation $c(d) = 3.50d$ represents the total cost of renting DVDs as a function of the number of DVDs rented. Which of the following tables contains values that satisfy this function?

A

Cost of Renting DVDs				
d	1	2	3	4
$c(d)$	3.50	7.00	10.50	14.00

B

Cost of Renting DVDs				
d	1	2	3	4
$c(d)$	3.50	6.00	9.00	12.50

C

Cost of Renting DVDs				
d	1	2	3	4
$c(d)$	3.50	7.50	10.50	14.50

D

Cost of Renting DVDs				
d	1	2	3	4
$c(d)$	7.00	10.50	14.00	17.50

48. What is the value of $f(3)$ if $f(x) = -4x + 2$?

F -14 H -6
G -10 J 14

49. EXTENDED RESPONSE For song downloads, Kaylee pays $15 per month plus $1.75 per song.

a. Write an equation using function notation that gives the total cost as a function of the number of songs downloaded.

b. How much would her total bill be if she downloaded 15 songs?

c. Make a function table to show her total bill if she downloads 15, 20, 25, and 30 songs.

50. Which of the following sets of ordered pairs does *not* represent a function?

A {(1, 2), (2, 3), (3, 4), (4, 5)}
B {(2, 3), (2, 4), (2, 5), (2, 6}
C {(1, 6), (2, 6), (3, 6), (4, 6)}
D {(5, 4), (4, 3), (3, 2), (2, 1)}

Spiral Review

51. RIDES The table shows the results of a survey about the favorite type of ride at an amusement park. Construct a circle graph using the information in the table. (Lesson 7-8)

Type of Ride	Percent
Roller coaster	48
Bumper cars	29
Ferris wheel	10
Carousel	3
Tilt-a-whirl	10

52. BANKING Suppose Marcus invests $750 at an annual rate of 6.25%. How long will it take until Marcus earns $125 in simple interest? (Lesson 7-7)

Find each product or quotient. (Lessons 3-3 and 3-4)

53. $\frac{8}{9} \cdot \frac{27}{28}$ **54.** $\frac{3}{4}\left(-\frac{1}{3}\right)$ **55.** $-\frac{7}{8} \cdot \frac{2}{5}$ **56.** $2 \cdot \frac{7}{12}$

57. $\frac{2}{9} \div \frac{1}{4}$ **58.** $-\frac{1}{2} \div \frac{5}{6}$ **59.** $-\frac{3}{5} \div -\frac{5}{9}$ **60.** $6\frac{2}{3} \div 5$

Skills Review

Evaluate each expression if $a = 12$ and $b = 8$. (Lesson 1-2)

61. $4a - b$ **62.** $3b - 2a$ **63.** $3a + 2b$ **64.** $a + 5b$

8-2 Sequences and Equations

Why?

Giant pandas spend more than 10 hours a day searching for food and eating. The table shows the average amount of bamboo a giant panda eats.

Number of Days	Total Amount of Bamboo (lb)
1	30
2	60
3	90
4	120

Source: Smithsonian National Zoo

a. As the number of days increases by 1, how much does the amount of bamboo increase?

b. How much bamboo would a giant panda eat in 5 days?

Then
You have already used variables to represent patterns. (Lesson 1-2)

Now
- Describe sequences using words and symbols.
- Find terms of arithmetic sequences.

New Vocabulary
sequence
term
arithmetic sequence
common difference

Math Online
glencoe.com
- Extra Examples
- Personal Tutor
- Self-Check Quiz
- Homework Help

Describe Sequences A **sequence** is an ordered list of numbers. Each number is called a **term** of the sequence. When the difference between any two consecutive terms is the same, the sequence is called an **arithmetic sequence**. The difference is called the **common difference**.

EXAMPLE 1 Describe an Arithmetic Sequence

Describe each sequence using words and symbols.

a. 6, 7, 8, 9, …

+1 +1 +1

Term Number (n)	1	2	3	4
Term (t)	6	7	8	9

+1 +1 +1

The difference of the term numbers is 1.
The common difference of the terms is 1.

The terms have a common difference of 1. A term is 5 more than the term number. So, the equation that describes the sequence is $t = n + 5$.

b. 4, 8, 12, 16, …

+1 +1 +1

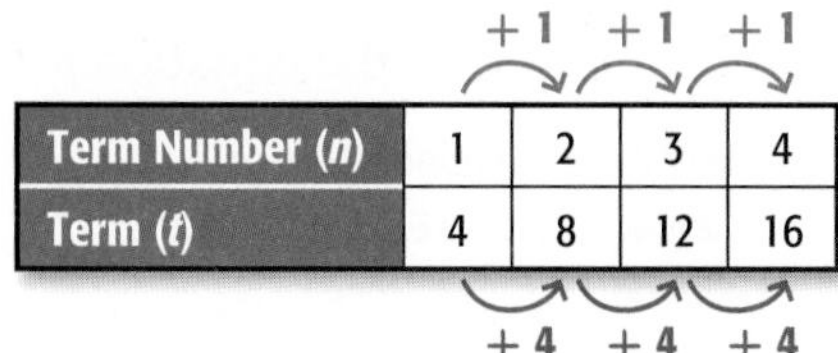

Term Number (n)	1	2	3	4
Term (t)	4	8	12	16

+4 +4 +4

The difference of the term numbers is 1.
The common difference of the terms is 4.

The terms have a common difference of 4. A term is 4 times the term number. So, the equation that describes the sequence is $t = 4n$.

Check Your Progress

1A. 10, 11, 12, 13, …

1B. 5, 10, 15, 20, …

Personal Tutor glencoe.com

StudyTip

Functions An arithmetic sequence is a function. The equation describing the sequence is the function rule and the term numbers are the inputs.

Finding Terms You can use rules to extend patterns and find other terms.

EXAMPLE 2 Find a Term in an Arithmetic Sequence

Write an equation that describes the sequence 7, 10, 13, 16, … . Then find the 15th term of the sequence.

Term Number (n)	1	2	3	4
Term (t)	7	10	13	16

+1 +1 +1 — The difference of the term numbers is 1.

+3 +3 +3 — The common difference of the terms is 3.

The terms have a common difference of 3. This difference is 3 times the difference of the term numbers. This suggests that $t = 3n$. However, you need to add 4 to get the value of t. So, $t = 3n + 4$ describes the sequence.

Check If $n = 2$, then $t = 3(2) + 4$ or 10. ✓

Use the equation to find the 15th term. Let $n = 15$.

$t = 3n + 4$ **Write the equation.**

$t = 3(15) + 4$ or 49 **Replace n with 15.**

So, the 15th term is 49.

Check Your Progress

2. Write an equation that describes the sequence 5, 8, 11, 14, … . Then find the 20th term of the sequence.

Personal Tutor glencoe.com

Real-World Link

About 12.8 million people work in the restaurant industry in the U.S. The only other employer in the country that is larger is the U.S. government.

Source: National Restaurant Association

Real-World EXAMPLE 3 Find a Term in an Arithmetic Sequence

RESTAURANTS The diagram shows the number of square tables needed to seat 4, 6, or 8 people at a restaurant. How many tables are needed to seat 16 people?

Make a table to organize your sequence and find a rule.

Number of Tables (t)	1	2	3
Number of People (p)	4	6	8

The difference of the term numbers is 1.

The common difference of the terms is 2.

The pattern in the table shows the equation $p = 2t + 2$.

$p = 2t + 2$ **Write the equation.**

$16 = 2t + 2$ **Replace p with 16. Solve for t.**

$7 = t$ **Simplify.**

So, 7 tables are needed to seat a party of 16.

Check Your Progress

3. **RESTAURANTS** How many tables shaped like hexagons are needed for 22 people?

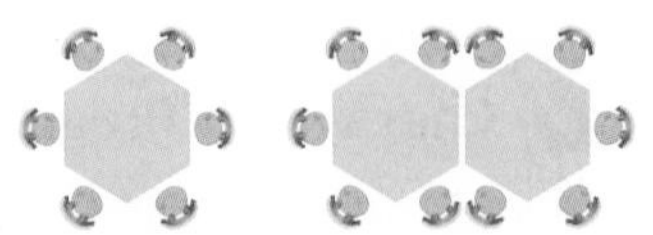

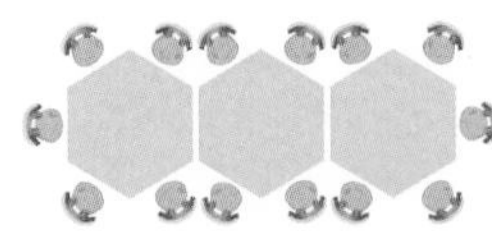

Personal Tutor glencoe.com

Check Your Understanding

Example 1 p. 401

Describe each sequence using words and symbols.

1. 2, 3, 4, 5, … **2.** 7, 8, 9, 10, …

3. 3, 6, 9, 12, … **4.** 7, 14, 21, 28, …

Example 2 p. 402

Write an equation that describes each sequence. Then find the indicated term.

5. 10, 11, 12, 13, …; 10th term **6.** 6, 12, 18, 24, …; 11th term

7. 4, 7, 10, 13, …; 23rd term **8.** 2, 6, 10, 14, …; 14th term

Example 3 p. 402

9. GEOMETRY Suppose each side of a square has a length of 1 foot. Determine which figure will have a perimeter of 60 feet.

Figure 1

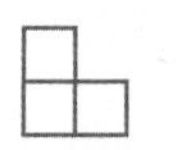
Figure 2

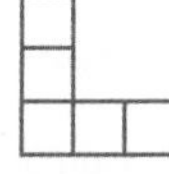
Figure 3

Practice and Problem Solving

 = **Step-by-Step Solutions** begin on page R11.
Extra Practice begins on page 810.

Example 1 p. 401

Describe each sequence using words and symbols.

10. 3, 4, 5, 6, … **11.** 8, 9, 10, 11, … **12.** 14, 15, 16, 17,

13. 15, 16, 17, 18, … **14.** 2, 4, 6, 8, … **15.** 8, 16, 24, 32, …

16. 12, 24, 36, 48, … **17.** 20, 40, 60, 80, … **18.** 6, 16, 26, 36, ...

Example 2 p. 402

Write an equation that describes each sequence. Then find the indicated term.

19. 16, 17, 18, 19, …; 23rd term **20.** 14, 15, 16, 17, …; 16th term

21. 4, 8, 12, 16, …; 13th term **22.** 11, 22, 33, 44, …; 25th term

23. 7, 10, 13, 16, …; 20th term **24.** 7, 9, 11, 13, …; 33rd term

25. 1, 5, 9, 13, …; 89th term **26.** 3, 8, 13, 18, …; 70th term

27 **CONSTRUCTION** A building frame consists of beams in the form of triangles, as shown in the diagram. The frame of a new office building will use 27 beams.

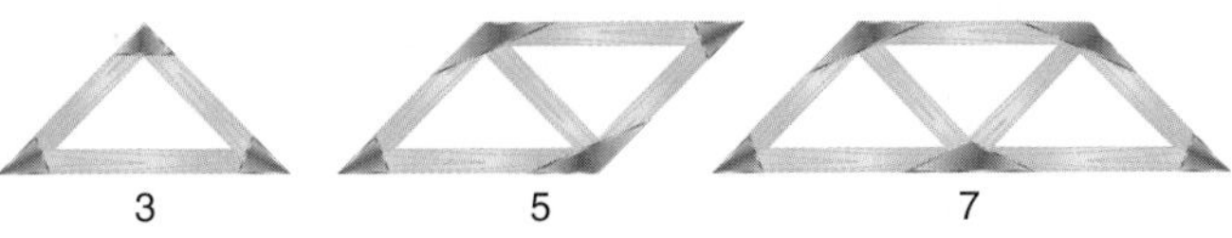

a. Use the pattern to find a rule that describes the sequence.

b. Find the number of triangles that will be formed for the frame.

Example 3 p. 402

28. GEOMETRY Study the pattern at the right. Which figure will have 40 squares?

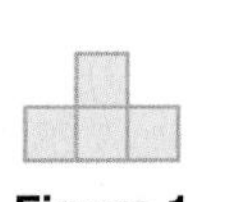
Figure 1

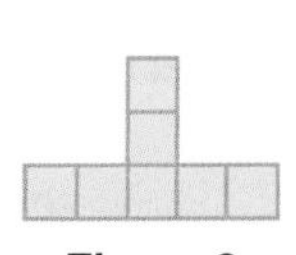
Figure 2

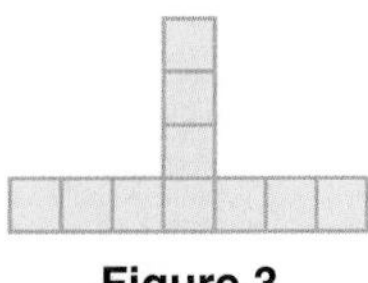
Figure 3

Real-World Link

A person on jumping stilts can leap up to 6 feet in the air and travel up to 20 miles per hour.

Source: Xtreme Importz

29. **EXTREME SPORTS** The table shows how far a person on *jumping stilts* can travel. Write an equation that describes the relationship between distance traveled d and number of steps taken s. Then determine how far a person could travel in 12 steps.

Number of Steps (s)	1	2	3	4	5
Distance Traveled in Feet (d)	9	18	27	36	45

Describe each sequence using words and symbols.

30. 3, 5, 7, 9, …
31. 4, 6, 8, 10, …
32. 1, 4, 7, 10, …
33. 4, 9, 14, 19, …

34. **MULTIPLE REPRESENTATIONS** In this problem, you will investigate sequences and graphs. Use the graph shown.

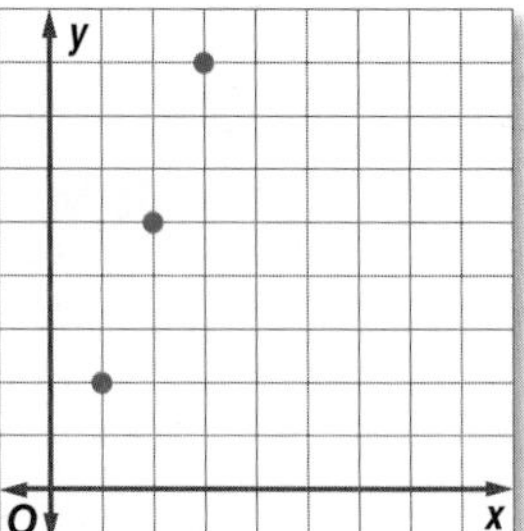

   a. **TABULAR** Make a table of ordered pairs (x, y).

   b. **VERBAL** Describe how the values of y change as the values of x increase by 1.

   c. **ALGEBRAIC** Write an equation relating x and y. Then find x when y is 101.

35. **THEATERS** One section of a movie theater has 26 seats in the first row, 35 seats in the second row, 44 seats in the third row, and so on. If there are 10 rows of seats, how many seats are in the section?

36. **ALGEBRA** The expression $1 + 2n(n + 2)$ describes a pattern of numbers, where n represents a number's position in the sequence. Write the first four terms of the sequence.

H.O.T. Problems
Use Higher-Order Thinking Skills

37. **OPEN ENDED** Write an arithmetic sequence whose common difference is -8.

38. **CHALLENGE** Use an arithmetic sequence to find the number of multiples of 6 between 41 and 523. Justify your reasoning.

39. **WRITING IN MATH** Explain how arithmetic sequences could be used to make real-world predictions.

40. **FIND THE ERROR** Edward and Alliyah are writing a rule for finding the nth term of the sequence 14, 18, 22, 26, … . Is either of them correct? Explain.

Edward	Alliyah
The difference is 4, so the rule is $t = 4n$.	*The difference is 4 and each term is 10 more than four times the term number, so the rule is* $t = 4n + 10$.

41. **WRITING IN MATH** The equation $y = x + 6$ represents the function {(1, 7), (2, 8), (3, 9), (4, 10)}. Write a rule using n and t to describe the arithmetic sequence 7, 8, 9, 10, … . Then compare the input values and the output values of both the function and the sequence.

Standardized Test Practice

42. In a study skills program, students increase the number of minutes they study each week. The first week, they study 40 minutes per day. Each week after that, they increase their study time by 10%, until they are studying 72 minutes per day. In what week of the program should the students begin studying over 60 minutes per day?

A Week 3
B Week 4
C Week 5
D Week 6

43. GRIDDED RESPONSE The table below shows temperatures in degrees Celsius and the equivalent temperature in degrees Fahrenheit.

Celsius	Fahrenheit
40	104
30	86
20	68

If the pattern in the table continues, what will be the temperature in degrees Fahrenheit if the temperature outside is 15 degrees Celsius?

44. Which of the following equations describes the arithmetic sequence 7, 13, 19, 25, 31, ... , where n is the term number?

F $t = 4n + 3$
G $t = 6n + 1$
H $t = 7n$
J $t = 5n + 2$

45. EXTENDED RESPONSE The table below follows a rule.

x	y
1	5
2	7
3	9
■	■
■	■

a. Complete the table by filling in the four missing numbers.

b. Based on the table, write a rule that represents the relationship between x and y.

Spiral Review

46. ENERGY The circle graph shows the source of the United States' energy. (Lesson 7-8)

a. What is the biggest source of energy for the United States?

b. If the petroleum contributed 40 quadrillion Btu, find the amount of Btu contributed by hydropower.

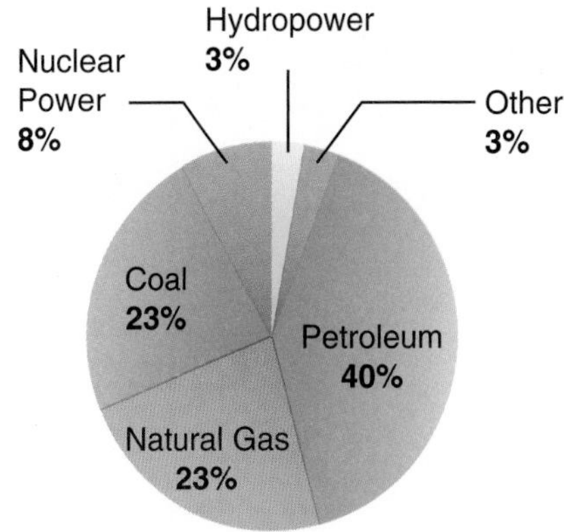

Find the simple interest to the nearest cent. (Lesson 7-7)

47. \$2,400 at 7% for $1\frac{1}{2}$ years

48. \$350 at $3\frac{1}{2}$% for 3 years

Skills Review

Evaluate each expression if $x = -2$ and $y = 3$. (Lesson 1-2)

49. $y + 7$

50. $4x$

51. $2y - 1$

52. $5x + 3y$

8-3 Representing Linear Functions

Why?

In a tennis match, Venus Williams served a tennis ball at a speed of nearly 60 meters per second.

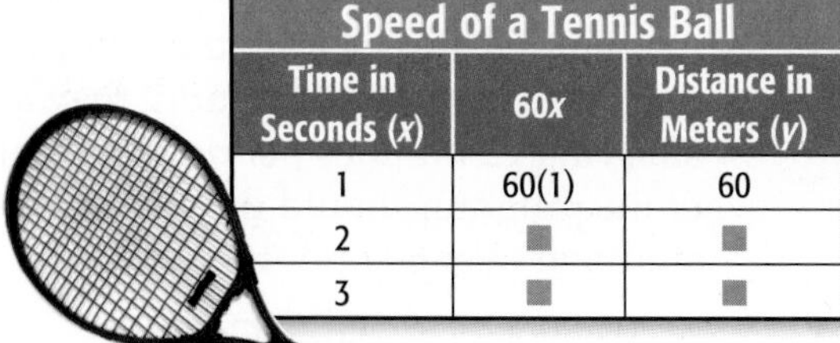

Speed of a Tennis Ball		
Time in Seconds (x)	$60x$	Distance in Meters (y)
1	60(1)	60
2	■	■
3	■	■

a. Complete the table to find the number of meters the tennis ball would travel in 1, 2, and 3 seconds at that speed.

b. On grid paper, graph the ordered pairs (time, distance). Then connect the points.

c. Do the data represent a function?

d. Write an equation representing the relationship between time x and number of meters y.

Then
You used functions to describe relationships between two quantities. (Lesson 1-5)

Now
- Solve linear equations with two variables.
- Graph linear equations using ordered pairs.

New Vocabulary
linear equation
x-intercept
y-intercept
discrete data

Math Online
glencoe.com
- Extra Examples
- Personal Tutor
- Self-Check Quiz
- Homework Help
- Math in Motion

Solve Linear Equations An equation such as $y = 60x$ is called a linear equation. A **linear equation** is an equation whose graph is a line. A linear equation is also a function because each member of the domain (x-value) is paired with exactly one member of the range (y-value).

The solution of an equation with two variables consists of two numbers, one for each variable, that make the equation true. One way to find solutions is to make a table.

EXAMPLE 1 Use a Table of Ordered Pairs

Find four solutions of $y = 7x$. Write the solutions as ordered pairs.

Step 1 Choose four values for x and substitute each value into the equation. We chose −1, 0, 1, and 2.

Step 2 Evaluate the expression to find the value of y.

Step 3 Write the solutions as ordered pairs.

Step 1		Step 2	Step 3
x	$y = 7x$	y	(x, y)
−1	$y = 7(-1)$	−7	(−1, −7)
0	$y = 7(0)$	0	(0, 0)
1	$y = 7(1)$	7	(1, 7)
2	$y = 7(2)$	14	(2, 14)

Four solutions of $y = 7x$ are (−1, −7), (0, 0), (1, 7), and (2, 14).

Check Your Progress

Find four solutions of each equation. Write the solutions as ordered pairs.

1A. $y = x + 2$

1B. $y = 3x - 1$

1C. $y = -2x + 5$

1D. $y = -4x - 6$

Personal Tutor glencoe.com

When solving real-world problems, check that the solution makes sense in the context of the original problem.

Real-World EXAMPLE 2 Use Function Equations

Real-World Link

Cell phone companies spend about $1.4 billion on mobile messaging and display advertisements.

Source: Jupiter Research

CELL PHONES Games cost $8 to download onto a cell phone. Ring tones cost $1. Find four solutions of $8x + y = 20$ in terms of the numbers of games x and ring tones y Darcy can buy with $20. Explain each solution.

Step 1 Rewrite the equation by solving for y.

$$8x + y = 20 \quad \text{Write the equation.}$$
$$8x - 8x + y = 20 - 8x \quad \text{Subtract } 8x \text{ from each side.}$$
$$y = 20 - 8x \quad \text{Simplify.}$$

Step 2 Choose four x values and substitute them into $y = 20 - 8x$.

x	$y = 20 - 8x$	y	(x, y)
1	$y = 20 - 8(1)$	12	(1, 12)
2	$y = 20 - 8(2)$	4	(2, 4)
$\frac{1}{4}$	$y = 20 - 8\left(\frac{1}{4}\right)$	18	$\left(\frac{1}{4}, 18\right)$
5	$y = 20 - 8(5)$	−20	(5, −20)

Step 3 Explain each solution.

(1, 12) → She can buy 1 game and 12 ring tones.

(2, 4) → She can buy 2 games and 4 ring tones.

$\left(\frac{1}{4}, 18\right)$ → This solution does not make sense in the situation because there cannot be a fractional number of games.

(5, −20) → This solution does not make sense in the situation because you cannot have a negative number of ring tones.

Check Your Progress

2. FOOD Michael and his two friends have a total of $9 to spend on tacos and burritos for lunch. A burrito x costs $2 and a taco y costs $1. Find two solutions of the equation $2x + y = 9$ to find how much food they can buy with $9. Explain each solution.

Personal Tutor glencoe.com

StudyTip

Graphs In this book, graphs of discrete functions will be represented by dashed lines. Continuous functions will be represented by solid lines.

The solutions in Example 2 represent discrete data because only whole number values are appropriate. **Discrete data** have space between possible data values. Continuous data can take on any value, so there is no space between data values for a given domain. You can determine if data that model real-world situations are discrete or continuous by considering whether all numbers are reasonable as part of the domain.

Discrete Data	Continuous Data
the number of fish a tank can hold	the size of fish in a fish tank
the number of shirts you can buy	the amount of money a shirt costs

Graph Linear Equations The x-coordinate of the point at which the graph crosses the x-axis is the **x-intercept**. The y-coordinate of the point at which the graph crosses the y-axis is the **y-intercept**.

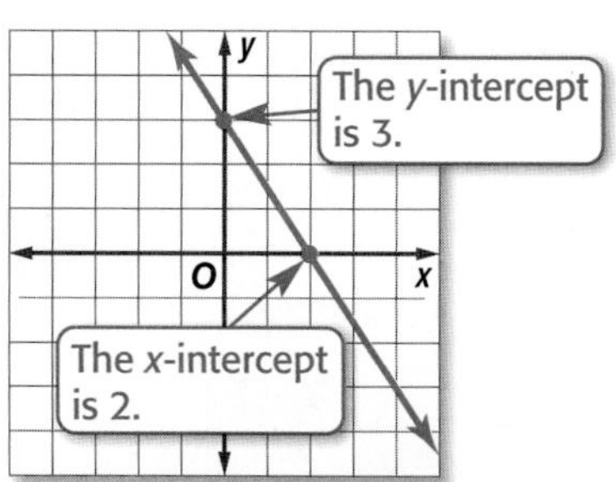

Since two points determine a straight line, a simple method of graphing a linear equation is to find the points where the graph crosses the x-axis and the y-axis and connect them.

StudyTip

An equation written in the form $Ax + By = C$, is written in standard form.

Math *in Motion*, Interactive Lab glencoe.com

EXAMPLE 3 Graph a Linear Function

Graph $-2x + y = 4$.

Step 1 Find the x-intercept.

To find the x-intercept, let $y = 0$.

$-2x + y = 4$	**Write the equation.**
$-2x + (0) = 4$	**Replace y with 0.**
$-2x = 4$	**Divide each side by -2.**
$x = -2$	

Since $x = -2$ when $y = 0$, graph the ordered pair $(-2, 0)$.

Step 2 Find the y-intercept.

To find the y-intercept, let $x = 0$.

$-2x + y = 4$	**Write the equation.**
$-2(0) + y = 4$	**Replace x with 0.**
$y = 4$	

Since $y = 4$ when $x = 0$, graph the ordered pair $(0, 4)$.

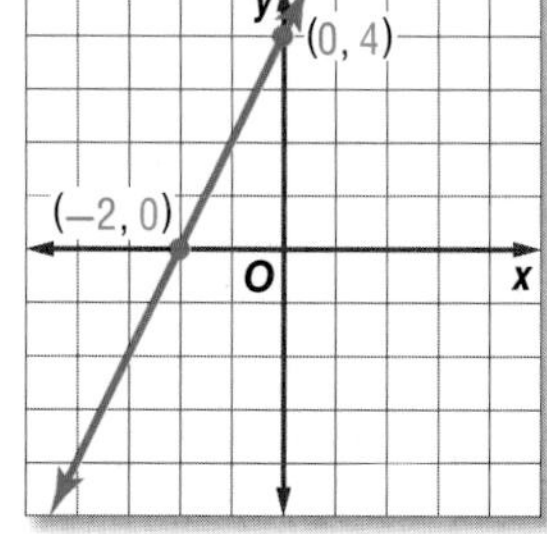

Step 3 Connect the points with a line.

Check Check by solving the equation for y.

$-2x + y = 4$	**Write the equation.**
$-2x + 2x + y = 4 + 2x$	**Add $2x$ to each side.**
$y = 4 + 2x$	**Simplify.**

If $x = -1$, $y = 4 + 2(-1)$ or 2. Notice that the graph of $(-1, 2)$ is on the line. ✔

StudyTip

Although only two points are needed to determine a straight line, always graph at least three to make certain you are correct.

Check Your Progress

Graph each equation.

3A. $y = x - 4$ **3B.** $x + y = 2$

Personal Tutor glencoe.com

Check Your Understanding

Example 1 p. 406

Copy and complete each table. Use the results to write four ordered pair solutions of the given equation.

1. $y = x + 7$

x	$y = x + 7$	y
−1	■	■
0	■	■
1	■	■
2	■	■

2. $y = 2x - 3$

x	$y = 2x - 3$	y
−2	■	■
0	■	■
2	■	■
4	■	■

Find four solutions of each equation. Write the solutions as ordered pairs.

3. $y = x + 5$ **4.** $y = -4x$ **5.** $y = 3x + 6$ **6.** $-x + y = 7$

Example 2 p. 407

7. **WAGES** The amount of money Toshiko earns y for working x hours at the library is given by the linear equation $y = 10x$. Find two solutions of this equation. Explain each solution.

Example 3 p. 408

Graph each equation by plotting ordered pairs.

8. $y = x - 1$ **9.** $y = 3x$ **10.** $y = 2x + 4$ **11.** $x + y = 6$

= **Step-by-Step Solutions** begin on page R11. **Extra Practice** begins on page 810.

Practice and Problem Solving

Example 1 p. 406

Copy and complete each table. Use the results to write four ordered pair solutions of the given equation. Write the solutions as ordered pairs.

12. $y = x - 2$

x	$y = x - 2$	y
−1	■	■
0	■	■
1	■	■
2	■	■

13. $y = -2x$

x	$y = -2x$	y
−1	■	■
0	■	■
1	■	■
2	■	■

14. $y = 5x + 1$

x	$y = 5x + 1$	y
−2	■	■
−1	■	■
0	■	■
1	■	■

15. $y = -2x + 8$

x	$y = -2x + 8$	y
−1	■	■
0	■	■
2	■	■
4	■	■

Find four solutions of each equation. Write the solutions as ordered pairs.

16. $y = 8x$ **17** $y = -2x$ **18.** $y = x + 7$ **19.** $y = -x + 3$

20. $y = 2x + 5$ **21.** $y = -3x - 4$ **22.** $x + y = -3$ **23.** $2x + y = 9$

Example 2 p. 407

24. **GEOMETRY** The circumference of a circle C with a radius of r units is approximately given by the linear equation $C \approx 6.3r$. Find two solutions of this equation. Explain each solution.

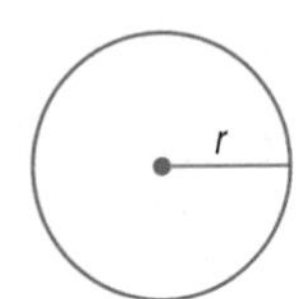

25. **FAIRS** Regular rides cost 3 tickets and children's rides cost 1 ticket. Find three solutions of $3x + y = 12$ to find the number of regular rides x and children's rides y a family can go on with 12 tickets. Explain each solution.

Example 3 p. 408

Graph each equation.

26. $y = 5x$ **27.** $y = -3x$ **28.** $y = x + 4$ **29.** $y = -x + 5$

30. $y = 2x + 2$ **31.** $y = 4x - 1$ **32.** $x + y = -6$ **33.** $x - y = 1$

Real-World Career

Marine Mammal Trainer Marine mammal trainers train dolphins, whales, seals, sea lions, walruses, and other marine animals.

A marine mammal trainer needs a Bachelor of Science degree in either zoology, biology, psychology, or marine biology.

34. **MULTIPLE REPRESENTATIONS** In this problem, you will investigate functions. A whale can swim half a mile per minute.

a. **TABULAR** Make a table to find the number of miles a whale can swim in 5, 10, 15, and 20 minutes.

b. **VERBAL** As the input values increase by 5, does the difference between the output values increase by the same amount? If not, what is the difference between the output values?

c. **GRAPHICAL** Graph the ordered pairs (time, distance). Then draw a line through the points.

d. **ALGEBRAIC** Write an equation to represent the relationship between time x and number of miles y. Explain your reasoning.

35 **GEOMETRY** A rectangle that is x inches long and y inches wide and has a perimeter of 16 inches.

a. Write an equation to represent this situation.

b. Find three ordered pairs (width, length) that satisfy the equation.

c. Graph the ordered pairs. Then draw a line through the points.

d. Is the ordered pair (−4, 12) a solution of the equation? Does it make sense in the context of the situation? Explain.

36. **FINANCIAL LITERACY** Mara has $440 to pay a painter to paint her bedroom. The painter charges $55 per hour. The equation $y = 440 - 55x$ represents the amount of money left after x number of hours worked by the painter. What does the solution (7, 55) represent?

H.O.T. Problems Use Higher-Order Thinking Skills

37. **OPEN ENDED** Write a linear equation that has (3, −2) as a solution. Then find another solution of the equation.

38. **CHALLENGE** Name a linear equation that is *not* a function.

39. **WRITING IN MATH** Explain why a linear function has infinitely many solutions. Then determine which representation shows all the solutions of a linear function: a table, a graph, or an equation. Explain.

40. **REASONING** Consider the arithmetic sequence 3, 10, 17, 24, 31, 38, … .

a. Make a scatter plot of the ordered pairs (term number, term). Do the points seem to lie on a line?

b. In which quadrant(s) would this graph make sense? Explain.

c. If you connect the points, you are including points where the x–value is 1.5, 2.7, or 5.9. Can an arithmetic sequence have these values for x? Explain your reasoning.

41. **WRITING IN MATH** Explain how a linear equation represents a function. Then describe four different real-world representations of a linear function that can be used to express the same relationship.

Standardized Test Practice

42. Which equation represents the table of values for the ordered pairs shown?

x	−1	0	1	2
y	12	9	6	3

A $x - y = -13$ C $x + y = 11$
B $3x - y = 9$ D $3x + y = 9$

43. Which of the following equations represents the line graphed below?

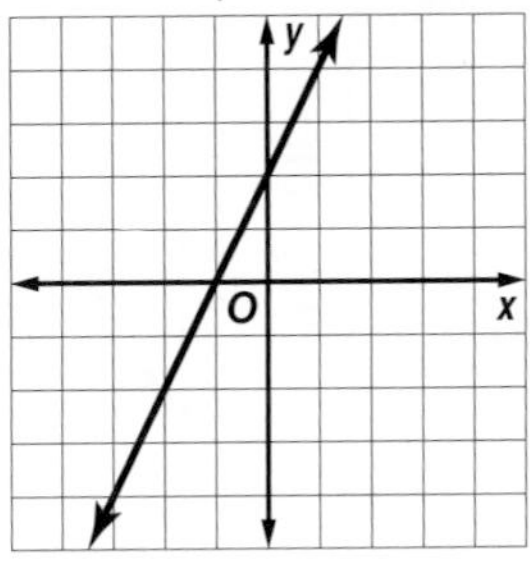

F $2x - y = -2$ H $x - y = 0$
G $x - y = -2$ J $x + 3y = -5$

44. Which of the following sets of ordered pairs represents a linear function?

A $\{(2, 2), (1, -1), (0, 0), (-1, -1)\}$
B $\{(4,-4), (3, -3), (2, -2), (1, -1)\}$
C $\{(-4, -4), (0, 0), (4, -4), (-8, -8)\}$
D $\{(-3, -5), (-1, 0), (0, 1), (3, -5)\}$

45. EXTENDED RESPONSE The equation $y = 64x$ describes the weight y of x cubic feet of water.

a. Describe the linear relationship between number of cubic feet and weight of water in words.
b. Make a table of values that describe the relationship.
c. Choose two ordered pairs from your table and describe what they mean.
d. Graph the ordered pairs in your table and draw a line through them.

Spiral Review

Write an equation that describes each sequence. Then find the indicated term. (Lesson 8-2)

46. 7, 8, 9, 10, …; 14th term

47. 6, 10, 14, 18, …; 23rd term

Determine whether each relation is a function. Explain. (Lesson 8-1)

48. $\{(0, 6), (-3, 9), (4, 9), (-2, 1)\}$

49. $\{(-0.1, 5), (0, 10), (-0.1, -5)\}$

50. FOOD The circle graph shows the results of a survey about favorite pizza toppings. (Lesson 7-8)

a. What is the favorite pizza topping of the people surveyed?
b. If 250,000 people were surveyed, how many would you expect to rate mushrooms as their favorite topping?

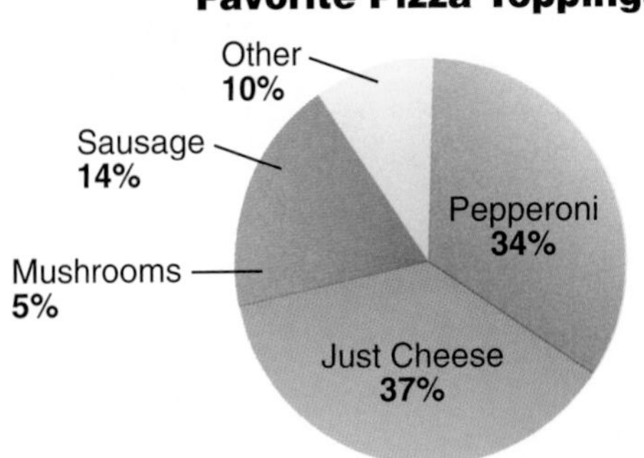

Skills Review

Evaluate each expression. (Lesson 3-4)

51. $\frac{18 - 10}{8 - 4}$

52. $\frac{16 - 7}{1 - 3}$

53. $\frac{46 - 22}{2008 - 2004}$

54. $\frac{31 - 25}{46 - 21}$

8-4 Rate of Change

Then

You have already used a rate to compare two quantities. (Lesson 5-2)

Now

- Find rates of change.
- Solve problems involving rates of change.

New Vocabulary

rate of change

Math Online

glencoe.com

- Extra Examples
- Personal Tutor
- Self-Check Quiz
- Homework Help

Why?

In *kitesurfing*, riders are pulled through the water and up in the air by a kite. The graph shows the change in height of a kitesurfer from 0 to 14 seconds.

a. Between which two consecutive points did the height of the kitesurfer increase the most? How do you know?

b. What is happening to the kitesurfer's height between points C and D?

c. Find the ratio of the change in height to the time for each section of the graph. Which section is the steepest?

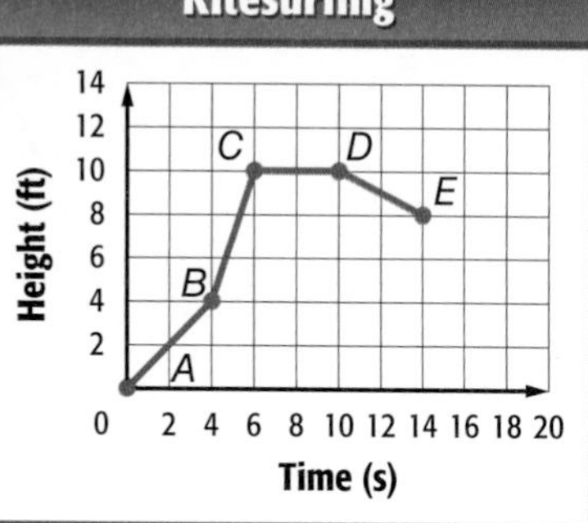

Rate of Change A **rate of change** is a rate that describes how one quantity changes in relation to another quantity. The rate of change of the height of the kitesurfer to the time from point B to point C is $\frac{10 - 4}{6 - 4}$ or 3 feet per second.

Real-World EXAMPLE 1 Find Rate of Change from a Graph

AIRPLANES The graph shows the changes in height and distance of a small airplane. Find the rate of change from point B to point C.

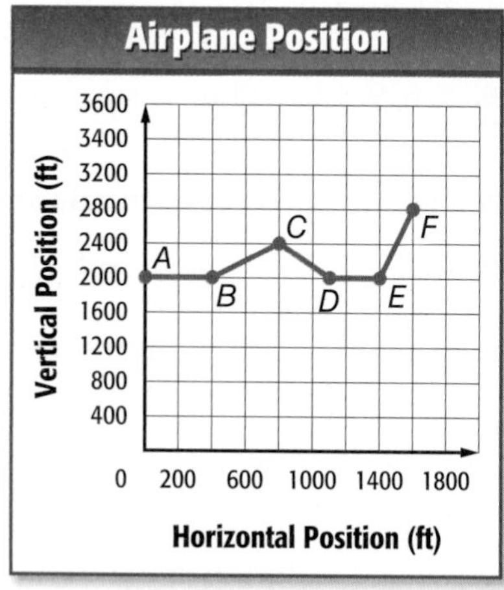

The vertical position changed from 2000 feet to 2400 feet. The horizontal position changed from 400 feet to 800 feet.

$$\frac{\text{change in vertical position}}{\text{change in horizontal position}} = \frac{2400 - 2000}{800 - 400}$$

$$= \frac{400}{400}$$

$$= 1$$

The rate of change is an increase of 1 foot in the vertical position for every 1 foot in the horizontal position.

Check Your Progress

1. **AIRPLANES** Find the rate of change from point E to point F in the airplane position graph above.

Personal Tutor glencoe.com

Real-World EXAMPLE 2 Find Rate of Change from a Table

MONEY The table shows the amount of money that Joshua earns working at a state park for the summer. Find the rate of change.

Time (h)	Income ($)
x	y
8	76.00
9	85.50

Understand You know the income Joshua receives. You need to find the rate of change.

Plan To find the rate of change, divide the change in income by the change in time.

Solve

$$\text{rate of change} = \frac{\text{change in income}}{\text{change in time}}$$

$$= \frac{\$85.50 - \$76.00}{9\text{ h} - 8\text{ h}}$$ **Income goes from \$76 to \$85.50. Time goes from 8 to 9 hours.**

$$= \$9.50/\text{h}$$ **Simplify.**

The rate of change is \$9.50/h or an increase of \$9.50 per hour.

Check Check by using the rate of change to find Joshua's income if he works 10 hours.

\$9.50/hour × 10 hours = \$95.00. Since 76 + 9.5 = 85.5 and 85.5 + 9.5 = 95, the solution is reasonable. ✓

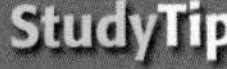

StudyTip

***x* and *y* Values** In rates of change, the x values always increase. The y values may increase, decrease, or remain the same.

Check Your Progress

2. EXERCISE Alondra uses a pedometer to keep track of the number of steps she takes on a walk. Find the rate of change.

Time (min)	Number of Steps
x	y
3	288
5	480

Personal Tutor glencoe.com

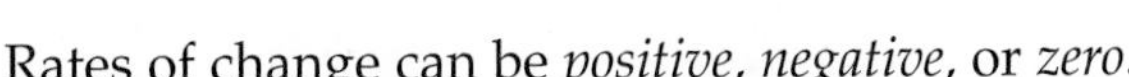

Rates of change can be *positive*, *negative*, or *zero*.

Real-World EXAMPLE 3 Negative Rate of Change

ROLLER COASTERS The biggest drop of the *Kingda Ka* roller coaster is described in the table. Find the rate of change. Interpret its meaning.

Time (s)	Height (ft)
x	y
0	456
1	268
2	80

$$\text{rate of change} = \frac{\text{change in height}}{\text{change in time}}$$

$$= \frac{268\text{ ft} - 456\text{ ft}}{1\text{ s} - 0\text{ s}}$$ **Height goes from 456 to 268 feet. Time goes from 0 to 1 second.**

$$= -188\text{ ft/s}$$ **Simplify.**

The rate of change can be described as a decrease of 188 feet per second.

Real-World Link

The Kingda Ka roller coaster, in Jackson, New Jersey, is 3118 feet long, travels up to 128 miles per hour, and lasts about half a minute.

Source: Coaster Grotto

Check Your Progress

3. MONEY The table shows the amount y remaining in Lauren's savings account after x weeks. Find the rate of change. Interpret its meaning.

Time (wk)	Amount ($)
x	y
4	160
6	120

Personal Tutor glencoe.com

Real-World Link

The music on a CD is imprinted on a spiral track that is 3.52 miles long.

Real-World EXAMPLE 4 Compare Rates of Change

TECHNOLOGY **A CD spins at different speeds when playing songs on the inside tracks (near the center of the CD) than when playing songs on the outside tracks, as shown in the table. Compare the rates of change.**

Number of Seconds x	Number of Revolutions y	
	Inside Tracks	Outside Tracks
0	0	0
2	16	6
4	32	12

Source: *Encyclopedia Americana*

Songs on Inside Tracks

$$\text{rate of change} = \frac{\text{change in } y}{\text{change in } x}$$

$$= \frac{32 - 16}{4 - 2}$$

$$= 8 \text{ revolutions/s}$$

Songs on Outside Tracks

$$\text{rate of change} = \frac{\text{change in } y}{\text{change in } x}$$

$$= \frac{12 - 6}{4 - 2} = \frac{6}{2}$$

$$= 3 \text{ revolutions/s}$$

A CD spins at a faster rate when playing songs on the inside tracks than when playing songs on the outside tracks. A steeper line on the graph indicates a greater rate of change.

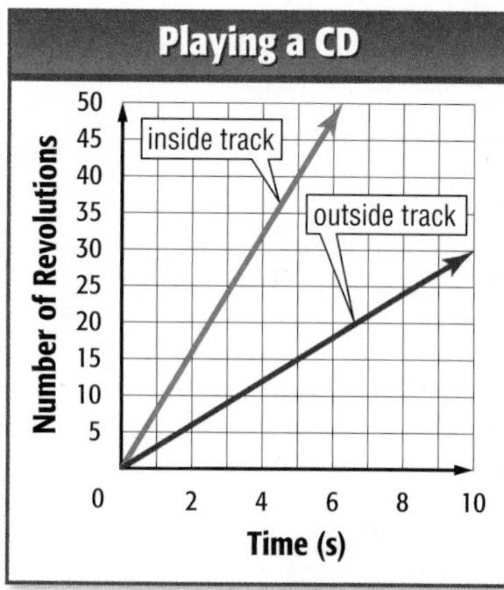

Check Your Progress

4. **GEOMETRY** The perimeter of a regular hexagon changes as its side lengths increase by 1 inch. Compare this rate of change with the rate of change for a square whose side lengths increase by 1 inch.

Personal Tutor glencoe.com

ReadingMath

Graphs In the Concept Summary, the rate of change in each graph can be read a different way. In the first graph, as time increases, the temperature increases. In the second graph, as time increases, temperature decreases. In the third graph, as time increases, temperature does not change.

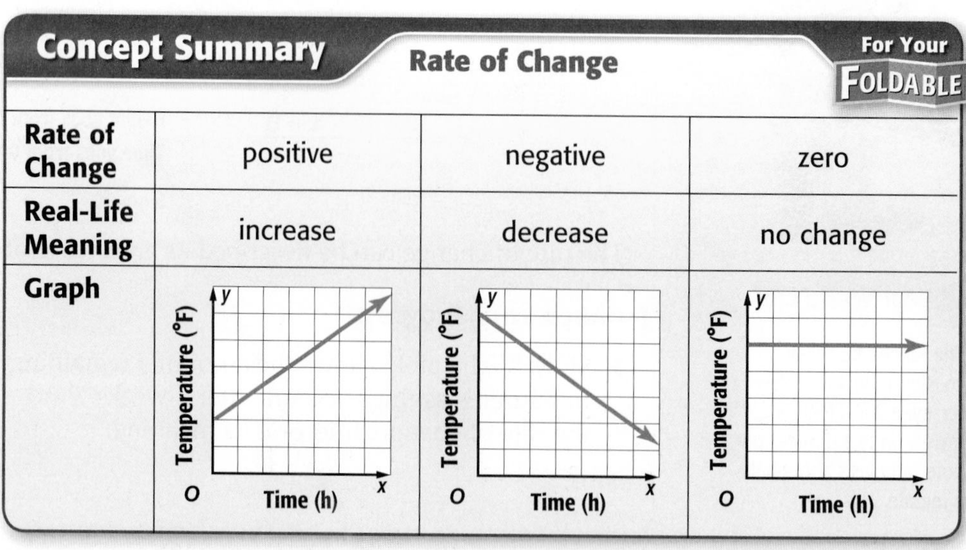

Concept Summary
Rate of Change
For Your FOLDABLE
Rate of Change
positive
negative
zero
Real-Life Meaning
increase
decrease
no change
Graph
Temperature (°F)
y
x
O
Time (h)

Check Your Understanding

Examples 1 and 2
pp. 412–413

Find the rate of change for each linear function.

1.

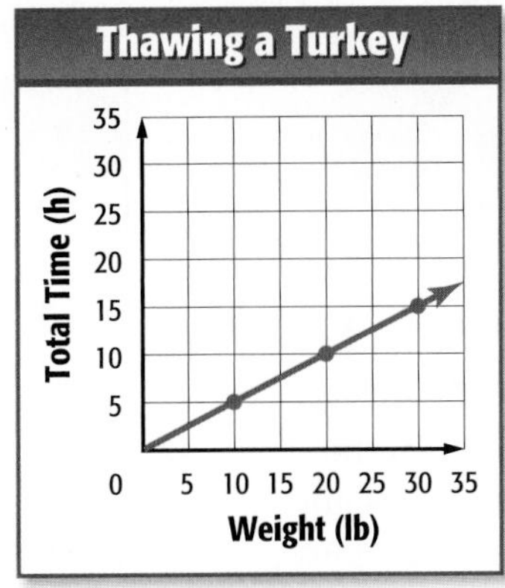

2.

Filling a Pool	
Time (min)	Depth (in.)
x	y
5	12
10	24
15	36
20	48

Examples 3 and 4
pp. 413–414

3. **HIKING** The table shows the altitude of a group of hikers during their hike. Find the rate of change.

Hiking	
Time (min)	Altitude (ft)
x	y
20	170
30	162
40	154

4. **HOT-AIR BALLOONS** The table shows the relationship between time and altitude of two hot-air balloons. Compare the rates of change for the balloons from 3 to 5 seconds.

Time (s)	Altitude (ft)	
	Balloon 1	Balloon 2
3	22.7	30.4
4	30.9	36.0
5	39.1	41.6

Practice and Problem Solving

● = **Step-by-Step Solutions** begin on page R11.
Extra Practice begins on page 810.

Examples 1 and 2
pp. 412–413

Find the rate of change for each linear function.

5.

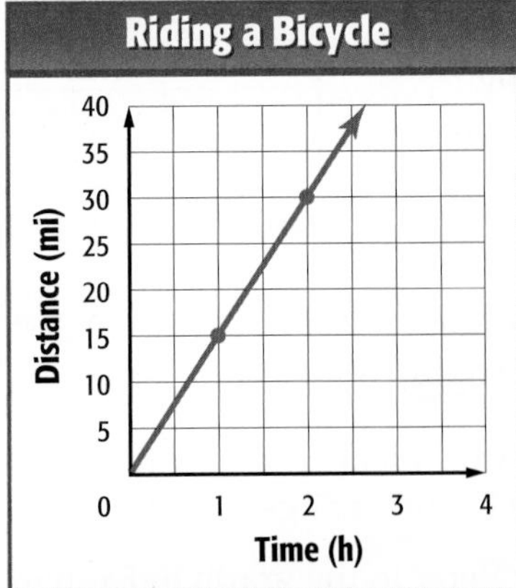

6.

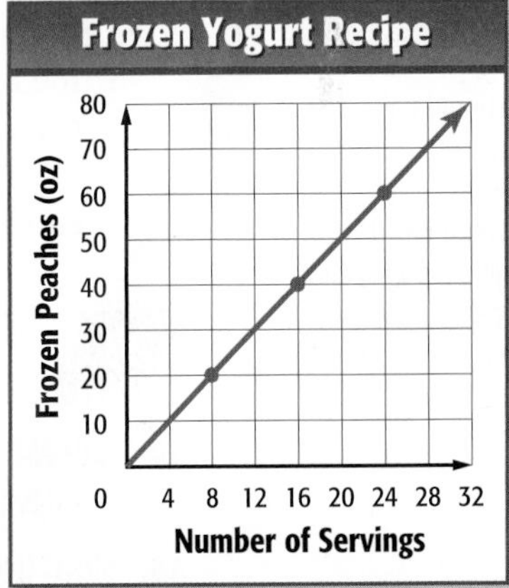

7.

Jumping on a Trampoline	
Time (min)	Number of Jumps
x	y
1	20
2	30
3	40
4	50

8.

Sending Video Messages	
Number of Messages	Cost ($)
x	y
4	1.00
6	1.50
8	2.00
10	2.50

Find the rate of change for each linear function.

Examples 3 and 4
pp. 413–414

9. **TEMPERATURE** The table shows the outside temperature after a number of hours. Find the rate of change.

Time (h)	Temperature (°F)
x	y
1	48
2	45
3	42
4	39

10. **LATE FEES** The table shows late fees for DVDs and video games at a video store. Compare the rates of change.

Days Late	Late Fees ($)	
	DVDs	Video Games
0	0	0
2	3	4
4	6	8
5	7.50	10

11. **INSECTS** The graph shows the speed at which mosquitoes and bees beat their wings. Compare the rates of change.

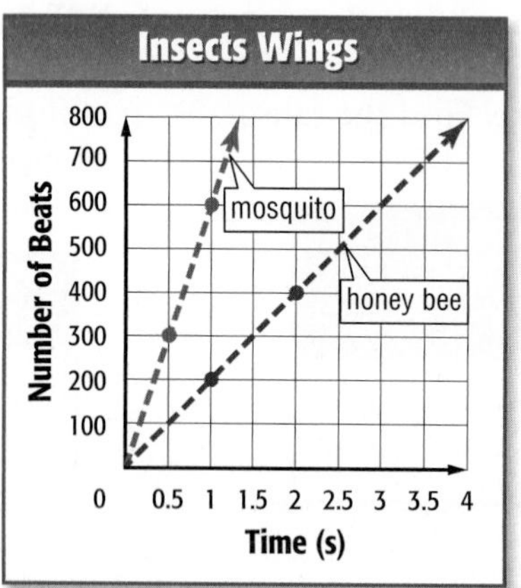

12. **SALES** The graph shows the number of books sold at a book fair. Find the rate of change.

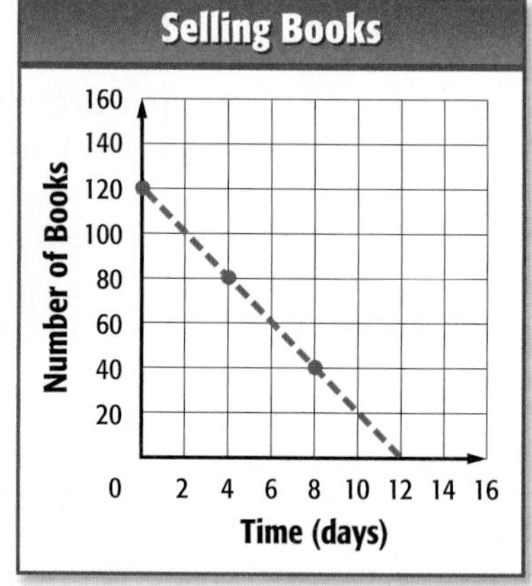

13. **PUPPIES** The table shows the number of times puppies should be fed at different ages. Find the three rates of change represented in the table and describe what they mean.

Total Puppy Feedings

Days	Age		
	< 6 Months	6–12 Months	> 12 Months
1	3	2	1
2	6	4	2
3	9	6	3

Source: Foster & Smith, Inc.

Real-World Link

A puppy should gain about 0.08–0.16 ounce per day per pound of their expected adult weight. For example, a puppy that grows to be 50 pounds should gain 2 to 4 ounces per day.

Source: Foster & Smith, Inc.

H.O.T. Problems

Use Higher-Order Thinking Skills

14. **OPEN ENDED** Describe a real-world relationship between two quantities that involve a positive rate of change.

15. **WRITING IN MATH** Refer to the graph in Exercise 11. What is the connection between the steepness of the lines and the rates of change?

16. **CHALLENGE** A person starts walking, then runs, and then sits down to rest. Sketch a graph of the situation to represent the different rates of change. Label the x-axis "Time" and the y-axis "Distance".

17. **REASONING** Explain why a horizontal line represents a rate of change of zero, but a vertical line cannot be used to represent a rate of change.

18. **WRITING IN MATH** Use the data about airplane position in Example 1 to explain how rate of change affects the graph of the airplane's position.

Standardized Test Practice

For Exercises 19 and 20, use the graph.

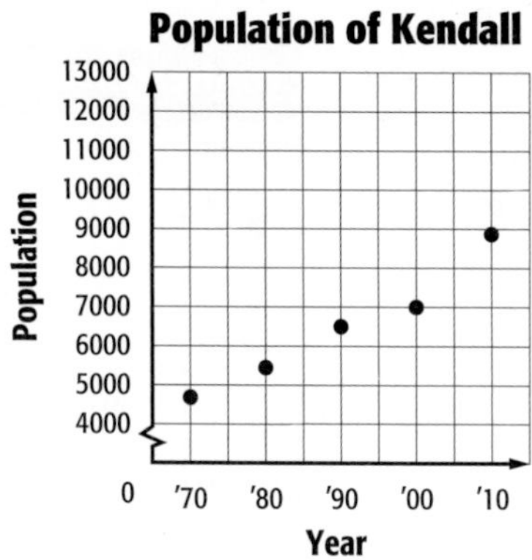

19. In which of the following decades was there the greatest population change?

A 1970–1980 C 1990–2000

B 1980–1990 D 2000–2010

20. In which of the following decades was there the least population change?

F 1970–1980 H 1990–2000

G 1980–1990 J 2000–2010

21. The table shows a relationship between time and altitude of a hot-air balloon. Which is the *best* estimate for the rate of change for the balloon from 1 to 5 seconds?

Time (s)	Altitude (ft)
1	6.3
2	14.5
3	22.7
4	30.9
5	39.1

A 7.6 ft/s C 8.2 ft/s

B 7.8 ft/s D 8.8 ft/s

22. SHORT RESPONSE Find the rate of change for the linear function represented in the table. Round to the nearest hundredth.

Time (min)	Distance (mi)
30	32.5
45	48.75
60	65.0
75	81.25
90	97.5

Spiral Review

Find four solutions of each equation. Show each solution in a table of ordered pairs. (Lesson 8-3)

23. $y = x - 7$

24. $y = 2x - 3$

25. $3x - y = 10$

Describe each sequence using words and symbols. (Lesson 8-2)

26. 4, 6, 8, 10, ...

27. 1, 4, 7, 10, ...

28. 3, 7, 11, 15, ...

29. GEOMETRY Study the pattern at the right. Write an expression that describes the pattern. Which figure will have 45 squares? (Lesson 8-2)

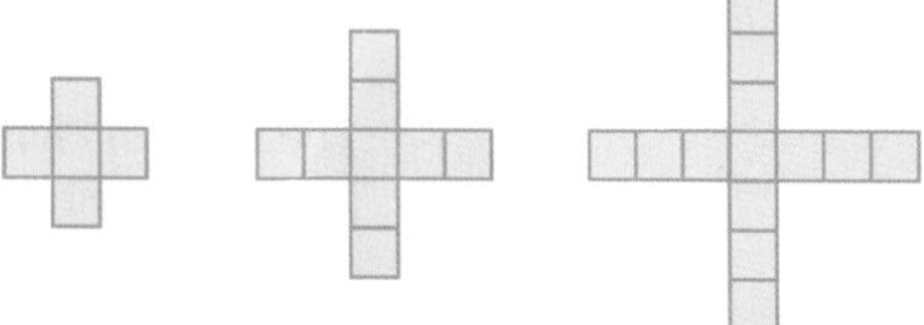

30. Find 95% of 256 using the percent equation. (Lesson 7-5)

31. RAIN A raindrop falls from the sky at about 17 miles per hour. How many feet per second is this? Round to the nearest foot per second. (Lesson 6-3)

Skills Review

Rewrite $y = kx$ by replacing k with each given value. (Lesson 1-2)

32. $k = 5$

33. $k = -2$

34. $k = 0.25$

35. $k = \frac{1}{3}$

8-5 Constant Rate of Change and Direct Variation

Then

You have already identified proportional and nonproportional relationships in tables and graphs. (Lesson 6-4)

Now

- Identify proportional and nonproportional relationships by finding a constant rate of change.
- Solve problems involving direct variation.

New Vocabulary

linear relationship
constant rate of change
direct variation
constant of variation

Math Online

glencoe.com

- Extra Examples
- Personal Tutor
- Self-Check Quiz
- Homework Help

Why?

A cat's heart beats about twice as fast as a human heart. The graph shows the average heartbeat of a cat.

a. Choose any two points on the graph and find the rate of change.

b. Repeat part **a** with a different pair of points. What is the rate of change?

c. **MAKE A CONJECTURE** What is the rate of change between any two points on the line?

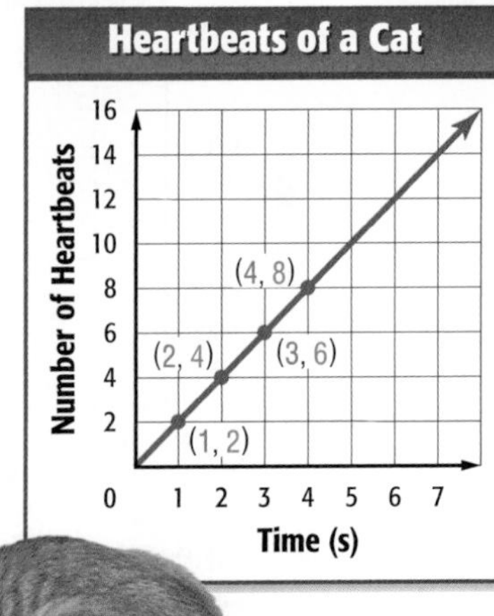

Constant Rates of Change The graph above is a straight line. A relationship that has a straight-line graph is called a **linear relationship**. Notice in the graph above that as time increases by 1, the number of heartbeats increases by 2.

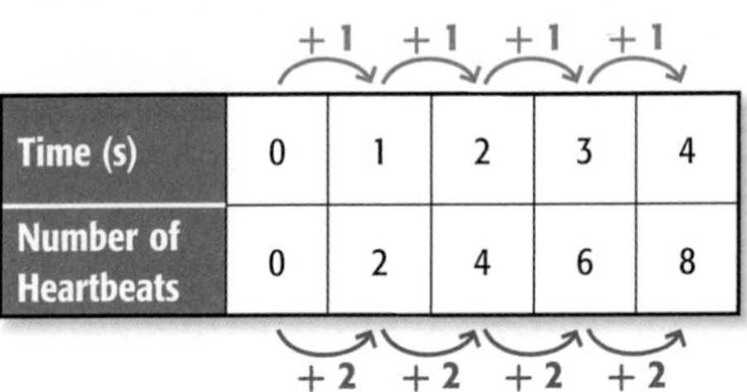

Time (s)	0	1	2	3	4
Number of Heartbeats	0	2	4	6	8

$$\text{rate of change} = \frac{\text{change in heartbeats}}{\text{change in time}}$$
$$= \frac{2}{1}$$

The rate of change is 2 heartbeats per second.

The rates of change between any two data points in a linear relationship are the same or constant. This is called a **constant rate of change**.

Constant Rate of Change

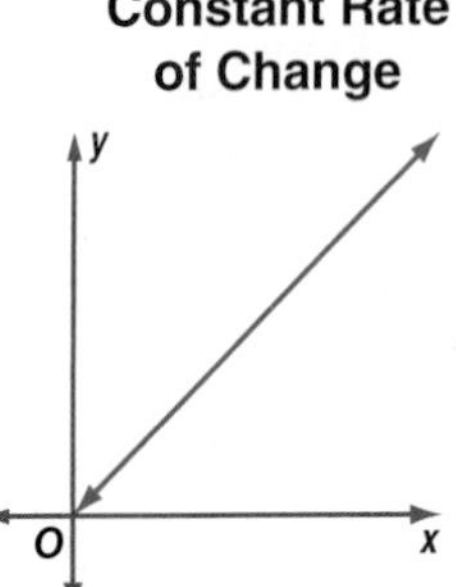

Constant Rate of Change

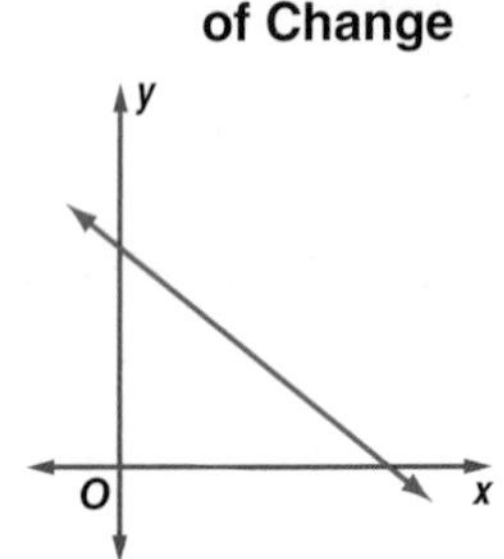

Not a Constant Rate of Change

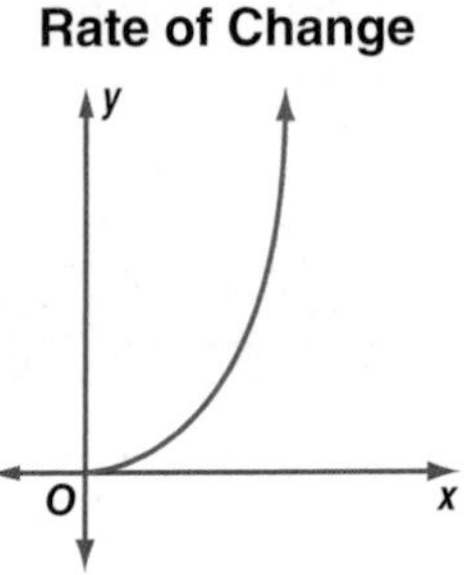

Real-World EXAMPLE 1 Find a Constant Rate of Change

DESIGNS A circular design on an Internet advertisement has two circles, one that is decreasing in size and one that is increasing in size. Find the constant rate of change for the radius of circle 1 in the graph shown. Then interpret its meaning.

Circle Designs

Radius of Circle (cm)

Number of Seconds

Circle 1

Circle 2

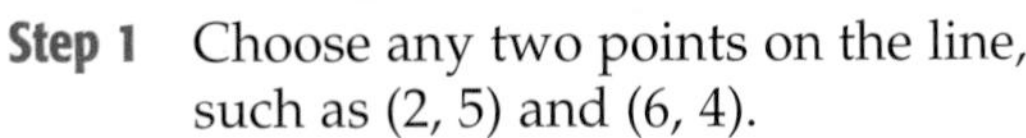

Step 1 Choose any two points on the line, such as (2, 5) and (6, 4).

(2, 5) → 2 seconds, radius 5 cm
(6, 4) → 6 seconds, radius 4 cm

Step 2 Find the rate of change between the points.

$$\text{rate of change} = \frac{\text{change in radius}}{\text{change in time}}$$

$$= \frac{4\text{ cm} - 5\text{ cm}}{6\text{ s} - 2\text{ s}}$$ **The radius goes from 5 cm to 4 cm. The time goes from 2 s to 6 s.**

$$= -\frac{1\text{ cm}}{4\text{ s}}$$ **Simplify.**

$$= -0.25\text{ cm/s}$$ **Express this as a unit rate.**

The rate of change –0.25 cm/s means that the radius of the circle is decreasing at a rate of 0.25 centimeter per second.

Check Your Progress

1. **DESIGNS** Use the graph above. Find the constant rate of change for circle 2. Then interpret its meaning.

Personal Tutor glencoe.com

> **Watch Out!**
> It does not matter which point you use first to find the difference in a rate of change problem. But, it is important that you *always* begin with that point in both the numerator and the denominator.

Some linear relationships are also *proportional*. That is, the ratio of each non-zero y-value compared to the corresponding x-value is the same.

Time x	1	2	3
Distance y	4	8	12

$$\frac{\text{distance } y}{\text{time } x} = \frac{4}{1} = 4 \quad \frac{8}{2} = 4 \quad \frac{12}{3} = 4$$

The ratios are equal, so the linear relationship is proportional.

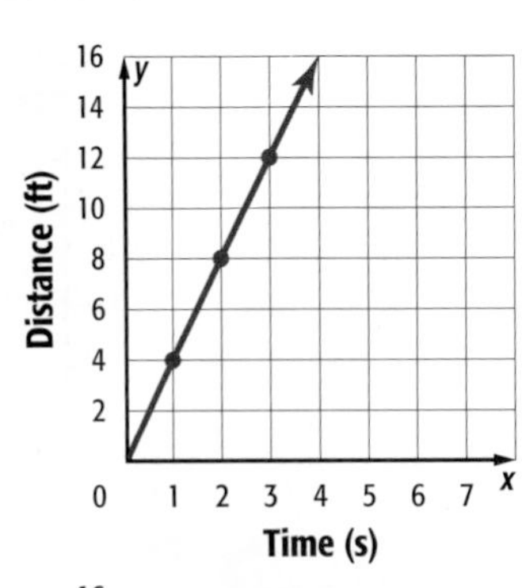

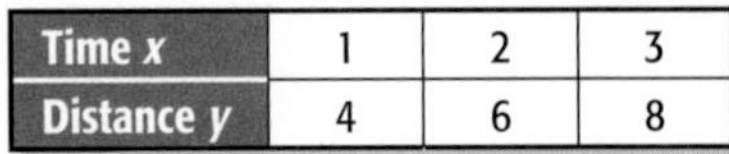

Time x	1	2	3
Distance y	4	6	8

$$\frac{\text{distance } y}{\text{time } x} = \frac{4}{1} = 4 \quad \frac{6}{2} = 3 \quad \frac{8}{3} = 2\frac{2}{3}$$

The ratios are *not* equal, so the linear relationship is nonproportional.

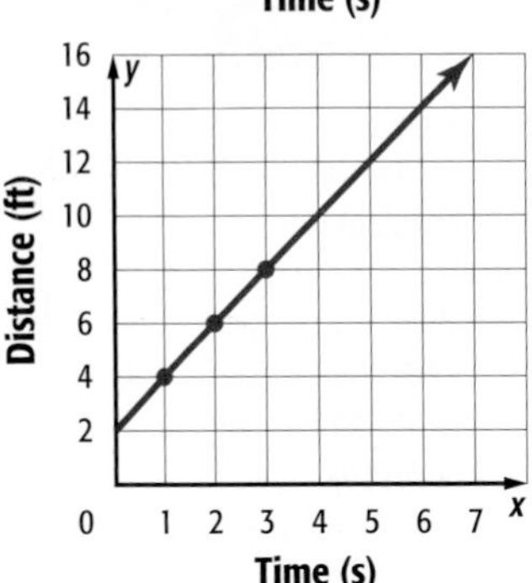

> **Review Vocabulary**
> **proportional relationship** a relationship in which the ratios of related terms are equal *Example:* $\frac{4}{1} = \frac{8}{2} = \frac{12}{3}$ are ratios that represent a proportional relationship. (Lesson 6-4)

Real-World Link

Trail mix is a popular snack because it is delicious and energy rich. It is usually made from a combination of nuts and dried fruits.

EXAMPLE 2 Use Graphs to Identify Proportional Linear Relationships

TRAIL MIX The graph shows the cost of different amounts of trail mix. Determine if there is a proportional linear relationship between the cost and the weight of the trail mix.

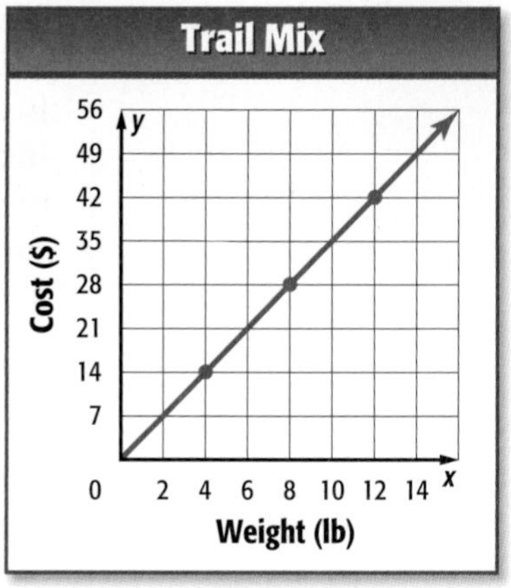

To determine if the quantities are proportional, find $\frac{\text{cost } y}{\text{weight } x}$ for points on the graph.

$\frac{\$14}{4 \text{ lb}} = \$3.50/\text{lb}$

$\frac{\$28}{8 \text{ lb}} = \$3.50/\text{lb}$

$\frac{\$42}{12 \text{ lb}} = \$3.50/\text{lb}$

Since the ratios are the same, the cost of the trail mix is proportional to the weight of the trail mix.

Check Your Progress

2. **DOG WALKING** Tyler charges his customers \$10 per week plus \$5 every time he walks their dogs. Determine whether the relationship between total weekly cost and the number of times the dog is walked is proportional.

Personal Tutor glencoe.com

Direct Variation In the example above, the cost and the weight both vary, but the ratio of the cost to the weight remains constant at \$3.50 per pound.

When the ratio of two variable quantities is constant, their relationship is a **direct variation**. The graph of a direct variation always passes through the origin and represents a proportional linear relationship.

Key Concept Direct Variation

For Your FOLDABLE

Words A direct variation is a relationship in which the ratio of y to x is a constant, k. We say y varies directly with x.

Symbols $y = kx$, where $k \neq 0$

Example $y = 2x$

Graph

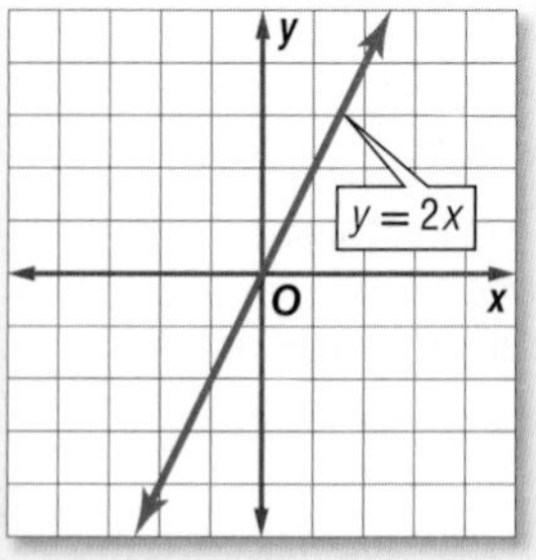

ReadingMath

Directly Proportional Since k is a constant rate of change in a direct variation, we can say the following:

- *y varies directly with x.*
- *y is directly proportional to x.*

In the equation $y = kx$, k is called the **constant of variation** or *constant of proportionality*. The direct variation $y = kx$ can be written as $k = \frac{y}{x}$. In this form, you can see that the ratio of y to x is the same for any corresponding values of y and x. In other words, x and y vary in such a way that they have a constant ratio, k.

Real-World EXAMPLE 3 Use Direct Variation to Solve Problems

TECHNOLOGY **The time it takes to burn amounts of information on a CD varies directly with the amount of information.**

Amount of Information (megabytes)	Time (s)
x	y
2.5	10
3.75	15
10	40
25	100

a. Write an equation that relates the amount of information and the time it takes.

Step 1 Find the value of k using the equation $y = kx$. Choose any point in the table. Then solve for k.

$y = kx$ **Direct variation equation**

$10 = k(2.5)$ **Replace y with 10 and x with 2.5.**

$4 = k$ **Simplify.**

Step 2 Use k to write an equation.

$y = kx$ **Direct variation equation**

$y = 4x$ **Replace k with 4.**

b. Predict how long it will take to fill a 700-megabyte CD.

$y = 4x$ **Write the direct variation equation.**

$y = 4(700)$ **Replace x with 700.**

$y = 2800$ **Simplify.**

It will take 2800 seconds to fill the CD.

Check Your Progress

3. UNIT COST The graph shows the cost y of x pounds of apples at the local grocery store.

a. Write an equation that relates cost and weight of the apples.

b. Predict how much 5 pounds of apples would cost.

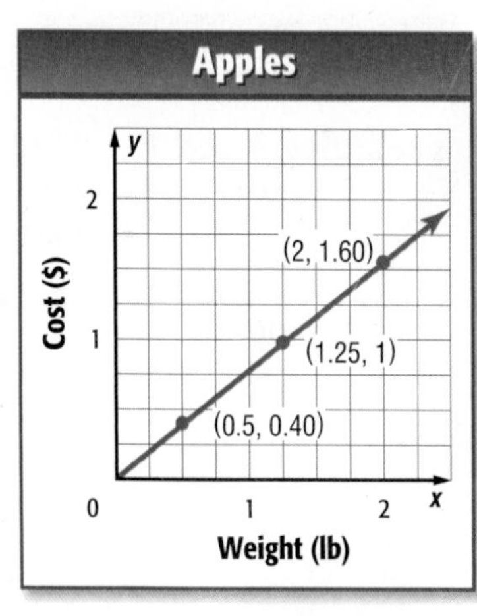

StudyTip

Graphs The graph of a direct variation is always a line that goes through the origin.

Personal Tutor glencoe.com

Concept Summary

For Your FOLDABLE

Proportional Linear Relationships

Words Two quantities a and b have a proportional linear relationship if they have a constant ratio and a constant rate of change.

Symbols $\frac{a}{b}$ and $\frac{\text{change in } b}{\text{change in } a}$ is constant.

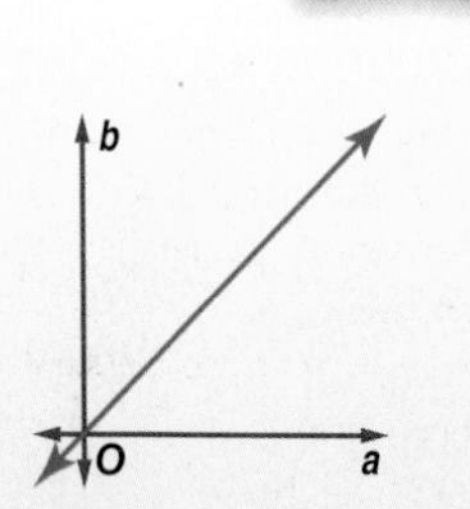

Check Your Understanding

Examples 1 and 2
pp. 419–420

Find the constant rate of change for each linear function and interpret its meaning. Then determine whether a proportional linear relationship exists between the two quantities. Explain your reasoning.

1.

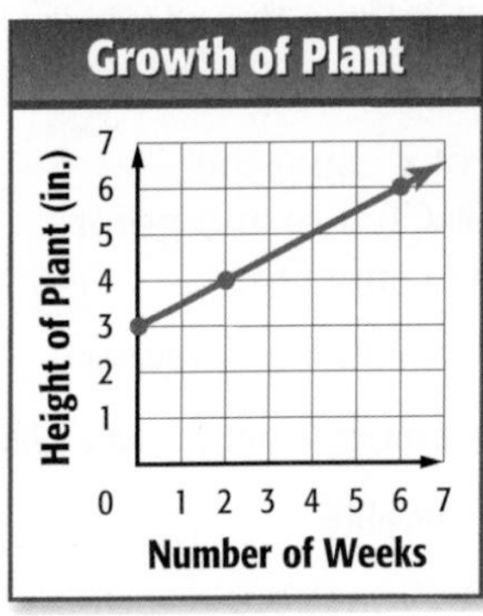

2.

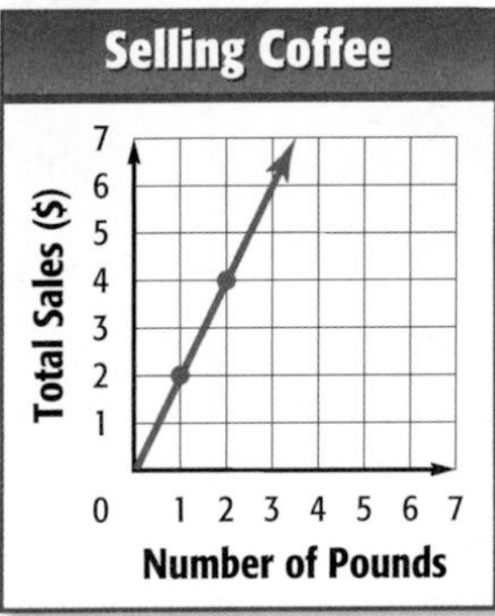

Example 3
pp. 421

3. PHYSICAL SCIENCE The length that a spring stretches varies directly with the amount of weight attached to it. When an 8-ounce weight is attached, a spring stretches 2 inches.

a. Write a direct variation equation relating the weight x and the amount of stretch y.

b. Estimate the stretch of the spring when it has a 20-ounce weight attached.

Practice and Problem Solving

= **Step-by-Step Solutions** begin on page R11.
Extra Practice begins on page 810.

Examples 1 and 2
pp. 419–420

Find the constant rate of change for each linear function and interpret its meaning. Then determine whether a proportional linear relationship exists between the two quantities. Explain your reasoning.

4.

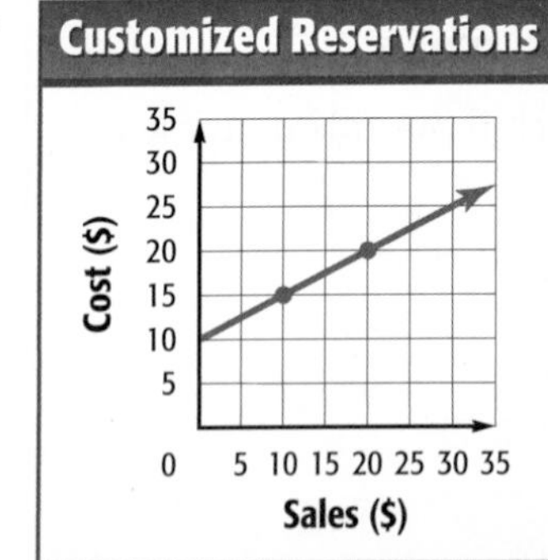

5

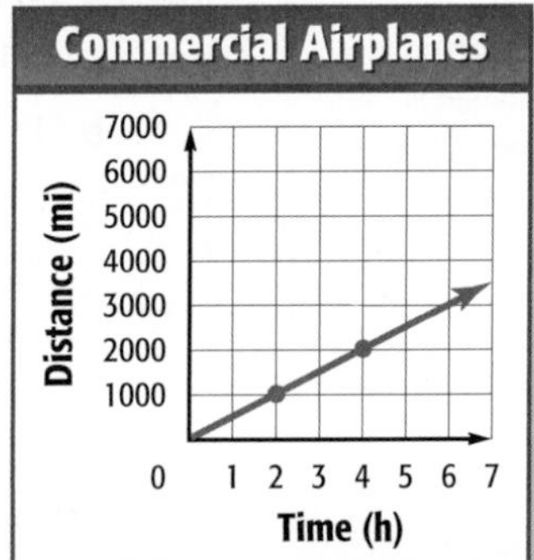

6.

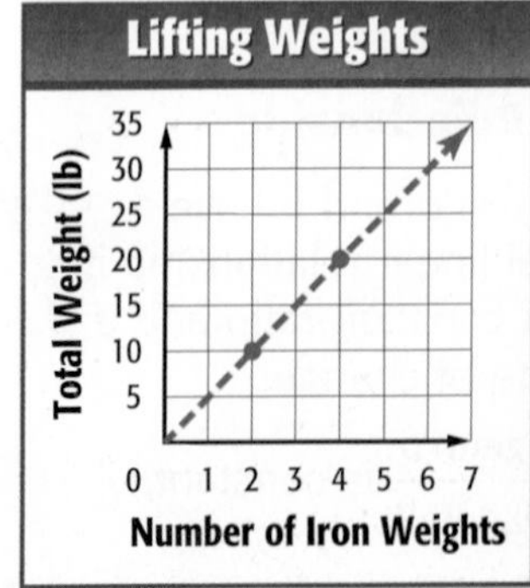

7.

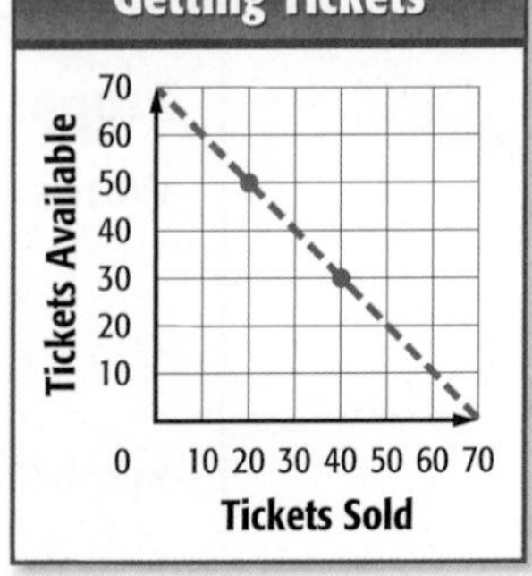

Example 3 p. 421

8. **FOOD COSTS** The cost of cheese varies directly with the number of pounds bought. Suppose 2 pounds cost $8.40.

   a. Write an equation relating the weight and the cost of the cheese.

   b. Find the cost of 3.5 pounds of cheese.

Real-World Link

Orca whales typically dive to a depth of between 100 and 200 feet. However, the deepest dive ever recorded by an orca whale is 900 feet.

Source: Sea World

9. **PRESSURE** Water pressure is measured in pounds per square inch (psi), which varies directly with the depth of the water.

Depth (ft)	Pressure (psi)
x	y
33	14.7
66	29.4
99	44.1
132	58.8

   a. Write an equation that relates the depth and the water pressure. Round to the nearest tenth.

   b. Use the information at the left to find the water pressure at the depth of the deepest dive by an orca whale.

CLEANING **Use the graph at the right to determine whether each statement is true or false. Explain your reasoning.**

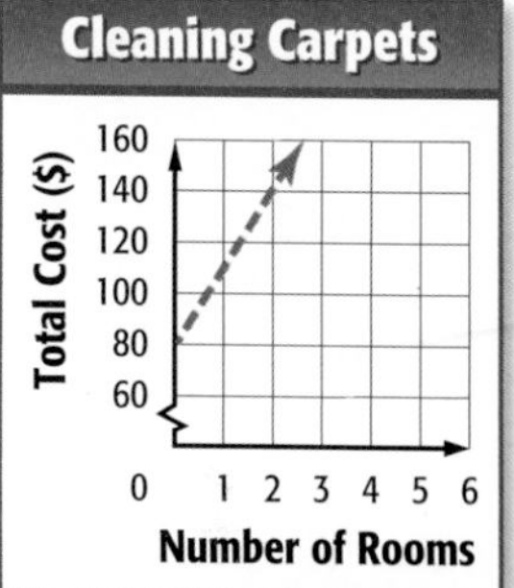

10. There is not a constant rate of change.

11. The two quantities are not proportional.

12. A proportional linear relationship exists.

13. The total cost varies directly with the number of rooms.

H.O.T. Problems

Use **H**igher-**O**rder **T**hinking Skills

14. **OPEN ENDED** Graph a line that shows a 3-unit increase in y for every 1-unit increase in x. State the rate of change.

15. **FIND THE ERROR** Keyshawn and Ramiro are determining whether the points with coordinates (2, 12) and (6, 18) represent a proportional linear relationship. Is either of them correct? Explain your reasoning.

Keyshawn

$\frac{y}{x} = \frac{12}{2} = 6 \quad \frac{18}{6} = 3$

Since the ratios are not the same, the relationship is not proportional.

Ramiro

$\frac{\text{change in } y}{\text{change in } x} = \frac{18 - 12}{6 - 2}$

$= \frac{6}{4} \text{ or } 1.5$

Yes, the relationship is proportional.

16. **WRITING IN MATH** Determine whether the following statement is *always*, *sometimes*, or *never* true. Justify your reasoning. *A linear relationship that has a constant rate of change is a proportional relationship.*

17. **CHALLENGE** In the equation $d = rt$, d is the distance, r is the rate, and t is the time. Suppose the rate is constant. Explain how the distance changes if the time is increased.

18. **WRITING IN MATH** Write about two real-world quantities that have a proportional linear relationship. Describe how you could change the situation to make the relationship between the quantities nonproportional.

Standardized Test Practice

19. Find the constant rate of change for the linear function shown in the graph.

Animal Race

Time (s)

4.5 4.0 3.5 3.0 2.5 2.0 1.5 1.0 0.5

0 1 2 3 4 5 6 7 8 9

Distance (in.)

A 2

B $\frac{1}{2}$

C $-\frac{1}{2}$

D -2

20. Which of the following is *not* true about the graph of a function with a constant rate of change?

F The graph can show a positive relationship.

G The graph can show a negative relationship.

H The relationship between x and y values can be proportional.

J The graph will look like a curve.

21. The cost of cookies at a bake sale is shown.

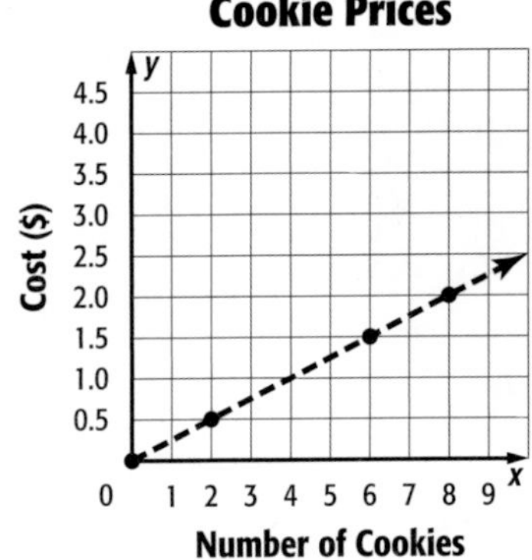

Which of the following is the *best* prediction for the cost of 21 cookies?

A $5.00

B $5.25

C $5.50

D $5.75

22. EXTENDED RESPONSE At an amusement park, the cost of admission varies directly with the number of tickets purchased. Suppose one ticket costs $12.75.

a. Write an equation that could be used to find the cost of any number of admission tickets.

b. Find the cost of 15 tickets.

Spiral Review

23. WATER PARKS Admission costs to a water park are shown in the table. (Lesson 8-4)

a. What is the rate of change?

b. What does the rate mean in this situation?

Water Park Costs	
Number of People	Total Cost ($)
x	y
3	36
4	48
5	60

Find four solutions of each equation. Write the solutions as ordered pairs. (Lesson 8-3)

24. $y = 12x$

25. $y = \frac{1}{2}x$

26. $y = -3x + 5$

Skills Review

Subtract. (Lesson 2-3)

27. $-11 - 13$

28. $15 - 31$

29. $-26 - (-26)$

30. $9 - (-16)$

CHAPTER 8

Mid-Chapter Quiz

Lessons 8-1 through 8-5

Determine whether each relation is a function. Explain. (Lesson 8-1)

1. $\{(0, 5), (1, 2), (1, -3), (2, 4)\}$
2. $\{(-6, 3.5), (-3, 4.0), (0, 4.5), (3, 5.0)\}$
3. $\{(3, 2), (-3, 2), (4, 4), (-4, 4)\}$

4. **MULTIPLE CHOICE** The relation $\{(2, 11), (-9, 8), (14, 1), (5, 5)\}$ is *not* a function when which ordered pair is added to the set? (Lesson 8-1)

   A (8, -9) C (0, 0)

   B (6, 11) D (2, 18)

Describe each sequence using words and symbols. (Lesson 8-2)

5. 4, 5, 6, 7, …
6. 9, 18, 27, 36, …
7. 3, 5, 7, 9, …
8. 2.5, 5, 7.5, 10, 12.5, …

Write an equation that describes each sequence. Then find the indicated term. (Lesson 8-2)

9. 12, 13, 14, 15, …; 15th term
10. 6, 11, 16, 21, …; 30th term
11. 7, 12, 17, 22, …; 20th term

12. **SHORT RESPONSE** Write an expression that could be used to find the value of the term in the nth position in the table below. (Lesson 8-2)

Position	1	2	3	4	n
Value	5	14	23	32	■

Find four solutions of each equation. Write the solutions as ordered pairs. (Lesson 8-3)

13. $y = x + 8$
14. $y = -5x$
15. $y = 2x + 3$
16. $y = -4x - 7$

17. **CURRENCY** The equation $y = 1.35x$ describes the approximate number of U.S. dollars y that are equal to x euros. (Lesson 8-3)

    a. What does the solution (3, 4.05) mean?

    b. About how many U.S. dollars is 15 euros?

Graph each equation. (Lesson 8-3)

18. $y = x + 2$
19. $y = 3x - 4$

Find the rate of change for each linear function. (Lesson 8-4)

20.

Saving Money	
Number of Weeks	Amount Saved ($)
x	y
3	25
4	40
5	55
6	70

21.

Water in Bucket	
Time (s)	Amount of Water (oz)
x	y
2	80
4	48
6	16

22. **MONEY** Find the constant rate of change for the linear function shown below and interpret its meaning. (Lesson 8-5)

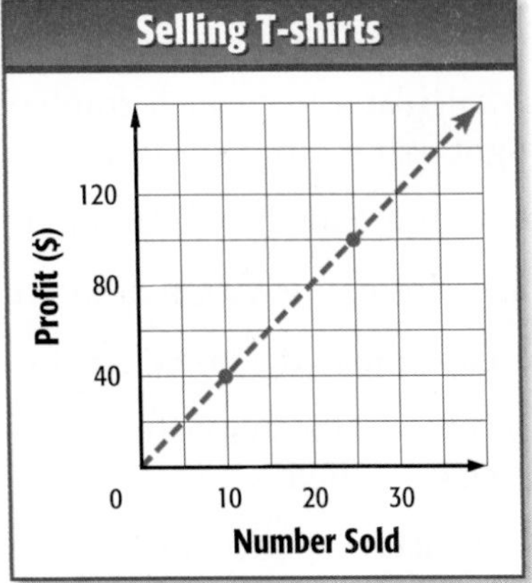

Determine whether a proportional linear relationship exists between the two quantities shown in each of the functions indicated. Explain your reasoning. (Lesson 8-5)

23. Exercise 20
24. Exercise 22

25. **CRAFTS** The cost of craft beads varies directly with the number purchased. The cost of 25 beads is $10. How much do 8 beads cost? (Lesson 8-5)

EXPLORE

8-6

Graphing Technology Lab
Slope and Rate of Change

In this activity, you will investigate the relationship between slope and rate of change.

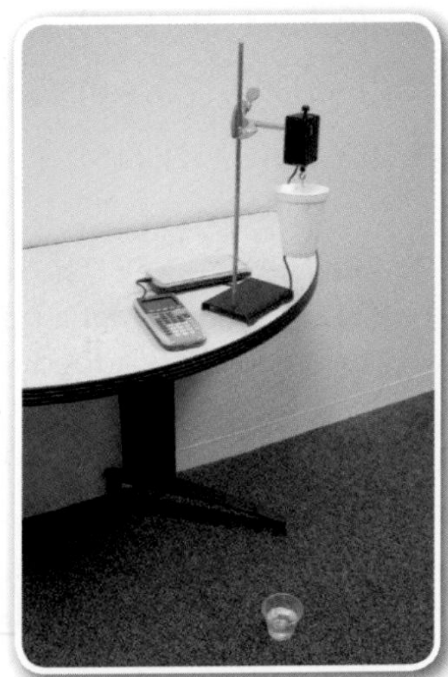

Set Up the Lab

- Connect the force sensor to the Calculator Based Lab (CBL). Then use the unit-to-unit link cable to connect the CBL to the calculator. Place the sensor in a ring stand as shown.
- Make a small hole in the bottom of a paper cup. Straighten a paper clip and use it to create a handle to hang the cup on the force sensor. Place another cup on the floor below.
- Set the device to collect data 100 times at intervals of 0.1 second.

ACTIVITY

Step 1 Hold your finger over the hole in the cup. Fill the cup with water.

Step 2 Begin collecting data as you begin to allow the water to drain.

Step 3 Make the hole in the cup larger. Then repeat Steps 1 and 2 for a second trial.

Analyze the Results

1. Use the calculator to create a graph of the data for Trial 1. The graph will show the weight of the cup y as a function of time x. Describe the graph.
2. Create the graph for Trial 2. Compare the steepness of the two graphs.
3. What happens to the weight of the cup as the time increases?
4. Did the cup empty at a faster rate in Trial 1 or Trial 2? Explain.
5. Describe the relationship between the steepness of the line and the rate at which the cup was emptied.
6. **MAKE A CONJECTURE** Suppose you emptied a cup using a hole half the size of the original hole. Then you emptied a cup using a hole twice the size of the original hole. What would graphs of the data for these trials look like? Explain.
7. Water is emptied at a constant rate from containers that have different shapes. A graph of the water level in each of the containers as a function of time is shown below. Make a sketch of each container.

a. 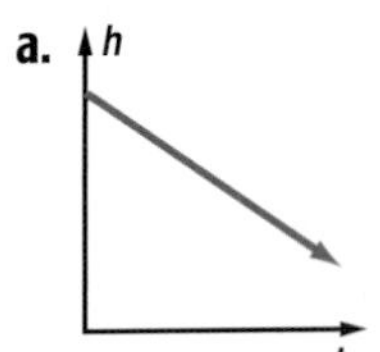

b. 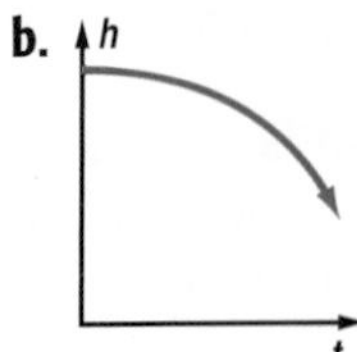

c. 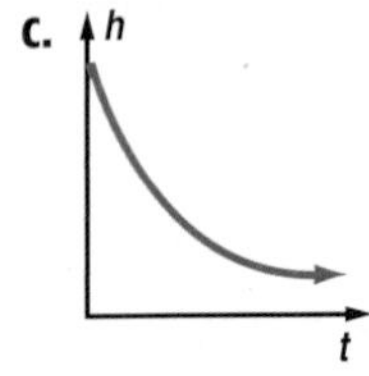

8-6 Slope

Why?

The inflatable water slide shown at the right is 15 feet high and 18 feet long.

a. Write the rate of change comparing the height of the slide to the length as a fraction in simplest form.

b. Find the rate of change of a slide that has the same length but is 5 feet higher than the slide at the right. Is this slide steeper or less steep than the original?

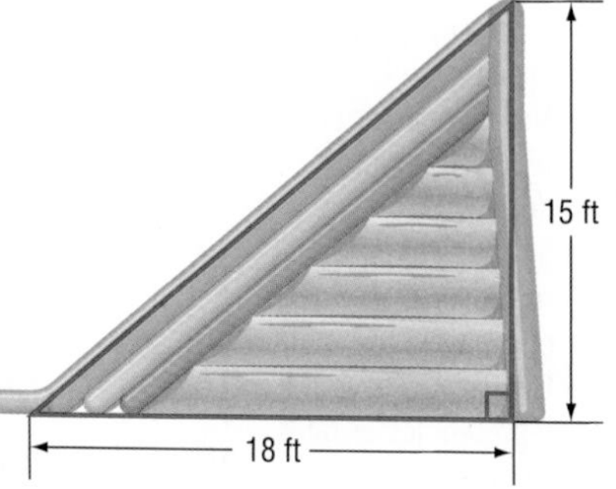

Then

You have already used graphs to find a constant rate of change and interpret its meaning. (Lesson 8-5)

Now

- Find the slope of a line.
- Use slope to describe a constant rate of change.

New Vocabulary

slope

Math Online

glencoe.com

- Extra Examples
- Personal Tutor
- Self-Check Quiz
- Homework Help

Slope **Slope** is the ratio of the *rise*, or the vertical change, to the *run*, or the horizontal change of a line. It describes the steepness of the line. In linear functions, no matter which two points you choose, the slope, or rate of change, of the line is always constant

$$\text{slope} = \frac{\text{rise}}{\text{run}} \quad \begin{matrix} \leftarrow \text{vertical change} \\ \leftarrow \text{horizontal change} \end{matrix}$$

Real-World EXAMPLE 1 Use Rise and Run to Find Slope

SKATEBOARDING Find the slope of a portable skateboard ramp that rises 15 inches for every horizontal change of 54 inches.

$\text{slope} = \frac{\text{rise}}{\text{run}}$ **Write the formula for slope.**

$= \frac{15 \text{ in.}}{54 \text{ in.}}$ **rise = 15 in., run = 54 in.**

$= \frac{5}{18}$ **Simplify.**

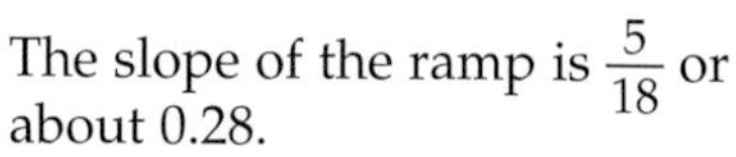

The slope of the ramp is $\frac{5}{18}$ or about 0.28.

Check Your Progress

1A. RAMPS What is the slope of a wheelchair ramp that rises 2 inches for every horizontal change of 24 inches? Write as a fraction in simplest form.

1B. KITES Arthur is flying a kite in the park. The kite is a horizonal distance of 20 feet from Arthur's position and a vertical distance of 70 feet. Find the slope of the kite string.

Personal Tutor glencoe.com

Since slope is a rate of change, it can be positive (slanting upward), negative (slanting downward), or zero.

StudyTip

Slopes
- Positive slopes represent a rate of increase.
- Negative slopes represent a rate of decrease.
- Steeper slopes represent greater rates of change.
- Less steep slopes represent lesser rates of change.

EXAMPLE 2 Use a Graph to Find Slope

Find the slope of each line.

a.

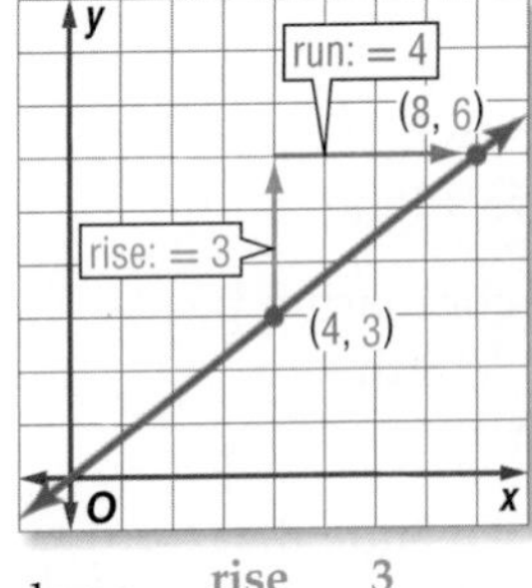

slope $= \frac{\text{rise}}{\text{run}} = \frac{3}{4}$

b.

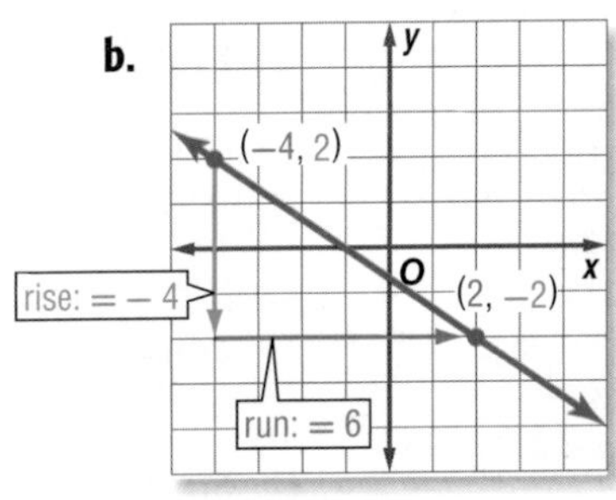

slope $= \frac{\text{rise}}{\text{run}} = \frac{-4}{6}$ or $-\frac{2}{3}$

Check Your Progress

2A.

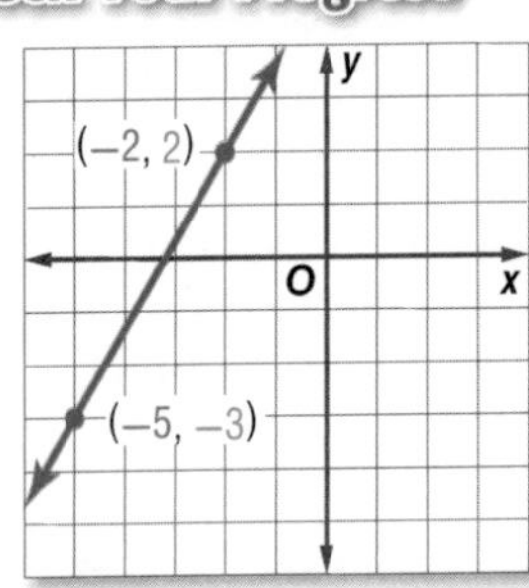

2B.

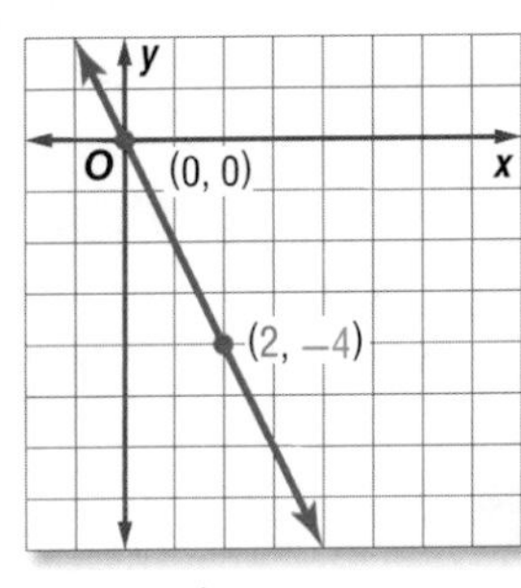

Personal Tutor glencoe.com

Slope and Constant Rate of Change The slope is the same for any two points on a straight line. It represents a constant rate of change.

Key Concept Slope

For Your FOLDABLE

Words The slope m of a line passing through points at (x_1, y_1) and (x_2, y_2) is the ratio of the difference in y-coordinates to the corresponding difference in x-coordinates.

Symbols $m = \frac{y_2 - y_1}{x_2 - x_1}$, where $x_2 \neq x_1$

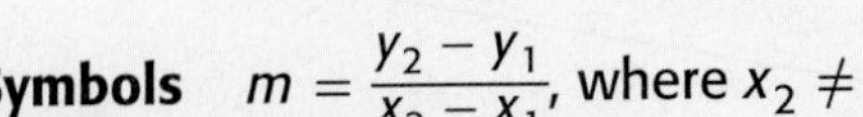

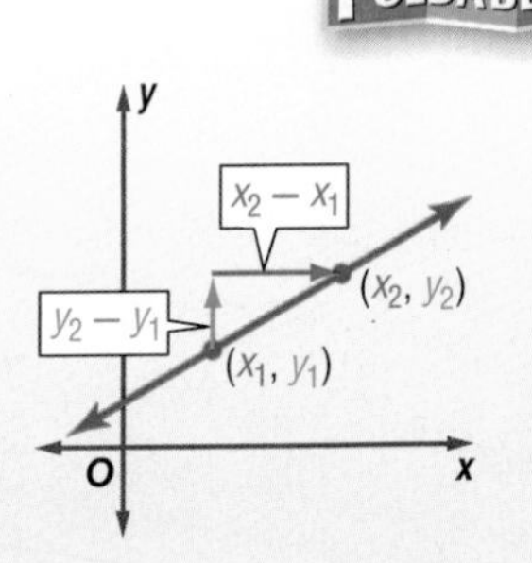

ReadingMath

Multiple Representations
There are different methods for finding the slope of a line.

$m = \frac{\text{rise}}{\text{run}}$

$m = \frac{\text{change in } y}{\text{change in } x}$

$m = \frac{y_2 - y_1}{x_2 - x_1}$

EXAMPLE 3 Use Coordinates of Points to Find Slope

Find the slope of the line that passes through $R(-2, 3)$ and $S(4, -1)$.

$m = \frac{y_2 - y_1}{x_2 - x_1}$ — **Definition of slope**

$m = \frac{-1 - 3}{4 - (-2)}$ — **$(x_1, y_1) = (-2, 3)$, $(x_2, y_2) = (4, -1)$**

$m = \frac{-4}{6}$ or $-\frac{2}{3}$ — **Simplify.**

Check Your Progress

3A. $A(-4, 3), B(1, 2)$

3B. $C(1, -5), D(8, 3)$

Personal Tutor glencoe.com

StudyTip

Choosing Points
- Any two points on a line can be chosen as (x_1, y_1) and (x_2, y_2).
- The coordinates of both points must be used in the same order.

EXAMPLE 4 Zero and Undefined Slopes

Find the slope of the line that passes through each pair of points.

a. $A(-3, 4)$, $B(2, 4)$

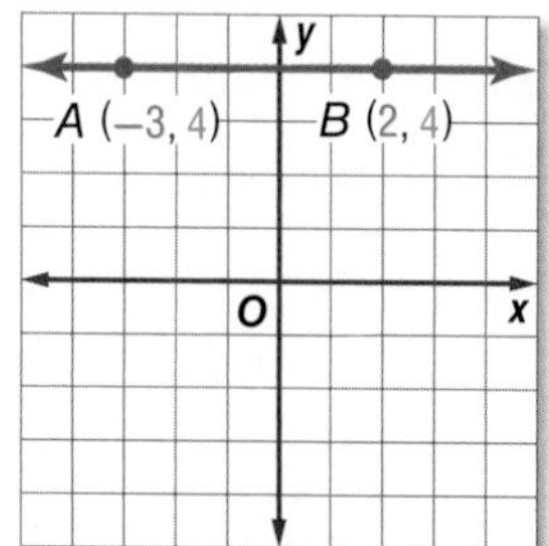

$m = \frac{y_2 - y_1}{x_2 - x_1}$ **Definition of slope**

$m = \frac{4 - 4}{2 - (-3)}$ **$(x_1, y_1) = (-3, 4)$, $(x_2, y_2) = (2, 4)$**

$m = \frac{0}{5}$ or 0 **Simplify.**

b. $T(1, 3)$, $U(1, 0)$

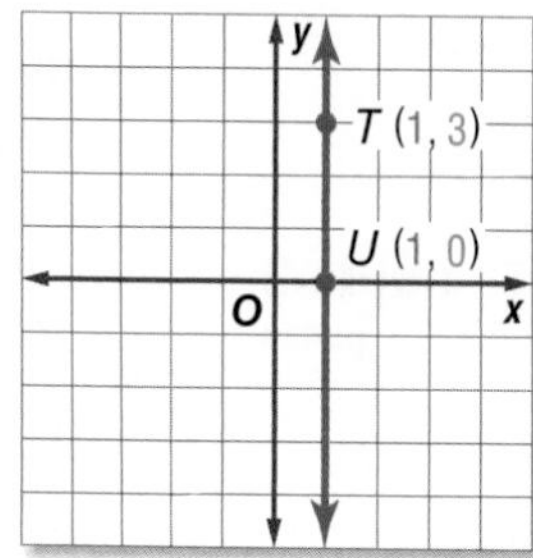

$m = \frac{y_2 - y_1}{x_2 - x_1}$ **Definition of slope**

$m = \frac{0 - 3}{1 - 1}$ **$(x_1, y_1) = (1, 3)$, $(x_2, y_2) = (1, 0)$**

$m = \frac{-3}{0}$ (crossed out) **Division by 0 is undefined.**

The slope is undefined.

Check Your Progress

4A. $E(-1, 7)$, $F(5, 7)$

4B. $G(2, 4)$, $H(2, -1)$

Personal Tutor glencoe.com

Check Your Understanding

Example 1 p. 427

1. **CLIMBING** Libby and a group of friends are rock climbing. She is climbing up the side of a cliff that rises 18 inches for every horizontal change of 3 inches. Find the slope of the cliff.

Example 2 p. 428

Find the slope of each line.

2.

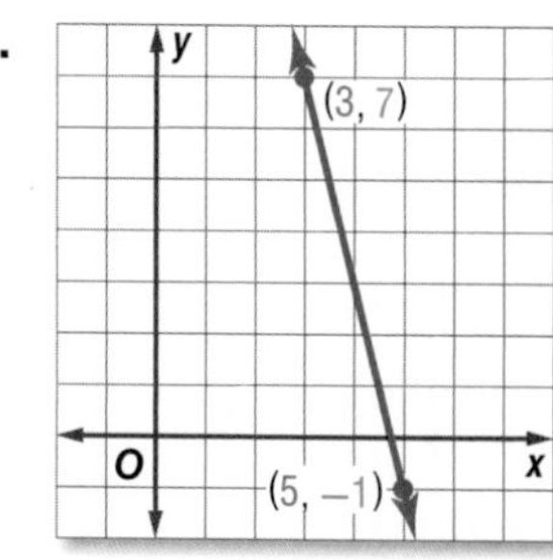

3.

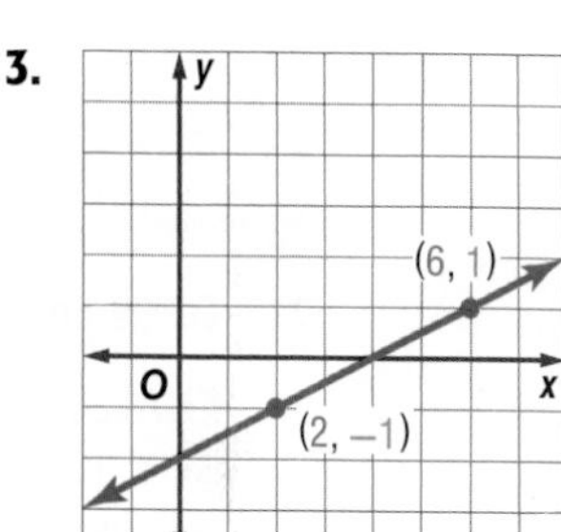

Examples 3 and 4 pp. 428–429

Find the slope of the line that passes through each pair of points.

4. $N(1, 2)$, $P(5, 0)$

5. $F(-2, -4)$, $G(7, 1)$

6. $A(4, -5)$, $B(9, -5)$

7. $Y(8, 3)$, $Z(-6, 3)$

Practice and Problem Solving

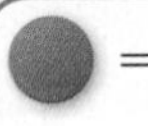 = **Step-by-Step Solutions** begin on page R11. **Extra Practice** begins on page 810.

Example 1 p. 427

8. **SKIING** Find the slope of a snowboarding beginner hill that decreases 24 feet vertically for every 30-foot horizontal increase.

9. **DRIVING** Find the slope of a road that rises 5 feet vertically for every 45-foot horizontal increase.

Example 2 p. 428

Find the slope of each line.

10. 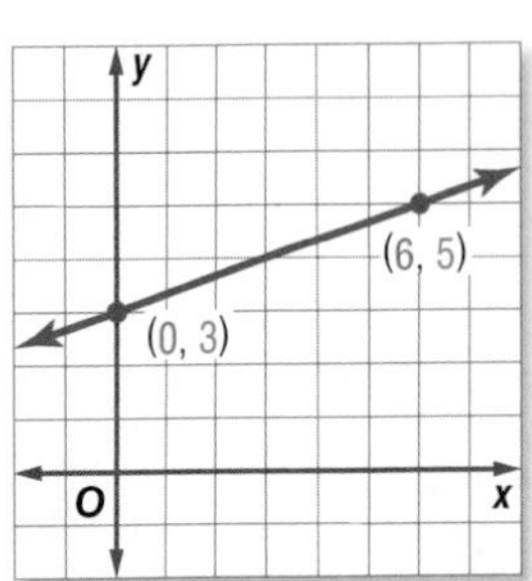

11. 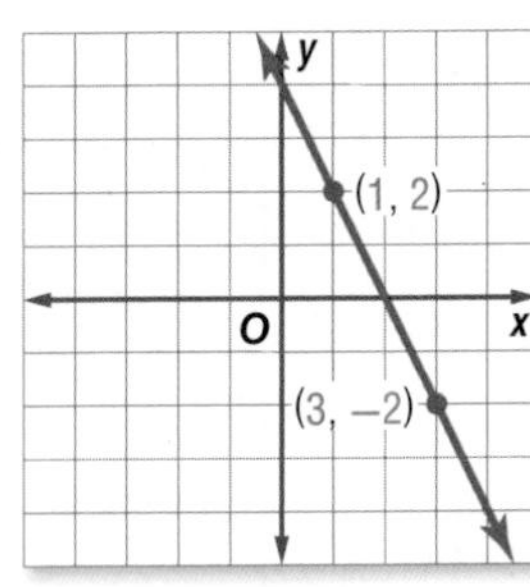

12. 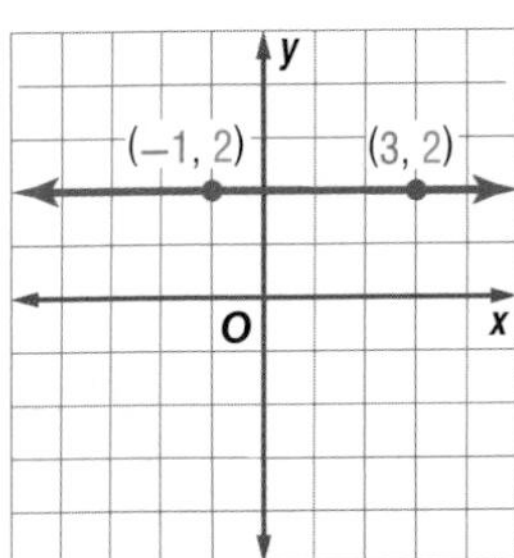

13. 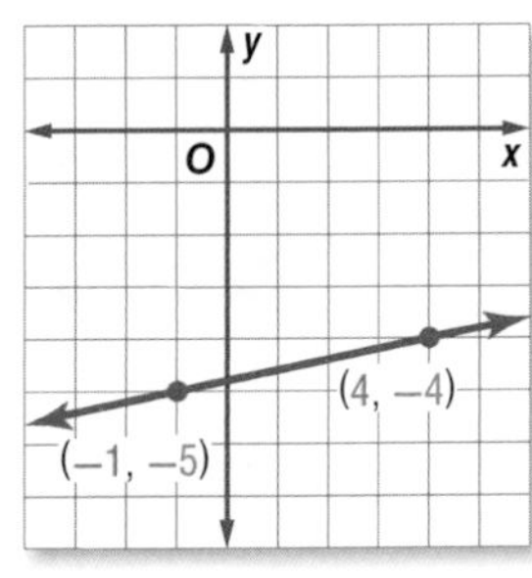

Examples 3 and 4 pp. 428–429

Find the slope of the line that passes through each pair of points.

14. $A(3, 2), B(10, 8)$
15. $R(5, 1), S(0, 4)$
16. $L(5, -6), M(9, 6)$
17. $J(-1, 3), K(-1, 7)$
18. $C(-8, 6), D(1, 6)$
19. $V(5, -7), W(-3, 9)$

20. **MULTIPLE REPRESENTATIONS** In this problem, you will investigate ordered pairs. Use the table shown.

x	y
−1	−6
0	−8
1	−10
2	−12

a. **NUMERICAL** What is the slope of the line represented by the data in the table?

b. **GRAPHICAL** Graph the points on a coordinate plane. Connect the points with a line.

c. **VERBAL** What does the point $(0, -8)$ represent?

H.O.T. Problems

Use **H**igher-**O**rder **T**hinking Skills

21. **OPEN ENDED** Name two points on a line that has a slope of $\frac{5}{8}$.

22. **CHALLENGE** The terms in arithmetic sequence A have a common difference of 3. The terms in arithmetic sequence B have a common difference of 8. In which sequence do the terms form a steeper line when graphed as points on a coordinate plane? Justify your reasoning.

23. **REASONING** *True* or *false*? As the constant of variation increases in a direct variation, the slope of the graph becomes steeper. Explain.

24. **WRITING IN MATH** Explain whether a horizontal line, such as $y = 1$, *always*, *sometimes*, or *never* has 0 slope.

Standardized Test Practice

25. The table shows the number of degrees Celsius for certain Fahrenheit temperatures.

Degrees Fahrenheit	Degrees Celsius
5	−15
14	−10
41	5
77	25

What is the slope of the line that fits these data?

A $-\frac{9}{5}$ C $\frac{5}{9}$

B $-\frac{5}{9}$ D $\frac{9}{5}$

26. A wheelchair ramp rises a vertical distance of 2.5 feet over a horizontal distance of 30 feet. Find the slope of the ramp.

F $\frac{1}{12}$ G $\frac{1}{11}$ H $\frac{1}{9}$ J $\frac{1}{8}$

27. Which of the following is true concerning the slope of the line below?

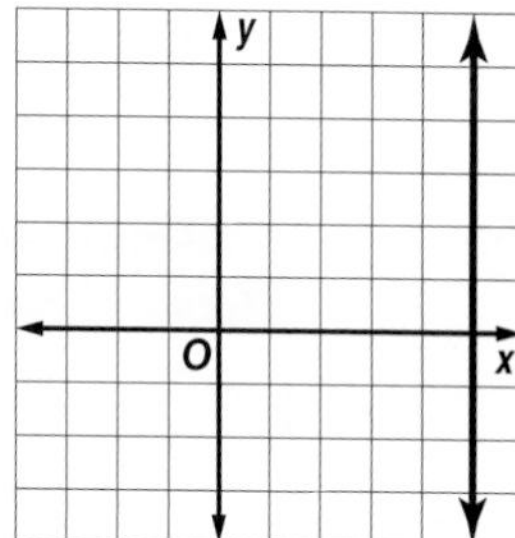

A The slope is −1.

B The slope is zero.

C The slope is 1.

D The slope is undefined.

28. SHORT RESPONSE What is the slope of the line that passes through the points (−3, 5) and (6, −1)?

Spiral Review

29. Determine whether a proportional linear relationship exists between the two quantities shown in the graph. Explain your reasoning. (Lesson 8-5)

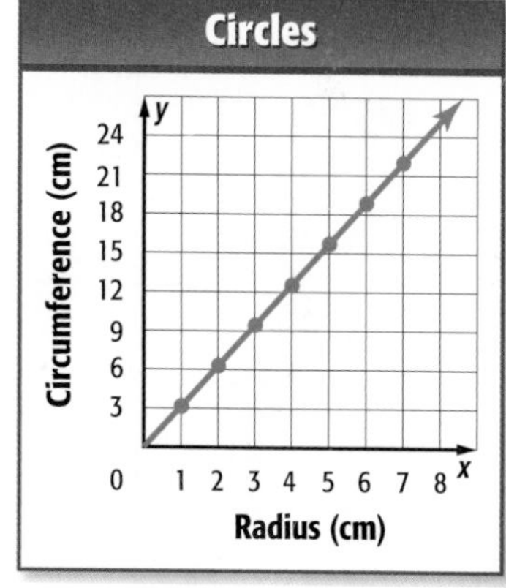

30. FRUIT The table below shows the cost of pears and oranges. Compare the rates of change. (Lesson 8-4)

Weight (lb) x	Cost (\$) y	
	Pears	Oranges
0	0	0
2	2.40	2.20
5	6.00	5.50

Solve each equation. (Lesson 4-3)

31. $-123 = x - 183$

32. $-205 + t = -118$

33. $471 = -196 + n$

Skills Review

Solve each equation for y. (Lesson 5-1)

34. $x + y = 6$

35. $3x + y = 1$

36. $-x + 5y = 10$

37. $\frac{y}{2} - 7x = 5$

EXPLORE **8-7**

Algebra Lab

Proportional and Nonproportional Linear Relationships

Math Online glencoe.com

Math *in Motion*, Animation

Objective
Identify direct proportional and nonproportional linear relationships.

In this lab, you will shade squares on grid paper to develop two different linear functions.

ACTIVITY

Step 1 Using grid paper, draw the two patterns shown.

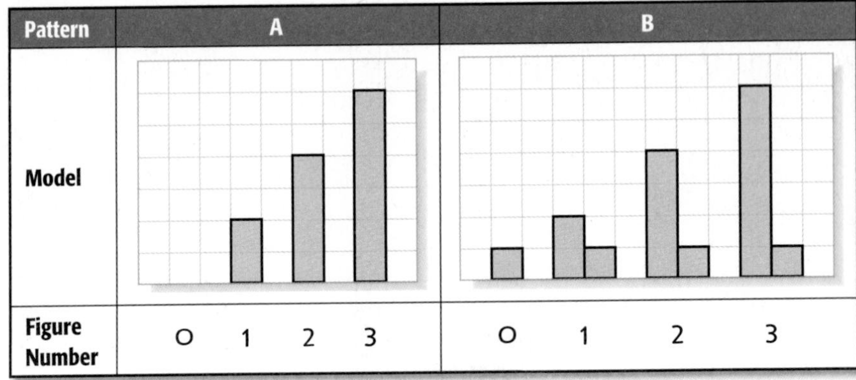

Pattern	A				B			
Model								
Figure Number	0	1	2	3	0	1	2	3

Step 2 Let x represent the figure number and y represent the number of shaded squares in each figure. Copy and complete the table below for each pattern. Then graph and label each set of data on separate coordinate planes.

Pattern ____		
x	Process	y
0	■	■
1	■	■
2	■	■
3	■	■
4	■	■
5	■	■
x	■	■

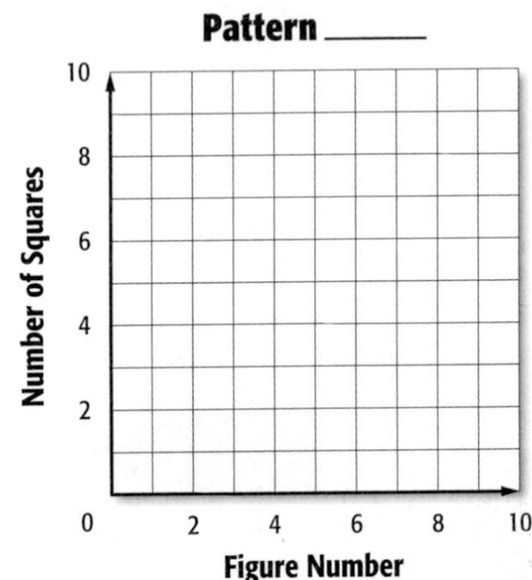

Analyze the Results

1. Describe how the physical models of patterns A and B are alike. Describe how they are different.
2. Compare the processes shown in the tables for patterns A and B.
3. How are the graphs of patterns A and B alike? How are they different?
4. Which pattern represents a direct proportional relationship or direct variation? Which represents a nonproportional relationship? Explain. How can you tell this from the data shown in the table? from the graph?

8-7 Slope-Intercept Form

Why?

A landscaping company charges a $20 fee to mow a lawn plus $8 per hour. The equation $y = 8x + 20$ represents this situation, where x is the number of hours it takes to mow the lawn and y is the total cost of mowing the lawn.

Number Hours, x	Total Cost, y
1	■
2	■
3	■

a. Copy and complete the table to find the total cost of mowing the lawn.

b. Use the table to graph the equation. In which quadrant does the graph lie? Explain.

c. Find the y-coordinate of the point where the graph crosses the y-axis and the slope of the line. How are they related to the equation?

Then

You have already graphed linear equations using ordered pairs. (Lesson 8-3)

Now

- Determine slopes and y-intercepts of lines.
- Graph linear equations using the slope and y-intercept.

New Vocabulary

slope-intercept form

Math Online

glencoe.com

- Extra Examples
- Personal Tutor
- Self-Check Quiz
- Homework Help
- Math in Motion

Find Slope and y-Intercept An equation with a y-intercept that is *not* 0 represents a nonproportional relationship. An equation of the form $y = mx + b$, where m is the slope and b is the y-intercept, is in **slope-intercept form**.

Key Concept — Slope-Intercept Form

For Your FOLDABLE

Words The slope-intercept form of an equation is $y = mx + b$, where m is the slope and b is the y-intercept.

Symbols $y = mx + b$

Example $y = 2x + 1$ (2 → slope; 1 → y-intercept)

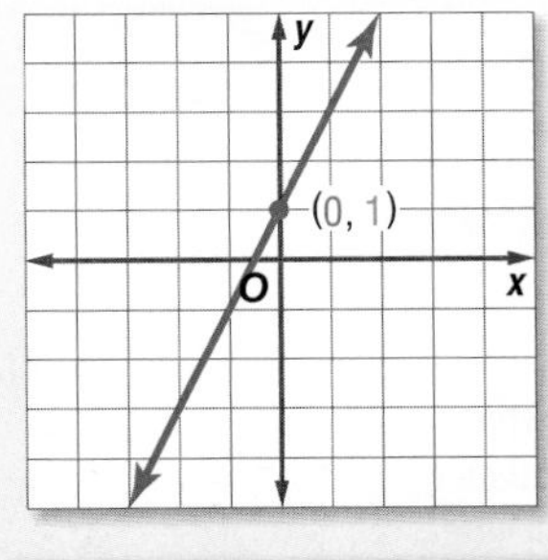

Math *in Motion,* BrainPOP® glencoe.com

EXAMPLE 1 Find the Slope and y-intercept

State the slope and the y-intercept of the graph of $y = \frac{1}{4}x - 6$.

$y = \frac{1}{4}x - 6$ — **Write the equation.**

$y = \frac{1}{4}x + (-6)$ — **Write the equation in the form $y = mx + b$.**

The slope is $\frac{1}{4}$ and the y-intercept is -6.

Check Your Progress

State the slope and the y-intercept of the graph of each equation.

1A. $y = 3x + 1$

1B. $y = -2x$

Personal Tutor glencoe.com

EXAMPLE 2 Find Slope and y-intercept

State the slope and the y-intercept of the graph of $-4x + y = 2$.

First, write the equation in slope-intercept form.

$-4x + y = 2$	**Write the original equation.**
$+4x \quad = \quad +4x$	**Add 4x to each side.**
$y = 2 + 4x$	**Simplify.**
$y = 4x + 2$	**Write in slope-intercept form.**
$y = mx + b$	$m = 4, b = 2$

The slope of the graph is 4 and the y-intercept is 2.

Check Your Progress

State the slope and the y-intercept of the graph of each equation.

2A. $6x + y = -3$

2B. $y - 5 = -x$

2C. $y - 5x = 10$

2D. $x = 2 + y$

Personal Tutor glencoe.com

Graph Equations You can use the slope-intercept form of an equation to graph a line.

EXAMPLE 3 Graph an Equation

Graph $y = -\frac{2}{3}x + 3$ using the slope and y-intercept.

Step 1 Find the slope and y-intercept.

slope $= -\frac{2}{3}$

y-intercept $= 3$

Step 2 Graph the y-intercept point at (0, 3).

Step 3 Write the slope as $\frac{-2}{3}$. Use it to locate a second point on the line.

$m = \frac{-2}{3}$ ← **Change in y: down 2 units** ← **Change in x: right 3 units**

Another point on the line is at (3, 1).

Step 4 Draw a line through the two points and extend the line.

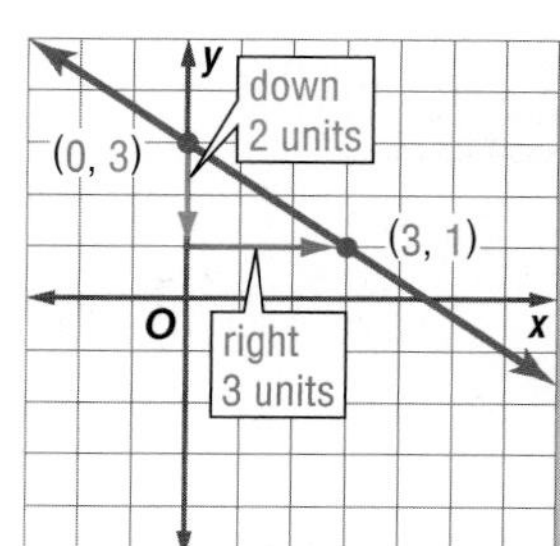

Check Your Progress

3. Graph $y = -x - 2$ using the slope and y-intercept.

Personal Tutor glencoe.com

StudyTip

Constant of Proportionality The direct variation equation $y = kx$ is a special case of a linear equation of the form $y = mx + b$. The constant of proportionality k is the slope and the y-intercept b is 0.

Real-World EXAMPLE 4 Analyze the Slope and y-Intercept

GECKOS **A typical leopard gecko is 3 inches long at birth and grows at a rate of about $\frac{1}{3}$ inch per week for the first few months. The total length of a leopard gecko y after x weeks can be represented by $y = \frac{1}{3}x + 3$.**

Real-World Link

Leopard geckos shed their skin every 2 to 4 weeks. The frequency of their shedding is determined by their age and the growth rate.

Source: Golden Gate Geckos

a. Graph the equation.

Step 1 First, find the slope and the y-intercept.

slope $= \frac{1}{3}$

y-intercept $= 3$

Step 2 Plot the point at (0, 3). Then go up 1 and right 3 and plot another point.

Step 3 Connect these points and extend the line.

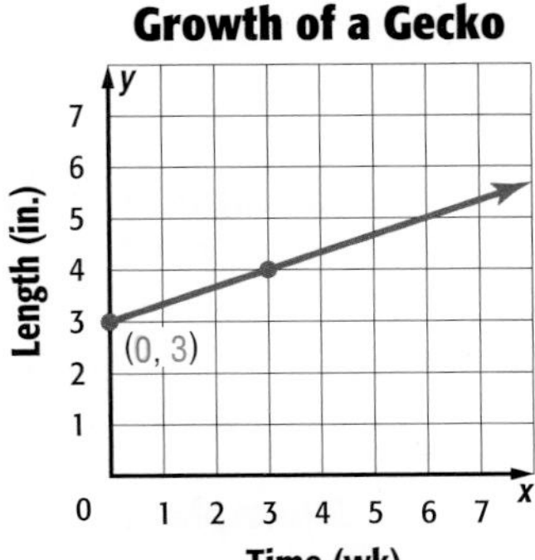

b. Describe what the y-intercept and the slope represent.

The y-intercept 3 represents the length of the gecko at birth. The slope $\frac{1}{3}$ represents the growth rate in inches per week, which is the rate of change.

Check Your Progress

4. **WRITING** Jack has written 30 pages of his novel. He plans to write 12 pages per week until he has completed his novel. The total number of pages written y can be represented by $y = 12x + 30$, where x is the number of weeks.

   A. Graph the equation.

   B. Describe what the y-intercept and the slope represent.

Personal Tutor glencoe.com

Check Your Understanding

Examples 1 and 2 pp. 433–434

State the slope and the y-intercept of the graph of each equation.

1. $y = 2x + 6$
2. $y = \frac{3}{4}x - 1$
3. $7x + y = 0$
4. $4x + y = 3$

Example 3 p. 434

Graph each equation using the slope and y-intercept.

5. $y = \frac{1}{3}x + 1$
6. $y = -x + 2$
7. $y = 2x - 4$
8. $y = -0.75x - 3$

Example 4 p. 435

9. **KITES** A kite flying 60 feet in the air is falling. The altitude of the kite can be represented by $y = -x + 60$, where x is the time in seconds.

   a. Graph the equation.

   b. Describe what the y-intercept and the slope represent.

Practice and Problem Solving

● = **Step-by-Step Solutions** begin on page R11.
Extra Practice begins on page 810.

Examples 1 and 2 pp. 433–434

State the slope and the y-intercept of the graph of each equation.

10. $y = x + 8$ **11.** $y = -\frac{5}{2}x - 2$ **12.** $y = \frac{1}{3}x$ **13.** $y = -9x$

14. $-x + y = 5$ **15.** $4x + y = 0$ **16.** $-9x + y = -5$ **17.** $y - 6 = \frac{1}{2}x$

Example 3 p. 434

Graph each equation using the slope and y-intercept.

18. $y = x - 2$ **19.** $y = 3x + 4$ **20.** $y = \frac{1}{4}x + 1$

21. $y = \frac{3}{2}x - 3$ **22.** $y = -2x - 6$ **23.** $y = -\frac{4}{3}x + 5$

Example 4 p. 435

24. **FINANCIAL LITERACY** To replace a set of brakes, an auto mechanic charges \$40 for parts plus \$50 per hour. The total cost y can be given by $y = 50x + 40$ for x hours.

a. Graph the equation using the slope and y-intercept.

b. State the slope and y-intercept of the graph of the equation and describe what they represent.

StudyTip

Real-World Data Look at the problem to determine how to draw a graph of the data. In Exercise 25, it is not reasonable for the time or the altitude to be negative. So, the graph should be in Quadrant I only.

25 **BIRDS** The altitude in feet y of an albatross who is slowly landing can be given by $y = 300 - 50x$, where x represents the time in minutes.

a. Graph the equation using the slope and y-intercept.

b. State the slope and y-intercept of the graph of the equation and describe what they represent.

26. **BAKING** Sam has 15 teaspoons of chopped nuts. She uses $1\frac{1}{2}$ teaspoons for each muffin. The total amount of nuts that she has left y after making x muffins can be given by $y = -\frac{3}{2}x + 15$, as shown in the graph.

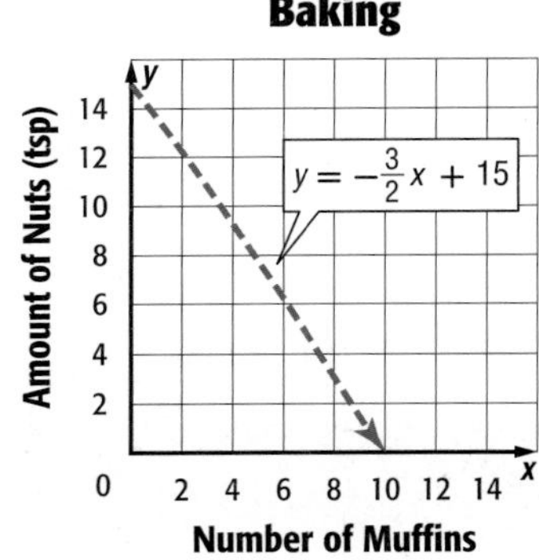

a. State the slope and y-intercept of the graph of the equation and describe what they represent.

b. Name the x-intercept and describe what it represents.

27. **PHOTOS** The table shows the prices of sitting fees and each 5×7 portrait for two photography studios.

Studio	Sitting Fee (\$)	5 × 7 Price (\$)
Lifetime Photos	18	15
Family Photos	12	15

a. Write an equation to represent y the total cost of having your photo taken and buying x number of 5×7's for each studio.

b. Graph each equation on the same coordinate plane.

c. Will the lines ever intersect? Explain.

d. Compare the slopes of each line.

Real-World Link

The amusement park with greatest number of rides is Cedar Point located in Sandusky, Ohio. The park has 20 children's rides and 47 major rides.

Source: National Amusement Park Historical Society

Graph each equation using the slope and y-intercept.

28. $x - 2y = 8$ **29.** $3x + 4y = 12$ **30.** $y = 6$ **31.** $x + 4y = 0$

32. **MULTIPLE REPRESENTATIONS** In this problem, you will investigate graphs of equations. The Math Club is planning a trip to an amusement park. The table shows the bus prices to travel to each park and the admission price per student.

Park	Bus Fee ($)	Admission Price Per Student ($)
Wild Waves	200	16.50
Coaster Haven	250	14.50

a. **ALGEBRAIC** Write an equation to represent the total cost y for x students at each park.

b. **GRAPHICAL** Graph the two equations on the same coordinate plane. For how many students is the cost of both trips the same? Explain.

c. **NUMERICAL** If 20 students decide to take the trip, which trip will cost less? If 28 students decide to take the trip, which trip will cost less?

d. **VERBAL** Is it possible to determine the answers to part **c** by examining the graph you made in part **a**? Explain your reasoning.

For Exercises 33–35, use the graph shown at the right.

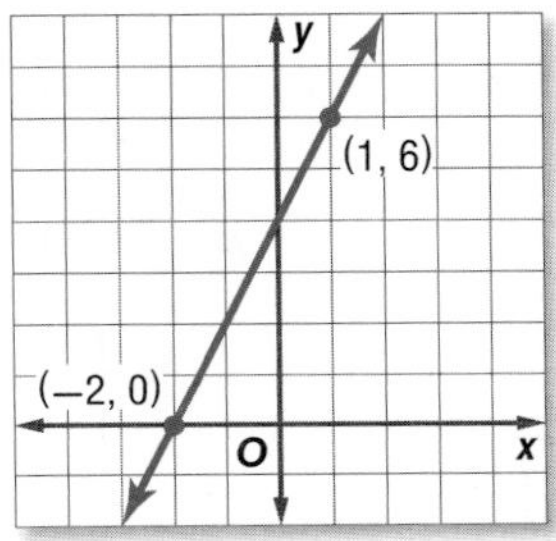

33. What is the slope of the line shown?

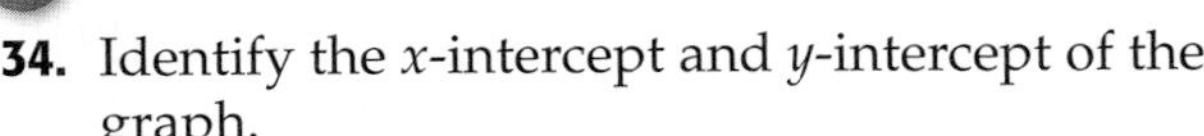

34. Identify the x-intercept and y-intercept of the graph.

35. What is the equation of the line? Write in slope-intercept form.

H.O.T. Problems

Use **H**igher-**O**rder **T**hinking Skills

36. **OPEN ENDED** Describe a line that has a y-intercept but no x-intercept. Identify the slope of the line.

37. **WRITING IN MATH** Write a real-world problem that can be represented by an equation in the form $y = mx + b$. Solve by graphing.

38. **CHALLENGE** Suppose the graph of a line has a negative slope and a positive y-intercept. Through which quadrants does the line pass? Justify your reasoning.

39. **REASONING** Describe what happens to the graph of $y = 3x + 4$ when the slope is changed to $\frac{1}{3}$.

40. **FIND THE ERROR** Maricruz and Francesco are finding the slope and y–intercept of $x - 2y = 3$. Is either of them correct? Explain your reasoning.

Maricruz	Francesco
slope = 1	slope = $\frac{1}{2}$
y-intercept = 3	y-intercept = $\frac{3}{2}$

41. **WRITING IN MATH** Describe the steps you take to graph an equation using the slope and y-intercept.

Standardized Test Practice

42. A line has a slope of $\frac{4}{5}$ and a y-intercept of 10. Which of the following represents the equation of the line?

A $4x - 5y = 10$

B $5x - 4y = 10$

C $4x - 5y = -50$

D $5x - 4y = -50$

43. Which *best* represents the graph of $y = 3x - 1$?

F
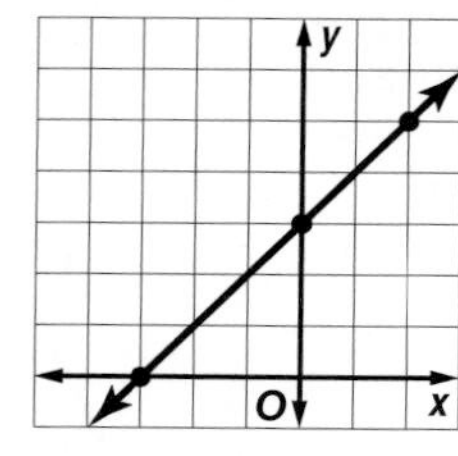

H
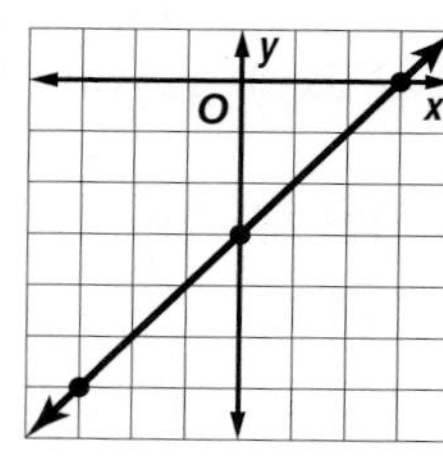

G
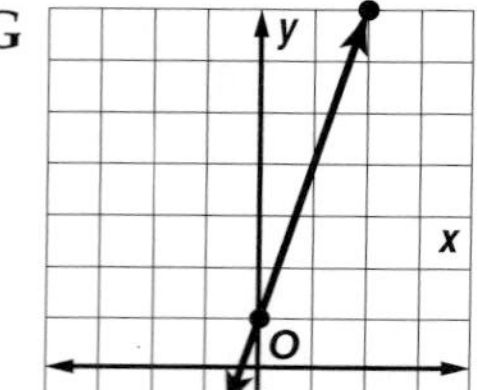

J
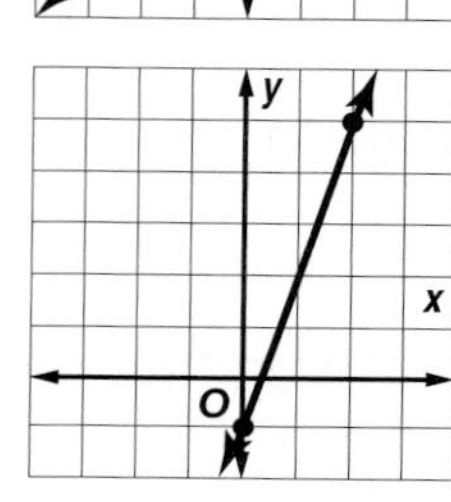

44. The cost of renting a cotton candy machine is \$30 plus \$5 for each hour. The total cost of renting a cotton candy machine is represented by $C(h) = 5h + 30$, where h is number of hours. What does the slope represent?

A number of hours

B cost of each hour

C cost per bag of cotton candy

D cost of renting the machine for no hours

45. EXTENDED RESPONSE Snow Mountain Ski Resort offers a special season pass at the beginning of each ski season. The pass costs \$35, and an additional \$25 is charged each time you ski. The total cost can be represented by $y = 35 + 25x$, where x is number of times you ski.

a. Graph the equation.

b. Explain what the y–intercept and the slope represent.

Spiral Review

46. CARS The cost of gas varies directly with the number of gallons bought. Marty bought 18 gallons of gas for \$49.50. Write an equation that could be used to find the unit cost of a gallon of gas. Then find the unit cost. (Lesson 8-6)

47. BIRDSEED Find the constant rate of change for the linear function in the table and interpret its meaning. (Lesson 8-5)

Amount of Birdseed (lb)	Total Cost (\$)
x	y
4	11.20
8	22.40
12	33.60

Find each sum in simplest form. (Lesson 3-6)

48. $\frac{1}{2} + \frac{7}{10}$

49. $\frac{1}{6} + \frac{3}{12}$

50. $\frac{7}{8} + \frac{2}{10}$

Skills Review

Simplify each expression. (Lesson 1-2)

51. $-3(5) - 7$

52. $4 + (-5)(6)$

53. $(-10 - 8) \div (-9)$

54. $(32 - 12) \div 4 \times 7$

55. $(14 - 6) \times 8 \div 2$

56. $(56 - 28) \div (7 \times 2)$

EXTEND 8-7

Graphing Technology Lab
Family of Linear Graphs

Math Online glencoe.com
- Other Calculator Keystrokes
- Graphing Technology Personal Tutor

A **family of functions** is a set of functions that is related in some way. The family of linear functions has the parent function $y = x$.

A TI-Nspire calculator allows you to enter a function and manipulate the graph. This is useful for investigating families of linear functions because you can easily compare characteristics such as slopes and y-intercepts.

ACTIVITY

Graph $y = 2x + 6$ in the default viewing window and move the line to see how the equation relates to the graph.

Step 1 Graph $y = 2x + 6$ in the default viewing window.

- Open a new **Graphs & Geometry** window by using the following keystrokes. (home) 2
- Enter the function $2x + 6$ in *f1(x)* and press (enter). When a new window is open, the cursor is automatically in the *f(x)* box. The graph and its equation will appear on the coordinate plane.
- Press (tab) twice to move the cursor to the graph screen.

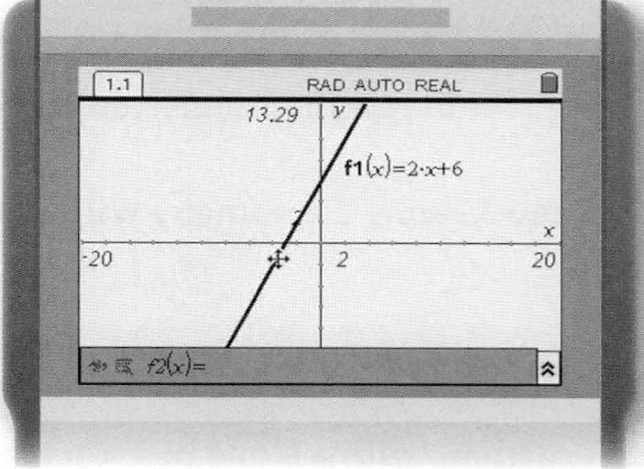

Step 2 Grab the line and translate the line up and down.

- Use the NavPad to move the cursor over the line in the third quadrant near the x-axis until it blinks with a ✣ sign on it. Press (ctrl)(click) to grab the line. Use the NavPad to translate the line up and down.

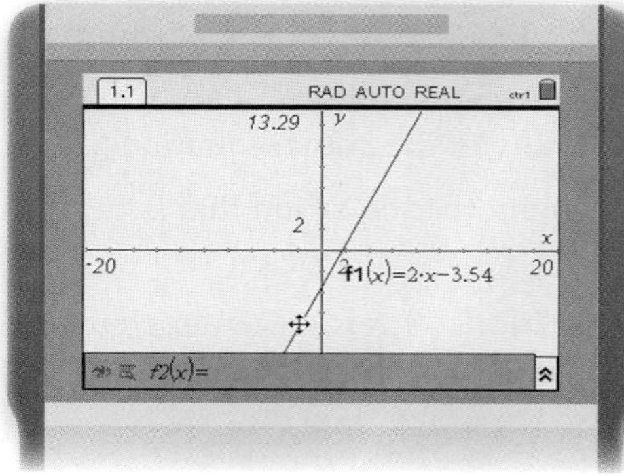

When graphing, be sure to use a viewing window that shows both the x- and y-intercepts of the graph of a function. If you need to set your own minimum and maximum values for the axes and the scale factor, use the **WINDOW** option from the menu button.

Exercises

Graph $y = 3x - 4$. Grab the line and move it up and down as instructed on the previous activity.

1. What changes in the equation? What remains the same?
2. How does adding or subtracting a constant c to a linear function affect its graph?
3. Move the line until it crosses the y-axis as near to 5 as possible. Write the equation of the line.
4. Write an equation of the line that is parallel to $y = 3x - 4$ and passes through the origin.

Open a new Graphs & Geometry window. Graph $y = 3x - 4$. Use the NavPad to move over the line until a blinking ⟲ appears on it. Grab the line as instructed in the Activity and rotate the line.

5. What changes in the equation? What remains the same?
6. How does changing the coefficient for x affect the graph of a linear function?
7. Without graphing, determine whether the graph of $y = 0.5x$ or the graph of $y = 1.5x$ has a steeper slope. Explain.
8. Rotate the graph until the x coefficient is negative. How does changing the sign of the coefficient of x affect the graph of a linear function?

Open a new Graphs & Geometry window. Graph the following lines.

$f1(x) = -2x$ $\qquad$ $f2(x) = -2x + 1$ $\qquad$ $f3(x) = \frac{1}{2}x + 1$

9. Describe the similarities and differences between the graphs.

Three functions with a slope of 1 are graphed in the standard viewing window, as shown.

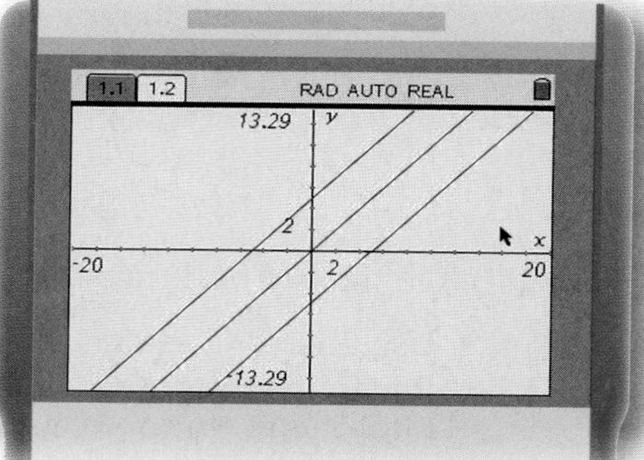

10. Write an equation for each, beginning with the left-most graph.
11. Does the equation $y = -x - 5$ have any special relationships with any of the above equations?
12. Write another equation that is part of this family of functions.
13. **GARDENING** A garden center charges $75 per cubic yard for topsoil. The delivery fee is $25.
    a. Write the equation that represents the charges.
    b. Describe the change in the graph if the delivery fee is changed to $35.
    c. How does the graph change if the price of a cubic yard of topsoil is increased to $80 when the delivery fee is $35.?
    d. What are the prices of a cubic yard of topsoil and delivery if the graph has slope 70 and y-intercept 40?

8-8 Writing Linear Equations

Then

You have already graphed linear equations using the slope and y-intercept. (Lesson 8-6)

Now

- Write equations given the slope and y-intercept, a graph, a table, or two points.
- Use linear equations to solve problems.

New Vocabulary

point-slope form

Math Online

glencoe.com

- Extra Examples
- Personal Tutor
- Self-Check Quiz
- Homework Help

Why?

Olivia purchased a music subscription for downloading music from the Internet. The initial fee was \$10. She then paid a cost of \$1 per song.

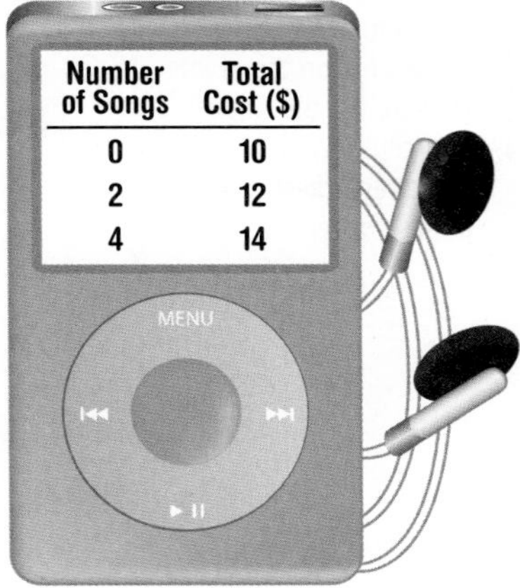

Number of Songs	Total Cost (\$)
0	10
2	12
4	14

a. Graph the ordered pairs (number of songs, total cost). Draw a line through the points.

b. Find the slope and the y-intercept of the line. What do these values represent?

Write Equations in Slope-Intercept Form One way to write a linear equation is to substitute the values for the slope and y-intercept in $y = mx + b$. Sometimes, you may need to find the y-intercept and slope from a graph.

EXAMPLE 1 Write the Equation of a Line in Slope-Intercept Form

Write an equation in slope-intercept form for each line.

a. The slope is $\frac{1}{2}$, and the y-intercept is −5.

$y = mx + b$ — **Slope-intercept form**

$y = \frac{1}{2}x + (-5)$ — **Replace m with $\frac{1}{2}$ and b with −5.**

$y = \frac{1}{2}x - 5$ — **Simplify.**

b.

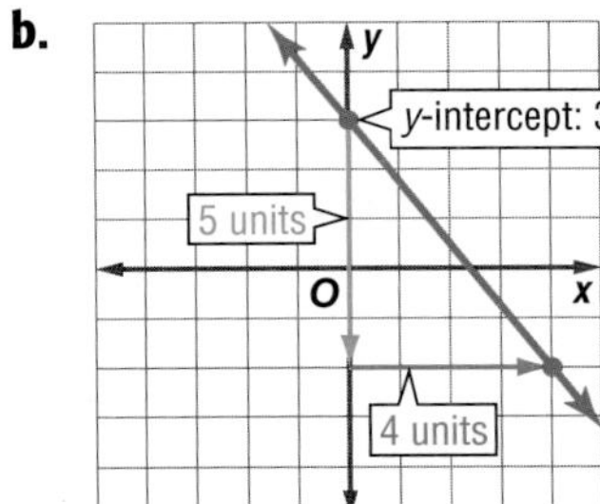

The y-intercept is 3. From (0, 3), you move down 5 units and right 4 units to another point on the line. So, the slope is $-\frac{5}{4}$.

$y = mx + b$ — **Slope-intercept form**

$y = -\frac{5}{4}x + 3$ — **Replace m with $-\frac{5}{4}$ and b with 3.**

Check Your Progress

1A. slope = 2, y-intercept = $\frac{1}{3}$

1B.

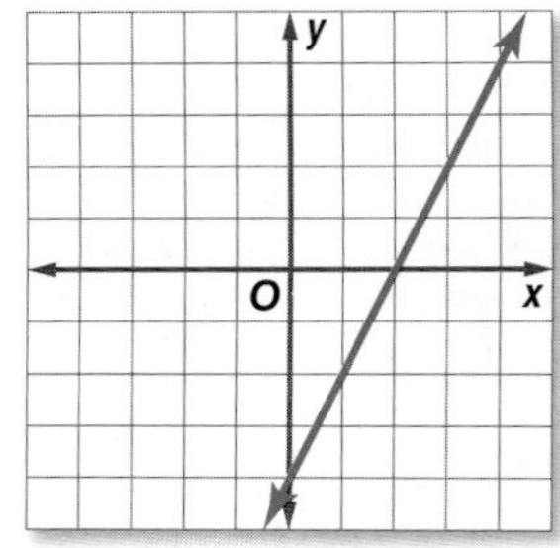

Personal Tutor glencoe.com

Equations of Lines The point-slope form of an equation can vary depending on the points chosen, but the slope-intercept form is the same regardless of the points chosen.

You can also write an equation for a line if you know the coordinates of two points on the line. An equation in the form $y - y_1 = m(x - x_1)$ where m represents the slope and (x_1, y_1) represents a point on the line, is called **point-slope form** of a line.

EXAMPLE 2 Write an Equation Given Two Points

Write an equation for the line that passes through (4, 8) and (−2, 5).

Step 1 Find the slope m.

$m = \frac{y_2 - y_1}{x_2 - x_1}$ **Definition of slope**

$m = \frac{5 - 8}{-2 - 4}$ $(x_1, y_1) = (4, 8), (x_2, y_2) = (-2, 5)$

$m = \frac{-3}{-6}$ or $\frac{1}{2}$ **Simplify.**

Step 2 Use the slope and the coordinates of either point to write the equation in point-slope form.

$y - y_1 = m(x - x_1)$ **Point-slope form**

$y - 8 = \frac{1}{2}(x - 4)$ **Replace (x, y) with (4, 8) and m with $\frac{1}{2}$.**

The equation is $y - 8 = \frac{1}{2}(x - 4)$. In slope-intercept form, $y = \frac{1}{2}x + 6$.

StudyTip

Check Equation To check, substitute the coordinates of the other point in the equation.

$y - 8 = \frac{1}{2}(x - 4)$

$5 - 8 \stackrel{?}{=} \frac{1}{2}(-2 - 4)$

$5 - 8 \stackrel{?}{=} \frac{1}{2}(-6)$

$-3 = -3$ ✓

✓ Check Your Progress

2. Write an equation for the line that passes through (3, 0) and (6, −3).

Personal Tutor glencoe.com

EXAMPLE 3 Write an Equation from a Table

Write an equation of the line in point-slope form that passes through the points shown in the table at the right.

x	y
4	−2
8	1
12	4
16	7

Step 1 Find the slope m. Use the coordinates of any two points.

$m = \frac{y_2 - y_1}{x_2 - x_1}$ **Definition of slope**

$m = \frac{1 - (-2)}{8 - 4}$ $(x_1, y_1) = (4, -2), (x_2, y_2) = (8, 1)$

$m = \frac{3}{4}$ **Simplify.**

Step 2 To write the equation, use the slope and the coordinates of any point.

$y - y_1 = m(x - x_1)$ **Point-slope form**

$y - 1 = \frac{3}{4}(x - 8)$ **Replace (x, y) with (8, 1) and m with $\frac{3}{4}$.**

The equation is $y - 1 = \frac{3}{4}(x - 8)$.

StudyTip

Alternative Strategy If a table includes the y-intercept, simply use this value and the slope to write an equation.

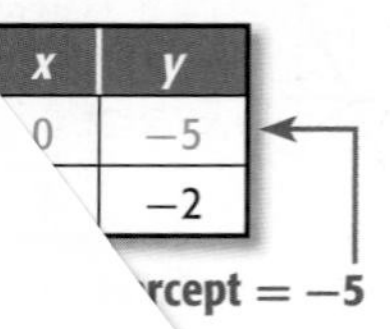

x	y
0	−5
[illegible]	−2

...rcept = −5

✓ Check Your Progress

3. Write an equation of the line in point-slope form that passes through the points shown.

x	−3	0	2	3
y	9	3	−1	−3

Personal Tutor glencoe.com

Solve Problems You can write an equation to describe the relationship between two quantities and to make predictions.

Real-World Link

Since the boiling point of water is lower at higher altitudes, cooks have to make adjustments in cooking times. A general rule is that for every 1000 feet above 2000 feet elevation, increase the cooking time by 5%.

Source: Kuhn Rikon Switzerland

Real-World EXAMPLE 4 Write an Equation to Make a Prediction

BOILING POINT The boiling point of water at sea level, or at altitude 0 feet, is 212°F. The boiling point decreases 1°F for every 540-foot increase in altitude. Estimate the boiling point for an altitude of 4000 feet.

Understand You know the rate of change of boiling point temperature to altitude and the temperature at altitude 0 feet. You need to estimate the boiling point for an altitude of 4000 feet.

Plan First, find the slope and y-intercept. Then write an equation to show the relationship between altitude x and temperature y. Use the equation to find the boiling point.

Solve The boiling point decreases 1°F for every 540-foot increase in altitude.

- Find the slope m and the y-intercept b.

$$m = \frac{\text{change in } y}{\text{change in } x} \quad \begin{array}{l}\leftarrow \text{change in boiling point}\\ \leftarrow \text{change in altitude}\end{array}$$

$$= \frac{-1}{540} \quad \begin{array}{l}\leftarrow \text{decrease of 1°F}\\ \leftarrow \text{increase of 540 ft}\end{array}$$

$$\approx -0.002$$

At $x = 0$, $y = 212$. So, the y-intercept b equals 212.

- Write the equation.

$y = mx + b$ **Slope-intercept form**

$y = -0.002x + 212$ **Replace m with -0.002 and b with 212.**

- Find the boiling point.

$y = -0.002x + 212$ **Write the equation.**

$= -0.002(4000) + 212$ **Replace x with 4000.**

$= 204$ **Simplify.**

At an altitude of 4000 feet, the boiling point of water is about 204°F.

Check Since $4000 \div 540$ is about 8, the boiling point would drop 8×1 or 8°F. 212°F − 8°F = 204°F. So, the answer is reasonable. ✓

Problem-SolvingTip

Use a Table Translate the words into a table of values to better understand what the slope represents. For every increase of 540 feet in altitude, the temperature decreases by 1°F.

Altitude (ft)	Boiling Point (°F)
x	y
0	212
540	211
1,080	210

Check Your Progress

4. **PIANO LESSONS** The cost of 7 half-hour piano lessons is $179. The cost of 11 half-hour lessons is $267.
   a. Write a linear equation that shows the cost y for x half-hour lessons.
   b. Use the equation from part **a.** to find the cost of 3 half-hour lessons.

Personal Tutor glencoe.c

Concept Summary — Write a Linear Equation

For Your FOLDABLE

From Slope and y-Intercept	• Substitute the slope m and y-intercept b in $y = mx + b$.
From a Graph	• Find the y-intercept b and the slope m from the graph. • Substitute the slope and y-intercept in $y = mx + b$.
From Two Points	• Use the coordinates of the two points to find the slope. • Substitute the slope and coordinates of one of the points in $y - y_1 = m(x - x_1)$
From a Table	• Use the coordinates of any two points to find the slope. • Substitute the slope and coordinates of one of the points in $y - y_1 = m(x - x_1)$

Check Your Understanding

Example 1 p. 441

Write an equation in slope-intercept form for each line.

1. slope = 2, y-intercept = 4
2. slope = 0, y-intercept = 1
3. slope = $-\frac{3}{4}$, y-intercept = 0
4. slope = $\frac{1}{3}$, y-intercept = −6

5. 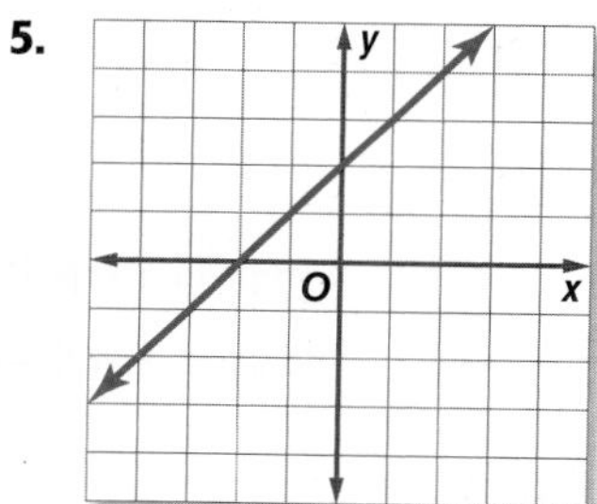

6. 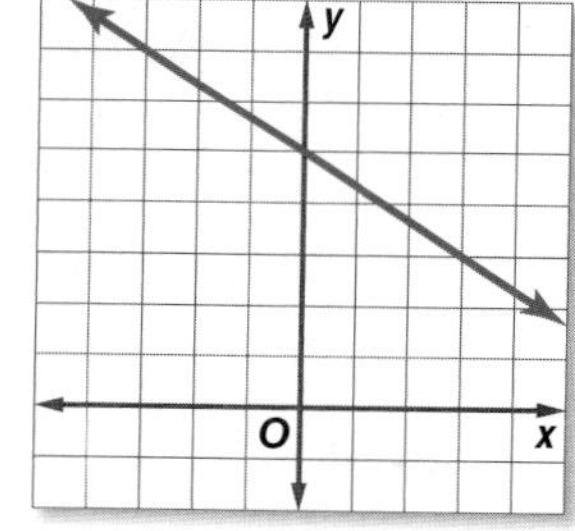

7. 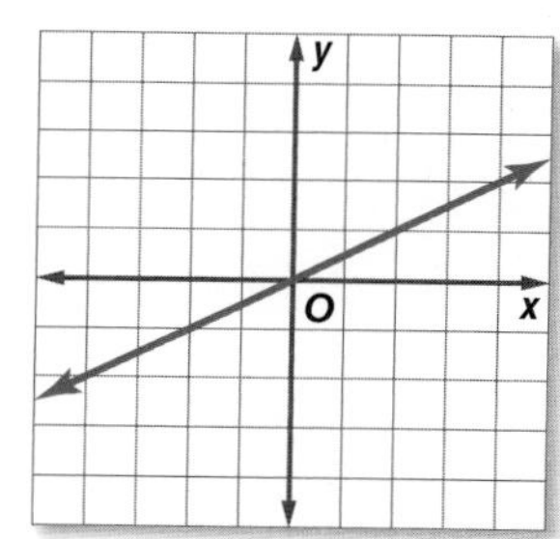

Example 2 p. 442

Write an equation for the line that passes through each pair of points.

8. (3, 3) and (6, 4)
9. (2, 5) and (3, −4)
10. (−1, 2) and (5, −10)

Example 3 p. 442

Write an equation in point-slope form to represent each table of values.

11.

x	−1	0	1	2
y	−6	−2	2	6

12.

x	−6	−3	3	6
y	−1	0	2	3

Example 4 p. 443

13. **SOUND** Use the table that shows the distance that a rip current travels through the ocean.

Time (s) x	Distance (ft) y
0	0
1	2.4
2	4.8
3	7.2

a. Write an equation in slope-intercept form to represent the data in the table. Describe what the slope means.

b. Estimate how far the rip current travels in 1 minute.

Practice and Problem Solving

 = **Step-by-Step Solutions** begin on page R11.
Extra Practice begins on page 810.

Example 1 p. 441

Write an equation in slope-intercept form for each line.

14. slope = 3, y-intercept = 2

15. slope = 1, y-intercept = −4

16. slope = 0, y-intercept = 1

17. slope = 2, y-intercept = 0

18. slope = $\frac{1}{4}$, y-intercept = −3

19. slope = 0, y-intercept = −7

20. slope = $-\frac{2}{3}$, y-intercept = 0

21. slope = $-\frac{5}{3}$, y-intercept = −6

22.

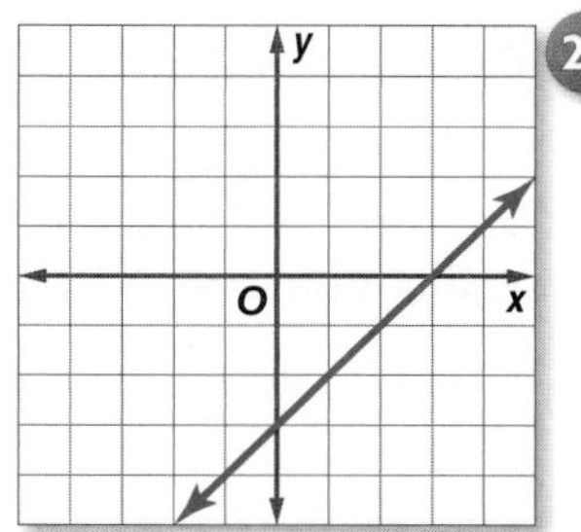

23.

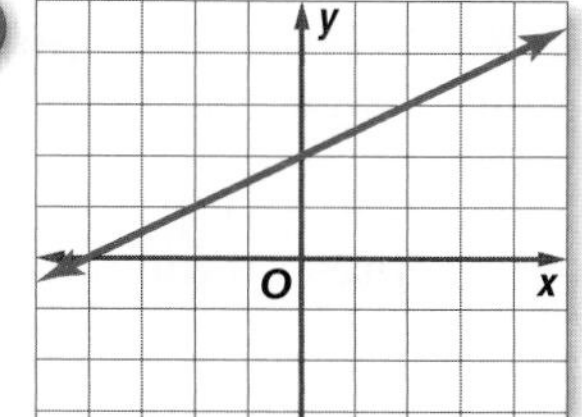

24.

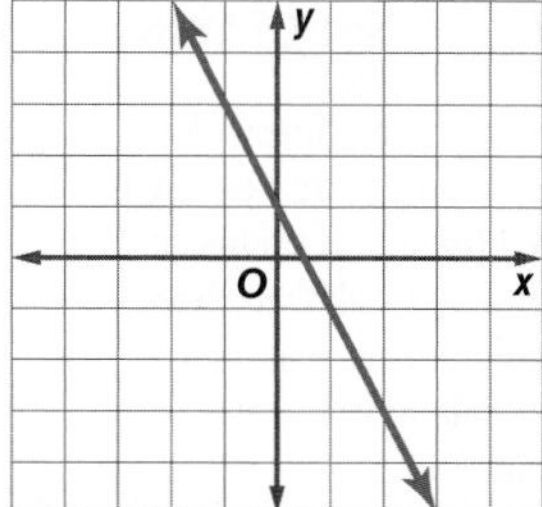

25.

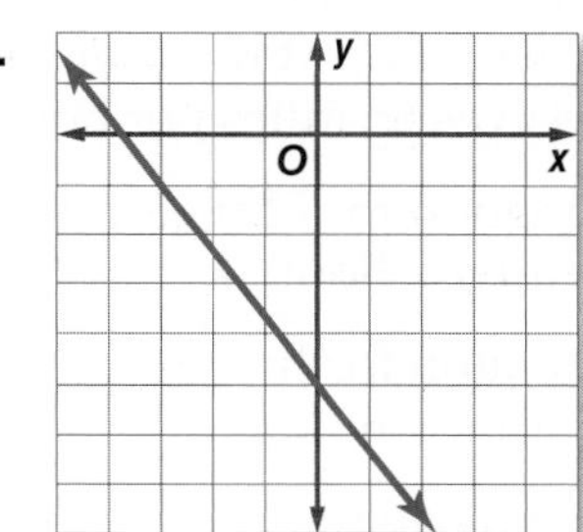

26.

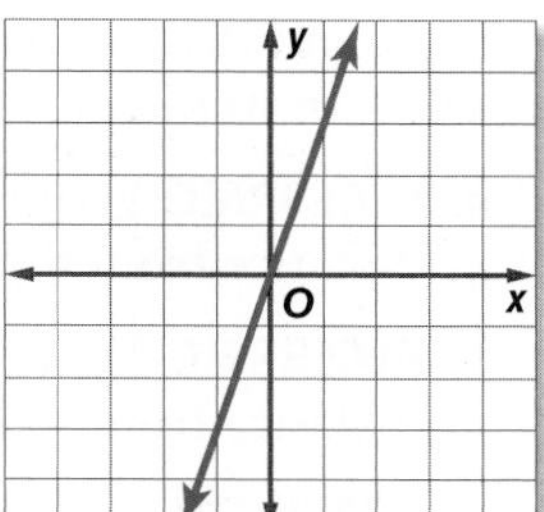

27.

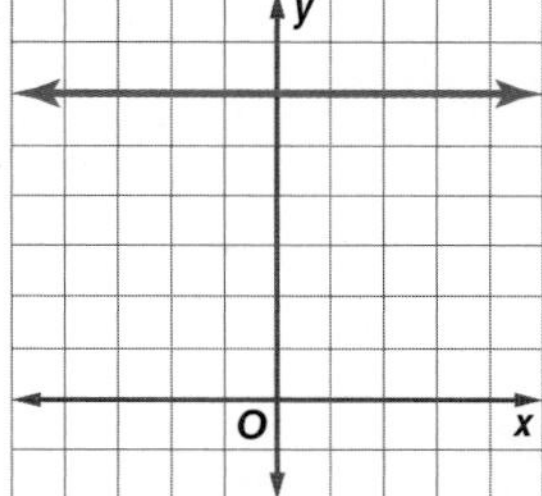

Example 2 p. 442

Write an equation for the line that passes through each pair of points.

28. (1, 2) and (3, 4)

29. (2, −2) and (4, −1)

30. (1, 4) and (2, 8)

31. (3, −6) and (5, −6)

32. (4, −4) and (8, −10)

33. (−1, 9) and (2, −6)

Example 3 p. 442

Write an equation in point-slope form to represent each table of values.

34.

x	−1	1	2	3
y	4	6	7	8

35.

x	−2	−1	1	2
y	−4	−2	2	4

36.

x	−4	−3	0	1
y	−3	−4	−7	−8

37.

x	−5	5	10	15
y	−1	3	5	7

Example 4 p. 443

38. **COMPUTERS** Compuworks, a computer repair company, charges a fee and an hourly charge. After two hours, the repair bill is \$110, and after three hours, it is \$150.

a. Write an equation in slope-intercept form to represent the data.

b. Describe what the slope and y-intercept mean.

c. How much would it cost for 1.5 hours of work?

Real-World Link

You can see a firework explosion before hearing it because sound travels at about 742 miles per hour, but light travels at about 670 million miles per hour.

Source: Zambelli Fireworks

39. **FIREWORKS** The table at the right shows the heights of different firework displays.

Shell Radius (in.)	Height of Fireworks (m)
3	120
4	150
5	180
6	210

Source: Skylighter

a. Write an equation in slope-intercept form to represent the data. Describe what the slope means.

b. Predict the height of fireworks that have shells with a radius of 9 inches.

40. **PAINTING** Mr. Awan has budgeted \$860 to have his dining room painted. The estimated cost for materials is \$100. The painter charges \$35 per hour and estimates that the work will take about 20 hours to complete. Has Mr. Awan budgeted enough money to paint the dining room? Explain.

41. **SNORKELING** The table at the right shows the cost of a snorkeling trip.

Number of People	Total Cost (\$)
3	104.85
5	174.75

a. Write an equation in slope-intercept form to represent the data. Describe what the slope means.

b. How much would it cost for 9 people to go on the snorkeling trip?

42. **MULTIPLE REPRESENTATIONS** Emilee and Justyne are traveling on the same highway to a family reunion at a park. Emilee starts out 225 miles from the park and drives 70 miles per hour. At the same time, Justyne starts out 200 miles from the park and drives 65 miles per hour.

a. **ALGEBRAIC** Write an equation for Emilee's trip where y is the total distance from the park after x hours.

b. **ALGEBRAIC** Write an equation for Justyne's trip where y is the total distance from the park after x hours.

c. **GRAPHICAL** Graph the two equations on the same coordinate plane.

d. **VERBAL** Do you think that Emilee will overtake Justyne before they reach the park? Explain your reasoning.

H.O.T. Problems

Use Higher-Order Thinking Skills

43. **OPEN ENDED** Choose two points in the second quadrant. Write an equation in point-slope form for the line that passes through the points.

44. **REASONING** *True or false*: The equation for a horizontal line has both an x and y term. Explain your reasoning.

45. **FIND THE ERROR** Daniel and Kayla are writing an equation for the line that passes through (–4, 0) and (0, 5). Is either of them correct? Explain.

Daniel	Kayla
$m = \frac{5}{4}$ and $b = 5$, so the equation is $y = \frac{5}{4}x + 5$.	$m = \frac{4}{5}$ and the y-intercept is 0, so the equation is $y = \frac{4}{5}x$.

46. **CHALLENGE** Write an equation in *slope-intercept* form for the line that passes through the points (3, 2) and (9, 4).

47. **WRITING IN MATH** Summarize how to find the slope and y-intercept of a linear function from a(n) equation, table, and graph.

Standardized Test Practice

48. Which of the following equations describes the data in the table below?

Number of DVDs	2	4	6	8
Cost ($)	16	22	28	34

A $3x + y = 10$
B $x - 3y = -10$
C $3x - y = -10$
D $x + y = 10$

49. Which of the following equations names the line that passes through the points (0, −6) and (2, 3)?

F $y = \frac{9}{2}x - 6$
G $y = \frac{2}{9}x - 6$
H $y = -\frac{2}{9}x - 6$
J $y = -\frac{9}{2}x - 6$

50. Which of the following equations describes the line graphed below?

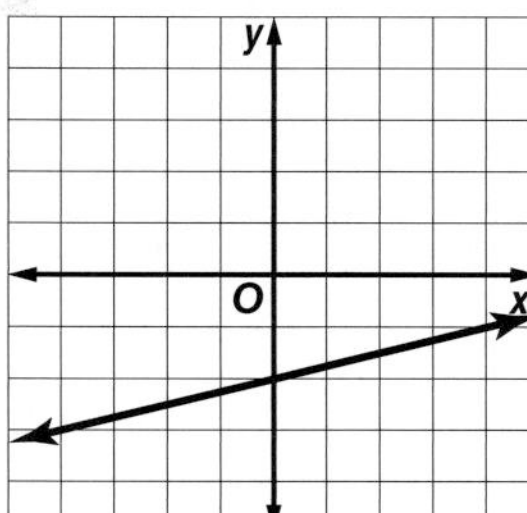

A $y = 4x - 2$
B $y = \frac{1}{4}x - 2$
C $y = \frac{1}{3}x - 2$
D $y = \frac{1}{4}x + 2$

51. SHORT RESPONSE Jan is 320 miles from home and is driving home at a speed of 65 mph. Write an equation to determine her distance from home at any point during her trip.

Spiral Review

Graph each equation using the slope and y-intercept. (Lesson 8-7)

52. $y = \frac{3}{4}x + 2$ **53.** $x + y = -3$ **54.** $x + y = 0$

Find the slope of each line. (Lesson 8-6)

55.

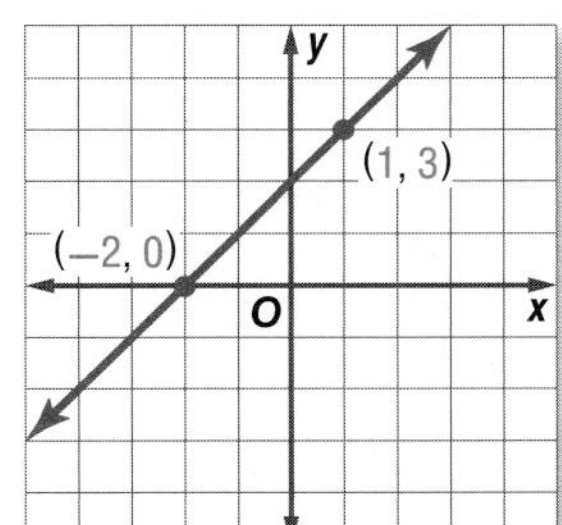

56.

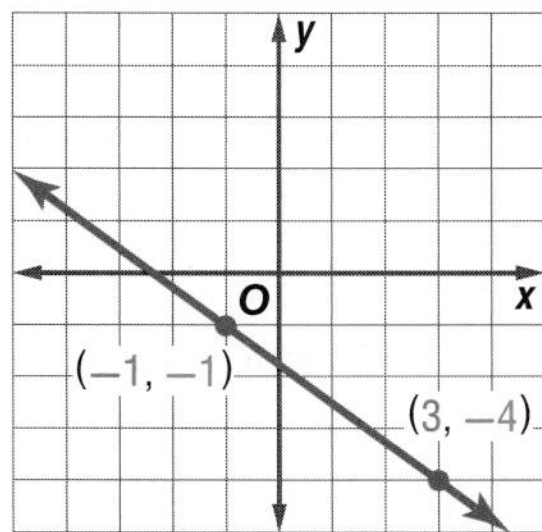

57. BIRDS If 12 of the 75 animals in a pet store are parakeets, what percent are parakeets? (Lesson 7-3)

Skills Review

Determine whether a scatter plot of the data for the following might show a *positive*, *negative*, or *no* relationship. Explain your answer. (Lesson 1-6)

58. size of household and amount of water bill

59. temperature and heating costs

60. speed and distance traveled

8-9 Prediction Equations

Then

You have already analyzed trends in data displayed in scatter plots. (Lesson 1-6)

Now

- Draw lines of fit for sets of data.
- Use lines of fit to make predictions about data.

New Vocabulary

line of fit

Math Online

glencoe.com

- Extra Examples
- Personal Tutor
- Self-Check Quiz
- Homework Help

Why?

The scatter plot shows the number of NCAA Women's softball teams.

a. Use the two points that are labeled to find the slope of the line drawn. Describe what the slope means.

b. Looking at the line, what would you expect the number would be in 2006? Based on the graph, is the actual number greater or less than your prediction?

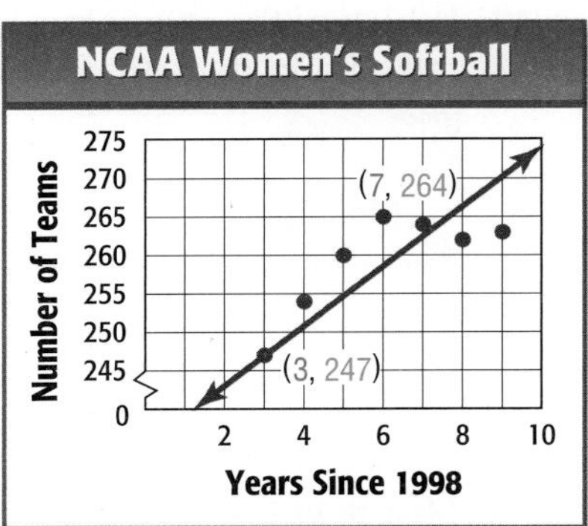

Source: National Collegiate Athletic Association

Lines of Fit When real-world data are collected, the points graphed usually do not form a straight line, but may approximate a linear relationship. A **line of fit** is a line that is very close to most of the data points. The line drawn in the graph shown above is a line of fit.

Real-World EXAMPLE 1 Make a Prediction from a Line of Fit

TESTS The table shows the number of college-bound students that took the ACT in different years.

Students Taking ACT				
Year	2000	2002	2004	2006
Number (thousands)	1065	1116	1171	1206

Source: ACT, Inc.

a. Make a scatter plot and draw a line of fit for the data.

Graph each of the data points. Draw a line that fits most of the data.

b. Use the line of fit to predict the number of students who will take the ACT in 2015.

Extend the line so that you can estimate the y-value for an x-value of 2015. The y-value for 2015 is about 1480 (thousand). So, we can predict that about 1480 (thousand) students will take the ACT in 2015.

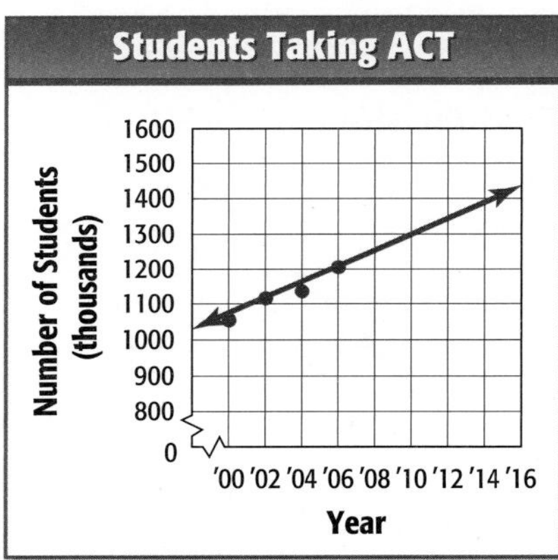

Check Your Progress

1. ENTERTAINMENT The table shows the percent of U.S. households that have a digital video camera.

Year	1999	2000	2001	2002	2003	2004	2005	2006
Percent	3	7	10	14	17	19	21	23

A. Make a scatter plot of the data and draw a line of fit.

B. Predict the percent of households with a digital video camera in 2015.

Personal Tutor glencoe.com

StudyTip

Estimation Drawing a line of fit using the method in this lesson is an estimation. Therefore, it is possible to draw different lines to approximate the same data.

Prediction Equations Predictions can be made from the equation of a line of fit.

Real-World EXAMPLE 2 Make Predictions from an Equation

MOVIES The scatter plot shows the number of movie theater screens in the U.S. for several years following 1999.

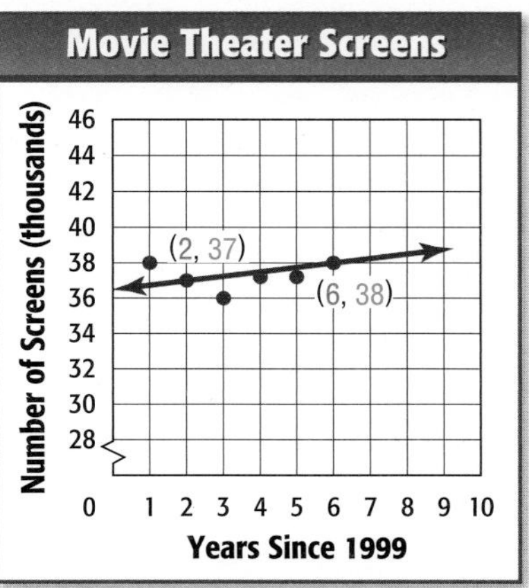

a. Write an equation in slope-intercept form for the line of fit that is drawn.

Step 1 Use two points on the line to find the slope. These may or many not be original data points.

$m = \frac{y_2 - y_1}{x_2 - x_1}$ **Definition of slope**

$m = \frac{38 - 37}{6 - 2}$ **Use $(x_1, y_1) = (2, 37)$ and $(x_2, y_2) = (6, 38)$.**

$m = 0.25$ **Simplify.**

Step 2 Use the slope and the coordinates of either point to write the equation of the line in point-slope form.

$y - y_1 = m(x - x_1)$ **Point-slope form**

$y - 38 = 0.25(x - 6)$ **Replace (x_1, y_1) with (6, 38) and m with 0.25.**

Step 3 Solve the point-slope equation for y.

$y - 38 = 0.25(x - 6)$ **Point-slope equation**

$y - 38 = 0.25x - 1.5$ **Distributive Property**

$y - 38 + 38 = 0.25x - 1.5 + 38$ **Add 38 to each side.**

$y = 0.25x + 36.5$ **Simplify.**

The equation for the line of fit is $y = 0.25x + 36.5$.

b. Predict the number of movie theater screens in 2013.

$y = 0.25x + 36.5$ **Write the equation of the line of fit.**

$y = 0.25(14) + 36.5$ **Since 2013 − 1999 = 14, replace x with 14.**

$y = 40$ **Simplify.**

There should be about 40,000 movie theater screens in 2013.

Real-World Link

In 1987, there were 2084 drive-in, or outdoor, movie screens in the U.S. By 2005, there were only 648.

Source: Motion Picture Association of America

Check Your Progress

2. SWIMMING The scatter plot shows the winning Olympic times in the men's 100-meter butterfly for several years following 1964.

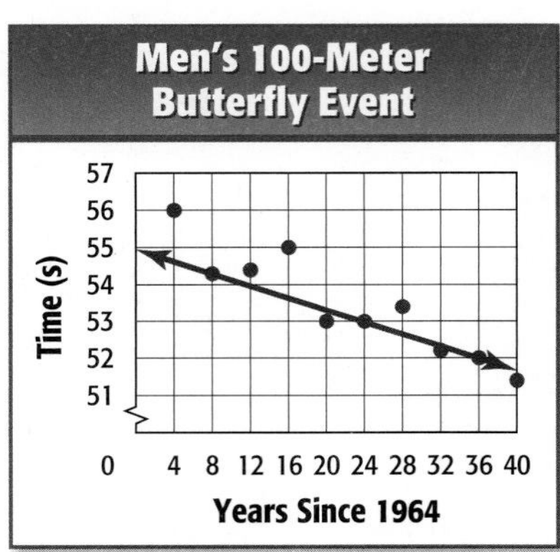

A. Write an equation in slope-intercept form for the line of fit that is drawn.

B. Predict the winning time in the men's 100-meter butterfly in 2012.

Personal Tutor glencoe.com

Check Your Understanding

Example 1
p. 448

1. **NEWSPAPERS** The table shows the number of Sunday newspapers in the U.S.

   a. Make a scatter plot of the data and draw a line of fit.

   b. Use the line of fit to predict the number of Sunday newspapers in the U.S. in 2014.

Year	Number of Sunday Newspapers
2001	913
2002	913
2003	917
2004	915
2005	917
2006	907

Example 2
p. 449

2. **ONLINE SHOPPING** Use the line of fit drawn that shows the amount of book sales that were purchased online for several years following 1997.

   a. Write an equation in slope-intercept form for the line of fit.

   b. Use the equation to predict the sales of books online in 2013.

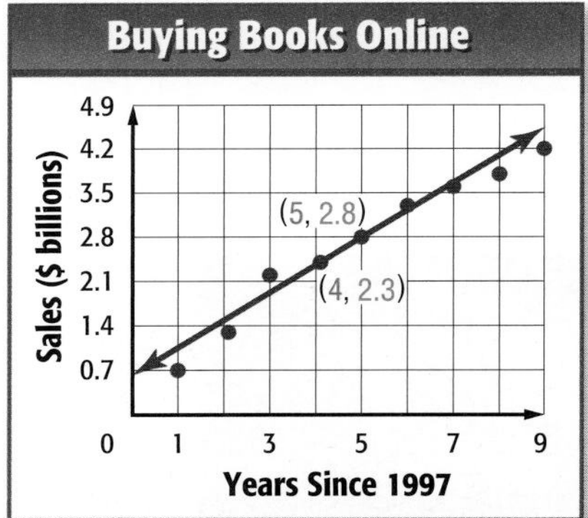

Practice and Problem Solving

 = **Step-by-Step Solutions** begin on page R11.
Extra Practice begins on page 810.

Example 1
p. 448

 **SPORTS** The table shows the sales of athletic equipment.

   a. Make a scatter plot of the baseball and softball sales and draw a line of fit.

   b. Use the line of fit to predict the sales of baseball and softball equipment in 2015.

   c. Make a scatter plot of the tennis sales and draw a line of fit.

   d. Use the line of fit to predict the sales of tennis equipment in 2015.

Year	Equipment Sales ($ millions)	
	Baseball and Softball	Tennis
1999	329	338
2000	319	383
2001	316	371
2002	334	358
2003	340	343
2004	346	362
2005	356	373

Example 2
p. 449

4. **FAN CLUBS** Use the line of fit drawn that shows the number of fan clubs in the U.S. for several years following 1999.

   a. Write an equation in slope-intercept form for the line of fit.

   b. Use the equation to predict the number of fan clubs in the U.S. in 2012.

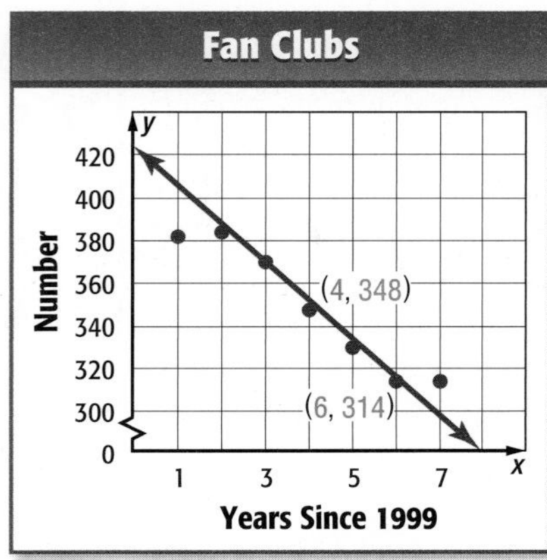

5. **SPACE** Use the line of fit drawn that shows the amount of money the government spent on space and other technology for several years following 1998.

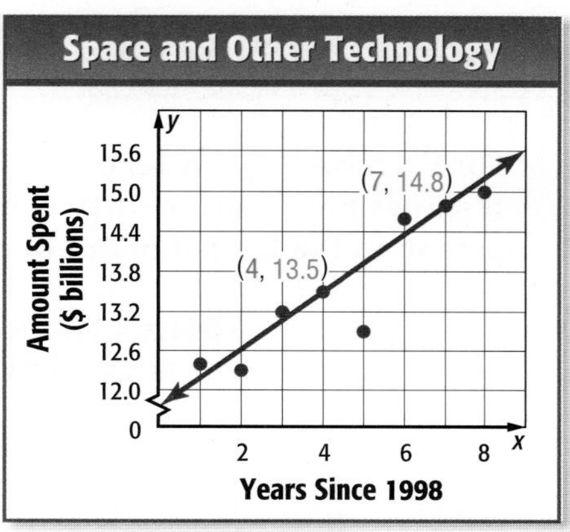

a. Write an equation in slope-intercept form for the line of fit. Round to the nearest tenth.

b. Use the equation to predict the amount that the government will spend on space and other technology in 2016.

StudyTip

A line of fit can be a good predictor if the relationship between the two variables is a strong positive or strong negative relationship.

6. **LIFE EXPECTANCY** According to the U.S. Census, the life expectancy of someone born in 2000 is 77.0 years. The life expectancy of someone born in 2010 is 78.5 years.

a. Write an equation in slope-intercept form for the line of fit of the data.

b. Use the equation to predict the life expectancy of a person born in 2015.

c. What are some limitations in using a line to predict life expectancy?

7. **MULTIPLE REPRESENTATIONS** In this problem, you will compare two graphs. Use the table that shows the populations of Illinois and Pennsylvania for several years following 1999.

Years Since 1999	Population (millions)	
	Illinois	Pennsylvania
1	12.4	12.3
2	12.5	12.3
3	12.6	12.3
4	12.7	12.4
5	12.7	12.4
6	12.8	12.4

a. **GRAPHICAL** Make a scatter plot and draw a line of fit for each set of data. What does the slope of each line represent?

b. **VERBAL** Which state's population appears to be growing at a faster rate? Explain. If the two lines were to intersect, what does the point of intersection represent?

c. **ALGEBRAIC** Write an equation for each line of fit. Round to the nearest thousandth. Use the equations to verify your answer in part **b**.

d. **NUMERICAL** Estimate the population of both states in 2015.

H.O.T. Problems

Use Higher-Order Thinking Skills

8. **OPEN ENDED** Find a set of real-world data in which the data change over a span of at least 5 years. Make a scatter plot of the data and draw a line of fit. Write an equation for the line of fit that you drew.

9. **WRITING IN MATH** Explain when it is reasonable to make a prediction from a line of fit. Then explain when it is not reasonable.

10. **CHALLENGE** Refer to part **c** of Exercise 7. Suppose you use (Years Since 1998, Population) to write the equations. Is the slope or y-intercept of the graphs of the equations the same? Explain.

11. **REASONING** Describe a scatter plot in which it would not be useful to draw a line of fit.

12. **WRITING IN MATH** Explain how a line can be used to make predictions. Include a description of a line of fit and an explanation of how a line can represent a set of data that is not exactly linear.

Standardized Test Practice

For Exercises 13 and 14, use the scatter plot. It shows how much money is in circulation per person in the United States in different years.

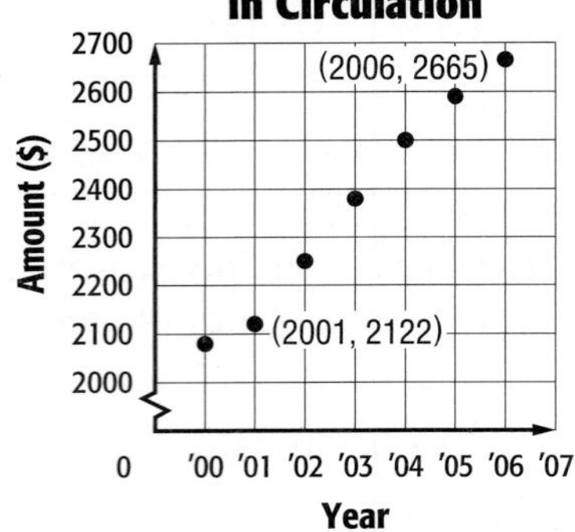

13. Which statement *best* describes the relationship shown on the scatter plot?

A There is no relationship between the data.
B The amount of money in circulation per person increased over time.
C The amount of money in circulation per person decreased over time.
D The amount of money in circulation per person stayed the same over time.

14. Which of the following is the best prediction for the amount of money in circulation per person in 2020?

F \$3605
G \$4050
H \$4185
J \$5240

15. Which of the following shows the equation for a line of fit that contains the data points (4, 534) and (10, 138)?

A $y = -66x + 798$
B $y = -66x - 798$
C $y = 66x + 798$
D $y = 66x - 798$

16. Given the equation $y = 24x + 55$ for a line of fit, which of the following sets of points could be data points?

F (−3, −127) and (2, 103)
G (−3, 127) and (−2, 103)
H (3, 127) and (2, 103)
J (3, 127) and (−2, 103)

17. EXTENDED RESPONSE The table shows the heat index at different humidity levels for a temperature of 90°F.

Heat Index at 90°F					
Humidity (%)	0	10	30	50	70
Heat Index (°F)	83	85	90	96	106

a. Make a scatter plot of the data.
b. Write an equation in slope-intercept form for a line of fit.
c. Predict the heat index when the humidity is at 100%.

Spiral Review

Write an equation in slope-intercept form for each line. (Lesson 8-8)

18. slope $= -\frac{1}{3}$, y-intercept $= 8$

19. slope $= \frac{2}{5}$, y-intercept $= 0$

Graph each equation using the slope and y-intercept. (Lesson 8-7)

20. $y = x + 5$

21. $y = -x + 6$

22. $y = 2x - 3$

Skills Review

Evaluate each expression if $x = 7$, $y = 3$, and $z = 9$. (Lesson 1-2)

23. $\frac{xy}{3} + 2$

24. $2x + 3z + 5y$

25. $5z - 3x - 2y$

8-10 Systems of Equations

Why?

Hannah and Luis each open a savings account. They make different initial deposits and weekly deposits, as shown in the table.

Savings Accounts

Person	Initial Deposit	Weekly Deposit
Hannah	\$0	\$5
Luis	\$30	\$2

a. Write an equation to represent the amount of money in each person's account. Let y = the amount of money in an account. Let x = the number of weeks.

b. Make a table of values that satisfies each equation. Then graph both equations on the same coordinate plane.

c. What are the coordinates of the point where the two lines meet? What does this point represent?

Then
You have already solved linear equations by graphing.
(Lesson 8-2)

Now
- Solve systems of linear equations by graphing.
- Solve systems of linear equations by substitution.

New Vocabulary
system of equations
substitution

Math Online
glencoe.com
- Extra Examples
- Personal Tutor
- Self-Check Quiz
- Homework Help

Solve Systems by Graphing A **system of equations** is a collection of two or more equations with the same set of variables. The equations $y = 5x$ and $y = 30 + 2x$ together are a system of equations. The solution of this system is (10, 50) because the ordered pair is a solution of both equations.

$y = 5x$	**Write the equations.**	$y = 30 + 2x$
$50 \stackrel{?}{=} 5(10)$	**Replace (x, y) with (10, 50).**	$50 \stackrel{?}{=} 30 + 2(10)$
$50 = 50$ ✓	**Simplify.**	$50 \stackrel{?}{=} 30 + 20$
		$50 = 50$ ✓

One way to solve a system of equations is to graph the equations on the same coordinate plane. The coordinates of the point where the graphs intersect is the solution of the system of equations.

EXAMPLE 1 Solve by Graphing

Solve the system of equations by graphing.

$y = x$
$y = -3x + 4$

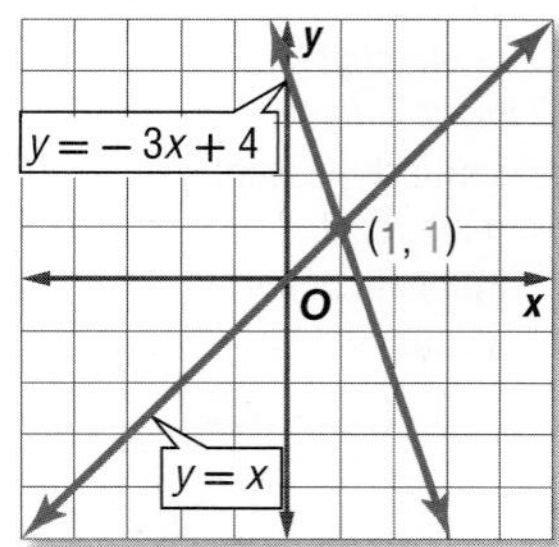

The graphs appear to intersect at (1, 1). Check this estimate by replacing x with 1 and y with 1.

Check

$y = x$	$y = -3x + 4$
$1 \stackrel{?}{=} 1$	$1 \stackrel{?}{=} -3(1) + 4$
$1 = 1$ ✓	$1 = 1$ ✓

The solution of the system of equations is (1, 1).

✓ Check Your Progress

1. Solve the system of equations by graphing.
$y = x + 2$
$y = -2x - 4$

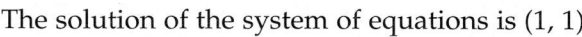

Personal Tutor glencoe.com

Test-Taking Tip

When answering extended response questions, make sure your solutions answer all parts of the question. Double check that you covered all parts.

STANDARDIZED TEST EXAMPLE 2

EXTENDED RESPONSE **Marjorie and Bryan are selling magazine subscriptions. Marjorie sells 3 times as many subscriptions as Bryan. Bryan sells 12 fewer subscriptions than Marjorie.**

a. Write a system of equations to represent this situation.

Let y represent Marjorie's sales and x represent Bryan's sales.

$y = 3x$ **Marjorie sells 3 times as many subscriptions as Bryan.**

$y = x + 12$ **Bryan sells 12 fewer subscriptions than Marjorie.**

b. Solve the system by graphing. Explain what the solution means.

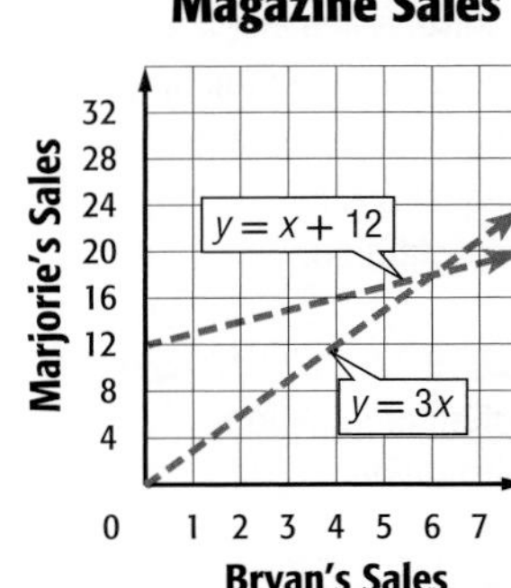

Graph each equation on the same coordinate grid. The equations intersect at (6, 18).

So, the solution to the system is $x = 6$ and $y = 18$. This means that Marjorie sells 18 subscriptions and Bryan sells 6 subscriptions.

Check Your Progress

2. A doctor's office has twice as many new patients as existing patients. The number of new patients is 22 more than the number of existing patients.

A. Write a system of equations to represent this situation.

B. Solve the system by graphing. Explain what the solution means.

Personal Tutor glencoe.com

A system of equations can have no solution or infinitely many solutions.

Study Tip

Slopes/Intercepts When the graphs of a system of equations have:
- different slopes → exactly 1 solution
- same slope, different y-intercepts → no solution
- same slope, same y-intercept → infinitely many solutions

EXAMPLE 3 No Solution and Infinitely Many Solutions

Solve each system of equations by graphing.

a. $y = -x + 1$
$y = -x - 3$

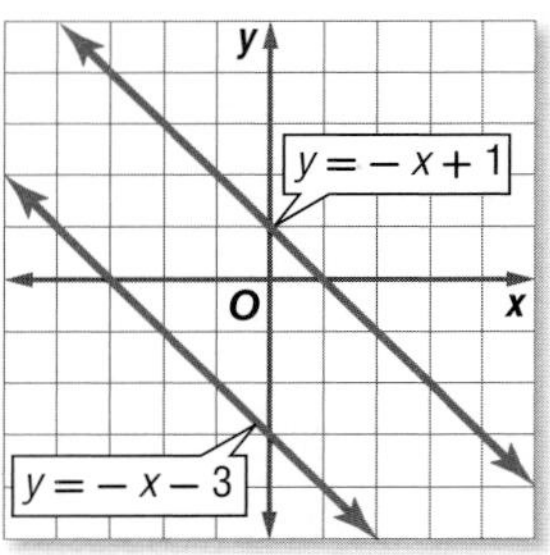

The graphs appear to be parallel lines. Since there is no coordinate pair that is a solution to both equations, there is no solution of this system of equations.

b. $y = 2x + 4$

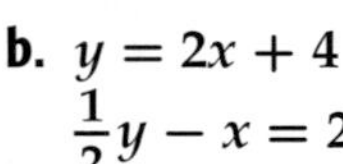
$\frac{1}{2}y - x = 2$

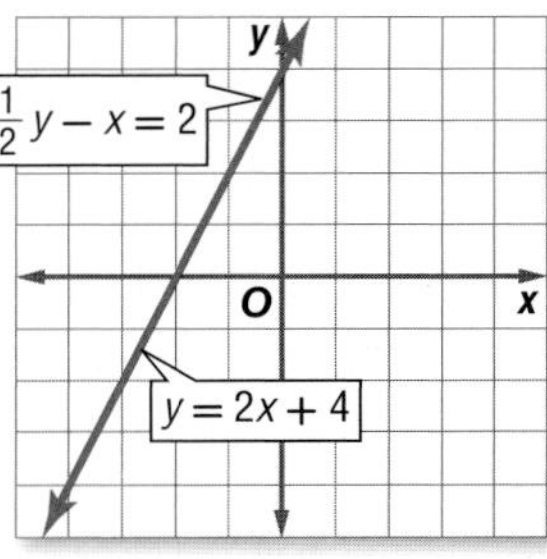

Both equations have the same graph. Any ordered pair on the graph will satisfy both equations. Therefore, there are infinitely many solutions of this system of equations.

Check Your Progress

3A. $y = x + 4$
$y = x$

3B. $y = \frac{1}{2}x - 1$
$x - 2y = 2$

Personal Tutor glencoe.com

Solve Systems by Substitution You can also use algebraic methods to solve a system of equations. One method is called **substitution**.

EXAMPLE 4 Solve by Substitution

> **StudyTip**
> **Systems** The variable y must have the same value in both equations.

Solve the system of equations by substitution.

$y = x + 2$
$y = 5$

Replace y with 5 in the first equation.

$y = x + 2$	**Write the first equation.**
$5 = x + 2$	**Replace y with 5.**
$3 = x$	**Solve for x.**

The solution of this system of equations is (3, 5). You can check the solution by graphing. The graphs appear to intersect at (3, 5).

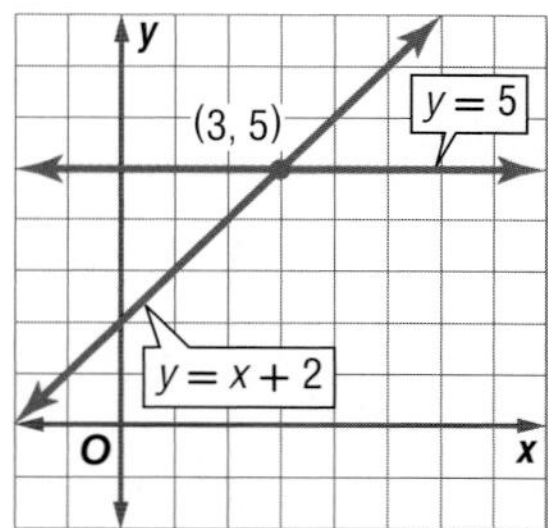

Check Your Progress

4A. $y = 5$
$y = x + 4$

4B. $y = 7 - x$
$x = 3$

Personal Tutor glencoe.com

Concept Summary — Systems of Equations

For Your FOLDABLE

Graph	Intersecting Lines	Parallel Lines	Same Line
Number of Solutions	one solution	no solutions	infinitely many

Check Your Understanding

Examples 1 and 3 pp. 453–454

Solve each system of equations by graphing.

1. $y = -x$
$y = x - 4$

2. $y = x + 1$
$x + y = 7$

3. $y = \frac{3}{2}x - 1$
$3x - 2y = 2$

Example 2 p. 454

4. EXTENDED RESPONSE Amanda pays an annual fee of \$100 to belong to a gym, plus a monthly fee of \$10. Maria pays only a monthly fee of \$20.

a. Write a system of equations to represent this situation.

b. Solve the system by graphing. Explain what the solution means.

Example 4 p. 455

Solve each system of equations by substitution.

5. $y = x - 2$
$y = 3$

6. $y = x + 4$
$x = 0$

7 $y = 2x + 3$
$y = 1$

Practice and Problem Solving

● = **Step-by-Step Solutions** begin on page R11.
Extra Practice begins on page 810.

Examples 1 and 3 pp. 453–454

Solve each system of equations by graphing.

8. $y = 2x$
$y = x + 1$

9. $y = x$
$y = -x + 4$

10. $y = -x + 1$
$y = x - 5$

11. $y = \frac{3}{4}x$
$3x - 4y = 0$

12. $y = \frac{1}{2}x + 1$
$y = \frac{1}{2}x - 2$

13. $x + y = -3$
$x - y = -3$

Example 2 p. 454

14. **NUMBER SENSE** The sum of two numbers is 5, and the difference of the numbers is 3.

a. Write a system of equations to represent this situation.

b. Solve the system of equations by graphing. Explain what the solution means.

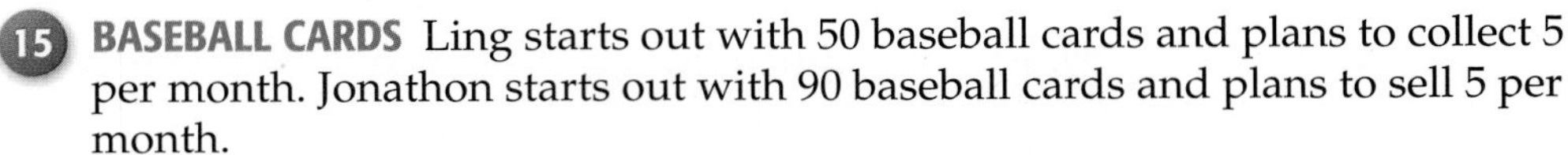

15 **BASEBALL CARDS** Ling starts out with 50 baseball cards and plans to collect 5 per month. Jonathon starts out with 90 baseball cards and plans to sell 5 per month.

a. Write a system of equations to represent this situation.

b. Solve the system of equations by graphing. Explain what the solution means.

Example 4 p. 455

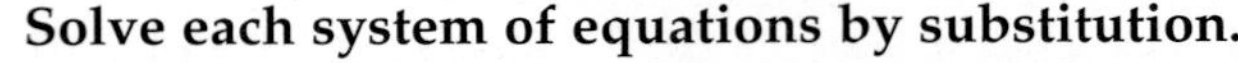

Solve each system of equations by substitution.

16. $y = x + 2$
$x = 1$

17. $y = x + 4$
$y = 0$

18. $y = 2x - 3$
$y = 5$

19. $y = -x - 4$
$x = 2$

20. $x + y = 2$
$x = -3$

21. $x - y = 6$
$y = -1$

22. **FINANCIAL LITERACY** The cost of 2 bagels and 2 cans of orange juice is \$4.40. The cost of 3 bagels and 4 cans of orange juice is \$7.80.

a. Write a system of equations to represent this situation.

b. Solve the system of equations by substitution. Explain what the solution means.

Real-World Link

Three of the rarest and most valuable baseball cards are from 1951. They are priced between \$30,000 and \$35,000 each.

Source: USA Today

H.O.T. Problems

Use **H**igher-**O**rder **T**hinking Skills

23. **OPEN ENDED** Write a system of equations that has the solution (1, 7). Write the equations in slope-intercept form.

24. **WRITING IN MATH** Describe the three ways that two lines can be related. Can a system of linear equations have exactly two solutions? Explain.

CHALLENGE **Solve each system of equations by substitution.**

25. $y = 2x + 6$
$y = x$

26. $x + 4y = 33$
$y = -3x$

27. $5x + y = 8$
$y = x - 4$

28. **REASONING** Describe when it is better to use substitution to solve a system of equations rather than graphing.

29. **WRITING IN MATH** Describe the graph of a system of equations if the system has 1 solution, no solution, or infinitely many solutions.

Standardized Test Practice

30. Amy took three times as many pictures as Jennifer. Jennifer has 16 fewer pictures than Amy. Which system of equations can be used to find the number of pictures each person took?

A $a = 3j$
$a = j + 16$

B $a = 3j$
$a = j - 16$

C $j = 3a$
$j = a + 16$

D $j = 3a$
$j = a - 16$

31. Refer to Exercise 30. How many pictures did each person take?

F Amy took 8 pictures and Jennifer took 24 pictures.

G Amy took 24 pictures and Jennifer took 8 pictures.

H Amy took 16 pictures and Jennifer took 6 pictures.

J Amy took 6 pictures and Jennifer took 16 pictures.

32. Which of the following is the solution of the system of equations graphed below?

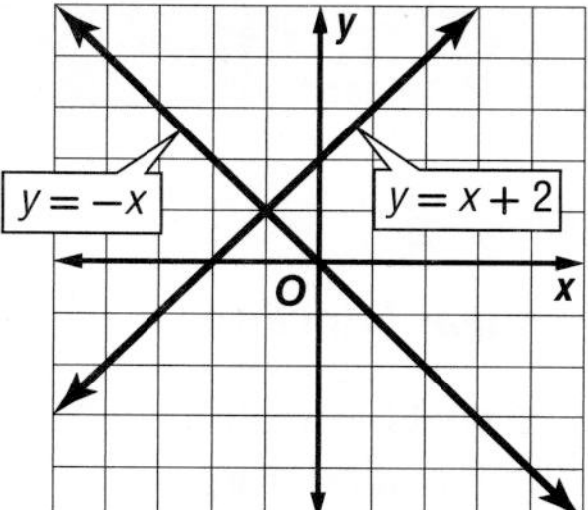

A (1, 1)

B (1, −1)

C (−1, 1)

D (−1, −1)

33. GRIDDED RESPONSE In one season, Tyra made 4 times as many goals as Kailey. Kailey made 9 fewer goals than Tyra. How many goals did Tyra make?

Spiral Review

34. EARTH SCIENCE The table shows the latitude and the average temperature in July for various U.S. cities. (Lesson 8-9)

a. Make a scatter plot of the data and draw a line of fit.

b. Write an equation for the line of fit you drew in part **a**.

c. Use the equation from part **b** to estimate the average July high temperature for a location with latitude 50° north. Round to the nearest degree Fahrenheit.

City	Latitude (°N)	Average July High Temperature (°F)
Chicago, IL	41	73
Dallas, TX	32	85
Denver, CO	39	74
New York, NY	40	77
Duluth, MN	46	66

Write an equation in slope-intercept form for each table of values. (Lesson 8-8)

35.

x	−1	0	1	2
y	−7	−3	1	5

36.

x	−3	−1	1	3
y	7	5	3	1

Skills Review

Multiply. (Previous Course)

37. $4 \cdot 4 \cdot 4$

38. $9 \cdot 9 \cdot 9 \cdot 9$

39. $2 \cdot 2 \cdot 2 \cdot 2 \cdot 2$

40. $5 \cdot 5 \cdot 5$

Study Guide and Review

Math Online glencoe.com
- STUDY TO GO
- Vocabulary Review

Chapter Summary

Key Concepts

Representing Linear Functions (Lesson 8-3)

- A solution of a linear equation is an ordered pair that makes the equation true.
- A linear equation can be represented by a set of ordered pairs, a table of values, or a graph.

Rate of Change and Slope (Lessons 8-4 through 8-6)

- A change in one quantity in relation to another quantity is called the rate of change.
- A quantity that increases over time has a positive rate of change. If it decreases over time, it has a negative rate of change. If it does not change over time, it has a zero rate of change.
- Two quantities a and b have a proportional linear relationship if $\frac{a}{b}$ is constant and $\frac{\text{change in } a}{\text{change in } b}$ is constant.
- Slope can be used to describe rate of change.
- Slope is the ratio of the rise, or the vertical change, to the run, or the horizontal change.

Write and Use Equations (Lessons 8-7 through 8-9)

- In the slope-intercept form $y = mx + b$, m is the slope and b is the y-intercept.
- You can write a linear equation by using the slope and y-intercept, two points on a line, a graph, a table, or a verbal description.

Systems of Equations (Lesson 8-10)

- The solution of a system of equations is the ordered pair that satisfies all equations in the system.

FOLDABLES® Study Organizer

Be sure the Key Concepts are noted in your Foldable.

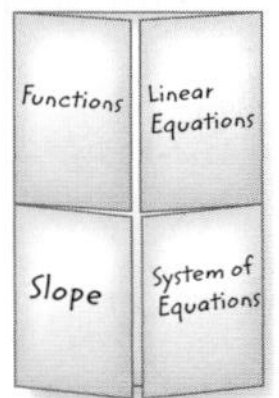

Key Vocabulary

arithmetic sequence (p. 401)
common difference (p. 401)
constant of variation (p. 420)
constant rate of change (p. 418)
dependent variable (p. 395)
direct variation (p. 420)
discrete data (p. 407)
family of functions (p. 439)
function notation (p. 396)
independent variable (p. 395)
line of fit (p. 448)
linear equation (p. 406)
linear relationship (p. 418)
point-slope form (p. 442)
rate of change (p. 412)
sequence (p. 401)
slope (p. 427)
slope-intercept form (p. 433)
substitution (p. 455)
system of equations (p. 453)
term (p. 401)
vertical line test (p. 396)
***x*-intercept** (p. 407)
***y*-intercept** (p. 407)

Vocabulary Check

Choose the term from the list above that best matches each phrase.

1. the ratio of the vertical change to the horizontal change of a line
2. an equation written in the form $y = mx + b$
3. an ordered list of numbers
4. a description of how one quantity changes in relation to another quantity
5. b in the equation $y = mx + b$
6. the rate of change between any two data points is the same
7. in a linear equation, a variable for the input
8. k in the equation $y = kx$
9. a set of equations with the same variables
10. a line that is close to most of the data points in a scatter plot

Lesson-by-Lesson Review

8-1 Functions (pp. 395–400)

11. Determine whether the relation {(5, 3), (−5, 4), (4, 2), (4, 1)} is a function. Explain.

12. **GASOLINE** Use the table that shows the cost of gas in different years. Is the relation a function? Explain.

Year	'02	'04	'06
Cost ($)	1.36	1.82	2.26

EXAMPLE 1

Determine whether the relation shown in the table is a function. Explain.

x	9	11	13	17	21
y	7	3	−1	−5	−7

Yes, it is a function since each domain value is paired with only one range value.

8-2 Sequences and Equations (pp. 401–405)

13. Describe the sequence 6, 12, 18, 24, … using words and symbols.

Write an equation that describes each sequence. Then find the indicated term.

14. 6, 10, 14, 18, …; 47th term

15. 7, 14, 21, 28, …; 50th term

16. **GEOMETRY** Which figure in the pattern below will have 99 squares?

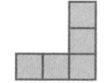

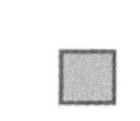
Figure 1

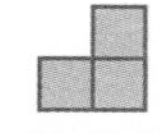
Figure 2

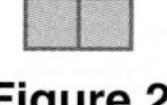
Figure 3

EXAMPLE 2

Write an equation that describes the sequence 9, 18, 27, 36, … . Then find the 16th term of the sequence.

Term Number (n)	1	2	3	4
Term (t)	9	18	27	36

The common difference is 9. Each term is 9 times the term number. So, $t = 9n$.

$t = 9n$ **Write the equation.**

$t = 9(16)$ or 144 **Replace n with 16.**

The 16th term of the sequence is 144.

8-3 Representing Linear Functions (pp. 406–411)

Find four solutions of each equation. Write the solution as ordered pairs.

17. $y = -5x$

18. $y = 4x$

19. $y = x + 9$

20. $x + y = -1$

Graph each equation.

21. $y = -2x$

22. $y = x + 5$

23. **SNACKS** Each small smoothie x costs $1.50, and each large smoothie y costs $3. Find two solutions of $1.5x + 3y = 12$ to determine how many of each type Lisa can buy with $12.

EXAMPLE 3

Find four solutions of $y = -x + 1$. Write the solutions as ordered pairs.

Choose four values for x and substitute each value into the equation. Then solve for y.

x	$y = -x + 1$	y	(x, y)
−1	$y = -(-1) + 1$	2	(−1, 2)
0	$y = -0 + 1$	1	(0, 1)
1	$y = -1 + 1$	0	(1, 0)
2	$y = -2 + 1$	−1	(2, −1)

Four solutions: (−1, 2), (0, 1), (1, 0), and (2, −1).

CHAPTER 8 Study Guide and Review

8-4 Rate of Change (pp. 412–417)

24. Find the rate of change for the linear function shown below.

Time (h) x	0	4	8
Money Earned ($) y	0	31	62

25. **ENTERTAINMENT** The table shows the total cost of tickets. Compare the rates of change.

Number of People x	Total Cost ($) y	
	Adults	Children
2	36	25
4	72	50
6	108	75

EXAMPLE 4

The table shows the time and water level of a pool. Find the rate of change.

Time (min) x	0	4	8
Water Level (ft) y	5	4	3

$$\text{rate of change} = \frac{\text{change in water level}}{\text{change in time}}$$
$$= \frac{5 \text{ ft} - 4 \text{ ft}}{0 \text{ min} - 4 \text{ min}}$$
$$= \frac{1 \text{ ft}}{-4 \text{ min}} \text{ or } -\frac{1}{4} \text{ ft/min}$$

The rate of change is $-\frac{1}{4}$ ft/min.

8-5 Constant Rate of Change and Direct Variation (pp. 418–424)

26. **WEATHER** The temperature one day is shown in the graph. Find the constant rate of change and interpret its meaning.

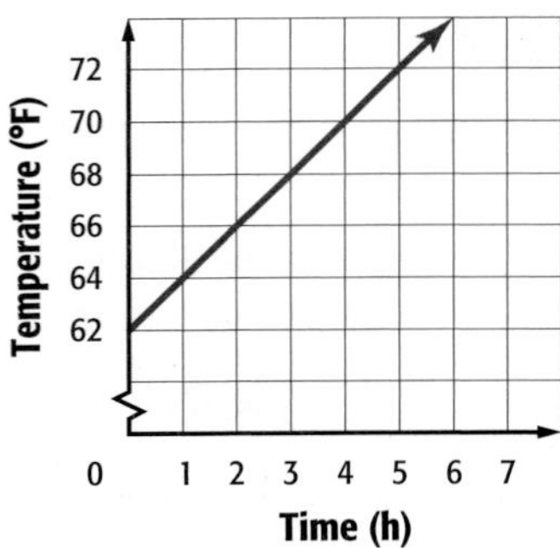

EXAMPLE 5

Find the constant rate of change for the linear function shown at the right and interpret its meaning.

Year	Population (1000s)
x	y
2001	688
2005	722

$$\text{rate of change} = \frac{\text{change in population}}{\text{change in years}}$$
$$= \frac{722 - 688}{2005 - 2001}$$
$$= \frac{34}{4} \text{ or } 8.5$$

The rate of change 8.5 means that the population increased at a rate of 8.5 thousand people per year.

8-6 Slope (pp. 427–431)

Find the slope of the line that passes through each pair of points.

27. $F(0, 1), G(6, 4)$

28. $R(-8, -2), S(4, 9)$

29. $A(-3, 7), G(5, -1)$

30. $P(6, -4), S(-1, 10)$

31. **ANIMALS** A lizard is crawling up a hill that rises 5 feet for every horizontal change of 30 feet. Find the slope.

EXAMPLE 6

Find the slope of the line that passes through $C(6, 1)$ and $D(0, -3)$.

$m = \frac{y_2 - y_1}{x_2 - x_1}$ — **Definition of slope**

$m = \frac{-3 - 1}{0 - 6}$ — **$(x_1, y_1) = (6, 1)$, $(x_2, y_2) = (0, -3)$**

$m = \frac{-4}{-6}$ or $\frac{2}{3}$ — **Simplify.**

MIXED PROBLEM SOLVING
For mixed problem-solving practice, see page 850.

8-7 Slope-Intercept Form (pp. 433–438)

State the slope and the y-intercept of the graph of each equation.

32. $y = 4x + 7$

33. $y = -\frac{4}{3}x$

34. $5x + y = 0$

35. $-x + y = -8$

Graph each equation using the slope and y-intercept.

36. $y = -x + 4$

37. $y = 2x - 6$

38. $y = \frac{3}{2}x - 3$

39. $y = -\frac{1}{4}x + 5$

40. BALLOONS A balloon is rising above the ground. The height in feet y of the balloon can be given by $y = 7 + 2x$, where x represents the time in seconds. State the slope and y-intercept of the graph of the equation. Describe what they represent.

EXAMPLE 7

State the slope and y-intercept of the graph of $y = 4x - 1$. Then graph the equation.

$y = 4x - 1$	**Write the original equation.**
$y = 4x + (-1)$	**Write the equation in the form $y = mx + b$.**
$y = mx + b$	**$m = 4, b = -1$**

The slope of the graph is 4, and the y-intercept is -1.

Now graph the equation. Write the slope as $\frac{4}{1}$. Plot the point at $(0, -1)$. Then go up 4 and right 1. Connect the points and extend the line.

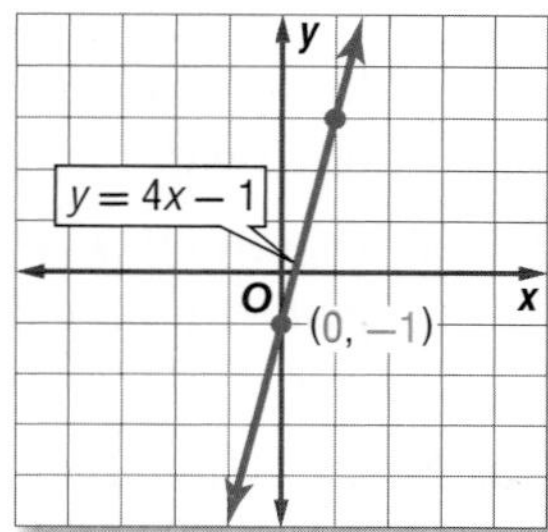

8-8 Writing Linear Equations (pp. 441–447)

Write an equation in slope-intercept form for each line.

41. slope $= -2$, y-intercept $= 5$

42. slope $= \frac{3}{4}$, y-intercept $= -1$

43. slope $= 4$, y-intercept $= 0$

Write an equation in point-slope form for the line passing through each pair of points.

44. $(3, 3), (7, -1)$

45. $(1, 5), (2, 8)$

46. BIRTHDAYS Diem's parents wants to rent the local movie theatre for her birthday party. It costs \$100 plus \$30 per hour to rent the movie theater.

a. Write an equation in slope-intercept form that shows the cost y for renting the theater for x hours.

b. Find the cost of renting the theater for 4 hours.

EXAMPLE 8

Write an equation in point-slope form for the line that passes through (5, 9) and (2, 0).

Step 1 Find the slope m.

$m = \frac{y_2 - y_1}{x_2 - x_1}$	**Definition of slope**
$m = \frac{9 - 0}{5 - 2}$	**$(x_1, y_1) = (5, 9)$, $(x_2, y_2) = (2, 0)$**
$m = 3$	**Simplify.**

Step 2 Use the slope and the coordinates of either point.

$y - y_1 = m(x - x_1)$	**Point-slope form**
$y - 9 = 3(x - 5)$	**Replace (x, y) with (5, 9) and m with 3.**

An equation of the line through (5, 9) and (2, 0) is $y - 9 = 3(x - 5)$.

CHAPTER 8

Study Guide and Review

8-9 Prediction Equations (pp. 448–452)

47. HOUSING The table shows the changes in the median price of existing homes.

Year	Median Price ($1000s)
2000	139.0
2002	158.1
2003	170.0
2004	195.2
2005	219.0
2006	221.9

a. Make a scatter plot and draw a line of fit for the data.

b. Use the line of fit to predict the median price for an existing home for the year 2015.

48. MUSIC The table shows the changes in the average concert ticket prices.

Year	'03	'04	'05	'06	'07	'08
Ticket Cost ($)	45	48	56	60	65	80

a. Make a scatter plot and draw a line of fit for the data.

b. Use the line of fit to predict the average price of a concert ticket in 2015.

EXAMPLE 9

Make a scatter plot and draw a line of fit for the table showing the attendance at home games for the first four games of a high school football season. Then use the line of fit to predict the attendance for the seventh home game.

Game	Attendance
1	1100
2	1200
3	1300
4	1500

Plot the points and draw a line as close to the points as possible. For an x value of 7, the y value is about 19. So, a prediction for the attendance is approximately 1900 people.

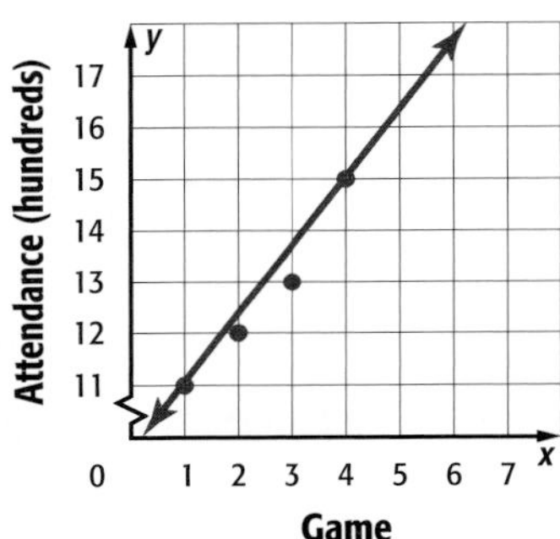

8-10 Systems of Equations (pp. 453–457)

Solve each system of equations by graphing.

49. $y = x$
$y = \frac{1}{2}x - 1$

50. $y = x + 2$
$y = 3x$

Solve each system of equations by substitution.

51. $y = x + 3$
$x = 1$

52. $y = 2x + 6$
$y = 0$

53. NUMBER SENSE The sum of two numbers is 9, and the difference of the numbers is 1. Write a system of equations to represent this situation. Then solve the system to find the numbers.

EXAMPLE 10

Solve the system of equations by graphing.

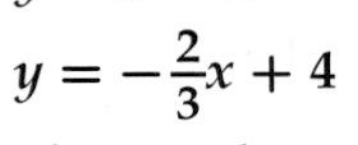

$y = x - 1$

$y = -\frac{2}{3}x + 4$

The graphs appear to intersect at (3, 2).

The solution of the system of equations is (3, 2).

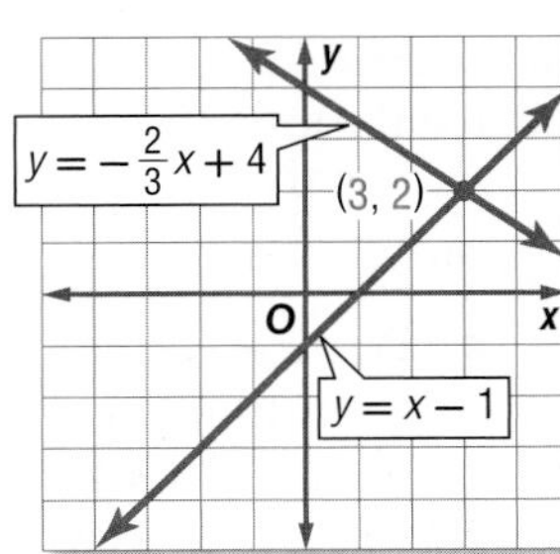

Practice Test

Math Online glencoe.com
Chapter Test

1. Is the relation {(7, 0), (9, 3), (11, 1), (13, 0)} a function? Explain.

2. Write an equation that describes the sequence 2, 7, 12, 17, … . Then find the 10th term.

3. **MULTIPLE CHOICE** Which of the following equations shows the relationship between the side length and perimeter of a regular pentagon?

Side Length (s)	Perimeter (p)
1	5
2	10
3	15
4	20

A $p = 5 + s$
C $s = 5p + 5$
B $p = 5s$
D $s = 5 + p$

Find four solutions of each equation. Write the solutions as ordered pairs.

4. $y = x + 7$
5. $y = -4x$

Graph each equation by plotting ordered pairs.

6. $y = x - 6$
7. $y = -2x + 3$

8. **LANDSCAPING** Find the rate of change for the linear function shown below. Then determine whether a proportional linear relationship exists between the two quantities. Explain your reasoning.

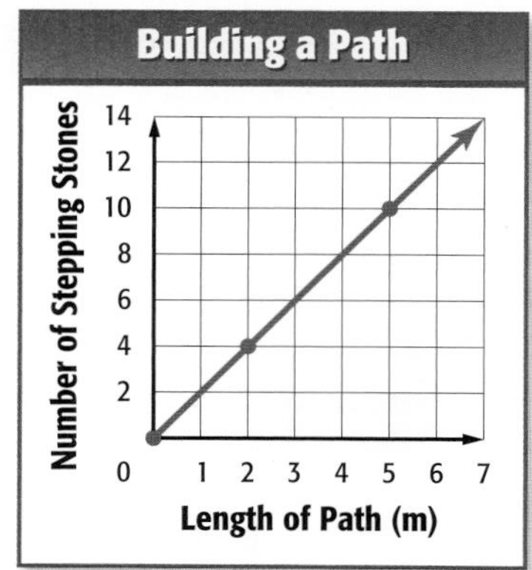

State the slope and y-intercept of the graph of each equation. Then graph each equation using the slope and y-intercept.

9. $y = 3x - 1$
10. $4x + 3y = 6$

11. **FUNDRAISING** The total profit for a school fundraiser varies directly with the number of potted plants sold. Suppose the school earns $57.60 if 12 plants are sold.

   a. Write an equation that could be used to find the profit per plant sold.

   b. Find the total profit if 65 plants are sold.

12. **MULTIPLE CHOICE** A stylist earns $10 an hour plus $3 per hair cut. Which equation represents the stylist's hourly earnings?

F $y = 3x + 10$
H $y = 3x - 10$
G $y = 10x + 3$
J $y = 10x - 3$

13. **RECYCLING** Use the graph below to write an equation in slope-intercept form for the line. What does the slope of the line represent?

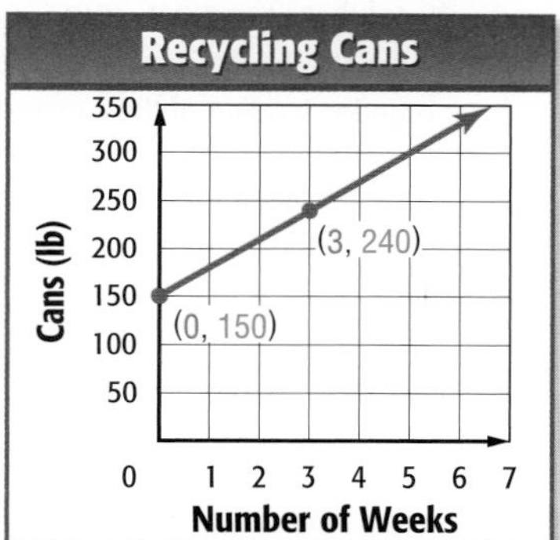

Find the slope of the line that passes through each pair of points.

14. $A(8, 5), B(7, 9)$
15. $R(11, 6), S(9, -1)$

16. **FESTIVALS** The table shows the attendance for an annual jazz festival.

Year	People
2008	1400
2009	1520
2010	1650
2011	1740

   a. Make a scatter plot and draw a line of fit.

   b. Use the line of fit to predict jazz festival attendance in 2014.

Solve each system of equations by graphing.

17. $y = x + 4$
$y = -x + 2$

18. $y = x$
$y = 2x - 1$

Solve each system of equations by substitution.

19. $y = x + 3$
$x = 4$

20. $y = 2x - 5$
$y = -1$

Extended-Response Questions

Extended-response questions are often called *open-ended* or *constructed-response* questions. These types of questions typically have multiple parts. You must answer all parts to receive full credit.

Strategies for Solving Extended-Response Questions

Step 1

As with short answer, extended-response questions are typically graded using a rubric. The following is an example of an extended-response scoring rubric.

Extended-Response Scoring Rubric		
Credit	**Score**	**Criteria**
Full	4	**Full Credit:** A correct solution is given that is supported by well-developed, accurate explanations.
Partial	3, 2, 1	**Partial Credit:** A generally correct solution is given that may contain minor flaws in reasoning or computation or an incomplete solution. The more correct the solution, the greater the score.
None	0	**No Credit:** An incorrect solution is given indicating no mathematical understanding of the concept, or no solution is given.

Step 2

In solving extended-response questions, remember to…

- explain your reasoning or state your approach to solving the problem.
- show all of your work or steps.
- solve each part of the question.
- check your answer if time permits.

EXAMPLE

Read the problem. Identify what you need to know. Then use the information in the problem to solve. Show your work.

The table at the right shows the altitude of a hot air balloon at different times since it began descending to land.

a. Write a linear equation in slope-intercept form to represent the data.

b. Describe what the slope means.

c. Predict how long it will take for the balloon to land.

Time (min)	Altitude (ft)
0	260
2	220
4	180
6	140

Read the problem carefully.

Example of a 4-point response:

a. Let x = time and y = altitude of the balloon.

Use the points (0, 260) and (2, 220) to find the slope.

$m = \frac{260 - 220}{0 - 2}$ or -20

When $x = 0$, $y = 260$, so the y-intercept is 260. The equation of the line is $y = -20x + 260$.

b. The slope gives the change in altitude for the change in time. It is negative because the altitude is decreasing. So, the hot air balloon is descending at a rate of 20 feet per minute.

c. The hot air balloon will have landed when its altitude reaches 0 feet. Let $y = 0$ in the equation and solve for x.

$y = -20x + 260$

$0 = -20x + 260$

$20x = 260$

$x = 13$

So, it will take the balloon 13 minutes to land.

Exercise

Read the question. Identify what you need to know. Then use the information in the question to solve. Show your work.

1. James works as an electrician. He charges a fixed amount per service call plus an hourly fee depending on how long the job takes. His total fee for jobs of different lengths are shown in the graph.

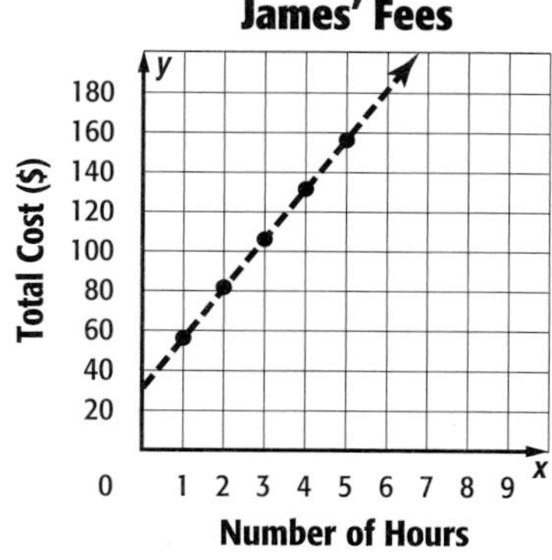

a. Write a linear equation to represent the data.

b. Describe what the slope means.

c. How much would a job cost if it takes James 7 hours?

CHAPTER 8

Standardized Test Practice

Cumulative, Chapters 1 through 8

Multiple Choice

Read each question. Then fill in the correct answer on the answer document provided by your teacher or on a sheet of paper.

1. Find the area of the rectangle below.

A 32 ft^2 C 58 ft^2

B 34 ft^2 D 66 ft^2

2. The table shows a relationship between x and y. Which equation is true for each ordered pair in the table?

x	y
0	0
1	42.50
2	85
3	127.5

F $y = 37.5x + 5$ H $y = 27.5x + 15$

G $y = 42.5x$ J $y = 37.5x + 10$

3. Bethany is riding her bike. After 20 seconds she has traveled 360 feet, and after 1 minute she has traveled 1080 feet. What is her rate of change?

A 18 feet per minute

B 16 feet per minute

C 18 feet per second

D 16 feet per second

Test-Taking Tip

Question 7 Be sure your answer choice reflects the correct units.

4. Janelle invested \$2000 in a savings account that pays 7.5% simple interest. How long will it be before she has \$2750 in the account?

F 10 years H 5 years

G 8 years J 4 years

5. What is the solution of the equation $-24 + 3n = -45$?

A $n = 23$ C $n = -7$

B $n = 7$ D $n = -23$

6. What is the slope of the line that passes through the points $(-4, 6)$ and $(3, -5)$?

F $\frac{11}{7}$ H $-\frac{7}{11}$

G $\frac{7}{11}$ J $-\frac{11}{7}$

7. The linear equation $y = 8.50x$ describes Jamie's wages y when she works x hours. Which of the following best describes the real world meaning of the ordered pair $(12, 102)$?

A She earns \$102 for working 12 hours.

B She earns \$12 for working 102 hours.

C She earns \$102 for working 102 hours.

D She earns \$12 for working 12 hours.

8. Find the equation of the line in the graph.

F $y = x + 2$

G $y = x - 2$

H $y = -x + 2$

J $y = -x - 2$

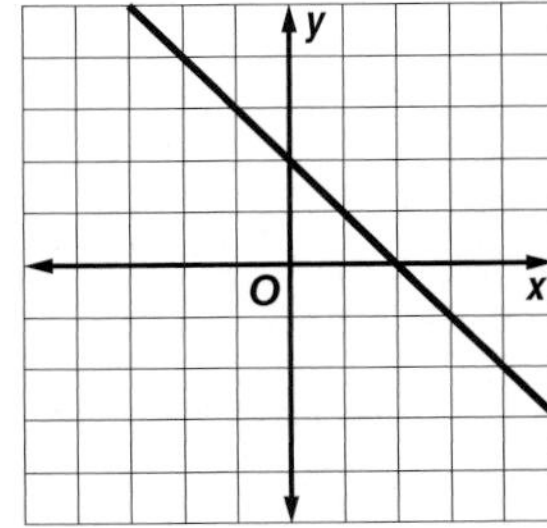

Short Response/Gridded Response

Record your answers on the answer sheet provided by your teacher or on a sheet of paper.

9. **GRIDDED RESPONSE** The equation $y = 464 - 8x$ describes the altitude y, in feet, of a plane x seconds after it begins its descent. How many seconds will it take the plane to land?

10. Is the relation shown in the graph a function? Explain.

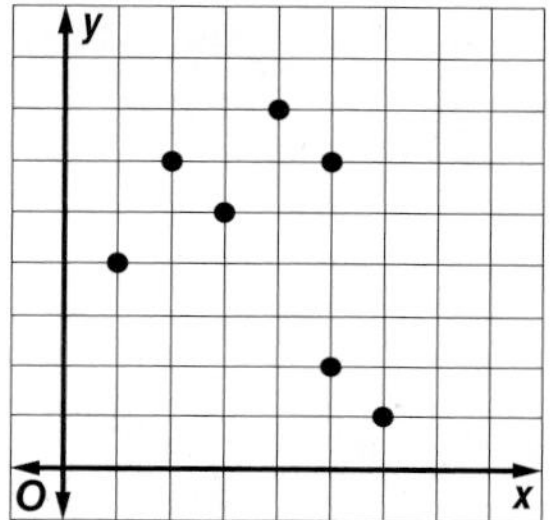

11. Use the sequence 2, 5, 8, 11, 14, … to answer parts **a** and **b**.
    - **a.** Describe the sequence of numbers using words and symbols.
    - **b.** What is the 24th term of the sequence?

12. At a sport's store, 20 out of 112 sets of golf clubs are junior clubs. Express the ratio of sets of junior golf clubs to the total number of sets of golf clubs as a fraction in simplest form. Explain its meaning.

13. Find the solution of the following system of equations. Show your work.

$$4x + 2y = -6$$
$$y = -5x$$

14. H&R Rentals charges $40 to rent a moving van, plus $0.15 per mile.
    - **a.** Write an equation, in slope-intercept form, to show the total cost of renting a van and driving it x miles.
    - **b.** Suppose Marcos rents a moving van and his total cost is $53.20. How many miles did he drive the van?

Extended Response

Record your answers on a sheet of paper. Show your work.

15. The table shows Margo's average keyboarding speed, in words per minute, for five weeks.

Margo's Keyboarding Speed	
Week	**Speed (wpm)**
1	18
2	21
3	26
4	28
5	35

- **a.** Make a scatter plot of Margo's average keyboarding speed versus the number of weeks she has been taking the class. Draw a line of fit.
- **b.** Use the line of fit to predict how fast Margo will be able to type after the sixth week of class.

Need Extra Help?															
If you missed Question...	1	2	3	4	5	6	7	8	9	10	11	12	13	14	15
Go to Lesson or Page...	5-1	8-1	8-4	7-7	4-5	8-6	8-3	8-8	8-7	8-10	8-1	8-9	6-1	8-8	8-9

CHAPTER 9 Powers and Nonlinear Functions

Then

In Chapter 8, you learned about linear functions.

Now

In Chapter 9, you will:

- Write and evaluate expressions with exponents.
- Write and compare numbers in scientific notation.
- Use quadratic functions to solve problems.

Why?

SPACE Our solar system has eight main planets. The distance between the Sun and these planets can be millions of miles. When writing numbers that are extremely small or large, you can use scientific notation.

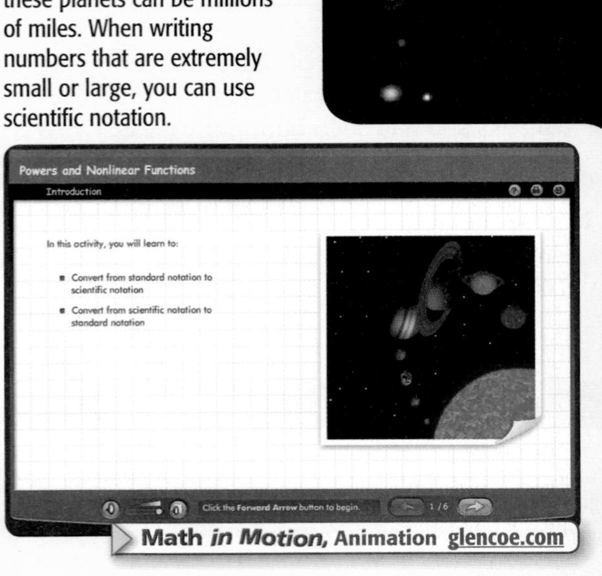

Powers and Nonlinear Functions
Introduction
In this activity, you will learn to:
Convert from standard notation to scientific notation
Convert from scientific notation to standard notation
Click the Forward Arrow button to begin.
1 / 6
Math in Motion, Animation glencoe.com

Get Ready for Chapter 9

Diagnose Readiness You have two options for checking Prerequisite Skills.

Text Option Take the Quick Check below. Refer to the Quick Review for help.

QuickCheck

Evaluate each expression if $a = 4$, $b = 7$, and $c = 5$. (Lesson 1-2)

1. $3a + 2c$
2. $\frac{ac}{2}$
3. $\frac{ab}{-4} + 2b$
4. $4b - 3c$
5. $-5b + 6a$
6. $\frac{3}{4}(2bc)$
7. **SALES** Carlita is buying a computer that costs \$900. She makes a down payment of \$180 and plans to pay the balance in 6 equal installments. How much will each installment be?

QuickReview

EXAMPLE 1

Evaluate $\frac{xy}{z}$ if $x = 3$, $y = 8$, and $z = 6$.

$\frac{xy}{z} = \frac{3(8)}{6}$ — **Replace x with 3, y with 8, and z with 6.**

$= \frac{24}{6}$ — **Multiply 3 and 8.**

$= 4$ — **Divide.**

Find each sum or difference. (Lessons 2-2 and 2-3)

8. $-27 + (-13)$
9. $11 + (-18)$
10. $-9 + 31$
11. $22 + (-16)$
12. $42 - 58$
13. $-15 - 4$
14. $6 - (-17)$
15. $-24 - (-28)$
16. **CARS** The value of Chris' car fell \$2365 in the last two years. If the original value was \$14,681, what is the value of the car now?

EXAMPLE 2

Find $-25 - (-36)$.

$-25 - (-36) = -25 + 36$ — **To subtract -36, add 36.**

$= 11$ — **Simplify.**

Find each product. (Lesson 3-3)

17. $7 \cdot 10$
18. $1.25 \cdot 100$
19. $16.78 \cdot 10$
20. $0.675 \cdot 1000$
21. $56 \cdot 0.1$
22. $162 \cdot 0.001$
23. $97.18 \cdot 0.01$
24. $0.316 \cdot 0.01$
25. **HOTELS** A hotel costs \$159 plus 10% in taxes and fees for each night. The amount of taxes and fees is found by multiplying the cost of the hotel by 10% or 0.1. What is the cost of taxes and fees for one night?

EXAMPLE 3

Find 3.76×0.01.

3.76	←2 decimal places
× 0.01	←2 decimal places
0.0376	←4 decimal places

The product is 0.0376.

Online Option **Math Online** Take a self-check Chapter Readiness Quiz at glencoe.com.

Get Started on Chapter 9

You will learn several new concepts, skills, and vocabulary terms as you study Chapter 9. To get ready, identify important terms and organize your resources. You may wish to refer to **Chapter 0** to review prerequisite skills.

FOLDABLES® Study Organizer

The Tools of Algebra Make this Foldable to help you organize your Chapter 9 notes about powers and nonlinear functions. Begin with five sheets of plain $8\frac{1}{2}$" by 11" paper.

1. **Stack** 5 sheets of paper $\frac{3}{4}$ inch apart.

2. **Roll** up the bottom edges. All tabs should be the same size.

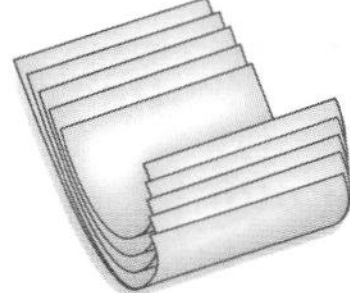

3. **Crease** and staple along the fold.

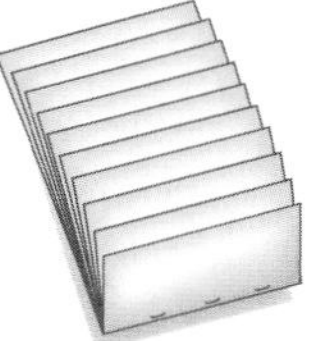

4. **Label** the tabs with topics from the chapter.

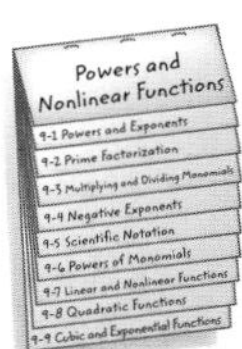
Powers and Nonlinear Functions

New Vocabulary

English		Español
exponent	• p. 471 •	exponente
power	• p. 471 •	potencia
base	• p. 471 •	base
prime number	• p. 476 •	número primo
composite number	• p. 476 •	número compuesto
prime factorization	• p. 477 •	factorización prima
factor tree	• p. 477 •	árbol de factores
monomial	• p. 477 •	monomio
factor	• p. 477 •	factorizar
standard form	• p. 493 •	forma estándar
scientific notation	• p. 493 •	notación cientifica
nonlinear function	• p. 504 •	función no lineal
quadratic function	• p. 510 •	función cuadrática
parabola	• p. 510 •	parábola
cubic function	• p. 516 •	función cúbica
exponential function	• p. 517 •	función exponencial

Review Vocabulary

constant rate of change • p. 418 • tasa constante de cambio consistent ratio of vertical change to horizontal change

factors • p. 856 • factores two or more numbers are multiplied to form a product

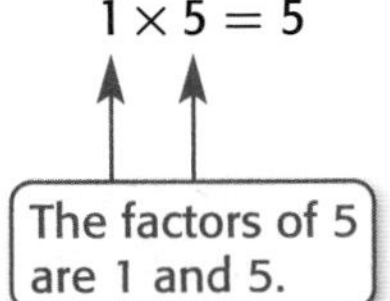
$1 \times 5 = 5$
The factors of 5 are 1 and 5.

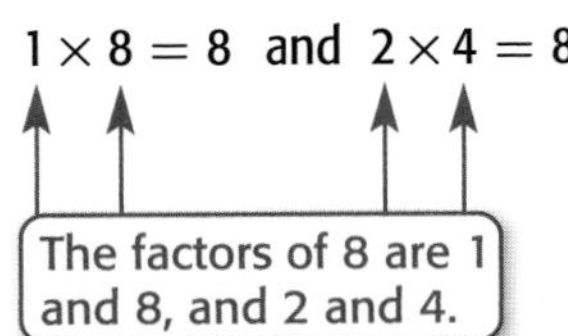
$1 \times 8 = 8$ and $2 \times 4 = 8$
The factors of 8 are 1 and 8, and 2 and 4.

function • p. 33 • función relation in which each member of the domain is paired with exactly one member of the range

Multilingual eGlossary glencoe.com

Math Online glencoe.com

- Study the chapter online
- Explore **Math in Motion**
- Get extra help from your own **Personal Tutor**
- Use **Extra Examples** for additional help
- Take a **Self-Check Quiz**
- **Review Vocabulary** in fun ways

9-1 Powers and Exponents

Why?

Elephants at the National Zoo in Washington, D.C., eat 125 pounds of hay each day. Each elephant weighs more than 10,000 pounds.

a. Using 5 as a factor, write a multiplication expression that equals 125.

b. Using 10 as a factor, write a multiplication expression that equals 10,000.

Then
You evaluated expressions without exponents. (Lesson 2-4)

Now
- Write expressions using exponents.
- Evaluate expressions containing exponents.

New Vocabulary
exponent
power
base

Math Online
glencoe.com
- Extra Examples
- Personal Tutor
- Self-Check Quiz
- Homework Help
- Math in Motion

Use Exponents An expression like $5 \cdot 5 \cdot 5$ with equal factors can be written using an exponent. An **exponent** tells how many times a number is used as a factor. A number that is expressed using an exponent is called a **power**. The number that is multiplied is called the **base**. So, $5 \cdot 5 \cdot 5$ equals the power 5^3.

base → 5^3 ← **exponent**; ↑ **power**

Read and Write Powers		
Power	**Words**	**Factors**
5^1	5 to the first power	5
5^2	5 to the second power or 5 squared	$5 \cdot 5$
5^3	5 to the third power or 5 cubed	$5 \cdot 5 \cdot 5$
5^4	5 to the fourth power or 5 to the fourth	$5 \cdot 5 \cdot 5 \cdot 5$
⋮	⋮	⋮
5^n	5 to the *n*th power or 5 to the *n*th	$\underbrace{5 \cdot 5 \cdot 5 \cdot \ldots \cdot 5}_{n \text{ factors}}$

Any number, except 0, raised to the zero power is defined as 1.

$1^0 = 1$ $2^0 = 1$ $3^0 = 1$ $4^0 = 1$ $5^0 = 1$ $x^0 = 1, x \neq 0$

EXAMPLE 1 Write Expressions using Exponents

Write each expression using exponents.

a. $4 \cdot 4 \cdot 4 \cdot 4 \cdot 4$

The base 4 is a factor 5 times.
So, the exponent is 5.
$4 \cdot 4 \cdot 4 \cdot 4 \cdot 4 = 4^5$

b. $(-8) \cdot (-8) \cdot (-8)$

The base -8 is a factor 3 times.
So, the exponent is 3.
$(-8) \cdot (-8) \cdot (-8) = (-8)^3$

Check Your Progress

1A. $6 \cdot 6 \cdot 6 \cdot 6$ **1B.** $(-2)(-2)(-2)$ **1C.** $\left(\frac{1}{2}\right)\left(\frac{1}{2}\right)\left(\frac{1}{2}\right)$

Personal Tutor glencoe.com

EXAMPLE 2 Write Expressions using Exponents

Watch Out!

Negative Bases $(-3)^2$ is not the same as -3^2.

- $(-3)^2 = (-3)(-3) = 9$
- $-3^2 = (-1)(3^2) = -9$

Write each expression using exponents.

a. $y \cdot y \cdot y \cdot y \cdot y \cdot y$

The base y is a factor 6 times. So, the exponent is 6.

$y \cdot y \cdot y \cdot y \cdot y \cdot y = y^6$

b. $(k + 2)(k + 2)(k + 2)(k + 2)$

The base $(k + 2)$ is a factor 4 times. So, the exponent is 4.

$(k + 2)(k + 2)(k + 2)(k + 2) = (k + 2)^4$

c. $5 \cdot r \cdot r \cdot s \cdot s \cdot s \cdot s$

$5 \cdot r \cdot r \cdot s \cdot s \cdot s \cdot s = 5 \cdot (r \cdot r) \cdot (s \cdot s \cdot s \cdot s)$ **Group factors with like bases.**

$= 5 \cdot r^2 \cdot s^4$ or $5r^2s^4$ $r \cdot r = r^2, s \cdot s \cdot s \cdot s = s^4$

Check Your Progress

2A. $x \cdot x \cdot x \cdot x \cdot x$ **2B.** $(c - d)(c - d)$ **2C.** $9 \cdot f \cdot f \cdot f \cdot f \cdot g$

Personal Tutor glencoe.com

Evaluate Expressions Since powers represent repeated multiplication, they need to be included in the rules for order of operations.

Concept Summary Order of Operations

For Your FOLDABLE

Step 1	Simplify the expressions inside grouping symbols first.
Step 2	Evaluate all powers.
Step 3	Do all multiplications or divisions in order from left to right.
Step 4	Do all additions or subtractions in order from left to right.

Real-World EXAMPLE 3 Evaluate Expressions

BEACH VOLLEYBALL The playing area for beach volleyball includes the playing court and the free zone. Evaluate each expression to find the area of the playing court and the free zone.

a. The playing court is a rectangle with an area of 2^7 square meters.

$2^7 = 2 \cdot 2 \cdot 2 \cdot 2 \cdot 2 \cdot 2 \cdot 2$ **2 is a factor 7 times.**

$= 128$ **Simplify.**

The area of the playing court is 128 square meters.

b. The area of the free zone is $2^2 \cdot 3^2 \cdot 5$ square meters.

$2^2 \cdot 3^2 \cdot 5 = 2 \cdot 2 \cdot 3 \cdot 3 \cdot 5$ **Evaluate powers.**

$= 180$ **Multiply.**

The area of the free zone is 180 square meters.

Real-World Link

Beach volleyball began in Santa Monica, California, in the 1920s. It is estimated that about 800 million people play volleyball worldwide.

Source: U.S. Beach-Volleyball

Check Your Progress

3. PHYSICS A tennis ball is dropped from the top of a building. After 8 seconds, the tennis ball hits the ground. The distance in meters the ball traveled is represented by $4.9(8)^2$. How far did the ball drop?

Personal Tutor glencoe.com

StudyTip

Exponents An exponent goes with the number, variable, or quantity in parentheses immediately preceding it.

EXAMPLE 4 Evaluate Algebraic Expressions

Evaluate $x^2 + y^3$ if $x = 6$ and $y = -2$.

$x^2 + y^3 = 6^2 + (-2)^3$ — Replace x with 6 and y with -2.

$= 36 - 8$ — Evaluate powers; $6^2 = (6 \cdot 6)$ or 36; $(-2)^3$

$= 28$ — Subtract.

Check Your Progress

Evaluate each expression if $a = 5$, $b = -2$, and $c = \frac{3}{4}$.

4A. $10 + b^2$ **4B.** $(a + b)^3$ **4C.** $2 - c^2$

Personal Tutor glencoe.com

Check Your Understanding

Examples 1 and 2 pp. 471–472

Write each expression using exponents.

1. $2 \cdot 2 \cdot 2 \cdot 2 \cdot 2 \cdot 2$ **2.** $d \cdot d \cdot d \cdot d \cdot d \cdot d$ **3.** $\left(-\frac{1}{4}\right)\left(-\frac{1}{4}\right)\left(-\frac{1}{4}\right)$

4. $4 \cdot m \cdot m \cdot m \cdot q \cdot q \cdot q$ **5.** $(y - 3)(y - 3)(y - 3)$ **6.** $(a + 1)(a + 1)$

Example 3 p. 472

7. **INSECTS** The longhorn beetle can have a body length of over 2^4 centimeters. How many centimeters long is this?

8. **E-MAIL** Theo sends an e-mail to three friends. Each friend forwards the e-mail to three friends. Each of those friends forwards it to three friends, and so on. Find the number of e-mails sent during the fifth stage as a power. Then find the value of the power.

Example 4 p. 473

Evaluate each expression if $a = 3$, $b = -4$, and $c = 3.5$.

9. $a^3 + 2$ **10.** $3(b - 1)^2$ **11.** $c^2 + b^2$

Practice and Problem Solving

= **Step-by-Step Solutions** begin on page R11. **Extra Practice** begins on page 810.

Examples 1 and 2 pp. 471–472

Write each expression using exponents.

12. $11 \cdot 11 \cdot 11 \cdot 11$

13 $3 \cdot 3 \cdot 3 \cdot 3 \cdot 3$

14. $(-8)(-8)(-8)(-8)(-8)(-8)$

15. $(-14) \cdot (-14) \cdot (-14)$

16. $\left(-\frac{1}{5}\right)\left(-\frac{1}{5}\right)\left(-\frac{1}{5}\right)\left(-\frac{1}{5}\right)$

17. $(-1.5)(-1.5)(-1.5)$

18. $ab \cdot ab \cdot ab \cdot ab$

19. $5 \cdot p \cdot p \cdot p \cdot q \cdot q \cdot q$

20. $3 \cdot 7 \cdot m \cdot m \cdot n \cdot n \cdot n \cdot n$

21. $8(c + 4)(c + 4)$

22. $(n - 5)(n - 5)(n - 5)$

23. $(2x + 3y)(2x + 3y)$

Example 3 p. 472

24. **VOLCANOES** The longest chain of active volcanoes is in the South Pacific. This chain is more than $3 \cdot 10^4$ miles long and has approximately $3^5 \cdot 5$ volcanoes.

a. How long is the chain of volcanoes?

b. How many volcanoes are there?

...le 4
p. 473

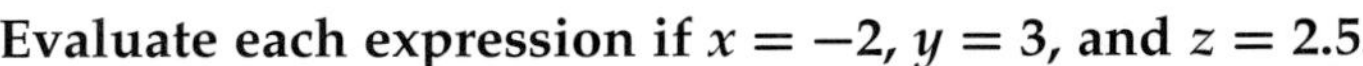

Evaluate each expression if $x = -2$, $y = 3$, and $z = 2.5$.

25. y^4	**26.** z^3	**27.** $7x^2$
28. xy^3	**29.** $z^2 + x$	**30.** $y^0 + 9$
31. $2y + z^3$	**32.** $x^2 + 2y - 3$	**33.** $y^2 - 3x + 8$
34. $4(y + 1)^4$	**35.** $3(2z + 4)^2$	**36.** $5(x^3 + 6)$

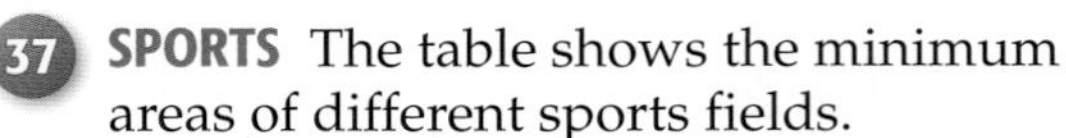

37 **SPORTS** The table shows the minimum areas of different sports fields.

Sport	Minimum Field Area (ft²)
Field Hockey	$2^6 \cdot 10^3$
Men's Lacrosse	$3^2 \cdot 7 \cdot 10^3$
Women's Soccer	$2^4 \cdot 5^2 \cdot 7 \cdot 13$

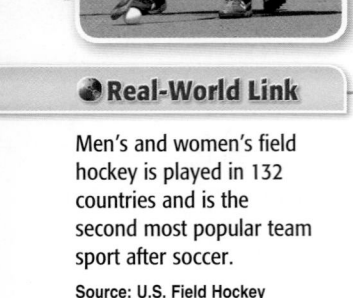

a. Find the minimum area of each playing field.

b. Order the areas from least to greatest.

c. How much greater is the area of a field hockey field than the area of a men's lacrosse field?

Real-World Link

Men's and women's field hockey is played in 132 countries and is the second most popular team sport after soccer.

Source: U.S. Field Hockey

Evaluate each expression.

38. 9^2	**39.** 11^3	**40.** $\left(-\frac{2}{3}\right)^3$	**41.** $(-5)^4$
42. $(-2)^7$	**43.** $2 \cdot 4^4$	**44.** $6^3 \cdot 4$	**45.** $3^5 \cdot 10$
46. $2^0 \cdot 10$	**47.** $7^3 \cdot 2^2$	**48.** $5 \cdot 2^4$	**49.** $(4.5)^4 \cdot 2$

Replace each ● with <, >, or = to make a true statement.

50. 2^5 ● 5^2	**51.** 3^6 ● 6^3	**52.** 2^6 ● 8^2	**53.** 8^3 ● 4^5

54. **MULTIPLE REPRESENTATIONS** In this problem, you will explore volume of a cube. The volume of a cube equals the side length cubed.

a. **SYMBOLIC** Write an equation showing the relationship between side length s and volume V of a cube.

b. **TABULAR** Make a table of values showing the volume of a cube with side lengths of 1, 2, 4, 8, and 16 centimeters.

c. **ANALYTICAL** Use your table to make a conjecture about the change in volume when the side length of a cube is doubled. Justify your response by writing an algebraic expression.

H.O.T. Problems

Use Higher-Order Thinking Skills

55. **OPEN ENDED** Use exponents to write two numerical expressions. Then find the product of the expressions.

56. **CHALLENGE** Determine whether x^3 is *always*, *sometimes*, or *never* a positive number for $x \neq 0$. Explain your reasoning.

57. **REASONING** Suppose the population of the United States is about 230 million. Is this number closer to 10^7 or 10^8? Explain.

58. **CHALLENGE** Explain why $5^0 = 1$. (*Hint*: Find a pattern in 5^4, 5^3, 5^2, and 5^1 to predict 5^0.)

59. **WRITING IN MATH** Describe the advantages of using exponents to represent numeric values.

Standardized Test Practice

60. Marta observed that a bacterium cell doubled every 3 minutes.

Time (min)	Number of Bacteria
0	2^0
3	2^1
6	2^2
9	2^3
12	2^4

Which expression represents the number of cells after one half hour?

A 2^{10} **C** 2^{20}

B 2^{15} **D** 2^{30}

61. GRIDDED RESPONSE Suppose a certain forest fire doubles in size every 8 hours. If the initial size of the fire was 1 acre, how many acres will the fire cover in 3 days?

62. Which of the following is equivalent to $4^3 \cdot 5^2$?

F $12 \cdot 25$

G $3 \cdot 3 \cdot 3 \cdot 3 \cdot 2 \cdot 2 \cdot 2 \cdot 2 \cdot 2$

H $4 \cdot 4 \cdot 4 \cdot 5 \cdot 5$

J $4 \cdot 4 \cdot 4 \cdot 5 \cdot 5 \cdot 5$

63. Evaluate $\left(\frac{4}{5}\right)^2$.

A $\frac{8}{25}$ **C** $\frac{8}{10}$

B $\frac{16}{25}$ **D** $1\frac{3}{5}$

Spiral Review

Solve each system of equations by substitution. (Lesson 8-10)

64. $y = x + 10$
$y = 2$

65. $y = x - 5$
$x = 0$

66. $y = 3x - 4$
$y = 1$

67. MULTIPLE REPRESENTATIONS In this problem, you will investigate the approximate barometric pressure at various altitudes. (Lesson 8-9)

a. GRAPHICAL Make a scatter plot of the data and draw a line of fit.

b. ALGEBRAIC Write an equation for the line of fit you drew in part **a.** Use it to estimate the barometric pressure at 60,000 feet. Is the estimation reasonable? Explain.

c. VERBAL Do you think that a line is the best model for this data? Explain.

Altitude (ft)	Barometric Pressure (in. mercury)
0	30
5000	25
10,000	21
20,000	14
30,000	9
40,000	6
50,000	3

Source: *New York Public Library Science Desk Reference*

Choose the greatest number in each set. (Lesson 7-2)

68. $\left\{\frac{2}{5}, 0.45, 35\%, 3 \text{ out of } 8\right\}$

69. $\left\{\frac{3}{4}, 0.70, 78\%, 4 \text{ out of } 5\right\}$

70. $\left\{19\%, \frac{3}{16}, 0.155, 2 \text{ to } 15\right\}$

71. $\left\{89\%, \frac{10}{11}, 0.884, 12 \text{ to } 14\right\}$

Skills Review

List all the whole number factors for each number. (Previous Course)

72. 7 **73.** 15 **74.** 18 **75.** 40

9-2 Prime Factorization

Why?

Isaiah wants to arrange 6 pictures of his family and friends on the wall.

a. Express 6 as the product of two whole numbers.

b. Suppose he adds two more pictures. Describe the possible rectangular arrangements of 8 pictures.

c. In how many different ways could he arrange 5 pictures in a rectangle? Explain.

Then

You simplified expressions by multiplying. (Lesson 2-4)

Now

- Write the prime factorizations of composite numbers.
- Factor monomials.

New Vocabulary

prime number
composite number
prime factorization
factor tree
monomial
factor

Math Online

glencoe.com

- Extra Examples
- Personal Tutor
- Self-Check Quiz
- Homework Help

Write Prime Factorizations Numbers that have whole number factors can be represented using rectangles.

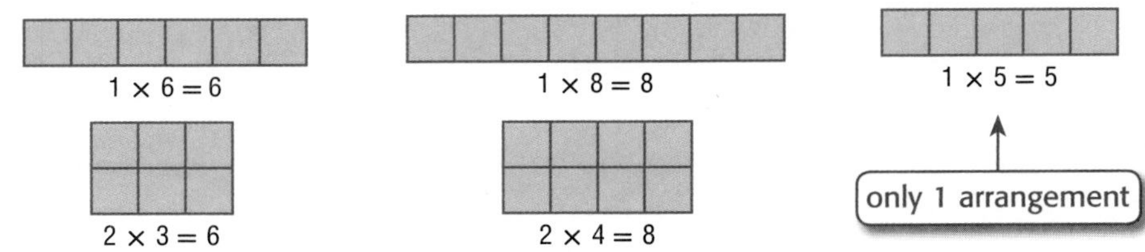

A **prime number** is a whole number that has exactly two unique factors, 1 and itself. So, 5 is a prime number. A **composite number** is a whole number that has more than two factors. The numbers 6 and 8 are composite numbers. The numbers 0 and 1 are neither prime nor composite.

Prime Number	Factors	Composite Number	Factors
2	1, 2	4	1, 2, 4
3	1, 3	6	1, 2, 3, 6
5	1, 5	8	1, 2, 4, 8

EXAMPLE 1 Identify Prime and Composite Numbers

Determine whether each number is *prime* or *composite*.

a. 15

Find factors of 15 by listing whole number pairs whose product is 15.

$15 = 1 \times 15$ $\quad$ $15 = 3 \times 5$

The factors of 15 are 1, 3, 5, and 15. Since the number has more than two factors, it is a composite number.

b. 23

$23 = 1 \times 23$

The number 23 has only two factors, 1 and 23. So, 23 is a prime number.

Check Your Progress

1A. 37 **1B.** 22

Personal Tutor glencoe.com

> **StudyTip**
>
> **Prime Factorization** There is exactly one prime factorization for a number. The order in which the prime factors are written does not matter.

When a composite number is expressed as the product of prime factors, it is called the **prime factorization** of the number. One way to find the prime factorization of a number is to use a **factor tree**.

Step 1 Write the number that you are factoring at the top.

Step 2 Choose any pair of whole number factors of the number.

Step 3 Continue to factor any number that is not prime.

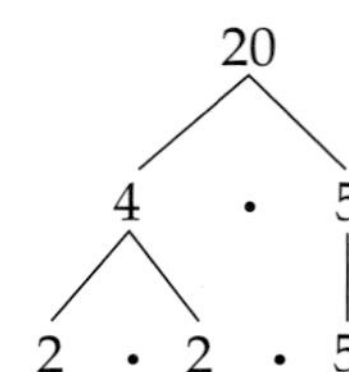

The factor tree is complete when you have a row of prime numbers. The prime factorization of 20 is $2 \cdot 2 \cdot 5$ or $2^2 \cdot 5$.

STANDARDIZED TEST EXAMPLE 2

What is the prime factorization of 90?

A $2 \cdot 5 \cdot 9$

B $2 \cdot 3^2 \cdot 5$

C $3 \cdot 6 \cdot 5$

D $2 \cdot 3 \cdot 15$

Read the Test Item

You are asked to express 90 as a product of prime factors. Construct a factor tree to find all of the prime factors.

Solve the Test Item

Choose any pair of whole number factors of 90.

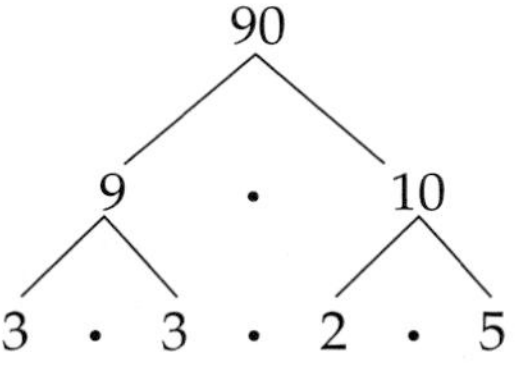

Continue to factor any number that is not prime.

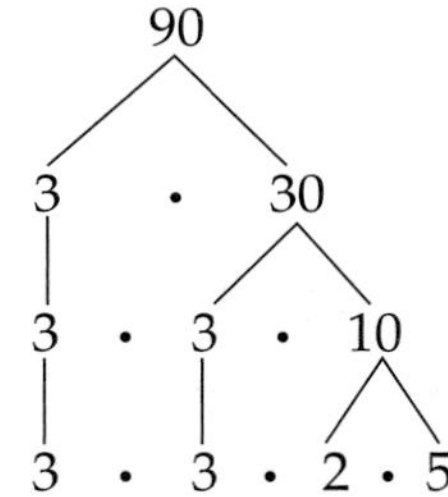

The prime factorization of 90 is $2 \cdot 3 \cdot 3 \cdot 5$ or $2 \cdot 3^2 \cdot 5$. The answer is B.

Check Your Progress

2. What is the prime factorization of 24?

F $3 \cdot 8$

G $2^3 \cdot 3$

H $2 \cdot 3 \cdot 4$

J $2 \cdot 12$

Personal Tutor glencoe.com

> **StudyTip**
>
> **Factors** In the equation $8 \times 10 = 80$, 8 and 10 are factors of 80. In the equation $8 \cdot x = 8x$, 8 and x are factors of $8x$.

Factor Monomials A **monomial** is a number, a variable, or a product of numbers and/or variables.

Monomials	Not Monomials
$80, x, 8x$	$x + 5, x^2 - y^2$

To **factor** a number means to write it as a product of its factors. A monomial can also be factored as a product of prime numbers and variables with no exponent greater than 1. Negative coefficients can be factored using -1 as a factor.

EXAMPLE 3 Factor Monomials

Factor each monomial.

a. $11r^4s$

$11r^4s = 11 \cdot r \cdot r \cdot r \cdot r \cdot s$ $\quad$ $r^4 = r \cdot r \cdot r \cdot r$

b. $-28a^2b^3$

$-28a^2b^3 = -1 \cdot 2 \cdot 2 \cdot 7 \cdot a^2 \cdot b^3$ $\quad$ $-28 = -1 \cdot 2 \cdot 2 \cdot 7$

$= -1 \cdot 2 \cdot 2 \cdot 7 \cdot a \cdot a \cdot b \cdot b \cdot b$ $\quad$ $a^2 \cdot b^3 = a \cdot a \cdot b \cdot b \cdot b$

Check Your Progress

3A. $10xy$ **3B.** $12q^2r^3$ **3C.** $-18mn^4$

Personal Tutor glencoe.com

Check Your Understanding

Example 1 p. 476

Determine whether each number is *prime* or *composite*.

1. 26 **2.** 19 **3.** 35

Example 2 p. 477

4. **MULTIPLE CHOICE** There are 30 students going on a field trip to an art museum. What is the prime factorization of 30?

A $2^2 \cdot 3 \cdot 5$

B $2 \cdot 3 \cdot 5$

C $3^2 \cdot 5$

D $2^2 \cdot 5$

Example 3 p. 478

ALGEBRA **Factor each monomial.**

5. $14a^3$ **6.** $-9rs^2t$ **7.** $20x^2y$

Practice and Problem Solving

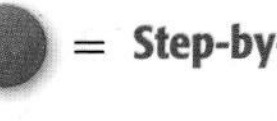 = **Step-by-Step Solutions** begin on page R11. **Extra Practice** begins on page 810.

Example 1 p. 476

Determine whether each number is *prime* or *composite*.

8. 8 **9.** 11 **10.** 29 **11.** 26

12. 13 **13.** 41 **14.** 57 **15.** 63

Example 2 p. 477

Write the prime factorization of each number. Use exponents for repeated factors.

16. 22 **17** 243 **18.** 105 **19.** 56

20. 104 **21.** 196 **22.** 450 **23.** 198

Example 3 p. 478

ALGEBRA **Factor each monomial.**

24. $15y$ **25.** $6n^2$ **26.** $-9p^4$ **27.** $-11n^3$

28. $18ab$ **29.** $20qrs$ **30.** $24f^2g$ **31.** $-35c^3d^2$

32. **GEOMETRY** The surface area of a cube is given by the expression $6s^2$, where s is the side length. Factor $6s^2$.

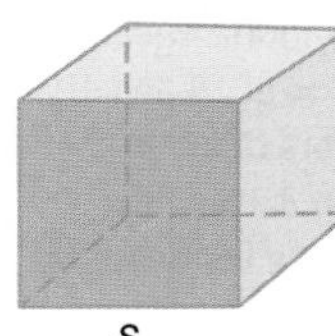

33. **REPTILES** The graph shows the maximum age in years of the world's longest-lived reptiles. Write the prime factorization of the age of each reptile.

a. boa constrictor

b. American alligator

c. box turtle

d. Galapagos tortoise

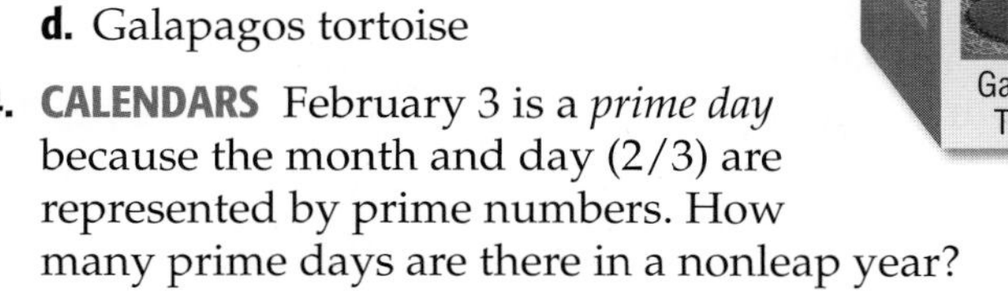

34. **CALENDARS** February 3 is a *prime day* because the month and day (2/3) are represented by prime numbers. How many prime days are there in a nonleap year?

35. **GEOMETRY** Determine how many rectangles with different whole-number dimensions can be drawn for each given area.

a. 30 square inches   b. 28 square feet

36. **TECHNOLOGY** *Mersenne primes* are prime numbers in the form $2^n - 1$. In 2006, a group of scientists used special software to discover the greatest prime number so far, $2^{32,582,657} - 1$. Write the prime factorization of each number, or write *prime* if the number is a Mersenne prime.

a. $2^5 - 1$ b. $2^6 - 1$ c. $2^7 - 1$ d. $2^8 - 1$

37. **CODES** Prime numbers are used to help keep messages sent over the Internet private. One step in the process involves multiplying two prime numbers to produce a key N. Determine which number could be N: 27, 29, 31, or 33.

38. **PACKAGING** A beverage company is developing the packaging for a case of soda that contains 36 cans. List the arrangement of the cans that could be used for the package. (*Hint*: The cans can be stacked as well as arranged in a rectangular pattern one-can high.)

Math History Link

Marin Mersenne (1588–1648) Mersenne Primes were first described by the French monk Marin Mersenne. His formula $2^n - 1$ continues to be used by mathematicians today to locate large primes.

Source: The Great Internet Mersenne Prime Search

H.O.T. Problems
Use Higher-Order Thinking Skills

39. **OPEN ENDED** Write three different monomials that have factors of 2, 5, and a^2.

40. **WRITING IN MATH** Explain how you can use the prime factorization of 30 to generate the factors of 30.

41. **FIND THE ERROR** Felipe and Ledell are describing prime numbers. Is either of them correct? Explain your reasoning.

Felipe	Ledell
All odd numbers are prime.	No prime numbers are even.

42. **CHALLENGE** Write five numbers that are divisible by 6 and find the prime factors of the numbers. Then write a rule to describe when a number is divisible by 6.

43. **WRITING IN MATH** Suppose n represents a prime number. Is $2n$ *always*, *sometimes*, or *never* prime? Explain your reasoning.

Standardized Test Practice

44. What is the prime factorization of $12x^3y^2$?

A $12 \cdot x \cdot x \cdot x \cdot y \cdot y$

B $2 \cdot 2 \cdot 3 \cdot x \cdot x \cdot x \cdot y \cdot y$

C $3 \cdot 4 \cdot x^3 \cdot y^2$

D $12 \cdot x^3 \cdot y^2$

45. SHORT RESPONSE Draw and label all of the rectangles with different whole-number dimensions that have an area of 36 square feet.

46. Which number is *not* a prime factor of 90?

F 2

G 3

H 5

J 9

47. Which is the least whole number that is divisible by 2, 3, 7, and 11?

A 452

B 456

C 462

D 468

Spiral Review

Evaluate each expression if $x = 2$, $y = 4$ and $z = -3$. (Lesson 9-1)

48. xz^3

49. $y^0 - 10$

50. $x^2 + 3x - 1$

51. $5(y - 1)^2$

52. $4z^4$

53. $z^2 + x^2$

54. GAMES Emilio has 17 video games and plans to purchase 2 new games every month. LaShaun has 47 video games and plans to donate 3 games to a charity every month. After how many months will the two have the same number of video games? (Lesson 8-10)

55. PIZZA Koto delivers pizzas on weekends. Her average tip is \$1.50 for each pizza that she delivers. How many pizzas must she deliver to earn at least \$20 in tips? (Lesson 5-4)

Write and solve an equation to find each number. (Lesson 4-3)

56. The sum of a number and 9 is −2.

57. The sum of −5 and a number is −15.

58. GEOMETRY The area of a trapezoid is the product of one half the height and the sum of both bases. (Lesson 4-6)

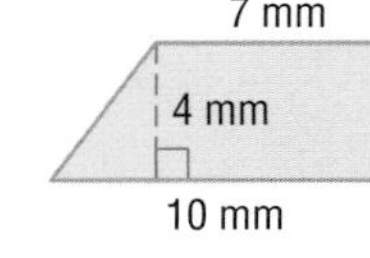

a. If h is the height, b_1 is one base, and b_2 is the second base, write an expression for the area of the trapezoid.

b. Find the area of the trapezoid shown at the right.

Skills Review

For each expression, use parentheses to group whole numbers together and powers with like bases together. (Lesson 1-3)

Example: $a \cdot 4 \cdot a^3 \cdot 2 = (4 \cdot 2)(a \cdot a^3)$

59. $3 \cdot a^4 \cdot 5$

60. $n \cdot p \cdot p^2 \cdot n^3$

61. $b \cdot 5 \cdot 10 \cdot b^4$

62. $t \cdot t^4 \cdot (-7) \cdot t^4 \cdot (-2)$

63. $2 \cdot c^2 \cdot b^3 \cdot d^3 \cdot c^3 \cdot (-8)$

64. $12 \cdot 15 \cdot a \cdot 9 \cdot a^5 \cdot c^3$

9-3 Multiplying and Dividing Monomials

Then

You used the Commutative and Associative Properties of Multiplication to simplify expressions.
(Lesson 1-3)

Now

- Multiply monomials.
- Divide monomials.

Math Online

glencoe.com

- Extra Examples
- Personal Tutor
- Self-Check Quiz
- Homework Help
- Math in Motion

Why?

The table shows the estimated number of songs on an MP3 player with different gigabytes (GB) of capacity.

Capacity (GB)	Estimated Number of Songs
5	$1000 = 10^3$
40	$10{,}000 = 10^4$

Complete.

a. $10 \times ■ = 1000 \quad \rightarrow 10^1 \times ■ = 10^3$

b. $10 \times ■ = 10{,}000 \quad \rightarrow 10^1 \times ■ = 10^4$

c. Examine the exponents of the factors and the exponents of the products above. Then write a rule for determining the exponent of the product when you multiply powers with the same base. Test your rule by multiplying $10^3 \cdot 10^4$ using a calculator.

Multiply Monomials Recall that exponents are used to show repeated multiplication. You can use the definition of exponent to help find a rule for multiplying powers with the same base.

$$10^1 \cdot 10^3 = \overbrace{(10)}^{\text{1 factor}} \cdot \overbrace{(10 \cdot 10 \cdot 10)}^{\text{3 factors}} = 10^4$$

$$\underbrace{(10) \cdot (10 \cdot 10 \cdot 10)}_{\text{4 factors}}$$

Key Concept — Product of Powers Property

For Your FOLDABLE

Words Multiply powers with the same base by adding their exponents.

Symbols $a^m \cdot a^n = a^{m+n}$ **Example** $2^4 \cdot 2^3 = 2^{4+3}$ or 2^7

EXAMPLE 1 Multiply Powers

Find each product. Express using exponents.

a. $4^3 \cdot 4^5$

$4^3 \cdot 4^5 = 4^{3+5}$ — Product of Powers Property; the common base is 4.

$= 4^8$ — Add the exponents.

b. $6 \cdot 6^4$

$6 \cdot 6^4 = 6^1 \cdot 6^4$ — $6 = 6^1$

$= 6^{1+4}$ — Product of Powers Property; the common base is 6.

$= 6^5$ — Add the exponents.

Watch Out!

Common Misconception When multiplying powers, do not multiply the bases. $6^1 \cdot 6^4 = 6^5$, not 36^5.

Check Your Progress

1A. $5^2 \cdot 5^3$

1B. $12^3 \cdot 12$

Personal Tutor glencoe.com

EXAMPLE 2 Multiply Monomials

Find each product.

a. $b^2 \cdot b^2$

$b^2 \cdot b^2 = b^{2+2}$ **Product of Powers Property; the common base is *b*.**

$= b^4$ **Add the exponents.**

b. $2x^3 \cdot 8x^4$

$2x^3 \cdot 8x^4 = 2 \cdot 8 \cdot x^3 \cdot x^4$ **Commutative Property of Multiplication**

$= 2 \cdot 8 \cdot x^{3+4}$ **Product of Powers Property; the common base is *x*.**

$= 2 \cdot 8 \cdot x^7$ **Add the exponents.**

$= 16x^7$ **Multiply.**

Check Your Progress

2A. $y^6 \cdot y^3$

2B. $(5a^2)(-3a^4)$

Personal Tutor glencoe.com

Divide Monomials There is also a property for quotients of powers.

$$\frac{4^5}{4^2} = \frac{\overbrace{4 \cdot 4 \cdot 4 \cdot 4 \cdot 4}^{\text{5 factors}}}{\underbrace{4 \cdot 4}_{\text{2 factors}}} = \frac{4 \cdot 4 \cdot 4 \cdot \cancel{4}^{1} \cdot \cancel{4}^{1}}{\cancel{4}_{1} \cdot \cancel{4}_{1}} = \underbrace{4 \cdot 4 \cdot 4}_{\text{3 factors}} \text{ or } 4^3$$

Key Concept Quotient of Powers Property

For Your FOLDABLE

Words Divide powers with the same base by subtracting their exponents.

Symbols $a^m \div a^n = a^{m-n}$

Example $3^6 \div 3^2 = 3^{6-2}$ or 3^4

Math *in Motion,* BrainPOP® glencoe.com

Watch Out!

Dividing Monomials When dividing powers, remember that the denominator cannot equal zero. So, in Example 3b, $c \neq 0$.

EXAMPLE 3 Divide Powers

Find each quotient.

a. $\frac{8^5}{8^3}$

$\frac{8^5}{8^3} = 8^{5-3}$ **Quotient of Powers Property; the common base is 8.**

$= 8^2$ **Subtract the exponents.**

b. $\frac{c^7}{c^2}$

$\frac{c^7}{c^2} = c^{7-2}$ **Quotient of Powers Property; the common base is *c*.**

$= c^5$ **Subtract the exponents.**

Check Your Progress

3A. $\frac{3^9}{3^2}$

3B. $\frac{b^7}{b^6}$

Personal Tutor glencoe.com

Real-World Link

Jupiter is the largest planet in the solar system. If Jupiter were hollow, more than 1200 Earths could fit inside.

Source: NASA

Real-World EXAMPLE 4 Use Powers to Compare Values

PLANETS The table shows the approximate diameters of Earth, Mars, and Neptune. About how many times as great is Neptune's diameter than Earth's diameter?

Planet	Approximate Diameter (mi)
Earth	2^{13}
Mars	2^{12}
Neptune	2^{15}

Write a division expression.

$\frac{2^{15}}{2^{13}} = 2^{15-13}$ **Quotient of Powers Property**

$= 2^2$ or 4 **Simplify.**

So, Neptune's diameter is about 4 times as great as Earth's diameter.

Check Your Progress

4. **PLANETS** About how many times as great is the diameter of Earth than the diameter of Mars?

Personal Tutor glencoe.com

Check Your Understanding

Examples 1 and 2 pp. 481–482

Find each product. Express using exponents.

1. $2^4 \cdot 2^6$
2. $8^5 \cdot 8$
3. $x^{10} \cdot x^6$
4. $-w^2(5w^7)$

Example 3 p. 482

Find each quotient. Express using exponents.

5. $\frac{4^5}{4^3}$
6. $7^9 \div 7$
7. $\frac{r^8}{r^4}$
8. $b^{11} \div b^2$

Example 4 p. 483

9. **CANYONS** The Grand Canyon is approximately 2^9 kilometers long. *Mariner Valley* is a canyon on Mars that is approximately 2^{12} kilometers long. About how many times longer is Mariner Valley than the Grand Canyon?

Practice and Problem Solving

● = **Step-by-Step Solutions** begin on page R11. **Extra Practice** begins on page 810.

Examples 1 and 2 pp. 481–482

Find each product. Express using exponents.

10. $5^6 \cdot 5^2$
11.  $(-2)^3 \cdot (-2)^2$
12. $a^7 \cdot a^2$
13. $(t^3)(t^3)$
14. $(10x)(4x^7)$
15. $6p^7 \cdot 9p^7$
16. $m^5 \cdot (-4m^6)$
17. $(-8s^3)(-3s^4)$

Example 3 p. 482

Find each quotient. Express using exponents

18. $\frac{5^{10}}{5^2}$
19. $\frac{7^6}{7}$
20. $\frac{a^8}{a^7}$
21. $\frac{k^{12}}{k^9}$
22. $(-1.5)^8 \div (-1.5)^3$
23. $8^{15} \div 8^9$
24. $r^{20} \div r^6$
25. $(-n)^6 \div (-n)^4$

Example 4 p. 483

26. **SOUND** Sound intensity is measured in *decibels*. The decibel scale is based on powers of ten as shown.

Sound	Decibels	Intensity
rock concert	110	10^{11}
vacuum cleaner	80	10^8
normal conversation	60	10^6
whispering	20	10^2

a. How many times as intense is a rock concert as normal conversation?

b. How many times as intense is a vacuum cleaner as a person whispering?

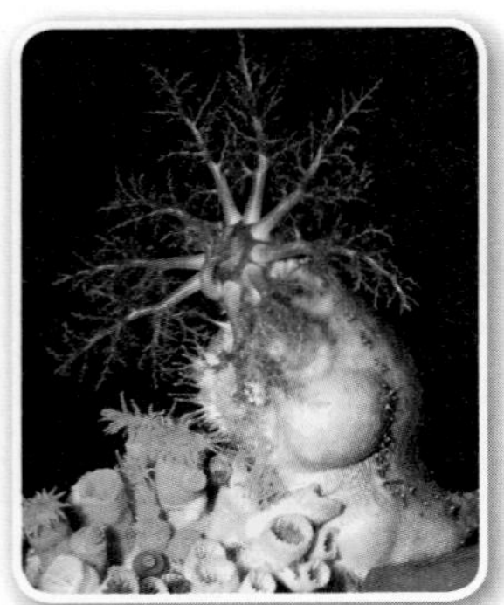

Real-World Link

Sea cucumbers are members of the echinoderm family that live on or near the ocean floor. The smallest species of sea cucumbers are about 10 millimeters long.

Source: OceanLink

27. **RUNNING** A person weighing 5^3 pounds can experience forces 5 times their body weight while running. Find $5^3 \cdot 5$ to find the number of pounds exerted on a person's foot while running.

28. **SEA CUCUMBERS** The largest sea cucumbers are more than 10^2 times longer than the smallest sea cucumbers. Use the information at the left to determine the approximate length of the largest sea cucumbers.

29. **HEALTH** A nurse draws a sample of blood. A cubic millimeter of the blood contains 22^5 red blood cells and 22^3 white blood cells. Compare the number of red blood cells to the number of white blood cells as a fraction. Explain its meaning.

Find each missing exponent.

30. $(5^{\blacksquare})(5^2) = 5^3$
31. $(9^{10})(9^{\blacksquare}) = 9^{15}$
32. $a^{12} \cdot a^{\blacksquare} = a^{19}$
33. $\frac{6^{\blacksquare}}{6^4} = 6^8$
34. $\frac{x^7}{x^{\blacksquare}} = 1$
35. $c^{10} \div c^{\blacksquare} = c^8$

36. **MULTIPLE REPRESENTATIONS** In this problem, you will investigate area and volume. The formulas $A = s^2$ and $V = s^3$ can be used to find the area of a square and the volume of a cube, respectively, with side length s.

a. **TABULAR** Copy and complete the table shown.

b. **VERBAL** How are the area and volume each affected if the side length is doubled? tripled?

c. **VERBAL** How are the area and volume each affected if the side length is squared? cubed?

Side Length (units)	Area of Square (units^2)	Volume of Cube (units^3)
s	s^2	s^3
$2s$	■	■
$3s$	■	■
s^2	■	■
s^3	■	■

Find each product or quotient. Express using exponents.

37. $ab^5 \cdot 8a^2b^5$
38. $10x^3y \cdot (-2xy^2)$
39. $\frac{n^3(n^5)}{n^2}$
40. $\frac{s^7}{s \cdot s^2}$

H.O.T. Problems

Use Higher-Order Thinking Skills

41. **OPEN ENDED** Write two algebraic expressions whose quotient is x^5.

42. **FIND THE ERROR** Addison and Noah are multiplying $(4a^2)(4a^3)$. Is either of them correct? Explain your reasoning.

Addison	Noah
$(4a^2)(4a^3) = 16a^{2+3}$ $= 16a^5$	$(4a^2)(4a^3) = 4a^{2+3}$ $= 4a^5$

43. **CHALLENGE** Use the Quotient of Powers Property and the equation $\frac{a^n}{a^n} = 1$ to show that a nonzero number raised to the zero power equals 1.

44. **REASONING** *True* or *false*. For any integer a, $(-a)^2 = -a^2$. If true, explain your reasoning. If false, give a counterexample.

45. **WRITING IN MATH** Explain how to use division of powers to divide large numbers.

Standardized Test Practice

46. In the metric system, one meter is equal to 10^2 centimeters. One kilometer is 10^3 meters. How many centimeters are in one kilometer?

A 1000
B 10,000
C 100,000
D 1,000,000

47. Which of the following expressions has the same value as $6a^3$?

F $6 \cdot a \cdot a \cdot a$
G $6 + a + a + a$
H $6 + a \cdot a \cdot a$
J $6 \cdot 6 \cdot 6 \cdot a \cdot a \cdot a$

48. Which of the following expressions is equivalent to the product of $5a^3$ and $3a^8$?

A $8a^{11}$
B $8a^{24}$
C $15a^{11}$
D $15a^{24}$

49. SHORT RESPONSE The formula $A = \frac{1}{2}bh$ can be used to find the area of a triangle with base b and height h. Write an expression in simplest form to represent the area of the triangle shown below.

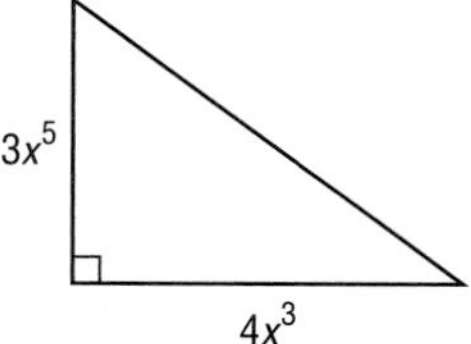

Spiral Review

Write the prime factorization of each number. Use exponents for repeated factors. (Lesson 9-2)

50. 49 **51.** 81 **52.** 100 **53.** 150

ALGEBRA Factor each monomial. (Lesson 9-2)

54. $18x^4$ **55.** $54a^6$ **56.** $-11m^3$ **57.** $-33t^5$

58. MILEAGE Which numbers in the table can be expressed as whole numbers raised to a power? Name the cities and express the numbers as powers. (Lesson 9-1)

Miles to Kentucky Dam	
City	**Miles**
Chicago	400
Evansville	100
Lexington	250
Louisville	200
Nashville	125
Paducah	25
St. Louis	225

59. REAL ESTATE A commission is a fee paid to a salesperson based on a percent of sales. Suppose a real estate agent earns a 3% commission. What commission would be earned for selling a house for $230,000? (Lesson 7-5)

Find each sum or difference. Write in simplest form. (Lesson 3-6)

60. $\frac{3}{5} + \frac{3}{10}$ **61.** $-\frac{1}{2} + \frac{3}{8}$ **62.** $-6\frac{2}{3} - \frac{8}{9}$

Skills Review

Evaluate each expression if $a = -3$, $b = 7$, and $c = 5$. (Lesson 2-5)

63. $\frac{1}{ab}$ **64.** $\frac{1}{(b)(b)}$ **65.** $\frac{c}{75}$ **66.** $\frac{ac}{5ab}$

9-4 Negative Exponents

Then

You evaluated expressions containing positive exponents. (Lesson 9-1)

Now

- Write expressions using negative exponents.
- Evaluate numerical expressions containing negative exponents.

Math Online

glencoe.com

- Extra Examples
- Personal Tutor
- Self-Check Quiz
- Homework Help

Why?

Saturn has seven major rings that are made of ice crystals: rings A, B, C, D, E, F, and G.

a. How many times as great is the maximum thickness than the minimum thickness of ring G?

b. How many times greater is the maximum thickness than the minimum thickness of ring A?

Thickness of Saturn's Rings

Ring	Minimum (km)	Maximum (km)
G	100	1000
A	0.1	1

c. How many times greater is the maximum thickness of ring A than the minimum thickness of ring G? Write your answer as a fraction with an exponent in the denominator.

Negative Exponents Negative exponents are the result of repeated division. Extending the pattern below shows that $\frac{1}{100}$ or $\frac{1}{10^2}$ can be defined as 10^{-2}.

Exponential Form	Standard Form
$10^3 = 10 \cdot 10 \cdot 10$	1000
$10^2 = 10 \cdot 10$	100
10^1	10
10^0	1
10^{-1}	$\frac{1}{10}$
10^{-2}	$\frac{1}{100}$

÷ 10
÷ 10
÷ 10
÷ 10
÷ 10

You can apply the Quotient of Powers rule and the definition of a power to $\frac{x^3}{x^5}$ and write a general rule about negative exponents.

Method 1 Quotient of Powers

$$\frac{x^3}{x^5} = x^{3-5} = x^{-2}$$

Method 2 Definition of Power

$$\frac{x^3}{x^5} = \frac{\overset{1}{\cancel{x}} \cdot \overset{1}{\cancel{x}} \cdot \overset{1}{\cancel{x}}}{\underset{1}{\cancel{x}} \cdot \underset{1}{\cancel{x}} \cdot \underset{1}{\cancel{x}} \cdot x \cdot x} = \frac{1}{x \cdot x} \text{ or } \frac{1}{x^2}$$

Since $\frac{x^3}{x^5}$ cannot have two different values, you can conclude that $x^{-2} = \frac{1}{x^2}$.

Key Concept — **Negative and Zero Exponents** — For Your FOLDABLE

Symbols For $a \neq 0$ and any whole number n, $a^{-n} = \frac{1}{a^n}$.
For $a \neq 0$, $a^0 = 1$.

Example $8^{-2} = \frac{1}{8^2}$ $\quad x^0 = 1, x \neq 0$

Watch Out!

Remember 2^{-3} equals $\frac{1}{2^3}$ *not* -6.

EXAMPLE 1 Write Expressions using Positive Exponents

Write each expression using a positive exponent.

a. 2^{-3}

$2^{-3} = \frac{1}{2^3}$ **Definition of negative exponent**

b. m^{-4}

$m^{-4} = \frac{1}{m^4}$ **Definition of negative exponent**

Check Your Progress

1A. 3^{-5} **1B.** y^{-3} **1C.** 2^0

Personal Tutor glencoe.com

EXAMPLE 2 Use Negative Exponents

Write each fraction as an expression using a negative exponent other than −1.

a. $\frac{1}{4^2}$

$\frac{1}{4^2} = 4^{-2}$ **Definition of negative exponent**

b. $\frac{1}{100}$

$\frac{1}{100} = \frac{1}{10^2}$ **Definition of exponent**

$= 10^{-2}$ **Definition of negative exponent**

Check Your Progress

2A. $\frac{1}{6^3}$ **2B.** $\frac{1}{25}$ **2C.** $\frac{1}{27}$

Personal Tutor glencoe.com

Negative exponents are often used in science when dealing with very small numbers. Usually the number is a power of 10.

Real-World Link

Geckos cannot blink because they have a transparent scale that covers each eye rather than eyelids. Most gecko species live for about three years, but some can live for up to 20 years.

Source: Wellington Zoo

Real-World EXAMPLE 3 Use Negative Exponents

REPTILES Geckos have tiny hairs on the bottom of their feet that are about 0.000001 meter long. Write the decimal as a fraction and as a power of 10.

$0.000001 = \frac{1}{1{,}000{,}000}$ **Write the decimal as a fraction.**

$= \frac{1}{10^6}$ **$1{,}000{,}000 = 10^6$**

$= 10^{-6}$ **Definition of negative exponent**

Therefore, 0.000001 is $\frac{1}{1{,}000{,}000}$ as a fraction and 10^{-6} as a power of 10.

Check Your Progress

3. FISH The slowest moving fish is a sea horse. It swims at a maximum speed of 0.0001 mile per minute. Write the decimal as a fraction and as a power of ten.

Personal Tutor glencoe.com

StudyTip

Order of Operations Remember to follow the order of operations when evaluating expressions.

Evaluate Expressions Algebraic expressions containing negative exponents can be written using positive exponents and then evaluated.

EXAMPLE 4 Algebraic Expressions with Negative Exponents

Evaluate $4a^{-5}$ if $a = -2$.

$4a^{-5} = 4 \cdot (-2)^{-5}$ — Replace a with -2.

$= 4 \cdot \frac{1}{(-2)^5}$ — Definition of negative exponent

$= 4 \cdot \frac{1}{-32}$ — Find $(-2)^5$.

$= \overset{1}{\cancel{4}} \cdot \frac{1}{-\underset{8}{\cancel{32}}}$ — Simplify.

$= \frac{1}{-8}$ — Simplify.

Check Your Progress

Evaluate each expression if $m = 4$ and $n = 3$.

4A. m^{-2} **4B.** $6mn^{-4}$

Personal Tutor glencoe.com

Check Your Understanding

Example 1 p. 487

Write each expression using a positive exponent.

1. 6^{-2} **2.** $(-2)^{-3}$ **3.** x^{-5} **4.** b^{-7}

Example 2 p. 487

Write each fraction as an expression using a negative exponent other than -1.

5. $\frac{1}{2^6}$ **6.** $\frac{1}{8^2}$ **7.** $\frac{1}{9}$ **8.** $\frac{1}{36}$

Example 3 p. 487

9. BASEBALL When a baseball is hit, it comes in contact with the bat for less than 0.001 of a second. Write 0.001 using a negative exponent other than -1.

Example 4 p. 488

Evaluate each expression if $x = -4$ and $y = 2$.

10. y^{-7} **11.** x^{-3} **12.** 3^x **13.** $8y^{-4}$

Practice and Problem Solving

Example 1 p. 487

Write each expression using a positive exponent.

14. 11^{-6} **15** 7^{-1} **16.** $(-4)^{-5}$ **17.** $(-5)^{-4}$

18. a^{-2} **19.** k^{-8} **20.** b^{-15} **21.** r^{-20}

Example 2 p. 487

Write each fraction as an expression using a negative exponent other than -1.

22. $\frac{1}{9^4}$ **23.** $\frac{1}{10^3}$ **24.** $\frac{1}{7^6}$ **25.** $\frac{1}{6^5}$

26. $\frac{1}{4}$ **27.** $\frac{1}{49}$ **28.** $\frac{1}{144}$ **29.** $\frac{1}{125}$

Example 3 p. 487

Write each decimal using a negative exponent.

30. **SPACE** The minimum thickness of Saturn's A ring is one tenth kilometer.

31. **SCIENCE** The diameter of a typical atom is 0.00000001 centimeter.

Example 4 p. 488

Evaluate each expression if $n = 3$, $p = -2$, and $q = 6$.

32. n^{-5}
33. $(pq)^{-2}$
34. p^{-3}
35. $-q^{-1}$
36. 9^{p}
37. 2^{-q}
38. $6n^{-3}$
39. $4pq^{-2}$
40. $7^{p}q^{2}$

41. **SCIENCE** The table at the right shows the average lengths of different objects.

Object	Length (m)
pinhead	10^{-3}
cell	10^{-4}
virus	10^{-7}
atom	10^{-10}

Source: NASA

a. How many times as long is a virus than an atom?

b. About how many viruses would fit across a pinhead?

c. A football field is about 10^2 meters long. How many times as long is this than a cell?

42. **SCIENCE** The shortest period of time ever measured directly was a light burst of a laser lasting about 0.000000000000001 second. Write this decimal as a fraction and as a power of ten.

Real-World Link

Milk is an excellent source of calcium. People ages 9–13 need about 1300 milligrams of calcium per day to strengthen bones and improve dental health.

Source: Milk Matters

43. **PHYSICAL SCIENCE** The pH of a substance is a measure of its acidity. The pH scale ranges from 0 to 14, with a pH of 7 being neutral. As the pH decreases, the substance is more acidic. The table shows the pH of several common substances.

	Substance	pH	Hydrogen Ion Concentration
acids	coffee	5	10^{-5}
	milk	6	10^{-6}
neutral	pure water	7	10^{-7}
bases	egg whites	8	10^{-8}
	baking soda	9	10^{-9}

Source: Vision Learning

a. Which substance in the table has the greatest hydrogen ion concentration? How many times as great is that hydrogen ion concentration than that of egg whites?

b. Which substance has a hydrogen ion concentration of *one millionth*?

c. As the pH increases by 1, describe what happens to the concentration of hydrogen ions.

d. How many times as great is the hydrogen ion concentration of coffee as the hydrogen ion concentration of pure water?

44. **SAND** A grain of sand has a volume of about $\frac{1}{10{,}000}$ cubic millimeters.

a. Write this number using a negative exponent.

b. An empty bottle used to create sand art can hold about 10^{10} grains of sand. What is the approximate volume of the sand art bottle?

c. If one cubic centimeter is equal to 10^3 cubic millimeters, how many cubic centimeters of sand will the bottle hold?

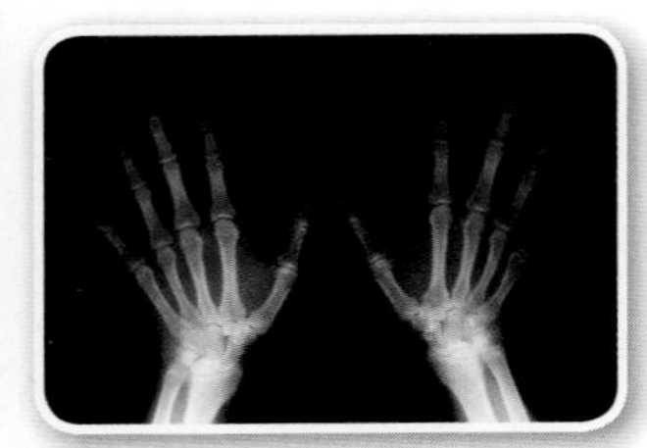

Real-World Link

The wavelengths of X-rays are between 1 and 10 nanometers.

Source: *Biology*, Raven

45. **MULTIPLE REPRESENTATIONS** In this problem, you will explore negative exponents. When using powers of 10, $10^{-1} = \frac{1}{10}$ or 0.1.

   a. **TABULAR** Copy and complete the table shown.

Power	Fraction	Decimal
10^{-1}	$\frac{1}{10}$	0.1
10^{-2}	■	■
10^{-3}	■	■
10^{-4}	■	■
10^{-5}	■	■

   b. **VERBAL** Do you notice a pattern between the negative powers of 10 and their decimal equivalents? Explain.

   c. **VERBAL** Write a verbal rule that could be used to find the decimal equivalent of any negative power of 10.

   d. **NUMERICAL** Use the rule from part **c** to find the value of 10^{-12}.

46. **SCIENCE** A nanometer is equal to a billionth of a meter. Use the information at the left to express the greatest wavelength of an X-ray in meters. Write the expression using a negative exponent.

Use the Product of Powers and Quotient of Powers rules to simplify each expression.

47. $a^3 \div a^9$

48. $n^{-4} \cdot n^{-2}$

49. $\dfrac{x^{-3}y^4}{x^{-2}y^2}$

50. $\dfrac{28b^4c^3}{4b^8c^{-5}}$

H.O.T. Problems

Use Higher-Order Thinking Skills

51. **OPEN ENDED** Write a power that has a negative exponent and show the steps you take to write the power as a fraction.

52. **FIND THE ERROR** Jeannette and Mahala are evaluating the expression $2 \cdot 4^{-2}$. Is either of them correct? Explain your reasoning.

Jeannette	Mahala
$2 \cdot 4^{-2} = 2 \cdot \frac{1}{16}$ $= \frac{2}{16}$ or $\frac{1}{8}$	$2 \cdot 4^{-2} = (2 \cdot 4)^{-2}$ $= 8^{-2}$ or $\frac{1}{64}$

53. **REASONING** Consider the following sets of numbers:

$$\text{Set 1: } 2^{-2}, (-2)^{-2}, (-2)^2, 2^2$$

$$\text{Set 2: } 2^{-3}, (-2)^{-3}, (-2)^3, 2^3$$

   a. Simplify each expression in Set 1. Which expressions, if any, are equal?

   b. Simplify each expression in Set 2. Which expressions, if any, are equal?

   c. Explain why the number of equal expressions is different for each list.

   d. Finish the conjecture: $2^{-x} = (-2)^{-x}$, if and only if ________.

   e. Finish the conjecture: $(-2)^x = 2^x$, if and only if ________.

54. **CHALLENGE** Compare and contrast x^{-n} and x^n where $x \neq 0$. Then give a numerical example to show the relationship.

55. **REASONING** Investigate the fraction $\frac{1}{2^n}$. Does it increase or decrease as the value of n increases? Explain.

56. **WRITING IN MATH** Explain the difference between the expressions $(-3)^4$ and 3^{-4}.

Standardized Test Practice

57. DNA contains the genetic code of an organism. The length of a DNA strand is about 10^{-7} meter. Which of the following represents the length of the DNA strand as a decimal?

A 0.00001 m
B 0.000001 m
C 0.0000001 m
D 0.00000001 m

58. When simplified, 2^{-5} is equal to which of the following?

F -32
G $-\frac{1}{32}$
H $\frac{1}{32}$
J 32

59. Which of the following shows the expressions 4^0, 4^{-2}, 4^2, and 4^{-3} in order from least to greatest?

A $4^{-3}, 4^{-2}, 4^2, 4^0$

B $4^0, 4^{-2}, 4^{-3}, 4^2$

C $4^2, 4^0, 4^{-2}, 4^{-3}$

D $4^{-3}, 4^{-2}, 4^0, 4^2$

60. **SHORT RESPONSE** It takes light 5.3×0.000001 seconds to travel one mile. Write 0.000001 as a fraction and as a power of 10.

Spiral Review

61. **ARTS AND CRAFTS** When a piece of paper is cut in half, the result is two smaller pieces of paper. When the two smaller pieces are stacked and then cut, the result is four pieces of paper. The number of resulting sheets of paper after c cuts is 2^c. (Lesson 9-3)

a. How many more pieces of paper are there if a piece of paper is cut and stacked 8 times than when a piece of paper is cut and stacked 5 times?

b. A stack of 500 sheets of notebook paper is about 1 inch thick. How thick would your stack be if you were able to make 10 cuts?

ALGEBRA **Factor each monomial.** (Lesson 9-2)

62. $-105x^2yz^5$
63. $-r^2st$
64. $53fg$

State the slope and y-intercept of each equation. (Lesson 8-7)

65. $2x + y = -3$
66. $5x + 4y = 20$
67. $y = 4$

68. **FOOD** The results of a survey about favorite hamburger condiments are shown in the table at the right. Which condiment was chosen by the most people? Explain. (Lesson 7-1)

Hamburger Condiment	
Condiment	**Part**
mustard	22%
ketchup	$\frac{2}{5}$
relish	0.2

Skills Review

Find each product. (Lesson 3-3)

69. 25×0.001
70. 107×0.0001
71. 3.8×0.01
72. 18×100
73. 76×1000
74. $134 \times 100{,}000$

CHAPTER 9

Mid-Chapter Quiz

Lessons 9-1 through 9-4

Write each expression using exponents. (Lesson 9-1)

1. $8 \cdot n \cdot n \cdot n$
2. $(x - 1)(x - 1)$

Evaluate each expression. (Lesson 9-1)

3. 6^3
4. $2^5 \cdot 3$
5. $3^4 \cdot 5$
6. $4^3 \cdot 2^2 \cdot 3$

7. **MULTIPLE CHOICE** The number of acres consumed by a forest fire triples every two hours. Which of the following expressions represents the number of acres consumed after 1 day? (Lesson 9-1)

Hours	2	4	6	8	10
Acres Consumed	3^1	3^2	3^3	3^4	3^5

A 3^{10} acres
B 3^{12} acres
C 3^{18} acres
D 3^{24} acres

Evaluate each expression if $x = -3$. (Lesson 9-1)

8. $x^3 - 4$
9. $6(x + 1)^2$

Write the prime factorization of each number. Use exponents for repeated factors. (Lesson 9-2)

10. 42
11. 99
12. 64

13. **MULTIPLE CHOICE** The kitchen floor shown is to be tiled.

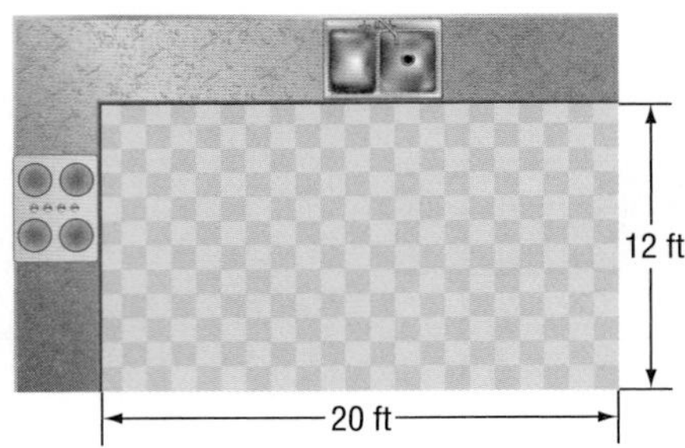

If the tiles are only available in dimensions that are prime numbers, which set of tile dimensions could *not* be used to tile the floor? (Lesson 9-2)

F 2 ft by 2 ft
G 2 ft by 3 ft
H 2 ft by 5 ft
J 3 ft by 3 ft

ALGEBRA Factor each monomial. (Lesson 9-2)

14. $7a^3$
15. $-12xy^2$

Find each product or quotient. Express using exponents. (Lesson 9-3)

16. $8^4 \cdot 8^5$
17. $c^2 \cdot c^6$
18. $\frac{5^9}{5^3}$
19. $\frac{x^7}{x^2}$
20. $(3n)(6n^2)$
21. $2y^3 \cdot 7y^3$

22. **TRAVEL** The table compares the number of people in Wyoming who drive to work to the number of people who walk to work in a recent year. How many times as many people drive than walk to work? (Lesson 9-3)

Mode of Transportation	Number of People
drove	10^5
walked	10^3

Source: U.S. Census Bureau

Write each expression using a positive exponent. (Lesson 9-4)

23. 2^{-5}
24. $(-6)^{-2}$
25. q^{-11}

Write each fraction as an expression using a negative exponent other than −1. (Lesson 9-4)

26. $\frac{1}{5^3}$
27. $\frac{1}{9}$
28. $\frac{1}{16}$

For Exercises 29 and 30, use the table that shows wavelengths. (Lesson 9-4)

Wave	Wavelength (m)
radio waves	10^0
microwaves	10^{-2}
X-rays	10^{-10}

Source: NASA

29. Write the wavelength of an X-ray as a fraction without an exponent and in words.
30. How many times as long is the wavelength of a radio wave as that of a microwave?

Evaluate each expression if $a = -3$ and $b = 5$. (Lesson 9-4)

31. b^{-2}
32. a^{-4}
33. 6^a

9-5 Scientific Notation

Why?

The width of Earth is 12.76 million meters. The width of a plant cell is a trillion times smaller with a width of 12.76 millionths of a meter.

a. Write 12.76 million in numbers.

b. Write 12.76 millionths as a decimal.

Then

You have already compared and ordered integers. (Lesson 2-1)

Now

- Express numbers in standard form and in scientific notation.
- Compare and order numbers written in scientific notation.

New Vocabulary

standard form
scientific notation

Math Online

glencoe.com

- Extra Examples
- Personal Tutor
- Self-Check Quiz
- Homework Help

Scientific Notation Numbers that do not contain exponents are written in **standard form**. However, when you deal with very large numbers like 12,760,000 or very small numbers like 0.00001276 it is difficult to keep track of the place value. A number that is expressed as a product of a factor and a power of 10 is written in **scientific notation**.

Key Concept — Scientific Notation (For Your FOLDABLE)

Words A number is expressed in scientific notation when it is written as the product of a factor and a power of 10. The factor must be greater than or equal to 1 and less than 10.

Symbols $a \times 10^n$, where $1 \leq a < 10$ and n is an integer.

Examples $3{,}500{,}000 = 3.5 \times 10^6$

$0.00004 = 4 \times 10^{-5}$

EXAMPLE 1 Express Numbers in Standard Form

Express each number in standard form.

a. 2×10^3

$2 \times 10^3 = 2 \times 1000$ — $10^3 = 1000$

$= 2000$ — Move the decimal point 3 places to the right.

b. 6.8×10^5

$6.8 \times 10^5 = 6.8 \times 100{,}000$ — $10^5 = 100{,}000$

$= 680{,}000$ — Move the decimal point 5 places to the right.

c. 3.25×10^{-4}

$3.25 \times 10^{-4} = 3.25 \times 0.0001$ — $10^{-4} = 0.0001$

$= 0.000325$ — Move the decimal point 4 places to the left.

StudyTip

Powers of Ten The exponent tells you how many places to move the decimal point.

Check Your Progress

1A. 4×10^2 **1B.** 5.94×10^7 **1C.** 1.3×10^{-3}

Personal Tutor glencoe.com

StudyTip

Sign of Exponents
- If a number in standard form is greater than or equal to 1, then the exponent is *positive*.
- If a number is between 0 and 1, the exponent is *negative*.

EXAMPLE 2 Express Numbers in Scientific Notation

Express each number in scientific notation.

a. 4,000,000

$4{,}000{,}000 = 4 \times 1{,}000{,}000$ — **The decimal point moves 6 places.**

$= 4 \times 10^6$ — **The exponent is positive.**

b. 5800

$5800 = 5.8 \times 1000$ — **The decimal point moves 3 places.**

$= 5.8 \times 10^3$ — **The exponent is positive.**

c. 0.072

$0.072 = 7.2 \times 0.01$ — **The decimal point moves 2 places.**

$= 7.2 \times 10^{-2}$ — **The exponent is negative.**

Check Your Progress

2A. 900 **2B.** 18,900 **2C.** 0.000064

Personal Tutor glencoe.com

Real-World Link

The Amazon Rainforest includes areas in Brazil, Venezuela, Colombia, Ecuador, and Peru. More than 20 percent of the world's oxygen is produced in this rain forest. More than half of the world's estimated 10 million species of plants, animals, and insects live in the tropical rain forests.

Source: Raintree Nutrition

Real-World EXAMPLE 3 Solve Problems using Scientific Notation

RAINFOREST Scientists estimate that there are over 3.5×10^6 ants per acre in the Amazon Rainforest. If the Amazon Rainforest covers approximately 1 billion acres, find the total number of ants. Write in scientific notation.

Understand You know there are about 3.5×10^6 ants per acre and there are about 1 billion acres. You need to know the total number of ants.

Plan Write 1 billion in scientific notation. Multiply the number of ants per acre by the number of acres to find the total number of ants.

Solve 1 billion $= 1 \times 10^9$

total number of ants = number per acre × number of acres

$= (3.5 \times 10^6) \times (1 \times 10^9)$

You can use a calculator to find the product.

3.5 [2nd] [EE] 6 [X] 1 [2nd] [EE] 9 [ENTER] 3.5E15

So, there are about 3.5×10^{15} ants in the Amazon Rainforest.

Check Check using mental math.

$(3.5 \times 10^6)(1 \times 10^9)$

$= (3.5 \times 1)(10^6 \times 10^9)$ — **Commutative Property**

$= 3.5 \times 10^{15}$ ✓ — **Product of Powers Property**

Check Your Progress

3. INSECTS About 1×10^6 fruit flies weigh 1.3×10^2 pounds. How much does one fruit fly weigh? Write in scientific notation.

Personal Tutor glencoe.com

Compare and Order Numbers To compare and order numbers in scientific notation, first compare the exponents. With positive numbers, the number with a greater exponent is greater. If the exponents are the same, compare the factors.

Real-World EXAMPLE 4 Order Numbers in Scientific Notation

EARTH SCIENCE The table shows different geologic time periods. Order the time periods from oldest to youngest.

Geologic Time Periods	
Period	**Number of Years Ago**
Jurassic	2.08×10^8
Silurian	4.38×10^8
Tertiary	6.64×10^7
Triassic	2.45×10^8

Source: U.S. Geological Survey

Step 1 Order the numbers according to their exponents.

The Tertiary period has an exponent of 7. So, it is the youngest period.

Step 2 Order the numbers with the same exponent by comparing the factors.

$4.38 > 2.45 > 2.08$

Silurian, Triassic, Jurassic

So, $4.38 \times 10^8 > 2.45 \times 10^8 > 2.08 \times 10^8$.

The time periods ordered from oldest to youngest is Silurian, Triassic, Jurassic, and Tertiary.

Check Your Progress

4. **EARTH SCIENCE** Approximately 1.372×10^7 square kilometers of Antarctica and about 1.834×10^6 square kilometers of Greenland are covered by an ice cap. Which land mass has a greater area covered by ice?

Personal Tutor glencoe.com

Check Your Understanding

Example 1 p. 493

Express each number in standard form.

1. 4.16×10^3
2. 3.2×10^{-2}
3. 1.075×10^5

Example 2 p. 494

Express each number in scientific notation.

4. 1,600,000
5. 135,000
6. 0.008

Example 3 p. 494

7. **ROADS** The U.S. has the most miles of road in the world at about 4×10^6 miles. Japan has about 7.3×10^5 miles. How many more miles of roads does the U.S. have than Japan? Write in scientific notation.

Example 4 p. 495

Order each set of numbers from least to greatest.

8. 3.4×10^2, 3.5×10^2, 3.7×10^{-2}, 400
9. 6.5×10^3, 6.12×10^5, 6.01×10^4, 6.1×10^{-2}

Practice and Problem Solving

● = **Step-by-Step Solutions** begin on page R11. **Extra Practice** begins on page 810.

Example 1 p. 493

Express each number in standard form.

10. 6.89×10^4 **11.** 1.5×10^{-4} **12.** 2.3×10^{-5}

13. 9.51×10^{-3} **14.** 3.062×10^6 **15.** 7.924×10^2

16. MONEY A dollar bill is approximately 1.09×10^{-2} centimeter thick.

17. E-MAILS It is estimated that more than 1.71×10^4 billion e-mails are sent each day around the world. Most of these are spam and viruses.

Example 2 p. 494

Express each number in scientific notation.

18. 700,000 **19** 32,000,000 **20.** 0.045

21. 0.000918 **22.** 1,000,000 **23.** 0.006752

Example 3 p. 494

24. WEATHER Each minute, there are approximately 6×10^3 flashes of lightning around the world. The air around a lightning bolt is heated to about 5.4×10^4 degrees Fahrenheit, which is about five times hotter than the Sun.

a. About how many flashes of lightning are there in a day? Write in scientific notation and in standard form.

b. About how hot is the Sun in degrees Fahrenheit? Write in scientific notation and in standard form.

Example 4 p. 495

Order each set of numbers from least to greatest.

25. $2.4 \times 10^2, 2.45 \times 10^{-2}, 2.45 \times 10^2, 2.4 \times 10^{-2}$

26. 2.81×10^4, 2805, $2.08 \times 10^5, 3.2 \times 10^4, 3.024 \times 10^2$

27. $5.9 \times 10^6, 5.9 \times 10^4, 5.01 \times 10^5, 5.1 \times 10^{-3}$

28. 9,562,301, $9.05 \times 10^{-6}, 9.5 \times 10^6$, 905,000

29. GOLD A sheet of gold leaf is approximately 1.25×10^{-5} centimeter thick.

a. Write the value of the thickness as a decimal.

b. Use the formula $V = \ell wh$ to find the volume in cubic meters of a sheet of gold that is 2 meters wide and 5 meters long.

Real-World Link

One ounce of gold can be beaten out to 300 square feet. The thinnest sheets of gold are just a few atoms thick.

Source: The Physics Factbook

30. SYRUP List the states in the table from least to greatest production of maple syrup.

State	Amount of Syrup Produced (L)
Maine	1.10×10^6
New Hampshire	3.14×10^5
New York	9.65×10^5
Vermont	1.89×10^6
Wisconsin	3.79×10^5

Source: *Book of World Records*

31. TRAFFIC In a recent year, route U.S. 59 in the Houston area averaged approximately 338,510 vehicles per day. About how many vehicles was this during the entire year? Write the number in scientific notation. Verify your solution by using estimation.

32. SPACE The Moon travels around Earth at a speed of about 3.68×10^3 kilometers per hour. If the Moon orbits Earth every 27.3 days, about how far does it travel in one orbit around Earth?

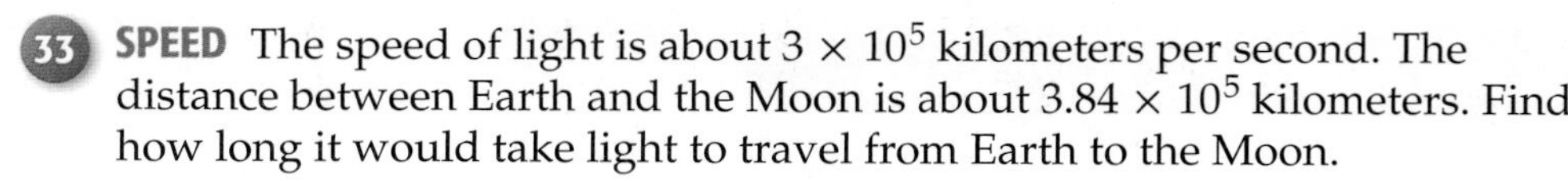

33 **SPEED** The speed of light is about 3×10^5 kilometers per second. The distance between Earth and the Moon is about 3.84×10^5 kilometers. Find how long it would take light to travel from Earth to the Moon.

Replace each ● with <, >, or = to make a true statement.

34. 5.72×10^8 ● 5.8×10^8

35. 35,400 ● 35.4×10^3

36. 0.042 ● 4.2×10^{-3}

37. 5×10^5 ● 5,000,000

Real-World Link

The hump on the back of the rhino's neck is made up of muscle that is needed to raise its head. This muscle can be $\frac{1}{4}$ of its total body weight.

38. **ANIMALS** The table shows the weights of various marine and land mammals.

Mammal	Weight (pounds)
African Elephant	1.44×10^4
Blue Whale	2.87×10^5
Fin Whale	9.92×10^4
Right Whale	8.82×10^4
White Rhinoceros	7.94×10^3

a. Order the animals' weights from heaviest to lightest.

b. Which animal is about 10 times lighter than a right whale?

c. About how many times heavier is the blue whale than the African elephant?

d. Estimate the combined weight of the fin whale, right whale, and white rhinoceros. Write the combined weight in scientific notation and in standard form.

Evaluate each expression. Express in scientific notation and in standard form.

39. $(6.3 \times 10^5) + (2.7 \times 10^7)$

40. $(8.5 \times 10^{-3}) - (4.8 \times 10^{-5})$

41. $(6.2 \times 10^2)(9.1 \times 10^3)$

42. $\dfrac{16.4 \times 10^{-5}}{3.2 \times 10^{-7}}$

H.O.T. Problems

Use **H**igher-**O**rder **T**hinking Skills

43. **OPEN ENDED** Write two numbers in scientific notation with different exponents. Then find the sum, difference, product, and quotient of the two numbers. Write the answers in scientific notation.

44. **CHALLENGE** A *googol* is a number that is 1 followed by 100 zeros. A *centillion* is a number that is 1 followed by 303 zeros. Write each of these numbers in scientific notation.

45. **REASONING** Miami is the second most populous city in Florida.

a. Which number better describes the population of Miami, 3.8×10^4 or 3.8×10^6? Explain.

b. Express Miami's population in another form.

c. Which notation is best to use when describing population? Explain.

46. **CHALLENGE** Which number is twice as great as 3×10^2: 6×10^2, 3×10^4, or 6×10^4? Explain.

47. **WRITING IN MATH** Your friend thinks 7.8×10^3 is greater than 6.5×10^2 because $7.8 > 6.5$. Explain why your friend's reasoning is incorrect.

Standardized Test Practice

48. The slowest land mammal is the three-toed sloth that moves 0.07 mile per hour. Which expression represents this number?

A 7×10^{-3}
B 7×10^{-2}
C 7×10^{2}
D 7×10^{3}

49. The distance from Earth to the Sun is about 9.6×10^7 miles. Which of the following represents this distance?

F 9,600,000
G 96,000,000
H 960,000,000
J 9,600,000,000

50. SHORT RESPONSE The weight of a fruit fly is about 1.3×10^{-4} pound. How much would one million fruit flies weigh?

51. EXTENDED RESPONSE A 45-acre farm in Florida produces 183.2 tons of avocados per year. One ton is 2000 pounds.

a. Write an expression, in scientific notation, for the number of pounds of avocados produced per year.

b. Write an expression, in scientific notation, for the average number of pounds of avocados produced per acre.

c. Find the average number of pounds produced per acre. Round to the nearest whole number.

Spiral Review

52. MEDICINE Which type of molecule in the table has a greater mass? How many times greater is it than the other type? (Lesson 9-4)

Molecule	Mass (kg)
penicillin	10^{-18}
insulin	10^{-23}

Find each product or quotient. Express using exponents. (Lesson 9-3)

53. $a \cdot a^5$

54. $(n^4)(n^4)$

55. $-3x^2(4x^3)$

56. $\frac{3^8}{3^5}$

57. MEASUREMENT Use the scale drawing shown. (Lesson 5-1)

a. What is the area of the lawn?

b. Suppose you want to fertilize the lawn. If one bag of fertilizer covers 2500 square feet, how many bags of fertilizer should you buy?

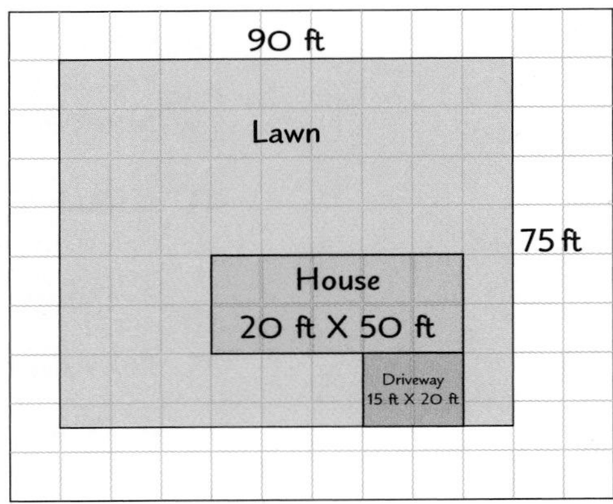

58. RANCHING The largest ranch in the world is in the Australian Outback. It is about 12,000 square miles, which is five times the size of the largest United States ranch. Write and solve a multiplication equation to find the size of the largest United States ranch. (Lesson 4-4)

Skills Review

Write each expression using exponents. (Lesson 9-1)

59. $4 \cdot 4 \cdot 4 \cdot 4 \cdot 4$

60. $(6 \cdot 6 \cdot 6) \cdot 6$

61. $3 \cdot 2 \cdot 3 \cdot 2 \cdot 2$

9-6 Powers of Monomials

Then

You multiplied and divided monomials. (Lesson 9-3)

Now

- Find the power of a power.
- Find the power of a product.

Math Online

glencoe.com

- Extra Examples
- Personal Tutor
- Self-Check Quiz
- Homework Help

Why?

Refer to Squares A, B, C, and D shown below.

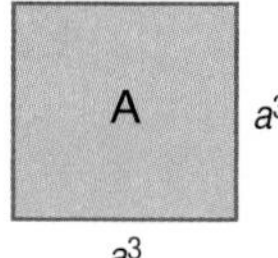

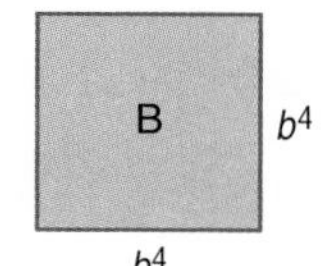

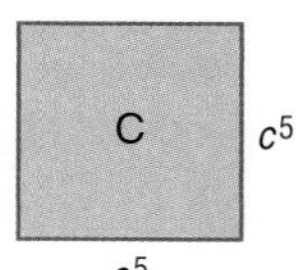

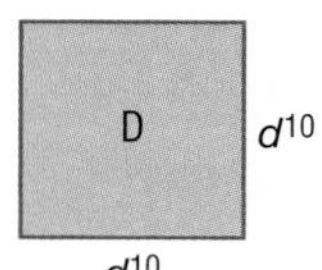

a. Use the squares to complete the table. The first one is done for you.

Square	Area = Side Length Squared	Area using Product of Powers
A	$(a^3)^2$	$a^3 \cdot a^3 = a^{3+3}$ or a^6
B	■	■
C	■	■
D	■	■

b. What is the relationship between the exponents in column 2 and the final exponent in column 3?

Power of a Power You can use the property for finding the *product* of powers to find a property for finding the *power* of a power.

$$(4^2)^5 = \overbrace{(4^2)(4^2)(4^2)(4^2)(4^2)}^{\text{5 factors}}$$
$$= 4^{2+2+2+2+2} \quad \leftarrow \text{Product of Powers Property}$$
$$= 4^{10}$$

Key Concept Power of a Power Property — For Your FOLDABLE

Words To find the power of a power, multiply exponents.

Symbols $(a^m)^n = a^{m \cdot n}$

Examples $(7^4)^2 = 7^{4 \cdot 2}$ or 7^8 $\qquad (n^3)^5 = n^{3 \cdot 5}$ or n^{15}

Watch Out!

Power of a Power When finding the power of a power, do not add the exponent. $(9^6)^2 = 9^{12}$, not 9^8.

EXAMPLE 1 Find the Power of a Power

Simplify.

a. $(9^6)^2$

$(9^6)^2 = 9^{6 \cdot 2}$ Power of a Power

$= 9^{12}$ Simplify.

b. $(q^4)^5$

$(q^4)^5 = q^{4 \cdot 5}$ Power of a Power

$= q^{20}$ Simplify.

Check Your Progress

1A. $(4^3)^7$ **1B.** $(n^4)^4$ **1C.** $(2^3)^{-2}$

Personal Tutor glencoe.com

Watch Out!

Power of a Product When finding the power of a product, do not multiply the constant by the exponent. $(2x^3)^5 = 32x^{15}$ not $10x^{15}$.

Power of a Product The Power of a Power Property can be extended to find the power of a product.

$$(2x^3)^5 = \underbrace{(2x^3)(2x^3)(2x^3)(2x^3)(2x^3)}_{\text{5 factors}}$$

$= 2^5 \cdot (x^3)^5$	Associative and Commutative Properties of Multiplication
$= 2^5 \cdot (x^3) \cdot (x^3) \cdot (x^3) \cdot (x^3) \cdot (x^3)$	Power of a Power
$= 2^5 \cdot x^{3+3+3+3+3}$	Product of Powers Property
$= 32 \cdot x^{15}$ or $32x^{15}$	Simplify.

Notice the power of each factor in the final power above.

Key Concept **Power of a Product Property** *For Your FOLDABLE*

Words To find the power of a product, find the power of each factor and multiply.

Symbols $(ab)^m = a^m b^m$, for all numbers a and b and any integer m

Examples $(7x^4)^2 = 7^2 \cdot (x^4)^2$ or $49x^8$

EXAMPLE 2 Find the Power of a Product

Simplify.

a. $(5r^7)^2$

$(5r^7)^2 = 5^2 \cdot (r^7)^2$	Power of a Product
$= 5^2 \cdot r^{7 \cdot 2}$	Power of a Power
$= 25r^{14}$	Simplify.

b. $(2x^6y^3)^4$

$(2x^6y^3)^4 = 2^4 \cdot (x^6)^4 \cdot (y^3)^4$	
$= 2^4 \cdot (x^{6 \cdot 4}) \cdot (y^{3 \cdot 4})$	
$= 16x^{24}y^{12}$	Simplify.

Check Your Progress

2A. $(6w^4)^5$ **2B.** $(-4s^5t^7)^3$ **2C.** $(3x^{-2}y^4)^2$

Personal Tutor glencoe.com

EXAMPLE 3 Find the Power of a Product

GEOMETRY Express the volume of the cube as a monomial.

$V = s^3$	Formula for volume of a cube
$= (2xy^5)^3$	Replace s with $2xy^5$.
$= 2^3x^3(y^5)^3$	Power of a Product
$= 8x^3y^{15}$	Simplify.

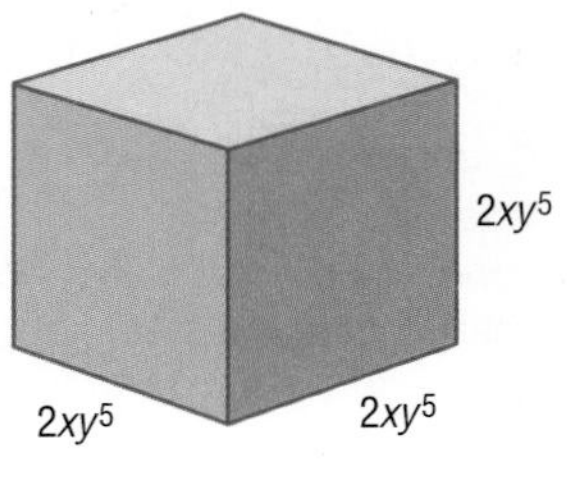

Check Your Progress

3. GEOMETRY Find the area of a square with sides of length $7a^8b^2$.

Personal Tutor glencoe.com

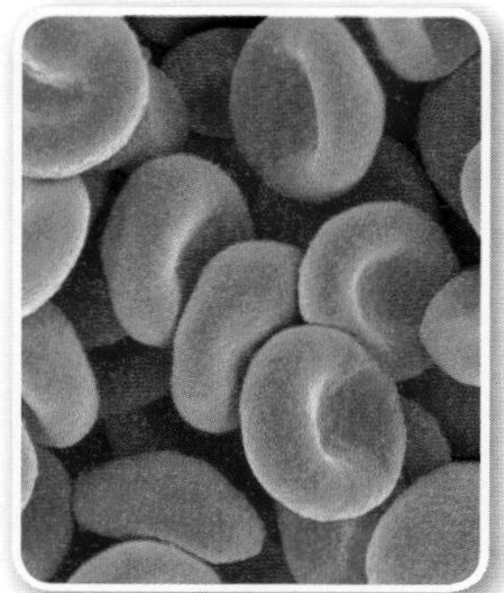

Real-World Link

Red blood cells make up about 40% of total blood volume and carry oxygen to the body. A few drops of whole blood contains approximately one billion red blood cells.

Source: University of Iowa Blood Center

Real-World EXAMPLE 4 Find the Power of a Product

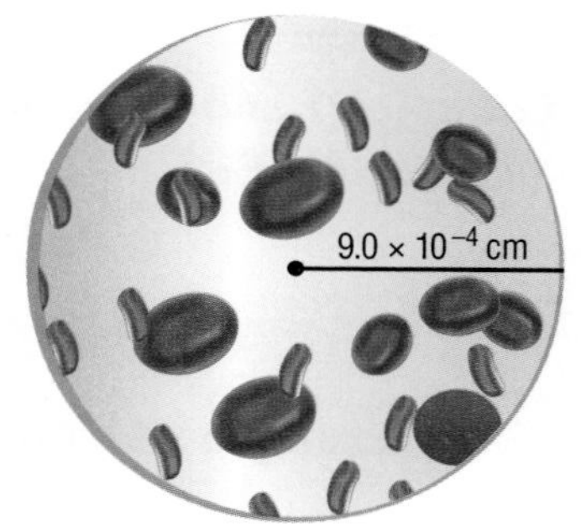

BIOLOGY Ebony wanted to view red blood cells under a microscope with a 1000X magnification. The radius of the circular field she could view is $9 \cdot 10^{-4}$ centimeters. Find the area of the viewing field. Use the formula $A = 3.14r^2$ where A is the area of a circle and r is the radius. Express your answer in scientific notation.

$A = 3.14 \cdot r^2$ — **Write the equation.**

$= 3.14 \cdot (9 \cdot 10^{-4})^2$ — **Replace r with $9.0 \cdot 10^{-4}$.**

$= 3.14 \cdot 9^2 \cdot (10^{-4})^2$ — **Power of a Product**

$= 3.14 \cdot 81 \cdot 10^{-8}$ — **Simplify.**

$= 2.5434 \times 10^{-6}$ — **Multiply. Write the answer in scientific notation.**

The area of the field is 2.5434×10^{-6} square centimeters.

Check Your Progress

4. **BIOLOGY** The radius of a grain of pollen from a certain flower is 2.5×10^{-3} millimeters. Use the formula $V = \frac{4}{3}(3.14)r^3$ to find the volume of the pollen. Express your answer in scientific notation rounded to the nearest hundredth.

Personal Tutor glencoe.com

Concept Summary: Powers

For Your FOLDABLE

Product of Powers	Quotient of Powers	Powers of Powers
Add exponents. $a^m \cdot a^n = a^{m+n}$	Subtract exponents. $a^m \div a^n = a^{m-n}$	Multiply exponents. $(ab)^m = a^m b^m$

Check Your Understanding

Examples 1 and 2 pp. 499–500

Simplify.

1. $(6^2)^4$
2. $(a^5)^5$
3. $(r^6)^{-2}$
4. $(3x^4)^3$
5. $(4m^2n)^2$
6. $(-2f^3g^4)^7$

Example 3 p. 500

7. **GEOMETRY** Express the area of the square at the right as a monomial.

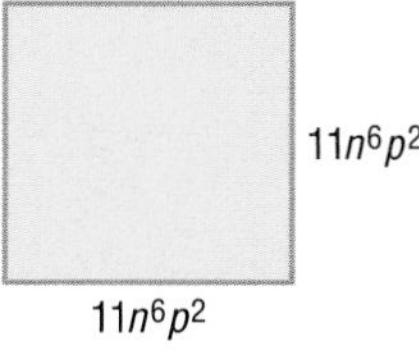

Example 4 p. 501

8. **BUILDINGS** The Lincoln Center for the Performing Arts in New York City takes up most of a square block that measures 2.5×10^2 meters on each side. Find the area of the block that houses the Lincoln Center. Express your answer in scientific notation and in standard notation.

Practice and Problem Solving

● = **Step-by-Step Solutions** begin on page R11. **Extra Practice** begins on page 810.

Examples 1 and 2 pp. 499–500

Simplify.

9. $(2^3)^2$
10. $(5^4)^6$
11. $(3^5)^3$
12. $(6^2)^9$
13. $(b^7)^4$
14. $(f^6)^{-3}$
15. $(2y^2)^8$
16. $(7x^{-5})^2$
17. $(3st^3)^4$
18. $(10y^5z)^3$
19. $(-4n^2p^4)^5$
20. $(-5a^8b^3)^4$

Example 3 p. 500

GEOMETRY Express each measure as a monomial.

21. area of square

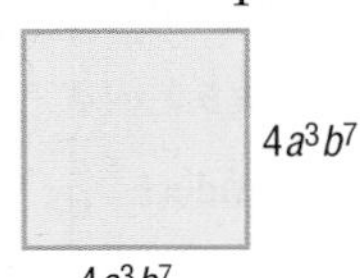

22. volume of cube

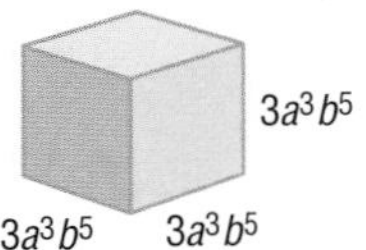

Example 4 p. 501

23. **SPACE** The diameter of Saturn at the equator is 6.027×10^4 kilometers. Use the formula $A = 3.14r^2$ to find the area of the cross section of the planet at the equator. Round to the nearest hundred million.

24. **COMPUTERS** A square microchip for a certain computer measures 1.6×10^{-2} meters on each side. Find the area of the microchip.

Simplify. Express your answer in scientific notation. Round to the nearest hundredth.

25. **STARS** There are approximately $10(4^9)^2$ stars in the Milky Way Galaxy.

26. **EARTH** The surface area of Earth is approximately $11.74[(2^3)^4]^2$ square miles.

Simplify.

27. $[(3^2)^4]^3$
28. $(0.4n^3)^2$
29. $\left(\frac{1}{2}t^4v^3\right)^4$
30. $(-2w)^{-4}(4w^2)^4$

31. **MULTIPLE REPRESENTATIONS** In this problem, you will explore functions.
    a. **ALGEBRAIC** Simplify $(2x^2)^2$ and $[(2x^2)^2]^2$.
    b. **GRAPHICAL** Using a graphing calculator, graph the functions $y = 2x^2$, $y = (2x^2)^2$, and $y = [(2x^2)^2]^2$. Sketch the graphs.
    c. **VERBAL** Compare the graphs. What is similar about the graphs? What is different?

H.O.T. Problems

Use Higher-Order Thinking Skills

32. **OPEN ENDED** Write three expressions that each are equivalent to x^8: one using the Product of a Power Property; one using the Quotient of a Power Property; and one using the Power of a Power Property.

33. **CHALLENGE** Solve $(8^{2x})^3 = 8^{30}$. Explain your reasoning.

34. **REASONING** Compare each pair of monomials. If the pair is not equivalent, explain.
    a. $-6q^3$ and $(-6q)^3$
    b. $(bc)^4$ and b^4c^4

35. **WRITING IN MATH** Summarize the steps you use to find the power of a power and the power of a product.

Standardized Test Practice

36. SHORT RESPONSE Find the area of the circle shown below using scientific notation. Use the formula $A = 3.14r^2$.

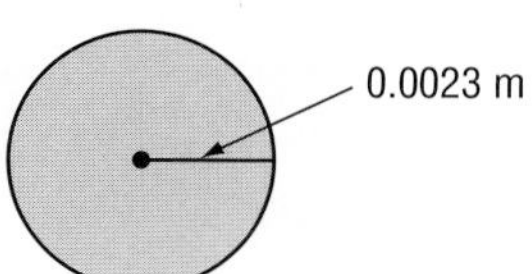

37. The length of the side of a square is $4x^3y^2$ units. Use the formula $A = s^2$ to find the area of the square.

A $4x^5y^4$

B $8x^6y^4$

C $16x^6y^4$

D $64x^6y^4$

38. Simplify the expression $(4x^3y^2)^4$.

F $16x^7y^6$

G $16x^{12}y^8$

H $256x^7y^6$

J $256x^{12}y^8$

39. A star known as Ross 154 is approximately $(3.0 \times 10^8)^2$ miles from Earth. Which of the following represents this distance?

A 6×10^{10}

B 6×10^{16}

C 9×10^{10}

D 9×10^{16}

Spiral Review

Express each number in standard form. (Lesson 9-5)

40. 4.24×10^2

41. 5.72×10^4

42. 3.347×10^{-1}

Express each number in scientific notation. (Lesson 9-5)

43. 2,000,000

44. 499,000

45. 0.006

46. BIRDS A mockingbird uses about 5^{-4} Joules of energy to sing a song. Write the amount of energy the bird uses as an expression using a positive exponent and as a decimal. (Lesson 9-4)

47. BUSINESS To make a profit, stores sell an item for more than it paid for the item. The increase in price is called the *markup*. Suppose a store purchases paint brushes for $8 each. Find the markup if the brushes are sold for 15% over the price the store paid for them. (Lesson 7-5)

Skills Review

Copy and complete each table to find the coordinates of three points through which the graph of each function passes. (Lesson 8-3)

48. $y = 5x + 1$

x	$5x + 1$	(x, y)
0	■	■
1	■	■
2	■	■

49. $y = 3x^2 - 2$

x	$3x^2 - 2$	(x, y)
0	■	■
1	■	■
2	■	■

50. $y = 2x^3 + 3$

x	$2x^3 + 3$	(x, y)
0	■	■
1	■	■
2	■	■

9-7 Linear and Nonlinear Functions

Then

You have already represented linear functions using graphs, equations, and tables. (Lesson 8-3)

Now

- Determine whether a function is linear or nonlinear from a graph.
- Determine whether a function is linear or nonlinear from an equation or a table.

New Vocabulary

nonlinear function

Math Online

glencoe.com

- Extra Examples
- Personal Tutor
- Self-Check Quiz
- Homework Help

Why?

The graph shows the number of cell phone subscribers in the United States in recent years.

a. Does the number of cell phone subscribers increase by a constant amount each year? Explain.

b. Does the graph represent a linear relationship? Explain.

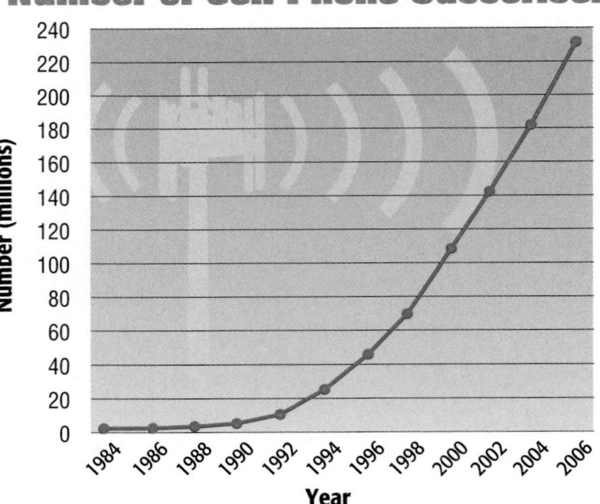

Graphs of Nonlinear Functions In Lesson 8-3, you learned that linear functions have graphs that are straight lines. These graphs represent constant rates of change. **Nonlinear functions** are functions that do not have constant rates of change. Therefore, their graphs are *not* straight lines.

EXAMPLE 1 Identify Functions Using Graphs

Determine whether each graph represents a *linear* or *nonlinear* function. Explain.

a.

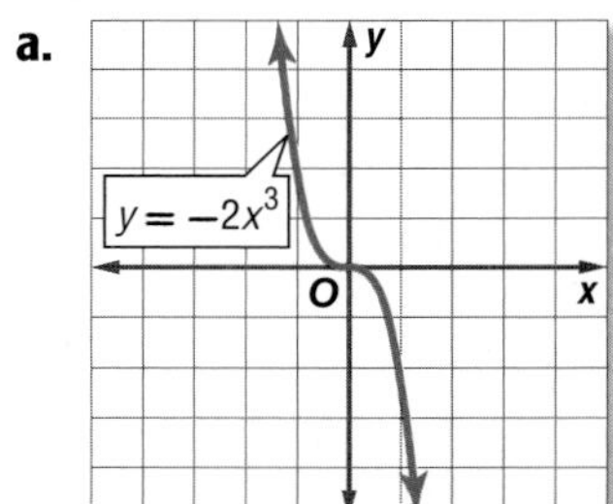

This graph is a curve, not a straight line. So, it represents a nonlinear function.

b.

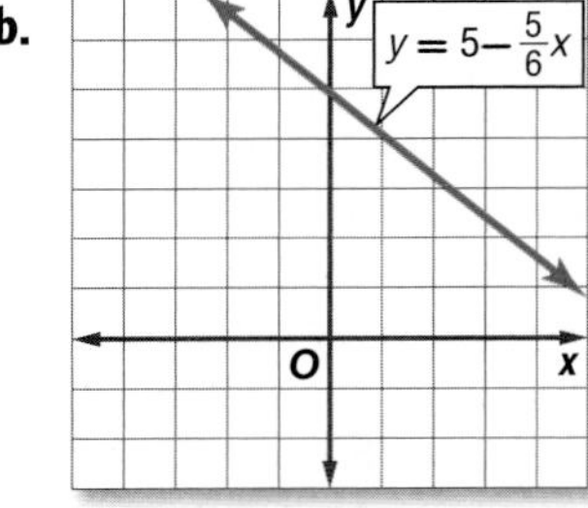

This graph is a line. So, it represents a linear function.

Check Your Progress

1A.

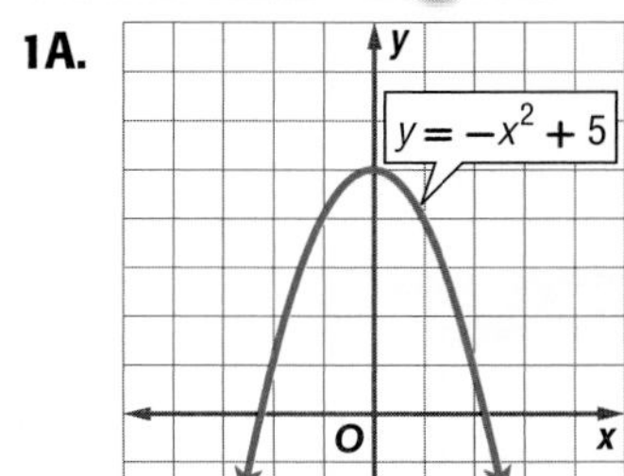

1B.

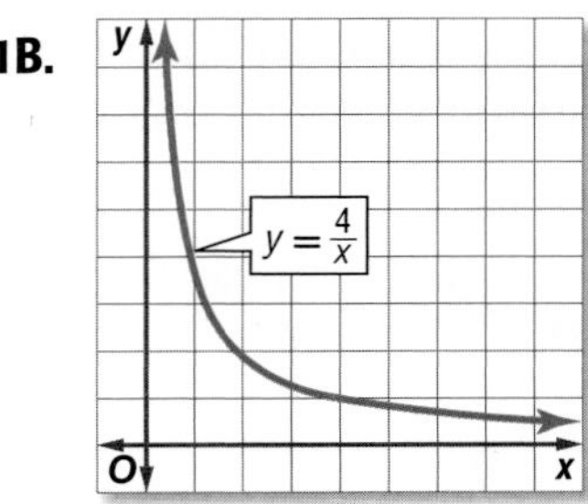

Personal Tutor glencoe.com

Vocabulary Review

constant rate of change a consistant ratio of vertical change to horizontal change

Example slope $= \frac{1}{2}$ (Lesson 8-5)

Equations and Tables Recall that the equation for a linear function can be written in the form $y = mx + b$, where m represents the constant rate of change. Therefore, you can determine whether a function is linear from its equation.

EXAMPLE 2 Identify Functions Using Equations

Determine whether each equation represents a *linear* or *nonlinear* function. Explain.

a. $3x + y = 7$

This equation represents a linear function because it can be written as $y = -3x + 7$.

b. $y = 2x^2$

This is nonlinear because x is squared and the equation cannot be written in the form $y = mx + b$.

Check Your Progress

2A. $y = \frac{1}{5}x$

2B. $y = 2^x$

Personal Tutor glencoe.com

A nonlinear function does not increase or decrease at the same rate. You can check this by using a table.

The tables represent the functions in Example 2.

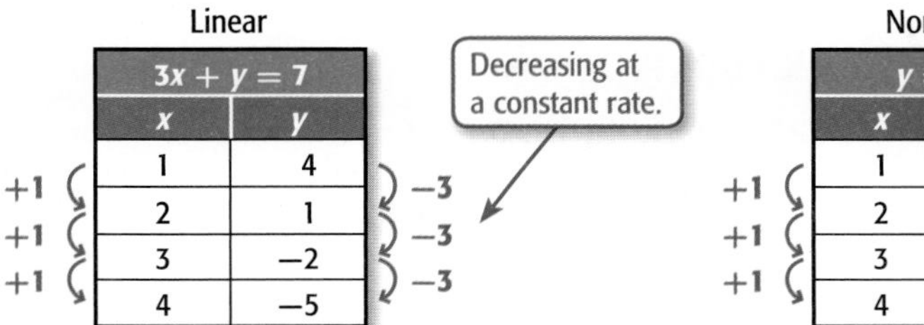

Linear

$3x + y = 7$	
x	y
1	4
2	1
3	−2
4	−5

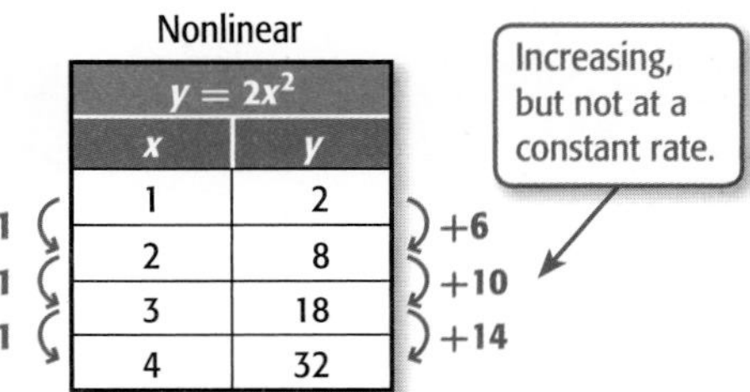

Nonlinear

$y = 2x^2$	
x	y
1	2
2	8
3	18
4	32

Problem-Solving Tip

Make a Graph You can plot a few points on a coordinate plane to determine whether the data in a table represent a linear function.

EXAMPLE 3 Identify Functions Using Tables

Determine whether each table represents a *linear* or *nonlinear* function. Explain.

a.

x	y
2	1
4	2
6	3
8	4

x: +2, +2, +2; y: +1, +1, +1

As x increases by 2, y increases by 1. So, this is a linear function.

b.

x	y
1	1
2	4
3	9
4	16

x: +1, +1, +1; y: +3, +5, +7

As x increases by 1, y increases by a greater amount each time. So, this is a nonlinear function.

Check Your Progress

3A.

x	y
1	−5
2	−20
3	−45
4	−80

3B.

x	y
3	10
6	14
9	18
12	22

Personal Tutor glencoe.com

Real-World Career

Aerospace Technician Aerospace technicians construct, test, and maintain aircraft and space vehicles. They use software to produce drawings and models used to construct aircraft. A minimum of a two-year degree in engineering or aerospace technology is usually required.

Source: Virginia Department of Education

Real-World EXAMPLE 4 Real-Life Functions

SPACE The table shows the flight data for a model rocket launch. Describe whether the data for the ascent more closely represent a *linear* or *nonlinear* function.

Ascent		Descent	
Time (s)	Height (m)	Time (s)	Height (m)
0	0	7	140
1	38	8	130
2	74	9	120
3	106	10	110
4	128	11	100
5	138	12	90
6	142	13	80

Understand You need to determine whether the data represent a linear or nonlinear function.

Plan Find the change in height for each second. Make a table.

Solve Subtract to find the changes in height.

s	0	1	2	3	4	5	6
m	0	38	74	106	128	138	142

+38 +36 +31 +22 +10 +4

As the seconds increase by 1, the height of the rocket changes by a different amount each time. So, this is a nonlinear function.

Check If you were to graph the function, you would see that the points do not lie on a straight line.

Check Your Progress

4. **SPACE** Use the data for the model rocket launch shown above. Describe whether the data for the descent more closely represent a *linear* or *nonlinear* function. Explain.

Personal Tutor glencoe.com

Check Your Understanding

Example 1 p. 504

Determine whether each graph, equation, or table represents a *linear* or *nonlinear* function. Explain.

1. 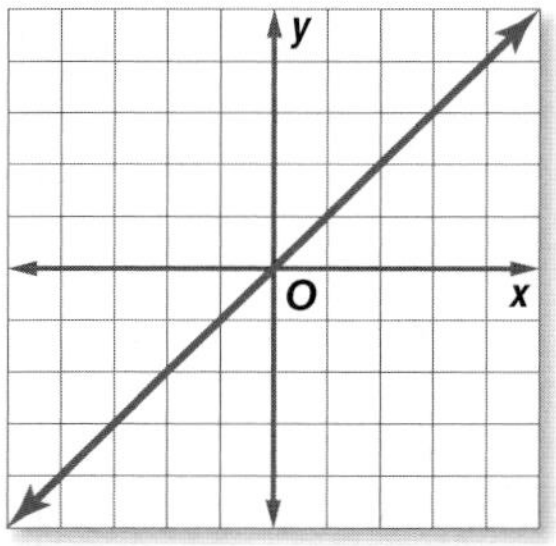

2. 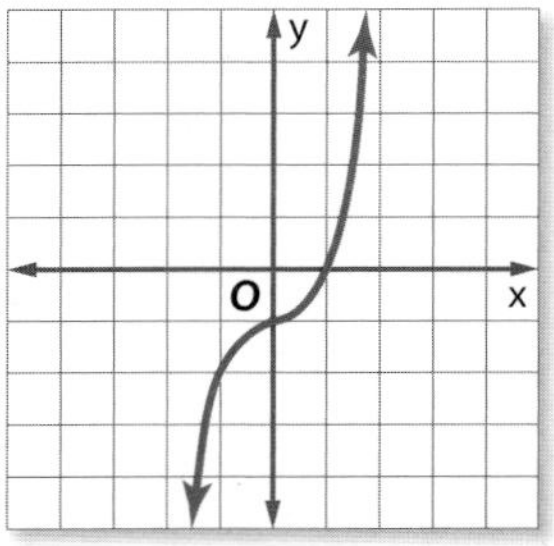

Examples 2 and 3 p. 505

3. $y = -x - 12$

4. $5xy = 25$

5.

x	y
1	6
2	9
3	12
4	15

6.

x	y
1	1
2	8
3	27
4	64

Example 4 p. 506

7. **FINANCIAL LITERACY** The amount of money in Juana's savings account for each of the last six months is \$100, \$100.50, \$101.00, \$101.51, \$102.02, and \$102.53. Do these data represent a *linear* or *nonlinear* function? Explain.

Practice and Problem Solving

● = **Step-by-Step Solutions** begin on page R11.
Extra Practice begins on page 810.

Example 1 p. 504

Determine whether each graph, equation, or table represents a *linear* or *nonlinear* function. Explain.

8.

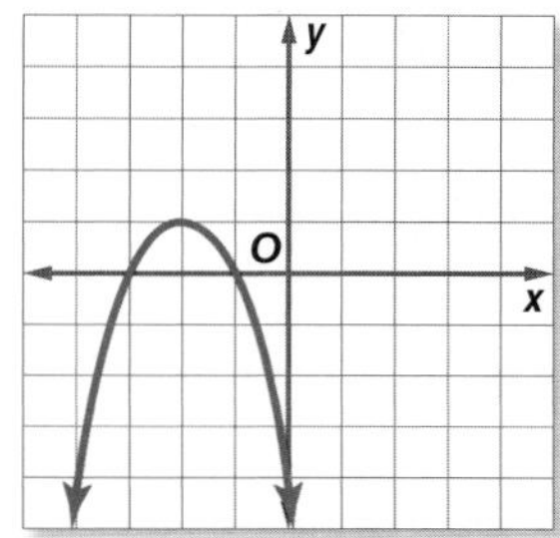

9.

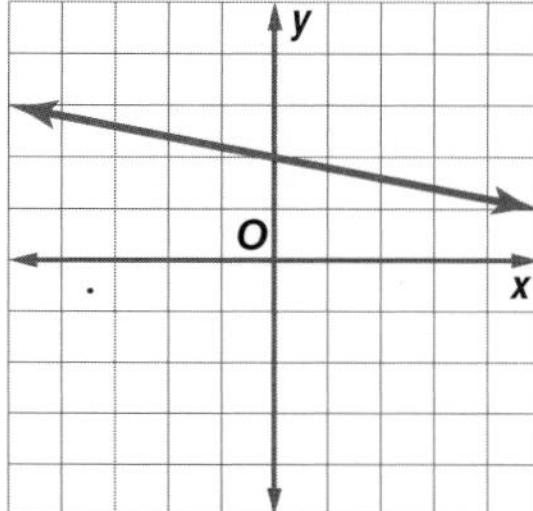

10.

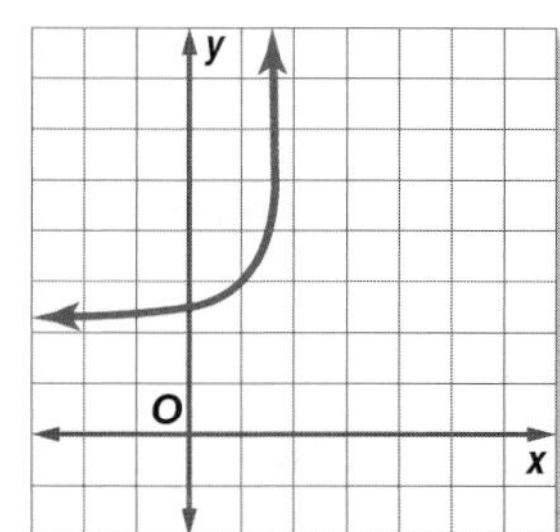

11.

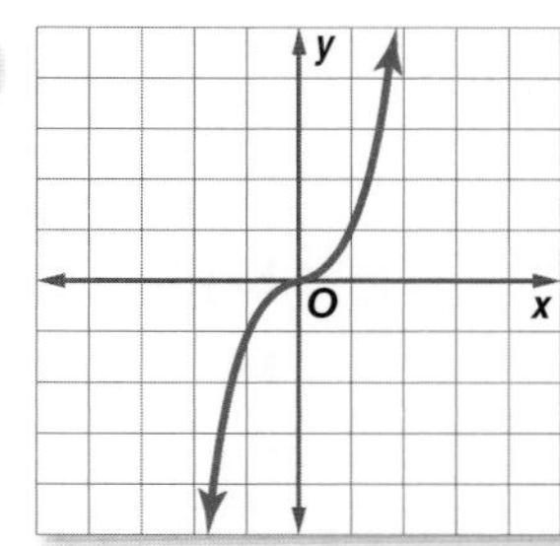

Example 2 p. 505

12. $3x = y$

13. $y = 7$

14. $xy = -4$

15. $y = \frac{10}{x}$

Example 3 p. 505

16.

x	y
1	21
2	16
3	11
4	6

17.

x	y
−2	5
0	8
2	10
4	15

18.

x	y
1	9
2	11
3	14
4	18

19.

x	y
10	1
11	3
12	5
13	7

Example 4 p. 506

20. HEALTH The chart below shows the average height in centimeters of a teenage boy. Do these data represent a linear or nonlinear function? Explain.

Age	13	14	15	16	17	18
Height (cm)	156	163	170	173	175	176

21. BASEBALL The graph shows the average price of a baseball ticket in recent years. Would you describe the change in price as a linear function? Explain.

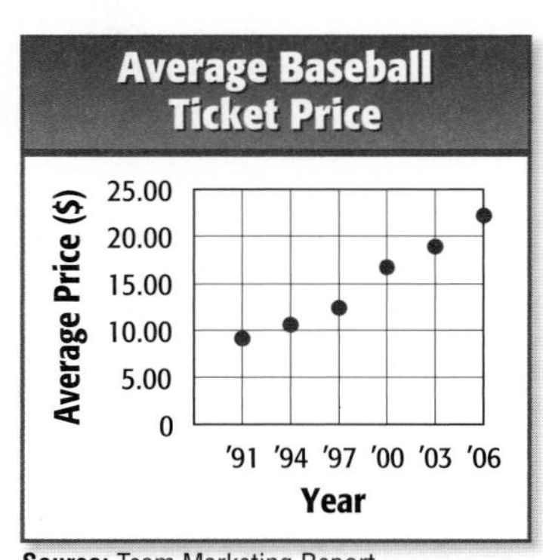

Source: Team Marketing Report

22. FINANCIAL LITERACY Adam puts \$15 into his savings account every month. Suzanne tries to double the amount of money in her bank account every month. Not including interest, which person's monthly balance represents a linear function? Explain why the other person's balance is best represented by a nonlinear function.

StudyTip

The area of a circle is the amount of surface *inside* the circle. The circumference of a circle is the distance *around* the circle.

23. **MULTIPLE REPRESENTATIONS** In this problem, you will investigate area and circumference of circle S. The formula for finding the area A of a circle given the radius r is $A = 3.14r^2$. The formula for finding the circumference of a circle is $C = 6.28r$.

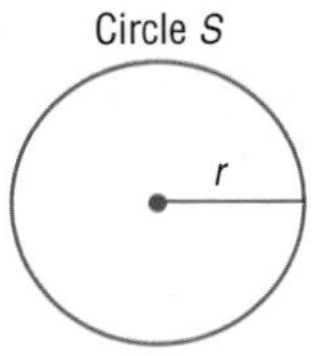

a. **TABULAR** Copy and complete the table at the right.

b. **GRAPHICAL** Graph the points whose ordered pairs are (r, C).

c. **GRAPHICAL** Graph the points whose ordered pairs are (r, A).

d. **ANALYTICAL** Does either graph represent a linear relationship? If so, name the slope of the line.

Radius r	Circumference C	Area A
1	■	■
2	■	■
3	■	■
4	■	■
5	■	■

Determine whether each situation represents a *linear* or *nonlinear* function. Explain your reasoning.

24. Amanda earns \$7.80 per hour.

25. As the number of inches increases by 1, the number of centimeters increases by about 2.54.

26. The population of a city increases by 3% each year.

27. **MONEY** Ben does yard work to earn extra money. Mrs. Rodriquez pays him \$10 per hour. Mrs. Benson pays him \$100 per weekend. For which situation is Ben's pay a linear function of the number of hours he works? Explain.

28. **GEOMETRY** The table below shows the corresponding width for the possible different lengths of a rectangle with a fixed area of 20 square feet. Graph these data on a coordinate plane. Do the data represent a linear function? Justify your solution.

Length (ft)	1	2	3	4	5	6	8	10
Width (ft)	20	10	$6\frac{2}{3}$	5	4	$3\frac{1}{3}$	$2\frac{1}{2}$	2

H.O.T. Problems

Use Higher-Order Thinking Skills

29. **WRITING IN MATH** Describe your preferred method for determining whether a function is linear or nonlinear given its equation.

30. **OPEN ENDED** Describe a real-world situation that represents a linear function.

31. **WHICH ONE DOESN'T BELONG?** Identify the equation that does not belong with the other three. Explain your reasoning.

$y = \frac{1}{3}x - x$ | $xy = 3$ | $5x + y = 6$ | $x = 5y$

32. **CHALLENGE** Are all straight lines linear functions? Explain.

33. **WRITING IN MATH** Describe the different representations that are possible for a function. Explain how you can use each representation to determine whether a function is linear.

Standardized Test Practice

34. Which graph represents a linear function?

A

C

B

D

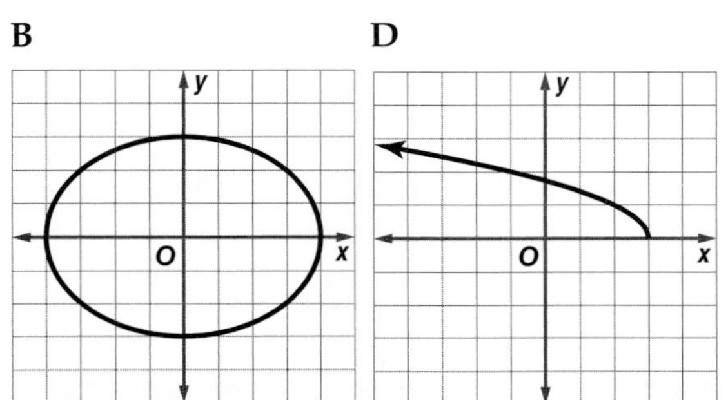

35. EXTENDED RESPONSE Does the equation for the area of a square represent a linear or nonlinear function? Use a table to explain your reasoning.

36. Which of the following equations represents a nonlinear function?

F $y = 3x + 9$

G $y = 9$

H $x = -2$

J $xy = -3$

37. Which of the following equations describes the data in the table?

x	−6	−4	−2	0	2
y	32	12	0	−4	0

A $y = 2x - 4$

B $y = x^2 - 4$

C $y = 4x + 8$

D $x^2 + y = 8$

Spiral Review

Simplify. (Lesson 9-6)

38. $4(x^3)^4$

39. $(-6p^2)^3$

40. $(2yw^5)^6$

Express each number in scientific notation. (Lesson 9-5)

41. 80,000,000

42. 697,000

43. 0.059

44. OCEANS Rank the oceans in the table by area from least to greatest. (Lesson 9-5)

Ocean	Area (sq mi)
Arctic	5.44×10^6
Atlantic	3.18×10^7
Indian	2.89×10^7
Pacific	6.40×10^7

45. MONEY If Simone wants to leave a tip of about 15% on a dinner check of $23.85, how much should she leave? (Lesson 7-4)

46. COOKING Gabriel used $3\frac{3}{4}$ cups of sugar to make $2\frac{1}{2}$ batches of cookies. How much sugar is needed for one batch of cookies? (Lesson 4-4)

Skills Review

Use a table to graph each function. (Lesson 8-3)

47. $y = x + 5$

48. $y = -2x - 6$

49. $x + y = -8$

50. $x - 3y = -12$

51. $y = -\frac{2}{3}x + 9$

52. $y = \frac{1}{2}x - 4$

9-8 Quadratic Functions

Then

You used linear functions to solve problems. (Lesson 8-7)

Now

- Graph quadratic functions.
- Use quadratic functions to solve problems.

New Vocabulary

quadratic function
parabola

Math Online

glencoe.com

- Extra Examples
- Personal Tutor
- Self-Check Quiz
- Homework Help

Why?

People on *zero-gravity flights* experience weightlessness, similar to what astronauts experience in space flight. Weightlessness occurs because the plane is flown in a *parabolic* path, climbing and diving at slopes of about 45 degrees.

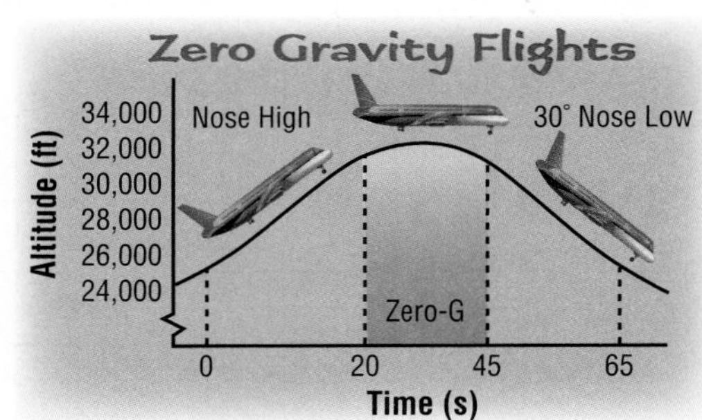

a. Describe the graph.

b. How is this graph different from linear graphs?

Graph Quadratic Functions The function that describes the flight path above is an example of a quadratic function. A **quadratic function** can be written in the form $y = ax^2 + bx + c$, where $a \neq 0$. The graph of a quadratic function is called a **parabola**.

Key Concept Quadratic Function — For Your FOLDABLE

Words A quadratic function can be described by an equation of the form $y = ax^2 + bx + c$, where $a \neq 0$.

Example $y = x^2 - 4$

Graph

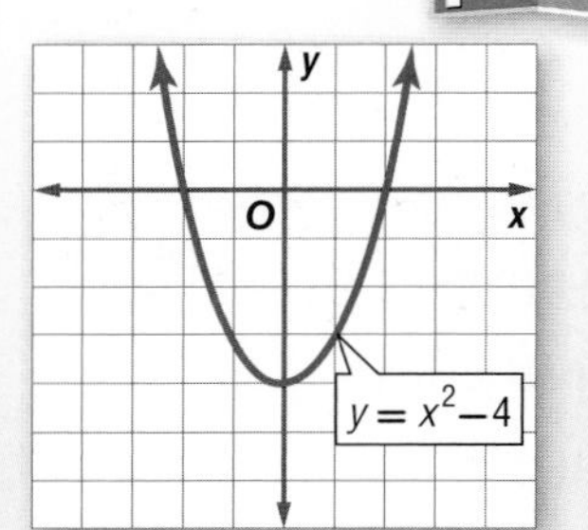

EXAMPLE 1 Graph Quadratic Functions

Graph $y = x^2 - 2$.

Make a table of values, plot the ordered pairs, and connect the points with a curve.

x	$y = x^2 - 2$	(x, y)
−2	$y = (-2)^2 - 2 = 2$	(−2, 2)
−1	$y = (-1)^2 - 2 = -1$	(−1, −1)
0	$y = (0)^2 - 2 = -2$	(0, −2)
1	$y = (1)^2 - 2 = -1$	(1, −1)
2	$y = (2)^2 - 2 = 2$	(2, 2)

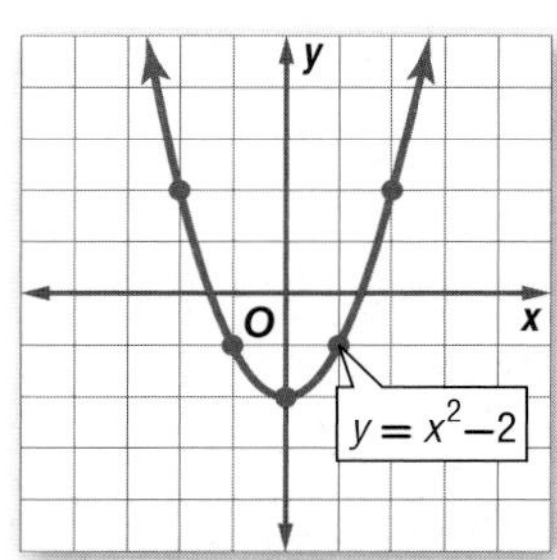

Check Your Progress

1. Graph $y = \frac{1}{4}x^2$.

Personal Tutor glencoe.com

Watch Out!

Squares Square x before finding the opposite.

EXAMPLE 2 Graph Quadratic Functions

Graph $y = -x^2$.

Make a table of values, then plot the ordered pairs.

x	$y = -x^2$	(x, y)
-2	$y = -(-2)^2 = -4$	$(-2, -4)$
-1	$y = -(-1)^2 = -1$	$(-1, -1)$
0	$y = -(0)^2 = 0$	$(0, 0)$
1	$y = -(1)^2 = -1$	$(1, -1)$
2	$y = -(2)^2 = -4$	$(2, -4)$

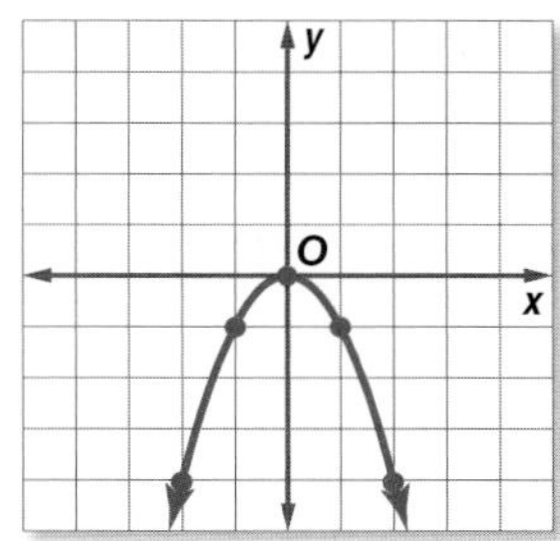

Check Your Progress

2. Graph $y = 2x^2 + 1$.

Personal Tutor glencoe.com

Use Quadratic Functions You can use quadratic functions to model and analyze real-world situations.

EXAMPLE 3 Use Quadratic Functions

BASEBALL Georgia threw a baseball into the air. The equation that gives the ball's height in meters h as a function of time t is $h = -4.9t^2 + 16t + 1.4$.

a. Graph this equation and interpret your graph. What was the height of the ball after 3 seconds?

Make a table of values, then plot the ordered pairs.

t	$h = -4.9t^2 + 16t + 1.4$	(t, h)
0	$1.4 = -4.9(0)^2 + 16(0) + 1.4$	$(0, 1.4)$
1	$12.5 = -4.9(1)^2 + 16(1) + 1.4$	$(1, 12.5)$
2	$13.8 = -4.9(2)^2 + 16(2) + 1.4$	$(2, 13.8)$
3	$5.3 = -4.9(3)^2 + 16(3) + 1.4$	$(3, 5.3)$

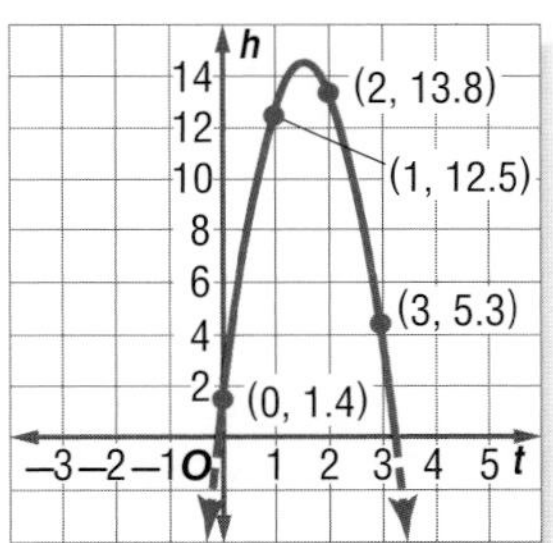

- The maximum height of the ball occurred between 1 and 2 seconds.
- The ball was released at 1.4 meters (0 seconds).
- The ball hit the ground between 3 and 4 seconds (0 meters).
- At 3 seconds, the ball was 5.3 meters above the ground (3, 5.3).

b. What values of the domain and range are unreasonable? Explain.

Unreasonable values for the domain and range would be any negative numbers because neither time nor height can be negative.

Review Vocabulary

Domain the set of x-values, or input, of a function (Lesson 1-4)
Range the set of y-values, or output, of a function (Lesson 1-4)

Check Your Progress

3. **FRAMES** Mei is building a picture frame, and she wants the width to be $\frac{2}{3}$ the length. Graph the equation that models the area of the framed picture. What is the area of the picture if the width is 6 inches?

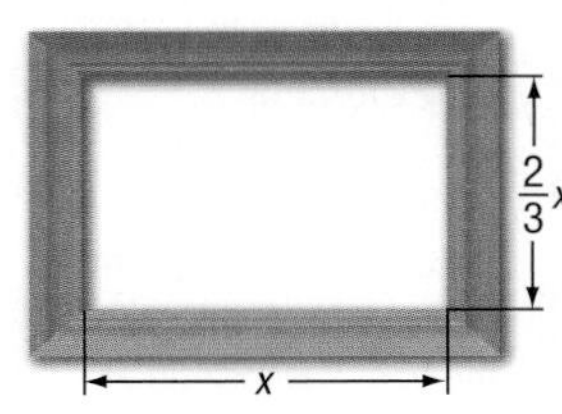

Personal Tutor glencoe.com

Check Your Understanding

Examples 1 and 2
pp. 510 and 511

Graph each function.

1. $y = 2x^2$
2. $y = -4x^2$
3. $y = \frac{1}{2}x^2$
4. $y = 2x^2 - 2$
5. $y = \frac{1}{4}x^2 + 1$
6. $y = -3x^2 - 3$

Example 3
p. 511

7. **SPORTS** A soccer ball is kicked straight up into the air. The height of the ball h after t seconds can be modeled by the equation $h = -16t^2 + 40t + 2$.
   a. Graph this equation and interpret the graph. What is the height of the ball after 2 seconds?
   b. What values of the domain and range are unreasonable? Explain.

Practice and Problem Solving

● = **Step-by-Step Solutions** begin on page R11.
Extra Practice begins on page 810.

Examples 1 and 2
pp. 510 and 511

Graph each function.

8. $y = x^2$
9. $y = x^2 + 1$
10. $y = -3x^2$
11. $y = 4x^2$
12. $y = x^2 + 2$
13. $y = x^2 + 4$
14. $y = -x^2 + 3$
15. $y = 2x^2 + 4$
16. $y = 2x^2 - 3$
17. $y = -\frac{1}{2}x^2 + 1$
18. $y = \frac{1}{2}x^2 + 1$
19. $y = -2x^2 + 5$

Example 3
p. 511

20. **SPACE** Refer to the graph at the beginning of the lesson.
    a. Estimate the maximum height of the aircraft during the parabolic maneuver. Round to the nearest thousand.
    b. Describe the altitude of the aircraft between 20 and 45 seconds, the time in which zero gravity is achieved.

21. **CARS** The function $d = \frac{1}{2}at^2$ represents the distance d that a race car will travel over an amount of time t given the rate of acceleration a.
    a. Suppose a car is accelerating at a rate of 7 feet per second each second. Graph this function on the coordinate plane.
    b. Use your graph to find the time it would take the car to travel 125 feet.

22. **CRAFTS** Meghan has 30 inches of ribbon to make a rectangular border for a scrapbook page.
    a. Write and graph a function to represent the area A of the section inside the border.
    b. What should the dimensions of the section be to enclose the maximum area inside the border? (*Hint*: Find the coordinates of the maximum point of the graph.)

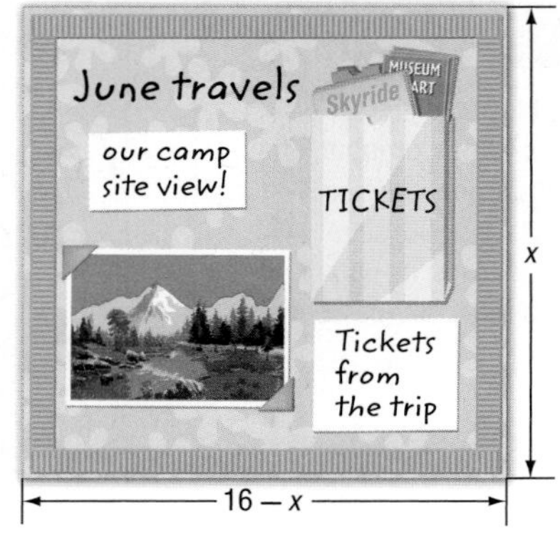

StudyTip

Vertex Form A quadratic equation can also be written in *vertex form* or $y = a(x - h)^2 + k$. In this form, the maximum or minimum point of the graph called the vertex is located at the point (h, k).

23. **MULTIPLE REPRESENTATIONS** In this problem, you will examine a system of equations. A square has side length s.

a. **SYMBOLIC** Write equations to represent the perimeter P and the area A of the square in terms of side length s.

b. **TABULAR** Make a table showing the perimeter and area of the square for side lengths 0, 1, 2, 3, and 4 units.

c. **GRAPHICAL** Graph the points whose ordered pairs are (side length, perimeter) and (side length, area). Describe the graphs.

d. **ANALYTICAL** Are there any values for s that have the same numerical value for A and P? How can you tell from the table or graph?

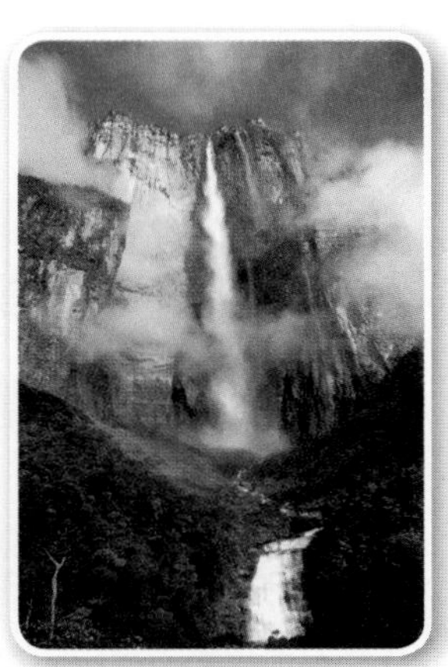

Real-World Link

The tallest waterfall in the world is Angel Falls in Venezuela. It is 979 meters (3212 feet) tall and 107 meters (350 feet) wide.

Source: World Waterfall Database

24. **WATERFALLS** The function $d = -16t^2 + h$ models the distance d in feet a drop of water falls t seconds after it begins its descent from the top of a waterfall of height h.

Tallest U.S. Waterfalls

Waterfall	State	Height (meters)
Olo'upena Falls	Hawaii	900
Pu'uka'oku Falls	Hawaii	840
Waihilau	Hawaii	792
Colonial Creek Falls	Washington	788
Johannesburg Falls	Washington	751

Source: World Waterfall Database

a. Choose two waterfalls from the table and graph the function for each waterfall.

b. Use your graph to estimate the time it will take a drop of water to reach the river at the base of each waterfall.

H.O.T. Problems Use Higher-Order Thinking Skills

25. **OPEN ENDED** Sketch the graph of a quadratic function that has x-intercepts at 2 and 6.

26. **REASONING** Consider the two equations $y = x^2 + 1$ and $y = -x^2 + 1$.

a. Make a table of values for each equation including 7 ordered pairs. Be sure to include $x = 0$ in each table.

b. Graph the two functions on the same coordinate plane.

c. Compare and contrast both graphs.

d. Make a conjecture about the value of a in the equation $y = ax^2 + bx + c$ and the direction that the curve opens.

e. Make a conjecture about the value of c and the placement of the graph on the y-axis.

CHALLENGE The graph of quadratic functions may have one maximum or one minimum point. The *maximum point* of a graph is the point with the greatest y-value coordinate. The *minimum point* is the point with the least y-value coordinate. Graph each equation. Find the coordinates of each point.

27. the maximum point of the graph of $y = -x^2 + 7$

28. the minimum point of the graph of $y = x^2 - 6$

29. **WRITING IN MATH** Describe the relationships between the different representations of quadratic functions and explain how to translate among these representations.

Standardized Test Practice

30. Which graph represents the function $y = -2x^2 - 3$?

A
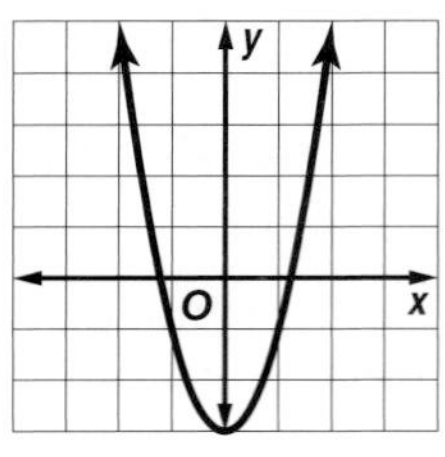

C
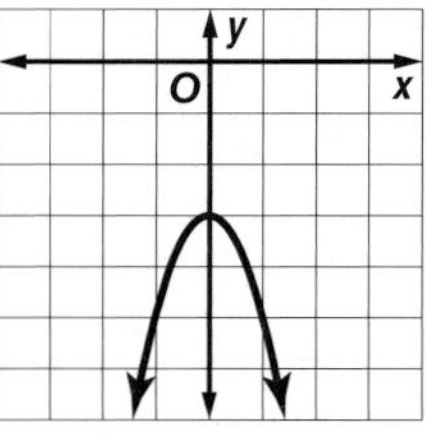

B
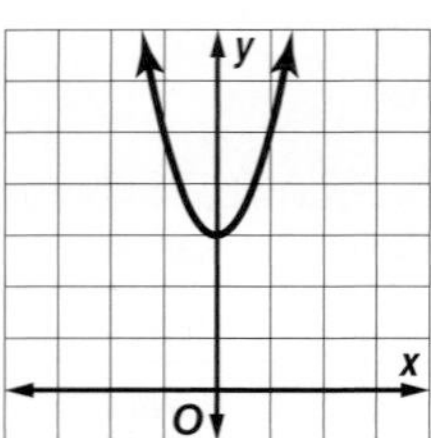

D
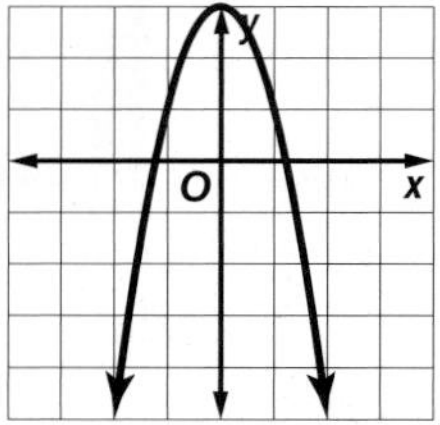

31. What are the x-intercepts of the graph of $y = x^2 - 4$?

F $-3, 3$

G $-2, 2$

H 0

J $-1, 3$

32. The equation $d = \frac{s^2}{20}$, modeled below, can be used to determine the stopping distance d in feet of a car moving at a speed s feet per second. Which of the following is the *best* estimate for the stopping distance of a car traveling at a speed of 15 feet per second?

A 10 ft

B 11 ft

C 12 ft

D 13 ft

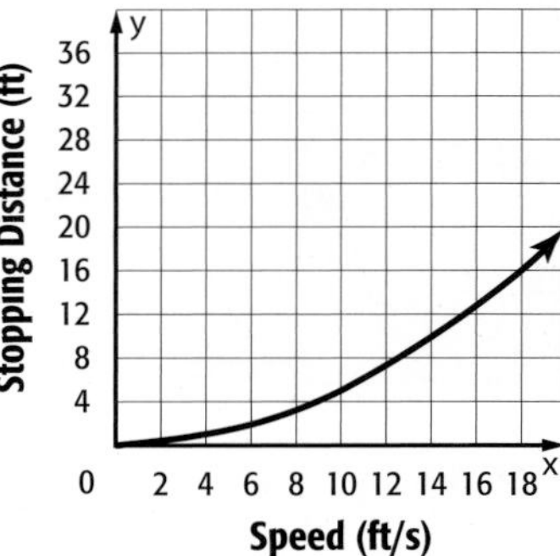

33. EXTENDED RESPONSE The equation $y = \frac{1}{2}x^2 + 3$ is a quadratic function.

a. Graph the function.

b. What is the minimum value of the function?

Spiral Review

Determine whether each table represents a *linear* or *nonlinear* function. Explain. (Lesson 9-7)

34.

x	y
−4	13
−2	0
0	4
2	0

35.

x	y
8	19
9	22
10	25
11	28

36.

x	y
9	−2
11	−8
13	−14
15	−20

Simplify. (Lesson 9-6)

37. $(7^4)^2$

38. $(3^4)^0$

39. $(2c^5d)^3$

40. $(10x^3y^4)^2$

41. SCHOOL A biology class had 28 students. Four students transferred out of the class to take chemistry. Find the percent of change in the number of students in the biology class. Round to the nearest tenth. (Lesson 7-6)

Skills Review

Evaluate each expression. (Lesson 9-1)

42. 2^4

43. 6^3

44. $3 \cdot 4^2$

45. $2 \cdot 4^3$

EXTEND 9-8

Graphing Technology Lab: Family of Quadratic Functions

Math Online glencoe.com
- Other Calculator Keystrokes
- Graphing Technology Personal Tutor

You can use a graphing calculator to investigate families of quadratic functions. The family of quadratic functions has the parent function $y = x^2$.

ACTIVITY

Graph $y = x^2$ and $y = x^2 - 5$ in the standard viewing window and describe how the graphs are related.

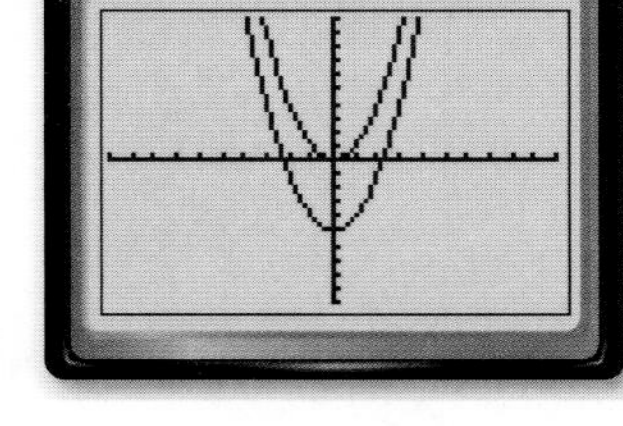

Step 1 Clear any existing equations from the [Y=] list by pressing [Y=] [CLEAR].

Step 2 Enter $y = x^2$ in [Y1] and $y = x^2 - 5$ in [Y2] and graph:

[Y=] [X,T,θ,n] [x^2] [ENTER]

[Y=] [X,T,θ,n] [x^2] [–] 5 [Zoom] 6

Press [Trace] and move along each function using the right and left arrow keys. Move from one function to another using the up and down arrow keys.

The graphs are similar in that they are both parabolas. However, the graph of $y = x^2$ has its vertex at (0, 0), and the graph of $y = x^2 - 5$ has its vertex at (0, –5).

Analyze the Results

For Exercises 1–3, graph $y = x^2$, $y = x^2 + 4$, and $y = x^2 - 6$ on the same screen.

1. Compare and contrast the graphs.
2. How does adding or subtracting a constant c from a quadratic function affect its graph?
3. The three parabolas at the right are graphed in the standard viewing window and have the same shape as the graph of $y = x^2$. Write an equation for each, beginning with the lowest parabola.

For Exercises 4–6, graph $y = x^2$, $y = 0.5x^2$, and $y = 4x^2$ on the same screen.

4. Compare and contrast the graphs.
5. How does changing the coefficient of x^2 affect the graph of a parabolic function?
6. Without graphing, determine whether the graph of $y = 0.2x^2$ or the graph of $y = 1.2x^2$ is more narrow. Explain.

7. **MAKE A CONJECTURE** Compare and contrast the graph of $y = x^2 + 1$ and the graph of each function listed below.

   a. $y = x^2 - 1$ **b.** $y = 2x^2 + 1$ **c.** $y = -x^2 + 1$

9-9 Cubic and Exponential Functions

Then

You have already graphed linear functions. (Lesson 8-3)

Now

- Graph cubic functions.
- Graph exponential functions.

New Vocabulary

cubic function
exponential function

Math Online

glencoe.com

- Extra Examples
- Personal Tutor
- Self-Check Quiz
- Homework Help

Why?

Dog cages like the one shown come in different sizes that are $4x$ units long, $3x$ units wide, and $3x$ units high. The volume V of the dog cage can be found by multiplying the length, width, and height.

a. Write an equation to represent the volume of the cage.

b. In the first quadrant of the coordinate plane, graph the volume as a function of side length. Use 0, 0.5, 1, 1.5, and 2 for the values of x.

c. Is the function linear, quadratic, or neither? Explain.

Cubic Functions A **cubic function** is a function that can be described by an equation of the form $y = ax^3 + bx^2 + cx + d$, where a $\neq 0$. You can make a table of values to graph a cubic function.

EXAMPLE 1 Graph Cubic Functions

Graph each cubic function.

a. $y = 2x^3$

Make a table of values, plot the ordered pairs, and connect the points with a curve.

x	$y = 2x^3$	(x, y)
-1.2	$y = 2(-1.2)^3 \approx -3.5$	$(-1.2, -3.5)$
-1	$y = 2(-1)^3 = -2$	$(-1, -2)$
0	$y = 2(0)^3 = 0$	$(0, 0)$
1	$y = 2(1)^3 = 2$	$(1, 2)$
1.2	$y = 2(1.2)^3 \approx 3.5$	$(1.2, 3.5)$

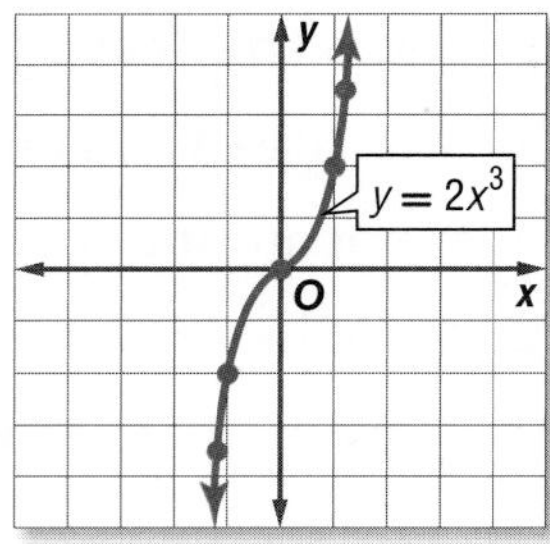

b. $y = x^3 + 3$

x	$y = x^3 + 3$	(x, y)
-1.5	$y = (-1.5)^3 + 3 \approx -0.4$	$(-1.5, -0.4)$
-1	$y = (-1)^3 + 3 = 2$	$(-1, 2)$
0	$y = (0)^3 + 3 = 3$	$(0, 3)$
1	$y = (1)^3 + 3 = 4$	$(1, 4)$
1.5	$y = (1.5)^3 + 3 \approx 6.4$	$(1.5, 6.4)$

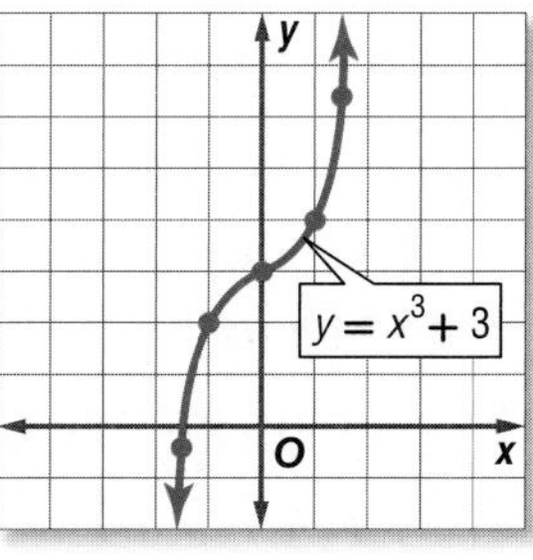

Check Your Progress

1A. $y = -4x^3$

1B. $y = x^3 + 4$

Personal Tutor glencoe.com

EXAMPLE 2 Investigate Cubic Functions

GEOMETRY The equation $y = \frac{1}{3}(3.14)x^3$ represents the volume of the cone. Graph the equation in the first quadrant.

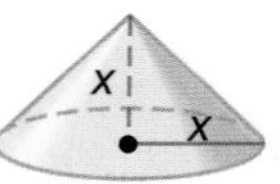

x	$y = \frac{1}{3}(3.14)x^3$	(x, y)
0	$y = \frac{1}{3}(3.14)(0)^3 = 0$	(0, 0)
1	$y = \frac{1}{3}(3.14)(1)^3 \approx 1.0$	(1, 1)
2	$y = \frac{1}{3}(3.14)(2)^3 \approx 8.4$	(2, 8.4)
3	$y = \frac{1}{3}(3.14)(3)^3 \approx 28.3$	(3, 28.3)

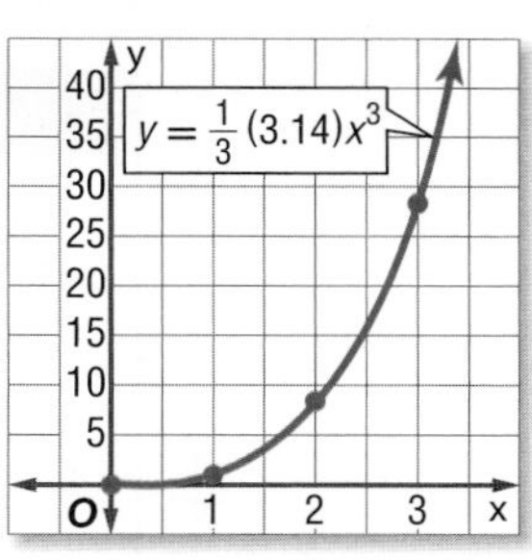

Check Your Progress

2. The volume V of a cube with side length s equals the cube of the side length. Write an equation for the volume and graph the equation in the first quadrant.

Personal Tutor glencoe.com

Exponential Functions In equations of linear, quadratic, and cubic functions, the variable was a base. In exponential functions, the variable is an exponent. An **exponential function** is a function that can be described by an equation of the form $y = a^x + c$, where $a \neq 0$ and $a \neq 1$.

Real-World EXAMPLE 3 Graph Exponential Functions

TEXT MESSAGES Hannah sends a text message to two of her friends. Each of her two friends forwards the text to two friends. Each of those friends forwards it to two friends, and so on. The function $N = 2^x$ represents the total number of text messages sent, where x is the stage of text messages.

a. Make a table of values.

x	$N = 2^x$	(x, N)
1	$N = 2^1 = 2$	(1, 2)
2	$N = 2^2 = 4$	(2, 4)
3	$N = 2^3 = 8$	(3, 8)
4	$N = 2^4 = 16$	(4, 16)
5	$N = 2^5 = 32$	(5, 32)
6	$N = 2^6 = 64$	(6, 64)

b. Graph the values.

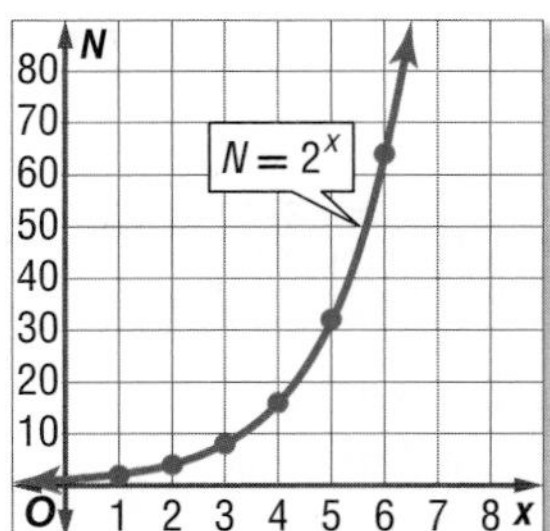

c. In what stage will the number of text messages sent be 64?

Use the graph. The x value that corresponds to the N value of 64 is 6. So, 64 messages will be sent at the 6th stage.

Real-World Link

The world record for hitting the most balls back and forth in 60 seconds was set in 1993 with 173.

Source: USA Table Tennis

Check Your Progress

3. **TABLE TENNIS** A table tennis tournament has 64 players. Half of the players are eliminated after each round. The function $y = 64\left(\frac{1}{2}\right)^x$ represents the total number of players remaining after each round, where x is number of rounds played. Graph the function. After how many rounds will there be a champion?

Personal Tutor glencoe.com

EXAMPLE 4 Graph Exponential Functions

Graph $y = 2^x + 1$.

First, make a table of ordered pairs. Then graph the ordered pairs.

x	$y = 2^x + 1$	(x, y)
-2	$y = 2^{-2} + 1 = 1.25$	$(-2, 1.25)$
-1	$y = 2^{-1} + 1 = 1.5$	$(-1, 1.5)$
0	$y = 2^0 + 1 = 2$	$(0, 2)$
1	$y = 2^1 + 1 = 3$	$(1, 3)$
2	$y = 2^2 + 1 = 5$	$(2, 5)$

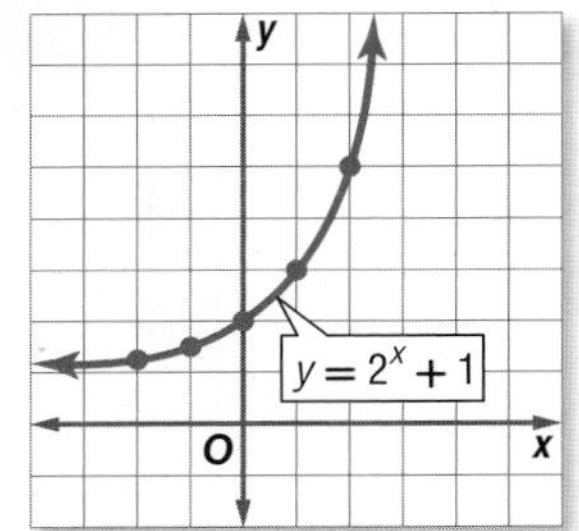

Check Your Progress

4. Graph $y = 3^x - 4$.

Personal Tutor glencoe.com

Concept Summary Functions

For Your FOLDABLE

Linear	Nonlinear		
	Quadratic	**Cubic**	**Exponential**
$y = mx + b$	$y = ax^2 + bx + c$	$y = ax^3 + bx^2 + cx + d$	$y = a^x + c$

Check Your Understanding

Examples 1 and 2 pp. 516–517

Graph each function.

1. $y = x^3$ **2.** $y = x^3 + 1$ **3.** $y = \frac{1}{3}x^3$ **4.** $y = -2x^3$

Example 3 p. 517

5. FINANCIAL LITERACY The amount of money spent at an amusement park continues to increase. The total $T(x)$ in millions of dollars can be estimated by the function $T(x) = 12(1.12)^x$, where x is the number of years after it opened in 2005.

a. Make a table of values showing the amount of money spent after the park has been open 1, 2, 3, and 4 years.

b. Graph the function.

c. What does the y-intercept represent in this problem?

Example 4 p. 518

Graph each function.

6. $y = 3^x$ **7** $y = 2^x - 3$ **8.** $y = 3^x + 3$

Practice and Problem Solving

● = **Step-by-Step Solutions** begin on page R11.
Extra Practice begins on page 810.

Examples 1 and 2 pp. 516–517

Graph each function.

9. $y = -x^3$
10. $y = 3x^3$
11. $y = 4x^3$
12. $y = x^3 - 2$
13. $y = x^3 - 1$
14. $y = -x^3 + 1$
15. $y = -x^3 + 3$
16. $y = 2x^3 - 1$

Example 3 p. 517

17. **MONEY** Jax opened a savings account with an interest rate of 5%. The balance of his account is represented by the function $y = 1000(1.05)^x$, where x represents the number of years the money has been in the account.

   a. Graph the function. Identify the y-intercept. Explain its meaning.

   b. In how many years will the balance be greater than $2000?

18. **BACTERIA** The population of bacteria in a culture increases according to the function $y = 500(2.1)^{0.01t}$, where t represents the number of hours.

   a. Estimate the number of bacteria after 10 hours. Graph the function.

   b. Identify the y-intercept and explain what it represents.

Example 4 p. 518

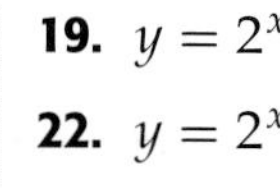

Graph each function.

19. $y = 2^x + 2$
20. $y = 4^x$
21. $y = 3^x + 1$
22. $y = 2^x - 1$
23. $y = 3^x - 2$
24. $y = 2^x + 4$

Real-World Link

Bacteria cells grow exponentially, but at different rates. In the time that it takes a single *Nitrosomonas* cell to double in population, a single *E. coli* bacterium can produce a population of more than 35 trillion cells.

Source: Fritz Industries

Identify each function as *linear, quadratic, cubic,* or *exponential.*

25. $y = 3x^2 + 4x$
26. $5x + 2y = 10$
27. $y = 2.4^x$
28. 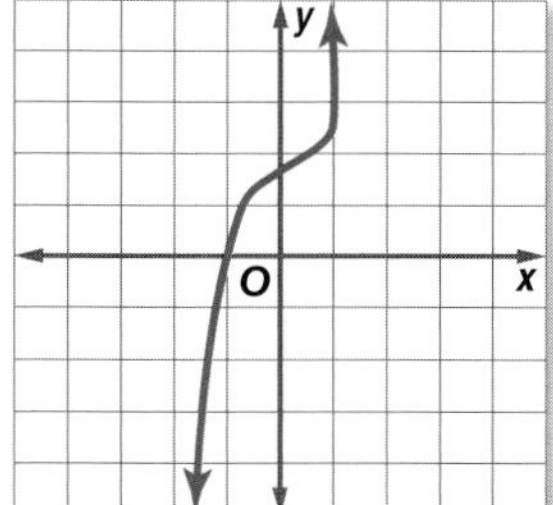

29. 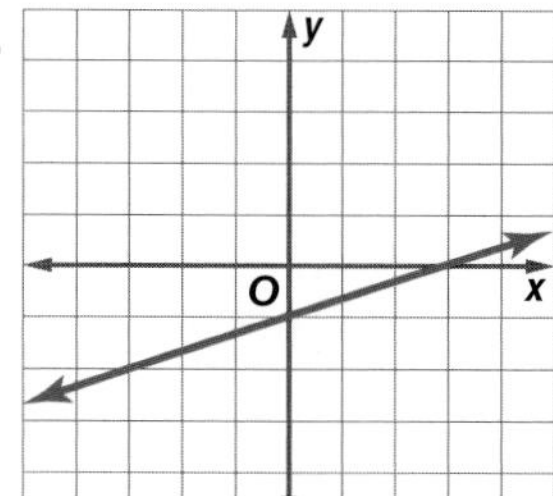

30. 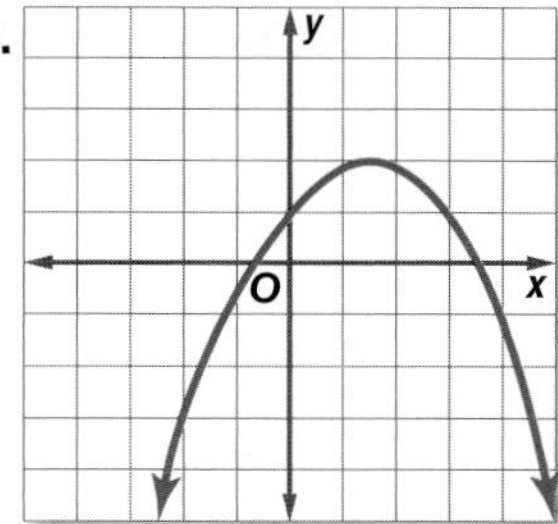

H.O.T. Problems

Use **H**igher-**O**rder **T**hinking Skills

31. **OPEN ENDED** Write a linear function and an exponential function. Then compare and contrast the rates of change represented by the functions.

32. **CHALLENGE** Analyze the graph of an exponential function by describing how each change in $y = a^x + c$ affects the graph of the function.

   a. c increases

   b. for $a > 1$, a decreases

33. **WHICH ONE DOESN'T BELONG?** Identify the equation that is not the same as the other three. Explain your reasoning.

$y = 5 + 3^x$	$y + 3^x = 5$	$y + 3x = 5$	$3^x + y = 5$

34. **REASONING** Graph $y = x^2$ and $y = x^3$ in the first quadrant on the same coordinate plane. Explain which graph shows faster growth.

35. **WRITING IN MATH** Describe the general shape of the exponential function $y = a^x$.

Standardized Test Practice

36. Which of the following equations represents the graph?

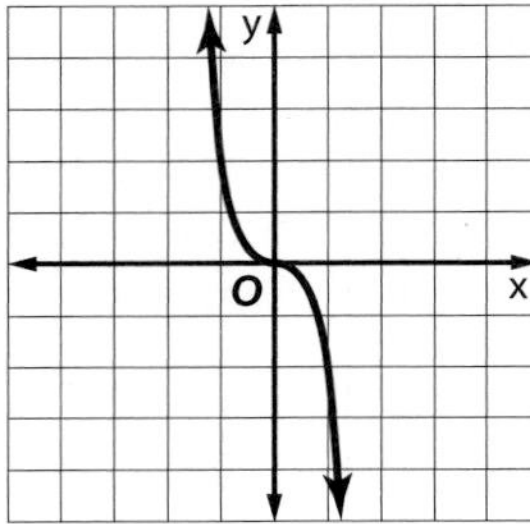

A $y = -2x^3$

B $y = 2x^3$

C $y = -2x^2$

D $y = 2x^2$

37. Which of the following equations represents a cubic function?

F $y = 5^x$

G $y = 5x$

H $y = 5x^2$

J $y = 5x^3$

38. Of the following, which describes the equation $y = 5^x - 4$?

A quadratic function

B linear function

C exponential function

D cubic function

39. EXTENDED RESPONSE Use the function $y = 3^x - 1$.

a. Make a table of values.

b. Graph the ordered pairs.

c. Does the equation represent a linear, quadratic, cubic, or exponential function? Explain your reasoning.

Spiral Review

40. CONSTRUCTION Use the figure at the right. A dog pen is being built with a 100-foot roll of chain link fence. (Lesson 9-8)

a. Write and graph an equation to represent the area A of the pen.

b. What dimensions of the dog pen would enclose the maximum area inside the fence? (*Hint*: Find the maximum point of the graph.)

Determine whether each graph represents a *linear* or *nonlinear* function. Explain. (Lesson 9-7)

41.

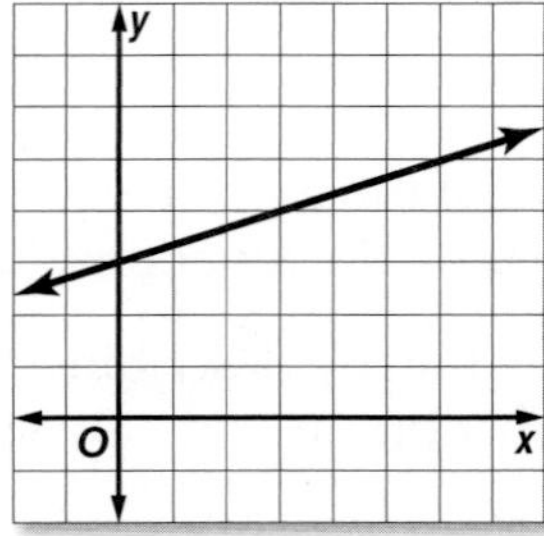

42.

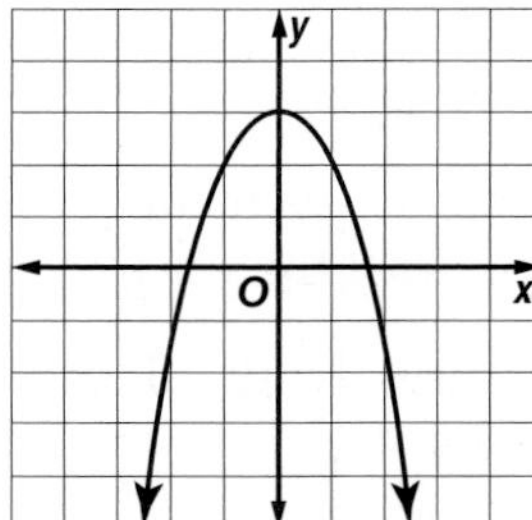

43.

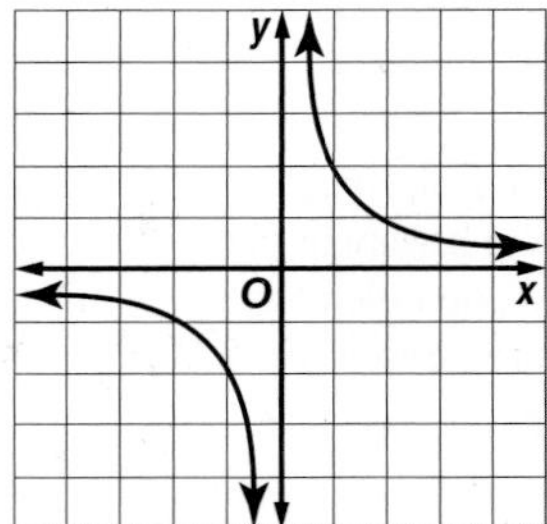

Express each number in standard form. (Lesson 9-5)

44. 3.08×10^{-4}

45. 1.4×10^2

46. 8.495×10^5

Skills Review

Multiply. (Previous Course)

47. $4 \times 4 \times 4 \times 4$

48. $2 \times 2 \times 2 \times 2 \times 2$

49. 9×9

50. $3 \times 3 \times 3$

EXTEND 9-9

Algebra Lab
A Half-Life Simulation

Math Online glencoe.com
Math *in Motion*, Animation

Objective
Students will make predictions about half-lives.

The *half-life* of a radioactive element such as uranium is the time that it takes for one-half a quantity of the element to decay.

ACTIVITY

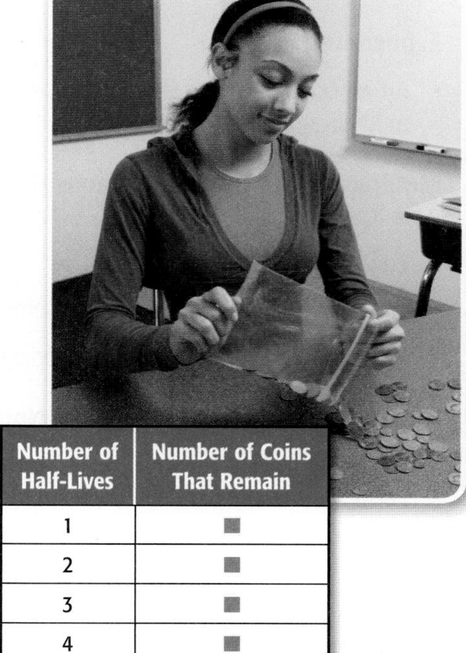

Step 1 Place 100 coins in a bag. Shake the bag and empty the coins on a table. This simulates one half-life.

Step 2 Remove all the coins that are tails up. In a table like the one shown, record the number of coins that remain.

Step 3 Place the remaining coins in the bag and shake it again. Then empty the coins on a table. This represents another half life.

Step 4 Remove all the coins that are tails up. Count the number of coins that remain and record it in the table.

Step 5 Repeat the activity until fewer than 5 coins remain.

Number of Half-Lives	Number of Coins That Remain
1	■
2	■
3	■
4	■
5	■

Analyze the Results

1. On a coordinate plane, let the x-axis represent the number of half-lives and the y-axis represent the number of coins that remain. Plot the points (number of half-lives, number of remaining coins) from your table.
2. Describe the graph of the data.

The expressions below represent the average number of coins that remain after three simulations of the half-life activity.

1 Half-Life	2 Half-Lives	3 Half-Lives
$100\left(\frac{1}{2}\right) = 100\left(\frac{1}{2}\right)^1$	$100\left(\frac{1}{2}\right)\left(\frac{1}{2}\right) = 100\left(\frac{1}{2}\right)^2$	$100\left(\frac{1}{2}\right)\left(\frac{1}{2}\right)\left(\frac{1}{2}\right) = 100\left(\frac{1}{2}\right)^3$

3. Use the expressions to predict how many coins remain after three half-lives. Compare this number to the number in the table. Explain any differences.
4. Suppose you started with 50 coins. Predict how many coins would remain after 4 half-lives.

Study Guide and Review

Math Online glencoe.com
• STUDY TO GO
• Vocabulary Review

Chapter Summary

Key Concepts

Exponents (Lesson 9-1)

- An exponent is a way of writing repeated multiplication.

Multiplying and Dividing Monomials (Lesson 9-3)

- Powers with the same base can be:
  - multiplied by adding their exponents.
  - divided by subtracting their exponents.

Negative Exponents and Scientific Notation (Lessons 9-4 and 9-5)

- For $a \neq 0$ and any whole number n, $a^{-n} = \frac{1}{a^n}$.
- A number in scientific notation is the product of a number between 1 and 10 and a power of 10.

Powers of Monomials (Lesson 9-6)

- To find the power of a power, multiply exponents.
- To find the power of a product, find the power of each factor and multiply.

Linear and Nonlinear Functions (Lesson 9-7)

- Nonlinear functions do not have constant rates of change.

Graphing Quadratic and Cubic Functions (Lessons 9-8 and 9-9)

- Quadratic and cubic functions can be graphed by plotting points.

FOLDABLES® Study Organizer

Be sure the Key Concepts are noted in your Foldable.

Key Vocabulary

base (p. 471)
composite number (p. 476)
cubic function (p. 516)
exponent (p. 471)
exponential function (p. 517)
factor (p. 477)
factor tree (p. 477)
monomial (p. 477)
nonlinear function (p. 504)
parabola (p. 510)
power (p. 471)
prime factorization (p. 477)
prime number (p. 476)
quadratic function (p. 510)
scientific notation (p. 493)
standard form (p. 493)

Vocabulary Check

Determine whether each statement is *true* or *false*. If *false*, replace the underlined word or number to make a true statement.

1. The exponent of a number raised to the first power can be omitted.
2. A number is in scientific notation when it does not contain exponents.
3. In 5^7, the number 7 is the base.
4. The number 49 is an example of a(n) prime number.
5. The equation $y = 5^x + 2$ is an example of a(n) exponential function.
6. The graph of a(n) cubic function is called a parabola.
7. To multiply powers with the same base, add the exponents.
8. A(n) nonlinear function has a constant rate of change.
9. To simplify $(n^6)^5$, the first step is to add 6 and 5.
10. The graph of a(n) quadratic function is symmetric.

CHAPTER 9

Study Guide and Review

Lesson-by-Lesson Review

9-1 Powers and Exponents (pp. 471–475)

Write each expression using exponents.

11. $6 \cdot 6 \cdot 6 \cdot 6 \cdot 6$

12. 4

13. $x \cdot x \cdot x$

14. $f \cdot f \cdot g \cdot g \cdot g \cdot g$

Evaluate each expression.

15. 3^5

16. $2 \cdot 4^3$

Evaluate each expression if $w = -\frac{3}{4}$, $x = 4$, $y = 1$, and $z = -5$.

17. $x^2 - 6$

18. $w^3 + y^2$

19. $2(y + z^3)$

20. w^4x^2yz

21. **TEETH** Adult humans have 2^5 teeth. How many teeth do adults have?

EXAMPLE 1

Write $a \cdot a \cdot b \cdot b \cdot b \cdot b \cdot b$ using exponents.

Group the factors with like bases. Then write using exponents.

$a \cdot a \cdot b \cdot b \cdot b \cdot b \cdot b = (a \cdot a) \cdot (b \cdot b \cdot b \cdot b \cdot b)$

$= a^2b^5$

EXAMPLE 2

Evaluate $(a + 2b)^2$ if $a = 3$ and $b = -2$.

$(a + 2b)^2 = [3 + 2(-2)]^2$ — $a = 3$ and $b = -2$

$= (-1)^2$ — Simplify inside the brackets.

$= 1$ — Simplify.

9-2 Prime Factorization (pp. 476–480)

Write the prime factorization of each number. Use exponents for repeated factors.

22. 34

23. 40

24. 63

25. 225

Factor each monomial.

26. $18x$

27. $10r^2$

28. $32pq$

29. $-25ab^2$

30. **PHOTOGRAPHS** Jacy has 24 photographs to put in a rectangular arrangement. In how many different numbers of rows and columns can she display them if each row has the same number of photographs? Name each arrangement.

EXAMPLE 3

Write the prime factorization of 52. Use exponents for repeated factors.

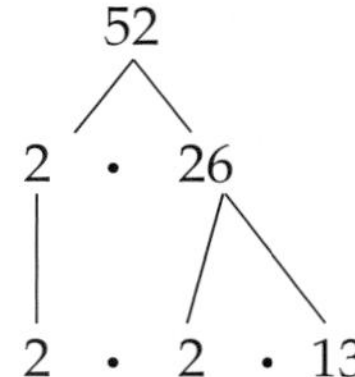

The prime factorization of 52 is $2 \cdot 2 \cdot 13$ or $2^2 \cdot 13$.

EXAMPLE 4

Factor $-21gh^3$.

$-21gh^3 = -1 \cdot 3 \cdot 7 \cdot g \cdot h^3$ — $-21 = -1 \cdot 3 \cdot 7$

$= -1 \cdot 3 \cdot 7 \cdot g \cdot h \cdot h \cdot h$ — $g \cdot h^3 = g \cdot h \cdot h \cdot h$

CHAPTER 9

Study Guide and Review

9-3 Multiplying and Dividing Monomials (pp. 481–485)

Find each product or quotient. Express using exponents.

31. $3^5 \cdot 3^2$

32. $(-7) \cdot (-7)^4$

33. $m^3 \cdot m^6$

34. $x^8 \cdot x$

35. $(2h^7)(6h)$

36. $(5a^3)(-6a^4)$

37. $\frac{9^6}{9^5}$

38. $\frac{k^{10}}{k^4}$

39. **PLANETS** Venus is about 10^8 kilometers from the Sun. Saturn is about 10^9 kilometers from the Sun. About how many times farther from the Sun is Saturn than Venus?

EXAMPLE 5

Find each product or quotient.

a. $4t^5 \cdot 2t^8$

$4t^5 \cdot 2t^8 = (4 \cdot 2)(t^5 \cdot t^8)$ — **Commutative Property of Multiplication**

$= (8)(t^{5+8})$ — **The common base is *t*.**

$= 8t^{13}$ — **Add exponents.**

b. $\frac{n^{15}}{n^9}$

$\frac{n^{15}}{n^9} = n^{15-9}$ — **The common base is *n*.**

$= n^6$ — **Subtract exponents.**

9-4 Negative Exponents (pp. 486–491)

Write each expression using a positive exponent.

40. 9^{-4}

41. $(-10)^{-2}$

42. m^{-5}

Write each fraction as an expression using a negative exponent other than −1.

43. $\frac{1}{6^3}$

44. $\frac{1}{64}$

45. $\frac{1}{125}$

46. **MEASUREMENT** If 1 millimeter is equal to 10^{-3} meter and 1 nanometer is equal to 10^{-9} meter, how many nanometers are in 1 millimeter? Write using a positive exponent.

EXAMPLE 6

Write $\frac{1}{32}$ as an expression using a negative exponent other than −1.

$\frac{1}{32} = \frac{1}{2 \cdot 2 \cdot 2 \cdot 2 \cdot 2}$ — **Find the prime factorization of 32.**

$= \frac{1}{2^5}$ — **Definition of exponent**

$= 2^{-5}$ — **Definition of negative exponent**

9-5 Scientific Notation (pp. 493–498)

Express each number in standard form.

47. 5.82×10^3

48. 9×10^{-2}

49. 3.4×10^{-4}

50. 1.705×10^5

Express each number in scientific notation.

51. 379

52. 26,880

53. 0.0014

54. 0.000561

55. **SPACE** The mass of the Sun is 1.98892×10^{15} exagrams. Express in standard form.

EXAMPLE 7

Express 0.0049 in scientific notation.

$0.0049 = 4.9 \times 0.001$ — **The decimal point moves 3 places.**

$= 4.9 \times 10^{-3}$ — **The exponent is negative.**

MIXED PROBLEM SOLVING
For mixed problem-solving practice, see page 851.

9-6 Powers of Monomials (pp. 499–503)

Simplify.

56. $(2^6)^3$

57. $(r^2)^8$

58. $(3x^7)^2$

59. $(-2n^4)^6$

60. $(4a^9b)^4$

61. $(5w^5x^8)^3$

62. GEOMETRY Find the area of the square shown below.

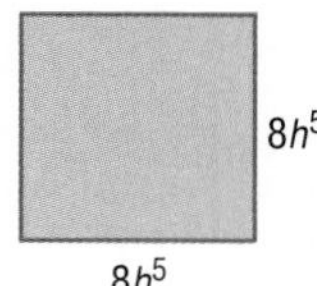

EXAMPLE 8

Simplify.

a. $(8^5)^3$

$(8^5)^3 = 8^{5 \cdot 3}$ — Power of a Power

$= 8^{15}$ — Simplify.

b. $(3rs^4)^5$

$(3rs^4)^5 = 3^5 \cdot r^5 \cdot (s^4)^5$ — Power of a product

$= 3^5 \cdot r^5 \cdot s^{4 \cdot 5}$ — Power of a power

$= 243r^5s^{20}$ — Simplify.

9-7 Linear and Nonlinear Functions (pp. 504–509)

Determine whether each graph, equation, or table represents a *linear* or *nonlinear* function. Explain.

63.

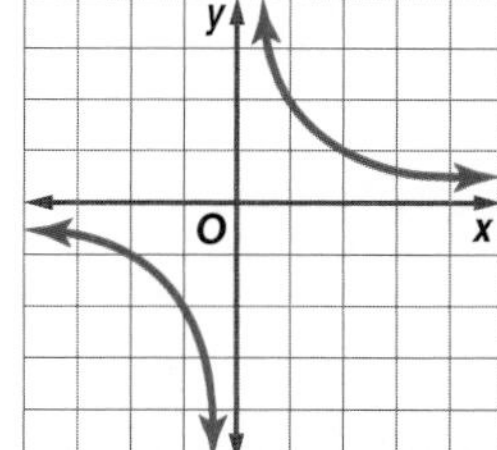

64.

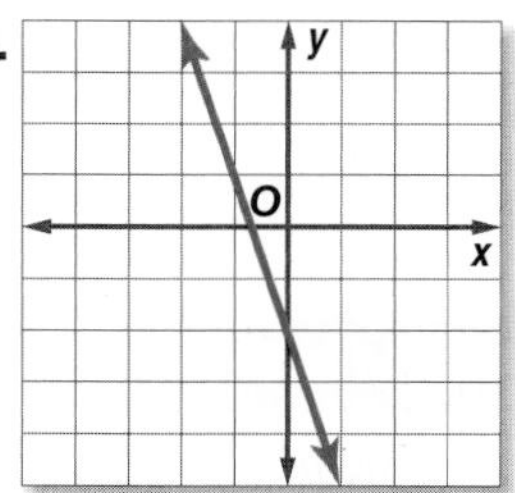

65. $y = \frac{3}{4}x$

66. $y = \frac{5}{x} + 1$

67.

x	y
1	8
2	6
3	4
4	2

68.

x	y
6	10
7	20
8	30
9	40

69. SCHOOLS A school district's spending on students over the last five years is represented by the equation $y = 325x^2 + 0.2x + 1427$. Is this equation linear? Explain.

EXAMPLE 9

Determine whether each graph, equation, or table represents a *linear* or *nonlinear* function. Explain.

a.

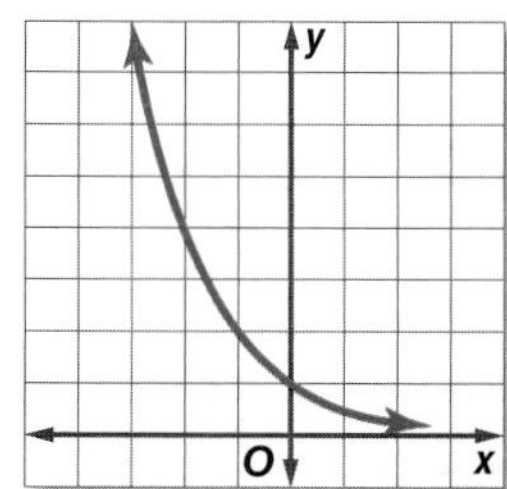

The graph is a curve, not a straight line. So, it represents a nonlinear function.

b. $2y = 8x + 1$

This equation represents a linear function because it can be written as $y = 4x + \frac{1}{2}$.

c.

	x	y	
+2	−6	5	+5
+2	−4	10	+10
+2	−2	20	+20
	0	40	

The table represents a nonlinear function because the rate of change is not constant.

CHAPTER 9 Study Guide and Review

9-8 Quadratic Functions (pp. 510–514)

Graph each function.

70. $y = 3x^2$

71. $y = -2x^2$

72. $y = x^2 - 4$

73. $y = -x^2 + 1$

74. $y = 2x^2 + 2$

75. $y = \frac{1}{2}x^2 - 3$

76. GEOMETRY The volume of a cylinder with a height of 8 inches can be found using the equation $V = 8(3.14)r^2$ where r is the radius of the cylinder. Graph the equation

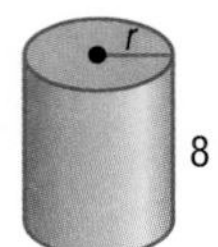

EXAMPLE 10

Graph $y = x^2 + 3$.

x	$x^2 + 3$	(x, y)
−2	$(-2)^2 + 3 = 7$	(−2, 7)
−1	$(-1)^2 + 3 = 4$	(−1, 4)
0	$(0)^2 + 3 = 3$	(0, 3)
1	$(1)^2 + 3 = 4$	(1, 4)
2	$(2)^2 + 3 = 7$	(2, 7)

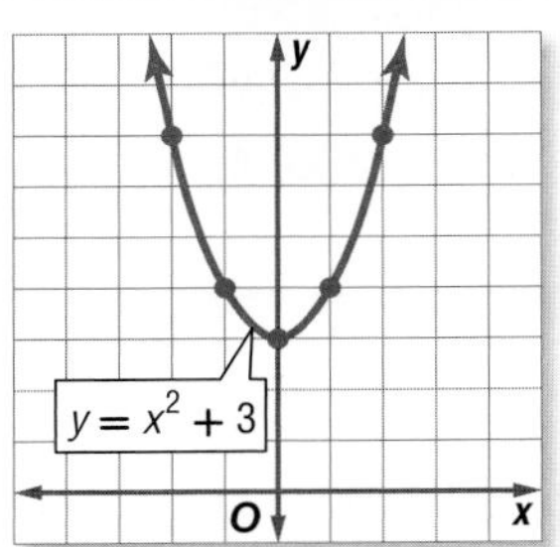

9-9 Cubic and Exponential Functions (pp. 516–520)

Graph each function.

77. $y = x^3 + 1$

78. $y = -3x^3$

79. $y = x^3 - 4$

80. $y = 2^x + 2$

81. $y = 3^x$

82. $y = 2^x - 1$

83. LIFE SCIENCE Starting from a single bacterium in a dish, the number of bacteria after t cycles of reproduction is 2^t. A bacterium reproduces every 30 minutes. If there are 1000 bacteria in a dish now, how many will there be in 1 hour?

EXAMPLE 11

Graph $y = -2^x$. The equation $y = -2^x$ is the same as $y = (-1)(2^x)$.

x	$y = -2^x$	(x, y)
−2	$y = -2^{-2}$	(−2, −0.25)
−1	$y = -2^{-1}$	(−1, −0.5)
0	$y = -2^0$	(0, −1)
1	$y = -2^1$	(1, −2)
2	$y = -2^2$	(2, −4)

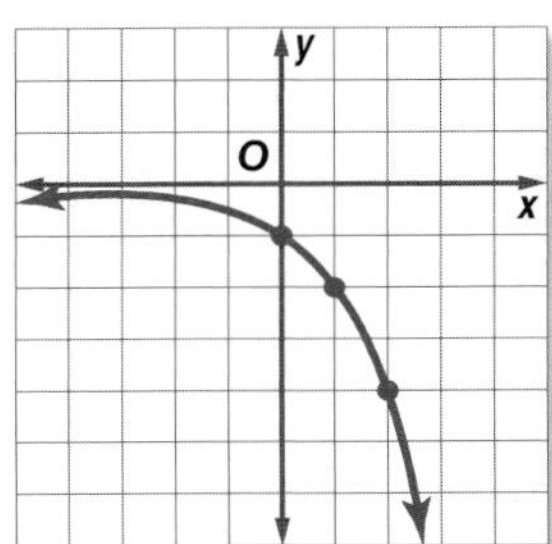

Practice Test

Math Online glencoe.com
Chapter Test

Write each expression using exponents.

1. $(-2)(-2)(-2)$
2. $3 \cdot b \cdot b \cdot b \cdot b \cdot b$

Evaluate each expression if $c = 2$ and $d = -3$.

3. $c^3 + 5$
4. $c - 4d^2$

5. **MULTIPLE CHOICE** When the United States had 48 states, the stars on the flag were in a 6×8 rectangular arrangement. Which of the following rectangular arrangements of the 48 stars would be impossible?

   A 2×24
   B 3×16
   C 4×12
   D 5×10

Factor each monomial.

6. $-15rs$
7. $26b^3c^2d$

8. **MULTIPLE CHOICE** Which of the following expressions represents the area of the square?

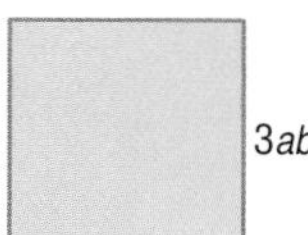

   F $3ab^2$
   G $3a^2b^2$
   H $9ab^2$
   J $9a^2b^2$

Find each product or quotient. Express using exponents.

9. $9^2 \cdot 9^6$
10. $k \cdot k^5$
11. $(-2)^5 \div (-2)^2$
12. $(11y^3)(-3y^7)$

13. **MEASUREMENT** How many square centimeters is equivalent to one square millimeter? Write as an expression with a positive exponent.

Write each expression using a positive exponent.

14. 10^{-10}
15. m^{-5}
16. Write $\frac{1}{121}$ as an expression using a negative exponent other than -1.

Write each number in standard form.

17. 3.4×10^{-5}
18. 7.29×10^3

Write each number in scientific notation.

19. 50,300
20. 0.008

VOLCANOES The table shows the greatest amounts of lava in cubic meters per second that erupted from six volcanoes in the last century.

Volcano	Date	Eruption Rate
Mount St. Helens	1980	2×10^4
Ngauruhoe	1975	2×10^3
Hekla	1970	4×10^3
Agung	1963	3×10^4
Bezymianny	1956	2×10^5
Hekla	1947	2×10^4
Santa Maria	1902	4×10^4

21. Order the volcano eruption rates from greatest to least eruption rate.
22. How many times as great was the Santa Maria eruption rate than the Mount St. Helens eruption rate?

Simplify.

23. $(4^2)^6$
24. $(p^7)^3$
25. $(3c^4d^6)^4$

Determine whether each equation or table represents a *linear* or *nonlinear* function. Explain.

26. $y = x^2 + 1$
27. $3x + 2y = 9$

28.

x	y
1	5
2	16
3	27
4	38

29.

x	y
7	−1
8	0
9	1
10	2

Graph each function.

30. $y = -\frac{1}{3}x^2$
31. $y = 2x^2 - 1$
32. $y = 2x^3 + 1$
33. $y = 2^x - 2$

CHAPTER 9

Preparing for Standardized Tests

Eliminate Unreasonable Answers

You can eliminate unreasonable answers to help you find the correct choice when solving multiple choice test items. Doing so may save you time by narrowing down the list of possible correct answers.

Strategies for Eliminating Unreasonable Answers

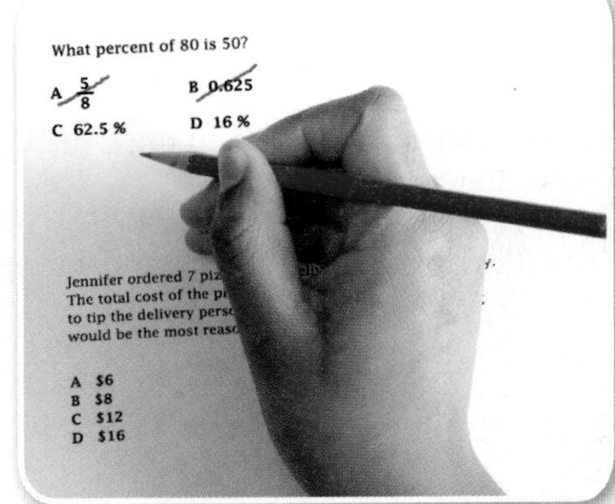

Step 1

Read the problem carefully to determine exactly what you are being asked to find.

Ask yourself:

- What am I being asked to solve?
- In what format (fraction, integer, decimal, percent, or graph) will the correct answer be?
- What units (if any) will the correct answer have?

Step 2

Carefully look over each possible answer choice and evaluate for reasonableness.

- Identify any answer choices that are clearly incorrect and eliminate them.
- Eliminate any answer choices that are not in the proper format.
- Eliminate any answer choices that do not have the correct units.

Step 3

Solve the problem and choose the correct answer from those remaining. Check your answer.

EXAMPLE

Read the problem. Identify what you need to know. Then use the information in the problem to solve.

The formula for the area A of a triangle with height h and base b is $A = \frac{1}{2}bh$. Write an expression to represent the area of the triangle below.

A $2x^2 + 4x$

B $32x^3$

C $16x^2$

D $16x^3$

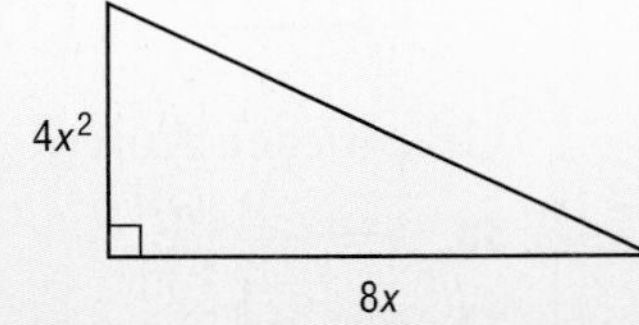

Finding the area of the triangle will result in multiplying a monomial by a monomial. You know by the product of powers property that the answer will have an x^3 term. Answer choices A and C are not in the proper form for the correct answer, so they can be eliminated.

So, the correct answer choice will be either B or D. Use the formula to multiply the monomials and find the correct answer.

$A = \frac{1}{2}bh$ **Write the formula.**

$A = \frac{1}{2}(8x)(4x^2)$ **Replace *b* with 8*x* and *h* with 4*x*².**

$A = (4x)(4x^2)$ **Multiply.**

$A = 16x^3$ **Multiply then simplify.**

The area of the triangle is $16x^3$. The correct answer is D.

Exercises

Read each question. Eliminate any unreasonable answers. Then use the information in the question to solve.

1. The table shows the surface areas, in square kilometers, of the oceans.

Ocean	Area (km²)
Pacific	156,000,000
Atlantic	77,000,000
Indian	69,000,000
Southern	20,000,000
Arctic	14,000,000

Which of the following represents the area of the Arctic Ocean in scientific notation?

A 0.14×10^8

B 1.4×10^7

C 1.4×10^8

D 14×10^6

2. The graph below represents a function.

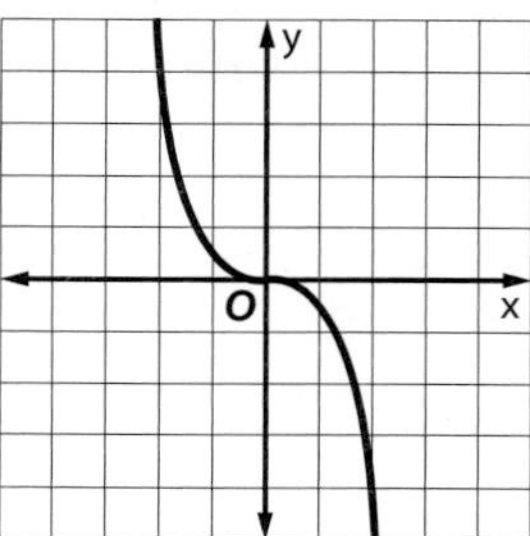

Which of the following equations represents the function?

F $y = 2^x - 1$

G $y = -2^x$

H $y = -\frac{1}{2}x^3$

J $y = \frac{1}{2}x^3$

CHAPTER 9

Standardized Test Practice

Cumulative, Chapters 1 through 9

Multiple Choice

Read each question. Then fill in the correct answer on the answer document provided by your teacher or on a sheet of paper.

1. In 2007, the population of the United States was about 3×10^8. Which of the following represents this number in standard form?

A 3,000,000

B 30,000,000

C 300,000,000

D 3,000,000,000

2. Michelle deposits $1200 in a savings account with a 6% interest rate. How much simple interest will she have earned after three years?

F $1,416 **H** $216

G $1,396 **J** $196

3. The area of a triangle is one half the product of its base and its height. Which of the following expressions represents the area of the right triangle shown?

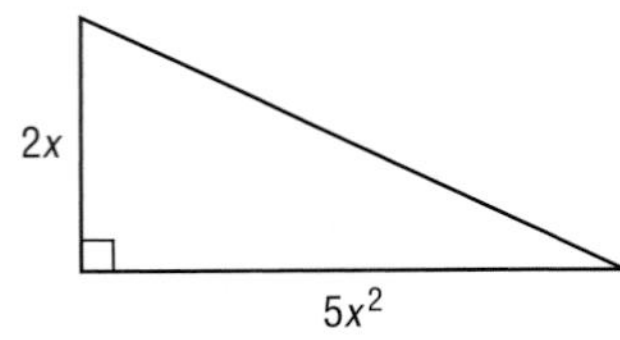

A $5x^3$ **C** $10x^3$

B $7x^2$ **D** $5x^2$

4. Find the slope of the line that passes through points $R(4, 1)$ and $S(-2, 6)$?

F $-\frac{5}{6}$ **H** $\frac{2}{3}$

G $-\frac{2}{3}$ **J** $\frac{5}{6}$

5. Which term does *not* describe the function graphed below?

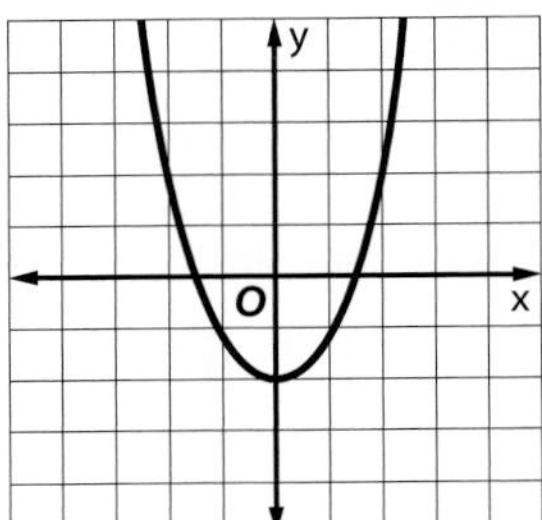

A linear **C** parabola

B nonlinear **D** quadratic

6. Write b^{-2} using a positive exponent.

F $\frac{1}{2b}$ **H** $\frac{1}{b+2}$

G $\frac{1}{2^b}$ **J** $\frac{1}{b^2}$

7. The radius of the Moon is about 1.7×10^6 meters. Use the formula $V = \frac{4}{3}\pi r^3$ to find the volume of the moon.

A 2.1×10^9 m^3

B 3.7×10^9 m^3

C 3.7×10^{18} m^3

D 2.1×10^{19} m^3

8. Which of the following expressions is equivalent to $3p - 4r + 4p - 5r$?

F $-p - r$ **H** $7p - r$

G $-p - 9r$ **J** $7p - 9r$

Test-TakingTip

Question 6 If time permits, you can check your answer using substitution. For example, let $b = 5$. Using a calculator, $5^{-2} = 0.04$, which is the same as $\frac{1}{25}$.

Short Response/Gridded Response

Record your answers on the answer sheet provided by your teacher or on a sheet of paper.

9. Describe the transformation shown on the coordinate plane below.

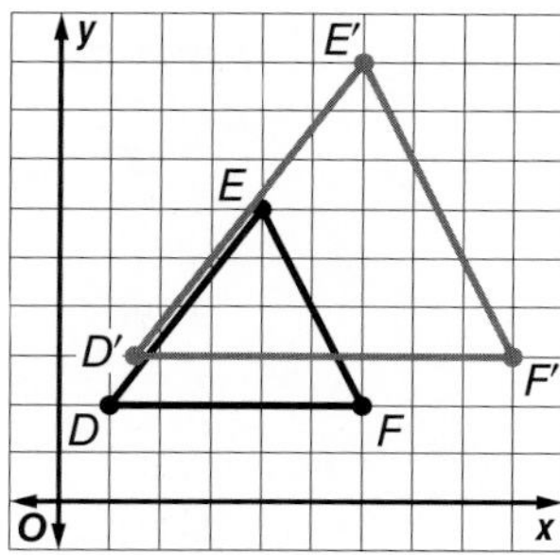

10. GRIDDED RESPONSE In a single elimination tournament, the function $y = 32\left(\frac{1}{2}\right)^x$ can be used to find the number of teams y left in the tournament after x rounds. If there are 32 teams in the tournament, how many rounds will be needed to determine the champion?

11. The width of a cotton fiber is about 0.001 centimeter.

a. Write the width of a cotton fiber as a fraction.

b. Write the width of a cotton fiber using a negative exponent.

c. How many cotton fibers would need to be stacked on top of each other to reach a height of 5 centimeters?

12. GRIDDED RESPONSE Write $\frac{8}{25}$ as a decimal.

13. Suppose a wild fire is doubling in size every 6 hours. If the initial size of the fire was 1 acre, how many acres will the fire cover in 2 days?

Extended Response

Record your answers on a sheet of paper. Show your work.

14. The quadratic function $y = -16x^2 + 128x$ models the height y of a toy rocket x seconds after it is launched.

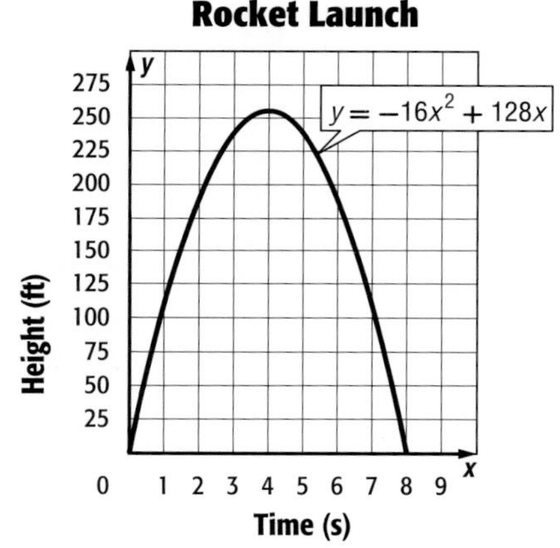

a. How long does it take the rocket to reach its peak height?

b. What is the peak height reached by the toy rocket?

c. How long does it take the rocket to land?

Need Extra Help?														
If you missed Question...	1	2	3	4	5	6	7	8	9	10	11	12	13	14
Go to Lesson or Page...	9-5	7-7	9-3	8-6	9-8	9-4	9-6	4-2	6-8	9-9	9-4	7-2	9-1	9-8

CHAPTER 10

Real Numbers and Right Triangles

Then

In Chapter 3 you learned how all fractions and some decimals are classified as rational numbers.

Now

In Chapter 10, you will:

- Identify irrational numbers and classify real numbers.
- Classify triangles.
- Solve problems using the Pythagorean Theorem and the Distance Formula.

Why?

RECREATION Looking for a sport that requires great skill, balance, and flexibility? Then give windsailing a try. Windsailing is a cross between sailing and surfing. Windsail designers use mathematics such as square roots when trying to create the best sail.

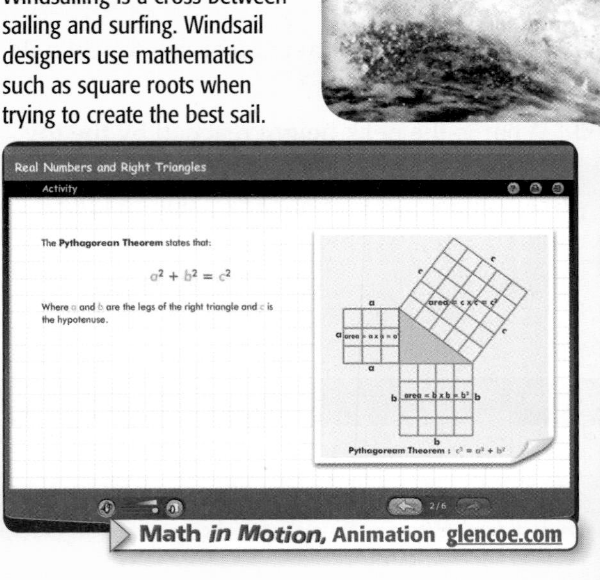

Real Numbers and Right Triangles
Activity
The Pythagorean Theorem states that:
$a^2 + b^2 = c^2$
Where a and b are the legs of the right triangle and c is the hypotenuse.
Pythagorean Theorem : $c^2 = a^2 + b^2$
2/6

Math in Motion, Animation glencoe.com

Get Ready for Chapter 10

Diagnose Readiness You have two options for checking Prerequisite Skills.

Text Option Take the Quick Check below. Refer to the Quick Review for help.

QuickCheck

Replace each ● with <, >, or = to make a true statement. (Concepts and Skills Bank)

1. 5.19 ● 5.187 **2.** 19.45 ● 19.5

3. 24.56 ● 24.56 **4.** 9.734 ● 9.73

5. 16.892 ● 16.9 **6.** 42.641 ● 42.64

7. **FIELD HOCKEY** Order the following winning averages of five field hockey teams from greatest to least.

0.523, 0.546, 0.601, 0.594, 0.509

QuickReview

EXAMPLE 1

Replace ● with <, >, or = to make 34.29 ● 34.3 a true statement.

34.29 **Line up the decimal points.**

34.3 **The digits in the tenths place are not the same.**

2 tenths < 3 tenths

So, 34.29 < 34.3.

Solve each equation. (Lesson 4-4)

8. $9a = 72$ **9.** $36 = 3m$

10. $16x = 64$ **11.** $144 = 16b$

12. $90 = 18c$ **13.** $120 = 40f$

14. **COOKIES** A batch of cookies contains 2 cups of sugar. How many batches of cookies contain 16 cups of sugar?

EXAMPLE 2

Solve $14w = 56$.

$14w = 56$ **Write the equation.**

$\frac{14w}{14} = \frac{56}{14}$ **Divide each side by 14.**

$w = 4$ **Simplify.**

Evaluate each expression. (Lesson 9-1)

15. $(8 - 3)^2 + (12 - 9)^2$

16. $(5 - 2)^2 + (8 - 4)^2$

17. $(-4 - 6)^2 + [(-2 - (-3)]^2$

18. $(-3 - 1)^2 + (5 - 3)^2$

19. $[(7 - (-4)]^2 + [(6 - (-2)]^2$

20. **BIOLOGY** Suppose a virus splits into two viruses every 45 minutes. How many viruses are there after 5 hours 15 minutes?

EXAMPLE 3

Evaluate $(6 - 2)^2 + (9 - 7)^2$.

$(6 - 2)^2 + (9 - 7)^2$

$= 4^2 + 2^2$ **Simplify the expressions inside parentheses first.**

$= 16 + 4$ **Evaluate 4^2 and 2^2.**

$= 20$ **Simplify.**

Online Option **Math Online** Take a self-check Chapter Readiness Quiz at glencoe.com.

Get Started on Chapter 10

You will learn several new concepts, skills, and vocabulary terms as you study Chapter 10. To get ready, identify important terms and organize your resources. You may wish to refer to **Chapter 0** to review prerequisite skills.

FOLDABLES® Study Organizer

Using Real Numbers Make this Foldable to help you organize information about real numbers and right triangles. Begin with three plain sheets of $8\frac{1}{2}$" × 11" paper.

1. **Fold** to make a triangle. Cut off the extra paper.

2. **Repeat** Step 1 twice. You now have three squares.

3. **Stack** the three squares and staple along the fold.

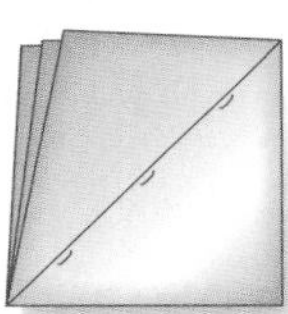

4. **Label** each section with a lesson number.

New Vocabulary

English		Español
square root	• p. 537 •	raíz cuadrada
irrational numbers	• p. 543 •	número irracional
real numbers	• p. 543 •	número reales
acute angle	• p. 551 •	ángulo agudo
right angle	• p. 551 •	ángulo recto
obtuse angle	• p. 551 •	ángulo obtuso
straight angle	• p. 551 •	ángulo llano
acute triangle	• p. 552 •	triángulo acutángulo
obtuse triangle	• p. 552 •	triángulo obtusángulo
right triangle	• p. 552 •	triángulo rectángulo
congruent	• p. 552 •	congruentes
scalene triangle	• p. 552 •	triángulo escaleno
isosceles triangle	• p. 552 •	triángulo isósceles
equilateral triangle	• p. 552 •	triángulo equilátero
hypotenuse	• p. 558 •	hipotenusa
Pythagorean Theorem	• p. 558 •	Teorema de Pitágoras
converse	• p. 560 •	recíproca
Distance Formula	• p. 565 •	Fórmula de la distancia

Review Vocabulary

integer • p. 61 • enteros whole numbers and their opposites

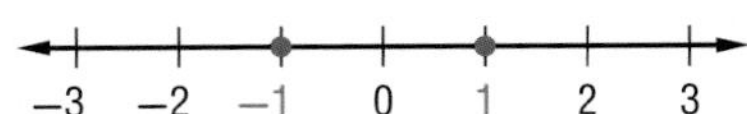

rational number • p. 128 • número racional a number that can be written in the form $\frac{a}{b}$, where a and b are integers and $b \neq 0$

Multilingual eGlossary glencoe.com

Math Online glencoe.com

- Study the chapter online
- Explore **Math in Motion**
- Get extra help from your own **Personal Tutor**
- Use **Extra Examples** for additional help
- Take a **Self-Check Quiz**
- **Review Vocabulary** in fun ways

EXPLORE 10-1

Algebra Lab

Squares and Square Roots

Math Online glencoe.com
Math *in Motion,* Animation

Numbers raised to the second power are called *squares*. You can use a geometric model to discover the reason they are called squares.

ACTIVITY 1

Use algebra tiles to evaluate 6^2.

- The power 6^2 is the product 6×6. The product can be represented by a square with one factor as the length and the other as the width.
- Arrange tiles in a 6-by-6 square.
- Since $6 \times 6 = 36$, $6^2 = 36$.

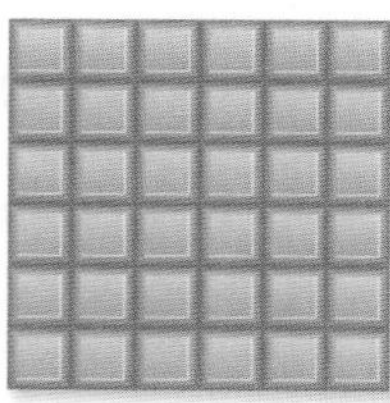

The opposite of squaring a number is finding a *square root*. These are inverse operations. To find the square root of a number, find two equal factors whose product is that number. The positive square root of a number is the *principal square root*. The symbol for the principal square root is $\sqrt{}$.

ACTIVITY 2

Use algebra tiles to find $\sqrt{25}$.

- You know that a square number can be represented by the area of a square. To find the square root of 25, arrange 25 tiles into a square.
- Twenty-five tiles can be arranged in a 5-by-5 square. So, $25 = 5 \times 5$, or 5^2.
- The length of each side of the square is 5 units. So, the principal square root of 25 is 5.

Analyze the Results

Model each power. Apply what you learned to evaluate it.

1. 3^2
2. 5^2
3. 7^2
4. 8^2
5. 10^2
6. 12^2
7. Explain why n^2 is called *n squared.*

Model each square root. Apply what you learned to evaluate it.

8. $\sqrt{4}$
9. $\sqrt{16}$
10. $\sqrt{81}$
11. $\sqrt{49}$
12. $\sqrt{100}$
13. $\sqrt{121}$
14. What part of the model represents the square root of the area of the square?

A **perfect square** is a number with a whole number square root but most whole numbers are not perfect squares. You can estimate the square roots of numbers that are not perfect squares.

ACTIVITY 3

Use algebra tiles to estimate the principal square root of 50.

Step 1 Arrange 50 tiles into the largest square possible. The largest square possible has 49 tiles, with 1 left over.

Step 2 Add tiles until you have the next larger square. So, 14 new tiles are needed for the next square.

Step 3 The square root of 49 is 7 and the square root of 64 is 8. Therefore, the square root of 50 is between the whole numbers 7 and 8.

Since 50 is closer to 49 than to 64, you can expect that $\sqrt{50}$ is closer to 7 than to 8.

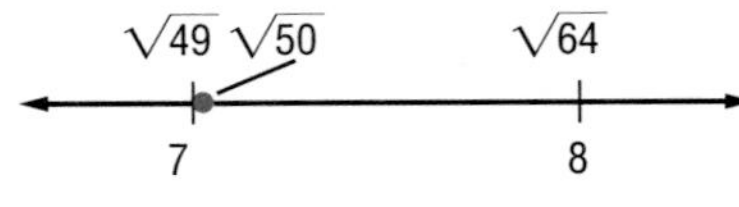

Check the estimate with a calculator. 2nd [$\sqrt{\ }$] 50 ENTER 7.071067812 ✓

Analyze the Results

Use a number line to model each square root. Apply what you learned to estimate the principal square root.

15. $\sqrt{20}$ **16.** $\sqrt{44}$ **17.** $\sqrt{58}$ **18.** $\sqrt{69}$ **19.** $\sqrt{91}$ **20.** $\sqrt{111}$

21. MAKE A CONJECTURE Describe how to use the guess and check strategy to find the square root of a number between 400 and 500.

10-1 Squares and Square Roots

Then

You have already evaluated expressions containing squares of numbers. (Lesson 8-1)

Now

- Find square roots.
- Estimate square roots.

New Vocabulary

perfect square
square root
radical sign

Math Online

glencoe.com

- Extra Examples
- Personal Tutor
- Self-Check Quiz
- Homework Help

Why?

The Space Needle in Seattle, Washington, was built on a square lot with an area of 14,400 square feet.

a. What number when multiplied by itself equals 144? 14,400?

b. What are the dimensions of the square lot?

Find Square Roots A number like 144 is a **perfect square** because it is the square of an integer. The opposite of squaring a number is finding the square root.

Key Concept Square Root — For Your FOLDABLE

Words A **square root** of a number is one of its two equal factors.

Symbols If $x^2 = y$, then x is a square root of y.

A **radical sign**, $\sqrt{}$, is used to indicate a positive square root. Every positive number has both a positive and a negative square root.

$$\sqrt{36} = 6 \qquad -\sqrt{36} = -6 \qquad \pm\sqrt{36} = \pm 6 \text{ or } 6, -6$$

A negative number like −36 has no real-number square root because the square of a number cannot be negative.

EXAMPLE 1 Find Square Roots

Find each square root.

a. $\sqrt{9}$

$\sqrt{9} = 3$ — Find the positive square root of 9; $3^2 = 9$.

b. $-\sqrt{64}$

$-\sqrt{64} = -8$ — Find the negative square root of 64; $8^2 = 64$.

c. $\pm\sqrt{4}$

$\pm\sqrt{4} = \pm 2$ — Find both square roots of 4; $2^2 = 4$.

d. $\sqrt{-81}$

There is no real square root because no number times itself is equal to −81.

Check Your Progress

1A. $\sqrt{49}$ **1B.** $-\sqrt{16}$ **1C.** $\pm\sqrt{100}$ **1D.** $\sqrt{-49}$

Personal Tutor glencoe.com

Math History Link

Aryabhata (476–550) Aryabhata, an Indian mathematician, published a very good approximation of the square roots of 2 and 3. He also wrote a summary of astronomy and Hindu mathematics.

Estimate Square Roots When integers are not perfect squares, you can estimate square roots mentally by using perfect squares.

EXAMPLE 2 Estimate Square Roots

Estimate each square root to the nearest integer.

a. $\sqrt{33}$

The first perfect square less than 33 is 25. $\sqrt{25} = 5$

The first perfect square greater than 33 is 36. $\sqrt{36} = 6$

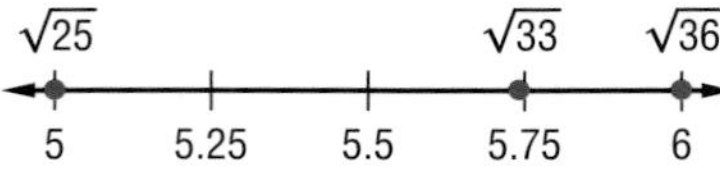

The square root of 33 is between the integers 5 and 6. Since 33 is closer to 36 than to 25, you can expect $\sqrt{33}$ to be closer to 6 than to 5.

b. $-\sqrt{129}$

The first perfect square less than 129 is 121. $\sqrt{121} = 11$

The first perfect square greater than 129 is 144. $\sqrt{144} = 12$

$-\sqrt{144}$ $-\sqrt{129}$ $-\sqrt{121}$

−12 −11.75 −11.50 −11.25 −11

The negative square root of 129 is between the integers −11 and −12. Since 129 is closer to 121 than to 144, you can expect $-\sqrt{129}$ to be closer to −11 than −12.

Check Your Progress

2A. $\sqrt{60}$ **2B.** $-\sqrt{23}$

Personal Tutor glencoe.com

StudyTip

Choose a Form Express a number as a square root if an exact answer is needed. Express a number as a decimal if an approximation is sufficient.

EXAMPLE 3 Use a Calculator to Find a Square Root

Use a calculator to find $\pm\sqrt{40}$ to the nearest tenth.

2nd [√‾] 40 ENTER 6.32455532 **Use a calculator.**

$\sqrt{40} \approx 6.3$ **Round to the nearest tenth.**

So, $\pm\sqrt{40}$ is $\approx \pm 6.3$ because you must find both square roots.

Check for Reasonableness Since $6^2 = 36$, the answer is reasonable. ✓

$\sqrt{36}$ $\sqrt{40}$ $\sqrt{49}$

6 6.25 6.5 6.75 7

Check Your Progress

Use a calculator to find each square root to the nearest tenth.

3A. $\sqrt{14}$ **3B.** $\sqrt{79}$

Personal Tutor glencoe.com

EXAMPLE 4 Use a Calculator to Find a Negative Square Root

Use a calculator to find $-\sqrt{18}$ to the nearest tenth.

[(–)] [2nd] [$\sqrt{\ }$] 18 [ENTER] 4.24264069 **Use a calculator.**

$-\sqrt{18} \approx -4.2$ **Round to the nearest tenth.**

Check for Reasonableness **Since $4^2 = 16$, the answer is reasonable. ✓**

$-\sqrt{25}$ $-\sqrt{18}$ $-\sqrt{16}$

–5 –4.75 –4.5 –4.25 –4

Check Your Progress

Use a calculator to find each square root to the nearest tenth.

4A. $-\sqrt{27}$

4B. $-\sqrt{92}$

4C. $\pm\sqrt{67}$

4D. $-\sqrt{135}$

Personal Tutor glencoe.com

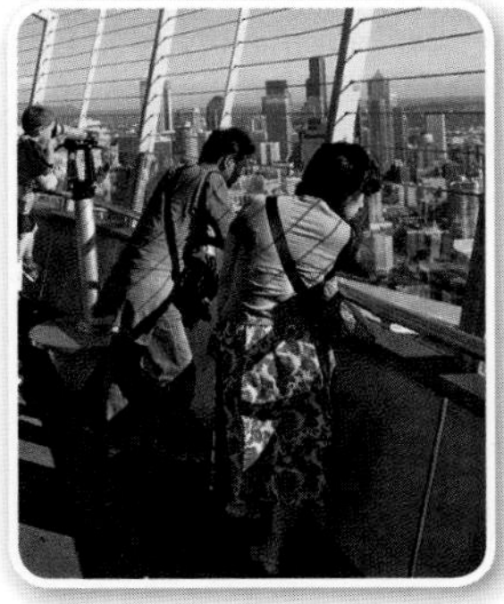

Real-World Link

The Space Needle has 848 steps from the bottom of the basement to the top of the Observation Deck.

Source: Space Needle Fun Facts

When finding square roots in real-world situations, use the positive, or *principal*, square root when a negative answer does not make sense.

Real-World EXAMPLE 5 Use Square Roots to Solve Problems

RECREATION On a clear day, the number of miles a person can see to the horizon can be found using the formula $d = 1.22 \cdot \sqrt{h}$, where d is the distance to the horizon in miles and h is the person's distance from the ground in feet. The observation deck of Seattle's Space Needle is 520 feet high. How far to the horizon can a person standing on the observation deck see? Round to the nearest tenth.

ESTIMATE **The distance is between $1 \cdot \sqrt{400}$ and $1 \cdot \sqrt{900}$. So, it's between 20 and 30.**

$d = 1.22 \cdot \sqrt{h}$ **Write the equation.**

$= 1.22 \cdot \sqrt{520}$ **Replace h with 520.**

$\approx 1.22 \cdot 22.8$ **Use a calculator.**

≈ 27.8 **Simplify.**

The approximate distance to the horizon is 27.8 miles to the nearest tenth.

Check for Reasonableness **$20 < 27.8 < 30$ ✓**

Check Your Progress

5. RECREATION Spring Port Ledge Lighthouse in Maine is approximately 55 feet high. Estimate and then calculate about how far a person who is standing on the observation deck can see on a clear day. Round to the nearest tenth of a mile.

Personal Tutor glencoe.com

Check Your Understanding

Example 1 p. 537

Find each square root.

1. $\sqrt{16}$
2. $-\sqrt{100}$
3. $\pm\sqrt{81}$

Example 2 p. 538

Estimate each square root to the nearest integer. Do not use a calculator.

4. $\sqrt{27}$
5. $-\sqrt{48}$
6. $\pm\sqrt{39}$

Examples 3 and 4 pp. 538–539

Use a calculator to find each square root to the nearest tenth.

7. $\sqrt{21}$
8. $-\sqrt{56}$
9. $\pm\sqrt{37}$

Example 5 p. 539

10. **RECREATION** A baseball diamond is actually a square with an area of 8100 square feet. Most baseball teams cover their diamond with a tarp to protect it from the rain. The sides are all the same length. How long is the tarp on each side?

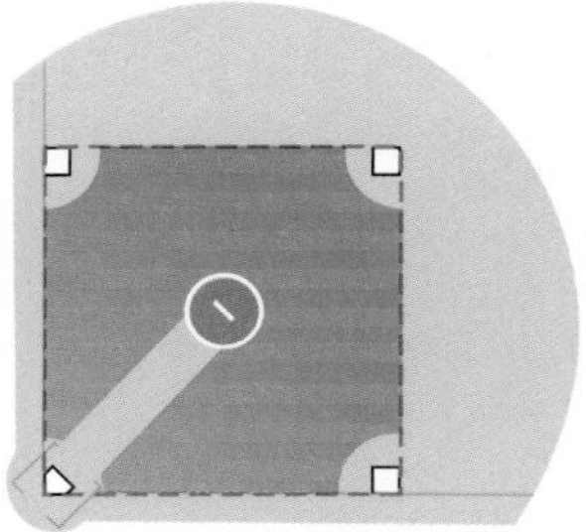

Practice and Problem Solving

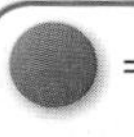 = **Step-by-Step Solutions** begin on page R11. **Extra Practice** begins on page 810.

Example 1 p. 537

Find each square root.

11. $\sqrt{36}$
12. $\sqrt{9}$
13. $-\sqrt{169}$
14. $-\sqrt{144}$
15. $\pm\sqrt{-25}$
16. $\pm\sqrt{1}$

Example 2 p. 538

Estimate each square root to the nearest integer. Do not use a calculator.

17. $\sqrt{83}$
18. $\sqrt{34}$
19. $-\sqrt{102}$
20. $-\sqrt{14}$
21. $\pm\sqrt{78}$
22. $\pm\sqrt{146}$

Examples 3 and 4 pp. 538–539

Use a calculator to find each square root to the nearest tenth.

23. $\sqrt{7}$
24. $\sqrt{32}$
25. $\pm\sqrt{71}$
26. $-\sqrt{48}$
27. $-\sqrt{155}$
28. $\sqrt{162}$
29. $\sqrt{310}$
30. $\pm\sqrt{215}$

Example 5 p. 539

31. **RECREATION** Cedar Point in Ohio is known as the "Roller Coaster Capital of the World." The table shows the heights of their tallest coasters. Use the formula from Example 5 to determine how far a rider can see from the highest point of each ride. Round to the nearest tenth.

Cedar Point Attractions

Roller Coaster	Height (ft)
Magnum XL-200	205
Mean Streak	161
Millennium Force	310
Top Thrill Dragster	420
Wicked Twister	215

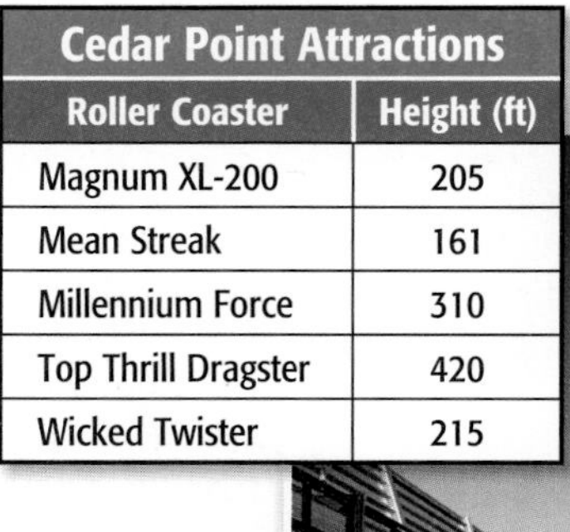

a. Millennium Force

b. Mean Streak

c. How much farther can a rider see on the Top Thrill Dragster than on the Magnum XL-200?

Real-World Link

Chess is thought to be the oldest skill game in the world. Viswanathan Anand was named the 2007 World Chess Champion in Mexico City, Mexico. The top prize for the tournament was $390,000.

32. **RECREATION** According to the *Guinness Book of World Records*, the smallest chessboard has an area of 1.44 square inches. The largest chessboard has an area of approximately 297.6 square feet. Find the side length of each board. Round to the nearest tenth, if necessary.

    a. smallest board **b.** largest board

33. **GEOMETRY** The area of the square at the right is given. Find the length of a side to the nearest tenth. Then find its approximate perimeter.

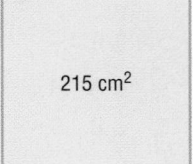

215 cm^2

34. **GARDENS** The recommended space needed for a certain plant is 2 square feet. You want to plant 32 of these plants in a square garden. What is the length of each side of the garden if you follow the recommendations?

Complete each of the following mentally.

35. Which is greater, $\sqrt{79}$ or 8? Explain your reasoning.

36. Which is less, -2 or $-\sqrt{6}$? Explain your reasoning.

37. Between which two consecutive whole numbers on a number line does $\sqrt{85}$ lie?

38. Order $\sqrt{77}$, -8, $-\sqrt{83}$, 9, -10, $-\sqrt{76}$, $\sqrt{65}$ from least to greatest.

39. **MULTIPLE REPRESENTATIONS** In this problem, you will investigate the graphs of $y = x^2$ and $y = \sqrt{x}$.

    a. TABULAR Copy and complete each of the tables shown. Choose an appropriate domain

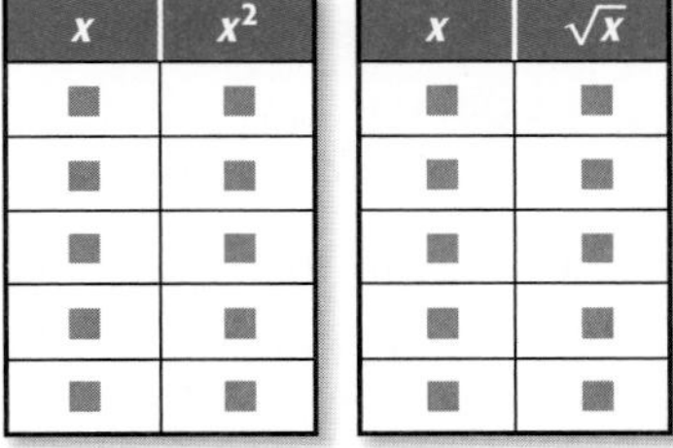

x	x^2
■	■
■	■
■	■
■	■
■	■

x	$\sqrt{x}$
■	■
■	■
■	■
■	■
■	■

    b. GRAPHICAL Using the ordered pairs from each table, graph $y = x^2$ and $y = \sqrt{x}$.

    c. VERBAL Describe each graph for $0 < x < 1$, for $x = 1$, and for $x > 1$.

    d. ANALYTICAL Which has the greater average rate of change? Explain.

H.O.T. Problems Use Higher-Order Thinking Skills

40. **OPEN ENDED** Find a square root that lies between 17 and 18.

41. **CHALLENGE** Addition and subtraction are *inverse operations* because one operation undoes the other operation. Use inverse operations to evaluate the following.

    a. $(\sqrt{246})^2$ **b.** $(\sqrt{811})^2$ **c.** $(\sqrt{732})^2$

    d. Describe the inverse operation of squaring a number.

42. **NUMBER SENSE** What are the possibilities for the ending digit of a number that has a whole number square root? Explain your reasoning. Then write three numbers between 1000 and 2000 that are *not* perfect squares.

43. **WRITING IN MATH** Describe the difference between an exact value and an approximation when finding square roots of numbers that are not perfect squares. Give an example of each.

Standardized Test Practice

44. Which point on the number line best represents $\sqrt{210}$?

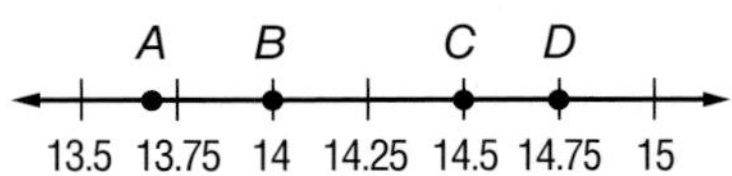

A A **C** C
B B **D** D

45. SHORT RESPONSE The area of each square is 25 square units. Find the perimeter of the figure shown below.

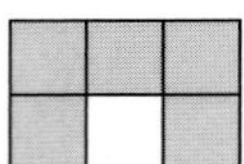

46. The new gymnasium at Oakdale Middle School has a hardwood floor in the shape of a square. If the area of the floor is 62,500 square feet, what is the length of one side of the square floor?

F 200 ft **H** 250 ft
G 225 ft **J** 275 ft

47. A surveyor determined the distance across a field was $\sqrt{1568}$ feet. What is the approximate distance?

A 25.6 ft **C** 39.6 ft
B 30.6 ft **D** 42.6 ft

Spiral Review

Graph each equation. (Lesson 9-9)

48. $y = 3x^3$ **49.** $y = -3x^3$ **50.** $y = 2^x - 4$

51. BASEBALL The equation $h = -16t^2 + 8t + 4$ can be used to model the height h of a baseball t seconds after it is hit. (Lesson 9-8)

a. Graph the equation on a coordinate plane.

b. After how many seconds will the ball hit the ground?

Solve each problem using the percent equation. (Lesson 7-5)

52. 9 is what percent of 25?

53. 48 is 64% of what number?

54. GEOMETRY Suppose that two-fifths of the rectangle at the right is shaded. (Lesson 5-1)

a. What is the area of the shaded region?

b. Is it better to use the decimal or fractional form of two-fifths in this situation? Explain.

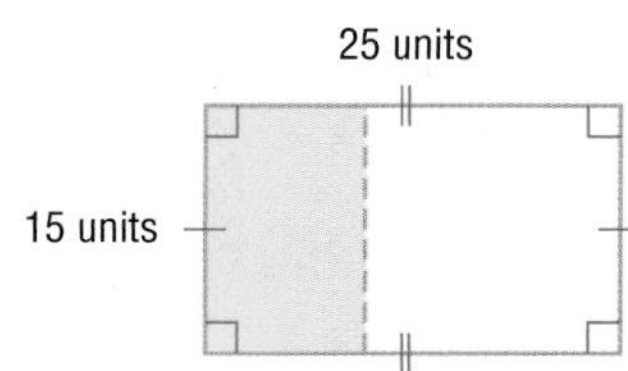

Skills Review

Identify all sets to which each number belongs. (Lesson 3-2)

55. 4 **56.** -7 **57.** $-2\frac{5}{8}$ **58.** $\frac{6}{3}$

59. 15.8 **60.** 9.0202020... **61.** 1.2345... **62.** 30.151151115...

10-2 The Real Number System

Why?

A roller coaster drops 100 feet at a 95-degree angle. The ride is 0.84 mile long and covers 5.5 acres.

a. Express the length of the ride as a fraction in simplest form.

b. Express the number of acres that the ride covers as an improper fraction.

Then
You have already compared fractions and decimals.
(Lesson 3-1)

Now
- Identify and compare numbers in the real number system.
- Solve equations by finding square roots.

New Vocabulary
irrational numbers
real numbers

Math Online
glencoe.com
- Extra Examples
- Personal Tutor
- Self-Check Quiz
- Homework Help

Identify and Compare Real Numbers Recall that *rational numbers* are numbers that can be written as fractions. Examples of rational numbers are given below.

$$1\frac{2}{5} = \frac{7}{5} \qquad -4 = -\frac{4}{1} \qquad 0.15 = \frac{15}{100}$$

$$0.\overline{3} = \frac{1}{3} \qquad \sqrt{25} = \frac{5}{1}$$

An **irrational number** is a number that cannot be written as a fraction. When written as decimals, irrational numbers neither terminate nor repeat.

Key Concept Irrational Number — For Your FOLDABLE

Words An irrational number is a number that cannot be expressed as $\frac{a}{b}$, where a and b are integers and $b \neq 0$.

Examples $\pi \approx 3.14159...$ $\qquad -\sqrt{5} \approx -2.2360679...$

The sets of rational numbers and irrational numbers together make up the set of **real numbers**. The diagram shows the relationship among the real numbers.

Real Numbers

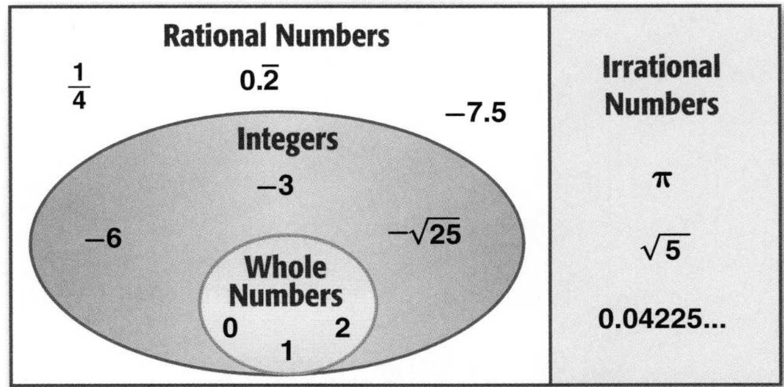

Computations with an irrational number and a rational number (other than zero) produce an irrational number.

StudyTip

Square Roots If an integer is not a perfect square, then its square root is an irrational number.

EXAMPLE 1 Classify Real Numbers

Name all of the sets of numbers to which each real number belongs. Write *whole, integer, rational,* or *irrational.*

a. $\frac{21}{7}$ Since $\frac{21}{7} = 3$, this number is a whole number, an integer, and a rational number.

b. -2.5 Since $-2.5 = -\frac{10}{4}$, this number is a rational number.

c. $0.\overline{2}$ Since $0.\overline{2} = 0.22222\ldots$ or $\frac{2}{9}$, this number is a rational number.

d. $\sqrt{38}$ Since $\sqrt{38} = 6.16441400\ldots$ It is not the square root of a perfect square so it is irrational.

Check Your Progress

1A. 0.7

1B. $\sqrt{100}$

1C. $\frac{9}{5}$

1D. -6

Personal Tutor glencoe.com

EXAMPLE 2 Compare Real Numbers

Review Vocabulary

Integers Integers are the whole numbers and their opposites. …, −3, −2, −1, 0, 1, 2, 3, … (Lesson 2-1)

Replace ● with <, >, or = to make $3\frac{1}{3}$ ● $\sqrt{15}$ a true statement.

Express each number as a decimal. Then compare the decimals.

$3\frac{1}{3} = 3.33333333\ldots$

$\sqrt{15} = 3.87298334\ldots$

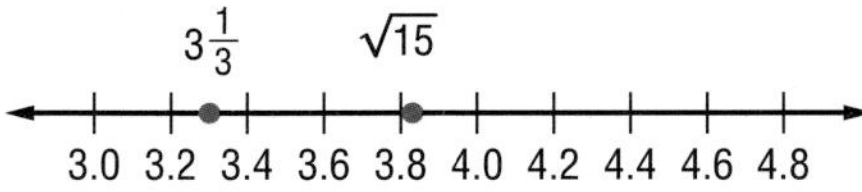

Since 3.333… is less than 3.872…, $3\frac{1}{3} < \sqrt{15}$.

Check Your Progress

2. Replace ● with <, >, or = to make $7\frac{2}{5}$ ● $\sqrt{57}$ a true statement.

Personal Tutor glencoe.com

EXAMPLE 3 Order Real Numbers

Order $8\frac{4}{5}$, $\sqrt{64}$, $8.\overline{3}$, $\sqrt{76}$ from least to greatest.

Express each number as a decimal. Then order the decimals.

$8\frac{4}{5} = 8.8$

$\sqrt{64} = 8$

$8.\overline{3} = 8.33333333\ldots$

$\sqrt{76} = 8.71779788\ldots$

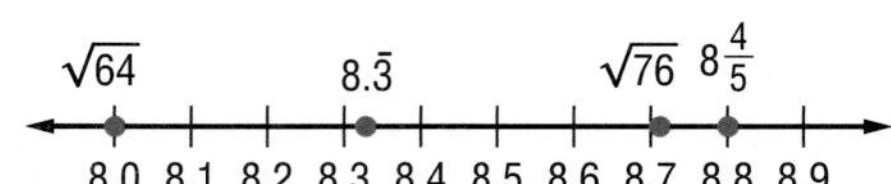

From least to greatest, the order is $\sqrt{64}$, $8.\overline{3}$, $\sqrt{76}$, and $8\frac{4}{5}$.

Check Your Progress

3. Order $\sqrt{30}$, 5.6, $\frac{15}{3}$, and $5\frac{2}{3}$ from greatest to least.

Personal Tutor glencoe.com

Solve Equations By the definition of a square root, if $x^2 = y$, then $x = \pm\sqrt{y}$. You can use this relationship to solve equations involving squares.

EXAMPLE 4 Solve Equations

Solve each equation. Round to the nearest tenth, if necessary.

a. $a^2 = 36$

$a^2 = 36$	**Write the equation.**
$a = \pm\sqrt{36}$	**Definition of square root**
$a = 6$ and -6	**Check $6 \cdot 6 = 36$ and $(-6) \cdot (-6) = 36$**

The solutions are 6 and −6.

b. $2n^2 = 170$

$2n^2 = 170$	**Write the equation.**
$n^2 = 85$	**Divide each side by 2.**
$n = \pm\sqrt{85}$	**Definition of square root**
$n \approx 9.2$ and -9.2	**Use a calculator.**

The solutions are approximately 9.2 and −9.2.

StudyTip

Check Reasonableness Check the results by calculating 9^2 and $(-9)^2$. $9^2 = 81$ $(-9)^2 = 81$ Since 81 is close to 85, the solutions are reasonable.

Check Your Progress

4A. $363 = 3d^2$ **4B.** $y^2 = 30$

Personal Tutor glencoe.com

In most real-world situations, a negative square root does not make sense. Consider only the positive, or *principal*, square root.

Real-World EXAMPLE 5 Use Real Numbers to Solve Problems

HANG GLIDING The *aspect ratio* of a hang glider allows it to glide through the air. The formula for the aspect ratio R is $R = \frac{s^2}{A}$, where s is the wingspan and A is the area of the wing. What is the wingspan of a hang glider if its aspect ratio is 4.5 and the area of the wing is 50 square feet?

$R = \frac{s^2}{A}$	**Write the formula.**
$4.5 = \frac{s^2}{50}$	**Replace *R* with 4.5 and *A* with 50.**
$225 = s^2$	**Multiply each side by 50.**
$\sqrt{225} = s$	**Consider the positive square root.**
$15 = s$	**Simplify.**

The wingspan of the hang glider is 15 feet.

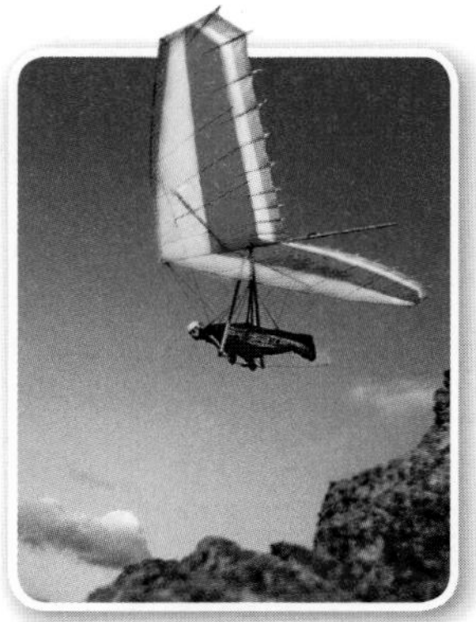

Real-World Link

The longest hang gliding trip lasted 36 hours. The highest altitude reached by a hang glider was 18,000 feet.

Source: HowStuffWorks

Check Your Progress

5. SEISMIC WAVES A *tsunami* is caused by an earthquake on the ocean floor. The speed of a tsunami can be measured by the formula $\frac{s^2}{d} = 9.61$, where s is the speed of the wave in meters per second and d is the depth of the ocean in meters where the earthquake occurs. What is the speed of a tsunami if an earthquake occurs at a depth of 632 meters? Round to the nearest tenth.

Personal Tutor glencoe.com

Check Your Understanding

Example 1 p. 544

Name all of the sets of numbers to which each real number belongs. Write *whole, integer, rational,* or *irrational*.

1. 10 **2.** $\frac{1}{5}$ **3.** $\sqrt{35}$ **4.** $-\frac{14}{2}$

Example 2 p. 544

Replace each ● with <, >, or = to make a true statement.

5. $\sqrt{6}$ ● $2\frac{3}{8}$ **6.** $-5.\overline{2}$ ● $-\sqrt{29}$ **7.** $-\sqrt{42}$ ● $-6\frac{2}{3}$

Example 3 p. 544

Order each set of numbers from least to greatest.

8. $-\frac{5}{4}, -\sqrt{4}, -1.5, -1\frac{3}{4}$ **9.** $\sqrt{110}, 10\frac{1}{5}, 10.\overline{5}, 10.15$

Example 4 p. 545

ALGEBRA Solve each equation. Round to the nearest tenth, if necessary.

10. $x^2 = 16$ **11.** $3m^2 = 222$ **12.** $42 = 1.4r^2$

Example 5 p. 545

13. **PIZZA** The formula $A \approx 3.14r^2$ can be used to determine the area of a circle where A is the area and r is the distance from the center of the circle to the outside edge. If the area of the largest pizza ever made was approximately 11,818 square feet, about how far is the distance from the center of the pizza to the outside edge? Round to the nearest tenth.

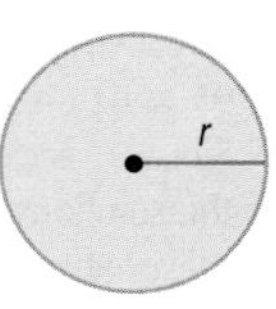

Practice and Problem Solving

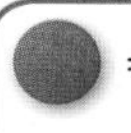 = **Step-by-Step Solutions** begin on page R11. **Extra Practice** begins on page 810.

Example 1 p. 544

Name all of the sets of numbers to which each real number belongs. Write *whole, integer, rational,* or *irrational*.

14. 4 **15.** $\frac{3}{5}$ **16.** $-\frac{7}{2}$ **17.** $\sqrt{26}$

18. $-\frac{36}{4}$ **19.** 8.2 **20.** 0.55555… **21.** $-\sqrt{81}$

22. 6.01 **23.** $\frac{42}{7}$ **24.** $\sqrt{144}$ **25.** $0.\overline{18}$

Example 2 p. 544

Replace ● with <, >, or = to make a true statement.

26. $\sqrt{17}$ ● 4.2 **27.** $5.\overline{15}$ ● $\sqrt{26}$ **28.** $3\frac{5}{6}$ ● $\sqrt{10}$

29. $\sqrt{2.56}$ ● $1\frac{3}{5}$ **30.** $-\sqrt{0.25}$ ● $-\frac{1}{2}$ **31.** $-7\frac{5}{8}$ ● $-\sqrt{55}$

Example 3 p. 544

Order each set of numbers from least to greatest.

32. $2.\overline{71}, 2\frac{3}{4}, \sqrt{5}, \frac{5}{2}$ **33.** $\sqrt{64}, 8\frac{1}{7}, 8.\overline{14}, \frac{15}{2}$

34. $-\sqrt{11}, -3.\overline{3}, -3.4, -\frac{16}{5}$ **35.** $-\frac{5}{6}, -5, -\sqrt{26}, -\frac{31}{6}$

Example 4 p. 545

ALGEBRA Solve each equation. Round to the nearest tenth, if necessary.

36. $y^2 = 64$ **37** $130 = n^2$ **38.** $5p^2 = 315$

39. $2d^2 = 162$ **40.** $190.5 = 1.5b^2$ **41.** $0.1x^2 = 0.169$

Example 5 p. 545

Real-World Link

The women's pole vault world record was set at the World Championships in 2005. A height of 5.01 meters, which is about 16 feet, was cleared.

Source: International Association of Athletics Federations

42. **TRACK AND FIELD** The height h in feet that a pole vaulter can reach can be estimated using the formula $h = \frac{v^2}{64}$, where v is the velocity of the athlete in feet per second. Use the information at the left to determine about how fast the record holder was running.

43 **PHYSICS** The formula $h = 16t^2$ describes the time t in seconds that it takes for an object to fall from a height of h feet. A thrill ride has a 60-foot tall freefall drop. How long does it take for the ride to complete its freefall? Round to the nearest tenth.

Determine whether each statement is *always, sometimes,* or *never* true. Explain your reasoning.

44. An integer is a rational number.

45. A real number can be written as a repeating decimal.

46. An irrational number can be written as a terminating decimal.

47. A whole number is an integer.

Tell whether each expression is *rational or irrational.* Explain.

48. $4 \times \sqrt{2}$ **49.** $\sqrt{49} - 15$ **50.** $\sqrt{10} \div 2$ **51.** $9 \cdot \pi$

52. **CARS** In the formula $s = \sqrt{30fd}$, s is the speed of a car in miles per hour, d is the distance the car skidded in feet, and f is friction. The table shows different values of f. At an accident scene, a car made 100-foot skid marks before hitting another car. If the speed limit was 55 miles per hour, was the car speeding before applying the brakes on a dry, concrete road? Explain.

Road Conditions	Type of Surface	
	Concrete	Asphalt
Wet	0.4	0.5
Dry	0.8	1.0

H.O.T. Problems

Use Higher-Order Thinking Skills

53. **OPEN ENDED** Find a rational number and an irrational number that are between 6.2 and 6.5. Include the decimal approximation of the irrational number to the nearest hundredth.

REASONING **Tell whether each expression is *true* or *false*. If false, give a counterexample.**

54. All whole numbers are integers.

55. All square roots are irrational numbers.

56. All rational numbers are integers.

57. **CHALLENGE** What is the value of x to the nearest tenth if $x^2 - 4^2 = \sqrt{15^2}$?

58. **WHICH ONE DOESN'T BELONG?** Identify the number that does not belong with the other three. Explain your reasoning.

$50.\overline{1}$	$-\frac{50}{2}$	-50.1	$\sqrt{50}$

59. **WRITING IN MATH** Explain the relationship between the area of a square and the length of its sides. Give an example of a square whose side length is rational and an example of a square whose side length is irrational.

Standardized Test Practice

60. For what value of x is $\frac{1}{\sqrt{x}} < \sqrt{x} < x$?

A -2

B $\frac{1}{4}$

C $\frac{1}{2}$

D 2

61. The formula $h = 16t^2$ describes the time t in seconds that it takes for an object to fall from a height of h feet. How long would it take a basketball to hit the ground from a height of 50 feet?

F 1.77 s

G 10.36 s

H 28.28 s

J 200 s

62. Which of the following is an example of an irrational number?

A -8

B $\frac{3}{4}$

C $\sqrt{10}$

D $\sqrt{16}$

63. GRIDDED RESPONSE The area of a triangle that has three equal sides can be found using the expression $\frac{s^2\sqrt{3}}{4}$, where s is the length of one side. What is the area in square inches of the triangle below to the nearest tenth?

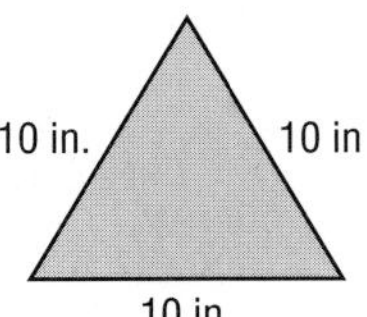

Spiral Review

Estimate each square root to the nearest integer. Do not use a calculator. (Lesson 10-1)

64. $\sqrt{79}$

65. $\sqrt{95}$

66. $-\sqrt{54}$

67. $-\sqrt{125}$

68. $\pm\sqrt{200}$

69. $\pm\sqrt{396}$

Graph each function. (Lesson 9-9)

70. $y = 4x^3$

71. $y = -2x^3 - 3$

72. $y = -\frac{1}{3}x^3 + 2$

73. $y = \frac{3}{4}x^3 - 4$

74. FORESTRY The table shows the percent of forest land in different states. (Lesson 7-3)

a. For each state in the table, how many square miles of land are covered by forests? Round to the nearest square mile.

b. Which state has the greatest amount of forest land?

STATE	PERCENT OF LAND COVERED BY FOREST	AREA OF STATE (SQUARE MILES)
Illinois	11.0%	55,584
Kentucky	49.1%	39,728
Michigan	44.7%	56,804
New York	56.1%	47,214
Ohio	27.3%	40,948

Source: U.S. Census Bureau

Skills Review

Solve each equation. (Lesson 5-5)

75. $18 + 57 + x = 180$

76. $x + 27 + 54 = 180$

77. $85 + x + 24 = 180$

78. $x + x + x = 180$

79. $2x + 3x + 4x = 180$

80. $2x + 3x + 5x = 180$

Geometry Lab
Angles in a Triangle

Math Online glencoe.com
Math *in Motion*, Animation

There is a relationship among the measures of the angles of a triangle.

ACTIVITY

Step 1 Use a straightedge or ruler to draw a triangle on a piece of paper. Then cut out the triangle and label the vertices *X*, *Y*, and *Z*.

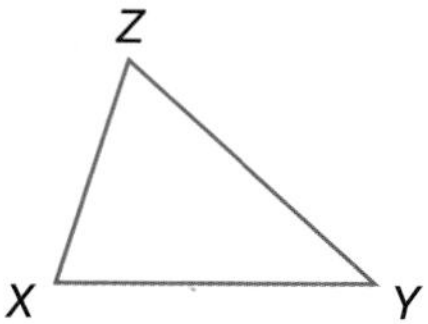

Step 2 Fold the triangle as shown so that point *Z* lies on side *XY* as shown. Label the back of ∠*Z* as ∠2.

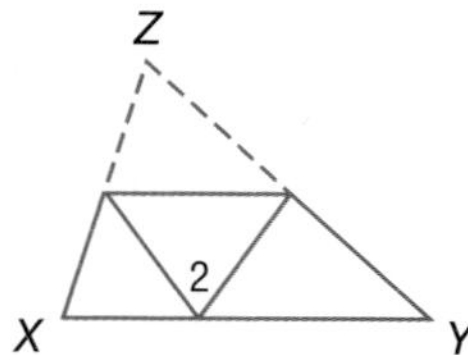

Step 3 Fold again so point *X* meets point *Z*. Label the back of ∠*X* as ∠1.

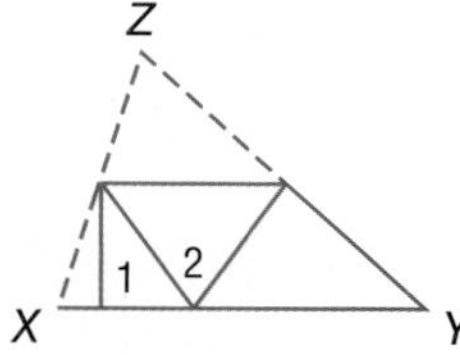

Step 4 Fold so point *Y* meets point *Z*. Label the back of ∠*Y* as ∠3.

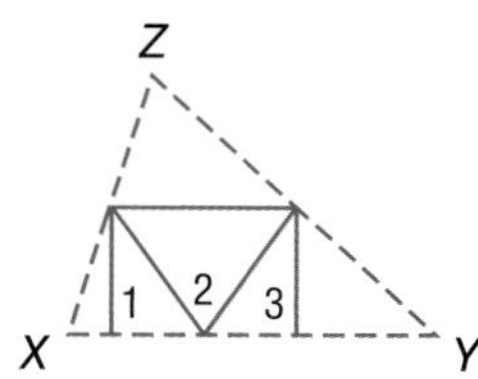

Analyze the Results

1. What kind of figure is formed by angles 1, 2, and 3?

2. Repeat the activity using a different triangle. Label the vertices *R*, *S*, and *T*. What kind of figure is formed by angles *R*, *S*, and *T*?

3. Make a table like the one shown at the right. Use a protractor to measure the angles of triangles *XYZ* and *RST*. Find the sum of the angle measures.

Triangle	Angle Measures	Sum of Angle Measures
XYZ		
RST		

4. **MAKE A CONJECTURE** Without measuring, make a conjecture about the sum of the measures of the angles of any triangle.

10-3 Triangles

Then

You solved equations by adding or subtracting. (Lesson 4-3)

Now

- Find the missing angle measure of a triangle.
- Classify triangles by properties and attributes.

New Vocabulary

line segment
triangle
vertex
congruent

Math Online

glencoe.com
- Extra Examples
- Personal Tutor
- Self-Check Quiz
- Homework Help

Why?

Geodesic domes are structures that are almost spherical. They are stronger and are able to cover more space than any other type of structure without internal supports.

a. Name the geometric figures that form the surface of the geodesic dome shown.

b. Describe the side lengths of the geometric figures.

Find Angle Measures A **line segment** is part of a line containing two endpoints and all of the points between them. A **triangle** is formed by three line segments that intersect only at their endpoints. A **vertex** is the point where the segments of a triangle intersect.

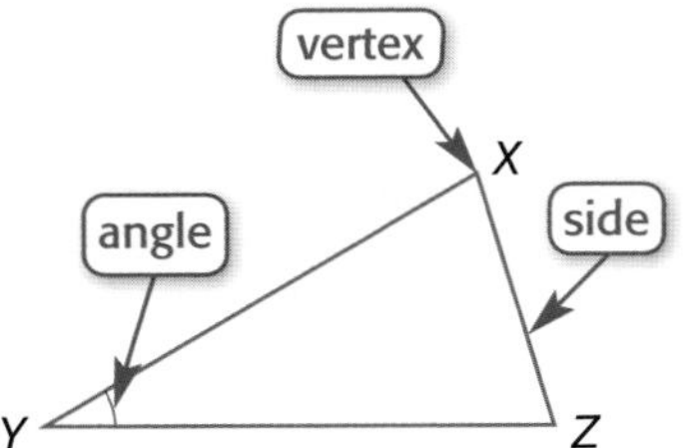

sides: $\overline{XY}$, $\overline{YZ}$, $\overline{XZ}$
vertices: X, Y, Z
angles: $\angle X$, $\angle Y$, $\angle Z$

Triangle *XYZ*, written $\triangle XYZ$, is shown at the right. When finding the measures of angles in a triangle, the notation $m\angle X$ means the measure of angle *X*.

Key Concept — Angles of a Triangle

For Your FOLDABLE

Words The sum of the measures of the angles of a triangle is 180°.

Symbols $x + y + z = 180$

Model

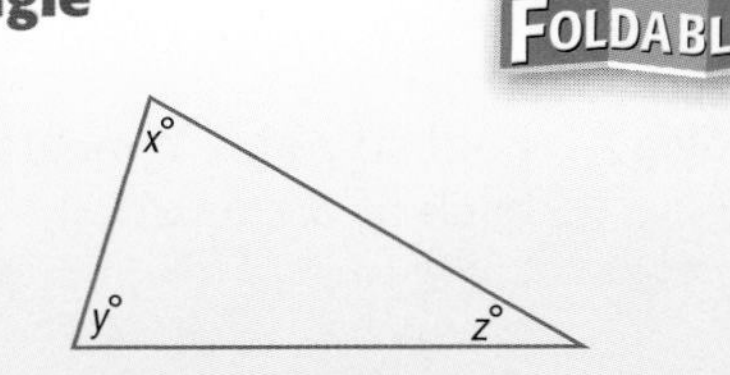

EXAMPLE 1 Find Angle Measures

Find the value of *x* in $\triangle PQR$.

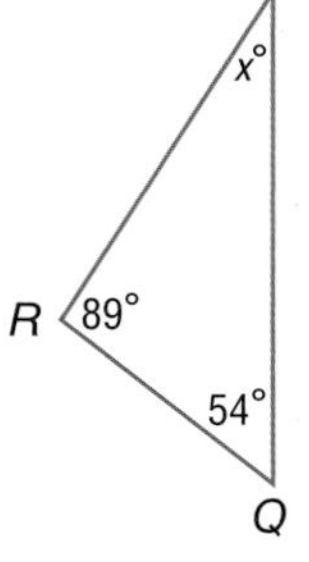

$m\angle P + m\angle Q + m\angle R = 180$	Write an equation.
$x + 54 + 89 = 180$	Substitution
$x + 143 = 180$	Simplify.
$x + 143 - 143 = 180 - 143$	Subtract 143 from each side.
$x = 37$	Simplify.

So, $m\angle P = 37°$.

Check Your Progress

1. Find the $m\angle E$ in $\triangle DEF$ if $m\angle D = 62°$ and $m\angle F = 39°$.

Personal Tutor glencoe.com

EXAMPLE 2 Use Ratios to Find Angle Measures

The measures of the angles of $\triangle ABC$ are in the ratio 1:3:8. What are the measures of the angles?

Words	The sum of the measures of the angles is 180°.
Variables	Let x represent the measure of the first angle, $3x$ the measure of a second angle, and $8x$ the measure of the third angle.
Equation	$x + 3x + 8x = 180$

$x + 3x + 8x = 180$	**Write the equation.**
$12x = 180$	**Combine like terms.**
$\frac{12x}{12} = \frac{180}{12}$	**Divide each side by 12.**
$x = 15$	**Simplify.**

Since $x = 15$, $3x = 3(15)$ or 45, and $8x = 8(15)$ or 120.

The measures of the angles are 15°, 45°, and 120°.

StudyTip

Check Solutions
15 + 45 + 120 = 180. So, the answer is correct.

Check Your Progress

2. The measures of the angles of a triangle are in the ratio 1:3:6. What are the measures of the angles?

Personal Tutor glencoe.com

Classify Triangles Angles can be classified by their degree measure.

Key Concept Types of Angles

For Your FOLDABLE

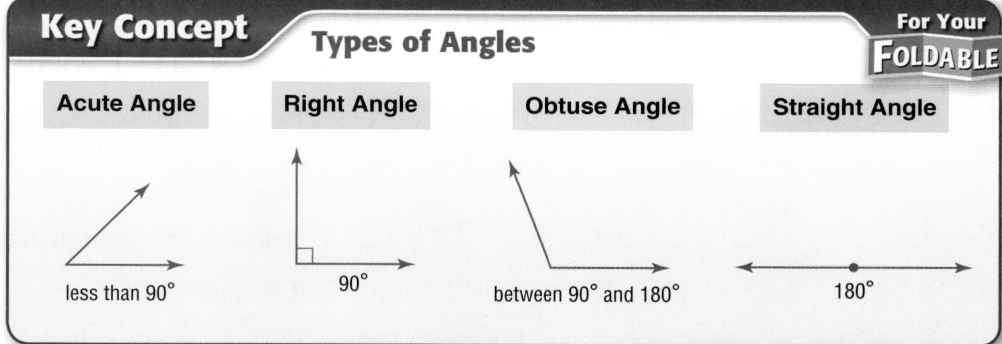

Real-World EXAMPLE 3 Classify Angles

ANIMALS *Smilodons* were saber-toothed cats that lived 11,000 years ago. Smilodons had jaws that opened to an angle of about 120°. What type of angle is formed by the jaws of a smilodon?

S 120°

The measure of $\angle S$ is greater than 90°, so $\angle S$ is obtuse. The jaws of a smilodon form an obtuse angle.

Real-World Link

Smilodons were about 4–5 feet long, 3 feet tall, and weighed about 440 pounds. Their saber-like teeth were up to 7 inches long.

Source: Enchanted Learning

3. ANIMALS Modern lions can open their jaws to an angle of about 65°. What type of angle is formed by the jaws of a lion?

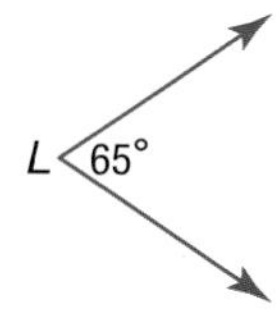

Personal Tutor glencoe.com

Triangles can be classified by their angles.

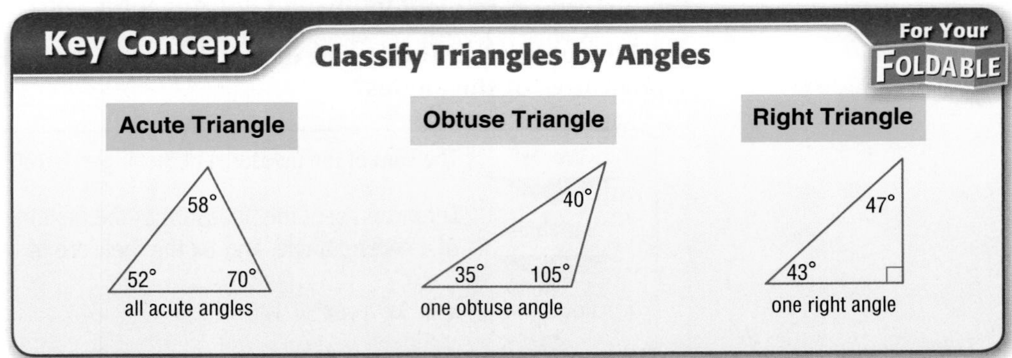

Triangles can also be classified by their sides. **Congruent** sides are sides that have the same length.

ReadingMath

Congruent Segments Tick marks on the sides of a triangle indicate that those sides are congruent. Sides without tick marks are not congruent.

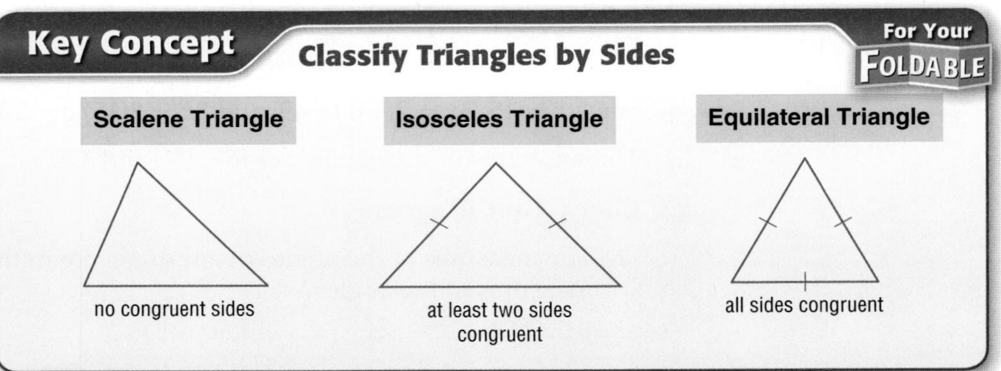

EXAMPLE 4 Classify Triangles

Classify each triangle by its angles and by its sides.

a.

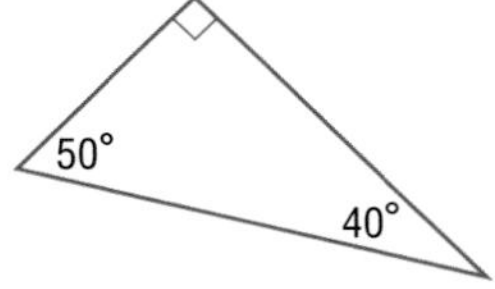

Angles: The triangle has a right angle.

Sides: The triangle has no congruent sides.

The triangle is a right scalene triangle.

b.

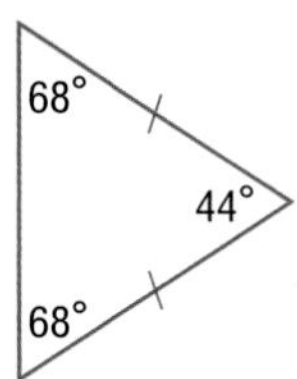

Angles: The triangle has all acute angles.

Sides: The triangle has two congruent sides.

The triangle is an acute isosceles triangle.

Check Your Progress

4A.

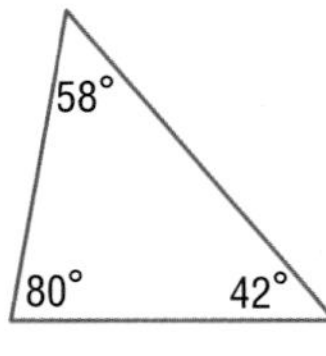

4B.

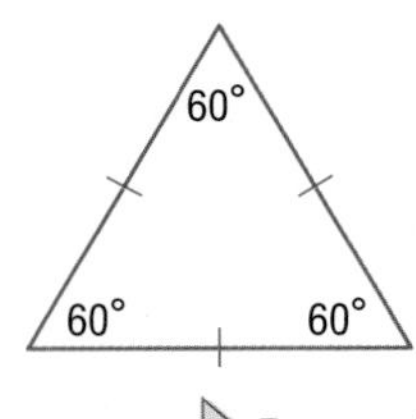

Personal Tutor glencoe.com

Check Your Understanding

Examples 1 and 4 pp. 550, 552

Find the value of x in each triangle. Then classify each triangle by its angles and by its sides.

1. 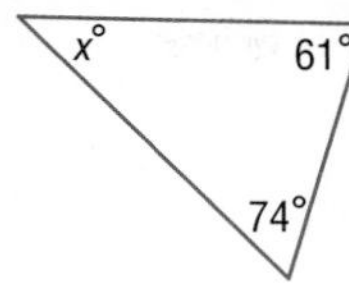

2. 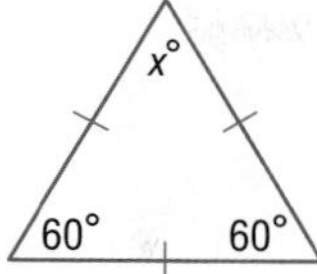

3. 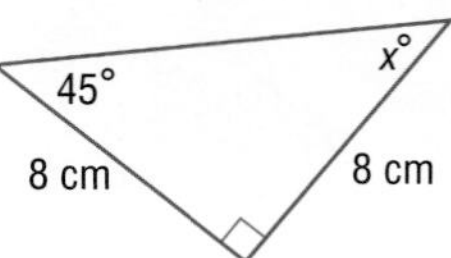

Example 2 p. 551

4. The measures of the angles of a triangle are in the ratio 2:3:5. What are the measures of the angles?

Example 3 p. 551

5. **ANIMALS** A hippopotamus can open its jaws to an angle of about 180°. What type of angle is formed by the jaws of a hippo?

Practice and Problem Solving

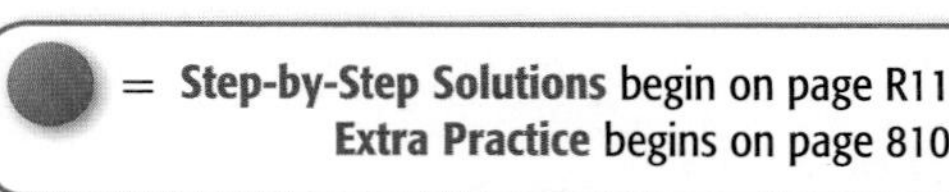
= **Step-by-Step Solutions** begin on page R11.
Extra Practice begins on page 810.

Examples 1 and 4 pp. 550, 552

Find the value of x in each triangle. Then classify each triangle by its angles and by its sides.

6. 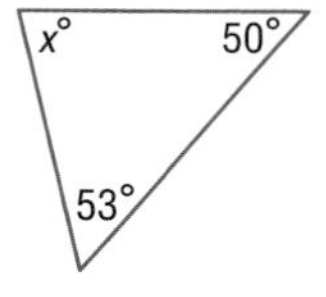

7. 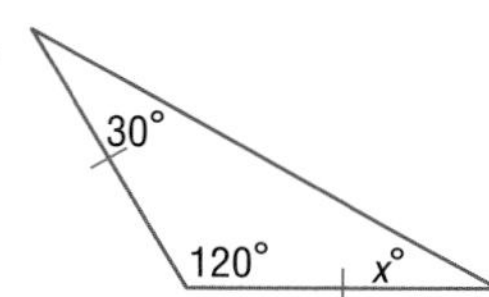

8. 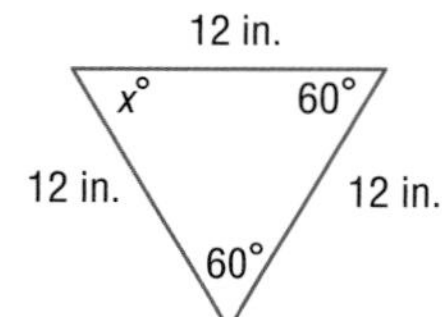

9. 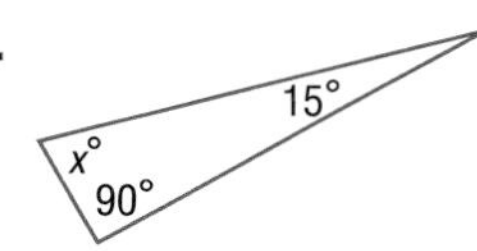

10. 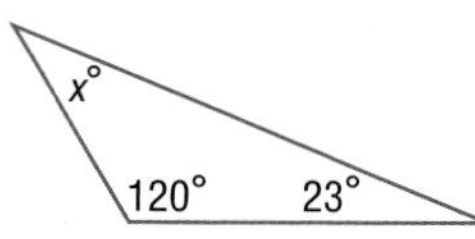

11. 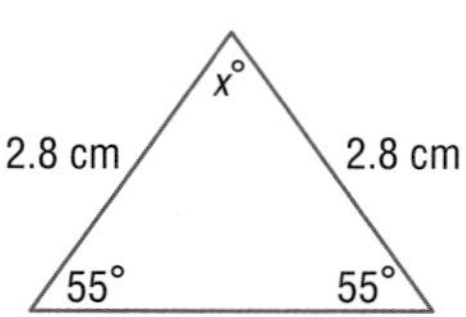

Example 2 p. 551

12. Determine the measures of the angles of $\triangle ABC$ if the measures of the angles are in the ratio 1:1:16.

13. Determine the measures of the angles of $\triangle TUV$ if the measures of the angles are in the ratio 1:6:8.

Example 3 p. 551

TIME What type of angle is formed by the hands on a clock at each time?

14. 3:00 **15.** 6:00

16. 10:30 **17.** 4:30

Classify each angle as *acute, obtuse, right,* or *straight.*

18. 30° **19.** 86° **20.** 90°

21. 145° **22.** 116° **23.** 55°

24. 42° **25.** 125° **26.** 92°

27. ASTRONOMY The Big Dipper may be the best known group of stars in the sky. The figures below show how the Big Dipper probably looked 100,000 years ago, how it looks today, and how it will look 100,000 years from now.

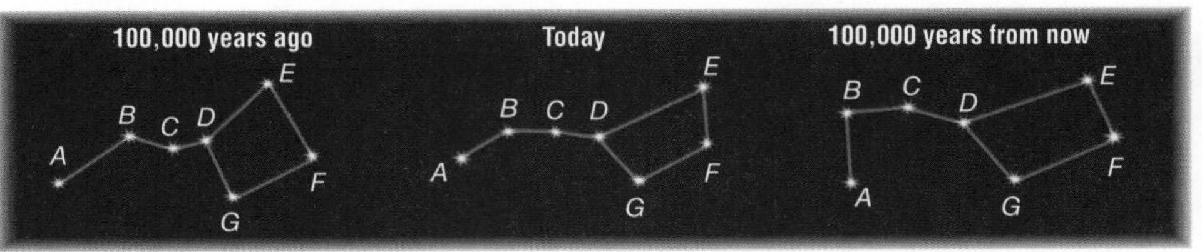

a. Which angle was acute 100,000 years ago and will be obtuse 100,000 years from now?

b. Identify an angle that appears to be a right angle.

Real-World Link

According to a recent survey, the average age of a skateboarder is 14.

Source: The Skate Park Association of the United States

28. SKATEBOARDING Tony Hawk was the first skateboarder to perform the 900 during the X-Games. He rode off a ramp and spun 900° in mid-air while on a skateboard. How many revolutions did Tony make performing the 900?

ALGEBRA Find the measures of the angles in each triangle.

29.

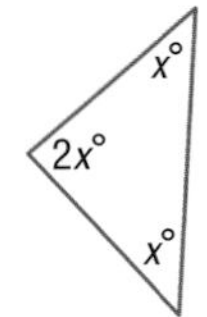

30.

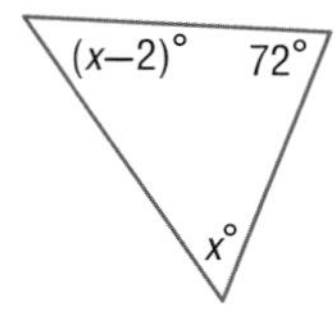

31.

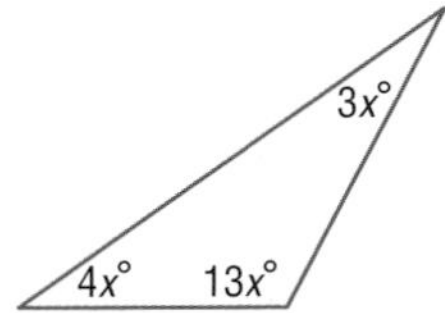

ALGEBRA The measures of the sides of a triangle are given. Classify each triangle by its sides.

32. $2x, 3x, 4x$

33. y, y, y

34. $3x, 3x, 2x$

H.O.T. Problems

Use Higher-Order Thinking Skills

35. OPEN ENDED Sketch each triangle. If it is not possible to sketch the triangle, write *not possible.*

a. acute scalene

b. obtuse and *not* scalene

c. right equilateral

d. obtuse equilateral

36. FIND THE ERROR Miguel says that an equilateral triangle is sometimes an obtuse triangle. Jane says that an equilateral triangle is always an acute triangle. Is either of them correct? Explain your reasoning.

37. CHALLENGE Find the value of x and y in the figure.

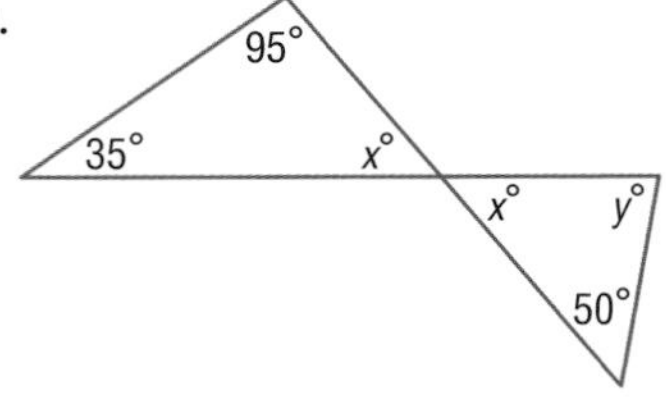

38. REASONING What is the relationship between the measures of two acute angles of any right triangle? Explain.

39. WRITING IN MATH *True* or *false*? Every triangle has at least 2 acute angles. Justify your reasoning.

Standardized Test Practice

40. How would you find the value of x?

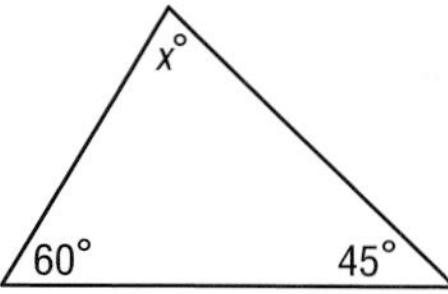

A Subtract 60 from 105.
B Add 180 to 105.
C Add 45 to 105.
D Subtract 105 from 180.

41. Which of the following best describes the triangle with the given measures?

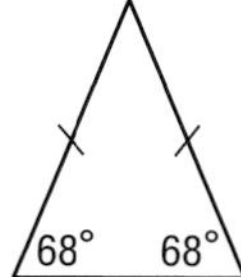

F acute isosceles triangle
G acute scalene triangle
H obtuse isosceles triangle
J acute equilateral triangle

42. Which of the following is an obtuse triangle?

A

C 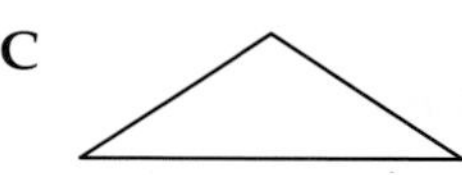

B 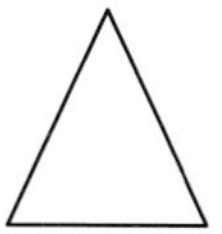

D 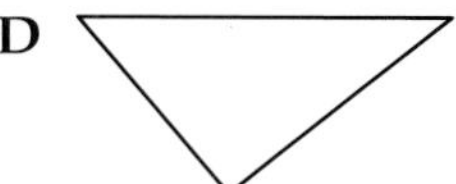

43. EXTENDED RESPONSE Use the triangle shown below.

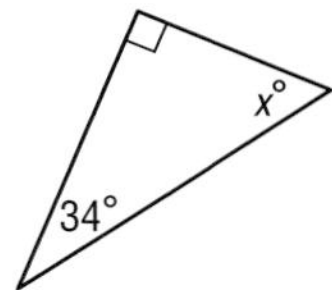

a. Find the value of x.
b. Explain how you found the value of x in part **a**.
c. Classify the triangle by its angles and by its sides.

Spiral Review

ALGEBRA Solve each equation. Round to the nearest tenth, if necessary. (Lesson 10-2)

44. $m^2 = 81$ **45.** $196 = y^2$ **46.** $168 = 2p^2$ **47.** $\frac{f^2}{2} = 51$

Estimate each square root to the nearest integer. Do not use a calculator. (Lesson 10-1)

48. $-\sqrt{52}$ **49.** $-\sqrt{17}$ **50.** $\sqrt{38}$ **51.** $\sqrt{140}$

52. WEATHER The time a storm will hit an area can be predicted using $d \div s = t$ where d is the distance in miles an area is from the storm, s is the speed in miles per hour of the storm, and t is the travel time in hours of the storm. Suppose it is 11:00 A.M. and a storm is heading toward a town at a speed of 30 miles per hour. The storm is about 150 miles from the town. What time will the storm hit? (Lesson 3-4)

Skills Review

Find the value of each expression. (Lesson 9-1)

53. 11^2 **54.** 3^2 **55.** 16^2 **56.** 17^2

CHAPTER 10

Mid-Chapter Quiz

Lessons 10-1 through 10-3

Find each square root. (Lesson 10-1)

1. $\pm\sqrt{49}$

2. $\sqrt{144}$

3. $\sqrt{64}$

4. $-\sqrt{121}$

5. **GARDENING** A square vegetable garden has an area of 169 square feet. If $\sqrt{A} = s$, where s is the length of one side and A is the area, how many feet of fencing is needed to enclose the garden? (Lesson 10-1)

Estimate each square root to the nearest whole number. Do not use a calculator. (Lesson 10-1)

6. $\sqrt{51}$

7. $-\sqrt{88}$

8. $\sqrt{17}$

9. $\pm\sqrt{111}$

10. $\pm\sqrt{41}$

11. $-\sqrt{1000}$

12. **MULTIPLE CHOICE** Which statement is *not* true? (Lesson 10-1)

A $6 < \sqrt{39} < 7$

B $9 < \sqrt{89} < 10$

C $-8 < -\sqrt{56} < -7$

D $-4 < -\sqrt{17} < -5$

Name all of the sets of numbers to which each real number belongs. Write *whole*, *integer*, *rational*, or *irrational*. (Lesson 10-2)

13. 0.3

14. $-\sqrt{49}$

15. $15.\overline{1}$

16. $\frac{56}{8}$

17. **GEOMETRY** Use $A = \pi r^2$ to find the radius of a circle with an area of 28.26 square inches. In the formula, A represents the area of a circle, r represents the radius, and $\pi \approx 3.14$. (Lesson 10-2)

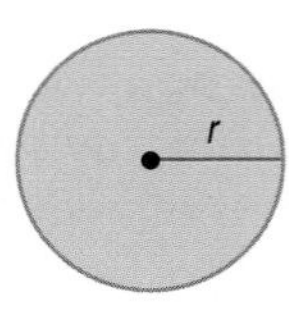

ALGEBRA Solve each equation. Round to the nearest tenth, if necessary. (Lesson 10-2)

18. $k^2 = 74$

19. $131 = n^2$

20. $2b^2 = 98$

21. $3.6r^2 = 518.4$

Replace each ● with <, >, or = to make a true statement. (Lesson 10-2)

22. $\sqrt{7}$ ● 2.5

23. $1\frac{4}{5}$ ● $\sqrt{3.24}$

24. **HORIZON** On a clear day, the number of miles a person can see to the horizon can be found using the formula $d = 1.22 \cdot \sqrt{h}$, where d is the distance in miles to the horizon and h is the person's distance from the ground. If a person is standing on the roof of a skyscraper at a height of 1200 feet, about how many miles could he see? (Lesson 10-2)

Classify each angle measure as *acute, obtuse, right,* or *straight.* (Lesson 10-3)

25. 77°

26. 180°

27. 90°

28. 165°

Classify each triangle by its angles and by its sides. (Lesson 10-3)

29.

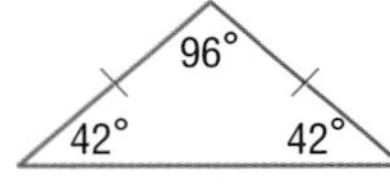

30.

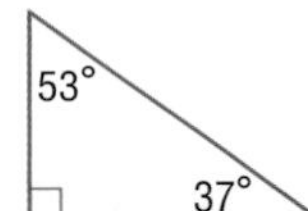

31.

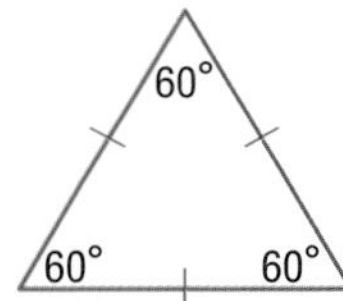

32.

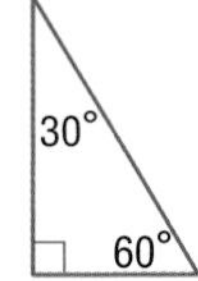

33. **MULTIPLE CHOICE** Refer to the figure shown. Lloyd lives in Salsburg, does his grocery shopping in Richmond, and sees a doctor in Thornville. What is the measure of the angle formed when Lloyd travels from Salsburg to Richmond and then to Thornville? (Lesson 10-3)

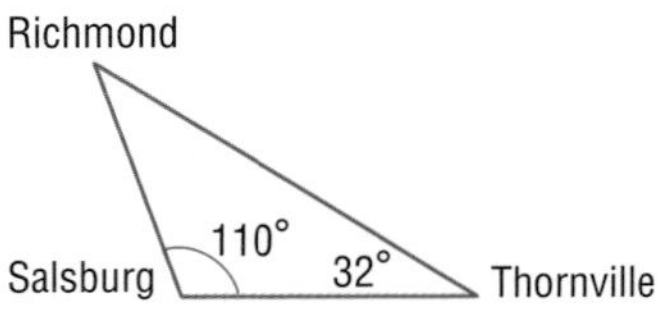

F 28°

G 38°

H 48°

J 180°

Algebra Lab
The Pythagorean Theorem

Math Online glencoe.com
Math in Motion, Animation

You can use grid paper to investigate the relationship that exists among the sides of a right triangle. Each square □ represents 1 square unit.

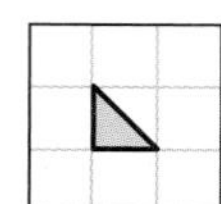

$A = \frac{1}{2}(1)$ or $\frac{1}{2}$ unit^2

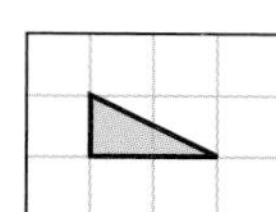

$A = \frac{1}{2}(2)$ or 1 unit^2

ACTIVITY

In each diagram shown below, a square is attached to each side of a right triangle.

Triangle 1

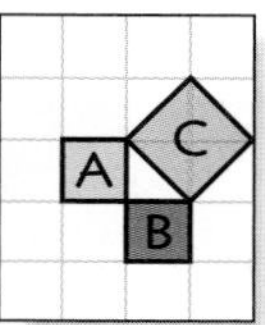

Triangle 2

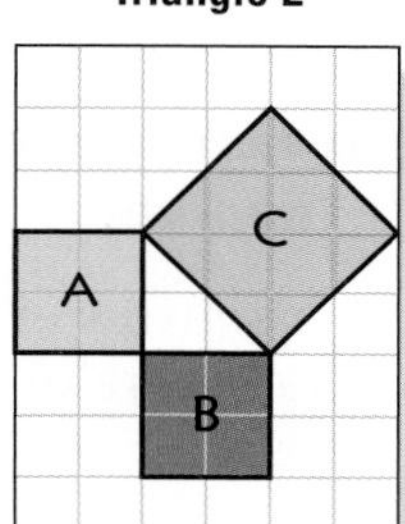

Triangle 3

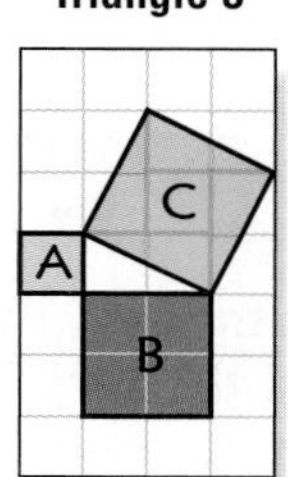

Triangle 4

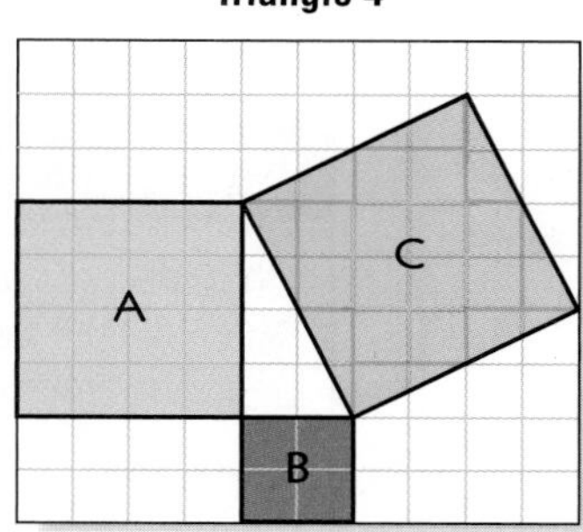

Copy and complete the table. Then find the area of each square that is attached to the triangle. Record the results in your table.

Triangle	Area of Square A (units^2)	Area of Square B (units^2)	Area of Square C (units^2)
1			
2			
3			
4			

Analyze the Results

1. How does the sum of the areas of square A and square B compare to the area of square C?
2. Draw a right triangle on centimeter grid paper. Count to find the measures of the shorter sides and use the relationship you discovered to calculate the measure of the longest side. Measure to verify your answer.
3. Refer to the diagram at the right. If the lengths of the sides of a right triangle are whole numbers such that $a^2 + b^2 = c^2$, the numbers a, b, and c are called a **Pythagorean Triple**. Tell whether each set of numbers is a Pythagorean Triple. Explain.

a c b

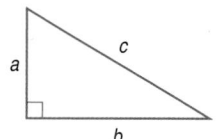

a. 3, 4, 5 **b.** 5, 7, 9 **c.** 6, 9, 12

10-4 The Pythagorean Theorem

Then

You have already found missing measures of similar triangles. (Lesson 6-7)

Now

- Use the Pythagorean Theorem to find the length of a side of a right triangle.
- Use the converse of the Pythagorean Theorem to determine whether a triangle is a right triangle.

New Vocabulary

legs
hypotenuse
Pythagorean Theorem
solving a right triangle
converse

Math Online

glencoe.com

- Extra Examples
- Personal Tutor
- Self-Check Quiz
- Homework Help
- Math in Motion

Why?

You are probably not thinking about right triangles as you speed down a water slide. But the figure at the right shows how they are related.

a. What parts of the water slide make up the right triangle?

b. Which side of the triangle is the longest side?

Use the Pythagorean Theorem In a right triangle, the sides adjacent to the right angle are called the **legs**. The side opposite the right angle is the **hypotenuse**. It is the longest side of a right triangle.

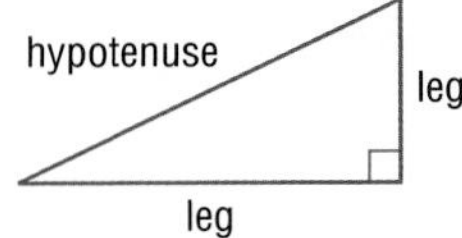

The **Pythagorean Theorem** describes the relationship between the lengths of the legs and the hypotenuse for any right triangle.

Key Concept — Pythagorean Theorem

For Your FOLDABLE

Words In a right triangle, the sum of the squares of the lengths of the legs is equal to the square of the length of the hypotenuse.

Symbols $a^2 + b^2 = c^2$

Model

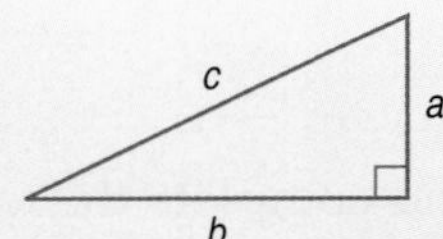

Math *in Motion*, Interactive Lab glencoe.com

EXAMPLE 1 Find the Hypotenuse Length

Find the length of the hypotenuse of the right triangle.

$a^2 + b^2 = c^2$	Pythagorean Theorem
$8^2 + 15^2 = c^2$	Replace *a* with 8 and *b* with 15.
$64 + 225 = c^2$	Evaluate 8^2 and 15^2.
$289 = c^2$	Add 64 and 225.
$\pm\sqrt{289} = c$	Definition of square root
$17 = c$	Use the principal square root.

8 mm
c
15 mm

The length of the hypotenuse is 17 millimeters.

Check Your Progress

1. Find the length of the hypotenuse of a right triangle if the legs are 24 inches and 45 inches long.

Personal Tutor glencoe.com

STANDARDIZED TEST EXAMPLE 2

GRIDDED RESPONSE **A volleyball court is 30 feet wide and 60 feet long. A player serves the ball from one corner of the court to the opposite corner. How far is this? Round to the nearest tenth.**

Problem-SolvingTip

Make a Model You can draw and label a model of the volleyball court to help you determine which values to substitute for the variables in $a^2 + b^2 = c^2$.

Read the Test Item

The sides of the court and the diagonal form a right triangle. Find the measure of the hypotenuse.

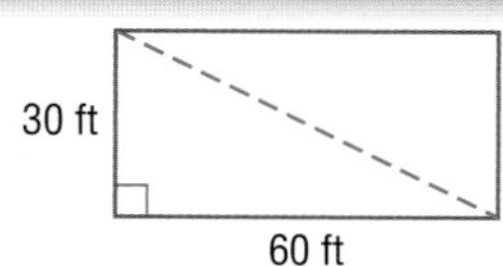

Solve the Test Item

$a^2 + b^2 = c^2$	**Pythagorean Theorem**
$30^2 + 60^2 = c^2$	**Replace a with 30 and b with 60.**
$900 + 3600 = c^2$	**Evaluate 30^2 and 60^2.**
$4500 = c^2$	**Add.**
$\pm\sqrt{4500} = c$	**Definition of square root**
$67.1 \approx c$	**Use the principal square root.**

Fill in the Answer Grid

The distance from corner to corner is about 67.1 feet.

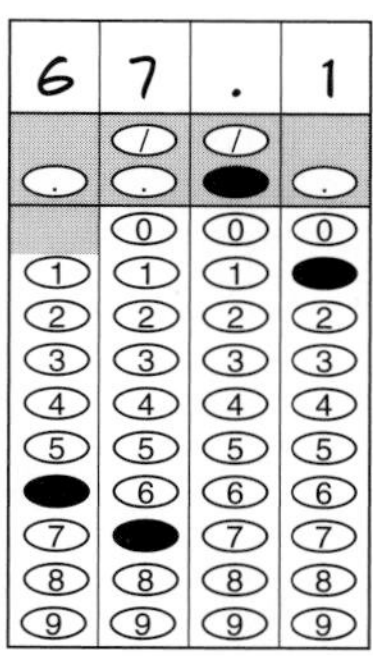

Check Your Progress

2. **GRIDDED RESPONSE** A doorway is 2.7 feet wide and 8.4 feet high. What is the longest piece of drywall that can be taken through the doorway? Round to the nearest tenth.

Personal Tutor glencoe.com

If you know the lengths of two sides of a right triangle, you can use the Pythagorean Theorem to find the length of the third side. This is called **solving a right triangle**.

Real-World EXAMPLE 3 Solve a Right Triangle

ADVERTISEMENTS **A balloon advertising the opening of a store is tethered to the ground as shown. About how many feet above the ground is the balloon?**

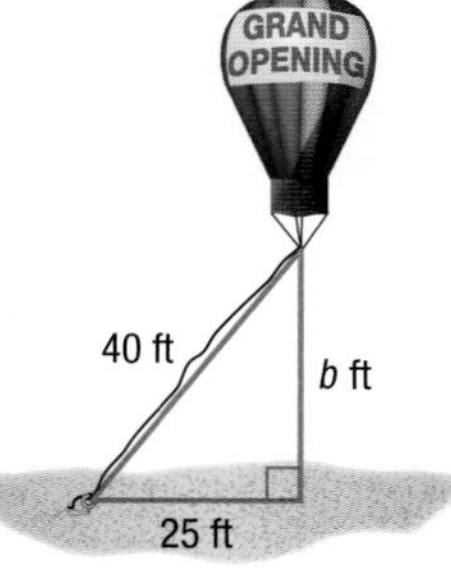

$a^2 + b^2 = c^2$	**Pythagorean Theorem**
$25^2 + b^2 = 40^2$	**Replace a with 25 and c with 40.**
$625 + b^2 = 1600$	**Evaluate 25^2 and 40^2.**
$625 - 625 + b^2 = 1600 - 625$	**Subtract 625 from each side.**
$b^2 = 975$	**Simplify.**
$b = \pm\sqrt{975}$	**Definition of square root**

2nd [√] 975 ENTER 31.22498999 **Use a calculator.**

The balloon is about 31.2 feet above the ground.

StudyTip

Square Roots In Lesson 10-1, you learned that a number has both a positive and negative square root. Since you are finding the lengths of sides of triangles, calculate the positive square root only.

Check Your Progress

3. **LADDERS** A 15-foot ladder is leaning against a house. The base of the ladder is 3.5 feet from the house. About how many feet does the ladder reach on the side of the house?

Personal Tutor glencoe.com

Use the Converse of the Pythagorean Theorem The Pythagorean Theorem is written in if-then form. If you reverse the statements after *if* and *then*, you have formed the **converse of the Pythagorean Theorem**.

Pythagorean Theorem	If a triangle is a right triangle, then $c^2 = a^2 + b^2$.
Converse	If $c^2 = a^2 + b^2$, then a triangle is a right triangle.

Not all converses are true. However, the converse of the Pythagorean Theorem *is* true. Use the converse to determine whether a triangle is a right triangle.

EXAMPLE 4 Identify a Right Triangle

The measures of three sides of a triangle are given. Determine whether each triangle is a right triangle.

a. 6 cm, 8 cm, 10 cm

$a^2 + b^2 = c^2$	Pythagorean Theorem
$6^2 + 8^2 \stackrel{?}{=} 10^2$	$a = 6, b = 8, c = 10$
$36 + 64 \stackrel{?}{=} 100$	Evaluate.
$100 = 100$	Simplify.

The triangle is a right triangle.

b. 4 in., 5 in., 6 in.

$a^2 + b^2 = c^2$	Pythagorean Theorem
$4^2 + 5^2 \stackrel{?}{=} 6^2$	$a = 4, b = 5, c = 6$
$16 + 25 \stackrel{?}{=} 36$	Evaluate.
$41 \neq 36$	Simplify.

The triangle is *not* a right triangle.

Check Your Progress

4A. 8 in., 9 in., 12 in.

4B. 15 mm, 20 mm, 25 mm

Personal Tutor glencoe.com

Check Your Understanding

Example 1 p. 558

Find the length of the hypotenuse of each right triangle. Round to the nearest tenth, if necessary.

1. 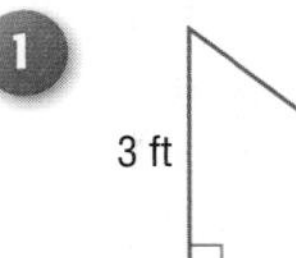

2. 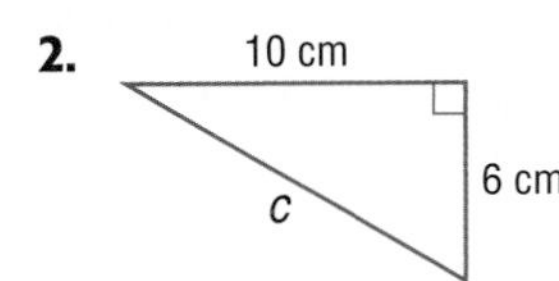

Example 2 p. 559

3. **GRIDDED RESPONSE** A gymnastics tumbling mat is a square that measures 40 feet on each side. During the floor routine, a gymnast makes a tumbling pass along the length of the diagonal of the mat. How many feet long is their tumbling pass? Round to the nearest tenth.

Example 3 p. 559

4. **SHADOWS** Marian's shadow is 94 inches long. The distance from the top of Marian's head to the end of her shadow is 115 inches. How many inches tall is Marian? Round to the nearest tenth.

Example 4 p. 560

The lengths of three sides of a triangle are given. Determine whether each triangle is a right triangle.

5. $a = 9, b = 12, c = 15$
6. $a = 6, b = 10, c = 12$
7. $a = 12, b = 14, c = 20$
8. $a = 15, b = 20, c = 25$

Practice and Problem Solving

● = **Step-by-Step Solutions** begin on page R11.
Extra Practice begins on page 810.

Example 1 p. 558

Find the length of the hypotenuse of each right triangle. Round to the nearest tenth, if necessary.

9.

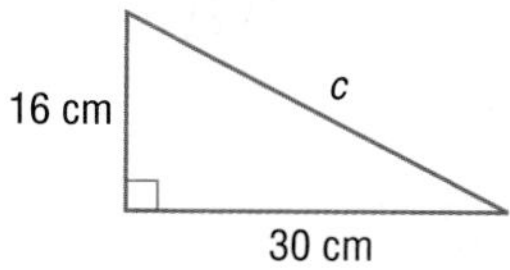

10.

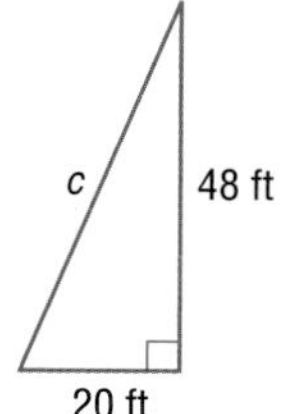

11.

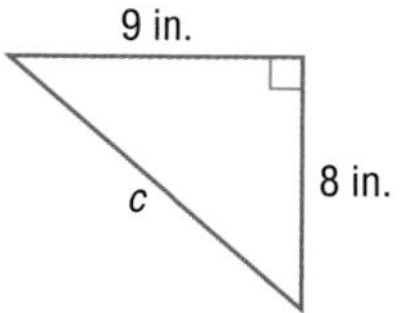

12.

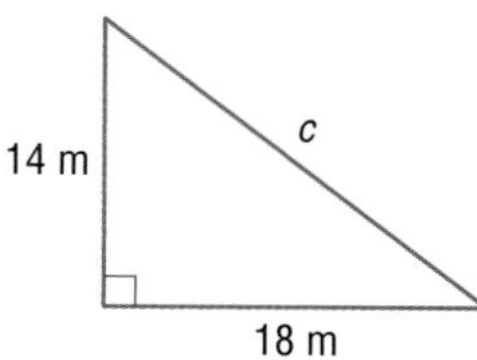

Example 2 p. 559

13. MAPS It is 68 miles from Columbia to Augusta and 130 miles from Augusta to Charleston. How many miles is it from Charleston to Columbia to the nearest tenth?

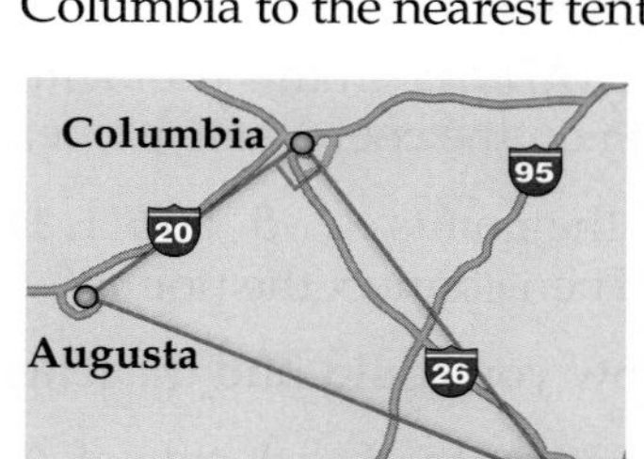

14. MODELS Inez is building a model sailboat. How long must she cut a piece of wood to make a brace for the mainsail?

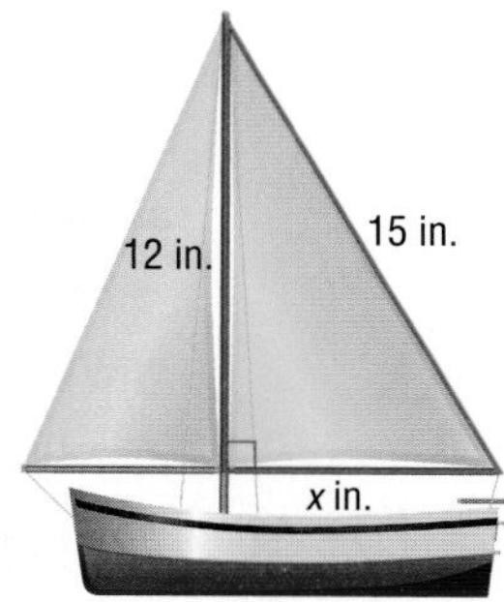

Example 3 p. 559

15 **DESIGN** A square courtyard in the center of an office building measures 35 feet on each side. The designer of the building is constructing a walkway diagonally through the center of the courtyard. What is the length of the walkway to the nearest tenth?

Example 4 p. 560

The lengths of three sides of a triangle are given. Determine whether each triangle is a right triangle.

16. $a = 4, b = 9, c = 12$

17. $a = 7, b = 24, c = 25$

18. $a = 12, b = 16, c = 20$

19. $a = 16, b = 30, c = 32$

HOME THEATERS Projection screens for home movie theaters are described according to the measure of the diagonal. Find each missing dimension. Round to the nearest inch.

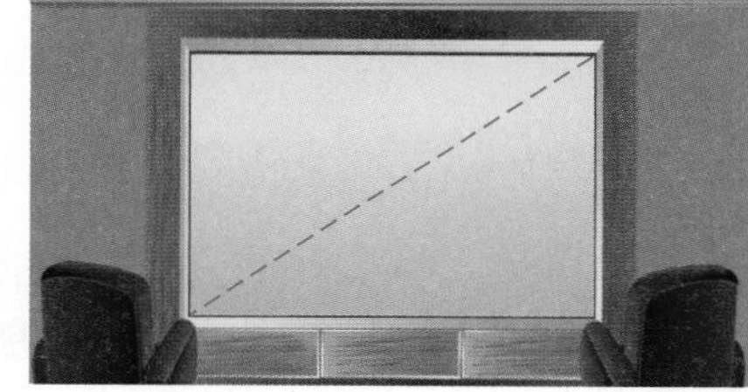

20. A screen is 80 inches long and 45 inches high. What is the diagonal of the screen?

21. An 84-inch screen is 73 inches long. What is the height?

22. A 120-inch screen is 59 inches high. What is the length?

If c is the measure of the hypotenuse, find each missing measure. Round to the nearest tenth, if necessary.

23. $a = 9$ m, $b = ?$, $c = 12$ m

24. $a = ?$, $b = 21$ cm, $c = 35$ cm

25. $a = ?$, $b = 16$ ft, $c = 29$ ft

26. $a = 30$ in., $b = ?$, $c = 40$ in.

27. $a = 8.1$ mi, $b = 3.5$ mi, $c = ?$

28. $a = 10.4$ yd, $b = 16.9$ yd, $c = ?$

29. $a = ?$, $b = \sqrt{123}$ ft, $c = 22$ ft

30. $a = \sqrt{127}$ m, $b = ?$, $c = 31$ m

Real-World Link

There are three main parasailing methods. The most popular method is called Winchboat parasailing. This is where the launch and recovery of the parasailor take place directly on the boat.

Source: Parasailing USA

31. RECREATION Parasailing is a popular water activity where a person with a parachute is attached to a boat by a long rope.

a. In Myrtle Beach, the maximum height off the water a parasailor can be is 500 feet. If the rope is 800 feet long, about how many feet from the boat is the parasailor to the nearest tenth?

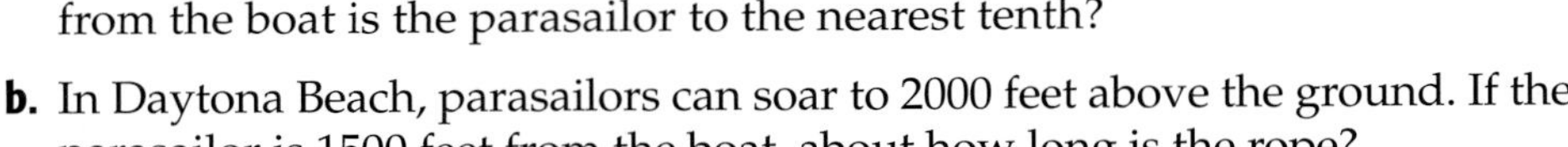

b. In Daytona Beach, parasailors can soar to 2000 feet above the ground. If the parasailor is 1500 feet from the boat, about how long is the rope?

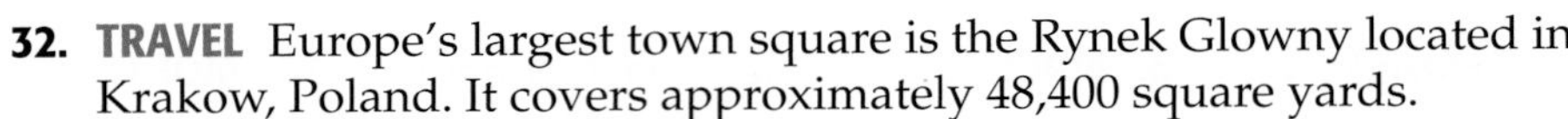

32. TRAVEL Europe's largest town square is the Rynek Glowny located in Krakow, Poland. It covers approximately 48,400 square yards.

a. How many feet long is a side of the square?

b. To the nearest foot, what is the diagonal distance across Rynek Glowny?

33. MULTIPLE REPRESENTATIONS In this problem, you will investigate the Pythagorean Theorem on the coordinate plane.

a. GRAPHICAL Graph the points $A(-5, 1)$, $B(1, 1)$, and $C(1, -3)$ on a coordinate plane. Then connect the points.

b. VERBAL Explain how you could find the length of segment $\overline{AC}$.

c. NUMERICAL Find the length of each side of $\triangle ABC$ to the nearest tenth.

d. NUMERICAL What are the perimeter and area of $\triangle ABC$?

H.O.T. Problems

Use Higher-Order Thinking Skills

34. OPEN ENDED The hypotenuse of a right triangle is 23 centimeters long. Find possible measures for the legs of the triangle. Round to the nearest hundredth. Justify your answer.

35. CHALLENGE In the figure, $\overline{BD}$ is the diagonal of the base and $\overline{FD}$ is the diagonal of the figure. Find FD to the nearest tenth.

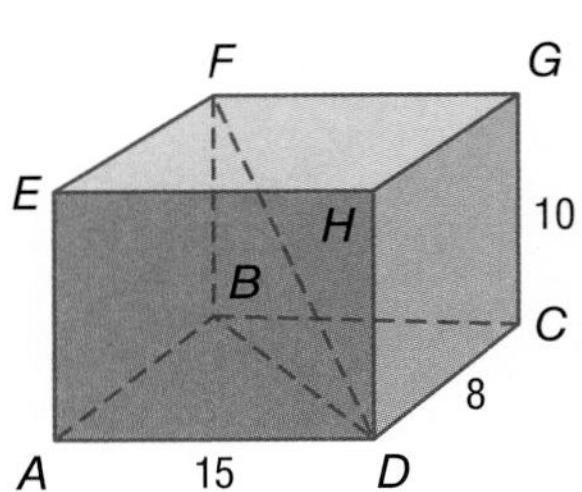

36. FIND THE ERROR The legs of a right triangle are 13 and 22 inches. Needa says that the hypotenuse is 17.7 inches long and Monifa says that it is 25.6 inches. Is either of them correct? Explain.

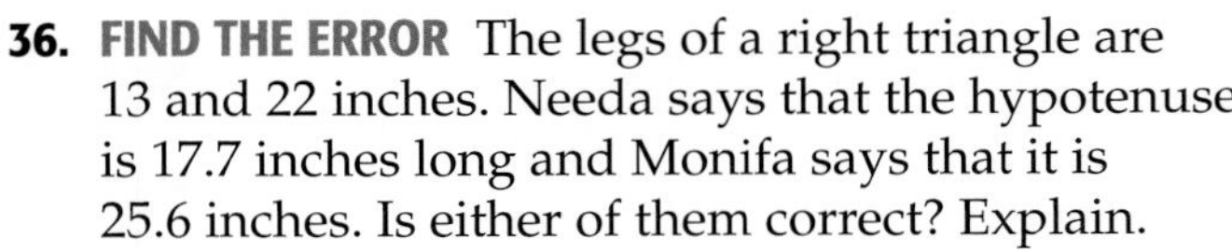

37. REASONING Find the length of the legs x of the isosceles right triangle shown. Describe your steps.

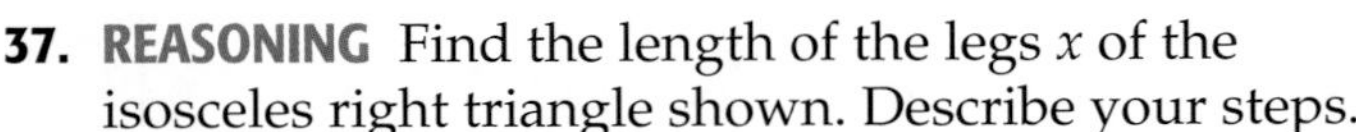

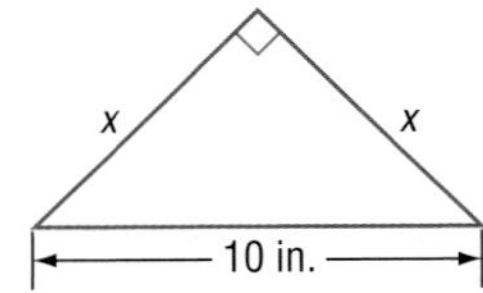

38. WRITING IN MATH Explain how you can use the measures of a triangle to determine whether a triangle is a right triangle. Give an example.

Standardized Test Practice

39. Find the amount of edging needed to enclose the triangular flower bed.

A 8 yd
B 12 yd
C 18 yd
D 24 yd

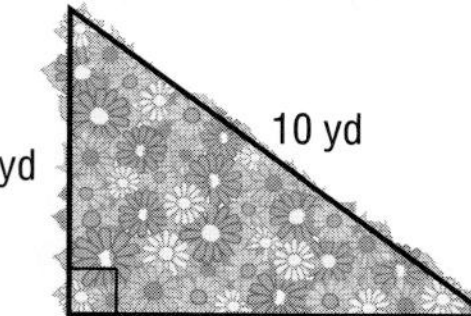

40. SHORT RESPONSE Is the triangle formed by the buildings below a right triangle? Explain your reasoning.

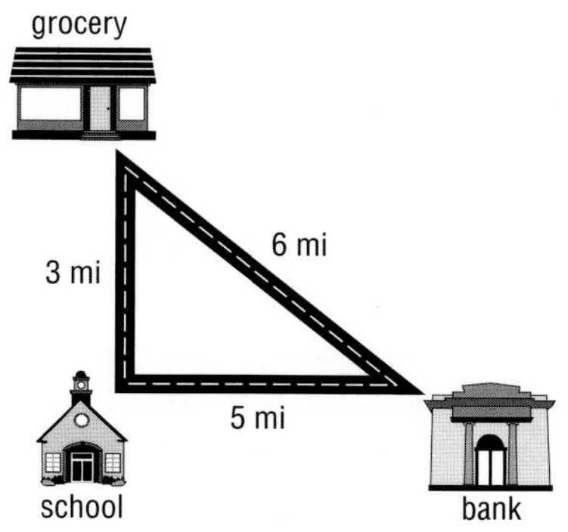

41. Which of the following *cannot* be the measures of sides of a right triangle?

F 6 cm, 8 cm, 10 cm
G 14 cm, 18 cm, 20 cm
H 10 cm, 24 cm, 26 cm
J 20 cm, 21 cm, 29 cm

42. EXTENDED RESPONSE Use the figure below.

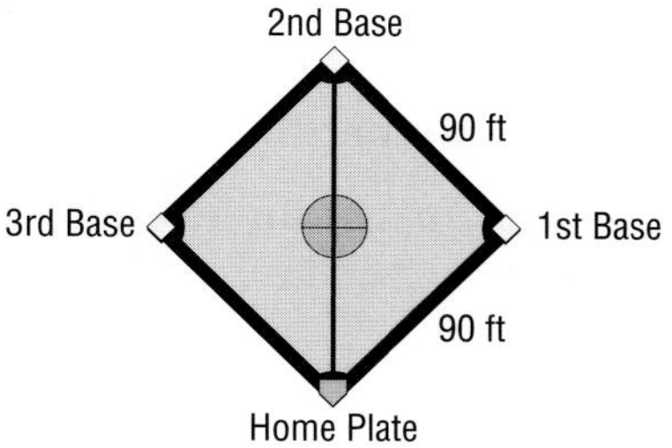

a. Explain how to determine the distance from home plate to second base.

b. About how many feet is it from home plate to second base? Show and justify all your steps.

Spiral Review

Find the value of x in each triangle. Then classify each triangle by its angles and by its sides. (Lesson 10-3)

43. 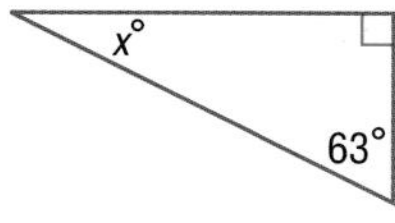

44. 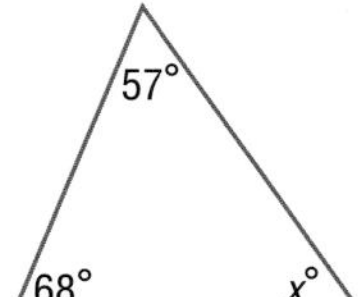

45. 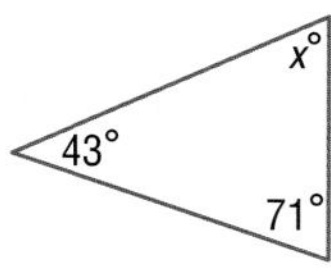

Name all of the sets of numbers to which each real number belongs. Write *whole, integer, rational,* or *irrational.* (Lesson 10-2)

46. -5

47. $0.\overline{4}$

48. $\sqrt{63}$

49. 7.4

50. GRADES Tobias' average for five quizzes is 86. He wants to have an average of at least 88 for six quizzes. What is the lowest score he can receive on his sixth quiz to obtain this average? (Lesson 4-6)

Skills Review

Simplify each expression. (Lesson 9-1)

51. $(10 - 3)^2 + (4 - 8)^2$

52. $[7 - (-2)]^2 + (5 - 3)^2$

53. $(-4 - 6)^2 + (-1 - 5)^2$

EXTEND 10-4

Algebra Lab

Graphing Irrational Numbers

Math Online glencoe.com

Math *in Motion*, Animation

You have already graphed integers and rational numbers on a number line. You can use right triangles to graph irrational numbers like $\sqrt{5}$ or $\sqrt{34}$ on a number line.

ACTIVITY

Graph $\sqrt{34}$ on a number line.

Step 1 Find two square numbers with a sum of 34. Since $34 = 9 + 25$ or $3^2 + 5^2$, you can use the numbers 3 and 5.

Step 2 Use the numbers to draw a right triangle.

- First, draw a number line on grid paper.
- Next, draw a right triangle with legs that measure 3 units and 5 units. Notice that this triangle can be drawn in two ways.

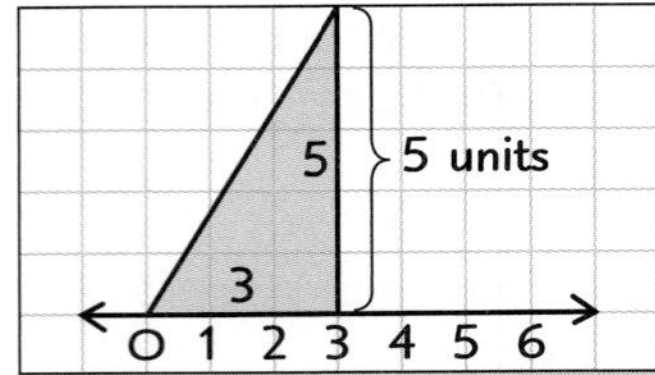

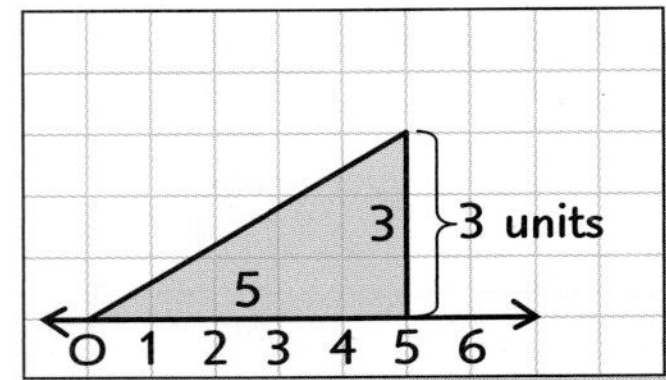

Step 3 Graph $\sqrt{34}$.

- Open your compass to the length of the hypotenuse.
- With the tip of the compass at 0, draw an arc that intersects the number line. Label the intersection point A.
- The distance from 0 to A is $\sqrt{34}$ units. From the graphs, $\sqrt{34} \approx 5.8$.

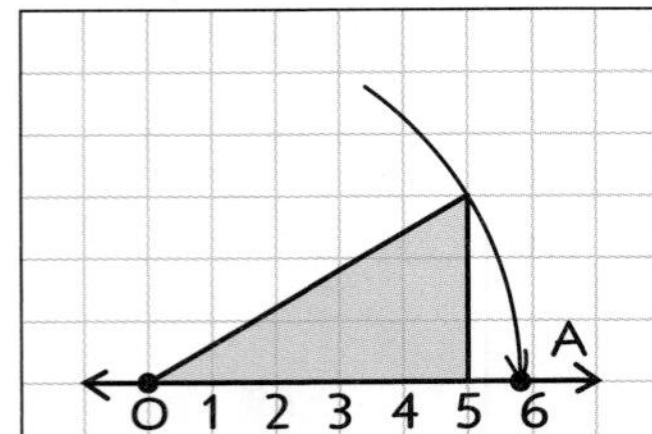

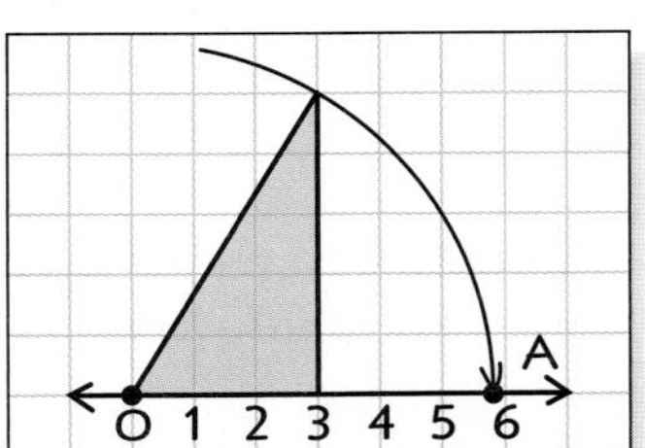

Analyze the Results

Use a compass and grid paper to graph each irrational number on a number line.

1. $\sqrt{5}$ **2.** $\sqrt{20}$ **3.** $\sqrt{97}$ **4.** $\sqrt{45}$

5. Describe two different ways to graph $\sqrt{61}$.

6. Explain how the graph of $\sqrt{2}$ can be used to locate the graph of $\sqrt{3}$.

10-5 The Distance Formula

Then

You found the slope of a line passing through two points on a coordinate plane. (Lesson 8-6)

Now

- Use the Distance Formula to find the distance between two points on a coordinate plane.
- Apply the Distance Formula to solve problems about figures on the coordinate plane.

New Vocabulary

Distance Formula

Math Online

glencoe.com

- Extra Examples
- Personal Tutor
- Self-Check Quiz
- Homework Help

Why?

Dakota is biking on the trail shown on the map at the right. His brother timed his ride from point B to point A.

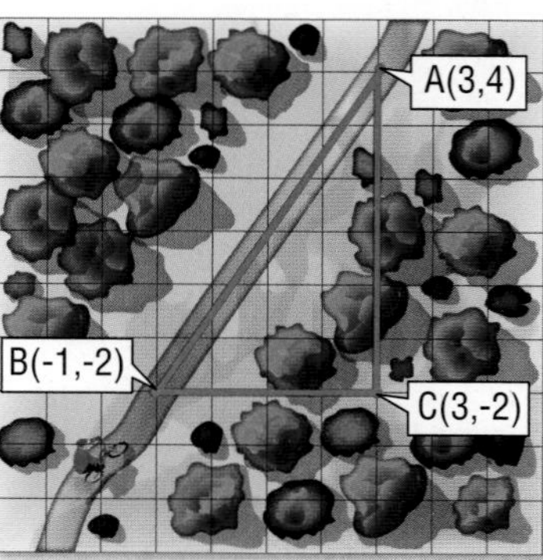

a. What type of triangle is formed by points A, B, and C?

b. How can you find the length of $\overline{BC}$ without counting the number of grids? $\overline{AC}$?

c. How can you find the distance between points A and B?

Find the Distance between Points The figure above shows that you can find the distance between points A and B by drawing a right triangle. You can also use the **Distance Formula**, which is based on the Pythagorean Theorem.

Key Concept — Distance Formula

For Your FOLDABLE

Symbols The distance d between two points with coordinates (x_1, y_1) and (x_2, y_2) is given by $d = \sqrt{(x_2 - x_1)^2 + (y_2 - y_1)^2}$.

Model

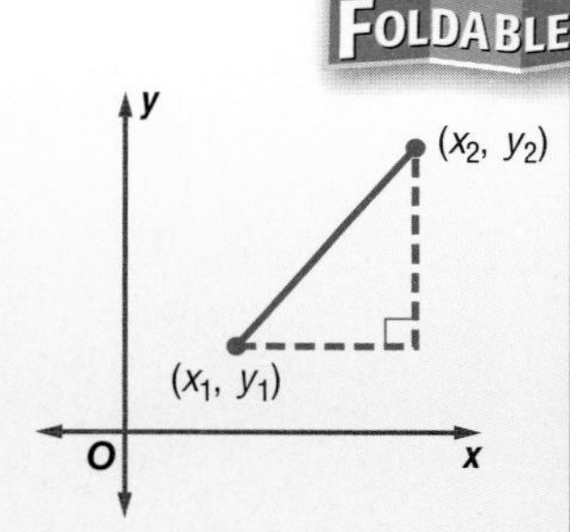

EXAMPLE 1 Find the Distance Between Two Points

Find the distance between $A(-2, 1)$ and $B(4, 3)$.

$d = \sqrt{(x_2 - x_1)^2 + (y_2 - y_1)^2}$	Distance Formula
$AB = \sqrt{[4 - (-2)]^2 + (3-1)^2}$	$(x_1, y_1) = (-2, 1)$, $(x_2, y_2) = (4, 3)$
$AB = \sqrt{6^2 + 2^2}$	Simplify.
$AB = \sqrt{36 + 4}$	Evaluate 6^2 and 2^2.
$AB = \sqrt{40}$	Add 36 and 4.
$AB \approx 6.3$	Simplify.

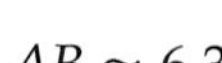

So, the distance between points A and B is about 6.3 units.

Check Your Progress

1. Find the distance between $F(5, -6)$ and $G(1, 2)$. Round to the nearest tenth, if necessary.

Personal Tutor glencoe.com

StudyTip

Midpoint To find the coordinates of the midpoint of a line segment, use the formula $M = \left(\frac{x_1 + x_2}{2}, \frac{y_1 + y_2}{2}\right)$. The midpoint of $\overline{AB}$ in Example 1 is at the point $\left(\frac{-2 + 4}{2}, \frac{1 + 3}{2}\right)$ or (1, 2).

Real-World Link

The District of Columbia is 67 square miles and is divided into four quadrants. The U.S. Capitol building is at the point where the quadrants meet. Numbered streets run north and south. Lettered streets run east and west.

Source: Washington, D.C., Convention and Tourism Corp.

Real-World EXAMPLE 2 Use the Distance Formula

RECREATION The Yeager family is visiting Washington, D.C. One unit on the coordinate system of their map is 0.05 mile. Find the distance between the Department of Defense at (−2, 9) and the Madison Building at (3, −3).

Department of Defense

U.S. Capitol

Madison Building

Understand You know the coordinates of the two locations and that each unit represents 0.05 mile You need to find the distance between the two points.

Plan Use the Distance Formula to find the distance between the two points. Then multiply to find the distance in miles.

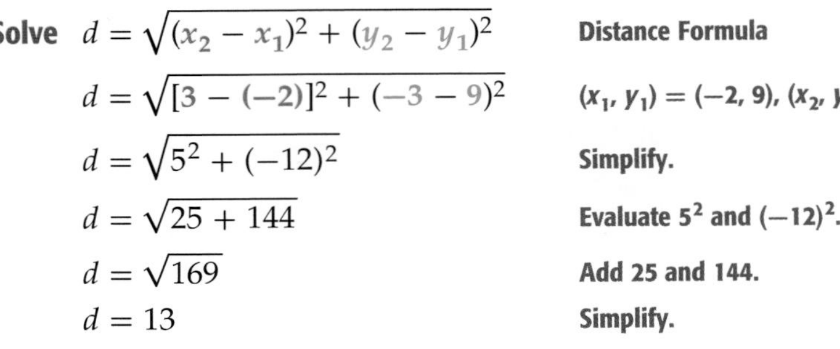

Solve

$d = \sqrt{(x_2 - x_1)^2 + (y_2 - y_1)^2}$	**Distance Formula**
$d = \sqrt{[3 - (-2)]^2 + (-3 - 9)^2}$	**$(x_1, y_1) = (-2, 9)$, $(x_2, y_2) = (3, -3)$**
$d = \sqrt{5^2 + (-12)^2}$	**Simplify.**
$d = \sqrt{25 + 144}$	**Evaluate 5^2 and $(-12)^2$.**
$d = \sqrt{169}$	**Add 25 and 144.**
$d = 13$	**Simplify.**

The distance between the two buildings is 13 units on the map. Since each unit is equal to 0.05 mile, the distance between the two buildings is $0.05 \cdot 13$ or 0.65 mile.

Check The distance is slightly greater than 12 units or 0.6 mile. So, the answer is reasonable. ✓

Check Your Progress

2. RECREATION Find the distance between the Madison Building at (3, −3) and the U.S. Capitol at (0, 0) to the nearest hundredth.

Personal Tutor glencoe.com

Concept Summary Formulas

For Your FOLDABLE

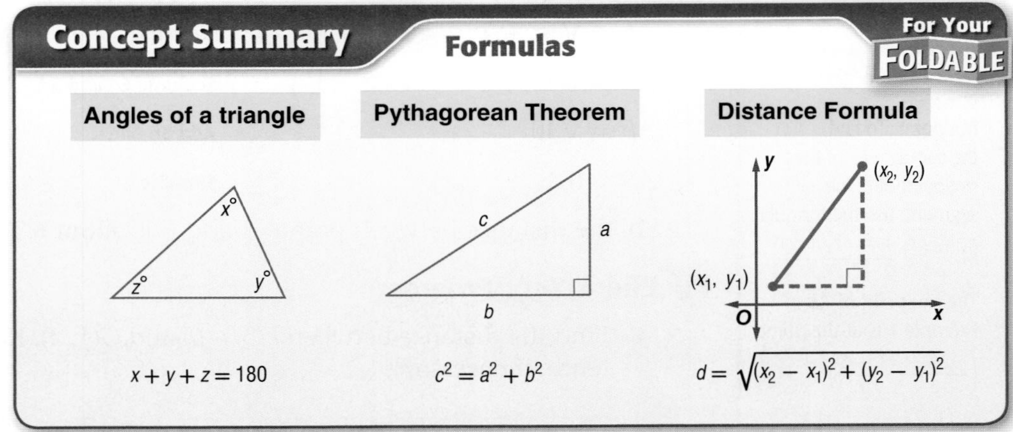

Apply the Distance Formula If you know the coordinates of points of a figure, you can draw conclusions and solve real-world problems on the coordinate plane.

StudyTip

Substitution You can use either point as (x_1, y_1). The distance will be the same.

EXAMPLE 3 Find the Perimeter

GEOMETRY Classify $\triangle JKL$ by its sides. Then find its perimeter to the nearest tenth.

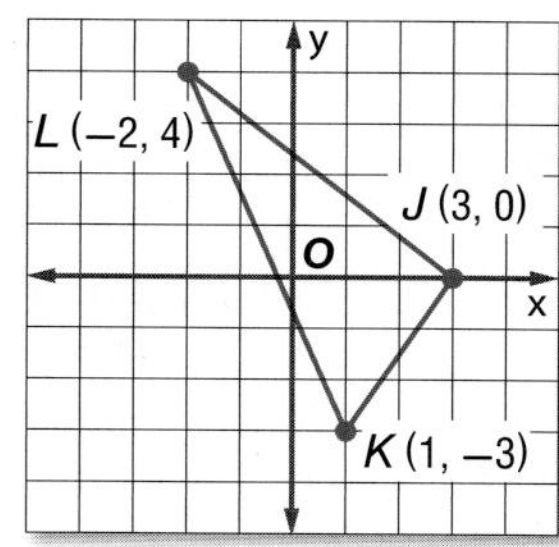

Step 1 Use the Distance Formula to find the length of each side of the triangle.

Side $\overline{JK}$

$\overline{JK}$ has endpoints $J(3, 0)$ and $K(1, -3)$.

$$d = \sqrt{(x_2 - x_1)^2 + (y_2 - y_1)^2}$$
$$JK = \sqrt{(1 - 3)^2 + (-3 - 0)^2}$$
$$JK = \sqrt{(-2)^2 + (-3)^2}$$
$$JK = \sqrt{4 + 9} \text{ or } \sqrt{13}$$

Side $\overline{KL}$

$\overline{KL}$ has endpoints $K(1, -3)$ and $L(-2, 4)$.

$$d = \sqrt{(x_2 - x_1)^2 + (y_2 - y_1)^2}$$
$$KL = \sqrt{(-2 - 1)^2 + [4 - (-3)]^2}$$
$$KL = \sqrt{(-3)^2 + 7^2}$$
$$KL = \sqrt{9 + 49} \text{ or } \sqrt{58}$$

Side $\overline{LJ}$

$\overline{LJ}$ has endpoints $L(-2, 4)$ and $J(3, 0)$.

$$d = \sqrt{(x_2 - x_1)^2 + (y_2 - y_1)^2}$$
$$LJ = \sqrt{[3 - (-2)]^2 + (0 - 4)^2}$$
$$LJ = \sqrt{5^2 + (-4)^2}$$
$$LJ = \sqrt{25 + 16} \text{ or } \sqrt{41}$$

None of the sides are congruent. So, $\triangle JKL$ is scalene.

Step 2 Round each side length to the nearest tenth. Then add the lengths of the sides to find the perimeter.

$$JK + KL + LJ = \sqrt{13} + \sqrt{58} + \sqrt{41}$$
$$\approx 3.6 + 7.6 + 6.4$$
$$\approx 17.6$$

The perimeter is about 17.6 units.

Watch Out!

To simplify the sum of square roots, do not add the numbers inside the square root symbol. Simplify first, then add the rounded numbers.

Check Your Progress

3. GEOMETRY Classify $\triangle XYZ$ with vertices $X(-2, 8)$, $Y(-3, 1)$, and $Z(3, 3)$ by its sides. Then find its perimeter to the nearest tenth.

Personal Tutor glencoe.com

Check Your Understanding

Example 1 p. 565

Find the distance between each pair of points. Round to the nearest tenth, if necessary.

1. $G(1, 5), H(9, 5)$ **2.** $R(0, -4), S(-2, 6)$

Example 2 p. 566

3. ARCHAEOLOGY An archaeologist creates a coordinate plane to record where artifacts were discovered. A unit on the grid represents 5 feet. Find the distance between two artifacts if one artifact was found at $(-3, 1)$ and the other was found at $(-6, -5)$ on the grid. Round to the nearest tenth.

Example 3 p. 567

4. GEOMETRY Classify $\triangle MNP$ shown at the right by its sides. Then find its perimeter. Round to the nearest tenth.

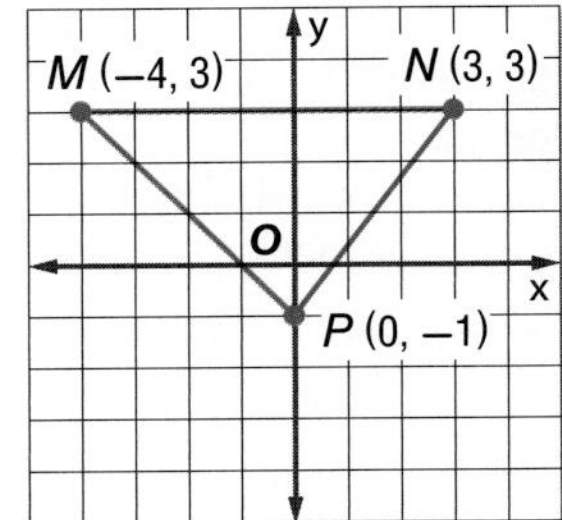

5. GEOMETRY Classify $\triangle ABC$ with vertices $A(4, 0)$, $B(-1, 6)$, and $C(7, -2)$ by its sides. Then find its perimeter to the nearest tenth.

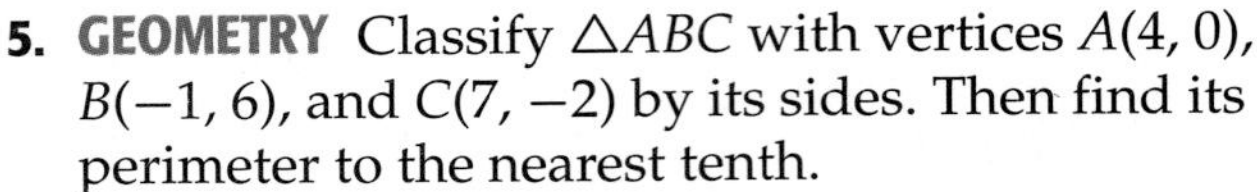

Practice and Problem Solving

● = **Step-by-Step Solutions** begin on page R11. **Extra Practice** begins on page 810.

Example 1 p. 565

Find the distance between each pair of points. Round to the nearest tenth, if necessary.

6. $F(1, 3), G(6, 3)$  **7** $C(5, 1), D(8, 4)$ **8.** $H(-3, -1), J(8, 0)$

9. $L(-2, 2), M(4, 7)$ **10.** $A(-2, 7), B(0, 5)$ **11.** $P(9, -3), Q(4, 4)$

Example 2 p. 566

12. MAPS Ashland, Kentucky, has a longitude of 82° W and a latitude of 38° N. Bowling Green, Kentucky, is located at 86° W and 35° N. Each degree is about 46.6 miles at this longitude/latitude. Find the distance between Ashland and Bowling Green.

13. AIRPORTS A distance of 3 units on a coordinate plane equals an actual distance of 1 mile. The locations of two airports on a map are at (121, 145) and (218, 401). Find the actual distance between these airports to the nearest mile.

Example 3 p. 567

GEOMETRY Classify each triangle by its sides. Then find the perimeter of each triangle. Round to the nearest tenth.

14.

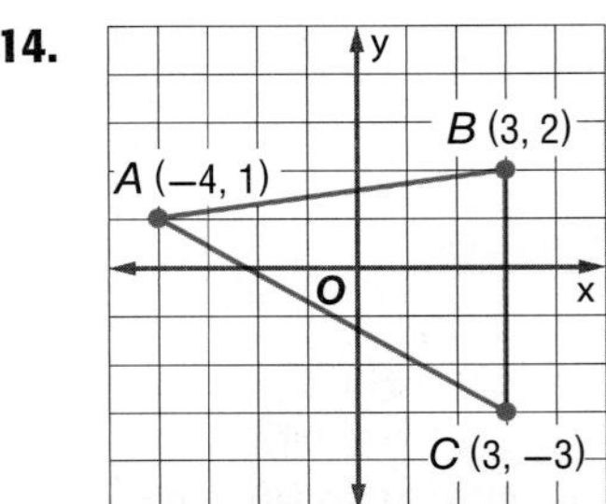

15.

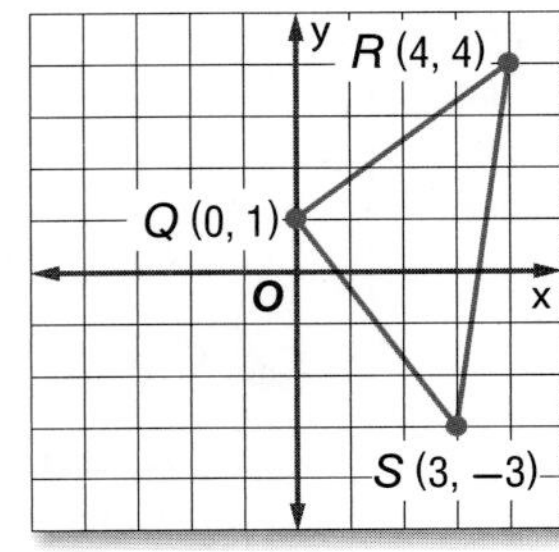

16. $H(-4, 8), J(1, 5), K(-4, 1)$ **17.** $L(0, 7), M(3, -4), N(-6, -7)$

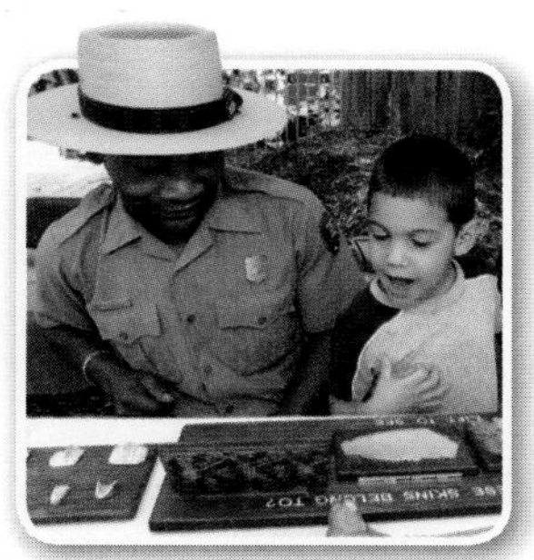

Real-World Career

Park Ranger
Responsibilities for a ranger in a nature park may include using maps to maintain trails. The coordinate plane is involved in the process of using maps.

This career may require a 4-year degree in parks management, or natural sciences.

18. **PARKS** Rosa is looking at a map of the park that is laid out on a coordinate system. Rosa is at (1, −1). The shelter house is at (−2, −4) and the fossil exhibit is at (3, 2). Is Rosa closer to the shelter house or the fossil exhibit?

19. **RECREATION** Darnell's first dart lands 2 inches to the right and 7 inches below the bull's-eye. What is the distance between the bull's-eye and where his first shot hit the target? Round to the nearest tenth of an inch.

Find the area of each rectangle.

20. 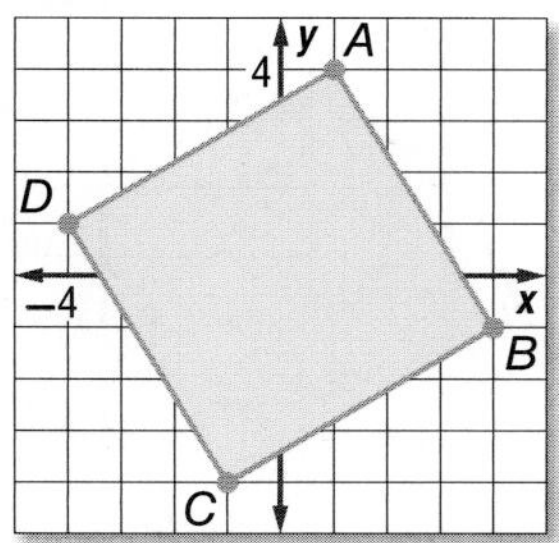

21. 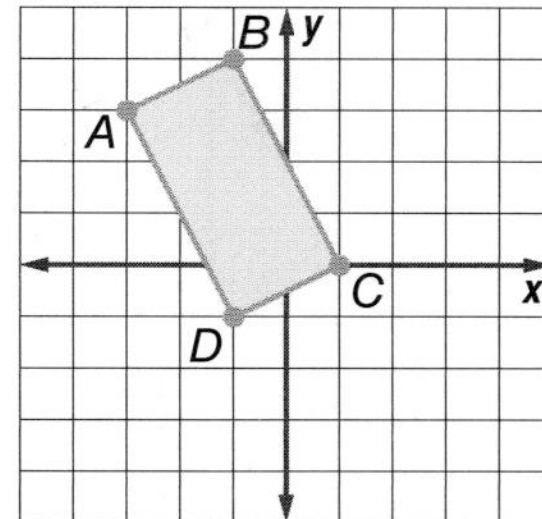

22. **RECREATION** Tomas is hiking the Appalachian Trail in Pennsylvania between Hamburg and Harrisburg at an average rate of 4 miles per hour. Hamburg is located at 40.56°N and 76.01°W, while Harrisburg is located at 40.30°N and 76.84°W. At this longitude/latitude, each degree is about 57.8 miles. Find about how long it will take him to hike the trail.

Find the distance between each pair of points. Round to the nearest tenth, if necessary.

23. $Q\left(5\frac{1}{4}, 3\right), R\left(2, 6\frac{1}{2}\right)$

24. $A\left(-2\frac{1}{2}, 0\right), B\left(-8\frac{3}{4}, -6\frac{1}{4}\right)$

25. $S(6.5, 3.2), T(-5.1, 9.3)$

26. $N(-0.4, -4.8), P(1.8, -8.8)$

H.O.T. Problems

Use Higher-Order Thinking Skills

27. **OPEN ENDED** Name the coordinates of the endpoints of a line segment that is neither horizontal nor vertical and has a length of 5 units.

28. **CHALLENGE** Find x if the distance between (1, 2) and (x, 7) is 13 units.

29. **REASONING** In a golf tournament, Joan's ball landed 2 feet to the left and 3 feet short of the cup. Carolina's ball landed 1 foot to the right and 4 feet beyond the cup. Which of the following techniques would you use to determine who is closer to the cup? Justify your selection(s). Then use the technique(s) to solve the problem.

mental math	number sense	estimation

30. **CHALLENGE** Plot the point (1, 2) on the coordinate plane.
    a. Graph eight points that are 5 units away from (1, 2) on the plane.
    b. Connect the points with a smooth curve.
    c. What figure is formed? Explain.

31. **WRITING IN MATH** Compare the steps for using the Pythagorean Theorem and for using the Distance Formula to find the distance between two points on the coordinate plane.

Standardized Test Practice

32. Which expression could be used to find the distance between points *A* and *B*?

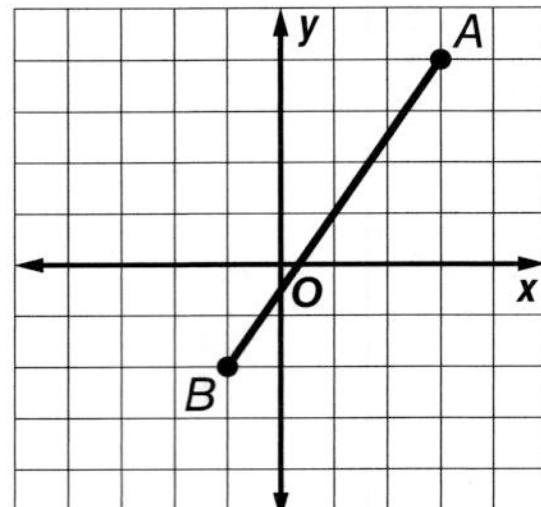

A $\sqrt{(3 + 1)^2 + (4 + 2)^2}$

B $\sqrt{(3 - 1)^2 + (4 - 2)^2}$

C $\sqrt{(-2 + 1)^2 + (4 + 3)^2}$

D $\sqrt{(3 - 1)^2 - (4 - 2)^2}$

33. What is the distance between *S* and *T* in quadrilateral *RSTU*? Round to the nearest tenth.

F 4.5
G 5.4
H 5.7
J 10.8

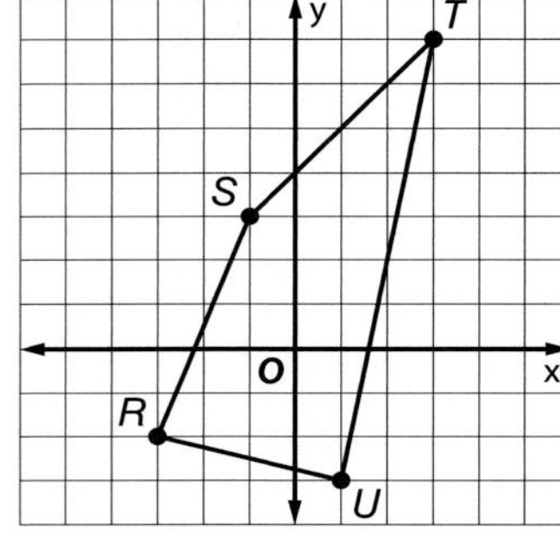

34. EXTENDED RESPONSE Ayana is hiking the trail shown on the graph below.

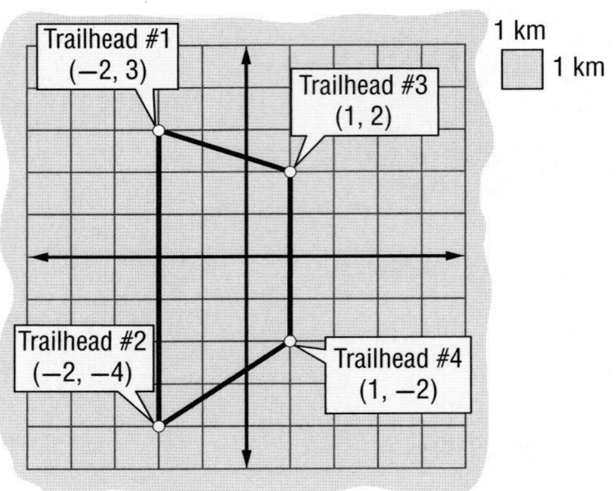

a. To the nearest tenth kilometer, find the total distance Ayana hikes if she completes the entire trail.

b. If there are paths between Trailheads 2 and 3 and between Trailheads 1 and 4, which path is shorter?

35. To the nearest tenth unit, what is the distance between (1, −5) and (5, −1)?

A 5.6
B 5.7
C 6.4
D 6.7

Spiral Review

Find the length of the hypotenuse in each right triangle. Round to the nearest tenth, if necessary. (Lesson 10-4)

36.

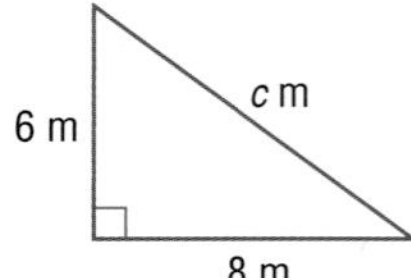

37.

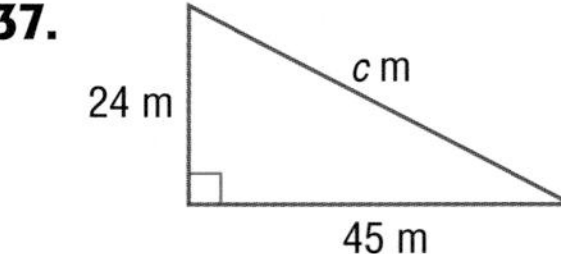

38.

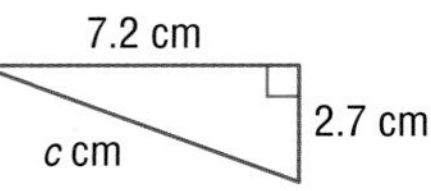

39. ALGEBRA The measures of the angles of a triangle are in the ratio 1:3:5. What is the measure of each angle? (Lesson 10-3)

Skills Review

Estimate each square root to the nearest integer. Do not use a calculator. (Lesson 10-1)

40. $\sqrt{45}$

41. $-\sqrt{139}$

42. $\pm\sqrt{170}$

10-6 Special Right Triangles

Why?

The diagram at the right shows the first step in making a paper airplane.

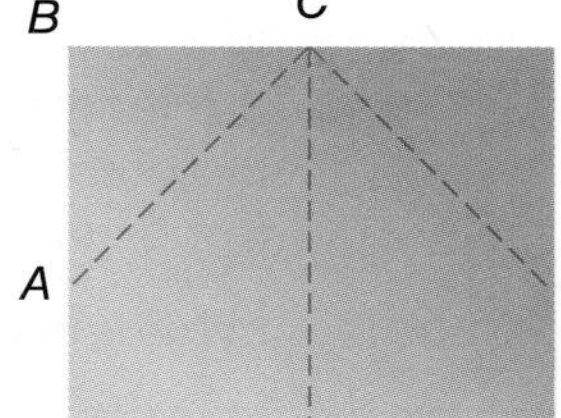

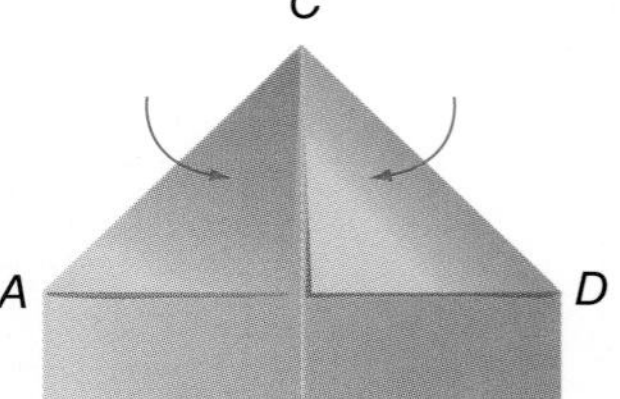

a. Suppose $AB = 5.5$ inches. If $AB = BC$, use the Pythagorean Theorem to find AC to the nearest tenth.

b. Use a calculator to find $AB \cdot \sqrt{2}$ and $BC \cdot \sqrt{2}$ to the nearest tenth. Compare the results to the answer in part **a**.

Then

You used the Pythagorean Theorem to find missing measures in right triangles. (Lesson 10-4)

Now

- Find missing measures in 45°-45°-90° triangles.
- Find missing measures in 30°-60°-90° triangles.

Math Online

glencoe.com

- Extra Examples
- Personal Tutor
- Self-Check Quiz
- Homework Help

Find Measures in 45°–45°–90° Triangles Study the triangles shown. The corresponding angles have the same measure and the corresponding sides are proportional with a scale factor of 2. This and other examples suggest that all 45°-45°-90° triangles are similar. In addition to the Pythagorean Theorem, you can use similar triangles to find missing measures in a 45°-45°-90° triangle.

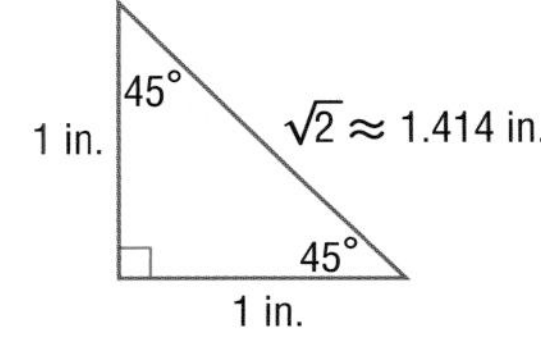

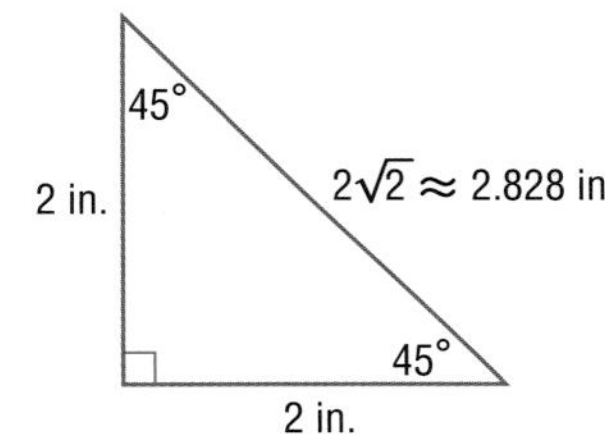

EXAMPLE 1 Find the Hypotenuse in a 45°–45°–90° Triangle

Triangle *ABC* and triangle *DEF* are 45°-45°-90° triangles. Find the length of the hypotenuse in △*DEF*.

The scale factor from △*ABC* to △*DEF* is $\frac{3}{1}$ or 3. Use the scale factor to find the hypotenuse.

$x = 3 \cdot \sqrt{2}$ **Multiply the length of $\overline{AC}$ by the scale factor, 3.**

$= 3\sqrt{2}$

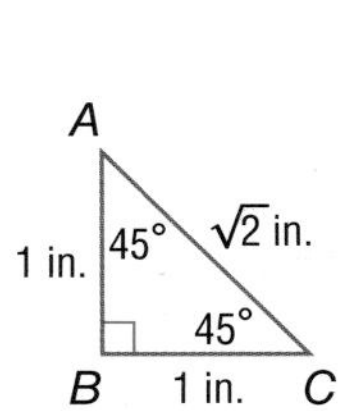

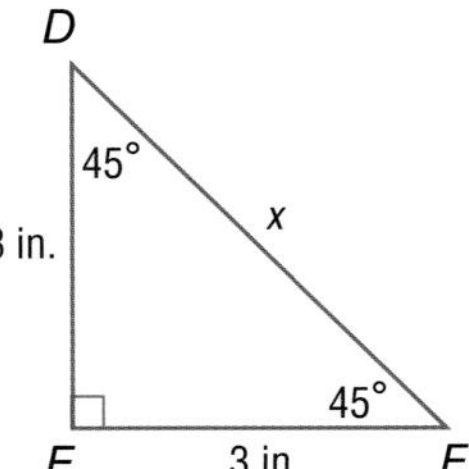

So, the hypotenuse of △*DEF* measures $3\sqrt{2}$ inches.

Check Your Progress

Find the length of each hypotenuse.

1A.

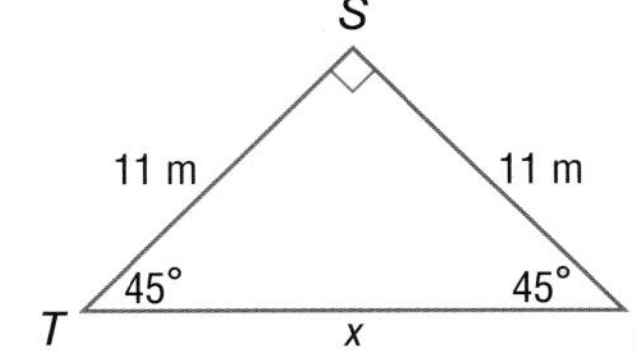

1B.

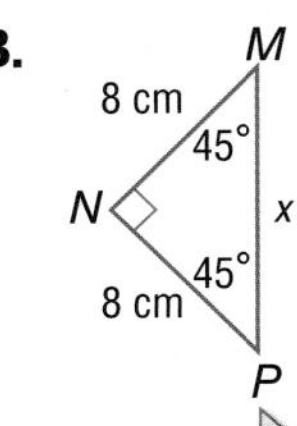

Personal Tutor glencoe.com

Example 1 suggests the following relationship for 45°-45°-90° triangles.

45°-45°-90° Triangles

For Your FOLDABLE

Words In a 45°-45°-90° triangle, the length of the hypotenuse is $\sqrt{2}$ times the length of a leg.

Model

$a\sqrt{2}$, 45°, a, 45°, a

Symbols hypotenuse = leg $\cdot \sqrt{2}$

Find Measures in 30°-60°-90° Triangles Study the triangles shown. Just as 45° -45°-90° triangles are similar, 30°-60°-90° triangles are similar.

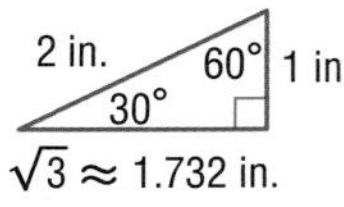

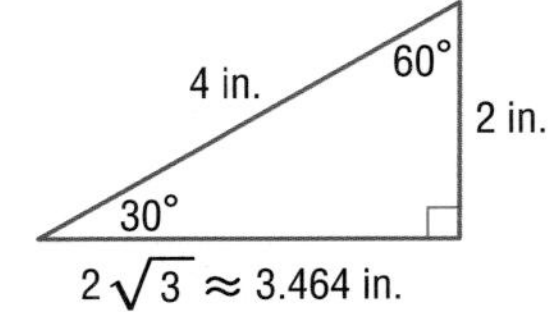

EXAMPLE 2 Find Missing Measures in a 30°-60°-90° Triangle

$\triangle ABC$ and $\triangle DEF$ are 30°-60°-90° triangles. Find the exact length of the missing measures.

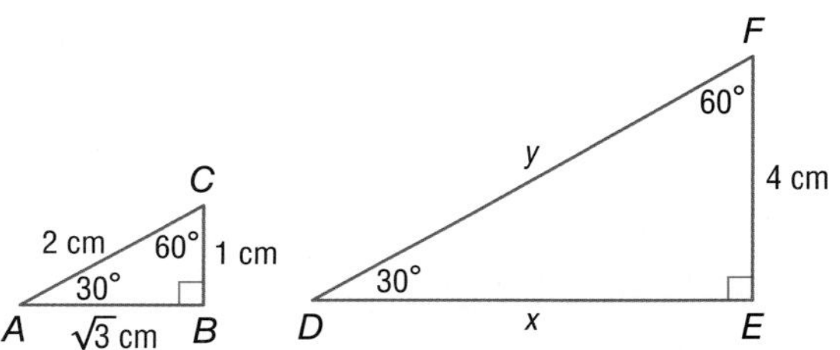

> **StudyTip**
>
> **Alternative Method** You can also use a proportion to find the missing measures of corresponding sides. In Example 2, you can use $\frac{2}{1} = \frac{x}{4}$.

The scale factor from $\triangle ABC$ to $\triangle DEF$ is $\frac{4}{1}$ or 4.

Use the scale factor to find the missing measures.

$x = 4 \cdot 2$ **Multiply the length of $\overline{AC}$ by the scale factor.**

$= 8$

So, x is 8 centimeters.

$y = 4 \cdot \sqrt{3}$ **Multiply $\overline{AB}$ by the scale factor.**

$= 4\sqrt{3}$

So, y is $4\sqrt{3}$ centimeters.

2. Triangle RST is a 30°-60°-90° triangle. Find the exact length of the missing measures.

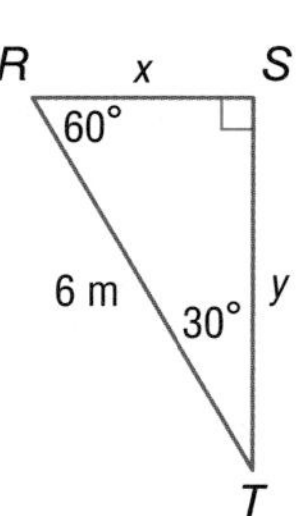

Personal Tutor glencoe.com

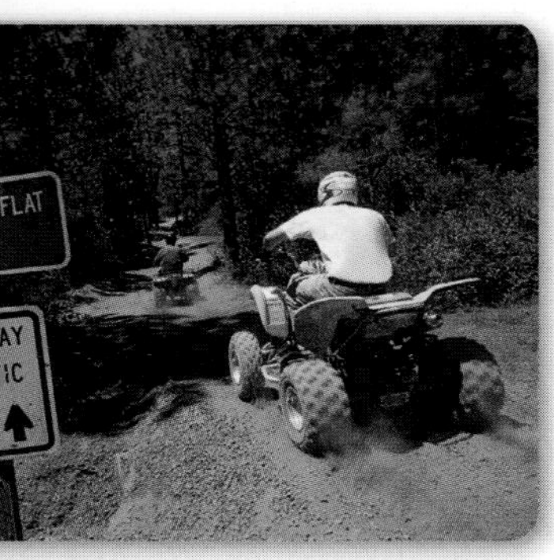

Real-World Link

(ATV) The all-terrain vehicle was initially developed in Japan as a means of transportation through mountainous areas. When ATVs were first sold in the U.S. in the early 1970s, they had 3 wheels.

Source: American Honda Motor Company

Example 2 suggests the following relationship for 30°-60°-90° triangles.

Key Concept 30°-60°-90° Triangles

For Your FOLDABLE

Words In a 30°-60°-90° triangle,
- the length of the hypotenuse is 2 times the length of the shorter leg, and
- the length of the longer leg is $\sqrt{3}$ times the length of the shorter leg.

Model

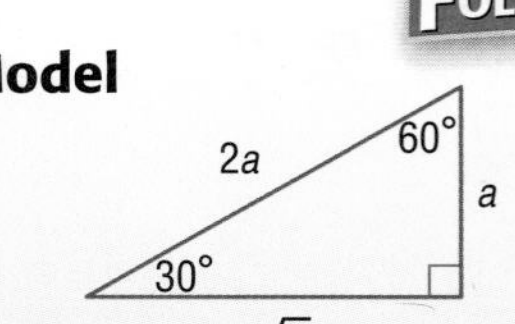

Symbols hypotenuse = 2 · shorter leg
longer leg = $\sqrt{3}$ · shorter leg

In real-world situations, use the decimal value for the square root.

Real-World EXAMPLE 3 Use Special Right Triangles

RAMPS A ten foot ramp for loading an all-terrain vehicle onto a truck makes a 30°-angle with the ground. Find *y*, the distance from the truck to the end of the ramp. Round to the nearest tenth.

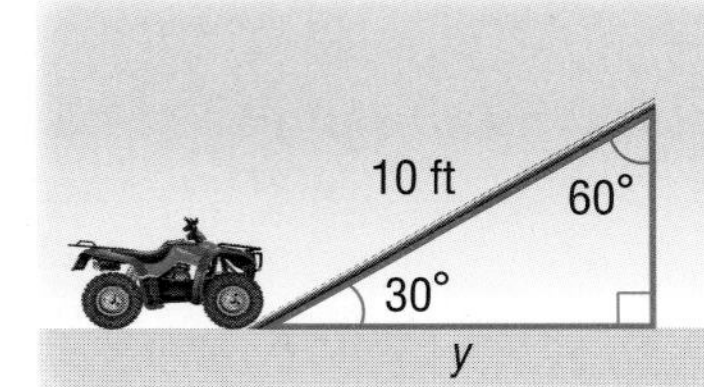

Understand You know the hypotenuse of the 30°-60°-90° triangle. You need to find the distance from the truck to the end of the ramp.

Plan To find *y*, use the relationship between the shorter leg and the longer leg in a 30°-60°-90° triangle. The shorter leg is half the length of the hypotenuse or 5 feet.

Solve longer leg = $\sqrt{3}$ · shorter leg — **Relationship between sides**

$y = \sqrt{3} \cdot 5$ — **Substitution**

To find the decimal value of *y* use a calculator.

[2nd] [√‾‾] 3 [ENTER] [×] 5 [ENTER] 8.660254038

$\sqrt{3} \cdot 5 \approx 8.7$ — **Round to the nearest tenth.**

The distance from the truck to the end of the ramp is about 8.7 feet.

Check Use the Pythagorean Theorem to check the solution. Since $5^2 + 8.7^2 = 100.69$ and $100.69 \approx 100$, the answer is correct. ✓

ReadingMath

Exact Values Since 8.6 is rounded, it is an approximate value, while $5\sqrt{3}$ is an exact value.

Check Your Progress

3. **KICKBALL** A kickball field is in the shape of a square. The distance from first base to second base is 60 feet. Find *x* the distance from home base to second base. Round to the nearest tenth.

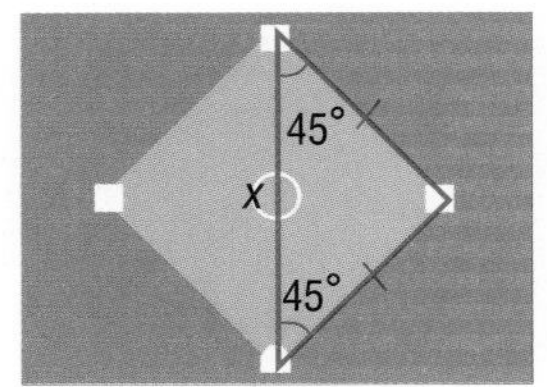

Personal Tutor glencoe.com

Check Your Understanding

Examples 1 and 2
pp. 571–572

Find each missing measure.

1.

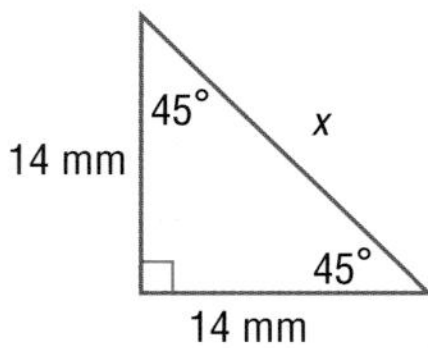

2.

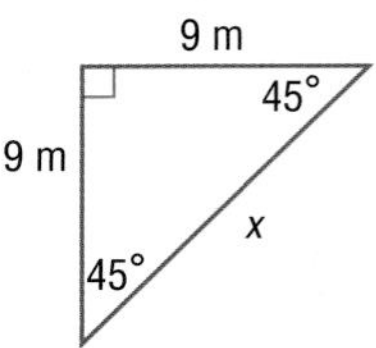

3.

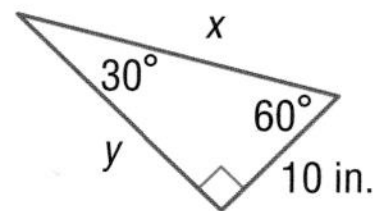

4.

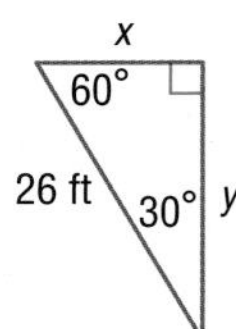

5. In a 45°-45°-90° triangle, a leg is 2 centimeters long. Find the exact length of the hypotenuse.

6. In a 30°-60°-90° triangle, the shorter leg is 9 feet long. Find the exact length of the hypotenuse and the length of the longer leg.

Example 3
p. 573

AIR HOCKEY An air hockey table is 40 inches wide, as shown at the right. Find each measure. Round to the nearest tenth, if necessary.

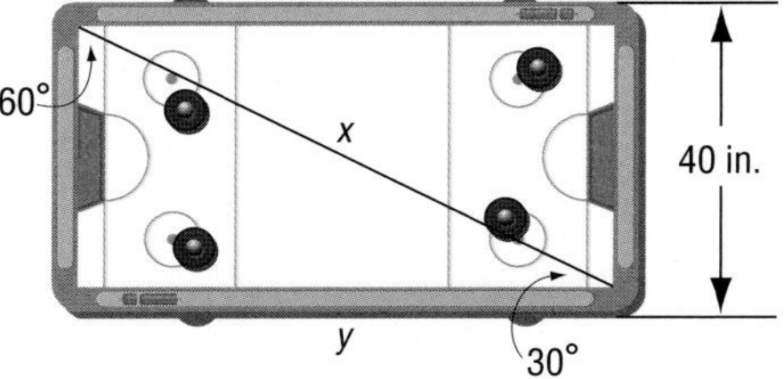

7. length of table

8. distance the puck can go from one corner of the table to another

Practice and Problem Solving

● = **Step-by-Step Solutions** begin on page R11.
Extra Practice begin on page 810.

Examples 1 and 2
pp. 571–572

Find each missing measure.

9.

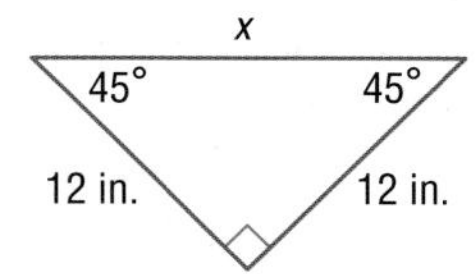

10.

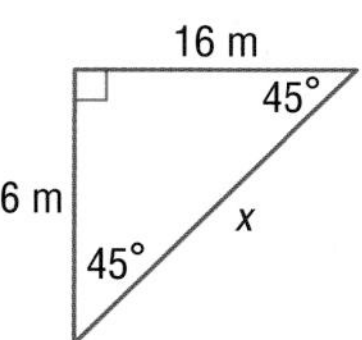

11.

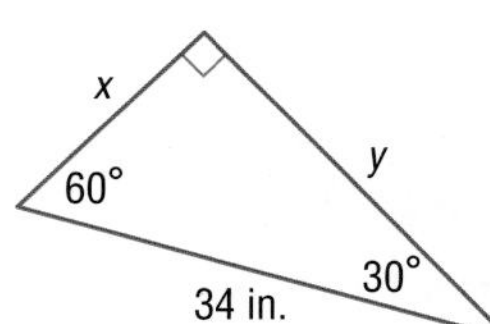

12.

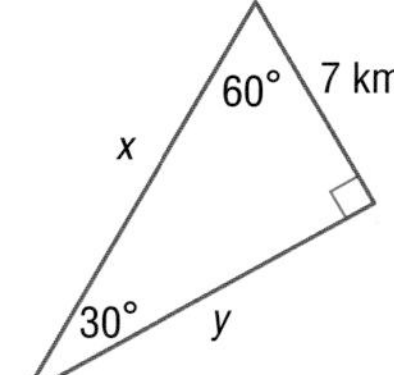

13.

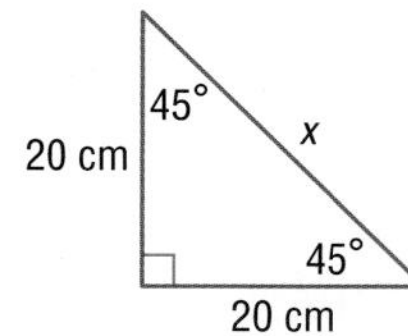

14.

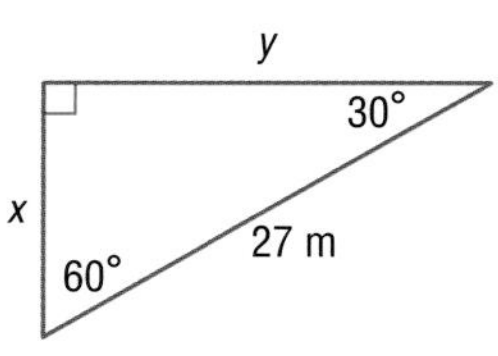

15. In a 45°-45°-90° triangle, a leg is 15 inches long. Find the exact length of the hypotenuse.

16. In a 30°-60°-90° triangle, the hypotenuse is 26 inches long. Find the exact length of the shorter leg and the length of the longer leg.

Example 3
p. 573

Real-World Link

Visitors to the Tate Modern museum in the United Kingdom will be able to sit in canvas sacks and slide from the fifth floor to the ground, accelerating up to 30 miles per hour.

Source: Tate Modern

17. SLIDES A slide forms a 60° angle with its ladder. If the slide is 12 feet tall, how long is it?

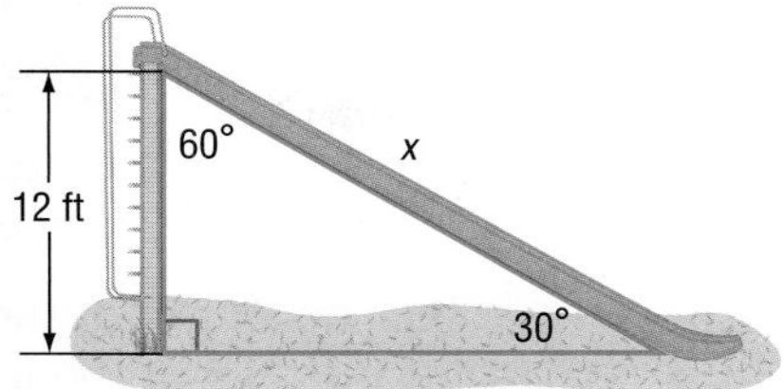

18. MAZES The corn maze below is a square. If the diagonal is $420\sqrt{2}$ feet long, what is the perimeter of the maze?

19. MULTIPLE REPRESENTATIONS In this problem, you will explore 45°-45°-90° triangles.

a. TABULAR Find the hypotenuse of isosceles right triangles with the following leg lengths: 1, 2, 3, 4, and 5. Round to the nearest tenth. Record the results in a table.

b. GRAPHICAL Graph the points (leg length, hypotenuse) on a coordinate plane. Describe the pattern of the points.

c. SYMBOLIC Write an equation that you could use to approximate the hypotenuse y if you know the side length x of a 45°-45°-90° triangle.

20. SIGNS What is the perimeter of the school crossing sign shown at the right? Round to the nearest tenth.

Find each missing measure.

21.

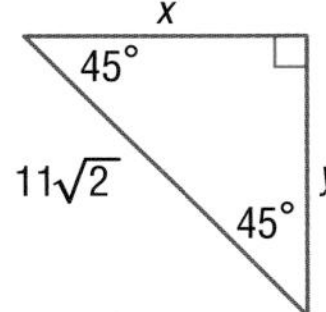

22.

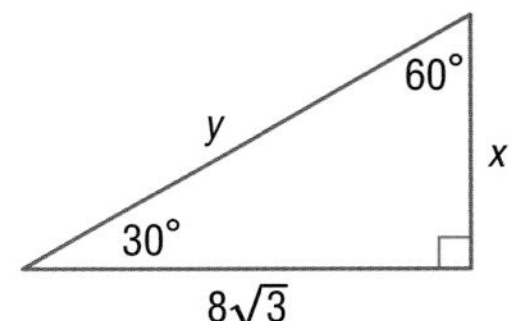

H.O.T. Problems

Use **H**igher-**O**rder **T**hinking Skills

23. OPEN ENDED Draw a 30°-60°-90° triangle. Measure the length of the shorter leg. Then use the measure to find the length of the longer leg and the hypotenuse. Check your answers by measuring the other two sides.

24. FIND THE ERROR Ava says that the length of $\overline{JK}$ in the figure is 16 inches. Deondre says the length is about 22.6 inches. Is either of them correct? Explain.

L
32 in.
60°
30°
J
K

25. REASONING A triangle has lengths 21 centimeters, 42 centimeters, and 36.4 centimeters. Is it a 30°-60°-90° triangle? Explain.

26. CHALLENGE In a 45°-45°-90° triangle, the ratio of the smaller angles to the larger angle is 1:2. Is the ratio of the leg lengths to the hypotenuse also 1:2? Explain your reasoning.

27. WRITING IN MATH Describe the relationships among the side lengths of a 30°-60°-90° triangle.

Standardized Test Practice

28. Mr. Govin installed the gate in his backyard. What is the value of x?

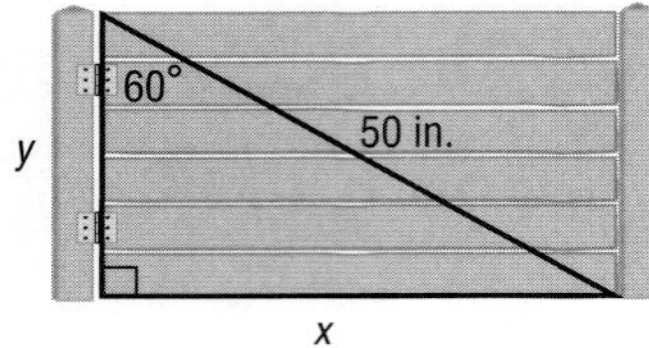

A 25 in.

B $25\sqrt{2}$ in.

C $25\sqrt{3}$ in.

D $50\sqrt{3}$ in.

29. Each leg of a 45°-45°-90° triangle has a length of 18 feet. What is the length of the hypotenuse?

F $\frac{\sqrt{2}}{18}$ ft

G 18 ft

H $18\sqrt{2}$ ft

J $18\sqrt{3}$ ft

30. Which of the following is *not* true about 30°-60°-90° triangles?

A The length of the hypotenuse is the product of the length of the leg and $\sqrt{2}$.

B The length of the hypotenuse is the product of the length of the shorter length and 2.

C The length of the longer leg is the product of the length of the shorter length and $\sqrt{3}$.

D The length of shorter leg is equal to half the length of the hypotenuse.

31. EXTENDED RESPONSE Use the triangle shown at the right.

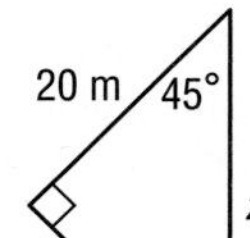

a. Find each missing measure.

b. Explain how you found each value.

Spiral Review

Find the distance between each pair of points. Round to the nearest tenth, if necessary. (Lesson 10-5)

32. $J(5, -4), K(-1, 3)$

33. $C(-7, 2), D(6, -4)$

34. $S(-9, 0), T(6, -7)$

35. $M(0, 0), N(-7, -8)$

If c is the measure of the hypotenuse, find each missing measure. Round to the nearest tenth, if necessary. (Lesson 10-4)

36. $a = 9, b = ?, c = 41$

37. $a = ?, b = 35, c = 37$

38. $a = ?, b = 12, c = 19$

39. $a = 7, b = ?, c = 14$

40. GEOMETRY If each side of the figures below has a length of 1 foot, find the perimeter of a figure with 9 pentagons. (Lesson 3-8)

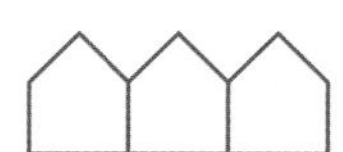

Skills Review

Solve each equation. (Lessons 4-3 and 4-4)

41. $2x = 180$

42. $x + 72 = 90$

43. $2x = 90$

44. $95 + x = 180$

45. $x + 55 = 90$

46. $3x = 180$

47. $x + 79 = 180$

48. $50 + x = 90$

Study Guide and Review

Math Online glencoe.com
- STUDY TO GO
- Vocabulary Review

Chapter Summary

Key Concepts

Squares and Square Roots (Lesson 10-1)

- Perfect squares are squares of integers.
- A square root of a number is one of two equal factors of the number.

The Real Number System (Lesson 10-2)

- Numbers that cannot be written as terminating or repeating decimals are called irrational numbers.
- The set of rational and irrational numbers together make up the set of real numbers.

Triangles (Lesson 10-3)

- An acute angle measures less than 90°, a right angle measures 90°, and an obtuse angle has a measure between 90° and 180°.
- Triangles can be classified by their angles as acute, obtuse, or right and by their sides as scalene, isosceles, or equilateral.

The Pythagorean Theorem (Lesson 10-4)

- In a right triangle with legs a and b and hypotenuse c, $c^2 = a^2 + b^2$.
- If you know the lengths of two sides of a right triangle, you can use the Pythagorean Theorem to find the length of the third side.

The Distance Formula (Lesson 10-5)

- The distance d between two points with coordinates (x_1, y_1) and (x_2, y_2) is given by $d = \sqrt{(x_2 - x_1)^2 + (y_2 - y_1)^2}$.

FOLDABLES® Study Organizer

Be sure the Key Concepts are noted in your Foldable.

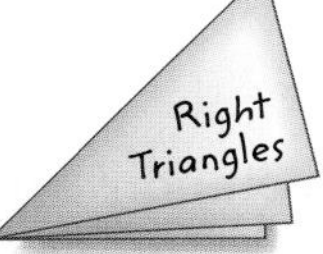

Key Vocabulary

congruent (p. 552)
converse (p. 560)
Distance Formula (p. 565)
hypotenuse (p. 558)
irrational numbers (p. 543)
legs (p. 558)
line segment (p. 550)
perfect square (p. 537)
Pythagorean Theorem (p. 558)
radical sign (p. 537)
real numbers (p. 543)
solving a right triangle (p. 559)
square root (p. 537)
triangle (p 550)
vertex (p. 550)

Vocabulary Check

Choose the term that best matches each statement or phrase.

1. a square of a whole number
2. a triangle with no congruent sides
3. decimals that do not repeat or terminate
4. the sides of a right triangle that are adjacent to the right angle
5. a triangle with angle measures 73°, 30°, and 77°
6. the side opposite the right angle in a triangle
7. sides of a figure that have the same length
8. the point at which two sides of a triangle intersect
9. used to indicate a positive square root
10. part of a line containing two endpoints and all the points between them

CHAPTER 10 Study Guide and Review

Lesson-by-Lesson Review

10-1 Squares and Square Roots (pp. 537–542)

Find each square root.

11. $\sqrt{169}$ **12.** $-\sqrt{25}$

13. $\pm\sqrt{1}$ **14.** $\sqrt{484}$

Estimate each square root to the nearest integer. Do not use a calculator.

15. $\sqrt{15}$ **16.** $-\sqrt{52}$

17. $-\sqrt{90}$ **18.** $\sqrt{415}$

19. CLOCKS The *period* of a pendulum is the time it takes to make one complete swing. The period P of a pendulum is given by the formula $P = 2\pi\sqrt{\frac{\ell}{32}}$, where ℓ is the length of the pendulum. If a clock's pendulum is 8 feet long, find the period. Use 3.14 for π.

EXAMPLE 1

Find $\pm\sqrt{256}$.

$\pm\sqrt{256} = \pm 16$ **Find both square roots of 256; $16^2 = 256$.**

EXAMPLE 2

Estimate $\sqrt{70}$ to the nearest integer.

The first perfect square less than 70 is 64.

$\sqrt{64} = 8$

The first perfect square greater than 70 is 81.

$\sqrt{81} = 9$

$\sqrt{64}$ $\sqrt{70}$ $\sqrt{81}$
8 8.25 8.5 8.75 9

Since 70 is closer to 64 than 81, $\sqrt{70}$ is closer to 8 than 9.

10-2 The Real Number System (pp. 543–548)

Name all of the sets of numbers to which each real number belongs. Write *whole*, *integer*, *rational*, or *irrational*.

20. 18 **21.** $\frac{6}{11}$ **22.** $\sqrt{74}$ **23.** $4.\overline{5}$

Replace each ● with <, >, or = to make a true statement.

24. $6.\overline{25}$ ● $\sqrt{39}$ **25.** $-\sqrt{70}$ ● $-8\frac{1}{5}$

26. $-11\frac{1}{9}$ ● $-\sqrt{124}$ **27.** $\sqrt{68}$ ● $8.\overline{4}$

Solve each equation. Round to the nearest tenth, if necessary.

28. $d^2 = 100$ **29.** $4y^2 = 5.76$

30. GARDENS The formula $A \approx 3.14\, r^2$ can be used to determine the area of a circle where A is the area and r is the distance from the center of the circle to the outside edge. If the area of a circular garden is 700 square feet, about how far is the distance from the center of the garden to the outside edge? Round to the nearest tenth.

EXAMPLE 3

Replace ● with <, >, or = to make $\sqrt{12}$ ● $\frac{10}{3}$ a true statement.

$\sqrt{12} = 3.46410162\ldots$ $\frac{10}{3} = 3.333\ldots$

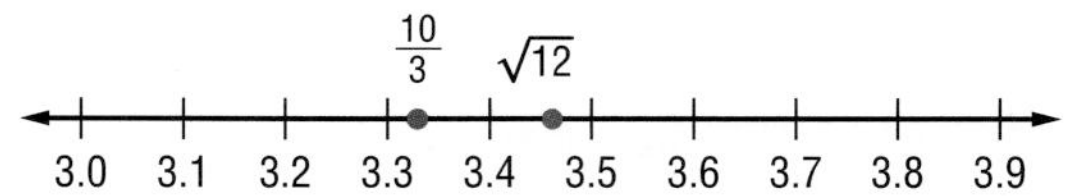

Since $\sqrt{12}$ is to the right of $\frac{10}{3}$, $\sqrt{12} > \frac{10}{3}$.

EXAMPLE 4

Solve $4n^2 = 44$.

$4n^2 = 44$ **Write the equation.**

$n^2 = 11$ **Divide each side by 4.**

$n = \pm\sqrt{11}$ **Definition of square root**

$n \approx 3.3$ and -3.3 **Use a calculator.**

The solutions are approximately 3.3 and -3.3.

MIXED PROBLEM SOLVING
For mixed problem-solving practice, see page 852.

10-3 Triangles (pp. 550–555)

Find the value of x in each triangle. Then classify each triangle by its angles and by its sides.

31. 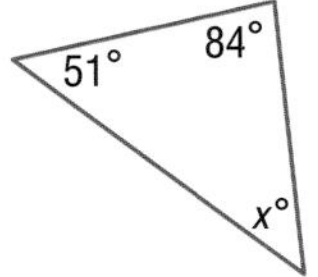

32. 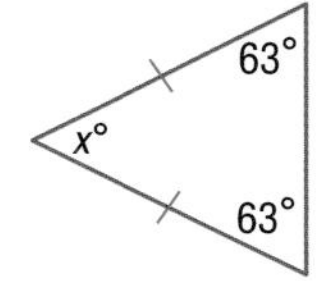

33. 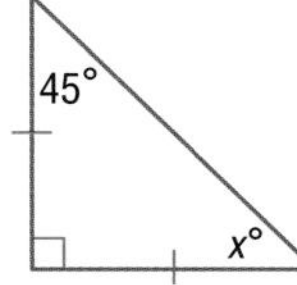

34. 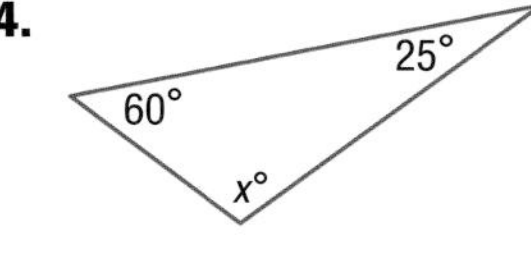

35. SIGNS Classify the yield sign by its angles and by its sides.

EXAMPLE 5

Find the value of x in the triangle. Then classify the triangle by its angles and by its sides.

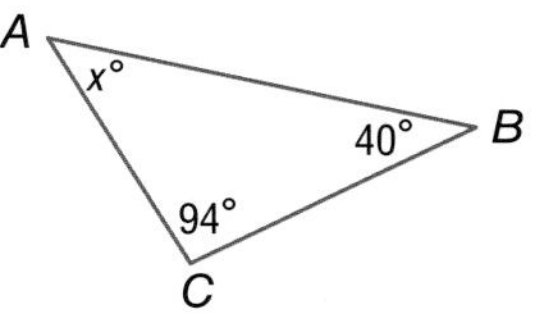

Step 1 Find the missing angle measure.

$x + 40 + 94 = 180$	**Write an equation.**
$x + 134 = 180$	**Simplify.**
$x + 134 - 134 = 180 - 134$	**Subtract 134 from each side.**
$x = 46$	**Simplify.**

Step 2 Classify the triangle.

Angles: The triangle has an obtuse angle.

Sides: The triangle has no congruent sides.

So, the triangle is obtuse scalene.

10-4 The Pythagorean Theorem (pp. 558–563)

Find the missing length of each triangle. Round to the nearest tenth, if necessary.

36. 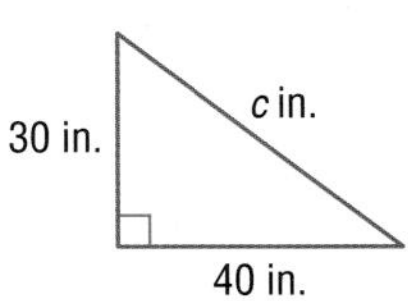

37. 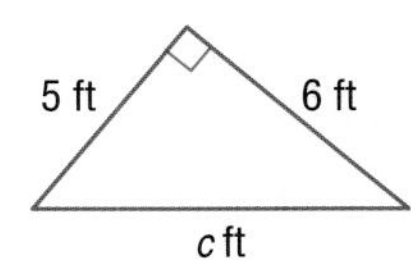

38. 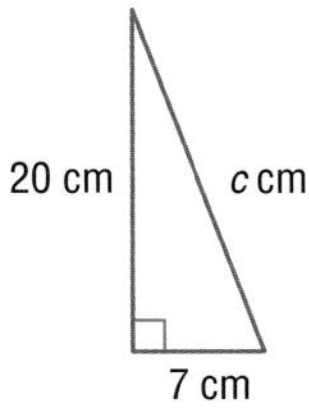

39. 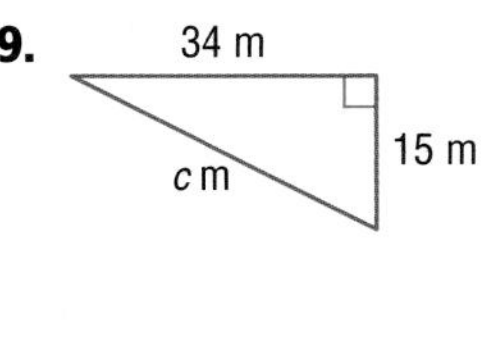

40. SOFTBALL On a fast pitch softball diamond, the bases are 60 feet apart. What is the distance from home plate to second base in a straight line to the nearest tenth of a foot?

EXAMPLE 6

Find the missing measure of the right triangle. Round to the nearest tenth, if necessary.

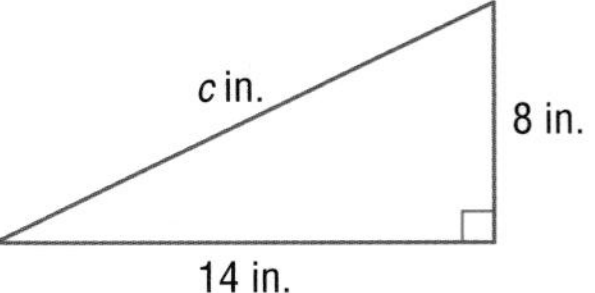

$a^2 + b^2 = c^2$	**Pythagorean Theorem**
$8^2 + 14^2 = c^2$	**Replace *a* with 8 and *b* with 14.**
$64 + 196 = c^2$	**Evaluate 8^2 and 14^2.**
$260 = c^2$	**Add 64 and 196.**
$\pm\sqrt{260} = c$	**Definition of square root**
$16.1 \approx c$	**Use the principal square root.**

So, c is about 16.1 inches.

Study Guide and Review

10-5 The Distance Formula (pp. 565–570)

Find the distance between each pair of points. Round to the nearest tenth, if necessary.

41. $G(0, 0)$, $H(3, 4)$

42. $B(-2, 7)$, $C(-5, 7)$

43. $J(9, -5)$, $K(0, 4)$

44. $M(-8, 1)$, $N(7, -6)$

45. **GEOMETRY** Find the perimeter of $\triangle DEF$. Round to the nearest tenth.

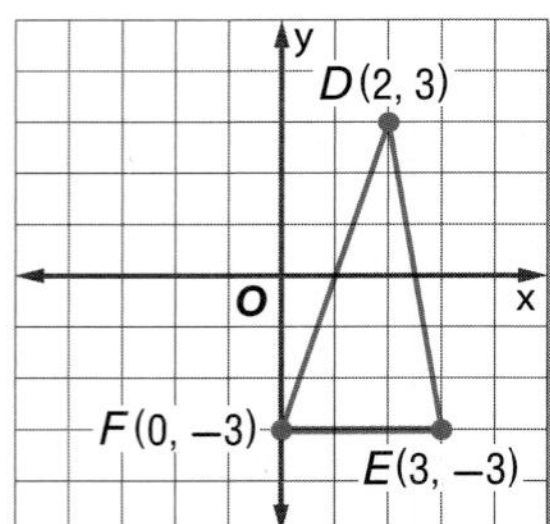

46. **LAKES** A distance of 2 units on a coordinate plane equals an actual distance of 1 mile. Suppose the locations of two lakes on a map are at (26, 15) and (9, 20). Find the actual distance between these lakes to the nearest mile.

EXAMPLE 7

Find the distance between $S(5, -3)$ and $T(-1, 2)$.

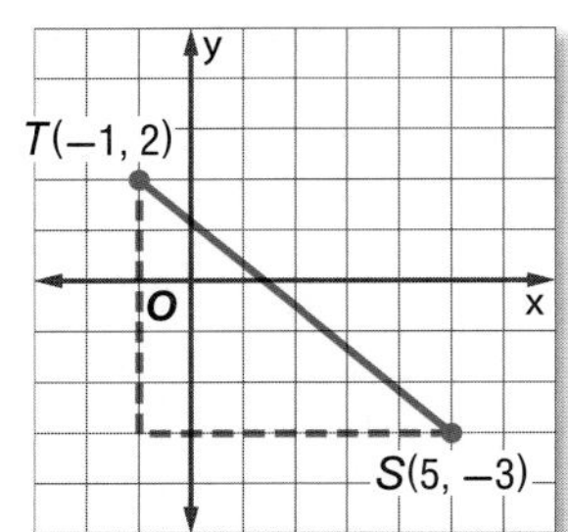

$d = \sqrt{(x_2 - x_1)^2 + (y_2 - y_1)^2}$	**Distance Formula**
$ST = \sqrt{(-1 - 5)^2 + [2 - (-3)]^2}$	**$(x_1, y_1) = (5, -3)$, $(x_2, y_2) = (-1, 2)$**
$ST = \sqrt{(-6)^2 + 5^2}$	**Simplify.**
$ST = \sqrt{36 + 25}$	**Evaluate 6^2 and 5^2.**
$ST = \sqrt{61}$	**Add 36 and 25.**
$ST \approx 7.8$	**Simplify.**

The distance between S and T is about 7.8 units.

10-6 Special Right Triangles (pp. 571–576)

Find each missing measure.

47.

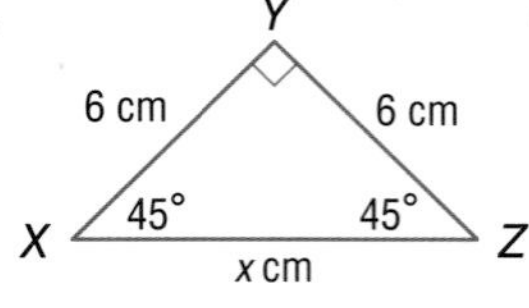

48.

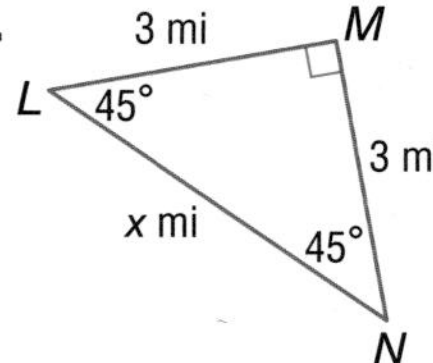

49. **50.**

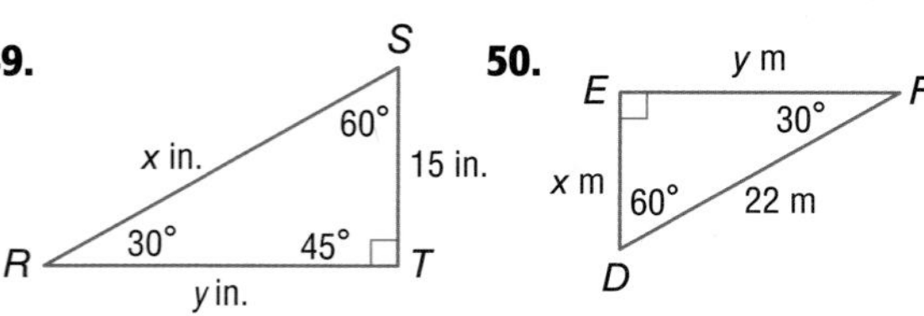

51. **BAKERY** A 9-inch square cake is cut in half along the diagonal. What is the length of the cut?

EXAMPLE 8

$\triangle ABC$ and $\triangle DEF$ are 45°-45°-90° triangles. Find the missing measure.

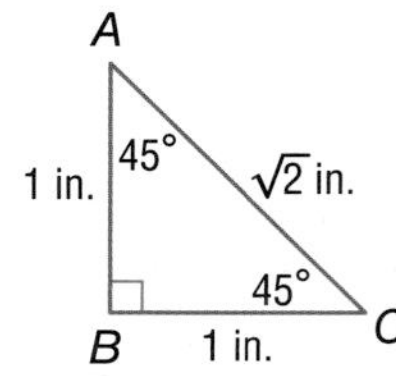

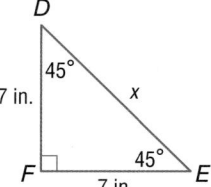

The scale factor from $\triangle ABC$ to $\triangle DEF$ is $\frac{7}{1}$ or 7. Use the scale factor to find the hypotenuse.

$x = 7 \cdot \sqrt{2}$ or $7\sqrt{2}$

So, the hypotenuse is $7\sqrt{2}$ inches.

Practice Test

Math Online glencoe.com
Chapter Test

Find each square root.

1. $\sqrt{256}$
2. $-\sqrt{400}$
3. Without using a calculator, estimate $-\sqrt{102}$ to the nearest integer.

Replace each ● with <, >, or = to make a true statement.

4. $4.\overline{9}$ ● $4\frac{9}{10}$
5. -6.8 ● $-\sqrt{45}$

ALGEBRA Solve each equation. Round to the nearest tenth, if necessary.

6. $t^2 = 49$
7. $72 = 6p^2$
8. **SPORTS** The formula $h = 16t^2$ describes the time t in seconds that it takes for an object to fall from a height of h feet. How long would it take a baseball dropped out of a window from a height of 50 feet to hit the ground?
9. **MULTIPLE CHOICE** The area of the square at the right is 256 square units. Which of the following is the value of x?

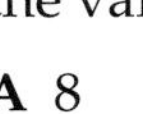

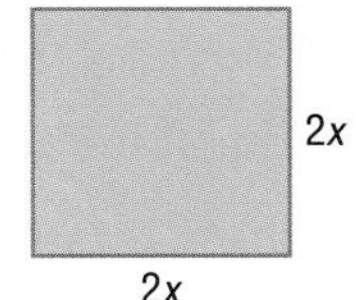

A 8 C 16

B 11.3 D 64

Classify each angle as *acute*, *obtuse*, *right*, or *straight*.

10. 45°
11. 95°
12. 180°

Find the value of x in each triangle. Then classify each triangle by its angles and by its sides.

13.

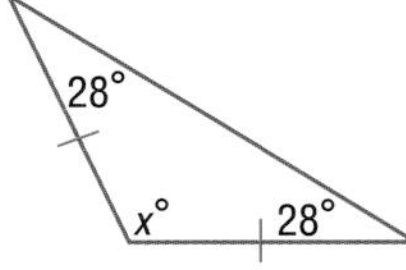

14.

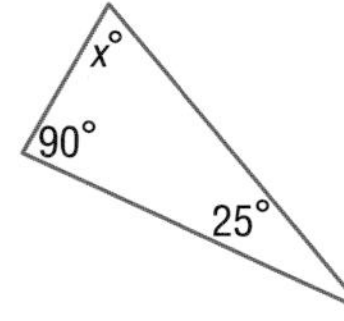

The lengths of three sides of a triangle are given. Determine whether each triangle is a right triangle.

15. $a = 15, b = 36, c = 39$
16. $a = 9, b = 15, c = 17$

Find each missing measure. Round to the nearest tenth, if necessary.

17.

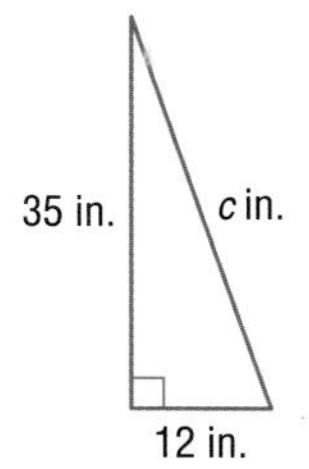

18.

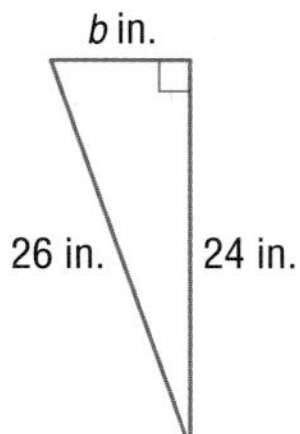

19. **LADDER** A building has a 12-foot-high window. If the bottom of a ladder is 5 feet away from the building, will a 15-foot ladder reach the window? Explain.

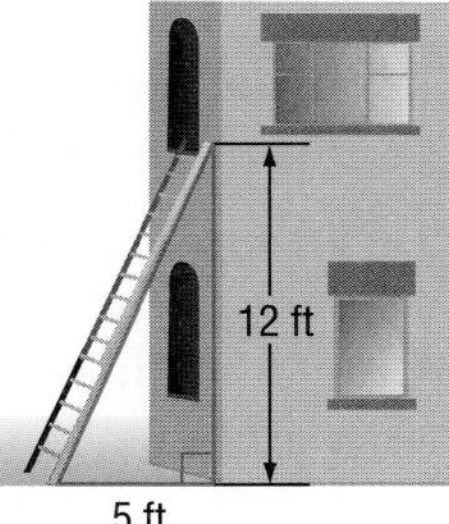

Find the distance between each pair of points. Round to the nearest tenth, if necessary.

20. $Q(-6, 4), R(6, -8)$
21. $C(5, 9), D(-7, 3)$
22. **MULTIPLE CHOICE** What is the perimeter of $\triangle GHJ$ shown?

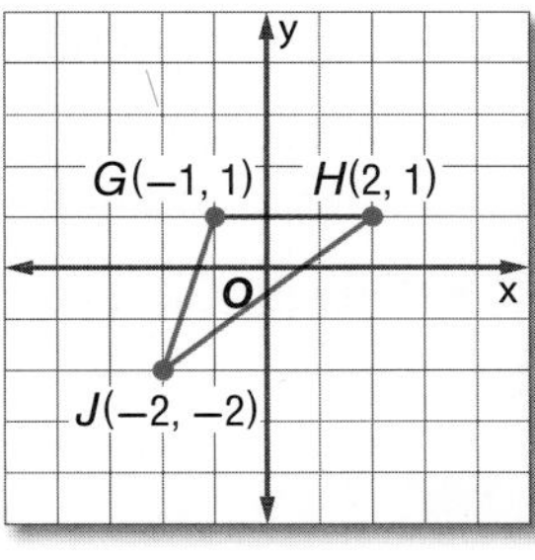

F 7.2 units

G 11.2 units

H 15.7 units

J 16.7 units

Find each missing measure.

23.

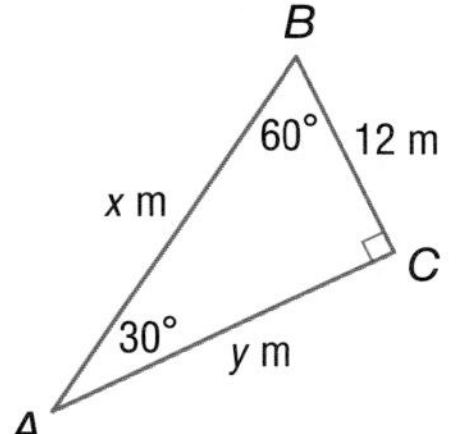

24.

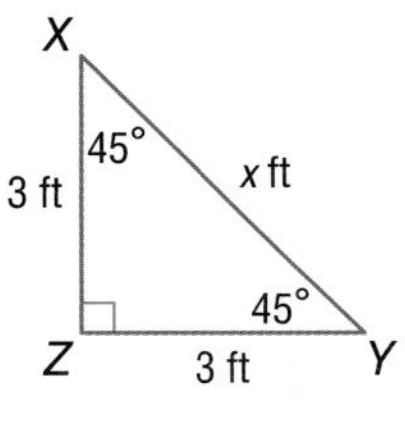

25. **CRAFTS** A square piece of fabric 8 inches on each side is folded in half diagonally. What is the length of the fold?

CHAPTER 10

Preparing for Standardized Tests

Use a Formula

A *formula* is an equation that shows a relationship among certain quantities. Many standardized test problems will require using a formula to solve them.

Strategies for Using a Formula

Formulas and Measures

Midpoint on a coordinate plane		$M = \left(\frac{x_1 + x_2}{2}, \frac{y_1 + y_2}{2}\right)$
Distance on a coordinate plane		$d = \sqrt{(x_2 - x_1)^2 + (y_2 - y_1)^2}$
Perimeter of a rectangle		$P = 2\ell + 2w$ or $P = 2(\ell + w)$
Circumference of a circle		$C = 2\pi r$ or $C = \pi d$
Area	rectangle	$A = \ell w$
	parallelogram	$A = bh$
	triangle	$A = \frac{1}{2}bh$
	trapezoid	$A = \frac{1}{2}h(b_1 + b_2)$
	circle	$A = \pi r^2$
Surface Area	cube	$S = 6s^2$
	prism	$S = Ph + 2B$
	cylinder	$S = 2\pi rh + 2\pi r^2$
	regular pyramid	$S = \frac{1}{2}P\ell + B$
	cone	$S = \pi r\ell + \pi r^2$
	cube	$V = s^3$

Step 1

Become familiar with common formulas and their uses. You may or may not be given access to a formula sheet to use during the test.

- **If you are given a formula sheet,** be sure to practice with the formulas before taking the test so you know how to apply them.
- **If you are not given a formula sheet,** study and practice with common formulas. Your teacher can provide you with the formulas you need to know. The list may include the perimeter, circumference, area, and volume formulas, the distance formula, the Pythagorean Theorem, the midpoint formula, and others.

Step 2

Choose the appropriate formula and solve.

- **Ask Yourself:** What quantities are given in the problem statement?
- **Ask Yourself:** Is there a formula I know that relates these quantities?
- **Solve:** Substitute known quantities into the formula and solve for the unknown quantity.
- **Check:** Check your answer if time permits.

EXAMPLE

Read the problem. Identify what you need to know. Then use the information in the problem to solve.

Carla is making a map of her hometown on a coordinate grid. She plots her school at (6, 2) and the zoo at (−4, −5) as shown. If each unit on the map represents 1 mile, what is the distance between the school and the zoo? Round to the nearest tenth.

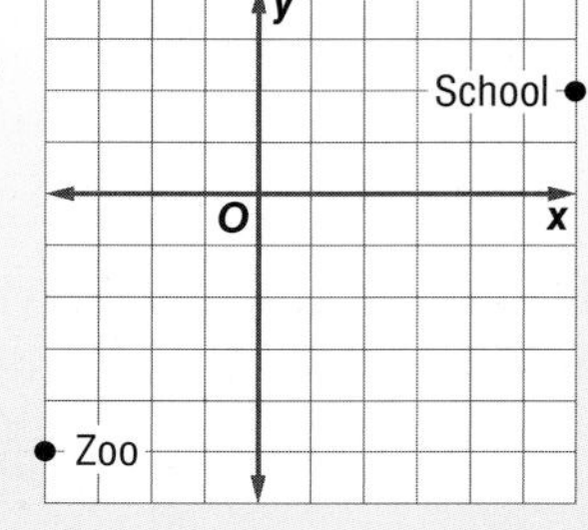

A 10.7 mi

B 11.5 mi

C 11.9 mi

D 12.2 mi

Read the problem carefully. You are given the coordinates of two points on a map and asked to find the distance between them. The Distance Formula relates the distance between two points on a coordinate grid.

Use the Distance Formula to find the distance between the school and the zoo.

$d = \sqrt{(x_2 - x_1)^2 + (y_2 - y_1)^2}$	**Write the Distance Formula.**
$d = \sqrt{((-4) - 6)^2 + ((-5) - 2)^2}$	**Replace x_1 with −4, 6 for x_2, −5 for y_1, and 2 for y_2.**
$d = \sqrt{(-10)^2 + (-7)^2}$	**Simplify.**
$d = \sqrt{100 + 49}$	**Simplify.**
$d = \sqrt{149} \approx 12.2$	**Use a calculator.**

Each unit on Carla's map represents 1 mile. So, the distance from the school to the zoo is about 12.2 miles.

The correct answer is D.

Exercises

Read each question. Identify what you need to know. Then use a formula to solve the questions.

1. Triangle *XYZ* has the vertices shown below. What is the perimeter of the triangle? Round to the nearest tenth.

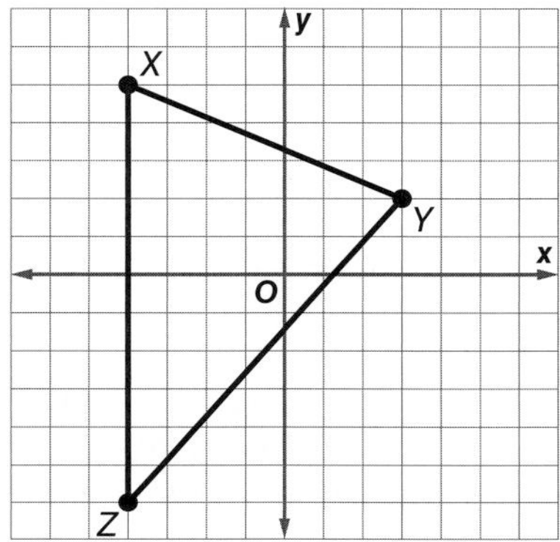

A 26.7 units

B 29.2 units

C 34.5 units

D 48.6 units

2. Alejandro is flying a kite on a breezy day. The kite is attached to the end of a piece of string that is 120 feet long. Using the diagram below, what is the current height of the kite? Round to the nearest foot.

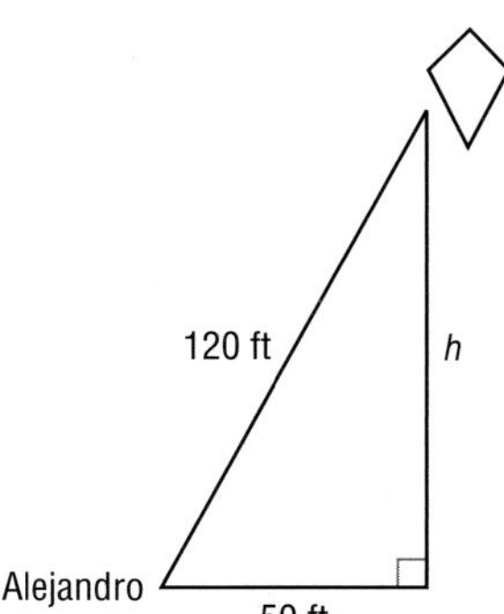

F 90 ft

G 98 ft

H 109 ft

J 130 ft

CHAPTER 10

Standardized Test Practice

Cumulative, Chapters 1 through 10

Multiple Choice

Read each question. Then fill in the correct answer on the answer document provided by your teacher or on a sheet of paper.

1. Between which two consecutive whole numbers on a number line does $\sqrt{52}$ lie?

A 7 and 8

B 8 and 9

C 9 and 10

D 10 and 11

2. Which of the following numbers is *not* rational?

F $-\frac{1}{40}$

G $40.\overline{40}$

H 40.1

J $\sqrt{40}$

3. What is the length of the diagonal of the rectangular picture frame below? Round to the nearest tenth.

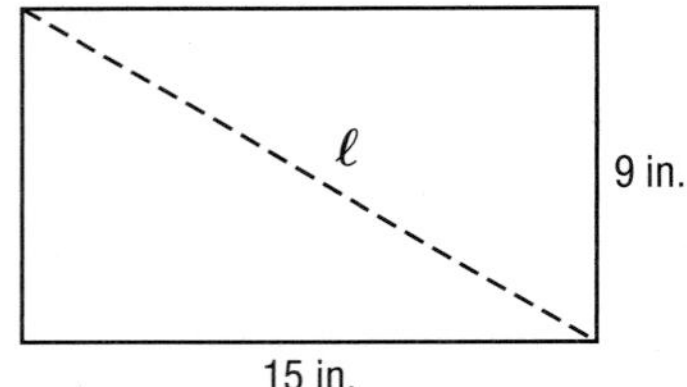

A 16.3 in.

B 16.7 in.

C 17.2 in.

D 17.5 in.

Test-TakingTip

Question 3 Many standardized tests provide a Formula Sheet. Find and use the Pythagorean Theorem if this sheet is available.

4. Tyrone earns a 6.5% commission on his weekly sales. How much commission will he earn for his sales for the week shown in the table? Round to the nearest cent.

Weekly Sales	
Day	**Sales**
Monday	\$1,520
Tuesday	\$1,945
Wednesday	\$0
Thursday	\$2,010
Friday	\$2,485
Saturday	\$0
Sunday	\$1,625

F \$574.14

G \$623.03

H \$655.49

J \$703.68

5. Solve for x in the triangle below.

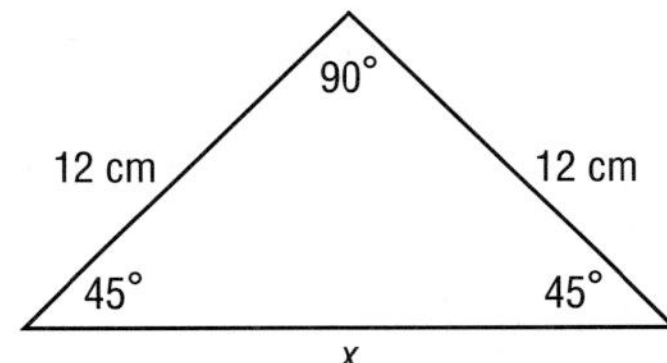

A $12\sqrt{3}$ cm

B $6\sqrt{3}$ cm

C $12\sqrt{2}$ cm

D $6\sqrt{2}$ cm

6. The average speed of the 2007 Tour de France winner was about 39 kilometers per hour. That is about the same rate as which of the following?

F 24 mph

G 45 mph

H 63 mph

J 71 mph

Short Response/Gridded Response

Record your answers on the answer sheet provided by your teacher or on a sheet of paper.

7. Classify the triangle by its angles and its sides.

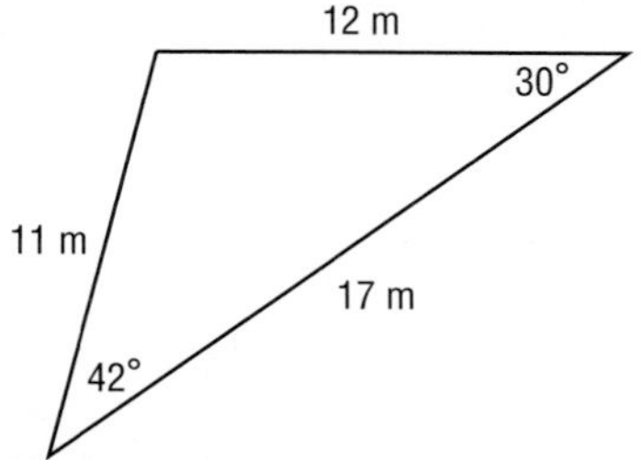

8. The table shows the approximate diameters of Mars and Neptune. About how many times as great is the diameter of Neptune than the diameter of Mars?

Planet	Approximate Diameter (mi)
Mars	2^{12}
Neptune	2^{15}

9. GRIDDED RESPONSE Reggie has a square tarp with an area of 196 square feet. What are the lengths of the sides of the tarp in feet?

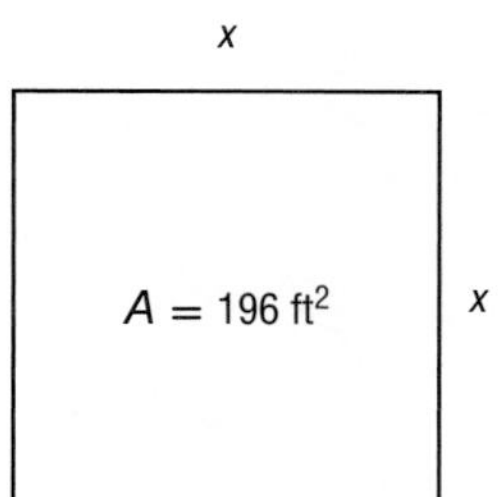

10. GRIDDED RESPONSE All of the coats in a clothing store were on sale for 30% off the original price. Carmen paid $41.30 for a coat after the discount. What was the original price of the coat in dollars?

Extended Response

Record your answers on a sheet of paper. Show your work.

11. Triangle *RST* has the vertices shown on the coordinate grid. Use the triangle to answer each question.

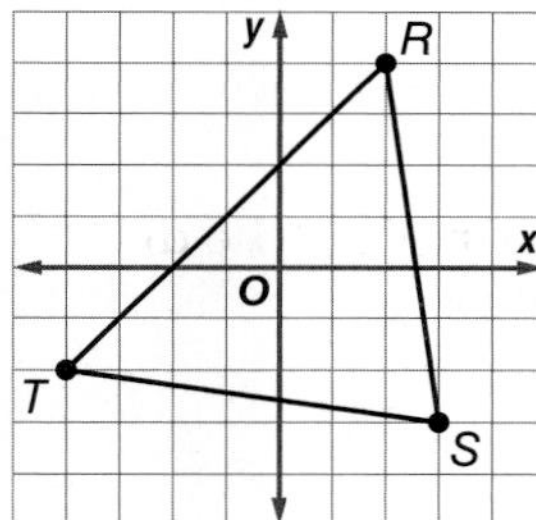

a. What are the lengths of $\overline{RS}$, $\overline{ST}$, and $\overline{RT}$? Round to the nearest hundredth.

b. Classify the triangle by its angles and sides.

c. What is the perimeter of the triangle? Round to the nearest tenth.

Need Extra Help?											
If you missed Question...	1	2	3	4	5	6	7	8	9	10	11
Go to Lesson or Page...	10-1	10-2	10-4	7-5	10-6	6-3	10-3	9-3	10-1	7-6	10-5

CHAPTER 11 Distance and Angle

Then

In Chapter 10 you worked with angles, triangles, and the Pythagorean Theorem.

Now

In Chapter 11, you will:

- Identify the relationship of parallel and intersecting lines.
- Identify properties of congruent triangles.
- Find the area of polygons, irregular figures, and circles.

Why?

NATURE There are about 24,000 different species of butterflies in the world. They range in size from an $\frac{1}{8}$ inch up to 12 inches. Did you know that butterflies taste with their feet? Their taste sensors are located in the feet. By standing on their food, they can taste it!

Distance and Angle
Activity
∠M and ∠P are corresponding angles.
Use the protractor to measure ∠M. Place your pointer anywhere on the protractor to drag it. Use the red button to rotate the protractor.
m∠M = °
Check Answer
M
B
P
Q
5/6

Math *in Motion,* Animation glencoe.com

Get Ready for Chapter 11

Diagnose Readiness You have two options for checking Prerequisite Skills.

Text Option Take the Quick Check below. Refer to the Quick Review for help.

QuickCheck

Solve each equation. (Lesson 5-5)

1. $x + 25 = -67$
2. $x + 18 = 106$
3. $3x - 22 = 83$
4. $2x - 16 = -128$
5. $4x + 140 = 420$
6. $5x + 160 = 220$
7. **RUNNING** Darius ran the same number of miles each day from Monday through Friday, and 2 miles on Saturday. If he ran a total of 27 miles during the week, how many miles did he run each weekday?

Find each product. Round to the nearest tenth, if necessary. (Lesson 0-3)

8. $2.7(5)$
9. $9.2(1.8)$
10. $2.5(16.7)$
11. $\frac{1}{2}(6)(3.2)$
12. $3(7.17)(5.1)$
13. $2(1.7)(2.62)$
14. **FOOD** Nicole poured 32 bowls of soup volunteering in a soup kitchen. If each bowl contained 16.5 ounces of soup, how much soup did Nicole pour?

Find each sum. (Lesson 3-6)

15. $3\frac{2}{3} + 5\frac{4}{5}$
16. $6\frac{3}{8} + 2\frac{1}{2}$
17. $3\frac{1}{3} + 2\frac{3}{4}$
18. $3\frac{2}{3} + 2\frac{5}{9}$
19. $5\frac{1}{4} + 2\frac{5}{6}$
20. $4\frac{1}{2} + 2\frac{2}{3}$
21. **RECYCLING** The class collected $12\frac{5}{6}$ pounds of bottles and $8\frac{1}{8}$ pounds of aluminum cans. How many pounds of bottles and aluminum cans did the class collect?

QuickReview

EXAMPLE 1

Solve $7x - 2 = -72$.

$7x - 2 = -72$	Write the equation.
$7x - 2 + 2 = -72 + 2$	Add 2 to each side.
$7x = -70$	Simplify.
$\frac{7x}{7} = \frac{-70}{7}$	Divide each side by 7.
$x = -10$	

EXAMPLE 2

Find 0.5(3)(6.25). Round to the nearest tenth.

$0.5(3)(6.25)$

$= [0.5(3)](6.25)$	Order of operations
$= (1.5)(6.25)$	$0.5 \times 3 = 1.5$
$= 9.375$	Multiply.
≈ 9.4	Round to the nearest tenth.

EXAMPLE 3

Find $1\frac{3}{4} + 4\frac{5}{6}$.

$1\frac{3}{4} + 4\frac{5}{6} = \frac{7}{4} + \frac{29}{6}$	Write the mixed numbers as improper fractions.
$= \frac{7}{4} \cdot \frac{3}{3} + \frac{29}{6} \cdot \frac{2}{2}$	Rename each fraction using the LCD.
$= \frac{21}{12} + \frac{58}{12}$	Simplify.
$= \frac{79}{12}$ or $6\frac{7}{12}$	Add the numerators.

Online Option Math Online Take a self-check Chapter Readiness Quiz at glencoe.com.

Get Started on Chapter 11

You will learn several new concepts, skills, and vocabulary terms as you study Chapter 11. To get ready, identify important terms and organize your resources. You may wish to refer to **Chapter 0** to review prerequisite skills.

FOLDABLES® Study Organizer

Distance and Angle Make this Foldable to help you study the topics of distance and angle. Begin with a piece of notebook paper.

1. **Fold** the construction paper in half lengthwise. Label the chapter title on the outside.

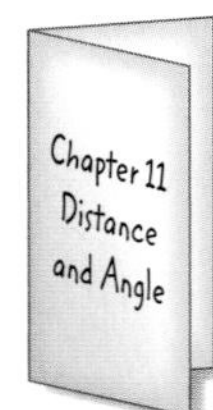

2. **Fold** the sheets of notebook paper in half lengthwise. Then fold top to bottom twice.

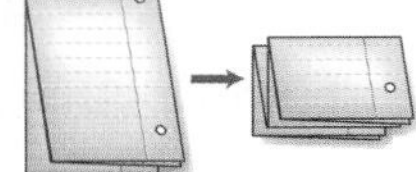

3. **Open** the notebook paper. Cut along the second folds to make four tabs.

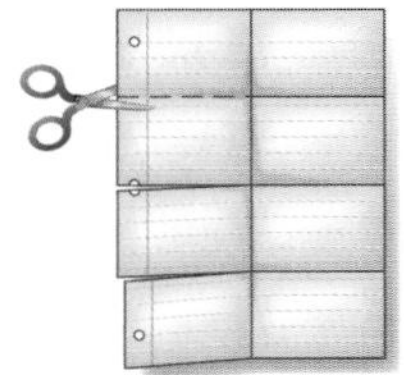

4. **Glue** the uncut notebook paper side by side onto the construction paper. Label each tab as shown.

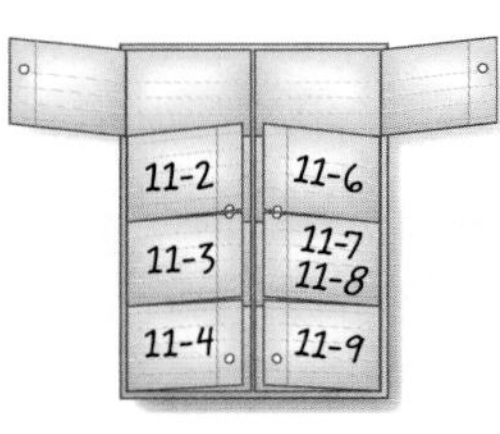

Math Online glencoe.com

- Study the chapter online
- Explore **Math in Motion**
- Get extra help from your own **Personal Tutor**
- Use **Extra Examples** for additional help
- Take a **Self-Check Quiz**
- **Review Vocabulary** in fun ways

New Vocabulary

English		Español
adjacent angles	• p. 589 •	ángulos adyacentes
complementary angles	• p. 589 •	ángulos complementarios
supplementary angles	• p. 589 •	ángulos supplementarios
perpendicular lines	• p. 589 •	rectas perpendiculars
parallel lines	• p. 590 •	rectas paralelas
transversal	• p. 590 •	transversal
congruent	• p. 598 •	congruentes
quadrilateral	• p. 612 •	cuadrilateral
polygon	• p. 617 •	polígono
interior angle	• p. 618 •	ángulo interiors
radius	• p. 631 •	radio
diameter	• p. 631 •	diámetro
circumference	• p. 631 •	circunferencia
π (pi)	• p. 631 •	pi

Review Vocabulary

exponents • p. 471 • exponente in a power, the exponent is the number of times the base is used as a factor

$5^3 \leftarrow$ **exponent**

ray • rayo a ray extends indefinitely in one direction

Multilingual eGlossary glencoe.com

11-1

Angle and Line Relationships

Why?

Then

You have already found the missing angle measure of a triangle. (Lesson 10-3)

Now

- Examine relationships between pairs of angles.
- Examine relationships of angles formed by parallel lines and a transversal.

New Vocabulary

perpendicular lines
vertical angles
adjacent angles
complementary angles
supplementary angles
parallel lines
transversal
alternate interior angles
alternate exterior angles
corresponding angles

Math Online

glencoe.com

- Extra Examples
- Personal Tutor
- Self-Check Quiz
- Homework Help

Step 1 Draw two different pairs of intersecting lines and label the angles formed as shown.

Step 2 Find and record the measure of each angle.

Step 3 Color angles that have the same measure.

a. For each set of intersecting lines, identify the pairs of angles that have the same measure.

b. What is true about the sum of the measures sharing a side?

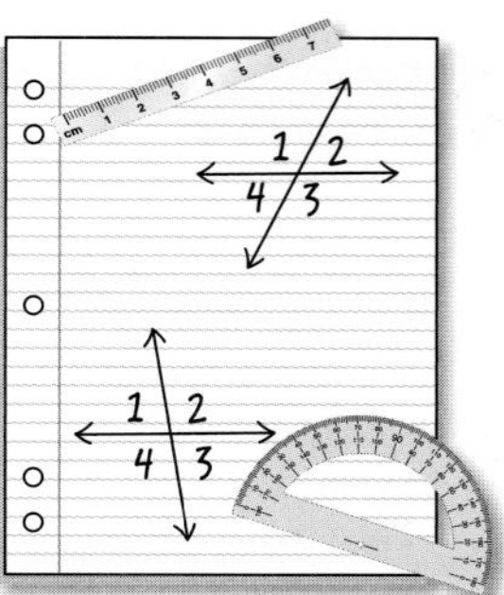

1 2
4 3
1 2
4 3

Angle Relationships Pairs of angles can be classified by their relationship to each other.

Key Concept Pairs of Angles — For Your FOLDABLE

Words	Models	Symbols
When two lines intersect, they form two pairs of opposite angles, called **vertical angles**. Vertical angles are congruent.	$\angle 1$ and $\angle 2$, $\angle 3$ and $\angle 4$	$\angle 1 \cong \angle 2$ $\angle 3 \cong \angle 4$
Two angles that have the same vertex between them, share a common side, and do not overlap are called **adjacent angles**.	A, B, C, 5, 6	$m\angle ABC = m\angle 5 + m\angle 6$
If the sum of the measures of two angles is 90°, the angles are called **complementary angles**.	7, 8	$m\angle 7 + m\angle 8 = 90°$
If the sum of the measures of two angles is 180°, the angles are called **supplementary angles**.	9, 10	$m\angle 9 + m\angle 10 = 180°$

A special case occurs when two lines intersect to form a right angle. These lines are **perpendicular lines**.

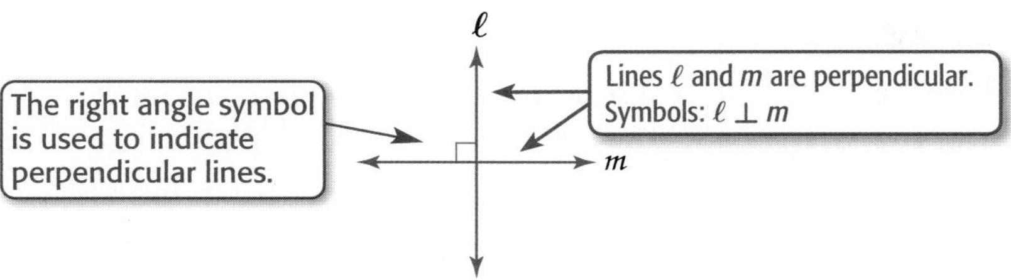

The right angle symbol is used to indicate perpendicular lines.
Lines ℓ and m are perpendicular. Symbols: ℓ ⊥ m
ℓ
m

Real-World EXAMPLE 1 Find a Missing Angle Measure

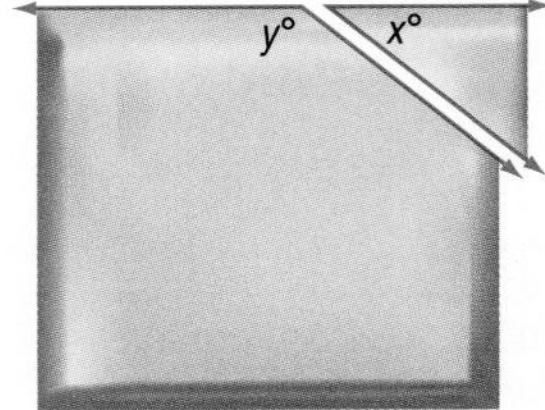

y°
x°

TILING Jun is cutting a piece of tile in the shape of a rectangle.

a. Classify the relationship between $\angle x$ and $\angle y$.

The angles are supplementary. The sum of their measures is 180°.

b. If $m\angle y = 135°$, what is the measure of $\angle x$?

$m\angle x + 135 = 180$ **Write the equation.**

$m\angle x + 135 - 135 = 180 - 135$ **Subtract 135 from each side.**

$m\angle x = 45$ **Simplify.**

So, $m\angle x = 45°$.

> **StudyTip**
>
> **Angles** Complementary angles and supplementary angles can either be adjacent angles or separate angles.

Check Your Progress

1. Angles R and S are complementary. If $m\angle R = 65.3°$, what is the measure of $\angle S$?

Personal Tutor glencoe.com

Parallel Lines Two lines in a plane that never intersect are called **parallel lines**. A line that intersects two or more other lines in a plane is called a **transversal**.

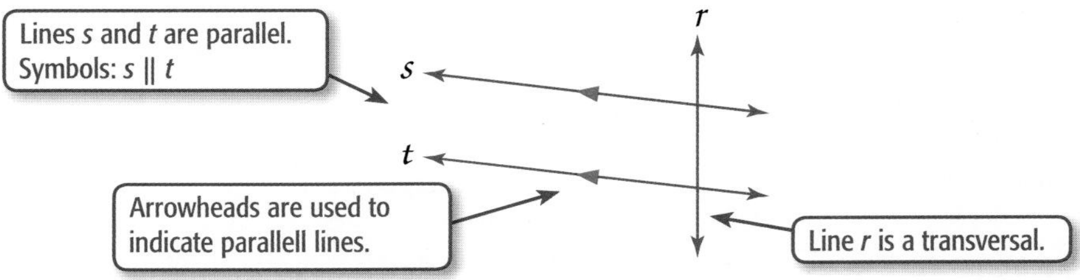

Lines s and t are parallel.
Symbols: s || t
r
s
t
Arrowheads are used to indicate parallell lines.
Line r is a transversal.

When parallel lines are cut by a transversal, special pairs of angles are congruent.

> **ReadingMath**
>
> **Symbols** The symbol for *is congruent to* is ≅. So, *angle 3 is congruent to angle 5* is written as $\angle 3 \cong \angle 5$.

Key Concept Special Angle Relationships

For Your FOLDABLE

When a transversal intersects two parallel lines, eight angles are formed.

- *Interior angles* lie inside the parallel lines. $\angle 3$, $\angle 4$, $\angle 5$, $\angle 6$
- *Exterior angles* lie outside the parallel lines. $\angle 1$, $\angle 2$, $\angle 7$, $\angle 8$

The following pairs of angles are congruent.

- **Alternate interior angles** are on opposite sides of the transversal and inside the parallel lines. $\angle 3 \cong \angle 5$, $\angle 4 \cong \angle 6$
- **Alternate exterior angles** are on opposite sides of the transversal and outside the parallel lines. $\angle 1 \cong \angle 7$, $\angle 2 \cong \angle 8$
- **Corresponding angles** are in the same position on the parallel lines in relation to the transversal. $\angle 1 \cong \angle 5$, $\angle 2 \cong \angle 6$, $\angle 3 \cong \angle 7$, $\angle 4 \cong \angle 8$

ReadingMath

Symbols The symbol $\parallel$ means *is parallel to*.

EXAMPLE 2 Find Measures of Angles Formed by Parallel Lines

In the figure at the right, $a \parallel b$ and q and r are transversals.

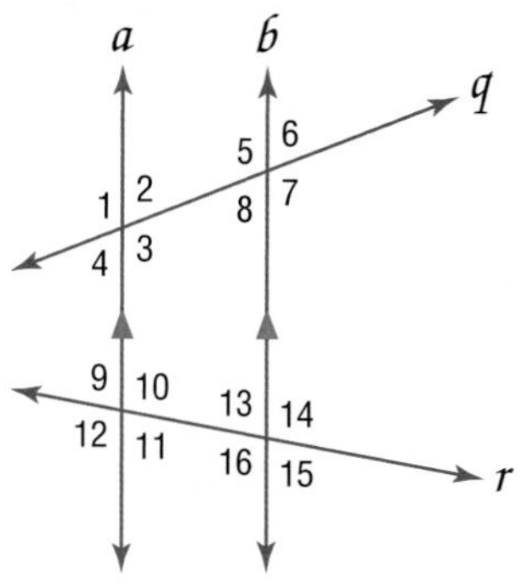

a. Classify the relationship between ∠3 and ∠5.

Since ∠3 and ∠5 are alternate interior angles, they are congruent.

b. If $m\angle 1 = 120°$, find $m\angle 5$ and $m\angle 3$.

Since ∠1 and ∠5 are corresponding angles, they are congruent. So, $m\angle 5 = 120°$.

Since $m\angle 5$ and $m\angle 3$ are congruent, $m\angle 3 = 120°$.

Check Your Progress

2. Classify the relationship between ∠11 and ∠15. If $m\angle 15 = 78.5°$, find $m\angle 11$ and $m\angle 9$.

Personal Tutor glencoe.com

Vocabulary Review

Ray A ray is a part of a line that extends indefinitely in one direction. In the figure at the right, $\overrightarrow{BA}$, $\overrightarrow{BC}$, and $\overrightarrow{BD}$ are rays.

EXAMPLE 3 Use Algebra to Find Missing Angle Measures

ALGEBRA In the figure at the right, $m\angle ABD = 164°$. Find the measures of ∠ABC and ∠CBD.

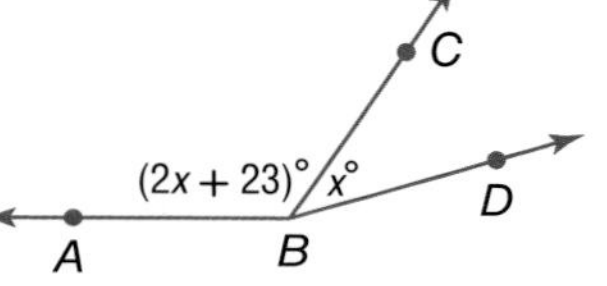

Angles *ABC* and *CBD* are adjacent angles that have a total measure of 164°.

Step 1 Find the value of *x*.

$m\angle ABC + m\angle CBD = 164$	**Adjacent angles**
$(2x + 23) + x = 164$	**Replace $m\angle ABC$ with $2x + 23$ and replace $m\angle CBD$ with x.**
$3x + 23 = 164$	**Combine like terms.**
$-23 = -23$	**Subtract 23 from each side.**
$3x = 141$	**Simplify.**
$\frac{3x}{3} = \frac{141}{3}$	**Divide each side by 3.**
$x = 47$	**Simplify.**

Step 2 Replace *x* with 47 to find the measure of each angle.

$m\angle ABC = 2x + 23$
$= 2(47) + 23$ or 117

$m\angle CBD = x$
$= 47$

So, $m\angle ABC = 117°$ and $m\angle CBD = 47°$.

Check Your Progress

ALGEBRA Angles *RVT* and *UVW* are vertical angles, with $m\angle RVT = 3x$ and $m\angle UVW = 5x - 36$.

3A. Find the value of *x*.

3B. Find $m\angle UVW$.

Personal Tutor glencoe.com

Check Your Understanding

Example 1 p. 590 — **Classify the pairs of angles shown. Then find the value of x in each figure.**

1.

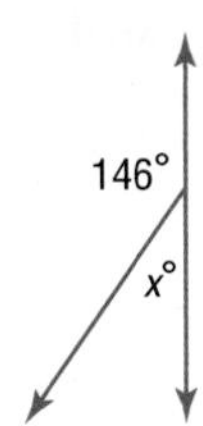

2.

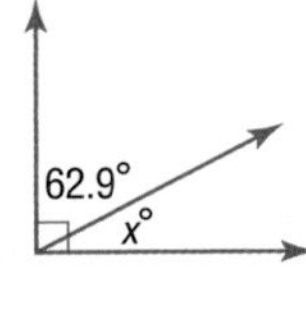

3.

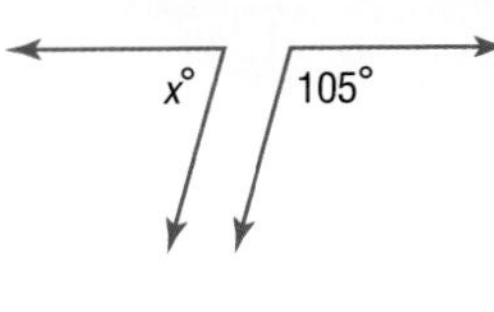

Example 2 p. 591 — **In the figure at the right, $r \parallel s$ and w is a transversal. If $m\angle 1 = 128°$, find the measure of each angle. Explain your reasoning.**

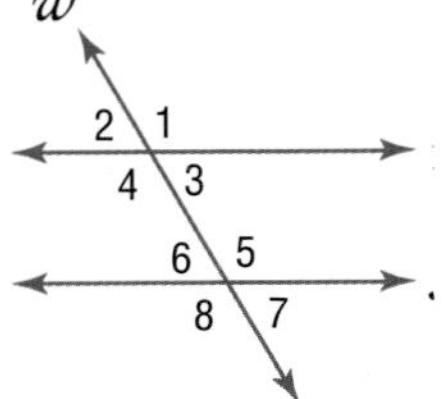

4. $\angle 5$ **5.** $\angle 8$ **6.** $\angle 2$

7. **GYMNASTICS** The balance beam shown below is parallel to the floor.

a. Classify the relationship between $\angle x$ and $\angle y$.

b. If $m\angle y = 117°$, find the value of x.

Example 3 p. 591 — **ALGEBRA The measure of $\angle Q$ is $6x + 16.8$ and the measure of $\angle R$ is $2x$.**

8. If $\angle Q$ and $\angle R$ are supplementary, what is the value of x? What is the measure of each angle?

9. If $\angle Q$ and $\angle R$ are complementary, what is the value of x? What is the measure of each angle?

Practice and Problem Solving

= **Step-by-Step Solutions** begin on page R11. **Extra Practice** begins on page 810.

Example 1 p. 590 — **Classify the pairs of angles shown. Then find the value of x in each figure.**

10.

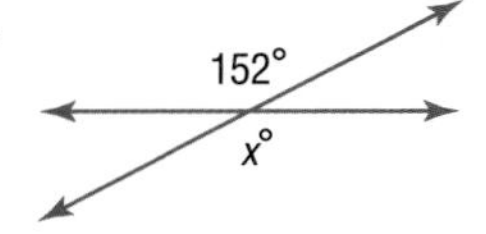

11.

58.9°
x°

12.

67.4°
x°

13. **ARCHITECTURE** Look at the semicircular window.

a. Classify the relationship between $\angle 1$ and $\angle 2$.

b. If $m\angle 2$ is 24°, find $m\angle 1$.

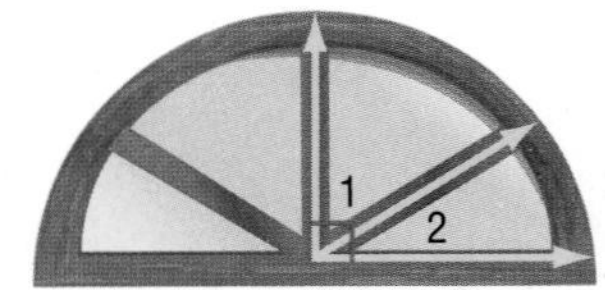

Example 2 p. 591

In the figure at the right, $f \parallel g$ and t is a transversal. If $m\angle 6 = 52.6°$, find the measure of each angle. Explain your reasoning.

14. $\angle 3$

15. $\angle 5$

16. $\angle 2$

17. $\angle 8$

18. $\angle 4$

19. $\angle 1$

Example 3 p. 591

20. If $m\angle FGH = 165°$, find $m\angle 1$.

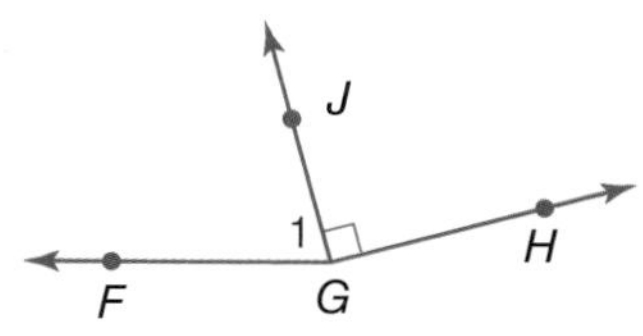

21. If $m\angle QRS = 158°$, find $m\angle TRS$.

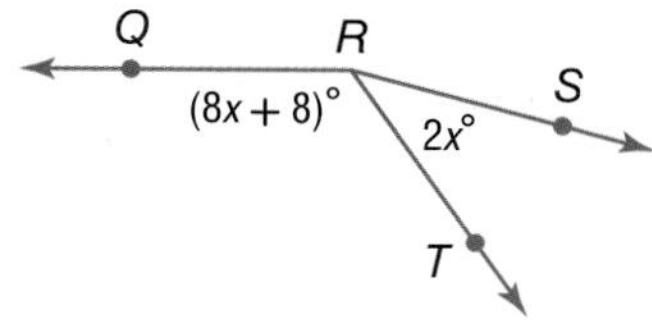

22. ALGEBRA Angles M and P are complementary. If $m\angle M = 3x - 45$ and $m\angle P = 2x + 15$, what is the value of x? What is the measure of each angle?

23. ALGEBRA The measure of $\angle C$ is $4x - 24.6$ and the measure of $\angle D$ is $x + 11.3$. If the angles are supplementary, find the value of x and $m\angle D$.

ALGEBRA In the figure below, $a \parallel b$ and z is a transversal.

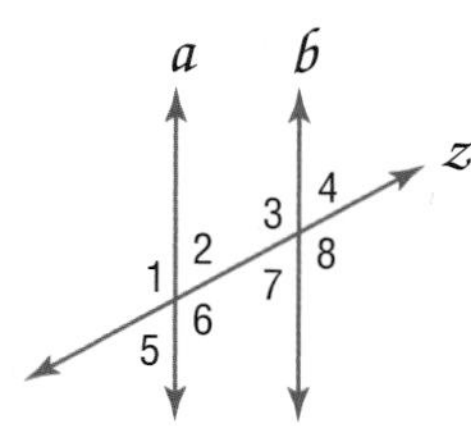

24. If $m\angle 1 = 5x$ and $m\angle 5 = 2x - 2$, find the value of x, $m\angle 1$, and $m\angle 5$.

25. If $m\angle 4 = 15x$ and $m\angle 2 = 11x + 20$, find the value of x, $m\angle 4$, and $m\angle 2$.

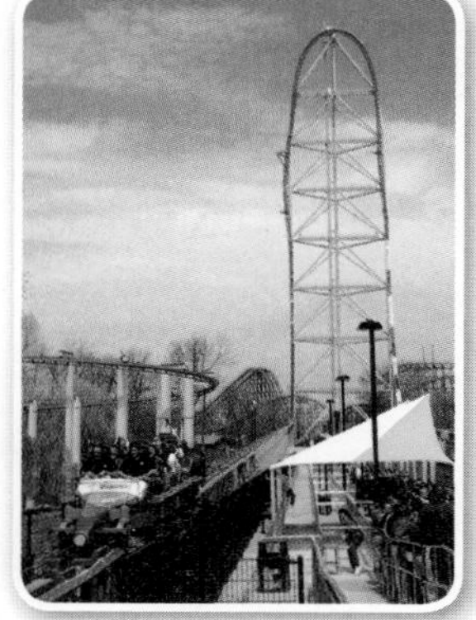

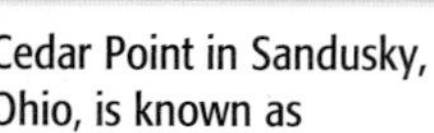

Cedar Point in Sandusky, Ohio, is known as "America's Roller Coast" because it has more roller coasters than any other amusement park.

Source: Cedar Point

26. ROLLER COASTERS In the roller coaster tower at the right, the measure of $\angle 1$ is 6° less than twice the measure of $\angle 2$. Find the measures of angles 1 and 2.

27. ALGEBRA The sum of the measures of $\angle A$, $\angle B$, and $\angle C$ is 180°, and $m\angle B = m\angle C$. If $m\angle A = 110°$, find $m\angle B$ and $m\angle C$.

28. ALGEBRA Angles U and V are supplementary angles. The ratio of their measures is 7:13. Find the measure of each angle.

29. ALGEBRA Use the pair of angles shown.

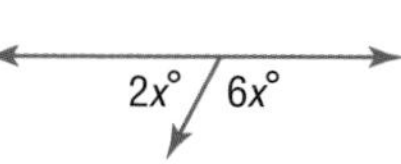

a. Write an equation to find the value of x. Then find the value of x.

b. What are the measures of the two angles?

Real-World Link

Every second lightning strikes somewhere on the surface of Earth about 100 times.

30. **GEOMETRY** Lines p, q, r, and s form the quadrilateral shown at the right. Can you conclude that opposite sides of the quadrilateral are parallel? Explain your reasoning.

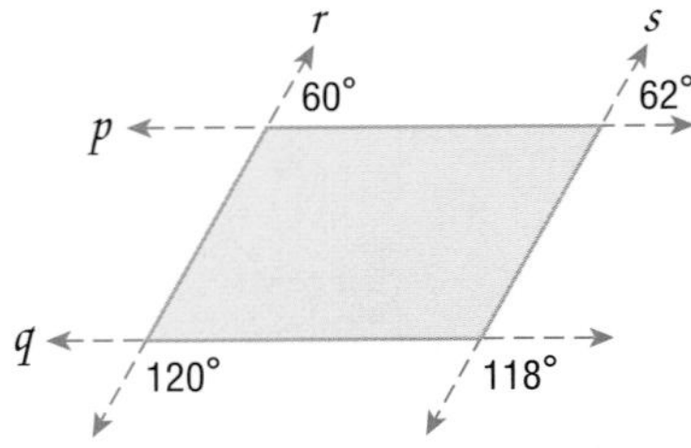

31. **TIME** Use the clock at the right that shows 6 o'clock and 10 seconds.

a. Find $m\angle WXY$ and $m\angle YXZ$.

b. Find the time that will show when $m\angle WXY = m\angle YXZ = 90°$.

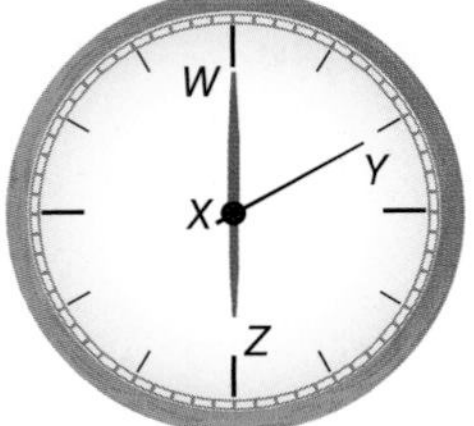

32. **MULTIPLE REPRESENTATIONS** In this problem, you will investigate parallel lines on the coordinate plane. Line f passes through points at (0, 2) and (2, 3). Line g passes through points at (0, −3) and (2, −2). Line h passes through points at (1, 0) and (2, −2).

a. **GRAPHICAL** Graph the three lines on the same coordinate plane. Label each line.

b. **VERBAL** Describe the angles that are formed by the lines.

c. **ANALYTICAL** Describe the relationship between the slopes of parallel lines.

H.O.T. Problems

Use Higher-Order Thinking Skills

33. **OPEN ENDED** Draw a pair of complementary adjacent angles. Label the measures of the angles.

34. **FIND THE ERROR** Elias and Taylor calculated the value of x for the missing angle shown at the right. Is either of them correct? Explain your reasoning.

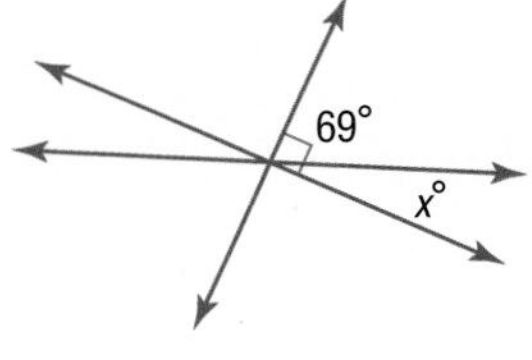

Elias	Taylor
$69 + x = 180$	$69 + x = 90$
$x = 111$	$x = 21$

35. **CHALLENGE** Lines ℓ and m shown at the right are parallel and are cut by transversals j and k. Describe how the following pairs of angles are related: 1 and 2, 3 and 4, 5 and 6. Then make a conjecture about how the interior angles on the same side of a transversal are related.

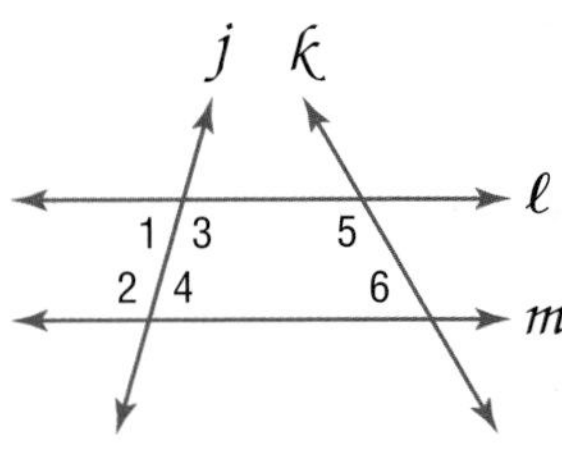

36. **WRITING IN MATH** Suppose two parallel lines are cut by a transversal and an exterior angle measures 90°. What can you conclude about the measures of the other seven angles that are formed? Explain your reasoning.

Standardized Test Practice

37. Lines a and b are parallel in the figure below. Find the value of x.

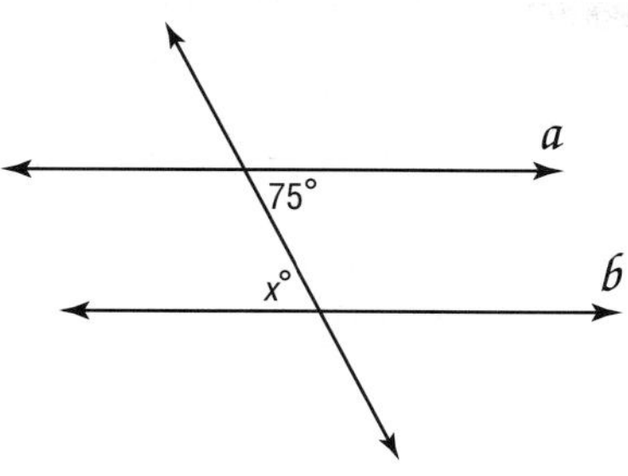

A 110
B 105
C 75
D 15

38. Which angles are *not* supplementary?

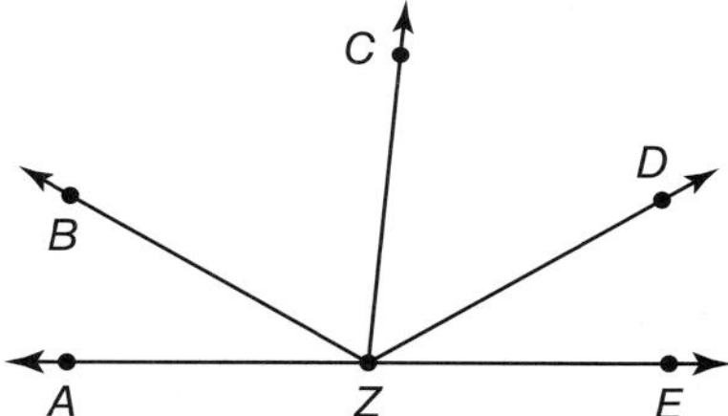

F $\angle EZC$ and $\angle CZA$
G $\angle BZC$ and $\angle CZD$
H $\angle AZB$ and $\angle BZE$
J $\angle DZE$ and $\angle AZD$

39. In the figure below, the two angles are congruent. Find the value of x.

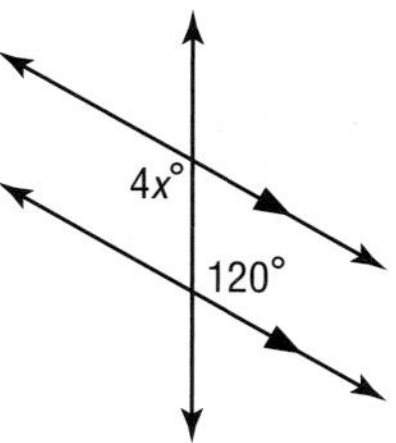

A 30
B 40
C 116
D 124

40. EXTENDED RESPONSE The hedge shears have two different angles shown. Find the value of x. Explain your reasoning.

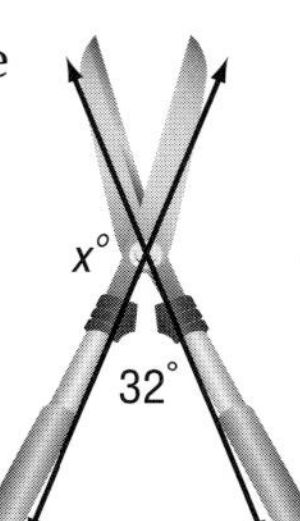

Spiral Review

41. GEOMETRY In a 30°-60°-90° triangle, the shorter leg is 12 feet long. Find the length of the hypotenuse and the length of the longer leg. (Lesson 10-6)

42. ARCHAEOLOGY Two artifacts are found at a dig. A coordinate plane was set up. One artifact was found at (1, 5), and the other artifact was found at (3, 1). How far apart were the two artifacts? Round to the nearest tenth of a unit if necessary. (Lesson 10-5)

43. ANIMALS In 2000, there were 356 endangered species in the U.S. Five years later, 389 species were considered endangered. What was the percent of change? Then state whether the percent of change is an *increase* or *decrease*. Round to the nearest tenth. (Lesson 7-6)

ALGEBRA Solve each inequality. (Lesson 5-4)

44. $5m < 5$

45. $\frac{a}{-2} > 3$

46. $-4x \geq -16$

Skills Review

Use a protractor to draw an angle having each measure. (Previous Course)

47. 35°

48. 65°

49. 85°

50. 155°

51. 180°

EXPLORE **11-2**

Geometry Lab: Investigating Congruent Triangles

Math Online glencoe.com
Math *in Motion*, Animation

In this lab, you will investigate whether it is possible to show that two triangles are congruent without showing that all six pairs of corresponding parts are congruent.

ACTIVITY 1

Step 1 Draw a triangle on a piece of patty paper. Copy the sides of the triangle onto another piece of patty paper and cut them out.

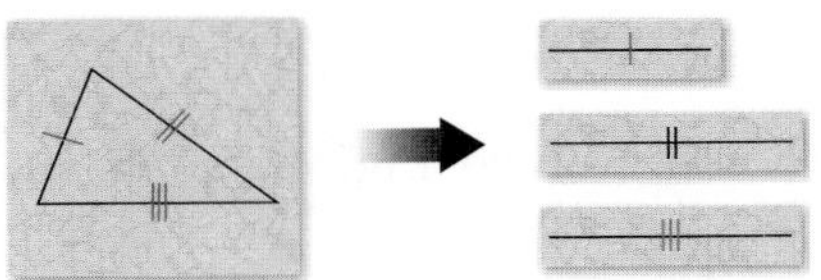

Step 2 Arrange and tape the pieces together so that they form a triangle.

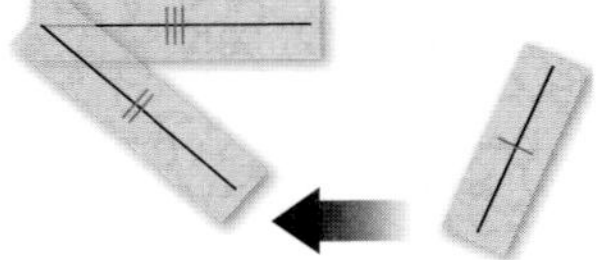

Analyze the Results

1. Is the triangle you formed congruent to the original triangle? Explain.
2. Repeat Activity 1 and try to form another triangle. Is it congruent to the original triangle?
3. **MAKE A CONJECTURE** Based on this activity, can three pairs of congruent sides be used to show that two triangles are congruent?

ACTIVITY 2

Step 1 Draw a triangle on a piece of patty paper. Copy each angle of the triangle onto separate pieces of patty paper. Extend each ray of each angle to the edge of the patty paper.

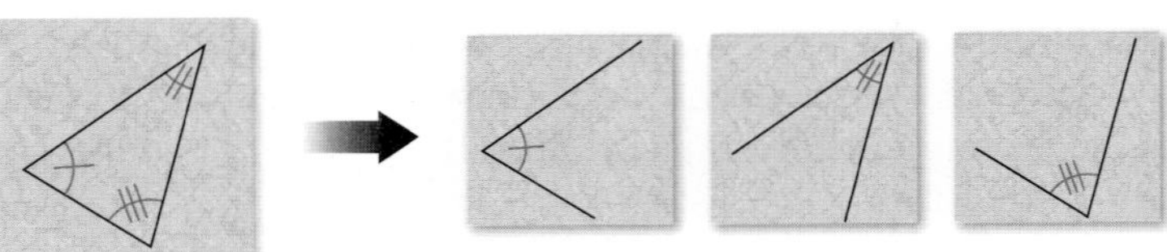

Step 2 Arrange and tape the pieces together so that they form a triangle.

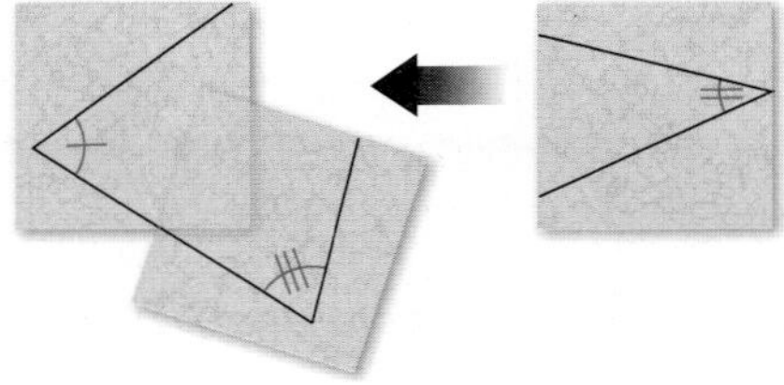

Analyze the Results

4. Is the triangle you formed congruent to the original triangle? Explain.
5. Repeat Activity 2 and try to form another triangle. Is it congruent to the original triangle?
6. **MAKE A CONJECTURE** Based on this activity, can three pairs of congruent angles be used to show that two triangles are congruent?

ACTIVITY 3

Step 1 Draw a triangle on a piece of patty paper. Copy two sides of the triangle and the angle between them onto separate pieces of patty paper and cut them out.

Step 2 Arrange and tape the pieces together so that the two sides are joined to form the rays of the angle. Then tape these joined pieces onto a piece of construction paper and connect the two rays to form a triangle.

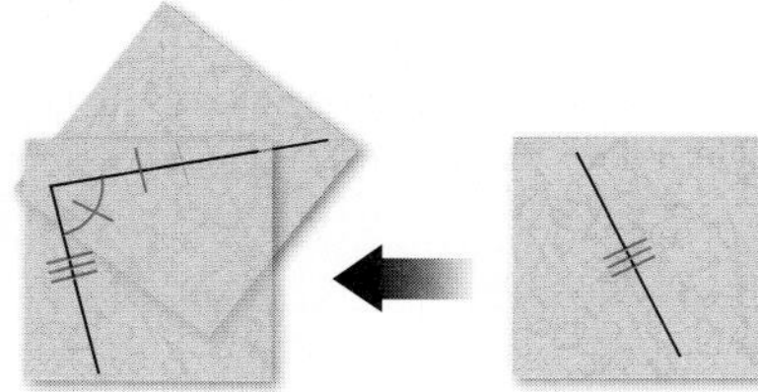

Analyze the Results

7. Is the triangle you formed congruent to the original triangle? Explain.
8. Repeat Activity 3 and try to form another triangle. Is it congruent to the original triangle?
9. **MAKE A CONJECTURE** Based on this activity, can two pairs of congruent sides and the pair of congruent angles between them be used to show that two triangles are congruent?
10. **EXTENSION** Use patty paper to investigate and determine whether each of the following can be used to show that two triangles are congruent.
    - two pairs of congruent sides and a pair of congruent angles *not* between them
    - two pairs of congruent angles and the pair of congruent sides between them
    - two pairs of congruent angles and a pair of congruent sides *not* between them

11-2 Congruent Triangles

Then
You identified triangles with congruent sides. (Lesson 9-3)

Now
- Identify corresponding parts of congruent triangles.
- Identify congruent triangles.

New Vocabulary
congruent
corresponding parts

Math Online
glencoe.com
- Extra Examples
- Personal Tutor
- Self-Check Quiz
- Homework Help

Why?

The flag of the United Kingdom is shown below. Consider the four large triangles appearing on the top and the bottom of the flag.

a. Measure the sides of the four triangles. What is true about the lengths?

b. Use a protractor to measure the angles of the four triangles. What is true about the measures?

c. **MAKE A CONJECTURE** Suppose the triangles were cut out and laid on top of one another so that the parts with the same measures were matched up. What is true about the triangles?

Corresponding Parts of Congruent Triangles Figures that have the same size and shape are **congruent**. In the figure below, triangle *ABC* is congruent to triangle *DEF*.

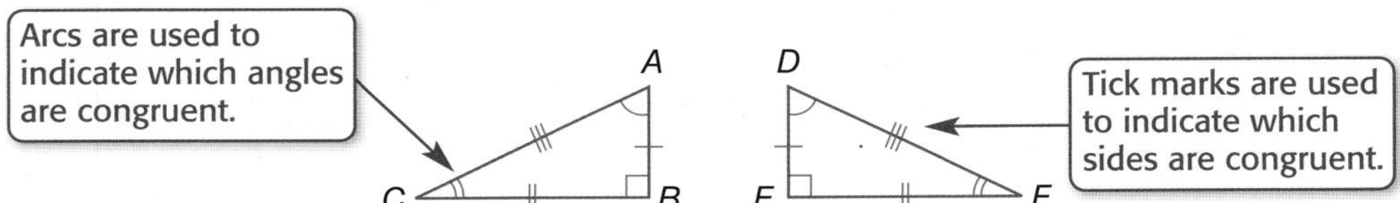

The parts of congruent triangles that *match* or correspond, are called **corresponding parts**.

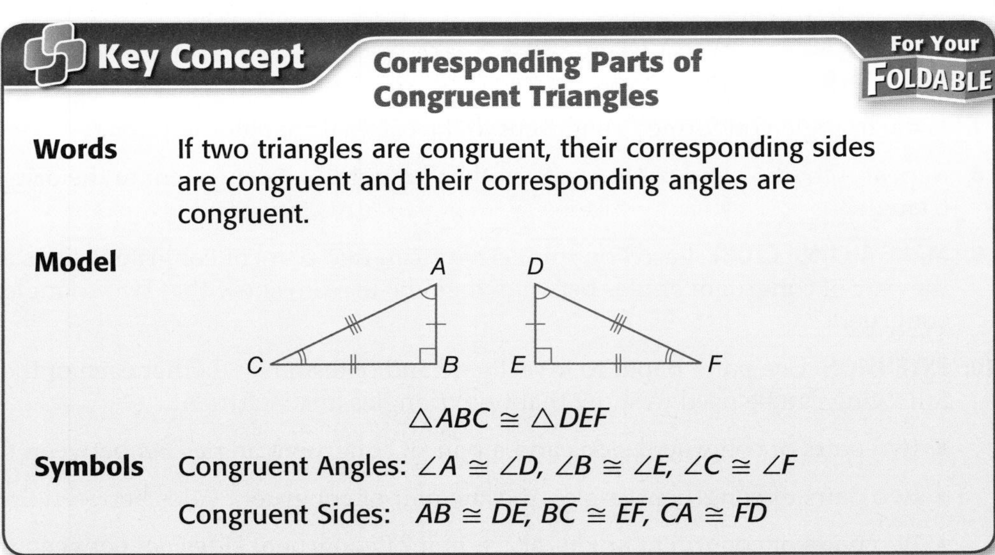

Key Concept
Corresponding Parts of Congruent Triangles
For Your FOLDABLE
Words
If two triangles are congruent, their corresponding sides are congruent and their corresponding angles are congruent.
Model
A
B
C
D
E
F
△ABC ≅ △DEF
Symbols
Congruent Angles: ∠A ≅ ∠D, ∠B ≅ ∠E, ∠C ≅ ∠F
Congruent Sides: AB ≅ DE, BC ≅ EF, CA ≅ FD

StudyTip

Congruence Statements You can write the following congruence statements for the congruent triangles at the right.

$\triangle RSQ \cong \triangle UVT$
$\triangle SQR \cong \triangle VTU$
$\triangle SRQ \cong \triangle VUT$
$\triangle RQS \cong \triangle UTV$
$\triangle QSR \cong \triangle TVU$
$\triangle QRS \cong \triangle TUV$

In a *congruence statement*, the letters are written so that corresponding vertices appear in the same order.

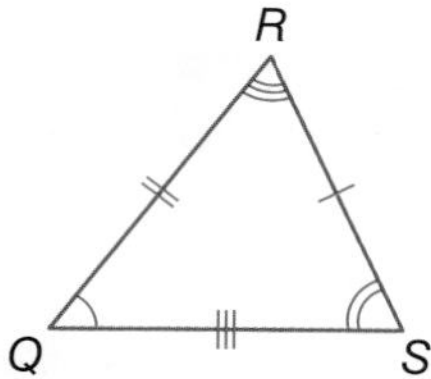

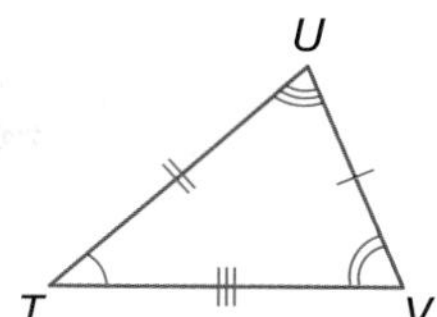

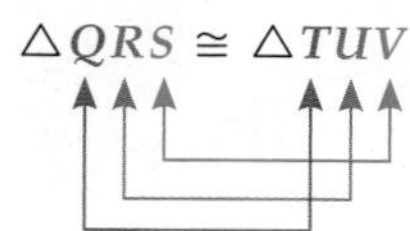

Vertex *Q* corresponds to vertex *T*.
Vertex *R* corresponds to vertex *U*.
Vertex *S* corresponds to vertex *V*.

EXAMPLE 1 Name Corresponding Parts

a. Name the corresponding parts in the congruent triangles below. Then complete the congruence statement $\triangle JKL \cong \triangle$ __?__.

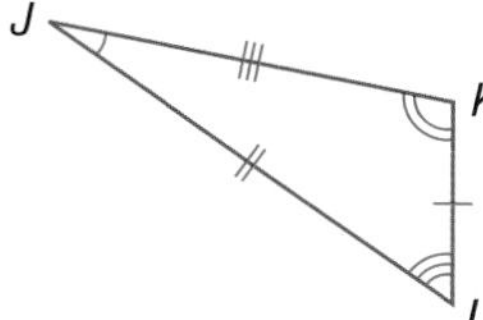

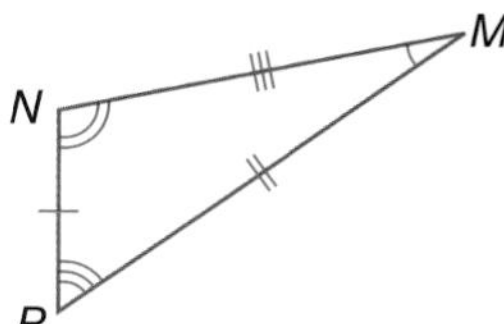

Use the matching arcs and tick marks to identify the corresponding parts.

Corresponding angles: $\angle J \cong \angle M$, $\angle K \cong \angle N$, $\angle L \cong \angle P$

Corresponding sides: $\overline{JK} \cong \overline{MN}$, $\overline{KL} \cong \overline{NP}$, $\overline{LJ} \cong \overline{PM}$

The congruence statement is $\triangle JKL \cong \triangle MNP$.

b. If $\triangle ABC \cong \triangle HGF$, name the corresponding parts. Then complete the congruence statement $\triangle BAC \cong \triangle$ __?__.

Use the order of the vertices in the congruence statement $\triangle ABC \cong \triangle HGF$ to identify the corresponding parts.

Corresponding angles: $\angle A \cong \angle H$, $\angle B \cong \angle G$, $\angle C \cong \angle F$

Corresponding sides: $\overline{AB} \cong \overline{HG}$, $\overline{BC} \cong \overline{GF}$, $\overline{CA} \cong \overline{FH}$

The congruence statement is $\triangle BAC \cong \triangle GHF$.

StudyTip

Draw a Figure When given only a congruence statement, it is helpful to draw the two triangles in the same position. Then it will be easier to visualize and confirm the corresponding angles and sides.

A B C H G F

Check Your Progress

Name the corresponding parts in each pair of congruent triangles. Then complete the congruence statement.

1A.

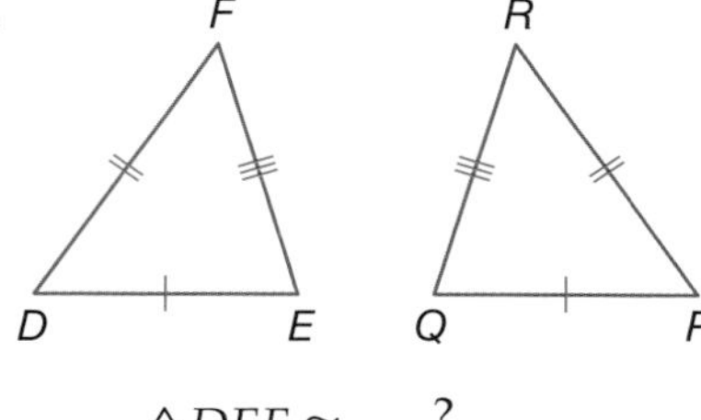

$\triangle DEF \cong$ __?__

1B. $\triangle TUV \cong \triangle CDE$

$\triangle VTU \cong$ __?__

Personal Tutor glencoe.com

You can use corresponding parts to find the measures of angles and sides in a figure that is congruent to a figure with known measures.

Real-World EXAMPLE 2 Find Missing Measures

SIGNS In the signs below, $\triangle RST \cong \triangle VWX$.

a. If $m\angle W = 60°$, what is the measure of $\angle S$?

$\angle W$ and $\angle S$ are corresponding angles, so they are congruent.

$m\angle S = 60°$ — **Congruent angles have equal measures.**

b. The length of $\overline{ST}$ is 30 inches. What is the length of $\overline{WX}$?

$\overline{ST}$ and $\overline{WX}$ are corresponding sides, so they are congruent.

$WX = 30$ inches — **Congruent sides have equal lengths.**

Check Your Progress

2. QUILTING In the quilt design shown, $\triangle ABC \cong \triangle ADE$. What is the measure of $\angle BCA$? What is the length of AD?

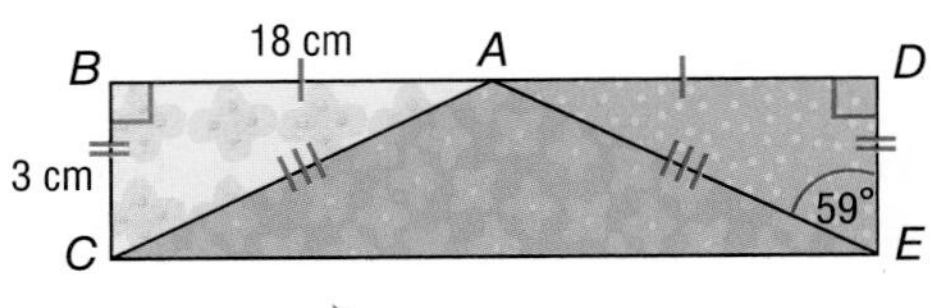

Personal Tutor glencoe.com

Identify Congruent Triangles In congruent triangles, all corresponding angles are congruent and all corresponding sides are congruent.

> **StudyTip**
>
> **Identify Congruent Parts** To write a congruence statement, note which angles are congruent and which segments have the same length. Then write corresponding vertices in the same order.

EXAMPLE 3 Identify Congruent Triangles

Determine whether the triangles shown at the right are congruent. If so, name the corresponding parts and write a congruence statement.

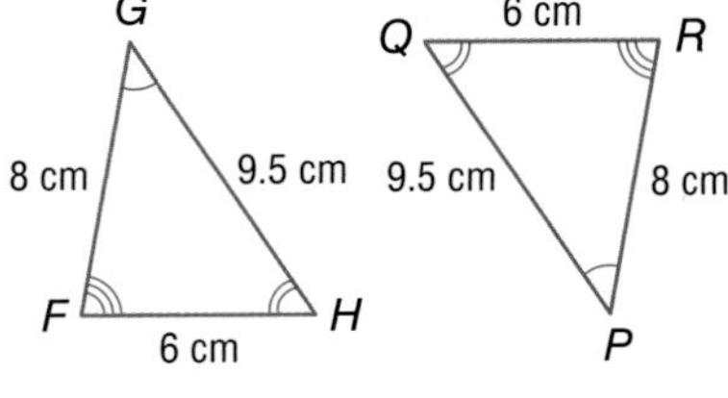

The arcs indicate that $\angle G \cong \angle P$, $\angle H \cong \angle Q$, and $\angle F \cong \angle R$.

The side measures indicate that $\overline{GH} \cong \overline{PQ}$, $\overline{HF} \cong \overline{QR}$, and $\overline{FG} \cong \overline{RP}$.

Since all pairs of corresponding angles and sides are congruent, the triangles are congruent. One congruence statement is $\triangle GHF \cong \triangle PQR$.

Check Your Progress

3. Determine whether the triangles shown are congruent. If so, name the corresponding parts and write a congruence statement.

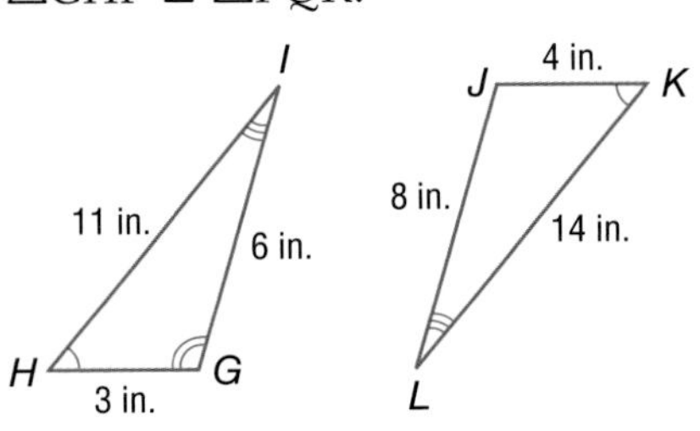

Personal Tutor glencoe.com

Check Your Understanding

Example 1 p. 599 **Name the corresponding parts in each pair of congruent triangles. Then complete the congruence statement.**

1. 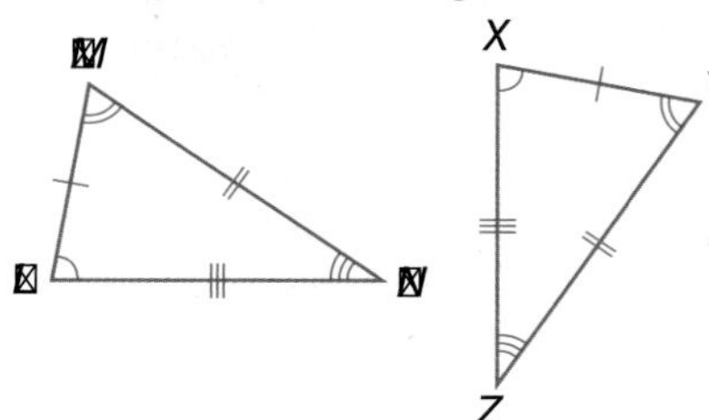

$\triangle YZX \cong$ ___?___

2. $\triangle RST \cong \triangle GJK$
$\triangle STR \cong$ ___?___

Example 2 p. 600

3. KITES In the umbrella kite shown at the right, $\triangle JLK \cong \triangle NLM$.

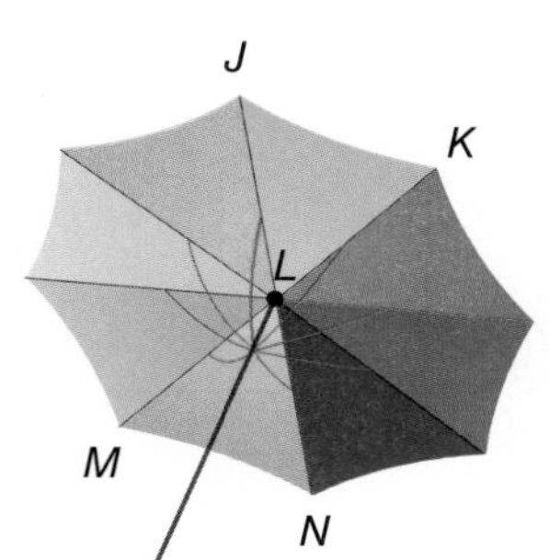

a. If $m\angle JKL = 67.5°$, what is $m\angle NML$?

b. If $NL = 14$ inches and $MN = 12$ inches, what is JK?

Example 3 p. 600 **Determine whether the triangles shown are congruent. If so, name the corresponding parts and write a congruence statement.**

4. 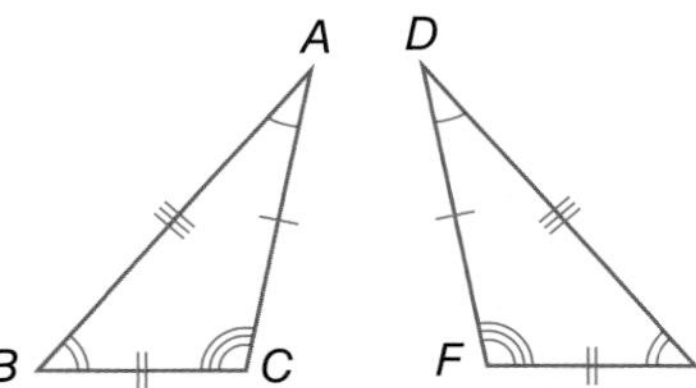

5. 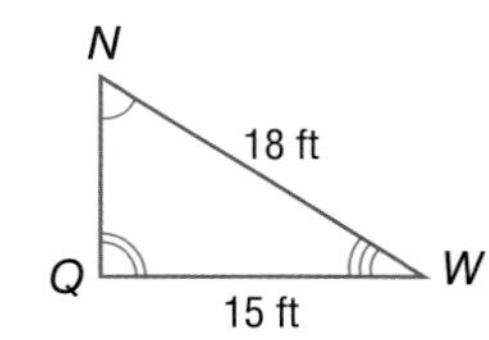

Practice and Problem Solving

● = **Step-by-Step Solutions** begin on page R11.
Extra Practice begins on page 810.

Example 1 p. 599 **Name the corresponding parts in each pair of congruent triangles. Then complete the congruence statement.**

6. 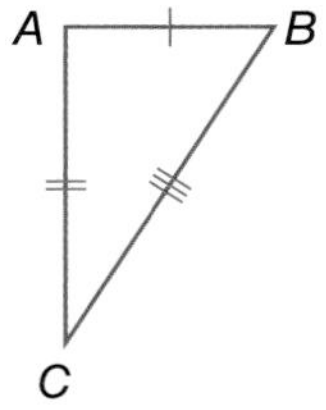

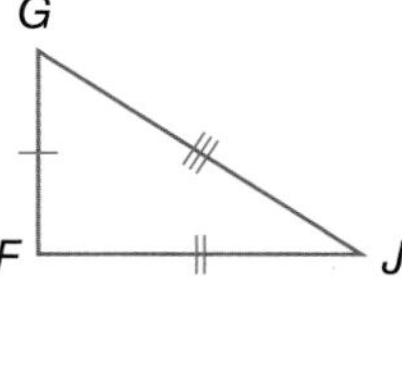

$\triangle GJF \cong$ ___?___

7 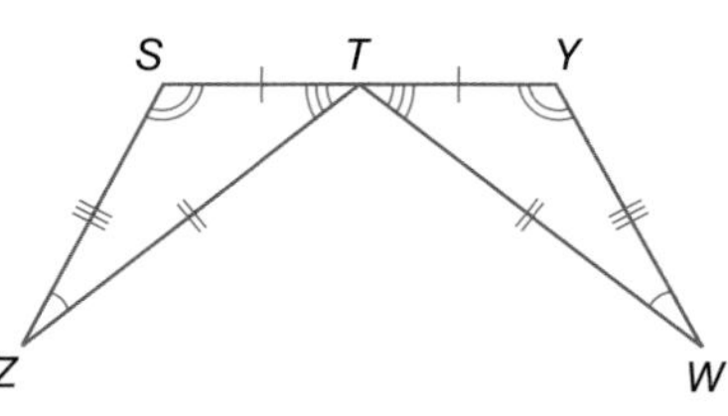

$\triangle STZ \cong$ ___?___

8. $\triangle KLM \cong \triangle WXY$
$\triangle LKM \cong$ ___?___

9. $\triangle NPQ \cong \triangle DEF$
$\triangle EFD \cong$ ___?___

Example 2 p. 600

10. **STAINED GLASS** In the stained glass window at the right, $\triangle ABC \cong \triangle ADC$.

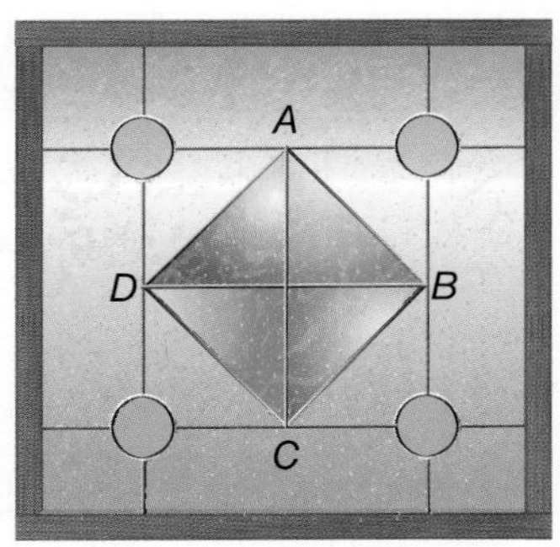

 a. If $BC = 26$ centimeters, what is DC?

 b. If $m\angle ADC = 90°$, and $m\angle DCA = 45°$, what is $m\angle BAC$?

11. **ALGEBRA** Triangle STU is congruent to triangle MNP. In the figures, $TU = 15.4$ millimeters, $US = 13$ millimeters, and $ST = 18.1$ millimeters. If $MN = 2x + 1$, what is the value of x?

Example 3 p. 600

Determine whether the triangles shown are congruent. If so, name the corresponding parts and write a congruence statement.

12.

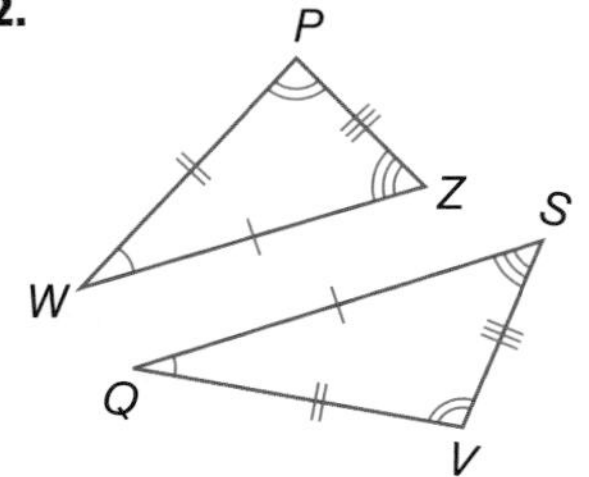

13.

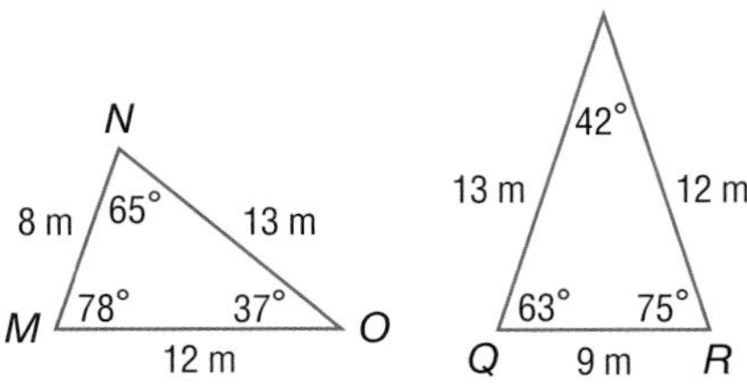

14.

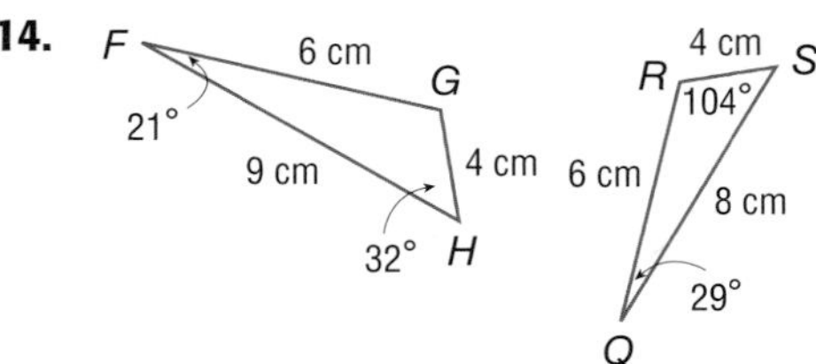

15.

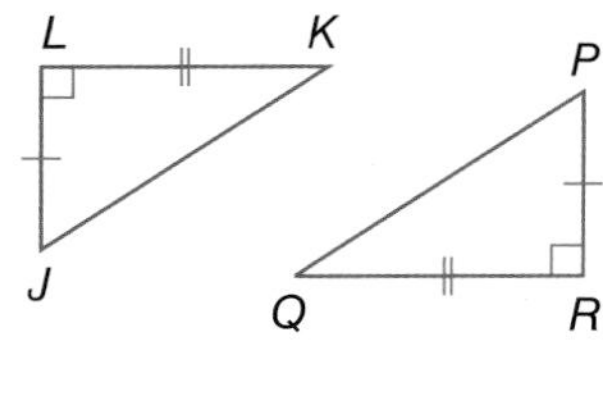

16. **ART** The structure shown at the right, *Moondog*, created by Tony Smith, is located at the Hirshhorn Museum and Sculpture Garden in Washington, D.C. If $\triangle ABC \cong \triangle DFG$, name all corresponding sides.

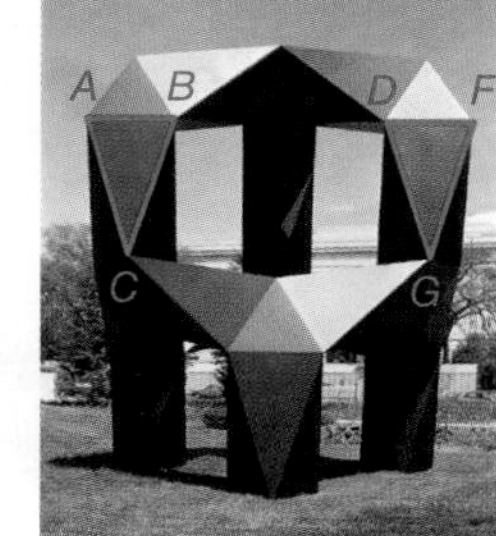

17. **SOCIAL STUDIES** The Khafre pyramid shown at the left is composed of triangles. If $\triangle MNO \cong \triangle NPO$, $m\angle M = 60°$, and $m\angle OPN = 50°$, find $m\angle O$.

Real-World Link

The average weight of an individual block of stone that makes up the Khafre pyramid in Egypt is 2.5 tons. Some of the outside stones weigh up to 7 tons.

Source: PBS

18. **GEOMETRY** Hexagon $ABCDEF$ has six congruent sides. Segments $\overline{CA}$, $\overline{CF}$, and $\overline{CE}$ are drawn on the hexagon, forming four triangles. Make a conjecture about which triangles are congruent. Test your conjecture by measuring the sides and angles of the triangles.

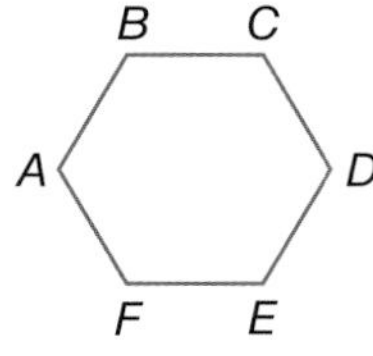

19. In the figure below, $\triangle FGI \cong \triangle LKM$. Find the value of x. Round to the nearest tenth.

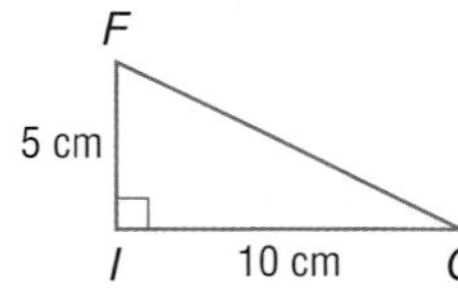

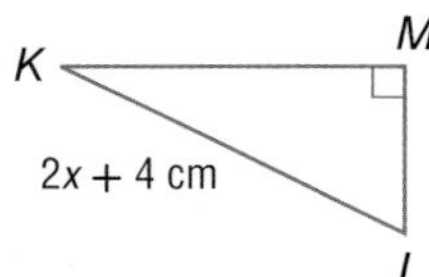

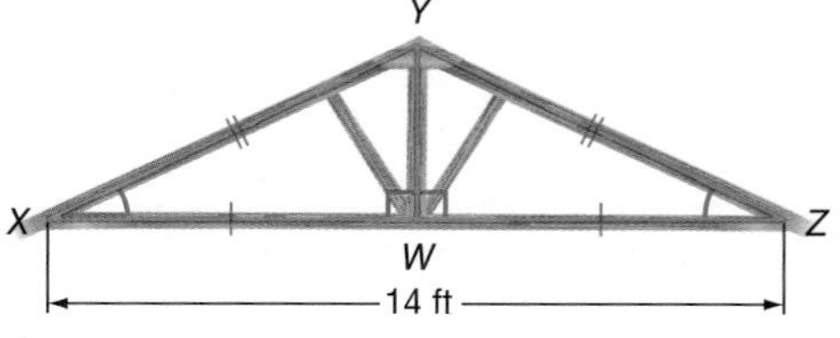

20. **CONSTRUCTION** The roof at the right contains two congruent triangles.

   a. Write a congruence statement involving the triangles. Then write a congruence statement involving a pair of corresponding sides.

   b. What is the distance from the right side of the roof at the base to the center? Name the side.

   c. If the length of $\overline{XW}$ is 4 feet longer than $\overline{WY}$, what is the length of $\overline{WY}$?

21. **BUTTERFLIES** Butterfly wings are triangular in shape. Using the art of the butterfly as a model, draw two sets of congruent triangles, label the vertices, and write a congruence statement for each.

Real-World Career

Biological Scientists Biological scientists study the behavior, genetics, and diseases of living things. Since the job requires collecting and analyzing data, knowledge of arithmetic, algebra and statistics are important.

A four-year degree in a biological field is the minimum level of education required for this career.

ALGEBRA Find the value of x for each pair of congruent triangles.

22. 23.

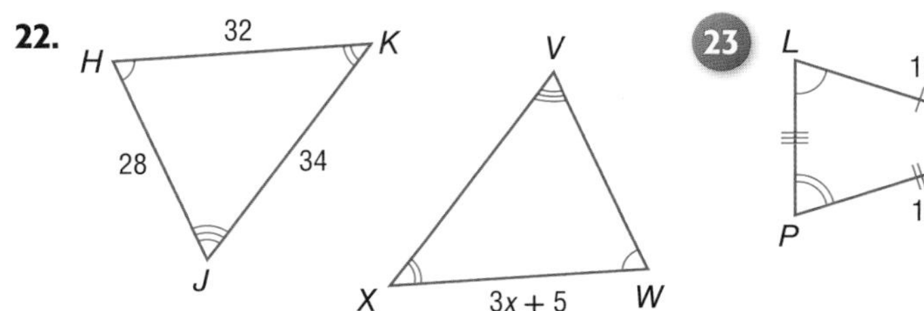

H.O.T. Problems

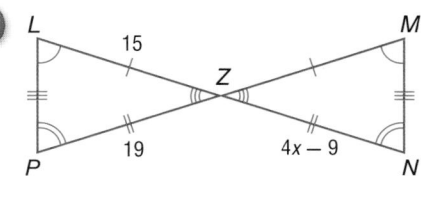

Use **H**igher-**O**rder **T**hinking Skills

24. **OPEN ENDED** Find a real-world example of congruent triangles. Explain how you know the triangles are congruent.

25. **FIND THE ERROR** Jade and Fernando are writing a congruence statement for the congruent triangles at the right. Is either of them correct? Justify your reasoning.

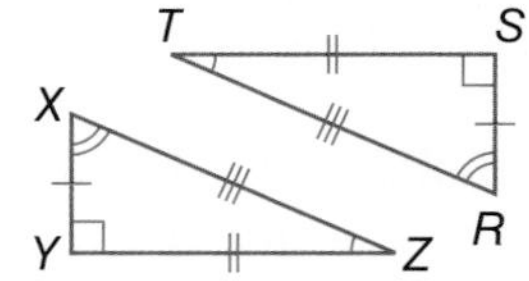

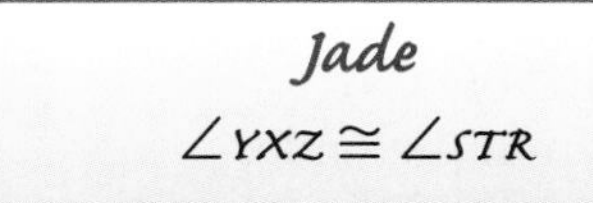

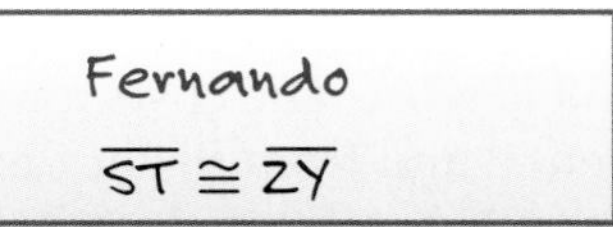

26. **CHALLENGE** Determine whether each statement is *true* or *false*. If true, explain your reasoning. If false, give a counterexample.

   a. Triangles *EFG* and *KLM* are congruent. So, their perimeters are equal.

   b. Triangles *WXY* and *HJK* are congruent. So, their areas are equal.

   c. The perimeter of $\triangle ABC$ is 24 millimeters and the perimeter of $\triangle RST$ is 24 millimeters. So, triangles *ABC* and *RST* are congruent.

27. **CHALLENGE** Describe the scale factor of two congruent triangles.

28. **WRITING IN MATH** Describe structures that are built using congruent triangles. Explain why you think congruent triangles are used.

Standardized Test Practice

29. In the road sign below, $\triangle ABC \cong \triangle DCB$, AC is 2.5 meters long, BC is 1 meter long, and AB is 2.7 meters long. What is the length of $\overline{BD}$?

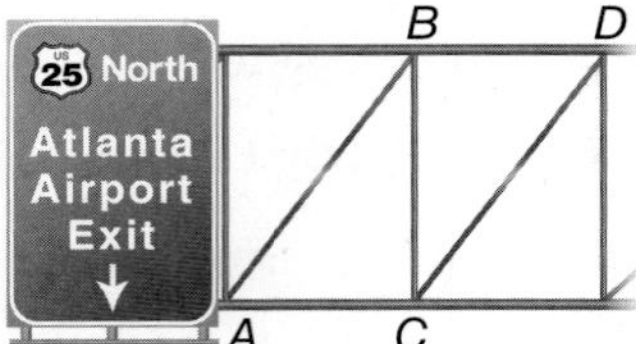

A 1 meter **C** 2.7 meters
B 2.5 meters **D** 2 meters

30. Guy–wires connected to a telephone pole create two congruent triangles $\triangle PQR$ and $\triangle SQR$. Find the length of $\overline{QS}$.

F 12 ft
G 24 ft
H 48 ft
J 65 ft

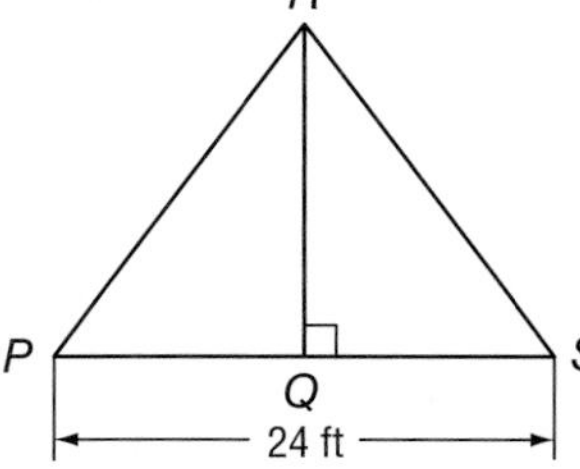

31. Which of the following statements is *not* true if $\triangle FGH \cong \triangle LMN$?

A $\overline{FH} \cong \overline{LN}$
B $\overline{GH} \cong \overline{MN}$
C $\angle G \cong \angle M$
D $\angle F \cong \angle N$

32. EXTENDED RESPONSE Determine whether the triangles shown below are congruent. If so, name the corresponding parts and write a congruence statement.

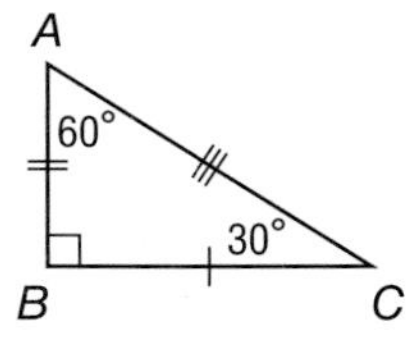

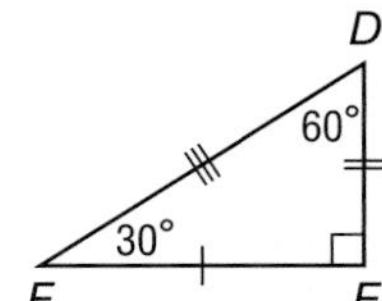

Spiral Review

33. Angles R and S are complementary. Find $m\angle R$ if $m\angle S = 65.7°$ (Lesson 11-1)

34. In a 45°-45°-90° triangle, a leg is 4 centimeters long. Find the length of the hypotenuse to the nearest tenth. (Lesson 10-6)

35. BIRDS A mockingbird uses about 5^{-4} Joules of energy to sing a song. Write the amount of energy the bird uses as an expression using a positive exponent. (Lesson 9-4)

36. CAR RENTAL The costs for renting a car from Able Car Rental and from Baker Car Rental are shown in the table. For what mileage does Baker have the better deal? Use the inequality $30 + 0.05x > 20 + 0.10x$. Explain why this inequality works. (Lesson 5-5)

Rental Car Costs

	Cost per Day	Cost per Mile
Able Car Rental	$30	$0.05
Baker Car Rental	$20	$0.10

Skills Review

Graph each ordered pair on a coordinate plane. (Lesson 2-6)

37. $F(4, 5)$ **38.** $G(2, -2)$ **39.** $C(-1, 4)$ **40.** $A(-3, -2)$

11-3 Rotations

Then

You drew translations and reflections on the coordinate plane. (Lesson 2-7)

Now

- Define, identify, and draw rotations.
- Determine if a figure has rotational symmetry.

New Vocabulary

rotation
center of rotation
rotational symmetry

Math Online

glencoe.com

- Extra Examples
- Personal Tutor
- Self-Check Quiz
- Homework Help

Why?

Jasmine gets on the Ferris wheel at point *A*. The Ferris wheel stops when she is at point *B* to let more people get on the ride.

a. Does the size and shape of the Ferris wheel change as the wheel spins?

b. Describe the movement involved as Jasmine travels from point *A* to point *B*.

Rotations A **rotation** is a transformation in which a figure is turned around a fixed point. This point is called the **center of rotation**. Rotations are also called *turns*. A rotated image has the same size and shape as the original figure.

The images below show clockwise rotations of a figure with a center of rotation at point *A*.

Original Figure	Angle of Rotation		
	90°	180°	270°
A, center of rotation	A′	A′	A′

EXAMPLE 1 Rotate a Figure about a Point

Draw the figure at the right after a 90° clockwise rotation about point *C*.

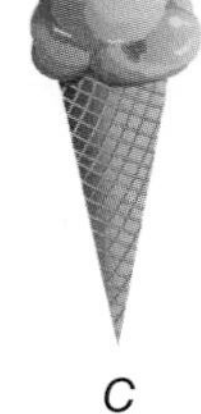

Since point *C* is the center of rotation, it remains in the same position. The figure moves one quarter turn clockwise.

Check Your Progress

Draw the figure in Example 1 after each rotation.

1A. 270° clockwise rotation about point *C*.

1B. 180° counterclockwise about point *C*.

Personal Tutor glencoe.com

ReadingMath

Notation The notation W' is read *W prime.* It represents the point W after the rotation.

EXAMPLE 2 Rotate a Figure about a Point

Quadrilateral $WXYZ$ has vertices $W(-4, -1)$, $X(-2, 0)$, $Y(-1, -3)$, and $Z(-2, -4)$. Graph the figure and its image after a clockwise rotation of 180° about vertex X. Then give the coordinates of the vertices for quadrilateral $W'X'Y'Z'$.

Step 1 Graph the original figure. Then graph vertex W' after a 180° rotation about vertex X. Note that $m\angle WXW' = 180°$ and $WX = XW'$.

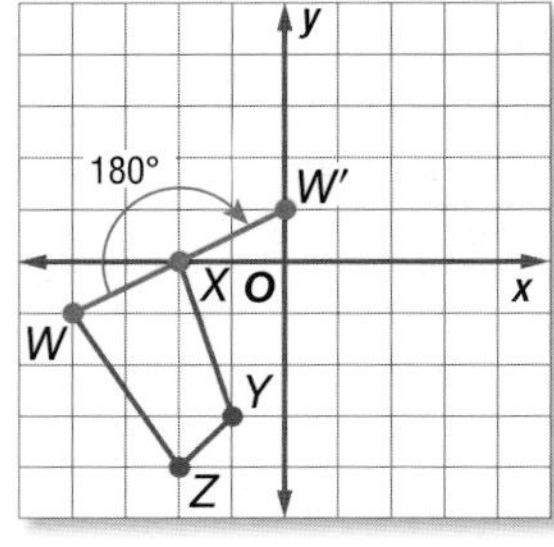

Step 2 Graph the remaining vertices after 180° rotations about vertex X. Connect the vertices to form quadrilateral $W'X'Y'Z'$.

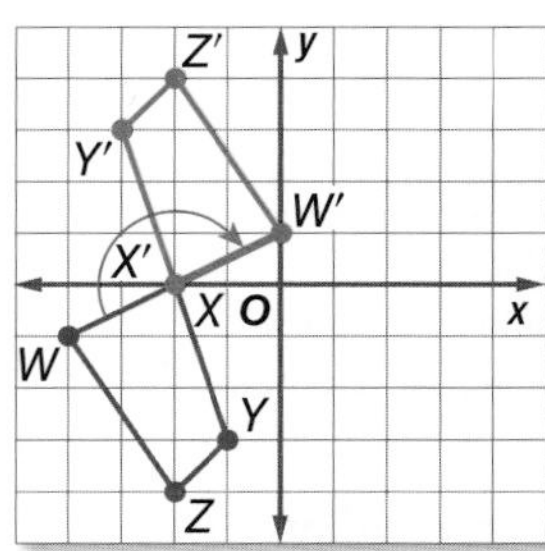

So, the coordinates of the vertices of quadrilateral $W'X'Y'Z'$ are $W'(0, 1)$, $X'(-2, 0)$, $Y'(-3, 3)$, and $Z'(-2, 4)$.

Check Your Progress

2. Graph quadrilateral $WXYZ$ and its image after a counterclockwise rotation of 270° about vertex Y. Then give the coordinates of the vertices for quadrilateral $W'X'Y'Z'$.

Personal Tutor glencoe.com

Vocabulary Link

Rotate
Everyday Use to take turns
Math Use to turn around a fixed point

The diagrams below show three clockwise rotations of a figure about the origin.

90° Rotation

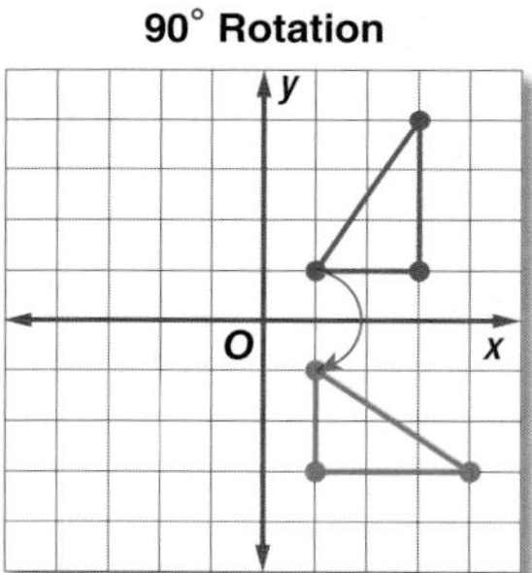

180° Rotation

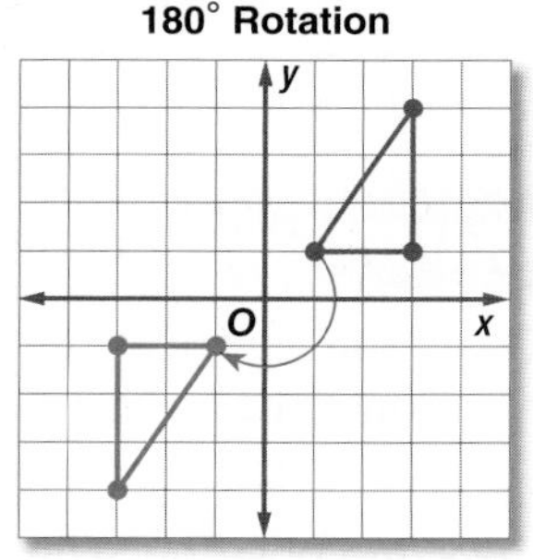

270° Rotation

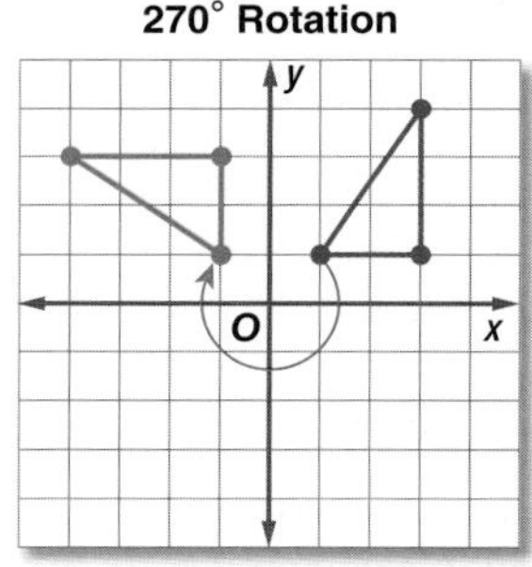

The following diagrams show three counterclockwise rotations about the origin.

270° Rotation

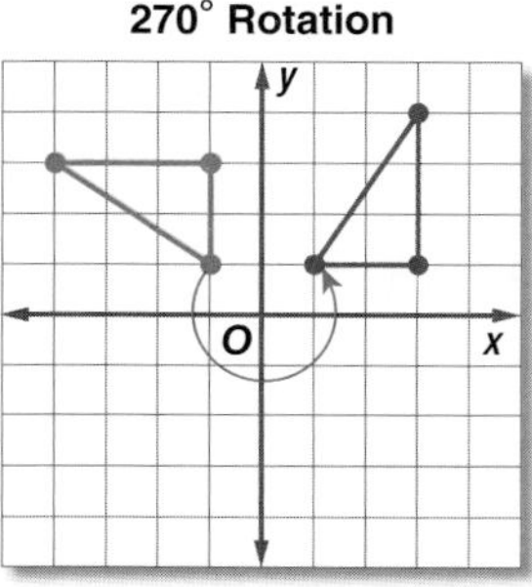

180° Rotation

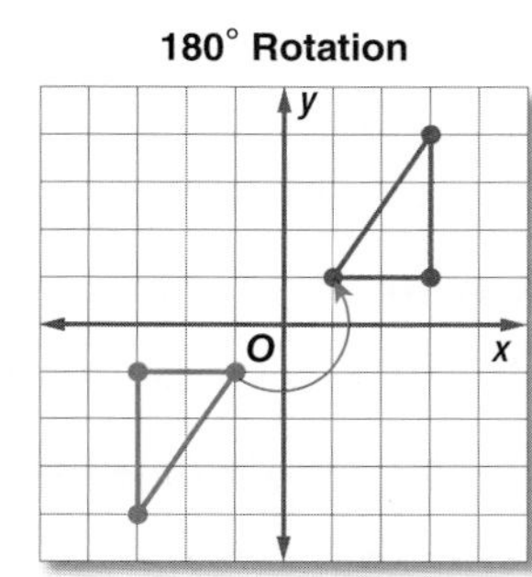

90° Rotation

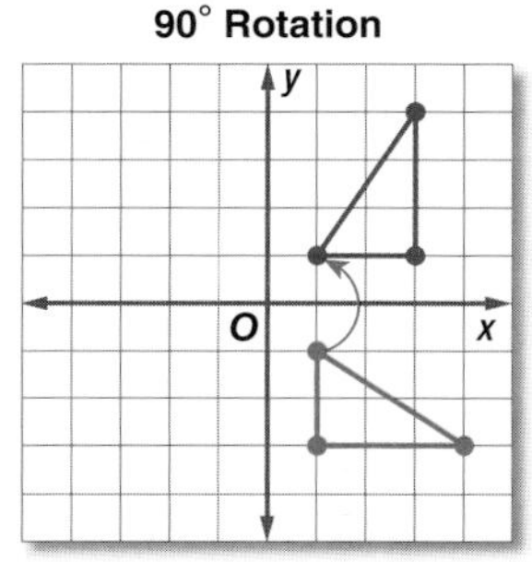

EXAMPLE 3 Rotations about the Origin

A triangle has vertices $A(2, -5)$, $B(4, -4)$, and $C(2, -1)$. Graph the triangle and its image after a rotation of 270° clockwise about the origin.

Step 1 Graph $\triangle ABC$ on a coordinate plane. A 270° degree rotation is the same as three 90° rotations or $\frac{3}{4}$ of a complete circle. Then graph vertex C' after a 270° clockwise rotation about the origin.

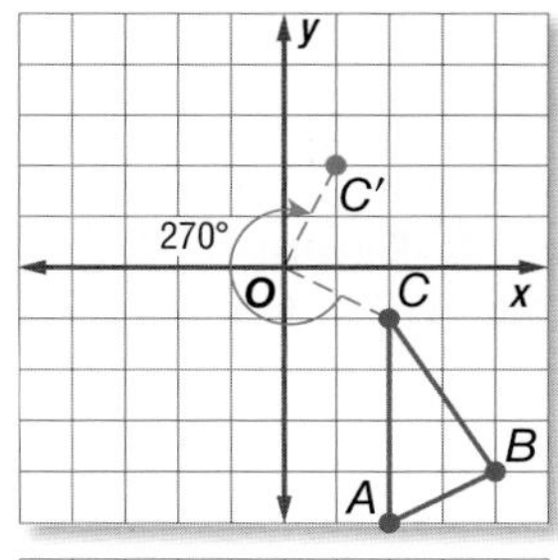

Step 2 Graph the remaining vertices after 270° rotations about the origin. Then connect the vertices to form $\triangle A'B'C'$.

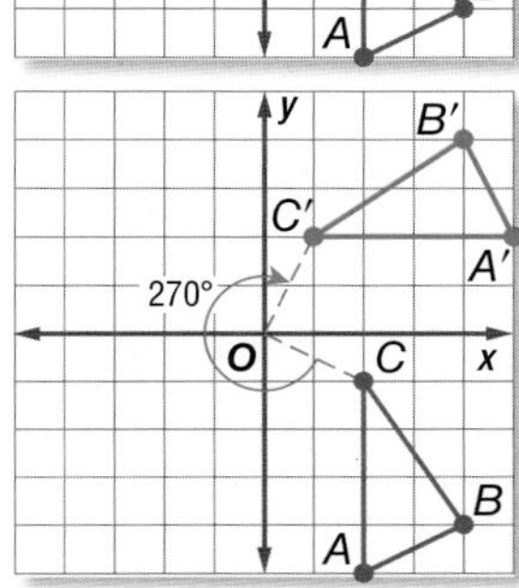

So, the coordinates of the vertices of $\triangle A'B'C'$ are $A'(5, 2)$, $B'(4, 4)$, and $C'(1, 2)$.

Check Your Progress

3. A figure has vertices $J(3, 2)$, $K(6, 4)$, $M(6, 7)$, and $N(3, 7)$. Graph the figure and its image after a rotation of 90° clockwise about the origin.

Personal Tutor glencoe.com

Rotational Symmetry If a figure can be rotated less than 360° about its center so that the image matches the original figure, the figure has **rotational symmetry.**

StudyTip

Rotational Symmetry One complete turn of a figure measures 360° because there are 360° in a circle. A figure that matches itself only after a 360° turn does not have rotational symmetry.

Real-World EXAMPLE 4 Rotational Symmetry

NATURE Determine whether the snowflake shown has rotational symmetry. If it does, describe the angle of rotation.

The snowflake can match itself in six positions.

The pattern repeats in 6 even intervals.

So, the angle of rotation is $360° \div 6$ or 60°.

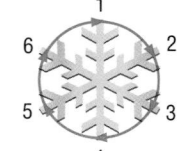

Check Your Progress

4. **NATURE** Determine whether the flower has rotational symmetry. If it does, describe the angle of rotation.

Personal Tutor glencoe.com

Check Your Understanding

Example 1 p. 605

1. Draw the figure at the right after a 270° clockwise rotation about point A.

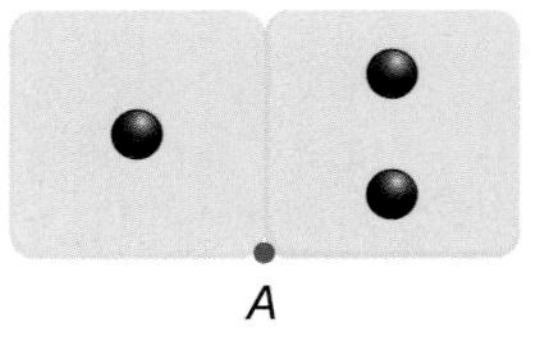

Example 2 p. 606

2. Triangle JKL has vertices $J(1, 4)$, $K(1, 1)$, and $L(5, 1)$.
   a. Graph the figure and its image after a clockwise rotation of 270° about vertex J.
   b. Give the coordinates of the vertices for triangle $J'K'L'$.

Example 3 p. 607

3. A figure has vertices $D(1, 1)$, $F(2, 3)$, $G(5, 3)$, and $H(5, 1)$. Graph the figure and its image after a rotation of 180° around the origin.

Example 4 p. 607

4. **WINDMILL** Determine whether the blades of the windmill shown at the right have rotational symmetry. If they do, describe the angle of rotation.

Practice and Problem Solving

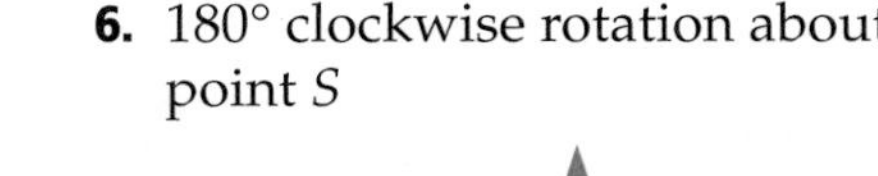
= **Step-by-Step Solutions** begin on page R11.
Extra Practice begins on page 810.

Example 1 p. 605

Draw each figure after the rotation described.

5. 90° clockwise rotation about point R

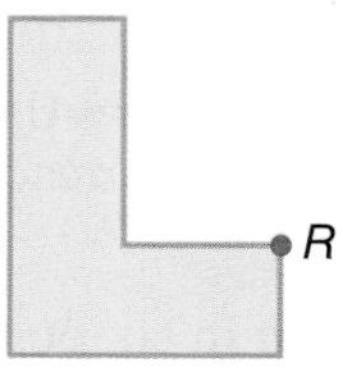

6. 180° clockwise rotation about point S

Example 2 p. 606

7. Triangle GHJ is shown at the right.
   a. Graph the figure after a clockwise rotation of 90° about vertex G.
   b. Give the coordinates of the vertices for $\triangle G'H'J'$.

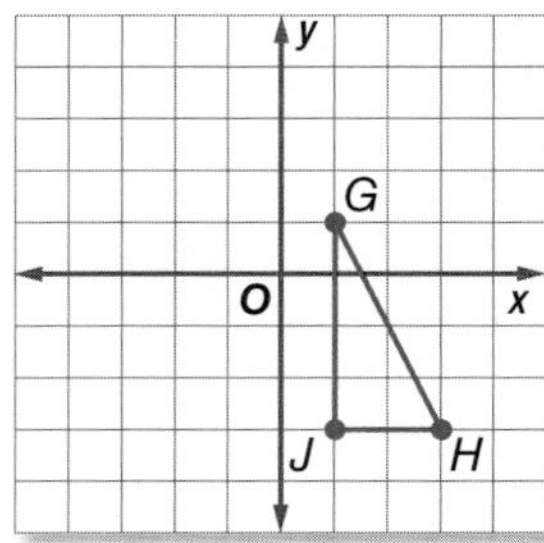

8. Trapezoid $ABCD$ has vertices $A(-3, 1)$, $B(-3, 4)$, $C(1, 4)$, and $D(-1, 1)$.
   a. Graph the figure and its image after a clockwise rotation of 180° about vertex D.
   b. Give the coordinates of the vertices for trapezoid $A'B'C'D'$.

Example 3 p. 607

9. A figure has vertices $K(1, -1)$, $L(3, -4)$, $M(1, -5)$, and $N(-1, -4)$. Graph the figure and its image after a rotation of 270° counterclockwise about the origin.

10. A triangle has vertices $P(-3, 1)$, $Q(0, 4)$, and $R(1, -1)$. Graph the triangle and its image after a rotation of 180° about the origin.

Example 4 p. 607

11. **SHAPES** Describe the rotational symmetry of the star in Exercise 6.

SWIMMING Determine whether each synchronized swimming formation has rotational symmetry. If it does, describe the angle of rotation.

12. 

13. 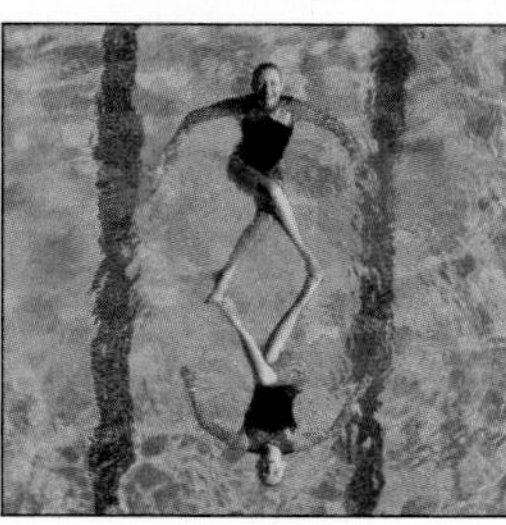

14. 

15. **DECORATING** Pat is placing the tile at the right in different orientations on a wall. Describe three rotations that would change the orientation of the tile. Use drawings to show what the tile looks like after each rotation.

16. **MULTIPLE REPRESENTATIONS** In this problem, you will investigate rotations. A figure is graphed on the coordinate plane. One of its vertices has coordinates (x, y).

 a. **SYMBOLIC** Write the coordinates of the vertex after each rotation.
 - 90° clockwise about the origin
 - 180° about the origin
 - 270° clockwise about the origin

 b. **ANALYTICAL** The vertex of the figure is located in Quadrant II. If the figure is rotated 180° about the origin, in which quadrant will the corresponding vertex of the image be located?

Real-World Link

Sixty percent of a synchronized swimming routine is underwater. Synchronized swimmers at the highest level can swim up to 75 meters underwater without coming up for air.

Source: Meredith College

H.O.T. Problems

Use **H**igher-**O**rder **T**hinking Skills

17. **OPEN ENDED** Sketch a figure that has rotational symmetry. Describe the angle of rotation.

18. **REASONING** If two polygons are similar, could they also be congruent? Explain your reasoning.

19. **CHALLENGE** Triangle RST is rotated 270° clockwise around the origin. Its rotated image has vertices $R'(-5, 1)$, $S'(2, 0)$, and $T'(-3, -6)$. What are the coordinates of triangle RST?

20. **REASONING** Will a geometric figure and its rotated image *sometimes, always,* or *never* have the same perimeter? Explain your reasoning.

21. **WRITING IN MATH** Describe the similarities and differences between reflections and rotations.

Standardized Test Practice

22. Triangle XYZ was rotated about the origin to $\triangle X'Y'Z'$. Which of the following describes the rotation?

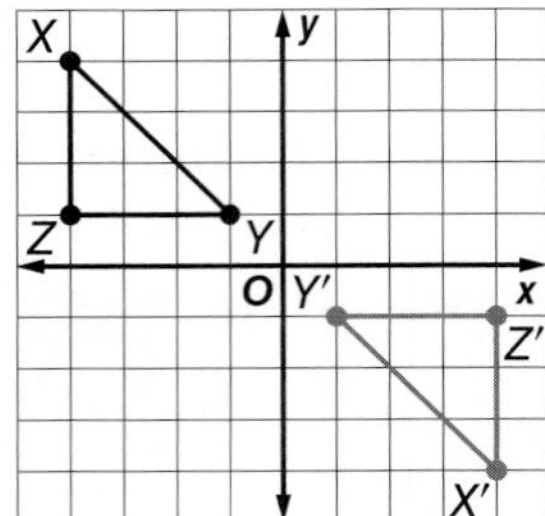

A 90° clockwise about the origin
B 90° counterclockwise about the origin
C 180° clockwise about the origin
D 270° clockwise about the origin

23. Triangle ABC has vertices $A(1, 2)$, $B(-1, -1)$, and $C(2, 0)$. If the triangle is rotated clockwise 90° about the origin, which of the following would be the vertices of $\triangle A'B'C'$?

F $A'(2, 1)$, $B'(-1, -1)$, $C'(0, 2)$
G $A'(-1, 2)$, $B'(1, 1)$, $C'(-2, 0)$
H $A'(-2, 1)$, $B'(1, -1)$, $C'(0, 2)$
J $A'(2, -1)$, $B'(-1, 1)$, $C'(0, -2)$

24. EXTENDED RESPONSE Figure $ABCD$ is shown.

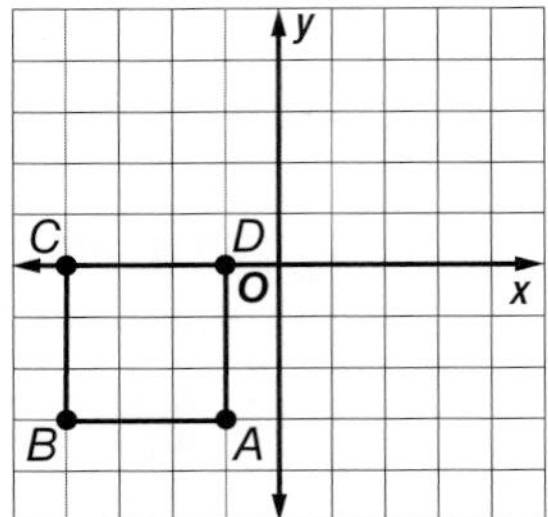

a. Graph the figure after a clockwise rotation of 270° about the origin.
b. Give the coordinates of the vertices for square $A'B'C'D'$.

25. Which of the following is true about the image shown below?

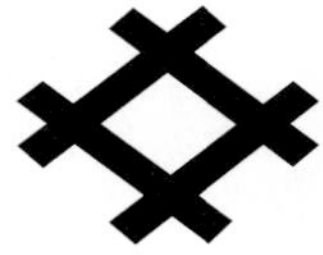

A The figure does not have rotational symmetry.
B The figure has angle of rotation of 90°.
C The figure has angle of rotation of 180°.
D The figure has angle of rotation of 270°.

Spiral Review

26. ARCHITECTURE Refer to the diagram of the roof truss. In the diagram, $\triangle TRU \cong \triangle SRU$. (Lesson 11-2)

a. Find the distance from the right metal plate connector to the center web.
b. What is the measure of the angle formed by the top left chord and the bottom chord?

top left chord
R
center web
metal plate connector
T
30°
S
U
32 ft
bottom chord

27. If $m\angle B = 17°$ and $\angle A$ and $\angle B$ are complementary, find $m\angle A$. (Lesson 11-1)

28. Find the distance between $F(-3, -6)$ and $G(4, 1)$. Round to the nearest tenth, if necessary. (Lesson 10-5)

Skills Review

Solve each equation. (Lesson 4-5)

29. $4x + 157 = 243$ **30.** $2x + 89 = 351$ **31.** $6x - 72 = 138$

EXPLORE

11-4

Geometry Lab
Angles in Polygons

Math Online glencoe.com
Math *in Motion*, Animation

Objective
Use parallel lines to investigate the sum of the measures of the angles in a triangle and similar triangles.

In Lesson 11-1, you identified special pairs of angles that are formed when parallel lines are cut by a transversal. In this lab, you will use the angle relationships of those angles to discover the sum of the measures of the angles in a triangle.

ACTIVITY Angles in a Triangle

Step 1 Draw a pair of parallel lines.

Step 2 Draw a transversal as shown. Label ∠1 and ∠2.

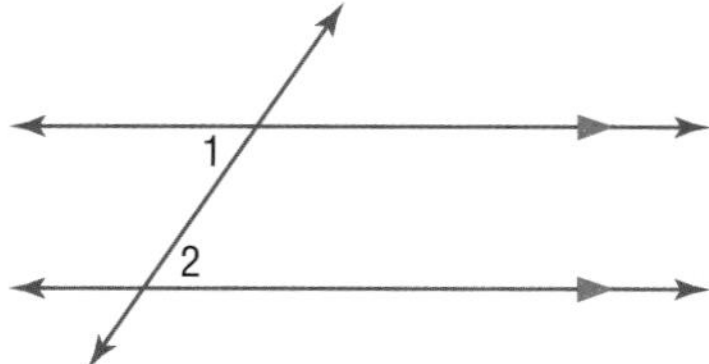

Step 3 Draw a second transversal as shown. Label ∠3 and ∠4. Label the triangle formed by these lines *ABC*.

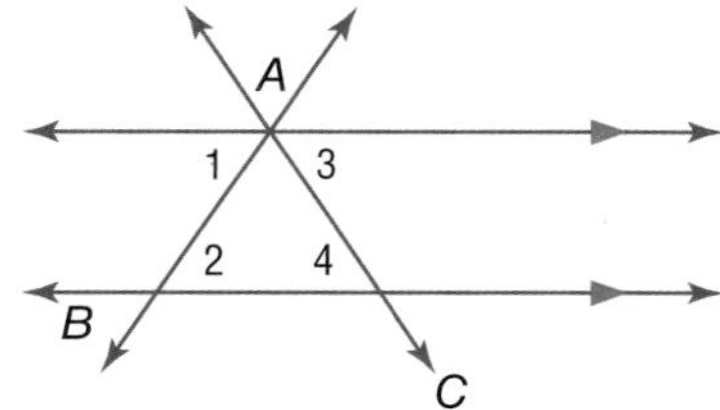

Analyze the Results

1. Classify the relationship between ∠1 and ∠2. What is true about these pairs of angles? Classify the relationship between ∠3 and ∠4. What is true about these pairs of angles?
2. What type of angle is formed by ∠1, ∠3, and ∠*BAC*? What is the sum of the measures of ∠1, ∠3, and ∠*BAC*?
3. What can you conclude about the sum of the measures of the angles in △*ABC*? Explain your reasoning.
4. **MAKE A CONJECTURE** Based on this activity, what is the sum of the measures of the angles of any triangle?
5. **MAKE A CONJECTURE** The quadrilateral at the right is separated into two triangles. Based on the activity above, what do you think is the sum of the angle measures of this quadrilateral? Explain.

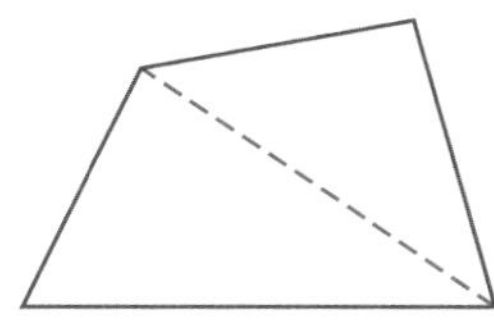

11-4 Quadrilaterals

Then

You found missing angle measures of a triangle. (Lesson 10-3)

Now

- Find missing angle measures of a quadrilateral.
- Classify quadrilaterals.

New Vocabulary

quadrilateral

Math Online

glencoe.com

- Extra Examples
- Personal Tutor
- Self-Check Quiz
- Homework Help
- Math in Motion

Why?

The *dance pad* shown is an electronic mat used in a video game. In the game, a player must move his or her feet to a set pattern, stepping on the panels of the mat according to the beat of a song.

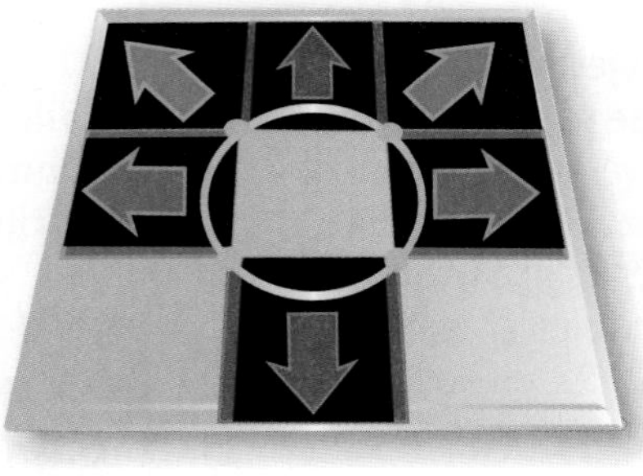

a. Describe the shapes of the panels that have arrows on them.

b. Describe the shape of the entire dance pad.

Find Angle Measures A **quadrilateral** is a closed figure with four sides and four angles. The segments that form a quadrilateral intersect only at their endpoints. Squares, rectangles, and trapezoids are examples of quadrilaterals.

Quadrilaterals | ***Not* Quadrilaterals**

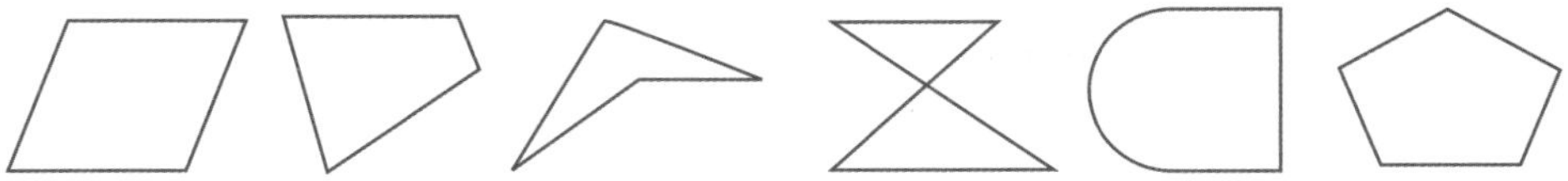

As with triangles, a quadrilateral can be named by its vertices. When you name a quadrilateral, begin at any vertex and name the vertices in order. Quadrilateral *ABCD* is shown below.

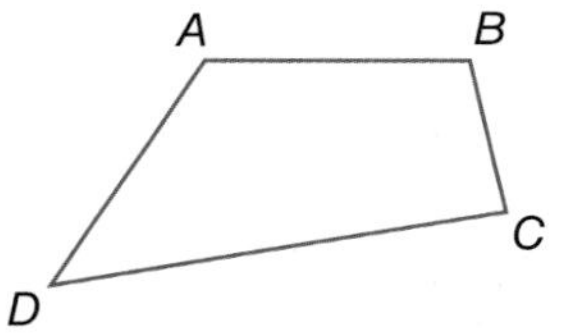

- sides: $\overline{AB}$, $\overline{BC}$, $\overline{CD}$, $\overline{DA}$
- vertices: A, B, C, D
- angles: $\angle A$, $\angle B$, $\angle C$, $\angle D$

Notice how a quadrilateral can be separated into two triangles. The sum of the measures of the angles of a triangle is 180°. So, the sum of the measures of the angles of a quadrilateral is 2(180°) or 360°.

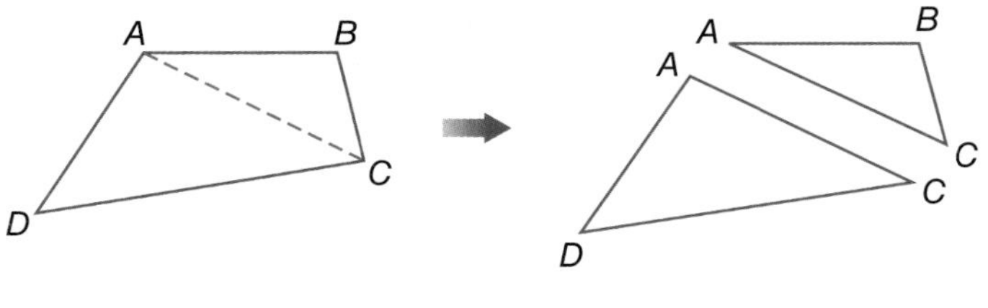

Key Concept | **Angles of a Quadrilateral** | For Your FOLDABLE

Words The sum of the measures of the angles of a quadrilateral is 360°.

Symbols $a + b + c + d = 360$

Model

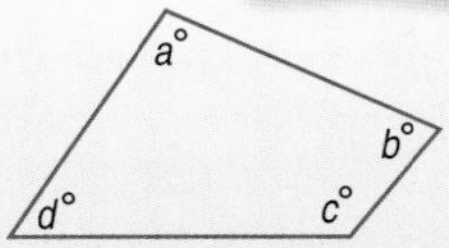

EXAMPLE 1 Find Angle Measures

Find the value of x in the quadrilateral. Then find each missing angle measure.

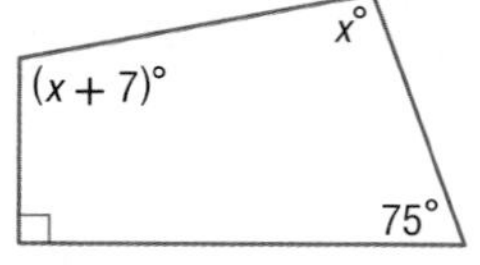

$(x + 7) + x + 75 + 90 = 360$	**The sum of the angle measures is 360°.**
$2x + 172 = 360$	**Combine like terms.**
$2x + 172 - 172 = 360 - 172$	**Subtract 172 from each side.**
$2x = 188$	**Simplify.**
$x = 94$	**Divide each side by 2.**

So, the missing angle measures are 94° and 94 + 7 or 101°.

StudyTip

Check Your Work To check the answer, find the sum of the measures of the angles. Since 75 + 90 + 101 + 94 = 360, the answer is correct.

Check Your Progress

1. In quadrilateral $EFGH$, $m\angle E = 3x°$, $m\angle F = 70°$, $m\angle G = x°$, and $m\angle H = 82°$. Find the value of x. Then find each missing angle measure.

Personal Tutor glencoe.com

Math *in Motion,* Animation glencoe.com

Classify Quadrilaterals Quadrilaterals can be classified by the relationship of their sides and angles, as shown in the diagram below.

Quadrilateral

Trapezoid
quadrilateral with exactly one pair of parallel sides

Parallelogram
quadrilateral with both pairs of opposite sides parallel and congruent

Rectangle
parallelogram with 4 right angles

Rhombus
parallelogram with 4 congruent sides

Square
parallelogram with 4 congruent sides and 4 right angles

StudyTip

Classifying Quadrilaterals The diagram at the right begins with the most general quadrilaterals and ends with the most specific. The name that *best* describes the quadrilateral is the one that is most specific.

Real-World EXAMPLE 2 Classify Quadrilaterals

BASKETBALL The free-throw lane used during International Basketball Federation competitions is shown. Classify the quadrilateral using the name that *best* describes it.

The quadrilateral has exactly one pair of opposite sides that are parallel. It is a trapezoid.

Check Your Progress

2. Classify the quadrilateral in Example 1 using the name that *best* describes it.

Personal Tutor glencoe.com

Check Your Understanding

Example 1 p. 613 **Find the value of x in each quadrilateral. Then find the missing angle measures.**

1. 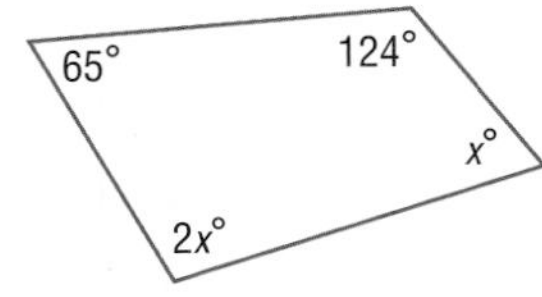

2. 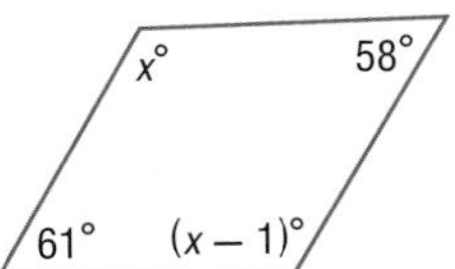

Example 2 p. 613 **Classify each quadrilateral using the name that *best* describes it.**

3. 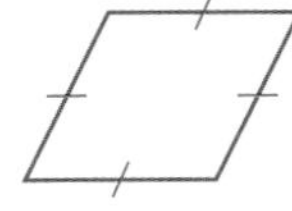

4. 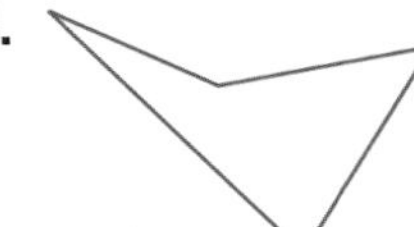

5. **SOCCER** Classify the quadrilaterals that are found on the soccer field.

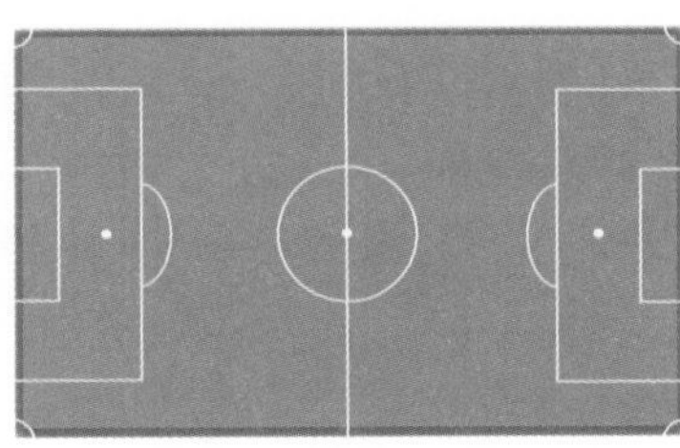

Practice and Problem Solving

● = **Step-by-Step Solutions** begin on page R11. **Extra Practice** begins on page 810.

Example 1 p. 613 **Find the value of x in each quadrilateral. Then find the missing angle measures.**

6. 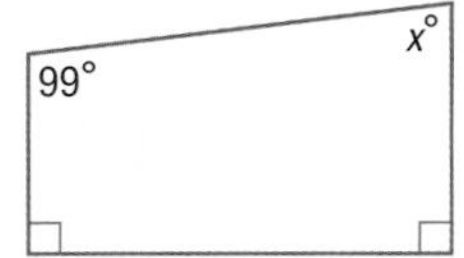

7. 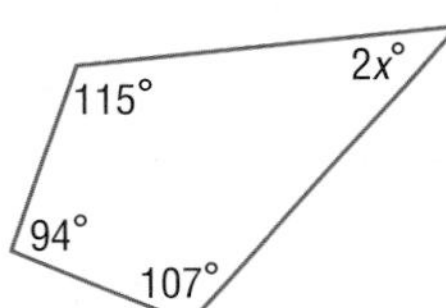

8. 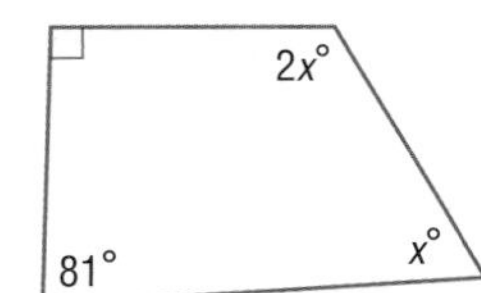

9. 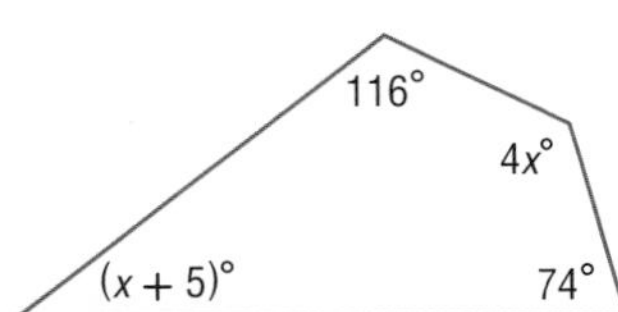

Example 2 p. 613

10. **CLOCK** Classify the quadrilateral that forms the face of the clock. Use the name that *best* describes it.

11. **PICTURE FRAMES** Classify the quadrilaterals that are found in the picture frames.

Classify each quadrilateral using the name that *best* describes it.

12. 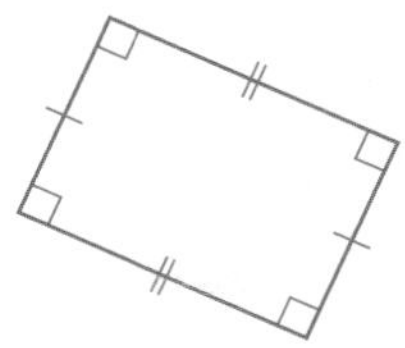

13. 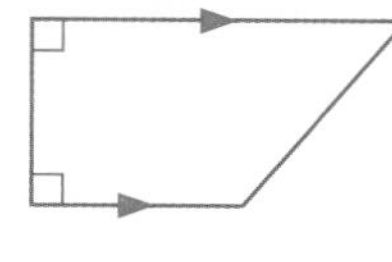

14. 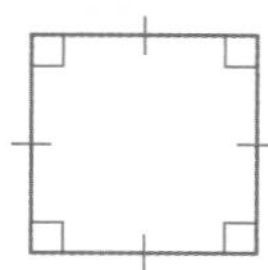

15. 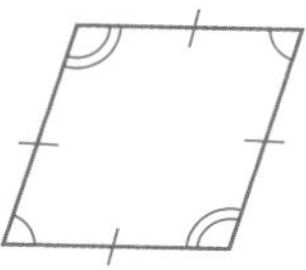

16. 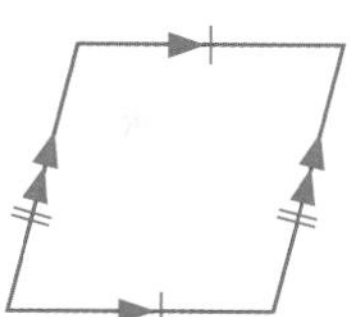

17. 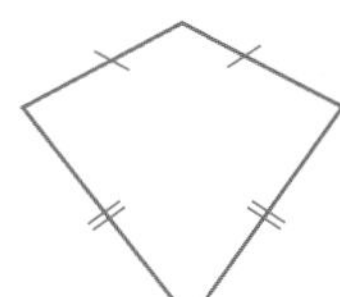

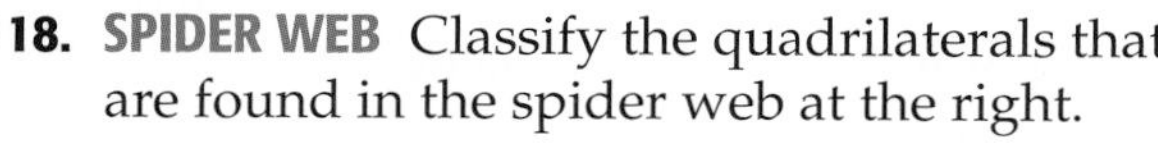

18. **SPIDER WEB** Classify the quadrilaterals that are found in the spider web at the right.

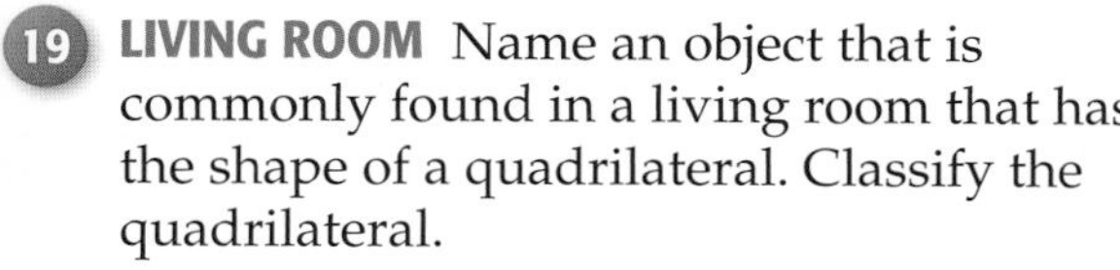

19. **LIVING ROOM** Name an object that is commonly found in a living room that has the shape of a quadrilateral. Classify the quadrilateral.

20. **SCHOOL SUPPLIES** Identify a school supply that is shaped like a quadrilateral. What characteristics did you use to identify the item?

Real-World Link

The combined length of thread in a spider's web is about 20 to 60 meters. On an equal weight basis, spider silk is twice as strong as steel.

Source: Belleza Spiders

Determine whether each statement is *sometimes, always,* or *never* true.

21. A square is a rectangle.
22. A trapezoid is a rhombus.
23. A parallelogram is a square.
24. A rhombus is a square.

25. **MULTIPLE REPRESENTATIONS** In this problem, you will use algebra to help you draw a figure. A quadrilateral has angle measures $x°$, $x°$, 70°, and 70°.

    a. **ALGEBRAIC** Write an equation that can be used to find the missing angle measures. Then find the measures.

    b. **GEOMETRIC** Sketch and label two different quadrilaterals that fit the description above. What types of quadrilaterals did you draw?

H.O.T. Problems

Use Higher-Order Thinking Skills

26. **OPEN ENDED** Use a map of the United States to find two states that appear to be shaped like quadrilaterals. Classify the quadrilaterals.

27. **CHALLENGE** Can a quadrilateral have two angles that are twice as large as the other two angles? Explain your reasoning or give an example.

28. **WHICH ONE DOESN'T BELONG?** Which quadrilateral does not belong with the other three? Explain your reasoning.

trapezoid	square	parallelogram	rectangle

29. **WRITING IN MATH** Describe the characteristics that a square shares with a rectangle and with a rhombus.

Standardized Test Practice

30. Find the value of x in the figure below.

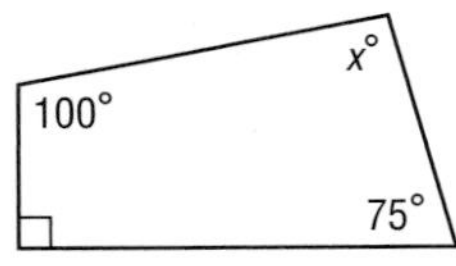

A 75°
B 90°
C 95°
D 105°

31. GRIDDED RESPONSE Mrs. Farias used the parallelogram below to design a pattern for a paving stone. She will use the paving stone for a sidewalk. Find x.

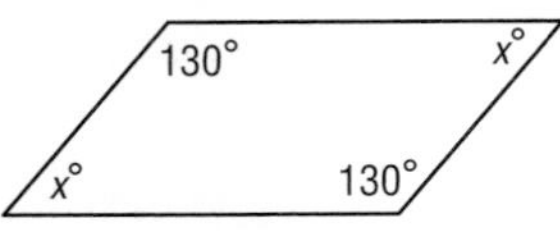

32. Which figure is *best* described as a square?

F 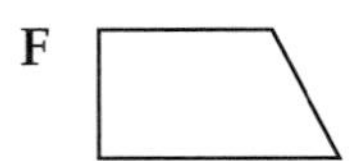

H 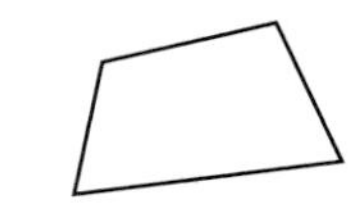

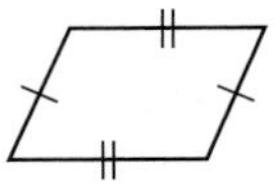

G 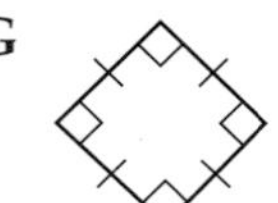

J

33. Find the value of x in the parallelogram below.

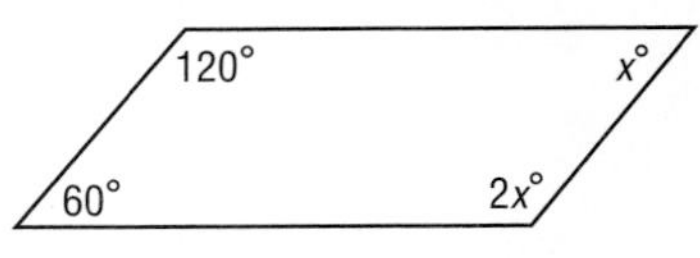

A 50°
B 55°
C 60°
D 65°

Spiral Review

34. Quadrilateral $FGHJ$ has vertices $F(-1, -1)$, $G(-3, -4)$, $H(3, -4)$, $J(2, -2)$. Graph the figure and its image after a 180° counterclockwise rotation about point H. (Lesson 11-3)

Complete each congruence statement if $\triangle LMN \cong \triangle QRS$. (Lesson 11-2)

35. $\angle Q \cong$ _?_

36. $\overline{QS} \cong$ _?_

37. $\overline{LM} \cong$ _?_

38. $\angle M \cong$ _?_

39. ALGEBRA Angles P and Q are complementary. If $m\angle P = (x + 3)°$ and $m\angle Q$ is twice $m\angle P$, write an equation that can be used to find the value of x. (Lesson 11-1)

40. SKYSCRAPERS On a clear day, the number of miles a person can see to the horizon can be found using the formula $d = 1.22 \cdot \sqrt{h}$ where d is the distance to the horizon in miles and h is the person's distance from the ground in feet. Suppose you are standing in the observation area of the Willis Tower in Chicago. About how far can a person see if the deck is 1353 feet above the ground? Round to the nearest tenth. (Lesson 9-1)

41. Evaluate $|25.3| - |-3.7|$. (Lesson 2-1)

Skills Review

Solve each equation. (Lesson 4-5)

42. $3x + 60 = 120$

43. $4x + 24 = 36$

44. $10x + 100 = 300$

45. $2x + 25 = 79$

11-5 Polygons

Why?

Puzzles pieces from two advanced level puzzles were accidentally mixed together.

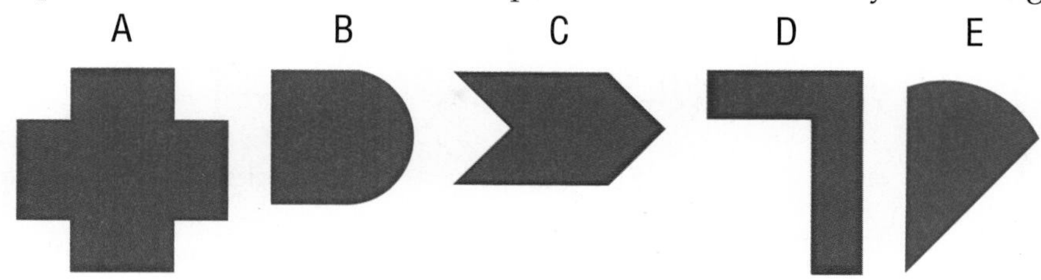

a. Pieces A, C, and D were from puzzle one. The other two pieces were from puzzle two. Describe a difference between the shapes of the two groups of puzzle pieces.

b. Draw a sample of another puzzle piece that could be a part of each puzzle.

Then

You have already classified quadrilaterals. (Lesson 11-4)

Now

- Classify polygons.
- Determine the sum of the measures of the interior angles of a polygon.

New Vocabulary

polygon
diagonal
interior angle
regular polygon
tessellation

Math Online

glencoe.com

- Extra Examples
- Personal Tutor
- Self-Check Quiz
- Homework Help

Classify Polygons A **polygon** is a simple, closed figure formed by three or more line segments called *sides*. The segments that form a polygon intersect only at their endpoints. The figures below are *not* polygons.

Non Polygons

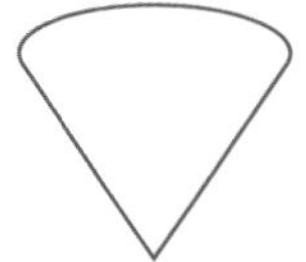

The figure has a curve.

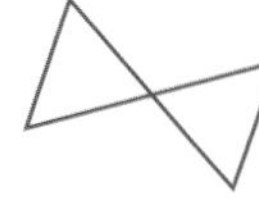

The sides overlap.

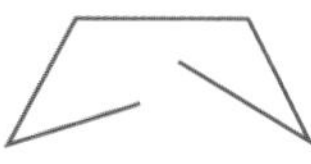

The figure is not closed.

Polygons can be classified by the number of sides they have, as shown in the table.

Polygon	Number of Sides	Polygon	Number of Sides
triangle	3	heptagon	7
quadrilateral	4	octagon	8
pentagon	5	nonagon	9
hexagon	6	decagon	10

EXAMPLE 1 Classify Polygons

Determine whether the figure is a polygon. If it is, classify the polygon. If it is not a polygon, explain why.

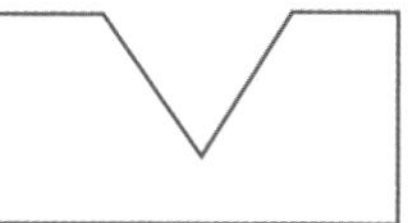

The figure has 7 sides that only intersect at their endpoints. It is a heptagon.

Check Your Progress

1A.

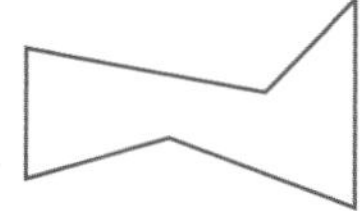

1B.

Personal Tutor glencoe.com

StudyTip

***n*-gon** A polygon with n sides is called an n-gon. For example, an octagon can also be called an 8-gon.

Find Angle Measures of a Polygon A **diagonal** is a line segment that joins two nonconsecutive vertices in a polygon. In the figures below, all possible diagonals from one vertex are shown.

quadrilateral	pentagon	hexagon	octagon
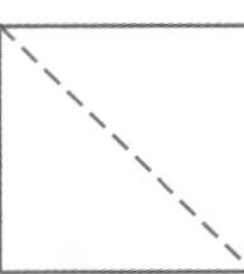	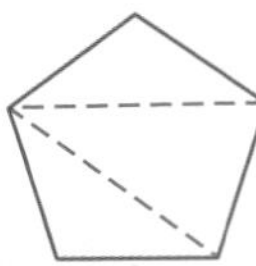	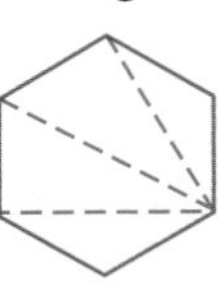	
4 sides 2 triangles	5 sides 3 triangles	6 sides 4 triangles	8 sides 6 triangles

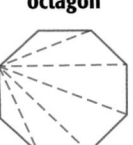

Notice that the number of triangles is 2 less than the number of sides. You can use this relationship to find the sum of the interior angle measures of a polygon. An **interior angle** is an angle formed at a vertex of a polygon.

Key Concept **Interior Angles of a Polygon** For Your FOLDABLE

Words The sum of the degree measures of the interior angles of the polygon is $(n - 2)180$.

Symbols $(n - 2)180$

Model

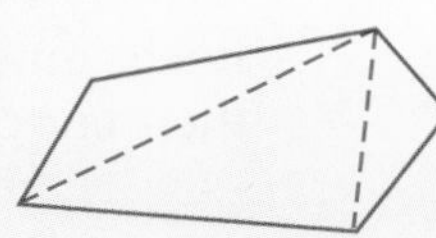

STANDARDIZED TEST EXAMPLE 2

Find the sum of the measures of the interior angles of a nonagon.

A 540° **B** 810° **C** 1260° **D** 1620°

Read the Test Item

The sum of the measures of the interior angles is $(n - 2)180$. Since a nonagon has 9 sides, $n = 9$.

Solve the Test Item

$(n - 2)180 = (9 - 2)180$ **Replace *n* with 9.**

$= 7(180)$ or 1260 **Simplify.**

The sum of the measures of the interior angles of a nonagon is 1260°. The answer is C.

Check

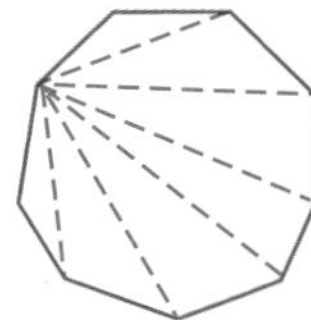

All possible diagonals from one vertex of a nonagon are shown at the right. You can see that 7 triangles are formed. So, $7 \cdot 180 = 1260$ is correct. ✓

Check Your Progress

2. Find the sum of the measures of the interior angles of a 13-gon.

F 990° **G** 1170° **H** 1980° **J** 2340°

Personal Tutor glencoe.com

A **regular polygon** is a polygon that has all sides congruent and all angles congruent. Since the angles are congruent, their measures are equal.

Real-World Link

The modern soccer ball is made of hexagonal and pentagonal panels because when they are sewn together and inflated, they make a near perfect sphere.

Source: SoccerBallWorld

Real-World EXAMPLE 3 Measure of One Interior Angle

SOCCER The surface of the soccer ball contains 12 regular pentagons and 20 regular hexagons. What is the measure of one interior angle of a hexagon?

Step 1 Find the sum of the measures of the interior angles of a hexagon.

$(n - 2)180 = (6 - 2)180$ **A hexagon has 6 sides. Replace n with 6.**

$= 4(180)$ or 720 **Simplify.**

The sum of the measures of the interior angles is 720°.

Step 2 Divide the sum of the measures by 6 to find the measure of one angle.

$720 \div 6 = 120$

The measure of one interior angle of a hexagon is 120°.

Check Your Progress

3. What is the measure of an interior angle of one of the pentagonal panels in the soccer ball?

Personal Tutor glencoe.com

Tessellations A repetitive pattern of polygons that fit together with no overlaps or holes is called a **tessellation**. The sum of the measures of the angles where the vertices meet in a tessellation is 360°.

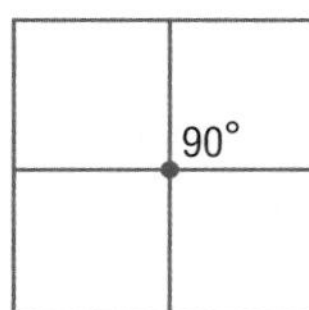

$4 \times 90° = 360°$

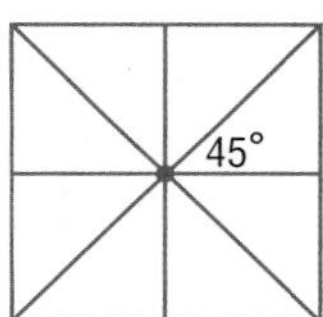

$8 \times 45° = 360°$

EXAMPLE 4 Find Tessellations

Determine whether or not a tessellation can be created using only regular hexagons. If not, explain.

The measure of each angle in a regular hexagon is 120°.

The sum of the measures of the angles where the vertices meet must be 360°. So, solve $120°n = 360$.

$120n = 360$ **Write the equation.**

$\frac{120n}{120} = \frac{360}{120}$ **Divide each side by 120.**

$n = 3$ **Simplify.**

Since 120° divides evenly into 360°, a regular hexagon can be used to make a tessellation.

Check Your Progress

Determine whether or not a tessellation can be created using each regular polygon. If not, explain.

4A. pentagon

4B. octagon

Personal Tutor glencoe.com

Check Your Understanding

Example 1
p. 617

Determine whether the figure is a polygon. If it is, classify the polygon. If it is not a polygon, explain why.

1. 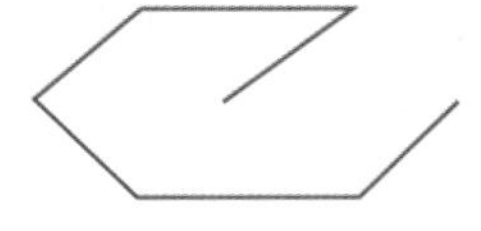

2. 

3. 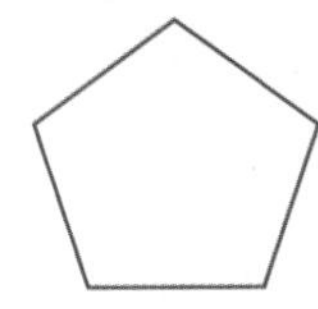

Example 2
p. 618

4. **MULTIPLE CHOICE** The sum of the measures of the interior angles of a certain regular polygon is 1800°. How many sides does this polygon have?

A 9 sides **B** 10 sides **C** 11 sides **D** 12 sides

Example 3
p. 619

5. **KALEIDOSCOPE** The kaleidoscope image at the right is a regular polygon with 14 sides. What is the measure of one interior angle of the polygon? Round to the nearest tenth.

Example 4
p. 619

6. Determine whether or not a tessellation can be created using only equilateral triangles. If not, explain.

Practice and Problem Solving

 = **Step-by-Step Solutions** begin on page R11.
Extra Practice begins on page 810.

Example 1
p. 617

Determine whether the figure is a polygon. If it is, classify the polygon. If it is not a polygon, explain why.

7. 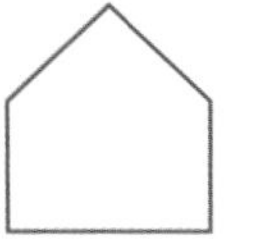

8. 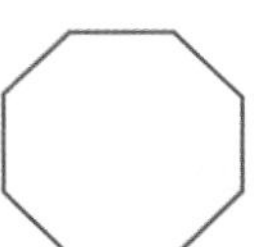

9. 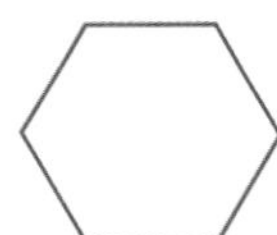

10. 

11. 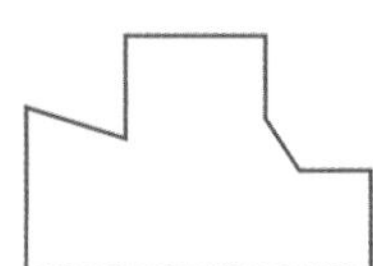

12. 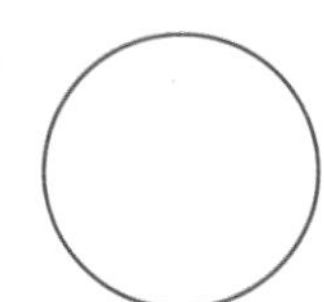

Example 2
p. 618

Find the sum of the measures of the interior angles of each polygon.

13. decagon **14.** 11-gon **15.** 16-gon **16.** 24-gon

Example 3
p. 619

17. **NATURE** The individual cells of a honeycomb are hexagons. What is the measure of an interior angle of a honeycomb?

18. **ARCHITECTURE** The dome in a state capitol building is octagonal. What is the measure of an interior angle of an octagon?

Example 4
p. 619

Determine whether or not a tessellation can be created using each regular polygon. If not, explain.

19. quadrilateral **20.** 12-gon **21.** 15-gon **22.** 20-gon

Real-World Link

M.C. Escher was an artist who often used tessellations to create fabulous works of art. The painting above was created using this method.

Identify the polygon given the sum of the interior angle measures.

23. 1080° **24.** 2340° **25.** 3240° **26.** 5040°

27. ART Refer to the painting at the left. How did the artist use tessellations to create the image?

TESSELLATIONS You can create a tessellation using a translation.

a. Draw a square. Then draw a triangle inside the top of the square.
b. Translate or slide the triangle from the top to the bottom of the square.
c. Repeat this pattern unit to create a tessellation.

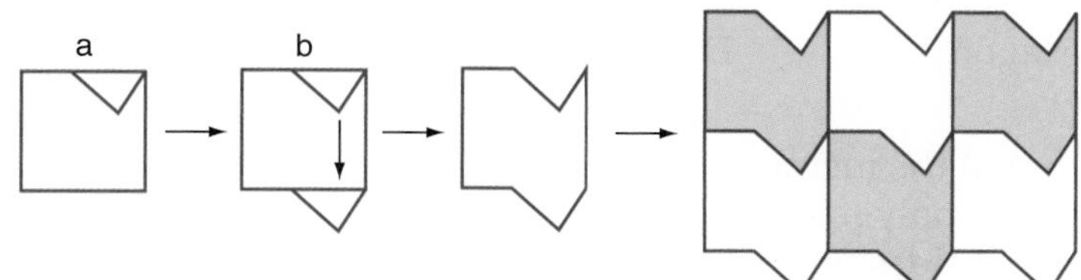

Use a translation to create a tessellation for each pattern shown.

28. 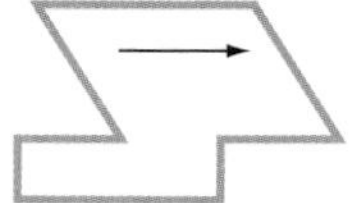

29. 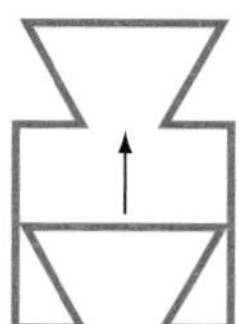

30. 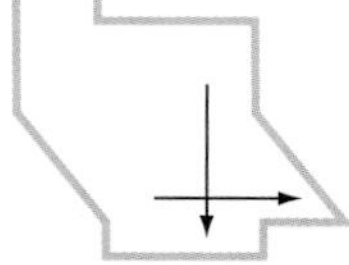

When a side of a polygon is extended, an *exterior angle* is formed. In any polygon, the sum of the measures of the exterior angles, one at each vertex, is 360°. Find the measure of an exterior angle of each regular polygon.

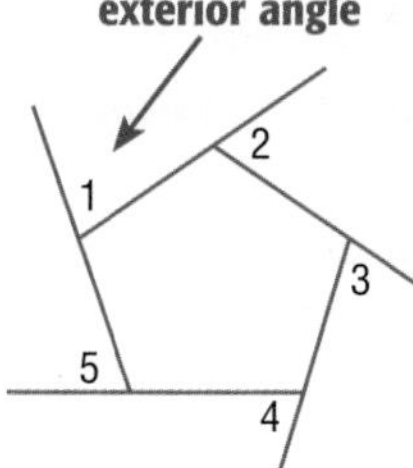

31 triangle **32.** octagon

33. decagon **34.** 15-gon

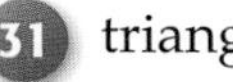

H.O.T. Problems
Use Higher-Order Thinking Skills

35. OPEN ENDED Use two types of polygons to create a tessellation that is different from the tessellations shown in this lesson. Describe the polygons and the transformation that you used.

36. REASONING If the number of sides of a regular polygon increases by 1, what happens to the sum of the measures of the interior angles?

37. FIND THE ERROR Jacinta says that it is possible to use a trapezoid to create a tessellation. Robert says this is impossible because the interior angles of a trapezoid are not congruent. Is either of them correct? Explain your reasoning.

38. CHALLENGE Create a tessellation using regular octagons and squares.

39. WRITING IN MATH Describe the difference between a regular polygon and a polygon that is irregular. Then explain the process used to find the interior angle measure of a regular polygon.

Standardized Test Practice

40. Which term identifies the shaded part of the design shown?

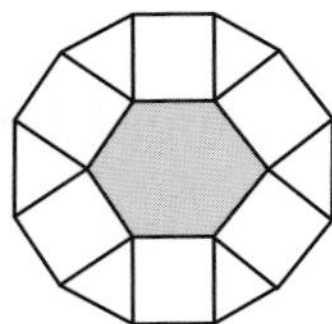

A heptagon
B hexagon
C octagon
D pentagon

41. The sum of the measures of the interior angles of a polygon is 2160°. Find the number of sides of the polygon.

F 10
G 14
H 16
J 18

42. A landscape architect is looking for a brick paver shape that will tessellate. Which shape by itself will allow her to tessellate a patio area?

A 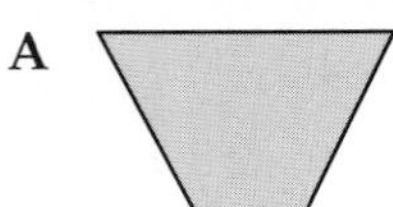

C 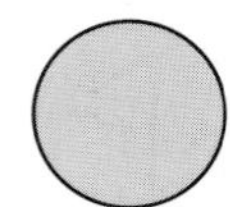

B 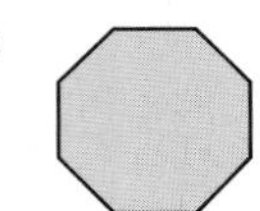

D 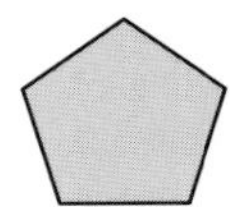

43. GRIDDED RESPONSE What is the measure in degrees of an interior angle of a regular polygon with 20 sides?

Spiral Review

Determine whether each statement is *always, sometimes* or *never* true. (Lesson 11-4)

44. A square is a rhombus.

45. A parallelogram is a rectangle.

46. A rectangle is a square.

47. A parallelogram is a quadrilateral.

48. A figure has vertices $A(-3, 2)$, $B(-1, 1)$, $C(-2, -3)$, and $D(-4, -2)$. Graph the figure and its image after a rotation of 180° around the origin. (Lesson 11-3)

49. Determine whether the triangles shown are congruent. If so, name the corresponding parts and write a congruence statement. (Lesson 11-2)

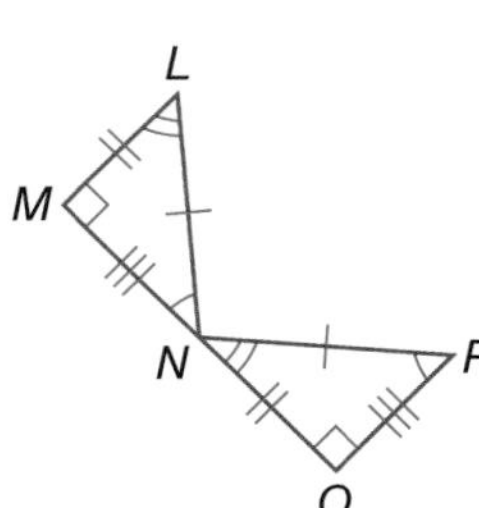

50. GARDENING Suppose you plant a square garden with an area of 300 square feet. What is the minimum amount of fencing needed to enclose the garden if the fencing only comes in whole-foot sections? (Lesson 5-1)

51. ALGEBRA Solve $x - 3.4 \geq 6.2$. Graph the solution on a number line. (Lesson 5-4)

Find each quotient. (Lesson 2-5)

52. $-69 \div 23$

53. $48 \div (-8)$

54. $-24 \div (-12)$

55. $-50 \div 5$

Skills Review

Simplify each expression. (Lesson 1-2)

56. $(5 - 2)180$

57. $(7 - 2)180$

58. $(10 - 2)180$

59. $(9 - 2)180$

CHAPTER 11

Mid-Chapter Quiz

Lessons 11-1 through 11-5

1. If $m\angle Y = 23°$ and $\angle Y$ and $\angle Z$ are complementary, what is $m\angle Z$? (Lesson 11-1)

2. Angles *G* and *H* are supplementary. If $m\angle G = x + 11$ and $m\angle H = x - 13$, what is x, $m\angle G$, and $m\angle H$? (Lesson 11-1)

In the figure, $m \parallel \ell$ and t is a transversal. If $m\angle 3 = 140°$, find the measure of each angle. Explain your reasoning. (Lesson 11-1)

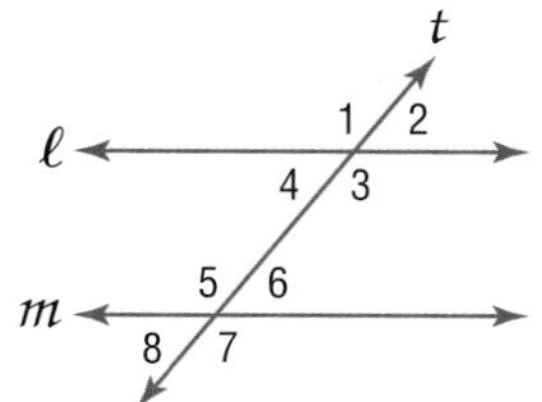

3. $m\angle 5$ **4.** $m\angle 7$

5. UMBRELLAS Each of the eight sections in the umbrella at the right is congruent, with the spokes having equal length. (Lesson 11-2)

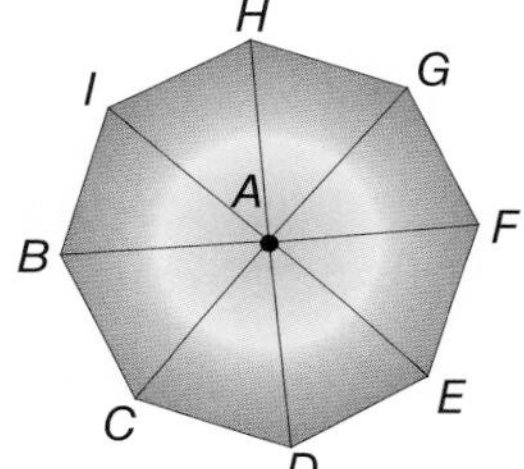

a. If $EF = 11$ inches and $EA = 14$ inches, what is CD?

b. Write a congruence statement involving $\triangle ABI$ and another triangle.

6. MULTIPLE CHOICE Which is *not* a true congruence statement for the congruent triangles shown below? (Lesson 11-2)

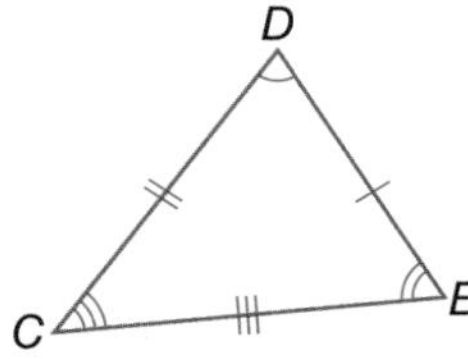

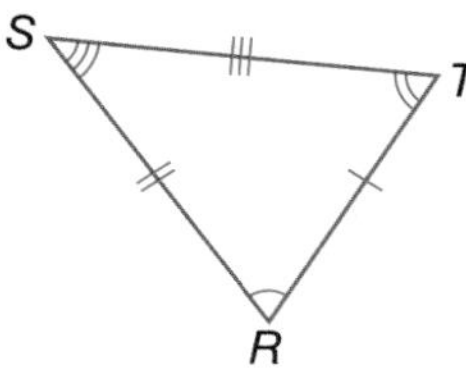

A $\triangle BCD \cong \triangle TSR$ **C** $\angle DBC \cong \angle RST$

B $\overline{CD} \cong \overline{SR}$ **D** $\triangle BDC \cong \triangle TRS$

7. Triangle *GHJ* has vertices $G(-2, 4)$, $H(-3, 0)$, and $J(-4, 3)$. Graph the figure and its image after a clockwise rotation of 90° about vertex *H*. Give the coordinates of the vertices for $\triangle G'H'J'$. (Lesson 11-3)

8. Determine whether the hubcap shown at the right has rotational symmetry. If it does, describe the angle of rotation. (Lesson 11-3)

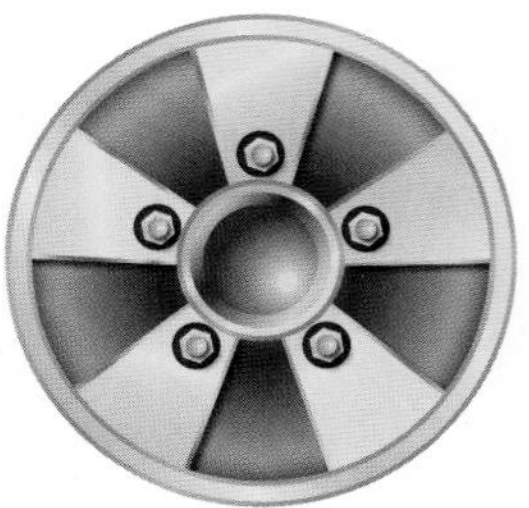

9. ALGEBRA Find the value of *x* in the quadrilateral at the right. Then find the missing angle measures. (Lesson 11-4)

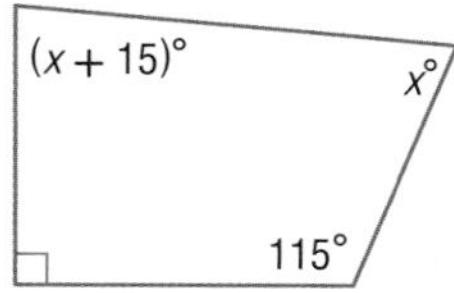

10. MULTIPLE CHOICE Which figure is a parallelogram? (Lesson 11-4)

F

H

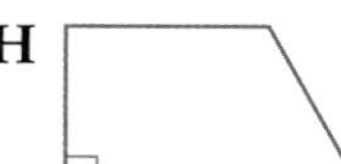

G

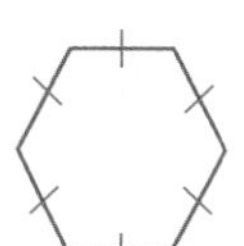

J

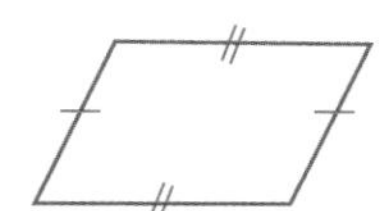

PARKING SPACES Classify each quadrilateral using the name that *best* describes it. (Lesson 11-4)

11.

12.

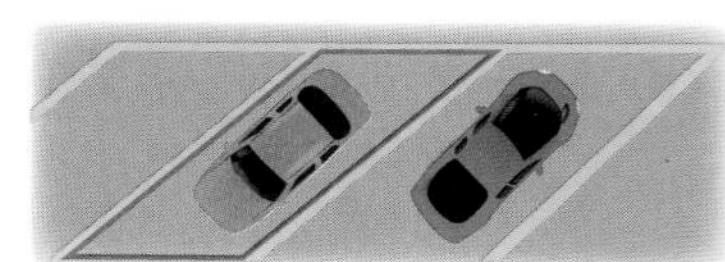

13. GRIDDED RESPONSE The outline of the star in the Houston Astros' logo is a regular pentagon. What is the measure in degrees of one interior angle in the pentagon? (Lesson 11-5)

Find the sum of the measures of the interior angles of each polygon. (Lesson 11-5)

14. octagon **15.** 14-gon

16. Find the measure of one interior angle of a regular nonagon. (Lesson 11-5)

11-6

Then
You found the area of rectangles. (Lesson 5-1)

Now
- Find areas of parallelograms.
- Find areas of triangles and trapezoids.

New Vocabulary
base
altitude

Math Online
glencoe.com
- Extra Examples
- Personal Tutor
- Self-Check Quiz
- Homework Help

Why?

Maya draws a rectangle on grid paper. She cuts off a triangle from the left and moves it to the right side of the figure.

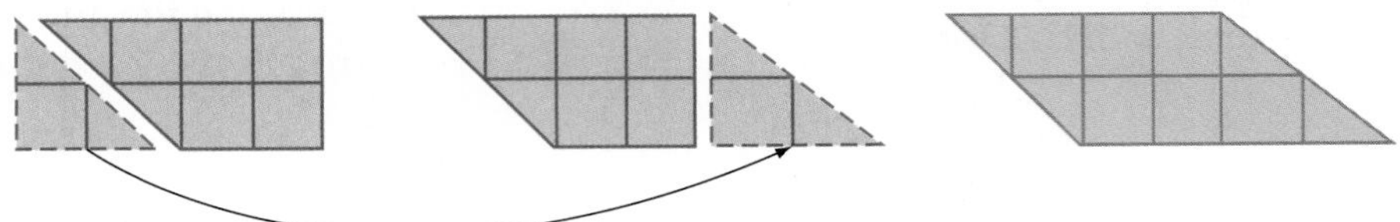

a. What is the area of the original rectangle?

b. Describe the new figure. What is its area?

c. What parts of a parallelogram could you use to find its area?

Area of Parallelograms The **base** of a parallelogram is any side of the parallelogram. The *height* is the length of an **altitude**, a line segment perpendicular to the base with endpoints on the base and the side opposite the base.

Key Concept — Area of Parallelogram
For Your FOLDABLE

Words The area A of a parallelogram in square units is $A = bh$, where b is the base of the parallelogram and h is the height.

Symbols $A = bh$

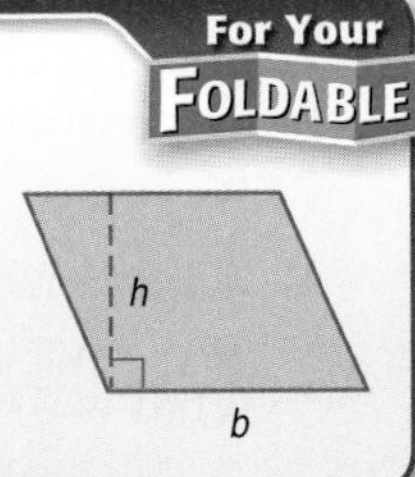

EXAMPLE 1 Find Areas of Parallelograms

Find the area of each parallelogram.

a.

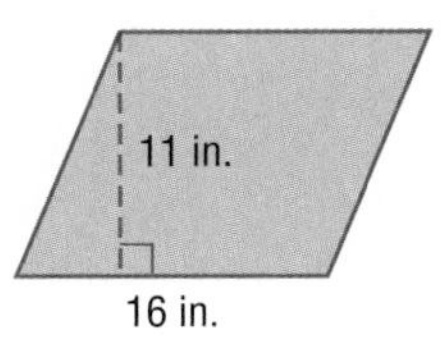

$A = bh$ Area of a parallelogram

$= 16 \cdot 11$ $b = 16$ and $h = 11$

$= 176$ Multiply.

The area is 176 square inches.

b.

7.5 cm

6.2 cm

$A = bh$ Area of a parallelogram

$= 6.2 \cdot 7.5$ $b = 6.2$ and $h = 7.5$

$= 46.5$ Multiply.

The area is 46.5 square centimeters.

Check Your Progress

1A.

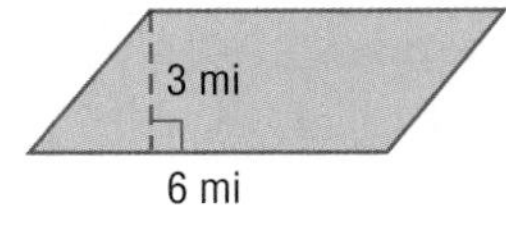

1B.

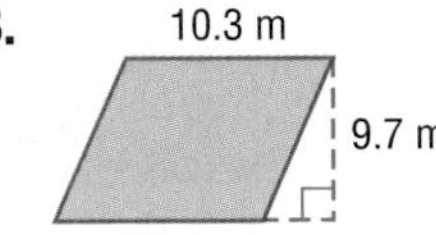

Personal Tutor glencoe.com

> **StudyTip**
>
> **Altitudes** An altitude can be outside a parallelogram or triangle.

Area of Triangles and Trapezoids The parallelogram below is separated into two congruent triangles by a diagonal.

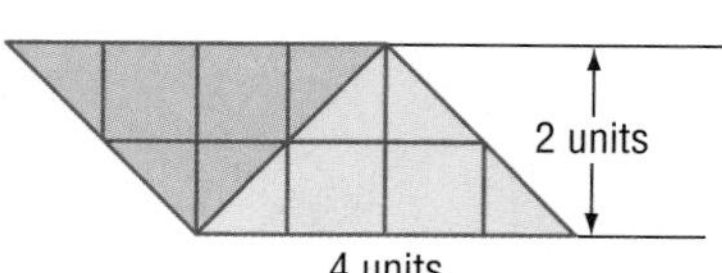

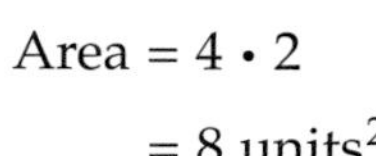

Area = 4 • 2
= 8 units^2

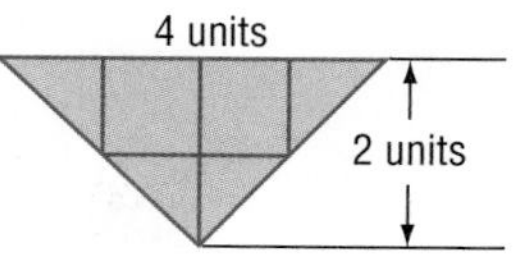

Area = $\frac{1}{2}$ • (4 • 2)
= 4 units^2

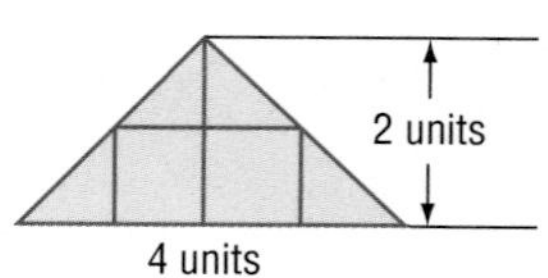

Area = $\frac{1}{2}$ • (4 • 2)
= 4 units^2

The area of each triangle is one-half the area of the parallelogram.

Key Concept — Area of Triangle

For Your 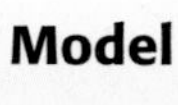

Words The area A of a triangle in square units is $A = \frac{1}{2}bh$, where b is the base of the triangle and h is the height.

Symbols $A = \frac{1}{2}bh$

Model

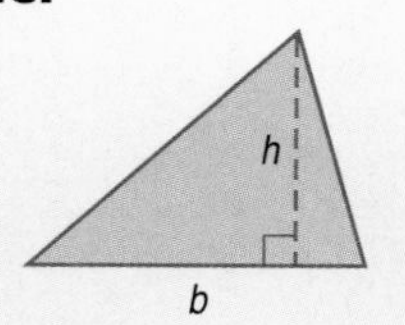

EXAMPLE 2 Find Areas of Triangles

Find the area of each triangle.

> **StudyTip**
>
> **Alternative Method** Multiplication is commutative and associative. So, you can also find $\frac{1}{2}(8)(5.2)$ by first multiplying $\frac{1}{2} \cdot 8$ and then multiplying the result by 5.2.

a.

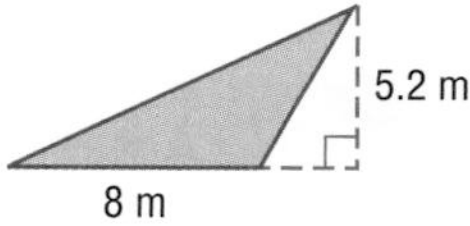

Estimate $\frac{1}{2} \cdot 8 \cdot 5 = 4 \cdot 5$ or 20

$A = \frac{1}{2}bh$	**Area of a triangle**
$= \frac{1}{2}(8)(5.2)$	**$b = 8$ and $h = 5.2$**
$= \frac{1}{2}(41.6)$	**Multiply 8 • 5.2.**
$= 20.8$	**Simplify.**

Check $20.8 \approx 20$ ✓

The area is 20.8 square meters.

b.

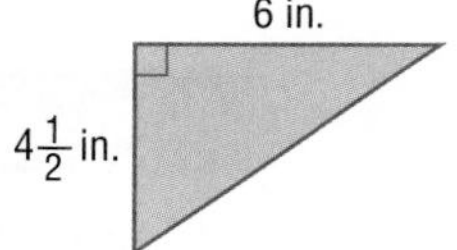

Estimate $\frac{1}{2} \cdot 6 \cdot 5 = 3 \cdot 5$ or 15

$A = \frac{1}{2}bh$	**Area of a triangle**
$= \frac{1}{2}(6)\left(4\frac{1}{2}\right)$	**$b = 6$ and $h = 4\frac{1}{2}$**
$= \frac{1}{2}\left(\frac{6}{1}\right)\left(\frac{9}{2}\right)$	**Rewrite terms as improper fractions.**
$= \frac{54}{4}$ or $13\frac{1}{2}$	**Multiply.**

Check $13\frac{1}{2} \approx 15$ ✓

The area is $13\frac{1}{2}$ square inches.

Check Your Progress

2A. 2 cm, 10 cm

2B. 3 ft, $5\frac{1}{3}$ ft

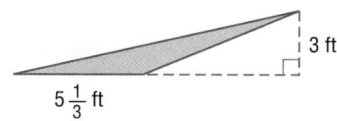

Personal Tutor glencoe.com

A trapezoid has two bases. The height of a trapezoid h is the distance between the bases. Notice that diagonal $\overline{HK}$ separates the trapezoid below into two triangles, $\triangle GHK$ and $\triangle HJK$. The area of the trapezoid is the sum of the areas of the two triangles.

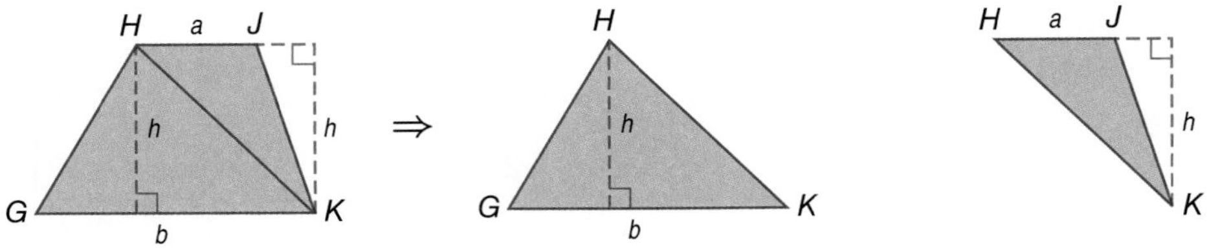

area of trapezoid $GHJK$	$=$	$\underbrace{\text{area of } \triangle GHK}$	$+$	$\underbrace{\text{area of } \triangle HJK}$
	$=$	$\frac{1}{2}bh$	$+$	$\frac{1}{2}ah$
	$=$	$\frac{1}{2}ah + \frac{1}{2}bh$	**Commutative Property**	
	$=$	$\frac{1}{2}h(a + b)$	**Distributive Property**	

Key Concept — Area of Trapezoid

For Your FOLDABLE

Words The area A of a trapezoid equals half the product of the height h and the sum of the bases $b_1 + b_2$.

Symbols $A = \frac{1}{2}h(b_1 + b_2)$

Model

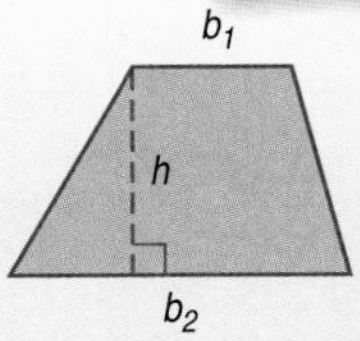

Watch Out!

Do not assume that the bases of a trapezoid are always the "top" and "bottom" sides of the figure.

Real-World EXAMPLE 3 Find Areas of Trapezoids

GEOGRAPHY The state of Nevada is shaped like a trapezoid as shown. Estimate the area of Nevada.

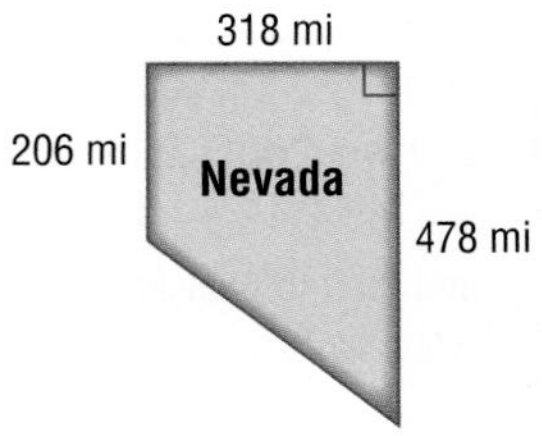

The height is 318 miles, and the bases are 206 miles and 478 miles.

$A = \frac{1}{2}h(a + b)$	**Area of a trapezoid**
$= \frac{1}{2} \cdot 318(206 + 478)$	**Replace *h* with 318, *a* with 206, and *b* with 478.**
$= \frac{1}{2} \cdot 318 \cdot 684$	**Simplify inside the parentheses.**
$= 108{,}756$	**Simplify.**

The area of Nevada is approximately 108,756 square miles.

Check Your Progress

3. Find the area of a trapezoid that has a height of 12.4 centimeters and bases of 10.5 centimeters and 7 centimeters.

Personal Tutor glencoe.com

Check Your Understanding

Examples 1–3 pp. 624–626

Find the area of each figure.

1. 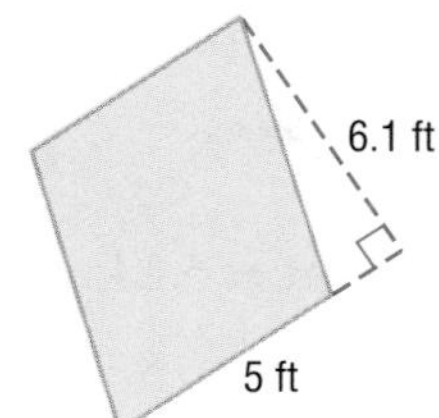

2. 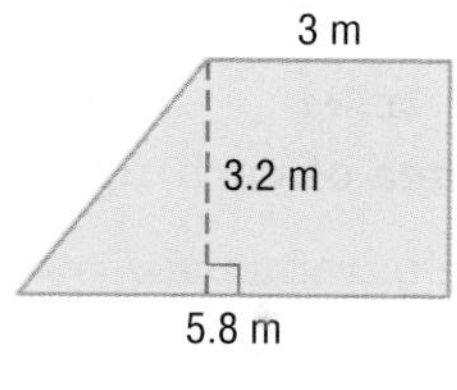

3. 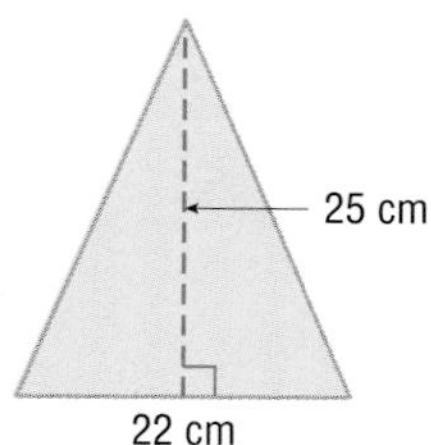

4. parallelogram: base = 9.4 m, height = 7.6 m
5. trapezoid: height = 16 in., bases = 3.1 in., 7.6 in.
6. **LACROSSE** A lacrosse goal with net is shown at the right. The goal is 6 feet wide, 6 feet high, and 7 feet deep. What is the area of the triangular region of the ground inside the net?

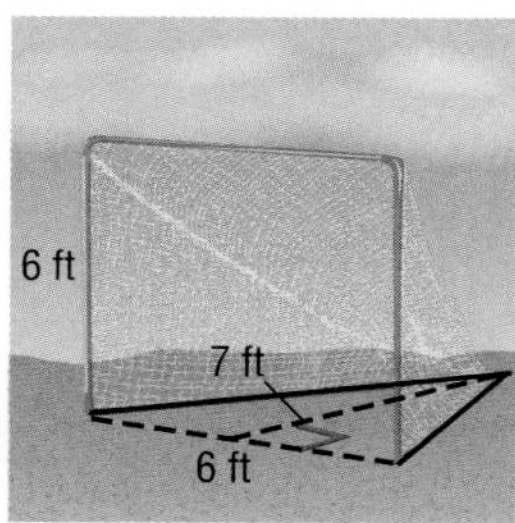

Practice and Problem Solving

● = **Step-by-Step Solutions** begin on page R11.
Extra Practice begins on page 810.

Examples 1–3 pp. 624–626

Find the area of each figure.

7. 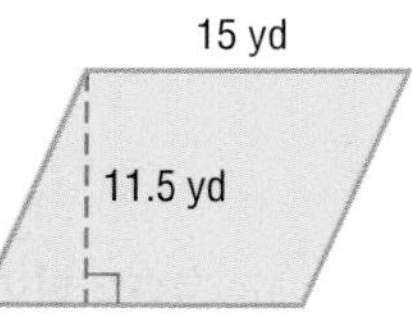

8. 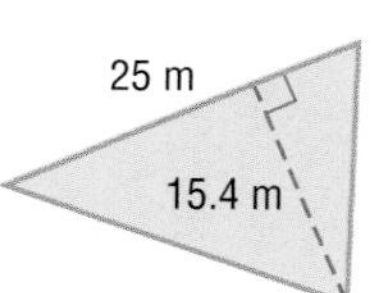

9. 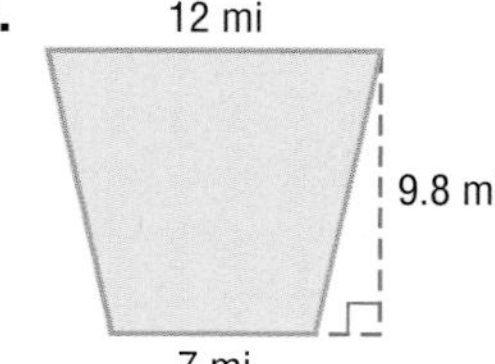

10. 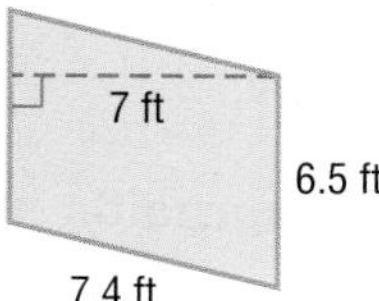

11. 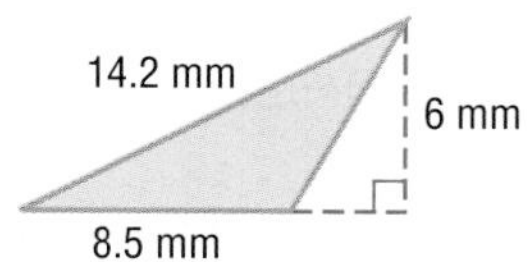

12. 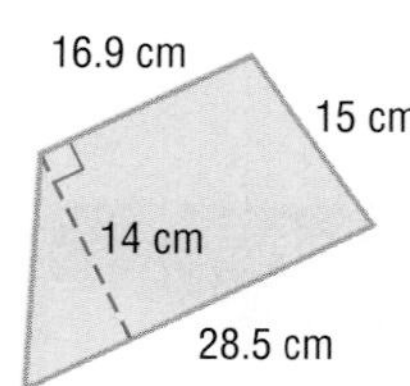

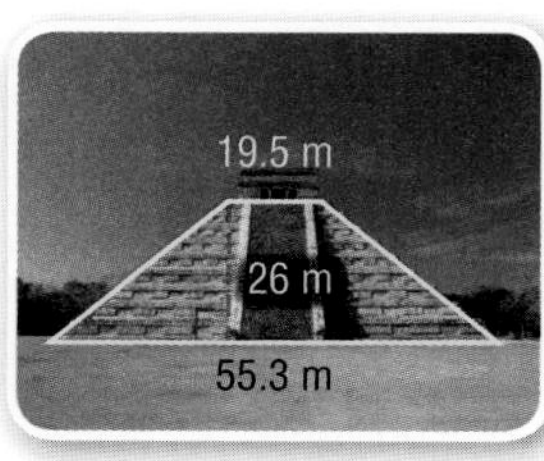

Real-World Link

The Mayan Pyramid of Kukulkan in the Yucatan Peninsula, Mexico, was built around 800 A.D. Each face of the pyramid has a stairway with ninety-one steps.

Source: World-Mysteries

13. **HISTORY** Each face of the Mayan Pyramid of Kukulkan is a trapezoid with the approximate dimensions shown at the left. What is the approximate area of one face of the structure?

Find the area of each figure.

14. parallelogram: height = 9 mm, base = 5 mm
15. trapezoid: height = 6 cm, bases = 4 cm, 7 cm
16. triangle: base = 12 ft, height = 4.4 ft

17. **HOPSCOTCH** The first six steps of a hopscotch pattern are shown at the right.

  a. What is the area of triangle 5?

  b. What is the area formed by triangles 3 and 4?

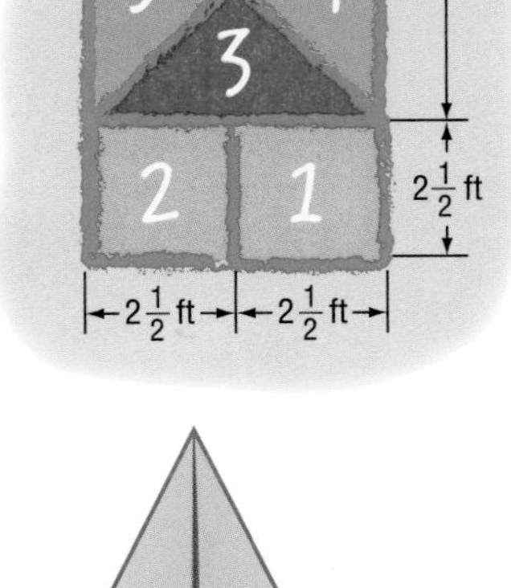

Find the area of each figure.

18. parallelogram: base = 8.2 km, height = 5.2 km

19. trapezoid: height = 2.4 m, bases = 7.9 m, 8.1 m

20. **SAILS** The sail at the right is formed from two congruent triangles. Find the area of the entire sail.

35 ft

28 ft

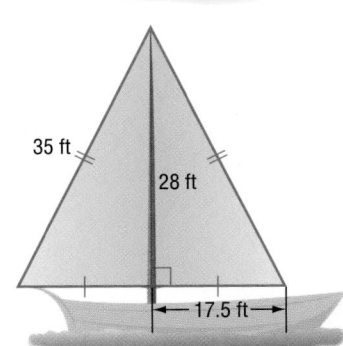

17.5 ft

21. Find the base of a parallelogram with a height of 14.3 centimeters and an area of 128.7 square centimeters.

22. Suppose a triangle has an area of 32 square kilometers and a base of 12.8 kilometers. What is the height?

23. A trapezoid has an area of 26 square feet. What is the measure of the height if the bases measure 1.3 feet and 3.2 feet? Round to the nearest tenth.

StudyTip

Estimation When finding areas of polygons, it is helpful to use estimation to determine whether your answer is reasonable.

Find the area of each figure.

24.

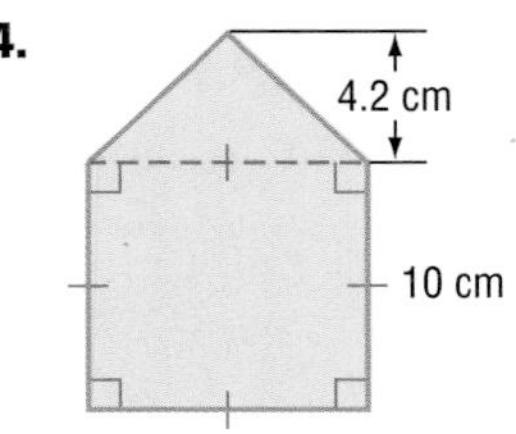

25.

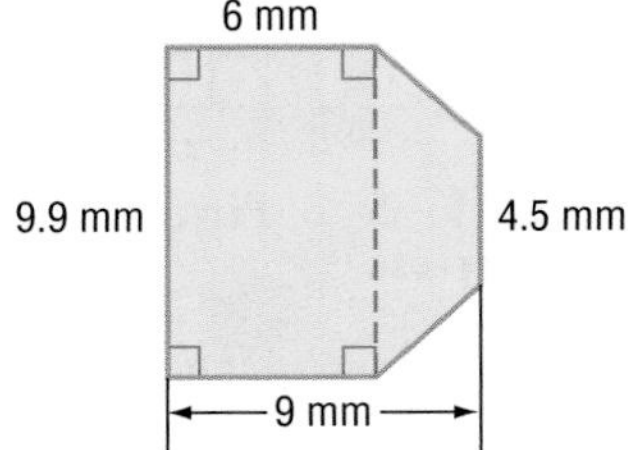

Find the area of each figure with the vertices shown.

26. rectangle: $A(-3, 4)$, $B(5, 4)$, $C(5, -1)$, $D(-3, -1)$

27. parallelogram: $H(-2, -1)$, $I(0, 2)$, $J(5, 2)$, $K(3, -1)$

28. triangle: $E(-2.5, 2)$, $F(3, -2.5)$, $G(3, 2)$

Find the perimeter and area of each trapezoid.

29.

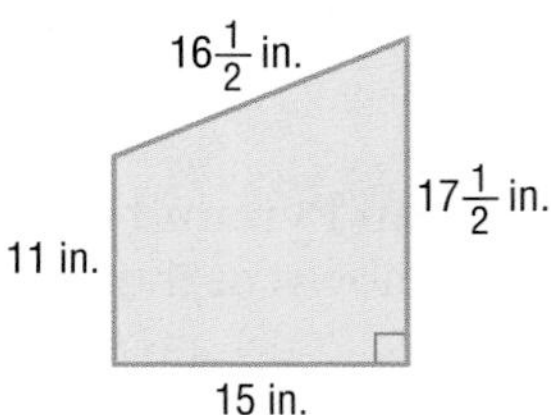

30.

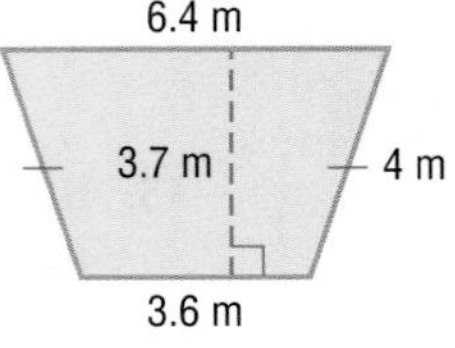

Vocabulary Review

Perimeter the distance around a geometric figure (Lesson 5-1)

31. **PAINTING** One wall of Tyve's room is in the shape of a trapezoid. The base of the wall is 12 feet wide. The top is 8 feet wide. The wall is 9 feet high. There is a rectangular window 2 feet by 3 feet. If she wants to paint the wall, what is the area that she needs to paint?

StudyTip

Measurement When converting measurements, remember that there will be a greater number of smaller units than larger units.

32. **MEASUREMENT** Find the area of the parallelogram at the right in square centimeters. Round to the nearest tenth. (*Hint*: 1 in. ≈ 2.54 cm)

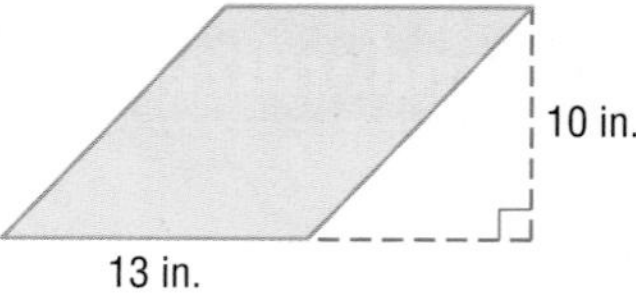

33 **MEASUREMENT** Find the area of the trapezoid at the right in square yards. Round to the nearest tenth. (*Hint*: 1 $yd^2 = 9\ ft^2$)

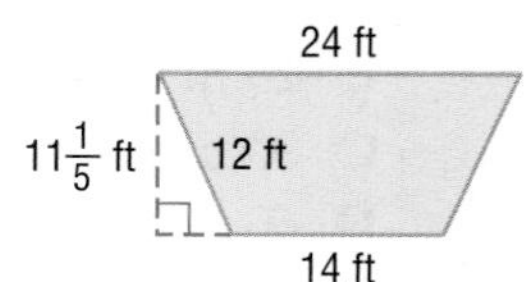

34. **MULTIPLE REPRESENTATIONS** In this problem, you will explore the area of a figure when a dilation occurs.

a. **GEOMETRIC** Sketch and label a triangle with a base of 6 units and a height of 3 units.

b. **TABULAR** Make a table like the one shown at the right. Find the area of each triangle.

c. **ANALYTICAL** Describe how the area of a triangle changes if the dimensions are doubled and if the dimensions are tripled.

Triangle	Base (in.)	Height (in.)	Area (in^2)
1	6	3	■
2	12	6	■
3	18	9	■

H.O.T. Problems

Use Higher-Order Thinking Skills

35. **OPEN ENDED** Give an example of a triangle and a parallelogram that have the same area. Describe the bases and heights of each figure. State the area.

36. **FIND THE ERROR** Cameron and Lydia found the area of the trapezoid shown at the right. Is either of them correct? Explain your reasoning.

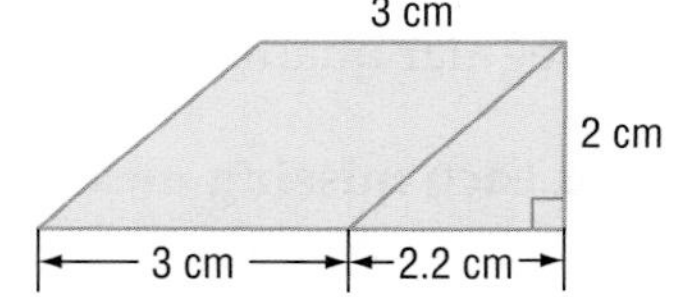

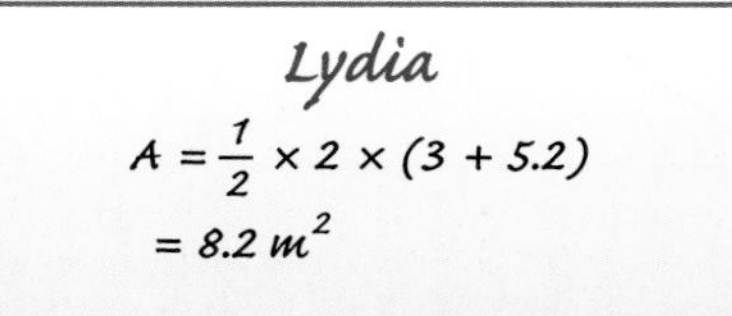

Cameron
$A = \frac{1}{2} \times 2 \times (3 \times 5.2)$
$= 15.6\ m^2$
Lydia
$A = \frac{1}{2} \times 2 \times (3 + 5.2)$
$= 8.2\ m^2$

37. **CHALLENGE** Refer to the parallelogram *QRST* shown at the right.

a. Describe a parallelogram that has the same area but a different perimeter than *QRST*. State the perimeter and area of the new parallelogram.

b. Describe a parallelogram that has the same perimeter but a different area than *QRST*. State the perimeter and area of the new parallelogram.

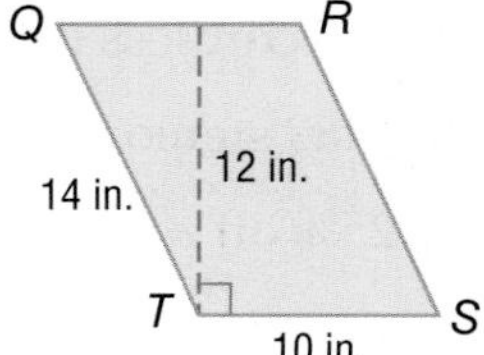

38. **REASONING** Describe another method for finding the area of the trapezoid in Exercise 36.

39. **WRITING IN MATH** Describe how the formula for the area of a trapezoid is related to the formula for the area of a triangle.

Standardized Test Practice

40. Find the area of the figure below.

A 17 m^2
B 25 m^2
C 30 m^2
D 60 m^2

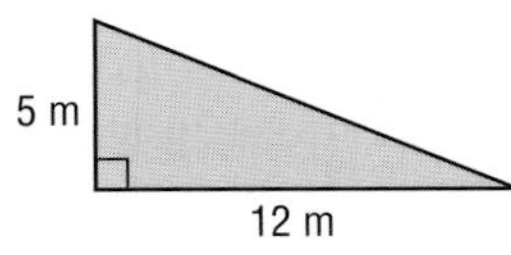

41. EXTENDED RESPONSE Mrs. Watts wants to fertilize her lawn for spring. The fertilizer that she wants to buy indicates one bag will fertilize 1000 square feet.

a. List the steps that Mrs. Watts will need to follow to determine the number of bags she will need to purchase.

b. Suppose Mrs. Watts' lawn measures 90 feet by 150 feet. Calculate the number of bags of fertilizer she will need.

42. Square X has an area of 9 square feet. The sides of square Y are twice as long as the sides of square X. Find the area of square Y.

F 18 ft^2
G 36 ft^2
H 9 ft^2
J 6 ft^2

43. An architect designed a room that was in the shape of a trapezoid. She wants to figure the amount of carpet that will be needed for the room. Use the room's dimensions below to determine the amount of carpet needed.

A 432.5 ft^2
B 487.5 ft^2
C 585 ft^2
D 607.5 ft^2

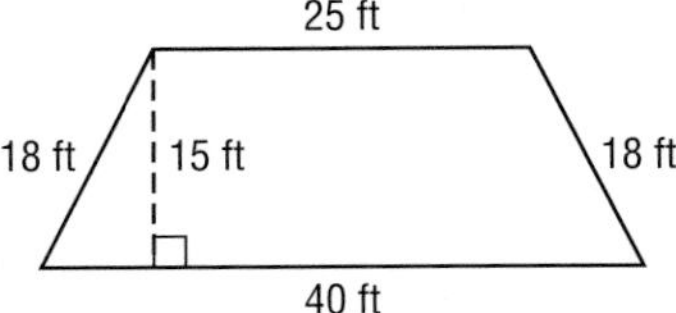

Spiral Review

Find the measure of an interior angle of each polygon to the nearest tenth. (Lesson 11-4)

44. regular pentagon

45. regular heptagon

46. regular quadrilateral

47. regular 12-gon

Find each missing measure. (Lesson 11-4)

48.

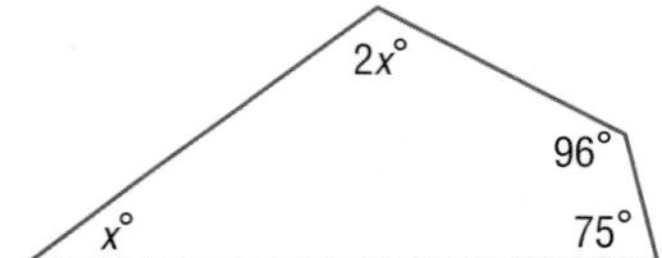

49.

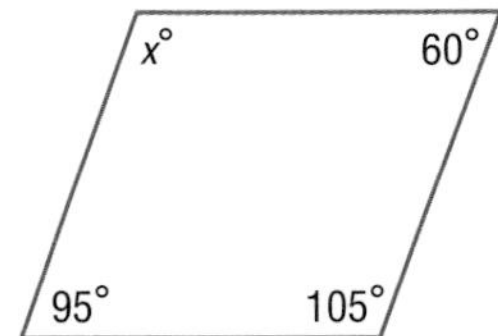

Find the discount to the nearest cent. (Lesson 7-5)

50. \$85 cordless phone, 20% off

51. \$489 stereo, 15% off

52. 25% off a \$74 baseball glove

53.

Skills Review

Use a calculator to find each product. Round to the nearest tenth. (Previous Course)

54. $3.14 \cdot 6$

55. $2 \cdot 3.14 \cdot 5.4$

56. $3.14 \cdot 2.2$

57. $3.14 \cdot 4.3$

11-7

Circles and Circumference

Then

You have already found the perimeter of rectangles. (Lesson 5-1)

Now

- Find the circumference of circles.
- Solve problems involving circumference.

New Vocabulary

circle
center
radius
diameter
chord
circumference
π (pi)

Math Online

glencoe.com

- Extra Examples
- Personal Tutor
- Self-Check Quiz
- Homework Help
- Math in Motion

Why?

Collect three different large circular objects.

a. Use a tape measure to find d, the distance across each object through its center and C, the distance around each object. Round to the nearest millimeter. Record your measures in a table.

b. For each object, find the ratio $\frac{C}{d}$.

c. Write an equation showing the relationship between C and d.

Circumference of Circles A **circle** is the set of all points in a plane that are the same distance from a given point in the plane.

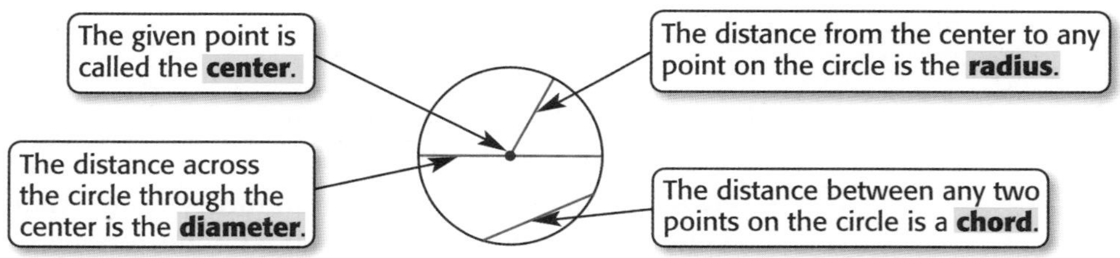

The given point is called the **center**.
The distance from the center to any point on the circle is the **radius**.
The distance across the circle through the center is the **diameter**.
The distance between any two points on the circle is a **chord**.

The **circumference** of a circle is the distance around the circle. In every circle, the ratio of the circumference to the diameter is equal to 3.1415926… . The Greek letter **π (pi)** stands for this number. So, $\frac{C}{d} = \pi$ or $C = \pi d$.

Key Concept — **Circumference of a Circle** — For Your FOLDABLE

Words The circumference C of a circle is equal to its diameter times π, or 2 times its radius times π.

Symbols $C = \pi d$ or $C = 2\pi r$

Model

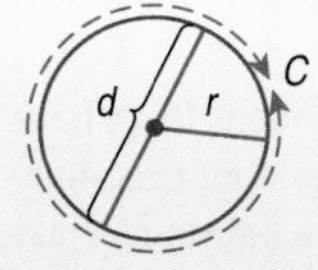

Math *in Motion*, Interactive Lab glencoe.com

EXAMPLE 1 Find the Circumference of a Circle

Find the circumference of the circle. Round to the nearest tenth.

$C = \pi d$ **Circumference of a circle**

$= \pi \cdot 3$ **Replace d with 3.**

$= 3\pi$ **Simplify. This is the *exact* circumference.**

3 [×] [2nd] [π] [ENTER] 9.424777961 **Use a calculator.**

3 in.

The circumference is about 9.4 inches.

Check Your Progress

1. Find the circumference of a circle with a diameter of $3\frac{3}{4}$ feet. Round to the nearest tenth.

Personal Tutor glencoe.com

StudyTip

Estimating If an exact answer is needed, leave the answer in terms of π. The exact answer to Example 2 is 10.6π. If an estimate is sufficient, use a calculator to find a decimal approximation. Round to the indicated place value.

EXAMPLE 2 Find the Circumference of a Circle

Find the circumference of the circle. Round to the nearest tenth.

$C = 2\pi r$	**Circumference of a circle**
$= 2 \cdot \pi \cdot 5.3$	**Replace r with 5.3.**
≈ 33.3	**Simplify. Use a calculator.**

The circumference is about 33.3 centimeters.

Check Your Progress

2. Find the circumference of the circle with a radius of 7 millimeters. Round to the nearest tenth.

Personal Tutor glencoe.com

Use Circumference to Solve Problems You can use the circumference of a circle to find the diameter or the radius.

Real-World EXAMPLE 3 Solve Problems

POOLS Bernard works at a community center that has a circular swimming pool with a circumference of 40 meters. He would like to use a rope to divide the pool down the center. What should be the length of the rope?

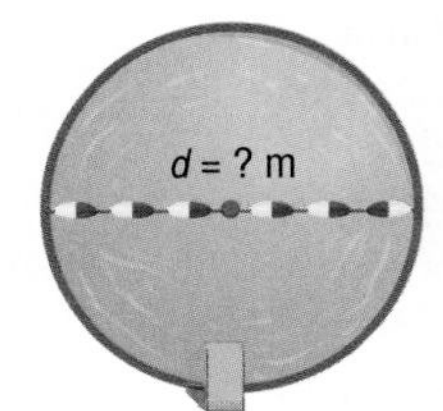

Understand You know the circumference of the pool. You need to find the diameter of the pool.

Plan Use the formula for the circumference of a circle to find the diameter.

Solve

$C = \pi d$	**Circumference of a circle**
$40 = \pi \cdot d$	**Replace C with 40.**
$\frac{40}{\pi} = d$	**Divide each side by π.**
$12.7 \approx d$	**Simplify. Use a calculator.**

So, the length of the rope should be about 12.7 meters.

Check Check the reasonableness of the solution by replacing d with 12.7 in $C = \pi d$.

$C = \pi d$	**Circumference of a circle**
$= \pi \cdot 12.7$	**Replace d with 12.7.**
≈ 39.9	**Simplify. Use a calculator.**

Since this circumference is close to the original circumference, 40 meters, the solution is reasonable.

Check Your Progress

3. MUSIC A CD has a diameter of 120 millimeters. A Universal Media Disc (UMD) has a diameter of 60 millimeters. Compare the circumferences of the discs.

Personal Tutor glencoe.com

Math History Link

π (pi)
Approximations for pi have been around for over 4000 years. One Babylonian tablet c.1900 B.C. used 3.125 as an estimate. One of the latest estimates was published in a book, *π to Five Million Places.* The book is 558 pages long and consists of the number pi computed to 5 million decimal places.

Check Your Understanding

Examples 1 and 2 pp. 631–632

Find the circumference of each circle. Round to the nearest tenth.

1. 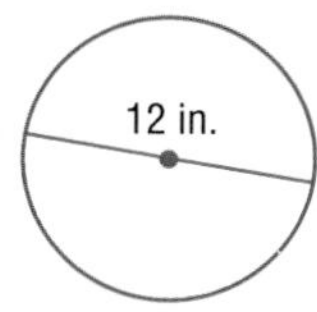

2. 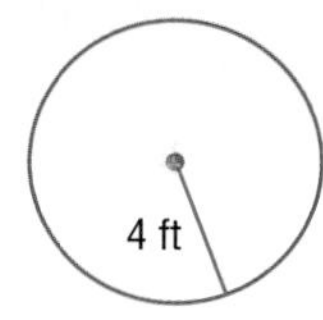

3. 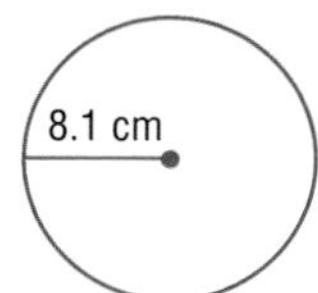

4. diameter = 10.5 centimeters

5. radius = 3.6 kilometers

Example 3 p. 632

6. **HISTORY** An ancient *timber circle* discovered in the United Kingdom was thought to have been built more than 4000 years ago. Known as "Seahenge," this ancient circle has a circumference of about 21.3 meters. What is the radius of the circle to the nearest tenth?

Practice and Problem Solving

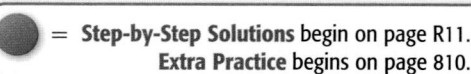

= **Step-by-Step Solutions** begin on page R11. **Extra Practice** begins on page 810.

Examples 1–3 pp. 631–632

Find the circumference of each circle. Round to the nearest tenth.

7 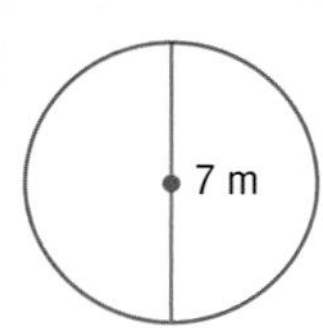

8. 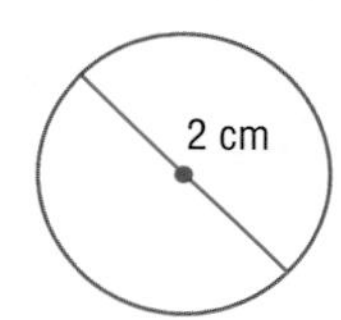

9. 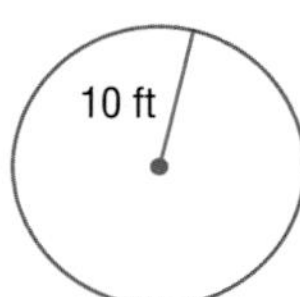

10. 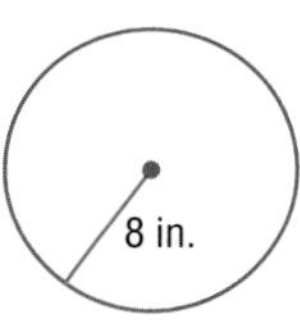

11. 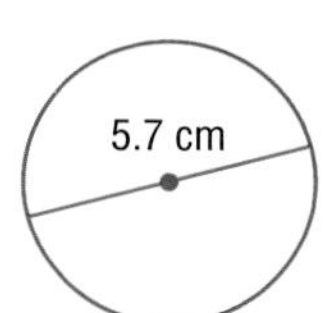

12. 

13. diameter = 9.4 meters

14. radius = 13.7 millimeters

15. radius = $11\frac{1}{4}$ feet

16. diameter = $14\frac{3}{5}$ yards

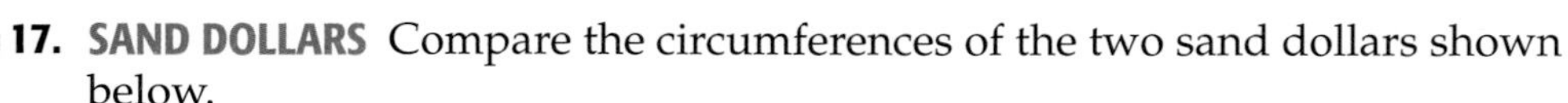

17. **SAND DOLLARS** Compare the circumferences of the two sand dollars shown below.

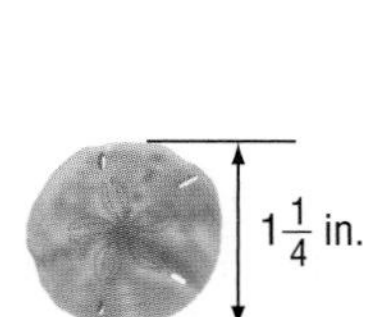

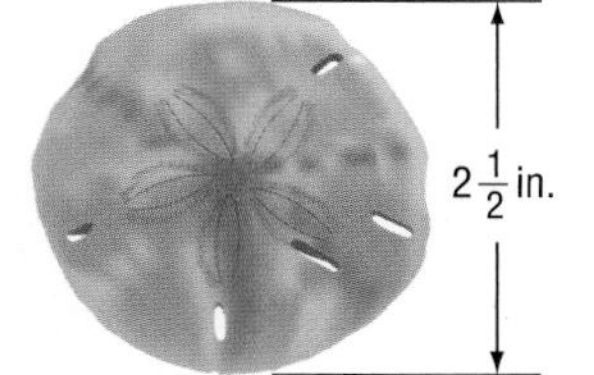

Real-World Link

The life span of a sand dollar ranges from 1 to 15 years. Sand dollars cannot survive if exposed to air for more than 2 or 3 hours.

Source: San Francisco State University

18. **SCIENCE** The circumference of the Moon is about 6790 miles. What is the distance to the center of the Moon in kilometers? Round to the nearest kilometer. (*Hint*: 1 mile ≈ 1.6 kilometers)

19. **CLOCKS** The tower at Philadelphia City Hall contains four clocks that have a radius of about 3.96 meters. Find how far the minute hand travels after each number of rotations around the clock face. Round to the nearest hundredth.

a. 2 rotations **b.** $\frac{1}{2}$ rotation **c.** $5\frac{3}{4}$ rotations

20. **CAROUSELS** The world's largest carousel in Spring Green, Wisconsin, has a diameter of 80 feet. How far does a rider travel on an outside horse after 10 revolutions? Round to the nearest foot.

21. **FOUNTAINS** A circular fountain at a park has a radius of 4 feet. The mayor wants to build a fountain that is quadruple the radius of the current fountain. Find the circumference of the new fountain. Round to the nearest tenth.

Real-World Link

The world's largest rideable motorcycle is 20 feet 4 inches long and weighs 6500 pounds.

Source: *Guinness World Records*

22. **MOTORCYCLES** The world's largest rideable motorcycle travels one mile after about 272.5 rotations of a tire.

    a. To the nearest tenth, how many feet does the motorcycle travel after one rotation of a tire? What does this measure represent? (*Hint*: 1 mile = 5280 feet)

    b. To the nearest tenth, how many feet tall are the tires?

23. **MULTIPLE REPRESENTATIONS** In this problem, you will investigate the relationship between the radius and circumference of a circle.

    a. **TABULAR** Make a table of values like the one at the right. Find the circumference of a circle having each radius. Round to the nearest tenth.

    b. **GRAPHICAL** Use your table to graph the circumference C of a circle as a function of the radius r.

    c. **ANALYTICAL** Describe the slope of the graph. How is the slope related to the formula for finding circumference?

Radius (in.)	Circumference (in.)
1	■
2	■
3	■
4	■
5	■

24. **GEOLOGY** The Beaverhead Crater is the largest impact crater in the United States. If the circumference of the crater is about 188.5 kilometers, what is the diameter of the crater?

H.O.T. Problems

Use Higher-Order Thinking Skills

25. **OPEN ENDED** Find a circular object in your home. Measure the diameter and use that value to calculate the circumference. Use a tape measure to check your calculation.

26. **REASONING** A *variable* is a quantity with a value that changes. In the formula for the circumference of a circle, identify any variables.

27. **CHALLENGE** Three congruent circles are inside a rectangle as shown at the right. Which is greater, the length of the rectangle ℓ, or the circumference of one circle? Explain your reasoning.

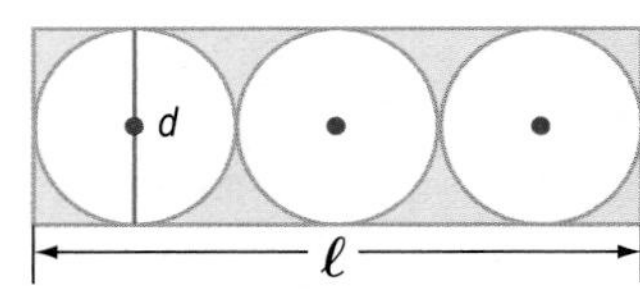

28. **REASONING** Explain why the formula $C = \pi d$ can be used to calculate the length of a circular bracelet. If the bracelet is opened up and laid out flat, can the same formula be used to find the length? Explain.

29. **WRITING IN MATH** Describe the relationship between circumference and radius. How does the circumference change if the radius is increased? How does the radius change if the circumference is decreased?

Standardized Test Practice

30. A plate has a radius of 5 inches. Which equation could be used to find the circumference of the plate in inches?

A $C = 2(10\pi)$
B $C = 5\pi$
C $C = 10\pi$
D $C = 2\pi$

31. A bicycle tire has a diameter of 24 inches. Find the circumference of the tire to the nearest tenth of an inch.

F 75.4 in.
G 83.5 in.
H 87.9 in.
J 91.5 in.

32. A planter has a circumference of 37.6 inches. Which measure is *closest* to the diameter of the planter?

A 10 in. **C** 12 in.
B 11 in. **D** 14 in.

33. EXTENDED RESPONSE The circle shown below has a diameter of 15 feet.

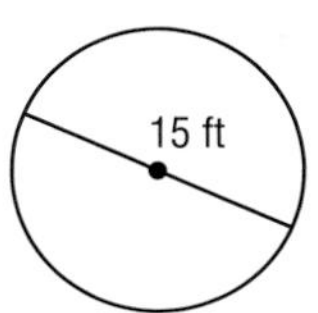

a. Write an expression to find the circumference of the circle.
b. What is the circumference to the nearest foot?

Spiral Review

Find the area of each figure described. (Lesson 11-6)

34. parallelogram: base, 10 yd; height 5 yd

35. trapezoid: height 3 ft; bases, 15 ft and 20 ft

Find the sum of the measures of the interior angles of each polygon. (Lesson 11-5)

36. triangle **37.** hexagon **38.** pentagon

39. INTERNET SHOPPING For every order submitted, an online bookstore charges a \$5 shipping fee plus a charge of \$2 per pound on the weight of the items being shipped. The total shipping charges y can be represented by $y = 2x + 5$, where x represents the weight of the order in pounds. Graph the equation. (Lesson 8-3)

40. ALGEBRA Solve $2x - 7 > 5x + 14$. (Lesson 5-5)

41. TOURS It costs \$15 per person to take a tour of an underground cave. (Lesson 4-6)

a. Write an equation to determine the total cost for any number of people to take the tour.
b. What is the total cost if 12 people take the tour?

Skills Review

Find each product. (Lesson 9-1)

42. $3.14 \cdot 4^2$ **43.** $3.14 \cdot \left(\frac{14}{2}\right)^2$ **44.** $3.14 \cdot (4.3)^2$

11-8 Area of Circles

Then

You have already found the circumference of a circle. (Lesson 11-7)

Now

- Find areas of circles.
- Find areas of sectors.

New Vocabulary

sector
central angle

Math Online

glencoe.com

- Extra Examples
- Personal Tutor
- Self-Check Quiz
- Homework Help

Why?

The trampoline at the right has a diameter of 16 feet. The circular area is shown on the grid paper below.

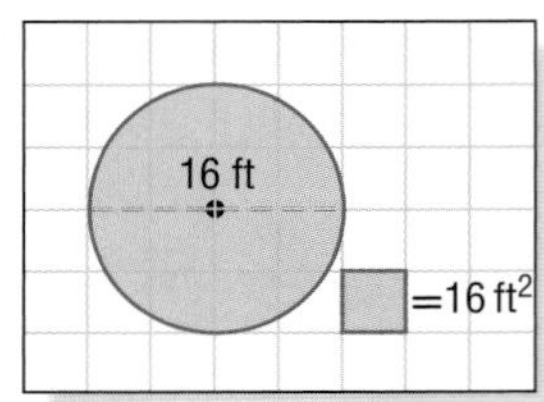

a. Approximately how many squares are shaded?

b. If one grid square represents 16 square feet, what is the approximate area of the trampoline?

Area of Circles A circle can be separated into parts and be arranged to form a figure that looks like a parallelogram.

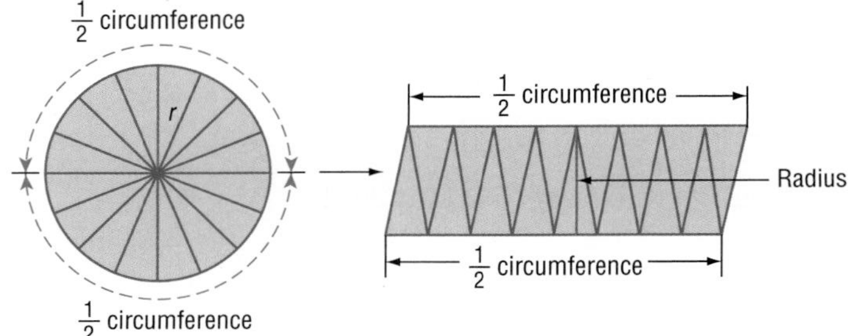

Since the areas of both figures are approximately the same, you can use the formula for the area of a parallelogram to find the area of a circle.

$A = bh$	**Area of a parallelogram**
$A = \left(\frac{1}{2} \times C\right)r$	**The base equals half the circumference of the circle; the height equals the radius of the circle.**
$A = \left(\frac{1}{2} \times 2\pi r\right)r$	**Replace C with $2\pi r$.**
$A = \pi \times r \times r$	**Simplify.**
$A = \pi r^2$	**Replace $r \times r$ with r^2.**

Key Concept — Area of a Circle

For Your FOLDABLE

Words The area A of a circle in square units is $A = \pi r^2$, where r is the length of the radius.

Symbols $A = \pi r^2$

Model

EXAMPLE 1 Find Areas of Circles

Find the area of each circle. Round to the nearest tenth.

a.

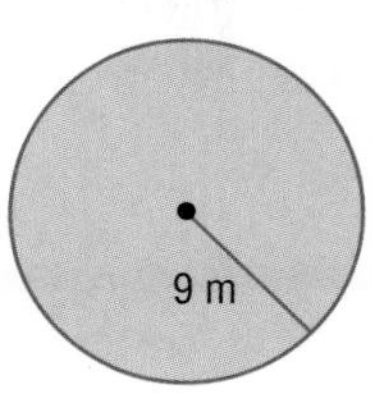

Estimate $A \approx 3 \cdot 9^2$ **or 243**

$A = \pi r^2$	**Area of a circle**
$= \pi \cdot 9^2$	**Replace *r* with 9.**
$= \pi \cdot 81$	**Evaluate 9^2.**
≈ 254.5	**Use a calculator.**

Check $254.5 \approx 243$ ✓

The area is about 254.5 m^2.

b.

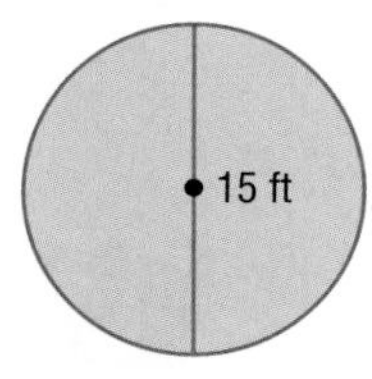

Estimate $A \approx 3 \cdot 8^2$ **or 192**

$A = \pi r^2$	**Area of a circle**
$= \pi \cdot 7.5^2$	$r = \frac{15}{2}$ **or 7.5**
$= \pi \cdot 56.25$	**Evaluate 7.5^2.**
≈ 176.7	**Use a calculator.**

Check $176.7 \approx 192$ ✓

The area is about 176.7 ft^2.

StudyTip

Estimation To estimate the area of a circle, multiply the square of the radius by 3.

Check Your Progress

1A.

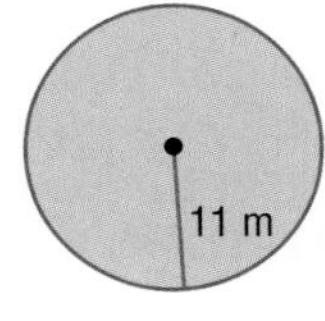

1B.

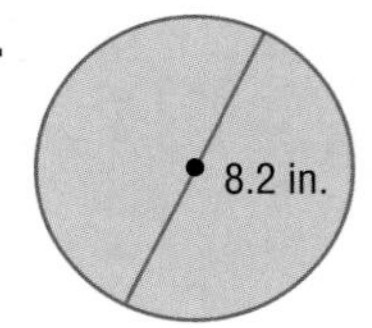

Personal Tutor glencoe.com

Real-World Link

The National Collegiate Athletic Association added women's hockey as a sanctioned sport in 1993. Women's ice hockey became an Olympic event in 1998. The U.S. team won the first gold medal.

Source: SportsKnowHow

Real-World EXAMPLE 2 Use Area of Circles to Solve Problems

HOCKEY A hockey rink is divided into three parts. The center part, called the *neutral zone*, is a rectangle with a center circle, as shown at the right. What is the area of the neutral zone around the center circle?

54 ft
85 ft
15 ft

Find the area of the rectangle minus the area of the circle.

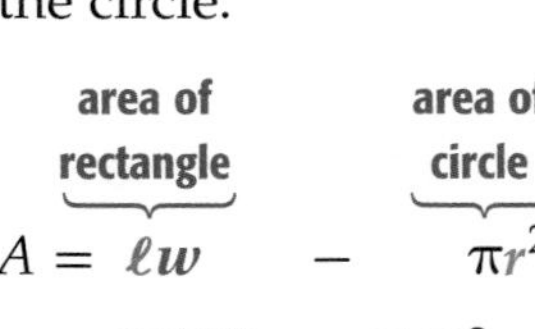

$$A = \ell w - \pi r^2$$

$= 85(54) - \pi(15)^2$	**Replace ℓ with 85, *w* with 54, and *r* with 15.**
$= 4590 - 225\pi$	**Simplify.**
≈ 3883.1	**Use a calculator.**

So, the area of the neutral zone around the circle is about 3883.1 square feet.

Check Your Progress

2. HOCKEY A *face-off circle* on a hockey rink is 30 feet across. At its center is a red spot 2 feet in diameter. What is the area of the face-off circle that is *not* red? Round to the nearest tenth.

Personal Tutor glencoe.com

Area of Sectors The shaded region in the circle at the right is called a **sector**. Its area depends on the radius of the circle and on the measure of the central angle. A **central angle** is an angle with a vertex at the center of a circle and with sides that intersect the circle.

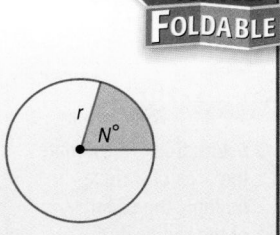

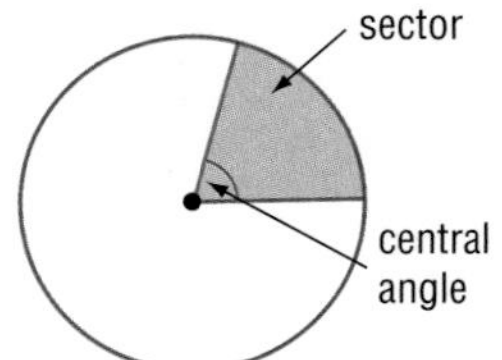

StudyTip

Units As with areas of circles, the area of a sector is given in square units.

Key Concept Area of a Sector

For Your FOLDABLE

Words The area A of a sector is $\frac{N}{360}(\pi r^2)$, where N is the degree measure of the central angle of the circle and r is the radius.

Symbols $A = \frac{N}{360}(\pi r^2)$

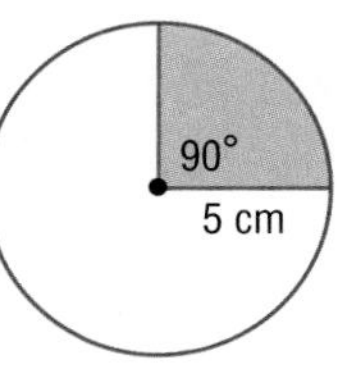

EXAMPLE 3 Find the Area of a Sector

Find the area of the shaded sector in the circle at the right. Round to the nearest tenth.

$A = \frac{N}{360}(\pi r^2)$	**Area of a sector**
$= \frac{90}{360}(\pi)(5^2)$	**Replace N with 90 and r with 5.**
$= \frac{1}{4}(\pi)(25)$	**Simplify.**
$\approx 19.6 \text{ cm}^2$	**Use a calculator.**

90°
5 cm

Check Your Progress

3. The diameter of a circle is 16 inches. It has a sector with a central angle of 120°. What is the area of the sector to the nearest tenth?

Personal Tutor glencoe.com

Check Your Understanding

Example 1 p. 637

Find the area of each circle. Round to the nearest tenth.

1.

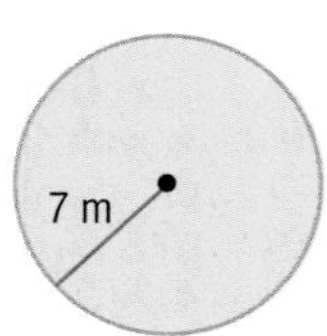

2.

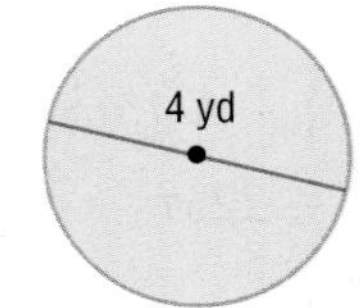

3.

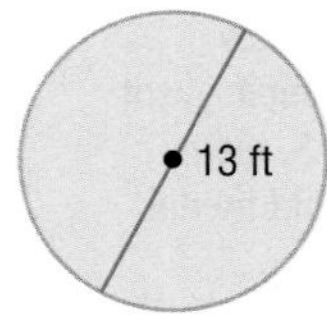

4. radius = 3.6 kilometers

5. diameter = 10.5 centimeters

Example 2 p. 637

6. TRAMPOLINES Refer to the trampoline at the beginning of the lesson.

a. Find the area of the trampoline. Compare to the estimate.

b. The safety padding around the jumping surface is 1 foot wide. What is the area of the jumping surface? (*Hint*: Draw and label a diagram.)

Example 3
p. 638

Find the area of each shaded sector. Round to the nearest tenth.

7.

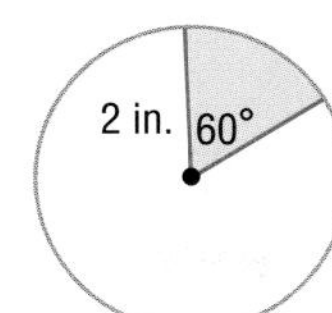

8.

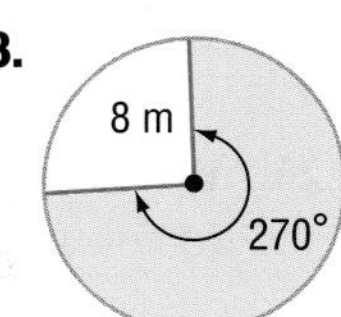

Practice and Problem Solving

● = **Step-by-Step Solutions** begin on page R11.
Extra Practice begins on page 810.

Example 1
p. 637

Find the area of each circle. Round to the nearest tenth.

9.

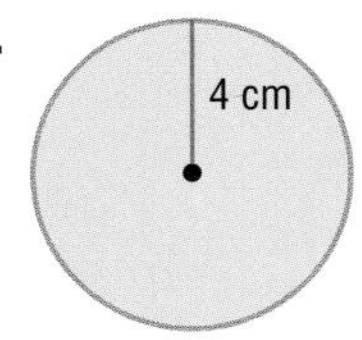

10.

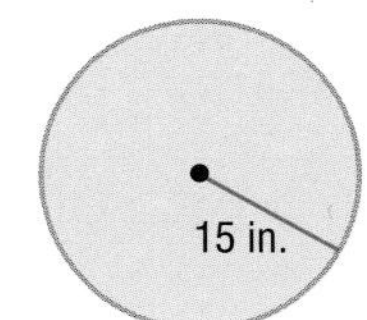

11.

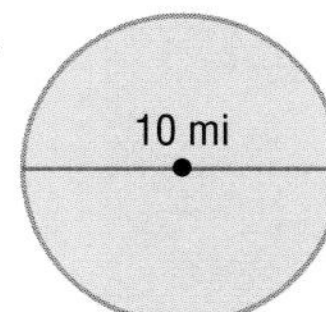

12.

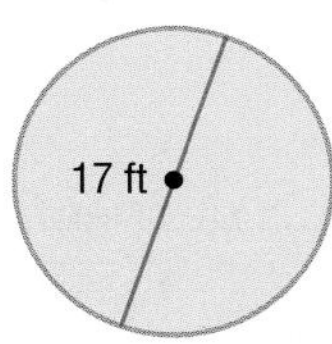

13.

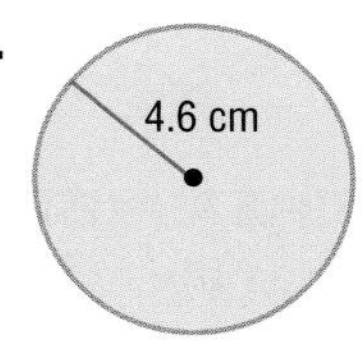

14.

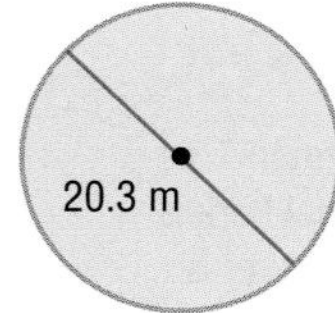

15. radius = 9.6 feet

16. diameter = 24.8 meters

17. diameter = $11\frac{1}{2}$ yards

18. radius = $3\frac{2}{3}$ miles

Example 2
p. 637

LAWN CARE Lauren has a sprinkler positioned in her lawn that directs a 12-foot spray in a circular pattern. About how much of the lawn does the sprinkler water if there is a rectangular flower bed 3 feet by 6 feet that is also in the path of the spray? Round to the nearest tenth.

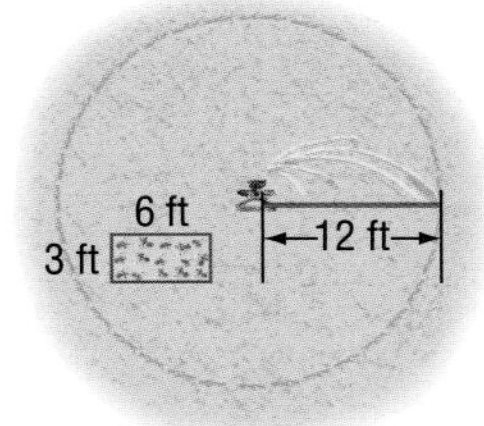

20. **CDS** What is the area of the CD shown at the right? Round to the nearest tenth.

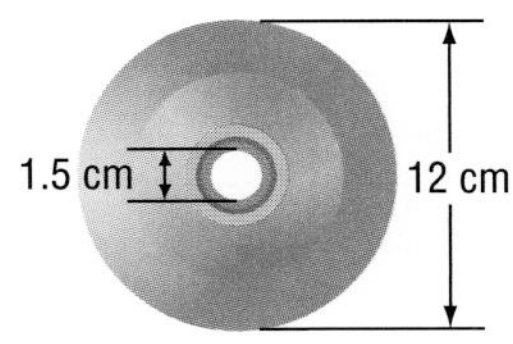

Example 3
p. 638

Find the area of each shaded sector. Round to the nearest tenth.

21.

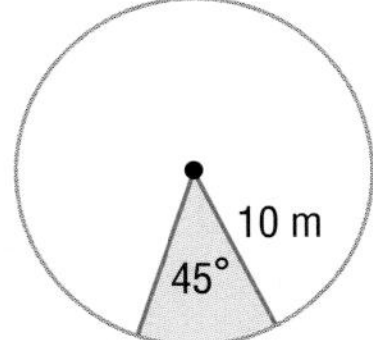

22.

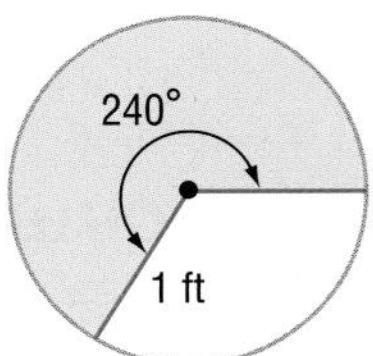

23. The diameter of a circle is 6 millimeters. It has a sector with a central angle of 72°. What is the area of the sector to the nearest tenth?

24. The diameter of a circle is 11 yards. It has a sector with a central angle of 20°. What is the area of the sector to the nearest tenth?

25. Find the radius of a circle if its area is 50 square inches. Round to the nearest inch.

26. What is the diameter of a circle if its area is 35.6 square centimeters? Round to the nearest tenth.

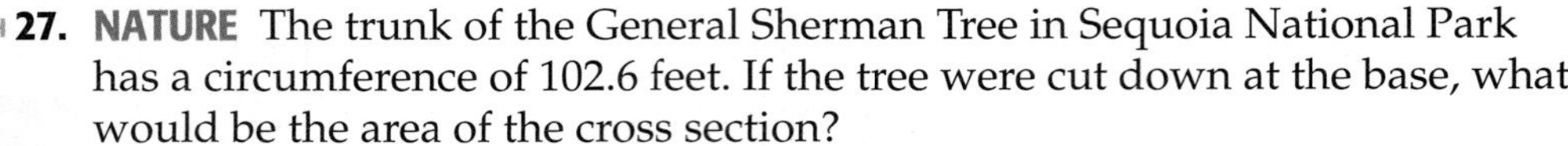

27. **NATURE** The trunk of the General Sherman Tree in Sequoia National Park has a circumference of 102.6 feet. If the tree were cut down at the base, what would be the area of the cross section?

Real-World Link

The General Sherman Tree in Sequoia National Park is nearly 275 feet tall and is estimated to be 2500 to 3000 years old.

Source: Illinois State University's Fell Arboretum

Find the distance around and the area of each figure. Round to the nearest tenth.

28. semicircle

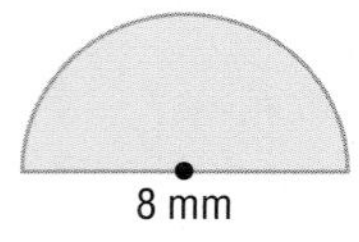

29. semicircle

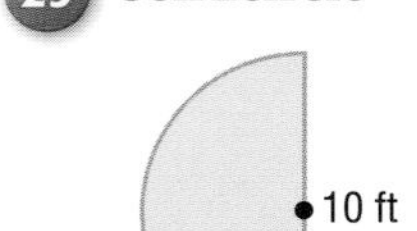

30. quarter circle

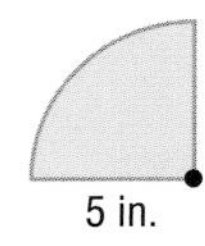

31. **MULTIPLE REPRESENTATIONS** In this problem, you will investigate the area of a circle as the radius changes.

a. **TABULAR** Make a table like the one at the right. Find the area of each circle to the nearest tenth.

Radius (cm)	Area (cm^2)
3	■
6	■
12	■
24	■
48	■

b. **ANALYTICAL** Describe how the area of a circle changes when the radius is doubled.

c. **LOGICAL** Predict the area of a circle that has a radius of 96 centimeters. Explain your reasoning. Then verify your prediction by finding the area.

H.O.T. Problems

Use Higher-Order Thinking Skills

32. **OPEN ENDED** Draw and label a circle that has an area between 800 square centimeters and 820 square centimeters. Label the length of the radius and state the area of the circle to the nearest tenth.

33. **WRITING IN MATH** Describe the difference between the circumference and area of a circle and explain how the formulas for circumference and area of a circle are related.

34. **CHALLENGE** The radius of circle B is 2.5 times the radius of circle A. If the area of circle A is 8 square yards, what is the area of circle B?

35. **REASONING** If the measures of the area and circumference of a circle have the same numerical values, what is the radius of the circle? Explain.

36. **WRITING IN MATH** Describe how you can find the area of a circle given the radius, diameter, or circumference.

Standardized Test Practice

37. Find the area of a circle with a diameter of 22 millimeters. Round to the nearest tenth.

A 380.1 mm^2
B 319.5 mm^2
C 189.9 mm^2
D 69.1 mm^2

38. A sprinkler is set to cover the area shown. Find the area of the grass being watered if the sprinkler reaches a distance of 20 feet.

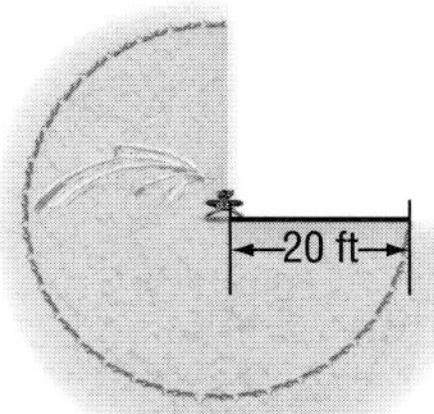

F 78.5 ft^2
G 314.2 ft^2
H 942.5 ft^2
J 1256.6 ft^2

39. The Blackwells have a circular pool with a radius of 10 feet. They want to install a 3–foot sidewalk around the pool. What will be the area of the walkway?

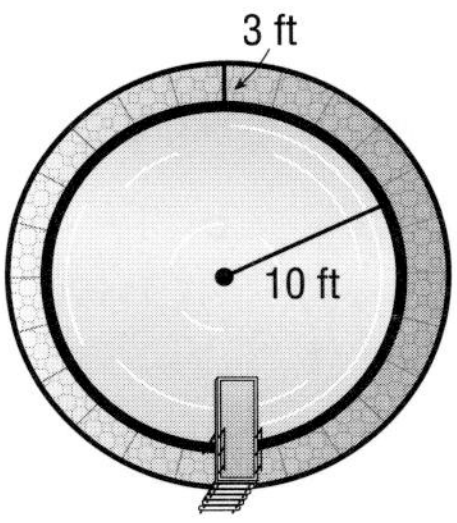

A 216.8 ft^2
B 285.9 ft^2
C 314.2 ft^2
D 442.2 ft^2

40. EXTENDED RESPONSE The area of a circle is 327.6 square centimeters.

a. Write an algebraic expression in terms of r that could be used to find the radius of the circle.

b. Find the radius to the nearest tenth.

Spiral Review

Find the circumference of each circle. Round to the nearest tenth. (Lesson 11-7)

41. radius: 8 in.
42. radius: 12.5 ft
43. diameter: 21 cm

Find the area of each figure. (Lesson 11-6)

44.

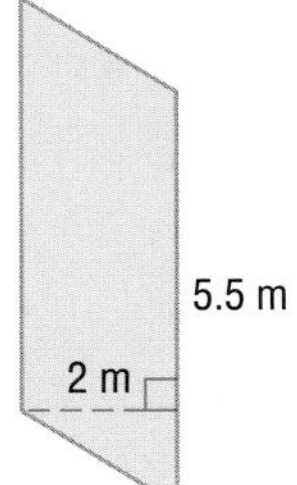

45.

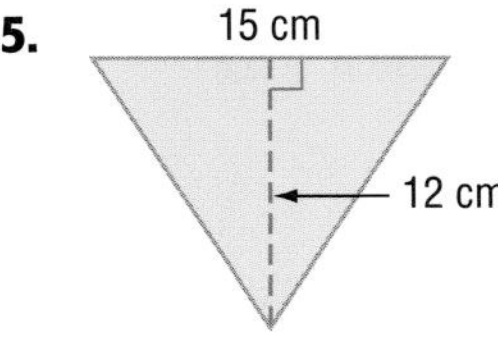

46.

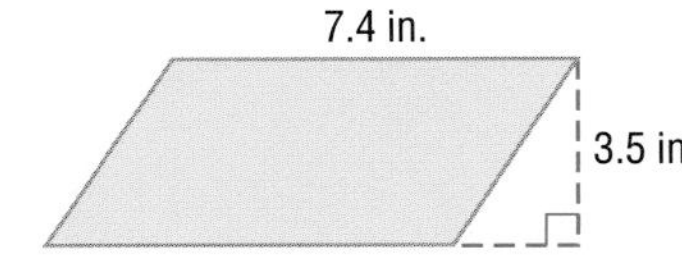

47. Find the product of $\frac{5}{9}$ and $\frac{8}{25}$. (Lesson 3-3)

Skills Review

Find each sum. (Previous Course)

48. 211 + 23.9
49. 512.41 + 21.3
50. 587 + 65.9
51. 52.89 + 85.56

11-9 Area of Composite Figures

Then

You found the areas of parallelograms, triangles, and trapezoids. (Lesson 11-6)

Now

- Find the area of composite figures.
- Solve problems involving the area of composite figures.

New Vocabulary

composite figure

Math Online

glencoe.com

- Extra Examples
- Personal Tutor
- Self-Check Quiz
- Homework Help

Why?

The flag of Ohio is the only state flag that is not in the shape of a rectangle.

a. Name different polygons that make up the flag.

b. Describe how these polygons could be used to find the total area of the flag.

Area of Composite Figures A **composite figure** is made up of two or more shapes. To find the area of a composite figure, decompose the figure into shapes with areas you know how to find. Then find the sum of those areas.

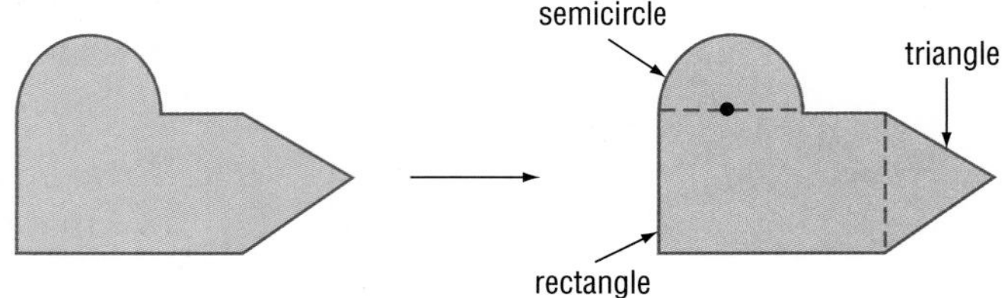

EXAMPLE 1 Find the Area of Composite Figures

Find the area of the composite figure.

Separate the figure into a rectangle and a triangle. Find the sum of the areas of the figures.

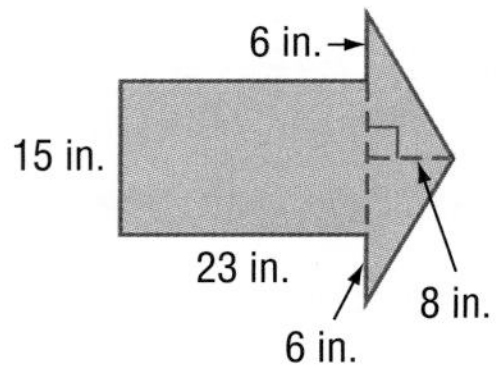

$A = bh$	**Area of a rectangle**
$= 15 \cdot 23$	$b = 15$ **and** $h = 23$
$= 345$	**Simplify.**

$A = \frac{1}{2}bh$	**Area of a triangle**
$= \frac{1}{2} \cdot 27 \cdot 8$	$b = 6 + 15 + 6$ **or 27 and** $h = 8$
$= 108$	**Simplify.**

The area of the composite figure is 345 + 108 or 453 square inches.

Check Your Progress

Find the area of each composite figure. Round to the nearest tenth, if necessary.

1A.

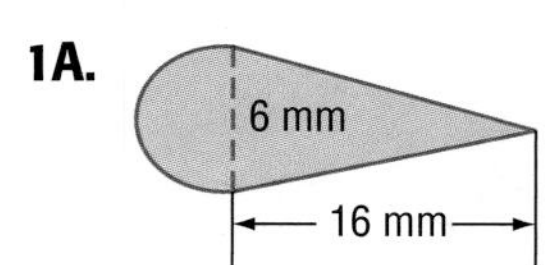

1B.

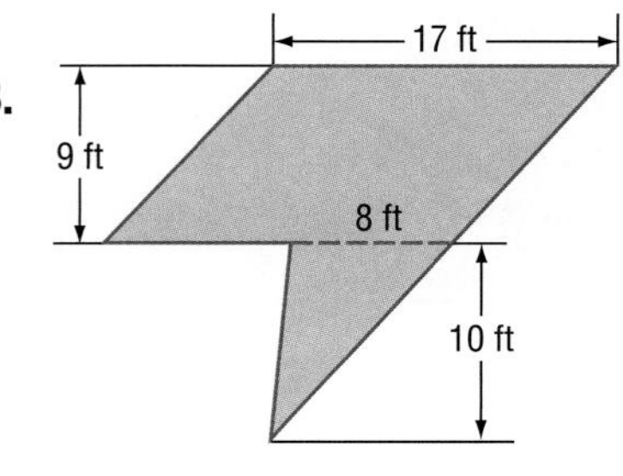

Personal Tutor glencoe.com

Concept Summary — Area Formulas

For Your FOLDABLE

Parallelogram	Triangle	Trapezoid	Circle
$A = bh$	$A = \frac{1}{2}bh$	$A = \frac{1}{2}h(a + b)$	$A = \pi r^2$

Problem-Solving Tip

Solve a Simpler Problem Sometimes it is necessary to break a problem into parts. You can use colored pencils to shade each separate figure that makes up the composite figure a different color. Find the area of each part, then combine the solutions of the parts.

Solve Problems Involving Area Often the first step in a multi-step problem is to find the area of a composite figure.

Real-World EXAMPLE 2 Solve Multi-Step Problems

CONSTRUCTION A ceiling tile covers 3 square feet. How many tiles are needed to cover the octagonal ceiling shown at the right?

18 ft, 5 ft, 20 ft, 5 ft, 10 ft

Understand You know the area of one ceiling tile and the dimensions of the ceiling.

Plan First, find the area of the ceiling by separating the figure into a rectangle and two trapezoids. Then, find the number of ceiling tiles.

Solve

$A = \frac{1}{2}h(a + b)$ — **Area of a trapezoid**

$= \frac{1}{2}(5)(10 + 18)$ — **$h = 5$, $a = 10$, $b = 18$**

$= 70$ — **Simplify.**

10 ft, 5 ft, 18 ft, 20 ft

The area of both trapezoids is $70 \cdot 2$ or 140 square feet.

$A = bh$ — **Area of a rectangle**

$= 18 \cdot 20$ — **$b = 18$, $h = 20$**

$= 360$ — **Simplify.**

The total area of the ceiling is $360 + 140$ or 500 square feet.

Each ceiling tile covers 3 square feet, so the total number of tiles needed is $500 \div 3 \approx 166.7$, or about 167 tiles.

Check The area of the figure should be a little less than the area of a rectangle that is 30 feet by 18 feet. Since the area of a 30-by-18-foot rectangle is 540 square feet, the answer of 500 square feet is reasonable.

Study Tip

Estimate Area One way to estimate the area of a composite figure is to draw a rectangle around the figure, as shown below. The area of the figure should be slightly less than the area of the rectangle.

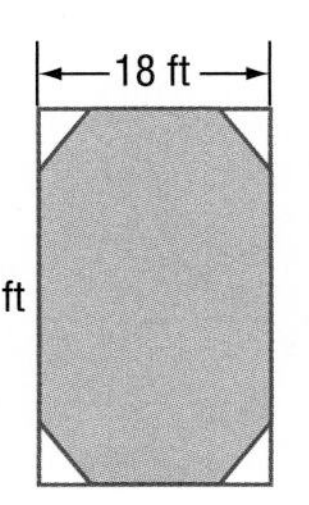

Check Your Progress

2. **REMODELING** The L-shaped counter shown is to be replaced with a new countertop that costs $52 per square foot. What will be the cost of the new countertop?

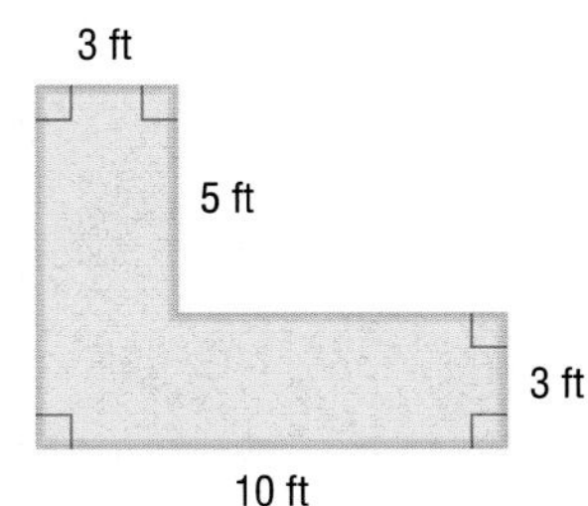

Personal Tutor glencoe.com

Check Your Understanding

Example 1 p. 642

Find the area of each figure. Round to the nearest tenth, if necessary.

1.

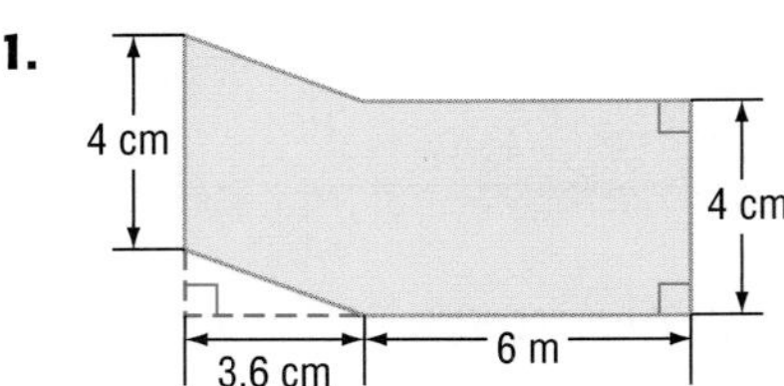

2.

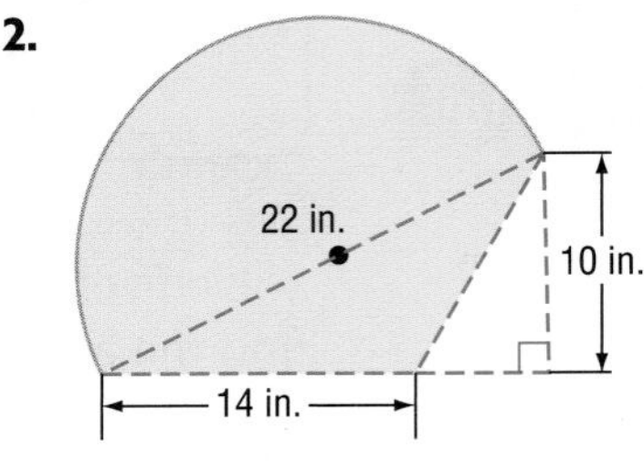

3.

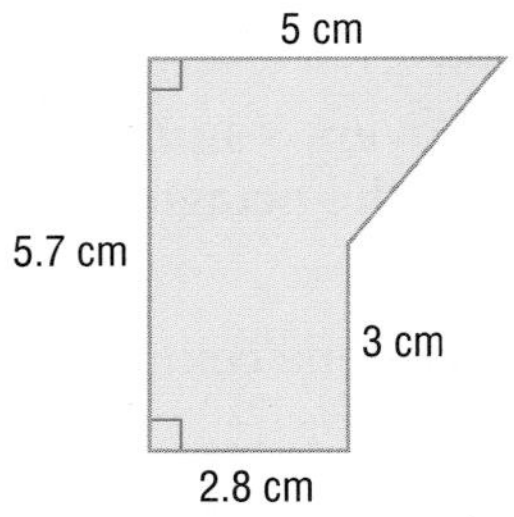

5 cm
5.7 cm
3 cm
2.8 cm

4.

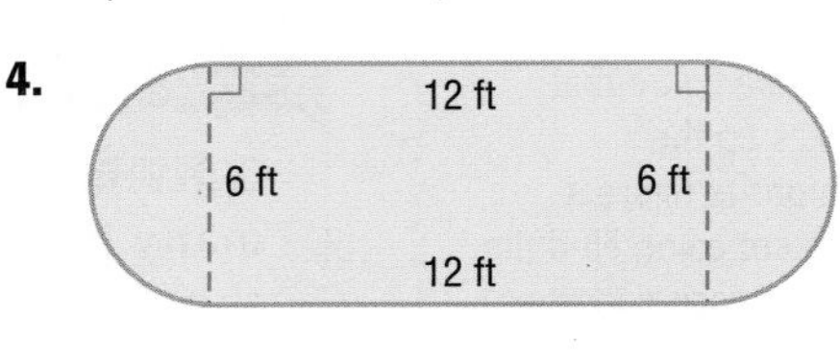

Example 2 p. 643

5. **LANDSCAPING** The Jamesons hired a landscaper to brick their walkway. One case of bricks costs $25 and covers 6 square feet.

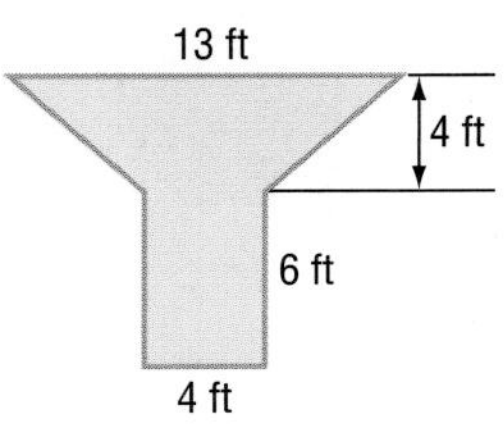

a. What is the area of the walkway?

b. How many cases are needed to cover the walkway with bricks? What would be the cost?

Practice and Problem Solving

● = **Step-by-Step Solutions** begin on page R11.
Extra Practice begins on page 810.

Example 1 p. 642

Find the area of each figure. Round to the nearest tenth, if necessary.

6.

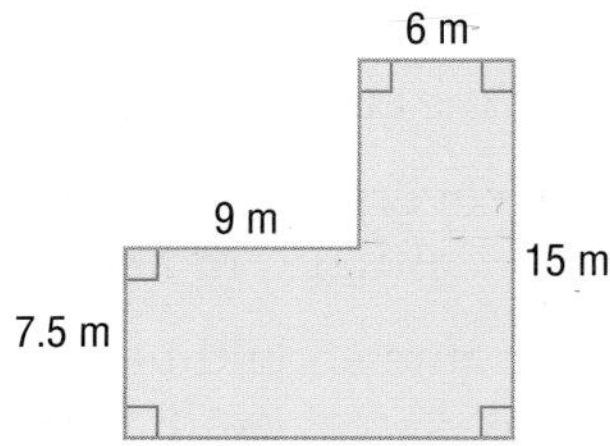

7

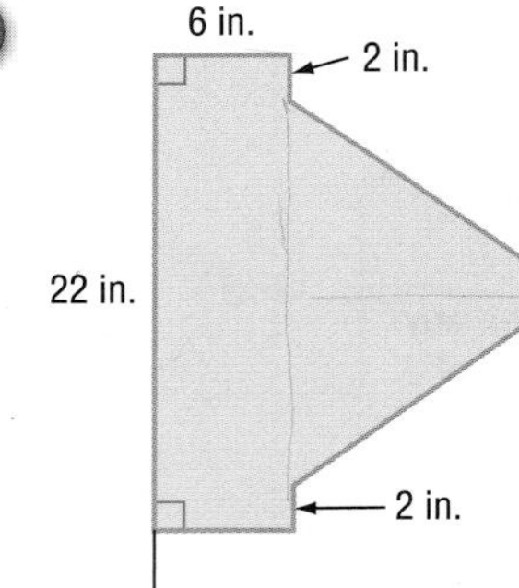

8.

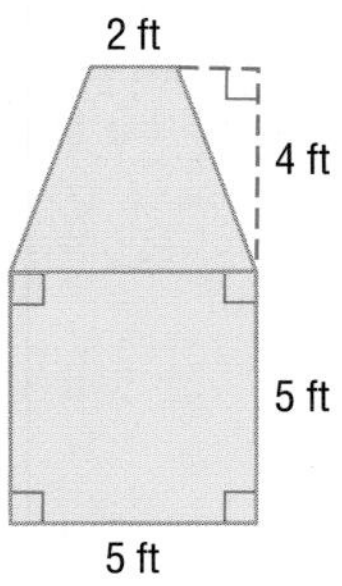

2 ft
4 ft
5 ft
5 ft

9.

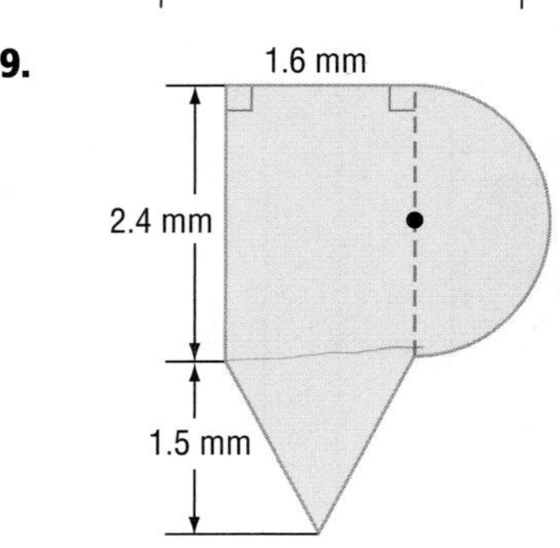

ReadingMath

Composite Figures Composite figures are also sometimes called *complex figures* or *irregular figures*.

Find the area of each figure. Round to the nearest tenth, if necessary.

10.

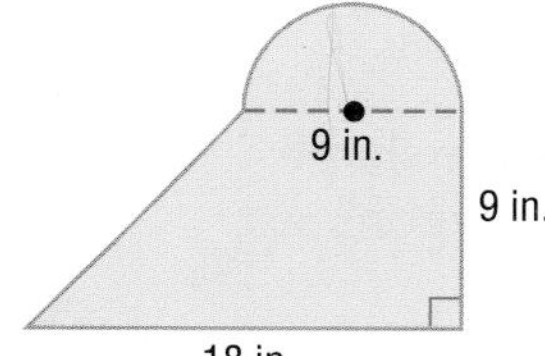

11.

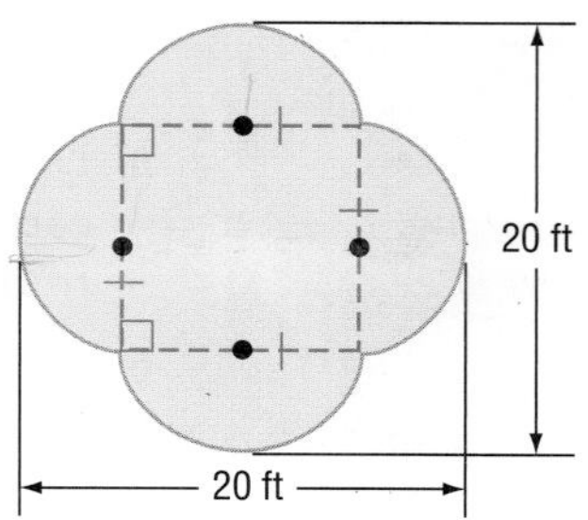

Example 2 p. 643

12. **POOLS** Refer to the swimming pool shown at the right.

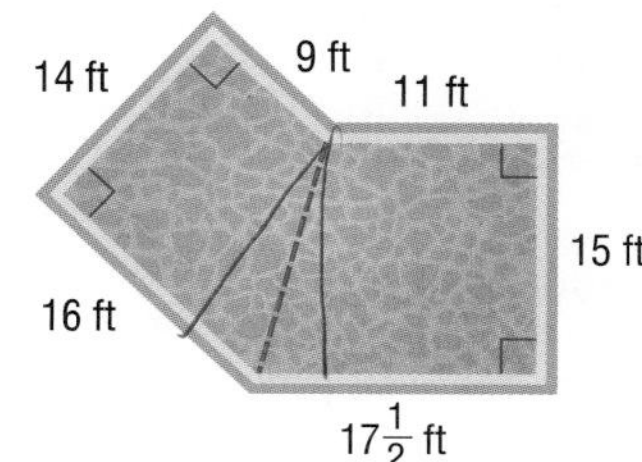

a. What is the area of the pool's floor?

b. A pool cover costs \$1.70 per square foot. How much would a cover for the pool cost?

13. **FINANCIAL LITERACY** Mr. Reyes wants to carpet his family room.

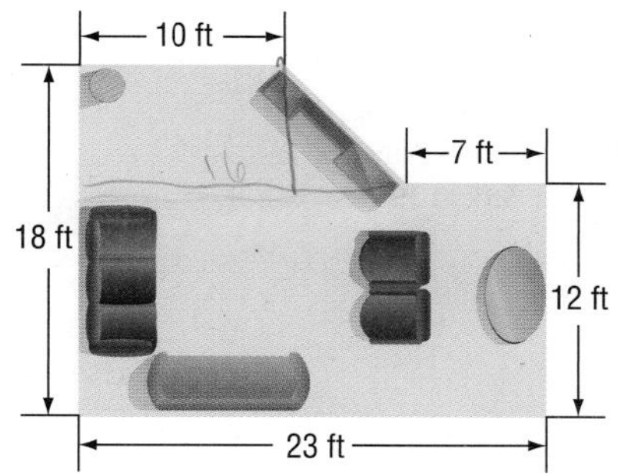

a. What is the area of the space to be carpeted?

b. If carpet costs \$2.25 per square foot, how much would it cost to carpet Mr. Reyes' family room if there is no leftover carpet?

14. What is the area, to the nearest tenth, of a figure that is formed using a rectangle 10 feet long and 7 feet wide and a semicircle with a diameter of 7 feet?

15. A figure is formed using a semicircle and a triangle that has a base of 8 inches and a height of 12 inches. Find the area of each figure to the nearest tenth.

a. The diameter of the semicircle equals the base of the triangle.

b. The diameter of the semicircle equals the height of the triangle.

16. **CARPENTRY** Mr. Reyes wants to put ceiling molding in the room described in Exercise 13. How many feet of molding will he need? Round to the nearest tenth.

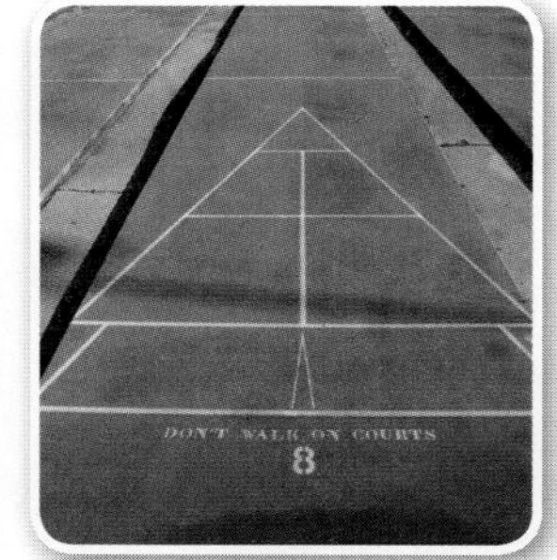

Real-World Link

A shuffleboard court is 52 feet long and 6 feet wide. Each player uses four discs to try to get 50, 75, or 100 points.

Source: Yahoo! Education

17 **SHUFFLEBOARD** Find the area of the part of the shuffleboard court shown.

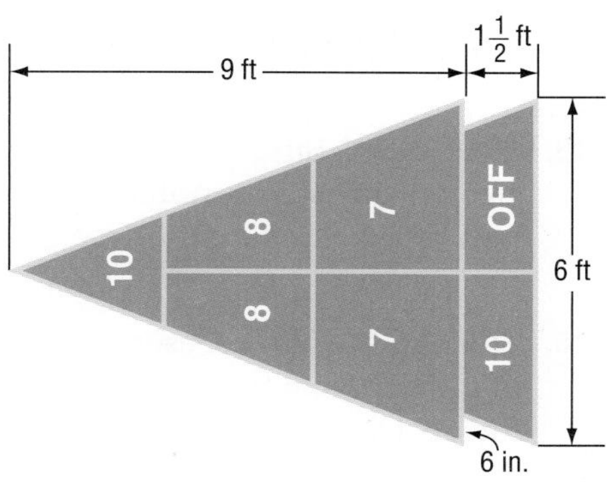

Real-World Link

The Denali National Park covers 6 million acres and contains Mt. McKinley, the highest mountain on the North American Continent. The park was originally named Mt. McKinley National Park in 1917.

18. **NATURE** A diagram of the Denali National Park Wilderness area in Alaska is shown at the right.

    a. Describe how the irregular shape can be separated into polygons to find its area.

    b. Estimate the number of square miles contained in the Denali National Park Wilderness area.

    c. Estimate the perimeter of the Wilderness area.

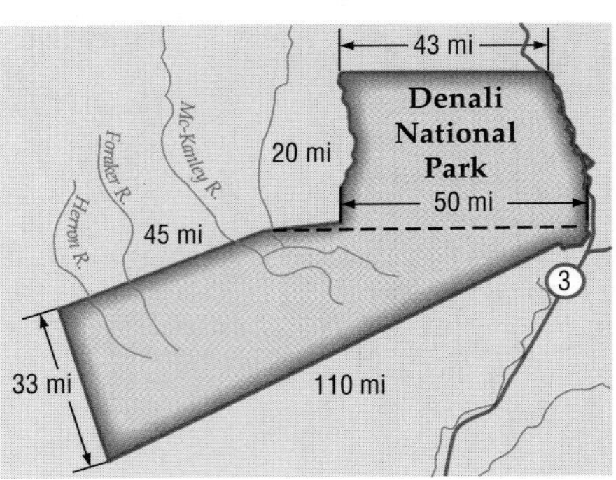

Find the area of each shaded region. Round to the nearest tenth, if necessary.

19.

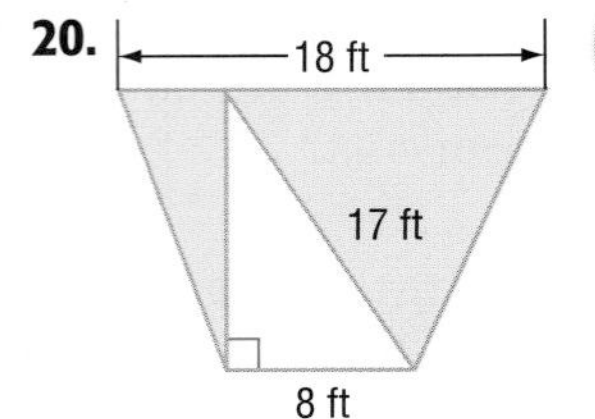

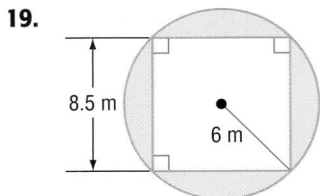

20.

21.

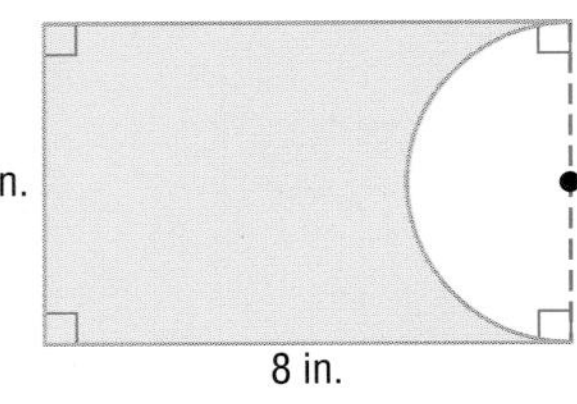

22. **MULTIPLE REPRESENTATIONS** In this problem, you will investigate areas of composite figures.

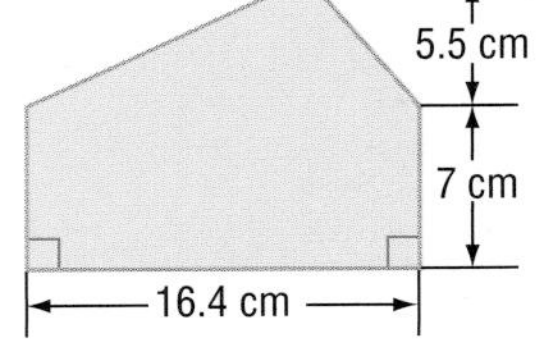

    a. **VERBAL** Describe the figure using names of geometric figures whose areas you can find.

    b. **SYMBOLIC** List the formulas that you can use to find the area of the composite figure.

    c. **ANALYTICAL** Make a conjecture about how the area of the composite figure changes if each dimension given is doubled. Then test your conjecture by doubling the dimensions and finding the area.

H.O.T. Problems

Use Higher-Order Thinking Skills

23. **OPEN ENDED** Describe real-life composite figures whose areas can be found by separating the figures into geometric figures that have area formulas.

24. **CHALLENGE** Reena created the flower at the right by placing semicircles around a regular pentagon. Using the measurements shown, find the area of the flower. Round to the nearest tenth.

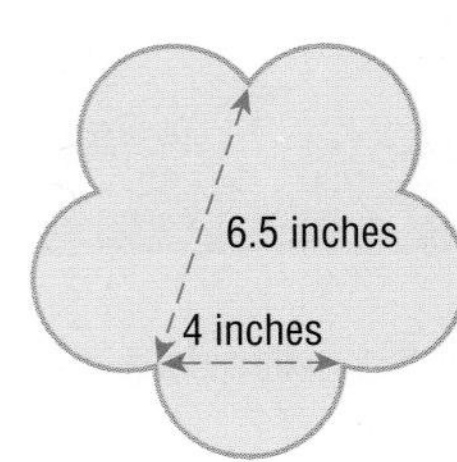

25. **REASONING** Suppose a composite figure has a curved side that is not a semicircle. Describe how you could estimate the area.

26. **WRITING IN MATH** Describe how circles and polygons help you find the area of a composite figure. Give an example by drawing and labeling a composite figure. Describe the figures that can be used to find the area and then find the area.

Standardized Test Practice

27. GRIDDED RESPONSE In the diagram, a patio that is 4 feet wide surrounds a swimming pool. What is the area of the patio in square feet? Round to the nearest tenth.

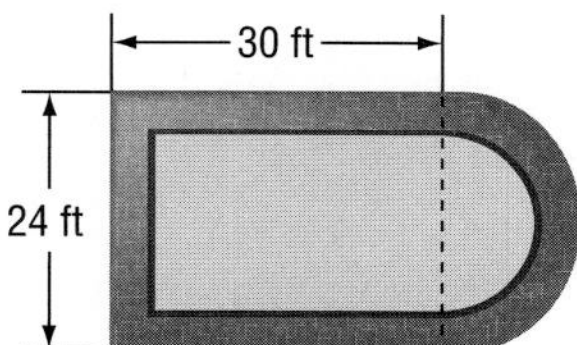

28. Find the area of the shaded region. Use 3.14 for π. Round to the nearest hundredth.

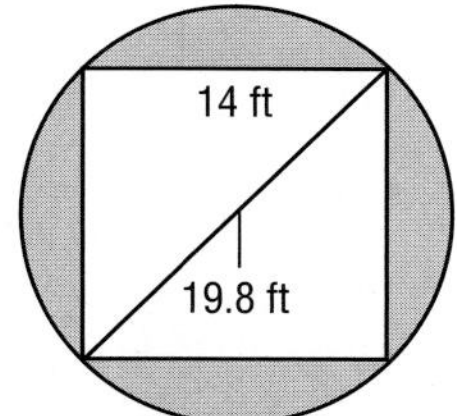

A 392.04 ft^2 C 111.75 ft^2
B 503.75 ft^2 D 258.17 ft^2

29. The Lin family is buying a cover for their swimming pool shown below. The cover costs $3.19 per square foot. How much will the cover cost?

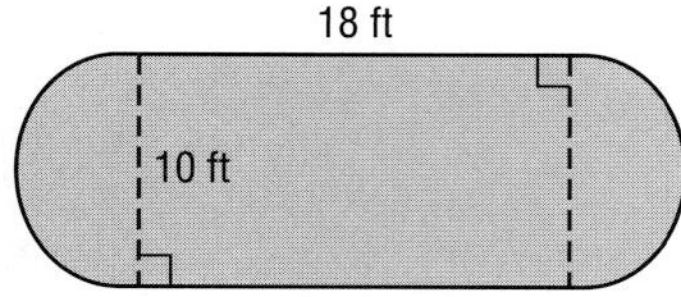

F $219.27 H $699.47
G $258.54 J $824.74

30. Using the floor plan below, find the difference between the areas of each bedroom.

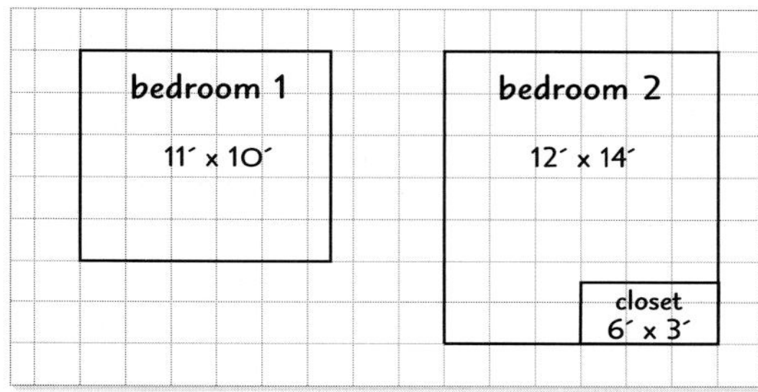

A 27 ft^2 C 58 ft^2
B 40 ft^2 D 168 ft^2

Spiral Review

Find the circumference and area of each circle. Round to the nearest tenth. (Lessons 11-7 and 11-8)

31. radius 6 cm **32.** diameter $9\frac{1}{2}$ ft **33.** radius 20 in.

Find the area of each figure described. (Lesson 11-6)

34. triangle: base, 9 in.; height, 6 in. **35.** trapezoid: height, 3 cm; bases, 4 cm, 8 cm

36. CROSS COUNTRY Jeremy can run $3\frac{1}{3}$ miles in 25 minutes. How many minutes would it take him to run 5 miles at this same rate? (Lesson 6-5)

Skills Review

Classify each polygon according to its number of sides. (Lessons 11-5)

37. 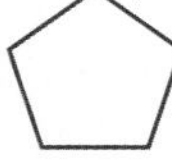

38. 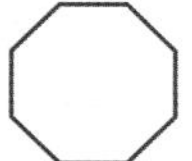

39. 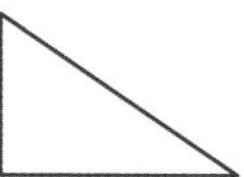

40. 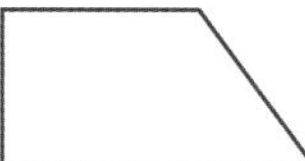

EXTEND

11-9 Spreadsheet Lab: Changes in Scale

Math Online glencoe.com
- Graphing Calculator Personal Tutor

In Lesson 6-6, you learned that when measurements have the same units the *scale factor* is the ratio of the length on a scale drawing or model to the corresponding length on the real figure. You can use a spreadsheet to investigate how the perimeter and area are affected when dimensions of a figure are changed by a scale factor.

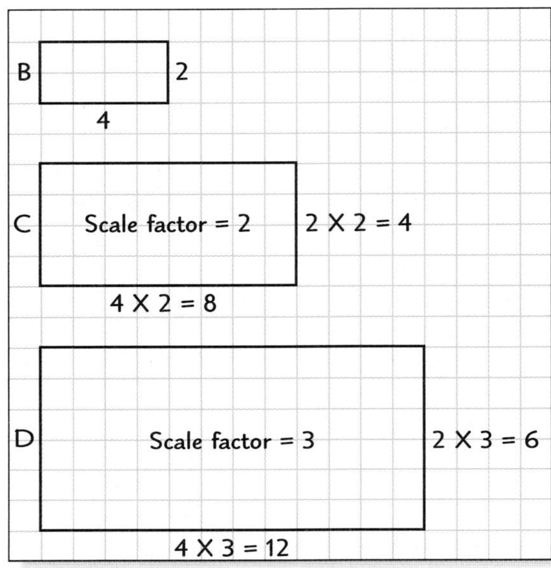

ACTIVITY

Step 1 In cells A1 through A7, enter the labels *Width, Length, Perimeter, Area, Scale Factor, Ratio of Perimeters,* and *Ratio of Areas*. Columns B, C, D, E, and F will be used for five similar rectangles.

Step 2 Enter a 2 in cell B1 and a 4 in cell B2 to represent the width and length of a rectangle.

Step 3 Enter the formula = 2*(B1 + B2) for the perimeter of the rectangle in cell B3. Copy the formula into the other cells in row 3.

Step 4 Write a formula to find the area of the rectangle. Copy the formula in the cells in row 4.

Step 5 Enter a scale factor of 2 in cell C5.

Perimeter, Area, and ....

◇	A	B	C
1	Width	2	4
2	Length	4	8
3	Perimeter	12	24
4	Area	8	32
5	Scale Factor		2
6	Ratio of Perimeters		
7	Ratio of Areas		

Sheet 1 | Sheet 2

Step 6 Enter the formula = B1*C5 in cell C1 and enter = B2*C5 in cell C2. These formulas find the dimensions of rectangle C based on the dimensions of rectangle B and the scale factor you entered. Enter similar formulas in the cells for columns D, E, and F.

Step 7 Type the formula = C3/B3 in cell C6, type = D3/B3 in cell D6, and so on. This formula will find the ratio of the perimeter of each of the other rectangles to the perimeter of rectangle B.

Step 8 Write a formula for the ratio of the area of rectangle C to the area of rectangle B. Enter the formula in cell C7. Enter similar formulas in the cells in row 7.

Step 9 Use Columns D, E, and F to find the perimeters, areas, and ratios for rectangles with scale factors of 3, 4, and 5.

Perimeter, Area, and Changes in Scale

◇	A	B	C	D	E	F
1	Width	2	4	6	8	10
2	Length	4	8	12	16	20
3	Perimeter	12	24	36	48	60
4	Area	8	32	72	128	200
5	Scale Factor		2	3	4	5
6	Ratio of Perimeters		2	3	4	5
7	Ratio of Areas		4	9	16	25
8						

Sheet 1 | Sheet 2 | Sheet 3

Analyze the Results

1. Compare the ratios in rows 5, 6, and 7 of columns C, D, and E. What do you observe?
2. What happens to the perimeter of a rectangle if the dimensions are doubled? multiplied by 4? multiplied by n? Change the original dimensions of the rectangle and the scale factors in the spreadsheet to verify your conclusion.
3. Describe the effect on the area of a rectangle if its dimensions are doubled. multiplied by 4? multiplied by n? Change the original dimensions of the rectangle and the scale factors in the spreadsheet to verify your conclusion.
4. Change the scale factors in cells C5, D5, E5, and F5 to 0.1, 0.2, 0.3 and 0.5. Describe the ratios of the perimeters and areas for these reductions.
5. Use the perimeter formula $P = 2(\ell + w)$ and the area formula $A = \ell w$ to explain the effects of changing the dimensions of a rectangle proportionally.

Study Guide and Review

Math Online glencoe.com
- STUDY TO GO
- Vocabulary Review

Chapter Summary

Key Concepts

Angle and Line Relationships (Lesson 11-1)

- Two angles are complementary if the sum of their measures is 90°.
- Two angles are supplementary if the sum of their measures is 180°.
- If two parallel lines are cut by a transversal, the following pairs of angles are congruent: corresponding angles, alternate interior angles, alternate exterior angles.

Congruent Triangles and Rotations
(Lessons 11-2 and 11-3)

- Figures that have the same size and shape are congruent.
- A rotation is a transformation in which a figure is turned around a fixed point.

Polygons (Lesson 11-5)

- Polygons are classified by their sides.
- If a polygon has n sides, then the sum of the interior angle measures is $(n - 2)180$.

Formulas (Lessons 11-6 through 11-9)

- Area of a parallelogram: $A = bh$
- Area of a triangle: $A = \frac{1}{2}bh$
- Area of a trapezoid: $A = \frac{1}{2}h(b_1 + b_2)$
- Circumference of a circle: $C = 2\pi r$
- Area of a circle: $A = \pi r^2$

FOLDABLES® Study Organizer

Be sure the Key Concepts are noted in your Foldable.

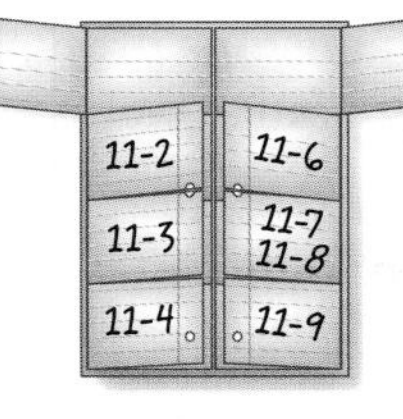

Key Vocabulary

- **adjacent angles** (p. 589)
- **alternate exterior angles** (p. 590)
- **alternate interior angles** (p. 590)
- **center of rotation** (p. 605)
- **central angle** (p. 638)
- **circumference** (p. 631)
- **complementary angles** (p. 589)
- **composite figure** (p. 642)
- **corresponding angles** (p. 590)
- **corresponding parts** (p. 598)
- **diameter** (p. 631)
- **interior angle** (p. 618)
- **parallel lines** (p. 590)
- **perpendicular lines** (p. 589)
- **quadrilateral** (p. 612)
- **radius** (p. 631)
- **regular polygon** (p. 619)
- **rotation** (p. 605)
- **rotational symmetry** (p. 607)
- **sector** (p. 638)
- **supplementary angles** (p. 589)
- **transversal** (p. 590)
- **vertical angles** (p. 589)

Vocabulary Check

Determine whether each statement is *true* or *false*. If false, replace the underlined word or phrase to make a true statement.

1. Two lines in a plane that never intersect are called <u>parallel lines</u>.
2. A <u>composite figure</u> has congruent sides and congruent angles.
3. The <u>altitude</u> can be any side of a parallelogram.
4. An <u>interior angle</u> is an angle formed at a vertex of a polygon.
5. A <u>rhombus</u> is an example of a quadrilateral.
6. The <u>radius</u> of a circle is the section formed by a central angle.
7. <u>Complementary angles</u> are two angles with a sum of 180°.
8. If two parallel lines are cut by a transversal, corresponding angles are <u>equal</u>.
9. A <u>tessellation</u> is a transformation in which a figure is turned around a fixed point.

Lesson-by-Lesson Review

11-1 Angle and Line Relationships (pp. 589–595)

In the figure below, $m \parallel n$ and r is a transversal. If $m\angle 4 = 112°$, find the measure of each angle. Explain your reasoning.

10. $\angle 6$

11. $\angle 2$

12. $\angle 8$

13. $\angle 1$

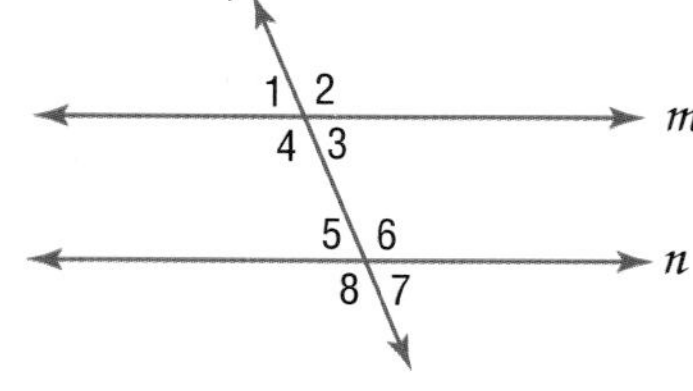

14. **DOORS** A door can swing open 180°. The door is open at an angle of 99°. What is the measure of the angle between the door and the door jam?

EXAMPLE 1

In the figure below, $a \parallel b$ and g is a transversal. If $m\angle 2 = 61°$, find $m\angle 7$ and $m\angle 4$. Explain your reasoning.

Since $\angle 2$ and $\angle 7$ are alternate exterior angles, they are congruent. So, $m\angle 7 = 61°$.

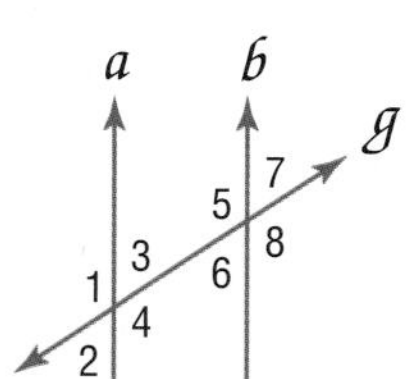

Since $\angle 2$ and $\angle 4$ are supplementary, the sum of their measures is 180°.

$m\angle 2 + m\angle 4 = 180$	**Supplementary angles**
$61 + m\angle 4 = 180$	**Replace $m\angle 2$ with 61.**
$m\angle 4 = 119$	**Subtract 61 from each side.**

So, $m\angle 4 = 119°$.

11-2 Congruent Triangles (pp. 598–604)

15. Name the corresponding parts in the congruent triangles below. Then complete the congruence statement $\triangle FGH \cong$ ___?___.

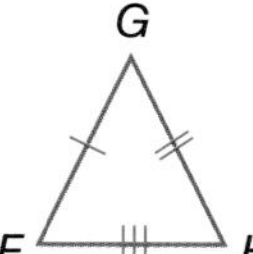

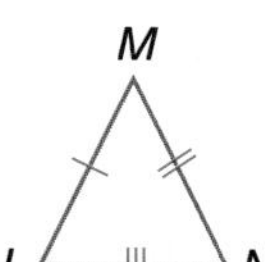

16. **SOFTBALL** On a softball diamond, the triangle formed by home plate, first base, and second base is congruent to the triangle formed by home plate, third base, and second base. If it is 65 feet from home plate to first base, how far is it from third base to home plate?

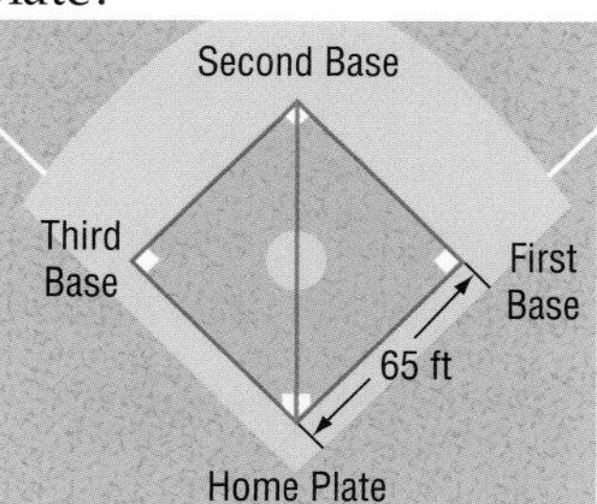

EXAMPLE 2

Name the corresponding parts in the congruent triangles shown below. Then complete the congruence statement $\triangle STU \cong \triangle$___?___.

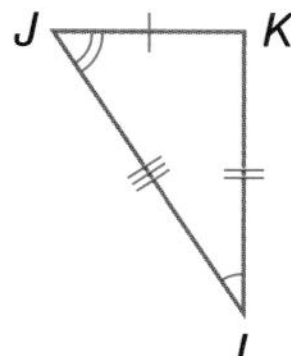

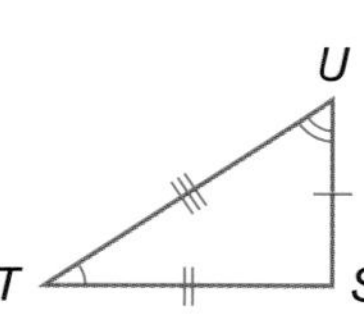

Use the matching arcs and tick marks to identify the corresponding parts.

Corresponding angles:

$\angle J \cong \angle U$, $\angle K \cong \angle S$, $\angle L \cong \angle T$

Corresponding sides:

$\angle \overline{JK} \cong \angle \overline{US}$, $\angle \overline{KL} \cong \angle \overline{ST}$, $\angle \overline{LJ} \cong \angle \overline{TU}$

The congruence statement is $\triangle STU \cong \triangle KLJ$.

CHAPTER 11

Study Guide and Review

11-3 Rotations (pp. 605–610)

17. Triangle ABC has vertices $A(2, 0)$, $B(4, -1)$, and $C(1, -3)$. Graph the figure and its image after a clockwise rotation of 180° about vertex A. Give the coordinates of the vertices for triangle $A'B'C'$.

Graph each figure and its image after a clockwise rotation about the origin.

18. triangle GHJ with vertices $G(0, -1)$, $H(3, 3)$, and $J(2, -3)$; 270° clockwise rotation

19. quadrilateral $NPQR$ with vertices $N(1, 1)$, $P(2, 3)$, $Q(4, 2)$, and $R(4, -2)$; 90° clockwise rotation

20. **SIGNS** Determine whether the shape of the sign shown at the right has rotational symmetry. If it does, describe the angle of rotation.

EXAMPLE 3

Triangle TVW has vertices $T(-1, 0)$, $V(0, 3)$, and $W(2, 2)$. Graph the figure and its image after a clockwise rotation of 270° about vertex T. Give the coordinates of the vertices for triangle $T'V'W'$.

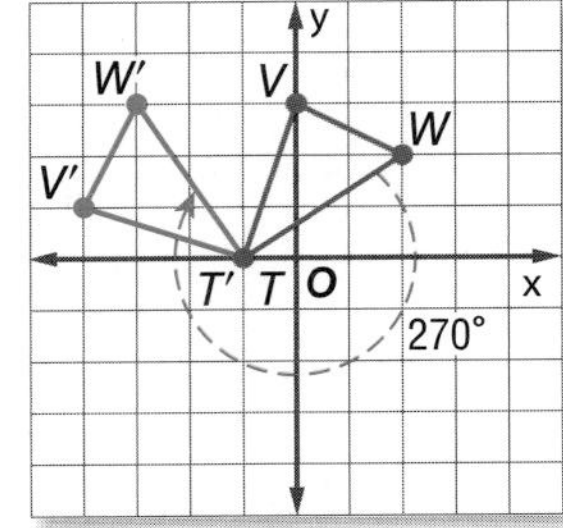

The coordinates of the vertices are $T'(-1, 0)$, $V'(-4, 1)$, and $W'(-3, 3)$.

11-4 Quadrilaterals (pp. 612–616)

Find the value of x in each quadrilateral. Then find the missing angle measures.

21.

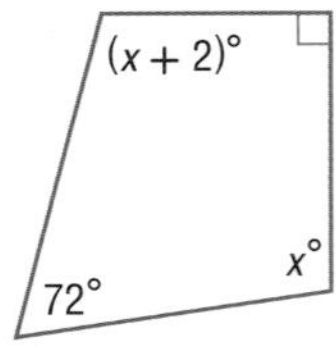

22.

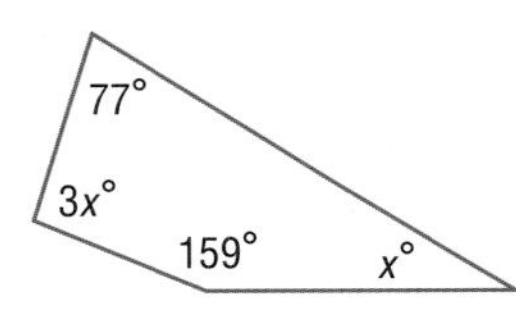

23. **WINDOWS** Classify the quadrilaterals shown in the window at the right. Use the name that *best* describes it.

EXAMPLE 4

Find the value of x in the quadrilateral below. Then find the missing angle measures.

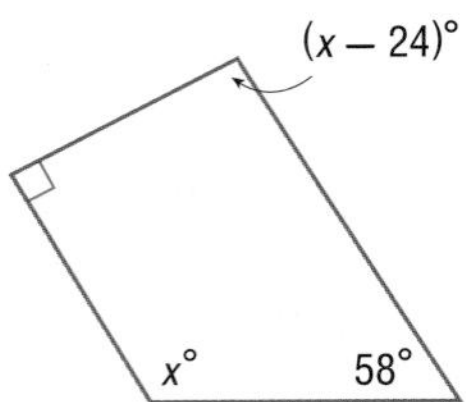

$x + 90 + (x - 24) + 58 = 360$	**Write an equation.**
$2x + 124 = 360$	**Simplify.**
$2x = 236$	**Subtract 124 from each side.**
$x = 118$	**Simplify.**

So, the missing angle measures are 118° and 118 – 24 or 94°.

MIXED PROBLEM SOLVING
For mixed problem-solving practice, see page 843.

11-5 Polygons (pp. 617–622)

Determine whether the figure is a polygon. If it is, classify the polygon and state whether it is regular. If it is not a polygon, explain why.

24. 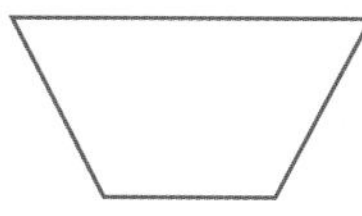

25. 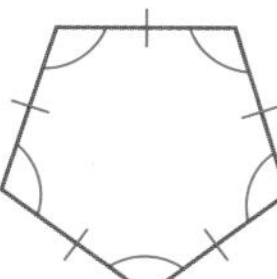

Find the measure of an interior angle of each regular polygon.

26. hexagon

27. 18-gon

28. **STREET SIGNS** What is the measure of each interior angle of the stop sign?

EXAMPLE 5

Find the measure of one interior angle in a regular heptagon.

Step 1 Find the sum of the measures of the angles. A heptagon has 7 sides. So, $n = 7$.

$(n - 2)180 = (7 - 2)180$ **Replace n with 7.**

$= 5(180)$ or $900°$ **Simplify.**

The sum of the measures of the interior angles is 900°.

Step 2 Divide the sum by 7 to find the measure of one angle.

$900 \div 7 \approx 128.6$

So, the measure of one interior angle in a heptagon is about 128.6°.

11-6 Area of Parallelograms, Triangles, and Trapezoids (pp. 624–630)

Find the area of each figure.

29. 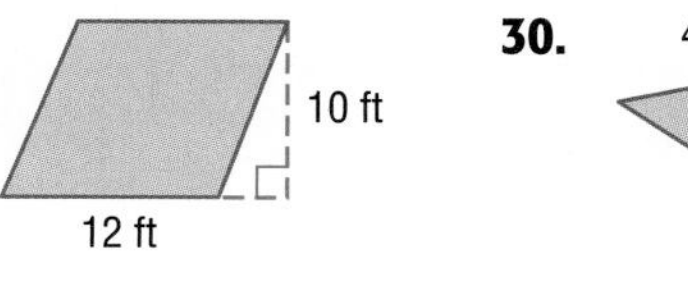

30. 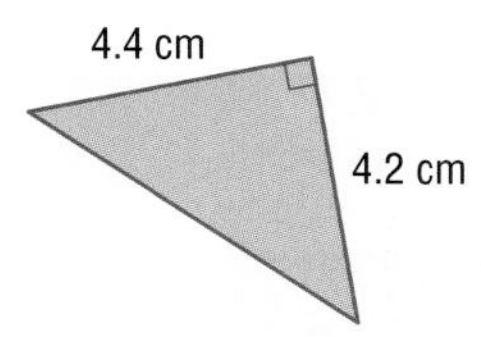

31. **DECORATING** Mrs. Jackson wants to lay tile in the area in front of her fireplace, as shown in the diagram. The tile costs $2.99 per square foot. About how much will it cost Mrs. Jackson to tile the area?

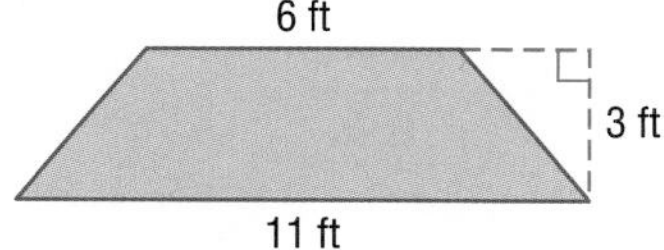

EXAMPLE 6

Find the area of the triangle.

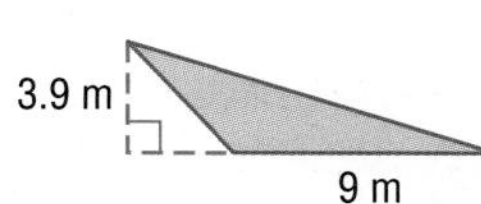

$A = \frac{1}{2}bh$ **Area of a triangle**

$= \frac{1}{2}(9)(3.9)$ **Replace b with 9 and h with 3.9.**

$= 17.55 \text{ m}^2$ **Simplify.**

EXAMPLE 7

Find the area of a trapezoid that has bases that measure 3 centimeters and 10.5 centimeters and a height that measures 5.2 centimeters.

$A = \frac{1}{2}h(a + b)$ **Area of a trapezoid**

$= \frac{1}{2}(5.2)(3 + 10.5)$ **Substitution**

$= 35.1 \text{ cm}^2$ **Simplify.**

CHAPTER 11 Study Guide and Review

11-7 Circles and Circumference (pp. 631–635)

Find the circumference of each circle. Round to the nearest tenth.

32. 

33. 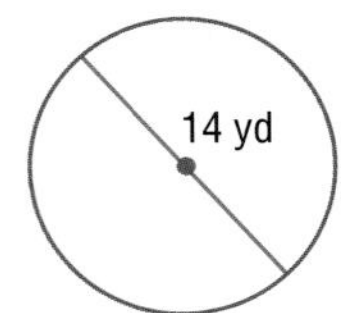

34. diameter = $5\frac{1}{3}$ ft

35. radius = 13.5 mm

36. **FERRIS WHEEL** The first Ferris wheel had a diameter of 250 feet. What was the distance, to the nearest whole foot, that riders traveled if they stayed on the ride for 15 rotations?

EXAMPLE 8

Find the circumference of the circle. Round to the nearest tenth.

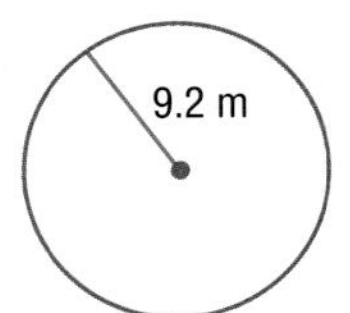

$C = 2\pi r$	**Circumference of a circle**
$= 2 \cdot \pi \cdot 9.2$	**Replace *r* with 9.2.**
≈ 57.8	**Simplify.**

So, the circumference of the circle is about 57.8 meters.

11-8 Area of Circles (pp. 636–641)

Find the area of each circle. Round to the nearest tenth.

37. 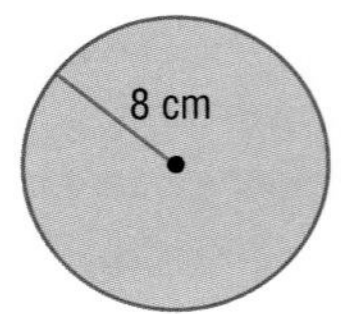

38. 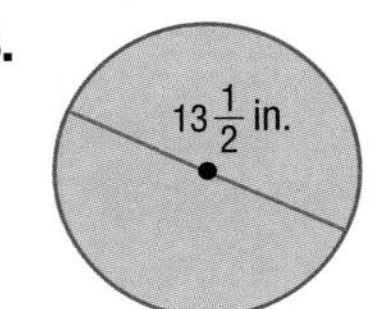

39. **BAKERY** A blueberry pie has a diameter of 9 inches. What is the area of the top crust? Round to the nearest tenth.

EXAMPLE 9

Find the area of the circle. Round to the nearest tenth.

$A = \pi r^2$	**Area of a circle**
$= \pi \cdot 3.5^2$	**$d = 7$, so $r = 3.5$**
≈ 38.5	**Simplify.**

So, the area of the circle is about 38.5 square inches.

11-9 Area of Composite Figures (pp. 642–647)

40. Find the area of the composite figure.

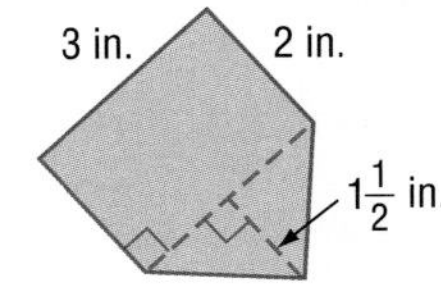

41. **MUSEUM** The floor plan of a new museum is shown. What is the area of the museum rounded to the nearest whole foot?

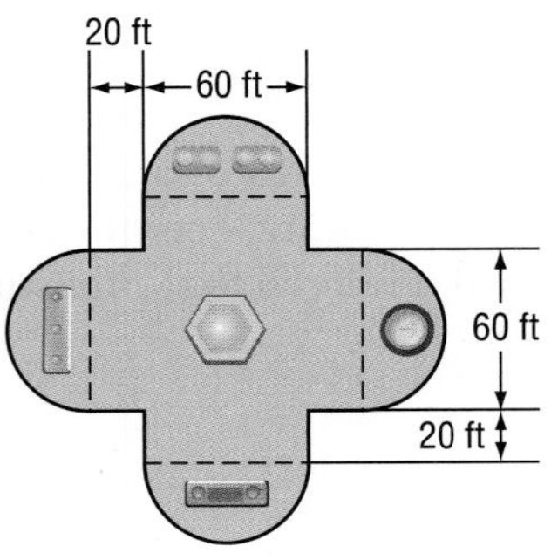

EXAMPLE 10

Find the area of the composite figure.

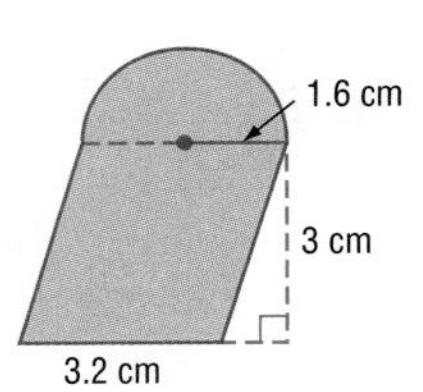

Find the areas of the semicircle and of the parallelogram. Then add.

Area of Semicircle

$A = \frac{1}{2}\pi r^2$

$= \frac{1}{2}\pi(1.6)^2$

$\approx 4.0 \text{ cm}^2$

Area of Parallelogram

$A = bh$

$= (3.2)(3)$

$= 9.6 \text{ cm}^2$

So, the area of the figure is 4.0 + 9.6 or about 13.6 square centimeters.

Practice Test

Find the value of x in each figure.

1. 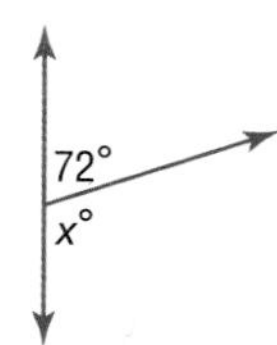

2. 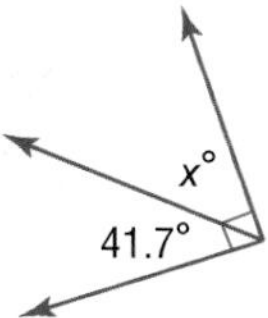

3. Name the corresponding parts in the congruent triangles. Then complete the congruence statement.

$\triangle CAB \cong$ ___?___

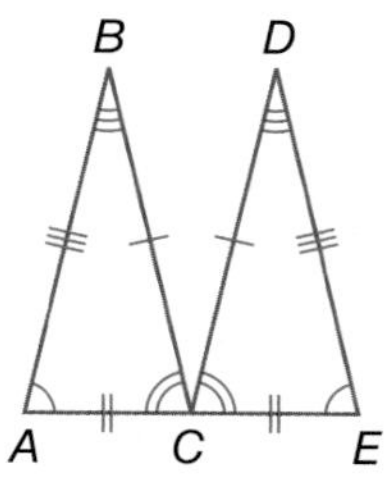

4. A figure has vertices $G(1, -1)$, $H(3, -4)$, $J(1, -5)$, and $K(-1, -4)$. Graph the figure and its image after a clockwise rotation of 90° about vertex J.

5. **MULTIPLE CHOICE** What are the coordinates of point R after a clockwise rotation of 270° about the origin?

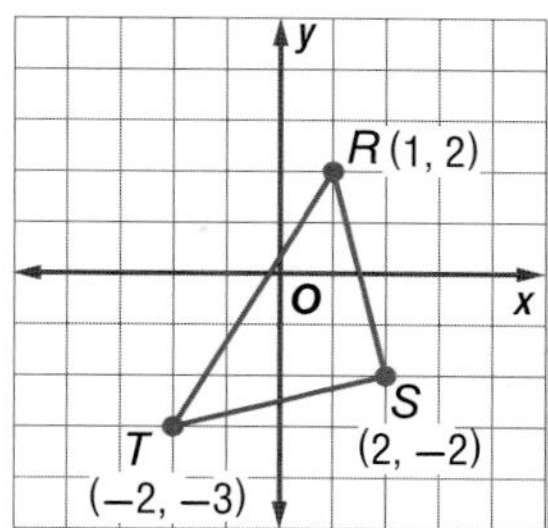

A $R'(-1, -2)$
B $R'(2, 1)$
C $R'(-2, 1)$
D $R'(2, -1)$

6. Find the value of x in the quadrilateral at the right. Then find the missing angle measures.

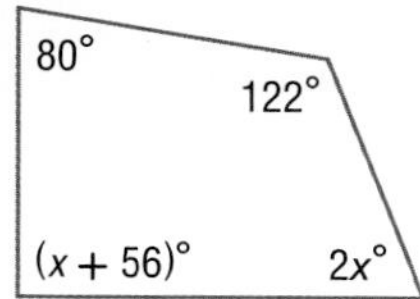

7. **MULTIPLE CHOICE** What is the measure of each interior angle of a regular 16-gon?

F 157.5°
G 205.7°
H 2520°
J 2880°

Classify each polygon. Then determine whether it is *regular* or *not regular*.

8. 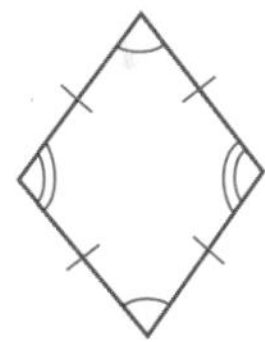

9. 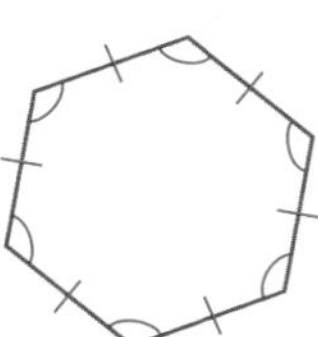

Find the area of each figure.

10. 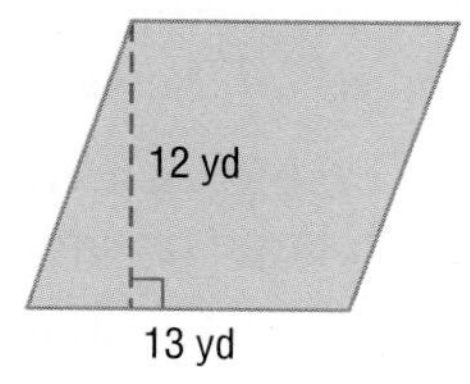

11. 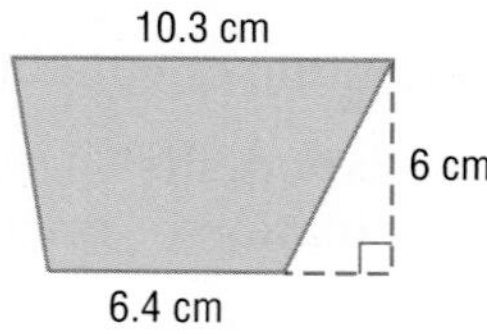

12. **BANNERS** A triangular banner has a base of 1 foot and a height of $1\frac{1}{2}$ feet. What is the area of the banner?

Find the circumference and area of each circle. Round to the nearest tenth.

13. 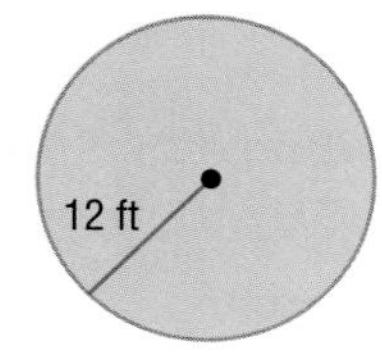

14. 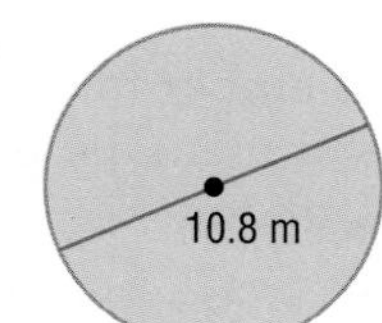

15. **PIZZA** The pizza at the right has a diameter of 15 inches. If one eighth of the pizza is eaten, what is the area of the pizza that is left? Round to the nearest tenth.

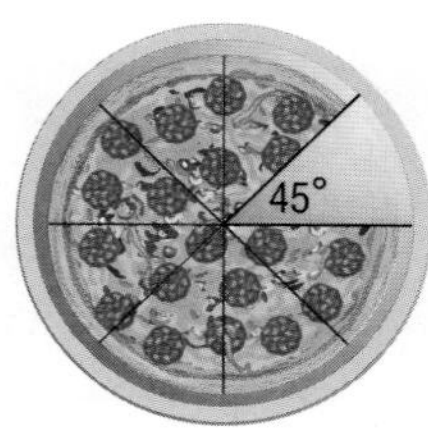

16. In the diagram, 1 square unit equals 5 square feet. What is the area of the shaded figure?

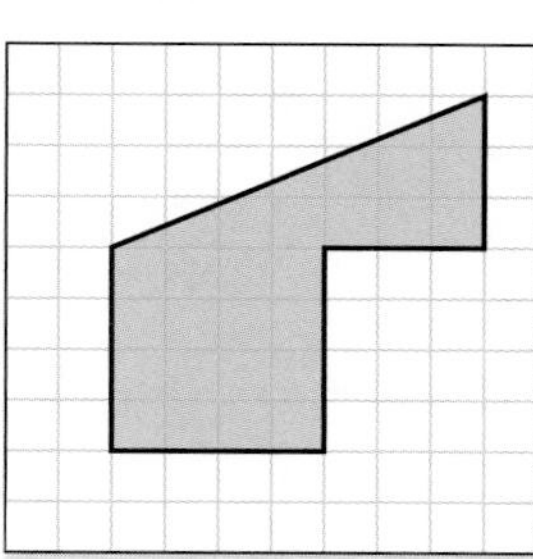

CHAPTER 11 Preparing for Standardized Tests

Using a Scientific Calculator

Scientific calculators are powerful problem-solving tools. Some problems can be solved faster or easier using a scientific calculator. Other problems that you encounter may have steps or computations that require the use of a scientific calculator.

Strategies for Using a Scientific Calculator

Step 1

Familiarize yourself with the various functions of a scientific calculator as well as when they should be used:

- **Exponents**—Used to solve problems involving scientific notation or calculations with large or small numbers.
- **Pi**—used to solve circle problems involving circumference and area.
- **Square roots**—Used to solve problems involving distance on a coordinate plane or the Pythagorean theorem.

Step 2

Use your calculator to solve the problem.

- Remember to work as efficiently as possible. Some steps may be done faster mentally or by hand, while others should be completed using your calculator.
- If time permits, check your answer.

EXAMPLE

Solve the problem below. Responses will be graded using the short-response scoring rubric shown.

What is the area of the shaded sector below? Round your answer to the nearest tenth.

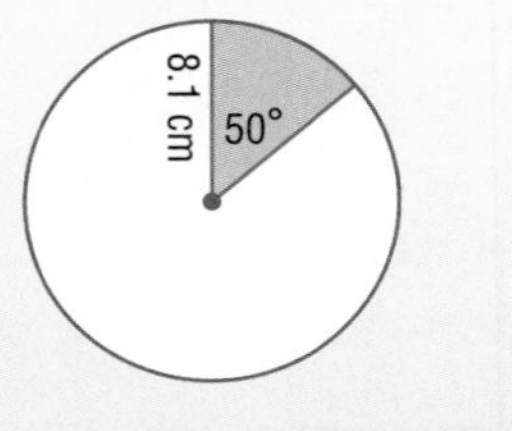

Scoring Rubric	
Criteria	**Score**
Full Credit: The answer is correct and a full explanation is provided that shows each step.	2
Partial Credit: • The answer is correct, but the explanation is incomplete. • The answer is incorrect, but the explanation is correct.	1
No Credit: Either an answer is not provided or the answer does not make sense.	0

Read the problem carefully. You are given the radius of a circle and the central angle of a sector and asked to find the area of the sector. Use the formula $A = \frac{N}{360}(\pi r^2)$ to find the area of the sector. Show your work to receive full credit.

Example of a 2-point response:

The formula $A = \frac{N}{360}(\pi r^2)$ gives the area of a sector with radius r and central angle N.

$A = \frac{N}{360}(\pi r^2)$

$A = \frac{50}{360}[\pi(8.1)^2]$

$A = \frac{5}{36}[\pi(65.61)]$

$A = 9.1125\pi$

$A \approx 28.6$

The area of the sector is about 28.6 square centimeters.

The steps, calculations, and reasoning are clearly stated. The student also arrives at the correct answer. So, this response is worth the full 2 points.

Exercises

Solve each problem. Show your work. Responses will be graded using the short-response scoring rubric given on page 656.

1. Heather's bicycle tire has a diameter of 22 inches. How far will she travel if the tire completes 50 revolutions? Round your answer to the nearest inch and to the nearest foot.

2. Find the area of the shaded region. Use 3.14 for π. Round to the nearest hundredth.

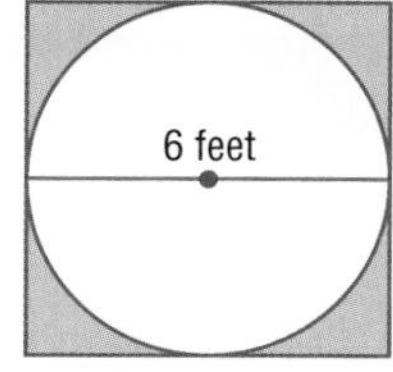

3. A metal washer is made by cutting a circular disk from the center of a larger disk. What is the area of the metal washer below? Round to the nearest tenth.

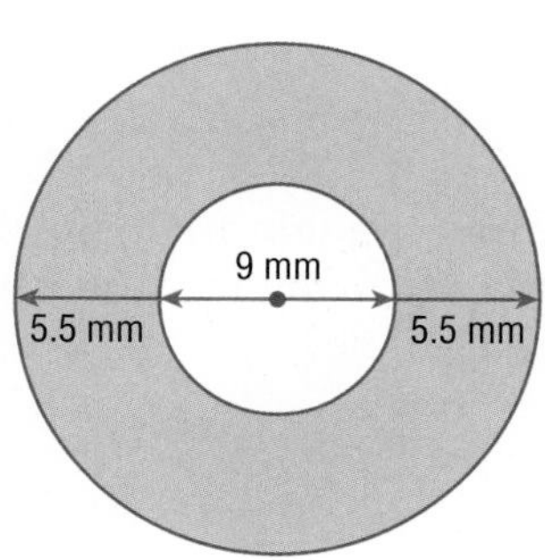

4. The diameter of a circle is 12 inches. Find the area of a sector of the circle with a central angle that measures 110°. Use 3.14 for π. Round to the nearest hundredth.

CHAPTER 11

Standardized Test Practice

Cumulative, Chapters 1 through 11

Multiple Choice

Read each question. Then fill in the correct answer on the answer document provided by your teacher or on a sheet of paper.

1. What is the value of x in the figure below?

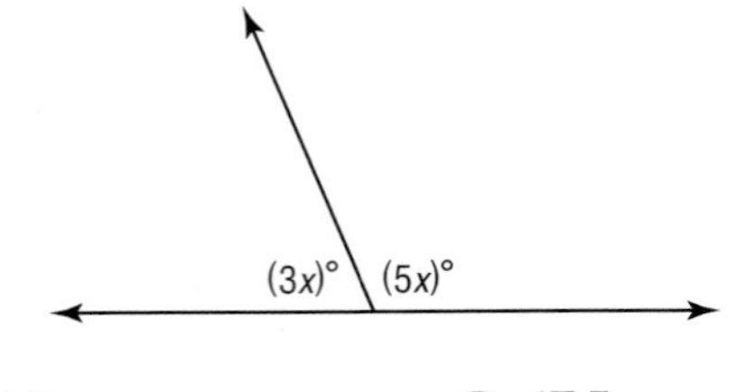

A 16.5
B 22.5
C 67.5
D 112.5

2. If right triangle DEF is congruent to right triangle LMN and $m\angle EDF = 42°$, what is $m\angle MLN$?

F 42°
G 48°
H 60°
J 90°

3. Which missing value for x would result in the relation shown in the table being a function?

x	2	5	1	■
y	11	4	8	10

A 1
B 2
C 3
D 5

4. The angles of a triangle are in the ratio 5:6:19. Which of the following measures is *not* an angle of the triangle?

F 30°
G 36°
H 114°
J 118°

5. Which of the following is *not* a factor of 3,003?

A 7
B 11
C 13
D 17

6. What is the sum of the interior angles of the figure below?

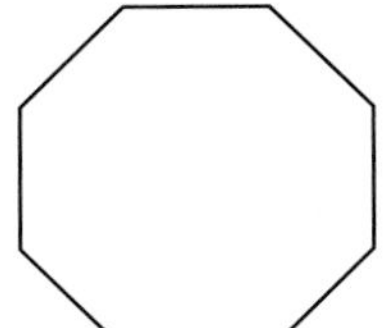

F 720°
G 900°
H 940°
J 1080°

7. Which of the following terms *best* describes a quadrilateral with opposite sides parallel and congruent?

A parallelogram
B rectangle
C rhombus
D square

8. Find the area of the circle below.

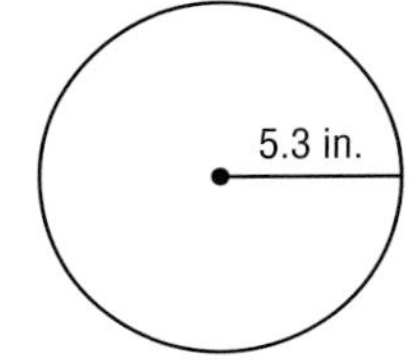

F 33.3 in^2
G 57.6 in^2
H 71.4 in^2
J 88.2 in^2

9. Lisa is going on a 2-week camping trip this summer. How many seconds are there in 2 weeks?

A 1,209,600 s
B 1,420,500 s
C 1,516,100 s
D 1,812,300 s

Test-TakingTip

Question 7 Sometimes a multiple choice item may have more than one *reasonable* answer, but you need to choose the *best* one.

Short Response/Gridded Response

Record your answers on the answer sheet provided by your teacher or on a sheet of paper.

10. GRIDDED RESPONSE Find the area of the composite figure below. Express your answer in square centimeters and round to the nearest tenth if necessary.

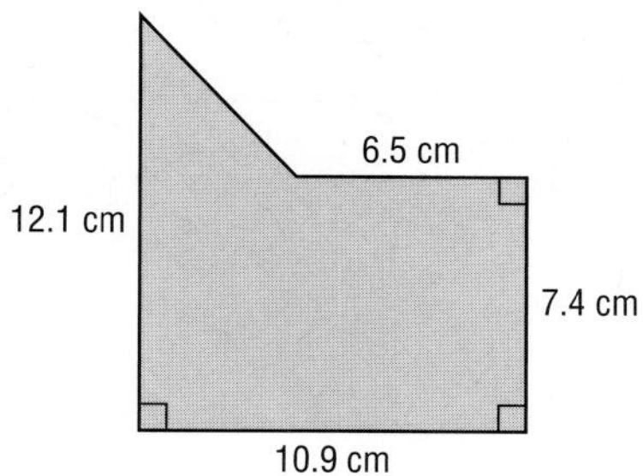

11. Triangle *EFG* is shown on the coordinate grid below.

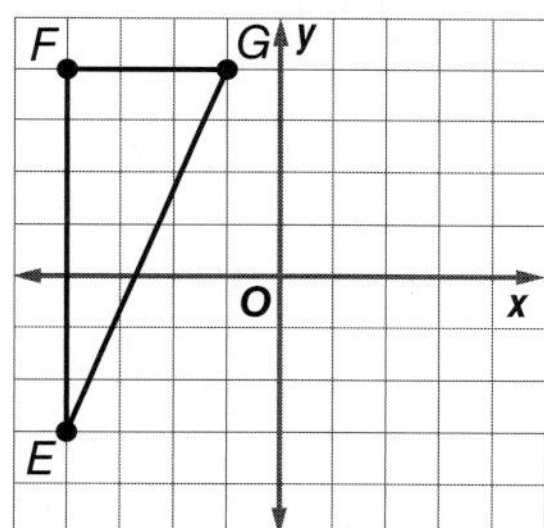

What are the coordinates of the vertices after a clockwise rotation of 90° about vertex *E*?

12. What is the distance between points *M*(2, 4) and *N*(-5, -3) on a coordinate plane? Round to the nearest tenth.

13. GRIDDED RESPONSE What is the angle of rotational symmetry in degrees for the figure shown below?

Extended Response

Record your answers on a sheet of paper. Show your work.

14. A large pizza from Santa Ana's Pizzeria has a 14-inch diameter as shown below.

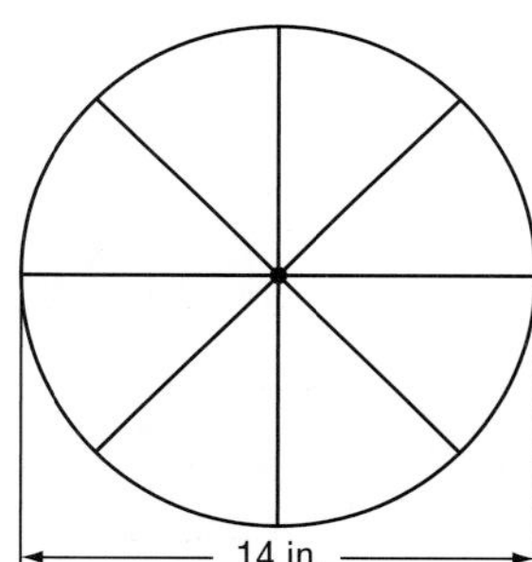

a. How many square inches of dough are needed to make a large pizza? Round to the nearest tenth.

b. What is the circumference, to the nearest tenth, of a large pizza?

c. Suppose the large pizza is cut into 8 equal slices. What is the central angle of each slice?

Need Extra Help?

If you missed Question...	1	2	3	4	5	6	7	8	9	10	11	12	13	14
Go to Lesson or Page...	11-1	11-2	8-1	10-3	9-2	11-4	11-5	11-8	6-3	11-9	11-3	10-5	11-3	11-5,7,8

CHAPTER 12 Surface Area and Volume

Then

In Chapter 11, you explored the properties of two-dimensional figures.

Now

In Chapter 12, you will:

- Describe three-dimensional figures.
- Find volumes and surface areas of three-dimensional figures.
- Examine properties of similar solids.

Why?

ART The world of art and design relies heavily on the use of three-dimensional figures. Artist Dale Chihuly uses blown glass to construct sculptures. His designs are geometric masterpieces!

Get Ready for Chapter 12

Diagnose Readiness You have two options for checking Prerequisite Skills.

Text Option Take the Quick Check below. Refer to the Quick Review for help.

QuickCheck

Determine whether each figure is a polygon. If it is, classify the polygon. (Lesson 11-5)

1. 

2. 

3. **SIGNS** Classify the shape of the sign shown.

Find each product. (Lesson 3-3)

4. $3(6)(12)$

5. $\frac{1}{3}(21)(5)$

6. $\frac{4}{3}(16)(3)$

7. $\frac{5}{4}(24)(11)$

8. **MUSIC** Suppose you practice the cello for $\frac{2}{3}$ hour every day. How many hours do you practice every week?

Determine whether each pair of ratios forms a proportion. (Lesson 6-4)

9. $\frac{1}{3}$ and $\frac{2}{3}$

10. $\frac{2}{5}$ and $\frac{10}{25}$

11. $\frac{2}{9}$ and $\frac{3}{18}$

12. $\frac{3}{4}$ and $\frac{39}{52}$

13. **WATER** Determine whether the set of numbers in the table are proportional. Explain your reasoning.

Bottles	1	2	3	4
Cost ($)	1.25	2.50	3.75	5.00

QuickReview

EXAMPLE 1

Determine whether the figure is a polygon. If it is, classify the polygon.

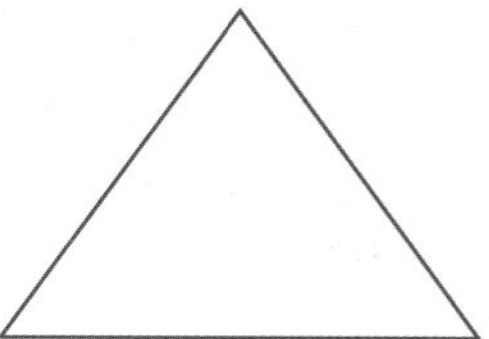

The polygon has 3 sides. It is a triangle.

EXAMPLE 2

Find $\frac{1}{3}(24)(5.8)$.

$\frac{1}{3}(24)(5.8) = \left[\frac{1}{3}(24)\right](5.8)$ **Associative Property**

$= (8)(5.8)$ **Simplify.**

$= 46.4$ **Simplify.**

EXAMPLE 3

Determine whether $\frac{2}{7}$ and $\frac{12}{42}$ form a proportion.

$\frac{2}{7} = \frac{12}{42}$ **Write a proportion.**

$2(42) = (7)12$ **Find the cross products.**

$84 = 84$ **Simplify.**

Since $\frac{2}{7} = \frac{12}{42}$, $\frac{2}{7}$ and $\frac{12}{42}$ form a proportion.

Online Option **Math Online** Take a self-check Chapter Readiness Quiz at glencoe.com.

Get Started on Chapter 12

You will learn several new concepts, skills, and vocabulary terms as you study Chapter 12. To get ready, identify important terms and organize your resources. You may wish to refer to **Chapter 0** to review prerequisite skills.

FOLDABLES® Study Organizer

Surface Area and Volume Make this Foldable to help you organize your Chapter 12 notes about surface area and volume. Begin with a sheet of 11″ × 17″ paper.

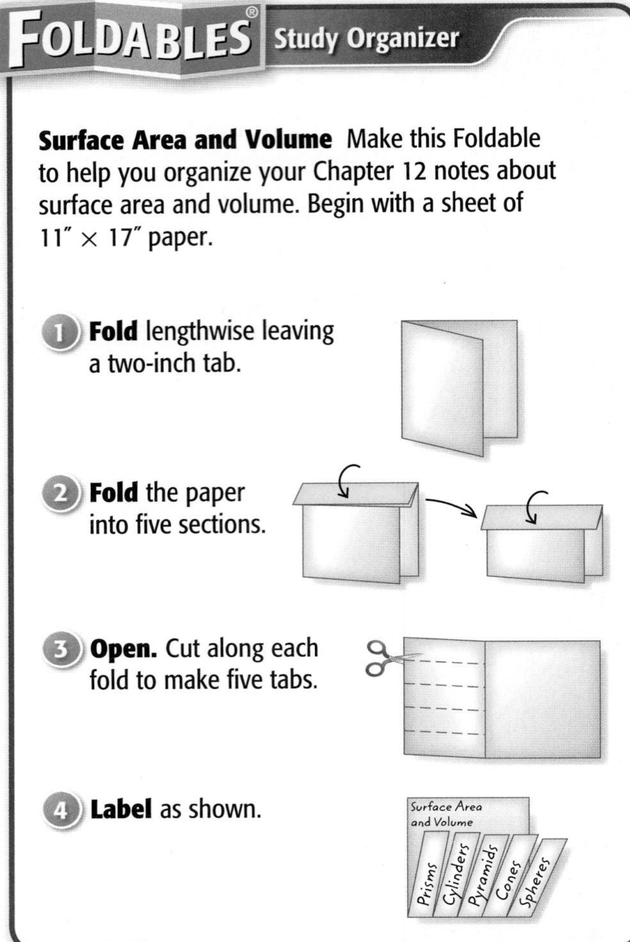

1. **Fold** lengthwise leaving a two-inch tab.
2. **Fold** the paper into five sections.
3. **Open.** Cut along each fold to make five tabs.
4. **Label** as shown.

Math Online glencoe.com

- Study the chapter online
- Explore **Math in Motion**
- Get extra help from your own **Personal Tutor**
- Use **Extra Examples** for additional help
- Take a **Self-Check Quiz**
- **Review Vocabulary** in fun ways

New Vocabulary

English		Español
plane	• p. 664 •	plano
solid	• p. 664 •	sólido
edge	• p. 664 •	arista
vertex	• p. 664 •	vértice
face	• p. 664 •	cara
prism	• p. 665 •	prisma
base	• p. 665 •	base
pyramid	• p. 665 •	pirámide
cylinder	• p. 665 •	cilindro
cone	• p. 665 •	cono sección
cross section	• p. 666 •	transversal
volume	• p. 671 •	volumen
sphere	• p. 684 •	esfera
nets	• p. 690 •	redes
lateral face	• p. 691 •	cara lateral
lateral area	• p. 691 •	área lateral
surface area	• p. 691 •	área
slant height	• p. 702 •	altura oblicua
similar solids	• p. 709 •	sólidos semejantes

Review Vocabulary

dilation • p. 307 • homotecia a transformation that alters the size of a figure by a scale factor, but not its shape

2 in. 4 in. 8 in. 16 in.

Multilingual eGlossary glencoe.com

EXPLORE **12-1**

Geometry Lab
Drawing Three-Dimensional Figures

Math Online glencoe.com
Math *in Motion,* Animation

Different views of a stack of cubes are shown. A point of view is called a **perspective.** You can build or draw three-dimensional figures using different perspectives.

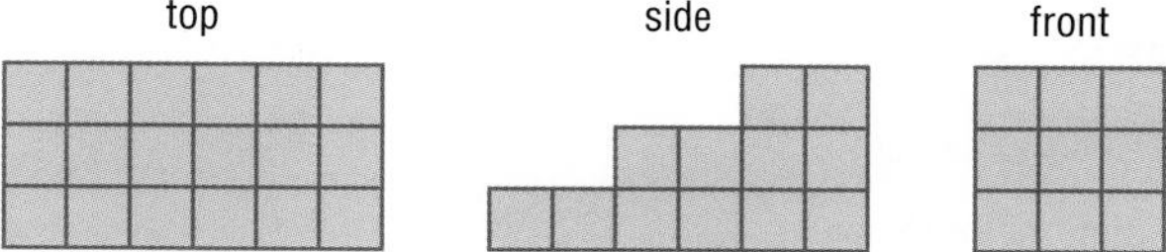

ACTIVITY

Build the figure that has the views shown above. Then use isometric dot paper to draw the model.

Step 1 Use the top view. The top view shows the shape of the base. It is a 3-by-6 rectangle.

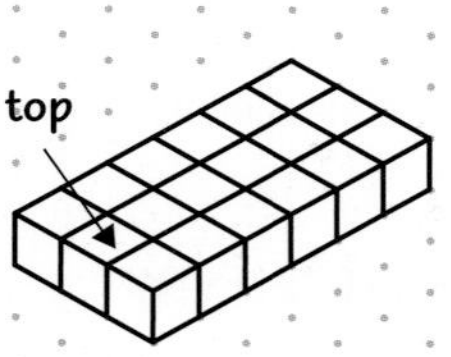

Step 2 The front view is a 3-by-3 square. This shows that the overall height and width of the figure is 3 units. The side view shows that the height increases in steps from 1 unit to 3 units.

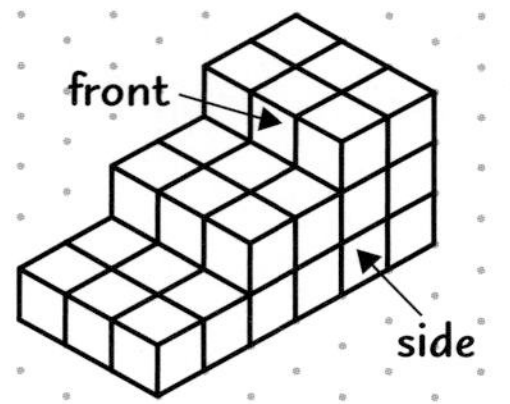

Exercises

The top view, a side view, and the front view of a three-dimensional figure are shown. Use cubes to build each figure. Then draw your model on isometric dot paper.

1. top 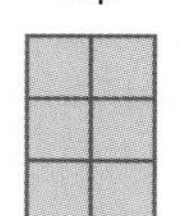 side 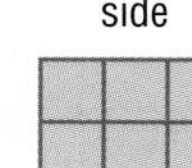 front 

2. top 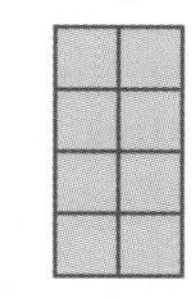 side 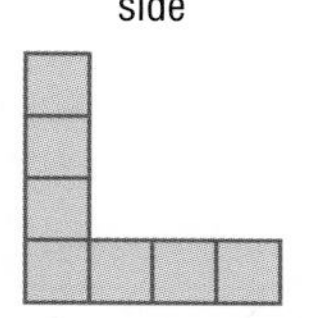 front 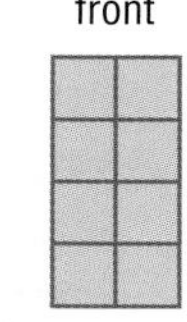

3. top 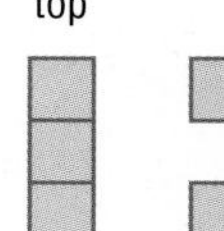 side 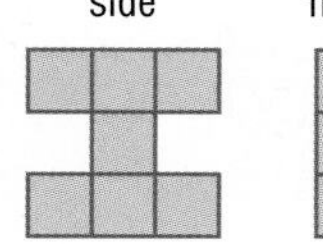 front

4. top 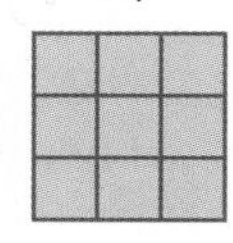 side front 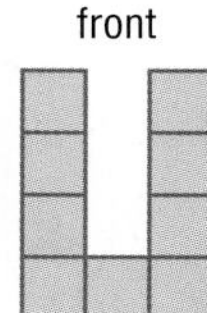

Draw and label the top view, a side view, and the front view for each figure.

5. 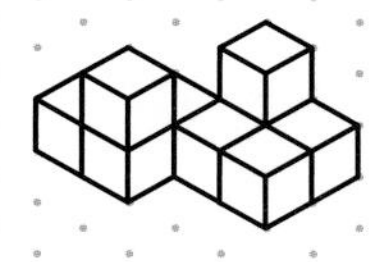

6. 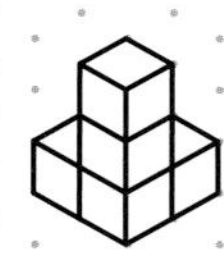

12-1 Three-Dimensional Figures

Then

You have already modeled three-dimensional figures. (Lesson 12-1a)

Now

- Identify three-dimensional figures.
- Describe and draw vertical, horizontal, and angled cross sections of three-dimensional figures.

New Vocabulary

plane
solid
polyhedron
edge
vertex
face
prism
base
pyramid
cylinder
cone
cross section

Math Online

glencoe.com

- Extra Examples
- Personal Tutor
- Self-Check Quiz
- Homework Help

Why?

a. If you observe the figures from directly above, what geometric figure(s) do you see?

b. If you view the figures directly from the front, what geometric figure(s) do you see?

c. Explain how you can see different two-dimensional figures when looking at a three-dimensional figure.

Identify Three-Dimensional Figures A two-dimensional figure has two dimensions—length and width. A **plane** is a two-dimensional flat surface that extends in all directions. There are different ways that planes may be related in space.

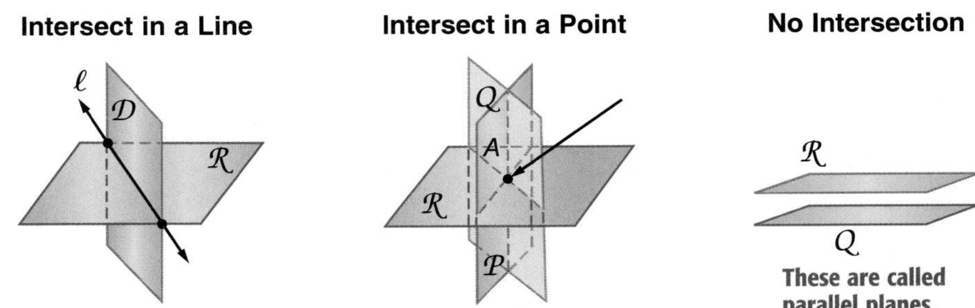
Intersect in a Line
ℓ
D
R
Intersect in a Point
Q
A
R
P
No Intersection
R
Q
These are called parallel planes.

A three-dimensional figure has three dimensions—length, width, and depth (or height). Intersecting planes can form three-dimensional figures or **solids**. A **polyhedron** is a solid with flat surfaces that are polygons.

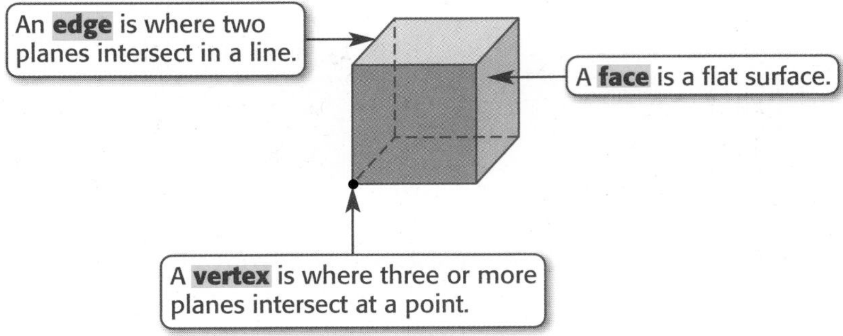
An edge is where two planes intersect in a line.
A face is a flat surface.
A vertex is where three or more planes intersect at a point.

> **Watch Out!**
>
> **Bases** In a *rectangular* prism, the bases do not have to be the top and bottom. Any two parallel rectangles are bases. In a *triangular* pyramid, any face is a base.

A **prism** is a polyhedron with two parallel, congruent faces called **bases** that are polygons. A **pyramid** is a polyhedron with one base that is any polygon. Its other faces are triangles. Prisms and pyramids are named by the shape of their bases.

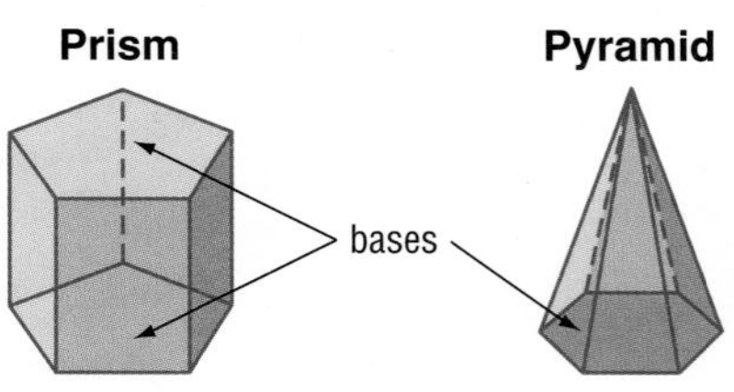

Concept Summary — Polyhedrons

For Your FOLDABLE

Polyhedron	triangular prism	rectangular prism	triangular pyramid	rectangular pyramid
Number of Bases	2	2	1	1
Polygon Base	triangle	rectangle	triangle	rectangle
Figure				

> **Vocabulary Link**
>
> **Cone**
>
> **Everyday Use** Lupe had an ice cream *cone* for dessert.
>
> **Math Use** A solid with one base that is a circle connected to a vertex by a curved side is a *cone*.

There are solids that are *not* polyhedrons. A **cylinder** is a three-dimensional figure with congruent, parallel bases that are circles connected with a curved side. A **cone** has one circular base and a vertex connected by a curved side.

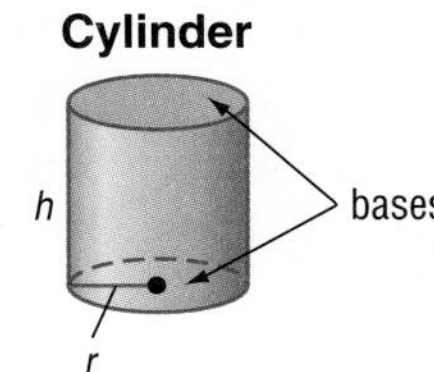

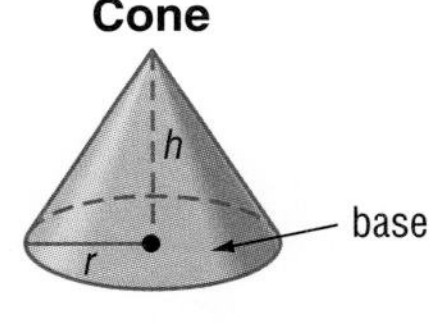

EXAMPLE 1 Identify Three-Dimensional Figures

Identify the figure. Name the bases, faces, edges, and vertices.

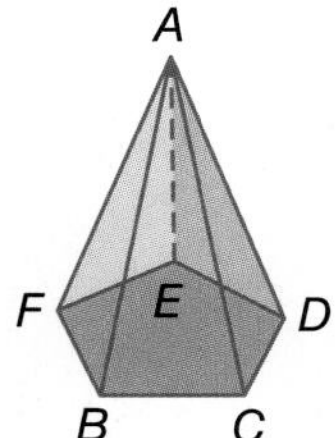

This figure has one pentagonal base, *BCDEF*, so it is a pentagonal pyramid.

faces: *ABF, AFE, AED, ADC, ACB, BCDEF*

edges: $\overline{AB}, \overline{AC}, \overline{AD}, \overline{AE}, \overline{AF}, \overline{BC}, \overline{CD}, \overline{DE}, \overline{EF}, \overline{FB}$

vertices: *A, B, C, D, E, F*

Check Your Progress

1A.

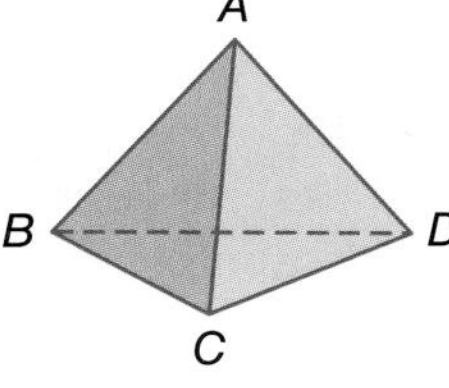

1B.

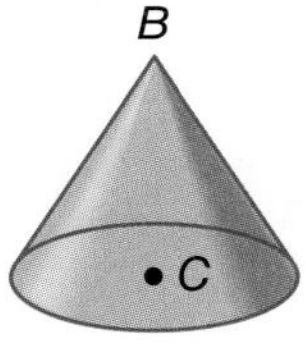

Personal Tutor glencoe.com

Cross Sections Interesting shapes can occur when a plane intersects, or slices, a three-dimensional figure. The intersection of the figure and the plane is called a **cross section** of the figure.

EXAMPLE 2 Cross Section of a Cone

Draw and describe the shape resulting from the following vertical, angled, and horizontal cross sections of a cone.

Vertical Slice

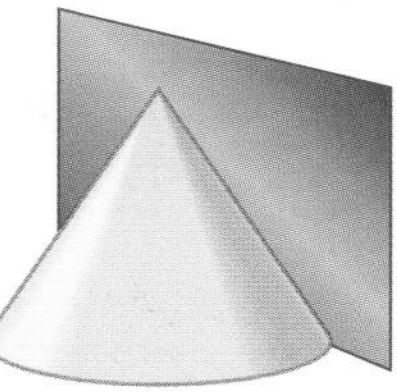

The cross section is a triangle.

Angled Slice

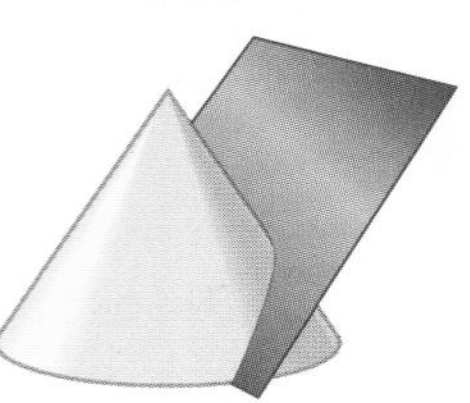

The cross section is a parabola.

Horizontal Slice

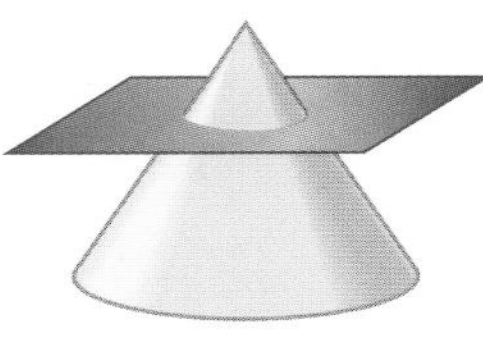

The cross section is a circle.

> **StudyTip**
>
> **Parabola** The shape that results from the angled slice through a cone is called a *parabola*. You learned about parabolas in Chapter 9.

Check Your Progress

2. Draw and describe the shape resulting from a vertical, angled, and horizontal cross section of a triangular pyramid.

Personal Tutor glencoe.com

Real-World EXAMPLE 3 Describe and Draw

CRYSTALS A fluorite crystal is shown at the right. Draw the top view and side view. Then draw and describe the shape resulting from a vertical cross section of the figure.

The crystal is a rectangular prism with two square pyramids attached. The vertical cross sections will look similar to the side view.

Top View

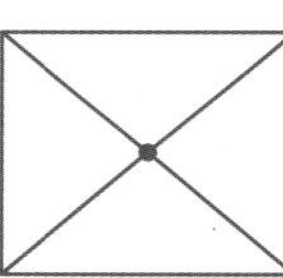

Side View

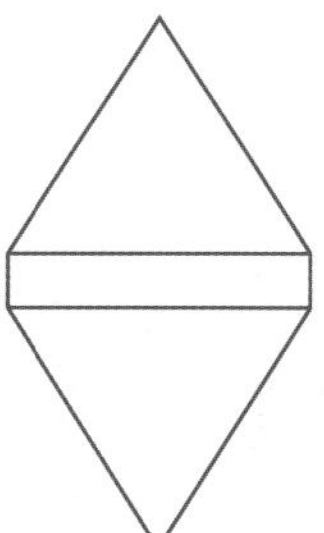

Vertical Cross Section

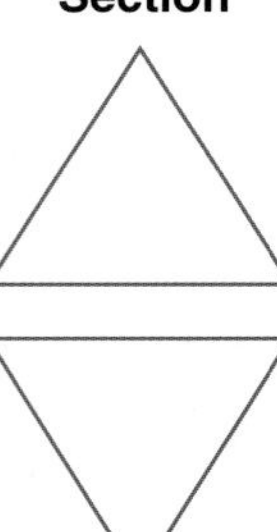

> **Math History Link**
>
> **Euclid (c. 325–283 B.C.)** The kind of solid geometry presented here is part of *Euclidean geometry.* Euclid was a Greek mathematician best known for his series of 13 books called *Elements*.

Check Your Progress

3. CAMPING Lemar and his brother went camping for the weekend. They set up the tent at the right. Draw the top view and side view. Then draw and describe the shape resulting from a vertical cross section of the figure.

Personal Tutor glencoe.com

Check Your Understanding

Example 1 p. 665

Identify each figure. Name the bases, faces, edges, and vertices.

1. 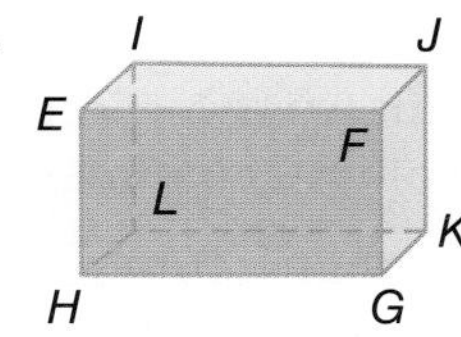

2. 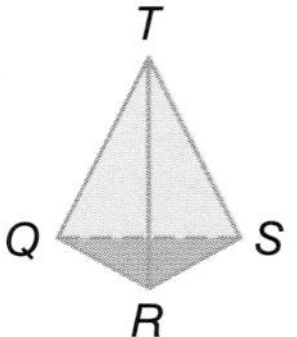

Example 2 p. 666

Draw and describe the shape resulting from each cross section.

3. 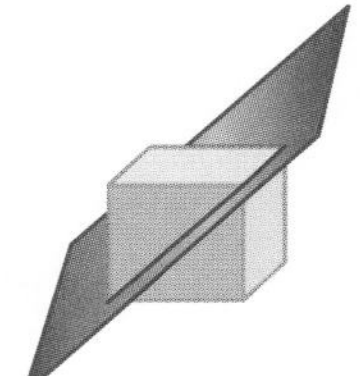

4. 

5. 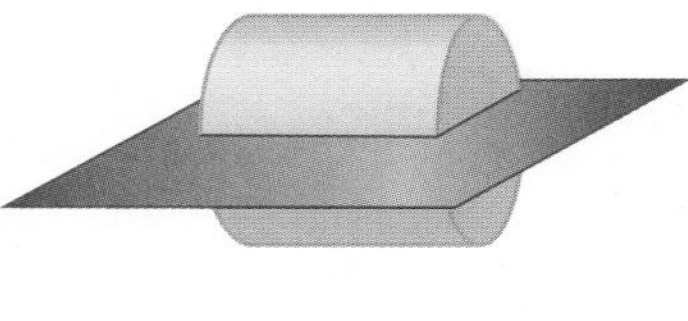

Example 3 p. 666

6. **PAPERWEIGHT** A glass paperweight in the shape of a pyramid is placed on a desk. Draw the top view and side view. Then draw and describe the shape resulting from a vertical cross section of the figure.

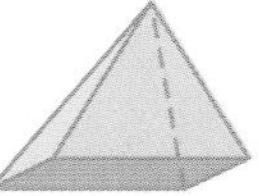

Practice and Problem Solving

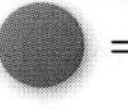 = **Step-by-Step Solutions** begin on page R11. **Extra Practice** begins on page 810.

Example 1 p. 665

Identify each figure. Name the bases, faces, edges, and vertices.

7. 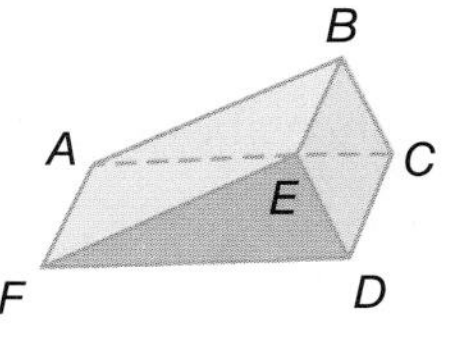

8. 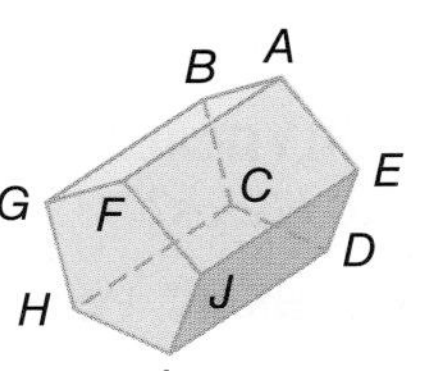

9. 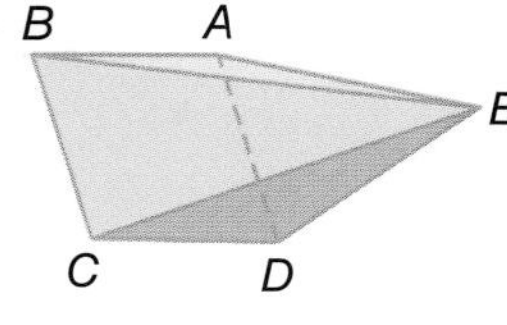

10. 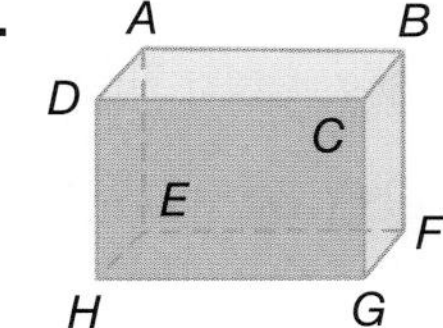

11. 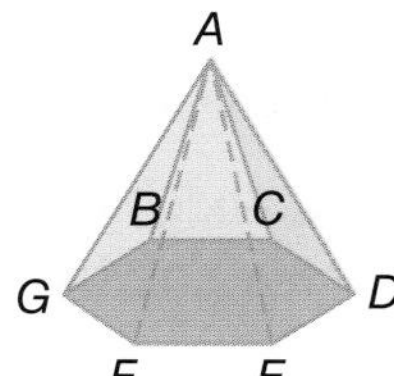

12. 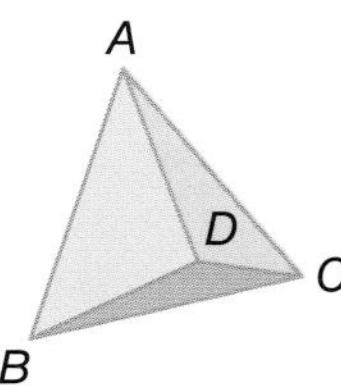

Example 2 p. 666

Draw and describe the shape resulting from each cross section.

13. 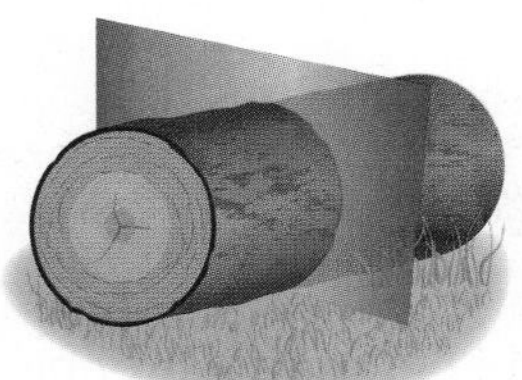

14. 

15. 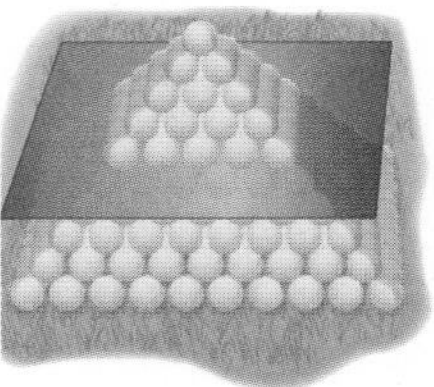

Example 3 p. 666

16. **BOXES** For an art project, Jorge is using the box shown at the right. Draw the top view and side view. Then draw and describe the shape resulting from an angled cross section of the figure.

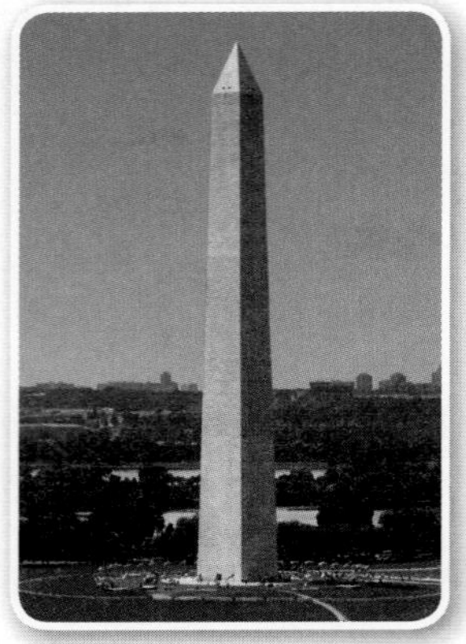

Real-World Link

The Washington Monument is in the shape of an *obelisk*. An obelisk is a four-sided tapered shaft that has a pyramid at the top.

17. **MONUMENTS** Use the information about the Washington Monument shown at the left.
    a. Sketch the pyramid shaped top of the monument and label the vertices.
    b. Identify the bases, faces, and edges.
    c. Describe the shapes that result from vertical, angled, and horizontal cross sections of the pyramid top of the monument.
    d. Draw and describe the shape that would result from an angled cross section through the top and base of the monument.

ART For each sculpture shown below, draw the top view and side view. Then draw and describe the shape resulting from a vertical cross section of the figure.

18. 

19. 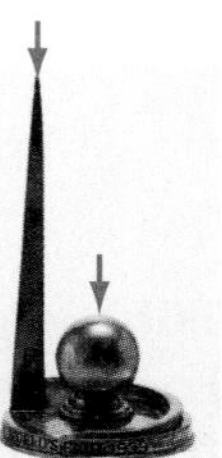

20. 

21. **MULTIPLE REPRESENTATIONS** In this problem, you will investigate Euler's Formula on polyhedra.
    a. **TABULAR** Draw each figure. Then, copy and complete the table shown.

Name	Triangular Pyramid	Square Pyramid	Pentagonal Pyramid
Vertices	4	5	■
Faces	■	■	6
Edges	■	8	■

    b. **ANALYTICAL** What do you notice about the number of vertices, faces, and edges?
    c. **ALGEBRAIC** Write an equation that compares the sum of the number of vertices V and the number of faces F to the number of edges E.

H.O.T. Problems

Use Higher-Order Thinking Skills

22. **OPEN ENDED** Choose a solid object from your home. Draw and describe the shape resulting from a vertical, angled, and horizontal cross section of it.

REASONING For Exercises 23–25, determine whether each statement is *always*, *sometimes*, or *never true*.

23. The bases of cylinders have different radii.
24. Two planes intersect in a single point.
25. Three planes do not intersect in a point.
26. **CHALLENGE** A triangular pyramid has 6 edges. A square pyramid has 8 edges. Write a formula that gives the number of edges e for a pyramid with an n-sided base.
27. **REASONING** Explain how you would classify the polyhedron shown at the right.
28. **WRITING IN MATH** Are cylinders polyhedrons? Explain.

Standardized Test Practice

29. Which of the following is *not* considered an edge of the triangular prism?

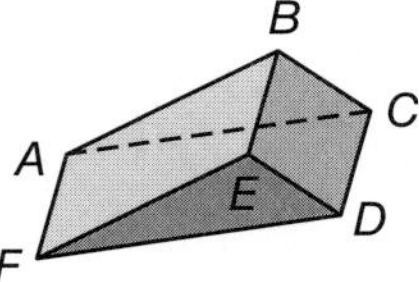

A $\overline{AF}$

B $\overline{BE}$

C $\overline{DE}$

D $\overline{AE}$

30. What three-dimensional figure has one vertex?

F cone

G cylinder

H triangular prism

J triangular pyramid

31. EXTENDED RESPONSE Draw and describe the shape resulting from a vertical cross section of the figure shown.

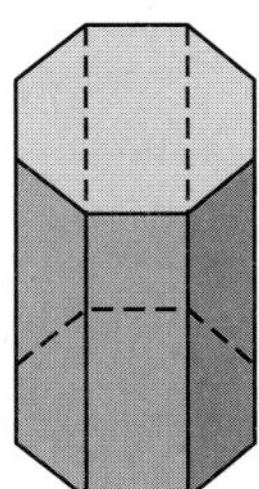

32. Which of the following real-world objects resembles a rectangular prism?

A bowl

B box

C soup can

D stop sign

Spiral Review

Find the area of each figure. Round to the nearest tenth, if necessary. (Lesson 11-9)

33.

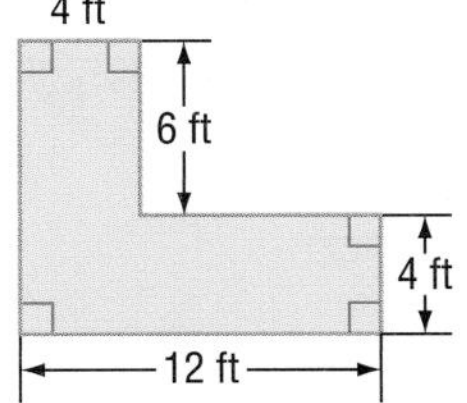

34.

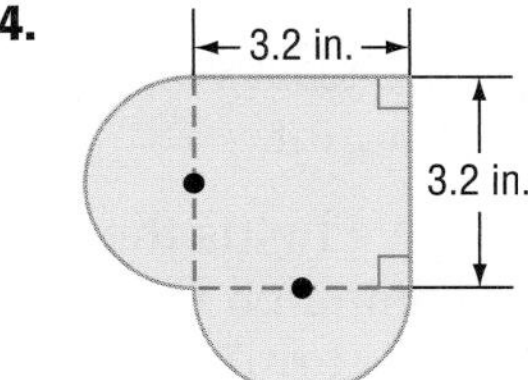

35.

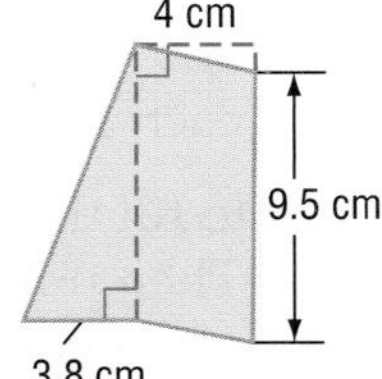

36. BAND During a football game, a marching band can be heard within a radius of 1.7 miles. What is the area that can hear the band? Round to the nearest tenth. (Lesson 11-8)

Skills Review

Find the area of each figure. (Lesson 11-6)

37.

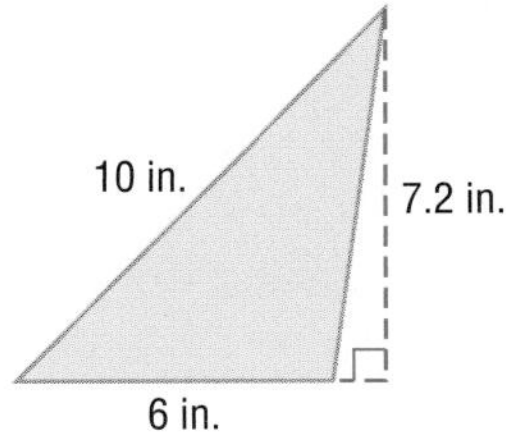

38.

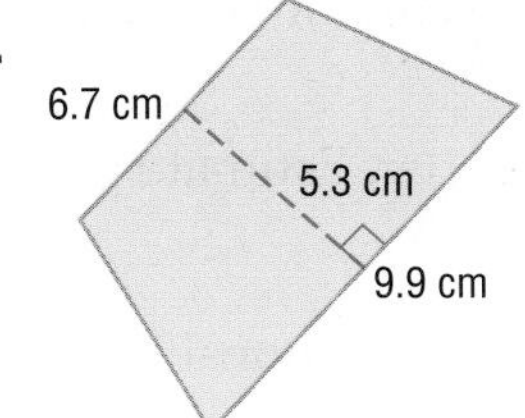

39.

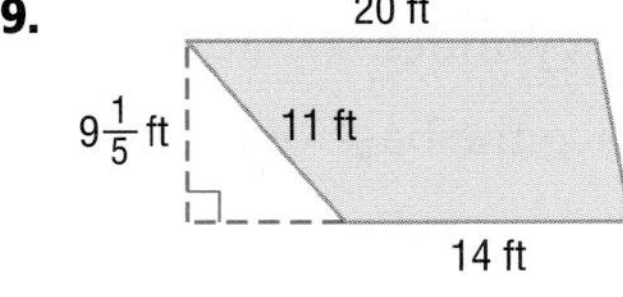

Geometry Lab
Volume

Math Online glencoe.com
Math *in Motion*, Animation

ACTIVITY

Step 1 Use three 5×8 index cards to make three containers each with a height of 5 inches as shown.

- Make one with a square base that has 2-inch sides.
- Make one with a triangular base that has sides of 2 inches, 3 inches, and 3 inches.
- Make one with a circular base that has an 8-inch circumference.

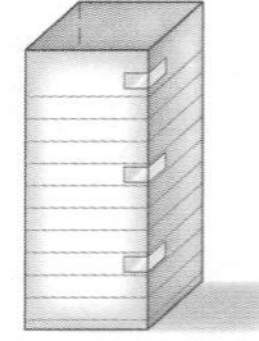

square base with 2-inch sides

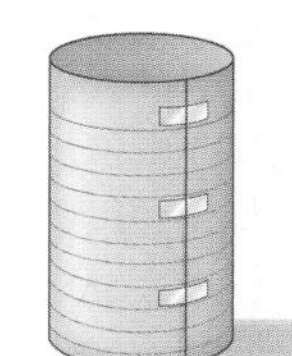

circular base with 8-inch circumference

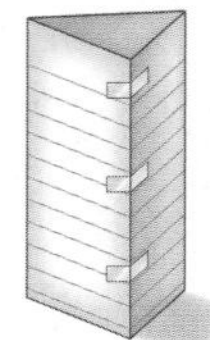

triangular base with sides 2 inches, 3 inches, and 3 inches

Step 2 Tape one end of each container to another card as a bottom, but leave the top open.

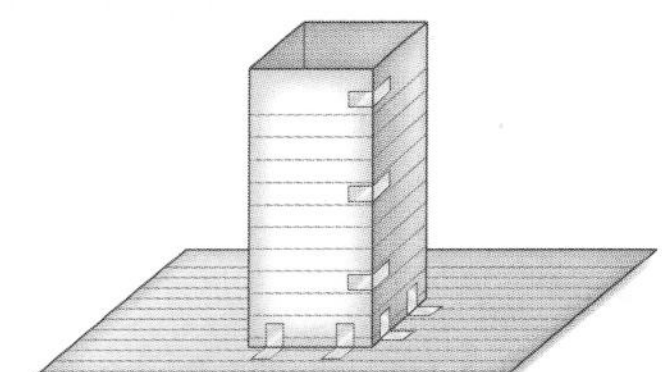

Step 3 Estimate which container holds the most (has the greatest volume) and which holds the least (has the least volume), or whether each container holds the same amount.

Step 4 Use rice to fill the container that you believe holds the least amount. Then put the rice into another container. Does the rice fill this container? Continue the process until you find which, if any, container has the least volume and which has the greatest.

Analyze the Results

1. Which container holds the most? the least?
2. How do the heights of the three containers compare? What is each height?
3. Compare the perimeters of the bases of each container with the circumference of the cylinder. What is each base perimeter?
4. Trace the base of each container onto grid paper. Estimate the area of each base.
5. Which container has the greatest base area?
6. **MAKE A CONJECTURE** Does there appear to be a relationship between the area of the base and the volume? Explain.

12-2 Volume of Prisms

Why?

You can investigate the amount of space a prism occupies by examining its cross sections. Look at the block of pottery clay at the right. It measures 4 inches long by 2 inches wide by 10 inches tall.

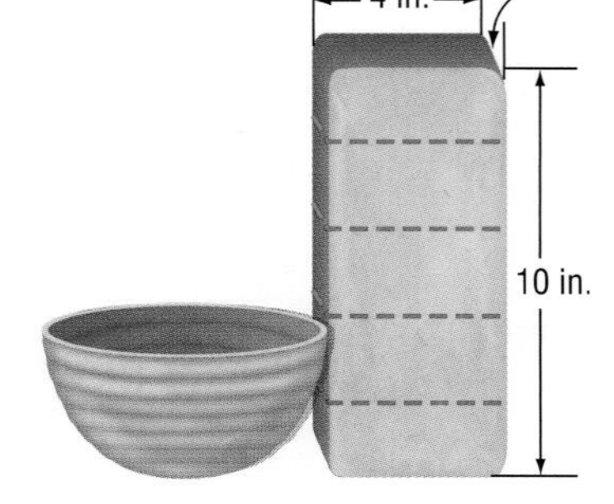

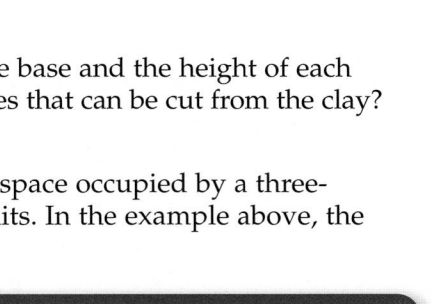

a. If the piece of clay is cut into 5 equal pieces horizontally, what are the dimensions of each piece?

b. How many one-inch cubes could be cut from each piece? from the entire piece of clay?

c. **MAKE A CONJECTURE** How are the area of the base and the height of each piece related to the number of one-inch cubes that can be cut from the clay?

Then

You have already found the areas of rectangles and triangles. (Lesson 5-1)

Now

- Find volumes of prisms.
- Find volumes of composite figures.

New Vocabulary

volume

Math Online

glencoe.com

- Extra Examples
- Personal Tutor
- Self-Check Quiz
- Homework Help

Volumes of Prisms **Volume** is the measure of space occupied by a three-dimensional region. It is measured in cubic units. In the example above, the *volume* of the block of clay is 80 cubic inches.

Key Concept — Volume of a Prism

For Your FOLDABLE

Words The volume V of a prism is the area of the base B times the height h.

Symbols $V = Bh$

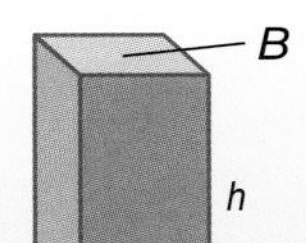

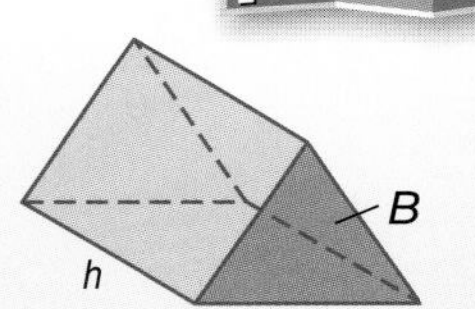

EXAMPLE 1 Find the Volume of a Rectangular Prism

Find the volume of the rectangular prism.

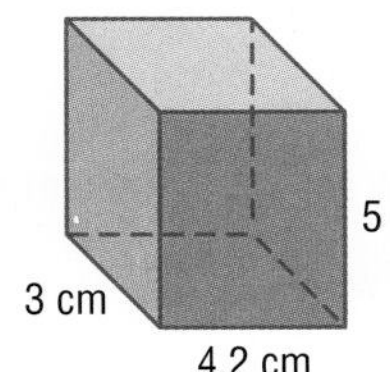

$V = Bh$	Write the formula for volume of a prism.
$= (\ell w)h$	The base is a rectangle, so $B = \ell w$.
$= (4.2 \cdot 3) \cdot 5$ or 63	Replace ℓ with 4.2 cm, w with 3 cm, and h with 5 cm.
	Simplify.

The volume is 63 cubic centimeters.

Check Your Progress

1. Find the volume of a rectangular prism with a length of 10 feet, a width of 13 feet, and a height of 21 feet.

Personal Tutor glencoe.com

EXAMPLE 2 Volume of a Triangular Prism

Find the volume of the triangular prism.

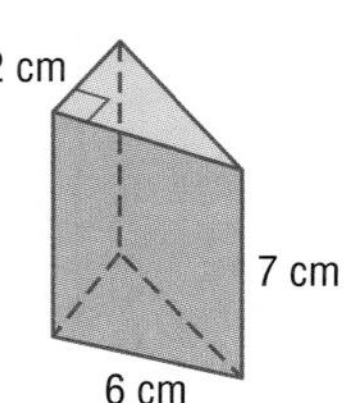

> **Watch Out!**
>
> **Height** Be sure not to confuse the height of the triangle with the height of the prism.

$V = Bh$ — **Write the formula for volume of a prism.**

$= \left(\frac{1}{2} \cdot 6 \cdot 2\right)h$ — **The base is a triangle, so $B = \frac{1}{2}bh$, $b = 6$ and $h = 2$.**

$= 6 \cdot 7$ — **The height of the prism is 7 cm.**

$= 42$ — **Simplify.**

The volume is 42 cubic centimeters.

Check Your Progress

Find the volume of each triangular prism.

2A.

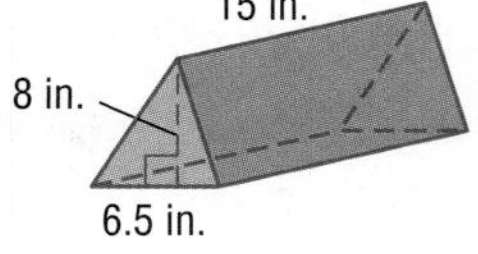

2B.

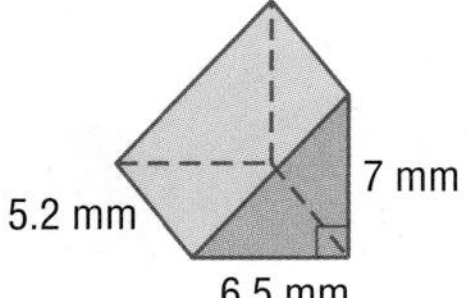

Personal Tutor glencoe.com

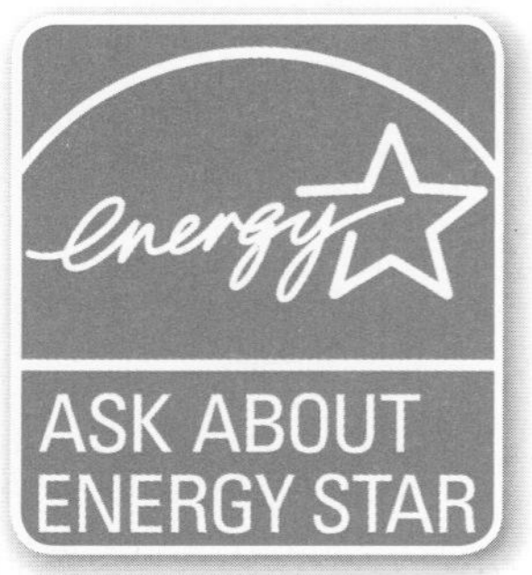

> **Real-World Link**
>
> Air conditioners have earned the energy star rating from the Department of Energy. In 2006, Americans avoided producing greenhouse gas emissions equal to those made by 25 million cars.
>
> **Source:** U.S. Department of Energy

Real-World EXAMPLE 3 Find the Missing Length

HOME IMPROVEMENT A room air conditioner can cool a room with a volume of 1600 cubic feet. If a room has a height of 8 feet and a length of 16 feet, what is the maximum width of the room?

Estimate $1600 \div (10 \times 16) = 10$

$V = Bh$ — **Write the formula for the volume of a prism.**

$V = \ell wh$ — **Replace B with ℓw.**

$1600 = 16 \cdot w \cdot 8$ — **Replace V with 1600, ℓ with 16, and h with 8.**

$1600 = 16 \cdot 8 \cdot w$ — **Commutative Property**

$1600 = 128w$ — **Simplify.**

$12.5 = w$ — **Divide each side by 128.**

The room is 12.5 feet wide.

Check for Reasonableness $12.5 \approx 10$ ✓

Check Your Progress

3. **POOLS** A children's rectangular pool holds 17.5 cubic feet of water. What is the width of the pool if its length is 3.5 feet and its height is 1 foot?

Personal Tutor glencoe.com

Volumes of Composite Figures You can find the area of composite three-dimensional figures by breaking them into smaller pieces.

> **Problem-Solving Tip**
>
> **Eliminating Choices** You can eliminate A and B as answers because the volume of the rectangular prism is 36 ft^3, so the volume of the whole figure must be greater.

STANDARDIZED TEST EXAMPLE 4

Find the volume of the ramp.

A 30 ft^3 **C** 66 ft^3

B 36 ft^3 **D** 96 ft^3

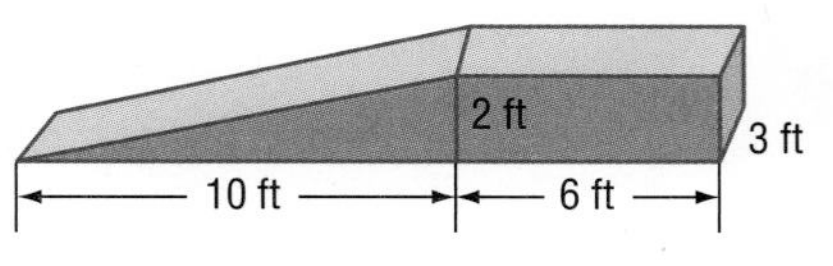

Read the Test Item

The solid is made up of a triangular prism and a rectangular prism. The volume of the solid is the sum of both volumes.

Solve the Test Item

Triangular Prism	**Rectangular Prism**
$V = Bh$	$V = \ell wh$
$V = \frac{1}{2} \cdot 10 \cdot 2 \cdot 3$	$V = 6 \cdot 3 \cdot 2$
$V = 30$	$V = 36$

The volume of the figure is 30 + 36 or 66 cubic feet. The answer is *C*.

Check Your Progress

4. Find the volume of the figure at the right.

F 60 cm^3 **H** 98 cm^3

G 73.5 cm^3 **J** 107.5 cm^3

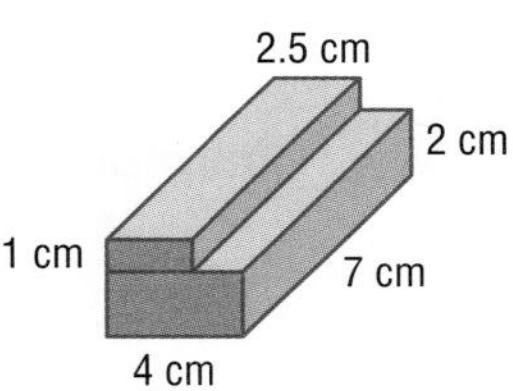

Personal Tutor glencoe.com

Check Your Understanding

Examples 1 and 2 pp. 671–672

Find the volume of each figure.

1.

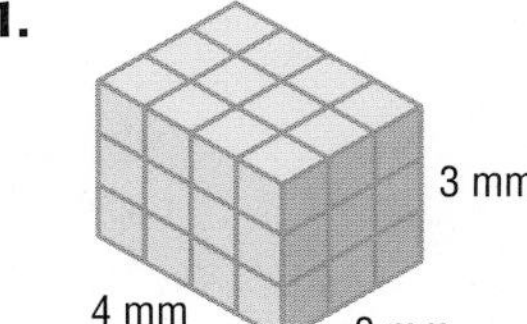

2.

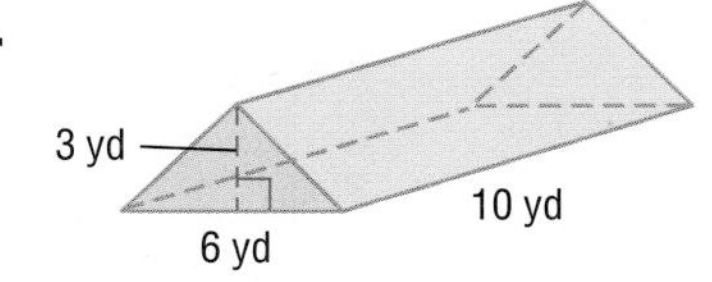

3.

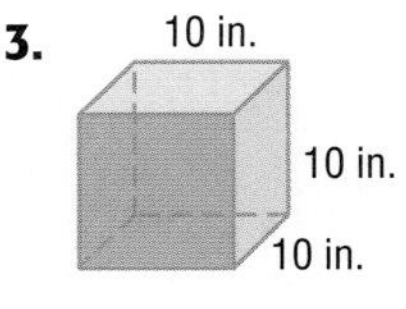

Example 3 p. 672

4. GARDENING A window box has a length of 8.5 inches and a height of 9 inches. If the volume of the box is 2295 cubic inches, what is the width of the box?

Example 4 p. 673

5 MULTIPLE CHOICE Find the volume of the figure at the right.

A 400 in^3 **C** 56 in^3

B 552 in^3 **D** 840 in^3

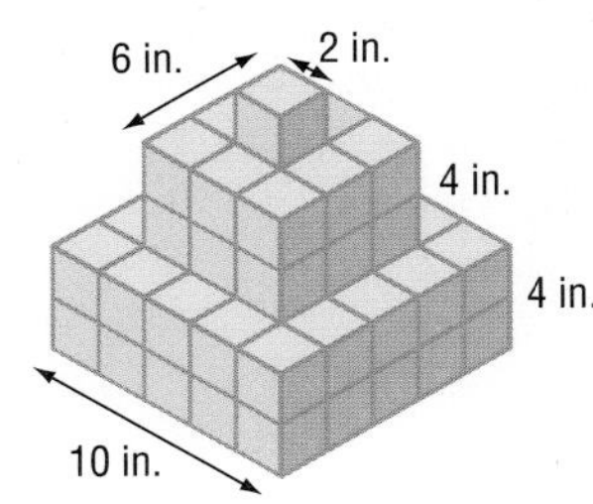

Practice and Problem Solving

● = **Step-by-Step Solutions** begin on page R11.
Extra Practice begins on page 810.

Examples 1 and 2
pp. 671–672

Find the volume of each figure.

6.

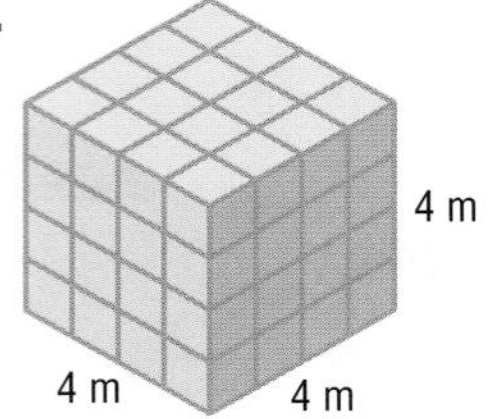

7.

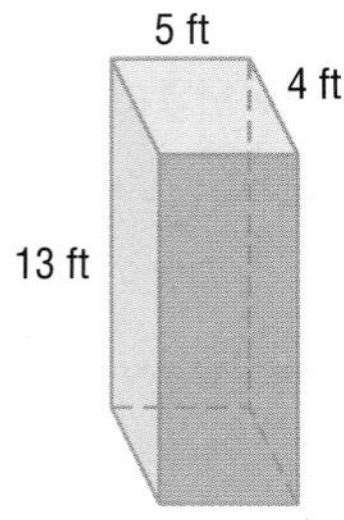

8.

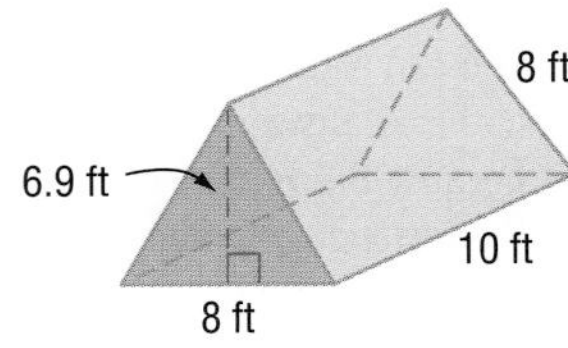

9.

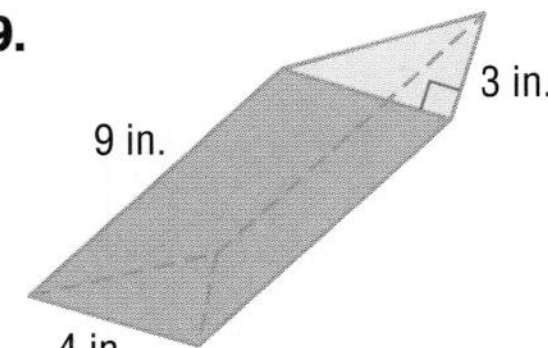

10.

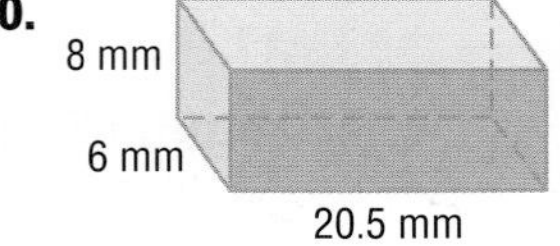

11.

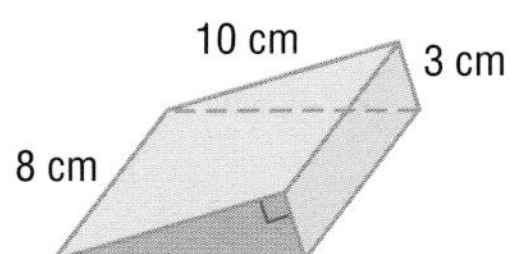

12. triangular prism: base of triangle 6.2 yards, height of triangle 20 yards, height of prism 14 yards

13. rectangular prism: height 4 inches, width $1\frac{1}{2}$ inches, length $\frac{1}{4}$ inch

Example 3
p. 672

14. Find the length of a rectangular prism with a width of 4 feet, a height of 6 feet, and a volume of 84 cubic feet.

15. Find the height of a triangular prism with a base length of 10 yards, a base height of 20 yards, and a volume of 600 cubic yards.

Example 4
p. 673

16. **DOG HOUSES** Josh and his mom are building the dog house shown below. Find the volume of the dog house.

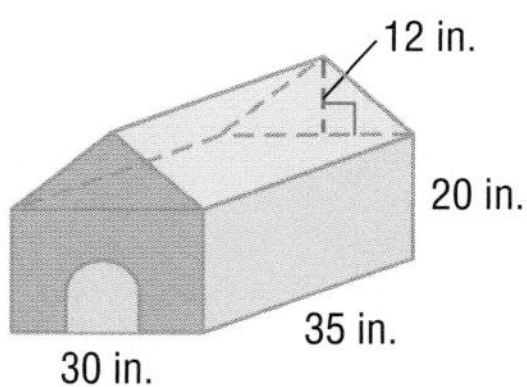

17. **STEPS** Morgan is building a model of some steps with 6 inch foam blocks. What is the total volume of the blocks?

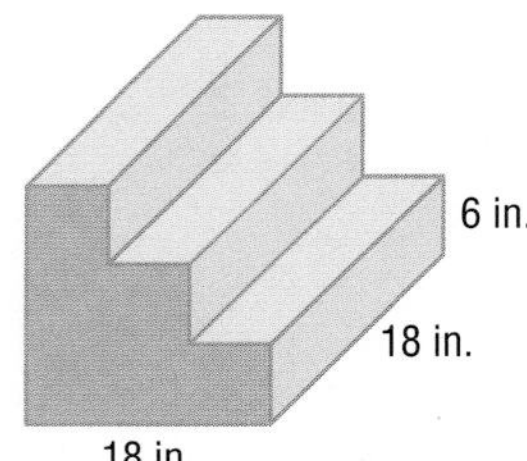

18. **CRAFTS** Jill is mailing a candle that is in the shape of a triangular prism as shown. She put the candle in a rectangular box that measures 3 inches by 5 inches by 7 inches and places foam pieces around the candle. Find the volume of the foam pieces needed to fill the space between the candle and the box.

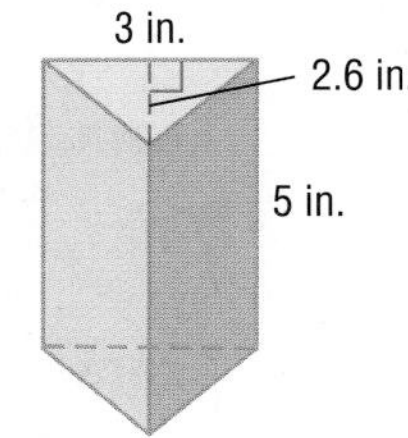

19. The height of a triangular prism is 8 feet and it has a volume of 200 cubic feet. If the base has a length of 5 feet, what is the height of the base?

20. **PLANTING** Ben wants to buy enough potting soil to fill a window box that is 42 inches long, 8 inches wide, and 6 inches high. If one bag of potting soil contains 576 cubic inches, how many bags should he buy?

Find the volume of each figure.

21.

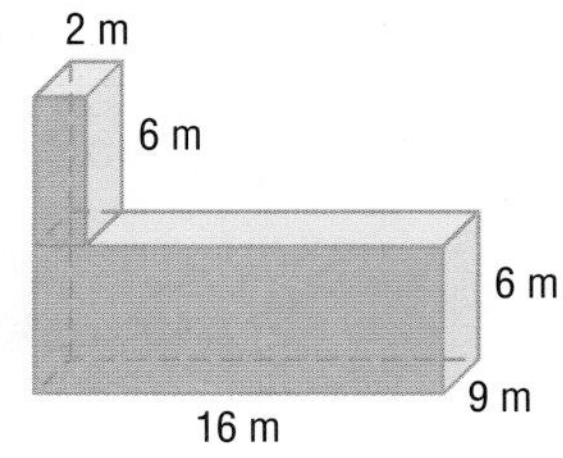

22.

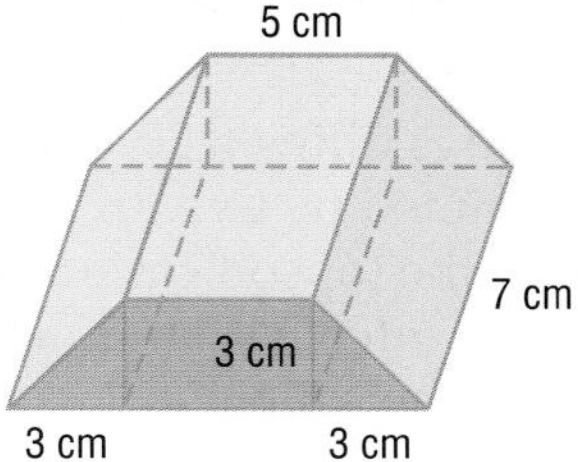

23.

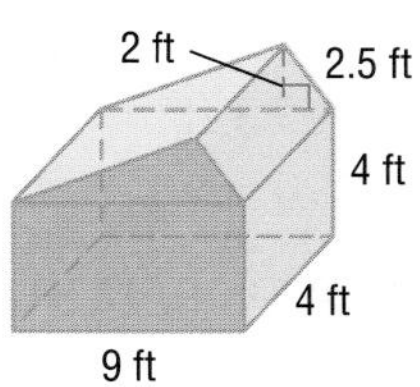

24.

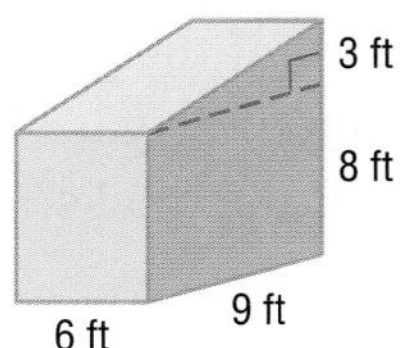

25. **CANDY** A chocolate bar is in the shape of a trapezoidal prism as shown at the right. Find the volume of the chocolate bar.

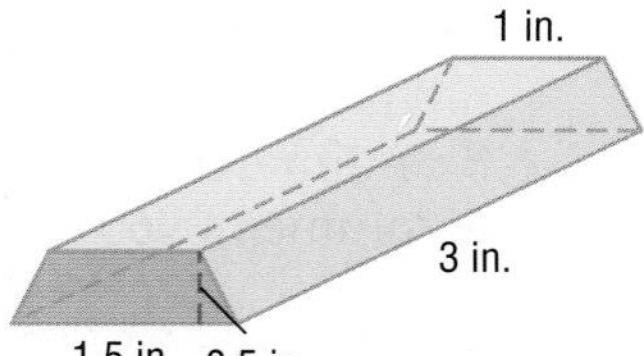

26. **FINANCIAL LITERACY** A movie theater sells different sizes of popcorn as shown in the table. If the containers are rectangular prisms, find the ratio of cost to the volume of each bag of popcorn. Which size of popcorn is the best buy?

Popcorn Sizes Available

Size	Length (in.)	Width (in.)	Height (in.)	Price ($)
Small	5	4	8	4.50
Medium	7	5	10	5.75
Large	10	6	12	6.50

Real-World Link

In the United States, 16 billion quarts of popcorn are consumed each year. That's about 16 gallons per person per year. Most popcorn is grown in Indiana, Nebraska, Ohio, Illinois, Iowa, Kentucky and Missouri.

Source: The Popcorn Board

27. **GEOMETRY** Refer to the figure at the right.

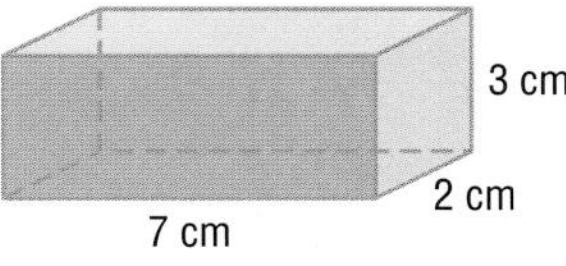

a. What is the volume of the figure?

b. How does the volume change if one of the dimensions is doubled? two dimensions? three dimensions? Explain.

c. Repeat the above steps and triple each dimension. What do you notice?

d. Without calculating, find the volume of the prism if each of the dimensions is multiplied by 6.

H.O.T. Problems

Use Higher-Order Thinking Skills

28. **REASONING** Without calculating, compare the volumes of the prisms shown. Explain.

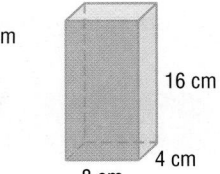

29. **OPEN ENDED** Find the dimensions of any triangular prism that has a volume of 44 cubic inches.

CHALLENGE **Use dimensional analysis to make each conversion.**

30. $5 \text{ yd}^3 = \blacksquare \text{ ft}^3$

31. $945 \text{ ft}^3 = \blacksquare \text{ yd}^3$

32. $2 \text{ m}^3 = \blacksquare \text{ cm}^3$

33. **WRITING IN MATH** Explain how doubling the length, width, and height of a box changes the volume of the box.

Standardized Test Practice

34. Mr. Toshio is filling a 20-foot by 35-foot garden framed by two levels of bricks with topsoil. If the topsoil costs $9 per cubic foot, what other information is needed to find s, the cost of the soil?

A The area of the garden.
B The perimeter of the garden.
C The price per cubic yard of soil.
D The height of the bricks.

35. GRIDDED RESPONSE How many centimeters tall is a rectangular prism with a length of 8 centimeters, width of 10 centimeters, and a volume of 960 cubic centimeters?

36. What is the volume of the prism below?

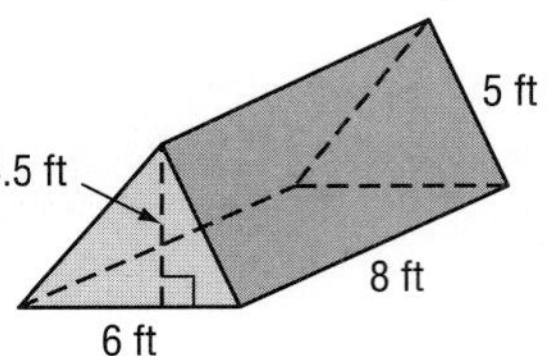

F 67.5 ft^3 **H** 108 ft^3
G 90 ft^3 **J** 216 ft^3

37. Which of the following is the best estimate for the volume of a shoe box with sides that measure 15.75 inches, 9.25 inches, and 8 inches?

A 10 in^3 **C** 1000 in^3
B 100 in^3 **D** $10{,}000 \text{ in}^3$

Spiral Review

Identify each figure. Name the bases, faces, edges, and vertices. (Lesson 12-1)

38.

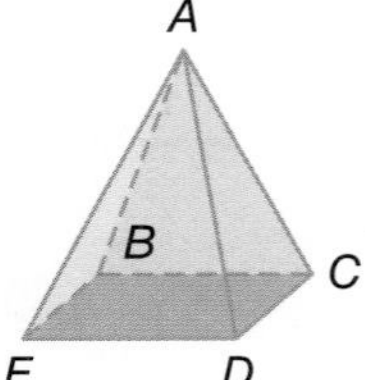

39.

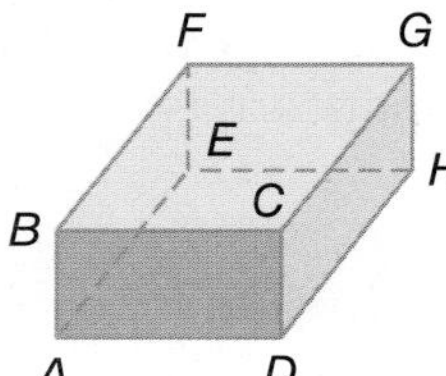

Find the area of each figure. Round to the nearest tenth. (Lesson 11-9)

40.

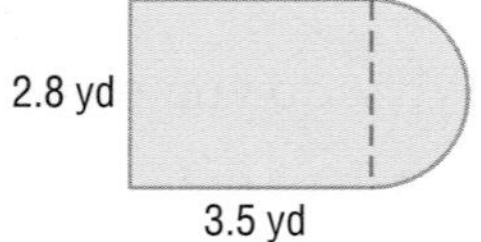

41.

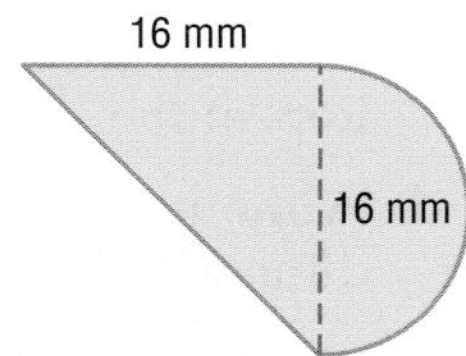

42. SOCCER Tomás wants to spend less than $100 for a new soccer ball and shoes. The ball costs $24. Write and solve an inequality that gives the amount that Tomás can spend on shoes. (Lesson 5-4)

Skills Review

Find the area of each circle. Round to the nearest tenth. (Lesson 11-8)

43.

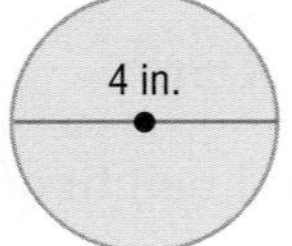

44.

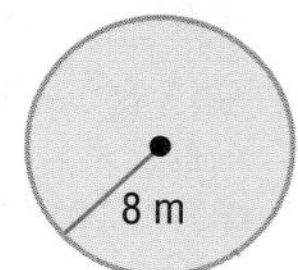

45.

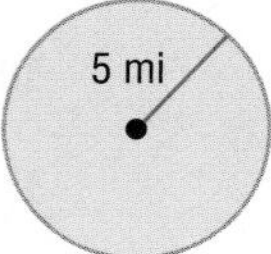

12-3 Volume of Cylinders

Why?

Marisol has a stack of dimes and another stack of quarters.

a. How would you find the value in dollars of the dimes? the quarters?

b. How much money would Marisol have if she had 30 dimes and 25 quarters?

c. **MAKE A CONJECTURE** How do the answers to part **a** and part **b** relate to finding the volume of a cylinder?

Then

You have already found the areas of circles. (Lesson 11-8)

Now

- Find the volumes of circular cylinders.
- Find the volumes of composite figures involving circular cylinders.

Math Online

glencoe.com

- Extra Examples
- Personal Tutor
- Self-Check Quiz
- Homework Help

Volumes of Cylinders Like prisms, the volume of a cylinder is the product of the area of the base and the height.

Key Concept — Volume of a Cylinder

For Your FOLDABLE

Words The volume V of a circular cylinder with radius r is the area of the base B times the height h.

Model

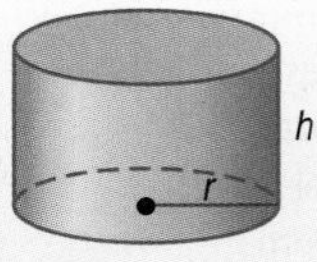

Symbols $V = Bh$, where $B = \pi r^2$ or $V = \pi r^2 h$

EXAMPLE 1 Volume of a Cylinder

Find the volume of each cylinder. Round to the nearest tenth.

a. radius of base 3 in., height 12 in.

3 in.
12 in.

$V = Bh$	Volume of a cylinder
$V = \pi r^2 h$	Replace B with πr^2.
$\approx \pi \cdot 3^2 \cdot 12$	Replace r with 3 and h with 12.
$\approx 339.1 \text{ in}^3$	Use a calculator.

b. diameter of base 14 cm, height 20 cm

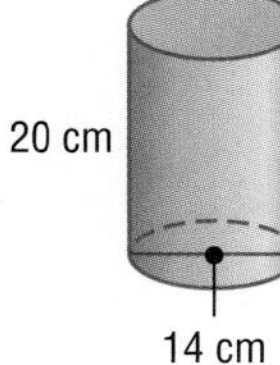

$V = Bh$	Volume of a cylinder
$V = \pi r^2 h$	Replace B with πr^2.
$\approx \pi \cdot 7^2 \cdot 20$	Replace r with 7 and h with 20.
$\approx 3078.8 \text{ cm}^3$	Use a calculator.

Check Your Progress

1A.

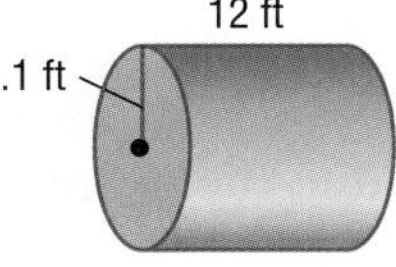

1B.

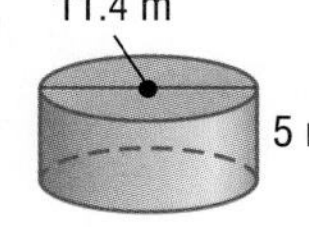

Personal Tutor glencoe.com

EXAMPLE 2 Height of a Cylinder

The volume of the cylinder is 618 cubic meters. Find the height of the cylinder. Round to the nearest tenth.

4 m

h

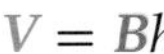

$V = Bh$	**Volume of a cylinder**
$V = \pi r^2 h$	**Replace B with πr^2.**
$618 = \pi \cdot 4^2 \cdot h$	**Replace V with 618 and r with 4.**
$618 = 16\pi h$	**Simplify.**
$12.3 \approx h$	**Divide each side by 16π. Round to the nearest tenth.**

The height of the cylinder is about 12.3 meters.

Check Your Progress

2. Find the height of a cylinder with a diameter of 10 yards and a volume of 549.5 cubic yards. Round to the nearest tenth.

Personal Tutor glencoe.com

Real-World Link

The Hirshshorn Museum and Sculpture Garden is part of the National Smithsonian Museum. It opened in 1974 and displays modern and contemporary art. It is located on the National Mall in Washington, D.C.

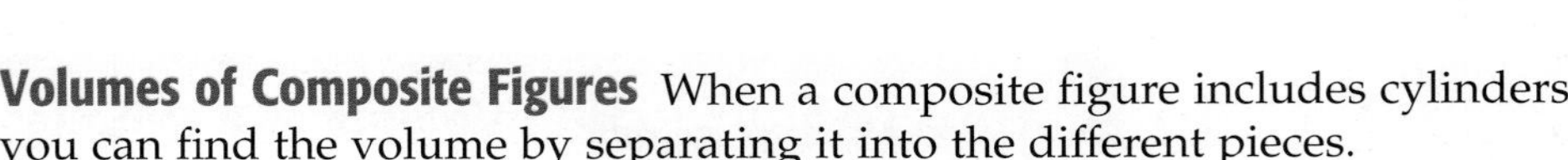

Volumes of Composite Figures When a composite figure includes cylinders, you can find the volume by separating it into the different pieces.

Real-World EXAMPLE 3 Volume of a Composite Figure

SCULPTURE An art museum is placing a sculpture on the stone pedestal shown at the right. Find the volume of the pedestal.

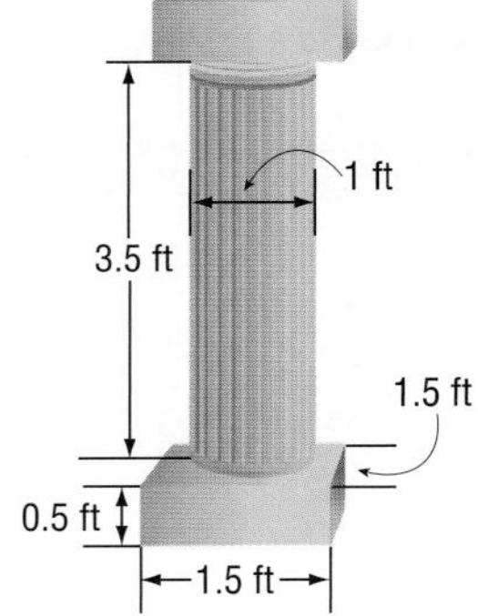

Find the volume. The volume of the pedestal is the sum of two rectangular prisms and a cylinder.

Step 1 Find the volume of the prisms.

$V = Bh$	**Volume of a prism**
$V = 1.5 \cdot 1.5 \cdot 0.5$	**The length and width are each 1.5 ft. The height is 0.5 ft.**
$= 1.125 \text{ ft}^3$	**Simplify.**

So, the volume of the two square bases is $2 \cdot 1.125$ or 2.25 ft^3.

Step 2 Find the volume of the cylinder.

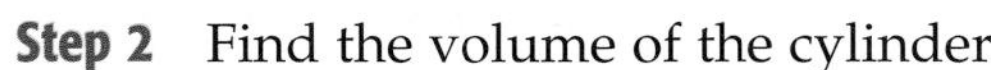

$V = \pi r^2 h$	**Volume of a cylinder**
$= \pi \cdot (0.5)^2 \cdot 3.5$	**Replace r with 0.5 and h with 3.5.**
$\approx 2.75 \text{ ft}^3$	**Use a calculator.**

Step 3 Find the volume of the composite figure.

$2.25 + 2.75 = 5$	**Add the volumes.**

So, the total volume of the pedestal is 5 ft^3.

Check Your Progress

3. Find the volume of the plastic building brick shown at the right. Round to the nearest tenth.

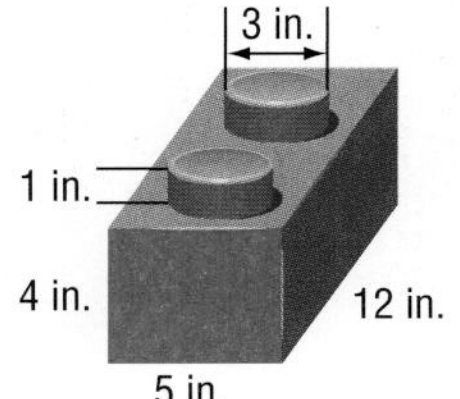

Personal Tutor glencoe.com

Check Your Understanding

Example 1 p. 677

Find the volume of each cylinder. Round to the nearest tenth.

1. 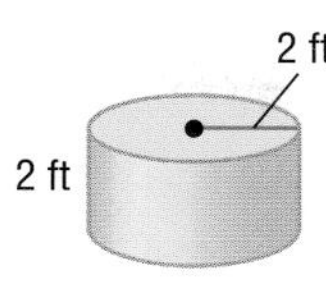

2. diameter of base: 33.2 mm
height: 60 mm

Example 2 p. 678

Find the height of each cylinder. Round to the nearest tenth.

3. volume = 283 in^3
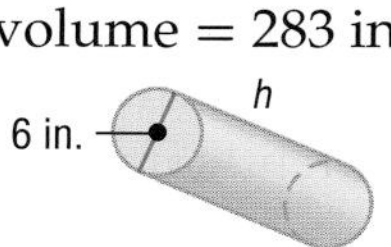

4. volume: 5700 m^3
diameter of base: 22 m

Example 3 p. 678

5. **CRAFTS** An oak peg like the one shown at the right was used in a toy truck. Find the volume of the peg. Round to the nearest tenth.

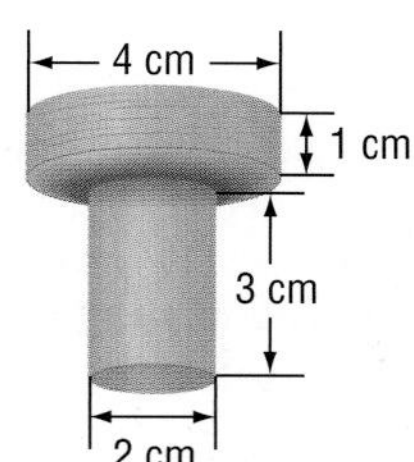

Practice and Problem Solving

● = **Step-by-Step Solutions** begin on page R11.
Extra Practice begins on page 810.

Examples 1 and 2 pp. 677 and 678

Find the volume of each figure. Round to the nearest tenth.

6. 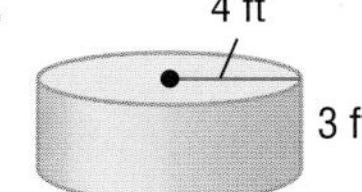

7. 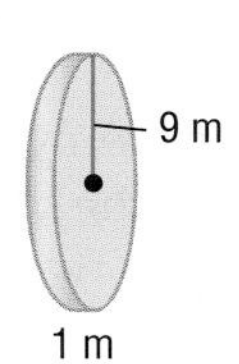

8. 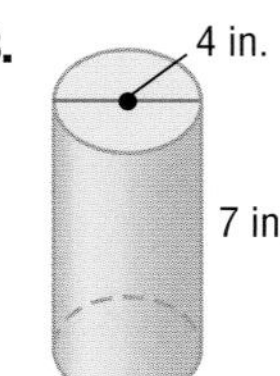

9. radius: 2.2 cm
height: 3 cm

10. diameter: 5 yd
height: 11 yd

11. diameter: 4.6 m
height: 6.1 m

Find the height of each cylinder. Round to the nearest tenth.

12. volume: 41.5 yd^3
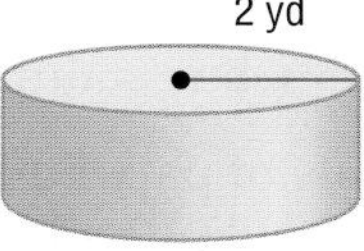

13. volume: 9.7 m^3
diameter: 1.5 m

Real-World Link

A size 303 can measures $3\frac{3}{16}$ inches in diameter. The can number indicates the measure of the diameter.

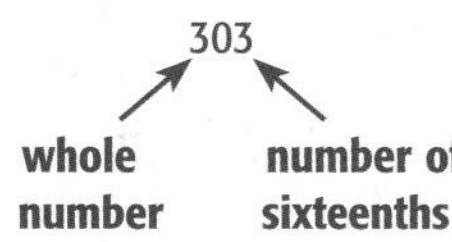

14. **CANDLES** A scented candle is in the shape of a cylinder that is 8 inches tall. The diameter of the candle is 3.5 inches. Find the volume of the candle. Round to the nearest tenth.

15. **FOOD** Use the information at the left to find the volume of soup in a size 400 can that is 5.5 inches tall.

Example 3 p. 678

Find the volume of each figure. Round to the nearest tenth.

16. 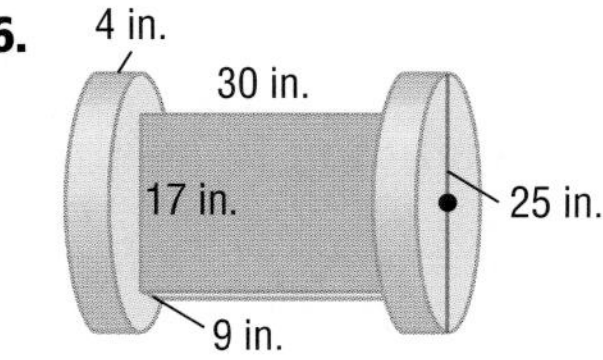

17. 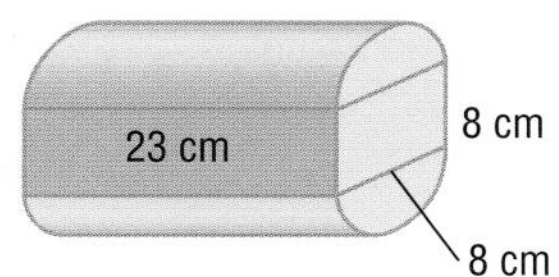

18. **MAIL** Find the volume of the mailbox below. Round to the nearest tenth.

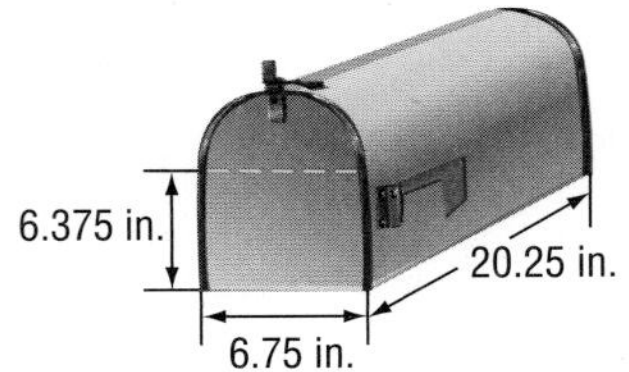

19. **TOWELS** A roll of paper towels has the dimensions shown. Find the volume of the roll. Round to the nearest tenth.

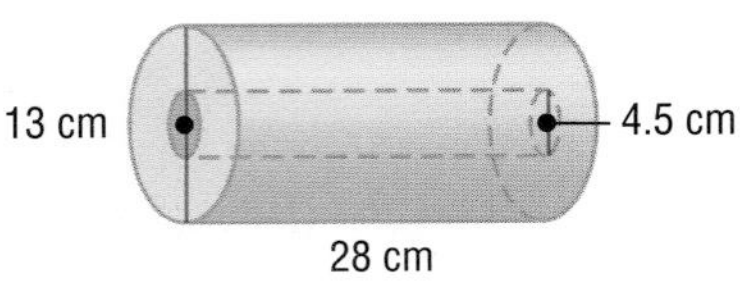

Watch Out!

In Exercise 19 do not subtract the diameter of the small tube from the roll of paper towels. You need to find the volume of each before you subtract.

20. **MULTIPLE REPRESENTATIONS** In this problem, you will examine how changing the size of a cylinder affects its volume. Round to the nearest tenth, if necessary.

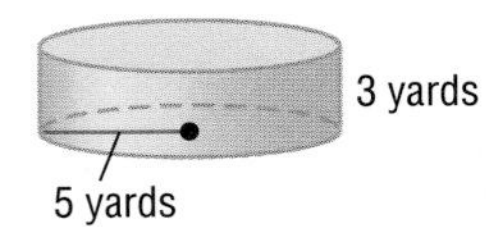

a. **TABULAR** Copy and complete the table for the cylinder shown.

Cylinder	Radius	Height	Volume (yd^3)
Original	5	3	■
Multiply radius by 3	$5 \cdot 3 =$ ■	3	■
Multiply height by 3	5	$3 \cdot 3 =$ ■	■
Multiply both by 3	■	■	■

b. **ANALYTICAL** Compare the original volume to the other volumes.

c. **ALGEBRAIC** Write an equation to find the volume of a cylinder after a dilation d.

21. **CONSERVATION** A cylindrical rain barrel is 38 inches tall and has a diameter of 28 inches. If the volume of one gallon of water is 231 cubic inches, how many gallons of water will the rain barrel hold? Round to the nearest tenth.

22. **FIREFIGHTING** Kamilah's uncle, a fire captain, said the diameters of fire hoses range from 1.5 inches to 6 inches and the hoses are 50 feet long. Find the approximate minimum and maximum volumes of a fire hose in cubic feet.

H.O.T. Problems

Use Higher-Order Thinking Skills

23. **OPEN ENDED** Write two real-world examples where you would want to change the dimensions of cylinders, but maintain the volume.

24. **CHALLENGE** Two equal-sized sheets of paper are rolled along the length and along the width, as shown. Which cylinder do you think has the greater volume? Explain.

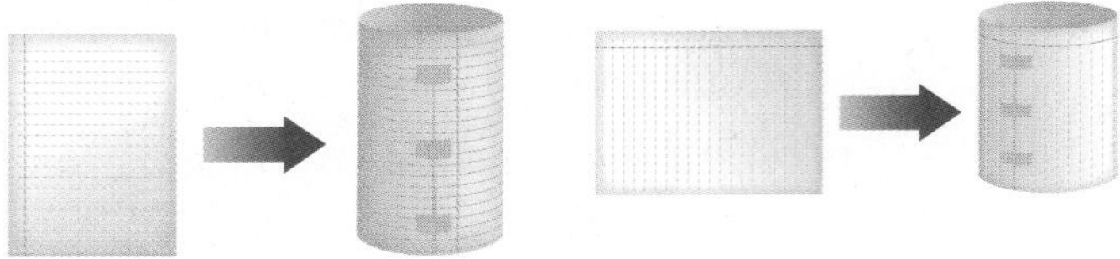

25. **NUMBER SENSE** Find the ratios of the volume of cylinder A to cylinder B.

a. Cylinder A has the same radius but twice the height of cylinder B.

b. Cylinder A has the same height but twice the radius of cylinder B.

26. **WRITING IN MATH** Explain how the formula for the volume of a cylinder is similar to the formula for the volume of a rectangular prism.

Standardized Test Practice

27. Find the maximum amount of water that can fill the trough shown.

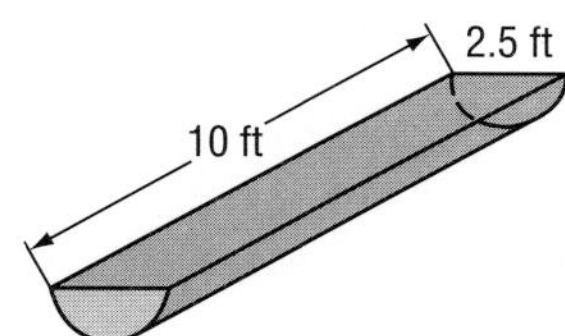

A 20.5 ft^3 C 48 ft^3
B 24.5 ft^3 D 49 ft^3

28. What is the volume of a cylinder with a radius of 8 inches and a height of 1 foot? Round to the nearest tenth.

F 25.1 in^3 H 301.6 in^3
G 201.1 in^3 J 2412.7 in^3

29. EXTENDED RESPONSE The smallest canister in a set of cylindrical canisters has a height of 11 inches and a radius of 1.5 inches.

a. Find the volume of the smallest canister.
b. The medium canister has the same height as the smallest but the radius is tripled. What is the volume of the medium canister?

30. A cylindrical diesel tank is 1.25 meters high and has a radius of 0.60 meter. If the tank can only be filled to an 85% capacity to allow for expansion and contraction of the fuel, what is the maximum volume of fuel? Round to the nearest hundredth.

A 4.71 m^3 C 1.41 m^3
B 2.01 m^3 D 1.20 m^3

Spiral Review

Find the volume of each figure. (Lesson 12-2)

31.

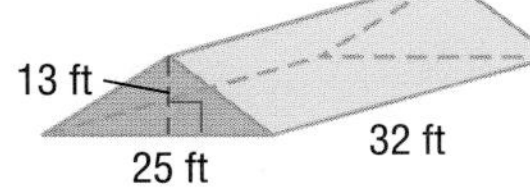

32.

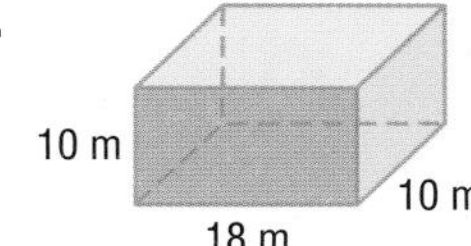

33.

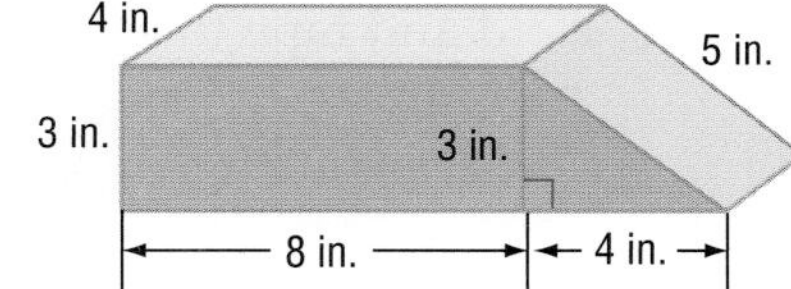

Identify each figure. Name the bases, faces, edges, and vertices. (Lesson 12-1)

34.

35.

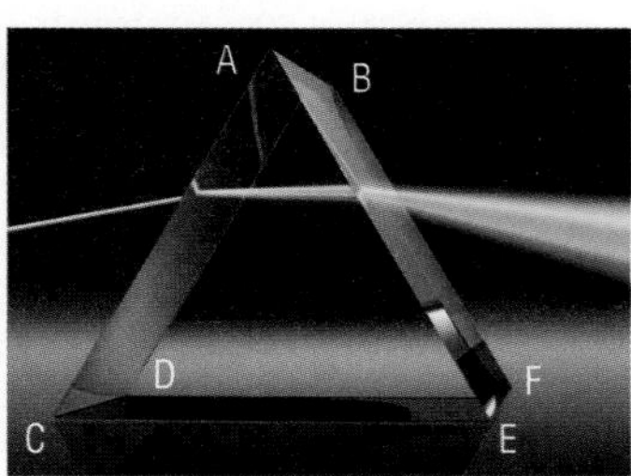

Solve each problem using the percent equation. Round to the nearest tenth. (Lesson 7-5)

36. 12 is what percent of 78?

37. 5% of what number is 8?

38. What is 42% of 45?

Skills Review

Find each product. (Lesson 3-3)

39. $\frac{1}{3} \cdot 7 \cdot 15$

40. $\frac{4}{3} \cdot 2 \cdot 5 \cdot 20$

41. $\frac{1}{3} \cdot \frac{5}{6} \cdot 12$

Geometry Lab
Volume of a Pyramid

Math Online glencoe.com
Math *in Motion*, Animation

In Lesson 12-2, you learned how to find the volume of a prism. You can use this idea to develop a method to find the volume of a pyramid.

ACTIVITY

Step 1 Draw and cut out five 2-inch squares. Then tape them together as shown.

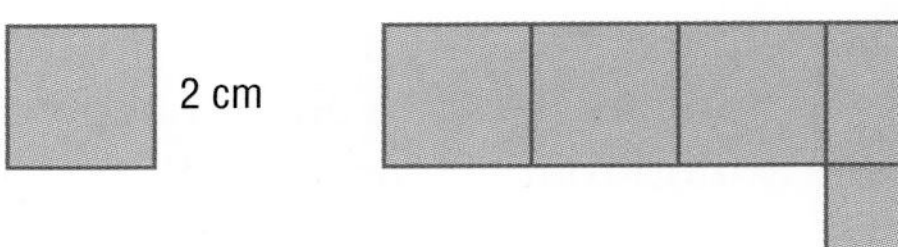

Step 2 Fold and tape to form a cube with an open top.

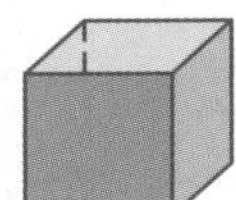

Step 3 Draw and cut out 4 isosceles triangles with the measurements shown. Then tape them together.

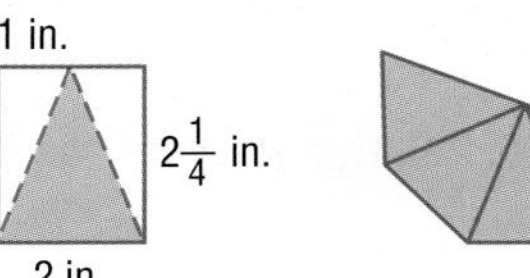

Step 4 Fold and tape to form an open square pyramid.

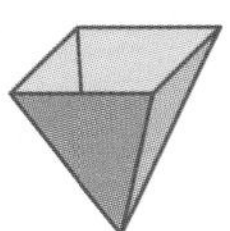

Analyze the Results

1. Compare the base areas and the heights of the prism and the pyramid.
2. Fill the pyramid with rice, sliding a ruler across the top to level the amount. Pour the rice into the cube. Repeat until the prism is filled. How many times did you fill the pyramid in order to fill the cube?
3. What fraction of the prism volume does one pyramid fill?
4. If the results of the activity apply to all pyramids, write a formula that relates the volume V of one pyramid to the dimensions of the prism.

Find the volume of each pyramid.

5. 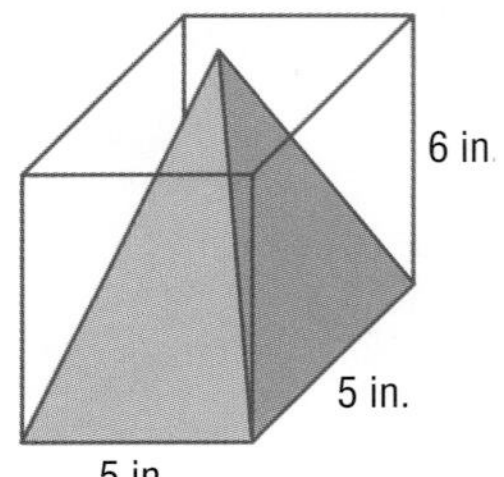

6. 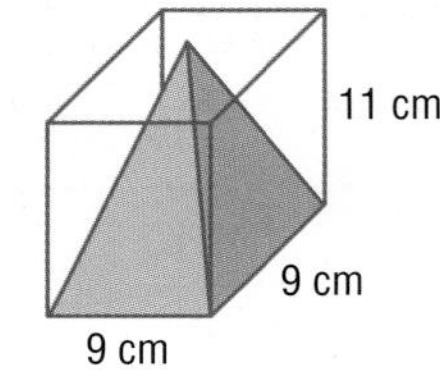

12-4

Volume of Pyramids, Cones and Spheres

Then

You have already found the volume of prisms and cylinders. (Lessons 12-2 and 12-3)

Now

- Find the volumes of pyramids and cones.
- Find the volumes of spheres.

New Vocabulary

sphere

Math Online

glencoe.com

- Extra Examples
- Personal Tutor
- Self-Check Quiz
- Homework Help

Why?

An ice cream shop serves ice cream in cones and cups. Both the cone and the cup have a height of 6 inches and a radius of 2 inches.

a. Do the cone and the cup hold the same amount of ice cream?

b. If the ice cream in the containers were to melt, would each container have the same amount of melted ice cream? Why or why not? Explain.

Volume of Pyramids A pyramid has one-third the volume of a prism with the same base and height. The height of a pyramid is the perpendicular distance from the vertex to the base.

Key Concept — Volume of a Pyramid

For Your FOLDABLE

Words The volume V of a pyramid is one-third the area of the base B times the height h.

Symbols $V = \frac{1}{3}Bh$

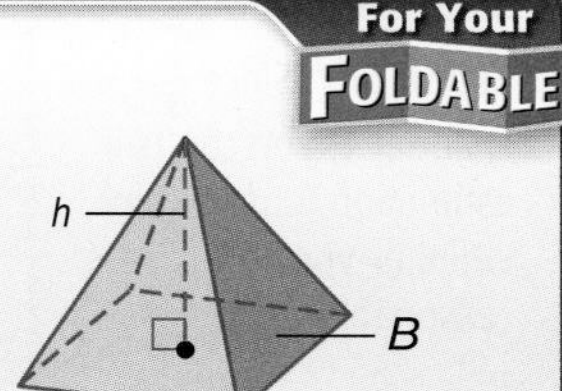

EXAMPLE 1 Volume of a Pyramid

Find the volume of the pyramid. Round to the nearest tenth, if necessary.

5 cm
6 cm
8 cm

$V = \frac{1}{3}Bh$	Volume of a pyramid
$V = \frac{1}{3}\left(\frac{1}{2} \cdot 8 \cdot 6\right)h$	The base is a triangle so $B = \frac{1}{2} \cdot 8 \cdot 6$.
$= \frac{1}{3} \cdot 24 \cdot 5$	The height of the pyramid is 5 cm.
$= 40$	Simplify.

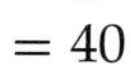

The volume of the pyramid is 40 cubic centimeters.

Check Your Progress

1. Find the volume of a pyramid with a base area of 90 square feet and a height of 12 feet.

Personal Tutor glencoe.com

The volumes of a cone and a cylinder are related in the same way as the volumes of a pyramid and a prism are related.

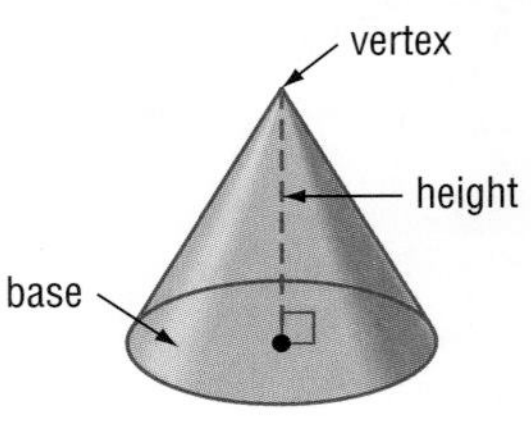

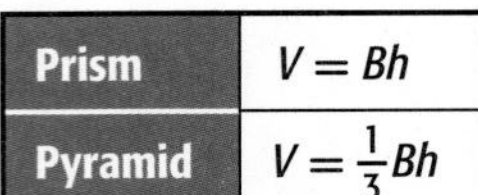

Prism	$V = Bh$
Pyramid	$V = \frac{1}{3}Bh$

Cylinder	$V = \pi r^2 h$
Cone	$V = \frac{1}{3}\pi r^2 h$

The volume of a cone is $\frac{1}{3}$ the volume of a cylinder with the same base and height.

Key Concept — Volume of a Cone

For Your FOLDABLE

Words The volume V of a cone with radius r is one third the area of the base πr^2 times the height h.

Symbols $V = \frac{1}{3}Bh$, where $B = \pi r^2$ or

$V = \frac{1}{3}\pi r^2 h$

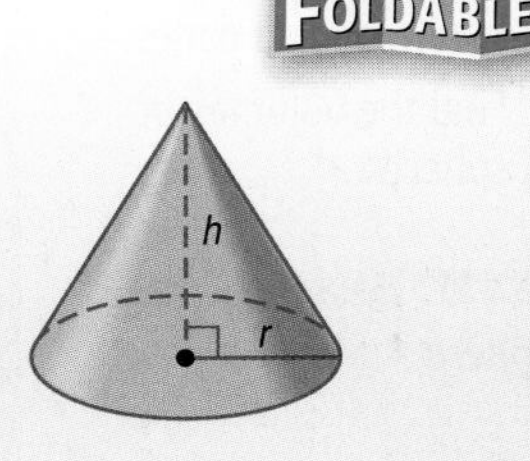

StudyTip

Estimating Since $\frac{1}{3} \cdot \pi$ is about 1, you can use estimation to determine whether your solution is reasonable.

$V \approx r^2h$

$\approx 5^2 \cdot 7$

$\approx 25 \cdot 7$ or 175

EXAMPLE 2 Volume of a Cone

Find the volume of the cone. Round to the nearest tenth.

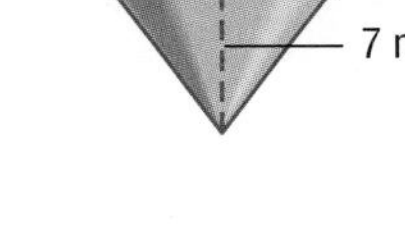

$V = \frac{1}{3}\pi r^2 h$	**Volume of a cone**
$V = \frac{1}{3} \cdot \pi \cdot 5^2 \cdot 7$	**Replace *r* with 5 and *h* with 7.**
$\approx 183.3 \text{ mm}^3$	**Simplify. Round to the nearest tenth.**

Check Your Progress

2A. radius 6 ft, height 20 ft

2B. radius $1\frac{1}{2}$ yd, height 9 yd

Personal Tutor glencoe.com

Volume of a Sphere A **sphere** is a set of points in space that are a given distance r from the center. Suppose a sphere with radius r is placed inside a cylinder with the same radius r and height $2r$. The volume of the sphere is $\frac{2}{3}$ of the volume of the cylinder. The volume of the cylinder is shown below.

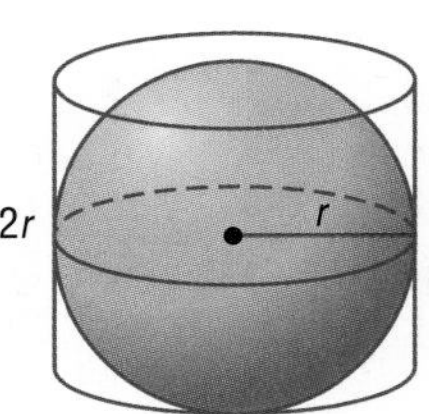

$V = \pi r^2 h$	**Volume of a cylinder**
$= \pi r^2(2r)$	**Replace *h* with 2*r*.**
$= 2\pi r^3$	**Simplify.**

Since the sphere is $\frac{2}{3}$ the size of the cylinder, you can find the volume of the sphere.

$V = \left(\frac{2}{3}\right)2\pi r^3$	**The sphere is $\frac{2}{3}$ the size of the cylinder.**
$= \frac{4}{3}\pi r^3$	**Simplify.**

Key Concept — Volume of a Sphere

For Your FOLDABLE

Words The volume V of a sphere with radius r is four-thirds times π times the radius cubed.

Symbols $V = \frac{4}{3}\pi r^3$

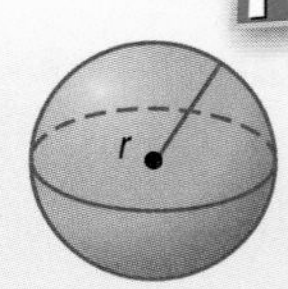

Math *in Motion*, Animation glencoe.com

EXAMPLE 3 Volume of a Sphere

Find the volume of the sphere. Round to the nearest tenth.

8 in.

$V = \frac{4}{3}\pi r^3$ — **Write the formula for the volume of a sphere.**

$= \frac{4}{3} \cdot \pi \cdot 8^3$ — **Replace r with 8.**

$\approx 2144.7 \text{ in}^3$ — **Simplify.**

Check Your Progress

3. Find the volume of a sphere with a radius of 10 feet. Round to the nearest tenth.

Personal Tutor glencoe.com

Real-World EXAMPLE 4 Volume of a Sphere

ICE CREAM A spherical scoop of ice cream with a diameter of 6.3 centimeters is placed in a bowl. Find the volume of the ice cream. Then find how long it would take the ice cream to melt if it melts at a rate of 2.1 cubic centimeters every minute.

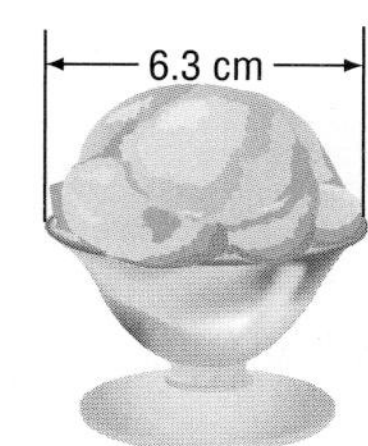

Understand You know the scoop of ice cream has a diameter of 6.3 centimeters and melts at a rate of 2.1 cubic centimeters per minute.

Plan Find the volume of the ice cream and how long it will take to melt.

Solve $V = \frac{4}{3}\pi r^3$ — **Write the formula for volume of a sphere.**

$= \frac{4}{3}\pi \cdot 3.15^3$ — **Since $d = 6.3$, replace r with 3.15.**

$\approx 130.9 \text{ cm}^3$ — **Simplify.**

Estimate $130 \div 2 = 65$

Use a proportion.

$$\frac{2.1 \text{ cm}^3}{1 \text{ min}} = \frac{130.9 \text{ cm}^3}{x \text{ min}}$$

$$2.1x = 130.9$$

$$x \approx 62.3$$

So, it will take approximately 62 minutes or about 1 hour for the ice cream to melt.

Check 62.3 is close to the estimate of 65. The answer is correct. ✓

Check Your Progress

4. TOYS A beachball has a diameter of 18 inches. A hand pump will inflate the ball at a rate of 325 cubic inches per minute. How long will it take to inflate the ball? Round to the nearest tenth.

Personal Tutor glencoe.com

Real-World Link

Ice cream and related frozen desserts are consumed by more than 90% of households in the United States.

Source: Mintel

Check Your Understanding

Examples 1–3 pp. 683–685

Find the volume of each figure. Round to the nearest tenth, if necessary.

1.

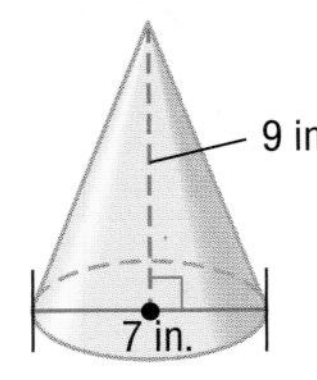

2.

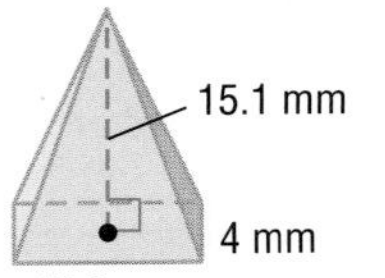

3.

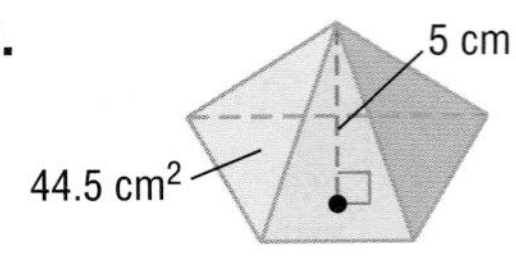

4.

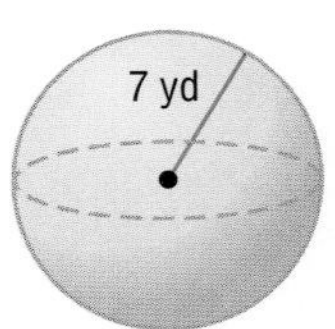

Example 4 p. 685

5. **JEWELRY** Amber purchased a necklace that contained an 8 millimeter diameter round pearl. Find the volume of the pearl to the nearest tenth.

Practice and Problem Solving

= **Step-by-Step Solutions** begin on page R11.
Extra Practice begins on page 810.

Examples 1–3 pp. 683–685

Find the volume of each figure. Round to the nearest tenth, if necessary.

6.

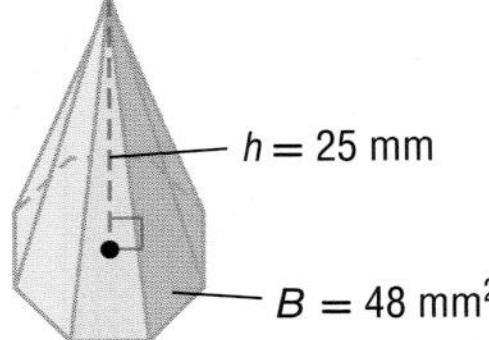

7.

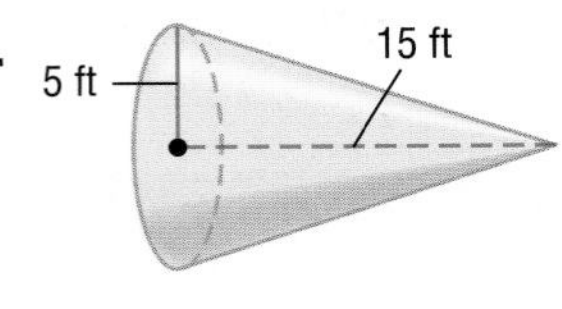

8.

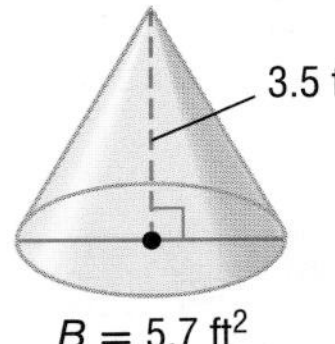

9.

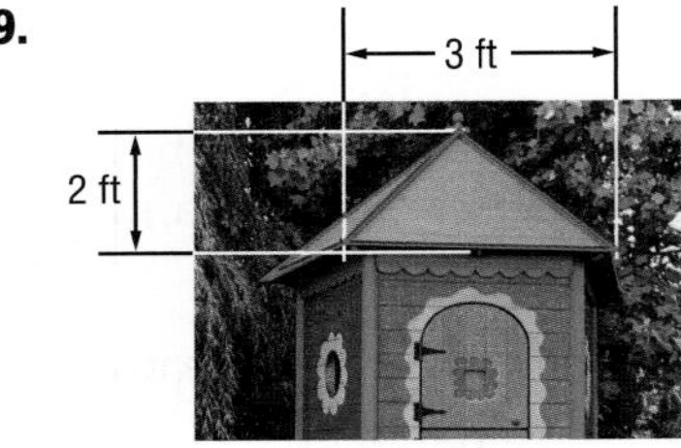

10.

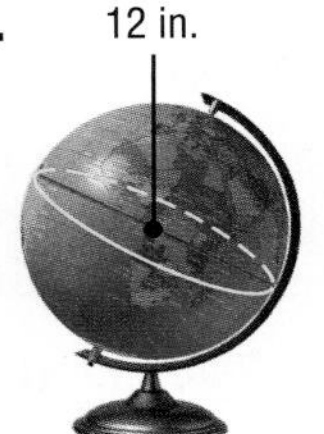

11.

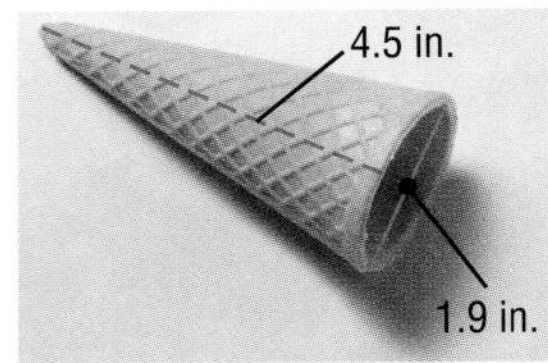

12. pentagonal pyramid: base area 52 in^2, height 12 in.

13. rectangular pyramid: length 3.5 feet, width 2 feet, height 4.5 feet

14. cone: diameter 6.7 mm, height 2.1 mm

15. sphere: radius 7.2 km

16. sphere: diameter 1.8 mm

Example 4 p. 685

17 **WEATHER** A cone-shaped icicle 2.5 feet long with a diameter of 1.5 feet has formed at the edge of a roof.

a. Find the amount of ice in the icicle to the nearest tenth of a cubic foot.

b. The icicle melts at a rate of 0.1 cubic foot every 5 minutes. How long will it take for the icicle to melt?

Find the height of each figure. Round to the nearest tenth, if necessary.

18. square pyramid: volume 873.18 m^3, length 12.6 m

19. cone: volume 306.464 ft^3, diameter 8 ft

Find the volume of each figure. Round to the nearest tenth, if necessary.

20.

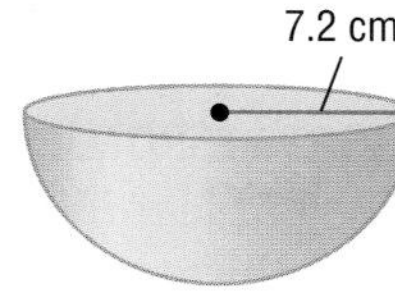

21.

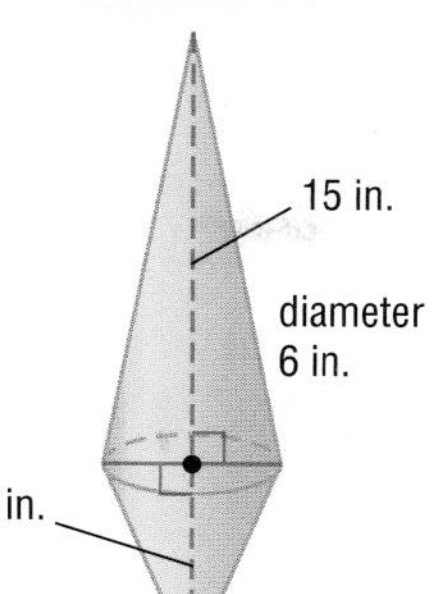

22.

Real-World Link

Pure silver's weight is about 10.5 grams per cubic centimeter. Its chemical symbol is Ag and it has a melting point over 1700°F.

Source: Silver Users Association

23. SILVER The solid silver bead shown at the left is made of two cones. It measures 11 centimeters in diameter and is 10 centimeters from top to bottom. The center portion is 2 centimeters tall. Use the information at the left to find the approximate mass in grams of the bead. Round to the nearest tenth.

24. BASKETBALL The volume of a mini-basketball is about 230 cubic inches. What is its radius? Round to the nearest inch.

25. PACKAGING Three golf balls are packaged in a box 13.1 centimeters long, 4.5 centimeters wide, and 4.5 centimeters tall. If each ball is 4.3 centimeters in diameter, find the volume of the empty space in the box, rounded to the nearest tenth.

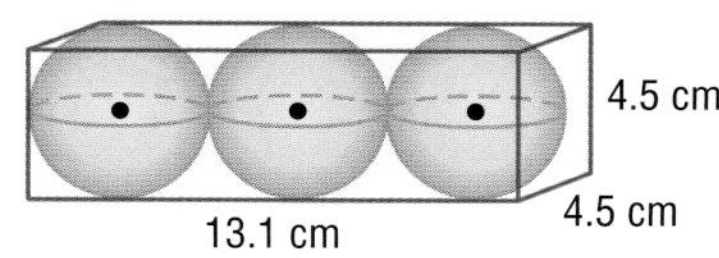

H.O.T. Problems

Use Higher-Order Thinking Skills

26. OPEN ENDED Name the dimensions of a cone whose volume is between 45 cubic units and 55 cubic units.

27. CHALLENGE A cone contains 93 cubic units of water. All the water is poured equally into three congruent cylindrical containers to a level of 4.5 units. What is the diameter of each container? Round to the nearest tenth.

28. WHICH ONE DOESN'T BELONG? Find the expression that does not represent the height of a three-dimensional figure. Explain your reasoning.

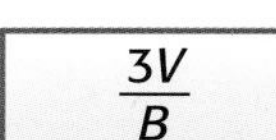

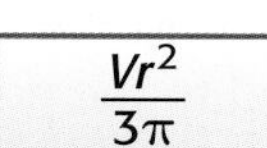

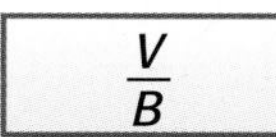

29. REASONING *True* or *false*? The volume of a rectangular-based pyramid and a cone with the same height and equal areas of the base are equal. Explain your reasoning.

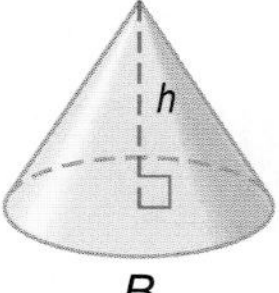

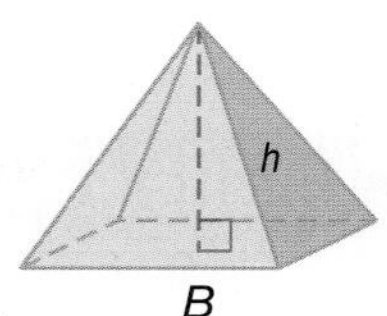

30. CHALLENGE Which has a greater effect on the volume of a cone: changing the radius or changing the height? Explain.

31. WRITING IN MATH Write a real-world example where you would want to change a cylinder to a cone. Describe how the dimensions of the cone would be affected if the volumes remain the same.

Standardized Test Practice

32. What is the volume of a square pyramid with a height of 12 centimeters and base edges of 4 centimeters?

 A 48 cm^3 C 96 cm^3
 B 64 cm^3 D 192 cm^3

33. Find the volume of the cone. Use 3.14 for π. Round to the nearest hundredth.

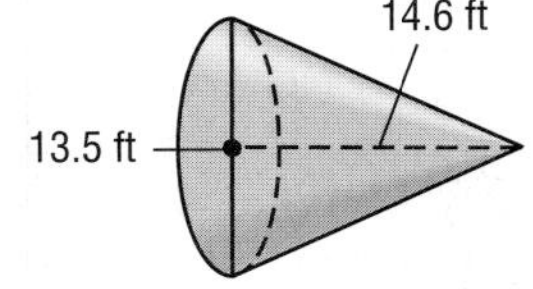

 F 8355.07 ft^3 H 696. 26 ft^3
 G 2785.02 ft^3 J 618.89 ft^3

34. **SHORT RESPONSE** The volume of a cone is 3768 cubic millimeters. The radius of the base is 5 millimeters Find the height of the cone.

35. A sphere has a radius of 5 meters. Which of the following is closest to the volume of the sphere in cubic meters?

 A 65.4 m^3 C 130.8 m^3
 B 104.7 m^3 D 523.3 m^3

Spiral Review

Find the volume of each figure. Round to the nearest tenth, if necessary. (Lessons 12-2 and 12-3)

36.
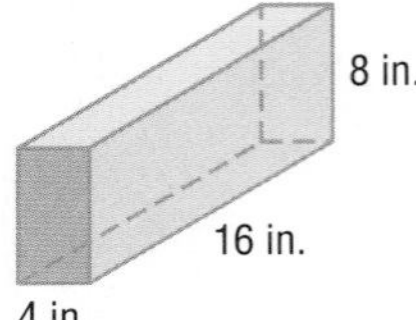

37.
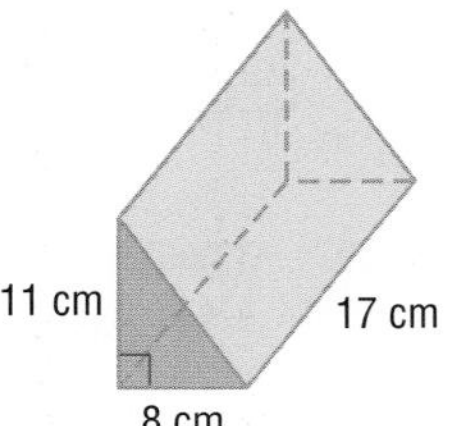

38.
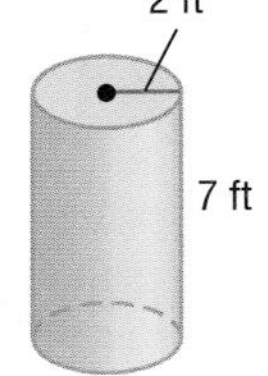

39. **TRAVEL** Loretta drives due north for 22 miles and then east for 11 miles. How far is Loretta from her starting point? Round to the nearest tenth of a mile. (Lesson 10-4)

40. **FUNCTIONS** Copy and complete the table. Use the results to write four solutions of $y = x + 5$. Write the solutions as ordered pairs. (Lesson 8-1)

x	$x + 5$	y
−3	−3 + 5	■
−1	■	■
0	■	■
1	■	■

41. **SNACKS** Manuel has $15 to buy snack mix that costs $5.75 per pound. Write and solve an equation to find the amount of snack mix he can buy if he spends all $15. Round to the nearest tenth. (Lesson 4-6)

Skills Review

Find the area of each figure. Round to the nearest tenth, if necessary. (Lessons 11-6 and 11-8)

42.
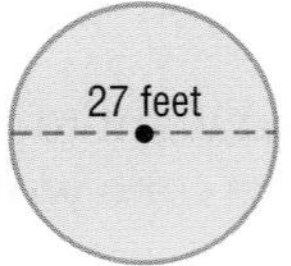

43.
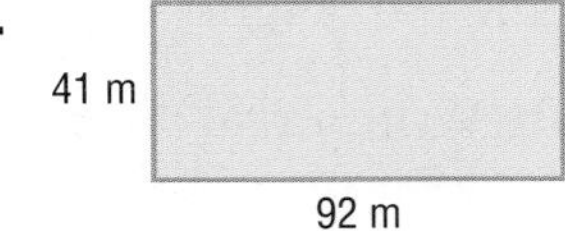

44.
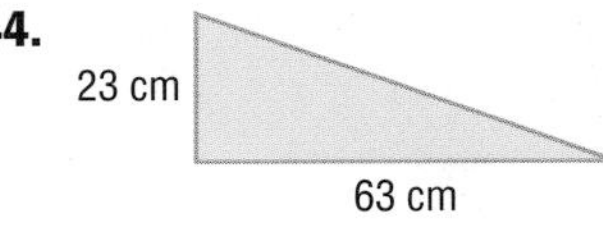

CHAPTER 12

Mid-Chapter Quiz

Lessons 12-1 through 12-4

Identify each figure. Name the bases, faces, edges, and vertices. (Lesson 12-1)

1. 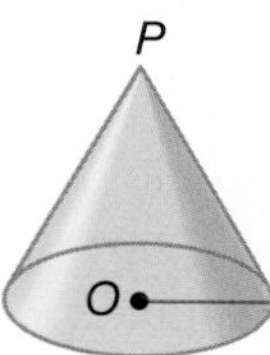

2. 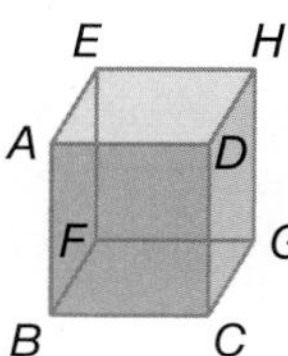

3. 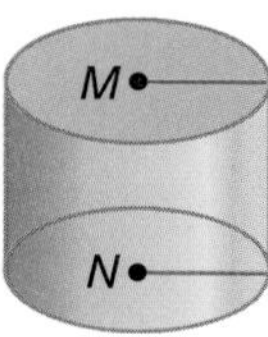

4. 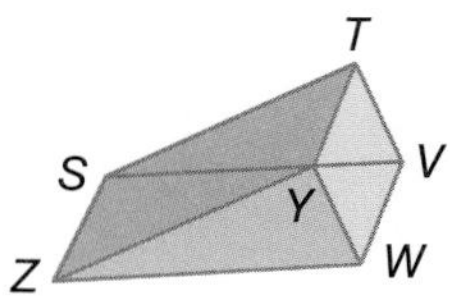

Draw and describe the shape resulting from each cross section. (Lesson 12-1)

5. 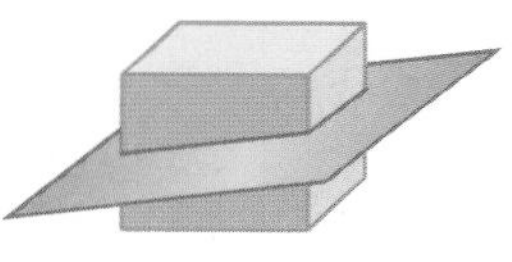

6. 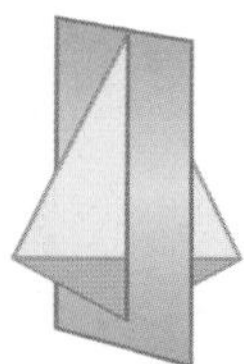

7. **MULTIPLE CHOICE** How much water can fit into an aquarium with a length 15 inches, width 12 inches, and height 16 inches? (Lesson 12-2)

A $1\frac{2}{3}$ ft^3 **C** 2400 ft^3

B $2\frac{2}{5}$ ft^3 **D** 2880 ft^3

8. What is the volume of the prism shown at the right? (Lesson 12-2)

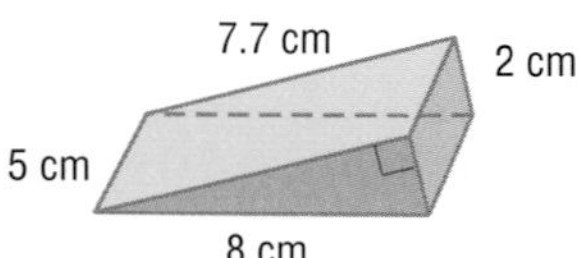

9. **MULTIPLE CHOICE** A can of lemonade concentrate has a diameter of 3 inches and a height of $4\frac{1}{2}$ inches. If the concentrate dissolves in $3\frac{1}{2}$ cans of water, how much water must be added? (Lesson 12-3)

F 31.8 in^3 **H** 127.2 in^3

G 111.3 in^3 **J** 381.7 in^3

Find the volume of each figure. Round to the nearest tenth, if necessary. (Lessons 12-2 and 12-3)

10. 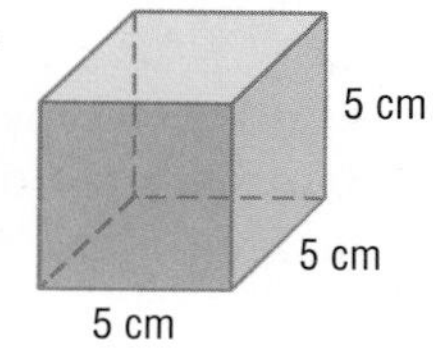

11. 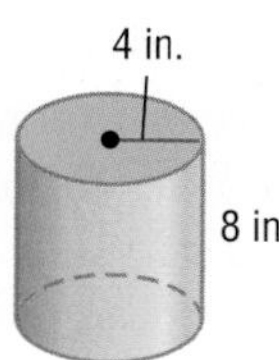

12. **HIGHWAY MAINTENANCE** Salt and sand mixtures are often used on icy roads. When the mixture is dumped from a truck into the staging area, it forms a cone-shaped mound with a diameter of 10 feet and a height of 6 feet. (Lesson 12-4)

    a. What is the volume of the salt-sand mixture to the nearest cubic foot?

    b. How many square feet of roadway can be salted using the mixture in part **a** if 500 square feet can be covered by 1 cubic foot of salt?

13. **MULTIPLE CHOICE** Mrs. McCullough is purchasing balloons for a party. Each spherical balloon is inflated with helium. How much helium is in the balloon if the balloon has a radius of 9 centimeters? (Lesson 12-4)

A 339.3 cm^3 **C** 1357.2 cm^3

B 381.7 cm^3 **D** 3053.6 cm^3

Find the volume of each figure. Round to the nearest tenth, if necessary. (Lesson 12-4)

14. 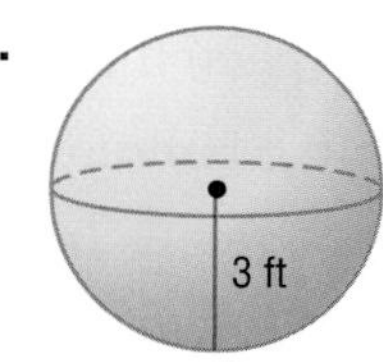

15. 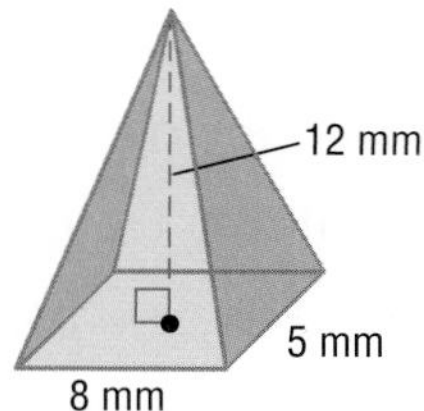

16. **FOOD** A baker is using a cake-decorating bag in the shape of a cone. How much frosting can fit in the bag if its diameter is 4 inches and height is 9 inches? Round to the nearest tenth. (Lesson 12-4)

Geometry Lab
Surface Area of Prisms

Math Online glencoe.com
Math in Motion, Animation

Nets are two-dimensional patterns of three-dimensional figures. When you construct a net, you are decomposing the three-dimensional figure into separate shapes. You can use a net to find the area of each surface of a three-dimensional figure such as the prism at the right.

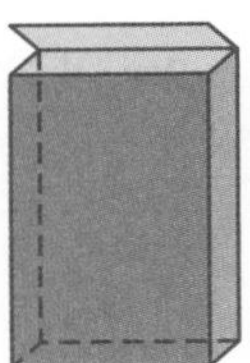

ACTIVITY

Find the surface area of a rectangular prism.

Step 1 Use an empty box with a tuck-in lid. Label the top, bottom, front, back, and side faces using a marker.

Step 2 Measure each face. Copy the table. Record the results in it. Then find and record the area of each face.

Face	Length	Width	Area
top	■	■	■
bottom	■	■	■
front	■	■	■
back	■	■	■
left	■	■	■
right	■	■	■

Step 3 Open the lid. Cut each of the 4 vertical edges. Open the box and lay it flat to form a net.

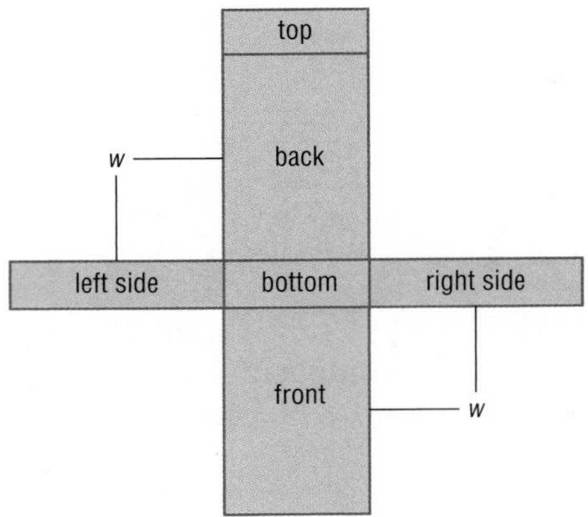

Step 4 Find the sum of the areas of the faces. This is the *surface area* of the prism.

Exercises

1. Classify the two-dimensional shape(s) that make up the net of the prism.
2. Find the perimeter of the top or the bottom base.
3. Multiply the perimeter of the base by the height of the box.
4. Add the product from Exercise 3 to the sum of the areas of the two bases.
5. Compare your answer from Exercise 4 to the answer in Step 4.
6. What do you observe about the areas of opposite faces?
7. **MAKE A CONJECTURE** Write a formula for the surface area S of a rectangular prism with length ℓ, width w, and height h.
8. Find several objects that are rectangular prisms. Measure the dimensions of the objects. Find the surface area of the objects using the formula from Exercise 7.

12-5 Surface Area of Prisms

Then

You have already found the area of two-dimensional figures. (Lesson 11–6)

Now

- Find lateral area and surface area of prisms.
- Find surface area of real-world objects shaped like prisms.

New Vocabulary

lateral face
lateral area
surface area

Math Online

glencoe.com

- Extra Examples
- Personal Tutor
- Self-Check Quiz
- Homework Help

Why?

The dimensions of a fish tank are given.

a. How many faces does the tank have?

b. Find the area of each face and the sum of the areas.

c. Find the volume of the fish tank.

d. Is the sum of the areas the same as the volume?

Prisms If you open up a box or prism and lay it flat, the result is a net. A net allows you to see all the surfaces. The surfaces of prisms have two characteristics.

- A prism has two parallel bases.
- Faces that are *not* bases are called **lateral faces**.

The **lateral area** is the sum of the areas of the lateral faces. The **surface area** is the sum of the lateral area plus the area of the bases. In the figures below, the lateral faces are shown in blue. The bases are shown in yellow.

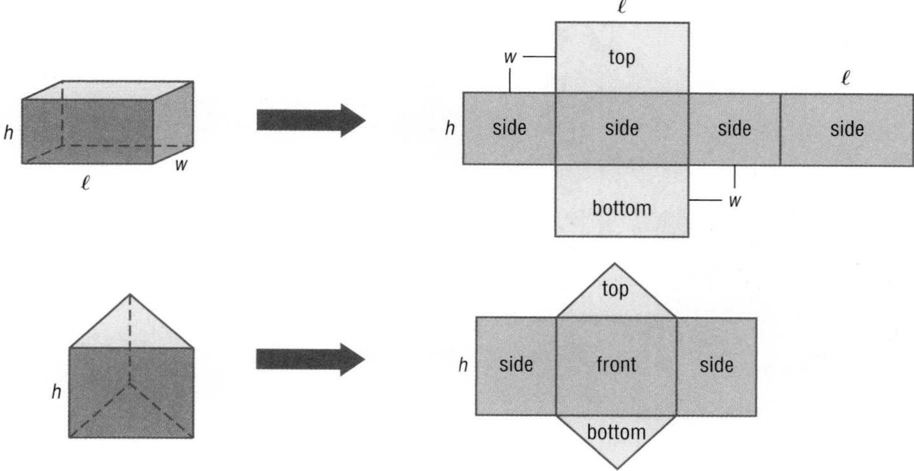

Key Concept

For Your FOLDABLE

Lateral Area of Prisms

Words The lateral area L of a prism is the perimeter of the base P times the height h.

Symbols $L = Ph$

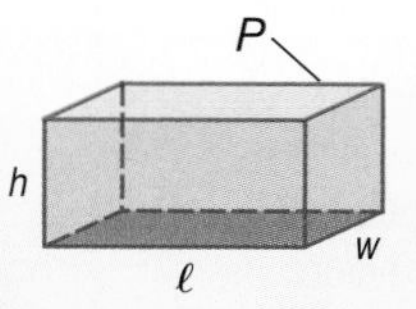

Surface Area of Prisms

Words The surface area S of a prism is the lateral area L plus the area of the two bases $2B$.

Symbols $S = L + 2B$ or $S = Ph + 2B$

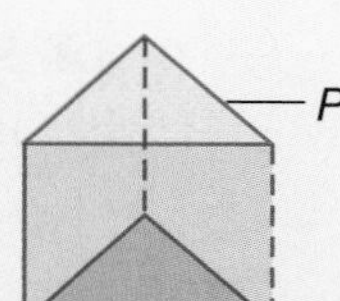

StudyTip

Nets Another way to find the surface area of any prism is to draw its net on grid paper and find the area of each face.

EXAMPLE 1 Surface Area of Prisms

Find the lateral and surface area of each prism.

a. Find the lateral area.

$L = Ph$

$L = (2\ell + 2w)h$

$= (2 \cdot 5.5 + 2 \cdot 3)15.1$

$= 256.7 \text{ in}^2$

Find the surface area.

$S = L + 2B$

$S = L + 2\ell w$

$= 256.7 \text{ in}^2 + 2 \cdot 5.5 \text{ in.} \cdot 3 \text{ in.}$

$= 256.7 + 33 \text{ or } 289.7 \text{ in}^2$

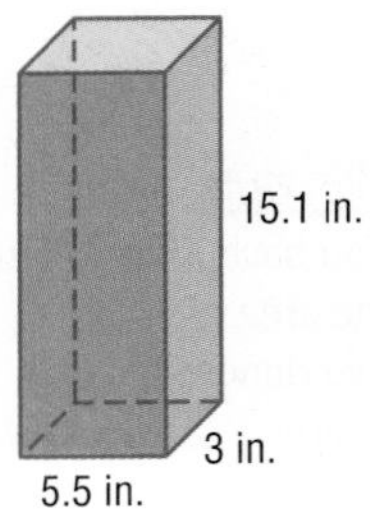

b. Find the lateral area. The lateral area is made up of faces that are *not* parallel.

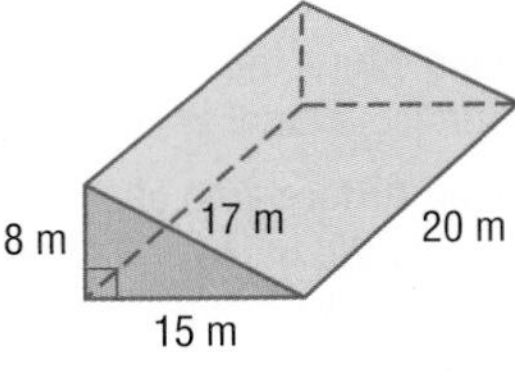

$L = Ph$ — **Formula for the lateral area**

$L = (8 + 15 + 17)20$ — ***P* is the perimeter of the triangular base.**

$= 40 \cdot 20 \text{ or } 800 \text{ m}^2$ — **Simplify.**

Find the surface area.

$S = L + 2B$ — **Formula for the surface area**

$= 800 + 2\left(\frac{1}{2} \cdot 15 \cdot 8\right)$ — $B = \frac{1}{2}bh$

$= 800 + 120 \text{ or } 920 \text{ m}^2$ — **Simplify.**

Check Your Progress

1A.

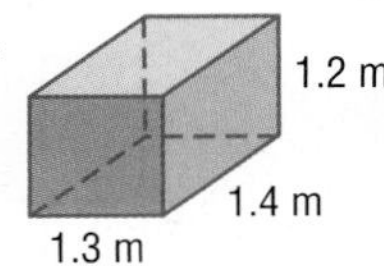

1B.

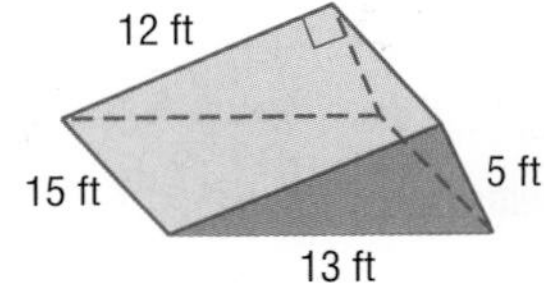

Personal Tutor glencoe.com

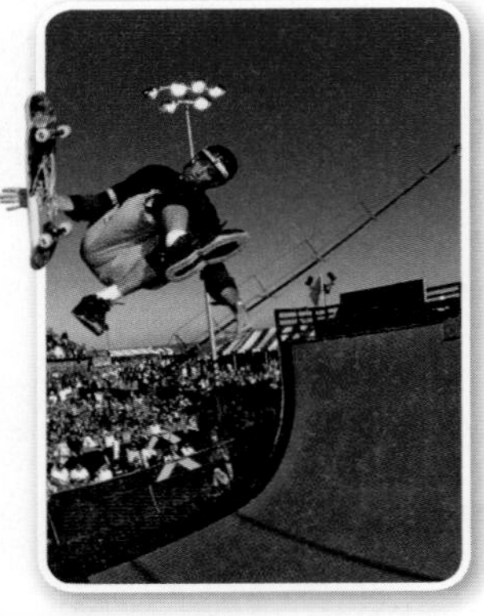

Real-World Link

Tony Hawk was the first skateboarder to complete the *900*–two and one-half rotations in the air before landing on the pipe.

Real-World EXAMPLE 2 Surface Area of Prisms

PAINT Marcus wants to paint a skateboard launch ramp like the one shown at the right. Find the amount of paint he will need if 1 quart covers 100 square feet.

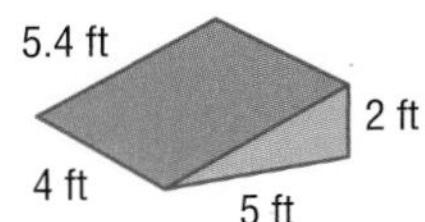

Find the lateral area.

$L = Ph$

$= (5.4 + 2 + 5)4$

$= 49.6 \text{ ft}^2$

Find the surface area.

$S = L + 2B$

$= 49.6 + 2\left(\frac{1}{2} \cdot 2 \cdot 5\right)$

$= 49.6 + 10$

$= 59.6 \text{ ft}^2$

Since $59.6 < 100$, Marcus will use less than 1 quart of paint.

Check Your Progress

2. CRAFTS Lucia is covering boxes with fabric to sell at a craft fair. The boxes are rectangular prisms and measure 10 inches wide, 14 inches long, and 5 inches high. Find the amount of fabric she will need to cover 1 box.

Personal Tutor glencoe.com

Check Your Understanding

Example 1 p. 692

Find the lateral and surface area of each prism.

1. 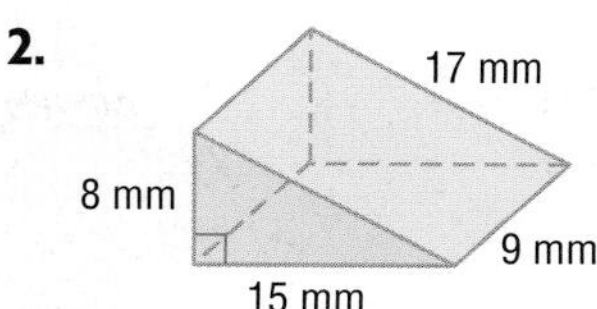

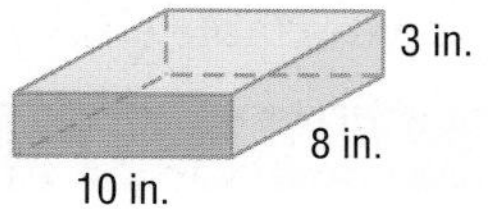

2. 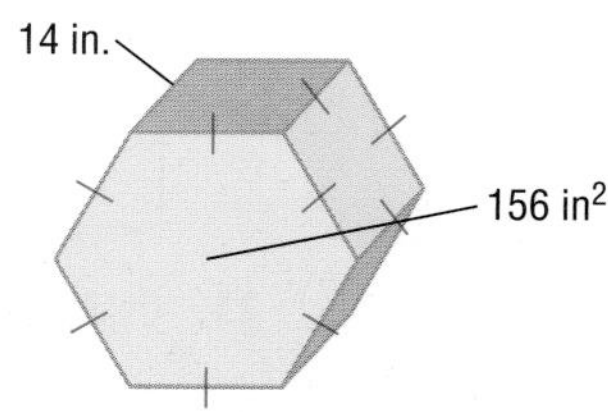

3. rectangular prism: length $5\frac{1}{2}$ yd, width $9\frac{1}{2}$ yd, height 12 yd

Example 2 p. 692

4. **PACKAGING** Find the amount of cardboard needed to make a box for a single slice of pizza. The box is in the shape of a triangular prism as shown.

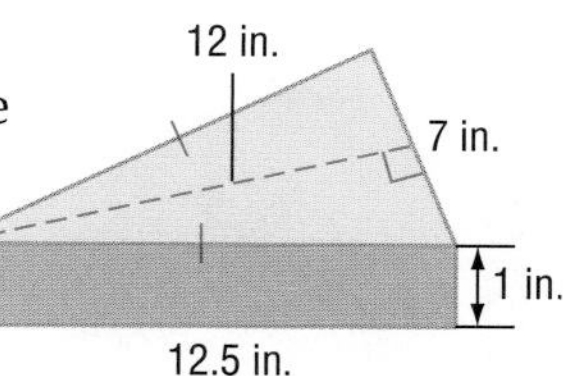

Practice and Problem Solving

● = **Step-by-Step Solutions** begin on page R11. **Extra Practice** begins on page 810.

Example 1 p. 692

Find the lateral and surface area of each prism.

5. 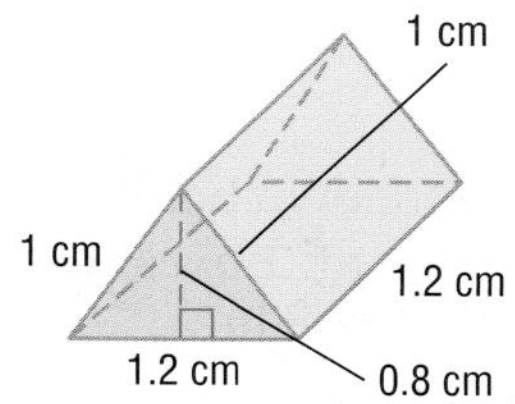

6. 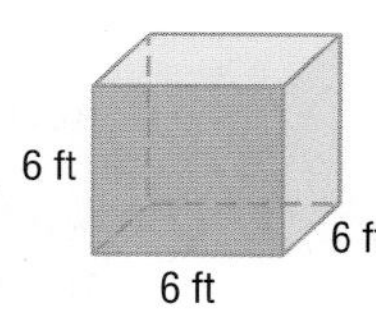

7. 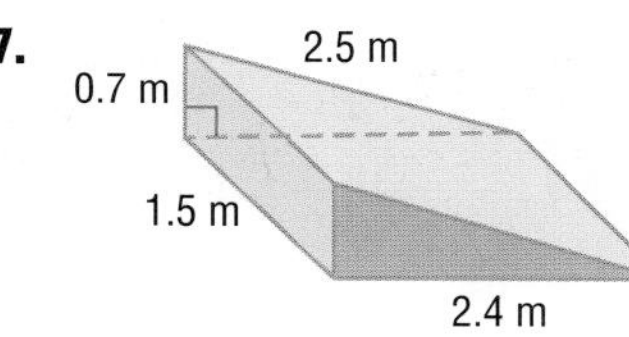

8. 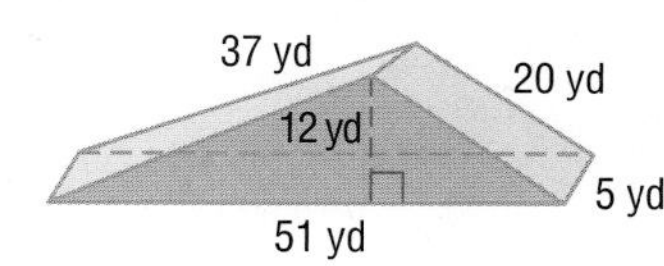

9. rectangular prism: length 2.2 cm, width 1.4 cm, height 9 cm

10. rectangular prism: length $6\frac{1}{2}$ in., width $8\frac{3}{4}$ in., height 10 in.

11. **JUICE** Find the amount of paper used to cover the juice box at the right.

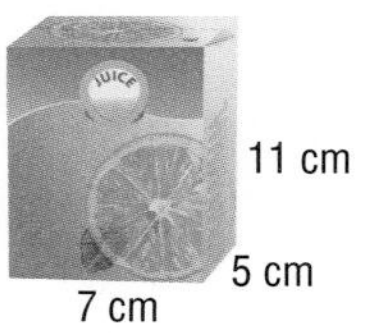

Example 2 p. 692

12. **PAINTING** Hinto is planning to paint the walls of a bedroom that is 20 feet long, 15 feet wide, and 8 feet high. If Hinto has 1 gallon of paint that covers 400 square feet, how many additional gallons of paint does he need?

Find the surface area of each prism.

13. 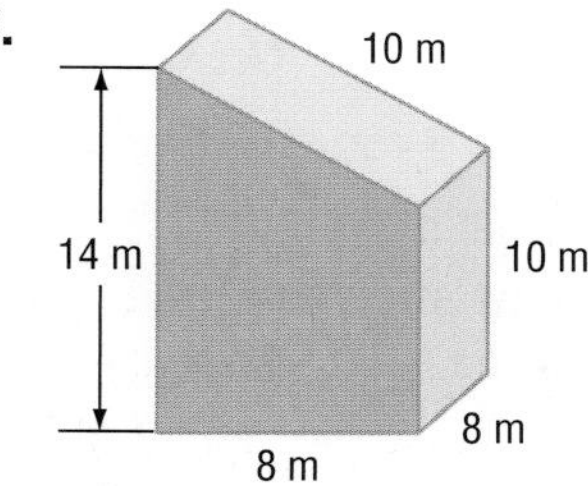

14. 14 in.

156 in²

 **15** **MULTIPLE REPRESENTATIONS** In this problem, you will explore how dilations affect surface area. Use the cube at the right.

a. **NUMERICAL** Find the surface area of the cube.

b. **TABULAR** Copy and complete the table shown.

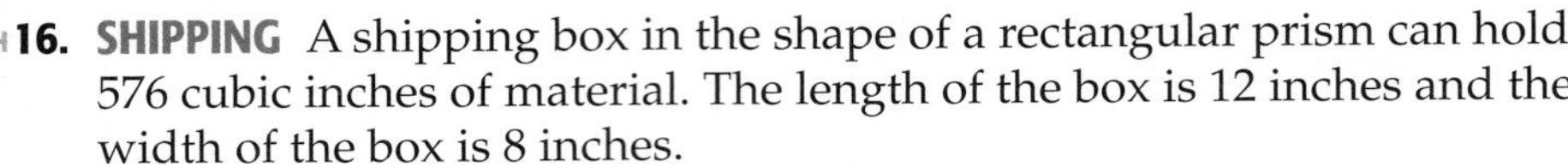

Scale Factor of Dilation	Original	× 2	× 3
Length of Side	■	■	■
Surface Area	■	■	■

c. **ANALYTICAL** Compare the original surface area to the surface area after a dilation by a scale factor of 2, then by a scale factor of 3.

d. **VERBAL** Does this same relationship exist with any rectangular prism? Explain your reasoning.

Real-World Link

In 2007, the world's cardboard manufacturers produced over 150 billion square meters of cardboard, enough to cover the state of Louisiana, and then some.

16. **SHIPPING** A shipping box in the shape of a rectangular prism can hold 576 cubic inches of material. The length of the box is 12 inches and the width of the box is 8 inches.

a. Find the surface area of the box.

b. Another box with the same volume has dimensions of 16 inches by 9 inches by 4 inches. Which box has the greater surface area? Explain.

c. Predict the shape of a rectangular prism with the same volume but with a greater surface area than either of the boxes above. Then, test your prediction.

17. **FINANCIAL LITERACY** The boxes shown below are puzzle boxes that are made from wood. If the wood to make the boxes costs $1.30 per square inch, which box would cost more to make? Explain.

Box 1

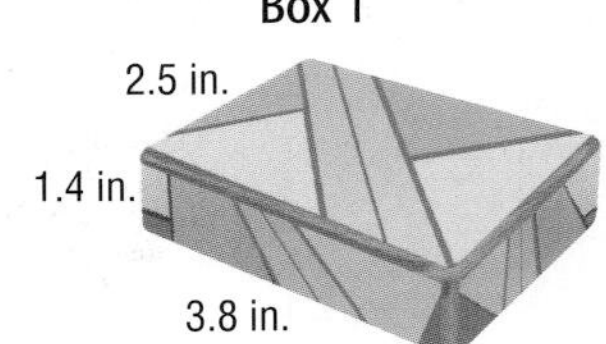

Box 2

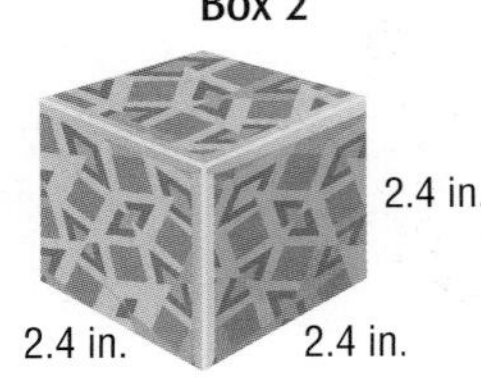

H.O.T. Problems

Use Higher-Order Thinking Skills

18. **OPEN ENDED** Name the dimensions of a cube with a surface area between 10 and 20 square inches.

19. **CHALLENGE** A box manufacturer wants to make a box that has a volume of 216 cm^3 and uses the least amount of cardboard possible. Find the dimensions of that box.

20. **FIND THE ERROR** Serena and Rhianna want to find the surface area of the cube shown. Is either of them correct? Explain.

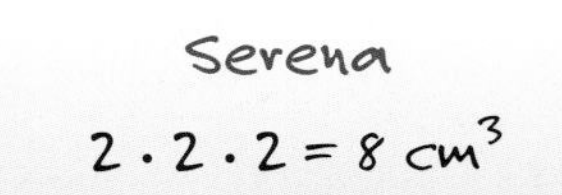

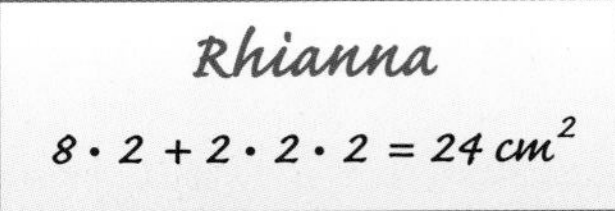

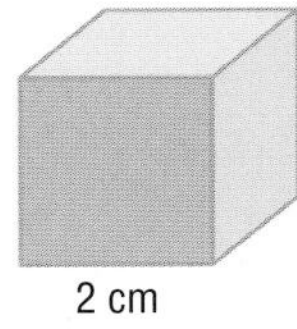

21. **CHALLENGE** Sketch two prisms such that one has a greater volume and the other has a greater surface area.

22. **WRITING IN MATH** Explain the difference between surface area and volume.

Standardized Test Practice

23. A rectangular cardboard box has the same volume as another rectangular box that measures 6 inches by 14 inches by 20 inches, but with less surface area. Which size box would *not* meet those requirements?

A 7 in. by 10 in. by 24 in.

B 7 in. by 12 in. by 20 in.

C 10 in. by 12 in. by 14 in.

D 5 in. by 16 in. by 21 in.

24. SHORT RESPONSE The side measures of a rectangular prism are tripled. What is the ratio, written as a fraction, of the surface area of the original prism to the surface area of the larger prism?

25. Find the surface area of the figure below to the nearest whole number.

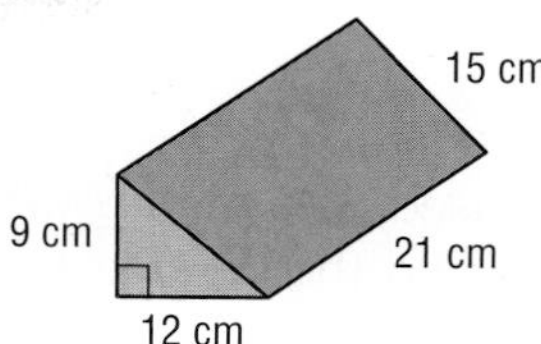

F 1065 cm^2

G 945 cm^2

H 777 cm^2

J 864 cm^2

26. How much cardboard is needed to make a rectangular box of cereal that measures 10 inches by 8 inches by 2 inches?

A 116 in^2

B 160 in^2

C 232 in^2

D 248 in^2

Spiral Review

Find the volume of each figure. Round to the nearest tenth. (Lesson 12-4)

27.

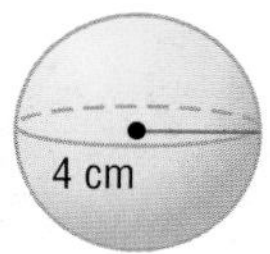

28.

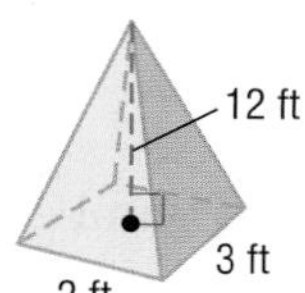

29.

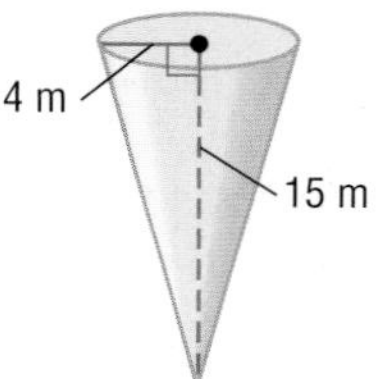

30. ANIMALS A water trough as shown at the right is used to provide water for farm animals. Find the volume of the trough. Round to the nearest tenth. (Lesson 12-3)

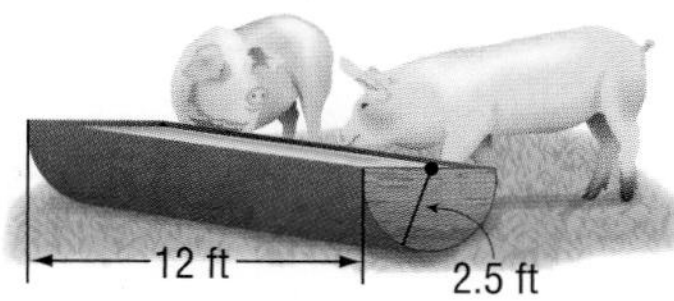

Evaluate each expression if $a = 3$, $b = 5$, and $c = 4$. (Lesson 9-1)

31. b^4

32. c^3

33. $2(3c + 7)^2$

34. $4(2a - c^3)^2$

Skills Review

Find the area of each figure. Round to the nearest tenth. (Lesson 11-9)

35.

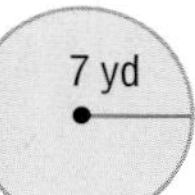

36.

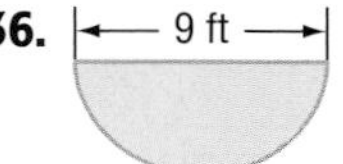

37.

25 m

30 m

Geometry Lab
Surface Area of Cylinders

Math Online glencoe.com
Math *in Motion,* Animation

In Lesson 12-5, you used nets to decompose prisms to find surface areas. You can also use nets to help find the surface area of cylinders.

ACTIVITY

Step 1 Use an empty cylinder-shaped container that has a lid. Measure and record the height of the container. Outline the bases on blank paper and cut them out.

Step 2 Wrap paper around the curved surface of the cylinder and tape it in place. Draw a line from the bottom of the cylinder to the top along the edge of the paper. Draw a line around the circumference of the cylinder.

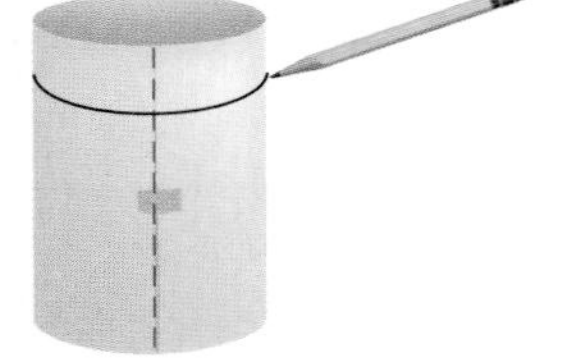

Step 3 Unroll the paper. Cut the paper along the lines marked.

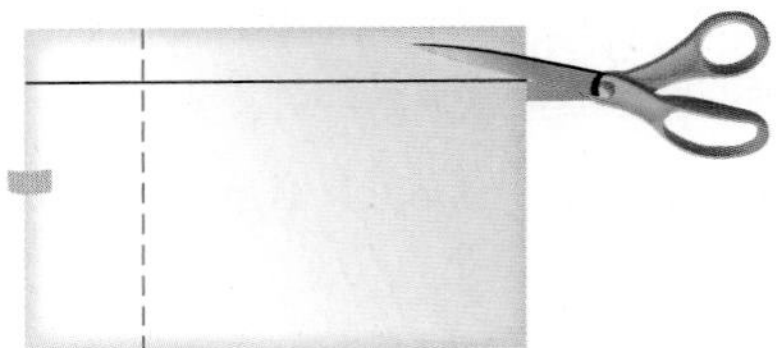

Step 4 Tape the bases and the side together so that they can be re-folded to make the original cylinder. This figure is the net of the cylinder.

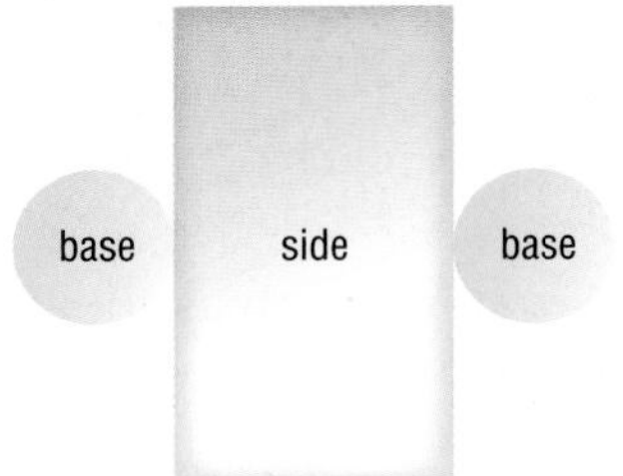

Analyze the Results

1. Classify the two-dimensional shapes that make up the net of the container.
2. Find the area of each shape. Then find the sum of these areas.
3. Find the diameter of the top of the container and use it to find the perimeter or circumference of that face.
4. Multiply the circumference by the height of the container. What does this product represent?
5. Add the product from Exercise 4 to the sum of the areas of the two circular bases.
6. Compare your answers from Exercises 2 and 5.
7. **MAKE A CONJECTURE** Describe a method for finding the area of all the surfaces of a cylinder given the diameter of one of its bases and its height.

12-6 Surface Area of Cylinders

Then

You have already found the lateral areas and surface areas of prisms. (Lesson 12-5)

Now

- Find lateral and surface areas of cylinders.
- Compare surface areas of cylinders.

Math Online

glencoe.com

- Extra Examples
- Personal Tutor
- Self-Check Quiz
- Homework Help
- Math in Motion

Why?

Seki is working in the craft room as a counselor at day camp this summer. She is planning a project to make storage containers out of used tin cans. She uses the label of each can as a template for the decorative wrap.

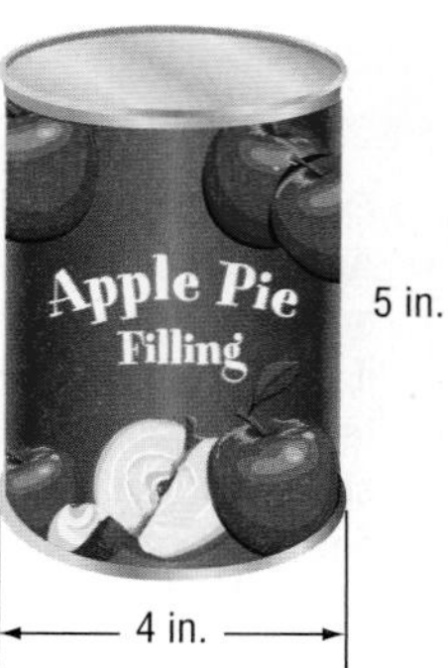

a. What shape is the label if Seki makes a vertical cut through the label and unrolls it?

b. What is the height of the label?

c. What is the length of the label?

d. What is the area of the label?

Surface Area of Cylinders You found the surface areas of prisms by adding the lateral area and the area of the two bases, $S = L + 2B$. You can find surface areas of cylinders in the same way. If you unroll a cylinder, its net is a rectangle (lateral area) and two circles (bases).

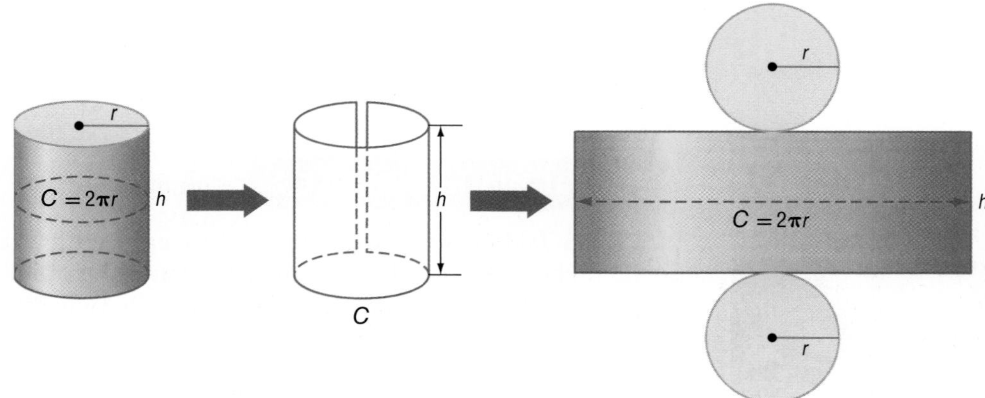

Key Concept — For Your FOLDABLE

Lateral Area of Cylinders

Words The lateral area L of a cylinder with radius r and height h is the circumference of the base ($2\pi r$) times the height h.

Symbols $L = 2\pi rh$

Model

circumference of base $= 2\pi r$

h

area of bases $= \pi r^2$

Surface Area of Cylinders

Words The surface area S of a cylinder is the lateral area L plus the area of the two bases ($2\pi r^2$).

Symbols $S = L + 2B$ or $S = 2\pi rh + 2\pi r^2$

Math *in Motion*, Interactive Lab glencoe.com

EXAMPLE 1 Surface Area of a Cylinder

Find the lateral area and the surface area of each cylinder. Round to the nearest tenth.

StudyTip

Exact vs. Approximate Answers Leave your answers in terms of π if exact answers are required. If approximate answers are sufficient, find a decimal approximation.

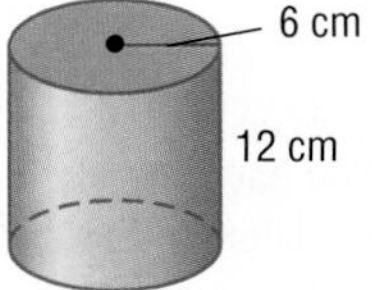

a. Lateral Area

$L = 2\pi rh$

$= 2 \cdot \pi \cdot 6 \cdot 12$

$= 144\pi \text{ m}^2$ **exact**

$\approx 452.4 \text{ m}^2$ **approximate**

Surface Area

$S = L + 2\pi r^2$

$= 144\pi + 2\pi(6)^2$

$= 216\pi \text{ m}^2$ **exact**

$\approx 678.6 \text{ m}^2$ **approximate**

b. diameter of 20 inches and height of 5.6 inches

The diameter is 20 inches so the radius r is $\frac{20}{2}$ or 10 inches.

Lateral Area

$L = 2\pi rh$

$= 2 \cdot \pi \cdot 10 \cdot 5.6$

$= 112\pi \text{ in}^2$

$\approx 351.9 \text{ in}^2$

Surface Area

$S = L + 2\pi r^2$

$= 112\pi + 2\pi(10)^2$

$= 112\pi \text{ in}^2 + 200\pi \text{ in}^2$

$\approx 980.2 \text{ in}^2$

Check Your Progress

1. Find the lateral area and surface area of a cylinder with a radius of 7 centimeters and a height of 12 centimeters. Round to the nearest tenth.

Personal Tutor glencoe.com

Real-World EXAMPLE 2 Compare Surface Areas of Cylinders

CRAFTS Isabel is making two candles, each with a radius of 1.5 inches. One candle is 4 inches tall and the other is 8 inches tall. Is the surface area of the larger candle twice that of the smaller candle? Explain.

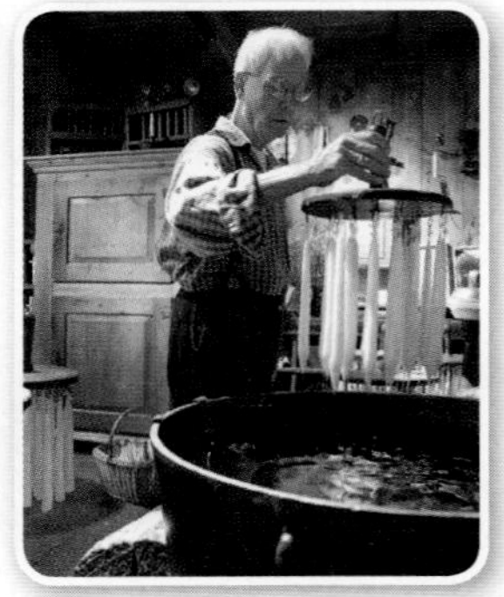

Real-World Link

Archaeologists have found remains of candles from as far back as 3000 B.C. Paraffin wax began to be popular in the 19th century.

Find the surface areas of both candles.

Surface Area of Candle A

$S = L + 2\pi r^2$

$S = 2\pi rh + 2\pi r^2$

$= 2\pi \cdot 1.5 \cdot 4 + 2\pi \cdot (1.5)^2$

$= 12\pi + 4.5\pi$

$= 16.5\pi$

Surface Area of Candle B

$S = L + 2\pi r^2$

$S = 2\pi rh + 2\pi r^2$

$= 2\pi \cdot 1.5 \cdot 8 + 2\pi \cdot (1.5)^2$

$= 24\pi + 4.5\pi$

$= 28.5\pi$

The surface area of candle B is *not* twice the surface area of candle A. The lateral area of candle B is twice the lateral area of candle A, but the base areas are equal.

Check Your Progress

2. CANNED FOODS Which can has a greater surface area: a tuna fish can with diameter 8 centimeters and height 4 centimeters or a soup can with a diameter 4 centimeters and height 8 centimeters?

Personal Tutor glencoe.com

Check Your Understanding

Example 1 p. 698

Find the lateral and surface area of each cylinder. Round to the nearest tenth.

1. 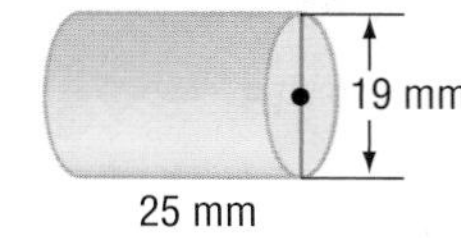

2. 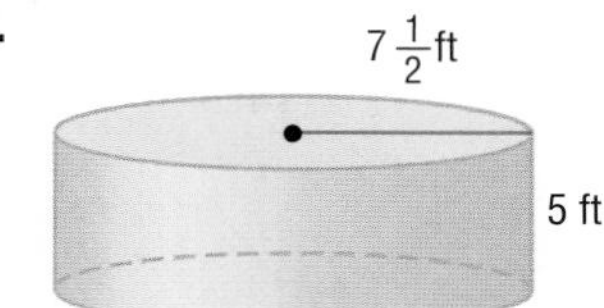

3. diameter of 4.2 meters and a height of 7.8 meters

Example 2 p. 698

4. **LABELS** A case of frozen juice contains 12 cans like the one shown at the right. Find the area of the labels for one case of canned juice to the nearest tenth.

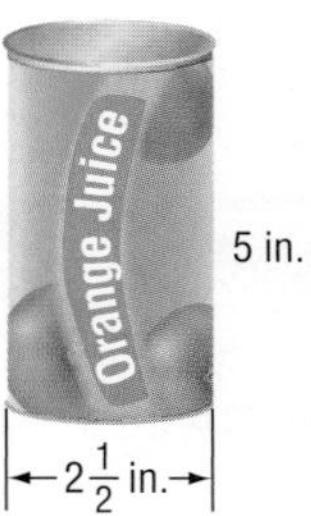

Practice and Problem Solving

= **Step-by-Step Solutions** begin on page R11.
Extra Practice begins on page 810.

Example 1 p. 698

Find the lateral and surface area of each cylinder. Round to the nearest tenth.

5. 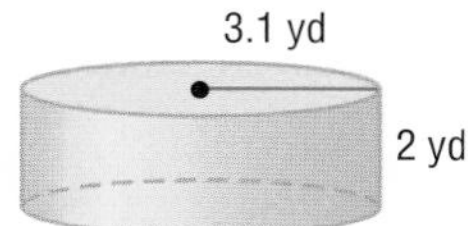

6. 

7. 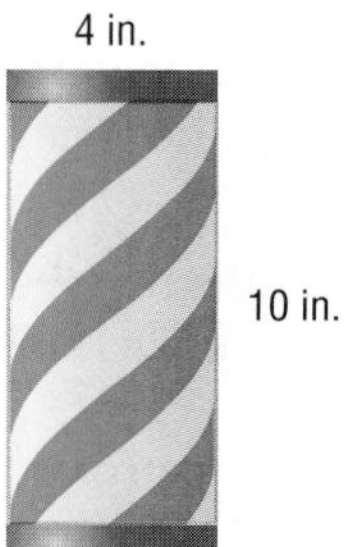

8. 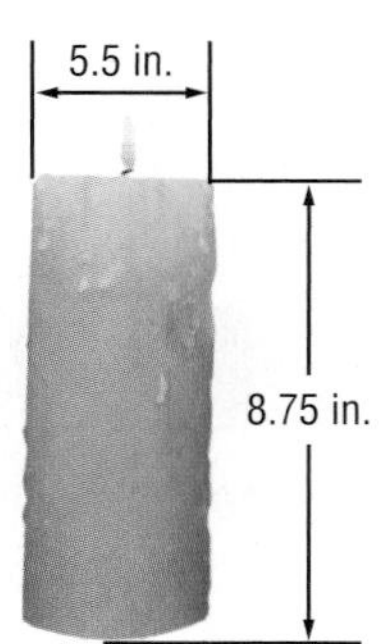

9. radius of $3\frac{2}{3}$ feet and a height of $5\frac{1}{4}$ feet
10. diameter of 9 meters and a height of 7.3 meters

Example 2 p. 698

11. **MAIL** The mailing box and mailing tube shown at the right have about the same volume. Which one has a smaller surface area? Explain.

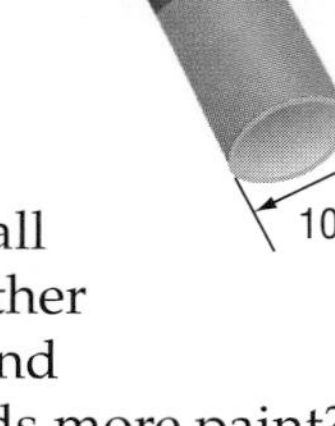

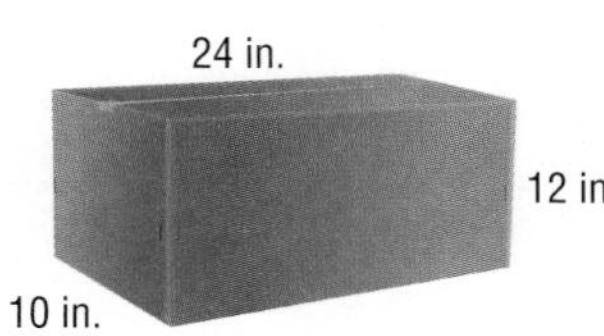

12. **HOME DECOR** Taryn is painting pillars. One pillar is 0.75 meter tall and 0.15 meter in diameter. Another pillar is 0.30 meter in diameter and 0.38 meter tall. Which pillar needs more paint?

13. **PLUMBING** Find the surface area (exterior and interior) of the pipe shown. Round to the nearest tenth of a square inch.

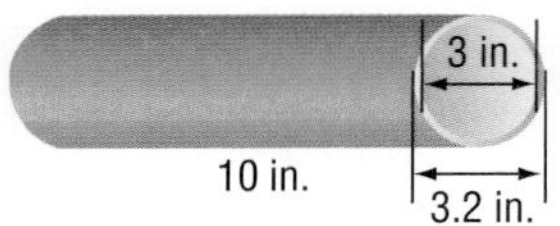

Find the surface area of each figure. Round to the nearest tenth.

14. 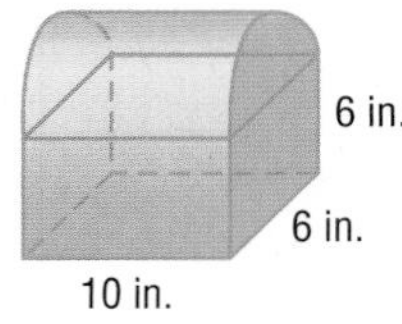

15. 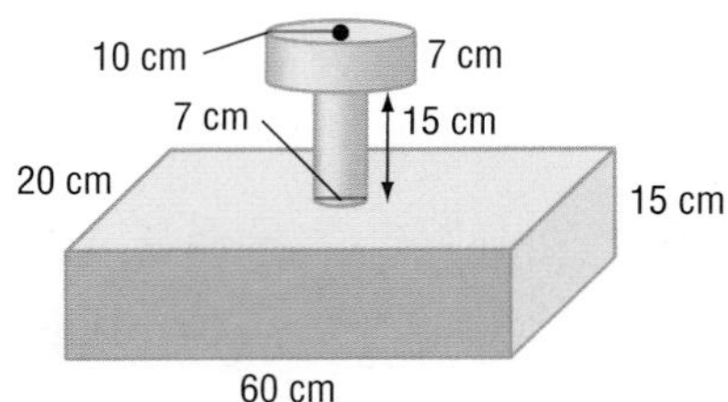

16. **MONEY** A penny's mass is exactly half the mass of a nickel. Using the information below, determine the ratio between their surface areas. Round to the nearest tenth.

Coin	Penny	Nickel
Weight	2.5 g	5.0 g
Diameter	19.05 mm	21.21 mm
Thickness	1.55 mm	1.95 mm

Real-World Link

The average life span of a coin is 30 years. Coins that are found to be too old or damaged for circulation are melted down to create new coins.

Source: U.S. Mint

17. **FINANCIAL LITERACY** The Student Council is planning a movie night. They will sell popcorn in one of the open-top containers shown. The cost depends on the amount of cardboard used to make each container. Which container should they buy? Use volume and surface area measurements to explain your choice.

18. Find the height of a cylinder if the surface area is 402 square centimeters and the radius is 4 centimeters. Round to the nearest tenth.

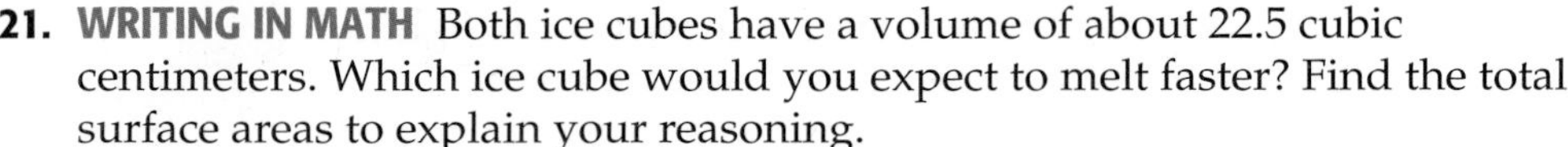

H.O.T. Problems Use Higher-Order Thinking Skills

19. **OPEN ENDED** Create a prism that has approximately the same surface area as the cylinder shown at the right.

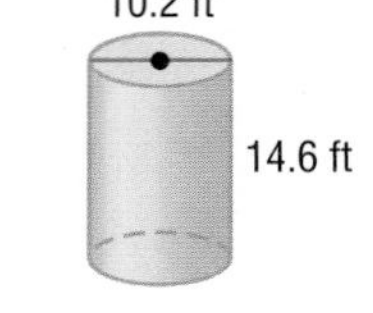

20. **CHALLENGE** A cylinder has a radius of 10 centimeters and a height of 6 centimeters. Without calculating, explain whether multiplying the radius or the height by a scale factor of 0.5 would have a greater effect on the surface area of the cylinder.

21. **WRITING IN MATH** Both ice cubes have a volume of about 22.5 cubic centimeters. Which ice cube would you expect to melt faster? Find the total surface areas to explain your reasoning.

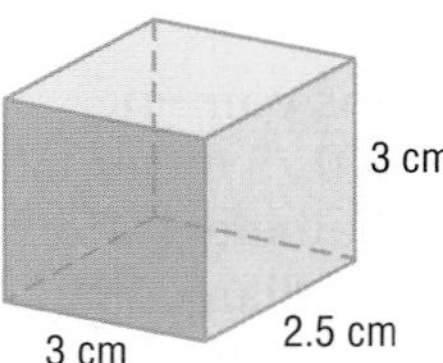

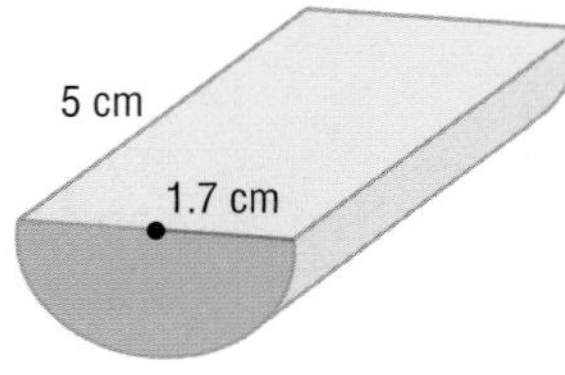

Standardized Test Practice

22. Find the amount of paper needed for the label on the can. Use 3.14 for π.

A 21.2 in^2
B 25.5 in^2
C 29.1 in^2
D 42.4 in^2

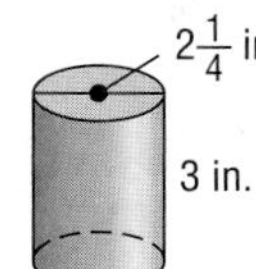

23. A cylinder has a surface area of 5652 square millimeters. If the diameter of the cylinder is 30 millimeters, what is the height? Use 3.14 for π.

F 10 mm
G 45 mm
H 59 mm
J 105 mm

24. GRIDDED RESPONSE What is the surface area in square meters of a cylinder with a radius of 3.4 meters and a height of 2.8 meters? Round to the nearest hundredth. Use 3.14 for π.

25. A cylindrical plastic bar is to have the same surface area as a cylindrical metal bar with a radius of 1 inch and a height of 4 inches. Which of the following dimensions meet these requirements? Use 3.14 for π.

A radius: 4 in., height: 2 in.
B radius: 2 in., height: 2 in.
C radius: 2 in., height: 4 in.
D radius: 2 in., height: 0.5 in.

Spiral Review

Find the lateral and surface area of each figure. (Lesson 12-5)

26.

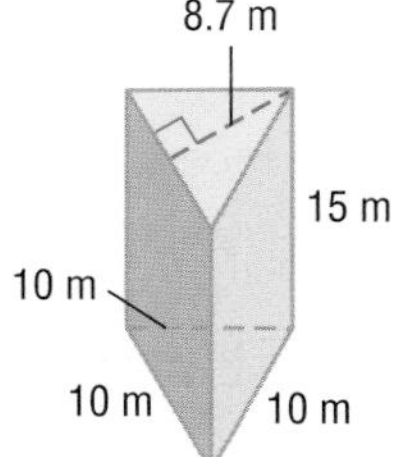

27.

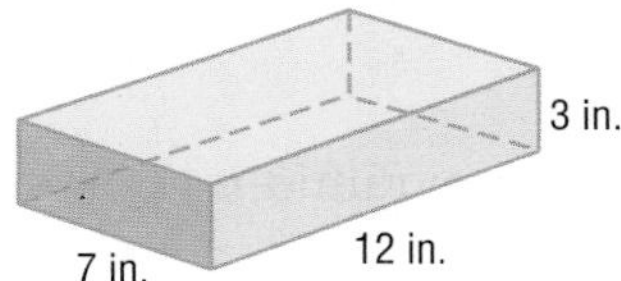

28.

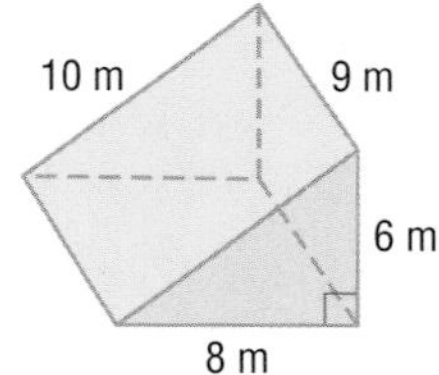

29. HISTORY The Pyramid of Cestius in Rome, Italy, is 27 meters high and has a square base 22 meters on a side. What is its volume? Use an estimate to check your answer. (Lesson 12-4)

Classify each angle as *acute*, *obtuse*, *right*, or *straight*. (Lesson 11-1)

30. 50° **31.** 180° **32.** 43° **33.** 114°

Find the simple interest earned to the nearest cent. (Lesson 7-7)

34. \$1500 at 7.5% for 5 years
35. \$750 at 12.25% for 10 years
36. \$625 at 5.75% for 8 years
37. \$10,150 at 4.5% for 20 years

Skills Review

Find each product. (Lesson 3-3)

38. 2.45^2
39. $4 \cdot \frac{1}{2} \cdot 3 \cdot 8$
40. $3.14(5.4)^2$
41. $7.5 \cdot \frac{1}{2} \cdot 15\frac{2}{3}$
42. $3.14\left(1\frac{1}{4}\right)^2$
43. $\frac{2}{5} \cdot 2\frac{1}{3}$

12-7 Surface Area of Pyramids and Cones

Then

You have already found the surface areas of prisms and cylinders. (Lessons 12-5 and 12-6)

Now

- Find lateral areas and surface areas of pyramids.
- Find lateral areas and surface areas of cones.

New Vocabulary

regular pyramid
slant height

Math Online

glencoe.com

- Extra Examples
- Personal Tutor
- Self-Check Quiz
- Homework Help

Why?

On a recent trip to Paris, Michelle visited the Louvre Museum.

a. How many faces are on the pyramid?

b. What are the shapes of the faces?

c. Do any of the shapes appear congruent?

d. How would you find the amount of glass needed to make the pyramid?

Surface Areas of Pyramids A **regular pyramid** is a pyramid whose base is a regular polygon. The lateral faces of a regular pyramid are congruent isosceles triangles that intersect at the vertex. The altitude or height of each lateral face is called the **slant height** of the pyramid.

Regular Pyramid

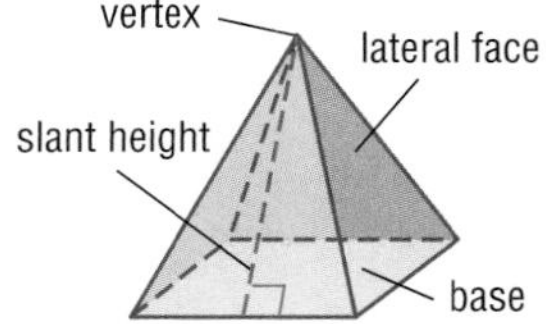

Net of Regular Pyramid

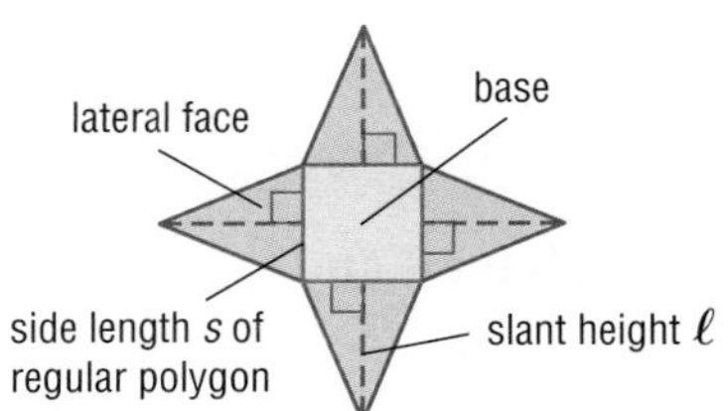

The lateral area of a pyramid is the sum of the areas of its lateral faces, which are all triangles.

$L = 4\left(\frac{1}{2}s\ell\right)$ **Area of the lateral faces**

$L = \frac{1}{2}(4s)\ell$ **Commutative Property of Multiplication**

$L = \frac{1}{2}P\ell$ **Replace $4s$ with P, the perimeter of the base.**

The total surface area is the lateral surface area plus the area of the base.

Key Concept

For Your FOLDABLE

Lateral Area of Pyramids

Words The lateral area L of a regular pyramid is half the perimeter P of the base times the slant height ℓ.

Symbols $L = \frac{1}{2}P\ell$

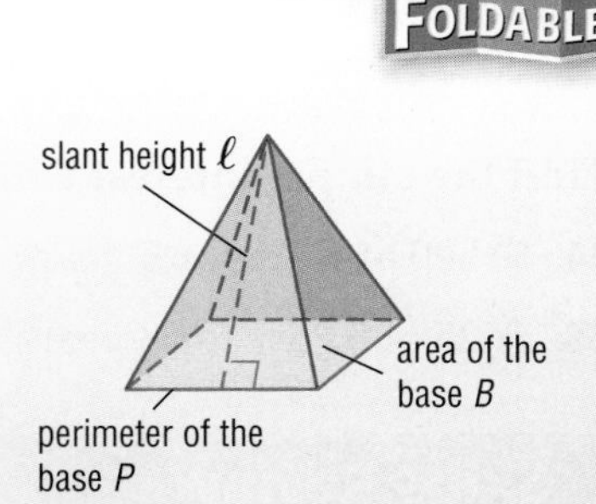

Surface Area of Pyramids

Words The total surface area S of a regular pyramid is the lateral area L plus the area of the base B.

Symbols $S = L + B$ or $S = \frac{1}{2}P\ell + B$

StudyTip

Height of a Pyramid and Slant Height The height of a pyramid is the perpendicular distance from the top point to the base of the pyramid. The slant height of the pyramid is the altitude of one of the triangles that forms the sides of the pyramid.

EXAMPLE 1 Surface Area of a Pyramid

Find the lateral and total surface area of the regular pentagonal pyramid.

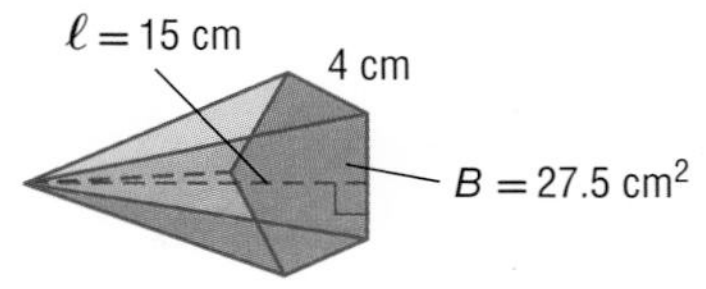

Find the lateral area.

$L = \frac{1}{2}P\ell$ **Write the formula.**

$L = \frac{1}{2}(4 \cdot 5)15$ **Replace *P* with 4 • 5 and ℓ with 15.**

$= 150 \text{ cm}^2$ **Simplify.**

Find the surface area.

$S = L + B$ **Write the formula.**

$S = 150 + 27.5$ **Replace *L* with 150 and *B* with 27.5.**

$= 177.5 \text{ cm}^2$ **Simplify.**

The lateral surface area is 150 cm^2 and the total surface area is 177.5 cm^2.

Check Your Progress

1. Find the lateral and surface area of a square pyramid with a base side length of 6 centimeters and a slant height of 18.4 centimeters.

Personal Tutor glencoe.com

Real-World EXAMPLE 2 Lateral Area of a Pyramid

HISTORY If the length of each side of the parachute shown at the left is 12 yards and the height of the pyramid is 12 yards, find the amount of cloth needed to make the parachute.

Real-World Link

In 1483, Leonardo Da Vinci sketched the first design for a parachute in the margin of a notebook. In 2000, Adrian Nicholas constructed and tested the parachute using Da Vinci's plans. It weighed 187 pounds, and it safely carried him down from an altitude of 8000 feet.

Step 1 Find the slant height ℓ of the square pyramid. The slant height of the pyramid is the hypotenuse of a right triangle with one leg the height of the pyramid and the other leg half the measure of one of the base sides.

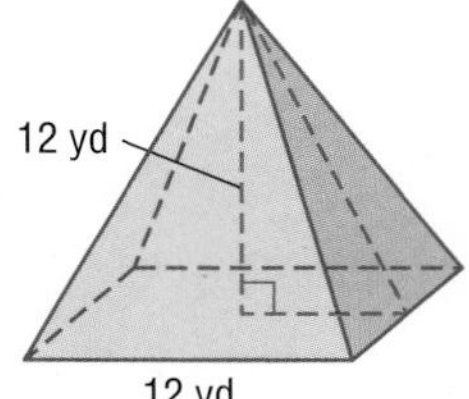

$a^2 + b^2 = c^2$ **Use the Pythagorean Theorem.**

$12^2 + 6^2 = \ell^2$ **Replace *a* with 12, *b* with 6, and *c* with ℓ.**

$180 = \ell^2$ **Simplify.**

$13.4 \approx \ell$ **Definition of square root**

Step 2 Since the parachute is open on the bottom, to find the amount of cloth needed for the parachute, find the lateral area of the pyramid.

$L = \frac{1}{2}P\ell$ **Write the formula for the lateral area.**

$L = \frac{1}{2}(12 \cdot 4)13.4$ **Replace *P* with 12 • 4 and ℓ with 13.4.**

$= 321.6 \text{ yd}^2$ **Simplify.**

So, 321.6 square yards of cloth are needed for the parachute.

Check Your Progress

2. MUSEUMS The pyramid at the Louvre Museum is a square pyramid with a height of 20.5 meters and sides of 35 meters. Find the amount of glass on the pyramid.

Personal Tutor glencoe.com

Surface Area of Cones You can also find surface areas of cones. The net of a cone shows the regions that make up the cone.

Model of Cone

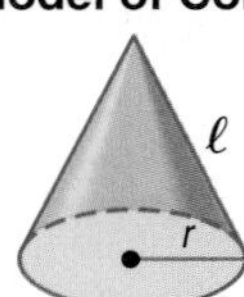

Net of Cone

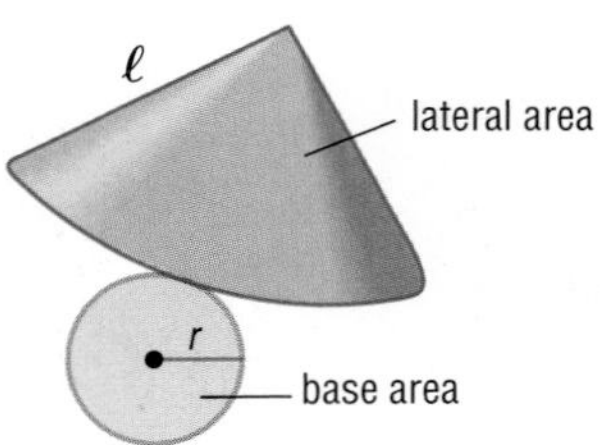

> **StudyTip**
>
> **Slant Height of a Cone** If you are not given the slant height of a cone, you can find it by using the Pythagorean Theorem. You need to know the radius and the height of the cone.
>
> $a^2 + b^2 = c^2$
>
> $\ell^2 = h^2 + r^2$
>
> 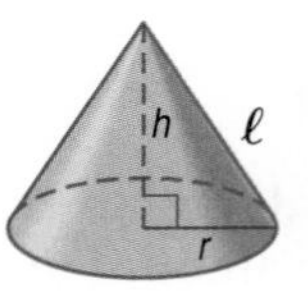
>

The lateral area of a cone with slant height ℓ is one-half the circumference of the base, $2\pi r$, times ℓ. So, $L = \frac{1}{2} \cdot 2\pi r \cdot \ell$ or $L = \pi r\ell$. The base of the cone is a circle with area πr^2.

Key Concept

For Your FOLDABLE

Lateral Area of Cones

Words The lateral area L of a cone is π times the radius times the slant height ℓ.

Symbols $L = \pi r\ell$

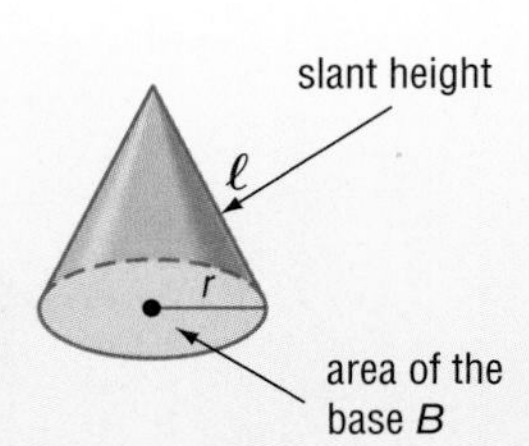

Surface Area of Cones

Words The surface area S of a cone with slant height ℓ and radius r is the lateral area plus the area of the base.

Symbols $S = L + \pi r^2$

EXAMPLE 3 Surface Area of a Cone

Find the lateral and surface area of the cone. Round to the nearest tenth.

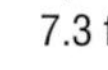

Find the lateral area.

$L = \pi r\ell$ **Lateral area of a cone**

$L = \pi(7.3)(11.6)$ **Replace r with 7.3 and ℓ with 11.6.**

≈ 266.0 **Simplify.**

Find the surface area.

$S = L + \pi r^2$ **Surface area of a cone**

$S = 266 + \pi(7.3)^2$ **Replace r with 7.3.**

$\approx 433.4 \text{ ft}^2$ **Simplify.**

The surface area of the cone is about 433.4 square feet.

Check Your Progress

3. Find the lateral and surface areas of a cone with a radius of $4\frac{1}{4}$ yards and a slant height of 7 yards. Round to the nearest tenth.

Personal Tutor glencoe.com

Check Your Understanding

Examples 1 and 3 pp. 703–704

Find the lateral and surface area of each figure. Round to the nearest tenth.

1.

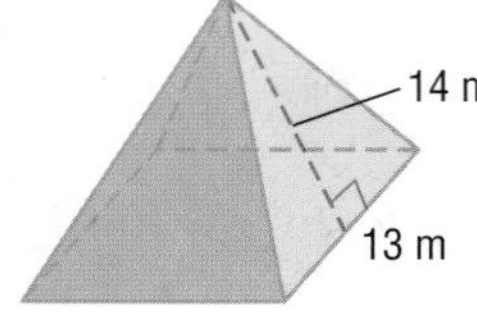

2.

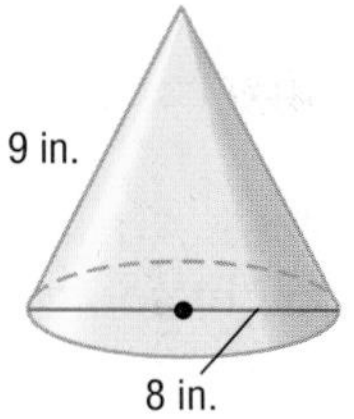

3.

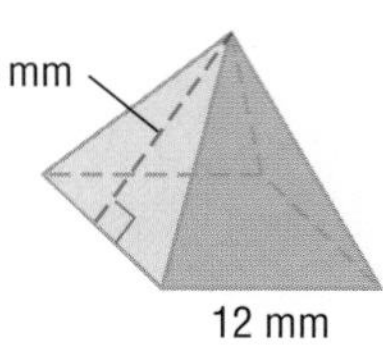

Example 2 p. 703

4. **GLASS** The Luxor Hotel in Las Vegas, Nevada, is a square pyramid made from glass with a base length of 646 feet and a height of 350 feet. Find the surface area of the glass on the Luxor.

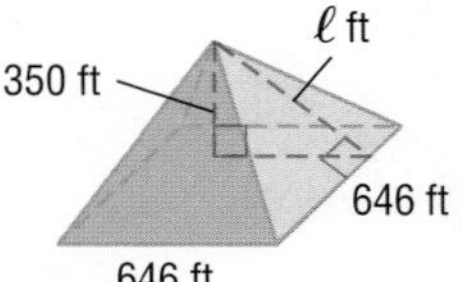

Practice and Problem Solving

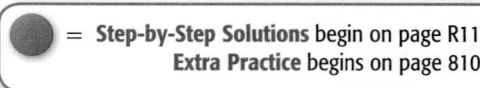

Example 1 p. 703

Find the lateral and surface area of each figure. Round to the nearest tenth.

5.

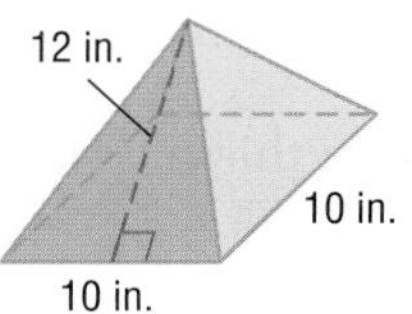

6.

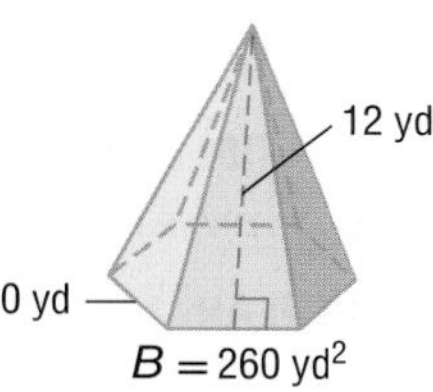

7.

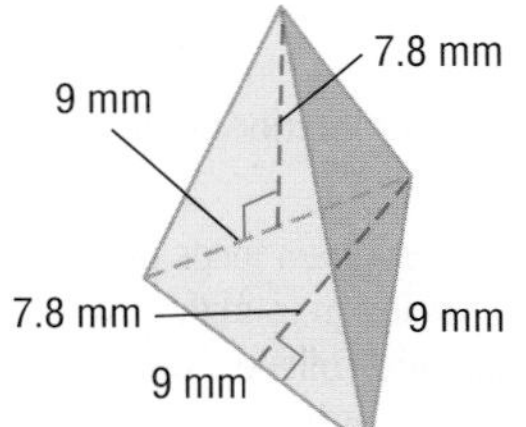

8. triangular pyramid: base side length 6 in., base area $15\frac{3}{5}$ in^2, slant height 8 in.

Example 2 p. 703

9. **JEWELRY** Brianne is making a necklace using the bead shown at the right. Find the approximate surface area of the silver bead.

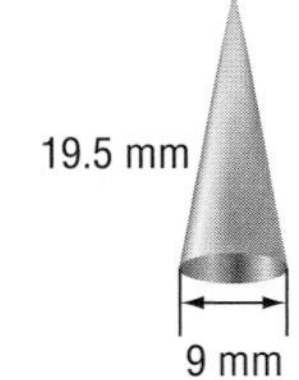

10. **ARCHITECTURE** The Transamerica Building in San Francisco is shaped like a square pyramid. It has a slant height of about 856 feet, and each side of its base is 145 feet. Find the lateral area of the building.

Example 3 p. 704

Find the lateral and surface area of each figure. Round to the nearest tenth.

11.

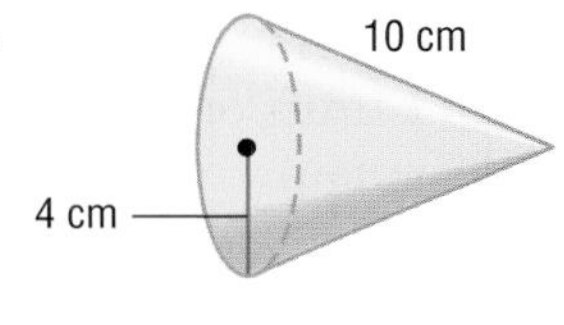

12.

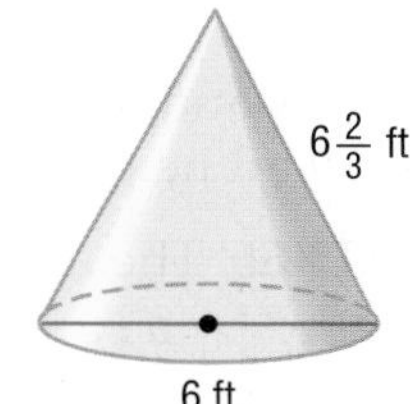

13.

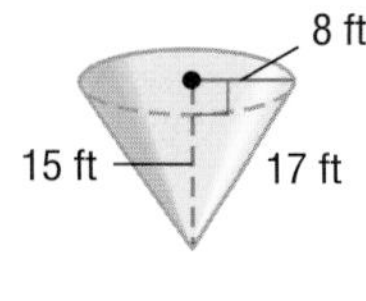

14. cone: diameter 19 cm, slant height 30 cm

Find the surface area of each figure. Round to the nearest tenth.

15.

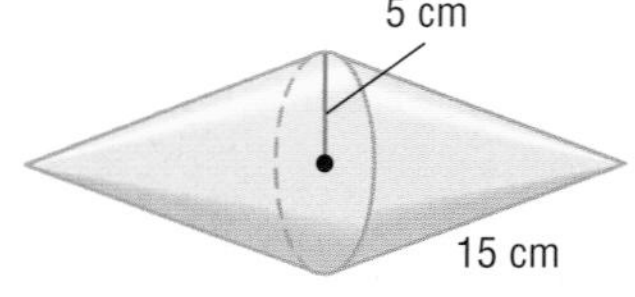

16.

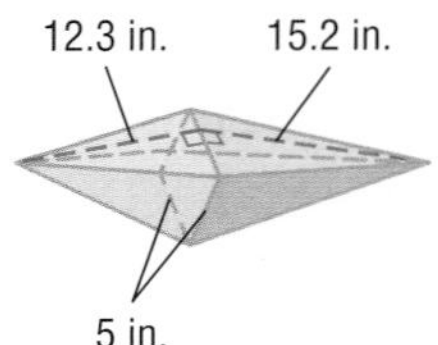

17. **COSTUMES** Adrienne is making costumes for the school play. She needs to make eight conical medieval hats. She wants each hat to be 18 inches tall and the bases of each to be 22 inches in circumference. How much material will she use to make the hats?

Real-World Career

Fabric Designers
There are many opportunities in the field of fashion, including fabric design. Fabric designers may develop fabrics for clothing or interior home design.

Fabric designers focus on art and design in high school and college. Some fashion institutes have two-year programs primarily for fabric designers.

Draw the figure represented by each net. Then find the lateral and total surface area of each figure.

18.

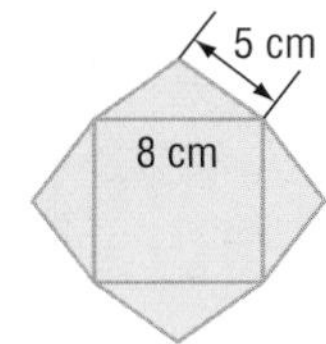

19.

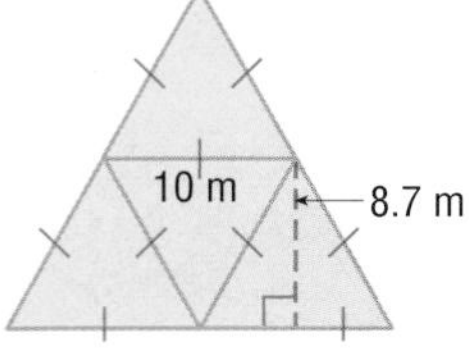

20.

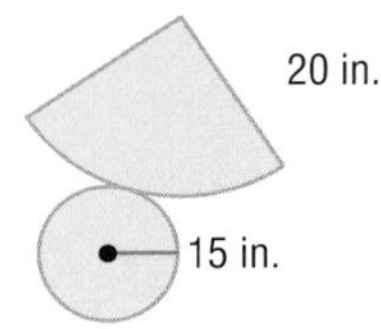

21. **TENTS** A rectangular piece of canvas is 50 feet by 60 feet. Delsin wants to make a tent in the shape of a cone that has a diameter of 30 feet. Find the slant height of the largest tent with a floor he can build using the canvas.

22. **MULTIPLE REPRESENTATIONS** In this problem, you will examine surface areas of pyramids.

a. **TABULAR** Find the surface area of a square pyramid with a side length of 1 and a slant height of 10. Then complete the table.

side length	1	2	3	4
slant height	10	20	30	40
surface area	■	■	■	■

b. **ANALYTICAL** What happens to the surface area if the base length and slant height are doubled, tripled, or multiplied by 4?

c. **VERBAL** Predict the surface area of a pyramid with the side length of 5 and slant height of 50. Then check your prediction.

H.O.T. Problems

Use Higher-Order Thinking Skills

23. **OPEN ENDED** Draw a cone with a surface area that is between 100 and 150 square units.

24. **CHALLENGE** Which has a greater surface area: a square pyramid with a base of x units and a slant height of ℓ units or a cone with a diameter of x units and a slant height of ℓ units? Explain your reasoning.

25. **CHALLENGE** The dimensions of the prism shown are increased by a scale factor and the total surface area of the new prism is 240,000 square meters. What was the scale factor used to create the new prism?

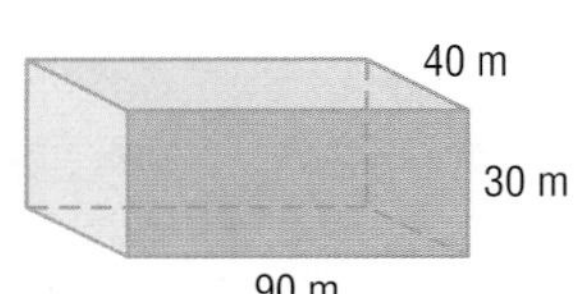

26. **WRITING IN MATH** Explain how to find the slant height of a cone if you are given the radius and the height of the cone.

Standardized Test Practice

27. Find the surface area of a cone with a radius of 6 centimeters and slant height of 15 centimeters. Round to the nearest tenth.

A 141.3 cm^2
B 264.4 cm^2
C 395.8 cm^2
D 565.2 cm^2

28. Find the area of the ice cream cone covered by the wrapper.

F 9.4 in^2
G 11.2 in^2
H 15.5 in^2
J 20.0 in^2

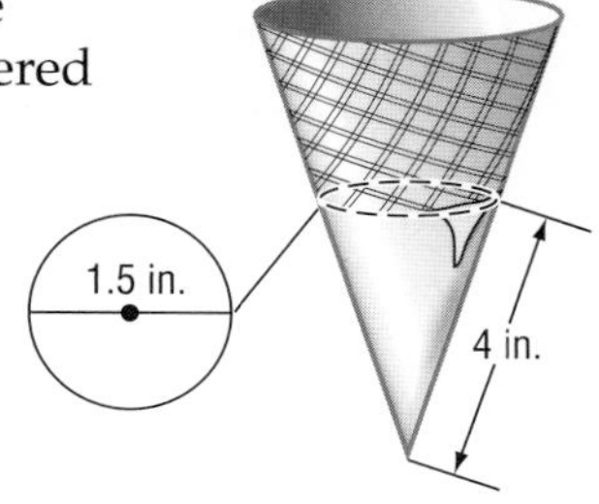

29. A square pyramid has a base with sides measuring 7 meters. If the surface area of the pyramid is 189 square meters, which of the following is the slant height of the pyramid?

A 8 m
B 10 m
C 14 m
D 16 m

30. **GRIDDED RESPONSE** A square pyramid has a surface area of 39 square feet. After a dilation, the surface area is 351 square feet. What was the scale factor of the dilation?

Spiral Review

Find the lateral and surface area of each three-dimensional figure. Round to the nearest tenth. (Lessons 12-5 and 12-6)

31. cylinder: radius 6 inches, height $10\frac{1}{4}$ inches

32. cylinder: diameter 5.7 meters, height 2.3 meters

33. equilateral triangular prism: base height 10.4 feet, base length 12 feet, prism height $3\frac{1}{3}$ feet

34. **DESSERTS** Find the volume of an ice cream cone that has a radius of 28 mm and a height of 117 mm. (Lesson 12-4)

35. **GEOMETRY** Mrs. Morales used the parallelogram at the right as a pattern for a paving stone for her sidewalk. If $m\angle 1$ is 130°, find $m\angle 2$. (Lesson 11-5)

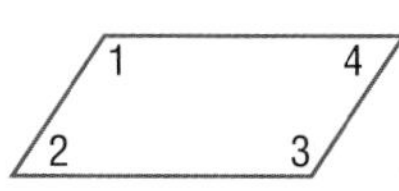

36. **RETAIL** Find the discount for a \$45 shirt that is on sale for 20% off. (Lesson 7-4)

Skills Review

Solve each proportion. (Lesson 6-5)

37. $\frac{x}{4} = \frac{3}{5.6}$

38. $\frac{2}{3} = \frac{14}{n}$

39. $\frac{n}{20} = \frac{15}{50}$

40. $\frac{14}{32} = \frac{x}{8}$

41. $\frac{3}{2.2} = \frac{7.5}{y}$

42. $\frac{30}{14} = \frac{m}{1.54}$

Geometry Lab
Similar Solids

Math Online glencoe.com
Math *in Motion*, Animation

Objective
Investigate similar solids.

A model train is an exact replica of a real train, but much smaller. The dimensions of the model and the original are proportional. Therefore, these two objects are *similar solids*.

You can use sugar cubes or centimeter blocks to investigate similar solids.

ACTIVITY 1

- If each edge of a cube is 1 unit long, then each face is 1 square unit and the volume of the cube is 1 cubic unit.
- Make a cube that has sides twice as long as the first cube.

1 unit
1 unit
1 unit

First Cube

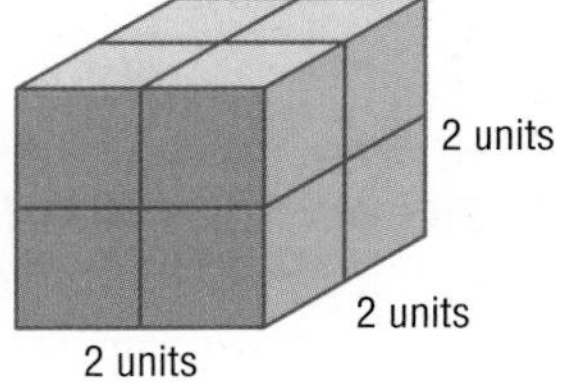

Second Cube

Analyze the Results

1. What is the area of one face of the first cube? second cube?

2. What is the volume of the first cube? second cube?

ACTIVITY 2

Build a cube that has sides three times as long as the first cube.

Analyze the Results

3. How many small cubes did you use?

4. What is the area of one face of the cube?

5. What is the volume of the cube?

6. Copy and complete the tables below.

Scale Factor	Side Length	Volume
1	■	■
2	■	■
3	■	■

Scale Factor	Side Length	Area of a Face	Surface Area
1	■	■	■
2	■	■	■
3	■	■	■

Vocabulary Review

Scale Factor number of times you increase or decrease linear dimensions

7. What happens to the area of a face when the length of a side is doubled? tripled?

8. Consider the unit cube. If the scale factor is x, what is the area of one face? the surface area?

9. What happens to the volume of a cube when the length of a side is doubled? tripled?

10. Consider a unit cube with side length s. Let the scale factor be x. Write an equation for the cube's volume V.

11. **MAKE A CONJECTURE** What are the surface area and the volume of a cube if the sides are 4 times as long as the original cube? Check your conjecture.

12-8 Similar Solids

Why?

ART In the 1800s, Edgar Degas created a wax sculpture of a young ballet dancer. The approximate height of the sculpture is 39 inches. Donise wants to purchase a model of the sculpture that is 13 inches tall.

a. Find the scale of the model.

b. If the base of the model is a rectangle that measures $4\frac{1}{2}$ inches by $4\frac{5}{8}$ inches, what would you expect the base measurements to be for the actual sculpture?

Then
You have already learned about similarity and two-dimensional figures. (Lesson 6-7)

Now
- Identify similar solids.
- Examine properties of similar solids.

New Vocabulary
similar solids

Math Online

glencoe.com
- Extra Examples
- Personal Tutor
- Self-Check Quiz
- Homework Help

Identify Similar Solids The model and the sculpture have the same shape but are different sizes. The ratio of their corresponding parts is 1 : 3. Two figures are **similar solids** if they have the same shape and their corresponding linear measures are proportional.

EXAMPLE 1 Identify Similar Solids

Determine whether the pair of solids is similar.

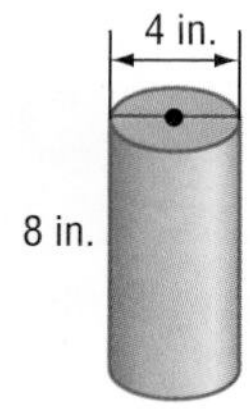

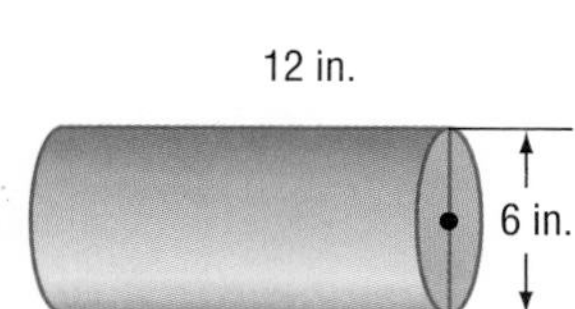

$\frac{4}{6} \stackrel{?}{=} \frac{8}{12}$ **Compare the diameters and the heights.**

$4 \cdot 12 \stackrel{?}{=} 8 \cdot 6$ **Find the cross products.**

$48 = 48$ **Simplify.**

The corresponding measurements are proportional. So, the cylinders are similar.

Check Your Progress

Determine whether each pair of solids is similar.

1A.

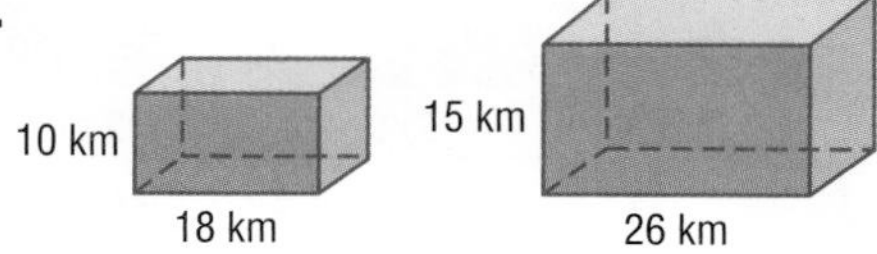

1B.

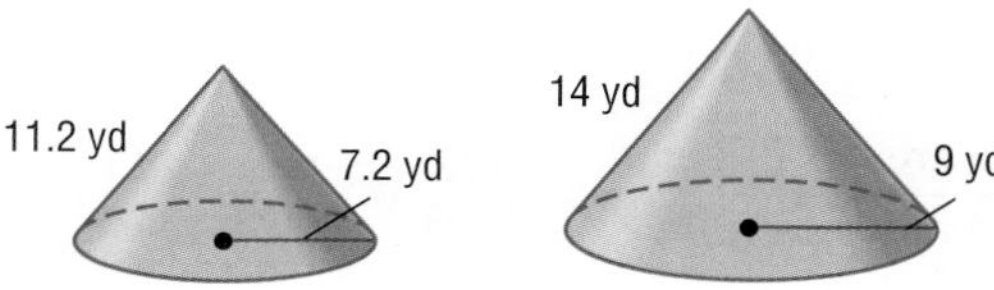

Personal Tutor glencoe.com

If you know solids are similar, you can find missing measures.

EXAMPLE 2 Find Missing Measures

The triangular pyramids at the right are similar. Find the side length of pyramid A.

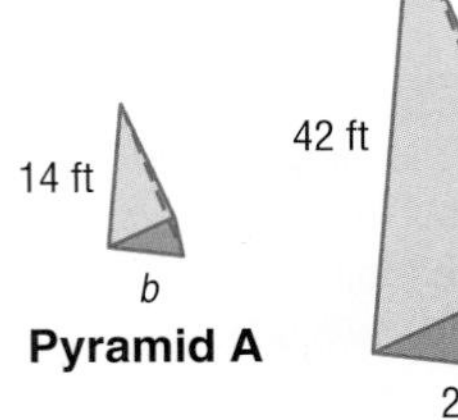

Since the two pyramids are similar, the ratios of their corresponding linear measures are proportional.

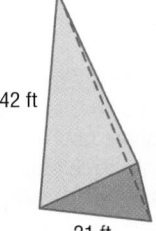

$$\frac{\text{side length of pyramid A}}{\text{side length of pyramid B}} = \frac{\text{side length of pyramid A}}{\text{side length of pyramid B}}$$

$\frac{14}{42} = \frac{b}{21}$ **Substitute the known values.**

$14 \cdot 21 = 42b$ **Find the cross products.**

$294 = 42b$ **Simplify.**

$7 = b$ **Divide each side by 42.**

The side length of pyramid A is 7 feet.

Check Your Progress

2. The triangular prisms are similar. Find the height of prism A.

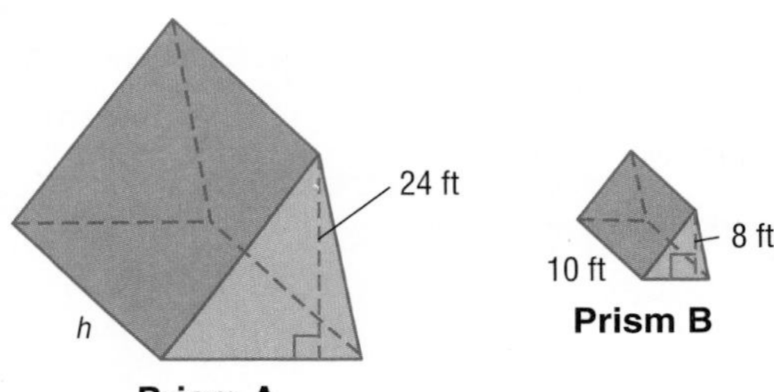

Personal Tutor glencoe.com

Properties of Similar Solids Corresponding linear measures of similar figures are proportional. Are corresponding surface areas and volumes also proportional? The two prisms below are similar with a scale factor of $\frac{1}{3}$.

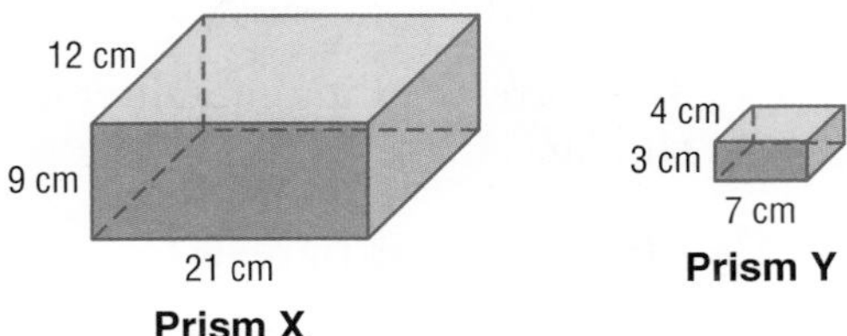

Scale Factor: $\frac{1}{3}$	Surface Area	Volume
Prism X	$(2 \cdot 21 + 2 \cdot 12)\, 9 + 2 \cdot 21 \cdot 12$ or 1098 cm^2	$21 \cdot 9 \cdot 12$ or 2268 cm^3
Prism Y	$(2 \cdot 7 + 2 \cdot 4)\, 3 + 2 \cdot 7 \cdot 4$ or 122 cm^2	$7 \cdot 3 \cdot 4$ or 84 cm^3

Notice the pattern in the following ratios.

$$\frac{\text{surface area of Prism Y}}{\text{surface area of Prism X}} = \frac{122}{1098} \text{ or } \frac{1}{9} \qquad \frac{1}{9} = \left(\frac{1}{3}\right)^2$$

$$\frac{\text{volume of Prism Y}}{\text{volume of Prism X}} = \frac{84}{2268} \text{ or } \frac{1}{27} \qquad \frac{1}{27} = \left(\frac{1}{3}\right)^3$$

StudyTip

Scale Factor and Dilation The characteristics of scale factors that applied to two-dimensional figures also apply to three-dimensional solids.

- If the scale factor > 1, the model is an enlargement.
- If the scale factor $= 1$, the figure is congruent to the original figure.
- If the scale factor < 1, the model is a reduction.

This and similar examples suggest the following relationships about the surface area and volume of similar solids.

Key Concept

Ratio of Surface Areas of Similar Solids

Words If two solids are similar with a scale factor of $\frac{a}{b}$, then the surface areas have a ratio $\left(\frac{a}{b}\right)^2$.

Symbols $\frac{\text{surface area of Solid } A}{\text{surface area of Solid } B} = \left(\frac{a}{b}\right)^2$ or $\frac{a^2}{b^2}$

Model

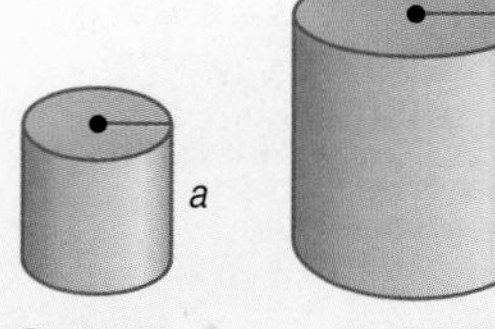

Ratio of Volumes of Similar Solids

Words If two solids are similar with a scale factor of $\frac{a}{b}$, then the volumes have a ratio $\left(\frac{a}{b}\right)^3$.

Symbols $\frac{\text{volume of Solid } A}{\text{volume of Solid } B} = \left(\frac{a}{b}\right)^3$ or $\frac{a^3}{b^3}$

EXAMPLE 3 Find Surface Areas of Similar Solids

A cube has a surface area of 600 square centimeters. If the dimensions are doubled, what is the surface area of the new cube?

Understand You know that the cubes are similar and the scale factor of the side lengths $\frac{a}{b}$ is $\frac{1}{2}$. You need to find the surface area of the new cube.

Plan Set up a proportion to find the surface area of the new cube.

Solve The surface areas of the cubes have a ratio of $\frac{a^2}{b^2}$ or $\frac{1^2}{2^2}$.

$\frac{\text{surface area of original cube}}{\text{surface area of new cube}} = \frac{a^2}{b^2}$ **Write a proportion.**

$\frac{600}{S} = \frac{1^2}{2^2}$ **Substitute the known values. Let $S =$ the surface area of the new cube.**

$\frac{600}{S} = \frac{1}{4}$ $\frac{1^2}{2^2} = \frac{1}{2} \cdot \frac{1}{2}$ **or** $\frac{1}{4}$

$600 \cdot 4 = S \cdot 1$ **Find the cross products.**

$2400 = S$ **Multiply.**

The surface area of the new cube is 2400 square centimeters.

Check If the surface area of the first cube is 600 square centimeters, the measure of each side is 600 ÷ 6 or 10 centimeters. The new cube measures 20 centimeters on each side and has a surface area of 2400 square centimeters. The answer is correct. ✓

StudyTip

Scale Factor When the lengths of all dimensions of a solid are multiplied by a scale factor x, then the surface area is multiplied by x^2 and the volume is multiplied by x^3.

Check Your Progress

3. A cone has a surface area of 160 square inches. If the dimensions are reduced by a factor of $\frac{1}{2}$, what is the surface area of the new cone?

Personal Tutor glencoe.com

Real-World Link

Spaceship Earth at Epcot Center in Orlando, Florida, is a geosphere with a diameter of 165 feet. The volume of the sphere is about 2,400,000 cubic feet.

Real-World EXAMPLE 4 Volumes of Similar Solids

BUILDINGS A model of the Spaceship Earth building has a diameter of 3 feet. Use the information at the left to find the volume of the model to the nearest tenth.

The model and the actual building are similar and the scale factor of the diameters is $\frac{3}{165}$ or $\frac{1}{55}$. The volumes of the spheres have a ratio of $\left(\frac{a}{b}\right)^3$ or $\left(\frac{1}{55}\right)^3$. Set up a proportion to find the volume of the model.

$\frac{\text{volume of model}}{\text{volume of geosphere}} = \frac{a^3}{b^3}$ — **Write the ratio of volumes.**

$\frac{m}{2{,}400{,}000} = \frac{1^3}{55^3}$ — **Replace *a* with 1, *b* with 55, and volume of geosphere with 2,400,000.**

$m \cdot 55^3 = 2{,}400{,}000 \cdot 1^3$ — **Find the cross products.**

$m \approx 14.4$ — **Divide each side by 55^3.**

The volume of the Spaceship Earth model is about 14.4 cubic feet.

Check Your Progress

4. **SPORTS** Baseballs and softballs are similar in shape. The scale between a baseball and a softball is 1 inch : 1.3 inches. The volume of a baseball is about 12.8 in^3. What is the volume of a softball to the nearest tenth?

Personal Tutor glencoe.com

Check Your Understanding

Example 1 p. 709 — **Determine whether each pair of solids is similar.**

1.

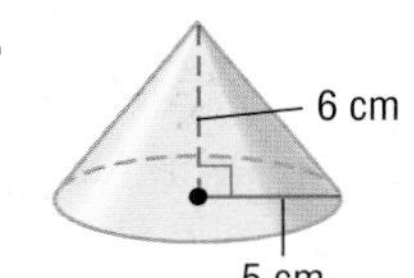

2.

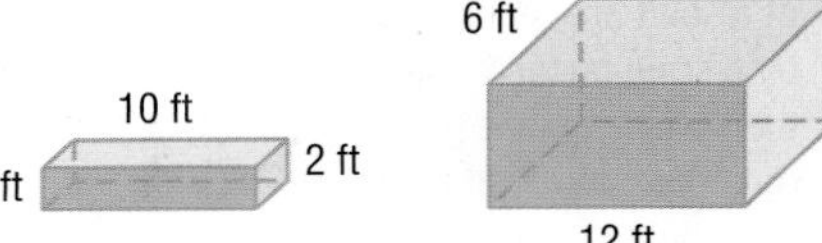

Example 2 p. 710 — **Find the missing measure for each pair of similar solids.**

3.

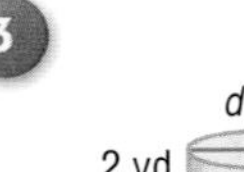

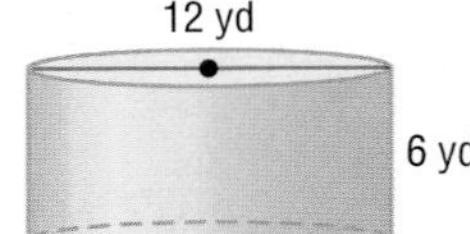

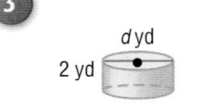

4.

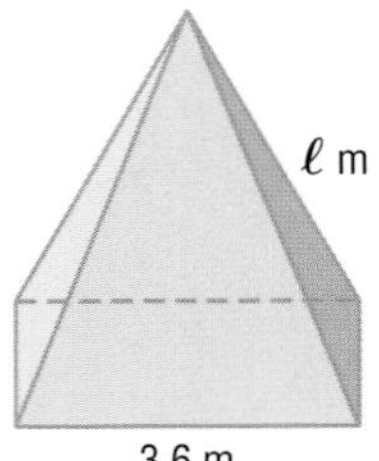

Example 3 p. 711

5. A pyramid has a surface area of 50 square feet. If the dimensions are tripled, what is the surface area of the new pyramid?

Example 4 p. 712

6. **CEREAL** A prototype for a new cereal box is 10 centimeters tall, 3 centimeters long, and 2 centimeters wide. The actual box will be 28 centimeters tall.

   a. What is the scale factor between the prototype and the actual box?

   b. What is the width of the actual box?

   c. Find the volume of the actual box.

Practice and Problem Solving

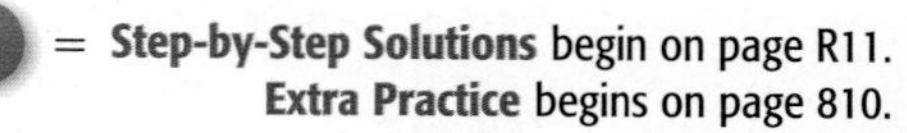

Example 1
p. 709

Determine whether each pair of solids is similar.

7. 8 m, 12 m; 12 m, 4 m

8. 2 in., 2 in., 2 in.; 7 in., 7 in., 7 in.

9.

10.

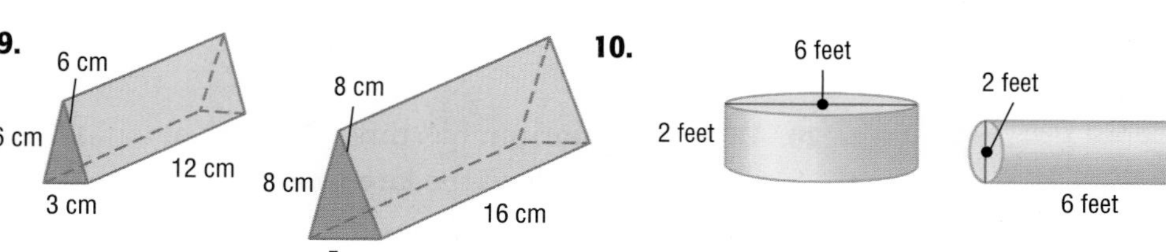

Example 2
p. 710

Find the missing measure for each pair of similar solids.

11.

12.

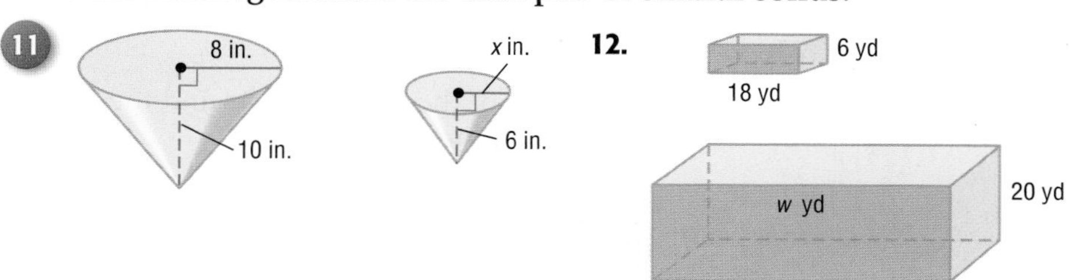

Example 3
p. 711

13. A rectangular prism has a surface area of 300 square millimeters. If the dimensions are quadrupled, what is the surface area of the new pyramid?

14. A pyramid has a surface area of 6400 square yards. If the new dimensions are $\frac{1}{8}$ the original size, what is the surface area of the new pyramid?

Example 4
p. 712

15. FOOD A wheel of cheese given to Thomas Jefferson in 1802 measured 48 inches in diameter and 17 inches in height and was similar to the cheese wheel pictured at the right.

a. What is the scale factor between the large and small cheese wheels?

b. Find the height of the smaller cheese wheel.

c. Find the volume of each wheel of cheese. Round to the nearest tenth.

d. How much greater is the volume of the large cheese wheel than the smaller cheese wheel?

16. The two cylinders are similar. What is the surface area of cylinder B if the volume of the cylinder A is 75 cubic meters? Round to the nearest tenth.

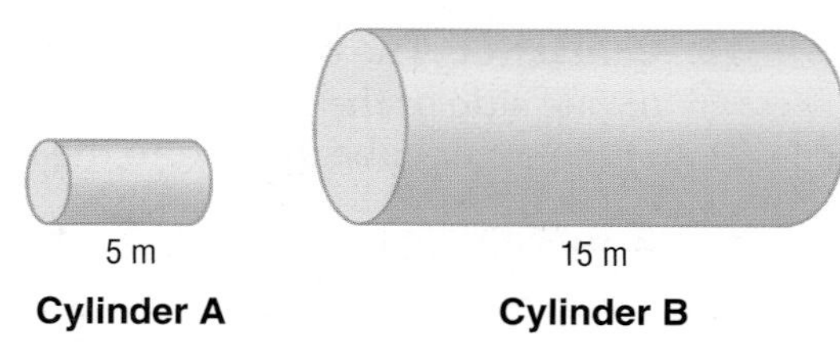

Real-World Link

A giant soccer ball pavilion greeted visitors to the 2006 Soccer World Cup semifinals in Dortmund, Germany. It housed exhibits on different aspects of soccer. The pavilion is 66 feet in diameter.

17. **SOCCER** Refer to the information at the left. The diameter of a regulation soccer ball is about 8.7 inches. If the volume of a regulation soccer ball is about 344.8 cubic inches, find the volume of the pavilion.

Find the missing measure for each pair of similar solids. Round to the nearest tenth if necessary

18.

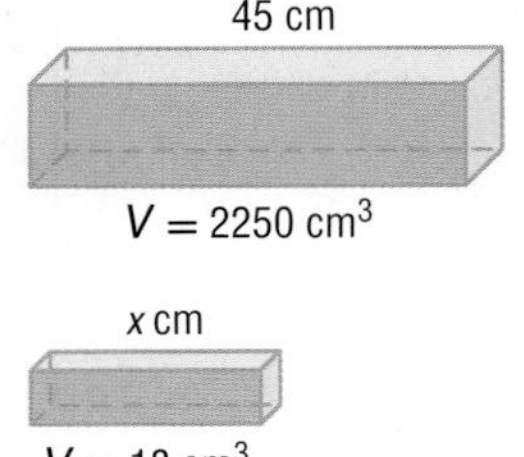

19.

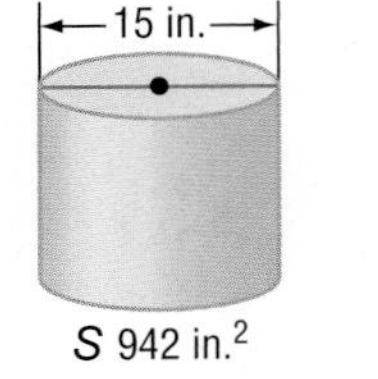

20. **FISH** An aquarium has three cylindrical tanks. The dimensions of the largest are $1\frac{1}{2}$ the size of the one shown, while the dimensions of the smallest are $\frac{3}{4}$ the size. Determine the volumes of the three aquariums. Round to the nearest tenth.

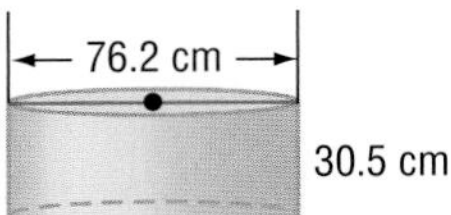

21. Use the two similar prisms at the right.

a. Write the ratio of the surface areas and the ratio of the volumes.

b. Find the surface area of prism B.

c. Find the volume of prism A.

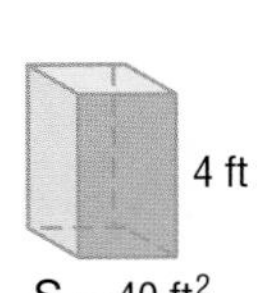

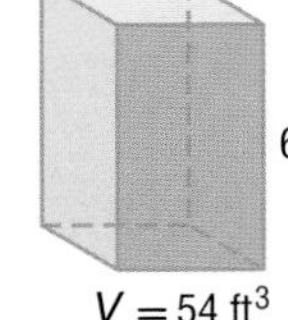

H.O.T. Problems

Use Higher-Order Thinking Skills

22. **OPEN ENDED** Find a scale for a model plane or car. Then determine the ratio of their surface areas and volumes.

23. **REASONING** True or false? *All spheres are similar.* If false, provide a counter example. Explain your reasoning.

24. **FIND THE ERROR** Fred and Cassandra are finding the ratio of the surface areas of a building given that the model of the building was built on a scale of 1 centimeter to 5 meters. Is either of them correct? Explain your reasoning.

Fred

$$\frac{a^2}{b^2} = \frac{1^2}{500^2} = \frac{1}{250{,}000}$$

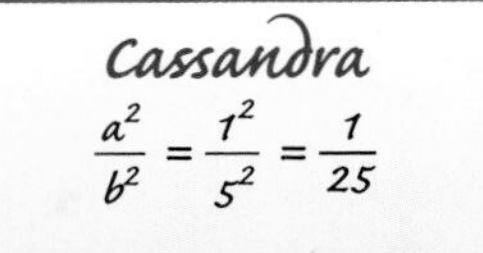

25. **NUMBER SENSE** Describe what happens to the surface area of a cone if its radius and slant height are doubled.

26. **CHALLENGE** The ratio of the surface areas of two cubes is 1:12. The length of one side of the smaller cube is 7 centimeters. Find the length of the side of the other cube. Round to the nearest tenth, if necessary.

27. **WRITING IN MATH** Write a real-world problem about similar solids that involves the purchase of a souvenir of a building. Then solve.

Standardized Test Practice

28. The two prisms below are similar. What is the ratio of the volume of the smaller prism to the larger prism?

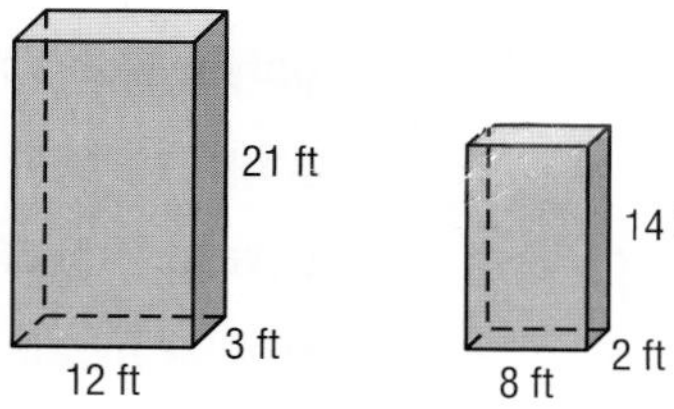

A 8 : 27
B 4 : 9
C 2 : 3
D 1 : 2

29. Which prism in the table is *not* similar to the other three?

Prism	Length	Width	Height
A	7.5	6	5
B	10.5	9.8	7
C	15	12	10
D	4.5	3.6	3

F A
G B
H C
J D

30. Find the missing measure in the two similar solids.

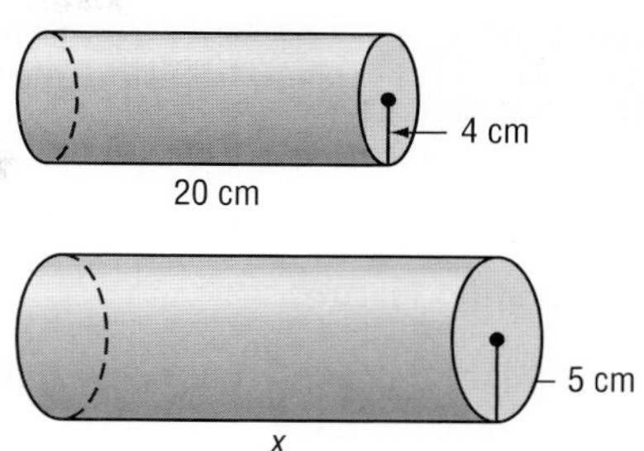

A 16 cm
B 25 cm
C 30 cm
D 35 cm

31. GRIDDED RESPONSE A prism has a surface area of 220 square feet. If the dimensions of a second prism are $\frac{1}{4}$ the original prism, what is the surface area in square feet of the second prism?

Spiral Review

Find the lateral and surface area of each figure. Round to the nearest tenth. (Lessons 12-6 and 12-7)

32.

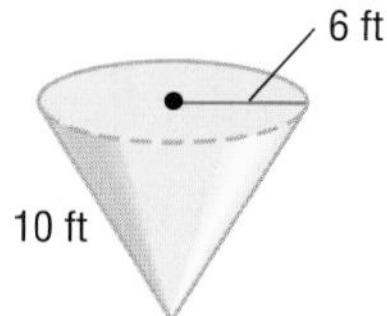

33.

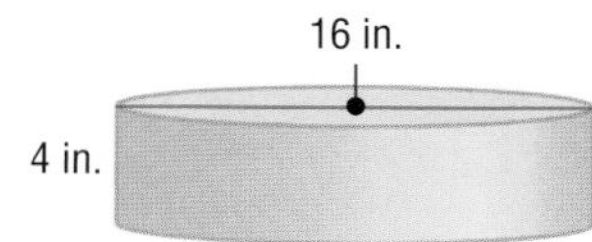

34. SCIENCE A standard funnel is shaped like a cone, and a buchner funnel is shaped like a cone with a cylinder attached to the base. Which funnel has the greater volume? (Lesson 12–4)

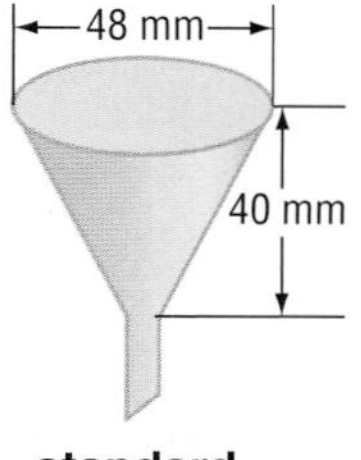

35. GARDENS Marty is designing two flower beds shaped like equilateral triangles. The lengths of each side of the flower beds are 8 feet and 20 feet, respectively. What is the ratio of the area of the larger flower bed to the smaller flower bed? (Lesson 10-6)

Skills Review

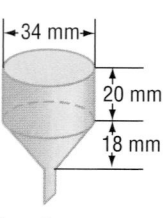

Evaluate each expression. (Lesson 1-1)

36. $\frac{21 + 43 + 59}{3}$

37. $\frac{11(96) + 219 + 10(15)}{3}$

38. $\frac{478 - 136 + 12 - 18}{4}$

CHAPTER 12

Study Guide and Review

Math Online glencoe.com

- STUDY TO GO
- Vocabulary Review

Chapter Summary

Key Concepts

Three-Dimensional Figures (Lesson 12-1)

- Prisms, pyramids, cylinders, cones, and spheres are three-dimensional figures.
- Prisms and pyramids are polyhedrons and are named by the shape of their bases.
- Cylinders, cones, and spheres are not polyhedrons.

Volume (Lessons 12-2 through 12-4)

- rectangular prism: $V = Bh$ or lwh
- cylinder: $V = \pi r^2 h$
- pyramid: $V = \frac{1}{3}Bh$
- cone: $V = \frac{1}{3}\pi r^2 h$
- sphere: $V = \frac{4}{3}\pi r^3$

Surface Area (Lessons 12-5 through 12-7)

- The surface area of a three-dimensional figure is the sum of the lateral area plus the area of the base(s).
- rectangular prism: $S = Ph + 2B$
- cylinder: $S = 2\pi r^2 + 2\pi rh$
- pyramid: $S = L + B$
- cone: $S = \pi rl + \pi r^2$

Similar Solids (Lesson 12-8)

- Solids are similar if they have the same shape and their corresponding linear measures are proportional.

FOLDABLES® Study Organizer

Be sure the Key Concepts are noted in your Foldable.

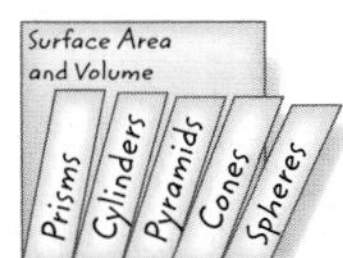

Key Vocabulary

base (p. 665)
cone (p. 665)
cross section (p. 666)
cylinder (p. 665)
edge (p. 664)
face (p. 664)
lateral area (p. 691)
lateral face (p. 691)
net (p. 690)
perspective (p. 663)
plane (p. 664)
polyhedron (p. 664)
prism (p. 665)
pyramid (p. 665)
regular pyramid (p. 702)
similar solids (p. 709)
slant height (p. 702)
solid (p. 664)
sphere (p. 684)
surface area (p. 691)
vertex (p. 664)
volume (p. 671)

Vocabulary Check

State whether each sentence is *true* or *false*. If *false*, replace the underlined term to make a true sentence.

1. A three-dimensional figure with two bases that are parallel circles is called a cylinder.
2. The volume of a prism is the sum of the areas of its lateral faces.
3. Figures that have the same shape and corresponding linear measures that are proportional are called similar solids.
4. An edge is where two planes intersect in a line.
5. A cone is a polyhedron with one base that is any polygon.
6. The intersection of a solid and a plane is called a cross section of the solid.
7. A cone has two bases.
8. The surface area of a rectangular prism is the sum of the areas of its faces.

Lesson-by-Lesson Review

12-1 Three-Dimensional Figures (pp. 664–669)

Identify each figure. Name the bases, faces, edges, and vertices.

9.

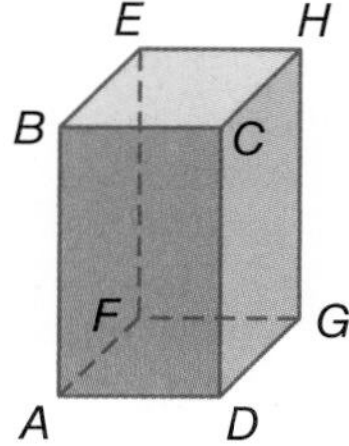

10.

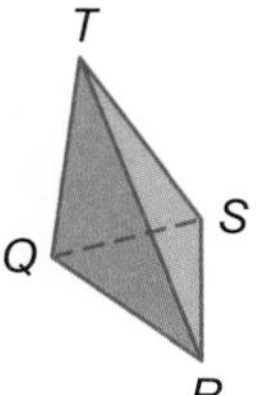

11.

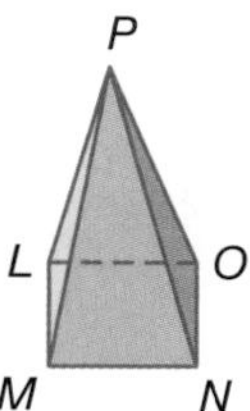

12.

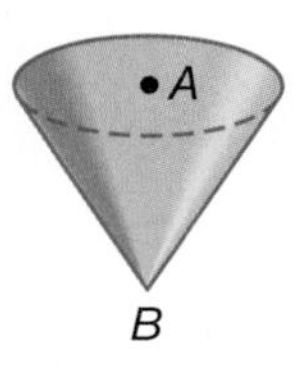

13. INSTRUMENTS Draw the top view and side view of the drum. Then draw and describe the shape resulting from a vertical cross section of the figure.

EXAMPLE 1

Identify the figure. Name the bases, faces, edges, and vertices.

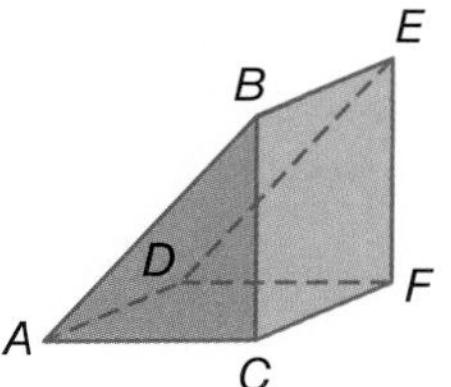

There are two triangular bases, so the solid is a triangular prism.

bases: *ABC, DEF*

faces: *ABED, BCFE, ACFD, ABC, DEF*

edges: $\overline{AB}, \overline{BC}, \overline{AC}, \overline{DE}, \overline{EF}, \overline{DF}, \overline{AD}, \overline{BE}, \overline{CF}$

vertices: *A, B, C, D, E, F*

12-2 Volume of Prisms (pp. 671–676)

Find the volume of each prism.

14.

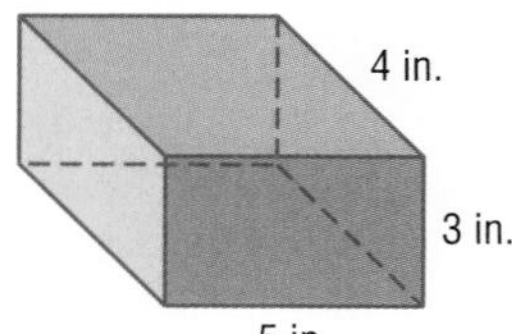

15.

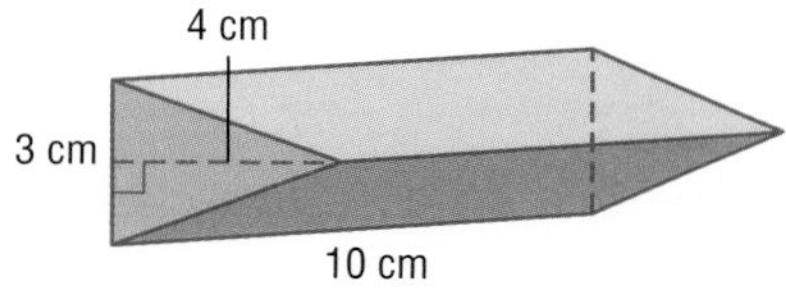

16. BOXES A shipping box is 11 inches long, 8.5 inches wide, and 5.5 inches high. What is the volume of the box?

EXAMPLE 2

Find the volume of the rectangular prism.

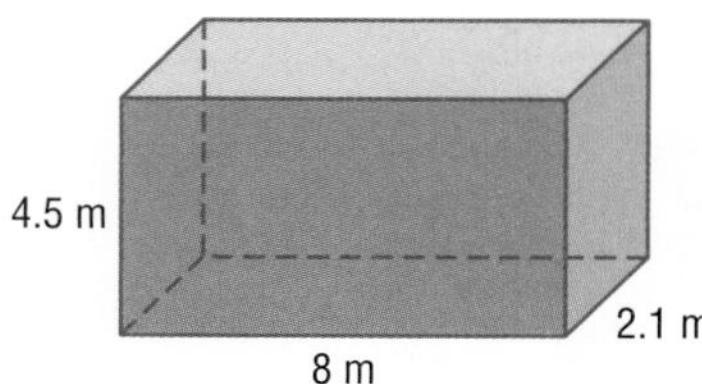

$V = \ell wh$	**Volume of a prism**
$V = 8 \cdot 2.1 \cdot 4.5$	**Replace ℓ with 8, w with 2.1, and h with 4.5.**
$V = 75.6 \text{ m}^3$	**Simplify.**

The volume of the prism is 75.6 cubic meters.

CHAPTER 12 Study Guide and Review

12-3 Volume of Cylinders (pp. 677–681)

Find the volume of each cylinder. Round to the nearest tenth, if necessary.

17.

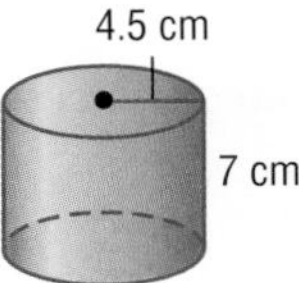

18.

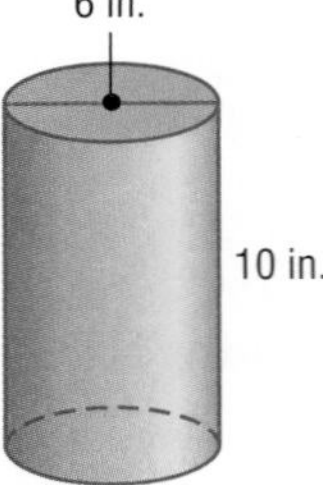

19. BEVERAGES A 12-ounce can of soda measures $4\frac{3}{4}$ inches high with a radius of $1\frac{1}{8}$ inches. Find the amount of soda that can fit in the can. Round to the nearest tenth.

EXAMPLE 3

Find the volume of the cylinder. Round to the nearest tenth, if necessary.

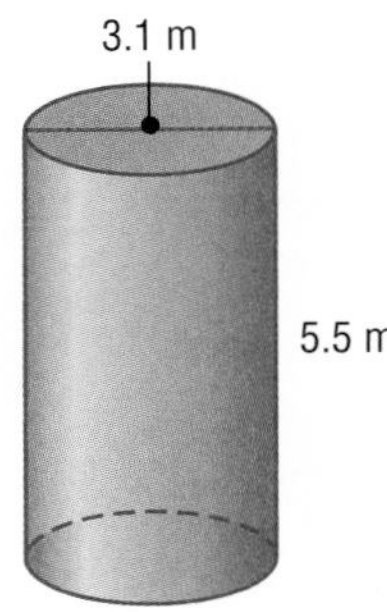

$V = \pi r^2 h$	**Volume of a cylinder**
$V = \pi(1.55)^2(5.5)$	**Replace *r* with 1.55 and *h* with 5.5.**
$V \approx 41.5 \text{ m}^3$	**Simplify.**

12-4 Volume of Pyramids, Cones, and Spheres (pp. 683–688)

Find the volume of each figure. Round to the nearest tenth, if necessary.

20.

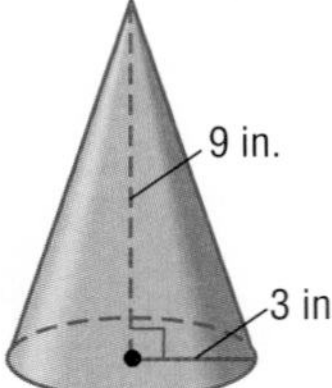

21.

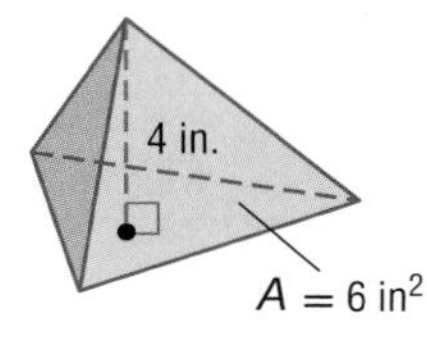

22.

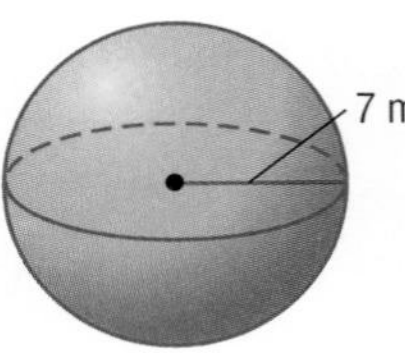

23.

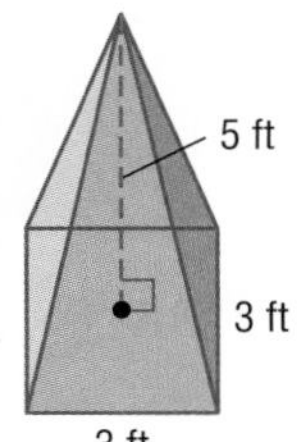

24. STORAGE Mr. Owens built a conical storage shed with a base 14 feet in diameter and a height of 11 feet. What is the volume of the shed?

EXAMPLE 4

Find the volume of the cone. Round to the nearest tenth, if necessary.

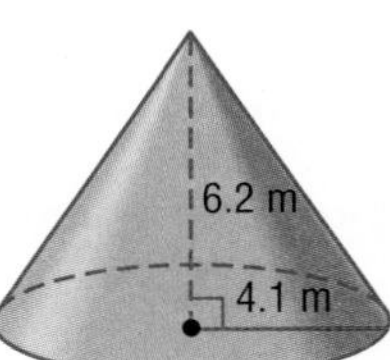

$V = \frac{1}{3}\pi r^2 h$	**Volume of a cone**
$= \frac{1}{3}\pi(4.1)^2(6.2)$	**Replace *r* with 4.1 and *h* with 6.2.**
$\approx 109.1 \text{ m}^3$	**Simplify.**

MIXED PROBLEM SOLVING
For mixed problem-solving practice, see page 843.

12-5 Surface Area of Prisms (pp. 691–695)

Find the lateral and surface area of each prism. Round to the nearest tenth, if necessary.

25.

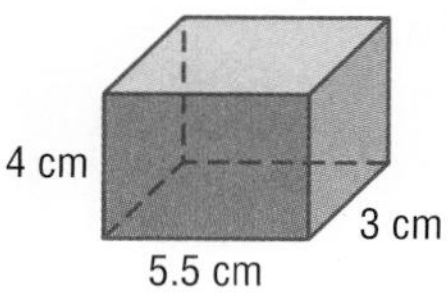

26.

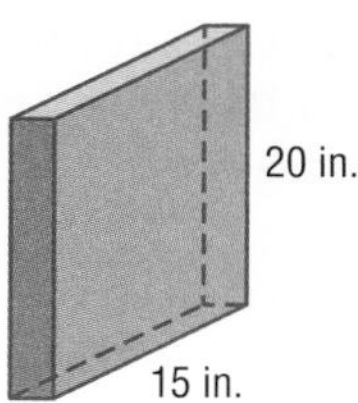

27.

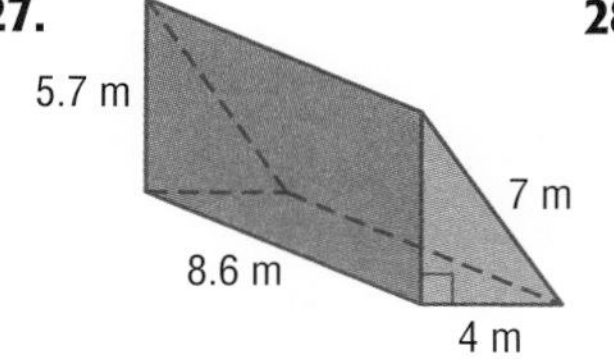

28.

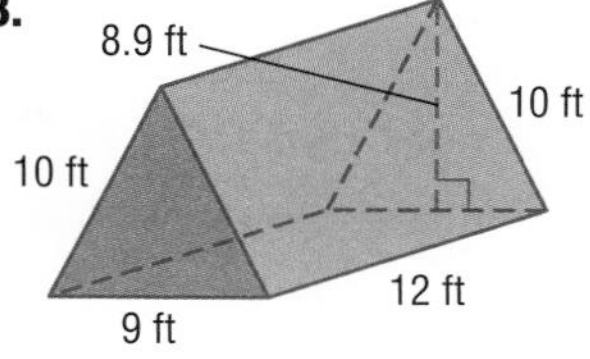

29. GIFTS Sarah is wrapping a gift that is 12 inches long, 6 inches wide, and 4 inches high. How many square inches of paper are needed to cover the gift?

EXAMPLE 5

Find the lateral and surface area of the prism.

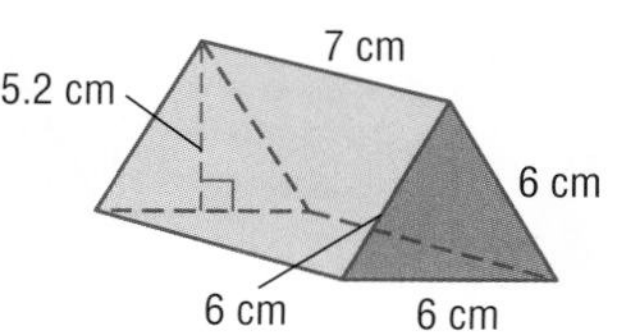

$L = Ph$	**Lateral area of a prism**
$= (6 + 6 + 6)7$	**$P =$ the perimeter of the base.**
$= 126 \text{ cm}^2$	**Simplify.**
$S = L + 2B$	**Surface area of a prism**
$= 126 + 2\left(\frac{1}{2} \cdot 6 \cdot 5.2\right)$	**$B = \frac{1}{2}bh$**
$= 157.2 \text{ cm}^2$	**Simplify.**

12-6 Surface Area of Cylinders (pp. 697–701)

Find the lateral and surface area of each cylinder. Round to the nearest tenth, if necessary.

30.

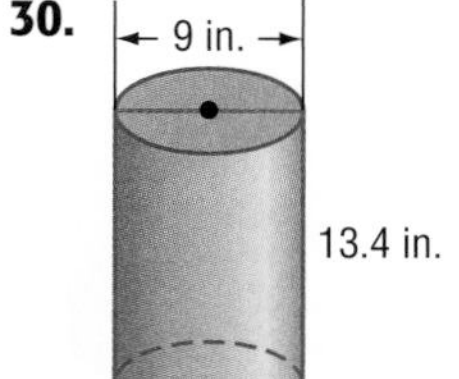

31.

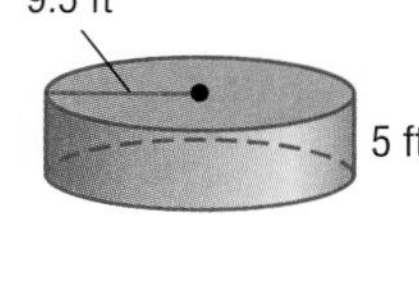

32.

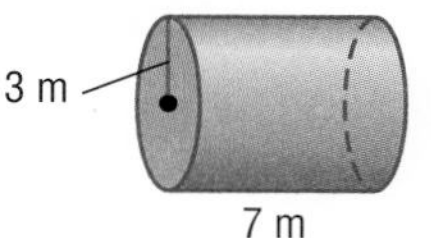

33.

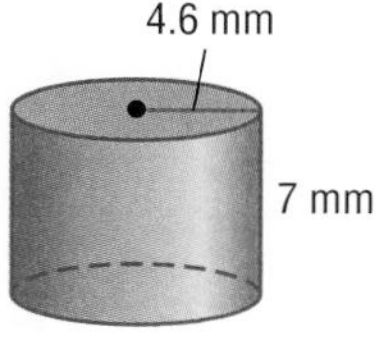

34. TELEVISION Coaxial cable is used to transmit cable television programming. The cable is covered by rubber sheathing. A typical coaxial cable has a diameter of 3 inches. How much rubber sheathing is there in 100 feet of cable?

EXAMPLE 6

Find the lateral and surface areas of the cylinder below. Round to the nearest tenth, if necessary.

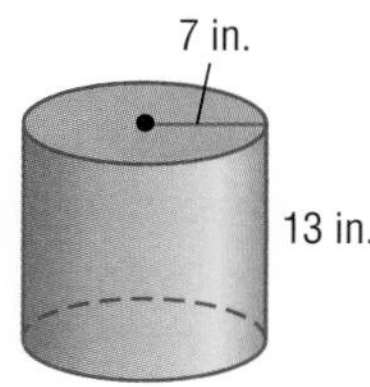

Lateral Area

$L = 2\pi rh$	**Lateral area of a cylinder**
$= 2 \cdot \pi \cdot 7 \cdot 13$	**Replace r with 7 and h with 13.**
$\approx 571.8 \text{ in}^2$	**Simplify.**

Surface Area

$S = L + 2\pi r^2$	**Surface area of a cylinder**
$= 571.8 + 2\pi(7)^2$	**Replace L with 571.8 and r with 7.**
$\approx 879.7 \text{ m}^2$	**Simplify.**

Study Guide and Review

12-7 Surface Area of Pyramids and Cones (pp. 702–707)

Find the lateral and surface areas of each figure. Round to the nearest tenth, if necessary.

35.

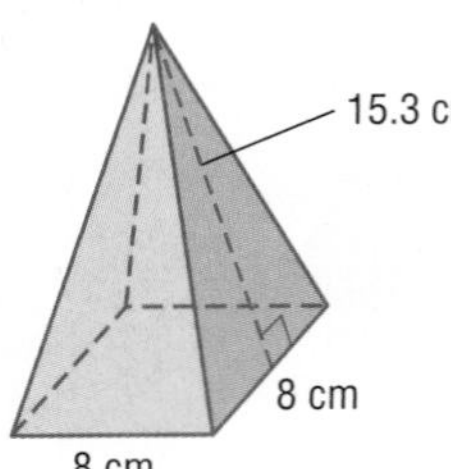

36.

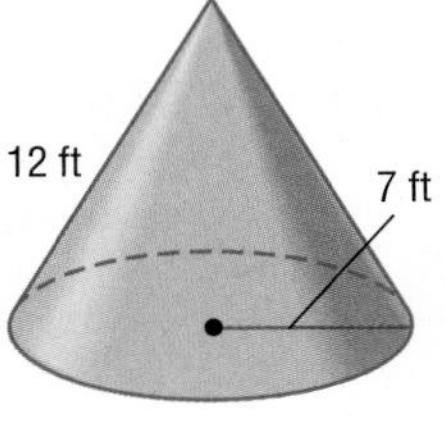

37.

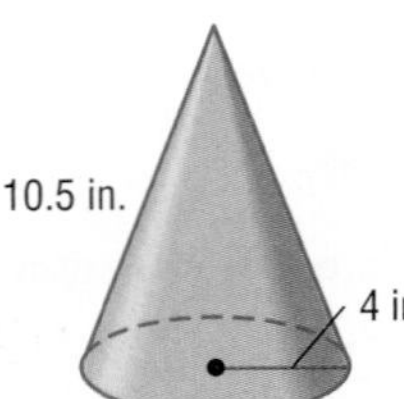

38.

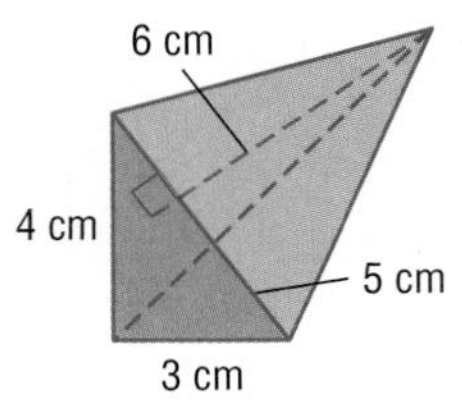

39. **ROOFS** A pyramid-shaped roof has a slant height of 18 feet and its square base is 55 feet wide. How many square feet of roofing material is needed to cover the roof?

EXAMPLE 7

Find the lateral and surface area of the square pyramid. Round to the nearest tenth, if necessary.

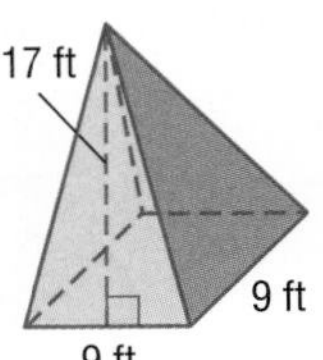

Lateral Area

$L = \frac{1}{2}P\ell$	**Lateral area of a pyramid**
$= \frac{1}{2}(4 \cdot 9)(17)$	**Substitute.**
$= 306 \text{ ft}^2$	**Simplify.**

Surface Area

$S = L + B$	**Surface area of a pyramid**
$= 306 + 9^2$	**Substitute.**
$= 387 \text{ ft}^2$	**Simplify.**

12-8 Similar Solids (pp. 709–715)

Find the missing measure for each pair of similar solids.

40.

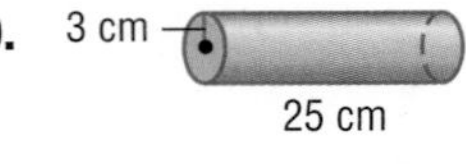

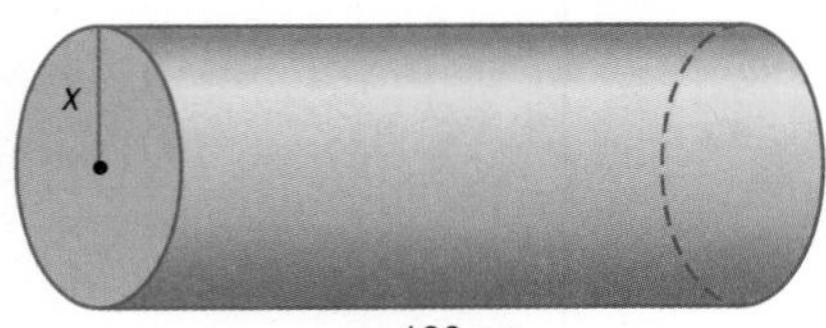

41.

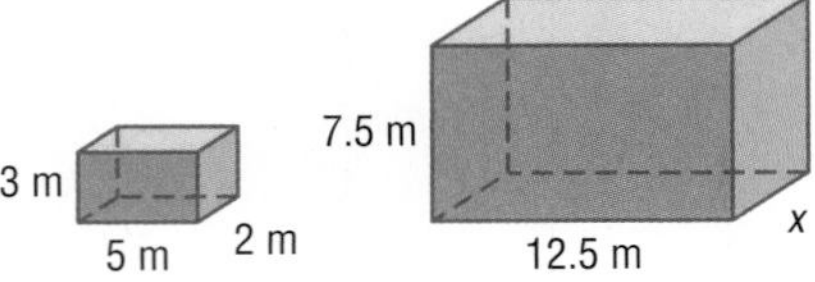

42. A prism has a surface area of 160 square feet. If the dimensions are reduced by a factor of $\frac{1}{4}$, what is the surface area of the new prism?

EXAMPLE 8

Find the missing measure for the pair of similar solids.

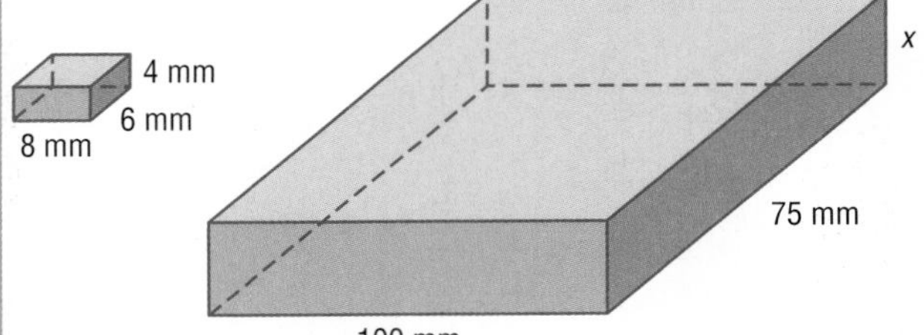

$\frac{4}{x} = \frac{8}{100}$	**Write the proportion.**
$4 \cdot 100 = 8x$	**Find the cross products.**
$400 = 8x$	**Simplify.**
$\frac{400}{8} = \frac{8x}{8}$	**Divide each side by 8.**
$50 = x$	**Simplify.**

CHAPTER 12 Practice Test

Math Online glencoe.com
Chapter Test

1. What is the shape resulting from a horizontal cross section of a cylinder?

Find the volume of each solid. Round to the nearest tenth, if necessary.

2. cube: length 2.5 in.
3. triangular prism: base of triangle 12 cm, altitude of triangle 8 cm, height of prism 4 cm
4. cone: radius 6 m, height 8 m
5. cylinder: radius 6.5 ft, height 9.6 ft

6. **MULTIPLE CHOICE** A rectangular prism has a volume of 32 cubic centimeters. If the length is 2 centimeters and the width is 2 centimeters, what is the height of the prism?

   A 2 cm C 8 cm

   B 4 cm D 18 cm

7. **BREAKFAST** The inside of the cereal bowl is in the shape of a hemisphere (half of a sphere). Find the maximum amount of milk that can fit in the bowl. Round to the nearest hundredth.

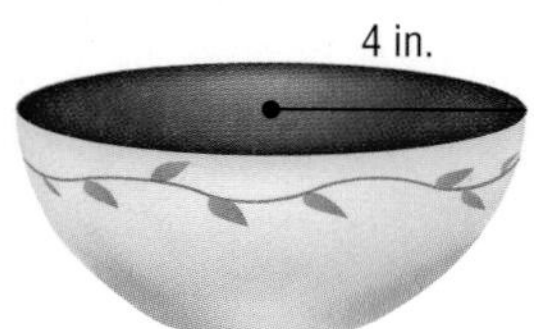

8. **COOKING** Sara is pouring 10 cartons of chicken broth into a large pot that holds 250 cubic inches. If each carton is 6 inches tall, 4 inches wide, and 1 inch thick, will all 10 cartons of broth fit in the pot? Explain.

9. **FURNITURE** Find the surface area of the ottoman that will be reupholstered, not including the bottom. Round to the nearest tenth.

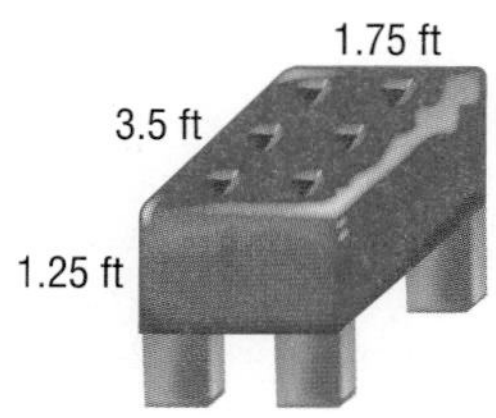

Find the lateral and surface area of each solid. Round to the nearest tenth, if necessary.

10. 11.

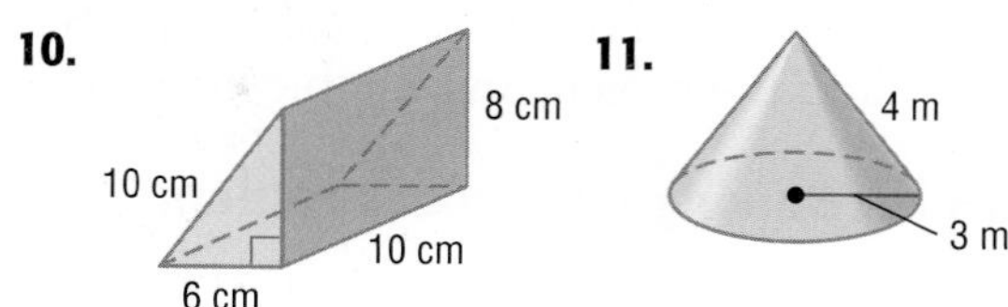

12. 13.

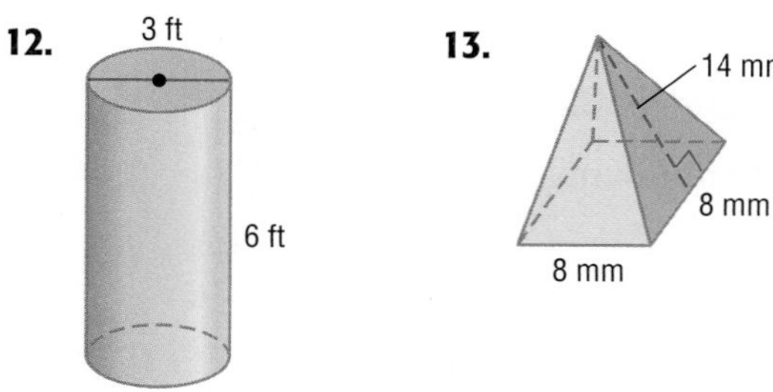

14. **MULTIPLE CHOICE** Chen is creating a model of a building. The building is in the shape of a pyramid. The model has a square base with sides of 10 inches and a slant height of 12 inches. What is the surface area of his model?

   F 100 in^2 H 580 in^2

   G 340 in^2 J 1200 in^2

15. A cube has a volume of 8 cubic inches. If the dimensions are doubled, what is the volume of the new cube?

16. Are the rectangular prisms described in the table similar? Explain your reasoning.

Prism	Length (m)	Width (m)	Height (m)
A	2	5	7
B	3	7.5	10.5

17. The model of Duane's room is 6 inches long, 5 inches wide, and 4 inches high. The scale is 2 feet to 1 inch. Find the length of his actual room.

18. **CAMPING** The sides and the floor of a pyramid tent are made of a water resistant fabric. The slant height of the tent is about 14 feet. Its square base is 15 feet on each side. How much fabric was used in making the tent?

CHAPTER 12 Preparing for Standardized Tests

Solve a Simpler Problem

There is often more than one way to solve a problem. Two different methods can both result in the correct answer, but one method may be more efficient.

Strategies for Solving a Simpler Problem

Step 1

Read the problem statement carefully.

Ask yourself:

- What information am I given?
- What am I being asked to solve?
- If the problem is a multi-step problem, can it be broken up into smaller, more manageable parts?

Step 2

Think of the most efficient way to solve the problem.

- If a calculator is allowed, it may speed up the process.
- Are there any shortcuts you can use to solve the problem?

Step 3

Solve the problem. Check your answer.

EXAMPLE

Read the problem. Identify what you need to know. Then use the information in the problem to solve.

Angela is using 2-centimeter blocks to build steps in a model. What is the total volume of the steps shown at the right?

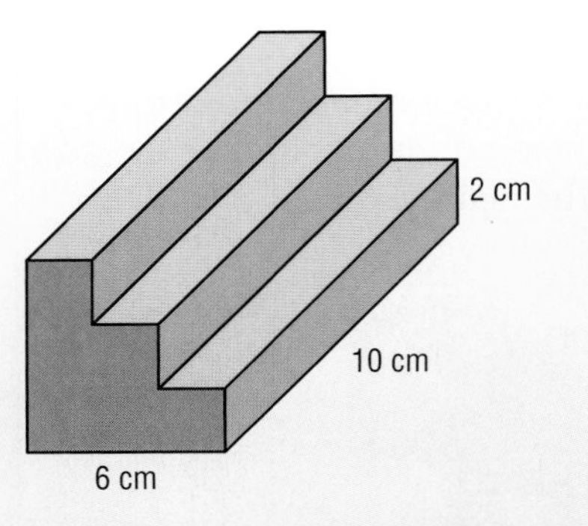

A 120 cm^3

B 180 cm^3

C 220 cm^3

D 240 cm^3

Read the problem carefully. It looks like a challenging volume problem. If you recognize that the steps are formed by stacking 6 square prisms on top of each other, it becomes much simpler to solve.

Divide the figure into 6 square prisms by drawing dashed lines. Then find the volume of one prism. Multiply by 6 to find the total volume.

Volume of one prism:

$V = 2 \times 2 \times 10$

$V = 40 \text{ cm}^3$

Total volume of steps:

$V = 6 \times 40$

$V = 240 \text{ cm}^3$

So, the total volume of the steps is 240 cubic centimeters. The correct answer is D.

Exercises

Read each problem. Identify what you need to know. Then use the information in the problem to solve.

1. What is the volume of the figure when a 10-millimeter diameter cylinder is removed?

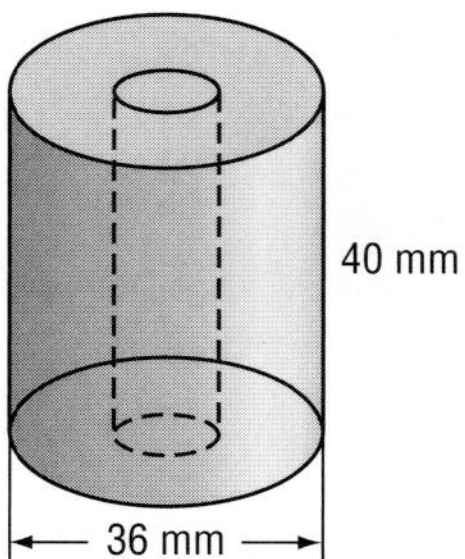

A $26{,}712.8 \text{ mm}^3$

B $37{,}573.4 \text{ mm}^3$

C $43{,}661.2 \text{ mm}^3$

D $48{,}985.3 \text{ mm}^3$

2. Neil and his father are building a tool shed against the side of a barn in the shape of the model shown below. What is the total volume of the shed?

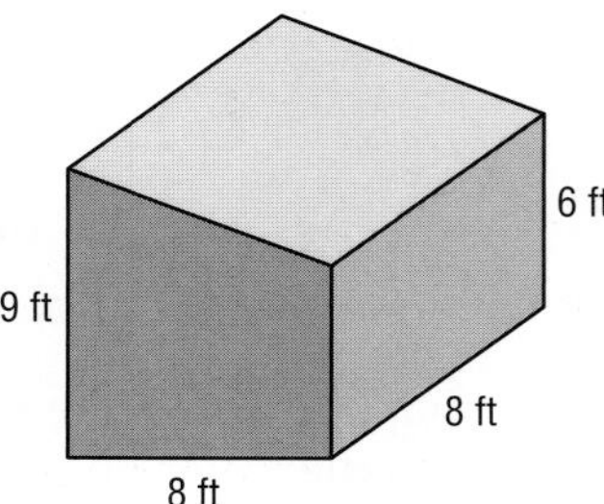

A 425 ft^3

B 464 ft^3

C 480 ft^3

D 512 ft^3

CHAPTER 12 Standardized Test Practice

Cumulative, Chapters 1 through 12

Multiple Choice

Read each question. Then fill in the correct answer on the answer document provided by your teacher or on a sheet of paper.

1. Classify the triangle below by its angles.

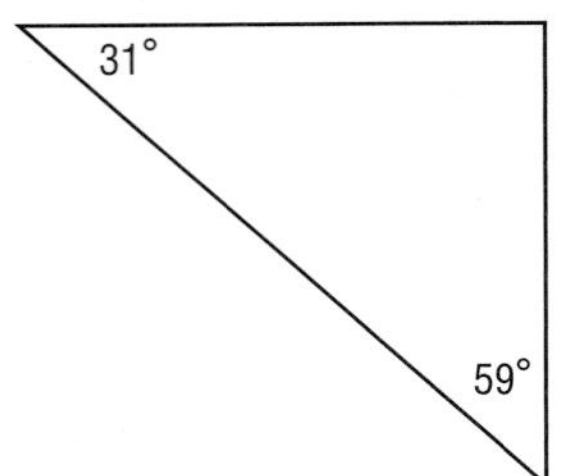

A acute

B obtuse

C right

D scalene

2. A water garden hose is shaped like a cylinder with a diameter of 1.25 inches and a length of 22 feet. What is the volume of the hose? Round your answer to the nearest tenth.

F 300 in^3

G 318 in^3

H 324 in^3

J 350 in^3

3. What is the volume of a hexagonal pyramid that has a base area of 42 square meters and a height of 5.4 meters?

A 75.6 m^3

B 78.4 m^3

C 226.8 m^3

D 234.9 m^3

4. A paper cup is shaped like a cone with a height of 12 centimeters, a diameter of 10 centimeters, and a slant height of 13 centimeters. How much paper is needed to make each cup?

F 162.2 cm^2

G 188.5 cm^2

H 204.2 cm^2

J 211.9 cm^2

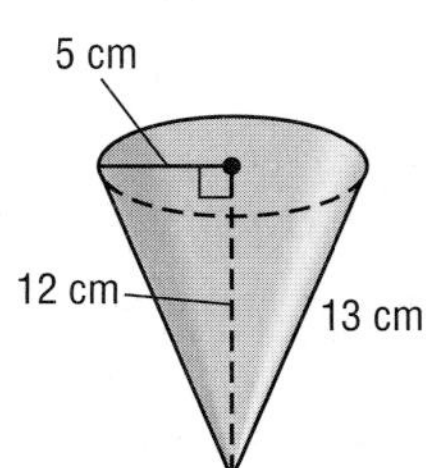

5. What is the constant rate of change shown in the graph below?

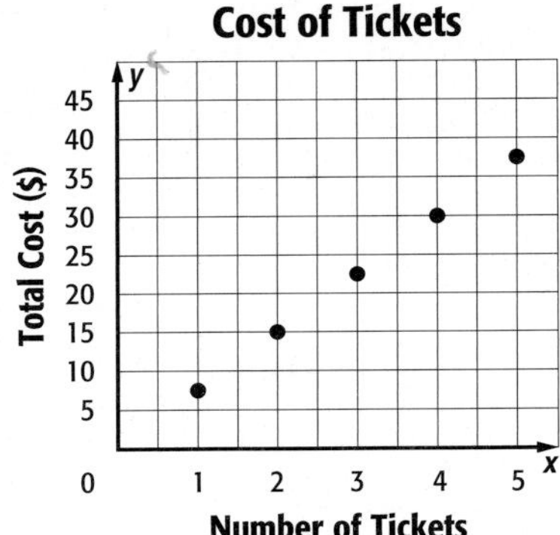

A $6.75 per ticket

B $7.50 per ticket

C $8.00 per ticket

D 7.5 tickets per dollar

6. A circular ice-skating rink has a radius of 65 feet. What is the circumference of the rink, rounded to the nearest tenth?

F 204.2 ft

G 261.3 ft

H 408.4 ft

J 522.6 ft

7. How much wrapping paper would Nicole need to completely cover the gift box shown below?

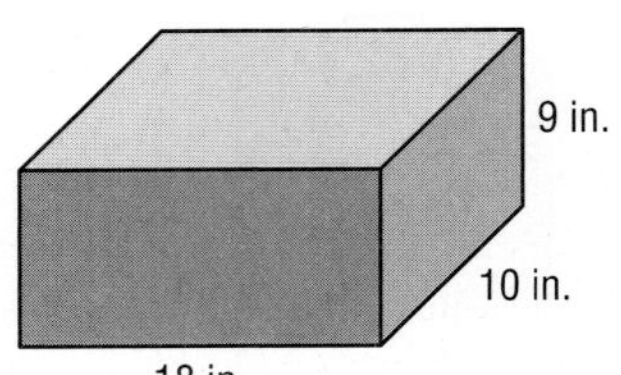

A 1620 in^2

B 1240 in^2

C 978 in^2

D 864 in^2

Test-TakingTip

Question 7 Remember to read each test item carefully. In this question you will need to find the surface area, not the volume of the prism.

Short Response/Gridded Response

Record your answers on the answer sheet provided by your teacher or on a sheet of paper.

8. **GRIDDED RESPONSE** The two triangular prisms below are similar solids. What is the value of x?

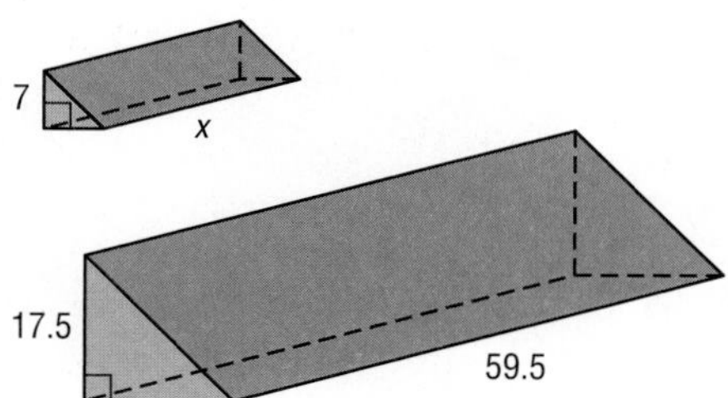

9. A plumber has two drainpipes shaped like cylinders. The first pipe is 8 inches long and has a diameter of 1.25 inches. The second pipe is 10 inches long and has a diameter of 1 inch. Which pipe has the greater outer surface area?

10. Find the volume of the rectangular prism shown below.

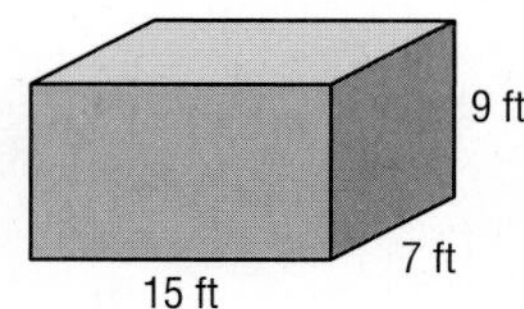

11. **GRIDDED RESPONSE** Terrance placed a 12-ft ladder against the side of his house to paint the shutters. The base of the ladder is 4 feet from the house. How high up the side of the house does the ladder reach? Round to the nearest tenth of a foot.

12. **GRIDDED RESPONSE** Estimate $\sqrt{125}$ to the nearest integer without using a calculator.

Extended Response

Record your answers on a sheet of paper. Show your work.

13. Use the rectangular prism below to answer each question.

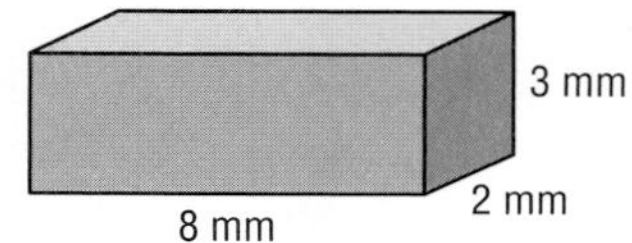

   a. What is the volume of the rectangular prism?

   b. What is the surface area of the rectangular prism?

   c. Double the dimensions of the prism. What is the new volume?

   d. What is the new surface area of the rectangular prism when the dimensions are doubled?

   e. What effect does doubling the dimensions have on the volume and surface area of a prism?

Need Extra Help?

If you missed Question...	1	2	3	4	5	6	7	8	9	10	11	12	13
Go to Lesson or Page...	10-3	12-3	12-4	12-7	8-5	11-7	12-5	12-8	12-6	12-2	10-4	10-1	12-2, 12-5

CHAPTER 13

Statistics and Probability

Then

In Chapter 2, you found the mean, or average, of a data set.

In Chapter 6, you have displayed data in circle graphs.

Now

In Chapter 13, you will:

- Find and use measures of central tendency and measures of variation.
- Display and interpret data.
- Find probabilities of simple events and independent and dependent events.

Why?

GAMES Statistics and statistical displays are frequently used to describe the results of the Olympic Games. In the 2006 Winter Olympics, the U.S. won 9 gold medals, 9 silver medals, and 7 bronze medals.

Statistics and Probabilities

Activity

Find the mean value

Add up the number of silver medals won by the top five countries.

12 + 9 + 7 + 6 + 10 = ☐

Check Answer

Data from 2006 Olympics

Rank	Nations	Gold	Silver	Bronze	Total
1	Germany	11	12	6	29
2	United State	9	9	7	25
3	Austria	9	7	7	23
4	Russia	8	6	8	22
5	Canada	7	10	7	24
6	Sweden	7	2	5	14
7	South Korea	6	3	2	11
8	Switzerland	5	4	5	14
9	Italy	5	0	6	11
10	France	3	2	4	9
11	Netherlands	3	2	4	9

3/7

Math *in Motion*, Animation glencoe.com

Get Ready for Chapter 13

Diagnose Readiness You have two options for checking Prerequisite Skills.

Text Option

Take the Quick Check below. Refer to the Quick Review for help.

QuickCheck

Find the mean for each data set. (Lesson 2-5)

1. \$8, \$22, \$16
2. 12.5 cm, 13 cm, 15.9 cm, 9.6 cm, 17.5 cm
3. **AQUARIUM** The table shows the number of visitors to the aquarium each month. Find the mean number of visitors.

Visitors to the Aquarium (thousands)			
3	11	5	4
5	3	6	3
12	2	2	4

Simplify. (Previous Course)

4. $\frac{6 \cdot 5}{5 \cdot 7}$
5. $\frac{9 \cdot 4}{2 \cdot 3}$
6. $\frac{7 \cdot 10 \cdot 9}{18 \cdot 15 \cdot 1}$
7. $\frac{8 \cdot 9 \cdot 10}{3 \cdot 4 \cdot 5}$
8. **SURVEYS** Seven-eighths of students surveyed said they drink one glass of milk per day. Three-fourths of these drink a glass of milk with dinner. What fraction of students drink their glass of milk with dinner?

Find each sum, difference, or product. (Lessons 3-3 and 3-6)

9. $\frac{2}{3} \times \frac{9}{10} \times \frac{4}{5}$
10. $\frac{7}{8} \times \frac{2}{3} \times \frac{9}{14}$
11. $\frac{9}{10} + \frac{2}{5} - \frac{3}{10}$
12. $\frac{1}{3} + \frac{1}{6} - \frac{1}{9}$
13. **GARDENING** Hannah finished filling a $\frac{7}{8}$ gallon watering can by pouring $\frac{1}{4}$ of a gallon of water into the can. How much water was already in the can?

QuickReview

EXAMPLE 1

Find the mean temperature of 65°, 70°, 75°, 70°, and 65°.

$$\frac{65 + 70 + 75 + 70 + 65}{5} = \frac{345}{5}$$

$$= 69 \quad \textbf{Simplify.}$$

The average temperature is 69°.

EXAMPLE 2

Simplify $\frac{9 \cdot 8 \cdot 7}{3 \cdot 2 \cdot 1}$.

$$\frac{9 \cdot 8 \cdot 7}{3 \cdot 2 \cdot 1} = \frac{\overset{3}{\cancel{9}} \cdot \overset{4}{\cancel{8}} \cdot 7}{\underset{1}{\cancel{3}} \cdot \underset{1}{\cancel{2}} \cdot 1}$$ **Divide out common factors.**

$$= 3 \cdot 4 \cdot 7 \text{ or } 84$$ **Simplify.**

EXAMPLE 3

Find $\frac{7}{12} + \frac{5}{8} - \frac{1}{12}$.

$$\frac{7}{12} + \frac{5}{8} - \frac{1}{12}$$

$$= \left(\frac{7}{12} \cdot \frac{2}{2}\right) + \left(\frac{5}{8} \cdot \frac{3}{3}\right) - \left(\frac{1}{12} \cdot \frac{2}{2}\right)$$ **Rename using the LCD, 24.**

$$= \frac{14}{24} + \frac{15}{24} - \frac{2}{24}$$ **Simplify.**

$$= \frac{27}{24}$$ **Simplify the numerators.**

$$= 1\frac{3}{24} \text{ or } 1\frac{1}{8}$$ **Simplify.**

Online Option

Math Online Take a self-check Chapter Readiness Quiz at glencoe.com.

Get Started on Chapter 13

You will learn several new concepts, skills, and vocabulary terms as you study Chapter 13. To get ready, identify important terms and organize your resources. You may wish to refer to **Chapter 0** to review prerequisite skills.

FOLDABLES® Study Organizer

Statistics and Probability Make this Foldable to help you organize your Chapter 13 notes about statistics and probability. Begin with 10 sheets of grid paper.

1. **Fold** each sheet of paper in half from top to bottom.
2. **Unfold** and cut four columns from the left side of each sheet, from the top to the crease.
3. **Stack** the sheets and staple to form a booklet.
4. **Label** each page with a lesson number and title.

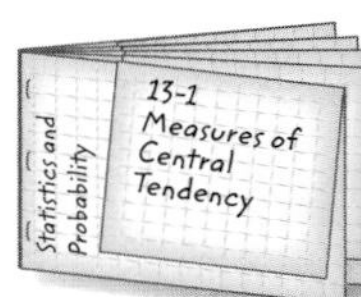

Math Online glencoe.com

- Study the chapter online
- Explore **Math in Motion**
- Get extra help from your own **Personal Tutor**
- Use **Extra Examples** for additional help
- Take a **Self-Check Quiz**
- **Review Vocabulary** in fun ways

New Vocabulary

English		Español
measures of central tendency	p. 730	medidas de tendencia central
median	p. 730	mediana
mode	p. 730	moda
stem-and-leaf plot	p. 737	diagrama de tallo y hojas
measures of variation	p. 743	medidas de varación
range	p. 743	amplitud
outlier	p. 745	valor atípico
box-and-whisker plot	p. 750	diagrama de caja y patillas
histogram	p. 757	histograma
probability	p. 765	probabilidad
sample	p. 771	muestra

Review Vocabulary

circle graph • p. 367 • gráfica circular type of statistical graph used to compare parts to a whole

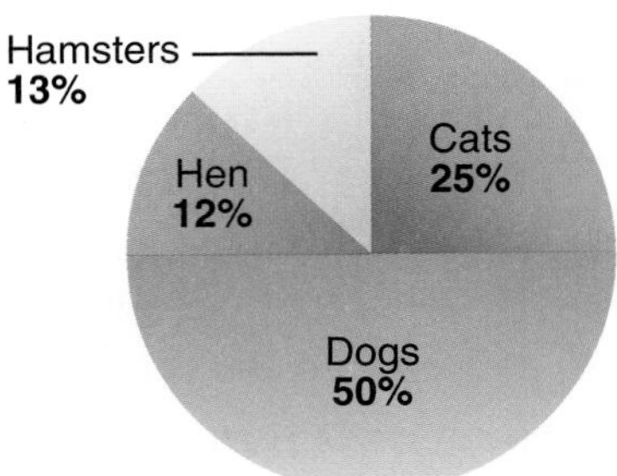

mean • p. 92 • media the sum of the data divided by the number of items in the data set, also called the average

Multilingual eGlossary glencoe.com

EXPLORE 13-1

Algebra Lab

Analyzing Data

Math Online glencoe.com
Math *in Motion,* Animation

Objective
Examine data and find a number to describe it.

Often, it is useful to describe or represent a set of data by using a single number. The table shows the speeds in miles per hour of first serves in a tennis match.

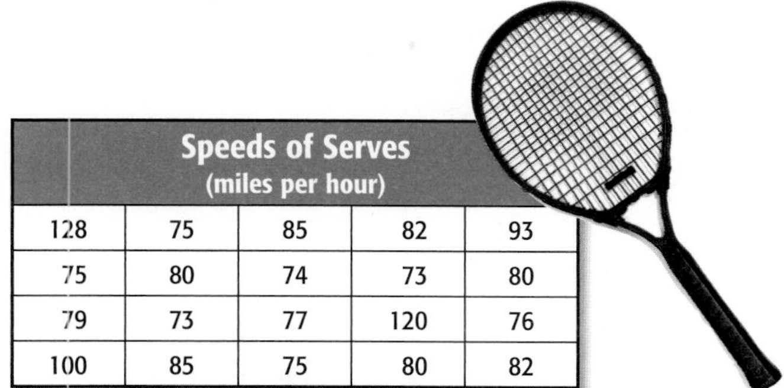

Speeds of Serves (miles per hour)				
128	75	85	82	93
75	80	74	73	80
79	73	77	120	76
100	85	75	80	82

The average of the data set is 84.6. This number is much higher than most of the data, so it is not a good number to describe the data.

A number that could be used to describe this data set is 80. Some reasons for choosing this number are listed below.

- It occurs three times. Only one other number appears as many times.
- If the numbers are arranged from least to greatest, 80 falls in the center of the data set.

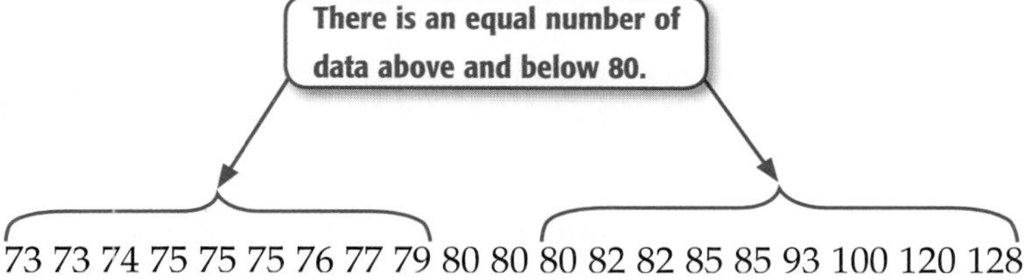

Collect the Data

Collect a group of data. Use one of the suggestions below, or use your own method. Display your data in a table.

- Research data about a sports team in your area, such as batting averages, number of wins per season, or number of goals per game.
- Find a graph or table of data in the newspaper or a magazine. Some examples include weather data and population data.
- Conduct a survey to gather some data about your classmates.
- Count the number of raisins in a number of small boxes.

Analyze the Results

1. Choose a number that best describes all of the data in the set.
2. Explain what your number means, and explain which method you used to choose your number.
3. Describe how your number might be useful in real life.

13-1 Measures of Central Tendency

Then

You have already found the mean or average of a data set. (Lesson 2-5)

Now

- Use the mean, median, and mode as measures of central tendency.
- Choose an appropriate measure of central tendency and recognize measures of statistics.

New Vocabulary

measures of central tendency
mean
median
mode

Math Online

glencoe.com

- Extra Examples
- Personal Tutor
- Self-Check Quiz
- Homework Help

Why?

The cartoon shows a basketball team getting ready to play a game.

a. Does the average of the heights of the players describe the overall height of the team? Explain.

b. If the player with the number 8 jersey was not on the team, would the average of the heights of the remaining players describe the overall height of the team? Explain.

Measures of Central Tendency When you have a list of numerical data, it is often helpful to use one or more numbers to describe the whole data set. **Measures of central tendency** describe the center of the data.

Key Concept — Measures of Central Tendency (For Your FOLDABLE)

mean	sum of the data divided by the number of items in the data set
median	middle number of the data ordered from least to greatest, or the mean of the middle two numbers
mode	number or numbers that occur most often

Real-World EXAMPLE 1 Find Measures of Central Tendency

GOVERNMENT **Find the mean, median, and mode of the data in the table.**

HOUSE OF REPRESENTATIVES			
California	53	New York	29
Florida	25	North Carolina	13
Illinois	19	Ohio	18
Michigan	15	Pennsylvania	19
New Jersey	13	Texas	32

$$\text{mean} = \frac{\text{total number of representatives}}{\text{number of states}}$$

$$= \frac{53 + 25 + \ldots + 32}{10}$$

$$= \frac{236}{10} \text{ or } 23.6$$

The mean number of representatives is 23.6.

To find the median, order the numbers from least to greatest.

13, 13, 15, 18, 19, 19, 25, 29, 32, 53

$$\frac{19 + 19}{2} = 19$$

There is an even number of items. Find the mean of the two middle numbers.

The median number of representatives is 19.

Since both 13 and 19 appear twice, the modes are 13 and 19 representatives.

Check Your Progress

1. SNAKES The reptile collection at a local zoo has snakes that measure 62, 48, 37, 45, 50, 65 , 48, 54, 48, 52, 40, and 51 centimeters long. Find the mean, median, and mode of the data.

Personal Tutor glencoe.com

STANDARDIZED TEST EXAMPLE 2

Karen talked on the phone for 25 minutes, 39 minutes, 28 minutes, and 20 minutes over the last 4 days. If she talks for 34 minutes on the next day, which of the following statements would be true?

A The mean increases and the median decreases.

B The median is unchanged and the mean increases.

C Both the mean and the median decrease.

D Both the mean and the median increase.

Read the Test Item

You need to determine which statement would be true if Karen talks for 34 minutes on the fifth day.

Solve the Test Item

The mean of the four days is $\frac{25 + 39 + 28 + 20}{4}$ or 28. The mean of the five days is $\frac{25 + 39 + 28 + 20 + 34}{5}$ or 29.2. Since the mean increased, you can eliminate answer choice C.

Find the median to check the other answer choices. Arrange the numbers from least to greatest, first with four days and then with five days.

20, 25, 28, 39 — **The median is 26.5.**

20, 25, 28, 34, 39 — **The median is 28.**

The median also increased. The answer is D.

Test-TakingTip

Number Sense Eliminate possibilities using number sense. Since the new data value is greater than the least value in the data set, the mean will *not* decrease. So, you can eliminate choice C.

Check Your Progress

2. Nikko earned the following amounts raking leaves: \$15, \$10, \$12, \$8, and \$17. If he drops the lowest amount he has earned raking leaves, which of the following statements would be true?

F The mean decreases and the median increases.

G The mean and the median both increase.

H The mean is unchanged and the median increases.

J The mean and the median both decrease.

Personal Tutor glencoe.com

Choose Appropriate Measures Different circumstances determine which measures of central tendency are most appropriate.

Concept Summary — Using Mean, Median, and Mode

For Your FOLDABLE

Measure	Most Useful When...
mean	• the data have no extreme values (values that are much greater or much less than the rest of the data)
median	• the data have extreme values • there are no big gaps in the middle of the data
mode	• data have many repeated numbers

Real-World EXAMPLE 3 Choose an Appropriate Measure

WEATHER The table shows daytime high temperatures for the previous week. Which measure of central tendency best represents the data? Then find the measure of central tendency.

Day	Temperature (°F)
Sun.	84
Mon.	83
Tues.	89
Wed.	90
Thurs.	91
Fri.	85
Sat.	80

Since the set of data has no extreme values or numbers that are identical, the mean would best represent the data.

Mean: $\frac{84 + 83 + \ldots + 80}{7} = \frac{602}{7}$ or 86

The temperature 86°F is the measure of central tendency that best represents the data.

Check Your Progress

3. **EXERCISE** The table shows the number of sit-ups Pablo had done in one minute for the past 7 days. Which measure of central tendency best represents the data? Justify your selection and then find the measure of central tendency.

Day	Number of Situps
Sun.	40
Mon.	37
Tues.	45
Wed.	19
Thurs.	49
Fri.	50
Sat.	46

Personal Tutor glencoe.com

You can also use measures of central tendency to show different points of view.

Real-World EXAMPLE 4 Different Points of View

RIDES The average wait times for 10 different rides at an amusement park are 65, 21, 17, 52, 25, 17, 11, 22, 60, and 44 minutes. Which measure of central tendency would the amusement park advertise to show that the wait times for its rides are short? Explain.

Mean: $\frac{65 + 21 + \ldots + 44}{10} = \frac{334}{10}$ or 33.4

Median: 11, 17, 17, 21, $\underbrace{22, 25,}$ 44, 52, 60, 65

$\frac{22 + 25}{2}$ or 23.5

Mode: 17

The amusement park would want to advertise a short wait time. So, the amusement park would want to use the mode, 17 minutes.

Real-World Link

There are more than 400 amusement parks around the world. The park with the greatest annual attendance is Magic Kingdom in Orlando, Florida. It averages almost 17 million visitors per year.

Source: Coaster Grotto

Check Your Progress

4. **SCORES** Maggie had the following times on her runs in the 100-meter dash: 11.6, 11.8, 12.7, 12.6, 11.9, and 12.0. Which measure of data would she want to use to describe her performance?

Personal Tutor glencoe.com

Check Your Understanding

Example 1 p. 730

1. **RACING** The table shows the winning speeds of the Indianapolis 500 in recent years. Find the mean, median, and mode of the data. Round to the nearest tenth.

Year	Driver	Speed (mph)
2001	Helio Castroneves	131
2002	Helio Castroneves	166
2003	Gil de Ferran	157
2004	Buddy Rice	139
2005	Dan Wheldon	158
2006	Sam Hornish Jr	157
2007	Dario Franchitti	152

Source: Indianapolis 500

2. **EXERCISE** Last week, Sarah spent 35, 30, 45, 30, 40, 37, and 28 minutes exercising. Find the mean, median, and mode. Round to the nearest whole number.

Example 2 p. 731

3. **MULTIPLE CHOICE** One week, a store sold 53, 61, 46, 59, 61, 55, and 49 board games. Suppose the store sells 83 board games on the eighth day. Which measure of central tendency would change the most?

   A mean
   B median
   C mode
   D all measures were affected equally

Example 3 p. 732

4. **BOOKS** The different eighth grade homerooms read 38, 45, 26, 51, 42, 38, 50, and 58 books for a reading competition. Which measure of central tendency best represents the data? Justify your selection and then find the measure of central tendency.

Example 4 p. 732

5. **EXERCISE** Refer to the data in Exercise 2. Which measure of central tendency could Sarah use to show that she exercises for large amounts of time each day? Explain.

Practice and Problem Solving

 = **Step-by-Step Solutions** begin on page R11.
Extra Practice begins on page 810.

Example 1 p. 730

Find the mean, median, and mode for each set of data. If necessary, round to the nearest whole number.

6. the minutes spent biking each day in one week: 45, 30, 65, 90, 74, 60, 35
7. the price, in dollars, of digital cameras: 250, 200, 320, 235, 265, 200

Example 2 p. 731

8. **BASKETBALL** Sasha scored the following point totals in 7 games this basketball season: 12, 18, 20, 8, 15, 18, and 14.
   a. If she drops her lowest score, which measure of central tendency would increase more, the mean or median? Explain.
   b. In her next basketball game, Sasha scores 13 points. Which measure of central tendency would decrease more, the mean or median? Explain.

Example 3 p. 732

9. The table shows the football teams with the most all-time wins. Which measure of central tendency best represents the data? Justify your selection and then find the measure of central tendency.

University	All-Time Football Wins*
Michigan	869
Notre Dame	824
Texas	820
Nebraska	817
Ohio State	798

Source: College Football Data Warehouse
*As of 2007.

Example 4
p. 732

10. **ATTENDANCE** The table shows the attendance at an art museum. Which measure of central tendency would the museum use to show it has a large number of visitors? Explain.

Day	Attendance
Tuesday	214
Wednesday	189
Thursday	214
Friday	248
Saturday	220
Sunday	253

11. **TEST SCORES** Matthew's math test scores this semester were 80, 76, 94, 90, 88, 92, 88, and 96. Which measure of central tendency might Matthew want to use to describe his test scores? Explain.

12. **FINANCIAL LITERACY** The hourly wages of employees in a small store are \$7, \$24, \$8, \$10, \$6, \$8, and \$8. Which measure of central tendency might the store use to attract people to work there? Explain.

13. **SIBLINGS** The graph shows the number of siblings that Ms. Delgado's students have. Which measure of central tendency best represents the data? Explain.

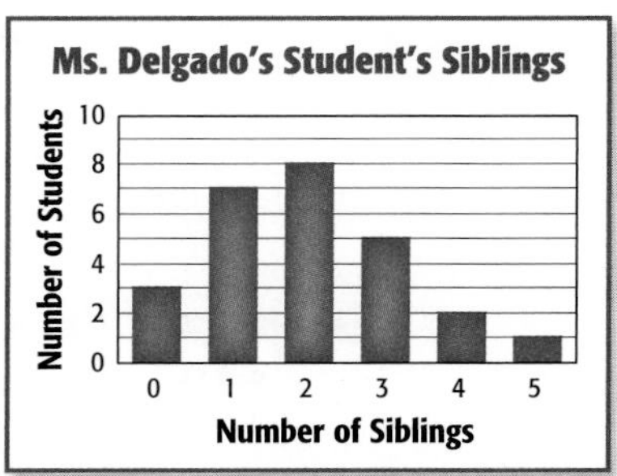

14. **ICE SKATING** Winona needs to average 5.8 points from 14 judges to win the competition. The mean score of 13 judges was 5.9. What is the lowest score she can receive from the 14th judge and still win?

Real-World Link

In the women's individual Olympic figure skating competitions, the United States has won gold medals 7 times since 1924. Sarah Hughes was 16 when she won the gold medal in 2002.

15. **COLLECT DATA** Survey your classmates to find their height. Display the results of your survey in a line plot. Then find the mean, median, and mode of your data. Which measure of central tendency would you use to represent the overall height?

H.O.T. Problems

Use Higher-Order Thinking Skills

16. **OPEN ENDED** Construct a data set with a mean of 4 and a median that is *not* 4.

17. **REASONING** Is it *always*, *sometimes*, or *never* possible for the mean, median, and mode to be equal? Justify your reasoning.

18. **CHALLENGE** The ages of the players on an intramural volleyball team are 29, 25, 26, 31, 28, 23, 21, and 25.
    a. Suppose another player joins the team. What must the age of the new player be so that the mean age is 27? Explain your reasoning.
    b. What must the age of the new player be so that the median age is 25? Explain your reasoning.

19. **CHALLENGE** A real estate guide lists the "average" home prices for counties in your state. Do you think the mean, median, or mode would be the most useful average for homebuyers? Explain.

20. **REASONING** Can a data set have more than one mode? median? Explain.

21. **WRITING IN MATH** Use the Internet to find some real-world data. Which measure of central tendency best represents the data you found? Justify your selection and then find the measure of central tendency.

Standardized Test Practice

22. The number of books read by the students in each reading class this year is shown in the table. Which measure of central tendency would the school use to show they read a lot of books?

104	90
162	134
110	97
145	126

A mode
B median
C mean
D cannot be determined

23. The high temperatures, in degrees Fahrenheit, for one week are 79°, 81°, 77°, 81°, 82°, 75°, and 76°. If the temperature on the eighth day is 80°, which of the following would be true?

F The mode will change.
G The mean will increase and the median will remain the same.
H The median will increase and the mean will remain the same.
J Both the mean and the median will increase.

24. Jamal said that the number that best represented the following set of data is 27. Which measure of central tendency is he referring to?

28, 32, 21, 25, 33, 32, 20, 26

A mean
B median
C mode
D all of the above

25. EXTENDED RESPONSE Shane and Chien have the bowling scores shown.

Game #	Shane	Chien
1	124	125
2	135	132
3	109	128
4	116	130
5	141	125

a. After the sixth game, the mean of Shane's scores is 130. What was Shane's score for his sixth game?

b. What must Chien bowl in the sixth game in order to have the same average as Shane?

Spiral Review

Determine whether each pair of solids is similar. (Lesson 12-8)

26.

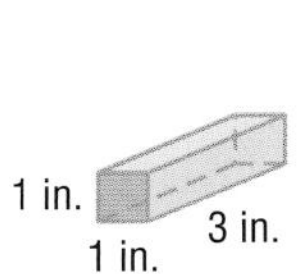

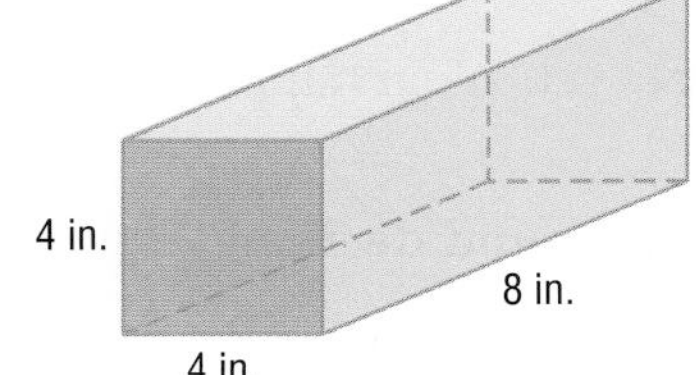

27.

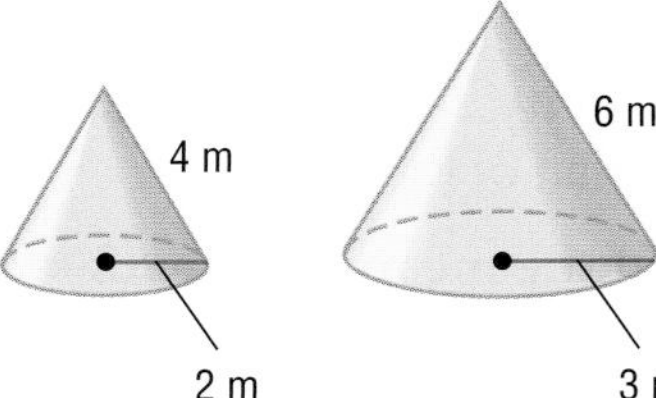

28. ARCHITECTURE The small tower of a historic house is shaped like a regular hexagonal pyramid as shown at the right. How much roofing will be needed to cover this tower? (*Hint*: Do not include the base of the pyramid.) (Lesson 12-7)

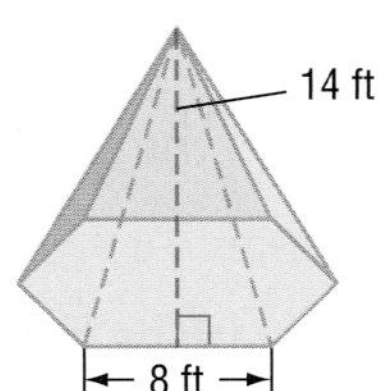

Skills Review

Make a line plot for the data set. (Previous Course)

29. ages of basketball players: 21, 20, 19, 16, 16, 18, 17, 19, 18, 20, 16, 18, and 20

Graphing Technology Lab
Mean and Median

Math Online glencoe.com
- Other Calculator Keystrokes
- Graphing Calculator Personal Tutor

A graphing calculator can quickly and accurately determine measures of central tendency for large data sets. You can use a graphing calculator to find the mean and median of a set of data.

ACTIVITY

SPORTS The number of wins for the last 15 seasons for the local soccer team is shown at the right.

12	9	13	11	12
7	8	7	11	14
8	6	2	10	13

Find the mean and median number of wins.

Step 1 Enter the data.

- Clear any existing lists.

KEYSTROKES: [STAT] [ENTER] [▲] [CLEAR] [ENTER]

- Enter the number of wins as L1.

KEYSTROKES: 12 [ENTER] 9 [ENTER] ... 13 [ENTER]

Step 2 Find the mean and median.

- Display a list of statistics for the data.

KEYSTROKES: [STAT] [▶] [ENTER] [ENTER]

Use the down arrow key to locate Med. The median number of wins is 10 and the mean number of wins is about 9.5.

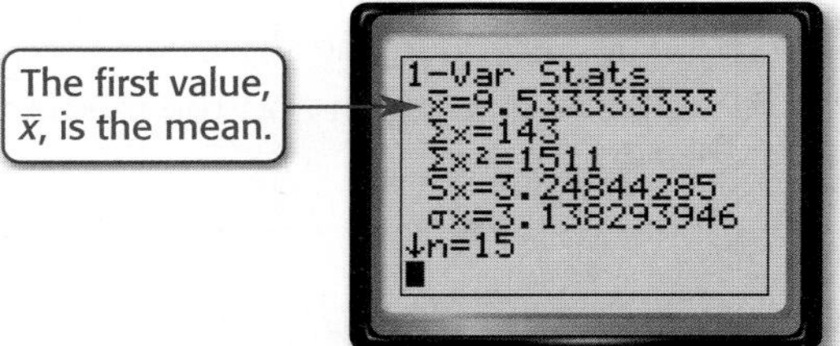

Analyze the Results

1. In the opening Activity, remove the number 2 from the data set. What happens to the mean and median of the data set?

Clear list L1 and find the mean and median of each data set. Round decimal answers to the nearest hundredth if necessary.

2. 7.6, 6.3, 8.9, 2.4, 6.1, 9.5
3. −54, −47, −42, −59, −46, −68, −52
4. 3.8, −3.4, −5.8, 7.6, −6.9, 8.7, 4.3, 2.5, 1.8, 7.2, −2.7, 5.9, −1.7, 54
5. Look back at the medians found. When is the median a member of the data set?
6. Refer to Exercise 4.
   a. Which statistic better represents the data, the mean or the median? Explain.
   b. Suppose the number 54 should have been 5.4. Recalculate the mean and median. Is there a significant difference between the first pair of values and the second?
   c. When there is an error in one of the data values, which statistic is less likely to be affected? Why?

13-2 Stem-and-Leaf Plots

Then

You have already displayed data in circle graphs. (Lesson 7-8)

Now

- Display data using stem-and-leaf plots.
- Interpret data in a stem-and-leaf plot.

New Vocabulary

stem-and-leaf plot
stems
leaves
back-to-back stem-and-leaf plot

Math Online

glencoe.com
- Extra Examples
- Personal Tutor
- Self-Check Quiz
- Homework Help

Why?

The average winning speeds, in miles per hour, for the Daytona 500 are shown. Write each number on a sticky note. Then group the numbers using the intervals: 130–139, 140–149, 150–159, 160–169, and 170–179.

DAYTONA 500 Winning Speeds (mph), 1979–2008

142	134	156	151	160	148
156	162	148	143	135	154
177	156	148	149	173	162
143	157	138	154	170	148
149	144	155	172	166	176

Source: Daytona International Speedway

a. Is there an equal number of speeds in each group? Explain.

b. What is an advantage of displaying data in groups?

Display Data In a **stem-and-leaf plot**, numerical data are listed in ascending or descending order. The greatest place values of the data are used for the **stems**. The least place value forms the **leaves**.

EXAMPLE 1 Draw a Stem-and-Leaf Plot

DOGS The table shows the average weight in pounds of different breeds of adult dogs. Display the data in a stem-and-leaf plot.

Adult Dog Weight (lb)			
10	21	15	9
17	6	4	18
9	26	20	24
11	13	8	25

Step 1 The stems are the greatest place values of the data. List the stems 0, 1, and 2 in order in the *Stem* column. Write the ones digits to the right of the corresponding stems in the *Leaf* column.

Stem	Leaf
0	9 6 4 9 8
1	0 5 7 8 1 3
2	1 6 0 4 5

Step 2 Order the leaves from least to greatest and write a *key* that explains how to read the stems and leaves. Include a title.

Adult Dog Weight (lb)

Stem	Leaf
0	4 6 8 9 9
1	0 1 3 5 7 8
2	0 1 4 5 6

2|6 = 26 lb

Write each data leaf as many times as it appears.

Check Your Progress

1. EXERCISE The numbers of minutes Will spent exercising are shown. Display the data in a stem-and-leaf plot.

Exercising Time (min)				
42	15	65	30	45
40	20	28	45	60
38	23	39	30	10

Personal Tutor glencoe.com

Interpret Data Stem-and-leaf plots are useful in analyzing data because you can see all the data values, including the least, greatest, median, and mode.

StudyTip

Median
The data values are already ordered in a stem-and-leaf plot. There are 21 data values in Example 2. The middle value is data value #11 or 38.

EXAMPLE 2 Interpret Data

SOFTBALL The stem-and-leaf plot shows the number of wins for various softball teams in one season.

Softball Wins

Stem	Leaf
1	8
2	2 3 4 4 4 7
3	1 5 6 8 9 9
4	0 4 7 7 9
5	0 0 1

2|3 = 23 wins

a. Find the median and mode.
The median, the middle value, is 38. The most frequent value is 24.

b. What is the difference between the least and greatest number of wins?
The difference is 51 − 18 or 33.

Check Your Progress

DOGS Refer to the stem-and-leaf plot of adult dog weight in Example 1.

2A. Find the median and mode.

2B. What is the difference between the least weight and greatest weight?

Personal Tutor glencoe.com

A **back-to-back stem-and-leaf plot** can be used to compare two sets of data. The back-to-back stem-and-leaf plot below compares daily high temperatures.

The leaves for one set of data are on one side of the stem.

Daily High Temperatures in June

Portland, OR	Stem	Chicago, IL
9 9 8 7 7	5	
2 2 2 1 1 1 1 1 1 0 0	6	7
8 8 7 6 6 6 6 6 5 5 4		
1 0 0	7	0 0 1 1 5 5 6 7 7 7 9 9
	8	1 2 2 3 3 4 4 4 5 5 5 5 5 7 7 9
	9	0

1|7 = 71°F *7|9 = 79°F*

The leaves for the other set of data are on the other side of the stem.

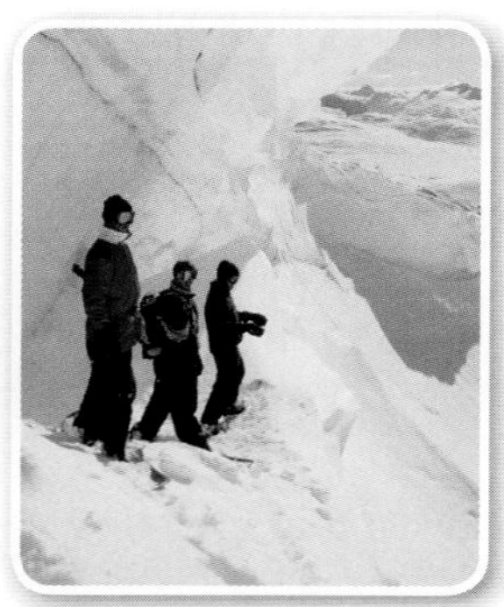

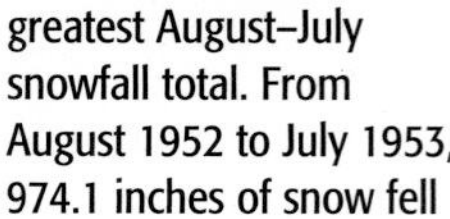

Real-World Link

Thompson Pass, Alaska, holds the record for the greatest August–July snowfall total. From August 1952 to July 1953, 974.1 inches of snow fell there.

Source: National Climatic Data Center

Real-World EXAMPLE 3 Back to Back Stem-and-Leaf Plots

SNOWFALL The monthly snowfall accumulations for nine months in Syracuse and Buffalo, New York, are shown.

Snowfall Accumulations

Syracuse	Stem	Buffalo
9 4 1	0	1 1 3 9
7	1	2 2 8
7 6	2	3 7
5 4	3	
0	4	

7|1 = 17 in. *1|2 = 12 in.*

a. In general, which city has the lower amount of monthly snowfall? Explain.
Buffalo; it experiences lower numbers and does not have snowfall over 27 inches.

b. Which city has more varied amounts of snowfall? Explain.
The data for Syracuse is more spread out, while the data for Buffalo is clustered. So, Syracuse has the more varied amount of snowfall.

Check Your Progress

TEMPERATURES Refer to the stem-and-leaf plot about temperatures.

3A. In general, which city had the higher daily temperatures? Explain.

3B. Which city has more varied temperatures? Explain.

Personal Tutor glencoe.com

Check Your Understanding

Example 1 p. 737

Display each set of data in a stem-and-leaf plot.

1.

Science Test Scores				
99	80	84	72	79
73	76	80	81	76

2.

Height of Plants (in.)				
40	51	68	57	55
50	57	51	67	41
67	57	48	58	67

Example 2 p. 738

3. **NUTRITION** The average amount of pasta that people in different countries consume each year is shown in the stem-and-leaf plot below.

Pasta Consumption (lb)

Stem	Leaf
0	3 4 5 8 8 9 9
1	0 1 4 5 5 5 5 5 9
2	0 0 8
3	
4	
5	9

2 | 8 = 28 lb

a. Find the median of the data.

b. Find the mode of the data.

c. What is the difference between the greatest amount of pasta and least amount of pasta consumed?

Example 3 p. 738

4. **STATES** The maximum allowable speed limits in various western and eastern states are shown in the back-to-back stem-and-leaf plot below.

Western States	Stem	Eastern States
5	5	
5 5	6	5 5 5 5 5 5 5 5 5
5 5 5 5 5 5 5 5 0 0	7	0 0 0
0 | 7 = 70 mph		*6 | 5 = 65 mph*

a. In general, which region has higher maximum speed limits? Explain.

b. Which region has more varied speed limits? Explain.

Practice and Problem Solving

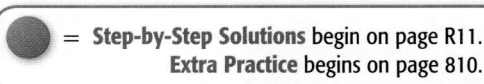

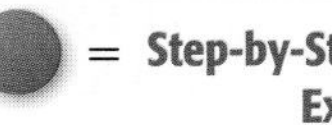
= **Step-by-Step Solutions** begin on page R11.
Extra Practice begins on page 810.

Example 1 p. 737

Display each set of data in a stem-and-leaf plot.

5. The ages of people attending a spinning class: 18, 27, 35, 22, 23, 30, 21, 40, 26, 33, 49, 55, 27, 22, 20, 39, and 42.

6. The number of students in each class in an elementary school that own a video game system: 20, 22, 18, 29, 30, 24, 19, 32, 27, 28, 30, 21, and 25.

7. The cost of a new pair of tennis shoes: \$50, \$65, \$80, \$96, \$65, \$70, \$76, \$80, \$60, \$75, \$78, and \$59.

8. Number of applicants at annual talent shows: 88, 90, 102, 104, 67, 109, 97, 98, 86, 88, 86, 100, 103, 99, 90, and 106.

Example 2 p. 738

9. **SCHOOL** The stem-and-leaf plot at the right shows the number of pages of a novel each student in Mrs. Switlick's class has read.

a. Find the median of the data.

b. Find the mode of the data.

c. What is the difference between the greatest number of pages read and the least number of pages read?

Pages Read by Students

Stem	Leaf
3	1 5 5 7 7 9
4	2 4 4 6 6 8 8
5	0 0 0 3 4 6 6 7
6	1 6 7 8 8
7	
8	8

6 | 1 = 61 pages

10. **NUTRITION** The stem-and-leaf plot shows the number of Calories per servings of various dry cereals.

a. Find the median of the data.

b. Find the mode of the data.

c. What is the difference between the greatest number of Calories and the least number of Calories for these cereals?

Calories in Cereal

Stem	Leaf
7	0 2
8	1 5
9	0 3
10	0 0 0 1 1 1 2 2 2 8 8 9
11	0 0 1 1 2 3 4 4 5
12	0 5 6 6
13	0
14	0 1

11 | 0 = 110 Calories

Example 3 p. 738

11. **GIRL SCOUTS** The number of boxes of cookies sold by each scout is shown in the back-to-back stem-and-leaf plot below.

Cookie Sales

Troop 60	Stem	Troop 122
9 9 5 4 3 2 0	3	0 1 3 5 8 9
5 5 4 4 2	4	4
3 0	5	0 4 5 7
	6	0 1 8

0 | 5 = 50 boxes *6 | 0 = 60 boxes*

a. Overall, which troop sold more cookies? Explain.

b. Which troop has more varied sales? Explain.

Real-World Link

Girl Scout cookies have been a part of the American culture for over 80 years. Individual troops receive about 17% of the purchase price of each box of cookie sold.

Source: Girl Scout

12. **FOOTBALL** The back-to-back stem-and-leaf plot at the right shows the points scored by two high school football teams in their games this season.

a. Which measure of central tendency would the Rams want to use to show they have a greater number of points scored than the Stallions? Explain.

b. Which measure(s) of central tendency would the Stallions want to use to show they have a greater number of points scored than the Rams? Explain.

High School Football Points

Rams	Stem	Stallions
9 8 7	0	
9 8 5 4	1	0 0 5 9
4 4 0	2	0 1 2 5 8
	3	1

9 | 0 = 9 points *1 | 9 = 19 points*

13. **TRACK** The times in seconds of the top 10 finishers of the 100-meter race are 14.3, 18.0, 19.9, 18.7, 16.2, 17.7, 16.5, 19.2, 17.9 and 17.4. Make and analyze a stem-and-leaf plot of the data. Draw two conclusions about the times.

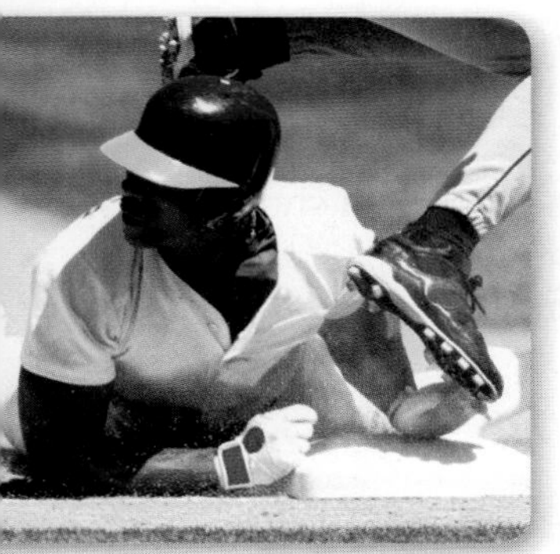

Real-World Link

Ricky Henderson played professional baseball from 1979 to 2003. He holds the career record for the greatest number of stolen bases, 1,406.

Source: MLB

14. **BASEBALL** The table below shows the number of stolen bases by team for each league during a recent season of professional baseball.

American League	139	144	96	78	72	103	78	
	112	123	52	81	131	88	57	
National League	96	200	138	68	55	56	69	119
	109	64	86	97	100	105	65	137

Display the data in a back-to-back stem-and-leaf plot. Then analyze and compare the data.

15. **MULTIPLE REPRESENTATIONS** In this problem, you will use a stem-and-leaf plot to analyze data. The test scores of 36 students are shown in the table at the right.

Test Scores (total score of 80)					
80	13	23	34	56	56
73	79	6	19	73	62
68	20	78	78	32	59
36	48	75	79	39	62
46	59	76	49	64	76
78	63	59	67	80	38

a. **TABULAR** Display the data in a stem-and-leaf plot.

b. **NUMERICAL** Find the mean, median, and mode of the data.

c. **GRAPHICAL** Turn the stem-and-leaf plot 90° counterclockwise. Sketch the graph that is formed by the peaks of the stem-and-leaf plot.

d. **ANALYTICAL** Describe the data by the number of peaks in the graph. Is it symmetric? Explain your reasoning.

e. **VERBAL** A set of data can be described as being *skewed* if a graph of the data is not symmetric. Is this set of data skewed? Explain your reasoning.

16. **COLLECT DATA** Display the hand lengths, in inches, of the students in your class in a frequency table.

a. Make a stem-and-leaf plot of the data.

b. Analyze the data. Then write a summary explaining your data.

H.O.T. Problems

Use Higher-Order Thinking Skills

17. **OPEN ENDED** Create a stem-and-leaf plot that has a median of 22.

18. **CHALLENGE** Create a stem-and-leaf plot that has at least 15 pieces of data in which the minimum value is 19 and the median is 43.

19. **FIND THE ERROR** Paolo and J'vonté are finding the median plant height from the stem-and-leaf plot at the right. Paolo says the median is 64. J'vonté says the median is 31. Is either of them correct? Explain your reasoning.

Plant Heights (cm)

Stem	Leaf
0	6 8
1	1
2	8
3	6 8
4	0 9

2 | 8 = 28 cm

20. **CHALLENGE** Use the stem-and-leaf plot about plant heights at the right. Add two data values that do not affect the median but add 3 to the mean.

21. **WRITING IN MATH** Refer to the information on page 737 about Daytona 500 winning speeds. Explain how a stem-and-leaf plot can help you analyze the data.

Standardized Test Practice

22. **GRIDDED RESPONSE** What is the median number of hours shown in the stem-and-leaf plot below?

Marathon Times (h)

Stem	Leaf
1	3 4 4
2	0 5 7 7
3	1 1 1 2 8
4	0 0 1

2|8 = 2.8 h

23. What are the stems for the following data?
22, 67, 43, 45, 7, 33, 29

A {2, 3, 4, 6}
B {1, 2, 3, 4, 5, 6, 7}
C {0, 2, 3, 4, 6}
D {0, 1, 2, 3, 4, 5, 6}

24. What is the mode of the set of data shown in the stem-and-leaf plot below?

Package Weights (oz)

Stem	Leaf
10	0 3 5 7 7
11	0 0 2 2
12	4 6 6 6 6
13	0 0 0 1
14	0

11|2 = 11.2 oz

F 0
G 10 and 12
H 12.6
J 13.0

25. Which of the following statements is true about the stem-and-leaf plot in Exercise 24?

A The mode is 1.40.
B The mean is about 11.9.
C The median is 12.3.
D Twice as many packages weigh between 10 and 13 pounds than 13 to 14 pounds.

Spiral Review

26. **TESTS** Which measure of central tendency best summarizes the test scores shown below? Explain. (Lesson 13-1)
97, 99, 95, 89, 99, 100, 87, 85, 89, 92, 96, 95, 60, 97, 85

Find the missing measure for each pair of similar solids. (Lesson 12-8)

27.

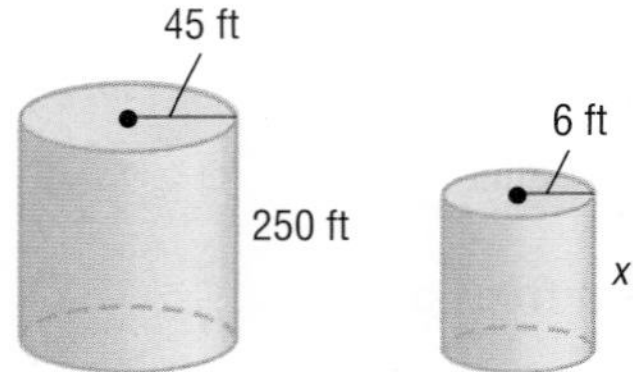

28.

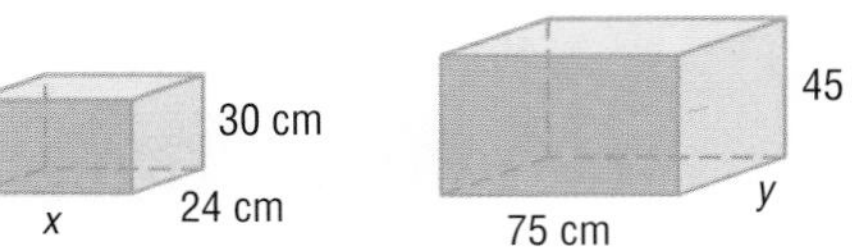

29. **FLOORING** A square room has a floor area of 324 square feet. The homeowners plan to cover the floor with 6-inch square tiles. How many tiles will be in each row on the floor? (Lesson 10-2)

Skills Review

Find the median of each set of data. Round to the nearest tenth, if necessary.
(Lesson 13-1)

30. 41, 37, 43, 43, 36

31. 2, 8, 16, 21, 3, 8, 9, 7, 6

32. 14, 6, 8, 10, 9, 5, 7, 13

33. 7.5, 7.1, 7.4, 7.6, 7.4, 9.0, 7.9, 7.1

13-3 Measures of Variation

Then

You have already found measures of central tendency. (Lesson 13-2)

Now

- Find measures of variation.
- Uses measures of variation to interpret and analyze data.

New Vocabulary

measures of variation
range
quartiles
lower quartile
upper quartile
interquartile range
outlier

Math Online

glencoe.com

- Extra Examples
- Personal Tutor
- Self-Check Quiz
- Homework Help

Why?

One of the main reasons people attend amusement parks is to ride roller coasters. The maximum speeds of ten of the world's fastest steel roller coasters are shown.

a. Find the difference between the fastest and slowest speeds.

b. Graph the data on a line plot.

c. Write a sentence describing how the data are distributed on the line plot.

Roller Coaster	Speed (mph)
American Eagle	66
Dodonpa	107
Furious Baco	84
Goliath	85
Kingda Ka	128
Millennium Force	93
Steel Dragon 2000	95
Titan	85
Top Thrill Dragster	120
Tower of Terror	100

Measures of Variation **Measures of variation** are used to describe the distribution of the data. One measure of variation is the range. The **range** of a set of data is the difference between the greatest and the least values of the set. It describes whether the data are spread out or clustered together.

EXAMPLE 1 Find Range

Find the range for each set of data.

a. **Study Time**

Stem	Leaf
1	0 0 2 3 6 7 8
2	2 5 6 9 9
3	1 2 8

2|6 = 26 minutes

The greatest value is 38 minutes, and the least value is 10 minutes. So, the range is 38 − 10 or 28 minutes.

b. The age in years of Mrs. Tyznik's grandchildren: 27, 8, 5, 19, 21, 10, 4, and 21.

The greatest value is 27 years and the least value is 4 years. So, the range is 27 − 4 or 23 years.

Check Your Progress

1A. The cost in dollars of DVDs: 20, 25, 15, 16, 10, and 9

1B. **Weight of Letters**

Stem	Leaf
3	3 8 9 7 8
4	4 4 6 6 8 9
5	0 2 6 8

3|3 = 33 grams

Personal Tutor glencoe.com

In a set of data, the **quartiles** are the values that divide the data into four equal parts. Recall that the median of a set of data separates the set in half.

StudyTip

Quartiles One half of the data lies between the lower quartile and the upper quartile.

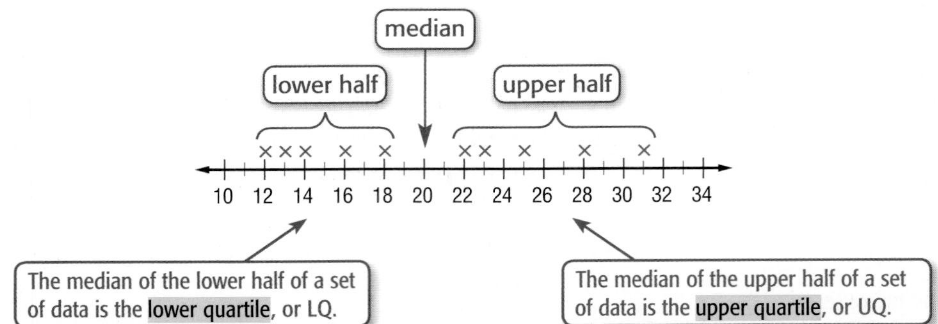

The upper and lower quartiles can be used to find another measure of variation called the **interquartile range**.

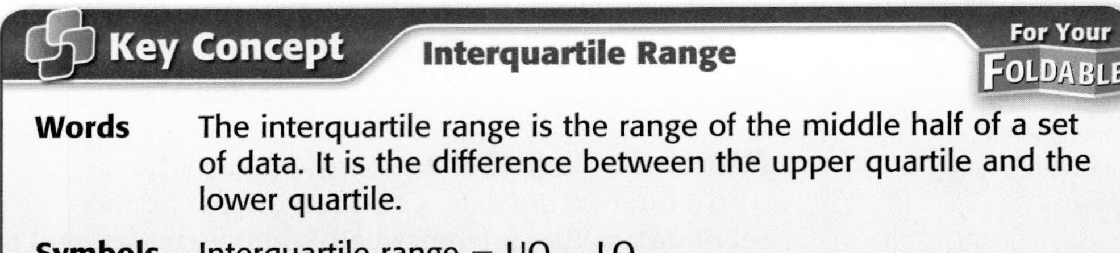

Key Concept **Interquartile Range** For Your FOLDABLE

Words The interquartile range is the range of the middle half of a set of data. It is the difference between the upper quartile and the lower quartile.

Symbols Interquartile range = UQ − LQ

EXAMPLE 2 Identify Measures of Variation

OLYMPICS Find the measures of variation for the data in the table.

U.S. Summer Olympic Silver Medals 1972–2004

34	32	31
31	35	0
39	24	61

Step 1 **Range:** 61 − 0 or 61 medals

Step 2 **Median, Upper Quartile, and Lower Quartile:**
Order the data from least to greatest.

lower quartile, median, upper quartile

0 24 31 31 32 34 35 39 61

$\frac{24 + 31}{2} = 27.5$ $\frac{35 + 39}{2} = 37$

The interquartile range is 37 – 27.5 or 9.5. The median is 32, the lower quartile is 27.5, and the upper quartile is 37.

StudyTip

Statistics A small interquartile range means that the data in the middle of the set are close in value. A large interquartile range means that the data in the middle are spread out, or vary.

Check Your Progress

2. **CONTESTS** Find the measures of variation for the data in the table.

Marley's Paper Airplane Tosses (ft)			
40	45	49	25
44	39	53	38

Personal Tutor glencoe.com

StudyTip

Outliers A data value that is much larger or much smaller than the median is an outlier.

Real-World Link

The cheetah is the fastest land animal in the world. It can sprint 70 miles per hour for only 200 to 300 yards at a time. It can run 3–4 miles at about 45 miles per hour.

Data that are more than 1.5 times the value of the interquartile range beyond the quartiles are called **outliers**.

EXAMPLE 3 Find Outliers

ANIMAL Find any outliers in the data set.

Animal Speeds	
Animal	**Speed (mph)**
Squirrel	12
Turkey (lower quartile →)	15
Elephant	25
Cat (median →)	30
Reindeer	32
Rabbit (upper quartile →)	35
Cheetah	70

Step 1 Find the interquartile range.

$35 - 15 = 20$

Step 2 Multiply the interquartile range, 20, by 1.5.

$20 \times 1.5 = 30$

Step 3 Subtract 30 from the lower quartile and add 30 to the upper quartile.

$15 - 30 = -15$ $\quad\quad 35 + 30 = 65$

The only outlier is 70 because it is greater than 65.

Check Your Progress

3. **MOVIES** Find any outliers in the data set.

Movie Running Time (min)			
105	120	155	115
96	100	110	120

Personal Tutor glencoe.com

Use Measures of Variation Measures of variation can be used to interpret and compare data.

EXAMPLE 4 Use Measures of Variation

EXERCISE The table shows the number of Calories burned for 30 minutes of each activity. Use measures of variation to describe the data at the right.

Number of Calories Burned for 30 Minutes	
Swimming (fast crawl)	261
Soccer	234
Racquetball	228
Football	222
Basketball	210
Tennis	183
Downhill Skiing	150
Volleyball	84

Find the measures of variation.

The range is $261 - 84$ or 177.

The median is $\frac{222 + 210}{2}$ or 216.

The upper quartile is 231.

The lower quartile is 166.5.

The interquartile range is $231 - 166.5$ or 64.5.

In one fourth of the activities, you will burn 166.5 Calories or less. In one fourth of the activities, you will burn 231 Calories or more. The number of Calories burned for half of the activities is in the interval 166.5 to 231.

Check Your Progress

4. **VIDEO GAMES** Use measures of variation to describe the data at the right.

Cost of Video Games ($)			
22.79	49.99	34.00	59.99
44.76	32.50	29.25	24.95

Personal Tutor glencoe.com

Real-World EXAMPLE 5 Measures of Variation

COUNTIES **The number of counties for certain western and northeastern states are shown.**

Number of Counties by Region

Western States		Northeastern States
	0	3 5 8
7 5	1	0 4 4 6
9 3	2	1 4
9 6 3	3	
4	4	
6 3	5	
	6	2

3|2 = 23 counties *2|4 = 24 counties*

a. Compare the western states' range with the northeastern states' range.
The range for the western region is 56 − 15, or 41 counties and the range for the northeastern region is 62 − 3, or 59. So, the number of counties in the northeast vary more than in the west.

b. Do the data for either region contain an outlier?

	Western States	Northeastern States
Lower Quartile:	23	8
Upper Quartile:	44	21
Interquartile Range:	44 − 23 = 21	21 − 8 = 13
Multiply by 1.5:	21 • 1.5 = 31.5	13 • 1.5 = 19.5
Determine Outliers:	23 − 31.5 = −8.5 ✗ 44 + 31.5 = 75.5 ✗	8 − 19.5 = −11.5 ✗ 21 + 19.5 = 40.5 ✓

Since 62 is greater than 40.5, 62 is an outlier for the northeastern states' data.

c. How does the outlier affect the measures of central tendency for the northeast region?

Calculate the mean, median, and mode without the outlier, 62.

	without the outlier	with the outlier
Mean:	$\frac{3 + 5 + \ldots + 24}{9} \approx 12.78$	$\frac{3 + 5 + \ldots + 62}{10} = 17.7$
Median:	14	14
Mode:	14	14

When the outlier is not included, the mean increased by 17.7 − 12.78, or 4.92, while the median and mode did not change.

Check Your Progress

5. MAGAZINES The number of magazines sold by each student in Homeroom 102 and Homeroom 104 is shown in the stem-and-leaf plot.

A. Compare Homeroom 104's range with Homeroom 102's range. Does either homeroom have an outlier in the data set?

B. How does the outlier affect the range for the number of magazines sold in that homeroom?

Magazines Sold

Homeroom 102	Stem	Homeroom 104
9 8 7 7 5 2	0	1 2 4 8 9 9
6 6 4 4 3	1	0 0 1 1 2 5 6 6
8 4 3	2	0 5 6
5 5 3	3	
0	4	
	5	1

3|2 = 23 magazines *2|5 = 25 magazines*

Personal Tutor glencoe.com

Real-World Link

Each year about 33 million 12- to 19-year-olds spend around 175 billion dollars on magazines.

Source: Study of Entertainment and Health

Check Your Understanding

Examples 1–3 pp. 743–745

Find the measures of variation and any outliers for each set of data.

1. The number of minutes spent bike riding are: 120, 80, 170, 100, 120, 110, 180, and 35.

2. **Animal Life Span**

Stem	Leaf
0	6 7
1	2 5 5 6
2	0

2 | 0 = 20 years

Examples 4 and 5 pp. 745–746

3. **FOOD** The number of Calories in a serving of certain fruits and vegetables is shown at the right.
   a. Compare the fruits' range with the vegetables' range.
   b. Determine any outliers. How do the outliers affect the measures of central tendency for the number of Calories in fruits?

Calories Per Serving

Fruits	Stem	Vegetables
	3	0 5 5
	4	0 0 0 5
0 0 0 0	5	0
0 0 0	6	
	7	
0	8	

0 | 6 = 60 Calories *3 | 5 = 35 Calories*

Practice and Problem Solving

= **Step-by-Step Solutions** begin on page R11.
Extra Practice begins on page 810.

Examples 1–3 pp. 743–745

Find the measures of variation and any outliers for each set of data.

4. **Computer Game Sales**

Day	Number Sold
Monday	89
Tuesday	90
Wednesday	80
Thursday	100
Friday	92
Saturday	104
Sunday	150

5. **Number of Sunny Days Per Month**

Stem	Leaf
0	
1	5 7 8 8
2	0 1 2 3 5 5 7
3	0

2 | 2 = 22 days

6. **Popcorn Sales at Movie Time Theatre**

Year	2003	2004	2005	2006	2007	2008	2009	2010
Sales (thousands)	0.66	0.43	1.25	0.2	0.53	0.6	0.58	0.48

Examples 4 and 5 pp. 745–746

7. **SCIENCE** The table shows the top scores on a science test.
   a. What is the median score for each room?
   b. Which room has a greater range of scores?
   c. Of the top scores shown, does either set of data have an outlier? If so, how does the outlier affect the range?

Science Scores

Room 100	Room 110
94	82
64	79
88	85
100	91
91	97
106	109
97	103
88	100
97	82

Examples 4 and 5
pp. 745–746

8. **SCORES** The table shows the number of points Cami scored per basketball game.

Cami's Basketball Stats

2008	Stem	2007
	0	6 7 8 9 9
8 8 6 6 4 4 4	1	0 0 2 4 6 7 8
6 6 4 4 2 2	2	1 7
	3	
0	4	

4|2 = 24 points   2|1 = 21 points

a. In which season did Cami have a greater range of points?

b. Select the appropriate measure of central tendency to describe the number of points she scored each season. Justify your response.

c. The score 40 is considered an outlier. How does the inclusion of the outlier for the number of points scored in the 2008 season affect the range?

9. **WEATHER** The table shows average monthly temperatures.

a. Which city has a greater range of temperatures?

b. Find the measures of variation for each city.

c. Compare the medians and the interquartile ranges of the average temperatures.

d. Select the appropriate measure of central tendency to describe the average high temperature for Augusta. Justify your response.

e. Describe the average temperatures of Antelope and Augusta, using both the measures of central tendency and variation

Average Monthly Highs

Month	Antelope, MT	Augusta, ME
January	21	28
February	30	32
March	42	41
April	58	53
May	70	66
June	79	75
July	84	80
August	84	79
September	72	70
October	58	58
November	37	46
December	24	34

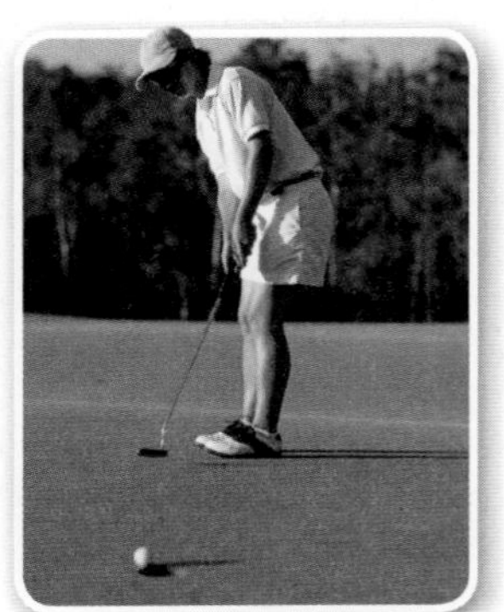

Real-World Link

In golf, the difficulty of a particular hole is given by its par, the number of strokes expected to reach the hole. A score of +3, read as 3 above par, means it took 3 additional tries to put the ball in the hole. A score of −2, read as 2 below par, means it took 2 less tries than expected to reach the hole.

10. **GOLF** Rondell's golf scores at the end of the high school golf tournament were −1, −2, 4, −6, 3, −1, and −3. Find the measures of variation of his golf scores. Explain what the measures of variation tell you about the data.

11. **ANALYZE DATA** Find a real-world data set with at least eight values that include one or more outliers. Display the data in a table. Find the mean and the interquartile range of the data set. Then remove the outlier(s) from the data set and find the mean and the interquartile range. Describe any differences in the values.

H.O.T. Problems

Use Higher-Order Thinking Skills

12. **OPEN ENDED** Write two lists of data that each have at least 15 values and a median of 10. One should have a range of 25 and the other a range of 6.

13. **REASONING** The range is *always, sometimes,* or *never* affected by outliers. Justify your reasoning.

14. **REASONING** *True or false.* The interquartile range is affected by high and low values of a data set. Explain your reasoning.

15. **REASONING** *True or false.* Of mean, median, and mode, the mean will always be most affected by the outliers. If false, give a counterexample.

16. **WRITING IN MATH** Explain how outliers affect the calculation of measures of variation in a data set.

Standardized Test Practice

17. The table shows the total number of wins by the team that won the women's NCAA basketball tournament for the past 15 years.

Total Wins				
34	39	29	31	34
34	34	32	37	39
33	36	35	33	31

Which of the following statements is *not* supported by these data?

A Less than half of the teams won more than 34 games and half won less than 34 games.

B The range of the data is 10 games.

C About one fourth of the teams won 32 or fewer games.

D An outlier of the data is 29 games.

18. Refer to the table in Exercise 17. What is the interquartile range of the data?

F 4 **H** 8

G 5 **J** 10

19. What is the lower quartile of the following set of data?

37, 12, 7, 8, 10, 5, 14, 19, 7, 15, 11

A 7 **C** 11

B 7.5 **D** 15

20. SHORT RESPONSE Find the measures of variation and any outliers for the set of data.

30, 62, 35, 80, 12, 24, 30, 39, 53, 38

Spiral Review

21. MONEY Display the data representing the cost of DVDs $12, $15, $18, $21, $14, $37, $27, $9 in a stem-and-leaf plot. (Lesson 13-2)

Find the mean, median, and mode for each set of data. Round to the nearest tenth. (Lesson 13-1)

22.

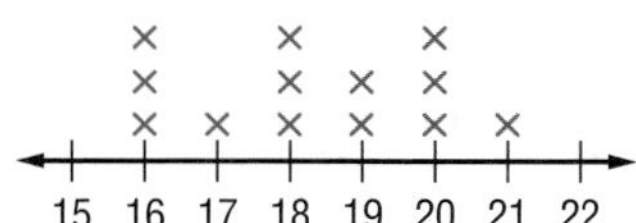

23.

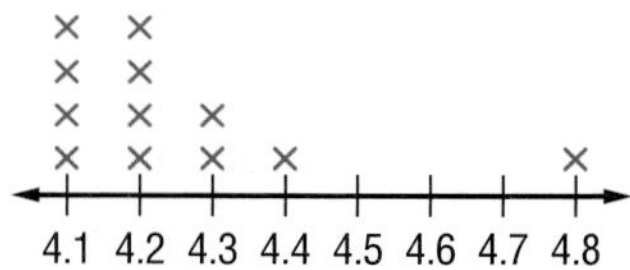

Determine whether each pair of solids is similar. (Lesson 12-8)

24.

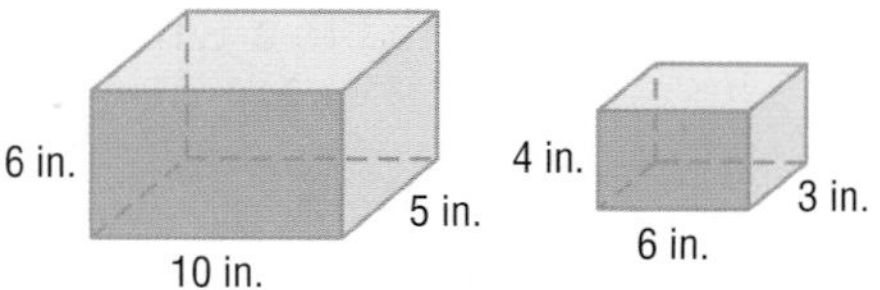

25.

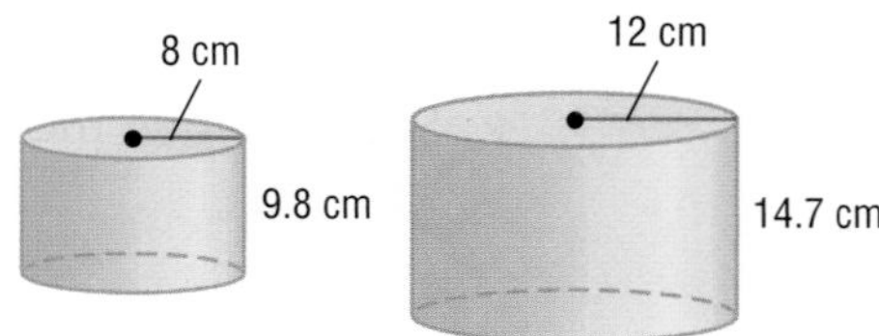

Find the volume of each cone. Round to the nearest tenth. (Lesson 12-4)

26. radius, 7 cm; height, 9 cm

27. diameter, 8.4 yd; height, 6.5 yd

Skills Review

Order each set of decimals from least to greatest. (Previous Course)

28. 1.0, 1.1, 0.9, 0.5, 1.9, 10.9, 0.1

29. 7.8, 8.7, 6.7, 6.8, 7.0, 6.9

13-4 Box-and-Whisker Plots

Then

You have already interpreted and analyzed data. (Lessons 13-1 through 13-3)

Now

- Display data in a box-and-whisker plot.
- Interpret data in a box-and-whisker plot.

New Vocabulary

box-and-whisker plot

Math Online

glencoe.com

- Extra Examples
- Personal Tutor
- Self-Check Quiz
- Homework Help

Why?

The table shows the number of wins per season for the Boston Red Sox and the Florida Marlins.

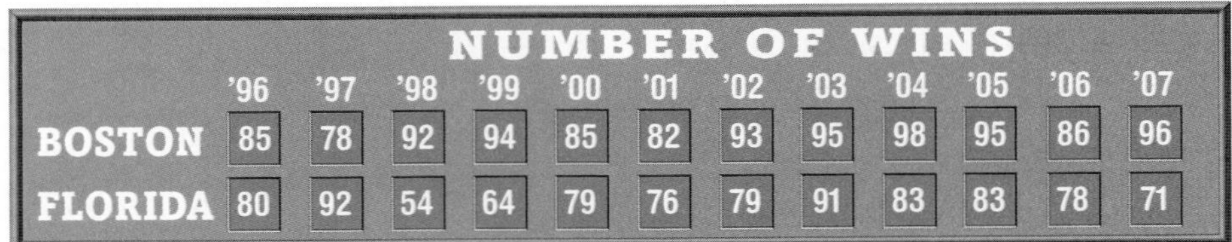

NUMBER OF WINS	'96	'97	'98	'99	'00	'01	'02	'03	'04	'05	'06	'07
BOSTON	85	78	92	94	85	82	93	95	98	95	86	96
FLORIDA	80	92	54	64	79	76	79	91	83	83	78	71

Source: Major League Baseball

a. What is the least value in the data set for Boston? Florida?

b. What is the lower quartile of the data for Boston? Florida?

c. What is the median of the data for Boston? Florida?

d. What is the upper quartile of the data for Boston? Florida?

e. What is the greatest value in the data for Boston? Florida?

Display Data A **box-and-whisker plot**, or box plot, uses a number line to show the distribution of a set of data. It divides a set of data into four parts using the median and quartiles. A *box* is drawn around the quartile values, and *whiskers* extend from each quartile to the minimum and maximum values that are not outliers.

EXAMPLE 1 Draw a Box-and-Whisker Plot

BASEBALL Use the data in the table above to draw a box-and-whisker plot for Boston's number of wins.

Step 1 Draw a number line that includes the least and greatest numbers in the data.

Step 2 Mark the minimum and maximum values, the median, and the upper and lower quartile above the number line. Check for outliers. If an outlier exists, mark the greatest value that is not an outlier. Use an asterisk (*) to indicate an outlier. It is not connected to a whisker.

Step 3 Draw the box and the whiskers.

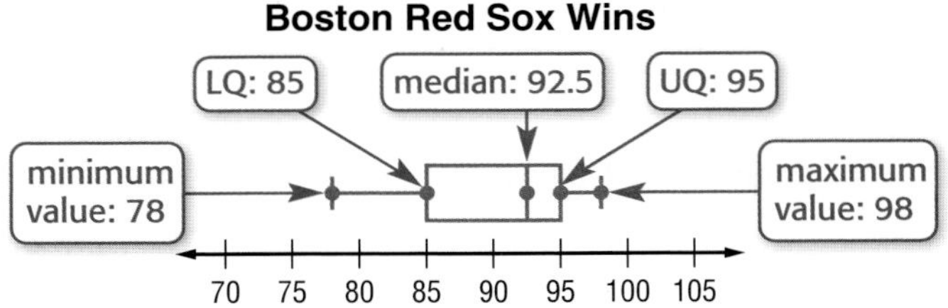

Check Your Progress

1. BASEBALL Use the data in the table above to draw a box-and-whisker plot for Florida's number of wins.

Personal Tutor glencoe.com

StudyTip

Five-number summary While the box-and-whisker plot separates data into four parts, it uses 5 numbers to do so: the minimum value, the lower quartile, the median, the upper quartile, and the maximum value.

Interpret Box-and-Whisker Plots Box-and-whisker plots separate data into four parts, excluding outliers. Even though the parts may differ in length, each part contains one-fourth or, 25%, of the data.

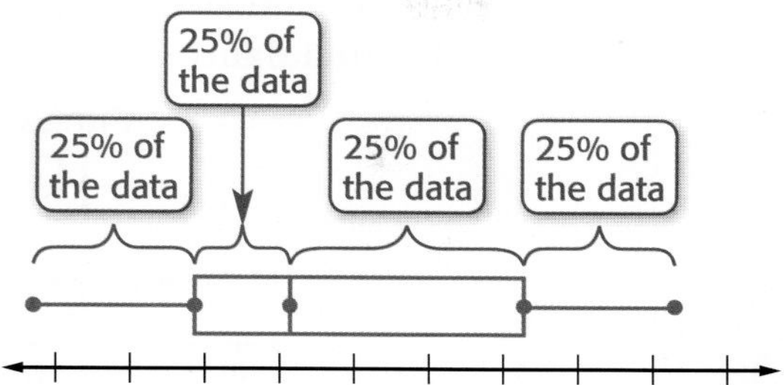

A long whisker or box indicates that the data have a greater range. The values of the data are spread out. A short whisker or box indicates the data have a lesser range. The values of the data in that part are concentrated.

Real-World EXAMPLE 2 Interpret Data

MUSIC The number of chart hits for the top female groups in the U.S. are displayed.

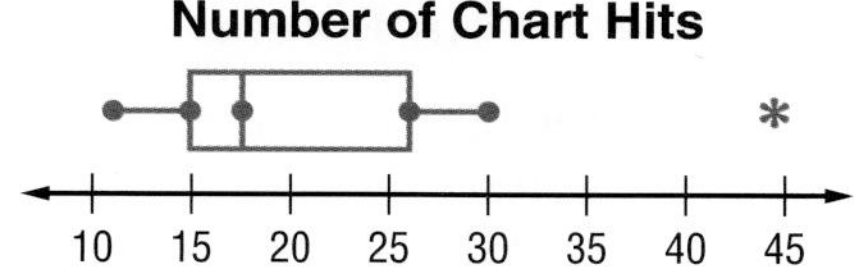

a. What percent of the groups had at least 18 chart hits?

Half of the groups, or 50%, had at least 18 chart hits.

b. What does the length of the box tell about the data? length of the whisker?

The median divides the data in the box into two unequal parts. The data between the lower quartile and the median are more clustered than the data between the median and the upper quartile.

The whiskers are the same length, so the data below the lower quartile and above the upper quartile have the same range.

Check Your Progress

2. **CALORIES** The number of Calories in one serving of various muffins are displayed in the box-and-whisker plot below.

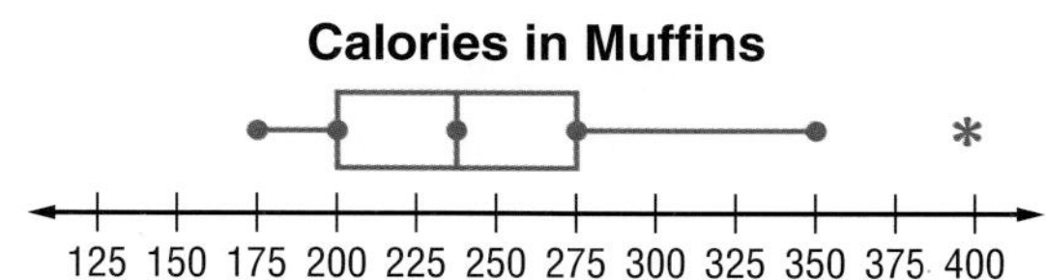

A. What percent of the muffins have more than 275 Calories per muffin?

B. What does the length of the box tell about the data? length of the whiskers?

Personal Tutor glencoe.com

Real-World EXAMPLE 3 Compare Data

FITNESS Two fitness clubs are analyzing their daily attendance for September. How does the daily attendance at the Athletic Club compare to the daily attendance at SuperFit?

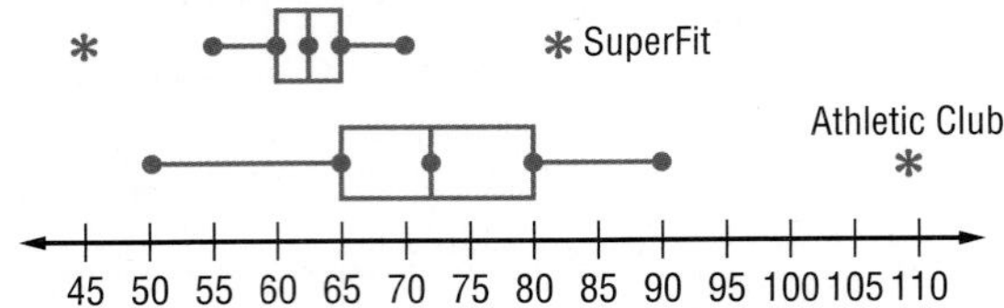

- The Athletic Club's highest and lowest daily attendance figures are both greater than SuperFit's corresponding attendance figures.
- The range of values from the Athletic Club is greater than the range of values from SuperFit.
- One fourth of the time, the attendance at The Athletic Club is greater than 80. The attendance at SuperFit is almost always less than 80.

These results suggest that the Athletic Club's daily attendance varies more than SuperFit's daily attendance. On average the Athletic Club has a higher daily attendance than SuperFit does.

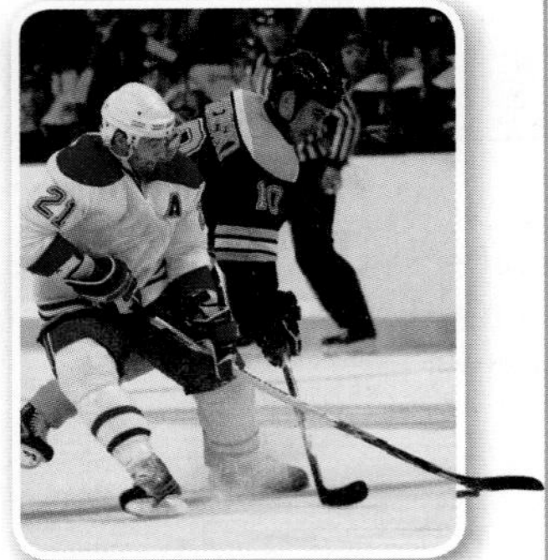

Real-World Link

The National Hockey League began in Montreal in 1917. The league expanded into the United States in 1924 with the Boston Bruins.

Check Your Progress

3. **HOCKEY** The number of goals scored in a recent regular season for players on the Philadelphia Flyers and the Pittsburgh Penguins is displayed. How does the number of goals for Philadelphia compare to the number of goals for Pittsburgh?

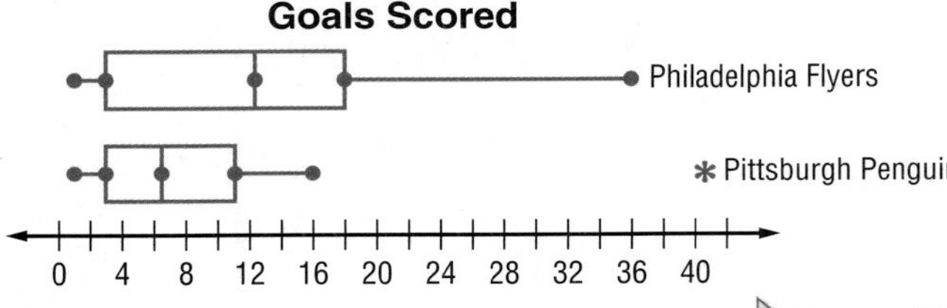

Personal Tutor glencoe.com

Check Your Understanding

Example 1 p. 750

1. **GEOGRAPHY** The heights in feet of several waterfalls in Africa and Asia are shown. Display the data in a box-and-whisker plot.

406	508	630	343	480	330	726
830	330	614	1100	885	1137	890

Source: *The World Almanac*

Example 2 p. 751

2. **EXERCISE** The number of Calories burned for 15 minutes of exercise for various activities is shown.

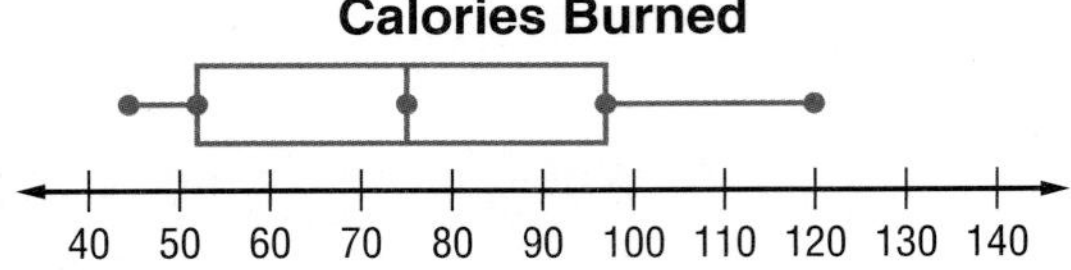

   a. What percent of the activities burn fewer than 98 Calories?

   b. What does the length of the box tell about the data?

Example 3
p. 752

3. **FOOTBALL** Two football teams are analyzing the number of points they scored in each game this season. How does the number of points scored by the Falcons compare to the number of points scored by the Cougars?

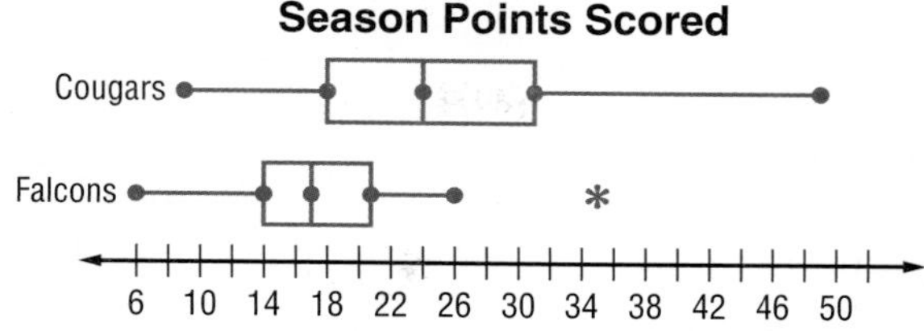

Practice and Problem Solving

= **Step-by-Step Solutions** begin on page R11.
Extra Practice begins on page 810.

Example 1
p. 750

Construct a box-and-whisker plot for each set of data.

4.

Number of Rainy Days Last Year for Various Cities		
173	176	185
182	172	120
190	173	182
182	180	173

5.

Price of Paintings ($)		
175	245	200
290	265	250
355	240	225
250	200	220

6. Miles per gallon, for city driving, of various automobiles: 28, 24, 23, 21, 27, 32, 19, 20, 18, 25, 25

7. Age of students in a pottery class: 23, 37, 34, 19, 28, 33, 26, 27, 35, 25, 21, 29, 28

Example 2
p. 751

8. **GAMES** The yearly games sales for a popular toy manufacturer are displayed.

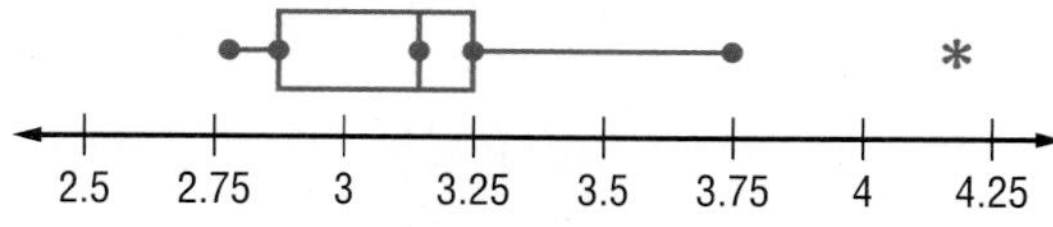

a. What percent of the years had sales that were more than $3.25 billion?

b. What does the length of the box tell about the data? the whiskers?

Real-World Link

The eye of a hurricane is a circular area that can range from 2–230 miles in diameter. Inside the eye, the wind is calm and the skies may be clear. The eyewall surrounds the eye and contains the most violent weather in a hurricane.

9. **HURRICANES** The top wind speeds for hurricanes in a recent year are shown below.

Top Wind Speeds for Hurricanes (miles per hour)

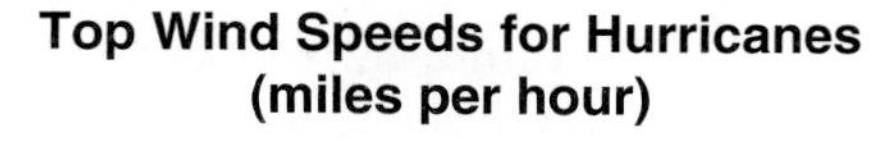

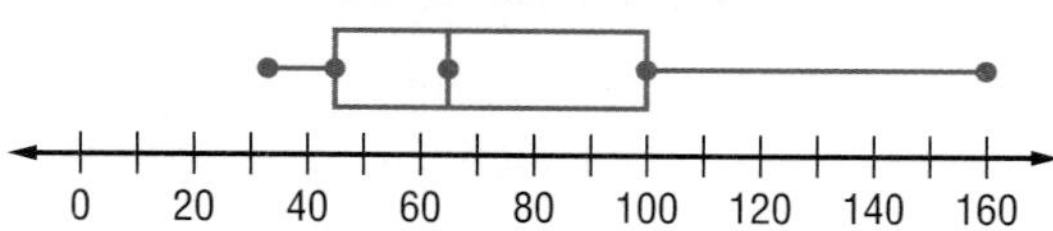

a. What percent of the hurricanes had top wind speeds less than 65 miles per hour?

b. What does the length of the box tell about the data? the whiskers?

Example 3
p. 752

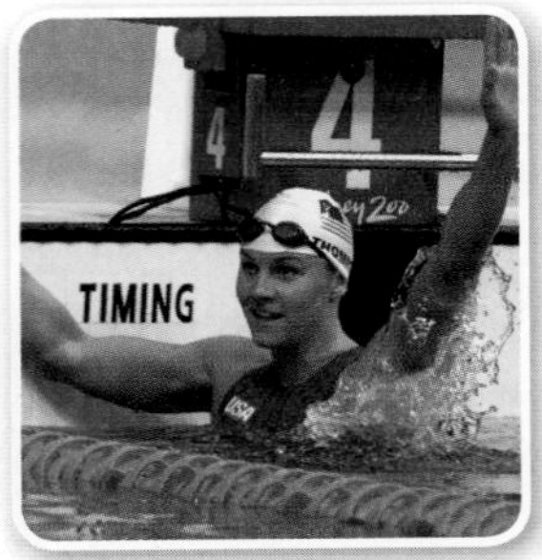

Real-World Link

Jenny Thompson competed in the 1992, 1996, 2000, and 2004 Olympics. She won a record total of 8 gold medals in swimming.

10. OLYMPICS Refer to the double box-and-whisker plot shown that shows the ages of the 2004 U.S. Men's and Women's Olympic Swimming Team.

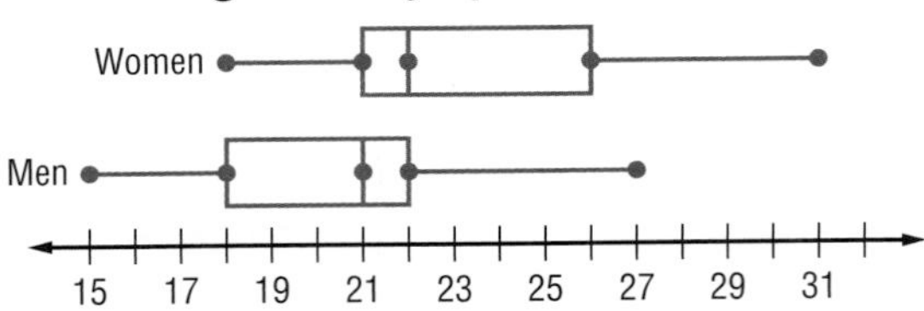

a. Which set of data has the greater range?

b. How many outliers are there in the data?

c. What percent of the women's team is 22 years or younger? men's?

d. Describe the ages of the women compared to the men.

Construct a box-and-whisker plot for each set of data.

11

DVD	A	B	C	D	E	F	G	H
Prices ($)	18.20	12.99	14.95	20.00	17.50	13.50	14.00	12.99

12.

Song	1	2	3	4	5	6	7	8
Length (min)	4.5	3.75	5.25	4.0	5.75	3.75	3.5	5.0

H.O.T. Problems

Use Higher-Order Thinking Skills

13. OPEN ENDED Create two data sets that contain from 10 to 20 values each that could have been used to generate the box-and-whisker plots shown below. Then write a few sentences that analyze the data.

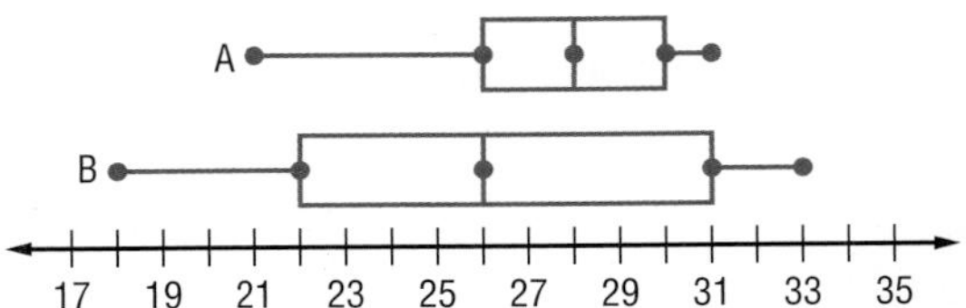

14. FIND THE ERROR Trina constructed a box-and-whisker plot showing her favorite NASCAR driver's yearly wins of 7, 14, 6, 1, 2, 2, 6, 3, 2, 8, 4, 7. Is she correct? Explain your reasoning.

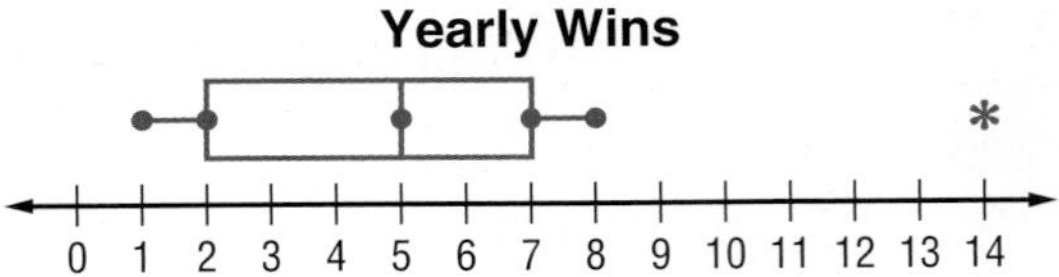

15. CHALLENGE The following data show the performance of a math class on a 50-point quiz: minimum: 28, lower quartile score: 30, median: 38, upper quartile score: 42, and maximum: 48.

a. Suppose there are 13 students in the class. Give a set of scores that would satisfy all the data shown.

b. Suppose six students have scores ranging from 38 to 42. How many students might there be in the class? Explain your reasoning.

16. WRITING IN MATH Explain how extending a whisker to include an outlier changes the look of a box-and-whisker plot.

Standardized Test Practice

For Exercises 17–19, use the box-and-whisker plot below of the following quiz scores for 25 students. {3, 3, 4, 4, 5, 5, 5, 5, 5, 5, 6, 6, 6, 6, 7, 8, 8, 9, 9, 9, 10, 10, 11, 11, 16}

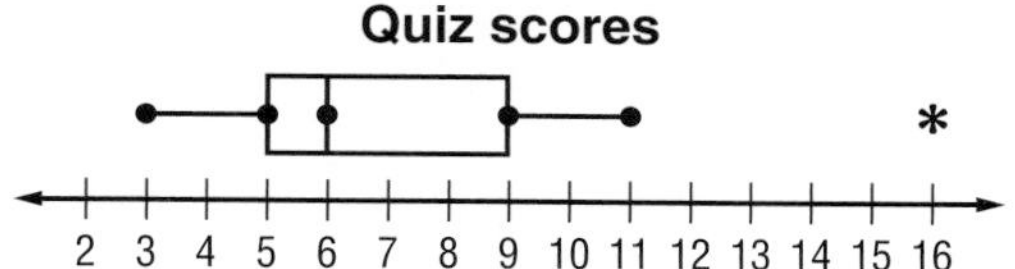

17. GRIDDED RESPONSE What is the interquartile range in the box-and-whisker plot?

18. Which of the following statements is *not* true about the box-and-whisker plot?

A The outlier is 16.

B Seventy-five percent of the data is less than or equal to 9.

C The median is 6.

D There are more data values in the interval 6 to 11 than in the interval 3 to 6.

19. Kenko was absent the day the quiz was given. If she scored a 12 on her quiz, how will the box-and-whisker plot at the left change?

F The lower quartile will change to 4.

G The median will change to 7.

H The right whisker will increase to 12.

J There will no longer be an outlier.

20. The five number summary required to make a box-and-whisker plot contains which of the following?

A mean, median, mode, range, outliers

B lower extreme, lower quartile, median, upper quartile, outliers

C range, lower quartile, median, upper quartile, mode

D lower extreme, lower quartile, median, upper quartile, upper extreme

Spiral Review

21. SPORTS Use the data in the stem-and-leaf plot. (Lessons 13-1 through 13-3)

Total Points Scored by Winners 1960–2008

Rose Bowl		Cotton Bowl
	0	7 7
8 7 7 7 7 7 7 4 4 4 4 3 0	1	0 0 0 0 2 3 3 4 4 7 7 7 7 9 9
8 8 7 7 7 4 3 3 2 2 1 1 1 0 0 0	2	0 1 3 4 4 4 7 8 8 8 9
8 8 8 8 7 4 4 4 4 2	3	0 0 0 1 1 1 5 5 5 5 6 6 8 8 8 8 8
9 6 5 5 4 2 2 2 1 1	4	1 5 6
	5	5
7|1 = 17 Points		*2|4 = 24 Points*

Source: *The World Almanac*

a. Find the range, median, upper quartile, lower quartile, interquartile range, and any outliers for each set of data.

b. Write a few sentences that compare the data.

Skills Review

22. PETS The table shows the number of pets owned by students in Mr. Hinkel's class. (Previous Course)

a. How many students were surveyed?

b. How many students have more than five pets?

Number of Pets	Frequency
0-2	13
3-5	7
6-9	5
10-12	0
13-15	1

EXTEND 13-4

Graphing Technology Lab
Box-and-Whisker Plots

Math Online glencoe.com
- Other Calculator Keystrokes
- Graphing Technology Personal Tutor

You can use a graphing calculator to create box-and-whisker plots.

EXAMPLE

The table shows the total precipitation (in inches) during each month in a recent year for two states.

State	J	F	M	A	M	J	J	A	S	O	N	D
Kentucky	5.72	2.22	3.69	4.63	4.44	4.24	4.61	4.63	8.55	5.05	3.17	3.01
Missouri	2.07	0.56	3.9	3.72	3.7	3.08	3.55	4.32	2.85	3.91	3.68	3.2

Source: NOAA

Make box-and-whisker plots for the precipitation levels in Kentucky and in Missouri.

Step 1 Enter the data.

- Clear any existing data.

KEYSTROKES: [STAT] [ENTER] [▲] [CLEAR] [ENTER]

- Enter the Kentucky data in L1 and the Missouri data in L2.

KEYSTROKES: *Review entering a list on page 47.*

Step 2 Format the graph.

- Turn on two statistical plots.

KEYSTROKES: *Review statistical plots on page 47.*

- For Plot 1, select the box-and-whisker plots and L1 as the Xlist.

KEYSTROKES: [▼] [▶] [▶] [▶] [ENTER] [▼] [2nd] [L1] [ENTER]

- Repeat for Plot 2, using L2 as the Xlist, to make a box-and-whisker plot for Missouri.

Step 3 Graph the box-and-whisker plots.

- Display the graph.

KEYSTROKES: [Zoom] 9

Press [TRACE]. Move from one plot to the other using the up and down arrow keys. The right and left arrow keys allow you to find the least value, greatest value, and quartiles.

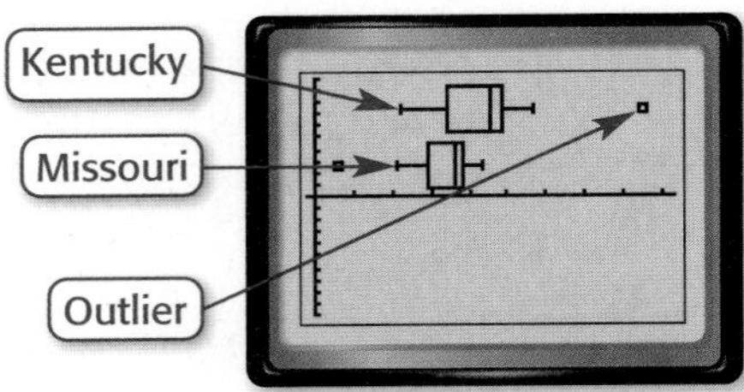

Exercises

1. What are the least, greatest, quartile, and median values for Kentucky and Missouri?
2. What is the interquartile range for Kentucky? Missouri?
3. Are there any outliers? How does the graphing calculator show them?
4. If the average monthly precipitation for Kentucky is 3.94 inches, estimate the percent of months that Kentucky had above average precipitation.
5. If the average monthly precipitation for Missouri is 3.2 inches, estimate the percent of months that Missouri had above average precipitation.
6. Based on the precipitation data, in which state would you prefer to live? Explain.

13-5 Histograms

Why?

A heliport pad is a landing and takeoff pad for helicopters. The number of heliports in each state in the United States are shown in a *frequency table.*

a. How many states have 151–200 heliports? more than 300 heliports?

b. What do you notice about the intervals?

Number of Heliports in Each State

Heliports	Tally	Frequency
1-50	卌 卌 卌 III	18
51-100	卌 卌 IIII	14
101-150	卌 III	8
151-200	II	2
201-250	II	2
251-300	III	3
301-350	I	1
351-400	I	1
401-450	I	1

Then

You have already displayed data in a stem-and-leaf plot. (Lesson 13-2)

Now

- Display data in a histogram.
- Interpret data in a histogram.

New Vocabulary

histogram

Math Online

glencoe.com

- Extra Examples
- Personal Tutor
- Self-Check Quiz
- Homework Help

Display Data A **histogram** uses bars to display numerical data that have been organized into equal intervals.

EXAMPLE 1 Draw a Histogram

RETAIL The table shows the ages of people who entered a store. Display the data in a histogram.

Age	Tally	Frequency
1–10	卌	5
11–20	卌 III	8
21–30	卌 卌 IIII	14
31–40	卌 卌 卌 III	18
41–50	卌 卌 卌 卌	20
51–60	卌 卌 III	13
61–70	卌 I	6

Step 1 Draw and label a horizontal and vertical axis as shown. Include a title.

Step 2 Show the intervals from the frequency table on the horizontal axis and an interval of 2 on the vertical axis.

Step 3 For each interval, draw a bar whose height is given by the frequency.

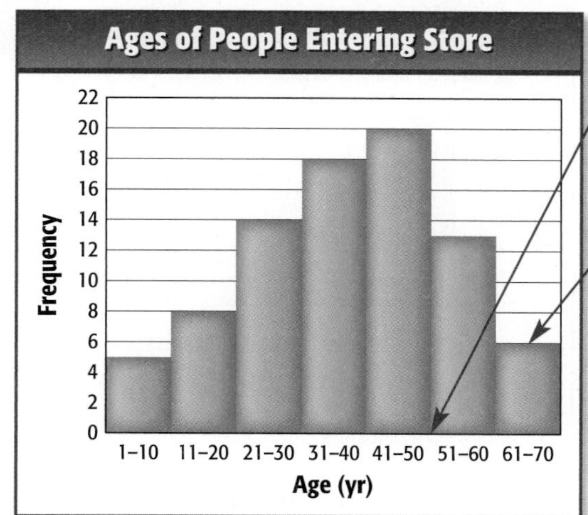

There is no space between bars.

Because the intervals are equal, all of the bars have the same width.

Check Your Progress

1. HELIPORTS Refer to the frequency table above about the number of heliports in each state. Display the data in a histogram.

Personal Tutor glencoe.com

Interpret Data A histogram gives a better visual display of data than a frequency table. Thus, it is easier to interpret data displayed in a histogram.

StudyTip

Break in Scale The symbol ⩘ means there is a break in the scale. The scale from 0 to 59 has been omitted.

EXAMPLE 2 Interpret Data

SCHOOL Refer to the histogram.

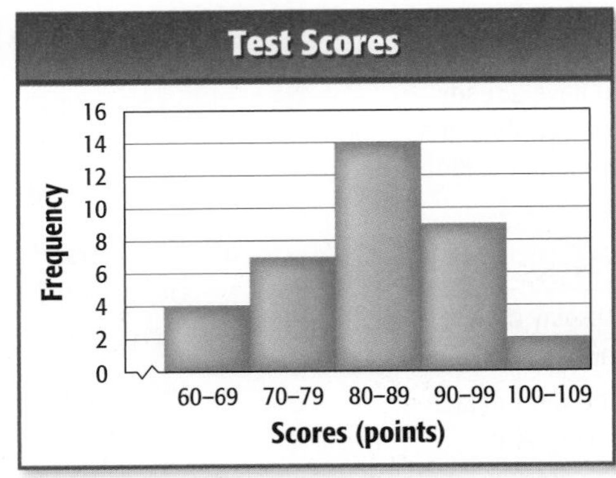

a. How many students scored at least 90 points on the test?

Since 9 students scored between 90 and 99 points and 2 received over one hundred points, $9 + 2$ or 11 students scored at least 90 points on the test.

b. What percent of the students scored 79 points or lower?

There were $4 + 7 + 14 + 9 + 2$ or 36 students who took the test. There were $4 + 7$ or 11 total students who scored 79 points or less. Since $\frac{11}{36}$ is about 30.56%, about 31% of the students scored 79 points or lower on the test.

Check Your Progress

2. SCHOOL Refer to the histogram about test scores above.

A. Is it possible to tell the score of the highest test?

B. In what range is a student most likely to score on the test?

Personal Tutor glencoe.com

Real-World EXAMPLE 3 Compare Data

BUILDINGS Use the histograms shown below. Which city has a greater number of buildings at least 600 feet tall?

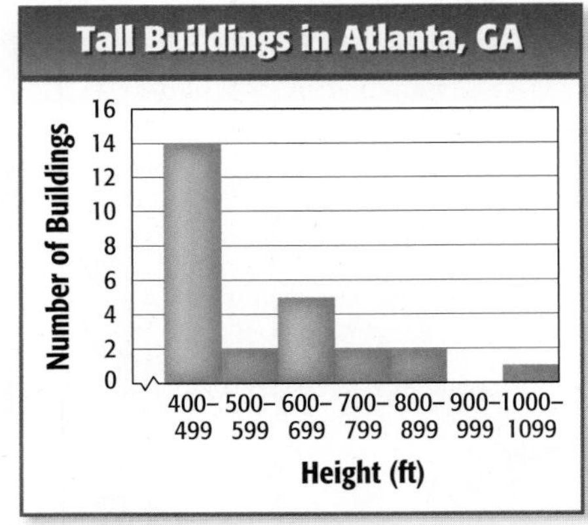

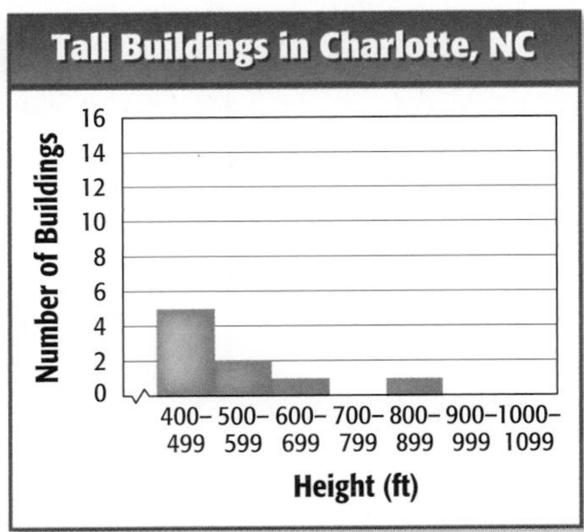

Atlanta has $5 + 2 + 2 + 1$ or 10 buildings at least 600 feet tall. Charlotte has $1 + 1$ or 2 buildings at least 600 feet tall. So, Atlanta has more buildings at least 600 feet tall.

StudyTip

No Data In both histograms, there is no bar for the 900–999 interval. When this happens, there is no data for that interval. In this situation, there are no buildings with a height of 900–999 feet in either city.

Check Your Progress

3. Compare the heights of the tall buildings in the two cities.

Personal Tutor glencoe.com

Check Your Understanding

Example 1 p. 757

Display each set of data in a histogram.

1.

Age of Indy 500 Winners		
Age	Tally	Frequency
21–25	||	2
26–30	~~||||~~ |	6
31–35	~~||||~~ |||	8
36-40	~~||||~~	5
41–45	||||	4
46–50	|||	3

2.

Weekly Time Spent Doing Chores		
Time (min)	Tally	Frequency
0–14	~~||||~~ |||	8
15–29	~~||||~~ ~~||||~~ |	11
30–44	~~||||~~ ~~||||~~ ||||	14
45–59	~~||||~~ |	6
60–74	||||	4

Example 2 p. 758

3. **TEMPERATURE** The histogram shown below shows the record high temperatures for several U.S. cities.

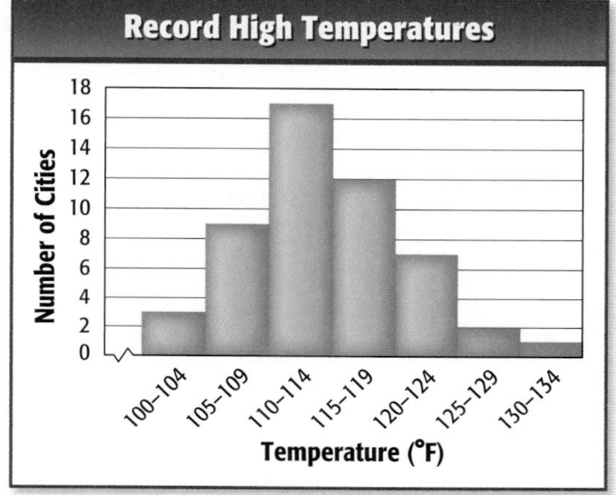

a. About what percent of U.S. cities had record high temperatures of 109° F or less?

b. How likely is it that a record high temperature is 125° F or higher?

c. What is the greatest temperature?

Example 3 p. 758

4. **ANALYZE GRAPHS** The histograms below show the winning times for two different women's swimming events from the summer Olympics.

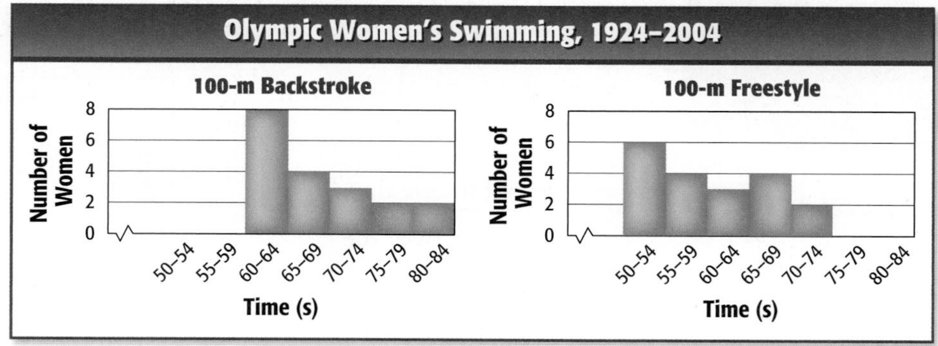

a. Which event has more winning times less than one minute?

b. How many Olympic Games were held from 1924 to 2004?

c. Compare the winning times of the two Olympic events.

Practice and Problem Solving

= **Step-by-Step Solutions** begin on page R11.
Extra Practice begins on page 810.

Example 1 p. 757

Display each set of data in a histogram.

5.

Calories of Fruit Bars		
Calories	**Tally**	**Frequency**
0–39	|||	3
40–79	~~||||~~ ||||	9
80–119	~~||||~~ |	6
120–159	||||	4
160–199	~~||||~~ ||	7

6.

Weekly Time Spent Reading		
Time (hr)	**Tally**	**Frequency**
0–1	~~||||~~ |	6
2–3	~~||||~~ ~~||||~~	10
4–5	~~||||~~ ||	7
6–7	|||	3
8–9	||||	4

Example 2 p. 758

7. **ANALYZE GRAPHS** The histogram shows the cost of different shoes at a store.

a. How many pairs of shoes cost $59.99 or less?

b. About what percent of the shoes cost $60 or more?

c. What price is a pair of shoes most likely to cost?

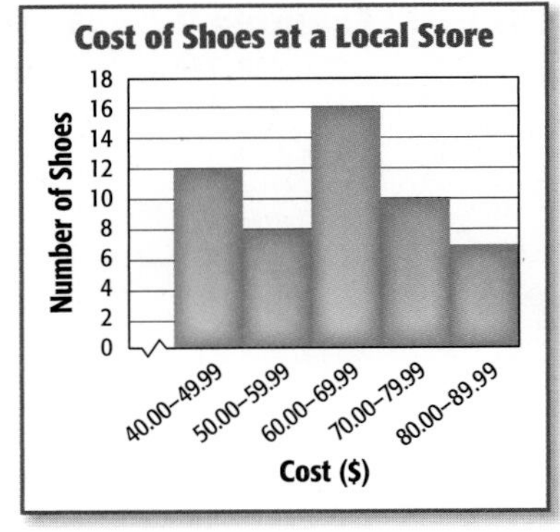

Example 3 p. 758

8. **HEIGHTS** The histograms below show the heights' of students in two homerooms. Compare the data in the histograms.

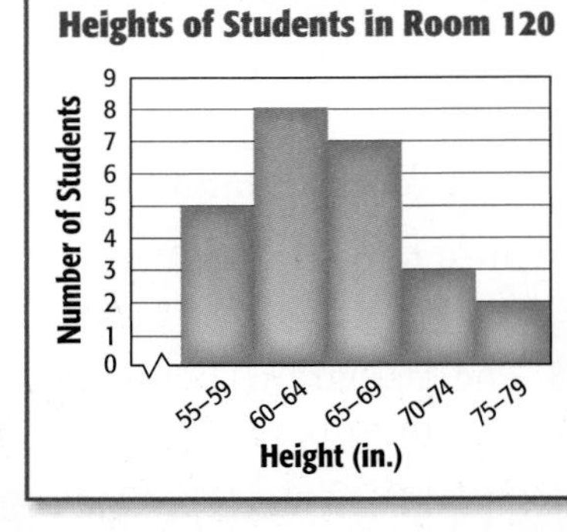

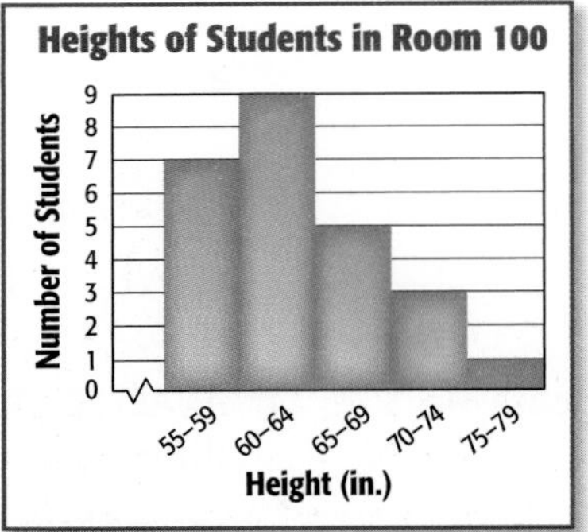

9. **INTERNET** The histogram shows the number of minutes students spend on the Internet in one day.

a. Discuss all of the information that you can collect from the histogram.

b. Are you able to find any measures of central tendency from the histogram? Explain your reasoning.

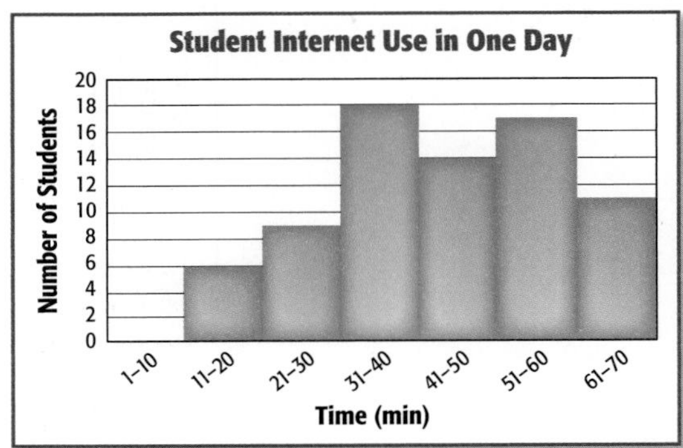

Real-World Link

According to a survey, 82% of teens age 12–14 and 92% of teens age 15–17 use the Internet, with approximately 11 million teens going online daily.

Source: Pew Internet & American Life Project

10. **FINANCIAL LITERACY** The table shows the average ticket prices to attend a game for each of the teams in the NHL.

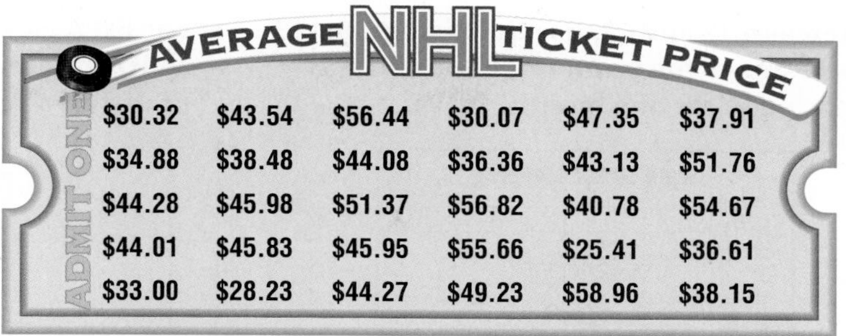

AVERAGE NHL TICKET PRICE

$30.32	$43.54	$56.44	$30.07	$47.35	$37.91
$34.88	$38.48	$44.08	$36.36	$43.13	$51.76
$44.28	$45.98	$51.37	$56.82	$40.78	$54.67
$44.01	$45.83	$45.95	$55.66	$25.41	$36.61
$33.00	$28.23	$44.27	$49.23	$58.96	$38.15

Source: Team Marketing Report

a. Select an appropriate interval for the data. Explain your reasoning. Then make a frequency table using the interval you selected.

b. Construct a histogram of the data.

11. **MEDALS** The histogram at the right shows the total number of medals won by the top ranking country per year for the Summer Olympics from 1896 to 2004.

a. Describe the data in the histogram.

b. Describe how the histogram was constructed incorrectly.

c. How could you change the histogram so that it is no longer constructed incorrectly?

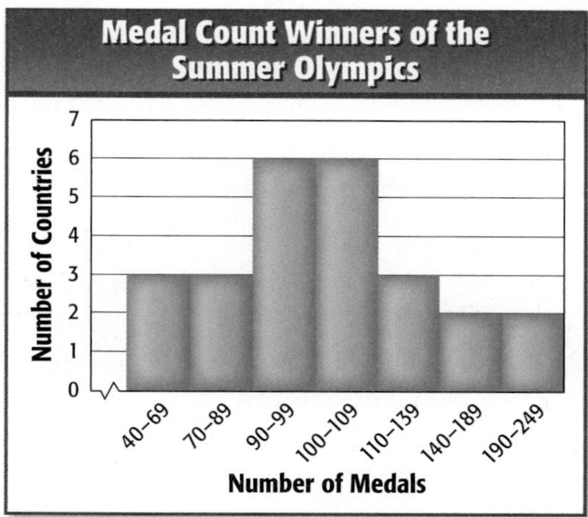

Real-World Link

Each individual Olympic city designs the medal for that year's competition. The medals must be at least 66 millimeters in diameter and at least 3 millimeters thick.

Source: Sports Illustrated

H.O.T. Problems

Use Higher-Order Thinking Skills

12. **OPEN ENDED** Construct a histogram from real-world data that has five intervals and exactly one gap.

13. **ANALYZE DATA** Create a set of data that, when plotted on a histogram, has a gap between 40 and 50, three items in the 20–29 interval, and the median value in the 30–39 interval. Display your data in a histogram.

14. **CHALLENGE** Eighty people were surveyed about the number of times they exercise in a month. The results are shown in the table. Construct a histogram based on these percents.

Exercise Habits	
Days per Month	**Percent of People**
1–5	10%
6–10	20%
11–15	15%
16–20	30%
21–25	20%
26–30	5%

15. **COLLECT THE DATA** Conduct a survey of your classmates to determine the number of text messages each person sends or receives during a typical week. Then make a frequency table and construct a histogram to represent the data. Compare the data in your histogram with a classmate's data.

16. **WRITING IN MATH** Compare a line plot and a histogram. What are the advantages? What are the disadvantages?

Standardized Test Practice

For Exercises 17 and 18, use the histogram below. It shows the number of hours Hugo slept each day for one month.

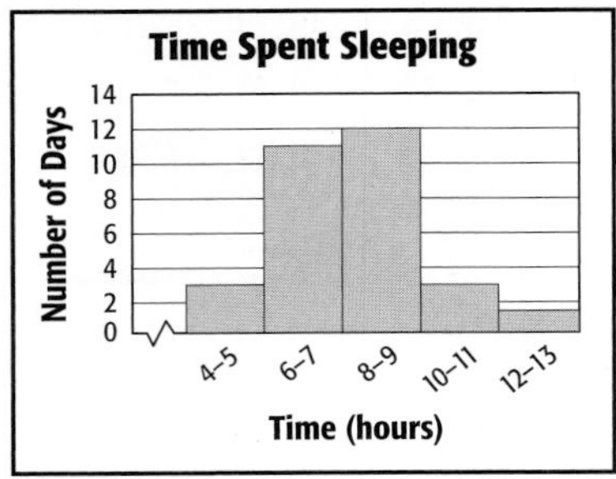

17. What percent of the nights did he sleep for 8 or more hours?

A 40% **C** 50%
B 47% **D** 53%

18. Which of the following can be concluded from the histogram above?

F The least number of hours Hugo slept is 4.
G The greatest number of hours Hugo slept is 13.
H Most of the nights Hugo slept for 8 or more hours.
J The mean number of hours he slept each night was 7.5.

For Exercises 19 and 20, use the data below that show the amount of playing time (in minutes) of different CDs from Janet's CD collection.

Playing Time of CDs (min)				
59	68	56	55	62
48	44	63	47	49
42	61	51	45	61
65	48	58	64	57
54	59	69	55	45

19. SHORT RESPONSE Complete the histogram of the data.

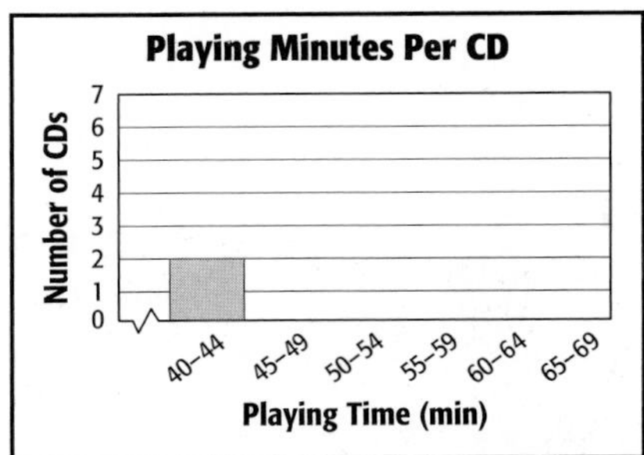

20. GRIDDED RESPONSE How many CDs have a playing time of at least 50 minutes?

Spiral Review

21. DANCE The ages of students in a dance class in years are 25, 30, 27, 35, 19, 23, 25, 22, 40, 34, and 20. Draw a box-and-whisker plot of the data. (Lesson 13-4)

22. CALORIES The stem-and-leaf plot shown shows the number of Calories found in a serving of yogurt. Find the range, interquartile range, and any outliers. (Lessons 13-3 and 13-2)

Calories in Yogurt

Stem	Leaf
4	0
5	0 1 1 5 7 7 7 8
6	7 7 9

5 | 7 = 57 Calories

Skills Review

Max surveyed his classmates to find the school activities they attended last weekend. The results are shown in the Venn diagram. (Previous Course)

School Musical
Basketball Game
23
9
10
8

23. How many students attended the musical? the basketball game?

24. How many students attended both? neither?

25. How many students participated in the survey?

EXTEND 13-5

Graphing Technology Lab
Histograms

Math Online glencoe.com
- Other Calculator Keystrokes
- Graphing Technology Personal Tutor

You can use a graphing calculator to make a histogram.

ACTIVITY

FOOTBALL The table below shows the total number of points scored in each Super Bowl. Make a histogram to show the point distribution.

Total Number of Points Scored in Each Super Bowl						
45	47	23	30	29	27	21
31	22	38	46	37	66	50
37	47	44	47	54	56	59
52	36	65	39	61	69	43
75	44	56	55	53	39	41
37	69	61	45	31	46	

Step 1 Enter the data.

- Clear any existing data in list L1.

KEYSTROKES: STAT ENTER ▲ CLEAR ENTER

- Enter the data in L1.

KEYSTROKES: *Review entering a list on page 47.*

Step 2 Format the graph.

- Turn on the statistical plot.

KEYSTROKES: 2nd [STAT PLOT] ENTER ENTER

- Select the histogram and L1 as the Xlist.

KEYSTROKES: ▼ ▶ ▶ ENTER ▼ 2nd [L1] ENTER

Step 3 Graph the histogram.

- Set the viewing window so the x-axis goes from 20 to 80 in increments of 5, and the y-axis goes from -5 to 15 in increments of 1. So, [20, 80] scl: 5 by [-5, 15] scl: 1. Then graph.

KEYSTROKES: WINDOW 20 ENTER 80 ENTER 5 ENTER -5 ENTER 15 ENTER 1 ENTER GRAPH

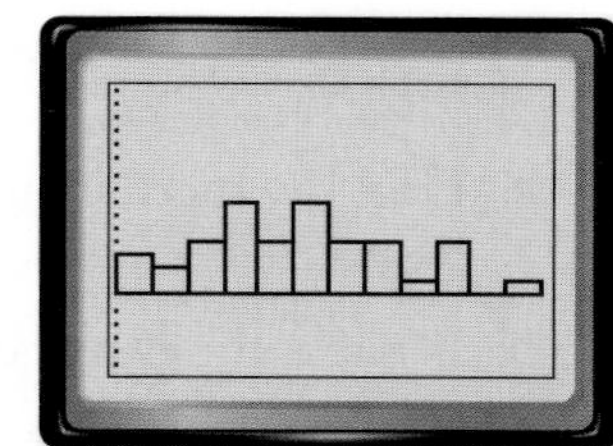

Analyze the Data

1. Press TRACE. Find the frequency of each interval using the right and left arrow keys.
2. Discuss why the domain is from 21 to 75 for this data set.
3. How does the graphing calculator determine the size of the intervals?
4. How many Super Bowls have had a point total of at least 35, but less than 60?
5. What percent of point totals falls in the range of Exercise 4?
6. Can you tell from the histogram how many Super Bowls had point totals of 48?
7. Make a stem-and-leaf plot of the data. How does the stem-and-leaf plot compare to the histogram you have graphed here? Which graph is easier to read?

CHAPTER 13

Mid-Chapter Quiz

Lessons 13-1 through 13-5

Find the mean, median and mode for each set of data. Round to the nearest whole number. (Lesson 13-1)

1. The number of miles several students bike each weekend: 5, 8, 6, 10, 12

2. The number of hours spent sleeping each week for several weeks: 45, 49, 41, 50, 53, 47, 45

3. The ages of people at a party: 2, 5, 7, 9, 10, 36, 37, 41, 42

4. **SCHOOL** The stem-and-leaf plot shows the test scores on Mr. Lisy's social studies test.

Test Scores

Stem	Leaf
4	5 8
5	1 5 7 7
6	2
7	2 2 3 5 8 9
8	1 1 1 1 7 8
9	2 7 9 9
10	0

5|1 = 51

a. Find the median and mode of the data. (Lesson 13-2)

b. What is the range of the data? (Lesson 13-3)

5. There are 22, 21, 24, 23, 26, 23, 28, 24, and 25 students in the homerooms at Morgan Middle School. Find the measures of variation for the data. (Lesson 13-3)

6. **MULTIPLE CHOICE** The number of minutes it took several students to complete a test are shown below. What is the interquartile range of the data? (Lesson 13-3)

45, 57, 55, 42, 48, 21, 39, 62, 45, 51

A 13   C 42

B 41   D 55

7. The total amount of monthly precipitation, in inches, for a city is 1.0, 1.2, 2.2, 3.6, 4.3, 4.6, 4.2, 4.5, 3.2, 2.6, 2.1, and 1.3. Display the data in a box-and-whisker plot. (Lesson 13-4)

8. **SKIING** The box-and-whisker-plot below shows the winning times for the women's downhill skiing event in the winter Olympics. (Lesson 13-4)

Women's Olympic Downhill

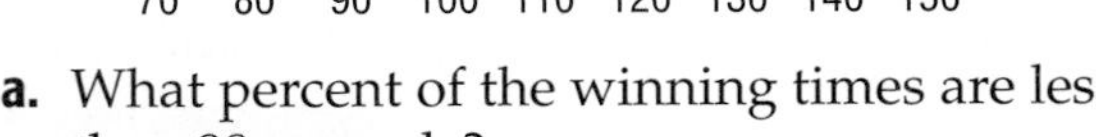

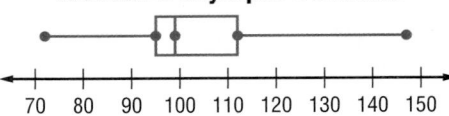

a. What percent of the winning times are less than 99 seconds?

b. What does the length of the box tell about the data? the whiskers?

9. **Display the data below in a histogram.** (Lesson 13-5)

# of Pets	Tally	Frequency
0–1	IIII	4
2–3	𝍸 𝍸	10
4–5	III	3
6–7	I	1

10. **MULTIPLE CHOICE** The histogram shows the prices for attending different sports activities at South High School. (Lesson 13-5)

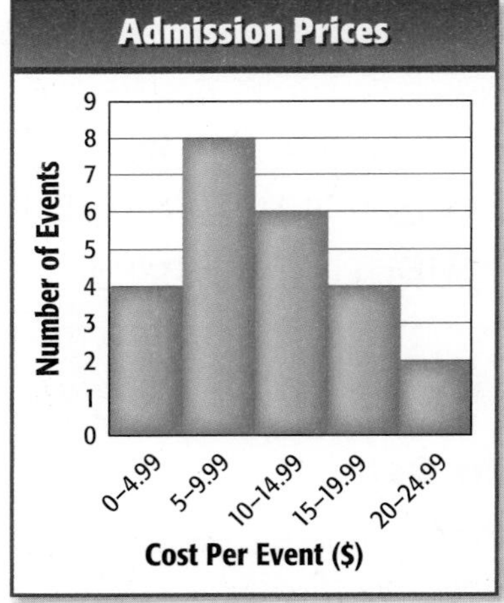

How many events cost $10 or more to attend?

F 3   H 8

G 5   J 12

11. Refer to the histogram above. How many different activities were sampled?

12. What fraction of events cost between $5 and $19.99 to attend?

13-6 Theoretical and Experimental Probability

Then

You have already simplified ratios. (Lesson 6-1)

Now

- Find the probability of simple events.
- Predict the actions of a larger group.

New Vocabulary

outcomes
simple event
probability
sample space
random
theoretical probability
experimental probability
odds in favor
odds against

Math Online

glencoe.com

- Extra Examples
- Personal Tutor
- Self-Check Quiz
- Homework Help
- Math in Motion

Why?

The table shows the number of gumballs of each flavor that is in a gumball machine.

Flavor	Number
blueberry	14
cherry	8
grape	18
lime	18
orange	16
strawberry	16

a. Write a ratio in simplest form that compares the number of grape gumballs to the total number of gumballs.

b. What percent of gumballs are grape?

c. Is there a better chance of getting a cherry or lime flavored gumball? Explain.

Probability of Simple Events In the activity above, there are 6 types of gumballs. These results are called **outcomes**. A **simple event** is one outcome or a collection of outcomes.

You can measure the chances of an event happening with probability. The **probability** of an event is a ratio that compares the number of favorable outcomes to the number of possible outcomes, assuming each outcome is equally likely to occur.

$$P(\text{event}) = \frac{\text{number of favorable outcomes}}{\text{number of possible outcomes}}$$

The probability of an event is always between 0 and 1, inclusive. The closer a probability is to 1, the more likely it is to occur.

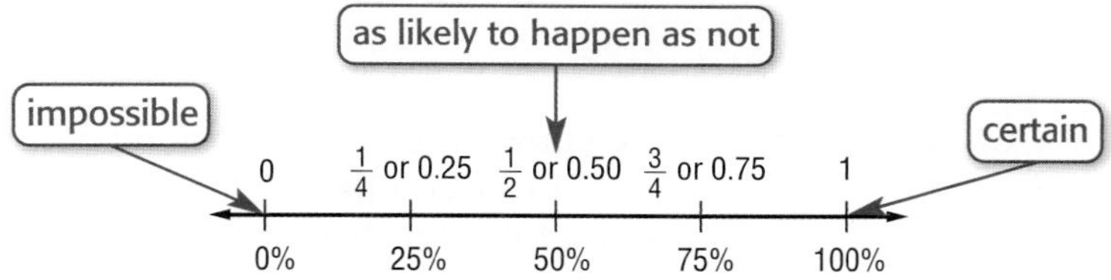

Math *in Motion*, Interactive Lab glencoe.com

EXAMPLE 1 Find Probability

A bag contains 7 pink, 2 white, and 5 blue marbles. A marble is selected without looking. Find the probability that a pink marble is chosen.

There are 14 marbles in all, and 7 of them are pink.

$$P(\text{pink}) = \frac{\text{number of pink marbles}}{\text{number of marbles in all}}$$

$$= \frac{7}{14} \text{ or } \frac{1}{2}$$

The probability of a pink marble being chosen is $\frac{1}{2}$ or 50%.

Check Your Progress

1. What is the probability of choosing a white marble? Express as a fraction and a percent.

Personal Tutor glencoe.com

The set of all possible outcomes is called the **sample space.** In Example 1, the sample space is {pink, white, blue}. When you roll a die, the sample space is {1, 2, 3, 4, 5, 6}. Outcomes occur at **random** if each outcome is equally likely to occur.

ReadingMath

***P*(prime)** is read as *the probability of rolling a prime number.*

EXAMPLE 2 Find Probability

A die is rolled and a coin is tossed. Find *P*(even, heads).

Make a table showing the sample space when rolling a die and tossing a coin.

	H	T
1	(1, H)	(1, T)
2	(2, H)	(2, T)
3	(3, H)	(3, T)
4	(4, H)	(4, T)
5	(5, H)	(5, T)
6	(6, H)	(6, T)

There are 3 outcomes (shown in green) in which an even number is rolled and heads is tossed.

So, $P(\text{even, heads}) = \frac{3}{12}$ or $\frac{1}{4}$.

This means there is a 25% chance of rolling an even number and tossing heads.

Check Your Progress

2. Find *P*(even, heads or tails).

Personal Tutor glencoe.com

The probabilities in Examples 1 and 2 are called theoretical probabilities. **Theoretical probability** is what *should* occur in an experiment. **Experimental probability** is what *actually* occurs when repeating a probability experiment many times. If the number of trials is very small, there can be a wide variation in results.

Math History Link

Gerolamo Cardano (1501–1576) was a mathematician who developed the classical definition of probability. His model described probability as a comparison of favorable outcomes to the number of possible outcomes.

EXAMPLE 3 Find Experimental Probability

The table shows the results of an experiment in which a die was rolled. Find the experimental probability of rolling a six for this experiment. Then compare the experimental probability with the theoretical probability.

Outcome	Tally	Frequency
1	IIII	4
2	~~IIII~~ I	6
3	II	2
4	~~IIII~~ III	8
5	III	3
6	~~IIII~~ I	6

$$\frac{\text{number of times six is rolled}}{\text{number of possible outcomes}} = \frac{6}{4+6+2+8+3+6} \text{ or } \frac{6}{29}$$

The experimental probability of rolling a six in this case is $\frac{6}{29}$ or about 21%. The theoretical probability of rolling a six is $\frac{1}{6}$ or about 17%. So, rolling a six in the experiment occurred more often than expected.

Check Your Progress

3. Find the experimental probability of rolling a four for the experiment above. Then compare it to the theoretical probability.

Personal Tutor glencoe.com

Another way to describe the chance of an event occurring is with odds. The **odds in favor** of an event is the ratio that compares the number of ways the event *can* occur to the number of ways that the event *cannot* occur. The **odds against** an event occurring is the ratio that compares the number of ways the event *cannot* occur to the number of ways that the event *can* occur.

StudyTip

Probability and Odds The probability of an event occurring can be written as a fraction or a percent.

$P(E) = \frac{3}{5}$ or 60%.

The odds in favor of the same event occurring is written as a ratio, 3:2.

EXAMPLE 4 Find the Odds

A die is rolled. Find the odds in favor of rolling a 1 or 2.

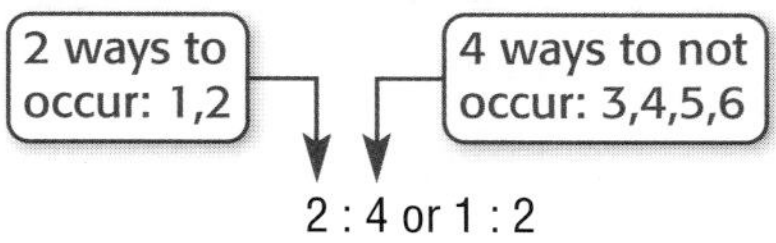

So, the odds in favor of rolling a 1 or a 2 is 1:2.

Check Your Progress

4. A die is rolled. Find the odds against rolling an even number.

Personal Tutor glencoe.com

Use a Sample to Make Predictions You can use past performance or a survey to predict future events.

Real-World EXAMPLE 5 Make A Prediction

HOBBIES The circle graph shows the results of a survey that asked teens, ages 13 to 19, what they would be doing if they were not online. Out of a similar group of 450 teens, predict how many would listen to music.

Teen Hobbies (Other Than Internet)

Watching TV 25%
Listening to music 26%
Reading 9%
Physical activity/Sports 27%
Writing/Drawing 5%
Other 8%

Understand You know that 26% listen to music. You need to know how many teens out of 450 listen to music.

Plan Use the percent proportion to find 26% of 450.

Review Vocabulary

Percent Proportion compares *part* of a quantity to the *whole* quantity as a fraction. (Lesson 7-3)

Solve The percent is 26%, and 450 is the whole. Let n represent the part.

part → / **whole →** $\frac{n}{450} = \frac{26}{100}$ } **percent**

$100 \cdot n = 26 \cdot 450$

$100n = 11{,}700$ **Find the cross products.**

$= 117$ **Mentally divide each side by 100.**

Check Estimate: 25% of 440 is 110. So, 117 is reasonable. ✓

Check Your Progress

5. HOBBIES Out of the 450 teens surveyed, how many would you expect to say they would be writing or drawing? Explain your reasoning.

Personal Tutor glencoe.com

Check Your Understanding

Example 1 p. 765 — **The spinner shown at right is spun once. Determine the probability of each outcome. Express each probability as a fraction and as a percent.**

1. $P(6)$ **2.** $P(\text{even})$

3. $P(\text{greater than } 6)$ **4.** $P(\text{less than } 5)$

Example 2 p. 766 — **5.** Two coins are tossed. What is the probability of both coins landing on heads?

Example 3 p. 766 — **6.** The table shows the results of an experiment in which a die was rolled. Find the experimental probability of rolling a 4. Then compare it to the theoretical probability.

Number	Frequency
1	||
2	|
3	||||
4	|||
5	|||
6	||

Example 4 p. 767 — **7.** A letter of the alphabet is chosen at random. Find the odds in favor of picking an A, E, I, O, or U.

Example 5 p. 767 — **8.** **COLORS** Without looking, Delores took a handful of multi-colored candies from a bag and found that 40% of the candies were yellow. Suppose that there were 375 candies in the bag. How many can she expect to be yellow?

 = **Step-by-Step Solutions** begin on page R11. **Extra Practice** begins on page 810.

Practice and Problem Solving

Example 1 p. 765 — **A dartboard like the one shown is divided into 20 equal sections. Determine the probability of each outcome if a dart is equally likely to land anywhere on the dartboard. Express each probability as a fraction and as a percent.**

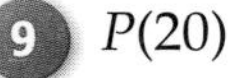

9. $P(20)$ **10.** $P(\text{less than } 8)$

11. $P(\text{odd})$ **12.** $P(\text{even})$

13. $P(\text{greater than } 16)$ **14.** $P(\text{multiple of } 3)$

Example 2 p. 766 — **15.** The spinners shown are each spun once.

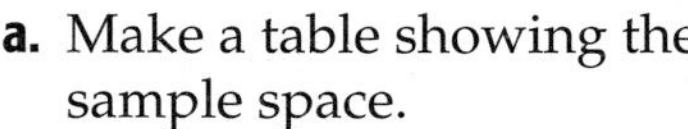

a. Make a table showing the sample space.

b. Find $P(\text{even sum})$.

c. Find $P(\text{two even numbers})$.

Example 3 p. 766 — **16.** The table shows the results of an experiment in which the spinners shown above were each spun 50 times.

Sum	Frequency
2	||
3	|||
4	卌 |||
5	卌 |
6	卌 卌 卌
7	卌 |||
8	卌 |||

a. What is the experimental probability of spinning a sum of 6?

b. What is the experimental probability of spinning a sum of 8?

c. What is the theoretical probability of spinning a sum of 8? Compare it to the experimental probability.

Example 4 p. 767

17. A jar contains 40 pennies, 18 nickels, 20 dimes, and 12 quarters. If a coin is selected at random, what are the odds in favor of picking a penny or dime?

18. In an assembly line, 56 out of 60 randomly selected parts have no faults. What are the odds in favor of a randomly selected part having a fault?

Example 5 p. 767

19. **ANALYZE GRAPHS** Refer to the graph that shows the results of a survey that asked youth about what is important to their personal success.

a. If 1200 youth were surveyed, how many would you expect to say friendships are a factor in their personal success?

b. Suppose 1500 youth were surveyed. How many would you expect to say immediate family is a factor in their personal success?

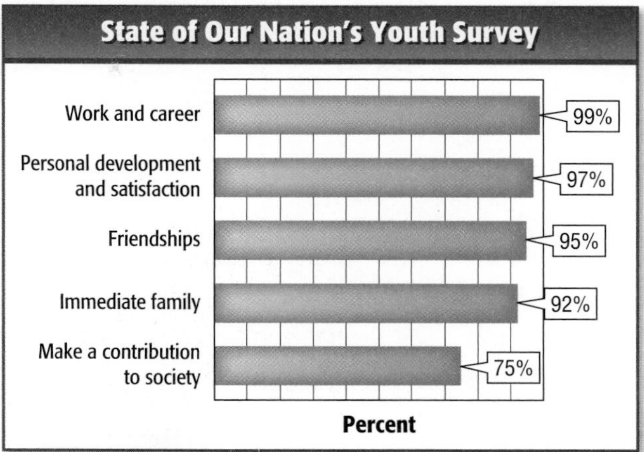

Real-World Link

The X games are held annually with an emphasis on action sports. The first X games were held in the summer of 1995 with competitions in Providence, Rhode Island, Newport, Rhode Island, and Mount Snow, Vermont.

20. **RAFFLE** In a raffle, one ticket is randomly chosen from 80 tickets to receive free haircuts for a year. If Wakim entered 6 tickets, what is the probability that he is *not* chosen to receive the free haircuts?

21. **X GAMES** Use the graph that shows the results of a class survey about students' favorite X Game sport.

a. What is the probability of skateboarding being someone's favorite sport?

b. Suppose 800 students in the school are surveyed. At this rate, predict how many will choose inline skating as their favorite sport.

c. Of the 800 surveyed, predict how many more students will prefer BMX than MotoX. Explain.

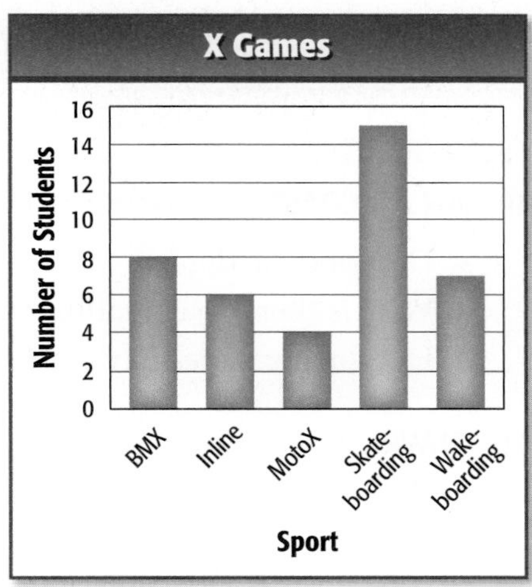

H.O.T. Problems

Use Higher-Order Thinking Skills

22. **OPEN ENDED** Give an example of a real-world situation in which the probability of an event is 25%.

23. **CHALLENGE** The experimental probability of a penny landing on tails is $\frac{9}{16}$. If the penny landed on heads 21 times, how many times was the coin tossed?

24. **CHALLENGE** A survey found that 95 households out of 150 have high speed internet and that 125 out of 200 have cable television. What is the probability that a household has both?

25. **WRITING IN MATH** Jasmine's sock drawer has 8 white pairs, 5 black pairs, 7 navy pairs, 3 khaki, and 12 other pairs. She randomly selected a pair of socks 20 times and selected a navy pair 6 times. Compare and contrast the theoretical and experimental probability.

Standardized Test Practice

26. The production records of a toy manufacturing company show that 5 out of every 75 toys have a defect. What is the probability that a randomly selected toy manufactured at the company will *not* have a defect?

A 7% C 70%
B 27% D 93%

27. If Luke spins a spinner like the one shown 400 times, how many times should he expect it to land on the space with a triangle?

F 100 H 300
G 200 J 400

28. EXTENDED RESPONSE Casey has a bag containing 3 white, 8 red, 2 blue, and 2 yellow marbles. She randomly picks a marble from the bag and replaces it. She repeats the experiment 250 times.

a. What is the probability that the marble Casey picks will be white?

b. Predict the number of times out of 250 Casey will pick a white marble.

29. The results of rolling a die are shown in the table. What is the experimental probability of rolling a 3?

Roll	1	2	3	4	5	6
Frequency	13	9	20	8	10	15

A 13% C 20%
B 17% D 27%

Spiral Review

30. ROLLER COASTERS Use the histogram shown. (Lesson 13-5)

a. Describe the data.
b. Why is there a jagged line in the vertical axis?
c. How many states have no roller coasters? Explain.

U.S. Roller Coasters

Number of States: 0, 2, 4, 6, 8, 10, 30

Number of Coasters: 0–9, 10–19, 20–29, 30–39, 40–49, 50–59, 60–69, 70–79

Source: *The Roller Coaster Database*

31. TRAVEL The box-and-whisker plots below show the average gas mileage for some cars. (Lesson 13-4)

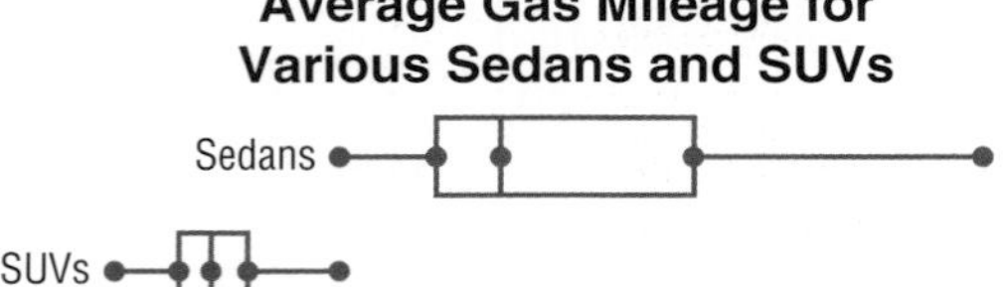

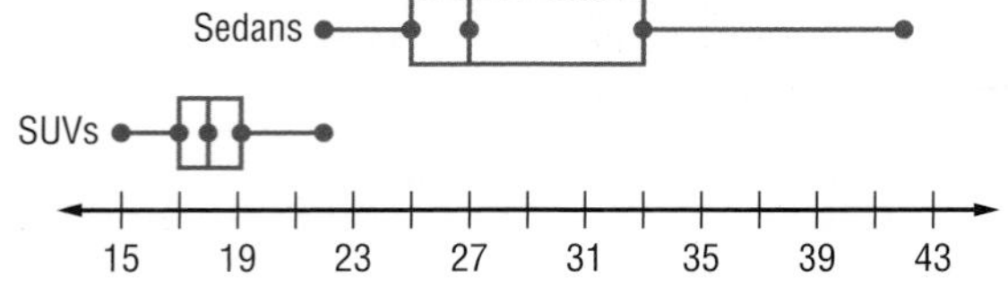

a. Which types of vehicles tend to be less fuel-efficient?
b. Compare the most fuel-efficient SUV to the least fuel-efficient sedan.

Skills Review

Find each product. (Previous Course)

32. $8 \times 2 \times 3$ **33.** $5 \times 10 \times 2 \times 9$ **34.** $20 \times 15 \times 20$ **35.** $12 \times 9 \times 12 \times 10$

13-7

Using Sampling to Predict

Then
You have already analyzed data in graphs. (Lessons 13-1 through 13-5)

Now
- Identify various sampling techniques.
- Determine the validity of a sample and predict the actions of a larger group.

New Vocabulary
sample
population
unbiased sample
simple random sample
stratified random sample
systemic random sample
biased sample
convenience sample
voluntary response sample

Math Online
glencoe.com
- Extra Examples
- Personal Tutor
- Self-Check Quiz
- Homework Help

Why?

The activities committee surveyed a group of students about a mascot for their new high school. The results are shown in the table.

Mascot	Number
Bisons	17
Cougars	4
Huskies	18
Knights	32
Other	3

a. About how many students would vote for the Huskies if the entire student body of 1600 voted? About how many would vote for the Knights?

b. Suppose the students surveyed were in the Spanish club. Do you think the results of the survey would fairly represent the student body? Explain.

c. How could you survey a part of the student population that would fairly represent all students? Give two examples.

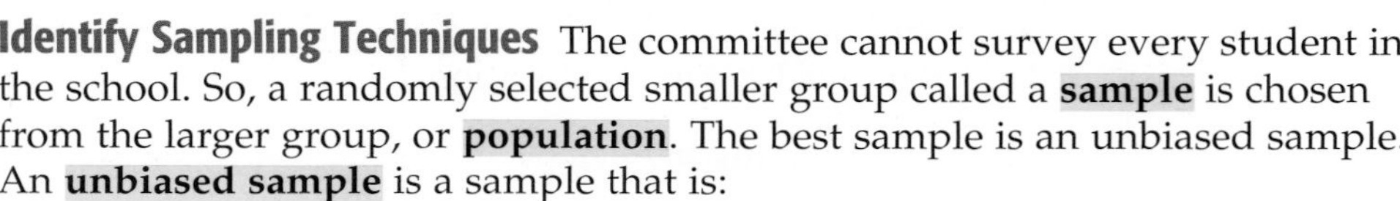

Identify Sampling Techniques The committee cannot survey every student in the school. So, a randomly selected smaller group called a **sample** is chosen from the larger group, or **population**. The best sample is an unbiased sample. An **unbiased sample** is a sample that is:

- representative of the larger population,
- selected at random or without preference, and
- large enough to provide accurate data. If a sample is too small, data accurately representing the larger population may not be available.

Concept Summary Unbiased Samples *For Your FOLDABLE*

Type	Definition	Example
Simple Random Sample	Each item or person in a population is as likely to be chosen as any other.	Thirty student ID numbers are randomly selected by a computer.
Stratified Random Sample	The population is divided into similar, nonoverlapping groups. A simple random sample is then selected from each group.	A population of election districts can be separated into urban, suburban, and rural strata.
Systematic Random Sample	The items or people are selected according to a specific time or item interval.	Every 20 minutes a customer is chosen, or every 10th customer in line is chosen.

A sample that is not representative of the population is called a **biased sample**. A biased sample usually favors certain parts of the population over others.

Concept Summary — Biased Samples

For Your FOLDABLE

Type	Definition	Example
Convenience Sample	Includes members of the population that are easily accessed.	The first 10 students in the cafeteria line.
Voluntary Response Sample	Involves only those who want to or can participate in the sampling.	The principal sent an email to graduating seniors asking them where to hold commencement. Seniors are asked to vote through an online poll.

EXAMPLE 1 Identify and Describe Samples

To determine the types of music their customers like, all the people attending a concert of a country music singer are surveyed. Identify the sample as *biased* or *unbiased* and describe its type. Explain your reasoning.

Since the customers at a country concert probably prefer country music, the sample is biased. The sample is a convenience sample since all of the people surveyed are in one location.

Check Your Progress

1. To determine which passengers' carry-on bags are to be inspected, every eighth person to check in will have his or her bags inspected. Identify the sample as *biased* or *unbiased* and describe its type. Explain your reasoning.

Personal Tutor glencoe.com

Validating and Predicting Samples Depending on the sampling method used, you can make predictions about larger populations.

Real-World Link

In the United States, there are approximately 73 million dogs that have owners.

Source: Humane Society

Real-World EXAMPLE 2 Using Sampling to Predict

PETS A pet store mailed a survey to residents to determine their favorite pets. Fifty people responded and the results are shown in the table. Is this sampling method valid? If so, how many people can you expect to choose dogs as their favorite pet in a city with 1585 people? Explain.

Pet	Number
dog	20
cat	16
fish	9
gerbil	5
no pets	0

This is a biased and voluntary response sample since it involves only those who want to participate in the survey. Therefore, this sampling method will not produce an accurate and valid prediction of the total number of dogs in the city.

Check Your Progress

2. WATCHES Of the 2000 watches made, the manufacturer tests the first 15 watches produced for defects. Of the watches, 3 were defective. Is this sampling method valid? If so, about how many of the 2000 are defective? Explain.

Personal Tutor glencoe.com

Real-World EXAMPLE 3 Using Sampling to Predict

TECHNOLOGY From a batch of 7500 computer chips produced, the manufacturer sampled every 150th chip at random for defects and found that 2 were defective. Is this sampling method valid? If so, find how many of the 7500 computer chips you can expect to be defective. Explain.

StudyTip

Alternate Method
Set up a proportion.

$\frac{2}{50} = \frac{n}{7500}$

$2 \cdot 7500 = 50 \cdot n$

$300 = n$

(Lesson 6-4)

This is a systematic random sample because the samples are selected according to a specific interval. So, this sampling method is reasonable and will produce a valid prediction.

Since every 150 chips were sampled, there were a total of 7500 ÷ 150 or 50 chips sampled and 2 were defective. Two out of 50, or 4%, were defective. So, find 4% of 7500.

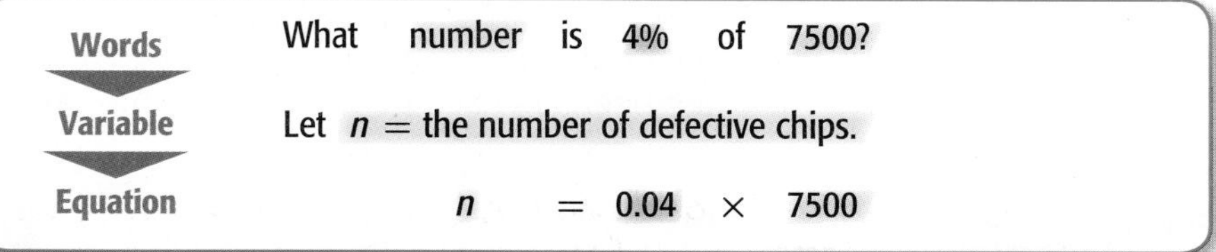

Words What number is 4% of 7500?
Variable Let n = the number of defective chips.
Equation $n = 0.04 \times 7500$

$n = 0.04 \times 7500$ **Write the equation.**

$= 300$ **Multiply.**

So, you would expect approximately 300 defective chips.

Check Your Progress

3. **FOOD** After finishing their meal, every fifth person that left the restaurant was surveyed about whether they ordered dessert after their meals. Out of 20 people, 12 said yes. Is this sampling method valid? If so, about how many of the 380 people who had dinner at the restaurant ate dessert?

Personal Tutor glencoe.com

Check Your Understanding

Example 1 p. 772

Identify each sample as *biased* or *unbiased* and describe its type. Explain your reasoning.

1. To determine how many students at a middle school bring their lunch from home, all the students on one school bus are surveyed.
2. To determine the theme for the homecoming dance, the homecoming committee surveys one classroom.
3. To determine shopping habits at a department store, one male and one female shopper are randomly selected and surveyed from each of their 75 stores.

Examples 2 and 3 pp. 772–773

4. **ANALYZE TABLES** The theater group took a survey about the type of popcorn they should sell during plays. They randomly surveyed 52 students at lunch. Their results are shown in the table. Is this sampling method valid? If so, how many of the boxes of popcorn should be caramel if they order 600 boxes?

Flavor	Number of Students
butter	18
cheese	14
caramel	13
plain	7

Practice and Problem Solving

● = **Step-by-Step Solutions** begin on page R11. **Extra Practice** begins on page 810.

Example 1
p. 772

Identify each sample as *biased* or *unbiased* and describe its type. Explain your reasoning.

5. To determine whether a new university library would be useful, all students whose student ID number ends in 2 are surveyed.

6. To determine the quality of cell phones coming off an assembly line, the manager chooses one cell phone every 20 minutes and checks it.

7. To determine the popularity of a musician, a magazine asks teenagers to log on to their website and participate in the survey.

8. To determine whether a candidate for governor is popular with the voters, 30% of citizens in each of the 254 counties are surveyed.

Examples 2 and 3
pp. 772–773

9. ANALYZE GRAPHS A school committee wanted to find out if students will recycle at school. The committee randomly surveyed 25% of the teenagers at a mall on a Saturday afternoon. The results are in the graph. Is this sampling method valid? If so, about how many of the 576 students at the school will participate in the program?

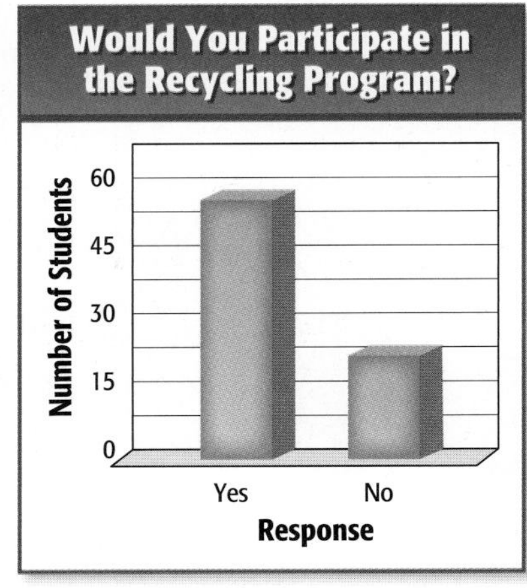

10. HEALTH Seven of the 28 students in math class have the flu. Is this sampling of the students who have the flu representative of the entire school? If so, how many of the 464 students who attend the school have the flu?

11. CONCERT As people leave a concert, every 10th person is surveyed. They are asked if they would buy a T-shirt. One hundred forty of 800 people surveyed said yes. Is this sampling method valid? If so, how many people would you expect to buy T-shirts at the next concert if 7000 attend? Explain your reasoning.

12. ANALYZE TABLES Every hour, twenty customers in a grocery store are randomly selected and surveyed on their milk preference. The results are shown in the table. After reviewing the data, the store manager decided that 40% of his total milk stock should be low-fat milk. Is this a valid conclusion? If not, what information should the store manager review to make a better conclusion?

Milk Preference	
Milk	**Number**
skim	88
low-fat	92
whole	60

Real-World Link

Approximately 170 billion pounds of milk are produced annually in the United States.

Source: National Agriculture Statistics Service

13. VIDEOS A video store is considering adding an international movie section. They randomly selected 300 customers, and 80 customers agree the international movie section is a good idea. Should the store add this section? Explain.

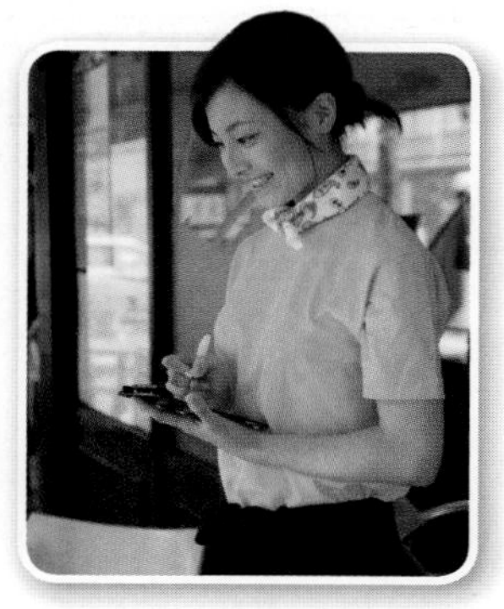

Real-World Link

The restaurant industry provides 12.8 million jobs nationally. This is the nation's second largest employer behind the government.

Source: National Restaurant Association

14. **FOOD** Ellis is planning on opening a restaurant in his community. He is conducting a survey to determine what type of food people in the community like.
    a. Describe an unbiased population sample that Ellis could survey to get unbiased results.
    b. Write two questions that Ellis could ask.
    c. After the survey is completed, how could Ellis use the results of the survey to determine what types of food he should serve in his restaurant?

15. **CARNIVALS** The student council is planning to have a school carnival.

CARNIVAL
Participate in a quick survey to help us decide what to have at the carnival.

    a. Describe an unbiased population sample they could survey to determine types of games and activities to have at the carnival.
    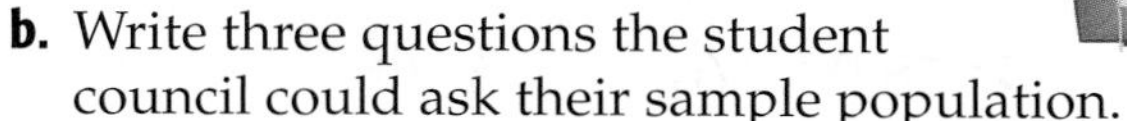
    b. Write three questions the student council could ask their sample population.
    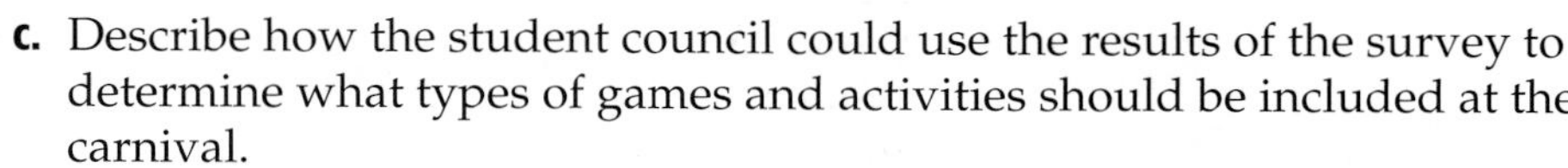
    c. Describe how the student council could use the results of the survey to determine what types of games and activities should be included at the carnival.

16. **GAMES** An online gaming site conducted a survey to determine the types of games people play online. The results are shown in the circle graph.

Games People Play Online

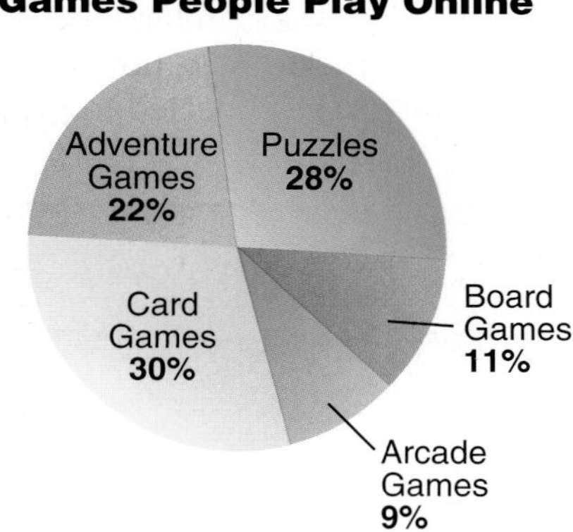

    a. If 2500 people participated in the study, how many of them would play arcade or board games?
    b. An article said 30% of Americans play card games online. Is this statement valid? Explain your reasoning.
    c. Describe how the study could have been conducted so that it represented all Americans and not just online gamers.

H.O.T. Problems

Use **H**igher-**O**rder **T**hinking Skills

17. **OPEN ENDED** Give an example of a unbiased survey. Then conduct the survey. Display the your results in a graph.
18. **CHALLENGE** Suppose you are a farmer and want to know if your corn crop is ready to be harvested. Describe an unbiased way to determine whether the crop is ready to harvest.
19. **REASONING** If someone were to conduct a survey in person, could the surveyor's tone of voice or how they ask the questions alter the response to the question? Explain.
20. **WRITING IN MATH** Why is sampling an important part of the manufacturing process? Illustrate your answer with an unbiased and biased sampling method you can use to check the quality of DVDs.

Standardized Test Practice

21. A real estate agent surveys people about their housing preferences at an open house for a luxury townhouse. Which is the best explanation for why the results of this survey might *not* be valid?

- **A** The survey is biased because the agent should have conducted the survey by telephone.
- **B** The survey is biased because the sample consisted of only people who already are interested in townhouses.
- **C** The survey is biased because the sample was a voluntary response sample.
- **D** The survey is biased because the agent should have conducted the survey at a single-family home.

22. GRIDDED RESPONSE One hundred people in a music store were surveyed about what type of music they prefer. If 35% of them said they prefer rock music, how many people out of 1500 can be expected to prefer rock music?

An online survey of about 38,000 children produced the results shown in the circle graph.

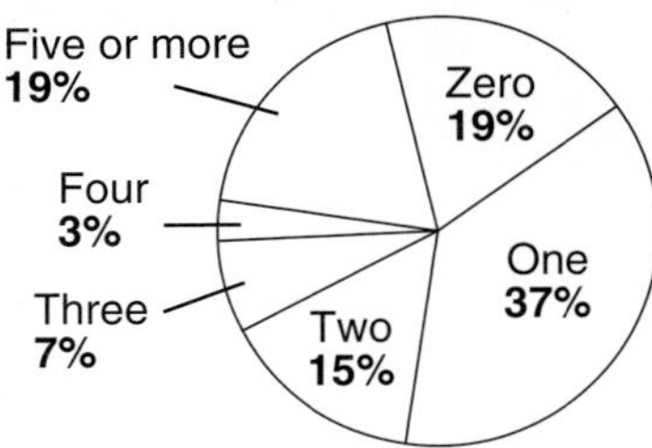

23. About how many of the children surveyed drink two cans of soda or less per day?

- **F** 5700
- **G** 8993
- **H** 14,060
- **J** 26,980

24. Based on the results, about how many children in a class of 30 would drink two or more cans of soda per day?

- **A** 5
- **B** 8
- **C** 13
- **D** 21

Spiral Review

25. What is the probability that a 12-month calendar is randomly turned to the month of January or April? (Lesson 13-6)

26. FOOTBALL Display the data about touchdowns in a histogram. (Lesson 13-5)

Touchdowns in a Season		
Amount	Tally	Frequency
80–96	~~||||~~ ~~||||~~	10
97–113	~~||||~~	5
114–130	||||	4
131–147	||	2
148–164		0
165–181			1

27. Find the percent of change from 32 feet to 79 feet. Round to the nearest tenth, if necessary. Then state whether the percent of change is a *percent of increase* or a *percent of decrease.* (Lesson 7-6)

Solve each problem using the percent equation. Round to the nearest tenth. (Lesson 7-5)

28. 7 is what percent of 32?

29. What is 28.5% of 84?

30. WEATHER During a 10-hour period, the temperature in Browning, Montana, changed at a rate of $-10°F$ per hour, starting at 44°F. What was the ending temperature? (Lesson 2-4)

Skills Review

Express each fraction as a percent. (Lesson 7-1)

31. $\frac{3}{4}$

32. $\frac{1}{5}$

33. $\frac{2}{3}$

34. $\frac{5}{6}$

13-8 Counting Outcomes

Then

You have already found the probability of simple events. (Lesson 13-6)

Now

- Use tree diagrams or the Fundamental Counting Principle to count outcomes.
- Use tree diagrams or the Fundamental Counting Principle to find the probability of an event.

New Vocabulary

tree diagrams
Fundamental Counting Principle

Math Online

glencoe.com

- Extra Examples
- Personal Tutor
- Self-Check Quiz
- Homework Help

Why?

An online store sells fish tanks in 4 sizes and 3 shapes. The choices are shown at the right.

Size	Shape
small	hexagon
medium	pentagon
large	rectangle
x-large	

a. Make a list of all of the possible fish tanks.

b. How many different fish tank choices are possible?

Counting Outcomes To solve the problem above, you can look at a simpler problem. Suppose there are only three size choices, small, medium, or large, and only two shape choices, pentagon or rectangle. You can draw a **tree diagram** to represent the possible outcomes.

EXAMPLE 1 Use a Tree Diagram to Count Outcomes

How many different fish tanks can be made from three size choices and two shape choices?

You can draw a diagram to find the number of possible fish tanks. List each size choice. Then pair each shape choice with each size.

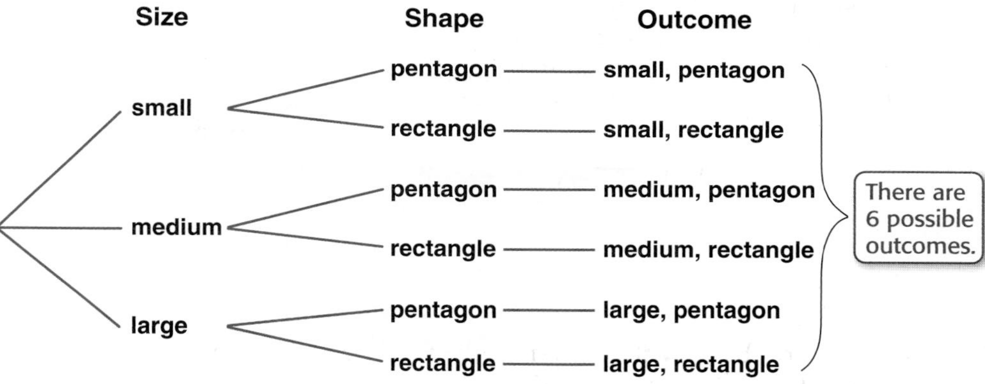

There are 6 possible outcomes.

Check Your Progress

1. Draw a tree diagram to find the number of different outfits that can be assembled using 5 shirts and 4 pairs of pants.

Personal Tutor glencoe.com

The **Fundamental Counting Principle** relates the number of outcomes to the number of choices.

Key Concept Fundamental Counting Principle

For Your FOLDABLE

Words If event M can occur in m ways and is followed by event N that can occur in n ways, then the event M followed by N can occur in $m \cdot n$ ways.

Example If there are 4 possible sizes for fish tanks and 3 possible shapes, then there are $4 \cdot 3$ or 12 possible fish tanks.

You can also use the Fundamental Counting Principle when there are more than two events.

EXAMPLE 2 Use the Fundamental Counting Principle

FOOD A pizza shop has regular, deep-dish, and thin pizza crusts, 2 different cheeses, and 5 toppings. How many different one-cheese and one-topping pizzas can be ordered?

The number of crust types	times	the number of cheeses	times	the number of toppings	equals	the number of possible outcomes.
3	$\times$	2	$\times$	5	$=$	30

So, 30 different pizzas can be ordered.

> **StudyTip**
>
> **Multiplying More than Two Factors** Remember, when you multiply, you can change the order of the factors. For example, in $3 \times 2 \times 5$ you can multiply 2×5 first, then multiply the product, 10, by 3 to get 30.

Check Your Progress

2. **ROUTES** When Shelly goes into her school, she can walk through 4 different doors. Once inside, she can go to her locker by using 4 different sets of stairs and then 3 different hallways. How many ways can Shelly get from outside the school to her locker?

Personal Tutor glencoe.com

Find the Probability of an Event When you know the number of outcomes, you can find the probability that an event will occur.

Real-World EXAMPLE 3 Find Probabilities

GAMES Emilio has 2 counters. Each counter has one side marked with an E and the other side marked with a J, for Jacob. Both counters are tossed. If one counter lands with E up and the other lands with J up, Emilio wins. Otherwise, Jacob wins. What is the probability that Emilio will win?

First find the number of outcomes.

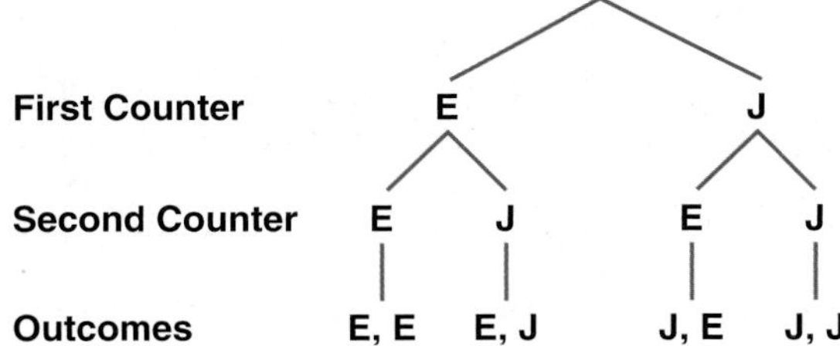

There are four equally likely outcomes with two favoring Emilio.

So, the probability of Emilio winning is $\frac{1}{2}$ or 50%.

> **Real-World Link**
>
> **Passwords** Approximately 80% of passwords are alphanumeric, which means they are a combination of letters and words.

Check Your Progress

3. **PASSWORDS** What is the probability of randomly choosing a 5-letter password for an Internet Web site that consists of only vowels if each vowel can be used more than once?

Personal Tutor glencoe.com

EXAMPLE 4 Find Probabilities

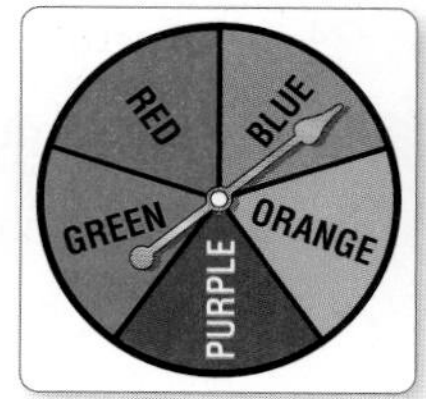

Lamar is going to spin each spinner once. What is the probability that he will spin red and the number 9?

Step 1 Find the number of possible outcomes.

Number of choices for first spinner	times	number of choices for second spinner	equals	the total number of outcomes.
5	$\times$	10	$=$	50

There are 50 outcomes.

Step 2 Find the probability. There is one way to spin red and the number 9. So, the probability is $\frac{1}{50}$ or 2%.

Check Your Progress

4. Three dice are rolled. What is the probability of rolling three 5s?

Personal Tutor glencoe.com

Check Your Understanding

Example 1 p. 777

1. Draw a tree diagram to find the number of tennis shoes available if they come in gray or white and are available in sizes 6, 7, or 8.

Example 2 p. 778

Use the Fundamental Counting Principle to find the total number of outcomes in each situation.

2. **CARS** The table shows the options a dealership offers for a model of a car.

Doors	Gears	Color
2-door	automatic	black
4-door	5-speed	tan
	4-speed	white
		red

3. **SCHOOL** Elisa can take 6 different classes first period, 4 different classes second period, 2 different classes third period, and 3 fourth period.

Examples 3 and 4 pp. 778–779

4. **ACCESSORIES** Amber has a denim and a black purse. Ebony has a black, a red, a denim, and a brown purse. Each girl picks a purse at random to bring to the mall. What is the probability the girls will bring the same color purse?

5. A spinner with 8 equal sections labeled 1–8 is spun twice. What is the probability that it will land on 6 after the first spin and on 6 after the second spin?

Practice and Problem Solving

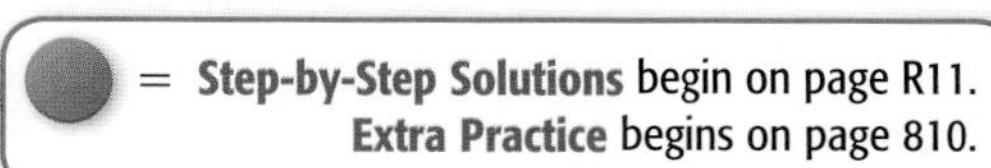

Example 1 p. 777

For each situation, draw a tree diagram to find the number of outcomes.

6. A pet store has male and female huskies with blue, green, and amber eyes.

7 There are 3 true-false questions on a quiz.

8. A coin is tossed and a die is rolled.

Example 2
p. 778

Use the Fundamental Counting Principle to find the total number of outcomes in each situation.

9. A month of the year is picked at random and a coin is tossed.

10. There are three choices for each of 5 multiple-choice questions on a science test.

11. A 3-digit password is created. Digits can repeat.

Examples 3 and 4
pp. 778–779

12. SANDWICHES The table shows the sandwich choices for lunch at a cafe. If a one-bread, one-meat, one-cheese sandwich is chosen at random, what is the probability that it will be turkey and Swiss on wheat bread?

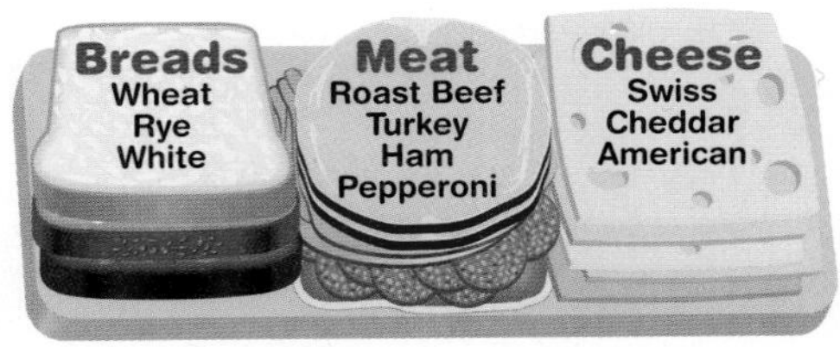

13. GAMES A game requires you to toss a 10-sided die and a 6-sided die to determine how to move on a game board. Find the following probabilities.

a. P(same number on both dice) **b.** P(odd, even) or P(even, odd)

14. GOVERNMENT Use the information at the left about Social Security numbers. What is the probability that the last two numbers are your age?

15. Each of the spinners at the right is spun once. Use a tree diagram to find the following probabilities.

a. P(at least one 2)

b. P (at least one 3)

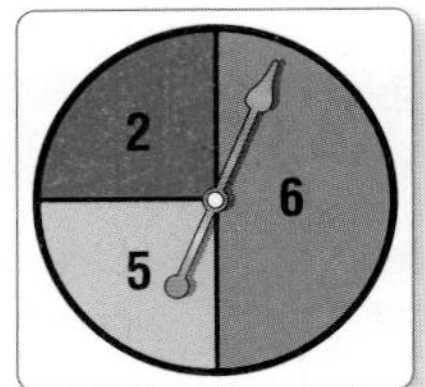

Real-World Link

The first three digits of a social security number are a geographic code. The next two digits are determined by the year and the state where the number is issued. The final four digits are random numbers.

H.O.T. Problems

Use Higher-Order Thinking Skills

16. OPEN ENDED Give an example of a real-world situation that has 16 outcomes.

17. FIND THE ERROR Cameron and Lisa are finding the number of different outfits they can make with 3 pairs of shoes, 4 pairs of pants, and 5 shirts. Is either of them correct? Explain your reasoning.

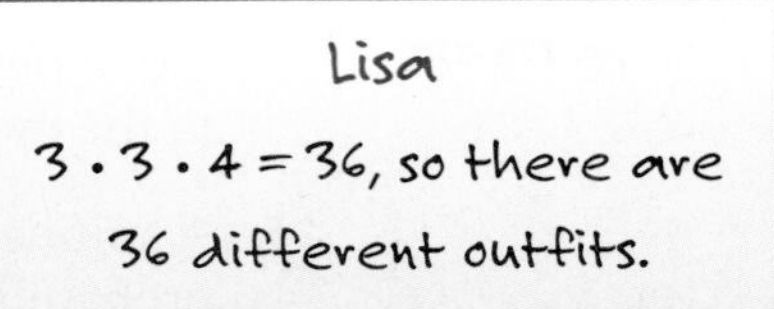

18. NUMBER SENSE Marcus has a choice of a white, grey, or black shirt to wear with a choice of tan, black, brown, or denim pants. Without calculating the number of possible outcomes, how many more outfits can he make if he buys a green shirt? Explain.

19. CHALLENGE Write an algebraic expression for the number of possible outcomes if the spinner at the right is spun x number of times.

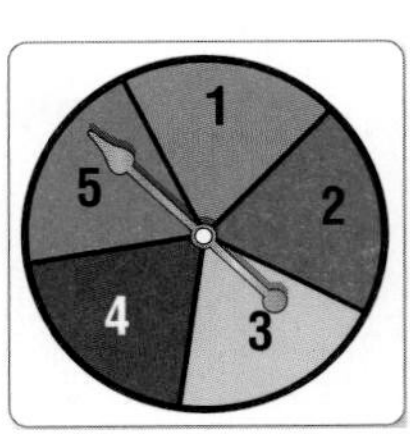

20. WRITING IN MATH Explain how a tree diagram might be more useful than the Fundamental Counting Principle when finding probability.

Standardized Test Practice

21. The spinner is spun twice. What is the probability that it will land on 2 after the first spin and on 5 after the second spin?

A $\frac{1}{64}$ C $\frac{1}{8}$

B $\frac{1}{16}$ D $\frac{5}{8}$

22. A bicycle lock has 4 rotating discs and each contains the digits 0–9. How many different lock combinations are possible?

F 3024 H 6561

G 5040 J 10,000

23. A restaurant offers a combo special with 3 different sandwiches, 2 different salads, and 5 different drinks. From how many different combos could Keisha choose?

A 10 C 30

B 11 D 50

24. **SHORT RESPONSE** Pilan can make outfits out of the clothes shown in the table. Draw a tree diagram to show all of the possible outcomes.

Shoes	Shirts	Pants
white	blue	tan
black	red	navy
	green	denim
	black	

Spiral Review

25. **MUSIC** Sixty-three of the 105 students in the band said that their favorite class was music. Is this sampling representative of the entire school? If so, how many of the 848 students who attend the school would say music is their favorite class? (Lesson 13–7)

26. **FOOD** Maresha took a random sample from a package of jellybeans without looking and found that 30% of the beans were red. Suppose there are 250 jellybeans in the package. How many can she expect to be red? (Lesson 13–6)

27. **NUTRITION** **Use the food data shown in the back-to-back stem-and-leaf plot.** (Lessons 13–2 and 13–3)

a. What is the greatest number of fat grams in each sandwich?

b. In general, which type of sandwich has a lower amount of fat? Explain.

c. Find the measures of variation and any outliers for the data.

Fat (g) of Various Burgers and Chicken Sandwiches

Chicken		Burgers
8	0	
9 8 5 5 3 3	1	0 5 9
0	2	0 6
	3	0 3 6
8|0 = 8 g		*2|6 = 26 g*

28. **POPULATION** Population density is a unit rate that gives the number of people per square mile. If the area of North Carolina is 48,711 square miles and its population is 8,856,505 people, what is the population density of North Carolina? Round to the nearest tenth. (Lesson 6–2)

Skills Review

Simplify. (Previous Course)

29. $\frac{4 \cdot 3}{2 \cdot 1}$ **30.** $\frac{6 \cdot 5 \cdot 4}{3 \cdot 2 \cdot 1}$ **31.** $\frac{9 \cdot 8}{3 \cdot 2}$ **32.** $\frac{7 \cdot 6 \cdot 5 \cdot 4}{4 \cdot 3 \cdot 2}$

EXTEND 13-8

Algebra Lab
Probability and Pascal's Triangle

Math Online glencoe.com
Math *in Motion,* Animation

Objective
Find probability of events.

ACTIVITY

Step 1 Copy and complete the tree diagram shown below listing all possible outcomes if you have two true and false questions on a test.

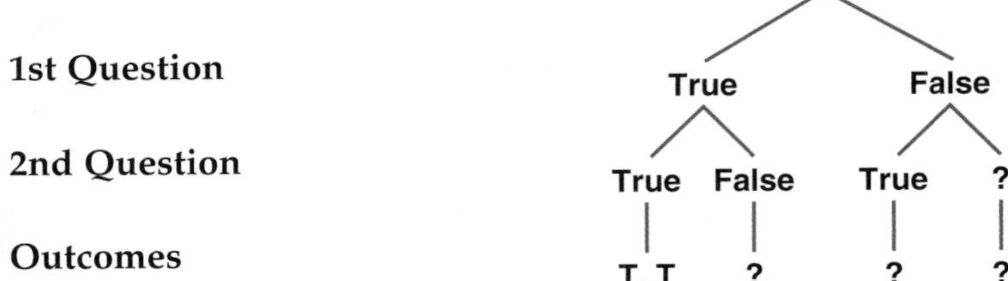

Step 2 Make another tree diagram showing the possible outcomes if there are three true and false questions on a test.

Step 3 Make another tree diagram showing the possible outcomes if there are four true and false questions on a test.

Analyze the Results

1. For two questions, how many outcomes are there? How many have one true and one false?

2. Find *P*(two true), *P*(one true, one false), and *P*(two false). Do not simplify.

3. For three questions, how many outcomes are there? How many have two true and one false? one true and two false?

4. Find *P*(three true), *P*(two true, one false), *P*(one true, two false), and *P*(three false). Do not simplify.

5. For four questions, how many outcomes are there? How many have three true and one false? two true and two false? one true and three false?

6. Find *P*(four true), *P*(three true, one false), *P*(two true, two false), *P*(one true, three false), and *P*(four false). Do not simplify.

Pascal was a French mathematician who lived in the 1600s. He is known for the triangle of numbers at the right, called Pascal's Triangle.

				1					Row 0
			1		1				Row 1
		1		2		1			Row 2
	1		3		3		1		Row 3
1		4		6		4		1	Row 4

7. Examine the rows of Pascal's Triangle. Explain how the numbers in each row are related to true and false questions. (*Hint*: Row 2 relates to answering two questions.)

13-9 Permutations and Combinations

Then

You have already used tree diagrams or the Fundamental Counting Principle to find the probability of an event. (Lesson 13-8)

Now

- Use permutations.
- Use combinations.

New Vocabulary

permutations
combinations

Math Online

glencoe.com

- Extra Examples
- Personal Tutor
- Self-Check Quiz
- Homework Help

Why?

The list shows the classes you plan to take next year. You wonder how many different ways there are to arrange your schedule for the first three periods of the day.

Scheduling Options
- Algebra
- Biology
- Language Arts
- Spanish
- World History

a. Make a tree diagram that lists all of the possibilities for the three periods. Do not repeat any classes in each arrangement.

b. How many different choices did you have for the first period? the second period? the third period?

Use Permutations An arrangement or listing in which order is important is called a **permutation**. Algebra, Language Arts, and Biology is a permutation of Language Arts, Biology, and Algebra because the order is different.

You can use the Fundamental Counting Principle to find the number of possible permutations.

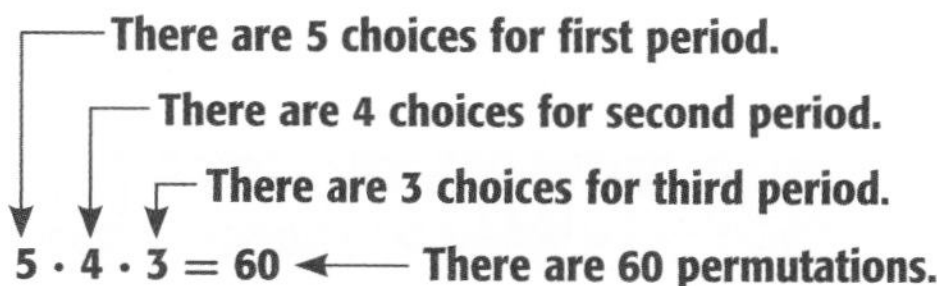

There are 5 choices for first period.
There are 4 choices for second period.
There are 3 choices for third period.
5 · 4 · 3 = 60
There are 60 permutations.

The notation $P(5, 3)$ represents the number of permutations of 5 things taken 3 at a time, as in 5 classes for 3 periods.

EXAMPLE 1 Use a Permutation

SOFTBALL How many ways can the first 3 batters of a batting order be arranged from a team of 12 players?

In a softball game, batting order is important. This arrangement is a permutation.

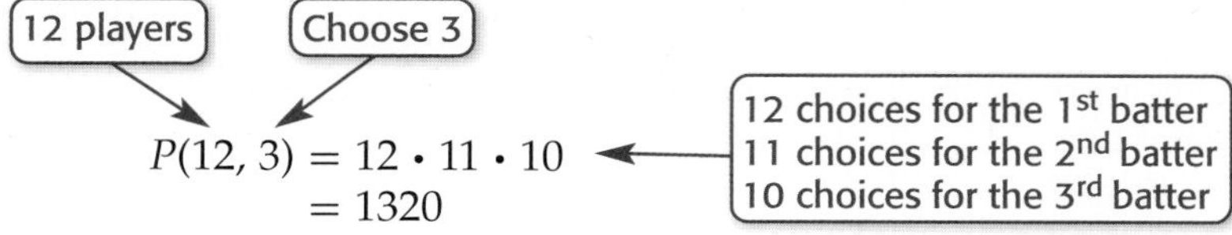

So, there are 1320 ways to arrange the first 3 batters.

Check Your Progress

1A. BANDS In how many ways can the first and second place in a Battle of the Bands be arranged if 8 bands participate?

1B. FOOD Four of ten different fruits are being selected to be placed in a row in a display window. In how many ways can they be placed?

Personal Tutor glencoe.com

ReadingMath

Permutation
Root Word: Permute
Permute means to change the order or arrangement of, especially to arrange the order in all possible ways.

EXAMPLE 2 Use a Permutation

How many zip codes can be made from the digits 2, 4, 5, 8, and 6 if each digit is used only once?

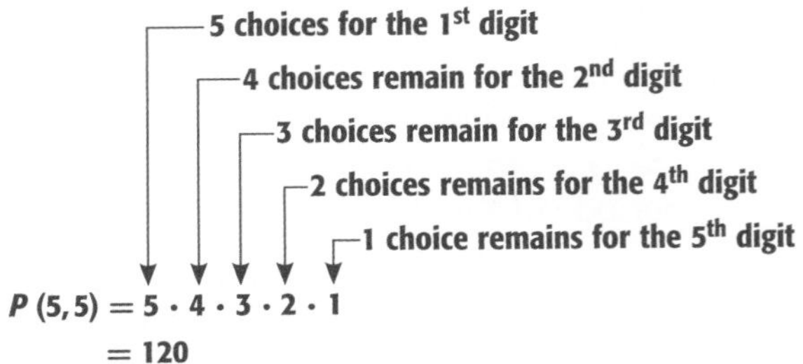

5 choices for the 1st digit
4 choices remain for the 2nd digit
3 choices remain for the 3rd digit
2 choices remains for the 4th digit
1 choice remains for the 5th digit

$$P(5, 5) = 5 \cdot 4 \cdot 3 \cdot 2 \cdot 1$$
$$= 120$$

Check Your Progress

2. How many 7-digit identification numbers can be made from the digits 1, 2, 3, 5, 6, 8, and 9 if each digit is used only once?

Personal Tutor glencoe.com

Use Combinations Sometimes order is not important. For example *chocolate, vanilla,* and *strawberry* is the same as *strawberry, chocolate,* and *vanilla* when you order ice cream. A **combination** is an arrangement or listing where order is *not* important.

Real-World EXAMPLE 3 Use a Combination

SHERBET How many ways can students choose two flavors of sherbet from orange, lemon, strawberry, and raspberry?

In choosing two flavors, order is not important. This arrangement is a combination.

Use the first letter of each flavor to list all of the permutations of the flavors taken two at a time. Then cross off arrangements that are the same as another one.

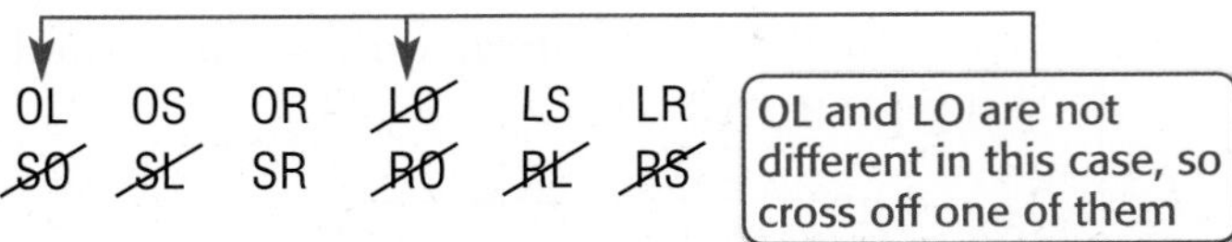

OL OS OR ~~LO~~ LS LR
~~SO~~ ~~SL~~ SR ~~RO~~ ~~RL~~ ~~RS~~
OL and LO are not different in this case, so cross off one of them

There are only six *different* arrangements. So, there are six ways to choose two flavors from a list of four flavors.

Real-World Link

Pizza Pepperoni is America's favorite pizza topping. 36% of all pizza orders contain pepperoni.

Check Your Progress

3A. STUDENT COUNCIL In how many ways can you choose two student council representatives from the students shown?

Student Council Candidates		
Jimmy	Evita	Debra
Molly	Julián	Candace

3B. PIZZA How many ways can a customer choose 3 pizza toppings from pepperoni, onion, sausage, green pepper, and mushroom?

Personal Tutor glencoe.com

You can find the number of combinations of items by dividing the number of permutations of the set of items by the number of ways each smaller set can be arranged.

$$C(4, 2) = \frac{4 \cdot 3}{2 \cdot 1} \text{ or } 6$$

From 4 flavors, take 2 at a time.

There are 2 • 1 ways to order 2 flavors.

Real-World EXAMPLE 4 Find Probability

CHESS The students listed are playing in a chess tournament. If in the first round each player plays every other player once, what is the probability that the first match played involves Abigail?

Lorenzo	Yasu
Abigail	Rashid
Irene	Destiny
Booker	William
Mato	Mercedes

Understand Abigail playing Yasu is the same as Yasu playing Abigail, so this is a combination.

Plan Find the combination of 10 people taken 2 at a time. This will give you the number of matches that take place during the tournament. Then find how many of the matches involve Abigail.

Solve $C(10, 2) = \frac{10 \cdot 9}{2 \cdot 1}$ or 45 **There are 45 ways to choose 2 people to play.**

Abigail plays each person once during the tournament. If there are 9 other people, Abigail is involved in 9 games. So the probability that Abigail plays in the first match is $\frac{9}{45}$ or $\frac{1}{5}$.

Check List all of the 2-player matches in the tournament. Check to see if there are 45 matches.

StudyTip

Checking Reasonableness of Results The number of combinations of 10 taken 2 at a time is less than the number of permutations of 10 taken 2 at a time because order does not matter in a combination. So, check that $C(10, 2) < P(10, 2)$.

Check Your Progress

4. **CHESS** Suppose Booker drops out of the chess tournament. What is the probability that the final first-round game involves Abigail and Mato?

Personal Tutor glencoe.com

Check Your Understanding

Examples 1 and 2 pp. 783–784

1. **MUSIC** A disc jockey has 12 songs he plans to play in the next hour. How many ways can he pick the next 3 songs if he does not repeat?

2. **CONSTRUCTION** A contractor can build 11 different model homes. She only has 4 lots. How many ways can she put a different house on each lot?

Example 3 p. 784

3. **BOOKS** Erica has to write 2 book reports this month. She has 6 books from which to choose. How many different ways can she pick 2 books?

Example 4 p. 785

4. **LOCKS** Omar knows the numbers to his locker combination are 24, 38, and 6. He cannot remember the correct order to the combination. What is the probability that Omar opens his locker on his first attempt?

Practice and Problem Solving

● = **Step-by-Step Solutions** begin on page R11. **Extra Practice** begins on page 810.

Examples 1 and 2 pp. 783–784

5. **PHOTOS** From among five photos, how can four of them be arranged on a shelf that holds four photographs?
6. **RACES** In a race with 7 runners, how many ways can the runners finish in first, second, and third place?
7. **GAMES** In the game Tic Tac Toe, players take turns placing an X or an O in any of the locations that are empty. How many different ways can the first 3 moves of the game occur?

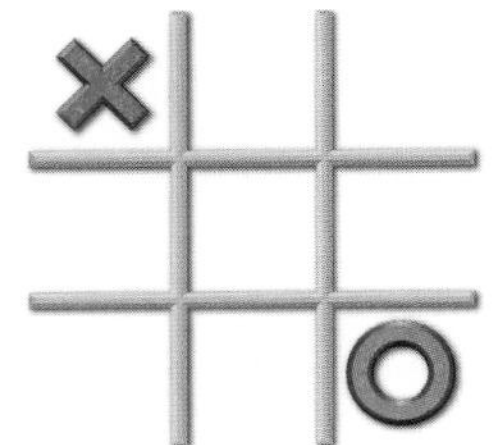

8. **SPELLING** There are 15 contestants in a spelling bee. How many different ways can the contestants finish first, second, third, and fourth?

Example 3 p. 784

9. **CHESS** The chess team wants to have practice two different days next week, from Monday through Friday. How many schedules can be made?
10. **FLAGS** A group of students are designing a new school flag using the colors blue, white, red, black, and gold. In how many ways can students choose 3 colors for the flag?

Example 4 p. 785

11. **MUSIC** Out of 10 songs on a CD, 3 are randomly selected to play without repeating a song. What is the probability that the first 3 songs on the CD are selected to play in any order?
12. **PHONE NUMBERS** Ivy knows that the last four digits of her friend's phone number contain the digits 0, 3, 5, and 6. She remembers that the first digit is 5. What is the probability that a randomly selected phone number with these digits will be her friend's?

Find each value.

13. $P(6, 4)$ 14. $P(5, 5)$ 15. $C(7, 7)$ 16. $C(12, 3)$

17. $P(14, 5)$ 18. $P(12, 4)$ 19. $C(25, 4)$ 20. $C(20, 15)$

Tell whether each situation is a *permutation* or *combination*. Then solve.

21. At a business meeting, each of the 12 people attending shakes hands with every other person exactly once. How many distinct handshakes are given?
22. A phone survey asks people to rank the activities shown according to how much time they spend listening to each activity on the radio. How many different rankings are possible?

Activities	
music	sports
weather	news
entertainment	

23. The school bus has 10 empty seats. At the next stop, 4 students get on the bus. How many ways can the 4 students arrange themselves in the empty seats, if each student takes one seat?
24. How many ways can 3 different movies be picked from 15 different movie choices?

LETTERS Each arrangement of the letters in the word *dream* is placed on a piece of paper. One paper is selected at random. Find each probability.

25. P(word begins with d) 26. P(word ends with am)

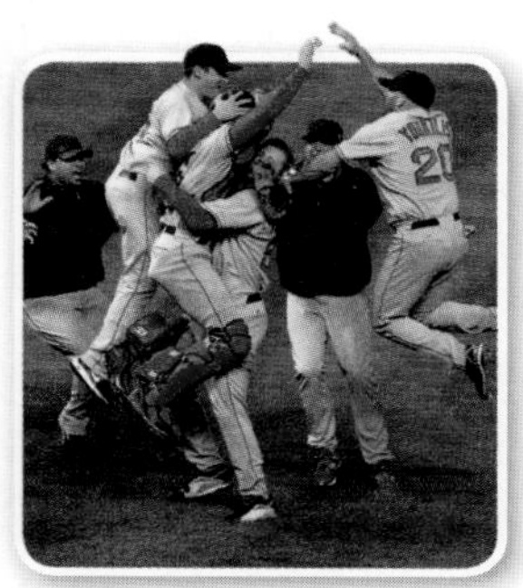

Real-World Link

The finals of the World Series is a best of seven series to determine the Major League Baseball champion. However, in 1903, 1919, 1920, and 1921, a best of nine series determined the winner.

Source: MLB

27. **SPORTS** In a best-of-three series, the first team to win 2 games wins the series. Two ways to win a best-of-three series are to win the first 2 games or lose the first game and win the next two games. How many ways are there to win a best-of-three series? a best-of-five series? a best-of-seven series?

28. **FOOD** At a restaurant, customers can choose three toppings for hamburgers from among ketchup, mustard, pickles, lettuce, onions, and cheese. Does ordering a hamburger with three toppings represent a combination or permutation? How many three-topping hamburgers are possible?

29. **TECHNOLOGY** Your cell phone uses a 4-digit personal identification number (PIN) to lock it from use.
    a. How many PINs can you choose to lock your cell phone?
    b. Use probability to explain how someone is unlikely to guess your PIN.

30. **MULTIPLE REPRESENTATIONS** In this problem, you will investigate points and line segments. In the figure at the right, each of the 4 non-collinear points are connected to every other point exactly once. There are a total of six line segments connecting the points.

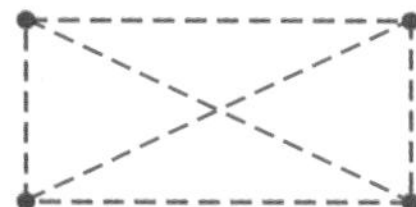

    a. **TABULAR** Copy and complete the table.
    b. **ANALYTICAL** Explain how you can find the number of segments mathematically.
    c. **NUMERICAL** Find $C(4, 2)$, $C(5, 3)$, $C(6, 4)$, and $C(7, 5)$. How do these results compare to the number of segments for 4, 5, 6, and 7 points?

Number of Points	Number of Segments
4	6
5	■
6	■
7	■

    d. **MAKE A PREDICTION** How many segments would be drawn between 10 points? 12 points? n points?

31. **BASKETBALL** There are 2730 ways for three teams to finish first, second, and third in their basketball league. How many teams are in the league?

32. **PARTIES** How many people were at a party if each person shook hands with every other person exactly once, and there were 300 handshakes?

H.O.T. Problems

Use **H**igher-**O**rder **T**hinking Skills

33. **OPEN ENDED** Describe a situation that could be represented by the expression $C(15, 5)$.

34. **CHALLENGE** *True* or *false*? The number of combinations of items is always greater than 1. If false, provide a counterexample.

35. **SELECT A TECHNIQUE** If $P(9, 9) = 362{,}880$, which technique would you use to find $P(10, 10)$? Justify your selection. Then find $P(10, 10)$.

mental math | number sense | estimation

36. **CHALLENGE** Is the value of $P(x, y)$ *sometimes*, *always*, or *never* greater than the value of $C(x, y)$? Explain. Assume x and y are positive integers and $x \geq y$.

37. **WRITING IN MATH** Give an example of a situation in which you would use a combination. Then, change the situation so that you need to use a permutation. Explain the difference between the situations.

Standardized Test Practice

38. When the Rockets win a basketball game, the 12 players on the team give each other high-fives. How many distinct high-fives are given?

A 24 C 66
B 36 D 96

39. Which situation is represented by $C(10, 5)$?

F the number of arrangements of 10 people in a line
G the number of ways to pick 5 out of 10 students for a project
H the number of ways to pick 5 out of 10 students to be first through fifth place in a spelling bee
J the number of ways 10 people can sit in a row of 5 chairs

40. How many seven-digit phone numbers are available if a digit can be used only once and the first digit cannot be zero or one?

A 40,320 C 1,209,600
B 483,840 D 2,097,152

41. SHORT RESPONSE Determine whether the following situation is a permutation or a combination. Then solve the problem.

There are 13 people running the 100-meter dash at a track meet. In how many ways can the runners finish in first, second, third, fourth, and fifth places?

Spiral Review

42. GAMES How many outcomes are possible for rolling three dice? (Lesson 13-8)

43. To determine what type of dessert people in a community like, Sara surveys 20% of the people who enter 3 different chocolate shops. Identify the sample as *biased* or *unbiased* and describe its type. Explain your reasoning. (Lesson 13-7)

44. ANIMALS The histogram shows the life spans of different animals. (Lesson 13-5)

a. How many years are there in each interval?

b. Which interval has the greatest number of animals?

c. How many of the animals in the histogram have a life span of more than 20 years?

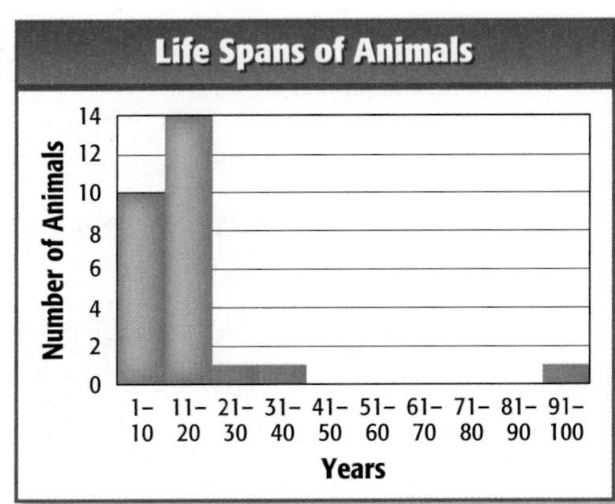

Source: *The World Almanac*

Solve each inequality. Then graph the solution on a number line. (Lesson 5-4)

45. $2a - 3 \geq 9$

46. $4c + 4 > 32$

47. $-2y + 3 < 9$

48. $-6m - 12 \leq 18$

Skills Review

Find each product. (Lesson 3-3)

49. $\frac{1}{6} \cdot \frac{1}{3}$

50. $\frac{2}{3} \cdot \frac{3}{6}$

51. $\frac{1}{3} \cdot \frac{1}{3} \cdot \frac{1}{3}$

52. $\frac{3}{8} \cdot \frac{2}{7} \cdot \frac{1}{6}$

EXPLORE
13-10 Graphing Technology Lab
Probability Simulation

Math Online glencoe.com
- Other Calculator Keystrokes
- Graphing Technology Personal Tutor

A simulation is an experiment that is designed to act out a given situation. You can use a random number generator on a graphing calculator to create data for the experiment. Repeating a simulation may result in different probabilities since the numbers generated are different each time.

ACTIVITY

Generate 30 random numbers from 1 to 10, simulating selecting a card from cards numbered 1 through 10 and replacing a card after it is drawn.

- Access the random number generator.
- Enter 1 as a lower bound and 10 as an upper bound for 30 trials.

KEYSTROKES: MATH ◀ 5 1 , 10 , 30) ENTER

Record all 30 numbers in a column on a separate sheet of paper.

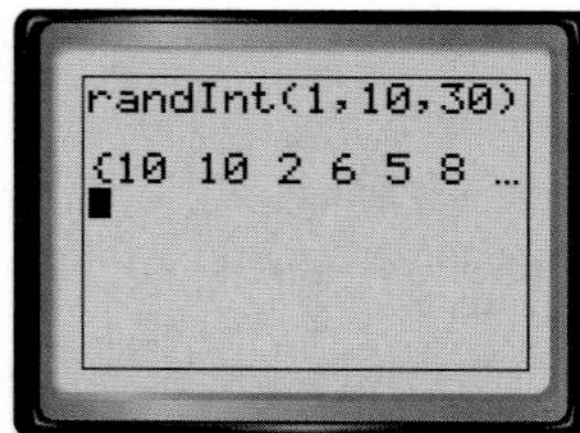

Analyze the Results

1. Record how often each number from 1 to 10 appeared.

a. Find the experimental probability of each number.

b. Compare the experimental probabilities with the theoretical probabilities.

2. Repeat the simulation of selecting 30 cards. Record this second set of numbers in a column next to the first set of numbers. Each pair of 30 numbers represents selecting two numbers. Find the sum for each of the 30 pairs of cards.

a. Find the experimental probability of each sum.

b. Compare the experimental probabilities with the theoretical probabilities.

3. Design an experiment to simulate 30 spins of a spinner that has equal sections labeled A, B, C, and D.

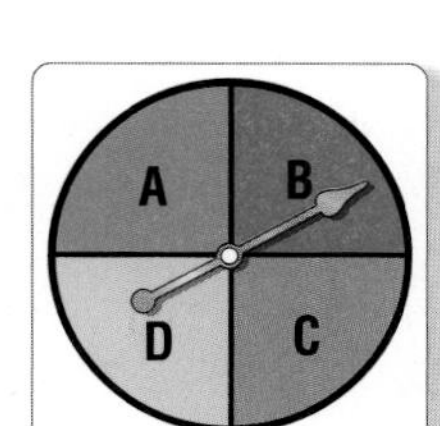

a. Find the experimental probability of each letter.

b. Compare the experimental probabilities with the theoretical probabilities.

4. Suppose you play a game where there are three containers, each with ten balls numbered 0 to 9. Pick three numbers and then use the random number generator to simulate the game. Score 2 points if one number matches, 16 points if two numbers match, and 32 points if all three numbers match. (Note: numbers can appear more than once.)

a. Play the game if the order of your number *does not* matter. Total your score for 10 simulations.

b. Now play the game if the order of the numbers *does* matter. Total your score for 10 simulations.

c. With which game did you score more points?

13-10

Probability of Compound Events

Then

You have already found simple probability. (Lesson 13-6)

Now

- Find the probability of independent and dependent events.
- Find the probability of mutually exclusive events.

New Vocabulary

compound events
independent events
dependent events
mutually exclusive events

Math Online

glencoe.com

- Extra Examples
- Personal Tutor
- Self-Check Quiz
- Homework Help
- Math in Motion

Why?

The 800-meter relay features 20 teams competing in a preliminary round of competition. Before the race, each team chooses a number from jar 1 to determine the heat in which they swim and a number from jar 2 to determine one of five lanes they occupy.

The tree diagram shows all of the different possibilities for a team's heat and lane numbers.

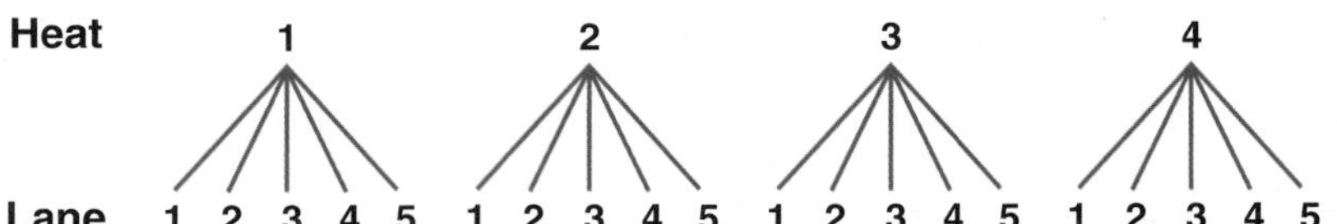

a. How many different outcomes are there for team placement in the competition?

b. Suppose Sabrina's team is the first to choose from the jars. What is P(third heat, lane 2)?

c. Find P(third heat) and P(lane 2). Multiply your answers and compare the product to your answer in part **b.**

Probabilities of Independent and Dependent Events In the above example, choosing the heat and the lane is a compound event. A **compound event** consists of two or more simple events. Since choosing the heat number does not affect choosing the lane number, the events are called independent events. In **independent events**, the outcome of one event does *not* influence the outcome of a second event.

$$P(\text{third heat}) = \frac{1}{4} \qquad P(\text{lane 2}) = \frac{1}{5}$$

There are 4 heats and one of them is third.

There are 5 lanes and one of them is lane 2.

Key Concept: Probability of Two Independent Events

For Your FOLDABLE

Words The probability of two independent events is found by multiplying the probability of the first event by the probability of the second event.

Symbols $P(A \text{ and } B) = P(A) \cdot P(B)$

Example $P(\text{third heat, lane 2}) = \frac{1}{4} \cdot \frac{1}{5}$ or $\frac{1}{20}$

EXAMPLE 1 Probability of Independent Events

The two spinners are spun. What is the probability that both spinners will show an odd number?

The events are independent since each spin does not affect the outcome of the next spin.

$P(\text{first spinner is odd}) = \frac{3}{5}$

$P(\text{second spinner is odd}) = \frac{1}{2}$

$P(\text{both spinners are odd})\ \frac{3}{5} \cdot \frac{1}{2} = \frac{3}{10}$

Check Your Progress

Use the spinners above to find each probability.

1A. $P(\text{both show a 2})$

1B. $P(\text{both are less than 4})$

Personal Tutor glencoe.com

Vocabulary Link

Dependent

Everyday Use Kayley's choice of coat was *dependent* on the weather.

Math Use Choosing two playing cards without replacement are *dependent* events.

If the outcome of one event affects the outcome of a second event, the events are called **dependent events**.

Key Concept Probability of Two Dependent Events

For Your FOLDABLE

Words If two events, *A* and *B*, are dependent, then the probability of both events occurring is the product of the probability of *A* and the probability of *B* after *A* occurs.

Symbols $P(A \text{ and } B) = P(A) \cdot P(B \text{ following } A)$

Math *in Motion,* BrainPOP® glencoe.com

EXAMPLE 2 Probability of Dependent Events

In a bag, there are 3 blue, 7 red, and 5 white marbles. Once a marble is selected, it is not replaced. Find the probability that two blue marbles are chosen.

Since the first marble is not replaced, the first event affects the second event. These are dependent events.

$P(\text{first marble is blue}) = \frac{3}{15}$ ← number of blue marbles / ← total number of marbles

$P(\text{second marble is blue}) = \frac{2}{14}$ ← number of blue marbles after one blue marble is removed / ← total number of marbles after one blue marble is removed

$P(\text{two blue marbles}) = \frac{3}{15} \cdot \frac{2}{14} = \frac{6}{210} \text{ or } \frac{1}{35}$

StudyTip

Check for Reasonableness Since at least one possibility is eliminated in a second event, the probability of 2 dependent events is greater than the probability of 2 similar independent events.

Check Your Progress

2A. $P(\text{two white marbles})$

2B. $P(\text{a red marble and then a white marble})$

Personal Tutor glencoe.com

Mutually Exclusive Events If two events cannot happen at the same time, they are said to be **mutually exclusive**. For example, when you spin two spinners like the one shown, you cannot spin a sum that is both 5 and even.

1
2
3
4
5

The probability of two mutually exclusive events is found by adding.

$P(5 \text{ or even}) = P(5) + P(\text{even})$

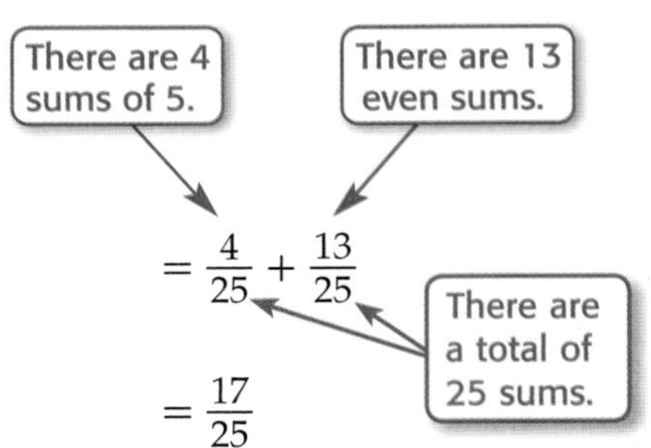

There are 4 sums of 5.
There are 13 even sums.
There are a total of 25 sums.

$= \frac{4}{25} + \frac{13}{25}$

$= \frac{17}{25}$

Second Spinner

First Spinner	+	1	2	3	4	5
	1	2	3	4	5	6
	2	3	4	5	6	7
	3	4	5	6	7	8
	4	5	6	7	8	9
	5	6	7	8	9	10

Key Concept: Probability of Mutually Exclusive Events

For Your FOLDABLE

Words The probability of one or the other of two mutually exclusive events can be found by adding the probability of the first event to the probability of the second event.

Symbols $P(A \text{ or } B) = P(A) + P(B)$

Example $P(5 \text{ or even}) = \frac{4}{25} + \frac{13}{25} \text{ or } \frac{17}{25}$

Real-World Career

Video Game Designer
A video game designer works with a team to create the latest video games. This job requires creativity, superior problem-solving skills, and a passion for video games.

Video game designers complete a two-year degree in either Game Design and Simulation Programming or Game Art and Design.

Real-World EXAMPLE 3 Probability of Mutually Exclusive Events

GAMES Teresa is playing a video game that involves rolling two dice. What is the probability that she will roll a sum of 4 or a sum that is an odd number?

The events are mutually exclusive because the number cubes cannot add up to 4 and be odd at the same time.

$$P(4 \text{ or odd}) = P(4) + P(\text{odd})$$
$$= \frac{3}{36} + \frac{18}{36}$$
$$= \frac{21}{36} \text{ or } \frac{7}{12}$$

The probability that the sum of the number cubes is 4 or odd is $\frac{7}{12}$.

Check Your Progress

3. What is the probability of drawing a ten or two from a standard deck of 52 playing cards?

Personal Tutor glencoe.com

Check Your Understanding

Example 1 p. 791

CARDS Two cards are drawn from a deck of cards numbered 1–10. After a card is selected, it is returned to the deck.

1. What is the probability of drawing an even-numbered card and then a card numbered greater than 8?
2. What is the probability of drawing an odd-numbered card and then a card numbered less than or equal to 5?

Example 2 p. 791

3. **FRUIT** A bowl of apples contains 6 red delicious, 7 granny smith, and 3 macintosh apples. After an apple is selected, it is not returned to the bowl. Find the probability of randomly choosing a macintosh and then a granny smith.

Example 3 p. 792

4. **CONTESTS** Rocko is drawing the winning ball of a contest. What is the probability that he will pick a ball that is purple or the number 2?

Practice and Problem Solving

● = **Step-by-Step Solutions** begin on page R11.
Extra Practice begins on page 810.

Example 1 p. 791

A penny is tossed and the spinner is spun. Find each probability.

5. P(heads and 4)
6. P(tails and odd)
7. P(heads and a number greater than 2)
8. P(tails and a number less than or equal to 5)

Example 2 p. 791

A card is drawn from a deck of eight cards with letters A, B, C, D, E, F, G, H. The card is *not* replaced and a second card is drawn. Find each probability.

9. P(B and F)
10. P(a vowel and D)

A jar contains 6 blue, 3 red, 5 green, and 2 yellow candies. Once a candy is drawn, it is *not* replaced. Find the probability of each outcome.

11. P(two green candies)
12. P(two red candies)
13. P(a yellow then a blue candy)
14. P(a blue then a green candy)

Example 3 p. 792

A card is drawn from the cards shown. Find the probability of each outcome.

1 2 4 5
7 8 9

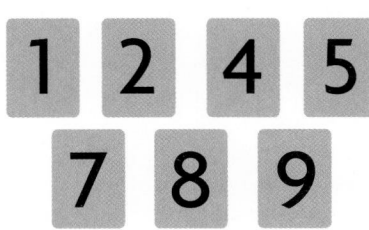

15. P(5 or even numbers)
16. P(2 or greater than or equal to 5)
17. P(3 or 7)
18. P(composite or prime number)

Real-World Link

Sales of video games, game consoles, and accessories were more than $12 billion in 2006. One video game alone made more than $115 million in its first week of sales.

A bag contains 3 red marbles, 4 green marbles, 2 yellow marbles, and 5 blue marbles. Once a marble is drawn, it is *not* replaced. Find the probability of each outcome.

19. three green marbles in a row

20. a blue marble, a yellow marble, and then a red marble

21 **CLOTHING** Your sock drawer contains 6 blue socks, 8 black socks, and 10 white socks. It is dark and you are getting dressed. What is the probability that you will randomly pick a pair of socks that match?

22. **VIDEO GAMES** A company claims its video game players have a defective rate of 4%. What is the probability that you and a friend both receive a defective player?

23. **SPORTS** A professional basketball player makes a field goal 49% of the time he shoots.

a. What is the probability that he will score on his next two attempts?

b. What is the probability that he will score on all of his next four attempts?

c. Would you say shooting multiple field goals are independent, dependent, or mutually exclusive events? Explain your reasoning.

24. **FOOD** A box contains 6 apple cinnamon granola bars, 4 peanut granola bars, and 8 oatmeal raisin granola bars. Luis is first in line and Rashona is second in line. If they are given a granola bar at random, what is the probability that they will receive the same kind of bar?

25. Draw a tree diagram to find the number of outcomes if the spinner at the right is spun twice. What is the probability of spinning red twice? Is this an independent event or dependent event?

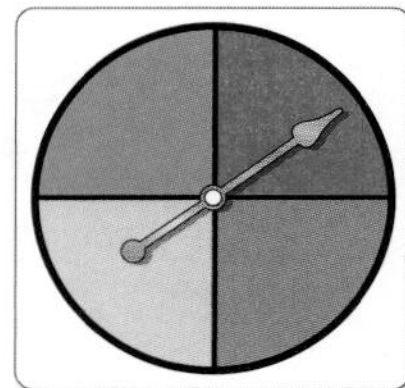

H.O.T. Problems

Use **H**igher-**O**rder **T**hinking Skills

26. **OPEN ENDED** Write a real-world example of two independent events.

27. **CHALLENGE** Seamus reaches into a box and randomly pulls out a marker. Without replacing it, he pulls out another marker. The probability that he first pulls out a red marker is 40%. If $P(\text{red, yellow})$ is 15%, find $P(\text{yellow})$.

28. **REASONING** Determine whether the following statement is *true* or *false*. If *false*, provide a counterexample.

If two events are dependent, then the probability of both events is less than 1.

29. **FIND THE ERROR** Dale and Shannon are finding the probability of rolling a difference of 0 or 1 on two number cubes. Is either of them correct? Explain.

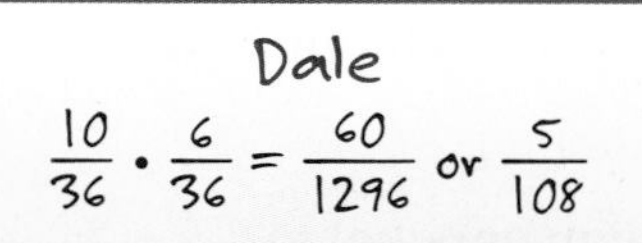

Dale
$\frac{10}{36} \cdot \frac{6}{36} = \frac{60}{1296}$ or $\frac{5}{108}$

Shannon

$\frac{10}{36} + \frac{6}{36} = \frac{16}{36}$ or $\frac{4}{9}$

30. **WRITING IN MATH** Compare and contrast independent events and dependent events.

Standardized Test Practice

31. A quarter is tossed and the spinner shown is spun. How do you find the probability of tossing heads and spinning an odd number?

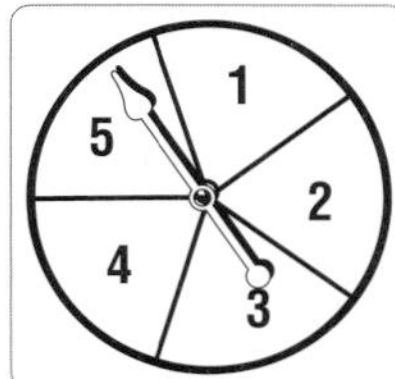

A $\frac{1}{2} \times \frac{3}{5}$

B $\frac{1}{2} + \frac{3}{5}$

C $\frac{1}{2} \times \frac{2}{5}$

D $\frac{1}{2} + \frac{2}{5}$

32. **GRIDDED RESPONSE** A jar contains 5 blue marbles, 4 red marbles, and 3 white marbles. Simone picks a marble at random. Without replacing it, she randomly chooses a second marble. Find P(red, white).

33. What is the probability of rolling a number cube three times and getting numbers less than 3 each time?

F $\frac{1}{216}$

G $\frac{1}{27}$

H $\frac{1}{8}$

J $\frac{27}{64}$

34. You buy a bag of assorted fruit that contains 3 apples, 5 oranges, and 2 bananas. You randomly choose a piece of fruit. Your friend then chooses a piece of fruit. What is the probability you both will get a banana?

A $\frac{1}{45}$

B $\frac{14}{45}$

C $\frac{20}{45}$

D $\frac{40}{45}$

Spiral Review

Tell whether each situation is a *permutation* or *combination*. Then solve. (Lesson 13-9)

35. How many ways can 6 cars line up for a race?

36. How many different flags can be made from the colors red, blue, green, and white if each flag has three vertical stripes and no colors can repeat?

37. How many ways can 4 shirts be chosen from 10 shirts to take on a trip?

38. How many ways can you buy 2 DVDs from a display of 15?

39. **FOOD SERVICE** Hastings Cafeteria serves toast, a muffin, or a bagel with coffee, milk, or orange juice. (Lesson 13-8)

a. How many different breakfasts of one bread and one beverage are possible?

b. What is the probability that a customer chooses a bagel with orange juice if a bread and beverage are equally likely to be chosen?

40. **SCIENCE** Use the information in the table about the length of days on each planet. (Lesson 13-3)

a. What is the median length of day for the planets?

b. Are there any outliers in the data? If so, name them.

c. Write a sentence describing how the lengths of days vary. Include a statement about any outliers in the data.

Planet	Length of Day* (Earth hours)
Mercury	1416
Venus	5832
Earth	24
Mars	25
Jupiter	10
Saturn	11
Uranus	17
Neptune	16

*The lengths are approximate
Source: *The World Almanac*

Algebra Lab: Simulations

Math Online glencoe.com
Math *in Motion,* Animation

You can use items such as a die, a coin, or a spinner to simulate many real-world situations. The items or combination of items should have the same number of outcomes as the number of possible outcomes of the situation.

ACTIVITY 1

Logan usually makes three out of every four free throws he attempts during a basketball game. Is spinning a spinner that is divided into four equal sections a good strategy to determine how many free throws he will make in his next game?

Since the probability of Logan making a free throw are 3 in 4, spinning a spinner with four equal sections where 3 sections are red and one section is black is a reasonable activity to simulate made free throws. The red sections will represent made free throws and the black section will represent a missed free throw.

Step 1 Spin the spinner and record the results. Write a ● for a made free throw and an X for a missed free throw.

Step 2 Repeat the simulation three times. Record your results in a table like the one shown below.

Simulation 1	●	●	●	X	●	X	●	●	●	X	X	●
Simulation 2												
Simulation 3												

Analyze the Results

1. Compare the results of the simulations to the theoretical probability. Explain any differences.

Use a simulation to act out the situation.

2. At the grand opening of a store, every person that comes to the store receives a free sports bottle. The bottle comes in six different colors.
   a. Use a die to simulate this situation. Let each number represent one color of the sports bottles. Conduct a simulation until you have one of each number.
   b. Based on your simulation, how many times must you enter the store in order to get all six colors of water bottles?

ACTIVITY 2

A student organization is selling raffle tickets to raise money for a trip. The odds in favor of winning an item in the raffle are 1 to 4. Conduct the following experiment to simulate the probability of two winning tickets being drawn in a row.

Step 1 Use red counters to represent a winning ticket and white counters to represent a losing ticket. There is one favorable outcome and four unfavorable outcomes. So, use 1 red counter and 4 white counters.

Step 2 Conduct a simulation for 30 tickets sold.

Step 3 Without looking, draw a counter from the bag and record its color. Replace the counter and draw a second counter.

Step 4 Repeat 30 times and record the results of the simulation in a table like the one shown below.

First ticket is a winner.	First ticket is a winner, second ticket is a loser.	Both tickets are winners.

Analyze the Results

3. Calculate the experimental probability that two tickets in a row will be winners.

4. How do the results in Activity 2 compare to the theoretical probability that two winners will be drawn in a row? (*Hint*: These are independent events.)

5. The odds in favor of winning a different raffle are 3 to 4. Calculate the theoretical probability that two winning tickets will be drawn in a row.

6. Refer to Exercise 5. To simulate the probability of drawing two winning tickets in a row, Marcus puts 40 red and blue marbles in a bag.

a. How many red and how many blue marbles should he use? Explain your reasoning.

b. Conduct a simulation for this situation. Compare the theoretical probability with the experimental probability.

7. A test has 12 multiple choice questions with answer choices A, B, C, or D for each question. The correct answers are A, B, B, A, C, A, D, A, B, A, D, C. You need to correctly answer 9 or more questions to pass the test.

a. Design a simulation that could be used to answer the questions on the test.

b. Is your simulation a good strategy for taking the test? Explain.

Study Guide and Review

Math Online glencoe.com
• STUDY TO GO
• Vocabulary Review

Chapter Summary

Key Concepts

Measures of Central Tendency (Lesson 13-1)

- Measures of central tendency describe the center of the data. The most common measures are mean, median, and mode.

Data Displays and Measures of Variation (Lessons 13-2 through 13-5)

- A stem-and-leaf plot is most often used when displaying data in a condensed form.
- The interquartile range is the range of the middle half of a set of data.
- A box-and-whisker plot separates data into four parts.
- A histogram displays data that have been organized into equal intervals.

Simple Probability and Probability of Compound Events (Lessons 13-6 and 13-10)

- The probability of an event is a ratio that compares the number of favorable outcomes to the number of possible outcomes.
- When the outcome of one event does not affect the outcome of a second event, these are called independent events.
- When the outcome of one event does affect the outcome of a second event, these are called dependent events.

Counting Outcomes and Permutations and Combinations (Lessons 13-7 and 13-8)

- The Fundamental Counting Principle states that an event M followed by an event N can occur in $m \times n$ ways if event M occurs in m ways and event N occurs in n ways
- Permutations: order is important
- Combinations: order is not important

FOLDABLES® Study Organizer

Be sure the Key Concepts are noted in your Foldable.

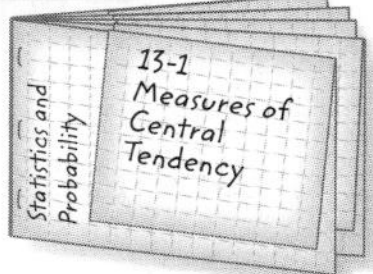

Key Vocabulary

box-and-whisker plot (p. 750)
compound events (p. 790)
dependent events (p. 791)
experimental probability (p. 766)
Fundamental Counting Principle (p. 777)
histogram (p. 757)
independent events (p. 790)
interquartile range (p. 744)
lower quartile (p. 744)
mean (p. 730)
measures of central tendency (p. 730)
measures of variation (p. 743)
median (p. 730)
mode (p. 730)
outcome (p. 765)
outlier (p. 745)
population (p. 771)
probability (p. 765)
quartile (p. 744)
range (p. 743)
sample (p. 771)
sample space (p. 766)
simple event (p. 765)
stem-and-leaf plot (p. 737)
theoretical probability (p. 766)
tree diagram (p. 777)
upper quartile (p. 744)

Vocabulary Check

Choose the term that best matches each statement or phrase. Choose from the list above.

1. sum of the data divided by the number of items in the data set
2. two or more simple events
3. drawing to represent possible outcomes
4. what should occur in an experiment
5. divides sets of data into four parts
6. number or numbers that occur more often
7. the difference between the greatest and least values in a set of data
8. uses bars to display numerical data that have been organized into equal intervals

Lesson-by-Lesson Review

13-1 Measures of Central Tendency (pp. 730–735)

Find the mean, median, and mode for each set of data. Round to the nearest tenth, if necessary.

9. number of students in each math class: 22, 23, 24, 22, 21
10. grams of fat per serving: 2, 7, 4, 5, 6, 4, 5, 6, 3, 5
11. inches of rain last week: 1.5, 2, 2.5, 2, 1.5, 2.5, 3
12. **MOVIES** At the movie theater, six movies are playing and their lengths are 138, 117, 158, 145, 135, and 120 minutes. Which measures of central tendency best represent the data? Justify your selections and then find the measure of central tendency.

EXAMPLE 1

Find the mean, median, and mode of 2, 3, 2, 4, 4, 6, 4, and 7.

Mean: $\frac{2+3+2+4+4+6+4+7}{8} = \frac{32}{8}$ or 4

Median:

2, 2, 3, 4, 4, 4, 6, 7 — **Arrange the numbers from least to greatest.**

$\frac{4+4}{2} = 4$ — **Find the middle number or the mean of the two middle numbers.**

Mode: 4 — **Find the data value(s) that occur most often.**

13-2 Stem-and-Leaf Plots (pp. 737–742)

Display each set of data in a stem-and-leaf plot.

13.

Heights of Football Players (in.)			
72	73	69	71
74	76	70	71
68	75	72	73

14.

Attendance at Key Club			
35	46	36	42
41	32	55	56
22	28	33	45
51	52	48	49

15. Frank's and Shandra's times for their last eight races are shown below.

Racing Times (min)

Frank	Stem	Shandra
9 8	0	
5 4 3 2 1	1	3 3 3 4 4 5
0	2	1 1

2 | 1 = 12 min   *1 | 3 = 13 min*

In general, which runner has a faster time? Explain.

EXAMPLE 2

The table below shows the low temperatures for two weeks for a certain city.

Low Temperatures for Two Weeks (°F)						
47	48	51	48	55	57	40
39	57	42	37	48	55	42

a. Display the data in a stem-and-leaf plot. The least number is 37, and the greatest number is 57. So, the stems are 3, 4, and 5.

Low Temperatures

Stem	Leaf
3	7 9
4	0 2 2 7 8 8 8
5	1 5 5 7 7

3 | 7 = 37°F

b. What is the median temperature?

The median is the mean of the two middle numbers of the set of data.

The median temperature is $\frac{48+48}{2}$ or 48°F.

Study Guide and Review

13-3 Measures of Variation (pp. 743–749)

Find the measures of variation and any outliers for each set of data.

16. The number of minutes spent reading each night: 31, 33, 32, 34, 35, 33

17. The number of fish in each fish tank: 6, 5, 7, 8, 5, 6, 7, 9, 8, 6

18. **GIFTS** Claire earned \$5, \$7, \$10, \$6, and \$8 doing errands for her neighbors. Find the measures of variation and any outliers for the set of data.

19. **GRADES** The scores Mr. Han's students earned on their last test are shown in the table. Use the measures of variations to describe the data in the table.

Test Scores			
99	88	81	89
77	58	92	80
83	82	74	84
76	73	99	74
82	87	82	74
86	76	85	92

EXAMPLE 4

During a baking contest, each baker had 26, 20, 21, 24, 23, 22, 21, 27, 23, 24, and 25 cookies to sample. Find the measures of variation and any outlier for the data.

Range: 27 - 20 or 7 cookies

Median, Upper Quartile, Lower Quartile:

List the data from the least to greatest.

lower quartile — median — upper quartile

{20, 21, 21, 22, 23, 23, 24, 24, 25, 26, 27}

The median is 23, the lower quartile is 21, and the upper quartile is 25.

Interquartile Range: 25 - 21 or 4.

Outliers:

Multiply the interquartile range by 1.5.

$4 \times 1.5 = 6$

Subtract 6 from the lower quartile and add 6 to the upper quartile.

$21 - 6 = 15$ $\qquad 25 + 6 = 31$

Since there are no values less than 15 or greater than 31, there are no outliers.

13-4 Box-and-Whisker Plots (pp. 750–755)

20. The box-and-whisker plot below shows the winning times in minutes for the men's marathon in the Summer Olympic Games from 1928 to 2004.

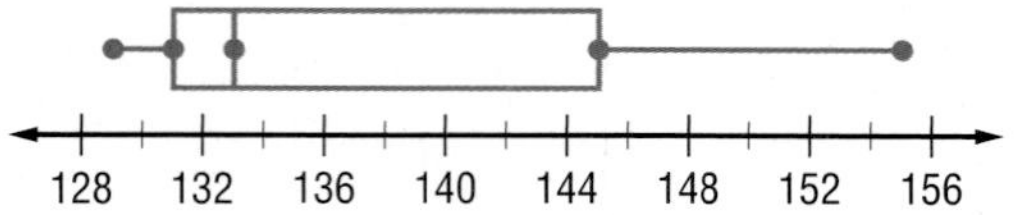

a. Find the percent of winning marathon times that were under 131.

b. Write a sentence describing what the length of the box-and-whisker plot tells about the winning times for the men's marathon.

EXAMPLE 5

Use the box-and-whisker plot shown to find the percent of New York City marathons that were held on days that had a high temperature greater than 72.5°F.

NYC Marathon High Temperatures (°F)

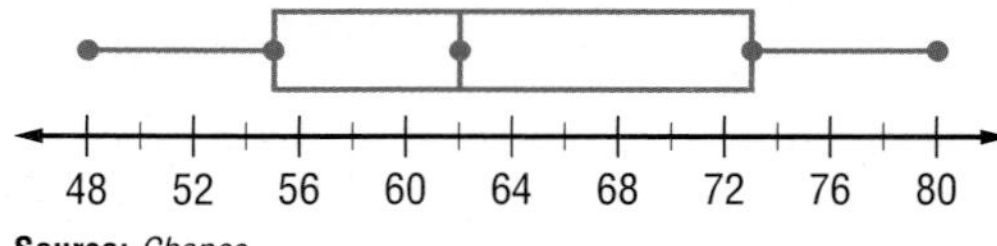

Source: *Chance*

Each of the four parts represents 25% of the data, so 25% of the marathons had a high temperature greater than 72.5°F.

MIXED PROBLEM SOLVING
For mixed problem-solving practice, see page 746.

13-5 Histograms (pp. 757–762)

21. The results of a class survey are shown in the table. Display the results in a histogram.

Number of Siblings		
Number of Siblings	Tally	Frequency
0–1	𝍸 𝍸 𝍸 III	18
2–3	𝍸	5
4–5	II	2

22. **U.S. PRESIDENTS** The frequency table shows the ages at which 43 U.S. Presidents began their terms in office. Display the data in a histogram.

Age	Frequency
42–51	14
52–61	23
62–71	6

EXAMPLE 6

Display the set of data in a histogram.

Times for 100-Meter Freestyle		
Times (S)	Tally	Frequency
0–29.9		0
30–59.9	𝍸 𝍸	10
60–89.9	𝍸 𝍸 𝍸 III	18
90–119.9	𝍸 II	7

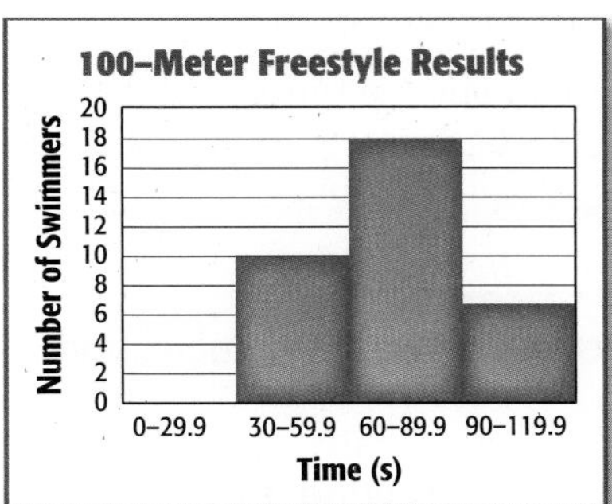

13-6 Theoretical and Experimental Probability (pp. 765–770)

MARBLES There are 2 blue marbles, 5 red marbles, and 8 green marbles in one bag. One marble is selected at random. Find the probability of each outcome.

23. *P*(red)
24. *P*(blue or green)
25. *P*(*not* blue)
26. *P*(yellow)

EXAMPLE 7

Suppose a number cube is rolled. Find the probability of rolling an even number.

$$P(\text{even}) = \frac{\text{number of favorable outcomes}}{\text{number of possible outcomes}}$$
$$= \frac{3}{6} \text{ or } \frac{1}{2}$$

13-7 Using Sampling to Predict (pp. 771–776)

27. To determine the weekly top ten songs, the local radio station asks people to log onto their Web site and vote for their favorite song. Identify the sample as *biased* or *unbiased* and describe its type. Explain.

28. **FOOD** Forty-five out of 60 people at a steakhouse said their favorite meal was steak. Is this sampling representative of the entire town? If so, how many of the 13,000 residents would say steak was their favorite meal?

EXAMPLE 8

Is polling students on the football team about their favorite sports a biased or unbiased sample? Then describe the type of sample.

Biased, convenience sample
Students who play football are more likely to choose football as their favorite sport.

13-8 Counting Outcomes (pp. 777–781)

Find the number of possible outcomes for each situation.

29. Two coins are tossed.
30. Customers can choose vanilla or chocolate ice cream and strawberry, chocolate, or caramel topping.
31. A number cube is rolled three times.
32. A coin is tossed and a spinner with four equal sections is spun.
33. James can choose blue or black socks, tan or black pants, and a red, blue, or green shirt.

EXAMPLE 9

A die is rolled and a coin is tossed. Find the number of possible outcomes.

Outcomes of Die		Outcomes of Coin Toss		
6	×	2	=	12

There are 12 possible outcomes.

13-9 Permutations and Combinations (pp. 783–788)

34. **RACES** In how many ways can five runners come in first, second, or third place?
35. **LOCKS** How many different combinations can be made with three distinct numbers on a lock with 40 numbers?
36. **PRIZES** How many different ways can you select eight prizes out of twenty choices?
37. **PARTIES** How many different sets of people could you have if you chose five people out of the 22 people in your class to invite to your birthday party?

EXAMPLE 10

How many ways can 8 runners place first, second, and third in a race?

The order is important, so this is a permutation.

$P(8, 3) = 8 \cdot 7 \cdot 6$

$= 336$

There are 336 ways for 8 runners to place first, second, and third.

13-10 Probability of Compound Events (pp. 790–795)

A box contains 8 red markers, 5 green markers, and 5 white markers. Once a marker is pulled from the box, it is not replaced. Find each probability.

38. P(red, then green)
39. P(red, then red)
40. P(white, then red)
41. P(green, then green)
42. P(white, then white)

EXAMPLE 11

A card is drawn from a deck of eight cards numbered from 1 to 8 and not replaced. Find the probability of drawing a 3 and then a 6.

$P(3, 6) = P(3 \text{ on } 1^{st} \text{ draw}) \cdot P(6 \text{ on the } 2^{nd} \text{ draw})$

$= \frac{1}{8} \cdot \frac{1}{7}$

$= \frac{1}{56}$

The probability of drawing a 3 then a 6 is $\frac{1}{56}$.

CHAPTER 13

Practice Test

Math Online glencoe.com
Chapter Test

Find the mean, median, and mode for the following data sets. Round to the nearest hundredth if necessary.

1. the ages of the students at a picnic:
7, 8, 7, 9, 8, 10, 11, 7

2. shoe sizes:
4, 5, 6, 5, 4, 7, 6

3. **TEST SCORES** Use the table that shows Ms. Fernandez' 1st period test grades.

Science Test Grades							
77	78	88	89	67	65	87	99
93	92	76	79	70	85	83	81

a. Display the data in a stem-and-leaf plot.

b. What is the median test grade?

c. In which interval do most of the grades occur?

4. **MULTIPLE CHOICE** The data show times for a race at a track meet. Find the range.

Track Times

Stem	Leaf
3	1 3 4 4 5 7
4	6 8 8 9 9
5	2 3 4

4 | 6 = 46 s

A 23 C 25

B 24 D 84

5. Refer to the stem-and-leaf plot above. Find the remaining measures of variation.

6. Display the following data in a box-and-whisker plot.

15	19	26	14	17
13	20	21	29	18

7. What percent of the data shown above is below 21?

8. Display the data shown in a histogram.

Length of Time Spent Reading		
Time (min)	Tally	Frequency
0–14	~~||||~~ ~~||||~~ ||	12
15–29	~~||||~~ ||	7
30–44	||||	4
45–59	||	2

A bowl contains 8 red apples, 4 green apples, and 8 yellow apples. Suppose one apple is selected at random. Find each probability as a fraction.

9. *P*(red) **10.** *P*(green)

11. *P*(yellow) **12.** *P*(red or green)

13. Identify each of the following as *biased* or *unbiased* and describe its type.

a. To determine the quality of TVs coming off an assembly line, every 5th TV is checked.

b. To determine if a new book store should be built, every tenth person exiting the library is surveyed.

14. **MULTIPLE CHOICE** Find the number of possible outcomes for a choice of bologna, salami, or ham and rye, wheat, sourdough, or white bread.

F 2 H 12

G 7 J 37

Tell whether each situation is a *permutation* or a *combination*. Then solve.

15. The number of ways 6 notebooks can be arranged on a bookshelf.

16. An envelope contains 12 different names. The number of ways you can choose 3 of the names.

A card is randomly drawn from the cards shown and not replaced. A second card is drawn. Find the probability of each outcome.

1	2	3	4
5	6	7	8

17. *P*(2, odd)

18. *P*(even, odd)

19. *P*(multiple of 3, multiple of 4)

20. *P*(5, odd)

CHAPTER

13 Preparing for Standardized Tests

Make an Organized List

One strategy for solving problems on standardized tests is to organize data in a table or list. Organizing data in a table or list can make it is easier to analyze the data and answer questions about the data.

Strategies for Making an Organized List

Step 1

Read the problem quickly to gain a general understanding of it.

- **Ask yourself:** "What information or data is given in the problem?"
- **Ask yourself:** "Would the problem be easier to comprehend and solve if the data were organized in a table or list?"

Step 2

Create an organized list.

- Choose the best format for organizing and displaying the data given in the problem statement.
- Fill the list or table with the data.
- Be sure to include all of the data values.
- Label the list or table as appropriate.

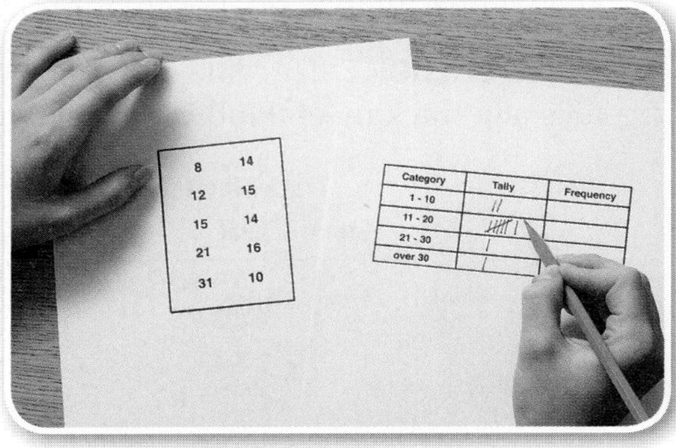

Step 3

Solve the problem.

- Use your organized list to solve the problem.
- If time permits, check your answer.

EXAMPLE

Read the problem. Identify what you need to know. Then use the information in the problem to solve.

A deli offers customers two choices of bread: white or wheat; two choices of cheese: cheddar or Swiss; and three types of meat: ham, turkey, or salami. If a sub sandwich with one type of bread, cheese, and meat is chosen at random, what is the probability that it contains ham or cheddar cheese?

A $\frac{3}{4}$ **B** $\frac{2}{3}$ **C** $\frac{1}{2}$ **D** $\frac{2}{5}$

Read the problem carefully. You are given the bread, cheese, and meat choices for sub sandwiches at a deli. You need to find the probability that a randomly selected sandwich contains ham or cheddar cheese. A *tree diagram* is a type of organized list that shows you all of the possible sandwich combinations.

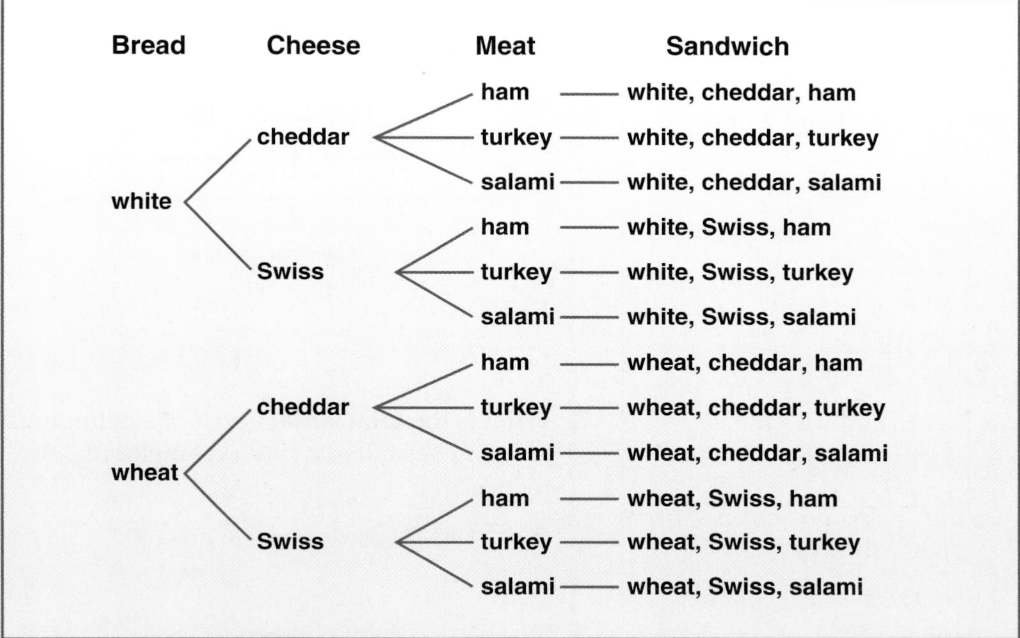

With the deli choices organized in the tree diagram, it is easy to see which sandwiches contain ham or cheddar cheese. There are 12 total sandwich combinations and 8 of them contain ham or cheddar cheese. So, the probability is $\frac{8}{12}$ or $\frac{2}{3}$. The correct answer is B.

Exercises

Read each problem. Identify what you need to know. Then use the information in the problem to solve.

1. A store has gray, white, or black shoes in sizes 8, 9, 10, 11, and 12. If a pair of shoes is chosen at random, what is the probability that the shoes are either gray, white, size 11 or size 12?

A $\frac{1}{12}$

B $\frac{3}{4}$

C $\frac{4}{5}$

D $\frac{5}{6}$

2. A car manufacturer offers two kinds of transmissions: standard or automatic; two different interiors: leather or cloth; and six different colors: black, red, white, gray, yellow, or blue. If a car is chosen at random, what is the probability that it is red and has leather interior?

F $\frac{2}{3}$

G $\frac{5}{12}$

H $\frac{3}{4}$

J $\frac{1}{12}$

CHAPTER 13 Standardized Test Practice

Cumulative, Chapters 1 through 13

Multiple Choice

Read each question. Then fill in the correct answer on the answer document provided by your teacher or on a sheet of paper.

1. Which point is 4 units to the left and 2 units below the origin?

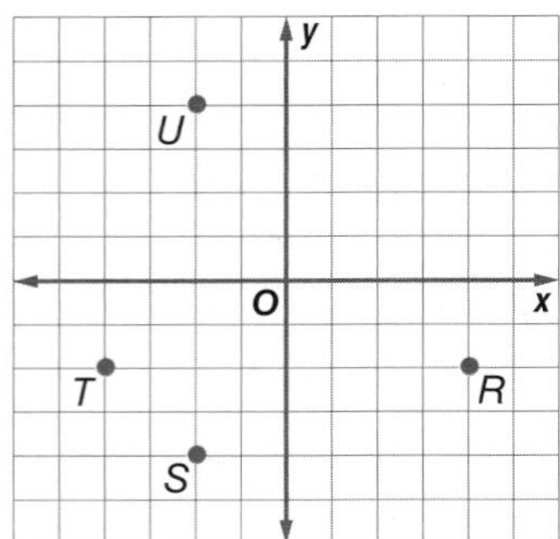

A *R* C *T*

B *S* D *U*

2. What is the median age shown in the stem-and-leaf plot below?

Customer Ages (years)

Stem	Leaf
1	8 9 9
2	1 2 3 3 3 4 5 5 6 7 9
3	0 0 1 4 6 6 8
4	2 3 3 4 6 7 9
5	
6	0
7	1 4

3|4 = 34 years

F 29 years H 31 years

G 30 years J 34 years

3. What is the range of the ages shown in the stem-and-leaf plot in Exercise 2?

A 18 years C 56 years

B 42 years D 74 years

4. In the box-and-whisker plot shown below, what percent of the cars get 24 miles per gallon or better?

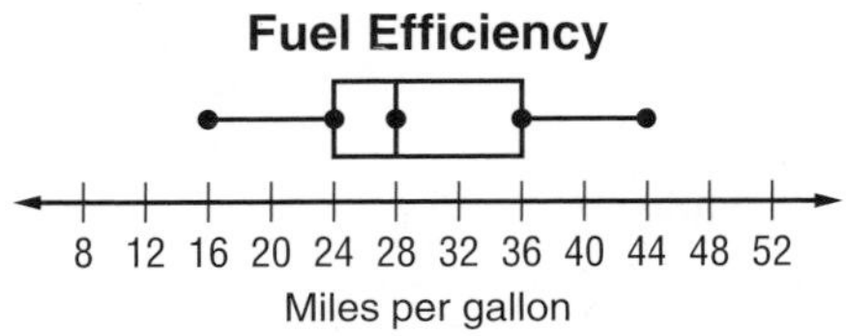

F 15% H 50%

G 25% J 75%

5. What is the total surface area of a cylinder that is 5.4 inches tall and has a diameter of 3.8 inches?

A 87.1 in.^{2} C 61.2 in.^{2}

B 73.3 in.^{2} D 55.4 in.^{2}

6. A random sample of 120 shoppers is interviewed as they leave a shopping center. Twenty-four of them say that they used a store coupon for their purchase. If this sample is representative of the population, how many of 650 customers would you expect to use a coupon?

F 130 H 135

G 140 J 150

7. Mickey has 3 pennies, 5 nickels, 2 dimes, and 8 quarters in her backpack. If she selects one coin at random, what is the probability that it is *not* a quarter?

A $\frac{1}{3}$ C $\frac{1}{2}$

B $\frac{4}{9}$ D $\frac{5}{9}$

Test-TakingTip

Question 7 Compare the number of *successes* to the total number of possible outcomes. In this problem, the number of success would be 10, the number of coins that are not quarters.

Short Response/Gridded Response

Record your answers on the answer sheet provided by your teacher or on a sheet of paper.

8. The table shows the pizza choices at Angelo's Pizzeria.

Angelo's Pizzeria	
Crust	**Toppings**
Hand-Tossed	Pepperoni
Deep Dish	Sausage
Pan	Green Peppers
	Mushrooms
	Olives

a. If a 1-topping pizza is selected at random from the menu, what is the probability that it will be a hand-tossed pizza with pepperoni?

b. What is the probability that a randomly selected 1-topping pizza will be a deep dish pizza?

9. Write a congruence statement for the triangles shown below.

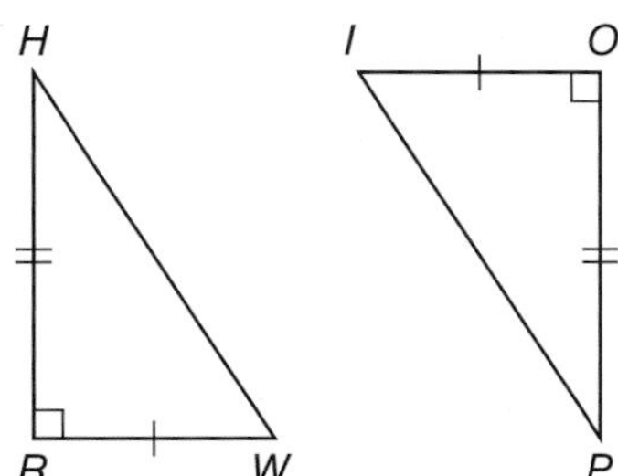

10. **GRIDDED RESPONSE** Coach Gonzalez has 9 batters playing in today's softball game. How many ways can she arrange the first 3 batters of the lineup, assuming order is important?

11. **GRIDDED RESPONSE** Gina has a jar of marbles that contains 7 blue marbles, 5 red marbles, and 8 green marbles. Once a marble is selected, it is not replaced. If she selects 2 marbles at random, find the probability that two red marbles are chosen.

12. This season, Antoine scored 13, 15, 9, 10, 14, and 11 points.

a. Calculate the mean, median, and mode for the set of data. Show your work.

b. Which measure of central tendency best describes Antoine's scoring production this season? Explain.

Extended Response

Record your answers on a sheet of paper. Show your work.

13. The table below shows the high temperatures over the past 20 days.

Daily High Temperatures (°F)				
72	74	76	76	75
79	81	84	83	76
73	72	67	69	70
74	76	78	79	80

a. Make a stem-and-leaf plot of the data.

b. What is the range of the daily high temperatures over the 20-day period?

c. Does there appear to be an outlier in the set of data? Explain.

Need Extra Help?													
If you missed Question...	1	2	3	4	5	6	7	8	9	10	11	12	13
Go to Lesson or Page...	2-6	13-2	13-3	13-4	12-6	13-7	13-6	13-8	11-2	13-9	13-10	13-2	13-3

Looking Ahead to Algebra 1

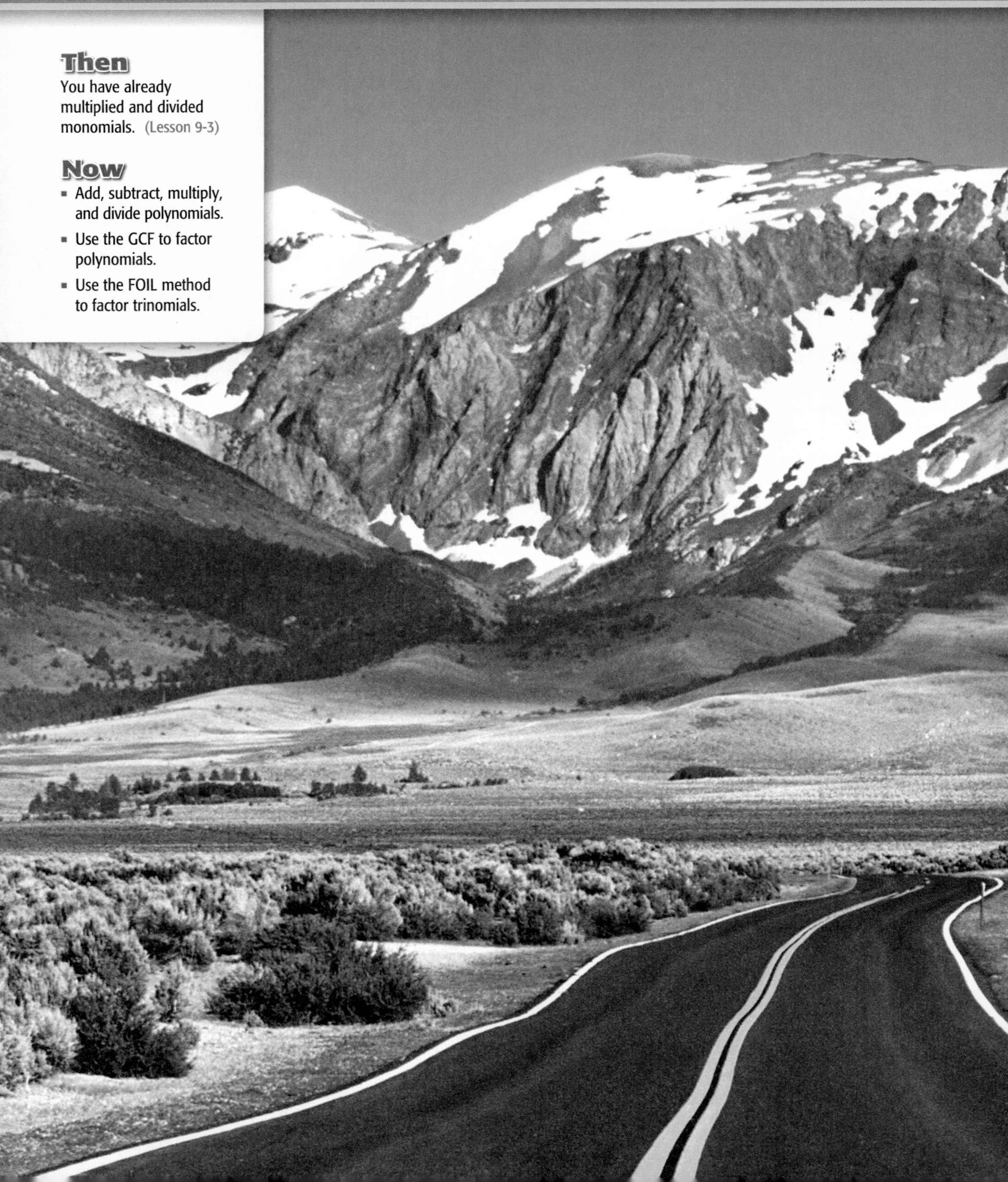

Then

You have already multiplied and divided monomials. (Lesson 9-3)

Now

- Add, subtract, multiply, and divide polynomials.
- Use the GCF to factor polynomials.
- Use the FOIL method to factor trinomials.

Let's Look Ahead

Polynomials

Why?

The value of x pencils that cost 5 cents is $5x$ cents.

a. Write an expression to represent the value in cents of y spiral notebooks that cost 25 cents each.

b. If erasers cost 5 cents each, write an expression to represent the total value of x pencils, y spiral notebooks, and z erasers.

Then
You have already simplified algebraic expressions. (Lesson 4-2)

Now
- Use models to simplify polynomials.
- Use polynomials to represent real-world problems.

New Vocabulary
polynomial

Math Online

glencoe.com
- Extra Examples
- Personal Tutor
- Self-Check Quiz
- Homework Help

Polynomials You can use algebra tiles to model monomials and **polynomials**, which are algebraic expressions that contain more than one monomial.

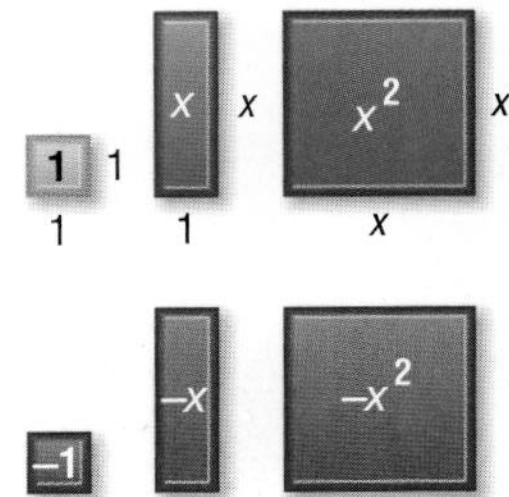

A tile that is 1 unit by 1 unit represents the integer 1. A tile that is 1 unit by x units is represented by the variable x. A tile that is x units by x units represents the expression x^2. Red tiles with the same shapes are used to represent -1, $-x$, and $-x^2$.

EXAMPLE 1 **Model Polynomials**

Use algebra tiles to model $x^2 - 3x + 4$.

$x^2 - 3x + 4 = x^2 + (-3x) + 4$ — **Use the definition of subtraction to write the polynomial as the sum of terms.**

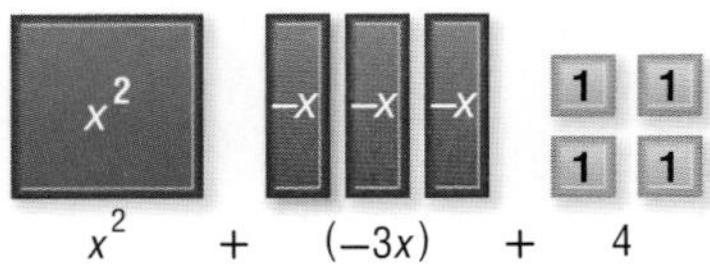

Check Your Progress

Use algebra tiles to model each polynomial.

1A. $5x - 6$

1B. $2x^2 + 4x + 3$

Personal Tutor glencoe.com

Recall that *like terms* are terms that contain the same variable. When working with polynomials, like terms contain the same variable and exponent.

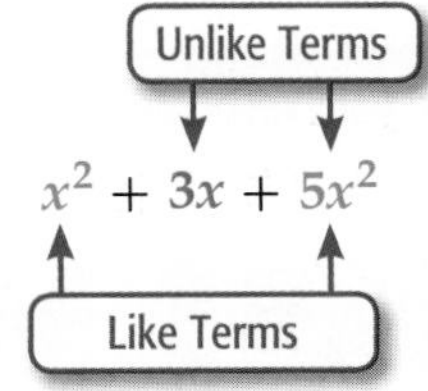

When simplifying polynomials, it is customary to write the result in *standard form*. That is, write the powers of the variable in decreasing order from left to right.

Standard form	Not standard form
$-3x^2 + 4x + 1$	$4x - 3x^2 + 1$

Vocabulary Review

Monomial An expression that is a number, a variable, or a product of numbers and/or variables. (Lesson 9–2)

EXAMPLE 2 Simplify Polynomials

Simplify $-3 + 4x + x^2 + x$.

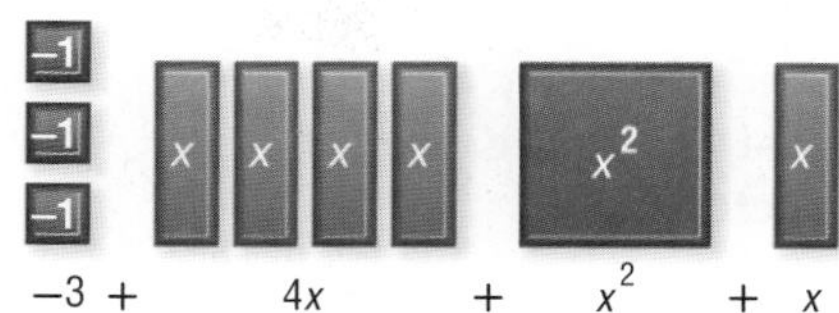

Group tiles with the same shape. Then write a polynomial.

Combine like terms.

So, $-3 + 4x + x^2 + x = x^2 + 5x + (-3)$ or $x^2 + 5x - 3$.

Check Your Progress

Simplify each polynomial. Use models if needed.

2A. $-x + 1 - 2x + 5$

2B. $x^2 + 4 + 3x + x^2$

Personal Tutor glencoe.com

When a positive tile and a negative tile of the same shape are paired, the result is a *zero pair*. Remove any zero pairs when simplifying polynomials.

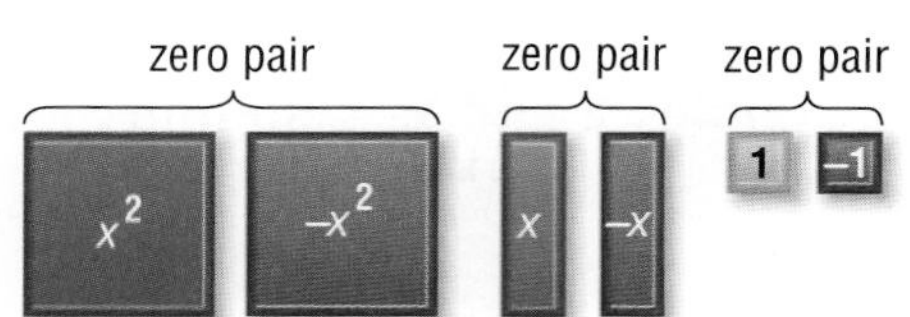

EXAMPLE 3 Remove Zero Pairs to Simplify Polynomials

Simplify $2x^2 - 3x - x^2$.

$2x^2 - 3x - x^2 = 2x^2 + (-3x) + (-x^2)$ **Write the polynomial as the sum of terms.**

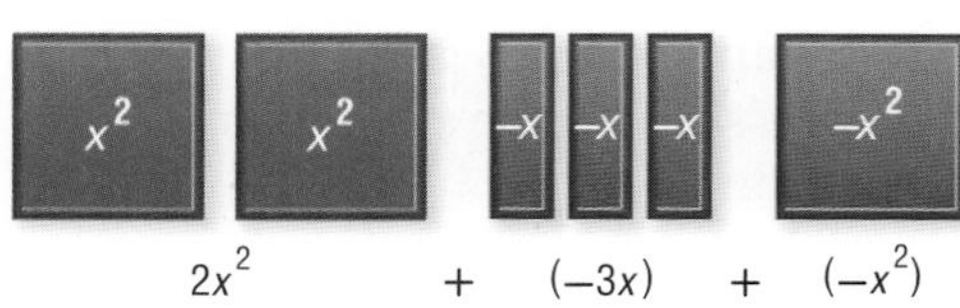

Group tiles with the same shape. Then write a polynomial.

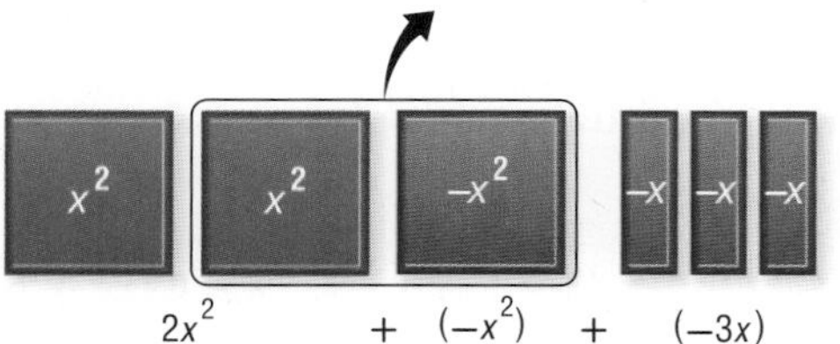

Remove the zero pairs and combine like terms.

So, $2x^2 - 3x - x^2 = x^2 - 3x$.

Check Your Progress

Simplify each polynomial. Use models if needed.

3A. $6 - 2x^2 + 3x^2$

3B. $-2x^2 + 5 - 1 + x^2$

Personal Tutor glencoe.com

Use Polynomials Polynomials can represent real-world problems.

Real-World EXAMPLE 4 Use a Polynomial

GEOMETRY Write and simplify a polynomial expression for the perimeter of the rectangle.

$P = 4y + 4y + 7 + 7$ **Definition of perimeter**

$= (8y + 14)$ in. **Combine like terms.**

Check Your Progress

4. **CRAFTS** Charlie bought a model airplane kit for \$5, three small jars of paint for \$$x$ each, and a bottle of glue for \$2. Write and simplify a polynomial expression to represent the total amount of money he spent.

Personal Tutor glencoe.com

Check Your Understanding

Example 1 p. LA2

Use algebra tiles to model each polynomial.

1. $3x^2 + 4x$
2. $-x^2 + 2x + 5$

Examples 2 and 3 p. LA3

Simplify each polynomial. Use models if needed.

3. $2x + 3 + 3x$
4. $x^2 - 3x + x$

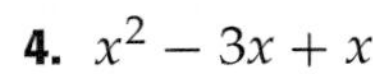

5. $1 + 2x + 4 + x$
6. $x^2 - 6x + 4x$
7. $3x^2 + 5x - x^2 + 2x$
8. $-2x^2 + 7 + 3x - x^2 - 5x$

Example 4 p. LA4

9. **GEOMETRY** Write and simplify a polynomial expression for the perimeter of the parallelogram.

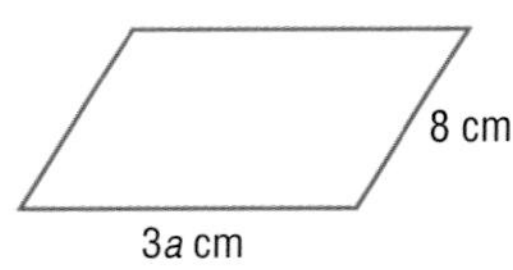

Practice and Problem Solving

Example 1 p. LA2

Use algebra tiles to model each polynomial.

10. $5x - 1$
11. $x^2 + 6$
12. $2x^2 - x + 3$
13. $-x^2 + 6x - 4.$

Examples 2 and 3 p. LA3

Simplify each polynomial.

14. $5 + x + 2$
15. $-x^2 + 4x - x^2$

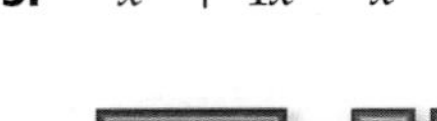

Examples 2 and 3 p. LA3

Simplify each polynomial. Use models if needed.

16. $x^2 - x + 4 + 2x - 3$

17. $-x^2 + 3x + x^2 - 2x$

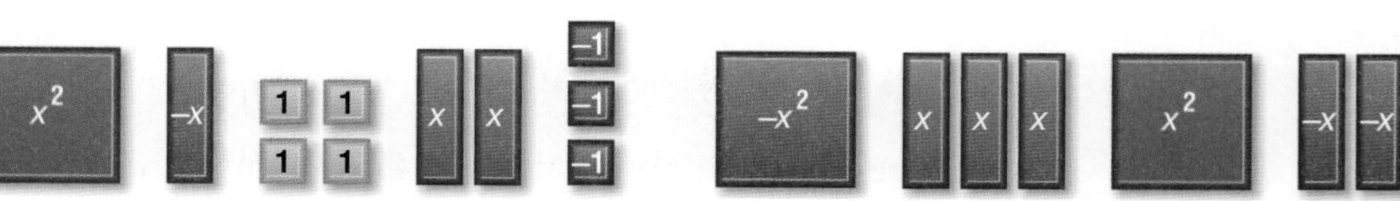

18. $x^2 + 7 + 3x^2$

19. $9 - 5x - 2x + 2$

20. $-2x^2 + x - 3x^2 - 6$

21. $7x - 8 + 2x + 4x^2 + x^2$

22. $-6x + 4 + 2x + 7$

23. $5 - x^2 + 7x + 4x^2$

24. $4x^2 + 5x + 1 - 4x^2 - 6x$

25. $6 + 2x^2 - 3x - 4x - x^2$

Example 4 p. LA4

26. **GEOMETRY** Write and simplify a polynomial expression for the perimeter of the trapezoid.

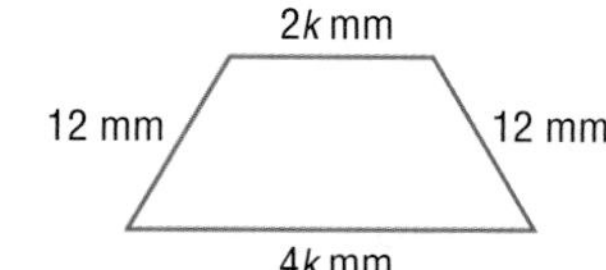

27. **EXERCISE** Taylor jogged x miles after school. Seth jogged twice the distance that Taylor jogged. Rashida jogged 4 miles. Write and simplify a polynomial expression to represent the total number of miles that the three students jogged.

28. **PIZZA** For a party, one classroom ordered 4 large pizzas, 2 medium pizzas, and 6 subs. Another classroom ordered 6 large pizzas and 8 subs. If ℓ represents the cost of a large pizza, m represents the cost of a medium pizza, and s represents the cost of a sub, write an expression in simplest form for the total amount of money that the two classrooms spent on food.

29. **JOBS** Michael and Olivia each earn x dollars per lawn that they mow and y dollars per hour for babysitting. The table shows how many lawns each mowed and how many hours each babysat. Write an expression in simplest form for the total amount of money they earned together.

	Number of Lawns	Number of Hours Babysitting
Michael	8	6
Olivia	9	4

Simplify each polynomial.

30. $1.5t^2 - 7.6 + 4t - 2t^2 + t - 5.1$

31. $10 + 6c^3 - c^2 - 2c^3 + 2c^2 - 4c^2$

32. $\frac{1}{2}n^2 - 6n - \frac{3}{4}n + 7n$

33. $4a^2 + 2b - b^2 - 6b + 3a^2 - 5a$

H.O.T. Problems

Use **H**igher-**O**rder **T**hinking Skills

34. **OPEN ENDED** Write a polynomial with four terms that simplifies to $x^2 + 3x$.

35. **CHALLENGE** Determine whether $x^2 + 6x = 7x^2$ is *always*, *sometimes*, or *never* true. Explain your reasoning.

36. **REASONING** Explain how you can tell from a model whether a polynomial will have at most one term, two terms, or three terms when it is simplified.

37. **WRITING IN MATH** Describe how you can use models to simplify polynomial expressions.

Adding Polynomials

Why?

Lauren has 38 phone calls and 47 text messages on her cell phone bill. Javier has 25 phone calls and 52 text messages on his cell phone bill.

a. The expression 38 calls + 47 texts represents the items on Lauren's phone bill. Write an expression for the items on Javier's phone bill.

b. Write an expression for the total number of calls and text messages.

Then

You have already solved equations by adding. (Lesson 4-3)

Now

- Add polynomials.
- Find perimeter by adding polynomials.

Math Online

glencoe.com

- Extra Examples
- Personal Tutor
- Self-Check Quiz
- Homework Help

Adding Polynomials You can use models to add polynomials.

EXAMPLE 1 Add Polynomials

Add. Use models if needed.

a. $(3x + 4) + (2x + 1)$

Step 1 Model each polynomial.

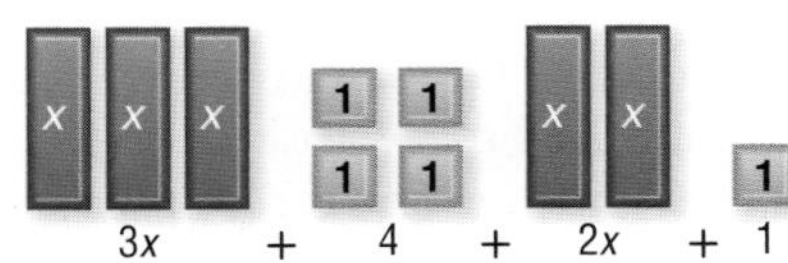

Step 2 Combine the tiles that have the same shape.

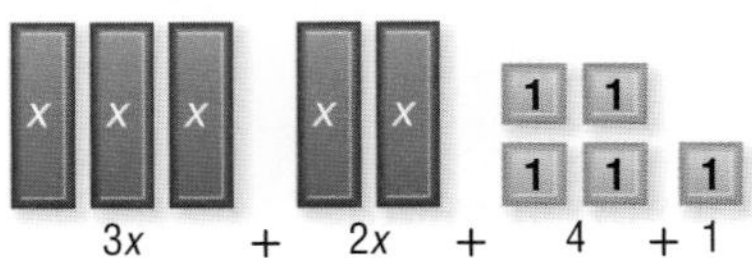

Step 3 Write the polynomial for the combined tiles.

$(3x + 4) + (2x + 1) = 5x + 5$

b. $(2x^2 - 4x + 2) + (-2x + 2)$

$2x^2 - 4x + 2$	
$\underline{\quad\quad -2x + 2}$	**Arrange like terms in columns.**
$2x^2 - 6x + 4$	**Add.**

So, $(2x^2 - 4x + 2) + (-2x + 2) = 2x^2 - 6x + 4$.

Check Your Progress

1A. $(x^2 + x - 3) + (3x^2 - 4)$

1B. $(-2x^2 - x + 1) + (-x^2 - 3x)$

Personal Tutor glencoe.com

EXAMPLE 2 Use Zero Pairs to Add Polynomials

Add $(-2x^2 + 3x + 2) + (x^2 - x + 4)$.

Model the polynomials.

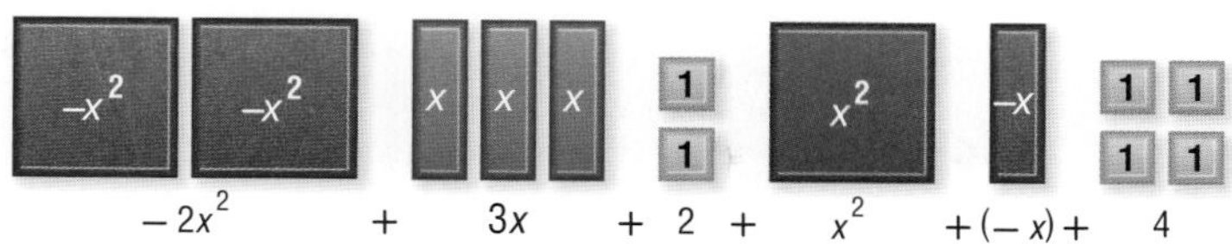

Group tiles with the same shape. Then remove any zero pairs.

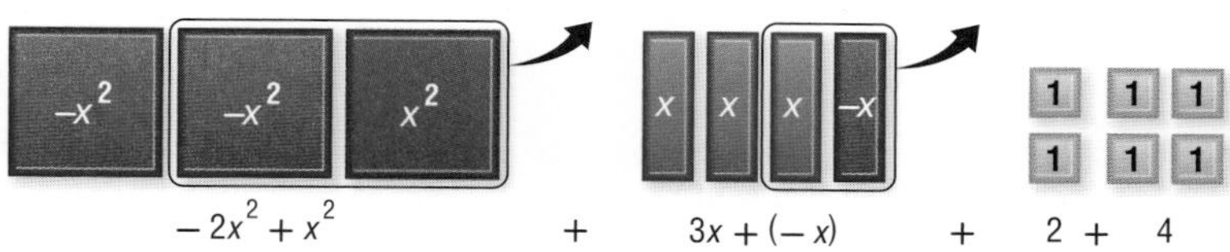

So, $(-2x^2 + 3x + 2) + (x^2 - x + 4) = -x^2 + 2x + 6$.

Check Your Progress

Add. Use models if needed.

2A. $(-3x^2 + 2x) + (3x^2 + 4x)$

2B. $(x^2 - 4x - 1) + (-2x^2 + 5x - 3)$

Personal Tutor glencoe.com

Find Perimeter Polynomials can be used to find perimeter.

Real-World EXAMPLE 3 Use Polynomials to Find Perimeter

GEOMETRY The lengths of the sides of golden rectangles are in the ratio 1:1.62. So, the length of a golden rectangle is approximately 1.62 times greater than its width.

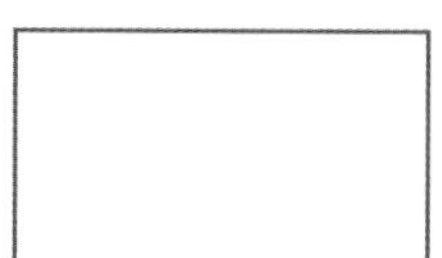

$1.62x$

a. Write the formula for the perimeter of a golden rectangle.

$P = 2\ell + 2w$ — **Formula for the perimeter of a rectangle**

$P = 2(1.62x) + 2x$ — **Replace ℓ with $1.62x$ and w with x.**

$P = 3.24x + 2x$ or $5.24x$ — **Simplify.**

The formula is $P = 5.24x$, where x is the measure of the width.

b. Find the perimeter of a golden rectangle if its width is 8.3 centimeters.

$P = 5.24x$ — **Perimeter of a golden rectangle**

$= 5.24(8.3)$ or 43.492 — **Replace x with 8.3 and simplify.**

The perimeter of the golden rectangle is 43.492 centimeters.

Real-World Link

The golden ratio has been incorporated in architecture from Ancient Greece to modern architecture.

Check Your Progress

3. GEOMETRY A rectangle has side lengths of $(x^2 - 5x)$ units and $(2x^2 + x)$ units.

A. Write the formula for the perimeter of the rectangle.

B. Find the perimeter of the rectangle if the value of x is 5.4 units.

Personal Tutor glencoe.com

Check Your Understanding

Examples 1 and 2 pp. LA6–LA7

Add. Use models if needed.

1. $(x + 5) + (2x + 3)$
2. $(x^2 - 4x) + (x^2 - 5x)$
3. $(2x^2 + 6) + (-x^2 + 3x - 1)$
4. $(x^2 - 7x + 2) + (-3x^2 + x + 4)$

Example 3 p. LA7

5. **GEOMETRY** Use the figure at the right.
   a. Write the formula for the perimeter of the figure.
   b. Find the perimeter of the figure if $x = 4$.

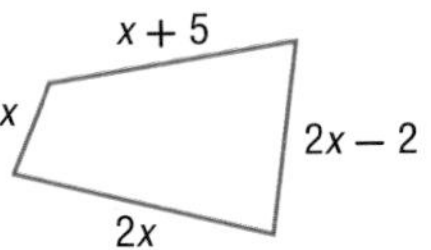

Practice and Problem Solving

Examples 1 and 2 pp. LA6–LA7

Add. Use models if needed.

6. $(7x + 5) + (x + 2)$
7. $(-x + 3) + (-5x + 6)$
8. $(3x^2 - 7x + 1) + (x^2 - x)$
9. $(6x^2 + 5x) + (-9x + 4)$
10. $(-4x^2 - 2x) + (-x^2 + 2x + 10)$
11. $(-3x^2 + x - 1) + (x^2 + x - 4)$

Example 3 p. LA7

GEOMETRY For each of the figures, write a formula for the perimeter of the figure. Then find the perimeter of each figure if $x = 0.8$.

12.

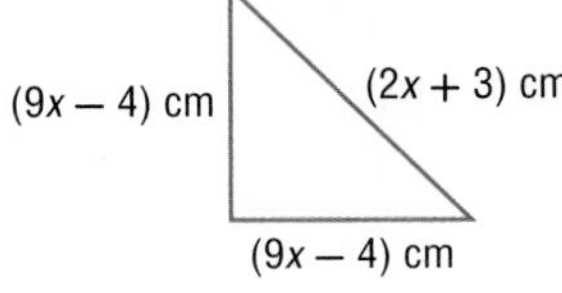

13.

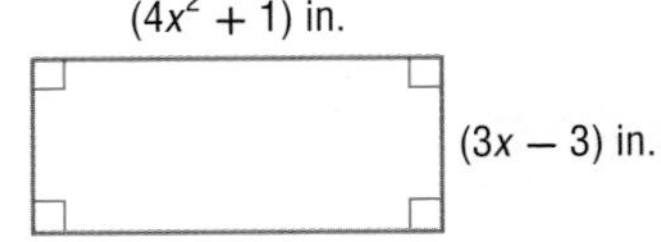

14. **NEWSPAPERS** Anna and Cole each earn x cents per newspaper that they deliver, plus tips. Anna delivered 55 newspapers and earned $12 in tips. Cole delivered 68 newspapers and earned $15 in tips.
    a. Write a polynomial expression to represent Anna's total earnings.
    b. Write a polynomial expression to represent Cole's total earnings.
    c. Write a polynomial expression to represent their total earnings.
15. **GEOMETRY** The angle measures of a triangle are $(x + 15)°$, $(2x - 20)°$, and $2x°$. What are the actual angle measures of the triangle?

Add. Use models if needed.

16. $(-2x^3 + 3x^2 + 6x + 1) + (x^3 - 10x^2 - 6x + 9)$
17. $(b^2 + a + 7 - 2b) + (4a^2 + 6b + 4b^2 - 8)$

H.O.T. Problems — Use Higher-Order Thinking Skills

18. **OPEN ENDED** Write two polynomials that have a sum of $3x^2 - 8x$.
19. **CHALLENGE** What polynomial would you add to $3y^2 - 4y + 2$ to have a sum of $2y^2 + 5$?
20. **REASONING** Explain how algebra tiles represent like terms and zero pairs.
21. **WRITING IN MATH** Explain how to add polynomials without using numbers.

Subtracting Polynomials

Then

You have already subtracted algebraic equations. (Lesson 4-3)

Now

- Subtract polynomials.
- Solve real-world problems by subtracting polynomials.

Math Online

glencoe.com

- Extra Examples
- Personal Tutor
- Self-Check Quiz
- Homework Help
- Math in Motion

Why?

A store had 14 video game systems and then received x more video game systems.

a. Write an expression for the total number of video game systems.

b. The store sold six of the systems. Write and simplify an expression to represent the number of systems the store has left.

Subtracting Polynomials Just as you add like terms, you can also subtract like terms.

EXAMPLE 1 Subtract Polynomials

Subtract. Use models if needed.

a. $(5x + 4) - (3x + 2)$

Step 1 Model the polynomial $5x + 4$.

Step 2 To subtract $3x + 2$, remove three x-tiles and two 1-tiles.

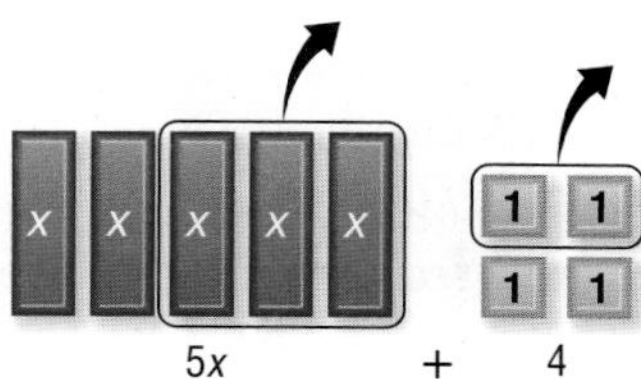

Step 3 Write the polynomial for the remaining tiles.

$(5x + 4) - (3x + 2) = 2x + 2$

b. $(x^2 - 4x - 6) - (-x - 3)$

Step 1 Model the polynomial $x^2 - 4x - 6$. $\quad x^2 - 4x - 6 = x^2 + (-4x) + (-6)$

Step 2 To subtract $(-x - 3)$, remove 1 negative x-tile and 3 negative 1-tiles.

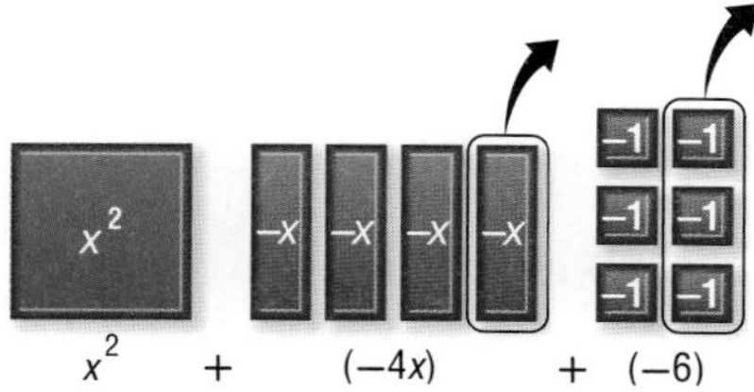

Step 3 Write the polynomial for the remaining tiles.

$(x^2 - 4x - 6) - (-x - 3) = x^2 - 3x - 3$

Check Your Progress

1A. $(7x - 5) - (2x - 1)$

1B. $(2x^2 + 6x - 4) - (x^2 + 2x - 4)$

Personal Tutor glencoe.com

Math *in Motion,* Interactive Lab glencoe.com

EXAMPLE 2 Add Zero Pairs to Subtract Polynomials

Find $(2x^2 + 3x + 2) - (x^2 - 2x)$.

Step 1

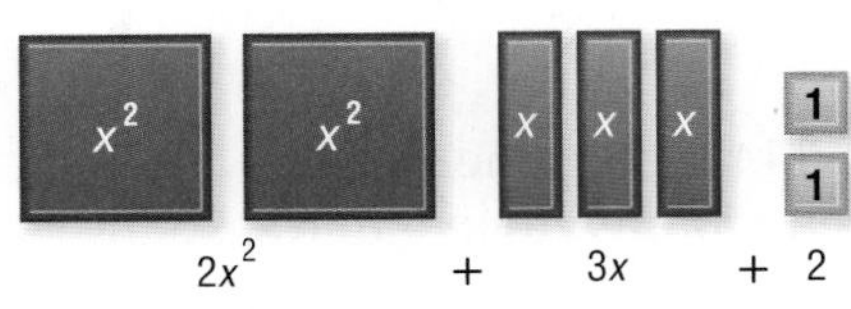

Model the polynomial.

Step 2

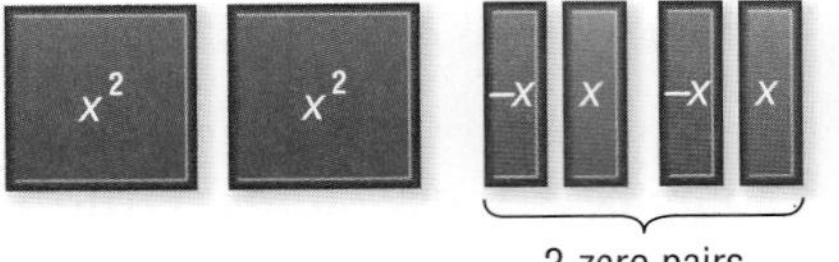

Since there are no negative x-tiles to remove, add 2 zero pairs of x-tiles.

Step 3

Remove 1 x^2-tile and 2 negative x-tiles.

So, $(2x^2 + 3x + 2) - (x^2 - 2x) = x^2 + 5x + 2$.

Check Your Progress

2. Find $(x - 5) - (2x - 1)$.

Personal Tutor glencoe.com

Solve Problems You can solve real-world problems by subtracting polynomials.

Real-World EXAMPLE 3 Subtract Polynomials to Solve Problems

EXERCISE The expression $8x + 50$ represents the total amount of money the soccer team earned from selling x T-shirts.

a. If the team had to pay $(2x + 24)$ dollars in expenses, write an expression to represent their profit.

Total − Expenses $= (8x + 50) - (2x + 24)$	**Subtract.**
$= 8x + 50 - 2x - 24$	**Distributive Property**
$= 6x + 26$	**Simplify.**

b. If the soccer team sold 54 T-shirts, what was their profit?

$6x + 26 = 6(54) + 26$	**Replace x with 54.**
$= 324 + 26$ or 350	**Simplify.**

So, the soccer team made \$350 profit.

Watch Out!

When subtracting $(2x + 24)$, subtract both $2x$ and 24, which is written as $-2x - 24$.

Check Your Progress

3. MONEY After working x hours on Monday, Kay earns $9x$ dollars. On Tuesday, she earns earns $(7x + 3)$ dollars.

a. Write an expression to represent how much more she earned on Monday.

b. If she worked for 5 hours each day, how much more did she earn on Monday?

Personal Tutor glencoe.com

Check Your Understanding

Examples 1 and 2 pp. LA9, LA10

Subtract. Use models if needed.

1. $(6x + 5) - (3x + 1)$
2. $(3x^2 - 4x + 2) - (x^2 - 2x)$
3. $(x^2 + 9x - 4) - (x^2 - 2x + 1)$
4. $(5x^2 + 7) - (x^2 + 2x + 4)$

Example 3 p. LA10

5. **SHIPPING** The cost of shipping an item that weighs x pounds from Charlotte to Chicago is shown in the table.

Shipping Company	Cost ($)
Atlas Service	$4x + 2.80$
Mid-Atlantic Service	$3x + 1.25$

   a. Write an expression to represent how much more Atlas charges than Mid-Atlantic for shipping an item.

   b. If an item weighs 2 pounds, how much more does Atlas charge for shipping it?

Practice and Problem Solving

Examples 1 and 2 pp. LA9, LA10

Subtract. Use models if needed.

6. $(3x + 7) - (x + 5)$
7. $(2x^2 - 4x) - (x^2 - x)$
8. $(x^2 + 8x - 9) - (3x - 1)$
9. $(-4x^2 + x + 7) - (-2x^2 + x + 2)$
10. $(5x + 6) - (x^2 + 2x)$
11. $(-4x^2 + x + 5) - (x^2 + 2x + 3)$

Example 3 p. LA10

12. **EXERCISE** The expression $5x + 2$ represents the number of miles Celeste rode her bike, and $10x$ represents the number of miles that Kimiko rode her bike in x hours.

    a. Write an expression to show how many more miles Kimiko rode than Celeste.

    b. If they each rode for 2 hours, how many more miles did Kimiko ride?

13. **CARS** A car accelerates for t seconds. The expression $2t + t^2$ represents the distance the car travels in meters. Another car has twice the acceleration and travels $(2t + 2t^2)$ meters in t seconds. After 10 seconds, how much farther does the second car travel?

14. **MEASUREMENT** What is the difference in the areas of the rectangles shown?

    $A = 2x^2 + 3x - 4$

    $A = x^2 - 5x + 2$

H.O.T. Problems
Use Higher-Order Thinking Skills

15. **OPEN ENDED** Write two polynomials that have a difference of $4x + 1$.

16. **CHALLENGE** Suppose A and B represent polynomials. If $A + B = 3x^2 + 2x - 2$ and $A - B = -x^2 + 4x - 8$, find A and B.

17. **WRITING IN MATH** Explain how to subtract $x^2 + 5x + 7$ from $2x^2$.

Multiplying a Binomial by a Monomial

Then

You have already multiplied monomials. (Lesson 9-3)

Now

- Use models to multiply a binomial by a monomial.
- Multiply a binomial by a monomial to solve equations.

New Vocabulary

binomial

Math Online

glencoe.com

- Extra Examples
- Personal Tutor
- Self-Check Quiz
- Homework Help

Why?

A square patio has a side length of x meters. If you increase the length by 3 meters, what is the area of the new patio?

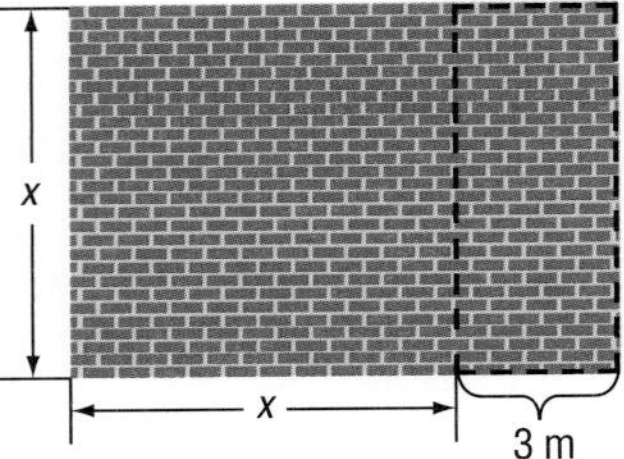

a. Write an expression to represent the new side length of the patio.

b. Write an expression to represent the area of the new patio.

Binomials You can use algebra tiles to model **binomials**, a polynomial with two terms.

The algebra tiles shown form a rectangle with a width of x and a length of $x + 2$. They represent the product $x(x + 2)$.

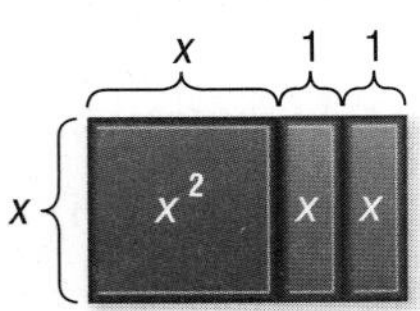

The area of the rectangle represents the product. Since the rectangle consists of one x^2-tile and two x-tiles, $x(x + 2) = x^2 + 2x$.

EXAMPLE 1 Multiplying a Binomial by a Monomial

Use algebra tiles to find $x(x + 3)$.

Step 1 Make a rectangle with a width of x and a length of $x + 3$. Use algebra tiles to mark off the dimensions on a product mat.

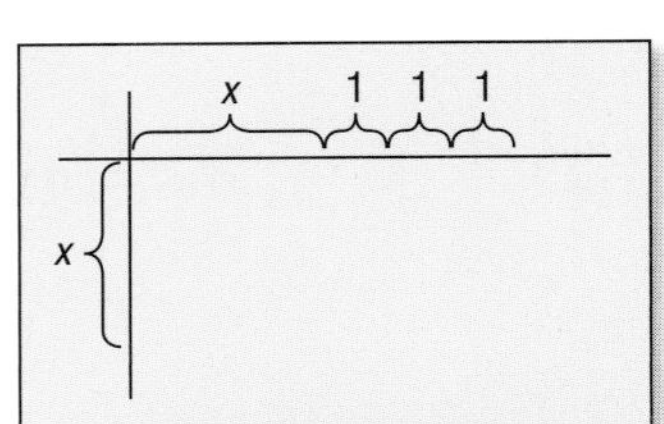

Step 2 Using the marks as a guide, fill in the rectangle with algebra tiles.

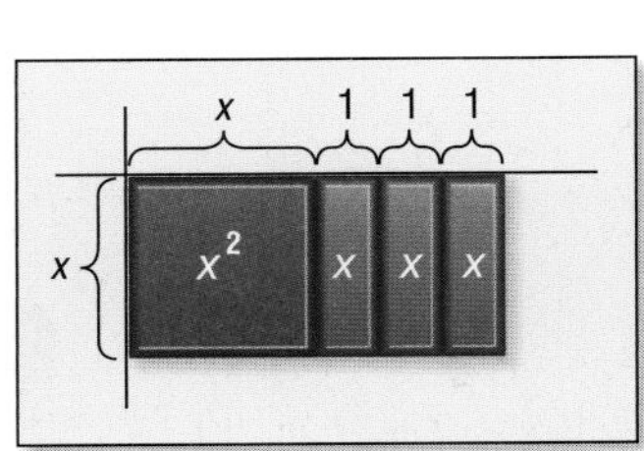

Step 3 The area of the rectangle is $x^2 + x + x + x$. In simplest form, the area is $x^2 + 3x$.

So, $x(x + 3) = x^2 + 3x$.

Check Your Progress

1. Use algebra tiles to find $2x(2x + 4)$.

Personal Tutor glencoe.com

In Example 1, each term in the binomial is multiplied by the monomial. This and other similar examples suggest that the Distributive Property can be used to multiply a binomial by a monomial.

Key Concept: Multiplying a Binomial by a Monomial

For Your FOLDABLE

Words To multiply a binomial by a monomial, use the Distributive Property.

Symbols $a(b + c) = ab + ac$

Model

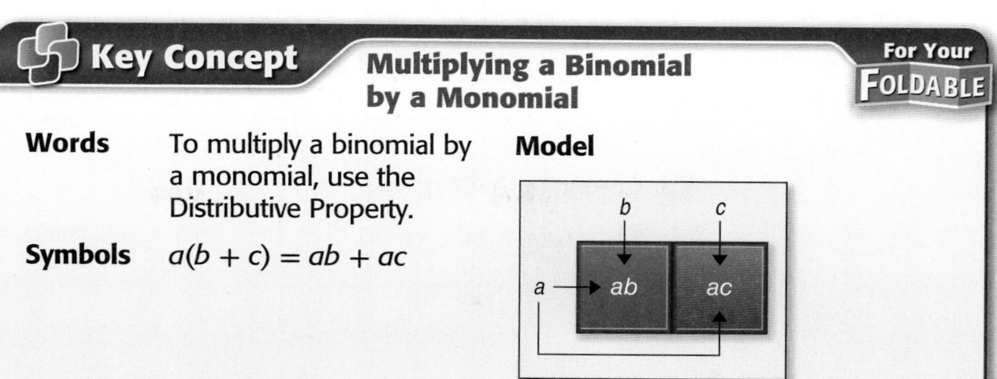

EXAMPLE 2 Multiplying a Binomial by a Monomial

Multiply. Use models if needed.

a. $x(x + 5)$

Method 1 Use a model.

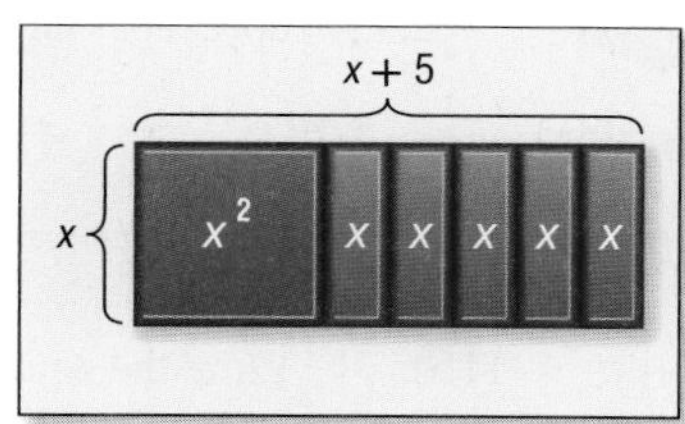

So, $x(x + 5) = x^2 + 5x$.

Method 2 Use the Distributive Property.

$x(x + 5) = x(x) + x(5)$ **Distributive Property**

$= x^2 + 5x$ **Simplify.**

So, $x(x + 5) = x^2 + 5x$.

b. $x(2x + 3)$

$x(2x + 3) = x(2x) + x(3)$

$= 2x^2 + 3x$

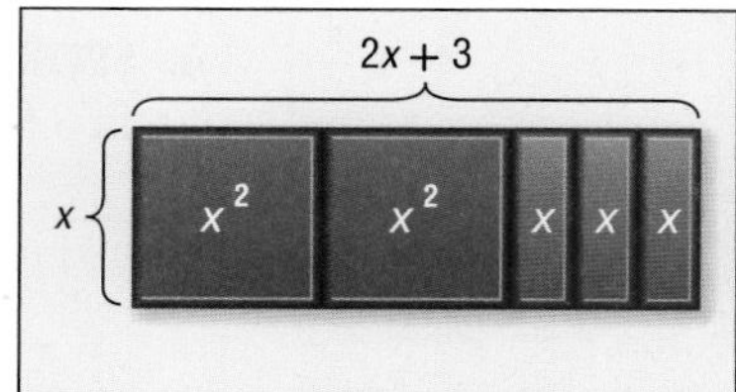

c. $3x^3(2x^2 - 5x)$

$3x^3(2x^2 - 5x) = 3x^3(2x^2) - 3x^3(5x)$ **Distributive Property**

$= 6x^5 - 15x^4$ **Simplify.**

StudyTip

Like Terms Recall that to multiply powers with the same base, add their exponents.

Check Your Progress

2A. $7(2x + 5)$ **2B.** $4x(3x^2 - 7)$ **2C.** $y^3(3y^2 + 2y)$

Personal Tutor glencoe.com

EXAMPLE 3 Simplifying Expressions

Simplify $5x(x + 2) - 2x(x)$.

$$5x(x + 2) - 2x(x) = 5x(x) + 5x(2) - 2x(x)$$ **Distributive Property**

$$= 5x^2 + 10x - 2x^2$$ **Multiply.**

$$= 3x^2 + 10x$$ **Simplify.**

Check Your Progress

3A. $-6a(a + 1) + 4a(2a)$

3B. $3n^2(n - 4) - 4(n^2 - 7)$

Personal Tutor glencoe.com

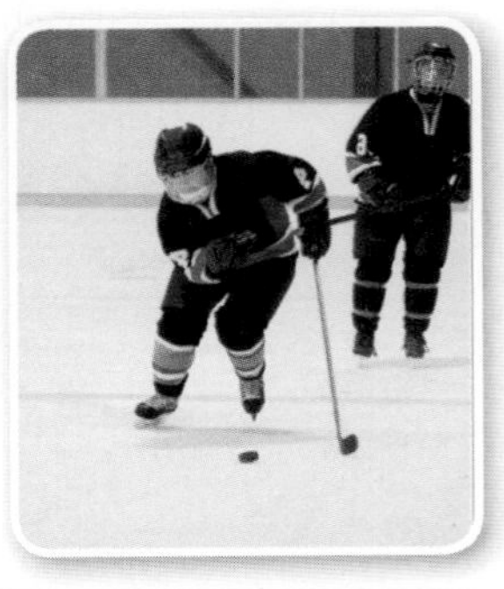

Real-World Link

The first youth hockey program was established in 1963 in Chicago, Illinois.

Source: *Chicago Encyclopedia*

Real-World EXAMPLE 4

SPORTS The length of a hockey rink is 115 feet longer than its width. If the perimeter of the rink is 570 feet, what are the dimensions of the rink?

Words	Perimeter equals twice the sum of the length and width.
Variable	Let w represent width. So, $w + 115$ represents the length.
Expression	$570 \quad = \quad 2 \quad (w + w + 115)$

$570 = 2(w + w + 115)$ **Write the equation.**

$570 = 2(2w + 115)$ **Combine like terms.**

$570 = 4w + 230$ **Distributive Property**

$340 = 4w$ **Subtract 230 from each side.**

$85 = w$ **Divide each side by 4.**

So, the width of the rink is 85 feet and the length is 85 + 115 or 200 feet.

Check Your Progress

4. SPORTS The perimeter of a tennis court is 228 feet. The length of the court is 6 feet more than twice the width. What are the dimensions of the court?

Personal Tutor glencoe.com

Check Your Understanding

Examples 1 and 2 p. LA13

Find each product. Use models if needed.

1. $2(2x + 1)$

2. $x(2x + 2)$

3. $3a(a - 1)$

4. $-3n(5 - 2n)$

5. $4z(z^2 - 2z)$

6. $3(8y^3 + 3)$

Example 3 p. LA14

Simplify.

7. $7x(x + 2) + 3x(x - 5)$

8. $4(y^2 - 8) + 10 - 2(y + 12)$

9. $-3(2a - 4) + 9 - 3(a + 1)$

10. $x(x + 3) + 5x + x(x + 5) + 9$

Example 4 p. LA14

11. **SPORTS** One of the world's largest swimming pools is the Orthlieb Pool in Casablanca, Morocco. It is 30 meters longer than 6 times its width. If the perimeter of the pool is 1110 meters, what are the dimensions of the pool?

Practice and Problem Solving

Examples 1 and 2 p. LA13

Find each product. Use models if needed.

12. $2x(x + 4)$

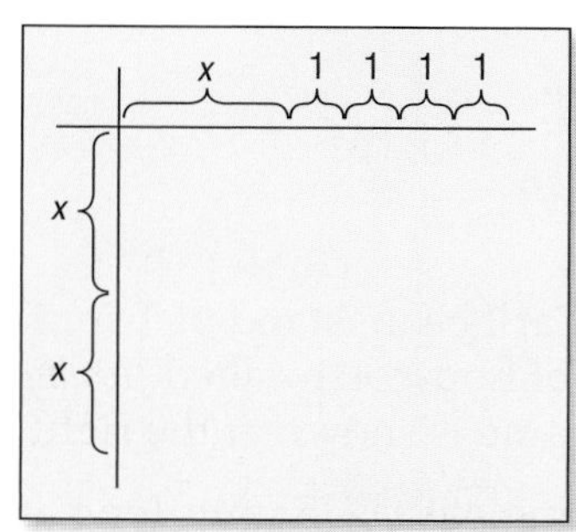

13. $2x(2x + 2)$

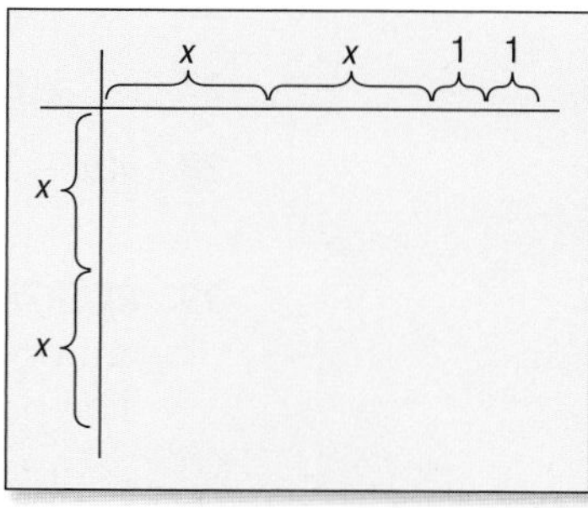

14. $4x(-2x + 7x^2)$

15. $-5y(6 - 2y)$

16. $5d(d^2 + 3)$

17. $0.25x(4x - 8)$

Example 3 p. LA14

Simplify.

18. $2a(a^2 - 5a) - 6(a^3 + 3a)$

19. $5t^2(t + 2) - 5t(4t^2 - 3t)$

20. $3n^2(n - 4) + 6n(3n^2 + n - 7) - 4(n - 7)$

Example 4 p. LA14

21. **FOOTBALL** The perimeter of an arena football field is 188 yards. The perimeter of an NFL football field is 346 yards. Use the information in the table to find the length and width of each field.

FOOTBALL FIELDS

Measure	Arena	NFL
Width	$a - 38$	$n + 3$
Length	a	$2n + 20$

22. **MONUMENTS** The rectangular Reflecting Pool extends from the Lincoln Memorial to the World War II memorial in Washington, D.C. The length of the pool is 25 feet more than 12 times its width.

a. If the perimeter of the pool is 4392 feet, find the dimensions of the pool.

b. Suppose the width of the pool is 83.5 times the depth. Find the volume of the pool.

23. **SPORTS** A billiards table has a perimeter of 24 feet. The length of a billiards table is twice as long as its width. If the width of the billiards table is $4x - 1$, what are the dimensions of the table?

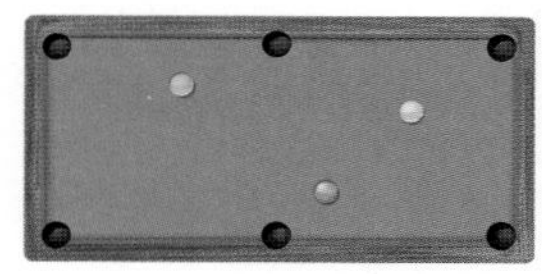

StudyTip

Properties of Equality You can use the same properties of equality from Chapter 4 to solve equations with polynomials.
For any numbers a, b, and c, if $a = b$ then
$a + c = b + c$
$a - c = b - c$
$ac = bc$ and
$\frac{a}{c} = \frac{b}{c}$ when $c = 0$

ALGEBRA **Solve $w(w + 12) = w(w + 14) + 12$.**

$w(w + 12) = w(w + 14) + 12$	**Write the equation.**
$w(w) + w(12) = w(w) + w(14) + 12$	**Distributive Property**
$w^2 + 12w = w^2 + 14w + 12$	**Simplify.**
$12w = 14w + 12$	**Subtract w^2 from each side.**
$-2w = 12$	**Subtract $14w$ from each side.**
$w = -6$	**Divide each side by -2.**

Solve each equation.

24. $4(y^2 - 3y) - 8 = 2y(2y + 4) + 32$

25. $n(n - 7) = n^2 + 3(-2n + 1) - 1$

26. $a(a + 2) + 3a = a(a - 3) + 8$

27. $c(c + 8) - c(c + 3) - 23 = 3c + 11$

28. $b(b + 10) + 6 = b(b + 6) - 2$

29. **RECREATION** On the Caribbean island of Trinidad, children play a form of hopscotch called *Jumby*. The pattern for this game is shown at the right.

a. Suppose each rectangle is $y + 5$ units long and y units wide. Write an expression in simplest form for the area of the pattern.

b. If y represents 10 inches, find the area of the pattern.

H.O.T. Problems

Use Higher-Order Thinking Skills

30. **OPEN ENDED** Write three different multiplication problems for which the product is $6a^2 + 8a$.

31. **CHALLENGE** The product of $x^2 - 2x + 1$ and a monomial is $4x^3 - 8x^2 + 4x$. What is the monomial? Explain your reasoning.

32. **FIND THE ERROR** Carmen and Paul are simplifying $5(p^2 + 2p - 2) - 4p(p - 1)$. Is either of them correct? Explain your reasoning.

Carmen	Paul
$5(p^2 + 2p - 2) - 4p(p - 1) =$	$5(p^2 + 2p - 2) - 4p(p - 1) =$
$5p^2 + 10p - 10 - 4p^2 + 4p =$	$5p^2 + 10p - 10 - 4p^2 - 4p =$
$p^2 + 14p - 10$	$p^2 + 6p - 10$

33. **REASONING** Explain how to multiply $-2n$ and $3 - 5n$.

34. **WRITING IN MATH** Describe how you can use models to multiply a monomial by a polynomial.

Multiplying Two Binomials

Why?

The local youth club has a stage for karaoke competitions. The main stage is a square. The technical crew needs to increase the length by 2 feet and the width by 1 foot for their equipment. A plan for the stage is shown below.

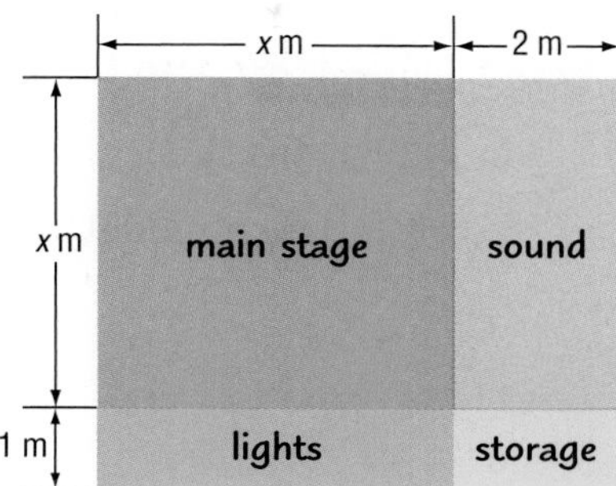

a. Write expressions for the area of each part of the stage.

b. Write an expression for the total area of the stage.

Then
You have already multiplied a binomial by a monomial. (Looking Ahead 4)

Now
- Multiply two binomials by using models.
- Multiply two binomials by using the Distributive Property.

New Vocabulary
FOIL Method

Math Online

glencoe.com
- Extra Examples
- Personal Tutor
- Self-Check Quiz
- Homework Help

Multiply Binomials As with multiplying a monomial by a binomial, you can use algebra tiles to multiply two binomials. In the model, the length and width of a rectangle represent the two binomials. The area represents the product.

EXAMPLE 1 Modeling Multiplication of Binomials

Find $(x + 2)(x + 1)$.

Step 1 Make a rectangle with a width of $x + 1$ and a length of $x + 2$.

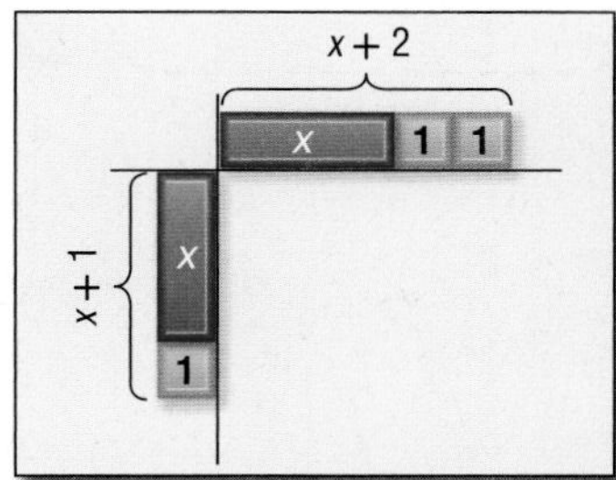

Step 2 Fill in the rectangle with algebra tiles. There is one x^2-tile, three x-tiles, and two 1-tiles.

So, $(x + 2)(x + 1) = x^2 + 3x + 2$.

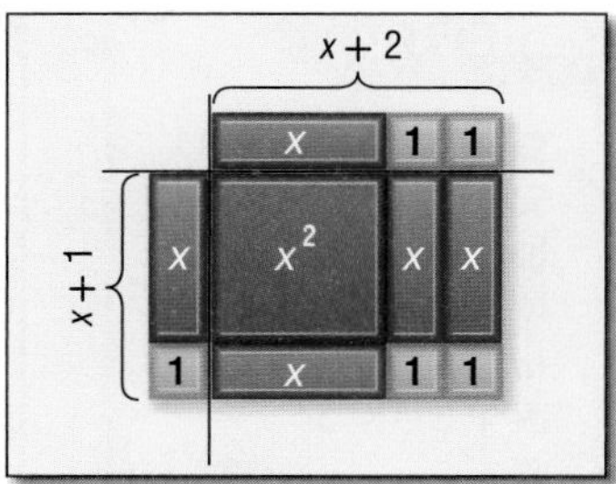

Check Your Progress

Multiply. Use a model.

1A. $(x + 3)(x + 4)$

1B. $(2x + 1)(x + 3)$

Personal Tutor glencoe.com

Distributive Property The Distributive Property can also be used to find the product of two binomials. The figure at the right shows the rectangle from Example 1 separated into four parts. Notice that each term from the first parentheses $(x + 2)$ is multiplied by each term from the second parentheses $(x + 1)$.

EXAMPLE 2 Use the Distributive Property

Find $(2x + 4)(x + 5)$.

Method 1 Use a model.

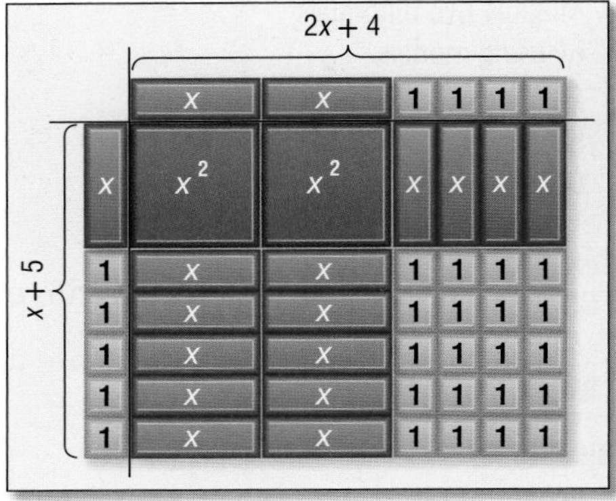

Method 2 Use the Distributive Property.

$$(2x + 4)(x + 5) = 2x(x + 5) + 4(x + 5) \quad \textbf{Distributive Property}$$
$$= 2x^2 + 10x + 4x + 20 \quad \textbf{Distributive Property}$$
$$= 2x^2 + 14x + 20 \quad \textbf{Simplify.}$$

So, $(2x + 4)(x + 5) = 2x^2 + 14x + 20$.

> **Watch Out!**
>
> **Negative Signs** If one or both of the binomials involve negatives or subtraction, remember to distribute the negatives.

Check Your Progress

Multiply. Use models if needed.

2A. $(3x + 2)(2x - 3)$

2B. $(x + 6)(2x + 4)$

Personal Tutor glencoe.com

A shortcut version of the Distributive Property is the **FOIL method**.

> **StudyTip**
>
> **Special Products** Some pairs of binomials have products that follow a specific pattern.
>
> $(a + b)^2 = a^2 + 2ab + b^2$
>
> $(a - b)^2 = a^2 - 2ab + b^2$
>
> $(a + b)(a - b) = a^2 - b^2$

Key Concept — FOIL Method (For Your FOLDABLE)

	Words	Symbols
		$(3x + 1)(x + 2)$
F	Multiply the **FIRST** terms.	$3x \cdot x$ or $3x^2$
O	Multiply the **OUTER** terms.	$3x \cdot 2$ or $6x$
I	Multiply the **INNER** terms.	$1 \cdot x$ or x
L	Multiply the **LAST** terms.	$1 \cdot 2$ or 2

So, $(3x + 1)(x + 2) = 3x^2 + 6x + x + 2$ or $3x^2 + 7x + 2$.

Real-World EXAMPLE 3 Use FOIL to Multiply Binomials

GEOMETRY A shipping box is shaped like a rectangular prism. The volume V is equal to the area of the base B times the height h. Express the volume of the prism as a polynomial.

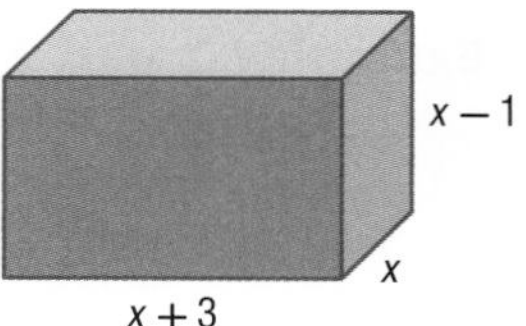

First, find the area of the rectangular base.

$B = \ell w$ — **Formula for area of a rectangle**

$= (x + 3)(x)$ — **Replace ℓ with $x + 3$ and w with x.**

$= x^2 + 3x$ — **Distributive Property**

To find the volume, multiply the area of the base by the height.

$V = Bh$ — **Formula for volume of a prism**

$= (x^2 + 3x)(x - 1)$ — **Replace B with $x^2 + 3x$ and h with $x - 1$.**

F O I L

$= x^2 \cdot x - x^2 \cdot 1 + 3x \cdot x - 3x \cdot 1$ — **Use the FOIL method.**

$= x^3 - x^2 + 3x^2 - 3x$ — **Multiply.**

$= x^3 + 2x^2 - 3x$ — **Simplify.**

Check Your Progress

3. **GIFTS** Express the volume of the gift box at the right as a polynomial.

Personal Tutor glencoe.com

Check Your Understanding

Example 1 p. LA17

Multiply. Use a model.

1. $(x + 2)(x + 3)$
2. $(x + 3)(x + 4)$
3. $(x + 1)(2x + 1)$

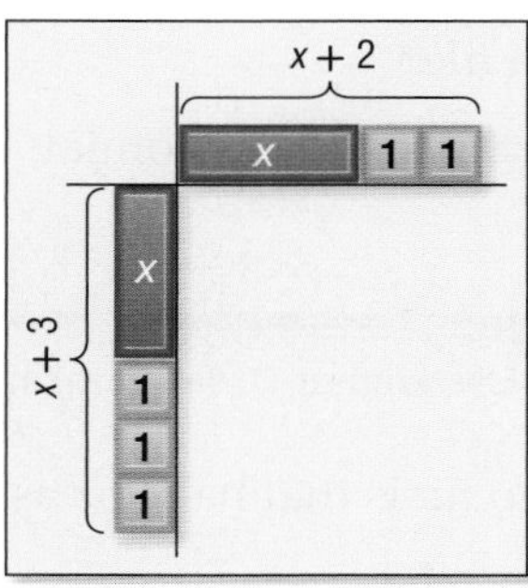

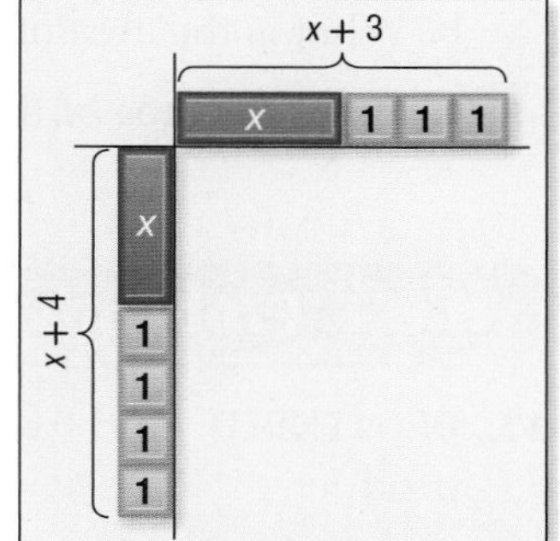

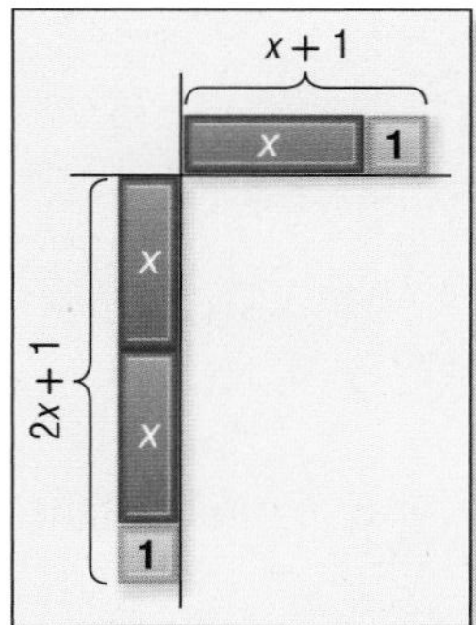

Multiply.

4. $(y + 4)(y - 2)$
5. $(m - 3)(m + 1)$
6. $(x - 5)(x - 2)$
7. $(3n + 2)(n - 2)$
8. $(2x - 5)(x - 4)$
9. $(4b - 3)(2b + 4)$

Example 3 p. LA19

10. **CEREAL** A cereal box has a length of $2x$ inches, a width of $x - 2$ inches, and a height of $2x + 5$ inches. Express the volume as a polynomial.

Practice and Problem Solving

Examples 1 and 2 pp. LA17–LA18

Multiply.

11. $(x + 4)(x + 8)$
12. $(r - 3)(r - 7)$
13. $(z + 6)(z - 4)$
14. $(2a + 5)(a - 7)$
15. $(5n + 2)(n - 3)$
16. $(2x + 5)(5x + 3)$
17. $(2x + 7)(x - 3)$
18. $(3a - b)(2a + b)$
19. $(n - 11)(n - 5)$
20. $(5n - 2p)(5n + 2p)$
21. $(3a + 1)(3a + 1)$
22. $(4h - 3)(3h + 2)$

Example 3 p. LA19

23. **GEOMETRY** A parcel box has a length of $3y$ centimeters, a width of $y + 3$ centimeters, and a height of $2y - 2$ centimeters. Express the volume of the package as a polynomial.

24. **INTEREST** Lauren deposited money into a savings account. The account earns an interest rate of r%. For each dollar deposited, the amount in the account after two years is given by the formula $(1 + r)(1 + r)$. Find this product.

Find each product.

25. $(y + 3)(y^2 - 4)$
26. $(x^2 + 3)(3x^2 - 1)$
27. $(2y^2 + 1)(y + 1)$
28. $(x + 2y)(x + 3y)$
29. $(2a - 5b)(a + 2b)$
30. $(m^3 - 2m)(m + 3)$
31. $(x - 4)(3x + 2)$
32. $(2y - 4z)(3y - 6z)$
33. $(x^2 + 1)(x - 2)$

34. **ENVELOPES** The mailing envelope shown has a mailing label on the front.
    a. Find the area of the label.
    b. Find the area of the envelope not covered by the label.

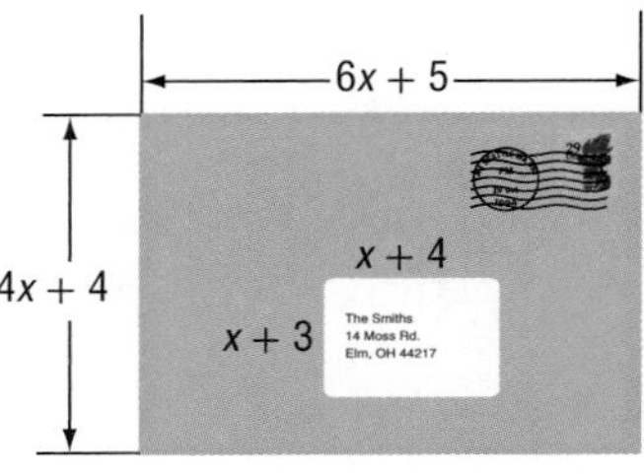

35. **GEOMETRY** A square has sides of length s. A rectangle is 5 inches longer and 4 inches wider than the square. Express the area of the rectangle as a polynomial.

36. **GEOMETRY** The model at the right represents the square of a binomial.
    a. What product does this model represent?
    b. What is the area of each tile?
    c. Write the area of the square as a polynomial.

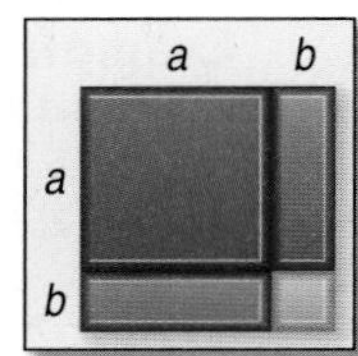

H.O.T. Problems Use Higher-Order Thinking Skills

37. **OPEN ENDED** Find two binomials that have $6x$ as one of the terms in their product.

38. **CHALLENGE** Find $(x + 1)^2$, $(x + 2)^2$, and $(x + 3)^2$. Is there a pattern in the products of binomials? If so, use the pattern to find $(x + 6)^2$ and $(x + n)^2$.

39. **REASONING** Does the product of two binomials always have three terms? If so, explain why. If not, give a counterexample.

40. **WRITING IN MATH** Compare and contrast the procedure for multiplying two binomials and the procedure for multiplying a binomial by a monomial.

Dividing a Polynomial by a Monomial

Why?

Student Council is selling milkshakes at lunch as a fundraiser. Each milkshake requires $\frac{1}{8}$ gallon of ice cream. They had $6\frac{1}{2}$ gallons of ice cream. Then their advisor brought them 5 more gallons.

a. Find the number of milkshakes that can be sold with the amount of ice cream they now have.

b. Describe two ways to find the number of milkshakes.

Then

You have already solved equations by dividing. (Lesson 4-4)

Now

- Divide polynomials by monomials.
- Solve problems using division of polynomials.

Math Online

glencoe.com

- Extra Examples
- Personal Tutor
- Self-Check Quiz
- Homework Help

Dividing Polynomials By Monomials To divide a polynomial by a monomial, divide each term of the polynomial by the monomial.

Key Concept: Dividing Polynomials

For Your FOLDABLE

Words To divide a polynomial by a monomial, divide each term of the polynomial by the monomial.

Symbols $\frac{a+b}{c} = \frac{a}{c} + \frac{b}{c}$

EXAMPLE 1 Divide a Polynomial by a Monomial

Divide.

a. $(9b^2 - 15b) \div (3b)$

$(9b^2 - 15b) \div (3b) = \frac{9b^2 - 15b}{3b}$ Write as a rational expression.

$= \frac{9b^2}{3b} - \frac{15b}{3b}$ Divide each term by $3b$.

$= \frac{9}{3} \cdot \frac{b^2}{b} - \frac{15}{3} \cdot \frac{b}{b}$ Associative Property

$= 3 \cdot b^{2-1} - 5 \cdot 1$ Quotient of Powers and Identity Properties

$= 3b - 5$ Simplify.

b. $(6x^2 + 4x) \div (2x)$

$(6x^2 + 4x) \div (2x) = \frac{6x^2 + 4x}{2x}$ Write as a rational expression.

$= \frac{6x^2}{2x} + \frac{4x}{2x}$ Divide each term by $2x$.

$= \frac{6}{2} \cdot \frac{x^2}{x} + \frac{4}{2} \cdot \frac{x}{x}$ Associative Property

$= 3 \cdot x^{2-1} + 2 \cdot 1$ Quotient of Powers and Identity Properties

$= 3x + 2$ Simplify.

Check Your Progress

1A. $(10x^2y^2 + 5xy) \div (5xy)$

1B. $(27x^2 - 21y^2) \div 3$

Personal Tutor glencoe.com

Real-World Link

There are about 110 pandas in captivity. Only seven of these are in the United States. The majority of the other pandas in captivity are found in China.

Source: Panda Bear Facts

Solve Problems You can use division to solve real-world problems.

Real-World EXAMPLE 2 Solve Problems

ZOOS Six friends visited the zoo to see the new panda exhibit. The group paid for admission and an additional \$12 for parking. The total cost of the visit can be shown by the expression \$6x + \$12. What was the cost of the visit for one person?

$$(6x + 12) \div 6 = \frac{6x + 12}{6} \quad \text{Write as a rational expression.}$$

$$= \frac{6x}{6} + \frac{12}{6} \quad \text{Divide each term by 6.}$$

$$= x + 2 \quad \text{Simplify.}$$

So, each person paid $x + 2$ dollars for admission to the zoo.

Check Your Progress

2. **GARDENS** The model at the right represents a garden. The total area of the garden shown can be represented by the expression $8x^2 + 12x$. If the width of the garden is $4x$, what is the length of the garden?

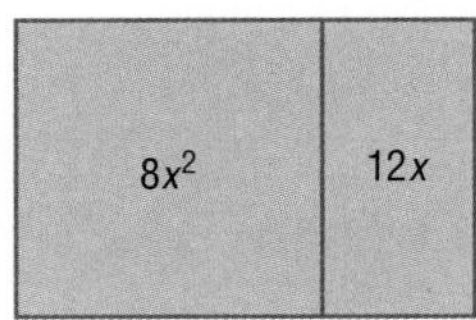

Personal Tutor glencoe.com

Check Your Understanding

Example 1 p. LA21

Divide.

1. $(5abc + c) \div c$
2. $(14ab + 28b) \div (14b)$
3. $(16x + 24xy) \div (8x)$
4. $(25st - 35s) \div (5s)$
5. $(30mn - 9m) \div (3m)$
6. $(42q + 56) \div 7$
7. $(18x^2 + 32x) \div (2x)$
8. $(20k^2 - 35k) \div (5k)$
9. $(20b^3 + 40b) \div (20b)$
10. $(4x^3 + 2x^2 - 6x) \div (2x)$

Example 2 p. LA22

11. **GEOMETRY** The area of the rectangle below can be shown by the expression $(16x + 4y)$ square feet. If the width of the rectangle is 4 feet, what is the length of the rectangle?

$A = 16x + 4y$

12. **GEOMETRY** The perimeter of the square below can be shown by the expression $12x^2 + 24x$. What is the length of each side of the square?

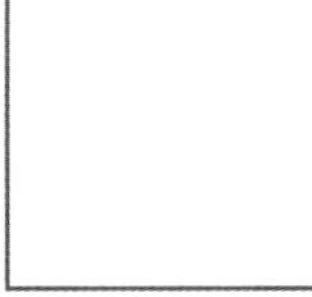

$P = 12x^2 + 24x$

Practice and Problem Solving

Example 1 p. LA21

Divide.

13. $(16x + 4y^2) \div 4$
14. $(15a + 3b^2) \div 3$
15. $(21a^2b - 14a^2) \div (7a^2)$
16. $(36st^2 - 9st) \div (9st)$
17. $(13r^2s - 26r^2) \div (13r^2)$
18. $(32np + m^2np^2) \div (np)$
19. $(42c^3d^2 + 56c^2d - 14c) \div (7c)$
20. $(81m^3n^2 - 45m^2n - 27n) \div (9n)$

Example 2 p. LA22

21. **CONSTRUCTION** A contractor built a basement for a new home. The volume of the rectangular basement shown at the right is represented by the expression $3x^3 + 6x^2 - 30x$. If the height of the basement is $3x$, what is the area of its base?

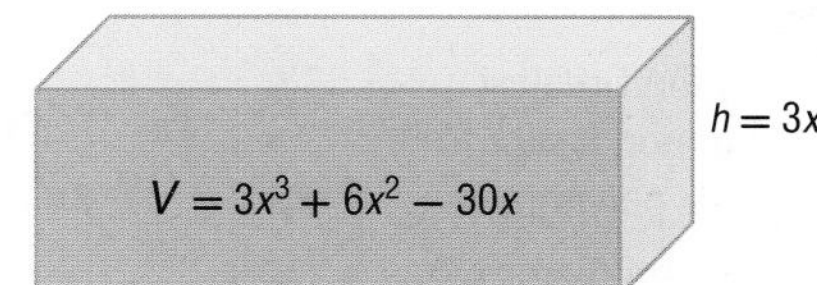

22. **BASKETBALL** The length of a basketball court is $6x^2 + 20x$. If the length is $2x$ times the width, what is the width of the court?

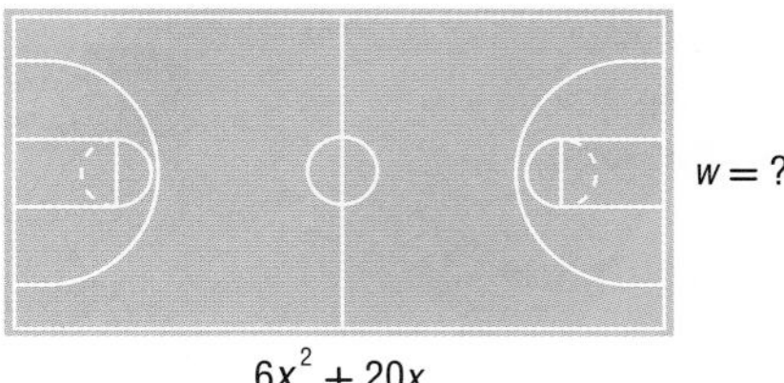

H.O.T. Problems

Use Higher-Order Thinking Skills

23. **OPEN ENDED** Write a polynomial and a monomial that have a quotient of $x^2 + 3x + 5$.

24. **CHALLENGE** The length and width of a rectangle are represented by $2x$ and $9 - 4x$. If x must be an integer, what are the only possible measures for the area of this rectangle?

25. **FIND THE ERROR** George and Marilyn are finding the quotient of $(15x^3 + 12x^2 - 24x) \div (3x)$. Is either of them correct? Explain your reasoning.

George	Marilyn
$= \dfrac{15x^3 + 12x^2 - 24x}{3x}$	$= \dfrac{15x^3 + 12x^2 - 24x}{3x}$
$= \dfrac{15x^3}{3x} + \dfrac{12x^2}{3x} - \dfrac{24x}{3x}$	$= \dfrac{15x^3}{3x} + \dfrac{12x^2}{3x} - \dfrac{24x}{3x}$
$= 5x^2 + 4x - 8$	$= 5x^3 + 4x^2 - 8x$

26. **REASONING** Explain why dividing a polynomial by a monomial could be called factoring.

27. **WRITING IN MATH** Explain why $2y^2 + 3$ *cannot* be divided by $2y$ without using fractions.

Using the GCF to Factor Polynomials

Then

You have already factored monomials. (Lesson 9-2)

Now

- Use the Greatest Common Factor to factor polynomials.
- Solve problems by using the Greatest Common Factor to factor polynomials.

New Vocabulary

factoring
factored form

Math Online

glencoe.com

- Extra Examples
- Personal Tutor
- Self-Check Quiz
- Homework Help

Why?

A large area of an art classroom is shown at the right. The total area is represented by $A = 3x^2 + 9x$.

a. The teacher wants to create three rectangular work stations in the classroom that are x feet long. Each station has the same width. Write an expression for the width.

b. If x is equal to 8 feet, what are the dimensions and area of the classroom?

Paint/Sketch Station
Sculpture Station
Ceramic Station

Factoring Sometimes, you know the product and are asked to find the factors. This process is called **factoring**. You can use algebra tiles to model factoring.

EXAMPLE 1 Using Algebra Tiles to Model Factoring

Use algebra tiles to factor $2x + 8$.

Step 1 Model the polynomial $2x + 8$.

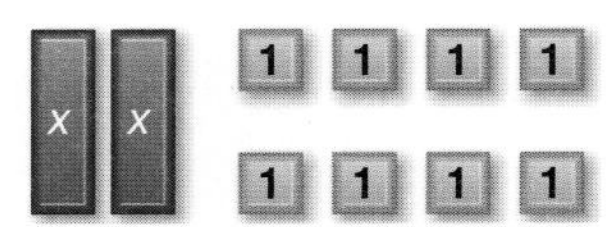

Step 2 Arrange the tiles into a rectangle. The total area of the tiles represents the product. Its length and width represent the factors. The rectangle has a width of 2 and a length of $x + 4$.

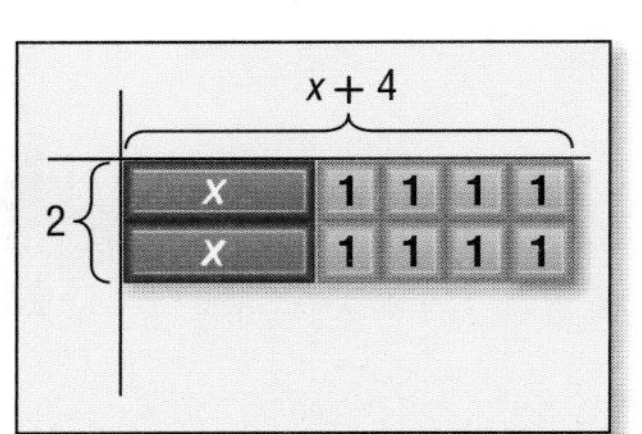

So, $2x + 8 = 2(x + 4)$.

Check Your Progress

Factor. Use algebra tiles.

1A. $3x + 9$ **1B.** $4x + 10$ **1C.** $x^2 + 5x$ **1D.** $3x^2 + 4x$

Personal Tutor glencoe.com

A polynomial is in **factored form** when it is expressed as the product of polynomials. To factor $8y^2 + 10y$, first find the greatest common factor (GCF) of each term.

$8y^2 = 2 \cdot 2 \cdot (2) \cdot (y) \cdot y$ **Write the prime factorization of $8y^2$ and $10y$.**

$10y = (2) \cdot 5 \cdot (y)$ **Circle the common factors.**

So, the GCF of $8y^2$ and $10y$ is $2y$.

Write each term as a product of the GCF and its remaining factors. Then use the Distributive Property.

$8y^2 + 10y = 2y(4y) + 2y(5)$

$= 2y(4y + 5)$ **Distributive Property**

StudyTip

Greatest Common Factor (GCF) Recall that the GCF is the greatest of the factors common to two or more numbers or terms.

EXAMPLE 2 Using GCF to Factor Polynomials

Factor $30x^2 + 12x$.

First, find the GCF of $30x^2$ and $12x$.

$30x^2 = \;\;\; 2 \cdot 3 \cdot 5 \cdot x \cdot x$ **Write the prime factorization of $30x^2$ and $12x$.**

$12x = 2 \cdot 2 \cdot 3 \cdot \;\;\; \cdot x$ **Circle the common factors.**

The GCF of $30x^2$ and $12x$ is $6x$. Write each term as a product of the GCF and its remaining factors.

$30x^2 + 12x = 6x(5x) + 6x(2)$

$= 6x(5x + 2)$ **Distributive Property**

Check Your Progress

Factor each polynomial using the GCF.

2A. $15ab^2 - 25abc$

2B. $18x^2y + 12xy^2 + 6xy$

Personal Tutor glencoe.com

Solve Problems You can solve some real-world problems by factoring.

Real-World EXAMPLE 3

ART **Kiyoshi is planning to mat a square painting with a mat that is 6 inches wide. Let x represent the length and the width of the painting. Write an expression in factored form that represents the area of the mat.**

Step 1 Find the total area.

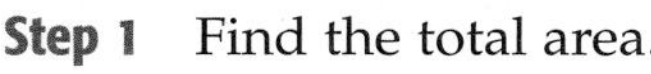

$A = \ell w$

$= (x + 12)(x + 12)$

$= x^2 + 24x + 144$

Step 2 Find the area of the mat alone.

Area of mat = Total Area − Area of painting

$= x^2 + 24x + 144 - x^2$

$= 24x + 144$

Step 3 Factor.

$24x = 2 \cdot 2 \cdot 2 \cdot \;\;\; 3 \;\;\; \cdot x$ **Write the prime factoriztion of $24x$ and 144.**

$144 = 2 \cdot 2 \cdot 2 \cdot 2 \cdot 3 \cdot 3$ **Circle the common factors. The GCF is 24.**

$24x + 144 = 24(x) + 24(6)$

$= 24(x + 6)$ **Distributive Property**

So, the area of the mat is $24(x + 6)$ inches.

Check Your Progress

3. **GEOMETRY** If the area of a rectangle is $16y^2 + 48y$, write the expression in factored form.

Personal Tutor glencoe.com

Check Your Understanding

Example 1 p. LA24

Use algebra tiles to factor each binomial.

1. $24y + 18y^2$
2. $x^2 + 2x$
3. $5 + 5y^2$

Example 2 p. LA25

Factor each polynomial using the GCF.

4. $3x + 6$
5. $2x^2 + 4x$
6. $12a^2b + 6a$

Example 3 p. LA25

7. **GEOMETRY** The diagram represents a walkway that is 2 meters wide surrounding a rectangular garden. Write an expression in factored form that represents the area of the walkway.

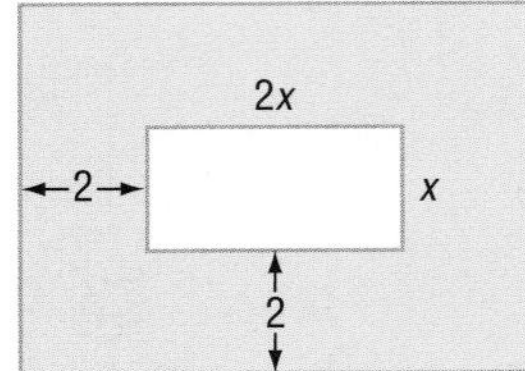

Practice and Problem Solving

Examples 1 and 2 pp. LA24, LA25

Factor each polynomial. If the polynomial cannot be factored, write *cannot be factored*.

8. $9x + 15$
9. $7a^2b^2 + 3ab^3$
10. $36mn - 11mn^2$
11. $14mn^2 - 2mn$
12. $24xy + 18xy^2 - 3y$
13. $12axy - 14ay + 20ax$

Example 3 p. LA25

14. The figure at the right shows a walkway built around a statue with a square base.

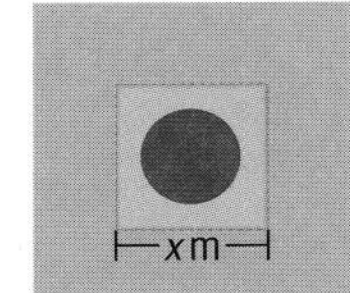

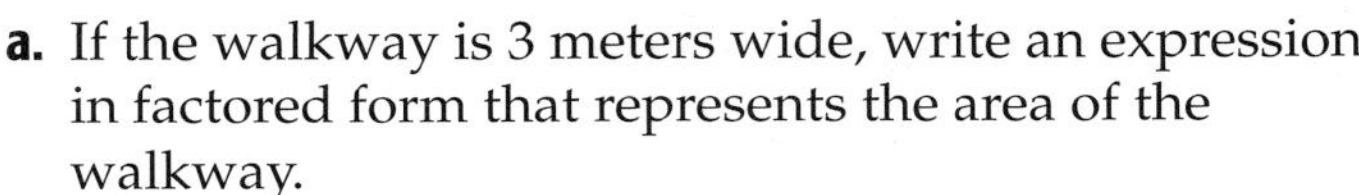

   a. If the walkway is 3 meters wide, write an expression in factored form that represents the area of the walkway.

   b. If the base of the statue is 3 meters wide, find the area of the walkway.

15. **MARINE BIOLOGY** In a pool at an aquarium, a dolphin jumps out of the water traveling at 20 feet per second. Its height h, in feet, above the water after t seconds is given by the formula $h = 20t - 16t^2$. Write an expression in factored form that represents the height.

H.O.T. Problems — Use Higher-Order Thinking Skills

16. **CHALLENGE** The area of a circle is found using the formula $A = \pi r^2$, where r is the radius of the circle. Using π, write an expression in factored form that represents the area of the shaded region at the right.

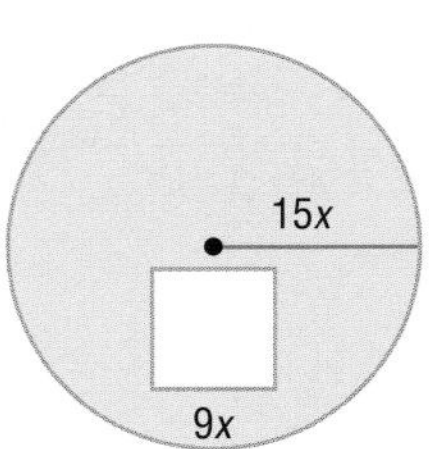

17. **REASONING** Explain what it means to factor a polynomial.

18. **OPEN ENDED** Write a polynomial that can be factored. Write another polynomial that *cannot* be factored.

19. **WRITING IN MATH** Write a few sentences explaining how the Distributive Property is used to factor polynomials. Include at least two examples.

Factoring Trinomials

Why?

Alvin has enough bricks to make a patio with a perimeter of 34 feet and area of 72 square feet.

a. How could you find the dimensions of the patio?

b. Find the dimensions.

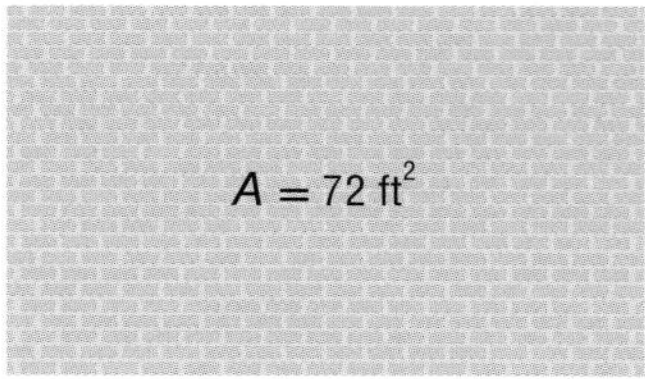

$P = 34$ ft

You can also use algebra tiles to factor trinomials. A **trinomial** is a polynomial with three terms. Trinomials that are in the form $x^2 + bx + c$ can sometimes be factored into two binomials.

Then

You have already factored polynomials. (Looking Ahead 7)

Now

- Factor trinomials in the form $x^2 + bx + c$.
- Solve real-world problems by factoring trinomials.

New Vocabulary

trinomial

Math Online

glencoe.com

- Extra Examples
- Personal Tutor
- Self-Check Quiz
- Homework Help

EXAMPLE 1 Factor Using Models

Factor.

a. $x^2 + 3x + 2$

Model the polynomial.

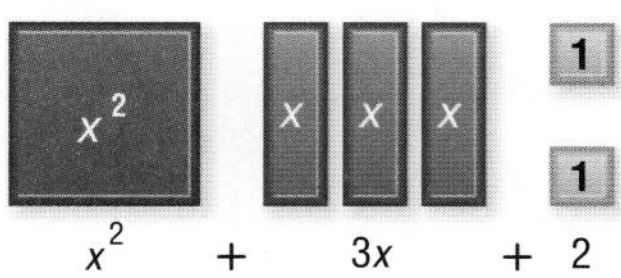

Try to form a rectangle with the tiles.

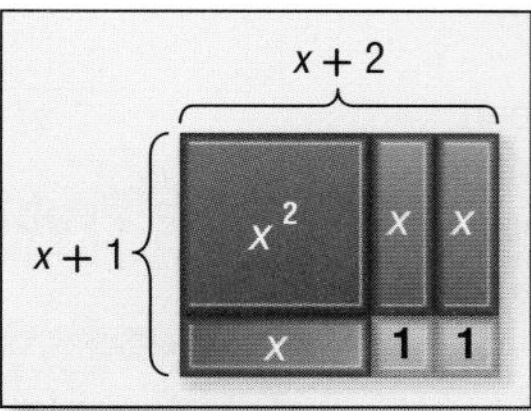

The rectangle has a width of $x + 1$ and a length of $x + 2$.

So, $x^2 + 3x + 2 = (x + 1)(x + 2)$.

b. $x^2 + 3x + 5$

Model the polynomial.

Try to form a rectangle with the tiles.

You cannot form a rectangle with the tiles.

So, $x^2 + 3x + 5$ cannot be factored.

Check Your Progress

Factor each trinomial.

1A. $x^2 + 7x + 10$

1B. $x^2 + 7 + x$

Personal Tutor glencoe.com

Use the FOIL method and guess-and-check to factor trinomials.

EXAMPLE 2 Factor Trinomials Using the Foil Method

Factor $x^2 + 6x + 8$.

Step 1 The x^2 term is the product of **F**irst terms. It only has one set of factors that are valid for two binomials.

$$x^2 + bx + c = (x + \blacksquare)(x + \blacksquare)$$

Step 2 The **L**ast two terms in each binomial must have a product of 8. Try several factor pairs of 8 until the sum of the products of the **O**uter and **I**nner terms is $6x$.

Try 1 and 8. $(x + 1)(x + 8) = x^2 + 8x + 1x + 8$

$= x^2 + 9x + 8$ **$9x$ is not the correct term.**

Try 2 and 4. $(x + 2)(x + 4) = x^2 + 4x + 2x + 8$

$= x^2 + 6x + 8$ ✓

So, $x^2 + 6x + 8 = (x + 2)(x + 4)$.

Check Your Progress

2A. $x^2 + 5x - 6$

2B. $x^2 + 8x + 15$

Personal Tutor glencoe.com

Real-World Link

In tennis, the service court fits inside the singles width of the court (27 feet) and extends back from the net 21 feet. The service court is divided in half with a center line marking the left and right service courts.

Source: SportsKnowHow

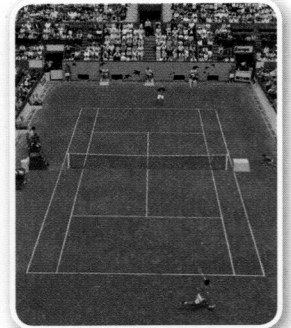

Solve Problems You can solve real-world problems by factoring trinomials.

Real-World EXAMPLE 3 Solve Problems By Factoring

SPORTS A doubles tennis court has an area represented by $x^2 + 78x + 1080$.

a. What are the dimensions of the court in terms of x?

$x^2 + 78x + 1080 = (x + \blacksquare)(x + \blacksquare)$ **Write the expression.**

The factor pair of 1080 that has a sum of 78 is 18 and 60.

So, $x^2 + 78x + 1080 = (x + 18)(x + 60)$

The dimensions are $(x + 18)$ feet by $(x + 60)$ feet.

b. If the service court occupies 1134 ft² and $x = 18$ feet, how much of the court is not considered service court area?

$(x + 18)(x + 60) = (18 + 18)(18 + 60) - 1134$ **Subtract the service area from the total area.**

$= (36)(78) - 1134$

$= 2808 - 1134$ or 1674 **Subtract.**

So, 1674 square feet is *not* considered service court area.

Check Your Progress

3. SPORTS The platform on which table tennis is played has an area that can be represented by $y^2 - 2y - 3$.

A. What are the dimensions of the platform in terms of y?

B. If $y = 8$ feet, what is the total platform area?

Personal Tutor glencoe.com

Check Your Understanding

Examples 1 and 2
pp. LA27, LA28

Factor.

1. $x^2 + 5x + 6$

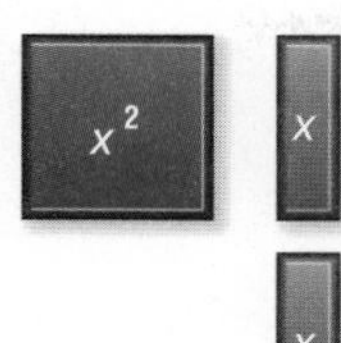

2. $x^2 + 4x + 3$

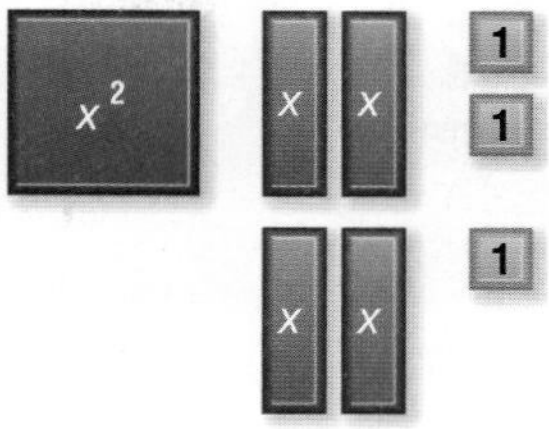

3. $x^2 + x + 2$

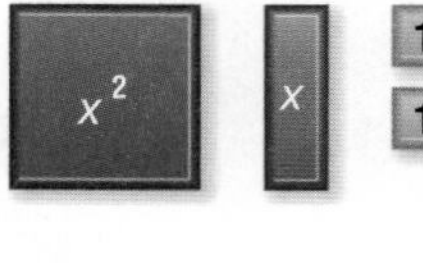

4. $y^2 + 9y + 20$

5. $a^2 - 5a + 4$

6. $x^2 + 3x - 10$

Example 3
p. LA28

7. **GEOMETRY** Refer to the figure at the right.

a. Find the area of the shaded region. Express in factored form.

b. If $x = 10$, what is the area?

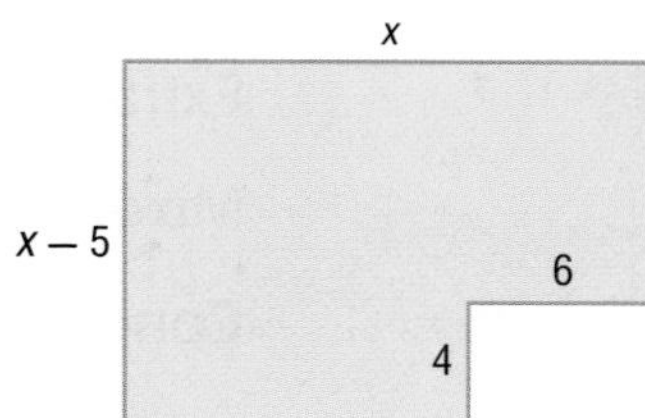

Practice and Problem Solving

Examples 1 and 2
pp. LA27, LA28

Factor.

8. $m^2 - 4m - 21$

9. $b^2 - 9b + 20$

10. $x^2 - x - 12$

11. $b^2 + 5b + 4$

12. $a^2 + 3a + 5$

13. $x^2 - 8x + 15$

14. $d^2 - 11d + 28$

15. $c^2 + 2c - 3$

16. $m^2 - 2m - 24$

17. Express $x^2 + 24x + 95$ as the product of two binomials.

Example 3
p. LA28

18. **GEOMETRY** Refer to the figure.

a. Find the area of the shaded region. Express the area in factored form.

b. If $\ell = 8$, what is the area?

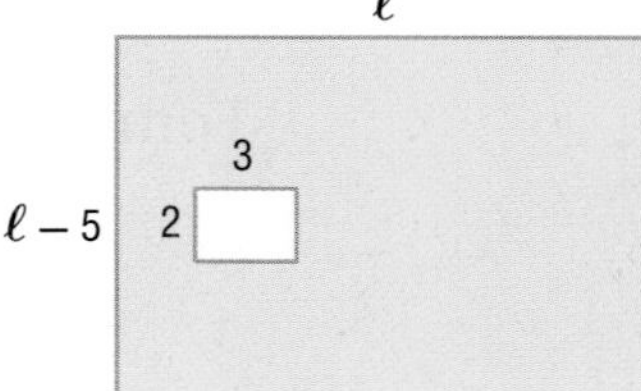

19. A square has an area of $x^2 + 6x + 9$ square inches. What is the perimeter?

H.O.T. Problems

Use **H**igher-**O**rder **T**hinking Skills

20. **CHALLENGE** The volume of a rectangular prism is $x^3 + 4x^2 + 3x$. Find the length, width, and height of the prism if each one can be written as a monomial or binomial. (*Hint*: Use the formula $V = \ell wh$.)

21. **REASONING** Complete the trinomial $x^2 + 6x +$ ____ with a positive integer so that the resulting trinomial can be factored. (*Hint*: There is more than one solution.)

22. **CHALLENGE** The binomials $x^2 - 1$, $x^2 - 4$, $x^2 - 36$, and so on, can all be factored into the product of two binomials. What pattern do you notice? Make and test a conjecture about the factors $a^2 - b^2$.

23. **WRITING IN MATH** Explain why the trinomial $x^2 + x + 5$ *cannot* be factored.

Student Handbook

Built-In Workbooks

Reference

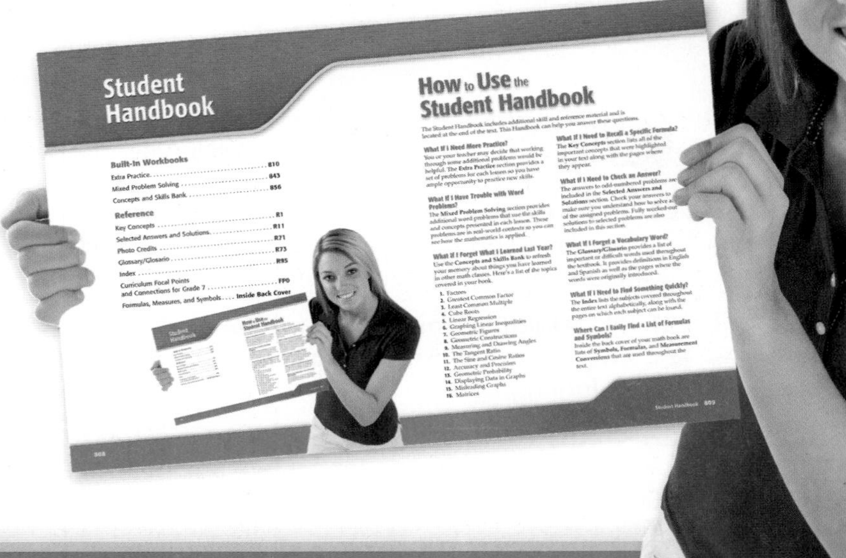

How to Use the Student Handbook

The Student Handbook includes additional skill and reference material and is located at the end of the text. This Handbook can help you answer these questions.

What If I Need More Practice?

You or your teacher may decide that working through some additional problems would be helpful. The **Extra Practice** section provides a set of problems for each lesson so you have ample opportunity to practice new skills.

What If I Have Trouble with Word Problems?

The **Mixed Problem Solving** section provides additional word problems that use the skills and concepts presented in each lesson. These problems are in real-world contexts so you can see how the mathematics is applied.

What If I Forget What I Learned Last Year?

Use the **Concepts and Skills Bank** to refresh your memory about things you have learned in other math classes. Here's a list of the topics covered in your book.

1. Factors
2. Greatest Common Factor
3. Least Common Multiple
4. Cube Roots
5. Linear Regression
6. Graphing Linear Inequalities
7. Geometric Figures
8. Geometric Constructions
9. Measuring and Drawing Angles
10. The Tangent Ratio
11. The Sine and Cosine Ratios
12. Accuracy and Precision
13. Geometric Probability
14. Displaying Data in Graphs
15. Misleading Graphs
16. Matrices

What If I Need to Recall a Specific Formula?

The **Key Concepts** section lists all of the important concepts that were highlighted in your text along with the pages where they appear.

What If I Need to Check an Answer?

The answers to odd-numbered problems are included in the **Selected Answers and Solutions** section. Check your answers to make sure you understand how to solve all of the assigned problems. Fully worked-out solutions to selected problems are also included in this section.

What If I Forget a Vocabulary Word?

The **Glossary/Glosario** provides a list of important or difficult words used throughout the textbook. It provides definitions in English and Spanish as well as the pages where the words were originally introduced.

What If I Need to Find Something Quickly?

The **Index** lists the subjects covered throughout the entire text alphabetically, along with the pages on which each subject can be found.

Where Can I Easily Find a List of Formulas and Symbols?

Inside the back cover of your math book are lists of **Symbols, Formulas,** and **Measurement Conversions** that are used throughout the text.

Extra Practice

Lesson 1-1 Words and Expressions (pp. 5–9)

Evaluate each expression.

1. $8 + 7 + 12 \div 4$

2. $20 \div 4 - 5 + 12$

3. $(25 \cdot 3) + (10 \cdot 3)$

4. $36 \div 6 + 7 - 6$

5. $30 \cdot (6 - 4)$

6. $(40 \cdot 2) - (6 \cdot 11)$

7. $\frac{86 - 11}{11 + 4}$

8. $\frac{12 + 84}{11 + 13}$

9. $\frac{5 \cdot 5 + 5}{5 \cdot 5 - 15}$

10. $(19 - 8)4$

11. $75 - 5(2 \cdot 6)$

12. $81 \div 27 \times 6 - 2$

Write a numerical expression for each verbal phrase.

13. three increased by nine

14. fifteen divided by three

15. six less than ten

Lesson 1-2 Variables and Expressions (pp. 11–16)

ALGEBRA Evaluate each expression if $a = 2$, $b = 4$, and $c = 3$.

1. $ba - ac$

2. $4b + a \cdot a$

3. $11 \cdot c - ab$

4. $4b - (a + c)$

5. $7(a + b) - c$

6. $8a + 8b$

7. $\frac{8(a + b)}{4c}$

8. $36 - 12c$

9. $\frac{9(b + a)}{c - 1}$

10. $abc - bc$

11. $28 - bc + a$

12. $a(b - c)$

Translate each phrase into an algebraic expression.

13. nine more than a

14. eleven less than k

15. three times p

16. the product of some number and five

17. twice Shelly's score decreased by 18

18. the quotient of 16 and n

Lesson 1-3 Properties (pp. 18–23)

Name the property shown by each statement.

1. $1 \cdot 4 = 4$

2. $6 + (b + 2) = (6 + b) + 2$

3. $9(6n) = (9 \cdot 6)n$

4. $8t \cdot 0 = 0 \cdot 8t$

5. $0(13n) = 0$

6. $7 + t = t + 7$

ALGEBRA Simplify each expression.

7. $(12 + x) + 9$

8. $31 + (15 + c)$

9. $d + (8 + 19)$

10. $2 \cdot (6 \cdot m)$

11. $(5 \cdot p) \cdot 3$

12. $9(4f)$

Lesson 1-4 Order Pairs and Relations (pp. 25–30)

Refer to the coordinate plane shown at the right. Write the ordered pair that names each point.

1. R
2. P
3. W
4. C
5. D
6. F

Express each relation as a table. Then determine the domain and range.

7. {(3, 6), (4, 9), (5, 1)}
8. {(2, 1), (4, 4), (6, 7), (4, 3)}

Lesson 1-5 Words, Equations, Tables, and Graphs (pp. 33–37)

Copy and complete each function table. Then state the domain and range of the function.

1. Each ticket cost $7.

Number of Tickets	Total Cost ($)
4	■
8	■
15	■
20	■

2. Natalie has 3 less than twice as many CDs as Kilan.

Kilan's CDs	Natalie's CDs
Input (x)	Output (y)
5	■
8	■
13	■
21	■

Lesson 1-6 Scatter Plots (pp. 40–46)

Determine whether a scatter plot of the data for the following might show a *positive, negative,* or *no* relationship. Explain your answer.

1. speed of airplane and miles traveled in three hours
2. weight and shoe size
3. outside temperature and heating bill
4. **GAMES** The number of pieces in a jigsaw puzzle and the number of minutes required for a person to complete it is shown below.

Number of Pieces	100	60	500	750	1000	800	75
Time (min)	35	20	175	315	395	270	25

a. Make a scatter plot of the data.

b. Does the scatter plot show any relationship? If so, is it positive or negative? Explain your reasoning.

c. Suppose Dave purchases a puzzle having 650 pieces. Predict the length of time it will take him to complete the puzzle.

Lesson 2-1 Integers and Absolute Value (pp. 61–66)

Replace each ● with <, >, or = to make a true sentence.

1. -4 ● -8
2. -6 ● 3
3. 0 ● -5
4. -12 ● -9
5. 12 ● -25
6. 3 ● -7
7. 0 ● -2
8. -15 ● 12
9. 5 ● -7
10. $|6|$ ● -2
11. -2 ● $|-3|$
12. $|-7|$ ● $|-4|$

Order the integers in each set from least to greatest.

13. {−1, 2, −5}
14. {0, −2, 8, 5, −9}
15. {100, −34, −86, 21, 0}
16. {−1, 16, −43, 8, 27, −40}
17. {0, −23, 75, −15, 24}
18. {−6, 6, −5, 18}

Evaluate each expression.

19. $|-3| + |9|$
20. $|-18| - |5|$
21. $|12 + 7|$
22. $-|6|$
23. $|-8| + |4|$
24. $-|-20|$
25. $|-6| \cdot |8|$
26. $-|12| \cdot |9|$
27. $-||-16| + |-22||$

Lesson 2-2 Adding Integers (pp. 69–74)

Find each sum.

1. $5 + (-6)$
2. $-17 + 24$
3. $15 + (-29)$
4. $-6 + 13$
5. $50 + (-14)$
6. $-21 + (-4)$
7. $30 + (-7)$
8. $(-3) + (-10)$
9. $-15 + 26$
10. $-11 + 15 + (-6)$
11. $23 + (-64)$
12. $-1 + 14 + (-13)$
13. $33 + (-18) + 7$
14. $-75 + (-13)$
15. $26 + 14 + (-71)$
16. $12 + (-20) + 16$
17. $100 + (-54) + (-17)$
18. $11 + (-22) + (-33)$

Lesson 2-3 Subtracting Integers (pp. 76–80)

Find each difference.

1. $8 - 17$
2. $-15 - 3$
3. $10 - 21$
4. $20 - (-5)$
5. $5 - (-9)$
6. $-12 - (-7)$
7. $-19 - (-6)$
8. $-16 - (-23)$
9. $-56 - 32$
10. $-49 - (-52)$
11. $-6 - 9 - (-7)$
12. $-6 - (-10) - 7$
13. $17 - 33$
14. $-21 - 19$
15. $12 - (-24)$
16. $-35 - (-18)$
17. $-54 - 27$
18. $32 - (-18)$

ALGEBRA Evaluate each expression if $x = 6$, $y = -8$, $z = -3$, and $w = 4$.

19. $y - z$
20. $3 - z$
21. $y - 5$
22. $x - y$
23. $14 - y - x$
24. $6 + x - z$
25. $y + z + w$
26. $w - z + 11$

Lesson 2-4 Multiplying Integers (pp. 83–88)

Find each product.

1. $-4(2)$
2. $-8(-5)$
3. $13(-4)$
4. $-5 \cdot 6 \cdot 10$
5. $-6(-2)(-14)$
6. $18(-3)(6)$
7. $4(-10)(-3)$
8. $-9(3)(2)$
9. $12(-8)$

ALGEBRA Simplify each expression.

10. $-3 \cdot 5x$
11. $7(-8m)$
12. $-10(-3k)$
13. $-4y(-8z)$
14. $(-2r)(-3s)$
15. $6(-2m)(3n)$

ALGEBRA Evaluate each expression.

16. $-6t$, if $t = 15$
17. $7p$, if $p = -9$
18. $-4k$, if $k = -16$
19. aw, if $a = 0$ and $w = -72$
20. dk, if $d = -12$ and $k = 11$
21. st, if $s = -8$ and $t = -10$
22. $3hp$, if $h = 9$ and $p = -3$
23. $-5bc$, if $b = -6$ and $c = 2$
24. $-4wx$, if $w = -1$ and $x = -8$

Lesson 2-5 Dividing Integers (pp. 90–95)

Find each quotient.

1. $-36 \div 9$
2. $112 \div (-8)$
3. $-72 \div 2$
4. $-26 \div (-13)$
5. $-144 \div 6$
6. $-180 \div (-10)$
7. $304 \div (-8)$
8. $-216 \div (-9)$
9. $80 \div (-5)$
10. $-105 \div 15$
11. $120 \div (-30)$
12. $-200 \div (-8)$
13. $42 \div (-6)$
14. $144 \div (-12)$
15. $-360 \div 9$
16. $-84 \div -6$
17. $125 \div (-5)$
18. $180 \div (-15)$
19. $-400 \div 20$
20. $72 \div (-9)$
21. $-156 \div (-2)$

ALGEBRA Evaluate each expression if $x = -5$, $y = -3$, $z = 2$, and $w = 7$.

22. $25 \div x$
23. $-42 \div w$
24. $3 \div y$
25. $2x \div z$
26. $-3x \div y$
27. $x \div (-1)$
28. $xyz \div 10$
29. $yz \div 2$
30. $\frac{3y}{-3}$
31. $\frac{6 - y}{y}$
32. $\frac{w}{-7}$
33. $\frac{w - x}{y}$

Lesson 2-6 Graphing in Four Quadrants (pp. 96–100)

Name the ordered pair for each point graphed at the right.

1. D
2. J
3. C
4. L
5. B
6. N
7. K
8. M

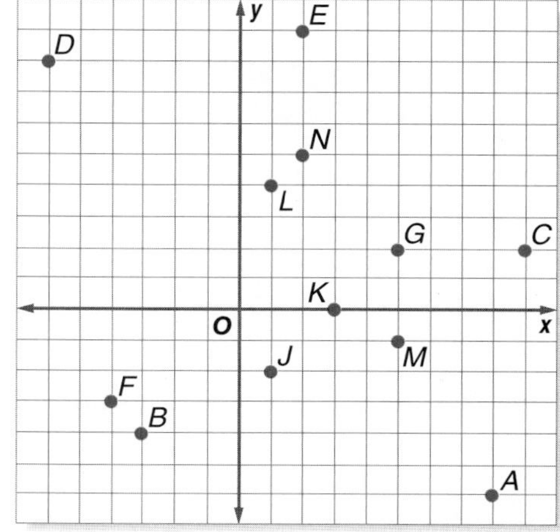

Graph and label each point on a coordinate plane. Name the quadrant in which each point is located.

9. $H(-2, -5)$
10. $P(1, 5)$
11. $R(-3, 1)$
12. $M(4, -2)$
13. $K(-4, 5)$
14. $G(3, -5)$

Lesson 2-7 Translations and Reflections on the Coordinate Plane (pp. 101–106)

1. The vertices of rectangle $ABCD$ are $A(-6, 5)$, $B(-2, 5)$, $C(-2, 2)$, and $D(-6, 2)$. What are the vertices of its image after a translation of 4 units to the right?
2. Rectangle $ABCD$ has vertices $A(2, 1)$, $B(5, 1)$, $C(5, 5)$, and $D(2, 5)$. Graph the rectangle and its image after a reflection over the x-axis.
3. A triangle has vertices $A(-14, 12)$, $B(6, 7)$, and $C(-5, 0)$. Its image has vertices $A'(-9, 8)$, $B'(11, 3)$, and $C'(0, -4)$. Describe the transformation.

Use the coordinate plane at the right.

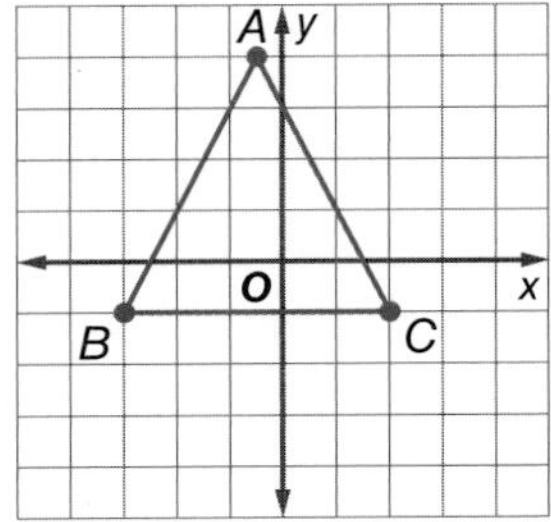

4. Find the coordinates of the vertices of the image of $\triangle ABC$ translated 3 units to the right and 4 units down.
5. Find the coordinates of the vertices of the image of $\triangle ABC$ reflected over the y-axis.

Lesson 3-1 Fractions and Decimals (pp. 121–127)

Write each fraction as a decimal. Use a bar to show a repeating decimal.

1. $\frac{6}{10}$
2. $\frac{4}{25}$
3. $-\frac{1}{8}$
4. $\frac{5}{6}$
5. $\frac{9}{20}$
6. $\frac{8}{11}$
7. $-\frac{3}{16}$
8. $\frac{6}{15}$

Replace each ● with <, >, or = to make a true sentence.

9. $\frac{7}{8}$ ● $\frac{5}{6}$
10. 0.04 ● $\frac{5}{9}$
11. $\frac{1}{3}$ ● $\frac{2}{7}$
12. $\frac{3}{5}$ ● $\frac{12}{20}$
13. $\frac{1}{2}$ ● 0.75
14. 0.3 ● $\frac{1}{3}$
15. $\frac{2}{3}$ ● 0.64
16. $\frac{2}{20}$ ● 0.10
17. $0.\overline{5}$ ● $\frac{5}{9}$
18. $2.\overline{1}$ ● $2\frac{1}{10}$
19. $3\frac{7}{8}$ ● 3.78
20. $-\frac{6}{7}$ ● $-\frac{5}{6}$

Lesson 3-2 Rational Numbers (pp. 128–133)

Write each number as a fraction.

1. $3\frac{4}{5}$
2. $-1\frac{2}{9}$
3. 15
4. $2\frac{3}{8}$
5. -13
6. $2\frac{6}{7}$
7. 36
8. $-1\frac{3}{5}$

Write each decimal as a fraction or mixed number in simplest form.

9. 0.6
10. 0.05
11. 0.38
12. 4.12
13. 0.375
14. -3.24
15. 0.222…
16. $-0.\overline{4}$

Lesson 3-3 Multiplying Rational Numbers (pp. 134–139)

Find each product. Write in simplest form.

1. $\frac{2}{5} \cdot \frac{3}{16}$
2. $3\frac{1}{4} \cdot \frac{2}{11}$
3. $\frac{3}{5}\left(-\frac{5}{12}\right)$
4. $\frac{5}{8} \cdot \frac{2}{3}$
5. $-\frac{9}{10} \cdot \frac{5}{24}$
6. $\frac{1}{7} \cdot \frac{21}{22}$
7. $\frac{4}{5} \cdot \frac{1}{8}$
8. $2\frac{2}{6} \cdot 6\frac{2}{7}$
9. $2\left(-\frac{7}{12}\right)$
10. $1\frac{3}{7}\left(-9\frac{4}{5}\right)$
11. $-\frac{6}{7}\left(-\frac{6}{7}\right)$
12. $\frac{6c}{10} \cdot \frac{2}{c}$
13. $\frac{p^3}{4} \cdot \frac{12}{p}$
14. $\frac{ab}{9} \cdot \frac{3}{b^2}$
15. $\frac{4x}{3y} \cdot \frac{12y^4}{x^2}$

Lesson 3-4 Dividing Rational Numbers (pp. 141–146)

Find the multiplicative inverse of each number.

1. $\frac{4}{7}$
2. $-\frac{5}{9}$
3. $\frac{1}{4}$
4. $5\frac{3}{8}$
5. 6
6. −18
7. $\frac{7}{10}$
8. 2.35
9. −1.4

Find each quotient. Write in simplest form.

10. $\frac{4}{5} \div \frac{2}{5}$
11. $-\frac{1}{3} \div \frac{6}{7}$
12. $\frac{4}{9} \div \frac{1}{5}$
13. $\frac{2}{3} \div \frac{1}{9}$
14. $\frac{4}{5} \div \left(-\frac{8}{15}\right)$
15. $\frac{1}{12} \div \frac{3}{4}$
16. $\frac{3}{4} \div \frac{15}{16}$
17. $16 \div 1\frac{7}{8}$
18. $2\frac{1}{6} \div \left(-1\frac{1}{5}\right)$
19. $-11 \div 3\frac{1}{7}$
20. $\frac{8}{45} \div \frac{10}{27}$
21. $-22 \div \left(-5\frac{1}{2}\right)$
22. $\frac{w}{5} \div \frac{w}{35}$
23. $\frac{ab}{12} \div \frac{b}{16}$
24. $\frac{21y}{8x^2} \div \frac{7y}{16x}$

Lesson 3-5 Adding and Subtracting Like Fractions (pp. 147–152)

Find each sum or difference. Write in simplest form.

1. $\frac{2}{7} + \frac{3}{7}$
2. $\frac{8}{15} - \frac{4}{15}$
3. $\frac{3}{7} + \frac{4}{7}$
4. $-\frac{8}{9} + \frac{1}{9}$
5. $\frac{5}{6} - \frac{1}{6}$
6. $\frac{7}{12} - \frac{5}{12}$
7. $\frac{5}{12} + \frac{11}{12}$
8. $-\frac{3}{14} - \frac{5}{14}$
9. $3\frac{1}{4} + \left(-\frac{3}{4}\right)$

ALGEBRA Find each sum or difference. Write in simplest form.

10. $\frac{n}{5} + \frac{3n}{5}$
11. $\frac{15}{k} - \frac{8}{k}, k \neq 0$
12. $12\frac{7}{8}s - 7\frac{3}{8}s$
13. $-6\frac{4}{9}t - 3\frac{2}{9}t$
14. $6\frac{1}{4}g + \left(-6\frac{3}{4}g\right)$
15. $7\frac{2}{5}n - \left(-4\frac{2}{5}n\right)$

Extra Practice

Lesson 3-6 Adding and Subtracting Unlike Fractions (pp. 153–158)

Find each sum or difference. Write in simplest form.

1. $\frac{1}{5}+\frac{2}{7}$
2. $\frac{4}{5}+\frac{7}{9}$
3. $\frac{1}{9}-\frac{7}{12}$
4. $\frac{8}{11}-\frac{4}{5}$
5. $\frac{7}{12}-\left(-\frac{4}{11}\right)$
6. $-\frac{9}{14}+\frac{15}{16}$
7. $-\frac{3}{8}+\left(-1\frac{5}{12}\right)$
8. $-\frac{2}{15}-3\frac{1}{5}$
9. $-5\frac{1}{3}+\left(-\frac{1}{6}\right)$
10. $3\frac{2}{5}+2\frac{4}{7}$
11. $-4\frac{1}{8}+2\frac{5}{9}$
12. $11\frac{3}{5}-\left(-6\frac{5}{8}\right)$

Lesson 4-1 The Distributive Property (pp. 171–176)

Use the Distributive Property to write each expression as an equivalent expression. Then evaluate the expression.

1. $2(4+5)$
2. $4(5+3)$
3. $3(7-6)$
4. $(2+5)9$
5. $(10-4)3$
6. $-6(1+3)$

Use the Distributive Property to write each expression as an equivalent algebraic expression.

7. $3(m+4)$
8. $(y+7)5$
9. $-6(x+3)$
10. $(p-4)5$
11. $-3(s-9)$
12. $5(x+y)$
13. $b(c+3d)$
14. $(a-b)(-5)$
15. $-6(v-3w)$
16. $5(x+12)$
17. $(m-6)(4)$
18. $-2(a-b)$
19. $(8-m)(-3)$
20. $8(p-3q)$
21. $(2x+3y)(4)$

Lesson 4-2 Simplifying Algebraic Expressions (pp. 178–183)

Identify the terms, like terms, coefficients, and constants in each expression.

1. $3+4x+x$
2. $5n+2-3n$
3. $6+1+7y$
4. $2c+c+8d$
5. $3a-9+b$
6. $2+6k+7-5k$

Simplify each expression.

7. $8k+2k+7$
8. $3+2b+b$
9. $t+2t$
10. $9(3+2x)$
11. $4(y+2)-2$
12. $(6+3e)4$
13. $4+9c+3(c+2)$
14. $5(7+2s)+3(s+4)$
15. $9(f+2)+14f$
16. $5a-9a$
17. $-6+4x+9-2x$
18. $6a+11+(-15)+9a$
19. $2(8w-7)$
20. $3(2d+5)+4d$
21. $2+4p-6(p-2)$
22. $-3(b+4)$
23. $-6+3s+11-5s$
24. $3(x-5)+7(x+2)$

Lesson 4-3 Solving Equations by Adding or Subtracting (pp. 184–189)

ALGEBRA Solve each equation. Check your solution and graph it on a number line.

1. $y + 49 = 26$
2. $d + 31 = -24$
3. $q - 8 = 16$
4. $x - 16 = 32$
5. $40 = a + 12$
6. $b + 12 = -1$
7. $21 = u + 6$
8. $-52 = p + 5$
9. $-14 = 5 - g$
10. $121 = k + (-12)$
11. $-234 = m - 94$
12. $110 = x + 25$
13. $f - 7 = 84$
14. $y - 864 = 652$
15. $475 + z = -18$
16. $x + 12 = -9$
17. $15 - h = 11$
18. $16 = p + 21$
19. $-13 + t = -2$
20. $86 = x + 43$
21. $y - 11 = -14$

Lesson 4-4 Solving Equations by Multiplying or Dividing (pp. 191–196)

Solve each equation. Check your solution.

1. $-y = -32$
2. $7r = -56$
3. $\frac{t}{-3} = 12$
4. $4 = \frac{s}{-14}$
5. $\frac{b}{47} = -2$
6. $64 = -4n$
7. $-144 = 12q$
8. $\frac{r}{11} = -12$
9. $-5g = -385$
10. $-16x = -176$
11. $-21 = \frac{y}{-4}$
12. $-372 = 31k$
13. $84 = \frac{k}{5}$
14. $-b = 19$
15. $\frac{v}{112} = -9$
16. $-3x = -27$
17. $\frac{p}{-12} = 4$
18. $5q = -100$
19. $\frac{d}{11} = -8$
20. $-9n = -45$
21. $125 = -25z$

Lesson 4-5 Solving Two-Step Equations (pp. 199–204)

Solve each equation. Check your solution.

1. $3t - 13 = 2$
2. $-8j - 7 = 57$
3. $9d - 5 = 4$
4. $6 - 3w = -27$
5. $\frac{k}{6} + 8 = 12$
6. $-4 = \frac{q}{8} - 19$
7. $15 - \frac{n}{7} = 13$
8. $44 = -4 + 8p$
9. $21 - h = -32$
10. $-19 = 11b - (-3)$
11. $6 = 20 + \frac{x}{3}$
12. $9 + 3a = -3$
13. $2x - 8 = 10$
14. $\frac{m}{4} - 6 = 10$
15. $-12 + 3p = 3$
16. $-18 = 6a - 6$
17. $\frac{t}{-3} + 11 = 23$
18. $3 + 2v = 11$
19. $16 = \frac{k}{3} - 11$
20. $-6g - 12 = -60$
21. $15 - 4c = -21$

Lesson 4-6 Writing Equations (pp. 205–209)

Translate each sentence into an equation. Then find each number.

1. Five less than three times a number is 13.
2. The product of 2 and a number is increased by 9. The result is 17.
3. Ten more than four times a number is 46.
4. The quotient of a number and −8, less 5, is −2.
5. Three more than two times a number is 11.
6. The quotient of a number and six, increased by 2 is −5.
7. The product of −3 and a number, decreased by 9 is 27.
8. A number is divided by 2. The sum of the result and 6 is −2.
9. The sum of 6 and a number divided by 3 is 7.

Lesson 5-1 Perimeter and Area (pp. 221–226)

Find the perimeter and area of each rectangle.

1. a rectangle 23 centimeters long and 9 centimeters wide
2. a 16-foot by 14-foot rectangle
3. a rectangle with a length of 31 meters and a width of 3 meters
4. a square with sides 7 meters long

Find the missing dimension of each rectangle.

	Length	Width	Area	Perimeter
5.	9 ft	■	126 ft^2	46 ft
6.	■	18 in.	108 in^2	48 in.
7.	13 yd	■	273 yd^2	68 yd
8.	■	12 cm	168 cm^2	52 cm
9.	■	3 m	162 m^2	114 m

10. The area of a triangle is 50 square meters. Its base is 10 meters. Find the height.
11. The area of a triangle is 96 square inches. Its height is 12 inches. Find the base.

Lesson 5-2 Solving Equations with Variables on Each Side (pp. 229–233)

Solve each equation. Check your solution.

1. $-7h - 5 = 4 - 4h$
2. $5t - 8 = 3t + 12$
3. $m + 2m + 1 = 7$
4. $2y + 5 = 6y + 25$
5. $3z - 1 = 23 - 3z$
6. $5a - 5 = 7a - 19$
7. $5x + 12 = 3x - 6$
8. $3x - 5 = 7x + 7$
9. $5c + 9 = 8c$
10. $3p = 4 - 9p$
11. $6z + 5 = 4z - 7$
12. $2a + 4.2 = 3a - 1.6$
13. $3.21 - 7y = 10y - 1.89$
14. $1.9s + 6 = 3.1 - s$
15. $12b - 5 = 3b$
16. $9 + 11a = -5a + 21$

Lesson 5-3 Inequalities (pp. 234–239)

Write an inequality for each sentence.

1. To pass his math class, Manuel needs to earn a grade of at least 70%.
2. Children who weigh less than 40 pounds must ride in an approved car seat.
3. To train for the upcoming marathon race, Jeff must run at least 30 miles each week.
4. There are at least 315 students at Glenwood Elementary School.
5. A cell phone bill increased by $10 is now more than $60 per month.
6. Citizens who are 18 years of age or older can vote.

For the given value, state whether the inequality is true or false.

7. $16 < 2x - 4; x = 7$
8. $3 - y \geq 6; y = -5$
9. $-5 \geq 3 - 4a; a = 2$
10. $\frac{1}{2}m - 1 < -3; m = -6$
11. $-8 + 3c < 4; c = 4$
12. $-5k + 20 > 0; k = 8$

Graph each inequality on a number line.

13. $x > -3$
14. $2 > n$
15. $y \leq 4$
16. $m < -2$
17. $q \geq 1$
18. $-1 \leq a$

Lesson 5-4 Solving Inequalities (pp. 241–247)

Solve each inequality. Check your solution.

1. $m + 9 < 14$
2. $k + (-5) < -12$
3. $-15 < v - 1$
4. $-7 + f \geq 47$
5. $r > -15 - 8$
6. $18 \geq s - (-4)$
7. $38 < r - (-6)$
8. $z - 9 \leq -11$
9. $-16 + c \geq 1$
10. $d + 1.4 < 6.8$
11. $-3 + x > 11.9$
12. $-0.2 \geq 0.3 + y$
13. $\frac{2}{3} \leq a - \frac{5}{6}$
14. $-7.42 \leq d - 5.9$
15. $6p < 78$
16. $\frac{m}{-3} > 24$
17. $-18 < 3b$
18. $-5k \geq 125$
19. $-75 > \frac{a}{5}$
20. $\frac{w}{6} < -5$
21. $8 < \frac{2}{3}c$
22. $\frac{m}{1.3} \geq 0.5$
23. $0.4y > -2$
24. $-\frac{1}{2}d \leq -5\frac{1}{2}$
25. The product of a number and -4 is greater than or equal to -20. What is the number?

Lesson 5-5 Solving Multi-Step Equations and Inequalities (pp. 248–253)

ALGEBRA Solve. Check your solution.

1. $6(m - 2) = 12$
2. $4(x - 3) = 4$
3. $5(2d + 4) = 35$
4. $w + 6 = 2(w - 6)$
5. $3(b + 1) = 4b - 1$
6. $7w - 6 = 3(w + 6)$
7. $4(k - 6) = 6(k + 2)$
8. $3x - 0.8 = 3x + 4$
9. $\frac{5}{9}g + 8 = \frac{1}{6}g + 1$
10. $2m + 1 < 9$
11. $-3k - 4 \leq -22$
12. $-2 > 10 - 2x$
13. $-6a + 2 \geq 14$
14. $3y + 2 < -7$
15. $\frac{d}{4} + 3 \geq -11$
16. $\frac{x}{3} - 5 < 6$
17. $-5g + 6 < 3g + 26$
18. $\frac{3(n + 1)}{7} \geq \frac{n + 4}{5}$

Lesson 6-1 Ratios (pp. 265–269)

Express each ratio as a fraction in simplest form.

1. 15 vans out of 40 vehicles
2. 6 pens to 14 pencils
3. 12 dolls out of 18 toys
4. 8 red crayons out of 36 crayons
5. 18 boys out of 45 students
6. 30 birds to 6 birds
7. 98 ants to 14 ladybugs
8. 140 dogs to 12 cats
9. 321 pennies to 96 dimes
10. 3 cups to 3 quarts

Lesson 6-2 Unit Rates (pp. 270–274)

Express each rate as a unit rate. Round to the nearest tenth, if necessary.

1. 343.8 miles on 9 gallons
2. $7.95 for 5 pounds
3. $52 for 8 tickets
4. $43.92 for 4 CDs
5. 450 miles in 8 hours
6. $3.96 for 12 cans of soda
7. $3.84 for 64 ounces
8. 200 yards in 32.3 seconds
9. **MONEY** Which costs more per notebook, a 4-pack of notebooks for $3.98 or a 5-pack of notebooks for $4.99? Explain.
10. **ANIMALS** A cheetah can run 70 miles in 1 hour. How many feet is this per second? Round to the nearest whole number.

Lesson 6-3 Converting Rates and Measurements (pp. 275–280)

Complete each conversion. Round to the nearest hundredth if necessary.

1. 4 pt ≈ ■ L
2. 70 in. ≈ ■ cm
3. 5 yd ≈ ■ m
4. 180 g ≈ ■ oz
5. 1500 m ≈ ■ yd
6. 13 kg ≈ ■ lb
7. 4 km ≈ ■ yd
8. 10 in ≈ ■ mm
9. 35 fl oz ≈ ■ mL
10. A wildfire in Montana destroyed 240,000 acres in 3 days. At this rate, how many acres were destroyed per hour?
11. A new energy-efficient automobile can travel 34.4 miles on one gallon of gasoline. At this rate, how many feet per quart does the car average?
12. If a two-year old child consumes 48 ounces of milk each day, how many gallons does the child consume in a year?
13. Convert 60 miles per hour to feet per minute.
14. Convert 40 kilometers per hour to miles per minute.
15. Convert 20 miles per gallon to kilometers per liter.
16. Convert 65 miles per hour to meters per second.

Extra Practice

Lesson 6-4 Proportional and Nonproportional Relationships (pp. 281–285)

Determine whether the set of numbers in each table is proportional. Explain.

1.

Baskets	2	4	6	8
Apples	10	20	30	40

2.

Days	5	10	15	20
Grass Height (inches)	4	6	8	10

3.

Cups of Flour	2	4	6	8
Cups of Sugar	1.5	3	4.5	6

4.

Men's Shoe Size	6	6.5	7	7.5
Women's Shoe Size	7.5	8	8.5	9

5.

Age (years)	5	10	15	20
Height (inches)	46	59	72	74

6.

Temperature (°C)	0	25	60	100
Temparature (°F)	32	77	140	212

7.

Visitors	1	2	3	4
Shoes Left by the Door	2	4	6	8

8.

Age (months)	8	10	14	20
Number of Teeth	2	6	12	16

Lesson 6-5 Solving Proportions (pp. 287–292)

Solve each proportion.

1. $\frac{7}{k} = \frac{49}{63}$

2. $\frac{s}{4.8} = \frac{30.6}{28.8}$

3. $\frac{6}{11} = \frac{19.2}{g}$

4. $\frac{8}{13} = \frac{b}{65}$

5. $\frac{x}{12} = \frac{26}{24}$

6. $\frac{21}{p} = \frac{3}{9}$

7. $\frac{6.5}{8} = \frac{w}{20}$

8. $\frac{10}{4.21} = \frac{7}{y}$

9. $\frac{12}{x} = \frac{1}{2.54}$

Write a proportion that could be used to solve for each variable. Then solve.

10. 6 plums at \$1
10 plums at *d*

11. 8 gallons at \$9.36
f gallons at \$17.55

12. 3 packages at \$53.67
7 packages at *m*

13. 10 cards at \$7.50
p cards at \$18

14. 12 cookies at \$3.00
16 cookies at *s*

15. 6 toy cars at \$4.50
c toy cars at \$6.75

Lesson 6-6 Scale Drawings and Models (pp. 294–299)

On a set of architectural drawings for a school, the scale is $\frac{1}{2}$ inch = 4 feet. Find the actual length of each room.

	Room	Drawing Distance (in.)
1.	Classroom	5
2.	Principal's Office	1.75
3.	Library	$7\frac{1}{2}$
4.	Cafeteria	$9\frac{1}{4}$
5.	Gymnasium	12.2
6.	Nurse's Office	1.3

Lesson 6-7 Similar Figures (pp. 301–306)

The figures are similar. Find each missing measure.

1.

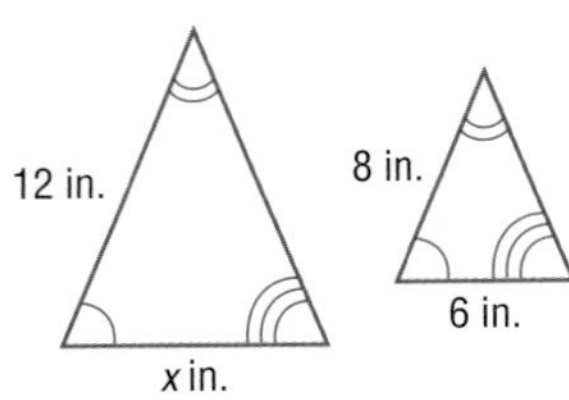

2.

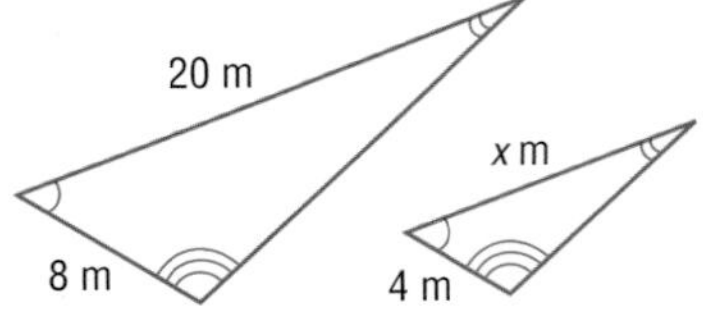

3.

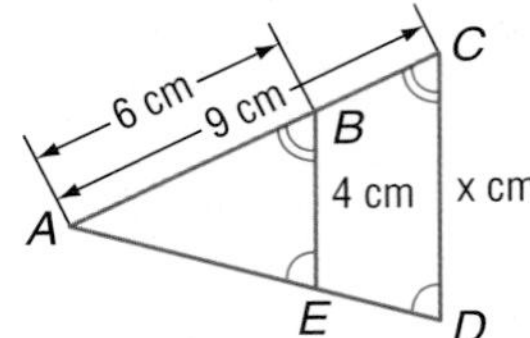

4.

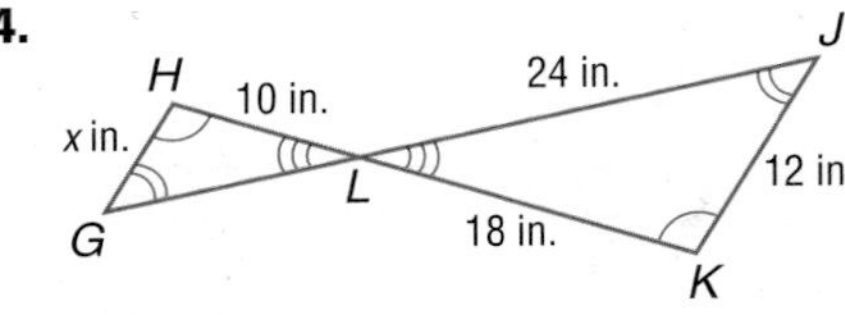

Lesson 6-8 Dilations (pp. 307–312)

1. A figure has vertices $A(-2, 1)$, $B(1, 3)$, $C(3, 2)$. Graph the figure and the image of the polygon after a dilation with a scale factor of 2.
2. A quadrilateral has vertices $K(3, 1)$, $L(1, 4)$, $M(6, 6)$, and $N(11, 1)$. Find the coordinates of the figure after a dilation with a scale factor of 0.5.
3. **BUSINESS** A $3\frac{1}{2}$ inch by 2 inch business card is being enlarged to $5\frac{1}{4}$ inches by 3 inches to use as an advertisement. What is the scale factor of the dilation?

Lesson 6-9 Indirect Measurement (pp. 313–317)

1. **FLAGS** A flag pole casts a 5 foot shadow while a nearby sign post casts a $1\frac{1}{4}$ foot shadow. Find the height of the flagpole if the sign post is 4 feet high.
2. A 56 inch tall woman casts a shadow that is 16 inches long. How tall is her child if his shadow is 7 inches long?
3. In the figure, $\triangle ABC \sim \triangle EDC$. Find the width of the Olen River $(\overline{AB})$.

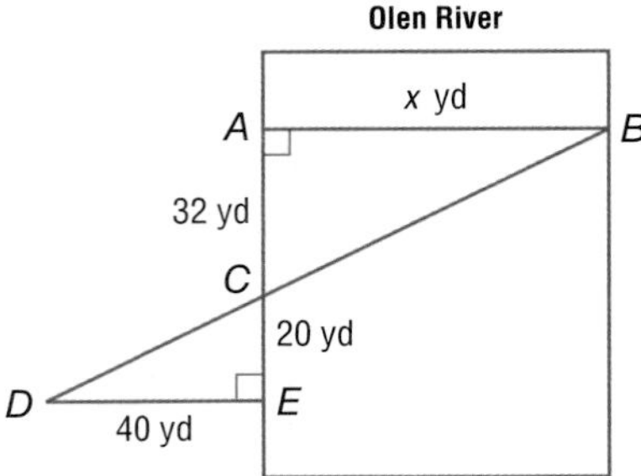

Lesson 7-1 Fractions and Percents (pp. 331–336)

Write each fraction as a percent. Round to the nearest hundredth.

1. $\frac{23}{25}$
2. $\frac{48}{60}$
3. $\frac{92}{96}$
4. $\frac{17}{50}$
5. $\frac{9}{25}$
6. $\frac{12}{8}$
7. $\frac{7}{40}$
8. $\frac{11}{33}$
9. $\frac{36}{27}$

Write each percent as a fraction or mixed number in simplest form.

10. 32%
11. 15%
12. $88\frac{1}{2}\%$
13. 250%
14. 21%
15. 64%
16. 25%
17. 131%
18. 72.5%
19. $66\frac{2}{3}\%$
20. 0.06%
21. 315%

Lesson 7-2 Fractions, Decimals, and Percents (pp. 337–342)

Write each percent as a decimal.

1. 72%
2. 7%
3. 380%
4. 0.03%

Express each decimal or fraction as a percent. Round to the nearest tenth, if necessary.

5. 0.19
6. 2.28
7. 0.009
8. 3.301
9. $\frac{5}{18}$
10. $\frac{3}{40}$
11. $\frac{13}{1000}$
12. $5\frac{3}{50}$

13. **SPORTS** In a recent survey, 370 students said gymnastics was their favorite sport to watch. If 3500 students were surveyed, what percent chose gymnastics as their favorite sport to watch? Round to the nearest percent.

Lesson 7-3 Using the Percent Proportion (pp. 345–350)

Use the percent proportion to solve each problem. Round to the nearest tenth, if necessary.

1. What is 81% of 134?
2. 52.08 is 21% of what number?
3. 11.18 is what percent of 86?
4. What is 120% of 312?
5. 140 is what percent of 400?
6. 430.2 is 60% of what number?
7. 32 is what percent of 80?
8. What is 15% of 125?
9. 22 is what percent of 110?
10. 9.4 is 40% of what number?
11. What is 41.5% of 95?
12. 17.92 is what percent of 112?

13. **FOOD** If 28 of the 50 soup cans on a shelf are chicken noodle soup, what percent of the cans are chicken noodle soup?

14. **SCHOOL** Of the students in a classroom, 60% are boys. If there are 20 students, how many are boys?

Extra Practice

Lesson 7-4 Find Percent of a Number Mentally (pp. 351–355)

Find the percent of each number mentally.

1. 40% of 60
2. 25% of 72
3. 50% of 96
4. $33\frac{1}{3}\%$ of 24
5. $37\frac{1}{2}\%$ of 80
6. 150% of 42
7. 200% of 125
8. $66\frac{2}{3}\%$ of 45

Estimate.

9. 60% of 49
10. 19% of 41
11. 82% of 60
12. 125% of 81
13. $\frac{1}{2}\%$ of 502
14. 31% of 19

Lesson 7-5 Using Percent Equations (pp. 357–362)

Solve each problem using a percent equation.

1. 9.28 is what percent of 58?
2. What number is 43% of 110?
3. 80% of what number is 90?
4. What number is 61% of 524?
5. 126 is what percent of 90?
6. 52% of what number is 109.2?
7. 62% of what number is 29.76?
8. 54 is what percent of 90?
9. Find 78% of 125.
10. What is 0.2% of 12?
11. 66% of what number is 49.5?
12. 36.45 is what percent of 81?

SHOPPING For each of the following, find the sales tax to the nearest cent.

13. $35 skirt, 7.5% sales tax
14. $108 lamp, 5.5% sales tax
15. $1585, 6% sales tax
16. $2934, 5.75% sales tax

Lesson 7-6 Percent of Change (pp. 364–369)

Find the percent of change. Round to the nearest tenth, if necessary. Then state whether each change is a *percent of increase* or a *percent of decrease*.

1. from $56 to $42
2. from $26 to $29.64
3. from $22 to $37.18
4. from $137.50 to $85.25
5. from $455 to $955.50
6. from $3 to $15
7. from $750.75 to $765.51
8. from $953 to $476.50
9. from $101.25 to $379.69
10. from $836 to $842.27
11. **BASEBALL CARDS** A baseball card collection contains 340 baseball cards. What is the percent of change if 25 cards are removed from the collection? Round to the nearest tenth.

Lesson 7-7 Simple and Compound Interest (pp. 370–374)

Find the simple interest to the nearest cent.

1. $1,100 at 5% for 3 years
2. $850 at 6% for 2 years
3. $12,500 at 3.5% for 4 years
4. $750 at $4\frac{1}{4}$% for 2 years
5. $140,700 at 5.4% for 6 years
6. $10,000 at $6\frac{1}{2}$% for 3 years

Find the total amount in each account to the nearest cent if the interest is compounded annually.

7. $600 at 5% for 3 years
8. $2500 at 4% for 2 years
9. $8240 at 10.5% for 4 years
10. $15,000 at 7.0% for 2 years
11. $10,500 at 3.75% for 3 years
12. $5075 at 6.25% for 1 year

13. **LOAN** Mariah borrowed $1200 to buy a new computer system. She paid the simple interest loan off in 24 monthly payments of $56 each. What was the interest rate on this loan?

14. **FINANCIAL LITERACY** Zachary took out a simple interest loan at $7\frac{1}{2}$% for 4 years when he borrowed $8000 to pay for a used car. How much will his monthly payments be?

15. **INVESTMENT** Eduardo's great aunt left him some money in her will. Rather than spend the money right away, Eduardo put it into a certificate of deposit for 3 years earning 5% simple interest annually. At the end of the 3 years, the certificate had earned $375. How much money did Eduardo inherit from his great aunt?

Lesson 7-8 Circle Graphs (pp. 376–381)

1. **FRUIT** During a recent lunch period, the cafeteria staff took a survey to find out which type of fruit the students preferred. The choices and results are listed in the table. Make a circle graph of the data.

Fruit	Number of Votes
Apples	144
Bananas	152
Grapes	98
Raisins	51
Oranges	205

2. **PROJECTS** For a project, students charted their activities during a week's time and reported their results in a circle graph.
   a. Miguel said he studied for 16.8 hours. How many hours are there in a week? How many degrees would Miguel need to use for the central angle on his circle graph for the Study sector?
   b. If Gabriel used 25° for the central angle for the sector labeled Piano, how many hours did he spend practicing the piano during that week? Round to the nearest hundredth.

3. **SPORTS** The circle graph at the right shows the results of a survey about favorite sports to watch on TV. If 300 students were surveyed, how many more preferred to watch baseball rather than basketball?

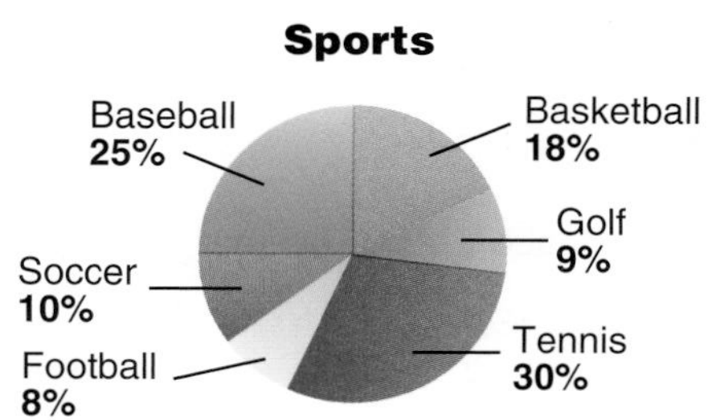

Lesson 8-1 Functions (pp. 395–400)

Determine whether each relation is a function.

1. $\{(4, -1), (2, -1), (0, 3), (5, -2)\}$

2.

x	6	8	6	7
y	4	3	9	1

3. 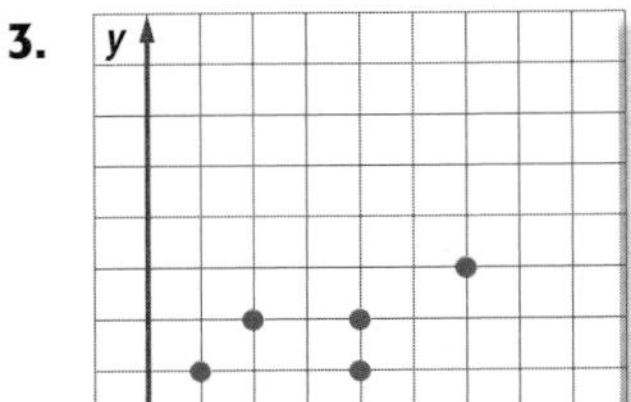

4. 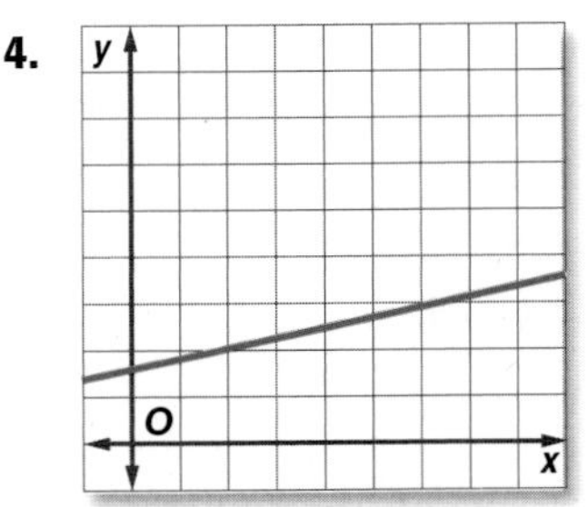

If $f(x) = 3x - 1$, find each function value.

5. $f(-1)$
6. $f(0)$
7. $f(2)$
8. $f(3)$

If $g(x) = -6x$, find each function value.

9. $g(-12)$
10. $g(15)$
11. $g(-16)$
12. $g(22)$

Lesson 8-2 Sequences and Equations (pp. 401–405)

Describe each sequence using words and symbols.

1. 25, 50, 75, …
2. 39, 40, 41, …
3. 4, 8, 12, …
4. 1.5, 2.5, 3.5, …
5. $\frac{1}{2}, 1, 1\frac{1}{2}, \ldots$
6. 4, 7, 10, …

Write an equation that describes each sequence. Then find the indicated term.

7. 5, 10, 15, …; 23rd term
8. 18, 17, 16, …; 13th term
9. 2, 1, 0, …; 7th term
10. 3, 5, 7, …; 99th term
11. 6, 13, 20, …; 38th term
12. $-6, -4, -2, \ldots$; 67th term
13. 0, 4, 8 …; 50th term
14. 5.5, 6, 6.5, …; 30th term

Lesson 8-3 Representing Linear Functions (pp. 406–411)

Find four solutions of each equation. Write the solutions as ordered pairs.

1. $x = 4$
2. $y = 0$
3. $x + y = 2$
4. $y = 2x - 6$
5. $x - y = 5$
6. $3x - y = 8$
7. $y = \frac{1}{2}x - 3$
8. $y = \frac{1}{3}x + 1$
9. $2x + y = -2$
10. $2x + 3y = 12$
11. $x + 2y = -4$
12. $2x - 4y = 8$

Graph each equation.

13. $y = x + 4$
14. $y = 4x$
15. $x + y = 3$
16. $y = x - 3$
17. $y = -2x + 5$
18. $2x + y = 6$

Lesson 8-4 Rate of Change (pp. 412–417)

Find the rate of change for each linear function.

1. 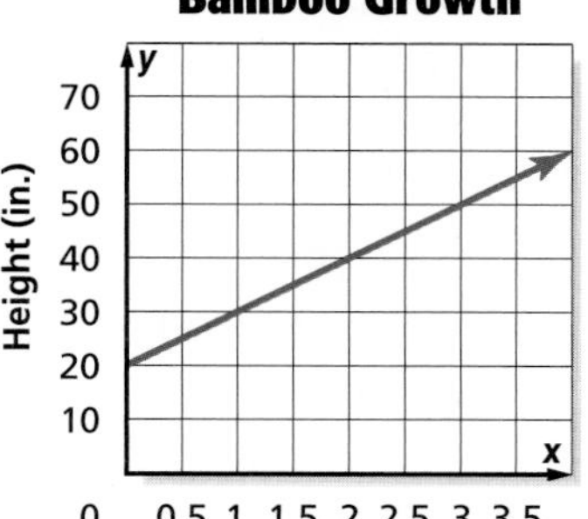

2. 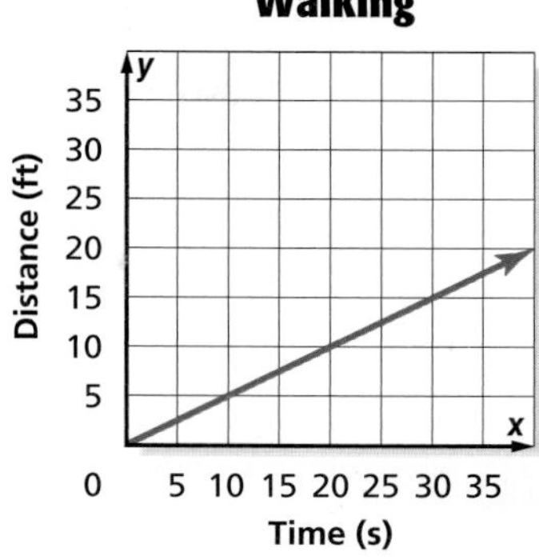

3.

Sales ($)	1240	1580	2250	2885
Commission ($)	49.60	63.20	90.00	115.40

4.

Time (min)	10	30	40	90
Candle Height (in.)	5.9	5.7	5.6	5.1

Lesson 8-5 Constant Rate of Change and Direct Variation (pp. 418–424)

Find the constant rate of change for each linear function and interpret its meaning.

1. 

2.

Time (min)	Volume (gal)
x	y
5	60
10	120
15	180
20	240

3. **JOBS** Lee works at a job where her pay varies directly as the number of hours she works. Her pay for 6.5 hours is $49.40. Write a direct variation equation relating Lee's pay x to the hours worked y. Then find her pay if she works 25 hours in a week.

Lesson 8-6 Slope (pp. 427–431)

Find the slope of each line.

1. 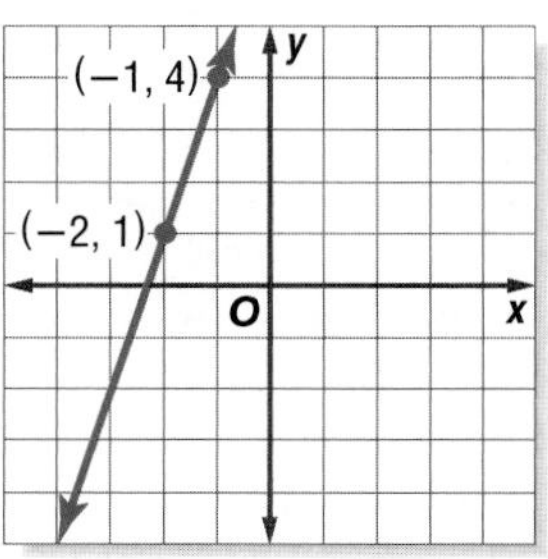

2. 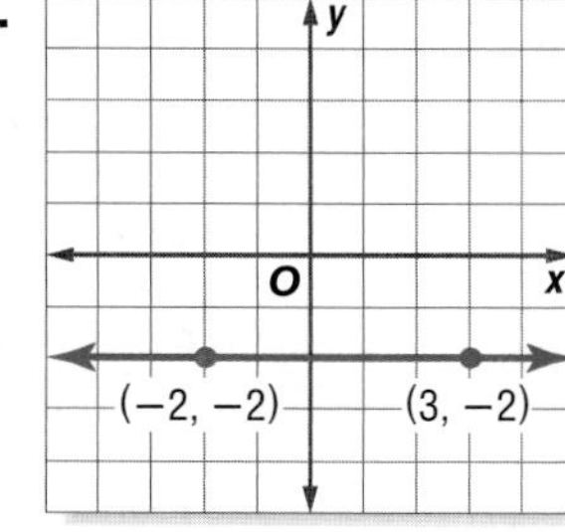

Find the slope of the line that passes through each pair of points.

3. $P(3, 8), Q(4, -3)$

4. $D(4, 5), E(-3, -9)$

5. $L(-1, 2), M(0, 5)$

6. $J(6, 2), K(6, -4)$

Lesson 8-7 Slope-Intercept Form (pp. 433–438)

State the slope and the y-intercept of the graph of each equation.

1. $y = x + 9$
2. $y = 2x - 5$
3. $y = -6x$
4. $y = \frac{3}{2}x$
5. $y = \frac{1}{3}x + 8$
6. $x + 2y = 12$

Graph each equation using the slope and y-intercept.

7. $y = 3x - 2$
8. $x - 3y = 9$
9. $y = \frac{1}{2}x + 4$
10. $y = -\frac{2}{3}x - 1$
11. $x - y = -4$
12. $2x + 4y = -4$

Lesson 8-8 Writing Linear Equations (pp. 441–447)

Write an equation in slope-intercept form for each line.

1. slope $= 3$, y-intercept $= -4$
2. slope $= \frac{3}{4}$, y-intercept $= 1$
3. slope $= -\frac{1}{2}$, y-intercept $= 0$
4. slope $= 0$, y-intercept $= -6$

Write an equation in slope-intercept form for the line passing through each pair of points.

5. $(4, 7)$ and $(0, 3)$
6. $(3, -6)$ and $(-1, 2)$
7. $(8, 7)$ and $(0, 0)$
8. $(1, 4)$ and $(3, -6)$
9. $(-2, 5)$ and $(3, 9)$
10. $(3, -1)$ and $(5, -1)$

Lesson 8-9 Prediction Equations (pp. 448–452)

1. **COMPUTER SALES** The table shows the amount of sales of computers in the U. S.
   a. Make a scatter plot of the computer sales and draw a line of fit.
   b. Use the line of fit to estimate the computer sales in 2010.
   c. Write an equation in slope-intercept form using the points (1994, 18.0) and (1998, 28.6).
   d. Use the equation to predict the amount of computer sales in 2020.

Year	Computer Sales ($ billions)
1990	8.9
1994	18.0
1996	23.6
1998	28.6
2000	35.1

2. **HOUSING** The scatter plot and line of fit show the number of housing units for the U. S.
   a. Write the equation in slope-intercept form for the line of fit. Round to the nearest tenth.
   b. Use the equation to estimate the number of housing units in 2014.

3. **JOBS** In 2006, 23.2% of the people between the ages of 65 and 74 were working. In 2000, only 19.6% of that age group was working.
   a. Write an equation in slope-intercept form for the line of fit for the data.
   b. Use the equation to predict the percent of that age group that will be working in 2015.

Lesson 8-10 Systems of Equations (pp. 453–457)

Solve each systems of equations by graphing.

1. $x + y = -2$, $2x - 3y = -9$
2. $y = 2x - 1$, $y = x + 1$
3. $3x - 2y = 11$, $-x + 6y = 7$
4. $-x + 2y = 6$, $x + 4y = 24$

Solve each systems of equations by substitution.

5. $2x - y = -2$, $4x - y = -6$
6. $2x + 2y = 3$, $x - 4y = -1$
7. $y = x - 4$, $4x + y = 26$
8. $x - 2y = -25$, $3x - y = 0$

Lesson 9-1 Powers and Exponents (pp. 471–475)

ALGEBRA Write each expression using exponents.

1. $8 \cdot 8 \cdot 8 \cdot 8$
2. 9
3. $(-6)(-6)(-6)(-6)(-6)$
4. $\left(\frac{1}{2}\right)\left(\frac{1}{2}\right)\left(\frac{1}{2}\right)\left(\frac{1}{2}\right)$
5. $(-0.4)(-0.4)(-0.4)$
6. $s \cdot s \cdot s \cdot s \cdot s$
7. $(y \cdot y \cdot y) \cdot (y \cdot y \cdot y \cdot y)$
8. $a \cdot b \cdot b$
9. $4 \cdot 4 \cdot 4 \cdot 4 \cdot x \cdot x \cdot x \cdot y$
10. $3q \cdot 3q \cdot 3q \cdot 3q \cdot 3q \cdot 3q$
11. $\underbrace{n \cdot n \cdot n \cdot \ldots \cdot n}_{17 \text{ factors}}$
12. $(x + y)(x + y)$

ALGEBRA Evaluate each expression if $m = 3$, $n = 2$, and $p = -4$.

13. $3m^2$
14. $n^0 + m$
15. 7^4
16. -5^3
17. p^3
18. $2(m - p)^2$
19. $-2n^3 + m$
20. $m - p^2$
21. $(m + n + p)^3$
22. $5p - m^2$
23. $(n + p)^4$
24. $(m - n)^8$

Lesson 9-2 Prime Factorization (pp. 476–480)

Determine whether each number is *prime* or *composite*.

1. 57
2. 369
3. 116
4. 125
5. 83
6. 99
7. 91
8. 79

Write the prime factorization of each number. Use exponents for repeated factors.

9. 21
10. 44
11. 51
12. 65
13. 30
14. 28
15. 117
16. 88
17. 54
18. 32
19. 300
20. 210

ALGEBRA Factor each monomial.

21. $40y$
22. $630a$
23. $187c^2$
24. $310p^2$
25. $510xy$
26. $1589cd$
27. $-18ab^2$
28. $-117x^3$
29. $105j^2k^5$

Extra Practice

Extra Practice

Lesson 9-3 Multiplying and Dividing Monomials (pp. 481–485)

ALGEBRA Find each product or quotient. Express using exponents.

1. $r^4 \cdot r^2$
2. $\frac{2^9}{2^3}$
3. $\frac{b^{18}}{b^5}$
4. $12^3 \cdot 12^8$
5. $x \cdot x^9$
6. $(2s^6)(4s^2)$
7. $w^3 \cdot w^4 \cdot w^2$
8. $(-2)^2(-2)^5(-2)$
9. $\frac{4^7}{4^6}$
10. $3(f^{17})(f^2)$
11. $(5k)^2 \cdot k^7$
12. $\frac{6m^8}{3m^2}$
13. $(3x^4)(-6x)$
14. $(4k^4)(-3k)^3$
15. $\left(\frac{42}{-6}\right)\left(\frac{g^{10}}{g^3}\right)$

Lesson 9-4 Negative Exponents (pp. 486–491)

Write each expression using a positive exponent.

1. y^{-9}
2. m^{-4}
3. 5^{-3}
4. 2^{-7}
5. 6^{-3}
6. a^{-11}

Write each fraction as an expression using a negative exponent.

7. $\frac{1}{p^4}$
8. $\frac{1}{b^9}$
9. $\frac{1}{5^3}$
10. $\frac{1}{7^4}$
11. $\frac{1}{15^2}$
12. $\frac{1}{25}$
13. $\frac{1}{c^7}$
14. $\frac{1}{64}$

Write each decimal using a negative exponent.

15. 0.01
16. 0.00001
17. 0.0001
18. 0.001
19. 0.1
20. 0.000001

Evaluate each expression if $x = 3$ and $y = -2$.

21. x^{-2}
22. 9^y
23. y^{-3}
24. x^{-3}
25. y^{-4}
26. $(xy)^{-2}$

Lesson 9-5 Scientific Notation (pp. 493–498)

Express each number in standard form.

1. 9.5×10^{-3}
2. 8.245×10^{-4}
3. 8.2×10^4
4. 9.102040×10^2
5. 4.02×10^3
6. 1.6×10^{-2}
7. 2.41023×10^6
8. 4.21×10^{-5}
9. 1.0012×10^{-3}
10. 8.604×10^2

Express each number in scientific notation.

11. 9040
12. 0.015
13. 6,180,000
14. 27,210,000
15. 0.00004637
16. 0.00546
17. 500,300,100
18. 0.0000032
19. 0.00047
20. 10,471,300

Lesson 9-6 Powers of Monomials (pp. 499–503)

Simplify.

1. $(3^2)^5$ **2.** $(a^4)^3$ **3.** $(k^2)^{-4}$

4. $(5^3)^3$ **5.** $(m^{-5})^{-2}$ **6.** $(4^3)^{-1}$

7. $(2a)^3$ **8.** $(-4c)^2$ **9.** $(xy^3)^5$

10. $(-3a^2b)^3$ **11.** $(-x^2yz^3)^6$ **12.** $(-2ab^3)^5$

13. $(-2m^4n^6)^3$ **14.** $(-abc^3)^2$ **15.** $(-xy^6)^5$

GEOMETRY **Express each measure as a monomial.**

16. Volume of a cube.

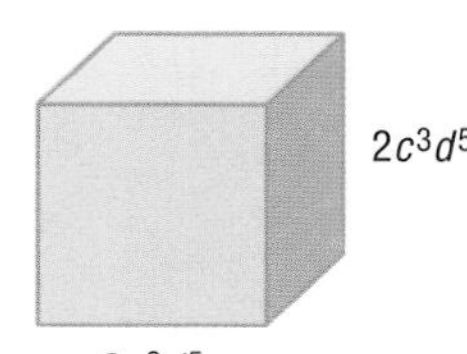

17. Area of square.

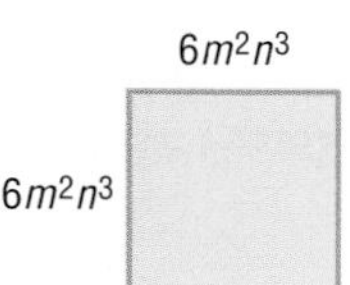

Lesson 9-7 Linear and Nonlinear Functions (pp. 504–509)

Determine whether each graph, equation, or table represents a *linear* or *nonlinear* function. Explain.

1.

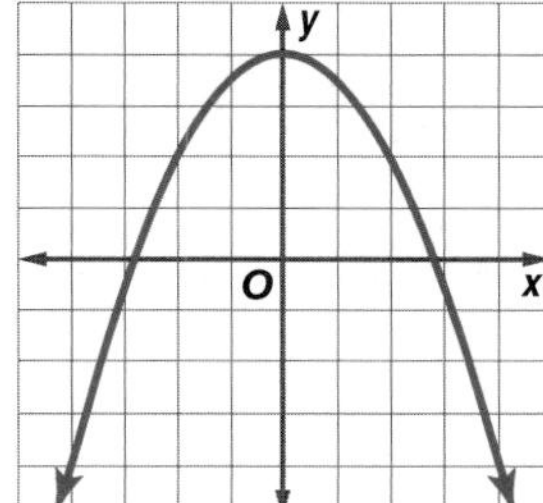

2.

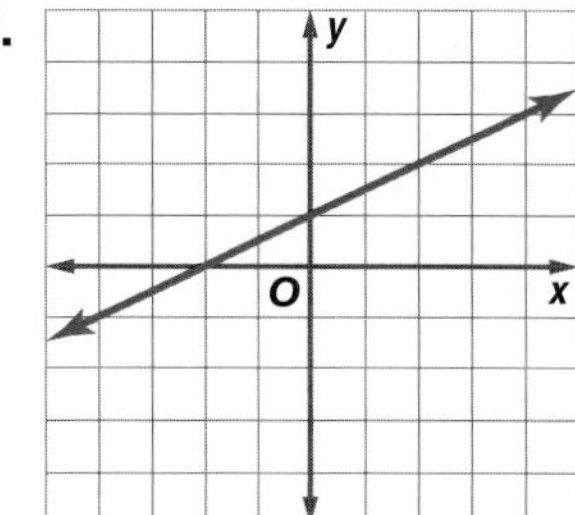

3.

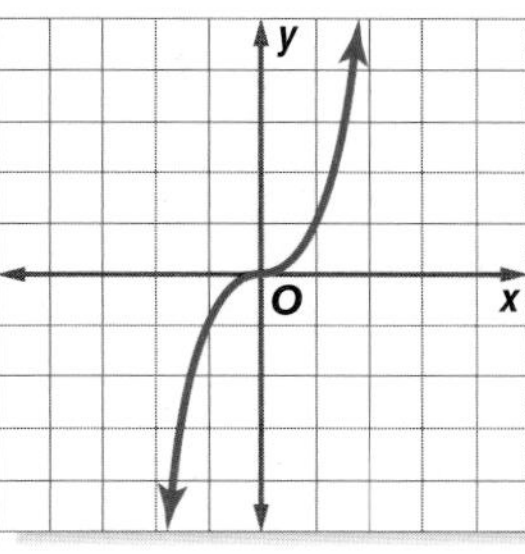

4. $y = -3x$ **5.** $y = 2x^3 - 5$ **6.** $-2x + 5y = 10$ **7.** $x = 7y$

8. $y = (-2)^x$ **9.** $y = \frac{6}{x}$

10.

x	y
2	5
4	7
6	9
8	11

11.

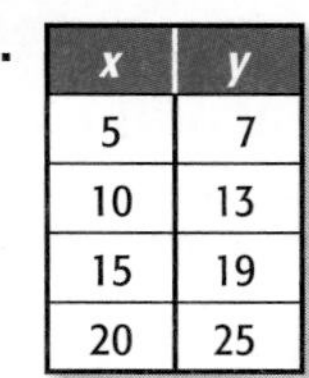

x	y
5	7
10	13
15	19
20	25

Lesson 9-8 Quadratic Functions (pp. 510–514)

Graph each function.

1. $y = 3x^2$ **2.** $y = -2x^2$ **3.** $y = \frac{1}{2}x^2$

4. $y = x^2 + 4$ **5.** $y = -0.5x^2 + 1$ **6.** $y = x^2 - 4x - 4$

7. $y = 5x^2 - 20x + 37$ **8.** $y = 3x^2 + 6x + 3$ **9.** $y = 2x^2 + 12x$

10. $y = x^2 - 6x + 5$ **11.** $y = x^2 + 6x + 9$ **12.** $y = -x^2 + 16x - 15$

13. $y = 4x^2 - 1$ **14.** $y = -2x^2 - 2x + 4$ **15.** $y = 6x^2 - 12x - 4$

Lesson 9-9 Cubic and Exponential Functions (pp. 516–520)

Graph each function. State the y-intercept.

1. $y = x^3$
2. $y = 0.3x^3$
3. $y = x^3 - 2$
4. $y = 3^x + 1$
5. $y = 2^x - 5$
6. $y = 2^{x+3}$
7. $y = 3^{x+1}$
8. $y = \left(\frac{2}{3}\right)^x$
9. $y = 5\left(\frac{2}{5}\right)^x$
10. $y = 5(3^x)$
11 $y = 4(5)^x$
12. $y = 2(5)^x + 1$
13. $y = \left(\frac{1}{2}\right)^{x+1}$
14. $y = \left(\frac{1}{3}\right)^x$
15. $y = \left(\frac{3}{4}\right)^x - 2$

Lesson 10-1 Squares and Square Roots (pp. 537–542)

Find each square root.

1. $\sqrt{36}$
2. $-\sqrt{81}$
3. $\sqrt{\frac{1}{4}}$
4. $-\sqrt{144}$
5. $\sqrt{25}$
6. $\sqrt{1.96}$
7. $\sqrt{100}$
8. $-\sqrt{0.49}$
9. $\sqrt{400}$

Estimate each square root to the nearest integer. Do not use a calculator.

10. $\sqrt{21}$
11. $-\sqrt{85}$
12. $\sqrt{7.3}$
13. $\sqrt{1.99}$
14. $-\sqrt{62}$
15. $\sqrt{74.1}$
16. $\sqrt{810}$
17. $-\sqrt{88.8}$
18. $\sqrt{1000}$

Use a calculator to find each square root to the nearest tenth.

19. $\sqrt{21}$
20. $\sqrt{99}$
21. $-\sqrt{60}$
22. $\sqrt{124}$
23. $-\sqrt{350}$
24. $\sqrt{18.6}$
25. $-\sqrt{42}$
26. $-\sqrt{84.2}$
27. $\sqrt{182}$

Lesson 10-2 The Real Number System (pp. 543–548)

Name all of the sets of numbers to which each real number belongs. Write *whole, integer, rational,* or *irrational.*

1. 15
2. 0
3. $\frac{3}{8}$
4. 0.666…
5. 1.75
6. $\sqrt{2}$
7. 5.14726…
8. $-\sqrt{36}$
9. 0.3535…

Replace each ● with <, >, or = to make a true statement.

10. $3\frac{3}{4}$ ● $\sqrt{15}$
11. $-\sqrt{41}$ ● -6.8
12. 5.2 ● $\sqrt{27.04}$
13. $-\sqrt{110}$ ● -10.5

ALGEBRA **Solve each equation. Round to the nearest tenth, if necessary.**

14. $x^2 = 14$
15. $y^2 = 25$
16. $34 = p^2$
17. $55 = h^2$
18. $225 = k^2$
19. $324 = m^2$
20. $d^2 = 441$
21. $r^2 = 25{,}000$
22. $10{,}000 = x^2$

Lesson 10-3 Triangles (pp. 550–555)

Find the value of x in each triangle. Then classify each triangle by its angles and by its sides.

1. 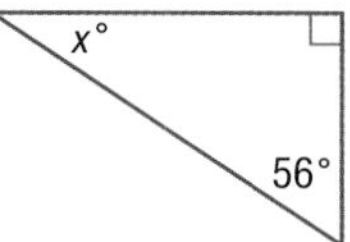

2. 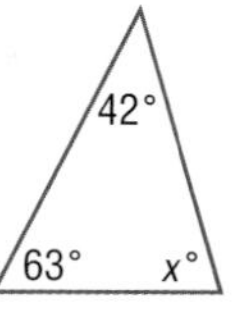

3. 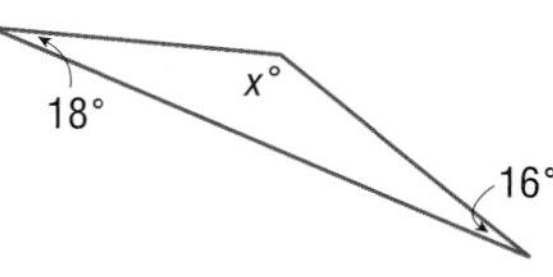

4. 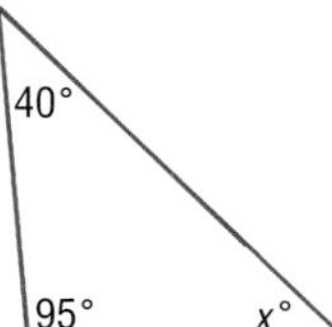

5. 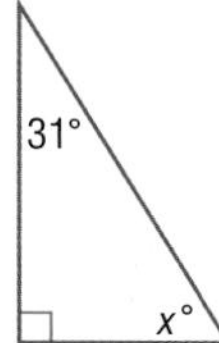

6. 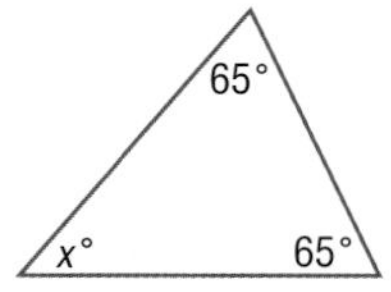

7. The measure of the angles of a triangle are in the ratio 1:2:3. What is the measure of each angle?
8. Determine the measures of the angles of $\triangle ABC$ if the measures of the angles of a triangle are in the ratio 1:1:2.
9. Suppose the measures of the angles of a triangle are in the ratio 1:9:26. What is the measure of each angle?

Lesson 10-4 The Pythagorean Theorem (pp. 558–563)

Find the length of the hypotenuse in each right triangle. Round to the nearest tenth, if necessary.

1. 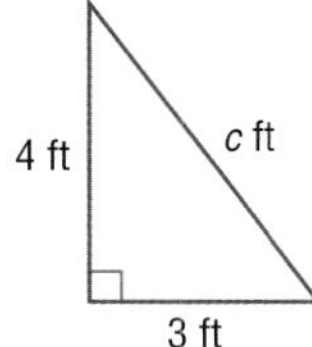

2. 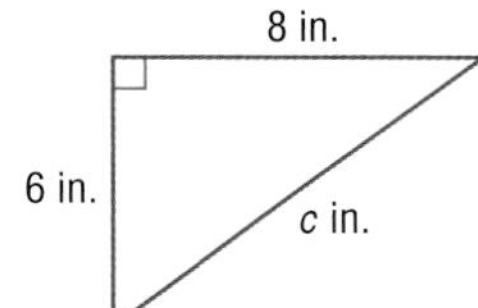

3. 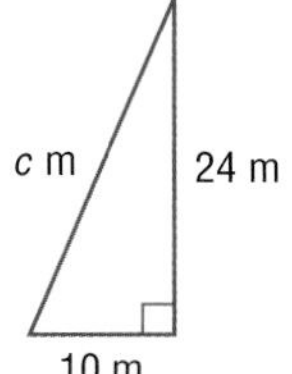

If c is the measurement of the hypotenuse, find each missing measure. Round to the nearest tenth, if necessary.

4. $a = 7$ m, $b = 24$ m
5. $a = 18$ in., $c = 30$ in.
6. $b = 10$ ft, $c = 20$ ft
7. $a = 3$ cm, $c = 9$ cm
8. $b = 8$ m, $c = 32$ m
9. $a = 32$ yd, $c = 65$ yd

Lesson 10-5 The Distance Formula (pp. 565–570)

Find the distance between each pair of points. Round to the nearest tenth, if necessary.

1. $A(2, 6), B(-4, 2)$
2. $C(-3, 9), D(2, 4)$
3. $E(6, -4), F(1, -6)$
4. $G(0, -1), H(9, -1)$
5. $I(-8, -3), J(2, 2)$
6. $K(3, 0), L(-7, -2)$
7. $M(3, 5), N(7, 1)$
8. $O(-6, 2), P(0, 8)$
9. $Q(4, -9), R(-2, 7)$
10. $S(13, -1), T(-5, -3)$

Lesson 10-6 Special Right Triangles (pp. 571–576)

Find each missing measure.

1.

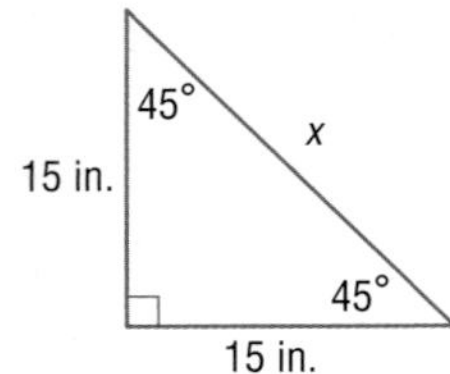

2.

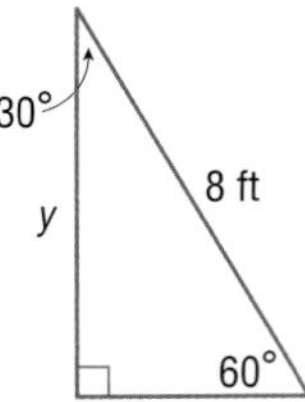

3.

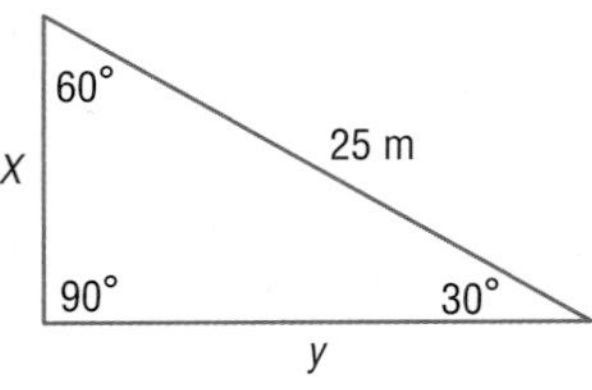

5. In a 45°-45°-90° right triangle, a leg is 36.4 inches long. Find the exact length of the hypotenuse.

6. In a 30°-60°-90° right triangle, the shorter leg is 17 centimeters long. Find the length of the hypotenuse and the length of the longer leg to the nearest tenth of a centimeter.

Lesson 11-1 Angle and Line Relationships (pp. 589–595)

In the figure at the right, $\ell \parallel m$ and p is a transversal. If the $m\angle 2$ is 38°, find the measure of each angle.

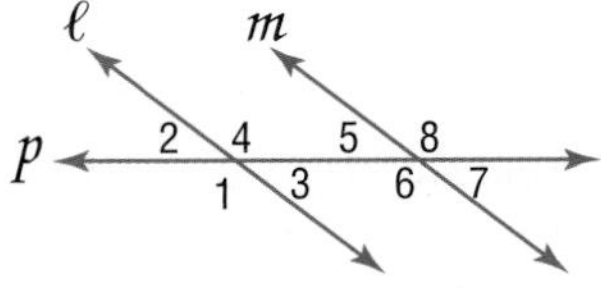

1. $\angle 1$ 

2. $\angle 4$

3. $\angle 3$

4. $\angle 6$

5. $\angle 5$

6. $\angle 8$

ALGEBRA Find the value of x in each figure.

7.

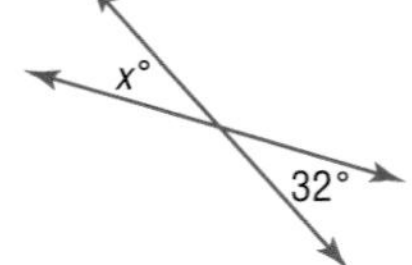

8.

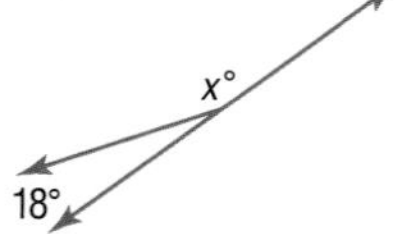

9.

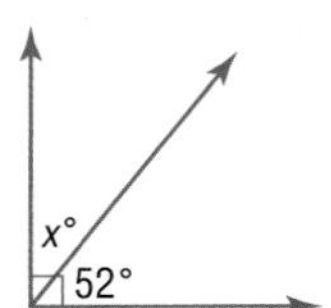

Lesson 11-2 Congruent Triangles (pp. 598–604)

Name the corresponding parts in each pair of congruent triangles. Then complete the congruence statement.

1.

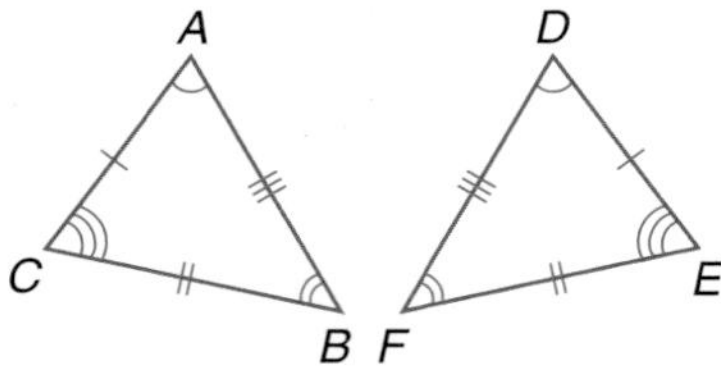

$\triangle ABC \cong \triangle$ _?_

2.

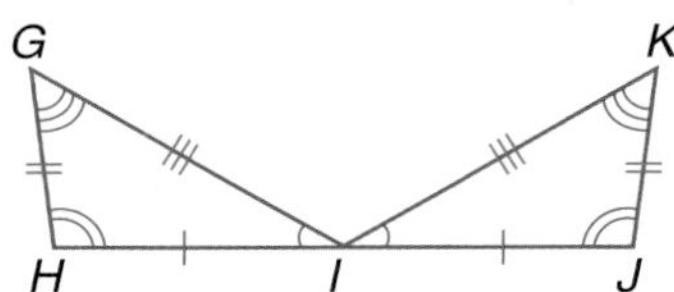

$\triangle GHI \cong \triangle$ _?_

Determine whether the triangles shown are congruent. If so, name the corresponding parts and write a congruence statement.

3.

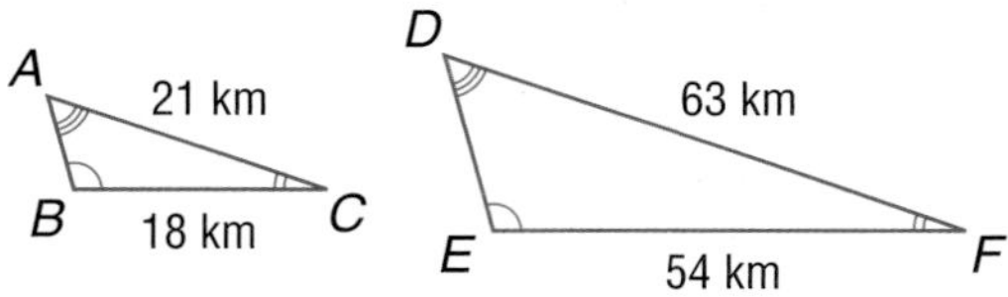

4.

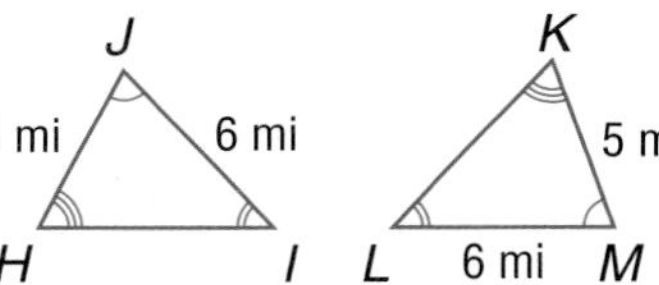

Lesson 11-3 Rotations (pp. 605–610)

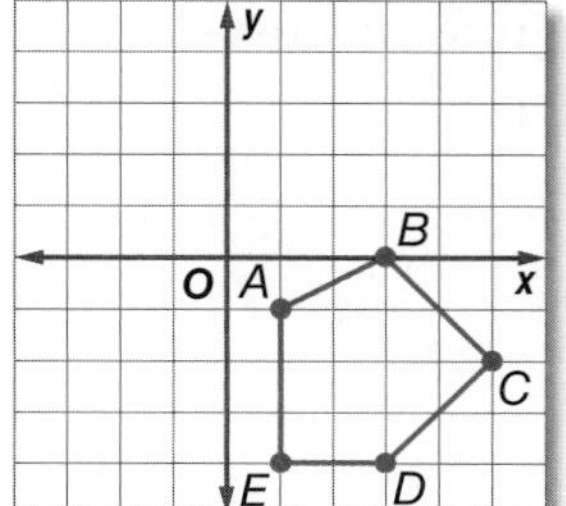

Figure *ABCDE* is shown.

1. Graph the image of the figure after a 90° counterclockwise rotation about the origin.
2. Find the coordinates of the vertices of the figure after a 180° rotation about the origin.
3. Graph the image of the figure after a 90° clockwise rotation about the origin.
4. **LETTERS** Determine whether the letter shown at the right has rotational symmetry. If it does, describe the angle of rotation.

J

Lesson 11-4 Quadrilaterals (pp. 612–616)

ALGEBRA Find the value of x. Then find the missing angle measures.

1. 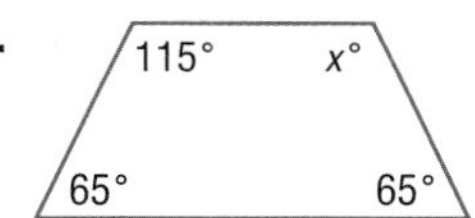

2. 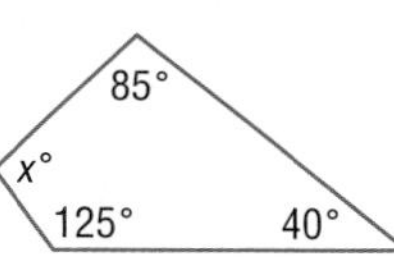

3. 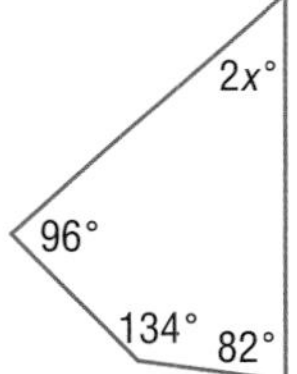

4. 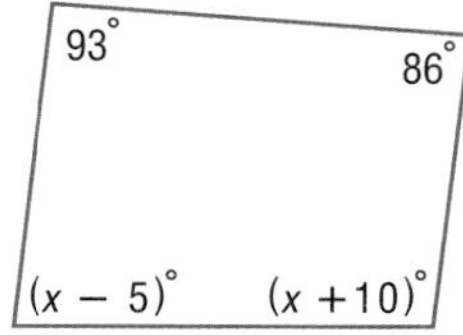

Lesson 11-5 Polygons (pp. 617–622)

Classify each polygon. Then determine whether it appears to be *regular* or *not regular*.

1. 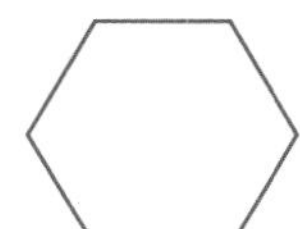

2. 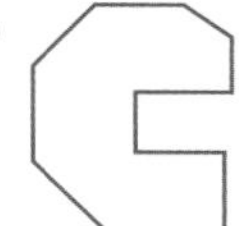

Find the sum of the measures of the interior angles of each polygon.

3. decagon
4. pentagon
5. nonagon
6. hexagon
7. octagon
8. 15-gon

Lesson 11-6 Area of Parallelograms, Triangles, and Trapezoids (pp. 624–630)

Find the area of each figure.

1. 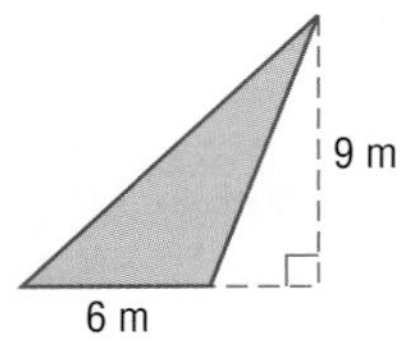

2. 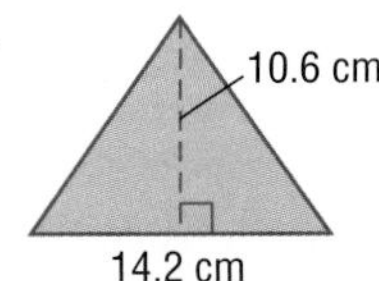

3. 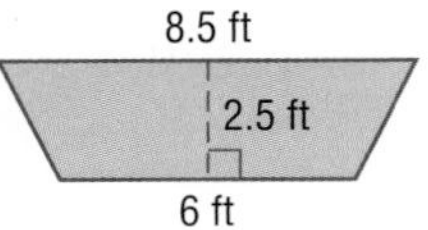

4. 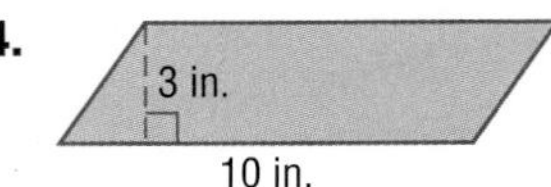

5. What is the height of a parallelogram with a base of 3.4 inches and an area of 32.3 square inches?

6. The bases of a trapezoid measure 8 meters and 12 meters. Find the measure of the height if the trapezoid has an area of 70 square meters.

Lesson 11-7 Circles and Circumference (pp. 631–635)

Find the circumference of each circle. Round to the nearest tenth.

1. 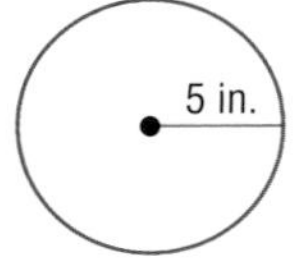

2. 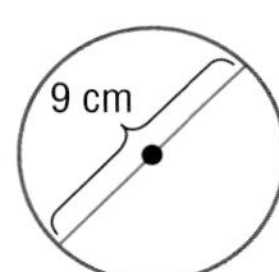

3. 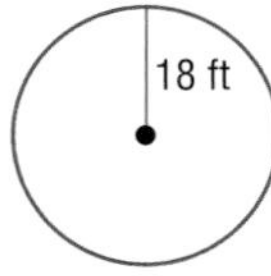

4. 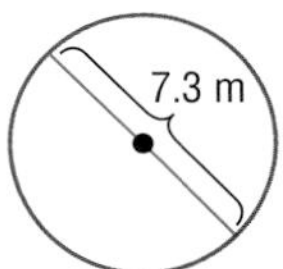

5. Find the diameter of a circle if its circumference is 18.5 feet. Round to the nearest tenth.

Lesson 11-8 Area of Circles (pp. 636–641)

Find the area of each circle. Round to the nearest tenth.

1. radius = 8.2 feet
2. diameter = 1.3 yards
3. diameter = 5.2 yards
4. radius = 4.8 centimeters
5. A circle has an area of 62.9 square inches. What is the radius of the circle? Round to the nearest tenth.

Lesson 11-9 Area of Composite Figures (pp. 642–647)

Find the area of each figure. Round to the nearest tenth.

1.

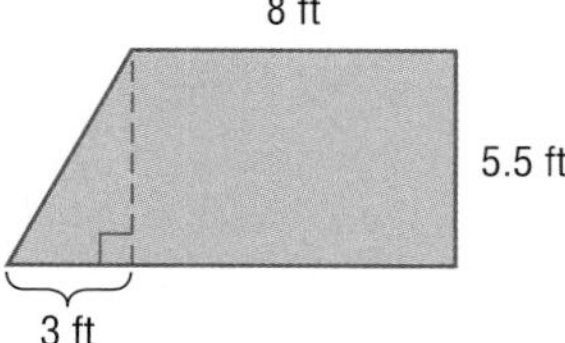

2.

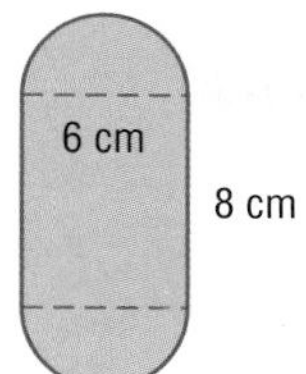

3.

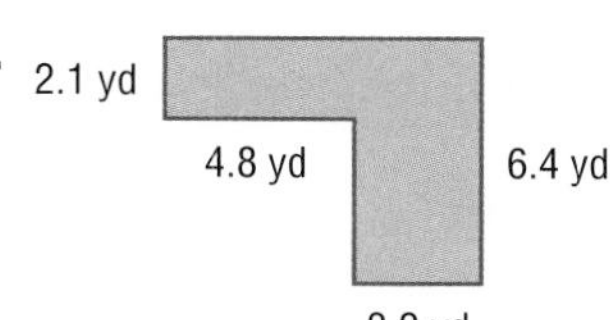

4.

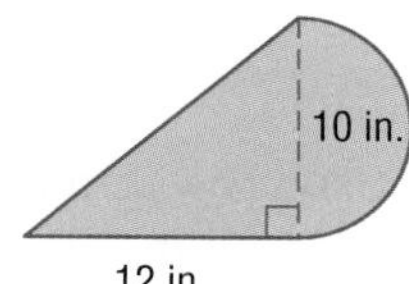

Lesson 12-1 Three-Dimensional Figures (pp. 664–669)

Identify each figure. Name the bases, faces, edges, and vertices.

1.

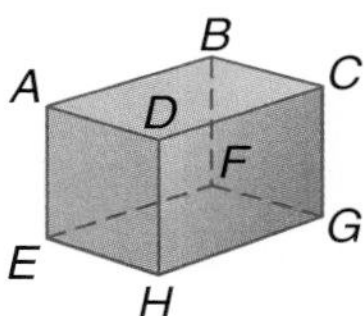

2.

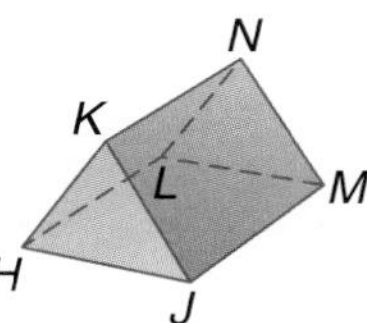

CROSS SECTIONS **Describe the shape resulting from each cross section.**

3.

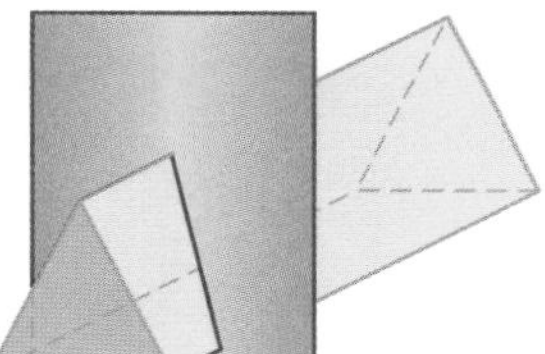

4.

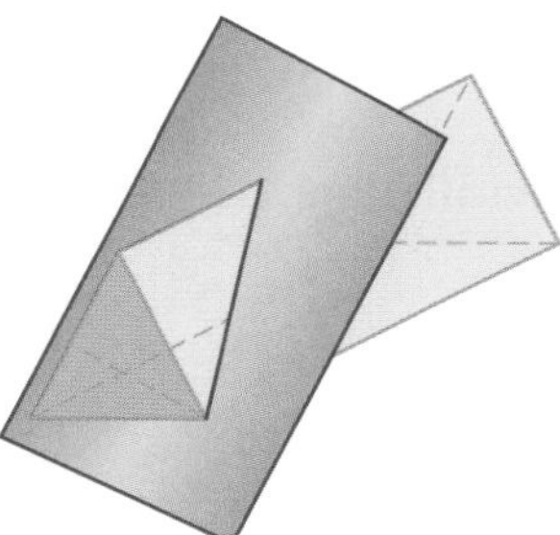

5.

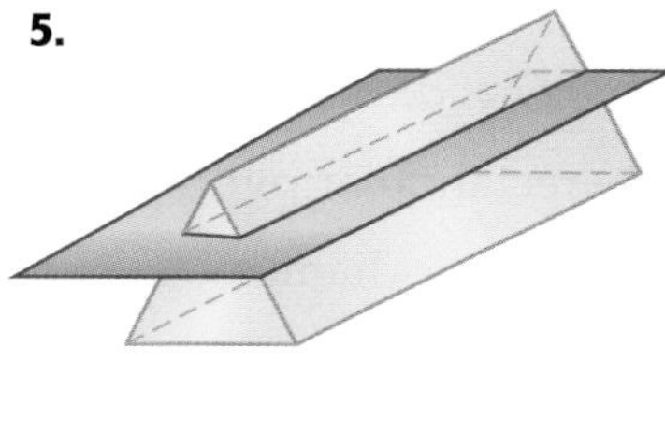

Lesson 12-2 Volume of Prisms (pp. 671–676)

Find the volume of each figure.

1.

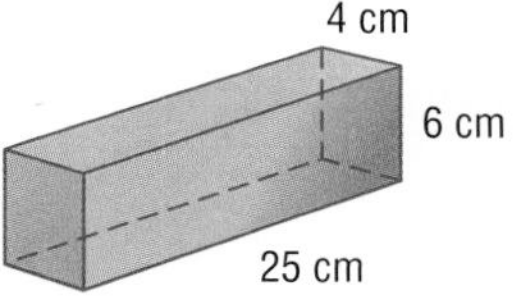

2.

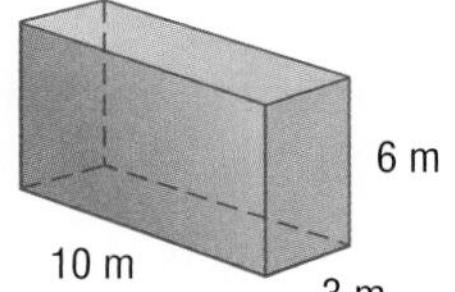

3.

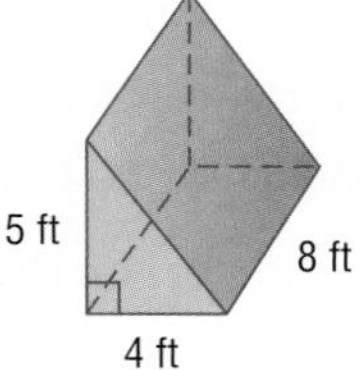

4. rectangular prism: length $2\frac{1}{2}$ yd, width 7 yd, height 12 yd

5. triangular prism: base of triangle 3.1 cm, altitude of triangle 1.7 cm, height of prism 5.0 cm. Round to the nearest tenth.

Lesson 12-3 Volume of Cylinders (pp. 677–681)

Find the volume of each cylinder. Round to the nearest tenth.

1. 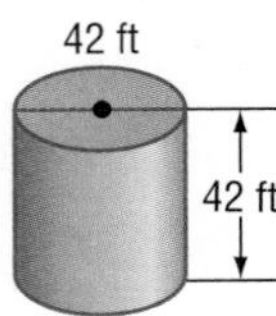

2. 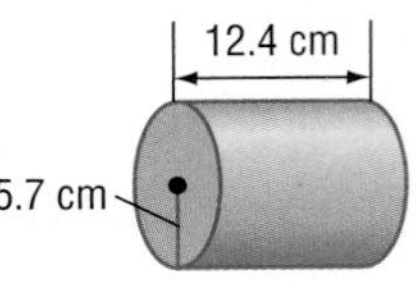

3. diameter: 6 mm
height: 13 mm

Find the height of each cylinder. Round to the nearest tenth.

4. volume: 128.41 cm^3
radius: 2.4 cm

5. volume: 208.81 cm^3
radius: 6.2 cm

6. volume: 2575.2 cm^3
radius: 8.1 cm

7. **PET CARE** Tina has an old fish tank in the shape of a cylinder. The tank is 2 feet in diameter and 6 feet high. How many cubic feet of water does it hold? Round to the nearest tenth.

Lesson 12-4 Volume of Pyramids, Cones, and Spheres (pp. 683–688)

Find the volume of each figure. Round to the nearest tenth, if necessary.

1. 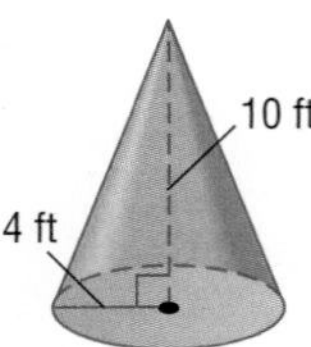

2. 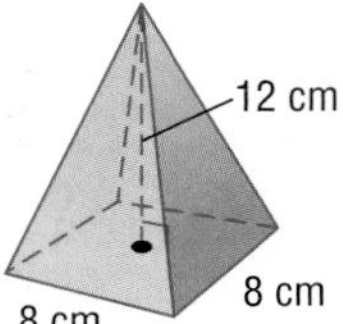

3. sphere: radius 1.2 cm
4. cone: diameter 10 yd, height 7 yd
5. rectangular pyramid: length 6 in., width 6 in., height 9 in.
6. sphere: diameter 14.4 ft
7. square pyramid: length $3\frac{1}{4}$ ft, height 12 ft
8. sphere: radius 3.1 cm

Lesson 12-5 Surface Area of Prisms (pp. 691–695)

Find the lateral area and surface area of each prism. Round to the nearest tenth.

1. cube: side length 14 in.
2. rectangular prism: length 5 cm, width 9 cm, height 3 cm
3. cube: side length 6 ft
4. cube: side length 4.9 m
5. rectangular prism: length 7.6 mm, width 8.4 mm, height 7.0 mm
6. triangular prism: right triangle 3 in. by 4 in. by 5 in., height of prism 10 in.

Lesson 12-6 Surface Area of Cylinders (pp. 697–701)

Find the lateral area and surface area of each cylinder. Round to the nearest tenth.

1. 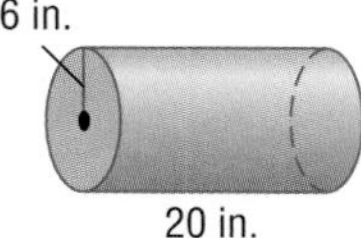

2. 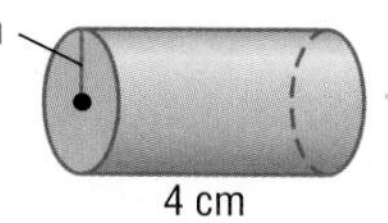

3. radius of 4.2 m and a height of 10 m

4. diameter of 8 m and a height of 12 m

5. diameter of 5 cm and a height of 5 cm

6. radius of 1.88 in. and a height of 3.9 in.

7. radius of 4.3 mm and a height of 12.1 mm

8. diameter of 4.5 yd and a height of 3.75 yd

Lesson 12-7 Surface Area of Pyramids and Cones (pp. 702–707)

Find the lateral area and surface area of each figure. Round to the nearest tenth.

1. 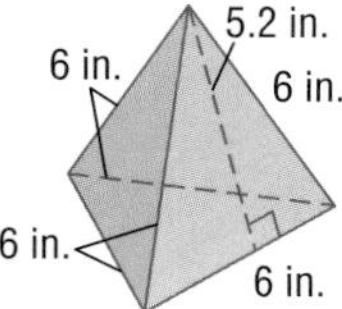

2. 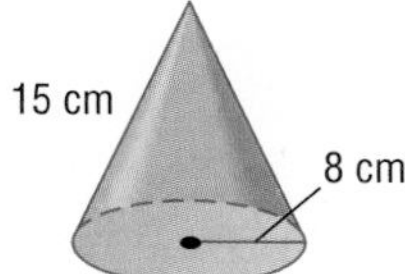

3. 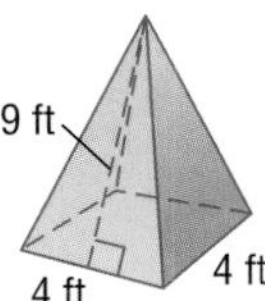

4. 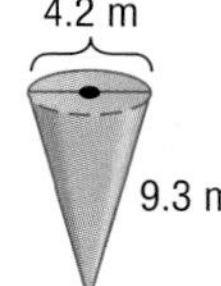

5. square pyramid: base side length 1.8 mm, slant height 3.0 mm

6. cone: radius 4 in., slant height 7 in.

7. cone: diameter 15.2 cm, slant height 12.3 cm

Lesson 12-8 Similar Solids (pp. 709–715)

Determine whether each pair of solids is similar.

1. 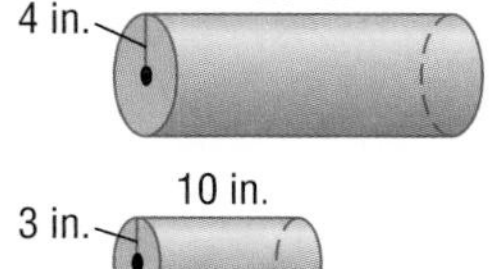

2. 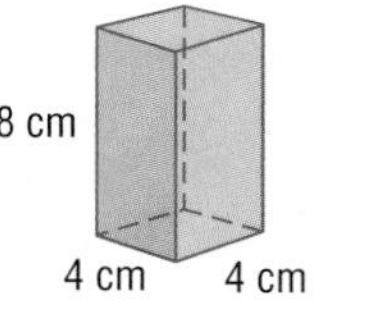

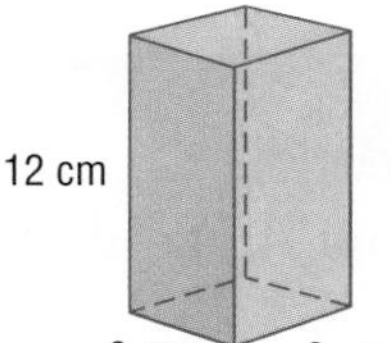

Find the missing measure of each pair of similar solids.

3. 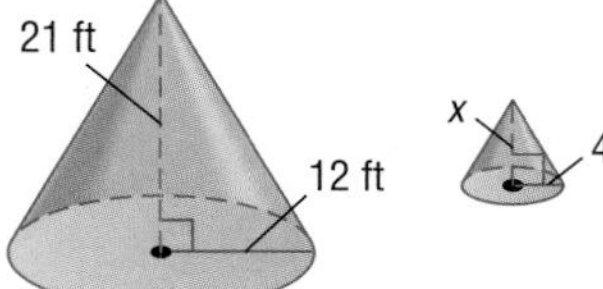

4. 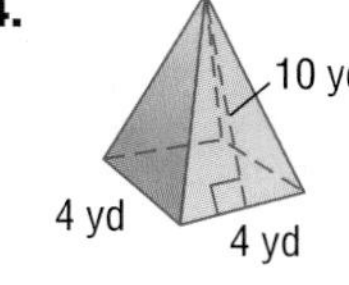

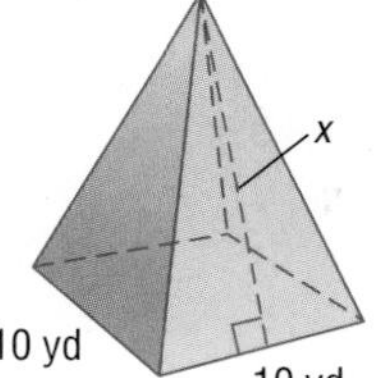

Lesson 13-1 Measures of Central Tendency (pp. 730–735)

Find the mean, median, and mode for each set of data. Round to the nearest tenth, if necessary.

1. quiz scores: 82, 79, 93, 91, 95
2. ages of people in an aerobics class: 23, 32, 19, 27, 41, 21, 26, 32, 23

3. **Students with Blue Eyes**

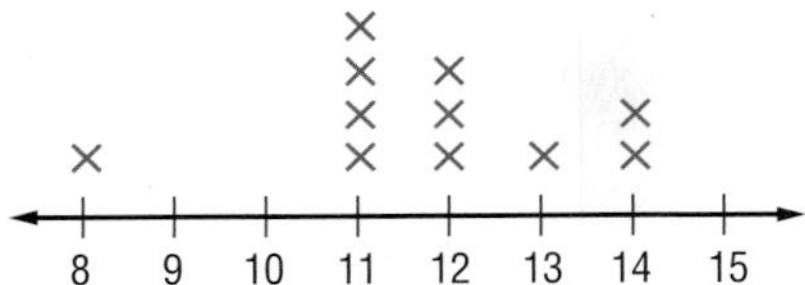

4. **Lengths of Snakes (meters)**

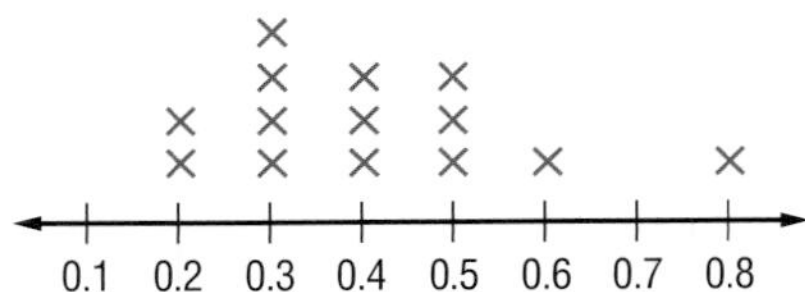

7. **POPULATION** The population of the Canadian provinces and territories in 2006 is shown in the table. Find the mean, median, and mode of the data. Round to the nearest tenth, if necessary.

Populations (thousands)						
3332.2	4292.2	1179.7	750.5	512.5	935.8	12,630.5
138.3	7636.7	989.0	42.2	31.1	29.8	

Lesson 13-2 Stem-and-Leaf Plots (pp. 737–742)

Display each set of data in a stem-and-leaf plot.

1. ages of people in a store: 3, 26, 35, 8, 21, 24, 30, 39, 35, 5, 38
2. prices of bicycles: \$172, \$198, \$181, \$182, \$193, \$171, \$179, \$186, \$181
3. number of vendors at craft shows: 17, 54, 37, 86, 24, 69, 77, 92, 21
4. science test scores: 73, 61, 89, 67, 82, 54, 93, 102, 59, 75, 83
5. **SPORTS** The stem-and-leaf plot shows the times in seconds of people in a race.
   a. What is the greatest value?
   b. In which interval do most of the values occur?
   c. What is the median value?

Race Times

Stem	Leaf
7	2 2 3 5 9
8	0 1 1 4 6 6 8 9
9	3 4 8

9 | 4 = 94 s

Lesson 13-3 Measures of Variation (pp. 743–749)

Find the range, interquartile range, and any outliers for each set of data.

1. cost of video games: 44, 37, 23, 35, 61, 95, 49, 96
2. dollars in savings account: 271, 891, 181, 193, 711, 791, 861, 818

3. **Ages of People in a Restaurant**

Stem	Leaf
2	0 1 1 2 4 7 9
3	3 3 6 8 8 8
4	2 4 5 7 9 9
5	2 9

3 | 6 = 36

4. **Typing Speeds (wpm)**

Stem	Leaf
4	0 2 2 3 4 5 6 6 7 8
5	1 2 5 5 5 9
6	4 7 8 8
7	0 0 1 4 9 9 9
8	1 7 9
9	0 0 1 3 5

8 | 7 = 87

Lesson 13-4 Box-and-Whiskers Plots (pp. 750–755)

Construct a box-and-whisker plot for each set of data.

1. Ages of home owners: 42, 23, 31, 27, 32, 48, 37, 25, 19, 26, 30, 41, 32, 29
2. Price in dollars, of MP3 players: 124, 327, 215, 278, 109, 225, 186, 134, 251, 308, 179
3. **VOLLEYBALL** Use the box-and-whisker plot shown.

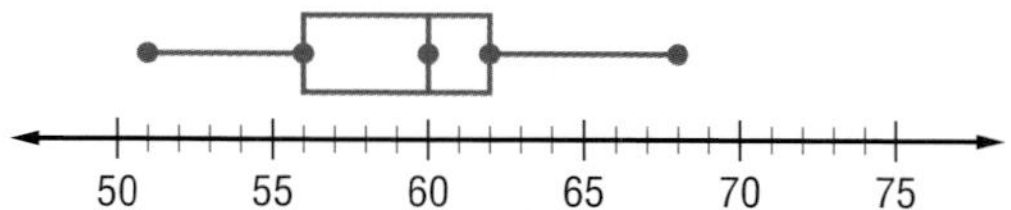

   a. What is the height of the tallest player?
   b. What percent of the players are between 56 and 68 inches tall?
   c. Explain what the length of the box-and-whisker plot tells us about the data.

Lesson 13-5 Histograms (pp. 757–762)

Display each set of data in a histogram.

1.

Weekly Exercise Time		
Time (h)	Tally	Frequency
0–2	𝍸 |||	8
3–5	||||	4
6–8	||	2
9–11	|||	3

2.

Weekly Grocery Bill		
Amount ($)	Tally	Frequency
0–49	𝍸		6
50–99	𝍸 𝍸 ||	12
100–149	𝍸 |||	8
150–199	||||	4
200–249	||	2

Lesson 13-6 Theoretical and Experimental Probability (pp. 765–770)

There are 4 blue marbles, 6 red marbles, 3 green marbles, and 2 yellow marbles in a bag. Suppose you select one marble at random. Find the probability of each outcome. Express each probability as a fraction and as a percent. Round to the nearest tenth of a percent if necessary.

1. P(green)
2. P(blue)
3. P(red)
4. P(yellow)
5. P(neither red nor green)
6. P(red or yellow)
7. P(not orange)
8. P(neither blue nor yellow)
9. P(not red)
10. P(neither green nor yellow)
11. Suppose two number cubes are rolled. What is the probability of rolling a sum greater than 8?
12. **COOKIES** A sample from a package of assorted cookies revealed that 20% of the cookies were sugar cookies. Suppose there are 45 cookies in the package. How many can be expected to be sugar cookies?

Extra Practice

Lesson 13-7 Using Sampling to Predict (pp. 771–776)

Identify each sample as *biased* or *unbiased* and describe its type. Explain your reasoning.

1. To determine the most popular kind of fish to eat, every third person coming out of a grocery store is interviewed.
2. To determine where to hold the senior prom, a survey is taken of the entire school.
3. To determine the popularity of blogging, members of the Blog Society are polled.
4. To determine whether a city's outerbelt should be widened, crews filmed traffic every six weeks at different times of day.

Lesson 13-8 Counting Outcomes (pp. 777–781)

Use the Fundamental Counting Principle to find the total number of outcomes in each situation.

1. Engagement rings come in silver, gold, and white gold. The diamond can weigh $\frac{1}{2}$ karat, $\frac{1}{3}$ karat, or $\frac{1}{4}$ karat. The diamond can have 4 possible shapes.
2. A dress can be long, tea-length, knee-length, or mini. It comes in 2 colors and the dress can have long sleeves or short sleeves.
3. The first digit of a 7-digit phone number is a 2. The last digit is a 3.
4. Three coins are tossed. What is the probability of three tails?
5. Two six-sided dice are rolled. What is the probability of getting an odd sum?
6. A ten-sided die is rolled and a coin is tossed. Find the probability of the coin landing on tails and the die landing on a number greater than 3.

Lesson 13-9 Permutations and Combinations (pp. 783–788)

1. **STUDENT COUNCIL** Seven people are running for four seats on student council. How many ways can the students be elected?
2. **LETTERS** How many ways can the letters of the word ISLAND be arranged?
3. **CANDLES** How many ways can five candles be arranged in three candlesticks?
4. **RACES** How many ways can six students line up for a race?
5. **GEOMETRY** Determine the number of line segments that can be drawn between any two vertices of a pentagon.

Lesson 13-10 Probability of Compound Events (pp. 790–795)

A card is drawn from a deck of cards numbered 6–19. Find each probability.

1. P(13 or even)
2. P(13 or less than 7)
3. P(even or odd)
4. P(14 or greater than 20)
5. P(even or less than 10)
6. P(odd or greater than 10)

Mixed Problem Solving

Chapter 1 The Tools of Algebra (pp. 2–57)

1. **FOOD** The Spanish Club is hosting a luncheon for all the students. The cost for each item is shown below. (Lesson 1-1)

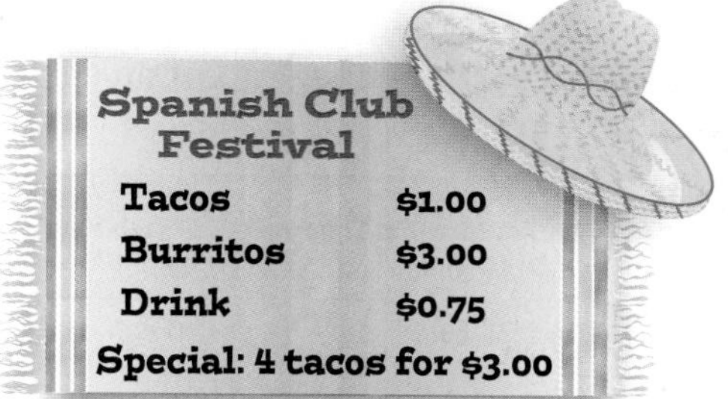

   a. Write an expression for the cost of a meal that includes one of each item.

   b. Inez is buying lunch for her friends. If she buys 3 tacos, 2 burritos and 3 drinks, what is the total cost?

2. **FOOTBALL** Admission to a high school football game is $8. At the game, raffle tickets are sold for $0.50 each. Write an expression showing the cost for admission and t tickets. Then find the total cost if 20 tickets are purchased. (Lesson 1-2)

3. **RIDES** The number of people who have ridden a ride and the number of ride tickets required at the fair are shown. (Lesson 1-2)

Rides	Number of People	Number of Tickets
Ferris wheel	f	3
Vortex	v	3
Hurricane	h	5

   a. Write an algebraic expression to show how many total ride tickets have been collected.

   b. Joseph rode the Ferris wheel twice, Vortex four times, and Hurricane three times. How many tickets did he use?

4. **HOMEWORK** After school, Morgan usually has a snack and works on her homework. Are the actions commutative? Explain. (Lesson 1-3)

5. **MOVIES** The cost of renting one movie from an online movie rental company is $3.50. (Lesson 1-4)

   a. Make a table of ordered pairs in which the x-coordinate represents the number of movies and the y-coordinate represents the cost for renting 1, 2, 3, 4, and 5 movies.

   b. Graph the ordered pairs. Then describe the graph.

6. **MULTIPLE REPRESENTATIONS** Pandas eat about 240 pounds of bamboo every three days. (Lesson 1-5)

   a. **ALGEBRAIC** Write an equation that can be used to find the pounds of bamboo p a panda will eat in any number of days d.

   b. **TABULAR** Make a function table to find the pounds of bamboo a panda will eat in 5, 7, 10, and 13 days.

   c. **GRAPHICAL** Graph the ordered pairs.

7. **SPORTS** The table shows the number of gold medals won by the United States Olympic team at the Summer Olympics from 1964 through 2004. (Lesson 1-6)

Year	1964	1968	1972	1976	1984
Number of Medals	36	45	33	34	83
Year	**1988**	**1992**	**1996**	**2000**	**2004**
Number of Medals	36	37	44	38	36

   a. Make a scatter plot of the data.

   b. Draw a conclusion about the type of relationship the data shows. Explain.

   c. Is it possible to predict the number of gold medals that will be won in 2016? Explain.

Chapter 2 Operations with Integers (pp. 58–115)

Mixed Problem Solving

1. **PLANETS** The maximum surface temperatures for different planets are shown in the table below. Order the temperatures from least to greatest. (Lesson 2-1)

Planet	Maximum Surface Temperature °C
Earth	58
Mars	−5
Mercury	427
Neptune	−214
Jupiter	−148
Uranus	−216

Source: NASA

2. **STOCKS** During a three-day period, a stock's price increased $2.03 the first day, then decreased $1.75 the second day and $0.89 the third day. Find the final price of the stock if the starting price was $97.86. (Lesson 2-2)

3. **BUSINESS SENSE** In January, Myers Construction Company had a net worth of −$12,061. By November of the same year, the company's net worth had increased by $33,385. What was the company's net worth in November? (Lesson 2-2)

4. **AVIATION** A commercial jet has a usual cruising altitude of 34,000 feet. Because of a storm, it climbs 2,500 feet and then descends 4,700 feet to its new cruising altitude. What is the jet's new cruising altitude? (Lesson 2-3)

5. **BANKING** Hasan wrote checks and made the deposits shown in his check registry. What was the change in his balance after these transactions? (Lesson 2-3)

CK #	Name	Amt	Dep
145	Car Payment	$177	
	Deposit		$130
146	Cell Phone	$38	
147	Gasoline	$76	
	Deposit		$82
148	Rent	$204	

6. **ENVIRONMENT** The erosion rate at Emerald Isle in North Carolina is about 2 feet per year. If erosion continues at the current rate, what integer represents the amount of erosion at Emerald Isle in 13 years? (Lesson 2-4)

7. **MOUNTAINS** Every 1000 feet above Earth's surface, air temperature decreases 5°F. How much would the temperature change from sea level to the highest point on each mountain? (Lesson 2-4)

Mountain	Highest Point (ft)
Mt. McKinley	20,320
Mt. St. Elias	18,008
Mt. Foraker	17,400

8. **AMUSEMENT PARK** After 6 seconds, the Dynamic Drop was −15 feet from the top of the ride. How many feet per second did the ride drop? (Lesson 2-5)

9. **GOLF** The first round scores for a golf tournament are shown in the table below. What was the average score for the round? (Lesson 2-5)

Player	Score
Sara	−4
Cherylynn	−6
Juanita	+2
Mia	−1

10. **HOCKEY** A hockey team scored a total of 5 goals in the first two periods of play. (Lesson 2-6)

    a. If x represents the number of goals in the first period, and y represents the number of goals in the second period, make a function table of possible values for x and y.

    b. Graph the ordered pairs and describe the graph.

11. **ANIMATION** Animators can use translations to show movement. Suppose the animator draws a ball with a center at (1, 5). Describe the translation if the next screen places the center of the ball at (0, −2). (Lesson 2-7)

1. **SCIENCE** A regular incandescent light bulb can have a life of 750 hours while a compact fluorescent bulb (CFB) can have a life of 10,000 hours. To the nearest thousandth, what part of the life of a CFB is the life of an incandescent bulb? (Lesson 3-1)

2. **COOKING** The side of a cookie box lists the following nutritional information about one cookie:

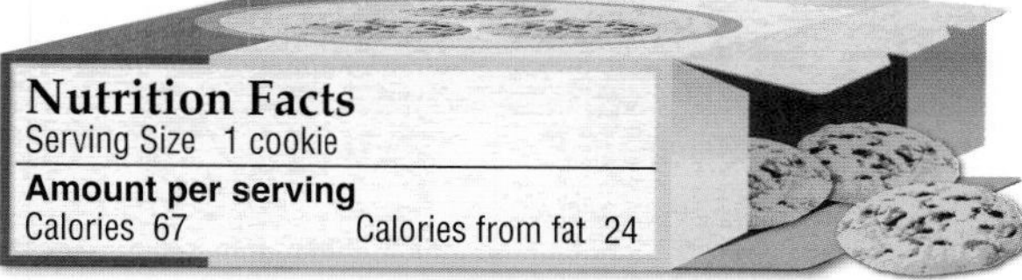

Nutrition Facts
Serving Size 1 cookie
Amount per serving
Calories 67
Calories from fat 24

To the nearest hundredth, what part of the total Calories are Calories from fat? (Lesson 3-1)

3. **BASKETBALL** At basketball practice yesterday, Zachary made $0.8\overline{6}$ of his free throw shots. What fraction of the shots did he miss? (Lesson 3-2)

4. **BASEBALL** A player's batting average is found by dividing the number of hits by the total number of times at bat. A player has a batting average of .225. Write the player's batting average as a fraction in lowest terms. (Lesson 3-2)

5. **RADIO STATIONS** In a recent survey $\frac{2}{3}$ of the people surveyed said they listen to Top 40 radio stations. Of these, $\frac{3}{4}$ said they listen to station WABC. What fraction of the people surveyed listen to station WABC? Write in simplest form. (Lesson 3-3)

6. **MUSIC** Gracie has a collection of 32 CDs, each in its own case. Each CD case is $\frac{5}{8}$ inch thick. (Lesson 3-3)

   a. If Gracie puts them on a stack on the floor, how tall would the stack be?

   b. Can she store the CDs on a 2-foot shelf in her room? Explain your reasoning.

7. **FOOD** A gallon of milk contains 128 ounces. Jung's favorite souvenir glass holds $10\frac{1}{2}$ ounces of milk. How many of these glasses of milk can Jung pour from one gallon? (Lesson 3-4)

8. **COINS** A nickel is approximately $\frac{2}{25}$-inch thick. How many nickels will fit in a 4-inch storage tube? (Lesson 3-4)

9. **MAIL** Mrs. Hamre wants to mail the packages shown. What is the total weight of the packages? Write in simplest form. (Lesson 3–5)

Package	Weight (oz)
1	$8\frac{7}{16}$
2	$5\frac{1}{16}$

10. **SEWING** Kelli wants to sew trim around the sleeves and hem of a dance costume. She needs a piece of trim that is $8\frac{1}{8}$ inches long for the sleeves and a piece that is $40\frac{5}{8}$ inches long for the hem. If the package contains 5 feet of trim, how much trim will she have left after making the costume? (Lesson 3-5)

11. **GARDENING** Kia wants to place a fence around her garden. The fencing material comes in a roll that is $40\frac{1}{2}$ feet long. If the width of the garden is $8\frac{7}{8}$ feet, what is the maximum length she can make her garden? (Lesson 3-6)

12. **CONSTRUCTION** Members of the Drama Club are building a set for a play. Rob needs a piece of wood $7\frac{3}{4}$ feet long for the front and a piece $4\frac{5}{8}$ feet long for the side. The wood comes in 12-foot lengths. After he cuts off the piece for the front, will Rob have enough left over for the side? Explain your reasoning. (Lesson 3-6)

1. **MUSEUMS** A group of 10 friends are taking a trip to the Rock and Roll Hall of Fame. The cost of admission and transportation per person is \$20. Lunch will cost each person \$8.00. Use mental math to find the total amount they will spend. Justify your answer by using the Distributive Property. (Lesson 4-1)

2. **GEOMETRY** The perimeter of a triangle can be found by adding the lengths of its sides. (Lesson 4-2)
   a. Write an expression for the perimeter of the triangle below.

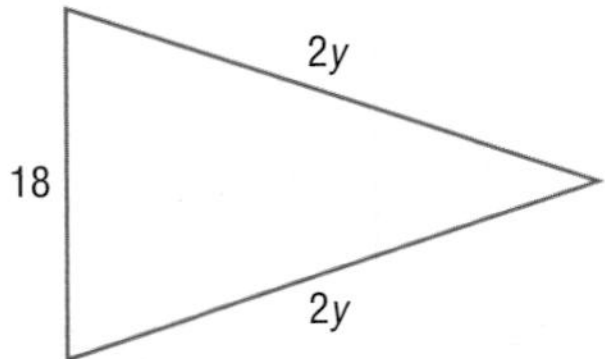

   b. If $y = 12$, what is the perimeter of the triangle?

3. **SHOPPING** Pilar purchased a new computer. She made an initial payment of \$70 and will make monthly payments of \$45 for x months. Write an expression to show the total amount Pilar will pay for the computer. (Lesson 4-2)

4. **TRAVEL** A city had an increase of 126,925 tourists from June to July. If there were 3.3 million tourists in July, how many were there in June? Write and solve an addition equation to find the number of tourists in June. (Lesson 4-3)

5. **SOCCER** During a recent season, Alex attempted 35 shots on goal and made 28 goals. Write and solve a subtraction equation to find the number of times he did not score. (Lesson 4-3)

6. **MONEY** Hector wants to donate an equal amount of money to four different charities. As part of a fundraiser, he raised a total of \$420 for charity. Write and solve a multiplication equation to find how much each charity will receive. (Lesson 4-4)

7. **TIME** Sebastian spent 1.25 hours practicing the piano on Wednesday. Write and solve a division equation to find how many minutes he spent practicing on Wednesday. (Lesson 4-4)

8. **MUSIC** An online music company advertises the rates shown in the table below. Sherita has \$30 to pay the membership fee and download songs. (Lesson 4-5)

Type of Fee	Cost (\$)
Membership	\$8.75
Song Download	\$0.85

   a. Solve the equation $0.85s + 8.75 = 30$ to find the number of songs she can download and have no money left over.
   b. If the membership fee increases to \$11.30, how many songs can she download?

9. **COLLECTIONS** Mathew and Aaron both collect baseball cards. Mathew has 66 cards in his collection which is 12 more than twice the number of cards in Aaron's collection. Solve the equation $12 + 2a = 66$ to find the number of cards in Aaron's collection. (Lesson 4-5)

10. **GEOMETRY** The perimeter of the figure below is 38 inches. Write and solve an equation to find the value of x. (Lesson 4-6)

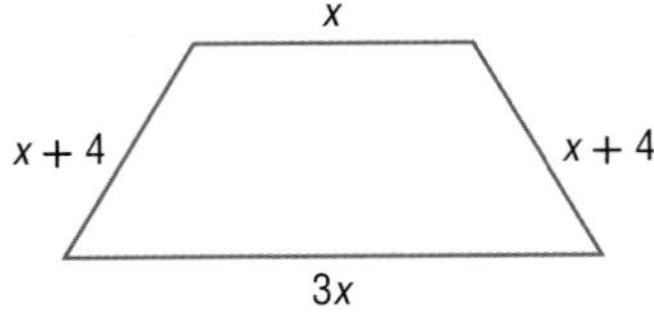

11. **CONSTRUCTION** The Trans-Pacific Express project involves laying 11,000 miles of telecommunications cable under the Pacific Ocean from China to the United States. The cost for the project will be approximately \$500 million. Write and solve an equation to find the cost per mile to the nearest dollar. (Lesson 4-6)

1. **ARCHITECTURE** The Reflecting Pool in between the Washington Monument and the Lincoln Memorial is 167 feet wide and 2029 feet long. What is the perimeter of the Reflecting Pool? (Lesson 5-1)

2. **FLOORS** The dining room in Mariah's house has a hardwood floor that is partially covered by a rug as shown below. If the rug measures 9 feet by 7.5 feet, how much of the hardwood floor is *not* covered by the area rug? (Lesson 5-1)

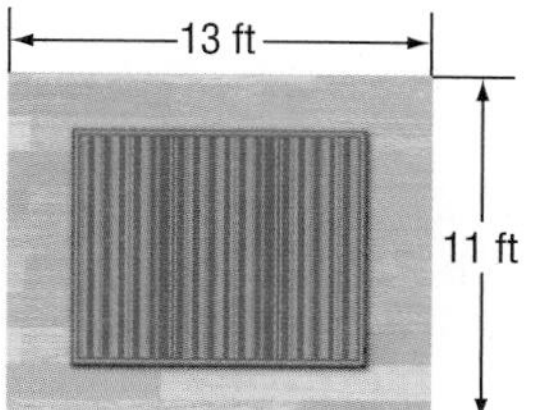

3. **SAVINGS** Theresa and Miranda are each saving money for a cruise. Theresa has already saved $500 and plans to deposit $40 each month. Miranda has $200 in her account and will deposit $60 each month. Write and solve an equation to find how many months it will take for them to have saved the same amount of money. (Lesson 5-2)

4. **CELL PHONES** The Happy Talk Cell Phone company offers two different rate plans for teens. Write and solve an equation to determine for what number of minutes will the costs be equal. (Lesson 5-2)

Happy Talk Cell Phone Rates		
	Monthly Fee	Cost per Minute
Option A	$25	$0.08
Option B	$15	$0.10

5. **RAINFALL** Mt. Waialeale in Hawaii receives the most amount of rainfall with an average rainfall of at least 397 inches. Write an inequality to describe the amount of rainfall. (Lesson 5-3)

6. **TICKETS** The Student Council is planning a dance. The expenses for the dance are shown below. If they sell tickets for $3.50, write and solve an inequality to find the minimum number of tickets they will have to sell to make a profit. (Lesson 5-4)

Autumn Dance Expenses	
Item	Cost($)
Gym Rental	75
DJ	165
Food and Drink	205
Cups, Napkins, etc.	35
Security	110

7. **MONEY** Jin earns $2350 per month plus $45 for each sale he completes. Write and solve an inequality to find how many sales he would have to make each month in order to earn at least $3000. (Lesson 5-4)

8. **SHOPPING** The table below shows the cost of school supplies at a local store.

Item	Cost	Number Purchased
notebook	$3.00	$2b$
box of pens	$1.50	b
package of paper	$2.00	$b + 1$

If Spencer spent a total of $30.50 before tax, how many of each item did he purchase? (Lesson 5-5)

9. **TRAINING** Mimi is planning to run a marathon. To prepare for the race, she will follow the schedule below. She plans on running 11 hours per week. How many hours will she run each day? (Lesson 5-5)

Running Schedule	
Day	Length of Time
Monday	x hours
Tuesday	2 hours more than Monday
Thursday	same as Monday
Saturday	3 times as much as Monday

1. **BASKETBALL** The league statistics for a school basketball league are shown in the table below. Copy and complete the table. Express the ratios as fractions in simplest form. (Lesson 6-1)

Team	Games	Win	Loss	W/L Ratio
Cougars	35	23	12	■
Bulldogs	36	30	6	■
Bears	33	22	11	■
Hawks	34	20	14	■
Sharks	36	26	10	■

2. **PET FOOD** Two stores have Munchies Dog Food on sale this week. Pets-a-lot sells a 17.6-pound bag for $11.99. Doggie Haven sells the 31.8-pound bag for $21.99. Which store has the lower price per pound of dog food? (Lesson 6-2)

3. **SPACE SHUTTLE** At an altitude of about 250 miles, the Space Shuttle can travel at approximately 17,500 mph in an orbit around Earth. In orbit, how many feet per second does the Shuttle travel? Round to the nearest foot. (Lesson 6-3)

4. **OLYMPICS** In the 2004 Olympics in Athens, Greece, Stefano Baldini from Italy won the 26.2 mile marathon in approximately 2 hours and 11 minutes. (Lesson 6–3)
   a. What was his average speed in miles per hour? Round to the nearest tenth.
   b. What was his average speed in kilometers per hour? Round to the nearest tenth.

5. **PIZZA** Is the cost of the pizza proportional to the size of the pizzas? Explain your reasoning. (Lesson 6-4)

Papino's Pizza

Diameter (inches)	6	10	12	15
Cost	$5	$9	$11	$15

6. **COOKING** A soup recipe uses $3\frac{1}{2}$ cups of water for 8 bowls of soup. Find the amount of water Jacinda needs if she wants to make 12 bowls of soup. (Lesson 6-5)

7. **MONEY** Jeff is preparing for a trip to China. The current exchange rate is 60 Chinese yuan for $8 US. (Lesson 6-5)
   a. How many Chinese yuan are in $1 US?
   b. If Jeff's hotel in China charges 482 yuan per night, how much would that be in US dollars?

8. **MODELS** Tony is making a scale model of his house for his granddaughter's doll's house. The front porch on Tony's house is 25 feet long. The front porch on the doll's house is 20 inches long. What is the scale of the model? (Lesson 6-6)

9. **PHOTOGRAPHS** A computer program can dilate photos so they print smaller or larger than the original. (Lesson 6-8)
   a. What scale factor should Jamie use to dilate a 4 inch by 6 inch photo so it will print as a 5 inch by 7.5 inch photo?
   b. What scale factor should he use to make a 2 inch by 3 inch print?

10. **SHADOWS** Georgio wants to know how tall the tree is that grows in his backyard. His brother Mario is 4 feet tall and casts a shadow that measures 1.5 feet long. At the same time, the tree casts a shadow 4.5 feet long. How tall is the tree? (Lesson 6-9)

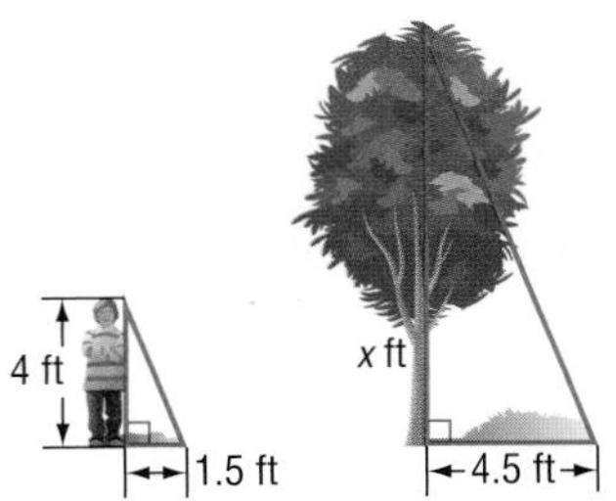

Percent (pp. 328–391)

1. **ALPHABET** The Hawaiian alphabet uses only 12 letters: A, E, H, I, K, L, M, N, O, P, U and W. What percentage of the English alphabet does the native Hawaiian alphabet use? Round to the nearest tenth of a percent. (Lesson 7-1)

2. **JOBS** A recent survey of students with part-time jobs showed that 33% worked in a grocery store, $\frac{7}{20}$ worked in fast food, and 0.27 worked in childcare. Which of these jobs has the largest percentage of student workers? (Lesson 7-2)

3. **EYES** The table below shows the eye colors of the students in Mr. Lehman's class. Use the percent proportion to find the percent of students with blue eyes. (Lesson 7-3)

Eye Color	Number of Students
blue	9
brown	5
green	6
hazel	4

4. **GRATUITY** The bill for the Morgan family at a restaurant was $62.14. Mr. Morgan would like to leave their server a 20% tip. About how much of a tip will the server receive? (Lesson 7-4)

5. **RECYCLING** In a community of 3200 homes, 72% of the households participate in the community recycling program. How many households is this? (Lesson 7-5)

6. **ADS** An ad for a portable DVD player was listed in this week's newspaper.

SPECIAL OFFER!
Portable DVD Player
Only $99.99
*After $20 in-store saving and $20 mail-in rebate.

What percent will a shopper save by buying during this sale? (Lesson 7-6)

7. **HOMES** The figures for home sales for 2010 and 2011 in Shore County are shown. (Lesson 7-6)

Shore County Home Sales

Year	Number of Homes Sold	Average Selling Price
2010	18,328	$220,988
2011	19,226	$219,748

a. Find the percent of change in the homes sold. Round to the nearest tenth.

b. Find the percent of change in the average selling price. Round to the nearest tenth.

8. **SAVINGS** Kira's parents started a college fund by depositing $5500 in a savings account at an interest rate of 5.75% for 3 years. Find the simple interest. (Lesson 7-7)

9. **MOVIES** Jacob surveyed his classmates to learn their favorite types of movies. Construct a circle graph of the data. (Lesson 7-8)

Type of Movie	Number of Students
animated	3
comedy	10
drama	7
horror	4
romance	4

10. **PODCASTS** The circle graph shows the results of a survey about the ages of people who listen to podcasts. Suppose 1500 people were surveyed. How many more people aged 18–34 listen to podcasts than people aged 45 or older? (Lesson 7-8)

Podcast Users

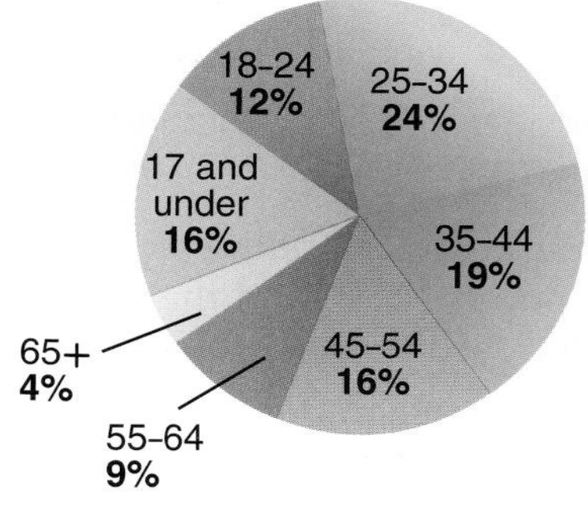

Source: Edison Media Research

1. **EXERCISE** Maria ran 17 miles at an average speed of 5 miles per hour. (Lesson 8-1)
   a. Use function notation to write an equation that gives the total distance she ran as a function of the total time.
   b. Use the function to find the total time she ran.

2. **SALARY** Julio makes \$215 per week at a shoe store plus an additional \$5 for every pair of shoes he sells. (Lesson 8-2)
   a. Write an equation that describes the relationship between Julio's total earnings e and the number of pairs of shoes he sells s.
   b. Determine how much money he would make if he sells 24 pairs of shoes.

3. **MOVIES** Lucia and her friends are going to Mario's Movies to rent m movies and buy b boxes of popcorn. Find three solutions of $5m + 1.25b = 20$. Explain each solution. (Lesson 8-3)

4. **AIRPLANES** The table below shows the altitude of an airplane during take off. Find the rate of change. Round to the nearest hundredth. (Lesson 8-4)

Airplane Flight

Time (seconds)	Altitude (feet)
0	0
30	113
60	226

5. **WATER** The amount of water used in a community varies directly with the population. About 18 million people living in Florida use 2.4 trillion gallons of water a year. (Lesson 8-5)
   a. Write a direct variation equation relating the population p and the amount of water used w.
   b. Estimate the amount of water that will be needed for 24 million people.

6. **CONSTRUCTION** A roofing contractor needed to find the slope of the roof on a house. The roof rises 24 inches for every horizontal change of 20 inches. Find the slope of the roof. (Lesson 8-6)

7. **FITNESS** A gym charges a \$59 initiation fee plus \$7.70 per week that a person attends. The total cost y can be given by $y = 7.7x + 59$ for x weeks. (Lesson 8-7)
   a. Graph the equation using the slope and y-intercept.
   b. State the slope and y-intercept of the graph of the equation.
   c. Describe what the slope and y-intercept represent.

8. **SCUBA DIVING** At a depth of 30 meters, a scuba diver noted the pressure was 58.8 pounds per square inch (psi). At a depth of 70 meters, the pressure was 117.6 psi. (Lesson 8-8)
   a. Write an equation in slope-intercept form to represent the data.
   b. Describe what the slope means.
   c. Find the diver's depth if the pressure is 107.3 psi.

9. **CONCERTS** The table below shows attendance at an annual Fourth of July concert. (Lesson 8-9)
   a. Make a scatter plot and draw a line of fit.
   b. Use the equation for the line of fit to predict the number of attendees in 2010.

Year	2002	2003	2004	2005	2006
Attendees	850	925	1000	1200	1350

10. **SHOPPING** Elena spent \$15 for 3 magazines and 2 puzzle books. At the same store, Christopher spent \$16 for 2 magazines and 4 puzzle books. (Lesson 8-10)
    a. Write a system of equations to represent this situation.
    b. Solve the system of equations. Explain what the solution means.

1. **BUSINESS** A popular beverage company has about $(2^3)(3)(5^4)$ locations worldwide. How many locations do they have? (Lesson 9-1)

2. **DISTANCE** The state of Hawaii is located approximately 2,400 miles from the California coast and 3,850 miles from Japan. Write the prime factorization of each distance. (Lesson 9-2)

3. **TRAVEL** The table compares the number of people who drove to work versus the number of people who walked to work in Wyoming in a recent year. How many times more people drove than walked to work? (Lesson 9-3)

Mode of Transportation	Number of People
Drove	10^5
Walked	10^3

4. **DIME** A dime has a thickness of approximately 0.001 meter. Write this decimal as a fraction and as a power of ten. (Lesson 9-4)

5. **PHYSICS** The length of an infrared light wave is approximately 0.0000037 meter. Write this number in scientific notation. (Lesson 9-5)

6. **PLANETS** The table shows the mass of the planets. Arrange the planets in order from least to greatest. (Lesson 9-5)

Mass of the Planets	
Planet	**Mass (kg)**
Mercury	3.303×10^{23}
Venus	4.869×10^{24}
Earth	5.976×10^{24}
Mars	6.421×10^{23}
Jupiter	1.900×10^{27}
Saturn	5.688×10^{26}
Uranus	8.686×10^{25}
Neptune	1.024×10^{26}

7. **HAIR** The average width of a human hair is 4×10^{-3} centimeter. If the cross section of the average hair is round, use the formula $A = 3.14 \cdot r^2$ to find the area of the cross section of a hair. Write your answer in scientific notation. (Lesson 9-6)

8. **SAVINGS** Contessa makes monthly deposits in her savings account. A record of her deposits is shown below. Do these data represent a *linear* or *nonlinear* function? Explain. (Lesson 9-7)

Month	Deposit
January	30
February	35
March	40
April	45
May	50

9. **FLOWER BED** Teresa is planning to make a rectangular flower garden. She wants the length of the garden to be 1.5 times the width. (Lesson 9-8)

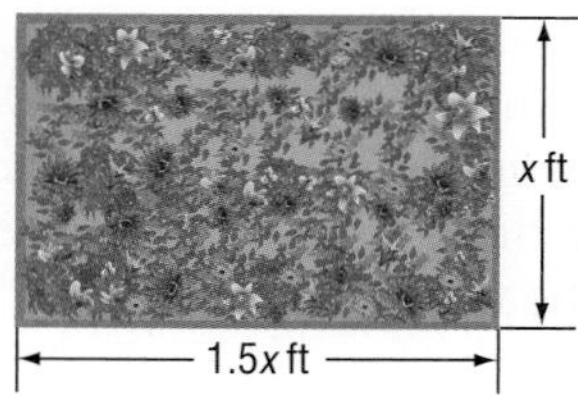

a. Write and graph a function that gives the area of the garden for different widths and lengths.

b. What is the area of the garden if it is 6 feet wide?

10. **MAIL** Monique sent a chain letter to three of her friends. Each of her three friends sends the letter to three of their friends. Each of those friends sends it to three friends, and so on. The function $N = 3^x$ represents the total number of letters sent, where x is the stage of letters. (Lesson 9-9)

a. Make a table showing the number of people that will have received the letter after 1, 2, 3, and 4 stages.

b. Graph the function.

c. After how many stages will 243 people receive the letter?

Chapter 10 Real Numbers and Right Triangles (pp. 532–585)

1. **HORIZON** The formula $d = \sqrt{\frac{h}{0.57}}$ where h represents the number of feet above sea level can be used to find the distance d in miles to the horizon. Ella, standing on a deck of a cruise ship, is approximately 80 feet above sea level. Estimate how far out to the horizon she can see. (Lesson 10-1)

2. **BELL TOWER** The Bell Tower in the Piazza San Marco in Venice, Italy is 99 meters tall. If a person drops a coin from the top of the tower, how many seconds would it take the coin to reach the ground? Use the formula $t = \sqrt{\frac{d}{4.9}}$ where t is the time in seconds and d is the distance in meters. Round your answer to the nearest tenth. (Lesson 10-1)

3. **EARTH** If you could cut Earth into two pieces at the equator, the area of a cross section of Earth would be approximately 127,796,483 square kilometers. Use the formula $A = \pi r^2$ to find the radius of Earth. (Lesson 10-2)

4. **TRAVEL** The table shows the distance between Cincinnati, Ohio, and two other cities. Suppose a triangle was formed by drawing a line between each pair of cities. Classify the triangle by its sides. (Lesson 10–3)

Distance from Cincinnati	
City	Distance (mi)
Minneapolis	692
Omaha	692

5. **POOL TABLE** The playing surface on a regulation 8-foot billiards table measures 88 inches by 44 inches. How far would a ball travel from one corner to the opposite corner? (Lesson 10-4)

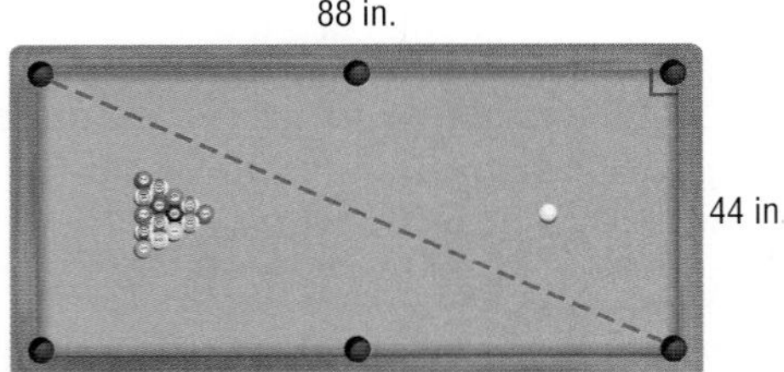

6. **TELEVISION** Katy learned that the 21-inch dimension on a television is the approximate diagonal of the screen. About how tall would the screen on a 21-inch television be if the screen is 16.5 inches wide? Round your answer to the nearest tenth. (Lesson 10-4)

7. **GAMES** In a popular strategy game a submarine has vertices (4, 8), a carrier ship has vertices (3, 1), and a battleship has vertices (11, 7). Is the submarine closer to the carrier or the battleship? Explain your reasoning. (Lesson 10-5)

8. **DISTANCE** Trey's home is at (4, 9) on the map. His friend Nicolas' home is at (6, 3) on the same map. If each unit on the map is 1 mile, how far do the two friends live from each other? Round to the nearest tenth. (Lesson 10–5)

9. **CONSTRUCTION** The diagram below shows the basic outline of the trusses that support a roof. The two sides of a truss meet in a 90° angle. How long will the base of the truss need to be if each of the sides measure 18 feet? Round your answer to the nearest tenth of a foot. (Lesson 10-6)

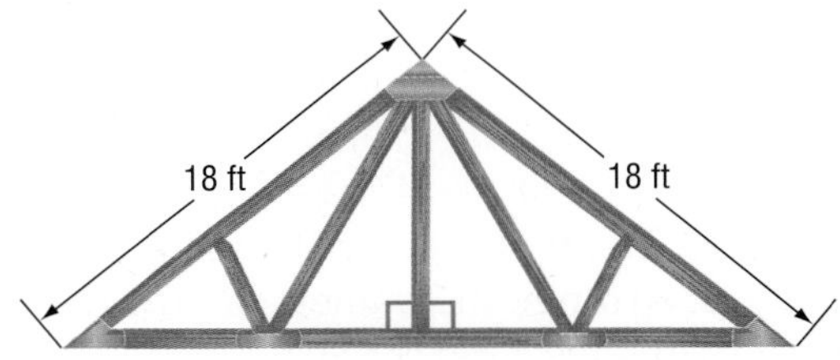

10. **UTILITIES** A telephone pole is braced by a cable attached at the top and anchored in the ground. If the angle between the cable and the ground is 60°, find the length of the cable to the nearest tenth. (Lesson 10-6)

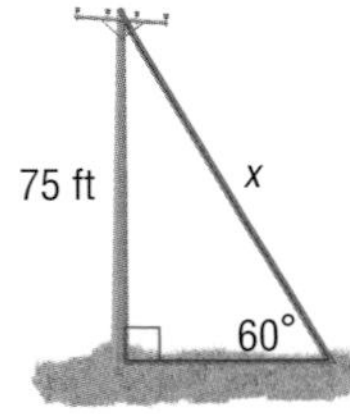

1. **FLOORING** A decorative floor has the tiling pattern shown below.

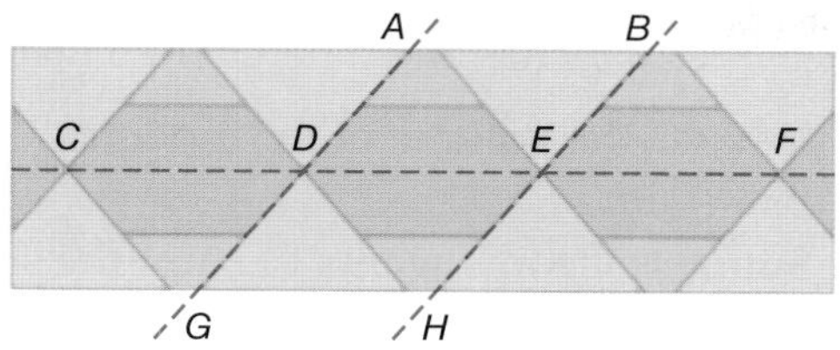

The horizontal lines are all parallel. Lines AG and BH are parallel and are transversals of the horizontal lines. If the measure of $\angle GDE = 130°$, find the measures of $\angle ADE$, $\angle ADC$, $\angle DAB$, $\angle ABE$, $\angle BEF$, and $\angle BED$. (Lesson 11-1)

2. **GEOMETRY** Triangle PQR is congruent to triangle MLK. In the figures, $m\angle P = 32°$ and $m\angle K = 56°$. What are the measures of the other angles? (Lesson 11-2)

3. **QUILTS** Determine whether the quilt pattern shown has rotational symmetry. If so, describe the angle of rotation. (Lesson 11-3)

4. **ALGEBRA** What is the measure of $\angle A$ in quadrilateral $ABCD$ if $m\angle B = 68°$, $m\angle C = 136°$, and $\angle D$ is a right angle? (Lesson 11-4)

5. **WINDOWS** Jamie is making a stained glass window from a tessellation of 12 isosceles triangles. What is the measure of $\angle x$ and $\angle y$? (Lesson 11-5)

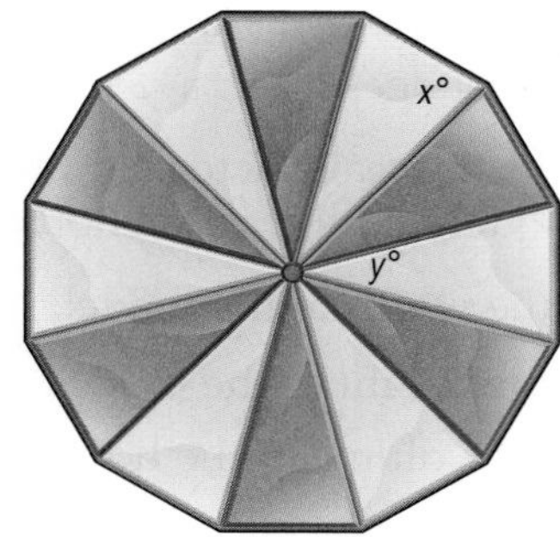

6. **SAILS** Mr. Johnson wants to replace the sail on his sailboat. What is the area of the sail shown below? (Lesson 11-6)

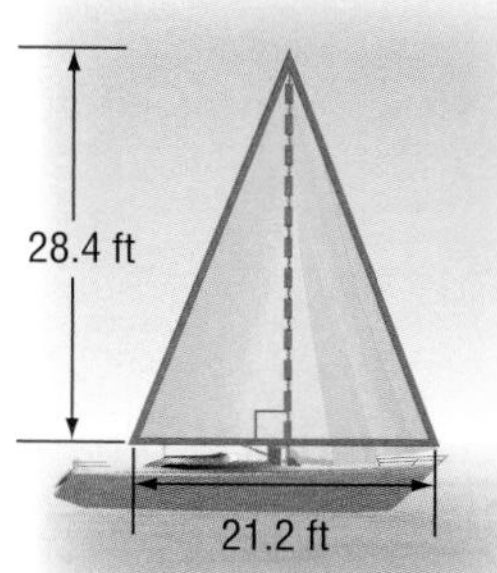

7. **FOUNTAINS** A landscape architect is designing a circular fountain to be placed in front of an office building. The fountain will have a decorative stone path around the outside of the fountain as shown.

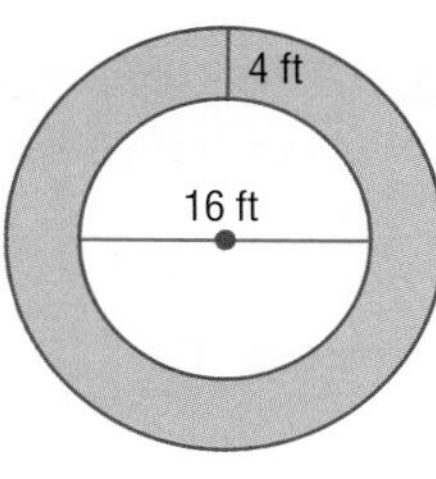

   a. What is the circumference of the space occupied by the fountain and the path to the nearest tenth? (Lesson 11-7)
   b. How many square feet of space are needed to build the fountain and the path? Round to the nearest tenth. (Lesson 11-8)
   c. If the stone for the path costs $1.75 per square foot, what is the total cost of stone for the walkway? (Lesson 11-9)

8. **FLOORING** Nigel wants to buy new outdoor carpeting for his wrap-around porch. A model of his porch is shown. (Lesson 11-9)

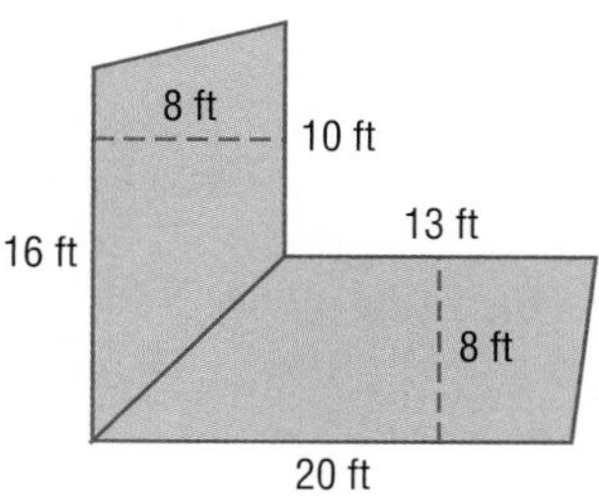

   a. Find the area of the porch.
   b. If the carpet costs $1.25 per square foot, how much will it cost to carpet the entire porch?

Mixed Problem Solving

1. **CAMPING** Jack uses the tent shown when he goes camping.

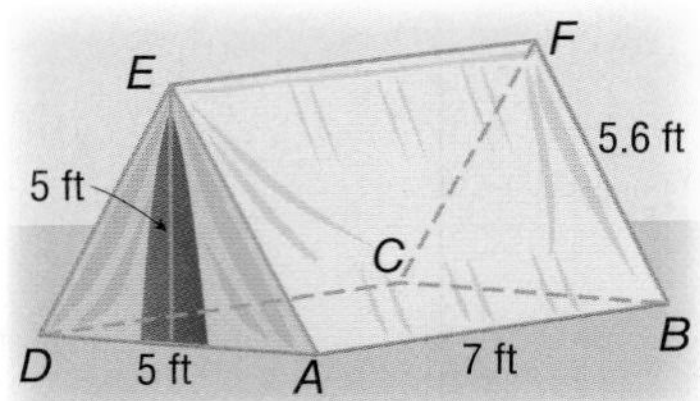

a. Identify the solid. Then name the bases, faces, edges, and vertices. (Lesson 12-1)

b. Find the volume of the tent. (Lesson 12-2)

c. How much fabric was needed to make the tent? (Lesson 12-5)

2. **GARDENING** Alexa wants to buy a greenhouse like the one shown below for her backyard. What is the volume of the greenhouse to the nearest foot? (Lesson 12-2)

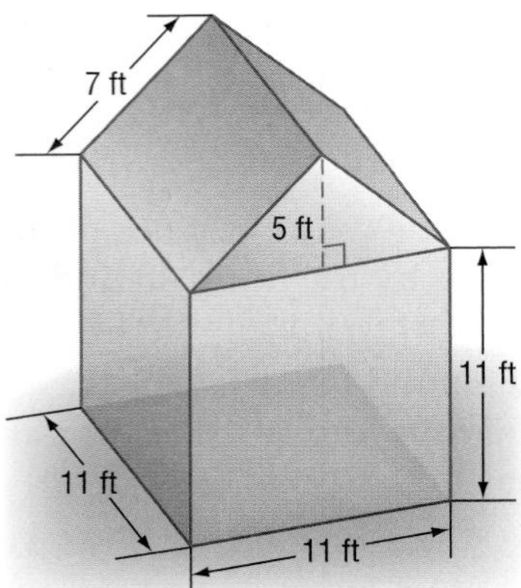

3. **AVIARY** The bird aviary shown is shaped like a cylinder with a flat top.

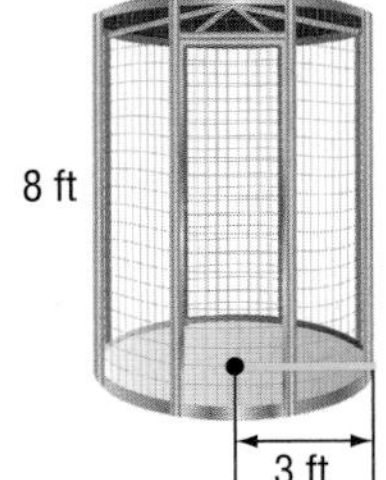

a. What is the volume of the aviary to the nearest tenth of a foot? (Lesson 12-3)

b. What is the surface area of the aviary to the nearest tenth of a foot? (Lesson 12-6)

4. **BALLOONS** A giant inflatable balloon used for a parade is spherical in shape. If the balloon has a diameter of 45 feet, how many cubic feet of helium are needed to fill the balloon? Round to the nearest tenth. (Lesson 12-4)

5. **ART** George is making a model of the Great Pyramid of Giza for a social studies project. The model has a square base with sides that measure 10 inches and a height of 6 inches.

a. What is the volume of the model? (Lesson 12-4)

b. Suppose George wants to paint the sides of the pyramid. If the slant height of the pyramid is 7.8 inches, what is the total area of the sides of the pyramid? (Lesson 12-7)

6. **SNOW SALT** The cone-shaped pile of snow salt shown has a base with a diameter of 12 meters and a height of 8 meters.

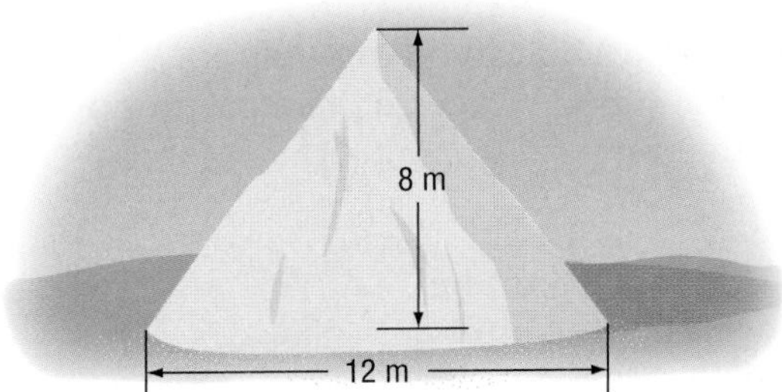

a. How many cubic meters of snow salt are in the pile? Round to the nearest tenth. (Lesson 12-4)

b. The pile of salt is covered with a tarp when the salt is not being used. What is the area of the smallest possible tarp needed to cover the entire pile of salt? Round to the nearest tenth. (Lesson 12-7)

7. **MODELS** Mr. Walker is renovating an old farmhouse. He wants to build a model of the farmhouse to use for his planning process. The main part of the original farm house is 25 feet wide, 30 feet deep and 20 feet tall. The model of the farm house will be 2 feet deep. (Lesson 12-8)

a. What is the scale factor between the model and the actual house?

b. Find the width of the model.

c. Find the height of the model.

d. Find the volume of the house.

e. Find the volume of the model.

1. **TESTS** Mrs. Thomas teaches two Pre-Algebra classes. The scores for the two classes are shown below.

First Period									
92	98	96	84	90	82	64	76	70	80
72	90	90	98	74	72	86	88	90	82
Second Period									
90	98	76	74	92	80	82	84	92	74
70	60	80	92	94	78	80	90	78	82

a. Find the mean, median and mode for each class. (Lesson 13-1)

b. Make a back-to-back stem-and-leaf plot for the data. (Lesson 13-2)

c. Find the measures of variation for the two classes. (Lesson 13-3)

2. **OLYMPICS** The table below shows the number of gold and silver medals won by athletes in past summer Olympic games. (Lesson 13-4)

United States Summer Olympic Medals								
Gold	35	40	44	37	36	83	34	33
Silver	39	24	32	34	31	61	35	31

a. Draw box-and-whisker plots for the gold and silver medals.

b. Which set of data has the greatest range?

c. In what percent of the years did the U.S. earn more than 40 gold medals?

3. **SUMMER WAGES** The table below shows the results of a survey about wages earned by students during the summer. (Lesson 13-5)

Hourly Rate ($)	Number of Students
6–7.99	6
8–9.99	29
10–11.99	23
12–13.99	12
14–15.99	10

a. Construct a histogram of the data.

b. About what percent of the students made $10 or more per hour?

c. Which hourly rate accounts for about 15% of the data?

4. **SOCKS** Of the 28 socks in Becca's sock drawer, 8 are black, 6 are brown, 4 are gray, and the rest are white. What is the probability she will pull a white sock out of the drawer without looking? Write in simplest form. (Lesson 13-6)

5. **PIZZA** The Walnut Springs Middle School cafeteria staff wants to know which types of pizza to serve. They survey every 5th student in the cafeteria line about their pizza preference. Is this sample biased or unbiased? Then describe the sample's type. (Lesson 13-7)

6. **TRAVEL** You want to go on a summer vacation. A travel agency offers a vacation package allows you to choose one hotel, one car, and two activities from the list below. How many ways can you plan your vacation? (Lesson 13-8)

Hotels	Cars	Activities
Sandy Beach	SUV	Parasailing
Ocean Waves	Convertible	Fishing
The Dunes		Surfboarding
		Snorkeling

7. **BASKETBALL** Coach Camaruca has 14 players on her junior varsity basketball team. She wants to try different groups of players to see how they work together. In how many ways can she pick her team of 5? (Lesson 13-9)

8. **ALPHABET** The Hawaiian alphabet consists of 12 letters: A, E, H, I, K, L, M, N, O, P, U, and W. Each letter is written on a block and placed in a bag. A block is randomly chosen from the bag and not replaced. Find the probability of each outcome. Write in simplest form. (Lesson 13-10)

a. P(vowel)

b. P(vowel or L)

c. three vowels in a row

d. three consonants in a row

e. one consonant followed by two vowels

Concepts and Skills Bank

1 Factors

Two or more numbers that are multiplied to form a product are called **factors.**

$$4 \times 9 = 36$$

factors → 4, 9; product → 36

So, 4 and 9 are factors of 36 because they each divide 36 with a remainder of 0. We can say that 36 is **divisible** by 4 and 9. However, 5 is not a factor of 36 because $36 \div 5 = 7$ with a remainder of 1.

Sometimes you can test for divisibility mentally. The following rules can help you determine whether a number is divisible by 2, 3, 5, 6, or 10.

Divisibility Rules		
A number is divisible by...	Examples	Reasons
2 if the ones digit is divisible by 2.	54	4 is divisible by 2.
3 if the sum of its digits is divisible by 3.	72	$7 + 2 = 9$, and 9 is divisible by 3.
5 if the ones digit is 0 or 5.	65	The ones digit is 5.
6 if the number is divisible by 2 and 3.	48	48 is divisible by 2 and 3.
10 if the ones digit is 0.	120	The ones digit is 0.

EXAMPLE 1

Determine whether each number is divisible by 2, 3, 5, 6, or 10.

a. 138

Number	Divisible?	Reason
2	yes	8 is divisible by 2.
3	yes	$1 + 3 + 8 = 12$, and 12 is divisible by 3.
5	no	The ones digit is 8, not 0 or 5.
6	yes	138 is divisible by 2 and 3.
10	no	The ones digit is not 0.

So, 138 is divisible by 2, 3, and 6.

b. 3050

Number	Divisible?	Reason
2	yes	0 is divisible by 2.
3	no	$3 + 0 + 5 + 0 = 8$, and 8 is not divisible by 3.
5	yes	The ones digit is 0.
6	no	3050 is divisible by 2, but not 3.
10	yes	The ones digit is 0.

So, 3050 is divisible by 2, 5, and 10.

You can also use the rules for divisibility to find the factors of a number.

EXAMPLE 2

List all the factors of 72.

Use the divisibility rules to determine whether 72 is divisible by 2, 3, 5, and so on. Then use division to find other factors of 72.

Number	72 Divisible by Number?	Factor Pairs
1	yes	1 • 72
2	yes	2 • 36
3	yes	3 • 24
4	yes	4 • 18
5	no	
6	yes	6 • 12
7	no	
8	yes	8 • 9
9	yes	9 • 8

Use division to find the other factor in each factor pair. $72 \div 2 = 36$

You can stop finding factors when the numbers start repeating.

So, the factors of 72 are 1, 2, 3, 4, 6, 8, 9, 12, 18, 24, 36, and 72.

Exercises

Use divisibility rules to determine whether each number is divisible by 2, 3, 5, 6, or 10.

1. 39 **2.** 135 **3.** 82 **4.** 120

5. 250 **6.** 118 **7.** 378 **8.** 955

9. 5010 **10.** 684 **11.** 10,523 **12.** 24,640

List all the factors of each number.

13. 75 **14.** 114 **15.** 57 **16.** 65

17. 90 **18.** 124 **19.** 102 **20.** 135

21. MUSIC The band has 72 students who will march during halftime of the football game. For one drill, they need to march in rows with the same number of students in each row.

a. Can the whole band be arranged in rows of 7? Explain.

b. How many different ways could students be arranged? Describe the arrangements.

22. CALENDARS Years that are divisible by 4, called *leap years,* are 366 days long. Also, years ending in "00" that are divisible by 400 are leap years. Use the rule given below to determine whether 2000, 2004, 2015, 2018, 2022, and 2032 are leap years.

If the last two digits form a number that is divisible by 4, then the number is divisible by 4.

2 Greatest Common Factor

Often, numbers have some of the same factors. The greatest number that is a factor of two or more numbers is called the **greatest common factor (GCF).**

EXAMPLE 1 Find the GCF

Find the GCF of 12 and 20.

Method 1 List the factors.

factors of 12: 1, 2, 3, (4), 6, 12

factors of 20: 1, 2, (4), 5, 10, 20

The common factors of 12 and 20: 1, 2, 4. The greatest one is 4.

Method 2 Use prime factorization.

$12 = (2) \cdot (2) \cdot 3$
$20 = (2) \cdot (2) \cdot 5$

The common prime factors of 12 and 20: 2, 2

The GCF is the product of the common prime factors.

$2 \cdot 2 = 4.$

The greatest common factor of 12 and 20 is 4.

EXAMPLE 2 Find the GCF

TRACK AND FIELD **There are 208 boys and 240 girls participating in a field day competition.**

a. What is the greatest number of teams that can be formed if each team has the same number of girls and each team has the same number of boys?

Find the GCF of 208 and 240.

$208 = (2) \cdot (2) \cdot (2) \cdot (2) \cdot 13$
$240 = (2) \cdot (2) \cdot (2) \cdot (2) \cdot 3 \cdot 5$

Write the prime factorization of each number. The common prime factors are 2, 2, 2, and 2.

The greatest common factor of 208 and 240 is $2 \cdot 2 \cdot 2 \cdot 2$ or 16. So, 16 teams can be formed.

b. How many boys and girls will be on each team?

$208 \div 16 = 13$ and $240 \div 16 = 15$

So, each team will have 13 boys and 15 girls

Factor Algebraic Expressions You can also find the GCF of two or more monomials by finding the product of their common prime factors.

EXAMPLE 3 Find the GCF of Monomials

Find the GCF of $16xy^2$ and $30xy$.

Completely factor each expression.

$16xy^2 = (2) \cdot 2 \cdot 2 \cdot 2 \cdot (x) \cdot (y) \cdot y$
$30xy = (2) \cdot 3 \cdot 5 \cdot (x) \cdot (y)$

Circle the common factors.

The GCF of $16xy^2$ and $30xy$ is $2 \cdot x \cdot y$

Exercises

Find the GCF of each set of numbers or monomials.

1. 6, 8
2. 12, 8
3. 3, 9
4. 24, 40
5. 21, 45
6. 16, 56
7. 28, 42
8. 21, 14
9. 20, 30
10. 12, 18
11. 42, 56
12. 30, 35
13. 9, 15, 24
14. 12, 24, 36
15. 6, 15, 24
16. 66, 90, 150
17. 20, 21, 25
18. 20, 28, 36
19. 18, $45mn$
20. $24t^2$, 32
21. $12x$, $40x^2$
22. $4st$, $10s$
23. $5ab$, $6b^2$
24. $7x^2$, $15xy$
25. $14b$, $56b^2$
26. $25k$, $35j$
27. $21x^2y$, $63xy^2$

28. The Venn diagram shows the factors of $10x$ and $18x^2$. What is the greatest common factor of the two monomials?

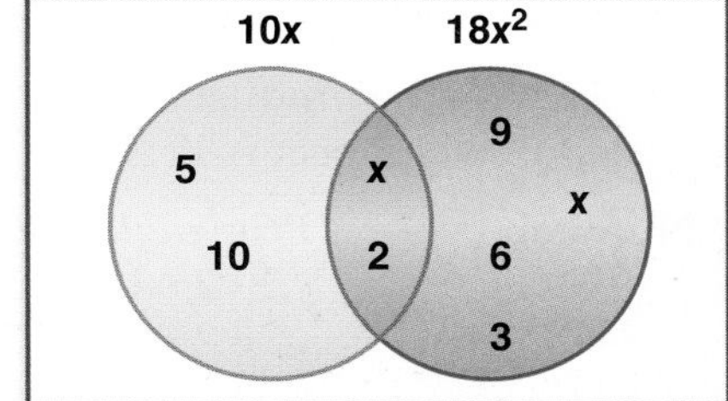

29. **FOOD** Marta is cutting a 16-inch and a 28-inch submarine sandwich for a party.

a. How long is the longest possible piece if she cuts them to be the same length?

b. How many total pieces are there?

30. **DESIGN** Lauren is covering the surface of an end table with equal-sized ceramic tiles. The table is 30 inches long and 24 inches wide.

a. What is the largest square tile that Lauren can use and not have to cut any tiles?

b. How many tiles will Lauren need?

31. **PARADES** In the parade, 36 members of the color guard are to march in front of 120 members of the high school marching band. Both groups are to have the same number of students in each row.

a. Find the greatest number of students that could be in each row.

b. How many rows will each group have?

32. **QUILTING** Suki is making a quilt from two different kinds of fabrics. One is 48 inches wide and the other is 54 inches wide. What are the dimensions of the largest squares she can cut from both fabrics so there is no wasted fabric?

33. **CARPENTRY** Tamika is helping her father make shelves to store her sports equipment in the garage. How many shelves measuring 12 inches by 16 inches can be cut from a 48-inch by 72-inch piece of plywood so that there is no waste?

34. **DECORATIONS** Terrell is cutting paper streamers to decorate for a party. He has a blue roll of paper 144 inches long, a red roll 192 inches long, and a yellow roll 360 inches long.

a. If he wants to have all colors the same length, what is the longest length that he can cut?

b. How many total streamers will he have?

3 Least Common Multiple

A **multiple** of a number is a product of that number and a whole number. Sometimes numbers have some of the same multiples. These are called **common multiples.**

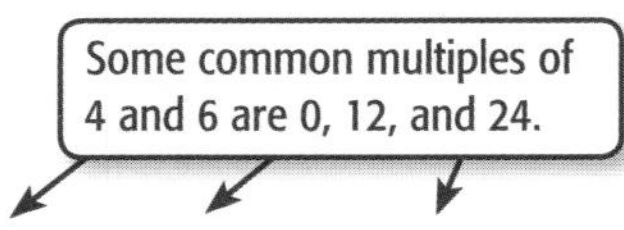

multiples of 4: 0, 4, 8, 12, 16, 20, 24, 28, . . .
multiples of 6: 0, 6, 12, 18, 24, 30, 36, 42, . . .

The least of the *nonzero* common multiples is called the **least common multiple (LCM).** So, the LCM of 4 and 6 is 12.

You can use prime factorization to find the least common multiple.

EXAMPLE 1 Find the LCM

Find the LCM of 108 and 240.

The LCM is the product of the prime factors, with each one raised to the *highest* power it occurs in *either* prime factorization.

Number	Prime Factorization	Exponential Form
108	$2 \cdot 2 \cdot 3 \cdot 3 \cdot 3$	$2^2 \cdot 3^3$
240	$2 \cdot 2 \cdot 2 \cdot 2 \cdot 3 \cdot 5$	$2^4 \cdot 3 \cdot 5$

The prime factors of both numbers are 2, 3, and 5. Multiply the greatest power of 2, 3, and 5 appearing in either factorization.

LCM = $2^4 \cdot 3^3 \cdot 5$ or 2160

So, the LCM of 108 and 240 is 2160.

EXAMPLE 2 Use the LCM to Solve Problems

ANIMALS The scales show the average weight of three different kinds of fish.

a. At what minimum weight will the scales show equal weights for each fish?

Find the LCM using prime factors.

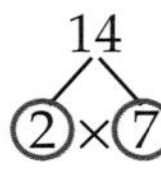

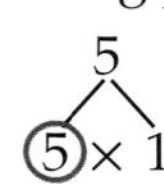

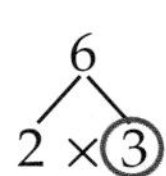

The scales will show the same weight at $2 \times 7 \times 5 \times 3$ or 210 pounds.

b. How many of each type of fish will show that weight?

You can find the number of fish by multiplying the prime factors that appear in the weight of the scale and *do not* appear in the weight of the fish.

Pink Salmon: 2×7 pounds   Scale: $2 \times 3 \times 5 \times 7$ pounds
You would need 3×5 or 15 pink salmon.

Bluegill: 1×5 pounds   Scale: $2 \times 3 \times 5 \times 7$ pounds
You would need $2 \times 3 \times 7$ or 42 bluegill.

Artic Grayling: 2×3 pounds   Scale: $2 \times 3 \times 5 \times 7$ pounds
You would need 5×7 or 35 Arctic grayling

The LCM of two or more monomials is found in the same way as the LCM of two or more numbers.

EXAMPLE 3 Find the LCM of Monomials

Find the LCM of $18xy^2$ and $10y$.

$18xy^2 = 2 \cdot 3^2 \cdot x \cdot y^2$

$10y = 2 \cdot 5 \cdot y$

$\text{LCM} = 2 \cdot 3^2 \cdot 5 \cdot x \cdot y^2$ **Multiply the greatest power of each prime factor.**
$= 90xy^2$

The LCM of $18xy^2$ and $10y$ is $90xy^2$.

Exercises

Find the least common multiple (LCM) of each pair of numbers or monomials.

1. 6, 8
2. 7, 9
3. 10, 14
4. $36ab$, $4b$
5. $5x^2$, $12y^2$
6. $14e^3$, $8e^2$
7. 4, 10
8. 20, 12
9. 2, 9
10. 16, 3
11. 15, 75
12. 21, 28
13. 14, 28
14. 20, 50
15. 18, 32
16. 24, 32
17. $20c$, $12c$
18. $16a^2$, $14ab$
19. $7x$, $12x$
20. $75n^2$, $25n^4$
21. $20ef$, $52f^3$

22. AUTO RACING One driver can circle a one-mile track in 30 seconds. Another driver takes 20 seconds. If they both start at the same time, in how many seconds will they be together again at the starting line?

23. LIGHTS Sahale decorated his house with two strands of holiday lights. The red lights strand blinks every 4 seconds, and the green lights strand blinks every 6 seconds. How many seconds will go by until both strands blink at the same time?

24. TIME If train A and train B both leave the station at 9:00 A.M., at what time will they next leave the station together?

Train Schedule	
Train	**Departs**
A	every 8 minutes
B	every 6 minutes

25. PRIZES A radio station is giving away 2 concert tickets to every sixteenth caller and a dinner for two to every twentieth caller. Which caller will receive both the concert tickets and the dinner?

26. SCIENCE The cycle for the appearance of cicadas and tent caterpillars is shown. Suppose the peak cycles of these two insects coincided in 1998. What will be the next year in which they will coincide?

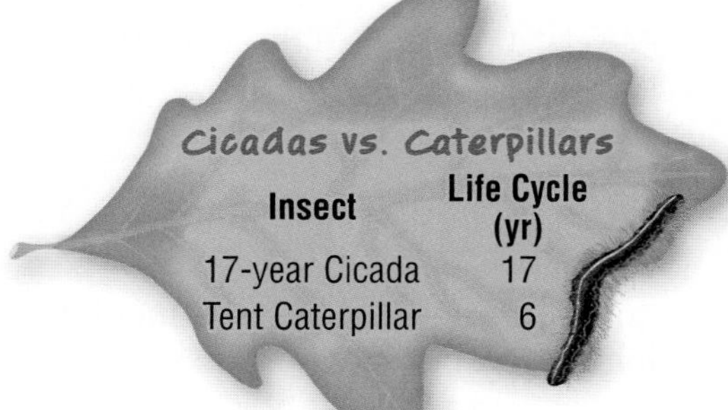

Cicadas vs. Caterpillars	
Insect	**Life Cycle (yr)**
17-year Cicada	17
Tent Caterpillar	6

4 Cube Roots

Numbers like 27, 125, and 1000 are perfect cubes because they are the cubes of integers.

$27 = 3 \times 3 \times 3$ or 3^3

$125 = 5 \times 5 \times 5$ or 5^3

$1000 = 10 \times 10 \times 10$ or 10^3

A **cube root** of a number is one of three equal factors of the number.

Key Concept — Cube Roots

For Your FOLDABLE

Words A cube root of a number is one of its three equal factors.

Symbols If $x^3 = y$, then $x = \sqrt[3]{y}$.

Examples Since $2 \times 2 \times 2$ or $2^3 = 8$, 2 is a cube root of 8.
Since $-6 \times (-6) \times (-6) = -216$, -6 is a cube root of -216.

Recall that a negative number has no real-number square root because the square of a number cannot be negative. The multiplication problem $(-6) \times (-6) \times (-6) = (-216)$ suggests that $\sqrt[3]{-216} = -6$

Every integer has exactly one cube root.

- The cube root of a positive number is positive.
- The cube root of zero is zero.
- The cube root of a negative number is negative.

EXAMPLE 1

Use a calculator to find each cube root to the nearest tenth.

a. $\sqrt[3]{150}$

[MATH] 4 150 [ENTER] 5.313292846 — Use a calculator.

Round to the nearest tenth.

$\sqrt[3]{150} \approx 5.3$ — **Check** Since $5^3 = 125$, the answer is reasonable.

b. $\sqrt[3]{-32}$

[MATH] 4 (–)32 [ENTER] –3.174802104 — Use a calculator.

Round to the nearest tenth.

$\sqrt[3]{-32} \approx -3.2$ — **Check** Since $(-3)^3 = -27$, the answer is reasonable.

You can also estimate cube roots mentally by using perfect cubes. The first twelve perfect cubes are shown at the right.

$1 = 1^3$ $8 = 2^3$ $27 = 3^3$	$64 = 4^3$ $125 = 5^3$ $216 = 6^3$	$343 = 7^3$ $512 = 8^3$ $729 = 9^3$	$1000 = 10^3$ $1331 = 11^3$ $1728 = 12^3$

EXAMPLE 2

Estimate each cube root to the nearest tenth. Do not use a calculator.

a. $\sqrt[3]{83}$

The first perfect cube less than 83 is 64. $\sqrt[3]{64} = 4$

The first perfect cube greater than 83 is 125. $\sqrt[3]{125} = 5$

Approximate the placement of $\sqrt[3]{83}$ on a number relative to 4 and 5. Since 83 is closer to 64 than 125, $\sqrt[3]{83}$ will be closer to 4.

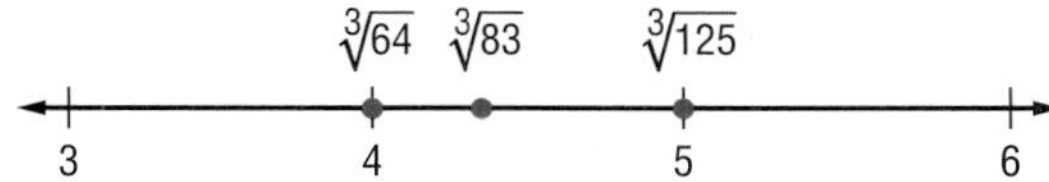

On the graph, it appears to be about 4.4. **Check** $4.4^3 = 4.4 \cdot 4.4 \cdot 4.4 = 85.184$

Since 83 is close to 85.184, the answer is reasonable.

b. $\sqrt[3]{-195}$

The first perfect cube less than −195 is −216. $\sqrt[3]{-216} = -6$

The first perfect cube greater than −195 is −125. $\sqrt[3]{-125} = -5$

Approximate the placement of $\sqrt[3]{-195}$ on a number relative to −6 and −5. Since −195 is closer to −216 than −125, $\sqrt[3]{-195}$ will be closer to −6.

$\sqrt[3]{-216}$ $\sqrt[3]{-195}$ $\sqrt[3]{-125}$

3 −6 −5 −4

On the graph, it appears to be about −5.8. **Check** $-5.8^3 = (-5.8) \cdot (-5.8) \cdot (-5.8) = -195.112$

Since −195 is close to −195.112, the answer is reasonable.

Exercises

Use a calculator to find each cube root to the nearest tenth.

1. $\sqrt[3]{20}$
2. $\sqrt[3]{35}$
3. $\sqrt[3]{-113}$
4. $\sqrt[3]{-92}$
5. $\sqrt[3]{563}$
6. $\sqrt[3]{854}$
7. $\sqrt[3]{1236}$
8. $\sqrt[3]{2024}$
9. $\sqrt[3]{4635}$

Estimate each cube root to the nearest tenth. Do not use a calculator.

10. $\sqrt[3]{74}$
11. $\sqrt[3]{39}$
12. $\sqrt[3]{499}$
13. $\sqrt[3]{576}$
14. $\sqrt[3]{-636}$
15. $\sqrt[3]{-879}$

16. Express the length of one side of a cube whose volume is 350 cubic meters. Round to the nearest tenth.

5 Linear Regression

Recall that a line of fit is a line that is very close to most of the data points in a scatter plot. A **linear regression equation** is an approximate equation describing a line of fit. You can use a graphing calculator to find a linear regression equation for a line of best-fit.

ACTIVITY 1

POSTAGE The table shows the cost to mail a first-class letter from 1978 to 2007. Find and graph a linear regression equation. Then predict the cost of a stamp in the year 2015.

Year	1978	1981	1985	1988	1991	1995	1999	2001	2002	2006	2007
Cost (¢)	15	20	22	25	29	32	33	34	37	39	41

Step 1 Find a regression equation.

- Clear any existing list.

  KEYSTROKES: STAT ENTER ▲ CLEAR ENTER

- Enter the years in **L1** and the cost in **L2**. Find the regression equation.

  KEYSTROKES: STAT ENTER 1978 ENTER . . . 2007 ENTER ▶ 15 ENTER . . . 41 ENTER STAT ▶ 4 ENTER

The equation is about $y = 0.82x - 1613.91$.

Step 2 Graph the regression equation.

- Use **STAT PLOT** to graph the scatter plot.

  KEYSTROKES: 2nd [STAT PLOT] ENTER ENTER

- Copy the equation to the **Y=** list and graph.

  KEYSTROKES: Y= VARS 5 ▶ ▶ 1 GRAPH

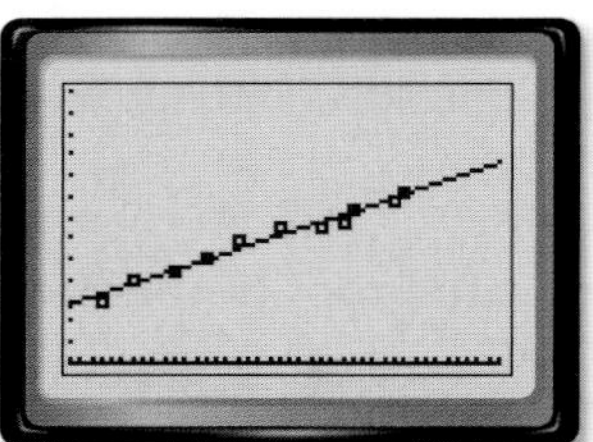

Step 3 Predict using the regression equation.

- Find y when $x = 2015$.

  KEYSTROKES: 2nd [CALC] 1 2015 ENTER

According to the regression equation, in 2015 a first-class stamp will cost about 47 cents.

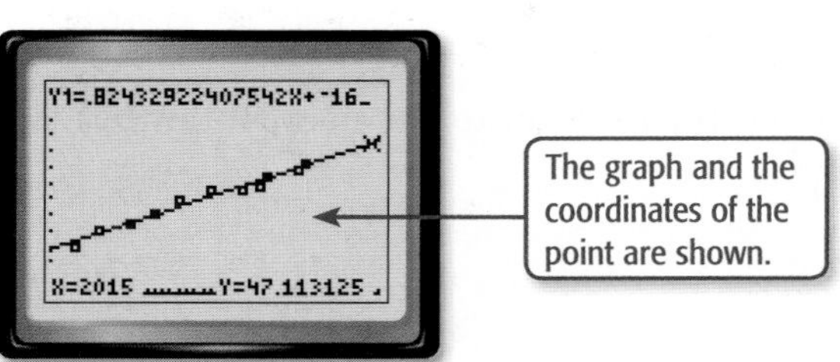

ACTIVITY 2

SWIMMING The table shows the winning times for the women's 4 by 100-meter freestyle swimming relay in the summer Olympics from 1964–2004. Find and graph a linear regression equation. Then predict the winning time in the year 2016.

Year	1964	1968	1972	1976	1980	1984	1988	1992	1996	2000	2004
Time (min)	4.06	4.04	3.92	3.75	3.71	3.72	3.68	3.66	3.35	3.6	3.6

Source: *ESPN Sport Almanac*

Step 1 **Find a regression equation.**

LinReg
y=ax+b
a=-.0135909091
b=30.69981818

- Clear any existing list.

  KEYSTROKES: STAT ENTER ▲ CLEAR ENTER

- Enter the years in **L1** and the times in **L2**. Find the regression equation.

  KEYSTROKES: STAT ENTER 1964 ENTER . . . 2004 ENTER ▶ 4.06 ENTER . . . 3.6 ENTER STAT ▶ 4 ENTER

The equation is about $y = -0.01x + 30.70$.

Step 2 **Graph the regression equation.**

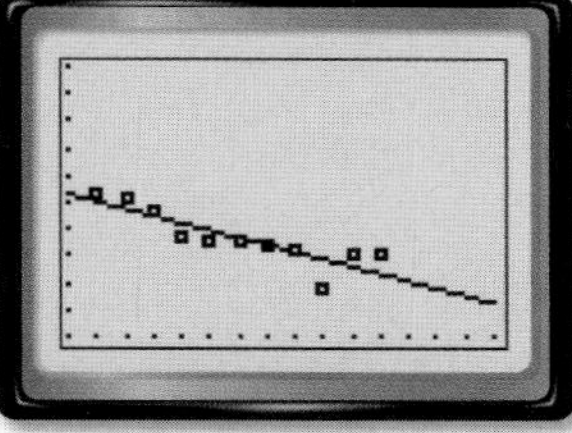

- Use **STAT PLOT** to graph the scatter plot.

  KEYSTROKES: 2nd [STAT PLOT] ENTER ENTER

- Copy the equation to the **Y=** list and graph.

  KEYSTROKES: Y= VARS 5 ▶ ▶ 1 GRAPH

Step 3 **Predict using the regression equation.**

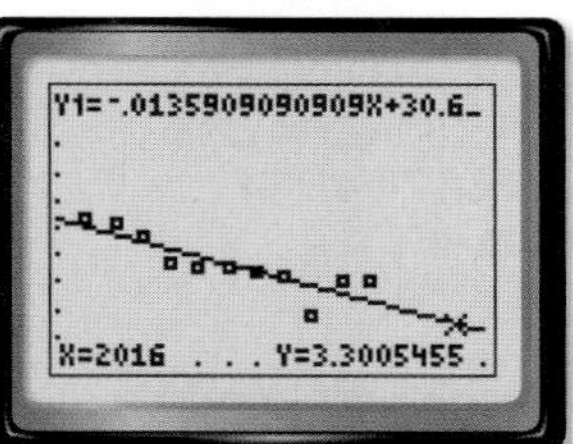

- Find y when $x = 2016$.

  KEYSTROKES: 2nd [CALC] 1 2016 ENTER

According to the regression equation, in 2016 the winning time for the 4×100 freestyle relay will be about 3.30 minutes.

Exercises

1. **CELL PHONES** The table below shows the number of cellular phone users in the United States from 1985 to 2007. Find and graph a linear regression equation. In what year will the number of cell phone users reach 300 million?

Cell Phone Users in the United States (millions)												
Year	1985	1987	1989	1991	1993	1995	1997	1999	2001	2003	2005	2007
Users	0.3	1.2	3.5	7.6	16.0	33.8	55.3	86.0	128.4	158.7	207.9	243.4

Concepts and Skills Bank

6 Graphing Linear Inequalities

Like a linear equation in two variables, the solution set of an inequality in two variables is graphed on a coordinate plane. The solution set for an inequality in two variables contains many ordered pairs when the domain and range are the set of real numbers. The graphs of all of these ordered pairs fill a region on the coordinate plane called a **half-plane**. An equation defines the **boundary** or edge for each half-plane.

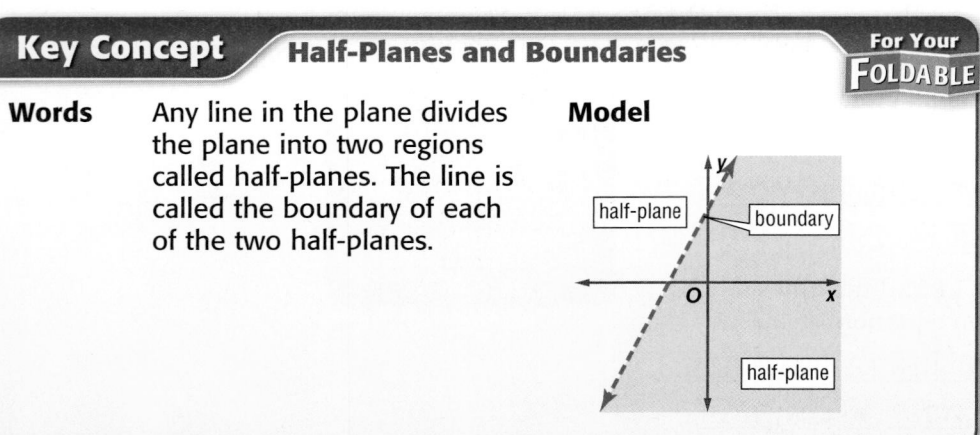

Key Concept — Half-Planes and Boundaries

For Your FOLDABLE

Words Any line in the plane divides the plane into two regions called half-planes. The line is called the boundary of each of the two half-planes.

Model

half-plane
boundary
y
O
x
half-plane

EXAMPLE 1

Graph $y \leq 2x - 1$.

Step 1 Graph $y = 2x - 1$. Since $y \leq 2x - 1$ means $y < 2x - 1$ or $y = 2x - 1$, the boundary is included in the solution set. The boundary should be drawn as a solid line.

Step 2 Select a point in one of the half-planes and test it. Let's use (0, 0).

$y \leq 2x - 1$	**Original inequality**
$0 \leq 2(0) - 1$	**$x = 0, y = 0$**
$0 \leq -1$	**false**

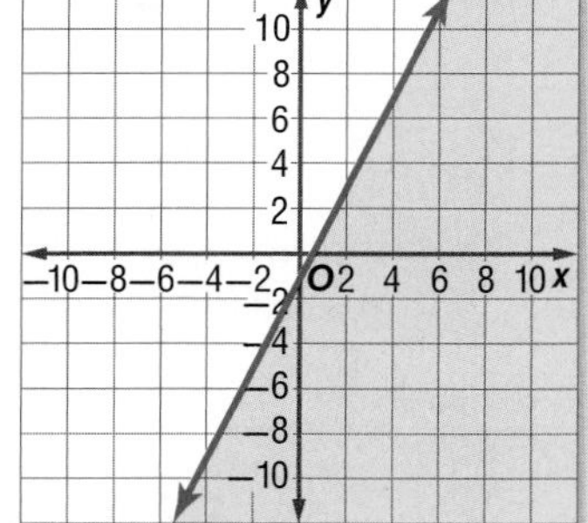

Since the statement is false, the half-plane containing the origin is *not* part of the solution. Shade the other half-plane.

Check Test a point in the other half-plane, such as (2, –2).

$y \leq 2x - 1$	**Original inequality**
$-2 \leq 2(2) - 1$	**$x = 2, y = -2$**
$-2 \leq 3$	**Simplify.** ✓

Since the statement is true, the half-plane containing (2, −2) should be shaded. The graph of the solution is correct.

In Example 1, the boundary was a solid line because $y < 2x - 1$ or $y = 2x - 1$. When the inequality symbol does not contain an "=", a dashed line on a coordinate plane indicates that the boundary is *not* part of the solution set.

EXAMPLE 2

Graph $y > \frac{1}{2}x + 2$.

Step 1 Graph $y = \frac{1}{2}x + 2$. Since $y > \frac{1}{2}x + 2$ does *not* contain an equality, the boundary is *not* included in the solution set.

The boundary should be drawn as a dashed line.

Step 2 Select a point in one of the half-planes and test it, such as (0, 0).

$y > \frac{1}{2}x + 2$ **Original inequality**

$0 > \frac{1}{2}(0) + 2$ **$x = 0, y = 0$**

$0 > 2$ **false**

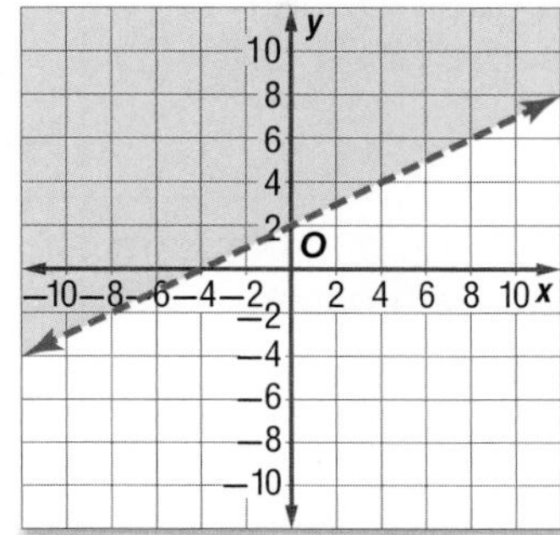

Since the statement is false, the half-plane containing the origin is *not* part of the solution. Shade the other half-plane.

Check Test a point in the other half-plane, such as (–8, 1).

$y > \frac{1}{2}x + 2$ **Original inequality**

$1 > \frac{1}{2}(-8) + 2$ **$x = -8, y = 1$**

$1 > -2$ **Simplify.** ✔

Since the statement is true, the half-plane containing (−8, 1) should be shaded. The graph of the solution is correct.

Concepts and Skills Bank

Exercises

Graph each inequality on a coordinate plane.

1. $y \leq x - 3$

2. $y \geq x - 1$

3. $y > \frac{1}{3}x$

4. $y < \frac{3}{4}x$

5. $y \geq -3x + 3$

6. $y \leq -x + 5$

7. $-4x + y < 6$

8. $4x + y \geq 5$

9. $2y \geq -9$

10. $-3y < -16$

7 Geometric Figures

The table below shows the basic geometric figures.

Key Concept Geometric Figures — For Your FOLDABLE

Definition	Model
A **point** is an exact location in space, represented by a dot.	*A* **Words** point *A*
A **line** is a set of points that form a straight path that goes in opposite directions without ending.	*M* *N* **Words** line *MN* or line *NM* **Symbols** $\overleftrightarrow{MN}$ or $\overleftrightarrow{NM}$, line ℓ
A **ray** is part of a line that has an endpoint and goes in one direction without ending.	*Y* *Z* **Words** ray *YZ* **Symbols** $\overrightarrow{YZ}$
A **line segment** is part of a line between two endpoints.	*C* *D* **Words** line segment *CD* or line segment *DC* **Symbols** $\overline{CD}$ or $\overline{DC}$
A **plane** is a flat surface that goes on forever in all directions.	*D* *F* *E* **Words** plane *DEF*, plane *FDE*, plane *EDF*, plane *DFE*, plane *FED*, plane *EFD*
An **angle** is formed by two rays with a common endpoint. The two rays that make up the angle are called the sides of the angle. The common endpoint is called the **vertex**.	*R* side *S* vertex side *T* **Words** angle *RST* or angle *TSR*, angle *S* **Symbols** $\angle RST$ or $\angle TSR$, $\angle S$

EXAMPLE Name Geometric Figures

Use the figure to name each of the following.

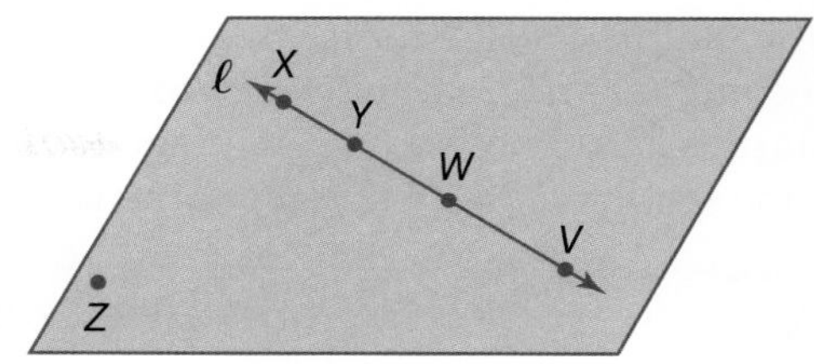

a. a line containing point X

There are four points on the line. Any two of the points can be used to name the line.

$\overleftrightarrow{XY}$ $\overleftrightarrow{YX}$ $\overleftrightarrow{XW}$ $\overleftrightarrow{WX}$ $\overleftrightarrow{XV}$ $\overleftrightarrow{VX}$

$\overleftrightarrow{YW}$ $\overleftrightarrow{WY}$ $\overleftrightarrow{YV}$ $\overleftrightarrow{VY}$ $\overleftrightarrow{WV}$ $\overleftrightarrow{VW}$

b. a line segment

There are 6 line segments: $\overline{XY}$, $\overline{XW}$, $\overline{XV}$, $\overline{YW}$, $\overline{YV}$, $\overline{WV}$

Exercises

For Exercises 1–5, refer to the figure at the right.

1. Name a line containing point Z.
2. Identify two rays.
3. What is another name for $\angle YRV$?
4. Name the plane shown.
5. List 3 line segments.

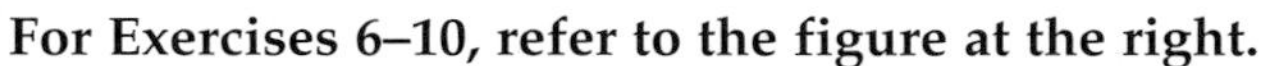

For Exercises 6–10, refer to the figure at the right.

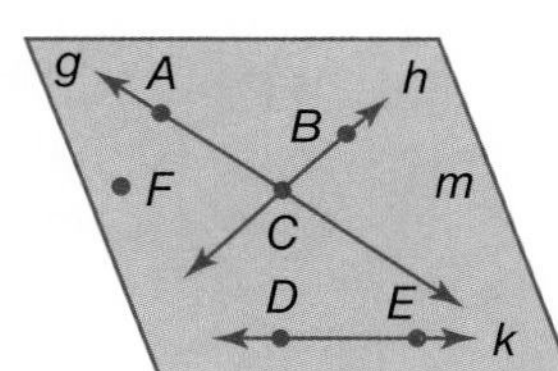

6. Name a line that contains point B.
7. Identify two angles.
8. List 3 line segments.
9. Name a point *not* contained in lines g, h, or k.
10. What is another name for line g?

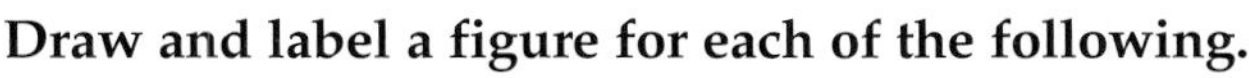

Draw and label a figure for each of the following.

11. point G
12. $\overline{XY}$
13. $\overrightarrow{XY}$ and $\overrightarrow{YZ}$
14. plane ABC

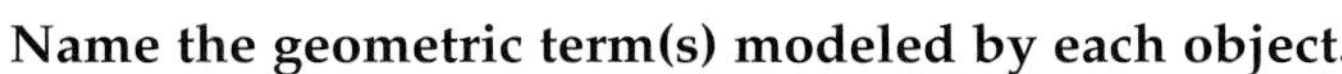

Name the geometric term(s) modeled by each object.

15. 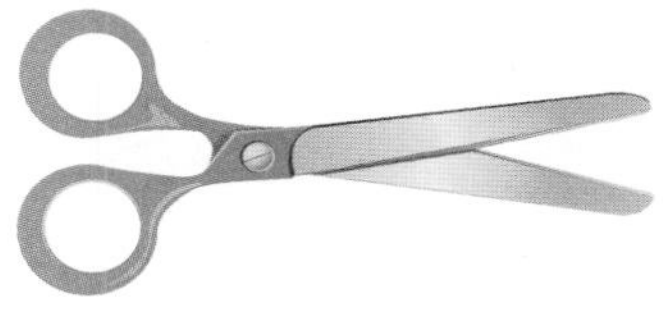

16. 

17. 

18. a napkin
19. the end of a pencil
20. woven threads in a blanket
21. Describe a real-life example of a plane containing points, lines, and angles.

Concepts and Skills Bank

8 Geometric Constructions

A compass is a drawing instrument used for drawing circles and arcs. A straightedge, such as a ruler, is used to draw segments. You can use a compass and a straightedge to construct basic elements of geometric figures. You know a line segment is part of a line with two endpoints. Line segments that have the same length are called **congruent segments**.

ACTIVITY 1 Construct Congruent Segments

Step 1 Draw $\overline{JK}$. Then use a straightedge to draw a line segment longer than $\overline{JK}$. Label it $\overline{LM}$.

Step 2 Place the compass at J and adjust the compass setting so you can place the pencil tip on K. The compass setting equals the length of $\overline{JK}$.

Step 3 Using this setting, place the compass tip at L. Draw an arc to intersect $\overline{LM}$. Label the intersection P.

$\overline{LP}$ is congruent to $\overline{JK}$.

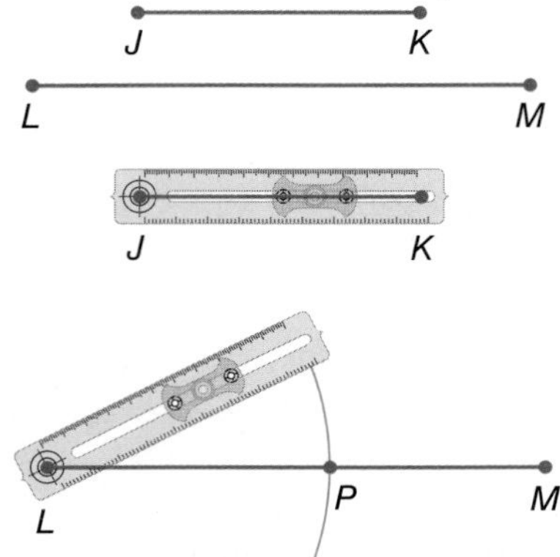

A **perpendicular bisector** is a perpendicular line that divides a line segment into two congruent segments.

ACTIVITY 2 Construct Perpendicular Bisectors

Step 1 Draw $\overline{AB}$. Then place the compass at point A. Using a setting greater than one half the length of $\overline{AB}$, draw an arc above and below $\overline{AB}$.

Step 2 Using this setting, place the compass at point B. Draw another set of arcs above and below $\overline{AB}$ as shown.

Step 3 Label the intersection of these arcs X and Y as shown.

Step 4 Draw $\overline{XY}$. Label the intersection of $\overline{AB}$ and this new line M.

$\overline{XY}$ is the perpendicular bisector of $\overline{AB}$.

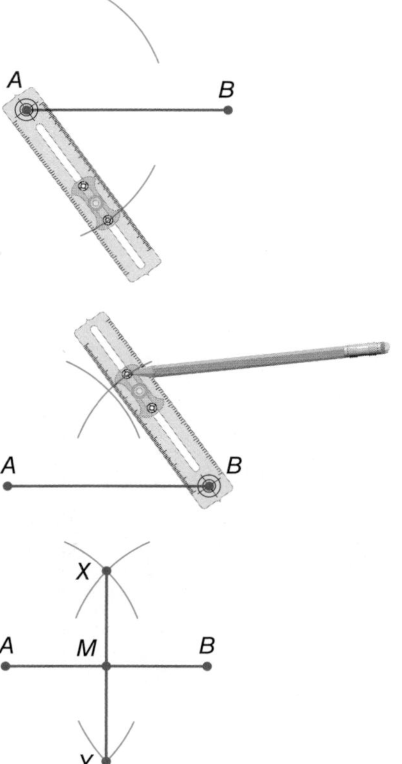

Two angles that have the same measure are **congruent angles**. You can use a protractor to construct congruent angles.

ACTIVITY 3 Construct Congruent Angles

Step 1 Draw $\angle ABC$.

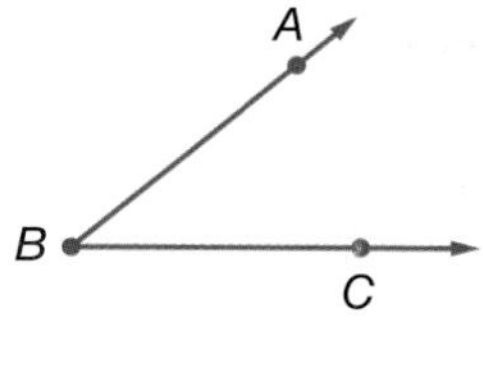

Step 2 Use a straightedge to draw $\overrightarrow{LK}$.

Step 3 With the compass at point *B*, draw an arc that intersects both sides of $\angle ABC$. Label the two points of intersection as *X* and *Y*.

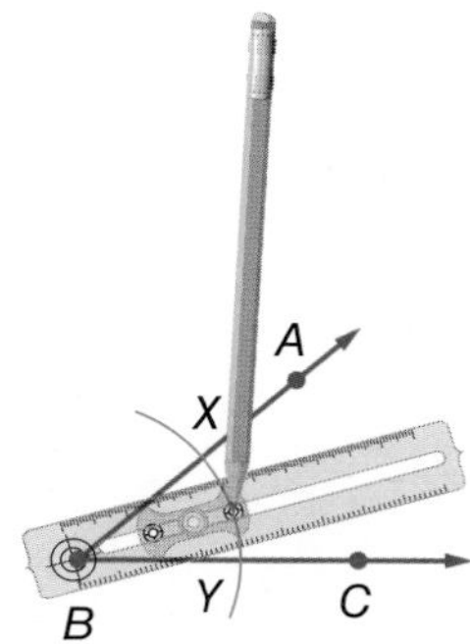

Step 4 With the same setting on your compass, place your compass at point *L*. Draw another arc. Label the intersection *R*.

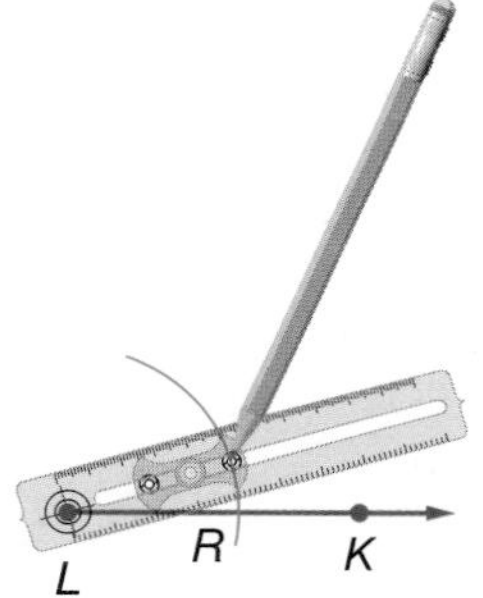

Step 5 Open your compass to the same width as the distance between points *X* and *Y*. Then place the compass at point *R*. Draw an arc that intersects the arc you drew in Step 4. Label this point of intersection *S*.

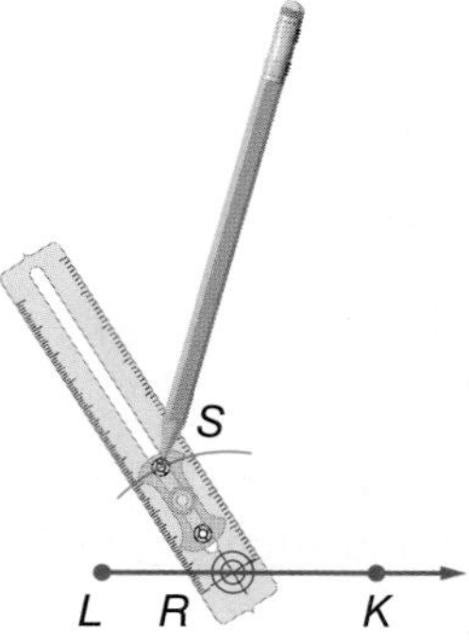

Step 6 Draw $\overrightarrow{LM}$ through point S. Angle *MLK* is congruent to $\angle ABC$.

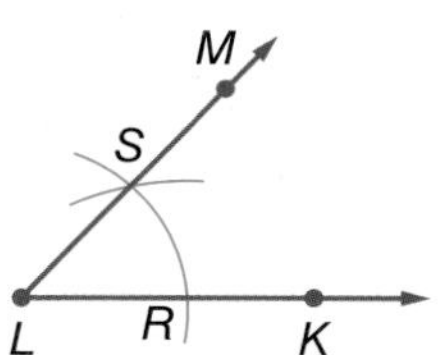

Concepts and Skills Bank

An **angle bisector** is a ray that divides an angle into two congruent angles.

ACTIVITY 4 Construct an Angle Bisector

Step 1 Draw $\angle JKL$.

Step 2 Place the compass at point K and draw an arc that intersects both sides of the angle. Label the intersections X and Y.

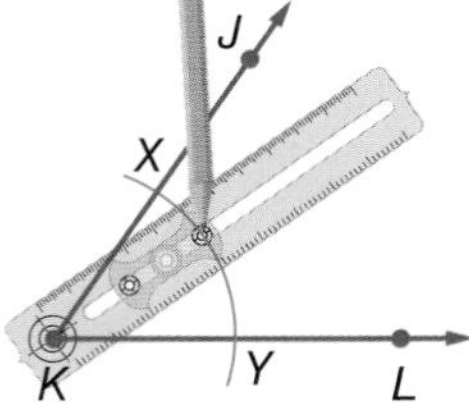

Step 3 With the compass at point X, draw an arc in the interior of $\angle JKL$.

Step 4 Using this setting, place the compass at point Y. Draw another arc.

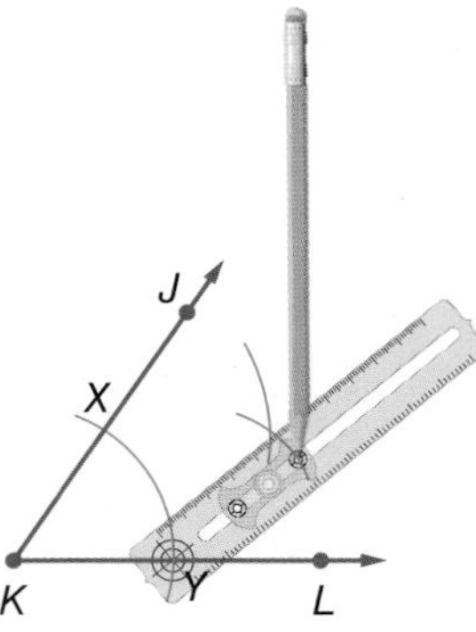

Step 5 Label the intersection of these arcs H. Then draw $\overrightarrow{KH}$.

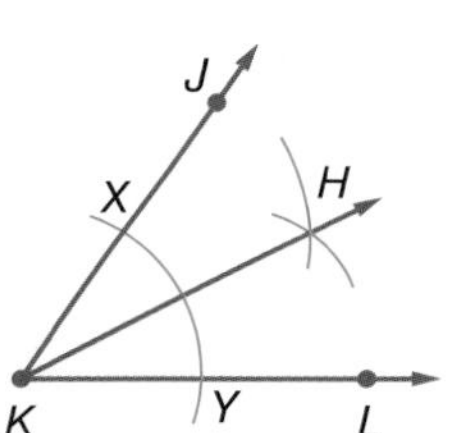

$\overrightarrow{KH}$ is the bisector of $\angle JKL$.

Exercises

Trace each segment. Then construct a segment congruent to it.

1.

2.

3.

Trace each segment. Then construct the segment's perpendicular bisector.

4.

5.

6.

Trace each angle. Then construct an angle congruent to it.

7.

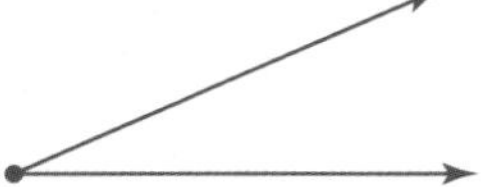

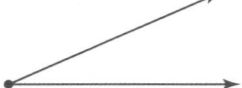

8.

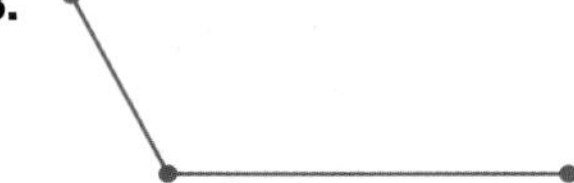

9.

10. Construct the angle bisector for Exercise 7.

11. Construct the angle bisector for Exercise 8.

12. Construct the angle bisector for Exercise 9.

⑨ Measuring and Drawing Angles

Recall a ray is part of a line that has an endpoint and goes in one direction without ending. Two rays that have the same endpoint form an angle. The common endpoint is called the vertex, and the two rays that make up the angle are called the **sides** of the angle.

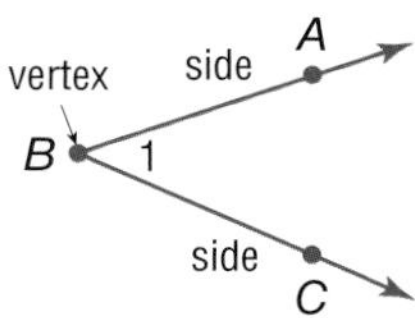

The symbol ∠ represents angle. There are several ways to name the angle shown above.

- Use the vertex and a point from each side. The vertex is always the middle letter. — $\angle ABC$ or $\angle CBA$
- Use the vertex only. — $\angle B$
- Use a number. — $\angle 1$

The most common unit of measure for angles is the **degree** (°). Angles are most often classified by their measures. The following are four types of angles.

Acute angle	Obtuse angle	Right angle	Straight angle
less than 90°	between 90° and 180°	exactly 90°	exactly 180°

You can use a **protractor** to measure angles.

EXAMPLE 1 Find Angle Measures

Use a protractor to find the measure of angle ∠CDE. Then classify each angle as *acute*, *obtuse*, *right*, or *straight*.

Step 1 Place the center point of the protractor's base on vertex D. Align the straight side with side $\overrightarrow{DE}$ so that the marker for 0° is on one of the rays.

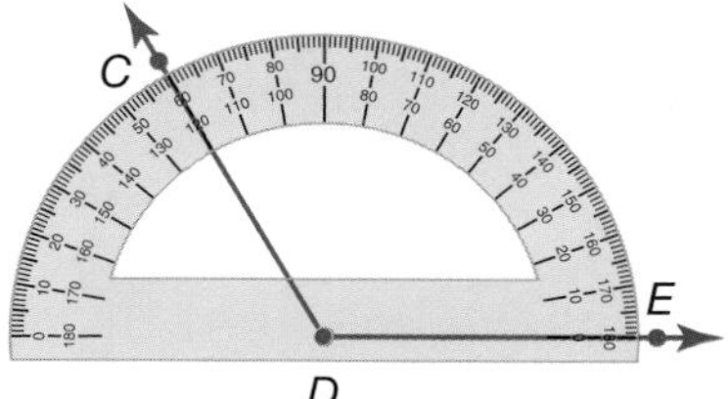

Step 2 Use the scale that begins with 0° at $\overrightarrow{DE}$. Read where the other side of the angle, $\overrightarrow{DC}$, crosses this scale.

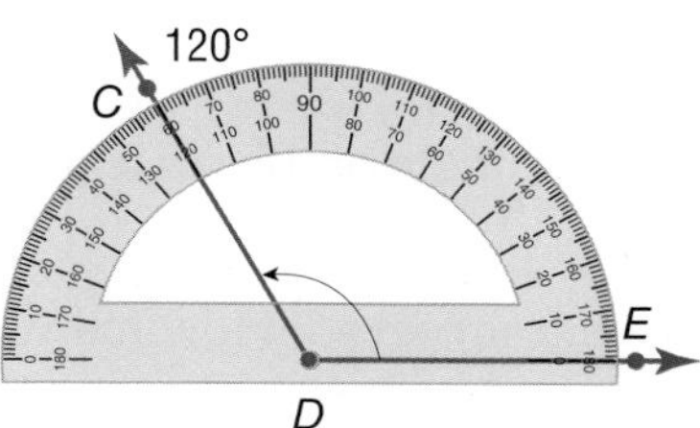

The measure of angle CDE is 120°. Using symbols, $m\angle CDE = 120°$.

The angle is between 90° and 180°, so it is an obtuse angle.

Concepts and Skills Bank

Protractors can also be used to draw an angle of a given measure.

EXAMPLE 2 Draw Angles

Draw ∠X having a measure of 85°.

Step 1 Draw a ray with endpoint X.

Step 2 Place the center point of the protractor on X. Align the mark labeled 0 with the ray.

Step 3 Use the scale that begins with 0. Locate the mark labeled 85. Then draw the other side of the angle.

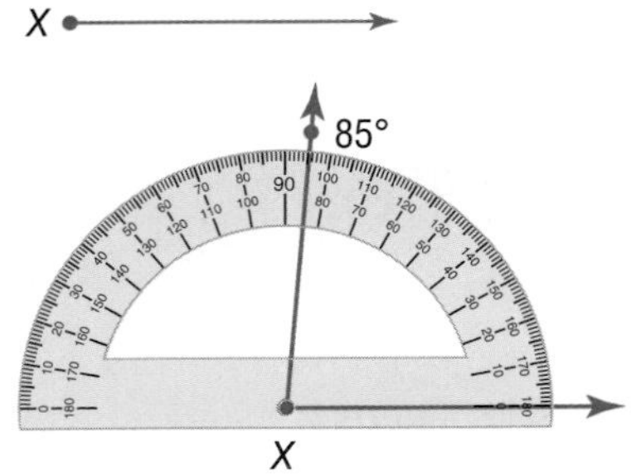

Exercises

Estimate the measure of each angle. Then use a protractor to find the actual measure. Classify each angle as *acute, obtuse, right,* or *straight.*

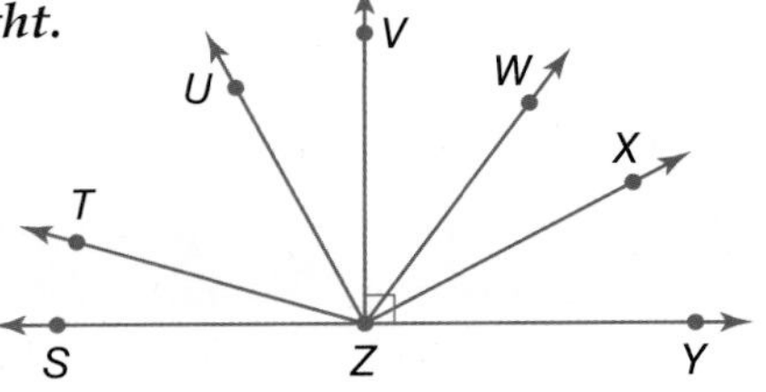

1. ∠*XZY**2.** ∠*SZT*

3. ∠*SZY*

4. ∠*UZX*

5. ∠*TZW*

6. ∠*XZT*

7. ∠*UZV*

8. ∠*WZU*

Estimate the measure of each angle. Then use a protractor to find the actual measure.

9.

10.

11.

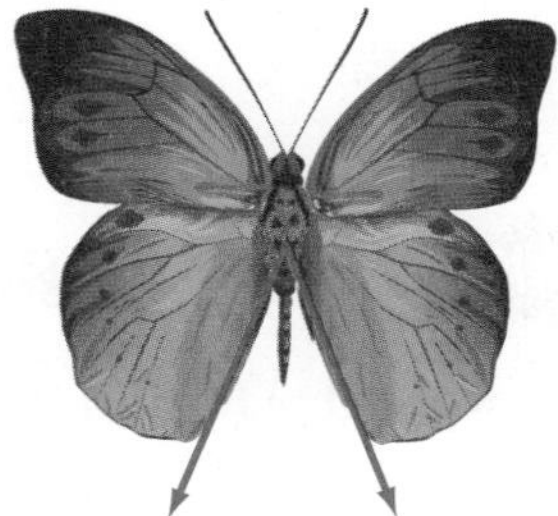

Use a protractor to draw an angle having each measurement. Then classify each angle as *acute, obtuse, right,* or *straight.*

12. 40°

13. 70°

14. 65°

15. 85°

16. 95°

17. 180°

18. 155°

19. 140°

20. 38°

21. 90°

22. 127°

23. 174°

24. SCHOOL The graph at the right shows the amount of money spent on school clothes.

a. Find the measure of each angle of the circle graph to the nearest degree.

b. Suppose 500 adults were surveyed. How many would you expect to spend between $250 and $349?

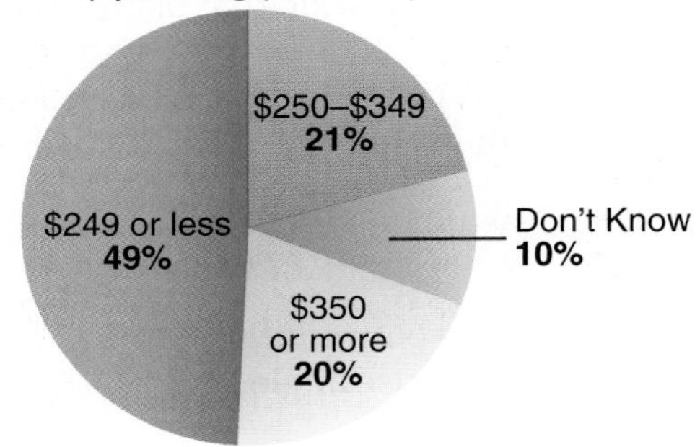

Concepts and Skills Bank

10 The Tangent Ratio

The study of the properties of triangles is called **trigonometry**. The word trigonometry means *angle measure.*

A **trigonometric ratio** is a ratio of the lengths of two sides of a right triangle. The **tangent** compares the measure of the leg opposite an angle with the measure of the leg adjacent to that angle. The symbol for the tangent of angle A is tan$A$.

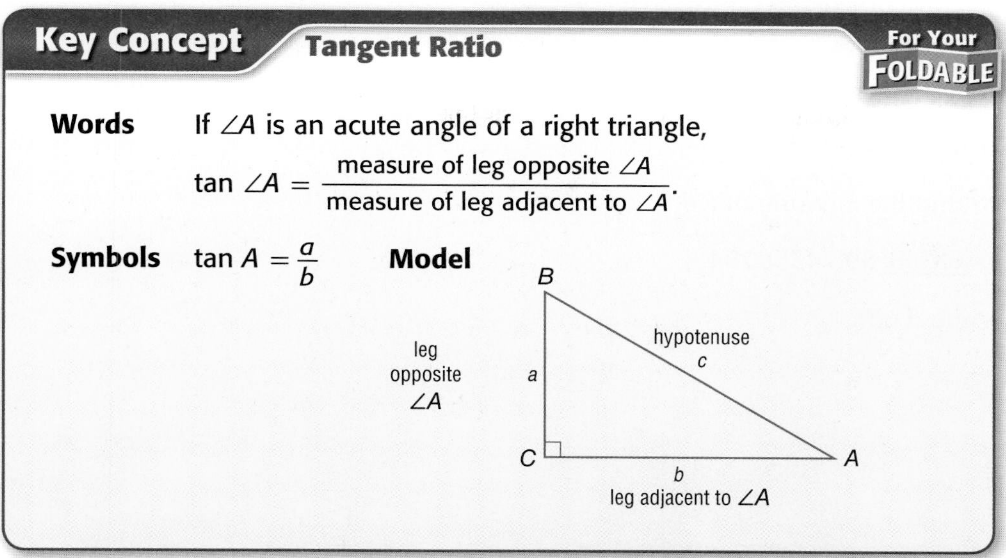

Key Concept Tangent Ratio — For Your FOLDABLE

Words If $\angle A$ is an acute angle of a right triangle,

$$\tan \angle A = \frac{\text{measure of leg opposite } \angle A}{\text{measure of leg adjacent to } \angle A}.$$

Symbols $\tan A = \frac{a}{b}$

Model

You can use the tangent ratio to solve real-world problems.

Real-World EXAMPLE 1 Find Length

ARCHITECTURE **The Leaning Tower of Pisa in Pisa, Italy, tilts about 5.2° from vertical. If the top of the tower is 55 meters from the ground, about how far has its top shifted from its original position?**

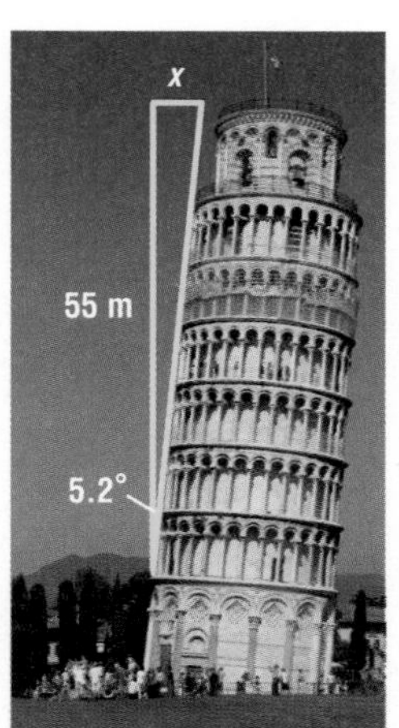

Angle A measures 5.2°, and you know the length of the side adjacent to $\angle A$. You want to find the length of the side opposite to $\angle A$.

Use the tangent ratio.

$\tan \angle A = \frac{\text{measure of leg opposite } \angle A}{\text{measure of leg adjacent to } \angle A}$ — **Write the tangent ratio.**

$\tan 5.2° = \frac{x}{55}$ — **Substitution**

$55(\tan 5.2°) = 55 \cdot \frac{x}{55}$ — **Multiply each side by 55.**

55 [{x}] [TAN] 5.2 [ENTER] 5.00539211 — **Use a calculator.**

$x \approx 5.0$ m — **Simplify.**

So, the top of the tower has shifted about 5.0 meters from its original position.

You can use the $\tan^{-1}$ function on your calculator to find the measure of an angle when you know the measures of the two legs.

EXAMPLE 2 Find Angle Measures

Find the measure of $\angle B$ to the nearest degree.

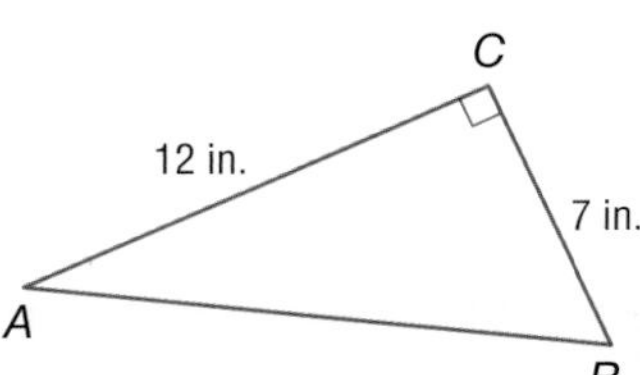

From the figure, you know the measures of the two legs. Use the definition of tangent.

$$\tan B = \frac{\text{opposite leg}}{\text{adjacent leg}}$$

$$\tan B = \frac{12}{7}$$

Use your calculator to find the measure of $\angle B$.

[2nd] [TAN^{-1}] 12 [÷] 7 [ENTER] 59.74356284

The measure of $\angle B$ is about 60°.

Concepts and Skills Bank

Exercises

Find each missing measure to the nearest tenth.

1.

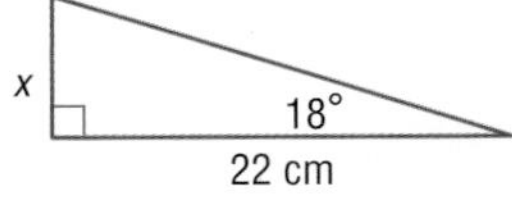

2.

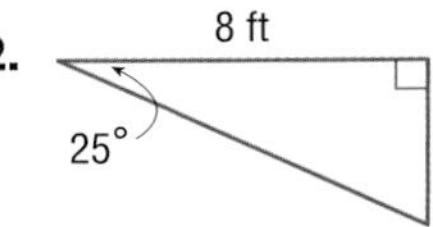

3.

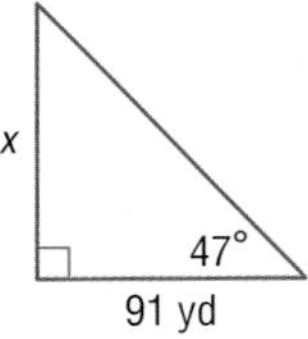

4.

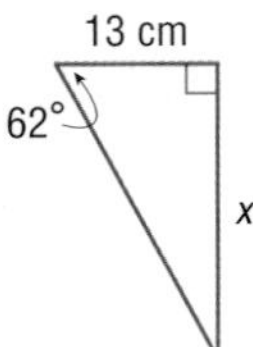

5.

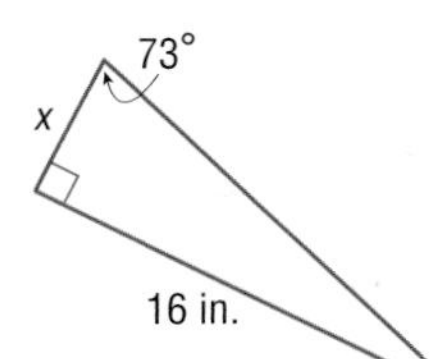

6.

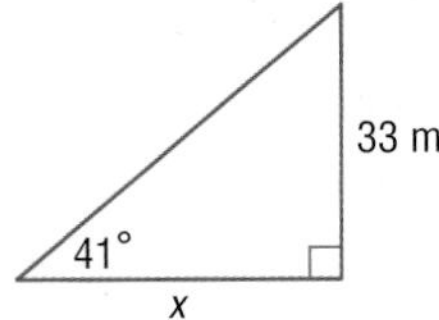

Find each missing angle measure to the nearest degree.

7.

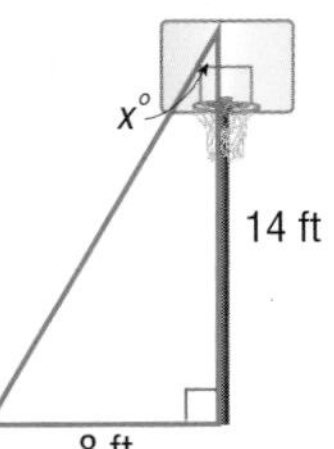

8.

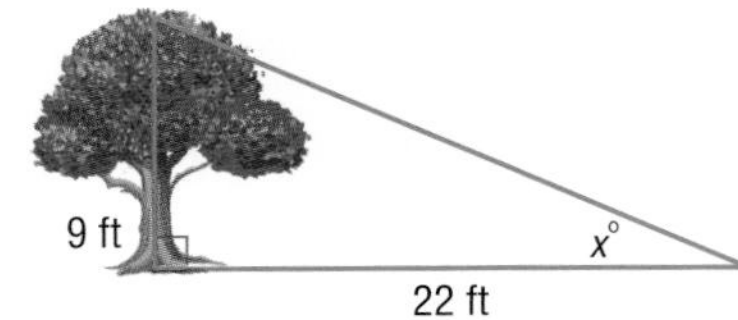

9.

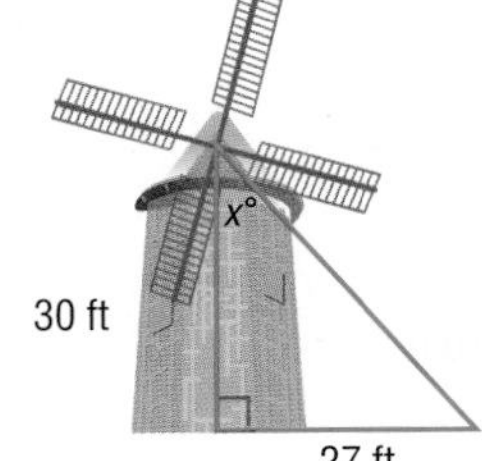

⑪ The Sine and Cosine Ratios

Some problems involving right triangles give the measures of one leg and the hypotenuse. You could use the Pythagorean Theorem to find the measure of the remaining leg or you could use the sine or cosine ratios.

For an acute angle of a right triangle,

- the **cosine** is the ratio of the measure of the leg adjacent to the acute angle to the measure of the hypotenuse.
- the **sine** is the ratio of the measure of the leg opposite the acute angle to the measure of the hypotenuse.

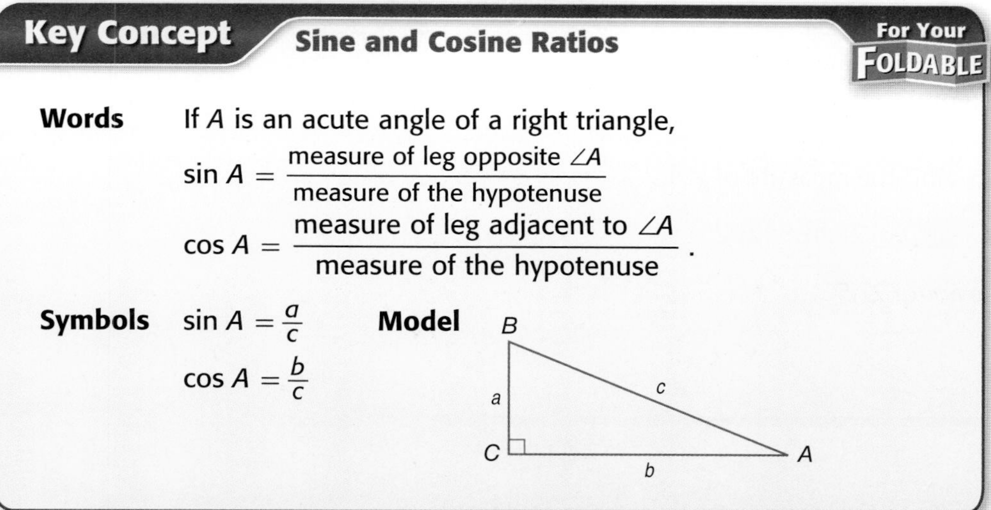

Key Concept Sine and Cosine Ratios — For Your FOLDABLE

Words If A is an acute angle of a right triangle,

$$\sin A = \frac{\text{measure of leg opposite } \angle A}{\text{measure of the hypotenuse}}$$

$$\cos A = \frac{\text{measure of leg adjacent to } \angle A}{\text{measure of the hypotenuse}}.$$

Symbols $\sin A = \frac{a}{c}$ $\cos A = \frac{b}{c}$

Model

You can use the sine and cosine ratios to find missing lengths of sides or angle measures in a right triangle.

EXAMPLE 1 Find Length

Use $\triangle ABC$ to find the length of BC.

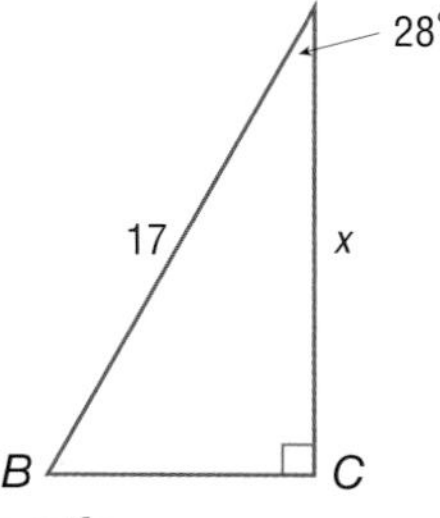

Angle A measures 28° and you know the length of the hypotenuse. You want to find the length of the side adjacent to ∠A.

Use the cosine ratio.

$\cos \angle A = \frac{\text{measure of leg opposite } \angle A}{\text{measure of the hypotenuse}}$	**Write the cosine ratio.**
$\cos 28° = \frac{x}{17}$	**Substitution**
$17(\cos 28°) = 17 \cdot \frac{x}{17}$	**Multiply each side by 17.**
17 [×] [cos] 28 [ENTER] 15.01010908	**Use a calculator.**
$x \approx 15$ m	**Simplify.**

So, the length of $\overline{BC}$ is about 15 meters.

You can use the $\sin^{-1}$ or $\cos^{-1}$ function on your calculator to find the measure of an angle when you know the measures of the two legs.

Real-World EXAMPLE 2 Find Angle Measures

SPORTS A parasailer is 30 feet above the water. The tow rope is 65 feet long. Find the angle of elevation between the parasailer and the surface of the water.

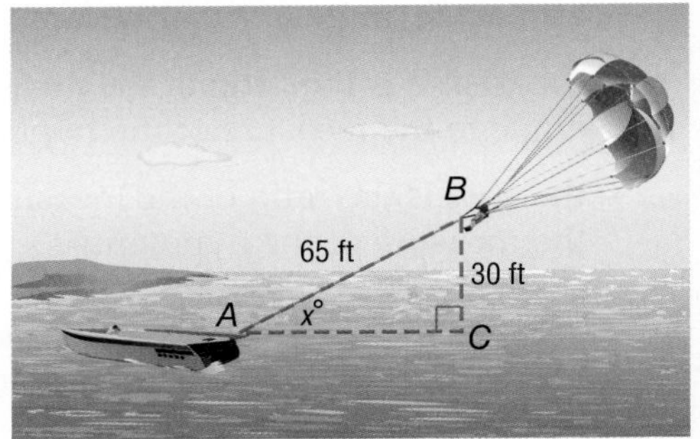

You know the side opposite the angle and the hypotenuse. You want to know the angle measure.

Use the sine ratio.

$\sin \angle A = \dfrac{\text{measure of leg opposite } \angle A}{\text{measure of the hypotenuse}}$ **Write the sine ratio.**

$\sin x° = \dfrac{30}{65}$ **Substitution**

Use your calculator to find the measure of $\angle A$.

2nd [$\sin^{-1}$] 30 ÷ 65 ENTER 27.48642625

The measure of ∠A is about 27.5°.

Exercises

Find each missing measure to the nearest tenth or nearest degree.

1. 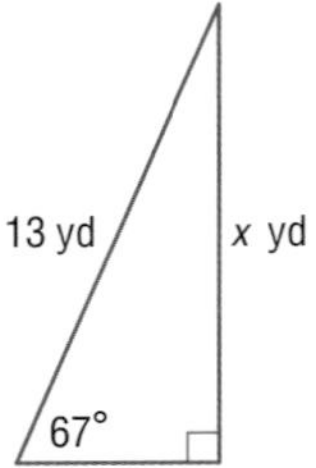

2. 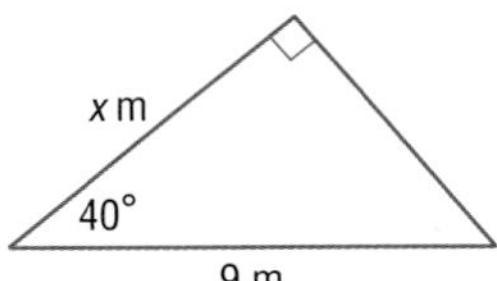

3. 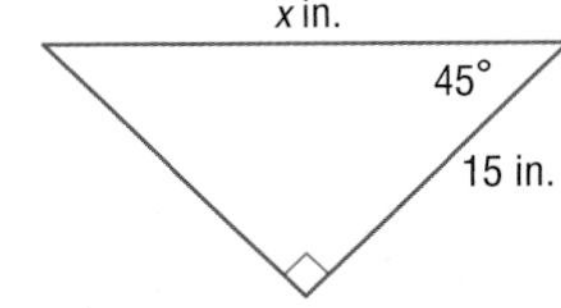

4. 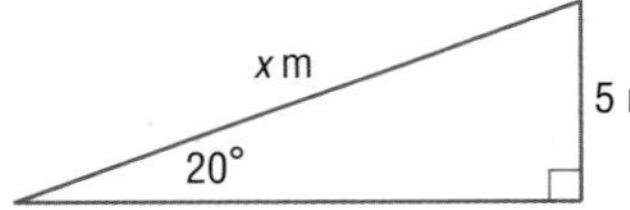

5. 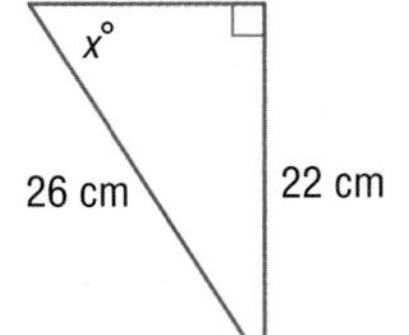

6. 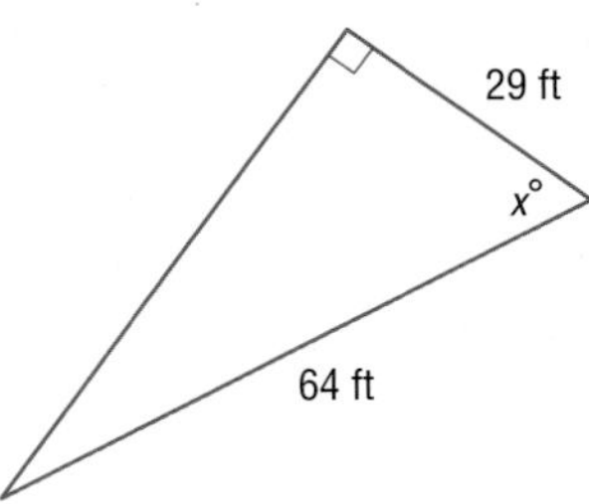

7. HOME IMPROVEMENT A painter props a 20-foot ladder against a house. The angle it forms with the ground is 65°. To the nearest foot, how far up the side of the house does the ladder reach?

8. TRANSPORTATION The end of an exit ramp from an interstate highway is 22 feet higher than the highway. If the ramp is 630 feet long, what angle does it make with the highway? Round to the nearest degree.

⑫ Accuracy and Precision

All measurements taken in the real world are approximations. The greater the care in which a measurement is taken, the more accurate it will be.

Accuracy is the degree of conformity of a measurement with the true value. **Precision** is the degree of perfection in which a measurement is made.

EXAMPLE 1 Determine Precision

Which is a more precise measurement of the line segment shown, 2 centimeters or 1.8 centimeters?

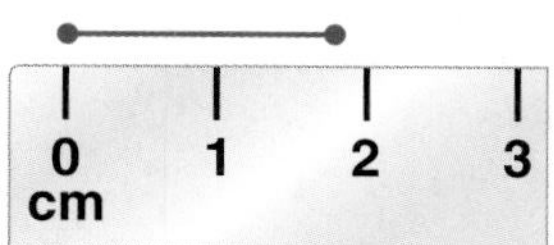

0 1 2 3
cm

Estimate 2 cm	**Estimate** 2 cm
measurement: 2 cm	measurement: 1.8 cm
precision: 1 cm	precision: 0.1 cm

So, 1.8 centimeters is a more precise measurement.

Exercises

Choose the correct term(s) to determine the degree of precision needed in each measurement situation.

1. In a pharmacy, the mass of one drop of a medicine is found to the nearest 0.01 (gram, kilogram).
2. The weight of a puppy in a veterinarian's office would be given to the nearest (tenth of a pound, tenth of an ounce).
3. On a cruise, the length of a tour on an island would be described in (minutes, hours).
4. A person making a sweater measures the fabric to the nearest (inch, eighth of an inch).

Estimate and then measure the length of each line segment to the nearest half, quarter, eighth, or sixteenth inch. State the measurement that is the most precise.

5.

6.

7. **GEOMETRY** Draw a line segment. Estimate the length of the line segment. Then measure using two different units. Which measuring unit gave the more precise measurement?
8. **CONSTRUCTION** A construction company is ordering cement to pour the garage floor for a new house. Would you say that precision or accuracy is more important in the completion of their order? Explain.

Concepts and Skills Bank

13 Geometric Probability

Recall that the probability of an event is defined as the ratio of the number of ways something can happen to the total possible outcomes. Probability can also be related to the area of a figure.

Key Concept **Probability and Area** For Your FOLDABLE

Words The probability of landing in a specific region of a target is the ratio of the area of the specific region to the area of the target.

Symbols $P(\text{specific region}) = \dfrac{\text{area of specific region}}{\text{area of the target}}$

You can find probability using an area model.

EXAMPLE 1 **Find Probability Using Area Models**

a. Find the probability that a coin dropped on the board will land within the shaded region.

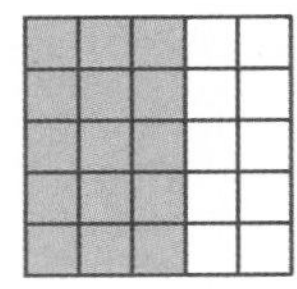

$P(\text{shaded region}) = \dfrac{\text{area of shaded region}}{\text{area of the target}}$

$= \dfrac{15}{25} \text{ or } \dfrac{3}{5}$

So, the probability is $\frac{3}{5}$, 0.6, or 60%.

b. To win a game at the school carnival, a thrown dart must land in the red section of the square board. What is the probability that a dart thrown at random will land in the red section?

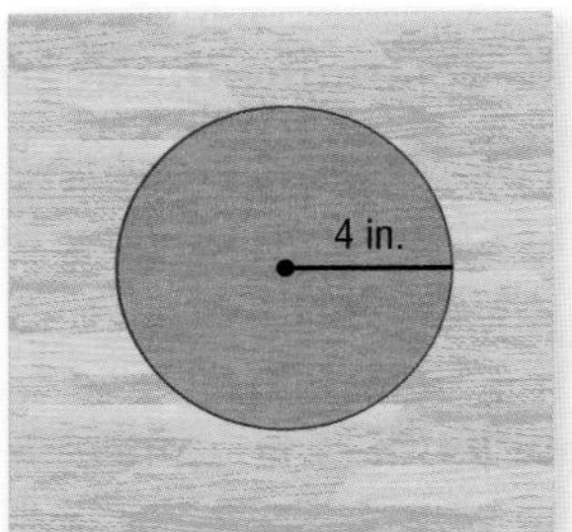

To find the probability of landing in the red section, you need to know the area of the red section and the area of the entire board.

Area of red section $= \pi \cdot r^2$ **Area of a circle**

$= \pi \cdot 4^2$ **Replace *r* with 4.**

$= 16\pi$ **Multiply 4 by 4.**

≈ 50.27 **Simplify.**

The area of the red circle is about 50.27 square inches.

Area of the square $= s^2$

$= 12^2 \text{ or } 144$

Now find the probability.

$P(\text{landing in the red section}) = \dfrac{\text{area of red section}}{\text{total area}}$

$\approx \dfrac{50.27}{144} \text{ or about } 0.35$

So, the probability of a dart landing in the red section is about 0.35 or 35%.

Exercises

Find the probability that a randomly thrown dart will land in the shaded region of each dartboard. Round to the nearest hundredth, if necessary.

1. 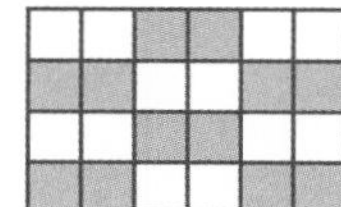

2. 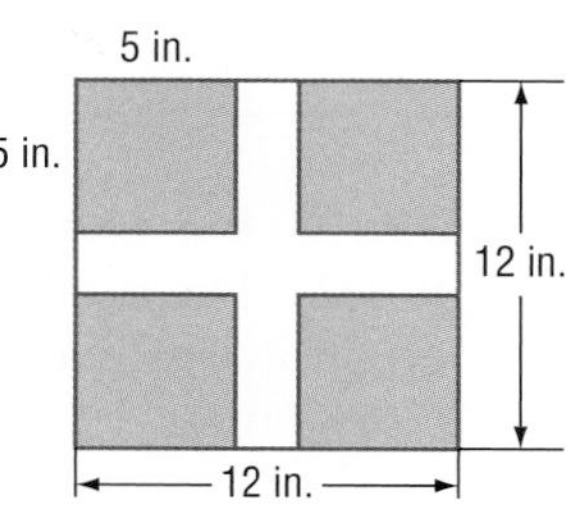

3. 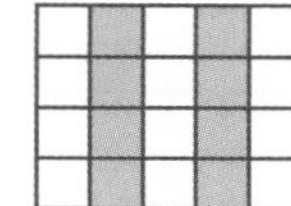

4. 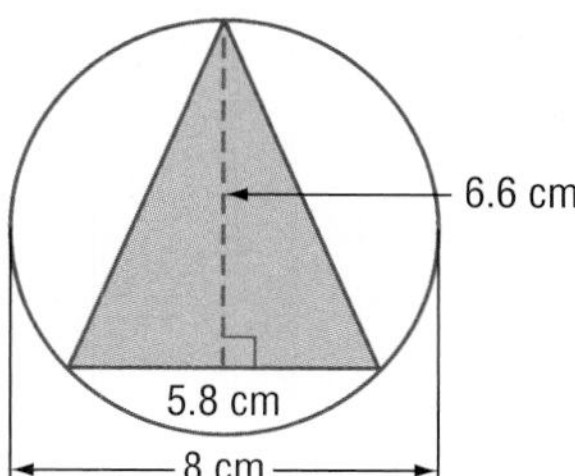

5. 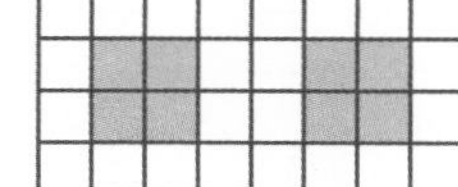

6. 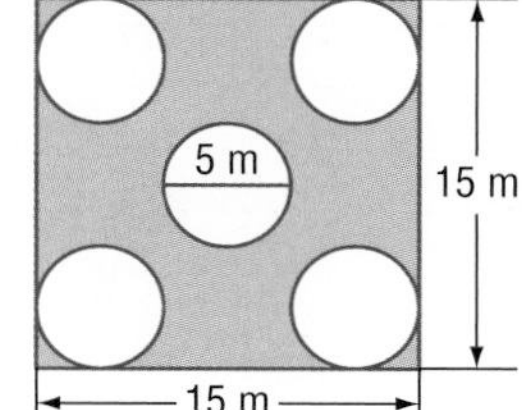

7. GAMES A popular game involves tossing a bean bag through a hole in a wooden board like the one shown. What is the probability of tossing the bean bag through the hole?

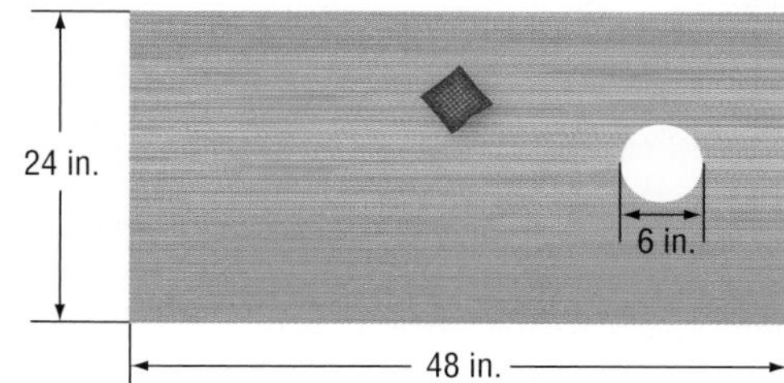

8. PATIOS A square patio is 10 feet by 10 feet. At the center of the patio is a rectangular rug 5 feet by 6 feet. If a coin is dropped somewhere on the patio, what is the probability that the coin lands on the rug?

9. PRIZES To win a prize, a coin must land on an odd number on the triangular board shown. It is equally likely that the coin will land anywhere on the board. What is the probability of winning a prize?

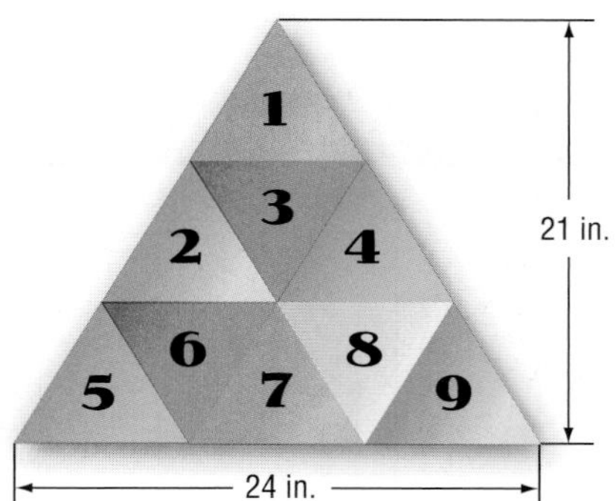

10. FOUNTAINS A circular flower garden has a diameter of 16 feet. At the center of the garden is a fountain 5 feet in diameter. If a coin is tossed at random into the garden, what is the probability that the coin will land in the fountain?

11. DARTS Draw a dartboard in which the probability of a dart landing in the shaded area is 60%.

12. GAMES It is equally likely that a thrown dart will land anywhere on the dartboard shown at the right. Find the probability of a randomly thrown dart landing inside the target region.

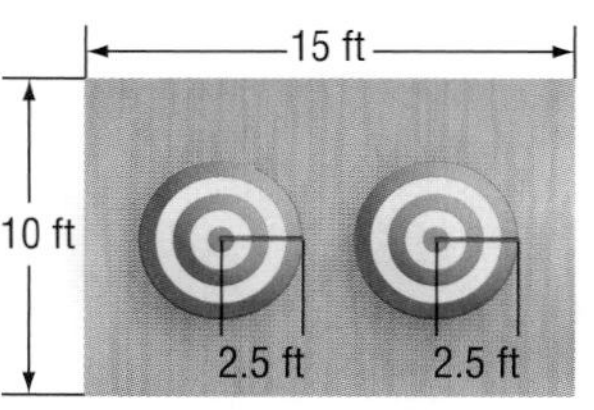

Concepts and Skills Bank

14 Displaying Data in Graphs

Statistics involves collecting, analyzing, and presenting information. The information that is collected is called data. Displaying data in graphs makes it easier to visualize the data.

- **Bar graphs** are used to compare the frequency of data. The bar graph at the right compares the amounts of recycled materials.

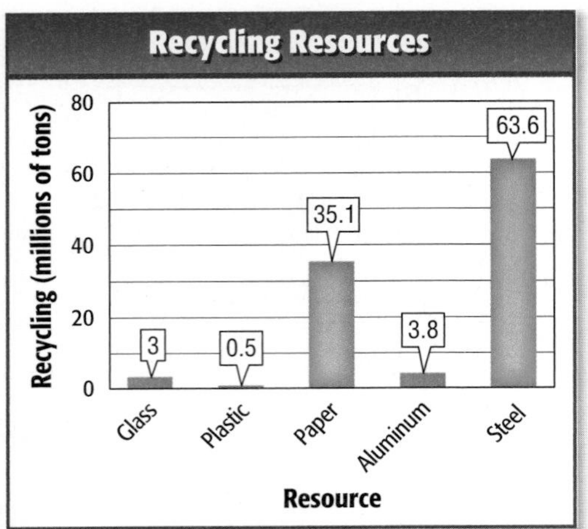

- **Double bar graphs** compare two sets of data. The double bar graph at the left shows movie preferences for men and women.

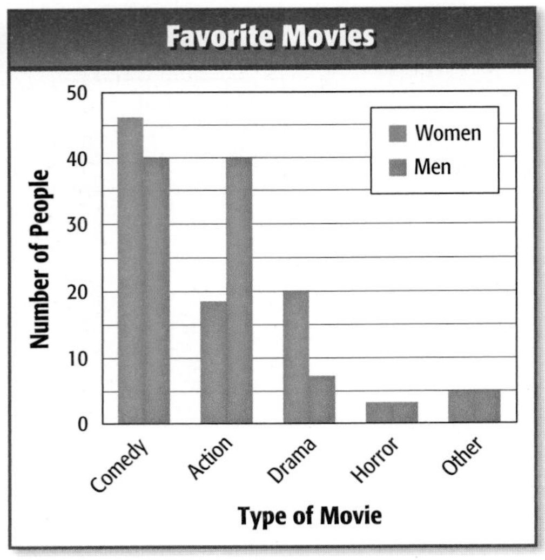

- **Line graphs** usually show how values change over a period of time. The line graph at the right shows the results of the women's Olympic Long Jump event from 1972 to 2004.

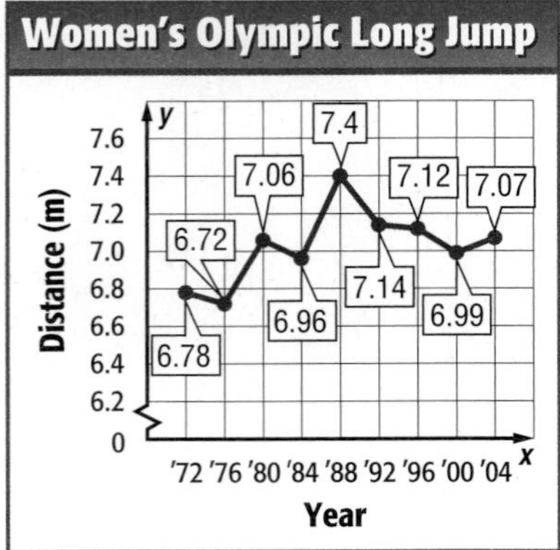

- **Double line graphs**, like double bar graphs, show two sets of data. The double line graph below compares the number of boys and the number of girls participating in high school athletics.

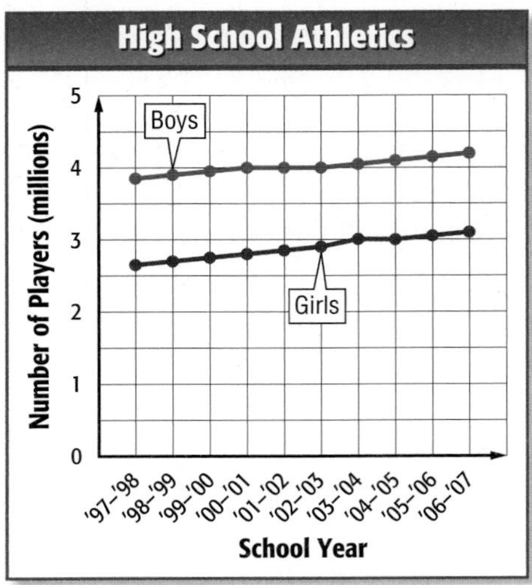

Source: *National Federation of State High School Association*

Concepts and Skills Bank

- **Circle graphs** show how parts are related to the whole. The circle graph below shows how electricity is generated in the United States.

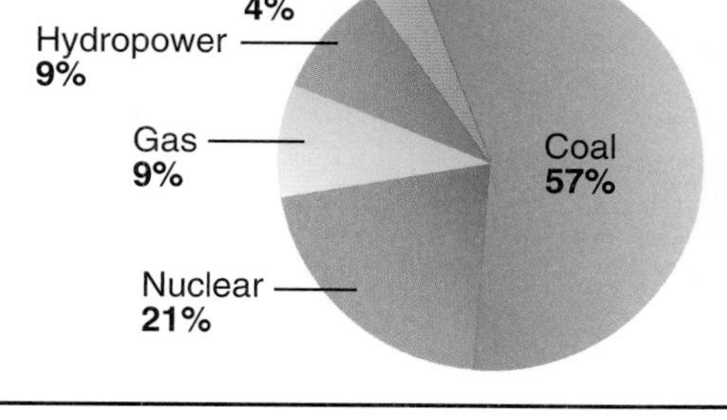

- **Line plots** organize data using a number line. The line plot below shows the number of Calories in a single serving of different brands of yogurt.

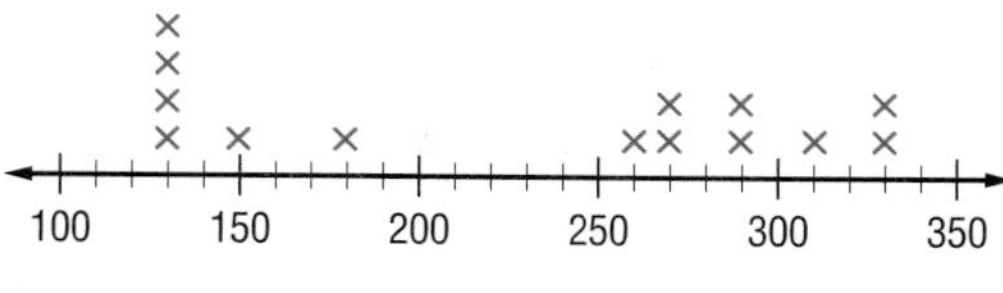

EXAMPLE 1 Select a Display

A newspaper wants to display the high temperature of the past week. Should they use a line graph, circle graph, or double bar graph?

Since the data would show change over time, a line graph would give the reader a clear picture of changes in temperature.

Exercises

Determine whether a bar graph, double bar graph, line graph, double line graph, line plot, or circle graph is the best way to display each sets of data. Explain.

1. the number of people who have different kinds of pets
2. the percent of students in class who have 0, 1, 2, 3, or more than 3 siblings
3. the number of teens who attended art museums, symphony concerts, rock concerts, and athletic events in 1990 compared to the number who attended the same events this year
4. the minimum wage every year from 1980 to the present
5. the number of boys and the number of girls participating in volunteer programs each year from 1995 to the present
6. The table below shows the number of events at recent Olympic games.

Olympic Year	1968	1972	1976	1980	1984	1988	1992	1996	2000	2004
Number of Events	172	196	199	200	223	237	257	271	300	296

7. The prices of lawn seats at events held at an amphitheater are shown in the chart below.

Lawn Seat Prices ($)							
23	29	31	34	19	28	35	37
22	28	26	33	35	39	26	27

Concepts and Skills Bank

15 Misleading Graphs

Two graphs that represent the same data may look quite different. If different vertical scales are used, each graph will give a different visual impression.

EXAMPLE 1 Misleading Graphs

TRAVEL The graphs show domestic traveler spending in the U.S.

Graph A

Traveler Spending in the U.S.

Amount (billions): $500, $400, $300, $0

Year: 1990, 1995, 2000

Graph B

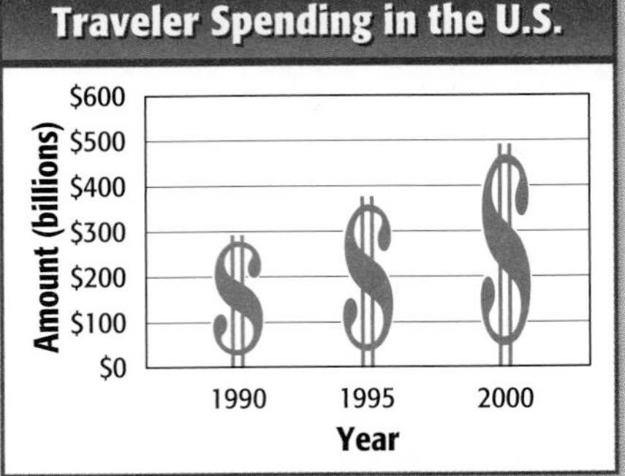

a. Why do the graphs look different?

The vertical scales differ.

b. Which graph appears to show a greater increase in the growth of traveler spending?

Graph A; the size of the dollar sign makes the increase appear more dramatic because both the height and width of the dollar sign are increasing.

When reading a statistical graph, you must interpret the information carefully and determine whether the inference made from the data is valid.

EXAMPLE 2 Accuracy of Predictions and Conclusions

SCOOTERS The graph displays units sold during the life of a scooter company. According to the graph, the number of scooters did not increase as fast in the 2000s as they did from 1970–2000. Determine whether this statement is accurate. Justify your reasoning.

No, the statement is not accurate. The horizontal scale is inconsistent; from 1970 to 2000, the interval is 10 years, but the interval is 1 year from 2000 to 2005. Also, the graph only goes through 2005. You would have to wait until 2010 to know the number of scooters sold and compare the rate of change from 2000 to 2010 to the rate of increase from 1970 to 2000.

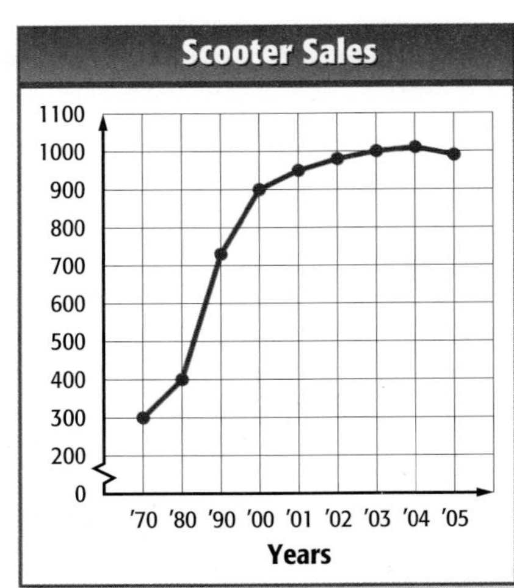

Exercises

1. **MOVIES** The graphs below show the dollars earned by the top five all time movies.

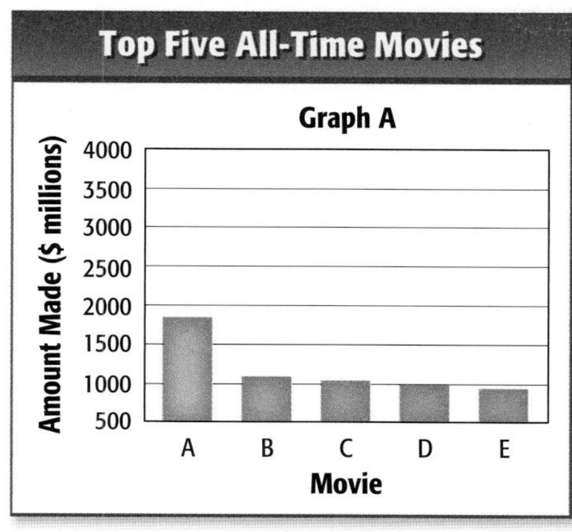

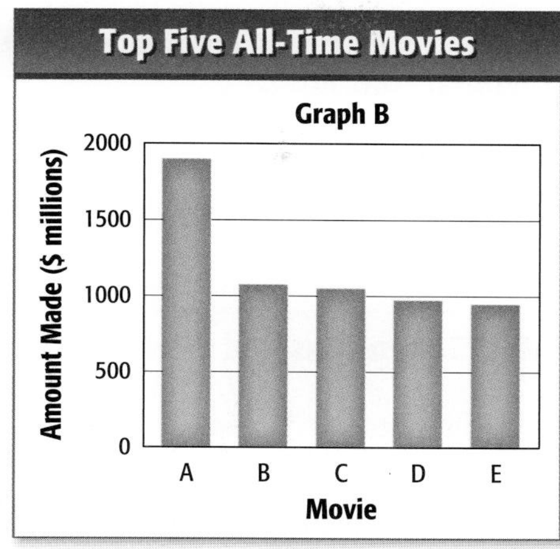

a. Which graph gives the impression that the top all-time movie made far more money than any other top all-time movie?

b. Which graph shows that the bottom four movies earned about the same amount?

2. **TRAVEL** The distance adults drive each week is shown in the graph below.

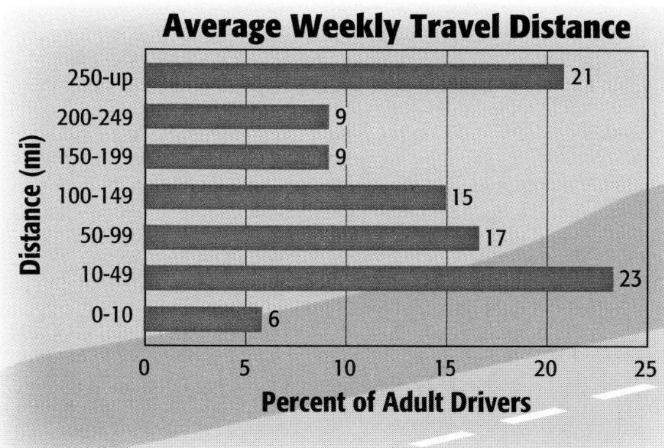

a. Is the graph misleading? Explain your reasoning.

b. Create a new graph. Change the scale to appropriately fit the data.

3. **HOCKEY** The graph at the right shows the number of wins after an 82-game schedule. According to the graph, Detroit had about 4 times as many wins as Chicago. Determine whether this statement is accurate. Justify your reasoning.

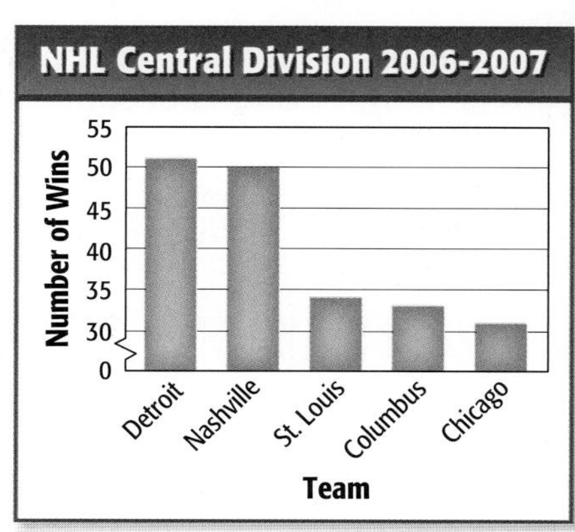

4. How can graphs be misleading? Give an example of a graph that is misleading and explain how to redraw the graph so it is not misleading.

16 Matrices

A rectangular arrangement of numerical data in rows and columns is called a **matrix**. Each number in a matrix is called an **element**.

$$\underbrace{\begin{bmatrix} 9 & 0 \\ 2 & 6 \\ 1 & 3 \end{bmatrix}}_{\text{2 columns}} \Big\} \text{3 rows}$$

- This matrix has 3 rows and 2 columns.
- The number 1 is an element in the third row and the first column.

A matrix is described by its **dimensions**, or the number of rows and columns, with the number of rows stated first. The dimensions of the matrix above are 3 by 2.

EXAMPLE 1 Identify Dimensions and Elements

State the dimensions of the matrix $\begin{bmatrix} 14 & ⑦ & 4 \\ 10 & 19 & 9 \end{bmatrix}$. Then identify the position of the circled element.

The matrix has 2 rows and 3 columns. So, the dimensions of the matrix are 2 by 3. The circled element is in the first row, second column.

$$\underbrace{\begin{bmatrix} 14 & ⑦ & 4 \\ 10 & 19 & 9 \end{bmatrix}}_{\text{3 columns}} \Big\} \text{2 rows}$$

Real-World EXAMPLE 2 Use a Matrix to Organize Data

TRACK AND FIELD The table shows the times of two athletes at a track and field meet.

Athlete	Event			
	100m	200m	400m	800m
Katherine	12.50	25.50	56.45	1:53.75
Brianne	12.25	25.55	57.65	1:54.35

a. Organize the information in a matrix.

$\begin{bmatrix} 12.50 & 25.50 & 56.45 & 1{:}53.75 \\ 12.25 & 25.55 & 57.65 & 1{:}54.35 \end{bmatrix}$ The matrix has 2 rows and 4 columns, like the information in the table.

b. Name the element in the first row, third column. What does it represent?

The element is 56.45. It is Katherine's time for the 400m event.

Key Concept Add or Subtract Matrices

For Your FOLDABLE

Words Two matrices with the same dimensions can be added or subtracted by adding or subtracting their corresponding elements.

Symbols
$$\begin{bmatrix} a & b \\ c & d \end{bmatrix} + \begin{bmatrix} e & f \\ g & h \end{bmatrix} = \begin{bmatrix} a+e & b+f \\ c+g & d+h \end{bmatrix}$$
$$\begin{bmatrix} a & b \\ c & d \end{bmatrix} - \begin{bmatrix} e & f \\ g & h \end{bmatrix} = \begin{bmatrix} a-e & b-f \\ c-g & d-h \end{bmatrix}$$

Example
$$\begin{bmatrix} 1 & 4 \\ 0 & 2 \end{bmatrix} + \begin{bmatrix} 5 & 5 \\ 1 & 3 \end{bmatrix} = \begin{bmatrix} 1+5 & 4+5 \\ 0+1 & 2+3 \end{bmatrix} \text{ or } \begin{bmatrix} 6 & 9 \\ 1 & 5 \end{bmatrix}$$

EXAMPLE 3 Add and Subtract Matrices

Add or subtract. If there is no sum or difference, write *impossible*.

a. $\begin{bmatrix} 7 & -6 & 2 \\ 4 & 6 & 8 \end{bmatrix} + \begin{bmatrix} 10 & 4 \\ -1 & 7 \end{bmatrix}$

The first matrix is 2 by 3. The second matrix is 2 by 2. Since the matrices do not have the same dimensions, it is impossible to add them.

b. $\begin{bmatrix} 0 & 2 \\ -3 & 4 \\ 1 & 4 \end{bmatrix} + \begin{bmatrix} 7 & 5 \\ 2 & 1 \\ -3 & 5 \end{bmatrix}$

$$\begin{bmatrix} 0 & 2 \\ -3 & 4 \\ 1 & 4 \end{bmatrix} + \begin{bmatrix} 7 & 5 \\ 2 & 1 \\ -3 & 5 \end{bmatrix} = \begin{bmatrix} 0+7 & 2+5 \\ -3+2 & 4+1 \\ 1+(-3) & 4+5 \end{bmatrix}$$

$$= \begin{bmatrix} 7 & 7 \\ -1 & 5 \\ -2 & 9 \end{bmatrix}$$

c. $\begin{bmatrix} 12 & 5 & 2 \\ 0 & 1 & -1 \end{bmatrix} - \begin{bmatrix} 3 & 5 & 1 \\ 9 & -6 & 0 \end{bmatrix}$

$$\begin{bmatrix} 12 & 5 & 2 \\ 0 & 1 & -1 \end{bmatrix} - \begin{bmatrix} 3 & 5 & 1 \\ 9 & -6 & 0 \end{bmatrix}$$

$$= \begin{bmatrix} 12-3 & 5-5 & 2-1 \\ 0-9 & 1-(-6) & -1-0 \end{bmatrix}$$

$$= \begin{bmatrix} 9 & 0 & 1 \\ -9 & 7 & -1 \end{bmatrix}$$

Exercises

1. WEATHER The table shows the high and low temperatures in a city for five days.

Temperature (°F)	Mon.	Tues.	Wed.	Thurs.	Fri.
High	76	79	71	69	74
Low	46	51	43	53	48

a. Organize the data in a matrix.

b. What are the dimensions of the matrix?

c. What is the element in the first row, fourth column? What does it represent?

Add or subtract. If there is no sum or difference, write *impossible*.

2. $\begin{bmatrix} 3 & 2 \\ -2 & 8 \end{bmatrix} + \begin{bmatrix} 5 & -2 \\ 5 & 1 \end{bmatrix}$

3. $\begin{bmatrix} 11 \\ 16 \\ 4 \end{bmatrix} - \begin{bmatrix} 7 \\ 19 \\ 2 \end{bmatrix}$

4. $\begin{bmatrix} 4 & 6 & 5 \end{bmatrix} - \begin{bmatrix} 6 & 14 & 7 \\ 5 & -3 & 8 \end{bmatrix}$

5. $\begin{bmatrix} 4 & 7 & 2 \\ 6 & 2 & -4 \end{bmatrix} + \begin{bmatrix} 1 & 3 & 6 \\ -4 & 5 & 7 \end{bmatrix}$

6. $\begin{bmatrix} -1 & 2 & 6 & 8 \end{bmatrix} + \begin{bmatrix} 3 & 6 & -2 & 4 \end{bmatrix}$

7. $\begin{bmatrix} 1 & 3 & 1 \\ 2 & 7 & 0 \end{bmatrix} + \begin{bmatrix} 3 & -7 \\ 15 & 9 \end{bmatrix}$

8. $\begin{bmatrix} 12 & 6 \\ -3 & 2 \end{bmatrix} - \begin{bmatrix} 3 & 11 \\ 4 & 9 \end{bmatrix}$

9. $\begin{bmatrix} 2 & -3 \\ 10 & 11 \\ 15 & -4 \end{bmatrix} + \begin{bmatrix} -1 & -6 \\ 2 & -3 \\ -7 & 3 \end{bmatrix}$

10. ACTIVITIES Janice practices trumpet 1.5 hours on Monday, 1 hour on Tuesday, and 0.5 hour on Wednesday. She practices soccer 2 hours on Monday, 1.5 hours on Tuesday, and 2 hours on Wednesday.

a. Organize this information in a matrix.

b. Name the element in the second row, third column and describe what it represents.

Key Concepts

Chapter 1

The Tools of Algebra

Order of Operations (p. 6)

Step 1 Evaluate the expressions inside grouping symbols.
Step 2 Multiply and/or divide in order from left to right.
Step 3 Add and/or subtract in order from left to right.

Substitution Property of Equality (p. 12)

If two quantities are equal, then one quantity can be replaced by the other.

Commutative Properties (p. 18)

For any numbers a and b, $a + b = b + a$.
For any numbers a and b, $a \cdot b = b \cdot a$.

Associative Properties (p. 18)

For any numbers a, b, and c, $(a + b) + c = a + (b + c)$.
For any numbers a, b, and c, $(a \cdot b) \cdot c = a \cdot (b \cdot c)$.

Numbers Properties (p. 19)

Property	Words	Symbols	Examples
Additive Identity	When 0 is added to any number, the sum is the number.	For any number a, $a + 0 = 0 + a = a$.	$5 + 0 = 5$ $0 + 5 = 5$
Multiplicative Identity	When any number is multiplied by 1, the product is the number.	For any number a, $a \cdot 1 = 1 \cdot a = a$.	$8 \cdot 1 = 8$ $1 \cdot 8 = 8$
Multiplicative Property of Zero	When any number is multiplied by 0, the product is 0.	For any number a, $a \cdot 0 = 0 \cdot a = 0$	$3 \cdot 0 = 0$ $0 \cdot 3 = 0$

Chapter 2

Operations with Integers

Absolute Value (p. 63)

The absolute value of a number is the distance the number is from zero on the number line. The absolute value of a number is always greater than or equal to zero.

Adding Integers with the Same Sign (p. 69)

To add integers with the same sign, add their absolute values. The sum is:
- positive if both integers are positive.
- negative if both integers are negative.

Adding Integers with Different Signs (p. 70)

To add integers with different signs, subtract their absolute values.
The sum is:
- positive if the positive integer's absolute value is greater.
- negative if the negative integer's absolute value is greater.

Additive Inverse Property (p. 71)

The sum of any number and its additive inverse is zero.

Subtracting Integers (p. 76)

To subtract an integer, add its additive inverse.

Multiplying Two Integers with Different Signs (p. 83)
The product of two integers with different signs is negative.

Multiplying Two Integers with the Same Sign (p. 84)
The product of two integers with the same sign is positive.

Dividing Integers with Different Signs (p. 90)
The quotient of two integers with different signs is negative.

Dividing Integers with the Same Sign (p. 91)
The quotient of two integers with the same sign is positive.

Chapter 3

Operations with Rational Numbers

Multiplying Fractions (p. 134)
To multiply fractions, multiply the numerators and multiply the denominators.

Inverse Property of Multiplication (p. 141)
The product of a number and its multiplicative inverse is 1.

Dividing Fractions (p. 142)
To divide by a fraction, multiply by its multiplicative inverse.

Adding Like Fractions (p. 147)
To add fractions with like denominators, add the numerators and write the sum over the denominator.

Subtracting Like Fractions (p. 148)
To add fractions with like denominators, add the numerators and write the sum over the denominator.

Adding Unlike Fractions (p. 153)
To add fractions with unlike denominators, rename the fractions with a common denominator. Then add and simplify as with like fractions.

Subtracting Unlike Fractions (p. 154)
To subtract fractions with unlike denominators, rename the fractions with a common denominator. Then subtract and simplify as with like fractions.

Chapter 4

Expressions and Equations

Distributive Property (p. 171)
$a(b + c) = ab + ac$ $\quad$ $(b + c)a = ba + ca$
$a(b - c) = ab - ac$ $\quad$ $(b - c)a = ba - ca$

Addition Property of Equality (p. 184)
If you add the same number to each side of an equation, the two sides remain equal.

Subtraction Property of Equality (p. 185)
If you subtract the same number from each side of an equation, the two sides remain equal.

Division Property of Equality (p. 191)
When you divide each side of an equation by the same nonzero number, the two sides remain equal.

Multiplication Property of Equality (p. 193)
When you multiply each side of an equation by the same number, the two sides remain equal.

Chapter 5

Multi-Step Equations and Inequalities

Perimeter (p. 221)

Words The perimeter of a rectangle is the sum of the measures of all four sides.

Symbols $P = \ell + \ell + w + w$

$P = 2\ell + 2w$ or $P = 2(\ell + w)$

Words The perimeter of a triangle is the sum of the measure of all three sides.

Symbols $P = a + b + c$

Area (p. 222)

Words The area of a rectangle is the product of the length and width.

Symbols $A = \ell w$

Addition and Subtraction Properties of Inequalities (p. 241)

When you add or subtract the same number from each side of an inequality, the inequality remains true.

Multiplication and Division Properties of Inequalities (p. 242)

When you multiply or divide each side of an inequality by the same positive number, the inequality remains true.

When you multiply or divide each side of an inequality by the same negative number, the inequality symbol must be reversed for the inequality to remain true.

Chapter 6

Ratio, Proportion, and Similar Figures

Ratio (p. 265)

Words A ratio is a comparison of two quantities by division.

Examples

Numbers			Algebra		
3 to 9	3:9	$\frac{3}{9}$	a to b	$a{:}b$	$\frac{a}{b}$

Property of Proportions (p. 287)

Words The cross products of a proportion are equal.

Symbols If $\frac{a}{b} = \frac{c}{d}$, then $ad = cb$. If $ad = cb$, then $\frac{a}{b} = \frac{c}{d}$ if $b \neq 0$ and $d \neq 0$.

Corresponding Parts of Similar Triangles (p. 301)

Words If two figures are similar, then

- the corresponding angles are congruent, or have the same measure, and
- the corresponding sides are proportional and opposite corresponding angles

Symbols $\triangle ABC \sim \triangle XYZ$: $\angle A \cong \angle X$, $\angle B \cong \angle Y$, $\angle C \cong \angle Z$ and $\frac{AB}{XY} = \frac{BC}{YZ} = \frac{AC}{XZ}$

Scale Factor (p. 308)

A dilation with a scale factor of k will be:

- an enlargement if $k > 1$,
- a reduction if $0 < k < 1$,
- the same as the original figure if $k = 1$.

Chapter 7

Percent

Percent (p. 331)

Words A percent is a part to whole ratio that compares a number to 100.

Examples 80% 80 out of 100 $\frac{80}{100}$

Percents and Decimals (p. 337)

To write a percent as a decimal, divide by 100 and remove the percent symbol. To write a decimal as a percent, multiply by 100 and add the percent symbol.

Percent Proportion (p. 345)

$\frac{a}{b} = \frac{p}{100}$, where a is the part, b is the whole or base and p is the percent.

Types of Percent Problems (p. 346)

Type	Example	Proportion
Find the Percent	1 is what percent of 5? or What percent of 5 is 1?	$\frac{1}{5} = \frac{p}{100}$
Find the Part	What number is 20% of 5?	$\frac{a}{5} = \frac{20}{100}$
Find the Base or Whole	1 is 20% of what number?	$\frac{1}{b} = \frac{20}{100}$

The Percent Equation (p. 358)

Type	Example	Equation
Find the Part	What number is 25% of 60? (What number = part)	$a = 0.25(60)$
Find the Percent	15 is what percent of 60? (what percent = percent)	$15 = p(60)$
Find the Whole	15 is 25% of what number? (what number = whole)	$15 = 0.25b$

Percent of Change (p. 364)

$$\text{percent of change} = \frac{\text{amount of change}}{\text{original amount}}$$

Chapter 8

Linear Functions and Graphing

Representing Functions (p. 408)

Equation	Ordered Pairs	Words	Table
$y = -2x$	(−1, 2), (0, 0), (1, −2)	The value of y is −2 times the corresponding value of x.	x / y: −1 / 2; 0 / 0; 1 / −2

Direct Variation (p. 420)

A direct variation is a relationship in which the ratio of y to x is a constant, k. We say y varies directly with x.

Proportional Linear Relationships (p. 421)

Two quantities a and b have a proportional linear relationship if they have a constant ratio and a constant rate of change.

Slope (p. 428)

Words The slope m of a line passing through points at (x_1, y_1) and (x_2, y_2) is the ratio of the difference in y-coordinates to the corresponding difference in x-coordinates.

Symbols $m = \frac{y_2 - y_1}{x_2 - x_1}$, where $x_2 \neq x_1$

Slope-Intercept Form (p. 433)

Words The slope-intercept form of an equation is $y = mx + b$, where m is the slope and b is the y-intercept.

Symbols $y = mx + b$

↑ slope ↑ y-intercept

Write a Linear Equation (p. 444)

From Slope and y-Intercept	• Substitute the slope m and y-intercept b in $y = mx + b$.
From a Graph	• Find the y-intercept b and the slope m from the graph. • Substitute the slope and y-intercept in $y = mx + b$.
From Two Points	• Use the coordinates of the two points to find the slope. • Substitute the slope and coordinates of one of the points in $y - y_1 = m(x - x_1)$
From a Table	• Use the coordinates of the two points to find the slope. • Substitute the slope and coordinates of one of the points in $y - y_1 = m(x - x_1)$

Chapter 9

Powers and Nonlinear Functions

Order of Operations (p. 472)

Step 1 Simplify the expressions inside grouping symbols first.

Step 2 Evaluate all powers

Step 3 Do all multiplications or divisions in order from left to right.

Step 4 Do all additions or subtractions in order from left to right.

Product of Powers Property (p. 481)

Words Multiply powers with the same base by adding their exponents.

Symbols $a^m \cdot a^n = a^{m+n}$

Quotient of Powers Property (p. 482)

Words Divide powers with the same base by subtracting their exponents.

Symbols $a^m \div a^n = a^{m-n}$

Negative and Zero Exponents (p. 486)

For $a \neq 0$ and any whole number n, $a^{-n} = \frac{1}{a^n}$.
For $a \neq 0$, $a^0 = 1$.

Scientific Notation (p. 493)

Words A number is expressed in scientific notation when it is written as the product of a factor and a power of 10. The factor must be greater than or equal to 1 and less than 10.

Examples $3{,}500{,}000 = 3.5 \times 10^6$ $0.00004 = 4.0 \times 10^{-5}$

Power of a Power Property (p. 499)

Words To find the power of a power, multiply exponents.

Symbols $(a^m)^n = a^{m \cdot n}$

Power of a Product Property (p. 500)

Words To find the power of a product, find the power of each factor and multiply.

Symbols $(ab)^m = a^m b^m$, for all numbers a and b and any integer m

Quadratic Function (p. 510)

A quadratic function can be described by an equation of the form $y = ax^2 + bx + c$, where $a \neq 0$.

Chapter 10

Real Numbers and Right Triangles

Square Root (p. 537)

Words A square root of a number is one of its two equal factors.

Symbols If $x^2 = y$, then x is a square root of y.

Irrational Number (p. 543)

An irrational number is a number that cannot be expressed as $\frac{a}{b}$, where a and b are integers and $b \neq 0$.

Angles of a Triangle (p. 550)

The sum of the measures of the angles of a triangle is 180°.

Pythagorean Theorem (p. 558)

Words In a right triangle, the sum of the squares of the lengths of the legs is equal to the square of the length of the hypotenuse.

Symbols $a^2 + b^2 = c^2$

Distance Formula (p. 565)

The distance d between two points with coordinates (x_1, y_1) and (x_2, y_2) is given by $d = \sqrt{(x_2 - x_1)^2 + (y_2 - y_1)^2}$.

45°-45°-90° Triangles (p. 572)

In a 45°-45°-90° triangle, the length of the hypotenuse is $\sqrt{2}$ times the length of a leg.

30°-60°-90° Triangles (p. 573)

In a 30°-60°-90° triangle,

- the length of the hypotenuse is 2 times the length of the shorter leg, and
- the length of the longer leg is $\sqrt{3}$ times the length of the shorter leg.

Chapter 11

Distance and Angle

Pairs of Angles (p. 589)

When two lines intersect, they form two pairs of opposite angles, called *vertical angles.* Vertical angles are congruent.

Two angles that have the same vertex, share a common side, and do not overlap are called *adjacent angles.*

If the sum of the measures of two angles is 90°, the angles are called *complementary angles.*

If the sum of the measures of two angles is 180°, the angles are called *supplementary angles.*

Names of Special Angles (p. 590)

When a transversal intersects two parallel lines, eight angles are formed.

- *Interior angles* lie inside the parallel lines.
- *Exterior angles* lie outside the parallel lines.

The following pairs of angles have the same measure, or are *congruent.*

- *Alternate interior angles* are on opposite sides of the transversal and inside the parallel lines.
- *Alternate exterior angles* are on opposite sides of the transversal and outside the parallel lines.
- *Corresponding angles* are in the same position on the parallel lines in relation to the transversal.

Corresponding Parts of Congruent Triangles (p. 598)

If two triangles are congruent, their corresponding sides are congruent and their corresponding angles are congruent.

Angles of a Quadrilateral (p. 612)

The sum of the measures of the angles of a quadrilateral is 360°.

Interior Angles of a Polygon (p. 618)

If a polygon has n sides, then $n - 2$ triangles are formed. The sum of the degree measures of the interior angles of the polygon is $(n - 2)180$.

Area of Parallelogram (p. 624)

Words The area A of a parallelogram in square units is $A = bh$, where b is the base of the parallelogram and h is the height.

Symbols $A = bh$

Area of Triangle (p. 625)

Words The area A of a triangle in square units is $A = bh$, where b is the base of the triangle and h is the height.

Symbols $A = \frac{1}{2}bh$

Area of Trapezoid (p. 626)

Words The area A of a trapezoid equals half the product of the height h and the sum of the bases $b_1 + b_2$.

Symbols $A = \frac{1}{2}h(b_1 + b_2)$

Circumference of a Circle (p. 631)

Words The circumference C of a circle is equal to its diameter times π, or 2 times its radius times π.

Symbols $C = \pi d$ or $C = 2\pi r$

Area of a Circle (p. 636)

Words The area A of a circle equals π times the square of its radius r.

Symbols $A = \pi r^2$

Area of a Sector (p. 638)

Words The area A of a sector is $A = \frac{N}{360}(\pi r^2)$, where N is the degree measure of the central angle of the circle and r is the radius.

Symbols $A = \frac{N}{360}(\pi r^2)$

Chapter 12

Surface Area and Volume

Volume of a Prism (p. 671)

Words The volume V of a prism is the area of the base B times the height h.

Symbols $V = Bh$

Volume of a Cylinder (p. 677)

Words The volume V of a circular cylinder with radius r is the area of the base B times the height h.

Symbols $V = Bh$, where $B = \pi r^2$ or $V = \pi r^2 h$

Volume of a Pyramid (p. 683)

Words The volume V of a pyramid is one-third the area of the base B times the height h.

Symbols $V = \frac{1}{3}Bh$

Volume of a Cone (p. 684)

Words The volume V of a cone with radius r is one-third the area of the base πr^2 times the height h.

Symbols $V = \frac{1}{3}Bh$, where $B = \pi r^2$ or $V = \frac{1}{3}\pi r^2 h$

Volume of a Sphere (p. 685)

Words The volume V of a sphere with radius r is four-thirds times π times the radius cubed.

Symbols $V = \frac{4}{3}\pi r^3$

Surface Area of Cylinders (p. 697)

Words The surface area S of a cylinder is the lateral area L plus the area of the two bases $(2\pi r^2)$.

Symbols $S = L + 2B$ or $S = 2\pi rh + 2\pi r^2$

Surface Area of Prisms (p. 691)

Words The surface area S of a prism is the lateral area L plus the area of the two bases $2B$.

Symbols $S = L + 2B$ or $S = Ph + 2B$

Lateral Area and Surface Area of a Pyramid (p. 702)

Words The lateral area L of a regular pyramid is half the perimeter P of the base times the slant height ℓ.

Symbols $L = \frac{1}{2}P\ell$

Words The total surface area S of a regular pyramid is the lateral area L plus the area of the base B.

Symbols $S = L + B$ or $S = \frac{1}{2}P\ell + B$

Lateral Area and Surface Area of a Cone (p. 704)

Words The lateral area L of a cone is π times the radius times the slant height ℓ.

Symbols $L = \pi r \ell$

Words The surface area S of a cone with slant height ℓ and radius r is the lateral area plus the area of the base.

Symbols $S = L + \pi r^2$

Ratio of Surface Area and Volume of Similar Solids (p. 711)

If two solids are similar with a scale factor of $\frac{a}{b}$, then the surface areas have a ratio $\left(\frac{a}{b}\right)^2$ and the volumes have a ratio $\left(\frac{a}{b}\right)^3$.

Chapter 13 Statistics and Probability

Measures of Central Tendency (p. 730)

mean sum of the data divided by the number of items in the data set

median middle number of the data ordered from least to greatest, or the mean of the middle two numbers

mode number or numbers that occur most often

Interquartile Range (p. 744)

Words The interquartile range is the range of the middle half of a set of data. It is the difference between the upper quartile and the lower quartile.

Symbols Interquartile range $= UQ - LQ$

Probability (p. 765)

Words The probability of an event is a ratio that compares the number of favorable outcomes to the number of possible outcomes.

Symbols $P(\text{event}) = \frac{\text{number of favorable outcomes}}{\text{number of possible outcomes}}$

Unbiased Samples (p. 771)

Simple Random Sample
a sample where each item or person in a population is as likely to be chosen as any other

Stratified Random Sample
a sample in which the population is divided into similar, nonoverlapping groups. A simple random sample is then selected from each group

Systematic Random Sample
a sample in which the items or people are selected according to a specific time or item interval

Convenience Sample
a sample which includes members of the population that are easily accessed

Voluntary Response Sample
a sample which involves only those who want to participate in the sampling.

Fundamental Counting Principle (p. 777)

If event M can occur in m ways and is followed by event N that can occur in n ways, then the event M followed by N can occur in $m \cdot n$ ways.

Probability of Two Independent Events (p. 790)

Words The probability of two independent events is found by multiplying the probability of the first event by the probability of the second event.

Symbols $P(A \text{ and } B) = P(A) \cdot P(B)$

Key Concepts

Probability of Two Dependent Events (p. 791)

Words If two events, A and B, are dependent, then the probability of both events occurring is the product of the probability of A and the probability of B after A occurs.

Symbols $P(A \text{ and } B) = P(A) \cdot P(B \text{ following } A)$

Probability of Mutually Exclusive Events (p. 792)

Words The probability of one or the other of two mutually exclusive events can be found by adding the probability of the first event to the probability of the second event.

Symbols $P(A \text{ or } B) = P(A) + P(B)$

Key Concepts

Selected Answers and Solutions

For Homework Help, go to Hotmath.com
Complete, step-by-step solutions of most odd-numbered exercises are provided free of charge.

Chapter 0 Start Smart: Preparing for Pre-Algebra

Pages P3–P4 Chapter 0 Pretest

1. 5 postcards and 5 letters **3.** 25 min **5.** 80 chairs **7.** 24.71 **9.** 81.85 **11.** 3568.87 **13.** 44.08 **15.** 26.95 **17.** 524.72 **19.** 47 **21.** 229.32 **23.** 221.652 **25.** 2.48 **27.** 5.67 **29.** 300.96 **31.** about 1.5 times **33.** Sample answer: the money earned is equal to 5 times the number of cars washed plus five; $55. **35.** Sample answer: the value of the term is equal to one less than five times the term number; 49. **37.** 101 cm **39.** inch **41.** mile **43.** 3.5 **45.** 135 **47.** $6\frac{2}{3}$ **49.** 6400 **51.** 430 **53.** 148 tickets

Page P6 Lesson 0-1

1. 16 pizzas **3.** $10 \times 10 \times 10$

Pages P9–P10 Lesson 0-2

1. 28 **3.** 27 and 29

5. 45 cards

Total Number of Cards	Amount Traded	Amount Received	New Total
55	8	5	52
52	6	4	50
50	5	3	48
48	12	9	45

7. $8 **9.** Sample answer: 125 candy bars and 105 pretzels **11.** 10 **13.** 12

15. 16 combinations

First Spin	Second Spin	First Spin	Second Spin
red	red	green	red
red	blue	green	blue
red	green	green	green
red	yellow	green	yellow
blue	red	yellow	red
blue	blue	yellow	blue
blue	green	yellow	green
blue	yellow	yellow	yellow

17. $119 **19.** Sample answer: 3 $5 bills, 2 $10 bills, and 7 $20 bills

Page P12 Lesson 0-3

1. 40.09 **3.** 20.411 **5.** 71.8 **7.** 2.918 **9.** $207.94 **11.** $12.39 **13.** 53.58 **15.** 105.35 **17.** 2.1 **19.** 3993 **21.** 1.15354 **23.** 7 **25.** 7.72 in.

Page P14 Lesson 0-4

1. Sample answer: the number of bracelets made is equal to 25 times the number of hours; 500 bracelets. **3.** Sample answer: the value of the term is equal to three more than the term number; 23.

5a.

Number of Tickets	1	2	3	4	5
Total Cost ($)	6.50	13.00	19.50	26.00	32.50

5b. Sample answer: The total cost is 6.5 times the number of tickets. **5c.** $52.00

7a.

Term Number	1	2	3
Number of Cubes	8	12	16

7b. Sample answer: The number of cubes is four more than four times the term number.

7c. 44 cubes

Page P16 Lesson 0-5

1. 33 in. **3.** 128 m **5.** $11\frac{1}{4}$ ft **7.** 76 mm **9.** 93 mm **11.** $3\frac{3}{16}$ in. **13a.** 8

13b.

Side Length	Perimeter
1 in.	4 in.
2 in.	8 in.
3 in.	12 in.
4 in.	16 in.

13c. it doubles; it triples; it quadruples

13d. 32 in.

Page P18 Lesson 0-6

1. in. **3.** mi **5.** cm **7.** km **9.** $1\frac{1}{2}$ in.; $1\frac{1}{4}$ in. **11.** 1.75 **13.** 63 **15.** 153 **17.** 750 **19.** 540 **21.** 7.4

Page P22 Lesson 0-7

1. 4 students **3.** Sample answer: Which homeroom had the greatest number of students with their driver's permit? Homeroom 16
5. Sample answer: 1300 pounds

7. Sample answer: There are about the same number of malls in New York and Ohio.

Pages P22–P23 *Chapter 0* *Posttest*

1. 7587 ft **3.** 204 squares **5.** 29 ways **7.** 8.8 **9.** 136.88 **11.** 603.247 **13.** 0.6 **15.** 280.94 **17.** 472.33 **19.** $35.60 **21.** 28.2 **23.** 3341.94 **25.** 555.744 **27.** 7.2 **29.** 15.84 **31.** 294.212963 **33.** about 2 times **35.** Sample answer: the bracelets made are equal to 10 times the number of people plus five; 105 bracelets. **37.** Sample answer: the value of the term is equal to one more than five times the term number; 51. **39.** 56 in. **41.** centimeter **43.** kilometer **45.** 4 **47.** 216 **49.** 6160 **51.** 165.6 **53.** 6.2 **55.** 430 **57.** 1.4 times

Chapter 1 The Tools of Algebra

Page 3 *Chapter 1* *Get Ready*

1. 10 **3.** 3.9 **5.** 7.5 **7.** $4.50 **9.** Sample answer: 1790 **11.** Sample answer: 7200 **13.** Sample answer: 200 mi **15.** 1 **17.** 4

19. (number line from 28 to 39, point at 32)

28 29 30 31 32 33 34 35 36 37 38 39

Pages 7–9 *Lesson 1-1*

1. 6×8 **3.** 26 **5.** 33 **7.** 36 **9.** 5
11. $75 + 6(25)$; $225

Number of Passengers	Expression	Cost ($)
25	75 + 6(25)	225
30	75 + 6(30)	255
35	75 + 6(35)	285
40	75 + 6(40)	315

13. $8 + 4$ **15.** 9×3 **17.** $15 - 10$ **19.** 10

21 $2[3 + 7(4)] = 2[3 + 28]$
$= 2[31]$
$= 62$

23. 76 **25.** 16 **27.** 28 **29.** 2

31 Words: $8 plus 6 times $0.75 per line
Expression: $8 + 6(0.75)$
$8 + 6(0.75) = 8 + 4.50$
$= \$12.50$

Number of Lines	Expression	Cost ($)
6	8 + 6(0.75)	12.50
10	8 + 10(0.75)	15.50
14	8 + 14(0.75)	18.50
18	8 + 18(0.75)	21.50

33a.

Term Number	Number of Toothpicks
1	4
2	6
3	8

33b. The number of toothpicks is two more than twice the term number. **35a.** around $(64 - 20)$
35b. $64 - (20 \div 4) + 6 = 64 - 5 + 6$
$= 59 + 6$
$= 65$

37a. 200(2.54); 508 cm **37b.** $508 + 350 = 858$ cm
37c. 90 in. $\approx$ 90(2.54) or about 230 cm; $858 \div 230 \approx 3.73$; about 4 packages of 90-in. trim
39. C **41.** B

Pages 14–16 *Lesson 1-2*

1. $c - 4$ **3.** $10h$ **5.** 25 **7.** 16 **9.** 12 **11a.** $16p$
11b. 80 fl oz

13 Words: twenty-four divided among some students
Variable: Let s represent some students.
Expression: $24 \div s$

15. $12n$ **17.** $10n - 4$ **19.** 4 **21.** 34 **23.** 18 **25.** 6 **27.** 18 **29.** 72 **31.** $\frac{w}{231}$ **33.** 13 **35.** 50 **37.** 58

39 Words: 42 pounds per bushel
Expression: $42b$
To find the number of pounds in 100 trees that each produce 6 bushels, find $100 \cdot 42 \cdot 6 = 25{,}200$ pounds.

41. $6a + c$; $a = 4$; $c = 8$; $32 total cost for 4 items costing $6 each and an $8 item **43.** $2n + 4$; $2(n + 2)$; Sample answer: The first expression represents the top multiplied by two to give the top and bottom, then add 4 for the sloped sides. The second expression represents half of the figure, then multiply by two to get the whole figure. **45.** C **47.** B **49.** 14 **51.** 15 **53.** 2 **55.** 9 **57.** 48 **59.** 43 points **61.** 23 **63.** 32

Pages 21–23 *Lesson 1-3*

1. no; $10 - 6 \neq 6 - 10$ **3.** Identity (×) **5.** Identity (+) **7.** Associative (×) **9.** $11 + k$ **11.** $40x$ **13.** $72a$ **15.** no; $13 + 15 = 28$ **17.** Identity (+) **19.** Commutative (+) **21.** Multiplicative (0) **23.** Commutative (+) **25.** $d + 28$

27 $(54 + p) + 16 = (54 + 16) + p$
$= 70 + p$

29. $72s$ **31.** $44t$ **33.** $72c$ **35.** $30 + t$ **37.** $1978 + f$

39 Words: the sum of two times a number and five added to six times a number

Expression: $2n + 5 + 6n$

Simplify: $2n + 5 + 6n = 8n + 5$

41. $8 + 6n + 9n + 1$; $15n + 9$ **43a.** $\$5x + \$61.25$ **43b.** \$79 **45.** false; $15 + (4 \cdot 6) = 39$ and $(15 + 4) \cdot 6 = 114$; Since $39 \neq 114$, the statement is false. **47a.** No; $2 - 3 = -1$ and -1 is not a whole number. **47b.** No; $1 + 1 = 2$ and 2 is not a member of the set. **47c.** The Closure Property for Multiplication states that because the product of two whole numbers is also a whole number, the set of whole numbers is closed under multiplication. **47d.** Yes. $0 \cdot 0 = 0$, $0 \cdot 1 = 0$, $1 \cdot 0 = 0$, and $1 \cdot 1 = 1$. **49.** D **51.** B **53.** $s + 200$ **55.** $h - 6$ **57a.** 71°F **57b.** 62°F **59.** 45 **61.** 24 **63.** 210 **65.** 405 **67.** 54 **69.** 336 **71.** 5117

Pages 28–30 ***Lesson 1-4***

1–4.

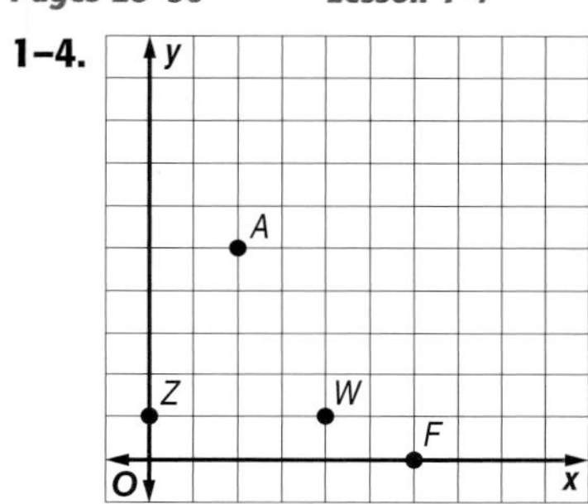

5. (3, 4) **7.** (5, 2)

9.

x	y
3	4
1	5
4	2

$D = \{1, 3, 4\}$, $R = \{2, 4, 5\}$

11a.

x	y	(x, y)
1	2	(1, 2)
2	4	(2, 4)
3	6	(3, 6)
4	8	(4, 8)

11b.

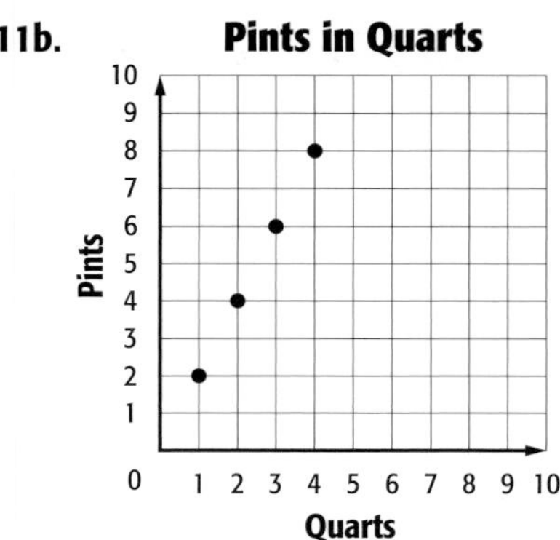

The points appear to lie in a line.

12–19.

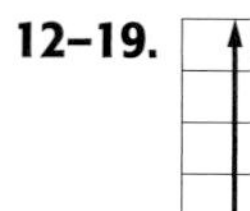

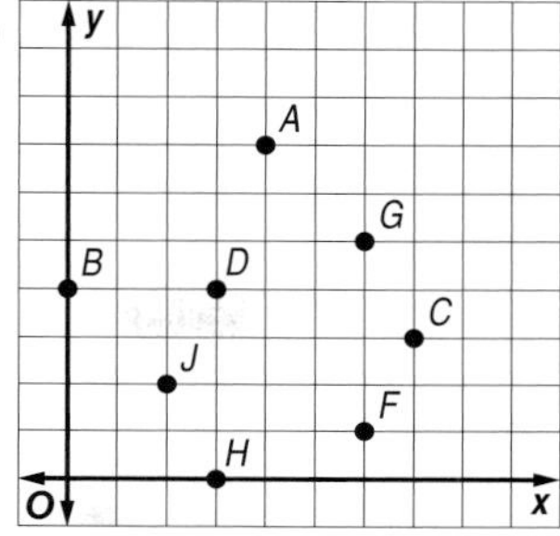

21. (6, 4) **23.** (5, 2) **25.** (7, 2) **27.** (4, 5)

29

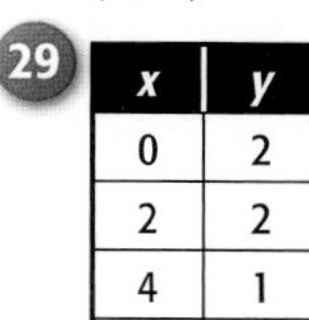

x	y
0	2
2	2
4	1
3	5

The domain is the set of x-values. $D = \{0, 2, 3, 4\}$
The range is the set of y-values. $R = \{1, 2, 5\}$

31.

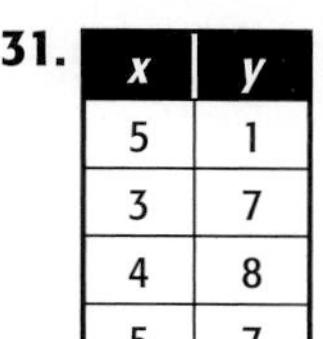

x	y
5	1
3	7
4	8
5	7

$D = \{3, 4, 5\}$, $R = \{1, 7, 8\}$

33a.

x	y	(x, y)
1	3	(1, 3)
2	6	(2, 6)
4	12	(4, 12)
6	18	(6, 18)

33b.

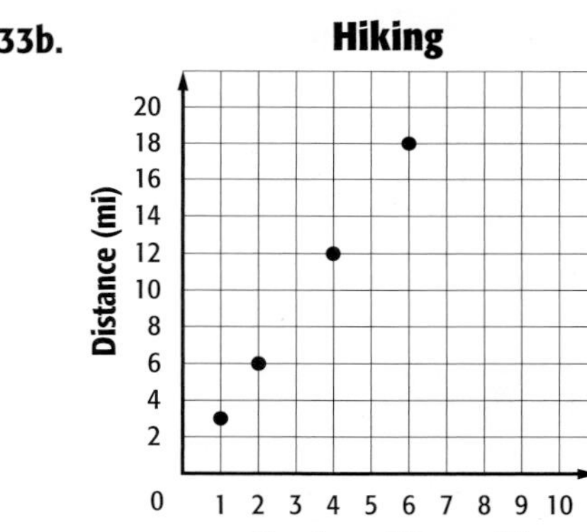

35 a.

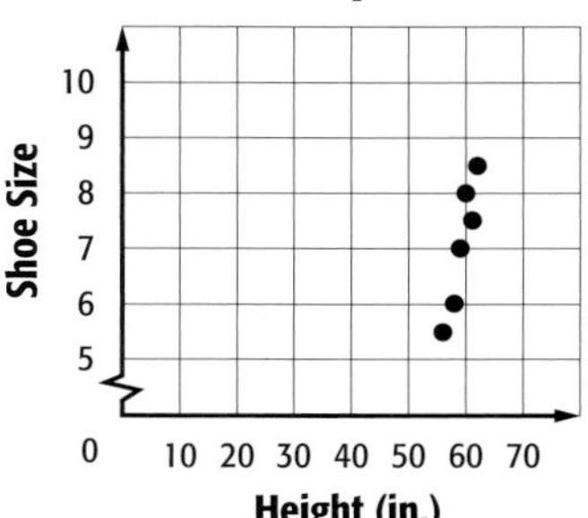

b. Sample answer: The graph is not linear like the other graph.

37a. {(1, 4), (2, 7), (3, 10), (4, 13)}

37b.

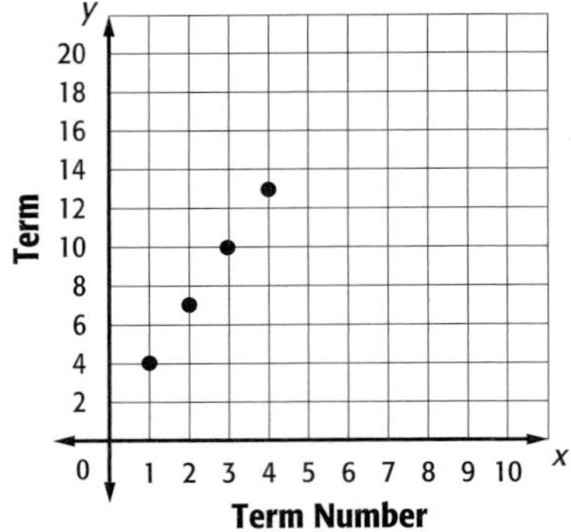

37c. The graph shows a positive, linear relationship. **37d.** 3 times the x value plus 1 to get the y-value; 61 **39.** Exercise 33; Sample answer: By connecting the points on the graph in Exercise 33, you could determine how far he will have hiked at any time other than hours. In Exercise 32, the points do not need to be connected because you wouldn't need to know how much a portion of a pizza would cost.
41. C **43.** A **45.** Commutative (×)
47. Identity (+) **49.** no; $(100 \div 10) \div 2 \neq 100 \div (10 \div 2)$ **51a.** $3s + 16$ **51b.** $91
53. 7 **55.** 16 **57.** 14

Pages 35–37 ***Lesson 1-5***

1.

Number of Touchdowns	Number of Points
Input (x)	**Output (y)**
1	6
2	12
5	30
7	42

D:{1, 2, 5, 7}; R:{ 6, 12, 30, 42}
3a. $z = 16p$
3b.

Input (p)	$16p$	Output (z)
5	16(5)	80
8	16(8)	128
11	16(11)	176
13	16(13)	208

3c.

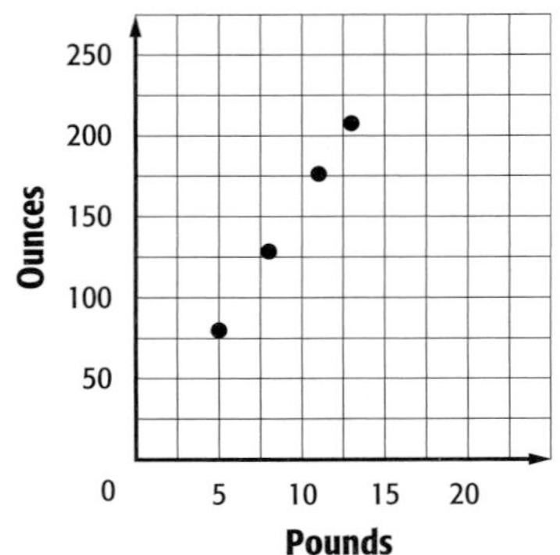

5.

Weight of Cat (lb)	Weight of Dog (lb)
Input (x)	**Output (y)**
3	7
6	10
9	13
12	16

D: {3, 6, 9, 12}; R: {7, 10, 13, 16}

7 Step 1: Create a function table showing the input and output. Step 2: The rule "five less than four times as many baseball cards." This translates into $4x - 5$. Use the rule to fill in the table. The domain is {3, 7, 11, 15}; the range is {7, 23, 39, 55}.

Ben's Cards	Casey's Cards
Input (x)	**Output (y)**
3	7
7	23
11	39
15	55

9 **a.** Words: area = side × side
Expression: $A = 10 \cdot s \cdot s$

b.

Input (x)	$10 \cdot s \cdot s$	Output (y)
6	$10 \cdot 6 \cdot 6$	360
12	$10 \cdot 12 \cdot 12$	1440
15	$10 \cdot 15 \cdot 15$	2250
24	$10 \cdot 24 \cdot 24$	5760

c. Plot the ordered pairs on a coordinate grid.

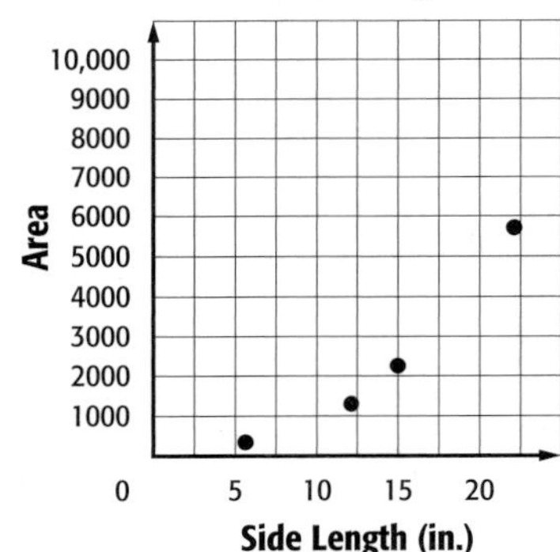

11a.

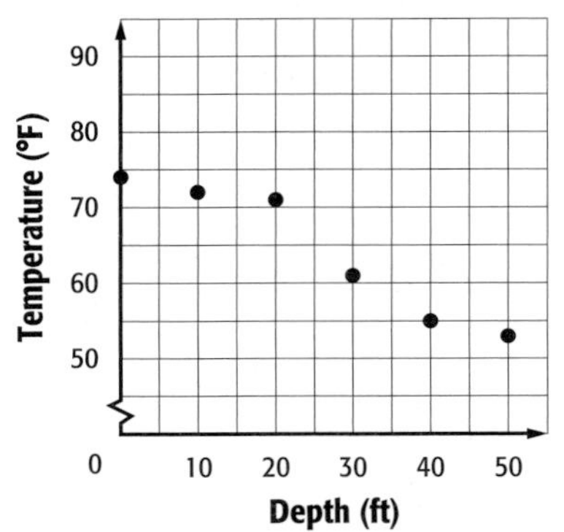

11b. No; the change in temperature is not constant so one equation cannot be used to find any temperature value. **11c.** yes; Sample answer: Each depth is paired with only one temperature. **13.** $2(x + 4)$ **15a.** 26, 32 **15b.** $3x + 2$ **17.** J **19.** (1, 7) **21.** (0, 4) **23.** (2, 1) **25.** $m + 12$ **27.** $37 + k$ **29.** $16y$ **31a.** $20n$ **31b.** 140 nickels **33.** 14 **35.** 15

Pages 42–46 *Lesson 1-6*

1 a.

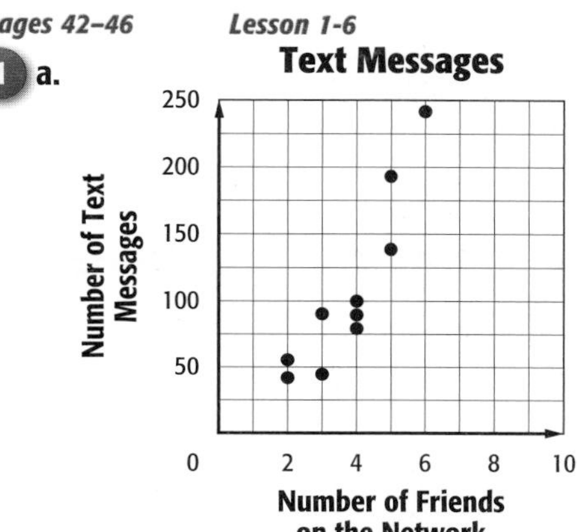

Let the x-axis represent the number of friends. Let the y-axis represent the number of text messages. Then graph the ordered pairs.

b. As the number of friends increases, the number of messages increases. There is a positive relationship between the number of friends and the number of messages.

c. By looking at the pattern in the graph, you can determine that a group of 8 friends will have around 300 messages.

3 Let the x-axis represent the year. Let the y-axis represent the winning times. Then graph the ordered pairs.

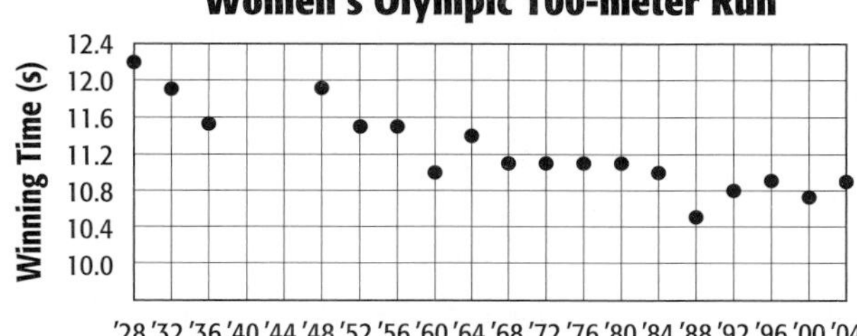

5. Negative; candle burn time increases as the height of the candle decreases. **7.** Positive; as the distance increases, the number of gallons increases. **9.** Positive; as the lawn size increases, the amount of water used increases.

11 a. Let the x-axis represent the month. Let the y-axis represent the temperature. Then graph the ordered pairs.

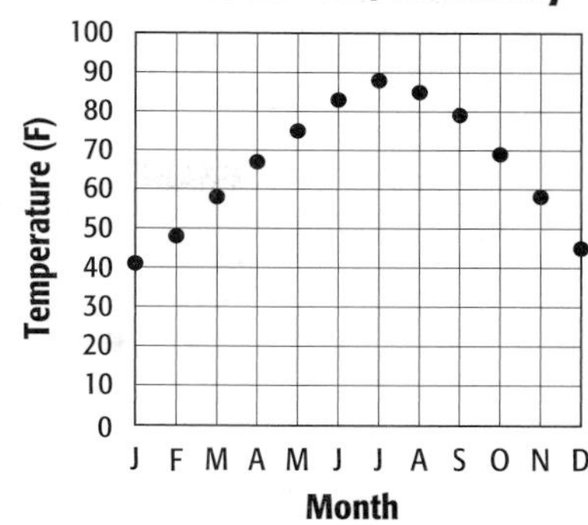

b. Sample answer: There is a positive relationship from January through July then a negative relationship for the rest of the year.

13a.

x	y
1	5
1	6
1	7
2	6
2	8
2	9
3	11
4	10
4	12

13b. Sample answer: The x-value could represent the number of people that went out to lunch and the y-value could represent the total cost of lunch.
13c. Sample answer: Overall, the data shows a positive relationship between the x and y values.

15a. Sample answer:

Length (in.)	Width (in.)
2	18
4	9
6	6
12	3
18	2

15b.

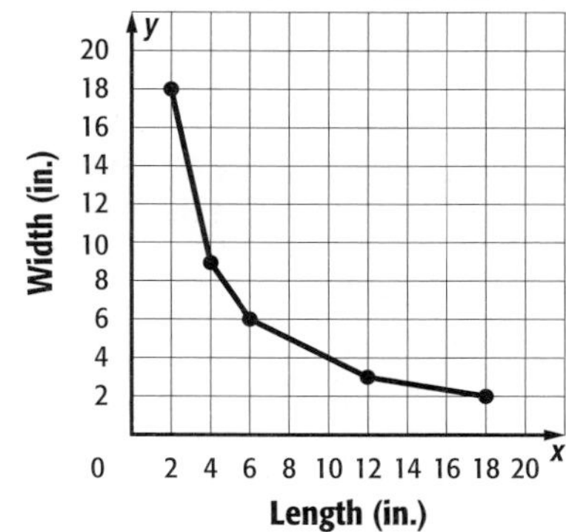

15c. about 5 in. **17.** negative **19.** Sample answer: Because both are warm weather activities, the scatter plot would show a positive relationship. They are both independent events, so they do not affect each other. **21.** No; Sample answer: The table does represent a function.

Sonya is incorrect because even though it shows a negative relationship, it is still a function. Melisa is incorrect because each x-value can be paired with only 1 y-value, but y-values can be paired with any number of x-values.
23. C **25.** B **27a.** $32.50 + 4.95c$

27b.

Number of Channels	Total Cost
0	32.50
1	37.45
2	42.40
3	47.35
4	52.30

27c.

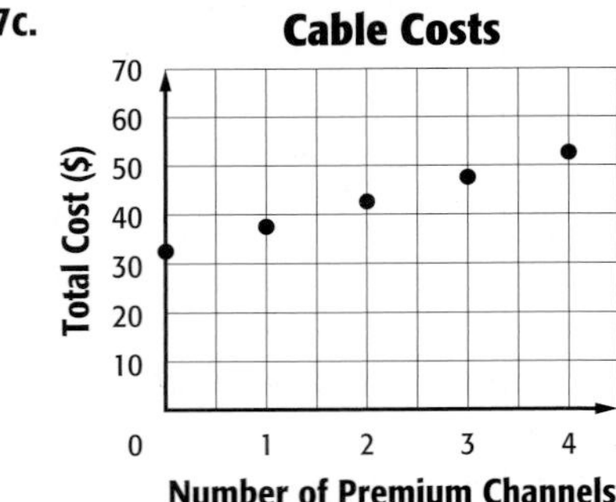

28–33.

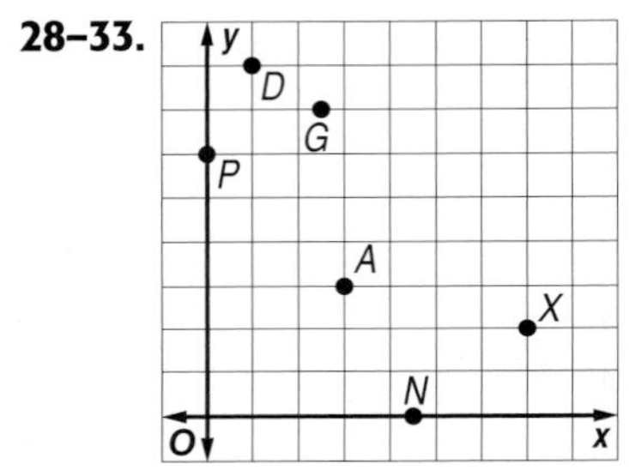

35. > **37.** = **39.** >

Pages 49–52 *Study Guide and Review*

1. true **3.** true **5.** true **7.** true **9.** false; scatter plot **11.** 45 **13.** \$130 **15.** 20 **17.** 33 **19.** $\frac{x}{36}$
21. Associative Property (×) **23.** $12 + v$

25.

x	y
2	3
2	6
2	5

D = {2}; R = {3, 5, 6}

27a.

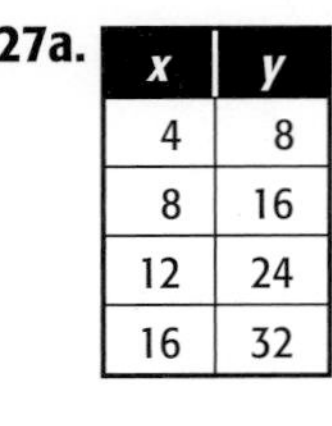

x	y
4	8
8	16
12	24
16	32

27b.

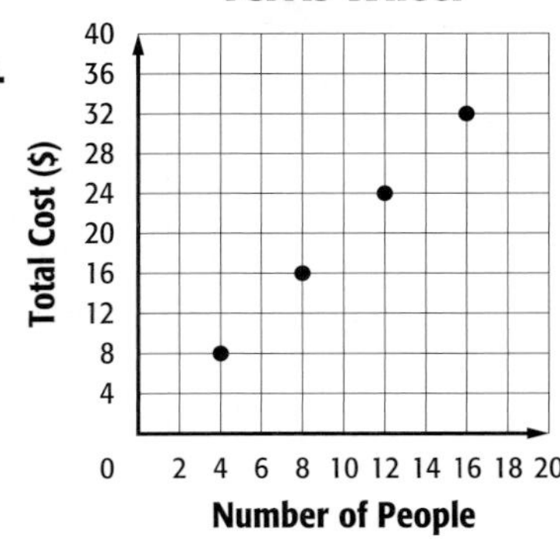

The points appear to fall in a line.

29.

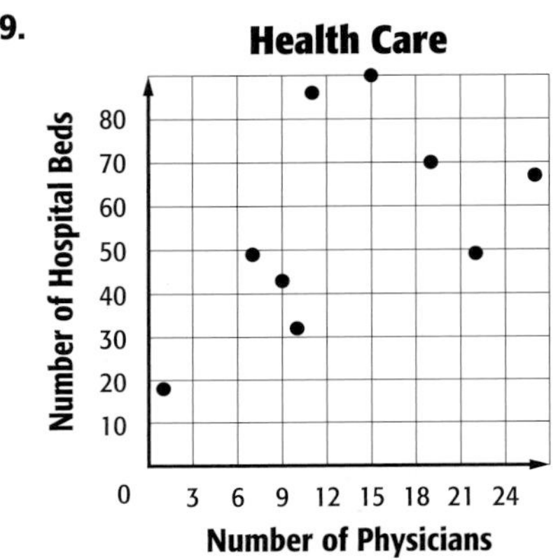

Chapter 2 Operations with Integers

Page 59 *Chapter 2* *Get Ready*

1. 18 **3.** 21 **5.** 40 **7.** 60 **9.** (1, 6) **11.** (2, 4) **13.** (3, 0) **15.** (5, 2)

Pages 64–66 *Lesson 2-1*

1. −500

−500 −400 −300 −200 −100 0 100

3. $2 > -5; -5 < 2$ **5.** $1 > -1; -1 < 1$ **7.** <
9. −80, −48, −45, −39, −34, −27, −23, −2,12
11. 17 **13.** 9 **15.** 18

17. −200

−300 −200 −100 0 100 200 300

19. 0

−6 −4 −2 0 2 4 6

21. $14 > -8; -8 < 14$ **23.** $-11 > -12; -12 < -11$
25. $4 > 0; 0 < 4$ **27.** $30 > 27; 27 < 30$ **29.** >
31. < **33.** > **35.** =

37 Graph each score on a number line.

−6 −4 −2 0 2 4 6

So, the scores from least to greatest are −4, −3, −2, −1, +1, +2, +3, +4, +5, +6

39. 17 **41.** −15

43 $-|-7| + |12| = -7 + 12$
$= 5$

45. 26 **47.** 7 **49.** 21 **51.** 23 **53.** Movie F; The absolute value of –8 is 8 and 8 is greater than any other value in the table. **55.** −272, −249, −201, −157, −101 **57.** Saturn; $-218 < -162$
59. 5, 0, −5, −13 **61.** 62, 20, −28, −35, −59
63. 232, 88, −72, −83, −94, −165
65. Always; The absolute value of a non-zero number is always positive. **67.** Sometimes; when $x = 0$, $|-0| = -|0|$, but when $x = 1$, $|-1| \neq -|1|$. **69.** 1 **71.** D **73.** −10, −9, −4, 0, 5, 7; Sample answer: Graph each integer on a number line. Then write the numbers as they appear from left to right. The integers −10, −9,

−4, 0, 5, and 7 are in order from least to greatest. **75.** Positive; As the number of hours increases, the number of homeruns increases. Thus, the scatter plot shows a positive relationship. **77a.** $y = 3x$

77b.

x	y
3	9
4	12
5	15
6	18

77c.

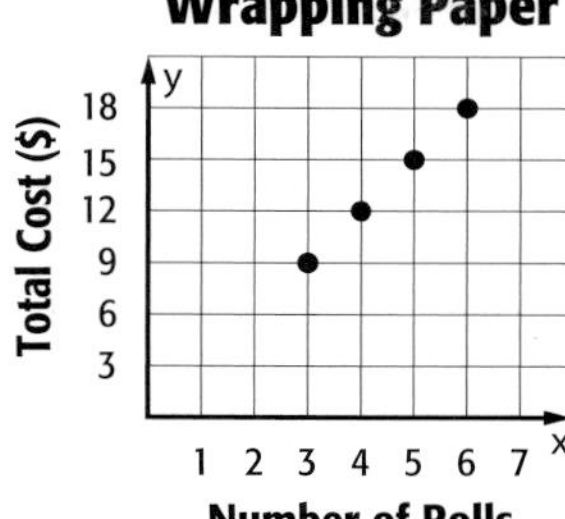

79. 438 **81.** 1025 **83.** 181

Pages 72–74 ***Lesson 2-2***

1. −11 **3.** −7 **5.** $-1500 + (-1250) = p$; −2750
7. −1 **9.** −10 **11.** 9 **13.** −5 **15.** 6
17. $6 + (-10) = x$; −4 yards **19.** 10 **21.** −16

23 $8 + (-11) + (-19) + 11 = (8 + 11) + (-11 + -19)$
$= 19 + (-30)$
$= -11$

25. −128 ft **27.** 11 **29.** 8 **31.** 35

33 **a.** Rock: $33 + 1 = 34\%$; Rap: $10 + 2 = 12\%$; Pop: $9 - 2 = 7\%$; Country: $14 - 1 = 13\%$
b. $1 + 2 + (-2) + (-1) = 0\%$

35. −20 **37.** Sample answer: At midnight the temperature was 0°F. From midnight to 3:00 A.M. the temperature dropped 5°. From 3:00 A.M. to 6:00 A.M. the temperature raised 4°. What was the temperature at 6:00 A.M.? **39.** Commutative Property (×) **41.** −22 + (28) equals positive 6, while the sum of the other expressions is negative 6. **43.** $4 - 3y$ **45.** A **47.** H
49. −54 **51.** no relationship **53.** Commutative Property (×) **55.** Identity Property (×)
57. 18 **59.** 51 **61.** 3 **63.** 49

Pages 78–80 ***Lesson 2-3***

1. −2 **3.** −24 **5.** 31 **7.** 4 **9.** −16 in. **11.** −9
13. −1 **15.** −7 **17.** 15

19 $-12 - (-11) = -12 + 11$
$= -1$

21. 10 **23.** −60 **25.** −$13 **27.** −1 **29.** −12
31. −20 **33.** 17 **35a.** +11,000; −3000; +4000; +26,000; −2000; −1000; +9000 **35b.** 44,000
37. −214

39 **a.** 30.59 − 33.30 = $−2.71; 31.04 − 30.59 = $0.45; 31.97 − 31.04 = $0.93; 30.15 − 31.97 = $−1.82
b. highest: $0.93;
lowest: −2.71; 0.93 −2.71 = 0.93 + 2.71
= $3.64

41. Sample answer: $4 - (-7)$; 11 **43.** false; $2 - (-2) = 4$ and $(-2) - 2 = -4$ **45.** C **47.** 5765
49. $4 + (-5) = -1$ **51.** > **53a.** $y = 6x$

53b.

Input (x)	$6x$	Output (y)
2	6(2)	12
4	6(4)	24
5	6(5)	30
7	6(7)	42

53c.

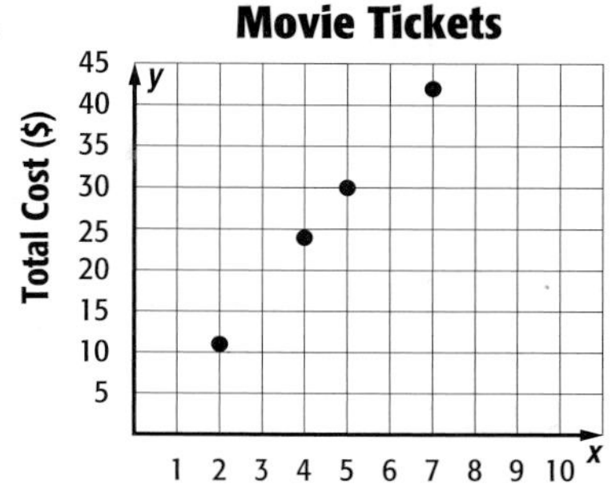

55. $r - 5$ **57.** $7 + n \div 8$ **59.** 56 **61.** 420

Pages 86–88 ***Lesson 2-4***

1. −42 **3.** 120 **5.** −$40 **7.** $-21ab$ **9.** −88

11 $3(-9) = -27$; The factors have different signs. The product is negative.

13. 75 **15.** 49 **17.** −336 **19.** −12°F **21.** $-30m$
23. $81mn$ **25.** $-36ab$ **27.** $-54efg$ **29.** −100
31. −72 **33.** $y = -2x$ **35.** $y = -4x$ **37.** < **39.** >

41 Team 1 answered 12 questions correctly, 3 questions incorrectly, and passed on 1 question. So, Team 1 earned $12(5) + 3(-8) + 1(-2)$ or 34 points. Team 2 answered 13 questions correctly, 2 questions incorrectly, and passed on 7 questions. So, Team 2 earned $13(5) + 2(-8) + 7(-2)$ or 35 points. Since $34 < 35$, Team 2 won.

43. Sample answer: 2 and −7 **45.** 22
47. false; Sample answer: $-3(-2)(-2) = -12$
49. Sample answer: The sign of the product will be negative if there is an odd number of negative integers; otherwise the sign will be positive. **51.** J **53.** 62.5 **55.** 6 **57.** −3
59a. $-54 > -70$ **59b.** −54°, −70°, −80°
61. 9 **63.** 13

Pages 92–95 ***Lesson 2-5***

1 $40 \div -10 = -4$; The quotient of two integers with different signs is negative.

3. $8.\overline{6}$ **5.** −16 **7.** −9 **9.** −4 **11.** −7 **13.** −3
15. 18 **17.** −15 **19.** 14 **21.** −110 **23.** 5
25. 14 **27.** −36 **29.** 12

31 To find the mean, add the amount of the transactions and divide by the number of transactions.

$$\frac{250 + (-60) + (-94) + 300 + (-186)}{5} = \frac{210}{5}$$
$$= 42$$

The mean transaction amount is $42.

33. −184.4°C **35.** −$300; Sample answer:

Every month, the company spends \$300 more than they earn. **37.** $<$ **39.** $>$

41 $22 = x \div (-34)$
Use guess and check.
$22 \stackrel{?}{=} (-714) \div (-34)$
$22 \stackrel{?}{=} 21$ too low
$22 \stackrel{?}{=} (-748) \div (-34)$
$22 \stackrel{?}{=} 22$ ✓
$x = -748$

43. −13 **45.** −40 ÷ 8; −5°F; Each hour the temperature dropped 5°F. **47a.** −244.8 m **47b.** −234.8 m; The mean would be 10 meters higher. **49.** Sample answer: −110 ÷ 5 = −22 **51.** 4, −1; Divide the previous term by −4 **53.** Sample answer: The Associate Property is not true for the division of integers because how the integers are grouped affects the solution. [24 ÷ (−6)] ÷ 2 = −2; 24 ÷ [(−6 ÷ 2) = −8; The Commutative Property is not true for the division of integers because the order of the integers effects the solution. −2 ÷ 10 = −0.2; 10 ÷ −2 = −5 **55.** J **57a.** −5°F **57b.** $-5h$

57c.

Input (x)	300m	Output (y)
1	5(1)	5
3	5(3)	15
5	5(5)	25

59. −5 **61.** −7 **63.** 3 **65.** −\$7 **67.** (5, 1) **69.** (5, 4)

Pages 98–100 ***Lesson 2-6***

1. (−5, 2) **3.** (5, 2) **5.** II **7.** III

9.

$x - y = 4$		
x	y	(x, y)
6	2	(6, 2)
5	1	(5, 1)
4	0	(4, 0)
−1	−5	(−1, −5)
−2	−6	(−2, −6)

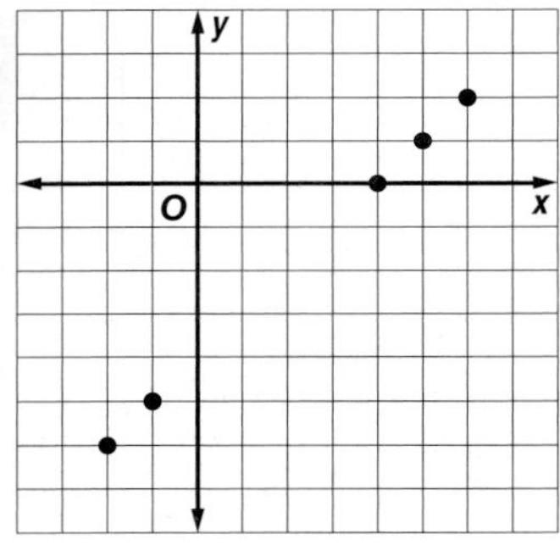

The points are along a diagonal line that crosses the y-axis at $y = 4$ and the x-axis at $x = 4$.
11. (−3, 1) **13.** (3, 5) **15.** (−4, −3) **17.** (5, −3) **19.** (0, 3)

20–31.

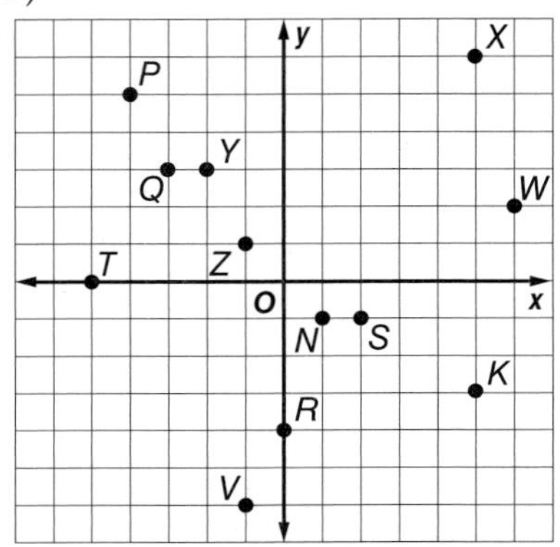

21 To graph (−2, 3), move left from the origin 2 units and then up 3 units. Draw a dot and label. The point is in quadrant II.

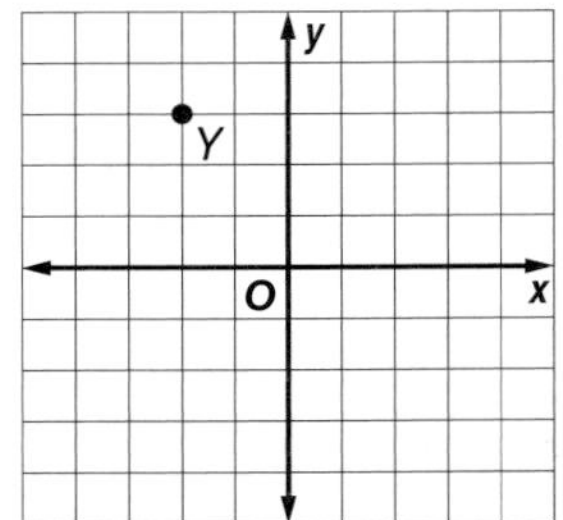

23. I **25.** IV **27.** none **29.** II **31.** IV

33.

$x - y = 10$		
x	y	(x, y)
30	20	(30, 20)
20	10	(20, 10)
10	0	(10, 0)
−20	−30	(−20, −30)
−30	−40	(−30, −40)

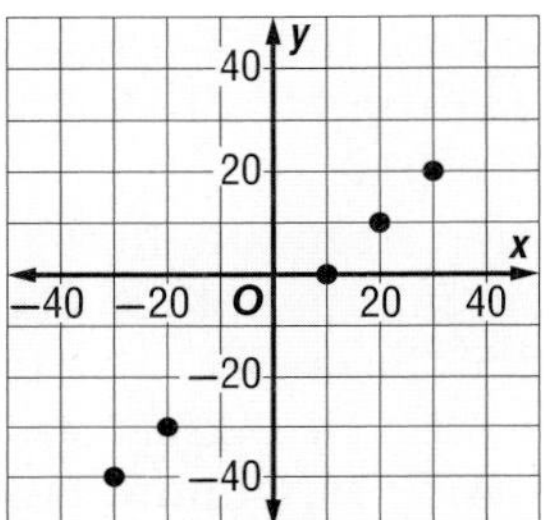

The points on the graph are in a line that slants downward to the left. The line crosses the x-axis at $x = 10$. **35.** IV **37.** IV

39 Graph each ordered pair on a coordinate plane.

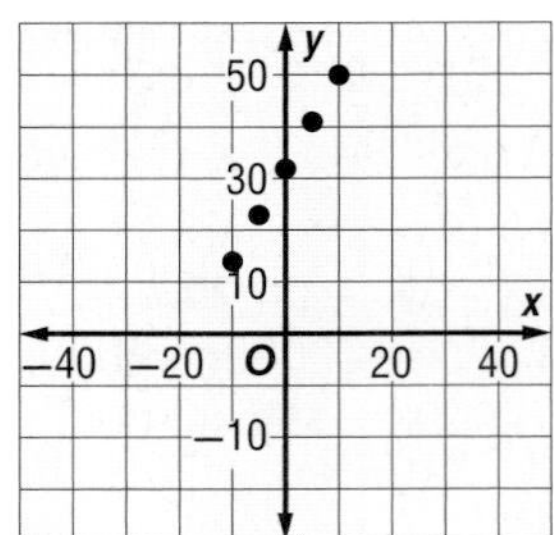

41.

Input	Rule: $x - 4$	Output
−2	−2 − 4	−6
−1	−1 − 4	−5
0	0 − 4	−4
1	1 − 4	−3
2	2 − 4	−2

43–46.

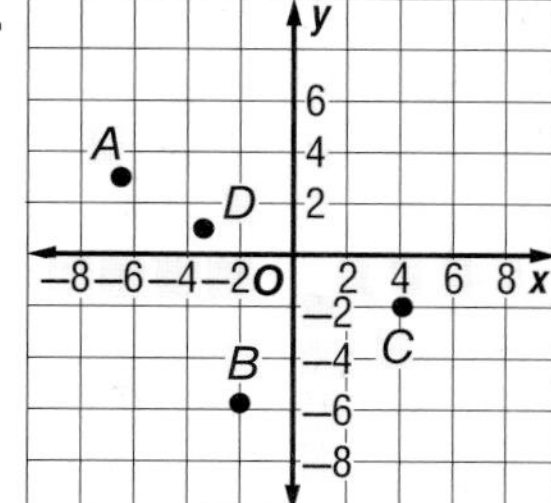

47. Sample answer: (−3, 1) **49a.** Sample answer: Never; both coordinates are positive.
49b. Sample answer: Sometimes; both (−2, 0) and (2, 0) lie on the x-axis. **51.** C **53.** D **55.** 3 **57.** −50 **59.** −11 s **61.** 6 **63.** *B* **65.** *D* **67.** *C*

Pages 104–106 ***Lesson 2-7***

1. A

3 original translation image

original		translation		image
$X(0, 4)$	+	(4, −5)	→	$X'(4, -1)$
$Y(-2, 0)$	+	(4, −5)	→	$Y'(2, -5)$
$Z(2, 0)$	+	(4, −5)	→	$Z'(6, -5)$

5.

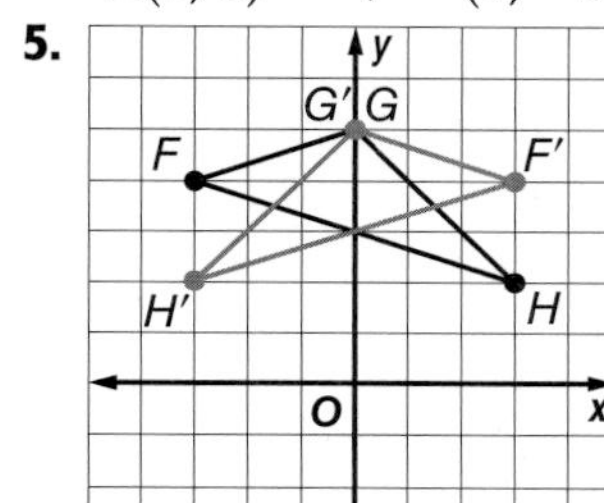

7.

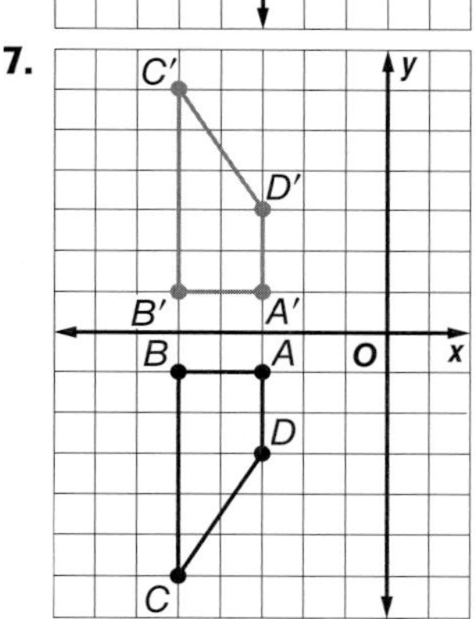

9. 5 units left and 3 units up

11 Since the letter T is flipped, it shows a reflection. It is reflected over the x-axis.

13. The translation is to the left 5 units and down 4 units. **15.** $S'(-14, 2)$, $T'(0, 9)$; The translation is to the left 6 units and up 2 units.

17. HTAM; y-axis **19.** No; For example, the vertices of △ABC are $A(1, 3)$, $B(7, 3)$, and $C(1, 6)$. The image's vertices reflected over the x-axis and y-axis would be $A'(-1,-3)$, $B'(-7, -3)$, and $C'(-1, -6)$. **21.** No single transformation is equivalent to the original two. The new figure is a reflection of the original but moved up and over to the right. **23.** D **25.** H **27.** (−4, 4) **29.** (−3, −2) **31.** 49 points **33.** $3n$ **35.** 0.2 **37.** 0.625

Pages 107–110 ***Study Guide and Review***

1. true **3.** false, flip **5.** true **7.** true **9.** false; mean **11.** true **13.** $0 > -5$; $-5 < 0$ **15.** = **17.** > **19.** 4 **21.** 13 **23.** −6 **25.** 4 **27.** 5 **29.** −13 **31.** −25 + (−50); −75 points **33.** 6 **35.** 11 **37.** −10 **39.** 20,602 ft **41.** −30 **43.** 48 **45.** −140 **47.** 4 **49.** −6 **51.** 8 **53.** 13.4 s

54–57.

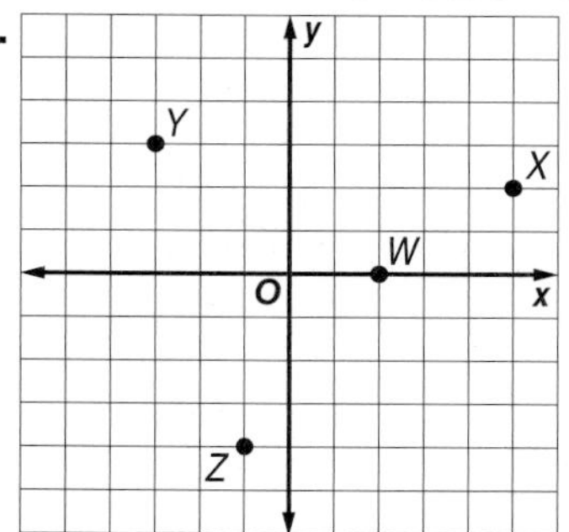

55. II **57.** III **59.** $A'(1, 3)$, $B'(4, 3)$, $C'(1, 1)$, and $D'(-2, 1)$ **61.** translation

Chapter 3 Operations with Rational Numbers

Page 117 ***Chapter 3*** ***Get Ready***

1. 0.6 **3.** 0.2 **5.** 0.5 **7.** −1.8 **9.** 29 bricks **11.** $\frac{4}{5}$ **13.** $\frac{3}{20}$ **15.** simplified **17.** −4 **19.** −7 **21.** +9 miles

Pages 124–127 ***Lesson 3-1***

1. 0.6

3 $-20\overline{)3.00}$ with quotient -0.15

5. $-0.\overline{6}$ **7.** 0.8 **9.** < **11.** > **13.** < **15.** test 1 **17.** 0.35 **19.** −0.1875 **21.** 0.36 **23.** −0.4375 **25.** $0.7\overline{3}$ **27.** $-0.\overline{2}$

29 955 ÷ 1028. Use a calculator:
955 ÷ 1028 ENTER **0.929**

31. < **33.** = **35.** < **37.** < **39.** more than; $\frac{1}{4} = 0.25$ and $0.28 > 0.25$ **41.** < **43.** > **45.** > **47.** $\frac{3}{32}, \frac{5}{16}, \frac{3}{8}, \frac{1}{2}, \frac{3}{4}$ **49.** $2\frac{3}{5}, 2\frac{2}{3}$, 2.67 **51.** $\frac{1}{13}, \frac{2}{25}$, 0.089

53 Place a bar over the part that repeats.
$0.999993\ldots = 0.9\overline{9}$

55. $-10.3\overline{4}$ **57.** Sample answer: A: $\frac{7}{9}$; B: $\frac{9}{10}$; C: $1\frac{1}{5}$; D: $1\frac{3}{10}$; E: $1\frac{3}{5}$; $\frac{7}{9} < \frac{9}{10}$ **59.** Sample answer: fractional form: customary measurement; decimal form: stock price **61.** $\frac{1}{3} = 0.\overline{3}$, $\frac{1}{6} = 0.1\overline{6}$, $\frac{1}{7} = 0.\overline{142857}$, $\frac{1}{9} = 0.\overline{1}$ **63.** $0.\overline{1}$, $0.\overline{23}$, and $0.\overline{75}$. Sample answer: When the denominator is a 9 or 99, the numerator repeats. **65.** C **67.** B **69.** 240 **71.** 9 **73.** 24 m below the surface **75.** 15 **77.** 13 **79.** thirty-four hundredths **81.** three tenths

Pages 130–133 ***Lesson 3-2***

1. $\frac{15}{4}$

3 $-1\frac{3}{4} = -\frac{1(4) + 3}{4}$
$= -\frac{4 + 3}{4}$
$= -\frac{7}{4}$

5. $-3\frac{85}{99}$ **7.** $2\frac{27}{50}$ **9.** rational **11.** $\frac{11}{6}$ **13.** $-\frac{87}{8}$ **15.** $3\frac{5}{8}$ **17.** $-5\frac{9}{25}$ **19.** $-1\frac{3}{10}$

21 $0.506 = \frac{506}{1000}$
$= \frac{506 \div 2}{1000 \div 2}$
$= \frac{253}{500}$

23. $-2\frac{5}{9}$ **25.** $\frac{16}{99}$ **27.** $-\frac{1}{11}$ **29.** integer, rational **31.** rational **33.** irrational **35.** Yes; $\frac{5}{8} = 0.625$, and $0.625 > 0.6$, so the bead will fit. **37.** > **39.** < **41.** > **43.** $\frac{652}{999}$ **45.** $\frac{163}{225}$ **47.** $9\frac{241}{990}$ **49a.** 3.1415927

3.14 3.1415927 $\frac{22}{7}$ 3.145

3.1 3.2

49b. $3.14 < \pi < \frac{22}{7}$ **49c.** Sample answer: If the diameter is a multiple of 7, use $\frac{22}{7}$. Otherwise, use 3.14.

51 To compare the numbers, rewrite the fractions as decimals. $3\frac{4}{11} = 3.\overline{36}$
On a number line, $-3.\overline{42}$ is to the left of -3.4 which is to the left of $3.\overline{36}$ which is to the left of 3.38. So, the order from least to greatest is $-3.\overline{42}$, -3.4, $3\frac{4}{11}$, 3.38.

53. $-1.95, -1\frac{13}{14}, -1.9, -1\frac{9}{11}$ **55.** $\frac{7}{9}; \frac{5}{8}; \frac{5}{8} < \frac{5}{7} < \frac{7}{9}$

57. Sample answer: Since $0.\overline{76} = 0.76767676...$ and $0.76 = 0.76000000...$, $0.\overline{76}$ is greater than 0.76. **59a.** true; Sample answer: Integers include all whole numbers and their opposites. Therefore, they belong to the set of rational numbers. **59b.** true; Sample answer: All whole numbers and their opposites belong to the set of integers. **59c.** false: Sample answer: $\frac{1}{2}$ is not an integer **59d.** true: Sample answer: All natural numbers are rational because they can be expressed as fractions. **61.** B **63.** C **65.** -0.625 **67.** -0.2

69.

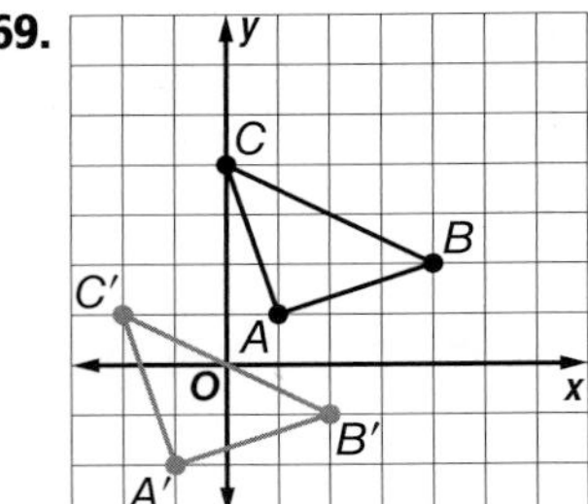

71.

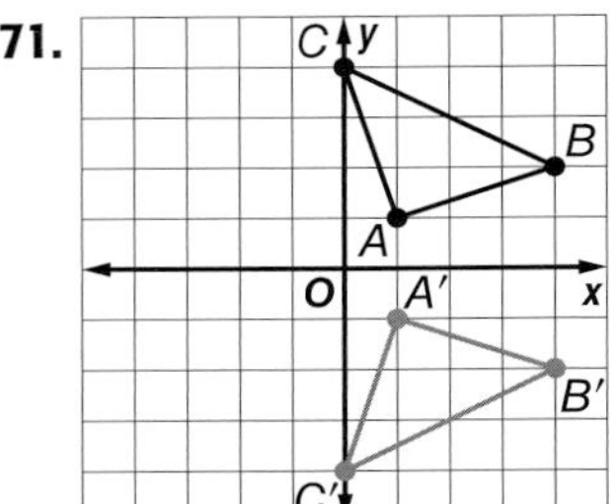

73. D = {0, 3, 4, 8}; R = {0, 2, 6, 12} **75.** 6050 m **77.** 0 **79.** -280

Pages 136–139 ***Lesson 3-3***

1 $\frac{7}{8} \cdot \frac{1}{2} = \frac{7 \cdot 1}{8 \cdot 2}$
$= \frac{7}{16}$

3. $-\frac{1}{8}$ **5.** 5 **7.** $-\frac{98}{75}$ or $-1\frac{23}{75}$ **9.** $-\frac{2}{5}$ **11.** $-\frac{28}{5}$ or $-5\frac{3}{5}$ **13.** 140 towns **15.** $\frac{3}{32}$ **17.** $\frac{8}{27}$ **19.** $\frac{1}{9}$ **21.** $\frac{2}{39}$ **23.** $-\frac{2}{3}$ **25.** -1 **27.** $\frac{5}{2}$ or $2\frac{1}{2}$ **29.** $-\frac{8}{5}$ or $-1\frac{3}{5}$ **31.** -10

33 $\frac{3}{5}(145) = \frac{3}{5} \cdot \frac{145}{1}$
$= \frac{3}{1} \cdot \frac{29}{1}$
$= \frac{3 \cdot 29}{1 \cdot 1}$
$= 87$ lb

35. $\frac{1}{21}$ **37.** -3 **39.** -2 **41.** 5 bags **43.** 12 **45.** 10 **47.** 4

51 **a.**

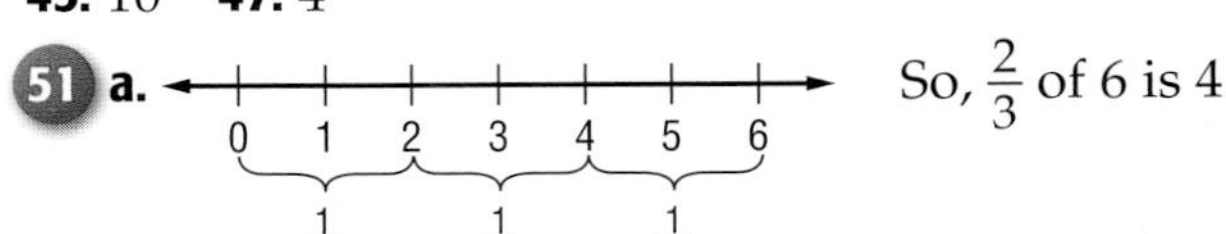

b.

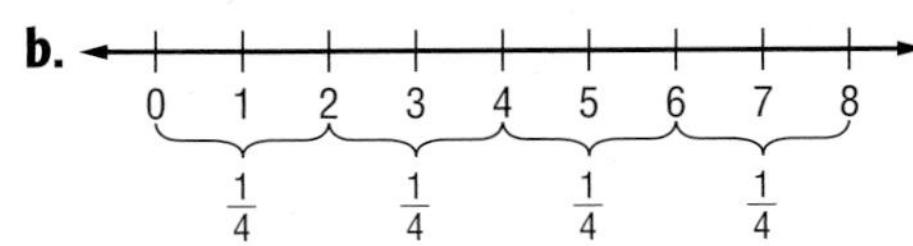

So, $\frac{3}{4}$ of 8 is 6.

c. 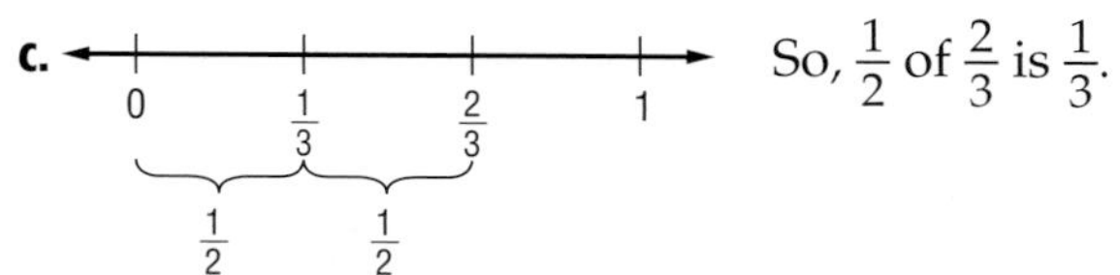

So, $\frac{1}{2}$ of $\frac{2}{3}$ is $\frac{1}{3}$.

d. 0 1 2; $\frac{1}{2}$ $\frac{1}{2}$

So, $\frac{1}{2}$ of 2 is 1.

e. 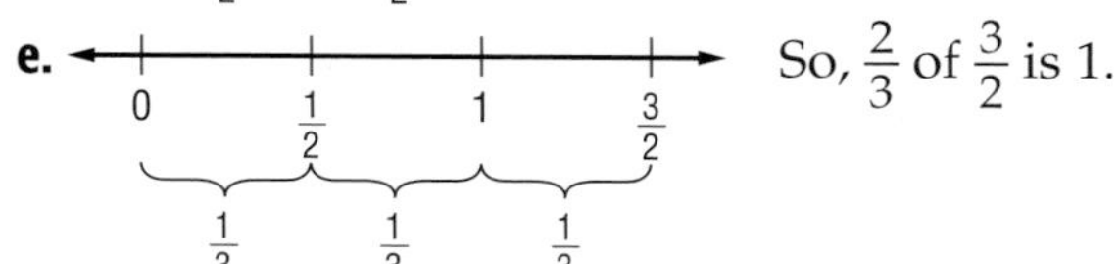

So, $\frac{2}{3}$ of $\frac{3}{2}$ is 1.

f. 0 $\frac{1}{2}$ $\frac{2}{3}$ 1 $\frac{4}{3}$ $\frac{5}{3}$ 2 $\frac{7}{3}$; $\frac{1}{7}$ $\frac{1}{7}$ $\frac{1}{7}$ $\frac{1}{7}$ $\frac{1}{7}$ $\frac{1}{7}$ $\frac{1}{7}$

So, $\frac{3}{7}$ of $\frac{7}{3}$ is 1.

g. When a number is multiplied by its reciprocal, the result is 1.

h. $\frac{a}{b} \cdot \frac{b}{a} = \frac{ab}{ba}$
$= \frac{1}{1}$
$= 1$

53. Marina; Kelly did not write the mixed numbers as improper fractions before multiplying. **55.** $-\frac{5}{6}$ **57.** always; Sample answer: When you multiply a fraction by a whole number or mixed number, you are finding a part of that whole number or mixed number. Since you are finding a part, the product will always be less than the whole number or mixed number. **59.** A **61.** B **63.** $4\frac{1}{50}$ **65.** $-5\frac{1}{8}$ **67.** $4\frac{5}{9}$ **69.** $>$ **71.** $=$ **73.** $<$ **75.** The group that goes to sleep at 12 A.M. or later. **77.** 88 **79.** 1440 **81.** -192 **83.** 84

Pages 144–146 ***Lesson 3-4***

1. $\frac{7}{6}$ **3.** $-\frac{1}{63}$ **5.** $-\frac{25}{2}$ or $-12\frac{1}{2}$ **7.** $-\frac{1}{18}$ **9.** $-\frac{16}{3}$ or $-5\frac{1}{3}$ **11.** $\frac{8b}{3}$ **13.** $\frac{5x}{2yz}$

15 The multiplicative inverse of $\frac{10}{19}$ is $\frac{19}{10}$ or $1\frac{9}{10}$.

17. $-\frac{7}{30}$ **19.** $-\frac{1}{54}$ **21.** $-\frac{5}{8}$ **23.** $\frac{1}{54}$ **25.** $-\frac{5}{3}$ or $-1\frac{2}{3}$ **27.** $\frac{57}{28}$ or $2\frac{1}{28}$ **29.** 12 costumes **31.** $\frac{1}{14}$ **33.** $\frac{15}{2}$ or $7\frac{1}{2}$

35 $405 \div \left(4\frac{1}{2}\right) = 4.0 \div \frac{9}{2}$
$= 405 \cdot \frac{2}{9}$
$= \frac{810}{9}$
$= 90$ mph

37. $-\frac{3}{25}$ **39.** $-\frac{3}{4}$ **41.** The quotient decreases; $\frac{3}{4} \div \frac{1}{8} = 6$. **43.** The quotient of $\frac{1}{2} \div \frac{3}{4}$ is $\frac{2}{3}$. The fraction $\frac{2}{3}$ is not a whole number. **45.** always **47.** C **49.** A **51.** $\frac{9}{8}$ or $1\frac{1}{8}$ **53.** $-\frac{3}{22}$ **55.** 1 **57.** $\frac{7}{250}$ mi^2 **59.** -72 **61.** 48 **63.** -68 **65.** $3n$ **67.** $-5n$

Pages 150–152 ***Lesson 3-5***

1. $1\frac{1}{3}$ **3.** $-\frac{1}{2}$ **5.** 10 **7.** $\frac{5}{7}$ **9.** $7\frac{6}{7}$ **11.** $-\frac{2x}{y}$ **13.** $-\frac{12c}{ab}$ **15.** $-\frac{1}{3}$ **17.** $\frac{3}{7}$ **19.** $13\frac{2}{3}$ **21.** $-\frac{17}{5}$ or $-3\frac{2}{5}$ **23.** $-\frac{2}{11}$

25 $-\frac{7}{20} - \frac{7}{20} = -\frac{7}{20} + \left(\frac{-7}{20}\right)$
$= \frac{-14}{20}$
$= -\frac{7}{10}$

27. $-2\frac{1}{9}$ **29.** $-6\frac{4}{9}$ **31.** $4\frac{1}{2}c$ **33.** $\frac{9m}{5}$ **35.** $\frac{5p}{14}$ **37.** $-2\frac{8}{9}$

39 a. $19\frac{4}{8} - 18\frac{3}{8} = 19 - 18\frac{4-3}{8}$
$= 1\frac{1}{8}$ lb
$20\frac{7}{8} - 17\frac{2}{8} = 20 - 17\frac{7-2}{8}$
$= 3\frac{5}{8}$ lb

b. $20\frac{7}{8} + 1\frac{3}{8} + 1\frac{3}{8} = 20 + 1 + 1 + \frac{7+3+3}{8}$
$= 22\frac{13}{8}$
$= 23\frac{5}{8}$ lb

41. $\frac{9pr}{2n}$ **43.** Sample answer: $\frac{1}{3} - 1 = -\frac{2}{3}$ **45.** No; he did not add the fraction part of the mixed numbers correctly. **47.** Sample answer: A recipe requires $\frac{3}{4}$ cup of milk and $1\frac{1}{4}$ cups of water. How much liquid is in the recipe? 2 cups. **49.** J **51.** $-6\frac{1}{3}$ **53.** -10 **55.** $\frac{2}{35}$ **57.** $\frac{5}{9}$ **59.** 17 yard line **61.** 60 **63.** 45 **65.** x^2y

Pages 155–158 ***Lesson 3-6***

1. $\frac{2}{3}$

3 $\frac{7}{8} + \left(-\frac{2}{7}\right) = \frac{49}{56} + \left(-\frac{16}{56}\right)$
$= \frac{49 + (-16)}{56}$
$= \frac{33}{56}$

5. $-9\frac{5}{6}$ **7.** $-1\frac{1}{36}$ **9.** $\frac{1}{24}$ **11.** $8\frac{1}{12}$ **13.** $\frac{7}{12}c$
15. $\frac{5}{21}$ **17.** $-1\frac{8}{33}$ **19.** $-\frac{23}{42}$ **21.** $-\frac{11}{30}$

23 $\frac{2}{3} - \frac{7}{15} = \frac{2}{3} + \frac{-7}{15}$
$= \frac{10}{15} + -\frac{7}{15}$
$= \frac{3}{15}$
$= \frac{1}{5}$

25. $12\frac{3}{10}$ **27.** $4\frac{1}{2}$ **29.** $-11\frac{7}{9}$ **31.** $4\frac{19}{48}$ **33.** $-17\frac{3}{14}$

35. $\frac{3}{4}c$ **37a.** Sample answer:

Perimeter of 20	
Length	Width
8	2
7	3
6	4
5	5

37b. (8, 2), (7, 3), (6, 4), (5, 5)

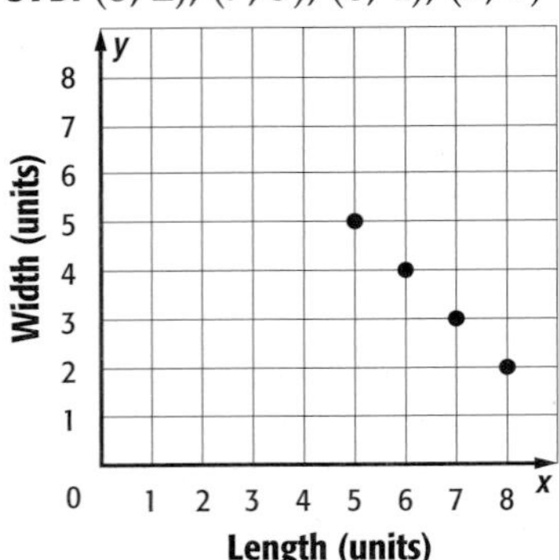

37c. $4\frac{1}{2}$ in.

41 $-19\frac{3}{8} - \left(-4\frac{3}{4}\right) = \frac{-155}{8} + \left(\frac{19}{4}\right)$
$= \frac{-155}{8} + \frac{38}{8}$
$= \frac{-117}{8}$
$= -14\frac{5}{8}$

43. $-59\frac{1}{6}$ **45.** $36\frac{67}{99}$ **47.** Sample answer: $\frac{2}{3} - \frac{5}{8} = \frac{1}{24}$

49. Sample answer: One way to respond is to use a diagram similar to the one given in the lesson opener to show that the fractions have to have the same "unit" (parts of a whole) to be added or subtracted. **51.** Sample answer: Fill the $\frac{1}{2}$-cup. From the $\frac{1}{2}$-cup, fill the $\frac{1}{3}$-cup. $\frac{1}{6}$ cup will be left in the $\frac{1}{2}$-cup because $\frac{1}{2} - \frac{1}{3} = \frac{1}{6}$. **53.** D **55.** H
57. $14\frac{1}{2}$ **59.** $\frac{3}{4}$ **61.** $\frac{2}{3}$ **63.** 6 **65.** $\frac{39}{1{,}000{,}000}$
67. 50

Pages 159–162 *Study Guide and Review*

1. false; rational **3.** false; unlike **5.** true
7. false; denominator **9.** false; add or subtract
11. 0.3 **13.** $-0\overline{83}$. **15.** 0.625 **17.** > **19.** >
21. < **23.** 5.3125 **25.** $-\frac{9}{20}$ **27.** $-\frac{14}{25}$ **29.** $-2\frac{1}{33}$
31. $10\frac{3}{11}$ **33.** rational **35.** rational **37.** $\frac{3}{20}$
39. $3\frac{1}{3}$ **41.** $6\frac{1}{4}$ in. **43.** $-\frac{1}{16}$ **45.** $\frac{5}{19}$ **47.** $-2\frac{11}{12}$
49. $\frac{2}{3}$ **51.** 8 days **53.** $-\frac{5}{12}$ **55.** $8\frac{2}{7}$ **57.** $11\frac{1}{2}c$
59. $-6\frac{1}{3}$ **61.** $\frac{17}{40}$ **63.** $9\frac{1}{8}$ **65.** $-3\frac{5}{12}$
67. $166\frac{17}{20}$ miles or 166.85 miles

Chapter 4 Expressions and Equations

Page 169 *Chapter 4* *Get Ready*

1. −9 **3.** 28 **5.** 88 **7.** $10.25 decrease
9. 4 + (−10) **11.** −19 + (−10) **13.** −1
15. 0 **17.** −21 **19.** −22 ft

Pages 173–176 *Lesson 4-1*

1 $7(9 + 3) = 7 \cdot 9 + 7 \cdot 3$
$= 63 + 21$
$= 84$

3. $7 \cdot 2 + 8 \cdot 2$; 30 **5.** $6.50; 5($1 + $0.30) = 5 · 1 + 5 · 0.30 **7.** $5p + 20$ **9.** $9a - 90$ **11.** $8 \cdot 8 + 8 \cdot 5$; 104 **13.** $5 \cdot 12 + 5 \cdot 7$; 95 **15.** $3 \cdot 15 - 3 \cdot 5$; 30 **17.** $-7 \cdot 16 - (-7)8$; −56 **19.** $65.70; 2($32.85) = 2($33 − $0.15) **21.** $4y + 28$

23 $(a + 9)6 = 6 \cdot a + 6 \cdot 9$
$= 6a + 54$

25. $5t - 30$ **27.** $-d + 10$ **29.** $-7x + 21$
31. 176 **33.** 779 **35.** 3000 **37.** 441 **39.** 1815
41. 2($8 + $7), 2($8) + 2($7); $30
43. $5 \cdot 4 + 5 \cdot \frac{1}{5}$; 21 **45.** $6 \cdot 4 + 6 \cdot \frac{2}{3}$; 28

47 $9(2 + \frac{1}{3}) = 9 \cdot 2 + 9 \cdot \frac{1}{3}$
$= 18 + 3$
$= 21$ yd

49. $-5e - 5f$ **51.** $-4j + 4k$ **53.** $8u - 8w$
55. Sample answer: $2(3 + 4) = 2 \cdot 3 + 2 \cdot 4$
57. no; $3 + (4 \cdot 5) = 23$, $(3 + 4)(3 + 5) = 56$
59. C **61.** C **63.** $\frac{1}{8}$ **65.** $\frac{17}{30}$ **67.** $20\frac{1}{8}$ yd
69. 120 **71.** 17 **73.** 9 + (−12) **75.** −10 + 3

Pages 180–183 *Lesson 4-2*

1. terms: $-2a$, $3a$, $5b$; like terms: $-2a$, $3a$; coefficients: −2, 3, 5; constant: none
3. terms: mn, $4m$, $6n$, $2mn$; like terms: mn, $2mn$; coefficients: 1, 4, 6, 2; constant: none

5 The terms are $3x$, $4x$, and $5y$. The like terms are $3x$ and $4x$ because they have the same variables. The coefficients are 3, 4, and 5. There is no constant in the expression.

7. $8x + 3$ **9.** $2x - 4$ **11.** $m + 5$ **13.** $2x + 7$ **15.** terms: $3a$, 2, $3a$, 7; like terms: $3a$, $3a$; coefficients: 3, 3; constants: 2, 7 **17.** terms: $3c$, $4d$, $5c$, 8; like terms: $3c$, $5c$; coefficients: 3, 4, 5; constant: 8 **19.** terms: $4x$, $4y$, $4z$, 4; like terms: none; coefficients: 4, 4, 4; constant: 4 **21.** $7a$ **23.** $-4m + 5$ **25.** $11p + 8$ **27.** $-3a - 6b$ **29.** $6x + 30$ **31.** $-18 + 3r$ **33.** $2y - 5$

35 first game: x; second game: $3x$; third game: $3x + 6$

Total $= x + 3x + 3x + 6$
$= 7x + 6$

37. $5x - 2y$ **39.** $-17m - 8n + y$ **41.** $-\frac{3}{4}m + \frac{3}{2}n$ **43.** $\frac{22}{15}a + \frac{14}{15}b$

45a. $7 + 4x + (-4) + (-2x) + 3$

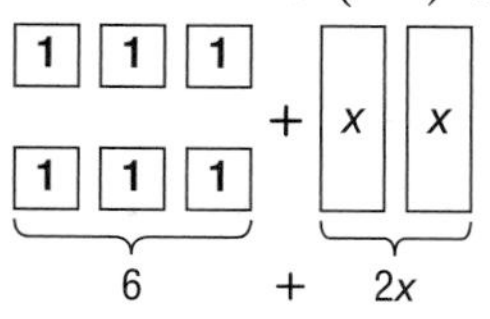

45b. $-8 + (-3x) + 2 + (-5x) + 2x + 4$

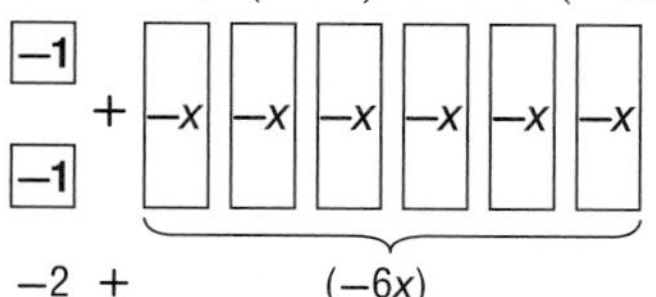

47. $6c - 8$

49 $16 \cdot (-31) + 16 \cdot 32$

$= 16(-31 + 32)$	Distributive Property
$= 16(1)$	Add –31 and 32.
$= 16$	Multiplicative Identity

51. $24 \cdot (-15) + 36 \cdot 15$

$= (-24 + 36)15$	Distributive Property
$= (12)15$	Simplify.
$= (10 + 2)15$	Distributive Property
$= 150 + 30$ or 180	Simplify.

53. $4x + 2$ **55.** $5x + xy + 2y + 10$ **57.** $-6x - 12$; the other expressions are equivalent to $-6x + 12$. **59.** Sample answer: The friend multiplied 5 by +2 instead of -2.

$4x - 2(x + 5) = 4x - 2x - 10$
$= 2x - 10$

61. G **63.** $9a - 9b - 6$ **65.** $-5a + 30$ **67.** 4(\$7 + \$3), 4(\$7) + 4(\$3); \$40 **69.** $-6 < -2$; $-2 > -6$ **71.** $0 > -9$; $-9 < 0$ **73.** $|15| < |18|$; $|18| > |15|$ **75.** 23 **77.** 3 **79.** 74

Pages 187–189 *Lesson 4-3*

1. 39 **3.** $\frac{1}{2}$ **5.** -32 **7.** 1.44 **9.** 3.5 **11.** $29.15 + x = 28.79$; -0.36 in.

13
$x - 24 = 73$
$x - 24 + 24 = 73 + 24$
$x = 97$

15. 5 **17.** 39 **19.** $-1\frac{1}{4}$ **21.** 134.9 **23.** -4.1 **25.** $-\frac{7}{8}$ **27.** $65 = m - 25$; \$90 **29.** -41 **31.** 0.1 **33.** -1 **35.** $\frac{7}{20}$ or 0.35 **37.** $97 - 9 = d$, $d - 13 = w$; Damon scored 88 points and Wes scored 75 points.

39 Words: The Gates of Arctic National Park is 2.78 million larger than Denali National Park.

Equation: $7.52 = 2.78 + x$
$7.52 = 2.78 + x$
$7.52 - 2.78 = 2.78 - 2.78 + x$
$4.74 = x$

4.74 million acres

41a. $y = 8 - x$ or $x + y = 8$

41b.

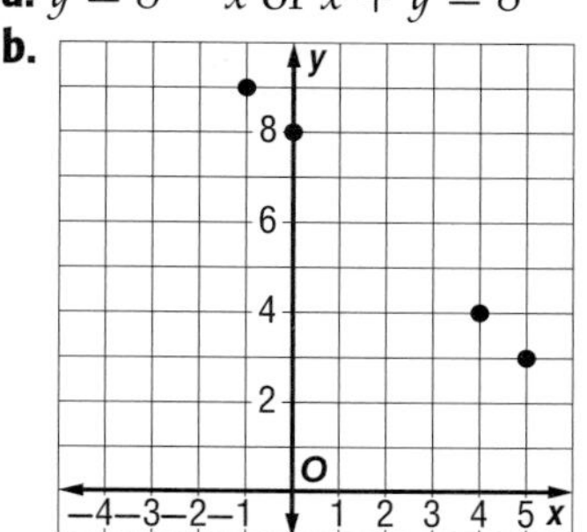

41c. Sample answer: Together, Melinda and Ariana sold 8 t–shirts for the school fundraiser. **41d.** The equation $x + y = 8$ has an infinite number of solutions. If $y = 3$, there is only one solution, 5. **43.** Sample answer: $x - (-1.2) = -1.2$ **45.** Sample answer: The small numbers allow you to use algebra tiles to solve the problem. $x - 5 = -9$; place an x–tile and 5 negative tiles on the left side of the mat and 9 negative tiles on the right side of the mat. Then remove 5 negative tiles from each side. The x-tile remains on the left side and four negative tiles remain on the right side. So, Jaime shot a -4 on the previous day. **47.** Sample answer: Peggy paid \$3.25 for a new notebook that was on sale. The sale price was \$2.70 less than the original price. Find the original price of the notebook. **49.** H **51.** B **53.** $6r - 30$ **55.** $3c - 32$ **57.** $8 - 4x$ **59.** Fate: $\frac{9}{500}$; McLendon-Chisholm: $\frac{1}{50}$; Rockwall: $\frac{21}{50}$; Royse City: $\frac{7}{100}$ **61.** -7 **63.** 4 **65.** 125

Pages 194–196 *Lesson 4-4*

1. -13 **3.** 12 **5.** -36 **7.** -8.5 **9.** 60 **11.** $\frac{x}{6} = 54$; 324 lb **13.** -12

15 $-72 = 3y$

$\frac{-72}{3} = \frac{3y}{3}$

$-24 = y$

CHECK: $-72 = 3(-24)$

$-72 = -72$

17. 16.5 **19.** 392 **21.** −64 **23.** 24 **25.** −80 **27.** $\frac{5}{6}x = 420$; 504 strawberries **29.** −4 **31.** −12 **33.** −64 **35.** $\frac{2}{3}$ **37.** 0.9 or $\frac{9}{10}$ **39.** $-\frac{7}{27}$ **41.** 70 bpm

43 **a.** Words: 50 miles per day for t days

Expression: $50 \cdot t$

The equation is $d = 50t$.

b. Multiply each input value by 50 to get the output value.

Time	1	2	3	4	5	6
Distance	50	100	150	200	250	300

c. Plot each ordered pair on a coordinate grid.

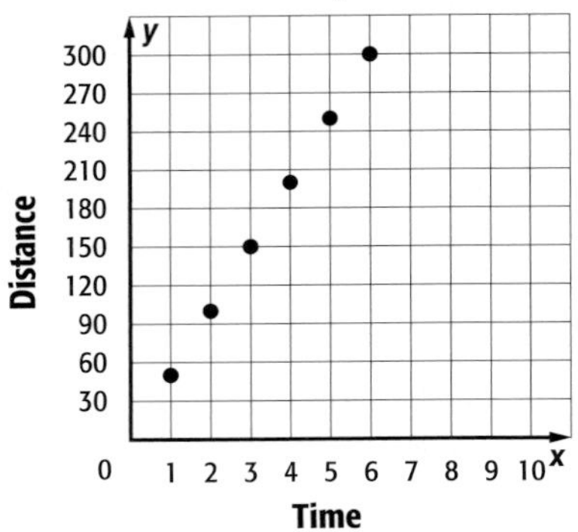

d. Using the graph, it looks like it should take 9 days to travel 450 miles.

e. $50t = 2500$

$\frac{50t}{50} = \frac{2500}{50}$

$t = 50$ It will take 50 days.

45. Sample answer: A shirt was on sale for half off. If the sale price was $15.60, find the original price; $31.20 **47.** Rachel; to undo division, you multiply. **49.** Sample answer: Yes; he can multiply each side of the equation by $\frac{1}{3}$ instead of dividing by 3. **51.** J **53a.** $c = \frac{64}{2}$ **53b.** $32 **55.** −3 **57.** −0.7 **59.** −8 **61.** $-\frac{1}{2}$ **63.** $5t + 15$ **65.** $12p + 4$ **67.** $-4x - 5$

69.

x	−2	−1	0	1
y	8	4	0	−4

71. 25 **73.** −19 **75.** 43

Pages 202–204 Lesson 4-5

1. 4 **3.** −4 **5.** −12 **7.** −3 **9.** 5 **11.** 1.2 **13.** C **15.** 4

17 $4d - 18 = -34$

$4d - 18 + 18 = -34 + 18$

$4d = -16$

$\frac{4d}{4} = -\frac{16}{4}$

$d = -4$

19. 45 **21.** 40 **23.** 8 **25.** 30 **27.** −11 **29.** 15 months **31.** 1.3 **33.** $\frac{5}{12}$ **35.** 30.4

37 $6s - 1.5s + 2.25s = 40.50$

$6.75s = 40.50$

$\frac{6.75s}{6.75} = \frac{40.50}{6.75}$

$s = 6$

39. 15.1 **41.** −14.5 **43.** Sample answer: You spent $7 at the bookstore and bought lunch for 2 days. You spent a total of $15. How much was lunch? $4 **45.** Sample answer: When solving equations you may have to combine like terms as a first step. Combining like terms is one way to simplify expressions. Also, after you use the addition property or the multiplication property, you will need to combine terms again. Throughout the entire equation solving process, you have to know how to simplify expressions. **47.** Sample answer: The problems both involve multiplying by 3 and adding 5. They are different because $3(2) + 5$ is an expression while $3x + 5 = 11$ is an equation in which you solve by subtracting 5 then dividing by 3. **49.** G **51.** 19 **53.** 56 **55.** −32 **57.** 3.83 **59.** $2x + 40$ **61.** 4 **63.** 8 **65.** −8

Pages 207–209 Lesson 4-6

1. $\frac{n}{3} - 8 = 16$ **3.** $5t - 12 = 98$ **5.** $2x - 14 = 96$; 41 cars

7 Words: Eighteen more than twice a number is 8.

Variables: Let x represent the number.

Equation: $18 + 2x = 8$

The equation is $18 + 2x = 8$.

$18 + 2x = 8$

$18 - 18 + 2x = 8 - 18$

$2x = -10$

$\frac{2x}{2} = \frac{-10}{2}$

$x = -5$

9. $3x - 3 = 48$, 17 teams **11.** $2g + 37 = 497$; 230 field goals **13.** $\frac{x}{3} + 8.50 = 13.25$; $14.25 **15.** $6d + 0.90 = 2.40$; $0.25

17 Words: Eight more than twice as many action movies is the number of animated movies.

Variables: Let n represent the number of action movies.

Equation: $8 + 2n = 24$

$$\begin{aligned} 8 + 2n &= 24 \\ 8 - 8 + 2n &= 24 - 8 \\ 2n &= 16 \\ \frac{2n}{2} &= \frac{16}{2} \\ n &= 8 \end{aligned}$$

19a. $47 + 15w$

Number of weeks	Amount of savings ($)
1	62
2	77
3	92
4	107

19b.

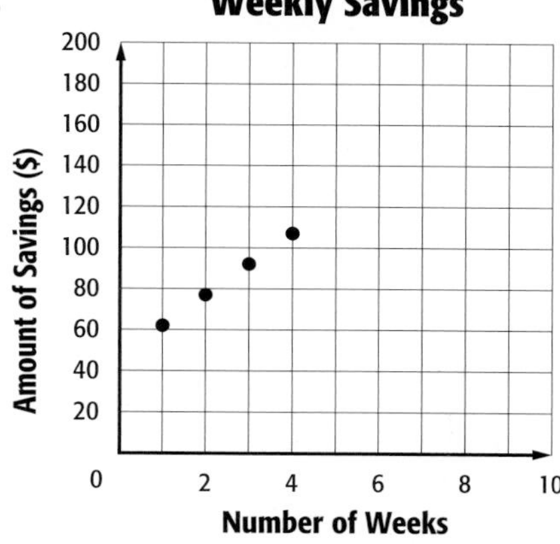

Draw a straight line through the plotted points. Then go to the point that has 212 as its y-value and find the corresponding x-value.
19c. $47 + 15w = 212$; 11 weeks **19d.** The scatter plot is less accurate and will take a longer time to find because the plot will be large. Solving an equation is quick and ends with an exact answer. **21.** 8, 10 ,12 **23.** 24 years old **25.** D **27a.** $60 + 6d = 180$ **27b.** \$20 per month **29.** -22 **31.** -10 **33.** 4 **35.** $n(t + h + s)$ **37.** 40 customers **39.** $8y - 16$ **41.** $18 + 2p$

Pages 210–212 Study Guide and Review

1. equivalent expressions **3.** constant **5.** solution **7.** Distributive Property **9.** inverse operations **11.** $7y + 21$ **13.** $-b + 9$ **15.** \$30; $5(2.50 + 3.50)$; $5 \cdot 2.50 + 5 \cdot 3.50$; **17.** $9x$ **19.** $8x + 2$ **21.** $5b + 6$ **23.** $x = 6$ **25.** $x = -1\frac{1}{20}$ **27.** $p = 18$ **29.** $x + 13 = 37$; 24 pages **31.** $x = 20$ **33.** $x = 5$ **35.** $x = -24$ **37.** 3 **39.** 9 **41.** $-\frac{1}{7}$ **43.** 35 books **45.** $2n - 6 = -22$; -8

Chapter 5 Multi-Step Equations and Inequalities

Page 219 Chapter 5 Get Ready

1. -6 **3.** 90 **5.** -7 **7.** \$75 **9.** -37 **11.** -73 **13.** 0 **15.** -40 **17.** 120 **19.** -4 **21.** $-36°$F

Pages 223–226 Lesson 5-1

1. 28 cm; 24 cm^2

3
$$\begin{aligned} A &= \ell w \\ 117 &= \ell(9) \\ \frac{117}{9} &= \frac{9\ell}{9} \\ 13 &= \ell \end{aligned}$$
She will use 13 squares along the length of the quilt

5. 22 cm, 24cm^2 **7.** 30in, 30 in^2 **9.** $50\frac{3}{4}$ ft

11
$$\begin{aligned} A &= \ell w \\ 432 &= 36w \\ \frac{432}{36} &= \frac{36w}{36} \\ 12 &= w \end{aligned}$$
The answer is 12 cm.

13. 527.52 in. **15a.** 74 ft **15b.** 3

17 $\ell = 4$ units; $w = 7$ units
$$\begin{aligned} A &= \ell w \\ &= 4(7) \\ &= 28 \text{ square units} \end{aligned}$$

19. $\frac{V}{A} = h$ **21.** $r - st = p$ **23.** $\frac{2A}{h} = b$
25. Sample answer: The dimensions of a lacrosse field are 330 ft by 180 ft. What is the perimeter and area of the field? $P = 1020$ ft
$A = 59{,}400$ ft^2
27. $\ell = 10$ ft, $w = 10$ ft **29.** C **31.** C **33.** $17 + 3x = 32$; 5 h **35.** 21 **37.** $12f = 132$; 11 ft **39.** $5280m = 10{,}560$; 2mi **41.** -8 **43.** -3

Pages 231–233 Lesson 5-2

1. 3 **3.** 4 **5.** 28 **7.** $50 + 1.99m = 3.99m$; 25 DVDs

9
$$\begin{aligned} 8 - v &= 7v \\ 8 - v + v &= 7v + v \\ 8 &= 8v \\ \frac{8}{8} &= \frac{8v}{8} \\ 1 &= v \end{aligned}$$

11. $\frac{3}{2}$ **13.** -3 **15.** $\frac{2}{5}$ **17.** $5 + 0.5s = s$; 10 songs **19.** -0.75 **21.** -0.5 **23.** -80 **25.** 100 text messages

27 Words: 118 miles shorter than 4 times the coastline of Texas is 983 miles longer than the coastline of Texas.
Variables: Let x represent the length of the coastline of Texas.
Equation: $4x - 118 = x + 983$
$$\begin{aligned} 4x - 118 &= x + 983 \\ 4x - 118 + 118 &= x + 983 + 118 \\ 4x &= x + 1101 \\ 4x - x &= 1101 \\ 3x &= 1101 \\ \frac{3x}{3} &= \frac{1101}{3} \\ x &= 367 \end{aligned}$$

Selected Answers and Solutions

The length of the coastline of Texas is 367 miles. The length of the coastline of Florida is 367 + 983 = 1350 miles.
29. Sample answer: Use the information in the figure to write and solve an equation.; $w = 55$ ft, $w + 40 = 95$ ft, $w + 45 = 100$ ft **31.** 7, 8, 9
33. She is incorrect. In the third step, the 4 should be negative, not positive; $x = -5$ **35.** B **37.** -2
39. 16 m; 12 m^2 **41.** 12 lessons **43.** $9b$ **45.** 2006: 0.36; 2003: 0.53; 2000: 0.34; 1997: 0.37 **47.** 53 **49.** 24

Pages 237–239 ***Lesson 5-3***

1. $x \leq 45$ **3.** $f > 8000$ **5.** true

7. 0 1 2 3 4 5 6 7 8 9 10

9. −3 −2 −1 0 1 2 3 4 5 6 7

11. $y > 5$

13 Words: children under the age of 2
Symbols: Let c represent the age of the children.
Inequality: $c < 2$

15. $s \leq 50$ **17.** $\ell \leq 15$ **19.** false **21.** false
23. true

25. −3 −2 −1 0 1 2 3 4 5 6 7

27. 1 2 3 4 5 6 7 8 9 10 11

29. −7 −6 −5 −4 −3 −2 −1 0 1 2 3

31. $x > 2$ **33.** $x \leq 0$ **35.** $x > -4$

37 Words: 4418 is at most 500 yards more than Brett Favre's passing yards.
Symbols: Let f represent Brett Favre's passing yards.
Inequality:

$$4418 \leq 500 + f$$
$$4418 \leq 500 + f$$
$$4418 - 500 \leq 500 - 500 + f$$
$$3918 \leq f$$

39. $-13, -14, -15$; −17 −15 −13 −11 −9 −7

41. Sample answer: $-\frac{1}{2}$ **43.** The Wilson family spent more than \$20.50 on groceries.

45. −5 −4 −3 −2 −1 0 1 2 3 4 5

47. D **49.** 4 **51.** -4 **53.** 180 min **55.** 1
57. $-1\frac{3}{8}$ **59.** $11\frac{3}{10}$ **61.** -21 **63.** -15 **65.** -1.8

Pages 244–247 ***Lesson 5-4***

1. $y \leq 5$

3
$$-7 < x + (-3)$$
$$-7 + 3 < x + (-3) + 3$$
$$-4 < x$$

CHECK: Try any number less than -4.
$$-7 < -5 + (-3)$$
$$-7 < -8\ ✓$$

5. $f < 4$ −1 0 1 2 3 4 5 6 7 8 9

7. $7.5t \geq 120$; at least 16 hours

9. $z \geq 3$ −2 −1 0 1 2 3 4 5 6 7 8

11. $a < 22$ **13.** $y \leq 10.7$ **15.** $p > 26$ **17.** $n \leq 1.5$
19. $c \leq 24.1$ **21.** $b > 15\frac{3}{4}$ **23.** $0.75y \leq 10$; at most 13 games.

25. $x > -9$ −14 −12 −10 −8 −6 −4

27. $m \leq -3$ −8 −7 −6 −5 −4 −3 −2 −1 0 1 2

29. $r \leq -4.5$ −9 −8 −7 −6 −5 −4 −3 −2 −1 0 1

31. $y > 18$ 12 13 14 15 16 17 18 19 20 21 22 23

33. $b \leq 2$ −3 −2 −1 0 1 2 3 4 5 6 7

35. $c \leq -\frac{9}{5}$ or $-1\frac{4}{5}$ −7 −6 −5 −4 −3 −2 −1 0 1 2 3

37 Words: $\frac{3}{4}$ of an hour plus the time on the history project is at most 3 hours
Symbols: Let h represent the time spent on the history project.
Inequality:

$$\frac{3}{4} + h \leq 3$$
$$\frac{3}{4} + h \leq 3$$
$$\frac{3}{4} - \frac{3}{4} + h \leq 3 - \frac{3}{4}$$
$$h \leq 2\frac{1}{4}$$

She spends at most $2\frac{1}{4}$ hours or 2 hours and 15 minutes.

39. $a < 3.9$ **41.** $c \leq -21$ **43.** $r \leq 3$ **45.** $t \leq 8.5$
47. $g \geq \frac{3}{2}$ **49.** $\frac{n}{-3} > \frac{5}{6}$; $n < -\frac{5}{2}$
51. $n - 15 \leq -8$; $n \leq 7$

53
$$4a - 7 \geq 21$$
$$4a - 7 + 7 \geq 21 + 7$$
$$4a \geq 28$$
$$\frac{4a}{4} \geq \frac{28}{4}$$
$$a \geq 7$$

CHECK: Try any number greater than 7.
$$4(8) - 7 \geq 21$$
$$32 - 7 \geq 21$$
$$25 \geq 21\ ✓$$

55. $b \geq 4$ **57.** $d \leq -20$ **59.** $g > -3$ **61.** $q \geq -2$

63. −4 −3 −2 −1 0 1 2 3 4 5 6

65. 5 6 7 8 9 10 11 12 13 14 15 16

67. $\frac{x}{-5} + 4 \leq 8$; $x \geq -20$; Sample answers: $-18, 0, 18$ because they are all greater than -20.

69. 50 **71.** Sometimes; if $x = 21$ and $y = 21$ then $x \leq y$ is true but $y > x$ is not true. **73.** B **75.** D **77.** $\ell \geq 45$ **79.** $b \geq 86$ **81.** 7.5 **83.** 13 **85.** −13 **87.** 2 **89.** 4 **91.** 8

Pages 251–253 ***Lesson 5-5***

1. 6 **3.** 11

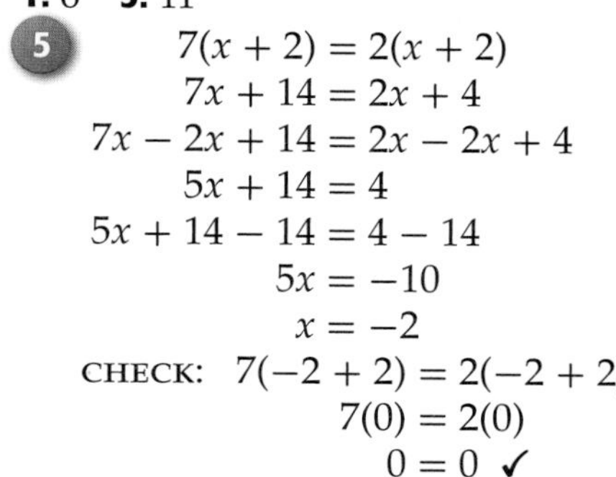

5

$$7(x + 2) = 2(x + 2)$$
$$7x + 14 = 2x + 4$$
$$7x - 2x + 14 = 2x - 2x + 4$$
$$5x + 14 = 4$$
$$5x + 14 - 14 = 4 - 14$$
$$5x = -10$$
$$x = -2$$

CHECK: $7(-2 + 2) = 2(-2 + 2)$
$$7(0) = 2(0)$$
$$0 = 0 \checkmark$$

7. null set; no solution **9.** Identity; all numbers **11.** B

13. $r \leq -8$

15. $p \geq 5$

17. $p \geq 7$

19. 2 **21.** 4 **23.** null set; no solution **25.** null set; no solution **27.** Identity; all numbers **29.** 6 granola bars and 6 magazines

31. $h \geq -2$

33. $s < -5$

35. $h \leq -7$

37. $t < 10$

39. 7 **41.** $x > 8$ **43.** 6

45 Words: three times her daily distance is at least 26.2 miles
Symbols: Let m represent the daily distance.
Inequality: $3m \geq 26.2$

$$3m \geq 26.2$$
$$\frac{3m}{3} \geq \frac{26.2}{3}$$
$$m \geq 8.73$$

So, she needs to increase her distance by 8.73 − 4 or 4.73 miles.

47. −16

$4(y - 3) = 2(3y + 10)$	Write the equation.
$4y - 12 = 6y + 20$	Distributive Property
$4y - 12 + 12 = 6y + 20 + 12$	Addition Property (=)
$4y = 6y + 32$	Simplify.
$4y - 6y = 6y + 32 - 6y$	Subtraction Property (=)
$-2y = 32$	Simplify.
$y = -16$	Division Property (=)

49. $w \geq 6.25$

$-1.2(w + 1.1) \leq 6.18$	Write the Inequality.
$-1.2w - 1.32 \leq 6.18$	Distributive Property
$-1.2w - 1.32 + 1.32 \leq 6.18 + 1.32$	Addition Property
$-1.2w \leq 7.5$	Simplify.
$\frac{-1.2w}{-1.2} \geq \frac{7.5}{-1.2}$	Division Property
$w \geq -6.25$	Simplify.

51. Sample answer: $10y - 3 \leq -12$
53. Neither girl is correct. Jada did not distribute the 5 in the second step. Liu did not switch the inequality sign when she divided by −2 in the last step. **55.** Sample answer: If you get an answer like 5 = 5 which is always true, all numbers are solutions. If you get an answer like 5 = 9 which is never true, there are no solutions. If you get an answer like $x = 8$, there is one solution. **57.** 12 **59.** D

61. $q < 24$

63. $x \leq -5$

65. $a \geq 2\frac{1}{2}$

67. $86 > w$ **69.** −0.625 **71.** $0.\overline{1}$ **73.** $-0.\overline{45}$ **75.** 3.6 **77.** \$7.40

Pages 254–256 ***Study Guide and Review***

1. inequality **3.** negative **5.** greater than **7.** product **9.** area **11.** 20 ft; 25 ft^2 **13.** 12 cm; 6 cm^2 **15.** 15 ft **17.** 1 **19.** 16 **21.** 0 **23.** $j \leq 15$ **25.** true **27.** true **29.** false

31. $x < 12$

33. $a \leq 18$

35. $z \geq 8$

37. $x > \frac{13}{5}$

39. $20x \leq 120$; $x \leq 6$ **41.** $n = 7$ **43.** $a = 4$ **45.** $m < 5$ **47.** $k < -6$

Chapter 6 Ratio, Proportion, and Similar Figures

Page 263 ***Chapter 6*** ***Get Ready***

1. $\frac{4}{5}$ **3.** $\frac{2}{3}$ **5.** $\frac{2}{3}$ **7.** $\frac{3}{4}$ students **9.** 120 **11.** 4 **13.** 4000 **15.** 13,000 **17.** $\frac{1}{5}$ **19.** 2.5 **21.** 2 **23.** 81 books

Pages 266–269 ***Lesson 6-1***

1 $\frac{12}{16} = \frac{12 \div 4}{16 \div 4}$
$$= \frac{3}{4}$$

This means that for every 3 boys there are 4 girls. **3.** $\frac{3}{7}$ **5.** $\frac{5}{8}$; For every 8 students, 5 participate in sports. **7.** $\frac{16}{3}$ **9.** $\frac{8}{1}$ **11.** $\frac{5}{9}$ **13.** $\frac{3}{4}$ **15.** $\frac{32}{1}$

17. $\frac{2}{3}$; For every 3 tables, 2 are booths or $\frac{2}{3}$ of the tables are booths.

19
$$\frac{4 \text{ ounces}}{2 \text{ pounds}} = \frac{4 \text{ ounces}}{32 \text{ ounces}} = \frac{4}{32} = \frac{4 \div 4}{32 \div 4} = \frac{1}{8}$$

21. $\frac{36}{7}$ **23a.** $\frac{7}{4}$ **23b.** $\frac{3}{4}$ **23c.** $\frac{11}{15}$ **23d.** $\frac{11}{4}$

25a. horse, cow, cat, hamster **25b.** horse; Sample answer: For every 27,272 grams a horse's heart will beat 1 time.

27 The turkey should be cooked at the ratio of 1 hour per four pounds which simplifies to $\frac{1}{4}$ hours per pound. The 18 pound turkey should be cooked for $18\left(\frac{1}{4}\right) = 4\frac{1}{2}$ hours. So it was not cooked long enough.

29. $\frac{27}{9} = \frac{45}{15}$ **31.** $\frac{6}{48} < \frac{14}{88}$ **33.** Sample answers: number of girls to boys in a class, number of apples to bananas in a fruit basket, a sale price of 3 for $8 **35.** 10 **37.** always; Sample answer: No matter what is used to measure the dimensions of the table, the actual dimensions will never change so the ratio will always be the same.

39. C **41.** B

43. $x \leq 9$
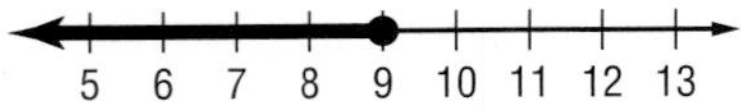

45. $y \leq -1$
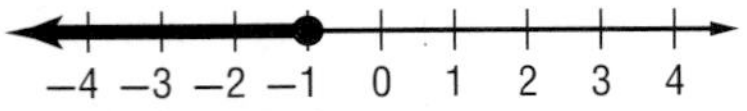

47. $n \leq -3$
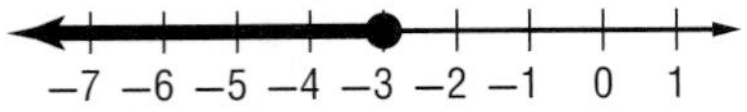

49. $a > 2500$ **51.** (2, −1) **53.** (3, 4) **55.** (4, −4) **57.** $1.44 **59.** $0.28

Pages 272–274 ***Lesson 6-2***

1. $24 per day **3.** 21.1 points per game **5.** 4.3 gallons per minute **7.** Mr. Nut; Sample answer: Barrel costs $0.34 cents per ounce, Mr. Nut costs $0.32 cents per ounce, and Chip's costs $0.35 cents per ounce.

9
$$\frac{156 \text{ students}}{6 \text{ classes}} = \frac{156 \div 6}{6 \div 6} = \frac{26}{1} = 26 \text{ students per class}$$

11. 59 miles per hour **13.** $77 per ticket **15.** 1 pizza for $6.50 **17.** Party Time **19.** $12.25 **21.** Jenny **23a.** Alicia ran 8 feet per second, Jermaine ran 6 feet per second **23b.** Alicia: 660 s or 11 min; Jermaine: 880 s or 14 min 40 s **23c.** His line would be just below Jermaine's because he ran at a slower rate than him.

25
$$\frac{525 \text{ grams}}{15 \text{ kilograms}} = \frac{525 \text{ grams}}{15000 \text{ grams}} = \frac{525 \div 15000}{15000 \div 15000} = \frac{0.035}{1} = \frac{0.035 \text{ grams salt}}{1 \text{ gram water}}$$

27. situation b; Sample answer: $\frac{120 \text{ mi}}{2 \text{ hr}} = 60$ mi/hr, $\frac{120 \text{ mi}}{3 \text{ hr}} = 40$ mi/hr **29.** $6.40; Sample answer: The unit rate for the 96 ounce container is $0.05 per ounce. So, 128 ounces would cost $0.05 × 128 or $6.40. **31.** B **33a.** x-small **33b.** Large; It has the least cost per ounce. **35.** $\frac{31}{15}$ **37.** $\frac{4}{1}$ **39.** 700 clicks **41.** 8 **43.** 144 **45.** 0.04 **47.** 3500

Pages 278–280 ***Lesson 6-3***

1. 28,800 acres **3.** 20.32 **5.** 425.25 **7.** 4.26 **9.** 138.72 grams per minute **11.** 1 billion cans per week **13.** 80.67 feet per second

15 $16 \text{ in.} \cdot \frac{2.54 \text{ cm}}{1 \text{ in.}} = 40.64 \text{ cm}$

17. 7.31 **19.** 1.85 **21.** 4.58 **23.** 3250.68 miles per hour **25.** 203.2 **27.** 2188 **29.** 6.02 **31a.** 2.60 meters per second **31b.** 12.27 miles per hour **33.** 7 yd/min, 500 m/h, 6 in./s **35.** 18 lb/min, 500 kg/h, 5 oz/s **37.** 72.79 liters per week **39.** <

41 $150 \text{ dollars} \cdot \frac{0.729 \text{ euro}}{1 \text{ dollar}} = 109.35 \text{ euro}$

43. 10,997.82 yen **45.** Sample answers: 600 cm/min, 360 m/h **47.** 500 ft/min; Sample answer: All of the other rates are equal to 60 mi/h. **49.** Identity Property; the conversion factor is equal to 1. **51.** C **53.** A **55.** $45.75 per ticket **57.** 24.2 miles per gallon **59.** The 6-pack of soda costs about $0.37 per can. The 12-pack of soda costs about $0.35 per can. So, the 12-pack is less expensive. **61.** $\frac{1}{4}$ **63.** $\frac{2}{5}$ **65.** −0.8 **67.** $7\frac{5}{14}$ **69.** $\frac{1}{4}$ **71.** $\frac{4}{5}$ **73.** $\frac{3}{13}$

Pages 283–285 ***Lesson 6-4***

1. No; The rates are not equal. **3.** $c = 3.19g$; $59.02

5 $\frac{1}{8}, \frac{2}{15}, \frac{3}{30} = \frac{1}{10}, \frac{4}{42} = \frac{2}{21}$

The rates are not equal, so the set of numbers is not proportional.

7. Yes; Each rate is equal to $\frac{1}{7}$ **9.** no; The rates are not equal. **11.** $p = \$18\ell$; \$126 **13.** no

Number of Rides	2	3	4	5	6
Cost	\$5.50	\$7.00	\$8.50	\$10.00	\$11.50

15 a. Yes; sample answer:

Hot Dog Packages	1	2	3	4
Hot Dogs	8	16	24	32

The hot dog packages to hot dogs ratio for hot dog packages of 1, 2, 3, and 4 units is $\frac{1}{8}$, $\frac{2}{16}$ or $\frac{1}{8}$, $\frac{3}{24}$ or $\frac{1}{8}$, and $\frac{4}{32}$ or $\frac{1}{8}$. Since these ratios are all equal to $\frac{1}{8}$, the number of hot dog packages is proportional to the number of hot dogs.

b. Yes; sample answer:

Hot Dogs	8	16	24	32
Hot Dog Buns	10	20	30	40

The hot dogs to hot dog buns ratio for hot dog packages of 1, 2, 3, and 4 units is $\frac{8}{10}$ or $\frac{4}{5}$, $\frac{16}{20}$ or $\frac{4}{5}$, $\frac{24}{30}$ or $\frac{4}{5}$, and $\frac{32}{40}$ or $\frac{4}{5}$. Since these ratios are all equal to $\frac{4}{5}$, the number of hot dogs is proportional to the number of hot dog buns.

17. Sample answer: If the ratio of red to pink flowers is $\frac{2}{8}$; $r = 0.25p$ for r red flowers and p pink flowers. At store A, there are always 2 red flowers for every 8 pink flowers in the bouquet. At store B, their signature bouquet will always have 3 more pink flowers than red flowers. The bouquet for store A is proportional while the bouquet for Store B is nonproportional. Store A: If $\frac{2}{10}$ of the flowers are red, $r = 0.25p$ for r red flowers and p pink flowers. Store B: $r = p - 3$ for r red flowers and p pink flowers. **19b.** The ratios should be close in value. **19c.** Sample answer: Pyramid of Khufu in Giza, Egypt; The Taj Mahal in India; The Lincoln Memorial in Washington, D.C. **21.** A **23.** D **25.** 10.16 **27.** 681 **29.** 23.3 miles per gallon **31.** 52.6 miles per day **33.** 12(\$15 + \$10 + \$8), 12(\$15) + 12(\$10) + 12(\$8); \$396 **35.** 18 **37.** 7 **39.** 7

Pages 290–292 *Lesson 6-5*

1. 24 **3.** 34 **5.** 29 **7.** 27 in.

9
$$\frac{6}{8} = \frac{z}{48}$$
$$6(48) = 8 \cdot z$$
$$288 = 8z$$
$$\frac{288}{8} = \frac{8z}{8}$$
$$36 = z$$

11. 24 **13.** 12 **15.** 14 **17.** 0.96 **19.** 23 **21.** 28 gal **23.** $d = 15.46t$; 927.6 ft; 1391.4 ft **25a.** 10 oz **25b.** 27 oz **25c.** 190.67 oz **27.** $\frac{s}{0.54} = \frac{4.55}{1.89}$; 1.3 **29.** $\frac{20}{4} = \frac{b}{20}$; 100 **31.** 0.8 **33.** 15 **35.** 7

37 a. $c = \frac{2.67}{3n}$ or $0.89n$; $c = \frac{9.50}{2n}$ or $4.75n$; $c = \frac{11.96}{4n}$ or $2.99n$; $c = \frac{10.35}{3n}$ or $3.45n$

b.

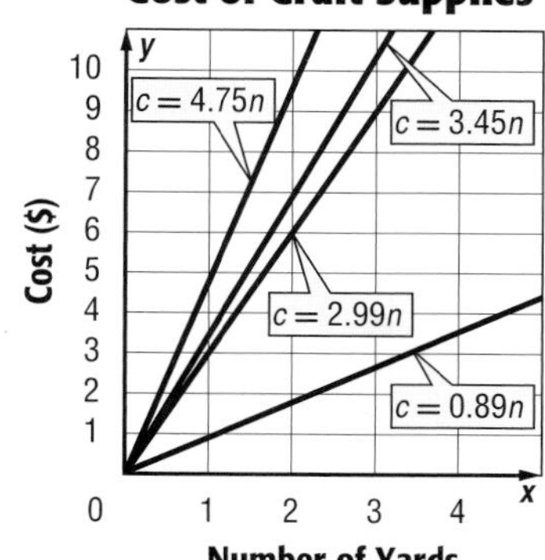

c. 18 inches is 1.5 feet which is $\frac{1}{2}$ yard.

The cost is: $c = 0.89\left(\frac{1}{2}\right) = \$0.45$

$$10 \text{ m} \cdot \frac{1 \text{ yard}}{0.914 \text{ m}} = 10.94 \text{ yard}$$
$$c = 4.75(10.94)$$
$$= \$51.97$$

39. 45 **41.** Trey; Sample answer: Morgan cross multiplied incorrectly. **43.** Sample answer: It is easier to use the constant of proportionality if you are solving the proportion for several unknowns. It is easy to find a unit rate and then multiply the amounts needed to find the equivalent amounts. No, because no matter how you set up and solve a proportion, you will also arrive at the same answer. **45.** J **47a.** $\frac{100}{258} = \frac{s}{645}$ **47b.** 250 students **47c.** $\frac{25}{258} = \frac{s}{220}$; about 21 students **49.** 12.7 **51.** 10.36 **53.** $\frac{8}{27}$ **55.** $\frac{9}{20}$

Pages 297–299 *Lesson 6-6*

1. 1 in. = 2 ft

3.

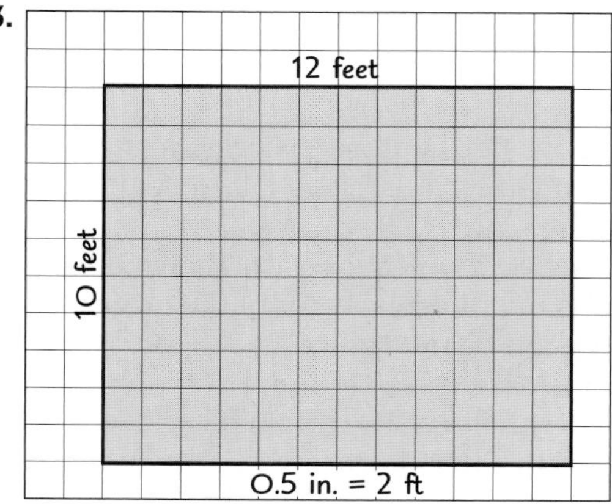

5 $\frac{255 \text{ ft}}{4 \text{ in.}} = \frac{x \text{ ft}}{1 \text{ in.}}$

$255 = 4x$

$63.75 = x$

$1 \text{ in.} = 63.75 \text{ ft}$

7a. $\frac{1}{2}$ in. by $\frac{5}{8}$ in. **7b.** $\frac{3}{4}$ in. by $\frac{3}{8}$ in.
7c. $\frac{1}{2}$ in. by $\frac{3}{8}$ in. **11** $\frac{6 \text{ in.}}{10 \text{ ft}} = \frac{6 \text{ in.}}{120 \text{ in.}}$

$= \frac{6 \div 6}{120 \div 6}$

$= \frac{1}{20}$

13. $\frac{1}{72}$ **15.** $\frac{16}{1}$ **17a.** Sample answer: $\frac{1}{4}$ in. = 3.0×10^7 mi **17b.**

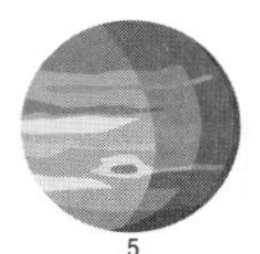

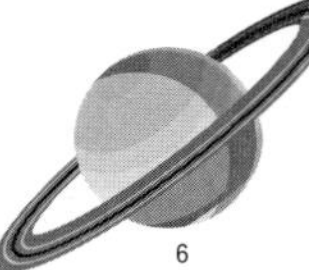

21. Always; Sample answer: A scale factor of $\frac{3}{1}$ means that 3 units is equal to 1 unit so the scale drawing or model will be larger than the actual object. **23.** Sample answer: The model is larger than the actual insect. The scale factor is $\frac{2.5}{1}$ which means the model is 2.5 times as large as the insect. **25.** B **27.** H **29.** 4 **31.** 16 **33.** 1.4 **35.** $h = 35d$; 105 in. **37.** 9 **39.** 18.4 **41.** 2.6

Pages 303–306 Lesson 6-7

1 $\frac{5}{x} = \frac{12}{8}$

$t(8) = 12x$

$40 = 12x$

$\frac{40}{12} = \frac{12x}{12}$

$3\frac{1}{3} = x$

3. 2 units

5 $\frac{25}{10} = \frac{x}{6}$

$25(6) = 10x$

$150 = 10x$

$\frac{150}{10} = \frac{10x}{10}$

$15 = x$

7. 2.125 cm **9.** 27 in.

11.

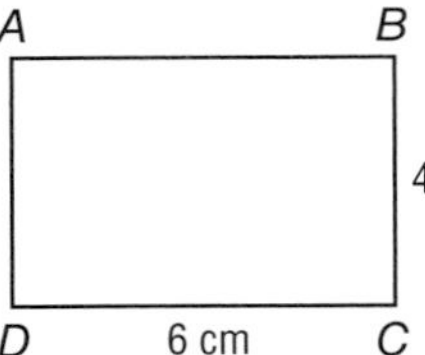

13. 102 m

15 **a.** $a + b + c$

b. The new sides will be the original sides multiplied by d. ad, bd, cd.

c. $ad + bd + cd$

d. $ad + bd + cd = d(a + b + c)$; This expression means that the perimeter of the original figure can be multiplied by d to find the perimeter of the new figure.

e. The perimeter of the original is $3 + 4 + 5 = 12$ and the scale factor is 2, so the perimeter of the new figure is $12(2) = 24$ inches.

f. If the figures are similar, then the perimeter is also similar. So, the perimeters are proportional.

17. Sometimes; Sample answer: Even though all rectangles have equal corresponding angles, the sides of one rectangle are not always proportional to the sides of another rectangle.
19. No; Sample answer: Both of them incorrectly set up the proportions. The proportion should be set up as $\frac{BC}{EF} = \frac{AB}{DE}$ or $\frac{16}{12} = \frac{x}{18}$. **21.** Triangle B is the original triangle. Since the scale factor is less than 1, the original triangle is being reduced, which means that the scaled triangle will be smaller. The measures of the sides of triangle A are less than the measures of the sides of triangle B, so triangle B must be the original triangle. **23.** J **25.** 8 **27.** 16 bags **29.** 4 **31.** 4

Pages 309–312 Lesson 6-8

1. $A'(-1, 0.5)$, $B'(0.5, 1)$, $C'(1.5, -1)$

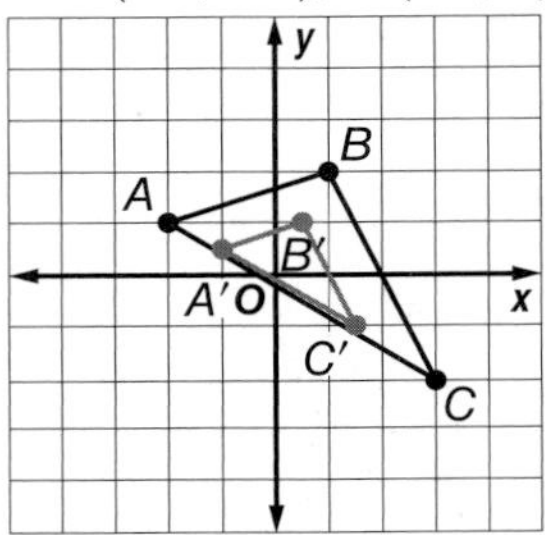

3 $M(0, 0)$, $N(3, -3)$, $O(0, -6)$, $P(-3, -3)$ yields $M'(0 \cdot 2.5, 0 \cdot 2.5)$, $N'(3 \cdot 2.5, -3 \cdot 2.5)$, $O'(0 \cdot 2.5, -6 \cdot 2.5)$, $P'(-3 \cdot 2.5, -3 \cdot 2.5) =$ $M'(0, 0)$, $N'(7.5, -7.5)$, $O'(0, -15)$, $P'(-7.5, -7.5)$ The correct answer choice is A.

5. $P'(4, 4)$, $Q'(8, 12)$, $R'(12, 4)$

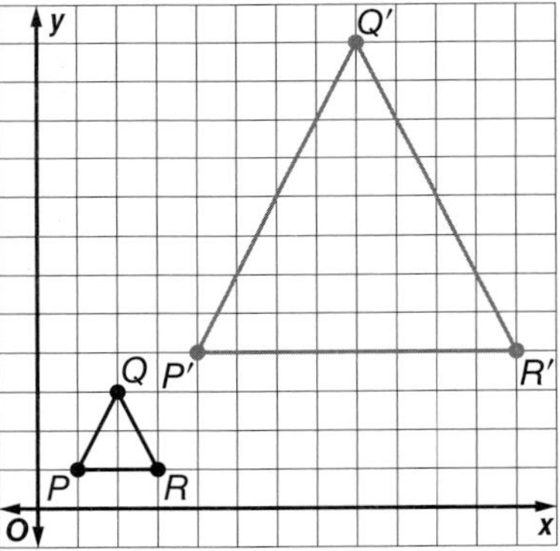

7. $X'(3, 3)$, $Y'(4.5, 6)$, $Z'(6, 0)$

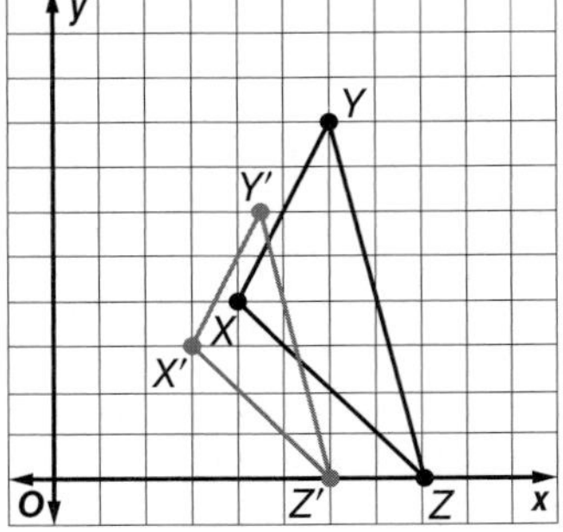

9. $G'(-2, 2)$, $H'(4, 2)$, $J'(6, -4)$, $K'(-4, -4)$

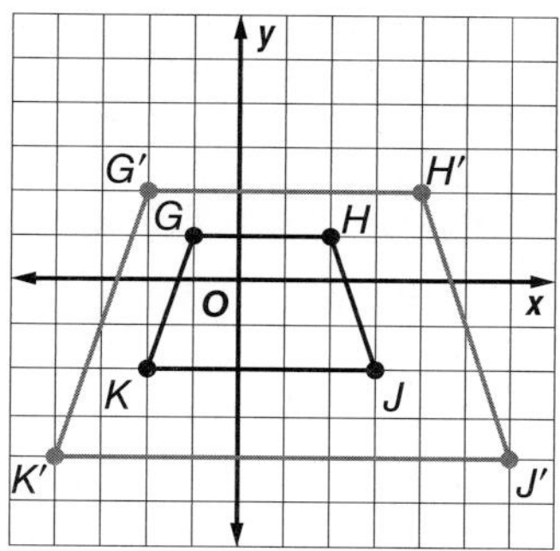

11. $R'(-3, 6)$, $S'(3, 12)$, $T'(3, 3)$

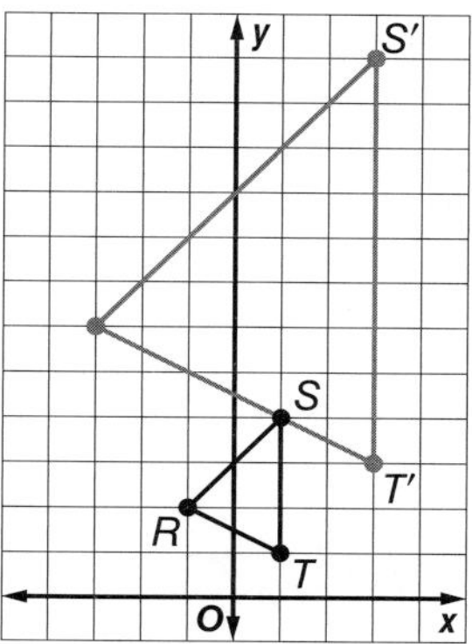

13. 200

15 a. 640 (1.5) = 960; 460(1.5) = 720 The image is 960 × 720 pixels.

b. $\frac{32}{640} = \frac{1}{20}$

c. $\frac{600}{480} = \frac{1.25}{1}$ The scale factor is 1.25.

17 The length of side DC is 3 units and the length of $D'C'$ is 2 units, so the scale factor is $\frac{2}{3}$. It is a reduction since the image is smaller than the original figure.

19. Sample answer:

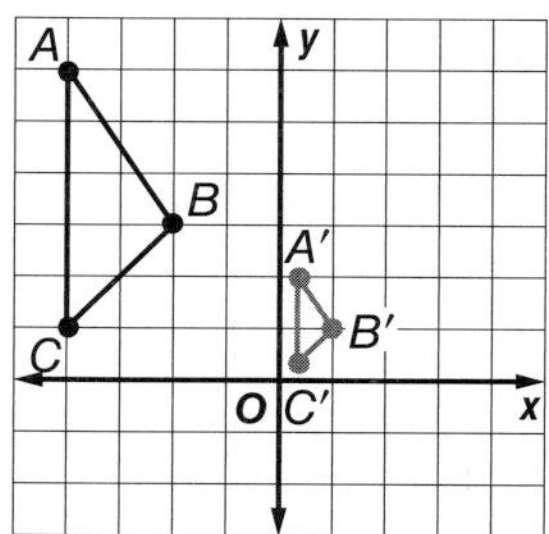

21a. $A'(0, 4)$, $B'(0, 0)$, $C'(8, 4)$ **21b.** $A'(0, 4)$, $B'(0, 0)$, $C'(8, 4)$ **21c.** $A'(0, 2.5)$, $B'(0, 0)$, $C'(5, 2.5)$ **21d.** $A'(0, 2.5)$, $B'(0, 0)$, $C'(5, 2.5)$ **23.** $C(2, 5)$, $C'(4, 7)$; Sample answer: Dilations involve multiplying the coordinates by the scale factor. In the points $C(2, 5)$ and $C'(4, 7)$, 2 is added to each coordinate. **25.** C **27.** B **29.** 2.1 **31.** 240 mi **33.** 9.6 **35.** 0.94

Pages 314–317 Lesson 6-9

1
$$\frac{8}{4.5} = \frac{x}{5.5}$$
$$8(5.5) = 4.5x$$
$$44 = 4.5x$$
$$\frac{44}{4.5} = \frac{4.5x}{4.5}$$
$$9.8 \text{ ft} \approx x$$

3. 45 ft **5.** 120 yd

7
$$\frac{157.5}{60} = \frac{5.25}{x}$$
$$157.5x = 60(5.25)$$
$$157.5x = 315$$
$$\frac{157.5x}{157.5} = \frac{315}{157.5}$$
$$x = 2 \text{ ft}$$

9a. Sample answer:

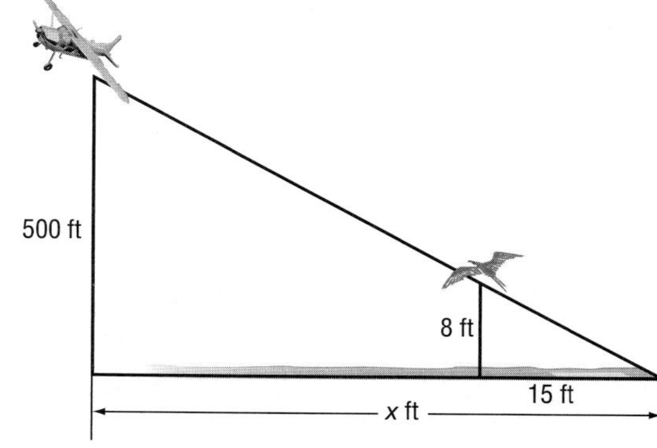

9b. $\frac{500}{8} = \frac{x}{15}$; 937.5 ft

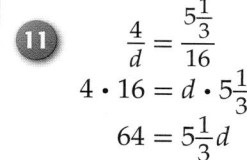

11
$$\frac{4}{d} = \frac{5\frac{1}{3}}{16}$$
$$4 \cdot 16 = d \cdot 5\frac{1}{3}$$
$$64 = 5\frac{1}{3}d$$
$$12 \text{ ft} = d$$

13. $\frac{37.5}{1.5} = \frac{150}{x}$; 6 ft **17.** False; Sample answer: You also need to know if the angles created by the two sides are congruent as well. **19.** D
21. 10 km
23.

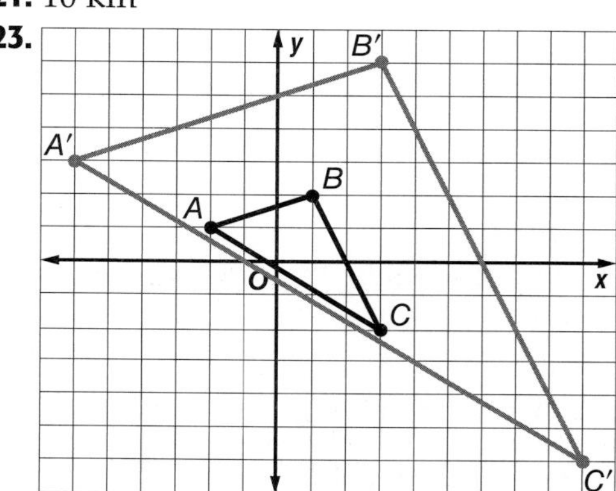

25. 18 pieces **27.** $0.58\overline{3}$ **29.** 0.6

Pages 318–322 Study Guide and Review

1. proportion **3.** constant of proportionality
5. similar figures **7.** dimensional analysis

9. $\frac{5}{12}$ **11.** $\frac{45}{1}$ **13.** $\frac{3}{4}$; For every 4 times at bat, Jean got 3 hits or on $\frac{3}{4}$ of his at bats, he got a hit. **15.** 80 meters per minute **17.** 0.6 miles per minute **19.** 17.78 **21.** 739.35 **23.** 8.51 **25.** 318.75 mi **27.** Yes; Each rate is equal to $\frac{1}{8}$. **29.** $c = 0.25r$, 11 rings cost \$2.75; 20 rings cost \$5 **31.** 7.2 **33.** 7 **35.** 120 ft **37.** 150 ft **39.** 7.2 in. **41.** 3.5 cm

43.

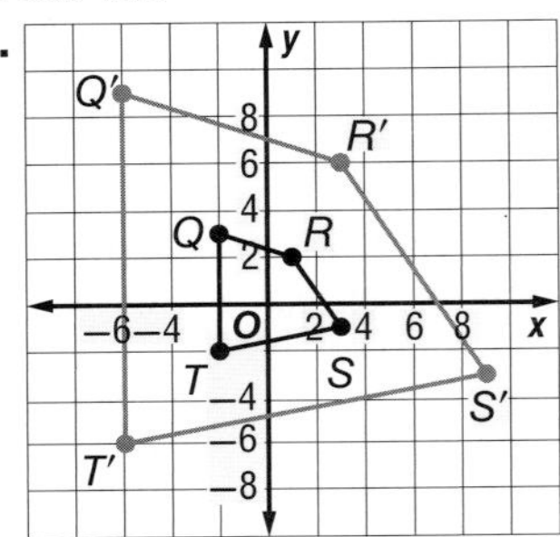

$Q'(-6, 9)$, $R'(3, 6)$, $S'(9, -3)$, $T'(-6, -6)$

45.

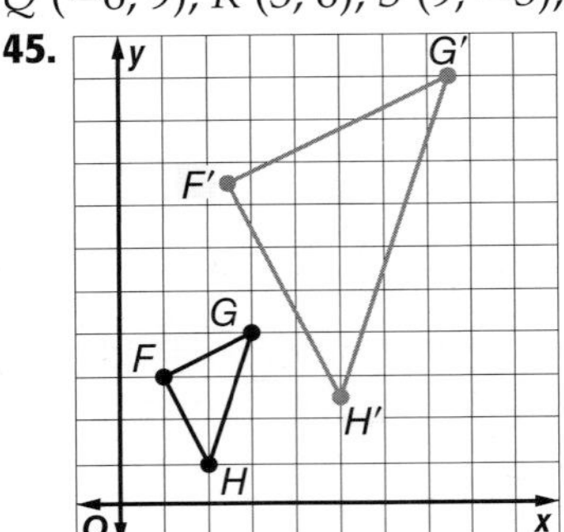

$F'(2.5, 7.5)$, $G'(7.5, 10)$, $H'(5, 2.5)$

47. 31 in. **49.** 70 m

Chapter 7 Percent

Page 329 ***Chapter 7*** ***Get Ready***

1. 0.25 **3.** 0.375 **5.** 0.67 **7.** 15 **9.** 44 **11.** 2 **13.** 5 **15.** 129 songs

Pages 334–336 ***Lesson 7-1***

1. 25 **3.** 112 **5.** 95% **7.** 94.44% **9.** 43.75% **11.** $\frac{61}{100}$ **13.** $\frac{31}{150}$ **15.** $\frac{29}{25}$ or $1\frac{4}{25}$ **17.** $\frac{3}{500}$ **19.** $\frac{5}{8}$ **21.** $\frac{3}{250}$

23
$$140\% = \frac{140}{100} = \frac{7}{5} = 1\frac{2}{5}$$

25. 25% **27.** 450% **29.** 87.5% **31.** 71.43% **33.** 62.5% **35.** 550% **37.** figure A: $\frac{1}{4}$, 25%; figure B: $\frac{7}{10}$, 70%; figure C: $\frac{3}{4}$, 75%; figure C

39
$$75\% = \frac{75}{100} = \frac{3}{4} \text{ and } \frac{36}{48} = \frac{3}{4}$$; So, they are equal.

41. < **43.** > **45.** Sample answer: 76%; $\frac{3}{4} = 0.75$ or 75% and $\frac{7}{9} \approx 77.8\%$. So, 76% is between 75% and 77.8%.

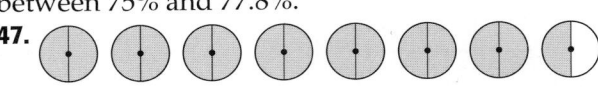

47. 750% **49.** 5; $\frac{5}{5+5} = 50\%$ **51.** B **53.** A **55.** 76 ft

57.

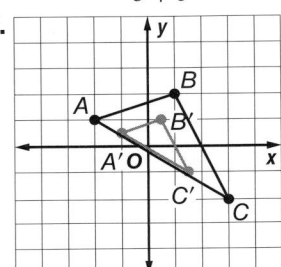

$A'(-1, 0.5)$, $B'(0.5, 1)$, $C'(1.5, -1)$

59. $y \geq -7$ **61.** a > 11 **63.** 1820 **65.** 0.333

Pages 340–342 ***Lesson 7-2***

1. 0.45 **3.** 0.006 **5.** 3% **7.** 60% **9.** dog **11.** 0.75 **13.** 0.03 **15.** 0.0091 **17.** 2.84

19 0.43 = 0.43 = 43% **21.** 28.6% **23.** 3.5% **25.** 80% **27.** 620% **29.** 366.7% **31.** 0.2% **33.** 0.02% **35.** action **37.** $\frac{5}{22}$, 2.2, 227% **39a.** 54 students **39b.** 100 students

41 $0.04x = 1$ is solved to show that there are 25 questions on the test. If Luther answered 1 incorrectly then he answered 24 questions correctly.

43. Sample answer: 9%; 0.09; $\frac{9}{100}$ **45.** Len; Carlita divided by 100 and she should have multiplied by 100 **47.** Sample answer: Greater than; by dividing by 100 and removing the % symbol, 0.005 > 0.0005. Since 0.005 > 0.0005, 0.5% > 0.0005. **49.** J **51.** 45 **53.** $\frac{1}{8}$ **55.** 80 m **57.** $8a + 4$ **59.** $3x + 6y$ **61.** $(3 \times 24) + (2 \times 12)$ is 96. So, the luggage is within the limit. **63.** 900 **65.** 0.42 **67.** 240

Pages 347–350 ***Lesson 7-3***

1. 25% **3.** 41.44 **5.** 400% **7.** 39%

9
$$\frac{21}{50} = \frac{p}{100}$$
$$21(100) = 50p$$
$$2100 = 50p$$
$$42 = p$$
42%

11. 55% **13.** 100 **15.** 166 **17.** 170 **19.** 375 **21.** 500 gumballs

23 **a.** $\frac{a}{2947} = \frac{33}{100}$
$100a = 2947(33)$
$100a = 97251$
$a = 972.51$
about 973 people

b. $\frac{a}{2947} = \frac{26}{100}$
$100a = 2947(26)$
$100a = 76622$
$a = 766.22$
about 766 people

25a. The percent is increased by a factor of 2 and the whole is halved. Each answer is 2. **25b.** 32% of 6.25 = 2 **27.** 300% **29.** 4.1 **31.** 2.2 **33.** 88.9%

35 **a.** Let x be the number of phones to add.
$$\frac{45 + x}{120 + x} = \frac{40}{100}$$
$$40(120 + x) = 100(45 + x)$$
$$4800 + 40x = 4500 + 100x$$
$$300 + 40x = 100x$$
$$300 = 60x$$
$$5 \text{ cell phones} = x$$
b. There will be 120 + 5 or 125 cell phones in stock.

37. 5% of 80, 25% of 80, 25% of 160; If the percent is the same but the whole is bigger, then the part is greater. If the whole is the same but the percent is greater, then the part is greater. **39.** 3
41. Always; Sample answer: To solve for x% of y, find $\frac{x}{100} = \frac{n}{y}$. So, $n = \frac{xy}{100}$. To solve for y% of x, find $\frac{y}{100} = \frac{n}{x}$. So, $n = \frac{xy}{100}$. **43.** C **45.** C
47. daily newspaper **49.** $1\frac{1}{5}$ **51.** $\frac{5}{6}$ **53.** $\frac{2}{3}$
55. $-\frac{3}{10}$ **57.** $\frac{11}{36}$ **59.** 3 **61.** 19

Pages 353–355 Lesson 7-4

1 $75\% \text{ of } 16 = \frac{3}{4}(16)$
$= 3(4)$
$= 12$

3. 3.7 **5.** 0.72 **7.** 12 exercises **9.** 10; 53% is about 50% or $\frac{1}{2}$, $\frac{1}{2} \cdot 20 = 10$ **11.** 36; 87% is about 90% or $\frac{9}{10}$, $\frac{9}{10} \cdot 40 = 36$ **13.** 44; 39 is about 40, 100% of 40 is 40 and 10% of 40 is 4, 40 + 4 = 44 **15.** 32 **17.** 70 **19.** 9 **21.** 6
23. 12.5 **25.** 0.3 **27.** \$7.50 **29.** 63; 73% is about 75% or $\frac{3}{4}$, $\frac{3}{4} \cdot 84 = 63$ **31.** 12.5; 49 is about 50, $\frac{1}{4} \cdot 50 = 12.5$ **33.** 0.5; 1% of 295 is about 3, $\frac{1}{6} \cdot 3 = 0.5$ **35.** 22; 276% is about 275%, 100% of 8 is 8 and 75% of 8 is 6, $8 \cdot 2 + 6 = 22$ **37.** 0.24; 1% of 30 is 0.3, $\frac{4}{5} \cdot 0.3 = 0.24$ **39.** 30; 194% is about 200%, 100% of 15 is 15, $15 \cdot 2 = 30$

41 $81\% \approx 80\% = \frac{4}{5}$
$\frac{4}{5}(6700) = 5360$ teens

43a. Sample answer: 155 mi **43b.** Sample answer: 90 mi **43c.** Sample answer: 65 mi
45. Sample answer: 60 and 600; These numbers are divisible by 3, making it easy to find $66\frac{2}{3}$% or $\frac{2}{3}$ of each number. **47.** Sample answer: One way is to use a fractional equivalent. 20% = $\frac{1}{5}$ and $\frac{1}{5} \cdot 60 = 12$. Another way is to find 10% of 60 and multiply the result by 2. $10\% \cdot 60 = 6$, $6 \cdot 2 = 12$ **49.** G **51a.** \$2.10; Sample answer: \$28 is about \$30. Since 1% of \$30 is \$0.30, then 7% of \$30 is $7 \cdot 0.30$ or \$2.10 **51b.** \$30.10; \$28 + \$2.10 = \$30.10 **51c.** \$4.50; Sample answer: \$30.10 is about \$30. 10% of \$30 is \$3 and 5% of 30 is \$1.50. \$3 + \$1.50 = \$4.50 **51d.** yes; Sample answer: The total bill with tax will be around \$30.10. The tip is \$4.50. \$30.10 + \$4.50 = \$34.60
53. $\frac{37}{1000}$ **55.** 4 **57.** 7.5 in. **59.** 0.25 **61.** 0.05
63. 0.35

Pages 360–362 Lesson 7-5

1. 30 **3.** $33\frac{1}{3}$% **5.** 275 **7.** 114 boxes

9 Part = Percent · Whole
$= 16\% \cdot 64$
$= 0.16(64)$
$= 10.24$

11. 20% **13.** 64 **15.** 78 **17.** 5.5% **19.** 39.9
21. 26% **23.** \$2500

25 **a.** $120\% \text{ of } 120 = 1.2(120)$
$= 144$ cars
$144 - 120 = 24$
They will need to add 24 cars.

b.

Percent	Equation	Number of Cars
5	1.05(120)	126
15	1.15(120)	138
25	1.25(120)	150
35	1.35(120)	162

27a. $s = 1.06 \cdot 500$; \$530
27b.

x	2010	2011	2012	2013	2014	2015
y	\$500.00	\$530.00	\$561.80	\$595.51	\$631.24	\$669.11

27c. Sample answer: No; The base amount is different each year so the part changes even though the percent remains constant. **29.** 96
31. Sample answer: The percent is greater than 100% because otherwise the part would be less than or equal to the whole. **33.** No; Suppose an item costs \$100. A 10% discount would be a discount of \$10, so the discounted price would be \$90. Adding a 10% sales tax adds \$9. So, \$90 + \$9 is not \$100. **35.** B **37.** A **39.** 48 **41.** 9
43. chocolate pieces: 2 c; peanuts: 3 c **45.** 26.4
47. 19.2 **49.** 37% **51.** 182% **53.** 5% **55.** 0.7%

Pages 366–369 Lesson 7-6

1. −20%; decrease

3 Step 1: Find the change in sales:
$900 - 1300 = -400$
Step 2: Divide the change by the original number:
$\frac{-400}{1300} = -0.308$
Step 3: Write the decimal as a percent:
$-0.308 = -30.8\%$; decrease

5. \$58.95

7 Step 1: Find the change in inches:
$26 - 14 = 12$
Step 2: Divide the change by the original number:
$\frac{12}{14} = 0.857$
Step 3: Write the decimal as a percent:
$0.857 = 85.7\%$; increase

9. −9.8%; decrease **11.** −12.5%; decrease **13.** 41.4%; increase **15.** 13.2%; increase **17.** \$76.80 **19.** \$33.00 **21.** \$18.85 **23.** \$598.80 **25.** \$7.50 **27.** 3088 calls **29.** −11.7%; decrease

31 Juliette's time will be 85% of Torie's since she is 15% faster.
85% of 74 is $0.85(74) = 62.9$ seconds
$62.9 < 74$

33a.

City	Population 2000	Population 2006	Amount of Change	%
Raleigh, NC	276,093	356,321	80,228	29%
Columbia, SC	116,278	119,961	3683	3%
Frankfort, KY	27,741	27,077	−664	−2%
Columbus, OH	711,470	733,203	21,733	3%

33b. Sample answer: The percents of change were the same but the amount of change for Columbus was much greater than the amount of change for Columbia. The percents were the same because the original amounts for each city were very different. **35.** Sample answer: from 80 mi to 160 mi; 100% increase **37a.** false; Sample answer: Suppose the cost of an item is \$25 and you want to mark it up 125% of the cost. So, multiply \$25 by 125% or 1.25. The new price is \$25 + \$31.25 or \$56.25. **37b.** true **39.** Sample answer: An internet service provider offers a plan for \$40 per month. If this month the plan is 45% off, what is the cost of the plan?; \$22 **41.** H **43.** \$855 **45.** 63 **47.** 14.52 **49.** 13 angelfish **51.** $\frac{2}{9}$ **53.** 0.875 **55.** 0.3

Pages 372–374 *Lesson 7-7*

1. \$567 **3.** \$117.81 **5.** 3.23% **7.** \$643.71 **9.** \$3512.55

11 $I = prt$
$= (620)(0.0625)(5)$
$= \$193.75$

13. \$242.40 **15.** \$919.80 **17.** \$4264.86 **19.** \$683.88 **21.** \$15,700.17 **23.** 42 months **25.** option A; Sample answer: After 3 years, the interest earned with option A is \$281.25. With option B, the interest earned is \$273.91. **27.** \$15,496.72 **29.** \$1123.16

31 #4 \$182.99(1.015) = \$185.73 − 50 = \$135.73
#5 \$135.73(1.015) = \$137.77 − 50 = \$87.77
#6 \$87.77(1.015) = \$89.09 − 50 = \$39.09
#7 \$39.09(1.015) = \$39.68
The bill amounts for the 5th and 7th months are \$137.77 and \$39.68.

33. Sample answer: \$2000 at 1%. Using the simple interest formula $I = \$2000 \cdot 0.01 \cdot 4$ or \$80 **35.** no; Sabino did not convert the time to years and Mya did not change the percent to a decimal correctly. **37.** Sample answer: With simple interest, the amount of money earned will be the same each year because it is always applied to the initial amount. With compound interest, the amount of interest will increase each year because it is being applied to the new total after the interest is added each year. **39.** J **41.** 6.5% compounded annually; Sample answer: The investment of \$500 at a 6.75% simple interest will earn \$168.75 and the investment at a 6.5% rate compounded annually will earn \$185.04. **43.** 3% **45a.** $\frac{1}{4} \times 7$ or 1.75 billion **45b.** $\frac{1}{5} \times 7$ or 1.4 billion **47.** $\frac{5a}{c}$ **49a.** $m = 8w - 1$ **49b.** 71 min **51.** 1.47

Pages 378–381 *Lesson 7-8*

1. Atmospheric Composition

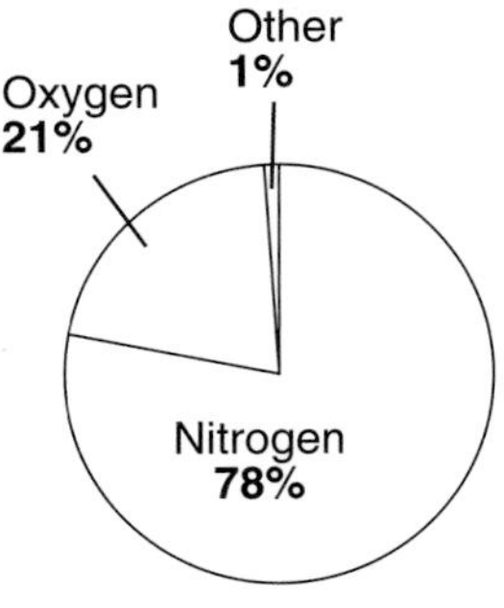

Source: *NASA*

3 Mystery: $0.18(600) = 108$
Historical Fiction: $0.10(600) = 60$
$108 - 60 = 48$ people

5. Layers of the Atmosphere (km)

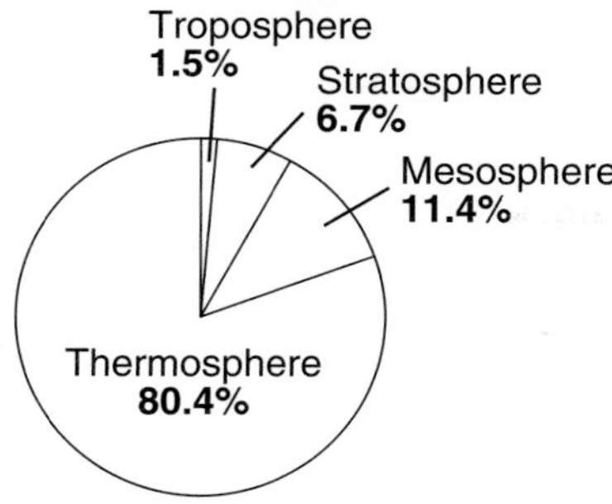

7 To find the angle measure for each category, multiply the entire circle, 360°, by the percent.

Type	Percent	Angle Measure in Graph
metal	8%	$0.08 \cdot 360 \approx 29°$
plastic	24%	$0.24 \cdot 360 \approx 86°$
food and yard waste	11%	$0.11 \cdot 360 \approx 40°$
rubber and leather	6%	$0.06 \cdot 360 \approx 21°$
other	21%	$0.21 \cdot 360 \approx 76°$
paper	30%	$0.30 \cdot 360 = 108°$

Use a compass to draw a circle and a radius. Then use a protractor to draw a 29° angle. Repeat for each of the remaining angles.

U.S. Landfill Composition

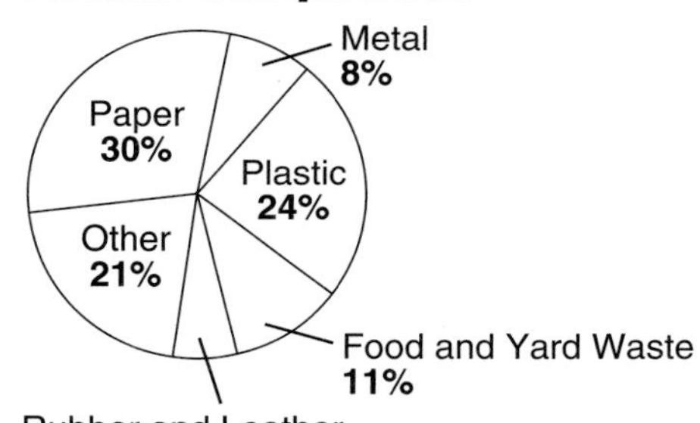

9. 45 households

11 $\frac{54}{360} = \frac{75}{x}$

$54x = 360(75)$

$54x = 27000$

$x = 500$

500 people were surveyed.

13a. Sampras's Grand Slam Wins

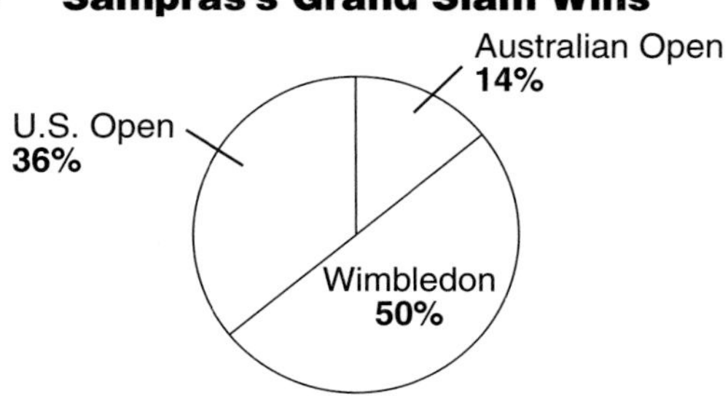

Graf's Grand Slam Wins

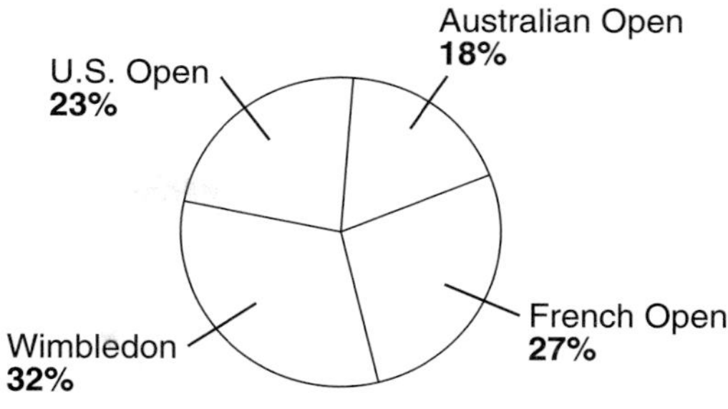

13b. No, even though they each have 5 U.S. Grand Slam wins, the total number of Grand Slam wins for each is different, therefore giving a different percentage for each. **15.** Sample answer: The percentages exceed 100% **17.** True; Sample answer: You can write a proportion comparing parts to wholes and use 360 as the whole in one of the ratios. **19.** Sample answer: Change each percent to a decimal. Then multiply each decimal by 360° to find the measure of each angle. Draw a circle and use a protractor to mark each angle. Label each section and add a title.
21. H

23.

Color	Number of Cars	Angle Measure
Red	5	18°
White	10	36°
Green	15	54°
Blue	30	108°
Other	40	144°

25. \$558.90 **27.** Domain: {4, −7, 0}, Range: {2, 9, −1} **29.** Domain: {18, 14, −6}, Range: {0, −9, 6}

Pages 382–385 Study Guide and Review

1. markup **3.** percent of change **5.** principal
7. percent **9.** $\frac{3}{10}$ **11.** $\frac{23}{25}$ **13.** $\frac{3}{500}$ **15.** $1\frac{2}{5}$
17. 25% **19.** 60% **21.** 33.33% **23.** 36% **25.** 80%
27. 0.072 **29.** 0.0048 **31.** 240% **33.** 77.5%
35. 20% **37.** 35 **39.** 18 **41.** 9 **43.** 10; 24% is about 25% or $\frac{1}{4}$, $\frac{1}{4} \cdot 40 = 10$ **45.** 0.5; 1% of 298 is about 3, $3 \div 6 = 0.5$ **47.** 200 free throws; 77% is about 80% or $\frac{4}{5}$ and 244 is about 250, $\frac{4}{5} \cdot 250 = 200$ **49.** 4 **51.** 150% **53.** No, the jersey is now 75% off the original price. If the jersey was originally \$100, after the first markdown it is \$50. Then the manager takes 50% off of \$50 making the jersey \$25, or 75% off the original price. **55.** 34.5%; increase
57. \$204 **59.** 25% **61.** \$6375.00 **63.** \$1898.44

65.

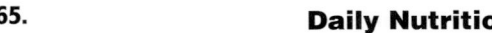

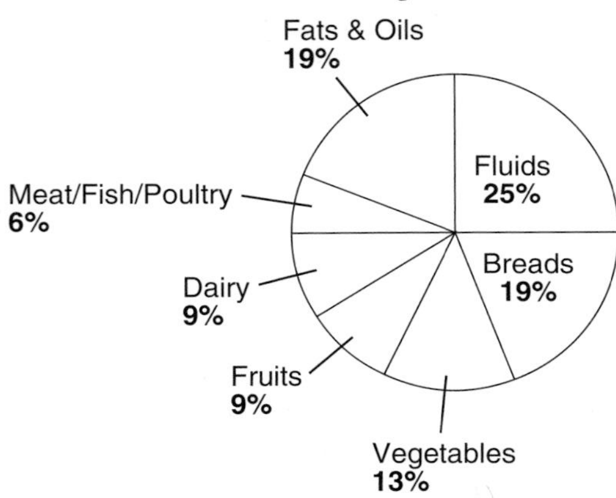

67. 2 championships

Chapter 8 Linear Functions and Graphing

Page 393 ***Chapter 8*** ***Get Ready***

1. 17 **3.** 20 **5.** 5th week **7.** A **9.** Sample answer: Walk north 10 blocks then west 5 blocks. **11.** 65 mph

Pages 397–400 ***Lesson 8-1***

1 This relation is a function because each element of the domain is paired with exactly one element of the range.
3. The graph represents a relation that is not a function because it does not pass the vertical line test. At least one input value has more than one output value. By examining the graph, you can see that when $x = 3$, there are two different y values. **5.** 14 **7.** 44 **9a.** $c(p) = 125 + 15p$
9b. 6 people **11.** This is not a function because 24 is paired with two range values, 16 and 17.
13. This is not a function because 3 is paired with 4 range values, 1, 3, 7, and 9. **15.** This relation is a function because each element of the domain is paired with exactly one element of the range.
17. This graph is a function because the vertical line test shows that it passes through no more than one point on the graph for each value of x.

19 $f(9) = 3(9) - 9$
$= 27 - 9$
$= 18$

21. −54 **23.** 0 **25.** −42 **27.** 74 **29.** −86 **31.** 254
33. −66 **35a.** $c(m) = 50 + 0.55m$ **35b.** 142 mi

37 **a.** {(1970, 37), (1980, 35), (1990, 34), (2000, 34)} Sample answer: The data represent a function because the x values or years are not repeated.
b. {(37, 1970), (35, 1980), (34, 1990), (34, 2000)} Sample answer: The inverse does not represent a function because the x-value of 34 has two different y-values.

39a. −3

39b.

Term Number	Term
1	36
2	33
3	30
4	27
5	24
6	21
7	18

The set of ordered pairs is a function because each input is paired with only one output.

39c.

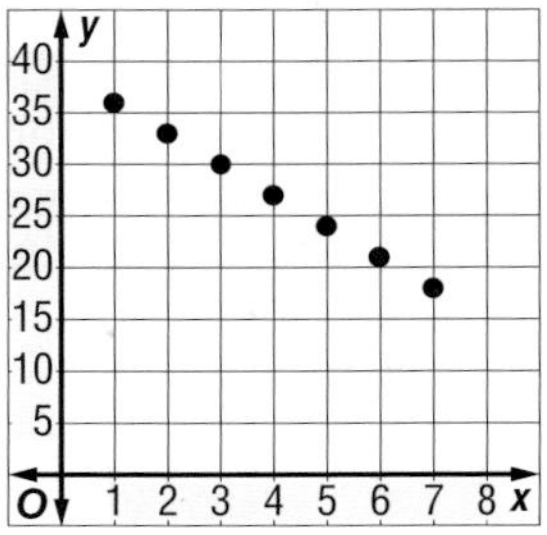

There is a negative relationship.
41. 138 **43.** 362 **45.** −982 **47.** A
49a. $c(s) = 15 + 1.75s$ **49b.** \$41.25

49c.

s	$c(s) + 15 = 1.75s$	$c(s)$
15	15 + 1.75(15)	41.25
20	15 + 1.75(20)	50.00
25	15 + 1.75(25)	58.75
30	15 + 1.75(30)	67.50

51.

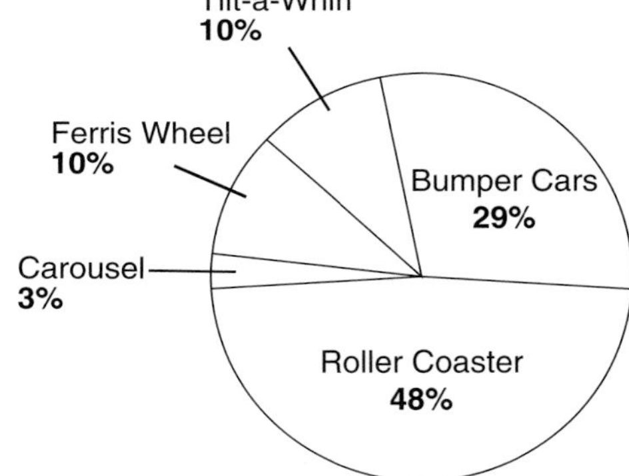

53. $\frac{6}{7}$ **55.** $-\frac{7}{20}$ **57.** $\frac{8}{9}$ **59.** $\frac{27}{25}$ or $1\frac{2}{25}$ **61.** 40 **63.** 52

Pages 403–405 ***Lesson 8-2***

1. The terms have a common difference of 1. A term is 1 more than the term number; $t = 1 + n$.
3. The terms have a common difference of 3. A term is 3 times the term number; $t = 3n$.
5. $t = 9 + n$; 19 **7.** $t = 3n + 1$; 70 **9.** 15 **11.** The terms have a common difference of 1. A term is 7 more than the term number; $t = 7 + n$. **13.** The terms have a common difference of 1. A term is 14 more than the term number; $t = 14 + n$.

15. The terms have a common difference of 8. A term is 8 times the term number; $t = 8n$.
17. The terms have a common difference of 20. A term is 20 times the term number; $t = 20n$.
19. $t = 15 + n$; 38 **21.** $t = 4n$; 52 **23.** $t = 3n + 4$; 64 **25.** $t = 4n - 3$; 353

27 **a.** Make a table to organize the sequence and find a rule. Each term is 2 more than the one before it. This would indicate that the rule should be $t = 2n$. However, you need to add 1 to get to the value of t. So, the rule is $t = 2n + 1$.

Term Number (n)	1	2	3	4
Term (t)	3	5	7	9

b. Use the rule to find to number of triangles if there are 27 beams.

$t = 2n + 1$
$27 = 2n + 1$
$26 = 2n$
$\frac{26}{2} = \frac{2n}{2}$
$13 = n$

There will be 13 triangles.

29. $d = 9s$; 108 ft

31 Make a table to organize the sequence. The terms have a common difference of 2. This would indicate that the rule should be $t = 2n$. However, you need to add 2 to get to the value of t. A term is 2 times the term number, plus 2. So, the rule is $t = 2n + 2$.

Term Number (n)	1	2	3	4	5
Term (t)	4	6	8	10	12

33. The terms have a common difference of 5. A term 5 times the term number, minus 1; $t = 5n - 1$. **35.** 665 seats **37.** Sample answer: 3, −5, −13, −21, . . . **39.** Sample answer: The pattern can be used to create an algebraic expression or equation that is a rule for the sequence. The expression or equation is then used to find values that continue the pattern, and it allows you to make predictions. **41.** $t = n + 6$; In both the function and the sequence, the input values are 1, 2, 3, 4 and the output values are 7, 8, 9, 10. **43.** 59

45a.

x	y
1	5
2	7
3	9
4	11
5	13

45b. $y = 2x + 3$

47. $252 **49.** 10 **51.** 5

Pages 409–411 *Lesson 8-3*

1.

x	$y = x + 7$	y
−1	$y = -1 + 7$	6
0	$y = 0 + 7$	7
1	$y = 1 + 7$	8
2	$y = 2 + 7$	9

(−1, 6), (0, 7), (1, 8), (2, 9)

3. Sample answer: (−1, 4), (0, 5), (1, 6), (2, 7)
5. Sample answer: (−1, 3), (0, 6), (1, 9), (2, 12)
7. Sample answer: (1, 10) means that she earns $10 for working 1 hour; (2, 20) means that she earns $20 for working 2 hours.

9.

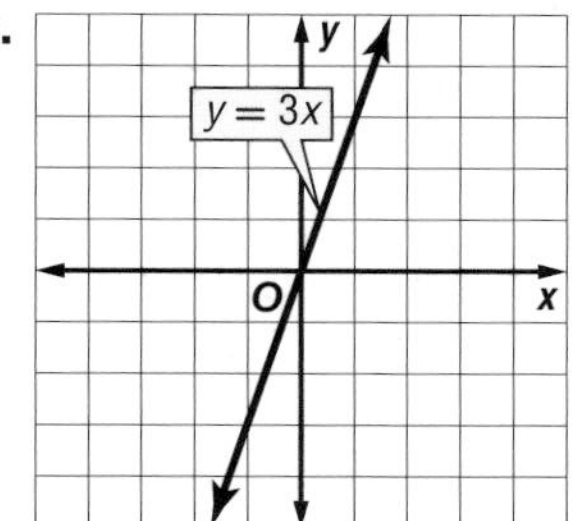

11.

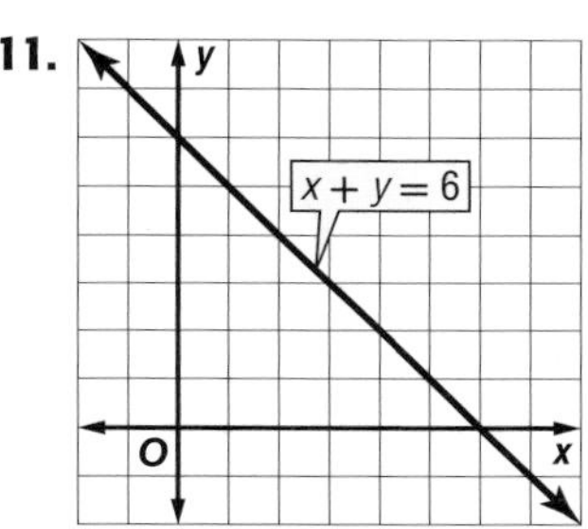

13.

x	$y = x + 7$	y
−1	$y = -2(-1)$	2
0	$y = -2(0)$	0
1	$y = -2(1)$	−2
2	$y = -2(2)$	−4

(−1, 2), (0, 0), (1, −2), (2, −4)

15.

x	$y = -2x + 8$	y
−1	$y = -2(-1) + 8$	10
0	$y = -2(0) + 8$	8
2	$y = -2(2) + 8$	4
4	$y = -2(4) + 8$	0

(−1, 10), (0, 8), (2, 4), (4, 0)

17 Choose four values for x and substitute each value into the equation. We chose −1, 0, 1, and 2. Evaluate the expression to find the value of y. Write the solution as ordered pairs. Sample answer: (−1, 2), (0, 0), (1, −2), (2, −4).

x	$2x$	y	(x, y)
−1	−2(−1)	2	(−1, 2)
0	−2(0)	0	(0, 0)
1	−2(1)	−2	(1, −2)
2	−2(2)	−4	(2, −4)

19. Sample answer: $(-1, 4)$, $(0, 3)$, $(1, 2)$, $(2, 1)$
21. Sample answer: $(-1, -1)$, $(0, -4)$, $(1, -7)$, $(2, -10)$ **23.** Sample answer: $(-1, 11)$, $(0, 9)$, $(1, 7)$, $(2, 5)$ **25.** Sample answer: $(1, 9)$ means they can ride 1 regular ride and 9 children's rides; $(2, 6)$ means they can ride 2 regular rides and 6 children's rides; $(3, 3)$ means they can ride 3 regular rides and 3 children's rides.

27.

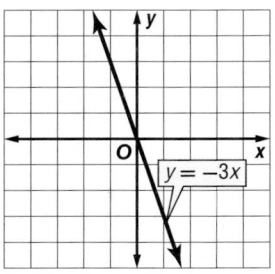

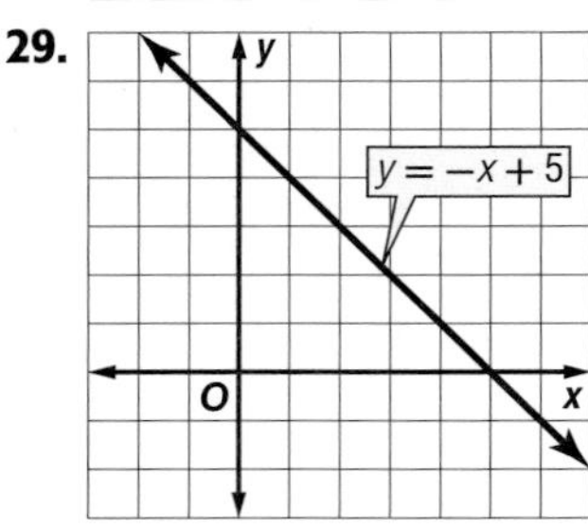

29.

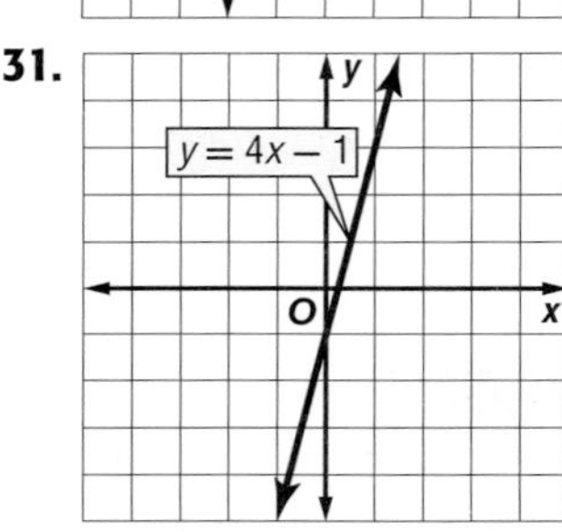

31.

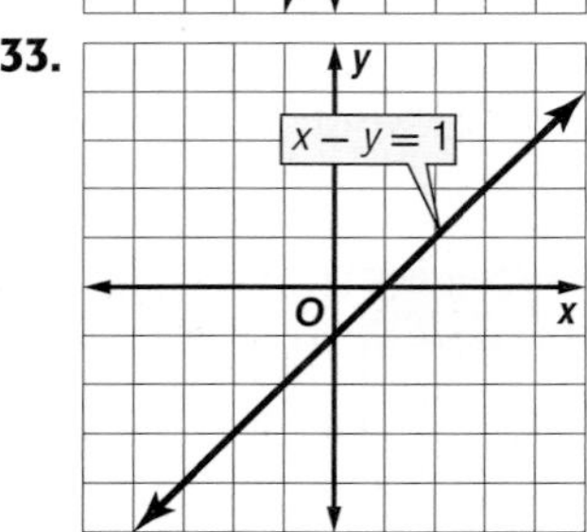

33.

33. graph: x − y = 1 (y, O, x)

35 a. $P = 2w + 2\ell$ Perimeter of a rectangle
$16 = 2x + 2y$ Replace P with 16, w with x, and ℓ with y.

x	$16 = 2x + 2y$	y	(x, y)
1	$16 = 2(1) + 2y$	7	(1, 7)
2	$16 = 2(2) + 2y$	6	(2, 6)
3	$16 = 2(3) + 2y$	5	(3, 5)

b. Choose three values for x and substitute each value into the equation. We chose 1, 2, and 3. Solve the equation to find the value of y. Write the solution as ordered pairs. Sample answer: $(1, 7)$, $(2, 6)$, $(3, 5)$.

c. Plot the ordered pairs on a coordinate plane.

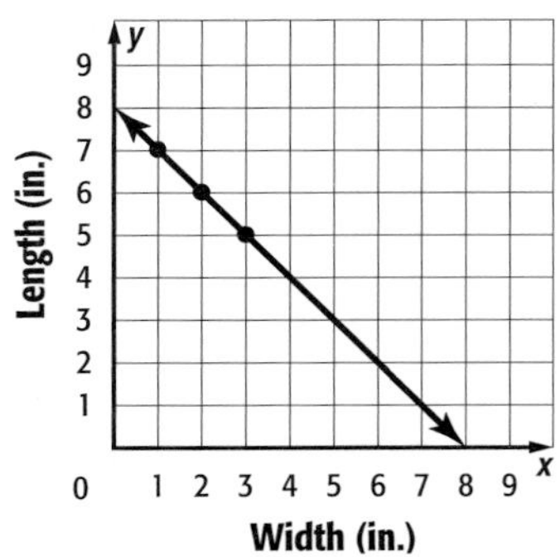

d. $(-4, 12)$: $16 \stackrel{?}{=} 2(-4) + 2(12)$
$16 \stackrel{?}{=} -8 + 24$
$16 \stackrel{?}{=} 16$ ✓

$(-4, 12)$ is a solution. It does not make sense because the width of a rectangle cannot have a negative value.

37. Sample answer: $y = x - 5$; $(1, -4)$
39. Sample answer: Infinitely many values can be substituted for x, or the domain. A table and a graph show some of the solutions. An equation represents all the solutions of a function. **41.** Sample answer: Linear equations use variables to show the relationship between the domain values and the range values of a function. Functions can be represented using a table, a graph, a verbal description, or an equation. **43.** F **45a.** Water weighs 64 pounds per cubic foot.

45b. Sample answer:

x	y
1	64
2	128
3	192
4	256

45c. Sample answer: $(1, 64)$ means that 1 cubic foot of water weighs 64 pounds; $(2, 128)$ means that 2 cubic feet of water weighs 128 pounds.
45d. Sample answer:

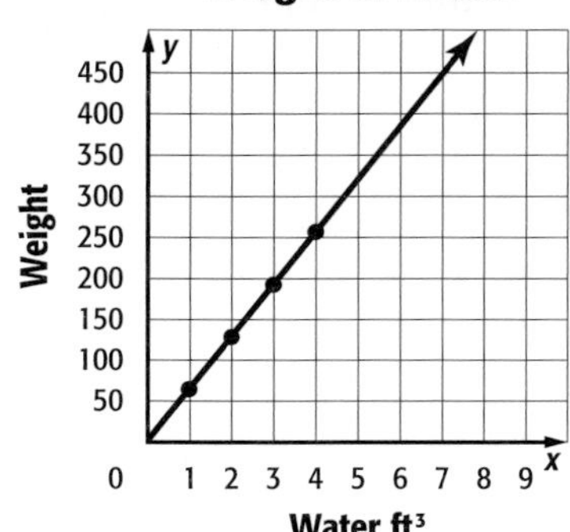

47. $t = 4n + 2$; 94 **49.** No; -0.1 in the domain is paired with 5 and -5 in the range. **51.** 2 **53.** 6

Pages 415–417 **Lesson 8-4**

1. increase of 0.5 h/lb **3.** −0.8 ft/min or decrease of 0.8 ft/min **5.** increase of 15 miles/h

7 rate of change $= \frac{\text{change in temperature}}{\text{change in time}}$
$= \frac{30 - 20}{2 - 1}$
$= \frac{10}{1}$

The rate of change is an increase of 10 jumps per minute.

9 rate of change $= \frac{\text{change in temperature}}{\text{change in time}}$
$= \frac{45 - 48}{2 - 1}$
$= \frac{-3}{1}$

The rate of change is a decrease of 3° per hour or −3°/h.

11. Mosquito: 600 beats/s; honey bee: 200 beats/s; the number of times a mosquito's wings beats increases at a faster rate than the number of times a honey bee's wings beat.
13. 3 times/day, 2 times/day, 1 time/day; A puppy less than 6 months old should be fed 3 times a day. A puppy between 6 and 12 months old should be fed twice a day. A puppy older than 12 months should be fed once a day.
15. Sample answer: The steeper the line, the greater the rate of change. **17.** Sample answer: For a horizontal line, as x increases, y does not change. So, the rate of change is 0. For a vertical line, x does not change. So, it cannot represent a rate of change. **19.** D **21.** C
23. Sample answer:

x	y
−1	−8
0	−7
1	−6
2	−5

25. Sample answer:

x	y
−1	−13
0	−10
1	−7
2	−4

27. The difference of the term numbers is 1. The terms have a common difference of 3. The term is 3 times the term number, minus 2. $t = 3n - 2$. **29.** $4n + 1$; figure 11
31. 25 ft/s **33.** $y = -2x$ **35.** $y = \frac{1}{3}x$

Pages 422–424 **Lesson 8-5**

1. $\frac{1}{2}$ in./wk; The plant grows $\frac{1}{2}$ inch per week. No; the ratio $\frac{\text{height}}{\text{time}}$ is not the same for every pair of values. **3a.** $y = 0.25x$ **3b.** 5 in.

5 rate of change $= \frac{\text{change in distance}}{\text{change in time}}$
$= \frac{2000 \text{ mi} - 1000 \text{ mi}}{4\,h - 2\,h}$
$= \frac{1000 \text{ mi}}{2\,h}$
$= 500$ mih

The airplane travels 500 miles per hour. To determine if the quantities are proportional, find $\frac{\text{distance } m}{\text{time } h}$ for points on the graph. $\frac{1000}{2} = 500$ mi/h; $\frac{2000}{4} = 500$ mi/h; $\frac{3000}{6} = 500$ mi/h
Since the ratio $\frac{\text{distance}}{\text{time}}$ is the same for every pair of values, the relationship is proportional.
7. −1 ticket available/ticket sold; For each ticket sold, there is 1 less ticket available to buy. No; the ratio $\frac{\text{tickets available}}{\text{tickets sold}}$ is not the same for every pair of values.

9 **a.** Use the equation $k = \frac{y}{x}$ to find the constant of variation; $k = \frac{14.7}{33}$ or 0.4; So, the equation is $y = 0.4x$ to the nearest tenth.
b. $y = 0.4x$
$y = 0.4(900)$
$y = 360$ psi

11. True; the ratio $\frac{\text{total cost}}{\text{number of rooms}}$ is not the same for every pair of values. **13.** False; the graph of a direct variation always passes through the origin. **15.** Keyshawn; Ramiro found the constant rate of change rather than finding the ratios $\frac{y}{x}$ for each point. **17.** It increases the distance. **19.** C **21.** B **23a.** \$12/person
23b. The cost of admission is \$12 per person.
25. Sample answer: $\left(-1, -\frac{1}{2}\right)$, (0, 0), $\left(1, \frac{1}{2}\right)$, (2, 1)
27. −24 **29.** 0

Pages 429–431 **Lesson 8-6**

1. 6 **3.** $\frac{1}{2}$

5 slope $= \frac{y_2 - y_1}{x_2 - x_1}$
$= \frac{1 - (-4)}{7 - (-2)}$
$= \frac{1 + 4}{7 + 2}$
$= \frac{5}{9}$

7. 0

9 slope $= \frac{\text{rise}}{\text{run}}$

$= \frac{5}{45}$

$= \frac{1}{9}$

11. -2 **13.** $\frac{1}{5}$ **15.** $-\frac{3}{5}$ **17.** undefined
19. -2 **21.** Sample answer: (2, 1) and (10, 6)
23. False; the graph of $y = 100x$ is as steep as the graph of $y = -100x$. **25.** C **27.** D
29. Yes; the ratio $\frac{\text{circumference}}{\text{radius}}$ is the same for every pair of values. **31.** 60 **33.** 667
35. $y = -3x + 1$ **37.** $y = 14x + 10$

Pages 435–438 Lesson 8-7

1 The equation $y = 2x + 6$ is written in the form $y = mx + b$. So, the slope is 2 and the y-intercept 6.
3. $-7; 0$

5.

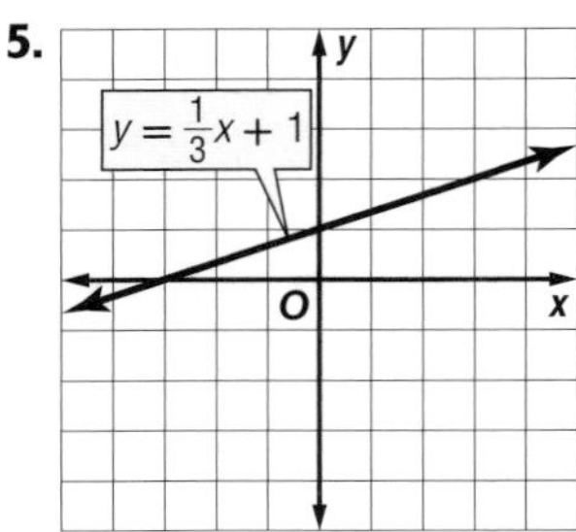

7.

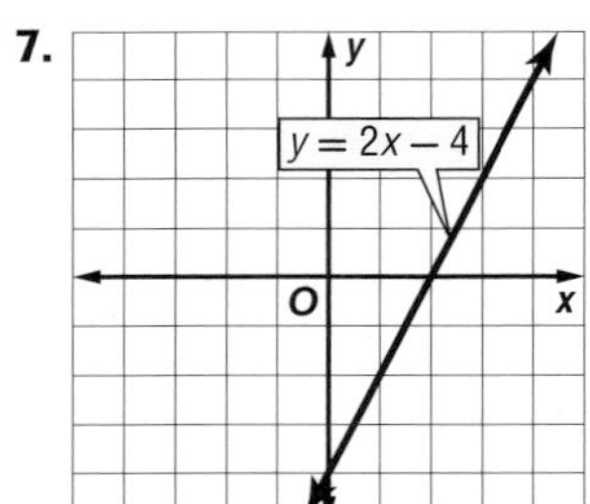

9a.

Flying a Kite

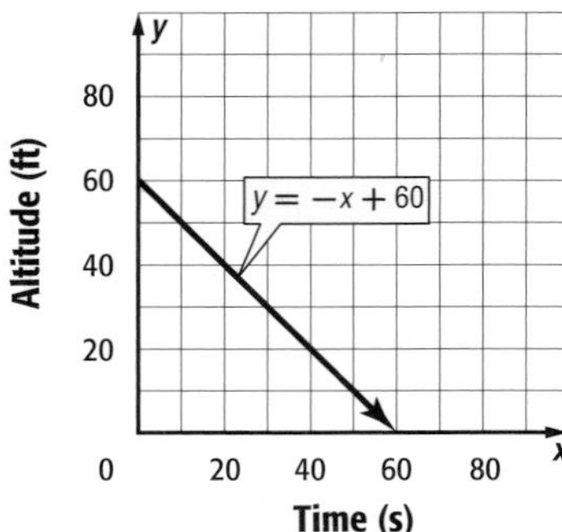

9b. The y-intercept 60 represents the initial height of the kite. The slope -1 represents the descent of 1 foot per second. **11.** $-\frac{5}{2}; -2$ **13.** $-9; 0$
15. $-4; 0$ **17.** $\frac{1}{2}; 6$

19.

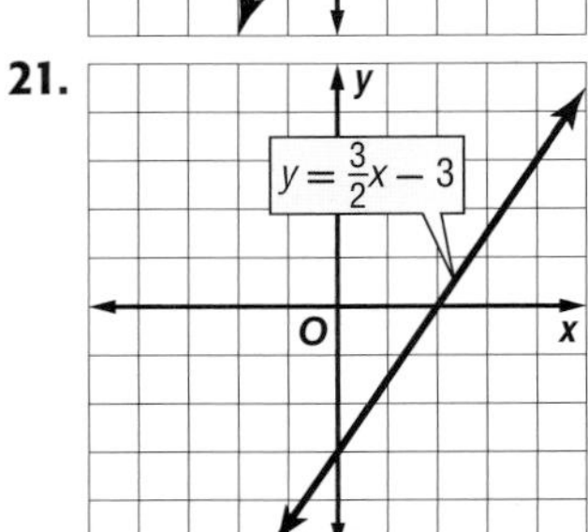

21.

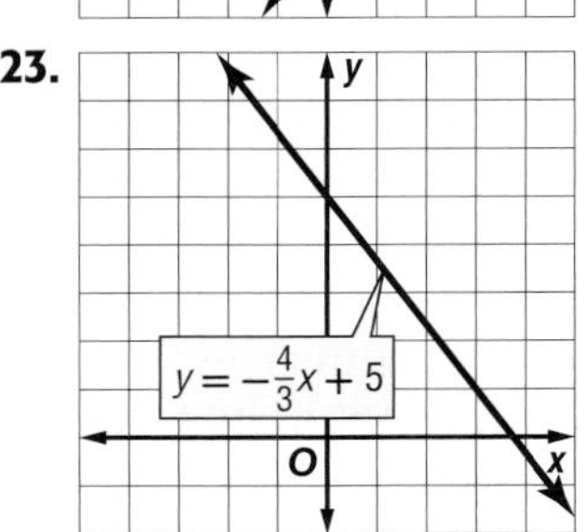

23.

y
$y = -\frac{4}{3}x + 5$
O
x

25 **a.** The slope is -50 and the y-intercept is 300. Plot the y-intercept and use the slope to find the next point. Connect the points with a line.

A Flying Albatross

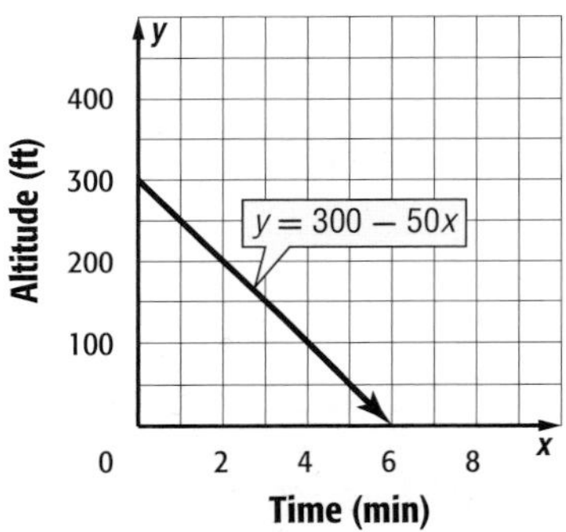

b. 300 is the y-intercept which represents the original height. The slope is -50 which represents descending at 50 ft/min.
27a. $y = 15x + 18$; $y = 15x + 12$
27b.

Total Cost ($)
100 90 80 70 60 50 40 30 20 10 0
y
Lifetime Photos
Family Photos
x
1 2 3 4 5 6 7 8 9 10
Number of Photos

27c. No; the lines are parallel and parallel lines do not intersect. **27d.** Each line has a slope of 15.

29.

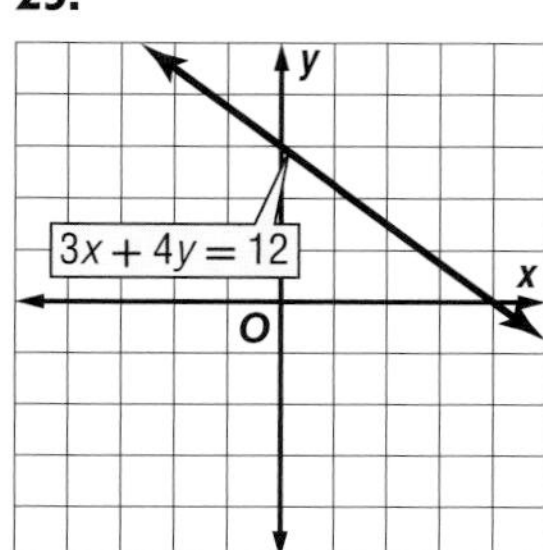

31.

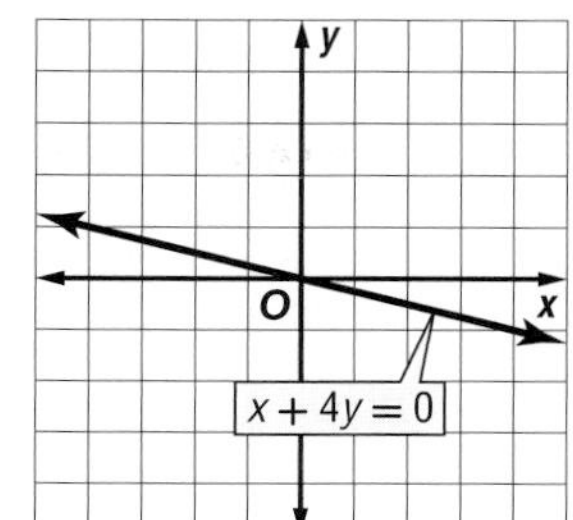

33 slope $= \dfrac{y_2 - y_1}{x_2 - x_1}$

$= \dfrac{6 - 0}{1 - (-2)}$

$= \dfrac{6}{3}$ or 2

35. $y = 2x + 4$ **37.** Sample answer: A photographer charges a \$50 sitting fee to come to your house to take family portraits and then \$15 for each 5×7 portrait. The total cost y can be represented by the equation $y = 50 + 15x$, where x represents the number of portraits ordered. A family has \$300 to spend on the portraits. How many portraits can they purchase?; 16 **39.** The graph becomes less steep. **41.** Sample answer: Graph the y-intercept point. Then use the slope to locate a second point on the line. Draw a line through the two points. **43.** J

45a.

Total Cost (\$) vs. Number of Times

45b. The y-intercept 35 is the initial cost of the pass and the slope 25 is the cost per time of skiing. **47.** \$2.80/lb; The birdseed costs \$2.80 per pound. **49.** $\frac{5}{12}$ **51.** -22 **53.** 2 **55.** 32

Pages 444–447 ***Lesson 8-8***

1. $y = 2x + 4$ **3.** $y = -\frac{3}{4}x$ **5.** $y = x + 2$

7. $y = \frac{1}{2}x$

9 First, find the slope.

$m = \dfrac{y_2 - y_1}{x_2 - x_1}$

$m = \dfrac{-4 - 5}{3 - 2}$

$m = \dfrac{-9}{1}$ or -9

Use the point (5, 2) to write an equation for the line in point-slope form.

$y - y_1 = m(x - x_1)$

$y - 5 = -9(x - 2)$

You can also use the point (3, –4) to write an equation in point-slope form.

$y - y_1 = m(x - x_1)$

$y + 4 = -9(x - 3)$

Slope either point-slope equation for y to write the equation in slope-intercept form.

$y + 4 = -9(x - 3)$

$y + 4 = -9x + 27$

$y = -9x + 23$

11. Sample answer: $y - 2 = 4(x - 1)$

13a. $y = 2.4x$; The speed of the rip current is 2.4 feet per second. **13b.** 144 ft **15.** $y = x - 4$

17. $y = 2x$ **19.** $y = -7$ **21.** $y = -\frac{5}{3}x - 6$

23 The slope is $\frac{1}{2}$ and the y-intercept is (0, 2). The equation is $y = \frac{1}{2}x + 2$.

25. $y = -\frac{4}{3}x - 5$ **27.** $y = 6$

29. $y + 2 = \frac{1}{2}(x - 2)$ or $y + 1 = \frac{1}{2}(x - 4)$ or $y = \frac{1}{2}x - 3$

31. $y + 6 = 0(x - 3)$ or $y + 6 = 0(x - 5)$ or $y = -6$

33. $y - 9 = -5(x + 1)$ or $y + 6 = -5(x - 2)$ or $y = -5x + 4$

35. Sample answer: $y + 2 = 2(x + 1)$

37. Sample answer: $y - 5 = \frac{2}{5}(x - 10)$

39 **a.** First, find the slope.

slope $= \dfrac{y_2 - y_1}{x_2 - x_1}$

$= \dfrac{150 - 120}{4 - 3}$

$= \dfrac{30}{1}$ or 30

Now, use the point-slope form.

$y - y_1 = m(x - x_1)$

$y - 120 = 30(x - 3)$

$y - 120 = 30x - 90$

$y - 120 + 120 = 30x - 90 + 120$

$y = 30x + 30$

The height of the fireworks increases 30 meters per 1-inch increase in shell radius.

b. $y = 30x + 30$

$y = 30(9) + 30$

$y = 270 + 30$

$y = 300$ m

41a. $y = 34.95x$; The slope \$34.95 is the cost per person. **41b.** \$314.55 **43.** Sample answer: (–1, 2) and (–4, 5); $y - 5 = -1(x + 4)$

45. Daniel; Kayla incorrectly calculated the slope by dividing the change in x by the change in y. Also, the y-intercept is 5, not 0. **47.** Sample answer: When using an equation in the form $y = mx + b$, the slope is m and the y-intercept is b. When using a table, choose two pairs of x- and y-coordinates to find the slope. The y-intercept is

the y value in the table when the corresponding x value is 0. When using a graph, choose two points on the line to find the slope. The y-intercept is the y-coordinate of the point where the graph crosses the y-axis. **49.** F
51. $d = 320 - 65s$
53.
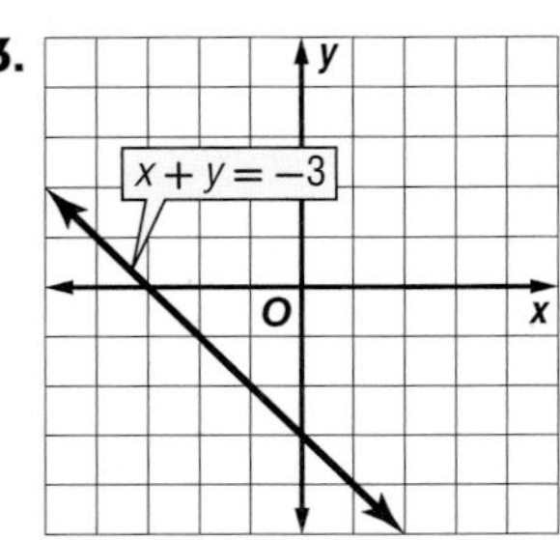

55. 1 **57.** 16%
59. negative; As the temperature decreases, heating costs increase.

Pages 450–452 **Lesson 8-9**

1a. Sample answer:

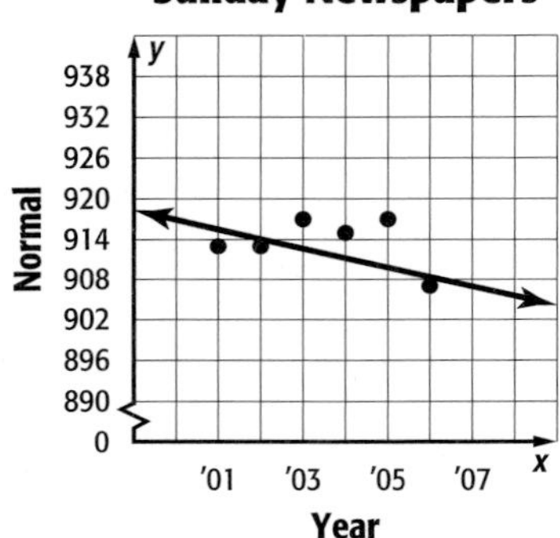

1b. Sample answer: 898

3 a. Plot the data points on a coordinate grid and draw a line that best fits the points. Sample answer:

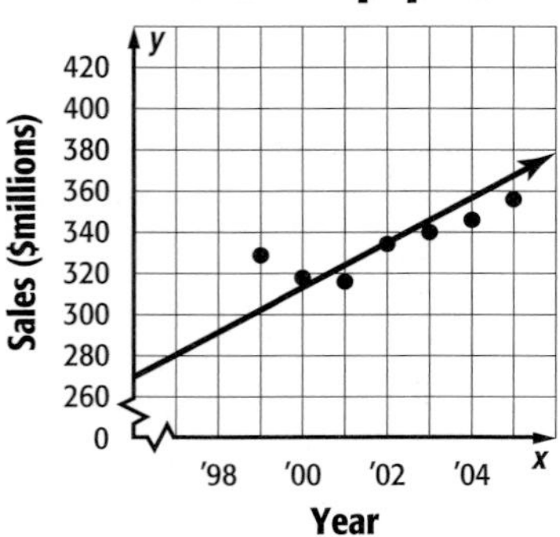

b. Extend the line to estimate the sales for 2015. Sample answer: $492 million

c. Plot the data points on a coordinate grid and draw a line that best fits the points. Sample answer:

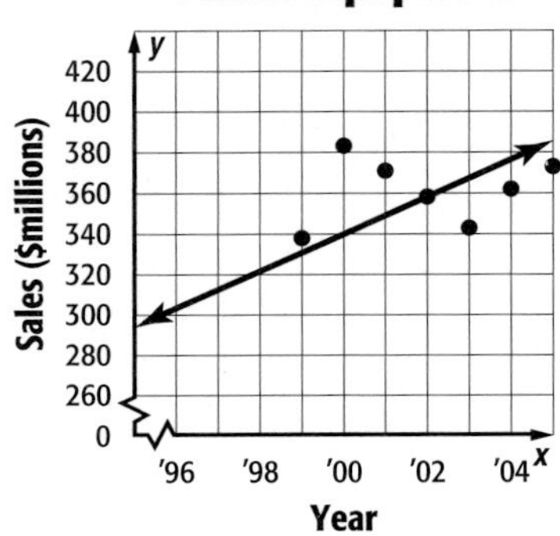

d. Extend the line to estimate the sales for 2015. Sample answer: $490 million

5 a. Use the points shown to find the slope of the line.

$$\text{slope} = \frac{y_2 - y_1}{x_2 - x_1} = \frac{14.8 - 13.5}{7 - 4} = \frac{1.3}{3} \approx 0.4$$

Now, use the point-slope form.

$$y - y_1 = m(x - x_1)$$
$$y - 13.5 = 0.4(x - 4)$$
$$y - 13.5 = 0.4x - 1.6$$
$$y - 13.5 + 13.5 = 0.4x - 1.6 + 13.5$$
$$y = 0.4x + 11.9$$

Sample answer: $y = 0.4x + 11.9$

b. $y = 0.4x + 11.9$
$y = 0.4(18) + 11.9$
$y = 7.2 + 11.9$
$y = 19.1$
Sample answer: $19.1 billion

7a.

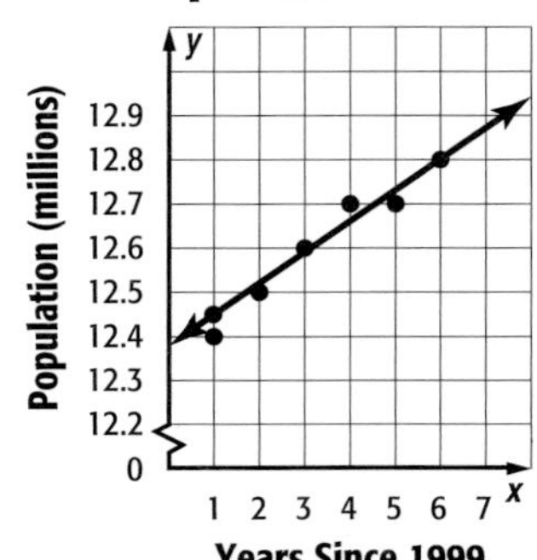

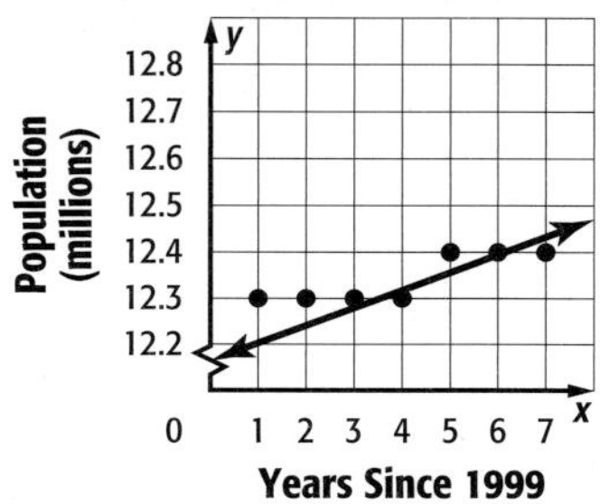

Sample answer: The slope of the line for the population of Illinois means that the population grows by 0.07 million (70,000) people every year. The slope of the line for the population of Pennsylvania means that the population grows by 0.03 million (30,000) people every year. **7b.** Illinois' population; the line of fit is steeper than the line of fit representing the growth of Pennsylvania's population. The intersection would represent the year in which the populations were equal. **7c.** Sample answer: Illinois: $y = 0.07x + 12.31$; Pennsylvania: $y = 0.03x + 12.22$; the slope of the Illinois equation is greater than the slope of the Pennsylvania equation. So, it is true that Illinois' population is growing at a faster rate. **7d.** Sample answer: Illinois 13.43 million; Pennsylvania 12.7 million **9.** Sample answer: It is reasonable to make a prediction when the data have either a positive or a negative relationship. It is not reasonable if there is no noticeable relationship between the data. **11.** Sample answer: a scatter plot in which the data do not appear to be linear **13.** B **15.** A

17a.

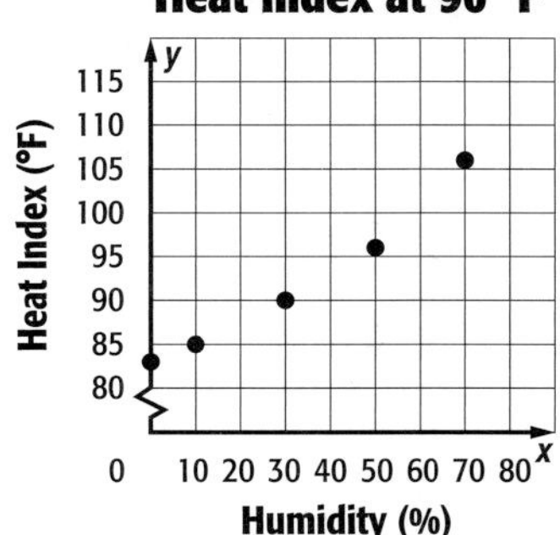

17b. Sample answer: $y = 0.35x + 81.5$

17c. 116.5°F **19.** $y = \frac{2}{5}x$

21.

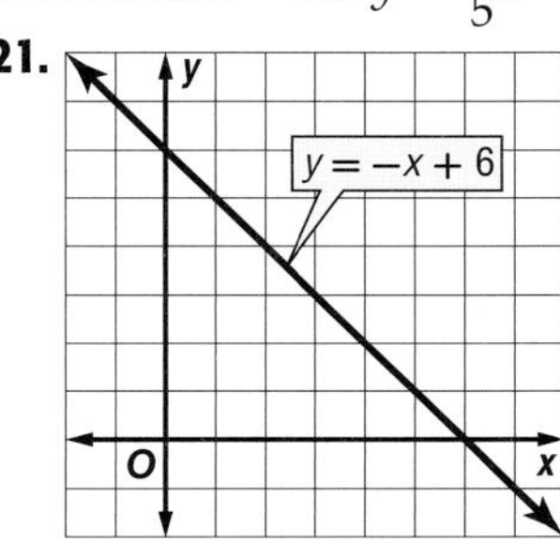

23. 9 **25.** 18

Pages 455–457 *Lesson 8-10*

1.

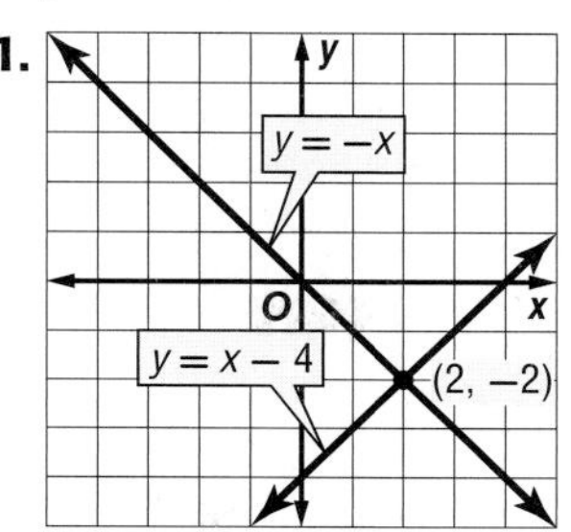

(2, −2)

3.

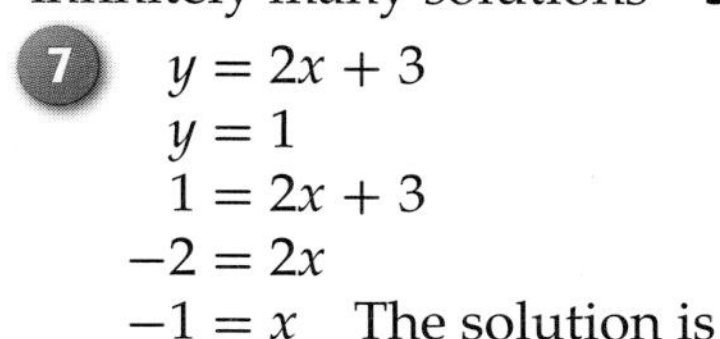

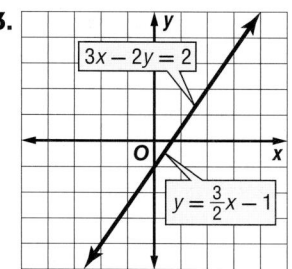

infinitely many solutions **5.** (5, 3)

7 $y = 2x + 3$
$y = 1$
$1 = 2x + 3$
$-2 = 2x$
$-1 = x$ The solution is $(-1, 1)$.

9.

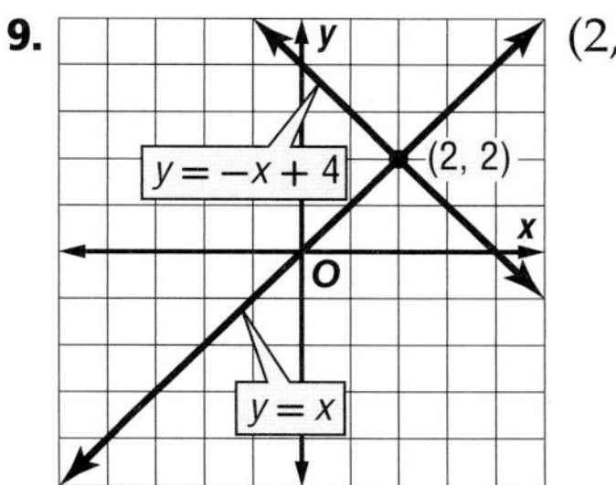

(2, 2)

11.

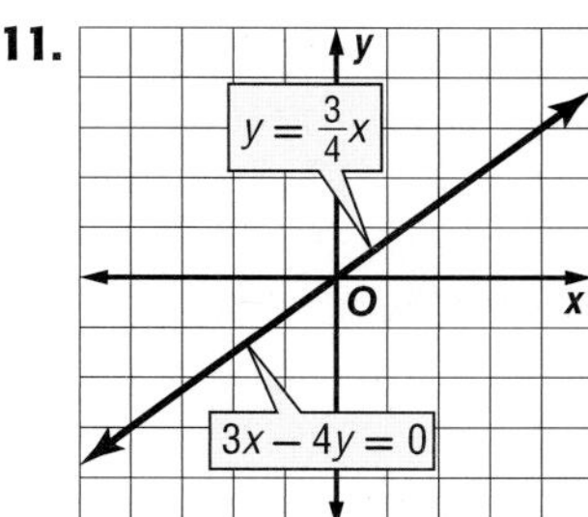

infinitely many solutions

13.

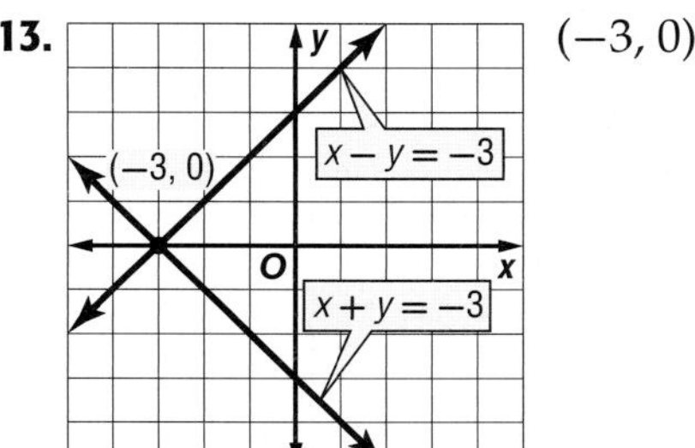

(−3, 0)

15 a. Let y represent the number of baseball cards after x months. Ling starts out with 50 baseball cards and *collects* 5 per month:

$y = 5x + 50$. Jonathon starts out with 90 baseball cards and *sells* 5 per month: $y = -5x + 90$.

b. Graph each equation on the same coordinate grid.

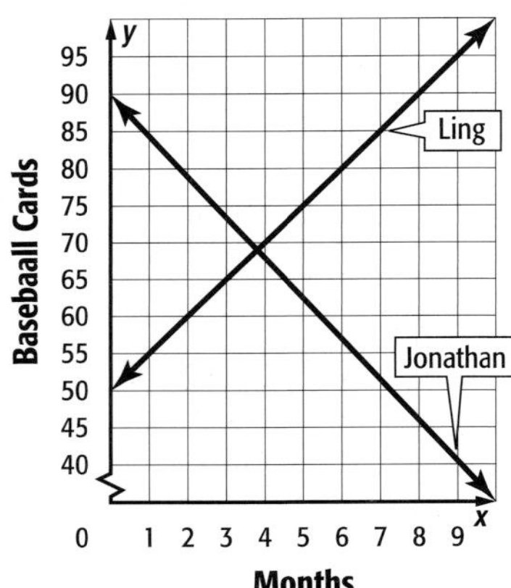

The equation intersects at (4, 70). The solution (4, 70) means that after 4 months, they will have the same number of cards (70).

17. (−4, 0) **19.** (2, −6) **21.** (5, −1) **23.** Sample answer: $y = x + 6$ and $y = 8x - 1$ **25.** (−6, −6) **27.** (2, −2) **29.** Sample answer: If a system of equations has 1 solution, the graphs are intersecting lines. If the system has no solution, the graphs are parallel lines. If the system has infinitely many solutions, then the graphs are the same line. **31.** G **33.** 12 **35.** $y = 4x - 3$ **37.** 64 **39.** 32

Pages 458–462 Study Guide and Review

1. slope **3.** sequence **5.** y-intercept **7.** independent variable **9.** system of equations **11.** No; The domain value 4 is paired with 2 range values, 1 and 2. **13.** The terms have a difference of 6. A term is 6 times the term number; $t = 6n$. **15.** $t = 7n$; 350 **17.** Sample answer: (−1, 5), (0, 0), (1, −5), (2, −10) **19.** Sample answer: (−1, 8), (0, 9), (1, 10), (2, 11)

21.

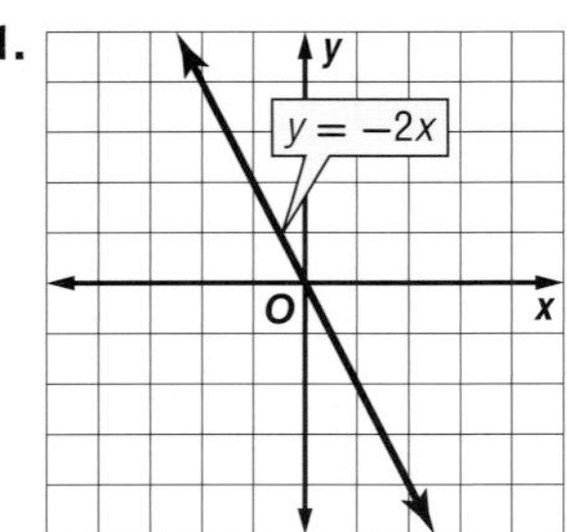

23. Sample answer: (0, 4) means she can buy 0 small smoothies and 4 large smoothies with \$12; (6, 1) means she can buy 6 small smoothies and 1 large smoothie with \$12. **25.** Adults: \$18/person; children: \$12.50/person; the cost for adults increases at a faster rate than the cost for children. **27.** $\frac{1}{2}$ **29.** −1 **31.** $\frac{1}{6}$ **33.** $-\frac{4}{3}$; 0 **35.** 1; −8

37.

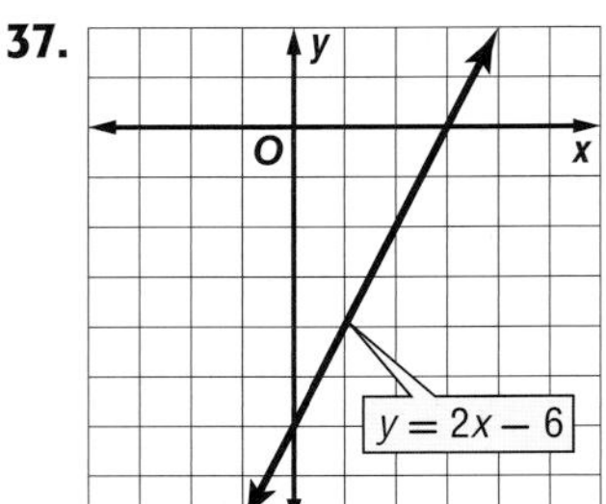

39.

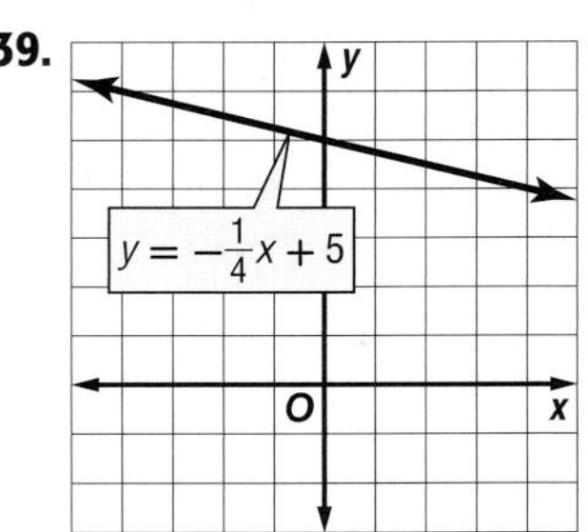

41. $y = -2x + 5$ **43.** $y = 4x$ **45.** $y - 5 = 3(x - 1)$

47a. Sample answer:

47b. Sample answer: \$362,000

49.

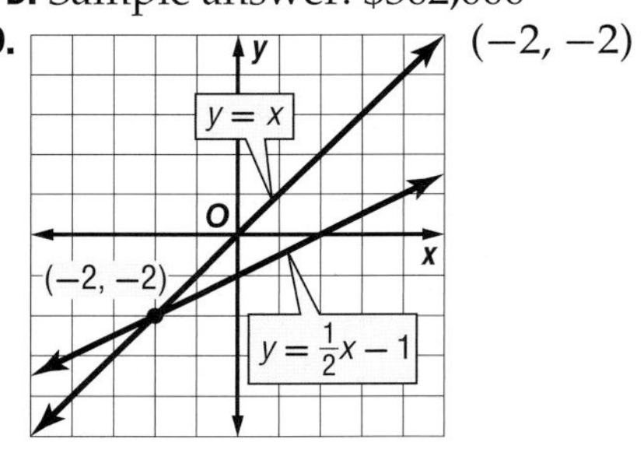

(−2, −2)

51. (1, 4) **53.** Sample answer: $x + y = 9$, $x - y = 1$; 5 and 4; $x = 5$; $y = 4$

Chapter 9 Powers and Nonlinear Functions

Page 469 Chapter 9 Get Ready

1. 22 **3.** 7 **5.** −11 **7.** \$120 **9.** −7 **11.** 6 **13.** −19 **15.** 4 **17.** 70 **19.** 167.8 **21.** 5.6 **23.** 0.9718 **25.** \$15.90

Pages 473–475 Lesson 9-1

1. 2^6 **3.** $\left(-\frac{1}{4}\right)^3$ **5.** $(y - 3)^3$

7. 16 cm **9.** 29 **11.** 28.25

13 The base 3 is a factor 5 times, so the exponent is 5, $3 \cdot 3 \cdot 3 \cdot 3 \cdot 3 = 3^5$ **15.** $(-14)^3$
17. $(-1.5)^3$ **19.** $5p^3q^3$ **21.** $8(c + 4)^2$
23. $(2x + 3y)^2$ **25.** 81 **27.** 28 **29.** 4.25
31. 21.625 **33.** 23 **35.** 243

37 **a.** Field Hockey: $2^6 \cdot 10^3 = 64 \cdot 1000 = 64{,}000 \text{ ft}^2$

Men's Lacrosse: $3^2 \cdot 7 \cdot 10^3 = 9 \cdot 7 \cdot 1000 = 63 \cdot 1000 = 63{,}000 \text{ ft}^2$

Women's Soccer:
$2^4 \cdot 5^2 \cdot 7 \cdot 13 = 16 \cdot 25 \cdot 7 \cdot 13 = 400 \cdot 7 \cdot 13 = 36{,}400 \text{ ft}^2$

b. The order from least to greatest is 36,400 ft^2; 63,000 ft^2; 64,000 ft^2
c. $64{,}000 - 63{,}000 = 1{,}000 \text{ ft}^2$

39. 1331 **41.** 625 **43.** 512 **45.** 2430 **47.** 1372
49. 820.125 **51.** > **53.** < **55.** Sample answer: $6^2, 4^3$; 2304 **57.** 10^8; Sample answer: $10^7 = 10{,}000{,}000$ and $10^8 = 100{,}000{,}000$. 100,000,000 is much closer to 230,000,000 than 10,000,000.
59. Using exponents is a more efficient way to describe and compare numbers.
61. 512 **63.** B **65.** (0, −5)

67a.

Barometric pressure

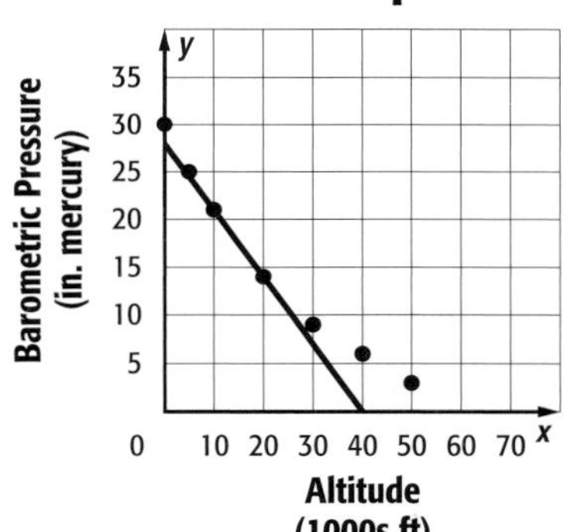

67b. Sample answer: Using (10,000, 21) and (30,000, 9), $y = -0.0006x + 27$; −9; this is not reasonable because barometric pressure cannot be negative. **67c.** No, the equation gives a negative value for barometric pressure, which is not possible. Also, the data in the scatter plot do not appear to be linear. **69.** 4 out of 5
71. $\frac{10}{11}$ **73.** 1, 3, 5, 15 **75.** 1, 2, 4, 5, 8, 10, 20, 40

Pages 478–480 *Lesson 9-2*

1. composite **3.** composite **5.** $2 \cdot 7 \cdot a \cdot a \cdot a$
7. $2 \cdot 2 \cdot 5 \cdot x \cdot x \cdot y$ **9.** prime **11.** composite
13. prime **15.** composite

17

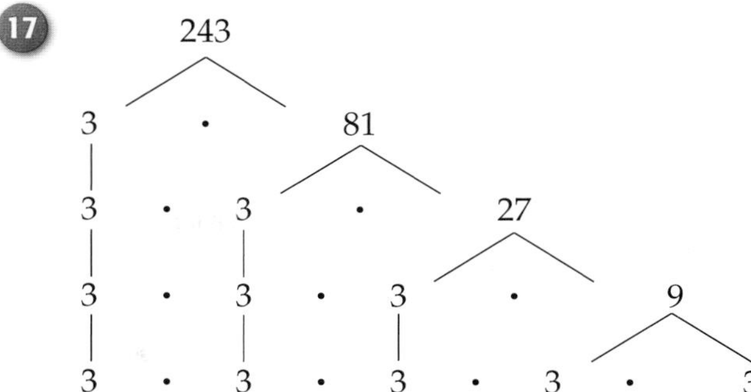

So, $243 = 3 \cdot 3 \cdot 3 \cdot 3 \cdot 3 = 3^5$
19. $2^3 \cdot 7$ **21.** $2^2 \cdot 7^2$ **23.** $2 \cdot 3^2 \cdot 11$
25. $2 \cdot 3 \cdot n \cdot n$ **27.** $-1 \cdot 11 \cdot n \cdot n \cdot n$
29. $2 \cdot 2 \cdot 5 \cdot q \cdot r \cdot s$ **31.** $-1 \cdot 5 \cdot 7 \cdot c \cdot c \cdot c \cdot d \cdot d$
33a. $2 \cdot 3 \cdot 5$ **33b.** $2 \cdot 5^2$ **33c.** $2^3 \cdot 3 \cdot 5$
33d. $2 \cdot 3 \cdot 5^2$ **35a.** 4 rectangles **35b.** 3 rectangles

37 $27 = 3 \cdot 9$ (not the product of two primes)
29 is prime, not the product of two primes
31 is prime, not the product of two primes
$33 = 3 \cdot 11$ which is the product of two primes
So, 33 could be N.

39. Sample answer: $10a^2$, $10a^2b$, $20a^2$ **41.** Both are incorrect; Filipe did not consider numbers like 9, 15, and 21 and Ledell did not consider 2 as a prime number. **43.** Sample answer: The number $2n$ is never prime. Since you are multiplying a number by 2, you have an additional set of factor pairs, 2 and n.

45.
1 ft × 36 ft
2 ft × 18 ft
3 ft × 12 ft
4 ft × 9 ft
6 ft × 6 ft

47. C **49.** −9 **51.** 45 **53.** 13 **55.** 14 pizzas
57. $-5 + n = -15$; −10 **59.** $(3 \cdot 5)(a^4)$ **61.** $(b \cdot b^4)(5 \cdot 10)$ **63.** $(-8 \cdot 2)(b^3)(c^2 \cdot c^3)(d^3)$

Pages 483–485 *Lesson 9-3*

1. 2^{10} **3.** x^{16} **5.** 4^2 **7.** r^4 **9.** 2^3 or 8 times
11 $(-2)^3 \cdot (-2)^2 = (-2)^{3+2} = (-2)^5$
13. t^6 **15.** $54p^{14}$ **17.** $24s^7$ **19.** 7^5
21. k^3 **23.** 8^6 **25.** $(-n)^2$

27 $5^3 \cdot 5 = 5^3 \cdot 5^1$
$= 5^{3+1}$
$= 5^4$
$= 5 \cdot 5 \cdot 5 \cdot 5$
$= 625$
5^4 or 625 lb
29. $\frac{484}{1}$; Sample answer: For every 484 red blood cells, there is 1 white blood cell. **31.** 5 **33.** 12 **35.** 2 **37.** $8a^3b^{10}$ **39.** n^6 **41.** Sample answer: x^7, x^2 **43.** Sample answer: By the Quotient of Powers, $\frac{a^n}{a^n} = a^{n-n}$ or a^0 for $a \neq 0$. Since $\frac{a^n}{a^n} = 1$, then $a^0 = 1$. So, any nonzero number raised to the zero power must equal 1.
45. Sample answer: Write the numbers as powers with the same base. Then subtract the exponents.
47. F
49. $A = \frac{1}{2}bh$
$A = \frac{1}{2}(4x^3)(3x^5)$
$A = \frac{1}{2} \cdot 4 \cdot 3 \cdot x^3 \cdot x^5$
$A = \frac{1}{2} \cdot 4 \cdot 3 \cdot x^8$
$A = 6x^8$
51. 3^4 **53.** $2 \cdot 3 \cdot 5^2$ **55.** $2 \cdot 3 \cdot 3 \cdot 3 \cdot a \cdot a \cdot a \cdot a \cdot a \cdot a$ **57.** $-1 \cdot 3 \cdot 11 \cdot t \cdot t \cdot t \cdot t \cdot t$
59. \$6900 **61.** $-\frac{1}{8}$ **63.** $-\frac{1}{21}$ **65.** $\frac{1}{15}$

Pages 488–491 *Lesson 9-4*

1. $\frac{1}{6^2}$ **3.** $\frac{1}{x^5}$ **5.** 2^{-6} **7.** 3^{-2}
9. 10^{-3} **11.** $-\frac{1}{64}$ **13.** $\frac{1}{2}$
15 $7^{-1} = \frac{1}{7^1}$
$= \frac{1}{7}$
17. $\frac{1}{(-5)^4}$ **19.** $\frac{1}{k^8}$ **21.** $\frac{1}{r^{20}}$ **23.** 10^{-3} **25.** 6^{-5}
27. 7^{-2} **29.** 5^{-3}
31 $0.00000001 = \frac{1}{100{,}000{,}000}$
$= \frac{1}{10^8}$
$= 10^{-8}$
33. $\frac{1}{144}$ **35.** $-\frac{1}{6}$ **37.** $\frac{1}{64}$ **39.** $-\frac{2}{9}$ **41a.** 10^3 or 1000 times **41b.** 10^4 or 10,000 **41c.** 10^6 or 1,000,000 times **43a.** coffee, 10^3 or 1,000 **43b.** milk **43c.** It is divided by 10. **43d.** 10^2 or 100 times

45a.

Power	Fraction	Decimal
10^{-1}	$\frac{1}{10}$	0.1
10^{-2}	$\frac{1}{100}$	0.01
10^{-3}	$\frac{1}{1000}$	0.001
10^{-4}	$\frac{1}{10000}$	0.0001
10^{-5}	$\frac{1}{100000}$	0.00001

45b. yes; Sample answer: As the exponents decrease, the number of zeros in the decimal places increase. **45c.** Sample answer: The number of zeros in the decimal equivalent is equal to one less than the absolute value of the negative exponent. For example, $10^{-3} = 0.001$.
45d. $10^{-12} = 0.000000000001$ **47.** a^{-6} or $\frac{1}{a^6}$
49 $x^{-3}y^4 \div x^{-2}y^2 = x^{-3-(-2)}y^{4-2} = x^{-1}y^2$ or $\frac{y^2}{x}$
51. Sample answer: $5^{-2} = \frac{1}{5^2} = \frac{1}{5 \cdot 5} = \frac{1}{25}$
53a. $2^{-2} = \frac{1}{4}$, $(-2)^{-2} = \frac{1}{4}$, $(-2)^2 = 4$, $2^2 = 4$; $2^{-2} = (-2)^{-2}$ and $(-2)^2 = 2^2$ **53b.** $2^{-3} = \frac{1}{8}$, $(-2)^{-3} = -\frac{1}{8}$, $(-2)^3 = -8$, $2^3 = 8$; none of the expressions are equal **53c.** Sample answer: When you square either a positive or a negative value, the answer is positive. When you cube a positive value, you get a positive and when you cube a negative value, you get a negative.
53d. x is an even number **53e.** x is an even number **55.** Sample answer: If $n = 3$, $\frac{1}{2^n} = \frac{1}{2^3}$ or $\frac{1}{8}$. If $n = 4$, $\frac{1}{2^n} = \frac{1}{2^4}$ or $\frac{1}{16}$.
57. C **59.** D **61a.** 224 pieces
61b. 1024 in. or $85\frac{1}{3}$ ft **63.** $-1 \cdot r \cdot r \cdot s \cdot t$
65. −2; −3 **67.** 0; 4 **69.** 0.025 **71.** 0.038
73. 76,000

Pages 495–498 *Lesson 9-5*

1. 4160 **3** $1.075 \times 10^5 = 1.075 \times 100{,}000$
$= 107{,}500$
5. 1.35×10^5 **7.** 3.27×10^6 mi **9.** 6.1×10^{-2}, 6.5×10^3, 6.01×10^4, 6.12×10^5 **11.** 0.00015
13. 0.00951 **15.** 792.4 **17.** 171,000,000,000
19 $32{,}000{,}000 = 3.2 \times 10{,}000{,}000$
$= 3.2 \times 10^7$
21. 9.18×10^{-4} **23.** 6.752×10^{-3} **25.** 2.4×10^{-2}, 2.45×10^{-2}, 2.4×10^2, 2.45×10^2 **27.** 5.1×10^{-3}, 5.9×10^4, 5.01×10^5, 5.9×10^6 **29a.** 0.0000125 cm
29b. $1.25 \text{ cm}^3 \times 10^{-6}$ **31.** 1.24×10^8

33 $(3.84 \times 10^5) \div (3 \times 10^5) = (3.84 \div 3) \times (10^5 \div 10^5)$
$= 1.28 \times 10^{5-5}$
$= 1.28 \times 10^0$
$= 1.28$ seconds

35. = **37.** < **39.** 2.763×10^7; 27,630,000 **41.** 5.642×10^6; 5,642,000 **43.** Sample answer: 2×10^2 and 4×10^3; sum: 4.2×10^3; difference: 3.8×10^3; product: 8×10^5; quotient: 5×10^{-2} **45a.** 3.8×10^6; 3.8×10^4 is only about 40,000 people, which is not very many for the second largest city in Florida **45b.** Sample answer: 3,800,000 or 3.8 million **45c.** Sample answer: 3.8 million is easier to read and understand than the standard form (3,800,000) or scientific notation (3.8×10^6) of the number.
47. Sample answer: 7.8×10^3 is greater than 6.5×10^2 because the exponent of 3 is greater than the exponent of 2. **49.** G **51a.** $(1.832 \times 10^2)(2 \times 10^3) = 3.664 \times 10^5$ **51b.** $(1.832 \times 10^2)(2 \times 10^3) / (4.5 \times 10^1) = 8.142 \times 10^3$ **51c.** 8142 lb **53.** a^6 **55.** $-12x^5$ **57a.** 5450 ft^2 **57b.** 3 bags **59.** 4^5 **61.** $2^3 \cdot 3^2$

Pages 501–503 ***Lesson 9-6***

1. 6^8 **3** $(r^6)^{-2} = r^{6 \cdot -2}$
$= r^{-12}$
$= \frac{1}{r^{12}}$

5. $16m^4n^2$ **7.** $121n^{12}p^4$ **9.** 2^6 **11.** 3^{15} **13.** b^{28} **15.** $256y^{16}$ **17.** $81s^4t^{12}$ **19.** $-1024n^{10}p^{20}$ **21.** $16a^6b^{14}$ **23.** 2.9×10^9 km^2

25 $10(4^9)^2 = 10(4^{9 \cdot 2})$
$= 10(4^{18})$
$= 10(68{,}719{,}476{,}736)$
$= 687{,}194{,}767{,}360$ or 6.87×10^{11} stars

27. 3^{24} **29.** $\frac{1}{16}t^{16}v^{12}$ **31a.** $4x^4$ and $16x^8$

31b.

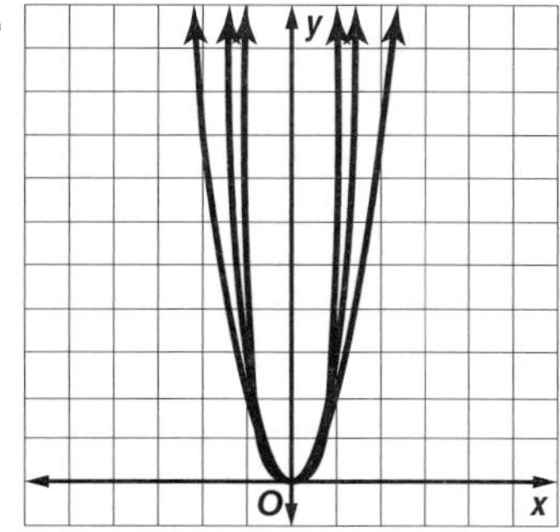

31c. Sample answer: All three graphs are shaped like "U". As the coefficient and exponent increase, the graph gets narrower and flatter on the bottom. **33.** $x = 5$; $(8^{2x})^3 = 8^{6x}$, so $6x = 30$ or $x = 5$ **35.** Sample answer: To find the power of a power, multiply exponents. To find the power of a product, find the power of each factor and multiply. **37.** C **39.** D **41.** 57,200

43. 2×10^6 **45.** 6×10^{-3} **47.** $1.20

49.

x	$3x^2 - 2$	(x, y)
0	$3(0)^2 - 2$	$(0, -2)$
1	$3(1)^2 - 2$	$(1, 1)$
2	$3(2)^2 - 2$	$(2, 10)$

Pages 506–509 ***Lesson 9-7***

1 The graph is a straight line, so it represents a linear function.
3. Linear; the equation is written in the form $y = mx + b$. **5.** Linear; as x increases by 1, y increases by 3. **7.** Nonlinear; the rate of change is not constant. **9.** Linear; graph is a straight line.
11 Nonlinear; graph is a curve
13. Linear; the equation is written in the form $y = mx + b$. **15.** Nonlinear; the equation cannot be written in the form $y = mx + b$.
17. Nonlinear; as x increases by 2, y increases by a different amount each time. **19.** Linear; as x increases by 1, y increases by 2. **21.** Nonlinear; the amount of change in price each year is not constant.

23a.

Radius r	Circumference C	Area A
1	6.28	3.14
2	12.56	12.56
3	18.84	28.26
4	25.12	50.24
5	31.4	78.5

23b.

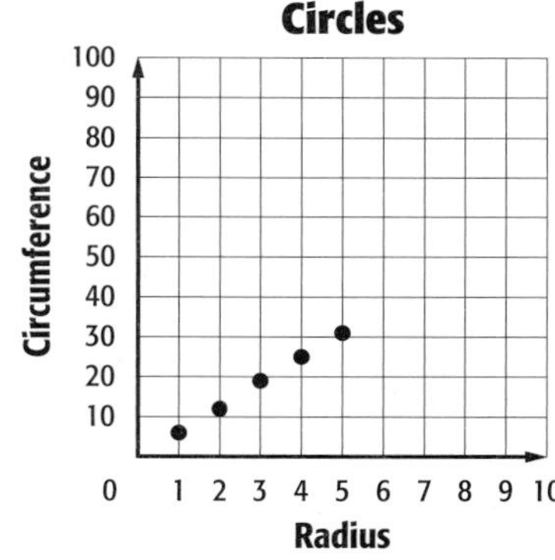

23c.

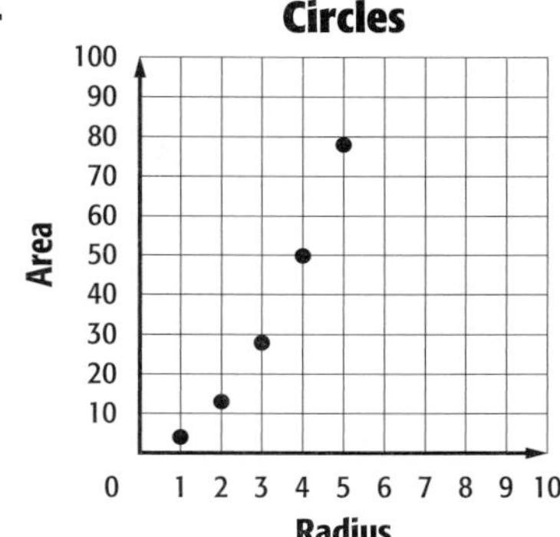

23d. Yes, the circumference is a linear relationship. The slope is 6.28.

25 Make a table of values.

	+1	+1	+1	
Inches	1	2	3	4
Centimeters	2.54	5.08	7.62	10.16
	+2.54	+2.54	+2.54	

Since the rate of change is constant, 2.54, the function is linear. **27.** Sample answer: Ben's pay from Mrs. Rodriquez is a linear relationship. As his hours increase, his pay increases by $10. His pay does not increase when he works for Mrs. Benson. **31.** $xy = 3$ because it is not a linear equation. **33.** Sample answer: Functions can be represented using graphs, equations, or tables. A graph that is a straight line represents linear function. An equation that can be written in the form $y = mx + b$ is a linear function. If a table of values shows a constant defined rate of change, the function is linear.

35. The equation for the area of a square $A = s^2$ represents a nonlinear function. You can construct a table of values and graph the points (s, A) to see that the graph is a curve.

Side	Area
1	1
2	4
3	9
4	16

37. B **39.** $-216p^6$ **41.** 8×10^7 **43.** 5.9×10^{-2}
45. About $3.60

47.

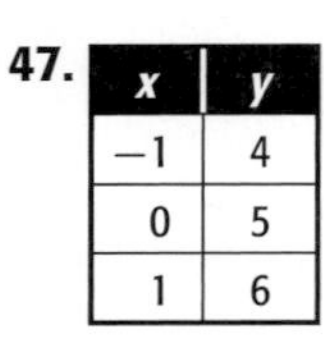

x	y
−1	4
0	5
1	6

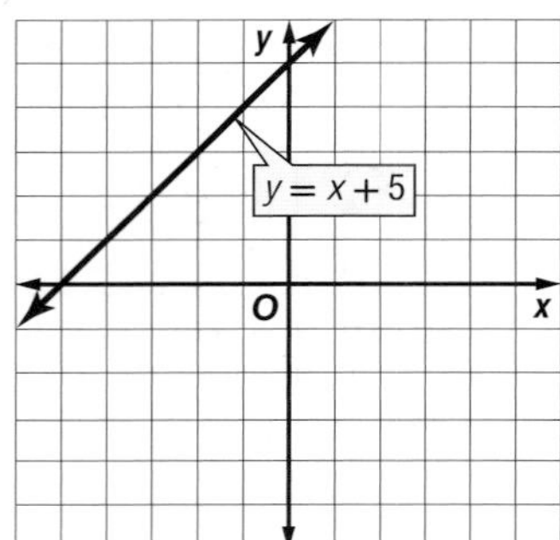

49.

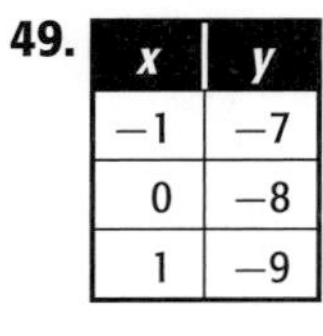

x	y
−1	−7
0	−8
1	−9

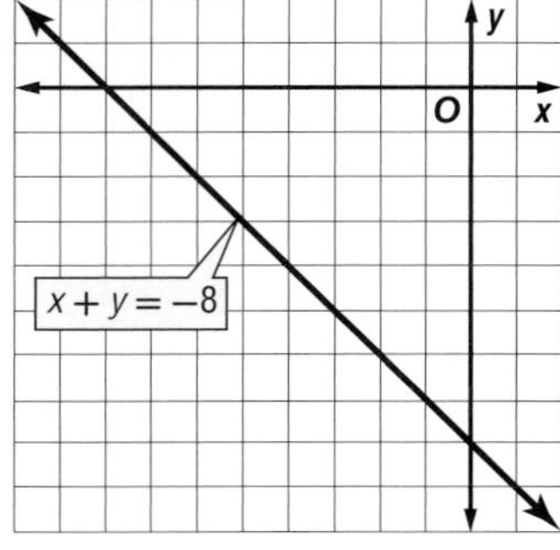

51.

x	y
−3	11
0	9
3	7

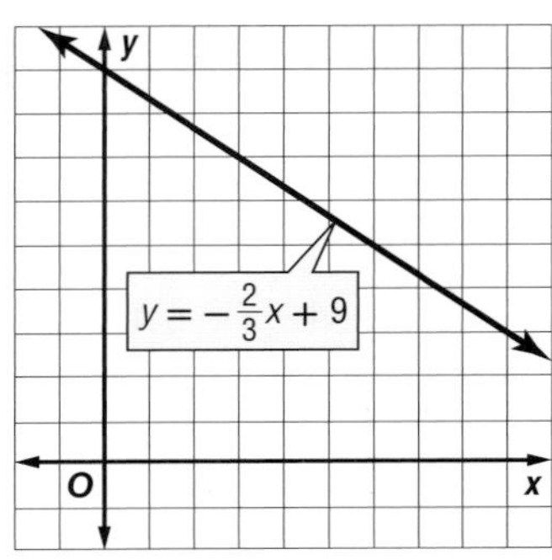

Pages 512–514 *Lesson 9-8*

1.

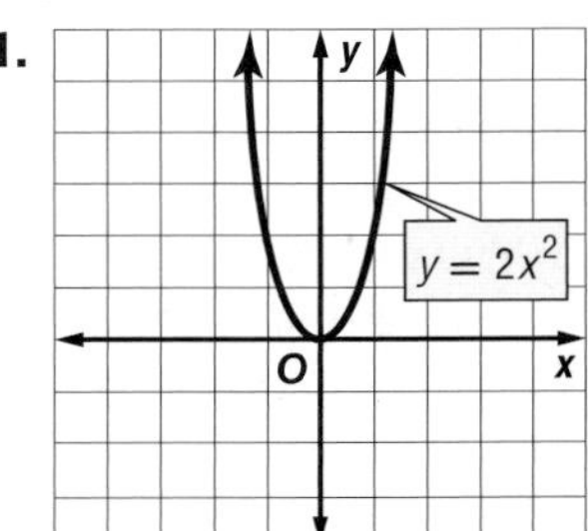

3.

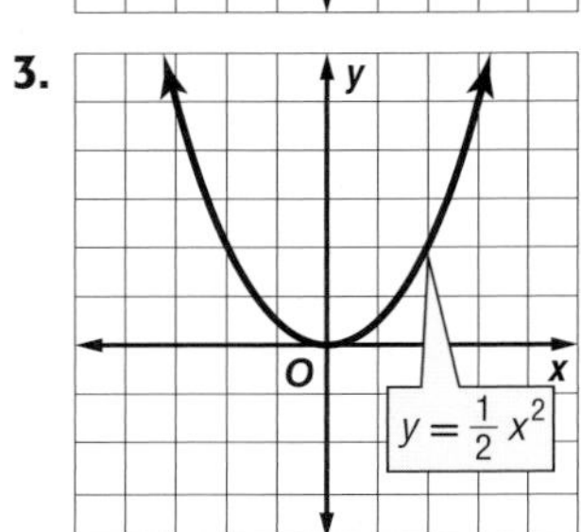

5.

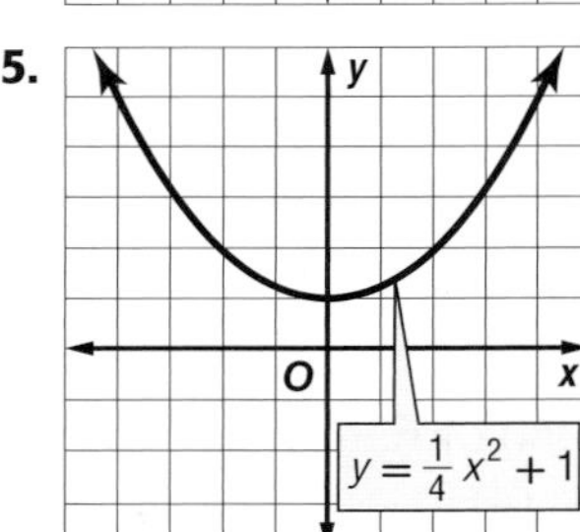

7a. 18 ft

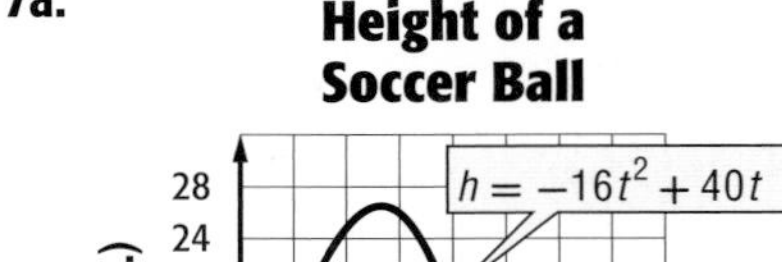

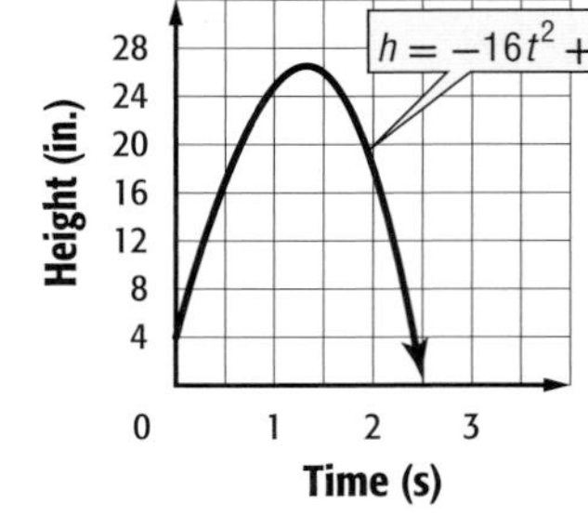

As the time increases up to about 1.5 seconds, the height of the ball increases. Then, the height of the ball decreases until it reaches 0 feet at about 2.5 seconds; 18 ft.

7b. Sample answer: Any negative values because height and time cannot be negative.

9 $y = x^2 + 1$
Make a table of values and plot the points. Then connect the points with a smooth curve.

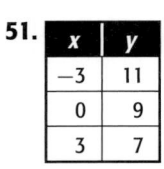

x	$y = x^2 + 1$	(x, y)
−2	$(-2)^2 + 1$	(−2, 5)
−1	$(-1)^2 + 1$	(−1, 2)
0	$(0)^2 + 1$	(0, 1)
1	$(1)^2 + 1$	(1, 2)
2	$(2)^2 + 1$	(2, 5)

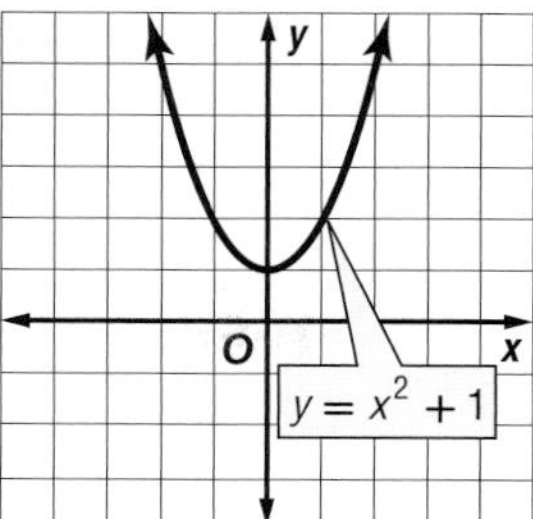

11.

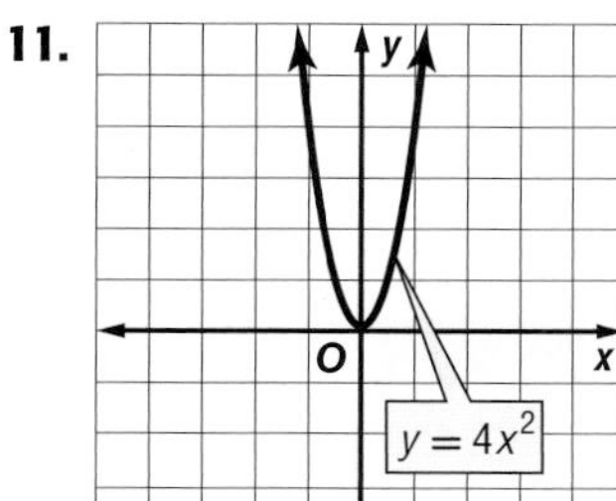

13.

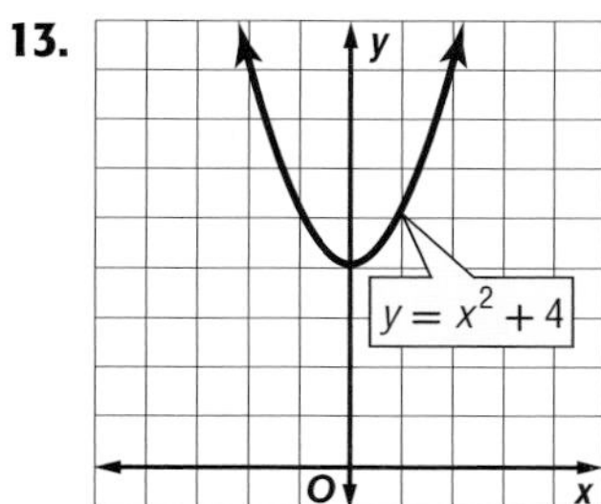

15.

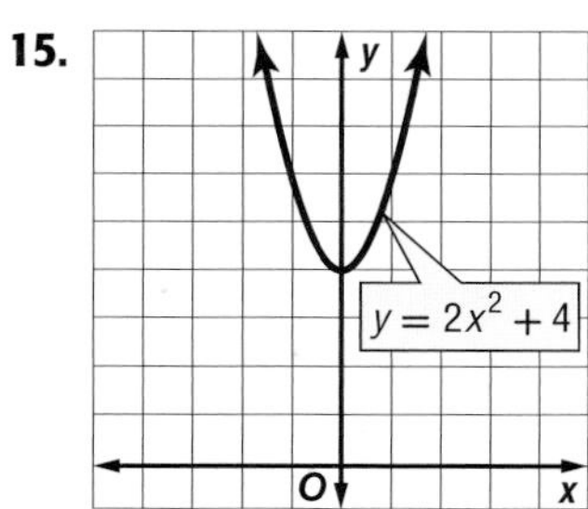

17.

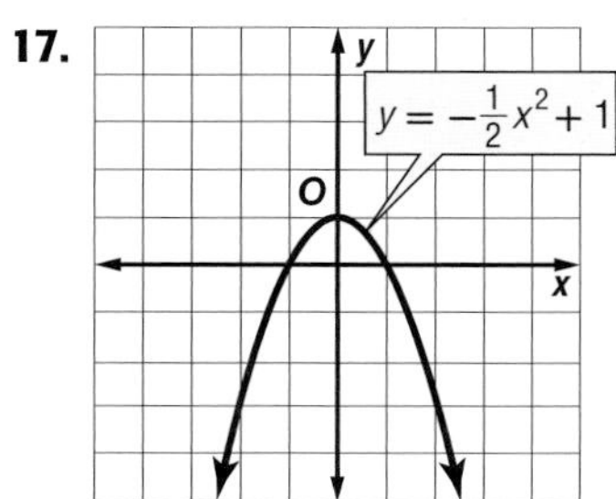

19.

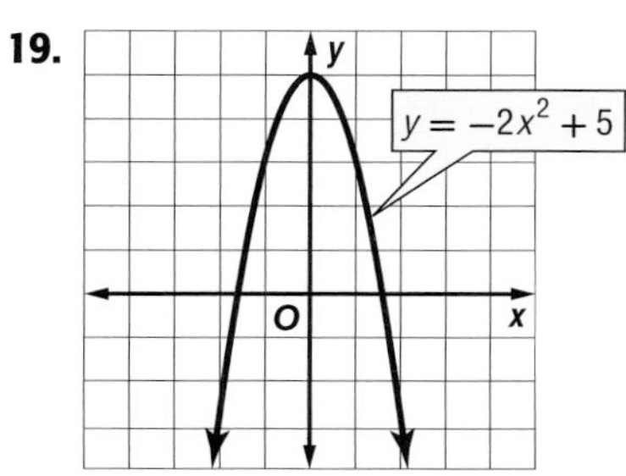

21a.

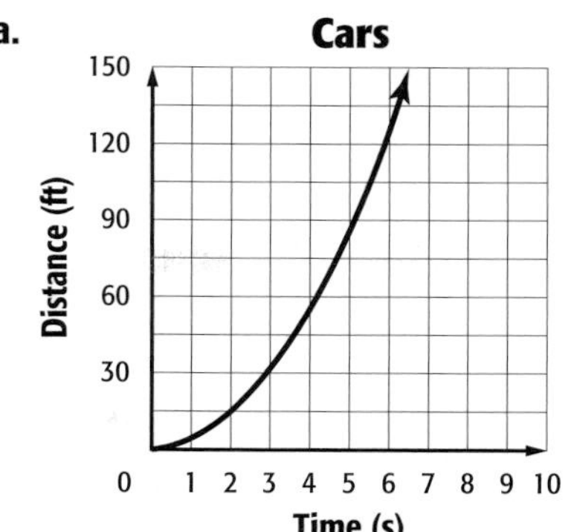

21b. about 6 seconds

23 **a.** $P = 4s$; $A = s^2$

b.

s	$P = 4s$	$A = s^2$
0	$4 \cdot 0 = 0$	$0^2 = 0$
1	$4 \cdot 1 = 4$	$1^2 = 1$
2	$4 \cdot 2 = 8$	$2^2 = 4$
3	$4 \cdot 3 = 12$	$3^2 = 9$
4	$4 \cdot 4 = 16$	$4^2 = 16$

c.

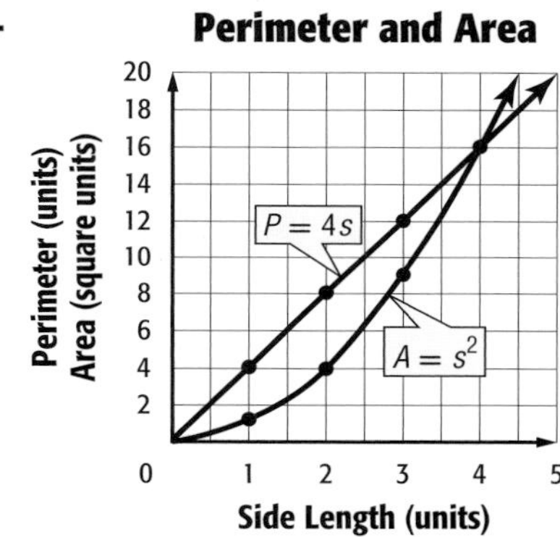

Graph the points (1, 4), (2, 8), (3, 12), and (4, 16). Then connect the points. Graph the points (1, 1), (2, 4), (3, 9), and (4, 16). Then connect the points. Since the points on the graph for the perimeter are in a straight line, it is a linear function. The points on the graph for the area follow a parabolic shape. So, it is a quadratic function.

d. Yes, when the side length is 4 units both the perimeter and area are 16. In the table, the values are the same. On the graph, it is where the two functions intersect.

25. Sample answer:

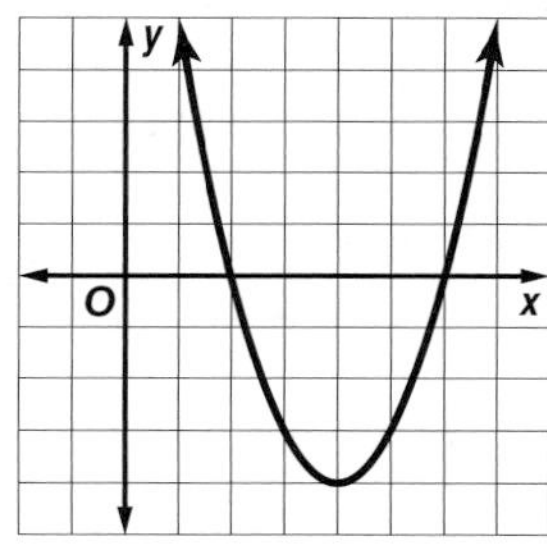

27. (0, 7)

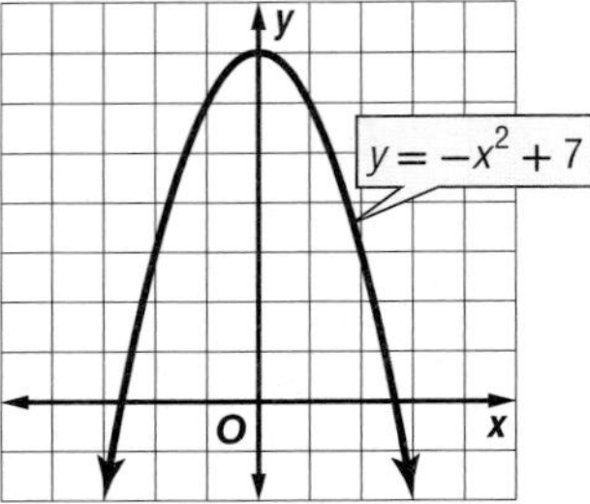

29. Sample answer: Formulas and tables can be used to make graphs. Tables and graphs can be used to write rules. To make a graph, use a rule to make a table of values. Then plot the points. To write a rule, find points that lie on the graph and make a table using the coordinates. Look for a pattern and write a rule that describes the pattern. **31.** G

33a.

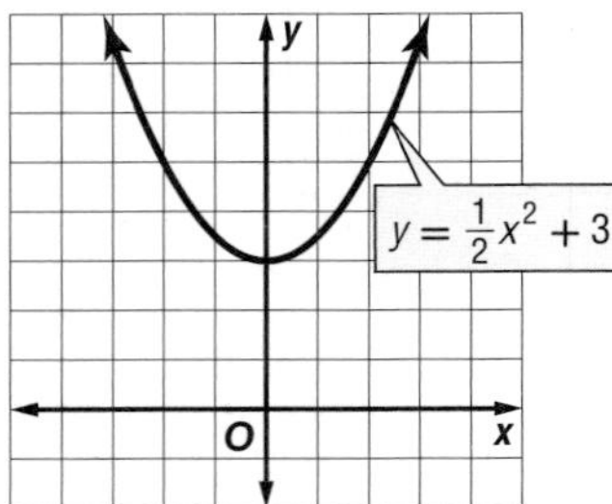

33b. (0, 3) **35.** Linear; rate of change is constant. **37.** 7^8 or 5,764,801 **39.** $8c^{15}d^3$ **41.** −14.3% **43.** 216 **45.** 128

Pages 518–520 ***Lesson 9-9***

1.

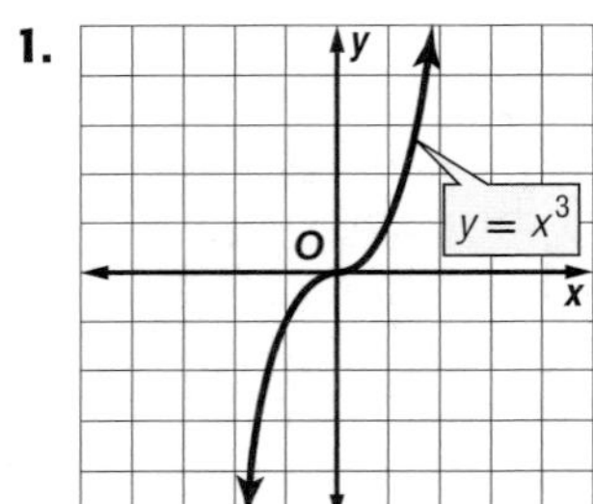

3.

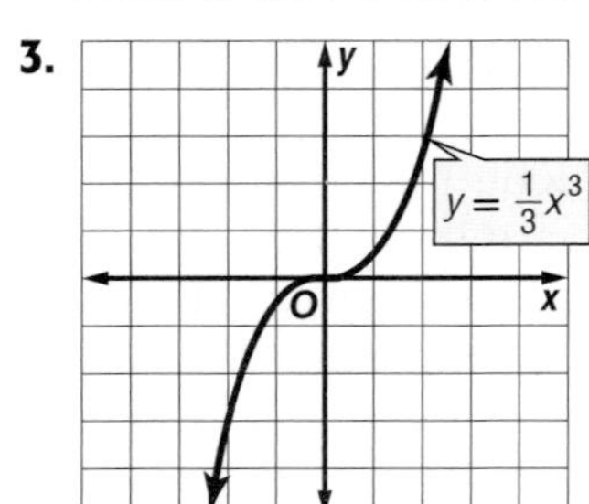

5a.

Years	Money Spent (in millions)
1	13.4
2	15.1
3	16.9
4	18.9

5b.

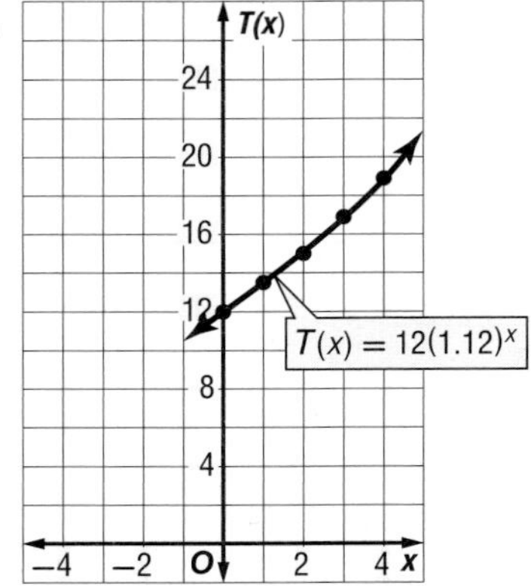

5c. $12 million sales in 2005

7

x	$y = 2^x - 3$	(x, y)
0	$y = 2^0 - 3 = -2$	(0, −2)
1	$y = 2^1 - 3 = -1$	(1, −1)
2	$y = 2^2 - 3 = 1$	(2, 1)
3	$y = 2^3 - 3 = 5$	(3, 5)

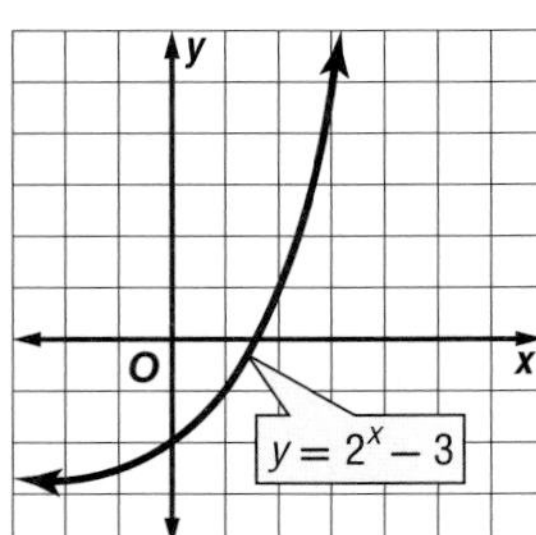

9.

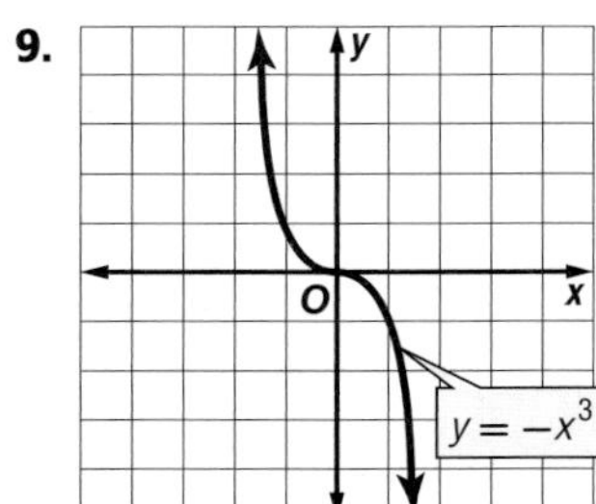

11.

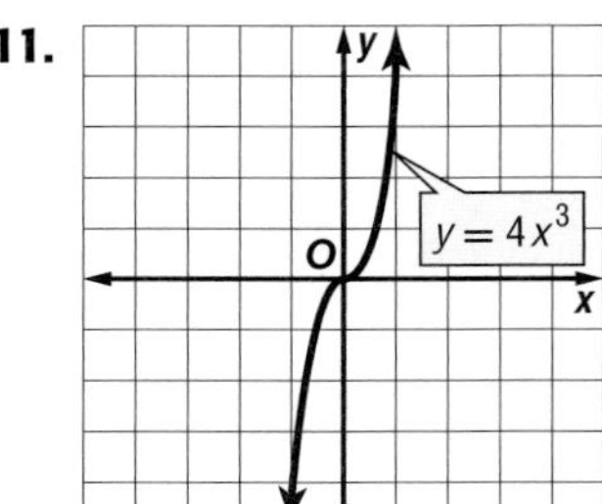

13.

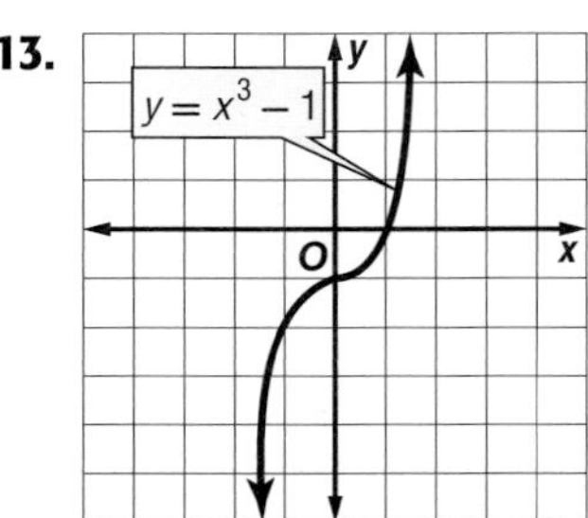

15.

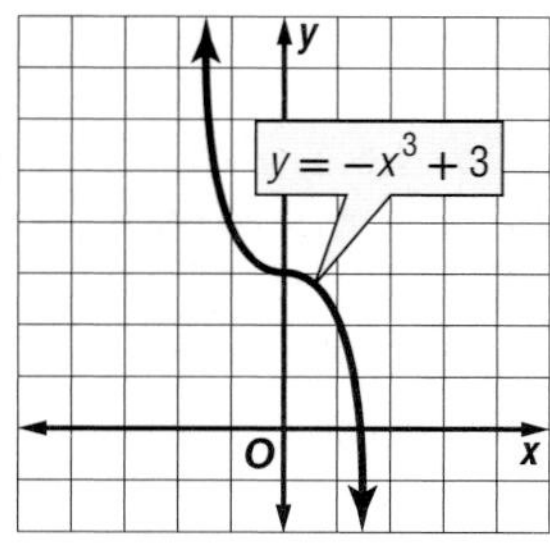

17 a.

x	$y = 1000(1.05)^x$	(x, y)
0	$y = 1000(1.05)^0 = 1000$	(0, 1000)
1	$y = 1000(1.05)^1 = 1050$	(1, 1050)
2	$y = 1000(1.05)^2 = 1102.50$	(2, 1102.50)
3	$y = 1000(1.05)^3 = 1157.63$	(3, 1157.63)

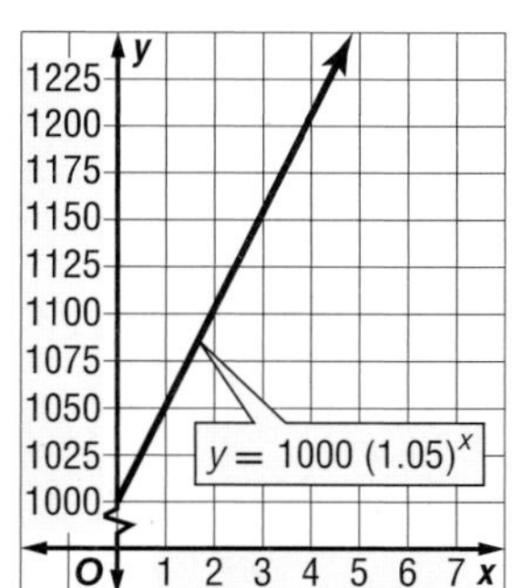

1000; The initial amount of money put into the account.

b. The balance will be greater than \$2000 after 15 years.

19.

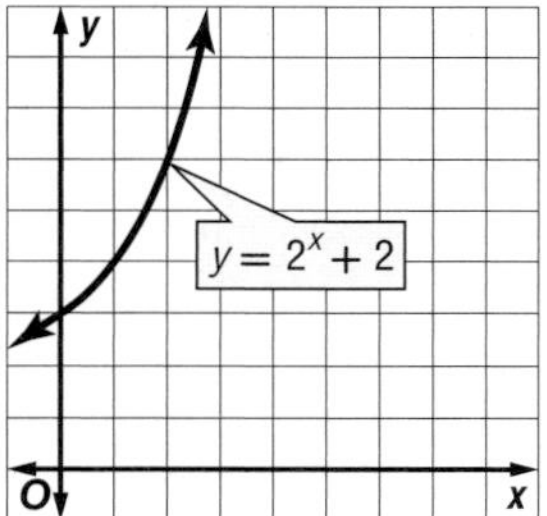

label: $y = 2^x + 2$

21.

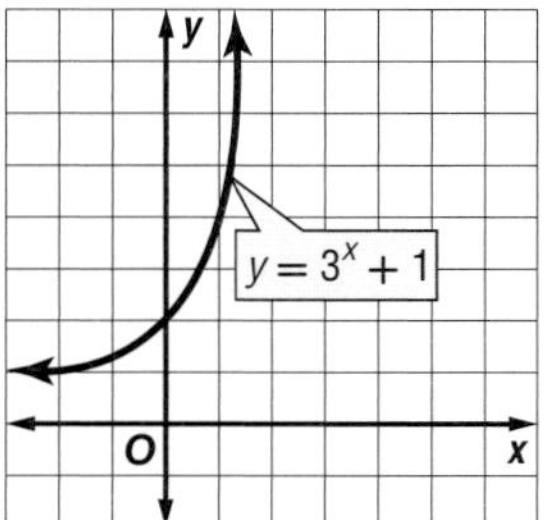

23.

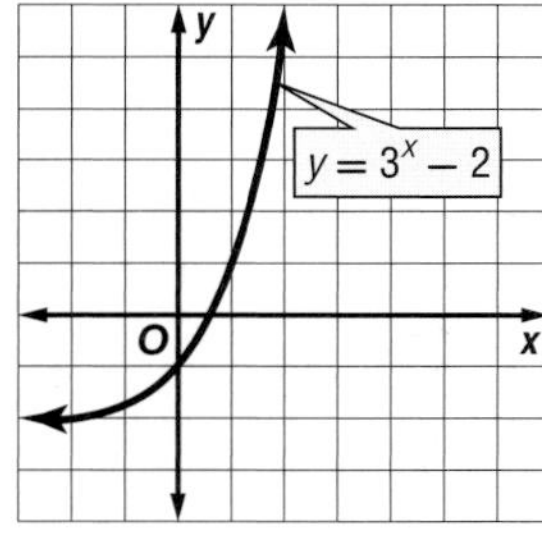

25. quadratic **27.** exponential **29.** linear **31.** Sample answer: $y = 3x$, $y = 3^x$; the linear function has a constant rate of change. The exponential function has a greater rate of change. **33.** $y + 3x = 5$; This equation represents a linear function and the others represent exponential functions. **35.** Sample answer: The graph begins almost flat, but then for increasing x values, it becomes more steep. **37.** J

39a.

x	y
−2	$-\frac{8}{9}$
−1	$-\frac{2}{3}$
0	0
1	2
2	8

39b.

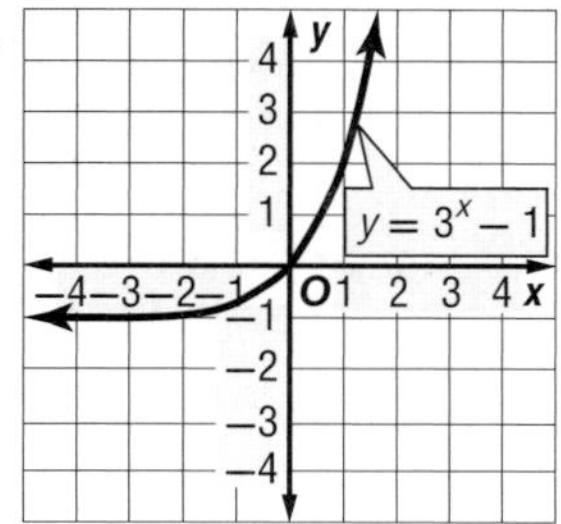

39c. Exponential; The variable is the exponent. **41.** Linear; graph is a straight line. **43.** Nonlinear; the graph is a curve. **45.** 140 **47.** 256 **49.** 81

Study Guide and Review

1. true **3.** false; exponent **5.** true **7.** true **9.** false; multiply **11.** 6^5 **13.** x^3 **15.** 243 **17.** 10 **19.** −248 **21.** 32 teeth **23.** $2^3 \cdot 5$ **25.** $3^2 \cdot 5^2$ **27.** $2 \cdot 5 \cdot r \cdot r$ **29.** $-1 \cdot 5 \cdot 5 \cdot a \cdot b \cdot b$ **31.** 3^7 **33.** m^9 **35.** $12h^8$ **37.** 9^1 or 9 **39.** about 10 times **41.** $\frac{1}{(-10)^2}$ **43.** 6^{-3} **45.** 5^{-3} **47.** 5820 **49.** 0.00034 **51.** 3.79×10^2 **53.** 1.4×10^{-3} **55.** 1,988,920,000,000,000 exagrams **57.** r^{16} **59.** $64n^{24}$ **61.** $125w^{15}x^{24}$ **63.** Nonlinear; the graph is a curve. **65.** Linear; the equation is written in the form $y = mx + b$. **67.** Linear; as x increases by 1, y decreases by 2. **69.** No; the equation cannot be written in $y = mx + b$ form.

71.

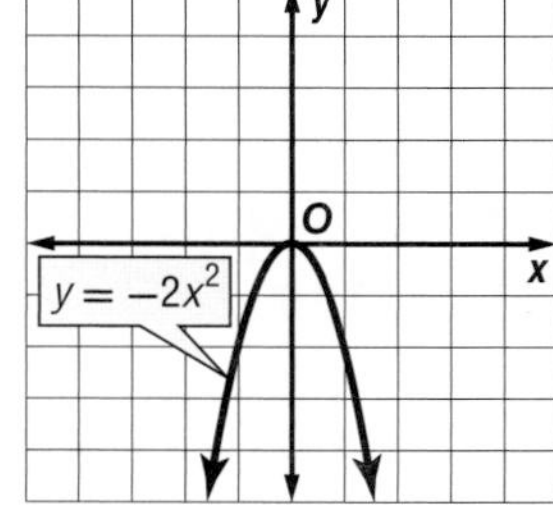

73.

y
$y = -x^2 + 1$
O
x

75.

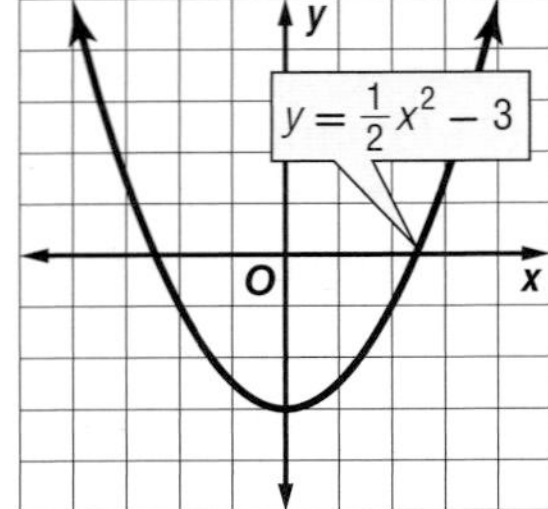

77.

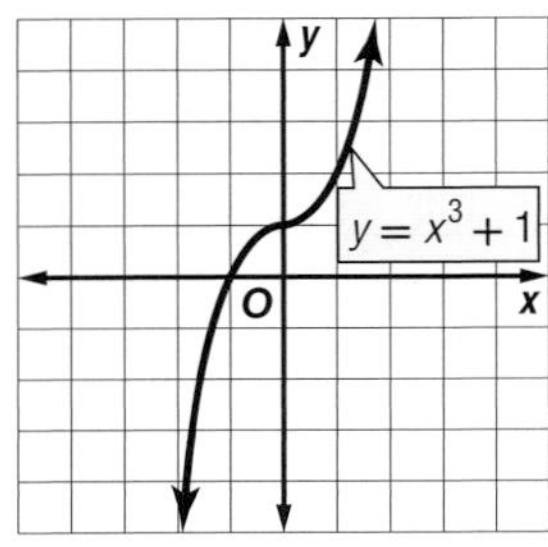

79.

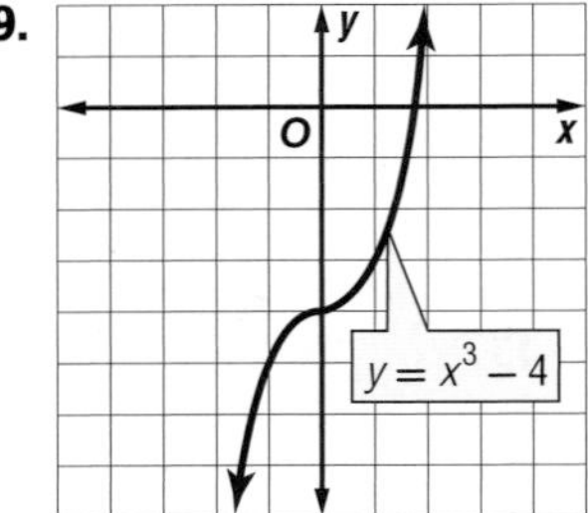

81.

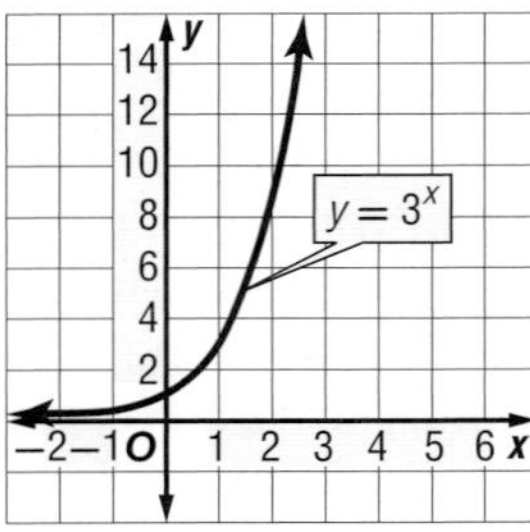

83. 4000

Chapter 10 Real Numbers and Right Triangles

Page 533 ***Chapter 10*** ***Get Ready***

1. > **3.** = **5.** < **7.** 0.601, 0.594, 0.546, 0.523, 0.509 **9.** 12 **11.** 9 **13.** 3 **15.** 34 **17.** 101 **19.** 185

Pages 540–542 ***Lesson 10-1***

1. 4 **3.** ±9 **5.** −7 **7.** 4.6 **9.** ±6.1 **11.** 6 **13.** −13 **15.** No real solution

17 The first perfect square integer less than 83 is 81. The first perfect square greater than 83 is 100. The square root of 83 is between 9 and 10. Since 83 is closer to 81, the square root of 83 is closer to 9.

19. −10 **21.** ±9 **23.** 2.6 **25.** ±8.4 **27.** −12.4 **29.** 17.6 **31a.** 21.5 mi **31b.** 15.5 mi **31c.** 7.5 mi

33

$$A = s^2 \qquad P = 4s$$
$$215 = s^2 \qquad = 4(14.7)$$
$$\sqrt{215} = s \qquad = 58.8 \text{ cm}$$
$$14.7 \text{ cm} \approx s$$

35. $\sqrt{79}$; Sample answer: $\sqrt{79}$ is between $\sqrt{64}$ and $\sqrt{81}$, which is greater than 8. **37.** 9 and 10

39a.

x	x^2
−2	4
−1	1
0	0
1	1
2	4

x	$\sqrt{x}$
4	2, −2
2.25	1.5, −1.5
1	1, −1
0.25	0.5, −0.5
0	0

39b.

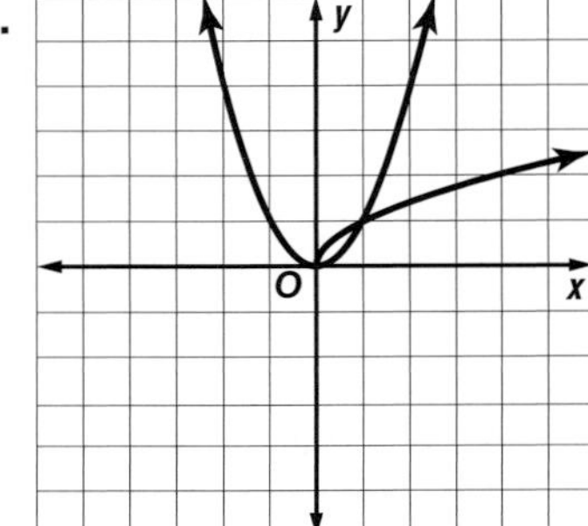

39c. Sample answer: For $0 < x < 1$, the graph for $y = x^2$ has a gradual climb then gets steeper as it approaches 1. The graph for $y = \sqrt{x}$ starts out steeper, then begins to flatten out. At $x = 1$, the graphs intersect. When $x > 1$, the first graph continues to get steeper while the second graph levels out. **39d.** Sample answer: The graph for $y = x^2$, it is steeper for most of the graph. **41a.** 246 **41b.** 811 **41c.** 732 **41d.** finding the square root of a number **43.** Sample answer: The exact value of a square root is given using the square root symbol, such as $\sqrt{13}$. An approximation is a decimal value, such as $13 \approx 3.6$. **45.** 60 units **47.** C

49.

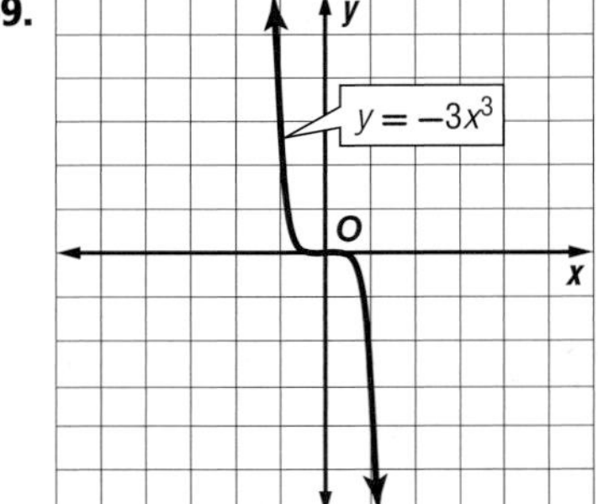

51a.

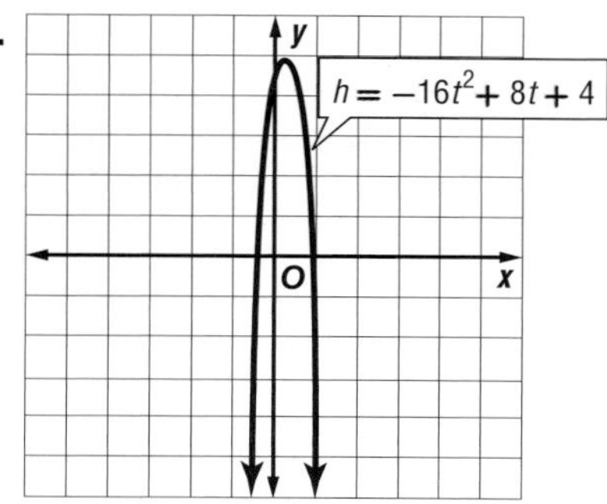

51b. about 0.85 s **55.** natural, whole, integer, rational **57.** rational **59.** rational **61.** not rational

Pages 546–548 Lesson 10-2

1. whole, integer, rational **3.** irrational **5.** > **7.** > **9.** 10.15, $10\frac{1}{5}$, $\sqrt{110}$, $10.\overline{5}$ **11.** 8.6, −8.6 **13.** 61.3 ft **15.** rational **17.** irrational **19.** rational **21.** integer, rational **23.** whole, integer, rational **25.** rational **27.** > **29.** = **31.** < **33.** $\frac{15}{2}$, $\sqrt{64}$, $8.\overline{14}$, $8\frac{1}{7}$ **35.** $-\frac{31}{6}$, $-\sqrt{26}$, −5, $-\frac{5}{6}$

37
$$130 = n^2$$
$$\pm\sqrt{130} = n$$
$$11.4 \text{ and } -11.4 \approx n$$

39. 9, −9 **41.** 1.3, −1.3

43
$$h = 16t^2$$
$$60 = 16t^2$$
$$3.75 = t^2$$
$$\pm\sqrt{3.75} = t$$
$$1.9 \text{ and } -1.9 = t$$

Since −1.9 seconds does not make sense, the answer is 1.9 seconds.

45. sometimes; Sample answer: $\frac{4}{9}$ can be written as $0.\overline{4}$, but $\frac{1}{2}$ is written as 0.5. **47.** always; Sample answer: All whole numbers are integers. **49.** Rational; $\sqrt{49}$ is rational. **51.** Irrational; π is irrational. **53.** Sample answer: 6.4; $\sqrt{40} \approx 6.32$ **55.** false; $\sqrt{16}$ is rational **57.** 5.6 or −5.6 **59.** If a square has an area that is not a perfect square, the lengths of the sides will be irrational. Sample answer: A square with an area of 25 square units has sides that are rational. A square with an area of 26 square units has sides that are irrational. **61.** F **63.** 43.3 **65.** 10 **67.** −11 **69.** ±20

71.

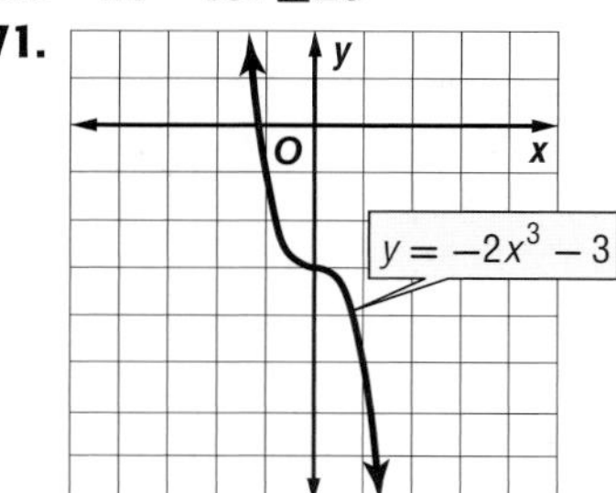

73.

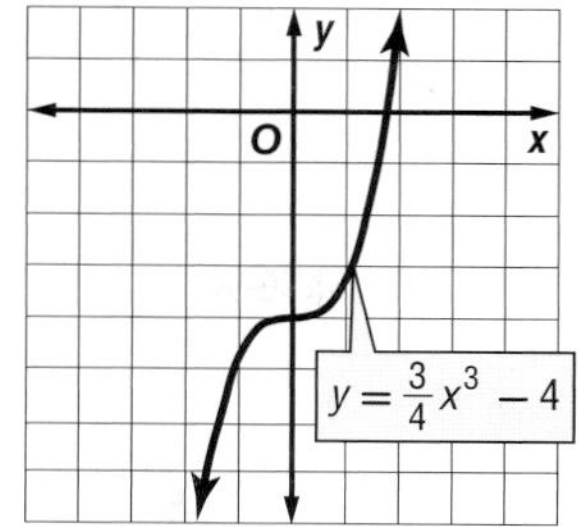

75. 105 **77.** 71 **79.** 20

Pages 553–555 Lesson 10-3

1. 45; acute scalene **3.** 45; right isosceles **5.** straight **7.** 30; obtuse isosceles **9.** 75; right scalene **11.** 70; acute isosceles **13.** 12°, 72°, 96°

15 At 6:00 the hands form an angle that measures 180°, so the angle is a straight angle.

17. acute **19.** acute **21.** obtuse **23.** acute **25.** obtuse **27a.** ∠E **27b.** ∠G from 100,000 years ago

29
$$x + x + 2x = 180$$
$$4x = 180$$
$$x = 45$$

So, the angle measures are 45°, 45°, and 90°.

31. 27°, 36°, 117° **33.** equilateral

35a.

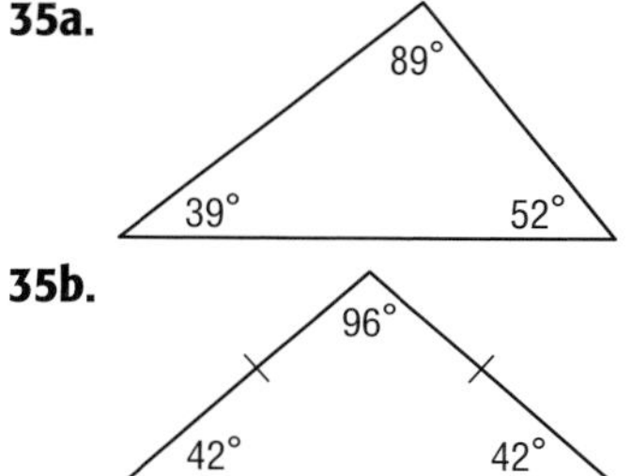

35b.

35c. not possible **35d.** not possible **37.** 50; 80 **39.** True; Sample answer: There can be at most one right or one obtuse angle. So, the sum of the two remaining angles must be ≤ 90°. **41.** F **43a.** 56 **43b.** 90 + 34 = 124, 180 − 124 = 56 **43c.** right scalene **45.** 14, −14 **47.** 10.1, −10.1 **49.** −4 **51.** 12 **53.** 121 **55.** 256

Pages 560–563 Lesson 10-4

1
$$a^2 + b^2 = c^2$$
$$3^2 + 4^2 = c^2$$
$$9 + 16 = c^2$$
$$25 = c^2$$
$$\pm\sqrt{25} = c$$
$$5 = c$$

The length of the hypotenuse is 5 feet.

3. 56.6 ft **5.** yes **7.** no **9.** 34 cm **11.** 12.0 in. **13.** 110.8 mi

15
$$a^2 + b^2 = c^2$$
$$35^2 + 35^2 = c^2$$
$$1225 + 1225 = c^2$$
$$2450 = c^2$$
$$\pm\sqrt{2450} = c$$
$$49.5 \approx c$$

The length of the walkway is 49.5 feet.

17. yes **19.** no **21.** 42 in.

23 $a^2 + b^2 = c^2$

$9^2 + b^2 = 12^2$

$81 + b^2 = 144$

$b^2 = 63$

$b = \pm\sqrt{63}$

$b \approx 7.9$ m

25. 24.2 ft **27.** 8.8 mi **29.** 19 ft **31a.** 624.5 ft
31b. 2500 ft
33a.

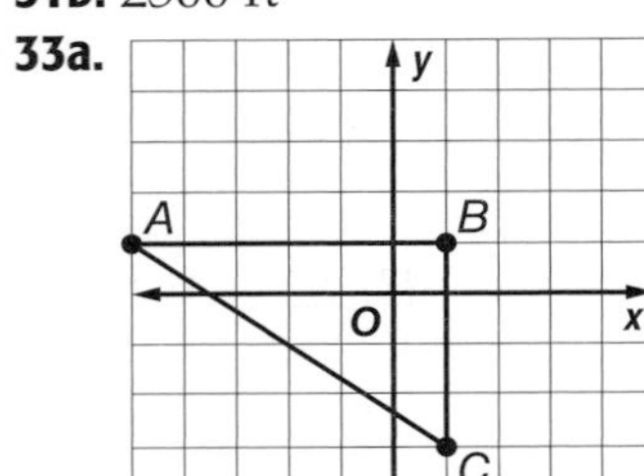

33b. Sample answer: Find the length of the legs using the units on the coordinate plane. Then use the Pythagorean Theorem to find the length of the hypotenuse. **33c.** $\overline{AB} = 6$ units, $\overline{BC} = 4$ units, $\overline{AC} \approx 7.2$ units **33d.** $P = 17.2$ units, $A = 12$ units^2 **35.** 19.7 units

37.

$c^2 = a^2 + b^2$	Pythagorean Theorem
$10^2 = x^2 + x^2$	Replace *c* with 10, *a* with *x* and *b* with *x*.
$100 = 2x^2$	Simplify.
$50 = x^2$	Divide each side by 2.
$\sqrt{50} = x$	Take the square root of each side.

So, the length of the legs is $\sqrt{50}$ or about 7.1 in.
39. D **41.** G **43.** 27; right scalene **45.** 66; acute scalene **47.** rational **49.** rational **51.** 65 **53.** 136

Pages 568–570 *Lesson 10-5*

1. 8 **3.** 33.5 ft **5.** scalene; 22.7

7 $d = \sqrt{(x_2 - x_1)^2 + (y_2 - y_1)^2}$

$= \sqrt{(8 - 5)^2 + (4 - 1)^2}$

$= \sqrt{(3)^2 + (3)^2}$

$= \sqrt{9 + 9}$

$= \sqrt{18}$

≈ 4.2

9. 7.8 **11.** 8.6 **13.** 91 mi **15.** isosceles; 17.1
17. scalene; 36.1

19 $a^2 + b^2 = c^2$

$2^2 + 7^2 = c^2$

$4 + 49 = c^2$

$53 = c^2$

$\pm\sqrt{53} = c$

$7.3 \text{ in.} \approx c$

21. 10 square units **23.** 4.8 **25.** 13.1
27. Sample answer: (1, 3) and (5, 6) **29.** Mental math; the cup is located at (0, 0) on a coordinate system with points at (−2, −3) and (1, 4). You can use mental math to find the distance between the cup and the two balls; $\sqrt{(-2)^2 + (-3)^2} = \sqrt{13}$; $\sqrt{(1)^2 + (4)^2} = \sqrt{17}$. Since $\sqrt{13} < \sqrt{17}$, Joan's ball is closer to the cup. **31.** Sample answer: To use the Pythagorean Theorem, connect the points. Then draw vertical and horizontal lines so that a right triangle is formed. Determine the lengths of the legs. Then use the Pythagorean Theorem formula to find the length of the hypotenuse. To use the Distance Formula, replace (x_1, y_1) and (x_2, y_2) in the formula with the coordinates of the two points. Then simplify.
33. H **35.** B **37.** 51 **39.** 20°, 60°, 100° **41.** −12

Pages 574–576 *Lesson 10-6*

1. $14\sqrt{2}$ mm **3.** $x = 20$ in.; $y = 10\sqrt{3}$ in.
5. $2\sqrt{2}$ cm **7.** 69.3 in. **9.** $x = 12\sqrt{2}$ in.
11. $x = 17$ in., $y = 17\sqrt{3}$ in. **13.** $20\sqrt{2}$ cm

15 In a 45°-45°-90° triangle, the length of the hypotenuse is $\sqrt{2}$ times the length of a leg. So, the length of the hypotenuse is $15\sqrt{2}$ inches.

17 In a 30°-60°-90° triangle, the length of the hypotenuse is 2 times the length of the shorter leg. So, the length of the hypotenuse is 2(12) or 24 feet.

19a.

Leg Length	Hypotenuse
1	1.4
2	2.8
3	4.2
4	5.7
5	7.1

19b. **45°-45°-90° Triangles**

Hypotenuse (units)

Leg length (units)

The points lie in a straight line.
19c. $y = x\sqrt{2}$ **21.** $x = 11$, $y = 11$ **25.** Yes; the legs are 21 cm and 36.4 cm and the hypotenuse is 42 cm. Since $21 \cdot 2 = 42$, the hypotenuse is twice the shorter leg. Since $21 \cdot \sqrt{3} \approx 36.4$, the longer leg is $\sqrt{3}$ times the length of the shorter leg.
27. The hypotenuse is twice the length of the shorter leg. The longer leg is $\sqrt{3}$ times the length of the shorter leg. **29.** H **31a.** $x = 20$, $y = 45$, $z = 20\sqrt{2}$ **31b.** First solve for y to determine the triangle is an isosceles triangle. Since the triangle is isosceles, both legs are congruent so find x. Use

the properties of 45-45-90 triangles to find z.
33. 14.3 **35.** 10.6 **37.** 12 **39.** 12.1 **41.** 90 **43.** 45 **45.** 35 **47.** 101

Pages 577–580 ***Study Guide and Review***

1. perfect square **3.** irrational numbers **5.** acute triangle **7.** congruent **9.** radical sign **11.** 13 **13.** ± 1 **15.** 4 **17.** -9 **19.** 3.14 s **21.** rational **23.** rational **25.** $<$ **27.** $<$ **29.** 1.2, -1.2 **31.** 45; acute scalene **33.** 45; right isosceles **35.** acute equilateral **37.** 7.8 ft **39.** 37.2 m **41.** 5 **43.** 12.7 **45.** 15.4 **47.** $6\sqrt{2}$ cm **49.** $x = 30$ in.; $y = 15\sqrt{3}$ in. **51.** $9\sqrt{2}$ in.

Chapter 11 Distance and Angle

Page 587 ***Chapter 11*** ***Get Ready***

1. -92 **3.** 35 **5.** 70 **7.** 5 mi **9.** 16.6 **11.** 9.6 **13.** 8.9 **15.** $9\frac{7}{15}$ **17.** $6\frac{1}{12}$ **19.** $8\frac{1}{12}$ **21.** $20\frac{23}{24}$ lb

Pages 592–595 ***Lesson 11-1***

1. supplementary angles; 34 **3.** supplementary angles; 75 **5.** 128°; ∠1 and ∠8 are alternate exterior angles, so they are congruent. **7a.** the angles are alternate interior angles **7b.** 117 **9.** $x = 9.15$, $m\angle Q = 71.7°$, $m\angle R = 18.3°$

11 Since the two angles form a 180° angle, they are supplementary angles.

$58.9 + m\angle x = 180$

$m\angle x = 121.1°$

13a. complementary angles **13b.** 66° **15.** 52.6°; ∠6 and ∠5 are corresponding angles, so they are congruent. **17.** 127.4°; ∠6 and ∠8 are supplementary angles, so the sum of their measures is 180°. **19.** 127.4°; Sample answer: ∠6 and ∠8 are supplementary angles, so $m\angle 8 = 127.4$. ∠8 and ∠1 are alternate exterior angles, so $m\angle 1 = 127.4$. **21.** 30° **23.** $x = 38.66$; $m\angle D = 49.96°$ **25.** $x = 5$, $m\angle 4 = 75°$, $m\angle 2 = 75°$ **27.** $m\angle B = 35°$, $m\angle C = 35°$

29 **a.** The two angles are supplementary, so $2x + 6x = 180$.

$$2x + 6x = 180$$
$$8x = 180$$
$$\frac{8x}{8} = \frac{180}{8}$$
$$x = 22.5$$

b. Since $x = 22.5$, substitute to find each angle measure.

$$2x = 2(22.5) = 45°$$
$$6x = 6(22.5) = 135°$$

31 **a.** $m\angle WXY = 60°$ because it is 1/3 of a straight angle which is 180°. $m\angle YXZ = 120°$ because $120° + 60° = 180°$

b. Sample answer: 6:00 and 15 seconds

33. Sample answer:

35. The sum of their measures is 180°. The interior angles on the same side of a transversal are supplementary. **37.** C **39.** A

41. hypotenuse: 24 ft., leg $12\sqrt{3}$ **43.** 9.3%; increase

45. $a < -6$ **47.** **49.**

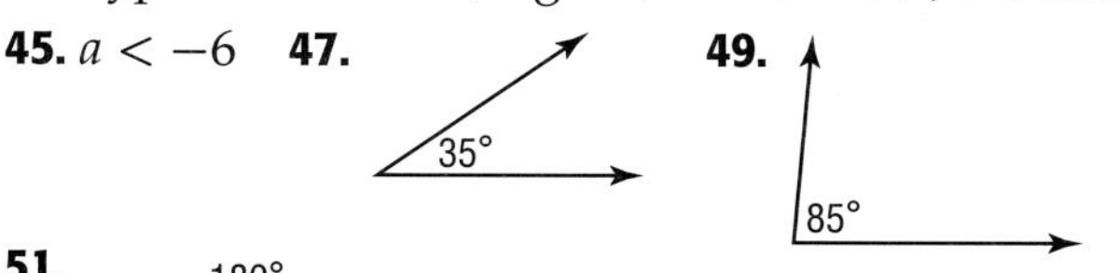

51.

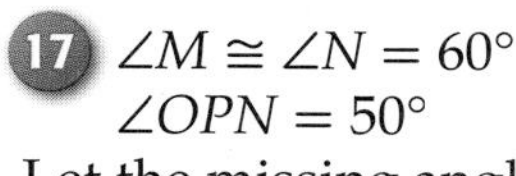

Pages 601–604 ***Lesson 11-2***

1. $\angle L \cong \angle X$, $\angle M \cong \angle Y$, $\angle N \cong \angle Z$, $\overline{LM} \cong \overline{XY}$, $\overline{MN} \cong \overline{YZ}$, $\overline{NL} \cong \overline{ZX}$; $\triangle MNL$

3a. 67.5° **3b.** 12 in. **5.** no

7 Use the matching arcs and tick marks to identify the corresponding angles and segments.

corresponding angles: $\angle S \cong \angle Y$, $\angle STZ \cong \angle YTW$, $\angle Z \cong \angle W$

corresponding sides: $\overline{ST} \cong \overline{YT}$, $\overline{TZ} \cong \overline{TW}$, $\overline{ZS} \cong \overline{WY}$

$\triangle STZ \cong \triangle YTW$

9. $\angle N \cong \angle D$, $\angle P \cong \angle E$, $\angle Q \cong \angle F$, $\overline{NP} \cong \overline{DE}$, $\overline{PQ} \cong \overline{EF}$, $\overline{QN} \cong \overline{FD}$; $\triangle PQN$

11. 8.55 mm **13.** no **15.** yes; $\angle J \cong \angle P$, $\angle L \cong \angle R$, $\angle K \cong \angle Q$, $\overline{JL} \cong \overline{PR}$, $\overline{JK} \cong \overline{PQ}$, $\overline{LK} \cong \overline{RQ}$; Sample answer: $\triangle JLK \cong \triangle PRQ$

17 $\angle M \cong \angle N = 60°$

$\angle OPN = 50°$

Let the missing angle be x

$$60° + 50° + x = 180°$$
$$110° + x = 180°$$
$$x = 70°$$

19. 3.6 cm

21. Sample answer:

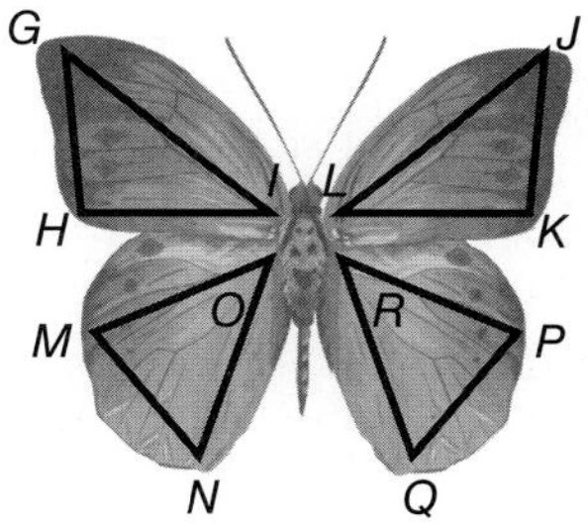

$\triangle GHI \cong \triangle JKL$, $\triangle MNO \cong \triangle PQR$

23 Since $\overline{PZ} \cong \overline{NZ}$, $18 = 4x - 9$
$19 = 4x - 9$
$28 = 4x$
$7 = x$

25. No; corresponding vertices must be written in the same order, so $\angle YXZ \cong \angle SRT$, not $\angle STR$, and $\overline{ST} \cong \overline{YZ}$ not $\overline{ZY}$ **27.** The scale factor is 1:1 or 1. **29.** B **31.** D **33.** 24.3° **35.** $\frac{1}{5^4}$

37–40.

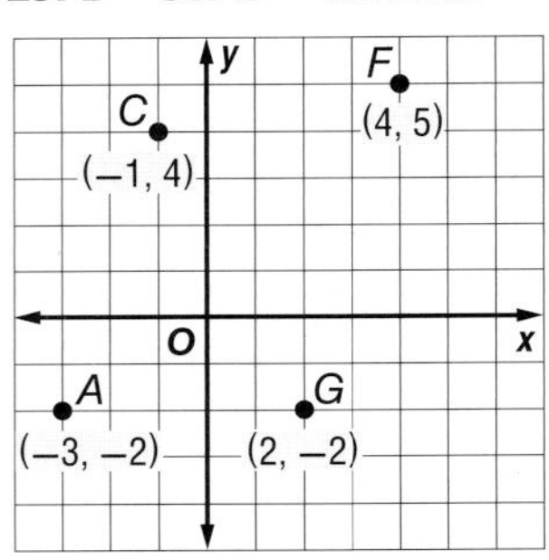

Pages 608–610 ***Lesson 11-3***

1.

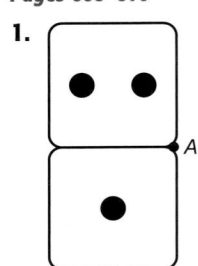

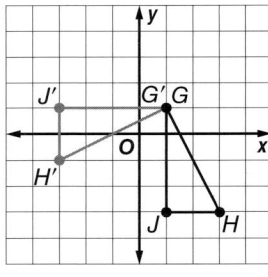

A

3.

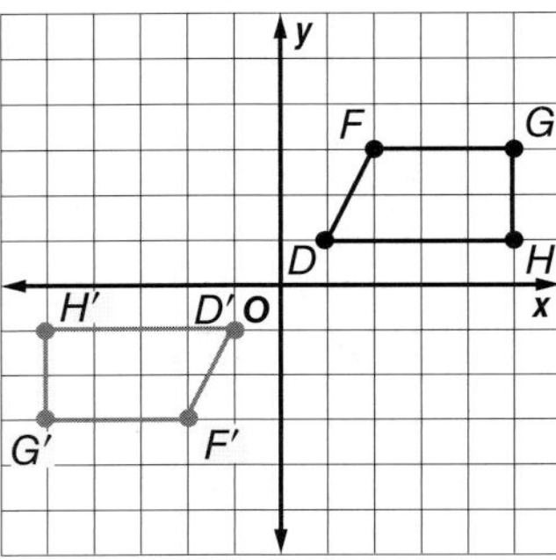

5.

R

7 **a.** Step 1: graph the original figure. Then graph the vertex H' after a 90° rotation about vertex G. Step 2: Graph the remaining vertex after a 90° rotation about vertex G. Connect the vertices to form $G'H'J'$.

b. $G'(1, 1)$, $H'(-3, -1)$, $J'(-3, 1)$

9.

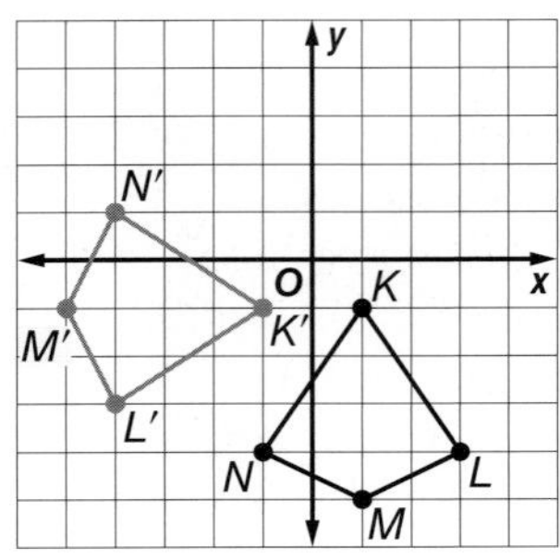

11. 72°

13 The formation has rotational symmetry. The angle of rotation is 180° because if the image is rotated 180° it looks like the original.

15. Sample answer:
90° clockwise; 270° clockwise; 180°

17. Sample answer:  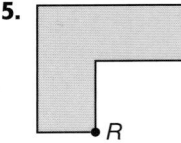 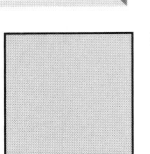 90°

19. $R(1, 5)$, $S(0, -2)$, $T(-6, 3)$ **21.** With both transformations, the original figure and the image are congruent. In a reflection, a figure is reflected across a line. In a rotation, a figure is rotated around a point. **23.** J **25.** C **27.** 73° **29.** 21.5 **31.** 35

Pages 614–616 ***Lesson 11-4***

1. 57; 57°, 114° **3.** rhombus **5.** rectangles

7 $115° + 94° + 107° + 2x = 360°$
$316° + 2x = 360°$
$2x = 44°$
$x = 22$

The missing angle is $2(22) = 44°$

9. 33; 132°, 38° **11.** square, rectangles
13. trapezoid **15.** rhombus **17.** quadrilateral

19 Sample answer: The top of a coffee table is shaped like a rectangle or a square.

21. always **23.** sometimes
25a. $x + x + 70 + 70 = 360$; 110°; 110°
25b. Sample answer:

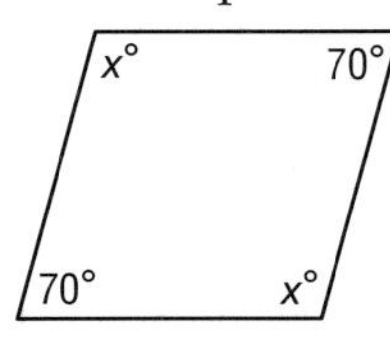

parallelogram

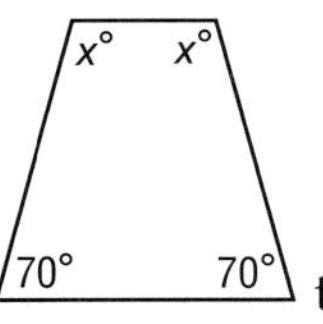

trapezoid

27. Yes; sample answer: quadrilateral with angles that measure 60°, 60°, 120°, and 120°.
29. A square and a rectangle both have four right angles. A square and a rhombus both have four congruent sides. **31.** 50 **33.** C **35.** $\angle L$
37. $\overline{QR}$ **39.** $2(x + 3) + (x + 3) = 90$
41. 21.6 **43.** 3 **45.** 27

Pages 620–622 ***Lesson 11-5***

1. The figure is not a polygon because it is an open figure. Two of the sides are not connected. **3.** The figure has 5 sides that only intersect at their endpoints. It is a pentagon. **5.** 154.3° **7.** The figure has 5 sides that only intersect at their endpoints. It is a pentagon. **9.** The figure has 6 sides that only intersect at their endpoints. It is a hexagon. **11.** The figure has 9 sides that only intersect at their endpoints. It is a nonagon. **13.** 1440° **15.** 2520°

17 A hexagon has 6 sides, so the sum of the measures of the interior angles is $(6-2)180 = 4(180)$ or 720. The measure of one interior angle is $720 \div 6 = 120°$.

19. yes **21.** no; Each interior angle of a regular 15-gon measures 156° and 360° is not evenly divisible by 156°. **23.** octagon **25.** 20-gon **27.** He used translations of the image of the bird and fish to make the tessellating pattern.

29.

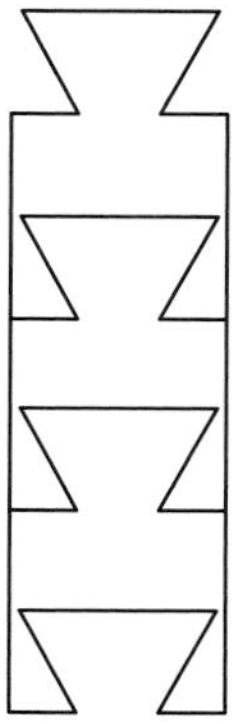

31 In a regular triangle, each interior angle has a measure of 60°. So, the exterior angle has a measure of 180° − 60° or 120°. **33.** 36°

35. Sample answer:

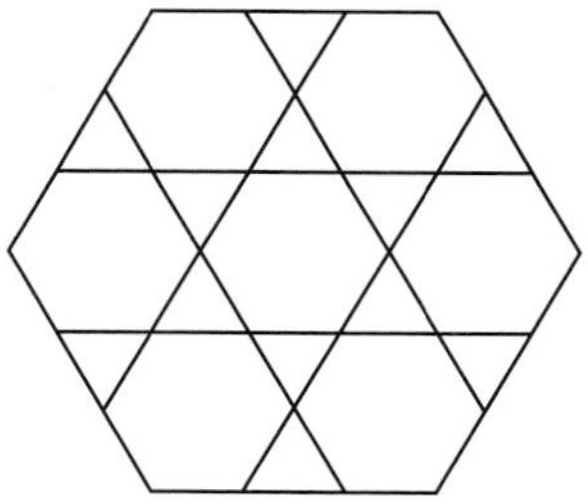

regular hexagons, equilateral triangles; translations, reflections, or rotations

37. Jacinta; the interior angles of a polygon do not have to be congruent in order to create a tessellation. The sum of their measures at a vertex must equal 360°. **39.** Sample answer: A regular polygon has all sides congruent and all angles congruent. A polygon that is not regular has different side lengths, different angle measures, or both. To find the interior angle measure of a regular polygon, subtract 2 from the number of sides, multiply the result by 180, then divide that result by the number of angles.

41. G **43.** 162° **45.** Sometimes **47.** Always

49. yes, $\angle M \cong O$, $\angle L \cong \angle PNO$, $\angle LNM \cong \angle P$, $\overline{NM} \cong \overline{PO}$, $\overline{ML} \cong \overline{ON}$, $\overline{LN} \cong \overline{NP}$, $\triangle LMN \cong \triangle NOP$

51. $x \geq 9.6$

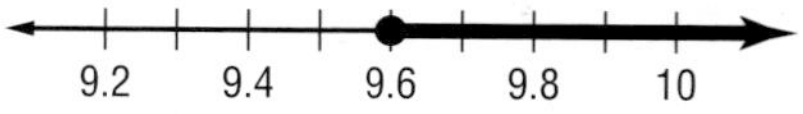

53. −6 **55.** −10 **57.** 900 **59.** 1260

Pages 627–630 ***Lesson 11-6***

1. 30.5 ft^2 **3.** 275 cm^2 **5.** 85.6 in^2

7
$$\begin{aligned} A &= bh \\ &= (15)(11.5) \\ &= 172.5 \text{ yd}^2 \end{aligned}$$

9. 93.1 mi^2 **11.** 25.5 mm^2 **13.** 972.4 m^2 **15.** 33 cm^2 **17a.** 6.25 ft^2 **17b.** 12.5 ft^2 **19.** 19.2 m^2 **21.** 9 cm **23.** 11.6 ft **25.** 81 mm^2 **27.** 15 units^2 **29.** 60 in.; 213.75 in^2

31
$$\begin{aligned} A &= \tfrac{1}{2}h(b_1 + b_2) - \ell w \\ &= \tfrac{1}{2}(9)(12 + 8) - (2)(3) \\ &= \tfrac{1}{2}(9)(20) - 6 \\ &= \tfrac{1}{2}(180) - 6 \\ &= 90 - 6 \\ &= 84 \text{ ft}^2 \end{aligned}$$

33
$$\begin{aligned} A &= \tfrac{1}{2}h(b_1 + b_2) \\ &= \tfrac{1}{2}(11.2)(24 + 14) \\ &= \tfrac{1}{2}(11.2)(38) \\ &= \tfrac{1}{2}(425.6) \\ &= 212.8 \text{ ft}^2 \end{aligned}$$

Divide by 9 to determine the area in square yards. $212.8 \div 9 \approx 23.6$ yd^2

35. Sample answer: a triangle with a base of 8 units and a height of 3 units has the same area as a parallelogram with a base of 4 units and a height of 3 units, 12 units^2. **37a.** Sample answer: a parallelogram with base 10 in., height 12 in., and sides 13 in.; 46 in., 120 in^2 **37b.** Sample answer: a parallelogram with base 11 in., height 11 in., and sides 13 in.; 48 in., 121 in^2

39. In the formula for the area of a triangle, $\frac{1}{2}$ is multiplied by the base and the height. The formula for the area of a trapezoid is the same, except that the sum of the bases is multiplied rather than a single base. **41a.** She would need to determine the area of the lawn that she is going to fertilize. After finding the total area, she will then divide that by 1000 square feet to determine how many bags she needs to buy. **41b.** $90 \cdot 150 = 13{,}500$. $13{,}500 \div 1000 = 13.5$ so she would need to buy 14 bags of fertilizer. **43.** B **45.** about 128.6° **47.** 150° **49.** $x = 100$ **51.** \$73.35 **53.** \$16.25 **55.** 33.9 **57.** 13.5

Pages 633–635 Lesson 11-7

1. 37.7 in. **3.** 50.9 cm **5.** 22.6 km

7 $d = 7$
$C = \pi d$
$= \pi(7)$
≈ 22.0 m

9. 62.8 ft **11.** 17.9 cm **13.** 29.5 m **15.** 70.7 ft
17. Sample answer: Small sand dollar: $C \approx 3.9$ in., large sand dollar: $C \approx 7.9$ in.; the circumference of the small sand dollar is half the circumference of the large sand dollar. **19a.** 49.76 m
19b. 12.44 m **19c.** 143.06 m

21 current fountain:
$C = \pi d$
$= \pi(8)$
≈ 25.1327 ft
new fountain:
$25.1237 \times 4 \approx 100.5$ ft

23a.

Radius (in.)	Circumference (in.)
1	6.3
2	12.6
3	18.8
4	25.1
5	31.4

23b. Sample answer:

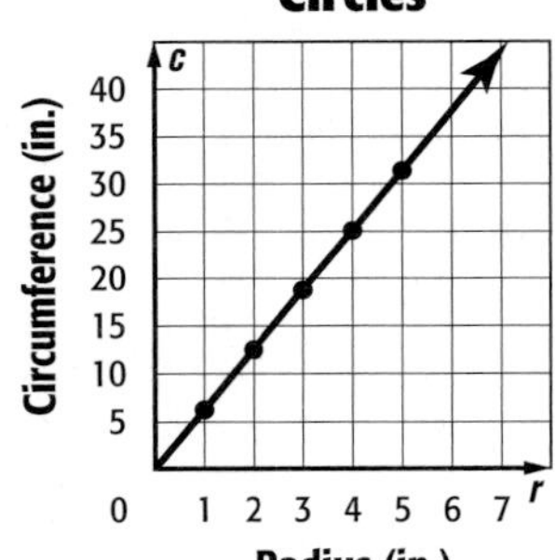

23c. 2π; Since the formula for the circumference of a circle, $C = 2\pi r$, is in the form of $y = mx$, 2π is the slope. **25.** Sample answer: A glass with a diameter of 7 centimeters has a circumference of about 22.0 centimeters. **27.** Circumference of one circle; ℓ equals $3d$ and the circumference of one circle is approximately $3.14d$
29. Circumference is 2π or about 6.3 times the radius. The circumference increases as the radius increases. The radius decreases as the circumference decreases. **31.** F **33a.** 15π
33b. 47 ft **35.** 52.5 ft^2 **37.** 720°
39.

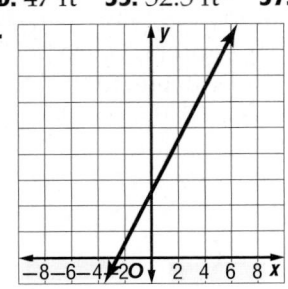

41a. $C = 15n$ **41b.** $180 **43.** 153.86

Pages 638–641 Lesson 11-8

1 $A = \pi r^2$
$= \pi(7)^2$
$= \pi(49)$
≈ 153.9 m^2

3. 132.7 ft^2 **5.** 86.6 cm^2 **7.** 2.1 in^2 **9.** 50.3 cm^2
11. 78.5 mi^2 **13.** 66.5 cm^2 **15.** 289.5 ft^2
17. 103.9 yd^2

19 $A = \pi r^2 - \ell w$
$= \pi(12)^2 - (3)(6)$
$= 144\pi - 18$
≈ 434.4 ft^2

21. 39.3 m^2 **23.** 5.7 mm^2 **25.** 4 in. **27.** 837.7 ft^2

29 $C = \frac{1}{2}\pi d + 10$ $A = \frac{1}{2}\pi r^2$
$= \frac{1}{2}\pi(10) + 10$ $= \frac{1}{2}\pi(5)2$
$= 5\pi + 10$ $= \frac{1}{2}\pi(25)$
≈ 25.7 ft ≈ 39.3 ft^2

31a.

Radius (cm)	Area (cm^2)
3	28.3
6	113.1
12	452.4
24	1809.6
48	7238.2

31b. The area is multiplied by 4.
31c. Sample answer: Since $96 = 48 \cdot 2$, the area should be $4 \cdot 7238.2$ or 28,952.8 cm^2; actual area $\approx$ 28,952.9 cm^2. **33.** Circumference measures the distance around a circle and is given in units. Area measures the surface enclosed by the circle and is given in square units. The formulas for both measures involves π and the radius. The formula for circumference is $C = 2\pi r$ and the formula for area is $A = \pi r^2$. **35.** 2 units; if $r = 2$, then $C = 2\pi(2)$ or 4π units and $A = \pi(2)^2$ or 4π units^2. **37.** A **39.** A **41.** 50.3 in. **43.** 66.0 cm
45. 90 cm^2 **47.** $\frac{8}{45}$ **49.** 533.71 **51.** 138.45

Pages 644–647 Lesson 11-9

1. 38.4 cm^2 **3.** 18.9 cm^2 **5a.** 58 ft^2
5b. 10 cases; $250

7 $A = \frac{1}{2}bh + \ell w$
$= \frac{1}{2}(18)(12) + (22)(6)$
$= \frac{1}{2}(216) + 132$
$= 108 + 132$
$= 240$ in^2

9. 7.3 mm^2 **11.** 257.1 ft^2 **13a.** 354 ft^2
13b. $796.50 **15a.** 73.1 in^2 **15b.** 104.5 in^2

17 $A = \frac{1}{2}(1.5)(6 + 5) + \frac{1}{2}(9)(6)$ or 35.25 ft^2

19. 40.8 m^2

21 $A = 5(8) - \frac{1}{2}\pi r^2$
$= 40 - 9.8$
$\approx 30.2 \text{ in}^2$

23. Sample answer: states, parks, shopping malls **25.** Sample answer: Use polygons to approximate the shape of the curved side. **27.** 429.7 **29.** J **31.** 37.7 cm; 113.1 cm^2 **33.** 125.7 in, 1256.6 in^2 **35.** 18 cm^2 **37.** pentagon **39.** triangle

Pages 650–654 Study Guide and Review

1. true **3.** false; base **5.** true **7.** false; supplementary angles **9.** false; rotation **11.** 112°; ∠4 and ∠2 are vertical angles, so they are congruent. **13.** 68°; ∠4 and ∠1 are a linear pair of angles, so they are supplementary. **15.** $\angle F \cong \angle L$, $\angle G \cong \angle M$, $\angle H \cong \angle N$; $\overline{FG} \cong \overline{LM}$, $\overline{GH} \cong \overline{MN}$, $\overline{HF} \cong \overline{NL}$; △LMN

17.

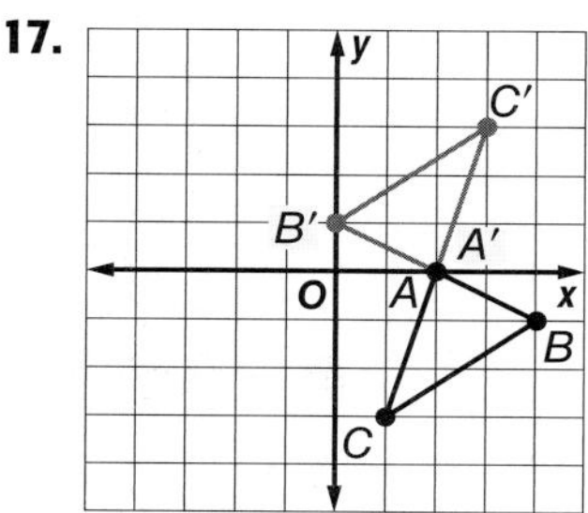

$A'(2, 0)$, $B'(0, 1)$, $C'(3, 3)$

19.

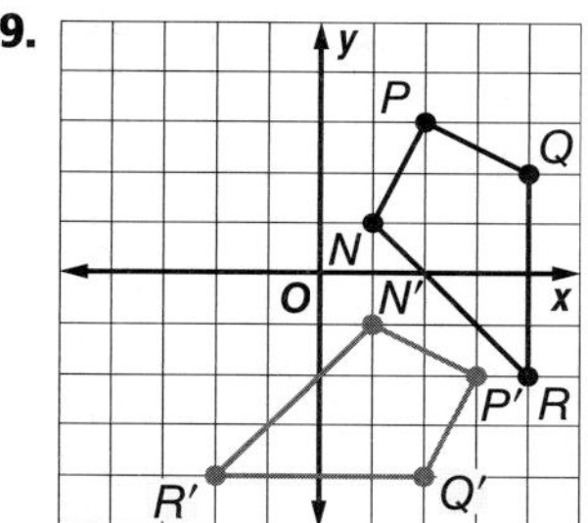

21. 98; 98°, 100° **23.** rhombus **25.** The figure has 5 sides that only intersect at their endpoints. It is a regular pentagon. **27.** 160° **29.** 120 ft^2 **31.** about \$76.25 **33.** 44.0 yd **35.** 84.8 mm **37.** 201.1 cm^2 **39.** 63.6 in^2 **41.** about 14,055 ft^2

Chapter 12 Surface Area and Volume

Page 661 Chapter 12 Getting Ready

1. yes; parallelogram **3.** octagon **5.** 35 **7.** 330 **9.** no **11.** no **13.** yes; $\frac{1}{1.25} = \frac{2}{2.5} = \frac{3}{3.75} = \frac{4}{5}$

Pages 667–669 Lesson 12-1

1. rectangular prism; EIJF and HLKG, or EILH and FGKJ, or IJKL and EFGH; EFGH, EFJI, GHLK, JFGK, EHLI, IJKL, E, F, G, H, I, J, K, L

3. parallelogram; **5.** rectangle;

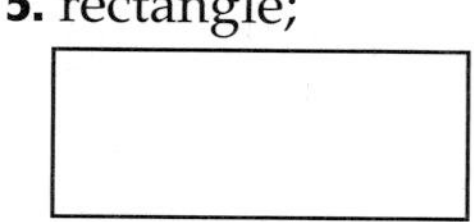

7. triangular prism, bases: DEF, ABC; faces: ABEF, BCDE, ACDF, FED, ABC; edges: $\overline{AB}$, $\overline{AC}$, $\overline{AF}$, $\overline{BC}$, $\overline{BE}$, $\overline{DF}$, $\overline{CD}$, $\overline{ED}$, $\overline{EF}$; vertices: A, B, C, D, E, F

9 This figure has one rectangular base, ABCD, so it is a rectangular pyramid; faces: BEC, BEA, AED, CED, ABCD; edges: $\overline{AB}$, $\overline{BC}$, $\overline{CD}$, $\overline{AD}$, $\overline{AE}$, $\overline{BE}$, $\overline{CE}$, $\overline{DE}$; vertices: A, B, C, D, E

11. hexagonal pyramid; base: BCDEFG; faces: ABC, ACD, ADE, AEF, AFG, AGB, BCDEFG; edges: $\overline{AB}$, $\overline{AC}$, $\overline{AD}$, $\overline{AE}$, $\overline{AF}$, $\overline{AG}$, $\overline{BC}$, $\overline{CD}$, $\overline{DE}$, $\overline{EF}$, $\overline{FG}$, $\overline{GB}$; vertices: A, B, C, D, E, F, G

13. circle; **15.** square;

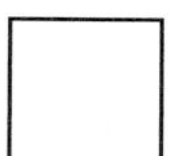

17a.

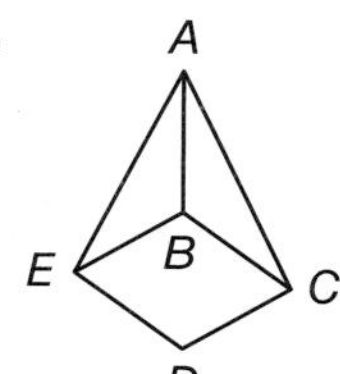

17b. base: BCDE; faces: ABC, BCDE, ABE, ACD, ADE; edges: $\overline{AB}$, $\overline{AC}$, $\overline{AD}$, $\overline{AE}$, $\overline{BC}$, $\overline{CD}$, $\overline{DE}$, $\overline{EB}$

17c. triangle, triangle or trapezoid, square

17d.

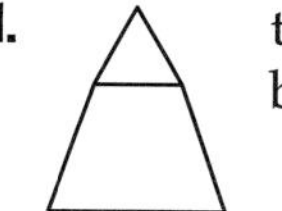

top-triangle
base-trapezoid

19. Top

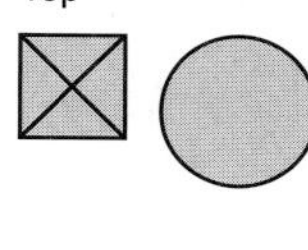

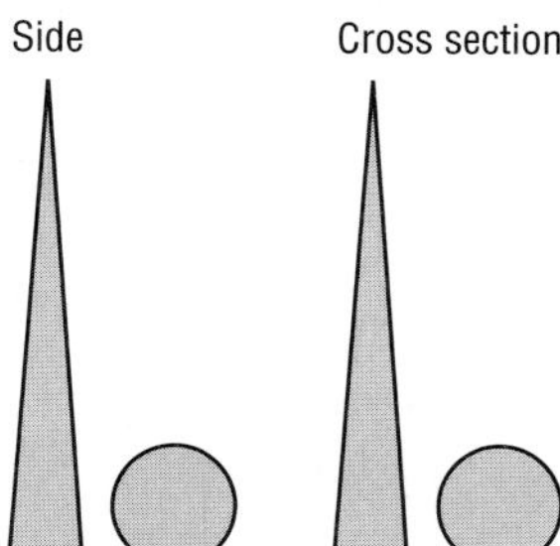

triangle and circle

21 **a.** The first figure is a triangular pyramid. It will have four triangular faces. The second figure is a square pyramid. It will have a square base and four triangular faces. The third figure is a pentagonal pyramid. It will have a pentagonal base and five triangular faces. From this information, you can complete the table.

Name	Triangular Pyramid	Square Pyramid	Pentagonal Pyramid
Figure			
Vertices	4	5	6
Faces	4	5	6
Edges	6	8	10

b. Sample answer: The number of vertices is equal to the number of faces. The number of edges increases by 2 as the number of vertices of the pyramid increases by 1.
c. The sum of the number of vertices V and the number of faces F is equal to the number of edges E plus 2. So, $V + F = E + 2$.
23. never **25.** sometimes **27.** Sample answer: Examine the base of the object; it is a hexagon. Since it has two parallel, congruent bases, it is also a prism. The figure is a hexagonal prism. **29.** D
31.  The shape resulting from a vertical cross section is a rectangle. **33.** 72 ft^2 **35.** 56.1 cm^2 **37.** 21.6 in^2 **39.** 156.4 ft^2

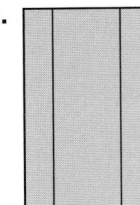

Pages 673–676 *Lesson 12-2*

1. 36 mm^3 **3.** 1000 in^3

5 The figure is made of three rectangular prisms.
V(figure) = V(rectangular prism 1) + V(rectangular prism 2) + V(rectangular prism 3)

$$\begin{aligned} V(\text{figure}) &= \ell wh + \ell wh + \ell wh \\ &= (10 \cdot 10 \cdot 4) + (6 \cdot 6 \cdot 4) + (2 \cdot 2 \cdot 2) \\ &= 400 + 144 + 8 \\ &= 552 \text{ cubic inches} \end{aligned}$$

The answer is B.
7. 260 ft^3 **9.** 54 in^3

11 $$\begin{aligned} V &= Bh \\ &= (\tfrac{1}{2} \cdot 3 \cdot 10)(8) \\ &= 15 \cdot 8 \\ &= 120 \text{ cm}^3 \end{aligned}$$

13. $1\frac{1}{2}$ in^3 **15.** 6 yd **17.** 3888 in^3
19. 10 ft **21.** 972 m^3 **23.** 180 ft^3

25 $V = Bh$
The base of the figure is a trapezoid.
So, $B = \frac{1}{2}(b_1 + b_2)h$.

$$\begin{aligned} V &= \left[\tfrac{1}{2}(b_1 + b_2)h\right]h \\ &= \left[\tfrac{1}{2}(1.5 + 1)0.5\right](3) \\ &= \tfrac{1}{2}(2.5)(0.5)(3) \\ &= 1.875 \text{ in}^3 \end{aligned}$$

27a. 42 cm^3 **27b.** The volume is multiplied by 2 because $V = \ell w(2h) = 2(\ell wh)$; The original volume is multiplied by 2^2 or 4 because $V = \ell(2w)(2h) = 4(\ell wh)$; The original volume is multiplied by 2^3 or 8 because $V = (2\ell)(2w)(2h) = 8(\ell wh)$. **27c.** When one dimension is tripled, the volume is multiplied by 3. When two dimensions are tripled, the original volume is multiplied by 3^2. When three dimensions are tripled, the original volume is multiplied by 3^3.
27d. 9072 cm^3 **29.** Sample answer: base length 4, base height 2, prism height 11. **31.** 35
33. Sample answer: The volume of the box is changed by 2^3 or 8. **35.** 12 **37.** C
39. rectangular prism; bases: $ABCD$ and $EFGH$, or $BCGF$ and $ADHE$, or $ABFE$ and $DCGH$; faces: $ABCD$, $EFGH$, $BCGF$, $ADHE$, $ABFE$, $DCGH$; edges: $\overline{AB}$, $\overline{DC}$, $\overline{HG}$, $\overline{EF}$, $\overline{AD}$, $\overline{BC}$, $\overline{EH}$, $\overline{FG}$, $\overline{AE}$, $\overline{BF}$, $\overline{DH}$, $\overline{CG}$ **41.** 228.5 mm^2
43. 12.6 in^2 **45.** 78.5 mi^2

Pages 679–681 *Lesson 12-3*

1. 25.1 ft^3 **3.** 10.0 in. **5.** 22.0 cm^3 **7.** 254.5 m^3

9 $$\begin{aligned} V &= Bh \\ &= \pi r^2 h \\ &= \pi(2.2)^2(3) \\ &\approx 45.6 \text{ cm}^3 \end{aligned}$$

11. 101.4 m^3 **13.** 5.5 m **15.** 69.1 in^3
17. 2628.1 cm^3

19 $$\begin{aligned} V &= \pi {r_1}^2 h_1 - \pi {r_2}^2 h_2 \\ &= \pi(6.5)^2(28) - \pi(2.25)^2(28) \\ &\approx 3271.2 \text{ cm}^3 \end{aligned}$$

21. 101.3 gal **23.** Sample answer: Changing the packaging size of a drink; changing the size of a garbage can to fit in a narrower but taller space. **25a.** 2:1 **25b.** 4:1 **27.** B **29a.** 77.8 in^3
29b. 699.8 in^3 **31.** 5200 ft^3 **33.** 120 in^3
35. triangular prism: bases ACE and BDF; faces: $ABFE$, $EFDC$, $ABCD$, ACE, BDF; edges: $\overline{AB}$, $\overline{EF}$, $\overline{CD}$, $\overline{AC}$, $\overline{CE}$, $\overline{EA}$, $\overline{BF}$, $\overline{FD}$, $\overline{DB}$; vertices: A, B, C, D, E, F
37. 160 **39.** 35 **41.** $3\frac{1}{3}$

Pages 686–688 *Lesson 12-4*

1. 115.5 in^3 **3.** 74.2 cm^3 **5.** 268.1 mm^3
7. 392.7 ft^3 **9.** 6 ft^3 **11.** 4.3 in^3 **13.** 10.5 ft^3
15. 1563.5 km^3

17 **a.** $$\begin{aligned} V &= \tfrac{1}{3}\pi r^2 h \\ &= \tfrac{1}{3}\pi(0.75)^2(2.5) \\ &\approx 1.5 \text{ ft}^3 \end{aligned}$$

b. Use a proportion.

$$\begin{aligned} \frac{0.1 \text{ ft}^3}{5 \text{ m}} &= \frac{1.5 \text{ ft}^3}{x \text{ m}} \\ 0.1x &= 5 \cdot 1.5 \\ \frac{0.1x}{0.1} &= \frac{7.5}{0.1} \\ x &= 75 \end{aligned}$$

It will take about 75 minutes or 1 hour 15 minutes for the icicle to melt.
19. 18.3 ft

21 $$\begin{aligned} V &= \tfrac{1}{3}\pi {r_1}^2 h_1 + \tfrac{1}{3}\pi {r_2}^2 h_2 \\ &= \tfrac{1}{3}\pi(3)^2(15) + \tfrac{1}{3}\pi(3)^2(7) \\ &\approx 207.3 \text{ in}^3 \end{aligned}$$

23. 4656.8 g **25.** 140.4 cm^3 **27.** 3 units
29. True. The volumes are equal if both heights and both bases are equal. Changing the shape of the base will not affect the volume.
31. Changing the shape of the packaging. If the volumes remained the same and they have congruent bases, the cone's height would be 3 times the height of the cylinder. **33.** H
35. D **37.** 748 cm^3 **39.** 24.6 mi **41.** $575p = 15$; 2.6 lb **43.** 3772 m^2

Pages 693–695 Lesson 12-5

1. 108 in^2; 268 in^2 **3.** 360 yd^2; 464.5 yd^2
5. 3.84 cm^2; 4.8 cm^2 **7.** 8.4 m^2; 10.08 m^2

9 $L = Ph$
$= (2\ell + 2w)h$
$= (2(2.2) + 2(1.4))(9)$
$= 64.8\ cm^2$

$S = L + 2B$
$= 64.8 + 2(2.2)(1.4)$
$= 70.96\ cm^2$

11. 334 in^2 **13.** 528 m^2

15 **a.** $S = 2(\ell h + wh + \ell w)$
$= 2(5 \cdot 5 + 5 \cdot 5 + 5 \cdot 5)$
$= 150\ in^2$

b.

Scale Factor of Dilation	Original	×2	×3
Length of Side	5	10	15
Surface Area	150	600	1350

$S = 2(\ell h + wh + \ell w)$
$= 2(10 \cdot 10 + 10 \cdot 10 + 10 \cdot 10)$
$= 600\ in^3$
$S = 2(\ell h + wh + \ell w)$
$= 2(15 \cdot 15 + 15 \cdot 15 + 15 \cdot 15)$
$= 900\ in^3$

c. Sample answer: The surface area after a dilation of scale factor of 2 is four times the original; The surface area after a dilation of a scale factor of 3 is 9 times the original.

d. Yes. The original formula for surface area is $S = Ph + 2B$. After doubling the sides, the new formula would be $S = (2P)(2h) + 2(4B)$ which simplifies to $S = 4(Ph + 2B)$. Remember, the B is four times as much because both base and height are doubled.

17. The rectangular prism would be more expensive to make. The surface area of the rectangular prism is 36.64 in^2, which would cost \$47.63 to make. The surface area of the cube is 34.56 in^2, which would cost \$44.93 to make.
19. The box is a cube that measures 6 cm on each side. **21.** See students' work. Sample answer: Prism A with dimensions 3 by 3 by 3 and Prism B with dimensions 10 by 2 by 1. Prism A has the larger volume while Prism B has the larger surface area. **23.** D **25.** J **27.** 268.1 cm^3
29. 251.3 m^3 **31.** 625 **33.** 722 **35.** 153.9 yd^2
37. 1240.9 m^2

Pages 699–701 Lesson 12-6

1. 1492.3 mm^2; 2059.3 mm^2
3. 102.9 m^2; 130.6 m^2

5 $L = 2\pi rh$
$= 2\pi(3.1)(2)$
$= 12.4\pi$
$\approx 39.0\ yd^2$

$S = L + 2\pi r^2$
$= 12.4\pi + 2\pi(3.1)^2$
$\approx 99.3\ yd^2$

7. 125.7 in^2; 150.8 in^2 **9.** 121.0 ft^2; 205.4 ft^2
11. The box has surface area of 1296 in^2 and the tube has a surface area of 1288.1 in^2. So the tube has a smaller surface area.

13 Exterior of Pipe:
$S = 2\pi rh$
$= 2\pi(1.6)(10)$
$= 32\pi$
$\approx 100.5\ in^2$

Interior of Pipe:
$S = 2\pi rh$
$= 2\pi(1.5)(10)$
$= 30\pi$
$\approx 94.2\ in^2$

So, the surface area of the pipe is about 100.5 + 94.2 or 194.7 in^2.
15. 6121.0 cm^2 **17.** The cylindrical popcorn container because it uses less material (surface area) and holds about the same amount of popcorn (volume). Rectangular Prism: $V = 275\ in^3$ S.A. $= 237.5\ in^2$; Cylinder: $V = 274.3\ in^3$ S.A. $= 211.1\ in^2$
19. Sample answer: surface area of the cylinder:
$S = 2\pi rh + 2\pi r^2$
$= 2(3)(5)15 + 2(3)5^2$
$= 150 + 450$
$= 600\ ft^2$
A cube measuring 10 ft would have a surface area of 600 square feet.
$S = 6s^2$
$600 = 6s^2$
$100 = s^2$
$s = 10$
21. The ice cube in the shape of a half cylinder would melt at a faster rate because it has a greater total surface area exposed to the air. **23.** G
25. D **27.** 114 in^2; 282 in^2 **29.** 4356 m^3
31. straight **33.** obtuse **35.** \$918.75 **37.** \$9135
39. 48 **41.** 58.75 **43.** $\frac{14}{15}$

Pages 705–707 Lesson 12-7

1. 364 m^2; 533 m^2 **3.** 192 mm^2; 336 mm^2

5 $L = \frac{1}{2}P\ell$
$= \frac{1}{2}(4 \times 10)(12)$
$= 240\ in^2$

$S = L + B$
$= 240 + 10(10)$
$= 340\ in^2$

7. 105.3 mm^2; 140.4 mm^2 **9.** 339.3 mm^2
11. 125.7 cm^2; 175.9 cm^2 **13.** 427.3 ft^2; 628.3 ft^2
15. 471.2 cm^2

17 $C = 2\pi r$
$22 = 2\pi r$
$3.5 = r$

$a^2 + b^2 = c^2$
$(3.5)^2 + (18) = c^2$
$12.25 + 324 = c^2$

Selected Answers and Solutions

$S = \pi r\ell$ $\quad 18.3 \approx c$

$= \pi(3.5)(18.3)$

≈ 201.2

$8(201.2) = 1609.8 \text{ in}^2$

19. lateral area = 130.5 m^2; surface area = 174 m^2 **21.** 48.7 feet **25.** 4 **27.** C **29.** B **31.** 386.4 in^2; 612.6 in^2 **33.** 120 ft^2; 244.8 ft^2 **35.** 50° **37.** $2\frac{1}{7}$ **39.** 6 **41.** 5.5

Pages 712–715 *Lesson 12-8*

1. yes

3 $\frac{2}{6} = \frac{d}{12}$

$2 \cdot 12 = 6 \cdot d$

$24 = 6d$

$\frac{24}{6} = \frac{6d}{6}$

$4 = d$

$d = 4$ yd

5. 450 ft^2 **7.** no **9.** no

11 $\frac{10}{6} = \frac{8}{x}$

$10 \cdot x = 6 \cdot 8$

$10x = 48$

$\frac{10x}{10} = \frac{48}{10}$

$x = 4.8$ in.

13. 4800 mm^2 **15a.** 4:1 **15b.** 4.25 inches **15c.** large wheel volume = 30,762.5 in^3; small wheel volume 480.7 in^3 **15d.** 64 times greater

17 $V = \frac{4}{3}\pi r^3$

$= \frac{4}{3}\pi(33)^3$

$\approx 150{,}536.3 \text{ ft}^3$

19. 3.3 in. **21a.** $\frac{4}{9}, \frac{8}{27}$ **21b.** 90 ft^2 **21c.** 16 ft^3 **23.** True. Spheres have only one measurement, the radius. **25.** The surface area quadruples. **29.** G **31.** 13.75 **33.** 201.1 in^2; 603.2 in^2 **35.** $\frac{25}{4}$ **37.** 475

Pages 716–720 *Chapter 12* *Study Guide and Review*

1. true **3.** true **5.** false; pyramid **7.** false; cylinder or prism **9.** rectangular prism; bases: *AFGD*, *BEHC*; faces: *ABCD*, *EFGH*, *BEFA*, *CHGD*; edges: $\overline{AB}, \overline{BC}, \overline{CD}, \overline{AD}, \overline{EF}, \overline{EH}, \overline{FG}, \overline{GH}, \overline{EB}, \overline{HC}, \overline{AF}, \overline{GD}$; vertices: *A, B, C, D, E, F, G, H* **11.** square pyramid; bases: *LMNO* faces: *LPM, LPO, NPO, NPM*; edges: $\overline{LM}, \overline{MN}, \overline{NO}, \overline{OL}, \overline{LP}, \overline{MP}, \overline{NP}, \overline{OP}$; vertices: *L, M, N, O, P*

13.

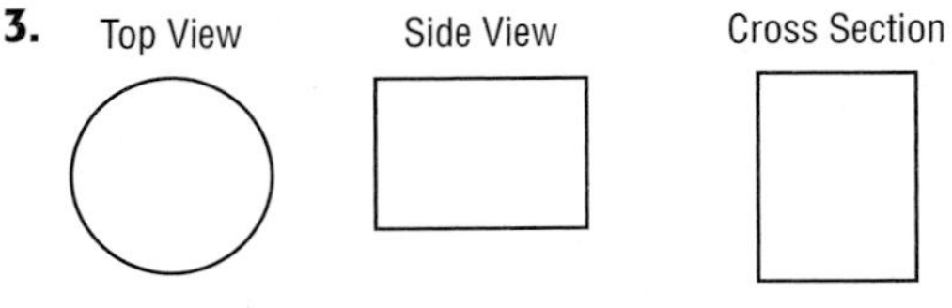

A vertical cross section of the drum is a rectangle.

15. 60 cm^3 **17.** 445.3 cm^3 **19.** 18.9 in^3 **21.** 8 in^3 **23.** 15 ft^3 **25.** 68 cm^2; 101 cm^2 **27.** 143.6 m^2; 166.4 m^2 **29.** 288 in^2 **31.** 298.5 ft^2; 865.5 ft^2 **33.** 202.3 mm^2; 335.3 mm^2 **35.** 244.8 cm^2; 308.8 cm^2 **37.** 131.9 cm^2; 182.2 cm^2 **39.** 1980 ft^2 **41.** 5 m

Chapter 13 Statistics and Probability

Page 727 *Chapter 13* *Get Ready*

1. \$15.33 **3.** 5000 visitors **5.** 6 **7.** 12 **9.** $\frac{12}{25}$ **11.** 1 **13.** $\frac{5}{8}$

Pages 733–735 *Lesson 13-1*

1. mean: 151.4 mph; median: 157 mph; mode: 157 mph **3.** A **5.** mean or median; Sample answer: The mean and median both show longer amounts of time.

7 mean

Add the data values and divide by the number of data values.

$\frac{250 + 200 + 320 + 235 + 265 + 200}{6} = \frac{1470}{6}$ or \$245

median

List the data values in ascending order.

200, 200, 235, 250, 265, 320

Since there is an even number of data values, find the mean of the two middle numbers.

$\frac{235 + 250}{2} = \frac{485}{2}$ or \$242.50

mode

Find the data value(s) which occur the most.

\$200

9. Median; 869 is an extreme value that shifts the mean upward.

11 mean

Add the data values and divide by the number of data values.

$\frac{50 + 76 + 94 + 90 + 88 + 92 + 88 + 96}{8} = \frac{704}{8}$ or \$88

median

List the data values in ascending order.

76, 80, 88, 88, 90, 92, 94, 96

Since there is an even number of data values, find the mean of the two middle numbers.

$\frac{88 + 90}{2} = \frac{178}{2}$ or 89

mode

Find the data value(s) which occur the most.

88

The measure of central tendency that makes his score look best is the median, 89.

13. mode; Sample answer: More students have 2 siblings than any other number. It is the best representation of the data.

15. Sample answer:

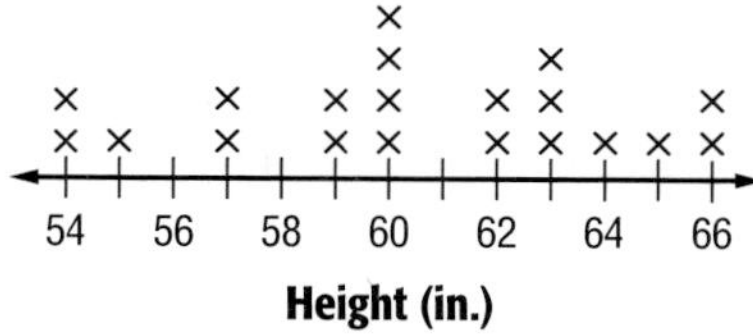

mean: 60.45; median: 60; mode: 60; any of them. Each measure is around 60 inches.

17. Sometimes; For example, the data set: \$3, \$5, \$5, \$5, and \$7. The mean is \$5, the median is \$5, and the mode is \$5.

19. Sample answer: The median home price would be useful because it is not affected by the cost of the very expensive homes. The cost of half the homes in the county would be greater than the median cost and half would be less.

21. Sample answer:

Completed Passes in the NFL

Player	Number of Passes
Jon Kitna	372
Peyton Manning	362
Carson Palmer	324
Steve McNair	295
Jake Plummer	175

Source: National Football League

The median; the mean is affected by the extreme value of 175, and there is no mode. The median is 324 passes.

23. J **25a.** 155 **25b.** 140 **27.** yes

29.

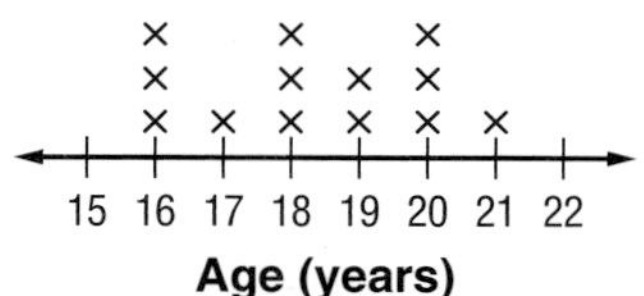

Pages 739–742 ***Lesson 13-2***

1. **Test Scores**

Stem	Leaf
7	2 3 6 6 9
8	0 0 1 4
9	9

8|1 = 81

3a. 14.5 lb; **3b.** 15lb **3c.** 56 lb

5 Step 1: Find the least and greatest number. Then identify the greatest place value in each number. The least number, 18, has 1 in the tens place. The greatest number, 55, has 5 in the tens place.

Step 2: Write the stems 1-5 in the Stem column from least to greatest. Write the ones digit to the right of each corresponding stem in the Leaf column.

Step 3: Order the leaves from least to greatest and write a key that explains how to read the stems in the Leaf column. Include a title.

Ages of People in Spinning Class

Stem	Leaf
1	8
2	0 1 2 2 3 6 7 7
3	0 3 5 9
4	0 2 9
5	5

2|8 = 28 years

7. **Tennis Shoe Cost**

Stem	Leaf
5	0 9
6	0 5 5
7	0 5 6 8
8	0 0
9	6

5|9 = \$59

9 **a.** List numbers in ascending order: 31, 35, 35, 37, 37, 39, 42, 44, 44, 46, 46, 48, 48, 50, 50, 50, 53, 54, 56, 56, 57, 61, 66, 67, 68, 68, 88 and determine the middle number; 50 pages

b. Observe data and determine which number occurs most often; 50 pages

c. 88 – 31 = 57 pages

11 **a.** Look at the overall data. Troop 60 members generally sold 30 to 40 boxes and Troop 122 members generally sold 50 to 60 boxes. So, Troop 122 sold more boxes.

b. Troop 122 sales are more spread out. Troop 60 are more clustered.

13. Sample answer: The person who came in first beat the other runners by almost 2 seconds. An average time is about 17.8 seconds.

100 Meter Times

Stem	Leaf
14	3
15	
16	2 5
17	4 7 9
18	0 7
19	2 9

18|0 = 18.0 s

15 **a.** Step 1: Find the least and greatest number. Then identify the greatest place value in each number. The least number, 6, has 0 in the tens place. The greatest number, 80, has 8 in the tens place.
Step 2: Write the stems 0-8 in the Stem column from least to greatest. Write the ones digit to the right of each corresponding stem in the Leaf column.
Step 3: Order the leaves from least to greatest and write a key that explains how to read the stems in the Leaf column. Include a title.

Test Scores

Stem	Leaf
0	6
1	3 9
2	0 3
3	2 4 6 8 9
4	6 8 9
5	6 6 9 9 9
6	2 2 3 4 7 8
7	3 3 5 6 6 8 8 8 9 9
8	0 0

b. mean: Find the sum of the data and divide by the total number of values.

$\frac{2003}{36} = 55.6$

median: Find the middle number of the data.

$\frac{62 + 59}{2} = 60.5$

mode: 59, 78

c.

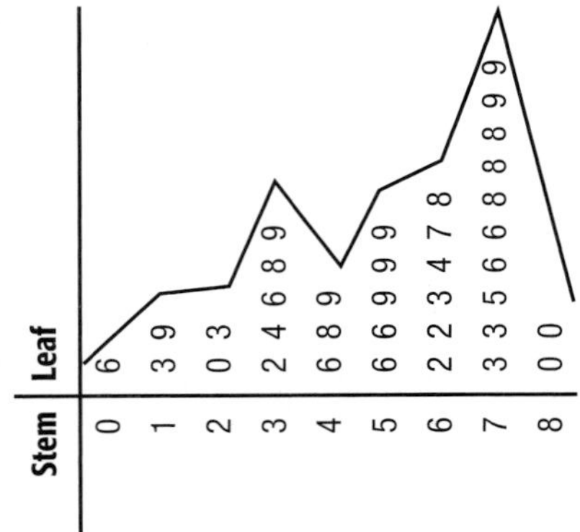

d. Sample answer: There are two main peaks in the graph with one valley. The graph is not symmetric since the shape is not the same on each side of the graph.

e. Sample answer: The data set is skewed. Most of the data is on the right side of the graph.

17. Sample answer:

Stem	Leaf
1	0 0 2 3 6 7 8
2	2 5 6 9 9
3	1 2 8

2|6 = 26

19. Neither; Sample answer: The median is 32. J'vonté divided wrong, and Paolo forgot to divide 64 by 2.

21. Sample answer: Stem-and-leaf plots can help you see how the winning speeds are distributed. You can identify the fastest and slowest winning speeds, the mean speed, the mode speed, and the median speed. **23.** D **25.** B **27.** $33\frac{1}{3}$ ft **29.** 36 **31.** 8 **33.** 7.5

Pages 747–749 ***Lesson 13-3***

1. R: 145; UQ: 145; LQ: 90; M: 115; IR 55; none
3a. The fruits' range is 30. The vegetables' range is 20. So, the number of Calories in fruits varies more than the number of Calories in vegetables.
3b. 80; With the outlier, the mean increased by 57.5–54.3 or 3.2, the median increased by 55–50 or 5, and the mode did not change.

5 The range is 30 − 15 or 15.
The median is $\frac{21}{2}$ or 21.5.
lower quartile: 15 17 18 18 10 21
LQ is 18
upper quartile: 22 23 25 25 27 30
UQ is 25
The interquartile range is 25 – 18 = 7
To find an outlier, multiply the interquartile range by 1.7.
$7 \cdot 1.5 = 10.5$
Add 10.5 to the lower quartile and the upper quartile to find any outliers
18 − 10.5 = 7.5   25 + 10.5 = 35.5
There are no outliers.
7a. Room 110: 91; Room 100: 94 **7b.** Room 100
7c. Room 100 has an outlier, 64. The outlier makes the range greater by 24 points.

9 **a.** Antelope, MT: 84 – 21 = 63
Augusta, ME: 80 – 28 = 52
Antelope has the greater range.
b. Antelope:
The median is $\frac{58 + 58}{2} = 58$.
lower quartile: 21 24 30 37 42 58
LQ is 33.5
upper quartile: 58 70 72 79 84 84
UQ is 75.5
The interquartile range is 75.5 − 33.5 = 42
Augusta:

The median is $\frac{53 + 58}{2} = 55.5$.
lower quartile: 28 32 34 41 46 53
LQ is 37.5
upper quartile: 58 66 70 75 79 80
UQ is 72.5
The interquartile range is 72.5 – 37.5 = 35

c. The median for Antelope is higher. The lower quartile for Augusta is higher and the upper quartile for Antelope is higher.
d. The median is the best measure of central tendency because it best describes the data.
e. The average temperature of the two cities is similar, although Antelope has the greater median and range. This means that the temperature varies more in Antelope than it does in Augusta.

11. Sample answer: Mean: 24.38; Interquartile Range: 6; Mean: 20.43; Interquartile Range: 6; Without the outlier, the mean price is lower and the prices vary less.

Colorado 1-day Lift Ticket Prices for Kids	
Ski Area	**Child Price**
Arapahoe Basin	$24
Aspen Highlands	$52
Howelsen	$10
Loveland	$23
Monarch	$19
Ski Cooper	$18
SolVista	$24
Wolf Creek	$25

13. Always; the range is the difference between the greatest and least value. An outlier is an extreme value, therefore will always effect the range.

15. true **17.** D **19.** A **21.**

Cost of DVDs

Stem	Leaf
0	9
1	2 4 5 8
2	1 7
3	7

3|7 = $37

23. 4.3; 4.2; 4.1 and 4.2 **25.** yes **27.** 120.1 yd^3
29. 6.7, 6.8, 6.9, 7.0, 7.8, 8.7

Pages 752–755 Lesson 13-4

1 Step 1: Draw a number line that includes the least and greatest numbers in the data.
Step 2: Mark the median, and the upper and lower quartile above the number line. Check for outliers. If an outlier exists, mark the greatest value that is not an outlier. Then mark the outlier with an asterisk.
Step 3: Draw the box and whiskers. Add a title.

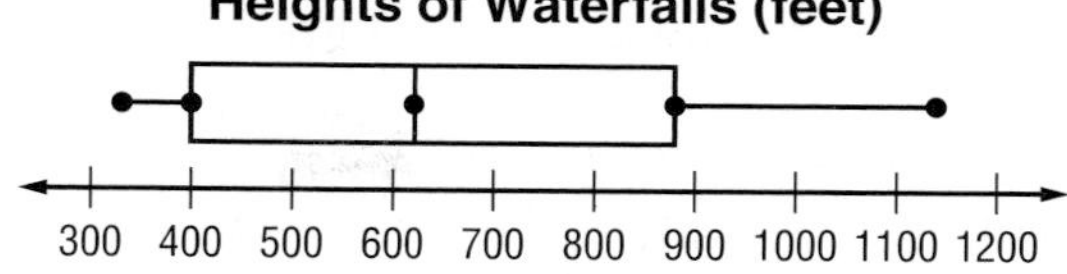

3. Sample answer: Half the time the Cougars scored between 18 and 31 points. Half the time the Falcons scored between 14 and 21 points. So, the Cougars usually scored more points per game than the Falcons.

5.

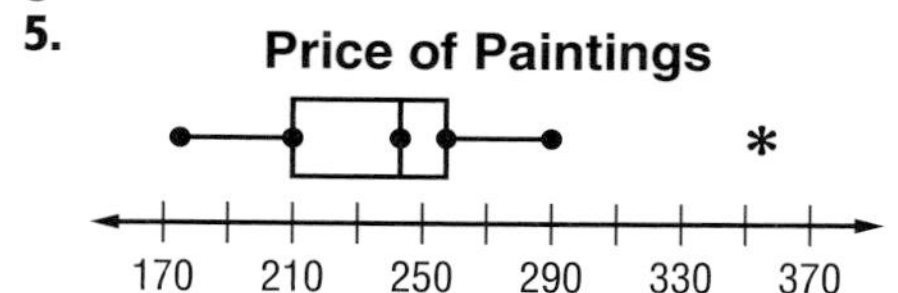

7.

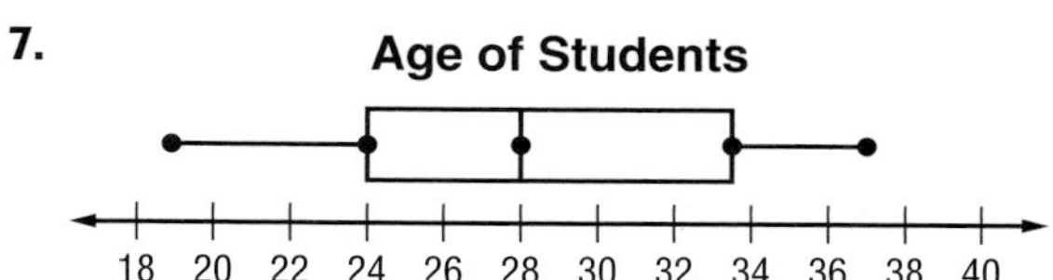

9 **a.** 50%
b. Sample answer: The median divides the data in the box into unequal parts. The left whisker is significantly shorter than the right whisker so the data below the lower quartile is more concentrated than the data above the upper quartile. The data below the upper quartile is more spread out.

11 Step 1: Draw a number line that includes the least and greatest numbers in the data.
Step 2: Mark the median, and the upper and lower quartile above the number line. Check for outliers. If an outlier exists, mark the greatest value that is not an outlier.
Step 3: Draw the box and whiskers. Add a title.

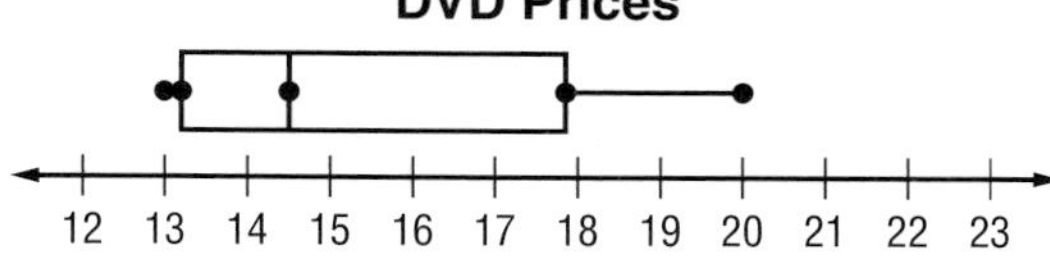

13. Sample answer: Data set A {21, 21, 22, 26, 26, 27, 28, 28, 28, 29, 30, 30, 30, 30, 31} Data set B {18, 18, 18, 22, 22, 22, 22, 26, 26, 26, 26, 31, 31, 31, 33}
15a. 28, 29, 30, 30, 31, 35, 38, 39, 41, 42, 42, 47, 48
15b. at least 21; Sample answer: since 6 students have scores ranging from 38 to 42, there are at most 4 students with scores between 38 and 42. If the measures of variation above are constant, there are at most 4 students with scores between 28 and 30, 30 and 38, 38 and 42, and 42 and 48. Therefore, 16 students are accounted for in these scores, plus the minimum, lower quartile, median,

upper quartile, and maximum scores gives a minimum of 21 students. **17.** 4 **19.** H **21a.** Rose Bowl: 39, 27, 38, 17.5, 20.5, no outliers; .Cotton Bowl: 48, 28, 35, 17, 18, no outliers **21b.** Sample answer: The winners of the Cotton Bowl scored more points on average than the winners of the Rose Bowl. The number of points scored by the Cotton Bowl winners varies greater than the number of points scored by the Rose Bowl winners. The Rose Bowl data in the middle are more spread out than the Cotton Bowl data.

Pages 759–762 Lesson 13-5

1 Step 1: Draw and label a horizontal and vertical axis as shown. Include a title.
Step 2: Show the intervals from the frequency table on the horizontal axis and the frequency on the vertical axis.
Step 3: For each interval, draw a bar whose height is given by the frequency.

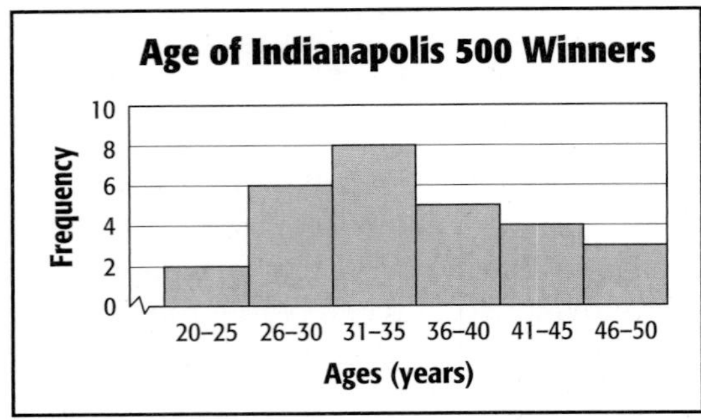

3a. about 24% **3b.** Not very likely. Only 3 temperatures out of 51are 125°F or higher. **3c.** This information cannot be determined from the data presented in the graph. We only know the highest temperature is between 130–134 degrees Fahrenheit and the lowest is between 100–104.

5.

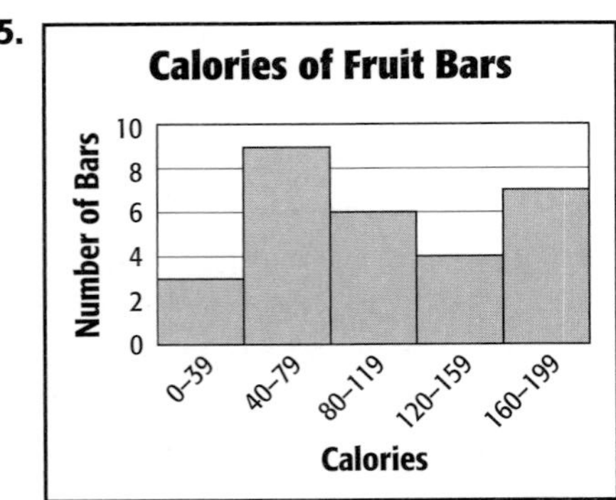

7 **a.** Find the value for each interval, then add.
$12 + 8 = 20$
b. Find the value for each interval, then add to find the total.
$16 + 10 + 7 = 33$
Write a ratio comparing the part to the base.
$\frac{33}{53} \approx 0.62$ or 62%
c. The tallest bar is 60–69.99, so that is the price most shoes are likely to cost.

9a. Sample answer: By counting the heights of the bars, there are 75 students represented in the graph. You can find the number of students in each 10-minute span. The students in the survey spent at least 11 minutes on the Internet. Eighty percent of students spent at least 31 minutes on the Internet. **9b.** Since exact data values are not listed, you are not able to find any measure of central tendency. However, the median is located in the 41-50 minute time period by counting the number of students in each time period.

11 **a.** There have been 25 summer Olympic games. In 52% of the games, the country that won the medal count won at least 100 medals.
b. The intervals along the x-axis are not equal. Because the intervals are not equal, the data is distributed unevenly.
c. The data could be rearranged into intervals of 50. Then the histogram would give a more accurate representation of the data.

13. Sample answer: 10, 12, 15, 19, 20, 25, 26, 31, 31, 31, 33, 36, 50, 52, 55, 57, 58, 59

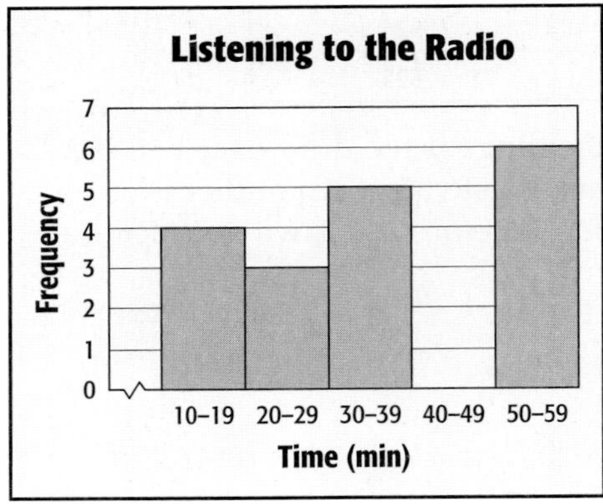

17. D

19.

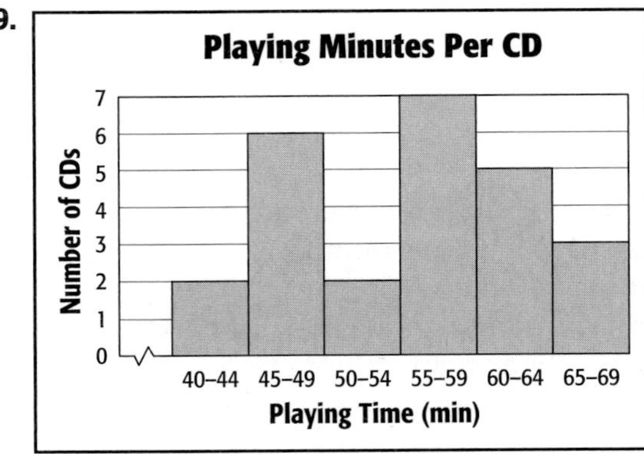

21. Student's Age

19 21 23 25 27 29 31 33 35 37 39 41 43

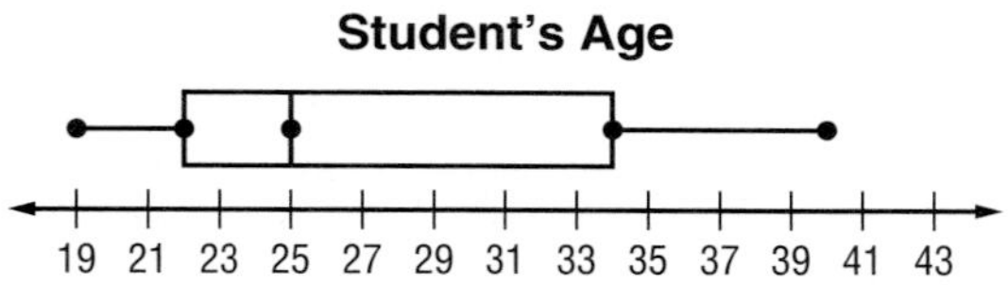

23. 32 students; 19 students **25.** 50 students

Pages 768–770 Lesson 13-6

1. $\frac{1}{10}$ or 10%. **3.** $\frac{2}{5}$ or 40%. **5.** $\frac{1}{4}$ or 25% **7.** 5:21

9 $P(20) = \frac{\text{number of sections labeled 20}}{\text{total number of sections}}$

$= \frac{1}{20}$

$= 5\%$

11. $\frac{1}{2}$ or 50% **13.** $\frac{1}{5}$ or 20%

15a.

(1, 1)	(1, 2)	(1, 3)	(1, 4)
(2, 1)	(2, 2)	(2, 3)	(2, 4)
(3, 1)	(3, 2)	(3, 3)	(3, 4)
(4, 1)	(4, 2)	(4, 3)	(4, 4)

15b. $\frac{1}{2}$ or 50% **15c.** $\frac{1}{4}$ or 25% **17.** 2:1 **19a.** 1140

19b. 1380

21 **a.** First, find the total number of students.
$8 + 6 + 4 + 15 + 7 = 40$
15 of the 40 students' favorite sport is skateboarding.
$\frac{15}{40} = \frac{3}{8}$ or 37.5%

b. Use the percent proportion.

$\frac{6}{40} = \frac{p}{100}$

$6 \cdot 100 = 40 \cdot p$

$600 = 40p$

$15 = p$

15% of the students chose inline skating.
Use the percent proportion to predict.

$\frac{p}{800} = \frac{15}{100}$

$p \cdot 100 = 800 \cdot 15$

$100p = 12{,}000$

$p = 120$

120 students should chose inline skating.

c. Use the percent proportion.
Students that prefer BMX:

$\frac{8}{40} = \frac{p}{100}$

$8 \cdot 100 = 40 \cdot p$

$800 = 40p$

$20 = p$

20% of the students chose inline skating.
Use the percent proportion to predict.

$\frac{p}{800} = \frac{20}{100}$

$p \cdot 100 = 800 \cdot 20$

$100p = 16{,}000$

$p = 160$

160 students should chose BMX. Students that prefer MotoX:

$\frac{4}{40} = \frac{p}{100}$

$4 \cdot 100 = 40 \cdot p$

$400 = 40p$

$10 = p$

20% of the students chose inline skating.
Use the percent proportion to predict.

$\frac{p}{800} = \frac{10}{100}$

$p \cdot 100 = 800 \cdot 10$

$100p = 8{,}000$

$p = 80$

80 students should chose MotoX.

$160 - 80 = 80$

So, 80 more students prefer BMX.

23. 48 **25.** Sample answer: The theoretical probability is $\frac{7}{35}$ or 20%. So, she would expect to select a navy pair of sock 4 out of 20 times. Since she selected a navy pair 6 times, her probability experimental of $\frac{6}{20}$ or 30%, exceeded her theoretical probability. **27.** F **29.** D
31a. Since SUVs average the least miles per gallon, they tend to be less fuel-efficient.
31b. The most fuel-efficient SUV and the least fuel-efficient sedan both average 22 miles per gallon. **33.** 900 **35.** 12,960

Pages 773–776 Lesson 13-7

1 This sample is biased and is a convenience sample because it is easy to survey students on one bus.
3. unbiased, stratified random sample; the shoppers are first divided into non-overlapping groups and then 1 male and 1 female is selected randomly from each group **5.** unbiased, simple random sample; the students are randomly selected

7 This sample is biased and is a voluntary response survey because those teenagers who are interested in participating in the survey are part of the sample.
9. This sampling method is not valid because it will include students who do not attend your

school. So, the results can not lead to a reasonable conclusion. **11.** Yes; this is a systematic random survey because the sample is selected according to an interval; 1225 **13.** No; only 80 out of 300 customers agree. From this random survey, you can predict that only about 27% of the customers would like a foreign movie section, so the store should not add such an area.

15 a. Sample answer: Send out a survey to all students is the school.
b. Sample Answer:
1. What is your favorite carnival game?
2. Which game would you play first?
3. If you could only play one game, which one would it be?

c. They could determine which games most students would play and which would raise the most money.

17. Sample answer: unbiased survey: Survey your classmates about their favorite sport.; See student's work. **19.** yes: Sample answer: If questions are asked in a neutral tone, then a more accurate answer can be expected. However, if the person asking the questions changes their tone of voice it can persuade someone to give an inaccurate response. **21.** B **23.** J **25.** $\frac{1}{6}$; 16.7% **27.** 146.9%; increase **29.** 23.9 **31.** 75% **33.** $66\frac{2}{3}$%

Pages 779–781 **Lesson 13-8**

1. 6 pairs

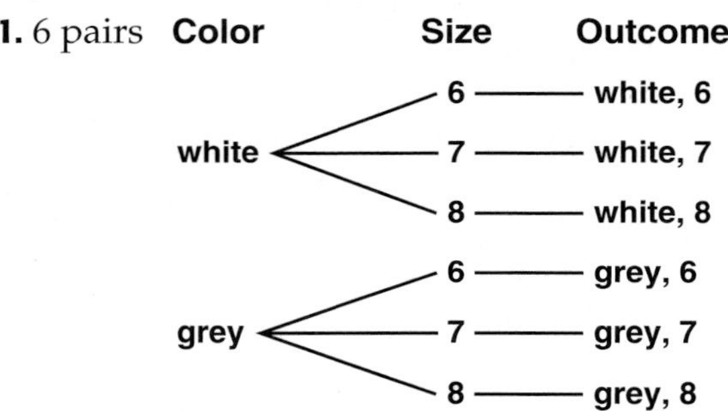

3. 144 **5.** $\frac{1}{64}$

7 8 outcomes

Question 1	Question 2	Question 3	Outcome
T	T	T	T, T, T
		F	T, T, F
	F	T	T, F, T
		F	T, F, F
F	T	T	F, T, T
		F	F, T, F
	F	T	F, F, T
		F	F, F, F

9. 24 **11.** 1000

13 a. P(same number on both dice) = P(1 on both dice) + P(2 on both dice) + P(3 on both dice) + P(4 on both dice) + P(5 on both dice) + P(6 on both dice)

P(1 on both dice)
$= \frac{1}{10} \cdot \frac{1}{6}$ or $\frac{1}{60}$

So, P(same number on both dice)
$= \frac{1}{60} + \frac{1}{60} + \frac{1}{60} + \frac{1}{60} + \frac{1}{60} + \frac{1}{60}$
$= \frac{6}{60}$ or $\frac{1}{10}$

b. P(odd, even) or P(odd, even)
$= \frac{5}{10} \cdot \frac{3}{6} + \frac{5}{10} \cdot \frac{3}{6}$
$= \frac{1}{4} + \frac{1}{4}$
$= \frac{1}{2}$

15a. $\frac{7}{16}$ **15b.** $\frac{1}{4}$ **17.** Neither girl is correct. Cameron added the possibilities and she should have multiplied them. Lisa multiplied the wrong possibilities. **19.** 5^x **21.** A **23.** C **25.** No; this is a biased sample. The students who participate in band are more likely to enjoy music, so more of them may say music is their favorite class, than compared to the entire school population. **27a.** 20; 36 **27b.** Chicken; whereas chicken sandwiches have 8–20 grams of fat, burgers have 10–36 grams of fat **27c.** chicken: R: 12; UQ: 18.5; LQ: 13; M: 15; IE: 5.5; outliers: none; burgers: R: 26; UQ: 31.5; LQ: 17; M: 23; IR: 14.5; outliers: none **29.** 6 **31.** 12

Pages 785–788 **Lesson 13-9**

1. 1,320 ways

3 Label her books A, B, C, D, E, and F
List all possibilities:

AB	AC	AD	AE	AF
~~BA~~	BC	BD	BE	BF
~~CA~~	~~CB~~	CD	CE	CF
~~DA~~	~~DB~~	~~DC~~	DE	DF
~~EA~~	~~EB~~	~~EC~~	~~ED~~	EF
~~FA~~	~~FB~~	~~FC~~	~~FD~~	~~FE~~

Cross out those that are the same. There are 15 possible ways.

5. 120 **7.** 504 ways **9.** 10 **11.** $\frac{1}{120}$ **13.** 360

15 $C(7, 7) = \frac{7 \cdot 6 \cdot 5 \cdot 4 \cdot 3 \cdot 2 \cdot 1}{7 \cdot 6 \cdot 5 \cdot 4 \cdot 3 \cdot 2 \cdot 1} = 1$

17. 240,240 **19.** 12,650 **21.** combination; 66 **23.** permutation; 5040 **25.** $\frac{1}{5}$

27 best of three: win first 2, win last 2, win first and last or C(3, 2): 3 ways best of five: $C(5, 3)$: $\frac{5 \cdot 4 \cdot 3}{3 \cdot 2 \cdot 1} = 10$ ways

best of seven: $C(7, 4)$: $\frac{7 \cdot 6 \cdot 5 \cdot 4}{4 \cdot 3 \cdot 2 \cdot 1} = 35$ ways

29a. 10,000 **29b.** Since there are 10,000 different PINs, there is only a $\frac{1}{10{,}000}$ chance that someone could guess the PIN. **31.** 15 teams **33.** Sample answer: the number of five-person committees that could be formed from a group of 15 people. **35.** mental math; $P(10, 10) = 10 \cdot P(9, 9)$ so $10 \times 362{,}880 = 3{,}628{,}800$. **37.** Sample answer: A combination could be used to find 9 batters from a baseball team with 20 players. A permutation would be used if the 9 batters are to bat in a particular order. in the first situation, the batters would hit in any order, while in the second they would hit in a determined order. **39.** G **41.** permutation; 154,440
43. biased, convenience sample; Sara is only surveying people who probably like chocolate and not other types of desserts

45. $a \geq 6$

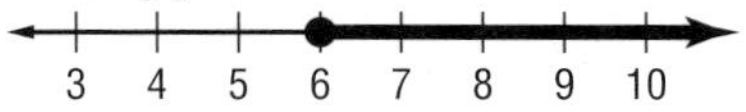

47. $y > -3$

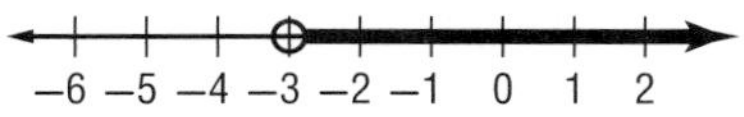

49. $\frac{1}{18}$ **51.** $\frac{1}{27}$

Pages 793–795 ***Lesson 13-10***

1. $\frac{1}{10}$ **3.** $\frac{7}{80}$

5 $P(\text{heads and } 4) = \frac{1}{2} \cdot \frac{1}{6} = \frac{1}{12}$

7. $\frac{1}{3}$ **9.** $\frac{1}{56}$ **11.** $\frac{1}{12}$ **13.** $\frac{1}{20}$ **15.** $\frac{4}{7}$ **17.** $\frac{1}{7}$

19. $\frac{1}{91}$

21 $P(\text{they match}) = P(\text{2 blue socks}) + P(\text{2 black socks}) + P(\text{2 white socks})$

$$= \frac{6}{24} \cdot \frac{5}{23} + \frac{8}{24} \cdot \frac{7}{23} + \frac{10}{24} \cdot \frac{9}{23}$$

$$= \frac{176}{552} \text{ or } \frac{22}{69}$$

23a. 0.2401 **23b.** 0.0576 **23c.** Independent; Making one field goal does not typically affect making another one.

25.

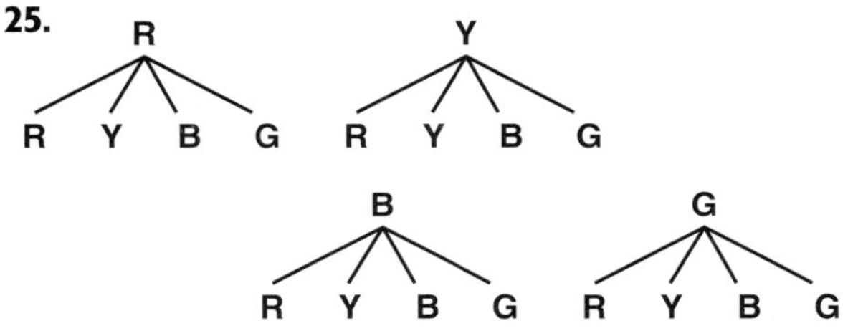

$\frac{1}{16}$; independent event **27.** 37.5% **29.** Shannon; rolling a difference of 0 or 1 are mutually exclusive events so the probability of rolling a difference of 0 must be added to the probability of rolling a difference of 1. **31.** A **33.** G
35. P; 720 ways **37.** C; 210 ways **39a.** 9
39b. $\frac{1}{9}$

Pages 798–802 ***Study Guide and Review***

1. mean **3.** tree diagram **5.** box-and-whisker plot **7.** range **9.** 22.4; 22; 22
11. 2.1; 2; 1.5, 2, 2.5

13. **Heights of Football Players**

Stem	Leaf
6	8 9
7	0 1 1 2 2 3 3 4 5 6

6|8 = 68 in.

15. Frank; whereas Frank's times range from 0.8-2.0 minutes, Shandra's times range from 1.3-2.1 minutes. **17.** R: 4; UQ: 8; LQ: 6; M: 6.5; IR 2; none **19.** The spread of the data is 41 points. The median is 82 points. One fourth of the students, earned 76 points or less. One fourth of the students earned 87.5 points or more. Half of the students earned between 76 and 87.5 points.

21.

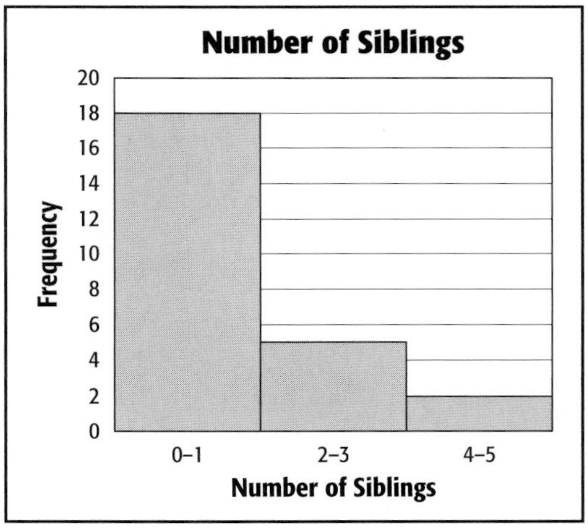

23. $\frac{1}{3}$ **25.** $\frac{13}{15}$ **27.** Biased, convenience sample; Diners at a steakhouse are more likely to choose steak as their favorite meal. **29.** 4 **31.** 216 **33.** 12 **35.** 59,280 combinations **37.** 26,334

39. $\frac{28}{153}$ **41.** $\frac{10}{153}$

Photo Credits

Cover i: (c) SW Productions/Brand X/CORBIS, (bkgd) Jason Reed/Photodisc/Getty Images; **v** Creatas/Punchstock; **P1** Jack Sullivan/Alamy; **P2** Jack Sullivan/Alamy; **P7** BananaStock/Jupiter Images; **P9** Bill Aron/PhotoEdit; **P10** Guy Motil/CORBIS; **P11** Robert W. Ginn/Alamy; **P12** Rick Rickman/NewSport/CORBIS; **P13** Matt King/Getty Images; **P15** Erik Dreyer/Taxi/Getty Images; **P16** H. Mark Weidman Photography/Alamy; **P21** Mira/Alamy; **2–3** Ron Chapple/Jupiter Images; **7** Ted Soqui/CORBIS; **11** Don Farrall/Getty Images; **12** Franck Guiziou/Hemis/CORBIS; **13** moodboard/CORBIS; **15** Anthony-Masterson/Getty Images; **18** ed-imaging; **22** Jason Reed/Reuters/CORBIS; **27** Stephen Frink/Getty Images; **29** Lee Foster/Alamy; **31** (l) G.K. & Vikki Hart/Getty Images, (r) Photodisc/PunchStock; **33** Stephen Frink/Science Faction/Getty Images; **34** vario images GmbH & Co. KG/Alamy; **36** Jeff Greenberg/PhotoEdit; **39** Mazer Creative Services; **41** ThinkStock LLC/Index Stock Imagery; **42** Dex Image/CORBIS; **44** Radius Images/Jupiter Images; **45** Superstudio/Iconica/Getty Images; **54** Colin Young-Wolff/PhotoEdit, Inc.; **58–59** Photo Network/Alamy; **62** Cosmo Condina/Stock Connection/Jupiter Images; **65** Erik Dreyer/Stone/Getty Images; **71** FLIP NICKLIN/MINDEN PICTURES; **73** Masterfile; **76** Brand X Pictures/PunchStock; **77** Cindy Charles/PhotoEdit; **79** David Stoecklein/CORBIS; **83** Stephen Frink/zefa/CORBIS; **84** © DoD/Roger-Viollet/The Image Work; **87** Visions of America/Joe Sohm/Getty Images; **90** Tyler Stableford/The Image Bank/Getty Images; **94** Peter Johnson/CORBIS; **97** Leonard de Selva/CORBIS; **99** Inti St Clair/Getty Images; **101** Bryan Rinnert/3Sight Photography; **105** AP Images; **116–117** David Papazian/Beateworks/CORBIS; (bkgd) Stockbyte/PunchStock; **121** Ed Kashi/IPN; **123** USDA; **124** Steve Hockstein/Star Ledger/CORBIS; **126** Eddie Gerald/Alamy; **132** British Museum, London, UK/Bridgeman Art Library; **136** KVLY; **137** AP Images; **143** Somos Images/CORBIS; **145** (l) Jeff Greenberg/PhotoEdit, (c and r) Larry Williams/CORBIS; **149** George Tiedemann/GT Images/CORBIS; **151** AP Images; **153** Dorling Kindersley/Getty Images; **155** Sonda Dawes/The Image Works; **157** Ian Shive/Aurora/Getty Images; **168–169** AP Images; **172** Gregg Newton/CORBIS; **175** Alan Schein Photography/CORBIS; **178** Duomo/CORBIS; **179** Pictorial Parade/Hulton Archive/Getty Images; **180** JupiterImages/Thinkstock/Alamy; **184** Thomas Northcut/Digital Vision/Getty Images; **186** Michael Newman/PhotoEdit; **188** Bill Heinsohn/Photographer's Choice/Getty Images; **191** Creatas/Jupiter Images; **192** Jeremy Woodhouse/Getty Images; **195** Howie Garber/Photographer's Choice/Getty Images; **199** Marvin E. Newman/Photographer's Choice/Getty Images; **201** Mario Tama/Getty Images; **203** Bob Daemmrich/PhotoEdit; **205** Edward Pond/Masterfile; **208** Kayte M. Deioma/PhotoEdit; **214** ed-imaging; **218–219** Picturenet/Blend Images/Getty Images; **221** David Madison/Getty Images; **222** The British Library/HIP/The Image Works; **224** Jeff Pape/New Bremen Giant Pumpkin Growers; **230** Bob Daemmrich/PhotoEdit; **232** Alaska Stock; **235** David McNew/Getty Images; **238** David Stluka/Getty Images; **243** Robert Harding/Getty Images; **245** Courtesy Ohio Expo Center; **246** Peter Dazeley/Photographer's Choice/Getty Images; **248** Tetra Images/Alamy; **249** Chris Farina/CORBIS; **252** AP Images; **258** Kelly Redinger/Design Pics/CORBIS; **262–263** Martin Beddall/Alamy; **267** (l) Buzzshotz/Alamy; (r) Martin Barraud/Riser/Getty Images; **268** Elsa/Getty Images; **270** AP Images; **271** Stuart Westmorland/Getty Images; **273** Karl Weatherly/CORBIS; **275** Nation Bill/CORBIS SYGMA; **277** Kevin Schafer/Riser/Getty Images; **279** Ralf-Finn Hestoft/CORBIS; **283** Michael Newman/PhotoEdit, Inc.; **289** moodboard/CORBIS; **291** Julie Hollern; **294** Steffen Schmidt/epa/CORBIS; **296** Dennis MacDonald/PhotoEdit; **297** Bob Thomas/CORBIS; **298** Thinkstock/Jupiter Images; **301** Digital Art/CORBIS; **301** Digital Art/CORBIS; **305** CORBIS; **307** David Young-Wolff/PhotoEdit; **310** Fendis/zefa/CORBIS; **315** Bettmann/CORBIS; **328–329** John Capella/Sports Imagery/Getty Images; **331** Photodisc/Getty Images; **332** The Stapleton Collection/The Bridgeman Art Library International; **333** Radius Images/Alamy; **335** Glowimages/Getty Images; **337** PureStock/Jupiter Images; **339** Don Mason/CORBIS; **341** SERGIO MORAES/Reuters/CORBIS; **345** Dasha Wright/Jupiter Images; **347** WireImage/Getty Images; **348** David Young-Wolff/PhotoEdit; **349** Joe Murphy/NBAE via Getty Image; **351** Comstock Images/Alamy; **352** ed-imaging; **353** Gabe Palmer/Alamy; **354** K. & H. Benser/zefa/CORBIS; **357** STAN HONDA/AFP/Getty Images; **359** Getty Images; **360** Brand X Pictures/Jupiter Images; **361** Bruno de Hogues/Sygma/CORBIS; **364** PhotoLink/Getty Images; **366** Robert Laberge/Getty Images; **367** Clive Rose/Getty Images; **368** Arthur Pollock/America 24-7/Getty Images; **370** David Young-Wolff/PhotoEdit; **373** ML Harris/Getty Images; **377** John Zoiner/Workbook Stock/Jupiter Images; **380** Clive Mason/Getty Images; **388** ed-imaging; **392–393** Art Wolfe/Getty Images; **396** State Central Artillery Museum, St. Petersburg, Russia Bridgeman Art Library; **399** Colin Young-Wolff/PhotoEdit;

401 Ablestock/Images.com/IPN; **402** Peter Mason/Getty Images; **404** AP Images; **406** CORBIS/PunchStock; **407** Michael Newman/PhotoEdit; **410** Lee Snider/The Image Works; **412** Darryl Leniuk/Getty Images; **413** STAN HONDA/AFP/Getty Images; **414** ed-imaging; **416** Colin Gray/Getty Images; **418** IPS Co., Ltd./Beateworks/CORBIS; **420** Envision/CORBIS; **423** GERARD LACZ/Animals Animals - Earth Scenes; **426** Horizons Companies; **427** Paul A. Souders/CORBIS; **429** Corey Rich/Aurora Photos; **433** Photo by Getty Images for John Deere; **435** Terry Whittaker/Photo Researchers, Inc.; **437** AP Images; **438** Clayton Sharrard/PhotoEdit; **439** Kevin Dodge/CORBIS; **443** Frances M. Roberts/Alamy; **446** Jon Hicks/CORBIS; **449** Benjamin Collier/Aurora Photos; **453** Kitt Cooper-Smith/Alamy; **456** David Young-Wolff/PhotoEdit; **468–469** Antonio M. Rosario/Stone/Getty Images; **471** ERWIN & PEGGY BAUER/Animals Animals - Earth Scenes; **472** AP Images; **474** AP Images; **476** BananaStock/Jupiter Images; **476** DreamPictures/Pam Ostrow/Blend; **479** The Granger Collection, New York; **481** RubberBall/Alamy; **483** Courtesy NSSDC Goddard Space Flight Center/NASA; **484** David Wrobel/Visuals Unlimited/Getty Images; **486** Brand X Pictures/PunchStock; **487** Art Wolfe/Stone/Getty Images; **489** Dana White/PhotoEdit; **490** Peter Miller/Iconica/Getty Images; **493** NASA; **494** MICHAEL & PATRICIA FOGDEN/Minden Pictures; **496** C Squared Studios/Getty Images; **497** Image Source/PunchStock; **501** Micro Discovery/CORBIS; **506** Hans Neleman/Stone/Getty Images; **513** James Marshall/CORBIS; **516** GK Hart/Vikki Hart/Getty Images; **517** Sean Justice/Riser/Getty Images; **519** S. Lowry/Univ Ulster/Stone/Getty Images; **521** ed-imaging; **528** ed-imaging; **532–533** Mark A. Johnson/CORBIS; **537** Brad Perks Lightscapes/Alamy; **538** Courtesy of Inter University Centre for Astronomy and Astrophysics/Dinodia; **539** dk/Alamy; **540** Cedar Point Amusement Park/Resort; **541** AP Images; **543** Cedar Point Amusement Park/Resort; **545** Peter Poby/zefa/CORBIS; **547** Getty Images; **550** Steve Vidler/SuperStock; **551** DEA PICTURE LIBRARY/Getty Images; **553** Martin Harvey/Alamy; **554** REUTERS/Mike Blake; **562** Chris A Crumley/Alamy; **566** Bilderbuch/Design Pics/CORBIS; **569** Jeff Greenberg/Alamy; **573** Todd Bigelow/Aurora Photos; **575** david pearson/Alamy; **579** Photodisc/Getty Images **582** ed-imaging; **586–587** Creatas Images/Jupiter Images; **593** (l) AP Images; **593** (r) Richard Cummins/SuperStock; **594** Comstock Images/PunchStock; **598** Tim Gainey/Alamy; **600** (l) Inmagine/Jupiter Images, (r) Darren Baker/Alamy; **602** (l) Robert Harding/Getty Images, (r) Courtesy National Gallery of Art, Washington DC; © 2010 Estate of Tony Smith/Artist Rights Society, NY; **603** UpperCut Images/Alamy; **605** Eric Nathan/PhotoShelter; **609** (l) Al Bello/Getty Images, (cl) Dennis O'Clair/Stone/Getty Images, (cr) Mike Powell/Getty Images, (r) Pete Saloutos/UpperCut Images/Getty Images; **613** John Kolesidis/Reuters/CORBIS; **615** (l) Oxford Scientific/Jupiter Images, (r) Hugh Lansdown/Alamy; **619** Comstock Images/Alamy; **621** The Granger Collection, New York; **627** Adam Crowley/Getty Images; **633** Stuart Westmorland/Riser/Getty Images; **634** Gregory Dunham; **636** plainpicture/Bornkessel, E./Jupiter Images; **637** David Stluka/Getty Images; **640** PureStock/Jupiter Images; **642** Stockbyte; **645** Hiep Vu/Masterfile; **646** Leo Keeler/Alamy; **652–653** Photodisc/Getty Images; **653** Photodisc Collection/Getty Images; **656** Bob Rowan; Progressive Image/CORBIS; **660–661** Andrea Jones/Alamy; **664** Kelly Redinger/Design Pics/CORBIS; **666** (l) Classic Image/Alamy, (r) Lester V. Bergman/CORBIS; **668** (t) Brand X Pictures/PunchStock, (bl) SuperStock, Inc./SuperStock/The Isamu Noguchi Foundation and Garden Museum, NY Artist Rights Society, NY, (bc) CORBIS/SuperStock, (br) Walter Bibikow/CORBIS; **670** ed-imaging; **672** EPA; **675** Brand X Pictures/Jupiter Images; **677** WidStock/Alamy; **678** James P. Blair/CORBIS; **679** Chloe Johnson/Alamy; **681** (l) Focal Depth/Alamy, (r) Matthias Kulka/zefa/CORBIS; **685** Frances Roberts/Alamy; **686** (l) Design and Photo © Barbara Butler; www.barbarabutler.com, (c) Getty Images/Photodisc, (r) FoodPhotography Eising/Stockfood; **687** Ed-Imaging; **692** Duomo/CORBIS; **694** Image Source Black/Alamy; **698** AP Images; **699** (t) Tannen Maury/epa/CORBIS, (b) Stockbyte Silver/Alamy; **700** CORBIS; **702** Jack Sullivan/Alamy; **703** The Granger Collection, New York; **706** Ron Chapple Stock/Alamy; **709** Rodolfo Arpia/Alamy; **712** Nik Wheeler/CORBIS; **713** Burke/Triolo Productions/Getty Images; **714** Tania Reinicke/Bilderberg/Aurora Photos; **722** Will & Deni McIntyre/CORBIS; **726–727** Elizabeth Kreutz/NewSport/CORBIS; **729** CORBIS/PunchStock; **732** Richard Klune/CORBIS; **734** Neal Preston/CORBIS; **738** Greg Von Doersten/Aurora/Getty Images; **740** William Perlman/Star Ledger/CORBIS; **741** Michael Zito/SportsChrome, Inc; **745** Hemera Technologies/Jupiter Images; **746** Geoff du Feu/Alamy; **748** Toshio Hoshi/Jupiter Images; **752** Bruce Bennett/Getty Images; **753** StockTrek/Getty Images; **754** Reuters/CORBIS; **760** Spencer Grant/PhotoEdit; **761** BOCOG - HANDOUT/epa/CORBIS; **765** Brian Hagiwara/Jupiter Images; **766** Roger-Viollet/The Image Works; **769** Jeff Gross/Getty Images; **772** David Young-Wolff/PhotoEdit; **774** Lester Lefkowitz/CORBIS; **775** MIXA/Getty Images; **778** Marc Romanelli/Photographer's Choice/Getty Images; **780** Raquel Ramirez/PhotoEdit; **784** Brand X Pictures; **787** Jeff Gross/Getty Images; **792** AP Images; **794** Zefa RF/Alamy; **796** ed-imaging; **804** ed-imaging; **808** Eclipse Studios; **LA0–LA1** TOM MARESCHAL/Alamy; **LA2** Burke/Triolo/Brand X Pictures/Jupiterimages, (inset) Ryan McVay/Getty Images, (inset) The McGraw-Hill Companies, Inc./Ken Cavanagh photographer; **LA6** Steve Hamblin/Alamy; **LA7** Ingolf Pompe/Aurora Photos; **LA9** David Young-Wolff/PhotoEdit; **LA14** Don Hammond/DesignPics/IPN; **LA21** Lew Robertson/Brand X/CORBIS; **LA22** John W. Banagan/Getty Images; **LA28** Christian Liewig/Liewig Media Sports/CORBIS

Glossary/Glosario

Math Online

A mathematics multilingual eGlossary is available at glencoe.com. The glossary includes the following languages:

Arabic	Haitian Creole	Portuguese	Tagalog
Bengali	Hmong	Russian	Urdu
Cantonese	Korean	Spanish	Vietnamese
English			

Cómo usar el glosario en español:

1. Busca el término en inglés que desees encontrar.
2. El término en español, junto con la definición, se encuentran en la columna de la derecha.

English

A

absolute value (p. 63) The distance a number is from zero on the number line.

accuracy (p. 879) The degree of conformity of a measurement with the true value.

acute angle (p. 551) An angle with a measure greater than 0° and less than 90°.

acute triangle (p. 552) A triangle that has three acute angles.

Addition Property of Equality (p. 184) If you add the same number to each side of an equation, the two sides remain equal.

additive inverses (p. 71) An integer and its opposite.

adjacent angles (p. 590) Two angles that have the same vertex, share a common side, and do not overlap.

algebra (p. 11) A branch of mathematics dealing with symbols.

algebraic expression (p. 11) An expression that contains sums and/or products of variables and numbers.

alternate exterior angles (p. 590) Nonadjacent exterior angles found on opposite sides of the transversal. In the figure below, ∠1 and ∠7, ∠2 and ∠8 are alternate exterior angles.

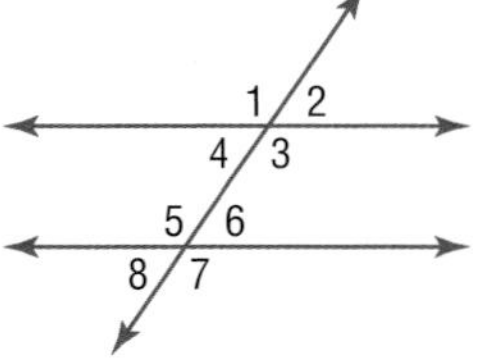

Español

valor absoluto Distancia que un número dista de cero en la recta numérica.

exactitud Grado de conformidad de una medida con el valor verdadero.

ángulo agudo Ángulo con una medida mayor que 0° y menor que 90°.

triángulo acutángulo Triángulo que posee tres ángulos agudos.

Propiedad de adición de la igualdad Si sumas el mismo número a ambos lados de una ecuación, los dos lados permanecen iguales.

inverso aditivo Un entero y su opuesto.

ángulos adyacentes Dos ángulos que poseen el mismo vértice, comparten un lado y no se traslapan.

álgebra Rama de las matemáticas que tiene que ver con signos.

expresión algebraica Expresión que contiene sumas y/o productos de números y variables.

ángulos alternos externos Ángulos exteriores no adyacentes que se encuentran en lados opuestos de una transversal. En la siguiente figura, ∠1 y ∠7, ∠2 y ∠8 son ángulos alternos externos.

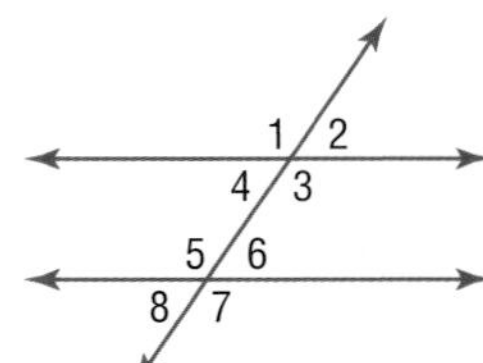

alternate interior angles (p. 590) Nonadjacent interior angles found on opposite sides of the transversal. In the figure on page R63, $\angle 4$ and $\angle 6$, $\angle 3$ and $\angle 5$ are alternate interior angles.

ángulos alternos internos Ángulos interiores no adyacentes que se encuentran en lados opuestos de una transversal. En la figura anterior, $\angle 4$ y $\angle 6$, $\angle 3$ y $\angle 5$ son ángulos alternos internos.

altitude (p. 624) A line segment that is perpendicular to the base of a figure with endpoints on the base and the side opposite the base.

altura Segmento de recta perpendicular a la base de una figura y cuyos extremos yacen en la base y en el lado opuesto de la base.

angle (p. 868) Two rays with a common endpoint form an angle. The rays and vertex are used to name an angle. The angle below is $\angle ABC$.

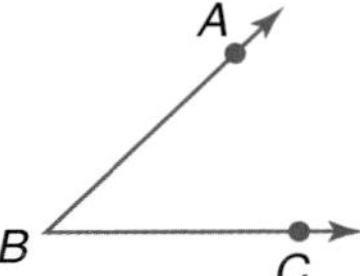

ángulo Dos rayos con un punto común forman un ángulo. Los rayos y el vértice se usan para identificar el ángulo. El siguiente ángulo es $\angle ABC$.

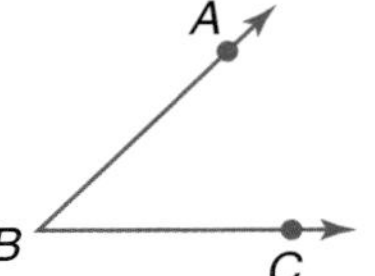

angle bisector (p. 872) A ray that divides an angle into two congruent angles.

bisectriz de un ángulo Semirrecta que divide un ángulo en dos ángulos congruentes.

area (p. 222) The measure of the surface enclosed by a geometric figure.

área Medida de la superficie que encierra una figura geométrica.

arithmetic sequence (p. 401) A sequence in which the difference between any two consecutive terms is the same.

sucesión aritmética Sucesión en que la diferencia entre dos términos consecutivos cualesquiera es siempre la misma.

Associative Property (p. 18) The way in which numbers are grouped when added or multiplied does not change the sum or product.

Propiedad asociativa La forma en que se suman o multiplican dos números no altera su suma o producto.

average (p. 92) The sum of data divided by the number of items in the data set, also called the mean.

promedio Suma de los datos dividida entre el número de elementos en el conjunto de datos. También llamado media.

B

back-to-back stem-and-leaf plot (p. 738) Used to compare two sets of data. The leaves for one set of data are on one side of the stem and the leaves for the other set of data are on the other side.

diagrama de tallo y hojas consecutivo Se usa para comparar dos conjuntos de datos. Las hojas de uno de los conjuntos de datos aparecen en un lado del tallo y las del otro al otro lado de éste.

bar graph (p. 882) A graphic form using bars to make comparisons of statistics.

gráfica de barras Tipo de gráfica que usa barras para comparar estadísticas.

bar notation (p. 122) In repeating decimals, the line or bar placed over the digits that repeat. For example, $2.\overline{63}$ indicates the digits 63 repeat.

notación de barra En decimales periódicos, la línea o barra que se escribe encima de los dígitos que se repiten. Por ejemplo, en $2.\overline{63}$ la barra encima del 63 indica que los dígitos 63 se repiten.

base (p. 471) In 2^4, the base is 2. The base is used as a factor as many times as given by the exponent (4). That is, $2^4 = 2 \times 2 \times 2 \times 2$.

base En 2^4, la base es 2. La base se usa como factor las veces que indique el exponente (4). Es decir, $2^4 = 2 \times 2 \times 2 \times 2$.

base (p. 624) The base of a parallelogram or a triangle is any side of the figure. The bases of a trapezoid are the parallel sides.

base La base de un paralelogramo o de un triángulo es cualquier lado de la figura. Las bases de un trapecio son los lados paralelos.

base (p. 665) The bases of a prism are any two parallel congruent faces.

biased sample (p. 771) A sample that is not representative of a population.

binomial (p. LA12) A polynomial with exactly two terms.

boundary (p. 866) A line or curve that separates the coordinate plane into regions.

box-and-whisker plot (p. 750) A diagram that divides a set of data into four parts using the median and quartiles. A box is drawn around the quartile values and whiskers extend from each quartile to the extreme data points.

base Las bases de un prisma son cualquier par de caras paralelas y congruentes.

muestra sesgada Una muestra que no es representativa de una población.

binomio Polinomio con exactamente dos términos.

frontera Recta o curva que divide el plano de coordenadas en regiones.

diagrama de caja y patillas Diagrama que divide un conjunto de datos en cuatro partes usando la mediana y los cuartiles. Se dibuja una caja alrededor de los cuartiles y se extienden patillas de cada uno de ellos a los valores extremos.

C

cell (p. 17) A box within a spreadsheet.

center (p. 631) The given point from which all points on the circle are the same distance.

center of rotation (p. 605) A fixed point around which shapes move in a circular motion to a new position.

central angle (p. 638) An angle whose vertex is the center of a circle and whose sides intersect the circle.

chord (p. 631) A segment with endpoints that are on a circle.

circle (p. 631) The set of all points in a plane that are the same distance from a given point called the center.

circle graph (p. 376) A type of statistical graph used to compare parts of a whole.

circumference (p. 631) The distance around a circle.

coefficient (p. 178) The numerical part of a term that contains a variable.

combination (p. 784) An arrangement or listing in which order is not important.

common difference (p. 401) The difference between any two consecutive terms in an arithmetic sequence.

common multiples (p. 860) Multiples that are shared by two or more numbers. For example, some common multiples of 4 and 6 are 0, 12, and 24.

Commutative Property (p. 18) The order in which numbers are added or multiplied does not change the sum.

celda Casilla dentro de una hoja de cálculos.

centro Punto dado del cual equidistan todos los puntos de un círculo.

centro de rotación Punto fijo alrededor del cual una figura gira con un movimiento circular hasta alcanzar una nueva posición.

ángulo central Ángulo cuyo vértice es el centro de un círculo y cuyos lados intersecan el círculo.

cuerda Segmento cuyos extremos están sobre un círculo.

círculo Conjunto de todos los puntos del plano que están a la misma distancia de un punto dado del plano llamado centro.

gráfica circular Tipo de gráfica estadística que se usa para comparar las partes de un todo.

circunferencia Longitud del contorno de un círculo.

coeficiente Parte numérica de un término que contiene una variable.

combinación Arreglo o lista en que el orden no es importante.

diferencia común Diferencia entre dos términos consecutivos cualesquiera de una sucesión aritmética.

múltiplos comunes Múltiplos compartidos por dos o más números. Por ejemplo, algunos múltiplos comunes de 4 y 6 son 0, 12 y 24.

Propiedad conmutativa La forma en que se suman o multiplican dos números no altera su suma o producto.

complementary angles (p. 589) Two angles are complementary if the sum of their measures is 90°.

composite figure (p. 644) A figure that is made up of two or more shapes.

composite number (p. 476) A whole number that has more than two factors.

compound event (p. 790) Two or more simple events.

compound interest (p. 371) Interest paid on the initial principal and on interest earned in the past.

cone (p. 665) A three-dimensional figure with one circular base. A curved surface connects the base and vertex.

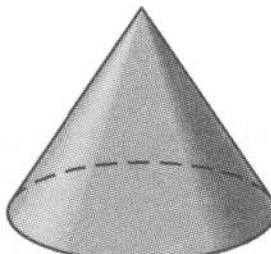

congruent (p. 301, 552, 598) Line segments that have the same length, or angles that have the same measure, or figures that have the same size and shape.

congruent angles (p. 871) Two angles that have the same measure.

congruent segments (p. 870) Line segments that have the same length.

constant (p. 178) A term without a variable.

constant of proportionality (p. 282) A constant ratio or unit rate of a proportion.

constant of variation (p. 420) The slope, or rate of change, in the equation $y = kx$, represented by k.

constant rate of change (p. 418) The rate of change between any two data points in a linear relationship is the same or constant.

convenience sample (p. 772) A sample which includes members of the population that are easily accessed.

converse (p. 560) The statement formed by reversing the phrases after *if* and *then* in an if-then statement.

converse of the Pythagorean Theorem (p. 560) The reversal of the *if* and *then* statement that forms the Pythagorean Theorem.

coordinate (p. 62) A number that corresponds with a point on a number line.

ángulos complementarios Dos ángulos son complementarios si la suma de sus medidas es 90°.

figura compleja Figura compuesta de dos o más formas.

número compuesto Número entero que posee más de dos factores.

evento compuesto Dos o más eventos simples.

interés compuesto Interés que se paga sobre el capital inicial y sobre el interés que se haya ganado en el pasado.

cono Figura tridimensional con una base circular, la cual posee una superficie curva que une la base con el vértice.

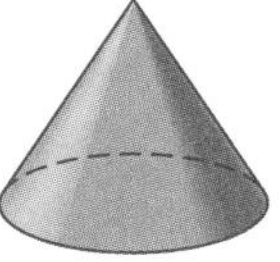

congruentes Segmentos de recta que tienen la misma longitud o ángulos que tienen la misma medida o figuras que poseen la misma forma y tamaño.

ángulos congruentes Ángulos que tienen la misma medida.

segmentos congruentes Segmentos de recta que tienen la misma longitud.

constante Término sin variables.

constante de proporcionalidad La razón constante o tasa unitaria de una proporción.

constante de variación La pendiente, o tasa de cambio, en la ecuación $y = kx$, representada por k.

tasa constante de cambio La tasa de cambio entre dos puntos cualesquiera en una relación lineal permanece constante o igual.

muestra de conveniencia Muestra que incluye miembros de una población fácilmente accesibles.

recíproca Un enunciado que se forma intercambiando los enunciados que vienen a continuación de *si-entonces* en un enunciado *si-entonces*.

recíproco del Teorema de Pitágoras El intercambio de las frases del enunciado *si-entonces* que forman el Teorema de Pitágoras.

coordenada Número que corresponde a un punto en la recta numérica.

coordinate plane (p. 25) Another name for the coordinate system.

coordinate system (p. 25) A coordinate system is formed by the intersection of two number lines that meet at right angles at their zero points, also called a coordinate plane.

corresponding angles (p. 590) Angles that have the same position on two different parallel lines cut by a transversal. In the figure, $\angle 1$ and $\angle 5$, $\angle 2$ and $\angle 6$, $\angle 3$ and $\angle 7$, $\angle 4$ and $\angle 8$ are corresponding angles.

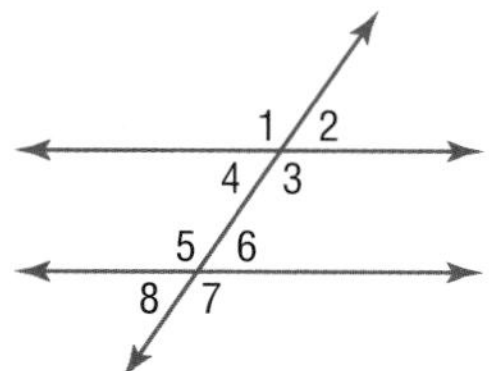

corresponding parts (pp. 301, 598) Parts of congruent or similar figures that match.

cosine (p. 877) For an acute angle of a right triangle, the ratio of the measure of the leg adjacent to the acute angle to the measure of the hypotenuse.

counterexample (p. 19) An example that shows a conjecture is not true.

cross products (p. 287) If $\frac{a}{c} = \frac{b}{d}$, then $ad = bc$. If $ad = bc$, then $\frac{a}{c} = \frac{b}{d}$.

cross section (p. 666) The intersection of a solid and a plane.

cube root (p. 862) A number that can be raised to the third power to create another number.

cubic function (p. 516) A function that can be described by an equation of the form $y = ax^3 + bx^2 + cx + d$, where $a \neq 0$.

cylinder (p. 665) A solid that has two parallel, congruent bases (usually circular) connected with a curved side.

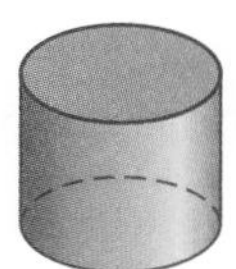

plano de coordenadas Otro nombre para el sistema de coordenadas.

sistema de coordenadas Un sistema de coordenadas se forma de la intersección de dos rectas numéricas perpendiculares que se intersecan en sus puntos cero. También llamado plano de coordenadas.

ángulos correspondientes Ángulos que tienen la misma posición en dos rectas paralelas distintas cortadas por una transversal. En la figura, $\angle 1$ y $\angle 5$, $\angle 2$ y $\angle 6$, $\angle 3$ y $\angle 7$, $\angle 4$ y $\angle 8$ son ángulos correspondientes.

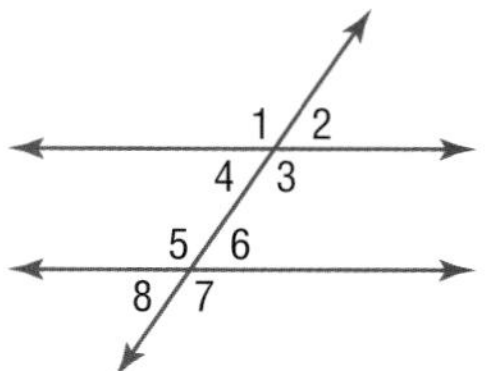

partes correspondientes Partes de figuras congruentes o semejantes que se corresponden mutuamente.

coseno Para un ángulo agudo de un triángulo rectángulo, la razón de la medida del cateto adyacente al ángulo agudo la medida de la hipotenusa.

contraejemplo Ejemplo que muestra que una conjetura no es verdadera.

productos cruzados Si $\frac{a}{c} = \frac{b}{d}$, entonces $ad = bc$. Si $ad = bc$, entonces $\frac{a}{c} = \frac{b}{d}$.

seccion transversal Intersección de un sólido con up plano.

raíz cúbica Número que se puede elevar a la tercera potencia para crear otro número.

función cúbica Función que puede describirse por una ecuación de la forma $y = ax^3 + bx^2 + cx + d$, donde $a \neq 0$.

cilindro Sólido que posee dos bases congruentes y paralelas (por lo general circulares) unidas por un lado curvo.

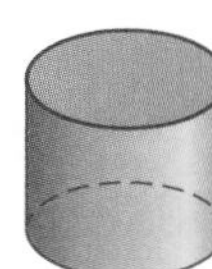

D

data (p. 882) Pieces of information, which are often numerical.

datos Información, la cual a menudo se presenta de manera numérica.

deductive reasoning (p. 20) The process of using facts, properties, or rules to justify reasoning or reach valid conclusions.

razonamiento deductivo Proceso de usar hechos, propiedades o reglas para justificar un razonamiento o para sacar conclusiones válidas.

defining a variable (p. 11) Choosing a variable and a quantity for the variable to represent in an equation.

definir una variable Seleccionar una variable y una cantidad para la variable que represente en la ecuación.

degree (p. 873) The most common unit of measure for angles.

grado La unidad de medida angular más común.

dependent events (p. 791) Two or more events in which the outcome of one event does affect the outcome of the other event(s).

eventos dependientes Dos o más eventos en que el resultado de uno de ellos afecta el resultado del otro o de los otros eventos.

dependent variable (p. 395) The variable in a relation with a value that depends on the value of the independent variable.

variable dependiente La variable en una relación cuyo valor depende del valor de la variable dependiente.

diagonal (p. 618) A line segment that joins two nonconsecutive vertices of a polygon.

diagonal Segmento de recta que une dos vértices no consecutivos de un polígono.

diameter (p. 631) The distance across a circle through its center.

diámetro La distancia a través de un círculo pasando por el centro.

dilation (p. 307) A transformation that alters the size of a figure but not its shape.

homotecia Transformación que altera el tamaño de una figura, pero no su forma.

dimensional analysis (p. 275) The process of including units of measurement when computing.

análisis dimensional Proceso que incorpora las unidades de medida al hacer cálculos.

dimensions (p. 886) The number of rows and columns in a matrix, with the number of rows stated first.

dimensiones El numero de filas y columnas que hay en una matriz, con el número de filas indicó primero.

direct variation (p. 420) A special type of linear equation that describes rate of change. A relationship such that as x increases in value, y increases or decreases at a constant rate.

variación directa Tipo especial de ecuación lineal que describe tasas de cambio. Relación en que a medida que x aumenta de valor, y aumenta o disminuye a una tasa constante.

discount (p. 366) The amount by which the regular price of an item is reduced.

descuento Cantidad por la que se reduce el precio normal de un artículo.

Distance Formula (p. 565) The distance between two points, with coordinates (x_1, y_1) and (x_2, y_2), is given by $d = \sqrt{(x_2 - x_1)^2 + (y_2 - y_1)^2}$.

Fórmula de la distancia La distancia entre dos puntos, con coordenadas (x_1, y_1) y (x_2, y_2), se calcula con $d = \sqrt{(x_2 - x_1)^2 + (y_2 - y_1)^2}$.

Distributive Property (p. 171) To multiply a sum by a number, multiply each number in parentheses by the number outside the parentheses.

Propiedad distributiva Para multiplicar una suma por un número, multiplica cada número en paréntesis por el número fuera del paréntesis.

divisible (p. 856) A number is divisible by another if, upon division, the remainder is zero.

divisible Un número es divisible entre otro si, al dividirlos, el residuo es cero.

Division Property of Equality (p. 191) When you divide each side of an equation by the same nonzero number, the two sides remain equal.

Propiedad de igualdad de la división Cuando divides ambos lados de una ecuación entre el mismo número no nulo, los dos lados permanecen iguales.

domain (p. 27) The domain of a relation is the set of all x-coordinates from each pair.

dominio El dominio de una relación es el conjunto de coordenadas x de todos los pares.

E

edge (p. 664) Where two planes intersect in a line.

arista Recta en donde se intersecan dos planos.

element (p. 886) Each entry in a matrix.

elemento Cada entrada de una matriz.

equation (pp. 34, 184) A mathematical sentence that contains an equals sign (=).

ecuación Enunciado matemático que contiene el signo de igualdad (=).

equilateral triangle (p. 552) A triangle with all sides congruent.

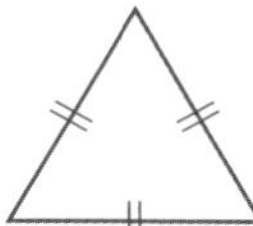

triángulo equilátero Un triángulo cuyos lados son todos congruentes.

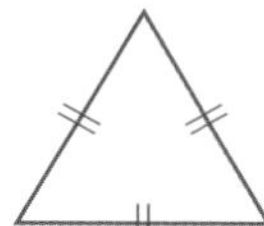

equivalent equations (p. 184) Two or more equations with the same solution. For example, $x + 4 = 7$ and $x = 3$ are equivalent equations.

ecuaciones equivalentes Dos o más ecuaciones con las mismas soluciones. Por ejemplo, $x + 4 = 7$ y $x = 3$ son ecuaciones equivalentes.

equivalent expressions (p. 171) Expressions that have the same value.

expresiones equivalentes Expresiones que tienen el mismo valor.

evaluate (p. 6) Find the numerical value of an expression.

evaluar Calcular el valor numérico de una expresión.

experimental probability (p. 766) What actually occurs in a probability experiment.

probabilidad experimental Lo que realmente sucede en un experimento probabilístico.

exponent (p. 471) In 2^4, the exponent is 4. The exponent tells how many times the base, 2, is used as a factor. So, $2^4 = 2 \times 2 \times 2 \times 2$.

exponente En 2^4, el exponente es 4. El exponente indica cuántas veces se usa la base, 2, como factor. Así, $2^4 = 2 \times 2 \times 2 \times 2$.

exponential function (p. 516) A function that can be described by an equation of the form $y = a^x$, where $a > 0$ and $a \neq 1$.

función exponencial Función que puede describirse mediate una ecuación de la forma $y = a^x$, donde $a > 0$ y $a \neq 1$.

exterior angles (p. 590) Four of the angles formed by the transversal and two parallel lines. Exterior angles lie outside the two parallel lines.

ángulos exteriores Cuatro de los ángulos formados por una transversal y dos rectas paralelas. Los ángulos exteriores yacen fuera de las dos rectas paralelas.

F

face (p. 664) A flat surface, the side or base of a prism.

cara plana, el lado o la base de un prisma.

factor (p. 478) To write a number as a product of its factors.

factorizar Escribir un número como el producto de sues propios factores.

factored form (p. LA24) A monomial expressed as a product of prime numbers and variables and no variable has an exponent greater than 1.

forma reducida Monomio escrito como el producto de números primos y variables y en el que ninguna variable tiene un exponente mayor que 1.

factoring (p. LA24) To express a polynomial as the product of monomials and polynomials.

factorización La escritura de un polinomio como producto de monomios y polinomios.

factors (p. 856) Two or more numbers that are multiplied to form a product.

factores Dos o más números que se multiplican para formar un producto.

factor tree (p. 477) A way to find the prime factorization of a number. The factors branch out from the previous factors until all the factors are prime numbers.

árbol de factores Forma de encontrar la factorización prima de un número. Los factores se ramifican de los factores anteriores hasta que todos los factores son números primos.

family of functions (p. 439) A set of functions that is related in some way.

familia de funciones Conjunto de funciones con características similares.

FOIL method (p. LA18) To multiply two binomials, find the sum of the products of the *F*irst terms, the *O*uter terms, the *I*nner terms, and the *L*ast terms.

método FOIL Para multiplicar dos binomios, busca la suma de los productos de los primeros (*F*irst) términos, los términos exteriores (*O*uter), los términos interiores (*I*nner) y los últimos términos (*L*ast).

formula (p. 221) An equation that shows a relationship among certain quantities.

fórmula Ecuación que muestra la relación entre ciertas cantidades.

frequency table (p. 757) A chart that indicates the number of values in each interval.

tabla de frecuencias Tabla que indica el número de valores en cada intervalo.

function (p. 33) A function is a special relation in which each element of the domain is paired with exactly one element in the range.

función Una función es una relación especial en que a cada elemento del dominio le corresponde un único elemento del rango.

function notation (p. 396) A way to name a function that is defined by an equation. In function notation, the equation $y = 3x - 8$ is written as $f(x) = 3x - 8$.

notación funcional Una manera de nombrar una función definida por una ecuación. En notación funcional, la ecuación $y = 3x - 8$ se escribe $f(x) = 3x - 8$.

function rule (p. 33) The operation performed on the input of a function.

regla de función Operación que se efectúa en el valor de entrada.

function table (p. 33) A table organizing the input, rule, and output of a function.

tabla de funciones Tabla que organiza las entradas, la regal y las salidas de una función.

Fundamental Counting Principle (p. 777) If event M can occur in m ways and is followed by event N that can occur in n ways, then the event M followed by event N can occur in $m \cdot n$ ways.

Principio fundamental de contar Si el evento M puede ocurrir de m maneras y lo sigue un evento N que puede ocurrir de n maneras, entonces el evento M seguido del evento N puede ocurrir de $m \cdot n$ maneras.

G

graph (p. 26) A dot at the point that corresponds to an ordered pair on a coordinate plane.

gráfica Marca puntual en el punto que corresponde a un par ordenado en un plano de coordenadas.

greatest common factor (GCF) (p. 858) The greatest number that is a factor of two or more numbers.

máximo común divisor (MCD) El número mayor que es factor de dos o más números.

H

half-plane (p. 866) The region of the graph of an inequality on one side of a boundary.

semiplano Región de la gráfica de una desigualdad en un lado de la frontera.

histogram (p. 757) A histogram uses bars to display numerical data that have been organized into equal intervals.

histograma Un histograma usa barras para exhibir datos numéricos que han sido organizados en intervalos iguales.

hypotenuse (p. 558) The side opposite the right angle in a right triangle.

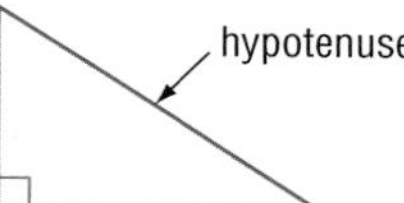

hipotenusa Lado opuesto al ángulo recto en un triángulo rectángulo.

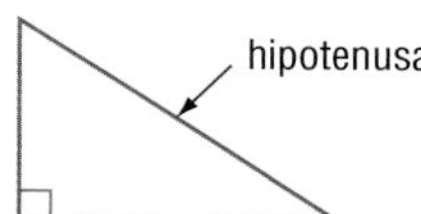

I

identity (p. 250) An equation that is true for every value of the variable.

identidad Ecuación que es verdadera para cada valor de la variable.

image (p. 101) Every corresponding point on a figure after its transformation.

imagen Todo punto correspondiente de una figura después de ser transformación.

independent events (p. 790) Two or more events in which the outcome of one event does not influence the outcome of the other event(s).

eventos independientes Dos o más eventos en que el resultado de uno de ellos no afecta el resultado del otro o de los otros eventos.

independent variable (p. 395) The variable in a function with a value that is subject to choice.

variable independiente La variable de una función sujeta a elección.

indirect measurement (p. 313) Using the properties of similar triangles to find measurements that are difficult to measure directly.

medición indirecta Uso de las propiedades de triángulos semejantes para hacer mediciones que son difíciles de realizar directamente.

inequality (pp. 62, 234) A mathematical sentence that contains $<$, $>$, $\neq$, $\leq$, or $\geq$.

desigualdad Enunciado matemático que contiene $<$, $>$, $\neq$, $\leq$, o $\geq$.

integer (p. 61) The whole numbers and their opposites.

$$\ldots, -3, -2, -1, 0, 1, 2, 3, \ldots$$

enteros Los números enteros y sus opuestos.

$$\ldots, -3, -2, -1, 0, 1, 2, 3, \ldots$$

interest (p. 370) The amount of money paid or earned for the use of money.

interés Cantidad que se cobra o se paga por el uso del dinero.

interior angle (p. 618) An angle inside a polygon.

ángulo interno ángulo ubicado dentro de un polígono.

interior angles (p. 590) Four of the angles formed by the transversal and two parallel lines. Interior angles lie between the two parallel lines.

ángulos interiores Cuatro de los ángulos formados por una transversal y dos rectas paralelas. Los ángulos interiores están entre las dos rectas paralelas.

interquartile range (p. 744) The range of the middle half of a set of data. It is the difference between the upper quartile and the lower quartile.

amplitud intercuartílica Amplitud de la mitad central de un conjunto de datos. Es la diferencia entre el cuartil superior y el inferior.

inverse operation (p. 184) Operation that undoes another, such as addition and subtraction.

operaciones inversas Operaciones que se anulan mutuamente, como la adición y la sustracción.

inverse proportion (p. 293) A relationship formed when the product of two variables is a constant.

proporción inversa Una relación formó cuando el producto de dos variables es una constante.

irrational number (p. 543) A number that cannot be expressed as $\frac{a}{b}$, where a and b are integers and b does not equal 0.

número irracional Número que no puede escribirse como $\frac{a}{b}$, donde a y b son enteros y b no es igual a 0.

isosceles triangle (p. 552) A triangle that has at least two congruent sides.

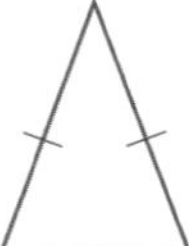

triángulo isósceles Triángulo que posee por lo menos dos lados congruentes.

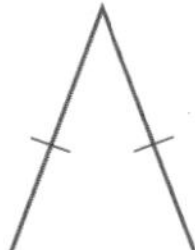

L

lateral area (p. 691) The sum of the areas of the lateral faces of a solid.

área lateral Suma de las áreas de las caras laterales de un sólido.

lateral faces (p. 691) The lateral faces of a prism, cylinder, pyramid, or cone are all the surfaces of the figure except the base or bases.

caras laterales Las caras laterales de un prisma, cilindro, pirámide o cono son todas las superficies de la figura, excluyendo la base o las bases.

least common multiple (LCM) (p. 860) The least of the nonzero common multiples of two or more numbers. The LCM of 4 and 6 is 12.

mínimo común múltiplo (mcm) El menor de los múltiplos comunes no nulos de dos o más números. El MCM de 4 y 6 es 12.

leaf (p. 737) In a stem-and-leaf plot, the next greatest place value of the data after the stem forms the leaves.

hojas En un diagrama de tallo y hojas, las hojas las forma el segundo valor de posición mayor después del tallo.

legs (p. 558) The sides that are adjacent to the right angle of a right triangle.

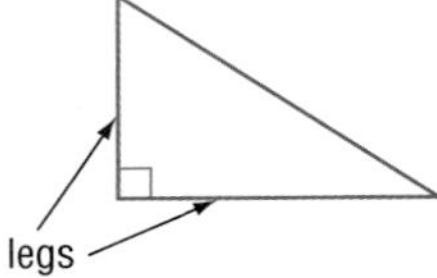

catetos Lados adyacentes al ángulo recto de un triángulo rectángulo.

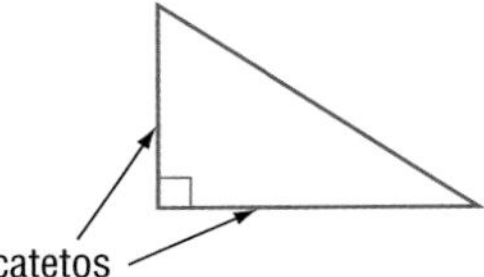

like fractions (p. 147) Fractions that have the same denominator.

fracciones semejantes Fracciones con el mismo denominador.

like terms (p. 178) Expressions that contain the same variables to the same power, such as $2n$ and $5n$ or $6xy^2$ and $4xy^2$.

términos semejantes Expresiones que tienen las mismas variables elevadas a los mismos exponentes, como $2n$ y $5n$ ó $6xy^2$ y $4xy^2$.

line (p. 868) A never-ending straight path.

recta Trayectoria rectilínea interminable.

line graph (p. 882) A type of statistical graph used to show how values change over a period of time.

gráfica lineal Tipo de gráfica estadística que se usa para mostrar cómo cambian los valores durante un período de tiempo.

line of fit (p. 448) On a scatter plot, a line drawn that is very close to most of the data points. The line that best fits the data.

recta de ajuste En una gráfica de dispersión, una recta que está muy cercana a la mayoría de los puntos de datos. La recta que mejor se ajusta a los datos.

line of symmetry (p. 101) Each half of a figure is a mirror image of the other half when a line of symmetry is drawn.

eje de simetría Cuando se traza un eje de simetría, cada mitad de una figura es una imagen especular de la otra mitad.

line plot (p. 883) A diagram that shows the frequency of data on a number line.

esquema lineal Diagrama que muestra la frecuencia de los datos sobre una recta numérica.

line segment (p. 550) Part of a line containing two endpoints and all the points between them.

segmento de recta Parte de una recta que contiene dos extremos y todos los puntos entre éstos.

linear equation (p. 406) An equation in which the variables appear in separate terms and neither variable contains an exponent other than 1. The graph of a linear equation is a straight line.

ecuación lineal Ecuación en que las variables aparecen en términos separados y en la cual ninguna de ellas tiene un exponente distinto de 1. La gráfica de una ecuación lineal es una recta.

linear regression equation (p. 864) One type of equation for a line of fit.

ecuación de la regresión linear Tipo de ecuación para una recta de regresión.

linear relationships (p. 418) Relationships that have straight line graphs.

relación lineal Relación que al ser graficada forma una línea recta.

lower quartile (p. 744) The median of the lower half of a set of data, indicated by LQ.

cuartil inferior Mediana de la mitad inferior de un conjunto de datos, se denota con CI.

M

markup (p. 365) The amount the price of an item is increased above the price of the store paid for an item.

margen de utilidad Cantidad de aumento en el precio de un artículo por encima del precio que paga la tienda por dicho artículo.

matrix (p. 886) A rectangular arrangement of numerical data in rows and columns.

matriz Disposición rectangular de números colocados en filas y columnas.

mean (pp. 92, 730) The sum of data divided by the number of items in the data set, also called the average.

media Suma de los datos dividida entre el número de elementos en el conjunto de datos. También llamada promedio.

measures of central tendency (p. 730) For a list of numerical data, numbers that can represent the whole set of data.

medidas de tendencia central Números que pueden representar todo el conjunto de datos en una lista de datos numéricos.

measures of variation (p. 743) Used to describe the distribution of statistical data.

medidas de variación Se usan para describir la distribución de datos estadísticos.

median (p. 730) In a set of data, the middle number of the ordered data, or the mean of the two middle numbers.

mediana En un conjunto de datos, el número central de los datos ordenados numéricamente o la media de los dos números centrales.

mode (p. 730) The number or numbers that occur(s) most often in a set of data.

moda Número o números de un conjunto de datos que aparecen más frecuentemente.

monomial (p. 477) An expression that is a number, a variable, or a product of numbers and/or variables.

monomio Expresión que es un número, una variable y/o un producto de números y variables.

Multiplication Property of Equality (p. 193) When you multiply each side of an equation by the same number, the two sides remain equal.

Propiedad de multiplicación de la igualdad Cuando multiplicas ambos lados de una ecuación por el mismo número, los dos lados permanecen iguales.

multiplicative inverse (p. 141) Two numbers whose product is 1.

inversos multiplicativos Dos números cuyo producto es igual a uno.

multiple (p. 860) The product of a number and a whole number.

múltiplo Producto de un número por un número entero.

mutually exclusive events (p. 792) Two or more events that cannot happen at the same time.

eventos mutuamente exclusivos Dos o más eventos que no pueden ocurrir simultáneamente.

Glossary/Glosario

N

negative number (p. 61) A number less than zero.

net (p. 690) A two-dimensional pattern for a three-dimensional figure.

nonlinear function (p. 504) A function with a graph that is not a straight line.

nonproportional (p. 281) A relationship in which two ratios are not equal.

null set or empty set (p. 250) A set with no elements shown by the symbol { } or Ø.

numerical expression (p. 5) A combination of numbers and operations such as addition, subtraction, multiplication, and division.

número negativo Número menor que cero.

redes Patrón bidimensional de una figura tridimensional.

función no lineal Función cuya gráfica no es una recta.

relación no proporcional Relación en la que dos razones no son iguales.

conjunto vacío Conjunto que carece de elementos y que se denota con el símbolo { } o Ø.

expresión numérica Combinación de números y operaciones, como adición, sustracción, multiplicación y división.

O

obtuse angle (p. 551) An angle with a measure greater than 90° but less than 180°.

obtuse triangle (p. 552) A triangle with one obtuse angle.

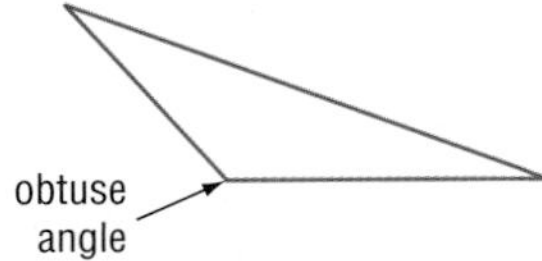

odds against (p. 767) The ratio that compares the number of ways the event *cannot* occur to the number of ways that the event *can* occur.

odds in favor (p. 767) The ratio that compares the number of ways the event *can* occur to the number of ways that the event *cannot* occur.

opposites (p. 71) Two numbers with the same absolute value but different signs.

order of operations (p. 6) The rules to follow when more than one operation is used in an expression.

1. Do all operations within grouping symbols first; start with the innermost grouping symbols.
2. Evaluate all powers before other operations.
3. Multiply and divide in order from left to right.
4. Add and subtract in order from left to right.

ordered pair (p. 25) A pair of numbers used to locate any point on a coordinate plane.

origin (p. 25) The point at which the number lines intersect in a coordinate system.

ángulo obtuso Ángulo que mide más de 90°, pero menos de 180°.

triángulo obtusángulo Triángulo que posee un ángulo obtuso.

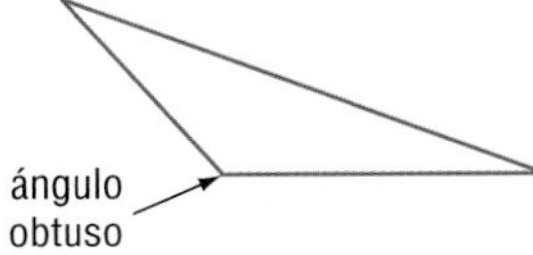

probabilidad en contra Como la razón entre el número de casos desfavorables y el número de caso favorables.

probabilidad a favor Como la razon entre el número de casos favorables y el número de casos desfavorables.

opuestos Dos números que tienen el mismo valor absoluto, pero que tienen distintos signos.

orden de las operaciones Reglas a seguir cuando se usa más de una operación en una expresión.

1. Primero ejecuta todas las operaciones dentro de los símbolos de agrupamiento
2. Evalúa todas las potencias antes que las otras aperaciones.
3. Multiplica y divide en orden de izquierda a derecha.
4. Suma y resta en orden de izquierda a derecha.

par ordenado Par de números que se usa para ubicar cualquier punto en un plano de coordenadas.

origen Punto de intersección de las rectas numéricas de un sistema de coordenadas.

outcome (p. 765) Possible result of a probability experiment.

outlier (p. 745) Data that are more than 1.5 times the interquartile range beyond the quartiles.

resultado Resultados posibles de un experimento probabilístico.

valores atípicos Datos que distan de los cuartiles más de 1.5 veces la amplitud intercuartílica.

P

parabola (p. 510) The graph of a quadratic function.

parallel lines (p. 590) Two lines in the same plane that do not intersect.

parallelogram (p. 613) A quadrilateral with opposite sides parallel and congruent.

part (p. 345) In a percent proportion, the number being compared to the whole quantity.

percent (p. 331) A ratio that compares a number to 100.

percent equation (p. 357) An equivalent form of the percent proportion, where % is written as a decimal.

Part = Percent × Whole

percent of change (p. 364) The ratio of the increase or decrease of an amount to the original amount.

percent of decrease (p. 364) The ratio of an amount of decrease to the previous amount, expressed as a percent. A negative percent of change.

percent of increase (p. 364) The ratio of an amount of increase to the original amount, expressed as a percent.

percent proportion (p. 345)

$$\frac{\text{part}}{\text{whole}} = \frac{\text{percent}}{100} \text{ or } \frac{a}{b} = \frac{p}{100}$$

perfect square (p. 537) Rational number whose square root is a whole number. 25 is a perfect square because $\sqrt{25} = 5$.

perimeter (p. 221) The distance around a geometric figure.

permutation (p. 783) An arrangement or listing in which order is important.

perpendicular bisector (p. 870) A perpendicular line that divides a line segment into two congruent segments.

parábola La gráfica de una función cuadrática.

rectas paralelas Dos rectas en el mismo plano que no se intersecan.

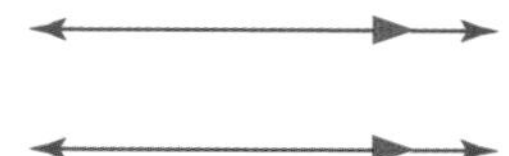

paralelogramo Cuadrilátero con lados opuestos congruentes y paralelos.

parte En una proporción porcentual, el número que se compara con la cantidad total.

por ciento Razón que compara un número con 100.

ecuación porcentual Forma equivalente a la proporción porcentual en la cual el % se escribe como decimal.

Parte = Por ciento × Entero

porcentaje de cambio Razón del aumento o disminución de una cantidad a la cantidad original.

porcentaje de disminución Razón de la cantidad de disminución a la cantidad original, escrita como por ciento. Un por ciento de cambio negativo.

porcentaje de aumento Razón de la cantidad de aumento a la cantidad original, escrita como por ciento.

proporción porcentual

$$\frac{\text{parte}}{\text{todo}} = \frac{\text{por ciento}}{100} \text{ or } \frac{a}{b} = \frac{p}{100}$$

cuadrados perfectos Números racionales cuyas raíces cuadradas son números racionales. 25 es un cuadrado perfecto porque $\sqrt{25} = 5$.

perímetro Longitud alrededor de una figura geométrica.

permutación Arreglo o lista en que el orden es importante.

bisector perpendicular Línea perpendicular que divide una línea segmento en dos segmentos congruentes.

perpendicular lines (p. 589) Lines that intersect to form a right angle.

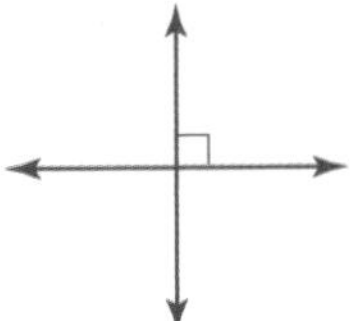

rectas perpendiculares Rectas que se intersecan formando un ángulo recto.

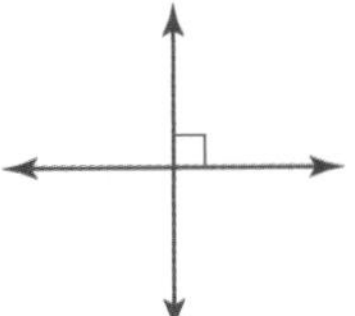

perspective (p. 663) A point of view.

perspective Un punto de vista.

pi, π (p. 631) The ratio of the circumference of a circle to the diameter of the circle. Approximations for π are 3.14 and $\frac{22}{7}$.

pi, π Razón de la circunferencia de un círculo al diámetro del mismo. 3.14 y $\frac{22}{7}$ son aproximaciones de π.

plane (p. 664) A two-dimensional flat surface that extends in all directions and contains at least three noncollinear points.

plano Superficie plana bidimensional que se extiende en todas direcciones y que contiene por lo menos tres puntos no colineales.

point (p. 868) An exact location in space that is represented by a dot.

punto Ubicación exacta en el espacio que se representa con un marca puntual.

point-slope form (p. 442) An equation of the form $y - y_1 = m(x - x_1)$, where m is the slope and (x_1, y_1) is a given point on a nonvertical line.

forma punto-pendiente Ecuación de la forma $y - y_1 = m(x - x_1)$, donde m es la pendiente y (x_1, y_1) es un punto dado de una recta no vertical.

polygon (p. 617) A simple closed figure in a plane formed by three or more line segments.

polígono Figura simple y cerrada en el plano formada por tres o más segmentos de recta.

polyhedron (p. 664) A solid with flat surfaces that are polygons.

poliedro Sólido con superficies planas que son polígonos.

polynomial (p. LA2) An algebraic expression that contains the sums and/or products of one or more monomials.

polinomio Expresión algebraica que contiene sumas y/o productos de uno o más monomios.

population (p. 771) A larger group used in statistical analysis.

población Grupo grande que se utiliza en análisis estadísticos.

positive number (p. 61) Any number that is greater than zero.

número positive Todo número mayor que cero.

power (p. 471) A number that is expressed using an exponent.

potencia Número que puede escribirse usando un exponente.

precision (p. 879) The degree of perfection in which a measurement is made.

precisión Grado de perfección el cual se hace una medida.

prime factorization (p. 477) A composite number expressed as a product of prime factors. For example, the prime factorization of 63 is $3 \times 3 \times 7$.

factorización prima Número compuesto escrito como producto de factores primos. Por ejemplo, la factorización prima de 63 es $3 \times 3 \times 7$.

prime number (p. 476) A whole number that has exactly two factors, 1 and itself.

número primo Número entero que sólo tiene dos factores, 1 y sí mismo.

principal (p. 370) The amount of money in an account.

capital Cantidad de dinero en una cuenta.

prism (p. 665) A polyhedron that has two parallel, congruent bases in the shape of polygons.

rectangular prism

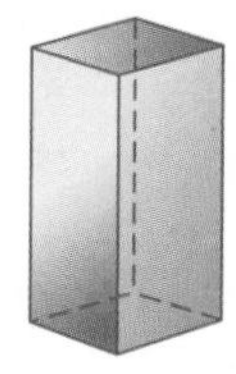

triangular prism

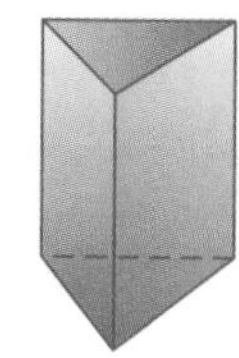

prisma Poliedro que posee dos bases congruentes y paralelas en forma de polígonos.

prisma rectangular

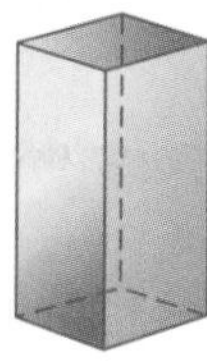

prisma triangular

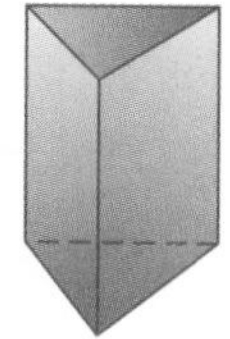

probability (p. 765) The ratio of the number of ways a certain event can occur to the number of possible outcomes.

$$P(\text{event}) = \frac{\text{number of favorable outcomes}}{\text{number of possible outcomes}}$$

probabilidad La razón del número de maneras en que puede ocurrir el evento al número de resultados posibles.

$$P(\text{evento}) = \frac{\text{número de resultados favorables}}{\text{número de resultados posíbles}}$$

properties (p. 18) Statements that are true for any numbers.

propiedades Enunciados que son verdaderos para cualquier número.

proportion (p. 287) A statement of equality of two or more ratios.

proporción Enunciado de la igualdad de dos o más razones.

proportional (p. 281) The ratios of related terms are equal.

proporcional Relación en la que la razón entre los términos relacionados permanece igual.

protractor (p. 873) An instrument used to measure angles.

transportador Instrumento que se usa para medir ángulos.

pyramid (p. 665) A polyhedron that has a polygon for a base and triangles for sides.

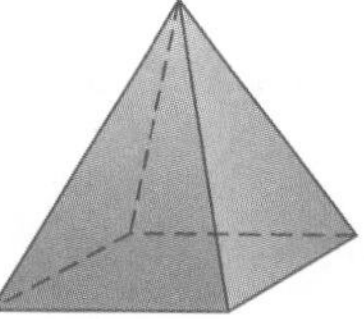

pirámide Poliedro cuya base es un polígono y cuyos lados son triángulos.

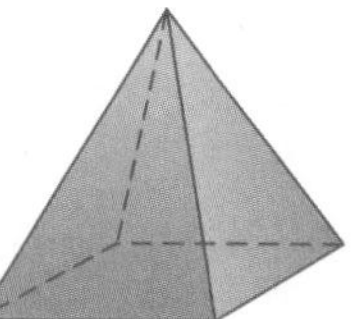

Pythagorean Theorem (p. 558) If a triangle is a right triangle, then the square of the length of the hypotenuse is equal to the sum of the squares of the lengths of the legs or $c^2 = a^2 + b^2$.

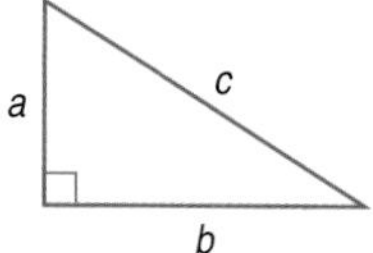

Teorema de Pitágoras Si un triángulo es rectángulo, entonces el cuadrado de la longitud de la hipotenusa es igual a la suma de los cuadrados de las longitudes de los catetos, o $c^2 = a^2 + b^2$.

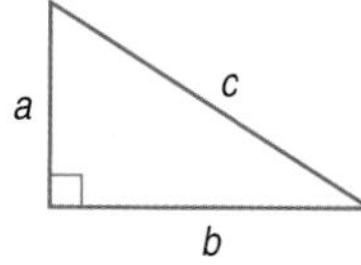

Pythagorean Triple (p. 557) The sides of right triangle represented by a, b, and c, when $a^2 + b^2 = c^2$.

Triple pitagórico Los lados de un triángulo rectángulo representados por a, b y c, cuando $a^2 + b^2 = c^2$.

Q

quadrant (p. 97) One of four regions into which the x-axis and y-axis separate the coordinate plane.

cuadrantes Las cuatro regiones en que los ejes x y y dividen el plano de coordenadas.

quadratic function (p. 510) A function that can be described by an equation of the form $y = ax^2 + bx + c$, where $a \neq 0$.

función cuadrática Función que puede describirse por una ecuación de la forma $y = ax^2 + bx + c$, donde $a \neq 0$.

quadrilateral (p. 612) A closed figure with four sides and four vertices, including squares, rectangles, and trapezoids.

quartile (p. 744) The values that divide a set of data into four equal parts.

cuadrilátero Figura cerrada de cuatro lados y cuatro vértices, incluyendo cuadrados, rectángulos y trapecios.

cuartiles Valores que dividen un conjunto de datos en cuatro partes iguales.

R

radical sign (p. 537) The symbol $\sqrt{}$ used to indicate a nonnegative square root.

radius (p. 631) The distance from the center to any point on the circle.

random (p. 766) Outcomes occur at random if each outcome is equally likely to occur.

range (p. 27) The range of a relation is the set of all y-coordinates from each ordered pair.

range (p. 743) A measure of variation that is the difference between the least and greatest values in a set of data.

rate (p. 270) A ratio of two measurements having different units.

rate of change (p. 412) A change in one quantity with respect to another quantity.

ratio (p. 265) A comparison of two numbers by division. The ratio of 2 to 4 can be stated as 2 out of 4, 2 to 4, 2 : 4, or $\frac{2}{4}$.

rational number (p. 128) A number that can be written as a fraction in the form $\frac{a}{b}$, where a and b are integers and $b \neq 0$.

ray (p. 868) A part of a line that extends indefinitely in one direction.

real numbers (p. 543) The set of rational numbers together with the set of irrational numbers.

reciprocal (p. 141) Another name for a multiplicative inverse.

rectangle (p. 613) A parallelogram with four right angles.

reflection (p. 101) A transformation where a figure is flipped over a line. Also called a flip.

regular polygon (p. 619) A polygon having all sides congruent and all angles congruent.

regular pyramid (p. 702) A pyramid whose base is a regular polygon.

relation (p. 27) A set of ordered pairs.

repeating decimal (p. 122) A decimal whose digits repeat in groups of one or more without end. Examples are 0.181818… and 0.8333….

signo radical El símbolo $\sqrt{}$ que se usa para indicar la raíz cuadrada no negativa.

radio Distancia del centro a cualquier punto de un círculo.

aleatorio Los resultados son aleatorios si todos son equiprobables.

rango El rango de una relación es el conjunto de coordenadas y de todos los pares.

amplitud Medida de variación que es la diferencia entre los valores máximo y mínimo de un conjunto de datos.

tasa Razón de dos medidas que tienen unidades distintas.

tasa de cambio Cambio de una cantidad con respecto a otra.

razón Comparación de dos números mediante división. La razón de 2 a 4 puede escribirse como 2 de cada 4, 2 a 4, 2 : 4 ó $\frac{2}{4}$.

número racional Número que puede escribirse como una fracción de la forma $\frac{a}{b}$ donde a y b son enteros y $b \neq 0$.

rayo Parte de una recta que se extiende indefinidamente en una dirección.

números reales El conjunto de los números racionales junto con el de números irracionales.

recíproco Otro nombre del inverso multiplicativo.

rectángulo Paralelogramo con cuatro ángulos rectos.

reflexión Transformación en que una figura se voltea a través de una recta.

polígono regular Polígono cuyos lados son todos congruentes y cuyos ángulos son también todos congruentes.

pirámide regular Pirámide cuya base es un polygon regular.

relación Conjunto de pares ordenados.

decimal periódico Decimal cuyos dígitos se repiten en grupos de uno o más. 0.181818… y 0.8333… son ejemplos de este tipo de decimales.

rhombus (p. 613) A parallelogram with four congruent sides.

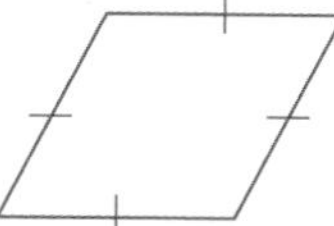

rombo Paralelogramo con cuatro lados congruentes.

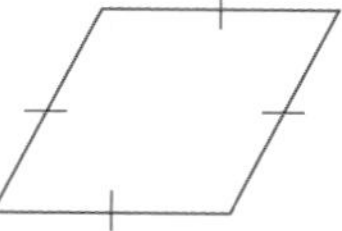

right angle (p. 551) An angle that measures 90°.

ángulo recto Ángulo que mide 90°.

right triangle (p. 552) A triangle with one right angle.

triángulo rectángulo Triángulo que tiene un ángulo recto.

rotation (p. 605) A transformation where a figure is turned around a fixed point. Also called a turn.

rotación Transformación en que una figura se hace girar alrededor de un punto fijo. También se llama vuelta.

rotational symmetry (p. 607) A figure has rotational symmetry if it can be turned less than 360° about its center and still look like the original.

simetría rotacional Una figura posee simetría rotacional si se puede girar menos de 360° en torno a su centro sin que esto cambia su apariencia con respecto a la figura original.

S

sample (p. 771) A subgroup or subset of a population used to represent the whole population.

muestra Subgrupo o subconjunto de una población que se usa para representarla.

sample space (p. 766) The set of all possible outcomes.

espacio muestral Conjunto de todos los resultados posibles.

scale (p. 294) The relationship between the measurements on a drawing or model and the measurements of the real object.

escala Relación entre las medidas de un dibujo o modelo y las medidas de la figura verdadera.

scale drawing (p. 294) A drawing that is used to represent an object that is too large or too small to be drawn at actual size.

dibujo a escala Dibujo que se usa para representar una figura que es demasiado grande o pequeña como para ser dibujada de tamaño natural.

scale factor (p. 295) The ratio of a length on a scale drawing or model to the corresponding length on the real object.

factor de escala Razón de la longitud en un dibujo a escala o modelo a la longitud correspondiente en la figura verdadera.

scale model (p. 294) A model used to represent an object that is too large or too small to be built at actual size.

modelo a escala Modelo que se usa para representar una figura que es demasiado grande o pequeña como para ser construida de tamaño natural.

scalene triangle (p. 552) A triangle with no congruent sides.

triángulo escaleno Triángulo que no tiene lados congruentes.

scatter plot (p. 40) The relationship between a set of data with two variables, graphed as ordered pairs on a coordinate plane.

gráfica de dispersión Es un diagrama que muestra la relación entre un conjunto de datos con dos variables, graficados como pares ordenados en un plano coordenadas.

scientific notation (p. 493) A number in scientific notation is expressed as $a \times 10n$, where $1 \leq a < 10$ and n is an integer. For example, $5{,}000{,}000 = 5.0 \times 10^6$.

notación científica Un número en notación científica se escribe como $a \times 10n$, donde $1 \leq a < 10$ y n es un entero. Por ejemplo, $5{,}000{,}000 = 5.0 \times 10^6$.

sector (p. 638) A pie-shaped part of a circle bound by the radius of a circle and an arc.

sector Una parte empanada-formada de un círculo limita por el radio de un círculo y de un arco.

selling price (p. 365) The amount a customer pays for an item.

sequence (p. 401) An ordered list of numbers, such as, 0, 1, 2, 3, or 2, 4, 6, 8.

sides (p. 868) The two rays that make up an angle.

similar figures (p. 301) Figures that have the same shape but not necessarily the same size.

similar solids (p. 709) Solids that have the same shape but not necessarily the same size.

simple event (p. 765) One outcome or a collection of outcomes.

simple interest (p. 370) The amount of money paid or earned for the use of money.

$I = prt$ (Interest = principal × rate × time)

simple random sample (p. 771) A sample where each item or person in the population is as likely to be chosen as any other.

simplest form (p. 179) An algebraic expression in simplest form has no like terms and no parentheses.

simplify (p. 20) To write an expression in a simpler form.

simplify the expression (p. 179) To use distribution to combine like terms.

sine (p. 877) For an acute angle of a right triangle, the ratio of the measure of the leg opposite the acute angle to the measure of the hypotenuse.

slant height (p. 702) The length of the altitude of a lateral face of a regular pyramid or cone.

slope (p. 427) The ratio of the rise, or vertical change, to the run, or horizontal change. The slope describes the steepness of a line.

$$\text{slope} = \frac{\text{rise}}{\text{run}}$$

slope-intercept form (p. 433) A linear equation in the form $y = mx + b$, where m is the slope and b is the y-intercept.

solid (p. 664) Three-dimensional figure.

solution (p. 184) A value for the variable that makes an equation true. For $x + 7 = 19$, the solution is 12.

precio de venta Cantidad de dinero que paga un consumidor por un artículo.

sucesión Lista ordenada de números, como 0, 1, 2, 3 ó 2, 4, 6, 8.

lados Los dos rayos que forman un ángulo.

figuras semejantes Figuras que tienen la misma forma, pero no necesariamente el mismo tamaño.

sólidos semejantes Sólidos que tienen la misma forma, pero no necesariamente el mismo tamaño.

evento simple Resultado o colección de resultados.

interés simple Cantidad que se paga o que se gana por usar el dinero.

$I = crt$ (Interés = capital × rédito × tiempo)

muestra aleatoria simple Muestra de una población que tiene la misma probabilidad de escogerse que cualquier otra.

forma reducida Una expresión algebraica reducida no tiene ni términos semejantes ni paréntesis.

reducir Escribir una expresión en forma más simple.

reducir la expresión Usar la distribución para combinar términos semejantes.

seno Es la razón entre la medida del cateto opuesto al ángulo agudo y la medida de la hipotenusa de un triángulo rectángulo.

altura oblicua En una pirámide regular o un cono, la longitud de la altura de una cara lateral.

pendiente Razón de la elevación o cambio vertical al desplazamiento o cambio horizontal. La pendiente describe la inclinación de una recta.

$$\text{pendiente} = \frac{\text{elevación}}{\text{desplazamiento}}$$

forma pendiente-intersección Una ecuación lineal de la forma $y = mx + b$, donde m es la pendiente y b es la intersección y.

sólido Figura tridimensional.

solución Valosss y que posee cuatro ángulos rectos.

solving a right triangle (p. 559) Using the Pythagorean Theorem to find the length of the third side of a right triangle, if the lengths of the other two sides are known.

resolver un triángulo rectángulo Uso del Teorema de Pitágoras para hallar la longitud de un tercer lado de un triángulo rectángulo, si se conocen las longitudes de los otros dos lados.

solving the equation (p. 184) The process of finding a solution to an equation.

resolver la ecuación Proceso de hallar una solución a una ecuación.

sphere (p. 684) The set of all points in space that are a given distance, *r*, from the center.

esfera El conjunto de todos los puntos en el espacio que se hallan a una distancia *r* del centro.

spreadsheet (p. 17) A table that performs calculations.

hoja de cálculos Tabla que realiza cálculos.

square (p. 613) A parallelogram with all sides congruent and four right angles.

cuadrado Paralelogramo cuyos lados son todos congruentes y que posee cuatro ángulos rectos.

square root (p. 537) One of the two equal factors of a number. The square root of 25 is 5 since $5^2 = 25$.

raíz cuadrada Uno de los dos factores iguales de un número. Una raíz cuadrada de 25 es 5 porque $5^2 = 25$.

standard form (p. 493) A number is in standard form when it does not contain exponents. The standard form for seven hundred thirty-nine is 739.

forma estándar Un número está en forma estándar si no contiene exponentes. Por ejemplo, la forma estándar de setecientos treinta y nueve es 739.

statistics (p. 882) The branch of mathematics that deals with collecting, organizing, and interpreting data.

estadística Rama de las matemáticas cuyo objetivo primordial es la recopilación, organización e interpretación de datos.

stem (p. 737) The greatest place value common to all the data values is used for the stem of a stem-and-leaf plot.

tallo Máximo valor de posición común a todos los datos que se usa como el tallo en un diagrama de tallo y hojas.

stem-and-leaf plot (p. 737) A system used to condense a set of data where the greatest place value of the data forms the stem and the next greatest place value forms the leaves.

diagrama de tallo y hojas Sistema que se usa para condensar un conjunto de datos, en que el valor de posición máximo de los datos forma el tallo y el segundo valor de posición máximo forma las hojas.

straight angle (p. 551) An angle with a measure equal to 180°.

ángulo llano Ángulo que mide 180°.

stratified random sample (p. 771) A sampling method in which the population is divided into similar, non-overlapping groups. A simple random sample is then selected from each group.

muestra aleatoria estratificada Método de muestreo en que la población se divide en grupos semejantes que no se sobreponen. Luego se selecciona una muestra aleatoria simple de cada grupo.

substitution (p. 455) Use algebraic methods to find an exact solution of a system of equations.

sustitución Usa métodos algebraicos para hallar una solución exacta a un sistema de ecuaciones.

Subtraction Property of Equality (p. 185) If you subtract the same number from each side of an equation, the two sides remain equal.

Propiedad de sustracción de la igualdad Si restas el mismo número de ambos lados de una ecuación, los dos lados permanecen iguales.

supplementary angles (p. 589) Two angles are supplementary if the sum of their measures is 180°.

suplementarios ángulos Dos ángulos son suplementarios si sus medidas suman 180°.

surface area (p. 691) The sum of the areas of all the surfaces (faces) of a 3-dimensional figure.

área de superficie Suma de las áreas de todas las superficies (caras) de una figura tridimensional.

system of equations (p. 453) A set of two or more equations with the same variables.

sistema de ecuaciones Sistema de ecuaciones con las mismas variables.

Glossary/Glosario

systematic random sample (p. 771) A sampling method in which the items or people are selected according to a specific time or item interval.

muestra aleatoria sistemática Muestra en que los elementos de las muestra se escogen según un intervalo de tiempo o elemento específico.

T

tangent (p. 875) For an acute angle of a right triangle, the ratio of the measure of the leg opposite the acute angle to the measure of the leg adjacent to the acute angle.

tangente La razón entre la medida del cateto opuesto al ángulo agudo y la medida del cateto adyacente al ángulo agudo de un triángulo rectángulo.

term (p. 178) When plus or minus signs separate an algebraic expression into parts, each part is a term.

término Cada una de las partes de una expresión algebraica separadas por los signos de adición o sustracción.

term (p. 401) Each number within a sequence is called a term.

término Cada número de una sucesión se llama término.

terminating decimal (p. 121) A decimal whose digits end. Every terminating decimal can be written as a fraction with a denominator of 10, 100, 1000, and so on.

decimal terminal Decimal cuyos dígitos terminan. Todo decimal terminal puede escribirse como una fracción con un denominador de 10, 100, 1000, etc.

tessellation (p. 619) A pattern formed by repeating figures that fit together without gaps or overlaps.

teselado Patrón formado por figuras repetidas que no se traslapan y que no dejan espacios entre sí.

theoretical probability (p. 766) What should occur in a probability experiment.

probabilidad teórica Lo que debería ocurrir en un experimento probabilístico.

transformation (p. 101) A movement of a geometric figure.

transformación Desplazamiento de una figura geométrica.

translation (p. 101) A transformation where a figure is slid from one position to another without being turned. Also called a slide.

translación Transformación en que una figura se desliza sin girar, de una posición a otra. También se llama deslizamiento.

transversal (p. 590) A line that intersects two parallel lines to form eight angles.

transversal Recta que interseca dos rectas paralelas formando ocho ángulos.

trapezoid (p. 613) A quadrilateral with exactly one pair of parallel sides.

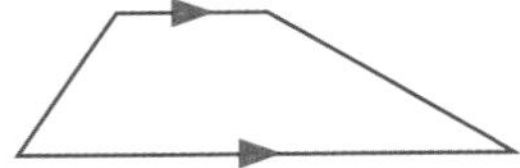

trapecio Cuadrilátero con sólo un par de lados paralelos.

tree diagram (p. 777) A diagram used to show the total number of possible outcomes.

diagrama de árbol Diagrama que se usa para mostrar el número total de resultados posibles.

triangle (p. 550) A figure having three sides.

triángulo Figura de tres lados.

trigonometric ratio (p. 875) A ratio of the lengths of sides of a right triangle.

razón trigonométrica Razón de las longitudes de las lados de un triángulo rectángulo.

trigonometry (p. 875) The study of the properties of triangles and trigonometric functions and their applications.

trigonometría Estudio de las propiedades de los triángulos y de las funciones trigonométricas y sus aplicaciones.

trinomial (p. LA27) A polynomial with three terms.

trinomio Polinomio de tres términos.

two-step equation (p. 199) An equation that contains two operations.

ecuación de dos pasos Ecuación que contiene dos operaciones.

U

unbiased sample (p. 771) A random sample that is representative of a larger sample.

muestra insesgada Muestra aleatoria que es representativa de una muestra más grande.

unlike fractions (p. 153) Fractions with different denominators.

fracciones con distinto denominador Fracciones cuyos denominadores son diferentes.

unit rate (p. 270) A rate simplified so that it has a denominator of 1.

tasa unitaria Tasa reducida que tiene denominador igual a 1.

upper quartile (p. 744) The median of the upper half of a set of data, indicated by UQ.

cuartil superior Mediana de la mitad superior de un conjunto de datos, denotada por CS.

V

variable (p. 11) A placeholder for any value.

variable Marcador de posición para cualquier valor.

vertex (p. 550) A vertex of a polygon is a point where two sides of the polygon intersect.

vértice El vértice de un polígono es un punto en que se intersecan dos lados del mismo.

vertex (p. 664) Where three or more planes intersect in a point.

vértice Punto en que se intersecan tres o más planos.

vertex (p. 868) The common endpoint of the rays forming an angle.

vértice Extremo común de los rayos que forman un ángulo.

vertical angles (p. 589) Two pairs of opposite angles formed by two intersecting lines. The angles formed are congruent. In the figure, the vertical angles are $\angle 1$ and $\angle 3$, $\angle 2$ and $\angle 4$.

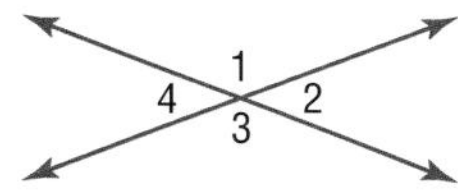

ángulos opuestos por el vértice Dos pares de ángulos opuestos formados por dos rectas que se intersecan. Los ángulos que resultan son congruentes. En la figura, los ángulos opuestos por el vértice son $\angle 1$ y $\angle 3$, $\angle 2$ y $\angle 4$.

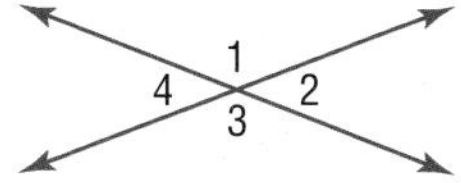

vertical line test (p. 396) If any vertical line drawn on the graph of a relation passes through no more than one point on the graph for each value of x in the domain, then the relation is a function.

prueba de la recta vertical Si todas las rectas verticales trazadas en la gráfica de una relación no pasan por más un punto para cada valor de x en el dominio, entonces la relación es una función.

volume (p. 671) The measure of space occupied by a solid region.

volumen Medida del espacio que ocupa un sólido.

voluntary response sample (p. 772) A sample which involves only those who want to participate in the sampling.

muestra de respuesta voluntaria Muestra que involucra sólo aquellos que quieren participar en el muestreo.

W

whole (p. 345) In a percent proportion, the whole quantity, or the number to which the part is being compared.

$$\frac{\text{part}}{\text{whole}} = \frac{\text{percent}}{100}$$

entero o base En una proporción porcentual, toda la cantidad o número al que se compara la parte.

$$\frac{\text{parte}}{\text{todo}} = \frac{\text{porciento}}{100}$$

X

x-axis (p. 25) The horizontal number line which helps to form the coordinate system.

x-coordinate (p. 25) The first number of an ordered pair.

x-intercept (p. 407) The x-coordinate of a point where a graph crosses the x-axis.

eje x Recta numérica horizontal que forma parte de un sistema de coordenadas.

coordenada x El primer número de un par ordenado.

intersección x La coordenada x de un punto en que una gráfica interseca el eje x.

Y

y-axis (p. 25) The vertical number line which helps to form the coordinate system.

y-coordinate (p. 25) The second number of an ordered pair.

y-intercept (p. 407) The y-coordinate of a point where a graph crosses the y-axis.

eje y Recta numérica vertical que forma parte de un sistema de coordenadas.

coordenada y El segundo número en un par ordenado.

intersección y La coordenada y de un punto en que una gráfica interseca el eje y.

Z

zero pair (p. 67) A positive tile paired with a negative tile.

par nulo Ficha positiva apareada con una negativa.

Index

A

Index

B

C

Index

D

Index

E

F

G

H

I

Index

J

K

L

Index

N

O

P

Index

Q

R

T

U

V

W

X

Y

Z

Curriculum Focal Points and Connections for Grade 7

The Curriculum Focal Points identify key mathematical ideas for this grade. They are not discrete topics or a checklist to be mastered; rather, they provide a framework for the majority of instruction at a particular grade level and the foundation for future mathematics study. The complete document may be viewed at www.nctm.org/focalpoints.

KEY

G7-FP1
Grade 7 Focal Point 1

G7-FP2
Grade 7 Focal Point 2

G7-FP3
Grade 7 Focal Point 3

G7-FP4C
Grade 7 Focal Point 4 Connection

G7-FP5C
Grade 7 Focal Point 5 Connection

G7-FP6C
Grade 7 Focal Point 6 Connection

G7-FP7C
Grade 7 Focal Point 7 Connection

G7-FP1 Number and Operations and Algebra and Geometry: **Developing an understanding of and applying proportionality, including similarity**

Students extend their work with ratios to develop an understanding of proportionality that they apply to solve single and multistep problems in numerous contexts. They use ratio and proportionality to solve a wide variety of percent problems, including problems involving discounts, interest, taxes, tips, and percent increase or decrease. They also solve problems about similar objects (including figures) by using scale factors that relate corresponding lengths of the objects or by using the fact that relationships of lengths within an object are preserved in similar objects. Students graph proportional relationships and identify the unit rate as the slope of the related line. They distinguish proportional relationships $\left(\frac{y}{x} = k, \text{ or } y = kx\right)$ from other relationships, including inverse proportionality $\left(xy = k, \text{ or } y = \frac{k}{x}\right)$.

G7-FP2 Measurement and Geometry and Algebra: **Developing an understanding of and using formulas to determine surface areas and volumes of three-dimensional shapes**

By decomposing two- and three-dimensional shapes into smaller, component shapes, students find surface areas and develop and justify formulas for the surface areas and volumes of prisms and cylinders. As students decompose prisms and cylinders by slicing them, they develop and understand formulas for their volumes (*Volume* = *Area of base* × *Height*). They apply these formulas in problem solving to determine volumes of prisms and cylinders. Students see that the formula for the area of a circle is plausible by decomposing a circle into a number of wedges and rearranging them into a shape that approximates a parallelogram. They select appropriate two- and three-dimensional shapes to model real-world situations and solve a variety of problems (including multistep problems) involving surface areas, areas and circumferences of circles, and volumes of prisms and cylinders.

G7-FP2 Number and Operations and Algebra: **Developing an understanding of operations on all rational numbers and solving linear equations**

Students extend understandings of addition, subtraction, multiplication, and division, together with their properties, to all rational numbers, including negative integers. By applying properties of arithmetic and considering negative numbers in everyday contexts (e.g., situations of owing money or measuring elevations above and below sea level), students explain why the rules for adding, subtracting, multiplying, and dividing with negative numbers make sense. They use the arithmetic of rational numbers as they formulate and solve linear equations in one variable and use these equations to solve problems. Students make strategic choices of procedures to solve linear equations in one variable and implement them efficiently, understanding that when they use the properties of equality to express an equation in a new way, solutions that they obtain for the new equation also solve the original equation.

Connections to the Focal Points

G7-FP4C **Measurement and Geometry:** Students connect their work on proportionality with their work on area and volume by investigating similar objects. They understand that if a scale factor describes how corresponding lengths in two similar objects are related, then the square of the scale factor describes how corresponding areas are related, and the cube of the scale factor describes how corresponding volumes are related. Students apply their work on proportionality to measurement in different contexts, including converting among different units of measurement to solve problems involving rates such as motion at a constant speed. They also apply proportionality when they work with the circumference, radius, and diameter of a circle; when they find the area of a sector of a circle; and when they make scale drawings.

G7-FP5C **Number and Operations:** In grade 4, students used equivalent fractions to determine the decimal representations of fractions that they could represent with terminating decimals. Students now use division to express any fraction as a decimal, including fractions that they must represent with infinite decimals. They find this method useful when working with proportions, especially those involving percents. Students connect their work with dividing fractions to solving equations of the form $ax = b$, where a and b are fractions. Students continue to develop their understanding of multiplication and division and the structure of numbers by determining if a counting number greater than 1 is a prime, and if it is not, by factoring it into a product of primes.

G7-FP6C **Data Analysis:** Students use proportions to make estimates relating to a population on the basis of a sample. They apply percentages to make and interpret histograms and circle graphs.

G7-FP7C **Probability:** Students understand that when all outcomes of an experiment are equally likely, the theoretical probability of an event is the fraction of outcomes in which the event occurs. Students use theoretical probability and proportions to make approximate predictions.

ormulas

Perimeter

are	$P = 4s$		
angle	$P = 2\ell + 2w$ or $P = 2(\ell + w)$		

Circumference

le	$C = 2\pi r$ or $C = \pi d$		

Area

are	$A = s^2$	triangle	$A = \frac{1}{2}bh$
angle	$A = \ell w$	trapezoid	$A = \frac{1}{2}h(b_1 + b_2)$
llelogram	$A = bh$	circle	$A = \pi r^2$

Surface Area

e	$S = 6s^2$	cylinder	$S = 2\pi rh + 2\pi r^2$
angular prism	$S = 2\ell w + 2\ell h + 2wh$	pyramid	$S = L + B$ or $S = \frac{1}{2}P\ell + B$

Volume

e	$V = s^3$	pyramid	$V = \frac{1}{3}Bh$
m	$V = \ell wh$ or Bh	cone	$V = \frac{1}{3}\pi r^2 h$ or $\frac{1}{3}Bh$
nder	$V = \pi r^2 h$ or Bh		

Pythagorean Theorem

t triangle	$a^2 + b^2 = c^2$		

Temperature

renheit to Celsius	$C = \frac{5}{9}(F - 32)$	Celsius to Fahrenheit	$F = \frac{9}{5}C + 32$

Slope

$$m = \frac{\text{rise}}{\text{run}} \text{ or } m = \frac{\text{change in } y}{\text{change in } x} \text{ or } m = \frac{y_2 - y_1}{x_2 - x_1}$$

easures

Metric	Customary
Length	
lometer (km) = 1,000 meters (m)	1 foot (ft) = 12 inches (in.)
eter = 100 centimeters (cm)	1 yard (yd) = 3 feet or 36 inches
ntimeter = 10 millimeters (mm)	1 mile (mi) = 1,760 yards or 5,280 feet
Volume and Capacity	
er (L) = 1,000 milliliters (mL)	1 cup (c) = 8 fluid ounces (fl oz)
oliter (kL) = 1,000 liters	1 pint (pt) = 2 cups
	1 quart (qt) = 2 pints
	1 gallon (gal) = 4 quarts
Weight and Mass	
logram (kg) = 1,000 grams (g)	1 pound (lb) = 16 ounces (oz)
am = 1,000 milligrams (mg)	1 ton (T) = 2,000 pounds
etric ton = 1,000 kilograms	

Measures (cont.)

Metric	Customary
Time	
1 minute (min) = 60 seconds (s)	1 week (wk) = 7 days
1 hour (h) = 60 minutes	1 year (yr) = 12 months (mo) or 52 weeks or 365 day
1 day (d) = 24 hours	1 leap year = 366 days
Metric to Customary	
1 meter ≈ 39.37 inches	1 kilogram ≈ 2.2 pounds
1 kilometer ≈ 0.62 mile	1 gram ≈ 0.035 ounce
1 centimeter ≈ 0.39 inch	1 liter ≈ 1.057 quarts

Symbols

Symbol	Meaning	Symbol	Meaning
Number and Operations			
$+$	plus or positive	$>$	is greater than
$-$	minus or negative	$<$	is less than
$a \cdot b$	a times b	$\geq$	is greater than or equal to
$a \times b$	a times b	$\leq$	is less than or equal to
ab or $a(b)$	a times b	$\approx$	is approximately equal to
$\div$	divided by	%	percent
$\pm$	plus or minus	$a : b$	the ratio of a to b, or $\frac{a}{b}$
$=$	is equal to	$0.7\overline{5}$	repeating decimal 0.75555 . . .
$\neq$	is not equal to		
Algebra and Functions			
$-a$	opposite or additive inverse of a	$\lvert x \rvert$	absolute value of x
a^n	a to the nth power	$\sqrt{x}$	principal (positive) square root of x
a^{-n}	$\frac{1}{a^n}$	$f(x)$	function, f of x
Geometry and Measurement			
$\cong$	is congruent to	$\perp$	is perpendicular to
$\sim$	is similar to	$\parallel$	is parallel to
$^\circ$	degree(s)	$\angle A$	angle A
$\overleftrightarrow{AB}$	line AB	$m\angle A$	measure of angle A
$\overrightarrow{AB}$	ray AB	$\triangle ABC$	triangle ABC
$\overline{AB}$	line segment AB	(a, b)	ordered pair with x-coordinate a and y-coordinate b
AB	length of $\overline{AB}$	O	origin
∟	right angle	π	pi $\left(\text{approximately } 3.14 \text{ or } \frac{22}{7}\right)$
Probability and Statistics			
$P(A)$	probability of event A		